Table 3
Derived SI units obtained by combining base units and units with special names

Quantity	Units	Quantity	Units
Acceleration	m/s²	Molar entropy	J/mol·K
Angular acceleration	rad/s²	Molar heat capacity	J/mol·K
Angular velocity	rad/s	Moment of force	Nm
Area	m²	Permeability	H/m
Concentration	mol/m³	Permittivity	F/m
Current density	A/m²	Radiance	W/m²·sr
Density, mass	kg/m³	Radiant intensity	W/sr
Electric charge density	C/m³	Specific heat capacity	J/kg·K
Electric field strength	V/m	Specific energy	J/kg
Electric flux density	C/m²	Specific entropy	J/kg·K
Energy density	J/m³	Specific volume	m³/kg
Entropy	J/K	Surface tension	N/m
Heat capacity	J/K	Thermal conductivity	W/m·K
Heat flux density	W/m²	Velocity	m/s
Irradiance	W/m²	Viscosity, dynamic	Pa·s
Luminance	cd/m²	Viscosity, kinematic	m²/s
Magnetic field strength	A/m	Volume	m
Molar energy	J/mol	Wavelength	m

Table 4
SI prefixes

Multiplication factor	Prefix[a]	Symbol
1 000 000 000 000 = 10^{12}	tera	T
1 000 000 000 = 10^{9}	giga	G
1 000 000 = 10^{6}	mega	M
1 000 = 10^{3}	kilo	k
100 = 10^{2}	hectot	h
10 = 10^{1}	deka[b]	da
0.1 = 10^{-1}	deci[b]	d
0.01 = 10^{-2}	centi[b]	c
0.001 = 10^{-3}	milli	m
0.000 001 = 10^{-6}	micro	μ
0.000 000 001 = 10^{-9}	nano	n
0.000 000 000 001 = 10^{-12}	pico	p
0.000 000 000 000 001 = 10^{-15}	femto	f
0.000 000 000 000 000 001 = 10^{-18}	atto	a

[a] The first syllable of every prefix is accented so that the prefix will retain its identity. Thus, the preferred pronunciation of kilometer places the accent on the first syllable, not the second.
[b] The use of these prefixes should be avoided, except for the measurement of areas and volumes and for the nontechnical use of centimeter, as for body and clothing measurements.

Water Reuse

Water Reuse
Issues, Technologies, and Applications

Metcalf & Eddy | AECOM

Written by

Takashi Asano
Professor Emeritus of Civil and Environmental Engineering
University of California at Davis

Franklin L. Burton
Consulting Engineer
Los Altos, California

Harold L. Leverenz
Research Associate
University of California at Davis

Ryujiro Tsuchihashi
Technical Specialist
Metcalf & Eddy, Inc.

George Tchobanoglous
Professor Emeritus of Civil and Environmental Engineering
University of California at Davis

New York   Chicago   San Francisco   Lisbon   London   Madrid   Mexico City
Milan   New Delhi   San Juan   Seoul   Singapore   Sydney   Toronto

The McGraw·Hill Companies

Library of Congress Cataloging-in-Publication Data

Water reuse : issues, technologies, and applications / written by
    Takashi Asano . . . [et al.]. — 1st ed.
       p.   cm.
    Includes index.
    ISBN-13: 978-0-07-145927-3 (alk. paper)
    ISBN-10: 0-07-145927-8 (alk. paper)
    1. Water reuse.   I. Asano, Takashi.
TD429.W38515   2007
628.1′62—dc22

                                                                2006030659

Copyright © 2007 by Metcalf & Eddy, Inc. All rights reserved. Printed in the United States of America. Except as permitted under the United States Copyright Act of 1976, no part of this publication may be reproduced or distributed in any form or by any means, or stored in a data base or retrieval system, without the prior written permission of the publisher.

3 4 5 6 7 8 9 0   DOC/DOC   0 1 3 2 1 0 9 8 7

ISBN-13: 978-0-07-145927-3
ISBN-10: 0-07-145927-8

Photographs: All of the photographs for this textbook were taken by George Tchobanoglous, unless otherwise noted.

The sponsoring editor for this book was Larry S. Hager and the production supervisor was Pamela A. Pelton. It was set in Times by International Typesetting and Composition. The art director for the cover was Brian Boucher.

Printed and bound by RR Donnelley.

This book is printed on acid-free paper.

McGraw-Hill books are available at special quantity discounts to use as premiums and sales promotions, or for use in corporate training programs. For more information, please write to the Director of Special Sales, McGraw-Hill Professional, Two Penn Plaza, New York, NY 10121-2298. Or contact your local bookstore.

Information contained in this work has been obtained by The McGraw-Hill Companies, Inc. ("McGraw-Hill") from sources believed to be reliable. However, neither McGraw-Hill nor its authors guarantee the accuracy or completeness of any information published herein, and neither McGraw-Hill nor its authors shall be responsible for any errors, omissions, or damages arising out of use of this information. This work is published with the understanding that McGraw-Hill and its authors are supplying information but are not attempting to render engineering or other professional services. If such services are required, the assistance of an appropriate professional should be sought.

This book is dedicated to Metcalf & Eddy's James Anderson, who died of cancer in March 2006 and was therefore unable to see this book through to publication. In addition, we would like to acknowledge Vera Anderson, Jim's wife, of whom Jim wrote the following: "I would like to thank my wife, Vera, whose support given to me through a difficult health crisis allowed me to complete what will probably be my last professional endeavor."

As Director of Technology, Jim was responsible for Metcalf & Eddy's research program and for the continued development of our textbooks. It was through his vision of the importance of water reuse in strategic water resources management that this book was brought to fruition. Jim also understood the need to train environmental engineering professionals and Metcalf & Eddy's commitment to do its part as originally conceived and carried out by Leonard Metcalf and Harrison P. Eddy nearly 100 years ago.

Steve Guttenplan
President
Metcalf & Eddy

ABOUT THE AUTHORS

Takashi Asano is a Professor Emeritus of the Department of Civil and Environmental Engineering at the University of California, Davis. He received a B.S. degree in agricultural chemistry from Hokkaido University in Sapporo, Japan, an M.S.E. degree in sanitary engineering from the University of California, Berkeley, and a Ph.D. in environmental and water resources engineering from the University of Michigan, Ann Arbor in 1970. His principal research interests are water reclamation and reuse and advanced water and wastewater treatment in the context of integrated water resources management. Professor Asano was on the faculty of Montana State University, Bozeman, and Washington State University, Pullman. He also worked for 15 years as a water reclamation specialist for the California State Water Resources Control Board in Sacramento, California, in the formative years of water reclamation, recycling, and reuse. He is a recipient of the 2001 Stockholm Water Prize and also a member of the European Academy of Sciences and Arts, the International Water Academy, and an honorary member of the Water Environment Federation. Professor Asano received an Honorary Doctorate from his *alma mater*, Hokkaido University in Sapporo, Japan, in 2004. He is a registered professional engineer in California, Michigan, and Washington.

Franklin L. Burton served as vice president and chief engineer of the western region of Metcalf & Eddy in Palo Alto, California, for 30 years. He retired from Metcalf & Eddy in 1986 and has been in private practice in Los Altos, California, specializing in treatment technology evaluation, facilities design review, energy management, and value engineering. He received his B.S. in mechanical engineering from Lehigh University and an M.S. in civil engineering from the University of Michigan. He was a coauthor of the third and fourth editions of the Metcalf & Eddy textbook *Wastewater Engineering: Treatment and Reuse*. He has authored over 30 publications on water and wastewater treatment and energy management in water and wastewater applications. He is a registered civil engineer in California and is a life member of the American Society of Civil Engineers, American Water Works Association, and Water Environment Federation.

Harold L. Leverenz is a research associate at the University of California, Davis. He received a B.S. in biosystems engineering from Michigan State University and an M.S. and Ph.D. in environmental engineering from the University of California, Davis. His professional and research interests include decentralized systems for water reuse, natural treatment processes, and ecological sanitation systems. Dr. Leverenz is a member of the American Ecological Engineering Society, the American Society of Agricultural and Biological Engineers, and the International Water Association.

Ryujiro Tsuchihashi is a technical specialist with Metcalf & Eddy, Inc. He received his B.S. and M.S. in civil and environmental engineering from Kyoto University, Japan, and a Ph.D. in environmental engineering from the University of California, Davis. The areas of his expertise include biological nutrient removal, molecular technologies in the detection of pathogenic organisms in the aquatic environment, health aspects of groundwater recharge, biological and various water reuse applications. He is a member of the American Society of Civil Engineers, International Water Association, and WateReuse Association.

George Tchobanoglous is a Professor Emeritus in the Department of Civil and Environmental Engineering at the University of California, Davis. He received a B.S. degree in civil engineering from the University of the Pacific, an M.S. degree in sanitary engineering from the University of California at Berkeley, and a Ph.D. from Stanford University in 1969. His research interests are in the areas of wastewater treatment and reuse, wastewater filtration, UV disinfection, aquatic wastewater management systems, wastewater management for small and decentralized wastewater management systems, and solid waste management. He has authored or coauthored over 350 technical publications including 13 textbooks and 4 reference works. The textbooks are used in more than 225 colleges and universities, as well as by practicing engineers. The textbooks have also been used extensively in universities worldwide both in English and in translation. He is a past president of the Association of Environmental Engineering and Science Professors. Among his many honors, in 2003 Professor Tchobanoglous received the Clarke Prize from the National Water Research Institute. In 2004, he was inducted into the National Academy of Engineering. In 2005, he received an Honorary Doctor of Engineering Degree from the Colorado School of Mines. He is a registered civil engineer in California.

Contents

Preface xxvii
Acknowledgments xxxiii
Foreword xxxvii

Part 1 Water Reuse: An Introduction 1

1 Water Issues: Current Status and the Role of Water Reclamation and Reuse 3

    Working Terminology 4
1-1 Definition of Terms 6
1-2 Principles of Sustainable Water Resources Management 6
    The principle of sustainability 7
    Working definitions of sustainability 7
    Challenges for sustainability 7
    Criteria for sustainable water resources management 7
    Environmental ethics 13
1-3 Current and Potential Future Global Water Shortages 15
    Impact of current and projected world population 15
    Potential global water shortages 19
    Water scarcity 19
    Potential regional water shortages in the continental United States 20
1-4 The Important Role of Water Reclamation and Reuse 23
    Types of water reuse 24
    Integrated water resources planning 24
    Personnel needs/sustainable engineering 27
    Treatment and technology needs 27
    Infrastructure and planning issues 28
1-5 Water Reclamation and Reuse and Its Future 30
    Implementation hurdles 31
    Public support 31
    Acceptance varies depending on opportunity and necessity 31
    Public water supply from polluted water sources 31
    Advances in water reclamation technologies 31
    Challenges for water reclamation and reuse 32
    Problems and Discussion Topics 32
    References 33

2 Water Reuse: Past and Current Practices 37

    Working Terminology 38
2-1 Evolution of Water Reclamation and Reuse 39
    Historical development prior to 1960 39
    Era of water reclamation and reuse in the United States-post-1960 41
2-2 Impact of State and Federal Statutes on Water Reclamation and Reuse 45
    The Clean Water Act 45
    The Safe Drinking Water Act 46
2-3 Water Reuse—Current Status in the United States 46
    Withdrawal of water from surface and groundwater sources 46
    Availability and reuse of treated wastewater 46
    Milestone water reuse projects and research studies 47
2-4 Water Reuse in California: A Case Study 47
    Experience with water reuse 47
    Current water reuse status 48

Contents

Water reuse policies and recycling regulations 51
Potential future uses of reclaimed water 52

2-5 Water Reuse in Florida: A Case Study 53
Experience with water reuse 54
Current water reuse status 54
Water reuse policies and recycling regulations 56
Potential future uses of reclaimed water 56

2-6 Water Reuse in Other Parts of the World 58
Significant developments worldwide 58
The World Health Organization's water reuse guidelines 59
Water reuse in developing countries 59

2-7 Summary and Lessons Learned 63
Problems and Discussion Topics 65
References 66

Part 2 Health and Environmental Concerns in Water Reuse 71

3 Characteristics of Municipal Wastewater and Related Health and Environmental Issues 73

Working Terminology 74

3-1 Wastewater in Public Water Supplies—*de facto* Potable Reuse 77
Presence of treated wastewater in public water supplies 78
Impact of the presence of treated wastewater on public water supplies 78

3-2 Introduction to Waterborne Diseases and Health Issues 78
Important historical events 79
Waterborne disease 80
Etiology of waterborne disease 81

3-3 Waterborne Pathogenic Microorganisms 83
Terminology conventions for organisms 83
Log removal 83
Bacteria 83
Protozoa 87
Helminths 89
Viruses 89

3-4 Indicator Organisms 92
Characteristics of an ideal indicator organism 92
The coliform group bacteria 93
Bacteriophages 93
Other indicator organisms 94

3-5 Occurrence of Microbial Pathogens in Untreated and Treated Wastewater and in the Environment 94
Pathogens in untreated wastewater 94
Pathogens in treated wastewater 97
Pathogens in the environment 102
Survival of pathogenic organisms 102

3-6 Chemical Constituents in Untreated and Treated Wastewater 103
Chemical constituents in untreated wastewater 103
Constituents added through domestic commercial and industrial usage 104
Chemical constituents in treated wastewater 108
Formation of disinfection byproducts (DBPs) 113
Comparison of treated wastewater to natural water 114
Use of surrogate parameters 115

3-7 Emerging Contaminants in Water and Wastewater 117
Endocrine disruptors and pharmaceutically active chemicals 117
Some specific constituents with emerging concern 118
New and reemerging microorganisms 120

3-8 Environmental Issues 120
Effects on soils and plants 121
Effects on surface water and groundwater 121
Effects on ecosystems 121
Effects on development and land use 122
Problems and Discussion Topics 122
References 124

4 Water Reuse Regulations and Guidelines   131

Working Terminology   132

4-1 Understanding Regulatory Terminology   134
  Standard and criterion   134
  Standard versus criterion   134
  Regulation   135
  Difference between regulations and guidelines   135
  Water reclamation and reuse   135

4-2 Development of Standards, Regulations, and Guidelines for Water Reuse   135
  Basis for water quality standards   136
  Development of water reuse regulations and guidelines   136
  The regulatory process   139

4-3 General Regulatory Considerations Related to Water Reclamation and Reuse   139
  Constituents and physical properties of concern in wastewater   139
  Wastewater treatment and water quality considerations   142
  Reclaimed water quality monitoring   145
  Storage requirements   146
  Reclaimed water application rates   147
  Aerosols and windborne sprays   147

4-4 Regulatory Considerations for Specific Water Reuse Applications   149
  Agricultural irrigation   149
  Landscape irrigation   150
  Dual distribution systems and in-building uses   151
  Impoundments   152
  Industrial uses   153
  Other nonpotable uses   153
  Groundwater recharge   154

4-5 Regulatory Considerations for Indirect Potable Reuse   155
  Use of the most protected water source   155
  Influence of the two water acts   155
  Concerns for trace chemical constituents and pathogens   156
  Assessment of health risks   157

4-6 State Water Reuse Regulations   157
  Status of water reuse regulations and guidelines   158
  Regulations and guidelines for specific reuse applications   158
  Regulatory requirements for nonpotable uses of reclaimed water   165
  State regulations for indirect potable reuse   167

4-7 U.S. EPA Guidelines for Water Reuse   169
  Disinfection requirements   169
  Microbial limits   178
  Control measures   178
  Recommendations for indirect potable reuse   178

4-8 World Health Organization Guidelines for Water Reuse   179
  1989 WHO guidelines for agriculture and aquaculture   180
  The Stockholm framework   180
  Disability adjusted life years   180
  Concept of tolerable (acceptable) risk   181
  Tolerable microbial risk in water   181
  2006 WHO guidelines for the safe use of wastewater in agriculture   182

4-9 Future Directions in Regulations and Guidelines   184
  Continuing development of state standards, regulations, and guidelines   184
  Technical advances in treatment processes   184
  Information needs   184

Problems and Discussion Topics   185
References   187

5 Health Risk Analysis in Water Reuse Applications   191

Working Terminology   192

5-1 Risk Analysis: An Overview   193
  Historical development of risk assessment   194
  Objectives and applications of human health risk assessment   194

Elements of risk analysis 194
Risk analysis: definitions and concepts 196

5-2 Health Risk Assessment 197
Hazard identification 198
Dose-response assessment 198
Dose-response models 200
Exposure assessment 204
Risk characterization 204
Comparison of human health and ecological risk assessment 205

5-3 Risk Management 205

5-4 Risk Communication 206

5-5 Tools and Methods Used in Risk Assessment 207
Concepts from public health 207
Concepts from epidemiology 208
Concepts from toxicology 209
National toxicology program cancer bioassay 213
Ecotoxicology: environmental effects 214

5-6 Chemical Risk Assessment 215
Safety and risk determination in regulation of chemical agents 215
Risks from potential nonthreshold toxicants 220
Risk considerations 224
Chemical risk assessment summary 225

5-7 Microbial Risk Assessment 225
Infectious disease paradigm for microbial risk assessment 225
Microbial risk assessment methods 227
Static microbial risk assessment models 227
Dynamic microbial risk assessment models 229
Selecting a microbial risk model 232

5-8 Application of Microbial Risk Assessment in Water Reuse Applications 234
Microbial risk assessment employing a static model 234
Microbial risk assessment employing dynamic models 239
Risk assessment for water reuse from enteric viruses 244

5-9 Limitations in Applying Risk Assessment to Water Reuse Applications 249
Relative nature of risk assessment 249

Inadequate consideration of secondary infections 249
Limited dose-response data 250
Problems and Discussion Topics 250
References 251

Part 3 Technologies and Systems for Water Reclamation and Reuse 255

6 Water Reuse Technologies and Treatment Systems: An Overview 257

Working Terminology 258

6-1 Constituents in Untreated Municipal Wastewater 260

6-2 Technology Issues in Water Reclamation and Reuse 260
Water reuse applications 262
Water quality requirements 262
Multiple barrier concept 263
Need for multiple treatment technologies 265

6-3 Treatment Technologies for Water Reclamation Applications 265
Removal of dissolved organic matter, suspended solids, and nutrients by secondary treatment 268
Removal of residual particulate matter in secondary effluent 269
Removal of residual dissolved constituents 271
Removal of trace constituents 271
Disinfection processes 271

6-4 Important Factors in the Selection of Technologies for Water Reuse 272
Multiple water reuse applications 273
Need to remove trace constituents 273
Need to conduct pilot-scale testing 276
Process reliability 276
Standby and redundancy considerations 279
Infrastructure needs for water reuse applications 280

6-5	Impact of Treatment Plant Location on Water Reuse 281	**7-5**	Membrane Bioreactor Processes for Secondary Treatment 328

6-5 Impact of Treatment Plant Location on Water Reuse   281
  Centralized treatment plants   282
  Satellite treatment facilities   282
  Decentralized treatment facilities   283

6-6 The Future of Water Reclamation Technologies and Treatment Systems   286
  Implication of trace constituents on future water reuse   287
  New regulations   287
  Retrofitting existing treatment plants   288
  New treatment plants   289
  Satellite treatment systems   289
  Decentralized treatment facilities and systems   289
  New infrastructure concepts and designs   290
  Research needs   291
  Problems and Discussion Topics   292
  References   293

7 Removal of Constituents by Secondary Treatment   295

  Working Terminology   296

7-1 Constituents in Untreated Wastewater   299
  Constituents of concern   299
  Typical constituent concentration values   299
  Variability of mass loadings   301

7-2 Technologies for Water Reuse Applications   304

7-3 Nonmembrane Processes for Secondary Treatment   307
  Suitability for reclaimed water applications   307
  Process descriptions   308
  Process performance expectations   310
  Importance of secondary sedimentation tank design   318

7-4 Nonmembrane Processes for the Control and Removal of Nutrients in Secondary Treatment   320
  Nitrogen control   320
  Nitrogen removal   321
  Phosphorus removal   324
  Process performance expectations   328

7-5 Membrane Bioreactor Processes for Secondary Treatment   328
  Description of membrane bioreactors   330
  Suitability of MBRs for reclaimed water applications   331
  Types of membrane bioreactor systems   332
  Principal proprietary submerged membrane systems   333
  Other membrane systems   338
  Process performance expectations   340

7-6 Analysis and Design of Membrane Bioreactor Processes   340
  Process analysis   340
  Design considerations   353
  Nutrient removal   358
  Biosolids processing   361

7-7 Issues in the Selection of Secondary Treatment Processes   361
  Expansion of an existing plant vs. construction of a new plant   362
  Final use of effluent   362
  Comparative performance of treatment processes   362
  Pilot-scale studies   362
  Type of disinfection process   362
  Future water quality requirements   363
  Energy considerations   363
  Site constraints   364
  Economic and other considerations   368
  Problems and Discussion Topics   368
  References   371

8 Removal of Residual Particulate Matter   373

  Working Terminology   374

8-1 Characteristics of Residual Suspended Particulate Matter from Secondary Treatment Processes   375
  Residual constituents and properties of concern   375
  Removal of residual particles from secondary treatment processes   385

8-2 Technologies for the Removal of Residual Suspended Particulate Matter   388

Technologies for reclaimed water applications 388
Process flow diagrams 390
Process performance expectations 390
Suitability for reclaimed water applications 392

8-3 Depth Filtration 392
Available filtration technologies 392
Performance of depth filters 398
Design considerations 407
Pilot-scale studies 415
Operational issues 417

8-4 Surface Filtration 417
Available filtration technologies 419
Performance of surface filters 422
Design considerations 423
Pilot-scale studies 425

8-5 Membrane Filtration 425
Membrane terminology, types, classification, and flow patterns 426
Microfiltration and ultrafiltration 430
Process analysis for MF and UF membranes 435
Operating characteristics and strategies for MF and UF membranes 436
Membrane performance 436
Design considerations 441
Pilot-scale studies 441
Operational issues 443

8-6 Dissolved Air Flotation 445
Process description 445
Performance of DAF process 448
Design considerations 448
Operating considerations 453
Pilot-scale studies 453

8-7 Issues in the Selection of Technologies for the Removal of Residual Particulate Matter 454
Final use of effluent 454
Comparative performance of technologies 455
Results of pilot-scale studies 455
Type of disinfection process 455
Future water quality requirements 455
Energy considerations 455
Site constraints 455
Economic considerations 455

Problems and Discussion Topics 456
References 459

9 Removal of Dissolved Constituents with Membranes 461

Working Terminology 462

9-1 Introduction to Technologies Used for the Removal of Dissolved Constituents 463
Membrane separation 463
Definition of osmotic pressure 463
Nanofiltration and reverse osmosis 465
Electrodialysis 466
Typical process applications and flow diagrams 467

9-2 Nanofiltration 467
Types of membranes used in nanofiltration 468
Application of nanofiltration 471
Performance expectations 471

9-3 Reverse Osmosis 473
Types of membranes used in reverse osmosis 473
Application of reverse osmosis 474
Performance expectations 474

9-4 Design and Operational Considerations for Nanofiltration and Reverse Osmosis Systems 475
Feedwater considerations 475
Pretreatment 477
Treatability testing 479
Membrane flux and area requirements 482
Membrane fouling 487
Control of membrane fouling 490
Process operating parameters 490
Posttreatment 492

9-5 Pilot-Plant Studies for Nanofiltration and Reverse Osmosis 499

9-6 Electrodialysis 501
Description of the electrodialysis process 501
Electrodialysis reversal 502
Power consumption 503
Design and operating considerations 506
Membrane and electrode life 507
Advantages and disadvantages of electrodialysis versus reverse osmosis 508

9-7 Management of Membrane Waste Streams   509
Membrane concentrate issues   509
Thickening and drying of waste streams   511
Ultimate disposal methods for membrane waste streams   515
Problems and Discussion Topics   519
References   522

10 Removal of Residual Trace Constituents   525

Working Terminology   526

10-1 Introduction to Technologies Used for the Removal of Trace Constituents   528
Separation processes based on mass transfer   528
Chemical and biological transformation processes   531

10-2 Adsorption   532
Applications for adsorption   532
Types of adsorbents   533
Basic considerations for adsorption processes   536
Adsorption process limitations   551

10-3 Ion Exchange   551
Applications for ion exchange   552
Ion exchange materials   554
Basic considerations for ion exchange processes   555
Ion exchange process limitations   559

10-4 Distillation   560
Applications for distillation   560
Distillation processes   560
Basic considerations for distillation processes   562
Distillation process limitations   563

10-5 Chemical Oxidation   563
Applications for conventional chemical oxidation   563
Oxidants used in chemical oxidation processes   563
Basic considerations for chemical oxidation processes   566
Chemical oxidation process limitations   567

10-6 Advanced Oxidation   567
Applications for advanced oxidation   568
Processes for advanced oxidation   569
Basic considerations for advanced oxidation processes   574
Advanced oxidation process limitations   577

10-7 Photolysis   578
Applications for photolysis   578
Photolysis processes   579
Basic considerations for photolysis processes   579
Photolysis process limitations   586

10-8 Advanced Biological Transformations   586
Basic considerations for advanced biological treatment processes   587
Advanced biological treatment processes   588
Limitations of advanced biological transformation processes   590
Problems and Discussion Topics   591
References   594

11 Disinfection Processes for Water Reuse Applications   599

Working Terminology   600

11-1 Disinfection Technologies Used for Water Reclamation   602
Characteristics for an ideal disinfectant   602
Disinfection agents and methods in water reclamation   602
Mechanisms used to explain action of disinfectants   604
Comparison of reclaimed water disinfectants   605

11-2 Practical Considerations and Issues for Disinfection   606
Physical facilities used for disinfection   606
Factors affecting performance   609
Development of the $C_R t$ Concept for predicting disinfection performance   616
Application of the $C_R t$ concept for reclaimed water disinfection   617
Performance comparison of disinfection technologies   618
Advantages and disadvantages of alternative disinfection technologies   618

11-3 Disinfection with Chlorine 622
  Characteristics of chlorine compounds 622
  Chemistry of chlorine compounds 624
  Breakpoint reaction with chlorine 626
  Measurement and reporting of disinfection process variables 631
  Germicidal efficiency of chlorine and various chlorine compounds in clean water 631
  Form of residual chlorine and contact time 631
  Factors that affect disinfection of reclaimed water with chlorine 633
  Chemical characteristics of the reclaimed water 635
  Modeling the chlorine disinfection process 639
  Required chlorine dosages for disinfection 641
  Assessing the hydraulic performance of chlorine contact basins 644
  Formation and control of disinfection byproducts 650
  Environmental impacts 654

11-4 Disinfection with Chlorine Dioxide 654
  Characteristics of chlorine dioxide 655
  Chlorine dioxide chemistry 655
  Effectiveness of chlorine dioxide as a disinfectant 655
  Byproduct formation and control 656
  Environmental impacts 657

11-5 Dechlorination 657
  Dechlorination of reclaimed water treated with chlorine and chlorine compounds 657
  Dechlorination of chlorine dioxide with sulfur dioxide 660

11-6 Disinfection with Ozone 660
  Ozone properties 660
  Ozone chemistry 661
  Ozone disinfection systems components 662
  Effectiveness of ozone as a disinfectant 666
  Modeling the ozone disinfection process 666
  Required ozone dosages for disinfection 669
  Byproduct formation and control 670
  Environmental impacts of using ozone 671
  Other benefits of using ozone 671

11-7 Other Chemical Disinfection Methods 671
  Peracetic acid 671
  Combined chemical disinfection processes 672

11-8 Disinfection with Ultraviolet Radiation 674
  Source of UV radiation 674
  Types of UV lamps 674
  UV disinfection system configurations 678
  Mechanism of inactivation by UV irradiation 682
  Factors affecting germicidal effectiveness of UV irradiation 684
  Modeling the UV disinfection process 690
  Estimating UV dose 691
  Ultraviolet disinfection guidelines 700
  Analysis of a UV disinfection system 708
  Operational issues with UV disinfection systems 708
  Environmental impacts of UV irradiation 711

  Problems and Discussion Topics 712
  References 718

12 Satellite Treatment Systems for Water Reuse Applications 725

  Working Terminology 726

12-1 Introduction to Satellite Systems 727
  Types of satellite treatment systems 728
  Important factors in selecting the use of satellite systems 730

12-2 Planning Considerations for Satellite Systems 730
  Identification of near-term and future reclaimed water needs 730
  Integration with existing facilities 731
  Siting considerations 731
  Public perception, legal aspects, and institutional issues 734
  Economic considerations 735
  Environmental considerations 735
  Governing regulations 735

12-3 Satellite Systems for Nonagricultural Water Reuse Applications 735

Reuse in buildings 736
Landscape irrigation 736
Lakes and recreational enhancement 736
Groundwater recharge 736
Industrial applications 737

12-4 Collection System Requirements 738
Interception type satellite system 738
Extraction type satellite system 738
Upstream type satellite system 739

12-5 Wastewater Characteristics 739
Interception type satellite system 740
Extraction type satellite system 740
Upstream type satellite system 741

12-6 Infrastructure Facilities for Satellite Treatment Systems 741
Diversion and junction structures 741
Flow equalization and storage 744
Pumping, transmission, and distribution of reclaimed water 745

12-7 Treatment Technologies for Satellite Systems 745
Conventional technologies 745
Membrane bioreactors 746
Sequencing batch reactor 746

12-8 Integration with Existing Facilities 748

12-9 Case Study 1: Solaire Building New York, New York 751
Setting 751
Water management issues 751
Implementation 752
Lessons learned 753

12-10 Case Study 2: Water Reclamation and Reuse in Tokyo, Japan 755
Setting 755
Water management issues 755
Implementation 756
Lessons learned 758

12-11 Case Study 3: City of Upland, California 760
Setting 760
Water management issues 760
Implementation 760
Lessons learned 761

Problems and Discussion Topics 761
References 762

13 Onsite and Decentralized Systems for Water Reuse 763

Working Terminology 764

13-1 Introduction to Decentralized Systems 766
Definition of decentralized systems 766
Importance of decentralized systems 767
Integration with centralized systems 770

13-2 Types of Decentralized Systems 770
Individual onsite systems 771
Cluster systems 771
Housing development and small community systems 772

13-3 Wastewater Flowrates and Characteristics 774
Wastewater flowrates 774
Wastewater constituent concentrations 778

13-4 Treatment Technologies 785
Source separating systems 786
In-building pretreatment 788
Primary treatment 788
Secondary treatment 792
Nutrient removal 797
Disinfection processes 802
Performance 804
Reliability 804
Maintenance needs 804

13-5 Technologies for Housing Developments and Small Community Systems 806
Collection systems 807
Treatment technologies 815

13-6 Decentralized Water Reuse Opportunities 816
Landscape irrigation systems 816
Irrigation with greywater 818
Groundwater recharge 818
Self-contained recycle systems 821
Habitat development 821

13-7 Management and Monitoring of Decentralized Systems 821
Types of management structures 821
Monitoring and control equipment 824

Problems and Discussion Topics 826
References 827

14 Distribution and Storage of Reclaimed Water 829

Working Terminology 830

14-1 Issues in the Planning Process 831
Type, size, and location of facilities 831
Individual reclaimed water system versus dual distribution system 832
Public concerns and involvement 833

14-2 Planning and Conceptual Design of Distribution and Storage Facilities 833
Location of reclaimed water supply, major users, and demands 834
Quantities and pressure requirements for major demands 834
Distribution system network 836
Facility design criteria 841
Distribution system analysis 845
Optimization of distribution system 847

14-3 Pipeline Design 856
Location of reclaimed water pipelines 856
Design criteria for reclaimed water pipelines 858
Pipeline materials 858
Joints and connections 860
Corrosion protection 861
Pipe identification 862
Distribution system valves 863
Distribution system appurtenances 863

14-4 Pumping Systems 866
Pumping station location and site layout 866
Pump types 867
Pumping station performance 870
Constant versus variable speed operation 870
Valves 871
Equipment and piping layout 872
Emergency power 872
Effect of pump operating schedule on system design 875

14-5 Design of Reclaimed Water Storage Facilities 877
Location of reclaimed water reservoirs 878
Facility and site layout for reservoirs, piping, and appurtenances 879
Materials of construction 881
Protective coatings—interior and exterior 881

14-6 Operation and Maintenance of Distribution Facilities 882
Pipelines 883
Pumping stations 884

14-7 Water Quality Management Issues in Reclaimed Water Distribution and Storage 884
Water quality issues 885
Impact of water quality issues 887
The effect of storage on water quality changes 887
Strategies for managing water quality in open and enclosed reservoirs 889

Problems and Discussion Topics 892
References 898

15 Dual Plumbing Systems 901

Working Terminology 902

15-1 Overview of Dual Plumbing Systems 902
Rationale for dual plumbing systems 902
Applications for dual plumbing systems 903

15-2 Planning Considerations for Dual Plumbing Systems 907
Applications for dual plumbing systems 907
Regulations and codes governing dual plumbing systems 908
Applicable health and safety regulations 908

15-3 Design Considerations for Dual Distribution Systems 908
Plumbing codes 908
Safeguards 908

15-4 Inspection and Operating Considerations 913

15-5 Case Study: Irvine Ranch Water District, Orange County, California 915
Setting 915
Water management issues 915
Implementation 916
Operational issues 918
Lessons learned 919

15-6 Case Study: Rouse Hill Recycled Water Area Project (Australia) 919

Setting 919
Water management issues 920
Implementation 920
Lessons learned 920

15-7 Case Study: Serrano, California 921
Setting 922
Water management issues 922
Implementation 923
Lessons learned 925
Problems and Discussion Topics 925
References 926

Part 4 Water Reuse Applications 927

16 Water Reuse Applications: An Overview 929

Working Terminology 930

16-1 Water Reuse Applications 930
Agricultural irrigation 931
Landscape irrigation 931
Industrial uses 931
Urban nonirrigation uses 933
Environmental and recreational uses 933
Groundwater recharge 933
Indirect potable reuse through surface water augmentation 933
Direct potable reuse 934
Water reuse applications in other parts of the world 934

16-2 Issues in Water Reuse 934
Resource sustainability 934
Water resource opportunities 935
Reliability of water supply 935
Economic considerations 935
Public policy 935
Regulations 936
Issues and constraints for specific applications 937

16-3 Important Factors in the Selection of Water Reuse Applications 937
Water quality considerations 937
Types of technology 939

Matching supply and demand 939
Infrastructure requirements 939
Economic feasibility (affordability) 940
Environmental considerations 941

16-4 Future Trends in Water Reuse Applications 941
Changes in regulations 942
Water supply augmentation 942
Decentralized and satellite systems 942
New treatment technologies 942
Issues associated with potable reuse 944
Problems and Discussion Topics 944
References 944

17 Agricultural Uses of Reclaimed Water 947

Working Terminology 948

17-1 Agricultural Irrigation with Reclaimed Water: An Overview 949
Reclaimed water irrigation for agriculture in the United States 950
Reclaimed water irrigation for agriculture in the world 952
Regulations and guidelines related to agricultural irrigation with reclaimed water 953

17-2 Agronomics and Water Quality Considerations 954
Soil characteristics 955
Suspended solids 958
Salinity, sodicity, and specific ion toxicity 959
Trace elements and nutrients 966
Crop selection 971

17-3 Elements for the Design of Reclaimed Water Irrigation Systems 971
Water reclamation and reclaimed water quantity and quality 977
Selection of the type of irrigation system 977
Leaching requirements 986
Estimation of water application rate 989
Field area requirements 997
Drainage systems 998
Drainage water management and disposal 1003
Storage system 1003
Irrigation scheduling 1008

17-4 Operation and Maintenance of Reclaimed Water Irrigation Systems 1008
Demand-supply management 1009
Nutrient management 1009
Public health protection 1011
Effects of reclaimed water irrigation on soils and crops 1011
Monitoring requirements 1014

17-5 Case Study: Monterey Wastewater Reclamation Study for Agriculture—Monterey, California 1015
Setting 1016
Water management issues 1016
Implementation 1016
Study results 1017
Subsequent projects 1021
Recycled water food safety study 1021
Lessons learned 1021

17-6 Case Study: Water Conserv II, Florida 1022
Setting 1023
Water management issues 1023
Implementation 1023
Importance of Water Conserv II 1027
Lessons learned 1027

17-7 Case Study: The Virginia Pipeline Scheme, South Australia—Seasonal ASR of Reclaimed Water for irrigation 1028
Setting 1028
Water management issues 1029
Regulatory requirements 1029
Technology issues 1029
Implementation 1030
Performance and operations 1032
Lessons learned 1035

Problems and Discussion Topics 1035
References 1038

18 Landscape Irrigation with Reclaimed Water 1043

Working Terminology 1044

18-1 Landscape Irrigation: An Overview 1045
Definition of landscape irrigation 1045
Reclaimed water use for landscape irrigation in the United States 1046

18-2 Design and Operational Considerations for Reclaimed Water Landscape Irrigation Systems 1047
Water quality requirements 1047
Landscape plant selection 1050
Irrigation systems 1054
Estimation of water needs 1054
Application rate and irrigation schedule 1065
Management of demand-supply balance 1065
Operation and maintenance issues 1066

18-3 Golf Course Irrigation with Reclaimed Water 1070
Water quality and agronomic considerations 1070
Reclaimed water supply and storage 1072
Distribution system design considerations 1075
Leaching, drainage, and runoff 1076
Other considerations 1076

18-4 Irrigation of Public Areas with Reclaimed Water 1076
Irrigation of public areas 1078
Reclaimed water treatment and water quality 1079
Conveyance and distribution system 1079
Aesthetics and public acceptance 1079
Operation and maintenance issues 1080

18-5 Residential Landscape Irrigation with Reclaimed Water 1080
Residential landscape irrigation systems 1080
Reclaimed water treatment and water quality 1081
Conveyance and distribution system 1081
Operation and maintenance issues 1082

18-6 Landscape Irrigation with Decentralized Treatment and Subsurface Irrigation Systems 1082
Subsurface drip irrigation for individual on-site and cluster systems 1082
Irrigation for residential areas 1086

18-7 Case Study: Landscape Irrigation in St. Petersburg, Florida 1086
Setting 1087
Water management issues 1087
Implementation 1087

        Project Greenleaf and resource management   *1089*
        Landscape irrigation in the city of St. Petersburg   *1091*
        Lessons learned   *1093*

18-8   Case Study: Residential Irrigation in El Dorado Hills, California   1093
        Water management issues   *1094*
        Implementation   *1094*
        Education program   *1096*
        Lessons learned   *1096*
        Problems and Discussion Topics   1097
        References   1099

19 Industrial Uses of Reclaimed Water   1103

        Working Terminology   1104

19-1   Industrial Uses of Reclaimed Water: An Overview   1105
        Status of water use for industrial applications in the United States   *1105*
        Water management in industries   *1107*
        Factors affecting the use of reclaimed water for industrial applications   *1108*

19-2   Water Quality Issues for Industrial Uses of Reclaimed Water   1109
        General water quality considerations   *1110*
        Corrosion issues   *1110*
        Indexes for assessing effects of reclaimed water quality on reuse systems   *1115*
        Corrosion management options   *1126*
        Scaling issues   *1127*
        Accumulation of dissolved constituents   *1129*

19-3   Cooling Water Systems   1132
        System description   *1132*
        Water quality considerations   *1132*
        Design and operational considerations   *1135*
        Management issues   *1138*

19-4   Other Industrial Water Reuse Applications   1141
        Boilers   *1141*
        Pulp and paper industry   *1147*
        Textile industry   *1150*
        Other industrial applications   *1154*

19-5   Case Study: Cooling Tower at a Thermal Power Generation Plant, Denver, Colorado   1155
        Setting   *1155*
        Water management issues   *1156*
        Implementation   *1158*
        Lessons learned   *1158*

19-6   Case Study: Industrial Uses of Reclaimed Water in West Basin Municipal Water District, California   1158
        Setting   *1158*
        Water management issues   *1158*
        Implementation   *1159*
        Lessons learned   *1161*
        Problems and Discussion Topics   1161
        References   1165

20 Urban Nonirrigation Water Reuse Applications   1169

        Working Terminology   1170

20-1   Urban Water Use and Water Reuse Applications: An Overview   1171
        Domestic potable water use in the United States   *1171*
        Commercial water use in the United States   *1172*
        Urban nonirrigation water reuse in the United States   *1172*
        Urban nonirrigation water reuse in other countries   *1172*

20-2   Factors Affecting the Use of Reclaimed Water for Urban Nonirrigation Reuse Applications   1175
        Infrastructure issues   *1175*
        Water quality and supply issues   *1176*
        Acceptance issues   *1179*

20-3   Air Conditioning   1179
        Description of air conditioning systems   *1179*
        Utilizing reclaimed water for air conditioning systems   *1181*
        Water quality considerations   *1181*
        Management issues   *1183*

20-4   Fire Protection   1183
        Types of applications   *1186*
        Water quality considerations   *1187*
        Implementation issues   *1187*
        Management issues   *1188*

20-5	Toilet and Urinal Flushing 1188		Operations and maintenance 1230
	Types of applications 1188		*Other considerations 1230*
	Water quality considerations 1188	**21-5**	Other Uses 1231
	Implementation issues 1192		*Snowmaking 1231*
	Satellite and decentralized systems 1193		*Animal viewing parks 1231*
	Management issues 1193	**21-6**	Case Study: Arcata, California 1231
20-6	Commercial Applications 1195		*Setting 1232*
	Car and other vehicle washing 1195		*Water management issues 1232*
	Laundries 1196		*Implementation 1232*
20-7	Public Water Features 1197		*Lessons learned 1233*
	Fountains and waterfalls 1197	**21-7**	Case Study: San Luis Obispo, California 1234
	Reflecting pools 1197		*Setting 1234*
	Ponds and lakes in public parks 1198		*Water management issues 1235*
20-8	Road Care and Maintenance 1198		*Implementation 1235*
	Dust control and street cleaning 1199		*Lessons learned 1238*
	Snow melting 1199	**21-8**	Case Study: Santee Lakes, San Diego, California 1238
	Problems and Discussion Topics 1200		*Setting 1239*
	References 1201		*Water management issues 1239*
			Implementation 1239
			Lessons learned 1241

21 Environmental and Recreational Uses of Reclaimed Water 1203

Working Terminology 1204

21-1 Overview of Environmental and Recreational Uses of Reclaimed Water 1205
Types of environmental and recreational uses 1206
Important factors influencing environmental and recreational uses of reclaimed water 1207

21-2 Wetlands 1210
Types of wetlands 1210
Development of wetlands with reclaimed water 1213
Water quality considerations 1216
Operations and maintenance 1216

21-3 Stream Flow Augmentation 1222
Aquatic and riparian habitat enhancement with reclaimed water 1222
Recreational uses of streams augmented with reclaimed water 1224
Reclaimed water quality requirements 1224
Stream flow requirements 1226
Operations and maintenance 1226

21-4 Ponds and Lakes 1228
Water quality requirements 1228

Problems and Discussion Topics 1242
References 1242

22 Groundwater Recharge with Reclaimed Water 1245

Working Terminology 1246

22-1 Planned Groundwater Recharge with Reclaimed Water 1248
Advantages of subsurface storage 1248
Types of groundwater recharge 1249
Components of a groundwater recharge system 1250
Technologies for groundwater recharge 1251
Selection of recharge system 1253
Recovery of recharge water 1254

22-2 Water Quality Requirements 1255
Water quality challenges for groundwater recharge 1255
Degree of pretreatment required 1255

22-3 Recharge Using Surface Spreading Basins 1256
Description 1256
Pretreatment needs 1257
Hydraulic analysis 1259

       Operation and maintenance issues   *1268*
       Performance of recharge basins   *1271*
       Pathogens   *1279*
       Examples of full-scale surface spreading facilities   *1280*

22-4 Recharge Using Vadose Zone Injection Wells   1282
       Description   *1282*
       Pretreatment needs   *1283*
       Hydraulic analysis   *1284*
       Operation and maintenance issues   *1285*
       Performance of vadose zone injection wells   *1286*
       Examples of operational full-scale vadose zone injection facilities   *1286*

22-5 Recharge Using Direct Injection Wells   1287
       Description   *1287*
       Pretreatment needs   *1288*
       Hydraulic analysis   *1288*
       Operation and maintenance issues   *1290*
       Performance of direct injection wells   *1291*
       Examples of full-scale direct aquifer injection facilities   *1292*

22-6 Other Methods Used for Groundwater Recharge   1293
       Aquifer storage and recovery (ASR)   *1293*
       Riverbank and dune filtration   *1294*
       Enhanced river recharge   *1295*
       Groundwater recharge using subsurface facilities   *1296*

22-7 Case Study: Orange County Water District Groundwater Replenishment System   1296
       Setting   *1297*
       The GWR system   *1297*
       Implementation   *1297*
       Lessons learned   *1298*
       Problems and Discussion Topics   1299
       References   1300

23 Indirect Potable Reuse through Surface Water Augmentation   1303

       Working Terminology   1304
23-1 Overview of Indirect Potable Reuse   1305
       De facto indirect potable reuse   *1305*
       Strategies for indirect potable reuse through surface-water augmentation   *1307*
       Public acceptance   *1308*

23-2 Health and Risk Considerations   1308
       Pathogen and trace constituents   *1308*
       System reliability   *1309*
       Use of multiple barriers   *1309*

23-3 Planning for Indirect Potable Reuse   1309
       Characteristics of the watershed   *1310*
       Quantity of reclaimed water to be blended   *1311*
       Water and wastewater treatment requirements   *1312*
       Institutional considerations   *1312*
       Cost considerations   *1313*

23-4 Technical Considerations for Surface-Water Augmentation in Lakes and Reservoirs   1314
       Characteristics of water supply reservoirs   *1314*
       Modeling of lakes and reservoirs   *1319*
       Strategies for augmenting water supply reservoirs   *1320*

23-5 Case Study: Implementing Indirect Potable Reuse at the Upper Occoquan Sewage Authority   1323
       Setting   *1323*
       Water management issues   *1323*
       Description of treatment components   *1323*
       Future treatment process directions   *1326*
       Water quality of the Occoquan Reservoir   *1327*
       Water treatment   *1328*
       Lessons learned   *1328*

23-6 Case Study: City of San Diego Water Repurification Project and Water Reuse Study 2005   1329
       Setting   *1330*
       Water management issues   *1330*
       Wastewater treatment mandates   *1330*
       Water Repurification Project   *1331*
       2000 Updated Water Reclamation Master Plan   *1332*
       City of San Diego Water Reuse Study 2005   *1332*
       Lessons learned   *1334*

23-7 Case Study: Singapore's NEWater for Indirect Potable Reuse   1334
       Setting   *1335*
       Water management issues   *1335*
       NEWater Factory and NEWater   *1335*

Implementation 1335
NEWater demonstration plant performance 1336
Project milestones 1336
Lessons learned 1337

23-8 Observations on Indirect Potable Reuse 1340
Problems and Discussion Topics 1341
References 1342

24 Direct Potable Reuse of Reclaimed Water 1345

Working Terminology 1346

24-1 Issues in Direct Potable Reuse 1346
Public perception 1347
Health risk concerns 1347
Technological capabilities 1347
Cost considerations 1348

24-2 Case Study: Emergency Potable Reuse in Chanute, Kansas 1348
Setting 1348
Water management issues 1349
Implementation 1349
Efficiency of sewage treatment and the overall treatment process 1349
Lessons learned 1351
Importance of the Chanute experience 1352

24-3 Case Study: Direct Potable Reuse in Windhoek, Namibia 1352
Setting 1353
Water management issues 1353
Implementation 1354
Lessons learned 1359

24-4 Case Study: Direct Potable Reuse Demonstration Project in Denver, Colorado 1361
Setting 1362
Water management issues 1362
Treatment technologies 1362
Water quality testing and studies 1364
Animal health effects testing 1371
Cost estimates on the potable reuse advanced treatment plant 1372
Public information program 1373
Lessons learned 1374

24-5 Observations on Direct Potable Reuse 1375
Problems and Discussion Topics 1376
References 1376

Part 5 Implementing Water Reuse 1379

25 Planning for Water Reclamation and Reuse 1381

Working Terminology 1382

25-1 Integrated Water Resources Planning 1384
Integrated water resources planning process 1385
Clarifying the problem 1386
Formulating objectives 1386
Gathering background information 1386
Identifying project alternatives 1388
Evaluating and ranking alternatives 1389
Developing implementation plans 1389

25-2 Engineering Issues in Water Reclamation and Reuse Planning 1392

25-3 Environmental Assessment and Public Participation 1392
Environmental assessment 1393
Public participation and outreach 1393

25-4 Legal and Institutional Aspects of Water Reuse 1393
Water rights law 1393
Water rights and water reuse 1395
Policies and regulations 1397
Institutional coordination 1397

25-5 Case Study: Institutional Arrangements at the Walnut Valley Water District, California 1397
Water management issues 1397
Lessons learned 1398

25-6 Reclaimed Water Market Assessment 1399
Steps in data collection and analysis 1399
Comparison of water sources 1399
Comparison with costs and revenues 1401
Market assurances 1402

25-7 Factors Affecting Monetary Evaluation of Water Reclamation and Reuse 1406
Common weaknesses in water reclamation and reuse planning 1407
Perspectives in project analysis 1408
Planning and design time horizons 1408
Time value of money 1409
Inflation and cost indices 1409

25-8 Economic Analysis for Water Reuse   1411
   Comparison of alternatives by present worth analysis   1412
   Measurement of costs and inflation   1412
   Measurement of benefits   1412
   Basic assumptions of economic analyses   1414
   Replacement costs and salvage values   1415
   Computation of economic cost   1417
   Project optimization   1420
   Influence of subsidies   1421

25-9 Financial Analysis   1422
   Construction financial plans and revenue programs   1422
   Cost allocation   1423
   Influence on freshwater rates   1423
   Other financial analysis considerations   1423
   Sources of revenue and pricing of reclaimed water   1424
   Financial feasibility analysis   1425
   Sensitivity analysis and conservative assumptions   1429
   Problems and Discussion Topics   1430
   References   1432

26 Public Participation and Implementation Issues   1435

   Working Terminology   1436

26-1 How Is Water Reuse Perceived?   1436
   Public attitude about water reuse   1436
   Public beliefs about water reuse options   1440

26-2 Public Perspectives on Water Reuse   1440
   Water quality and public health   1441
   Economics   1441
   Water supply and growth   1441
   Environmental justice/equity issues   1441
   The "Yuck" factor   1442
   Other issues   1442

26-3 Public Participation and Outreach   1443
   Why involve the public?   1443
   Legal mandates for public involvement   1443
   Defining the "public"   1444
   Approaches to public involvement   1444
   Techniques for public participation and outreach   1446
   Some pitfalls in types of public involvement   1448

26-4 Case Study: Difficulties Encountered in Redwood City's Landscape Irrigation Project   1450
   Setting   1450
   Water management issues   1450
   Water reclamation project planned   1450
   Lessons learned   1452

26-5 Case Study: Water Reclamation and Reuse in the City of St. Petersburg, Florida   1451
   Setting   1453
   Water and wastewater management issues   1453
   Development of reclaimed water system   1455
   Current status of water reclamation and reuse   1456
   Lessons learned   1456
   Access to city's proactive water reclamation and reuse information   1459

26-6 Observations on Water Reclamation and Reuse   1459
   Problems and Discussion Topics   1459
   References   1460

Appendixes

A Conversion Factors   1463
B Physical Properties of Selected Gases and the Composition of Air   1471
C Physical Properties of Water   1475
D Statistical Analysis of Data   1479
E Review of Water Reclamation Activities in the United States and in Selected Countries   1485
F Evolution of Nonpotable Reuse Criteria and Groundwater Recharge Regulations in California   1509
G Values of the Hantush Function $F(\alpha, \beta)$ and the Well Function $W(u)$ 1523
H Interest Factors and Their Use   1525

Indexes

   Name Index   1529
   Subject Index   1541

Preface

With many communities approaching the limits of their available water supplies, water reclamation and reuse has become a logical option for conserving and extending available water supply by potentially (1) substituting reclaimed water for applications that do not require drinking (potable) water, (2) augmenting existing water sources and providing an additional source of water supply to assist in meeting both present and future water needs, (3) protecting aquatic ecosystems by decreasing the diversion of freshwater as well as reducing the quantity of nutrients and other toxic contaminants entering waterways, (4) postponing and reducing the need for water control structures, and (5) complying with environmental regulations by better managing water consumption and wastewater discharges. The increasing importance and recognition of water reclamation and reuse have led to the need for specialized instruction of engineering and science students in their undergraduate and graduate levels, as well as practicing engineers and scientists, and a technical reference for project managers and government officials. Aside from the need for a textbook on water reuse applications and the technologies used to treat and distribute reclaimed water, there is also the need to address the special considerations of public health, project planning and economics, public acceptance, and the diverse uses of reclaimed water in society.

ORGANIZATION OF THE TEXTBOOK AND CONTENT

This textbook, *Water Reuse: Issues, Technologies, and Applications*, is an endeavor by the authors to assemble, analyze, and synthesize a vast amount of information on water reclamation and reuse. To deal with the amount of available material, the book is organized into five parts, each dealing with a coherent body of information which is described below.

Part 1: Water Reuse: An Introduction

It is important to understand the concept of sustainable water resources management as a foundation for water reclamation and reuse. Thus, in Part 1 of this textbook, current and potential future water shortages, principles of sustainable water resources management, and the important role of water reclamation and reuse are introduced briefly. The past and current practices of water reclamation and reuse are presented, which also serve as an introduction to the subsequent engineering and water reuse applications chapters.

Part 2: Health and Environmental Concerns in Water Reuse

Health and environmental issues related to water reuse are discussed in three related chapters in Part 2. The characteristics of wastewater are introduced, followed by a discussion of the applicable regulations and their development. Because health risk analysis is an important aspect of water reuse applications, a separate chapter is devoted to this subject including tools and methods used in risk assessment, chemical risk assessment, and microbial risk assessment.

Part 3: Water Technologies and Systems for Water Reclamation and Reuse

The various technologies and systems available for the production and delivery of reclaimed water are the subject of Part 3. Although design values are presented, detailed design is not the focus of these chapters. Rather, the focus is on the dependable performance of the processes and technologies. Detailed discussions are provided with respect to constituents of concern in water reuse applications including particulate matter, dissolved constituents, and pathogenic microorganisms. Another important aspect of water reclamation is related to meeting stringent water quality performance requirements as affected by wastewater variability and process reliability, factors which are emphasized repeatedly throughout this textbook.

Part 4: Water Reuse Applications

Because water quality and infrastructure requirements vary greatly with specific water reuse application, major water reuse applications are discussed in separate chapters in Part 4: nonpotable water reuse applications including agricultural uses, landscape irrigation, industrial uses, environmental and recreational uses, groundwater recharge, and urban nonpotable and commercial uses. Indirect and direct potable reuses are discussed with several notable projects. Groundwater recharge can be considered as a form of indirect potable reuse if the recharged aquifer is interconnected to potable water production wells.

Part 5: Implementing Water Reuse

In the final Part 5 of this textbook, the focus is on planning and implementation for water reuse. Integrated water resources planning, including reclaimed water market assessment, and economic and financial analyses are presented. As technology continues to advance and cost effectiveness and the reliability of water reuse systems becomes more widely recognized, water reclamation and reuse plans and facilities will continue to expand as essential elements in sustainable water resources management. Implementation issues in water reclamation and reuse are discussed including soliciting and responding to community concerns, development of public support through educational programs, and the development of financial instruments.

IMPORTANT FEATURES OF THIS TEXTBOOK

To illustrate the principles, applications, and facilities involved in the field of water reclamation and reuse, more than 350 data and information tables and 80 detailed worked examples, more than 500 illustrations, graphs, diagrams, and photographs are included. To help the readers of this textbook hone their analytical skills and mastery of the material, problems and discussion topics are included at the end of each chapter. Selected references are also provided for each chapter.

The International System (SI) of Units is used in this textbook. The use of SI units is consistent with teaching practice in most universities in the United States and in most countries throughout the world.

To further increase the utility of this textbook, several appendixes have been included. Conversion factors from SI Units to U.S. Customary Units and the reverse are presented in Appendixes A-1 and A-2, respectively. Conversion factors used commonly for the analysis and design of water and wastewater management systems are presented in Appendix A-3. Abbreviations for SI and U.S. Customary Units are presented in Appendixes A-4 and A-5, respectively. Physical characteristics of air and selected gases and water are presented in Appendixes B and C, respectively. Statistical analysis of data with an example is presented in Appendix D.

Milestone water reuse projects and research studies in the United States and a summary of water reclamation and reuse in selected countries of the world are presented in Appendixes E-1 and E-2, respectively. Evolution of nonpotable reuse criteria and groundwater recharge regulations in California is presented in Appendix F. Dimensionless well function W(u) values are presented in Appendix G. Finally, interest factors and their use are presented and illustrated in Appendix H.

With recent Internet developments, it is now possible to view many of the facilities discussed in this textbook through satellite images using one of the many search engines available on the Internet. Where appropriate, global positioning coordinates for water reuse facilities of interest are given to allow viewing of these facilities in their natural setting.

USE OF THIS TEXTBOOK

Enough material is presented in this textbook to support a variety of courses for one or two semesters or three quarters at either the undergraduate or graduate level. The specific topics to be covered will depend on the time available and the course objectives. Three suggested course plans are presented below.

Course Title: Survey of Water Reuse
Setting: 1 semester or 1 quarter, stand-alone class
Target: Upper division or MS, environmental science major
Course Objectives: Introduce important considerations influencing water reuse planning and implementation.
Sample outline:

Course Plan I

Topic	Chapters	Sections
Introduction to water reuse	1, 2	All
Wastewater characteristics	3	3-1, 3-2, 3-5 to 3-8
Regulations for water reuse	4	4-1 to 4-7
Public health protection and risk assessment	5	5-1 to 5-5, 5-9
Introduction to water reclamation technologies	6	All
Infrastructure for water reuse	12, 13, 14, 15	12-1, 12-2, 13-1, 13-2, 13-6, 14-1, 14-2, 15-1, 15-2
Overview of disinfection for reuse applications	11	11-1, 11-2
Introduction to water reuse applications	16	All
Perspectives on water reuse planning	25	25-1 to 25-4
Perspectives on public acceptance	26	26-1 to 26-3

Course Plan II

Course Title: Water Reuse Applications
Setting: 1 semester or 1 quarter class
Target: Upper division or MS, environmental engineering major
Course Objectives: Introduce nonconventional engineering aspects of water reuse including satellite, decentralized, and onsite treatment and reuse systems. An overview of various water reuse applications are introduced.

Sample outline:

Topic	Chapters	Sections
Introduction to water reclamation and reuse	1, 2	1-1 to 1-5, 2-1
Wastewater characteristics	3	3-1, 3-2, 3-5 to 3-8
Water reuse regulations and guidelines	4	4-1 to 4-4, 4-6 to 4-8
Public health protection and risk assessment	5	5-1 to 5-5, 5-8, 5-9
Introduction to water reclamation technologies	6	6-1 to 6-5
Overview of disinfection for reuse applications	11	11-1, 11-2
Introduction to water reuse applications	16	All
Reclaimed water use for irrigation	17, 18	17-1 to 17-3, 18-1 to 18-2, 18-4 to 18-5
Reclaimed water use for industrial processes	19	19-1 to 19-3
Urban nonirrigation, environmental, and recreational uses	20, 21	20-1, 20-2, 21-1
Indirect potable reuse by groundwater and surface water augmentation	22, 23	22-1 to 22-2, 22-7, 23-1 to 23-3, 23-8
Economic and financial analysis	25	25-6 to 25-9
Public participation and public acceptance	25, 26	25-3, 26-1 to 26-3

Course Plan III

Course Title: Advanced Treatment Technologies and Infrastructure for Water Reuse Applications
Setting: 1 semester or 1 quarter class
Target: MS level, environmental engineering major
Course Objectives: Introduce treatment technologies important in water reuse. Introduce reliability issues, concept of probability distribution in assessing disinfection performance, and future directions. The course will be a stand-alone class on advanced treatment, or part of a wastewater treatment class that covers both conventional and advanced technologies emphasizing water reclamation, recycling, and reuse.

This textbook is a useful supplement to a companion textbook, *Wastewater Engineering: Treatment and Reuse*, 4th ed., (Tchobanoglous, G., F.L. Burton, and H.D. Stensel) for the following topics:

Sample outline:

Topic	Chapters	Sections
Introduction to water reuse	1, 2	All
Wastewater characteristics	3	3-1, 3-2, 3-5 to 3-8
Introduction to water reclamation and reuse	6, 16	6-2 to 6-4, 16-1 to 16-4
Membrane filtration, membrane bioreactor	7, 8	7-5, 7-6, 8-5
Nanofiltration, reverse osmosis, and electrodialysis	9	9-1 to 9-4
Adsorption, Advanced oxidation	10	10-1, 10-2, 10-6, 10-7
Disinfection	11	11-1 to 11-3, 11-5, 11-6, 11-8
Alternative systems for water reuse	12, 13	12-1, 12-2, 13-1, 13-2, 13-6,
Infrastructure for water reuse	14, 15	14-1, 14-2, 15-1 to 15-3

Acknowledgments

This textbook, *Water Reuse: Issues, Technologies, and Application* is a tribute to the pioneering planners and engineers who were able to look ahead of their time and push forward the frontiers of water reclamation and reuse from obscure practice to a growing discipline in sustainable water resources management. Based on the widespread acceptance of water reuse and the development of new treatment technologies and applications, it is an appropriate time to produce a comprehensive textbook on the subject. A book of this magnitude, however, could not have been written without the assistance of numerous individuals, some are acknowledged below and others who remain in the background. The authors are particularly grateful to many individuals who contributed the information through personal contacts and the "grey" literature as well as conference and symposium proceedings.

The principal authors were responsible for writing, editing, coordinating, and also responding to reviewer's comments for this textbook. Individuals who contributed specifically to the chapters, listed in chapter order, included Dr. James Crook, environmental engineering consultant, who prepared Chapter 4, Water Reuse Regulations and Guidelines; Dr. Joseph Cotruvo, J. Cotruvo Associates, prepared chemical risk assessment, and Dr. Adam W. Olivieri, Eisenberg Olivieri & Associates and Mr. Jeffery A. Soller, Soller Environmental, prepared microbial risk assessment in Chapter 5, Health Risk Analysis in Water Reuse Applications; Mr. Max E. Burchett of Whitley Burchett & Associates prepared Chapter 14, Storage and Distribution of Reclaimed Water; Professor Audrey D. Levine of the University of South Florida prepared portions of Chapter 19, Industrial and Commercial Uses of Reclaimed Water; Professor Peter Fox of Arizona State University prepared Chapter 22, Groundwater Recharge with Reclaimed Water; Mr. William C. Lauer and Mr. Stephen E. Rogers prepared the publication on the Denver, CO, direct potable reuse demonstration project from which the case study in Chapter 24 was adapted; Mr. Richard A. Mills of California State Water Resources Control Board prepared Chapter 25, Planning for Water Reclamation and Reuse. The help and assistance of Mr. Pier Mantovani in the formative stage of the textbook preparation is also acknowledged. A significant contributor to preparation of this textbook was Ms. Jennifer Cole Aieta of Aieta Cole Enterprises who edited and provided insightful commentary for all of the chapters.

Other individuals who contributed, arranged in alphabetical order, are: Mr. Robert Angelotti, Upper Occoquan Sewage Authority, who reviewed portions of Chapter 23; Dr. Akissa Bahri of the International Water Management Institute in Ghana who reviewed Chapter 17; Mr. Harold Bailey, Padre Dam Municipal Water District, reviewed portions of Chapter 21 and provided several pictures used in Chapters 18 and 21; Drs. Jamie Bartram and Robert Bos, World Health Organization in Switzerland reviewed portions of Chapter 4; Mr. Matt Brooks, Upper Occoquan Sewage Authority, who reviewed portions of Chapter 23; Mr. Bryan Buchanan, City of Roseville, California, provided several photos used in Chapter 18; Ms. Katie DiSimone, City of San Luis Obispo, California, provided information for Chapter 21; Mr. Bruce Durham of Veolia, UK, provided materials for Chapter 24; Mr. Jeffery Goldberg, City of St. Petersburg, reviewed part of Chapter 18; Dr. Stephen Grattan, the University of California, Davis, reviewed Chapter 17; Ms. Lori Kennedy, University of California, Davis, who helped compiling information and drafted portions of Chapters 1, 2, and 25; Mr. Tze Weng Kok, Singapore Public Utilities Board reviewed portions of Chapter 24; Professor Naoyuki Funamizu of Hokkaido University in Japan reviewed portions of Chapter 5 and also provided water reuse pictures; Dr. Josef Lahnsteiner, WABAG in Austria and Dr. Günter G. Lempert, Aqua Services & Engineering (Pty) Ltd. in Namibia reviewed and contributed to Chapter 24; Messrs Gary Myers and John Bowman, Serrano El Dorado Owners' Association, California, provided materials used in Chapters 14 and 18; Professor Slawomir W. Hermanowicz of the University of California, Berkeley reviewed Chapter 1; Professor Audrey D. Levine of the University of South Florida reviewed Chapters 1 and 2; Dr. Loretta Lohman of Colorado State University Cooperative Extension reviewed Chapter 26; Professor Rafael Mujeriego of Technical University of Catalonia in Spain in numerous discussions over many years has contributed valuable insight; Dr. Kumiko Oguma of the University of Tokyo in Japan reviewed portions of Chapter 11 and provided information on microbial regrowth in UV disinfection; Professor Choon Nam Ong of the National University of Singapore reviewed portions of Chapter 24; Professor Gideon Oron, Ben-Gurion University of the Negev in Israel provided irrigation pictures; Mr. Erick Rosenblum, City of San Jose, California, reviewed portions of Chapter 26; Dr. Bahman Sheikh, water reclamation consultant, reviewed Chapters 17, 23, and 24; Messrs. Keiichi Sone and Toshiaki Ueno of the Tokyo Metropolitan Government in Japan provided several water reuse pictures used in Chapters 20 and 21; Professor H. David Stensel of the University of Washington reviewed Chapters 6 and 7; Mr. Tim Sullivan, El Dorado Irrigation District, California, provided information and reviewed portions of Chapter 18; Professor Kenneth Tanji of the University of California, Davis, reviewed Chapter 17; Mr. Thai Pin Tan of Singapore Public Utilities Board reviewed portions of Chapter 24 and provided the information; Professor Hiroaki Tanaka of Kyoto University in Japan reviewed microbial risk assessment sections of Chapter 5; Dr. R. Shane Trussell reviewed and provided valuable comments on membrane bioreactors in Chapter 7; Professor Gedaliah Shelef of the Israel Institute of Technology in Israel through numerous discussions over many years has contributed valuable insight on water reclamation and reuse; Professor Edward D. Schroeder of the University of California, Davis reviewed an early draft of Chapters 1 and 2; Dr. David York of Florida Department of Environmental Protection reviewed portions of Chapter 2. The collective efforts of these individuals were invaluable and greatly appreciated.

The assistance of the staff of Metcalf & Eddy in preparation of this textbook is also acknowledged. The efforts of Mr. James Anderson were especially important in making this book possible and in managing the resources made available by Metcalf & Eddy to the authors. Sadly, Mr. Anderson never saw the published version of this textbook; he passed away as the manuscript was nearing completion. It was his vision that water reclamation and reuse would become an important part of global water resources management. As Metcalf & Eddy's full time author, Dr. Ryujiro Tsuchihashi with

Ms. Kathleen Esposito took on the additional responsibility for the completion of this textbook, Ms. Dorothy Frohlich provided liaison between the authors and reviewers.

Members of the McGraw-Hill staff were also critical to the production of this textbook. Mr. Larry Hager was instrumental in the development of this textbook project. Mr. David Fogarty served as editing supervisor and helped keep all of the loose ends together. Ms. Pamela Pelton served as the production supervisor. Ms. Arushi Chawla served as project manager at International Typesetting and Composition.

Takashi Asano, Davis, CA
Franklin L. Burton, Los Altos, CA
Harold L. Leverenz, Davis, CA
Ryujiro Tsuchihashi, New York, NY
George Tchobanoglous, Davis, CA

Foreword

The history of Metcalf & Eddy textbooks is nearly as long as the firm's. A few years after the firm's founding 100 years ago, Leonard Metcalf and Harrison P. Eddy undertook the preparation of a book bringing together in a form convenient for ready reference the more important principles of theory and rules of practice in sewerage design and operation. The work was published in three volumes in 1914–1915 under the title *American Sewerage Practice*. Due to urging from academicians, a single-volume abridgement for use in engineering schools was published in 1922.

Since that time, Metcalf & Eddy books have undergone numerous revisions and printings. To meet global needs, Metcalf & Eddy textbooks have also been translated into Chinese, Italian, Japanese, Korean, and Spanish. To date, the books have been used in over 300 universities worldwide.

After the fourth edition, entitled *Wastewater Engineering: Treatment and Reuse*, was published in 2003, it became evident that global water issues and needs will make water reuse one of the crucial components of water resources management. For that reason, Metcalf & Eddy concluded that a proper response would be to launch a full textbook on the subject of water reuse. The new textbook, *Water Reuse: Issues, Technologies, and Applications*, is therefore focused on providing education for the building blocks needed to rationally manage our most critical resource—water.

Metcalf & Eddy believes it is essential to encourage wastewater and water supply professionals to elevate water reuse to a strategic level in their planning process so that this limited resource can be efficiently managed and properly preserved. It is envisioned that wastewater professionals will see this textbook as a road map to the implementation of complex water reuse projects. There is no other single source of information available today that combines a discussion of issues in water reuse, policy, up-to-date treatment technologies, real-life practical water reuse applications, as well as planning and implementation considerations. Metcalf & Eddy takes great pride in presenting the first textbook to address water reuse in such a comprehensive fashion. This book combined with the fourth edition represents the most complete treatise on the subject of wastewater today.

Metcalf & Eddy was able to assemble a team of authors that has no equal, consisting of Dr. Takashi Asano, the 2001 Stockholm Water Prize Laureate; Dr. George Tchobanoglous, a member of the National Academy of Engineering; and Franklin Burton, former Vice President and Chief Engineer in the western regional office of Metcalf & Eddy. New additions to the author team are Dr. Harold Leverenz, and Metcalf & Eddy's Dr. Ryujiro Tsuchihashi. Dr. Tsuchihashi also served as a full-time Metcalf & Eddy liaison to our California-based author team.

This textbook could not be completed without the contribution of many individuals, in addition to our principal authors. Other Metcalf & Eddy professionals (unless otherwise noted) who contributed as reviewers of chapters are: William Bent, Bohdan Bodniewicz, Anthony Bouchard (Consoer Townsend Envirodyne Engineers), Gregory Bowden, Timothy Bradley, Pamela Burnett, Theping Chen, William Clunie, Nicholas Cooper, Ashok Dhingra, Bruce Engerholm, Kathleen Esposito, Robert Jarnis, Gary Johnson (Connecticut Department of Environmental Protection), Mark Laquidara, Thomas McMonagle, Chandra Mysore, William Pfrang, Charles Pound, John Reidy, James Schaefer, Robert Scherpf, Betsy Shreve, Beverley Stinson, Brian Stitt, Patrick Toby (Consoer Townsend Envirodyne Engineers), Dennis Tulang, Larry VandeVenter, Stanley Williams (Turner Collie & Braden) and Alan Wong. Kathleen Esposito contributed to the coordination aspects of this project with the assistance of Dorothy Frohlich.

I would also like to acknowledge Mr. Larry Hager of the McGraw-Hill Professional Division who was instrumental in bringing the resources of McGraw-Hill to this project from inception to completion.

The new textbook could not have been launched without the enthusiastic support of Metcalf & Eddy's parent company, AECOM Technology Corporation. I thank Mr. Richard Newman, Chairman of the Board, and Mr. John Dionisio, President and Chief Executive Officer, for their support and vision.

Steve Guttenplan, PE
President

Part 1

WATER REUSE: AN INTRODUCTION

The social, economic, and environmental impacts of past water resources development and inevitable prospects of water scarcity are driving the shift to a new paradigm in water resources management. New approaches now incorporate the principles of sustainability, environmental ethics, and public participation in project development. With many communities approaching the limits of their available water supplies, water reclamation and reuse have become an attractive option for conserving and extending available water supply by potentially (1) substituting reclaimed water for applications that do not require high-quality drinking water, (2) augmenting water sources and providing an alternative source of supply to assist in meeting both present and future water needs, (3) protecting aquatic ecosystems by decreasing the diversion of freshwater, reducing the quantity of nutrients and other toxic contaminants entering waterways, (4) reducing the need for water control structures such as dams and reservoirs, and (5) complying with environmental regulations by better managing water consumption and wastewater discharges.

Water reuse is particularly attractive in the situation where available water supply is already overcommitted and cannot meet expanding water demands in a growing community. Increasingly, society no longer has the luxury of using water only once. Part 1 serves as an introduction to the general subject of water reuse. Current and potential water shortages, principles of sustainable water resources management, and the important role of water reclamation and reuse are discussed in Chap. 1. An overview of existing water reclamation and reuse applications and issues is presented in Chap. 2, which also serves as an introduction to the subsequent chapters.

1 Water Issues: Current Status and the Role of Water Reclamation and Reuse

	WORKING TERMINOLOGY 4
1-1	DEFINITION OF TERMS 6
1-2	PRINCIPLES OF SUSTAINABLE WATER RESOURCES MANAGEMENT 6
	The Principle of Sustainability 7
	Working Definitions of Sustainability 7
	Challenges for Sustainability 7
	Criteria for Sustainable Water Resources Management 7
	Environmental Ethics 13
1-3	CURRENT AND POTENTIAL FUTURE GLOBAL WATER SHORTAGES 15
	Impact of Current and Projected World Population 15
	Potential Global Water Shortages 19
	Water Scarcity 19
	Potential Regional Water Shortages in the Continental United States 20
1-4	THE IMPORTANT ROLE OF WATER RECLAMATION AND REUSE 23
	Types of Water Reuse 24
	Integrated Water Resources Planning 24
	Personnel Needs/Sustainable Engineering 27
	Treatment and Technology Needs 27
	Infrastructure and Planning Issues 28
1-5	WATER RECLAMATION AND REUSE AND ITS FUTURE 30
	Implementation Hurdles 31
	Public Support 31
	Acceptance Varies Depending on Opportunity and Necessity 31
	Public Water Supply from Polluted Water Sources 31
	Advances in Water Reclamation Technologies 31
	Challenges for Water Reclamation and Reuse 32
	PROBLEMS AND DISCUSSION TOPICS 32
	REFERENCES 33

WORKING TERMINOLOGY

Term	Definition
Agricultural water use	Water used for crop production and livestock uses.
Aquifer	Geological formations that contain and transmit groundwater.
Beneficial uses	The many ways water can be used, either directly by people, or for their overall benefit. Examples include municipal water supply, agricultural and industrial applications, navigation, fish and wildlife habitat enhancement, and water contact recreation.
Consumptive use	The part of water withdrawn that is evaporated, transpired, incorporated into products or crops, consumed by humans or livestock, or otherwise removed from the immediate water environment.
Direct potable reuse	See Potable reuse, direct.
Domestic water use	Domestic water use includes water for normal household purposes, such as drinking, food preparation, bathing, washing clothes and dishes, flushing toilets, and watering lawns and gardens.
Ecoefficiency	The efficiency with which environmental resources are used to produce a unit of economic activity.
Environmental ethics	A discipline of ethics that explores moral responsibility in relation to the environment.
Evapotranspiration	A collective term that includes loss of water from the soil by evaporation and by transpiration from plants.
Global hydrologic cycle	The annual accounting of the moisture fluxes over the entire globe in all of their various forms.
Groundwater	The subsurface water that occurs beneath the water table in soils and geologic formations that are fully saturated and supplies wells and springs.
Groundwater recharge	The infiltration or injection of natural waters or reclaimed waters into an aquifer, providing replenishment of the groundwater resource or preventing seawater intrusion.
Indirect potable reuse	See Potable reuse, indirect.
Industrial water use	Water used in industrial operations and processes. The principal industrial water users are thermal and atomic power generation.
Irrigation water use	Artificial application of water on lands to assist in the growing of crops and pastures or to maintain vegetative growth in recreational lands such as parks and golf courses.
Integrated water resources planning	A process that promotes the coordinated development and management of water, land, and related resources to maximize the resultant economic and social welfare in an equitable and sustainable manner.
Landscape irrigation	Irrigation systems for applications such as golf courses, public parks, playgrounds, school yards, and athletic fields.
Municipal water use	The water withdrawals made by the populations of cities, towns, and housing estates, and domestic and public services and enterprises. Also includes water used to provide directly for the needs of urban populations, which consume high-quality water from city water supply systems.
Nonpotable reuse	All water reuse applications that do not involve either indirect or direct potable reuse.
Per capita water use	The average amount of water used per person during a standard time period, usually per day.
Potable water	Water suitable for human consumption without deleterious health risks. The term drinking water is a preferable term better understood by the community at large.

Potable reuse, direct	The introduction of highly treated reclaimed water either directly into the potable water supply distribution system downstream of water a treatment plant, or into the raw water supply immediately upstream of a water treatment plant (see Chap. 24).
Potable reuse, indirect	The planned incorporation of reclaimed water into a raw water supply such as in potable water storage reservoirs or a groundwater aquifer, resulting in mixing and assimilation, thus providing an environmental buffer (see Chaps. 22 and 23).
Public water supply	Water withdrawn by public and private water suppliers and delivered to multiple users for domestic, commercial, industrial, and thermoelectric power uses.
Reclaimed water	Municipal wastewater that has gone through various treatment processes to meet specific water quality criteria with the intent of being used in a beneficial manner (e.g., irrigation). The term recycled water is used synonymously with reclaimed water, particularly in California.
Renewable water resources	The water entering a country's surface and groundwater systems. Not all of this water can be used because some falls in a place or time that precludes tapping it even if all economically and technically feasible storage and diversion structures were built.
Return flow	The water that reaches a ground- or surface-water source after release from the point of use and thus becomes available for further use.
Runoff	Part of the precipitation that appears in surface streams. It is the same as streamflow unaffected by artificial diversions, storage, or other works of man in or on the stream channels.
Sustainability	The principle of optimizing the benefits of a present system without diminishing the capacity for similar benefits in the future.
Sustainable development	Development that meets the needs of the present without compromising the ability of future generations to meet their own needs.
Transpiration	Water removed from soil that undergoes a change-of-state from liquid water in the stomata of the leaf to the water vapor of the atmosphere.
Wastewater	Used water discharged from homes, business, cities, industry, and agriculture. Various synonymous uses such as municipal wastewater (sewage), industrial wastewater, and stormwater.
Water reclamation	Treatment or processing of wastewater to make it reusable with definable treatment reliability and meeting appropriate water quality criteria.
Water reuse	The use of treated wastewater for a beneficial use, such as agricultural irrigation and industrial cooling.
Watershed	The natural unit of land upon which water from direct precipitation, snowmelt, and other storage collects and flows downhill to a common outlet where the water enters another water body such as a stream, river, wetland, lake, or the ocean.
Withdrawals	The water removed from the ground or diverted from a stream or lake for use.

The feasibility and reliability of providing adequate quantities and quality of water to meet societal needs is constrained by geographic, hydrologic, economic, and social factors. Projections of unprecedented global population growth, particularly in urban areas, have fueled concerns about water availability in increasingly complex environmental, economic, and social settings. Some of the important questions and concerns are: (1) how long can existing water sources be sustained? (2) how can we ensure the reliability of current and future water sources? (3) where will the next generation of water sources be found to meet the needs of growing populations and uses and provide for agriculture and industrial water requirements? and (4) how will conflicts between watershed interests in environmental preservation and beneficial uses of water sources be resolved? To address the social, economic, and environmental impacts of water resources development and avert the ominous prospects of water scarcity, there is a critical need to reexamine the way water resources systems are planned, constructed, and managed.

The emerging paradigm of sustainable water resources management emphasizes whole-system solutions to reliably and equitably meet the water needs of present and future generations. Understanding the concepts of sustainable water resources management as a foundation of water reclamation and reuse is of fundamental importance. Thus, the purpose of this introductory chapter is to provide a perspective on (1) a definition of terms including working terminology used in this chapter, (2) principles of sustainable water resources management, (3) current and potential future global water shortages, (4) the important role played by water reclamation and reuse, and (5) the future of water reclamation and reuse. The discussion in this chapter is designed to stimulate readers to think about future water resources development and management in more sustainable and comprehensive ways, incorporating water reclamation and reuse as one of the viable options.

1-1 DEFINITION OF TERMS

Several different terms are used to describe forms of water and wastewater and their subsequent treatment and reuse. To facilitate communication among different disciplines associated with water reclamation and reuse practices, it is important to establish a broad understanding of the terminology used in the field of water reclamation and reuse. Useful terminology related to water reclamation and reuse is presented as Working Terminology at the beginning of this chapter and every chapter in this textbook.

For the purpose of gaining broader public acceptance of water reuse, in 1995 the State of California amended the provisions of the existing Water Code substituting the term *recycled water* for *reclaimed water* and the term *recycling* for *reclamation* (State of California, 2003). *Water recycling* is defined to mean water, which as a result of treatment of wastewater, is suitable for a direct beneficial use or a controlled use that would not otherwise occur. However, because of the traditional usage of the word and the practice in water reclamation and reuse, the terms *reclaimed water* and *recycled water* are used synonymously in this textbook. It should be noted that the terminology given above may be considered *working definitions* that have evolved from water and wastewater treatment, several water reuse legislations and regulations, as well as in response to questions raised by reclaimed water users and the public at large.

1-2 PRINCIPLES OF SUSTAINABLE WATER RESOURCES MANAGEMENT

Historically, water resources management has focused on supplying water for human activities, with an intrinsic assumption that technological solutions would keep pace with steadily increasing water demands and progressively more stringent water quality requirements. Past water resources development was based on manipulating the natural hydrologic cycle by attempting to balance the inherent water availability in a region with societal needs for water in the context of the social and economic background of the region, population, and the extent of urbanization (Baumann et al., 1998; Thompson, 1999; Bouwer, 2000). Because of the social, economic, and environmental impacts of past development and the prospects of potential water shortages, a new paradigm for water resources development and management is evolving, based on the principles of sustainability and environmental ethics. Sustainability and environmental ethics are examined further in this section.

The Principle of Sustainability

The principle of sustainability, a cornerstone in the Brundtland Commission's report entitled *Our Common Future* (WCED, 1987), is defined as follows: "Humanity has the ability to make development sustainable to ensure that it meets the needs of the present without compromising the ability of future generations to meet their own needs." Sustainability is becoming a driving principle of political, economic, and social development and it has achieved considerable public acceptance; however, the debate still continues over just what is to be sustained, how, and for whom (Wilderer et al., 2004; Sikdar, 2005).

Working Definitions of Sustainability

Sustainability can be applied to a range of human activities (e.g., sustainable agriculture) or to human society as a whole. From an environmental perspective, human activities are not sustainable if they irreversibly degrade natural ecosystems that perform essential life-supporting functions. In economics, sustainability may be defined, for example, as ". . . nondeclining utilities (welfare) of a representative member of society for millennia into the future . . ." (Pezzey, 1992). Despite the lack of a common understanding of what sustainability is and the variable interpretations among different disciplines, there is a general understanding that a whole system, long-term view is needed to assess and approach sustainability, particularly in the case of water resources management. In this textbook, *sustainability* is defined as the principle of optimizing the benefits of a present system without diminishing the capacity for similar benefits in the future.

Challenges for Sustainability

The goal of sustainable water resources development and management is to meet water needs reliably and equitably for current and future generations by designing integrated and adaptable systems, optimizing water-use efficiency, and making continuous efforts toward preservation and restoration of natural ecosystems. The transition to a sustainable society poses a number of technological and social challenges. Technological innovations can help to improve what is called the ecoefficiency of human activities. Recognizing that water resources are finite, it is essential that the overall use of the resource be sustainable despite the increased efficiency of current and future technologies. Unless population and consumption growth rates are reduced, technological improvements may only delay the onset of negative consequences (Huesemann, 2003). Today, considerations for sustainability must include a number of aspects that vary both temporally and spatially, including energy and resource use and environmental pollution (Hermanowicz, 2005).

Criteria for Sustainable Water Resources Management

The emerging paradigm of sustainable water resources management has been interpreted in different ways by different stakeholders. The American Society of Civil Engineers (ASCE, 1998) proposed the following working definition for sustainable water resources systems: "Sustainable water resources systems are those designed and managed to fully contribute to the objectives of society, now and in the future, while maintaining their ecological, environmental, and hydrological integrity." In practice, the extent of sustainability in water resources management needs to be measurable with relevant criteria. Criteria often identified with sustainable water resources management are shown in Table 1-1.

Traditional approaches to water resources development have focused on modifying water storage and flow patterns by constructing dams and reservoirs and/or designing systems for interbasin transfers to secure water supplies (see Fig. 1-1). In many cases, developing additional water resources satisfies the first criterion in Table 1-1 (i.e., to meet basic human needs for water). However, in a growing number of cases, there is not enough water available to meet basic water needs, as evidenced by the rise in water scarcity in many regions of the world. New sources of water that can be developed cost-effectively are not available for many of the major urban areas of the developing world. Cost-effective sources of water have already been developed or are in the process of development, and, in most cases, water that has been harnessed has been fully allocated and in many cases overallocated.

Further, construction of dams and reservoirs is becoming less feasible due to consideration of ecological and social impacts, safety, and the cost of complying with environmental regulations. Thus, in many places, additional supplies of drinking water can be obtained only by reallocating water that is currently used by other sectors such as agriculture or by using alternative water sources such as saline or brackish water, stormwater, or reclaimed water. Under the principles of sustainable water resources management, demand management, such as water conservation, is used to meet basic water needs. It is argued by some that the need to develop new sources of water can be avoided by implementing measures for more efficient use of water (Vickers, 1991; Gleick, 2002). It might also be argued that multiple approaches are needed to ensure the sustainability of water resources management including water reclamation and reuse, water conservation, and other demand management as listed in Table 1-1.

Water Conservation

Water conservation has been viewed historically by the water industry as a standby or temporary measure that is utilized only during times of drought or other emergency water shortages. This limited view of the role of water conservation is changing; utilities that have pioneered the use of conservation have shown that it is a viable long-term supply option (Vickers, 2001). Water conservation can yield a number of benefits for the water utility, environment, and community. These benefits include reduced energy and chemical inputs for water treatment, downsized or postponed expansions of water facilities, and reduced costs and impacts of wastewater management.

Common conservation measures include customer education about water use, water-efficient fixtures, water-efficient landscaping, metering, economic incentives, and water-use restriction programs (Maddaus, 2001). In the United States, 42 percent of annual water use is, on the average, for indoor purposes and 58 percent for outdoor purposes (Mayer et al., 1999). Indoor residential water use can be reduced significantly by installing water-efficient fixtures, such as low-volume flush toilets. Typical indoor domestic uses of water in the United States with potential water savings with residential

Table 1-1
Criteria for sustainable water resource management[a]

Objective	Action
Meet basic human needs for water	Provide adequate quantity of water of a quality appropriate to protect public health without compromising enviromental quality.
Maintain long-term renewability	Replenish freshwater through return flows to the environment.
Preserve ecosystems	Manage the interface between societal activities and sensitive ecosystems; ensure that ecosystem water balance is maintained. Strive to achieve zero effluent discharge goals.
Promote efficient use of resources	Optimize the use of energy, material, water, and control the release of greenhouse gas emissions.
Encourage water conservation	Ensure that water users are informed of the advantages of water conservation; develop new ways to conserving water; implement incentives to promote water conservation.
Encourage water reclamation and reuse	Preserve high quality water sources for other uses; develop new ways of water reclamation and reuse; prevent environmental degradation by closed-loop management of treated wastewater.
Emphasize importance of water quality in multiple uses of water	Identify relationships between pollution prevention programs, effective management of industrial water use and wastewater treatment, and alternative uses of water. Strive to achieve zero effluent discharge goals.
Examine necessity and opportunity of water resources needs and build consensus	Involve public and private stakeholders in planning and decision-making, equitably distribute costs and benefits.
Design for resilience and adaptability	Develop design strategies that incorporate mechanisms to deal with uncertainty, risk, and changing societal values.

[a]Compiled, in part, from various sources including ASCE (1998); Gleick (1998 and 2000); Braden and van Ierland (1999); Loucks (2000); Asano (2002); Baron et al. (2002).

Figure 1-1

Shasta Dam, on the Sacramento River near Redding, CA, serves to control flood waters and store surplus winter runoff for irrigation in the Sacramento and San Joaquin Valleys, maintain navigation flows, provide flows for the conservation of fish and water for municipal and industrial use, protect the Sacramento-San Joaquin Delta from intrusion of saline ocean water, and generate hydroelectric power (Courtesy of U.S. Department of the Interior, Bureau of Reclamation). (Coordinates: 40.718 N, 122.420 W)

water conservation are shown in Table 1-2. Water conservation can reduce indoor water use by 32 percent on a per capita basis as shown in Table 1-2. In addition to indoor water uses, the water use efficiency for outdoor residential water applications such as landscape irrigation, washing cars, and other cleaning or recreational uses can also benefit from implementing water conservation practices.

Water Reclamation and Reuse

Water reclamation is the treatment or processing of wastewater to make it reusable with definable treatment reliability and meeting water quality criteria. Water reuse is the use of treated wastewater for beneficial uses, such as agricultural irrigation and industrial cooling. Treated municipal wastewater represents a more reliable and significant source for reclaimed water as compared to wastewaters coming from agricultural return flows, stormwater runoff, and industrial discharges. As a result of the Federal Clean Water Act and related wastewater treatment regulations, centralized wastewater treatment has become commonplace in urban areas of the United States (see Chap. 2, Sec. 2-2). New technologies in decentralized and satellite wastewater treatment have also been developed

Table 1-2
Typical single family home water use, with and without water conservation[a]

	Typical single family home water use			
	Without water conservation		With water conservation	
Water uses	L/capita·d[b]	Percent	L/capita·d[b]	Percent
Toilets	76.1	27.7	36.3	19.3
Clothes washers	57.2	20.9	40.1	21.4
Showers	47.7	17.3	37.9	20.1
Faucets	42.0	15.3	40.9	21.9
Leaks	37.9	13.8	18.9	13.8
Other domestic	5.7	2.1	5.7	3.1
Baths	4.5	1.6	4.5	2.4
Dish washers	3.8	1.3	3.8	2.0
Total	274.4	100	187.8	100

[a]Adapted from AWWA Research Foundation (1999).
[b]L/capita·d, liters per capita per day.

(see Chaps. 12 and 13). The emphasis of this textbook is, therefore, focused on planning and implementation of water reclamation and reuse from municipal wastewater. The benefits of water reclamation and reuse and factors driving its future are summarized in Table 1-3.

With many communities approaching the limits of their readily available water supplies, water reclamation and reuse has become an attractive option for conserving and extending available water supply by potentially (1) substituting reclaimed water for applications that do not require high-quality water supplies, (2) augmenting water sources and providing an alternative source of supply to assist in meeting both present and future water needs, (3) protecting aquatic ecosystems by decreasing the diversion of freshwater, reducing the quantity of nutrients and other toxic contaminants entering waterways, (4) reducing the need for water control structures, and (5) complying with environmental regulations by better managing water consumption and wastewater discharges.

Water reuse is attractive particularly in situations where the available water supply is already overcommitted and cannot meet expanding water demands in a growing community. Increasingly, society no longer has the luxury of using water only once. Examples of signs highlighting water conservation and reuse are shown on Fig. 1-2.

Water reuse offers an alternative water supply that is consistently available in urban areas, even during drought years, for various beneficial uses. However, because of its genesis from municipal wastewater (traditionally known as *sewage*), acceptance of

Table 1-3

Water reclamation and reuse: rationale, potential benefits, and factors driving its further use[a]

Rationale for water reclamation and reuse

- Water is a limited resource. Increasingly, society no longer has the luxury of using water only once
- Acknowledge that water recycling is already happening and do it more and better
- The quality of reclaimed water is appropriate for many nonpotable applications such as irrigation and industrial cooling and cleaning water, thus providing a supplemental water source that can result in more effective and efficient use of water
- To meet the goal of water resource sustainability it is necessary to ensure that water is used efficiently
- Water reclamation and reuse allows for more efficient use of energy and resources by tailoring treatment requirements to serve the end-users of the water
- Water reuse allows for protection of the environment by reducing the volume of treated effluent discharged to receiving waters

Potential benefits of water reclamation and reuse

- Conservation of fresh water supplies
- Management of nutrients that may lead to environmental degradation
- Improved protection of sensitive aquatic environments by reducing effluent discharges
- Economic advantages by reducing the need for supplemental water sources and associated infrastructure. Reclaimed water is available near urban development where water supply reliability is most crucial and water is priced the highest
- Nutrients in reclaimed water may offset the need for supplemental fertilizers, thereby conserving resources. Reclaimed water originating from treated effluent contains nutrients; if this water is used to irrigate agricultural land, less fertilizer is required for crop growth. By reducing nutrient (and resulting pollution) flows into waterways, tourism and fishing industries are also helped

Factors driving further implementation of water reclamation and reuse

- Proximity: Reclaimed water is readily available in the vicinity of the urban environment, where water resources are most needed and are highly priced
- Dependability: Reclaimed water provides a reliable water source, even in drought years, as production of urban wastewater remains nearly constant
- Versatility: Technically and economically proven wastewater treatment processes are available now that can provide water for nonpotable applications and can produce water of a quality that meets drinking water requirements
- Safety: Nonpotable water reuse systems have been in operation for over four decades with no documented adverse public health impacts in the United States or other developed countries
- Competing demands for water resources: Increasing pressure on existing water resources due to population growth and increased agricultural demand

Table 1-3
Water reclamation and reuse: rationale, potential benefits, and factors driving its further use[a] (*Continued*)

Factors driving further implementation of water reclamation and reuse

- Fiscal responsibility: Growing recognition among water and wastewater managers of the economic and environmental benefits of using reclaimed water
- Public interest: Increasing awareness of the environmental impacts associated with overuse of water supplies, and community enthusiasm for the concept of water reclamation and reuse
- Environmental and economic impacts of traditional water resources approaches: Greater recognition of the environmental and economic costs of water storage facilities such as dams and reservoirs
- Proven track record: The growing numbers of successful water reclamation and reuse projects throughout the world
- A more accurate cost of water: The introduction of new water charging arrangements (such as full cost pricing) that more accurately reflect the full cost of delivering water to consumers, and the growing use of these charging arrangements
- More stringent water quality standards: Increased costs associated with upgrading wastewater treatment facilities to meet higher water quality requirements for effluent disposal
- Necessity and opportunity: Motivating factors for development of water reclamation and reuse projects such as droughts, water shortages, prevention of seawater intrusion and restrictions on wastewater effluent discharges, plus economic, political, and technical conditions favorable to water reclamation and reuse

[a]Compiled from various sources including Asano (1998); Queensland Water Recycling Strategy (2001); Mantovani et al. (2001); Simpson (2006).

reclaimed water as an alternative water source has to overcome unique hurdles. In the United States and other developed countries, reclaimed water is treated using strict water quality control measures to ensure that it is nontoxic and free from disease causing microorganisms, but it does carry potential risks inherent in the use of any resource exposed to human waste. Concerns for health and safety must be addressed in the planning and implementation of water reclamation and reuse. It has been found that the success of water reclamation and reuse projects in many parts of the world has hinged on the pressures associated with the urgent necessity for water coupled with the opportunity to develop water reuse systems.

Environmental Ethics

Environmental ethics involves the application of moral responsibility in relation to management of the natural environment. Similar to the principle of sustainability, environmental ethics has emerged in response to serious environmental degradation resulting from societal activities such as over-allocation of natural resources. There are several theories of environmental ethics that are used to describe human obligations in the protection of natural systems. The anthropocentric (human-centered) perspective emphasizes environmental protection for the survival and well-being of humans alone. The ecocentric (nature-centered) perspective regards humans as only one element of the broader natural community, and bases moral responsibility on the intrinsic value and rights of nature.

(a)

(b)

Figure 1-2
Examples of signs highlighting (a) water conservation and (b) reuse.

Equitable Water Allocation

An ongoing water resources management debate questions whether society has an obligation to meet the basic water needs for all people and ecosystems. Because of the uneven geographic distribution of populations, water availability, and wealth, it is difficult to provide for equitable and balanced allocation of water resources. Balancing societal water needs with ecosystem requirements is even more challenging, considering the complex science-defining ecosystem needs, the widely varying perceptions of ecosystem value, and the dire social consequences of water scarcity (Harremoës, 2002).

Precautionary Principle

Another ethical question is whether human activities should proceed if there is a potential, but unproven risk to the environment or public health. The precautionary principle, introduced in European environmental policies in the late 1970s, has been providing both guidance and controversy in this area (Foster et al., 2000; Krayer von Krauss et al., 2005). The definition of precautionary principle used in the Third North Sea Conference in 1990 was, "To take action to avoid potentially damaging impacts of substances that are persistent, toxic, and liable to bioaccumulation even where there is no scientific evidence to prove a causal link between emissions and effects" (Harremoës et al., 2001). At its core, the precautionary principle embodies the idea of "better safe than sorry," but, undoubtedly, some people would argue that no progress will be made with this mindset.

Similar to sustainable development, the greatest difficulty with using the precautionary principle as a policy tool is its extreme variability in interpretation. The principle can be interpreted as calling for absolute proof of safety before any action is taken, or it may be interpreted as opening the door to cost-benefit analysis and discretionary judgment as stated in the Rio de Janeiro Declaration (United Nations, 1992; Foster et al., 2000). A challenging final question is: how to use uncertainty information in policy context? More research is required to answer this question (Krayer von Krauss et al., 2005).

1-3 CURRENT AND POTENTIAL FUTURE GLOBAL WATER SHORTAGES

The total volume of renewable freshwater in the global hydrologic cycle is several times more than is needed to sustain the current world population. However, only about 31 percent of the annual renewable water is accessible for human uses due to geographical and seasonal variations associated with the renewable water (Postel, 2000; Shiklomanov, 2000). On a global scale, annual withdrawals for irrigation are over 65 percent of the total withdrawn for human uses; 2,500 out of a total of 3,800 km^3. Withdrawals for industry are about 20 percent, and those for municipal use are about 10 percent (Cosgrove and Rijsberman, 2000).

Countries of North Africa and the Middle East, especially Egypt and the United Arab Emirates, are among the countries with the lowest freshwater availability (see Figs. 1-3 and 1-4). On the contrary, Iceland, Suriname, Guyana, Papua New Guinea, Gabon, Canada, and New Zealand are examples of the most water abundant countries, based on per capita water availability (WRI, 2000).

The implementation of water reclamation and reuse projects is driven mainly by existing and projected water shortages in specific water-poor countries. Other factors such as preventing saltwater intrusion into freshwater resources in coastal areas and prohibition of wastewater effluent disposal into sensitive environments will certainly influence water reuse decisions. The impacts associated with current and projected world population, water requirements, and potential global and regional water scarcity are considered briefly in the following discussion.

Impact of Current and Projected World Population

The world population in 2002 was estimated at 6.2 billion with an annual growth rate of 1.2 percent, or 77 million people per year. To put the recent growth in perspective, the world population in the year 1900 was only 1.6 billion and in 1950 it was 2.5 billion. It is projected that the world population in 2050 will be between 7.9 billion and 10.3 billion (United Nations, 2003).

Figure 1-3

Transporting water in buckets near the pyramid in Saqqara, Egypt (Coordinates: 29.871 N, 31.216 E). The limited availability of water infrastructure is common in many parts of the world.

The rate of growth in industrialized countries is well under one percent per year. In developing countries, however, the growth rate exceeds two percent per year, and in some parts of Africa, Asia, and the Middle East it exceeds three percent per year. As a result, over 90 percent of all future population increases will occur in the developing world (United Nations, 2003). Six countries currently account for half of the annual population growth: India, China, Pakistan, Nigeria, Bangladesh, and Indonesia. The population in the United States was estimated at about 285 million in 2001 and was growing at an annual rate of about one percent (U.S. Census Bureau, 2003).

Urbanization

In 1950, New York was the only city in the world with a population of more than 10 million. The number of cities with more than 10 million people increased to 5 in 1975 and 17 in 2001, and is expected to increase to 21 cities in 2015. The world's urban population reached 2.9 billion in 2000 and is expected to increase by 2.1 billion by 2030, just slightly below the world's total population increase (United Nations, 2002). The population of cities with 10 million inhabitants or more in 1950, 1975, 2001, and 2015 is listed in Table 1-4. It is projected that Asia and Africa will have more urban dwellers than any other continent of the world, and Asia will contain 54 percent of the world's urban population by 2030.

Although urbanization is more prominent in the developing world, urban populations in developed countries are also expanding. In the United States, the average annual

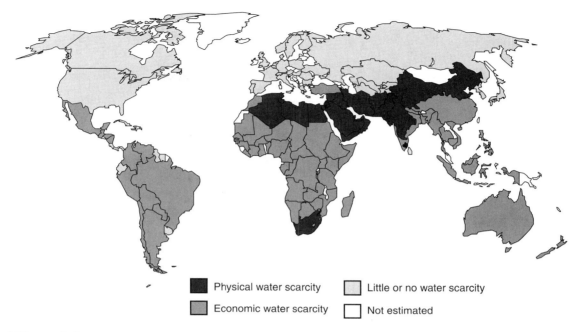

Figure 1-4
Projected global water scarcity in 2025 (Adapted from IWMI, 2000). In the global scale, countries of North Africa and the Middle East, Pakistan, India, and the northern part of China are projected to face severe water scarcity.

population growth in metropolitan areas (cities and suburbs) between 1990 and 1998 was 1.14 percent, while nonmetropolitan areas grew at a slower rate of 0.88 percent, reflecting population shifts from rural to urban areas. Of the country's total population in 1998, 28.1 percent lived in metropolitan areas with five million or more people. Among urban areas with five million or more people, the Los Angeles-Riverside-Orange County area and the San Francisco-Oakland-San Jose area in California grew most rapidly between 1990 and 1998—reflecting an annual increase of 1.08 percent, slightly lower than the growth rate of all U.S. metropolitan areas (Mackun and Wilson, 2000). Metropolitan areas in the United States with populations of five million or more are shown in Table 1-5.

Urbanization intensifies the pressures of population growth on water resources due to imbalances between water demands and the proximity of water sources. In addition, significant differences exist in water use patterns between rural, agricultural, and urban areas. Because of this, population growth and urbanization will pose significant challenges for water resources management throughout the world.

Irrigation Water Use
The expansion of the aerial extent of irrigated land-use due to population growth is one of the most important contributors to the increase of total water use in the world. In 1995, over 65 percent of the total global water withdrawal for human uses was for irrigation,

Table 1-4

The population of cities and metropolitan areas with 10 million inhabitants or more, for 1950, 1975, 2001, and 2015[a]

1950		1975		2001		2015	
City	Population, millions	City	Population, millions	City	Population, millions	City	Population, millions
New York	12.3	Tokyo	19.8	Tokyo	26.5	Tokyo	27.2
		New York	15.9	Sao Paulo	18.3	Dhaka	22.8
		Shanghai	11.4	Mexico City	18.3	Mumbai	22.6
		Mexico City	10.7	New York	16.8	Sao Paulo	21.2
		Sao Paulo	10.3	Mumbai	16.5	Delhi	20.9
				Los Angeles	13.3	Mexico City	20.4
				Calcutta	13.3	New York	17.9
				Dhaka	13.2	Jakarta	17.3
				Delhi	13.0	Calcutta	16.7
				Shanghai	12.8	Karachi	16.2
				Buenos Aires	12.1	Lagos	16.0
				Jakarta	11.4	Los Angeles	14.5
				Osaka	11.0	Shanghai	13.6
				Beijing	10.8	Buenos Aires	13.2
				Rio de Janeiro	10.8	Metro Manila	12.6
				Karachi	10.4	Beijing	11.7
				Metro Manila	10.1	Rio de Janeiro	11.5
						Cairo	11.5
						Istanbul	11.4
						Osaka	11.0
						Tianjin	10.3

[a]Adapted from United Nations (2002).

which includes both agricultural and nonresidential landscape applications. Irrigation consumes a large volume of water through evaporation from reservoirs, canals, and soil and through incorporation into and transpiration by crops. *Consumptive use* is the portion of withdrawn water that is evaporated, transpired, incorporated into products or crops, consumed by humans or livestock, or otherwise removed from the immediate water environment. Depending on the technology and management, consumptive use associated with irrigation can range from 30 to 90 percent of the total water withdrawn (Cosgrove and Rijsberman, 2000).

Applied water that is not consumed either recharges groundwater or contributes to drainage or return flows. This water can be—and often is—reused, but, because return flows tend to have higher salt concentrations and are likely to be contaminated with nutrients, sediments, pesticides, and other chemicals, beneficial reuse of this water has limited applications unless it is treated prior to use.

Table 1-5
Metropolitan areas in United States with population of 5 million or more: 1990 to 1998[a]

Metropolitan area	1998 population	Population change 1990 to 1998	
		Number	Percent
New York-Northern New Jersey-Long Island, NY-NJ	20,126,150	558,939	2.9
Los Angeles-Riverside-Orange County, CA	15,781,273	1,249,744	8.6
Chicago-Gary-Kenosha, IL-IN-WI	8,809,846	570,026	6.9
Washington-Baltimore-Northern Virginia, DC-MD-VA	7,285,206	558,811	8.3
San Francisco-Oakland-San Jose, CA	6,816,047	538,522	8.6
Philadelphia-Wilmington-Atlantic City, PA-NJ-DE	5,988,348	95,329	1.6
Boston-Worcester-Lawrence-Southern Maine and New Hampshire, MA-NH-ME	5,633,060	177,657	3.3
Detroit-Ann Arbor-Flint, MI	5,457,583	270,412	5.2

[a] Adapted from Mackun and Wilson (2000). Original source: U.S. Census Bureau, Population Estimates Program.

Domestic and Industrial Water Uses

Conversion of farmland into residential and industrial areas results in a decrease in agricultural water use and a concurrent increase in domestic and industrial water uses. A large share of the water used by households, services, and industry—up to 90 percent in areas where total water use is high—is returned as wastewater. While a large proportion of the water used in domestic and industrial water is collected as wastewater, water is in such a degraded state that treatment is required before it can be discharged or reused.

Potential Global Water Shortages

Globally, the water resources in various regions and countries are expected to face unprecedented pressures in the coming decades as a result of continuing population growth and uneven distributions of population and water. Although the number of persons served has increased, about 1.1 billion people, or about 18 percent of the world population lacked access to clean drinking water, and 2.4 billion did not have adequate sanitation services in 2000 (WHO, 2000). Surging populations throughout the developing world are intensifying the pressures on limited water supplies. The concentration of populations within urban areas further exacerbates the disparity between water demand and regional water availability.

Water Scarcity

A country is considered water-scarce when its annual supply of renewable freshwater is less than 1,000 m³ per capita (Falkenmark and Widstrand, 1992; Falkenmark and Lindh, 1993). Such countries can expect to experience chronic and widespread shortages of water that hinder their development and welfare. Globally, water scarcity is resulting in a host of crises, such as food shortages, regional water conflicts, limited economic development, and environmental degradation (Postel, 2000). These issues have put freshwater availability at the forefront of state, national, and international efforts in recent decades.

Two types of water-scarce countries can be identified: (1) the countries with physical water scarcity which will not have sufficient water to meet their future agricultural, domestic, industrial, and environmental needs even with the highest feasible efficiency and productivity of water use, and (2) the countries with economic water scarcity: countries that have sufficient water resources but lack the monetary resources needed to access or use these resources or face severe financial and development capacity problems. These countries will need to increase water supply by 25 percent or more over 1995 levels through additional storage and conveyance facilities to meet their water demands in 2025. The projected global water scarcity in 2025 is depicted on Fig. 1-4. Countries of North Africa and the Middle East, Pakistan, India, and the northern part of China are projected to face severe water scarcity (IWMI, 2000).

While the data presented on Fig. 1-4 provide a global perspective, it is difficult to apply that information on a regional or watershed scale. For example, about one-half of the population of China lives in the wet region of southern China, mainly in the Yangtze basin, while the other half lives in the arid north, mainly in the Yellow River basin. This is also true for India, where about 50 percent of the population lives in the arid northwest and southeast, while the remainder lives in fairly wet areas (IWMI, 2000). In many countries, the distance between available sources of water and population centers is too far to allow for moving water from the source to the needed area due to the lack of resources to construct, operate, and maintain the extensive infrastructure that would be required. In addition, there may be environmental, social, and economic constraints that limit the overall feasibility of transporting water. Thus, much more attention needs to be paid to the governance of water to ensure that sustainable water supplies will be available through the twenty-first century (Rogers, et al., 2006). The value of implementing water reclamation and reuse is recognized by many in the context of sustainable water resources management because municipal wastewater is produced at the doorstep of the metropolis where water is needed the most and priced the highest.

Potential Regional Water Shortages in the Continental United States

A comparison of the average regional consumptive use and renewable water supply in the United States is depicted on Fig. 1-5. The renewable water supply is the sum of precipitation and imports of water, minus the water not available for use through natural evapotranspiration and exports. Renewable water supply is a simplified upper limit to the amount of water consumption that could occur in a region on a sustained basis. Requirements to maintain minimum flows in streams leaving the region for navigation, hydropower, fish, and other instream uses limit the amount of the renewable supply available for use. Also, total development of a surface-water supply is never possible because the extent of evaporative losses increases as more reservoirs are constructed. Nevertheless, the renewable supply compared to consumptive use is an index of the degree to which the resource has already been developed (USGS, 1984; Adams, 1998).

Water resources regions having potential limitations in water supply with respect to adequacy and dependability are the Rio Grande Region, Missouri, Texas-Gulf, the Upper and Lower Colorado River Basin, Great Basin, and California as depicted on Fig. 1-5. From the water supply point of view, several major regions of the country are using water in excess of their presently sustainable water resources. Some areas are entirely dependent on groundwater mining. Other areas, where surface waters are used, have been able

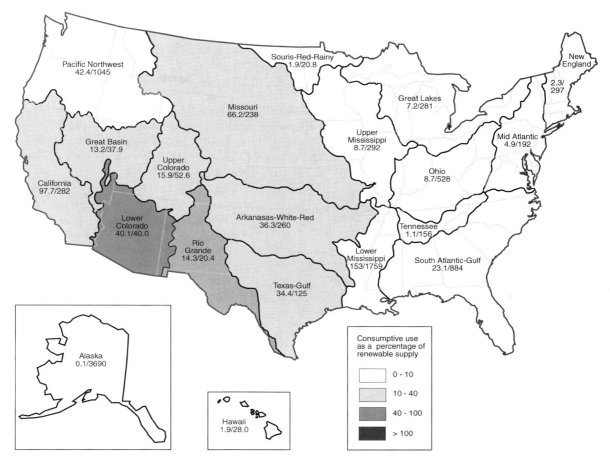

Figure 1-5

Comparison of average consumptive use and renewable water supply for the 20 water resources regions of United States (Adapted from USGS, 1948; updated using 1995 estimates of water use). The number in each water resource region is consumptive use/renewable water supply in 10^6 m³/d, respectively, or consumptive use as a percentage of renewable supply as shown in the legend.

to satisfy growing demands by means of the relatively high yields from normal and wet-year stream flows. Identified water resources issues from various regions are summarized below based on the U.S. Geological Survey Water-Supply Paper 2250 (USGS, 1984).

Central Great Plains

The Central Great Plains relies on water imported to the region. The main transbasin water diversions are the tunnels drilled through the Rockies to bring supplies of water from the Colorado River to the Great Plains. Irrigated agriculture is a main end use in this region, and this demand is increasing (although in some areas water use is shifting from agriculture to urban development). The biggest regional issue is the lack of surplus capacity in regional water supplies. For example, water from the Arkansas River serves multiple uses as it passes through the individual states. The resulting conflicts over allocation of limited groundwater and surface water supplies have led to a number of lawsuits in the region.

Eastern Midwest
The Eastern Midwest includes some of the largest river systems in the nation, and this region is also strongly affected by drought and flood. Drought brings on low flow and depletion of groundwater. Flooding causes crop and property damage, erosion, and sedimentation. In addition, agricultural runoff from the region is causing hypoxia (a reduction in aquatic oxygen concentration to levels where life cannot be sustained) within the Gulf of Mexico. However, floods help the fish population by diluting agricultural runoff and increasing the concentration of dissolved oxygen. Generally, the region has plenty of water, but the efficiency of water distribution varies seasonally, resulting in water shortages during droughts.

Great Lakes
The Great Lakes, while making up 95 percent of the fresh surface water in the United States, are a shared resource with Canada. The potential for degradation in water quantity, quality, associated ecosystems, and coastline is a concern for both nations. Regional needs include a serious consideration of sustainability, the development of a robust water management plan including groundwater supplies, and an assessment of water quality and ecosystem impacts on the 121 watersheds around the Great Lakes.

Metropolitan East Coast, New York City
Although many communities in this region have their own water supply systems, they are generally small compared to that for New York City. The quality of discharged effluent from these communities has improved significantly over time. In general, new institutional forms and changes are needed as growth is occurring and to cope with degraded water quality and growing water demand, along with needs for new infrastructure systems.

Mid-Atlantic
The Middle Atlantic region is an area with significant climate variability and large vulnerabilities. During the past few decades, the region has experienced both severe drought and flooding produced by winter storms and summer hurricanes. The region includes several metropolitan areas which rely on water systems that are highly sensitive to climate variation. A large portion of the population obtains water from private wells. As a result, water management in dry periods is a major issue for this region.

Rio Grande
Water shortage is a concern for the entire region, yet at the same time the region is experiencing rapid urban and population growth. With the expanding population in the region aquifers are being depleted rapidly. Conflicts are arising between Native American tribes and the rest of the community, resulting in legal battles in many cases. Rio Grande river water along the Mexican border is being allocated to agriculture, yet no drought management plan is in place. The ecology of the region is also threatened due to instream flows as low as 20 percent of historical levels. One potential answer to supply problems is increased efficiency of agricultural water use.

Southeast, including the Atlantic Coast

This region has abundant water, but water management policy is critical because of the strong pressure for further development in the region. Demographic impacts also play an important role in water management and use in this area because of the high population densities along the coast and because of large seasonal swings in population. Agriculture, forestry, and ecological systems are identified as the main areas of concern, especially with respect to water quality and availability. In addition, some health hazards are also associated with contaminated water resources.

The State of Florida receives about 1400 mm of rain on an annual basis, however most of the precipitation occurs over a three to four months period (rainy season). The remainder of the year is relatively dry. Water use patterns are inverse to rainfall with higher water usage occurring during the dry season (winter) and lower water usage occurring during the rainy season (summer). Shifts in land use patterns from agriculture to urbanization have resulted in an imbalance between water availability and water use. In addition, seasonal population shifts due to tourism and retirement communities impose further pressures on water resources during the dry season. Overdrafting of groundwater has also resulted in land subsidence. There is a critical need for alternative reliable water sources to meet water demands associated with population increases projected to occur in the future.

Preventing Crises and Conflict in the West

Chronic water supply problems in the West are some of the greatest challenges the United States will be facing in the coming decades. The U.S. Department of the Interior (2003) published a report entitled, *Water 2025: Preventing Crises and Conflict in the West*, which describes the issues that are driving major conflicts between water users in the West. The specific competing issues described in this report are (1) the explosive population growth in western urban areas, (2) the emerging need for water for environmental and recreational uses, and (3) the national importance of the domestic production of food and fiber from western farms and ranches. *Water 2025* provides a basis for a public discussion of the realities that face the West so that decisions can be made at the appropriate level in advance of water supply crises.

1-4 THE IMPORTANT ROLE OF WATER RECLAMATION AND REUSE

Water reclamation and reuse involves considerations of public health and also requires close examinations of infrastructure and facilities planning, wastewater treatment plant siting, treatment process reliability, economic and financial analyses, and water utility management involving effective integration of water resources and reclaimed water. Whether water reuse will be appropriate depends upon careful economic considerations, potential uses for the reclaimed water, public health protection, stringency of waste discharge requirements, and public policy where the desire to conserve rather than develop available water resources may override other obstacles. In addition, the varied interests of many stakeholders, including those representing the environment, must be considered.

Types of Water Reuse

The principal categories of water reuse applications for reclaimed water originating from treated municipal wastewater are shown in Table 1-6, in descending order of volume of use. The majority of water reuse projects are for nonpotable applications such as agricultural and landscape irrigation and industrial uses (see Figs. 1-6 and 1-7). Groundwater recharge can be designed for indirect potable reuse where groundwater is recharged with reclaimed water and replenishes portions of potable groundwater. The detailed discussions on the technical aspects of water reuse applications are given in Part 4 of this textbook.

Integrated Water Resources Planning

Integrated water resources planning is a process that promotes the coordinated development and management of water, land, and related resources to maximize the resultant economic and social welfare in an equitable and sustainable manner. A framework to compare competing interests, including those of future generations, does not currently exist in water management and planning. A new definition of sustainable water development is also

Table 1-6

Water reuse categories and typical applications

Category	Typical application
Agricultural irrigation	Crop irrigation
	Commercial nurseries
Landscape irrigation	Parks
	School yards
	Freeway medians
	Golf courses
	Cemeteries
	Greenbelts
	Residential
Industrial recycling and reuse	Cooling water
	Boiler feed
	Process water
	Heavy construction
Groundwater recharge	Groundwater replenishment
	Salt water intrusion control
	Subsidence control
Recreational/environmental uses	Lakes and ponds
	Marsh enhancement
	Streamflow augmentation
	Fisheries
	Snowmaking
Nonpotable urban uses	Fire protection
	Air conditioning
	Toilet flushing
Potable reuse	Blending in water supply reservoirs
	Blending in groundwater
	Direct pipe to pipe water supply

Figure 1-6
Irrigation with reclaimed water: (a) fodder, (b) vegetable crops, (c) golf course irrigation, Crete, Greece, and (d) landscape (front yard) irrigation.

needed that expands the traditional supply and demand approach and encompasses environmental and social issues. Suitable methodology to assess various aspects of sustainability is needed especially for detailed engineering analysis.

Although the immediate drivers behind water reuse may differ in each case, the overall goal is to close the hydrologic cycle on a much smaller, local scale. In this way, the used water (wastewater), after proper treatment, becomes a valuable resource literally "at the doorstep of the community" instead of being a waste to be disposed. In many cases, water reuse is practiced because other sources of water are not available due to physical, political, or economic constraints and further attempts to reduce consumption are not feasible. An important breakthrough in the evolution of sustainability for water resources was achieved when water reclamation and reuse were introduced as options to satisfy water demand. Water reclamation and reuse are also the most challenging options, technically and economically, because the source of water is normally of the lowest quality. As a result, extensive treatment is commonly applied, often beyond pure requirements

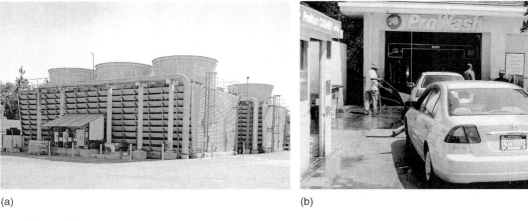

(a) (b)

(c) (d)

Figure 1-7

Nonirrigation use of reclaimed water: (a) evaporative cooling towers, (b) commercial car washing, (c) groundwater recharge, and (d) recreational impoundment.

stemming from the final water use, with a goal of alleviating health concerns to help make the water reuse option palatable to the public. The requirements for reclaimed water (e.g., advanced treatment and a separate distribution system), however, make water reuse costly, thus, limiting its wider use (Hermanowicz, 2005).

Substituting Reclaimed Water for Nonpotable Uses

A growing water resource management trend worldwide is to prioritize the use of water based on availability and quality. Preferentially, the emphasis is on preserving the highest quality water sources for drinking water supplies by using an alternative source such as reclaimed water for applications that have less significant health risks such as irrigating croplands and golf courses. Increasing water productivity for irrigation is an urgent need especially in regions of high water vulnerability. The integration of water reclamation and reuse into water resources management allows for preservation of higher quality water supplies by substituting reclaimed water for direct nonpotable applications.

Water Use Patterns

To assess the role of water reclamation and reuse and provide a framework for evaluating water reuse feasibility, it is important to correlate major water use patterns with potential water reuse applications. For example, in urban areas, industrial, commercial, and nonpotable urban water requirements account for the majority of water demand. In arid and semiarid regions, irrigation is the dominant component of water demand. Water requirements for irrigation applications tend to vary seasonally whereas industrial water needs are more constant. The degree of water reuse for a given watershed depends on the water demand patterns in commercial, industrial, and agricultural applications within the watershed. Seasonal variations in water reuse, needs for reclaimed water storage, and distribution facilities are discussed in Chap. 14.

Personnel Needs/ Sustainable Engineering

A dramatic change has occurred in the water resources development and management over the past three decades. Whereas twentieth century engineers and managers were trained to build dams, reservoirs, and water and wastewater treatment facilities, today's water professionals are confronted with the complex task of assessing the sustainability of water and its impact on society and the environment. In addition to considering technical and economic aspects of water management projects, today's water professionals are becoming the stewards of water resources for the current and future needs of humans and the environment.

For more than a quarter century, a recurring thesis in environmental and water resources engineering has been that improved municipal wastewater treatment could provide a treated effluent of such quality that it should not be wasted but put to beneficial use (see Fig. 1-8). This conviction coupled with the vexing problems of increasing water shortages and environmental pollution, provides a realistic framework for considering municipal wastewater as a water resource in many parts of the world. Water pollution control efforts have made treated effluent from municipal wastewater treatment plants a viable alternative for augmentation of the existing water supply, especially when compared to increasingly expensive and often environmentally destructive development of new water resources.

Treatment and Technology Needs

An important determinant of the potential applications and treatment requirements for water reuse is the quality of water resulting from various municipal uses. A conceptual comparison of the extent to which water quality changes through municipal applications is illustrated on Fig. 1-9. Water treatment technologies are applied to source water such as surface water, groundwater, or seawater to produce drinking water that meets applicable drinking water regulations and guidelines. Conversely, municipal water uses degrade water quality by absorbing and accumulating chemical or biological contaminants and other constituents. The quality changes necessary to upgrade the resulting wastewater then become the basis for wastewater treatment. In practice, treatment is carried out to the point required by regulatory agencies for protection of the environment, including aquatic ecosystems and preservation of beneficial uses of receiving waters.

As the quality of treated water approaches that of unpolluted natural water, the practical benefits of water reclamation and reuse become evident. The levels of treatment and the resultant water quality endow the water with economic value as a water resource.

Figure 1-8
Overview of Harford County, Maryland, Sod Run biological nutrient removal (BNR) wastewater treatment plant (Coordinates: 39.426 N, 76.219 W). The capacity of the plant is 76 × 10³ m³/d (20 Mgal/d).

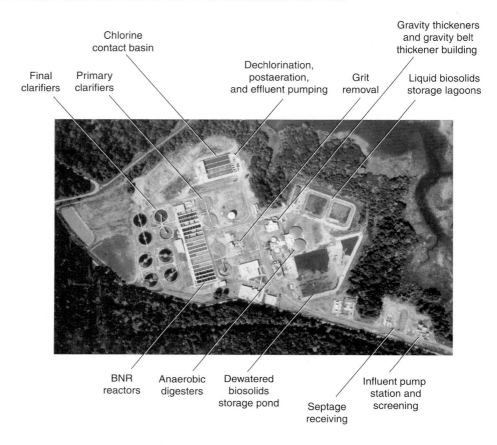

As more advanced technologies are applied for water reclamation, such as carbon adsorption, advanced oxidation, and membrane technologies (see Chaps. 9 and 10), the quality of reclaimed water can meet or exceed the conventional drinking water quality standards by all measurable parameters. This high quality water for indirect potable reuse was termed *repurified* water in the case of San Diego, California and *NEWater* in the case of Singapore (see Chap. 23). Today, technically proven water reclamation or water purification processes exist to provide water of almost any quality desired, including ultrapure water for precision industries and medical uses.

Infrastructure and Planning Issues

Often, reclaimed water system design is approached in the same way as conventional potable water system design. However, special issues arise from the water quality, reliability, variation in supply and demand, and other differences between reclaimed water and freshwater. Engineering issues for a water reclamation and reuse project generally fall into the following categories: (1) water quality, (2) public health protection, (3) wastewater treatment alternatives, (4) pumping, storage, and distribution system siting and design (see Fig. 1-10), (5) on-site conversions at water reuse sites, such as potable and reclaimed water plumbing separation, (6) matching of supply and demand for reclaimed water, and (7) supplemental and backup water supplies. Many aspects of these issues are addressed throughout this textbook.

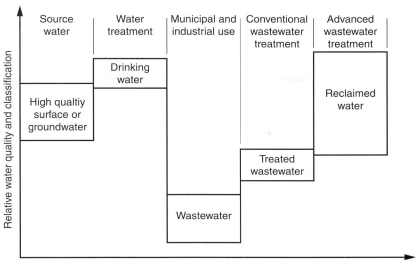

Figure 1-9
Water quality changes during municipal uses of water in a time sequence and the concept of water reclamation and reuse.

It is instructive to examine population growth patterns in the western United States and consider their implications on water reuse infrastructure and planning issues. The counties with the highest population growth rate, up to 60 percent above the average, were characterized by low-to-medium population density (around four people/km^2). In contrast, the counties with high population densities (large cities and densely populated suburbs) and those with very low population densities grew at a much lower rate, sometimes even losing people. Such high growth rates at relatively modest population densities result in significant challenges for water supply, wastewater disposal, and, more importantly, water reuse. At these population densities, individual solutions such as

(a)

(b)

Figure 1-10
Infrastructure is essential in successful water reuse applications: (a) Irrigation pumps and (b) storage reservoir.

wells for water supply, septic tanks and leach fields for wastewater treatment and disposal may no longer be feasible. Yet, traditional communal solutions involving pipelines and collection systems become very expensive due to long distances between individual users (Hermanowicz and Asano, 1999; Anderson, 2003).

Providing the municipal infrastructure is costly. The costs will be unevenly distributed (in the absence of subsidies) with less densely populated communities liable for much higher per capita expenses. Higher costs for larger, less densely populated communities combined with the demographic trend toward modest population densities are likely to strain financially future water projects. It must also be recognized that, until recently, most of the water reclamation and reuse projects have been implemented from centralized municipal wastewater treatment facilities with treatment and disposal requirements that were developed since the late 1970s.

To alleviate the needs for large infrastructure construction, concepts and technologies have advanced using satellite water and wastewater treatment, and decentralized and onsite systems. Topics related to water reclamation and reuse in satellite, decentralized, and onsite systems are discussed in detail in Chaps. 12 and 13.

Ultimately, after appropriate treatment, wastewater collected from cities must be returned to the land or water. The complex question of which contaminants in urban wastewater should be removed to protect the environment, to what extent, and where they should be placed must be answered in light of an analysis of local conditions, environmental and health risks, scientific knowledge, engineering judgment, economic feasibility, and public acceptance. Planning for water reuse is discussed in detail in Chap. 25.

1-5 WATER RECLAMATION AND REUSE AND ITS FUTURE

The social, economic, and environmental impacts of historic water resources development practices and the inevitable prospects of water scarcity are driving the shift to a new paradigm in water resources management. The new approach incorporates the principles of sustainability, environmental ethics, and public participation.

Sustainable water resources management emphasizes whole-system solutions to meet the water needs of present and future generations reliably and equitably. Achieving sustainable water resources management is dependent upon a clear understanding of the distribution and availability of water resources in the hydrologic cycle and the effect that human activities may have on the environment. Sustainable water resources management seeks to design integrated and adaptable systems, increasing efficiency of water use, and making continuous efforts toward protecting ecosystems (Baron et al., 2002).

Environmental ethics plays a significant role in sustainable water resources management by bringing equity into consideration in the context of societal needs and environmental stewardship. Public participation in planning and project development is essential to identify community priorities and concerns, which include not only equity but also growth impacts, cost, and public safety.

While the world's water problems may loom high, steady progress in water reclamation and reuse has been made since the 1970s. To make full use of the water resource created by reclaimed water, several challenges must be met. These include institutional and social obstacles such as regulatory developments and public acceptance. Technical and economic challenges also must be addressed. Important issues related to the future of water reclamation and reuse are summarized in the following paragraphs.

Implementation Hurdles

While water reclamation and reuse is a sustainable approach and can be cost-effective in the long run, the additional treatment of wastewater beyond secondary treatment for reuse and the installation of reclaimed water distribution systems can be costly and energy-intensive as compared to such water supply alternatives as imported water (interbasin transfer of water) or groundwater. Furthermore, institutional barriers as well as varying agency priorities can make it difficult to implement water reuse projects in some cases.

Public Support

The public's awareness of sustainable water resources management is essential; thus, planning should evolve through a community value-based decision-making model. It is important that water reuse is placed within the broader context of water resources management and other options such as desalting to address water supply and water quality problems. Community values and priorities are then identified to guide planning from the beginning in the formulation and selection of alternative solutions.

Acceptance Varies Depending on Opportunity and Necessity

To date the major emphasis of water reclamation and reuse has been on nonpotable applications such as agricultural and landscape irrigation, industrial cooling, and in-building applications such as toilet flushing in large commercial buildings. Indirect and direct potable reuse options raise more public concern and uncertainty. In any case, the value of water reuse is weighed within a context of larger public issues. Water reuse implementation continues to be influenced by diverse factors such as opportunity and necessity; drought and reliability of water supply; growth versus no growth; urban sprawl, traffic noise, and air pollution; and the perception of reclaimed water safety, aesthetics, political will, and public policy governing sustainable water resources management.

Public Water Supply from Polluted Water Sources

Due to land use practices and the increasing proportion of treated wastewater discharged into the nation's waters, freshwater sources of drinking water now contain many of the same constituents of public health concern that are found in reclaimed water. Much of the research that addresses direct and indirect potable water reuse is becoming equally relevant to *unplanned indirect potable reuse* (*de facto* indirect potable reuse) that occurs naturally when water sources containing wastewater discharges are used as a source for drinking water supply. Because of the research interest and public concerns, emerging pathogens and trace organic constituents including disinfection byproducts, pharmaceutically active compounds, and personal care products have been investigated and reported on extensively with regard to public water sources. However, the ramifications of many of these constituents in trace quantity are not well understood with respect to long-term health effects (see Chap. 5).

Advances in Water Reclamation Technologies

Cost-effective and reliable water reclamation technologies are vital to successful implementation of water reuse projects. Comprehensive research on advanced treatment technologies and their combinations, including membrane processes, advanced oxidation, and reliable disinfection is essential (see Fig. 1-11).

(a)   (b)

Figure 1-11
Advanced treatment system consisting of (a) reverse osmosis membrane process, and (b) ultraviolet disinfection system.

Challenges for Water Reclamation and Reuse

The incentives for a water reclamation and reuse program make perfect sense to technical experts—a new water source, water conservation, economic advantages, environmental benefits, government support, and the fact that the cost of wastewater treatment makes the product too valuable to "throw away" or dispose. So why hasn't the concept been embraced and supported wholeheartedly by the community? (Wegner-Gwidt, 1998). The human side of politics, public policy, and decision-making associated with technological advances are not always in concert with technical experts and technological advances. As technology continues to advance and the reliability and safety of water reuse systems is widely demonstrated and public policy and perception changes to embrace these technological advances, water reclamation and reuse will continue to expand as an essential element in sustainable water resources management.

PROBLEMS AND DISCUSSION TOPICS

1-1 What role has water played in the historic development and decline of civilizations such as Mesopotamia? Cite a minimum of three references and summarize your findings.

1-2 Review three articles that deal with renewable water resources and compare the definitions given in the articles to the definition given in the working terminology in this chapter. Discuss the reasons for any differences.

1-3 Discuss what temporal and geographic factors affect "renewable water resources" in the region in which you live.

1-4 What impact does the development of megacities have on renewable water resources?

1-5 Discuss briefly the geopolitical implications of the global distribution of water. Cite three references in your response.

1-6 A much quoted definition of sustainable development was presented in the Brundtland Commission's report *Our Common Future* (WCED, 1987, also available online). However, the question of what is to be sustained, how, and for whom, has been debated extensively for the past two decades. Discuss briefly the elements of sustainable water resources management with respect to equity and interdependence.

1-7 What is your answer to the opinion that water conservation practices are unnecessary because future generations will be able to work out new solutions for any water shortages, should they develop.

1-8 Using reclaimed water is technically, economically, and socially challenging because the source of water is municipal wastewater. Discuss the engineering, social, and economic factors that can be used to justify water reclamation and reuse.

1-9 The incentives for a water reuse program make perfect sense to technical experts— a new water source, water conservation, economic advantages, environmental benefits, government support, and the fact that the cost of wastewater treatment makes the product too valuable to "throw away" or dispose. So why hasn't the concept been embraced and supported wholeheartedly by the community?

1-10 Currently, in the United States, the highest rates of water reuse occur in California and Florida, even though these states have widely different precipitation patterns. Compare regional factors that influence the potential for implementing water reuse.

REFERENCES

Adams, D. B. (1998) "Regional Water Issues, Newsletter of the U.S. National Assessment of the Potential Consequences of Climate Variability and Change," *Acclimations*, 11–12. http://www.usgcrp.gov/usgcrp/Library/nationalassessment/newsletter/1998.12/frame7.html

Anderson, J. (2003) "The Environmental Benefits of Water Recycling and Reuse," *Water Sci. Technol: Water Supply*, **3**, 4, 1–10.

Asano, T. (2002) "Water from (Waste) Water—the Dependable Water Resource," *Water Sci. Technol.*, **45**, 8, 24–33.

Asano, T. (ed.) (1998) *Wastewater Reclamation and Reuse*, Water Quality Management Library, **10**, CRC Press, Boca Raton, FL.

ASCE (1998) *Sustainability Criteria for Water Resources Systems,* prepared by the Task Committee on Sustainability Criteria, Water Resources Planning and Management Division, American Society of Civil Engineers and the Working Group of UNESCO/IHP IV Project M-4.3, Reston, VA.

Baron, J. S., N. L. Poff, P. L. Angermeier, C. N. Dahm, P. H. Gleick, N. G. Hairston, R. B. Jackson, C. A. Johnston, B. D. Richter, and A. D. Steinman (2002) "Meeting Ecological and Social Needs for Freshwater," *Ecol. Appl.*, **12**, 5, 1247–1260.

Baumann, D. D., J. J. Boland, and W. M. Hanemann (1998) *Urban Water Demand Management and Planning*, McGraw-Hill, New York.

Bouwer, H. (2000) "Integrated Water Management: Emerging Issues and Challenges," *Agric. Water Mgmt.*, **45**, 217–228.

Braden, J. B., and E. C. van Ierland (1999) "Balancing: the Economic Approach to Sustainable Water Management," *Water Sci. Technol.*, **39**, 5, 17–23.

Cosgrove, W. J., and F. R. Rijsberman (2000) *World Water Vision: Making Water Everybody's Business*, Earthscan Publications, London, UK.

Falkenmark, M., and G. Lindh (1993) "Water and Economic Development," in P. H. Gleick (ed.), *Water in Crisis: A Guide to the World's Fresh Water Resources*, Pacific Institute for Studies in Development, Environment, and Security, Stockholm Environment Institute, Oxford University Press, New York.

Falkenmark, M., and M. Widstrand (1992) "Population and Water Resources: A Delicate Balance," *Population Bulletin*, Population Reference Bureau, Washington, D.C., **47**, 3, 2–35.

Foster, K. R., P. Vecchia, and M. H. Repacholi (2000) "Science and the Precautionary Principle," *Science*, **288**, 5468, 979–981.

Gleick, P. H. (1998) "Water in Crisis: Paths to Sustainable Water Use," *Ecol. Appl.*, **8**, 3, 571–579

Gleick, P. H. (2000) "The Changing Water Paradigm: A Look at Twenty-First Century Water Resources Development," *Water Inter.*, **25**, 1, 127–138.

Gleick, P. H. (2002) "Soft Water Paths," *Nature*, **418**, 373.

Hermanowicz, S. W., and T. Asano (1999) "Abel Wolman's "The Metabolism of Cities" Revisited: A Case for Water Recycling and Reuse," *Water Sci. Technol.*, **40**, 4–5, 29–36.

Hermanowicz, S. W. (2005) "Sustainability in Water Resources Management: Changes in Meaning and Perception," University of California Water Resources Center Archives. http://repositories.cdlib.org/wrca/wp/swr_v3

Harremoës, P., D. Gee, M. MacGarvin, A. Stirling, J. Keys, B. Wynne, and S. G. Vaz, (eds.) (2001) "Late Lessons from Early Warnings: the Precautionary Principle 1896–2000," *Environmental Issue Report*, No. 22, European Environment Agency, Copenhagen, Denmark.

Harremoës, P. (2002) "Water Ethics: a Substitute for Over-Regulation of a Scarce Resource. Water Scarcity for the 21st Century—Building Bridges Through Dialogue," *Water Sci. Technol.*, **45**, 8, 113–124.

Huesemann, M. W. (2003) "The Limits of Technological Solutions to Sustainable Development," *Clean Tech. Environ. Pollut.*, **5**, 1, 21–34.

IWMI (2000) "World Water Supply and Demand: 1990 to 2025," International Water Management Institute, Colombo, Sri Lanka.

Krayer von Krauss, M., M. B. A. van Asselt, M. Henze, J. Ravetz, and M. B. Beck (2005) "Uncertainty and Precaution in Environmental Management," *Water Sci. Technol.*, **52**, 6, 1–9.

Loucks, D. P. (2000) "Sustainable Water Resources Management," *Water Inter.*, **25**, 1, 3–10.

Mackun, P. J., and S. R. Wilson (2000) *Population Trends in Metropolitan Areas and Central Cities: 1990 to 1998*, Current Population Reports, P25–1133, U.S. Department of Commerce, U.S. Census Bureau, Washington, DC.

Maddaus, W. O. (2001) *Water Resources Planning: Manual of Water Supply Practices*, AWWA Manual M50, American Water Works Association, Denver, CO.

Mantovani, P., T. Asano, A. Chang, and D. A. Okun (2001) *Managing Practices for Nonpotable Water Reuse*, Project 97-IRM-6, Water Environment Research Foundation, Alexandria, VA.

Mayer, P. W., W. B. DeOreo, E. M. Opitz, J. C. Kiefer, W. Y. Davis, B. Dziegielewski, and J. O. Nelson, (1999) *Residential End Uses of Water*, American Water Works Research Foundation, Denver, CO.

Pezzey, J. (1992) "Sustainability: An Interdisciplinary Guide," *Environ. Values*, **1**, 4, 321–362.

Postel, S. L. (2000) "Entering an Era of Water Scarcity: the Challenges Ahead," *Ecol. Appl.*, **10**, 4, 941–948.

Queensland Water Recycling Strategy (2001) *Queensland Water Recycling Strategy: An Initiative of the Queensland Government*, The State of Queensland, Environmental Protection Agency, Queensland, Australia.

Rogers, P. P., M. R. Llamas, and L. Martínez-Cortina (eds.) (2006) *Water Crisis: Myth or Reality?* Taylor & Francis, London.

Sikdar, S. K. (2005) "Science of Sustainability," *Clean Tech. Environ. Pol.*, **7**, 1, 1–2.

Simpson, J. (2006) *Water Quality Star Rating—From Waste-d-Water to Pure Water*, Woombye, Qld, Australia.

Shiklomanov, I. A. (2000) "Appraisal and Assessment of World Water Resources," *Water Inter.* **25**, 1, 11–32.

State of California (2003) California Code—Water Code Section 13050, Subdivision (n). (http://www.leginfo.ca.gov)

Thompson, S. A. (1999) *Water Use, Management, and Planning in the United States*, Academic Press, San Diego, CA.

United Nations (1992) *Agenda 21: The United Nations Programme of Action from Rio de Janeiro*, New York.

United Nations (2002) *World Urbanization Prospects: The 2001 Revision—Data Tables and Highlights*, United Nations, Population Division, Department of Economic and Social Affairs, United Nations Secretariat, United Nations, New York.

United Nations (2003) *World Population Prospects: The 2002 Revision—Highlights*, United Nations Population Division, Department of Economic and Social Affairs, United Nations, New York.

U.S. Department of the Interior (2003) *Water 2025: Preventing Crises and Conflict in the West*, Washington, DC.

USGS (1984) *National Water Summary 1983—Hydrologic Events and Issues*, U.S. Geological Survey Water-Supply Paper 2250.

U.S. Census Bureau (2003) *Population Briefing National Population Estimates for July, 2001*, United States Census Bureau. http://www.census.gov/

Vickers, A. (1991) "The Emerging Demand-Side Era in Water Management," *J. AWWA*, **83**, 10, 38–43.

Vickers, A. (2001) *Handbook of Water Use and Conservation*, WaterPlow Press, Amherst, MA.

Wegner-Gwidt, J. (1998). Public Support and Education for Water Reuse, Chap. 31, 1417–1462, in T. Asano (ed.), *Wastewater Reclamation and Reuse,* Water Quality Management Library, **10**, CRC Press, Boca Raton, FL.

Wilderer, P. A., E. D. Schroeder, and H. Kopp (eds.) (2004) *Global Sustainability*, Wiley-VCH, Germany.

WCED (1987) *Our Common Future (The Brundtland Commision's Report)*, World Commission on Environment and Development, Oxford University Press, Oxford, UK.

WHO (2000) *Global Water Supply and Sanitation Assessment 2000 Report*, WHO/UNICEF Joint Monitoring Programme for Water Supply and Sanitation, World Health Organization, Geneva, Switzerland.

WRI (2000) *World Resources 2000–2001: The Fraying Web of Life*, World Resources Institute, Washington, DC.

2 Water Reuse: Past and Current Practices

	WORKING TERMINOLOGY 38
2-1	EVOLUTION OF WATER RECLAMATION AND REUSE 39
	Historical Development Prior to 1960 39
	Era of Water Reclamation and Reuse in the United States-Post-1960 41
2-2	IMPACT OF STATE AND FEDERAL STATUTES ON WATER RECLAMATION AND REUSE 45
	The Clean Water Act 45
	The Safe Drinking Water Act 46
2-3	WATER REUSE—CURRENT STATUS IN THE UNITED STATES 46
	Withdrawal of Water from Surface and Groundwater Sources 46
	Availability and Reuse of Treated Wastewater 46
	Milestone Water Reuse Projects and Research Studies 47
2-4	WATER REUSE IN CALIFORNIA: A CASE STUDY 47
	Experience with Water Reuse 47
	Current Water Reuse Status 48
	Water Reuse Policies and Recycling Regulations 51
	Potential Future Uses of Reclaimed Water 52
2-5	WATER REUSE IN FLORIDA: A CASE STUDY 53
	Experience with Water Reuse 54
	Current Water Reuse Status 54
	Water Reuse Policies and Recycling Regulations 56
	Potential Future Uses of Reclaimed Water 56
2-6	WATER REUSE IN OTHER PARTS OF THE WORLD 58
	Significant Developments Worldwide 58
	The World Health Organization's Water Reuse Guidelines 59
	Water Reuse in Developing Countries 59
2-7	SUMMARY AND LESSONS LEARNED 63
	PROBLEMS AND DISCUSSION TOPICS 65
	REFERENCES 66

WORKING TERMINOLOGY

Term	Definition
Beneficial uses	The many ways water can be used, either directly by people or for their overall benefit. Examples include municipal water supply, agricultural and industrial applications, navigation, fish and wildlife, habital enhancement, and water contact recreation.
Direct potable reuse	See Portable reuse, direct.
Imported water	Water from one hydrologic region is transferred to another hydrologic region. Examples include the California State Water Project and the Colorado River Project.
Indirect potable reuse	See Portable reuse, indirect.
Integrated water resources planning	A process that promotes the coordinated development and management of water, land, and related resources to maximize the resultant economic and social welfare in an equitable sustainable manner.
Nonpotable reuse	All water reuse applications that do not involve either direct or indirect potable reuse.
Planned water reuse	Deliberate direct or indirect use of reclaimed water, without relinquishing control over the water during its delivery.
Potable reuse, direct	The introduction of highly treated reclaimed water either directly into the potable water supply distribution system downstream of a water treatment plant, or into the raw water supply immediately upstream of a water treatment plant (see Chap. 24).
Potable reuse, indirect	The planned incorporation of reclaimed water into a raw water supply such as in potable water storage reservoirs or a groundwater aquifer, resulting in mixing and assimilation, thus providing an environmental buffer (see Chaps. 22 and 23).
Reclaimed water (also, recycled water)	Municipal wastewater that has gone through various treatment processes to meet specific water quality criteria with the intent of being used in a beneficial manner (e.g., irrigation). The term recycled water is used synonymously with reclaimed water, particularly in California (see Chap. 1, Sec. 1-1).
Sewer mining	The process of tapping into a sewer main and extracting wastewater locally, which can then be treated in a satellite treatment plant and reused for beneficial purposes.
Title 22 regulations	State of California regulations for how treated and recycled water is used and discharged is listed in Title 22 of the California Administrative Code. The statewide Water Recycling Criteria are developed by the Department of Health Services and enforced by the nine State Regional Water Quality Control Boards.
Water reclamation	Treatment or processing of wastewater to make it reusable with definable treatment reliability and water quality criteria (from Chap. 1).
Water recycling	The use of wastewater that is captured and redirected back into the same water use scheme such as in industry. However, the term *water recycling* is often used synonymously with water reclamation (see Chap. 1, Sec. 1-1).
Water reuse	The use of treated wastewater for a beneficial use, such as agricultural irrigation and industrial cooling.

In Chap. 1, it was noted that continued population growth, contamination of both surface water and groundwater, uneven distribution of water resources, and periodic droughts have forced water agencies to search for additional sources of water supply. The reuse of treated wastewater effluent was examined as an important element of future water resources management strategies. The purpose of this chapter is to provide an overview of the past and current practices of water reuse in the United States and in selected parts of the world and to discuss future trends. These practices will serve as a basis for developing more effective and sustainable water reuse practices in the future.

To provide the needed perspective on past and current water reuse practices, this chapter is organized in seven sections dealing with (1) the evolution of water reclamation and reuse, (2) the impact of federal statutes on water reclamation and reuse, (3) the current status of water reuse in the United States, (4) a case study of water reuse in California, (5) a case study of water reuse in Florida, (6) water reuse in other parts of the world, and (7) a summary of lessons learned in implementing water reuse.

2-1 EVOLUTION OF WATER RECLAMATION AND REUSE

The purpose of this section is to provide a brief overview of the evolution of water reclamation and reuse. Topics considered include (1) a brief historical review of water reuse prior to 1960, (2) significant water reclamation and reuse in the United States post 1960, and (3) significant developments worldwide. The year 1960 is used as a time division because significant water pollution control activities in the United States and the modern era of water reclamation and reuse both occurred after 1960. The impact of state and federal statutes on water reclamation and reuse is discussed in Sec. 2-2.

Key events that have contributed to the evolution of water reclamation and reuse up to about 1960 are summarized in Table 2-1. The reuse of wastewater is not new. For example, indications of the use of wastewater for agricultural irrigation extend back approximately 3000 years to the Minoan Civilization in Crete, Greece (Angelakis et al., 1999 and 2003). In modern times, the beginnings of water reclamation and reuse can be traced to the mid-nineteenth century with the introduction of wastewater systems for conducting household wastes away from urban dwellings into the nearest water courses. The considerable pollution of the Thames River as it passed through London, UK, not only caused nauseating conditions in the city but also was responsible for repeated epidemics of cholera among those served by a public water supply taken from the unsanitary Thames. The solution was the construction of a vast interceptor along the Thames, which, following the admonition of Sir Edwin Chadwick—*the rain to the river and the sewage to the soil*, carried the wastewater downstream for spreading on sewage farms. Such land disposal schemes were widely adopted by large cities in Europe as well as in the United States up to the early twentieth century (Metcalf and Eddy, 1928; Barty-King, 1992; Okun, 1997; Cooper, 2001).

Historical Development Prior to 1960

When the water supply link with disease became clearer, engineering solutions were implemented that included the development of alternative water sources using reservoirs and aqueduct systems, the relocation of water intakes to upstream of wastewater discharges, and the progressive introduction of water filtration during the 1850s and '60s (Barty-King, 1992; Cooper, 2001). Microbiological advances in the late nineteenth century precipitated the *Great Sanitary Awakening* (Fair and Geyer, 1954) and the advent of chlorine disinfection. The development of the activated sludge process around 1913 was a significant step toward advancement of wastewater treatment and, specifically, the development of biological wastewater treatment systems.

Table 2-1
Historic and milestone events related to the evolution of water reclamation and reuse worldwide through 1968[a]

Period	Location	Events
~ 3000 BC	Crete, Greece	Minoan civilization: use of wastewater for agricultural irrigation.
97 AD	Rome, Italy	The City of Rome has a water supply commissioner, Sexus Julius Frontinus.
1500 ~	Germany	Sewage farms are used for wastewater disposal.
1700 ~	United Kingdom	Sewage farms are used for wastewater disposal.
1800–1850	France, England, United States	Legal use of sewers for human waste disposal in Paris (1880), London (1815), and Boston (1833) instituted.
1850–1875	London, England	Cholera epidemic is linked to polluted well water by Snow.
1850–1875	England	Typhoid fever prevention theory developed by Budd.
1850–1875	Germany	Anthrax connection to bacterial etiology demonstrated by Koch.
1875–1900	France, England	Microbial pollution of water demonstrated by Pasteur. Sodium hypochlorite disinfection by Down to render water "pure and wholesome" advocated.
1890	Mexico City, Mexico	Drainage canals are built to take untreated wastewater to irrigate an important agricultural area north of the city, a practice that still continues today. Untreated or minimally treated wastewater from Mexico City is delivered to the Valley of Mexico where it is used to irrigate about 90,000 ha of agricultural lands, including vegetables.
1906	Jersey City, NJ	Chlorination of water supply.
1906	Oxnard, CA	The earliest reference related to a public health viewpoint of water quality requirements for the reuse of wastewater appears in the *Monthly Bulletin, California State Board of Health*, February, 1906 on the Oxnard septic tank system of sewage disposal.
1908	England	Disinfection kinetics elucidated by Chick.
1913–1914	United States and England	Activated sludge process is developed at the Lawrence Experiment Station in Massachusetts and demonstrated by Ardern and Lockett in England.
1926	United States	In Grand Canyon National Park treated wastewater is first used in a dual water system for toilet flushing, lawn sprinkling, cooling water, and boiler feed water.
1929	United States	The City of Pomona, CA initiated a project utilizing reclaimed water for irrigation of lawns and gardens.
1932–1985	San Francisco, CA	Treated wastewater is used for watering lawns and supplying ornamental lakes in Golden Gate Park.

Table 2-1
Historic and milestone events related to the evolution of water reclamation and reuse worldwide through 1968[a] (*Continued*)

Period	Location	Events
1955	Japan	Industrial water is supplied from Mikawajima wastewater treatment plant by Tokyo Metropolitan Sewerage Bureau.
1968	Namibia	Direct potable reuse begun at Windhoek's Goreangab Water Reclamation Plant.

[a]Adapted in part from Metcalf and Eddy (1928); Ongerth and Jopling (1977); Barty-King (1992); Okun (1997); Cooper (2001); Angelakis et al. (2003).

The earliest reference related to a public health viewpoint of water quality requirements for the reuse of wastewater appears in the *Monthly Bulletin, California State Board of Health*, February 1906, on the Oxnard septic tank system of sewage disposal. "Why not use it for irrigation and save the valuable fertilizing properties in solution, and at the same time completely purify the water? The combination of the septic tank and irrigation seems the most rational, cheap, and effective system for this state." (Ongerth and Jopling, 1977). In a 1915 U.S. Public Health Service Bulletin, it was noted that if effluent from a septic tank were disposed of in a shallow trench located 0.3 m below the soil surface the effluent ". . . may be used advantageously to cultivate an attractive hedge of roses or other shrubs or to cultivate a row of corn or other plants, the edible parts of which are produced well above the surface of the ground." (Lumsden et al., 1915).

One of the earliest cases of industrial reuse in the United States was the use of chlorinated wastewater effluent for steel processing at the Bethlehem Steel Company in Baltimore, Maryland, which was practiced from 1942 until the company ceased operations in the late 1990s (see Chap. 19, Sec. 19-3). In the 1960s, planned urban water reuse systems were developed in response to rapid urbanization in California, Colorado, and Florida.

Era of Water Reclamation and Reuse in the United States—Post 1960

Further technological advances in physical, chemical, and biological processing of water and wastewater during the first half of the twentieth century led to the contemporary era of water reclamation and reuse, which had its beginnings around 1960. Factors contributing to the development of water reclamation and reuse since 1960 include: (1) rapid population growth in the West, (2) increased development in humid climatic regions, particularly in the State of Florida, (3) more stringent wastewater treatment and effluent discharge regulations, (4) conducting water reuse demonstration projects, and (5) the development of water reclamation and reuse guidelines and regulations in many states. The impact of more stringent wastewater treatment and effluent discharge requirements is considered in Sec. 2-3. Milestone events related to the evolution of water reclamation and reuse in the United States since 1960 are summarized in Table 2-2.

Rapid Growth in the Arid West
Since the 1960s, rapid population growth in the arid west, the associated regulatory pressures related to water pollution control, and water shortages have encouraged the use of reclaimed water (see Fig. 2-1). For example, Colorado Springs, Colorado, is located at the eastern base of the Rocky Mountains in a water-short area. To reduce dependence on water from the western slopes of the mountains, in the early 1960s the city implemented a limited dual-distribution system in which reclaimed water was used to meet irrigation

Table 2-2

Milestone events related to the evolution of water reclamation and reuse in the United States—post-1960[a]

Period	Location	Events
1960	Sacramento, CA	California legislation encourages wastewater reclamation and reuse in the State Water Code.
1962	Los Angeles County, CA	A major groundwater recharge project by surface spreading is initiated at the Whittier Narrows spreading basin.
1965	San Diego County, CA	Santee recreational lakes, supplied with reclaimed water, are opened for swimming, and put-and-take fishing.
1972	Washington, DC	U.S. Clean Water Act to restore and maintain water quality is passed.
1975	Fountain Valley, CA	Groundwater recharge by direct injection of reclaimed water into aquifers is started by the Orange County Water District (known as Water Factory 21).
1977	Pomona, CA	Pomona Virus Study, conducted by Sanitation Districts of Los Angeles County, is published.
1977	Irving, CA	Irving Ranch Water District initiates a major landscape irrigation project with a dual water system delivering reclaimed water.
1977	St. Petersburg, FL	Another major urban water reuse system is initiated in St. Petersburg, Florida.
1978	Sacramento, CA	California Wastewater Reclamation Criteria (Title 22 regulations) are promulgated by the Department of Health Services to be enforced by nine Regional Water Quality Control Boards.
1982	Tucson, AZ	Initiates a metropolitan water reuse program mandating use of reclaimed water in golf courses, school grounds, cemeteries, and parks.
1984	Los Angeles, CA	Health Effects Study by Los Angeles County Sanitation Districts is published.
1987	Monterey, CA	Monterey Wastewater Reclamation Study for Agriculture by Monterey Regional Water Pollution Control Agency is published.
1987	Sacramento, CA	Report of the Scientific Advisory Panel on Groundwater Recharge with Reclaimed Wastewater is published by the State of California Interagency Water Reclamation Coordinating Committee.
1992	Washington, DC	U.S. Environmental Protection Agency and U.S. Agency for International Development publish *Guidelines for Water Reuse*.
1993	Denver, CO	Potable Water Reuse Demonstration Plant—Final report (pilot plant operation began in 1984) is published.
1996	San Diego, CA	City of San Diego Total Resource Recovery Health Effects Study is published by Western Consortium for Public Health.

Table 2-2
Milestone events related to the evolution of water reclamation and reuse in the United States—post-1960[a] (*Continued*)

Period	Location	Events
2003	Sacramento, CA	Water Recycling 2030: Recommendations of California's Recycled Water Task Force, Department of Water Resources and State Water Resources Control Board, State of California is published.
2004	Washington, DC	U.S. Environmental Protection Agency and U.S. Agency for International Development published *Guidelines for Water Reuse*.

[a]Adapted in part from Metcalf and Eddy (1928); Barty-King (1992); Ongerth and Jopling (1977); Okun (1997); Asano (1998); Cooper (2001); U.S. EPA (1992); State of California (2002a); U.S. EPA and U.S. AID (2004).

demands in addition to surface water from a nearby stream. This is one of the oldest operating systems in the United States in which reclaimed water is used for urban landscape irrigation. Current reclaimed water uses in Colorado Springs include parks, golf courses, cemeteries, and commercial properties, as well as the 280 MW Martin Drake Power Plant.

Development in Humid Climatic Regions

Water reclamation and reuse is taking on added significance in humid climatic regions where increased community development is putting considerable pressure on water resources and collection system services. In St. Petersburg, Florida, for example, the reclaimed water system has continued to expand and change in character. From its inception in the late 1970s, the St. Petersburg system has evolved from one of an alternative mode of wastewater effluent disposal to one of a fully operational reclaimed water supply. The growth in the use of reclaimed water has contributed significantly to the suppression of potable water demands over the past 20 years (see Chap. 26, Sec. 26-5). Also, Venice, Florida, which has a critical water supply problem and a high growth rate, constructed the East Side Wastewater Treatment Plant, which now provides reclaimed water for urban landscape irrigation.

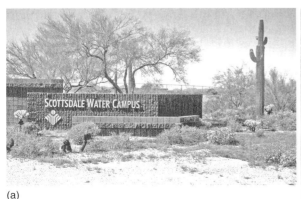
(a)

(b)

Figure 2-1
Era of water reclamation and reuse in the United States. Rapid population growth, regulatory pressure on water pollution, and water shortages have encouraged the use of reclaimed water: (a) Scottsdale Water Campus, AZ, and (b) St. Petersburg, FL.

Cold Weather Discharge Permits

One of the innovative approaches that the State of Georgia has implemented to encourage water reuse is allowing discharge of treated effluents during cold weather to the surface waters of the state. The state limits discharges of treated effluents during warm weather due to the impact on aquatic life. As a result, water reuse is encouraged during the summer months, whereas these cities would have little place to store the flows in the winter months.

Water Reclamation and Reuse Research, and Development of Regulations and Guidelines

Several water reclamation and reuse research and demonstration projects have provided valuable insight into treatment system design concepts and health risk assessment in water reuse. In 1977, a comprehensive research project, known as the *Pomona Virus Study* (SDLAC, 1977; see also Table E-1 in App. E), was completed at the Pomona Research Facility of the Sanitation Districts of Los Angeles County which evaluated various tertiary wastewater treatment systems for the removal of enteric viruses (Dryden et al., 1979; Chen et al., 1998). Following the completion of the project, the California Department of Health Services recommended specific design and operational requirements for treatment alternatives for water reclamation including in-line coagulation and flocculation, and direct filtration, both of which are more cost-effective filtration systems than a conventional treatment train consisting of chemical coagulation, flocculation, sedimentation, and filtration (see Fig. 2-2).

Figure 2-2

Pilot plant used to conduct the Pomona Virus Study in 1977. The study was completed at the Pomona Research Facility of the County Sanitation Districts of Los Angeles County. The overall objective of the study was to evaluate various tertiary wastewater treatment systems for the removal of enteric viruses.

The findings of the *Pomona Virus Study* were influential in the formulation of the State of California's 1978 *Wastewater Reclamation Criteria* (Title 22 regulations) which have been referenced widely in various states and also abroad (State of California, 1978). For example, in the State of California Water Code, it is noted that "It is the intention of the Legislature that the State undertakes all possible steps to encourage development of water reclamation facilities so that reclaimed water will be available to help meet the growing water requirements of the State." (Water Code Sections 13510–13512). In 2003, the State of California published a report, *Water Recycling 2030: Recommendations of California's Recycled Water Task Force,* which evaluated the current framework of state and local rules, regulations, ordinances, and permits to identify opportunities for and obstacles or disincentives to increasing the safe use of reclaimed water in the next 25-year horizon (State of California, 2003b).

2-2 IMPACT OF STATE AND FEDERAL STATUTES ON WATER RECLAMATION AND REUSE

The development of programs for planned reuse of wastewater began in the early part of the twentieth century. The State of California was a pioneer in promoting water reclamation and reuse, with the Board of Public Health adopting in 1918 its initial *Regulation Governing Use of Sewage for Irrigation Purposes*. The regulations prohibited the use of ". . . raw sewage, septic or Imhoff tank effluents, or similar sewage or water polluted by such sewage . . ." for the irrigation of tomato, celery, lettuce, berries, and other produce that is eaten raw (Ongerth and Jopling, 1977). The standards for treatment and reuse have continued to evolve for the purpose of protecting public health. Two U.S. federal statutes that have a significant impact on the quantity and quality of wastewater discharged and the potential for water reuse are the Water Pollution Control Act and its Amendments, now known also as the Clean Water Act (CWA), and the Safe Drinking Water Act (SDWA). The combined effectiveness of these two acts working in consonance will ultimately determine the quantity and quality of viable water sources available for water reuse. These two acts are discussed briefly in the following paragraphs.

The Clean Water Act

Discharges of untreated wastewater from municipalities, industries, and businesses caused widespread pollution of rivers, lakes, and coastal waters. In 1972, Congress responded to public outrage over the deplorable condition of the nation's waters by enacting the CWA. The CWA and its amendments determine the degree and type of wastewater treatment necessary to meet prescribed effluent standards—whether that effluent is to be reclaimed and reused, or discharged to a receiving body of water.

The CWA was the milestone event in water pollution control in the United States designed *to restore and maintain the chemical, physical, and biological integrity of the Nation's waters* with the ultimate goal of zero discharge of pollutants into navigable, fishable, and/or swimmable waters. Within the language of the CWA is the goal to achieve greater use of those systems that reclaim and reuse water by productively treating and recycling wastewater. Furthermore, the CWA ensures improvement in the general quality of wastewater through increasingly more stringent pretreatment standards of industrial discharges. As a result of the CWA, centralized wastewater treatment has become commonplace in urban areas and treated effluents have become readily available sources for water reuse (WEF, 1997; U.S. EPA, 1998).

The Safe Drinking Water Act

The SDWA was enacted in 1974 and has had a major impact on the way water treatment and distribution are mandated. Subsequent amendments have updated the SDWA to keep abreast of health concerns and technical advances. The purpose of the SDWA is to ensure that water supply systems serving the public meet minimum standards for the protection of public health. The SDWA was designed to achieve uniform safety and quality of drinking water in the United States by identifying contaminants and establishing maximum acceptable contaminant levels.

The SDWA, which provides regulations for potable water supplies, indirectly affects the quality of wastewater as well because many wastewaters are discharged into streams that are used for public water supplies (see Chap. 3, Sec. 3-1). A public water supply system must maintain a watershed control program that will minimize the potential for contamination by human enteric viruses and *Giardia lamblia* cysts (Clark and Summers, 1993).

2-3 WATER REUSE—CURRENT STATUS IN THE UNITED STATES

The current status of water reclamation and reuse in the United States is examined in this section. A closer look at water reuse practices in two states, California and Florida, is provided in the following two sections to further illustrate the extent and applications of water reuse, the driving factors, and the different policy approaches for promoting and regulating water reuse. California and Florida are also the major states to compile comprehensive inventories of water reuse projects by types of water reuse application.

Withdrawal of Water from Surface and Groundwater Sources

Conclusions drawn from estimates of water use in the United States are that approximately 1.5×10^9 m^3/d water withdrawals were made for all uses during 2000 (Hutson et al., 2004). California, Texas, and Florida accounted for one-fourth of all water withdrawals. States with the largest surface water withdrawals were California, which had large withdrawals for irrigation and thermoelectric power, and Texas, which had large withdrawals for thermoelectric power. States with the largest groundwater withdrawals were California, Texas, and Nebraska, all of which had large withdrawals for irrigation.

Availability and Reuse of Treated Wastewater

Information on the quantities of wastewater treated and released from publicly owned treatment facilities and returned directly to the hydrologic cycle, or released for beneficial reuse (reclaimed water) were reported by the U.S. Geological Survey (Solley et al., 1998). About 16,400 publicly owned treatment facilities released some 155×10^6 m^3/d of treated wastewater nationwide during 1995. In addition, only about two percent (4×10^6 m^3/d) of the treated wastewater was reclaimed for beneficial uses such as irrigation of golf courses and public parks. The States of Florida, California, and Arizona all reported large uses of reclaimed water. Data from 1995 is reported because the U.S. Geological Survey's latest publication, *Estimated Use of Water in the United States in 2000,* did not report reclaimed water, number of wastewater facilities, or wastewater returned. Quality of data was cited as the reason for the omission in this latest report (Hutson et al., 2004). However, the WateReuse Association (an organization promoting water reuse research and implementation), estimates that 9.8×10^6 m^3/d (2.6×10^3 Mgal/d) of municipal wastewater are reclaimed and reused currently, and reclaimed water use on a volume basis is growing at an estimated rate of 15 percent per year (WateReuse Association, 2005).

Water scarcity and wastewater discharge regulations have been the motivating factors in the development of water reclamation projects. Most water reuse sites are located in the arid and semiarid western and southwestern states where water supplies are limited. However, an increasing number of water reuse projects are being implemented in the humid regions of the United States due to the rapid growth and urbanization in these regions.

A number of milestone water reuse projects and research studies over the past century have led to the current knowledge of water reclamation and reuse. Selected milestone projects and research studies in the United States are shown in Table E-1 in App. E. These projects were selected either because of their pioneering water reuse applications, or their significant scientific and engineering impacts on later developments in water reclamation and reuse. The presentation of milestones is also a recognition of the pioneering planners and engineers who were able to look ahead of their time and push forward the frontiers of water reclamation and reuse from obscure practice to a growing discipline in sustainable water resources management.

Milestone Water Reuse Projects and Research Studies

2-4 WATER REUSE IN CALIFORNIA: A CASE STUDY

California, the most populous state (2004 population: 35.9 million) in the union, is a state where two-thirds of the population live in a semiarid and desert climate. As a result, efficient water use is critical to sustaining water availability. To meet the water demands associated with future growth, the State of California is working to develop a balanced portfolio of water resources. The future water resource portfolios include not only traditional dams and reservoirs but also an array of other types of facilities and management techniques, such as water transfers, water conservation, desalination, and water reclamation and reuse (State of California, 2005). In 1991, the State of California established a statewide goal to reclaim and reuse 1234×10^6 m^3/yr by the year 2010. Furthermore, it has been estimated that reclaimed water could free up enough freshwater to meet the household water demands of 30 to 50 percent of the additional 17 million Californians expected to live there in 2030. To achieve this potential, an investment of $11 billion will be needed (State of California, 2003b).

In many ways, California has been in the vanguard of water reclamation and reuse since its early days as a state. Water reclamation has been practiced in California as early as 1890 for agriculture. By 1910 at least 35 communities were using wastewater for farm irrigation, 11 without wastewater treatment, and 24 after septic tank treatment. Landscape irrigation in Golden Gate Park in San Francisco (see Table E-1 in App. E) began with untreated municipal wastewater, but minimal treatment was added in 1912.

Experience with Water Reuse

Table 2-3

Type and quantity of water reuse in the States of California and Florida[a]

	Water reuse quantity			
	California		Florida	
Type of water reuse	10^6 m³/yr	% of total	10^6 m³/yr	% of total
Agricultural irrigation	297	46	131	16
Landscape irrigation	137	21	379	45
Industrial use	34	5	122	15
Groundwater recharge	60	9	135	16
Seawater intrusion barrier	32	5	na	–
Recreational impoundment	41	6	na	–
Wildlife habitat	25	4	61	7
Geysers/energy production	3	1	na	–
Other uses or mixed type	19	3	6	1
Total	648	100	834	100

[a]Adapted from State of California (2002); State of Florida (2004).
na = not applicable

Wastewater treatment standards have continued to evolve and further protect public health, and by 1952, there were 107 communities in California using reclaimed water for agricultural and landscape irrigation.

Current Water Reuse Status

The first comprehensive statewide estimate of water reuse was made in 1970, when 216×10^6 m³ of recycled water were used. By the end of 2001, reclaimed water use in California had reached over 648×10^6 m³/yr (State of California, 2002).

Water Reuse Applications

Types and quantity of reclaimed water use are shown in Table 2-3. Agricultural and landscape irrigation is the dominant use of reclaimed water (67 percent of the total water reuse by volume). At least 20 varieties of food crops are grown with reclaimed water, including vegetables eaten uncooked such as lettuce, celery, and strawberries. Eleven nonfood crops, especially pasture and feed for animals, as well as nursery products, are irrigated with reclaimed water. Landscape irrigation is primarily for turf, including over 125 golf courses and many parks, schoolyards and freeway landscaping. Industrial and commercial uses include cooling towers in power stations, boiler feed water in oil refineries, carpet dying, and recycled newspaper processing. Reclaimed water is also used in office and commercial buildings for toilet and urinal flushing (CSWRCB, 2003; State of California, 2003b; Crook, 2004; Levine and Asano (2004).

In many groundwater basins in California, the rate of pumping exceeds the rate of natural replenishment. Artificial recharge of groundwater is practiced in some areas by percolating either stormwater captured from streams, imported water, or reclaimed water into aquifers. The most notable use of reclaimed water for this purpose is groundwater recharge in the Montebello Forebay, which has been in operation since 1962, located near Whittier in Los Angeles County (see Fig. 2-3; also Table E-1 in App. E). In coastal

Figure 2-3
Aerial view of Rio Hondo Spreading Grounds (Courtesy of County Sanitation Districts of Los Angeles County). These basins and the unlined portions of the rivers and creeks permit large volumes of reclaimed water to percolate into the aquifer. (Coordinates: 33.993 N, 118.105 W, view at altitude 4 km.)

areas, where excessive groundwater pumping has taken place, the groundwater levels have fallen to the extent that seawater has been drawn inland, contaminating aquifers. Reclaimed water has been injected into the aquifers along the coast to create barriers to the seawater, thus protecting the groundwater while, in part, also replenishing the drinking water aquifer. Highly treated reclaimed water from Orange County Water District's historic Water Factory 21 has been injected into coastal aquifers to act as a seawater intrusion barrier since 1976 (see Fig. 2-4). Other groundwater recharge facilities in Orange County are

Figure 2-4
Orange County Water District's Water Factory 21, CA. A view from effluent launders of chemical (lime) precipitation clarifiers looking toward administration building. Lime recalcining and chemical storage building is on the left; ammonia stripping towers are visible on the right (CA. 1976)

Figure 2-5
Aerial view of groundwater recharge facilities in Orange County, CA (Courtesy of Orange County Water District). Deep spreading basins (left) used to recharge Colorado River water and Santa Ana River spreading basins with finger levees (right) used to recharge groundwater with river water dominated by reclaimed water from upstream plants. (Coordinates: 33.856 N, 117.845 W, view at altitude 4 km)

shown on Fig. 2-5. A more recently constructed project also operates along the coast in Los Angeles County (State of California, 2003b). Construction began in 2004 on a new Groundwater Replenishment System (replacing and dismantling Water Factory 21), which is a joint project of the Orange County Water District and the Orange County Sanitation District. Replacement of Water Factory 21 with newer technology (e.g., microfiltration and reverse osmosis membrane systems), is a part of this project (see Chap. 22).

Geographic Distribution of Water Reuse Sites

Most of the reclaimed water use in California is in the Central Valley and the South Coastal Regions, amounting to 80 percent of the reclaimed water produced in California. The coastal areas from Santa Barbara County north and the desert and eastern Sierra Nevada regions use the remaining 20 percent. The uses of reclaimed water reflect the land uses in these regions. The Central Valley of California is dominated by agriculture, which is a readily accessible market that can use reclaimed water receiving relatively low levels of treatment (e.g., secondary treatment).

Urban uses of reclaimed water are dominant in the South Coastal Region (Counties of Ventura, Los Angeles, Orange, and San Diego and portions of San Bernardino and Riverside), where about half of the state's population resides. The dependence of the

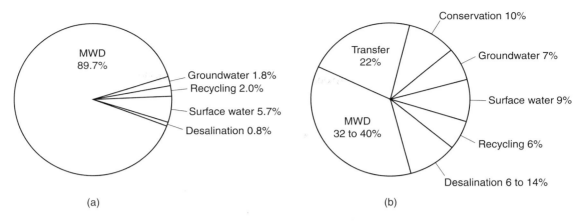

Figure 2-6
Comparison of regional water supply sources for San Diego County, CA the years 2002 and 2020 (a) 2002 and (b) 2020. The principal source of water is from the Metropolitan Water District (MWD) of Southern Californa (Adapted from San Diego County Water Authority, 2002).

south coastal area on expensive imported water has stimulated demand for alternative sources of water, such as reclaimed water. In fact, water and wastewater agencies in these regions were the first to use reclaimed water extensively. An exception to this trend is the City of San Diego. Despite a large metropolitan water demand, supplied mostly by the imported water, only limited water reclamation and reuse projects have been implemented. Water reuse was the victim of politics, planning limitations, and a lack of public support. However, it is anticipated that water reclamation and reuse will play an important role in San Diego in the future (see Chap. 23). Projections of regional water supply sources are that six percent of the water supply will come from water reclamation and reuse in the year 2020 as depicted on Fig. 2-6 (San Diego County Water Authority, 2002).

Size of Water Reclamation Systems

The measure of the size of a water reclamation system is the total annual reclaimed water deliveries from each wastewater treatment plant. System sizes range from less than 400 m³/yr (Terra Bella Sewer Maintenance District in Tulare County) to over 50×10^6 m³/yr (City of Los Angeles, Donald C. Tillman Water Reclamation Plant). Some agencies, either on their own or in cooperation with water districts or other water purveyors, have played a major role in developing the use of reclaimed water. Some of the districts operate more than one treatment plant producing reclaimed water. The 15 largest reclaimed water producing agencies in California are listed in Table 2-4. In 2002, there were over 200 water reclamation plants delivering reclaimed water throughout California, but nearly 60 percent of reclaimed water came from the 15 largest water reclamation and reuse agencies identified in Table 2-4.

Water Reuse Policies and Recycling Regulations

The California Department of Health Services (DHS) has the authority and responsibility to establish statewide health-related regulations for water reclamation and reuse. The *Wastewater Reclamation Criteria* (State of California, 1978) were widely used for over 20 years, the formative years of water reclamation and reuse, and were commonly

Table 2-4

The 15 largest reclaimed water producing agencies in California[a]

Rank	Agency	Number of plants	Reclaimed water deliveries, 10^6 m^3/yr	
			1987	2001
1	County Sanitation Districts of Los Angeles County	8	66	103
2	City of Los Angeles	2	4	50
3	City of Bakersfield	2	30	39
4	Eastern Municipal Water District	4	12	35
5	West Basin Municipal Water District	1	0	32
6	Irvine Ranch Water District	1	10	24
7	City of Santa Rosa	2	11	15
8	Monterey Regional Water Pollution Control Agency	1	0	15
9	Orange County Water District	1	3	14
10	City of Modesto	1	18	13
11	Inland Empire Utilities Agency	4	2	12
12	Las Virgenes Municipal Water District	1	5	8
13	East Bay Municipal Utility Distict	1	0	7
14	City of San Jose	1	0	7
15	South Tahoe Public Utility District	1	6	6
	Total	31	167	380

[a]Adapted from State of California (1990) and (2002).
Note: There were over 200 water reclamation plants in California delivering reclaimed water statewide in 2001, but 59 percent (380/648) of the reclaimed water came from the 15 largest water reclamation and reuse agencies as listed in this table.

known as *Title 22 regulations* because they were listed in Title 22, Division 4 of the California Code of Regulations. The current *Water Recycling Criteria* were adopted by DHS in 2000 (State of California, 2000). The water recycling criteria include water quality standards, treatment process requirements, operational requirements, and treatment reliability requirements (see detailed discussions in Chap. 4).

The State of California Water Code mandates nine Regional Water Quality Control Boards (RWQCBs) to establish water quality standards, to prescribe and enforce waste discharge requirements, and, in consultation with DHS, to prescribe and enforce water reclamation requirements. Thus, the regional boards enforce DHS's *Water Recycling Criteria*, and each water reclamation project must have a permit from the appropriate RWQCB conforming to DHS criteria.

Potential Future Uses of Reclaimed Water

Water planners are continually evaluating a variety of alternative water sources to determine the most cost-effective and feasible options available [e.g., *The California Water Plan Update 2005* (State of California, 2005)]. Public health concerns are increasing, not only with respect to reclaimed water but also with all sources of water including drinking water.

Table 2-5
Projections for reclaimed water use in California[a] ($\times 10^6$ m³/yr)

Application	Year			
	2002	2007	2010	2030
Planned nonpotable use	494–629	642–913	950–1234	1875–2283
Planned indirect potable use[b]	61–86	99–148	148–210	407–494
Total	555–715	741–1061	1098–1444	2282–2777

[a]Adapted from State of California (2003b).
[b]Planned indirect potable use includes groundwater recharge, a portion of recharged groundwater in seawater intrusion barriers, and surface water reservoir augmentation for domestic water supply.

However, technology is becoming more effective in removing pathogens and trace chemical constituents of concern. Evolving technology will make water reclamation and reuse, and alternative treatment methods such as membrane processes, more reliable and economical in the future. It is anticipated that the next areas for expanded reclaimed water use will be landscape irrigation, industrial reuse, groundwater recharge, and surface water augmentation.

It is difficult to predict exactly how reclaimed water will compare with alternative supply options in the long term. However, two comprehensive studies estimating future water reuse potential were conducted in regions covering the metropolitan areas of the southern California coastal region and the San Francisco Bay area (State of California, 2003b). Additional surveys were conducted in which wastewater agencies were polled regarding potential projects within their service areas. Based on these studies, projections of available wastewater, and the caveats of uncertainty, a range of projections for reclaimed water use is presented in Table 2-5. Planned nonpotable and planned indirect potable uses are listed separately in Table 2-5 because of the different public health concerns and public acceptance issues related to indirect potable reuse.

To put water reuse in perspective, a total of 635×10^6 m³ of reclaimed water was used in 2002 (the midrange of values listed in Table 2-5), which is approximately 10 percent of the amount of treated municipal wastewater produced in California in 2000, estimated to be about 6.2×10^9 m³/yr. In 2030, the amount of reclaimed water use is projected to be 2500×10^6 m³/yr, which is approximately 23 percent of the anticipated available municipal wastewater.

2-5 WATER REUSE IN FLORIDA: A CASE STUDY

The State of Florida receives on average over 1,270 mm of rainfall each year. While the state may appear to have an abundance of water, continuing population growth, primarily in the coastal areas, contribute to increased concerns about future water availability. Florida's population was approximately 17.4 million in 2004, the fourth largest in the United States after California, Texas, and New York, and the population growth rate between 1990 and 2000 was 23.5 percent (State of Florida, 2003a).

The major driving force for Florida to continue to pursue water reclamation and reuse is the state's rapid population growth, which is projected to reach about 20 million by

2020, and its associated water demand (York and Wadsworth, 1998). However, Florida was motivated initially to adopt water reclamation and reuse as a means to control wastewater discharge and associated environmental impacts such as coastal eutrophication. In recent years, Florida has risen to become a nationally recognized leader in water reuse along with California.

Experience with Water Reuse

Until the late 1960s, secondary wastewater treatment and discharge into surface water was common practice in Florida. With growing environmental awareness, however, municipalities and utilities in Florida were charged with managing wastewater in an environmentally sound and cost-effective manner. Most of Florida's streams are small, warm, and slow moving, and there are a number of environmentally sensitive lakes, estuaries, and coastal waters throughout the state. Regulations limit significantly the quantity and quality of effluent that may be discharged to surface waters to protect them from environmental degradation. As a result, a move toward land application and water reuse systems emerged in the 1970s and grew in size and scope during the 1980s (Young and York, 1996). Two state regulations, one in 1986 and one in 1990 were developed to further protect ecologically sensitive coastal areas (State of Florida, 2002 and 2003b). These regulations required full advanced wastewater treatment (AWT) to protect ecologically sensitive coastal areas, and surface discharge was essentially precluded unless AWT was provided. The specified limits for AWT were 5 mg/L for carbonaceous biochemical oxygen demand (CBOD) and total suspended solids (TSS), 3 mg/L for total nitrogen (TN), and 1 mg/L for total phosphorus (TP).

The City of Tallahassee initiated testing of spray irrigation systems with reclaimed water in 1961. Due to the success of these systems, they were expanded to 809 ha of major agricultural irrigation reuse. Another major irrigation reuse system was developed about 10 years later by the City of St. Petersburg. The development of this urban reclaimed water irrigation distribution system, which was the largest in the United States, was precipitated by two important events. The first was a 1972 decision by the city council to implement a recycling and deep injection well system for reclaimed water. The second was the Wilson-Grizzle Act, which required advanced wastewater treatment for the disposal of wastewater into environmentally sensitive bays (Johnson and Parnell, 1998). Other major water reuse projects that have been developed since 1972 include CONSERV II (an agricultural reuse project in Orlando and Orange counties), the Project APRICOT (Altamonte Spring's urban reuse system), and a wetlands project in Orlando (York and Wadsworth, 1998). The reclaimed water distribution system is quite extensive in Collier County and the City of Naples. Water reuse in St. Petersburg and CONSERV II are depicted on Fig. 2-7.

Current Water Reuse Status

Approximately 834×10^6 m^3 of reclaimed water was used in Florida for beneficial purposes in 2003. The total reuse capacity of domestic wastewater treatment facilities has increased from 500×10^6 m^3/yr in 1986 to $1,590 \times 10^6$ m^3/yr in 2003, which amounts to an increase of 233 percent. The current reuse capacity represents about 54 percent of the total permitted domestic wastewater treatment capacity in Florida (State of Florida, 2004). While Florida has been remarkably successful in implementing water reuse, it is interesting to note that over 1200×10^6 m^3/yr of wastewater effluent is disposed of using deep injection wells, ocean outfalls, and other surface water discharges.

 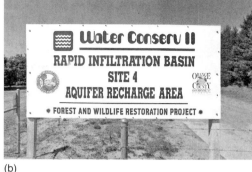

(a) (b)

Figure 2-7
Water reuse in Florida: (a) St. Petersburg—the reclaimed water system had continued to expand and change in character from an alternate mode of wastewater disposal to full operation as a water resource for irrigation and other uses of the city's Public Utilities Department, and (b) CONSERV II—Water Conserv II is the one of the largest water reuse projects with a combination of agricultural irrigation and rapid infiltration basins. (See also Table E-1 in App. E for details.)

Water Reuse Applications

Reclaimed water was used in 2003 to irrigate 154,234 residences, 427 golf courses, 486 parks, and 213 schools. A summary of water reclamation and reuse activities in Florida is shown in Table 2-3 jointly with California for comparison. Golf courses are important users of reclaimed water. In 2003, 184 water reuse systems included one or more golf courses within their list of reclaimed water customers (State of Florida, 2004).

Geographic Distribution of Water Reuse Sites

Water reclamation is practiced statewide with the largest reuse sites located in central Florida (Orlando-Lakeland area), the Tampa Bay area, southwestern Florida, and at some of the Atlantic coast counties such as Palm Beach, Volusia, and Brevard. Miami-Dade and Broward counties, the two most populous counties (a combined population of over three million), contain over 24 percent of Florida's population and generate 33 percent of the state's domestic wastewater. However, these two counties, located in the Miami-Ft. Lauderdale area, reclaim only 3.1 to 5.7 percent of their wastewater flow, respectively (State of Florida, 2004).

Size of Water Reclamation Systems

In Florida, 63 of its 67 counties reclaim effluent from wastewater treatment plants. The four counties that do not reclaim wastewater have populations that are less than 20,000. The amount of reclaimed water ranges from approximately 40,000 m^3/yr (Holmes County) to 124×10^6 m^3/yr (Orange County). The 15 largest reclaimed water-producing counties are listed in Table 2-6, and approximately 60 percent of all reclaimed water in Florida in 2003 came from these 15 counties. Overall, the amount of wastewater that is reclaimed for reuse averages 33.8 percent for these 15 counties as compared to 39.3 percent statewide. As noted in Table 2-6, the percent of wastewater that is reclaimed for the 15 counties

Table 2-6

The 15 largest reclaimed water producing counties in Florida[a]

County	WWTP flow[b], × 10³ m³/d	Reuse capacity, × 10³ m³/d	Reuse flow, × 10³ m³/d	Reuse flow/ WWTP flow, %	Annual Reuse flow, × 10⁶ m³/yr
Orange	345	639	339	98.3	124
Pinellas	383	492	186	48.4	68
Seminole	186	287	137	73.7	50
Lee	150	200	132	88.2	48
Hillsborough	546	352	115	26.7	42
Palm Beach	424	199	110	26.0	40
Collier	114	141	101	89.3	37
Polk	102	221	97	95.3	35
Volusia	119	125	69	57.5	25
Brevard	136	162	68	50.0	25
Leon	675	115	67	100.0	24
Osceola	68	149	67	98.8	24
Miami-Dade	1165	860	67	5.7	24
Okaloosa	63	113	63	100.9[c]	23
Manatee	104	142	62	59.5	23
Total—15 counties	3971	2786	1342	33.8	490
Total—67 counties	5627	4357	2211	39.3	807

[a]Adapted from State of Florida (2004).
[b]WWTP = Wastewater Treatment Plant.
[c]Percentage greater than 100 due to roundoff error.

ranges from 5.7 percent in Miami-Dade County to over 98 percent in Orange, Leon, Osceola, and Okaloosa counties.

Water Reuse Policies and Recycling Regulations

The Florida Legislature has established "... the encouragement and promotion of reuse of reclaimed water and water conservation ... ." as formal state objectives in Florida Statutes (F.S.) Section 403.064(1), and Section 373.250. Florida initiated a program to promote use of reclaimed water in 1987. In 1988, a water reuse provision, including mandatory reuse in Water Resource Caution Areas (WRCAs), was added to the Florida Administrative Code (FAC) Chapter 62-40. Chapter 62-610 contains the rules governing water reuse. Water Resource Caution Areas are areas that have critical water supply problems or are projected to have critical water supply problems within the next 20 years. Water reuse is required within these WRCAs, unless such reuse is not economically, environmentally, or technically feasible as determined by a water reuse feasibility study. Domestic wastewater facilities located within, discharging within, or serving a population within designated WRCAs are required to prepare water reuse feasibility studies before receiving a waste discharge permit (York and Wadsworth, 1998).

Potential Future Uses of Reclaimed Water

The Reuse Coordinating Committee along with the Water Reuse Work Group developed strategies for water reuse in Florida, which included a vision of water reuse in 2020. The vision statement included the following: (1) water reuse would be employed by all domestic wastewater treatment facilities having capacities of 380 m³/d and larger;

(2) statewide, on the order of 65 percent of all domestic wastewater would be reclaimed and used for beneficial purposes; (3) effluent disposal using ocean outfalls, other surface discharges, and deep injection wells would be limited to facilities that serve as backups to water reuse facilities; (4) groundwater recharge and indirect potable reuse projects would become common practice; (5) sewer mining would be common practice, particularly in larger urban areas, as a means for enabling effective use of reclaimed water; and (6) reclaimed water would be used widely to flush toilets in commercial facilities, industrial facilities, hotels and motels, and multiple-family residential units.

To achieve these visions, the State of Florida established the following 16 strategies for managing reclaimed water as a valuable resource (State of Florida, 2004). Highlights of these strategies follow.

- Encourage metering and volume-based rate structures. This strategy encourages municipalities, water, and wastewater agencies to meter and charge for reclaimed water service.
- Implement viable funding programs. Funding should be targeted at reuse projects featuring high potable quality water offsets or recharge fractions as a means for encouraging efficient and effective water use.
- Facilitate seasonal reclaimed water storage including aquifer storage and recovery (ASR). Storage represents a major concern particularly for projects emphasizing irrigation with reclaimed water where large seasonal fluctuations in use may occur.
- Encourage use of reclaimed water in lieu of other water sources in agricultural irrigation, landscape irrigation, industrial/commercial/institutional, and indoor water use sectors.
- Link water reuse to regional water supply planning (including integrated water resource planning). Water planning must fully consider the full range or alternative supplies, including reclaimed water.
- Develop integrated water education programs. This issue addresses the need to inform the public fully about the need for and issues involved with alternative water supplies.
- Encourage groundwater recharge and indirect potable reuse as they offer significant advantages for augmenting existing water supplies.
- Discourage effluent disposal to emphasize that large quantities of wastewater effluent are being wasted.
- Provide water use permitting incentives for utilities that implement water reuse programs.
- Encourage reuse in Southeast Florida. In this area, particularly Miami-Dade and Broward counties, the vast majority of treated wastewater is wasted. For significant gains in water reuse in the state, effluent disposal must be discouraged and water reuse encouraged. Sewer mining has been one method identified in implementing water reuse.
- Encourage use of supplemental water supplies from all sources including treated stormwater.
- Encourage efficient irrigation practices.
- Encourage interconnection of reuse systems to provide greater flexibility and reliability.
- Enable redirecting of existing reuse systems to more desirable reuse options as a means of motivating utilities.

- Use reclaimed water at government facilities. The state should lead by example in water reuse not only to conserve water but to also serve as an effective means of educating the public.
- Ensure continued safety of water reuse. This strategy addresses such topics as cross-connection control, control of pathogens and emerging contaminants, responsible utility management and oversight, and public education.

One of the objectives of water reuse planning in Florida is the removal of institutional and regulatory inconsistencies related to water. A key component is the development of "use-based" standards that are independent of the source of water used (State of Florida, 2004). In other words, Florida recognizes that "water is water" and alternative water resources, including reclaimed water, will play increasingly important roles in water management in the future. Water reuse is already recognized as a key component of wastewater management and water resource management. These water reuse strategies will ensure that water and wastewater agencies continue to pursue the State's objectives of encouraging and promoting water reuse.

2-6 WATER REUSE IN OTHER PARTS OF THE WORLD

Similar to the situation in the United States, the growing trends in water reclamation and reuse in the world are to consider water reuse practices as an essential component of integrated water resources management. The development of water reclamation and reuse in many countries is closely related to water scarcity, water pollution control measures, and obtaining alternative water resource. In cities and regions of the developed world, where wastewater collection and treatment have been the common practice, water reuse is practiced with proper attention to the environment, public health, and esthetic considerations.

Significant Developments Worldwide

The water reclamation and reuse activities in the countries belonging to the European Union (EU) are guided by the EU Water Framework Directives promulgated in 2000. In the European Communities Commission Directive (91/271/EEC), "Treated wastewater shall be reused whenever appropriate . . . ," and that ". . . disposal routes shall minimize the adverse effects on the environment . . ." (EEC, 1991). Most of the significant developments in water reclamation and reuse have occurred in arid regions of the world. Several Mediterranean countries in Europe, particularly in Portugal, Spain, southern provinces of France and Italy, Cyprus, and Greece, have been the vanguards in water reclamation and reuse using secondary or tertiary treated effluents. In addition, Israel, Tunisia and other Maghreb countries have well-established agricultural irrigation programs using reclaimed water (Mujeriego and Asano, 1991 and 1999; Angelakis et al., 1996, 1999, and 2003; Shelef and Azov, 1996; Marecos do Monte, 1998; Bonomo et al., 1999; Shelef, 2000; Brissaud et al., 2001; Sala et al., 2002; Jimenez and Asano, 2004; Bahri and Brissaud, 2004; Bixio et al., 2005; Lazarova and Bahri, 2005).

The drought that afflicted much of Australia in 2001–2003 resulted in water restrictions being imposed in Sydney, Melbourne, Canberra, Perth, and the Queensland Gold Coast. Over 500 municipal wastewater treatment plants now engage in the water reclamation of at least part of their treated effluent. Specific water reclamation and reuse targets have been established for major cities (Radcliffe, 2004; Anderson, 2005). For example, the Queensland Water Recycling Strategy is a whole government initiative aimed at maximizing water reclamation and reuse in an efficient, economic, and environmentally sustainable manner without adverse health effects.

Unique to the prevailing water reuse applications which are mostly in irrigation uses, Japan's water reclamation and reuse has focused on urban water applications such as in building water reuse for toilet flushing in commercial and office buildings, urban landscapes, stream flow augmentation, and even snow melting and heating and air conditioning using heat content of the reclaimed water (Japan Sewage Works Association, 2005; UNEP and GEC, 2005).

Some of the significant worldwide activities in water reuse that have occurred since 1960 are summarized in Table 2-7. In addition, a summary of water reclamation and reuse in leading countries of the world is shown in Table E-2 in App. E. A wide range of water reuse applications, which may be closely tied to local regulatory, environmental, and pressing water resources conditions, are presented in Table E-2. The majority of water reuse is for nonpotable applications such as agricultural and landscape irrigation, and industrial reuse. Some of the representative water reuse applications are shown on Fig. 2-8.

In Windhoek, Namibia, because of extreme drought conditions, extensive research was conducted in 1968 on direct potable reuse technology and an epidemiological study was conducted to assess the health effects of reclaimed water consumption (Isaäcson et al., 1987; Odendaal et al., 1998). Based on the findings from the research, highly treated wastewater has been commingled with other drinking water sources. In Singapore, water reclamation and reuse has been implemented as a source of raw water to supplement Singapore's water supply. Indirect and direct potable reuse including Singapore and Windhoek are discussed in detail in Chaps. 23 and 24, respectively. Technologies such as membrane bioreactors, membrane filtration, and ultraviolet disinfection are important in the production of high quality reclaimed water and are further discussed in Part 3.

The World Health Organization's Water Reuse Guidelines

In 1989, the World Health Organization (WHO) published *Health Guidelines for the Use of Wastewater in Agriculture and Aquaculture* (WHO, 1989) that provided guidance for less developed countries that had little or no experience with planned reuse of wastewater. In these countries, waste stabilization ponds and wastewater storage and treatment reservoirs are two possible treatment options prior to water reuse in agriculture. The WHO guidelines have been under revision since 2002 and revised guidelines are expected to be published in 2006 (Carr et al., 2004; also see Chap. 4, Sec. 4-8). The guidelines are intended to be used as the basis for the development of international and national approaches (including standards and regulations) to managing the health risks from hazards associated with wastewater use in agriculture and aquaculture, as well as providing a framework for national and local decision-making (WHO, 2005 and 2006).

Water Reuse in Developing Countries

Urban growth impacts on infrastructure in developing countries are extremely pressing (see Chap. 1, Sec. 1-2). In many cities of Asia, Africa, and Latin America, engineered wastewater collection systems and wastewater treatment facilities are nonexistent.

Table 2-7

Significant events related to water reclamation and reuse in the world[a]

Period	Location	Event
1962	La Soukra, Tunisia	Irrigation with reclaimed water for citrus plants and groundwater recharge to reduce saltwater intrusion into coastal groundwater.
1965	Israel	Use of secondary effluent for crop irrigation.
1969	Wagga Wagga, Australia	Landscape irrigation of sporting fields, lawns, and cemeteries.
1968	Windhoek, Namibia	Research on direct potable reuse and subsequent implementation.
1977	Tel-Aviv, Israel	Dan Region Project—Groundwater recharge via basins. Pumped groundwater is transferred via a 100-km-long conveyance system to southern Israel for unrestricted crop irrigation.
1984	Tokyo, Japan	Toilet flushing water for commercial buildings in the Shinjuku District using reclaimed water from the Ochiai Wastewater Treatment Plant operated by the Tokyo Metropolitan Sewerage Bureau.
1988	Brighton, UK	Inauguration of the Specialist Group on Wastewater Reclamation, Recycling and Reuse at the 14th Biennial Conference of the International Association on Water Pollution Research and Control (currently, the International Water Association, headquartered in London, UK).
1989	Girona, Spain	Golf course irrigation using reclaimed water from the Consorci de la Costa Brava wastewater treatment facility.
1999	Adelaide, South Australia	The Virginia Pipeline Project, the largest water reclamation project in Australia—irrigating vegetable crops using reclaimed water from the Bolivar Wastewater Treatment Plant (120,000 m^3/d).
2002	Singapore	NEWater-reclaimed water that has undergone significant purification using microfiltration, reverse osmosis, and ultraviolet disinfection. NEWater is used as a raw water source to supplement Singapore's water supply.

[a]Compiled from various sources including Metcalf and Eddy (1928); AWWA (1981); Ongerth and Ongerth (1982); Asano and Levine (1996); Baird and Smith (2002).

Figure 2-8
Some representative water reuse applications in various parts of the world: (a) fodder crop, Australia; (b) row crop, Israel (Courtesy of MEKOROT, Israel National Water Company); (c) Agave, Jordan (Courtesy of A. Bahri); and (d) constructed wetland, Costa Brava, Spain (Courtesy of L. Sala).

Where wastewater collection systems are available, they often discharge untreated wastewater to the nearest drainage channel or watercourse. For developing countries, particularly in arid areas, wastewater is simply too valuable to waste. It contains scarce water and valuable plant nutrients, and crop yields are higher when crops are irrigated with wastewater than with freshwater (Shende et al., 1988). Farmers use untreated wastewater out of necessity and it is a reality that cannot be denied or effectively banned (Buechler et al., 2002). Unfortunately, these are the realities in developing countries, and should not be confused with planned and regulated water reclamation and reuse. Major health concerns make it imperative to governments and the United Nations agencies to implement public health and environmental protection during the era of rapid urbanization in these developing countries.

Almost all water reuse in developing countries is for agricultural purposes. Some of the representative water reuse applications are shown on Fig. 2-9. Because alternative low-cost

Figure 2-9
Water reuse applications in developing countries:
(a) hand watering on vegetable crops with stream water dominated by untreated wastewater in Ghana (Courtesy of IWMI, Ghana), (b) drip irrigation of date palms in Aqaba, Jordan. (Coordinates: 29.563 N, 34.988 E)

(a) (b)

sources of water are generally not available for irrigation of high-value market crops near these cities, the common practice is to use untreated wastewater directly or to withdraw it from nearby streams that may be grossly polluted with untreated municipal and industrial wastewaters. One-tenth or more of the world's population consumes food grown with irrigation supplied by wastewater (Smit and Nasr, 1992). Wastewater and excreta are also used in urban agriculture which often supplies a large proportion of the fresh vegetables sold in many cities, particularly in less developed countries. For example, in Dakar, Senegal, more than 60 percent of the vegetables consumed in the city are grown in urban areas using a mixture of groundwater and untreated wastewater (Faruqui et al., 2002).

In most developing countries where wastewater is used for irrigation, it is used without adequate treatment (see Fig. 2-10). The consequence of contamination of food that is eaten uncooked is a high level of enteric diseases and has serious impacts on visitors to these regions. Thus, the protection of the public health, as well as the provision of additional water supply, is an incentive to the initiation of agricultural water reuse projects near the cities in developing countries. Collecting wastewater for treatment is a formidable and expensive task at present in many developing countries. Under these conditions, WHO is trying to develop realistic health guidelines for the use of wastewater in agriculture (Blumenthal et al., 2000; Mara, 2003; Carr et al., 2004; see also Chap. 4).

Water Lines, an international journal of appropriate technologies for water supply and wastewater treatment reported several water reuse practices in developing countries which included water reuse by a natural filtration system in a Vietnamese rural community (Takizawa, 2001) and sewage reclamation for industrial uses in Chennai (formally Madras), India (Kurian and Visvanathan, 2001).

(a)

(b)

(c)

(d)

Figure 2-10

Mexico City's untreated municipal wastewater and Mezquital Valley irrigation canal system. The complex hydraulic system was implemented to regulate water distribution according to crop water needs with nine dams (six with wastewater), three rivers, and 858 km channels that convey 60 m^3/s of untreated municipal wastewater produced by 19 million Mexico City residents: (a) view of the Grand Canal (facing upstream) used to transport untreated wastewater from Mexico City to agricultural areas some 28 km from the city. In addition to serving as a transport canal, the Grand Canal also serves as one of the world's largest oxidation ponds, (b) view (facing downstream) from one of the pumping stations used to lift water from the canal to agricultural areas through a series of distribution canals, (c) and (d) views of the distribution canals and agricultural lands irrigated with untreated wastewater. (Coordinates: from 19.778 N, 99.120 W to 19.579 N, 99.024 W)

2-7 SUMMARY AND LESSONS LEARNED

Several milestone water reuse projects and research studies in the twentieth century have led to the current knowledge of water reclamation and reuse. Selected milestone projects and research studies in the United States are shown in Table B-1 in App. B.

These projects were selected because some are pioneering projects as measured by their water reuse applications and others had significant scientific and engineering impacts on later developments in water reuse.

In the United States, two federal statutes, the CWA and the SDWA, have had a significant impact on the quantity and quality of wastewater discharges and the potential for water reuse. These regulations were enacted in the early 1970s and have encouraged water reclamation and reuse through more stringent discharge regulations and specific water reuse encouragement via federal and state grants and loans.

Historically, water reclamation and reuse sites tend to be located where water is the scarcest. Scarcity occurs in areas such as the arid and semiarid western and southwestern United States, including Arizona, California, Colorado, Nevada, Texas, and Utah, and humid regions where rapid growth is occurring such as Florida, Georgia, Maryland, and Missouri. Overall, the States of California and Florida have the most comprehensive water reclamation and reuse regulations and practices, most likely because these states have been actively involved with water reclamation and reuse for close to half a century.

The growing trend in water reclamation and reuse in the world is to consider water reuse practices as an essential component of sustainable, integrated water resources management. Similar to the situation in the United States, underlying the development of water reclamation and reuse in many countries is water scarcity, water pollution control measures, and obtaining alternative water resources. In cities and regions of developed countries, where wastewater collection and treatment have been the common practice, water reuse is practiced with appropriate attention to the environment, public health, and aesthetic considerations.

In many developing countries, however, confined wastewater collection system and wastewater treatment are often nonexistent, and untreated or partially treated wastewater often provides an essential water and fertilizer source. For developing countries, particularly in arid areas, wastewater is simply too valuable to waste and untreated wastewater is used out of necessity. Step-by-step implementation of public health and environmental protection to address major health concerns associated with food contaminated by raw wastewater is necessary to develop safe and effective water reclamation and reuse programs in developing countries.

There is a wide spectrum of challenges and solutions in implementing water reclamation and reuse, even in areas where public health standards are high. Regulations for water pollution control and environmental protection are in place and enforced rigorously, and there is little opportunity for year-round irrigation using reclaimed water. Some of the salient lessons learned in implementing water reuse in such areas follow.

- Motivating factors in water reclamation and reuse include water scarcity, wastewater effluent discharge regulations, and obtaining dependable alternative water sources.
- In all cases, reliable wastewater treatment is the foundation for successful water reclamation and reuse.

- As demand for water reuse has increased, locating treatment systems closer to the point of use has become more feasible, resulting in an increase in decentralized or satellite wastewater treatment and water reuse systems. Water reclamation and reuse by mining wastewater from sewer lines (sewer mining) in local areas or on-site water reuse systems have been implemented effectively with technology such as membrane bioreactors and ultraviolet disinfection systems.
- Based on studies on future reclaimed water use, it is anticipated that the next uses for large volumes of reclaimed water will be (1) landscape irrigation in urban areas, (2) industrial reuse, and (3) indirect potable reuse with groundwater recharge and surface water augmentation. In addition, the U.S. EPA has a program for artificial wetlands development using reclaimed water that may become more important in the future (see Chap. 21).
- Water reclamation and reuse are generally one part of a comprehensive water resources approach. Urban water supply sources consist of multiple water sources which may include (1) water transfer from agriculture uses to domestic uses, (2) imported water (interbasin transfer of water), (3) local surface water and groundwater, (4) water conservation, (5) water reclamation and reuse, and (6) seawater and blackish water desalination. A water source plan that is illustrative of the concept of multiple water sources is shown on Fig. 2-6.
- The development of a successful water reuse project is contingent on multiple factors, including nontechnical issues. Public perception and the political process are vital to incorporating water reclamation and reuse into a comprehensive water resource plan.
- Public health concerns with water, both reclaimed and potable, are increasing. Advances in treatment technology have made water reclamation and reuse safer, more reliable, and more economical, which is helping to address public health concerns. Newer technology, such as membrane treatment is important in developing safe and effective decentralized and on-site treatment facilities, which in turn may encourage greater use of reclaimed water.

PROBLEMS AND DISCUSSION TOPICS

2-1 Prepare a brief summary of water reuse opportunities in your community or region. What factors might affect the implementation of water reuse opportunities cited in your summary.

2-2 Based on a review of the literature, how would you explain the relative differences in the first four water reuse applications between Florida and California, as listed in Table 2-3.

2-3 Do you feel, based on a review of the literature, that desalination in coastal water-short areas will reduce the incentive to conserve and reuse water?

2-4 What impact has the synthesis and use of chemicals in consumer products in the twenty-first century had on the admonition of Sir Edwin Chandwick—*the rain to the river and the sewage to the soil*. Cite a minimum of three references.

2-5 How does the use of dual distribution systems (see Chap. 14) affect the economic viability of water reuse for urban applications.

2-6 What impact will the imposition of more stringent discharge requirements for wastewater treatment plants have on water reclamation and reuse?

2-7 While direct potable reuse of reclaimed municipal wastewater is, at present, limited to extreme situations, it has been argued that there should be a single water quality standard for potable water. If reclaimed water can meet this standard, it should be acceptable regardless of the source of water. Discuss pros and cons of this argument focusing on health risks as well as ethics and public acceptance issues.

2-8 Approximately 60 percent of reclaimed water was produced by the 15 largest water producing agencies in California in 2001 (see Table 2-5). These agencies have considerable experience in water reclamation and reuse. It can be argued that water reclamation and reuse contemplated by small communities should be discouraged because impact on overall water resources is insignificant, lack of expertise, and difficulty in local use area control. As a policy maker for the state water resources agency, what would be your position on water reclamation and reuse as it applies to small versus large communities (see also Chap. 13) in your geographic location?

2-9 Two examples of potable water reuse schemes are presented in Chap. 2. Considering the various necessities and opportunities which exist in different states and countries, develop a rational basis for adopting a direct or indirect potable water reuse option in sustainable water resources management. How may the public react to your potable water reuse proposal and on what basis should such a decision be made?

2-10 It may be argued that a direct or indirect potable reuse may be the most cost-effective option in large-scale water reuse in the future. It is also argued that water reclamation technologies have advanced to the point where any quality water can be produced reliably by a combination of treatment processes and operations. However, the future of planned direct or indirect potable reuse is uncertain. List pros and cons of direct or indirect potable reuse with respect to decision-making, engineering, public health protection, public perception and acceptance, and cost. Provide a rational basis of how to promote water reclamation and reuse and to what extent, in the context of integrated water resources management.

REFERENCES

Anderson, J. M. (2005) "Integrating Recycled Water into Urban Water Supply Solutions," in S. J. Khan, M. H. Muston, and A. I. Schäfer (eds.), *Integrated Concepts in Water Recycling*, 32–40, University of Wollongong, Australia.

Angelakis, A., T. Asano, E. Diamadopoulos, and G. Tchobanoglous (Issue eds.) (1996) "Wastewater Reclamation and Reuse 1995," *Water Sci. Technol.*, **23**, 10–11.

Angelakis, A. N., M. H. F. Marecos do Monte, L. Bontoux, and T. Asano (1999) "The Status of Wastewater Reuse Practice in the Mediterranean Basin: Need for Guidelines," *Water Res.*, **33**, 10, 2201–2217.

Angelekis, A. N., N. V. Paranychianakis, and K. P. Tsagarakis (Issue eds.) (2003) "Water Recycling in the Mediterranean Region," *Water Sci. Technol.*, **3**, 4.

Asano, T., and A. D. Levine (1996) "Wastewater Reclamation, Recycling and Reuse: Past, Present, and Future," *Water Sci. Technol.*, **33**, 10–11, 1–14.

Asano, T. (ed.) (1998) *Wastewater Reclamation and Reuse*, Water Quality Management Library, **10**, CRC Press, Boca Raton, FL.

AWWA (1981) *The Quest for Pure Water*, Vol. 1 and 2, 2nd ed., American Water Works Association, Denver, CO.

Bahri A., and F. Brissaud (2004) "Setting up Microbiological Water Reuse Guidelines for the Mediterranean," *Water Sci. Technol.*, **50**, 2, 39–46.

Baird, R. B., and R. K. Smith (2002) *Third Century of Biochemical Oxygen Demand*, Water Environment Federation, Alexandria, VA.

Barty-King, H. (1992) *Water: the Book, an Illustrated History of Water Supply and Wastewater in the United Kingdom*, Quiller Press, Ltd., London.

Bixio, D., J. De Koning, D. Savic, T. Wintgens, T. Melin, and C. Thoeye (2005) "Water Reuse in Europe," in S. J. Khan, M. H. Muston, and A. I. Schäfer (eds.), *Integrated Concepts in Water Recycling*, 80–92, University of Wollongong, Australia.

Blumenthal, U., D. D. Mara, A. Peasey, G. Ruiz-Palacios, and R. Stott (2000) "Guidelines for the Microbiological Quality of Treated Wastewater Used in Agriculture: Recommendations for Revising WHO Guidelines," *Bulletin of the World Health Organization*, **78**, 9, 1104–1116.

Bonomo, L., C. Nurizzo, R. Mujeriego, and T. Asano (eds.) (1999) "Advanced Wastewater Treatment, Recycling and Reuse," Proceedings volume, *Water Sci. Technol.*, **40**, 4–5.

Brissaud, F., J. Bontoux, R. Mujeriego, A. Bahri, C. Nurizzo, and T. Asano, (Issue eds.) (2001) "Wastewater Reclamation, Recycling and Reuse," Proceedings volume, *Water Sci. Technol.*, **43**, 10.

Buechler, S., W. Hertog, and R. Van Veenhuizen (2002) "Wastewater Use for Urban Agriculture," *Urban Agric. Mag.*, **8**, 12, 1–4.

Carr, R. M., U. J. Blumenthal, and D. D. Mara (2004) "Guidelines for the Safe Use of Wastewater in Agriculture: Revising WHO Guidelines," *Water Sci. Technol.*, **50**, 2, 31–38.

Chen, C. L., J. F. Kuo, and J. F. Stahl (1998) "The Role of Filtration for Wastewater Reuse," 219–262, in T. Asano (ed.) *Wastewater Reclamation and Reuse*, Water Quality Management Library, **10**, CRC Press, Boca Raton, FL.

Clark, R. M., and R. S. Summers (eds.) (1993) *Strategies and Technologies for Meeting SDWA Requirements*, Technomic Publishing Co., Inc., Lancaster, PA.

Cooper, P. F. (2001) "Historical Aspects of Wastewater Treatment," 11–54, in P. Lens, G. Zeeman, and G. Lettinga (eds.) *Decentralised Sanitation and Reuse: Concepts, Systems and Implementation*, IWA Publishing, London.

Crook, J. (2004) *Innovative Applications in Water Reuse: Ten Case Studies*, WateReuse Association, Alexandria, VA.

CSWRCB (2003) *Recycled Water Use in California*, Office of Water Recycling, California State Water Resources Control Board, Sacramento, CA. http://www.waterboards.ca.gov/recycling/docs/wrreclaim1.attb.pdf

Dryden, F. D., C. L. Chen., and M. W. Selna (1979) "Virus Removal in Advanced Wastewater Treatment Systems," *J. WPCF*, **51**, 8, 2098–2109.

EEC (1991) European Communities Commission Directive (91/271/EEC), The Council of the European Community.

Fair, G. M., and J. C. Geyer (1954) *Water Supply and Waste-Water Disposal*, John Wiley & Sons Inc., New York.

Faruqui, N., S. Niang, and M. Redwood (2002) "Untreated Wastewater Reuse in Market Gardens: A Case-Study of Dakar, Senegal," Paper presented at the International Water Management Institute Workshop on Wastewater Reuse in Irrigated Agriculture: Confronting the Livelihood and Environmental Realities, International Water Management Institute, Hyderabad, India.

Hutson, S. S., N. L. Barber, J. F. Kenny, D. S. Lumia, and M. A. Maupin (2004) *Estimated Use of Water in the United States in 2000*: Reston, VA, U.S. Geological Survey, Circular 1268.

Isaäcson, M., A. R. Sayed, and W. Hattingh (1987) *Studies on Health Aspects of Water Reclamation during 1974 to 1983 in Windhoek, South West Africa/Namibia*, Report WRC 38/1/87 to the Water Resources Commission, Pretoria, South Africa.

Japan Sewage Works Association (2005) *Sewage Works in Japan 2005: Wastewater Reuse*, Tokyo, Japan.

Jimenez, B., and T. Asano (2004) "Acknowledge All Approaches: The Global Outlook On Reuse," *Water 21*, **3**, 32–37.

Johnson, W. D., and J. R. Parnell (1998) "Wastewater Reclamation and Reuse in the City of St. Petersburg, Florida," 1037–1104, in T. Asano (ed.) *Wastewater Reclamation and Reuse*, Water Quality Management Library, **10**, CRC Press, Boca Raton, FL.

Kurian J., and C. Visvanathan (2001) "Sewage Reclamation Meets Industrial Water Demands in Chennai," *Water Lines*, **19**, 4, 6–9.

Lazarova V., and A. Bahri (eds.). (2005) *Water Reuse for Irrigation: Agriculture, Landscapes, and Turf Grass*, CRC Press, Boca Raton, FL.

Levine, A. D., and T. Asano (2004) "Recovering Sustainable Water from Wastewater," *Environ Sci. Technol.*, **38**, 11, 201A–208A.

Lumsden, L. L., C. W. Stiles, and A. W. Freeman (1915) *Safe Disposal of Human Excreta in Unsewered Homes*, Public Health Bulletin No. 68, United States Public Health Service, Government Printing Office, Washington, DC.

Mara, D. (2003) *Domestic Wastewater Treatment in Developing Countries*, Earthscan, London.

Marecos do Monte, M. H. F. (1998) "Agricultural Irrigation with Treated Wastewater in Portugal," Chap. 18, in T. Asano (ed.) *Wastewater Reclamation and Reuse*, Water Quality Management Library, **10**, CRC Press, Boca Raton, FL.

Metcalf, L., and H. P. Eddy (1928) *American Sewerage Practice*, Vol. **1**, Design of Sewers, McGraw-Hill Book Co., Inc., New York.

Mujeriego, R., and T. Asano (Issue eds.) (1991) "Wastewater Reclamation and Reuse," *Water Sci. Technol.*, **24**, 9.

Mujeriego, R., and T. Asano (1999) "The Role of Advanced Treatment in Wastewater Reclamation and Reuse," *Water Sci. Technol.*, **40**, 4–5, 1–9.

Odendaal, P. E., J. L. J. van der Westhuizen, and G. J. Grobler. (1998) "Water Reuse in South Africa," 1163–1192, in T. Asano (ed.) *Wastewater Reclamation and Reuse*, Water Quality Management Library, **10**, CRC Press, Boca Raton, FL.

Okun, D. A. (1997) "Distributing reclaimed water through dual systems," *J. AWWA*, **89**, 11, 52–64.

Ongerth, H. J., and W. F. Jopling (1977) "Water Reuse in California," Chap. 8, in H. I. Shuval (ed.) *Water Renovation and Reuse*, Academic Press, Inc., New York.

Ongerth, H. J., and J. E. Ongerth (1982) "Health Consequences of Wastewater Reuse," *Annu. Rev. Public Health*, **3**, 419–444.

Radcliffe, J. C. (2004) *Water Recycling in Australia*, Australia Academy of Technological Sciences and Engineering, Parkville, Victoria, Australia.

Sala, L., L. Mujeriego, M. Serra, and T. Asano (2002) "Spain Sets the Example," *Water 21*, **4**, 18–20.

San Diego County Water Authority (2002) *2000 Urban Water Management Plan*, San Diego, CA. http://www.sdcwa.org/news/plan2000.phtml

SDLAC (1977) *Pomona Virus Study Final Report,* prepared for California State Water Resources Control Board and U.S. Environmental Protection Agency, Sanitation Districts of Los Angeles County, Los Angeles, CA.

Shelef, G., and Y. Azov (1996) "The Coming Era of Intensive Wastewater Reuse in the Mediterranean Region," *Water Sci. Technol.*, **33**, 10–11, 115–125.

Shelef, G. (2000) Wastewater Treatment, Reclamation and Reuse in Israel, in *Efficient Use of Limited Water Resources: Making Israel a Model State*, Begin-Sadat (BESA) Center for Strategic Studies, Bar-Ilan University, Ramat Gan 52900 Israel.

Shende, B, C. Chakrabarti, R. P. Rai, V. J. Nashikkar, D. G. Kshirsagar, P. B. Deshbhratar, and A. S. Juwarkar (1988) "Status of Wastewater Treatment and Agricultural Reuse with Special Reference to Indian Experience and Research and Development Needs," 185–209, in M. B. Pescod and A. Arar (eds.) *Treatment and Use of Sewage Effluent for Irrigation*, Butterworths, London

Smit, J., and J. Nasr (1992) "Urban Agriculture for Sustainable Cities: Using Wastes and Idle Land and Water Bodies as Resources," *Environ. Urban.*, **4**, 2, 141–152.

Solley, W. B., R. R. Pierce, and H. A. Perlman (1998) *Estimated Use of Water in the United States in 1995*, U.S. Geological Survey Circular 1200, U.S. Geological Survey, Reston, VA.

State of California (1978) *Wastewater Reclamation Criteria, An Excerpt from the California Code of Regulations, Title 22, Division 4*, Environmental Health, Department of Health Services, Berkeley, CA.

State of California (1990) *California Municipal Wastewater Reclamation in 1987*, Office of Water Recycling, State Water Resources Control Board, Sacramento, CA.

State of California (2000) *Code of Regulations, Title 22, Division 4, Chap. 3 Water Recycling Criteria*, Sections 60301 *et seq.*, Sacramento, CA.

State of California (2002) *Statewide Recycled Water Survey*, Office of Water Recycling, State Water Resources Control Board, Sacramento, CA. http://www.waterboards.ca.gov/recycling/munirec.html

State of California (2003a) *California Code*—Water Code Section 13050, subdivision (n). http/www.leginfo.ca.gov

State of California (2003b) *Water Recycling 2030: Recommendations of California's Recycled Water Task Force*, Department of Water Resources, Sacramento, CA.

State of California (2005) *California Water Plan Update 2005*, Department of Water Resources. http://www.waterplan.water.ca.gov/cwpu2005/

State of Florida (2002) *2001 Reuse Inventory*, Florida Department of Environmental Protection, Tallahassee, FL.

State of Florida (2003a) *Florida Population*, Office of Economic and Demographic Research, State of Florida. http://www.state.fl.us/edr/population.htm

State of Florida (2003b) *Water Reuse for Florida: Strategies for Effective Use of Reclaimed Water*, Florida Department of Environmental Protection, Tallahassee, FL.

State of Florida (2004) *2003 Reuse Inventory*, Department of Environmental Protection, Division of Water Resources Management, Tallahassee, FL.

Takizawa, S. (2001) Water reuse by a natural filtration system in a Vietnamese rural community, *Water Lines*, **19**, 2–5.

UNEP, and GEC (2005) *Water and Wastewater Reuse: An Environmentally Sound Approach for Sustainable Urban Water Management*, United Nations Environment Programme and Global Environment Centre Foundation, Osaka, Japan. http://www.unep.or.jp/Ietc/Publications/Water_Sanitation/wastewater_reuse/index.asp

U.S. EPA, and U.S. AID (1992) *Manual—Guidelines for Water Reuse*, EPA/625/R-92/004, U.S. Environmental Protection Agency and U.S. Agency for International Development, Washington, DC.

U.S. EPA (1998) *Water Pollution Control—Twenty-five Years of Progress and Challenges for the New Millennium*, 833-F-98-003, Office of Water, U.S. Environmental Protection Agency, Washington, DC.

U.S. EPA, and U.S. AID (2004) *Guidelines for Water Reuse*, EPA/625/R-04/108, U.S. Environmental Protection Agency and U.S. Agency for International Development, Washington, DC.

WateReuse Association (2005). http://www.watereuse.org/aboutus.htm

WEF (1997) *The Clean Water Act: 25th Anniversary Edition,* Water Environment Federation Alexandria, VA.

WHO (1989) *Health Guidelines for the Use of Wastewater in Agriculture and Aquaculture,* Report of a WHO Scientific Group, Technical Report Series 778, World Health Organization, Geneva, Switzerland.

WHO (2005) *Meeting Report: Final Expert Review Meeting for the Finalization of the Third Edition of the WHO Guidelines for the Safe Use of Wastewater, Excreta and Greywater: 13–17 June, 2005,* World Health Organization, Geneva, Switzerland.

WHO (2006) *WHO Guidelines for the Safe Use of Wastewater, Excreta and Greywater,* Third Edition, Volume II, *Wastewater Use in Agriculture,* World Health Organization, Geneva, Switzerland.

York, D. W., and L. Wadsworth (1998) "Reuse in Florida: Moving Toward the 21st Century," *Florida Wat. Res. J.,* **11**, 31–33.

Young, H. W., and D. W. York (1996) "Reclaimed Water Reuse in Florida and the South Gulf Coast," *Florida Wat. Res. J.,* **11**, 32–36.

Part 2

HEALTH AND ENVIRONMENTAL CONCERNS IN WATER REUSE

While there is no reliable epidemiological evidence that the use of reclaimed water for any of its applications has caused a disease outbreak in the United States, potential transmission of infectious disease by pathogenic organisms is the most common concern in water reclamation and reuse. This concern is true particularly in developing countries where untreated or inadequately treated wastewater is used widely, unfortunately. In addition, the production, distribution, and use of reclaimed water that is regulated inadequately may result in a number of adverse environmental impacts.

In Part 2, health and environmental issues associated with water reuse are discussed in three related chapters. Characteristics of municipal wastewater and health and environmental issues are presented in Chap. 3. Waterborne pathogens, chemical constituents in wastewater and reclaimed water, and emerging contaminants, as well as environmental impacts are discussed in this chapter. The development and implementation of water reclamation and reuse regulations, which have played such an important role in the advancement of water reuse, are presented and discussed in Chap. 4. Applicable regulations and guidelines for various uses of reclaimed water are also discussed in Chap. 4. Health risk assessment is an emerging and potentially useful tool in evaluating the risk to human health due to microbiological, and the natural and anthropogenic chemical constituents of water, reclaimed water, and wastewater. Following a brief introduction to tools and methods used in health risk analysis that include concepts from public health, epidemiology, and toxicology, chemical and microbial risk assessment in water reuse applications are discussed in Chap. 5.

3 Characteristics of Municipal Wastewater and Related Health and Environmental Issues

WORKING TERMINOLOGY   74

3-1   WASTEWATER IN PUBLIC WATER SUPPLIES—*DE FACTO* POTABLE REUSE   77
Presence of Treated Wastewater in Public Water Supplies   78
Impact of the Presence of Treated Wastewater on Public Water Supplies   78

3-2   INTRODUCTION TO WATERBORNE DISEASES AND HEALTH ISSUES   78
Important Historical Events   79
Waterborne Disease   80
Etiology of Waterborne Disease   81

3-3   WATERBORNE PATHOGENIC MICROORGANISMS   83
Terminology Conventions for Organisms   83
Log Removal   83
Bacteria   83
Protozoa   87
Helminths   89
Viruses   89

3-4   INDICATOR ORGANISMS   92
Characteristics of an Ideal Indicator Organism   92
The Coliform Group Bacteria   93
Bacteriophages   93
Other Indicator Organisms   94

3-5   OCCURRENCE OF MICROBIAL PATHOGENS IN UNTREATED AND TREATED WASTEWATER AND IN THE ENVIRONMENT   94
Pathogens in Untreated Wastewater   94
Pathogens in Treated Wastewater   97
Pathogens in the Environment   102
Survival of Pathogenic Organisms   102

3-6   CHEMICAL CONSTITUENTS IN UNTREATED AND TREATED WASTEWATER   103
Chemical Constituents in Untreated Wastewater   103
Constituents Added through Domestic Commercial and Industrial Usage   104
Chemical Constituents in Treated Wastewater   108
Formation of Disinfection Byproducts (DBPs)   113
Comparison of Treated Wastewater to Natural Water   114
Use of Surrogate Parameters   115

3-7	EMERGING CONTAMINANTS IN WATER AND WASTEWATER 117
	Endocrine Disruptors and Pharmaceutically Active Chemicals 117
	Some Specific Constituents with Emerging Concern 118
	New and Reemerging Microorganisms 120
3-8	ENVIRONMENTAL ISSUES 120
	Effects on Soils and Plants 121
	Effects on Surface Water and Groundwater 121
	Effects on Ecosystems 121
	Effects on Development and Land Use 122
	PROBLEMS AND DISCUSSION TOPICS 122
	REFERENCES 124

WORKING TERMINOLOGY

Term	Definition
Abiotic reaction	Nonliving reaction in an ecosystem. The abiotic factors of the environment include light, temperature, and atmospheric gases (e.g., chemical oxidation, photolysis, volatilization, and sorption).
Advanced treatment	Removal of total dissolved solids and or trace constituents as required for specific water reuse applications. See Table 3-8 for the related treatment stages.
Anthropogenic compounds	Chemical compounds created by humans, often resistant to biodegradation.
Asymptomatic	Used to describe an individual who does not currently show symptoms of the disease being discussed. Asymptomatic individuals may develop symptoms of the disease at a later point in time if and when the disease onsets.
Biotic reaction	Produced or caused by living organisms. See also abiotic reaction.
Carcinogen	Cancer-causing substance or agent. Radiation and some chemicals and viruses are known carcinogens.
Coliform group of bacteria	Coliforms include several genera of bacteria belonging to the family Enterobacteriaceae, of which *Escherichia coli* is the most important member. The historical definition of this group is based on the method (lactose fermentation) used for its detection.
Cyst	In parasitology, a cyst is the resistant dormant stage of a single-celled organism which is passed out and encourages the propagation of the species (see Oocyst).
De facto indirect potable reuse	Many cities withdraw drinking water from rivers that contain varying amount of discharges from upstream cities and industries. Thus, indirect, unplanned, or *de facto* potable reuse of wastewater in domestic and public water supply is widespread and increasing.
Disinfection byproducts (DBPs)	Chemicals that are formed with the residual organic matter found in treated reclaimed water as a result of the addition of a strong oxidant (e.g., chlorine or ozone) for the purpose of disinfection.

Emerging contaminants	Constituents, which have been identified in water, that are being considered for regulatory action pending the development of additional information on health and the environmental impacts.
Endocrine-disrupting compounds (EDCs)	Synthetic and natural compounds that mimic, block, stimulate, or inhibit natural hormones in the endocrine systems of animals, including humans. The origins of EDCs include pesticides, pharmaceutically active chemicals (PhACs), personal care products (PCPs), herbicides, industrial chemicals, and disinfection byproducts.
Enteric	Intestinal, associated with human feces [e.g., enteric disease, diseases of the intestinal tract, generally causing diarrhea; or enteric bacteria (or virus) to describe pathogens that affect the intestinal tract].
Enterohemorrhagic	Causes bloody diarrhea.
Epidemiology	Medical science that involves the study of the incidence and distribution of diseases in large populations, and the conditions influencing the spread and severity of disease.
Etiology	A branch of medical science concerned with the causes and origins of diseases.
Fecal coliforms	Bacteria in the coliform group that inhabit the intestinal tract and are associated with fecal contamination. *E.coli*, the most common enteric bacterium, is commonly used as an indicator organism.
Gastrointestinal illness	A broad range of symptoms including vomiting, diarrhea, or nausea combined with abdominal cramps relating to both the stomach and the intestines.
Hemolytic uremic syndrome (HUS)	A disease in which red blood cells are destroyed and the kidneys fail.
Indicator organism	An organism whose presence or absence in an environment indicates the presence of other organisms of concern. For example, the coliform group of bacteria in water indicates the possible presence of pathogens.
In vitro	Biological studies which take place in isolation from a living organism such as in a test tube or petri dish.
In vivo	Biological studies which take place within a living biological organism.
Oocyst	Enteric protozoan parasites produce a cyst or oocyst. The oocyst is usually the infectious and environmental stage, and it contains sporozoites.
Personal care products (PCPs)	Products such as shampoo, hair conditioner, deodorants, and body lotion.
Pharmaceutically active compounds (PhACs)	Chemicals synthesized for medical purposes (e.g., antibiotics).
Pathogens	Disease-causing organisms capable of inflicting damage on a host it infects.
Public health	The science and practice of protecting and improving the health of a community through preventive medicine, health education, control of communicable diseases, application of sanitary measures, and monitoring of environmental hazards.
Sodicity	A parameter representing the amount of exchangeable sodium cation in water and relating to water infiltration in soil.

Tertiary treatment	Removal of residual suspended solids (after secondary treatment), usually by granular medium filtration, surface filtration, and membranes. Disinfection is also typically a part of tertiary treatment. Nutrient removal is often included in this definition. See Table 3-8 for the related treatment stages.
Thermotolerant coliforms (also known as fecal coliform)	A subset of the coliform group of bacteria found in the intestinal tract of humans and other warm-blooded animals. They can produce acid and gas from lactose at 44.0–44.5°C; hence the test for them is more specific than for total coliforms and selects a narrower range of organisms. *E.coli* is typically the major proportion of thermotolerant coliforms.
Total coliforms	All bacteria in the coliform group, including those not associated with the fecal matter of warm-blooded animals. Total coliform is commonly used as an indicator organism.
Trace organics	Organic compounds detected at very low (minute) levels by the use of sophisticated instrumentation capable of measuring concentrations in the range of 10^{-12} to 10^{-3} mg/L.
Vadose zone	Designation of the layer of the ground below the surface (unsaturated zone) but above the water (groundwater) table.

Reclaimed water derived from municipal wastewater (traditionally known as sewage) comes from a variety of sources including households, schools, offices, hospitals, and commercial and industrial facilities. The quantity and quality of wastewater derived from each source varies among communities, depending on the number and type of commercial and industrial establishments in the area, and the condition of the wastewater collection system such as the extent of infiltration and inflow, and, in the case of combined sewer systems, urban stormwater runoff. Thus, untreated municipal wastewater typically contains a variety of biological and chemical constituents that may be hazardous to human health and the environment. In many developing countries, the irrigation of vegetable crops with untreated or inadequately treated wastewater is a major source of enteric disease. The situation is different, however, in the United States and other industrialized countries where reliable wastewater treatment and health-related water reclamation and reuse regulations dictate the feasibility and acceptability of water reuse.

Health and environmental issues associated with water reclamation and reuse are related to wastewater treatment, reclaimed water quality, chemical and microbiological constituents that may be present in water, health risk assessment, and public perception and acceptance. Many issues related to nonpotable reclaimed water applications have been addressed successfully, and numerous agricultural and landscape irrigation projects and industrial cooling applications have been implemented throughout the world.

Characteristics of municipal wastewater and related health and environmental issues are presented in this chapter to serve as an introduction to water reuse regulations and guidelines (Chap. 4) and health risk analysis in water reuse applications (Chap. 5). The following topics are discussed in this chapter: (1) wastewater in public water supplies—*de facto* potable reuse, (2) introduction to waterborne diseases and health issues, (3) waterborne pathogenic microorganisms, (4) indicator organisms, (5) occurrence of microbial pathogens in untreated and treated wastewater and the environment, (6) chemical constituents in untreated and treated wastewater, (7) emerging contaminants in water and wastewater, and (8) environmental issues.

3-1 WASTEWATER IN PUBLIC WATER SUPPLIES—*DE FACTO* POTABLE REUSE

Many cities withdraw drinking water from surface water impoundments in protected upstream watersheds, which generally provide high quality surface water. In less desirable situations, drinking water is drawn from rivers that contain discharges from upstream cities and industries, as shown on Fig. 3-1 (see also Figure 23-1 in Chap. 23). Philadelphia, Cincinnati, New Orleans, and Los Angeles are examples of such cities. Other cities including New York, San Francisco, and Seattle have been able to develop protected upstream sources. Some cities are fortunate enough to have groundwater sources available, which are generally of high quality because they are protected from many environmental influences. However, many cities have overdrawn their groundwater sources and have been obliged

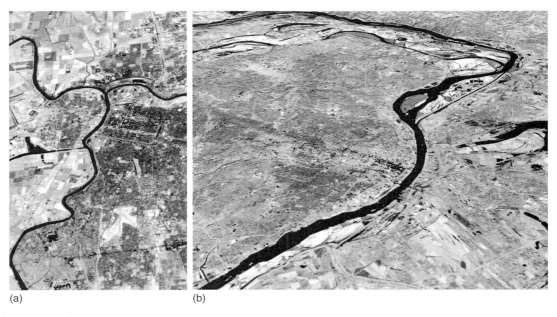

(a)   (b)

Figure 3-1

Unplanned and incidental (*de facto*) potable reuse occurs in many river systems in the United States: (a) Sacramento River at Sacramento, CA and (b) the Mississippi River near St. Louis, MO (Adapted from U.S. Geological Survey). In (a), river water containing treated wastewater discharges from the City of Sacramento, and other cities adjacent to the Sacramento River and its tributaries, is transported from the San Francisco Bay-Delta to southern California via the California Aqueduct as a source of potable water supply. In (b), the Mississippi River flows from the State of Minnesota to the Gulf coast; cities along its path use it as a source of potable water and for the discharge of treated wastewater.

to turn to surface waters containing varying amounts of treated wastewater for expanded drinking water supply. Thus, the indirect, unplanned, or *de facto* potable reuse of wastewater in domestic and public water supply is widespread and increasing.

Presence of Treated Wastewater in Public Water Supplies

Treated wastewater sometimes represents a significant portion of the total flow in many receiving waters. Notable examples include the Santa Ana River in southern California; the Platte River downstream from the City of Denver, Colorado; the Ohio River near the City of Cincinnati, Ohio; and the Occoquan Watershed located southwest of Washington, DC.

Most water reuse in these situations is incidental and unplanned, and goes largely unrecognized by the public and many professionals. Although these situations are beyond the scope of this textbook, which deals with formal and planned water reclamation and reuse, it must be recognized that where treated wastewater is present in a water supply source, what occurs is the *de facto* reuse of wastewater for potable purposes. In fact, the distinctions between the various types of water reuse are arbitrary and every degree of water reuse exists. "The distinction between inadvertent or unplanned and planned indirect potable reuse is, after all, one of intention or attention" (Dean and Lund, 1981).

Impact of the Presence of Treated Wastewater in Public Water Supplies

Because conventional wastewater treatment does not remove all of the known constituents from wastewater, and stormwater is not treated typically, concerns exist about the health risk to downstream water supplies. As the quantities of treated wastewater discharged into the nation's waters increase, much of the research that is focused on *unplanned indirect potable reuse* is becoming equally relevant to planned direct and indirect potable water reuse. Because of the research interest, advanced analytical techniques, and public concerns, emerging pathogens (i.e., pathogens that have been identified recently) including several enteric viruses, and trace organic constituents, including disinfection byproducts, PhACs, and PCPs have been reported in natural waters as well as in reclaimed water. Many of these compounds are suspected endocrine disruptors. The ramifications of many of these constituents in trace quantity are not well understood with respect to long-term health effects and the environmental impact.

It is important to note, however, that the great majority of planned water reuse applications in the United States and in the developed world are for nonpotable reuse, such as irrigation of agricultural lands and landscapes, and industrial applications. Thus, public health concerns related to the possible ingestion of reclaimed water are remote and not directly applicable to most water reuse applications.

3-2 INTRODUCTION TO WATERBORNE DISEASES AND HEALTH ISSUES

The potential transmission of infectious disease by pathogenic organisms is the most common concern in water reclamation and reuse. While there is no epidemiological evidence that the use of reclaimed water (i.e., appropriately treated municipal wastewater meeting strict water reclamation and reuse regulations) for any of its applications has caused a disease outbreak in the United States, the potential spread of infectious disease, particularly in developing countries, through untreated or inadequately treated municipal wastewater remains a public health concern.

Important Historical Events

Concerns over particular waterborne microorganisms have changed over the years due to improved sanitation, evolving microorganisms, the use of preventive medicine, and improved microbiological and epidemiological methods for identifying the microorganisms responsible for disease outbreaks. Historically, microorganisms were first identified as agents of waterborne disease during the cholera outbreak in England in the 1860s. In 1884, a pioneering German pediatrician and bacteriologist, Theodor Escherich isolated organisms, which he initially thought were the cause of cholera, from the stools of a cholera patient. Later it was found that similar organisms were also present in the intestinal tracts of every healthy individual. The organism isolated by Escherich was eventually named for him—*Escherichia coli* or *E. coli*. In 1892, the New York State Board of Health used the fermentation tube method, developed by Theobald Smith, for the detection of *E. coli* to demonstrate the connection between sewage contamination of the Mohawk River and the spread of typhoid fever (see Fig. 3-2). In the 1920s, typhoid fever was linked to the waterborne bacterium *Salmonella typhi*. *Giardia lamblia,* a waterborne protozoan, became a major concern in the 1960s; rotavirus and Norwalk virus were associated with a large number of disease outbreaks beginning in the 1970s; and *Cryptosporidium parvum,* also a protozoan, was first associated with

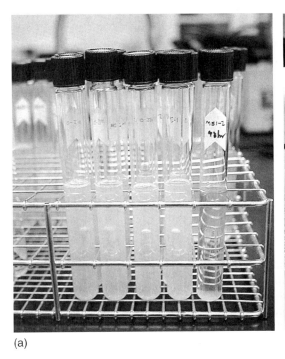
(a)

(b)

Figure 3-2
Detection of coliform group of bacteria by (a) multiple-tube fermentation technique, and (b) membrane filter technique.

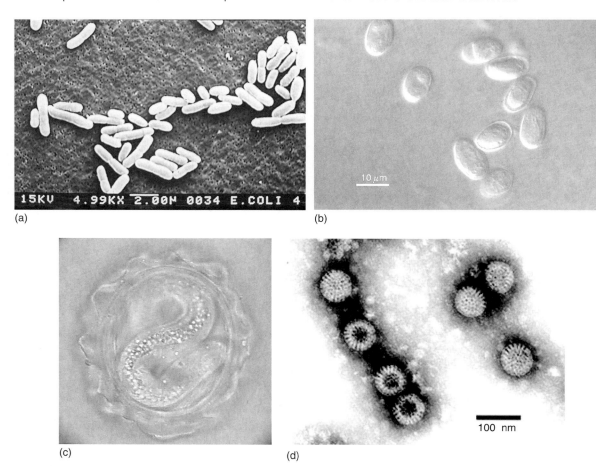

Figure 3-3

Microscopic pictures of representative pathogens: (a) *E. coli*, (b) protozoa, (c) helminths, and (d) virus. [Images courtesy of (a) A. Levine, (b) and (d) U.S. EPA, and (c) K. Nelson.]

waterborne outbreaks in the 1980s (Hunter, 1997; NRC, 1998; Crittenden et al., 2005). Microscopic pictures of representative pathogens are shown on Fig. 3-3.

Waterborne Diseases

Microorganisms associated with waterborne disease are primarily enteric pathogens, including enteric bacteria, protozoa, and viruses. These pathogens can survive in water and infect humans through ingestion of feces-contaminated water, person-to-person contact, or contaminated surfaces and food. A schematic representation of the routes of transmission for enteric disease is shown on Fig. 3-4.

Any potable water supply receiving human or animal wastes can be contaminated with disease-causing microbial agents. Even so-called pristine water supplies have been associated with disease outbreaks, presumably due to contamination from wildlife in protected watersheds (Cooper and Olivieri, 1998; NRC, 1998; Yates and Gerba, 1998).

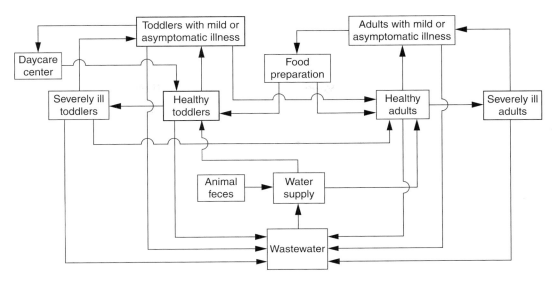

Figure 3-4
Conceptual framework for disease transmission and the roles of wastewater, water supply, and food preparation. (From Crittenden et al., 2005.)

As shown in Table 3-1, a diversity of pathogenic organisms, including bacteria, protozoa, cyanobacteria, helminths (intestinal worms), and viruses are potentially present in untreated municipal wastewater. The concentration of helminths is particularly high in untreated municipal wastewater in developing countries due to the high rates of infection in these areas.

Etiology of Waterborne Disease

In the United States, state and local public health departments are responsible for detecting disease outbreaks, monitoring, and conducting epidemiological investigations of suspected waterborne outbreaks. When an outbreak occurs and waterborne pathogens are suspected, epidemiological studies to obtain the information on the etiology (causes and origins) of waterborne disease are conducted to identify whether water is the vehicle of transmission.

For gastrointestinal illness, routine stool examinations by hospital laboratories typically include culturing for *Salmonella, Shigella,* and *Campylobacter* bacteria. At the specific request of a physician, many laboratories can also test for rotavirus, *Giardia,* and *Cryptosporidium.* Nevertheless, no specific agent is identified in many outbreaks, leaving the cause classified only as acute gastrointestinal illness (AGI) of unknown etiology. Before 1982, in fact, most waterborne outbreaks reported were listed as AGI (NRC, 1998). Improper collection of clinical and/or water samples and limitations of diagnostic techniques for many enteric pathogens can prevent accurate determination of the pathogen. Based on the clinical symptoms it appears that many of the AGI outbreaks may be due to viral agents, such as Norovirus (previously known as Norwalk-like virus) and related human Caliciviruses (NRC, 1998; Craun and Calderon, 1999; Huffman et al., 2003).

Table 3-1 Examples of major groups and genera of waterborne and water-based pathogens[a]

Group	Pathogen	Diseases and symptoms caused
Bacteria	Salmonella	Typhoid and diarrhea
	Shigella	Diarrhea
	Campylobacter	Diarrhea—leading cause in foodborne outbreaks
	Yersinia enterocolitica	Diarrhea
	Escherichia coli O157:H7 and other certain strains	Diarrhea, which can lead to hemolytic uremia syndrome in small children.
	Legionella pneumophila	Pneumonia and other respiratory infections
Protozoa	Naegleria	Meningoencephalitis
	Entamoeba histolytica	Amoebic dysentery
	Giardia lamblia	Chronic diarrhea
	Cryptosporidium parvum	Acute diarrhea, fatal for immunocompromised individuals
	Cyclospora	Diarrhea
	Microsporidia includes Enterocytozoon spp.	Chronic diarrhea and wasting, pulmonary, ocular, muscular, and renal disease
	Encephalitozoon spp.	
	Septata spp.	
	Pleistophora spp.	
	Nosema spp.	
Cyanobacteria (blue-green algae)	Microcystis	Diarrhea from ingestion of the toxins these organisms produce
	Anabaena	Microcystin toxin is implicated in liver damage
	Aphantiomenon	
Helminths	Ascaris lumbricoides	Ascariasis
	Trichuris trichiora	Trichuriasis (whipworm)
	Taenia saginata	Beef tapeworm
	Schistosoma mansoni	Schistosomiasis (affecting the liver, bladder, and large intestine)
Viruses	Enteroviruses (polio, echo, coxsackie)	Meningitis, paralysis, rash, fever, myocarditis, respiratory disease, and diarrhea
	Hepatitis A and E	Infectious hepatitis
	Human Caliciviruses	
	Noroviruses	Diarrhea/gastroenteritis
	Sapporo	Diarrhea/gastroenteritis
	Rotavirus	Diarrhea/gastroenteritis
	Astroviruses	Diarrhea
	Adenovirus	Diarrhea (types 40 and 41), eye infections, and respiratory disease
	Reovirus	Respiratory and enteric infections

[a]Adapted from Gerba (1996); Straub and Chandler (2003).

Similar to drinking water safety, available information on health issues and reclaimed water quality continues to expand, which in turn increases the ability to answer questions related to the safety of reclaimed water. Conclusions drawn from data gathered from actual water reuse applications in the United States and other developed countries are that the risk of transmission of infectious disease is minimal after proper treatment and when the applicable water reclamation and reuse regulations are met, as further discussed in Chaps. 4 and 5.

3-3 WATERBORNE PATHOGENIC MICROORGANISMS

The principal infectious agents that may be found in untreated municipal wastewater can be classified into four broad groups: bacteria, protozoa, helminths, and viruses. Many of the infectious agents reported in Table 3-1 are potentially present in untreated municipal wastewater. Waterborne gastroenteritis associated with drinking water and recreational water is shown in Tables 3-2 and 3-3, respectively. Important members of each of these groups are considered briefly in the following discussion.

Terminology Conventions for Organisms

According to convention, every biological species (except viruses) bears a Latinized name that consists of two words. The first word is the genus (e.g., *Giardia*), and the second word is the species (e.g., *lamblia*). The first letter of the genus name is capitalized, and both the genus and species are either italicized or underlined. After the full names of genus and species names (e.g., *Escherichia coli*) have been given, further reference to the organism may be abbreviated as *E. coli*. Many of these organisms can be further differentiated on the basis of antigenic recognition by antibodies of the immune system, a process called serotyping (Cohn et al., 1999). It should be noted that these conventions do not apply to viruses, which are not living.

Log Removal

Because microorganisms often exist in large numbers in excreta or municipal wastewater, their removal or inactivation in wastewater treatment processes is often expressed as *log removal*. With detectable levels of microorganisms, log removal represents the reduction associated with wastewater treatment or water reclamation processes. Log removal is defined as

$$\text{Log removal} = -\log\left(\frac{\text{conc}_{out}}{\text{conc}_{in}}\right) \quad (3\text{-}1)$$

For example, if the concentration of *Giardia lamblia* is reduced from 100/L in the influent to 1/L in the effluent by activated sludge treatment process, the log removal due to the treatment is

$$\text{Log removal} = -\log\left(\frac{1}{100}\right) = 2 \text{ or } 99\% \text{ removal}$$

Bacteria

Bacteria are microscopic organisms ranging from approximately 0.2 to 10 μm in length. They are distributed ubiquitously in nature and have a wide variety of nutritional requirements. Many types of harmless and beneficial bacteria colonize in the human intestinal tract and are routinely shed in the feces. Pathogenic bacteria are also present in the feces of infected individuals. Therefore, municipal wastewater can contain a wide

Table 3-2
Waterborne gastroenteritis outbreaks in the United States associated with drinking water, 1993–2002[a,b]

Etiologic agent	1993–1994 Outbreaks	1993–1994 Cases	1995–1996 Outbreaks	1995–1996 Cases	1997–1998 Outbreaks	1997–1998 Cases	1999–2000 Outbreaks	1999–2000 Cases	2001–2002 Outbreaks	2001–2002 Cases
Unknown	5	495	8	684	5	163	17	416	7	117
Noroviruses	0	0	2	742	1	1450	5	512	5	727
Giardia spp.	5	385	3	1546	4	159	6	52	3	18
Cryptosporidium spp.	5	403,271	0	0	2	1432	2	10	1	10
Campylobacter jejuni	3	223	0	0	0	0	2	117	1	13
Escherichia coli O157:H7	1	2	1	33	3	164	4	60	1	2
Salmonella spp.	1	625	0	0	0	0	2	208	0	0
Shigella spp.	2	263	2	93	1	83	0	0	0	0
Other bacteria	1[c]	11[c]	1[d]	60[d]	0	0	0	0	0	0
More than two bacterial agents	0	0	0	0	0	0	1[e]	781[e]	1	12
Chemical agents (total)	8	93	7	90	3	44	2	3	5	39
Total	31	405,368	24	3248	19	3495	41	2159	24	938

[a] Data from Blackburn et al. (2004); Lee et al. (2002); Barwick et al. (2000); Levy et al. (1998); Kramer et al. (1996).
[b] Outbreaks of gastroenteritis reported to CDC. Outbreaks of meningoencephalitis are not included.
[c] Non-O1 *Vibrio cholerae*.
[d] *Plesiomonas shigelloides*.
[e] *C. jejuni* and *E. coli* O157:H7.
[f] *C. jejuni* and *Yersinia enterocolitica*.

Table 3-3
Waterborne gastroenteritis outbreaks in the United States associated with recreational water, 1993–2002[a,b]

Etiologic agent	1993–1994 Outbreaks	1993–1994 Cases	1995–1996 Outbreaks	1995–1996 Cases	1997–1998 Outbreaks	1997–1998 Cases	1999–2000 Outbreaks	1999–2000 Cases	2001–2002 Outbreaks	2001–2002 Cases
Unknown	1	12	4	65	3	939	6	95	7	141
Noroviruses	0	0	1	55	2	48	3	202	5	146
Giardia spp.	4	141	1	77	0	0	1	18	1	2
Cryptosporidium spp.	6	693	6	8512	9	538	16	1394	11	1474
Campylobacter jejuni	0	0	0	0	0	0	1	6	0	0
Escherichia coli O157:H7	1	166	6	52	3	39	5	61	4	78
Salmonella spp.	0	0	1	3	0	0	0	0	0	0
Shigella spp.	4	737	3	190	1	9	3	46	2	78
Other bacteria	0	0	0	0	0	0	0	0	0	0
More than two bacterial agents	0	0	0	0	0	0	1[c]	38[c]	0	0
Total	16	1749	22	8954	18	1573	36	1860	30	1919

[a]Data from Yoder et al. (2004); Lee et al. (2002); Barwick et al. (2000); Levy et al. (1998); Kramer et al. (1996).
[b]Outbreaks of gastroenteritis reported to CDC. Outbreaks of meningoencephalitis, dermatitis, keratitis, leptospirosis, and Pontiac fever (caused by *Legionella pneumophila*) are not included.
[c]*Cryptosporidium parvum* and *Shigella sonnei*.

variety and concentration range of bacteria, including those pathogenic to humans (Schroeder and Wuertz, 2003).

Enteric bacteria are associated with human and animal feces and may be transmitted to humans through fecal-oral transmission routes (refer to Fig. 3-4). Most illnesses due to enteric bacteria cause acute diarrhea, and certain bacteria tend to produce particularly severe symptoms. Classical waterborne bacterial diseases such as dysentery, typhoid, and cholera, while still important in developing countries, have dramatically decreased in the United States since the 1920s (Craun, 1991). However, *Campylobacter*, nontyphoid *Salmonella*, and pathogenic *E. coli* have been estimated to cause three million illnesses per year in the United States (Bennett et al., 1987). As measured by hospitalization rates during waterborne disease outbreaks (i.e., the percentage of illnesses requiring hospitalization), the most severe illnesses are due to pathogenic *E. coli* (14 percent), *Shigella* (5.4 percent), and *Salmonella* (4.1 percent) (Gerba et al., 1994). Hence, enteric bacterial pathogens remain an important cause of waterborne disease in the United States. It is estimated that enteric bacteria caused 14 percent of all waterborne disease outbreaks in the United States from 1970 to 1990 (Craun, 1991). Enteric bacteria of particular concern are discussed below (Cohn et al., 1999; AWWA, 1999; Schroeder and Wuertz, 2003).

Shigella

Shigella infects humans and primates and causes shigellosis bacillary dysentery. *S. sonnei* causes the bulk of waterborne infections, although all four subgroups (*S. dysenteriae, S. flexneri, S. boydii,* and *S. sonnei*) have been isolated during different disease outbreaks (Moyer, 1999). Waterborne shigellosis is most often the result of contamination from one identifiable source, such as an improperly disinfected well. The survival of *Shigella* in water and their response to water treatment is similar to that of the coliform bacteria. Therefore, systems that control coliforms effectively protect against *Shigella*.

Salmonella

Over 2,200 known serotypes of *Salmonella* exist, all of which are pathogenic to humans. Most cause gastrointestinal illness; however, a few can cause other types of disease, such as typhoid (*S. typhi*) and paratyphoid (*S. paratyphi*) fevers. The latter two species infect only humans; while the others are carried by both humans and animals. At any time, about 0.1 percent of the population is excreting *Salmonella* (mostly as a result of infections caused by contaminated foods).

Escherichia coli

E. coli is a member of the fecal coliform group of bacteria found in the intestinal tracts of humans and warm-blooded animals, and is normally harmless (see Fig. 3-3a). This organism in water indicates fecal contamination. Some strains of *E. coli* are, however, pathogenic and cause gastroenteritis. A particular strain, *E. coli* O157:H7, causes acute bloody diarrhea and abdominal cramps (enterohemorrhagic), and in some cases (two to seven percent of infections) have resulted in hemolytic uremic syndrome (HUS), in which red blood cells are destroyed and the kidneys fail. One of the highest mortality rates of all waterborne diseases is due to HUS.

Known microbial reservoirs for *E. coli* O157:H7 are healthy cattle. Transmission can occur by ingestion of undercooked beef or raw milk, and by drinking contaminated water (NRC, 1998). Drinking water was identified as the source of an outbreak of *E. coli* O157:H7 in a Missouri community in 1989, which involved 243 cases (i.e., a person with the disease) that included 32 hospitalizations, and four deaths. Unchlorinated well water and breaks in the water distribution system were considered to be contributing factors. Another waterborne outbreak of *E. coli* O157:H7 involved 80 cases in Oregon in 1991 and was attributed to recreational water contact in a lake (Oregon Health Division, 1992; CDC, 1993).

Yersinia enterocolitica

Yersinia enterocolitiea can cause acute gastrointestinal illness, and is carried by humans, pigs, and a variety of other animals. The organism is found commonly in surface waters and has been isolated occasionally from groundwater and drinking water. *Yersinia* can grow at temperatures as low as 4°C and has been isolated in untreated surface waters more frequently during colder months than warmer months.

Campylobacter jejuni

Campylobacter jejuni can infect humans and a variety of animals and is the most common bacterial cause of gastrointestinal illness requiring hospitalization, and a major cause of foodborne illness. The natural habitat of *Campylobacter* is the intestinal tract of warm-blooded animals, and it is found commonly in wastewater and surface waters.

Protozoa

Protozoa are single-celled organisms that lack a cell wall, but do possess a flexible covering called a pellicle (see Fig. 3-3*b*). Typically they are larger than bacteria and, unlike algae, cannot photosynthesize. Protozoa are common in fresh and marine water, and some can grow in soil and other locations (Cohn et al., 1999). The enteric protozoan parasites produce cysts or oocysts that aid in their survival in wastewater and under adverse conditions in the aquatic environment. Important pathogenic protozoa include *Giardia lamblia, Cryptosporidium parvum,* and *Entamoeba histolytica.*

Giardia lamblia

Waterborne giardiasis, caused by the protozoan *G. lamblia*, is recognized as the most common protozoan infection in the United States and remains a major public health concern (Craun, 1986; Kappus et al., 1992). The reported incidence of waterborne giardiasis, a gastrointestinal disease manifested by diarrhea, fatigue, and cramps, has increased in the United States since 1971 (Craun, 1986). According to the Giardia Surveillance data for the period from 1998 to 2002, the total number of reported cases ranged from about 19,700 to 24,200 per year (Hlavsa et al., 2005a). Between 1993 and 2002, there were 21 outbreaks of giardiasis associated with drinking water, and seven associated with recreational water. Because *G. lamblia* is endemic in wild and domestic animals, infection can result from water supplies that have no wastewater contribution. The disease cycle for *G. lamblia* is illustrated on Fig. 3-5. Densities of *G. lamblia* cysts in untreated wastewater have been reported in a range between 10^1 and 10^4 cysts/L (Sykora et al., 1991; Rose et al., 1996; Chauret et al., 1999; Caccio et al., 2003) and as high as 3375 cysts/L. In addition, *G. lamblia* has been detected in treated wastewater effluent and is much more resistant to disinfection with chlorine than is bacteria. Ultraviolet irradiation has been found effective for inactivating *G. lamblia* and *G. muris* (Craik et al., 2000; Linden et al., 2002).

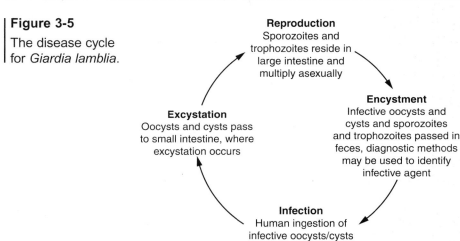

Figure 3-5
The disease cycle for *Giardia lamblia*.

Cryptosporidium parvum

C. parvum was first described as a human pathogen in 1976 (Juranek, 1995). Two *Cryptosporidium* species, *C. parvum* and *C. hominis*, which was formerly recognized as a genotype of *C. parvum*, are known to infect humans. Other species including *C. canis*, *C. felis*, *C. meleagridis*, and *C. muris* may also infect immunocompromised persons (CDC, 2005a). In the environment, *Cryptosporidium* is in the form of an oocyst, which is about 4 to 6 μm in diameter and capable of surviving until it is ingested by an animal. Once it reaches the intestinal tract of an animal, sporozoites in an oocyst initiate infection, causing a gastrointestinal disorder, that is, cryptosporidiosis.

Cryptosporidiosis causes severe diarrhea; no pharmaceutical cure exists at present. Average infection rates in the United States, as measured by oocyst excretion in a population, range from 0.6 to 20 percent (Fayer and Ungar, 1986; Lisle and Rose, 1995). The disease can be particularly hazardous for people with compromised immune systems (Current and Garcia, 1991).

According to the CDC's *Surveillance for Waterborne Disease Outbreaks*, there were 10 outbreaks of cryptosporidiosis associated with drinking water, and 49 associated with recreational water between 1993 and 2002 (Hlavsa et al., 2005b). In 1993, a massive outbreak of cryptosporidiosis occurred in Milwaukee, WI, causing approximately 400,000 illnesses and at least 50 fatalities. Deterioration of raw water quality by either animal or human wastes and decreased effectiveness of water treatment processes due to stormwater inflow were attributed to the outbreak, but the original source of *Cryptosporidium* was not identified definitively (MacKenzie et al., 1995; Kramer et al., 1996).

Cryptosporidium has been found in secondary effluent samples at various levels, typically between 10^1 and 10^3 oocysts/L (Madore et al., 1987; Peeters et al., 1989; Villacorta-Martinez et al., 1992; Rose et al., 1996; Robertson et al., 2000). Low concentrations of oocysts have been detected in reclaimed waters that were treated with conventional secondary treatment followed by filtration and chlorination; some of the detected oocysts were determined to be infective (Korick et al., 1990; Gennaccaro et al., 2003; Ryu et al., 2005). Chlorination is not effective for inactivating *Cryptosporidium*. Alternatively, ultraviolet (UV) irradiation has been proven to be effective to inactivate *Cryptosporidium* oocysts (Clancy et al., 2000; Craik et al., 2001).

Entamoeba histolytica

When ingested, *E. histolytica* can cause amoebic dysentery, with symptoms ranging from acute bloody diarrhea and fever to mild gastrointestinal illness. Occasionally, the organism can cause ulcers and then invade the bloodstream, causing more serious effects. However, most infected individuals do not have clinical symptoms. In contrast to the case for *G. lamblia* and *C. parvum,* animals are not reservoirs for *E. histolytica*, so the potential for source water contamination is relatively low, especially if municipal wastewater treatment practices are adequate. About 3000 cases of amebiasis occur typically in the United States each year, and waterborne disease outbreaks caused by *E. histolytica* are infrequent (CDC, 1985).

Helminths

The term *helminths* is used to describe a group of mostly parasitic worms (see Fig. 3-3c). Worldwide, helminths are one of the principal causative agents of human disease, collectively on the order of 4.5 billion illnesses per year. Over the last century, helminth infections in the United States decreased dramatically, because of more extensive sanitation facilities, and improved wastewater treatment facilities and food handling practices. However, due to increased levels of immigration to the United States of persons from countries where parasitic worms are endemic, helminths and helminth ova (eggs) are found increasingly in untreated municipal wastewater in the United States (Tchobanoglous et al., 2003; Maya et al., 2006).

Ascaris lumbricoides

The infectious disease caused by *A. lumbricoides* (an intestinal roundworm) is known as ascariasis. In its moderate form ascariasis is characterized by digestive and nutritional problems, abdominal pain, vomiting, and the passage of live worms in stools or vomit. More serious cases involving the liver can cause death. Transmission is through the ingestion of salads and vegetables contaminated with helminth ova from human feces. Worldwide, especially in moist tropical areas, the prevalence of this type of infection can exceed 50 percent. In the United States, ascariasis is most common in the south.

Schistosoma mansoni

Schistosomiasis, caused by *S. mansoni,* is a debilitating infection where worms inhabit veins of the host and chronic infection affects the liver or urinary system. Humans, domestic animals, and rats serve as the primary hosts and snails act as a necessary intermediate host. Larvae found in water, incubated and released from snails, are able to penetrate through human skin. Eggs are excreted via urine or feces and the cycle begins again as larvae develop in water and reinfect snails. Schistosomiasis is prevalent in Africa, the Arabian Peninsula, South America, the Middle East, Asia, and parts of India.

Viruses

Viruses are obligate intracellular parasites able to multiply only within a host cell and are host-specific. Viruses occur in various shapes and range in size from 0.01 to 0.3 μm in cross-section and are composed of a nucleic acid core surrounded by an outer coat of protein. Enteric viruses are obligate human pathogens, which mean they replicate only in the human host (see Fig. 3-3d). Their simple structure, a protein coat surrounding a core of genetic material (DNA or RNA), allows for prolonged survival in the environment. There are more than 120 identified human enteric viruses. Some of the better understood viruses include the enteroviruses (polio-, echo-, and coxsackieviruses), hepatitis A virus, rotavirus, and human caliciviruses (e.g., Noroviruses).

Most enteric viruses cause gastroenteritis or respiratory infections, but some may cause other diseases as well, including encephalitis, neonatal disease, myocarditis, aseptic meningitis, and jaundice (Gerba et al., 1985, 1996; Frankel-Conrat et al., 1988; Wagenkneckt et al., 1991; see also Table 3-1). Some common enteric viruses that have caused, or could potentially cause, waterborne diseases are discussed below (Cohn, et al., 1999).

Hepatitis A

Although all enteric viruses are potentially transmitted by drinking water, evidence of this route of infection is strongest for hepatitis A virus (HAV). The HAV causes infectious hepatitis, an illness characterized by inflammation and necrosis of the liver. Symptoms include fever, weakness, nausea, vomiting, diarrhea, and sometimes jaundice.

Noroviruses and Other Caliciviruses

The pathogenic viruses classified as caliciviruses are not well quantified as they do not grow in culture. Viruses in this group are generally identified by molecular technologies such as reverse-transcriptase polymerase chain reaction (RT-PCR), and electron microscopy. Human caliciviruses (HuCVs) have generally been named after the location of the first outbreak (i.e., Norwalk agent, Snow Mountain agent, Hawaii agent, Montgomery County agent, and so on) (Gerba et al., 1985). The family of caliciviruses (*Caliciviridae*) is divided into four genera, of which Noroviruses and Sapoviruses have been associated with human diseases. Noroviruses, which were previously recognized as Norwalk-like viruses, or small round structured viruses, are considered to be responsible for a vast majority of nonbacterial gastroenteritis (Karim and LeChevallier, 2004).

Based on current estimates, over 90 percent of nonbacterial gastroenteritis outbreaks of unidentified etiology may be due to HuCVs. Between 1997 and 2000, for example, fecal specimens from 284 nonbacterial outbreaks were examined by the Center for Disease Control and Prevention (CDC), of which 93 percent were attributed to Noroviruses (Fankhauser et al., 2002). Information on several documented waterborne outbreaks of calicivirus is shown in Table 3-4. With advances in molecular methods for identification and quantification of previously unidentifiable viruses, a strategy for the detection of the caliciviruses in various water matrices is being refined (Huffman et al., 2003; Karim and LeChevallier, 2004).

Rotaviruses

Rotaviruses cause acute gastroenteritis, primarily in children. Almost all children have been infected at least once by the age of five years; and in developing countries, rotavirus infections are a major cause of infant mortality. Rotaviruses are spread by fecal-oral transmission and have been found in municipal wastewater, lakes, rivers, groundwater, and even tap water (Gerba et al., 1985; Gerba, 1996).

Table 3-4 Documented waterborne calicivirus outbreaks[a]

Year	Location	Water source	Estimated no. of primary cases	Viral genotype
2000	Italy	Municipal	344	GGII
1999	France	Municipal	~6	GGII
1998	Switzerland	Groundwater	>1750	GGI, GGII
1998	Finland	Municipal	1700 to 3100	GGII
1998	Wisconsin	Lake[d]	18	Serum Ab positive[b]
1998	Ohio	Lake[d]	30	Serum Ab positive[b]
1996	Florida	Well	594	Serum Ab positive[b]
1995	Wisconsin	Municipal	148	SRSV
1995	Alaska	Shallow well	433	GGII
1994	United Kingdom Bristol/South Wales	Municipal	130	GGI GGII
1988	Idaho	Well	339	Serum Ab positive[b]
1987	Pennsylvania, Delaware, New Jersey	Well[c]	5000	Serum Ab positive[b]
1986	South Dakota	Well	135	Serum Ab positive[b]
1986	California	Lake[d]	41	Serum Ab positive[b]
1986	New Mexico	Stream	36	Serum Ab positive[b]
1978	Washington	Municipal (cross-connection)	>1600	Serum Ab positive[b]
1977	Ohio	Swimming pool	103	Serum Ab positive[b]
1976	Colorado	Spring	418	Immune, electron microscopy

[a]Adapted from Huffman et al. (2003).
[b]Fourfold increase in serum antibody titer compared to control sera.
[c]Noncommunity well used to manufacture ice.
[d]Recreational water-related outbreak.

Enteroviruses

The enteroviruses include polioviruses, coxsackieviruses, and echoviruses. Enteroviruses are found in wastewater and surface water, and sometimes in drinking water. In 1952, a polio outbreak with 16 cases of paralytic disease was attributed to a drinking water source, but since then, no well-documented case of waterborne disease caused by poliovirus has been reported in the United States (Craun, 1986). Poliovirus vaccine and large-scale vaccination programs have eradicated paralytic poliomyelitis from the Western Hemisphere (Gerba, 1996). Vaccination with oral poliovirus vaccine (OPV) was discontinued in the United States in 2000. In 2005, however, four unvaccinated children in Minnesota were infected by poliovirus, raising concerns regarding transmission of poliovirus to other communities with low levels of vaccination, and the potential for an outbreak in the United States (CDC, 2005b).

Coxsackieviruses, and to a lesser extent echoviruses, cause a large variety of illnesses, some very serious in humans, including the common cold, aseptic meningitis, and heart disease. Symptoms can include fever and gastrointestinal problems.

Adenoviruses

There are 47 known types of adenoviruses, but only types 40 and 41 are important causes of gastrointestinal illness, especially in children. Other types of adenoviruses are responsible for upper respiratory illness, including the common cold. However, all types may be shed in the feces, and may be spread by the fecal-oral route. Although adenoviruses have been detected in wastewater, surface water, and drinking water, data on their occurrence in water are limited. Drinking water outbreaks implicating these viruses have not been reported and, therefore, their significance as waterborne pathogens is uncertain. Adenoviruses are relatively resistant to disinfectants and may not readily be inactivated or removed by traditional treatment methods (Cohn et al., 1999).

3-4 INDICATOR ORGANISMS

The number and variety of microbial constituents that may be present in municipal wastewater are considerable. Routine monitoring for all possible microbial constituents, especially viruses, is either impossible or impractical. In addition, the time required to complete most identification analyses precludes their utility as a water quality control tool. Thus, tests for surrogate microorganisms (known as indicator organisms) that are present when pathogens are present have been used to estimate the presence of pathogens.

Characteristics of an Ideal Indicator Organism

An ideal indicator organism should have the following characteristics (Cooper and Olivieri, 1998; Maier et al., 2000; NRC, 2004):

1. The indicator organism must be present when fecal contamination is present.
2. The numbers of indicator organisms present should be equal to or greater than those of the target pathogenic organism (e.g., pathogenic viruses)
3. The indicator organism must exhibit the same or greater survival characteristics in treatment processes and the environment as the target pathogen organism for which it is a surrogate.
4. The indicator organism must not reproduce outside of the host organism (i.e., the culturing procedure itself should not produce a serious health threat to laboratory workers).
5. The isolation and quantification of the indicator organism must be faster than that of the target pathogen (i.e., the procedure must be less expensive and it must be easier to cultivate the indicator organisms than the target pathogen).
6. The organism should be a member of the intestinal microflora of warm-blooded animals.

As noted above, one of the ideal characteristics of an indicator organism is that it must be present when the target pathogen is present. Unfortunately, the target pathogen(s) may not be present during the entire year, because the shedding of pathogenic organisms is not uniform throughout the year. Thus, it is important that the indicator organism be present when fecal contamination is present, if public health is to be protected. To date, no ideal indicator organism has been found.

The Coliform Group of Bacteria

The intestinal tract of humans contains a large population of rod-shaped bacteria known collectively as the coliform group of bacteria (see Figs. 3-2 and 3-3a). Each person excretes from 100 to 400 billion coliform bacteria per day, in addition to other kinds of bacteria. Thus, the presence of coliform bacteria in environmental samples has, over the years, been taken as an indication that pathogenic organisms associated with feces (e.g., viruses) may also be present, and the absence of coliform bacteria has been taken as an indication that the water is also free from disease-producing organisms.

Fecal coliform are indicative of fecal contamination and associated health risks; however, the measurement and control of total coliforms (rather than only fecal coliforms) during disinfection is considered to be a more stringent treatment goal. Fecal coliform bacteria are classified as the coliform group of bacteria that are able to ferment lactose at 44.5°C and produce indole from tryptophan. Most organisms identified using the fecal coliform test are *E. coli* that originate from warm-blooded animals; however, some other nonfecal thermotolerant bacteria may also be present. Organisms identified with the total coliform test must be able to grow at 35°C in the presence of bile salts and produce acid and gas during the fermentation of lactose (*Standard Methods*, 2005). Water quality standards have used either (total or fecal) or both measures, depending on the type of water use (NRC, 1998). While coliform bacteria serve well as indicators of bacterial pathogens, they may not predict the inactivation or removal of enteric protozoa, viruses, and helminths.

Standards for drinking water quality have been based upon the total coliform count, which is quite conservative as the standard is low (≤ 1 coliform/100 mL) regardless of the type of coliform. The U.S. EPA has proposed fecal coliform to be the standard indicator bacteria for reclaimed water. However, some regulatory agencies, for example, California Department of Health Services, are more conservative, and require total coliform measurement for the compliance with the standard/criteria for reclaimed water (see Chap. 4).

Bacteriophages

Bacteriophages are viruses that infect bacteria. They have been used as models or surrogates for human viruses in basic genetic research as well as water quality assessment (Grabow, 2001). Coliphages are viruses that infect *E. coli* (see Fig. 3-6). The presence of coliphages in water, therefore, is taken as an indication of the presence of their host *E. coli*, which is excreted by animals and humans. Coliphages may serve as better indicators for human enteric viruses than bacterial indicators, because coliphages more closely resemble human enteric viruses in size, shape, and resistance to treatment processes. In a comparison of untreated and treated wastewater, river water, treated river water, and treated lake water, Havelaar et al. (1993) found significant correlations between levels of coliphage and levels of enteric viruses in all but the untreated and treated wastewater samples. The conclusion reached from an analysis of these data was that other unknown factors may complicate the use of coliphages as indicators when evaluating recent wastewater inputs into a water body (NRC, 1998).

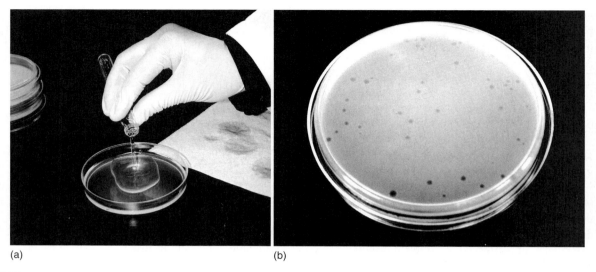

(a)   (b)

Figure 3-6
Test procedure for the determination of coliphage MS-2 viruses that infect *Escherichia coli:* (a) sample containing coliphage is poured onto a preformed lawn (growth) of *E. coli* in a petri dish, and (b) each clear spot on the petri dish after incubation for 12 h is counted as an individual coliphage.

Other Indicator Organisms

Other microorganisms that have been used or proposed for use as indicators of fecal contamination are summarized in Table 3-5. Indicator organisms that have been used to establish performance criteria for various water uses are reported in Table 3-6.

3-5 OCCURRENCE OF MICROBIAL PATHOGENS IN UNTREATED AND TREATED WASTEWATER AND IN THE ENVIRONMENT

The presence of pathogenic microorganisms in various types of wastewater is discussed in this section. The types of wastewater considered includes (1) untreated wastewater, (2) primary effluent, (3) secondary effluent, (4) tertiary effluent, and (5) effluent produced by advanced wastewater treatment (AWT). Because all forms of wastewater have been used in various water reuse applications, the information presented is useful when assessing associated health risks in water reuse applications, which are discussed in Chap. 5.

Pathogens in Untreated Wastewater

The occurrence and concentration of pathogenic microorganisms in untreated municipal wastewater depends on a number of factors that are not entirely predictable such as overflows of untreated wastewater (see Fig. 3-7). Important variables include the source and original use of the water, the general health of the population, the existence of disease carriers for particular infectious agents, excretion rates of infectious agents, duration of the infection, and the ability of infectious agents to survive outside their hosts under various environmental conditions (NRC, 1998). In the following discussion it is

Table 3-5
Specific organisms or groups of organisms that have been used, or proposed for use, as indicators of fecal contamination[a]

Indicator organism	Characteristics
Total coliform bacteria	Species of gram-negative rods which ferment lactose with gas production (or produce a distinctive colony within 24 ± 2 to 48 ± 3 h incubation on a suitable medium) at $35 \pm 0.5°C$. However, there are strains that do not conform to this definition. The total coliform group includes four genera in the Enterobacteriaceae family. These are *Escherichia, Citrobactor, Enterobacter,* and *Klebisella.* Of the group, the *Escherichia* genus (*E. coli* species) appears to be most representative of fecal contamination.
Fecal coliform bacteria	The fecal coliform bacteria group is the group of gram-negative rods that have the ability to produce gas (or colonies) at an elevated incubation temperature ($44.5 \pm 0.2°C$ for 24 ± 2 h).
Klebisella spp.	The total coliform population includes the genera *Klebisella*. The thermotolerant *Klebisella* are also included in the fecal coliform group. This group is cultured at $35 \pm 0.5°C$ for 24 ± 2 h.
E. Coli	The *E. coli* is one of the coliform bacteria populations and is more representative of fecal sources than other coliform genera.
Bacteroides	Bacteroides, an anaerobic organism, has been proposed as a human specific indicator.
Fecal Streptococci	This group has been used in conjunction with fecal coliform to determine the source of recent fecal contamination (man or farm animals). Several strains appear to be ubiquitous and cannot be distinguished from the true fecal streptococci under usual analytical procedures, which detract from their use as an indicator organisms.
Enterococci	Two strains of fecal streptococci, *S. faecalis* and *S. faecium*, are the most human specific members of the fecal streptococcus group. By eliminating the other strains through the analytical procedures, the two strains known as enterococci can be isolated and enumerated. The enterococci are generally found in lower numbers than other indicator organisms, however, they exhibit better survival in seawater.
Clostridium perfringens	This organism is a spore-forming anaerobic-persistent bacteria, and the characteristics make it a desirable indicator where disinfection is employed, where pollution may have occurred in the past, or where the interval before analysis is protracted.
P. aeruginosa and *A. hydrophila*	These organisms may be present in wastewater in large numbers. Both can be considered aquatic organisms and can be recovered in water in the absence of immediate sources of fecal pollution.

[a] Adapted from Tchobanoglous et al. (2003).

Table 3-6
Indicator organisms used in establishing performance criteria for various water uses and types[a]

Water type or use	Indicator organism
Drinking water	Total coliform
Freshwater recreation	Fecal coliform
	E. coli
	Enterococci
Saltwater recreation	Fecal coliform
	Total coliform
	Enterococci
Shellfish growing areas	Total coliform
	Fecal coliform
Agricultural irrigation (for reclaimed water)	Total coliform
Wastewater effluent Disinfection	Total coliform
	Fecal coliform
	MS2 coliphage

[a] Adapted from Tchobanoglous et al. (2003).

assumed that the principal sources of pathogenic organisms in wastewater are from municipal wastewater from residential, commercial, and industrial sources. Additional information on the sources of wastewater in a collection system is presented in the next section, Pathogens in Treated Wastewater.

Reported microorganism concentrations in untreated municipal wastewater are shown in Table 3-7, along with an estimate of the median infectious dose. Note that a wide range of concentrations of pathogenic microorganisms are encountered in the field, and the median infectious dose, N_{50}, corresponds to the typical dose needed to cause disease

(a)

(b)

Figure 3-7
Pathogens in the environment: (a) stormwater drain at the swimming beach, and (b) health warning indicating bacterial levels exceed health standards. (Photos courtesy of Orange County Sanitation District, CA.)

Table 3-7
Microorganism concentrations found in untreated wastewater and the corresponding median infectious dose[a]

Organism	Concentration in raw wastewater, MPN/100 mL[b]	Median infectious dose number (N_{50})
Bacteria		
Bacteroides	10^7–10^{10}	
Coliform, total	10^7–10^9	
Coliform, fecal[c]	10^5–10^8	10^6–10^{10}
Clostridium perfringens	10^3–10^5	1–10^{10}
Enterococci	10^4–10^5	
Fecal streptococci	10^4–10^6	
Pseudomonas aeruginosa	10^3–10^6	
Shigella	10^0–10^3	10–20
Salmonella	10^2–10^4	
Protozoa		
Cryptosporidium parvum oocysts	10^1–10^5	1–10
Entamoeba histolytica cysts	10^0–10^5	10–20
Giardia lamblia cysts	10^1–10^4	< 20
Helminth		
Ova	10^0–10^3	
Ascaris lumbricoides		1–10
Virus		
Enteric virus	10^3–10^4	1–10
Coliphage	10^2–10^4	

[a] Adapted in part from; Feacham et al. (1983); NRC (1996); Crook (1992).
[b] MPN = most probable number.
[c] *Escherichia coli* (enteropathogenic).

in humans (see Fig. 3-8). There is also a wide person-to-person variation in the N_{50} dose, depending on the overall health of the individual, genetic factors, the age of the person, and whether the immune system is compromised, which is represented by reporting the N_{50} dose as a range of values. The subject of median infectious dose is considered further in Chap. 5.

The occurrence and concentration of pathogenic microorganisms in treated municipal wastewater depends on a number of factors including (1) the number of organisms in the untreated wastewater, (2) the level of treatment, (3) the treatment technologies employed, and (4) the regulatory requirements. A discussion on the level of treatment and the available treatment technologies is presented first, followed by information on the pathogens in treated wastewater. Treatment technologies are discussed in great detail in Part 3 of this textbook.

Pathogens in Treated Wastewater

Treatment Levels and Technologies

Methods of treatment in which the application of physical forces predominate are known as *unit operations*. Methods of treatment in which the removal of contaminants is brought about by chemical or biological reactions are known as *unit processes*. At the

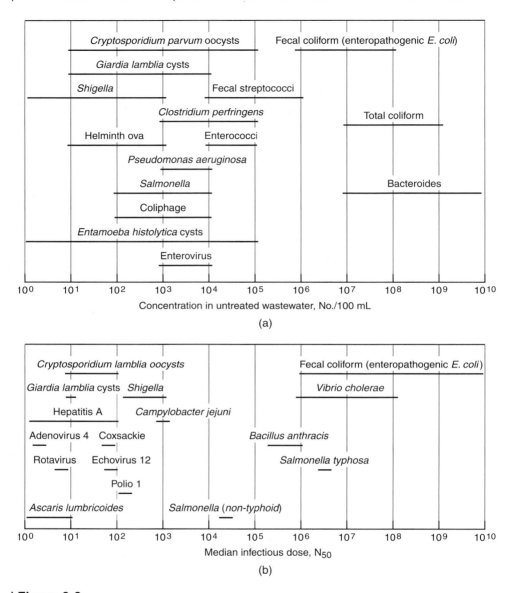

Figure 3-8

Reported microorganism concentrations in untreated municipal wastewater and median infectious dose. (Adapted from Crittenden et al., 2005.)

present time, unit operations and processes are grouped together to provide various levels of treatment known as preliminary, primary, advanced primary, secondary, tertiary, and advanced treatment (see Table 3-8 and Fig. 3-9). In preliminary treatment, gross solids that may damage equipment are removed by screening. In primary treatment, a physical operation, usually sedimentation, is used to remove the floating and settleable

Table 3-8
Classification of stages used for wastewater treatment and water reclamation[a]

Treatment level[b]	Description
Preliminary	Removal of wastewater constituents such as rags, sticks, floatables, grit, and grease that may cause maintenance or operational problems with the treatment operations, processes, and ancillary systems.
Primary	Removal of a portion of the suspended solids and organic matter from the wastewater.
Advanced primary	Enhanced removal of suspended solids and organic matter from the wastewater; typically accomplished by chemical addition or filtration.
Secondary	Removal of biodegradable organic matter (in solution or suspension) and suspended solids. Disinfection typically is also included in the definition of conventional secondary treatment.
Secondary with nutrient removal	Removal of biodegradable organics, suspended solids, and nutrients (nitrogen, phosphorus, or both nitrogen and phosphorus).
Tertiary	Removal of residual suspended solids (after secondary treatment), usually by granular medium filtration, surface filtration, and membranes. Disinfection is also typically a part of tertiary treatment. Nutrient removal is often included in this definition.
Advanced	Removal of total dissolved solids and or trace constituents as required for specific water reuse applications.

[a]Adapted, in part, from Crites and Tchobanoglous (1998).
[b]See also Fig. 3-9 for treatment process diagrams.

materials found in wastewater (see Fig. 3-9a). In secondary treatment, biological and chemical processes are used to remove most of the organic matter (see Fig. 3-9b and also Fig. 3-10). Disinfection is typically a part of secondary treatment. Nutrient removal is also often included in this step (see Fig. 3-9c). In tertiary treatment residual suspended solids are removed to enhance the disinfection process, usually by filtration. In advanced treatment (see Fig. 3-9d), additional combinations of unit operations and processes are used to remove constituents that are not reduced significantly by conventional secondary and tertiary treatment for specific water reuse applications (Tchobanoglous et al., 2003).

Pathogens in Primary Effluent
Primary treatment does little to remove microbiological pathogens from wastewater. However, some protozoa and parasite ova and cysts will settle out during primary treatment, and some particulate-associated microorganisms may be removed with settleable matter. Estimated microorganism removals during primary treatment are reported in Table 3-9.

Pathogens in Secondary Effluent
Secondary treatment reduces pathogens but does not eliminate them from the effluent, even with disinfection (see Fig. 3-9b). Typical log removal of microorganisms by various wastewater treatment processes is shown in Table 3-9. Based on the data presented in Table 3-9, it can be concluded that wastewater discharges may contribute enteric

Figure 3-9
Various municipal wastewater treatment operations and processes: (a) primary treatment, (b) secondary treatment, (c) tertiary treatment, and (d) advanced treatment.

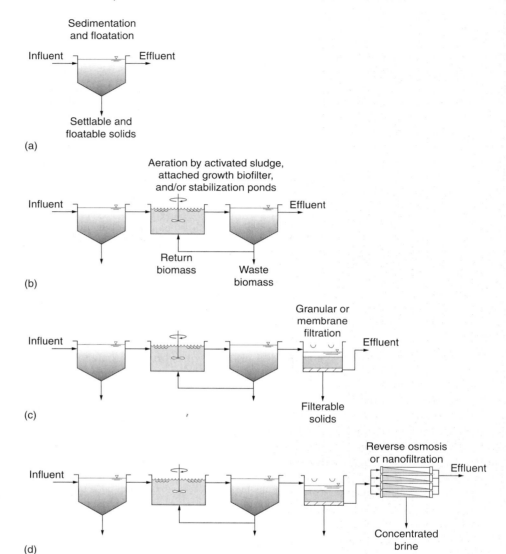

pathogens to natural waters, many of which may be used downstream of the wastewater effluent discharge as a source of water for potable purposes (see Chap. 1).

Pathogens in Tertiary and Advanced Wastewater Treatment Effluent

The concentration of microorganisms in the effluent from advanced treatment processes is dependent on the specific microorganism and the form of advanced treatment (e.g., chemical treatment, granular medium filtration, membrane filtration). Reclaimed water derived from tertiary and advanced wastewater treatment processes is deemed safe for unrestricted landscape irrigation (see Fig. 3-11b).

Figure 3-10
City of San Diego, CA, aquaculture facility (ca. 1996) employed water hyacinths in place of conventional secondary treatment with either activated sludge process or trickling filters. (a) empty plug-flow basin with stepped influent feed distribution piping and aeration system and (b) view of process in operation with full coverage of water hyacinths.

Table 3-9
Typical microorganism log removal by wastewater treatment processes[a]

	Removal of organism for given treatment process, log units					
	Primary	Secondary		Tertiary		Advanced
Organism	Plain sedimentation	Activated sludge	Trickling filter	Depth filtration	Microfiltration[b]	Reverse osmosis[c]
Fecal coliforms	<0.1–0.3	0–2	0.8–2	0–1	1–4	4–7
Salmonella	<0.1–2	0.5–2	0.8–2	0–1	1–4	4–7
Mycobacterium tuberculosis	0.2–0.4	0–1	0.5–2	0–1	1–4	4–7
Shigella	<0.1	0.7–1	0.8–2	0–1	1–4	4–7
Campylobacter	1	1–2		0–1	1–4	4–7
Cryptosporidium parbum	0.1–1	1		0–3	1–4	4–7
Entamoeba histolytica	0–0.3	<0.1	<0.1	0–3	2–6	>7
Giardia lamblia	<1	2		0–3	2–6	>7
Helminth ova	0.3–1.7	<0.1	1	0–4	2–6	>7
Enteric viruses	<0.1	0.6–2	0–0.8	0–1	0–2	4–7

[a] Adapted in part from Crook (1992).
[b] Wide range of values due to differences in performance of membranes from different manufacturers and imperfections or failure of the membrane (see Example 8-4 in Chap. 8).
[c] In theory, reverse osmosis should remove all organisms, however, due to imperfections or failure of the membrane some organisms may pass through with the permeate stream (see Example 8-4 in Chap. 8).

(a) (b)

Figure 3-11
Reclaimed water is used safely in various irrigation applications: (a) irrigation of grape vines, and (b) landscapes.

Pathogens in the Environment

In receiving waters, natural processes tend to reduce the concentrations of enteric microorganisms due to dilution and die-off. The natural inactivation or die-off rate is usually reported in terms of the time required for a 90 percent reduction in the viability of the microbial population. Many factors influence the inactivation rate, including the amount of particulate matter, oxygen, salinity, and UV light the water is exposed to. However, temperature, as discussed below, appears to play the most significant role.

Survival of Pathogenic Organisms

It is known that enteric pathogens generally survive longer at lower temperatures. Survival rates of some pathogens at selected temperatures are shown in Table 3-10. The data given in Table 3-10 are incomplete and should be used as a rough guide only, as numerous exceptions have been reported in the literature.

Table 3-10
Survival of enteric pathogens and indicator bacteria in freshwaters[a]

Microorganism	Time reported for 90 percent reduction in viable concentrations
Coliforms	0.83 to 4.8 d at 10 to 20°C, avg. 2.5 d
E. coli	3.7 d at 15°C
Salmonella	0.83 to 8.3 d at 10 to 20°C
Yersinia	7 d at 5 to 8.5°C
Giardia	14 to 143 d at 2 to 5°C
	3.4 to 7.7 d at 12 to 20°C
Enteric viruses	1.7 to 5.8 d at 4 to 30°C

[a]Adapted from Feachem et al. (1983); Korhonen and Martikainen (1991); Kutz and Gerba (1988); McFeters and Terzieva (1991).

3-6 CHEMICAL CONSTITUENTS IN UNTREATED AND TREATED WASTEWATER

Chemical constituents of wastewater are classified typically as inorganic and organic. Inorganic chemical constituents of concern include dissolved constituents, nutrients, nonmetallic constituents, metals, and gases. Organic constituents of interest in wastewater are classified as aggregate and individual. Aggregate organic constituents are comprised of a number of individual compounds that cannot be distinguished separately. Both aggregate and individual organic constituents are of great significance in the treatment and reuse of wastewater.

To understand the chemical characteristics of wastewater, it is important to know the sources of the chemical constituents found in wastewater. Untreated wastewater contains known and unknown inorganic and organic constituents that are (1) present naturally in the water supply source, (2) present in the treated water as supplied at the tap, (3) added from residential, commercial, industrial, and other human activities in the water and wastewater service area; (4) added from stormwater in combined collection systems and from infiltration into the collection system; (5) formed in the collection system as result of abiotic and biotic reactions; and (6) added to the wastewater in collection systems for odor or corrosion control. Each of the sources of chemical constituents in wastewater is considered briefly in the following discussion.

Chemical Constituents in Untreated Wastewater

Constituents in Natural Water

Natural waters contain both inorganic and organic constituents. Inorganic constituents in natural waters are derived from the dissolution of the rocks and minerals, which have been in contact with the water. Concentrations of inorganic constituents are increased by the natural evaporation process that removes some of the surface water and leaves the inorganic substance behind in the water. The principal inorganic cation constituents of most natural waters are calcium (Ca^{2+}), magnesium (Mg^{2+}), potassium, (K^+), and sodium (Na^+). The corresponding major anions are bicarbonate (HCO_3^-), sulfate (SO_4^{2-}), and chloride (Cl^-). Many trace inorganic constituents occur in natural water in varied concentrations, depending on the geologic characteristics of the region, and human and agricultural activities in the watershed.

In addition to inorganic constituents, most natural waters, especially surface waters, also contain a variety of natural organic matter (NOM), the breakdown products of these compounds, and a vast array of microorganisms. Groundwater generally does not contain measurable concentrations of organic compounds. However, some groundwaters, which have been in contact with peat bogs or other organic materials found in the subsurface or which are under the influence of surface water, do contain a variety of organic compounds, most of which have not been identified. The concentration of organic matter in natural waters will vary widely, depending on the source (e.g., reservoirs versus aquifers).

Typically NOM is composed of humic materials from plants and algae, microorganisms and their metabolites, and high molecular weight aliphatic and aromatic hydrocarbons. These organics are typically benign, although some are nuisance constituents such as odoriferous metabolites that can cause aesthetic concerns such as taste and odor. A few of the high molecular weight aliphatic and aromatic hydrocarbons may have adverse health effects. In addition, humic materials may serve as precursors in the formation of trihalomethanes (THMs) and other organohalogen oxidation byproducts during disinfection.

Constituents in Public Water Supplies

With the exception of one or two large water supply sources, which are of pristine quality, most public water supplies are treated to remove specific inorganic and organic constituents to meet regulatory requirements, as specified in the U.S. EPA Drinking Water Standards. Residual inorganic chemicals found in drinking water represent a varying degree of health concerns. Some are known or suspected carcinogens, such as arsenic, lead, and cadmium. Several inorganic chemicals are essential to human nutrition at low doses, yet demonstrate adverse health effects at higher doses. These include aluminum, chromium, copper, manganese, molybdenum, nickel, selenium, zinc, and sodium. Additional constituents often leach into treated drinking water from contact with piping or plumbing materials, such as lead, copper, zinc, and asbestos.

The aggregate residual organic compounds in treated water are generally of little concern. Other organic constituents such as acrylamide or epichlorohydrin, components of coagulants (e.g., polyacrylamide), which can leach out during water treatment, may be present. In addition, it has been found that undesirable components of pipe coatings, linings, and joint adhesives, such as polynuclear aromatic hydrocarbons (PAHs), epichlorohydrin, and solvents can also leach into the treated water. Disinfection byproducts (DBPs), discussed below, can be formed in the distribution system before arriving at the tap.

Constituents Added through Domestic, Commercial, and Industrial Usage

When public water supplies are used for domestic, commercial, and industrial purposes, a wide variety of known *and* unknown constituents are added to the water that ends up as wastewater. Data on the increase in the mineral content of wastewater resulting from water use, and the variation of the increase within a collection system, are especially important in evaluating the reuse potential of wastewater. Typical data on the incremental increase in mineral content that can be expected in municipal wastewater resulting from domestic use are reported in Table 3-11. Increases in the mineral content of wastewater may be due in part from addition of highly mineralized water from private wells and groundwater infiltration, and from industrial use. Domestic and industrial water softeners also contribute significantly to the increase in mineral content and, in some areas, may represent the major source. Occasionally, water added from private wells and groundwater infiltration (because of its high quality) will serve to dilute the mineral concentration in wastewater. The amount of salt added to wastewater from water softeners can be estimated using the following equation.

$$\text{Salt in blended effluent, kg/m}^3 = \frac{\begin{pmatrix}\text{fraction of homes} \\ \text{with water softeners}\end{pmatrix}\begin{pmatrix}\text{salt added to each water} \\ \text{softener per year, kg/yr}\end{pmatrix}}{(365 \text{ d/yr})(\text{average daily flow rate per home, m}^3\text{/d})} \quad (3\text{-}2)$$

Organic compounds are normally composed of a combination of carbon, hydrogen, and oxygen, together with nitrogen and sulfur, in some cases. The organic matter in wastewater

Table 3-11 Typical mineral increase from domestic water use[a]

Constituent	Increment range, mg/L[b,c]
Anions:	
Bicarbonate (HCO_3)	50–100
Carbonate (CO_3)	0–10
Chloride (Cl)	20–50
Sulfate (SO_4)	15–30
Cations:	
Calcium (Ca)	6–16
Magnesium (Mg)	4–10
Potassium (K)	7–15
Sodium (Na)	40–70[d]
Other constituents	
Aluminum (Al)	0.1–0.2
Boron (B)	0.1–0.2
Fluoride (F)	0.2–0.4
Manganese (Mn)	0.2–0.4
Silica (SiO_2)	2–10
Total alkalinity (as $CaCO_3$)	60–120
Total dissolved solids (TDS)	150–380

[a] From Tchobanoglous et al., 2003.
[b] Based on 460 L/capita·d (120 gal/capita·d), which is classified as medium strength wastewater.
[c] Values do not include commercial and industrial additions.
[d] Excluding the addition from domestic water softeners.

consists typically of proteins (40 to 60 percent), carbohydrates (25 to 50 percent), and oils and fats (8 to 12 percent). Urea, the major constituent of urine, is another important organic compound found in fresh wastewater. Because urea decomposes rapidly, it is seldom found in other than fresh wastewater. Along with the proteins, carbohydrates, fats and oils, and urea, wastewater typically contains varying amounts of a large number of different synthetic organic chemicals (SOCs), with structures ranging from simple to extremely complex. While many individual chemical compounds are known, the vast majorities of these compounds are unknown and are usually reported as aggregate constituents.

It is interesting to note that prior to about 1940, most of the municipal wastewater in the United States was generated from domestic sources. After 1940, as industrial development in the United States grew significantly, increasing amounts of commercial and industrial wastewater have been and continue to be discharged to municipal collection systems. The amounts of SOCs generated by commercial and industrial activities have increased, and some 10,000 new organic chemicals are developed each year. Many of these chemicals are now found in the wastewater from most municipalities and communities. The addition of new chemicals will continue to make the complete characterization of wastewater an unachievable goal.

Constituents Added from Stormwater in Combined Collection Systems and Infiltration

In addition to the constituents added through usage, other usually unknown, inorganic and organic constituents are often added to the wastewater from stormwater inflow and infiltration. In addition, combined sewers are used in many parts of the country. Constituents of concern in stormwater include oils, grease, tars, and metals from roadway runoff; pesticides and herbicides; fertilizers; animal feces; and decayed humic materials.

Infiltration is an ongoing problem with wastewater collection systems, especially as collection systems age and become less watertight. A constituent of great concern in coastal areas is salinity, principally in the form of sea- or brackish water. Dissolved humic substances are constituents of concern, which are difficult to treat and can interfere with the disinfection process.

Constituents Formed in the Collection System as a Result of Abiotic and Biotic Reactions

In some collection systems with long travel times, typically greater than 6 h, a number of abiotic and biologically mediated reactions occur as wastewater is transported to a centralized location for treatment. The formation of hydrogen sulfide under anoxic conditions is a well known example of a biological reaction that occurs in collection systems. However, little is known about the exact nature of most of the transformations that occur under anoxic and anaerobic conditions as wastewater is transported.

Constituents Added to the Wastewater in Collection Systems for Odor or Corrosion Control

In some collection systems with long travel times, chemicals are added to control the formation of odors and to mitigate corrosion. In some cases pure oxygen is added to suppress anoxic and anaerobic reactions leading to the formation of hydrogen sulfide.

Composition of Untreated Wastewater

The influent to a wastewater treatment plant contains a mixture of the constituents discussed above, which varies with the day of the week, the month of the year, and seasonally. Typical data on the composition of untreated domestic wastewater as found in wastewater collection systems are reported in Table 3-12. The data presented in Table 3-12 for medium-strength wastewater are based on an average flow of 460 L/capita·d (120 gal/capita·d) and include constituents added by commercial, institutional, and industrial sources. Typical concentrations for low-strength and high-strength wastewater, which reflect different amounts of infiltration, are also given. Because there is no "typical" wastewater, it must be emphasized that the data presented in Table 3-12 should only be used as a guide.

The aggregate constituents, biochemical oxygen demand (BOD), chemical oxygen demand (COD), and total organic carbon (TOC), are used to characterize the bulk of the organic matter in wastewater (see Table 3-12). Within these categories there are a number of trace SOCs that are unknown. Volatile organic chemicals (VOCs), is the category used to characterize some of these compounds. It should be noted that the term SOCs is now often used as a regulatory rather than a chemical description for these compounds.

Table 3-12
Typical composition of untreated domestic wastewater[a]

Contaminants	Unit	Range	Typical[b]
Solids, total (TS)	mg/L	390–1230	720
Dissolved, total (TDS)	mg/L	270–860	500
Fixed	mg/L	160–520	300
Volatile	mg/L	110–340	200
Suspended solids, total (TSS)	mg/L	120–400	210
Fixed	mg/L	25–85	50
Volatile	mg/L	95–315	160
Settleable solids	mg/L	5–20	10
Biochemical oxygen demand (BOD) 5 d, 20°C	mg/L	110–350	190
Total organic carbon (TOC)	mg/L	80–260	140
Chemical oxygen demand (COD)	mg/L	250–800	430
Nitrogen (total as N)	mg/L	20–70	40
Organic	mg/L	8–25	15
Free ammonia	mg/L	12–45	25
Nitrites	mg/L	0–trace	0
Nitrates	mg/L	0–trace	0
Phosphorus (total as P)	mg/L	4–12	7
Organic	mg/L	1–4	2
Inorganic	mg/L	3–10	5
Chlorides[c]	mg/L	30–90	50
Sulfate[c]	mg/L	20–50	30
Oil and grease	mg/L	50–100	90
Volatile organic compounds (VOCs)	mg/L	<100–>400	100–400
Total coliform	no./100 mL	10^6–10^9	10^7–10^8
Fecal coliform	no./100 mL	10^3–10^7	10^4–10^5
Cryptosporidum oocysts	no./100 mL	10^{-1}–10^2	10^{-1}–10^1
Giardia lamblia cysts	no./100 mL	10^{-1}–10^3	10^{-1}–10^2

[a]Adapted from Tchobanoglous et al. (2003).
[b]Typical wastewater composition is based on an approximate flow rate of 460 L/capita·d (120 gal/capita·d).
[c]Values should be increased by amount of constituent present in domestic water supply.

Because the actual compounds that compose aggregate parameters such as BOD, COD, TOC, TDS, and TSS are unknown, there is a degree of concern about using treated wastewater in indirect potable reuse applications. However, advanced treatment methods and new and improved analytical methods have helped to mitigate these concerns.

Chemical Constituents in Treated Wastewater

The required water quality for reclaimed water varies with each reuse application. The focus of the following discussion is to consider the constituents that are present in treated wastewater after varying degrees of treatment as discussed previously (see Table 3-8). Information on the chemical constituents in treated wastewater is of importance in assessing the potential health risks associated with the use of reclaimed water.

Constituents Remaining after Primary Treatment

As noted previously, primary treatment is used to remove floating and settleable materials found in wastewater. As a result of the removal of settleable materials there are measurable reductions in BOD, TSS, TOC, along with some metals that are associated with TSS. Performance data on the removals that can be achieved with primary treatment and the constituents remaining are presented in Table 3-13. The data in Table 3-13 were collected at a 3800 m^3/d water reclamation facility employing fine screens in place of conventional sedimentation facilities and water hyacinths in place of conventional secondary treatment employing activated sludge or trickling filters (see Fig. 3-10). Generalized information on the constituents remaining after primary treatment is reported in Table 3-14.

Constituents Remaining after Secondary Treatment

Biological and chemical processes are used in secondary treatment to remove most of the organic matter and the TSS. Performance data on the removals that can be achieved with secondary treatment and the constituents remaining are presented in Tables 3-13 and 3-14. The data in Table 3-13 are for the water reclamation plant described previously. Generalized information on the constituents remaining after secondary treatment with a variety of process combinations are reported in Table 3-14.

Constituents Remaining after Tertiary Treatment

The principal application of tertiary treatment is for the removal of residual TSS remaining after secondary sedimentation, typically by cloth or media filtration. Typical performance data on the removals that can be achieved with tertiary treatment and the constituents remaining are presented in Tables 3-13 and 3-14. The data in Table 3-13 are for the water reclamation plant described previously. Generalized information on the constituents remaining after tertiary treatment is reported in Table 3-14.

Constituents Remaining after Advanced Wastewater Treatment

As noted previously, AWT is used to remove residual suspended solids and other constituents that are not reduced significantly by conventional secondary treatment. Performance data on the removals that can be achieved with advanced treatment and the constituents remaining are presented in Tables 3-13 and 3-14. The data in Table 3-13 are for the water reclamation plant described previously. Generalized information on the constituents remaining after AWT is reported in Table 3-14.

The ability of AWT processes to remove many trace chemical contaminants is well established (see Chap. 10). Several pilot and demonstration potable water reuse studies have shown that AWT can produce water that exceeds U.S. EPA primary, and some secondary drinking water standards. A comparison of the quality of water produced by San Diego's Aqua III pilot plant, Tampa's Hookers Point AWT pilot plant, and Denver's Potable Reuse Demonstration Project to U.S. EPA drinking water standards is shown in Table 3-15.

Table 3-13
Removal of wastewater constituents in a water reclamation facility[a]

	Raw Conc.	Primary effluent Conc.	%R[c]	Secondary effluent Conc.	% R	Tertiary effluent Conc.	% R	AWT effluent[b] Conc.	% R	Overall % R
Conventional[d]										
CBOD	185	149	19	13	74	4.3	5	NA		98
TSS	219	131	40	9.8	55	1.3	4	NA		99+
TOC	91	72	21	14	64	7.1	8	0.6	7	99+
TS	1452	1322	9	1183	10	1090	6	43	72	97
Turb. (NTU)	100	88	12	14	74	0.5	14	0.27	0	99+
Ammonia-N	22	21	5	9.5	52	9.3	1	0.8	39	96
Nitrate-N	0.1	0.1	0	1.4	0	1.7	0	0.7	0	0
TKN	31.5	30.6	3	13.9	53	14.2	0	0.9	41	97
Phosphate-P	6.1	5.1	16	3.4	28	0.1	54	0.1	0	98
Nonconventional										
Arsenic	0.0032	0.0031	3	0.0025	19	0.0015	30	0.0003	40	92
Boron	0.35	0.38	0	0.42	0	0.31	13	0.29	3	17
Cadmium	0.0006	0.0005	17	0.0012	0	0.0001	67	0.0001	0	83
Calcium	74.4	72.2	3	66.7	7	70.1	0	1.0	88	99
Chloride	240	232	3	238	0	284	0	15	90	94
Chromium	0.003	0.004	0	0.002	32	0.001	24	0.001	28	83
Copper	0.063	0.070	0	0.043	33	0.009	52	0.011	0	83
Iron	0.60	0.53	11	0.18	59	0.05	22	0.04	2	94
Lead	0.008	0.008	0	0.008	0	0.001	93	0.001	0	91
Magnesium	38.5	38.1	1	39.3	0	6.4	82	1.5	13	96
Manganese	0.065	0.062	4	0.039	37	0.002	57	0.002	0	97
Mercury	0.0003	0.0002	33	0.0001	33	0.0001	0	0.0001	0	67
Nickel	0.007	0.010	0	0.004	33	0.004	11	0.001	45	89
Selenium	0.003	0.003	0	0.002	16	0.002	0	0.001	64	80
Silver	0.002	0.003	0	0.001	75	0.001	0	0.001	0	75
Sodium	198	192	3	198	0	211	0	11.9	91	94
Sulfate	312	283	9	309	0	368	0	0.1	91	99+
Zinc	0.081	0.076	6	0.024	64	0.002	27	0.002	0	97

[a] Adapted from Western Consortium for Public Health (1992). Primary treatment consisted of a rotary drum screen followed by disk screens, secondary treatment was with water hyacinths, tertiary treatment consisted of lime precipitation and depth filtration, and AWT comprised reverse osmosis, air stripping, and carbon adsorption.
[b] AWT = Advanced wastewater treatment
[c] % R = percent removed
[d] Raw and primary effluent results are BOD not CBOD.
All units are mg/L unless otherwise noted.

Table 3-14
Typical range of effluent quality after secondary treatment

Constituent	Unit	Untreated wastewater[a]	Conventional activated sludge[b]	Conventional activated sludge with filtration[b]	Activated sludge with BNR[c]	Activated sludge with BNR and filtration[c]	Membrane bioreactor	Activated sludge with microfiltration and reverse osmosis
Total suspended solids (TSS)	mg/L	120–400	5–25	2–8	5–20	1–4	≤2	≤1
Colloidal solids	mg/L		5–25	5–20	5–10	1–5	≤1	≤1
Biochemical oxygen demand (BOD)	mg/L	110–350	5–25	<5–20	5–15	1–5	<1–5	≤1
Chemical oxygen demand (COD)	mg/L	250–800	40–80	30–70	20–40	20–30	<10–30	≤2–10
Total organic carbon (TOC)	mg/L	80–260	10–40	8–30	8–20	1–5	0.5–5	0.1–1
Ammonia nitrogen	mg N/L	12–45	1–10	1–6	1–3	1–2	<1–5	≤0.1
Nitrate nitrogen	mg N/L	0–trace	10–30	10–30	2–8	1–5	<10[d]	≤1
Nitrite nitrogen	mg N/L	0–trace	0–trace	0–trace	0–trace	0–trace	0–trace	0–trace
Total nitrogen	mg N/L	20–70	15–35	15–35	3–8	2–5	<10[d]	≤1
Total phosphorus	mg P/L	4–12	4–10	4–8	1–2	≤2	<0.3[e]–5	≤0.5
Turbidity	NTU		2–15	0.5–4	2–8	0.3–2	≤1	0.01–1
Volatile organic compounds (VOCs)	µg/L	<100–>400	10–40	10–40	10–20	10–20	10–20	≤1

Metals	mg/L	1.5–2.5	1–1.5	1–1.4	1–1.5	1–1.5	trace	≤?
Surfactants	mg/L	4–10	0.5–2	0.5–1.5	0.1–1	0.1–1	0.1–0.5	≤1?
Totals dissolved solids (TDS)	mg/L	270–860	500–700	500–700	500–700	500–700	500–700	≤5–40
Trace constituents	µg/L	10–50	5–40	5–30	5–30	5–30	0.5–20	≤0.1
Total coliform	No./100 mL	10^6–10^9	10^4–10^5	10^3–10^5	10^4–10^5	10^4–10^5	<100	~0
Protozoan cysts and oocysts	No./100 mL	10^1–10^4	10^1–10^2	0–10	0–10	0–1	0–1	~0
Viruses	PFU/100 mL[f]	10^1–10^4	10^1–10^3	10^1–10^3	10^1–10^3	10^1–10^3	10^0–10^3	~0

[a]From Table 3-12.
[b]Conventional secondary is defined as activated sludge treatment with nitrification.
[c]BNR is defined as biological nutrient removal for the removal of nitrogen and phosphorus.
[d]With anoxic stage.
[e]With coagulant addition.
[f]Plaque forming units.

Table 3-15 Comparison of U.S. EPA Drinking Water Standards with water quality parameters for three reclaimed waters[a]

Constituent[b]	U.S. EPA drinking water standards	Reclaimed water[c]		
		San Diego	Tampa	Denver
Physical				
TOC	–	0.27	1.88	0.2
TDS	500	42	461	18
Turbidity (NTU)	–	0.27	0.05	0.06
Nutrients				
Ammonia-N	–	0.8	0.03	5
Nitrate-N	–	0.6	0	0.1
Phosphate-P	–	0.1	0	0.02
Sulfate	250	0.1	0	1
Chloride	250	15	0	19
TKN	–	0.9	0.34	5
Metals				
Arsenic	0.05	<0.0005	0[d]	ND[e]
Cadmium	0.005	<0.0002	0[d]	ND
Chromium	0.1	<0.001	0[d]	ND
Copper	1.0	0.011	0[d]	0.009
Lead	[f]	0.007	0[d]	ND
Manganese	0.05	0.008	0[d]	ND
Mercury	0.002	<0.0002	0[d]	ND
Nickel	0.1	0.0007	0.005	ND
Selenium	0.05	<0.001	0[d]	ND
Silver	0.05	<0.001	0[d]	ND
Zinc	5.0	0.0023	0.008	0.006
Boron	–	0.29	0	0.2
Calcium	–	<2.0	–	1.0
Iron	0.3[g]	0.37	0.028	0.02
Magnesium	–	<3.0	0	0.1
Sodium	–	11.9	126	4.8

[a]Adapted from CH2M Hill (1993), Lauer et al. (1991), Western Consortium for Public Health (1992).

[b]NTU = nephelometric turbidity units; TDS = total dissolved solids; TKN = total Kjeldahl nitrogen; TOC = total organic carbon. All reported values with the exception of turbidity are expressed in mg/L.

[c]San Diego physical and nutrient concentration values are arithmetic means. Any nondetected observations were assumed to be present at the corresponding detection limit. Metal concentration values are geometric means determined through probit analysis. Tampa values are arithmetic means of detected values. Denver values are geometric means of detected values.

[d]Not detected in seven samples.

[e]Not detected in more than 50 percent of samples.

[f]Lead is regulated according to a treatment standard.

[g]Noncorrosive limit for iron.

Table 3-16
Reported removal ranges for selected emerging constituents of concern

Constituent[a]	Removal range, percent		
	Secondary treatment	Microfiltration	Reverse osmosis
N-nitrosdimethylamine (NDMA)	50–75	50–75	50–75
17β-Estradiol			50–100
Alkyphenols ethoxylates (APEOs)	40–80	40–80	40–80

[a] Significant variations have been observed in the concentrations of these constituents in the influent wastewater.

Removal of Trace Constituents

The removal of trace constituents occurs in both conventional and AWT processes, but the levels to which individual constituents are removed are not well defined. Typical performance data for the removal of nonconventional constituents are reported in Table 3-16 for complete secondary treatment, microfiltration, and reverse osmosis. From a review of the data presented in Table 3-16, it can be concluded that the treatment performance for nonconventional constituents in trace concentrations is quite variable and not well defined.

Impact of Constituents Remaining after Treatment

The impact of the constituents that remain after various treatment processes can be of profound importance with respect to the long-term protection of public health and the environment. The ability to measure the emerging constituents listed in Table 3-16 in the 10^{-9} or 10^{-12} g/L range is new as of a few years ago. Their health and environmental impacts are mostly unknown at the present time.

Formation of Disinfection Byproducts (DBPs)

The chemical oxidation processes, such as chlorination, that are used to disinfect wastewater effluents produce DBPs. Most DBPs are primarily dissolved organohalogens derived from the oxidative breakdown of organic substances in water (Bellar et al., 1974; Rook, 1974; Cooper et al., 1983; Bauman and Stenstrom, 1990; Rebhun et al., 1997). The DBPs may be grouped generally into trihalomethanes, haloacetonitriles, haloketones, haloacetic acids, chlorophenols, aldehydes, trichloronitromethane, chloral hydrate, and cyanogen chloride. Among them, trihalomethanes and haloacetic acids are by far the most common DBPs and often present at higher concentrations than the other less frequently detected DBPs (Krasner et al., 1989). Another DBP that has been found in wastewater and reclaimed water is N-nitrosodimethylamine (NDMA), a potent carcinogen. Occurrence of NDMA in reclaimed water is further discussed in Sec. 3-7.

Chlorine disinfection has been the most common disinfection method for wastewater (see Chap. 11). The extent of DBP formation by chlorination depends on pH, temperature, reaction time, free and combined chlorine concentrations, ammonia concentration, DBP precursor concentration, and precursor type (Stevens et al., 1989; Reckhow et al., 1990; Rebhun et al., 1997). Organic matters that are highly aromatic, and contain chlorine reactive sites such as phenol, 2, 4-pentanedione, organic nitrogen, meta-dihydroxybenzene, and various acetyl moieties, are thought to be precursors of DBPs (Stevens et al., 1989). Even though organic matters in wastewater effluent tend to be less aromatic than many NOMs,

DBP formation is observed in most chlorinated wastewater effluent because of high dissolved organic carbon (DOC).

The likelihood of harm caused by DBPs is derived primarily from the direct ingestion of chlorinated water (Canter et al., 1998; Hildesheim et al., 1998; Waller et al., 1998). In crop irrigation, consumers may be indirectly exposed to DBPs through food chain transfer and/or contamination of the underlying groundwater. Disinfection byproducts, however, are subject to volatilization in the ambient environment and are readily degradable through chemical and biological reactions. Chlorinated reclaimed water is stored typically prior to use for irrigation and the DBPs formed during chlorination decay typically during storage. After land application, the processes of degradation continue in the soil. Because DBPs are not expected to accumulate in the soil they are of little concern in agricultural irrigation. It is also unlikely that DBPs are a serious threat to the groundwater underneath the irrigated fields, considering the transient time of water and degradation of DBPs in the vadose zone (Thomas et al., 2000; Chang, 2002). Formation of DBPs in reclaimed water is of greater concern when indirect or direct potable reuse is considered.

Comparison of Treated Wastewater to Natural Water

Natural water, or water that has been in contact with the environment for a long period of time, will attain chemical and biological signatures related to the mineral surfaces of the soils, the microorganisms found in the aquatic and terrestrial environment, and anthropogenic compounds. Because of the long retention time in the environment, a number of abiotic and biotic transformations occur which are mediated by physical, chemical, and biological reactions. Ultimately, these reactions result in the accumulation of a wide variety of naturally synthesized organic compounds and breakdown products of biological origin in the water body. As noted previously these organic compounds found in natural waters are collectively known as NOM. Thus, drinking water obtained from surface water sources will contain background concentrations of these NOM and anthropogenic chemicals.

The organisms found in wastewater treatment are primarily derived from organisms found in natural aquatic systems as well as in excreta. The microbial metabolic reactions occurring in biological wastewater treatment processes are representative of various biological reactions that occur in nature. However, while the reactions are similar to those that occur in nature, the reaction rates in the engineered system are designed to increase, through process optimization, to the extent required to meet wastewater discharge permits. The increased reaction rates necessary for accelerated biological wastewater treatment also impact the constituents found in reclaimed water. The easily biodegradable organic constituents found in wastewater are assimilated readily by the biomass, whereas many of the more difficult to treat constituents may not or only be degraded partially. As a result, effluent from wastewater treatment will contain much of the natural chemical signature as well as the chemical and biological characteristics resulting from conventional biological treatment. For water reclamation and reuse, additional treatment process may include granular media filtration, membrane filtration, reverse osmosis, conventional (low-level) chemical oxidation, advanced (high-level) chemical oxidation, or UV exposure. Each of these processes and operations has a unique impact on the residual constituents in the reclaimed water. However, with the exception of reverse osmosis, advanced oxidation, and extensive advanced treatment processes, much of the natural chemical signature including that added during biological treatment will remain intact.

Following discharge to the environment, natural processes and systems will assimilate the compounds, and the reclaimed water will once again take on the chemical signature of the environment. The degree of treatment and environmental factors will control the rate and extent of assimilation. Anthropogenic constituents can be found in systems with short retention time in the environment such as water flowing in streams with low environmental reaction rates (Barber et al., 1996; Kolpin et al., 2004; Lin et al., 2006). The use of several of these organic compounds to determine the origins of the organic contamination in the Mississippi River from municipal and industrial wastewater sources is illustrated in Table 3-17. Examples of natural systems which assimilate residual constituents more rapidly include shallow open water systems, wetlands, and vadose zone percolation. Conversely, reclaimed water from advanced treatment processes may be of much higher quality than the receiving water in the environment, resulting in degradation of the reclaimed water quality.

Use of Surrogate Parameters

Because reclaimed water may contain hundreds of compounds that can be traced to natural and human origins, rigorous analytical methods are needed to characterize even a portion of these chemicals quantitatively. To overcome the limitations associated with expensive sampling and time-consuming laboratory work, several surrogate parameters have been developed to assess the chemical makeup of treated effluent and reclaimed water, and the potential degree of environmental assimilation. The primary surrogates for aggregate organic trace constituents in water, depending on concentration, include assimilable organic carbon (AOC), biodegradable dissolved organic carbon (BDOC), total organic carbon (TOC), and chemical oxygen demand (COD). However, the difficulty with these aggregate organic parameters is that they do not relate any information regarding the specific compounds that makeup the measured parameter. Any measurable AOC, BDOC, TOC, or COD is an indication that an unknown organic suite of chemical compounds is present in the reclaimed water.

Because of the uncertainties associated with the unknown chemicals in reclaimed water, particularly in the indirect and direct potable reuse situations, it would be useful if one or more surrogates were available (see Chaps. 23 and 24). Unfortunately, surrogates that may be used to characterize the safety and suitability of reclaimed water for human and environmental exposure do not exist at present, due in part, to the limitations of health and environmental risk analysis, as reviewed in Chap. 5. Environmental systems are extremely complex. For example, some trace constituents in treated wastewater effluents are known or suspected to cause abnormalities in fish in receiving waters. Fish are increasingly recognized as an excellent model for such tests, in that the aquatic environment may provide early warnings of the effects that these chemicals will have on human health (Klime, 1998). Because the observed abnormalities can also be caused by naturally occurring compounds or other factors such as temperature, methods must be developed to assess the effects of different chemical compounds and mixtures of compounds. Although numerous surrogates have been evaluated including caffeine and other medicines, more research will be needed to resolve the many issues involved (NRC, 1998).

Table 3-17
Organic compounds measured to evaluate wastewater contamination of the Mississippi River, 1987–1992[a]

Contaminant	Abbreviation	Compounds and sources
Dissolved organic carbon	DOC	All natural and synthetic organic compounds, regional-scale natural sources
Fecal coliform bacteria	None	Bacteria derived from human and animal fecal wastes, from wastewater effluents, and feedlot and agricultural runoff
Methylene-blue-active substances	MBAS	Composite measure of synthetic and natural anionic surfactants; predominantly from municipal wastewater discharges
Linear alkylbenzenesulfonate	LAS	Complex mixture of specific anionic surfactant compounds used in soap and detergent products; primary source is domestic wastewater effluent
Nonionic surfactants	NP, PEG	Complex mixture of compounds derived from nonionic surfactants that includes nonylphenol (NP) and polyethylene glycol (PEG) residues; from wastewater and industrial sources
Adsorbable organic halogen	AOX	Adsorbable halogen-containing organic compounds, including by-products from chlorination of DOC and synthetic organic chemicals (SOCs), solvents, and pesticides; from multiple natural and anthropogenic sources
Fecal sterols	None	Natural biochemical compounds found predominantly in human and livestock wastes; primary source is domestic wastewater and feedlot runoff
Polynuclear aromatic hydrocarbons	PNA	Complex mixture of compounds, many of which are priority pollutants; from multiple sources associated with combustion of fuels
Caffeine	None	Specific component of beverages, food products, and medications specifically for human consumption; most significant source is domestic wastewater effluent
Ethylenediaminetetraacetic acid	EDTA	Widely used synthetic chemical for complexing metals; from a variety of domestic, industrial, and agricultural sources
Volatile organic compounds	VOCs	A variety of chlorinated solvents and aromatic hydrocarbons; predominantly from industrial and fuel sources
Semivolatile organic compounds	TTT, THAP	Wide variety of compounds including priority pollutants and chemicals such as trimethyltriazinetrione (TTT) and trihaloalkylphosphates (THAP); predominantly from industrial sources

[a]Adapted from U.S. Geological Survey (1996).

3-7 EMERGING CONTAMINANTS IN WATER AND WASTEWATER

The term *emerging contaminants* is used for chemicals and microorganisms that have been identified in water only recently and are under consideration to be regulated. The potential environmental impacts of emerging contaminants such as PhACs, endocrine disruptors, and new and reemerging pathogenic microorganisms are discussed in this section. These constituents are inherent to municipal wastewater (Rebhun et al., 1997; Hale et al., 2000; Huffman et al., 2003); however, knowledge of their occurrence in reclaimed water is limited. When reclaimed water is applied on land or discharged to aquatic environments, these chemical constituents and pathogens may be inadvertently released, potentially resulting in an adverse impact on the environment (Bouwer et al., 1998). The fate and transport in the vadose zone and in groundwater, and the risks associated with the unintentional transfer of these chemicals and pathogens to humans, are virtually unknown.

Endocrine Disruptors and Pharmaceutically Active Chemicals

For over several decades, scientists have reported that certain synthetic and natural compounds could mimic, block, stimulate, or inhibit natural hormones in the endocrine systems of animals. These substances are now collectively known as endocrine-disrupting compounds (EDCs), and have been linked to a variety of adverse effects in both humans and wildlife. Chemicals classified as EDCs have a wide variety of origins including pharmaceuticals, personal care products, household chemicals, pesticides and herbicides, industrial chemicals, disinfection byproducts, naturally occurring hormones, and metals. Chemicals that are classified as PhACs are synthesized for medical purposes, such as antibiotics, anti-inflamatories, X-ray contrast media, and antidepressants. Some PhACs, such as contraceptives and steroids, are also EDCs (NRC, 1999).

These synthetic and naturally occurring chemicals have been discovered in various surface and groundwaters, some of which have been linked to ecological impacts at trace concentrations (Kolpin et al., 2004). The majority of EDCs and PhACs are more polar than traditional contaminants and several have acidic or basic functional groups. These properties, coupled with occurrence at trace levels (i.e., <1 µg/L), create unique challenges for both removal processes and analytical detection. Reports of EDCs and PhACs in water have raised substantial concern among the public and regulatory agencies; however, little is known about the fate of these compounds during drinking water and wastewater treatment. A substantial number of studies have shown that conventional drinking water and wastewater treatment plants can not completely remove many EDCs and PhACs.

Oxidation with chlorine and ozone can result in transformation of some compounds with reactive functional groups under the conditions employed in water and wastewater treatment plants. Advanced treatment technologies, such as activated carbon, advanced oxidation, and reverse osmosis, appear viable for the removal of many trace contaminants including EDCs and PhACs (see Chap. 10). Future research needs include more detailed fate and transport data, standardized analytical methodology, removal kinetics, predictive models, and determination of the toxicological relevance of trace levels of EDCs and PhACs in water (Snyder et al, 2003).

Some Specific Constituents with Emerging Concern

During the past decade, a variety of water contaminants have indeed become much more prominent in the minds of public health officials, environmental engineers, and scientists. This situation is an illustration of how the intersection of sensitive new analytical techniques, modern industrial products, and improved understanding of science and engineering lead to the emergence of new contaminants (Alvarez-Cohen and Sedlak, 2003). In this section, a brief discussion on nitrosodimethylamine (NDMA); 1, 4-dioxane; perchlorate; methyl tertiary-butyl ether (MTBE); and other oxygenates is provided.

N-Nitrosodimethylamine

N-nitrosodimethylamine is a member of a family of extremely potent carcinogens, the *N*-nitrosamines. Until recently, concerns about NDMA mainly focused on the presence of NDMA in food, consumer products, and polluted air. However, current concern focuses on NDMA as a drinking water contaminant resulting from reactions occurring during chlorination or via direct industrial contamination. Because of the relatively high concentrations of NDMA formed during wastewater chlorination, the intentional and unintentional indirect potable reuse of reclaimed water is a particularly important area of concern. Although UV irradiation can effectively remove NDMA, there is considerable interest in the development of less expensive alternative treatment technologies. These alternative technologies include approaches for removing organic nitrogen-containing NDMA precursors prior to chlorination and the use of sunlight photolysis, and *in situ* bioremediation to remove NDMA and its precursors (Mitch et al., 2003; Sedlak and Kavanaugh, 2006).

Effluents from conventional and AWT plants can contain relatively high concentrations of NDMA. In addition, NDMA is often present in untreated municipal wastewater prior to chlorination. For example, NDMA concentrations as high as 105,000 ng/L have been reported in effluents from printed circuit board manufacturers using NDMA-contaminated dimethyldithiocarbamate to remove metals (Orange County Sanitation District, 2002). These industrial inputs resulted in concentrations of NDMA of approximately 1500 ng/L in the untreated wastewater. As a result of removal processes that occur during secondary treatment, NDMA concentrations in unchlorinated secondary effluent are typically less than 20 ng/L, although industrial inputs can result in larger spikes in NDMA influent and effluent concentrations. Chlorination of secondary wastewater effluent typically results in the formation of between 20 and 100 ng/L NDMA. In water reclamation plants receiving secondary wastewater effluent, NDMA concentrations in microfiltration effluent may increase by approximately 30 to 50 ng/L as a result of chlorination before the membrane to prevent biological fouling (Mitch et al., 2003).

Facilities with advanced treatment capabilities typically use MF-RO and/or UV treatment. This treatment train has been shown to be effective in removing NDMA and NDMA precursors (Sedlak and Kavanaugh, 2006).

1, 4-Dioxane

1, 4-dioxane has been reported as a water contaminant in a wide variety of locations, due to its widespread occurrence in industrial and commercial products, high aqueous solubility, and resistance to biodegradation.

As early as 1975, Kraybill (1977) reported the detection of 1, 4-dioxane in drinking water in the United States. Johns et al. (1998) identified 1, 4-dioxane as a frequent contaminant of the lower Mississippi River. In a survey of natural waters in Japan, Abe (1999) found 1, 4-dioxane at concentrations from 1.9 to 94.8 µg/L in 83 of 95 river, ocean, and groundwater samples. The 1, 4-dioxane was believed to originate from 1, 1, 1-trichloroethane (TCA) contaminated groundwater and chemical and municipal treatment plant effluents.

1, 4-dioxane is classified as a probable human carcinogen. It is used as a stabilizer for chlorinated solvents, particularly TCA, and it is formed as a byproduct during the manufacture of polyester and various polyethoxylated compounds. Improper disposal of industrial waste and accidental solvent spills have resulted in the contamination of groundwater with 1, 4-dioxane. Volatilization and sorption are not significant attenuation mechanisms due to 1, 4-dioxane's complete miscibility with water. The low 1, 4-dioxane removal efficiency in conventional wastewater treatment processes contributes to its presence in aquatic environments. At present, advanced oxidation processes (AOPs) are the only proven technology for their removal (Adams et al., 1994; Zenker et al., 2004). Chemical and energy costs for many AOPs, however, may be substantial, thus their use is not widespread.

Ultraviolet light is also used commonly as part of an AOP. Because 1, 4-dioxane is a relatively weak absorber of UV light, it is degraded poorly by direct photolysis. Ultraviolet light can be used in combination with H_2O_2, however, to produce hydroxyl radicals that react with 1, 4-dioxane. Ultraviolet light, in combination with a TiO_2 catalyst, has also been demonstrated to degrade 1, 4-dioxane. Hill et al. (1997) achieved greater than 99 percent reduction in 1, 4-dioxane using wavelengths greater than 300 nm. Ethylene diformate was observed as the most significant oxidation byproduct. Hydrogen peroxide can also be used in combination with ferrous ion (Fenton's reagent) to degrade 1, 4-dioxane. There are several different AOPs that are commercially available for the treatment of 1, 4-dioxane that use combinations of H_2O_2, O_3, and UV light (Mohr, 2001). Additional discussion on AOPs is presented in Chap. 10.

Perchlorate

Perchlorate (ClO_4^-) is a highly oxidized (+7) chlorine oxyanion manufactured for use as the oxidizer in solid propellants for rockets, missiles, explosives, and pyrotechnics (Gullick et al., 2001; Logan, 2001). Perchlorate release into the environment has occurred primarily in association with its manufacture and use in solid rocket propellant. When released into groundwater, perchlorate can spread over large distances because it is highly soluble in water and adsorbs poorly to soil. Two proven techniques to remove perchlorate from drinking water are anaerobic biological reactors and ion exchange.

Perchlorate contamination of the environment may affect agricultural plants as well as naturally occurring flora (U.S. EPA, 2002). Chlorate has been used as a defoliant, and therefore, it is not surprising that perchlorate can also be taken up by plants. The accumulation of perchlorate in plants is of concern for several reasons (Hutchinson et al., 2000). Perchlorate can be toxic to some plants. If the perchlorate accumulates in, and is not degraded by, a plant, it may be released back into the environment when the plant dies, which could be toxic to other plants or wildlife. Perchlorate accumulation in food plants could present another route of human exposure to perchlorate. Perchlorate-contaminated water, such as Lake Mead or the Colorado River, is presently used for irrigating food crops. Recently, perchlorate accumulation has been found in crops irrigated with contaminated water and subsequently used as animal feed, resulting in significant concentrations in dairy products (Urbansky et al., 2000).

Methyl Tertiary-Butyl Ether and Other Oxygenates
The production and use of fuel oxygenates has increased dramatically in the United States since the early 1990s due to federal and state regulations aimed to improve air quality. Currently, methyl tertiary-butyl ether (MTBE) is the most widely used oxygenate in gasoline, followed by ethanol. Widespread use of oxygenates in gasoline has been accompanied by widespread release of these materials into the environment. Accidental gasoline releases from underground storage tanks and pipelines are the most significant point sources of oxygenates in groundwater. Because of their polar characteristics, oxygenates migrate through aquifers with minimal retardation, raising great concerns nationwide of their potential for reaching drinking water sources (Deeb et al., 2003).

New and Reemerging Microorganisms

Within the past decade there has been an increase in the number of disease outbreaks in the United States and in many other parts of the world, some caused by a number of endemic contagious diseases that were thought to have been controlled or eliminated (only smallpox to date). For example, the bacteria *Legionella pneumophila,* the causative agent in Legionnaire's disease, has been found in wastewater and reclaimed water. The high incidence of tuberculosis reported in Africa is an example of the reemergence of a disease that was thought to be under control or essentially eliminated. The significance of the identification of new disease organisms, disease outbreaks, and the reemergence of old diseases is that the concern for public health must remain the primary objective of wastewater management including water reclamation and reuse (Tchobanoglous et al., 2003).

3-8 ENVIRONMENTAL ISSUES

The production, distribution, and use of reclaimed water may result in a number of environmental impacts. An environmental impact statement (EIS) must be submitted for federally funded projects and is required by some state laws if certain criteria are applicable (Kontos and Asano, 1996). The EIS criteria, as listed below, are useful in identification of the potential effects that may result from a water reuse project:

- The project may significantly alter land use.
- The project is in conflict with any land use plans or policies.
- Wetlands will be adversely impacted.
- Endangered species or their habitat will be affected.
- The project is expected to displace populations or alter existing residential areas.
- The project may adversely affect a flood plain or important farmlands.
- The project may adversely affect parklands, preserves, or other public lands designated to be of scenic, recreational, archaeological, or historical value.

- The project may have a significant adverse impact upon ambient air quality, noise levels, surface or groundwater quality or quantity.
- The project may have adverse impacts on water supply, fish, shellfish, wildlife, and their actual habitats.

Potential impacts from water reclamation systems range from aesthetic (placement of tanks and reservoirs), sociocultural (disturbance of above ground and below ground cultural resources), physical (location of underground pipelines), environmental (beneficial uses of groundwater and surface waters), sensory (offensive odors), to health and safety issues (aerosols, air pollution, chemicals, pathogens).

Effects on Soils and Plants

Constituents in reclaimed water such as nutrients, salts, and organic and inorganic compounds may all affect soil and plants when applied to soil for irrigation. In addition, reclaimed water may contain microorganisms that could alter the native microbial community or be pathogenic to vegetation. Excessive or insufficient watering for local plant uptake and soil drainage requirements may also impact soil and vegetation. A water balance for soil and vegetation under irrigation can be used to estimate the proper amounts of reclaimed water that may be applied. Guidelines for the application of reclaimed water with chemical constituents, salinity, and nutrients are discussed in detail in Chap. 17.

Effects on Surface Water and Groundwater

The discharge of salts, nutrients, and pathogens to surface water and groundwater may impact water quality and the beneficial use of these waters. Surface waters may be contaminated from reclaimed water runoff or direct discharge. Runoff waters may be controlled by water application at proper rates or facilities for the catchment of excess flow. High efficiency irrigation methods, such as drip and subsurface techniques, are preferred for preventing runoff. Low-permeability soils under irrigation may require drainage systems to prevent waterlogging. Water flowing onto the irrigation site as a result of rainfall should be controlled to avoid saturating soils. Irrigation immediately before, during, or after rainfall events is not recommended. In addition to the dangers associated with the discharge of pathogens to the environment, excess nutrients may result in algal blooms in receiving waters. Algal blooms can affect drinking water systems, aquatic life, and may limit the use of the water for other beneficial uses. Toxic compounds produced by some algae may be of particular concern.

Groundwater may be impacted by the leaching of irrigation water into underlying unconfined aquifers. Of the compounds likely to be present in reclaimed water, nitrate is among the most well-known groundwater contaminant because of its mobility with water through soil. A variety of other contaminants may also be present, especially when industrial discharges are present to the wastewater system. The movement of constituents contained in applied reclaimed water depends on many properties of the site, including soil properties and the site hydrogeology. Thus, a groundwater monitoring program should be implemented for aquifers that may be impacted by reclaimed water applications.

Effects on Ecosystems

Water reclamation programs can have an adverse impact on ecosystems associated with rivers and terrestrial systems where the water flow and receiving water quality are modified. Therefore, careful consideration should be given to the application of reclaimed water to areas that contain or are located near sensitive ecological resources. The application of reclaimed water may selectively induce the growth of one species over another or alter the structure and characteristics of a given ecosystem. In some cases, the flow in a river or stream may be composed of a large fraction, or entirely of, wastewater effluent. The implementation of a water reclamation and reuse program may have a large impact on downstream water use under these conditions. In some cases, reclaimed water may be used beneficially for stream-flow augmentation, where minimum flows are required to protect the habitat of aquatic organisms or support downstream activities (see Chap. 21). Limits on water quality and quantity may be implemented to protect sensitive and important species.

Wastewater effluent diversion to water reuse may affect environmental water quality in a number of ways. In some cases, wastewater effluent discharges are considered a source of pollution in ecological systems and diversion is considered to be an improvement. For example, there have been reports of effects in the reproductive systems of aquatic organisms living in the vicinity of wastewater outfalls. However, there are also examples where it is necessary to augment stream flows with reclaimed water and an ecosystem has adapted well to the affected stream flows over a period of time.

Application of reclaimed water on soil for irrigation may indirectly increase the amount of stormwater runoff due to the increased moisture content present in the soil. Irrigation and groundwater recharge systems both have the potential to increase the level of the groundwater. Both increased stormwater runoff and an increase in the level of the groundwater may result in increased runoff in impacted river systems.

Effects on Development and Land Use

Reclaimed water projects have the potential to promote urban growth and changes in patterns of development. In areas constrained by a limited water supply, reclaimed water may be used as a reliable source for nonpotable uses. Residential, industrial, municipal, and agricultural developments have all been made possible due to the implementation of a water reuse system. Parks, golf courses, nurseries, and gardens are examples of outdoor water users that may be feasible when an adequate and reliable water supply is made available. However, changes in the pattern of water use may have a negative effect on these developments whose viability is dependent on reclaimed water. Because land-use changes that may result from a water reuse project are difficult to predict in advance and because sometimes there is opposition to new development, it is important to involve the public in the decision-making process from inception of the project (see Chap. 26).

PROBLEMS AND DISCUSSION TOPICS

3-1 An increasing number of communities use water sources that contain a significant wastewater component. Review the current literature and prepare a brief synopsis of the articles on the growing knowledge of the potential impact of trace contaminants in the nation's freshwater sources used for drinking water supply. Cite a minimum of three references.

3-2 So-called "emerging" pathogens and contaminants in water and wastewater have changed with time. For examples, there were major concerns for nonbiodegradable (hard) detergent in 1960s; disinfection byproducts in 1970s; *Legionellae* and protozoan (*Giardia* and *Cryptosporidium*) in air, drinking water, and wastewater; and trace contaminants (pharmaceuticals and endocrine disruptors in the 1990s and the early 21st century) with much debate and little consensus. Review the current literature (a minimum of three articles) and prepare a brief synopsis of the articles on emerging contaminants of interest and discuss the significance of these constituents in drinking water, wastewater, reclaimed water, and the aquatic environment. How should municipalities deal with future "emerging" pathogens and other contaminants in water reclamation and reuse?

3-3 Obtain a list of the influent and effluent characteristics for your local wastewater treatment plant. How do the values compare with the values given in Tables 3-12 and 3-14? Discuss any major differences.

3-4 Determine the amount of mineral increase from domestic water use by comparing the TDS in the public water supply in your community to the TDS in the effluent from the corresponding wastewater treatment plant. How does the value obtained compare to the value given in Table 3-11?

3-5 Indirect potable reuse schemes where treated wastewater is first discharged into aquatic environments or aquifers and blended with natural water is widespread. The scheme may be planned or unplanned where unplanned indirect potable water reuse has been practiced for centuries in many parts of the world. Assess the pros and cons of future indirect potable reuse possibilities in light of increasing knowledge of analytical chemistry, toxicology, public health, economics, and public perception and acceptance. How can the increased knowledge be used to rationalize and increase acceptance of planned indirect potable reuse?

3-6 Compare reclaimed water to natural water and discuss significant differences in terms of concentration, constituents, variability, and public perception.

3-7 In 1968, a direct potable water reclamation system from municipal wastewater was pioneered in Windhoek, Namibia, to supplement the potable water supply to the city. Review the following paper and discuss implications of the sage words of Dr. Lucas van Vuuren, one of the pioneers of the Windhoek water reclamation system, "Water should not be judged by its history, but by its quality."

Harrhoff, J., and B. Van der Merwe (1996) "Twenty-five Years of Wastewater Reclamation in Windhoek, Namibia," *Water Sci. Technol.*, 33, 10–11, 25–35.

3-8 Discuss briefly the management alternatives for chemicals and/or pharmaceuticals that cannot be degraded during conventional biological treatment. Summarize the pros and cons of each management option.

3-9 Discuss the relative significance of the sources of *N*-nitrosodimethylamine (NDMA) from dietary intake, industrial chemical use, and disinfection byproducts. Review three or more articles and explain how this impacts water reclamation and reuse.

REFERENCES

Abe, A. (1999) "Distribution of 1, 4-Dioxane in Relation to Possible Sources in the Water Environment," *Sci. Total Environ.,* **227**, 41–47.

Adams, C. D., P. A. Scanlan, and N. D. Secrist (1994) "Oxidation and Biodegrability Enhancement of 1, 4-Dioxane using Hydrogen Peroxide and Ozone," *Environ. Sci. Tech.,* **28**, 1812–1818.

Alvarez-Cohen, L., and D. L. Sedlak (2003)" Introduction—Emerging Contaminants in Water," *Environ. Eng. Sci.,* **20**, 5, 387–388.

AWWA (1999) *Waterborne Pathogens,* AWWA M48, American Water Works Association, Denver, CO.

Barber, L. B., J. A. Leenheer, W. E. Pereira, T. I. Noyes, G. K. Brown, C. F. Tubor, and J. H. Writer (1996) "Organic Contamination of the Mississippi River from Municipal and Industrial Wastewater," in R. H. Mead (ed.) *Contaminants in the Mississippi River 1987–1999,* U.S. Geological Survey Circular 1133, Denver, CO. Accessible at: http://pubs.usgs.gov/circ/circ1133/organic.html

Barwick, R. S., D. A. Levy, G. F. Craun, M. J. Beach, and R. L. Calderon (2000) "Surveillance for Waterborne-Disease Outbreaks—United States, 1997–1998," Centers for Disease Control and Prevention (CDC), *Morbidity and Mortality Weekly Report (MMWR), Surveillance Summaries,* **49**, SS-4, 1–35.

Bauman, L. C., and M. K. Stenstrom (1990) "Removal of Organohalogen and Organohalogen Precursors in Reclaimed Wastewater I," *Water Res.,* **24**, 8, 957–964.

Bellar, T. A., J. J. Lichtenberg, and R. C. Kroner (1974) "The Occurrence of Organohalides in Chlorinated Drinking Water," *J. AWWA,* **66**, 22, 703–706.

Bennett, J. V., S. D. Homberg, M. F. Rogers, and S. L. Solomon (1987) "Infectious and Parasitic Diseases," *Amer Prev. Med.,* **3**, 102–114.

Blackburn, B. G., G. F. Craun, J. S. Yoder, V. Hill, R. L. Calderon, N. Chen, S. H. Lee, D. A. Levy, and M. J. Beach (2004) "Surveillance for Waterborne-Disease Outbreaks Associated with Drinking Water—United States, 2001–2002," Centers for Disease Control and Prevention (CDC), *Morbidity and Mortality Weekly Report, Surveillance Summaries,* **53**, SS-8, 23–45.

Bouwer, H., P. Fox, and P. Westerhoff (1998) "Irrigating with Treated Effluent—How Does this Practice Affect Underlying Groundwater?" *Water Environ. Tech.,* **10**, 9, 115–118.

Caccio, S. M., M. De Giacomo, F. A. Aulicino, and E. Pozio (2003) "*Giardia* Cysts in Wastewater Treatment Plants in Italy," *Appl. Environ. Microbiol.,* **69**, 6, 3393–3398.

Cantor, K. P., C. F. Lynch, M. E. Hildwsheim, M. Dosemeci, J. Lubin, M. Alavanji, and G. Cruan (1998) "Drinking Water Sources and Chlorination Byproducts, I. Risk of Bladder Cancer," *Epidemiol.,* **9**, 21–28.

CDC (1993) "Surveillance for Waterborne Disease Outbreaks—United States, 1991–1992," Centers for Disease Control and Prevention, *Morbidity and Mortality Weekly Report, Surveillance Summaries,* **42**, 1–22.

CDC (1985) *Water-Related Disease Outbreaks: Annual Summary 1984,* Centers for Disease Control, Atlanta, GA.

CDC (2005*a*) "Parasites and Health, Cryptosporidiosis," Centers for Disease Control and Prevention, Atlanta, GA, Accessible at: http://www.dpd.cdc.gov/dpdx/HTML/Cryptosporidiosis.htm

CDC (2005*b*) "Poliovirus Infections in Four Unvaccinated children—Minnesota, August–October 2005," Centers for Disease Control and Prevention, *Morbidity and Mortality Weekly Report, Surveillance Summaries,* **54**, 1–3. http://www.cdc.gov/mmwr/preview/mmwrhtml/mm54d1014a1.htm

Chang, A. C., G. Pan, A. L. Page, and T. Asano (2002) *Developing Human Health-related Chemical Guidelines for Reclaimed Water and Sewage Sludge Applications in Agriculture,* Prepared for World Health Organization, Geneva, Switzerland.

Chauret, C., S. Springthorpe, and S. Sattar (1999) "Fate of *Cryptosporidium* Oocysts, *Giardia* Cysts, and Microbial Indicators during Wastewater Treatment and Anaerobic Sludge Digestion," *Canadian J. of Microbiol.*, **45**, 3, 257–262.

CH2M Hill (1993) *Tampa Water Resources Recovery Project Pilot Studies*, Vol. 1 Final Report, Tampa, FL.

Clancy, J. L., Z. Bukhari, T. M. Hargy, J. R. Bolton, B. W. Dussert, and M. M. Marshall (2000) "Using UV to Inactivate *Cryptosporidium*," *J. AWWA*, **92**, 9, 97–104.

Cohn, P. D., M. Cox, and P. S. Berger (1999) "Health and Aesthetic Aspects of Water Quality," in *Water Quality & Treatment, A Handbook of Community Water Supplies,* American Water Works Association, McGraw-Hill, Inc., New York.

Cooper, R. C., and A. W. Olivieri (1998) "Infectious Disease Concerns in Wastewater Reuse," Chap. 12, in: T. Asano, (ed), *Wastewater Reclamation and Reuse*, Water Quality Management Library **10**, CRC Press, Boca Raton, FL.

Cooper, W. J., J. T. Villate, E. M. Ott, R. Slifker, and F. Z. Parsons (1983) "Formation of Organohalogen Compounds in Chlorinated Secondary Wastewater Effluent," 483–497, in: R. L. Jolley et al. (eds) *Water Chlorination: Environemtnal Impact and Health Effects,* Vol. 4. Ann Arbor Science, Ann Arbor, MI.

Craik, S. A., G. R. Finch, J. R. Bolton, and M. Belosevic (2000) "Inactivation of *Giardia muris* Cysts Using Medium-Pressure Ultraviolet Radiation in Filtered Drinking Water," *Water Res.*, **34**, 18, 4325–4332.

Craik, S. A., D. Weldon, G. R. Finch, J. R. Bolton, and M. Belosevic (2001) "Inactivation of *Cryptosporidium parvum* Oocysts Using Medium- and Low-Pressure Ultraviolet Radiation," *Water Res.*, **35**, 6, 1387–1398.

Craun, G. F. (1986) *Waterborne Diseases in the United States*, CRC Press, Boca Raton, FL.

Craun, G. F. (1991) "Statistics of Waterborne Disease in the United States," *Water Sci. Technol.*, **24**, 2, 10–15.

Craun, G. F., and R. L. Calderon (1999) *Waterborne Disease Outbreaks: Their Causes, Problems, and Challenges to Treatment Barriers*, Chap. 1, "Waterborne Pathogens", AWWA Manual M48, 1st ed., American Water Works Association, Denver, CO.

Crites, R., and G. Tchobanoglous (1998) *Small and Decentralized Wastewater Management Systems*, WCB/McGraw-Hill, Boston, MA.

Crittenden, J. C., R. R. Trussell, D. W. Hand, K. J. Howe, and G. Tchobanoglous (2005) *Water Treatment: Principles and Design*, 2nd ed., John Wiley & Sons, Inc., Hoboken, NJ.

Crook, J. (1992) "Water Reclamation," 559–589, in R. Meyers (ed.) *Encyclopedia of Physical Science and Technology,* **17**, Academic Press, San Diego, CA.

Current, W. L., and L. S. Garcia (1991) "Cryptosporidiosis," *Clin. Microbiol. Rev.*, **4**, 3, 325–358.

Dean R. B., and E. Lund (1981) *Water Reuse: Problems and Solutions*, Academic Press, London.

Deeb, R. A., K. H Chu, T. Shih, S. Linder, I. Suffet, M. C. Kavanaugh, and L. Alvarez-Cohen (2003) "MTBE and Other Oxygenates: Environmental Sources, Analysis, Occurrence, and Treatment," *Environ. Eng. Sci.*, **20**, 5, 443–447.

Fankhauser, R. L., S. S. Monroe, J. S. Noel, C. D. Humphrey, J. S. Bresee, U. D. Parashar, T. Ando, and R. I. Glass (2002) "Epidemiologic and Molecular Trends of 'Norwalk-like Viruses' Associated with Outbreaks of Gastroenteritis in the United States," *J. Infect. Dis.*, **186**, 1, 1–7.

Fayer, R. G., and B. L. P. Ungar (1986) "*Cryptosporidium* spp. and Cryptosporidiosis," *Microbiol. Rev.*, **50**, 4, 458–483.

Feachem, R G., D. J. Bradley, H. Garelick, and D. D. Mara (eds.) (1983) "*Entamoeba histolytica* and Amebiasis," 337–347 in *Sanitation and Disease: Health Aspects of Excreta and Wastewater Management*, John Wiley and Sons, New York.

Frankel-Conrat H., P. C. Kimball, and J. A. Levy (1988) *Virology*, 2nd ed., Prentice Hall, Englewood Cliffs, NJ.

Gennaccaro, A. L., M. R. McLaughlin, W. Quintero-Betancourt, D. E. Huffman, and J. B. Rose (2003) "Infectious *Cryptosporidium parvum* Oosysts in Final Reclaimed Effluent," *App. Environ. Microbiol.*, **69**, 8, 4983–4984.

Gerba, C. P., S. N. Singh, and J. B. Rose (1985) "Waterborne Viral Gastroenteritis and Hepatitis," *CRC Cri. Rev. Environ. Contr.*, **15**, 213–236.

Gerba, C. P., J. B. Rose, and C. N. Haas (1994) "Waterborne Disease: Who is at Risk?" *Proceedings of the American Water Works Association's Water Quality Technology Conference*, San Francisco, American Water Works Association, Denver, CO.

Gerba, C. P. (1996) "Pathogens in the Environment," 279–299, in I. L. Pepper, C. P. Gerba, and M. L. Brusseau, (eds.), *Pollut. Sci.*, Academic Press, New York.

Grabow, W. O. K. (2001) "Bacteriophages: Update on Application as Models for Viruses in Water," *Water SA*, **27**, 2, 251–268.

Gullick, R. Q., M. W. Lechvallier, and T. A. S. Barhorst (2001) "Occurrence of Perchlorate in Drinking Water Sources," *J. AWWA*, **93**, 1, 66–76.

Hale, R. C., C. L. Smith, P. O. de Fur, E. Harvey, E. O. Bush (2000) "Nonylphenols in Sediments and Effluents Associated with Diverse Wastewater Outfalls Environ.," *Toxicol. Chem.* **19**, 4, 946–952.

Harrhoff, J., and B. Van der Merwe (1996) "Twenty-five Years of Wastewater Reclamation in Windhoek, Namibia," *Water Sci. Technol.*, **33**, 10–11, 25–35.

Havelaar, A. H., M. Van Olphen, and Y. C. Drost (1993) "F-specific RNA Bacteriophages are Adequate Model Organisms for Enteric Viruses in Fresh Water," *Appl. Environ. Microbiol.*, **59**, 2956–2962.

Hill, R. R., G. E. Jeffs, and D. R. Roberts (1997) "Photocatalytic Degradation of 1, 4-dioxane in Aqueous Solution," *J. Photochem. Photobiol., A: Chemistry*, **108**, 55–35.

Hildesheim, M. E., K. P. Cantor, C. F. Lynch, M. Dosemeci, J. Lubin, M. Alavanji, and G. Cruan (1998) "Drinking Water Sources and Chlorination Byproducts, II. Risk of Colon and Rectal Cancers," *Epidemiol.*, 9:29–35.

Hlavsa, M. C., J. C. Watson, and M. J. Beach (2005a) "*Giardia* Surveillance—United States, 1998–2002," Centers for Disease Control and Prevention (CDC), *Morbidity and Mortality Weekly Report, Surveillance Summaries*, **54**, SS01, 9–16.

Hlavsa, M. C., J. C. Watson, and M. J. Beach (2005b) "Cryptosporidiosis Surveillance—United States, 1999–2002," Centers for Disease Control and Prevention (CDC), *Morbidity and Mortality Weekly Report, Surveillance Summaries*, **54**, SS01, 1–8.

Huffman, D. E., K. L. Nelson, and J. B. Rose (2003) "Calicivirus-An Emerging Contaminant in Water: State of the Art," *Environ. Eng. Sci.,* **20**, 5, 503–515.

Hunter, P. R. (1997) *Waterborne Disease—Epidemiology and Ecology,* John Wiley & Sons, Chichester, UK.

Hutchinson, S. L., S. Susarla, N. L. Wolfe, and S. C. McCutcheon (2000) "Perchlorate Accumulation from Contaminated Irrigation Water and Fertilizer in Leafy Vegetables," *Proccedings Second International Conference on Remediation of Chlorinated and Recalcitrant Compounds,* May 2000, Monterey, CA.

Johns, M. M., W. E. Marshall, and C. A. Toles (1998) "Agricultural By-products as Granular Activated Carbons for Adsorbing Dissolved Metals and Organics," *J. Chem. Technol. Biotechnol.*, **71**, 131–140.

Juranek, D. D. (1995) "Cryptosporidiosis—Source of Infection and Guidelines for Prevention," *Clin. Infect. Dis.*, **21**, Suppl. 1, S57–S61.

Kappus. K. K., D. D. Juranek, and J. M. Roberts (1992) "Results of Testing for Intestinal Parasites by State: Diagnostic Laboratories, United States, 1987," *Morbidity and Mortality Weekly Report, Surveillance Summaries*, **40**, SS-4, 25.

Karim, M. R., and M. W. LeChevallier (2004) "Detection of Noroviruses in Water: Current Status and Future Directions," *J. Water Sup. Res. Technol.—AQUA*, **53**, 6, 359–380.

Klime, D. E. (1998) *Endocrine Disruption in Fish*, Kluwer Academic Publishers, Norwell, MA.

Kolpin, D. W., M. Skopec, M. T. Meyer, E. T. Furlong, and S. D. Zaugg (2004), "Urban Contribution of Pharmaceuticals and Other Organic Wastewater Contaminants to Streams during Differing Flow Conditions," S*ci. Total Environ.,* **328**, 1–3, 119–130.

Kontos, N., and Asano, T. (1996), "Environmental Assessment for Wastewater Reclamation and Reuse Projects," *Wat. Sci. Tech.*, **33**, 10–11, 473–486.

Korhonen, L. K., and P. J. Martikainen (1991) "Survival of *Escherichia coli* and *Campylobacter jejuni* in Untreated and Filtered Lake Water," *J. Appl. Bacteriol.*, **71**, 379–382.

Korick, D. G., J. R. Mead, M. S. Madore, N. A. Sinclair, and C. R. Sterling (1990) "Effects of Ozone, Chlorine Dioxide, Chlorine and Monochloramine on *Cryptosporidium parvum* Oocysts Viability," *Appl. Environ. Microbiol.,* **56**, 5, 1423–1428.

Kramer, M. H., B. L. Herwaldt, G. F. Craun, R. L. Calderon, and D. D. Juranek (1996) "Surveillance for Waterborne-Disease Outbreaks—United States, 1993–1994," Centers for Disease Control and Prevention (CDC), *Morbidity and Mortality Weekly Report, Surveillance Summaries*, **45**, SS-1, 1–33.

Krasner, S. W., M. J. McGuire, J. G. Jacangelo, N. L. Patania, K. M. Reagan, and E. M. Aieta (1989) "The Occurrence of Disinfection By-products in U.S. Drinking Water," *J. AWWA*, **81**, 41–53.

Kraybill, H.R. (1977), "Global Distribution of Carcinogenic Pollutants in Water," *Ann. New York Acad. Sci.*, **298**, 80–89.

Kutz, S. M., and C. P. Gerba (1988) "Comparison of Virus Survival in Freshwater Sources," *Wat. Sci. Tech.*, **20**, 11/12, 467–471.

Lauer, W. C., S. E. Rogers, A. M. La Chance, and M. K. Nealy (1991) "Process Selection for Potable Reuse Health Effect Studies," *J. AWWA*, **83**, 11, 52–63.

Lee, S. H., D. A. Levy, G. F. Craun, M. J. Beach, and R. L. Calderon (2002); "Surveillance for Waterborne-Disease Outbreaks—United States, 1999–2000," Centers for Disease Control and Prevention (CDC), *Morbidity and Mortality Weekly Report, Surveillance Summaries*, **51**, SS-8, 1–47.

Levy, D. A., M. S. Bens, G. F. Craun, R. L. Calderon, and B. L. Herwaldt (1998); "Surveillance for Waterborne-Disease Outbreaks—United States, 1995–1996," Centers for Disease Control and Prevention (CDC), *Morbidity and Mortality Weekly Report, Surveillance Summaries*, **47**, SS-5, 1–33.

Lin, A. Y-C., M. H. Plumlee, and M. Reinhard (2006) "Natural Attenuation of Pharmaceuticals and Alkylphenol Polyethoxylate Metabolites during River Transport: Photochemical and Biological Transformation," *Env. Toxicol. and Chem.*, **25**, 6, 1458–1464.

Linden, K. G., G. Shin, G. Faubert, W. Cairns, and M. D. Sobsey (2002) "UV Disinfection of *Giardia lamblia* Cysts in Water," *Environ. Sci. and Tech.*, **36**, 11, 2519–2522.

Lisle, J. T., and J. B. Rose (1995) "*Cryptosporidium* Contamination of Water in the USA and UK: A Mini-review," *J. Inter. Wat. SRT-Aqua,* **42**, 4, 1–15.

Logan, B. E. (2001) "Assessing the Outlook for Perchlorate Remediation," *Environ. Sci. Technol.*, **35**, 482A–487A

MacKenzie, W. R., N. J. Hoxie, M. E. Proctor, M. S. Gradus, K A. Blair, D. E. Peterson, J. J. Kazmierczak, D. G. Addiss, K. R. Fox, J. B. Rose, and J. P. Davis (1995) "A Massive Outbreak in Milwaukee of *Cryptosporidium* Infection Transmitted through the Public Water Supply," *New Eng. J. Med.,* **331**, 3, 161–167.

Madore, M. S., J. B. Rose, C. P. Gerba, M. J. Arrowood, and C. R. Sterling (1987) "Occurrence of *Cryptosporidium* Oocysts in Sewage Effluents and Selected Surface Waters," *J. Parasitol.*, **73**, 4, 702–705.

Maier, R. M., I. L. Pepper, and C. P. Gerba (2000) *Environmental Microbiology*, Academic Press, San Diego, CA.

Maya, C., B. Jimenez, and J. Schwartzbrod (2006) "Comparison of Techniques for the Detection of Helminth Ova in Drinking Water and Wastewater," *Water Environ. Res.* **78**, 2, 118–124.

McFeters, G. A, and S. I. Terzieva (1991) "Survival of *Escherichia coli* and *Yersinia enterocolitica* in Stream Water: Comparison of Field and Laboratory Exposure," *Microb. Ecol.*, **22**, 65–74.

Mitch, W. A., J. O. Sharp, R. R. Trussell, R. L. Valentine, L. Alvarez-Cohen, and D. L. Sedlak (2003) "*N*-Nitrosodimethylamine (NDMA) as a Drinking Water Contaminate A Review," *Environ. Eng. Sci.*, **20**, 5, 389–404.

Mohr, T. K. G. (2001) *Solvent Stabilizers,* Santa Clara Valley Water District, San Jose, CA.

Moyer, N. P. (1999) "Shigella," Chap. 17, 115–117, in *Waterborne Pathogens*, AWWA M48, American Water Works Association, Denver, CO.

NRC (1996) *Use of Reclaimed Water and Sludge in Food Crop Production*, National Research Council, National Academy Press., Washington, DC.

NRC (1998) *Issues in Potable Reuse: The Viability of Augmenting Drinking Water Supplies with Reclaimed Water*, National Research Council, National Academy Press, Washington, DC.

NRC (1999) *Hormonally Active Agents in the Environment*, National Research Council, National Academy Press, Washington, DC.

NRC (2004) *Indicators for Waterborne Pathogens,* National Research Council, National Academy Press, Washington, DC.

Orange County Sanitation District (2002) "Industrial sampling and IRWD sampling," Presentation at the NDMA Workshop: *Removal and/or Destruction of NDMA and NDMA Precursors in Wastewater Treatment Processes, March 14, 2002,* West Basin Municipal Water District, Carson, CA.

Oregon Health Division (1992) *A Large Outbreak of Cryptosporidiosis in Jackson County, Communicable Disease Summary*, Oregon Health Division, Portland, OR, **41**, 14.

Peeters, J. E., E. A. Mazas, W. J. Masschelein, L. Villacorta Martinez de Maturana, and E. DeBacker (1989) "Effect of Drinking Water with Ozone or Chlorine Dioxide on Survival of *Cryptosporidium parvum* Oocysts," *Appl. Environ. Microbiol.*, **55**, 6, 1519–1522.

Rebhun, M., L. Heller-Grossman, and J. Manka (1997) "Formation of Disinfection Byproducts during Chlorination of Secondary Effluent and Renovated Water," *Water Environ. Res.*, **69**, 6, 1154–1162.

Reckhow, D. A., P. C. Singer, and R. L. Malcolm (1990) "Chlorination of Humic Materials: Byproducts Formation and Chemical Interpretations," *Environ. Sci. Technol.*, **24**, 11, 478–482.

Robertson, L. J., C. A. Paton, A. T. Campbell, P. G. Smith, M. H. Jackson, R. A. Gilmour, S. E. Black, D. A. Stevenson, and H. V. Smith (2000) "*Giardia* Cycts and *Cryptosporidium* Oocysts at Sewage Treatment Works in Scotland, UK," *Water Res.*, **34**, 8, 2310–2322.

Rook, J. J. (1974) "Formation of Haloforms During Chlorination of Natural Waters," *J. Soc. Water Treat. Exam.,* **23**, 234–243.

Rose, J. B., J. Dickson, S. R. Farrah, and R. P. Carnahan (1996) "Removal of Pathogenic and Indicator Microorganisms by a Full-scale Water Reclamation Facility*,*" *Water Res.*, **30**, 11, 2785–2797.

Ryu, H., A. Alum, and M. Abbaszadegan (2005) "Microbial Characterization and Population Changes in Nonpotable Reclaimed Water Distribution Systems," *Environ. Sci. Technol.*, **39**, 22, 8600–8605.

Schroeder, E. D., and S. Wuertz (2003) "Chap. 3 Bacteria," 57–68, in D. Mara and N. Horan (eds.) *The Handbook of Water and Wastewater Microbiology,* Elsevier Science, Amsterdam, The Netherlands.

Sedlak, D., and M. Kavanaugh (2006) Removal and Destruction of NDMA and NDMA Precursors during Wastewater Treatment, *WateReuse Foundation, Alexandria, VA.*

Snyder, S. A., P. Westerhoff, Y. Yoon, and D. L. Sedlak (2003) "Pharmaceuticals, Personal Care Products, and Endocrine Disruptor in Water: Implications for the Water Industry," *Environ. Eng. Sci.,* **20**, 5, 449–469.

Standard Methods for the Examination of Water and Wastewater (2005) 21st edition, Prepared and published jointly by American Public Health Association, American Water Works Association, and Water Environment Federation, Washington, DC.

Stevens, A. A., L. A. Moore, and R. S. Miltner (1989) "Formation and Control of Nontrihalomethanes Disinfection Byproducts," *J. AWWA*, **81**, 8, 54–60.

Straub, T. M, and D. P. Chandler (2003) "Toward a Unified System for Detecting Waterborne Pathogens," *J. Microbiol. Meth.*, **53**, 185–197.

Sykora, J. L., C. A. Sorber, W. Jakubowski, L. W. Casson, P. D. Gavaghan, M. A. Shapiro, and M. J. Schott (1991) "Distribution of *Giardia* cysts in Wastewater," *Wat. Sci. Technol.*, **24**, 2, 187–192.

Tchobanoglous, G., F. L. Burton, and H. D. Stensel (2003) *Wastewater Engineering: Treatment and Reuse*, 4th ed., McGraw-Hill, New York.

Thomas, J. M., W. A. McKay, E. Cole, J. E. Landmeyer, P. M. Bradley (2000) "The Fate of Haloacetic Acids and Trihalomethanes in an Aquifer Storage and Recovery Program, Las Vegas, Nevada," *Ground Water*, **38**, 4, 605–614.

Urbansky, E. T., M. L. Magnuson, C. A. Kelty, and S. K. Brown (2000) "Perchlorate Uptake by Salt Cedar *(Tamarix ramosissima)* in the Las Vegas Wash Riparian Ecosystem," *Sci. Tot. Environ.*, **256**, 227–232.

U.S. EPA (2002) *Perchlorate Environmental Contamination: Toxicological Review and Risk Characterization (External Review Draft), NCEA-I-O503*, Office of Research and Development, U.S. Environmental Protection Agency, Washington, DC.

U.S. Geological Survey (1996) Meade, R.H. (ed.) *"Contaminants in the Mississippi River, 1987–92"*, USGS Circular 1133, Denver, CO.

Villacorta-Martinez De Maturana, L., M. E. Ares-Mazas, D. Duran-Oreiro, and M. J. Lorenzo-Lorenzo (1992) "Efficacy of Activate Sludge in Removing *Cryptosporidium parvum* Oocysts from Sewage," *Appl. Environ. Microbiol.*, **58**, 11, 3514–3516.

Wagenknecht, L. E., J. M. Roseman, and W. H. Herman (1991) "Increased Incidence of Insulin-dependent Diabetes mellitus following an Epidemic of Coxsackievirus B5," *Am. J. of Epidemiol.*, **133**, 1024–1031.

Waller, K., S. H. Swan, G. DeLorenze, and B. Hopkins (1998) "Trihalomethane in Drinking Water and Spontaneous Abortion," *Epidemiol.*, **9**, 2, 134–140.

Western Consortium for Public Health (1992) *The City of San Diego Total Resource Recovery Project Health Effects Study, Final Summary Report*, Oakland, CA.

Yates, M. V., and Gerba, C. P. (1998) "Microbial Considerations in Wastewater Reclamation and Reuse," Chap. 10, in T. Asano (ed.), *Wastewater Reclamation and Reuse*, Water Quality Management Library **10**, CRC Press, Boca Raton, FL.

Yoder, J. S., B. G. Blackburn, G. F. Craun, V. Hill, D. A. Levy, N. Chen, S. H. Lee, R. L. Calderon, and M. J. Beach (2004) "Surveillance for Waterborne-Disease Outbreaks Associated with Recreational Water—United States, 2001–2002," Centers for Disease Control and Prevention (CDC), *Morbidity and Mortality Weekly Report, Surveillance Summaries*, **53**, SS-8, 1–21.

Zenker, M. J., R. C. Borden, and M. A. Barlaz *(2004) "*Biodegradation of 1, 4-Dioxane using Trickling Filter," *J. Environ. Eng. Div.* ASCE, **130**, 9, 926–931.

4 Water Reuse Regulations and Guidelines

WORKING TERMINOLOGY   132

4-1   UNDERSTANDING REGULATORY TERMINOLOGY   134
Standard and Criterion   134
Standard versus Criterion   134
Regulation   135
Difference between Regulations and Guidelines   135
Water Reclamation and Reuse   135

4-2   DEVELOPMENT OF STANDARDS, REGULATIONS, AND GUIDELINES FOR WATER REUSE   135
Basis for Water Quality Standards   136
Development of Water Reuse Regulations and Guidelines   136
The Regulatory Process   139

4-3   GENERAL REGULATORY CONSIDERATIONS RELATED TO WATER RECLAMATION AND REUSE   139
Constituents and Physical Properties of Concern in Wastewater   139
Wastewater Treatment and Water Quality Considerations   142
Reclaimed Water Quality Monitoring   145
Storage Requirements   146
Reclaimed Water Application Rates   147
Aerosols and Windborne Sprays   147

4-4   REGULATORY CONSIDERATIONS FOR SPECIFIC WATER REUSE APPLICATIONS   149
Agricultural Irrigation   149
Landscape Irrigation   150
Dual Distribution Systems and In-building Uses   151
Impoundments   152
Industrial Uses   153
Other Nonpotable Uses   153
Groundwater Recharge   154

4-5   REGULATORY CONSIDERATIONS FOR INDIRECT POTABLE REUSE   155
Use of the Most Protected Water Source   155
Influence of the Two Water Acts   155
Concerns for Trace Chemical Constituents and Pathogens   156
Assessment of Health Risks   157

4-6   STATE WATER REUSE REGULATIONS   157
Status of Water Reuse Regulations and Guidelines   158
Regulations and Guidelines for Specific Reuse Applications   158
Regulatory Requirements for Nonpotable Uses of Reclaimed Water   165
State Regulations for Indirect Potable Reuse   167

4-7 U.S. EPA GUIDELINES FOR WATER REUSE 169
 Disinfection Requirements 169
 Microbial Limits 178
 Control Measures 178
 Recommendations for Indirect Potable Reuse 178

4-8 WORLD HEALTH ORGANIZATION GUIDELINES FOR WATER REUSE 179
 1989 WHO Guidelines for Agriculture and Aquaculture 180
 The Stockholm Framework 180
 Disability Adjusted Life Years 180
 Concept of Tolerable (Acceptable) Risk 181
 Tolerable Microbial Risk in Water 181
 2006 WHO Guidelines for the Safe Use of Wastewater in Agriculture 182

4-9 FUTURE DIRECTIONS IN REGULATIONS AND GUIDELINES 184
 Continuing Development of State Standards, Regulations, and Guidelines 184
 Technical Advances in Treatment Processes 184
 Information Needs 184

PROBLEMS AND DISCUSSION TOPICS 185

REFERENCES 187

WORKING TERMINOLOGY

Term	Definition
Coliform group	All aerobic and facultative anaerobic, gram-negative, non-spore-forming, rod-shaped bacteria that ferment lactose with gas formation within 48 h at 35°C. The coliform group consists of several genera of bacteria belonging to the family Enterobacteriaceae, mostly of intestinal origin (see also Chap. 3).
Criteria	Standards, rules, or tests on which a judgment or decision can be based. Sometimes used interchangeably with "standards," "rules," "requirements," or "regulations."
Criterion	A constituent or other parameter concentration or level or a narrative statement upon which scientific judgment may be based. Sometimes used interchangeably with "standard," "rule," "requirement," or "regulation."
Direct potable reuse	Introduction of reclaimed water directly into a drinking water distribution system, without intervening storage or additional treatment (e.g., pipe-to-pipe).
Fecal coliforms	Bacteria in the coliform group that inhabit the intestinal tract and are associated with fecal contamination. *Escherichia coli*, the most common enteric bacterium, is commonly used as an indicator organism (see also Chap. 3).
Guidelines	Recommended or suggested standards, criteria, rules, or procedures that are voluntary, advisory, and nonenforceable.

Indicator organism	A nonpathogenic microorganism used to detect the possible presence of pathogenic microorganisms in water or other medium. An ideal indicator organism has attributes such as survival and transport similar to pathogens and is present in greater numbers than pathogens (see also Chap. 3).
Indirect potable reuse	Augmentation of a raw water supply with reclaimed water followed by an environmental buffer. The mixture of raw and reclaimed water typically receives additional treatment before distribution as drinking water.
Morbidity	A disease or the incidence of disease within a population. The morbidity rate is a ratio that measures the incidence and prevalence of a specific disease. Within the framework of a given time period, it gives the number of people who are afflicted with that disease per standard unit of population.
Mortality	Mortality refers to death. The number of individuals that die as a result of a specific disease or group of diseases each year. When expressed as a rate, it is the number of deaths during that time period per standard unit of population.
Most probable number (MPN)	A statistical determination of coliform organism density per 100 mL employing liquid culture medium in test tubes and serial dilutions. It is not an actual enumeration.
Multiple-barrier concept	To limit the presence of pathogens and harmful chemicals in reclaimed water, multiple barriers including source control, various unit operation and process combinations, and design and operation of the reclaimed water distribution system to increase reliability of treatment operations and processes and to provide consistent water quality.
Pathogens	Disease-causing organisms capable of inflicting damage on a host it infects.
Personal care products (PCPs)	Products such as shampoo, hair conditioner, deodorants, and body lotion.
Pharmaceutically active compounds (PhACs)	Chemicals synthesized for medical purposes (e.g., antibiotics) (see Chap. 3).
Standard	Standard applies to any enforceable rule, principle, or measure established by a regulatory authority. Often synonymous with numerical water quality limits.
Total coliforms	All bacteria in the coliform group, including those not associated with the fecal matter of warm-blooded animals. Total coliform is commonly used as an indicator organism (see Chap. 3).
Regulations	Criteria, standards, rules, or requirements that have been legally adopted and are enforceable by government agencies.
Stakeholder	A person, persons, community, business, regulatory agency, or organization with a concern and interest in some issue.
Use area	A location with defined boundaries where reclaimed water is used for one or more beneficial purposes, such as a golf course or other irrigation site, impoundment, or a building.
Vector	An organism, such as a mosquito or tick that carries disease-causing microorganisms from one host to another.
Water reclamation	The act of treating wastewater to make it acceptable for beneficial reuse. For the purposes of this chapter, water reuse criteria, standards, regulations, or guidelines implicitly include water reclamation requirements.

The development and implementation of water reclamation and reuse regulations and guidelines were important milestones in the advancement of water reuse in the United States and around the world. The purpose of this chapter, which deals with the regulations and guidelines for the reuse of municipal wastewater, is to present information on (1) regulatory terminology, (2) the development of standards, regulations, and guidelines for water reuse, (3) general regulatory considerations for water reclamation and reuse, (4) regulatory considerations for specific nonpotable water reuse applications, and (5) regulatory considerations for indirect and direct potable reuse. The information presented in the first five sections is intended to serve as background for the three sections that follow them, which deal with (1) state regulations; (2) U.S. Environmental Protection Agency (U.S. EPA) guidelines for water reuse; and (3) World Health Organization (WHO) guidelines for water reuse. The chapter concludes with a discussion of future directions in establishing regulations and guidelines.

4-1 UNDERSTANDING REGULATORY TERMINOLOGY

Before discussing the basis for the development of water reuse regulations and guidelines, it is useful to define and examine the basis of some commonly used regulatory terminology. Terms routinely used in regulatory documentation concerning water reclamation and reuse include standard, criterion, criteria, regulation, and guideline. The meanings of these terms are examined in the following discussion.

Standard and Criterion

The following definitions of a *standard* and *criterion* are excerpted from the seminal publication, *Water Quality Criteria* by McKee and Wolf (1963). The term *standard* applies to any rule, principle, or measure established by an authority. Because the standard has been established by an authority, it tends to be quite rigid, official, or quasi-legal. An authoritative origin does not necessarily mean that a standard is fair, equitable, or based on sound scientific knowledge, for it may have been established somewhat arbitrarily and tempered by a cautious factor of safety. When scientific data are being accumulated to serve as yardsticks of water quality without regard for legal authority, the term *criterion* is most applicable. Unlike a standard, a criterion generally does not connote authority other than that of fairness and equity; nor does it imply an ideal condition. To be useful, a criterion should be capable of quantitative evaluation by acceptable analytical procedures. Without numerical criteria, vague descriptive qualitative terms are subject to legal interpretation or administrative decisions.

Standard versus Criterion

In a classical sense, *criterion* should not be used interchangeably with, or as a synonym for, *standard*. A criterion typically represents a constituent concentration or level associated with a degree of environmental effect. A criterion may be a narrative statement, for example, needed treatment processes to produce acceptable reclaimed water. Criteria usually are developed solely on the basis of available data and scientific judgment, often without consideration of technical or economic feasibility, and serve as a basis for standards. In practice, the terms *standards* and *criteria* (standard and criterion) are often used interchangeably by states, and regulations contain enforceable standards or criteria. Standards usually—but not always—infer numerical limits, while criteria may include both numerical limits and narrative statements prescribing other than numerical limits. The term *requirement* is also used to describe an administrative decision by a regulatory body and may include a standard, criterion, or other requisite, such as signage or other use area operational or management controls.

Regulation

A standard, criterion, or guideline becomes a regulation when adopted officially by a regulatory body, such as a state legislature or a water pollution control agency. It is important to note that regulations are mandatory and enforceable by governmental agencies. For example, once adopted by the individual states the U.S. EPA drinking water standards then become enforceable drinking water regulations. Water reuse regulations usually include wastewater treatment process requirements, treatment reliability requirements, reclaimed water quality criteria, and use area controls such as cross-connection control provisions, signage, and setback distances. Developing regulations for water reclamation and reuse is challenging and often controversial. Currently, there are no federal regulations governing water reclamation and reuse practices in the United States, typically regulations are developed at the state level. Thus, state regulatory agencies impose water reclamation and reuse regulations and guidelines to reduce threats to public health and the environment.

Difference between Regulations and Guidelines

Understanding the difference between regulations and guidelines is important. Whereas regulations are legally adopted, enforceable, and mandatory, guidelines are advisory, voluntary, and nonenforceable but can be incorporated in water reuse permits and, thus, become enforceable requirements. Some states prefer the use of guidelines to provide flexibility in regulatory requirements depending on project-specific conditions, which can result in differing requirements for similar uses within a state and lead to inequities in water reuse permits if guidelines are not uniformly imposed. As reclaimed water use becomes more pronounced in states having guidelines, most states eventually progress to development and imposition of regulations.

Water Reclamation and Reuse

The treatment of wastewater and its subsequent beneficial use commonly is called water reclamation and reuse, although different terms are used in various states or regions. For example, some frequently used terms include *water reuse, water recycling, water purification, reclaimed water, recycled water, reuse water, repurified water,* and *NEWater*. Throughout this textbook, the term *water reclamation* refers to treatment or processing of municipal wastewater to make it reusable; *reclaimed water* refers to treated municipal wastewater that is used for beneficial purposes; the term *water reuse* refers to the use of reclaimed water.

4-2 DEVELOPMENT OF STANDARDS, REGULATIONS, AND GUIDELINES FOR WATER REUSE

The focus of the early history of public health in the environmental field was to provide safe water supply and safe disposal of wastewater. With the latter, the first efforts were directed at eliminating indiscriminate discharges of untreated wastewater to the environment and at providing wastewater treatment. These efforts progressed to (1) providing higher levels of treatment—in particular, biological oxidation to restore receiving waters to aerobic conditions; (2) disinfection of effluents to protect against microbial health hazards from public contact with recreational waters; and (3) reducing contamination of potable water supplies. Standards for acceptable performance evolved from these practices—standards that represented good practice, that could be attained by well-designed and operated wastewater treatment plants, and that were validated by indications that the resulting conditions were no longer producing epidemic disease. Hence, water quality standards evolved as part of the process to control major public health hazards associated with drinking water supply and municipal wastewater disposal. To understand more fully the complexities involved in developing standards and implementing regulations it is useful to review (1) the basis for water quality standards, (2) the development of water reuse standards, and (3) the steps involved in the regulatory process.

Basis for Water Quality Standards

Water quality standards ultimately must express quality factors in numbers that will establish a desired boundary condition. McGauhey (1968) listed the following bases that may be used: (1) established or ongoing practice; (2) attainability, either easily or reasonably attainable technologically and economically; (3) educated guess, making use of best information available; (4) epidemiological and toxicological data; (5) human exposure; and (6) data from mathematical models or treatment process effectiveness. Since the above listing was put forth, a significant body of information has become available as a result of ongoing research studies, from the operation of pilot and demonstration plants, and long-term full-scale treatment plant operational data. In addition, the science of risk assessment and risk analysis has progressed significantly and risk analysis is used extensively in developing water quality standards, especially for many of the trace constituents. Because of the importance of health risk assessment, an entire chapter, Chap. 5, is devoted to the subject. Also included in Chap. 5 is an analysis of the use of epidemiological and toxicological studies to assess the safety of reclaimed water.

In practice, the supporting basis for reclaimed water quality standards usually includes consideration of several of the above-mentioned factors. Considering the factors involved in establishing the standards, it is evident that standards should be dynamic in nature and subject to revision as new information becomes available. As a practical matter, however, standards promulgation can be a time-consuming process, often involving several years of effort, and once standards are established and adopted, they are difficult to change.

Development of Water Reuse Regulations and Guidelines

Although agricultural irrigation with low quality wastewater was practiced in some areas of the United States in the late 1800s, there were no significant regulations or restrictions on the practice until the early part of the twentieth century. As urban areas began to encroach on sewage farms and as the scientific basis of disease became more widely understood, concern about the health risks associated with irrigation using wastewater grew among public health officials. This led to the establishment of regulations and guidelines for the use of wastewater for agricultural irrigation, which was the first reclaimed water application to be regulated.

Currently, water reuse regulations and guidelines are based on a variety of considerations, including the factors identified in Table 4-1. The protection of public health is

Table 4-1
Factors affecting water reuse guidelines and regulations and water quality requirements

Factor	Description
Public health protection	Water reuse guidelines and regulations are directed principally at public health protection. For nonpotable reclaimed water applications, criteria generally address only microbiological and environmental concerns. Health risks associated with both pathogenic microorganisms and chemical constituents need to be addressed where reclaimed water is to be used for potable water supply augmentation.
Use area controls	Reclaimed water quality requirements are based on proper controls and safety precautions implemented at areas where the water is used. Depending on reclaimed water quality and type of use, controls may include warning signs, color-coded pipes and appurtenances, fencing, confinement of the water to approved areas of use, cross-connection control provisions, and other public health protection measures.
Use requirements	Many industrial uses and some other applications have specific physical and chemical water quality requirements that are not related to health considerations. Similarly, the effect of individual constituents or parameters on crops or other vegetation, soil, and groundwater or other receiving water is an important consideration for reclaimed water irrigation applications. Physical, chemical, and/or microbiological quality may limit user or regulatory acceptability of reclaimed water for specific uses. Numerous guidelines with suggested or recommended water quality limits are available. Water quality requirements not associated with public health or environmental protection are seldom included in water reuse criteria by regulatory agencies.
Environmental considerations	The natural flora and fauna in and around reclaimed water use areas and receiving waters should not be adversely impacted by the reclaimed water.
Aesthetics	For high level nonpotable uses, e.g., urban irrigation and toilet flushing, the reclaimed water should be no different in appearance than potable water, i.e., clear, colorless, and odorless. For recreational impoundments, reclaimed water should not promote algal growth.
Economics	Although regulatory agencies take into account the costs that regulations impose on reclaimed water producers and users, they are prone to set standards thought to be safe and do not lower health or environmental standards for the sole purpose of making projects economically attractive.
Political realities	Regulatory decisions regarding water reclamation and reuse may be influenced by public policy, public acceptance, technical feasibility, and financial considerations.

achieved by eliminating or reducing the concentrations of microbial and chemical constituents of concern through wastewater treatment and/or by limiting public or worker exposure to the water via design and operational controls. A variety of use area controls have been developed and implemented to further protect public health. Examples of signs that have been used to alert the public that reclaimed water is being used are illustrated on Fig. 4-1. Where reclaimed water is used for agricultural crop irrigation (see Fig. 4-2) water quality requirements are of critical importance. Environmental protection of natural flora and fauna is always of concern in water reuse applications. Because of the importance of public acceptance, special attention to water quality is required in water reuse applications such as recreational impoundments and toilet flushing. During development

(a)   (b)

Figure 4-1
Examples of signs highlighting (a) water conservation and (b) water reuse.

of water reuse regulations and guidelines, regulatory agencies also consider other factors such as economics, technical feasibility, enforceability, and political realities (see Table 4-1).

Regulations and guidelines may take the form of (1) process specifications or level of treatment (such as requiring granular medium filtration); (2) reclaimed water quality specifications (such as turbidity and coliform limits); and (3) design and operational controls (such as treatment reliability requirements, cross-connection control provisions, setback distances, and operator certification requirements). The ideal method for establishing regulations and guidelines involves a scientific determination of environmental benefits and health risks, a technical/engineering decision of costs to meet various water quality objectives, and a regulatory/political decision that weighs benefits and costs.

Figure 4-2
Irrigation of grape vines with reclaimed water.

The regulatory process leading to adoption of water reuse regulations typically proceeds in a logical stepwise fashion: (1) beneficial uses of reclaimed water are identified; (2) a regulatory agency forms a technical advisory committee (TAC) to help develop draft standards; (3) draft standards are developed and provided to stakeholders and others for review and comments; (4) comments and suggested revisions are received by the regulatory agency; (5) the regulatory agency evaluates all comments and may revise the draft standards, often with assistance and advice from the TAC; (6) public hearings are held to present draft standards and receive additional comments; (7) final revisions are made by the regulatory agency and standards are promulgated; (8) public notice of proposed regulation adoption is made; and (9) standards/regulations are adopted. For various reasons, these steps are not always followed in exactly this progression, but the above process is a common procedure. The range of factors that must be considered is discussed in the following section. Considerations for specific applications are considered in Sec. 4-4.

The Regulatory Process

4-3 GENERAL REGULATORY CONSIDERATIONS RELATED TO WATER RECLAMATION AND REUSE

General regulatory considerations related to water reclamation and reuse encompass a wide range of concerns including the constituents in water, wastewater treatment technologies, monitoring requirements, storage requirements, reclaimed water application rates, aerosols and windborne sprays, and cross-connections. In addition to these general concerns, there are additional concerns related to specific water reuse applications and, more recently, about the use of reclaimed water for indirect potable reuse. General regulatory considerations are addressed in this section; specific reuse applications and indirect potable reuse are addressed in Secs. 4-4 and 4-5, respectively. The material presented in this and the following two sections is meant to serve as background material for understanding the basis for the development of the various water reuse regulations and guidelines presented and discussed in Secs. 4-6, 4-7, and 4-8.

The constituents found in municipal wastewater were presented and discussed in detail in Chap. 3. In what follows, the microbial and chemical constituents and physical properties of water that are of concern in water reuse applications, as delineated in Table 4-2, are considered briefly.

Constituents and Physical Properties of Concern in Wastewater

Microbiological Constituents

As discussed in Chap. 3, untreated municipal wastewater contains pathogenic microorganisms and a wide range of chemical constituents, many of which are known to cause illness or disease upon contact, inhalation, or ingestion. In developing countries, adverse health effects, including disease outbreaks associated with the use of untreated or improperly treated wastewater for irrigation, are well documented. In considering the potential adverse health and environmental risks involved in the use of reclaimed water originating from municipal wastewater, controls are needed to assure that microbial and chemical constituents are removed or reduced to safe levels in reclaimed water. The classes of microorganisms of concern are identified in Table 4-2. It should be noted that the occurrence and concentration of pathogenic microorganisms in untreated municipal

Table 4-2
Microbiological and chemical constituents and physical properties of concern in water reclamation and reuse

Factor	Description
Microbiological constituents	
Bacteria	Although untreated wastewater may contain large numbers of bacterial pathogens, bacteria have not been shown to represent a major public health threat in adequately treated reclaimed water. Research studies and operating experience over the last 50 years or more indicate that conventional secondary or tertiary treatment, when coupled with a high level of disinfection to reduce either total or fecal coliform organisms to low levels, effectively eliminates bacterial pathogens or reduces them to insignificant levels in reclaimed water.
Protozoa	Several waterborne disease outbreaks around the world have been attributed to the protozoan parasites *Giardia lamblia* and *Cryptosporidium parvum* in drinking water and recreational water. Although no giardiasis or cryptosporidiosis cases related to water reuse projects have been confirmed, protozoa have emerged as major waterborne causes of diseases.
Helminths	Helminths (worms) represent a significant health threat in developing countries where untreated or poorly treated wastewater is used for irrigation, but are of lesser concern in industrialized countries where secondary or higher levels of treatment readily remove the organisms. Helminths have not been shown to present a health problem at any reclaimed water use site in the United States.
Viruses	Untreated municipal wastewater contains a myriad of pathogenic viruses. Some viruses are more resistant to disinfection than coliforms and there is little direct correlation between coliform level and virus concentration in reclaimed water. However, there is no consensus among public health experts regarding the health significance of low levels of viruses in reclaimed water. A significant body of information exists indicating that enteric viruses are removed, destroyed, or inactivated to low or nondetectable levels via wastewater treatment processes including filtration and disinfection (Crook, 1989; *Engineering Science,* 1987; SDLAC, 1977). Difficulties with the identification of specific viruses is discussed in Chap. 3.
Chemical constituents	
Biodegradable organics	Biodegradable organics can create aesthetic and nuisance problems. Organics provide food for microorganisms, adversely affect disinfection processes, make water unsuitable for some industrial or other uses, and may cause acute or chronic health effects if reclaimed water is used for potable purposes.
Total organic carbon, TOC	Total organic carbon (TOC) is the most common monitoring parameter for gross measurement of organic content in reclaimed water used for potable purposes. TOC is used as a measure of treatment process effectiveness. TOC analysis, however, is not a useful predictive tool for indicating very low levels of some health-significant chemicals, and identification of one or more wastewater constituents that can be used as surrogates for nonregulated chemicals is needed.
Nitrates	When applied at excessive levels on land, the nitrate form of nitrogen will readily leach through the soil and may cause groundwater concentrations to exceed drinking water standards. Nitrate and nitrite are also of concern when reclaimed water is used for potable reuse.

Heavy metals	Some heavy metals such as cadmium, copper, molybdenum, nickel, and zinc may accumulate in crops to levels that are toxic to consumers of the crops. Heavy metals in reclaimed water that has received at least secondary treatment are generally within acceptable levels for most uses; however, if industrial wastewater pretreatment programs are not enforced, certain industrial wastewaters discharged to a municipal wastewater collection system may contribute significant amounts of heavy metals.
Hydrogen ion concentration, pH	The pH of wastewater affects disinfection efficiency, coagulation, metal solubility, and alkalinity of soils. Normal pH range in municipal wastewater is 6.5 to 8.5, but some industrial wastes may have pH levels well outside of this range.
Trace constituents	Pharmaceutically active compounds (PhACs), endocrine disrupting compounds (EDCs), personal care products, and other trace constituents have been implicated in adverse effects to frogs, fish, and other aquatic animals. Although a number of trace constituents are removed via conventional treatment, low concentrations of some of them may be present in wastewater effluent. The health risks associated with low concentrations of many of these compounds are unknown; however, they may present a health concern if reclaimed water is used for potable purposes or if reclaimed water used for irrigation or other uses makes its way into groundwater or surface supplies.
Disinfection byproducts	The reaction of chemical oxidants such as chlorine and ozone with organics in water can create a wide range of disinfection byproducts (DBPs), some of which may be harmful to human health if ingested over the long term. The principal DBPs of concern in drinking water are the trihalomethanes, haloacetic acids, bromate, and haloacetonitriles (see Chap. 11).
Total dissolved solids, TDS	A measure of the total ionic constituents in water. High TDS concentrations are of concern in a number of reuse applications including agricultural and landscape irrigation, and industrial applications (see Chaps. 17, 18, and 19).
Physical properties	
Total suspended solids, TSS	Suspended solids can shield microorganisms from disinfectants and react with disinfectants such as chlorine or ozone to lessen disinfection effectiveness. Where ultraviolet (UV) radiation is used as the disinfection process, it is essential to reduce particulate matter to low levels in the wastewater prior to disinfection. The TSS can affect the performance of reuse facilities such as drip irrigation systems.
Turbidity	Turbidity is used as a surrogate measure of suspended solids. Unfortunately, the measurement of turbidity often does not reflect the presence of large particles that can shield microorganisms in the disinfection process.
Temperature	Because wastewater temperatures are generally higher than the ambient environment, the temperature of the water may affect certain reuse applications including accelerated biological growth and scaling in pipes and appurtenances.

wastewater depend on a number of factors, and it is not possible to predict with any degree of assurance what the general characteristics of a particular wastewater will be with respect to infectious agents.

The potential transmission of infectious disease by pathogenic microorganisms is the most common concern associated with nonpotable reuse of treated municipal wastewater. Due to the progress in public health engineering and preventive medicine, waterborne disease outbreaks of epidemic proportions have, to a great extent, been controlled in the United States. However, the potential for disease transmission through the water route has not been eliminated. With a few exceptions, the pathogens of past epidemic history are still present in municipal wastewater today and the status is more one of severance of the transmission chain than a total eradication of the disease agent. Although there have not been any confirmed cases of infectious disease resulting from the use of properly treated reclaimed water in the United States, the potential spread of infectious diseases through inappropriate water reuse remains a public health concern. For nonpotable applications of reclaimed water, regulations and guidelines generally are based on the control of pathogenic organisms, while potable water augmentation requires control of both microbial and chemical constituents. Regulations and guidelines become more stringent and restrictive as the degree of human contact with reclaimed water increases. Health and environmental issues associated with water reuse are discussed further in Chap. 5.

Chemical Constituents

The principal chemical constituents of concern in water reuse applications are also listed in Table 4-2. In general, the chemical constituents identified in Table 4-2 are related to specific reuse applications. Where there is a potential for indirect potable reuse there is concern over the composition of the organic matter, measured as BOD, TOC, or COD, whose specific composition is unknown. In a recent comprehensive study, a concerted effort was undertaken to identify the individual constituents that comprise the organic fraction (Leenheer, 2003). Based on the results of the analytical studies, it was found that the majority of the residual organic matter in treated effluent was comprised of cell fragments of the terpenoid family. For example, some vitamins, hormones, and biological polymers are terpenoids.

Physical Properties

The principal physical properties of concern in water reuse applications are identified in Table 4-2. Total suspended solids and turbidity are related to the treatment processes used to treat wastewater. Turbidity is commonly used both as a surrogate measure of TSS and as a parameter for process control. Unfortunately, turbidity measurements themselves provide little information on the particle size distribution of the particles measured as turbidity, which is important for evaluating disinfection efficacy. Temperature is a factor that is site-specific and its significance will depend on the water reuse application (e.g., snowmaking, snowmelting, and rate of pipe corrosion).

Wastewater Treatment and Water Quality Considerations

To provide assurance that reclaimed water will be produced that is essentially free of measurable levels of pathogens and health-significant chemicals, it is necessary to prescribe both treatment unit processes and operations and acceptable water quality limits. A combination of treatment and quality requirements known to produce reclaimed water of acceptable quality obviates the need to monitor the finished water for some chemical and microbial contaminants. Expensive and time-consuming monitoring of product water for certain constituents of interest, for example, some health-significant chemical constituents or pathogenic microorganisms such as enteric viruses or parasites, may be eliminated without compromising health protection.

Type of Wastewater Treatment

Where expected exposure is incidental or not likely, a low level of wastewater treatment is usually acceptable and undisinfected or disinfected secondary treated effluent may be allowed depending on the type of use. In most states, the definition of secondary treatment means that neither the BOD nor TSS exceed 30 mg/L. A few states use the term "oxidized wastewater" to define secondary treated wastewater, where oxidized wastewater is defined as wastewater in which the organic matter has been stabilized, is nonputrescible, and contains dissolved oxygen. Most state regulations do not require a specific type of secondary treatment (e.g., conventional activated sludge, extended aeration activated sludge, lagoon systems), and various types of secondary treatment may be acceptable. Where public exposure to reclaimed water used for nonpotable applications is expected to occur, tertiary treatment usually is required. Types of acceptable tertiary treatment may include sand filtration, multi-media filtration, membranes, or other methods shown to be effective in reducing particulate and organic matter.

Economic Considerations

It is sometimes argued that inclusion of treatment process requirements in regulations may result in economic infeasibility, that treatment process requirements will stifle the development and implementation of innovative treatment techniques, and, thus, that selection of treatment processes to meet established water quality limits should be left to project proponents. While regulatory agencies do consider economic and technical feasibility during regulation development, their primary responsibility is public health and environmental protection. Thus, regulatory agencies do not compromise health and welfare in the interest of making water reuse projects more economically attractive. States with comprehensive regulations allow alternative methods of treatment provided they are demonstrated—in the opinion of the regulatory agency—to be as effective as those specified in the regulations (Crook, 1998; Crook et al., 2002).

BOD, TSS, and Turbidity Requirements

Most states specify wastewater treatment processes and reclaimed water quality limits for TSS and/or turbidity, total or fecal coliforms, and disinfection. States that have regulations for potable reuse also include limits on chemical constituents that include, but are not limited to, the U.S. EPA drinking water standards. For uses of reclaimed water that require a high quality product water, BOD and TSS limits as low as 5 mg/L are specified in some states. These limits are applicable where filtration or other tertiary treatment processes are used to remove some objectionable constituents and prepare the water for disinfection. Daily sampling for BOD and TSS, using composite samples is usually required, although less frequent sampling is allowed in some states. Not all states include limits for BOD and TSS, and several states specify turbidity requirements in lieu of TSS. Turbidity limits generally are required only for tertiary-treated reclaimed water where human contact is expected or likely. Where necessary, most states require that turbidity be monitored continuously. The compliance point for turbidity usually is just prior to disinfection.

Where specified, limits on turbidity in reclaimed water after filtration range from 1 to 10 NTU, with 2 NTU being a common requirement. California specifies different turbidity requirements depending on the type of tertiary treatment. Where media filtration is the tertiary treatment process, turbidity after filtration cannot exceed an average of 2 NTU within any 24-h period, cannot exceed 5 NTU more than 5 percent of the time within a 24-h period, and cannot exceed 10 NTU at any time. Where membranes are used in lieu of media filtration, turbidity cannot exceed 0.2 NTU more than 5 percent of the time within a 24-h period and cannot exceed 0.5 NTU at any time. The rationale for California's turbidity requirements is discussed in Sec. F-1 in Appendix F.

The Use and Limitations of Indicator Organisms

Because it is impractical to monitor reclaimed water for all pathogenic organisms of concern, surrogate parameters are accepted universally (see Chap. 3, Sec. 3-4). Currently, either total or fecal coliform organisms are the preferred indicator organisms for monitoring reclaimed water in the United States. Regulatory decisions regarding the selection of which coliform group to use are somewhat subjective. Where low levels of coliform organisms are required to indicate the absence of pathogenic bacteria, there is no consensus among microbiologists that the total coliform analysis is superior to the fecal coliform analysis. The use of total coliforms provides an added safety factor that appeals to regulatory agencies that adhere to a conservative approach to water reuse. Other indicator organisms such as enterococci, *E. coli*, *Clostridium perfringens,* and coliphage have been proposed but for various reasons are not recommended or required in any existing water reuse regulations or guidelines in the United States, with the one exception that *E. coli* is used as the indicator in Colorado's reclaimed water regulations.

As analytical detection and identification techniques have improved through the years, it has become apparent that coliforms, by themselves, are inadequate indicators of the presence or concentration of some pathogens, particularly viruses and parasites, as many pathogens have been shown to be more resistant to wastewater treatment than classical microbial indicators such as coliforms. In addition, concerns for pathogenic organisms such as *Giardia lamblia* and *Cryptosporidium parvum,* which may originate from nonhuman sources, have led to questioning the use of indicators that arise primarily from human fecal inputs. Therefore, it is improper to infer that a high level of either total or fecal coliform removal—by itself—is indicative of high levels of pathogen removal from reclaimed water.

Disinfection Requirements

Where chlorine is used as the disinfectant, several states require continuous monitoring of chlorine residual and specify both the chlorine residual and contact time that must be met, particularly for reclaimed water uses where human contact with the water is likely to occur. Required chlorine residuals and disinfection contact times differ substantially from state to state, ranging from 1 to 5 mg/L and 15 to 90 min at peak flow, respectively. Where UV is used for disinfection, most states do not specify UV dosage or design or operating conditions, although some state regulations require compliance with the *Ultraviolet Disinfection Guidelines for Drinking Water and Water Reuse* (NWRI, 2003).

While the need to maintain a chlorine residual in reclaimed water distribution systems to prevent odors, slimes, and bacterial regrowth was recognized early in the development of dual water systems (Okun, 1979); only in the last decade or so have regulatory agencies begun to require such residuals. A few states now require maintenance of a chlorine residual (typically 0.5 or 1.0 mg/L) in distribution systems carrying reclaimed water.

Reclaimed Water Quality Monitoring

Necessary decisions involving monitoring water quality include selection of water quality parameters, numerical limits, sampling method and frequency, and the monitoring compliance point. It is impractical to monitor reclaimed water for all toxic chemicals and pathogenic organisms of concern. Thus, surrogate parameters are necessary and widely accepted. In addition to the previously described issue concerning indicator organisms, important issues include the need to monitor for viruses, and the appropriate parameter for measurement of particulates.

Suspended Solids Monitoring

Because particulate matter has a direct influence on disinfection effectiveness, turbidity or TSS measurements are useful parameters to monitor in wastewater immediately prior to disinfection. Suspended solids measurements typically are performed daily on a 24-h composite sample and produce an average value for the sampling period. A common argument in support of monitoring for suspended solids is that the required sampling frequency for most other parameters is daily on either grab or composite samples, and, therefore, more frequent monitoring for particulate matter is unjustified. Monitoring for TSS is appropriate for reclaimed water that receives only secondary treatment, since the effluent is not intended to be completely free of measurable levels of all pathogens.

Turbidity Monitoring

It is clear that continuously monitored turbidity is superior to daily suspended solids measurements as an aid to treatment performance. Reliable instrumentation is available for continuous online measurement of turbidity, and turbidity monitoring has found wide application as a water quality parameter at water reclamation facilities. Low turbidity or suspended solids values by themselves do not indicate that reclaimed water is devoid of pathogenic microorganisms, and turbidity or suspended solids measurements are not used as an indicator of microbiological quality but rather as a quality criterion for reclaimed water prior to disinfection.

Monitoring Compliance Point

The location of the monitoring point for indicator organism regulatory compliance has been an issue in some states. One viewpoint is that the reclaimed water should meet microbial water quality limits at the point of use. Arguments in favor of this position generally center on the possible regrowth of microorganisms between the treatment plant and the point of reuse. However, restrictive coliform requirements ensure that pathogenic bacteria are destroyed during disinfection and any bacterial regrowth would only be that of nonpathogenic organisms. Viruses require living cells to invade and replicate themselves and do not increase in concentration in the open environment. Similarly, parasites such as *Giardia* and *Cryptosporidium* require a host to reproduce. Many regulatory agencies subscribe to the rationale that any degradation that may occur during storage and distribution would be no different than that which would occur with the use of other water. This approach is not meant to imply that subsequent water quality control should be ignored. Depending on the treatment provided, use of the reclaimed water, and storage and distribution system characteristics, it may be appropriate to maintain a chlorine residual to reduce slime growths in distribution systems, help eliminate musty odors, and provide an added disinfection safety factor. In most states, the monitoring compliance point for indicator organisms is immediately after disinfection.

Groundwater Monitoring

Groundwater monitoring is often required when reclaimed water is used for irrigation or for impoundments that are not sealed to prevent seepage. In general, the groundwater monitoring programs require that one well be placed hydraulically upgradient of the water reuse site to assess background and incoming groundwater conditions within the aquifer in question and one or more wells be placed hydraulically down gradient of the reuse site to monitor compliance with groundwater quality requirements (see Fig. 4-3). Groundwater monitoring programs associated with reclaimed water irrigation generally focus on water quality in shallow aquifers. Sampling parameters and frequency of sampling are considered generally on a case-by-case basis.

Storage Requirements

Current regulations and guidelines regarding storage requirements are based primarily upon the need to limit or prevent surface water discharge and are not related to storage required to meet diurnal or seasonal variations in supply and demand. Storage requirements vary from state to state and are dependent generally upon geographic location, climate, and site conditions. A minimum storage volume equal to 3 d of the average design flow is typical in water-short states with warm climates, while more than 200 d of storage are required in some northern states because of the high number of nonirrigation days due to high rainfall or freezing temperatures.

Most states that specify storage requirements do not differentiate between operational and seasonal storage. The majority of states that have storage requirements in their regulations

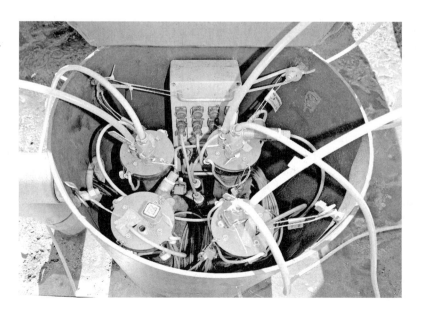

Figure 4-3
Groundwater monitoring (wellhead sampler) and instrumentation.

or guidelines require that a water balance be performed on the water reuse system, taking into account all inputs and outputs of water to the system based on a specified rainfall recurrence interval.

Reclaimed Water Application Rates

Most state regulations do not include requirements or recommendations regarding reclaimed water irrigation application rates, as these are based on plant or crop irrigated and site-specific conditions. Of the states that do recommend application rates, the maximum recommended hydraulic loading rate typically is 50 mm/wk (2.5 in./wk).

Aerosols and Windborne Sprays

Exposure to reclaimed water in aerosols and wind-borne sprays has often been cited as a public health concern (see Fig. 4-4). Aerosols are particles suspended in air that range in size from 0.01 to 50 μm. Pathogen levels in aerosols caused by spraying of reclaimed water are a function of their concentration in the applied water, droplet size, and the aerosolization efficiency of the spray process. Typically, one percent or less of the sprayed water is aerosolized. The possibility of disease transmission depends on several factors, including degree of wastewater treatment, extent of aerosol or windblown spray formation and travel, proximity to populated areas or areas accessible to the public, prevailing climatic conditions, and design of the irrigation system. Infection or disease may be contracted directly by inhalation or indirectly by aerosols containing infectious organisms that are deposited on surfaces such as food, vegetation, and clothes. The infective dose of some pathogens is lower for respiratory tract infections than for infections via the gastrointestinal tract; thus, for some pathogens, inhalation may be a more likely route for disease transmission than either contact or ingestion.

Pathogen Survival

Some pathogenic organisms, such as enteroviruses and *Salmonella*, appear to survive the wastewater aerosolization process much better than indicator organisms (Teltsch et al., 1980). If pathogens are present in aerosols, they generally remain viable and travel

(a)

(b)

Figure 4-4
Public health concerns related to the exposure to reclaimed water in aerosols and windborne sprays: (a) from agricultural irrigation and (b) from landscape irrigation.

farther with increased wind velocity, increased relative humidity, lower temperature, and lower solar radiation. Aerosols can be transmitted for several hundred meters under optimum conditions.

Little research has been conducted in the last two decades on aerosol formation and pathogen transport resulting from spray irrigation with wastewater or reclaimed water. Studies directed at residents in communities subjected to aerosols from municipal wastewater treatment plants have not detected any definitive correlation between exposure to aerosols and disease (Camann et al., 1980; Fannin et al., 1980; Johnson et al., 1980). While bacteria and viruses have been found in aerosols emitted by spray irrigation systems using untreated and poorly treated wastewater, there have not been any documented disease outbreaks resulting from spray irrigation with disinfected reclaimed water, and studies indicate that the health risk associated with aerosols from spray irrigation sites using disinfected reclaimed water is low (U.S. EPA, 1980; U.S. EPA, 1981).

Limiting Exposure
The general practice is to limit exposure to aerosols and airborne sprays produced from reclaimed water that is not heavily disinfected through design or operational controls. Design features include setback distances, which are sometimes called buffer zones; windbreaks such as trees or walls around irrigated areas; low-pressure irrigation systems and/or spray nozzles with large orifices to reduce the formation of fine mist; low-profile sprinklers; and surface or subsurface methods of irrigation. Operational measures include spraying only during periods of low wind velocity, not spraying when wind is blowing toward sensitive areas subject to aerosol drift or windblown spray, and irrigating at off-hours when the public or employees would not be in areas subject to aerosols or spray.

Setback Distances
Windblown spray of reclaimed water droplets may present a greater potential health hazard than that from aerosols. The intent of setback distances is to prevent excessive human contact with the reclaimed water or to prevent potential contamination of potable water supply sources. Although predictive models have been developed to estimate microorganism concentrations in aerosols or larger water droplets resulting from spray irrigation of wastewater, setback distances are somewhat arbitrarily determined by regulatory agencies based on experience and engineering judgment.

Many states have established setback distances between reclaimed water use areas and surface waters, potable water supply wells, or areas accessible to the public. Setbacks are usually required where reclaimed water is used for spray irrigation, cooling water in towers, and other areas where spray or mist is formed. Setbacks may also be required at irrigation or impoundment sites to prevent percolated reclaimed water from reaching potable water supply wells. Setback distances vary depending on the quality of reclaimed water, type of reuse, method of application, and purpose of the setback, for example, to avoid human contact with the water or protect potable water sources from contamination. Setback distances, where required, vary considerably from state to state, and range from 15 m (50 ft) to as much as 240 m (800 ft). Some states do not require setback distances from irrigated areas to areas accessible to the public if a high level of treatment and disinfection is provided. Setback distances required in California are given in Sec. F-1 in Appendix F.

4-4 REGULATORY CONSIDERATIONS FOR SPECIFIC WATER REUSE APPLICATIONS

In addition to the factors discussed in Sec. 4-2, there are a number of considerations for specific reuse applications, including those for (1) agricultural irrigation, (2) landscape irrigation, (3) dual distribution systems and in-building uses, (4) impoundments, (5) industrial uses, (6) miscellaneous nonpotable uses, and (7) groundwater recharge. These considerations are discussed briefly in the following sections. Regulatory aspects of indirect and direct potable reuse, which are evolving and challenging water reuse applications, are discussed separately in Sec. 4-5.

Agricultural Irrigation

The major health concern associated with using reclaimed water for agricultural irrigation is the potential for contamination of food crops and resulting adverse effects to consumers of the crops (see Chap. 17). Important considerations for agricultural irrigation, as discussed below, include: (1) the direct and indirect contamination of crops, (2) the survival of pathogenic constituents, (3) processing of crops before distribution, (4) the uptake of trace chemical constituents, and (5) specifying the level of treatment.

Crop Contamination
Wastewater containing microbial pathogens can contaminate crops directly by contact during irrigation or indirectly as a result of soil contact. Spray irrigation of food crops that grow above the ground surface and are eaten uncooked, requires more stringent requirements because of the direct contact between the reclaimed water and the crops. Spray or surface irrigation of root crops, such as carrots, beets, and onions also results in direct contact between the crop and reclaimed water. Indirect contamination of crops can occur by blowing dust or by workers, birds, and insects that convey organisms from irrigation water or soil to the edible portion of the crop.

Concern for Pathogens
Organisms contaminating food crops may remain viable on food surfaces. Many pathogens can survive for extended periods on plants and in soil, and simply providing extensive time periods between irrigation and crop harvest, or providing commercial storage before public sale cannot be relied upon to destroy all pathogens.

Crop Processing
If reclaimed water used to irrigate food crops is not highly treated to destroy pathogens, physical or chemical commercial processing should be performed before the crops are sold for human consumption. Transmission of infectious organisms may occur by handling crops that are contaminated and from selling or distributing the crops before processing.

Trace Constituents
Trace chemical constituents may be of concern due to the potential for uptake through the roots from the applied water or the soil and by foliar uptake. Some constituents are

known to accumulate in particular crops, thus presenting potential health hazards to both grazing animals and/or humans; however, it has been postulated that most trace organic compounds are too large to pass through the semipermeable membrane of plant roots (U.S. EPA, 1981; NRC, 1996).

Level of Treatment

The level of treatment depends on water quality, type of crop irrigated (food, nonfood, eaten uncooked, cooked before consumption), method of irrigation employed (spray, surface, or subsurface), and degree of contact between the crop and reclaimed water.

Landscape Irrigation

Landscape irrigation involves the irrigation of golf courses, parks, cemeteries, school grounds, freeway medians, residential lawns, and similar areas (see Chap. 18). Depending on the area being irrigated, its location relative to populated areas, and the extent of public access or use of the grounds, water quality requirements and operational controls placed on the system may differ. Considerations for landscape irrigation, as discussed below, include: (1) level of public access, (2) accumulation of trace chemical constituents, and (3) use area controls.

Public Access

Landscape irrigation frequently takes place in urban areas or on grounds frequented by the public where control over the use of the reclaimed water is more critical than where public access is limited or prohibited (see Fig. 4-5). For example, the irrigation of landscaped areas where children congregate such as parks and playgrounds, may result in contact or ingestion of turf or soil. Irrigation of areas not subject to public access have limited potential for creating public health problems, whereas the need to reduce the level of microbial pathogens in the irrigation water becomes more important as the expected level of direct or indirect human contact with reclaimed water increases.

(a) (b)

Figure 4-5

Examples of unrestricted public access to the areas using highly treated reclaimed water: (a) golf course irrigation in California and (b) artificial stream (known as *Seseragi* in Japanese) in Sapporo, Japan (Courtesy of City of Sapporo, Japan).

Trace Constituents
Unsupported concerns have been voiced in recent years that trace chemical constituents such as PhACs and EDCs may be present in irrigation water. These constituents may migrate to groundwater used as drinking water supply sources or accumulate to health-significant levels in turf or soil and be ingested inadvertently or contacted by children.

Use Area Controls
Use area controls need to be imposed at open access landscape irrigation sites as an added safety precaution to protect both children and adults who frequent the irrigation sites. Useful controls may include signs warning the public that the area is irrigated with reclaimed water, protecting drinking water fountains from direct contact with the irrigation water, eliminating the potential for ponding of reclaimed water, confining the reclaimed water and spray to the designated irrigation site(s), and irrigating only during off-hours. All use areas should be subjected to routine surveillance by the producer or user of the reclaimed water to ensure adherence to all use area controls.

Dual Distribution Systems and In-building Uses

Increasing use of reclaimed water for multiple uses (e.g., residential and other irrigation, ornamental fountains, toilet and urinal flushing, car washes, and commercial laundries) in urban areas has resulted in the development of several large dual water systems that distribute both potable water and reclaimed water to the same service area. Important considerations include (1) identification of reclaimed water lines and appurtenances, (2) cross-connection control, (3) reclaimed water quality, and (4) distribution system design and construction.

Identification Considerations
Identification of reclaimed water lines and appurtenances is usually accomplished by color-coding and labeling. Proper identification of building piping used to transport reclaimed water is necessary for maintenance activities and for avoiding cross-connections. Identification of pipelines and plumbing systems for reclaimed water service is discussed in Chaps. 14 and 15.

Cross-Connection Control
Reclaimed water used inside buildings for toilet and urinal flushing or for fire protection presents cross-connection control concerns. Although such uses do not result in frequent human contact with the reclaimed water, inadvertent cross-connections to potable water systems have occurred; thus, highly disinfected reclaimed water is needed for those uses to reduce the potential for disease transmission upon inadvertent ingestion of small quantities of reclaimed water.

Regulations often address identification of transmission and distribution lines and appurtenances via color-coding, taping, or other means; separation of reclaimed water and potable water lines; allowable pressures; surveillance; and backflow prevention devices (see Chaps. 14 and 15). At use areas that receive both potable and reclaimed water, backflow prevention devices are usually required on the potable water supply line to each site to reduce the potential of contaminating the potable drinking water system in the event of a cross-connection at a use area. Direct connections between reclaimed water and potable water lines are not allowed in any state.

California's Water Recycling Criteria require compliance with the California Department of Health Services cross-connection control regulations (State of California, 2000). Those regulations require that water systems serving residences through a dual-water system that uses reclaimed water for landscape irrigation must, as a minimum, be protected by a double-check valve assembly backflow preventer. The same requirement applies to a public water system in buildings using reclaimed water in a separate piping system within buildings for fire protection. A reduced-pressure principle backflow–prevention device is required as a minimum to protect the potable system at sites other than those mentioned earlier. An air gap separation is required where a public water system is used to supplement a reclaimed water supply.

California's criteria for dual-plumbed systems within buildings include the following requirements:

- Internal use of reclaimed water within any individually owned residential unit, including multiplexes or condominiums is prohibited
- Submission of a report that includes a detailed description of the intended use area, plans and specifications, and cross-connection control provisions and testing procedures
- Testing for possible cross-connections at least every four years
- Notification of any incidence of backflow from the reclaimed water system into the potable water system within 24 h of discovery
- Conformance to the Department of Health Services (DHS) cross-connection control regulations
- Facilities that produce or process food products or beverages can use reclaimed water internally only for fire suppression systems

Reclaimed Water Quality
Where there is likely to be any human contact with the reclaimed water, advanced treatment and a high level of disinfection are needed to minimize health risks. Chemical constituents in highly treated effluent are generally not a problem for most types of nonpotable urban reuse. Nutrients in reclaimed water—particularly nitrogen and phosphorous—may stimulate biofilm growth, but, if necessary, can be controlled or removed by advanced wastewater treatment processes (see Chap. 7).

Distribution System Features
Distribution system design and construction features include monitoring, storage, use area controls, and management (see Chap. 14).

Impoundments

Impoundments may serve a variety of functions from aesthetic, noncontact uses, to boating, fishing, and swimming. As with other uses of reclaimed water, the level of treatment needed varies with the intended use of the water and increases as the potential for human contact increases. The important considerations for impoundments include: (1) impoundment use and (2) water quality issues.

Impoundment Use
Reclaimed water impoundments may be categorized according to whether they are used for aesthetic or recreational purposes. Recreational impoundments can be subdivided

into either restricted (nonbody contact only) or nonrestricted (body contact allowed) impoundments. Nonbody contact includes activities such as boating and fishing where there is only incidental contact with the reclaimed water, while body contact includes full body immersion.

Water Quality Issues
The water quality requirements for an impoundment vary depending on the intended use. Nonrestricted recreational impoundments (and reflecting pools and decorative fountains) should not contain chemical substances that are toxic following inadvertent ingestion or be irritating to the eyes or skin, and the water should be safe from a microbiological standpoint. Other concerns are temperature, pH, chemical constituents, aquatic growths, and clarity. Clarity is important for several reasons, including safety, visual appeal, and recreational enjoyment. Reclaimed water used for recreational impoundments where fishing and boating are allowed should not contain high levels of pathogenic microorganisms or heavy metals that accumulate in fish to levels that present health risks to the consumers of the fish. If fish, shellfish, or plants exposed to reclaimed water are used for human consumption, both the microbiological and chemical quality of the source water should be thoroughly assessed for possible bioaccumulation of toxic contaminants through the food chain.

Industrial Uses

Reclaimed water from conventional wastewater treatment processes is of adequate quality for many industrial applications that can tolerate water of less than potable quality. Reclaimed water also has the important advantage of being a reliable supply even in drought years. Industrial uses of reclaimed water include cooling, process water, stack scrubbing, boiler feed, wash water, transport of material, and as an ingredient in a nonfood-related product (see Chap. 19). Regulatory considerations for reuse of water in industrial applications include: (1) generation of aerosols and (2) safety of manufactured products.

Aerosols
Pathogenic microorganisms in reclaimed water used in cooling towers present potential hazards to workers and to the public in the vicinity of cooling towers from aerosols and windblown spray. In practice, however, biocides and other chemicals are usually added to all cooling waters onsite to prevent slimes and otherwise inhibit microbiological activity, which has the secondary effect of eliminating or greatly diminishing the potential health hazard associated with aerosols or windblown spray.

Safety of Manufactured Products
The suitability of reclaimed water for use in industrial processes depends on the particular use and the potential for worker or public contact with the water. Low quality reclaimed water should not be used in the manufacture of—or incorporated into—products subject to contact or ingestion upon sale to the public unless processing is sufficient to assure that microbial pathogens or health-significant chemical constituents are eliminated or reduced to acceptable limits.

Other Nonpotable Uses

Less common uses of reclaimed water include flushing sanitary sewers, street cleaning, dust control, soil compaction, making concrete, snowmaking, snowmelting, decorative fountains, commercial laundries, commercial car washes, equipment washing, and fire protection systems (see Chap. 20). While each application must be evaluated on a case-by-case basis, the common regulatory considerations include: (1) the amount of human contact and (2) potential environmental impacts.

Health Considerations
The expected degree of human contact with reclaimed water determines the appropriate level of disinfection. Minimal disinfection of reclaimed water is needed for uses where there is little or no expected human contact with the water, such as flushing sanitary sewers or making concrete, whereas uses such as snowmaking and vehicle washing are likely to result in contact with the reclaimed water, thus necessitating a considerably higher level of disinfection.

Environmental Impacts
Environmental impacts from reclaimed water depend on the specific application and should be evaluated on a case-by-case basis. For example, additional treatment, such as nutrient removal, may be needed for reclaimed water used to make snow at ski resorts in the event that the snowmelt runs off into the pristine environment or percolates to potable water supplies.

Groundwater Recharge

The regulatory considerations related to groundwater recharge depend on: (1) characterization of the aquifer, (2) recharge of nonpotable aquifers, (3) recharge of potable aquifers, (4) design of a soil-aquifer treatment (SAT) process, and (5) issues associated with direct aquifer injection (see Chap. 22).

Aquifer Characterization
Health concerns pervade almost all recharge projects, because rarely are the boundaries between potable and nonpotable aquifers well defined. In the event that distinct boundaries cannot be identified, a conservative assumption is that the aquifer will be used for potable purposes and appropriate treatment levels for recharging a potable aquifer should be used.

Recharge of Aquifers Used for Nonpotable Purposes
Where reclaimed water is recharged into nonpotable aquifers and there is no possibility of the water migrating to potable aquifers, health concerns are mitigated, although the reclaimed water, upon extraction, is subject to the appropriate water quality requirements for the subsequent use of the water.

Recharge of Aquifers Used for Potable Purposes
Groundwater recharge of potable aquifers is problematic, as many utilities distribute drinking water from potable water supply wells with little or no treatment. As a consequence, it is necessary for reclaimed water to meet all drinking water standards—and water quality limits for potentially toxic unregulated chemical contaminants and microbial pathogens—prior to extraction.

Soil-Aquifer Treatment
Surface spreading provides additional treatment of the reclaimed water as it percolates through the vadose zone. In some cases, all applicable standards may be met prior to mixing with the native groundwater. If SAT is intended to replace conventional media

filtration as one of the required wastewater treatment processes prior to reuse, controls or limits may be placed on the percolation rate and required depth of the vadose zone. Extracted water would have to meet all reclaimed water quality requirements specified for its subsequent use, which may result in the need for disinfection following SAT.

Direct Aquifer Injection

Injection directly into a confined aquifer provides little opportunity for additional water quality improvement in the subsurface, resulting in the need to meet all water quality limits prior to injection via incorporation of advanced wastewater treatment processes. Regulatory considerations relating to groundwater recharge of potable aquifers are complex and are discussed in detail in Chap. 22.

4-5 REGULATORY CONSIDERATIONS FOR INDIRECT POTABLE REUSE

Planned indirect potable reuse involves the use of reclaimed water to augment surface water that is used as a source of drinking water supply. The water typically receives additional treatment prior to distribution as drinking water. In direct potable reuse, by contrast reclaimed water is introduced directly into a drinking water distribution system. However, direct potable reuse is not practiced in the United States at present.

Use of the Most Protected Water Source

The use of natural waters derived from the most protected source as water supply is practiced traditionally as much as practicable; thus, there are relatively few formal and planned indirect potable reuse projects. The principle of using protected water sources has guided the selection of potable water supplies for almost 150 yr in the United States and was well-stated in the *1962 Public Health Service Drinking Water Standards*: "The water supply should be taken from the most desirable source which is feasible, and effort should be made to prevent or control pollution of the source." (U.S. Department of Health, Education, and Welfare, 1962). This finding was reaffirmed by U.S. EPA in 1975 in its *Primary Drinking Water Regulations*, ". . . priority should be given to selection of the purest source. Polluted sources should not be used unless other sources are economically unavailable . . ." (U.S. EPA, 1975). Public health concerns related to potable reuse centers on water quality, treatment reliability, aesthetics, and the difficulty of identifying and estimating human exposures to the potentially toxic chemicals and pathogens that may be present. To some extent the assessment of possible health risks can rely on the vast body of knowledge that has been developed for drinking water supplies using conventional source water containing substantial discharges from municipal wastewater treatment plants.

Influence of the Two Water Acts

The Safe Drinking Water Act (SDWA) requirements assure a safe drinking water when relatively uncontaminated, protected water sources are used. Conversely, the federal Clean Water Act (CWA) is intended to eliminate pollution and maintain the physical, chemical, and biological integrity of the nation's waters, but its water quality limits are not reflective of drinking water standards (see Section 2-3 in Chap. 2). Thus, the provisions of the CWA and SDWA are insufficient to address all the public health concerns associated with municipal wastewater constituents since neither the CWA nor the SDWA establish standards for all of the potentially harmful constituents that may be present in wastewater. The level of wastewater contribution that triggers additional constituent controls has not been identified in law or in the literature. The threshold level in any particular case depends on a number of factors, including: the industrial, commercial, research, and medical contributions to the municipal wastewater that may present unique problems; the wastewater treatment processes utilized; and the natural barriers to contaminant transport that exist between the waste discharge and the drinking water system surface intake or well. Thus, reclaimed water used for planned indirect potable reuse may have to meet additional water quality criteria for known or suspected microbial and chemical constituents of concern than those that normally apply to drinking water and wastewater discharges.

Indirect potable reuse is practiced where treated wastewater is discharged into a water course, a raw water reservoir, or an underground aquifer and withdrawn downstream or down gradient at a later time for treatment and subsequent distribution as drinking water. Potable reuse, on its face, is less desirable than using a higher quality source water for drinking, and reclaimed water is inherently suspect as a source water supply as untreated municipal wastewater contains potentially harmful contaminants, including pathogens, heavy metals, and organic compounds. Reclaimed water used for potable reuse ultimately must meet all physical, chemical, radiological, and microbiological drinking water standards. However, drinking water standards are not intended to apply to contaminated source waters that may contain unregulated constituents that are known or suspected to be harmful upon ingestion. Thus, the drinking water standards cannot be relied on as the sole standard of safety.

Concerns for Trace Chemical Constituents and Pathogens

Most chemical constituents found in treated municipal wastewater are present at concentrations that are of concern only with chronic exposure. Thus, these constituents are of particular importance where treated wastewater discharges persist for extended periods of time. However, constituents found in treated wastewater at concentrations that are high relative to those considered safe for potable use may present a health-risk due to acute exposure, even at lower discharge rates.

Quality standards have been established for many inorganic constituents and treatment and analytical technology has demonstrated the capability to identify, quantify, and control these substances. Similarly, available technology is capable of eliminating pathogenic agents from contaminated waters. On the basis of available information, there is no indication that health risks from using highly treated reclaimed water for potable purposes are greater than those from using existing water supplies (NRC, 1994). However, unanswered questions remain with organic constituents, due mainly to their potential large numbers and unresolved health-risk potential resulting from long-term ingestion of low concentrations.

Studies have been made on the chemical and microbiological characteristics of reclaimed water, although they are limited in number and scope. Several studies have indicated that reclaimed water can meet drinking water standards and often exceed such standards. Such findings lead some experts to conclude that reclaimed water is acceptable as a drinking water source. Other experts disagree stating, for example, that: (1) disinfection of reclaimed water may create different and unidentified disinfection byproducts than those found in conventional water supplies; (2) less than 25 percent by weight of the organic compounds in reclaimed water have been identified and the health effects of only a few of the individual constituents have been determined; (3) the health effects of mixtures of two or more of the thousands of compounds potentially present in reclaimed water are not characterized easily; and (4) throughout the whole process, there is increased reliance on technology and management (see Chap. 3).

Assessment of Health Risks

The assessment of health risks associated with indirect potable reuse is not definitive due to limited chemical and toxicological data and inherent limitations in the available toxicological and epidemiological methods. The results of epidemiological studies directed at drinking water have generally been inconclusive, although the hypothesis that there may be a health risk is still present. Recognizing the limitations of epidemiological studies because of the many confounding variables, health-related studies do provide a basis for concern for potable use of reclaimed water. In addition, the limited data and extrapolation methodologies used in toxicological assessments (see Chap. 5) provide a source of limitations and uncertainties in the overall risk characterization. In these circumstances, the readers are reminded of the precautionary principle discussed in Chap. 1.

A multiple barrier system using demonstrated treatment technologies is essential to assure that reclaimed water used to augment drinking water supplies is at least as safe and reliable as other alternative supplies. Existing treatment technology is able to produce reclaimed water that meets all current drinking water standards. However, in consideration of the source water, meeting drinking water standards does not necessarily indicate that the water is safe. Intensive water quality monitoring and contingency plans for response to system failures should be a part of a conservative regulatory approach. Monitoring programs should be adequate to verify the performance of treatment processes and to detect potentially harmful regulated and unregulated contaminants. Monitoring is a particular concern for membrane processes, where development of online water quality monitoring is needed to detect contaminant breakthrough via leaking seals, imperfections or holes in membranes, or improper operating conditions (see Chaps. 8 and 9).

4-6 STATE WATER REUSE REGULATIONS

There are no federal regulations governing water reclamation and reuse in the United States; thus, regulations are developed and implemented at the state government level. The lack of federal regulations has resulted in differing standards among states that have developed water reuse regulations. In the 1990s, several states adopted or revised their respective regulations, and it was common practice to base water reuse regulations on those of states that had comprehensive regulations, guidelines, and background information to support them. The *Guidelines for Water Reuse* (U.S. EPA, 1992; 2004) were also used as a resource by states that had limited or no regulations or guidelines. Since the guidelines were published, interest in water reuse has increased in several states that previously did not have water reuse regulations.

At present, no states have regulations that cover all potential uses of reclaimed water, but several states have extensive regulations that prescribe requirements for a wide range of end uses of the reclaimed water. Other states have regulations or guidelines that focus on land treatment of wastewater effluent, emphasizing additional treatment or effluent disposal rather than beneficial reuse, even though the effluent may be used for irrigation of agricultural sites or public access lands.

Status of Water Reuse Regulations and Guidelines

The status and summary of water reclamation and reuse regulations and guidelines in the United States as of 2004 have been documented in the *Guidelines for Water Reuse* (U.S. EPA, 2004) and are provided in Table 4-3. The absence of state regulations and guidelines for specific reuse applications does not necessarily prohibit those applications; many states evaluate specific types of water reuse on a case-by-case basis. Based on the data in Table 4-3, 25 states have adopted regulations regarding the use of reclaimed water, 16 states have guidelines or design standards, and nine states have no regulations or guidelines. These data are somewhat misleading, as they include regulations and guidelines directed at land disposal of effluent or land application of wastewater intended primarily as a disposal mechanism rather than for beneficial reuse.

Regulations and Guidelines for Specific Reuse Applications

The number of states with regulations or guidelines for each type of reuse is summarized in Table 4-4, which has been adapted from the *Guidelines for Water Reuse*. As indicated in Table 4-4, agricultural and landscape irrigation represent the reclaimed water uses most commonly regulated, and many states have implemented regulations that apply only to those types of use. As noted above, these data include state regulations that pertain to land disposal of effluent or land application of wastewater intended primarily as a disposal mechanism rather than beneficial reuse. The standards in states having the most reuse experience tend to be more stringent than those in states with fewer reuse projects. States that have water reuse regulations or guidelines typically set standards for reclaimed water quality and specify minimum treatment requirements; although a few states, such as Texas and New Mexico, do not prescribe treatment processes and rely solely on water quality limits.

Variations amongst State Regulations

In the past, most state water reuse regulations were developed in response to a need to regulate a growing number of water reuse projects in the particular state. Recently, some states that currently have few reuse projects have taken a proactive approach, and have adopted criteria which tend to encourage implementation of projects. Arizona, California, Florida, and Texas, which have had comprehensive criteria for a number of years, have revised their water reuse regulations within the last ten years to reflect additional reclaimed water uses, advances in wastewater treatment technology, and increased knowledge in the areas of microbiology and public health protection. The variations and inconsistencies among state regulations are illustrated in Table 4-5, which includes

Table 4-3
Summary of state water reuse regulations and guidelines for nonpotable reuse applications[a]

State	Regulations	Guidelines	No regulations or guidelines	Unrestricted nonpotable urban uses	Restricted nonpotable urban uses	Agricultural irrigation of food crops	Agricultural irrigation of nonfood crops	Unrestricted recreational impoundments	Restricted recreational impoundments	Environmental uses	Industrial uses
Alabama		•					•				
Alaska	•				•	•	•				
Arizona	•			•	•	•	•	•	•		
Arkansas	•						•				
California	•			•	•	•	•		•		•
Colorado	•				•		•				•
Connecticut			•								
Delaware	•			•	•	•	•				
Florida	•			•	•	•	•		•	•	•
Georgia		•			•		•				
Hawaii	•			•	•	•	•				
Idaho	•			•	•	•	•				
Illinois	•				•						
Indiana	•			•	•		•				
Iowa	•			•	•	•	•				
Kansas					•						
Kentucky			•								
Louisiana			•								
Maine			•								
Maryland		•		•	•						
Massachusetts		•			•						
Michigan			•								
Minnesota			•								
Mississippi	•					•	•				
Missouri		•		•	•						
Montana	•					•	•				
Nebraska	•				•		•				
Nevada	•			•	•	•	•	•	•		
New Hampshire			•								

(Continued)

Table 4-3
Summary of state water reuse regulations and guidelines for nonpotable reuse applications[a]

State	Regulations	Guidelines	No regulations or guidelines	Unrestricted nonpotable urban uses	Restricted nonpotable urban uses	Agricultural irrigation of food crops	Agricultural irrigation of nonfood crops	Unrestricted recreational impoundments	Restricted recreational impoundments	Environmental uses	Industrial uses
New Jersey		•			•		•				•
New Mexico	•	•		•	•	•	•				
New York		•		•	•		•				•
North Carolina	•			•			•				
North Dakota		•			•	•	•				
Ohio	•			•	•		•				
Oklahoma	•			•		•	•				
Oregon	•			•	•		•	•	•		
Pennsylvania		•					•				
Rhode Island			•								
South Carolina		•					•				
South Dakota	•						•				
Tennessee	•				•		•				
Texas	•			•	•	•	•	•	•	•	•
Utah	•			•	•	•	•	•	•		•
Vermont		•					•				
Virginia			•								
Washington	•			•	•	•	•	•	•	•	•
West Virginia	•				•		•				
Wisconsin					•		•				
Wyoming	•			•		•	•				

[a] Adapted from U.S. EPA (2004).

Table 4-4

Number of states with water reuse regulations or guidelines for different types of use[a]

Type of use	Number of states with regulations or guidelines	Description
Unrestricted urban water reuse	28	Irrigation of areas in which public access is not restricted, such as parks, playgrounds, school yards, and residences. Toilet flushing, air conditioning, fire protection, construction, cleansing, ornamental fountains, and aesthetic impoundments.
Irrigation	28	
Toilet flushing	10	
Fire protection	9	
Construction	9	
Landscape impoundment	11	
Street cleaning	6	
Restricted urban water reuse	34	Irrigation of areas in which public access can be controlled, such as golf courses, cemeteries, and highway medians.
Agricultural irrigation of food crops	21	Irrigation of food crops which are intended for human consumption. Food crop is processed. Food crop is consumed uncooked.
Agricultural irrigation of nonfood crops	40	Irrigation of fodder, fiber, and seed crops, pasture land, commercial nurseries, and sod farms.
Unrestricted recreational water reuse	7	An impoundment of water in which no limitations are imposed on body-contact water recreational activities.
Restricted recreational water reuse	9	An impoundment of reclaimed water in which recreation is limited to fishing, boating, and other noncontact recreational activities.
Environmental water reuse	3	Reclaimed water used to create manmade wetlands, enhance natural wetlands, and to sustain stream flows.
Industrial water reuse	9	Reclaimed water used in industrial facilities primarily for cooling system makeup water, boiler feedwater, process water, and general washdown and cleansing.
Groundwater recharge	5	Using infiltration basins, percolation ponds or injection wells, reclaimed water is used to recharge groundwater aquifers.
Indirect potable reuse	5	The intentional discharge of highly treated reclaimed water into surface waters or groundwater that will be used as a source of potable water supply.

[a]Adapted from U.S. EPA (2004).

Table 4-5
Examples of state water reuse regulations for selected nonpotable applications

State	Fodder crop irrigation[a]		Processed food crop irrigation[b]		Food crop irrigation[c]		Restricted recreational impoundments[d]	
	Quality limits	Treatment required	Quality limits	Treatment required	Quality limits	Treatment required	Quality limits	Treatment required
Arizona	1,000 fecal coli/100 mL	Secondary	Not covered	Not covered	No detect. fecal coli/100 mL 2 NTU	Secondary Filtration Disinfection	No detect. fecal coli/100 mL 2 NTU	Secondary Filtration Disinfection
California	Not specified	Oxidation	Not specified	Oxidation	2.2 total coli/100 mL 2 NTU	Oxidation Coagulation[e] Filtration Disinfection	2.2 total coli/100 mL	Oxidation Disinfection
Colorado	Not covered	Not covered	Not covered	Not covered	Not covered	Not covered	Not covered	Not covered
Florida	200 fecal coli/100 mL 20 mg/L CBOD 20 mg/L TSS	Secondary Disinfection	No detect. fecal coli/100 mL 20 mg/L CBOD 5 mg/L TSS	Secondary Filtration Disinfection	Use prohibited	Use prohibited	No detect. fecal coli/100 mL 20 mg/L CBOD 5 mg/L TSS	Secondary Filtration Disinfection
New Mexico (Policy)	1000 fecal coli/100 mL 75 mg/L TSS 30 mg/L BOD	Not specified	Not covered	Not covered	Use prohibited	Use prohibited	100 fecal coli/100 mL 30 mg/L BOD 30 mg/L TSS	Not specified

State	Quality	Treatment	Quality	Treatment	Quality	Treatment	Quality	Treatment
Utah	200 fecal coli/100 mL, 25 mg/L BOD, 25 mg/L TSS	Secondary Disinfection	No detect. fecal coli/100 mL, 10 mg/L BOD, 2 NTU	Secondary Filtration Disinfection	No detect. fecal coli/100 mL, 10 mg/L BOD, 2 NTU	Secondary Filtration Disinfection	200 fecal coli/100 mL, 25 mg/L BOD, 25 mg/L TSS	Secondary Disinfection
Texas	200 fecal coli/100 mL, 20 mg/L BOD, 15 mg/L CBOD	Not specified	200 fecal coli/100 mL, 20 mg/L BOD, 15 mg/L CBOD	Secondary Filtration Disinfection	Use prohibited	Use prohibited	20 fecal coli/100 mL, 3 NTU, 5 mg/L BOD or CBOD	Not specified
Washington	240 total coli/100 mL	Oxidation Disinfection	240 total coli/100 mL	Oxidation Disinfection	2.2 total coli/100 mL, 2 NTU	Oxidation Coagulation Filtration Disinfection	2.2 total coli/100 mL	Oxidation Disinfection

[a] In some states more restrictive requirements apply where milking animals are allowed to graze on pasture irrigated with reclaimed water.

[b] Physical or chemical processing sufficient to destroy pathogenic microorganisms. Less restrictive requirements may apply where there is no direct contact between reclaimed water and the edible portion of the crop.

[c] Food crops eaten raw where there is direct contact between reclaimed water and the edible portion of the crop.

[d] Recreation is limited to fishing, boating, and other nonbody contact activities.

[e] Not needed if filter effluent turbidity does not exceed 2 NTU, the turbidity of the influent to the filters is continually measured, the influent turbidity does not exceed 5 NTU for more than 15 min and never exceeds 10 NTU, and there is the capability to automatically activate chemical addition or divert the wastewater should the filter influent turbidity exceed 5 NTU for more than 15 min.

examples of several states' reclaimed water regulations for uses ranging from fodder crop irrigation to toilet and urinal flushing in buildings. The reader is referred to the *Guidelines for Water Reuse* (U.S. EPA, 2004) for a complete tabulation of all state water reuse regulations. Some of the notable variations among state regulations are highlighted below:

Coliform Bacteria Limits   Most states use fecal coliform organisms as the indicator organism for microbial pathogens in reclaimed water, while a few states use total coliform. Fecal or total coliform limits depend on use of the water and are highly variable among states. Arizona, Florida, and some other states' regulations are similar to, or based on, the *Guidelines for Water Reuse* and use fecal coliform organisms as the indicator organism. In those states regulations typically require that reclaimed water has no detectable fecal coliform/100 mL for high level nonpotable applications and does not exceed 200 fecal coliform/100 mL for uses where human contact is minimal.

States that use total coliform as the indicator organism require that the number of total coliform organisms not exceed 2.2/100 mL for high level uses and either 23 or 240/100 mL for uses where there is no or minimal human contact with the water. Higher single sample maximum coliform limits are allowed in several states. Regulatory compliance varies in different states, but usually is based on median or geometric mean values over a given time period. Coliform samples are usually required to be collected on a daily basis during peak flow conditions to represent the most demanding treatment facility operating conditions. Less frequent coliform sampling is allowed in some states. Several states require that coliform analyses be conducted using the multiple tube fermentation technique with the results expressed as the most probable number (MPN), while others allow use of the membrane filter (MF) technique. A few states do not specify which enumeration technique to use, and some states allow the use of either the MPN or MF methods. While the presence of coliforms can still be taken as a sign of fecal contamination, the absence of coliforms should not be viewed as an indication that the water is uncontaminated.

Limits and Monitoring for Pathogenic Organisms   At present, no states have set limits on pathogenic organisms for any nonpotable reuse application, but at least two states require monitoring for specific pathogens under certain circumstances—Florida and California. In an effort to learn more about the possible presence of protozoan pathogens in reclaimed water that receives tertiary treatment and a high level of disinfection, Florida's reuse rules contain parasite monitoring requirements. Facilities with capacities of 3.78×10^3 m^3/d (1.0 Mgal/d) and larger are required to sample their reclaimed water for *Giardia* and *Cryptosporidium* at least once every two years. Smaller facilities must sample at least once every five years. Samples are required to be taken following the disinfection process.

California requires that reclaimed water used for nonrestricted recreational impoundments be monitored for enteric viruses, *Giardia*, and *Cryptosporidium* if tertiary treatment does not include a sedimentation process between the chemical coagulation and filtration processes. Monthly sampling is required for the first year of operation, and quarterly sampling is required during the second year of operation. Sampling may be discontinued after the second year of operation with approval of the California DHS.

Treatment Facility Reliability Some states have adopted treatment reliability requirements to ensure that inadequately treated reclaimed water is not reused. Generally, requirements consist of alarms warning of power failure or failure of essential unit processes, automatic standby power sources, emergency storage or disposal provisions, and the provision that each treatment process be equipped with multiple units or a backup unit. Reliability requirements for California and Florida are presented below as examples.

CALIFORNIA REQUIREMENTS California's *Water Recycling Criteria* provide design and operational considerations covering alarms, power supply, emergency storage and disposal, wastewater treatment processes, and chemical supply, storage, and feed facilities. For treatment processes, several reliability features are acceptable. For example, for all biological treatment processes one of the following is required: (1) alarm (failure and power loss) and multiple units capable of producing oxidized wastewater (i.e., secondary treatment) with one unit not in operation; (2) alarm (failure and power loss) and short-term (at least 24 h) storage or disposal provisions and standby replacement equipment; or (3) alarm (failure and power loss) and long-term (at least 20 d) storage or disposal provisions. Similar reliability requirements apply to other treatment processes (California Department of Health, 1973).

FLORIDA REQUIREMENTS Florida requires Class I reliability as defined by the U.S. EPA (U.S. EPA, 1974) at water reclamation facilities where filtration and high-level disinfection are provided. Class I reliability requires multiple treatment units or backup units and a secondary power source. In addition, a minimum of one day of storage is required to store reclaimed water of unacceptable quality. Florida also requires staffing at the water reclamation facility 24 h/d, 7 d/wk or 6 h/d, 7 d/wk as long as reclaimed water is delivered to the reuse system only during periods when a qualified operator is present. Operator presence can be reduced to 6 h/d if additional reliability features are provided.

Voluntary versus Mandatory Water Reuse

In almost all states, water reuse is voluntary and not mandated by governmental agencies. An exception is Florida, where a mandatory reuse program has been established that is actively enforced (Florida Department of Environmental Protection, 1999). The policy requires the state's water management districts to identify water resource caution areas having water supply problems that have become critical or are anticipated to become critical within the next 20 yr. State legislation requires preparation of water reuse feasibility studies for treatment facilities located within the water resource caution areas. A reasonable amount of reclaimed water use from municipal wastewater treatment facilities is required within the designated water resource caution areas unless water reuse is not economically, environmentally, or technically feasible.

Water reuse regulations focus on public health implications of using reclaimed water; thus, water quality criteria not related to health protection usually are not included in water reuse regulations. Most states with extensive water reuse experience have comparable, conservatively based water quality criteria or guidelines. Arguments for less restrictive standards are most often predicated upon a lack of documented health hazards rather than upon any certainty that hazards are small or nonexistent. In the absence of definitive

Regulatory Requirements for Nonpotable Uses of Reclaimed Water

epidemiological data and a unified interpretation of scientific and technical data on pathogen exposures, selection of water quality limits will continue to be somewhat subjective and inconsistent among the states. Regulatory requirements for some nonpotable uses of reclaimed water not included in Table 4-5 are discussed below:

Wetlands

In most cases, the primary intent in applying reclaimed water to wetlands is to provide additional treatment of effluent prior to discharge or reuse, although wetlands are sometimes created solely for environmental enhancement. In such cases, secondary treatment is usually acceptable as influent to the wetland system. Very few states have regulations that specifically address the use of reclaimed water for creation of artificial wetlands or the restoration or enhancement of natural wetlands. Where there are no regulations, regulatory agencies prescribe requirements on a case-by-case basis. In addition to state requirements, natural wetlands, which are considered waters of the United States, are protected under EPA's National Pollutant Discharge Elimination System (NPDES) Permit and Water Quality Standards programs. Constructed wetlands built and operated for the purpose of wastewater treatment generally are not considered waters of the United States.

In the few states that have adopted regulations for reclaimed water use in wetlands, requirements vary based on the type of wetland system and degree of public access. For example, the State of Washington requires that reclaimed water discharged to natural wetlands where there is no expected human contact with the water must meet Class D reclaimed water standards (secondary treatment and not more than 240 total coliforms/100 mL). Discharges to natural or constructed wetlands providing human-contact recreational or educational beneficial uses must meet Class A reclaimed water standards (tertiary treatment and not more than 2.2 total coliforms/100 mL in the reclaimed water). Reclaimed water discharged to any wetland system in Washington cannot exceed the following water quality limits: 20 mg/L BOD, 20 mg/L TSS, 3 mg/L total Kjeldahl nitrogen (as N), and 1 mg/L total phosphorus (as P).

Industrial Uses Other than Cooling

Due to the myriad of industrial processes that use water, regulatory agencies generally prescribe water reuse requirements for industrial applications other than cooling on an individual case basis. For example, Florida regulations address the use of reclaimed water for food processing at industrial facilities. Florida's water reuse rule specifically prohibits the use of reclaimed water in the manufacture or processing of food or beverages for human consumption where the reclaimed water will be incorporated into, or come in contact, with the food or beverage product. Similarly, Washington standards do not allow the use of reclaimed water for food preparation and prohibit its use in food or drink for humans. While many industrial uses require water of higher chemical quality than that typically present in reclaimed water, (e.g., computer chip manufacturing requires reverse osmosis treatment to produce ultrapure wash water), water reuse regulations are intended to provide health protection and only include requirements to attain that end.

Miscellaneous Nonpotable Uses

While all states that have water reuse regulations or guidelines include criteria for crop and/or landscape irrigation, some include requirements for less common uses of reclaimed water, such as flushing sanitary sewers, street cleaning, dust control, soil compaction, making concrete, snowmaking, decorative fountains, commercial laundries, commercial car washes, equipment washing, and fire protection systems. For these and similar uses, the various state standards impose wastewater treatment process requirements, reclaimed water quality limits, and design and operational requirements reflective of the degree of human exposure to the water that are in concert with other more common uses of reclaimed water. For example, secondary treatment with a minimal level of disinfection is acceptable for uses where there is little or no expected human contact with the water, such as flushing sanitary sewers or making concrete. Conversely, uses such as snowmaking and vehicle washing are likely to result in contact with the reclaimed water, and tertiary treatment with a high level of disinfection is usually required.

State Regulations for Indirect Potable Reuse

There are no planned direct potable reuse projects in the United States, and no state has developed regulations allowing such use (see Chap. 24). From a regulatory standpoint, few states have addressed the challenge of developing regulations for indirect potable reuse (see Chap. 23). California and Florida are in the forefront of developing discrete criteria relating to planned indirect potable reuse of reclaimed water. Some of the other states rely on U.S. EPA's Underground Injection Control regulations to protect potable groundwater basins, while some states prohibit indirect potable reuse altogether. There are no federal regulations that specifically address indirect or direct potable reuse of reclaimed water.

State of California

The existing California *Water Recycling Criteria* include general requirements for groundwater recharge of domestic water supply aquifers by surface spreading. The regulations state that reclaimed water used for groundwater recharge of domestic water supply aquifers by surface spreading "shall be at all times of a quality that fully protects public health" and that DHS recommendations "will be based on all relevant aspects of each project, including the following factors: treatment provided; effluent quality and quantity; spreading area operations; soil characteristics; hydrogeology; residence time; and distance to withdrawal." Until more definitive criteria are adopted, proposals to recharge groundwater by either surface spreading or injection will be evaluated on a case-by-case basis. California has prepared draft criteria for groundwater recharge (the most recent being in 2004), which are presented in Appendix F in Section F-2.

State of Florida

Florida's water reuse rules pertaining to groundwater recharge and indirect potable reuse are summarized in Table 4-6. The rules address rapid-rate infiltration basin systems and absorption field systems, both of which may result in groundwater recharge. Although not specifically designated as indirect potable reuse systems, groundwater recharge projects located over potable aquifers could function as an indirect potable reuse system. If more than 50 percent of the wastewater applied to the systems is collected after percolation, the systems are considered to be effluent disposal systems and not beneficial reuse. Loading to these systems is limited to 230 mm/d (9 in./d). For systems having higher loading rates or a more direct connection to an aquifer than normally encountered, reclaimed water must receive secondary treatment, filtration, disinfection, and must meet primary and secondary drinking water standards.

Table 4-6

State of Florida water reuse rules for groundwater recharge and indirect potable reuse[a]

Type of use	Water quality limits	Treatment required
Groundwater recharge via rapid infiltration basins (RIBs)	200 fecal coli/100 mL 20 mg/L $CBOD_5$ 20 mg/L TSS 12 mg/L NO_3 (as N)	Secondary Disinfection
Groundwater recharge via RIBs in unfavorable conditions	No detectable fecal coli/100 mL 20 mg/L $CBOD_5$ 5.0 mg/L TSS Primary[b] and secondary drinking water standards 10 mg/L total N	Secondary Filtration Disinfection
Groundwater recharge or injection to groundwaters having TDS < 3000 mg/L	No detectable total coli/100 mL 20 mg/L $CBOD_5$ 5.0 mg/L TSS 3.0 mg/L TOC 0.2 mg/L TOX[c] 10 mg/L total N Primary[b] and secondary drinking water standards	Secondary Filtration Disinfection Multiple barriers for control of pathogens & organics Pilot testing required
Groundwater recharge or injection to groundwaters having TDS 3000–10,000 mg/L	No detectable total coli/100 mL 20 mg/L $CBOD_5$ 5.0 mg/L TSS 10 mg/L total N Primary drinking water standards[b]	Secondary Filtration Disinfection
Indirect potable reuse: discharge to Class I surface waters (used for public water supply)	No detectable total coli/100 mL 20 mg/L $CBOD_5$ 5.0 mg/L TSS 3.0 mg/L TOC 10 mg/L total N Primary[b] and secondary drinking water standards WQBELs[d] may apply	Secondary Filtration Disinfection

[a]Adapted from Florida Department of Environmental Protection (1999).
[b]Except for asbestos.
[c]TOX = Total organic halogen.
[d]WQBELs are water quality based effluent limitations to ensure that water quality standards in a receiving body of water will not be violated.

The Florida regulations include requirements for planned indirect potable reuse by injection into water supply aquifers and augmentation of surface supplies. A minimum horizontal separation distance of 150 m (500 ft) is required between reclaimed water injection wells and potable water supply wells. The injection regulations pertain to G-I, G-II, and F-I groundwaters, all of which are classified as potable aquifers. Reclaimed water must meet G-II groundwater standards prior to injection. G-II groundwater standards are, for the most part, primary and secondary drinking water standards. Florida considers discharges to Class I surface waters (public water supplies) as indirect potable reuse. Discharges less than 24 h travel time upstream from Class I waters are also considered as indirect potable reuse. Outfalls for surface water discharges cannot be located within 150 m (500 ft) of existing or approved potable water intakes within Class I surface waters. Pilot testing is required prior to implementation of injection or surface water augmentation projects.

Other States

In some states, regulations addressing indirect potable reuse are independent from the state's water reuse regulations. For example, the use of reclaimed water for groundwater recharge in Arizona is regulated under statutes and administrative rules administered by the Arizona Department of Environmental Quality (ADEQ) and the Arizona Department of Water Resources (ADWR). Several different permits are required by these agencies prior to implementation of a groundwater recharge project. In general, ADEQ regulates groundwater quality and ADWR manages groundwater supply. All aquifers in Arizona currently are classified for drinking water protected use, and the state has adopted National Primary Drinking Water Maximum Contaminant Levels (MCLs) as aquifer water quality standards. These standards apply to all groundwater in saturated formations that yield more than 20 L/d (5 gal/d) of water. Any groundwater recharge project involving injection of reclaimed water into an aquifer is required to demonstrate compliance with aquifer water quality standards at the point of injection.

4-7 U.S. EPA GUIDELINES FOR WATER REUSE

In recognition of the increasing role of water reuse as an integral component of the nation's water resources management—and to facilitate the orderly planning, design, and implementation of water reuse projects—the U.S. EPA, in conjunction with the U.S. Agency for International Development (U.S. AID), published *Guidelines for Water Reuse* in 1992 (U.S. EPA, 1992). The U.S. EPA took the position that national water reuse standards were not necessary and comprehensive guidelines, coupled with flexible state regulations, would foster increased consideration and implementation of water reuse projects. The guidelines were updated in 2004 (U.S. EPA, 2004) to include technological advances, research data, and other information generated in the last decade. The guidelines address various aspects of water reuse and include recommended treatment processes, reclaimed water quality limits, monitoring frequencies, setback distances, and other controls for various water reuse applications. The suggested guidelines for wastewater treatment and reclaimed water quality are presented in Table 4-7.

Disinfection Requirements

It is recommended in the guidelines that, regardless of the type of reclaimed water use, some level of disinfection be provided to avoid adverse health consequences from inadvertent contact or accidental or intentional misuse of a water reuse system. Two different levels of disinfection are recommended for nonpotable uses of reclaimed water. Reclaimed water used for applications where no direct public or worker contact with the water is expected should be disinfected to achieve a fecal coliform concentration not

Table 4-7
U.S. EPA suggested guidelines for reuse of municipal wastewater[a]

Types of reuse	Treatment	Reclaimed water quality[b]	Reclaimed water monitoring	Setback distances[c]	Comments
Urban Reuse					
All types of landscape irrigation (e.g., golf courses, parks, cemeteries) also vehicle washing, toilet flushing, use in fire protection systems and commercial air conditioners, and other uses with similar access or exposure to the water	• Secondary[d] • Filtration[e] • Disinfection[f]	• pH = 6–9 • ≤ 10 mg/L BOD[g] • ≤ 2 NTU[h] • No detectable fecal coli/100 mL[i,j] • ≥ 1 mg/L Cl$_2$ residual[k]	• pH—weekly • BOD—weekly • Turbidity—continuous • Coliform—daily • Cl$_2$ residual—continuous	• 15 m (50 ft) to potable water supply wells	• Consult recommended agricultural (crop) limits for metals • At controlled-access irrigation sites where design and operational measures significantly reduce the potential of public contact with reclaimed water, a lower level of treatment, for example, secondary treatment and disinfection to achieve ≤ 14 fecal coli/100 mL, may be appropriate • Chemical (coagulant and/or polymer) addition prior to filtration may be necessary to meet water quality recommendations • The reclaimed water should not contain measurable levels of pathogens[l] • Reclaimed water should be clear and odorless • A higher chlorine residual and/or a longer contact time may be necessary to assure that viruses and parasites are inactivated or destroyed • A chlorine residual of 0.5 mg/L or greater in the distribution system is recommended to reduce odors, slime, and bacterial regrowth • Provide treatment reliability

	**Restricted access area irrigation**
Sod farms, silviculture sites, and other areas where public access is prohibited, restricted, or infrequent	• Secondary[d] • Disinfection[f]	• pH = 6–9 • ≤ 30 mg/L BOD[g] • ≤ 30 mg/L TSS • ≤ 200 fecal coli/100 mL[i,m,n] • ≥ 1 mg/L Cl_2 residual[k]	• pH—weekly • BOD—weekly • TSS—daily • Coliform—daily • Cl_2 residual—continuous	• 90 m (300 ft) to potable water supply wells • 30 m (100 ft) to areas accessible to the public (if spray irrigation)	• Consult recommended agricultural (crop) limits • If spray irrigation, TSS less than 30 mg/L may be necessary to avoid clogging of sprinkler heads • Provide treatment reliability
	**Agricultural reuse—food crops not commercially processed[p]**
Surface or spray irrigation of any food crop, including crops eaten raw	• Secondary[d] • Filtration[e] • Disinfection[f]	• pH = 6–9 • ≤ 10 mg/L BOD[g] • ≤ 2 NTU[h] • No detectable fecal coli/100 mL[i,j] • ≥ 1 mg/L Cl_2 residual[k]	• pH—weekly • BOD—weekly • Turbidity—continuous • Coliform—daily • Cl_2 residual—continuous	• 15 m (50 ft) to potable water supply wells	• Consult recommended agricultural (crop) limits • Chemical (coagulant and/or polymer) addition prior to filtration may be necessary to meet water quality recommendations • The reclaimed water should not contain measurable levels of pathogens[l] • A higher chlorine residual and/or a longer contact time may be necessary to assure that viruses and parasites are inactivated or destroyed • High nutrient levels may adversely affect some crops during certain growth stages • Provide treatment reliability

(Continued)

Table 4-7
U.S. EPA suggested guidelines for reuse of municipal wastewater[a] (*Continued*)

Types of reuse	Treatment	Reclaimed water quality[b]	Reclaimed water monitoring	Setback distances[c]	Comments
Agricultural reuse—food crops commercially processed[o]					
Surface irrigation of orchards and vineyards	• Secondary[d] • Disinfection[f]	• pH = 6–9 • ≤ 30 mg/L BOD[g] • ≤ 30 mg/L TSS • ≤ 200 fecal coli/100 mL[i,m,n] • ≥ 1 mg/L Cl$_2$ residual[k]	• pH—weekly • BOD—weekly • TSS—daily • Coliform—daily • Cl$_2$ residual—continuous	• 90 m (300 ft) to potable water supply wells • 30 m (100 ft) to areas accessible to the public (if spray irrigation)	• Consult recommended agricultural (crop) limits • If spray irrigation, TSS less than 30 mg/L may be necessary to avoid clogging of sprinkler heads • High nutrient levels may adversely affect some crops during certain growth stages • Provide treatment reliability
Agricultural reuse—Nonfood crops					
Pasture for milking animals; fodder, fiber, and seed crops	• Secondary[d] • Disinfection[f]	• pH = 6–9 • ≤ 30 mg/L BOD[g] • ≤ 30 mg/L TSS • ≤ 200 fecal coli/100 mL[i,m,n] • ≥ 1 mg/L Cl$_2$ residual[k]	• pH—weekly • BOD—weekly • TSS—daily • Coliform—daily • Cl$_2$ residual—continuous	• 90 m (300 ft) to potable water supply wells • 30 m (100 ft) to areas accessible to the public (if spray irrigation)	• Consult recommended agricultural (crop) limits • If spray irrigation, TSS less than 30 mg/L may be necessary to avoid clogging of sprinkler heads • High nutrient levels may adversely affect some crops during certain growth stages • Milking animals should be prohibited from grazing for 15 d after irrigation ceases. A higher level of disinfection, for example, to achieve ≤ 14 fecal coli/100 mL, should be provided if this waiting period is not adhered to • Provide treatment reliability

Recreational impoundments

Incidental contact (e.g., fishing and boating) and full body contact with reclaimed water allowed	• Secondary[d] • Filtration[e] • Disinfection[f]	• pH = 6–9 • ≤ 10 mg/L BOD[g] • ≤ 2 NTU[h] • No detectable fecal coli/100 mL[i,j] • ≥ 1 mg/L Cl$_2$ residual[k]	• pH—weekly • BOD—weekly • Turbidity—continuous • Coliform—daily • Cl$_2$ residual—continuous	150 m (500 ft) to potable water supply wells (minimum) if bottom not sealed	• Dechlorination may be necessary to protect aquatic species of flora and fauna • Reclaimed water should be nonirritating to skin and eyes • Reclaimed water should be clear and odorless • Nutrient removal may be necessary to avoid algae growth in impoundments • Chemical (coagulant and/or polymer) addition prior to filtration may be necessary to meet water quality recommendations • The reclaimed water should not contain measurable levels of pathogens[l] • A higher chlorine residual and/or a longer contact time may be necessary to assure that viruses and parasites are inactivated or destroyed • Fish caught in impoundments can be consumed • Provide treatment reliability

Landscape impoundments

Aesthetic impoundments where public contact with reclaimed water is not allowed	• Secondary[d] • Disinfection[f]	• ≤ 30 mg/L BOD[g] • ≤ 30 mg/L TSS • ≤ 200 fecal coli/100 mL[m,n,o] • ≥ 1 mg/L Cl$_2$ residual[k]	• pH—weekly • TSS—daily • Coliform—daily • Cl$_2$ residual—continuous	• 150 m (500 ft) to potable water supply wells (minimum) if bottom not sealed	• Nutrient removal processes may be necessary to avoid algae growth in impoundments • Dechlorination may be necessary to protect aquatic species of flora and fauna • Provide treatment reliability

(Continued)

Table 4-7
U.S. EPA suggested guidelines for reuse of municipal wastewater[a] (Continued)

Types of reuse	Treatment	Reclaimed water quality[b]	Reclaimed water monitoring	Setback distances[c]	Comments
Construction uses					
Soil compaction, dust control, washing aggregate, making concrete	• Secondary[d] • Disinfection[f]	• ≤ 30 mg/L BOD[g] • ≤ 30 mg/L TSS • ≤ 200 fecal coli/100 mL [m,n,o] • ≥ 1 mg/L Cl$_2$ residual[k]	• BOD—weekly • TSS—daily • Coliform—daily • Cl$_2$ residual—continuous		• Worker contact with reclaimed water should be minimized • A higher level of disinfection, for example, to achieve ≤ 14 fecal coli/100 mL, should be provided where frequent worker contact with reclaimed water is likely • Provide treatment reliability
Industrial reuse					
Once-through cooling	• Secondary[d]	• pH = 6–9 • ≤ 30 mg/L BOD[g] • ≤ 30 mg/L TSS • ≤ 200 fecal coli/100 mL [i,m,n] • ≥ 1 mg/L Cl$_2$ residual[k]	• pH—weekly • BOD—weekly • TSS—weekly • Coliform—daily • Cl$_2$ residual—continuous	• 90 m (300 ft) to areas accessible to the public	• Windblown spray should not reach areas accessible to users or the public
Recirculating cooling towers	• Secondary[d] • Disinfection[f] (chemical coagulation and filtration[e] may be needed)	• Variable, depends on recirculation ratio • pH = 6–9 • ≤ 30 mg/L BOD[g] • ≤ 30 mg/L TSS • ≤ 200 fecal coli/100 mL [i,m,n] • ≥ 1 mg/L Cl$_2$ residual[k]	• pH—weekly • BOD—weekly • TSS—weekly • Coliform—daily • Cl$_2$ residual—continuous	• 90 m (300 ft) to areas accessible to the public. May be reduced or eliminated if high level of disinfection is provided	• Windblown spray should not reach areas accessible to the public • Additional treatment by user is usually provided to prevent scaling, corrosion, biological growths, fouling, and foaming • Provide treatment reliability
Other industrial uses	Depends on specific use	Depends on specific use	Depends on specific use	Depends on specific use	

Types of reuse	Treatment	Reclaimed water quality	Reclaimed water monitoring	Setback distances	Comments

Environmental reuse

Wetlands, marshes, wildlife habitat, stream augmentation	• Variable • Secondary[d] and disinfection[f] (min.)	• Variable, but not to exceed: • ≤ 30 mg/L BOD[g] • ≤ 30 mg/L TSS • ≤ 200 fecal coli/100 mL[i,m,n]	• BOD—weekly • TSS—daily • Coliform—daily • Cl$_2$ residual—continuous		• Dechlorination may be necessary to protect aquatic species of flora and fauna • Possible effects on groundwater should be evaluated • Receiving water quality requirements may necessitate additional treatment • The temperature of the reclaimed water should not adversely affect ecosystem • Provide treatment reliability

Groundwater recharge

By spreading or injection into aquifers not used for potable water supply	• Site-specific and use-dependent • Primary (min.) for spreading • Secondary[d] (min.) for injection	• Site-specific and use-dependent	• Depends on treatment and use	• Site-specific	• Facility should be designed to ensure that no reclaimed water reaches potable water supply aquifers • For spreading projects, secondary treatment may be needed to prevent clogging • For injection projects, filtration and disinfection may be needed to prevent clogging • Provide treatment reliability

Indirect potable reuse

Groundwater recharge by spreading into potable aquifers	• Secondary[d] and disinfection[f] • May also need filtration[e] and/or advanced wastewater treatment[p]	• Meet drinking water standards after percolation through vadose zone	• Includes, but not limited to, the following: • pH—daily • Coliform—daily • Cl$_2$ residual—continuous	• 150 m (500 ft) to extraction wells. May vary depending on treatment provided and site-specific conditions	• The depth to groundwater (i.e., thickness of the vadose zone) should be at least 2 m (6 ft) at the maximum groundwater mounding point • The reclaimed water should be retained underground for at least 6 mo prior to withdrawal

(Continued)

Table 4-7
U.S. EPA suggested guidelines for reuse of municipal wastewater[a] (Continued)

Types of reuse	Treatment	Reclaimed water quality[b]	Reclaimed water monitoring	Setback distances[c]	Comments
Indirect potable reuse (Continued)					
			• Drinking water standards—quarterly • Other[q]—depends on constituent • BOD—weekly • Turbidity—continuous		• Recommended treatment is site-specific and depends on factors such as type of soil, percolation rate, thickness of vadose zone, native groundwater quality, and dilution • Monitoring wells are necessary to detect the influence of the recharge operation on the groundwater • The reclaimed water should not contain measurable levels of pathogens after percolation through the vadose zone[k] • Provide treatment reliability
Groundwater recharge by injection into potable aquifers	• Secondary[d] • Filtration[e] • Disinfection[f] • Advanced wastewater treatment[p]	• Includes, but not limited to, the following: • pH = 6.5–8.5 • ≤ 2 NTU[h] • No detectable total coli/100 mL[i,j] • ≥ 1 mg/L Cl$_2$ residual[k] • ≤ 3 mg/L TOC • ≤ 0.2 mg/L total organic halogen (TOX) • Meet drinking water standards	• Includes, but not limited to, the following: • pH—daily • Turbidity—continuous • Total coliform—daily • Cl$_2$ residual—continuous • Drinking water standards—quarterly • Other[q]—depends on constituent	• 600 m (2000 ft) to extraction wells. May vary depending on site-specific conditions	• The reclaimed water should be retained underground for at least 9 mo prior to withdrawal • Monitoring wells are necessary to detect the influence of the recharge operation on the groundwater • Recommended quality limits should be met at the point of injection • The reclaimed water should not contain measurable levels of pathogens at the point of injection[l] • A higher chlorine residual and/or a longer contact time may be necessary to assure virus inactivation • Provide treatment reliability

Augmentation of surface supplies	• Secondary[d] • Filtration[e] • Disinfection[f] • Advanced wastewater treatment[p]	• Includes, but not limited to, the following: • pH = 6.5–8.5 • ≤ 2 NTU[h] • No detectable total coli/100 mL[i,j] • ≥ 1 mg/L Cl$_2$ residual[k] • ≤ 3 mg/L TOC • Meet drinking water standards	• Includes, but not limited to, the following: • pH—daily • Turbidity—continuous • Total coliform—daily • Cl$_2$ residual—continuous • Drinking water standards—quarterly • Other[q]—depends on constituent	• Site-specific	• Recommended level of treatment is site-specific and depends on factors such as receiving water quality, time and distance to point of withdrawal, dilution and subsequent treatment prior to distribution for potable uses • The reclaimed water should not contain measurable level of pathogens[l] • A higher chlorine residual and/or a longer contact time may be necessary to assure virus inactivation • Provide treatment reliability

[a] Adapted from U.S. EPA (2004).
[b] Unless otherwise noted, recommended quality limits apply to reclaimed water at the point of discharge from the treatment facility.
[c] Setbacks are recommended to protect potable water supply sources from contamination and to protect humans from unreasonable health risks due to exposure to reclaimed water.
[d] Secondary treatment processes include activated sludge processes, trickling filters, rotating biological contactors, and may include stabilization pond systems. Secondary treatment should produce effluent in which both the BOD and SS do not exceed 30 mg/L.
[e] Filtration means the passing of wastewater through natural undisturbed soils or filter media such as sand and/or anthracite, filter cloth, or passing the wastewater through microfilters or other membrane processes.
[f] Disinfection means the destruction, inactivation, or removal of pathogenic microorganisms by chemical, physical, or biological means. Disinfection may be accomplished by chlorination, ozonation, other chemical disinfectants, UV radiation, membrane processes, or other processes. The use of chlorine as defining the level of disinfection does not preclude the use of other disinfection processes as acceptable means of providing disinfection for reclaimed water.
[g] As determined from the 5-d BOD test.
[h] The recommended turbidity limit should be met prior to disinfection. The average turbidity should be based on a 24-h time period. The turbidity should not exceed 5 NTU at any time. If TSS is used in lieu of turbidity, the TSS should not exceed 5 mg/L.
[i] Unless otherwise noted, recommended coliform limits are median values determined from the bacteriological results of the last 7 d for which analyses have been completed. Either the membrane filter or fermentation tube technique may be used.
[j] The number of fecal coliform organisms should not exceed 14/100 mL in any sample.
[k] Total chlorine residual should be met after a minimum contact time of 30 mins.
[l] It is advisable to fully characterize the microbiological quality of the reclaimed water prior to implementation of a reuse program.
[m] The number of fecal coliform organisms should not exceed 800/100 mL in any sample.
[n] Some stabilization pond systems may be able to meet this coliform limit without disinfection.
[o] Commercially processed food crops are those that, prior to sale to the public or others, have undergone chemical or physical processing sufficient to destroy pathogens.
[p] Advanced wastewater treatment processes include chemical clarification, carbon adsorption, reverse osmosis and other membrane processes, air stripping, ultrafiltration, and ion exchange.
[q] Monitoring should include inorganic and organic compounds, or classes of compounds, that are known or suspected to be toxic, carcinogenic, teratogenic, or mutagenic and are not included in the drinking water standards.

exceeding 200/100 mL for the following reasons: most bacterial pathogens will be destroyed or reduced to low or insignificant levels in the water; the concentration of viable viruses will be reduced somewhat; disinfection of secondary effluent to this coliform level is readily achievable at minimal cost; and significant health-related benefits associated with disinfection to lower, but not pathogen-free, levels are not obvious.

For uses where direct or indirect contact with reclaimed water is likely or expected, and for dual water systems where there is a potential for cross-connections with potable water lines, the guidelines recommend a high level of disinfection to produce reclaimed water having no detectable fecal coliform organisms/100 mL. This more restrictive disinfection level is intended for use in conjunction with tertiary treatment and other water quality limits, such as a turbidity of ≤ 2 NTU in the wastewater prior to disinfection. The combination of treatment and water quality limits has been shown to be capable of producing reclaimed water that is essentially free of measurable levels of bacterial and viral pathogens.

Microbial Limits

The guidelines include limits for fecal coliform organisms but do not include parasite or virus limits. Parasites such as helminths have not been shown to be a problem at water reuse operations in the United States at the treatment levels and reclaimed water limits recommended in the guidelines, although there has been considerable interest in recent years regarding the occurrence and significance of *Giardia* and *Cryptosporidium* in reclaimed water. Where filtration and a high level of disinfection are recommended to produce reclaimed water that is essentially free of measurable levels of pathogens, the guidelines indicate that it may be necessary to provide chemical addition prior to filtration to assure removal or inactivation of parasites and viruses. While enteric viruses are a concern in reclaimed water, virus limits are not recommended in the guidelines for the following reasons: a significant body of information exists indicating that viruses are inactivated or removed to low or immeasurable levels via appropriate wastewater treatment; there is a limited number of facilities having the personnel and equipment necessary to perform the analyses; there is no consensus among public health experts regarding the health significance of low levels of viruses in reclaimed water; and no cases of viral disease resulting from the reuse of wastewater have been documented in the United States.

Control Measures

As with state water reuse criteria, the guidelines are directed principally at health protection and include various control measures. For example, for nonpotable urban uses of reclaimed water, the guidelines recommendations include the following: clear, colorless, and odorless product water; a setback distance of 15 m (50 ft) from irrigated areas to potable water supply wells; maintenance of a chlorine residual of at least 0.5 mg/L in the distribution system; treatment reliability and emergency storage or disposal of inadequately treated water; and cross-connection control devices on potable water service lines and color-coded or taped reclaimed water lines and appurtenances. Similar design and operational recommendations are included in the guidelines for the other reclaimed water applications.

Recommendations for Indirect Potable Reuse

Whereas the water quality requirements for nonpotable water uses are tractable and not likely to change significantly in the future, the number of water quality constituents to be monitored in drinking water and, hence, reclaimed water intended for indirect potable reuse will increase and quality requirements are likely to become more restrictive. Consequently, the authors of the guidelines determined that it would not be prudent to suggest a complete list on reclaimed water quality limits for all constituents of concern.

In addition to some specific wastewater treatment and reclaimed water quality recommendations, the guidelines provide some general recommendations to indicate the extensive treatment and water quality requirements that are likely to be imposed where planned indirect potable reuse is contemplated. The guidelines do not advocate direct potable reuse and do not include recommendations for such use. Some of the pertinent topics related to potable reuse are discussed in detail in the National Research Council report, *Issues in Potable Reuse: The Viability of Augmenting Drinking Water Supplies with Reclaimed Water* (NRC, 1998).

It is explicitly stated in the *Guidelines for Water Reuse* that the recommended treatment unit processes and water quality limits presented in the guidelines "are not intended to be used as definitive water reclamation and reuse criteria. They are intended to provide reasonable guidance for water reuse opportunities, particularly in states that have not developed their own criteria or guidelines." (U.S. EPA, 2004).

4-8 WORLD HEALTH ORGANIZATION GUIDELINES FOR WATER REUSE

As discussed in Chap. 1, within the next 50 yr, it is estimated that more than 40 percent of the world's population will live in countries facing water stress or water scarcity. Most population growth is expected to occur in urban and periurban areas of developing countries. Given the fact that only about 10 percent of all wastewater produced in developing countries receives any treatment, the challenge to public health and environmental protection is enormous.

Over the years, the World Health Organization (WHO) has provided guidance for the safe use of wastewater. In 1971, WHO sponsored a meeting of experts on water reuse, which culminated in a 1973 report recommending health criteria and treatment processes for various wastewater applications (WHO, 1973). The 1973 criteria were revised in 1989; and the most recent, third edition of the WHO *Guidelines* has been published in 2006 (WHO, 2006).

In general, the WHO guidelines are significantly less restrictive than water reuse regulations or guidelines adopted by various states of the United States. The intentions of international technical cooperation organizations such as WHO, the Food and Agriculture Organization of the United Nations (FAO), and of multilateral development agencies such as the World Bank and the United Nations Development Programme (UNDP) are to introduce at least some level of treatment of wastewater and to achieve positive disease transmission interruption or exposure prevention prior to food crop irrigation. The WHO guidelines satisfy that intent and can be considered appropriate as an interim measure in the context of socio-economic realities of many countries, until they have developed the capacity to produce higher quality reclaimed water.

1989 WHO Guidelines for Agriculture and Aquaculture

In 1989, WHO published *Health Guidelines for the Use of Wastewater for Agriculture and Aquaculture* (WHO, 1989). The guidelines were based on the premise that the main health risks with wastewater use are associated with helminth infections and, therefore, a high degree of helminth egg removal is necessary for the safe use of wastewater in agriculture and aquaculture. Waste stabilization ponds were identified as the method of choice in meeting the guidelines in warm climates, where land is available at reasonable cost, more sophisticated treatment methods are not affordable, and adequate technical backup support is lacking. Based on helminth removal, the guidelines recommend a pond retention time of 8 to 10 d, with at least twice that time required in warm climates to reduce fecal coliforms to the guideline level of 1000/100 mL. However, based on actual field experience at some existing full-scale and demonstration stabilization pond systems, it has been found that the desired reductions of helminths and fecal coliform organisms may be difficult to achieve in practice. However, following sound planning, design, and operation and maintenance, stabilization ponds may increase the possibility of meeting the desired helminth and coliform reductions. Comprehensive manuals and publications are available on the subjects related to stabilization ponds (U.S. EPA, 1983; Arthur, 1983; Mara and Pearson, 1998; Mara, 2003).

The Stockholm Framework

In 2001, at a WHO expert meeting in Stockholm, Sweden, a framework was developed that facilitates an integrated approach combining risk assessment and risk management to control water-related diseases. This approach harmonizes the process of developing health-based guidelines and standards in terms of water- and sanitation-related microbial hazards and provides the conceptual framework for all WHO water-related guidelines. The Stockholm framework involves: (1) the assessment of health risks prior to the setting of health-based targets and the development of guideline values, (2) the definition of basic control approaches, and (3) the evaluation of the impact of these combined approaches on public health (Bartram et al., 2001; WHO, 2006).

The framework allows countries to adjust guidelines to local, social, cultural, economic, and environmental circumstances and to compare the associated health risks with the risks that may result from microbial exposures through wastewater use, drinking water, and contact with recreational or occupational waters. This approach requires that diseases be managed from an integrated health perspective and not in isolation. This, in turn, implies that determination of acceptable risk, or tolerable risk, needs to be put into the context of actual disease rates in a population related to all the exposures that lead to a particular disease, including other water and sanitation-related exposures. Different countries may, therefore, set different health targets, based on their own contexts. Furthermore, disease outcome from one exposure pathway, or from one illness to another, can be compared by using a common measure, such as disability adjusted life years, discussed in the following section.

Disability Adjusted Life Years

The disability adjusted life years (DALYs) is a summary measure of population health and the loss of DALYs is an indicator for the burden of disease due to a specific illness or risk factor. The DALYs is an attempt to measure the time lost through disability or death from a particular disease, by comparing it to a long life free of disability in the absence of the disease. The DALYs are calculated by adding the years of life lost (YLLs) to premature death to the years lived with a disability (YLDs). Years of life lost are calculated from age-specific mortality rates and the standard life expectancies of a given population. Years lived with a disability are calculated from the number of cases multiplied by the average duration of the disease and a severity factor, which ranges from one (death) to zero (perfect health), based on the disease. For example, watery diarrhea has a severity factor ranging from 0.09 to 0.12, depending on the age group (Murray and Lopez, 1996; Prüss and Havelaar, 2001). Disability adjusted life years are an important tool for comparing health outcomes because they account not only for acute health effects but also for delayed and chronic effects, including morbidity and mortality (Bartram, Fewtrell & Stenström, 2001). When risk is described in DALYs, different health outcomes can be compared (e.g., cancer can be compared to giardiasis) and risk management decisions can be prioritized in a cost-effective way (Aertgeerts and Angelakis, 2003).

Concept of Tolerable (Acceptable) Risk

The following criteria can be used to judge whether a risk is acceptable (Hunter and Fewtrell, 2001):

- The risk falls below an arbitrary, defined probability.
- The risk falls below some level that is already tolerated.
- The risk falls below an arbitrary, defined attributable fraction of total disease burden in the community.
- The cost of reducing the risk would exceed the costs saved when the "costs of suffering" are also factored in.
- The money would be better spent on other, more pressing public health problems.
- Public health professionals say that the risk is acceptable.
- The general public say that the risk is acceptable (or more likely, do not say that it is not acceptable).
- Politicians say that the risk is acceptable.

Tolerable risks are not necessarily static. As tools for managing water-related disease transmission improve, the levels of risk that are tolerable may decrease. Tolerable risks can therefore be set with the idea of continuous improvement. For example, smallpox and polio were eradicated because it was technologically feasible to do so, not because of the continually decreasing global burden of disease attributed to these pathogens.

Tolerable Microbial Risk in Water

For water-related exposures, WHO has determined that a disease burden of 10^{-6} DALYs (i.e., one micro-DALY) per person per year from a disease (caused by either a chemical or an infectious agent) transmitted through drinking water is a tolerable risk (WHO, 2003). This level of health burden is equivalent to a mild illness (e.g., watery diarrhea) with a low fatality rate (e.g., 1 in 100,000) at an approximately 1 in 1000 annual risk of disease to an individual, which is equivalent to a 1 in 10 risk over a lifetime (WHO, 1996).

Tolerable risk can be looked at in the context of total risk from all exposures; risk management decisions can then be used to address the greatest risks first. For example, if 99 percent of cases of salmonellosis were related to food, then halving the number of cases attributed to drinking water would have very little impact on the disease burden. For water-related exposures to microbial contaminants, the incidence of diarrhea or gastrointestinal disease is often used to represent all waterborne infectious diseases.

2006 WHO Guidelines for the Safe Use of Wastewater in Agriculture

Following a final expert review meeting held during June 13–17, 2005, in Geneva, Switzerland, the third edition of the WHO *Guidelines for the Safe Use of Wastewater, Excreta and Greywater,* published in 2006, was an extensive update of the previous two editions (1973 and 1989, respectively), expanded to include new scientific evidence and contemporary approaches to risk management encompassing the Stockholm Framework, discussed earlier. The Guidelines are presented in four separate volumes: Vol. I—*Policy and Regulatory Aspects;* Vol. II—*Wastewater in Agriculture;* Vol. III—*Wastewater and Excreta Use in Aquaculture;* and Vol. IV—*Excreta and Greywater Use in Agriculture.* The *Guidelines* are intended to be used as the basis for the development of international and national approaches (including standards and regulations) to managing the health risks from hazards associated with wastewater use in agriculture and aquaculture, as well as providing a framework for national and local decision-making (WHO, 2005 and 2006).

Assessment of Health Risks

Three types of evaluations were used to assess risk in the *Guidelines*: microbial and chemical laboratory analysis, epidemiological studies, and quantitative microbial risk assessment (QMRA). Wastewater contains a variety of different pathogens, many of which are capable of survival in the environment (in the wastewater, on the crops, or in the soil) long enough to be transmitted to humans. In places where wastewater is used without adequate treatment, the greatest health risks are usually associated with intestinal helminths. Another conclusion reached from the QMRA evaluation was that the risk for rotavirus transmission was estimated to be higher than the risks associated with *Campylobacter* or *Cryptosporidium* infections.

Health-Based Targets

Health-based targets define a level of health protection that is relevant to each hazard. A health-based target can be based on a standard metric of disease, such as a DALY (e. g., 10^{-6} DALYs), or it can be based on an appropriate health outcome, such as the prevention of the transmission of vector-borne (from an organism, such as a mosquito or tick, that carries disease-causing microorganisms from one host to another) diseases resulting from exposures associated with wastewater use in agriculture. To achieve a health-based target, health protection measures are developed. Usually a health-based target can be achieved through a combination of health protection measures targeted at different components of the system to achieve the tolerable risk of 10^{-6} DALYs.

The WHO's health-based target for wastewater use in agriculture is shown in Table 4-8. The health-based targets for rotavirus are based on QMRA conclusions that the pathogen reduction required to achieve 10^{-6} DALY for different exposures. To develop health-based targets for helminth infections, epidemiological evidence was used. This evidence demonstrated that excess helminth infections (for both product consumers and

Table 4-8
Health-based targets for wastewater use in agriculture[a]

Exposure scenario	Health-based target (DALY per person per year)	Log pathogen reduction needed	Number of helminth eggs/L
Unrestricted irrigation	≤10^{-6} [b]		
Lettuce		6	≤1 [c,d]
Onion		7	≤1 [c,d]
Restricted irrigation	≤10^{-6} [b]		
Highly mechanized		3	≤1 [c,d]
Labor intensive		4	≤1 [c,d]
Localized (drip) irrigation	≤10^{-6} [b]		
High-growing crops[e]		2	No recommendation
Low-growing crops[e]		4	≤1 [c]

[a] Adapted from WHO (2006).
[b] Rotavirus reduction. The health-based target can be achieved, for unrestricted and localized irrigation, by a 6–7 log unit pathogen reduction (obtained by a combination of wastewater treatment and other health protection measures); for restricted irrigation, it is achieved by a 2–3 log unit pathogen reduction.
[c] When children under 15 yr are exposed, additional health-protection measures should be used (e.g., treatment to ≤0.1 eggs/L, protective equipment such as gloves or shoes/boots, or chemotherapy.)
[d] Arithmetic mean should be determined throughout the irrigation season. The mean value of ≤1 eggs/L should be obtained for at least 90 percent of samples in order to allow for the occasional high-value sample (i.e., with >10 eggs/L). With some wastewater treatment processes (e.g., waste stabilization ponds), the hydraulic retention time can be used as a surrogate to assure compliance with ≤1 eggs/L.
[e] No crops to be picked up from the soil.

farmers) could not be measured when wastewater of a quality of ≤1 helminth egg/L was used for irrigation. This level of health protection could also be met by treatment of wastewater; by a combination of wastewater treatment and washing of produce to protect consumers of raw vegetables; or by wastewater treatment and the use of personal protective equipment (shoes, gloves) to protect workers. When children less than 15 yr old are exposed in the fields, either additional wastewater treatment (to achieve a wastewater quality of ≤0.1 helminth egg/L) or the addition of other health protection measures (e.g., antihelminthic treatment such as chemotherapy) should be considered.

Health Protection Measures

A variety of health protection measures can be used to reduce health risks to consumers, workers and their families, and local communities. Strategies for managing health risks to achieve health targets include wastewater treatment to achieve appropriate microbiological quality guidelines, crop restriction, wastewater application methods, control of human exposure, chemotherapy (e.g., antiparasitic pills), and immunization. Phased implementation of microbial water quality standards may be necessary as treatment is gradually introduced or upgraded over a period of time, for example, 10 to 15 yr. For a maximum public health effect, the guidelines should be coimplemented with other health interventions such as hygiene promotion, provision of adequate drinking water and sanitation, and other health care measures (Carr et al., 2004; WHO, 2006).

4-9 FUTURE DIRECTIONS IN REGULATIONS AND GUIDELINES

As noted previously, there are currently no federal regulations in the United States that specifically address or control water reclamation and reuse and, where they exist, water reuse regulations have been established at the state level. While the U.S. EPA has published guidelines for water reuse (U.S. EPA, 1992; 2004), there is no indication that national regulations similar in scope to those for drinking water will be promulgated in the foreseeable future. Thus, the burden of developing water reuse standards and permitting reclaimed water projects will continue to rest with the individual states. As a result, the current lack of uniformity among state water reuse regulations will continue.

Continuing Development of State Standards, Regulations, and Guidelines

Historically, water-short states such as Arizona, California, Florida, and Texas have taken the lead in developing comprehensive water reclamation and reuse standards, which provide direction and encourage water reuse in those states. Other states typically have adopted less-comprehensive and—in some cases—less rigorous standards. As water shortages have begun to pervade many regions of the country previously thought to be water-rich, water reuse is becoming an integral component of water resources management throughout the nation. As a consequence, water reuse regulations have recently been developed in states where water reuse was not previously considered to be an economically viable alternative source of water.

Technical Advances in Treatment Processes

Technical advances in wastewater treatment processes and microbial and chemical contaminant detection methodology, coupled with decreased costs as the technologies mature, undoubtedly will be reflected in future regulations. For example, membranes already are beginning to replace conventional media filtration at many water reclamation facilities as a more effective means of tertiary treatment; UV irradiation is now the favored method of disinfection at many facilities; and polymerase chain reaction (PCR) analytical techniques—while not yet required in any water reuse monitoring regulations for microbial pathogens—may ultimately be required to monitor the quality of reclaimed water used for some applications. As research and demonstration projects validate the effectiveness of new or improved treatment processes over time, regulatory agencies will no longer consider them "alternative treatments" that require extensive data collection and documentation of treatment effectiveness at each individual site prior to approval. Such processes will be incorporated in state regulations and become part of the standard requirements.

Information Needs

While current water reclamation and reuse regulations and practices have not been shown to present unreasonable health risks, information is needed in several areas to assist regulatory agencies in improving or verifying the effectiveness of their criteria, including the following:

- Better indicator organisms to estimate the presence or absence of microbial pathogens in reclaimed water.
- Less expensive methodology to determine the presence and viability of microbial pathogens.
- More rapid, online monitoring of indicator organisms and pathogens.
- Better parameters for determination of disinfection pretreatment effectiveness.
- Advances in risk assessment methodology to make it a more useful tool during regulation development.
- Determination of treatment effectiveness and reliability for removal of potentially hazardous microbial and chemical constituents in reclaimed water.
- Toxicological assessment of PhACs, EDCs, and other potential health-significant chemical contaminants in reclaimed water.
- Better surrogates for chemical constituents of health concern.
- Development of online biomonitoring methods for reclaimed water intended for potable reuse in lieu of current toxicological assessment methodology.

Regulations will be modified as new information becomes available. The many discrepancies among state water reuse regulations indicate that some state regulations provide a greater degree of health protection than others. Determination of the actual health risks resulting from compliance with different requirements and a more uniform interpretation of acceptable risk are needed. Resolution of these issues ultimately should result in more uniform state regulations.

Revision and improvement of water reuse regulations often is a long-term process, sometimes requiring years to achieve. It is, however, an ongoing task that is necessary to advance the state-of-the-art of water reuse. Regulatory agencies have the responsibility and authority to produce standards that are protective of public health and the environment. Most state agencies do not have the financial capability to conduct needed research to resolve knowledge gaps and rely heavily on academic institutions, research organizations, consulting engineering firms, and others to assist in information development. This situation is not likely to change, and the water reuse industry will continue to work in concert with regulatory agencies to help craft reasonable and scientifically sound water reuse regulations.

PROBLEMS AND DISCUSSION TOPICS

4-1 The U.S. EPA has determined that guidelines for drinking water standards should be designed to ensure that human populations are not subjected to the risk of infection by enteric disease greater than 10^{-4} for a yearly exposure (Regli et al., 1991; Macler and Regli, 1993). Shuval et al. (1997) estimated the additional cost of treating wastewater to no detectable fecal coliform organisms/100 mL (U.S. EPA suggested guidelines), rather than to 1000 fecal coliform organisms/100 mL (WHO suggested guidelines) and the cost of each case of disease avoided. The total cost for rotavirus disease was

$3.5 million and for hepatitis A $35 million. The additional health benefit that might result from a further reduction of risk achieved by adhering to the U.S. EPA-suggested guidelines appears to be insignificant in relation to the additional costs associated with the expensive technology required to treat wastewater to such a rigorous standard.

Based on the assessment given above, one may argue that such costs can never be justified, and the money would be better spent on primary health care facilities such as building new hospitals. Discuss pros and cons of more stringent standards in water reclamation and reuse considering various factors involved in implementing a water reuse project. Considerations may include local water scarcity, the value of water, public health, public perception and acceptance, and economics.

4-2 The California Department of Health Services (DHS) has the authority to develop water reuse criteria for groundwater recharge, which are enforced by the California Regional Water Quality Control Boards (RWQCBs). Draft DHS recharge regulations only apply to planned groundwater recharge projects and not to wastewater disposal projects. Recharges resulting from disposal of wastewater to percolation ponds or to rivers where the water eventually recharges potable groundwater basins are not subject to the DHS groundwater recharge criteria. However, it occurs on a large scale in California, particularly in the Central Valley region of the state. Such "unplanned" recharges are subject to water quality limits imposed by the RWQCBs—limits that are considerably less restrictive than those in the DHS draft groundwater recharge regulations.

The inconsistency described above results in much more restrictive regulations applied to planned versus unplanned or incidental groundwater recharge projects. Discuss: (1) if DHS should only be concerned with the quality of groundwater that is extracted for potable supply and thus, not worry about water quality at the point of recharge; and (2) how, if you deem appropriate, imposition of a common set of criteria could be imposed on all projects that result in augmentation of potable groundwater supplies. Considerations may include changes in regulatory authority, groundwater quality, antidegradation policy, soil aquifer treatment, and public health protection.

4-3 For reclaimed water uses where human contact with the water is expected or likely, most states require either turbidity or suspended solids limits after filtration to enhance the potential for effective disinfection of the water. While turbidity is superior to suspended solids as a measure of particulate removal by filtration, neither parameter correlates well with pathogen removal via filtration or the ability of the disinfection process to destroy certain pathogens.

Discuss the potential use of parameters other than turbidity or suspended solids to serve as measures of particulate removal and thus, disinfection process effectiveness—for example, particle counting. Considerations may include online monitoring capability, analytical instrumentation, practicality, meaning of data obtained, relationship to treatment provided and disinfection achieved, and suggested limits.

4-4 There are no federal regulations relating to water reuse in the United States, and water reclamation and reuse regulations or guidelines are developed at the state level.

As a result, there are considerable differences among the various state regulations. The U.S. EPA has published *Guidelines for Water Reuse* (U.S. EPA, 2004) that includes suggested reclaimed water treatment processes, water quality limits, and other recommendations; however, these guidelines have no enforcement status.

Discuss the pros and cons of developing and adopting national water reclamation and reuse regulations similar in scope to the U.S. EPA drinking water standards. Considerations may include the effort required to develop national regulations (time, cost, expertise needed), enforcement, conflict with existing regulations, local conditions, state's rights issues, public confidence, and consistency.

4-5 Treated wastewaters sometimes represent a significant portion of the total flow in many receiving waters including rivers, streams, and reservoirs. Thus, regulation of municipal wastewater discharge and nonpoint source pollution will become intimately and formally connected with the regulation of drinking water in the future.

Propose indirect potable reuse regulations for a community and compare and analyze the situation described above with reference to your proposed regulations in terms of engineering reliability, consistency in water quality, public health protection, and public acceptance.

4-6 Compare and discuss similarity and differences between the U.S. EPA's 2004 *Guidelines for Water Reuse* and the WHO's 2006 *Guidelines for the Safe Use of Wastewater in Agriculture* in terms of: (1) use of sciences such as microbiology, epidemiology, quantitative microbial risk assessment, and health statistics, (2) concept of tolerable risk, (3) tolerable microbial risk in water and wastewater, (4) health protection measures, and (5) relation to water pollution control and environmental protection. Summarize the principal reasons why the guidelines are different.

REFERENCES

Aertgeerts, R., and A. Angelakis (eds.) (2003) *State of the Art Report: Health Risks in Aquifer Recharge Using Reclaimed Water*, SDE/WSH/03.08, WHO, Geneva, Switzerland, and WHO Regional Office for Europe, Copenhagen, Denmark.

Arthur, J. (1983) *Notes on the Design and Operation of Waste Stabilization Ponds in Warm Climates of Developing Countries*, World Bank Technical Paper Number 7, The World Bank, Washington, DC.

Bartram, J., L. Fewtrell, and T. A Stenström (2001) "Water Quality: Guidelines, Standards and Health," Chap. 1, in L. Fewtrell and J. Bartram (eds.) *Water Quality: Guidelines, Standards and Health; Assessment of Risk and Risk Management For Water-Related Infectious Disease*, IWA Publishing, London, UK.

California Department of Health (1973) *Development of Reliability Criteria for Water Reclamation Operations*, California Department of Health Services, Water Sanitation Section, Berkeley, CA.

Camann, D. E., D. E. Johnson, H. J. Harding, and C. A. Sorber (1980) "Wastewater Aerosol and School Attendance Monitoring at an Advanced Wastewater Treatment Facility: Durham Plant, Tigard, Oregon," 160–179, in H. Pahren and W. Jakubowski (eds.) *Wastewater Aerosols and Disease*, EPA-600/9-80-028, U.S. Environmental Protection Agency, Cincinnati, OH.

Carr, R. M., U. J. Blumenthal, and D. D. Mara (2004) "Health Guidelines for the Use of Wastewater in Agriculture: Developing Realistic Guidelines," in C. Scott, N. Faruqui, and L. Raschid-Sally (eds.) *Wastewater Use in Irrigated Agriculture: Confronting the Livelihood and Environmental Realities*, CAB International, London.

Crook, J. (1989) "Viruses in Reclaimed Water," 231–237, in *Proceedings of the 63rd Annual Technical Conference,* Florida Pollution Control Association, and Florida Water & Pollution Control Operators Association, St. Petersburg Beach, FL.

Crook, J. (1998) "Water Reclamation and Reuse Criteria," Chap. 14, in T. Asano (ed.) *Wastewater Reclamation and Reuse,* **10**, CRC Press, Boca Raton, FL.

Crook, J. (2002) "The Ongoing Evolution of Water Reuse Criteria," in *Proceedings of the AWWA/WEF 2002 Water Sources Conference,* Las Vegas, NV.

Crook, J., R. H. Hultquist, R. H. Sakaji, and M. P. Wehner (2002) "Evolution and Status of California's Proposed Criteria for Groundwater Recharge with Reclaimed Water," in *Proceedings of the AWWA Annual Conference and Exposition,* American Water Works Association, Denver, CO.

Engineering-Science (1987) *Monterey Wastewater Reclamation Study for Agriculture: Final Report,* prepared for the Monterey Regional Water Pollution Agency by Engineering-Science, Berkeley, CA.

Fannin, K. F., K. W. Cochran, D. E. Lamphiear, and A. S. Monto (1980) "Acute Illness Differences with Regard to Distance from the Tecumseh, Michigan Wastewater Treatment Plant," 117–135, in H. Pahren and W. Jakubowski (eds.) *Wastewater Aerosols and Disease,* EPA-600/9-80-028, Health Effects Research Laboratory, U.S. Environmental Protection Agency, Cincinnati, OH.

Florida Department of Environmental Protection (1999) *Reuse of Reclaimed Water and Land Application,* Chap. 62–610, *Florida Administrative Code.* Florida Department of Environmental Protection, Tallahassee, FL.

Hunter, P. R., and L. Fewtrell (2001) "Acceptable Risk," Chap. 10, in L. Fewtrell and J. Bartram (eds.) *Water Quality: Guidelines, Standards and Health; Assessment of Risk and Risk Management For Water-Related Infectious Disease,* IWA Publishing, London.

Johnson, D. E., D. E. Camann, D. T. Kimball, R. J. Prevost, and R. E. Thomas (1980) "Health Effects from Wastewater Aerosols at a New Activated Sludge Plant: John Egan Plant, Schaumburg, Illinois," 136–159, in H. Pahren and W. Jakubowski (eds.), *Wastewater Aerosols and Disease,* EPA-600/9-80-028, U.S. Environmental Protection Agency, Cincinnati, OH.

Leenheer, J. (2003) *Comprehensive Characterization of Dissolved and Colloidal Organic Matter in Waters Associated with Groundwater Recharge at the Orange County Water District,* U.S. Geological Survey, Denver, CO.

Macler, B. A., and S. Regli (1993) "Use of Microbial Risk Assessment in Setting U.S. Drinking Water Standards," *Intern. J. Food Microbiol.,* **18**, 4, 245–256.

Mara, D. (2003) *Domestic Wastewater Treatment in Developing Countries,* Earthscan, London.

Mara, D., and H. Pearson (1998) *Design Manual for Waste Stabilization Ponds in Mediterranean Countries,* Lagoon Technology International Ltd., Leeds, UK.

McGauhey, P. H. (1968) *Engineering Management of Water Quality,* McGraw-Hill, New York.

McKee, J. E., and H. W. Wolf (eds.) (1963) *Water Quality Criteria,* 2nd ed., Publication 3-A, California State Water Resources Control Board, Sacramento, CA.

Murray, C. J. L., and A.D. Lopez (1996) *The Global Burden of Disease,* Vol. 1, Global Burden of Disease and Injury Series, Harvard School of Public Health, Cambridge, MA.

NRC (1994) *Ground Water Recharge Using Waters of Impaired Quality,* National Research Council, National Academy Press, Washington, DC.

NRC (1996) *Use of Reclaimed Water and Sludge in Food Crop Production,* National Research Council, National Academy Press, Washington, DC.

NRC (1998) *Issues in Potable Reuse: The Viability of Augmenting Drinking Water Supplies with Reclaimed Water*, National Research Council, National Academy Press, Washington, DC.

NWRI (2003) *Ultraviolet Disinfection Guidelines for Drinking Water and Water Reuse*. Report Number NWRI-2003-06, National Water Research Institute, Fountain Valley, CA.

Okun, D.A. (1979) *Criteria for Reuse of Wastewater for Nonpotable Urban Water Supply Systems in California*. Report prepared for the California Department of Health Services, Sanitary Engineering Section, Berkeley, CA.

Prüss, A., and A. Havelaar (2001) "The Global Burden of Disease Study and Applications in Water, Sanitation, and Hygiene," Chap. 3, in L. Fewtrell and J. Bartram (eds.) *Water Quality: Guidelines, Standards and Health; Assessment of Risk and Risk Management for Water-Related Infectious Disease*, IWA Publishing, London, UK.

Regli, S., J. B. Rose, C. N. Haas, and C. P. Gerba (1991) "Modeling the Risk from *Giardia* and Viruses in Drinking Water," *J. AWWA*, **83**, 11, 76–84.

SDLAC (1977) *Pomona Virus Study: Final Report*. Sanitation Districts of Los Angeles County, California State Water Resources Control Board, Sacramento, CA.

Shuval, H. I., Y. Lampert, and B. Fattal (1997) "Development of a Risk Assessment Approach for Evaluating Wastewater Reuse Standards for Agriculture," *Water Sci. Technol.*, **35**, 11–12, 15–20.

State of California (2000) *Cross-Connection Control by Water Users. Health and Safety Code*, Division 104, Part 12, Chap. 5, Art. 2, California Department of Health Services, Sacramento, CA.

Teltsch, B., S. Kidmi, L. Bonnet, Y. Borenzstajn-Roten, and E. Katzenelson (1980) "Isolation and Identification of Pathogenic Microorganisms at Wastewater-Irrigated Fields: Ratios in Air and Wastewater," *App. Environ. Microbiol.*, **39**, 1184–1195.

U.S. Department of Health, Education, and Welfare (1962) *Public Health Service Drinking Water Standards*, Public Health Service, U.S. Department of Health, Education, and Welfare, Washington, DC.

U.S. EPA (1974) *Design Criteria for Mechanical, Electric, and Fluid System and Component Reliability. MCD-05*. EPA-430/99-74-001, Office of Water Program Operations, U.S. Environmental Protection Agency, Washington, DC.

U.S. EPA (1975) *National Interim Primary Drinking Water Regulations, Fed. Reg.* **40**(248): 59566–59588.

U.S. EPA (1980) "Wastewater Aerosols and Disease," in H. Pahren and W. Jakubowski (eds.) *Proceedings of a Symposium*, 1979, EPA-600/9-80-028, Health Effects Research Laboratory, U.S. Environmental Protection Agency, Cincinnati, OH.

U.S. EPA (1981) *Process Design Manual for Land Treatment of Municipal Wastewater*, EPA-625/1-81-013, Center for Environmental Research Information, U.S. Environmental Protection Agency, Cincinnati, OH.

U.S. EPA (1983) *Design Manual: Wastewater Stabilization Ponds*. EPA-625/1-83-015, Office of Research and Development, U.S. Environmental Protection Agency, Washington, DC.

U.S. EPA (1992) *Manual: Guidelines for Water Reuse*, EPA-625/R-92-004, U.S. Environmental Protection Agency and U.S. Agency for International Development, Washington, DC.

U.S. EPA (2004) *Guidelines for Water Reuse*, EPA-625/R-04-108, U.S. Environmental Protection Agency and U.S. Agency for International Development, Washington, DC.

WHO (1973) *Reuse of Effluents: Methods of Wastewater Treatment and Health Safeguards*, Report of a WHO Meeting of Experts, Technical Report Series No. 17, World Health Organization, Geneva, Switzerland.

WHO (1989) *Health Guidelines for the Use of Wastewater in Agriculture and Aquaculture*, Report of a WHO Scientific Group, Technical Report Series 778, World Health Organization, Geneva, Switzerland.

WHO (1996) *Guidelines for Drinking-Water Quality,* Vol. 2, *Health Criteria and Other Supporting Information.* 2nd ed., World Health Organization, Geneva, Switzerland.

WHO (2003) *Guidelines for Drinking Water Quality*, 3rd ed., World Health Organization, Geneva, Switzerland.

WHO (2005) *Meeting Report: Final Expert Review Meeting for the Finalization of the Third Edition of the WHO Guidelines for the Safe Use of Wastewater, Excreta and Greywater, 2005,* World Health Organization, Geneva, Switzerland.

WHO (2006) *WHO Guidelines for the Safe Use of Wastewater, Excreta and Greywater,* 3rd ed., Vol. 2, *Wastewater Use in Agriculture,* World Health Organization, Geneva, Switzerland.

5 Health Risk Analysis in Water Reuse Applications

WORKING TERMINOLOGY 192

5-1 RISK ANALYSIS: AN OVERVIEW 193
- Historical Development of Risk Assessment 194
- Objectives and Applications of Human Health Risk Assessment 194
- Elements of Risk Analysis 194
- Risk Analysis: Definitions and Concepts 196

5-2 HEALTH RISK ASSESSMENT 197
- Hazard Identification 198
- Dose-Response Assessment 198
- Dose-Response Models 200
- Exposure Assessment 204
- Risk Characterization 204
- Comparison of Human Health and Ecological Risk Assessment 205

5-3 RISK MANAGEMENT 205

5-4 RISK COMMUNICATION 206

5-5 TOOLS AND METHODS USED IN RISK ASSESSMEMT 207
- Concepts from Public Health 207
- Concepts from Epidemiology 208
- Concepts from Toxicology 209
- National Toxicology Program Cancer Bioassay 213
- Ecotoxicology: Environmental Effects 214

5-6 CHEMICAL RISK ASSESSMENT 215
- Safety and Risk Determination in Regulation of Chemical Agents 215
- Risks from Potential Nonthreshold Toxicants 220
- Risk Considerations 224
- Chemical Risk Assessment Summary 225

5-7 MICROBIAL RISK ASSESSMENT 225
- Infectious Disease Paradigm for Microbial Risk Assessment 225
- Microbial Risk Assessment Methods 227
- Static Microbial Risk Assessment Models 227
- Dynamic Microbial Risk Assessment Models 229
- Selecting a Microbial Risk Model 232

5-8 APPLICATION OF MICROBIAL RISK ASSESSMENT IN WATER REUSE APPLICATIONS 234
Microbial Risk Assessment Employing a Static Model 234
Microbial Risk Assessment Employing Dynamic Models 239
Risk Assessment for Water Reuse from Enteric Viruses 244

5-9 LIMITATIONS IN APPLYING RISK ASSESSMENT TO WATER REUSE APPLICATIONS 249
Relative Nature of Risk Assessment 249
Inadequate Consideration of Secondary Infections 249
Limited Dose-Response Data 250

DISCUSSION TOPICS AND PROBLEMS 250

REFERENCES 251

WORKING TERMINOLOGY

Term	Definition
Acute toxicity	A toxicity effect occurring shortly after a single exposure event.
Anthropogenic	Human-induced or resulting from human activities. The term is used in this textbook in the context of chemical compounds or biological emissions that are produced as a result of human activities often used to refer to environmental changes.
Cancer potency	The upper 95 percent confidence limit slope of the dose/response relationship when graphed (see Fig. 5-2) for a carcinogen as the dose approaches zero.
Chemical carcinogen	A chemical that has been shown to produce tumors in either experimental animals or in humans.
Chemical noncarcinogen	A chemical that can produce adverse effects other than tumors in experimental animals or humans.
Chronic toxicity	A toxic effect occurring from exposure over a long period of time (e.g., a year).
Drinking water equivalent level (DWEL)	The concentration of a chemical in water at which no adverse noncancer health effect is anticipated over a person's lifetime, assuming a typical adult weight of 70 kg and a daily water consumption of 2 L.
Ecological risk assessment	Evaluation of available toxicological and ecological information for the purpose of estimating the probability that some undesired ecological event will occur.
Ecotoxicology	The study of the fate and effects of toxic substances on ecosystems.
Exposure	Contact with a chemical or physical agent by ingestion, inhalation, or dermal routes.
Hazard	The intrinsic capacity of a substance to cause harm.
Health risk assessment	An evaluation of the potential for adverse health effects to occur as a result of actual or potential exposures to chemicals.

Maximum contaminant level (MCL)	Enforceable drinking water standards applicable to public water supplies. They are set as close to the MCLG as feasible using the best available analytical and treatment technology and taking cost into consideration.
Maximum contaminant level goal (MCLG)	A nonenforceable regulatory health goal set at a level at which no known or anticipated adverse effect on health occurs with an adequate margin of safety.
Reference dose (RfD)	An estimate (with uncertainty spanning perhaps an order of magnitude) of a daily dose (usually expressed in [(mg/kg)/d]) to the human population, including sensitive subgroups, that is likely to be without an appreciable risk of deleterious effects over a lifetime of exposure.
Risk	The probability that an organism exposed to a specified hazard will have an adverse response.
Risk analysis	Risk analysis consists of three principal elements: (1) risk assessment, (2) risk management, and (3) risk communication.
Risk assessment	The qualitative or quantitative characterization and estimation of potential adverse health effects associated with exposure of individuals or populations to hazardous materials and situations.
Risk communication	Interactive exchange of information and opinions concerning risk and risk management among risk assessors, risk managers, consumers, and other interested parties about the nature, magnitude, significance, or control of a risk.
Risk management	The process of evaluating and, if necessary, controlling sources of exposure and risk. Sound environmental risk management means weighing many different attributes of a decision and developing alternatives.
Vector	An organism, such as a mosquito or tick that carries disease-causing microorganisms from one host to another (see also Chap. 4).

The need to quantify the health risks associated with exposure to environmental and occupational toxicants has generated an interdisciplinary methodology referred to as health risk analysis. Health risk analysis is potentially a useful tool for comparing the risk to human health due to exposure to microbiological, natural, and anthropogenic constituents in water and wastewater. Health risk analysis in water reuse originated from U.S. drinking water regulations. Although health risk analysis is still in its infancy in water reuse, knowledge has been accumulating rapidly in recent years. The purpose of this chapter is to introduce and discuss health risk analysis, and how it relates to water reuse applications. The topics introduced in this chapter include (1) an overview of risk analysis, (2) health risk assessment, (3) risk management, (4) risk communication, (5) tools and methods in risk assessment, (6) chemical risk assessment, (7) microbial risk assessment, (8) microbial risk assessment in water reuse applications, and (9) limitations in applying risk analysis to water reuse applications.

5-1 RISK ANALYSIS: AN OVERVIEW

Although the focus in this chapter is human health risk analysis, to understand risk analysis it is useful to consider (1) the historical development of risk assessment, (2) the objectives and applications of health risk assessments, (3) the steps involved in a health risk analysis, and (4) definitions and concepts pertaining to risk analysis.

Historical Development of Risk Assessment

Risk assessment as a formal discipline emerged in the 1940s and 1950s paralleling the onset of the nuclear industry and its regulatory activities. Safety hazard analyses (a type of risk assessment) have been used since the 1950s in the nuclear, petroleum refining, chemical, and aerospace industries. Human health risk assessment began in the 1980s with the publication of *Carcinogenic Risk Assessment Guidelines* (U.S. EPA, 1986) and continued to grow propelled by the "Superfund"—Comprehensive Environmental Response, Compensation, and Liability Act (CERCLA)—and the Resource Conservation and Recovery Act (RCRA) programs. Growing public interest in ecological resources with reference to sustainable development has provided further impetus for evaluating ecological risks (Kolluru et al., 1996). Selected milestones in the evolution of risk assessment are reported in Table 5-1.

Objectives and Applications of Human Health Risk Assessment

For given conditions of exposure, risk assessment is used to provide (Hallenbeck and Cunningham, 1986):

- A characterization of the type of health effects expected
- An estimate of the probability (risk) of occurrence of these health effects
- An estimate of the number of cases with these health effects
- A suggested acceptable concentration of a hazard in air, water, or food

The outputs of risk assessment are necessary for making informed regulatory decisions regarding worker exposures, plant emissions and effluents, ambient air and water exposures, chemical residues in foods, waste disposal sites, consumer products, and naturally occurring contaminants. Risk assessment and risk management are an integral part of contemporary regulatory activities of federal and state regulatory agencies. Key objectives and the advantages and concerns of risk assessments are listed in Table 5-2.

Elements of Risk Analysis

Considerable uncertainty pervades health risk analyses because of multifactorial causation, disease occurrence in unexposed populations (background noise), and long latency periods, the cause-effect relationship being at best tenuous. For example, humans are all exposed to thousands of chemicals every day, most of which are not likely to cause disease at the low concentrations to which they are generally exposed. However, discovering a true hazard may be difficult because some diseases, especially cancers, have a long latency of 10 to 20 or more years. Ecological risks may be even more difficult to assess because the effects may not be evident except in retrospect, if at all, because of natural fluctuations, instability, and resilience of ecosystems.

Risk analysis consists of three principal elements: (1) risk assessment, (2) risk management, and (3) risk communication. Risk assessment is the qualitative or quantitative characterization and estimation of potential adverse health effects associated with exposure of individuals or populations to hazardous materials and situations (Hoppin, 1993). In risk management, policy alternatives are examined in light of the results of risk assessment and, if required, appropriate control options are selected and implemented, including regulatory measures. Risk communication is the interactive exchange of information and opinions concerning risk and risk management among risk assessors, risk managers, consumers, and other interested parties (Charnley et. al., 1997; WHO, 1999). Each of these elements is considered further in Secs. 5-2, 5-3, and 5-4, respectively.

Table 5-1
Selected milestones in the development of quantitative risk assessment[a]

Year	Milestone
1938	Federal Food, Drug, and Cosmetic Act
1940s–50s	Development and application of probabilistic techniques in atomic energy and aerospace operations (HAZOP, failure mode, fault-tree techniques, and procedures)
	Food, Drug, and Cosmetic Act Amendments—Delaney clause (which prohibits the use of any food additive or animal drug if it is found to induce cancer in humans or animals, exemplifies the zero-risk ideal)
1958–1975	WASH-1400 Reactor Safety Study (Rasmussen), U.S. Nuclear Regulatory Commission
1976	Publication of the U.S. EPA carcinogenic risk assessment guidelines (first quantification of chemical cancer risks following radiation cancer risks)
1980s	Renewed emphasis on protecting human health, especially from carcinogenic risks, e.g., U.S. EPA water quality criteria based on 10^{-7} to 10^{-5} risk, linearized multistage dose-response model
1980	Supreme Court ruling that OSHA should prove health benefit of lowering benzene limit from 10 ppm
1981	First publication of the Society for Risk Analysis journal *Risk Analysis*
1983	National Research Council (NRC) Report: *Risk Assessment in the Federal Government: Managing the Process*
1985	California Department of Health Services: "Guidelines for chemical carcinogens: risk assessment and their scientific rationale"

1986	U.S. EPA formalized risk assessment guidelines:
	    Guidelines for carcinogen risk assessment
	    Guidelines for developmental toxicity risk assessment
	    Guidelines for exposure assessment
	Superfund Public Health Evaluation Manual
	Mounting importance of risk communications in risk management (SARA Title III, 1986)

1987	Publication of the U.S. EPA report *Unfinished Business: A Comparative Assessment of Environmental Problems*
1988	California Department of Health Services: Guidelines and safe use determination procedures for the *Safe Drinking Water and Toxic Enforcement Act of 1986* (Proposition 65)
1989	Publication of the U.S. EPA report *Risk Assessment Guidance for Superfund (RAGS), Human Health Evaluation Manual; Environmental Evaluation Manual*
1990	U.S. EPA Science Advisory Board publication of *Reducing Risk: Setting Priorities and Strategies for Environmental Protection* U.S. EPA risk management programs under the *Clean Air Act Amendments of 1990* (air toxics, accidental release prevention)
1990s	OSHA *Process Safety Management (PSM) Standard* (1992). Growing emphasis on noncancer (e.g., reproductive) effects; increasing use of pharmacokinetic models, toxicity equivalence, (e.g., dioxins, PAHs) New guidelines for exposure; developmental, reproductive, and neurotoxicities; carcinogenic risk; and indoor exposures (air quality) Increasing attention to ecological/environmental impacts International harmonization of risk issues through WHO, UNEP, OECD, etc. Expanded use of risk and cost-benefit criteria in environmental decision-making

[a] Adapted from Paustenback (1989), Kolluru et al. (1996).

Table 5-2
Risk assessments: objectives, advantages, and limitations[a]

Item	Objectives, advantages, and limitations
Objectives	• Obtain perspective on different sources and nature of risk—gain insights into risks across sources, space, and time • Identify worst risks as well as investment-sensitive and time-sensitive risks • Seek a systematic framework for optimal resource allocation to avoid or control risks • Estimate the likelihood of an adverse effect on humans, wildlife, or ecological systems posed by a specific level of exposure to chemical or microbial agents
Advantages	• Bottom line public health and safety concerns addressed with a common language • Systematic framework for prioritizing problems, allocating resources, and avoiding future problems • Scientific underpinnings for risk management
Limitations	• No broad consensus on the purpose, the approach, or the results; inadequate data, speculative and myopic nature of assumptions • Few qualified professionals with needed range of skills; risk assessors, engineers, and economists talk different languages • Multiple clients, diverse interests, unrealistic expectations, credibility problems

[a] Adapted from Kolluru et al. (1996).

Risk Analysis: Definitions and Concepts

Risk assessment can be defined broadly as the process of estimating the probability of occurrence of an event and the probable magnitude of adverse effects on safety, health, ecology, or finances over a specified time period. Perhaps the most widely cited definition of health risk assessment as applied to human health is the one given by the National Research Council in 1983 (NRC, 1983).

> We use *risk assessment* to mean the characterization of the potential adverse health effects of human exposures to environmental hazards. Risk assessments include several elements: description of the potential adverse health effects based on an evaluation of results of epidemiologic, clinical, toxicologic, and environmental research; extrapolation from those results to predict the type and estimate the extent of health effects in humans under given conditions of exposure; judgments as to the number and characteristics of persons exposed at various intensities and durations; and summary judgments on the existence and overall magnitude of the public-health problem. Risk assessment also includes characterization of the uncertainties inherent in the process of inferring risk.

The purpose of risk assessment is to provide complete information to risk managers, specifically, policy makers and regulators, so that they can make decisions based on the best information available. Factors other than those addressed in a risk assessment include societal concerns such as equity, control, and trust can, however, influence decisions about risk. More specifically, a *health risk assessment* is a written document wherein all the pertinent scientific information, regarding toxicology, epidemiology, human experience, environmental fate, and exposure are assembled, critiqued, and interpreted in the absence of moral judgments.

The NRC emphasized that the processes of risk assessment and risk management should be separate activities, under the umbrella of risk analysis. This was because many assessments were laden with value judgments and subjective views of the risk, not just scientific information (Paustenbach, 1989). To encourage this separation of assessment and management by scientists, policymakers, and the public, the NRC report contained the following definition of risk management (NRC, 1983):

> The Committee uses the term *risk management* to describe the process of evaluating alternative regulatory actions and selecting among them. Risk management, which is carried out by regulatory agencies under various legislative mandates, is an agency decision-making process that entails consideration of political, social, economic, and engineering information with risk-related information to develop, analyze, and compare regulatory options and to select the appropriate regulatory response to a potential chronic health hazard. The selection process necessarily requires the use of value judgments on such issues as the acceptability of risk and the reasonableness of the costs of control.

5-2 HEALTH RISK ASSESSMENT

The focus of this section is on the health risk assessment component of the overall risk analysis process. The health risk assessment component is the qualitative or quantitative characterization and estimation of potential adverse health effects associated with exposure of individuals or populations to hazardous materials and situations. Health risk assessment can be divided into four major steps including (1) hazard identification, (2) dose-response assessment, (3) exposure assessment, and (4) risk characterization (NRC, 1983). The interrelationship of these steps is illustrated on Fig. 5-1. Health risk assessment includes chemical and microbial risk assessment, which are further discussed in Secs. 5-6 and 5-7.

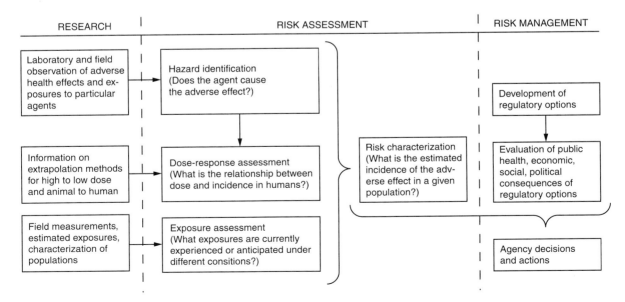

Figure 5-1
Elements of risk assessment and risk management. (Adapted from NRC, 1983.)

Hazard Identification

Hazard identification, defined as the process of determining whether exposure to an agent can cause an increase in the incidence of a health condition (such as cancer, birth defect), is the most easily recognized in the actions of regulatory agencies. In water reuse, the principal hazards are chemical and microbial constituents in reclaimed water.

Chemical Constituents

Chemical hazard identification involves characterizing the nature and strength of the evidence of causation. Although the question of whether a chemical constituent can cause cancer or other adverse health effects is theoretically a yes-no question, there are few chemicals on which human data are definitive. Therefore, the question is often restated in terms of effects in laboratory animals or other test systems, for example, "Does the agent induce cancer in test animals?" Positive answers to such questions are taken typically as evidence that an agent may pose a cancer risk for any exposed humans. Information from short-term *in vitro* (test tube) tests and on structural similarity to known chemical hazards may also be considered.

Microbial Constituents

For microbial agents, the purpose of hazard identification is to identify the microorganisms or the microbial toxins of concern. Hazards can be identified from relevant data sources such as scientific literature, databases, and solicitation of expert opinion. Relevant information for the hazard identification often includes review of clinical studies, epidemiological studies and surveillance, laboratory animal studies, investigations of the characteristics of microorganisms, interaction between microorganisms and their environment, and studies on analogous microorganisms and situations.

Dose-Response Assessment

Dose-response assessment is the process of characterizing the relationship between the dose of an agent administered or received and the incidence of an adverse health effect in exposed populations and then estimating the incidence of the effect as a function of human exposure to the agent (see Fig. 5-2a).

Chemical Constituents

Factors considered in developing dose-response relationships for chemical constituents include intensity of exposure, age pattern of exposure, and possibly other variables that might affect response, such as gender, lifestyle, and other modifying factors. A dose-response assessment usually requires extrapolation from high dose to low dose and extrapolation from animal test results to estimate human effects. The extrapolation methods used to predict incidence and the statistical and biologic uncertainties in these methods must be delineated and justified carefully (see Figs. 5-2b and 5-3a). A detailed discussion on chemical dose-response extrapolation is given in Sec. 5-6.

Microbial Constituents

For the microbial risk assessment, the dose-response assessment provides a quantitative or qualitative description of the likelihood, severity and/or duration of adverse effects that may result from exposure to a microorganism or its toxin. Dose-response relationships

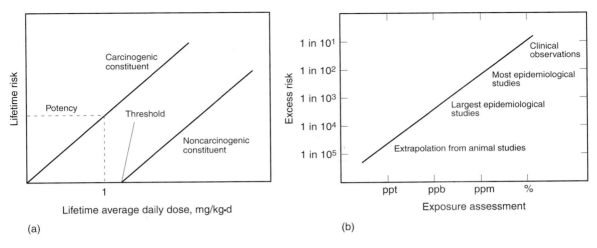

Figure 5-2
Definition illustrations for risk assessment: (a) dose-response curves for carcinogenic and noncarcinogenic constituents (as shown, it is assumed that dose-response curve for a carcinogenic constituent has no threshold value) and (b) relative sensitivity of epidemiological studies in defining excess risk. (Adapted from NRC, 1993.)

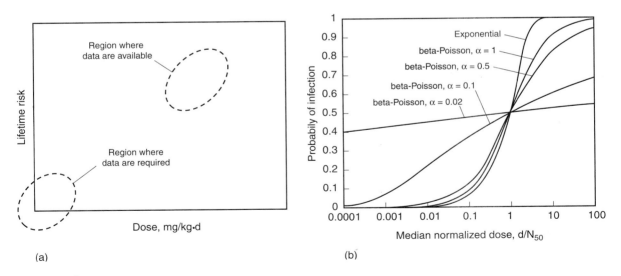

Figure 5-3
Definition sketch for dose-response curves: (a) illustration of where data are available and where data are required and (b) comparison of exponential and beta-Poisson dose-response functions. (Adapted from Haas and Eisenberg, 2001.)

Dose-Response Models

can be developed for different end points, such as infection or illness, depending on the microorganism of interest. In the absence of appropriate dose-response data, risk assessment tools such as expert elicitations could be used to consider factors such as infectivity that may be necessary to describe hazard characterizations.

To predict what will happen at extremely low concentrations using dose-response data obtained at high concentration values, the reported dose-response data are fit to models that relate the probability of infection to the mean dose ingested. In some cases, illness as an end-point is also investigated; however, the conditional modeling of illness given infection has proven to be difficult (Teunis and Havelaar, 1999). Typical dose-response models that have been proposed and used for human exposure include (1) single-hit models, (2) multistage model, (3) linear multistage model, (4) multi-hit model, (5) beta-Poisson, and (6) probit model. The characteristics of these models are summarized in Table 5-3. The single-hit exponential, multistage, and beta-Poisson models, used most frequently in both chemical and microbial risk assessment, are described in the following sections, using infection as an endpoint.

Single-Hit Models

The simplest form of the single-hit model is:

$$P_{inf}(n_p, r) = 1 - (1 - r)^{n_p} \qquad (5\text{-}1)$$

where P_{inf} = the probability of infection which is a function of n and r
n_p = number of pathogens ingested
r = the nonzero probability that an ingested pathogen will survive all barriers and colonize the host

Table 5-3 Models used to assess nonthreshold effects of toxic constituents[a]

Model[b]	Description
Single-hit	A single exposure can lead to the development of a tumor
Multi-stage	The formation of a tumor is the result of a sequence of biological events
Linear multi-stage	Modification of the multistage model. The model is linear at low doses with a constant of proportionality that statistically will produce less than five percent chance of underestimating risk
Multi-hit	Several interactions are required before cell becomes transformed
beta-Poisson	The model is based on similar assumptions to the exponential model except that the third assumption (that the probability of infection per ingested organism is constant) is relaxed. In the beta-Poisson model, the probability of surviving and reaching a host site ("r" in the exponential model) is beta distributed, and thus the model contains the two parameters (α and β) of the beta distribution
Probit	Tolerance of exposed population is assumed to follow a lognormal (probit) distribution

[a] Adapted from Cockerham and Shane (1994); Pepper et al. (1996).
[b] In all of the models cited above, it is assumed that exposure to the toxic constituent will always produce an effect regardless of the dose.

The exponential single-hit model, which is derived from the single-hit model given above, is based on the following assumptions (Haas et al., 1999):

- Microorganisms are distributed in water randomly and thus, follow the Poisson distribution as explained below
- For infection to occur, at least one pathogen must survive within the host
- The probability of infection per ingested or inhaled organism is constant

The mathematical function used to describe the relationship between risk and dose for the single-hit exponential model is:

$$P_{inf}(r, d) = 1 - \exp(-rd) \qquad (5\text{-}2)$$

where P_{inf} = the probability of infection which is a function of r and d
 r = empirical parameter assumed to be constant for any given host and given pathogen picked to fit the data
 d = mean ingested dose

The dose-response relationship for many protozoans and viruses tend to follow this model. The biological implication of this model is that differential susceptibility in the challenged population tends to be weak, that is, members of the challenged population are equally likely to become infected (McBride et al., 2002).

Multi-stage Model

The mathematical formulation used to describe the relationship between risk and dose for the multi-stage model is:

$$P_{inf}(r_i, d, s) = 1 - \exp[-\sum_{i=0}^{s}(r_i\, d^s)] \qquad (5\text{-}3)$$

where P_{inf} = the probability of infection which is a function of r, d, and n
 r_i = positive empirical parameters picked to fit the data
 d = mean ingested dose
 s = number of stages

Beta-Poisson Model

The beta-Poisson model is based on similar assumptions to the exponential model except that the third assumption (that the probability of infection per ingested organism is constant) is relaxed. This model allows the probability of infection per ingested or inhaled organism to vary with the population. In the beta-Poisson model, the probability of surviving and reaching a host site in the exponential model is beta distributed, and thus the model contains the two parameters (α and β) of the beta distribution. The most commonly used approximation to the beta-Poisson model has the following two approximate forms, depending on the how the dose term is defined:

$$P_{inf}(d, \alpha, \beta) \simeq 1 - \left(1 + \frac{d}{\beta}\right)^{-\alpha} \qquad (5\text{-}4)$$

where P_{inf} = the probability of infection which is a function of d, α, and β
$\quad$ d = mean ingested dose
$\quad$ β = a slope parameter, which holds when $\beta \geq 1$ and $\alpha \leq \beta$
$\quad$ α = a slope parameter

and

$$P_{inf}(d, \alpha, N_{50}) \simeq 1 - \left[1 + \frac{d}{N_{50}}(2^{1/\alpha} - 1)\right]^{-\alpha} \tag{5-5}$$

where N_{50} = the median dose
$\quad$ other terms are defined as above.

The median dose is given by the following expression:

$$N_{50} = \frac{\beta}{2^{1/\alpha} - 1} \tag{5-6}$$

Often it is necessary to convert interchangeably between annual and daily probability of infection, as shown in the following expression.

$$P_{yr} = 1 - (1 - P_d)^n \tag{5-7}$$

where P_{yr} = acceptable annual risk of infection caused by a pathogenic organism
$\quad$ P_d = acceptable daily (single) exposure risk caused
$\quad$ n = number of exposure events per year (events/yr)

For more rigorous discussion of the beta-Poisson model refer to Haas et al. (1999). Unfortunately, in this approximation to the beta-Poisson model, α does not have an obvious physical interpretation. What can be said is that it is a shape parameter governing the steepness of the dose-response curve; the larger its value the steeper the curve (McBride, et al., 2002).

New methods for dose-response assessment relying on Bayesian approaches have begun to appear in the literature over the last several years. A key feature of Bayesian approaches is the notion of using an empirically derived probability distribution for a population parameter. The Bayesian approach permits the use of objective data or subjective opinion in specifying a prior distribution (Messner et al., 2001; Englehardt, 2004; Englehardt and Swartout, 2004).

Model Coefficients

Coefficients for the various models discussed above are given in Table 5-4 for a variety of microorganisms. The relationship between the single-hit exponential and the beta-Poisson models is illustrated on Fig. 5-3b. The effect of the slope parameter on the beta-Poisson dose-response relationship is also depicted on Fig. 5-3b. The beta-Poisson is linear at low doses and is always shallower than the exponential model. However, as $\alpha \to \infty$, the beta-Poisson model approaches the exponential model (Haas et al., 1999). The use of these coefficients is illustrated in Example 5-1.

Table 5-4
Summary of dose-response parameters for exponential and beta-Poisson models from various enteric pathogen ingestion studies

		Model		
	Exponential	beta-Poisson		
Constituent	r	α	β	Reference
Virus				
Echovirus 12		0.374	186.69	Regli et al. (1991)
Rotavirus		0.253	0.422	Ward et al. (1986)
Poliovirus 1	0.009102	0.1097	1524	Regli et al. (1991)
Poliovirus 3		0.409	0.788	Regli et al. (1991)
Bacteria				
Salmonella	0.00752			Regli et al. (1991)
		0.33	139.9	
Shigella flexneri		0.2	2000	
Escherichia coli		0.1705	1.61 × 10⁶	Regli et al. (1991)
Campylobacter jejuni		0.145	7.589	Black et al. (1988)
		0.039	55	
Vibrio cholerae		0.097	13,020	
Protozoa				
Cryptosporidium	0.004191			Regli et al. (1991)
Giardia lamblia	0.02			Regli et al. (1991)

EXAMPLE 5-1. Application of the beta-Poisson Dose-Response Model.

A drinking water source contaminated with *Campylobacter jejuni* contains 1200 organisms/100 mL. Using the beta-Poisson model, estimate the probability of infection for an individual who ingests 250 mL of the drinking water. The coefficients for the beta-Poisson model for *C. jejuni* have been determined to be $\alpha = 0.145$ and $\beta = 7.589$ (see Table 5-4).

Solution

1. Calculate the dose obtained from ingestion of the drinking water.

   Dose = (1200 org/100 mL)(250 mL) = 3000 organisms

2. Estimate the probability of infection using Eq. (5-4).

$$P_{inf} \simeq 1 - \left(1 + \frac{d}{\beta}\right)^{-\alpha} \simeq 1 - \left(1 + \frac{3000 \text{ org}}{7.589}\right)^{-0.145} = 0.58$$

Comment

As shown in the above computation, ingestion of 3000 *C. jejuni* cells is expected to result in infection in 58 percent of individuals. A portion of the infected individuals may further develop a clinical illness (a disease with clinical signs and symptoms that are recognizable).

(a)   (b)

Figure 5-4
Examples of exposure to reclaimed water in the urban environment: (a) child running in park irrigated with reclaimed water in southern California (Courtesy of A. Bahri) and (b) children playing in an artificial stream (known as *Seseragi* in Japanese) in the urban environment where reclaimed water [microfiltration and reverse osmosis followed by low dose chlorination (~0.1 mg/L)] is used for water features and even for body contact recreation (Courtesy of Tokyo Metropolitan Government).

Exposure Assessment

Exposure assessment is the process of measuring or estimating the intensity, frequency, and duration of human exposures to an agent currently present in the environment, or of estimating hypothetical exposures that might arise from the release of new chemicals into the environment. In its most complete form, the magnitude, duration, schedule, and route of exposure; the size, nature, and classes of the human populations exposed; and the uncertainties in all estimates must be quantified. Exposure assessment is often used to identify feasible prospective control options and to predict the effects of available control technologies on exposure (see Fig. 5-4).

For microbial risk assessment, exposure assessment describes the magnitude and/or probability of actual or anticipated human exposure to pathogenic microorganisms or microbiological toxins. For microbiological agents, exposure assessments may be based on the potential contamination in water by a particular agent or its toxins, and on other exposure pattern information (e.g., the frequency and/or duration of exposure).

Factors that must be considered for exposure assessment include the frequency of human exposure to the pathogenic agents and the associated concentrations of those pathogens over time. Another factor that may be considered in the assessment is the pattern of consumption. Consumption patterns may be related to socioeconomic status, ethnicity, seasonality, age (population demographics), regional differences, and/or consumer preferences and behavior. Other factors to be considered include the potential impact of environmental conditions and/or treatment system reliability, if appropriate.

Risk Characterization

Risk characterization is the process of estimating the incidence of a health effect under the various conditions of human exposure described in exposure assessment. In addition, risk characterization may require compiling all of the data necessary for a given model and running simulations. It is performed by combining the exposure and dose-response assessments. The summary effects of the uncertainties in the preceding steps are described in this step.

Risk characterization represents the integration of the hazard identification, dose-response assessment, and exposure assessment components to obtain a risk estimate. The risk characterization process results in a qualitative or quantitative estimate of the likelihood and severity of the adverse effects that may occur in a given population, including a description of the uncertainties associated with these estimates.

Risk characterization depends on available data and expert interpretation of the data. The weight of evidence integrating quantitative and qualitative data may permit only a qualitative estimate of risk. The degree of confidence in the final estimation of risk depends on the variability, uncertainty, and assumptions identified in all previous steps (WHO, 1999).

Comparison of Human Health and Ecological Risk Assessment

Although the focus of this chapter has been on human health risk assessment, it is important to note that there are corresponding parallels to ecological risk assessment (ERA). Ecological risk assessment is a process that is used to evaluate the likelihood that adverse ecological effects may occur or are occurring as a result of exposure to one or more stressors (U.S. EPA, 1992). Ecological risk assessment is of particular importance when reclaimed water is used for augmentaion of aquatic systems, habitat ehancement, and other environmental uses (see also Chap. 21). A comparison of the major steps, typical end points, and applications between human health and ERA is presented in Table 5-5. As shown in Table 5-5, there is a wide range of applications for ERA, many of which are relevant to water reclamation and reuse.

5-3 RISK MANAGEMENT

In risk management, policy alternatives are weighed in light of the results of risk assessment and, if required, appropriate control options are selected and implemented including regulatory measures (Cothern 1992; Charnley et. al., 1997). Risk management is the process of evaluating and, if necessary, controlling sources of exposure and risk. Sound environmental risk management means weighing many different attributes of a decision and developing alternatives.

The scientific information provided by risk assessment is but one input to the process. Other criteria include politics, economics, competing risks and equity, and other social concerns. Although risk assessment is rooted in science, how useful its results are to risk management depends on the questions it is designed to answer, how it is conducted, and the way it is structured. Unfortunately, too many risk assessments prove to be of little value to risk managers because of inadequate planning (Kolluru, et al., 1996). While the intent of keeping risk assessment separate from management issues is to avoid prejudgment of the results by cost implications and value judgments (isolate science from politics and policy), the assessment and management phases often suffer from this disjunction in practice.

Table 5-5
Overview and comparison of human health and ecological/environmental risk assessments[a]

Human health	Ecological/environmental
Major steps	
1. Data analysis/hazard identification Quantities and concentrations of chemical, physical, and biological agents in environmental media at a site or study area; selection of chemicals of concern.	1. Problem formulation (hazard screening) Resident and transient flora and fauna, especially endangered or threatened species; aquatic, and terrestrial surveys; contaminants, and stresses of concern in study boundary.
2. Exposure assessment Pathways and routes, potential receptors including sensitive subgroups, exposure rates, and timing.	2. Exposure assessment Pathways, habitats, or receptor populations, especially valued and protected species; exposure point concentrations.
3. Dose-response or toxicity assessment Relationship between exposure or dose and adverse health effects.	3. Toxicity effects assessment Aquatic, terrestrial, and microbial tests, for example, LC_{50}, field studies.
4. Risk characterization Integration of toxicity and exposure data for qualitative or quantitative expression of health risks; uncertainty analysis.	4. Risk characterization Integration of field survey, toxicity and exposure data for characterizing significant ecological risks, causal relationship, uncertainty.
Typical endpoints	
Individual and population cancer risks, non-cancer hazards.	Ecosystem or habitat impacts, for example, population abundance, species diversity; global impacts.
Typical applications	
Hazardous-waste sites (Superfund, RCRA) Air, water, land permitting Food, drugs, cosmetics Facility expansion or closure	Environmental impact statements Natural Resource Damage Assessments (NRDA) Superfund/RCRA sites Facility siting, wetland studies Pesticide registration

[a]Adapted from Kolluru et al. (1996).

5-4 RISK COMMUNICATION

The risk communication component of risk analysis is the interactive exchange of information and opinions concerning risk and risk management among risk assessors, risk managers, consumers, and other interested parties (Charnley et. al., 1997; WHO, 1999).

Risk communication can be defined as the exchange of information among interested parties about the nature, magnitude, significance, or control of a risk. Interested parties include government agencies, corporations or industry groups, unions, the media, scientists, professional organizations, special interest groups, communities, and individual citizens.

Information about risks can be communicated through a variety of channels, ranging from media reports and warning labels on products to public meetings or hearings involving representatives from government, industry, the media, and the general public. Issues related to public communications and acceptances are discussed in Chap. 26.

5-5 TOOLS AND METHODS USED IN RISK ASSESSMENT

Understanding the characteristics of human health hazards and exposures associated with chemicals and microorganisms is important in the study of water reclamation and reuse. Issues related to public health and safety of water reuse applications have been based primarily on three areas of study. Within the medical field, there are infectious disease and toxicology. The third area is in public health under which epidemiology has focused on specific transmission routes, such as waterborne microbial agents. In this and the following two sections, the relevant concepts in public health, epidemiology, and toxicology are briefly reviewed.

Concepts from Public Health

Public health is defined broadly as the science and the art of preventing disease, prolonging life, promoting physical and mental health, and enhancing efficiency through organized community efforts geared toward a sanitary environment; the control of community infections; the education of the individual in principles of personal hygiene; the organization of medical and nursing service for the early diagnosis and treatment of disease; and the development of the social machinery to ensure to every individual in the community a standard of living adequate for the maintenance of health.

The mission of public health is to fulfill society's desire to create conditions so that people can be healthy (IOM, 1988). The goal of public health is the reduction of disease and the improvement of health in the community. The scientific basis for public health rests on the study of risks to the health of populations and the environment, and on the systems designed to deliver required services. Epidemiology and biostatistics are the scientific disciplines that underpin inquiry in all of public health (Scutchfield and Keck, 1997).

Measures of Disease

Three measures of disease are normally used in public health: incidence, prevalence, and mortality. *Incidence* (or *morbidity*) is the number of people who contract a disease during a specific period of time. *Cumulative incidence* (CI) is the proportion of people who became diseased during a specified period of time, and is calculated as:

$$\text{CI} = \frac{\text{number of new cases of a disease during a given period of time}}{\text{total population at risk}} \qquad (5\text{-}8)$$

Prevalence (P) is incidence (morbidity) plus the number of people who already had and still have the disease, and is calculated as:

$$P = \frac{\text{number of existing cases of a disease at a given point in time}}{\text{total population}} \qquad (5\text{-}9)$$

Mortality is the number of people who died during the specific period of time, usually expressed as number of deaths per year (Hennekens and Buring, 1987).

Environmental Public Health Indicators

Indicators that describe the public health consequences of environmental exposures are called *environmental public health indicators* (EPHIs). Numerous national and international organizations have recognized the compelling need for EPHIs. When combined with other information, such as environmental monitoring data and data from toxicological, epidemiological, or clinical studies, EPHIs can be an important key to improving understanding of the relationship between pollution or hazards and health outcomes.

Concepts from Epidemiology

Epidemiology in its broadest sense is the study of disease patterns in human populations. A crucial objective of epidemiology is to identify subgroups in the population who are at higher risk for disease than others. *Environmental epidemiology* is the study of the effects on human health of biological, chemical, and physical factors in the external environment. By examining specific populations or communities exposed to different ambient environments, environmental epidemiology is used to clarify the relationship between physical, biologic, or chemical factors and human health.

The Dynamics of Disease Transmission

Disease is the result of an interaction between the host, an infectious or other type of agent, and the environment that promotes the exposure. In many cases, a *vector* (an organism, such as a mosquito or tick, that carries disease-causing microorganisms from one host to another) is involved in the transmission of a disease. The relationship between host, agent, environment, and vector is illustrated as an epidemiologic triad on Fig. 5-5.

Study Design in Epidemiology

Unlike laboratory experiments, epidemiological studies generally involve observation of a population or individuals that cannot be controlled by the investigator. Therefore, factors potentially affecting the cause-effect relationship have to be examined carefully to extract useful information. Epidemiological studies can be divided in two basic types (1) observational studies and (2) experimental studies. A framework for epidemiological studies is shown on Fig. 5-6. The studies identified on Fig. 5-6 are described briefly in Table 5-6.

Issues in Environmental Epidemiology

Epidemiology is a useful tool to determine cause-effect relationship when the status of exposure and outcome is evident. However, the conventional epidemiological approach is encountering increasing difficulties in being able to identify causal inferences as both suspected causes and effects become more complex and subtle. This problem is more pronounced in environmental epidemiology. In environmental epidemiology, exposure to the study risk generally occurs in a large population. However, the status of exposure is difficult to identify because the exposure is usually at very low levels, and a number of other risk factors, confounders and/or effect modifiers, interact in a complex manner,

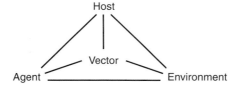

Figure 5-5

The epidemiologic triad of a disease. (Adapted from Gordis, 1996.)

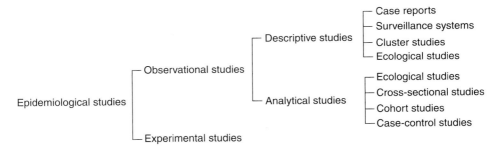

Figure 5-6

Categories of epidemiological studies. The various types of studies are described in Table 5-6.

while the expected effect is usually subtle, indirect, and chronic. Even for an acute effect such as infection with enteric viruses, it is usually difficult to separate the case from background prevalence. Furthermore, effects on local and global ecosystems are more difficult to identify, and are often overlooked by environmental epidemiologists (Pekkanen and Pearce, 2001).

It is a rare epidemiological study in which it is not concluded that more research is needed. Such a conclusion is disconcerting, given that epidemiological studies are expensive, typically costing one or two million dollars or more. Because of these problems, research focus in environmental epidemiology has been shifting to the individual and molecular levels. This shift leads essentially to rejection of epidemiological studies on environmental risk, and moves toward risk assessment based primarily on toxicological studies. However, the molecular and individual levels of study lack direct implication to the human population. Development of study designs that incorporate molecular and individual levels of study into population-based study are necessary to extrapolate from the individual and molecular levels to the human population (Pekkanen and Pearce, 2001).

Concepts from Toxicology

Toxicology is the study of the adverse effects of chemicals on living organisms (Klaassen, 2001) and is a multidisciplinary field that encompasses biology, chemistry, and environmental science. Environmental toxicology is the field of science that evaluates effects of toxic substances, released from human activities into the environment, on the biosphere and ecosystem, including humans. The toxic substances of interest are usually at low concentrations and widely distributed in the environment.

A toxic or potentially toxic substance cannot be administered deliberately to humans to test its toxicity. Therefore, various organisms, from mammals to bacteria, are usually used to examine the potential human health effects of toxicants, and the results are extrapolated to assess human health risk. Cells, nucleic acids and other components of an organism can be used to investigate the potential toxicity and mechanism of the toxic effects. Effects of toxicants on wildlife are examined using the living organisms in the environment of interest. A summary of tests for toxicity is shown in Table 5-7. *In vitro* and *in vivo* tests, the basis of toxicological testing, are described below.

Table 5-6
Principal types of epidemiological studies used to investigate cause and effect disease events and patterns[a]

Type of study	Description
Experimental	In experimental studies, baseline conditions are recorded first, and then exposed and nonexposed status is randomly assigned. Both groups are followed prospectively over time for the occurrence of disease or other outcome of interest. A major advantage in experimental studies is that unknown statistical confounders can be controlled by randomization. For ethical and other reasons, however, subjects cannot be assigned deliberately to receive a known risk. Therefore, experimental studies are used mostly in clinical trials in a treatment or preventive measure, but cannot be used in the study of health effects of toxic substances.
Observational	As a follow-up to anecdotal evidence and case histories, epidemiologists conduct two major types of observational studies to assess the relationship or association between suspected risk factors and disease: (1) descriptive and (2) analytical. These studies are used most commonly to monitor: (1) disease, study risk and other risks apart from the exposure of interest, (2) person and host characteristics, and (3) environmental conditions, without altering the conditions of the sample.
Descriptive (observational)	Descriptive studies are implemented when little information is available about a disease, exposure, or trait. In descriptive studies, current conditions are reported but no attempt is made to link any of the variables. Rather, descriptive studies generate hypotheses to be tested. Descriptive studies include case reports, cluster studies, ecologic studies, and surveillance systems.
Case report	A case report is a descriptive study of a single individual or small group of individuals. An association between an observed effect and a specific environmental exposure is studied based on detailed clinical evaluations and histories of the individual(s).
Cluster	A cluster study is a descriptive study of the population in a geographic area, occupational setting, or other small group in which the rate of a specific adverse effect is much higher than expected.
Ecological	In ecological studies, the relationship between two or more variables is examined at the population level. Ecological studies are most useful when large sets of data are available. Statistical confounders are not important in ecological studies due to a broad scale analysis. Ecological studies are generally categorized as descriptive, but the study could be analytic depending on the type of analysis used. Ecological studies are often used in attempt to assess the health effects of environmental pollutants.
Surveillance systems	Surveillance systems provide broad-scale information on specific populations for which epidemiologic analyses can be conducted.
Analytic (observational)	Analytic studies are conducted when sufficient information is available to form an *a priori* hypothesis. Analytic studies include case-control studies, cohort studies, and cross-sectional studies.
Case-control	Case-control studies are used to investigate the relationships between potential risk factor and disease by observing two groups of subjects: one with disease, trait, or condition of interest (case); the other without these conditions (control). Case-control studies usually depend on retrospective data. In a case-control study, selection of cases and controls is a crucial part of the study design. Cases must be selected so that the data can be generalized to all patients with the disease. Controls can be individuals without disease selected from the same group of people.

Table 5-6
Principal types of epidemiological studies used to investigate cause and effect disease events and patterns[a] (*Continued*)

Type of study	Description
Cohort	In a cohort study, a group of exposed individuals and a group of nonexposed individuals are observed and followed through time to evaluate changes in the incidence of disease. There are two particular types in cohort studies: prospective (historical) study and retrospective (concurrent) study. In the prospective cohort study, the exposure status at present time is identified and the samples are followed up to determine any future disease onset. In the retrospective approach, exposure status of cohort(s) in the past is identified and they are followed up until the present time. Questionnaires or laboratory tests are generally used in cohort studies to measure both exposure and outcome.
Cross-sectional	In cross-sectional studies, the status of exposure and the state and/or occurrence of disease are measured at a single point in time. A population is first defined, and presence or absence of exposure and presence or absence of disease for individuals is determined. Advantages in cross-sectional studies include (1) one-stop, one-time collection of data, (2) less expensive and more expedient to conduct, and (3) associations and correlation between variables can be easily evaluated. A case-control study is desirable when the disease occurrence is rare, because investigation of rare disease with a cohort study will require a tremendous number of people to be followed to generate enough cases for the study, and may not be practical.

[a]Adapted from NAS (1998); Gordis (1996); Sullivan and Krieger (2001).

In Vitro Tests

In vitro tests (biological studies which take place in isolation from a living organism such as in a test tube or Petri dish) are conducted using microorganisms, cells, or other specific components of organisms such as nucleic acids. With an *in vitro* assay, a part of the process to exert toxicity is examined. A classic example of an *in vitro* test is the Ames mutagenicity test. The Ames test examines the chemical substance's capability to cause mutations in *Salmonella typhimurium* (McCann et al., 1975). *In vitro* assays are used increasingly because they are less costly than other types of tests, can be conducted in a short time period, and are often very sensitive. However, it is difficult to establish a correlation between the observed effects and actual toxicity, such as carcinogenicity in *in vitro* assays. In addition, the results from *in vitro* tests cannot be applied directly to human toxicity because the processes of absorption, distribution, detoxication, and excretion in the human body are not duplicated in this type of test.

LD_{50} and LC_{50} The median lethal dose (LD_{50}) is the dose of a toxicant at which 50 percent of the population exposed under the defined conditions dies. The median lethal concentration (LC_{50}) is the concentration of a toxicant in the defined environment at which 50 percent of the population will be killed. A conceptual diagram of dose-response curve is shown on Fig. 5-7.

NOAEL and LOAEL The toxicity of a chemical from subchronic exposure (usually exposed for 30- to 90-d period) is examined to establish a "no observed adverse effect level" (NOAEL) and the "lowest observed adverse effect level" (LOAEL). Determination of NOAEL and LOAEL values is important for regulatory purposes. The U.S. EPA

Table 5-7

A summary of tests for toxicity[a]

Type of test	Description
I. Chemical and physical properties	For the compound in question, probable contaminants from synthesis as well as intermediates and waste products from synthetic processes.
II. Exposure and environmental fate	A. Degradation studies—hydrolysis, photodegradation, etc. B. Degradation in soil, water, etc., under various conditions. C. Mobility and dissipation in soil, water, and air. D. Accumulation in plants, aquatic animals, wild terrestrial animals, food plants, and animals, etc.

III. *In vivo* tests	A. Acute
	  1. LD_{50} and/or LC_{50}—oral, dermal, or inhaled
	  2. Eye irritation
	  3. Dermal irritation
	  4. Dermal sensitization
	B. Subchronic
	  1. 90-d feeding
	  2. 30- to 90-d dermal or inhalation exposure
	C. Chronic
	  1. Chronic feeding (including oncogenicity tests)
	  2. Teratogenicity
	  3. Reproduction
	D. Special tests
	  1. Neurotoxicity (delayed neuropathy)
	  2. Potentiation
	  3. Metabolism
	  4. Pharmacodynamics
	  5. Behavioral

IV. *In vitro* tests	A. Mutagenicity—prokaryote (Ames test) B. Mutagenicity—eukaryote (Drosophilia, mouse, etc.) C. Chromosome aberration (Drosophilia, sister chromatid exchange, etc.)
V. Effects on wildlife	Selected species of wild mammals, birds, fish, and invertebrates: acute toxicity, accumulation, and reproduction. A. Bioassay—determination of toxicity using organisms B. Biomonitoring—determination of effects on aquatic life

[a]Adapted from Hodgson and Levi (1997).

utilizes the NOAEL, incorporates a safety factor, then calculates the reference dose (RfD), which is used to establish "acceptable" levels of pollutants in regulations (Eaton and Klaassen, 2001). The concept of NOAEL and LOAEL is also shown on Fig. 5-7.

In Vivo Tests

In vivo tests (biological studies which take place within a living biological organism) are conducted using living organisms, such as mammals and fishes. The chemical compound of interest is administered to the experimental animals to qualitatively and quantitatively examine short-term acute toxicity or long-term chronic toxicity. Human toxicity is

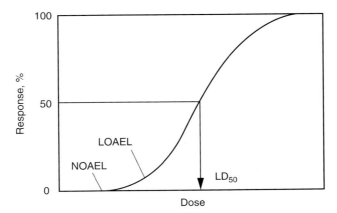

Figure 5-7
Conceptual diagram of a dose-response curve. LD_{50} is a dose at which 50 percent mortality is observed. The highest dose at which no observed adverse effect is observed (NOAEL), and the lowest dose at which any adverse effect is observed (LOAEL) are also indicated in the figure.

estimated using the results of *in vivo* tests with various assumptions. The major advantage of *in vivo* tests is that the similar exposure pathways for human toxicity can be examined with mammalian animals. The major disadvantages of *in vivo* tests are that a large number of animals is required, they are expensive to conduct, and are time consuming.

Whole Animal Tests for Carcinogenicity

The intent of whole animal tests is to identify chemicals that may be human carcinogens at low lifetime environmental or dietary doses, as indicated by their effects in animals at very high test doses. The human lifespan exceeds 70 yr whereas the test animals' is only about two years; so test doses must be extreme to reduce the possibility that the test would give false negatives. A false negative is a term used to indicate a test showed an incorrect negative result.

Confidence that an apparently negative result in a long-term carcinogenicity test does not represent a false negative is increased with increasing the numbers of animals in the study, increased longevity of the test animals, and a high quality of pathologic examination (see Fig. 5-8).

The national toxicology program (NTP) cancer bioassay generally involves high dose testing of individual chemicals in groups of 50 male and female inbred genetically similar mice and rats for 18 to 24 mo (their approximate lifetimes) at each dose. Typically two or three doses have been used: the maximum tolerated dose (MTD), half of the MTD, and another positive dose. Various control population tests are also employed: positive (known carcinogen), negative (vehicle only), as well as historical controls. Test exposures may be by augmented dietary or drinking water consumption, or force feeding

National Toxicology Program Cancer Bioassay

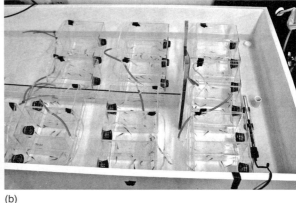

(a)   (b)

Figure 5-8
Examples of toxicity testing using fish bioassays: (a) laboratory setup used to conduct whole-effluent toxicity tests using a series of fish tanks and (b) using Japanese Medaka fish as a test species where mortality is the endpoint. (Courtesy of S. A. Lyon.)

in a vehicle like corn oil, or inhalation. The MTD is usually determined from a prior 90 d range-finding study; it is the highest dose that does not result in adverse effects other than up to about 10 percent weight loss. The goal is to select an MTD that will allow the animals to survive the subsequent carcinogenicity test so that the principal end result is cancers rather than noncancer chronic toxic effects.

Frequently doses must be adjusted during the cancer tests due to premature mortality or other confounding factors. It is necessary to use unusually high test doses because many carcinogens are relatively low potency and high doses provide a greater potential for carcinogenicity to be detected as statistically significantly distinct from the controls, considering that only 50 animals are in each specie and sex group. At termination, the health status of the animals and survival rate are assessed. The remaining animals are euthanized and sections from numerous organ tissues are examined by pathologists for indications of cancers. The types of cancers are diagnosed and the numbers and types of cancerous lesions are tabulated (NIH, 2004).

Ecotoxicology: Environmental Effects

Toxic substances affect not only humans but also ecosystems. The study of the fate and effects of toxic substances on ecosystems is called *ecotoxicology*. The science itself requires an understanding of ecologic principles and ecologic theory, and of how chemicals potentially affect individuals, populations, communities, and ecosystems. Measurements are accomplished with the use of either species-specific responses to toxicants or impacts at higher levels of organization. The ability to measure chemical transport and fate and exposure of organisms in ecotoxicologic testing is critical to the ultimate development of an ecologic risk assessment as outlined in Table 5-5 (Daughton and Ternes, 1999; Klassen and Watkins, 1999).

5-6 CHEMICAL RISK ASSESSMENT

Risk assessment is an integral part of the regulatory decision process particularly in the qualitative determination of the strength of evidence relating to carcinogenicity and establishing MCLGs and MCLs. The assessment of chemical risk is reviewed in this section.

Safety and Risk Determination in Regulation of Chemical Agents

Toxicity has been defined as the intrinsic quality of a chemical to produce an adverse effect. The toxicology of chemical substances found in drinking water can be divided into three broad classes: (1) acute or chronic toxicity, (2) carcinogenicity, and (3) reproductive, developmental, and neurotoxicity. The same substance may be capable of causing any and all of the effects depending upon the dose and individual's characteristics (Cotruvo, 1987; 1988). The distinguishing characteristics among these categories of effects lie:

1. In the probably unverifiable assumption that dose thresholds *exist* for chronic toxicity effects
2. In the also unverifiable assumption that dose thresholds *do not exist* (or have not been demonstrated) for (genotoxic) carcinogenic effects and neurotoxic effects
3. Where there are specific time periods during gestation when the fetus is at risk to insult from certain toxicants, often even at relatively low doses

In the case that dose thresholds exist for chronic toxicity effects, the nominal basis for standard setting is to achieve a total daily dose of the substance that is with practical certainty below the level at which any injury would result to any individual in the population during exposure or after the exposure has ceased. For toxicants assumed to be acting by nonthreshold mechanisms, it follows, at least hypothetically, that some finite risk may exist at any nonzero dose level. For reproductive and developmental toxicants, the timing of the exposure, even though brief, can be highly significant. Thus, standard setting objectives range from zero, which is not quantifiable and often not practically achievable (other than by banning the product), to a daily dose level that contributes only a negligible theoretical or hypothetical incremental increase in the lifetime risk of the effect to individuals and/or the population exposed.

Incremental Lifetime Risk

The U.S. EPA has defined incremental lifetime risk for cancer, above background, as follows:

$$\text{Incremental lifetime risk} = \text{CDI} \times \text{PF} \qquad (5\text{-}10)$$

The chronic daily intake (CDI) is computed as follows:

$$\text{CDI} = \frac{\text{average daily dose, mg/d}}{\text{body weight, kg}} \qquad (5\text{-}11)$$

where CDI = chronic daily intake over a 70 yr lifetime, mg/kg·d
PF = potency factor, (mg/kg·d)$^{-1}$

In its most general form, the total dose is defined as

$$\text{Total dose, mg/kg·d} = \begin{pmatrix} \text{constituent} \\ \text{concentration} \end{pmatrix} \begin{pmatrix} \text{intake} \\ \text{rate} \end{pmatrix} \begin{pmatrix} \text{exposure} \\ \text{duration} \end{pmatrix} \begin{pmatrix} \text{absorption} \\ \text{factor} \end{pmatrix} \quad (5\text{-}12)$$

Recommended standard values of weight and water ingestion for daily intake calculations have also been developed by the U.S. EPA. The average body weights used for an adult and child are 70 and 10 kg, respectively, and the corresponding rates of water ingestion are 2 and 1 L/d (U.S. EPA, 1986).

The potency factor (PF), often identified as the *slope factor*, is the slope of the dose-response curve at very low doses (see Fig 5-3*b*). In effect, the PF corresponds to the incremental risk above background resulting from a lifetime average dose of a toxicant. The U.S. EPA has selected the linear multi-stage model as the basis for assessing risk. Using this model and the best available data, the U.S. EPA has developed and maintained an extensive database on toxic substances known as the Integrated Risk Information System (IRIS). Typical toxicity data for several chemical constituents are reported in Table 5-8. The relative potency of the chemical constituents can be assessed by comparing the magnitude of the given values listed in Table 5-8 (e.g., for the oral route, the potency of arsenic is about 245 times that of chloroform). It is important to remember, however, that because of the numerous uncertainties involved in the development of the data base, including extrapolations of animal data to humans and from high experimental dosages to the low environmental dosages encountered in real life, the values given in the IRIS data base cannot be used to predict the incidence of human disease or the type of effects a given chemical constituent will have on an individual. Use of the data listed in Table 5-8 is illustrated in Example 5-2.

Table 5-8
Toxicity data for selected potential carcinogenic chemical constituents[a,b]

Chemical constituent	CASRN[c]	Potency factor, PF	
		Oral route, $(\text{mg/kg·d})^{-1}$	Inhalation route, $(\mu\text{g/kg·d})^{-1}$
Arsenic, inorganic	7440-38-2	1.5 E+0	3.0 E-2
Benzene	71-43-2	1.5 to 5.5 E-2	1.54 to 5.45 E-5
Bromate	15541-45-4	7 E-1	na
Chloroform	67-66-3	6.1 E-3	1.6 E-4
Dieldrin	60-57-1	1.6 E+1	3.2 E-2
Heptachlor	76-44-8	4.5 E+0	9.1 E-3
N-Nitrosodiethylamine	55-18-5	1.2 E+2	3.0 E-1
N-Nitrosodimethylamine	62-75-9	5.1 E+1	9.8 E-2
Vinyl chloride[d]	75-01-4	7.2 E-1	3.1 to 6.2 E-5

[a]Adapted from U.S. EPA IRIS database (1996) (http://www.epa.gov/iris).
[b]Because the data in the IRIS data base is being revised continuously, it is important to check the data base for the most current values.
[c]Chemical Abstracts Service Registry Number.
[d]Continuous lifetime exposure during adulthood.

EXAMPLE 5-2. Risk Assessment for Lifetime Consumption of Drinking Water Containing N-Nitrosodimethylamine.

Estimate the incremental lifetime risk for an adult associated with drinking groundwater containing 2.0 µg/L of N-Nitrosodimethylamine (NDMA). Determine the concentration that would be needed to limit the risk to 1 in 100,000.

Solution
1. Compute the CDI in mg/kg·d using Eq. (5-11).

$$CDI = \frac{(\text{average daily dose, mg/d})}{\text{body weight, kg}}$$

$$CDI = \frac{(2.0 \text{ µg/L})(2 \text{ L/d})(1 \text{ mg}/10^{-3} \text{ µg})}{70 \text{ kg}} = 0.57 \times 10^{-4} \text{ mg/kg·d}$$

2. Compute the incremental lifetime risk for drinking water consumption using Eq. (5-11) and Table 5-8.

    Incremental lifetime risk = CDI × PF

    The PF of NDMA from Table 5-8 for the oral route is $5.1 \times 10 \text{ (mg/kg·d)}^{-1}$

    Incremental lifetime risk = $(0.57 \times 10^{-4} \text{ mg/kg·d})[5.1 \times 10 \text{ (mg/kg·d)}^{-1}]$
    = 0.29×10^{-2}

    Thus, the calculated estimated probability of developing cancer as a result of lifetime consumption of water containing 2.0 mg/L of NDMA would be 2.9 per 1000 persons.

3. Determine the concentration of NDMA to limit the risk to 1 in 100,000.
    a. Estimate the CDI based on the determined risk and PF listed in Table 5-8.

    $$10^{-5} = (CDI)[5.1 \times 10 \text{ (mg/kg·d)}^{-1}]$$

    $$CDI = 1.96 \times 10^{-7} \text{ mg/kg·d}$$

    b. Estimate the concentration of NDMA by rearranging Eq. (5-12).

    $$\frac{(C, \text{ µg/L})(2 \text{ L/d})(1 \text{ mg}/10^{-3} \text{ µg})}{70 \text{ kg}} = 1.96 \times 10^{-7} \text{ mg/kg·d}$$

    $$C = 0.0069 \text{ µg/L}$$

Comment
NDMA is a member of a family of extremely potent carcinogens, the N-nitrosamines. Until recently, concerns about NDMA focused mainly on the presence of NDMA in food, consumer products, and polluted air. However, current concern focuses on NDMA as a drinking water contaminant resulting from reactions occurring during chlorination or via direct industrial contamination. Because of the relatively high concentrations of NDMA formed during wastewater chlorination, the fate of NDMA in planned or unplanned indirect potable reuse is a particularly important area of concern.

Noncarcinogenic Effects

In addition to the carcinogenic dose-response information, the U.S. EPA has developed reference doses (RfDs) for a number of constituents based on the assumption that thresholds exist for certain toxic effects (see Fig. 5-2b), such as cellular *necrosis* (localized death of living tissue), but may not exist for other toxic effects, such as carcinogenicity. In general, RfDs are established based on reported results from human epidemiological data, long-term animal studies, and other available toxicological information. The RfD values represent an estimate (with uncertainty spanning perhaps an order of magnitude) of the daily exposure to the human population (including sensitive subgroups) that is likely to be without an appreciable risk of deleterious effects during a lifetime. Values for constituent RfDs are available in the IRIS database.

The RfD is used as a reference point for gauging the potential effects of other doses. Usually, doses that are less than the RfD are not likely to be associated with health risks. As the frequency of exposure exceeds the RfD and the size of excess increases, the probability increases that adverse health effects may be observed in a human population. The RfD is derived using the following formula:

$$\text{RfD} = \frac{\text{NOAEL or LOAEL}}{(\text{UF}_1 \times \text{UF}_2 \ldots) \times \text{MF}} \tag{5-13}$$

where NOAEL = no observable adverse effect level
LOAEL = lowest observable adverse effect level
UF_1, UF_2 = uncertainty factors
MF = modifying factor

In the above equation, uncertainty factors are based on experimental species, effects, and duration of the study, while modifying factors represent professional assessments reflecting the confidence in the study. Taken together, these factors represent a safety (uncertainty) factor. The LOAEL is used only when a suitable NOAEL is unavailable.

The modified RfD for long-term or lifetime exposure for an adult can be computed by multiplying the experimental NOAEL (in milligrams per kilogram per day) by the reference weight of a typical adult (70 kg) and dividing by the safety (uncertainty) factor.

$$\text{RfD, mg/person} \cdot \text{d} = \frac{(\text{NOAEL, mg/kg} \cdot \text{d})(70 \text{ kg/person})}{\text{safety (uncertainty) factor}} \tag{5-14}$$

Because the RfD is intended to account for total daily intake of the toxicant, inhalation and food intake, as well as water, should be accounted for when attempting to arrive at the maximum drinking water level or the adjusted RfD for drinking water at the maximum drinking water level considering only health factors. Thus, in the optimum case when such information is available, the daily uptake from inhalation and the daily intake from food (if 100 percent uptake is assumed) should be subtracted from the RfD. Finally, for the determination of the acceptable drinking water concentration per liter, the common assumption is that adults consume 2 L of water per person per day, thus, the final value should be divided by a factor of 2:

Drinking water target, mg/L =

$$\frac{[(\text{RfD, mg/person} \cdot \text{d}) - (\text{inhalation, mg/person} \cdot \text{d}) - (\text{food, mg/person} \cdot \text{d})]}{2 \text{ (L/person} \cdot \text{d)}} \quad (5\text{-}15)$$

One of the uncertainties is the actual amount of water consumed by each individual. Studies have shown consistently that 2 L/d is about the 90th percentile of daily water consumption across the population in the United States. The range is from something less than 1 to 4 L at the 99th percentile (Ershow and Canter, 1989). Thus, the uncertainty range for water consumption is about a factor of 2. This uncertainty is minute compared to that associated with extrapolating animal toxicology to humans. The assumed daily consumption of 2 L of drinking water provides an additional small safety factor for most people.

The Relative Source Contribution The relative source contribution (RSC), the allocation of the actual or assumed contribution from each route of exposure, should be factored into the ultimate MCL. Typically, the U.S. EPA will assume an RSC of 20 percent from drinking water for inorganic chemicals with higher values for volatile organic chemicals (VOCs). Note that as the RSC value decreases, the drinking water standard will become more stringent, which is somewhat counter-intuitive because the burden on the drinking water supplier increases as the significance of the drinking water contribution to health risk decreases. For a NOAEL value of 20 mg/kg · d, a 70-kg person, and an uncertainty factor of 1000, the RfD value is:

$$\text{RfD, mg/person} \cdot \text{d} = \frac{(20 \text{ mg/kg} \cdot \text{d})(70 \text{ kg/person})}{1000} = 1.4 \text{ mg/person} \cdot \text{d}$$

The drinking water equivalent level (DWEL) calculated by using the default assumption of 20 percent of the daily dose allocated to drinking water is:

$$\text{DWEL} = \frac{(1.4 \text{ mg/person} \cdot \text{L})}{(2 \text{ L/person} \cdot \text{d})} \times 0.20 = 0.14 \text{ mg/L}$$

The validity of a RfD is dependent entirely on the quality of the experimental data and the judicious selection of the safety (uncertainty) factor, which is judgmental. Among the factors influencing the quality of the experimental data beyond the mechanics of the study are an understanding of the mechanism of action at low doses in humans; the selection of the appropriate animal model as the human surrogate; the number of animals at each dose, the number and range of the doses for acceptable statistical significance of results and the shape of the experimental dose-response curve; the actual detection of the most sensitive adverse effect (which could be biochemical change only, frank organ damage, or death); the length of the study (lifetime studies versus shorter-term studies); and the appropriate route of exposure (inhalation, gavage, ingestion in food or water, etc.). The quality of the experimental evidence determines the magnitude of the safety (uncertainty) factor to be applied. The lesser the understanding of the toxicology, the greater the uncertainty factor and the lower the RfD or MCL.

Safety (Uncertainty) Factors Safety factors are numbers that reflect the degree of uncertainty that must be considered when experimental data are extrapolated to the human population. When the quality and quantity of dose-response data are high, the uncertainty factor is low; when the data are inadequate or equivocal, the uncertainty factor must be larger.

The following general guidelines have been adopted by the National Academy of Sciences Safe Drinking Water Committee, and the guidelines or a variant were also used by the U.S. EPA in the development of drinking water standards, guidelines, and health advisories.

1. 10 Factor: Valid experimental results from studies on prolonged human ingestion with no indication of carcinogenicity.
2. 100 Factor: Experimental results of studies of human ingestion not available or scanty. Valid results from long-term feeding studies on experimental animals or, in the absence of human studies, on one or more species. No indication of carcinogenicity.
3. 1000 Factor: No long-term or acute human data. Scanty results on experimental animals. No indication of carcinogenicity.

Various modifying factors ranging from 1 to 10 are also used sometimes to reflect quality of data, significance of effect and/or other concerns, particularly for high-risk populations such as children. In summary, the larger the uncertainty factor, the less weight given to the experimental data, the RfD or MCL becomes predominantly driven by the uncertainty factor and is therefore less scientifically defensible. At least in theory, improved understanding of the toxicology should result in smaller uncertainty factors being required to reach a comfort level for the risk assessor.

Risks from Potential Nonthreshold Toxicants

Four principles for dealing with the assessment of hazards are outlined that involve chronic irreversible toxicity or the effects of long-term exposure. These principles (paraphrased as follows) were intended to apply primarily to cancer risks from substances whose mechanisms involve somatic mutations and may also be applicable to mutagenesis and teratogenesis:

1. Effects in animals, properly qualified, are applicable to humans. Large bodies of data indicate that exposures that are carcinogenic to animals are likely to be carcinogenic to humans, and vice versa.
2. Methods do not exist now to establish a threshold for long-term effects of toxic agents. Thresholds in carcinogenesis that would be applicable to a total population cannot be established experimentally.
3. The exposure of experimental animals to toxic agents in high doses is a necessary and valid method of discovering possible carcinogenic hazards in humans. High dosages, relative to expected human exposures, are given to animals under the experimental conditions because there is no choice but to use numbers of animals that are small relative to the exposed human populations. Biologically reasonable models can then be used to extrapolate the results to estimate risk at low doses.
4. Material should be assessed in terms of human risk rather than as safe or unsafe. Extrapolation techniques may permit the estimation of upper limits of risk to human populations. To do so, data are needed to estimate population exposure; valid, accurate, precise, reproducible animal assay procedures are required; and appropriate statistical methods are necessary.

Risk Extrapolation for Carcinogens

Many mathematical models have been developed to estimate potential risks to humans from low-dose exposures to carcinogens. Each model incorporates numerous unverifiable assumptions. Low-dose calculations are highly model dependent, widely differing results are obtained commonly, and none of the models have been demonstrated to apply at very low doses. Thus, the decision to use this approach and the choice of how to do the calculations are primarily matters of policy judgment. Among the choices that decision-makers must consider are which model(s) to employ, which assumptions to incorporate, and which acceptable risk to allow. A default linearized multistage model has been used typically for purposes of estimating nominal risks from lifetime (70 yr) exposure to carcinogens in drinking water. However, as a better understanding of the mechanisms modes and actions is achieved, biologically based models are being employed with nonlinear dose-response models.

The methodology used to extrapolate risk to low doses is to develop an extrapolation line based on the available data. Risk extrapolation, done by first selecting a point of departure (POD) from the observed data, is the starting point for extrapolation to lower doses. Unit risks for drinking water can be calculated from the slope factor [(mg/kg)/d] at any point of the extrapolation line by converting the slope to milligrams per liter terms (see Example 5-2).

Identification of Compounds Likely to be Carcinogenic to Humans

Risk assessment of potential human carcinogens is performed on substances that exceed an evidentiary threshold for cancer. Thus, a conclusion of a substance causing known or potential human cancer is based on scientific evidence of the substance exceeding the defined evidentiary threshold. The International Agency for Research on Cancer (IARC) and other organizations have provided guidelines for assessing the epidemiological and animal toxicological database leading to a conclusion of the strength of the evidence of carcinogenicity of numerous substances.

Chemical Risk Assessment in U.S. Drinking Water Regulations

The Safe Drinking Water Act (SDWA, 1974) requires that the U.S. EPA set drinking water standards after making a determination that a substance may have an adverse effect on health. Setting drinking water standards is a two step process involving (1) establishing a maximum contaminant level goal (MCLG), and (2) establishing a maximum contaminant level (MCL). The MCLG is developed through a risk assessment process in which the health risks that can occur from excess exposure to the toxicant are evaluated. The MCLG is the level of a contaminant in drinking water below which there is no known or expected risk to health. Maximum contaminant level goals allow for a margin of safety and are nonenforceable public health goals.

The MCL is the standard that is enforceable, and it is set as close as technologically feasible, taking costs in consideration, to the MCLG. Both the MCLG and the MCL for a substance are developed through a formal process defined in the SDWA that includes public notice in the *Federal Register* and a public comment period. Both the MCLG and the MCL may be legally challenged. Typically, the process of establishing a MCLG and a MCL can take three to five years and frequently it takes longer than five years.

Table 5-9
Generalized schematic of analyses and drinking water regulation development process

Step	Process or activity
Health Assessments	Nationwide Occurrence in Water Exposure Assessment Risk Assessment Cancer Classification Draft MCLG Toxicology Assessment
Technology/Economics Assessments	Analytical Methods Performance (PQL) National and Local Impact Draft MCL and MCLG Costs of Options Assessment
Draft MCLG and MCL	Publish in *Federal Register* as Proposed Public Comments Review/Revisions
Publish Final MCLG and MCL	MCLG and MCL values are published in *Federal Register* with designated effective date

A generalized drinking water regulation development process is shown in Table 5-9. The risk assessment process is an attempt to quantify human exposure from all routes including drinking water using animal toxicology data and human epidemiology (when available), to arrive at concentrations in drinking water at which exposure would result in no known or anticipated adverse effects on health, with a margin of safety. This process is complex requiring a number of informed judgments and assumptions. Frequently, the available data are incomplete or deficient. Applying risk assessment to decision-making, particularly for toxic substances that are considered to potentially act without a discernable threshold (interact directly with DNA, i.e. a genotoxic carcinogen), requires many policy choices and assumptions beyond the scientific data.

U.S. EPA's Qualitative Assessment of Carcinogens

In developing the current drinking water regulations, the U.S. EPA applied a qualitative weight of evidence scheme in which the following five groupings were defined to assess the potential for a contaminant to increase the risk of cancer in humans.

Group A—Human carcinogen: sufficient evidence in humans.

Group B—Probable human carcinogen: limited evidence in humans or no evidence in humans but sufficient evidence in animals.

Group C—Possible human carcinogen: limited or equivocal evidence in animals in the absence of human data.

Group D—Not classifiable as to human carcinogenicity: inadequate or no data available.

Group E—No evidence of carcinogenicity for humans: negative evidence in at least two species.

The above qualitative assessment of carcinogenicity has sometimes been applied in an inflexible manner and without full consideration being given to all of the data (positive and negative) and mechanistic considerations. The U.S. EPA has been developing new interpretive guidelines that are focused on weight-of-evidence with the intent of providing toxicologists more incentive to exercise scientific judgment rather than typically choosing the most conservative interpretation. For example, most researchers now believe that chloroform is not a cancer risk to humans at levels found typically in drinking water, although it is carcinogenic at doses high enough to cause cytotoxicity in the liver.

Three-Category Approach for Setting MCLGs

In setting the current drinking water regulations, the U.S. EPA employed a three-category approach for setting MCLGs, based on the qualitative carcinogenicity classification scheme discussed previously. Following this methodology, substances were grouped into three categories as follows:

Category I: Group A and B substances: Goal equals zero (aspirational goal).

Category II: Group C substances: The goal equals 10^{-5} to 10^{-6} (1 per 100,000 to 1 per 1,000,000) hypothetical excess cancer risk per 70 yr lifetime, or the goal equals the RfD value converted to DWEL with an additional safety factor applied to allow for an adequate margin of safety due to uncertainties in the substances carcinogenic potential to humans. (Category II, Group C substances include those substances where some limited but insufficient evidence of carcinogenicity exists from animal data.)

Category III: Group D and E substances: Goal is calculated using the RfD approach with a portion allocated to drinking water.

For a compound to be considered part of carcinogenicity Category II, Group C, the following conditions were imposed:

1. The studies involved a single species, strain, or experiment
2. The experiments were restricted by inadequate dosage levels, inadequate reporting
3. There was an increase in benign tumors only

Two approaches were used to set MCLGs—either (a) the goal was set based upon noncarcinogenic endpoints (the RfD) with an additional uncertainty (safety) factor of up to 10 applied; or (b) the goal was set based upon a nominal lifetime risk calculation in the range of 10^{-5} to 10^{-6} using a conservative calculation model (linearized multistage). The first approach is generally preferred; however, the second approach is used when valid noncarcinogenicity data are not available but adequate experimental data are available to perform the risk calculation.

The most controversial decisions in setting MCLGs and MCLs were related to the establishment of nonenforceable goals for substances to be regulated based upon the risk of human carcinogenicity (i.e., Category I, Groups A and B). The statutory directive (no known or anticipated adverse effect) and a brief statement in the legislative history that MCLGs for nonthreshold activity substances should be zero were the basis for the three options considered as possible MCLGs. The three options considered were

1. MCLG = zero
2. MCLG based upon a target calculated risk
3. The analytical detection limit

Little support for the analytical detection limit approach was received because detection limits are a consequence of the analytical techniques and practices, which are constantly changing (improving). The choice between the zero option and the finite risk option was made based upon the statutory direction and the legislative history. Setting the MCLG using the finite risk option occurred after discussing whether it was appropriate to set a target that hypothetically permitted some number of cancer deaths. The U.S. EPA also pointed out that setting an MCLG at zero did not imply that actual harm would occur at levels somewhat above zero, and MCLGs are ideal or aspirational goals, not standards.

Enforceable National Drinking Water Standards

The MCLs are set as close to the MCLGs as is feasible. Setting MCLs is usually not difficult or costly for Category II and Category III substances that have finite goals. However, a dilemma occurs when trying to establish MCLs for Category I substances with MCLGs of zero. The ideal MCL for an MCLG of zero is zero as well; however, MCLs consider feasibility, which MCLGs do not. The MCLs are determined based upon an assessment of a variety of factors including the feasibility of best availability technology (BAT), costs in a variety of water system conditions, the number of supplies affected, total national costs, and the reliability of analytical methods. Then the nominal residual lifetime risks that are theoretically associated with exposures at the technologically determined MCLs are examined.

Even though an assessment of a large number of factors is considered when setting an MCL, there are times when analytical methods are the limiting or deciding factor, especially where the MCL could potentially be set at a level below which the substance could be measured reliably. Thus the concept of practical quantification limits (PQLs) was developed. Practical quantification limits are defined as the lowest concentration that can be determined reliably within specified limits of precision and accuracy during routine laboratory operations in qualified laboratories. In cases where the analytical method is the limiting factor, the PQL is the lowest level that may be set as the MCL even though the risk assessment might lead to a lower value. As analytical capabilities improve it should be expected that the MCL would be lowered. However, procedurally, a new rulemaking process is undertaken, which includes a new risk assessment, regulatory assessment, a new proposal and public comment, and promulgation of a new regulation. Thus, even though analytical methods are changing continuously, existing MCLs are changed infrequently.

Risk Considerations

After the above aspects are considered, the MCLs for the probable carcinogens are proposed. The U.S. EPA then examines the putative risks at the proposed MCL level to determine whether this level would be acceptable from a safety standpoint. The upper bound generally considered acceptable (safe and protective of public health) is a risk range of 10^{-4} to 10^{-6} (1 death per 10,000 people to 1 death per 1,000,000 people) when calculated by a typically conservative linear multistage model [see Eq. (5-2)]. The lower bound risk can actually be zero. This approach used by the U.S. EPA is consistent with the concept expressed in the WHO's *Guidelines for Drinking-water Quality* (2004), which supports 10^{-5} as a general health protection-based guideline value for carcinogens in drinking water. It was also noted that country-to-country applications could vary by a factor of 10 on either side (i.e., 10^{-4} to 10^{-6}) for economic and other practical considerations.

The U.S. EPA has set aspirational MCLGs of zero for probable carcinogens, nonzero MCLGs for noncarcinogens based upon classical toxicology, and a related system that involves either additional safety factors or a nonthreshold risk model calculated target for substances that have equivocal evidence regarding their carcinogenicity.

Chemical Risk Assessment Summary

Legally enforceable drinking water standards, that is MCLs, are required to be set as near as technically and economically feasible to MCLGs. For noncarcinogens and substances with equivocal evidence of carcinogenicity, the MCL is usually the same as the MCLG. For probable carcinogens the MCL is set based on a variety of technological performance/cost factors, but also a reference risk rank is targeted between 10^{-4} and 10^{-6} (hypothetical incremental lifetime risk using a conservative model unlikely to have underestimated the risk). Standards falling in that range are concluded to be safe and protective of public health. The regulatory process requires publication of the proposed MCLGs and MCLs in the *Federal Register* along with their supporting rationales. A public comment period follows and each comment must be addressed and resolved before the final MCLGs and MCLs are established and implemented.

5-7 MICROBIAL RISK ASSESSMENT

Microbial risk assessment (MRA), also known as quantitative microbial risk assessment (QMRA), is an emerging field and that can potentially provide useful tools for analyzing microbial risk to human health. As discussed in Sec. 5-1, risk analysis consists of three principal components: (1) risk assessment; (2) risk management; and (3) risk communication. The focus of this section is on the first component of the risk analysis process. Topics considered in this section include: (1) microbial risk assessment is defined; (2) the implementation of MRA is described; (3) an overview of MRA methodologies is provided; and (4) a discussion of the lessons learned in the selection of MRA methodologies is presented. The application of MRA in water reuse applications is illustrated in the following section.

Microbial risk assessment (also known as pathogen risk assessment) is the process that is used to evaluate the likelihood of adverse human health effects that can occur following exposure to pathogenic microorganisms or to a medium in which pathogens occur (Cooper et al., 1986a; Cooper, 1991; Haas et al., 1999; ILSI, 2000). To the extent possible, the MRA process includes evaluation and consideration of quantitative information; however, qualitative information is also employed as appropriate (Cooper et al., 1986b; WHO, 1999; Ashbolt, et al., 2005). Many of the early MRAs employed the NRC conceptual chemical risk assessment framework (see Sec. 5-1) to provide a structure from which the assessments could be conducted (Haas, 1983a; Cooper et al., 1986b; ILSI, 1996; Regli et al., 1991; Rose et al., 1991a).

Infectious Disease Paradigm for Microbial Risk Assessment

Figure 5-9
Crowded beach. Complexities of person-to-person interactions for infection and disease create difficulties in identifying exposure pathways.

Complexities of Person-to-Person Interactions

As the field of MRA developed, it became clear that there are some complexities associated with modeling the infectious disease process that are unique to pathogens, such as person-to-person transmission of infection, and individual immunity (see Fig. 5-9). Thus, the conceptual framework for chemicals used in chemical risk assessment (static modeling) as discussed in Sec. 5-6 is not adequate for the assessment of risk of human infection following exposure to pathogens (via dynamic modeling). The fundamental difference between the two risk assessment techniques is that static models do not account for properties that are unique to a dynamic infectious disease process. For instance, in static models, the number of individuals that are assumed to be susceptible to infection is not time-varying, whereas in dynamic models that number is time-varying, which more closely resembles what occurs in nature. Additional comparisons of static and dynamic MRA models are described in Table 5-10. The fundamental difference between these two model approaches is that the risk characterization perspective is shifted away from an individual (static MRA) to a population-based perspective in the dynamic MRA investigations. Static and dynamic MRA models are considered further following a discussion of risk assessment methods (see Fig. 5-10).

Risk Analysis Framework

The U.S. EPA/ILSI (International Life Sciences Institute) framework for assessing the risk of human infection following exposure to water- and food-borne pathogens is comprised of three principal components (1) problem formulation, (2) analysis, and (3) risk characterization (Teunis and Havelaar, 1999; Soller et al., 1999; ILSI, 2000). The framework is similar conceptually to the NRC paradigm for human health risk assessments for exposure to chemicals (NRC, 1983) and the ecological risk assessment framework (U.S. EPA, 1992), as discussed in Sec. 5-2.

Table 5-10
Comparison of static and dynamic risk assessment models

Static risk assessment model	Dynamic risk assessment model
Static representation (not varying in time)	Dynamic representation (time-varying)
Direct exposure (environment-to-person)	Direct (environment-to-person) and indirect exposure (person-to-person)
Individual-based risk	Population-based risk
Potential for secondary transmission of infection or disease is negligible	Potential for secondary or person-to-person transmission of infection or disease exists
Immunity to infection from microbial agents is negligible	Exposed individuals may not be susceptible to infection or disease because they may already be infected or may be immune from infection due to prior exposure
Dose-response function is the critical health component	The dose-response function is important, however, factors specific to the transmission of infectious diseases may also be important

Microbial Risk Assessment Methods

Quantitative methods to characterize human health risks associated with exposure to pathogenic microorganisms started to appear in the published literature in the mid 1970s (Dudely et. al., 1976; Fuhs, 1975; Haas 1983a, b; Cooper, et. al., 1986b; Olivieri et. al., 1986). Since then, the field of MRA has been developing and maturing and dose-response relationships have been developed for various pathogenic microorganisms, and the dose-response information can now be applied for practical use (Haas et al., 1999; McBride et al., 2002).

Static Microbial Risk Assessment Models

Assessments using a static model for evaluating microbial risk are focused typically on estimating the probability of infection or disease to an individual as a result of a single-exposure event. These assessments generally assume that multiple or recurring exposures constitute independent events with identical distributions of contamination (Regli et al., 1991). Secondary transmission and immunity are assumed to be negligible or that they effectively cancel each other out. In this context, secondary transmission would increase the level of infection/disease in a community relative to a specific exposure to pathogens, and immunity would decrease the level of infection/disease in a community relative to a specific exposure to pathogens.

Model States

In static MRA models, as shown on Fig. 5-10a, it is assumed that the population may be categorized into two epidemiological states: (1) a susceptible state; and (2) an infected or diseased state. Susceptible individuals are exposed to the pathogen of interest and move into the infected/diseased state with a probability that is governed by the dose of the pathogen to which they are exposed and the infectivity of the pathogen. The solid lines on Fig. 5-10 are used to represent the movement of individuals from one epidemiological state to another, and the dotted lines represent the movement of pathogens. Although humans may be exposed to pathogens from a number of potential

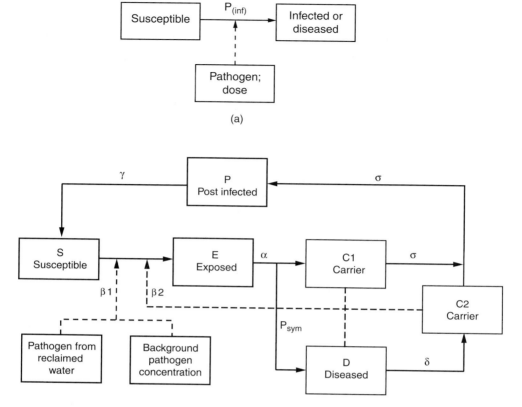

Figure 5-10
Conceptual models for microbial risk assessment: (a) static and (b) dynamic. The rate parameters for the dynamic model, shown as Greek symbols, are described in Table 5-11.

environmental sources, static models typically employ the assumption that susceptible individuals are exposed to pathogens from the specific pathway under consideration for the investigation and do not include the potential interaction and implications of multiple routes of exposure.

Probability of Infection or Disease

The probability that a susceptible individual becomes infected or diseased is a function of the dose of pathogens to which that individual is exposed. When individuals are exposed to pathogens from an environmental source, they move with a given probability to an infected or diseased state. This probability dose-response function is labeled P_{inf} on Fig. 5-10b. The dose is typically calculated by estimating two quantities: (1) the concentration of pathogens at the exposure site and (2) the volume of water ingested. This dose quantity is then input into the dose-response function and the probability that an exposed individual will become infected or diseased is estimated.

Required Health Effects Information

The critical health effects information required for the static model, therefore, is captured in the function that represents the probability of infection (P_{inf}), the pathogen-specific dose-response function. The probability of infection following exposure to a virulent pathogen depends on several host and pathogen-specific factors. The interaction between a pathogen and the host can be viewed as a series of conditional events, in which each event must occur to result in infection. The infection status depends on a number of factors such as (1) the number of organisms that enter the host; (2) the ability of the host to inactivate these organisms; (3) the number of organisms that can withstand the host's local immune defenses, adhere to mucosal surfaces, and multiply to infect the host; and (4) variation in pathogen virulence and host susceptibility (Eisenberg et al., 1996; Eisenberg et al., 2004). The probability of infection is often multiplied by the number of exposed individuals to estimate the expected number of infected individuals for the exposure scenario under consideration.

Dynamic Microbial Risk Assessment Models

In a dynamic risk assessment model, the population is assumed to be divided into a broader group of epidemiological states. Individuals move from state to state based on epidemiologically relevant data (such as duration of infection and duration of immunity). Only a portion of the population is in a susceptible state at any point in time, and only those in the susceptible state can become infected or diseased through exposure to microorganisms.

Movement from Susceptible to Exposed State

The probability that a susceptible person moves into an exposed state is governed by (1) the dose of pathogen to which they are exposed, (2) the infectivity of that pathogen, and (3) the number of infected/diseased individuals with whom they may come into contact. Infectivity as a function of dose (estimated using a dose-response function) is an important factor in estimating risk in static representations of the disease process. The dose-response function is also important in a dynamic MRA model; however, other factors such as person-to-person transmission, immunity, asymptomatic infection, and/or incubation period may also be as or more important.

Accounting for Additional Factors

Accounting for the additional factors, as cited above, when estimating risks associated with exposure to pathogenic microorganisms requires a more sophisticated mathematical model than the static model shown conceptually on Fig. 5-10a. When a dynamic disease transmission model is used, it is possible to account for attributes specific to the transmission of infectious diseases. Depending on the infectious disease processes that are important, the dynamic model may include more or less components, and therefore vary in complexity. For example, a dynamic model may account for person-to-person transmission, immunity, incubation, and asymptomatic infection, as is illustrated on Fig. 5-10b (Soller et al., 2004).

Epidemiological States

The population on Fig. 5-10 is separated into six epidemiological states. Rate parameters that are used to specify the movement between epidemiological states, given as Greek letters on Fig. 5-10b, are described in Table 5-11. The model shown on Fig. 5-10b is called a dynamic model because the number of people in each epidemiological state varies over time. The dynamic model is more comprehensive mathematically as

Table 5-11

Description of rate parameters for the representative dynamic model shown on Fig. 5-10

Symbol	Description
α	Rate of movement from an exposed state to a carrier (infectious and asymptomatic) state or a diseased state (infectious and symptomatic). $1/\alpha$ corresponds to the latency period prior to infection for the pathogen of interest.
σ	Rate of movement from a carrier state to a postinfection state. $1/\sigma$ corresponds to the duration of infectiousness, or equivalently, the duration of asymptomatic shedding of pathogens in feces.
δ	Rate of movement from a diseased state (infectious and symptomatic) to an asymptomatic (carrier) state. $1/\delta$ corresponds to the duration of symptoms during infection.
γ	Rate of movement from a postinfection state (not infectious, asymptomatic, and not susceptible to infection) to a susceptible state. $1/\gamma$ corresponds to the duration of immunity or protection from infection.
β_1	Rate of movement from a susceptible state to an exposed state due to exposure to pathogens from an environmental source (i.e., not person-to-person transmission). Function of the number of pathogens to which an individual is exposed and the infectivity of the pathogen of interest. The infectivity is described quantitatively through a dose-response function which is comprised of one or two dose-response parameters.
β_2	Rate of movement from a susceptible state to an exposed state due to exposure to pathogens from secondary (person-to-person or person-to-environment-to-person) transmission.
P_{sym}	Probability of a symptomatic response. Clinical data describing the proportion of infected individuals that develop symptoms.

compared to the static model. However, as discussed in the following section, for a specific set of assumptions the two models are essentially equivalent (Soller et. al., 2004).

Equivalence of Static and Dynamic Models
Comparing Figs. 5-10a and 5-10b, the two risk assessment models would be equivalent when:

- The background concentration of the pathogen (or equivalently the endemic level of infection/disease) in the population is zero or unimportant
- The duration of infection and disease approaches zero, and
- Infection and/or disease do not confer immunity or the duration of immunity approaches zero

In categorizing the epidemiological status of the population, individuals are considered infected if they are shedding pathogens in their feces or exhibit an immune response such as increased antibody levels. People are considered diseased if they exhibit any of the clinical symptoms related to the specific pathogen of interest, for example, diarrhea and/or vomiting. Individuals in the population move from one epidemiological state to another based on clinically observable data, exposure data, and a dose-response relation between the exposure and the probability of infection.

Deterministic or Stochastic Modeling
Dynamic MRA models can take two main forms: (1) deterministic, or (2) stochastic. In the deterministic form, the model is expressed as a set of differential equations that have defined parameters and starting conditions, which determine the rate of transfer of individuals from one epidemiologic state to another. This type of model is most suitable for large populations of individuals interacting randomly with one another (Eisenberg et al., 1996; Soller et al., 2003). In the stochastic form, the model incorporates probabilities at an individual level and is evaluated by an iterative process such as Markov Chain Monte Carlo analysis. Stochastic model forms are most suitable for small populations with heterogeneous mixing patterns (Koopman et al., 2002).

Microbial Risk Characterization Model Complexity
A variety of model types (from simple to complex) are available that can be used to characterize infectious disease transmission, and evaluate the potential for interventions. Different aspects of the disease transmission system are simulated through the model parameters, with each model capturing a unique set of transmission methods or pathways. Thus, it is unrealistic to presume that one model type is appropriate for all waterborne MRA. For exposures to microbes from reclaimed water applications, it has been demonstrated that the selection of an appropriate model form (static or dynamic) can be identified based on as few as three to four model parameters (Soller et al., 2004). It was also demonstrated that no one model form will be appropriate for all possible combinations of potential pathogens of interest and exposures.

The selection of a model type involves tradeoffs. Biological or demographic "realism" can be achieved, frequently through analytical complexity that distances the model from the available data. Further, each model form involves certain types of assumptions that may or may not be realistic or appropriate for a particular situation. With the perspective that different model types, and forms, and accompanying analytical approaches may be necessary for different applications, Koopman et al. (2001) suggest an analysis strategy involving a hierarchy of models from simple to increasingly complex models that could be traversed to make MRA analyses more realistic while remaining mathematically tractable. From the information available today it is anticipated that the issue of model complexity for MRAs will be an area of research that will receive substantial attention in the future.

Selecting a Microbial Risk Model

The most rigorous and scientifically defensible approach for modeling an infectious disease process in a population is with a dynamic mathematical model because infectious diseases behave dynamically. However, as noted previously, there may be conditions where the results from the static and dynamic models yield similar results. These conditions are of particular interest to risk assessors, because modeling the transmission of infectious diseases as a static process requires substantially less data and mathematical sophistication than modeling the process dynamically.

Evaluating Static and Dynamic Models

Recent work by Soller et al. (2004) evaluated the two types of models, and focused on identifying when it may be appropriate to use the static model and when it may be necessary to use the dynamic model for assessing risk for exposure to pathogens of public health concern from reclaimed water. The work by Soller et al. (2004) is discussed in detail in the following section.

The premises upon which the model evaluation results were based are as follows:

- Under the conditions in which the two models predict similar estimations of risk, the static model is more appropriate, as it is simpler yet yields similar results.
- Under the conditions in which the two models predict substantially different estimations of risk, the dynamic model is more appropriate, as one or more infectious disease processes impact the assessment enough to impact the assessment of risk.

Data Required

The data required to assess the risk associated with a given exposure to a pathogenic microorganism for the static and dynamic models are compared in Table 5-12. As shown, the static model requires substantially less data than the dynamic model. To differentiate between the conditions under which the static and dynamic models predict similar and substantially different estimations of risk for the specific exposure scenario under consideration, a series of numerical simulations were set up to explore the range of feasible parameter combinations. The exposure represented a range of pathogenic microorganisms via reclaimed water applications. Based on the simulation strategy selected, over 500,000 simulations were run. A sensitivity analysis was then conducted to determine which model parameters impacted the predicted difference in infection incidence between the two models.

Table 5-12 Parameters required for modeling static and dynamic disease processes[a]

Parameters	Static	Dynamic
Exposure related		
Concentration of pathogen	X	X
Volume of water ingested	X	X
Proportion of population exposed		X
Frequency of exposure		X
Pathogen related		
Dose response parameter(s)	X	X
Duration of incubation		X
Duration of infectiousness		X
Duration of disease		X
Duration of protection		X
Probability of symptomatic response		X
Person-to-person transmission potentials		X
Background concentration level		X

[a]Adapted from Soller et al. (2004).

Parameters Most Strongly Impacted

Based on the results of the sensitivity analysis, the parameters that most strongly impacted the difference in predicted incidence between the static and dynamic models, in order of decreasing importance are:

- Dose of pathogen
- Exposure intensity
- Dose-response parameters (α and β for the beta-Poisson model)
- Duration of infection

Important Caveats

Important caveats in the comparison of models (Soller et al., 2004) include:

- Extrapolation of the results to routes of exposure, pathogens, and/or other model variants not investigated, including levels of incidence difference, must be done with caution, as the results are only applicable within the bounds investigated.
- Microbial risk assessment is inherently agent specific, therefore, the cumulative effects of exposure to multiple pathogens were not addressed explicitly.
- The health outcome associated with infection and disease was gastroenteritis, as dose-response data predict this health outcome. There are a number of other more serious disease outcomes that are also associated with pathogenic microorganisms and characterizing the risk associated only with one outcome likely underestimates the true cumulative risk to public health. Developing a characterization methodology for other endpoints was beyond the scope of the investigation, however the likelihood for such health outcomes is important and should be considered during the risk management process.

The above results provide some insight and guidance to utilities and risk managers regarding the conditions under which a less complex MRA model may be employed for human exposure to pathogens via a reclaimed water pathway.

The Risk Manager's Role

When selecting a MRA model, the risk manager is urged to use caution, as risk issues are typically complex, and have impacts on both the development and the compliance with regulations. A discussion of the latter point is addressed in a paper by Olivieri et al. (2005) in which the costs and relative public health benefits of seasonally based municipal wastewater treatment plant effluent limits (i.e., secondary treatment with disinfection in the winter and tertiary treatment in the summer to protect public health in swimming and body contact sports) were investigated. Allowance for seasonal limits raises a significant water quality policy question with regard to the costs and relative benefits of providing tertiary treatment during the winter season in addition to the summer season. The assumed societal benefit of winter tertiary treatment was enhanced water quality for recreational purposes, and thus reduced risk to public health. Olivieri et al. (2005) estimated that between 4 million and 16 million recreation events would need to occur annually in northern California during the winter to justify the costs of providing winter tertiary treatment. The information presented in the paper could be used by water quality regulatory agencies to develop a risk-based policy to consider seasonal water quality limits for effluent discharge and water reuse applications.

5-8 APPLICATION OF MICROBIAL RISK ASSESSMENT IN WATER REUSE APPLICATIONS

Three typical examples are presented in this section to illustrate the application of both static and dynamic MRAs in water reuse applications: (1) a risk assessment employing a static microbial model, (2) a risk assessment employing a dynamic microbial model, and (3) an assessment of the health risks associated with enteric viruses in reclaimed water. It should be noted that the derivation of the required equations and the computational procedures that are used in the examples presented in this section are beyond the scope of this chapter. Nevertheless, where appropriate, the required equations are included, without detailed derivation, and the computational procedures are described to illustrate the methodology involved in MRA applications. Details on the derivation of the equations used and the computational procedures employed may be found in the cited references.

Microbial Risk Assessment Employing a Static Model

Static MRA methods, as noted in Sec. 5-7, have been used to evaluate the potential public health effects associated with drinking water containing a range of waterborne pathogens as well as reclaimed water. The methods employed in those assessments have varied from relatively straightforward assessments using point estimate values for model parameters to more complex assessments relying on stochastic (probabilistic) models. Some representative examples of risk assessments employing static models are summarized in Table 5-13. The use of a static MRA model is illustrated in Example 5-3.

Table 5-13
Examples of risk assessment employing static models

Model purposes, components, and findings	References
The enteroviruses (a subgroup of enteric viruses), for which a standard analytical method has been available for some time, could serve as an indicator of worst-case potential occurrence for any specific virus. The dose-response relation for rotavirus has been used to derive upper-limit risk estimates for viruses in water as rotavirus is the most infectious waterborne virus for which dose-response information is currently available.	Regli et al. (1991)
Enteric virus monitoring data in California from secondary and tertiary effluents were evaluated in conjunction with the State of California's *Water Recycling Criteria*. The analysis showed that annual risk of infection from exposure to chlorinated tertiary effluent containing 1 viral unit/100 L in recreational activities such as swimming or golfing is in the range of 10^{-2} to 10^{-7} while exposures resulting from foodcrop irrigation or groundwater recharge with reclaimed municipal wastewater is in the range of 10^{-6} to 10^{-11}.	Asano et al. (1992)
Uncertainties are accounted for in exposure assessments (lognormal distribution for volume ingested) and the dose-response relationship (95 percent confidence intervals about the maximum likelihood estimate for α and β) for viruses in drinking water by applying Monte Carlo simulation techniques.	Haas et al. (1993)
Point estimate values are used for the concentration of rotavirus in drinking water (0.004/L and 100/L) and an assumed 99.99 percent reduction of rotavirus through drinking water treatment. The volume of water ingested (2 L/d and 4 L/d) and beta-Poisson dose response parameters ($\alpha = 0.26$, $\beta = 0.42$) were also based on point estimate values. The probability of clinical illness was determined by multiplying the resulting probabilities of infection by 0.5. The probability of mortality was determined by multiplying the probability of illness by 0.01 percent for the general population and one percent for the elderly.	Gerba et al. (1996)
Two concepts related to safety of water reclamation and reuse are presented. The first is *reliability*, defined as the probability that the risk of infection from enteric viruses in reclaimed wastewater does not exceed an acceptable risk. The second is based on the *expectation* of the acceptable annual risk in which the exposure to enteric viruses may be estimated stochastically by numerical simulation (Monte Carlo methods). Because enteric virus concentrations in unchlorinated secondary effluents were found to vary over a wide range, characterizing their variability was found to be extremely important. The reliability criterion of meeting the less than 10^{-4} annual risk of infection (less than or equal to one infection per 10,000 population per year) at least 95 percent of the time was used to assess the safety of using reclaimed water in the four different exposure scenarios. The findings of this study served as an independent verification of the *California Water Recycling Criteria* (Title 22 regulations).	Tanaka et al. (1998)
A model for virus decay on lettuce and carrot crops has been derived as part of a comprehensive wastewater irrigation microbial risk assessment model under development. Results from the decay modeling indicated the presence of a very	Petterson and Ashbolt (2001)

(Continued)

Table 5-13
Examples of risk assessment employing static models (*Continued*)

Model purposes, components, and findings	References
persistent subpopulation of viruses evidenced by an initial rapid phase of decay followed by a very slow phase. In addition, virus counts fitted a negative binomial rather than Poisson distribution indicating overdispersion. Hence the data indicated that viruses were not uniformly distributed over the surfaces of both crops.	
Rose et al. evaluated the potential public health effects associated with drinking water contaminated with *Giardia lamblia* and *Cryptosporidium*. Point estimates were used to characterize the volume of water consumed daily (2 L), average levels of cysts in surface waters (0.22 to 104/100 L), reduction of cysts due to drinking water treatment (99.9 percent), and the dose-response relation. Annual risks were computed as described above and source water concentrations corresponding to annual risks of 1/10,000 were derived.	Rose et al. (1991a, b) Teunis et al. (1997) Perz et al. (1998) Teunis and Havelaar (1999) Makri et al. (2004)
Teunis et al. (1997) conducted an assessment of the risk of infection by *Cryptosporidium* and *Giardia lamblia* in drinking water from a surface water supply in which the major contributing factors to risk were each treated as stochastic variables. The stochastic variables investigated included the concentration of cysts (*Giardia lamblia*) and oocysts (*Cryptosporidium*) in raw water, the recovery of the detection method, the viability of recovered cysts or oocysts, the removal of organisms in the treatment process, and the daily consumption of unboiled tap water.	
Teunis and Havelaar (1999) conducted a case study in which the risk of human infection from *Cryptosporidium parvum* in drinking water was characterized. Exposure was assessed by splitting into different stages the route of the pathogens from river water to the consumed tap water. Assessment of the performance of a drinking water treatment process was modeled using spores from sulfite reducing clostridia as the surrogate organism. For dose-response assessment, the beta-Poisson model was employed. The dose-response relations for infection and illness were used to generate, via Monte Carlo methods, distributions for the risk of daily, annual, and lifetime infection and illness.	
Static models have been used to investigate the expected public health risk associated with exposure to pathogens from recreational exposure in freshwater. The city of Vacaville and the El Dorado Irrigation District in California each conducted investigations to evaluate the potential health risks posed from exposure to pathogens via contact with the respective treated effluents from recreational activities. For the city of Vacaville, the estimated risk values range from approximately 1 infection per 10^4 to 10^6 recreation events for secondary treatment to 1 infection per 10^6 to 10^7 events for tertiary treatment. For El Dorado Irrigation District, the median probability of infection to swimmers from exposure to tertiary treated effluent is estimated to be on the order of 5 infections per 10^7 exposures. The results of the investigation indicated that in the United States, the U.S. EPA acceptable health risk levels for illness (i.e., 8/1000) are more than met by the current performance of the municipal wastewater treatment plants during both the winter and summer seasons.	Olivieri and Soller (2001) HDR Engineering et al. (2001)

EXAMPLE 5-3. Derivation of Wastewater Treatment Requirements for Unrestricted Crop Irrigation with Reclaimed Water Based on Rotavirus.

Use microbial risk assessment procedures to derive the required reduction of rotavirus concentration through wastewater treatment for unrestricted crop irrigation with reclaimed water. Assume rotavirus is the organism of concern and an acceptable level of risk of disease is $P_{ill} = 10^{-3}$ illnesses/person·yr. Assume further that 10 percent of the persons infected will become ill. The following parameter values are to be used in the quantitative microbial risk analysis.

1. 100 g lettuce consumed every second day throughout the year
2. The concentration of rotaviruses is 1000 /L in untreated wastewater
3. Following irrigation, 10 mL of reclaimed water remain on 100 g lettuce
4. 2 log viral die-off between last irrigation and consumption, and
5. 1 log unit viral reduction occurs due to washing the lettuce with clean water prior to consumption (i.e., a total of 3 logs reduction due to die-off and washing)

A schematic diagram of the problem is as follows:

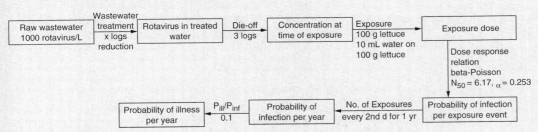

Use the schematic diagram above and work backward from the "Probability of illness per year" to determine the log reduction necessary during wastewater treatment.

Solution

1. Because the acceptable probability (risk) of illness per year is 10^{-3}, the acceptable risk probability of infection is:

   (10^{-3} infections/person·yr)/0.1 disease/infection = 1×10^{-2} infections/person·yr

2. Convert the acceptable probability of infection (1×10^{-2} infections/person·yr) to the probability of infection per person per exposure event using Eq. (5-7).

$$P_{yr} = 1 - (1 - P_d)^n$$

where n = number of exposure events
P_{yr} = acceptable level of risk (probability) of infection/person·year
P_d = acceptable probability of infection per person per day (single event)

Solving the above equation for the daily exposure, P_d, yields

$$P_d = 1 - (1 - P_{yr})^{1/n}$$

Substituting for P_{yr}, and noting that the exposure takes place every two days throughout the year (i.e., the value of n in the above equation is 365/2) the value of P_d is equal to

$$P_d = 1 - (1 - 10^{-2})^{1/(365/2)} = 5.5 \times 10^{-5}$$

3. Compute the corresponding dose of rotavirus using the β-Poisson dose-response equation based on the median ingested dose (N_{50}) [Eq. (5-5)].

   Solve the β-Poisson dose-response equation for dose (d). Note that P_{inf} probability of infection from single exposure event corresponds to P_d calculated above.

$$P_{inf}(d, \alpha, N_{50}) = 1 - \left[1 + \frac{d}{N_{50}}(2^{1/\alpha} - 1)\right]^{-\alpha}$$

   Solving the above expression for d yields

$$d = \frac{(1 - P_d)^{-1/\alpha} - 1}{N_{50}/(2^{1/\alpha} - 1)}$$

   Substituting the given values for the dimensionless "infectivity constants" for rotavirus, $N_{50} = 6.17$ and $\alpha = 0.253$, solve the above expression for the dose, d.

$$d = \frac{[1 - (5.5 \times 10^{-5})]^{-1/0.253} - 1}{6.17/(2^{1/0.253} - 1)} = 5 \times 10^{-4}$$

4. Compute the concentration at the time of exposure: the computed dose of 5×10^{-4} rotavirus is contained in the 10 mL remaining on the lettuce at the time of consumption. Thus, the concentration at the time of exposure, expressed in units of rotavirus per liter is 5×10^{-2} /L.

5. Compute the concentration of rotavirus in the treated wastewater taking into account the 3-log unit rotavirus reduction between the last irrigation and consumption, the maximum allowable concentration of rotaviruses, C, in the effluent of the treatment plant is

$$C = [(5 \times 10^{-2}) \times 10^3] = 50/L$$

6. Compute the required rotavirus log reduction, R, through wastewater treatment. The required reduction is by

$$\text{Log } R = \log (1000/L)/\log (50/L) = 1.77, \text{ round to } 2$$

Therefore the total required rotavirus reduction corresponds to approximately 2 log reduction achieved by wastewater treatment, plus the 2 log reduction due to die-off between the last irrigation and consumption, plus the 1 log reduction due to washing the lettuce in clean water immediately prior to consumption, is equal to (2 + 2 + 1); approximately a reduction of 5 logs overall.

Microbial Risk Assessment Employing Dynamic Models

The fundamental difference between the static and dynamic MRA models is that the risk characterization perspective is shifted away from an individual to a population-based perspective in the dynamic MRA model applications. In dynamic MRAs, the models are used to simulate the epidemiologic status of a population over time as well as environmental variables such as pathogen density. A conceptual model for health effects must be developed for each case under investigation. Risk characterization is implemented by integrating the exposure and health effects components (models) via a parameterization step, and by running Monte Carlo simulations. The resultant output from the simulations is distributions of predicted adverse health effects. Examples of risk assessment employing dynamic models are reported in Table 5-14. The use of a dynamic MRA model is illustrated in Example 5-4. Although the computational procedure utilized in Example 5-4 is beyond the scope of this textbook, the approaches utilized in this example are valid and illustrative of the dynamic microbial risk assessment models.

Table 5-14
Examples of risk assessment employing dynamic models

Model purposes, components, and findings	References
Dynamic microbial risk assessment methods have been used to characterize the potential public health effects associated with rotavirus in drinking water, obtain insight into the epidemic process related to drinking water treatment failures, characterize risks from microbiological contaminants associated with recreational activities, and estimate the bias associated with modeling the infectious disease process using a static model. In each investigation, a conceptual model for health effects was developed. Risk characterization was implemented by integrating the exposure and health effects components (models) via a parameterization step, and by running Monte Carlo simulations. The resultant outputs from the simulations are distributions of predicted adverse health effects.	Olivieri 1995a, b Eisenberg et al., 1996, and 2003. Soller et al., 1999, 2003, and 2006.
A dynamic model was constructed and possible parameter combinations were evaluated to find combinations consistent with surveillance data from the outbreak of disease incidence for the 1993 cryptosporidiosis outbreak in Milwaukee, Wisconsin. Evaluation of the model output indicated that a smaller outbreak likely occurred prior to the large reported outbreak. This finding suggested that had surveillance systems detected the earlier outbreak, up to 85 percent of the cases might have been prevented. Further analysis using the incidence data resulted in three inferred properties of the infection process (1) the mean incubation period was likely to have been between three and seven days; (2) there was a necessary concurrent increase in *Cryptosporidium* oocyst influent concentration and a decrease in treatment efficiency of the water treatment facility; and (3) the variability of the dose-response function in the model did not appreciably affect the simulated outbreaks.	Eisenberg et al., 1998

EXAMPLE 5-4. Risk Assessments Employing a Dynamic Model.

Use a dynamic MRA approach to assess the potential public health risk from enteric viruses that is associated with recreational activities in Newport Bay in southern California. The following analysis is adapted from Soller et al., 2006.

Analysis

1. A schematic diagram showing how different types of data were used for the investigation is presented below.

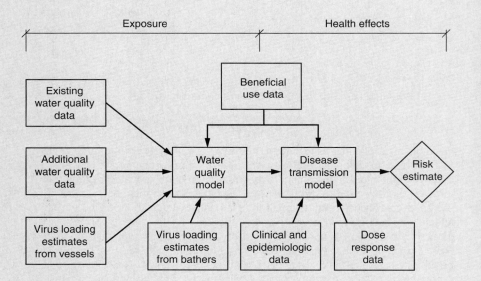

2. A model for disease transmission was developed to describe the epidemiologic status of individuals within a population, and how that status varies over time. As illustrated on the following page, four state variables in the dynamic disease transmission model (S, C, D, and P) are used to track the number of people that are in each of the epidemiologic states at any point in time. Rate parameters are used to determine the movement of the population from one state to another. The rate parameters include the rate of acquiring infection, the rate of recovery from infectious states, and the rate of decline in immunity. Rate parameters are determined through literature review directly or are functions of model parameters determined from the literature.

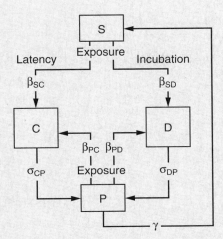

State variables

S: Susceptible to infection, not infectious, not symptomatic
C: Infections and not symptomatic
D: Infectious and symptomatic
P: Protected from infection, not infectious, not symptomatic

Rate parameters

β_{SC}: Rate at which individuals in state S move to state C
β_{SD}: Rate at which individuals in state S move to state D
β_{PC}: Rate at which individuals in state P move to state C
β_{PD}: Rate at which individuals in state S move to state D
σ_{CP}: Rate at which individuals in state C move to state P
σ_{DP}: Rate at which individuals in state D move to state P
γ : Rate at which individuals in state P move to state S

Two routes of transmission are considered—primary transmission by background exposure and/or recreational contact in Newport Bay, and secondary transmission which includes person-to-person transmission. The change in the fraction of the population in any state from one time period to the next is modeled as a first order differential equation. For example, the relative change in state S from one time period to the next due to primary infection is:

$$dS_1/dt = -\beta_{SC1} S - \beta_{SD1} S + \gamma P$$

Note that the numeric subscripts indicate that the route of transmission is (1) primary or (2) secondary (person-to-person transmission). Similarly, the relative change in state S from one time period to the next due to secondary infections is directly related to the number of individuals who are in states S, C, and D during that time period.

$$dS_2/dt = -(\beta_{SC2} + \beta_{SD2}) S (D + C)$$

The overall change in the number of susceptible individuals from one time period to the next is equal to $dS_1/dt + dS_2/dt$.

3. A series of Monte Carlo simulations were run with randomly selected model parameter values.

Model parameters	Parameter	Units	Range
Dose response parameters	DR_α	Unitless	0.125–0.5
	DR_β	Unitless	0.21–0.84
Probability of symptomatic response	P_{sym}	Unitless	0.1–0.45
Previous exposure factor	ε	Unitless	0.1–0.9
Reciprocal of incubation	$1/\tau_I$	d^{-1}	0.33–1.0
Reciprocal of latency	$1/\tau_L$	d^{-1}	0.143–0.333
Rate diseased move to postinfection state	σ_{dp}	d^{-1}	0.09–0.5
Rate carriers move to postinfection state	σ_{cp}	d^{-1}	0.05–0.125
Rate of susceptible re-establishment	γ	d^{-1}	0.0009–0.0027

Rate parameters dependant on model parameters

$\beta_{SC} = (P_{dose} + P_{contact}) \times (1 - P_{sym}) \times \tau_L$
$\beta_{SD} = (P_{dose} + P_{contact}) \times (P_{sym}) \times \tau_I$
$\beta_{PC} = (P_{dose} + P_{contact}) \times \varepsilon \times (1 - P_{sym}) \times \tau_L$
$\beta_{SD} = (P_{dose} + P_{contact}) \times \varepsilon \times (P_{sym}) \times \tau_I$

Intermediate variables used to compute rate variables

$P_{dose} = 1 - (1 + dose/DR_\beta)^{-DR_\alpha}$
$P_{contact} = 1.38 \times \sigma/N$
N = Population size — 1,200,000

4. Output from the disease transmission model is the number of people in each of the states as well as the average daily prevalence for the simulation, defined as the average proportion of the population that is symptomatic (in state D) during the whole simulation period. For any given simulation the number of individuals in each state changes until a steady state is achieved.

5. The number of individuals in each of the epidemiological states at the end of 1000 simulations was found to be very similar for background and background plus body contact recreation exposure conditions (see the following figure).

From the results of this analysis, it was found that approximately 99.9 percent of the population in the diseased state at any time is due to background exposure compared to approximately 0.1 percent due to recreational activities (REC-1). In these simulations, all members of the population are subject to background exposure, whereas only those who choose to recreate in the Newport Bay are subject to the incremental recreational exposure.

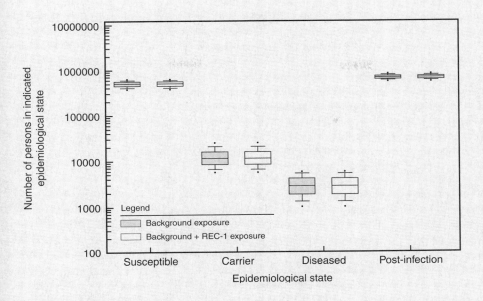

6. To provide an independent method of checking the disease transmission modeling results, a separate analytical approach was employed to evaluate enterococcus data collected in the most heavily used recreation site in Newport Bay. The approach involved applying a static based risk assessment approach utilizing the U.S. EPA enterococcus concentration—illness response function. The water quality results of 246 enterococcus samples were available from recreation sites for 1999 and 2000. The samples (48 percent of the total samples) that were below the detection limit of 10 MPN/100 mL were assumed to be present at the detection limit.

The enterococcus data were fit to a lognormal distribution using the method of maximum likelihood. Based on this distribution and the U.S. EPA equation relating enterococci density in marine water to illness rate (U.S. EPA, 1986), Monte Carlo simulations were used to estimate the expected distribution of disease attributable to REC-1 activities (see figure above).

Comparing the results of these simulations with the output from the disease transmission modeling simulations (see figure following—disease transmission model output), the levels of disease predicted by both the enterococcus data and the disease transmission model are below the U.S. EPA accepted marine levels. Furthermore, the levels of disease attributable to body contact recreation estimated by the disease transmission model are approximately an order of magnitude lower than those estimated using enterococcus data.

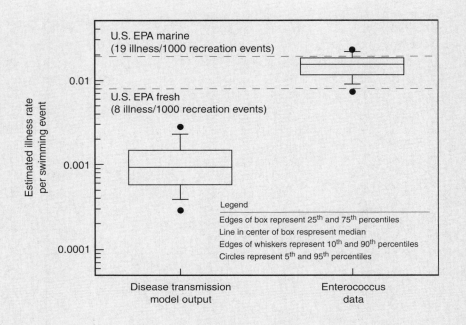

Comment
Often risks to human health are below levels that are practical to observe. In such cases simulation based approaches are invaluable because of their ability to evaluate the potential benefits and costs of proposed management options. Although MRA methods inherently do not characterize the cumulative risk associated with all pathogens potentially present in an environment, it is possible to construct a simulation-based model that captures the salient features of a class of pathogens of interest, and frame an investigation in a manner such that practical risk management decisions can be made.

Risk Assessment for Water Reuse from Enteric Viruses

In the United States, the constituents in reclaimed water that have received the most attention are pathogens, and specifically enteric viruses. The focus is on pathogens because of their low-dose infectivity, long-term survival in the environment, difficulties in monitoring them, and their low removal and inactivation efficacy in conventional wastewater treatment. Health risks associated with enteric viruses in reclaimed water are typically encountered in the following water reuse applications: (1) golf course irrigation, (2) food crop irrigation, (3) recreational impoundments, and (4) groundwater recharge. An analysis of the health risks associated with each of these water reuse applications is presented in Example 5-5.

EXAMPLE 5-5. Estimation of the Required Virus Reduction in Tertiary Treatment to Achieve Acceptable Infection Risk by the Exposure to Enteric Viruses in Reclaimed Water.

The distribution of rotavirus in the unchlorinated effluent from a secondary treatment process is lognormal, with a geometric mean value of -1.47 virus units/L (vu/L) and geometric standard deviation of 0.91 vu/L. If the acceptable annual risk of infection is 10^{-4} and rotavirus is the constituent of interest, use the beta-Poisson model to determine the reliability of the tertiary treatment process for rotavirus concentration log reductions of 3, 4, and 5 for each of the four exposure scenarios given in the following table. In this example, reliability is defined as the probability of time that the risk of infection from the ingestion of virus is equal to or less than the acceptable annual risk. Also determine the required level of virus reduction in tertiary treatment for 90, 95, and 99.9 percent reliability. Use the values for the dose response model parameters given by Rose and Gerba (1991a, b), where $\alpha = 0.232$ and $\beta = 0.247$.

Summary of exposure scenarios[a]

Item	Scenario			
	I	II	III	IV
Application	Golf course irrigation	Crop irrigation	Recreational impoundment	Groundwater recharge
Risk group receptor	Golfer	Consumer	Swimmer	Groundwater consumer
Exposure frequency	Twice per wk	Every day	40 d/yr—summer season only	Every day
Amount of water ingested in a single exposure, V, mL	1	10	100	1000
Virus reduction measures	Stop irrigation 1 d before golf play	Stop irrigation 2 weeks before harvest and shipment	No virus reduction	3 m vadose zone and 6 mo retention in aquifer
Environmental removal rate	0.5/d	0.5/d	none	0.69/d

[a]Adapted from Tanaka et al. (1998).

Solution
1. Estimate the acceptable daily virus exposure concentration for exposure Scenario I.
   a. The acceptable daily (single) exposure risk caused by virus ingestion can be estimated using Eq. (5-7), rearranged to solve for the daily exposure, P_d.
   $$P_d = 1 - (1 - P_{yr})^{1/n}$$
   For Scenario I, the exposure event occurs twice per week (52×2) with an ingestion of 0.001 L per exposure. Thus, the daily risk is computed as follows:
   $$P_d = 1 - (1 - 10^{-4})^{1/104} = 9.62 \times 10^{-7}$$
   b. Compute the acceptable daily exposure virus concentration, C_d, associated with the given acceptable daily risk, P_d, using the beta-Poisson dose-response equation Eq. (5-4). Note that the daily dose is given by $C_d \times V$. Use the values for the dose response model parameters, where $\alpha = 0.232$ and $\beta = 0.247$. The daily concentration for Scenario I is:
   $$C_d = [(1 - P_d)^{-1/\alpha} - 1]\beta/V$$
   $$= [(1 - 0.000000962)^{-1/0.232} - 1]\,0.247/0.001\,L = 1.024 \times 10^{-3}\ \text{vu/L}$$

2. Estimate the reliability, p, of the water reuse Scenario I to provide reclaimed water with a virus concentration equal to or less than the acceptable daily virus concentration computed in Step 1. A sample calculation is shown for Scenario I.
   a. Compute the fraction remaining after the virus reduction resulting from the environmental exposure, E, after 1 day (note that virus reduction measures, E = 1 d for Scenario I).
   $$E = (0.5\ d^{-1})(1\ d) = 0.5$$
   b. Compute the process reliability using the following expression, which represents the probability that the performance of the treatment process will be equal to or less than the acceptable daily virus concentration
   $$p = \Phi\{[\log C_d + R - \log E - (\mu)]/\sigma\}$$
   where Φ = standardized normal function
   R = log removal achieved by treatment process, unitless
   E = virus reduction resulting from the environmental exposure, unitless
   μ = geometric mean of the lognormal distribution whose random variable is logarithmically transformed with respect to viruses in unchlorinated secondary effluent
   σ = standard deviation of the lognormal distribution whose random variable is logarithmically transformed with respect to enteric viruses in unchlorinated secondary effluent

   Substituting the value of C_d, computed in Step 1b, and the values given in the problem statement, p is:
   $$p = \Phi\{[\log 0.001024 + 3 - \log 0.5 - (-1.47)]/0.91\} = \Phi\{1.96\}$$

c. Convert the standardized value obtained above to the percent reliability. The value of p may be determined using statistical tables or appropriate computer software. The computation in Excel may be determined using the standardized normal distribution function NORMSDIST.

$$p = \text{NORMSDIST}(1.96) = 0.975$$

3. Determine the corresponding values for different log removals for each of the four given scenarios. The required values, obtained using the procedure outlined above in Steps 1 and 2, are summarized in the following table:

Log removal of virus by tertiary treatment process	Reliability, p, for indicated reuse scenario, %			
	Golf course Irrigation	Crop irrigation	Recreational impoundment	Groundwater recharge
3	97.5	100.0	45.4	100.0
4	99.9	100.0	83.7	100.0
5	100.0	100.0	98.1	100.0

4. Determine the virus log reductions required for the secondary effluent to meet the acceptable virus concentration at 90, 95, and 99.9 percent reliability using the expression given in Step 2b and solving for R.
   a. The computation for golf course irrigation (Scenario I) at 90 percent reliability is as follows: (Note that the Excel function NORMSINV may be used to compute the inverse standardized normal function)

   $$R = \sigma \Phi^{-1}(p) + \mu + \log E - \log C_d$$
   $$= 0.91\, \Phi^{-1}(0.90) + (-1.47) + \log 0.5 - \log 0.001024 = 2.4$$

   b. The required log removal levels for the secondary effluent to meet the acceptable rotavirus concentration (i.e., the concentrations expected to result in one enteric virus infection per 10,000 population) at 90, 95, and 99.9 percent reliability for the four scenarios are summarized in the following table:

Scenario	Log reductions, R, necessary to meet indicated percent reliability		
	90	95	99.9
I. Golf course irrigation	2.4	2.7	4.0
II. Crop irrigation	0.02	0.35	1.66
III. Swimming	4.3	4.6	5.92
IV. Groundwater recharge	0.0	0.0	0.0

5. Determine the uncertainty in the annual risk of infection. Although beyond the scope of this example, the Monte Carlo Method may be used to simulate the distribution of the risk for each of the exposure scenarios. The method may be conducted using a spreadsheet model or proprietary risk analysis software. The required steps are illustrated in the following schematic diagram.

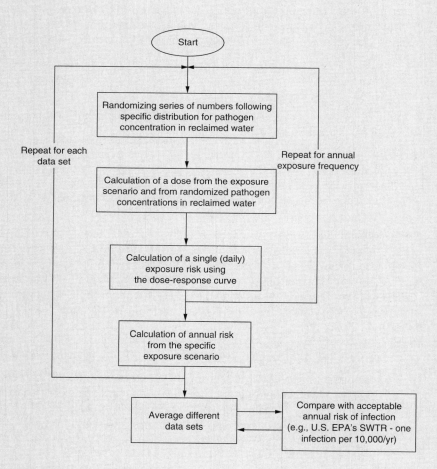

Comment

Because of the variable nature of wastewater treatment processes, a probability distribution is used to describe the effluent constituent concentrations. In the above analysis, only one organism (rotavirus) was considered; however, it is likely in practice that a number of organisms may be present and the analysis will need to be repeated for each organism and exposure scenario.

5-9 LIMITATIONS IN APPLYING RISK ASSESSMENT TO WATER REUSE APPLICATIONS

Although risk assessment is used in a variety of settings as an aid to decision making, a number of serious limitations exist with the application of risk assessment to water reuse. The principal limitations of risk assessment as applied to water reuse are (1) risk assessment is determined as a health risk relative to other things as opposed to being a definitive or absolute risk, (2) inadequate consideration of secondary infections in MRA, and (3) most importantly, the limited availability of exposure data and valid dose-response data.

Relative Nature of Risk Assessment

Because it is impossible to determine absolute risk with the present state of knowledge, relative health risk as opposed to absolute risk, as illustrated above, must be used to assess the safety of water reuse practices and to evaluate alternative water reuse applications. For instance, in a recently completed study by Englehardt et al. (2004) relative human and ecological risks involved in three different effluent discharge options were examined using a Bayesian analysis, based on available information and expert opinion. Arsenic, *Cryptosporidium parvum*, rotavirus, and NDMA were used as indicators of human health risks for the three alternative discharge options which were (1) injection to a deep subsurface aquifer located below a drinking water aquifer, (2) discharge to surface canals where the treated effluent can infiltrate into and become mixed with the existing natural groundwater, and (3) discharge to the ocean where body contact sports including swimming and beach activities may be involved.

The Bayesian approach used in this study is directly applicable to water reuse projects. However, regardless of the approach used, it is important to keep in mind that relative risk is not well understood by the public; thus, if risk assessment is to be used for project evaluation, it is imperative that the process be made as transparent as possible.

Inadequate Consideration of Secondary Infections

In chemical risk assessment, the basis for the analysis is the individual who is ingesting the chemical constituent of concern. Although the same approach can be used for MRA, as discussed above, more rigorous quantitative methodologies are required because pathogens can be transferred from person to person. Because of the possibility of disease transmission from person-to-person, secondary infection must be considered in assessing microbial risks, especially where large populations are exposed (e.g., swimming in a reclaimed water–dominated water body). To describe the transmission of enteric virus infection and disease within an exposed population, it has been proposed that consideration should be given to the quantification of persons who are (1) infectious and symptomatic, (2) infectious and not symptomatic, (3) not infectious and not symptomatic, and (4) not infectious and not symptomatic with short-term or partial immunity and their movement between these states (Eisenberg et al., 1996; Soller et al., 1999). Such an approach has been applied in assessing the risk associated with the land application of digested sludge (Eisenberg et al., 2004). However, this is not the typical approach, and because of the possibility of disease transmission from person-to-person, additional research should be done before MRA can be considered a routine undertaking.

Limited Dose-Response Data

Finally, the most serious limitation in the application of risk assessment to water reuse is the limited availability of dose-response data for most of the constituents of concern in water reuse. As noted previously, numerous uncertainties are involved in the development of the dose-response data including extrapolations of animal data to humans and mathematical extrapolations from high experimental dosages to the extremely low dosages encountered in most water reuse risk assessments. Because dose-response data serve as the basis for most mathematical modeling, the results obtained from modeling efforts must be used judiciously. Risk assessment in many water reuse applications will remain a qualitative exercise until valid dose-response data becomes widely available.

PROBLEMS AND DISCUSSION TOPICS

5-1 Using Table 5-3 as a general guide, find examples of each model application from literature for chemical and microbiological constituents that may be of importance in water reclamation and reuse. Discuss practical applications of dose-response models and limitations in water reclamation and reuse.

5-2 Use the exponential and beta-Poisson models to prepare a plot of probability of infection as a function of normalized dose for poliovirus 1 using the data in Table 5-4. Compare and comment on the two curves.

5-3 Verify the results shown on Fig. 5-3b for the exponential and beta-Poisson with alpha equal to 1.

5-4 Using the data given in Table 5-8, estimate the incremental cancer risk for an adult associated with drinking groundwater containing 2.0 µg/L of one of the following constituents (to be selected by the instructor): inorganic arsenic, benzene, bromate, chloroform, dieldrin, heptachlor, NDMA, or vinyl chloride. To limit the constituent exposure to an acceptable cancer risk of 1 in 100,000, determine the concentration of the constituent that can be allowed in extracted groundwater. Compare your computed value to the value given in the U.S. EPA IRIS data base (http://www.epa.gov/iris).

5-5 Review the current literature (a minimum of three articles should be reviewed and cited) and list health and regulatory factors affecting implementation of a water reuse project. What is the rationale for setting less stringent microbiological standards in developing countries where enteric diseases are rampant among the population?

5-6 Inactivation of pathogens in the environment occurs due to sunlight exposure, predation, and natural die-off. Because the amount of environmental removal is difficult to control or predict, a range of environmental removal values is used to assess the degree of treatment required. Solve the problem presented in Example 5-5 for golf course irrigation using environmental removal rates of 0.25, 0.5, and 0.75. Compute the log reductions necessary for 90, 95, and 99.9 percent reliability for the assumed environmental removal rates. Given that the results represent a limited sensitivity analysis, discuss the implications of the results with respect to the degree of treatment needed to compensate for environmental factors.

5-7 Review briefly the status of static and dynamic risk assessment in modeling microbial risk associated with water reuse.

5-8 In its present state of development, what are the principal limitations of microbial risk assessment? Cite three or more references.

REFERENCES

Asano, T., L. Y. C. Leong, M. G. Rigby, and R. H Sakaji (1992) "Evaluation of the California Wastewater Reclamation Criteria Using Enteric Virus Monitoring Data," *Water Sci. Technol.,* **26**, 7–8, 1513–1524.

Asbolt, N. J., S. R. Petterson, T-A. Stenström, C. Schonning, T. Westrell, and J. Ottson (2005) *Microbial Risk Assessment Tool*, Report 2005:7, Urban Water, Chalmers University of Technology, Gothenburg, Sweden.

Charnley, G., S. Newsome, and J. Foellmer (1997) *Risk Assessment and Risk Management in Regulatory Decision-Making*, The Presidential/Congressional Commission on Risk Assessment and Risk Management, **1–2**.

Cockerham, L. G., and B. S. Shane (1994) *Basic Environmental Toxicology*, CRC Press, Boca Raton, FL.

Cooper, R. C. (1991) "Public Health Concerns in Wastewater Reuse," *Water Sci Technol*, **4**, 9, 55–65.

Cooper, R. C., A. W. Olivieri, R. E. Danielson, and P. G. Badger (1986a) "Evaluation of Military Field-Water Quality," **6**, *Infectious Organisms of Military Concern Associated with Nonconsumptive Exposure: Assessment of Health Risks and Recommendations for Establishing Related Standards*, UCRL-21008.

Cooper, R. C., A. W. Olivieri, R. E. Danielson, P. G. Badger, R.C. Spear, and S. Selvin (1986b) "Evaluation of Military Field-Water Quality," **5**, *Infectious Organisms of Military Concern Associated with Consumption: Assessment of Health Risks and Recommendations for Establishing Related Standards*, UCRL-21008.

Cothern, C. R. (1992) *Comparative Environmental Risk Assessment,* Lewis Publishers, Boca Raton, FL.

Cotruvo, J. A. (1987). "Risk Assessment and Control Decisions for Protecting Drinking Water Quality," in I. A. Suffet and M. Malayiandi (eds.) *Organic Pollutants in Water, Advances in Chemistry Series, No. 214*, Am. Chem. Soc., Washington, DC.

Cotruvo, J. A. (1988) "Drinking Water Standards and Risk Assessment," *Reg. Toxicol. Pharmacol.,* **8**, 288–299.

Daughton, C. G., and T. A. Ternes (1999) "Pharmaceuticals and Personal Care Products in the Environment: Agents of Subtle Change?" *Environ. Health Perspect.*, **107**, Suppl. 6, 907–938.

Dudely, R. H., K. K. Hekimian, and B. J. Mechalas (1976) "A Scientific Basis for Determining Recreational Water Quality Criteria," *J. WPCF,* **48**, 12, 2761–2777.

Eaton, D. L., and C. D. Klaassen (2001) "Principles of Toxicology," in C. D. Klaassen (ed.) *Casarett and Doull's Toxicology: The Basic Science of Poisons*, 6th, McGraw-Hill, New York, NY.

Eisenberg, J. N., A. W. Olivieri, K. Thompson, E. Y. W. Seto, and J. I. Konnan (1996) "An Approach to Microbial Risk Assessment," *AWWA and WEF Water Reuse Conference Proceedings*, 735–744, American Water Works Assoc., Denver, CO.

Eisenberg, J. N. S., E. Y. W Seto, J. M. Colford, A. Olivieri, and R. C. Spear (1998) "An Analysis of the Milwaukee Cryptosporidiosis Outbreak Based on a Dynamic Model of the Infection Process," *Epidemiol.*, **9**, 3, 255–263.

Eisenberg, J. N. S., B. L. Lewis, T. C. Porco, A. H. Hubbard, and J. M. Colford (2003) "Bias Due to Secondary Transmission in Estimation of Attributable Risk from Intervention Trials," *Epidemiol.* **14**, 4, 442–450.

Eisenberg, J. N. S., J. A., Soller, J., Scott, D. Eisenberg, and J. Colford (2004) "A Dynamic Model to Assess Microbial Health Risks Associated with Beneficial Uses of Biosolids," *Risk Anal.*, **24**, 1, 221–236.

Englehardt, J. (2004) "Predictive Bayesian Dose-Response Assessment for Appraising Absolute Health Risk from Available Information," *Human Ecol Risk Assess.*, **10**, 1, 69–74.

Englehardt, J., and J. Swartout (2004) "Predictive Population Dose-Response Assessment for *Cryptosporidium parvum*: Infection Endpoint," *J. Toxicol. Environ. Health: Part A.* **67**, 8–10, 651–667.

Ershow, A. G., and K. P. Canter (1989) *"Total Water and Tapwater Intake in the United States: Population-Based Estimates of Quantities and Sources,"* National Cancer Institute, Order No. 263-MD-810264. Life Science Research Office, Fed. Am. Soc. Exper. Biol., Bethesda, MD.

Fuhs, G.W. (1975) "A Probabilistic Model of Bathing Beach Safety," *The Sci. Total Environ.*, **4**, 165–175.

Gerba, C. P., J. B. Rose, C. N. Haas, and K. D. Crabtree (1996) "Waterborne Rotavirus: A Risk Assessment," *Water Res.*, **30**, 12, 2929–2940.

Gordis, L. (1996) *Epidemiology*, W. B. Saunders Company, Philadelphia, PA.

Haas, C. N. (1983a) "Estimation of Risk Due to Low Doses of Microorganisms—A Comparison of Alternative Methodologies," *Am. J. Epidemiol.*, **118**, 4, 573–582.

Haas, C. N. (1983b) "Effect of Effluent Disinfection on Risks of Viral Disease Transmission Via Recreational Water Exposure," *J. WPCF*, **55**, 8, 1111–1116.

Haas, C. N., and J. N. S. Eisenberg (2001) "Risk Assessment," Chap. 8, 161–183, in L. Fewtrell and J. Bartram (eds.), *Water Quality: Guidelines, Standards and Health, Assessment of Risk and Risk Management for Water-Related Infectious Disease*, IWA Publishing, London.

Haas, C. N., J. B. Rose, C. P. Gerba, and S. Regli (1993) "Risk Assessment of Virus in Drinking Water," *Risk Anal.* **13**, 5, 545–52.

Haas, C. N., J. B. Rose, and C. P. Gerba (1999) *"Quantitative Microbial Risk Assessment."* John Wiley & Sons, Inc., New York.

Hallenbeck, W. H., and K. M. Cunningham (1986) *Quantitative Risk Assessment for Environmental and Occupational Health*, Lewis Publishers, Inc., Chelsea, MI.

Hennekens, C. H., and J. E. Buring (1987) *Epidemiology in Medicine*, Little, Brown and Company, Boston, MA.

HDR Engineering, EOA Inc., and Roberston-Bryan Inc. (2001) "Investigation of Effluent Coliform Bacteria Levels and CBOD/TSS Mass Loading for the Deer Creek Wastewater Treatment Plant," *Technical Report Prepared for El Dorado Irrigation District.*

Hodgson, E., and P. E. Levi (1997) *A Textbook of Modern Toxicology*, Appleton and Lange, Stamford, CT.

Hoppin, J. (1993) "Risk Assessment in the Federal Government: Questions and Answers," Center for Risk Analysis, Harvard School of Public Health, Boston, MA.

ILSI (1996) "A Conceptual Framework for Assessing the Risks of Human Disease following Exposure to Waterborne Pathogens," Risk Science Institute Pathogen Risk Assessment Working Group, International Life Sciences Institute, *Risk Anal.*, **16**, 841–848.

ILSI (2000) *Revised Framework for Microbial Risk Assessment*, International Life Sciences Institute, Risk Science Institute Pathogen Risk Assessment Working Group, ILSI Press, Washington, DC.

IOM (1988) *The Future of Public Health*, Institute of Medicine, The National Academy of Sciences, National Academy Press, Washington, DC.

Klaassen, C. D. (ed.) (2001) *Casarett and Doull's Toxicology: The Basic Science of Poisons*, 6th ed., McGraw-Hill, New York.

Klaassen, C. D., and J. B. Watkins III (1999) *Casarett and Doull's Toxicology: The Basic Science of Poisons, Companion Handbook,* 5th ed., McGraw-Hill, New York.

Kolluru, R., S. Bartell, R. Pitblado, and S. Stricoff (1996) *Risk Assessment and Management Handbook for Environmental, Health, and Safety Professionals,* McGraw-Hill, New York.

Koopman, J. S., S. E. Chick, C. P. Simon, C. S. Riolo, and G. Jacquez (2002) "Stochastic Effects on Endemic Infection Levels of Disseminating Versus Local Contacts," *Math. Biosci.,* **180**, 49–71.

Koopman, J. S., G. Jacquez, and S. E. Chick (2001) "New Data and Tools for Integrating Discrete and Continuous Population Modeling Strategies," *Ann. New York Acad. Sci.* **954**, 268–294.

McBride, G., D. Till, and T. Ryan (2002) *Pathogen Occurrence and Human Health Risk Assessment Analysis, Freshwater Microbiology Research Programme Report,* Ministry of Health, Wellington, New Zealand.

McCann J., E. Choi, E. Yamasaki, and B. N. Ames (1975) "Detection of Carcinogens as Mutagens in the *Salmonella*/Microsome Test: Assay of 300 Chemicals," *Proc. Nat. Acad. of Sci.,* **72**, 12, 5135–5139.

Makri, A, M. Goveia, J. Balbus, and R. Parkin (2004) "Children's Susceptibility to Chemicals: A Review by Developmental Stage," *J. Toxicol. Environ. Health—Part B—Crit. Rev.,* **7**, 6, 417–435.

Messner, M. J., C. L. Chappell, and P. O. Okhuysen (2001) "Risk Assessment for *Cryptosporidium*: a Hierarchical Bayesian Analysis of Human Dose Response Data," *Water Res.* **35**, 16, 3934–3940.

NAS (1998) *Issues in Potable Reuse: The Viability of Augmenting Drinking Water Supplies with Reclaimed Water,* 135–145. National Academy of Sciences, National Academy Press, Washington, DC.

NIH (2004) *National Toxicology Program—A National Toxicology Program for the 21st Century: A Roadmap for the Future,* National Institute of Environmental Health Sciences, National Institutes of Health, Research Triangle Park, NC.

NRC (1983) *Risk Assessment in the Federal Government: Managing the Process,* National Research Council, National Academy Press. Washington, DC.

Olivieri, A. W., R. C. Cooper, R. C. Spear, S. Selvin, R. E. Danielson, D. E. Block, and P. G. Badger (1986) "Risk Assessment of Waterborne Infectious Agents," *Envirosoft 86,* Computational Mechanics Publications.

Olivieri, A. W., R. C. Cooper, J. Konnan, J. Eisenberg, and E. Seto (1995a) *Mamala Bay Study—Infectious Disease Public Health Risk Assessment,* Prepared by EOA, Inc. for the Mamala Bay Study Commission, Honolulu, HI, Oakland, CA.

Olivieri, A. W., J. Eisenberg, J. Konnan, and E. Seto (1995b) *Microbial Risk Assessment for Reclaimed Water,* Prepared by EOA, Inc. and University of California at Berkeley School of Public Health for the Irvine Ranch Water District and the National Water Resource Association, Oakland, CA.

Olivieri, A. W., and J. Soller (2001) *Evaluation of the Public Health Risks Concerning Infectious Disease Agents Associated with Exposure to Treated Wastewater Discharged by the City of Vacaville, Easterly Wastewater Treatment Plant,* Prepared by EOA, Inc. for the City of Vacaville, Oakland, CA.

Olivieri, A. W., J. A. Soller, K. J. Olivieri, R. P. Goebel, and G. Tchobanoglous (2005) "Seasonal Tertiary Wastewater Treatment in California: An Analysis of Public Health Benefits and Costs," *Water Res.,* **39**, 13, 3035–3043.

Paustenbach, D. J. (ed.) (1989) *The Risk Assessment of Environmental and Human Health Hazards: A Textbook of Case Studies,* John Wiley & Sons, Inc., New York.

Pekkanen, J., and N. Pearce (2001) "Environmental Epidemiology: Challenges and Opportunities," *Environ. Health Perspect.,* **109**, 1, 1–5.

Pepper, I. L., C. P. Gerba, and M. L. Brusseau (eds.) (1996) *Pollution Science,* Academic Press, San Diego, CA.

Perz, J. F., F. K. Ennever, and S. M. Blancq (1998) "*Cryptosporidium* in Tap Water—Comparison of Predicted Risks with Observed Levels of Disease," *Am. J. Epidemiol.*, **147**, 3, 289–301.

Petterson, S. R., and N. J. Ashbolt (2001) "Viral Risk Associated with Wastewater Reuse: Modeling Virus Persistence on Wastewater Irrigated Salad Crops," *Water Sci. Technol.*, **43**, 12, 23–26.

Regli, S., J. B. Rose, C. N. Haas, and C. P. Gerba (1991) "Modeling the Risk from *Giardia* and Viruses in Drinking-Water," *J. AWWA,* **83**, 11, 76–84.

Rose, J. B., and C. P. Gerba (1991a) "Use of Risk Assessment for Development of Microbial Standards," *Water Sci. Technol.*, **24**, 2, 29–34.

Rose, J. B., C. N. Haas, and S. Regli (1991b) "Risk Assessment and Control of Water Borne *Giardiasis*," *Am. J. Public Health*, **81**, 6, 709–713.

The Safe Drinking Water Act, 42 U.S.C. s/s 300f et seq (1974). Revisions June 19, 1986; January 16, 1996.

Scutchfield, F. D., and C. W. Keck (1997) *Principles of Public Health Practice*, Delmar Publishing, Albany, New York.

Soller, J. A., J. N. S. Eisenberg, and A. W. Olivieri (1999) *Evaluation of Pathogen Risk Assessment Framework*, Prepared by EOA, Inc. for ILSI Risk Science Institute, Oakland, CA.

Soller, J. A., J. N. S. Eisenberg, J. DeGeorge, R. Cooper, G. Tchobanoglous, and A. W. Olivieri (2006), "A Public Health Evaluation of Recreational Water Impairment," *J. Water Health*, **4**, 1–19.

Soller, J. A., A. Olivieri, J. Crook, R. Parkin, R. Spear, G. Tchobanoglous, and J. N. S. Eisenberg (2003) "Risk-based Approach to Evaluate the Public Health Benefit of Additional Wastewater Treatment," *Environ. Sci. Technol.,* **37**, 9, 1882–1891.

Soller, J., A. Olivieri, J. N. S. Eisenberg, R. Sakaji, and R. Danielson (2004) *Evaluation of Microbial Risk Assessment Techniques and Applications, Water Environment Research Foundation*, Project 00-PUM-3, Final Project Report, Washington, DC.

Sullivan, J. B., and G. R. Krieger, *Clinical Environmental Health and Toxic Exposures*, 2nd ed., Lippincott Williams and Wilkins, Philadelphia, PA.

Tanaka, H., T. Asano, E. D. Schroeder, and G. Tchobanoglous (1998) "Estimating the Safety of Wastewater Reclamation and Reuse Using Enteric Virus Monitoring Data," *Water Environ. Res.*, **70**, 1, 39–51.

Teunis, P. F. M., G. J. Medema, L. Kruidenier, and A. H. Havelaar (1997) "Assessment of the Risk of Infection by *Cryptosporidium* or *Giardia* in Drinking Water from a Surface Water Source," *Water Res.,* **31**, 6, 1333–1346.

Teunis, P. F. M., and A. H. Havelaar (1999) "*Cryptosporidium* in Drinking Water: Evaluation of the ILSI/RSI Quantitative Risk Assessment Framework," RIVM report No. 284 550 006. Bilthoven, National Institute of Health and the Environment, The Netherlands.

U.S. EPA (1986) *Guidelines for Carcinogen Risk Assessment, Federal Register,* 51, 33, 992 (September 24, 1986).

U.S. EPA (1992) *Framework for Ecological Risk Assessment*, EPA/630/R-92/001, U.S. Environmental Protection Agency, Washington, DC.

Ward, R. L., D. L. Bernstein, C. E. Young, J. R. Sherwood, D. R. Knowlton, and G. M. Schiff (1986) "Human Rotavirus Studies in Volunteers: Determination of Infectious Dose and Serological Response to Infection," *J. Infect. Dis.*, **154**, 5, 871–880.

WHO (1999) *Principles and Guidelines for the Conduct of Microbiological Risk Assessment,* CAC/GL-30, World Health Organization, Geneva, Switzerland.

WHO (2004) *Guidelines for Drinking-water Quality,* 3rd ed., Vol. 1, *Recommendations,* World Health Organization, Geneva, Switzerland.

Part 3

TECHNOLOGIES AND SYSTEMS FOR WATER RECLAMATION AND REUSE

An infrastructure consisting of treatment facilities, storage reservoirs, pumping stations, and pipelines is needed for the production and delivery of the reclaimed water to the user. The infrastructure can range from very complex for large urban systems to small, low-technology systems for small communities or clusters of homes. Each water reuse system is unique because it has to be designed to meet local conditions. The technologies and systems used for water reclamation and reuse are determined by many factors including wastewater characteristics, reuse criteria and regulations, user characteristics, and system geography. *Process variability and reliability*, emphasized in the presentations in the Part 3 chapters, are also very important considerations in water reclamation and reuse, especially in indirect potable reuse applications. A fundamental understanding of these factors is required in selection of an appropriate system and its components.

Various technologies and systems available for the production and delivery of reclaimed water are the subject of Part 3. An overview of the factors considered in process and system selection is provided in Chap. 6. Removal of constituents by secondary treatment, removal of residual particulate matter, and removal of dissolved constituents are considered in Chaps. 7, 8, and 9, respectively. Removal of specific constituents is discussed in Chap. 10. Disinfection, a critical issue in water reuse, is presented in Chap. 11. Satellite and decentralized systems that are used for special and small community applications are introduced in Chaps. 12 and 13, respectively. Storage and distribution and plumbing systems for reclaimed water are discussed in Chaps. 14 and 15, respectively.

6 Water Reuse Technologies and Treatment Systems: An Overview

WORKING TERMINOLOGY   258

6-1   CONSTITUENTS IN UNTREATED MUNICIPAL WASTEWATER   260

6-2   TECHNOLOGY ISSUES IN WATER RECLAMATION AND REUSE   260
Water Reuse Applications   262
Water Quality Requirements   262
Multiple Barrier Concept   263
Need for Multiple Treatment Technologies   265

6-3   TREATMENT TECHNOLOGIES FOR WATER RECLAMATION APPLICATIONS   265
Removal of Dissolved Organic Matter, Suspended Solids, and Nutrients by Secondary Treatment   268
Removal of Residual Particulate Matter in Secondary Effluent   269
Removal of Residual Dissolved Constituents   271
Removal of Trace Constituents   271
Disinfection Processes   271

6-4   IMPORTANT FACTORS IN THE SELECTION OF TECHNOLOGIES FOR WATER REUSE   272
Multiple Water Reuse Applications   273
Need to Remove Trace Constituents   273
Need to Conduct Pilot-Scale Testing   276
Process Reliability   276
Standby and Redundancy Considerations   279
Infrastructure Needs for Water Reuse Applications   280

6-5   IMPACT OF TREATMENT PLANT LOCATION ON WATER REUSE   281
Centralized Treatment Plants   282
Satellite Treatment Facilities   282
Decentralized Treatment Facilities   283

6-6 THE FUTURE OF WATER RECLAMATION TECHNOLOGIES AND TREATMENT SYSTEMS 286
Implication of Trace Constituents on Future Water Reuse 287
New Regulations 287
Retrofitting Existing Treatment Plants 288
New Treatment Plants 289
Satellite Treatment Systems 289
Decentralized Treatment Facilities and Systems 289
New Infrastructure Concepts and Designs 290
Research Needs 291

PROBLEMS AND DISCUSSION TOPICS 292

REFERENCES 293

WORKING TERMINOLOGY

Term	Definition
Centralized wastewater management	The collection and drainage of wastewater, and sometimes stormwater, from a large, generally urban and suburban, area using an extensive network of pumps and piping for transport to a central location for treatment and reclamation, usually near the point of a convenient environmental discharge.
Constituents	Individual and aggregate components, elements, or biological entities such as total suspended solids (TSS), biochemical oxygen demand (BOD), *E. coli*, and ammonia nitrogen present and quantifiable in wastewater.
Conventional secondary treatment	Activated sludge treatment, commonly with nitrification, used for the removal of soluble organic matter and particulate constituents.
Decentralized wastewater management	Collection, treatment, and discharge/reuse of wastewater from individual homes, clusters of homes, isolated communities, industries, or institutional facilities, as well as from portions of existing communities at or near the point of wastewater generation.
Membrane bioreactor (MBR)	A process that combines a suspended growth activated sludge reactor with a membrane separation system; membrane separation is accomplished by either microfiltration or ultrafiltration and used in place of conventional gravity sedimentation.
Multiple barrier concept	The provision of multiple safeguards to maintain reliably the finished water quality; examples include source control, redundant systems, and treatment processes arranged sequentially.
Pilot-scale testing	The testing of unit operations or processes at a small-scale to establish the suitability of the treatment method in the treatment of a specific wastewater under specific environmental conditions and to obtain necessary data on which to base full-scale design.
Process reliability	The level of assurance that a process will achieve consistently the needed degree of constituent removal over the expected range of operating conditions.

Satellite treatment systems	Systems where wastewater in an upstream portion of the collection system is intercepted and diverted for treatment in a water reclamation facility located close to the point of reuse. Satellite treatment systems generally do not have solids-processing facilities; solids removed during treatment are returned to the collection system for processing in a central treatment plant located downstream.
Secondary effluent	Treated wastewater from a conventional biological treatment plant that typically meets average 30 d concentrations of 30 mg/L TSS and 30 mg/L BOD.
Sidestream	A portion of the wastewater flow that has been diverted from the main treatment process flow for specialized treatment.
Solids retention time (SRT)	The average period of time that biosolids remain in the activated sludge aeration tank.
Standby	A device or process that can be placed in service in an emergency or serve as a substitute.
Treatment process flow diagram, also known as treatment train	A combination of treatment operations and processes used to produce water meeting specified water quality goals or standards.
Unit operation	Method of treatment in which the application of physical forces predominates. Gravity sedimentation and filtration are common examples.
Unit process	Method of treatment in which constituent removal is brought about by chemical or biological reactions.

For many water reuse applications such as agricultural irrigation and industrial cooling water, effluent from secondary wastewater treatment plants was historically of sufficient quality. However, as quality goals for these and other water reuse applications have increased, spurred by the adoption of water reuse regulations (see Chap. 4), additional treatment has become necessary. Meanwhile, as regulations for effluent disposal have become more stringent, additional treatment has become necessary even for plants not practicing reuse. Thus, where feasible, it is reasonable to consider designing a treatment system suitable for potential future water reuse applications.

The technologies now used for water reclamation have evolved from operations and processes used for water and wastewater treatment. Even greater removals of measurable constituents are possible through recent technological advances. With the increased scientific knowledge developed over the past 10 yr concerning the potential impact of specific constituents found in water and wastewater, the focus on water quality in both drinking water and water reuse applications has intensified further, and especially so for indirect potable water reuse applications such as groundwater recharge and reservoir augmentation. Thus, in response to water quality issues and concerns, greater emphasis is now being given to technologies that provide higher levels of removal of suspended, colloidal, and dissolved solids; pathogenic organisms; and trace constituents. The technologies may include utilizing a combination of processes, applying newly developed processes, or modifying or upgrading existing systems to enhance process performance and reliability.

The purpose of this chapter is to provide an overview of the issues affecting the selection of treatment technologies for water reuse applications and to serve as an introduction to the nine technology chapters that follow. Issues considered in this chapter include (1) a review of the constituents of concern found in untreated municipal wastewater, (2) technology issues in water reclamation and reuse, (3) an overview of treatment technologies used for water reuse applications, (4) important factors in the selection of technologies for water reclamation, (5) the impact of treatment plant location on water reuse, and (6) the future of water reclamation technologies and treatment systems.

Although design values are presented in many of the chapters in Part 3, detailed design is not the focus of these chapters. Rather, the focus is on the performance of the processes and technologies presented and discussed with respect to constituents of concern in water reuse applications including particulate matter (turbidity, particle size, and particle size distribution), dissolved constituents (trace constituents, nutrients, and salts), and pathogenic microorganisms (bacteria, protozoa, helminths, and viruses).

6-1 CONSTITUENTS IN UNTREATED MUNICIPAL WASTEWATER

As discussed previously in Chap. 3, the constituents found in untreated wastewater, in addition to those present in the water supply, derive from the substances added to the water used for various domestic, commercial, and industrial uses. The principal classes of constituents and the corresponding physical properties found in untreated wastewater are identified in Table 6-1. What is apparent from a review of Table 6-1 is that several constituents or parameters are listed under multiple constituent and property classifications. Therefore, removal of a particular wastewater component may affect several water quality parameters. Alternately, various treatment processes may be used individually or combined to remove a given wastewater component. It should be noted that the constituents and physical properties of untreated wastewater are also dependent on the substances present in the source water and added to the water supply during potable water treatment.

While nearly all of the constituents found in untreated wastewater are potentially of concern, not all of the constituents are removed equally by all wastewater treatment processes. The classes of constituents that are removed by various treatment technologies are introduced in Sec. 6-3 and discussed in detail in the chapters where the corresponding technologies are presented.

6-2 TECHNOLOGY ISSUES IN WATER RECLAMATION AND REUSE

In water reclamation and reuse, the reuse applications will govern the type of treatment needed and the degree of reliability required for the treatment system. Because health and environmental concerns are primary issues in implementing water reuse, attention must be focused on developing approaches to ensure that water quality requirements are met consistently. As compared to conventional wastewater treatment systems that produce an effluent usually of a single quality for disposal, the challenge for water reclamation systems is greater because (1) the water quality goals will be more stringent

Table 6-1

Principal classes of constituents and physical properties found in untreated wastewater[a]

Parameter	Description of components
Constituent classification	
Suspended solids	Includes both suspended and colloidal matter. Typically made up of silt and clay, microorganisms, and particulate organic matter. Suspended solids can lead to the development of sludge deposits and anaerobic conditions, and, if inadequately treated, can affect disinfection efficiency.
Organic matter	Includes both dissolved and particulate matter. Composed principally of proteins, carbohydrates, and fats. Biodegradable organics are measured most commonly in terms of biochemical oxygen demand (BOD) and chemical oxygen demand (COD). Organic compounds selected on the basis of their known or suspected carcinogenicity, mutagenicity, teratogenicity, or high acute toxicity are often identified as priority pollutants. Organic compounds that tend to resist conventional methods of wastewater treatment are often classified as refractory organics. Typical examples include surfactants, phenols, and agricultural pesticides. Inadequate stabilization of organic matter can lead to the development of septic conditions and odors.
Inorganic matter	Inorganic constituents such as calcium, sodium, and sulfate are added to the original domestic water supply as a result of water use. Heavy metals are usually added to wastewater from commercial and industrial activities. Inorganic compounds selected on the basis of their known or suspected carcinogenicity, mutagenicity, teratogenicity, or high acute toxicity are often identified as priority pollutants. Specific inorganic constituents can greatly affect the uses to which reclaimed water is to be applied.
Pathogens	The principal classes of pathogenic organisms are bacteria, protozoa, helminths, and viruses.
Nutrients	The principal nutrients are nitrogen and phosphorus in various forms. Other inorganic constituents are also nutrients. When discharged to water bodies, nutrients can stimulate the growth of undesirable aquatic life. When applied to land, especially for groundwater recharge, excessive amounts can lead to groundwater contamination.
Trace constituents	Constituents found in extremely low concentrations including pesticides, pharmaceuticals, hormonally active agents, and residual personal care products. Some metals are often identified as trace constituents. Trace constituents may present health concerns if consumed (see Chap. 3).
Total dissolved solids (TDS)	Total dissolved solids are comprised of dissolved inorganic and organic matter. Total dissolved solids content can affect the suitability of reclaimed water for applications such as industrial reuse, agricultural irrigation, and groundwater recharge.
Physical property	
Turbidity	Particulate matter in water that scatters light may be quantified as turbidity. Turbidity is often used as a surrogate parameter for evaluating process performance and suitability for reuse.

(Continued)

Table 6-1
Principal classes of constituents and physical properties found in untreated wastewater[a] (*Continued*)

Parameter	Description of components
Physical property	
Color	Color can be used to assess the age or condition of the untreated wastewater (e.g., fresh or septic).
Odor	Results from the biological conversion of organic and inorganic constituents. Odorous compounds can also be discharged into the collection systems.
Temperature	A measure of how hot or cold is the wastewater. Temperature affects the rate of biological activity in treatment systems.
Transmittance	A measure of the amount of light, expressed as a percentage, that passes through a solution. Dissolved and colloidal constituents in water will absorb light and reduce the overall transmittance, which may affect the performance of UV disinfection processes.
Conductivity	A measure of the concentration of dissolved chemical elements.

[a] Numerical values for the various chemical constituents are presented in Table 3-12 and are also given in subsequent chapters.

with little or no margin for exceeding specified limits, (2) product water with different quality levels may be necessary to meet a variety of uses, and (3) fail-safe provisions are necessary to ensure public health protection.

To further examine technology issues in water reclamation and reuse it will be helpful to consider (1) the range of potential reuse applications (more detailed descriptions are provided in Part 4 of this textbook), (2) a general discussion of water quality for various reuse applications, (3) a description of the multiple barrier concept used to ensure water quality integrity, and (4) the utilization of multiple technologies for producing multiple grades of product water and for implementing the multiple barrier concept.

Water Reuse Applications

Seven general categories of water reuse applications are described in Table 6-2 and are listed in descending order of probable volume of use. The categories presented in Table 6-2 cover a wide variety of potential applications, many of which may not be applicable in specific cases. Each potential application has unique requirements related to product water quality (discussed below), volume of water required, rate of use, and time of use (continuous, intermittent, and seasonal).

Water Quality Requirements

Water quality requirements and regulations, as described in Chap. 4, are established by regulatory agencies for each reuse application and vary depending on the regulatory agency having jurisdiction. Ranges of water quality for the general water reuse categories identified in Table 6-2 are presented in Table 6-3 and are based on water quality requirements of the states of California and Florida. The purpose of Table 6-3 is to present some of the more common constituent limits for public health protection, but, for a given region and application, the number of constituents could be quite extensive and the water quality requirements different from those presented in this table. The constituent limits for treated effluent or water to be reused could also be stated as not-to-exceed values with some level of compliance. For example, the U.S. EPA recommends criteria for constituents that have

Table 6-2
Potential applications for water reuse[a]

General category	Potential application
Agricultural irrigation	Crop irrigation
	Commercial nurseries
Landscape irrigation	Public parks and school yards
	Roadway medians and roadside plantings
	Residential lawns
	Golf courses
	Cemeteries
	Greenbelts
	Industrial parks
Industrial	Cooling water
	Boiler feed water
	Process water
	Heavy construction (dust control, concrete curing, fill compaction, and cleanup)
Groundwater recharge	Groundwater replenishment
	Barrier against brackish or seawater intrusion
	Ground subsidence control
Recreation/environmental	Surface water augmentation
	Wetlands enhancement
	Fisheries
	Artificial lakes and ponds
	Snowmaking
Nonpotable urban uses	Toilet flushing
	Fire protection
	Air conditioning
	Sewer flushing
	Commercial car wash
	Driveway and tennis court wash down
Indirect potable use	Blending with public water supplies (surface water or groundwater)

[a]Adapted from Tchobanoglous et al. (2003).

been associated with specific acute and chronic health effects. These criteria could apply in special circumstances with a not to exceed limit of once in 3 yr or at a 99.9 percent level of compliance (U.S. EPA, 1994).

Multiple Barrier Concept

In selecting appropriate treatment operations and processes for water reuse applications, the provision of multiple barriers is an important consideration. The multiple barrier concept is utilized in potable water treatment and is based on the principle of establishing a series of barriers to preclude the passage of pathogens and harmful organic and inorganic contaminants into the water system to the greatest extent practicable (WEF, 1998). For water reuse, barriers may take the form of (1) source control programs designed to prevent the entrance of deleterious substances into the wastewater collection system that will inhibit treatment or may preclude reuse; (2) a combination of treatment processes wherein each provides a specific level of constituent reduction; and (3) an environmental buffer. Environmental buffers may be retention or storage ponds, dilution

Table 6-3
Ranges of water quality for water reuse applications in California and Florida[a,b]

Type of reuse	BOD, mg/L	TSS, mg/L	Total N, mg/L	Turbidity, NTU	Total coliforms, No./100 mL	Fecal coliforms No./100 mL
California[a]						
Agricultural irrigation						
Nonfood crop	≤ 30	≤ 30			< 23	
Food crop	≤ 10			≤ 2	< 2.2	
Landscape irrigation						
Restricted access	≤ 30	≤ 30			< 23	
Unrestricted access	≤ 10			≤ 2	2.2	
Industrial[c]					23	
Groundwater recharge[d]				≤ 2	≤ 2	
Recreational/environmental	≤ 10			≤ 2	< 2.2	
Nonpotable urban uses	≤ 10			≤ 2	< 2.2	
Indirect potable use[d]				≤ 2	< 2.2	
Florida[b]						
Agricultural irrigation						
Nonfood crop	≤ 20	≤ 5				200
Food crop	≤ 20	≤ 5				25 max[e]
Landscape irrigation						
Restricted access	≤ 20	≤ 5				25 max[e]
Unrestricted access	≤ 20	≤ 5				25 max[e]
Industrial[c]	≤ 20	≤ 20				200
Groundwater recharge						
Rapid infiltration basins	≤ 20					
Favorable conditions	≤ 20	≤ 20	≤ 12[f]			200
Unfavorable conditions		≤ 5	≤ 10			
Recharge or injection		≤ 5				
Recreational impoundments (restricted)	≤ 20	≤ 5				
Nonpotable urban uses	≤ 20	≤ 5				
Indirect potable use[d]	≤ 20	≤ 5	≤ 10			

[a] Adapted from the California Code of Regulations, Title 22, except as noted (State of California, 2001).
[b] Adapted from Florida Administrative Code (State of Florida, 1999).
[c] Industrial water quality varies based on the type of reuse and may require removal of specific constituents including TDS.
[d] Removal of specific constituents may also be required.
[e] Nondetect in 75 percent of samples.
[f] As NO_3-N.

Note: Blanks denote no values are given.

with fresh water, or soil-aquifer treatment that permits blending or equalizing water quality. The primary advantages of the multiple barrier approach are that (1) the public and the environment are provided a degree of protection even in the event one of the barriers should fail, (2) the probability that multiple processes will fail simultaneously is reduced significantly, and (3) a robustness to potential process upsets is provided because a greater number of barriers is used. Monitoring to ensure compliance with water reuse standards at several locations in a reuse system is an integral part of a multiple barrier system.

Need for Multiple Treatment Technologies

Where there are potentially several water reuse applications, each user may have specific requirements for the quality of reclaimed water. The use of multiple technologies may be appropriate to meet the requirements in instances where (1) multiple grades of product water are required that cannot be produced economically by a single process; (2) additional safeguards are needed for public health protection based on the results of the risk analysis, as described in Chap. 5; (3) changes in process operating parameters to meet one reuse application are not compatible with other uses; and (4) high levels of constituent removal are required to meet reuse criteria. In other cases where the use of multiple technologies cannot be justified at the water reclamation plant, the user of the reclaimed water may elect to employ additional treatment at or near the point of use. The multiple treatment technology approach also applies to wastewater treatment plants that must meet increasingly stringent discharge requirements.

6-3 TREATMENT TECHNOLOGIES FOR WATER RECLAMATION APPLICATIONS

Because of the importance of water quality in wastewater treatment and water reuse applications, different technologies are utilized, either singly or in combination, to achieve desired levels of constituent removal. Various levels of treatment were defined in Table 3-8 as preliminary, primary, advanced primary, secondary, secondary with nutrient removal, tertiary, and advanced. In this textbook, the focus of treatment technologies is on secondary, tertiary, and advanced treatment processes that meet the general water quality limits for reuse described in Table 6-3.

The principal unit operations and processes along with the constituents classes for which they are used in water reuse applications are listed in Table 6-4. The technologies listed in Table 6-4 are shown in a hierarchical arrangement according to application on Fig. 6-1. In reviewing the operations and processes shown on Fig. 6-1 it is clear that an almost endless number of treatment process flow diagrams can be developed, depending on the water quality requirements. The general categories of treatment technologies discussed in this textbook are those technologies used for the:

- Removal of dissolved organic matter, suspended solids, and nutrients, by secondary treatment (Chap. 7)
- Removal of residual particulate matter from secondary effluent (Chap. 8)
- Removal of residual dissolved constituents (Chap. 9)
- Removal of residual trace constituents (Chap. 10)
- Removal or inactivation of pathogens (disinfection) (Chap. 11)

Table 6-4
Unit operations and processes used for the removal of classes of constituents found in wastewater for reuse applications

Unit operation or process	Suspended solids	Colloidal solids	Organic matter (particulate)	Dissolved organic matter	Nitrogen	Phosphorus	Trace constituents	Total dissolved solids	Bacteria	Protozoan cysts and oocysts	Viruses	See Chap. no.
Secondary treatment	x			x								7
Secondary with nutrient removal	x			x	x	x						7
Depth filtration	x								x	x		8
Surface filtration	x	x	x						x	x		8
Microfiltration	x	x	x						x	x		8
Ultrafiltration	x	x	x						x	x	x	8
Dissolved air flotation	x	x	x									8
Nanofiltration			x	x			x	x	x	x	x	9
Reverse osmosis				x	x	x	x	x	x	x	x	9
Electrodialysis		x						x				9
Carbon adsorption				x			x					10
Ion exchange					x		x	x				10
Advanced oxidation			x	x			x		x	x	x	10
Disinfection				x					x	x	x	11

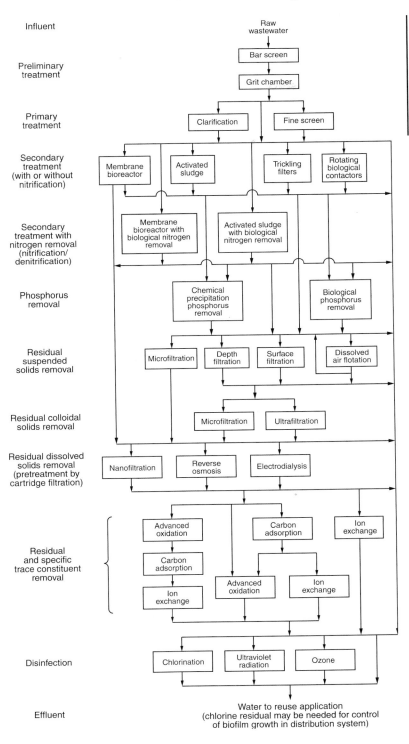

Figure 6-1
Matrix of alternative treatment processes that have been applied in wastewater reclamation and reuse.

Removal of Dissolved Organic Matter, Suspended Solids, and Nutrients by Secondary Treatment

Each of these categories is described briefly below and discussed in greater detail in subsequent chapters. In addition to treatment technologies, the types of water reuse systems, i.e., centralized, satellite, and decentralized, and infrastructure elements are introduced later in this chapter and discussed in Chaps. 12 through 15.

In secondary treatment, biological and chemical processes are used to remove most of the organic matter measured as BOD and TSS. Removal rates achieved by secondary treatment for BOD and TSS range from 85 to 95 percent. Typical biological processes used for secondary treatment, as illustrated on Fig. 6-1, include activated sludge, membrane bioreactors, trickling filters, and rotating biological contactors. A typical process flow diagram for secondary treatment with optional depth filtration and chlorine disinfection is shown on Fig. 6-2a. Views of secondary treatment processes are shown on Fig. 6-3.

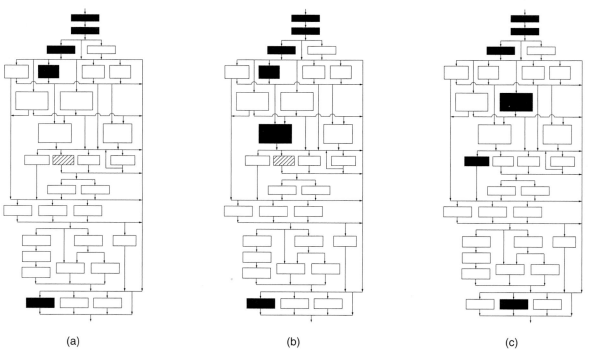

Figure 6-2
Typical treatment process flow diagrams based on the matrix of processes shown on Fig. 6-1: (a) conventional activated sludge treatment, optional depth filtration (shown cross hatched), and chlorination for agricultural reuse; (b) activated sludge with nitrification, chemical phosphorus removal, optional depth filtration, and chlorination for golf course irrigation; (c) activated sludge with nitrogen removal, microfiltration, and UV disinfection for landscape irrigation and industrial cooling tower applications; (d) membrane bioreactor with biological nitrogen and phosphorus removal and UV disinfection for ornamental water features; (e) activated sludge, microfiltration, electrodialysis, and disinfection with chlorine for removal of TDS for landscape irrigation; and (f) activated sludge with nitrification, microfiltration reverse osmosis, and UV/H_2O_2 advanced oxidation for indirect potable reuse through groundwater recharge or surface water augmentation.

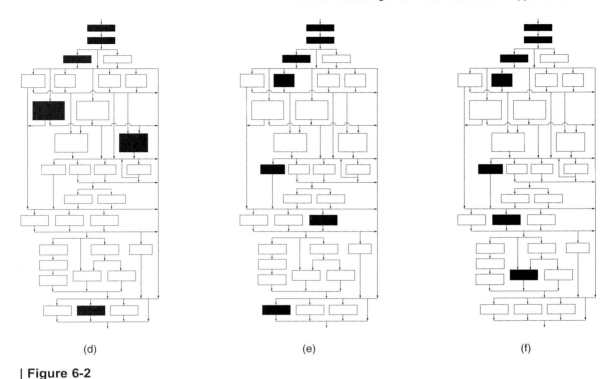

(d) (e) (f)

Figure 6-2
(*Continued*)

Increasing numbers of secondary treatment plants have incorporated anoxic or anaerobic reactors or treatment zones to aid in the biological removal of nitrogen and/or phosphorus. Typical process flow diagrams incorporating nutrient removal are shown on Figs. 6-2b–d. For small and medium size treatment plants, chemical precipitation is generally favored over biological processes for the removal of phosphorus. Chemical precipitation followed by filtration is also used where extremely low effluent phosphorus levels must be achieved.

Effluent from secondary treatment plants contains residual suspended and colloidal particulate matter that may require further removal. For example, residual particulate matter shields pathogenic organisms from disinfection by chlorine or ultraviolet (UV) light. In general, additional removal of residual particulate matter is required to optimize the disinfection processes. To meet reclaimed water standards such as California Recycled Water Criteria (commonly referred to as "Title 22") and for the protection of public health, typical treatment technologies employed include depth, surface, and membrane filtration and dissolved air flotation. Typical process flow diagrams incorporating some form of filtration for the removal of residual particulate matter are shown on Figs. 6-2a–f. A typical depth filter is shown on Fig. 6-4a and a typical microfiltration module is shown on Fig. 6-4b.

Removal of Residual Particulate Matter in Secondary Effluent

Figure 6-3

Views of processes used for secondary treatment: (a) plug-flow activated sludge, (b) membrane bioreactor with membrane being placed in reactor, (c) sequencing batch reactor, and (d) square covered tower trickling filter with odor control facilities in foreground.

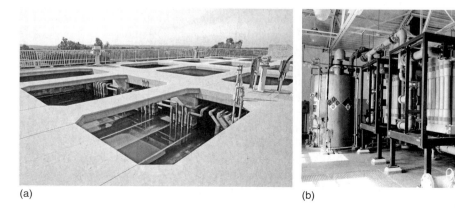

Figure 6-4

Views of filters: (a) granular-medium filters with fixed cast in place concrete washwater troughs and (b) microfiltration process with pressurized membrane modules.

Figure 6-5
View of a typical reverse osmosis unit for the removal of residual dissolved constituents.

Removal of Residual Dissolved Constituents

Dissolved constituents, such as salts, may be present in treated wastewater in amounts that limit or preclude water reuse in specific applications such as boiler makeup water. Dissolved inorganic constituents may cause scaling or corrosion in equipment and piping systems, especially in cooling tower systems. The dissolved constituents may result from: (1) high levels of minerals in the source water (Colorado River water with TDS levels over 500 mg/L is an example), (2) mineral pickup through domestic water use, ranging from 150 to 380 mg/L, (3) salt based water softeners, (4) discharges to the collection system by commercial and industrial facilities, and (5) chemicals added during the water reclamation process such as sodium hypochlorite and some coagulants. Although not very common, saline water may be introduced into the collection system in coastal areas. Demineralization may be accomplished by nanofiltration (NF), reverse osmosis (RO), or electrodialysis (ED). Typical process flow diagrams incorporating ED and RO for the removal of dissolved constituents are shown on Figs. 6-2e and f, respectively. A typical RO installation is shown on Fig. 6-5.

Removal of Trace Constituents

Specific reuse applications such as groundwater recharge, surface water augmentation, and industrial process water may require the removal of trace constituents. The need to remove trace constituents has to be determined on a case-by-case basis. When trace constituents of concern are not removed sufficiently by the conventional and tertiary treatment methods described above, advanced oxidation, adsorption, or ion exchange may be required, singly or in combination with NF or RO. A typical process flow diagram incorporating RO and advanced oxidation are shown on Figs. 6-2f. A typical ion exchange contactor is shown on Fig. 6-6.

Disinfection Processes

A major goal of water reclamation and reuse is to reduce the pathogen content to decrease the public health risks associated with exposure to reclaimed water. While disinfection

Figure 6-6

Views of ion exchange process used in water reuse applications: (a) stationary fixed bed exchange contactors and (b) fixed bed exchange contactors on a moveable platform for continuous operation in which one contactor is regenerated while the others are in service.

(a) (b)

requirements may vary depending on the specific water reuse application, disinfection is accomplished most commonly by the use of chlorine compounds and/or UV light. Because of high cost, ozone is used only in selected applications. When UV is used as the principal disinfectant, chlorine is often added to maintain a residual in the distribution system to control regrowth of microorganisms. Monitoring of the residual disinfectant should be done to maintain water quality in the distribution system (see Chaps. 11 and 14).

Disinfection is enhanced by the upstream removal of particulate matter that often shields pathogenic organisms from the disinfecting agent. The removal of particulate matter is especially critical for UV disinfection. Surface and membrane filtration are effective in conditioning effluents for disinfection. Use of one or more disinfectants following the treatment process and in conjunction with multiple barriers represents the best available technology to ensure public health protection and maximize process reliability. All of the processes shown on Fig. 6-2 with the exception of Fig. 6-2*f* incorporate a separate distinct disinfection step. In Fig. 6-2*f*, disinfection is achieved in the advanced oxidation step. Examples of chlorination and UV installations are shown on Figs. 6-7*a* and *b*, respectively.

6-4 IMPORTANT FACTORS IN THE SELECTION OF TECHNOLOGIES FOR WATER REUSE

To meet current and future reclamation requirements and regulations, the selection of technologies for water reuse will involve the careful consideration and evaluation of numerous factors. In selecting technologies for water reuse, consideration has to be given as to whether existing facilities are to be modified or upgraded, or an entirely new facility is to be constructed. In general, both physical and operational factors will have to be considered. The most important factors that must be evaluated in the selection of

(a) (b)

Figure 6-7
Views of disinfection systems: (a) chlorine contact basin with serpentine channels and (b) ultraviolet light disinfection unit with horizonal lamp placement perpendicular to flow with hand operated lamp cleaning mechanism.

technology for existing and new facilities are identified in Table 6-5. While each factor listed in Table 6-5 is important in its own right, some factors are considered further in the following discussion. These factors are also discussed in subsequent chapters.

The first factor presented in Table 6-5, *types of water reuse applications*, is particularly important where different water qualities are required to support multiple uses. Multiple processes that produce water of different qualities might be considered in lieu of a system that produces a single quality reclaimed water, albeit of the highest level required. For example, as shown on Fig. 6-8, a treatment process flow diagram comprised of biological nitrogen removal, surface filtration, nanofiltration, ion exchange, and UV disinfection could be used to provide three different qualities of reclaimed water. Alternatively, a given water quality could be produced for distribution and additional treatment provided at the point of use. For upgrading an existing secondary treatment plant where only a portion of the flow is reclaimed, an add-on or sidestream process that treats some of the secondary effluent might be considered. For example, to provide product water low in nitrogen and TSS at an existing secondary plant, a submerged attached growth process may be used to remove nitrogen from the flow to be reclaimed (see Chap. 7). For a new water reclamation plant, a membrane bioreactor with biological nutrient removal that produces a single grade of reclaimed water with very low nitrogen and TSS content might be the process of choice.

Multiple Water Reuse Applications

Another important factor deals with the concern for treating trace constituents that have been undetected previously and are now of concern. Examples of trace chemical constituents mentioned previously are N-nitrosodimethylamine (NDMA), methyl tertiary butyl ether (MTBE), pharmaceutically active substances, pesticides, and industrial chemicals, many of which are identified or suspected as endocrine disruptors (see Chap. 3). Some constituents may be removed in the course of normal biological treatment; other trace constituents may be generated in the wastewater treatment process, for example, NDMA may be formed during the disinfection process where chlorine is used.

Need to Remove Trace Constituents

Table 6-5

Important factors and issues that must be considered in technology selection

Factor/Issue	Comment	
	Upgrading an existing secondary treatment plant	New water reclamation plant
Type(s) of water reuse application	Different types of reuse place special constraints on the technologies to be used in terms of product water quality (see below); product water delivery schedule, i.e., continuous, intermittent, seasonal, and level of public health and environmental protection required. Where treatment upgrading is needed to support a single reuse application, process selection may be focused on "add-on" type processes. Where multiple grades of product water quality are needed, multiple processes might be considered.	For a new plant, integrated processes will most likely be the processes of choice instead of an "add-on" process. Where multiple grades of product water may be desired, there may be fewer opportunities for use of multiple processes because of practicality and cost.
Wastewater characteristics of the process feed stream	The physical and chemical characteristics of the secondary effluent affect the types of processes to be used and the effectiveness of disinfection.	The physical, chemical, and biological characteristics of the untreated wastewater affect the types of processes to be used, including pretreatment.
Reuse water quality goals (requirements)	The water quality goals limit process selection to those systems that are capable of meeting the constituent constraints. For example, if the goals limit total dissolved solids in the product water, some form of membrane treatment (most likely nanofiltration or reverse osmosis) will be required.	Same as for upgrading an existing secondary plant. As stated above, if more than one grade of product water is acceptable for reuse, where practicable a process train capable of producing multiple grades may enhance reuse opportunities.
Trace constituents	In special applications where trace constituents have been identified in treated effluent from an existing plant, pilot-plant studies should be conducted to determine the effectiveness of removal and the design parameters for a full-scale facility.	Unless a specific trace constituent problem has been identified, such as in the case of treated effluent from a plant in another part of the collection system, unknown trace constituents will have little effect on process selection, except for allowing space for the addition of a future process.
Compatibility with existing conditions	Selection of a new process may be influenced by compatibility with existing processes, hydraulic considerations, and site constraints.	For a plant on a new site, compatibility will be concerned mainly with site and environmental constraints discussed below.

Table 6-5

Important factors and issues that must be considered in technology selection (*Continued*)

Factor/Issue	Comment	
	Upgrading an existing secondary treatment plant	New water reclamation plant
Process flexibility	In the event of future changes in regulations or changes in the characteristics of the feed stream, the adaptability of the process in accommodating probable changes should be considered.	Same as for upgrading an existing secondary plant, and process flexibility should consider past experience in similar applications and results of pilot-plant studies.
Operating and maintenance requirements	Evaluation and implementation of new operating and maintenance (O&M) requirements. Additional training will be required for O&M of new process equipment. Supervisory control and data acquisition system upgrading may be required for additional process monitoring and control.	Evaluation of what replacement parts will be required and what their availability and cost will be. Service life of key components such as membranes should be considered as well. Automated process control systems will require specialized training.
Energy requirements	Assessment of the impact of the addition of more energy-using devices such as pumps and blowers on the existing electrical distribution system and standby power system. Determine how the peak demand and energy rate schedule will be affected.	The energy requirements, as well as probable future energy cost, must be known or estimated if cost-effective systems are to be designed. Evaluate whether standby power is required or whether the process can be interrupted safely if a power failure occurs.
Chemical requirements	The addition of new chemicals might have an effect on the treatment system, such as ozone on the membrane material. Evaluate whether the new chemicals will adversely affect product water quality, the formation of disinfection byproducts, and operation and maintenance of the system.	Same as for upgrading an existing secondary plant. In addition, determine if the use of chemicals will cause hazards in delivery to and use at the plant site.
Personnel requirements/ automation	New processes may necessitate special operational skills, thus requiring training of the existing staff or adding new personnel. The existing supervisory control system should be evaluated to determine the ease with which new processes can be integrated and what new system features are needed.	Staffing issues have to be determined such as the number of people, skill levels needed to operate and maintain the system, type of worker (full-time or part-time), degree of automation needed, and attended or unattended operation (24 h, 7 d, or less).

(*Continued*)

Table 6-5

Important factors and issues that must be considered in technology selection (*Continued*)

Factor/Issue	Comment	
	Upgrading an existing secondary treatment plant	New water reclamation plant
Environmental constraints	Additional noise and odor production could impact the surrounding area or violate regulations. Mitigation measures might include new or upgraded systems and enclosures. Special residuals processing or disposal may also be needed.	Environmental factors such as proximity to residential areas, traffic, potential odor generation, and noise may affect selection of the plant site, the type of process, and the need for equipment enclosure and automation. The residuals generated may also require special processing or disposal considerations.

The need for special treatment, therefore, may be limited to those situations where the trace constituents have been detected in treated effluent from an existing plant. In that case, the sources of the constituents in question need to be identified and controlled, where possible. If the sources cannot be identified or controlled and treatment is required, pilot-scale studies should be conducted to determine applicable and appropriate methods of treatment.

Need to Conduct Pilot-Scale Testing

Where the applicability or performance of a process for a given situation is unknown but the potential benefits of using the process are substantial, pilot-scale testing should be conducted. The purpose of conducting pilot-scale studies is to establish the suitability of the process for the intended use and to obtain necessary data including reliability characteristics and scaling parameters that will serve as a basis for full-scale design. Factors that should be considered in pilot-scale studies are presented in Table 6-6. Although the relative importance of the factors presented in Table 6-6 will depend on the specific conditions, it will be useful to consider the items in Table 6-6 as a preliminary checklist. Specific pilot-scale testing programs are discussed further in chapters dealing with various treatment technologies. Typical pilot-plant installations used for evaluating advanced treatment processes are shown on Fig. 6-9.

Process Reliability

Process reliability is defined as the probability of adequate performance for a specified period of time under specified conditions, where performance is determined by the ability to meet regulated effluent constituent concentrations. Process reliability may apply to an individual process or the effluent from a group of processes. The reason for the emphasis on reliability is that *reclaimed water treatment processes must perform reliably if the public is to have confidence that the practice of water reuse is acceptable, especially where indirect potable reuse may occur*. Monitoring of performance is an important part of process operation to ensure reliability. In facilities employing the multiple barrier approach, monitoring of each process stage will help ensure the integrity of the system. Component failures can be detected early so appropriate corrective measures can be applied. Further, regulations governing water reuse are often based on

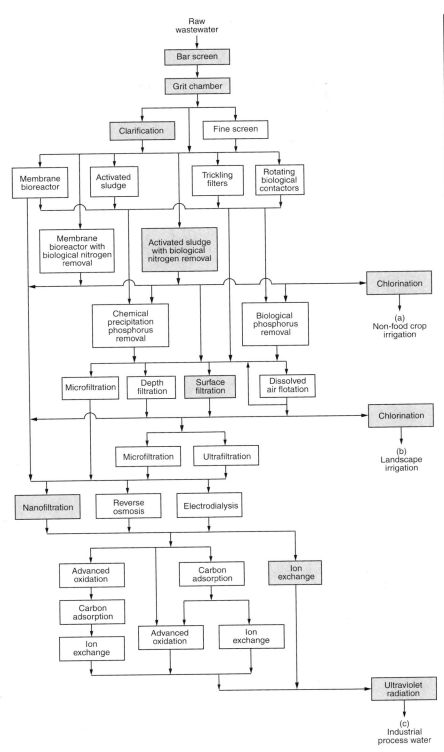

Figure 6-8

Typical process flow diagram of a treatment system that produces reclaimed water for multiple uses: (a) nonfood crop irrigation, (b) landscape irrigation, and (c) industrial process water.

Table 6-6

A checklist for the conduct of pilot-scale studies[a]

Item	Consideration
Reasons for conducting pilot testing	Test new process
	Test new application for existing process
	Simulation of another process (e.g., use of MS2 for enteric virus)
	Predict process performance, develop model parameters and treatment kinetic coefficients
	Document process performance
	Optimize system design
	Satisfy regulatory requirements
	Satisfy legal requirements (e.g., Title 22 requirements)
Pilot-plant size	Bench- or laboratory- scale model
	Pilot-scale tests
	Full (prototype) scale tests
Nonphysical design factors	Available time, money, and labor
	Degree of innovation and modification involved
	Quality of water to be treated
	Availability and location of facilities
	Complexity of process
	Similar testing experience
	Security concerns
	Demobilization requirements
Physical design factors	Scale-up factors
	Size of prototype
	Flow variations expected
	Variations in constituent concentrations
	Facilities and equipment required and setup
	Materials of construction
	Power requirements
Design of pilot testing program	Dependent variables including ranges and anticipated variability
	Independent variables including ranges and anticipated variability
	Data collection (data loggers or other means, communications, availability of phone service or wireless network)
	Time required
	Time of year (seasonal effects)
	Test facilities and appurtenances
	Test protocols including number of samples and sampling schedule
	Statistical design of experiments and analysis of experimental data

[a] Adapted from Tchobanoglous et al. (2003).

meeting water quality standards that are considered to be protective of public health with essentially 100 percent confidence.

Process reliability is important to water reuse because of its relationship to the risk of an adverse effect following exposure to reclaimed water. Process reliability is particularly important for acute exposure events, for example, where a failed disinfection process

 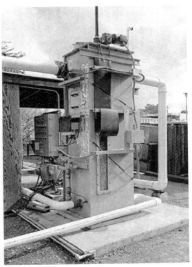

Figure 6-9
Views of typical pilot-plant installations: (a) upflow continuous backwash filter and (b) compressible medium filter.

(a) (b)

results in a temporary increase in the effluent concentration of pathogens. An increase in the effluent concentration of a toxic or pathogenic constituent, resulting from a process failure or a breakthrough event, increases the risk of illness for the exposed population. The mean design values specified during the engineering and design process are sufficient for determination of expected performance on a daily basis; however, the peak events that occur at intervals of once a year (99.7 percent) or once every 3 yr (99.9 percent) will be of greater concern from a risk management perspective (see Chap. 5). Factors that can influence reliability include:

- Amount of source wastewater available to meet the reclaimed water needs
- Range of constituent concentrations in the influent to the reclamation process and the range of quality that the treatment system can produce successfully
- Rate of change of influent water quality, especially if the influent source is a connection to the collection system subject to infiltration/inflow problems or industrial discharges
- Amount and type of instrumentation and automation
- Availability of skilled operating and maintenance personnel
- Mode of operation, i.e., continuous, intermittent, seasonal
- Availability of standby processes and equipment including an auxiliary electric power supply

Standby and Redundancy Considerations

In water reuse applications, the need for standby or redundant processes or equipment depends in part on the extent the supply of reclaimed water can be interrupted or the maximum time a process can be out of service without detrimental effects. If, for example, water is required to be delivered continuously, considerations have to be given to provide (1) standby equipment or processes if mechanical or structural failure occurs,

 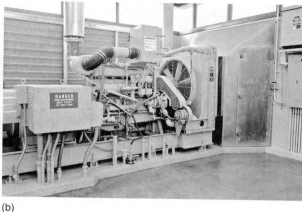

(a) (b)

Figure 6-10
Views of standby facilities: (a) redundant pumps at a reclaimed water distribution pumping station and (b) an engine-generator used to provide an emergency source of electric power.

(2) redundant equipment to permit periodic servicing of equipment in duty service (see Fig. 6-10a), and (3) backup electrical power sources or a standby generator (see Fig. 6-10b) in the event of a power failure.

The requirements for standby and redundancy have to be evaluated on a case-by-case basis. An example where standby service is needed is the furnishing of process water to an industry where service cannot be interrupted except for scheduled shutdowns. In this case, emergency power generation and redundant equipment are necessary to maintain service. Another example is the use of reclaimed water for fire protection for which an adequate water supply must be available at all times. If the supply of reclaimed water can be interrupted with minimal detrimental effects, the standby requirements can be reduced considerably. Facilities from alternative sources such as potable water may be considered to secure supply reliability in emergencies. Landscape irrigation is an example where the supply of water could be interrupted for a day or two, thus the need for auxiliary power or redundant equipment may not be necessary.

Infrastructure Needs for Water Reuse Applications

In a water reuse project, facility requirements go beyond providing treatment; a supporting infrastructure is needed. Comprehensive planning is required to ensure that all of the functional aspects of a complete system are met for delivering reclaimed water to the user. The infrastructure may consist of pumping stations, transmission and distribution pipelines, storage, and appurtenances such as diversion structures, service connections, and metering stations. Typical infrastructure components are summarized in Table 6-7 and described in more detail in Chap. 14. Typical above-ground storage reservoirs are shown on Fig. 6-11. The requirements and cost of the infrastructure have to be considered carefully in evaluating the overall project costs and economic benefits because the cost of providing a dual piping system could preclude installation of the system.

Table 6-7
Description of infrastructure components used in water reclamation systems

Component	Function and Description
Pumping station	Pumping stations may be required for the transport of untreated wastewater to the reclamation plant and for delivering reclaimed water to the point of use via the transmission and distribution system. For conveying untreated wastewater, the pumps should have solids-handling capabilities. Pumping stations for reclaimed water are similar in design to those used in water pumping applications. Booster stations may also be required in large systems with different pressure zones
Treated water storage	Storage is needed to compensate for the difference in the water production rate and the rate and time of use (demand). Storage facilities may consist of underground or aboveground structures, lakes or ponds, or aquifer storage
Transmission and distribution pipelines	Transmission lines are used to deliver water from the point of production to turnout locations that connect to the distribution network. Distribution pipelines are used to supply reclaimed water from the turnouts directly to the users
Diversion structures	For satellite facilities, diversion structures are used to intercept or divert untreated wastewater flows from the collection system to the satellite reclamation plant
Service connections	Service connections are the individual piping or plumbing that connects to the user from the distribution line. Service connections consist normally of a corporation stop or valve at the point of connection to the distribution main, service line to the user's property, and meter. Connections from the meter to the point of use on the property are usually the property owner's responsibility. Service connections should be marked distinctively, such as by color codes, to safeguard against possible improper use of nonpotable water and installation of a cross-connection
Metering	Metering is used to control the demand and to provide the basis for charging for the amount used. The same type of water meter used for domestic service is suitable for reclaimed water use except that it should be marked distinctively

6-5 IMPACT OF TREATMENT PLANT LOCATION ON WATER REUSE

Water reclamation can be accomplished at a principal wastewater treatment plant (the term "centralized" is used in this textbook) where all or a portion of the plant effluent is reclaimed, or in a satellite or decentralized facility designed especially for reclaiming wastewater close to the source(s) of generation and the point of use. The alternatives for

(a) (b)

Figure 6-11
Views of a storage reservoirs: (a) a concrete tank and (b) a steel storage tank.

locating a water reclamation facility are shown schematically on Fig. 6-12. Often the centralized plant is located remotely from potential areas of reuse, thereby limiting reuse opportunities because the costs of transporting reclaimed water to the points of reuse can be expensive. By constructing satellite facilities in upstream portions of the drainage area, wastewater from the local collection systems can be intercepted, treated, and distributed to areas of local reuse, thereby obviating the need to transport reclaimed water great distances from a centralized facility. Decentralized facilities, which are used to collect and treat wastewater for reuse independent of the primary collection system, are also an alternative for implementing local reuse (see Fig. 6-13b, c). The treatment plant location, therefore, has a major effect on (1) how a water reuse plan can be implemented, (2) selecting the size and type of treatment facility, and (3) the type and extent of ancillary facilities required. Satellite and decentralized wastewater management systems are discussed in more detail in Chaps. 12 and 13, respectively.

Centralized Treatment Plants

In a typical centralized treatment plant as shown on Fig. 6-12, the treatment plant is located at a low point in the drainage area, usually near the point of effluent disposal. At the time of selecting the original location of the treatment plant, the area surrounding the plant may have been relatively uninhabited. Over time, however, the surrounding land may be developed for residential, commercial, or industrial uses. In this environment, some local water reuse opportunities might be available such as landscape irrigation and supplying industries with process water, particularly if a high quality of water is produced to meet discharge requirements. If there is a substantial demand for reclaimed water, a sidestream reclamation process might be added to produce water for special applications. An example is the installation of an advanced treatment system to produce high quality reclaimed water for use as industrial process boiler feedwater or cooling water. The advantages and disadvantages of a centralized system are identified in Table 6-8.

Satellite Treatment Facilities

In a satellite system, as shown on Fig. 6-12, a water reclamation plant is located in the upper reaches of the service area close to potential applications such as groundwater recharge, agricultural irrigation, and recreational enhancement. For this system, untreated

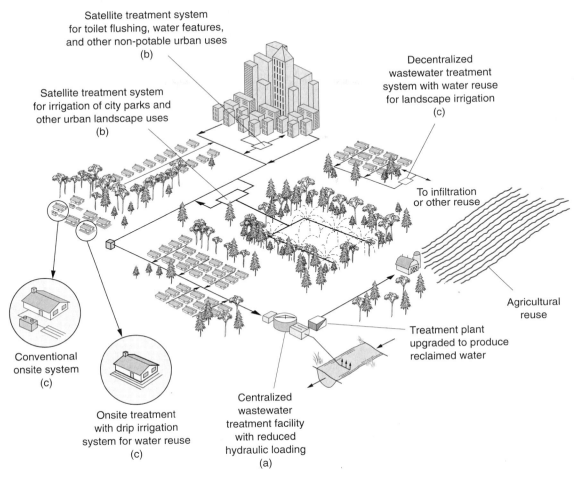

Figure 6-12
Schematic of types and locations of wastewater treatment facilities used for water reclamation and reuse: (a) centralized system, (b) satellite systems, and (c) decentralized systems.

wastewater is diverted from the collection system to the satellite plant. The satellite plant is designed for treating the wastewater to reclamation grade product water, and residuals produced in the treatment process are returned to the collection system for processing at the central plant. Satellite treatment plants may use processes similar to those used at a centralized treatment plant; however, the development of compact treatment facilities has made satellite treatment applications more feasible. The advantages and disadvantages of satellite systems are identified in Table 6-8. Satellite wastewater management systems are discussed in more detail in Chap. 12.

Decentralized Treatment Facilities

Decentralized wastewater management is defined in this text as the collection, treatment, and reuse of wastewater from individual homes, clusters of homes, isolated communities, industries, institutional facilities, or portions of existing communities at or

Figure 6-13
Views of wastewater management systems for water reuse applications: (a) centralized system, (b) satellite (extended aeration activated sludge) system, and (c) decentralized system (Courtesy of Orenco Systems, Inc.) Covers of six buried treatment units are shown on lower left at the third point.

near the point of wastewater generation and not connected to a centralized collection system. Decentralized systems are also illustrated on Fig. 6-12. The elements of a decentralized system comprise (1) wastewater pretreatment, (2) wastewater collection, (3) wastewater treatment, (4) reclaimed water production, (5) infrastructure for water reuse, and (6) biosolids disposal or reuse. It should be noted that not every decentralized system will incorporate all of these elements (Crites and Tchobanoglous, 1998). Advantages and disadvantages of a decentralized system are presented in Table 6-8.

Table 6-8
Advantages and disadvantages of centralized, satellite, and decentralized treatment facilities

Advantages	Disadvantages
Centralized treatment facilities	
• A suitable site for the treatment facilities may exist • Skilled operating, laboratory, and maintenance personnel are readily available • Multiple qualities of water, i.e., secondary effluent, filtered effluent, or low TSS effluent may be produced economically depending on the reuse requirements • Unit costs for consumables, such as electricity and chemicals, are lower because of volume discounts	• Large reuse markets in the vicinity of the plant may be limited due to the extent of surrounding residential and commercial development • Areas of reuse may be located remotely from the central plant, thus requiring a costly investment in infrastructure to provide service • Operating costs, especially for high service pumping, may be expensive for delivery to outlying reaches of the service area • Additional treatment processes required for producing reclamation-grade product water may not necessarily be compatible with existing wastewater treatment processes
Satellite treatment facilities	
• Opportunities for finding sites for local reuse of reclaimed water are enhanced • The supporting infrastructure and its cost can be reduced significantly as compared to a centralized system • Greater potential for having reuse applications adjacent to treatment system, thus minimizing transmission costs • Availability of land in the upper reaches of the service area may be better for locating satellite treatment and storage facilities • Diversion of untreated wastewater from the collection system reduces the hydraulic load on the collection system and central treatment • The overall cost of a distributed treatment system, i.e., using one or more satellite plants, may be more cost effective than an expanded centralized system • Energy consumption may be reduced by eliminating long distance and high pressure reclaimed water transport • Construction disruptions may be less, especially for pipelines in public streets	• Site selection may be controversial for treatment plant and storage locations in or near residential areas due to zoning, local land use ordinances, and public opposition • Availability of wastewater supply in the collection system may not correlate with water reuse demand • Requires additional monitoring equipment and telemetry for operation and control • Labor and monitoring requirements will be more costly with the addition of facilities in remote locations • May be more difficult to ensure reliability of water reclamation system and power supply • If chemicals are required for disinfection or other purposes, transport of hazardous materials through nonindustrial areas may be required • If membrane bioreactors are used, special chemicals and equipment may be required for membrane cleaning and replacement • Discharge of biosolids back to the collection system may lead to the formation of odors

(Continued)

Table 6-8

Advantages and disadvantages of centralized, satellite, and decentralized treatment facilities (*Continued*)

Advantages	Disadvantages
Decentralized treatment facilities	
• Can be used where there is no supporting infrastructure such as a collection and treatment system • Can be used to limit or control development in a given area • Adaptable to individual homes, clusters of homes, subdivisions, and isolated developments • Localized use of reclaimed water can be facilitated and implemented for landscape irrigation and groundwater recharge • Because systems are generally not complicated, highly skilled maintenance is generally not required	• Topographic and geologic features such as steep slopes, subsurface bedrock formations, and shallow soil water depth may limit use • Solids generated in decentralized treatment systems require periodic removal and further processing • Installation and management of many small systems may be difficult to accomplish because small communities have limited resources and expertise • Inadequately maintained systems are subject to failure • Designs must consider wider variations in wastewater flows and loadings

Decentralized treatment systems maintain both the liquid and solid fractions of wastewater near their point of origin, although residual solids may be transported to a centralized facility for further treatment and reuse. Typical situations where decentralized wastewater management can be considered for water reuse applications are:

- Where existing onsite systems must be improved or discontinued
- Where the community or facility is located remotely from an existing collection system
- Where localized reuse opportunities are available
- Where fresh water for domestic supply is in short supply
- Where the existing centralized collection and treatment system lacks capacity and funding for expansion.

Decentralized treatment facilities typically use septic tanks for primary treatment, intermittent and recirculating packed-bed filters, constructed wetlands, or compact treatment technologies for secondary treatment. Decentralized and onsite treatment systems are discussed in Chap. 13.

6-6 THE FUTURE OF WATER RECLAMATION TECHNOLOGIES AND TREATMENT SYSTEMS

The concept of sustainable water resources management, as discussed in Chap. 1, will be one of the driving forces in expanding the use of reclaimed water for conserving and extending existing water supplies. Other driving forces include increasingly stringent wastewater discharge requirements that incorporate considerations for environmental effects and health-based water quality standards and regulations.

To meet the challenge of expanded use, the facilities for water reuse must (1) have the capabilities of meeting existing and new reclaimed water regulations consistently and reliably, (2) be capable of being integrated into existing wastewater treatment and water reuse systems, where it is cost-effective to do so, (3) take advantage of new methods of treatment that have been developed for water reclamation applications, and (4) become affordable when compared to other water supply alternatives. The growth in water reuse systems will be tested by their ability to meet the challenges of the future.

Implication of Trace Constituents on Future Water Reuse

Trace constituents that have been identified or will be detected are of concern for the future use of reclaimed water. Some of the trace constituents are recognized as carcinogens, while others are suspected of interfering with the normal functioning of the endocrine system. The commercial, residential, and agricultural utilization for natural and synthetic products containing these compounds makes them ubiquitous in daily life, resulting in their eventual release into the environment. Discharges of treated wastewater have been cited as primary sources of trace constituents in the water cycle and particularly in water bodies used as sources for potable water. Thus, a future challenge to the wastewater industry and especially in water reuse applications is to develop (1) cost-effective methods (e.g., online monitors) of identifying the sources of these constituents so that they can be eliminated and (2) wastewater treatment processes that can remove or reduce these constituents.

Encouraging results have been obtained in the use of activated sludge with increased solids retention times (SRTs) of 11 to 13 d and coupled with nitrification/denitrification in the degradation of natural and synthetic estrogens (Andersen et al., 2003). Advanced treatment processes such as nanofiltration, reverse osmosis, advanced oxidation, and carbon adsorption provide a high level of treatment for the removal of these residual constituents. Soil aquifer treatment (SAT), adopted commonly for groundwater recharge, has been reported to be an effective process to reduce most trace constituents (Crites, 2000). Research into the occurrence and fate of trace constituents, therefore, must be forward looking in anticipation of changes in future regulations and their effect on water reuse applications (Esposito et al., 2005).

New Regulations

New regulations related to treated effluent disposal and the use of reclaimed water continue to evolve. As regulations governing the disposal of treated effluent become more stringent, requiring plant upgrading or alternative methods of disposal, water reuse becomes a more attractive alternative than strictly effluent disposal. Recent development of water reuse plans in relatively water-rich regions of the eastern U.S. is due largely to more stringent effluent discharge regulations. For environmental and other reasons, regulations have begun to limit the disposal of treated effluent to water bodies. Three examples are cited below.

- In the State College, Pennsylvania area, limits were placed on the discharge of highly treated effluent to Spring Creek, designated as a high-quality trout stream by the State of Pennsylvania. Temperature increases due to effluent discharge were determined to be detrimental to the trout population, inhibiting their ability to reproduce (Marcino, 2004).
- Most water reuse projects in Florida are driven by strict effluent disposal regulations. In St. Petersburg, an area-wide water reuse plan was initiated to ban effluent discharge unless wastewater is treated by an approved advanced wastewater treatment process.

- In San Jose, California, limitations on treated effluent discharges due to adverse effects of low TDS effluent on the salinity of South San Francisco Bay resulted in development of an extensive water recycling system for landscape irrigation.

New regulations for reclaimed water will continue to deal with the issues of removal of microbial pathogens, chemical constituents, and trace constituents. As research into the characteristics of water and wastewater becomes more extensive and potential impacts of these constituents on public health and the environment become better known, some changes in the regulations can be anticipated; others will evolve as more information becomes available. Future regulations for water reuse will relate to the detection and monitoring of pathogenic organisms, measurement and removal of residual solids, and identification and treatment of trace constituents. Regulations for trace contaminants such as those chemicals identified as carcinogens and endocrine disruptors will evolve where those substances have been identified as potential problems to public health and the environment and thus require mitigation. The situation where wastewater is present in public water supplies, i.e., *de facto* water reuse, is highlighted in Chaps. 3 and 23. For groundwater recharge, regulations may require limitations on both TDS and nitrates. When reclaimed water is used for indirect or direct potable reuse, multiple treatment technologies and multiple barriers are necessary to ensure treatment reliability and public health protection.

Because there are no federal regulations currently (2006) that govern concentrated residuals streams from membrane systems such as NF and RO, requirements for concentrate disposal are regulated by the states and vary widely. State regulations are often based on limited information and experience (Hightower and Keyes, 2005). Because of future uncertainties, the lack of consistent regulations and guidelines acts as a hindrance to the application of desalination and water reuse technologies. Long-term planning for concentrate disposal will be difficult and will depend greatly on future regulations that are enacted (Lynch et al., 2005).

Retrofitting Existing Treatment Plants

Many existing wastewater treatment plants are undergoing or will undergo retrofitting due to the need to replace aging and deteriorating equipment, increase capacity, improve performance, mitigate odor issues, and meet new water quality requirements. Space becomes more of a factor as many facilities have limited areas available for the addition of new processes. Compact treatment technologies for residual solids removal such as ballasted flocculation, high-rate clarification, cloth-media filters, and membrane filtration are attractive alternatives to conventional processes such as gravity sedimentation and media filtration. Optimization of the activated sludge process to enhance performance and to remove trace constituents offers increasing reliability to water reclamation processes. Development of new treatment technologies may allow expanded use of reclaimed water with increasing reliability. For example, the addition of fixed-film media to the aeration tanks [the integrated fixed-film activated sludge (IFAS) process] enhances nitrification in a relatively small aeration basin volume (Johnson et al., 2004). A new process, an oxygen-based membrane biofilm reactor (MBfR) discussed in Chaps. 7 and 10, may find application in reducing total nitrogen levels and other constituents such as perchlorate (Nerenberg, 2005). Improved disinfection using UV light in lieu of or in addition to chlorination may also necessitate the improved removal of residual suspended solids using one or more of the devices described in Sec. 6-3.

New Treatment Plants

During the coming years as additional demands are placed on the quality and quantity of global water supplies, greater emphasis will be placed on optimizing existing technologies and applying new concepts to maintain high levels of constituent control and removal. Conventional biological treatment technologies such as activated sludge will continue to improve and be used either singly or in combination with other processes. The use of membrane technologies for improved levels of constituent removal will grow at an expanding rate because of the high levels of water quality attainable, improved membrane designs, competition in the marketplace that will reduce costs, and the application for satellite and decentralized systems (introduced in Sec. 6-5). Additionally, more new treatment plant designs will include improved nutrient removal and recovery and enhanced residual suspended solids removal to facilitate improved disinfection and reuse.

As more water reuse applications are identified and reclaimed water quality requirements become more stringent, the removal of TDS becomes increasingly necessary. With the improved design of membranes for NF and RO, the high energy requirements for pumping have been reduced considerably. Devices for energy recovery from NF and RO systems are also entering the marketplace. Because of the improvements in NF and RO, many new reuse plants will be able to consider membrane treatment as a cost-effective process alternative. As the removal of TDS becomes increasingly necessary, considerations for the management of the waste concentrate (brine) will become increasingly important.

Satellite Treatment Systems

Satellite as well as decentralized systems are expected to be used increasingly as urban growth continues. In communities where development is occurring on the extremities of urbanized areas, adding new wastewater flows places a strain on the capacity of the existing collection, treatment, and disposal systems. By utilizing a satellite system concept, local reuse can be implemented and the hydraulic loads on the existing system lessened. Satellite systems can also be used in developed metropolitan areas for producing recycled water for toilet flushing and other nonpotable uses in apartment and office complexes in addition to local irrigation projects such as city parks (see Chap. 20).

Decentralized Treatment Facilities and Systems

Decentralized systems are flexible because they can be used for individual systems, cluster systems, housing developments, and commercial, institutional, and recreational facilities. Effluent from decentralized treatment systems can be used for a variety of applications similar to the general categories listed in Table 6-2. Landscape irrigation is the most common use. It is interesting to note that currently over 60 million people in the U.S. are served by decentralized collection and treatment systems and that more than one-third of the new homes built will be served by onsite or decentralized systems (Crites and Tchobanoglous, 1998).

Because water shortages will continue to occur as the population grows, local recycling using decentralized systems helps to offset demands for potable water by substituting reclaimed water in nonpotable applications. The membrane bioreactor technology will be important in advancing the concept of decentralized treatment because of its compact size and ability to meet stringent water quality regulations.

Figure 6-14
Examples of excessive energy loss due to free fall at clarifier weir structures.

New Infrastructure Concepts and Designs

Most of the wastewater system infrastructure in the United States was built during the period of the early 1970s to the mid-1990s. As this infrastructure has aged and is reaching the end of its useful life, the requirements for upgrading are becoming imminent. New infrastructure concepts and designs must be developed that will make wastewater management systems more energy efficient, more robust with respect to treatment performance, and more resistant to natural disasters and malicious attacks.

Energy Efficiency

The designs used for most wastewater management facilities were developed with little emphasis on energy-efficient operation. The formulas used for the design of collection systems were developed more than 75 yr ago. Based on numerous evaluations, it is clear that these formulas should be revisited, especially in light of new materials and methods of construction. The loss of energy at wastewater treatment facilities (see Fig. 6-14) must also be reduced through improved methods of hydraulic analysis. While significant strides have been made in the energy-efficient design of pumps and pumping stations, further improvements in energy efficiency need to be accomplished, especially because processes such as membrane treatment and UV disinfection will increase energy consumption. The aging infrastructure also includes power distribution systems that might not be able to handle increased demands of membrane and UV systems, the effects (high amperage and oscillation) of variable frequency drives, and the redundant equipment needed to ensure process reliability. The replacement of existing and the addition of new treatment processes will have to consider the overall aspects of how to make the treatment system more efficient. Examples of the energy impacts of different technologies on wastewater treatment based on volume of water processed are given in Table 6-9.

Robust Treatment Processes

Based on recent developments in biotechnology including genetic engineering and new methods of identifying microorganisms, significant new developments will affect the

Table 6-9
Energy impacts of technologies on wastewater treatment[a]

Technology	Energy impact[b]	
	kWh/Mgal	MJ/1000 m³
Fine pore diffusers for aeration (in lieu of coarse bubble diffusers)	−125 to −150	−120 to −140
Ultrafine pore diffusers	−180 to −220	−170 to −210
Dissolved oxygen control systems (as compared to manual control)	−50 to −100	−48 to −95
Energy-efficient blower control systems, i.e., inlet guide vanes, inlet butterfly valves, or adjustable-speed drives	−50 to −150	−48 to −140
Energy-efficient aeration blowers (as compared to blowers with inlet guide vanes)	−100 to −150	−95 to −140
Ultraviolet (UV) disinfection	+50 to +200	+48 to +190
Membranes		
Microfiltration	+200 to +400	+190 to +380
Reverse osmosis	+1000 to +2000	+950 to +1900

[a] Adapted in part from Burton (1998).
[b] Minus values correspond to energy savings. Plus values correspond to increased energy use.

implementation of biological wastewater treatment systems. Improvements in membrane technology will make it possible to remove essentially all of the constituents of concern from wastewater. The use of combined biological membrane technologies to treat specific constituents will become commonplace. In some cases membranes alone will be used to treat wastewater, without the need for biological treatment.

Security Concerns

The use of distributed, satellite, and decentralized treatment facilities that can be interconnected with intermediate storage and operated remotely will make the total system less prone to single failure. In an extreme case, a diverse system will provide greater safety in the event of terrorist activities. Although many chemical and biological sensors are now available, their sensitivity is such that they all produce far too many false positive responses to be of value in their current state of development. Based on ongoing applied and fundamental research, it is anticipated that robust water quality sensors will be available and used to detect the presence of foreign substances in wastewater at extremely low levels.

Research Needs

With the growing interest in water reuse applications, many issues are driving research efforts and include the rapid development in treatment technologies, improved ability to detect constituent content to very low levels, and concerns about potential health effects. Over the next 5 to 10 yr, continued research efforts in treatment technologies will include:

- Further evaluation of the health effects of constituents that have been identified only recently, and at extremely low concentrations; also in need of evaluation is the level of treatment required to render these constituents harmless
- Improved process control to optimize the performance of activated sludge systems

- Evaluation of SRTs in the activated sludge process or in MBR applications for the removal of trace constituents
- Further advancement of technologies such as IFAS and MBfR for the removal of nutrients and specific constituents
- Continued improvements in membrane operating performance and lowering membrane operating pressures and cost
- Methods to further control membrane fouling to improve membrane performance and reduce cleaning requirements
- Development of concentrate processing and disposal systems applicable to small installations or centralized brine processing
- Further evaluation of various disinfection systems including improved methods of monitoring to ensure that consistent high levels of disinfection performance are achieved

PROBLEMS AND DISCUSSION TOPICS

6-1 Reclaimed water will be used for a recreational impoundment. Based on a review of Fig. 6-1, what types of processes should be considered for a new treatment plant for this application? Explain your reasons.

6-2 A water reclamation plant is being planned for landscape irrigation and groundwater recharge. The TDS and total nitrogen in the influent wastewater is 850 and 36 mg/L, respectively. For the specified reuse, the TDS must be reduced to 500 mg/L or less, and the total nitrogen in the effluent must be equal to 10 mg/L or less as N. Develop a treatment process flow diagram to meet the required water quality limits. Explain your reasons for the selection of the various unit operations and processes.

6-3 Review two articles in the literature dealing with the concept of multiple barriers and summarize how the authors quantify the effectiveness of the multiple barrier systems.

6-4 It is often stated that various environmental buffers, such as travel through wetlands and/or streams, serve as an additional barrier before reuse for some constituents. Based on a brief review of the literature, are such barriers real or perceptional? Are there any limitations with taking credit for the use of an environmental buffer as a barrier?

6-5 Reclaimed water is to be used for industrial cooling water on a continuous year-round basis and for agricultural irrigation over a summer season with a duration of 4 mo. What type of infrastructure may be needed to meet the reclaimed water delivery schedule? In your answer, consider the need for any additional treatment, the possible need for winter storage to meet agricultural water demand, and the impact that discharge requirements may have on the reclaimed water supply (see also Chap. 23).

6-6 A satellite water reclamation plant is being planned to supply reclaimed water for a golf course located in a semiarid area in the southwestern United States. What are some of the standby and redundancy considerations for system design and operation that should be evaluated? State your assumptions for components of the system infrastructure.

6-7 Reclaimed water is being considered for landscape irrigation at a large urban park. The park will also contain playgrounds and athletic fields. Sources of reclaimed water being evaluated are effluent from a remote centralized conventional activated sludge plant and a proposed new satellite treatment plant located near the park (see Fig. 6-12 for a general system concept). For this application, (a) what types of water quality, treatment processes, and infrastructure would be required and (b) what are the advantages and disadvantages of the two types of reclaimed water systems?

6-8 A water reclamation plant that employs a membrane bioreactor with nitrogen removal will be used for groundwater recharge. Influent to the plant has been found to contain chloride concentrations in excess of 1000 mg/L. What measures can be employed to mitigate or control the high chloride concentration in the reclaimed water?

6-9 Groundwater recharge is proposed using effluent from an existing activated sludge treatment plant that has biological nutrient removal. An additional two-stage process using commercially available microfiltration and reverse osmosis membranes is being considered. You, as leader of the design team, are faced with the question—should pilot testing be done? State your reasons for or against undertaking a pilot testing program.

6-10 Given the concern with homeland security, what simple steps could be taken to enhance the integrity of the infrastructure for wastewater management?

REFERENCES

Andersen, A., H. Siegrist, B. Halling-Sorensen, and T. A. Ternes (2003) "Fate of Estrogens in a Municipal Sewage Treatment Plant," *Environ. Sci. Tech.,* **37**, 18, 4021–4026.

Burton, F. L. (1998) "Saving on Wastewater Treatment," *Energy Magazine,* **23**, 1, 17–20.

Crites, R. W., and G. Tchobanoglous (1998) *Small and Decentralized Wastewater Management Systems,* McGraw-Hill, New York.

Crites, R. W. (2000) "Soil Aquifer Treatment of Municipal Wastewater," *Proceedings of WEFTEC 2000,* Water Environment Federation, Alexandria, VA.

Esposito, K. M., P. J. Phillips, B. M. Stinson, R. Tsuchihashi, and J. Anderson (2005) "The Implication of Emerging Contaminants in the Future of Water Reuse," *Proceedings of the 2005 Annual Symposium,* WateReuse Association, Alexandria, VA.

Hightower, M. M. and C. G. Keyes, Jr. (2005) "Working Group Efforts to Establish Concentrate Management Guidelines for Desalination and Water Reuse," *Proceedings of WEFTEC 2005,* Water Environment Federation, Alexandria, VA.

Johnson, T. L., J. P. McQuarrie, and A. R. Shaw (2004) "Integrated Fixed-Film Activated Sludge (IFAS): The New Choice in Nitrogen Removal Upgrades in the United States," *Proceedings of WEFTEC 2004,* Water Environment Federation, Alexandria, VA.

Lynch, S. T., B. Rohwer, Z. Erdal, and A. Lynch (2005) "Brine/Concentrate Management Strategies for Southern California," *Proceedings of WEFTEC 2005,* Water Environment Federation, Alexandria, VA.

Marcino, S. A. (2004) "Water Reuse in the Northeast Region," *Proceedings of the 2003 Annual Symposium,* WateReuse Association, Alexandria, VA.

Nerenberg, R. (2005) "Membrane Biofilm Reactors for Water and Wastewater Treatment," *2005 Borchardt Conference: A Seminar on Advances in Water and Wastewater Treatment,* University of Michigan, Ann Arbor, MI.

State of California (2001) Code of Regulations, Title 22, Division 4, Chapter 3, *Water Recycling Criteria*, Sections 60301 et seq., June, 2001 Edition.

State of Florida (1999) *Reuse of Reclaimed Water and Land Application*, Chap. 62–610, Florida Administrative Code, Florida Department of Environmental Protection, Tallahassee, FL.

Tchobanoglous, G., F. L. Burton, and H. D. Stensel (2003) *Wastewater Engineering: Treatment and Reuse*, 4th ed., McGraw-Hill, New York.

U. S. EPA (1994) *Water Quality Handbook*, 2nd ed., EPA 823-B-94-005a, U.S. Environmental Protection Agency, Washington, DC.

WEF (1998), *Using Reclaimed Water to Augment Potable Water Resources*, A Special Publication, Water Environment Federation, Alexandria, VA.

7 Removal of Constituents by Secondary Treatment

	WORKING TERMINOLOGY 296
7-1	CONSTITUENTS IN UNTREATED WASTEWATER 299
	Constituents of Concern 299
	Typical Constituent Concentration Values 299
	Variability of Mass Loadings 301
7-2	TECHNOLOGIES FOR WATER REUSE APPLICATIONS 304
7-3	NONMEMBRANE PROCESSES FOR SECONDARY TREATMENT 307
	Suitability for Reclaimed Water Applications 307
	Process Descriptions 308
	Process Performance Expectations 310
	Importance of Secondary Sedimentation Tank Design 318
7-4	NONMEMBRANE PROCESSES FOR THE CONTROL AND REMOVAL OF NUTRIENTS IN SECONDARY TREATMENT 320
	Nitrogen Control 320
	Nitrogen Removal 321
	Phosphorus Removal 324
	Process Performance Expectations 328
7-5	MEMBRANE BIOREACTOR PROCESSES FOR SECONDARY TREATMENT 328
	Description of Membrane Bioreactors 330
	Suitability of MBRs for Reclaimed Water Applications 331
	Types of Membrane Bioreactor Systems 332
	Principal Proprietary Submerged Membrane Systems 333
	Other Membrane Systems 338
	Process Performance Expectations 340
7-6	ANALYSIS AND DESIGN OF MEMBRANE BIOREACTOR PROCESSES 340
	Process Analysis 340
	Design Considerations 353
	Nutrient Removal 358
	Biosolids Processing 361

7-7 ISSUES IN THE SELECTION OF SECONDARY TREATMENT PROCESSES 361
Expansion of an Existing Plant vs. Construction of a New Plant 362
Final Use of Effluent 362
Comparative Performance of Treatment Processes 362
Pilot-Scale Studies 362
Type of Disinfection Process 362
Future Water Quality Requirements 363
Energy Considerations 363
Site Constraints 364
Economic and Other Considerations 368

PROBLEMS AND DISCUSSION TOPICS 368

REFERENCES 371

WORKING TERMINOLOGY

Term	Definition
Activated sludge	Biological treatment process that involves the conversion of organic matter and/or other constituents in the wastewater to gases and cell tissue by a large mass of aerobic microorganisms maintained in suspension by mixing and aeration. The microorganisms form flocculent particles that are separated from the process effluent in a sedimentation tank (clarifier) and subsequently returned to the aeration process or wasted.
Aerobic (oxic) process	Biological treatment process that occurs in the presence of free dissolved oxygen. Oxygen is consumed by aerobic microorganisms to drive metabolic reactions.
Attached growth process	Biological treatment process in which the microorganisms responsible for the conversion of organic matter or other constituents in the wastewater to gases and cell tissue are attached to an inert medium, such as rocks, slag, or specially designed ceramic or plastic materials. Attached growth processes are also known as fixed-film processes.
Anaerobic process	Biological process that occurs in the absence of oxygen and oxidized compounds.
Anoxic process	Biological treatment process that occurs in the absence of free dissolved oxygen, where oxidized compounds such as nitrate and sulfate are used to drive metabolic reactions.
Atomic mass unit (amu)	A measure of molecular weight. An amu is also known as a Dalton (Da) and is equal to 1.66054×10^{-24} g.
Backpulse	A method of membrane cleaning in which the flow through the membrane is reversed at specified intervals using permeate or chlorinated permeate.
Biological nutrient removal	Removal of nitrogen and phosphorus by biological treatment.
Biological phosphorus removal (BPR)	Removal of phosphorus by accumulation in biomass and subsequent solids separation.
Carbonaceous biochemical oxygen demand (CBOD)	Biological conversion of organic matter in wastewater to cell tissue and various gaseous end products. It should be noted that the CBOD does not include the oxygen demand for oxidation of nitrogen compounds.

Concentrate (also called retentate)	The portion of the feed stream that contains salts rejected from the membrane system. The term can also refer to the rejected mixed liquor suspended solids (MLSS) in a membrane bioreactor (MBR) process, which is recycled or wasted to maintain a given biomass concentration in the biological treatment process.
Conventional treatment technologies	Technologies such as activated sludge and trickling filters that remove BOD and suspended solids from wastewater. Conventional treatment technologies are often coupled with anoxic and anaerobic processes for nutrient removal.
Denitrification	The biological process by which nitrate is reduced biologically to nitrogen gas under anoxic conditions.
Flow equalization	The damping of flowrate variations to obtain a constant or nearly constant flowrate, usually by means of a large storage tank. Influent flow is accumulated during peak flow periods for release during non peak periods, thus controlling the flowrate to downstream facilities.
Flux	The mass or volume rate of transfer through the membrane surface, usually expressed as kg/m$^2 \cdot$h or L/m$^2 \cdot$h. Flux is the prevalent term for referring to a membrane system's rate of water production.
Fouling	The accumulation of contaminants on the surface of or within the pores of the membrane that impedes the flow of permeate through the membrane.
Hybrid processes	Those processes that use a combination of suspended growth and attached growth biological treatment.
Membrane	A device, usually made of an organic polymer, that allows the passage of certain constituents but rejects others above a certain size or weight.
Membrane bioreactor (MBR)	A process that combines a suspended growth biological reactor with a membrane separation system; membrane separation is accomplished by either microfiltration or ultrafiltration membranes.
Membrane element	A single membrane unit containing a bound group of hollow-fiber membranes or a membrane plate.
Microfiltration (MF)	A membrane separation process used typically to remove relatively large particles from the feed stream; MF pore sizes range from approximately 0.05 to 2 μm.
Mixed liquor	The mixture of solids resulting from combining recycled sludge with influent wastewater in the bioreactor is termed *mixed liquor suspended solids* (MLSS) and *mixed liquor volatile suspended solids* (MLVSS). The total biomass solids concentration in a bioreactor is commonly measured as *total suspended solids* (TSS) and *volatile suspended solids* (VSS).
Module	A complete unit comprised of the membranes, the support structure for the membranes, and the permeate collection piping.
Nitrification	The two-step biological process by which ammonia (NH_4^+-N) is converted first to nitrite (NO_2^--N) and then to nitrate (NO_3^--N).
Permeability	A measure of membrane performance related to a specific flux of clean, deionized water through a new membrane.
Permeate (also called filtrate or product water)	The liquid stream that has passed through the membrane.
Pore size	The nominal size of a membrane's pores (typically measured in microns) that allows passage of permeate through the membrane wall while retaining constituents larger than the pore size on the membrane surface.

Residual streams	Waste streams produced by the water reclamation processes. Residual streams include waste sludge, waste washwater, concentrate, and chemical cleaning wastes.
Retentate (also called concentrate)	The portion of the feed stream that does not pass through the membrane.
Satellite treatment systems	The interception of wastewater in an upstream portion of the collection system for treatment in a water reclamation plant that is located close to the point of reuse. Satellite treatment plants generally do not have solids processing facilities; solids are returned to the collection system for processing in a central treatment plant located downstream.
Sequencing batch reactor (SBR)	A fill-and-draw type of reactor system involving a single complete-mix reactor in which all steps of the activated sludge process occur.
Sidestream treatment	Treatment of a portion of the wastewater flow in a separate process.
Solids retention time (SRT)	The average period of time during which the biomass has remained in a biological treatment system. The SRT is a critical parameter for the design of activated sludge processes.
Suspended growth processes	Biological treatment processes in which the microorganisms responsible for the conversion of organic matter or other constituents in the wastewater to gases and cell tissue are maintained in suspension within the liquid.
Transmembrane pressure (TMP)	The driving force required to filter solids from the liquid in a membrane separation process.
Ultrafiltration (UF)	A membrane separation process similar to MF except the membrane pore sizes range from approximately 0.005 to 0.1 μm. Generally, UF membranes are able to achieve higher levels of separation than MF, particularly for bacteria and viruses.

Municipal wastewater treatment systems are designed to meet a number of treatment objectives for effluent discharge and reuse. The term *secondary treatment* is used to describe the processes used for the removal of suspended solids, dissolved organic matter, nutrients, and pathogenic microorganisms. In this chapter, conventional nonmembrane biological treatment processes are described, but detailed design examples are not included as comprehensive coverage of such processes is provided in the companion text, *Wastewater Engineering: Treatment and Reuse* (Tchobanoglous et al., 2003). Add-on treatment processes that can be used for nutrient removal are also described. Because of the increasing importance of biological treatment processes employing membranes for water reuse, special emphasis is given to the description and design of these systems. Subjects discussed in this chapter include: (1) constituents of concern in untreated wastewater, (2) technologies for water reuse applications, (3) nonmembrane processes for secondary treatment,(4) nonmembrane processes for the control and removal of nutrients in secondary treatment, (5) membrane processes for secondary treatment, (6) analysis and design of MBRs, and (7) issues in the selection of secondary treatment processes. While effluent from secondary treatment is suitable for a number of reuse applications, higher levels of treatment are often needed for reuse applications requiring removal of residual suspended, dissolved, and trace constituents. The removal of residual matter after secondary treatment is considered in Chap. 8 and the removal of dissolved and trace constituents is considered in Chaps. 9 and 10, respectively.

7-1 CONSTITUENTS IN UNTREATED WASTEWATER

To design biological treatment and reclamation processes, information must be available on the influent wastewater constituents. Accordingly, the purpose of this section is to identify the constituents of concern and to present typical data on the concentration and variability of the constituents found in untreated wastewater. The product of the mean constituent concentration and corresponding mean flowrate over a given time interval is known as the mass loading for that time interval. The constituent mass loadings are important for the operation and design of some treatment processes. Because the variability in constituent mass loadings to a treatment facility is a function of the flowrate, the variability observed in influent flowrates must also be considered.

Constituents of Concern

Using the constituent classes introduced in Table 6-1 in Chap. 6, the constituents of concern in each class are identified in Table 7-1. In reviewing the information presented in Table 7-1 it is important to note that both *discrete* and *aggregate* measurements are used to quantify the constituents in untreated wastewater. Discrete measurements are used to quantify individual elements, specific compounds, or specific microorganisms. Constituents that are comprised of a number of individual components that are not distinguished separately are known as aggregate constituents. For example, the TSS test is used to measure the total mass of suspended particulate matter in a water sample; however, the mass of individual particles and the particle size distribution of the water sample cannot be determined using the TSS results. Similarly, in the measurement of BOD, COD, TOC, and oil and grease, the individual compounds that are measured in each test are unknown. *It is important to note that not knowing what compounds comprise the organic matter in both untreated and treated wastewater is often of concern in some reuse applications, particularly direct and indirect potable reuse.*

Several techniques are available that can be used to obtain more detailed information on the nature of the particulate and organic matter in treated wastewater. The approximate contribution of particles of varying size to the measured value of TSS can be obtained by serial filtration. Measurement of discrete particle sizes using optical techniques can be used to define the particle size distribution. Serial filtration and particle size distribution are discussed in detail in Sec. 8-1 in Chap. 8. Measurement of individual constituents can also be done, but is more difficult because of the complex and unknown nature of the chemical composition of untreated wastewater. Typical constituent concentration values and variability of wastewater parameters that affect mass loadings are considered in the following discussion.

Typical Constituent Concentration Values

The water quality issues for many water reuse applications are concerned with the removal of both aggregate and specific constituents; the constituents of concern are most commonly particulate matter (TSS and turbidity), organic matter (BOD, COD, and TOC), nutrients, and pathogenic organisms. Typical concentration values for the constituents used to characterize untreated wastewater are reported in Table 3-12 in Chap. 3. As shown in Table 3-12, the range of values reported for the individual constituents vary considerably. The range in the constituent values, with the exception of microorganisms,

Table 7-1
Constituents found in untreated wastewater: concerns and analytical tests

Parameter	Concerns	Analytical tests
Constituent classification		
Particulate matter	Can shield embedded bacteria thus interfering with disinfection effectiveness.	TSS, fixed and volatile SS, settleable solids
Organic matter	Organic substrate for microbial or algal growth; can exert a chlorine demand that can inhibit disinfection effectiveness; organic compounds can combine with chlorine to form chlorinated byproducts such as trihahomethanes (THMs).	BOD, COD, TOC, oil and grease
Inorganic matter	Inorganic constituents such as calcium, sodium, and sulfate are added to the original domestic water supply as a result of water use. Heavy metals are usually added to wastewater from commercial and industrial activities and are also often identified as priority pollutants.	Individual elements such as Ca^{2+}, Cl^-, Na^+, Fe^{3+}
Pathogens	Measure of microbial health risks due to enteric viruses, pathogenic bacteria, and protozoa.	Individual bacteria (e.g., total and fecal coliforms), protozoa, helminths, and viruses
Nutrients	Nutrient source for irrigation, may cause excessive amount of nitrates in groundwater, can contribute to microbial and algal growth. Organic and inorganic phosphorus is a nutrient source for irrigation; can contribute to microbial and algal growth.	Org. N, NH_4^+, NO_2^-, NO_3^-, Org. P, PO_4^{3-}
Trace constituents	Inorganic and organic compounds selected on the basis of their known or suspected carcinogenicity, mutagenicity, teratogenicity, or high acute toxicity.	Individual compounds such as NDMA
Total dissolved solids	High concentrations may lead to the degradation of local groundwater.	TDS, fixed and volatile dissolved solids
Physical property		
Color	Can be used to assess the age or condition of the untreated wastewater (e.g., fresh or septic); can also affect transmissibility and the performance of UV disinfection.	CU (color units), measured directly
Conductivity	Measure of the concentration of dissolved chemical elements.	µS/m, measured directly
Temperature	Will affect wastewater treatment kinetics.	°C, measured directly
Transmittance	Can affect the effectiveness of UV disinfection.	%, measured directly
Turbidity	Measure of particulate matter; is often used as a surrogate parameter for evaluating process performance and suitability for reuse.	NTU (nephelometric turbidity unit), measured directly

is due primarily to the amount of extraneous water that enters the collection systems through nonresidential discharges and infiltration. The observed range in the concentration of microorganisms will vary with the population that is infected and shedding within the community.

Variability of Mass Loadings

The variability of mass loadings is affected by the influent wastewater parameters of concern, influent flowrates, and the constituent concentrations.

Variability in Influent Wastewater Parameters

When selecting and sizing processes for the management and treatment of wastewater constituents, the variability of influent parameters must be considered. The design of pumping stations, flow equalization facilities, and unit treatment processes all require knowledge of the variability of the influent parameters. Peaking factors are used to estimate the maximum values that would be expected. One method used to characterize the variability of wastewater parameters and treatment processes is the use of the geometric standard deviation, s_g (see Appendix D). The value of s_g can be used to approximate an entire distribution of all expected values if a mean value is known or can be estimated. As discussed in Appendix D, the greater the numerical value of s_g, the greater the observed range in the measured values. Peaking factors are also related to s_g by specification of a frequency value. The peaking factor is calculated as the value at a given frequency divided by the mean value (see Fig. 7-1). For example, the peak day value, which

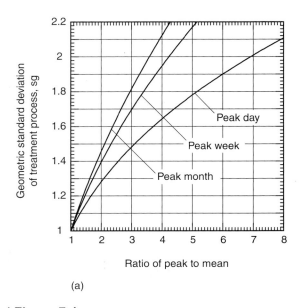 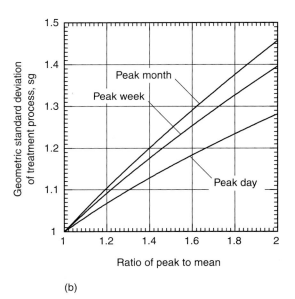

Figure 7-1

Relationship of s_g values to peaking factors for peak day, week, and month: (a) diagram for facilities with large peak-to-mean variations and (b) expanded portion of diagram (a) for facilities with small peak-to-mean variations.

corresponds to one event per year, is the value that occurs at a frequency of 99.7 percent [(364/365) × 100]. The variability of the influent wastewater flowrates and constituents is considered in the following sections.

Variability in Influent Flowrates

The influent flowrate to a treatment facility is dependent on factors such as the time of day, season, size and characteristics of the contributing population, and infiltration to and exfiltration from the collection system. Influent flowrate may be moderated by equalization occurring in the collection system or in specially designed facilities. The amount of variability is also correlated with the type of development. In large cities the wastewater flow is distributed more evenly because there is a greater diversity of lifestyles and a high amount of activity at night. In contrast, wastewater treatment facilities used for small residential communities are more likely to experience higher peak flow relative to mean flow values. Typical ranges of observed values for s_g for influent flow rates for small, medium, and large capacity wastewater treatment plants are given in Table 7-2. The relationship between s_g values and the peaking factors for peak day, week, and month can be determined using the curves given on Fig. 7-1. An example of the use of the s_g value and the curves given in Fig. 7-1 is illustrated in Example 7-1.

Variability in Constituent Concentrations

The variability of the constituents in wastewater must be considered carefully in the design of biological treatment processes, especially with respect to the design of the aeration facilities. Geometric standard deviation values for the variability observed in influent wastewater constituents, BOD, COD, and TSS are given in Table 7-2 in terms of the s_g value. The range of s_g values given in Table 7-2 corresponds to the range of values reported in the literature and in the authors' experience. The use of Fig. 7-1 for estimating peak expected values is shown in Example 7-1.

Table 7-2

Range of geometric standard deviations (s_g) for influent parameters observed at small, intermediate, and large wastewater treatment facilities

	Range of s_g for typical wastewater treatment facilities[a]					
	Small[b]		Intermediate[c]		Large[d]	
Parameter	Range	Typical	Range	Typical	Range	Typical
Flowrate	1.4–2.0	1.6	1.1–1.5	1.25	1.1–1.2	1.15
BOD	1.4–2.1	1.6	1.3–1.6	1.3	1.1–1.3	1.27
COD	1.5–2.2	1.7	1.4–1.8	1.4	1.1–1.5	1.30
TSS	1.4–2.1	1.6	1.3–1.6	1.3	1.1–1.3	1.27

[a] Excluding systems with large amounts of infiltration in the collection system.
[b] Flowrate of 4000–40,000 m³/d.
[c] Flowrate of 40,000–400,000 m³/d.
[d] Flowrate > 400,000 m³/d.

EXAMPLE 7-1. Estimation of Variability of Influent Wastewater Parameters.

Compute the expected maximum values for the influent parameters: flowrate, BOD, COD, and TSS for a small and large size wastewater treatment facility. Assume the following mean design values apply:

		Mean design values	
Parameter	Unit	Small	Large
Flowrate	m³/d	10,000	500,000
BOD	mg/L	250	250
COD	mg/L	600	600
TSS	mg/L	200	200

Determine the maximum value of the influent parameters for maximum day and maximum month. Comment on the importance of the results.

Solution

1. Select s_g values from Table 7-2 that correspond to the wastewater parameters of interest. In the absence of site- and regionally-specific information, use the typical s_g values given in Table 7-2 as follows:

Size of Facility	Parameter			
	Flowrate	BOD	COD	TSS
Small	1.6	1.6	1.7	1.6
Large	1.15	1.27	1.30	1.27

2. Locate the selected s_g value on Fig. 7-1 for a given frequency and determine the corresponding peaking factor.

    Using the s_g values determined in Step 1, the corresponding peaking factors for peak day and peak month can be found on Fig. 7-1a for small facility and Fig. 7-1b for large facility. The peaking factors are summarized in the following table:

	Small facility			Large facility		
		Peaking factor			Peaking factor	
Parameter	s_g	Day	Month	s_g	Day	Month
Flowrate	1.6	3.70	2.35	1.15	1.48	1.29
BOD	1.6	3.70	2.35	1.27	1.95	1.55
COD	1.7	4.40	2.65	1.30	2.20	1.62
TSS	1.6	3.70	2.35	1.27	1.95	1.55

3. Obtain the maximum values for a specified frequency. Multiply the peaking factor determined in Step 2 and the mean value from the table given in the problem statement

   a. For the peak day flowrate, the peaking factor is 3.70 and the mean design value is 10,000 m³/d:

   $$(3.70)(10{,}000 \text{ m}^3/\text{d}) = 37{,}000 \text{ m}^3/\text{d}$$

   b. The design values for the two facilities are summarized in the following table:

		Design values					
		Small facility			Large facility		
Parameter	Unit	Mean	Peak day	Peak month	Mean	Peak day	Peak month
Flowrate	m³/d	10,000	37,000	23,500	500,000	740,000	645,000
BOD	mg/L	250	925	587.5	250	487.5	387.5
COD	mg/L	600	2640	1590	600	1320	972
TSS	mg/L	200	740	470	200	390	310

Comment

As shown in the summary table presented in Step 3, the smaller facility must be designed to accommodate a larger range in influent wastewater parameters relative to the large facility.

In addition to the variability of the influent wastewater flowrate and constituents, the design, performance, and reliability of a wastewater treatment plant are also affected by the inherent variability in wastewater treatment processes and the variability caused by mechanical breakdown, design deficiencies, and operational failures. Changes in recycle flows, especially from biosolids processing facilities, can upset process performance. The inherent variability in wastewater treatment processes is considered in Secs. 7-3, 7-4, and 7-5, where the performance of various treatment processes is discussed. A discussion of the variability caused by mechanical breakdown, design deficiencies, and operational failures may be found in Tchobanoglous et al. (2003).

7-2 TECHNOLOGIES FOR WATER REUSE APPLICATIONS

For water reuse applications, a variety of treatment processes may be used depending on the type and degree of constituent removal required. Technologies that are discussed in this chapter and used in water reuse applications are presented in Table 7-3. Generalized process flow diagrams of commonly used processes are shown on Fig. 7-2. All of the treatment processes employ biological treatment for the removal of TSS and BOD.

Table 7-3
Technologies used commonly for the removal of suspended solids, dissolved organic matter, and nutrients in water reuse applications

Type	Common name	Use
Aerobic processes		
Flow-through suspended growth	Activated sludge variations, principally plug flow, complete mix, step feed, and oxidation ditch	Carbonaceous BOD and TSS removal, nitrification
Batch suspended growth	Sequencing batch reactor	Carbonaceous BOD and TSS removal, nitrification
Attached growth	Trickling filter	Carbonaceous BOD and TSS removal, nitrification
	Submerged attached growth	Carbonaceous BOD removal, nitrification
	Packed-bed reactor	Carbonaceous BOD removal, nitrification
Hybrid (combined suspended and attached growth processes)	Trickling filter/activated sludge, trickling filter/solids contact	Carbonaceous BOD and TSS removal, nitrification
Anoxic/aerobic processes		
Flow-through suspended growth	Modified Ludzack-Ettinger (MLE)	Denitrification
Batch suspended growth	Sequencing batch reactor (modified operation)	Denitrification
Attached growth	Upflow and downflow packed bed reactors	Denitrification
	Fluidized bed reactor	Denitrification
Anaerobic/aerobic processes		
Suspended growth	Phoredox, A^2/O, VIP	Carbonaceous BOD, TSS, and phosphorus removal
Batch suspended growth	Sequencing batch reactor (modified operation)	Phosphorus removal
Membrane systems	Membrane bioreactors and variations	Carbonaceous BOD, TSS, colloidal solids, and phosphorus removal; nitrification; denitrification

Filtration, discussed in Chap. 8, is often included as part of a treatment system to improve the removal of residual suspended solids and the effectiveness of disinfection. Chemical treatment is used in some cases to enhance physical separation of solids and for the removal of phosphorus. In many applications, especially for groundwater recharge and surface water enhancement, removal of nitrogen and/or phosphorus is an important requirement and may be integrated with biological treatment or accomplished by an add-on treatment step.

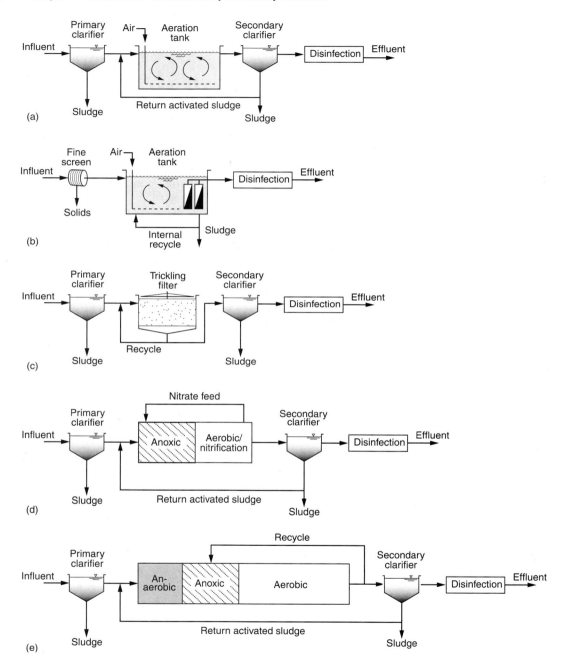

Figure 7-2

Generalized process flow diagrams for typical treatment processes: (a) activated sludge for TSS and BOD removal and nitrification, (b) membrane bioreactor for TSS and BOD removal, (c) trickling filter for TSS and BOD removal, (d) suspended growth biological treatment for nitrogen removal, and (e) suspended growth biological treatment for phosphorus removal.

The membrane bioreactor (MBR), a relatively new technology that continues to evolve rapidly, is especially well-suited for water reuse applications. Membrane bioreactors are being used for upgrading existing wastewater treatment plants and for producing high-grade product water suitable for many uses. Membrane bioreactors combine suspended growth treatment with membrane filtration, thereby removing a high percentage of particles and pathogens without the need for secondary clarification facilities. Membrane bioreactors require less space than traditional activated sludge systems because of the shorter hydraulic retention time in the bioreactor and the small footprint of the membrane separation unit. Membrane bioreactors are particularly adaptable to satellite and decentralized wastewater management systems because of their compact size. Satellite and decentralized systems are discussed in Chaps. 12 and 13, respectively.

7-3 NONMEMBRANE PROCESSES FOR SECONDARY TREATMENT

The main categories of aerobic biological processes used for water reclamation applications for the removal of suspended solids and organic matter are *suspended growth* (both flow-through and batch types), *attached growth*, and *hybrid* processes (see Working Terminology and Table 7-3).

In suspended growth processes, the microorganisms responsible for treatment are maintained in liquid suspension by mixing and aeration to maintain aerobic conditions. The principal suspended growth process is continuous flow activated sludge; the sequencing batch reactor (SBR) is a modification of the activated sludge process where all of the treatment occurs in batches in a single tank.

In attached growth processes, a medium, such as a fixed packing, rotating disks, or granular medium packing, is used to which microorganisms attach and form a biofilm. The biofilm microorganisms come in contact with the liquid and oxidize the organic matter. As the microorganisms grow, they slough from the surface of the medium and are removed subsequently by a solids separation device, usually a gravity settling tank and/or filter.

Several hybrid biological systems have also been developed that use a combination of attached growth and suspended growth reactors to meet specific conditions. Descriptions and discussions of the many treatment processes used for a variety of other applications are covered in detail in the companion text, Tchobanoglous et al. (2003). Those technologies suitable for water reuse applications are highlighted in this chapter.

Suitability for Reclaimed Water Applications

Many of the processes used in reclaimed water applications are termed conventional treatment as they use traditional treatment processes such as activated sludge, trickling filters, and gravity sedimentation. These processes are employed mainly in systems where large quantities of reclaimed water are produced especially for reuse applications such as agricultural irrigation, landscape irrigation, and groundwater recharge. The advantages of using conventional treatment technologies for water reclamation and reuse are: (1) the processes are familiar and well understood, (2) they can be automated to a fairly high degree, and (3) highly skilled operators may not necessarily be required except in the case where nutrient removal is included. The disadvantages of conventional technologies are that (1) they may require larger physical facilities, e.g., clarifiers, (2) they may be more susceptible to process upset, and (3) effluent quality, especially TSS and turbidity, may be more difficult to control and thus may not meet the water quality requirements for reuse without special process augmentation.

Conventional treatment technologies may be used in a centralized water reclamation plant that includes both liquid and solids processing facilities or in a satellite plant that *scalps* wastewater from the collection system for treatment and reuse and returns the residuals removed in the treatment process to the collection system for downstream processing. The removal of wastewater from the collection system for treatment and reuse is also called *sewer mining*. Membrane bioreactors, described in Sec. 7-5, are often used in sewer mining applications for satellite systems described in Chap. 12.

Process Descriptions

Various types of suspended growth, attached growth, and hybrid processes that are suitable for secondary treatment and reuse applications are described in this section. For many water reuse applications, chemical coagulation and filtration, described in Chap. 8, are often added following these processes to remove residual suspended solids and to further enhance the disinfection process.

Suspended Growth Processes

Activated sludge is the suspended growth process used most commonly for the biological treatment of municipal wastewaters. Several modifications or variations of the activated sludge process are used in reclaimed water applications. Plug flow, complete mix, and step-feed processes shown and described in Table 7-4 are utilized for medium and large size water reclamation plants. Views of typical plug flow and complete mix activated sludge reactors are shown on Figs. 7-3a and b. For smaller plants, the oxidation ditch and the SBR, also described in Table 7-4, are used more commonly because they are relatively simple to operate, can be adapted to influent variability, and can be used for nutrient removal. A view of a SBR is shown on Fig. 7-3c. Advantages and disadvantages of each process are presented in Table 7-5.

Attached Growth Processes

The most common aerobic attached growth process used is the trickling filter which is described in Table 7-6 and shown on Fig. 7-4. Although the trickling filter is simple in concept and simple to operate, the trickling filter by itself does not always produce a water quality in consistent conformance with reclaimed water treatment goals. When trickling filters are used for both BOD removal and nitrification, the filters have to be operated at low organic loadings as significant nitrification occurs only after the BOD concentrations are reduced appreciably. More often, trickling filters are used following secondary treatment as a tertiary step for nitrification. When the trickling filter is combined with the activated sludge process, as described in the following section on the hybrid process, a higher quality effluent can be produced that is more suitable for water reclamation purposes.

Table 7-4
Description of commonly used activated sludge processes for BOD and TSS removal and nitrification[a]

Process	Description
Flow-through suspended growth processes	
(a) Complete-mix activated sludge (CMAS) 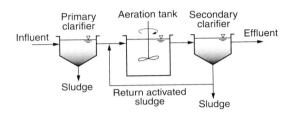	In the CMAS process settled wastewater and recycled activated sludge are introduced typically at several points in the aeration tank. The organic load on the aeration tank, mixed liquor suspended solids concentration, and oxygen demand are uniform throughout the tank. An advantage of the CMAS system is the dilution of shock loads. The CMAS process is relatively simple to operate but tends to have low organic substrate concentrations (i.e., low food to microorganism ratios) that encourage the growth of filamentous bacteria, causing sludge bulking problems. Sludge bulking, however, can be controlled by the use of a selector reactor prior to aerobic treatment.
(b) Conventional plug flow 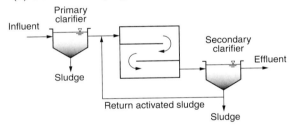	Settled wastewater and return activated sludge enter the front end of the aeration tank and are mixed by diffused air or mechanical aeration. Typically, three to five channels (passes) are used. The aeration system is designed to match the oxygen demand along the length of the tank by tapering the aeration rates, i.e., applying higher rates in the beginning and lower rates near the end of the tank.
(c) Step feed 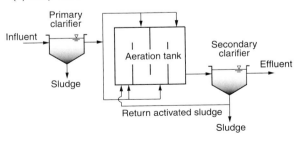	Step feed is a modification of the conventional plug flow process in which settled wastewater is introduced at three to four feed points in the aeration tank to equalize the food to microorganism ratio thus lowering peak oxygen demand. Flexibility of operation is one of the important features of this process because the apportionment of the wastewater feed can be changed to suit operating conditions. The step feed process has the capability of carrying a higher solids inventory, and thus a higher solids retention time (SRT) for the same volume as the conventional plug-flow process.

(Continued)

Table 7-4

Description of commonly used activated sludge processes for BOD and TSS removal and nitrification[a] (*Continued*)

Process	Description
Flow-through suspended growth processes	
(d) Oxidation ditch 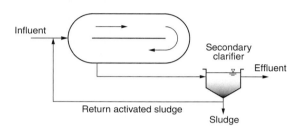	The oxidation ditch consists of a ring- or oval-shaped channel equipped with mechanical aeration and mixing devices. Screened wastewater enters the channel and is combined with return activated sludge. The tank configuration and aeration and mixing devices promote unidirectional channel flow, so that the energy used for aeration is sufficient to provide mixing in the system with a relatively long hydraulic retention time. As the wastewater leaves the aeration zone, the DO concentration decreases and denitrification may occur.
Batch suspended growth process	
(e) Sequencing batch reactor (SBR) 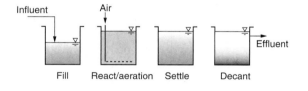	The SBR is a fill-and-draw type of reactor system involving a single complete-mix reactor in which all steps of the activated sludge process occur. For continuous flow, at least two basins are used so that one basin is in the fill mode while the other goes through react, solids settling, and effluent withdrawal. Mixed liquor remains in the reactor during all cycles, thereby eliminating the need for separate clarifiers. Sludge wasting occurs normally during the aeration period. By prolonging the aeration period, nitrification can occur.

[a] Adapted from Tchobanoglous et al. (2003).

Process Performance Expectations

Hybrid Process

The hybrid process most suited for water reclamation applications is the trickling filter/solids contact (TF/SC) process described in Table 7-6. The process consists of a trickling filter (either rock or plastic packing), an activated sludge aeration tank, and a final clarifier. The advantages of the TF/SC process are: (1) smaller activated sludge reactor (the aeration time ranges from 10 to 60 min.), (2) lower energy requirements as compared to conventional activated sludge, (3) good solids settling, and (4) high-quality effluent (Tchobanoglous et al., 2003). Depending on the length of time for aeration, nitrification can occur.

Where treated effluent is to be reused, it is important to know what typical mean effluent constituent values can be expected and the variability in those values. Information on

Figure 7-3
Views of typical suspended growth reactors: (a) plug flow, (b) complete mix, (c) sequencing batch reactor, and (d) membrane bioreactor.

the constituent values and variability is of importance in the selection of technologies that might be used to further process the treated effluent.

Effluent Constituent Values
The ranges of typical mean effluent constituent values that can be achieved with various biological treatment processes are reported in Table 7-7. In most cases the range of observed mean BOD and TSS values is due to the type of activated sludge process, the mode of operation [e.g., solids retention time (SRT) value], and the design of the secondary sedimentation facilities. It should be noted that the factors cited above will also affect the effluent particle size distribution as illustrated on Fig. 7-5. Particle size distribution is important as it affects the performance of the filtration and disinfection systems. As shown on Fig. 7-5, the type of activated sludge process and the SRT value will have a significant impact on the distribution of particles with diameters less than about 10 to 20 μm. For particles greater than about 20 μm, the design of the secondary sedimentation facilities will control the particle size distribution. Additional details on the

Table 7-5

Advantages and limitations of activated sludge processes for BOD removal and nitrification[a]

Process	Advantages	Limitations
Complete mix	Common, proven process that is adaptable to many types of wastewater Design is relatively uncomplicated Relatively easy to operate	Susceptible to filamentous sludge bulking that may cause excessive TSS and turbidity in reclaimed water and high chlorine demand Poor mixing and short-circuiting may adversely affect effluent quality
Conventional plug flow	Proven process Adaptable to many operating schemes including step feed, selector design, and anoxic/aerobic processes	More susceptible to process upset under shock loads May be difficult to match oxygen supply to oxygen demand, thus adversely affecting performance
Step feed	Distributes load to provide more uniform oxygen demand Adaptable to many operating schemes including anoxic/aerobic processes	More complicated design for process and aeration system Operation is more complex, particularly under changing operating conditions
Oxidation ditch	Highly reliable process; simple operation Economical process for small plants Adaptable to nutrient removal Lower biosolids production	Large structure, greater space requirement Some oxidation ditch process modifications are proprietary and license fees may be required Requires more aeration energy than conventional CMAS and plug flow treatment Plant capacity expansion is more difficult
Sequencing batch reactor (SBR)	Compact facility; final clarifiers and return activated sludge (RAS) pumping are not required Operation is flexible; nutrient removal can be accomplished by operational changes Quiescent settling enhances solids separation (low effluent TSS) Economical process for small plants	Process control is more complicated Batch discharge may require equalization prior to filtration and disinfection Redundant units required for continuous flow operation

[a] Adapted from Tchobanoglous et al. (2003).

methods of analysis for effluent particle size and particle size distributions are given in Sec. 8-1 in Chap. 8. The importance of the secondary sedimentation facilities is considered following the discussion of effluent variability.

Variability in Effluent Constituents

All physical, chemical, and biological treatment processes exhibit some measure of variability with respect to the performance that can be achieved. The observed variability of the treatment processes is due to (1) variability of the influent wastewater flowrate and constituents, (2) inherent variability of biological treatment processes due

Table 7-6

Description of commonly used attached growth and hybrid processes for BOD and TSS removal and nitrification[a]

Process	Description
Attached growth Trickling filter	The trickling filter is a nonsubmerged fixed film biological reactor using rock or plastic packing over which wastewater is distributed continuously. Virtually all new trickling filters are constructed with plastic packing. Influent wastewater is applied normally at the top of the packing through rotary distributor arms to provide a uniform application rate. A portion of the underflow from the filter is often recycled to the filter influent. Trickling filters can be used for BOD removal only, BOD removal and nitrification, or tertiary nitrification.
Hybrid Trickling filter/solids contact (TF/SC)	The TF/SC process consists of a trickling filter used in combination with the activated sludge solids contact process. Effluent from the trickling filter is fed directly to the solids contact reactor along with settled biosolids from the clarifier. A portion of the trickling filter effluent is often returned to the filter influent. The hydraulic retention time in the solids contact reactor ranges from 10 to 60 min. The process can be used for BOD removal only or for combined BOD removal and nitrification. Solids removal is enhanced by using a secondary clarifier with a flocculating center well.

[a] Adapted from Tchobanoglous et al. (2003).

to the presence of living microorganisms and the laws of chance, (3) variability caused by mechanical breakdown, design deficiencies, and operational failures, and (4) design limitations. The variability observed in the performance of various activated sludge processes with respect to BOD, TSS, and turbidity in the treated effluent is given in Table 7-8. Variability in terms of s_g values is illustrated graphically on Fig. 7-6 for BOD and TSS. The range of s_g values is representative of the values reported in the literature. Use of the data in Table 7-8 is illustrated in Example 7-2. Further, as illustrated on Fig. 7-6b (and on Fig. 7-7) and discussed in the following section, the physical characteristics of the secondary sedimentation facilities can have a significant impact on the observed performance of the activated sludge process.

(a) (b)

Figure 7-4
Views of trickling filter: (a) exterior view of a rock-media trickling filter and (b) interior view of a rock-media trickling filter.

Table 7-7
Typical range of effluent quality after secondary treatment[a]

Constituent	Unit	Untreated wastewater	Conventional activated sludge[b]	Activated sludge with BNR[c]	Membrane bioreactor
Total suspended solids (TSS)	mg/L	120–400	5–25	5–20	≤1
Biochemical oxygen demand (BOD)	mg/L	110–350	5–25	5–15	<1–5
Chemical oxygen demand (COD)	mg/L	250–800	40–80	20–40	<10–30
Total organic carbon (TOC)	mg/L	80–260	10–40	8–20	0.5–5
Ammonia nitrogen	mg N/L	12–45	1–10	1–3.0	<1–5
Nitrate nitrogen	mg N/L	0–trace	10–30	2–8	<10[d]
Nitrite nitrogen	mg N/L	0–trace	0–trace	0–trace	0–trace
Total nitrogen	mg N/L	20–70	15–35	3–8	<10[d]
Total phosphorus	mg P/L	4–12	4–10	1–2.0	0.5–2.0[d]
Turbidity	NTU		2–15	2–8	≤1
Volatile organic compounds (VOCs)	µg/L	<100–>400	10–40	10–20	10–20
Metals	mg/L	1.5–2.5	1–1.5	1–1.5	trace
Surfactants	mg/L	4–10	0.5–2	0.1–1	0.1–0.5
Totals dissolved solids (TDS)	mg/L	270–860	500–700	500–700	500–700
Trace constituents	µg/L	10–50	5–40	5–30	0.5–20
Total coliform	No./100 mL	10^6–10^9	10^4–10^5	10^4–10^5	<100
Protozoan cysts and oocysts	No./100 mL	10^1–10^4	10^1–10^2	0–10	0–1
Viruses	PFU/100 mL[e]	10^1–10^4	10^1–10^3	10^1–10^3	10^0–<10^3

[a] From Chap. 3, Tables 3-12 and 3-14.
[b] Conventional secondary is defined as activated sludge treatment with nitrification.
[c] BNR is defined as biological nutrient removal for the removal of nitrogen and phosphorus.
[d] With BNR process.
[e] Plaque forming units.

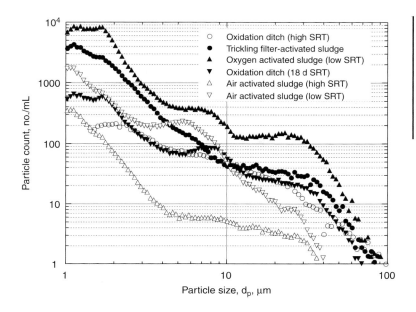

Figure 7-5
Effluent particle size distribution for several biological secondary treatment processes. (Courtesy of K. Bourgeous.)

Table 7-8
Typical range of effluent quality variability observed from secondary treatment processes[a]

Constituent	Unit	Range of effluent values	Geometric standard deviation, s_g[b] Range	Typical
Conventional activated sludge				
BOD	mg/L	5–25	1.3–2.0	1.5
TSS	mg/L	5–25	1.2–1.8	1.4
Turbidity	NTU	5–15[c]	1.2–1.6	1.4
Activated sludge with BNR				
BOD	mg/L	5–15	1.3–2.0	1.5
TSS	mg/L	5–20	1.2–1.8	1.4
Turbidity	NTU	2–8	1.2–1.6	1.4
Membrane bioreactor				
BOD	mg/L	<3	1.3–1.6	1.4
TSS	mg/L	≤1	1.3–1.9	1.5
Turbidity	NTU	≤1	1.1–1.4	1.3

[a] All of the reported distributions are log normal, M_g = geometric mean, s_g = geometric standard deviation.
[b] $s_g = P_{84.1}/P_{50}$.
[c] Turbidity values of less than 2 NTU have been observed in plants with deep clarifiers (e.g., sidewater depths of 5.5 to 6 m). Corresponding BOD and TSS values are in the range from 3 to 6 mg/L.

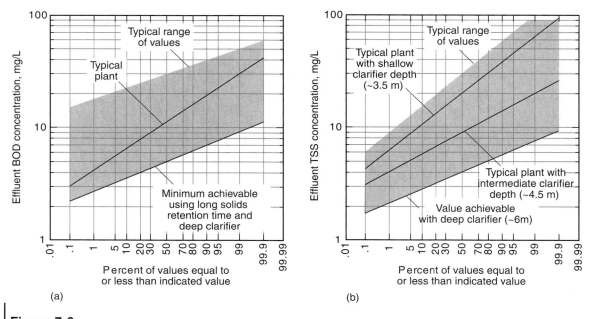

Figure 7-6
Variability of effluent BOD and TSS values from activated sludge processes: (a) BOD and (b) TSS.

EXAMPLE 7-2. Evaluation of Activated Sludge Process Reliability.

A conventional activated sludge process has been designed to have a mean effluent BOD and TSS value of 15 mg/L. Determine the maximum BOD and TSS values that are expected to occur with a frequency of (a) once per year and (b) once every 3 yr. If the effluent limit for both BOD and TSS is 30 mg/L, estimate how often the effluent limits will be exceeded annually.

Solution

1. Select s_g values for BOD and TSS from Table 7-8 that correspond to the effluent BOD and TSS for a conventional activated sludge process. From Table 7-8, use the typical s_g values of 1.5 and 1.4 for BOD and TSS, respectively.
2. Determine the probability distribution of the effluent BOD and TSS values.
   a. Using the s_g values, compute the BOD and TSS values corresponding to the plotting position on $P_{84.1}$ (see footnote b from Table 7-8).
      i. For BOD
      $$P_{84.1} = s_g \times P_{50} = 1.5 \times 15 \text{ mg/L} = 22.5 \text{ mg/L}$$
      ii. For TSS
      $$P_{84.1} = s_g \times P_{50} = 1.4 \times 15 \text{ mg/L} = 21 \text{ mg/L}$$

b. Estimate the distribution of effluent BOD and TSS values by plotting the $P_{84.1}$ and P_{50} values. As the effluent BOD and TSS values are expected to follow a log normal distribution, a straight line can be drawn through the $P_{84.1}$ and P_{50} values, as shown on the following plot.

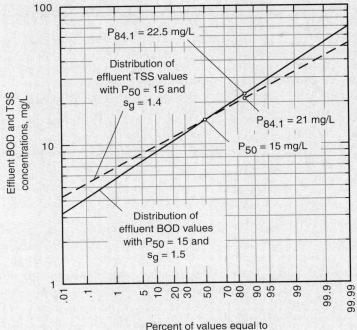

3. Compute the effluent BOD and TSS values expected to occur with the frequency of interest.
    a. The probability of occurrence of a given event with a frequency of once per year is $(1/365) \times 100 = 0.3\%$. Therefore, the percent of events occurring less than once per year is $100 - 0.3 = 99.7\%$. Using the plot developed in Step 2, the effluent BOD and TSS values corresponding to 99.7% are:
       i. For BOD
$$P_{99.7} = 45.8 \text{ mg/L}$$
       ii. For TSS
$$P_{99.7} = 37.8 \text{ mg/L}$$
    b. Similarly, the probability of occurrence of a given event with a frequency of once in 3 yr (i.e., 99.9%) is :
       i. For BOD
$$P_{99.9} = 52.6 \text{ mg/L}$$

ii. For TSS

$$P_{99.9} = 42.5 \text{ mg/L}$$

4. Estimate how often the annual effluent BOD and TSS values will exceed the effluent standard of 30 mg/L.
   a. From the plot presented in Step 2, the effluent BOD will exceed 30 mg/L approximately 4.5% of the time (~ 16 d/yr).
   b. From the plot presented in Step 2, the effluent TSS will exceed 30 mg/L approximately 2.0% of the time (~ 7 d/yr).

Comment
As found in step 4, the effluent BOD and TSS values will exceed the discharge limit of 30 mg/L about 4.5 and 2.0 percent of the time, respectively. If the BOD and TSS are not to exceed the effluent limits, then either the process will have to be designed for a lower mean value or some form of effluent filtration must be added to meet the discharge limits reliably. The impact of adding some form of filtration is considered in Chap. 8.

Importance of Secondary Sedimentation Tank Design

The importance of well-designed secondary sedimentation tanks can not be overemphasized in conventional secondary treatment. The importance of the design with respect to effluent quality in water reuse applications is illustrated on Fig. 7-7 in which the effluent particle size distribution for two clarifiers is plotted. The nature of the distribution

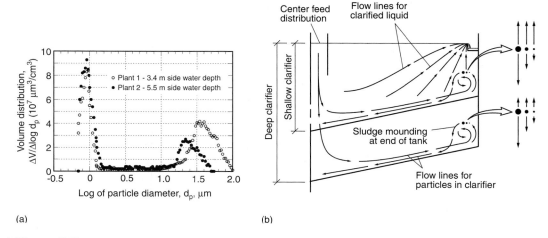

Figure 7-7
Particle removal performance in shallow and deep secondary clarifiers: (a) effluent particle size distribution and (b) flow lines in shallow and deep clarifiers. Note: In a shallow clarifier, the settling velocity of the equivalent small and medium size particles is less than the upward component of the fluid velocity. In a deep clarifier, the settling velocity of only the smallest particles is less than the upward component of the fluid velocity.

of particle sizes is bimodal as the volume distribution of particles has two peaks. As shown, the particle size data for small particles are similar for both treatment plants. However, the distribution of particle sizes for the medium and large particles is quite different, due primarily to the design of the sedimentation facilities, in particular the sedimentation tank sidewater depth. The mass percentage distribution between the two particle sizes will vary depending on the operating conditions of the biological process and the degree of flocculation achieved in the secondary settling facilities. The bimodal particle-size distribution has also been observed in water treatment plants. Other important factors that affect solids separation include flow distribution, tank inlet design, and weir placement, and loading, as discussed in Tchobanoglous et al. (2003).

Sidewater Depth

The importance of sedimentation tank sidewater depth can be understood by referring to Fig. 7-7b. In the shallow sedimentation tank, the upward velocity of the clarified effluent at the top of the sludge mound that forms at the end of the tank is greater than the downward velocity of both the small and medium sized particles. By comparison, in the deeper sedimentation tank, the upward velocity of the clarified effluent is only greater than that of the smaller sized particles. The importance of sidewater depth is also clearly illustrated on Fig. 7-7b in which the characteristics of effluent TSS settling velocities are given for shallow and deep clarifiers. Many existing activated sludge plants with shallow clarifiers will have difficulty meeting more stringent effluent standards without the addition of a follow-on process such as granular medium filtration. The significance of the particle size distribution with respect to subsequent processing is discussed in Sec. 8-1 in Chap. 8.

Other Physical Factors

Other physical factors that may affect the operation of the secondary sedimentation facilities include density flow, dead spaces, and wind driven circulation cells. As shown on Fig. 7-8, each of these conditions can lead to a reduction in the hydraulic detention time and to deterioration in the performance of the sedimentation tank with respect to the total mass of solids discharged and the corresponding particle size distribution. Additional details on the impact of these variables on sedimentation tank performance may be found in Tchobanoglous et al. (2003). To enhance the operation of downstream processes used for the removal of residual suspended solids for reuse applications, the design of sedimentation facilities is of critical importance. If effluent weirs are located at the tank perimeter in circular tanks or at the end walls in rectangular tanks, baffles should be provided to deflect the upward movement of particles away from the effluent weirs. Typical baffle arrangements used to deflect the upward velocity of the clarified effluent are found in Tchobanoglous et al. (2003).

Modification of Sedimentation Facilities

If greater levels of suspended solids removal are required that cannot be achieved consistently with existing sedimentation facilities, options are available for modifying the operation of the sedimentation facilities. Such modifications may include enhancing sedimentation tank performance by adding chemicals to the mixed liquor, modifying the inlet arrangement to dissipate energy and improve flow distribution, or by adding plate or tube settlers to an existing sedimentation tank. The installation of center flocculation wells may

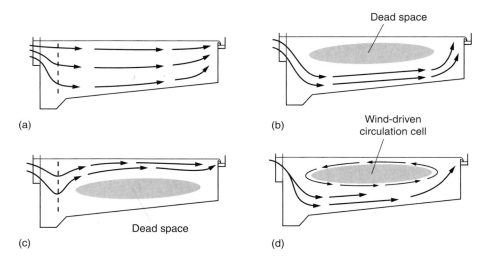

Figure 7-8

Effect of physical factors on flow patterns observed in rectangular sedimentation tanks: (a) ideal flow, (b) effect of density flow or thermal stratification (water in tank is warmer than influent), (c) effect of thermal stratification (water in tank is colder than influent), and (d) formation of wind-driven circulation cell.

be an option but it requires substantial modification to the clarifier mechanism. Each of the methods will improve performance, but using deep clarifiers with or without center flocculating wells will be the simplest to operate because no chemicals are involved. Filtration, as discussed in Chap. 8, is also an option for improving solids separation.

7-4 NONMEMBRANE PROCESSES FOR THE CONTROL AND REMOVAL OF NUTRIENTS IN SECONDARY TREATMENT

Nutrient removal is often required where reclaimed water is discharged to recreational and sensitive water bodies, used for groundwater recharge, or used for other reuse applications. The principal nutrients of concern are nitrogen and phosphorus. In selecting a technology for nutrient control and removal in water reuse applications, it is important to assess the characteristics of the untreated wastewater; the type of facility, if existing; and the level of nutrient removal required. The approaches used for nutrient control may involve the integration of nutrient removal with the main biological process, chemical addition, or adding a process for the removal of a specific nutrient. Technologies for (1) nitrogen control (nitrification), (2) nitrogen removal (nitrification/denitrification), (3) phosphorus removal, and (4) nitrogen and phosphorus removal are discussed below.

Nitrogen Control

Less than 30 percent of the total nitrogen in wastewater is removed by secondary treatment; therefore, additional treatment measures have to be undertaken for the control and removal of nitrogen. Nitrogen control can be accomplished by the conversion of ammonia to nitrate (nitrification); nitrogen removal can be done by nitrification/

denitrification. The term nitrification is used to describe the two-step biological process in which ammonia (NH_4^+-N) is oxidized to nitrite (NO_2^--N) and nitrite is oxidized to nitrate (NO_3^--N). The biological reduction of nitrate to nitric oxide (NO), nitrous oxide (N_2O), and nitrogen gas (N_2) is termed denitrification. Only a small amount of nitrogen is lost from the nitrification process, on the order of 5 to 20 percent, due to stripping or uptake. If nitrogen removal or reduction is required, a denitrification step must follow nitrification. Biological nitrogen removal is generally more cost effective and used more often than physical/chemical methods such as air stripping and ion exchange.

Nitrification can be accomplished biologically using either suspended or attached growth processes, although suspended growth is used most commonly. The principal reasons suspended growth is used are that (1) nitrification can be integrated conveniently in the design of the aerobic reactor and (2) operation of the process is relatively simple. The flow-through suspended growth processes in Table 7-4 can also be used for nitrification with modifications in their aeration configurations, aeration equipment design, solids retention time, and operating mode. The SBR can also be used for nitrification by modifying the operating cycle. The Biofor and Biostyr upflow submerged aerobic attached growth processes described in the following section can also be used for nitrification.

Nitrogen Removal

A variety of methods can be used for the removal of nitrogen and include (1) variations of the activated sludge process for nitrification/denitrification, (2) submerged attached growth processes, and (3) activated sludge with fixed film packing. Where low effluent nitrogen concentrations (e.g., less than 3 mg/L) are required, and an alternate nitrogen-free carbon source such as methanol is added to an anoxic/denitrification process that follows aerobic treatment.

Suspended Growth Processes
Several suspended growth processes can be used for biological nitrogen removal; the processes most commonly used are (1) Modified Ludzack-Ettinger (MLE), (2) step feed, (3) SBR, and (4) oxidation ditch. Each of these processes is described in Table 7-9. Each process represents modifications to the basic processes described in Table 7-4. The MLE process is a later development of the Ludzack-Ettinger process where the concept of anoxic-aerobic treatment was introduced for nitrogen removal. In MLE, internal recycle is added to increase the nitrate addition to the anoxic zone. In step feed a series of anoxic-aerobic compartments are used and wastewater feed is introduced to each of the anoxic compartments.

Submerged Attached Growth Processes
Two types of upflow submerged attached growth processes are the proprietary Biofor and Biostyr, shown in Table 7-10. Each can be used as an add-on process for BOD removal and nitrification, tertiary nitrification, or nitrogen removal. In general, the process consists of three phases: packing, biofilm, and liquid. The BOD and/or ammonia are oxidized as the applied wastewater flows past the biofilm that is attached to the packing. Oxygen is supplied by diffused aeration into the packing or by being predissolved into the influent wastewater. The type and size of packing is a major factor that affects the performance and operating characteristics of submerged attached growth processes. Designs differ by their packing configuration and inlet and outlet flow distribution and collection. No clarification is used with aerobic submerged attached growth processes, and excess solids

Table 7-9
Description of commonly used suspended growth process used for nitrogen removal[a]

Process	Description
(a) Modified Ludzack-Ettinger (MLE) 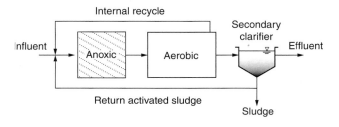	The MLE process is one of the most common methods used for biological nitrogen removal and can be adapted easily to existing activated sludge facilities. In the MLE process the initial contact of the wastewater and return activated sludge occurs in an anoxic zone. In a preanoxic configuration, as shown, the anoxic zone is located ahead of the aerobic zone. Nitrate produced in the aerobic zone is recycled to the preanoxic compartment. The amount of nitrate removal is limited by the practical levels of internal recycle to the preanoxic zone. The process is used more generally to achieve effluent total nitrogen concentrations between 5 and 10 mg/L.
(b) Step feed 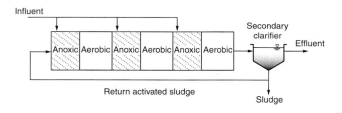	The step feed process is also applicable for meeting effluent total nitrogen concentrations of less than 10 mg/L. However, it is theoretically possible to achieve lower effluent nitrogen concentrations (3 to 5 mg/L) with internal recycle, such as in the MLE process, for the last pass of the anoxic-aerobic step feed process. The dissolved oxygen (DO) concentration in the aerobic zone must be controlled to minimize DO returned to the anoxic zone.
(c) Sequencing batch reactor (SBR) 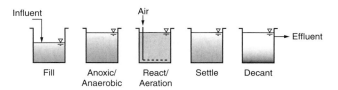	The SBR process provides a high degree of flexibility for nitrogen removal. Mixing during the fill period provides an opportunity for anoxic conditions for nitrate removal. During the aeration react period, the DO concentration may be cycled to provide anoxic operating periods. Batch decant reactor designs are slightly less flexible than the SBR processes for BOD removal and nitrification because they depend on internal recycle like the MLE for a major portion of the nitrate removal.

Table 7-9
Description of commonly used suspended growth process used for nitrogen removal[a] (*Continued*)

Process	Description
(d) Oxidation ditch	Several modifications of the oxidation ditch configuration are used for nitrogen removal. In the configuration shown, an anoxic zone is created at a point where the DO is depleted and nitrate is used for endogenous respiration by the mixed liquor. Because of the large tank volumes and long solids retention times (SRTs), sufficient capacity is available for nitrification and denitrification zones. Other process configurations use on-off operation of the aerators to create anoxic-aerobic conditions (Nitrox) or simultaneous nitrification/denitrification (Sym-Bio).

[a]Adapted from Tchobanoglous et al. (2003).

Table 7-10
Description of attached growth processes for nitrification and denitrification[a]

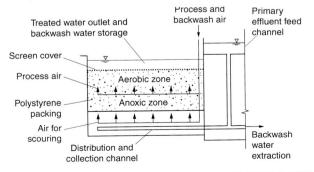

Process	Description
Biofor	The Biofor process is an upflow submerged aerobic attached growth process. The upflow reactor has a typical bed depth of 3 m. The packing is an expanded clay material with a density greater than 1.0 and a 2 to 4 mm size range. The Biofor process has been used for BOD removal and nitrification, tertiary nitrification, and denitrification.
Biostyr	The Biostyr process is an upflow process that uses 2 to 4 mm polystyrene beads having a specific gravity less than water. The bed can be operated entirely aerobic by providing air at the bottom or as an anoxic/anaerobic bed by providing air at an intermediate level. Nitrified effluent is recycled for anoxic/anaerobic operation. The Biostyr process has been applied for BOD removal only, BOD removal and nitrification, tertiary nitrification, and postdenitrification.

[a]Adapted from Tchobanoglous et al. (2003).

from biomass growth and influent suspended solids are trapped in the system and must be periodically removed. Most designs require a backwashing system much like that used at a water filtration plant to flush out accumulated solids, usually on a daily basis.

The major advantages of submerged attached growth processes are their relatively small space requirement, the ability to effectively treat dilute wastewaters, no sludge settling issues as in activated sludge process, and adaptability for incorporating nitrogen removal. Also for many processes, solids filtration occurs to produce a high-quality effluent. Their disadvantages include a more complex system in terms of instrumentation and controls, limitations of economies of scale for application to larger facilities, and generally a higher capital cost than activated sludge treatment.

Activated Sludge with Fixed Film Packing

Several synthetic packing materials have been developed for use in activated sludge processes. These packing materials may be suspended in the activated sludge mixed liquor or fixed in the aeration tank. A term used to describe these types of processes is an *integrated fixed film activated sludge process* (Sen et al., 1994). These processes are intended to enhance the activated sludge process by providing a greater biomass concentration in the aeration tank and thus offer the potential to reduce the basin size requirements. They are also used to improve volumetric nitrification rates and to accomplish denitrification in aeration tanks by having anoxic zones within the biofilm depth. Because of the complexity of the process and issues related to understanding the biofilm area and activity, the process designs are empirical and based on prior pilot-plant or limited full-scale results. Typical examples of suspended packing processes are Captor, Linpor, and Kaldnes and are described in Table 7-11.

Phosphorus Removal

Phosphorus removal is required only in reclaimed water applications where special circumstances such as the development of aquatic growths or biofouling of process equipment in industrial use are of concern. For decentralized applications, phosphorus removal is accomplished typically by filtration through reactive media or chemical precipitation. Biological phosphorus removal (BPR) is not well suited for decentralized applications because it adds a level of operating complexity that may not be practicable (see Chap. 13). Where phosphorus removal is required for water reuse applications, it can be accomplished by configuring an anaerobic contact zone or compartment ahead of the aerobic or anoxic zone or by adding chemical treatment.

Biological Phosphorus Removal

Certain phosphorus accumulating bacteria can be used to remove phosphorus under specific redox conditions as part of the activated sludge process. After the phosphorus uptake has occurred, these organisms must be removed or wasted from the process to accomplish phosphorus removal. Barnard (1975) used the term *Phoredox* to represent any process with an anaerobic/aerobic sequence to promote biological phosphorus removal (BPR). Various modifications to the basic Phoredox process are used for both biological phosphorus and nitrogen removal. Some of the process names that have evolved to designate specific process configurations include A/O (anaerobic/aerobic only) and A^2O (anaerobic/anoxic/aerobic). The A/O process (see Fig. 7-9a) is similar to the Phoredox process and was patented and marketed by Air Products and Chemicals, Inc.,

Table 7-11
Description of activated sludge processes with fixed-film packing for nitrogen removal[a]

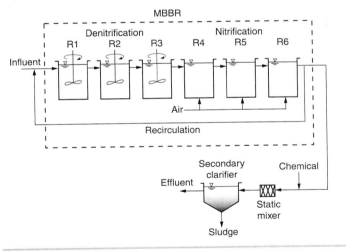

Process	Description
Captor and Linpor	In the Captor and Linpor processes foam pads are placed in the bioreactor in a free-floating fashion and retained by an effluent screen. The pad volume can account for 20–30% of the reactor volume. Mixing from the diffused aeration system circulates the pads in the system, but without additional mixing methods they may tend to accumulate at the effluent end of the aeration basin and float at the surface. An air knife is used to continuously clean the screen and a pump is used to return the packing material to the influent end of the reactor. Based on the results of full-scale and pilot-scale tests with the sponge packing installed, nitrification appears to occur at lower SRTs than those for activated sludge without internal packing.
Kaldnes	The process, termed a *moving bed biofilm reactor (MBBR)*, was developed by a Norwegian company, Kaldnes Miljøteknlogi. The process consists of adding small cylindrical shaped polyethylene carrier elements in aerated or nonaerated basins to support biofilm growth. The biofilm carriers are maintained in the reactor by a perforated plate at the tank outlet. Air agitation or mixers are applied in a manner to continuously circulate the packing. The packing may fill 25 to 50% of the tank volume. The MBBR does not require any return activated sludge flow or backwashing. A final clarifier is used to settle sloughed solids. For the anoxic-aerobic treatment mode, a six-stage reactor design is used.

[a] Adapted from Tchobanoglous et al. (2003).

along with the A²O process. The main difference between the Phoredox (A/O) process and the A²O process (see Fig. 7-9b) is that nitrification does not occur in the Phoredox (A/O) process. The A²O process is one of the basic types used for nitrate removal along with BPR. Views of anaerobic reactors for phosphorus removal are shown on Fig. 7-10. As shown on Fig. 7-9c, phosphorus removal has also been adapted to the MBR (discussed in Sec. 7-5) by the addition of an anaerobic zone.

Figure 7-9

Typical biological phosphorus removal processes: (a) Phoredox (A/O), (b) A²O, and (c) membrane bioreactor with anaerobic zone for phosphorus removal.

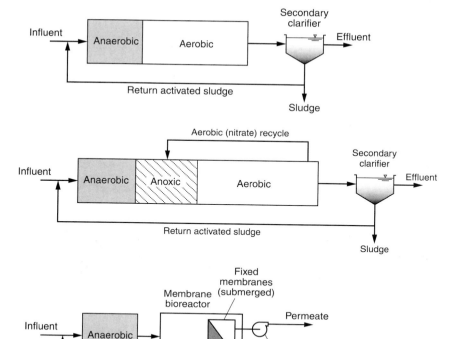

Figure 7-10

Views of biological phosphorus removal processes: (a) separate anaerobic stage and (b) continuous process with contiguous (from front to back) anaerobic, anoxic, and aerobic sections.

Phosphorus Removal by Chemical Addition

The removal of phosphorus from wastewater involves the incorporation of phosphate into the TSS and the subsequent removal of those solids. Phosphorus removal is brought about by the addition of salts of the multivalent metal ions that form precipitates of sparingly soluble phosphates. The multivalent metal ions that are used most commonly are calcium [Ca(II)], aluminum [Al(III)], and iron [Fe(III)]. Polymers have been used effectively in conjunction with alum and lime as flocculent aids.

The precipitation of phosphorus from wastewater can occur in a number of different locations within a process flow diagram (see Fig. 7-11). The general locations where

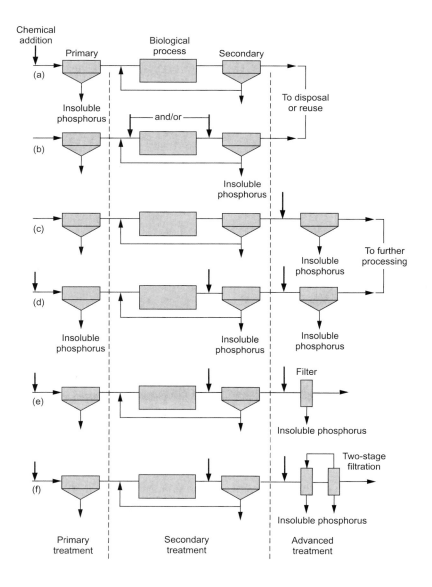

Figure 7-11

Alternative points of chemical addition for phosphorus removal: (a) before primary sedimentation, (b) before and/or following biological treatment, (c) following secondary treatment, and (d–f) at several locations in a process known as *split treatment*.

phosphorus can be removed may be classified as (1) preprecipitation—the addition of chemical for precipitation in the primary sedimentation tanks, (2) coprecipitation—the addition of chemicals to form precipitates that are removed with waste biological sludge, and (3) postprecipitation—the addition of chemicals to the secondary effluent for removal in subsequent sedimentation or filtration facilities. Chemicals can be added in more than one location, depending on the degree of removal of phosphorus required. The advantages and disadvantages of chemical addition at various locations in the wastewater treatment process are given in Table 7-12. For more information on the chemical removal of phosphorus, the companion text, Tchobanoglous et al. (2003), may be consulted.

Process Performance Expectations

In general, the performance of activated sludge processes that incorporate biological nutrient removal is better for the removal of BOD and TSS than conventional activated sludge processes.

Effluent Constituent Values

Typical effluent constituent values that can be achieved with biological nutrient removal (BNR) are reported in Table 7-7. In treatment plants that nitrify only, effluent ammonia concentrations as low as 1 mg N/L can be achieved. In treatment plants that both nitrify and denitrify, effluent total nitrogen concentrations in the range of 2 to 10 mg N/L can be achieved, depending on the chemical characteristics of the wastewater. Where postanoxic denitrification is accomplished using methanol, effluent nitrogen concentrations less than 3 mg N/L are attainable. It should be noted that treatment facilities located in cold climates have difficulty in nitrifying during wintertime. In BNR plants, effluent phosphorus levels on the order of 1 to 2 mg P/L can be achieved without chemical addition. With chemical addition, effluent phosphorus levels as low as 0.1 mg P/L or lower can be achieved. Generally, chemical precipitation of phosphorus is used in most small plants where phosphorus removal is required.

Variability in Effluent Constituents

The effluent variability observed in BNR processes is similar to that for the activated sludge processes (see Table 7-8). In many cases a BNR process will be selected over a conventional activated sludge process because more stable operation can be achieved. Also with chemical addition for phosphorus removal, the observed variability is generally reduced.

7-5 MEMBRANE BIOREACTOR PROCESSES FOR SECONDARY TREATMENT

One of the newer and most promising technologies for utilization in water reuse systems is the MBR. Membrane bioreactors combine biological treatment with an integrated membrane system to provide enhanced organics and suspended solids removal. Membranes function to replace sedimentation and depth filtration for separating the biomass in suspended growth systems from the treated water. By coupling a biological reactor with a membrane system as shown on Fig. 7-2b, conventional treatment operations can be eliminated such as gravity sedimentation and media filtration that might be used to produce an equivalent effluent. With an MBR, overall space requirements and facilities costs can be reduced. A smaller "footprint" allows MBR plants to be located in sites with limited area or completely enclosed in residential areas for satellite treatment applications.

Table 7-12
Advantages and disadvantages of chemical addition with metal salts and lime at various locations of a treatment plant for phosphorus removal[a]

Location of chemical addition	Advantages	Disadvantages
Addition of metal salts		
Prior to primary sedimentation	Increases BOD and TSS removal; lowest degree of metal leakage	Least efficient use of metal; polymer may be required for flocculation; sludge more difficult to dewater than primary sludge
Prior to secondary sedimentation	Lowest cost; lower chemical dosage than prior to primary sedimentation; less effect on pH than lime; available mixing in activated sludge aeration system is utilized; improves stability of activated sludge; polymer may not be required	Overdose of metal can cause pH toxicity; increases TDS in reclaimed water
Post secondary treatment with clarification	Low phosphorus content in effluent; most efficient metal use	Additional flocculation and sedimentation step required thus increasing capital cost; highest metal leakage; increases TDS in reclaimed water
Post secondary treatment with filtration (single or two stage)	Lowest phosphorus content in effluent	Length of filter run may be reduced with single-stage filtration; two-stage filtration more costly
Multiple locations	Provides greatest flexibility in optimizing phosphorus removal	More complicated chemical distribution and control system
Addition of lime		
Prior to primary sedimentation	Increased BOD and TSS removal thereby reducing load in aeration tanks; lime recovery demonstrated	Because excessively high pH interferes with the biological process, lime addition is limited to a pH of about 9.0; soluble phosphorus level is 2 to 3 mg/L
Prior to secondary sedimentation	Lower chemical dosage than prior to primary sedimentation; biological system breaks down complex phosphates to more readily precipitated orthophosphate form	Inert solids added to mixed liquor, reducing the percentage of volatile solids; high pH or returned solids may affect biological treatment performance
Post secondary treatment with clarification	Low phosphorus content in effluent; lime recovery demonstrated	Additional flocculation and sedimentation step required; with high pH or low alkalinity in treated wastewater, recarbonation is required
Post secondary treatment with filtration (single or two-stage)	Lowest phosphorus content in effluent	Length of filter run may be reduced with single-stage filtration; two-stage filtration is more costly; solids settling may be more difficult instage first-settling

[a] Adapted from WEF (1998).

Membranes were used initially for the desalination of brackish water and seawater and began to be used on wastewater experimentally in the 1970s. The first trials of using hollow fibers for the liquid/solids separation of activated sludge were done by Yamamoto et al. (1989). The researchers used the concept of immersing the membrane in a bioreactor and withdrawing permeate through the membrane by suction. Subsequent testing of a submerged MBR on domestic wastewater indicated that consistently high removals of turbidity, COD, and nitrogen could be obtained when operating under diurnal flow patterns typical in domestic households (Chiemchaisri et al., 1993). The submerged MBR concept is the basis of many of the MBR systems that have been developed and are used currently for wastewater treatment and water reuse applications.

Description of Membrane Bioreactors

Membrane bioreactors come in several different configurations because most MBRs are of proprietary design and have distinctively different features. Typical configurations are illustrated on Fig. 7-12. A common feature of most bioreactor systems is a low-pressure

Figure 7-12
General types of membrane bioreactors:
(a) with external pressure-driven membrane,
(b) integrated submerged, (c) with external submerged, and
(d) with external submerged rotating membrane.

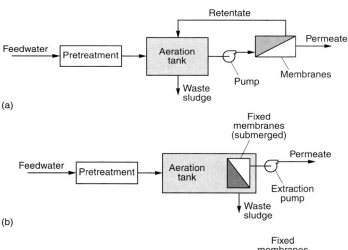
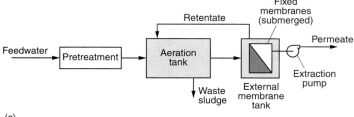
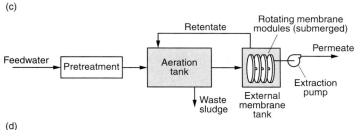

membrane system (e.g., MF or UF) that is used for liquid/solids separation. The advantages of membranes over conventional clarification are: (1) ancillary facilities are much smaller; (2) higher quality product water can be obtained; (3) increased MLSS produces a more stable sludge that is less susceptible to upsets; (4) in small MBR plants, return sludge systems can be eliminated or greatly reduced; and (5) systems are simpler to operate.

In the various types of MBR systems, the key component is the MF or UF membrane. The types of MF and UF membranes used commonly in MBRs are either hollow fiber or fixed plate and are described in Table 8-18 and shown on Fig. 8-28 in Chap. 8. The membranes may be either pressure driven or vacuum driven. Pressure-driven membranes are installed external to the bioreactor (see Fig. 7-12a) and the mixed liquor from the bioreactor is pumped to the membranes. Pressure-driven membranes are manufactured commonly in a tubular configuration and are referred to as external MBRs (EMBRs) (Trussell et al., 2005). To maintain permeability and improve performance, a pretreatment device such as a fine screen or a cloth-media filter is installed ahead of the membrane unit.

Vacuum-driven membranes may be immersed directly into the activated sludge reactor or in a separate membrane separation tank (see Figs. 7-12b–d). The membranes are subjected to a vacuum (less than 50 kPa) that draws water (permeate) through the membrane while retaining solids in the reactor or the membrane separation tank. To clean the exterior of the membranes, air is introduced below the membranes. As the air bubbles rise to the surface, scouring of the membrane surface occurs and rejected material is returned to the mixed liquor.

Membrane biological reactor technology promises to be one of the most important treatment devices for wastewater treatment and water reclamation and reuse. In a recent survey, over 1000 MBRs are in operation worldwide with many more proposed or under construction. MBRs have proliferated in Japan, which has approximately two-thirds of the world's total installations. Over 98 percent of MBR systems employ an aerobic biological reactor with membrane filtration (van der Roest et al., 2002).

Although most MBR installations using submerged membranes are less than 10 years old, the technology is advancing rapidly and more suppliers are entering the marketplace. Membrane bioreactors are well suited for satellite systems or for supplementing existing treatment plant capacity. Membrane bioreactors are also important in integrated water resource management for the production of high-quality reclaimed water to replace potable water now used for nonpotable purposes such as landscape irrigation and process cooling.

To introduce MBR technologies and their applications, the following subjects are considered in this section: (1) suitability of MBRs for reclaimed water applications, (2) types of MBR systems, (3) performance expectations, (4) proprietary submerged membrane systems, and (5) other membrane systems. Other applications of membrane systems used for the removal of residual particulate matter and dissolved constituents are discussed in Chaps. 8 and 9, respectively.

Suitability of MBRs for Reclaimed Water Applications

The principal advantages of MBR systems for water reuse applications are: (1) a high level of water quality is produced consistently and (2) because of their compact size,

MBRs can be sited close to the points of potential reuse. Because the membrane units have small pore sizes, usually ranging from 0.04 to 0.4 µm, highly clarified product water is produced that is low in BOD, TSS, turbidity, and bacteria, similar to effluent from secondary clarification followed by MF (see Chap. 8). The product water quality is suitable for a variety of reclaimed water applications and, after disinfection, can be utilized for a number of nonpotable, unrestricted uses (see Chap. 6). Membrane bioreactors can also incorporate treatment for nutrient removal in applications where nutrients, in particular nitrogen, must be controlled.

In planning water reuse facilities, one of the major considerations is the cost of the infrastructure necessary to store, transport, and deliver reclaimed water to the points of use. Because MBR facilities can be located strategically to intercept wastewater in upstream portions of the collection systems (see Chap. 12), wastewater can be withdrawn for treatment and reuse locally. The following benefits accrue from this concept: (1) reuse opportunities can be developed that previously were not feasible and (2) local needs for reclaimed water are met more economically because of reduced requirements for pipelines, pumping, and storage.

Types of Membrane Bioreactor Systems

Currrently, over 50 percent of the MBR systems used worldwide have the membranes submerged in the bioreactor, while the remaining systems have the membranes located external to the biological process (van der Roest, 2002; Judd and Judd, 2006). In submerged systems, extraction of the permeate through the membrane by suction from a permeate pump maintains the integrity of the floc particles thus improving membrane performance and solids separation.

Four of the commonly used MBR configurations are shown on Fig. 7-12. On Fig. 7-12a, an external membrane system is shown following the aeration tank. Mixed liquor from the aeration tank is pumped to and through the membrane unit. The membrane is backwashed periodically to remove material accumulated on the membrane surface. Submerged membrane filtration has been developed in different configurations; three basic types are shown on Figs. 7-12b–d. The types are (1) integrated immersed membranes installed in fixed modules (Fig. 7-12b), (2) fixed membrane modules installed in an external membrane separation vessel (Fig. 7-12c), and (3) rotating membrane modules (Fig. 7-12d). Each configuration type has certain distinct advantages:

- Type 1 can be installed in an existing aeration tank, thus saving space.
- Type 2 can be designed to utilize fine pore diffusers in the aeration tank to improve energy efficiency, and coarse bubble diffusers in the membrane compartment for membrane scouring and fouling control.
- Type 3 can utilize rotation of the membrane modules to assist in membrane fouling control and minimize air requirements for coarse bubble diffusers. The rotating bioreactor shown on Fig. 7-12d is usually installed in a separate membrane tank but it can also be adapted for installation in the bioreactor.

As compared to conventional suspended growth systems, MBRs have the following advantages: (1) because MBRs operate with higher suspended solids concentrations, the reactor hydraulic retention times are shorter, thus reducing the reactor size; (2) longer SRTs, on the order of two to three times those for conventional processes, result in less sludge production, more stable operation, and less chance for process upsets; and

(3) simultaneous nitrification-denitrification can be achieved through process control when longer SRTs are combined with lower dissolved oxygen (DO) concentrations in the bioreactor. Disadvantages of MBRs include (1) high capital costs for the membrane modules; (2) limited data on membrane life, thus a potential high recurring cost of periodic membrane replacement; (3) higher energy costs due to membrane scouring as compared to conventional suspended-growth processes; (4) potential membrane fouling that affects the ability to treat design flows; and (5) waste sludge from the membrane process may be more difficult to dewater.

Principal Proprietary Submerged Membrane Systems

Because most MBR installations are less than 10 yr old, there is a limited number of suppliers of the process equipment for reclaimed water; however, many others are entering or poised to enter the marketplace. There are four principal suppliers of proprietary submerged MBRs in the U.S.: Zenon Environmental, a part of GE Water and Process Technologies; Kubota Corporation; Mitsubishi Rayon Corporation; and USFilter, a subsidiary of Siemens. Each type of proprietary membrane installation discussed in this section differs in the type of membrane used and membrane arrangement. The characteristics of the various types of MBRs are summarized in Table 7-13 and described below. The first three suppliers, Zenon, Kubota, and Mitsubishi, market MBRs generally of the integrated type, although Zenon's membranes have been installed in a separate membrane tank in Traverse City, Michigan (Crawford et al., 2005). USFilter furnishes membrane modules for installation in a separate membrane vessel. The essential features of each membrane system are discussed below. Different methods of membrane fouling control, which varies with each type of proprietary MBR design, are discussed in a separate subsection.

Another supplier, Huber Technology, manufactures a rotating type of MBR. As Huber has not received approval in California to meet Title 22 requirements as of the writing of this text (2006), they are not considered as a principal MBR supplier for reuse applications. Description of the Huber system is provided for information purposes as it has been used in other locations. Huber has been evaluated in side-by-side pilot testing in Hawaii (Babcock, 2005).

Zenon Environmental

The Zenon system, called Zenogem, utilizes cassettes composed of tubular hollow-fiber membrane modules that are submerged in compartmentalized cells within an activated sludge bioreactor. Each cassette has overall dimensions of 0.91 m wide by 2.13 m long, and approximately 2.44 m high (see Fig. 7-13). The cassettes contain groups of 32 to 48 membrane modules; each module has a membrane surface area of 32 m^2. The configuration allows cleaning of the membrane cassettes without their removal from the bioreactor. Support facilities required include permeate pumps, chemical storage tanks, chemical feed pumps, and process controls. The membrane support system also includes an air-scour system and a backpulse water flushing system. The air-scour system consists of air compressors or blowers and coarse bubble diffusers located in the aeration basin. The air-scour system provides continuous agitation on the outside of the membranes to minimize solids deposition. The air supply for the air-scour system is typically provided in addition to the activated sludge process air. Additional information on membrane maintenance and cleaning is provided later in this chapter in the "Membrane Fouling Control and Cleaning" subsection.

Table 7-13
Characteristics of various proprietary MBR systems[a]

Manufacturer	Zenon	Kubota	Mitsubishi	USFilter	Huber
Membrane Type	Hollow fiber, UF	Plate, MF	Hollow fiber, UF	Hollow fiber, UF	Plate, UF
Configuration	Fixed, vertical	Fixed, vertical	Fixed, horizontal	Fixed, vertical	Rotating disks
Pore size, μm	0.04	0.4	0.04	0.04	0.04
Surface area/module, m^2	31.6	0.8	105	9.3	3
Modules/element	—	—	—	—	6 or 8
Estimated membrane life, yr	8	Unknown	Unknown	5–10	Unknown
Location	In basin or cell compartment	In basin or cell compartment	Throughout basin	Cell compartment	In basin or cell compartment
Pretreatment screening size, mm	1–2	≤3	1–2	1–2	≤3
Membrane aeration					
Type	Coarse bubble	Coarse bubble	Coarse bubble	Jet aeration	Coarse or fine bubble
Aeration cycle	10 s on/10 s off	9 s on/1 s off	Constant	Constant	Constant
Flux rate					
Average, $L/m^2 \cdot h$	17–25	17–25	8.5–17	17–25	16–22
Peak (≤6 h), $L/m^2 \cdot h$	<37	<73	<55	<51	<55
Maintenance cleaning (in addition to air scour)					
Type	Relax and Cl backpulse	Relax	Relax	Cl backpulse	None
Length of time, min	1, 2, or 3	1, 2, or 3	1, 2, or 3	—	—
Frequency	10 min/as needed	10 min	10 min	Weekly	—
Recovery cleaning					
Type	Chemical soak	Chlorine backwash	Chlorine backwash	Chemical soak	Chlorine backwash
Location	Drained cell	*In situ*	*In situ*	Drained cell	*In situ*
Frequency, mo	6	6–12	3	3	As needed
Biological parameters					
SRT, d	12–15	15	20	12–15	15–20
MLSS, mg/L	≤10,000	≤10,000	≤15,000	≤10,000	12,000–16,000

[a] Adapted in part from Crawford (2002); Wallis-Lage (2003); Adham and DeCarolis (2004); Babcock (2005); and Judd and Judd (2006).

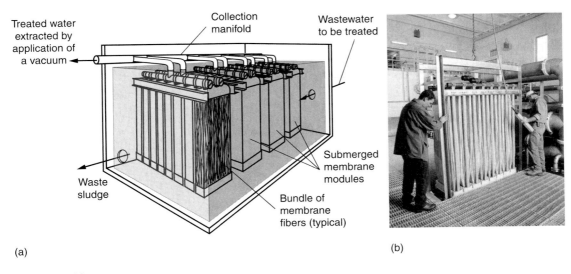

Figure 7-13
Typical Zenon submerged membrane bioreactor: (a) schematic of placement of bundles in an activated sludge reactor and (b) membrane bundle in position to be placed in a membrane bioreactor. (Courtesy of Zenon Environmental a division of GE Water and Process Technologies.)

Kubota
The membrane unit shown on Fig. 7-14 and manufactured by Kubota in Japan is marketed by Enviroquip, Inc. in the United States. The system utilizes one or more membrane panels; each panel has overall dimensions of 490 mm wide by 1000 mm high by 6 mm thick with an effective surface area of 0.8 m². The membrane panels are lined up adjacent to each other at 8 mm center-to-center dimensions to form a cassette. A cassette can contain up to 200 panels in a single stack or 400 panels if double stacked. The cassettes are installed in single file, perpendicular to the flow in multiple plug flow basins. In designs for smaller installations, the panels can be removed individually for cleaning and replacement. Each of the cassettes has a lower diffuser case and an upper membrane case. The diffuser case houses a coarse diffuser manifold that is designed to distribute air for membrane cleaning. The membrane case has a series of tubes that extract effluent from the membranes and connect to a collection manifold. For larger facilities, the membrane panels are fixed to a common header. The permeate support facilities are similar to those for the Zenon system.

Mitsubishi
The Mitsubishi MBR unit, manufactured in Japan and marketed in the U.S. by Ionics, a division of GE Water and Process Technologies, uses a 0.4 μm hollow fiber Sterapore membrane. The membranes are arranged horizontally and attached at both ends to permeate lines. Each membrane module has a total surface area of 105 m² and the modules can be stacked in up to three layers. A low-pressure air stream is used to maintain a turbulent flow pattern across the membrane fibers. Permeate is drawn through the membranes by a partial vacuum in a method similar to the Zenon and Kubota units. The membranes are "relaxed" periodically by turning off the permeate pump and allowing aeration to continue.

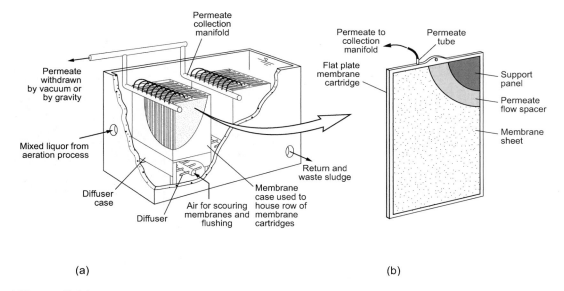

Figure 7-14
Typical Kubota submerged membrane bioreactor: (a) schematic of placement of membrane units in an activated sludge reactor and (b) view of a membrane cartridge.

USFilter

The USFilter MBR system, termed MemJet, uses hollow fiber, immersed membrane modules installed in an open basin separate from the aeration tank (see Fig. 7-15). The system utilizes multiple racks with up to 40 modules installed in a rack. Operation consists of introducing mixed liquor from the aeration tank and air at the bottom of each membrane module through a two-phase jet. The jet system provides both fluid transfer and air-scour energy to help keep the membrane from fouling. The air bubbles blend with the mixed liquor and rise through the membrane bundle, providing scouring energy to the membrane surface as well as fluidizing the membrane surface to prevent solids accumulation. The two-phase jet introduces fluid consistently to all membranes in the system by dividing the membrane modules into narrow fiber bundles that allow air and fluid to move up between the individual membrane fibers.

Advantages cited for the MemJet system are: (1) independent optimization of the biological and membrane processes is allowed, (2) chemical cleaning of the membranes is performed in-place without the necessity of removing the membrane modules from the membrane basin, and (3) all membranes are cleaned uniformly. The disadvantage of this system is that in-place cleaning requires all or a portion of the membrane system to go offline for a period of 4 to 6 h, thus interrupting the production of reclaimed water for short periods. In applications where continuous production of reclaimed water is not required, shutdowns for cleaning should not affect water reuse significantly. Where continuous production of reclaimed water is required, multiple units would have to be used.

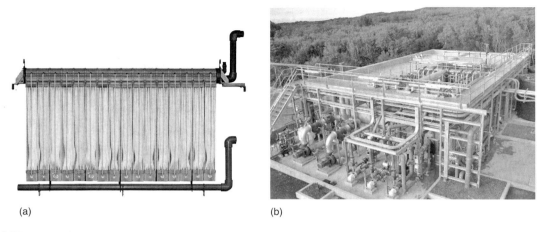

Figure 7-15
Views of a US Filter membrane technology: (a) filter module and (b) aerial view of small membrane bioreactor. (Courtesy of Siemans Water Technologies Corporation.)

Huber Technology

Huber Technology, a subsidiary of Huber Germany, markets the rotating MBR in the United States. The Huber VRM unit consists of UF membrane disks mounted on a support frame and submerged in either an aeration tank or a separate compartment or tank (see Fig. 7-16a). A stack of four membrane plates form a module, and six or eight membrane modules are arranged around a rotating hollow shaft to form a disk-shaped

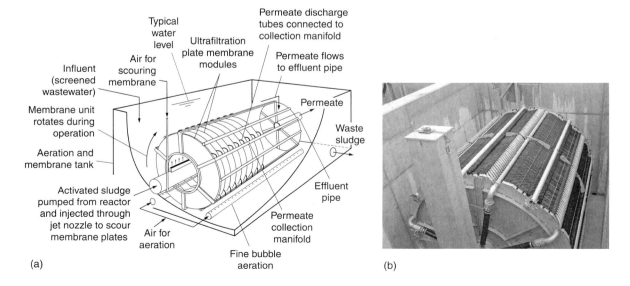

Figure 7-16
Huber Technology membrane bioreactor: (a) schematic diagram and (b) view of interior of the reactor showing rotating membrane elements. (Courtesy of Huber Technologies.)

membrane element (see Fig. 7-16b). Up to 60 elements are mounted on a rotating shaft and permeate is drawn through the membranes and effluent collectors by a permeate pump. Transmembrane pressure is maintained less than 30 kPa. Fouling is controlled by an air-scour system that helps remove solids buildup on the membrane elements. Coarse bubble aeration is used commonly for aeration and air scour. A separate backwash system using permeate is not required; the membranes are allowed to relax periodically by turning off the permeate pump.

Where the membranes are installed in a compartment separate from the aeration tanks, concentrated activated sludge is returned to the aeration tanks to control the MLSS concentration in the membrane tank. Activated sludge in turn is pumped from the aeration tank to the membrane tank through jet nozzles on the hollow shaft of the membrane unit, thus generating a scouring sludge flow between the membranes.

Other Membrane Systems

In some MBR configurations, largely for industrial applications, pressurized cross-flow membrane filters are used following the activated sludge process to achieve solids separation. Mixed liquor from the aeration tanks is pumped under pressure to external cross-flow membranes, similar to those used in water treatment operations for softening or demineralization. Retentate is cycled back to the aeration tank. This method of solids separation requires pumping energy to generate sufficient scouring velocity across the membrane surface for cleaning and to provide sufficient pressure for permeation. In industrial applications, the membrane modules are generally small, and pumping energy is less of a consideration.

Other types of membrane systems are entering the marketplace and are undergoing demonstration testing to satisfy requirements of approving authorities. These new products are based mainly on proprietary devices. Brief descriptions of some of these products are presented below.

Sequencing Batch Reactor/Cloth Filter/Microfiltration Process (AquaMB Process)

In the AquaMB Process, marketed by Aqua-Aerobic Systems, solids separation is accomplished by using a cloth filter and pressure-driven membrane filtration following biological treatment in a sequencing batch reactor (SBR) (see Fig. 7-17). The SBR

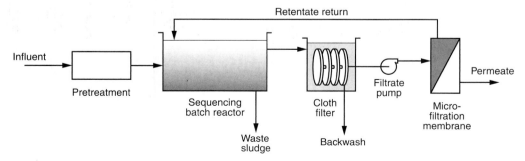

Figure 7-17

Membrane system using sequencing batch reactor, cloth filter, and microfiltration.

process, described in Table 7-4, utilizes a fill-and-draw reactor. After the treatment and settling, clarified liquid is decanted and discharged to a cloth filter for removal of fine suspended solids. The cloth filter (see Chap. 8) serves as a pretreatment step prior to membrane filtration. Effluent from the cloth filter is then pumped to a membrane separation unit for further removal of residual particulate matter. Either MF or UF can be used for solids separation. As described previously in this section, membrane separation may also be accomplished using vacuum or pressure type membranes.

Airlift Membrane Bioreactor

An airlift membrane bioreactor (AL-MBR) has been developed by Norit, a Netherlands company, for municipal wastewater reuse applications. Wastewater is treated in an aerated bioreactor and then discharged to a specially designed external membrane unit. The external membrane unit consists of a series of 3 m long vertical tubes that contain tubular membranes. The unit differs from the Zenon, Kubota, and Mitsubishi membrane systems in that the membranes are not immersed in mixed liquor. Mixed liquor from the bioreactor is introduced into the tubular modules, which are internally aerated. The filtration surface is a 0.03 mm pore size ultrafiltration polyvinylidene fluoride (PVDF) membrane on a tubular support. Low-pressure air and low-pressure sludge circulation are used to maintain a turbulent flow pattern along the membrane tubes and help prevent contaminant buildup (van der Roest, 2002). The system differs from that shown on Fig. 7-12a in that permeate is drawn through the membrane by applying a partial vacuum on the outside of the tube bundle. The membranes are back-pulsed periodically by pumping some of the permeate back through the membrane tubes to remove fouling material from the membrane surfaces.

Koch/Puron Membrane Bioreactor

Koch Membrane Systems has developed a hollow fiber membrane that is submerged in a tank separate from the biological reactor. The proprietary Puron membrane modules consist of bundles of long, hollow fibers (see Fig. 7-18). The fibers have pores of approximately

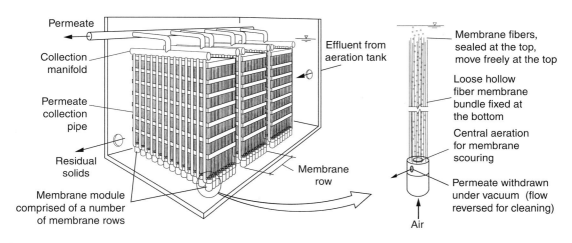

Figure 7-18
Koch Puron submerged hollow fiber membrane bioreactor. (Adapted from Koch Industries, Inc.)

0.05 μm in size. The membrane module uses a single-header design with the hollow fibers fixed only at the bottom that allows them to move freely along their entire length. The top ends of the fibers are sealed. Permeate is drawn under vacuum through the fiber walls from the outside to the inside. The bundles of fibers are mounted vertically in modules and compressed air is introduced in the center at the base of the bundle to scour the fibers and remove accumulated solids. The fibers are also backflushed periodically with permeate. A large-scale MBR plant (2000 m^3/d) is in operation in Belgium for treating wastewater from a malt production operation, and an MBR demonstration facility was placed in operation in 2003 for treating municipal wastewater (www.kochmembrane.com).

Process Performance Expectations

Typical effluent constituent concentrations from MBR processes and their variability are considered below. Additional performance data can be obtained from the manufactures.

Effluent Constituent Values

Because the membrane serves as barrier to the discharge of larger colloidal and suspended particles, the concentration of suspended solids is often unmeasurable (e.g., less than 1 mg/L) due to the limitations of the TSS test. The solids that accumulate on the membrane surface are important in the removal of soluble organic matter (Trussell et al., 2004).

Variability in Effluent Constituents

The observed variability in the degree of treatment achieved with membrane processes is due to (1) the variability in the influent wastewater constituent concentrations, (2) the concentration of soluble and colloidal COD, (3) the operation of the biological treatment process and especially the SRT (see Table 7-8), and (4) the presence of broken fibers or compromised membranes. The growth of biological films on the downstream piping and appurtenances will also contribute to the measured variability in the particulate matter.

7-6 ANALYSIS AND DESIGN OF MEMBRANE BIOREACTOR PROCESSES

The process analysis and design considerations for MBR systems are presented in this section. The process analysis is similar to that used for the design of suspended growth activated sludge systems, and the kinetic equations are basically the same. The kinetic equations and related coefficients are provided in this section for convenience of reference for use in examples and for homework problems.

Process Analysis

The process analysis for MBRs is based on water quality issues, kinetic equations, and kinetic coefficients developed for special applications. It is not the intent of this text to reproduce the theory of biological treatment design that is covered in detail in the companion text, Tchobanoglous et al. (2003). The basic kinetic equations are required, however, for determining the design of the suspended growth reactor used in most MBR designs. The basic kinetic equations and coefficients required in reactor design are covered in this section. The methodology and equations for determining biological solids production are also provided.

Key Wastewater Constituents

Wastewater characteristics are important in the design of activated sludge systems, particularly for biological nutrient-removal processes and for evaluating the capacity of an existing system. A partial listing of parameters used commonly in wastewater characterization and equations used in their computation are given in Table 7-14. A complete listing may be found in the companion text, Tchobanoglous et al. (2003).

Water Quality Issues

Water quality issues encompass both influent and effluent water quality. Characteristics of the feed stream to MBRs are important because they affect the design and performance of the bioreactor and the membranes, particularly if nutrient removal is required. Effluent (permeate) water quality requirements affect the need for chemical pretreatment or posttreatment.

Influent Water Quality Considerations Influent water quality characterization is important because (1) targeted constituents that need to be removed have to be identified and quantified and (2) contaminants that inhibit performance of the membranes and require pretreatment need to be determined.

Because of the fine pore sizes of the membranes, special attention has to be paid to the wastewater characteristics in the feed stream to the membrane units. Typical constituents that affect membrane performance are presented in Table 7-15. Some of the physical constituents such as settleable and suspended solids can be controlled by screening, grit removal, and primary clarification in conventional centralized treatment plants, but in satellite and decentralized facilities, fine screening and/or surface filters are needed to limit the buildup of material on the membranes. Biological byproducts from the suspended growth process, such as extracellular polymeric substances (EPS) and soluble

Table 7-14
Summary of equations used commonly in wastewater characterization[a]

Equation	Eq. No.	Definition of terms
$nbVSS = [1 - (bpCOD/pCOD)] \, VSS$	(7-1)	$bCOD$ = biodegradable COD, mg/L
$bpCOD = \dfrac{(bCOD/BOD)(BOD - sBOD)}{COD - sCOD}$	(7-2)	$bpCOD$ = biodegradable particulate COD, mg/L
		$iTSS$ = inert TSS, mg/L
$COD = bCOD + nbCOD$	(7-3)	$nbCOD$ = nonbiodegradable COD, mg/L
$bCOD = {\sim}1.6(BOD)$	(7-4)	$nbpCOD$ = nonbiodegradable particulate COD, mg/L
$nbCOD = nbsCOD + npbCOD$	(7-5)	$nbsCOD$ = nonbiodegradable soluble COD, mg/L
$bCOD = sbCOD + rbCOD$	(7-6)	$nbVSS$ = nonbiodegradable VSS, mg/L
$iTSS = TSS - VSS$	(7-7)	$pCOD$ = particulate COD, mg/L
		$rbCOD$ = readily biodegradable COD, mg/L
		$sBOD$ = soluble BOD, mg/L
		$sCOD$ = soluble COD, mg/L
		$sbCOD$ = slowly biodegradable COD, mg/L

[a]Adapted from Tchobanoglous et al. (2003).

Table 7-15
Wastewater constituents that affect the performance of membrane bioreactors

Type of constituent	Specific constituent	Effect on MBR
Physical	High concentration of TSS (>30 mg/L), hair, fibrous material, and other inert solids	Buildup on membrane surfaces that may cause reduced membrane efficiency, physical damage to membranes, and ability to maintain membrane cleaning. May decrease permeate quality.
	Temperature variations	Affects water viscosity and flux rate.
Chemical	High alkalinity Soluble iron	Membrane fouling that may require acid cleaning to remove chemical foulants.
	Oil and grease	Membrane fouling causing diminished performance and more frequent cleaning.
	Surfactants	Foaming that requires cleanup.
	Oxidants, e.g., ozone and chlorine	Attacks certain types of membrane material.
Biological	Dissolved and colloidal organic matter	Membrane fouling causing diminished performance and more frequent cleaning.
	Extracellular polymeric substances (EPS)	Clogs membrane pores resulting in diminished membrane performance and more frequent cleaning; also affects viscosity of sludge.

microbial products (SMP), may also cause fouling, especially at low SRTs (~2 d) (Trussell et al., 2004). Some of the biological foulants and trace constituents can be controlled by increasing the SRT in the bioreactor, thus promoting better adsorption of these constituents onto the biological floc particles. High MLSS concentrations can also cause severe membrane fouling.

Effluent Water Quality Considerations Because MF and UF membranes are effective in producing product water low in BOD, COD, TSS, and turbidity, other effluent water quality issues focus typically on nutrient values, virus levels, and total dissolved solids concentrations. In each case, development of the process train must consider elements such as the types of nutrient removal processes, disinfection systems, and posttreatment. Many process options are available as discussed in this and other chapters and have to be selected carefully for the specific application. For example, where total dissolved solids reduction is necessary, nanofiltration or reverse osmosis will be required following the MBR (see Chap. 9). If control of trace constituents such as NDMA or perchlorate is required, posttreatment with advanced oxidation may have to be considered (see Chap. 10).

Kinetic Equations
Kinetic relationships are used to model biomass growth and substrate utilization, and to define process performance. Important kinetic relationships used commonly in the analysis and design of suspended growth processes are given in Table 7-16. Derivation of the kinetic equations is given in Tchobanoglous et al. (2003).

Table 7-16
Summary of equations used commonly in the analysis of suspended growth processes[a]

Equation	Eq. No.	Definition of terms
$k_T = k_{20}\theta^{T-20}$	(7-8)	DO = dissolved oxygen concentration, ML^{-3}
$r_{su} = -\dfrac{kXS}{K_s + S}$	(7-9)	F/M = food-to-microorganism ratio
$\mu_m = kY$	(7-10)	f_d = fraction of cell mass remaining
		k = maximum rate of substrate utilization, T^{-1}
$r_{su} = -\dfrac{\mu_m XS}{Y(K_s + S)}$	(7-11)	k_d = endogenous decay coefficient, T^{-1}
		k_{dn} = endogenous decay coefficient for nitrifying organisms, T^{-1}
		k_T = reaction rate coefficient at temperature (T)
$r_g = Y\dfrac{kXS}{K_s + S} - k_d X$	(7-12)	k_{20} = reaction rate coefficient at 20°C
		K_n = half-velocity constant for nitrification, ML^{-3}
		K_o = half-saturation coefficient for DO, ML^{-3}
		K_o' = oxygen inhibition coefficient, ML^{-3}
$\mu = \dfrac{r_g}{X}$	(7-13)	K_s = half-velocity constant, ML^{-3}
		K_{s,NO_3} = half-velocity constant for denitrification, ML^{-3}
$SRT = \dfrac{VX}{(Q - Q_w)X_e + Q_w X_R}$	(7-14)	L_{org} = volumetric organic loading rate, $ML^{-3}T^{-1}$
$SRT = \dfrac{1}{\mu}$	(7-15)	μ = specific growth rate, T^{-1}
		μ_m = maximum specific growth rate, T^{-1}
		μ_n = specific growth rate for nitrification, T^{-1}
$\dfrac{1}{SRT} = -\dfrac{YkS}{K_s + S} - k_d$	(7-16)	μ_{nm} = maximum specific growth rate of nitrifying bacteria, T^{-1}
		N = NH_4-N concentration, ML^{-3}
$S = \dfrac{K_s[1 + (k_d)SRT]}{SRT(Yk - k_d) - 1}$	(7-17)	NO_3 = nitrate nitrogen concentration, ML^{-3}
		η = ratio of substrate utilization rate with nitrate versus oxygen a the electron acceptor
$X = \left(\dfrac{SRT}{\tau}\right)\left(\dfrac{Y(S_o - S)}{1 + (k_d)SRT}\right)$	(7-18)	P_X = solids, M
$(X_{VSS})(V) = (P_{X,VSS})SRT$	(7-19)	Q = flow rate, L^3T^{-1}
$(X_{TSS})(V) = (P_{X,TSS})SRT$	(7-20)	Q_w = waste sludge flow rate L^3T^{-1}
$R_o = Q(S_o - S) - 1.42 P_{X,bio}$	(7-21)	R_o = oxygen, MT^{-1}
		r_g = net biomass production rate, $ML^{-3}T^{-1}$
		r_{su} = soluble substrate utilization rate, $ML^{-3}T^{-1}$
$F/M = \dfrac{QS_o}{VX}$	(7-22)	S = concentration of growth limiting substrate in solution, ML^{-3}
		S_o = influent concentration, ML^{-3}
$L_{org} = \dfrac{(Q)(S_o)}{(V)}$	(7-23)	SRT = solids retention time, T
		TSS = total suspended solids, M
		τ = hydraulic retention time (V/Q), T
$\mu_n = \left(\dfrac{\mu_{nm}N}{K_n + N}\right)\left(\dfrac{DO}{K_o + DO}\right) - k_{dn}$	(7-24)	θ = temperature activity coefficient
		V = volume, L^3

(Continued)

Table 7-16
Summary of equations used commonly in the analysis of suspended growth processes[a] (*Continued*)

Equation	Eq. No.	Definition of terms
$r_{su} = \left(\dfrac{kXS}{K_s + S}\right)\left(\dfrac{NO_3}{K_{s,NO_3} + NO_3}\right)\left(\dfrac{K'_o}{K'_o + DO}\right)\eta$	(7-25)	VSS = volatile suspended solids, M X = biomass concentration, ML^{-3} X_e = concentration of biomass in the effluent, ML^{-3} X_R = concentration of biomass in the return line from clarifier, ML^{-3} X_{VSS} = volatile solids mass in reactor, ML^{-3} X_{TSS} = total solids mass in reactor, ML^{-3} Y = biomass yield, M of cell formed per M of substrate consumed Y_n = g biomass produced/g NH_4-N utilized

[a] Adapted from Tchobanoglous et al. (2003).
Note: Expressions for units are M = mass, L = length, and T = time.

The kinetic equations for the design of nitrification with activated sludge are also given in Table 7-16. Because nitrification kinetics are critical process design parameters, bench-scale or in-plant testing should be undertaken to evaluate the potential for site-specific toxicity and rate inhibition. Aeration tank volume requirements and SRT values are directly related to nitrification μ_m values.

Coefficients

Typical kinetic coefficients to be used with the equations in Table 7-16 for the removal of carbonaceous material (based on biodegradable COD) by heterotrophic bacteria in conventional suspended growth processes are given in Table 7-17. The μ_m and K_s values given in Table 7-17 are the default values recommended in the International Association of Water Pollution Research and Control (IAWPRC) Activated Sludge Model (ASM) 1 (Henze et al., 1987). The IAWPRC model is the commonly accepted standard model

Table 7-17
Activated sludge design coefficients for heterotrophic bacteria at 20°C[a]

Coefficient	Unit	Range	Typical value
μ_m	g VSS/g VSS·d	3.0–13.2	6.0
K_s	g bCOD/m^3	5.0–40.0	20.0
Y	g VSS/g bCOD	0.30–0.50	0.40
k_d	g VSS/g VSS·d	0.06–0.20	0.12
f_d	Unitless	0.08–0.20	0.15
θ values			
μ_m	Unitless	1.03–1.08	1.07
k_d	Unitless	1.03–1.08	1.04
K_s	Unitless	1.00	1.00

[a] Adapted from Henze et al. (1987); Barker and Dold (1997); and Grady et al. (1999).

Table 7-18
Activated sludge design nitrification kinetic coefficients at 20°C[a]

Coefficient	Unit	Range	Typical value
μ_{nm}	g VSS/g VSS · d	0.50–0.90	0.85
K_n	g NH_4-N/m^3	0.5–1.0	0.70
Y_n	g VSS/g NH_4-N	0.10–0.15	0.12
k_{dn}	g VSS/g VSS · d	0.05–0.17	0.17
θ values			
μ_n	unitless	1.06–1.123	1.072
K_n	unitless	1.03–1.123	1.053
k_{dn}	unitless	1.02–1.08	1.029

[a]Adapted from Melcer et al. (2003).

for the activated sludge process. Typical kinetic coefficients for activated sludge nitrification are given in Table 7-18.

Because the replacement of the sedimentation process with membranes alters the selection pressure upon the microbial population, differences in the biological kinetic constants have been noted. Factors such as changes in the shear forces acting on the floc particles, varying mixing, and mass transfer conditions also affect kinetic activity. Because limited information is available, kinetic coefficients to be used in process analysis should be confirmed by test data from comparable facilities or by pilot testing prior to design. The coefficients in Tables 7-17 and 7-18 should be used only as a guide.

Biological Solids Production

Included with the design of a suspended growth process is the determination of the biological solids (biosolids) production. Determination of the solids production is important for two reasons: (1) for process control and (2) for the design of subsequent solids processing facilities, if required. Two methods are used to determine biosolids production as described in the following paragraphs.

Estimating Biosolids Production Based on Published Data The first method is based on an estimate of an observed solids yield from published data from similar facilities, and the second is based on the actual activated sludge process design in which wastewater characterization is done and the various sources of solids are considered and accounted for. With the first method, the quantity of sludge produced daily (and thus wasted daily) can be estimated using Eq. (7-26) (Tchobanoglous et al., 2003):

$$P_{X,VSS} = Y_{obs}(Q)(S_o - S)(1 \text{ kg}/10^3 \text{ g}) \tag{7-26}$$

where $P_{X,VSS}$ = net waste activated sludge produced each day, kg VSS/d
Y_{obs} = observed yield, g VSS/g substrate removed
Q = influent flow, m^3/d
S_o = influent substrate concentration, g/m^3 (mg/L)
S = effluent substrate concentration, g/m^3 (mg/L)

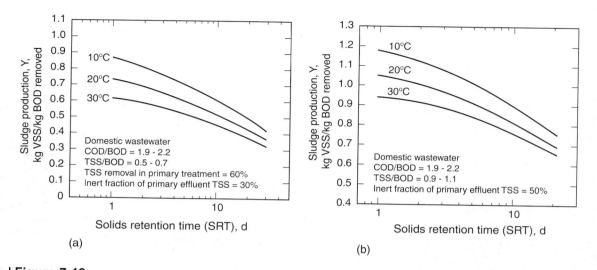

Figure 7-19

Net solids production vs. solids retention time (SRT) and temperature: (a) with primary treatment and (b) without primary treatment.

Observed volatile suspended solids (VSS) yield values, based on BOD, are illustrated on Fig. 7-19. The observed yield decreases as the SRT is increased due to biomass loss by increased endogenous respiration. The yield is lower with increasing temperature as a result of a higher endogenous respiration rate at higher temperature. The yield is higher when no primary treatment is used as more settleable organic matter and nonbiodegradable VSS (nbVSS) remains in the influent wastewater. The temperature correction value θ for endogenous respiration in the reaction rate equation, Eq. (7-8), is 1.04 between 20 and 30°C, and 1.12 between 10 and 20°C (Tchobanoglous et al., 2003).

Estimating Biosolids Production Using Kinetic Coefficients With sufficient wastewater characterization, a more accurate prediction of biosolids production can be made. The following equation accounts for the heterotrophic biomass growth, cell debris from endogenous decay, nitrifying bacteria biomass, and nonbiodegradable VSS, and can be used to estimate sludge production (Tchobanoglous et al., 2003).

$$P_{X,VSS} = \underbrace{\frac{QY(S_o - S)(1\ kg/10^3\ g)}{1 + (k_d)SRT}}_{\text{(A) Heterotrophic biomass}} + \underbrace{\frac{(f_d)(k_d)QY(S_o - S)SRT(1\ kg/10^3\ g)}{1 + (k_d)SRT}}_{\text{(B) Cell debris}} +$$

$$\frac{Q\ Y_n(NO_X)(1\ kg/10^3\ g)}{1 + (k_{dn})SRT} + Q(nbVSS)(1\ kg/10^3\ g) \qquad (7\text{-}27)$$

(C)   (D)

Nitrifying bacteria   Nonbiodegrdable

biomass   VSS in influent

where P_X, VSS = total mass of volatile suspended solids produced per day, kg VSS/d
Y = heterotrophic biomass yield, g biomass produced/g substrate utilized
k_d = endogenous decay coefficient, g VSS/g VSS·d
f_d = fraction of cell mass remaining, g/g
SRT = solids retention time, d
NO_x = concentration of $NH_4\text{-}N$ in the influent flow that is nitrified, g/m³
Y_n = nitrifier biomass yield, g biomass produced/g $NH_4\text{-}N$ utilized
k_{dn} = endogenous decay coefficient for nitrifying organisms, g VSS/g VSS·d
nbVSS = nonbiodegradable volatile suspended solids, g/m³

As previously mentioned, typical values of kinetic coefficients for the removal of carbonaceous material by heterotrophic organisms are given in Table 7-17 and typical coefficients for nitrification are given in Table 7-18.

The total mass of dry solids wasted per day must also include the influent inert TSS fraction (TSS includes VSS plus inorganic solids). Inorganic solids in the influent wastewater ($TSS_o - VSS_o$) contribute to the waste solids and are an additional solids production term that must be added to Eq. (7-27). The biomass terms in Eq. (7-27) (A, B, and C) contain inorganic solids and the VSS fraction of the total biomass is about 0.85, based on cell composition (Tchobanoglous et al., 2003). Thus, Eq. (7-27) is modified as follows to calculate the solids production in terms of TSS.

$$P_{X,TSS} = \frac{A}{0.85} + \frac{B}{0.85} + \frac{C}{0.85} + D + Q(TSS_o - VSS_o) \qquad (7\text{-}28)$$

(E)

Inert TSS

in influent

where $P_{X,TSS}$ = total mass of solids wasted per day, kg TSS/d
TSS_o = influent wastewater TSS concentration, g/m³
VSS_o = influent wastewater VSS concentration, g/m³

The daily mass of solids in the aeration tanks is determined from the SRT. The daily sludge production can be computed using Eq. (7-19) in Table 7-16. The total solids mass can be computed using Eq. (7-20) in Table 7-16. By selecting an appropriate MLSS concentration, the aeration reactor volume can be determined using Eq. (7-20).

EXAMPLE 7-3. Estimate the Solids Produced in a Conventional Activated Sludge Process as Compared to a Membrane Bioreactor.

Determine the solids produced in a conventional activated sludge reactor that treats 18,900 m³/d of primary effluent and compare the results to those produced in an MBR. The reactors are designed for BOD removal only.

The following wastewater characteristics apply:

Constituent	Concentration, g/m³
BOD	140
sBOD	70
COD	300
sCOD	132
rbCOD	80
TSS	70
VSS	60

Note: g/m³ = mg/L

The following design conditions and assumptions apply:
1. The SRT for the activated sludge process is 5 d and the SRT for the MBR is 15 d
2. The aeration basin mixed-liquor temperature is 12°C
3. Use the kinetic coefficients from Table 7-17
4. bCOD/BOD = 1.6

Solution

1. Develop wastewater characteristics and calculate the mass of VSS and TSS in the aeration basin of the activated sludge process
   a. Develop the wastewater characteristics
      i. Find bCOD using Eq. (7-4)

$$bCOD = S_o = 1.6 \, (BOD)$$
$$= 1.6(140 \text{ g/m}^3) = 224 \text{ g/m}^3$$

   ii. Find nbVSS using Eq. (7-1):

$$nbVSS = (1 - bpCOD/pCOD)VSS$$

$$\frac{bpCOD}{pCOD} = \frac{(bCOD/BOD)(BOD - sBOD)}{COD - sCOD}$$

$$\frac{bpCOD}{pCOD} = \frac{1.6(BOD - sBOD)}{COD - sCOD} = \frac{1.6[(140 - 70) \text{ g/m}^3]}{[(300 - 132) \text{ g/m}^3]} = 0.67$$

$$nbVSS = (1 - 0.67)(60 \text{ g VSS/m}^3) = 20 \text{ g/m}^3$$

iii. Find iTSS using Eq. (7-7)

   iTSS = TSS − VSS = (70 − 60) g/m³ = 10 g/m³

iv. Determine S, the concentration of growth limiting substrate in solution, using Eq. (7-17) in Table 7-16

$$S = \frac{K_s[1 + (k_d)SRT]}{SRT(Yk - k_d) - 1}$$

Note: $Yk = \mu_m$
Use μ_m, K_s, and k_d from Table 7-17: μ_m = 6.0 g/g·d; K_s = 20 g/m³;

$$k_d = 0.12 \text{ g/g·d}$$

Determine μ_m at T = 12°C using Eq. (7-8) and θ = 1.07

$$\mu_{m,T} = \mu_{m,\,12°C}\,\theta^{T-20} = 6.0 \text{ g/g·d}(1.07)^{12-20} = 3.5 \text{ g/g·d}$$

Determine k_d at T = 12°C using Eq. (7-8) and θ = 1.04

$$k_{d,T} = k_{d,12°C}\,\theta^{T-20} = 0.12 \text{ g/g·d}(1.04)^{12-20} = 0.088 \text{ g/g·d}$$

$$S = \frac{(20 \text{ g/m}^3)[1 + (0.088 \text{ g/g·d})(5 \text{ d})]}{(5 \text{ d})[(3.5 - 0.088) \text{ g/g·d} - 1]} = 1.8 \text{ g bCOD/m}^3$$

b. Determine mass of VSS and TSS in the activated sludge reactor.
   i. Determine the daily biomass production using parts (A) and (B) in Eq. (7-27). The term (C) = 0 as there is no nitrification, and the term (D) is not considered as the nbVSS is not part of the biological solids produced in the reactor. Use Y = 0.40 g VSS/g bCOD and f_d = 0.15

$$P_{X,\,VSS} = \frac{QY(S_o - S)(1 \text{kg}/10^3 \text{g})}{1 + (k_d)SRT}$$

$$+ \frac{(f_d)(k_d)QY(S_o - S)SRT(1 \text{ kg}/10^3 \text{ g})}{1 + (k_d)\,SRT}$$

$$= \frac{(18{,}900 \text{ m}^3/\text{d})(0.40 \text{ g/g})[(224 - 1.8) \text{ g/m}^3(1 \text{ kg}/10^3 \text{ g})]}{[1 + (0.088 \text{ g/g·d})(5 \text{ d})]}$$

$$+ \frac{(0.15 \text{ g/g})(0.088 \text{ g/g·d})(0.40 \text{ g/g})(18{,}900 \text{ m}^3/\text{d})[(224 - 1.8) \text{ g/m}^3](5 \text{ d})(1 \text{ kg}/10^3 \text{ g})}{[1 + (0.088 \text{ g/g·d})(5 \text{ d})]}$$

PX, VSS 5 (1166.5 1 77.0) kg/d 5 1243.5 kg VSS/d

ii. Determine mass of VSS in the aeration tank using Eq. (7-19) in Table 7-16

Mass of VSS in aeration tank = $(X_{VSS})(V) = P_{X,VSS}$ (SRT)

Determine $P_{X,VSS}$ using Eq. (7-27) for terms (A), (B), and (D). The term (D) is included as the nbVSS contributes to the overall VSS budget

From Eq. (7-27), $P_{X,VSS}$ is:

$$P_{X,VSS} = 1243.5 \text{ kg/d} + Q(nbVSS)(1 \text{ kg}/10^3 \text{ g})$$
$$= 1243.5 \text{ kg/d} + (18,900 \text{ m}^3/\text{d})(20 \text{ g/m}^3)(1 \text{ kg}/10^3 \text{ g})$$
$$= (1243.5 + 378) \text{ kg/d} = 1621.5 \text{ kg/d}$$

Calculate the mass of VSS in the aeration tank

$$\text{Mass of VSS in the aeration tank} = (P_{X,VSS})\text{SRT}$$
$$= (1621.5 \text{ kg/d})(5 \text{ d}) = 8107.5 \text{ kg}$$

iii. Determine the mass of TSS in the aeration tank using Eq. (7-20) in Table 7-16

$$\text{Mass} = (X_{TSS})(V) = P_{X,TSS}(\text{SRT})$$

From Eq. (7-28), $P_{X,TSS}$ is:

$$P_{X,TSS} = [(1243.5 \text{ kg/d})/0.85] + (378 \text{ kg/d}) + Q(TSS_o - VSS_o)$$
$$= 1463 \text{ kg/d} + 378 \text{ kg/d} + (18,900 \text{ m}^3/\text{d})(10 \text{ mg/L})(1 \text{ kg}/10^3 \text{ g})$$
$$= 2030 \text{ kg/d}$$

Calculate the mass of TSS in the aeration tank

$$\text{Mass of TSS in the aeration tank} = (P_{X,TSS})\text{SRT}$$
$$\text{MLTSS} = (2030 \text{ kg/d})(5 \text{ d}) = 10,150 \text{ kg}$$

2. Develop wastewater characteristics and calculate the mass of VSS and TSS for an MBR. Use a computation procedure similar to Step 1.
   a. Develop the wastewater characteristics

$$S_o = 224 \text{ g/m}^3 \text{ (see Step 1a)}$$

Compute S following the same procedure as Step 1b using SRT = 15 d

$$S = \frac{(20 \text{ g/m}^3)[1 + (0.088 \text{ g/g} \cdot \text{d})(15 \text{ d})]}{(15 \text{ d})[(3.5 - 0.088) \text{ g/g} \cdot \text{d} - 1]} = 1.3 \text{ g bCOD/m}^3$$

   b. Determine mass of VSS and TSS in the MBR.
      i. Substitute values in Eq. (7-27) for terms (A) and (B) and solve for the biological solids production component of $P_{X,VSS}$

$$P_{X, VSS} = \frac{(18,900 \text{ m}^3/\text{d})(0.40 \text{ g/g})[(224 - 1.3) \text{ g/m}^3(1 \text{ kg}/10^3 \text{ g})]}{[1 + (0.088 \text{ g/g} \cdot \text{d})(15 \text{ d})]}$$

$$+ \frac{(0.15 \text{ g/g})(0.088 \text{ g/g} \cdot \text{d})(0.40 \text{ g/g})(18,900 \text{ m}^3/\text{d})[(224 - 1.3) \text{ g/m}^3](15 \text{ d})(1 \text{ kg}/10^3\text{g})}{[1 + (0.088 \text{ g/g} \cdot \text{d})(15 \text{ d})]}$$

$$P_{X, VSS} = (725.7 + 143.7) \text{ kg/d} = 869.4 \text{ kg VSS/d}$$

ii. Determine mass of VSS in the bioreactor using Eq. (7-19) in Table 7-16.

Mass of VSS in the bioreactor = $(X_{VSS})(V) = P_X$ (SRT)

Determine $P_{X,VSS}$ using Eq. (7-27) for terms (A), (B), and (D). Again, the term (D) is included as the nbVSS contributes to the overall VSS budget.
From Eq. (7-27), $P_{X,VSS}$ is:

$$P_{X,VSS} = 869.4 \text{ kg/d} + Q(\text{nbVSS})(1 \text{ kg}/10^3 \text{ g})$$

$$= 869.4 \text{ kg/d} + (18{,}900 \text{ m}^3/\text{d})(20 \text{ g/m}^3)(1 \text{ kg}/10^3 \text{ g})$$

$$= (869.4 + 378) \text{ kg/d} = 1247.4 \text{ kg/d}$$

Mass of VSS in the bioreactor = $(P_{X,VSS})\text{SRT}$

$$= (1247.4 \text{ kg/d})(15 \text{ d}) = 18{,}711 \text{ kg}$$

iii. Determine mass of TSS in the bioreactor using Eq. (7-20) in Table 7-16.

Mass of TSS in the bioreactor = $(X_{TSS})(V) = P_{X,TSS}$ (SRT)

From Eq. (7-20), $P_{X,TSS}$ is:

$$P_{X,TSS} = [(869.4 \text{ kg/d})/0.85] + (378 \text{ kg/d}) + Q(\text{TSS}_o - \text{VSS}_o)$$

$$= 1022.8 \text{ kg/d} + 378 \text{ kg/d} + (18{,}900 \text{ m}^3/\text{d})(10 \text{ mg/L})(1 \text{ kg}/10^3 \text{ g})$$

$$= 1589.8 \text{ kg/d}$$

Mass of TSS = $(P_{X,TSS})\text{SRT} = (1589.8 \text{ kg/d})(15 \text{ d}) = 23{,}847 \text{ kg}$

3. Summarize results

Parameter	Units	Activated sludge	Membrane bioreactor
Daily biological solids production	kg VSS/d	1243.5	869.4
Mass of daily VSS produced in the aeration tank and bioreactor and influent of nonbiodegradable VSS[a]	kg VSS/d	1621.5	1247.8
SRT	d	5	15
Total mass of VSS in reactor	kg VSS	8107.5	18,711
Total mass of TSS in reactor	kg TSS	10,150	23,847

[a] Mass of VSS that has to be wasted daily.

Comment

Because of the longer SRT, the biological solids production in the MBR is reduced by 30 percent as compared to the activated sludge process. The total amount of VSS that must be wasted daily is also reduced by 23 percent. As a result, a similar downsizing of the solids processing facilities can be expected.

Process Variables

The principal process variables in the design and operation of MBR systems include temperature, pore size, membrane flux rate, membrane life, bioreactor suspended solids concentration, and solids and hydraulic retention times.

Temperature

The temperature of the incoming wastewater is a consideration in assessing membrane performance, as temperature affects the viscosity of the permeate and the concentrate (the biomass in the reactor). The pores of the membrane are very small and as the viscosity of the water increases with decreasing temperature, the driving force needed to achieve the required flux will increase, thus reducing permeability and the flux rate. The effects of temperature change are more pronounced during peak flow conditions, especially during wet weather. Where significant changes in peak flows and operating temperature are expected, pilot plant testing prior to design is valuable in assessing MBR performance, including fouling, under a range of operating conditions. Under test conditions, permeability over different periods of time can be correlated with temperature changes, which in turn permits the development of operating strategies for varying flow and temperature conditions.

Pore Size

Most membrane systems use either MF or UF membranes with pore sizes ranging from 0.04 to 0.4 μm; MF membranes have the larger pore size. Microfiltration membranes are used typically to remove relatively large particles such as suspended solids, emulsified oils, and macromolecules with an approximate molecular mass greater than 10^7 amu (see Fig. 8-1 in Chap. 8). In general, the UF membranes are able to achieve higher levels of separation, particularly for bacteria and viruses. Ultrafiltration membranes can generally separate macromolecules greater than 10^5 AMU.

Membrane Flux Rate

The membrane flux rate, defined as the mass or volume rate of transfer through the membrane surface (in terms of $L/m^2 \cdot h$) is an important design and operating parameter that affects the process economics and operating conditions. Typical flux rates for various proprietary systems are reported in Table 7-13. Lower flux rates are expected at higher MLSS concentrations and lower temperatures. The design flux rate is typically for the peak day and peak hour, which can be attenuated by peak flow management. As indicated in Table 7-13, the flux rate at peak hour can vary significantly, therefore, peak flowrates and membrane peak flux rates need to be evaluated carefully to ensure design hydraulic conditions can be maintained in the treatment facility without excessive membrane fouling, backups, overflows, or bypassing. Flow equalization, discussed in the Design Considerations subsection, is an option for controlling high peak flows where necessary.

Membrane Life

Because MBRs have been used for a relatively short time, limited data are available on the life of the membranes. Membrane life is affected by the characteristics of the wastewater being treated, the type and frequency of cleaning method employed, and especially the effectiveness of pretreatment. Estimated membrane life for some of the equipment suppliers is reported in Table 7-13.

Bioreactor Suspended Solids Concentration

By replacing gravity settling in secondary clarifiers with membrane systems, issues of filamentous sludge bulking and other floc settling and clarification problems are avoided. Aeration tank MLSS concentrations are no longer controlled by secondary clarifier solids loading limitations. The MBR systems can operate at much higher MLSS concentrations (up to 15,000 mg/L) than conventional activated sludge processes (Cote et al., 1998). Although high concentrations of MLSS have been reported, MLSS concentrations in the range of 8000 to 12,000 mg/L appear to be most cost effective when all factors are considered, especially membrane cleaning. Typically, MBRs are designed for MLSS concentrations of 8000 to 10,000 mg/L in aeration tanks and 10,000 to 12,000 mg/L in membrane tanks. Ranges of MLSS concentrations for various proprietary systems are given in Table 7-13. Rapid membrane fouling occurs at MLSS concentrations above 15,000 to 20,000 mg/L (Trussell et al., 2005).

Because of the higher MLSS concentrations in the bioreactor, solids management is an important consideration in limiting the caking of solids on the membrane surfaces. However, solids buildup, especially in separate tanks used to house the membranes, has to be controlled as membrane caking may intensify. In the USFilter system that uses a small membrane separation tank, mixed liquor solids in the membrane tank are recycled to the aeration tank. Recycle rates are typically four times the influent flow to limit solids accumulation (Wallis-Lage, 2003). If a separate membrane tank is used, a mass balance analysis can be used to determine the method of controlling the MLSS in the membrane tank.

Solids can be wasted from either the aeration basin or the separate membrane tank, if one is used. Scum removal also has to be provided as oil, grease, and foam will accumulate over time and should be removed to prevent odor generation and unsightliness. Oil and grease will also affect membrane performance. Solids wasting from the surface is an effective way to remove objectionable surface scum and maintain process control (Trussell et al., 2004).

Solids and Hydraulic Retention Times

Values for SRTs in the range of 10 to 20 d are customary as is a hydraulic retention time, τ, of about 4 h. As discussed below in design considerations, the alpha (α) factors for aeration also decrease as the MLSS values increase.

Design Considerations

Design considerations include pretreatment, air supply for the MBR, membrane fouling control and cleaning, peak flow management, solids production and management, nutrient removal, and biosolids processing.

Pretreatment

Because membrane systems in municipal wastewater treatment applications are sensitive to damage by debris and decreased efficiency of coarse bubble aeration devices, pretreatment is required to remove coarse solids such as plastics, leaves, and rags and fine particulate and colloidal matter including oils, fats, and hair from the wastewater. Pretreatment that can be used to prevent macro-fouling includes coarse screening with primary treatment or alternatively by fine screens (see Chap. 8). Because fine solids and especially hair can accelerate membrane clogging, fine screens or surface filters should be considered to protect the membranes. Screens with openings less than 1 up to 3 mm are becoming more common for MBR installations. Pretreatment has other benefits in reducing inert solids and organic matter loadings to the bioreactor. Primary treatment will remove about 30 percent of the organic matter and about 60 percent of the TSS. Fine screens will remove about 10 to 15 percent of the COD. The percentage of inert solids is also lower in primary effluent as compared to screened wastewater. As discussed in the section Biosolids Production and Management, the influent substrate concentration (as measured by BOD) and the inert solids concentration affect the total solids production and the wasting requirements from the bioreactor.

Air Supply

For MBR processes, the design of the aeration system requires careful consideration. The air supply for the MBR process consists of the aeration air to sustain the biological process and the air for cleaning the membranes. Turbulence induced by coarse bubble aeration creates a cross-flow velocity in the vicinity of the membrane module. The airflow rate or the airflow intensity (the airflow rate per unit area) can affect cleaning of the membranes, MLSS concentration, and flux rate. In experiments by Trussell et al. (2005), it was found that higher aeration rates increased the MLSS concentrations that resulted in significant specific flux decline. The flux decline was caused by the reversible accumulation of filtration cake at the membrane surface and not by adsorption of fouling components or by pore plugging. The specific flux was restored to within 1 to 2 percent of its initial value following dilution of the mixed liquor to its initial MLSS value.

Depending on the type(s) of aeration system used, the aeration requirements for fouling control may exceed the aeration for conventional biological treatment by two to four times. In testing two full-scale MBR plants, the following observations regarding aeration requirements were noted (Wagner et al., 2002):

- The alpha (α) correction factor for fine pore aerations systems varied depending on the MLSS concentration in the aeration tank (see Fig. 7-20). A similar relationship was observed when α was plotted against increasing viscosity of the mixed liquor.

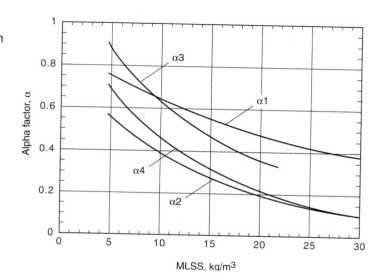

Figure 7-20
Effect of MLSS on alpha values for fine pore aeration in MBRs. [Sources: $\alpha 1$—Bratby et al. (2002) for coarse bubble aeration, $\alpha 2$—ibid. for fine bubble aeration, $\alpha 3$—Wagner et al. (2002), and $\alpha 4$—Thompson (2004).]

- The α value of fine bubble aeration systems in municipal full-scale MBRs at 12 g/L MLSS is in the same order of magnitude of 0.6 (±0.1) compared to conventional wastewater treatment plants at a lower MLSS and SRT.
- The cross-flow coarse bubble aeration system showed no dependence on α value and MLSS. The standard aeration efficiency of the cross-flow aeration system for fouling control, expressed as kWh/m^3 of air flow, was about one-third of that for the fine pore system which provides the biological oxygen supply.
- The aeration requirements for fouling control need to be carefully evaluated in determining the total capacity and energy requirements for the air supply.

The issue of air requirements is a matter of ongoing research and needs to be evaluated on a case-by-case basis. As shown on Fig. 7-20, α values plotted versus MLSS were reported to vary widely depending on the type of MBR used, type of aeration system, and specific characteristics of the MLSS (Bratby et al., 2002; Thompson, 2004; and Wagner et al., 2002).

The amount of air required for air scour depends on the type of membrane system used and is a critical parameter in respect to membrane fouling rate. As noted in Table 7-13, coarse bubble diffusers are used by three of the membrane manufacturers cited; another uses jet aeration, and the fifth uses either coarse or fine bubble aeration. As an example of air supply needs, in a project completed in early 2004 using Zenon membranes, the air supply for membrane scour was 0.56 m^3/h of air per m^3/d of influent flow (Crawford and Lewis, 2004).

Membrane Fouling Control and Cleaning

Membrane bioreactor technology, as cited above, has several advantages as compared to conventional biological treatment processes. One of the main disadvantages, however, is the necessity to clean the membranes routinely. Pretreatment functions mainly to maintain membrane integrity and to prevent hair and other wastewater constituents from changing the hydraulics of the membranes. The term fouling is used to describe the potential deposition and accumulation of constituents in the feed stream on the membrane. In the activated sludge reactor, rejected components in the mixed liquor biomass coat the outer layer of the membranes during effluent withdrawal. Accumulated cake and finer particles adsorb onto the membrane and cause an increase in pressure loss. The biomass not only contains biological flocs, formed by the conglomeration of microorganism within a floc matrix, but a whole range of soluble, insoluble, and colloidal compounds that are either contained in the incoming wastewater or result from bacterial metabolism. In one test facility it was found that the addition of alum to an MBR for phosphorus removal inhibited fouling by reducing the organic fouling material and improving floc structure and strength (Holbrook et al., 2004). Continuous membrane fouling control methods, therefore, are used during the operation of the MBR, with periodic more aggressive cleaning to maintain the filtration capacity of the membrane. Various methods of membrane fouling control and cleaning by manufacturers of proprietary membrane systems are described in the following paragraphs.

Zenon Environmental A method developed by Zenon Environmental Inc. to control fouling on the outside surfaces of the membrane fibers involves a three step process.

First, coarse bubble aeration is provided at the bottom of the membrane tank directly below the membrane fibers. The air bubbles flow upward between the vertically oriented fibers, causing the fibers to agitate against one another to provide mechanical scouring. Second, filtration is interrupted about every 10 to 20 min and the membrane fibers are backwashed with permeate for 30 to 45 s or relaxed (Giese and Larsen, 2000). Typically, a low concentration of chlorine (less than 5 mg/L) is maintained in the backflush water to inactivate and remove microorganisms that colonize the outer membrane surface. Third, about three times per week a strong sodium hypochlorite solution (about 100 mg/L) or citric acid is used in the backflush mode for 45 to 60 min in a procedure called *maintenance clean* that is similar to backwashing a filter. After the 45 min *in situ* cleaning, the system is flushed with permeate for 15 min. An additional permeate flush-to-drain operation is performed for 10 to 15 min to purge the system of free chlorine once the vacuum is initiated and permeate is extracted. The total system downtime during a maintenance clean is about 75 min.

The combination of air scour, backflushing, and maintenance cleaning is not completely effective in controlling membrane fouling, and the pressure drop across the membrane increases with time. Pressure drop is monitored to indicate fouling problems and cleaning needs. At a maximum operating pressure drop of ~60 kPa, the membranes are removed from the aeration basin for a recovery cleaning (Fernandez et al., 2000). During recovery cleaning, a membrane cassette is soaked in a tank containing a 1000 to 2000 mg/L sodium hypochlorite solution for about 24 h. Spare membranes are typically installed in the aeration tank during recovery cleaning so that there is no reduction in treatment capacity.

Mitsubishi and Kubota The Mitsubishi and Kubota membrane systems also use an air-scouring system for membrane cleaning. In addition to scouring air, a method of relaxing the membranes is employed for membrane cleaning. To relax the membranes, the membranes are taken out of service for 1 to 2 min at intervals of 8 to 15 min. The automation of the maintenance cleaning operation allows for remote operation (Wallis-Lage, 2003).

Both the Mitsubishi and Kubota units utilize in-place recovery cleaning. For the Mitsubishi unit, recovery cleaning is recommended at monthly intervals using a backpulse of 3000 mg/L chlorine solution. For the Kubota unit, a recovery cleaning is recommended every 6 mo. During recovery cleaning, flow is reversed for approximately 1 h. The flat plate membranes are not removed for cleaning, and an infrequent backflush with a 0.5 percent solution of hypochlorite has been shown to be effective for fouling control (Stephenson et al., 2000).

USFilter Jet mixing provides the cleaning mechanism for minimizing fouling during normal operation of the USFilter MBR system. Jet mixing is also important while cleaning the membranes in place. When cleaning is required, the membrane units are taken off line for a period of 4 to 6 h and cleaned chemically. The cleaning-in-place operation is accomplished without lifting or removing the membrane module, disconnection of the permeate piping system, or the use of high-pressure hoses. Because membranes are isolated from the biological process in the USFilter MBR system, oxidizing chemicals such as chlorine or acid will not affect the biomass. The principal advantage cited for the clean-in-place method is that all of the membranes on a common suction header are cleaned together to ensure uniformity of cleaning.

Huber Technology As noted in the previous description of the Huber Technology system, fouling is controlled principally by continuous rotational movement of the membrane disks and an air scour system that helps remove solids buildup on the membrane elements.

Peak Flow Management

Where MBRs are used as the mainstream treatment process, as compared to a sidestream or satellite application, peak flow management is an important consideration in the overall sizing and selection of membrane equipment. Individual membrane equipment manufacturers employ their own sizing criteria for their membrane equipment, and the criteria include the duration and magnitude of peak flow design conditions as well as average daily flow. As a general rule, peak flow should be limited to one and one-half times the average flow. In most applications, the peak flow will determine the selection of the type of membrane system and the number of modules required. Where there are extreme variations in flowrates, either upstream flow equalization will be needed to attenuate the flows or else a large number of additional membranes modules are required to accommodate peak flow (Crawford, 2002). Examples of on-line or off-line flow equalization are shown on Fig. 7-21. Because membrane systems and flow equalization can be expensive, a cost analysis should be made for alternative solutions in handling and treating peak flowrates.

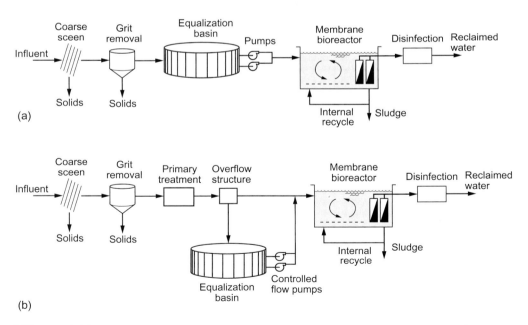

Figure 7-21

Typical process flow diagram incorporating flow equalization: (a) online and (b) off-line. Flow equalization can be applied after grit removal and after primary sedimentation.

Biosolids Production and Management

Because an MBR is a confined vessel with no gravity outlet to a clarifier as in the conventional activated sludge process, an understanding of biosolids generation and control is very important. If the solids in the bioreactor continue to increase and SRTs become longer, the process dynamics in the reactor undergoes gradual change. The aeration rate is affected by biosolids production relative to the amount of BOD removed and also by nitrification, if it was not accounted for in the process design. At long SRTs, some biomass loss occurs by more endogenous respiration that also affects aeration requirements. In essence, partial aerobic digestion occurs. If excessive solids buildup in the mixed liquor occurs, more frequent cleaning of the membranes may be needed. Solids production, therefore, has to be understood so that wasting of excess solids can be managed.

Nutrient Removal

Each of the MBR systems described in Sec. 7-5 can be modified or adapted for biological nutrient removal in configurations similar to those described in Sec. 7-4. For biological nutrient removal, additional zones or compartments are added ahead of the aerobic zone of the MBR to establish anoxic and/or anaerobic conditions conducive for nutrient removal. Chemicals can also be added in some applications to enhance nitrogen and phosphorus removal.

Nitrogen Removal

Process flow diagrams of how nitrogen removal can be incorporated with the MBR process are shown on Fig. 7-22 for integrated and separate stage membrane separation processes. On Fig. 7-22a, pretreated wastewater enters an anoxic reactor located upstream of the aerobic MBR unit. Mixed liquor from the MBR is recycled back to the anoxic tank. Recycle rates range typically from 2 to 5 times the influent flowrate and detention times in the anoxic tank range typically from 2 to 4 h.

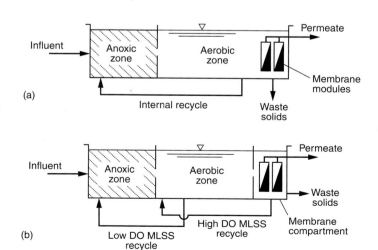

Figure 7-22

Membrane bioreactors with nitrogen removal: (a) integrated and (b) separate stage (with alternative solids recycle).

The process flow diagram for a separate-stage membrane system, shown on Fig. 7-22b, is similar to the integrated type except the recycle is taken from the membrane compartment instead of the aerobic reactor. Two recycle lines are shown for better process control, as discussed below. Recycle rates and anoxic tank detention times are similar to those in the integrated type.

The MLSS recycle is extracted typically at the location of the highest nitrate and DO concentrations. Recycling all of the return flow from the membrane zone or the aerobic zone to the anoxic zone may affect denitrification performance, however. Due to the intensive aeration and high DO in the membrane compartment, the residual DO in the return mixed liquor may inhibit denitrification in the anoxic zone. To mitigate the problem of nitrogen removal control, two recycle lines can be used: one for solids return from the membrane zone to the aerobic zone, and one for lower DO mixed liquor solids return to the anoxic reactor (see Fig. 7-22b). Two recycle lines can be used with either the integrated of separate stage configuration. Thus, positive control of the internal recirculation rates can ensure that the nutrient removal process is optimized.

Full-scale and pilot-plant MBR systems have been operated with the anoxic/aerobic MLE biological nitrogen removal process, similar to that shown in Table 7-9, with resulting effluent total nitrogen concentrations of <10 mg/L (Mourato et al., 1999; ReVoir et al., 2000; and Giese et al., 2000). Influent recycle flowrate ratios of 4 to 6 have been used in those studies to feed nitrate to a separate preanoxic tank. Reclaimed water with the total nitrogen of less than 10 mg/L as nitrogen is generally an acceptable value for groundwater recharge. An MBR with an anoxic section for nitrogen removal is shown on Fig. 7-23.

(a) (b)

Figure 7-23
Membrane bioreactor with an anoxic section for nitrogen removal: (a) overview of plant and (b) close-up of anoxic section.

Phosphorus Removal

Phosphorus removal can be accomplished typically by configuring an anaerobic contact zone or compartment ahead of the anoxic zone for BPR in addition to nitrogen removal. A typical process flow diagram for the biological removal of nitrogen and phosphorus is shown on Fig. 7-24. Typically, the hydraulic retention time, τ, in the anaerobic contact zone is from 0.5 to 1.0 h for fermentation of the readily biodegradable organic matter, as measured by readily biodegradable COD (rbCOD). The contents of the anaerobic zone are mixed to provide contact with the return sludge and the influent wastewater. The reaction kinetics for BOD removal, nitrification, and denitrification are similar to those used for conventional suspended growth processes.

Phosphorus removal can also be accomplished by the addition of metal salts to the MBR to enhance floc formation. Pilot testing is recommended to verify the effectiveness of chemical addition in phosphorus removal and to evaluate possible benefits in membrane fouling control. The reader is referred to Holbrook et al. (2004) for more information.

As described above for nitrogen removal, control of the solids recycle streams is important so that anaerobic and anoxic conditions can be maintained. Using the process flow diagram as shown on Fig. 7-24, the total phosphorus content of the reclaimed water can expected to be less than 2 mg/L. Where phosphorus limits of less than 1 mg/L are

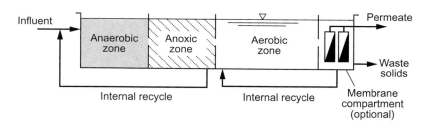

Figure 7-24
Typical process flow diagram for phosphorus removal using an MBR: (a) schematic and (b) plan view of full-scale plant.

required, additional phosphorus removal may be accomplished by one of three methods: (1) chemical addition of either alum or ferric salts to the anaerobic or aerobic zones, (2) adding acetate or other waste streams high in rbCOD, or (3) limiting recycle flows from other operations (such as dewatering) that are high in phosphorus (Stensel, 2003). The potential effects of chemical addition on the membranes, including fouling, must be considered, however. Other methods of phosphorus removal can be employed and should be evaluated on a case-by-case basis. Alternative methods of phosphorus removal are discussed in the companion text, Tchobanoglous et al. (2003).

Biosolids Processing

Although not a factor in satellite plants where solids are returned to the collection system, the characteristics of the biosolids produced by MBRs could affect thickening, dewatering, and digestion operations and processes at central treatment plants, particularly if the bioreactor is operated at high MLSS values (greater than 12,000 mg/L). Based on research by Merlo et al. (2005), biosolids from submerged MBRs are characterized by a higher frequency of small particles, higher amounts of colloidal material, lower levels of extracellular polymeric substances (EPS) and higher amounts of microorganisms due to the presence of nocardioform. Presence of these substances has been found to be important in sludge thickening and dewatering.

Limited information is available in the literature on the thickening of waste sludge from MBRs, but it has been noted that thickening of waste sludge from MBRs may be comparable to that of extended aeration plants (Crawford et al., 2000). Successful dewatering of biosolids from MBRs has been accomplished using aerobic digestion followed by a belt-filter press. Generally, waste MBR sludge cannot be gravity thickened, but other thickening methods may be suitable. Where solids from the MBR process constitutes a major part of the solids stream requiring processing, the thickening and dewaterability characteristics of MBR biosolids should be verified as part of a pilot-scale testing program.

Limited data are available on the dewatering characteristics of MBR biosolids. In an investigation by Fernandez et al. (2000), results of measurements of sludge volume index (SVI) and capillary suction time (CST) suggest that biosolids with high MLSS levels (~10,000 mg/L) are more viscous and may have an adverse impact on dewatering characteristics and membrane permeability. Other researchers suggest that as the particle size decreases, the resistance to dewaterability increases (Sanin and Vesilind, 1999). As the activated sludge solids undergo endogenous respiration, especially in MBRs with longer SRTs, smaller particle sizes often result (Merlo et al., 2005).

7-7 ISSUES IN THE SELECTION OF SECONDARY TREATMENT PROCESSES

Because of the wide variety of secondary treatment technologies and the number of variables that must be considered, selection of the appropriate technology can sometimes be a daunting task. Some of the issues that must be considered include: (1) expansion of an existing plant or construction of a new plant, (2) final use of the effluent, (3) comparative performance of various technologies, (4) results of pilot-plant studies, (5) type of disinfection process, (6) future water quality requirements, (7) energy considerations, (8) site constraints, and (9) economic and other considerations. In most situations a number of these factors, taken together, will govern the final selection.

Expansion of an Existing Plant vs. Construction of a New Plant

As discussed in Sec. 6-4, different considerations apply, depending on whether existing facilities are to be modified or upgraded, or an entirely new facility is to be constructed. In the former case, compatibility with existing processes is important to the extent the existing facilities can be utilized effectively. Integration of a new process into an existing plant can often be a challenging task, especially if effluent disposal must continue to be practiced. Several factors affecting modification of an existing plant include space for new facilities, plant hydraulic profile, piping modifications, operating considerations, and ancillary systems requirements, not the least of which is biosolids processing. For a new plant, most of the constraints cited above are eliminated and a different approach can be taken for process selection, one considering the application of new technologies.

Final Use of Effluent

The treated effluent must meet the intended reuse water quality requirements. The single most important effluent criterion is the disinfection requirement for most applications using secondary effluent. Because the level of disinfection attainable ultimately depends on suspended solids particle size distribution in the secondary effluent (discussed in Sec. 8-1), the method of solids separation is an important design consideration, be it gravity sedimentation or MF (for a new plant). Consideration should also be given to what future water reuse applications are likely and how those might impact process selection to meet different reuse standards.

Comparative Performance of Treatment Processes

The ranges of constituent removal required determine the types of processes considered to meet water quality standards for given water reuse applications. Examples of treatment levels achievable with various combinations of unit processes for secondary treatment are presented in Table 7-7.

Pilot-Scale Studies

For the selection of a secondary treatment process for removal of conventional levels of BOD and TSS, pilot studies are not necessary because of the maturity of the processes and the well-documented fundamentals of biological treatment (Tchobanoglous et al., 2003). If treatment using a membrane is being considered, pilot-plant testing is recommended because most MBR designs are proprietary and design parameters and performance can vary. Views and a schematic diagram of typical MBR pilot plants are shown on Fig. 7-25.

Type of Disinfection Process

As discussed above, selection of the solids separation process may be dictated by the type of disinfection process to be employed. For example, if UV disinfection is to be used, the presence of particulate matter in the effluent may require that deep clarifiers or membrane filtration be used (see Chap. 11).

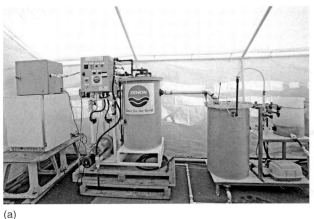

(a) (b)

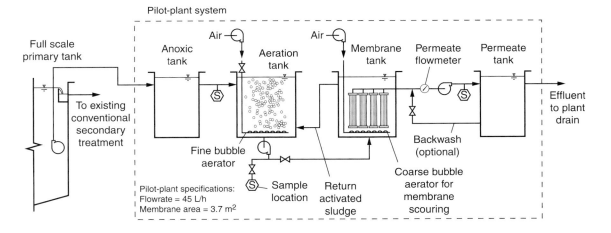

(c)

Figure 7-25
Typical MBR pilot plants: (a) view of MBR pilot plant with external membrane (b) view of MBR pilot plant with internal membrane, and (c) schematic flow diagram for pilot plant shown in (a) including sampling points. Sludge wasting is accomplished manually. (Courtesy O. Virgadamo and K. Bourgeous.)

Future Water Quality Requirements

In any application, careful consideration must be given to anticipated future water quality requirements. For example, if it is anticipated that nutrient removal may be required in the future, provisions for anoxic or anaerobic compartments should be considered in the design.

Energy Considerations

Operating costs are ongoing concerns for the operating agency; after labor cost, the cost of energy is the highest component of the operating costs. In conventional secondary treatment, most of the energy is used for (1) biological treatment by the activated sludge process or trickling filters, (2) pumping systems for the transfer of wastewater, liquid

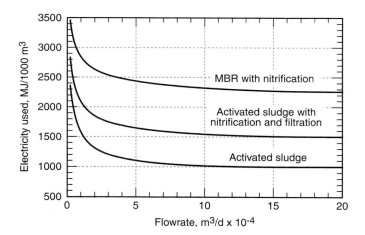

Figure 7-26
Comparison of electrical energy used for different types of activated sludge treatment processes as a function of flowrate. (Adapted from Burton, 1996.)

sludge, biosolids, and process water, and (3) equipment for the processing, dewatering, and drying of solids and biosolids. In activated sludge treatment, approximately one-half of the electricity used is for activated sludge aeration. The amount of electricity consumed varies generally according to plant size and type of treatment system (see Fig. 7-26). With the addition of nitrification, the total electricity consumption increases on the order of 20 to 30 percent as compared to conventional activated sludge.

Due to the limited data available and the different types of proprietary systems available, the energy consumption by MBRs is not well documented. As discussed in Sec. 7-6, MBRs require aeration to sustain the biological process and for membrane scouring for fouling control, and the aeration for fouling control may exceed the aeration for biological treatment by two to four times. The energy requirements for permeate also vary depending on the process configuration. Some of the energy requirements are offset by the elimination of return sludge pumping and some of the ancillary equipment such as clarifier drives used in conventional activated sludge systems. As part of the pilot-testing program and alternative equipment evaluation, quantification of the energy requirements should be made.

Site Constraints In many treatment plant sites, the area available for locating aeration tanks, trickling filters, clarifiers, and ancillary facilities may be limited. In such situations, the footprint of compact treatment technologies becomes an important factor in the technology selection and siting of the facilities. In a large (32,000 m^3/d) MBR plant built in Traverse City, Michigan, the space required for a separate membrane compartment was equivalent to 36 percent of the aeration tank volume, or significantly less than the space required for a secondary treatment facility using clarifiers (Crawford and Lewis, 2004). The space requirement for a conventional activated sludge plant as compared to an MBR plant is illustrated in Example 7-4.

EXAMPLE 7-4. Compare the Area Requirements (Footprint) of an Activated Sludge Plant to that of a Membrane Bioreactor Facility.

Compute the space requirements for a conventional activated sludge plant and an MBR treatment system using the wastewater characteristics and solids production determined in Example 7-3. The space requirements for preliminary and primary treatment are assumed to be the same for both plants and are not included in the example. The plants are designed to meet an effluent BOD of less than 30 g/m³. For the activated sludge process, secondary clarifiers are to be used. For an MBR, an allowance equal to 35 percent of the aeration tank volume is provided for installation of the membrane units in a separate compartment. The following design conditions apply:

Process parameters

Process unit/parameter	Units	Conventional activated sludge	Membrane bioreactor
Average flowrate	m³/d	18,900	18,900
Aeration tanks			
Solids retention time (SRT)	d	5	15
MLSS (X_{TSS})	g/m³	3000	10,000
Mass of TSS (Example 7-3)	kg/d	10,150	23,847
Depth of tank	m	5	5
Secondary clarifier			
Hydraulic application rate	m³/m² · d	22	N/A
Membrane compartment			
Area allowance for membrane installation			35 percent of the aeration tank volume

Solution

1. For a conventional activated sludge plant:
   a. Determine the aeration tank volume

   $$(V)(X_{TSS}) = 10{,}150 \text{ kg/d}$$
   $$\text{At } X_{TSS} = 3000 \text{ g/m}^3$$
   $$V = \frac{(10{,}150 \text{ kg})(10^3 \text{ g/kg})}{(3000 \text{ g/m}^3)} = 3383 \text{ m}^3$$

   b. Determine surface area of aeration tank
   $$A = V/d = (3383 \text{ m}^3)/5 \text{ m} = 677 \text{ m}^2$$

c. Determine aeration tank hydraulic retention time

$$\tau = \frac{V}{Q} = \frac{(3383 \text{ m}^3)(24 \text{ h/d})}{(18{,}900 \text{ m}^3/\text{d})} = 4.30 \text{ h}$$

d. Determine surface area of clarifiers

$$\text{Area} = \frac{(18{,}900 \text{ m}^3/\text{d})}{(22 \text{ m}^3/\text{m}^2 \cdot \text{d})} = 859 \text{ m}^2$$

e. Total surface area of aeration tanks plus clarifiers

$$\text{Area}_T = (677 + 859) \text{ m}^2 = 1536 \text{ m}^2$$

2. For an MBR:
   a. Determine the aeration tank volume

   $$(V)(X_{TSS}) = 23{,}847 \text{ kg/d}$$
   $$\text{At } X_{TSS} = 10{,}000 \text{ g/m}^3$$
   $$V = \frac{(23{,}847 \text{ kg})(10^3 \text{ g/kg})}{(10{,}000 \text{ g/m}^3)} = 2385 \text{ m}^3$$

   b. Determine surface area of aeration tank

   $$A = V/d = (2385 \text{ m}^3)/5 \text{ m} = 477 \text{ m}^2$$

   c. Determine aeration tank detention time

   $$\tau = \frac{V}{Q} = \frac{(2385 \text{ m}^3)(24 \text{ h/d})}{(18{,}900 \text{ m}^3/\text{d})} = 3.02 \text{ h}$$

   d. Determine surface area of membrane tank

   $$\text{Area} = 0.35 \times 477 \text{ m}^2 = 167 \text{ m}^2$$

   e. Total surface area of aeration tank plus membrane tank

   $$\text{Area}_T = (477 + 167) \text{ m}^2 = 644 \text{ m}^2$$

Comment

Although this example does not account for a number of structural features such concrete walls, influent and effluent channels, weir structures, and support facilities such as blower building and feedwater and sludge pumping stations, it does demonstrate some of the features that make an MBR facility compact in design. In this particular example, the surface area of an MBR is less than half of that required for conventional process units. Also in a conventional activated sludge plant, a circular configuration of secondary clarifiers, a preferred design in most plants, requires a larger footprint than a rectangular configuration. Finally, the effluent from an MBR Facility would be of much higher quality as a result of the membrane filtration aspect. A comparable activated sludge process would have to incorporate filtration of the clarified effluent.

Table 7-19
Comparison of the advantages and disadvantages of conventional activated sludge and membrane bioreactor treatment systems

Advantages	Disadvantages
Conventional activated sludge	
• Technology is well understood and various models are available • Many nonproprietary designs are available • Process capabilities are almost universally accepted • Different configurations such as step feed allow the process to be designed to maximize contact time between macromolecules and microorganisms • Air requirements are well understood • Skilled operation and maintenance personnel are widely available • The largest wastewater treatment facilities make use of conventional activated sludge, with plant capacities exceeding 2.5×10^6 m³/d • Slow rate of kinetic reaction	• Limitations in suspended solids removal necessitate a high level of disinfection • Greater sludge production (as compared to MBR) increases biosolids handling requirements and costs • Process performance dependent, in part, on design and operation of the secondary clarifier • Clarifier performance is reduced by development of filamentous organisms or poor settling sludge in aeration process • Subsequent filtration is required for effective UV disinfection • Membrane filtration is required to achieve effluent quality comparable to MBR • Large process footprint
Membrane bioreactor	
• Produces high quality effluent with greater reuse potential • Low suspended solids concentration and removal of large particles in product water enables effective disinfection to be accomplished • System is adaptable for adding nutrient removal processes • Provides longer retention of nitrifying bacteria, resulting in greater nitrification • Because of buffering effect of high MLSS values, effluent quality is not affected by influent constituent variations • Footprint is reduced substantially because clarifiers and/or filters are not required • Modular construction creates the potential for future expandability • Potential reduction in sludge volume due to longer SRT values • Can be readily adapted to satellite and large scale decentralized applications • Process is relatively easy to automate	• Long-term history of operation is not available • All MBR systems make use of proprietary equipment and designs; there is no standard configuration • Some proprietary processes have not been approved by regulatory authorities • Pretreatment required (usually fine screening) to avoid damaging and clogging membrane elements • Greater consumption of energy for effective operation • Peaking factors in excess of 1.5 usually require flow equalization or the addition of more process units • Mechanisms and control of membrane fouling still under investigation • Membrane replacement is relatively expensive • Pilot testing is often required for full-scale design • Limited availability of experienced operating personnel • Capacity of largest MBR facilities (to date) is less than 40,000 m³/d

Economic and Other Considerations

The final selection of an appropriate secondary treatment process must consider the factors discussed above as well as economic and other considerations such as compatibility with existing systems, skilled labor requirements, and the need for future expansion and upgrading. These considerations are discussed in the following paragraphs.

Economic Considerations

Capital and operation and maintenance (O & M) costs are an important consideration in most municipal settings. Typically a life cycle cost analysis, incorporating a sensitivity analysis for uncontrollable costs such as chemicals, is used to assess the feasibility of different technologies. Based on the results of a life cycle cost analysis, a more capital-intensive technology may be selected if the O & M costs are more reasonable and/or controllable. Finally, in preparing a life cycle cost analysis, it is also important to consider the potential revenue that may be derived by supplying reclaimed water to water reuse applications.

Other Considerations

As discussed in this chapter, several different types of technologies are available for secondary treatment. Although cost and the production of reclaimed water that meets user requirements are primary considerations, other factors influence process selection. Often the familiarity with existing and proven methods may govern process selection; however, the potential benefits of a new technology may lead to its selection. To distinguish a conventional activated sludge from an MBR process, a generalized comparison of the advantages and disadvantages of each is presented in Table 7-19. For an actual application, however, factors specific to the project have to be considered.

PROBLEMS AND DISCUSSION TOPICS

7-1 Prepare a probability plot of effluent daily BOD and TSS data from a local wastewater treatment plant. How does the geometric standard deviation compare to the values given on Figs. 7-6a and 7-6b?

7-2 Consider the nature of variability of wastewater treatment processes. Comment on why some processes have higher s_g values than other processes. Which factors do you think are important in minimizing s_g for a given process?

7-3 Obtain flow data from a local wastewater treatment or reclamation facility and determine the peaking factor for the peak day, week, and month. What are the corresponding s_g values?

7-4 A conventional secondary treatment process (suspended growth activated sludge followed by sedimentation in a shallow clarifier) has an effluent turbidity of 5 NTU with s_g value of 2.5. Determine what percent of the time the effluent turbidity will exceed 10 NTU and suggest process modifications that could improve performance.

7-5 Using the flowrate data presented below, estimate the size of an off-line equalization tank needed to normalize the flowrate variation. The maximum hourly flowrate to the treatment process should not exceed (a) 1.25, (b) 1.35, and (c) 1.45 times the average daily flow (a, b, or c to be selected by instructor).

Time period	Average flowrate during time period, m³/s	Time period	Average flowrate during time period, m³/s
M–1	0.275	N–1	0.465
1–2	0.220	1–2	0.465
2–3	0.165	2–3	0.420
3–4	0.130	3–4	0.380
4–5	0.105	4–5	0.355
5–6	0.120	5–6	0.355
6–7	0.140	6–7	0.360
7–8	0.245	7–8	0.395
8–9	0.385	8–9	0.430
9–10	0.450	9–10	0.440
10–11	0.485	10–11	0.390
11–N	0.480	12–M	0.340

7-6 The effluent from an existing activated sludge plant is proposed to be used in an ornamental pond but additional nitrification is required. It has been proposed that suspended packing be installed in the aeration tank to convert the process to an integrated fixed film nitrification process. Discuss the benefits and limitations of this proposal.

7-7 Prepare a summary table of the design considerations and process modifications that can improve the reliability of a secondary treatment process.

7-8 Explain in your own words how sidewater depth impacts the removal of particles by secondary clarification.

7-9 Explain in what geographic locations the situations described on Figs. 7-8b, 7-8c, and 7-8d would be most likely to occur.

7-10 Develop two MBR process flow diagrams for nutrient removal (a) to accomplish total nitrogen less than 10 mg/L and phosphorus less than 5 mg/L and (b) for total nitrogen less than 3 mg/L and phosphorus less than 1 mg/L.

7-11 As the design engineer for an MBR project, you have been asked to evaluate two different proprietary designs. What specific evaluation criteria would you use in the equipment selection process? Cite any reference documents that you may have used.

7-12 Comment on the suitability of MBRs for satellite water reclamation compared to conventional activated sludge processes.

7-13 Review a minimum of three articles in the current literature on the fouling of MBR membranes in the activated sludge process. Prepare a summary report on the operational issues and modifications that were implemented to control fouling.

7-14 Estimate the additional footprint required for the conventional activated sludge process design given in Example 7-4 for granular media filtration operating at a filtration rate of 100 L/m² · min

7-15 Estimate the energy requirements for the conventional activated sludge process and the MBR process described in Example 7-4.

7-16 An aeration system for an MBR was designed for an MLSS concentration of 12,000 g/m³. However, because of fouling limitations, the treatment process can only be operated at a MLSS concentration of 8000 g/m³. Estimate the additional flow that can be treated with the existing aeration system, assuming that adequate tankage is available. Refer to Tchobanoglous et al. (2003), if needed.

7-17 In the section "Other Membrane Systems," three alternative proprietary membrane systems are introduced. Select one of these systems and compile descriptive and design information in a format similar to that presented in Table 7-13. An alternative system that is not described in the textbook and is currently being marketed is also acceptable.

7-18 Develop a spreadsheet solution for the MBR process design procedure given in Examples 7-3 and 7-4. After confirming the results given in the examples, conduct a sensitivity analysis and determine the independent effect of (a) wastewater temperature, (b) reactor MLSS concentration, and (c) wastewater constituent concentrations on sludge production and reactor size. Use the following values in the spreadsheet model in your analysis and comment on the results and the implications on MBR process design:

Parameter	Unit	Low value	High value
Temperature	°C	5	20
Reactor MLSS	g/m³	5000	15,000
Wastewater constituents			
BOD	g/m³	98	420
sBOD	g/m³	49	210
COD	g/m³	210	900
sCOD	g/m³	92.4	396
rbCOD	g/m³	56	240
TSS	g/m³	49	210
VSS	g/m³	42	180

REFERENCES

Adham, S., and J. F. DeCarolis (2004) "Optimization of Various MBR Systems for Water Reclamation—Phase III," *Desalination and Water Purification Research and Development Report No. 103*, U.S. Department of the Interior, Bureau of Reclamation, Denver CO.

Babcock, R., Jr. (2005) "Comparison of Five Different Membrane Bioreactors," presented at the 27th Annual HWEA Conference, Honolulu, HI.

Barker, P. L., and P. L. Dold (1997) "General Model for Biological Nutrient Removal in Activated Sludge Systems: Model Presentation," *Water Environ. Res.*, **69**, 5, 961–984.

Barnard, J. L. (1975) "Biological Nutrient Removal without the Addition of Chemicals," *Water Res.*, **9**, 485–490.

Bratby, J. R., B. Gaines, M. Loyer, F. Luiz, and D. Parker (2002) "Merits of Alternative MBR Systems," *Proceedings of WEFTEC 2002*, Water Environment Federation, Alexandria, VA.

Burton, F. L. (1996) *Water and Wastewater Industries: Characteristics and Energy Management Opportunities*, CR-10691, Electric Power Research Institute, St. Louis, MO.

Chiemchaisri, C., K. Yamamoto, and S. Vigneswaran (1993) "Household Membrane Bioreactor in Domestic Wastewater Treatment," *Water Sci. Technol.*, **27**, 1, 171–178.

Cote, P., H. Buisson, and M. Praderie (1998) "Immersed Membranes Activated Sludge Process Applied to the Treatment of Municipal Wastewater," *Water Sci. Technol.*, **38**, 4–5, 437–442.

Crawford, G., D. Thompson, J. Lozier, G. Daigger, and E. Fleischer (2000) "Membrane Reactors—A Designer's Perspective," in *Proceedings of WEFTEC 2000*, Water Environment Federation, Alexandria, VA.

Crawford, G. (2002) "Competitive Bidding and Evaluation of Membrane Bioreactor Equipment—Three Large Plant Case Studies," in *Proceedings of WEFTEC 2002*, Water Environment Federation, Alexandria, VA.

Crawford, G. and R. Lewis (2004) "Exceeding Expectations," *Civil Engineering, ASCE*, **74**, 1, 62–67.

Crawford, G., G. Daigger, J. Fisher, S. Blair, and R. Lewis (2005) "Parallel Operation of Large Membrane Bioreactors in Traverse City," in *Proceedings of WEFTEC 2005*, Water Environment Federation, Alexandria, VA.

DeCarolis, J., S. Adham, B. Pearce, and L. Wasserman (2003) "Application of Various MBR Systems for Water Reclamation," in *Proceedings of WEFTEC 2003*, Water Environment Federation, Alexandria, VA.

Fernandez, A., J. Lozier, and G. Daigger (2000) "Investigating Membrane Bioreactor Operation for Domestic Wastewater Treatment: A Case Study," in *Proceedings of WEFTEC 2000*, Water Environment Federation, Alexandria, VA.

Giese, T. P. and M. D. Larsen (2000) "Pilot Testing New Technology at the Kitsap County Sewer District No. 5/City of Port Orchard, WA Joint Wastewater Treatment Facility," *Proceedings of WEFTEC 2000*, Water Environment Federation, Alexandria, VA.

Grady, C. P. L., Jr., G. T. Daigger, and H. C. Lim (1999) *Biological Wastewater Treatment,* 2nd ed., Marcel Dekker, Inc., New York.

Henze, M. C., P. L. Grady, W. Gujer, G. v. R. Marias, and T. Matsuo (1987) *Activated Sludge Model No. 1*, IAWPRC Scientific and Technical Reports, no. 1, IAWPRC, London.

Holbrook, R. D., M. J. Higgins, S. N. Murthy, A. N. Fonseca, D. Anabela, E. J. Fleischer, G. T. Daigger, T. J. Grizzard, N. G. Love, and J. T. Novak (2004) "Effect of Alum Addition on the Performance of Submerged Membranes for Wastewater Treatment," *Water Environ. Res.*, **76**, 7, 2699–2702.

Jefferson, B., P. Le Clech, S. Smith, A. Laine, and S. Judd (2000) "The Influence of Reactor Configuration on the Efficacy of Membrane Bioreactors for Domestic Waste Water Recycling," in *Proceedings of WEFTEC 2000*, Water Environment Federation, Alexandria, VA.

Judd S., and C. Judd (eds.) (2006) *The Membrane Book: Principles and Applications of Membrane Bioreactors in Water and Wastewater Treatment,* 1st ed., Elsevier, Amsterdam.

Lorenz, W., T. Cunningham, and J. P. Penny (2002) "Phosphorus Removal in a Membrane Reactor System: A Full-Scale Wastewater Demonstration Study," in *Proceedings of WEFTEC 2002,* Water Environment Federation, Alexandria, VA.

Merlo, R., R. S. Trussell, S. H. Hermanowicz, and D. Jenkins (2005) "Properties Affecting the Thickening and Dewatering of Submerged Membrane Bioreactor Sludge," in *Proceedings of WEFTEC 2005,* Water Environment Federation, Alexandria, VA.

Mourato, D., D. Thompson, C. Schneider, N. Wright, M. Devol, and S. Rogers (1999) "Upgrade of a Sequential Batch Reactor into a Zenogem Membrane Bioreactor," in *Proceedings of WEFTEC 1999,* Water Environment Federation Alexandria, VA.

ReVoir, G. J., II, D. R. Refling, and H. J. Losch (2000) "Wastewater Process Enhancements Utilizing Submerged Membrane Technology," in *Proceedings of WEFTEC 2000,* Water Environment Federation, Alexandria, VA.

Sanin, F. and P. A. Vesilind (1999) "A Comparison of Physical Properties of Synthetic Sludge with Activated Sludge," *Water Environment Research,* **71**, 2, Water Environment Federation, Alexandria, VA.

Sen, D., P. Mitta, and C. W. Randall (1994) "Performance of Fixed Film Media Integrated in Activated Sludge Reactors to Enhance Nitrogen Removal," *Water Sci. Technol.,* **30**, 11, 13–24.

Stensel, H. D. (2003) "MBR Processes for Nitrogen and Phosphorus Removal: Alternatives and Process Considerations," *Membrane Bioreactors Designed for Ultimate Nutrient Removal,* Enviroquip Workshop VII, Los Angeles, CA.

Stephenson, T., S. Judd, B. Jefferson, and K. Brindle (2000) *Membrane Bioreactors for Wastewater Treatment,* IWA Publishing, London, UK.

Tchobanoglous, G., F. L. Burton, and H. D. Stensel (2003) *Wastewater Engineering: Treatment and Reuse,* 4th ed., McGraw-Hill, New York.

Thompson, D. (2004) "ZeeWeed MBR Technical Workshop," WEFTEC 2004, Workshop #114, Water Environment Federation, Alexandria, VA.

Trussell, R. S., R. P Merlo, S. Hermowicz, and D. Jenkins (2004) "The Effect of Organic Loading on Membrane Fouling in a Submerged Membrane Bioreactor Treating Municipal Wastewater," *Proceedings of WEFTEC 2004,* Water Environment Federation, Alexandria, VA.

Trussell, R. S., S. Adham, and R. R. Trussell (2005) "Process Limits of Municipal Wastewater Treatment with the Submerged Membrane Bioreactor," *J. Environ. Eng., ASCE,* **131**, 410.

Van der Roest, H. F., D. P. Lawrence, and A. G. N. Bentem (2002) *Membrane Reactors for Municipal Wastewater Treatment,* STOWA Report, IWA Publishing, London, UK

Wagner, M., P. Cornel, and S. Krause (2002) "Efficiency of Different Aeration Systems in Full Scale Membrane Bioreactors," in *Proceedings of WEFTEC 2002,* Water Environment Federation, Alexandria, VA.

Wallis-Lage, C. (2003) "MBR Similarities and Differences between Manufacturers," *Proceedings of WEFTEC 2003,* Water Environment Federation, Alexandria, VA.

WEF (1998) *Design of Municipal Wastewater Treatment Plants, Manual of Practice 8,* Water Environment Federation, Alexandria, VA.

Yamamoto, K., M. Hiasa, T. Mahmood, and T. Matsuo (1989) "Direct Solid-Liquid Separation Using Hollow Fiber Membrane in an Activated Sludge Aeration Tank," *Water Sci. Technol.,* **21**, 43–54.

8 Removal of Residual Particulate Matter

	WORKING TERMINOLOGY 374
8-1	CHARACTERISTICS OF RESIDUAL SUSPENDED PARTICULATE MATTER FROM SECONDARY TREATMENT PROCESSES 375
	Residual Constituents and Properties of Concern 375
	Removal of Residual Particles from Secondary Treatment Processes 385
8-2	TECHNOLOGIES FOR THE REMOVAL OF RESIDUAL SUSPENDED PARTICULATE MATTER 388
	Technologies for Reclaimed Water Applications 388
	Process Flow Diagrams 390
	Process Performance Expectations 390
	Suitability for Reclaimed Water Applications 392
8-3	DEPTH FILTRATION 392
	Available Filtration Technologies 392
	Performance of Depth Filters 398
	Design Considerations 407
	Pilot-Scale Studies 415
	Operational Issues 417
8-4	SURFACE FILTRATION 417
	Available Filtration Technologies 419
	Performance of Surface Filters 422
	Design Considerations 423
	Pilot-Scale Studies 425
8-5	MEMBRANE FILTRATION 425
	Membrane Terminology, Types, Classification, and Flow Patterns 426
	Microfiltration and Ultrafiltration 430
	Process Analysis for MF and UF Membranes 435
	Operating Characteristics and Strategies for MF And UF Membranes 436
	Membrane Performance 436
	Design Considerations 441
	Pilot-Scale Studies 441
	Operational Issues 443
8-6	DISSOLVED AIR FLOTATION 445
	Process Description 445
	Performance of DAF Process 448
	Design Considerations 448
	Operating Considerations 453
	Pilot-Scale Studies 453

8-7 ISSUES IN THE SELECTION OF TECHNOLOGIES FOR THE REMOVAL OF RESIDUAL PARTICULATE MATTER   454
Final Use of Effluent   454
Comparative Performance of Technologies   455
Results of Pilot-Scale Studies   455
Type of Disinfection Process   455
Future Water Quality Requirements   455
Energy Considerations   455
Site Constraints   455
Economic Considerations   455

PROBLEMS AND DISCUSSION TOPICS   456

REFERENCES   459

WORKING TERMINOLOGY

Term	Definition
Backwash	The process of removing solids accumulated on or in a filtration medium by applying air and/or clean water in the opposing flow direction. Backwash water that contains the removed solids is called waste washwater.
Depth filtration	The removal of particulate matter suspended in a liquid by passing the liquid through a granular medium such as sand and/or anthracite coal.
Dissolved air flotation (DAF)	The removal of particulate matter by attaching the particles to a blanket of rising air bubbles in a specially designed flotation tank.
Feed stream (sometimes referred to as feedwater)	Water (or wastewater) being supplied to the membrane treatment process.
Flux	The mass or volume rate of transfer through the membrane surface, usually expressed as $m^3/m^2 \cdot h$ or $L/m^2 \cdot h$ ($gal/ft^2 \cdot d$). Flux is the prevalent term for referring to the rate of water production from a membrane system.
Fouling	The accumulation of material on the membrane surface resulting in the loss of performance.
Membrane	A device, usually made of an organic polymer, that allows the passage of water and certain constituents, but rejects others above a certain physical size or molecular weight.
Microfiltration (MF)	A membrane separation process used typically to remove relatively large particles from the feed stream; microfiltration pore sizes range from approximately 0.05 to 2 μm.
Permeate (sometimes called filtrate or product water)	The liquid stream that has passed through the membrane.
Pore size	The nominal size of a membrane's pores (typically measured in microns) that allows passage of permeate through the membrane wall while retaining selected contaminants on the membrane surface. Pore size is a classification system used typically to distinguish between types of membranes.

Residuals	Waste streams produced by water reclamation processes. For depth and surface filtration, the residual waste stream is filter waste washwater. For membrane systems, residual waste streams include waste washwater, concentrate, and chemical cleaning wastes.
Retentate (sometimes called concentrate)	The portion of the feed stream that does not pass through the membrane.
Surface filtration	The removal of particulate matter suspended in a liquid by passing the liquid through a thin septum, usually a cloth or metal medium.
Transmembrane pressure (TMP)	The driving force that transmits permeate through the membrane.
Ultrafiltration (UF)	A membrane separation process similar to MF except the membrane pore sizes can range from approximately 0.005 to 0.1 µm. Generally, UF membranes are able to achieve higher levels of separation than MF, particularly for bacteria and viruses.

For many water reuse applications, removal of residual particulate matter remaining after secondary biological treatment is required. Particulate matter contributes to turbidity, may be associated with undesirable chemical contaminants or pathogens, and may interfere with disinfection processes. The removal of these particles from wastewater effluents is the subject of this chapter; the removal of the residual dissolved constituents is considered in Chap. 9. Subjects considered in this chapter include: (1) characteristics of residual suspended particulate matter from secondary treatment processes, (2) an introduction to the technologies used for the removal of residual suspended particulate matter, (3) depth filtration, (4) surface filtration, (5) membrane filtration, in particular microfiltration (MF) and ultrafiltration (UF), (6) dissolved air flotation, and (7) issues in the selection of filter technologies for the removal of residual particulate matter.

8-1 CHARACTERISTICS OF RESIDUAL SUSPENDED PARTICULATE MATTER FROM SECONDARY TREATMENT PROCESSES

Residual particulate matter found in secondary effluent varies in size and composition depending on the type of treatment process employed. In general, secondary effluent contains subcolloidal, colloidal, and suspended particles, along with aggregate clumps of particles (often identified as flocculent particles). To understand the mechanisms of particle removal and the effect of particulate matter on a treatment process, it is necessary to review the nature and properties of these constituents. Because it is also important to understand the underlying principles, limitations, and biases of the analytical tests used to assess the performance of processes for residual particle removal, the principal tests used are reviewed in some detail in this section. The types and sizes of the particulate matter in treated effluent, the methods used for their quantification, and the treatment processes used for their removal are illustrated on Fig. 8-1.

Residual Constituents and Properties of Concern

Particles have a range of impacts on water quality depending upon their physical, chemical, and biological properties. As described in Table 8-1, the principal residual particulate constituents include microorganisms and colloidal and suspended particles. The principal properties of concern with respect to the removal of residual particulate matter are the

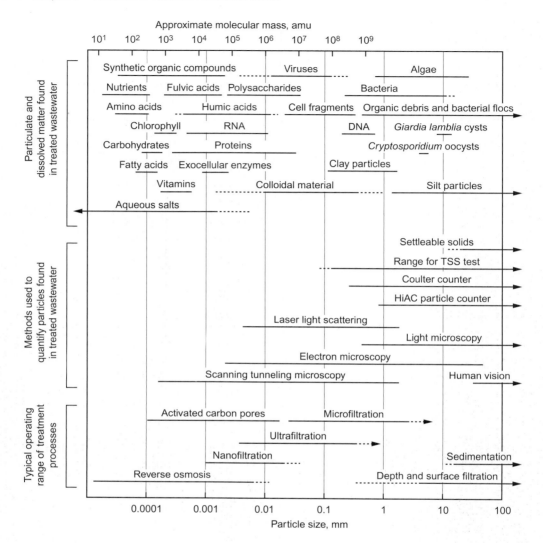

Figure 8-1
Size range of constituents in wastewater, methods used to quantify particulate matter, and operating range of particle removal processes.

distribution of particle sizes and turbidity. Other physical properties of interest such as absorbance and transmittance are considered in Chap. 11, which deals with disinfection.

Microorganisms

Pathogenic bacteria, protozoa, helminths, and viruses are the principal microorganisms in secondary effluent from conventional secondary biological treatment processes. Lagoon effluents will contain significant amounts of algae in addition to the microorganisms cited above. As noted in Table 7-7 in Chap. 7, the range of numbers

Residual constituent	Effect
Inorganic and organic suspended and colloidal solids	
Suspended solids	Can impact disinfection by shielding organisms
	Can clog sprinklers and drip irrigation tubing
Colloidal solids	May affect effluent turbidity; toxic constituents may be adsorbed on colloidal particles
Particulate organic matter	May shield bacteria during disinfection
Biological	
Bacteria	May cause disease
Protozoan cysts and oocysts	May cause disease
Viruses	May cause disease

Table 8-1
Typical residual constituents found in treated wastewater effluents and their effect on reuse applications[a]

[a]Adapted in part from Tchobanoglous et al. (2003).

of microorganisms found in untreated wastewater can be extremely large. As shown on Fig. 8-1, depth and surface filtration may effectively remove some portion of the protozoan cysts and oocysts of concern but are of limited value for the removal of bacteria and virus. The importance of the filtration process relevant to microorganisms is in the preparation of a water quality that is suitable for reliable disinfection.

Suspended Particles

Depending on the design of the secondary treatment facilities, the suspended solids found in treated effluent typically vary from about 1 to 200 μm. Light larger floc particles with sizes up to 500 μm and greater will also be present, depending on the operation of the biological treatment process and the design of the secondary clarifier. Suspended solids are of concern because they contribute to turbidity and can interfere with downstream treatment processes, such as disinfection, reverse osmosis (RO), and advanced oxidation. Large suspended particles can also interfere with spray nozzle and drip irrigation tubing used for irrigation applications.

Measurement of Suspended Solids The size range for colloidal particles reported in the literature varies from 0.001 to 0.003 μm for the lower size range to 1.0 to 2.0 μm for the upper size range. The size range for colloidal particles considered in this text is from 0.01 to 1.0 μm. As illustrated on Fig. 8-1, colloidal materials can include the smaller bacteria and a number of large viruses. The number of colloidal particles in treated wastewater is typically in the range of 10^8 to 10^{12}/mL. Colloidal materials are also important with respect to the measurement of turbidity and transmittance.

The total mass of suspended solids is determined by filtering a known volume of water through a membrane of known weight, with a nominal pore size of 1 μm. Because a filter is used to differentiate between total suspended solids (TSS) and total dissolved solids (TDS), the TSS test is somewhat arbitrary, depending on the pore size of the filter used for the test. Because filters with nominal pore sizes varying from 0.45 μm to about 2.0 μm have been used for the TSS test, it is difficult to compare TSS values reported in the literature. Also, because more TSS will be measured when a filter with a smaller pore size is used, it is important to note clearly the pore size of the filter used when comparing reported TSS values.

Limitations of Suspended Solids Measurements It is also important to note that the TSS test itself has no fundamental significance. The principal reasons that the test lacks a fundamental basis are:

1. The measured values of TSS are dependent on the type and pore size of the filter used in the analysis.
2. Depending on the sample size used for the determination of TSS, auto filtration, where the suspended solids that have been intercepted by the filter also serve as a filter, can occur. Auto filtration will capture smaller particles than otherwise possible and cause an apparent increase in the measured TSS value over the actual value.
3. Depending on the characteristics of the particulate matter, small particles may be removed by adsorption to material already retained by the filter.
4. Because the number and size distribution of the particles that comprise the measured value is unknown, TSS is a *lumped parameter*.

Distribution of Particle Sizes

Particle size distribution data may be used to optimize the performance of processes used for particle removal. Methods that have been used to characterize the distribution of particle sizes of residual colloidal and particulate matter from secondary treatment processes include (1) serial filtration, (2) electronic particle size counters, and (3) direct microscopic observation.

Serial Filtration Serial filtration may be used to determine an approximate particle size distribution of suspended solids based on mass (Levine et al., 1985). In the serial filtration method, a wastewater sample is passed sequentially through a series of membrane filters (see Fig. 8-2) with circular openings of known diameter (typically 12, 8, 5, 3, 1, and 0.1 μm), and the amount of particulate material retained in each filter is measured. Typical results from such a measurement are shown on Fig. 8-3. What is interesting to note in Fig. 8-3 is that a significant amount of colloidal material will be found between 0.1 and 1.0 μm. Although some information is gained on the size and distribution of the particles in the wastewater sample, little information is gained on the nature of the individual particles.

Electronic Particle Size Analyzers To understand more about the nature and distribution of particles in wastewater, nondestructive measurement of particle size and particle size distribution is now quite common. However, it should be noted that electronic particle sizing and counting techniques cannot be used reliably for determining the source or type of particle (e.g., distinguishing between a viable cyst, a nonviable cyst, or a similar size silt particle).

In electronic particle size counting, particles are counted by diluting a treated wastewater sample and then passing the diluted sample through a calibrated orifice or past laser beams. As the particles pass through the orifice, the conductivity of the fluid changes,

8-1 Characteristics of Residual Suspended Particulate Matter from Secondary Treatment Processes | 379

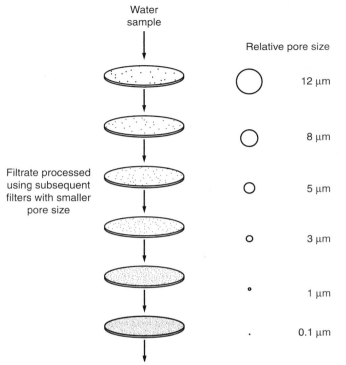

Figure 8-2
Definition sketch for the determination of the particle size distribution (by mass) using serial filtration with membrane filters.

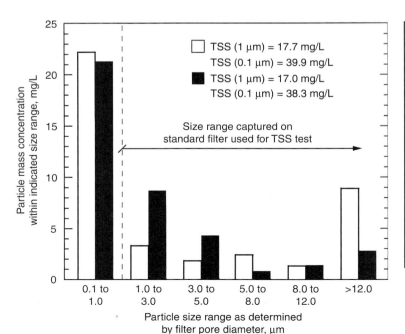

Figure 8-3
Typical data on the distribution of filterable solids obtained by serial filtration in treated trickling filter effluent. Note: large fraction of unmeasured solids between 0.1 and 1 μm using conventional TSS test (Adapted from Levine et al., 1985).

Table 8-2
Analytical techniques applicable to particle size analysis of wastewater constituents[a]

Technique	Typical particle size range, μm
Particle counters	
Conductivity difference	0.2 to >100
Equivalent light scattering	0.005 to >100
Light blockage	0.2 to >100
Microscopy	
Light	0.2 to >100
Transmission electron	0.2 to >100
Scanning electron	0.002 to 50
Image analysis	0.2 to >100

[a] Adapted from Levine et al. (1985).

due to the presence of the particle. The change in conductivity is correlated to the size of an equivalent sphere. In a similar fashion, as a particle passes by a laser beam, it reduces the intensity of the laser due to light scattering. The reduced intensity is correlated to the diameter of the particle.

Particle counters have sensors available in different sizes ranges, such as 1.0 to 60 μm or 2.5 to 150 μm, depending on the manufacturer and application. The typical size range quantifiable with different types of particle size analyzers is shown in Table 8-2. Particle counters that do not measure particles smaller than 1 μm may be a limitation in some cases. Particle counts are typically measured and recorded in about 10 to 20 size ranges (e.g., 2 to 5 μm) called channels (or bins) of the chosen sensor range. Channel sizes can be arithmetic, logarithmic, or arbitrary, depending on the measurement objective. Using a logarithmic scale, the upper channel limit is equal to the lower channel limit times a scaling factor. For disinfection studies channel sizes should be selected that represent size ranges of interest, for example, *Cryptosporidium* (2 to 5 μm) and *Giardia* (5 to 15 μm). With particle size counters that use large numbers of small channel sizes, the interpretation of the resulting data is more difficult. Where extremely small channel sizes are used, it is recommended that the data be aggregated into appropriate bin sizes (see Fig. 8-16 in Sec. 8-2). In addition to reporting particle number by size, the data can be reported in terms of surface area and volume; the volume fraction corresponding to each particle size range can also be computed, if needed (Standard Methods, 2005). Typical particle size counters are shown on Fig. 8-4.

In wastewater, it has been observed that the number of particles increases with decreasing particle diameter and that the frequency distribution typically follows a power law distribution of the form:

$$\frac{dN}{d(d_p)} = A(d_p)^{-\beta} = \frac{\Delta N}{\Delta(d_{pi})} \tag{8-1}$$

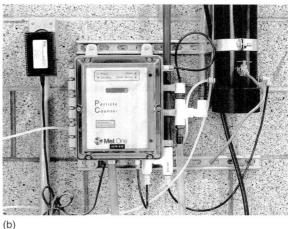

Figure 8-4
Views of particle size counters: (a) laboratory type and (b) online field type.

where dN = the particle number concentration with respect to the incremental change in particle diameter $d(d_p)$, number/mL·μm
$d(d_p)$ = incremental change in particle diameter, μm
A = power law density coefficient, unitless
d_p = arithmetic (or geometric) mean particle diameter, depending on counter channel configuration, μm
β = power law slope coefficient
ΔN = the particle number concentration in given channel, number/mL
$\Delta(d_{pi})$ = incremental channel size, μm

In effect the right-hand term in Eq (8-1) is used to normalize the data and allows for comparison between particle size distributions. Taking the log of both sides of Eq. (8-1) results in the following expression, which can be plotted to determine the unknown coefficients A and β

$$\log[\Delta N/\Delta(d_{pi})] = \log A - \beta \log(d_p) \qquad (8\text{-}2)$$

The value of A is determined when $d_p = 1$ μm. As the value of A increases, the total number of particles in each size classification increases. The slope β is a measure of the relative number of particles in each size range. Thus, if β is less than one, the particle size distribution is dominated by large particles, if β is equal to one all particle sizes are represented equally, and if β is greater than one the particle size distribution is dominated by small particles (Trussell and Tate, 1979). Because different slope values will be obtained, depending on the selection of the bin sizes, care must be exercised in interpreting the results. The analysis of data obtained from a particle size counter is shown in Example 8-1.

EXAMPLE 8-1. Analysis of Particle Size Information.

Determine the coefficients A and β in Eq. (8-1) for the following particle size data obtained using a particle counter with arithmetic channel settings.

Channel size, μm	Number
1–2	20,030
2–5	6688
5–15	1000
15–20	300
20–30	150
30–40	26.8
40–60	12.3
60–80	5.4
80–100	3.4
100–140	1.1

Solution

1. Set up a table to determine the information needed to plot the data.

Channel size, μm	Mean diameter[a], d_p, μm	ΔN, number/mL	Channel size interval, $\Delta(d_{pi})$	log (d_p)	log[$\Delta N/\Delta(d_{pi})$]
1–2	1.50	20,030	1	0.18	4.30
2–5	3.50	6688	3	0.54	3.35
5–15	10.00	1000	10	1.00	2.00
15–20	17.50	300	5	1.24	1.78
20–30	25.00	150	10	1.40	1.18
30–40	35.00	26.8	10	1.54	0.43
40–60	50.00	12.3	20	1.70	−0.21
60–80	70.00	5.4	20	1.85	−0.57
80–100	90.00	3.4	20	1.95	−0.77
100–140	120.00	1.1	40	2.08	−1.56

[a] Arithmetic mean diameter; for example, 1.5 = [(1 + 2)/2].

2. Prepare a plot of the log of the geometric mean particle diameter, d_p, versus the normalized number of particles for the corresponding bin size, $\log[\Delta N/\Delta(d_{pi})]$.

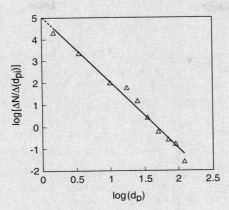

3. Determine A and β in Eq. (8-1).
   a. Determine A
      When $\log(d_p) = 0$, $d_p = 1$, and $A = 10^{5.05}$
   b. Determine β

$$-\beta = \frac{3.5 - (-1.0)}{0.5 - 2} = -3.0$$

$$\beta = 3.0$$

Comment
As the value of β is greater than one, the distribution is dominated by small particles, which is consistent with the actual data. It is important to note that the slope of the line of best fit through the plotted data will vary depending on the bin sizes selected for analysis. It should be noted that the line used to define β may not be linear depending on the characteristics of the suspension and the minimum and maximum particle sizes measured, a characteristic of the specific instrument used in the analysis. It should also be noted that the channel sizes of 2 to 5 μm and 5 to 15 μm were selected to determine if the number of *Cryptosporidium* or *Giardia* determined analytically can be correlated with particle size measurements.

Direct Observation For visualization of particles that are smaller than those visible to the unaided eye, microscopic techniques may be used. The use of microscopic observation allows for the determination of particle size counts, and in some cases, for more rigorous identification of a particle's origin than is possible with other analysis techniques. In microscopic observation, a measured volume of sample is placed in a particle counting cell and the individual particles may be counted, often with the use of a stain to enhance the particle contrast. The size range quantifiable using a variety of microscopic techniques is reported in Table 8-2. In general, microscopic counting of particles is impractical on a routine basis, given the number of particles per mL of wastewater. Nevertheless, this method can be used to qualitatively assess the nature and size of the particles in wastewater.

Turbidity

Turbidity, a measure of the light-transmitting properties of water, is another test used to indicate the clarity of treated wastewater with respect to colloidal and residual particulate matter. Turbidity in water is caused by the presence of suspended particles that reduce the clarity of the water. Turbidity is defined as "... an expression of the optical property that causes light to be scattered and absorbed rather than transmitted with no change in direction or flux level through the sample" (Standard Methods, 2005).

Turbidity Measurement Turbidity measurements require a light source (incandescent or light-emitting diode) and a sensor to measure the scattered light. As shown on Fig. 8-5, the scattered light sensor is located at 90 degrees to the light source. The measured turbidity increases as the intensity of the scattered light increases. Turbidity is expressed in nephelometric turbidity units (NTU). The spatial distribution and intensity of the scattered light, also illustrated on Fig. 8-5, will depend on the size of the particle relative to the wavelength of the light source (Hach, 1997). For particles less than one-tenth of the wave length of the incident light, the scattering of light is fairly symmetrical (see Fig. 8-5a).

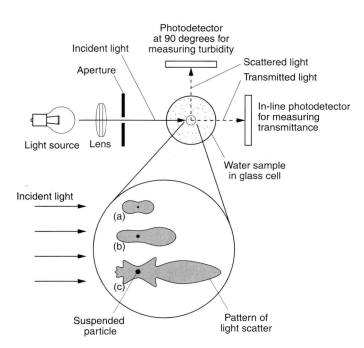

Figure 8-5

Definition sketch for the determination of turbidity and light scattering patterns for various size particles (Adapted in part from Hach, 1997).

Limitations of Turbidity Measurements As the particle size increases relative to the wave length of the incident light, the light reflected from different parts of the particle create interference patterns that are additive in the forward direction (see Figs. 8-5b and c). Also, the intensity of the scattered light varies with the wavelength of the incident light. For example, blue light is scattered more than red light. Based on these considerations, turbidity measurements tend to be more sensitive to particles in the size range of the incident light wavelength (0.3 to 0.7 μm for visible light). Thus, two filtered wastewater samples with nearly identical turbidity values could have very different particle size distributions. A further complication with turbidity measurements is that some particles will essentially adsorb most of the light, and only scatter a minimal amount of the incident light. Also, it should be noted that because of the light scattering characteristics of large particles, a few large particles would not be detected in the presence of many smaller particles. Because, there is no fundamental relationship between turbidity and the concentration of total suspended solids, turbidity alone is not a good measure of whether a wastewater can be disinfected effectively.

Removal of Residual Particles from Secondary Treatment Processes

Most treated wastewaters contain a wide variety of residual colloidal and suspended solids. In many wastewater reuse applications, removal of residual particulate matter is required, typically by some form of filtration. The characteristics of the residual solids remaining from various biological treatment processes are discussed in the following paragraphs.

Activated Sludge Processes

The total suspended solids, particle size distribution, and floc strength of the residual particles from secondary sedimentation facilities are considered in the following discussion, as these parameters influence the filtration processes used for their removal as well as other downstream processes.

Total Suspended Solids and Turbidity Typically, the TSS concentration in the effluent from activated sludge (and trickling filter) plants varies between 5 and 25 mg/L. In modern treatment plant designs employing nutrient removal and deep secondary clarifiers, the effluent TSS range is from 4 to 10 mg/L. Corresponding turbidity values can vary from less than 2 to 15 NTU.

Probability distributions for TSS and turbidity values from activated sludge processes are shown on Fig. 8-6. Although the variation in the value of the geometric standard deviation from plant to plant is not too great, there is considerable variation in the reported mean values observed from different activated sludge processes. The variations in the reported mean values are related to the operation of the activated sludge process, the design and operation of the secondary sedimentation facilities, and to a lesser extent the influent wastewater characteristics.

Particle Size Distribution Typical effluent particle size data from a variety of activated sludge treatment plants were reported on Fig. 7-5. The effluent particle size distribution shown on Fig. 7-5 is bimodal with respect to volume distribution, with distinct size ranges. Small particles varying in areal size (equivalent circular diameter) from

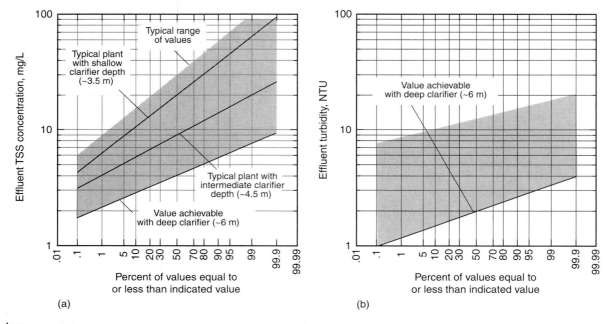

Figure 8-6
Variability of effluent TSS and turbidity values from activated sludge processes: (a) TSS and (b) turbidity.

0.8 to 1.2 μm, correspond primarily to bacteria and cell debris, and larger floc particles vary in size from about 5 to 100 μm. The mass fraction of the smaller particles varies from about 40 to 60 percent of the total. This percentage varies, however, depending on the operating conditions of the biological process and the degree of flocculation achieved in the secondary settling facilities. The observed particle size distribution is of importance, as it will influence the removal mechanisms that may be operative during the filtration process. For example, it is reasonable to assume that the removal mechanism for particles 1.0 μm in size may be different from that for particles in the size range from 10 to 100 μm.

Floc Strength Floc strength, which varies with the type of process and the mode of operation, is also important. For example, the residual floc from the chemical precipitation of biologically processed wastewater may be considerably weaker than the residual biological floc before precipitation. Further, the strength of the biological floc varies with solids retention time (SRT), increasing with longer SRT. The increased strength derives in part from the production of extracellular polymers as the SRT is lengthened. At extremely long SRTs (15 d and longer), it has been observed that the floc strength decreases due to floc breakup.

Microorganisms The removal of microorganisms in the activated sludge process depends on the type of process and the mode of operation (see Sec. 7-3 in Chap. 7) Typically, a reduction of 2 to 3 log can be expected for coliform organisms in most activated sludge processes. A 1.5 to 2 log reduction can be expected in viruses.

Trickling Filters

The average effluent TSS and BOD concentrations from trickling filter processes are typically higher than those of a well-operated activated sludge process, typically 20 to 30 mg/L (see Fig. 7-5 in Chap. 7). As shown on Fig. 7-5, the distribution of effluent suspended solids from trickling filters is similar to that for the activated sludge process, with low solids retention times.

Lagoons

In addition to some residual suspended solids, the bulk of the suspended solids in the effluent from lagoon treatment systems will be comprised of algae of various sizes (see Crites and Tchobanoglous, 1998). As shown on Fig. 8-7, the algae in lagoon effluents vary in species, size, and number concentration depending on the season and other factors. The slope of the probability distribution for suspended solids is similar to that for activated sludge.

Other Treatment Processes

An alternative biological treatment process involves the use of a trickling filter with a short detention time activated sludge reactor (see Chap. 7 for hybrid processes). The average TSS and BOD effluent concentrations and particle size distribution are similar to the activated sludge process (see Fig. 7-5).

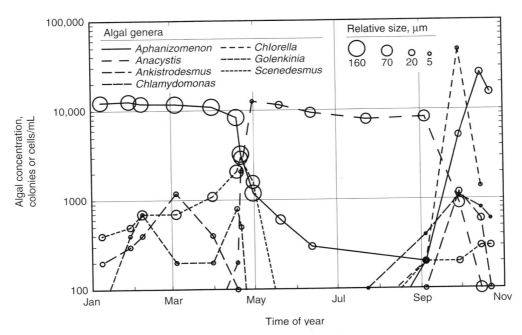

Figure 8-7
Variation in algal concentration in pond effluent at different times of the year (Adapted from Stowell, 1976).

Constructed wetlands have also been used in some water reuse applications. Effluent TSS concentrations vary from about 20 to 30 mg/L. In addition to suspended solids, the effluent from constructed wetlands also contains some recalcitrant particles produced in the process.

8-2 TECHNOLOGIES FOR THE REMOVAL OF RESIDUAL SUSPENDED PARTICULATE MATTER

The physical operations used most commonly are depth, surface, and membrane filtration. In some instances, dissolved air flotation (DAF) is used. Each of these technologies is introduced briefly in this section and discussed in more detail in the following four sections.

Technologies for Reclaimed Water Applications

The removal of particulate material suspended in a liquid can be accomplished by (1) depth filtration (passing the liquid through a filter bed comprised of a granular or compressible filter medium), (2) surface filtration (the removal of particulate material suspended in a liquid by mechanical sieving through a thin septum), (3) membrane filtration (passing the liquid through semipermeable membranes to exclude particles ranging in size from 0.005 to 2.0 μm), or (4) dissolved air flotation (attaching air bubbles to particulate matter to provide buoyancy so the particles can be removed by skimming). A schematic representation of each type of technology is presented on Fig. 8-8 and a brief introductory discussion is presented below.

Depth Filtration

Depth filtration (see Fig. 8-8a) was developed originally for the treatment of surface water for potable uses and later adapted for wastewater treatment applications. Depth filtration is used in reuse applications to achieve supplemental removal of suspended solids (including particulate BOD) from wastewater effluents for the following purposes: (1) to allow more effective disinfection; (2) as a pretreatment step for subsequent treatment steps such as carbon adsorption, membrane filtration, or advanced oxidation; and (3) to remove chemically precipitated phosphorus.

Surface Filtration

Surface filtration (see Fig. 8-8b) is used to remove the residual suspended solids from secondary effluents and stabilization pond effluents, and is being used as an alternative to depth filtration as pretreatment for membrane filtration. Surface filtration, a relatively new technology, involves a sieving action similar to a kitchen colander.

Membrane Filtration

Membrane filtration (see Fig. 8-8c) with MF and UF membranes is being used increasingly for water and wastewater applications. Microfiltration and UF membrane filters are also surface filtration devices but are differentiated on the basis of the sizes of the pores in the filter medium; the pore size can vary from 0.005 to 2.0 μm. In water reuse applications, MF and UF usually follow biological treatment and are used to remove particulate matter (including pathogens), organic matter, and some nutrients not removed by secondary clarification. Product water from MF and UF may be used

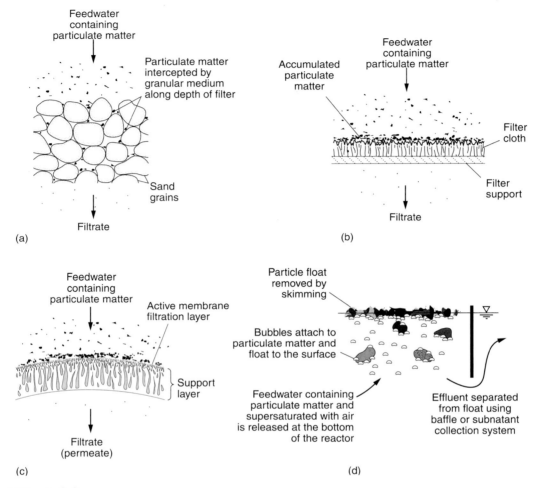

Figure 8-8
Definition sketch for the removal of particulate matter by (a) depth filtration, (b) surface filtration, (c) membrane filtration, and (d) dissolved air flotation.

directly for a variety of reuse applications (after disinfection) or used as pretreated feedwater for further treatment by nanofiltration (NF) or reverse osmosis (RO).

Dissolved Air Flotation

Flotation (see Fig. 8-8d) is a gravity separation process in which gas bubbles attach to solid particles to cause the density of the bubble-solid agglomerates to be lighter than water. In DAF, bubbles are produced by the reduction of pressure in a water stream saturated with air, similar to the bubble formation in a carbonated beverage when the top is removed. Dissolved air flotation has been used in water treatment as an alternative to sedimentation for treatment of nutrient-rich reservoir waters containing heavy algae blooms and for low-alkalinity, colored waters (AWWA, 1999). For water reuse

Figure 8-9
Typical process flow diagrams used for the removal of suspended particulate matter from secondary effluents: (a) depth and surface filtration, (b) membrane filtration, and (c) dissolved air filtration.

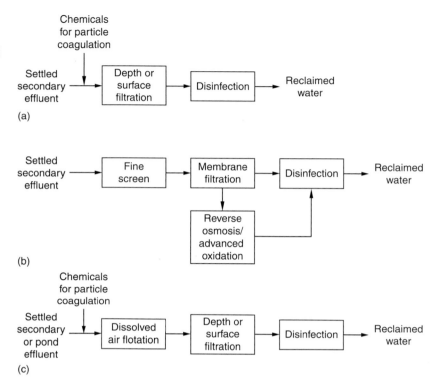

applications, DAF has been used principally for treating pond effluents containing algae and for low-density particles that are difficult to remove by gravity sedimentation, as a pretreatment step for depth or surface filtration.

Process Flow Diagrams

Typical process flow diagrams that can be used for the removal of residual suspended particulate matter from secondary effluents are shown on Fig. 8-9 for depth and surface filtration, membrane filtration, and dissolved air flotation. With depth and surface filtration, chemical addition is used commonly, often in combination with flocculation and sedimentation, as shown on Fig. 8-9a. For membrane filtration, a fine screen will often precede a membrane filter to mitigate the effects of solids that tend to clog the membrane. Dissolved air flotation is used often with chemical coagulation to help form flocculent particles that can be removed by flotation from secondary and pond effluents.

Process Performance Expectations

As discussed in Sec. 7-3 in Chap. 7, where treated effluent is to be reused, it is important to know what typical mean effluent constituent values can be expected and the variability in those values. Information on the constituent values and variability is of importance in the selection of technologies that might be used to further process the treated effluent.

The ranges of typical mean effluent constituent values that can be achieved with various particulate removal processes are reported in Table 8-3. The variability observed in the performance of various particulate removal processes with respect to TSS and turbidity in the treated effluent is discussed in the following three sections, which deal with depth, surface, and membrane filtration.

Table 8-3
Typical range of effluent quality after removal of residual particulate matter[a]

Constituent	Unit	Conventional activated sludge[b]	Conventional activated sludge with filtration[b]	Activated sludge with BNR and filtration[c]	Membrane bioreactor
Total suspended solids (TSS)	mg/L	5–25	2–8	1–4	≤1
Colloidal solids	mg/L	5–25	5–20	1–5	≤1
Biochemical oxygen demand (BOD)	mg/L	5–25	<5–20	1–5	<1–5
Chemical oxygen demand (COD)	mg/L	40–80	30–70	20–30	<10–30
Total organic carbon (TOC)	mg/L	10–40	15–30	1–5	0.5–5
Ammonia nitrogen	mg N/L	1–10	1–6	1–2	<1–5
Nitrate nitrogen	mg N/L	10–30	10–30	1–5	<10[d]
Nitrite nitrogen	mg N/L	0–trace	0–trace	0–trace	0–trace
Total nitrogen	mg N/L	15–35	15–35	2–5	<10[d]
Total phosphorus	mg P/L	4–10	4–8	≤2	<0.3[e]–5
Turbidity	NTU	2–15	0.5–4	0.3–2	≤1
Volatile organic compounds (VOCs)	µg/L	10–40	10–40	10–20	10–20
Metals	mg/L	1–1.5	1–1.4	1–1.5	trace
Surfactants	mg/L	0.5–2	0.5–1.5	0.1–1	0.1–0.5
Totals dissolved solids (TDS)	mg/L	500–700	500–700	500–700	500–700
Trace constituents	µg/L	5–40	5–30	5–30	0.5–20
Total coliform	No./100 mL	10^4–10^5	10^3–10^5	10^4–10^5	<100
Protozoan cysts and oocysts	No./100 mL	10^1–10^2	0–10	0–1	0–1
Viruses	PFU/100 mL[f]	10^1–10^3	10^1–10^3	10^1–10^3	10^0–10^3

[a]From Chap. 3, Tables 3-12 and 3-14.
[b]Conventional activated sludge is defined as activated sludge treatment with nitrification.
[c]BNR is defined as biological nutrient removal for the removal of nitrogen and phosphorus.
[d]With anoxic stage.
[e]With coagulant addition.
[f]Plaque-forming units.

Suitability for Reclaimed Water Applications

The technologies introduced in this section can be used to produce reclaimed water of varying quality depending on the elements of the process train. The filtration systems have different pore sizes and therefore can exclude different constituents, as shown on Fig. 8-1. Ultrafiltration membranes also have the ability to remove some bacterial cells, some colloidal material, and viruses. In wastewater reclamation, MF or UF might provide a level of treatment suitable for a variety of the reuse applications or they might be used in tandem with NF or RO where higher constituent removals of dissolved solids or dissolved organics are required. Nanofiltration and RO are discussed in Chap. 9.

8-3 DEPTH FILTRATION

Depth filtration is one of the oldest unit operations used in the treatment of potable water and is the most common method used for the filtration of effluents from wastewater treatment processes, especially in water reuse applications. In addition to providing supplemental removals of suspended solids (including particulate BOD) from wastewater effluents of biological and chemical treatment processes, depth filtration may be used as a pretreatment step for membrane filtration. Depth filtration is especially important as a conditioning step for effective disinfection.

Available Filtration Technologies

The principal types of depth filters that have been used for the filtration of wastewater are described in Table 8-4. As shown in Table 8-4, the filters can be classified in terms of their operation as semicontinuous or continuous. Filters that must be taken off-line periodically to be backwashed are classified operationally as semicontinuous. Filters in which the filtration and backwash operation occurs simultaneously are classified as continuous. Within each of these two classifications there are a number of different types of filters depending on bed depth (e.g., shallow, conventional, and deep bed), the type of filtering medium used (mono-, dual-, and multi-medium), whether the filtering medium is stratified or unstratified, the type of operation (downflow or upflow), and the method used for the management of solids (i.e., surface or internal storage). For the mono- and dual-medium semicontinuous filters, a further classification can be made based on the driving force (e.g., gravity or pressure), although most of the filters used commonly in reuse applications are gravity flow. Another important distinction that must be noted for the filters identified in Table 8-4 is whether they are proprietary or individually designed.

The five types of depth filters used most commonly for wastewater filtration at larger treatment plants [greater than 1000 m^3/d (0.25 Mgal/d)] are: (1) conventional downflow filters (mono-, dual-, and multi-medium), (2) deep-bed downflow filters, (3) deep-bed upflow continuous-backwash filters, (4) the pulsed bed filter, and (5) traveling bridge filters. Recent developments in filtration technology for reuse applications include a filter employing a synthetic filter medium and a two-stage filtration system which incorporates phosphorus removal. Pressure filters, which operate in the same manner as gravity filters, are used at smaller plants. Many of the filters are proprietary and are supplied by the manufacturer as a complete unit. Each of these eight filter types is described in Table 8-5. Views of several different types of filter installations are shown on Fig. 8-10.

Table 8-4
Comparison of principal types of granular medium filters

Type of filter	Type of filter operation	Filter bed details[a] Type	Filtering medium	Typical direction of flow	Backwash operation	Flowrate through filter	Solids storage location	Type of design	Remarks
Conventional	Semi-continuous	Mono-medium (stratified or unstratified)	Sand or anthracite	Downward	Batch	Constant/variable	Surface and upper bed	Individual	Rapid headloss buildup
Conventional	Semi-continuous	Dual-medium (stratified)	Sand and anthracite	Downward	Batch	Constant/variable	Internal	Individual	Dual-medium design used to extend length of filter run
Conventional	Semi-continuous	Multi-medium (stratified)	Sand, anthracite, and garnet	Downward	Batch	Constant/variable	Internal	Individual	Multi-medium design used for particle depth penetration
Deep bed	Semi-continuous	Mono-medium (stratified or unstratified)	Sand or anthracite	Downward	Batch	Constant/variable	Internal	Individual	Deep bed used to store solids and extend length of filter run
Deep bed	Semi-continuous	Mono-medium (stratified)	Sand	Upward	Batch	Constant	Internal	Proprietary	Deep bed used to store solids and extend length of filter run
Deep bed	Semi-continuous	Mono-medium (unstratified)	Sand	Upward	Continuous	Constant	Internal	Proprietary	Sand bed moves in countercurrent direction to fluid flow
Pulsed bed	Semi-continuous	Mono-medium (stratified)	Sand	Downward	Batch	Constant	Surface and upper bed	Proprietary	Air pulses used to break up surface mat and increase run length
Fuzzy filter	Semi-continuous	Mono-medium (unstratified)	Synthetic fiber	Upward	Batch	Constant	Internal	Proprietary	Perforated plate is used to retain the filter medium during backwash
Traveling bridge	Continuous	Mono-medium (stratified)	Sand	Downward	Semi-continuous	Constant	Surface and upper bed	Proprietary	Individual filter cells backwashed sequentially
Traveling bridge	Continuous	Dual-medium (stratified)	Sand and anthracite	Downward	Semi-continuous	Constant	Surface and upper bed	Proprietary	Individual filter cells backwashed individually
Pressure filters	Semi-continuous	Mono or dual medium	Sand and/or anthracite	Downward	Batch	Constant/variable	Surface and upper bed	Individual and proprietary	Used for small plants

[a] For filter bed depths, see Tables 8-8 and 8-9.

Table 8-5
Description of commonly used depth filters for reclaimed water applications[a]

Filter type	Description
(a) Conventional downflow 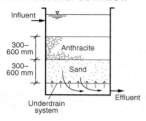	Wastewater containing suspended matter is applied to the top of the filter bed. Single-, dual-, or multi-medium filter materials are used. Typically sand or anthracite is used as the filtering material in single-medium filters. Dual-medium filters usually consist of a layer of anthracite over a layer of sand. Other combinations include: (1) activated carbon and sand, (2) resin beads and sand, and (3) resin beads and anthracite. Multi-medium filters typically consist of a layer of anthracite over a layer of sand over a layer of garnet or ilmenite. Other combinations include: (1) activated carbon, anthracite, and sand, (2) weighted spherical resin beads, anthracite, and sand, and (3) activated carbon, sand, and garnet.
(b) Deep-bed downflow 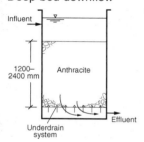	The deep-bed downflow filter is similar to the conventional downflow filter with the exception that the depth of the filter bed and the size of the filtering medium (usually anthracite) are greater than the corresponding values in a conventional filter. Because of the greater depth and larger medium size (i.e., sand or anthracite), more solids can be stored within the filter bed and the run length can be extended. The maximum size of the filter medium used in these filters depends on the ability to backwash the filter. In general, deep-bed filters are not fluidized completely during backwashing. To achieve effective cleaning, air scour plus water is used in the backwash operation.
(c) Deep-bed upflow continuous backwash 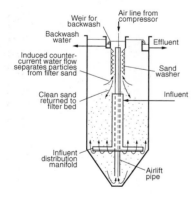	Wastewater to be filtered is introduced into the bottom of the filter where it flows upward through a series of riser tubes and is distributed evenly into the sand bed through the open bottom of an inlet distribution hood. The water then flows upward through the downward-moving sand. Clean filtrate exits from the sand bed, overflows a weir, and is discharged from the filter. At the same time sand particles, along with trapped solids, are drawn downward into the suction of an airlift pipe that is positioned in the center of the filter. A small volume of compressed air, introduced into the bottom of the airlift, draws sand, solids, and water upward through the pipe by creating a fluid with a density less than one. Impurities are scoured (abraded) from the sand particles during the turbulent upward flow. Upon reaching the top of the airlift, the dirty slurry spills over into the central reject compartment. A steady stream of clean filtrate flows upward, countercurrent to the movement of sand, through the washer section. The upflow liquid carries away the solids and reject water. Because the sand has a higher settling velocity than the removed solids, the sand is not carried out of the filter. The sand is cleaned further as it moves down through the washer. The cleaned sand is redistributed onto the top of the sand bed, allowing for a continuous uninterrupted flow of filtrate and reject water.

Table 8-5

Description of commonly used depth filters for reclaimed water applications[a] (*Continued*)

Filter type	Description
(d) Pulsed-bed 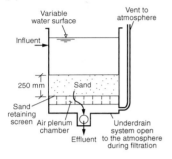	The pulsed-bed filter is a proprietary downflow gravity filter with an unstratified shallow layer of fine sand as the filtering medium. The shallow bed is used for solids storage, as opposed to other shallow-bed filters where solids are stored principally on the sand surface. An unusual feature of this filter is the use of an air pulse to disrupt the sand surface and thus allow penetration of suspended solids into the bed. The air pulse process involves forcing a volume of air, trapped in the underdrain system, up through the shallow filter bed to break up the surface mat of solids and renew the sand surface. When the solids mat is disturbed, some of the trapped material is suspended, but most of the solids are trapped within the filter bed. The intermittent air pulse causes a folding over of the sand surface, burying solids within the medium and regenerating the filter bed surface. The filter continues to operate with intermittent pulsing until a terminal headloss limit is reached. The filter then operates in a conventional backwash cycle to remove solids from the sand. During normal operation the filter underdrain is not flooded as it is in a conventional filter.
(e) Traveling bridge 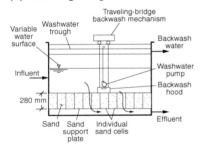	The traveling bridge filter is a proprietary continuous downflow, automatic backwash, low-head, granular medium-depth filter. The bed of the filter is divided horizontally into long independent filter cells. Each filter cell contains approximately 280 mm of medium. Treated wastewater flows through the medium by gravity and exits to the clearwell plenum via a porous-plate, polyethylene underdrain. Each cell is backwashed individually by an overhead traveling bridge assembly, while all other cells remain in service. Water used for backwashing is pumped directly from the clearwell plenum up through the medium and deposited in a backwash trough. During the backwash cycle, wastewater is filtered continuously through the cells that are not being backwashed. The backwash mechanism includes a surface wash pump to assist in breaking up of the surface matting and "mudballing" in the medium. Because the backwashing operation is performed on an "as needed" basis, the backwash cycle is termed semi-continuous.
(f) Synthetic medium 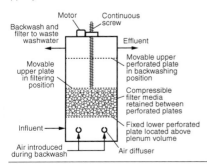	A synthetic medium filter, developed originally in Japan, is used for reclaimed water filtration. Unusual features of the filter are: (1) the porosity of the filter bed can be modified by compressing the filter medium and (2) the size of the filter bed is increased mechanically to backwash the filter. The filter medium, a highly porous synthetic material made of polyvaniladene, allows the influent to flow through the medium as opposed to flowing around the filtering medium, as in sand and anthracite filters. The porosity of the uncompacted quasi-spherical filter medium itself is estimated to be about 88 to 90 percent, and the porosity of the filter bed is approximately 94 percent. Filtration rates of 400 to 1200 $L/m^2 \cdot min$ have been pilot tested (Caliskaner and Tchobanoglous, 2000).

(*Continued*)

Table 8-5

Description of commonly used depth filters for reclaimed water applications[a] (*Continued*)

Filter type	Description
(f) Synthetic medium (*Continued*)	In the filtering mode, secondary effluent is introduced in the bottom of the filter. The influent wastewater flows upward through the filter medium, retained by two porous plates, and is discharged from the top of the filter. To backwash the filter, the upper porous plate is raised mechanically. While flow to the filter continues, air is introduced sequentially from the left and right sides of the filter below the lower porous plate, causing the filter medium to move in a rolling motion. The filter medium is cleaned by the shearing forces as the wastewater moves past the filter, and by abrasion as the filter medium rubs against itself. Wastewater containing the solids removed from the filter is diverted for subsequent processing. To put the filter back into operation after the backwash cycle has been completed, the raised porous plate is returned to its original position. After a short flushing cycle, the filtered effluent valve is opened, and filtered effluent is discharged.
(g) Two-stage 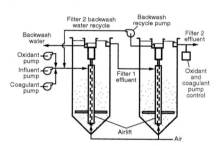	A proprietary two-stage filtration process is used for the removal of turbidity, total suspended solids, and phosphorus. Two deep-bed upflow continuous backwash filters are used in series to produce a high quality effluent. A large-size sand diameter is used in the first filter to increase the contact time and to minimize clogging. A smaller sand size is used in the second filter to remove residual particles from the first-stage filter. The waste washwater from the second filter, which contains small particles and residual coagulant, is recycled to the first filter to improve floc formation within the first-stage filter and the influent-to-waste ratio. Based on full-scale installations, the reject rate has been found to be less than five percent. Phosphorus levels equal to or less than 0.02 mg/L have been achieved in the final filter effluent.
(h) Pressure filters 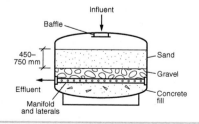	Pressure filters operate in the same manner as gravity filters and are used at smaller plants. The only difference is that in pressure filters, the filtration operation is carried out in a closed vessel under pressurized conditions achieved by pumping. Pressure filters normally are operated at higher terminal headlosses, resulting in longer filter runs and reduced backwash requirements. If, however, they are not backwashed on a regular basis, problems have been experienced with the formation of mudballs.

[a] Adapted from Tchobanoglous et al. (2003).

Figure 8-10
Views of typical filtration installations (a) view looking across banks of filters at a large installation. The physical size of the facility can be judged by the truck on access road between the filter banks, (b) view of complex piping system inside one of the filter galleries needed for the filters shown in (a) to function, (c) typical traveling bridge filter (empty) with individual cells exposed (see Table 8-5e), (d) deep denitrifying filter, (e) continuous backwash upflow filters (Courtesy of Austep, s.r.l., Italy), and (f) bank of small pressure filters used at small wastewater treatment plants.

Performance of Depth Filters

The critical question associated with the selection of any depth filter is whether it will perform as anticipated. Insight into the performance of depth filters can be gained from a review of the operational considerations and requirements, the operative particle removal mechanisms, and from performance data on the removal of turbidity, total suspended solids, and particle size alteration. Typical values of effluent quality and variability for depth filtration using granular media are presented in Table 8-6, along with comparable data for other filtration processes used for the removal particulate matter.

Operational Considerations

The principal operational consideration for a depth filter is the volume of water produced in a given time period. The volume of water produced is related to the

Table 8-6
Typical range of effluent quality variability observed from particulate removal processes

Parameter	Unit	Typical range of effluent values	Geometric standard deviation, s_g[a]	
			Range	Typical
Depth filtration following activated sludge process				
TSS	mg/L	2–8	1.3–1.5	1.4
Turbidity	NTU	0.5–4	1.2–1.4	1.25
Depth filtration following activated sludge with BNR				
TSS	mg/L	1–4	1.3–1.5	1.35
Turbidity	NTU	0.3–2	1.2–1.4	1.25
Surface filtration following activated sludge process				
TSS	mg/L	1–4	1.3–1.5	1.25
Turbidity	NTU	0.5–2	1.2–1.4	1.55
Microfiltration following activated sludge process				
TSS	mg/L	0–1	1.3–1.9	1.5
Turbidity	NTU	0.1–0.4	1.1–1.4	1.3
Ultrafiltration following surface filtration of activated sludge				
TSS	mg/L	0–1	1.3–1.9	1.5
Turbidity	NTU	0.1–0.4	1.1–1.4	1.3
Membrane bioreactor				
BOD	mg/L	<1–5	1.3–1.6	1.4
TSS	mg/L	<1–5	1.3–1.9	1.5
Turbidity	NTU	0.1–1	1.1–1.4	1.3

[a] Geometric standard deviation; $s_g = P_{84.1}/P_{50}$.

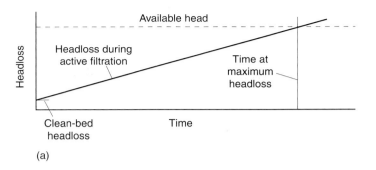
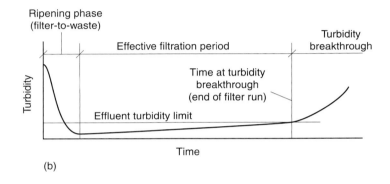

Figure 8-11
Definition sketch for length of filter run based on (a) headloss buildup and (b) effluent turbidity breakthrough.

development of headloss and filter performance, typically measured in terms of turbidity. The basic relationship between headloss development and effluent turbidity is shown on Fig. 8-11. Following a ripening period, after a backwash cycle, particles begin to accumulate in the filter medium, headloss gradually increases, and filter performance gradually decreases. The filter will continue to operate until headloss reaches some limiting headloss value or a turbidity breakthrough occurs.

The objective of a balanced filter design is to have the limiting headloss and turbidity breakthrough occurring at or near the same time. In small plants, the water filtered during the ripening period is wasted (usually returned to the plant inflow) until the acceptable effluent quality is reached. In large plants with many filters, the filter-to-waste cycle is omitted. Chemical addition has also been used to extend the time until turbidity breakthrough occurs, and to achieve a variety of other treatment objectives including the removal of specific contaminants such as phosphorus, metal ions, and humic substances. Chemicals commonly used in effluent filtration include a variety of organic polymers, alum, and ferric chloride.

Particle Removal Mechanisms

The principal particle removal mechanisms, believed to contribute to the removal of material within a granular-medium filter, are identified and described in Table 8-7.

Table 8-7
Principal mechanisms and phenomenon contributing to removal of material within a granular medium depth filter[a]

Mechanism/phenomenon	Description
1. Straining	
a. Mechanical	Particles larger than the pore space of the filtering medium are strained out mechanically.
b. Chance contact	Particles smaller than the pore space are trapped within the filter by chance contact.
2. Sedimentation	Particles settle on the filtering medium within the filter.
3. Impaction	Heavy particles do not follow the flow streamlines around the medium.
4. Interception	Many particles that move along in the streamline are removed when they come into contact with the surface of the filtering medium.
5. Adhesion	Particles become attached to the surface of the filtering medium as they pass by. Because of the force of the flowing water, some material is sheared away before it becomes firmly attached and is pushed deeper into the filter bed. As the bed becomes clogged, the surface shear force increases to a point at which no additional material can be removed. Some material may break through the bottom of the filter, causing the sudden appearance of turbidity in the effluent.
6. Flocculation	Flocculation can occur within the interstices of the filter medium. The larger particles formed by the velocity gradients within the filter are then removed by one or more of the above removal mechanisms.

7. Chemical adsorption
   a. Bonding
   b. Chemical interaction
8. Physical adsorption
   a. Electrostatic forces
   b. Electrokinetic forces
   c. van der Waals forces
	Once a particle has been brought into contact with the surface of the filtering medium or with other particles, either one of these mechanisms, or both, may be responsible for holding it there.

9. Biological growth	Biological growth within the filter will reduce the pore volume and may enhance the removal of particles with any of the above removal mechanisms (1 through 5).

[a] Adapted from Tchobanoglous et al. (2003).

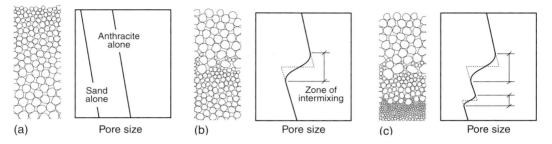

Figure 8-12
Definition sketch of typical pore size distribution according to the type of filter bed configuration (a) single-medium, (b) dual-medium, and (c) multi-medium.

Straining has been identified as the principal mechanism that is operative in the removal of the larger residual suspended solids remaining after secondary treatment (Tchobanoglous and Eliassen, 1970). Other mechanisms including interception, impaction, and adhesion are also operative even though their effects are small and, for the most part, masked by the straining action. The removal of the smaller particles found in wastewater is accomplished in two steps involving (1) the transport of the particles to or near the surface where they will be removed and (2) the removal of particles by one or more of the operative removal mechanisms. This two-step process has been identified as transport and attachment (O'Melia and Stumm, 1967).

Conventional downflow filters, dual- and multi-medium and deep-bed mono-medium depth filters (see Fig. 8-12) were developed to allow the suspended solids in the liquid to be filtered to penetrate further into the filter bed, and thus use more of the solids-storage capacity available within the filter bed. The deeper penetration of the solids into the filter bed also permits longer filter runs because the buildup of headloss is reduced. By comparison, in shallow mono-medium beds, most of the removal occurs in the upper few millimeters of the bed.

Removal of Turbidity

The results of long-term testing of seven different types of pilot-scale filters on the effluent from the same activated sludge process (SRT >15 d), without chemical addition, are shown on Fig. 8-13. Long-term data from other large-scale water reclamation plants are also shown. The principal conclusions to be reached from an analysis of the data presented on Fig. 8-13 are: (1) given a high quality filter influent (turbidity less than 5 to 7 NTU), all of the filters tested, including the large-scale plant, are capable of producing an effluent with an average turbidity of 2 NTU or less; (2) when the influent turbidity is greater than about 5 to 7 NTU, chemical addition is required with all of the filters to achieve an effluent turbidity of 2 NTU or less; and (3) effluent quality is directly related to influent quality if chemical addition is not used.

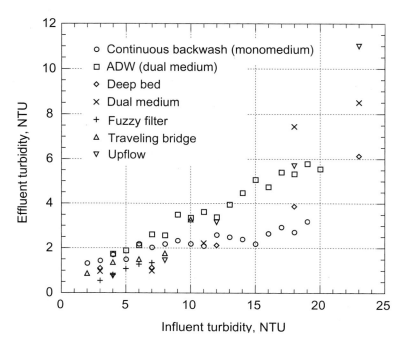

Figure 8-13
Performance data for seven different types of depth filters used for wastewater applications tested using the effluent from the same activated sludge plant at filtration a rate of 160 L/m²·min (4 gal/ft²·min) with the exception of the Fuzzy Filter that was operated at 800 L/m²·min (20 gal/ft²·min).

Removal of Total Suspended Solids

Keeping in mind the limitations associated with turbidity measurements, the following two relationships can be used to approximate TSS values from turbidity.

Settled secondary effluent

$$\text{TSS, mg/L} = (2.0 \text{ to } 2.4) \times (\text{turbidity, NTU}) \qquad (8\text{-}3)$$

Filter effluent

$$\text{TSS, mg/L} = (1.3 \text{ to } 1.6) \times (\text{turbidity, NTU}) \qquad (8\text{-}4)$$

Using the above equations, turbidity values of 5 to 7 NTU in the settled secondary effluent, which is the influent to the filter, correspond to TSS concentrations varying from about 10 to 17 mg/L, and an effluent turbidity of 2 NTU corresponds to TSS concentrations varying from 2.6 to 3.2 mg/L.

Variability in Turbidity and Total Suspended Solids Removal

In water reuse applications the variability of filter performance is of critical importance because there are specific effluent turbidity limits that must be met consistently. For example, as reported in Chap. 4, the turbidity standard for reclaimed water for unrestricted use in California is equal to or less than 2 NTU. Because the required turbidity value is written without a decimal point, a turbidity value of 2.49 NTU is reported as 2 NTU. The variability observed in the operating data from a large water reclamation facility is illustrated on Fig. 8-14 for the years 1995 and 1998. Comparing the mean turbidity and TSS values for 1995 and 1998, the TSS/turbidity

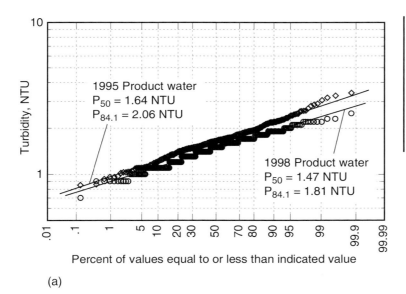

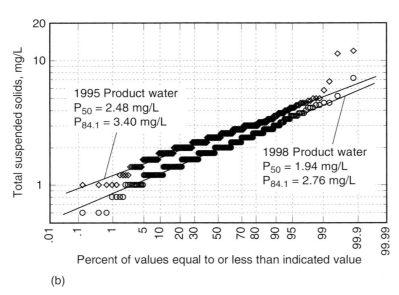

Figure 8-14
Probability distributions for filter performance for the filtration of settled activated sludge effluent: (a) turbidity and (b) total suspended solids.

ratios are 1.51 and 1.32, respectively, which is consistent with the range given in Eq. (8-4). The corresponding geometric standard deviations, s_g, for turbidity for 1995 and 1998 are 1.26 and 1.23, respectively. Similarly, the geometric standard deviations for TSS for 1995 and 1998 are 1.37 and 1.42, respectively. Both sets of values are consistent with the range of s_g values reported in the literature, as given in Table 8-6. Characterization of the variability in effluent constituents using the geometric standard deviation, s_g, is discussed in Appendix D. The greater the numerical value of s_g, the greater the observed range in the measured values. Use of the data in Table 8-6 is illustrated in Example 8-2.

EXAMPLE 8-2. Evaluation of the Effluent Variability of an Activated Sludge Process with Granular Medium Filtration.

An activated sludge process with granular medium filtration has been designed to have a mean effluent turbidity value of 2 NTU. Determine the maximum turbidity value that is expected to occur with a frequency of (a) once per year and (b) once every three years. If the effluent turbidity standard is 2.49 NTU, estimate how often the process will exceed the turbidity limit.

Solution

1. Select an s_g value from Table 8-6 that corresponds to the effluent turbidity for an activated sludge with filtration process. From Table 8-6, use the typical s_g value of 1.25.
2. Determine the probability distribution of the effluent turbidity values.
   a. Using the s_g value, compute the turbidity value corresponding to the plotting position on $P_{84.1}$ (see footnote b from Table 7-8).

   $$P_{84.1} = s_g \times P_{50} = 1.25 \times 2 \text{ NTU} = 2.5 \text{ NTU}$$

   b. Estimate the distribution of effluent turbidity values by plotting the $P_{84.1}$ and P_{50} values. As the effluent turbidity values are expected to follow a log normal distribution, a straight line can be drawn through the $P_{84.1}$ and P_{50} values, as shown on the following plot.

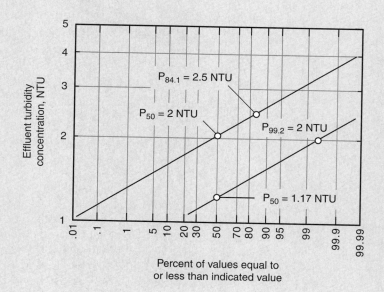

3. Compute the effluent turbidity value expected to occur with the frequency of interest.
   a. The probability of occurrence of a given event with a frequency of once per year is (1/365) × 100 = 0.3 percent. Using the plot developed in Step 2, an effluent turbidity value equal to or greater than 3.5 NTU will occur 0.3 percent of the time.
   b. Similarly, turbidity values equal to greater than 3.7 NTU will occur with a frequency of once in three years (i.e., 99.9 percent).
4. Estimate how often the combined treatment process will exceed the turbidity standard of 2.49 NTU. From the plot presented in Step 2, the effluent turbidity will exceed 2.49 NTU approximately 16 (100 − 84) percent of the time.

Comment
Recognition of the variability in performance is of importance in the design of filtration systems, especially where more stringent mean effluent turbidity values must be met. For example, if the turbidity standard had been 2.0 NTU at a reliability of at least 99.2 percent (three exceedances per year) the mean design value, as illustrated in the above figure, would have to be about 1.17 NTU, assuming that the geometric standard deviation remained constant and was equal to 1.25. To reach a mean turbidity value of 1.17 NTU would, in most cases, require the addition of chemicals.

Alteration of Particle Size

Although all of the filters listed in Table 8-5 can produce an effluent with an average turbidity of two or less, the effluent particle size distribution is different for each of the filters. Typical data on the removal of particle sizes from activated sludge effluent using depth filtration are shown on Fig. 8-15. As shown, the particle removal rate is essentially independent of the filtration rate up to about 240 L/m²·min. It is significant that most depth filters will pass some particles with diameters greater than 20 μm.

Depending on the quality of the settled secondary effluent, chemical addition has been used to improve the performance of effluent filters, with respect to turbidity. An example of the change in the distribution of particle sizes in the effluent from an activated sludge process following depth filtration without and with chemical coagulation is illustrated on Fig. 8-16. The original data, as collected, are shown on Fig. 8-16a. The data aggregated into selected bin sizes are shown on Fig. 8-16b, and, finally, the original data, plotted functionally according to the power law (see Example 8-1), are presented on Fig. 8-16c. As shown on Fig. 8-16a, filtration alone only affected the larger particles, whereas with chemical coagulation all of the particles were affected more or less uniformly. As shown on Fig. 8-16b, even though the number of particles in each size range was reduced by an order of magnitude, a significant number of particles remain in each size range.

Particles in the size range from 2 to 5 and 5 to 15 μm, which correspond to the approximate sizes of *Cryptosporidium* and *Giardia,* are important with respect to disinfection.

Figure 8-15
Particle size removal efficiency for a depth filter for effluent from an activated sludge plant.

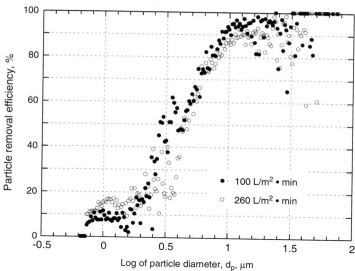

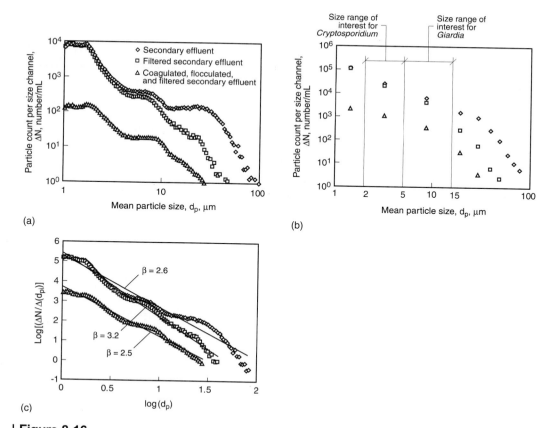

Figure 8-16
Effect of chemicals on filter particle size removal performance: (a) original data as collected (Courtesy of K. Bourgeous), (b) original data aggregated into selected bin sizes, and (c) original data, plotted functionally according to the power law (see Example 8-1).

Particles larger than about 10 to 15 μm are of importance because they are of sufficient size to shield microorganisms. Thus, depending on the disinfection method to be employed, it may be necessary to conduct pilot-plants studies, especially with chemicals, to assess the impact of the particles remaining after filtration on the disinfection process.

Removal of Microorganisms

Where chemicals are not used, the removal of coliform bacteria and viruses from biologically treated secondary effluent is on the order of 0 to 1.0 and 0 to 0.5 logs, respectively. The degree of removal depends on the solids retention time (SRT) at which the biological process is operated. For example, as shown on Fig. 8-17, as the SRT is increased fewer of the particles have one or more associated coliform bacteria. Typical data on the removal of the bacteriophage MS2 are illustrated on Fig. 8-18. As shown, the mean removal of MS2 across the effluent filters is about 0.3 log. However, what is of more interest is the distribution of the removal data. Based on the distribution shown on Fig. 8-18, which is also typical for the removal of coliform organisms, allowing a disinfection credit of one log of removal for filtration in water reuse applications may not be protective of public health. Where chemicals are used, the data on the removal for microorganisms is confounded statistically. In general, it is not possible to separate the effect of chemical addition from the performance of the filter.

Design Considerations

For new installations, extra care should be devoted to the design of the secondary settling facilities. With properly designed settling facilities resulting in an effluent with low TSS (typically 5 mg/L) and turbidity (less than 2 NTU), the decision on the type of filtration system used is often based on plant-related variables, such as the space available, duration of filtration period (seasonal versus year-round), the time available for construction, and costs. For existing plants that have variable suspended solids concentrations in the treated effluent, the type of a filter that can continue to function even when heavily loaded is an important consideration. The pulsed-bed filter and both downflow and upflow deep-bed coarse medium filters have been used in such applications.

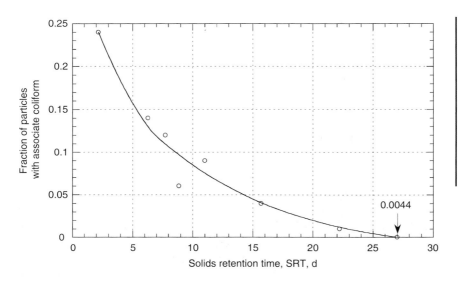

Figure 8-17

Number of particles with one or more associated coliform organisms as a function of the solids retention time for the activated sludge process (Darby et al., 1999).

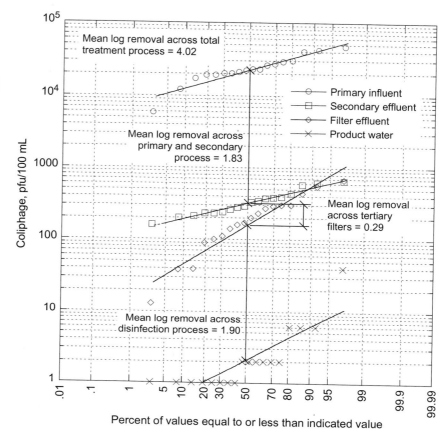

Figure 8-18
Removal of MS2 coliphage through a treatment process comprised of an activated sludge process, depth filtration, and chlorine disinfection.

Filter-Medium Characteristics

Grain size is the principal filter-medium characteristic that affects the filtration operation. Grain size affects both the clear-water headloss and the buildup of headloss during the filter run. If a filtering medium with too small a size is selected, much of the driving force will be wasted in overcoming the frictional resistance of the filter bed. On the other hand, if the size of the medium is too large, many of the small particles in the influent will pass directly through the bed. The size distribution of the filter material is usually determined by sieve analysis using a series of decreasing sieve sizes. The results of a sieve analysis are usually analyzed by plotting the cumulative percent passing a given sieve size on arithmetic-log or probability-log paper.

The effective size of a filtering medium is defined as the 10 percent size based on mass and is designated as d_{10}. For sand, it has been found that the 10 percent size by weight corresponds to the 50 percent size by count. The uniformity coefficient (UC) is defined as the ratio of the 60 percent size to the 10 percent size, UC = d_{60}/d_{10}. Sometimes it is advantageous to specify the 99 percent passing size and the one percent passing size to define more accurately the gradation curve for each filter medium. Additional information on filter medium characteristics is presented in the following section dealing with the design of filters.

Selection of Filter Medium

Selection of a filter medium (or media) typically involves the selection of the grain size as specified by the effective size, d_{10}; uniformity coefficient, UC; the 90 percent size; the specific gravity; solubility; hardness; and depth of the various materials used in the filter bed. Typical particle size distribution ranges for sand and anthracite filtering material are shown on Fig. 8-19. The 90 percent size designated, d_{90}, as read from a grain size analysis, is used commonly to determine the required backwash rate for depth filters. Typical sizes of filter materials for mono-, dual-, and multi-medium depth filters are given in Tables 8-8 and 8-9. Physical properties of filter materials used in depth filters are summarized in Table 8-10.

The degree of intermixing in the dual-medium and multi-medium beds depends on the density and size differences of the various media. To avoid extensive intermixing the settling rate of the filter mediums comprising the dual- and multi-medium filters must have essentially the same settling velocity. The following relationship can be used to establish the appropriate sizes (Kawamura, 2000).

$$\frac{d_1}{d_2} = \left(\frac{\rho_2 - \rho_w}{\rho_1 - \rho_w}\right)^{0.667} \qquad (8\text{-}5)$$

where d_1, d_2 = effective size of filter medium
ρ_1, ρ_2 = density of filter medium
ρ_w = density of water

The application of Eq. (8-5) is illustrated in Example 8-3.

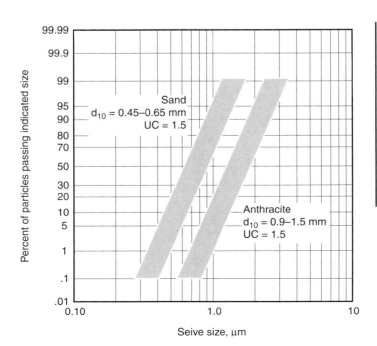

Figure 8-19

Typical particle size distribution ranges for sand and anthracite used in dual medium depth filters. Note that for sand the 10 percent size by weight corresponds to the 50 percent size by count.

Table 8-8
Typical design data for depth filters with mono-medium[a]

Characteristic	Unit	Value Range	Value Typical
Shallow bed (stratified)			
Anthracite			
Depth	mm	300–500	400
Effective size	mm	0.8–1.5	1.3
Uniformity coefficient	unitless	1.3–1.8	≤1.5
Filtration rate	L/m²·min	80–240	120
Sand			
Depth	mm	300–360	330
Effective size	mm	0.45–0.65	0.45
Uniformity coefficient	unitless	1.2–1.6	≤1.5
Filtration rate	L/m²·min	80–240	120
Conventional (stratified)			
Anthracite			
Depth	mm	600–900	750
Effective size	mm	0.8–2.0	1.3
Uniformity coefficient	unitless	1.3–1.8	≤1.5
Filtration rate	L/m²·min	80–400	160
Sand			
Depth	mm	500–750	600
Effective size	mm	0.4–0.8	0.65
Uniformity coefficient	unitless	1.2–1.6	≤1.5
Filtration rate	L/m²·min	80–240	120
Deep-bed (unstratified)			
Anthracite			
Depth	mm	900–2100	1500
Effective size	mm	2–4	2.7
Uniformity coefficient	unitless	1.3–1.8	≤1.5
Filtration rate	L/m²·min	80–400	200
Sand			
Depth	mm	900–1800	1200
Effective size	mm	2–3	2.5
Uniformity coefficient	unitless	1.2–1.6	≤1.5
Filtration rate	L/m²·min	80–400	200
Fuzzy filter			
Depth	mm	600–1080	800
Effective size	mm	25–30	28
Uniformity coefficient	unitless	1.1–1.2	1.1
Filtration rate	L/m²·min	600–1000	800

[a] Adapted in part from Tchobanoglous (1988) and Tchobanoglous et al. (2003).

Table 8-9
Typical design data for dual- and multi-medium depth filters[a]

Characteristic	Unit	Value[b] Range	Value[b] Typical
Dual-medium			
Anthracite ($\rho = 1.60$)			
Depth	mm	360–900	720
Effective size	mm	0.8–2.0	1.3
Uniformity coefficient	unitless	1.3–1.6	≤1.5
Sand ($\rho = 2.65$)			
Depth	mm	180–360	360
Effective size	mm	0.4–0.8	0.65
Uniformity coefficient	unitless	1.2–1.6	≤1.5
Filtration rate	L/m²·min	80–400	200
Multi-medium			
Anthracite (top layer of quad-media filter, $\rho = 1.60$)			
Depth	mm	240–600	480
Effective size	mm	1.3–2.0	1.6
Uniformity coefficient	unitless	1.3–1.6	≤1.5
Anthracite (second layer of quad-media filter, $\rho = 1.60$)			
Depth	mm	120–480	240
Effective size	mm	1.0–1.6	1.1
Uniformity coefficient	unitless	1.5–1.8	1.5
Anthracite (top layer of tri-media filter, $\rho = 1.60$)			
Depth	mm	240–600	480
Effective size	mm	1.0–2.0	1.4
Uniformity coefficient	unitless	1.4–1.8	≤1.5
Sand ($\rho = 2.65$)			
Depth	mm	240–480	300
Effective size	mm	0.4–0.8	0.5
Uniformity coefficient	unitless	1.3–1.8	≤1.5
Garnet ($\rho = 4.2$)			
Depth	mm	50–150	100
Effective size	mm	0.2–0.6	0.35
Uniformity coefficient	unitless	1.5–1.8	≤1.5
Filtration rate	L/m²·min	80–400	200

[a] Adapted from Tchobanoglous (1988) and Tchobanoglous et al. (2003).
[b] Anthracite, sand, and garnet sizes selected to limit the degree of intermixing. Use Eq. (8-5) for other values of density ρ.

Table 8-10
Typical properties of filter materials used in depth filtration[a]

Filter material	Specific gravity	Porosity, α	Sphericity[b]
Anthracite	1.4–1.75	0.56–0.60	0.40–0.60
Sand	2.55–2.65	0.40–0.46	0.75–0.85
Garnet	3.8–4.3	0.42–0.55	0.60–0.80
Ilmenite	4.5	0.40–0.55	
Fuzzy filter medium		0.87–0.89	

[a]Adapted in part from Cleasby and Logsdon (1999).
[b]Sphericity is defined as the ratio of surface area of an equal volume sphere to the surface area of the filter medium particle.

EXAMPLE 8-3. Determination of Filter Medium Sizes.

A dual-medium filter bed comprised of sand and anthracite is to be used for the filtration of settled secondary effluent. If the effective size of the sand in the dual-medium filter is to be 0.55 mm, determine the effective size of the anthracite to avoid significant intermixing.

Solution

1. Summarize the properties of the filter mediums
   a. For sand
      i. Effective size = 0.55 mm
      ii. Specific gravity = 2.65 (see Table 8-10)
   b. For anthracite
      i. Effective size = to be determined, mm
      ii. Specific gravity = 1.7 (see Table 8-10)
2. Compute the effective size of the anthracite using Eq. (8-5)

$$d_1 = d_2 \left(\frac{\rho_2 - \rho_w}{\rho_1 - \rho_w} \right)^{0.667}$$

$$d_1 = 0.55 \left(\frac{2.65 - 1}{1.7 - 1} \right)^{0.667}$$

$$d_1 = 0.97 \text{ mm}$$

Comment

Another approach that can be used to assess whether intermixing will occur is to compare the fluidized bulk densities of the two adjacent layers (e.g., upper 450 mm sand and lower 100 mm of anthracite).

Filter Bed Characteristics
The principal factors that must be considered in design are identified in Table 8-11. In the application of filtration to the removal of residual suspended solids, the nature of the particulate matter in the influent to be filtered, the filter bed configuration, the size of the filter material or materials, and the filtration flowrate are the most important of the process variables.

Selection of Filtration Technology
In selecting a filter technology, important issues that must be considered include: (1) anticipated feedwater quality, (2) type of filter to be used: proprietary or individually designed, (3) filtration rate, (4) filtration driving force, (5) number and size of filter units, (6) backwash water requirements, and (7) system redundancy. Each of these issues is described in Table 8-12.

The principal types of nonproprietary filter bed configurations used for wastewater filtration may be classified as mono-medium, dual-medium, or multi-medium beds. In conventional downflow filters, the distribution of grain sizes for each medium after backwashing is from small to large. Typical design data for mono-, dual-, and multi-medium filters were presented previously in Tables 8-8 and 8-9, respectively.

Table 8-11 Principal factors to be considered in the design of granular medium filters[a]

Factor	Significance
1. Effluent quality	Must meet specific reuse applications or fixed regulatory requirements.
2. Influent wastewater characteristics 　a. Suspended solids concentration 　b. Floc or particle size and distribution 　c. Floc strength 　d. Floc or particle charge 　e. Fluid properties	Affect the removal characteristics of a given filter-bed configuration. To a limited extent the listed influent characteristics can be controlled by the design engineer in the selection of the pretreatment and filtration system.
3. Filter medium characteristics 　a. Effective size, d_{10} 　b. Uniformity coefficient, UC 　c. Type, grain shape, density, and composition	Affect particle removal efficiency and headloss buildup.
4. Filter-bed characteristics 　a. Number of filtering mediums, i.e., mono-, dual-, or multi-medium 　b. Bed depth 　c. Stratification 　d. Degree of medium intermixing 　e. Porosity	Bed depth affects initial headloss, length of run. Degree of intermixing affects performance of filter bed. Porosity affects the amount of solids that can be stored within the filter.
5. Filtration rate	Affects filter size.

[a]Adapted in part from Tchobanoglous and Schroeder (1985) and Tchobanoglous et al. (2003).

Table 8-12

Important issues in selecting filter technology for water reuse applications[a]

Anticipated feed water quality	The anticipated effluent quality will impact the selection process, as some filters are more able to withstand periodic shock loadings. For example, wider variations in effluent quality would be expected where shallow clarifiers are used. More predictable effluent quality can be expected from deep clarifiers. In recent designs employing deep clarifiers (5 to 6 m sidewater depths), effluent turbidity values of less than 2 NTU are achieved consistently.
Type of filter: proprietary vs. individually designed	Currently available filter technologies are either proprietary or individually designed. With proprietary filters, the manufacturer is responsible for providing the complete filter unit and its controls, based on basic design criteria and performance specifications. In individually designed filters, the design engineer is responsible for working with several suppliers in developing the design of the system components. Contractors and suppliers then furnish the materials and equipment in accordance with the engineer's design.
Filtration rate	The filtration rate affects the real size of the filters that will be required. For a given filter application, the rate of filtration depends primarily on floc strength and the size of the filtering medium. For example, if the strength of the floc is weak, high filtration rates tend to shear the floc particles and carry much of the material through the filter. Filtration rates generally in the range of 80 to 320 $L/m^2 \cdot min$ will not affect the effluent quality when filtering settled activated sludge effluent.
Filtration driving force	Either the force of gravity or an applied pressure force can be used to overcome the frictional resistance to flow offered by the filter bed. Gravity filters of the type discussed in Table 8-5 are used most commonly for the filtration of treated effluent at large plants. Pressure filters operate in the same manner as gravity filters and are used at smaller plants. In pressure filters, the filtration operation is carried out in a closed vessel under pressurized conditions achieved by pumping.
Number and size the of filtration units	The number of filter units generally should be kept to a minimum to reduce cost of piping and construction, but it should be sufficient to assure that (1) backwash flowrates do not become excessively large and (2) when one filter unit is taken out of service for backwashing, the transient loading on the remaining units is not excessive. Transient loadings due to backwashing are not an issue with filters that backwash continuously. To meet redundancy requirements, a minimum of two filters should be used.
	The sizes of the individual filter units should be consistent with the sizes of equipment available for use as underdrains, wash-water troughs, and surface washers. Typically, width-to-length ratios for individually designed gravity filters vary from 1:1 to 1:4. A practical limit for the surface area on an individual depth filter (or filter cell) is about 100 m^2, although larger filters units have been built. For proprietary filters, use standard sizes that are available from manufacturers.
	The surface area of a depth filter is based on the peak filtration and peak plant flowrates. The allowable peak filtration rate is usually established on the basis of regulatory requirements. Operating ranges for a given filter type are based on past experience, the results of pilot-plant studies, manufacturers' recommendations, and regulatory constraints.

Table 8-12

Important issues in selecting filter technology for water reuse applications[a] (*Continued*)

Backwash water requirements	As noted in Table 8-4, depth filters operate in either a semicontinuous or continuous mode. In semicontinuous operation, the filter is operated until the effluent quality starts to deteriorate or the headloss becomes excessive, at which point the filter is taken out of service and backwashed to remove the accumulated solids. With filters operated in the semicontinuous mode, provision must be made for the backwash water needed to clean the filters. Typically, the backwash water is pumped from a filtered water clearwell or obtained by gravity from an elevated storage tank. The backwash storage volume should be sufficient to backwash each filter every 12 h. For filters that operate continuously, such as the upflow filter and the traveling bridge filter, the filtering and backwashing phases take place simultaneously. In the traveling bridge filter, the backwash operation can either be continuous or semicontinuous as required. For filters that operate continuously, there is no turbidity breakthrough or terminal headloss.
System redundancy	System redundancy is related to uninterruptible power and the need to provide standby capacity for routine maintenance. Most water reclamation plants in continuous service have emergency storage and onsite power generation to operate process equipment. In general, one standby filter as a minimum is recommended for standby service. Where the provision of standby facilities is not possible due to space or other limitations, the filters and related piping should be sized to handle periodic overloads during maintenance periods.

[a]Adapted in part from Tchobanoglous et al. (2003).

Pilot-Scale Studies

Although the clean water headloss can be estimated using well known equations (Tchobanoglous et al., 2003), it must be stressed that there is no generalized approach to the design of full-scale filters for the treatment of wastewater. The principal reasons are the inherent variability in the characteristics of the influent suspended solids to be filtered, the wide range of filter types that are available commercially, and the tolerance for variability in the product water. For example, changes in the degree of flocculation of the suspended solids in the secondary settling facilities significantly affects the particle sizes and their distribution in the effluent, which in turn affects the performance of the filter. Further, because the characteristics of the effluent suspended solids also vary with the organic loading on the process as well as with the time of day, filters must be designed to function under a rather wide range of operating conditions. The best way to ensure that the filter configuration selected for a given application will function properly and the effluent water quality is maintained within prescribed limits is to conduct pilot-scale studies (see Fig. 8-20).

The filter pilot plant shown on Fig. 8-20 was designed to determine if the maximum filtration rate of 200 L/m^2·min (5 gal/ft^2·min) currently allowed at the full-scale plant can be increased to 300 L/m^2·min (7.5 gal/ft^2·min) without compromising health and environmental concerns. If it can be demonstrated that the effluent characteristics at filtration rates of 200 and 300 L/m^2·min are essentially the same, the California Department of Health Services (DHS) will consider allowing the increased filtration rate at full-scale plants,

Figure 8-20
Views of filtration pilot plant: (a) filter columns fed from the source and (b) instrumentation used to monitor filter performance including turbidity and particle size counting.

subject to rigorous testing. For the effluents to be equivalent, the DHS has set forth the following preliminary unofficial criteria for equivalency for the Phase I pilot-plant testing.

1. Less than 10 percent increase in effluent turbidity
2. Less than 10 percent increase in 2 to 5 μm and 5 to 15 μm particles
3. Less than 10 percent decrease in log removal of MS2
4. Ability to disinfect effluent

The criteria for Phase II, in which full-scale tests at increased filter loading rates will be conducted at operating facilities, are still being discussed (as of 2006) but will likely include turbidity, particles, and disinfection, but not MS2. The DHS recognizes that the "accepted standard" cannot be current performance at 200 $L/m^2 \cdot min$ (5 $gal/ft^2 \cdot min$), because that will punish plants that currently produce high-quality secondary effluent with turbidity values varying from 1 to 3 NTU (FLEWR, 2005).

Because of the many variables that can be analyzed, care must be taken not to change more than one variable at a time so as to confound the results in a statistical sense. As

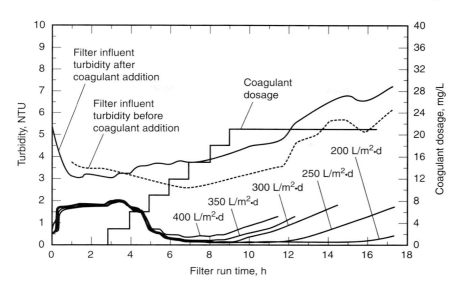

Figure 8-21
Effect of increasing coagulant dose in stepped increments, measured using the filter shown on Fig. 8-20. (Courtesy of G. Williams and K. Nelson).

shown in Fig. 8-21, the impact of increasing the chemical dosage is tracked clearly by the filter effluent turbidity at the test facility described above. Testing should be carried out at several intervals, ideally throughout a full year, to assess seasonal variations in the characteristics of the effluent to be filtered. Factors that should be considered in the conduct of pilot-scale tests were reported previously in Table 6-6 in Chap. 6.

The principal problems encountered in wastewater filtration are: (1) turbidity breakthrough, (2) mudball formation, (3) buildup of emulsified grease, (4) development of cracks and contraction of the filter bed, (5) loss of filter medium or media, and (6) gravel mounding. Because these problems can affect both the performance and operation of a filter system, care should be taken in the design phase to provide the necessary facilities to minimize their impact. These issues are considered further in Table 8-13.

Operational Issues

8-4 SURFACE FILTRATION

Surface filtration has been used in several applications including: (1) as a replacement for depth filtration to remove residual suspended solids from secondary effluents, (2) for the removal of suspended solids and algae from stabilization pond effluents, and (3) as a pretreatment operation before MF or UV disinfection.

Surface filtration, as shown on Fig. 8-8b, involves the removal of particulate material suspended in a liquid by mechanical sieving by passing the liquid through a thin septum

(i.e., filter material). Materials that have been used as filter septums include cloth fabrics of different weaves, woven metal fabrics, and a variety of synthetic materials.

Table 8-13
Summary of commonly encountered problems in depth filtration of wastewater and control measures for those problems[a]

Problem	Description/control
Turbidity breakthrough[b]	Unacceptable levels of turbidity are recorded in the effluent from the filter, even though the terminal head loss has not been reached. To control the buildup of effluent turbidity levels, chemicals and polymers have been added to the filter. The point of chemical or polymer addition must be determined by testing.
Mudball formation	Mudballs are an agglomeration of biological floc, dirt, and the filtering medium or media. If the mudballs are not removed, they will grow into large masses that often sink into the filter bed and ultimately reduce the effectiveness of the filtering and backwashing operations. The formation of mudballs can be controlled by auxiliary washing processes such as air scour or water surface wash concurrent with or followed by water wash. To avoid the formation of mudballs and the buildup of grease (see below) wastewater filters should be backwashed at least once per day, even though longer runs may be possible.
Buildup of emulsified grease	The buildup of emulsified grease within the filter bed increases the headloss and thus reduces the length of filter run. Both air scour and water surface wash systems help control the buildup of grease. In extreme cases, it may be necessary to steam clean the bed or to install a special washing system.
Development of cracks and contraction of filter bed	If the filter bed is not cleaned properly, the grains of the filter bed filtering medium become coated. As the filter compresses, cracks develop, especially at the sidewalls of the filter. Ultimately, mudballs may develop. This problem can be controlled by adequately backwashing and scouring.
Loss of filter medium or media (mechanical)	In time, some of the filter material may be lost during backwashing and through the underdrain system (where the gravel support has been upset or the underdrain system has been installed improperly). Loss of the filter material can be minimized through the proper placement of washwater troughs and underdrain system. Special baffles have also proven effective.
Loss of filter medium or media (operational)	Depending on the characteristics of the biological floc, grains of the filter material can become attached to it, forming aggregates light enough to be floated away during the backwashing operations. The problem can be minimized by the addition of an auxiliary air and/or water scouring system.
Gravel mounding	Gravel mounding occurs when the various layers of the support gravel are disrupted by the application of excessive rates of flow during the backwashing operation. A gravel support with an additional 50 to 75 mm (2 to 3 in.) layer of high density material, such as ilmenite or garnet, can be used to overcome this problem.

[a]Adapted from Tchobanoglous et al. (2003).
[b]Turbidity breakthrough does not occur with filters that operate continuously.

Membrane filters, MF and UF discussed in Sec. 8-5, are also surface filtration devices but are differentiated on the basis of the sizes of the pores in the filter medium. Cloth-medium surface filters typically have openings in the size range from 10 to 30 μm or larger; in MF and UF, the pore size can vary from 0.08 to 2.0 μm for MF and 0.005 to 0.2 μm for UF.

The principal types of cloth-medium surface filtration devices used in water reuse applications are the Cloth-Media Filter (CMF) the Discfilter (DF) and the diamond cloth-media filter (DCMF). Cartridge filters, also used for pretreatment prior to membrane filtration, particularly where RO is used, are discussed in Chap. 9. The operational features of the CMF and DF are described in Table 8-14.

Available Filtration Technologies

Table 8-14
Description of surface filters used in water reclamation applications

Type	Description
Cloth-Media Filter (CMF) 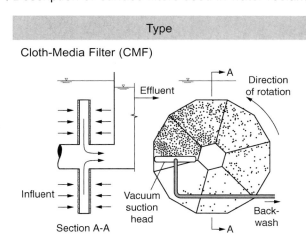	The CMF, marketed under the trademark AquaDisk by Aqua-Aerobic Systems, consists of several disks mounted vertically in a tank. Each disk is comprised of six equal segments. The CMF differs from the DF in that water flows by gravity from the exterior of the disks through the filter medium to an internal collection system. Two types of filter cloth can be used: (1) a needle felt cloth made of polyester or (2) synthetic pile fabric cloth.
Discfilter (DF) 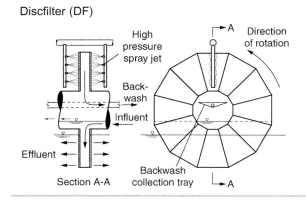	The DF, developed by Hydrotech and marketed in the United States by Veolia Water Systems, consists of a series of disks comprised of two vertically mounted parallel disks that are used to support the filter cloth. Each disk is connected to a central feed tube. The cloth screen material used can be of either polyester or Type 304 or 316 stainless steel. The filter mechanism can be furnished with a self-contained tank or for installation in a concrete tank. In cold climates or where odor control is a consideration, an enclosure can be provided for the disks.

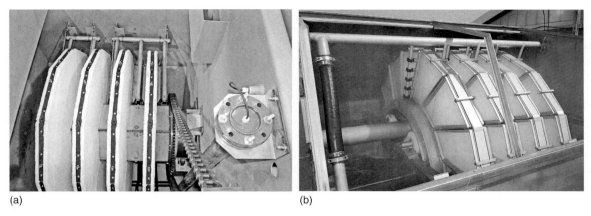

Figure 8-22
Views of surface filters: (a) Cloth-Media Disk Filter (Courtesy of Aqua Aerobic Systems, Inc.) and (b) Discfilter surface filter.

Cloth-Media Filter

In the Cloth-Media Filter (CMF), shown in Table 8-14 and on Fig. 8-22a, water enters the feed tank and flows through the filter cloth into a central collection tube or header. The resulting filtrate is collected in a central tube or filtrate header where it flows to final discharge over an overflow weir in the effluent channel. As solids accumulate on and in the cloth medium, resistance to flow or headloss increases. When the headloss through the cloth medium reaches a predetermined set-point, the disks are backwashed. After filtering to waste after the backwash cycle, the filter is put back into operation. When a backwash cycle is initiated, the disks remain submerged and rotate at 1 rev/min, allowing each segment to be cleaned. Solids are backwashed from both sides of the disk by liquid suction. Vacuum suction heads, located on either side of the CMF, draw filtrate water from the filtrate header back through the cloth media while the disk is rotating. This reversal of flow removes particles that have become entrapped on the surface and within the cloth medium. Typically, the backwash system uses less than three percent of the filtered water flow.

Over time, particles will accumulate in the cloth medium that cannot be removed by a typical backwash. This accumulation of particles leads to increased headloss across the filter, an increase in the backwash suction pressure, and shorter run times between backwashes. When the backwash suction pressure or operating time reaches predetermined setpoints, a high-pressure spray wash is initiated automatically. During the high-pressure spray wash, the disks rotate slowly at 1 rev/min while filtrate water is sprayed at a high pressure from the outside of the filter cloth. The high-pressure spray wash flushes the particles that have become lodged inside the cloth filter media in 2 revolutions of the disk. The time interval between high-pressure spray washes is a function of the feedwater quality. The CMF can be furnished with a self-contained tank or for installation in a concrete tank. Typical design data for a CMF are given in Table 8-15.

Table 8-15

Typical design information for surface filtration of secondary settled effluent using a Cloth-Media Filter[a]

Item	Unit	Typical value	Remarks
Nominal pore size	μm	10	Polyester three-dimensional needle felt cloths are employed as the filter material
Hydraulic loading rate	$m^3/m^2 \cdot min$	0.1–0.27	Depends on characteristics of suspended solids that must be removed
Headloss through screen	mm	50–300	Based on solids accumulation on or within the cloth
Disk submergence	% height	100	
	% area	100	
Disk diameter	m	0.90 or 1.80	Two sizes are available
Disk speed	rev/min	Stationary during normal operation	1 rev/min during backwash
Backwash and sludge wasting requirements	% throughput	4.5 at 0.1 $m^3/m^2 \cdot min$ 7.2 at 0.27 $m^3/m^2 \cdot min$	A function of hydraulic loading rate and feedwater quality

[a]Adapted from Tchobanoglous et al. (2003).

Discfilter

In the Discfilter (DF), shown on Fig. 8-22b and in Table 8-14, water enters through a central channel and flows outward through the filter cloth. Solids are retained within the filter discs while clean water flows to the outside of the disc into the collection tank. During normal operation 60 to 70 percent of the surface area of the DF is submerged and the disk rotates, depending on headloss, from 1 to 8.5 rev/min. The DF has the ability of operating in an intermittent or continuous backwash mode. When operating in a continuous backwash mode, the disks of the DF both produce filtered water and are backwashed simultaneously. When the DF is operating in an intermittent backwash mode, backwash spray jets are activated only when headloss through the filter reaches a preset level. Typically, the backwash system uses less than three to five percent of the filtered water flow. Typical design information for the DF is presented in Table 8-16.

Diamond Cloth-Media Filter

The diamond cloth-media filter (DCMF) is an innovative recent development. The DCMF is shown schematically and pictorially on Fig. 8-23. As shown on Fig. 8-23a, the cloth filter elements, which have a diamond-shaped cross-section, are cleaned by a vacuum sweep which moves back and forth along the length of the filter. Solids that settle to the bottom of the reactor below the filter element are removed periodically by a vacuum header. Using a diamond shape for the filter, it is possible to increase the cloth filter surface area per unit of aerial surface area. Because higher volumes for filtered water can be produced per unit area, the DCFM is used in new installations and as a replacement for existing sand filters as shown on Fig. 8-23a.

Table 8-16
Typical design information for surface filtration of secondary settled effluent using a Discfilter[a]

Item	Unit	Typical value	Remarks
Size of opening in screen material	μm	20–35	Stainless steel or polyester screen cloths are available in size ranging from 10 to 60 μm.
Hydraulic loading rate	$m^3/m^2 \cdot min$	0.25–0.83	Depends on characteristics of suspended solids that must be removed.
Headloss through screen	mm	75–150	Based on submerged surface area of drum.
Disc submergence	% height % area	70–75 60–70	Bypass should be provided when head loss exceeds 200 mm.
Disc diameter	m	1.75–3.0	Varies depending on screen design; 3 m is most commonly used size. Smaller sizes increase backwash requirements.
Backwash requirements	% throughput	2 at 350 kPa 5 at 100 kPa	

[a] Adapted in part from Tchobanoglous et al. (2003).

Performance of Surface Filters

In investigations of surface filters in comparison to granular-medium filters in filtering secondary effluent (Reiss et al., 2001; Olivier et al., 2003), it has been observed that surface filters outperformed granular-medium filters in removing turbidity and the number and size of particles. Typical values of effluent quality and variability of surface filtration compared to other filtration processes used for particulate removal are shown in Table 8-6.

Removal of Total Suspended Solids and Turbidity

To evaluate performance capabilities of surface filtration, a CMF pilot plant was tested using secondary effluent from an extended aeration-activated sludge process with a solids retention time greater than 15 d. Effluent TSS and turbidity values from the activated sludge process ranged from 3.9 to 30 mg/L and 2 to 30 NTU, respectively. Based on a long-term study, it was found, as shown on Fig. 8-24a, that both TSS and turbidity values of the filtered effluent were less than 1, 92 percent of the time (Reiss et al. 2001). The performance of the CMF as compared to depth filters, all tested with the same activated sludge effluent, is shown on Fig. 8-24b. As shown, the effluent turbidity from the CMF remained constant over a range of influent turbidity values that tested up to 30 NTU.

As with granular-medium filtration, the degree of removal of TSS from activated sludge effluent will depend on the SRT at which the process is operated.

Particle Size Removal

In comparative testing with a granular medium filter, the surface filter consistently outperformed the granular medium filter in respect to particle removal (see Fig. 8-25). The particle size reduction also had a significant impact on the inactivation of total coliform bacteria when used with UV disinfection (Olivier et al., 2003).

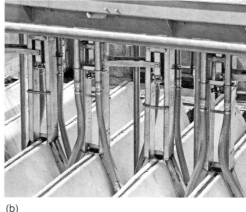
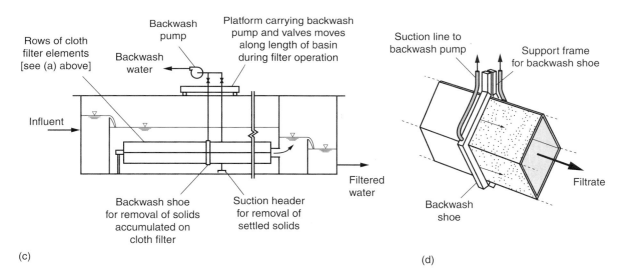

Figure 8-23
View of Diamond Cloth-Media Filter: (a) view of diamond filters installed in an existing sand filter basin, (b) close-up of solids removal mechanism, (c) definition sketch for operation of vacuum sweep, (d) cross-sectional view through filter element. [Photos (a) and (b) courtesy of Agua Aerobic Systems, Inc.]

Removal of Microorganisms

Where chemicals are not used, the removal of coliform bacteria and viruses from biologically treated secondary effluent is on the order of 0 to 1.0 and 0 to 0.5 log, respectively.

Design Considerations

Pilot studies are recommended in developing design and operating parameters for new installations. Useful data for design includes (1) the variability of the characteristics of the feed stream to be treated and (2) the amount of backwash water required for normal operation. The backwash water requirements are a function of the TSS in the feed

Figure 8-24

Performance data for Cloth-Media Filter for secondary effluent: (a) effluent turbidity as a function of influent turbidity at a filtration rate of 176 L/min · m² and (b) effluent probability distributions for turbidity and TSS.

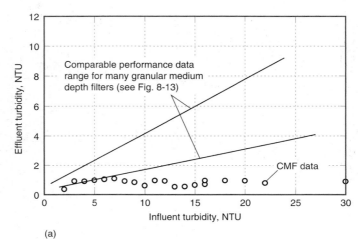

(a)

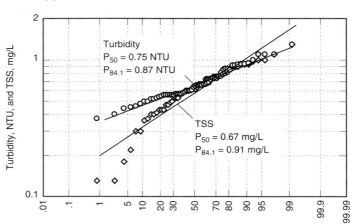

(b)

Figure 8-25

Comparison of particle sizes in effluent from secondary treatment, granular-medium filter, and Cloth-Media Filter. (Adapted from Olivier et al., 2003.)

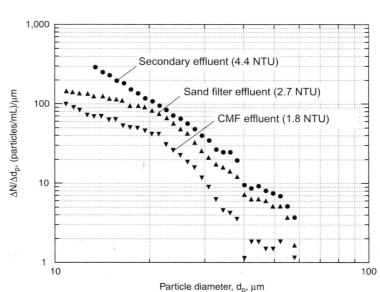

Figure 8-26
View of cloth filter pilot plant. It should be noted that the cloth filter disk is a full size mini-disk operational unit.

stream and the solids loading on the filters. If the secondary treatment system is effective in TSS removal, the volume of backwash water can be reduced substantially.

Because cloth-media surface filtration is a relatively new technology, little data are available on the life of the filter cloth. Where surface filtration is being considered, performance should be evaluated from operating installations using a similar type of cloth medium. One operating advantage cited for cloth-media filters is that the filter cloth can be removed and washed in a heavy-duty washing machine.

As with granular medium filtration, discussed above, there is no generalized approach to the design of full-scale filters for the treatment of wastewater. The discussion presented in the previous section on pilot-scale testing also applies to the cloth filter. A typical cloth-filter pilot plant is illustrated on Fig. 8-26. It should be noted that the single disk shown is full sized. In a larger installation, a number of disks would be arranged on the center shaft (see Fig. 8-22a).

Pilot-Scale Studies

8-5 MEMBRANE FILTRATION

Membrane filtration involves the passage a wastewater, usually from biological treatment, through a thin membrane (sometimes called a septum) for the purpose of removing particulate material, pathogens, organic matter, nutrients, and dissolved substances not removed by treatment processes. Membrane processes include microfiltration (MF), ultrafiltration (UF), nanofiltration (NF), reverse osmosis (RO), dialysis, and electrodialysis (ED). In this section, the focus is on the MF and UF membranes used for the filtration of secondary effluent in place of depth and surface filtration discussed previously. Nanofiltration, RO, and ED, processes used for the removal of dissolved solids, are considered in Chap. 9.

Figure 8-27
Definition sketch for operation of a membrane process.

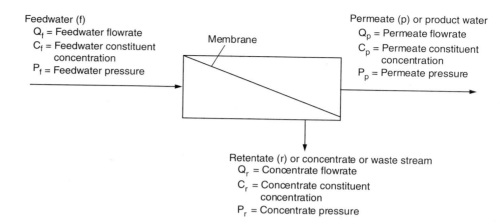

Membrane Terminology, Types, Classification, and Flow Patterns

Before discussing the MF and UF membranes used for water reclamation in greater detail, it will be useful to consider some terms that are used commonly in the membrane technology field, methods used to classify membranes, the types of membranes used, and the general flow patterns.

Membrane Process Terminology

Terms used commonly in the membrane technology field include feedwater, permeate, and retentate. These terms, defined previously in Working Terminology, are illustrated on Fig. 8-27. The influent water to be supplied to the membrane system for treatment is known as the *feedwater*. The liquid stream that has passed through the membrane is known as the *permeate*. The portion of the feed stream that does not pass through the membrane is known as the *retentate* (also referred to as concentrate or waste stream). Flux, the rate at which permeate flows through the membrane and expressed as $L/m^2 \cdot d$ or $kg/m^2 \cdot s$, is the principal measure of membrane performance.

Membrane Classification

Membrane processes can be classified in a number of different ways including: (1) the nature of the driving force for separation, (2) the separation mechanism, (3) the pore size of the membrane, (4) the nominal size of the separation achieved, and (5) the type of material from which the membrane is made. Although the focus of this section is on MF and UF, information on the different ways of classifying the membrane processes considered in this chapter and Chap. 9 are reported in Table 8-17 for comparative purposes. The types of membranes used and the materials of construction are considered further in the following section.

Types of Membranes and Materials

The principal types of membranes are: tubular, hollow fiber, spiral wound, plate and frame, and cartridge. Definition sketches for the various membranes are shown on Fig. 8-28 and more detailed descriptions are presented in Table 8-18. There are two basic flow patterns with membranes: (1) outside-in (see Fig. 8-28e) and (2) inside-out (see Fig. 8-28f). In most wastewater treatment applications where hollow fiber and membrane sheets are

Table 8-17
General characteristics of membrane processes

Membrane process	Membrane driving force	Typical separation mechanism	Typical pore size, μm	Typical operating range, μm	Materials
Microfiltration	Hydrostatic pressure difference or vacuum in open vessels	Sieve	Macropores (>50 nm)	0.008–2.0	Acrylonitrile, ceramic (various materials), polypropylene, polysulfone, polytetrafluoroethylene (PVDF), nylon, teflon
Ultrafiltration	Hydrostatic pressure difference or vacuum in open vessels	Sieve	Mesopores (2–50 nm)	0.005–0.2	Aromatic polyamides, ceramic (various materials) cellulose acetate, polypropylene, polysulfone, polyvinylidene fluoride (PVDF), teflon
Nanofiltration[a]	Hydrostatic pressure difference	Sieve + solution/diffusion + exclusion	Micropores (<2 nm)	0.001–0.01	Cellulosic, aromatic polyamide, polysulfone, polyvinylidene fluoride (PVDF), thin film composite
Reverse osmosis[a]	Hydrostatic pressure difference	Solution/diffusion + exclusion	Dense (<2 nm)	0.0001–0.001	Cellulosic, aromatic polyamide, thin film composite
Dialysis[a]	Concentration difference	Diffusion	Mesopores (2–50 nm)	—	Ion exchange resin cast as a sheet
Electrodialysis[a]	Electromotive force	Ion exchange	—	—	Ion exchange resin cast as a sheet

[a]Discussed in Chap. 9.

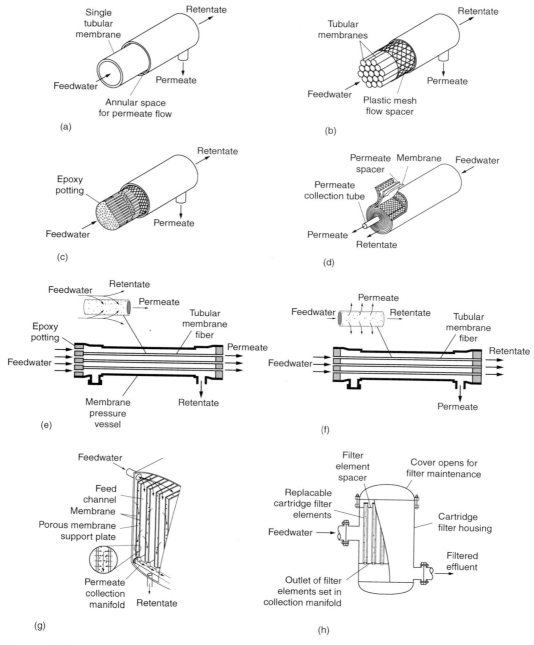

Figure 8-28

Definition sketch for types of membranes: (a) single tubular hollow fiber membrane, (b) bundle of tubular hollow fiber membranes in a container, (c) bundle of hollow fine fiber membranes in a container with flow from the inside to the outside of the fiber, (d) cutaway of a spiral wound thin film composite membrane module, (e) definition sketch for hollow fine-fiber membranes with flow from the outside to the inside fiber, (f) definition sketch for hollow fine-fiber membranes with flow from the inside to the outside, (g) parallel plate and frame membrane, and (h) cartridge filter with replaceable elements.

Table 8-18
Description of commonly used membrane types

Type	Description
Tubular (see Fig. 8-28a)	In the tubular configuration the membrane is cast on the inside of a support tube. A number of tubes (either singly or in a bundle) are then placed in an appropriate pressure vessel. The feedwater is pumped through the feed tube and product water is collected on the outside of the tubes. The retentate continues to flow through the feed tube. These units are used generally for water with high suspended solids or plugging potential. Tubular units are the easiest to clean, which is accomplished by circulating chemicals and pumping a "foamball" or "spongeball" through to mechanically wipe the membrane. Tubular units produce at a low product rate relative to their volume, and the membranes are generally expensive.
Hollow-fiber (see Fig. 8-28c)	The hollow-fiber membrane module consists of a bundle of hundreds to thousands of hollow fibers. The entire assembly is inserted into a pressure vessel. The feed can be applied to the inside of the fiber (inside-out flow) or the outside of the fiber (outside-in flow). Hollow-fiber membrane modules are commonly used in membrane bioreactors (MBRs) as described in Chap. 7.
Spiral wound (see Fig. 8-28d)	In the spiral wound membrane, a flexible permeate spacer is placed between two flat membrane sheets. The membranes are sealed on three sides. The open side is attached to a perforated pipe. A flexible feed spacer is added and the flat sheets are rolled into a tight circular configuration. Thin film composites are used most commonly in spiral wound membrane modules. The term spiral derives from the fact that the flow in the rolled-up arrangement of membranes and support sheets follows a spiral flow pattern.
Plate and frame (see Fig. 8-28g)	Plate and frame membrane modules are comprised of a series of flat membrane sheets and support plates. The water to be treated passes between the membranes of two adjacent membrane assemblies. The plate supports the membranes and provides a channel for the permeate to flow out of the unit.
Cartridge (see Fig. 8-28h)	Pleated cartridge filters are used most commonly in microfiltration applications, and usually are designed as disposable units. Pleated cartridge filters are used almost exclusively to concentrate virus from treated wastewater for analysis.

used, the flow is pattern is outside-in. With an outside-in flow pattern, the membrane can be backwashed with air, water, or a combination of both. The outside-in flow pattern is used for feedwater solutions with higher TSS and turbidities.

Most commercial membranes are produced as tubular, fine hollow fibers, or flat sheets. In general, three types of membranes are produced: symmetric, asymmetric, and thin film composite (TFC) (see Fig. 8-29). As shown on Fig. 8-29a and b symmetric membranes are the same throughout. Symmetric membranes can vary from microporous to nonporous (so called dense). Asymmetric membranes (see Fig. 8-29c) are cast in one process and consist of a very thin (less than 1 μm) layer and a thicker (up to 100 μm) porous layer that adds support and is capable of high water flux.

Thin-film composite membranes (see Fig. 8-29d) are made by bonding a thin cellulose acetate, polyamide, or other active layer (typically 0.15 to 0.25 μm thick) to a thicker

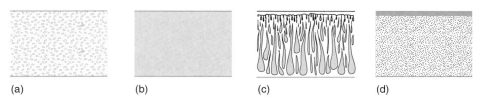

Figure 8-29
Types of membrane construction: (a) microporous symmetric membrane, (b) nonporous (dense) symmetric membrane, (c) asymmetric membrane, and (d) thin film composite (TFC), sometimes identified as an asymmetric membrane.

porous substrate, which provides stability. As reported in Table 8-17, membranes can be made from a number of different organic and inorganic materials. The membranes used for wastewater treatment are typically organic, although some ceramic membranes have been used. The choice of membrane and system configuration is based on minimizing membrane clogging and deterioration, typically based on pilot-plant studies.

Microfiltration and Ultrafiltration

Both MF and UF are used in water and wastewater treatment as alternatives to solids removal processes such as depth and surface filtration and combined processes involving chemical addition, flocculation, and gravity settling. In addition to removing suspended solids, MF and UF remove some large organic molecules, large colloidal particles, and many microorganisms (see Fig. 8-1). Microfiltration is used most commonly to reduce turbidity and some types of colloidal suspensions.

Ultrafiltration offers higher removals than MF, but operates at higher pressures. Some UF membranes with small pore sizes have also been used to remove dissolved compounds with high molecular weight, such as colloids, proteins, and carbohydrates. Product water from MF and UF membrane processes, after disinfection, may be used for a variety of reuse applications or as a pretreatment step to help prevent fouling of the less permeable NF, RO, and ED membranes (see Fig. 8-30). Advantages and disadvantages of MF and UF membranes as compared to conventional filtration are presented in Table 8-19.

Process Configurations

Two process configurations are used with membrane modules: pressurized and submerged.

Pressurized In the pressurized configuration, a pump is used to pressurize the feedwater and circulate it through the membrane (see Figs. 8-31a and c). The primary purpose of the pressure vessel (or tube) is to support the membrane and keep the feedwater and product water streams isolated. The vessel must also be designed to prevent leaks and pressure losses to the outside. Depending on the operating pressure and characteristics of the feedwater, materials used commonly include plastic and fiberglass tubes. Each module is generally 100 to 300 mm in diameter, 0.9 to 5.5 m long, and arranged in racks or

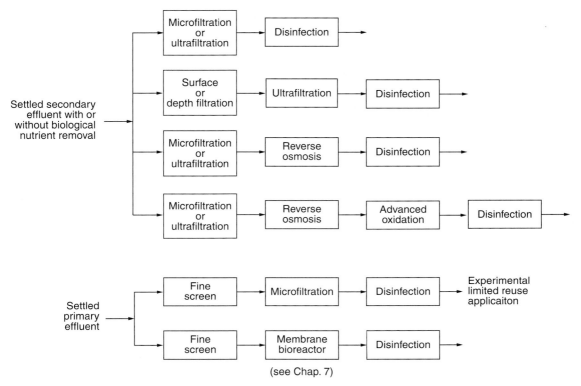

Figure 8-30
Typical process flow diagrams employing microfiltration and ultrafiltration with (a) settled secondary effluent and (b) settled primary effluent.

skids. Each module must be piped individually for feed- and permeate water. A typical pressurized MF membrane module system is shown on Fig. 8-32a.

Submerged (Vacuum) Type In the submerged system, the membrane elements are immersed in a feedwater tank and the permeate is withdrawn through the membrane by applying a vacuum, usually from the suction of a centrifugal pump (see Fig. 8-31e). Transmembrane pressure developed by the permeate pump causes clean water to be extracted through the membrane. Net positive suction head (NPSH) limitations of the permeate pump restrict the submerged membranes to a maximum transmembrane pressure of about 50 kPa and typically operate at a transmembrane pressure of 20 to 40 kPa (Crittenden et al., 2005). A typical submerged type membrane module is shown on Fig. 8-32c.

Operational Modes For Pressurized Configurations

Two different operational modes are used with pressurized MF and UF units. In the first operational mode known as *cross flow* (see Fig. 8-31a) the feedwater is pumped more-or-less tangentially, parallel to the membrane (see also Figs. 8-28e and f). The differential pressure across the membrane causes a portion of the feedwater to pass through the

Table 8-19
Advantages and disadvantages of microfiltration and ultrafiltration

Advantages	Disadvantages
• Can reduce the amount of treatment chemicals.	• Uses more electricity; high-pressure systems can be energy intensive.
• Smaller space requirements (footprint); membrane equipment requires 50 to 80 percent less space than conventional plants.	• May need pretreatment to prevent fouling; pretreatment facilities increase space needs and overall costs.
• Reduced labor requirements; can be automated easily.	• May require residuals handling and disposal of concentrate.
• New membrane design allows use of lower pressures; system cost may be competitive with conventional wastewater treatment processes.	• Requires replacement of membranes about every 5 yr.
	• Scale formation can be a serious problem. Scale-forming potential difficult to predict without field testing.
• Removes protozoan cysts, oocysts, and helminths ova; may also remove limited amounts of bacteria and viruses.	• Flux rate (the rate of feedwater flow through the membrane) gradually declines over time. Recovery rates may be considerably less than 100 percent.
	• Lack of a reliable low-cost method of monitoring performance.

membrane. Water that does not pass through the membrane is recirculated back to the membrane after blending with influent feedwater or is recirculated to a blending (or balancing) tank. In addition, a portion of the water that did not pass through the membrane is bled off for separate processing and disposal (see Fig. 8-31a). It should be noted that cross flow is the flow pattern in spiral-wound membranes.

In the second configuration, known as *dead-end* (also known as direct-feed or perpendicular-feed) and illustrated on Fig. 8-31c, there is no cross flow (or liquid waste stream) during the permeate production mode. All of the water applied to the membrane passes through the membrane. Particulate matter that cannot pass through the membrane pores is retained on the membrane surface. Dead-end filtration is used both for pretreatment and where the filtered water is to be used directly.

As constituents in the feedwater accumulate on the membranes (often termed membrane fouling), the pressure builds up on the feed side, the membrane flux (i.e., flow through membrane) starts to decrease, and the percent rejection [see Eq. (8-11)] also starts to decrease (see Fig. 8-33). When the performance has deteriorated to a given level, the membrane modules are taken out of service, backwashed, and periodically cleaned chemically (see Figs. 8-31b, d, and f and Fig. 8-33). Chemical cleaning is used to restore the membrane performance to its initial state, relative to the irreversible loss of membrane permeability that occurs during process operation (see Fig. 8-33). The degree of irreversible permeability loss depends on the membrane material and operating conditions, including (1) long-term aging of the membrane material, (2) mechanical compaction and deformation from high operating pressures, (3) hydrolysis reactions related to solution pH, and (4) reactions with specific constituents in the feedwater.

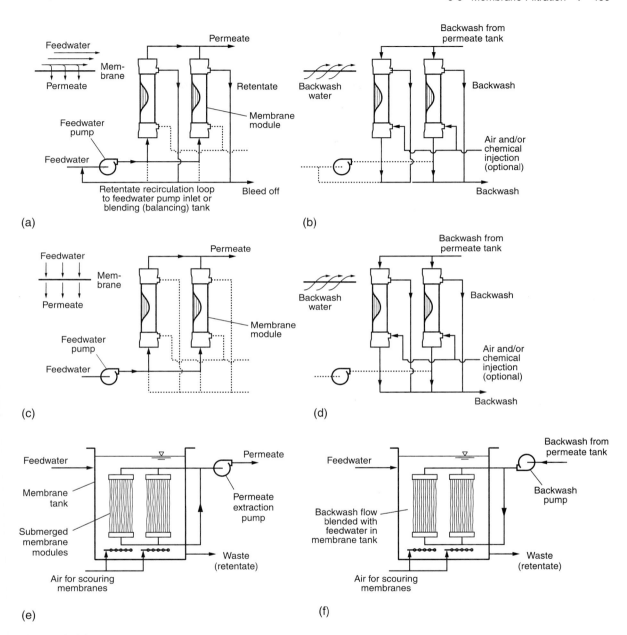

Figure 8-31

Definition sketch for membrane systems: (a) pressurized cross-flow membrane system (see insert), (b) backwashing pressurized cross-flow system, (c) pressurized dead end flow (see insert, see also Fig. 8-32a), (d) backwashing pressurized dead-end flow system, (e) submerged membrane with vacuum draw-off (see also Fig. 8-32d), and (f) backwashing submerged system.

Figure 8-32

Typical pressurized and submerged installations for the filtration of settled secondary effluent: (a) and (b) pressurized microfiltration, (c) ultrafiltration membrane module used in open vessel, and (d) mirofiltration membrane modules placed in open vessel.

Figure 8-33

Definition sketch for the performance of a membrane filtration system as function of time with and without proper cleaning.

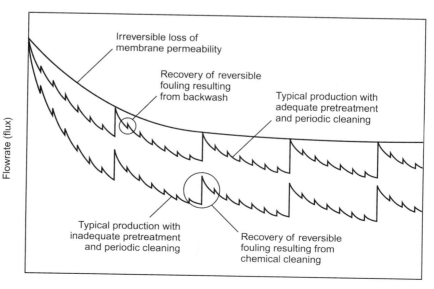

Process Analysis for MF and UF Membranes

Process analysis for membranes involves consideration of the operating pressure, permeate flow, the degree of recovery, degree of rejection, and mass balance.

Operating Pressure
For the cross-flow mode of operation, the transmembrane pressure is given by the following expression:

$$P_{tm} = \left[\frac{P_f + P_r}{2}\right] - P_p \qquad (8\text{-}6)$$

where P_{tm} = transmembrane pressure gradient, bar (Note: 1 bar = 10^5 Pa)
P_f = inlet pressure of the feed stream, bar
P_r = pressure of the retentate stream, bar
P_p = pressure of the permeate stream, bar

The overall pressure drop across the filter module for the cross-flow mode of operation is given by:

$$P = P_f - P_p \qquad (8\text{-}7)$$

where P = pressure drop across the module, bar

For the dead-end pressurized and submerged modes of operation, the transmembrane pressure is given by the following expression:

$$P_{tm} = P_f - P_p \qquad (8\text{-}8)$$

where P_{tm} = transmembrane pressure gradient, bar

Permeate Flow
The total permeate flow from a membrane system is given by:

$$Q_p = F_w A \qquad (8\text{-}9)$$

where Q_p = permeate flowrate, m³/h
F_w = transmembrane water flux rate, m/h (m³/m²·h)
A = membrane area, m²

As would be expected, the transmembrane water flux rate is a function of the quality and temperature of the feed stream, the degree of pretreatment, the characteristics of the membrane, and the system operating parameters.

Recovery
Recovery, r, is defined as the ratio of the net water produced to the gross water production during a filter run as follows:

$$r, \% = \frac{V_p}{V_f} \times 100 \qquad (8\text{-}10)$$

where V_p = net volume of the permeate, m³
V_f = volume of water fed to the membrane, m³

In computing the net volume of permeate, the amount of backwash water used should be taken into consideration.

Rejection

Rejection, R, is a measure of the fraction of material removed from the feed stream. It should be noted that there is a difference in the recovery, r, (which refers to the water) and rejection, R, (which refers to the solute). Rejection, R, expressed as a percentage, is given by the following expression:

$$R, \% = \frac{C_f - C_p}{C_f} \times 100 = \left(1 - \frac{C_p}{C_f}\right) \times 100 \tag{8-11}$$

where C_f = concentration in the feed stream, g/m³, mg/L
C_p = concentration in the permeate, g/m³, mg/L

Another commonly used approach is to express the rejection as log rejection as given below.

$$R_{log} = -\log(1-R) = \log\left(\frac{C_f}{C_p}\right) \tag{8-12}$$

Mass Balance

The corresponding flow and constituent mass balance equations for the pressurized cross-flow membrane are:

Flow balance: $\quad Q_f = Q_p + Q_r \tag{8-13}$

Constituent mass balance: $\quad Q_f C_f = Q_p C_p + Q_r C_r \tag{8-14}$

where Q_r = flowrate of retentate, m³/h, m³/s
C_r = retentate concentration, g/m³, mg/L

Operating Characteristics and Strategies for MF and UF Membranes

Typical characteristics of MF and UF membrane technologies used for water reclamation and reuse including operating pressures and flux rates are presented in Table 8-20. Three different operating strategies can be used to control the operation of a membrane process with respect to flux and the transmembrane pressure (TMP). The three modes, illustrated on Fig. 8-34, are: (1) constant flux in which the flux rate is fixed and the TMP is allowed to vary (increase) with time, (2) constant TMP in which the TMP is fixed and the flux rate is allowed to vary (decrease) with time, and (3) both the flux rate and the TMP are allowed to vary with time. Traditionally, the constant flux mode of operation has been used. However, based on the results of a study with various wastewater effluents (Bourgeous et al., 1999), it appears the mode in which both the flux rate and the TMP are allowed to vary with time may be the most effective mode of operation. It should be noted that the diagrams in Fig. 8-34 do not reflect the irreversible permeability loss, as described previously.

Membrane Performance

Microfiltration and UF have both been used extensively for the filtration of secondary effluents from biological processes. Perfomance expectations and process variability are considered below. Microfiltration has also been used with primary effluent.

Table 8-20
Characteristics of MF and UF membrane systems[a]

Parameter	MF	UF
Product particle size, μm	0.08–2.0	0.005–0.2
Retained compounds	Very small suspended particles, some colloids, most bacteria	Organics >1000 MW, pyrogens, viruses, bacteria, colloids
Flux rate, L/m² · d	400–1600	400–800
Operating pressure,[b] bar	0.07–1	0.7–7
Energy consumption, kWh/m³	0.4	3.0
Recovery, %	94–98	70–80
Material	Acrylonitrile, ceramic (various materials), polypropylene, polysulfone, polytetrafluorethylene, (PVDF), nylon, teflon	Aromatic polyamides, ceramic (various materials) cellulose acetate, polypropylene, polysulfone, polyvinylidene fluoride (PVDF), teflon
Configurations	Hollow fiber, spiral wound, plate and frame, tubular	Hollow fiber, spiral wound, plate and frame, tubular
Principal manufacturers	Dow, Koch, Pall, USFilter	Dow, Koch, Pall, Hydranautics, Zenon

[a]Adapted in part from Crites and Tchobanoglous (1998) and Paranjape et al. (2003).
[b]kPa × 10⁻² = bar, (1 bar = 10⁵ N/m²).
kPa × 0.145 = lb/in.²

Filtration of Secondary Effluent

Where MF and UF are used for the filtration of settled secondary effluent, constituent removals on the order of those shown in Table 8-21 may be expected. As noted in footnote b to Table 8-21, in addition to the normal variability that would be expected, there is significant variability in performance between UF membranes from different manufacturers. The performance of a UF membrane used for the removal of particles from secondary effluent following treatment by cloth (surface) filtration is shown on Fig. 8-35. For MF and UF systems where separation is solely based on pore size, increasing the flux through the membrane by increasing the operating pressure tends to

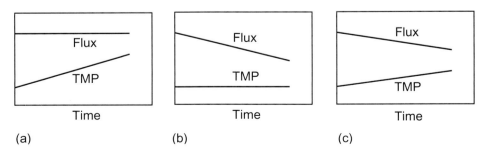

Figure 8-34
Three modes of membrane operation: (a) constant flux, (b) constant pressure, and (c) nonrestricted flux and pressure. (Adapted from Bourgeous et al., 1999.)

Table 8-21

Expected performance of microfiltration and ultrafiltration membranes on secondary effluent

Constituent	Rejection	Value Microfiltration	Value Ultrafiltration
TOC	%	45–65	50–75
BOD	%	75–90	80–90
COD	%	70–85	75–90
TSS	%	95–98	96–99.9
TDS	%	0–2	0–2
NH_4^+-N	%	5–15	5–15
NO_3^--N	%	0–2	0–2
PO_4^{3-}	%	0–2	0–2
SO_4^{2-}	%	0–1	0–1
Cl^-	%	0–1	0–1
Total coliform[a]	log	2–5	3–6
Fecal coliform[a]	log	2–5	3–6
Protozoa[a]	log	2–5	>6
Viruses[a]	log	0–2	2–7[b]

[a]The reported values reflect observed practice and integrity concerns. Also a wide range of performance differences occurs between membranes, as given in the following footnote.

[b]The low and corresponding mean removal values for four different UF membranes treating the same water were 2.5, 4.0, 5.3, and 6.1 and 3.8, 5.5, 6.5, and 7.5, respectively (Sakaji, 2006).

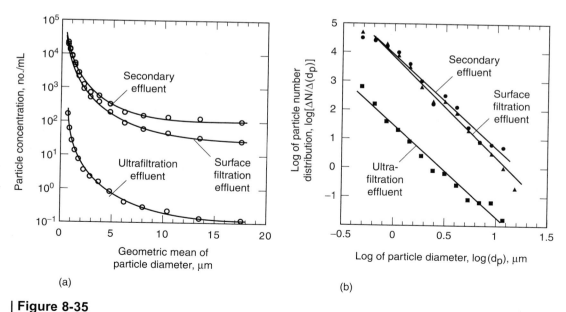

Figure 8-35

Typical performance data for UF membrane filtration system: (a) particle concentration versus diameter and (b) log of particle number versus log of particle diameter (Courtesy N. Tooker).

increase the permeate flow and decrease the permeate quality. Lower operating pressures are desirable because the degree of separation and product quality are improved. The effect of the pore size distribution of UF membranes on virus (coliphage QB) rejection under cross-flow conditions has been investigated (Urase et al., 1994). None of the five membranes tested provided a complete barrier for the model virus used in the study, contrary to expectations based on the nominal molecular weight cutoff size.

Typical values of effluent quality and variability of membrane filtration compared to other filtration processes used for particulate removal are shown in Table 8-6. As noted in Table 8-6, the s_g values for membrane filtration are actually larger than the corresponding values for depth and surface filtration. The reason the s_g values are larger is that the constituent concentrations are extremely low and slight perturbations can have a significant effect. Also because some of the measured values are near the method detection levels, the error in the detection method can contribute to the observed process variability. The importance of membrane integrity with respect to process performance and variability is examined in Example 8-4.

Filtration of Primary Effluent

Extensive MF testing on screened primary effluent using submerged membranes has been conducted at the Orange County Sanitation District's plant in Santa Ana, California, and the results are reported in Table 8-22 (Juby, 2003). As shown in the test data, the removal of TSS and microorganisms was exceptionally high and the removal of organic matter, although good, was about what was expected considering the permeability of MF membranes. Where low levels of organic matter are not necessary in certain reuse applications such as nonfood crop irrigation, MF of primary effluent may be an appropriate technology.

Table 8-22
Performance of microfiltration membranes on screened primary effluent at Orange County Sanitation District's Plant[a]

Parameter	Units	Pilot plant results		Demonstration plant results	
		Feedwater	Permeate	Feedwater	Permeate
TSS[b]	mg/L	39 (171)	<2 (158)	60 (92)[d]	4 (93)[d]
COD[b]	mg/L	274 (123)	138 (117)	253 (16)	111 (15)
BOD[b]	mg/L	124 (124)	65 (124)	105 (16)[d]	56 (89)[d]
Total coliform[c]	MPN/100 mL	2.4×10^6	5 to 7 log	5.0×10^7	4.9 log
Fecal coliform[c]	MPN/100 mL	2.4×10^6	5 to 7 log	6.8×10^6	4.7 log
Coliphage[c]	PFU/100 mL	8.4×10^5	2 log	1.6×10^3	1.7 log

[a]Adapted from Juby (2003).
[b]Average values; values in parentheses indicate number of samples.
[c]Typical values for feedwater; typical log removals for permeate.
[d]24 h composite samples.

EXAMPLE 8-4. Impact of Broken Fibers on Membrane Filter Effluent Quality.

Membrane filtration is used to treat secondary effluent for reuse applications. The effluent from the wastewater treatment plant, which serves as the influent to the membrane filter installation, has an effluent turbidity of 5 NTU and contains a heterotrophic plate count (HPC) of 10^6 microorganisms/L. The effluent from the membrane filters typically contains less that 10 microorganism/L and a turbidity of about 0.2 NTU. Using this information, what is the log rejection for microorganisms under normal operation with no broken fibers? If it, assumed that 6 out of 6000 (0.1 percent) membrane fibers have been broken during operation, determine the impact on the effluent microorganism count and turbidity. For the following analysis, neglect the water lost during the backwashing cycle.

Solution

1. Calculate the log rejection for microorganisms with no broken fibers using Eq. (8-12).

$$R_{log} = \log\left(\frac{C_f}{C_p}\right) = \log\left(\frac{10^6 \text{ org/L}}{10 \text{ org/L}}\right) = 5.0$$

2. Determine the log rejection for microorganisms assuming that 6 fibers have been broken.
   a. Prepare a mass balance diagram for the condition with the broken fibers.

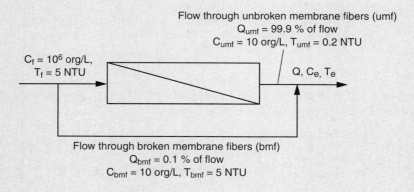

   b. Write mass balance equation for microorganisms in the effluent from the membrane and solve for effluent microorganism concentration.

$$C_e = \frac{C_{umf}Q_{umf} + C_{bmf}Q_{bmf}}{Q_e}$$

$$= \frac{(10 \text{ org/L})(0.999) + (10^6 \text{ org/L})(0.001)}{1} = 1010 \text{ org/L}$$

c. Calculate the log rejection for microorganisms for the condition with the broken fibers.

$$R_{log} = \log\left(\frac{C_p}{C_f}\right) = \log\left(\frac{10^6 \text{ org/L}}{1010 \text{ org/L}}\right) = 3.0$$

3. Calculate the impact on turbidity assuming that 6 fibers have been broken. Use the mass balance equation developed in Step 2 and solve for the effluent turbidity.

$$T_e = \frac{T_{umf}Q_{umf} + T_{bmf}Q_{bmf}}{Q_e}$$

$$= \frac{(0.2 \text{ NTU})(0.999) + (5 \text{ NTU})(0.001)}{1} = 0.205 \text{ NTU}$$

Comment

This example is used to demonstrate that a few broken fibers can have a significant impact on the microorganism count in the effluent (1010 versus 10/L) and the log removal (5 versus 3.0 log) and essentially no impact on the effluent turbidity (0.2 versus 0.205 NTU; the difference is not measurable). For this reason turbidity cannot be used as a surrogate measure for bacterial quality, and disinfection of MF effluent will be required to protect public health in sensitive applications.

Design Considerations

Design considerations for MF and UF systems include the character of the wastewater to be treated, membrane selection, pretreatment requirements, system configuration and pressure, flux rate, and membrane fouling control. Each of these design considerations is presented in Table 8-23.

Once membrane technology has been established as the process of choice, it is recommended that the least complicated membrane suitable for the intended application be selected; for example, MF membrane in lieu of a UF membrane. An initial evaluation for membrane selection can be based on the following criteria (Celenza, 2000):

- Microfiltration should be used for removing suspended solids and solutes of molecular weights greater than 300,000 and sizes ranging from 0.02 to 10 μm.
- Ultrafiltration should be used for removing suspended solids and solutes of molecular weights from 300 to 300,000 and sizes ranging from 0.0010 to 0.02 μm.

Pilot-Scale Studies

Because every wastewater is unique with respect to its chemistry, it is difficult to predict *a priori* how a given membrane process will perform. As a result, the selection of the best membrane for a given application is usually based on the results of pilot-scale studies. The elements that comprise a pilot plant include: (1) the pretreatment system, if used, (2) tankage for flow equalization and cleaning, (3) pumps, with appropriate controls, for pressurizing and backflushing the membrane, (4) the membrane test module, (5) adequate facilities for monitoring the performance of the test module, and (6) an appropriate

Table 8-23

Important design considerations and issues in selecting membrane technology

Issue	Significance
Wastewater characterization	The characterization of wastewater is important to: (1) ensure that the membrane system selected is compatible with the wastewater to be treated and (2) determine whether any special pretreatment is required. Constituents of special concern are suspended solids characteristics including molecular weight that might affect membrane performance; oil, grease, or floatable solids that might cause fouling; erosive substances that might affect membrane wear; and chemical constituents that could cause membrane deterioration. Typical constituents in the wastewater that cause membrane fouling or deterioration are given in Table 8-24. Those constituents that would be detrimental to membrane life and performance must be removed or reduced or, in the case of pH, conditioned to an inactive state. Fine particles, especially hair, that cause fouling of the membranes by forming a fibrous mat, need to be removed by pretreatment.
Membrane selection	Membranes used for the treatment of water and wastewater typically consist of a thin skin having a thickness of about 0.20 to 0.25 μm supported by a more porous structure of about 100 μm in thickness. Most commercial membranes are produced as flat sheets, fine hollow fibers, or in tubular form. The membranes used for wastewater treatment are typically made of organic materials (see Table 8-20). The choice of the type of membrane is based on minimizing membrane clogging and deterioration for the given application, based typically on pilot-plant studies.
Pretreatment	Membrane systems are very sensitive and care must be provided to optimize their performance. Depending on the wastewater to be treated, pretreatment may consist of fine screening, typically with 1 to 4 μm openings; surface filtration; or chemical neutralization, conditioning, or precipitation. Pilot testing of membrane systems under consideration can be very helpful in determining specific pretreatment requirements.
System configuration and operation	System configuration may include using either pressure-type or submerged membranes operated most commonly in a continuous flow mode (batch operation is possible, but difficult to manage). Various multistage arrangements are used where nanofiltration or reverse osmosis are incorporated in the process flowsheet (see Fig. 8-30). Operationally, the direct-flow (dead-end) method of operation is used most commonly.
System pressure	Pressure is selected to overcome membrane resistance and membrane fouling resistance. Lower pressures result in lower membrane compaction forces thereby improving flux restoration, minimizing flux deterioration, and increasing membrane life. The pressure requirements depend on membrane selection and whether a pressure-driven or submerged membrane system is used.
Flux rate	The primary factor affecting membrane performance is flux, which when related to the feed rate establishes the filter size and membrane area. The factors affecting flux rate include the applied pressure, fouling potential, and wastewater characteristics. Maintaining the flux rate is a function of fouling control that can include flushing rates to control cross flow velocity, back flushing, air scouring, and membrane cleaning. Typically, MF processes operate at flux rates ranging from 40 to 400 $L/m^2 \cdot min$, and UF processes operate at flux rates ranging from 0.15 to 0.60 $L/m^2 \cdot min$ (Celenza, 2000).
Membrane fouling	Membrane fouling affects pretreatment needs, cleaning requirements, operating conditions, cost, and performance. Typically, three approaches are used to control membrane fouling: (1) pretreatment of the feedwater, (2) membrane backflushing, and (3) chemical cleaning of the membrane. Pretreatment is used to reduce the TSS and bacterial content of the feedwater. Often feedwater is conditioned to limit chemical precipitation within the units. The commonly used method of eliminating the accumulated material from the membrane surface is backflushing with water and/or air. Chemical treatment is used to remove constituents that are not removed during conventional backwashing. Damage of the membranes due to deleterious constituents cannot be reversed.

system for backflushing the membranes. The information collected should be sufficient to allow for the design of the full-scale system and should include as minimum the following items (Tchobanoglous et al., 2003).

Membrane operating parameters:

    Pretreatment requirements including chemical dosages

    Transmembrane flux rate correlated to operating time

    Transmembrane pressure

    Washwater requirements

    Recirculation ratio

    Cleaning frequency including protocol and chemical requirements

    Posttreatment requirements

Typical water quality measurements may include:

    Turbidity

    Temperature

    Particle counts

Additional factors that should be considered in the conduction of pilot-plant studies are reported in Table 6-6 in Chap. 6. Typical pilot-test facilities used to evaluate various membrane treatment options in connection with the production of reclaimed water are shown on Fig. 8-36.

Operating issues deal mainly with membrane life, membrane performance, operating efficiency, and membrane clean-in-place frequency for flux maintenance.

Operational Issues

Membrane Life
Membrane life is important as excessive costs associated with the repair, replacement, and startup and operating time that occur with frequent membrane replacement could render the membrane system uneconomical. Membrane life on the order of 5 to 10 yr can be expected under normal circumstances, but a membrane life of less than 2 to 3 yr can affect the economics of membrane treatment significantly.

Membrane Performance
Membrane performance must be established carefully through the use of pilot-scale testing to ensure that proper design and operating parameters, including appropriate cleaning strategies, are selected. System upgrading or corrective actions due to inadequate pilot-scale testing can be expensive once the system has been installed and placed in full operation.

Operating Efficiency
Operating efficiency is a result of membrane life and stability. If the process does not have a minimum of 80 percent online operating efficiency, including planned maintenance, the system will not be viable or reliable and will be an economic liability. The factors having the greatest effect on operating performance are: (1) consistent membrane fouling

Figure 8-36

Views of membrane pilot plants: (a) enclosed pressurized microfiltration, (b) horizontal microfiltration submerged type, and (c) enclosed pressurized ultrafiltration.

(a)

(b)

(c)

involving frequent and long regeneration times, (2) membrane failure and replacement, (3) failure of high pressure pumps, and (4) failure of pumps and membranes resulting from abrasive waste components (Celenza, 2000).

Clean-in-Place Frequency

Flux maintenance affects all of the factors discussed above. If deterioration of the flux rate occurs frequently due to fouling, the operating efficiency is affected, frequent backflushing or chemical cleaning may be required, and deterioration of the membrane may result. Typical constituents in wastewater that cause membrane fouling or deterioration

Table 8-24
Typical constituents in wastewater that cause membrane fouling or deterioration[a]

Responsible wastewater constituents	Type of membrane fouling	Remarks
Metal oxides Organic and inorganic colloids Bacteria Microorganisms Concentration polarization	Fouling (cake formation sometimes identified as biofilm formation)	Damage to membranes can be limited by controlling these substances (for example, by fine screening).
Calcium sulfate Calcium carbonate Calcium fluoride Barium sulfate Metal oxide formation Silica	Scaling (precipitation)	Scaling can be reduced by limiting salt content, by pH adjustment, and by other chemical treatments such as the addition of antiscalants.
Acids Bases pH extremes Free chlorine Free oxygen	Damage to membrane	Membrane damage can be limited by controlling the amount of these substances in the feedwater. The extent of the damage depends on the nature of the membrane selected.

[a] Adapted from Tchobanoglous et al. (2003).

are described in Table 8-24. The overall operating costs, thus, may be affected due to increased labor cost, increased energy cost, and increased cost of repair and replacement. The importance of adequate pilot-scale testing, discussed above, is further underscored so that unforeseen operating conditions can be minimized.

8-6 DISSOLVED AIR FLOTATION

Dissolved air flotation (DAF) has been used in the wastewater field for the removal of oil and grease from industrial wastes; for the thickening of waste-activated sludge prior to digestion or further processing; for thickening of backwash water from depth filters, usually in large installations; for removing suspended particulate matter that is difficult to remove by conventional flocculation and sedimentation; and for the removal of algae from stabilization and storage pond effluents prior to filtration. In water treatment applications when compared to gravity sedimentation, DAF is a more efficient process for separating low density floc particles. Considerable research has been done in recent years in adapting high-rate DAF for drinking water applications. Much of the same technology can be used for reclaimed water.

Process Description

In DAF, air is dissolved under pressure in the water to be treated, according to Henry's law of dissolution. After pressurization the pressure is released to standard conditions, thus creating millions of microbubbles. The bubbles surround slow-settling particles and float them to the surface for removal (see Fig. 8-8d). As the float layer increases, it begins to slightly dewater and thicken.

Types of DAF

The two basic types of DAF processes used in water reclamation applications involve pressurization of the recycle-flow (see Fig. 8-37a) or pressurization of the full-flow (see Fig. 8-37b). As shown, the principal elements of a DAF system are the pressurization

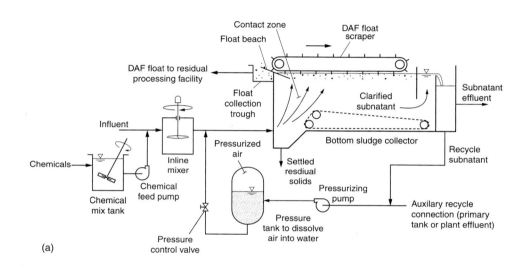

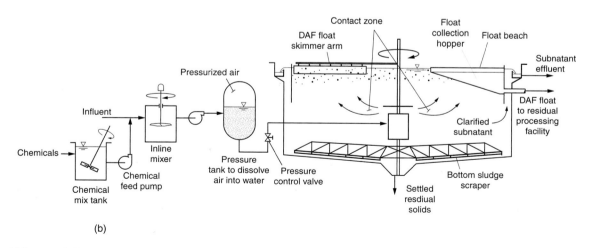

Figure 8-37
Definition sketch for dissolved air flotation systems with (a) pressurization of the recycle flow and (b) pressurization of the full flow. Either rectangular or circular tank types can be used with either pressurization system.

system and the dissolved air flotation tank. The pressurization system consists of a pressurization pump, air compressor, air pressurization tank, and pressure release valve. The flotation tank can be square, rectangular (see Fig. 8-38a), or circular (see Fig. 8-38b) and is equipped with surface skimmers for the removal of the float material, and bottom scrapers to remove settled solids. Baffles are provided to retain the float material while the clarified underflow passes over an effluent weir. The selection of the type of process to be used depends on the characteristics of the feed stream, size of the facility, and the results of pilot flotation studies. It should be noted that there other types of DAF processes such as the stacked DAF with integral media filter, used more commonly in the water treatment field.

Recycle-Flow versus Full-Flow Saturation
Recycle-flow DAF is most appropriate for systems or applications requiring the removal of nearly buoyant or neutrally buoyant floc particles. In this process, the total influent flows either initially through the flocculation tank or directly to the flotation tank if separate flocculation is not required. A portion of the clarified effluent (ranging from 5 to 20 percent) is recycled and delivered to an air saturation tank. The saturation tank may contain a packing material that breaks down the incoming water into small droplets. The upper atmosphere of the tank contains a pressurized volume of air that is dissolved readily into the water. The pressurized recycle water is introduced to the flotation tank through a pressure release device and is mixed with the flocculated water. In the pressure release device, the pressure is reduced to atmospheric pressure, releasing the air in the form of a blanket of fine bubbles (30 to 100 μm in diameter). Full-flow DAF is most appropriate for small systems. The process is similar to recycle-flow

(a)

(b)

Figure 8-38
Views of typical dissolved air flotation systems in operation: (a) rectangular and (b) circular.

DAF except that chemicals are injected into the suction of the feed pump and the entire influent flow is pressurized.

Removal of Thickened and Settled Sludge
The float layer containing air bubbles and the attached solids is removed periodically by a mechanical skimming device (see Figs. 8-37 and 8-38). Clarified water is removed from below the surface. The underflow from a DAF unit can be removed mechanically with scrapers or hydraulically with a manifold system. Because mechanical extraction results in a higher percentage of solids, it is usually favored over hydraulic extraction.

Performance of DAF Process

The performance of the DAF process depends on the characteristics of the wastewater from which suspended solids are to be removed, the type of secondary treatment process, and whether chemicals are used. The variability in process performance that can be expected from a DAF is similar to that for the activated sludge process (see Table 8-6).

Using pond effluent containing algae, it has been possible to achieve effluent TSS values in the range from less than 5 to 30 mg/L with chemical addition. The concentration of algae can vary widely depending on the time of year (see Fig. 8-7) and the location from which the effluent is withdrawn. As will be noted later, the chemical dose is typically independent of the solids concentration.

Design Considerations

Because flotation is very dependent on the type of surface of the particulate matter, laboratory and pilot-plant tests should be performed to yield the necessary design criteria. The design and performance of a DAF system depends on several factors, as presented in Table 8-25; typical design parameters are described in Table 8-26. Two of the principal factors, discussed below, are quantity of air and the need for chemical addition. Typical DAF installations used for the removal of algae are shown on Fig. 8-39.

Quantity of Air for Low Solids Concentrations
As compared to flotation thickening of waste activated sludge, the quantity of air required for flotation of reclaimed water with a low TSS concentration is significantly larger and is similar to the air required for flotation treatment of surface water in drinking water applications. The quantity of air required for treatment is independent of the TSS present because of the relatively low TSS concentration. The air/solids ratio becomes a variable only when the influent TSS is high (> 1000 mg/L), such as in treating stabilization pond effluent that has an extremely high algal content. In low TSS applications, it was found that for raw surface water with an influent TSS of 20 mg/L, approximately 380 mL air/g TSS is required (AWWA, 1999). The large air-to-solids ratio is required to ensure adequate collision between the floc particles and air bubbles to facilitate attachment before liquid/solids separation. The preliminary sizing of a DAF system for the removal of algae from water drawn from a winter storage reservoir is illustrated in Example 8-5.

Table 8-25
Principal factors to be considered in the design of dissolved air flotation systems[a]

Factor	Significance
Influent characteristics a. TSS b. pH c. Temperature d. Particle size	Influent characteristics, especially those cited, affect the need for and type of chemical treatment and the overall performance of the DAF system. Temperature is a factor mainly under cold weather conditions as it affects the flocculation time.
Coagulation a. Type of coagulant b. Coagulant dose c. Mixing requirements	Chemical coagulants create a structure that can easily absorb or entrap air bubbles. Coagulant selection and dose depend on the influent water quality constituents cited above. Thorough mixing is required and some form of in-line mixing is preferred.
Flocculation a. Flocculation time b. Degree of agitation c. Method of flocculation	Flocculation is effective for smaller floc particles, usually less than 100 μm. The flocculation time, which may range from 10 to 20 min, varies depending on the characteristics of the influent water, type of coagulant, and temperature. The method and degree of agitation is important in that excessive floc shear needs to be avoided.
Process efficiency	In practice, the efficiency of the DAF processes is dependent on the number and size of the air bubbles and the velocity of the bubbles. The size of bubbles that are formed in the pressurization tank are dependent on the operating pressure (see Table 8-26).
Air requirements	In low TSS applications, the quantity of air depends only on the quantity of water treated and is independent of the TSS concentration. For high TSS applications, i.e., >1000 mg/L, refer to the text for determining air/solids ratio.
Recycle rate	Recycle rate depends upon the amount of air that is saturated in the recycle stream. In surface water applications using DAF, the recycle rate has ranged from about 7 to 10 percent.
Hydraulic loading rate	Excessive hydraulic loading rates may cause effluent deterioration due to the carryover of air bubbles and the suspended particles.
Solids loading	The range of solids loadings should be verified by pilot testing. If the influent solids concentration increases substantially, a greater recycle rate or pressure may be required to achieve efficient flotation.
Removal of thickened solids and settled solids	In general, the thickened solids are removed with mechanical scrapers (see Fig. 8-37). Settled solids can be removed mechanically of hydraulically.

[a] Adapted in part from AWWA (1999) and Edzwald et al. (1999).

Table 8-26

Typical design factors for dissolved air flotation systems[a]

Factor	Unit	Low Solids concentration (<500 mg/L)	High Solids concentration (>1000 mg/L)
Hydraulic loading rate	m/h	8–20	10–20
Hydraulic detention time	min	5–30	5–30
Basin crossflow velocity	m/h	20–100	20–100
Contact zone detention time	s	30–240	30–240
Contact zone hydraulic loading rate[b]	m/h	35–90	35–90
Width-to-length ratio	unitless	1–2 to 1–4	1–2 to 1–4
Basin depth	m	1.5–3.0	1.5–3.0
Pressurization contact time	s	30–240	30–240
Pressurization tank operating pressure	kPa	400–600	450–600
Bubble size range	μm	10–100	10–100
Bubble size at 400 kPa	μm	50–60	50–60
Bubble size at 500 kPa	μm	40–50	40–50
Bubble size at 600 kPa	μm	30–40	30–40
Air loading[c]	g/m³	8–12	n.a.
Air-to-solids ratio	g/g	n.a., see text	See Eqs. (8-15) and (8-16)
Recycle ratio (recycle systems)	%	50–300	50–300

[a] Adapted in part from AWWA (1999); Couto et al. (2004); Crittenden et al. (2005); and Edzwald et al. (1999).
[b] Contact zone length in rectangular tanks is typically 15 to 25 percent of the total length. The contact zone in circular tanks extends approximately 1/3 of the radius measured from the center of the tank.
[c] Gram of air/m³ of influent wastewater.

(a)

(b)

Figure 8-39

Views of complete dissolved air flotation (DAF) installations: (a) circular DAF unit (foreground) used for the removal of algae from pond water (background). Pressurizing tank is shown on the left. Depth filter is located behind shed on the left, and (b) shallow circular DAF used in Europe for the removal of algae. Pressurizing tank is shown in the foreground. (Courtesy of Austep s.r.l., Italy.)

EXAMPLE 8-5. Removal of Algae by Flotation.

Prepare an initial estimate of the size of a full-flow dissolved air flotation process to remove algae from reclaimed water drawn from a winter storage pond. Use the average design values given in Table 8-26 for low solids concentration and assume that the following conditions apply. Also, if the operating pressure is 400 kPa, estimate the number of bubbles that could potentially be formed per mL of water at the point of release.

1. Reclaimed water flowrate = 4000 m³/d
2. Algal concentration in pond water = 75 mg/L
3. Concentration of alum to overcome alkalinity = 175 mg/L
4. Density of air at 20°C = 1.204 kg/m³ (see Appendix B)

Solution

1. Determine the required surface area. From Table 8-26 select a hydraulic loading rate of 14 m/h (14 m³/m²·h).

$$A = \frac{(4000 \text{ m}^3/\text{d})}{(14 \text{ m}^3/\text{m}^2 \cdot \text{h})(24 \text{ h/d})} = 11.9 \text{ m}^2$$

2. Select tank dimensions. From Table 8-26 use width to length ratio of 1 to 3. Solve for tank length L and width W.

$$L = \sqrt{(11.9 \text{ m}^2) \times 3} = 6.97 \text{ m, say } 6.0 \text{ m}$$

$$W = (1/3) \times 6.0 = 2.0 \text{ m}$$

3. Check detention time τ. From Table 8-26, assume a tank depth of 2.25 m.

$$\tau = \frac{V}{Q} = \frac{(2.0 \text{ m} \times 6.0 \text{ m} \times 2.25 \text{ m})(1440 \text{ min/d})}{(4000 \text{ m}^3/\text{d})} = 18.0 \text{ min, ok}$$

4. Check basin cross flow velocity v_{cf}.

$$v_{cf} = \frac{Q}{\text{Cross sectional area}} = \frac{(4000 \text{ m}^3/\text{d})}{(2.0 \text{ m} \times 2.25 \text{ m})(24 \text{ h/d})} = 37.0 \text{ m/h, ok}$$

5. Check contact zone hydraulic loading CZ_{hyd}. From Table 8-26, assume first 20 percent of the basin length serves as the contact zone.

$$CZ_{hyd} = \frac{Q}{A_{cz}} = \frac{(4000 \text{ m}^3/\text{d})}{[2.0 \text{ m} \times 6.0 \text{ m} \times (0.2)](24 \text{ h/d})} = 69.4 \text{ m/h, ok}$$

6. Compute the air flowrate Q_{air}. From Table 8-26, use an air volume of 10 g air/m³ of water.

$$V_{air} = \frac{(10 \text{ g/m}^3)(4000 \text{ m}^3/\text{d})}{(1.204 \text{ kg/m}^3)(10^3 \text{ g/kg})(1440 \text{ min/d})} = 0.023 \text{ m}^3/\text{min}$$

7. Compute the number of bubbles per mL of water at the point of release. From Step 6, the air per m³ of liquid is equal to 10 g/m³.

   a. Determine the density of air at 20°C and 400 kPa.

   $$\rho_{air,400\ kPa} = (400\ kPa/101.3\ kPa)(1.204\ kg/m^3\ air)$$
   $$= 4.754\ kg/m^3\ air$$

   b. Determine the volume of air.

   $$V_{air} = \frac{(0.010\ kg\ air/m^3\ water)}{(4.754\ kg/m^3\ air)} = 0.00210\ m^3\ air/m^3\ water$$

   c. Determine the volume of an air bubble.

   $$V_{bubble} = \frac{4 \times 3.14 \times [(55/2) \times 10^{-6}\ m]^3}{3} = 8.71 \times 10^{-14}\ m^3$$
   $$= 8.71 \times 10^{-8}\ mL$$

   d. Determine the number of air bubbles.

   $$\text{Number of bubbles} = \frac{(0.00210\ mL\ air/mL\ water)}{(8.71 \times 10^{-8}\ mL/bubble)}$$
   $$= 24{,}144\ bubbles/mL$$

Comment

It is important to note, again, that the air required at low concentrations is independent of the solids concentration up to a value of 1000 mg/L. In this example the TSS concentration is 250 mg/L (75 mg/L + 175 mg/L).

Quantity of Air for High Solids Concentrations

For high influent suspended solids concentrations, the relationship between the-air-to solids ratio and the solubility of air, the operating pressure, and the concentration of solids for a system in which all the flow is pressurized is given in Eq. (8-15).

$$\frac{A}{S} = \frac{1.3\ s_a(f P - 1)}{S_a} \tag{8-15}$$

where A/S = air-to-solids ratio, mL (air)/mg(solids)
s_a = air solubility, mL/L (values vary with temperature, see below)
f = fraction of air dissolved at pressure P, usually 0.5
P = pressure, atm
$= \dfrac{p + 101.35}{101.35}$

p = gage pressure, kPa
S_a = influent suspended solids, g/m³ (mg/L)

Temp., °C	s_a, mL/L
0	29.2
10	22.8
20	18.7
30	15.7

The corresponding equation for a system with only pressurized recycle is

$$\frac{A}{S} = \frac{1.3\, s_a(f P - 1)R}{S_a Q} \quad (8\text{-}16)$$

where R = pressurized recycle, m³/d
Q = influent flowrate, m³/d

In both of the foregoing equations, the numerator represents the weight of air and the denominator the weight of the solids. The factor 1.3 is the weight in milligrams of 1 mL of air and the term (−1) within the brackets accounts for the fact that the system is to be operated at atmospheric conditions.

Chemical Addition

In treating pond effluent, it should be noted that the chemical dose is independent of the algae concentration and only depends on the alkalinity that must be overcome to produce a sweep floc, which is used to intercept the suspended algae. Chemical dosages reported in the literature vary from 20 to 225 mg/L, depending on the specific chemical used and the alkalinity of the wastewater. Where extremely high chemical dosages (e.g., 175 to 225 mg/L) are used, the variability of the effluent tends to be somewhat lower than that for the filtration of activated sludge effluent. Chemical conditioning with polymers can enhance the performance of a DAF unit. The amount of conditioning agent required, the point of addition in the feed stream or recycle stream, and the method of intermixing should be determined for each application. Bench- or pilot-scale flotation tests help in determining the need for chemicals, type and amount of polymer required, and method of mixing.

Operating Considerations

In recycle-flow DAF systems, approximately 50 percent of the power costs of the flotation process is used for pumping the recycle against the saturator pressure. As a result, optimization of the recycle system is important in minimizing power costs. The use of a packed saturator as compared to one that is unpacked should be a design and operating consideration. A packed saturator has been shown to have a lower operating pressure requirement, but a possible disadvantage is that the packing may accumulate biological growth or other precipitates resulting in blockage of the packing (AWWA, 1999).

Pilot-Scale Studies

Because many factors interact to determine DAF performance, bench-and pilot-scale plant investigations are useful in determining expected performance and identifying design factors. Preliminary information for the design of pilot-scale studies can be obtained

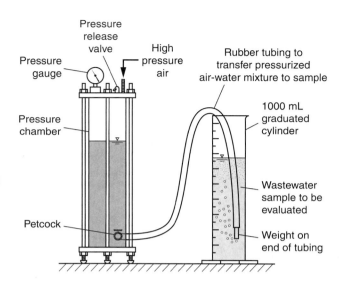

Figure 8-40
Bench-scale flotation cell used to determine DAF performance. The cell is pressurized and the gas is dissolved by shaking the cell. Once the air is dissolved, the supersaturated liquid is released into the graduated cylinder.

using bench-scale flotation cell (see Fig. 8-40). Ideally, pilot-scale units should operate with the same characteristics as the full-scale facility; the feed stream should be the same, gas bubbles created in the pilot plant should be of the same size, and the same operating pressures should be used. Equipment suppliers should be consulted to determine scale-up factors for translating pilot-plant results to a full-size installation.

8-7 ISSUES IN THE SELECTION OF TECHNOLOGIES FOR THE REMOVAL OF RESIDUAL PARTICULATE MATTER

Because of the wide variety of available filter technologies and the number of variables that must be considered, selection of the appropriate technology can sometimes be a daunting task. Some of the issues that must be considered include: (1) the final use of the effluent, (2) the comparative performance of filter technologies, (3) the results of pilot-scale studies, (4) type of disinfection process, (5) future water quality requirements, (6) energy considerations, (7) site constraints, and (8) economic considerations. In most situations a number of these factors, taken together, will govern the final selection.

Final Use of Effluent

Clearly, the filtered water must meet the intended reuse water quality requirements, for example, the State of California's Title 22 turbidity criterion of less than or equal to 2 NTU. It should be noted, however, that even though a number of filters can meet the turbidity criterion, there may be differences in the effluent variability and particle size distributions that will also impact the selection process. Thus, meeting the single effluent criterion is only one of many selection considerations. The marketability of a higher quality water for reuse applications will also be factor.

Treatment levels achievable with various combinations of unit processes used for the removal of residual solids are presented previously in Tables 8-3 and 8-6. Limited information is available for the performance of DAF in water reuse applications; removals similar to those achieved by enhanced sedimentation may be expected.

Comparative Performance of Technologies

Selection of a particular filter technology is often based on the results of pilot-scale studies, and especially so when coupled to the disinfection process to be used. For example, although two filters may produce a comparable effluent with respect to turbidity, the effluent particle size from one of the filters may be more suitable for disinfection with chlorine or UV. In another case, the quantity of backwash water that must be processed may be the determining factor. For example, consider two filters with backwash percentages of 4 and 12 percent of the total throughput capacity. If the total plant flowrate is to be filtered, the treatment plant may not have the hydraulic capacity to process the higher backwash percentage. Also, high backwash return rates increase the capital cost of the treatment facilities. In some cases, it may be necessary to provide special facilities for processing the backwash water to limit the return flow. In some larger installations technologies employing lamella plate settlers have been used. The quantity of backwash water may not be an issue for treatment plants that only filter a portion of the wastewater flow for reuse purposes.

Results of Pilot-Scale Studies

Selection of the filtration process may also be dictated by the type of disinfection process to be employed. For example, if UV disinfection is to be used, the particle size distribution achievable in the filter effluent (as determined by pilot testing) may influence the type of filtration system selected.

Type of Disinfection Process

In any application careful consideration must be given to anticipated future water quality requirements. For example, if it is anticipated that NF or RO may be needed in the future to remove specific trace constituents, MF may be a more suitable choice as compared to either depth or surface filtration because it is a superior method of pretreatment for NF and RO.

Future Water Quality Requirements

As the price of energy continues to increase, the energy requirement for effluent filtration becomes more of a factor in the selection and design of an appropriate filter technology than it has been in the past. Because it is anticipated that energy consumption and cost will continue to be important factors for the foreseeable future, it will be necessary to assess requirements and operating costs for different types of filtration (e.g., depth filtration versus surface filtration versus MF or UF).

Energy Considerations

At many treatment plant sites, the area available for locating effluent filtration facilities may be limited. In such situations, the footprint of the available filter technologies will become an important factor in the selection of a filter technology.

Site Constraints

Capital and operation and maintenance (O & M) costs are important considerations in municipal settings. Typically, a life cycle cost analysis, incorporating a sensitivity analysis for uncontrollable costs such as chemicals and energy, is used to assess the feasibility of different filter technologies. Based on the results of a life cycle cost analysis,

Economic Considerations

Chapter 8 Removal of Residual Particulate Matter

a more capital intensive technology may be selected if the O & M costs are more reasonable and/or controllable. Further, if chemical costs, which are difficult to control, are a significant part of the O & M costs, many municipalities might select a technology with more predictable costs even if it is somewhat more expensive. Finally, in preparing a life cycle cost analysis, it will also be important to consider the potential revenue that could be achieved by producing and marketing a higher quality water for reuse applications.

PROBLEMS AND DISCUSSION TOPICS

8-1 Given the following particle size data (A, B, C, or D to be selected by instructor), determine the coefficients A and β for the power law equation [Eq. (8-1)].

Bin size, mm	Number of particles, number/mL			
	A	B	C	D
1–2	20,000	30,000	50,000	10,000
2–5	12,000	12,000	12,000	2000
5–15	2000	2000	2000	800
15–20	200	250	200	150
20–30	100	100	100	125
30–40	40	40	40	90
40–60	15	15	15	50
60–80	8	7	7	20
80–100	6	4	4	12
100–140	3	1	1	6

8-2 Given the following values for the power law coefficients (water to be selected by instructor), estimate the number of particles in the size range between 2 to 5 and 5 to 15 μm.

Power law coefficient	Water sample			
	A	B	C	D
A	4.5	5.0	4.9	3.5
β	3.5	2.5	2.9	1.8

8-3 Obtain effluent turbidity and suspended solids (if available) data from your local wastewater treatment plant, and determine if the effluent variability as measured by the geometric standard deviation, s_g, is consistent with the information reported in Table 8-4.

8-4 Given the following granular medium filter effluent turbidity data collected at four different treatment plants, estimate the mean, the geometric standard deviation, s_g, and the probability of exceeding a turbidity reading of 2.5 NTU (treatment plant to be selected by instructor).

Turbidity, NTU			
Treatment plant			
A	B	C	D
1.7	1.7	1.0	1.2
1.8	1.1	1.8	1.4
2.2	0.9	1.5	1.5
2.0	1.4	1.1	1.6
	1.3	1.7	1.7
		1.3	1.9
			2.0
			2.1

8-5 Assuming the data for treatment plants A and B were collected at the same treatment plant at different times, what is the impact of using all of the data given for A and B as one data set versus using the individual data sets. In general, what are the advantages or disadvantages of collecting more turbidity samples?

8-6 Given the following sand size distribution (distribution to be selected by instructor), determine the effective size, d_{10}, and the uniformity coefficient, UC. If a layer of anthracite is to be added over 600 mm of sand, determine the effective size required to minimize intermixing.

Sieve number	Percent of sand retained			
	A	B	C	D
6–8	2	0	1	0.1
8–10	8	0.1	2	0.7
10–14	10	0.5	4	1.2
14–20	30	7.4	13	10
20–30	26	32	20	24
30–40	14	30	20	29
40–60	8	25	23	25
Pan	2	5	17	10

8-7 Using the performance data given in the following table for a microfiltration membrane, determine (water to be selected by instructor) the rejection and log rejection for each microorganism group.

	Microorganisms concentration, org/mL			
	Water A		Water B	
Microorganism	Feed stream	Permeate	Feed stream	Permeate
HPC	6.5×10^7	3.3×10^2	8.6×10^7	1.5×10^2
Total coliform	3.4×10^6	100	5×10^5	60
Enteric virus	7×10^3	6.6×10^3	2.0×10^3	9.1×10^2

8-8 A hollow-fiber membrane system with inside-to-outside flow is operated in a cross-flow arrangement. Each module contains 6000 fibers that have an inside diameter of 1.0 mm and a length of 1.25 m. Using this information determine:

a. The feed stream flow rate at the entrance to the module needed to achieve a cross-flow velocity of 1 m/s within the membrane fibers.
b. The permeate flow rate if the permeate flux of 100 L/m²·h is maintained.
c. The retentate cross-flow velocity at the exit from the membrane fibers.
d. The ratio of velocity of flow through the membrane surface to the average crossflow velocity within an individual membrane fiber.
e. The ratio of permeate flow to feed stream flow rate.

(This problem was adapted from Crittenden et al., 2005).

8-9 Membrane filtration, operated in a dead-end mode, is used to treat secondary effluent. If the heterotrophic microorganism plate count (HPC) in the effluent increased from 5 org/L under normal operation to 200 org/L, after an extended period of operation, estimate the number of broken fibers for the following conditions. The inflow rate and organism count are 4000 m³/d and 6.7×10^7 org/L, respectively. The membrane bundle contains 5000 individual fibers. If the influent and effluent turbidity values under normal operation are 4 and 0.25 NTU respectively, estimate the increase in the effluent turbidity assuming the increase could be measured.

8-10 Contrast the advantages and disadvantages between depth filtration, surface filtration, and MF.

8-11 Solve Example 8-5 assuming a circular dissolved flotation unit will be used. Based on a review of the current literature, which type unit (rectangular or circular) would you recommend?

8-12 A small community has decided to reclaim some treated wastewater that is stored in an open reservoir. Two different methods have been proposed to reclaim the water for reuse: (1) chemical coagulation with alum followed by flocculation, sedimentation, and granular medium filtration and (2) chemical coagulation followed by dissolved air flotation and granular medium filtration. The alum dose in both options is 175 mg/L, which is needed to overcome the alkalinity. Assuming the overall efficiency of the two processes is about the same, contrast the two options with respect to: (a) detention time, (b) surface-loading rates, (c) power input, and (d) efficiency. If you were the consultant responsible for process selection, which process would you recommend and why?

REFERENCES

AWWA (1999) *Water Quality and Treatment*, 5th ed., Chap. 8, 8.54–8.58, American Water Works Association, Denver, CO.

Bourgeous, K., G. Tchobanoglous, and J. Darby (1999) "Performance Evaluation of the Koch Ultrafiltration (UF) Membrane System for Wastewater Reclamation," Center For Environmental and Water Resources Engineering, Report No. 99-2, Department of Civil and Environmental Engineering, University of California, Davis, CA.

Caliskaner, O., and G. Tchobanoglous (2000) "Modeling Depth Filtration of Activated Sludge Effluent Using a Synthetic Compressible Filter Medium," Presented at the 73rd Annual Conference and Exposition on Water Quality and Wastewater Treatment, Water Environment Federation, Anaheim, CA.

Celenza, G. J. (2000) *Specialized Treatment Systems, Industrial Waste Treatment Process Engineering*, Vol. 3, Technomic Publishing, Lancaster, PA.

Cleasby, J. L., and G. S. Logsdon (1999) "Granular Bed and Precoat Filtration," Chap. 8, in R. D. Letterman (ed.) *Water Quality and Treatment: A Handbook of Community Water Supplies*, 5th ed., American Water Works Association, McGraw-Hill, New York.

Couto, H. J. B., M. V. Melo, and G. Massarani (2004) "Treatment of Milk Industry Effluent by Dissolved Air Flotation," *Braz. J. Chem. Eng.*, **21**, 1, 83–91.

Crites, R., and G. Tchobanoglous (1998) *Small and Decentralized Wastewater Management Systems*, McGraw-Hill, Boston.

Crittenden, J. C., R. R. Trussell, D. W. Hand, K. J. Howe, and G. Tchobanoglous (2005) *Water Treatment: Principles and Design*, 2nd ed., John Wiley & Sons Inc., New York.

Darby, J., R. Emerick, F. Loge, and G. Tchobanoglous (1999) *The Effect of Upstream Treatment Processes on UV Disinfection Performance*, Project 96-CTS-3, Water Environment Research Foundation, Washington, DC.

Edzwald, J. K., J. E. Tobiason, T. Amato, and L. J. Maggi (1999) "Integrating High Rate DAF Technology into Plant Design," *J. AWWA*, **91**, 12, 41–53.

FLEWR (2005) *Filter Loading Evaluation for Water Reuse*, Monterey Regional Water Pollution Control Agency and National Water Research Institute, Monterey, CA.

Hach (1997) *Hach Water Analysis Book*, 3rd ed., Hach Co., Loveland, OH.

Juby, G. (2003) "Non-Biological Microfiltration Treatment of Primary Effluent," *Wastewater Professional*, **39**, 2, 6–14.

Kawamura, S. (2000) *Integrated Design and Operation of Water Treatment Facilities*, 2nd ed., John Wiley & Sons Inc., New York.

Levine, A. D., G. Tchobanoglous, and T. Asano (1985) "Characterization of the Size Distribution of Contaminants in Wastewater: Treatment and Reuse Implications," *J. WPCF*, **57**, 7, 205–216.

Olivier, M., and D. Dalton (2002) "Filter Fresh: Cloth-Media Filters Improve a Florida Facility's Water Reclamation Efforts," *Water Environ. Technol.*, **14**, 11, 43–45.

Olivier, M., J. Perry, C. Phelps, and A. Zacheis (2003) "The Use of Cloth Media Filtration Enhances UV Disinfection through Particle Size Reduction," in *Proceedings 2003 WateReuse Symposium*, WateReuse Association, Alexandria, VA.

O'Melia, C. R., and W. Stumm (1967) "Theory of Water Filtration," *J. AWWA*, **59**, 11, 1393–1411.

Paranjape, S., R. Reardon, and X. Foussereau (2003) "Pretreatment Technology for Reverse Osmosis Membrane Used in Wastewater Reclamation Application—Past, Present, and Future—A Literature Review," in *Proceedings 76th Annual Technical Exhibition and Conference*, Water Environment Federation, Alexandria, VA.

Riess, J, K. Bourgeous, G. Tchobanoglous, and J. Darby (2001) *Evaluation of the Aqua-Aerobics Cloth Medium Disk Filter (CMDF) for Wastewater Recycling in California*, Center for Environmental and Water Resources Engineering, Report No. 01-2, Department of Civil and Environmental Engineering, University of California, Davis, CA.

Sakaji, R. H. (2006) "What's New for Membranes in the Regulatory Arena," Presented at Microfiltration IV, National Water Research Institute, Anaheim/Orange County, Orange, CA.

Standard Methods (2005) *Standard Methods for the Examination of Water and Waste Water*, 20th ed., American Public Health Association, Washington, DC.

Stowell, R (1976) *A Study of the Screening of Algae from Stabilization Ponds*, Masters Thesis, Department of Civil and Environmental Engineering, University of California, Davis, CA.

Tchobanoglous, G. (1988) "Filtration of Secondary Effluent for Reuse Applications," Proceedings of the Twelfth Sanitary Engineering Conference, University of Illinois, Urbana, IL.

Tchobanoglous, G., F. L. Burton, and H. D. Stensel (2003) *Wastewater Engineering: Treatment and Reuse*, 4th ed., McGraw-Hill, New York.

Tchobanoglous, G., and R. Eliassen (1970) "Filtration of Treated Sewage Effluent," *J. San. Eng. Div.*, ASCE, **96**, SA2, 243–265.

Tchobanoglous, G., and E. D. Schroeder (1985) *Water Quality: Characteristics, Modeling, Modification*, Addison-Wesley Publishing Company, Reading, MA.

Trussell, R. R., and C. H. Tate (1979) Measurement of Particle Size Distribution in Water Treatment, in *Proceedings Advances In Laboratory Techniques for Water Quality Control*, AWWA, Philadelphia, PA.

Trussell, R. R., and M. Chang (1999) "Review of Flow through Porous Media as Applied to Head Loss in Water Filters," *J. Environ. Eng.*, ASCE, **125**, 11, 998–1006.

Urase, T., K. Yamamoto, and S. Ohgaki (1994) "Effect of Pore Size Distribution of Ultrafiltration Membranes in Virus Rejection in Crossflow Conditions, *Water Sci. Technol.* **30**, 9, 199–208.

9 Removal of Dissolved Constituents with Membranes

WORKING TERMINOLOGY 462

9-1 INTRODUCTION TO TECHNOLOGIES USED FOR THE REMOVAL OF DISSOLVED CONSTITUENTS 463
Membrane Separation 463
Definition of Osmotic Pressure 463
Nanofiltration and Reverse Osmosis 465
Electrodialysis 466
Typical Process Applications and Flow Diagrams 467

9-2 NANOFILTRATION 467
Types of Membranes Used in Nanofiltration 468
Application of Nanofiltration 471
Performance Expectations 471

9-3 REVERSE OSMOSIS 473
Types of Membranes Used in Reverse Osmosis 473
Application of Reverse Osmosis 474
Performance Expectations 474

9-4 DESIGN AND OPERATIONAL CONSIDERATIONS FOR NANOFILTRATION AND REVERSE OSMOSIS SYSTEMS 475
Feedwater Considerations 475
Pretreatment 477
Treatability Testing 479
Membrane Flux and Area Requirements 482
Membrane Fouling 487
Control of Membrane Fouling 490
Process Operating Parameters 490
Posttreatment 492

9-5 PILOT-SCALE STUDIES FOR NANOFILTRATION AND REVERSE OSMOSIS 499

9-6 ELECTRODIALYSIS 501
Description of the Electrodialysis Process 501
Electrodialysis Reversal 502
Power Consumption 503
Design and Operating Considerations 506
Membrane and Electrode Life 507
Advantages and Disadvantages of Electrodialysis versus Reverse Osmosis 508

9-7 MANAGEMENT OF MEMBRANE WASTE STREAMS   509
Membrane Waste Stream Issues   509
Thickening and Drying of Waste Streams   511
Ultimate Disposal Methods for Membrane Waste Streams   515

PROBLEMS AND DISCUSSION TOPICS   519

REFERENCES   522

WORKING TERMINOLOGY

Term	Definition
Brine	Waste stream containing elevated concentrations of total dissolved solids.
Concentrate	See retentate.
Cross-flow filtration	Filtration technique in which the feed stream is pumped at high velocity parallel to the membrane surface to reduce the collection of retained species at the membrane surface.
Electrodialysis (ED)	A process that moves ions (charged molecular species) from one solution to another employing an electrical potential as the driving force and using a semipermeable membrane as a separator.
Electrodialysis reversal (EDR)	A modification of the ED process in which the polarity of the applied voltage is changed periodically, reversing the direction of ion movement to "electrically" flush the membranes to control membrane scaling and fouling.
Flux	The mass or volume rate of transfer through the membrane surface, usually expressed as $m^3/m^2 \cdot h$ or $L/m^2 \cdot h$ ($gal/ft^2 \cdot d$). Flux is the prevalent term for referring to the rate of water production from a membrane system.
Fouling	Deposition of material on the membrane surface resulting in the loss of performance.
Molecular weight cutoff (MWCO)	The designation for the size of materials retained by a membrane.
Nanofiltration (NF)	A pressure-driven membrane separation process that typically operates at pressures in the range of 5 to 10 bar and removes particle and dissolved material as small as approximately 0.001 μm.
Osmotic pressure	A natural pressure phenomenon exhibiting a force from a low concentration stream to a high concentration stream.
Recovery	Permeate or filtrate flow divided by the feedwater flowrate, expressed as a percentage.
Rejection or solute rejection	The fraction of solute (usually TDS) in the membrane feedwater that remains in the retentate (concentrate) stream. For ED/EDR systems, the term removal is used instead of rejection.
Retentate (also known as concentrate)	The portion of the process stream that contains salts and other constituents rejected from the membrane process.
Reverse osmosis (RO)	A high pressure [over 10 bar (1000 kPa)] membrane separation process used primarily for the removal of organic matter and salts from wastewater and for desalting brackish water and seawater.

Semipermeable membrane	A membrane that is permeable to some components in a feed solution and impermeable to other components.
Sodium adsorption ratio (SAR)	A measure of the sodicity of the soil; the SAR is the ratio of the sodium cation to the calcium and magnesium cations (see Chap. 17).
Thin film composite (TFC) membrane	Membranes that are composed of two or more materials cast on top of one another. Their separation and structural properties can be optimized independently for specific applications.
Transmembrane pressure	The pressure drop across a membrane.

For many water reuse applications, the removal of residual particulate matter by methods described in Chap. 8 provides an appropriate level of treatment. Where increased removal of dissolved organic and inorganic constituents is required, different technologies are required for their removal. Dissolved solids are little affected by the various treatment operations discussed previously. Many water reuse applications, however, are adversely affected by dissolved solids and membrane technologies such as nanofiltration (NF), reverse osmosis (RO), and electrodialysis (ED) are being used to increase removal of dissolved constituents. In this chapter, the following topics are discussed: (1) introduction to membrane technologies used for the removal of dissolved constituents, (2) NF, (3) RO, (4) ED, and (5) management of concentrate waste streams. Ion exchange, which can be used to remove specific constituents by exchanging them for other constituents, is considered in Chap. 10. Microfiltration (MF) and ultrafiltration (UF) are considered in Chap. 8.

9-1 INTRODUCTION TO TECHNOLOGIES USED FOR THE REMOVAL OF DISSOLVED CONSTITUENTS

With the increased use of reclaimed water for applications where quality and reliability are critical, such as indirect potable reuse and some industrial uses, the increased removal of dissolved solids and trace constituents may be required. The development of membrane technologies has provided practical means of achieving high removals reliably and at reasonable cost. The applicable membrane technologies NF, RO, and ED are introduced in this section.

Membrane Separation

The removal of dissolved constituents in reclaimed water can be accomplished by two basic membrane separation processes: pressure driven and electrically driven. The pressure-driven processes are NF and RO, and the electrically-driven process is ED. The general characteristics of membrane processes used for the removal of dissolved constituents from reclaimed water are presented Table 9-1. Comparative information on the MF and UF processes is given in Table 8-17 in Chap. 8.

Definition of Osmotic Pressure

Because the principle, osmosis, involved in NF and RO is different from MF and UF discussed in Chap. 8, the following definition of osmotic pressure is provided. Where two solutions having different solute concentrations are separated by a semipermeable membrane, a difference in chemical potential will exist across the membrane

Table 9-1
General characteristics of nanofiltration, reverse osmosis, and electrodialysis membrane processes[a]

Factor	Membrane process		
	Nanofiltration	Reverse osmosis	Electrodialysis
Membrane driving force	Hydrostatic pressure difference	Hydrostatic pressure difference	Electromotive force
Typical separation mechanism	Sieve, solution/diffusion, and exclusion	Solution/diffusion and exclusion	Ion selective membrane
Typical pore size	Micropores (<2 nm)	Dense (<2 nm)	na
Typical operating range, μm	0.001–0.01	0.0001–0.001	na
Molecular weight cut-off	300–1000	<300	na
Permeate description	Water, very small molecules, ionic solutes	Water, very small molecules, ionic solutes	Water, ionic solutes
Typical constituents removed	Small molecules, color, some hardness, bacteria, viruses, proteins	Very small molecules, color, hardness, sulfates, nitrate, sodium, other ions	Charged ionic solutes
Operating pressure[b]	350–550 kPa 3.5–5.5 bar (50–80 lb/in.^{2})	1200–1800 kPa 12–18 bar (175–260 lb/in.^{2})	na
Energy consumption[b]	0.6–1.2 kWh/m^3	1.5–2.5 kWh/m^3	1.1–2.6 kWh/m^3
Material	Cellulosic, aromatic polyamide, thin film composite[c]	Cellulosic, aromatic polyamide, thin film composite[c]	Ion exchange resin cast as a sheet
Configuration	Spiral wound, hollow fiber	Spiral wound, hollow fiber	Sheets

[a] Adapted from Tchobanoglous et al. (2003).
[b] Based on treating reclaimed water with a TDS concentration in the range from 1000 to 2500 mg/L (see also Table 9-11). Significantly higher operating pressures are required for seawater.
[c] With surface layer formed from different types of polyamide compounds. Support structure usually made of polysulfone.

Note: kPa × 10^{-2} = bar, (1 bar = 100 kPa = 10^5 N/m^2).
kPa × 0.145 = lb/in.^{2}.
na = not applicable.

(see Fig. 9-1a). Water will tend to diffuse through the membrane from the lower-concentration (higher-potential) side to the higher-concentration (lower-potential) side. In effect, the water passing through the membrane is trying to dilute the higher ionic concentration solution to equalize the concentrations on both sides of the membrane. This balancing pressure difference is termed the *osmotic pressure* (see Fig. 9-1b) and is a function of the solute characteristics and concentration and temperature. If a pressure gradient opposite in direction and greater than the osmotic pressure is imposed across the membrane, flow from the more concentrated to the less concentrated region will occur and is termed *reverse osmosis* (see Fig. 9-1c). The principle of RO is employed in both the NF and RO processes.

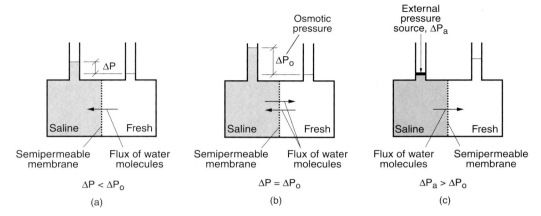

Figure 9-1
Definition sketch of (a) osmotic flow, (b) osmotic equilibrium, and (c) reverse osmosis.

Nanofiltration and Reverse Osmosis

Nanofiltration and RO (see Fig. 9-2) unlike MF and UF, discussed in Chap. 8, are capable of separating dissolved ions from the feed stream. As described in Chap. 8, MF and UF use pressure (and sometimes a vacuum) to provide convective flow of the liquid through the membrane, and NF and RO require hydrostatic pressure to overcome the osmotic pressure of the feed stream. Whereas MF and UF membranes reject constituents based on size and are rated in terms of pore size or porosity, NF and RO membranes are rated on the basis of salt rejection and flow. As shown on Fig. 8-1 in Chap. 8, NF and RO can remove particles of sizes less than 10^{-2} μm, which encompass dissolved and colloidal material such as aqueous salts, organic matter, pesticides, and herbicides.

In concept and operation, NF is much the same as RO; the key difference is the degree of removal of monovalent ions such as sodium and chlorides. Reverse osmosis removes monovalent ions in the 98 to 99+ percent range while removals with NF membranes vary between 50 and 90 percent, depending on the material and manufacture of the membrane. Nanofiltration has often been incorrectly categorized as a "loose RO" membrane. Nanofiltration has the ability to reject uncharged, dissolved materials and positively charged ions according the size and shape of the molecule. In contrast, "loose RO"—sometimes called low-pressure RO, is an RO membrane with reduced salt rejection. This type of membrane has been used in applications where moderate salt removal is acceptable (www.gewater.com). Nanofiltration is discussed in more detail in Sec. 9-2.

Reverse osmosis has been used in many applications including sea and brackish water desalination and industrial process water for the removal of dissolved constituents. Reverse osmosis typically removes 95 to 99.5 percent of the total dissolved solids and 95 to 97 percent of dissolved organic matter. With the development of new low-pressure membranes, RO is finding increased use for water reuse applications. Reverse osmosis is discussed in Sec. 9-3.

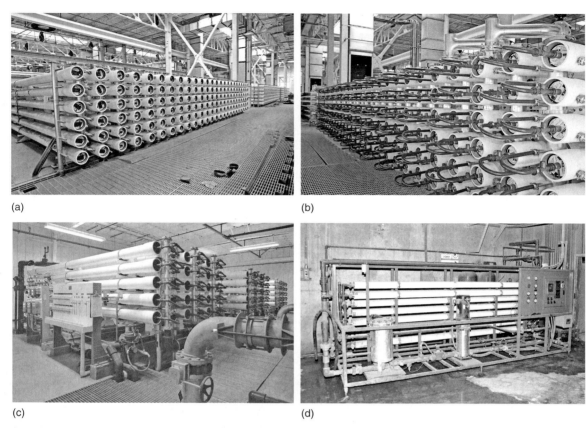

Figure 9-2
Views of various reverse osmosis installations: (a) and (b) views of large installation under construction used to treat activated sludge effluent following microfiltration, chemical addition, and cartridge filtration. Each bank of RO modules is designed to treat 19,000 m^3/d (5 Mgal/d). The capacity of the entire facility is 265,000 m^3/d (70 Mgal/d). It should be noted that the appearance of a nanofiltration installation would be the same. (c) view of facility designed to treat about 4000 m^3/d (1 Mgal/d) following microfiltration, and (d) skid mounted unit with integral cartridge filter and pump for point of use treatment.

Electrodialysis

Electrodialysis (ED) is an electrically driven process in which mineral salts and other species are transported through ion selective membranes from one solution into another under the driving force of direct electrical potential. Salts are in solution as ionized particles with positive and negative charges. When direct current is imposed on the solution, the positive ions migrate to the negative electrode, or cathode, and the negative ions migrate to the positive electrode, or anode (see Fig. 9-3). As compared to NF and RO, which transports water through the membrane leaving the salts behind, with ED, salt is gradually removed from solution leaving a dilute solution behind. Electrodialysis does not remove colloidal matter, matter that is not ionized, or bacteria (USBR, 2003).

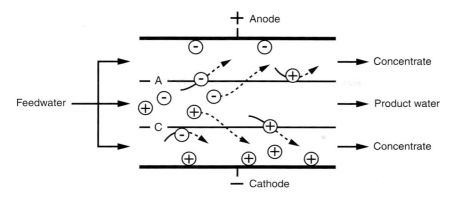

Figure 9-3
Conceptual diagram of electrodialysis process. Anions migrate toward positive pole (anode) and cations migrate toward negative pole (cathode). More detailed process diagrams are given on Figs. 9-14 and 9-15. Note that hydrogen gas (H_2) evolves from the cathode and oxygen gas (O_2) evolves from the anode to maintain charge neutrality during operation.

Electrodialysis and a modification of the electrodialysis process, termed electrodialysis reversal (EDR), used principally in the desalting brackish water, have also been used for wastewater applications. In a recent side-by-side test of RO and EDR it was found that comparable results can be achieved in demineralizing nonpotable reclaimed water (Lehman et al., 2004). Electrodialysis and EDR are discussed in Sec. 9-5.

Typical Process Applications and Flow Diagrams

Typical applications for the membrane processes used for the removal of dissolved constituents are reported in Table 9-2. Representative process flow diagrams using NF, RO, and ED are shown on Fig. 9-4. The diagrams differ mainly in the type of pretreatment required and desired level of constituent removal required. The process shown on Fig. 9-4d is a hybrid, combining ED and RO for desalination. Pretreatment customarily may consist of depth or surface filtration, MF, UF, or ED. Fine screens and cartridge filters may also be used in combination with other methods of pretreatment to minimize fouling of the membranes. Scaling and fouling of the membranes is of principal concern in selecting elements of the treatment system for specific applications.

9-2 NANOFILTRATION

Nanofiltration is used to remove particles in the 300 to 1000 molecular weight (MW) range, rejecting selected salts and most organics and microorganisms, operating at higher recovery rates and at lower pressures than RO systems. Even though most inorganic and organic constituents and microorganisms are removed, disinfection is required to ensure system reliability in the event of a leak or defect in the membrane (see Example 8-4 in Chap. 8).

Table 9-2
Typical applications for membrane technologies for the removal of dissolved constituents[a]

Applications	Process and process function	See Sec.
Nanofiltration (NF)		9-2, 9-4
Water softening	Used to reduce the concentration of multivalent ions contributing to hardness for specific water reuse applications.	
Water reuse	Used to reduce the TDS concentration of reclaimed water for specific applications. Also used in conjunction with reverse osmosis (see Fig. 9-6).	
Water reuse	Used to treat prefiltered effluent (typically with MF or UF) for indirect potable reuse applications such as groundwater injection.	
Reverse osmosis (RO)		9-3, 9-4
Desalination (desalting)	Used to remove dissolved constituents from both brackish and sea water.	
Water reuse	Used to treat prefiltered effluent (typically with cartridge filtration, MF, or UF) for indirect potable reuse applications such as groundwater injection.	
Two-stage treatment for boiler use	Two stages of RO are used to produce water suitable for high pressure boilers in industrial reuse applications.	
Electrodialysis (ED) and electrodialysis reversal (EDR)		9-6
Desalination (desalting)	Used to remove dissolved charged ionic constituents from brackish water.	
Water reuse	Used to treat prefiltered (usually with cartridge filters) brackish water with a low TDS concentration. ED/EDR removes only ionized compounds; dissolved organic compounds pass through ED/EDR systems and must be removed by other means.	
Water softening	Used to reduce the concentration of multivalent ions contributing to hardness for specific water reuse applications.	

[a] Adapted in part from Stephenson et al. (2000).

Types of Membranes Used in Nanofiltration

Two membrane configurations are used commonly for NF: spiral wound and hollow fiber (see Fig. 9-5). Most NF facilities use polyamide thin film composite (TFC) membranes in a spiral-wound configuration, but there are more than ten other types of membranes available including polyamide hollow fiber, polyvinyl acetate spiral wound, and asymmetric cellulose acetate (see Fig. 8-29c in Chap. 8) in a tubular configuration. Tubular configurations are seldom used as they have high pressure drops (200 to 280 kPa) and have low surface-to-volume ratios that result in large space requirements (Celenza, 2000).

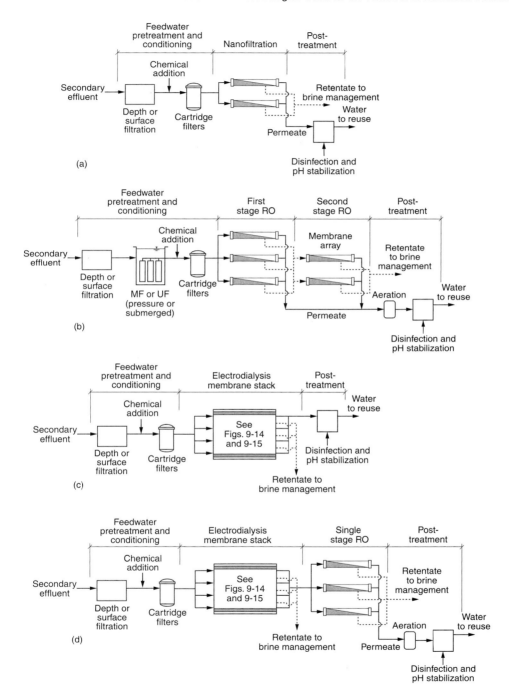

Figure 9-4

Typical process flow diagrams: (a) nanofiltration, (b) reverse osmosis, (c) electrodialysis, and (d) combined electrodialysis with reverse osmosis (for desalination).

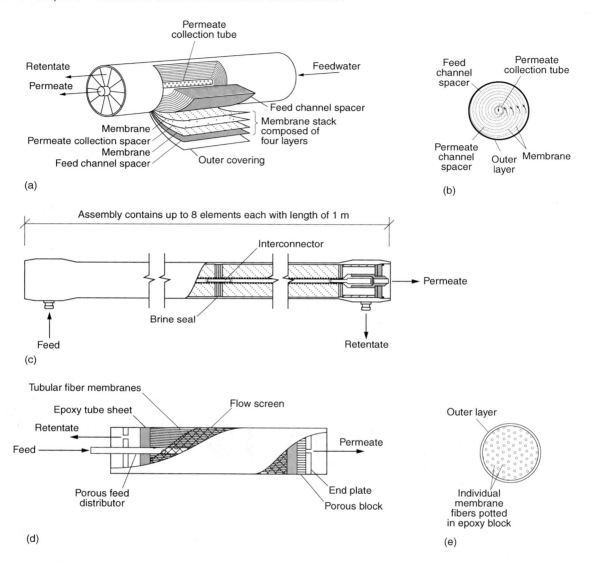

Figure 9-5

Typical membrane configurations used for nanofiltration and reverse osmosis: (a) spiral wound element construction, (b) cross section through spiral wound element (c) spiral wound membranes and vessel assembly, (d) hollow-fiber bundle, and (e) end view of hollow-fiber membrane bundle.

Nanofiltration membranes have pore sizes ranging from less than 0.001 to 0.003 μm, and they reject on two levels of selectivity. First, noncharged soluble organics are rejected on size and shape; the molecules are too large to pass through the pores. Second, charged soluble salts smaller than the membrane pores are rejected because the water is more soluble in the membrane than is a specific salt. Most NF membranes can be used to separate divalent anions from monovalent anions (Lien, 1998).

Application of Nanofiltration

Nanofiltration is used where the high salt rejection of RO is not necessary although NF is capable of removing hardness elements in water such as calcium and magnesium (see Table 9-2). Nanofiltration membranes are often used for removing salts to meet TDS requirements for groundwater recharge and for water softening applications.

Nanofiltration is also used in conjunction with RO in Australia in a two-stage process as shown on Fig. 9-6 for salinity reduction in high TDS-treated wastewater. In this process, NF is used to remove divalent and trivalent ions. The permeate from the NF, which contains mostly monovalent ions, is processed through RO while the divalent ions such as calcium and magnesium in the NF retentate are blended in the final product water to adjust the sodium adsorption ratio (SAR) for use as irrigation water. The removal of the divalent and trivalent ions enables the RO process to operate at higher recovery rates (Leslie et al., 2005).

Performance Expectations

The performance of NF with respect to the removal of specific constituents is site specific related primarily to the characteristics of the water to be treated, the type of membrane, and the operational strategies. Two issues related to process performance with respect to the removal of dissolved constituents are rejection rate and the degree of variability.

Rejection Rates

Nanofiltration membranes are used most commonly to reject high percentages of multivalent ions and divalent cations allowing monovalent ions to pass. Rejection rates for various constituents are given in Table 9-3 for NF and compared to rates for loose RO, although it should be noted that the distinction between NF and loose RO is not well defined in the literature.

Process Performance Variability

Typical values for the variability of the NF process are reported in Table 9-4, along with corresponding values for other membrane processes. Typical effluent values are not

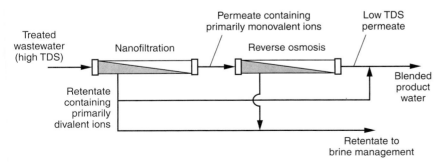

Figure 9-6
Schematic flow diagram for a two-stage process employing a nanofiltration membrane in conjunction with reverse osmosis for salinity reduction.

Table 9-3
Typical rejection rates for NF and RO[a]

Constituent	Unit	Rejection rate Nanofiltration	Loose RO
Total dissolved solids	%	40–60	
Total organic carbon	%	90–98	
Color	%	90–96	
Hardness	%	80–85	
Sodium chloride	%	10–50	70–95
Sodium sulfate	%	80–95	80–95
Calcium chloride	%	10–50	80–95
Magnesium sulfate	%	80–95	95–98
Nitrate	%	10–30	
Fluoride	%	10–50	
Arsenic (+5)	%	<40	
Atrazine	%	85–90	
Proteins	log	3–5	3–5
Bacteria[b]	log	3–6	3–6
Protozoa[b]	log	>6	>6
Viruses[b]	log	3–5	3–5

[a]Adapted in part from www.gewater.com and Wong (2003).
[b]Theoretically all microorganisms should be removed. The reported values reflect membrane integrity concerns (see Example 8-4 in Chap. 8).

Table 9-4
Typical effluent quality variability observed with processes used for the removal of constituents from reclaimed wastewater[a]

Constituent	Geometric standard deviation, s_g[c]	
	Range	Typical
Nanofiltration		
TDS	1.3–1.5	1.4
TOC	1.2–1.4	1.5
Turbidity	1.5–2.0	1.75
Reverse osmosis[d]		
TDS	1.3–1.8	1.6
TOC	1.2–2.0	1.8
Turbidity	1.2–2.2	1.8
Electrodialysis		
TDS	1.2–1.75	1.5

[a]Adapted in part from WCPH (1996).
[b]Typical effluent values are not given for the processes because they will vary widely, depend on the operating conditions and water quality requirements.
[c]s_g = geometric standard deviation; $s_g = P_{84.1}/P_{50}$.
[d]Because measured effluent values are typically near the constituent detection limits, the error in the detection method can contribute to the observed effluent variability.

given for the processes in Table 9-4 because of the wide range of values reported in the literature for different applications. In general, the range and mean geometric standard deviation for the NF process is somewhat higher than the values reported for depth and surface filtration. Additional comments on the reasons for the higher variability for the NF process are presented in the discussion of the variability for the RO process in the following section.

9-3 REVERSE OSMOSIS

Reverse osmosis is used to remove dissolved materials, commonly salts, under pressures ranging from 1200 to 1800 kPa for low TDS water (1000 to 2500 mg/L) to 5500 to 8500 kPa for seawater and at flux rates varying from about 12 to 200 $L/m^2 \cdot h$. Separation using RO membranes is effective generally at solute molecular weights below 300, and related solute sizes from 0.0001 to 0.001 μm (Celenza, 2000).

Types of Membranes Used in Reverse Osmosis

Reverse osmosis membranes are typically TFC membranes in a spiral wound configuration or hollow fiber with a pore size of approximately 0.0005 μm. Typical membrane configurations are as shown on Fig. 9-5 and their characteristics are summarized previously in Table 9-1. As shown on Fig. 9-5, the membranes are operated in a modified cross-flow mode without retentate recirculation. Additional operating information on hollow fiber and spiral wound membranes is given in Table 9-5. Spiral wound configurations have a low modular cost and are economical to operate because of low power consumption. Pressure drops range from 100 to 140 kPa. The area-to-volume ratio is between those of the tubular and hollow fiber membranes thus requiring careful examination of the space requirements. Plugging of the spiral wound elements can be minimized by filtering the influent to 1 to 10 μm, depending on the spacer construction (Celenza, 2000).

Spiral Wound Thin Film Composite Membranes
Conventional TFC membranes (see Fig. 8-29*d* in Chap. 8) using aromatic polyamide polymer are strongly hydrophobic and prone to high fouling rates during treatment of feedwaters with high concentrations of organic matter. Newer low fouling products are characterized by a low surface charge or coating that yields a surface similar to a

Table 9-5 Characteristics of commonly used RO membrane configurations[a]

Performance characteristic	Membrane type	
	Hollow fiber	Spiral wound
Resistance to mechanical damage	Good	Poor
Plugging potential	High	High
Mechanical cleaning	Poor	Poor
Area-to-volume ratio	High	Moderate
Power consumption	Good	Good
Membrane replacement costs	High	Low

[a]Adapted from Celenza (2000).

hydrophilic membrane with reduced affinity for dissolved organics. From field results with membranes operating in municipal wastewater reclamation systems, it has been found that the fouling rate is very low, comparable with that observed in RO operation with clean well water. The reduced fouling rate is attributed to a lower rate of adsorption of dissolved organics on the hydrophilic membrane surface (Wong, 2003; Freeman et al., 2002; Pearce et al., 2001).

Hollow Fiber Membranes

Hollow fiber membrane configurations have the highest surface-to-volume ratio and relatively low operating pressure drop (35 to 140 kPa), making them attractive in terms of energy consumption. Their backflushing capabilities make them relatively easy to clean. The small tube diameters, however, make the fiber prone to clogging thus requiring influent prefiltering of 20 to 100 μm (Celenza, 2000).

Application of Reverse Osmosis

In water reuse applications, RO is used for the removal of dissolved constituents remaining in wastewater after advanced treatment with depth filtration or MF or UF (see Table 9-2 and Fig. 9-4). Reverse osmosis membrane systems have been used to treat reclaimed municipal wastewater for groundwater recharge, surface water augmentation, cooling tower and evaporative cooler makeup water, and high pressure boiler feedwater.

Performance Expectations

As with NF discussed previously, RO process performance with respect to the removal of dissolved constituents is site specific related to the characteristics of the water to be treated, the type of membrane used, and the operational strategies.

Rejection Rates

Typical removal rates of constituents by RO are presented in Table 9-6. As reported in Table 9-6, the removal rates for most constituents are very high, especially when treating reclaimed water with relatively low TDS concentrations (1000–2500 mg/L). As the TDS increases, the observed removals move towards the lower end of the range given in Table 9-6.

Process Performance Variability

Typical values for the variability of the RO process are reported in Table 9-4, along with corresponding values for other membrane processes. In general, the ranges and mean geometric standard deviations for the RO process are larger than those for the other membrane processes. As noted in Chap. 8 in the discussion of the performance of MF and UF, as the constituent concentrations in the effluent get lower and lower, even slight perturbances can have a significant effect on the observed variability. In addition, the standard error in the detection methods can contribute to the variability for a given constituent. In trying to determine the geometric standard deviation for the RO process, a number of nondetect values will be reported. Although several different statistical methods are available for dealing with nondetect values, it is common practice, for the purpose of data analysis, to assign a value equal to one-half of the detection limit for the test to a nondetect value.

Table 9-6
Typical performance for reverse osmosis treatment[a]

Constituent	Unit	Rejection rate
Total dissolved solids	%	90–98
Total organic carbon	%	90–98
Color	%	90–96
Hardness	%	90–98
Sodium chloride	%	90–99
Sodium sulfate	%	90–99
Calcium chloride	%	90–99
Magnesium sulfate	%	95–99
Nitrate	%	84–96
Fluoride	%	90–98
Arsenic (+5)	%	85–95
Atrazine	%	90–96
Proteins	log	4–7
Bacteria[b]	log	4–7
Protozoa[b]	log	>7
Viruses[b]	log	4–7

[a]Adapted in part from www.gewater.com and Wong (2003).
[b]Theoretically all microorganisms should be removed. The reported values reflect integrity concerns (see Example 8-4 in Chap. 8).

9-4 DESIGN AND OPERATIONAL CONSIDERATIONS FOR NANOFILTRATION AND REVERSE OSMOSIS SYSTEMS

In the design of membrane systems for the removal of dissolved constituents, success of the system depends on careful analysis of the feedwater and selection of an appropriate pretreatment system. Membrane systems are extremely sensitive devices and considerable care must be taken to protect and optimize their useful life. Process design considerations for NF and RO systems are given in Table 9-7 and design and operations considerations are discussed in this section.

Pretreatment of the feedwater must be selected to allow successful performance of the specific membranes. The characteristics of the feedwater that must be considered include (Celenza, 2000):

Feedwater Considerations

- Suspended solids or turbidity to prevent clogging
- Organics that promote membrane degradation and destruction
- Iron, manganese, and other precipitates
- Oil, grease, or floatables to prevent fouling
- Erosive substances to minimize membrane wear
- In some cases, temperature within membrane operating limits
- pH to prevent membrane degradation

Table 9-7
Process design considerations for NF and RO[a]

Design consideration	Discussion
Feedwater characterization	Complete characterization of the feedwater is essential for identifying constituents that produce a high potential for membrane fouling. The effect of residual suspended solids in the influent to the membranes especially should be evaluated.
Pretreatment	Pretreatment must be evaluated to extend membrane life and issues such as flow equalization, pH control, chemical treatment, and residual solids removal should be considered.
Flux rate	Flux rate influences system costs by establishing the filter area, affecting polarization control, and affecting membrane life.
Recovery	Recovery rate affects solute rejection and membrane performance.
Membrane fouling	Parameters should be developed based on pilot-scale testing. Acid, antiscalants, and biocides are used to control membrane fouling, as are staging and operational conditions.
Membrane cleaning	Cleaning procedures and frequency need to be established.
Membrane life	The principal economic consideration that governs successful application of membrane technology.
Operating and maintenance costs	High pressure systems require significant energy costs, high capital costs for high pressure pumps, and high maintenance costs. associated with equipment wear. After membrane replacement, energy is the next major operating expense.
Recycle flows	Provisions for recycling a portion of the product water should be included as an operating consideration to control membrane velocity, influent concentration, and for equalizing influent flow variations.
Retentate and backwash disposal	Retentate and backwash characteristics need to be considered especially if chemicals are used in pretreatment or for membrane cleaning, and large volumes of waste require disposal.

[a] Adapted in part from Celenza (2000).

Factors that should be considered in designing NF and RO systems include the following (EPRI, 1999):

- A method of particle filtration. Because NF and RO are intended mainly to remove dissolved salts, they should, ideally, have turbidity levels less than 0.5 NTU in the influent.
- Presence and level of oxidants in the feedwater. Many thin film composite membranes are less tolerant of free chlorine or other oxidants such as ozone in the feedwater. Generally, membrane manufacturers recommend less that 1000 mg/L·h with free chlorine; membranes are warranted for up to 1000 h of contact with water containing 1 mg/L of free chlorine, or 200 h with 5 mg/L. Membranes made of polyvinylidene fluoride, which have a high resistance to ozone, are becoming available.
- Biological contamination. Biological contamination and large organic molecules should also be removed in a pretreatment system as they tend to accumulate and foul

the membrane surface and cause a decline in performance. To prevent biological fouling, in addition to particle filtration, an appropriate level of disinfection should be provided in accordance with recommendations of the membrane manufacturer.
- Membrane protection by chemical addition. Generally, only a few chemicals are needed to enhance membrane performance. Acid addition may be needed to adjust the feed pH to control scale formation if cellulose acetate membranes are used. Antiscalants may be required to prevent precipitation of slightly soluble salts.

Pretreatment

Membrane elements in NF and RO units can be fouled by colloidal matter and other constituents in the feed stream. For example, certain chemical constituents in the feed stream can increase membrane fouling because they are at or near their solubility limits and can precipitate, especially in RO applications. In most cases, pretreatment is absolutely necessary to remove or reduce these constituents. Several pretreatment options have been used singly or in combination and are listed in Table 9-8. Appropriate forms of pretreatment include depth or surface filtration, MF or UF, dissolved air flotation, as discussed in Chap. 8, and cartridge filters.

Cartridge filters, such as shown on Fig. 9-7, are almost always used as a further pretreatment step, both with reclaimed water as well as untreated seawater, following one or more of the pretreatment options cited above. Cartridge filters serve two important

Table 9-8
Methods of pretreatment for nanofiltration and reverse osmosis systems

Material to be removed	Method of pretreatment	Description or discussion
Iron and manganese	Ion exchange or chemical treatment	Removal of iron and manganese will decrease scaling potential.
Microorganisms	Disinfection	Disinfection of the feedwater may be accomplished using either chlorine, ozone, or UV irradiation to limit bacterial activity. Ultrafiltration can also be used to reduce the number of microorganisms.
Particulate matter	Depth or surface filtration, microfiltration, or ultrafiltration	Particulate matter can be removed by various methods of filtration (see Chap. 8). Fouling agents may pass through these filtration systems, thus the potential for membrane fouling should be verified by pilot-testing.
Particulate and colloidal matter	Cartridge filter (also ultrafiltration)	Cartridge filters are pressure-driven filters with pore sizes varying from 5 to 15 μm and are installed commonly ahead of RO membranes. Cartridge filters provide a final level of protection against the intrusion of relatively large solids into the RO system. The filters do not remove dissolved substances. Most cartridge

(Continued)

Table 9-8
Methods of pretreatment for nanofiltration and reverse osmosis systems (*Continued*)

Material to be removed	Method of pretreatment	Description or discussion
Particulate and colloidal matter (*Continued*)		filters are polypropylene wound cartridges from 800 to 1000 mm in length housed inside a vertical or horizontal stainless steel or fiberglass vessel. Generally, the pressure drop across a clean cartridge filter is between 0 and 35 kPa. As the solids accumulate and the pressure drop reaches a threshold range of 70 to 80 kPa, the cartridge has to be removed and replaced (Paranjape et al., 2003).
Scale formation	pH adjustment	To inhibit scale formation, pH adjustment of the feedwater within the range from 4.0 to 7.5 is required. A low pH enhances conversion of carbonate into bicarbonate species, which are much more soluble. Cellulose acetate RO membranes have an optimum pH range of 5 to 7, as they are prone to hydrolysis below a pH of 5. Newer polyamide RO membranes can be used over a broader pH range of 2 to 11 (Paranjape et al., 2003).
	Antiscalants	Antiscalants are polymeric compounds that either prevent scale formation entirely or permit formation of scales that can be removed easily during cleaning. Certain antiscalants, however, may increase the fouling of humic acids on RO membranes (Richard et al., 2001).
Sparingly soluble salts	Chemical treatment	Sparingly soluble salts such as silica can be removed by chemical treatment where reclaimed water is to be used for industrial purposes, i.e., removal of silica may be required to prevent precipitation on heat exchangers. Chemical treatment may include the addition of aluminum and iron oxides, zinc chloride, magnesium oxide, ozone (when ozone-resistant membranes are used), and ultra-high lime clarification. Lime clarification, however, may not be as effective as other pretreatment methods in removing materials that foul RO membranes, thus resulting in more frequent cleaning of the membranes (Gagliardo, 2000).

purposes: (1) as an inexpensive secondary barrier in the event of a failure in the pretreatment system and (2) to remove any particulate impurities originating from the chemicals added to precondition the feedwater to the NF, RO, or EDR membranes.

Common chemical pretreatments include pH adjustment, constituent precipitation, and the use of antiscalants. In some cases, dechlorination might be required to protect the membranes. It is important to note that the highest grade of chemicals (i.e., food grade) are not used typically. Often, the chemicals added to the feedstream will contain

(a) (b)

Figure 9-7
View of cartridge filters used to pretreat the feedstream to nanofiltration, reverse osmosis, and electrodialysis membranes (see Fig. 9-4): (a) a system of 10 cartridge filters, positioned back-to-back and connected to a central manifold located between them, is used to treat effluent from submerged microfiltration units to which chemicals have been added. The cartridge filters are used in conjunction with the reverse osmosis units shown in Figs. 9-2a and b and (b) cartridge filters (right foreground) used for pretreatment before UV and reverse osmosis. (Courtesy of Austep, r.s.l., Italy.)

particulate matter that must be removed to avoid damaging the membranes. For example, sulfuric acid used to lower the pH of the feedstream will often contain particulate matter that can be damaging to most membranes. Thus, as noted above, cartridge filters are used to protect against the presence of chemical impurities.

To assess the treatability of a given wastewater with NF and RO membranes, several indexes have been developed to determine the susceptibility of membranes to fouling. The three principal indexes are: (1) the silt density index (SDI), (2) the modified fouling index (MFI), and (3) the mini plugging factor index (MPFI). All three fouling indexes are determined using a simple laboratory dead-end membrane filter apparatus. The sample to be tested is passed through a 0.45 μm millipore filter with a 47-mm internal diameter at a constant pressure of 207 kPa (30 lb/in.²) gauge and various measurements are made. The specific measurements that are made depend on the index to be determined. The time to complete data collection for these tests varies from 15 min to 2 h, depending on the fouling nature of the water.

Treatability Testing

Silt Density Index

The most widely used index is the SDI (DuPont, 1977; ASTM, 2002). The SDI is defined as follows:

$$\text{SDI} = \frac{\%P_{207}}{t} = \frac{100\,[1 - (t_i/t_f)]}{t} \qquad (9\text{-}1)$$

where %P_{207} = percent plugging at 207 kPa (30 lb/in.²) feed pressure
 t = total time for running the test
 t_i = time to collect the initial sample of 500 mL
 t_f = time to collect final sample of 500 mL

The SDI is a static measurement of resistance, determined by taking samples at the beginning and end of the test. The SDI test cannot be used to assess the rate of change of resistance during the test. Approximate values for SDI are reported in Table 9-9. The calculation of the SDI is demonstrated in Example 9-1.

Modified Fouling Index

The MFI is determined using the same equipment and procedure used for the SDI, but the volume is recorded every 30 s over a 15-min filtration period (Schippers and Verdouw, 1980). Derived from a consideration of cake filtration, the MFI is defined as follows:

$$\frac{1}{Q} = (MFI)\, V + b \tag{9-2}$$

where Q = average flow, L/s
 MFI = modified fouling index, s/L²
 V = volume, L
 b = constant (intercept of linear portion of curve)

The value of the MFI is obtained as the slope of the straight-line portion of the curve obtained by plotting the inverse flow versus the cumulative volume (see Fig. 9-8a). Recommended MFI values are reported in Table 9-9.

Mini-Plugging Factor Index

The mini-plugging factor index (MPFI) is a measure of the change in flowrate as a function of time, as illustrated on Fig. 9-8b (Taylor and Jacobs, 1996). The equipment used for the MPFI test is the same as that used for the SDI and MFI tests. The MPFI is defined as the slope of the linear portion of the flowrate versus time curve

Table 9-9 Recommended values for fouling indexes[a]

Membrane process	Fouling index		
	SDI	MFI, s/L²	MPFI, L/s²
Nanofiltration	0–3	0–10	0–1.5 × 10⁻⁴
Reverse osmosis hollow fiber	0–2	0–2	0–3 × 10⁻⁵
Reverse osmosis spiral wound	0–3	0–2	0–3 × 10⁻⁵

[a]Adapted in part from Taylor and Wiesner (1999); and AWWA (1996).

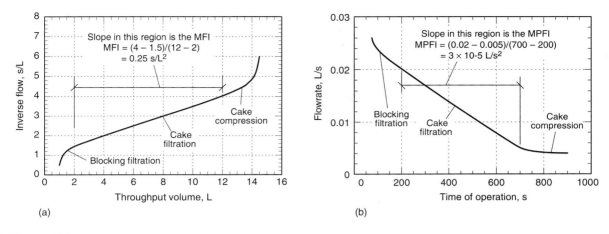

Figure 9-8
Typical plots used to determine fouling indexes: (a) modified fouling index (MFI) and (b) mini plugging factor index (MPFI).

(see Fig. 9-8b), which is ascribed to cake fouling. In equation form, the MPFI is expressed as follows:

$$Q = (MPFI)\, t + a \qquad (9\text{-}3)$$

where Q = average flow at 30 s intervals, L/s
MPFI = mini plugging factor index, L/s^2
t = time, s
a = constant (intercept of linear portion of curve)

Typical values for the MPFI are reported in Table 9-9. Because the MFI is based on throughput volume, it is thought to be a more sensitive index than the MPFI for characterization of fouling.

Limitations of Fouling Indexes

The SDI, MFI, and MPFI fouling indexes described above, and others currently in use, have serious limitations including: (1) a dead-end test is used to gather data to predict the fouling performance of a cross-flow membrane, (2) the test is conducted with a 0.45 mm filter which does not capture the effect of smaller colloidal particles (see Fig. 8-1 in Chap. 8), (3) the test is not representative of cake filtration, which occurs in cross-flow, and (4) the test is conducted under conditions of constant pressure with variable flux, where the opposite operational mode is normally used in practice. It should be noted that several other indexes, using MF or UF membranes in place of the millipore filter, to reflect the effect of smaller colloidal material and large dissolved organic material on fouling, are currently (2006) under development. In addition, a dynamic cross-flow test is under development (Adham, 2006).

EXAMPLE 9-1. Silt Density Index for Reverse Osmosis.

Determine the silt density index for a proposed feedwater from the following test data. If a spiral wound RO membrane is to be used, will pretreatment be required?

Test run time = 30 min
Initial 500 mL = 2 min
Final 500 mL = 10 min

Solution

1. Calculate the SDI using Eq. (9-1).

$$SDI = \frac{100[1 - (t_i/t_f)]}{t}$$

$$SDI = \frac{100[1 - (2/10)]}{30} = 2.67$$

2. Compare the SDI to the acceptable criteria.

The calculated SDI value of 2.67 is less than 3 (see Table 9-9); therefore, no further pretreatment is expected to be necessary.

Comment

As a practical matter, because the SDI value is close to 3 it may be prudent, depending on the variability of the feedwater and membrane characteristics, to consider some method of pretreatment to protect the membrane.

Membrane Flux and Area Requirements

A number of different models have been developed to determine the membrane surface area and the number of arrays required (see Fig. 9-4b). The basic equations used to develop the various models are described in the following paragraphs.

Water Flux Rate

Referring to Fig. 9-9, the flux of water through the membrane is a function of the pressure gradient:

$$F_w = k_w(\Delta P_a - \Delta \Pi) = \frac{Q_p}{A} \tag{9-4}$$

where F_w = water flux rate, L/m²·h
k_w = mass transfer coefficient for water flux (involving temperature, membrane characteristics, and solute characteristics), L/m²·h·bar

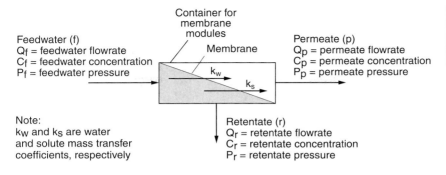

Figure 9-9
Definition sketch for a membrane process.

ΔP_a = average imposed pressure gradient, bar (Note: 1 bar = 10^5 Pa)

$$= \left[\frac{P_f + P_r}{2}\right] - P_p$$

$\Delta \Pi$ = osmotic pressure gradient, bar

$$= \left[\frac{\Pi_f - \Pi_r}{2}\right] - \Pi_p$$

where P_f = inlet pressure of feedwater, bar
P_r = pressure of retentate stream, bar
P_p = pressure of permeate water, bar
Π_f = osmotic pressure in feedwater, bar
Π_r = osmotic pressure in retentate, bar
Π_p = osmotic pressure in permeate, bar
Q_p = permeate stream flow, L/h
A = membrane area, m^2

Mass (Solute) Flux Rate

Some solute passes through the membrane in all cases. Solute flux can be described adequately by an expression of the form:

$$F_s = k_s \, \Delta C_s = \frac{(Q_p)(10^{-3} \, m^3/L) C_p}{A} \tag{9-5}$$

where F_s = mass flux of solute, g/m^2·h
k_s = mass transfer coefficient for solute, m/h
ΔC_s = solute concentration gradient across membrane, g/m^3

$$= \left[\frac{C_f + C_r}{2}\right] - C_p$$

C_f = solute concentration in feed stream, g/m^3
C_r = solute concentration in retentate (concentrate) stream, g/m^3
C_p = solute concentration in permeate stream, g/m^3
Q_p = permeate stream flow, L/h

Recovery Ratio

The permeate recovery ratio, r, which is the conversion of feedwater to permeate (product water), is defined as:

$$r, \% = \frac{Q_p}{Q_f} \times 100 \qquad (9\text{-}6)$$

where Q_p = permeate stream flow, L/h, m³/h, or m³/s
Q_f = feed stream flow, L/h, m³/h, or m³/s

As noted in Chap. 8, high recoveries are possible with low pressure membranes, but for NF and RO systems typical recoveries are in the range of 60 to 90 percent, depending on the feedwater quality. A recovery of 80 percent is often the practical limit because of severe reductions in operating efficiencies (IAEA, 2004).

The permeate recovery ratio affects the capital and operating cost of a membrane system. The volume of feedwater required for a given permeate capacity is determined directly by the design recovery ratio. Therefore, the size of the feedwater system, capacity of the pretreatment system, size of the high pressure pumps and supply piping are also functions of the recovery ratio. With increased recovery, the feedwater flow is reduced, the pressure may increase somewhat, but the brine will be more concentrated, which can make disposal more difficult.

An example of the effect of the permeate recovery ratio on feed pressure, power consumption, and feed flow is shown on Fig. 9-10 for an RO system operating at recovery rates between 60 and 90 percent. The feedwater flowrate depends only on the recovery ratio. The feed pressure is a complex function of recovery ratio, feedwater salinity, feedwater temperature, and specific permeate flux of the membrane. The power requirement of the high pressure pump is proportional to the flow and pressure. In the usual range of operating parameters, for an increase in recovery ratio, the decrease in

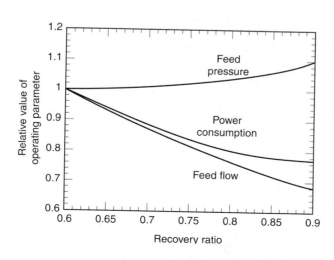

Figure 9-10
Effect of recovery on feed pressure, feed flow, and power consumption.

feedwater flow will have a greater effect on power consumption than an increase in feedwater pressure (Wilf, 1998). For RO, higher operating pressures are desirable because the degree of separation and the quality of the product are improved.

Rejection Efficiency

Rejection, or retention, is a measure of the fraction of solute or solid that is retained or does not pass through the membrane. It is calculated generally as a percentage, as follows:

$$R, \% = \left(\frac{C_f - C_p}{C_f}\right) \times 100 = \left(1 - \frac{C_p}{C_f}\right) \times 100 \tag{9-7}$$

where C_f = concentration in the feed stream, g/m³
C_p = concentration in the permeate, g/m³

The rejection efficiency of RO membranes for specific species can range from 85 to 99.5 percent and is quoted by the manufacturer for a standard set of feed conditions. When the rejection of microorganisms is considered, it is more convenient to express the rejection as log rejection, R_{log}, as given below:

$$R_{log} = -\log(1 - R) = \log\left(\frac{C_f}{C_p}\right) \tag{9-8}$$

Mass Balance Equations

The flow and constituent mass balance equations given in Chap. 8 for MF and UF also apply to NF and RO systems. The equations are:

Flow balance: $\qquad Q_f = Q_p + Q_r$ \hfill (9-9)

Constituent mass balance: $\quad Q_f C_f = Q_p C_p + Q_r C_r$ \hfill (9-10)

where Q_r = flowrate of retentate, m³/h, m³/s
C_r = retentate concentration, g/m³
Other terms are as defined previously

Use of the above equations to estimate the required surface area for TDS reduction is illustrated in Example 9-2.

EXAMPLE 9-2. Determine Membrane Performance for TDS Reduction in Reclaimed Water.

A wastewater with a TDS concentration of 1500 g/m³ is required to be reclaimed for groundwater recharge by means of surface spreading. RO treatment is required with product water having a TDS of no more than 200 g/m³. A thin film composite membrane is used having a mass transfer coefficient for water flux

k_w of 1.0 L/m²·h·bar and a mass transfer coefficient for solute k_s of 5×10^{-4} m/h. The flow rate is to be 150 m³/h. The net operating pressure ($\Delta P_a - \Delta \Pi$) will be 20 bar (2000 kPa). Assume the recovery rate will be 85 percent. Estimate the rejection rate and the concentration of the retentate stream.

Solution

1. Determine the membrane area required to produce 150 m³/h of water and the TDS concentration of the permeate.
2. Estimate membrane area using Eq. (9-4).

$$F_w = k_w(\Delta P_a - \Delta \Pi)$$

$$= (1.0 \text{ L/m}^2 \cdot \text{h} \cdot \text{bar})(20 \text{ bar}) = 20 \text{ L/m}^2 \cdot \text{h}$$

$$Q_p = F_w \times A, \quad Q = rQ_f = 0.85 Q_f$$

$$A = \frac{(0.85)(150 \text{ m}^3/\text{h})(10^3 \text{ L/m}^3)}{(20 \text{ L/m}^2 \cdot \text{h})} = 6375 \text{ m}^2$$

3. Estimate permeate TDS concentration using Eq. (9-5).

$$F_s = k_s \Delta C = \frac{Q_p C_p}{A}$$

Substituting the definition for ΔC from Eq. (9-5), and solving for C_p yields

$$C_p = \frac{k_s[(C_f + C_r)/2]A}{Q_p + k_s A}$$

Assume $C_r \approx 10 \, C_f$ and solve for C_p (Note: If the estimated C_r value and computed value of C_r, as determined below, are significantly different, the value of C_p must be recomputed).
Assume $Q_p = rQ_f$

$$C_p = \frac{(5 \times 10^{-4} \text{ m/h})[(1500 \text{ g/m}^3 + 15{,}000 \text{ g/m}^3)/2](6375 \text{ m}^2)}{(0.85)(150 \text{ m}^3/\text{h}) + (5 \times 10^{-4} \text{ m/h})(6375 \text{ m}^2)} = 201 \text{ g/m}^3$$

The permeate solute concentration is lower than necessary. The area could be reduced if blending were allowed.

4. Estimate the rejection rate using Eq. (9-7).

$$R, \% = \frac{C_f - C_p}{C_f} \times 100$$

$$R = \frac{(1500 \text{ g/m}^3 - 201 \text{ g/m}^3)}{(1500 \text{ g/m}^3)} \times 100 = 86.6\%$$

5. Estimate the retentate stream TDS by rewriting Eq. (9-10).

$$C_r = \frac{Q_f C_f - Q_p C_p}{Q_r}$$

$$C_r = \frac{(150 \text{ m}^3/\text{h})(1500 \text{ g/m}^3) - (0.85 \times 150 \text{ m}^3/\text{h})(201 \text{ g/m}^3)}{(0.15)(150 \text{ m}^3/\text{h})} = 8.9 \times 10^3 \text{ g/m}^3$$

Comment

If the permeate TDS concentration of the effluent were significantly below 200 g/m³, blending of feed and permeate could be used to reduce the required membrane area. In this example blending cannot be used.

Membrane Fouling

Membrane fouling is an important consideration in the design and operation of membrane systems as it affects pretreatment needs, cleaning requirements, operating conditions, cost, and performance. Membrane fouling will occur depending on the site-specific physical, chemical, and biological characteristics of the feedwater; the type of membrane; and operating conditions. As reported in Table 9-10, four general forms of fouling can occur: (1) particulate fouling, due to a buildup of the constituents in the feedwater on the membrane surface, (2) precipitation of inorganic salts resulting in the formation of inorganic scales, (3) organic fouling due to the presence of organic matter, and (4) biological fouling due to the presence of microorganisms in the feedwater. In addition, membranes can be damaged by the presence of chemical substances that can react with the membrane. Typical constituents in wastewater that can cause membrane fouling are also presented in Table 9-10.

Particulate Fouling Caused by Buildup of Solids

Three accepted mechanisms resulting in resistance to flow due to the accumulation of material (see Fig. 9-11) are: (1) pore narrowing, (2) pore plugging, and (3) gel/cake formation caused by concentration polarization (Ahn et al., 1998). The mechanisms of pore plugging and pore narrowing will only occur when the particulate matter in the feedwater is smaller than the pore size or the molecular weight cutoff. As the name describes, pore plugging occurs when particles the size of the pores become stuck in the pores of the membrane. Pore narrowing consists of solid material attaching to the interior surface of the pores which results in a narrowing of the pores. It has been hypothesized that once the pore size is reduced, concentration polarization is amplified further causing an increase in fouling (Crozes et al., 1997).

Gel/cake formation, caused by concentration polarization, occurs when the majority of the solid matter in the feed is larger than the pore sizes or molecular weight cutoff of the membrane. Concentration polarization can be described as the buildup of matter close to or on the membrane surface that causes an increase in resistance to solvent

Table 9-10

Typical types of membrane fouling and the constituents in wastewater that cause fouling and other constituents that can cause damage to the membranes[a]

Type of fouling	Responsible wastewater constituents	Remarks
Particulate fouling	Organic and inorganic colloids Clays and silts Silica Iron and manganese oxides Oxidized metals Metal salt coagulant products	Particulate fouling can be reduced by cleaning the membrane at regular intervals.
Scaling (precipitation of supersaturared salts)	Barium sulfate Calcium carbonate Calcium fluoride Calcium phosphate Stontium sulfate Silica	Scaling can be reduced by limiting salt content, by pH adjustment, and by other chemical treatments such as the addition of antiscalants.
Organic fouling	Natural organic matter (NOM) including humic and fulvic acids, proteins, and a polysaccharides Polymers used in treatment process	Effective pretreatment can be used to limit organic fouling.
Biofilm fouling	Dead microorganisms Living microorganisms Polymers produced by microorganisms	Biofilms are formed on the membrane surface by colonizing bacteria.
Damage to membrane	Acids Bases pH extremes Free chlorine Free oxygen	Membrane damage can be limited by controlling the amount of these substances in the feedwater. The extent of the damage depends on the nature of the membrane selected.

[a] In many cases, multiple types of fouling can occur simultaneously.

transport across the membrane. Some degree of concentration polarization will always occur in the operation of a membrane system. The formation of a gel or cake layer, however, is an extreme case of concentration polarization where a large amount of matter has actually accumulated on the membrane surface forming a gel or cake layer.

Scaling

As chemical constituents in the feedwater are removed at the surface of the membrane, their concentration increases locally. When the concentrations of the individual constituents increase beyond their solubility limits, a variety of different types of salts can

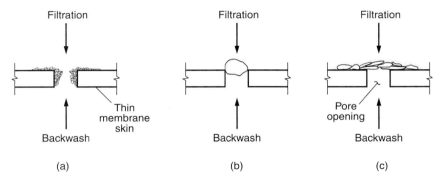

Figure 9-11
Modes of membrane fouling: (a) pore narrowing, (b) pore plugging, and (c) gel/cake formation caused by concentration polarization. (Adapted from Bourgeous et al., 1999.)

be precipitated, depending on the chemical characteristics and temperature of the feedwater (see Table 9-8). Chemical precipitation is especially critical in RO units used for desalination because of the high initial salt concentration in seawater. The chemical scale that forms on the membrane surface is of importance because it can reduce the water permeability of the membrane and potentially cause irreversible damage to the membrane.

Organic Fouling

Most treated wastewater contains, a variety of organic matter in varying concentrations. As noted in Table 9-10, organic foulants can include natural organic matter (NOM) that was present originally in the water supply, NOM produced during biological treatment, and organic polymers that may have been used in the wastewater treatment process. Polymers also include those used as filter aids in tertiary treatment and polymers recycled to the treatment process from dewatering activities. Because these polymeric materials are sticky, they can accumulate on the membrane surface and accelerate fouling by forming stable organic/inorganic particulate matter that can reduce the water permeability of the membrane.

Biological Fouling

Effluent from biological treatment systems presents a special problem, as the membranes are susceptible to fouling because of the biological activity that can occur. Because the concentration of organic matter and nutrients is elevated at the membrane surface, conditions are favorable for the growth of microorganisms. As microorganisms begin to colonize on the membrane surface, the water permeability of the membrane is reduced. When a membrane process is operated intermittently, the water permeability of the membrane can be reduced further if the microorganisms start to grow into the membrane pores. The growth of microorganisms is also of concern because of the production of extracellular polymers that can interact with other foulants, as described above.

Control of Membrane Fouling

Typically, four strategies are used to control membrane fouling: (1) pretreatment of the feedwater, (2) hydraulic flushing, (3) chemical treatment and conditioning, and (4) chemical cleaning of the membranes.

Feedwater Pretreatment

Pretreatment of the feedwater is used to reduce the TSS and bacterial content of the feedwater to limit fouling. Microfiltration or UF, depending on the type of membrane, is used most commonly for RO pretreatment. Use of NF for the removal of divalent and trivalent ions has been found to improve the performance of RO systems significantly. Similarly, the use of ED to reduce the total TDS by 50 to 60 percent has also been found to improve the performance of RO systems. Typical RO process flow diagrams were presented previously on Fig. 9-4. A system in which two membrane processes are used is identified as a *dual membrane system*. Where RO is combined with a nonmembrane process, the combination is known as an *integrated membrane system*.

Chemical Treatment and Conditioning

To limit the precipitation of sparingly soluble salts such as calcium carbonate, chemical precipitation can be used to remove the calcium as a pretreatment step or, as discussed above, NF can be used to remove calcium and magnesium salts. Alternatively, to maintain the calcium in solution, the pH can be adjusted to a value in range of 5.5 to 6 using sulfuric or hydrochloric acid. It is important to note that at these pH values most of the carbonate will pass through the membrane as carbon dioxide, which is why posttreatment is necessary (see Example 9-3). Another approach that is used commonly is the continuous addition of antiscalant and dispersants to inhibit scale formation. Sodium hexametaphosphate (SHMP) is a commonly used antiscalant, which makes it possible to have supersaturation without chemical precipitation by preventing crystal formation and growth (Crittenden et al., 2005).

Hydraulic Flushing

The most commonly used method of limiting the accumulation of particulate matter on the membrane is to maintain a cross-flow velocity across the membrane. Some membrane systems are designed to be flushed periodically to limit the accumulation of material on the membrane surface.

Chemical Cleaning

Chemical cleaning, often identified as cleaning-in-place (CIP), is used to remove constituents that are not removed with conventional cross-flow hydraulic flushing. Chemical precipitates can be removed by altering the chemistry of the feedwater and by chemical cleaning. Typically, both high and low pH solutions have been used in combination with detergents to loosen and remove materials that have accumulated on the membrane surface. Some recently developed membranes can withstand cleaning solutions varying from a pH of 1 to 13. High pH solutions are useful for the removal of biofouling and organic fouling. Low pH solutions are useful for the removal of calcium carbonate deposits.

Process Operating Parameters

Process operating considerations described in this section include permeate recovery ratio and rejection rate (discussed earlier), operating pressures and flux rates, energy requirements, split treatment and blending. Posttreatment, an important operating consideration, is discussed in the next section.

Operating Pressures and Flux Rates

Typical ranges of operating pressures and flux rates for NF and RO membranes are reported in Table 9-11. Two of the key variables that affect flux are temperature and operating pressure. Flux increases with higher temperature because fluid viscosity decreases. The flux through membranes increases by about three percent per degree Celsius (IAEA, 2004). As stated previously, as the pressure increases the flux increases linearly and the product quality increases.

Energy Consumption and Recovery

Typical energy consumption and product recovery values for various membrane systems are presented in Table 9-11. In reviewing Table 9-11, it is important to note that the reported operating pressure values for all of the membrane processes are considerably lower than comparable values from 5 to 10 yr ago. Operating pressures are expected to decrease further as new membranes are developed.

Because RO in particular produces a high-pressure concentrate stream, various methods have been developed or are under development to recover the energy lost in depressurizing the concentrate. Energy recovery devices are designed to recover most of this energy and transfer it to the feedwater to reduce the overall process energy. Typical devices that have been tried or used are (Beck, 2002):

- Reverse running pumps
- Pelton wheel turbines
- Flow work exchangers

Table 9-11
Typical operating parameters for nanofiltration and reverse osmosis[a]

Parameter	Unit	Nanofiltration	Reverse osmosis
Typical operating range	μm	0.001–0.01	0.0001–0.001
Flux rate	L/m²·h	10–35	12–20
Operating pressure			
1000–2500 mg/L TDS	kPa	350–550	1200–1800
Seawater TDS	kPa	500–1000	5500–8500
Energy consumption			
1000–2500 mg/L TDS	kWh/m³	0.6–1.2	1.5–2.5
Seawater TDS	kWh/m³	na	5–10

[a]Adapted from Crittenden et al. (2005); Taylor and Weisner (1999); and Tchobanoglous et al. (2003).
Note: kPa × 10^{-2} = bar (1 bar = 100 kPa = 10^5 N/m²).
　　　kPa × 0.145 = lb/in.².
na = not applicable.

A new device that has been tested and shows promise is the VARI-RO system. The system is an integrated variable-flow positive displacement pumping and energy recovery system for use in seawater and brackish water desalination by RO. In a pilot-scale testing program in Santa Barbara, California on a high pressure RO system, energy savings on the order of 30 percent were reported as compared to a commercially available system using a centrifugal pump, Pelton wheel turbine, and variable-frequency drive energy recovery system (Childs and Dabiri, 1998). The device may have application in RO applications for water reclamation where energy costs are high.

Split Treatment and Blending

Depending on the water quality objectives of the final product water, split treatment may be used where two or more flow streams undergo varying levels of treatment with different product water characteristics. The treated flow streams are then blended to meet the desired final product water requirements. The split treatment technique is useful and cost effective where a high quality product waste such as that produced by RO can be blended with a lesser quality water to meet the desired final product water requirements. Blending has the added advantages of (1) reducing the size of the higher cost treatment unit, (2) helping stabilize the low TDS and low alkalinity RO water, and (3) minimizing the need for posttreatment chemicals.

Posttreatment

Following NF and RO treatment, some posttreatment may be required. Typical posttreatments involve the addition of chemicals to adjust the stability of the treated water; in some reuse applications, the removal or addition of gases, and the addition of chemicals to meet disinfection requirements and to control the growth of microorganisms in pipelines. These subjects are considered briefly in the following discussion and in subsequent chapters.

Reverse Osmosis Effluent Stability

Depending on the level of dissolved solids removal, the product water from NF and RO processes may be corrosive to equipment and piping. Further, as the membranes become less permeable and increasingly greater amounts of minerals are removed, it is more likely that pH adjustment or other methods of corrosion control will be needed. Chemicals used commonly in the control of pH are given in Table 9-12.

Blending and/or Chemical Addition

In general, some form of blending and/or chemical addition may be required to stabilize the water. Techniques that have been used to reduce the corrosive characteristics of RO water include (1) blending with a less treated water, as discussed previously, (2) blending with brackish water from subsurface aquifers (typical blends are 60 to 70 percent RO water), (3) blending in surface water reservoirs, (4) for small systems, passing of RO effluent through beds of calcareous material such as dolomite or calcite, (5) remineralization with suitable chemicals such as sodium bicarbonate, calcium chloride, and sodium hypochlorite (expensive for large systems), (6) the addition of calcium carbonate ($CaCO_3$), and (7) the addition of lime [$Ca(OH)_2$]. The use of $Ca(OH)_2$ or $CaCO_3$ is generally favored because by properly controlling the $CaCO_3$ equilibrium, a thin protective film of calcium carbonate can be deposited on pipelines to limit corrosion, as discussed below. Facilities for the preparation of lime for injection in the RO product water are shown on Fig. 9-12.

Table 9-12
Chemicals used most commonly for the control of pH (neutralization)

Chemical	Formula	Molecular weight, g/mole	Equivalent weight	Availability Form	Availability Percent
Chemicals used to raise pH					
Calcium carbonate	$CaCO_3$	100.0	50.0	Powder crushed	90 to 98
Calcium hydroxide (lime)	$Ca(OH)_2$	74.1	37.1	Powder granules	90 to 94
Calcium oxide	CaO	56.0	28.0	Lump, pebble, ground	75 to 99
Dolomitic hydrated lime	$[Ca(OH)_2]_{0.6}[Mg(OH)_2]_{0.4}$		33.8	Powder	58 to 50 as $Ca(OH)_2$
Dolomitic quicklime	$(CaO)_{0.6}(MgO)_{0.4}$		24.8	Powder or other solid forms	35 to 46 as $MgCO_3$
Magnesium hydroxide	$Mg(OH)_2$	58.3	29.2	Light and heavy powder	90 to 96
Magnesium oxide	MgO	40.3	20.2	Powder	90 to 98
Sodium bicarbonate	$NaHCO_3$	84.0	84.0	Powder	95 to 99
Sodium carbonate (Soda ash)	Na_2CO_3	106.0	53.0	Powder	99.2
Sodium hydroxide (Caustic soda)	$NaOH$	40.0	40.0	Solid flake, ground flake	90 to 98
Sodium hydroxide (Caustic soda)	$NaOH$	40.0	40.0	Liquid	50
Chemicals used to lower pH					
Carbonic acid	H_2CO_3	62.0	31.0	CO_2	
Hydrochloric acid	HCl	36.5	36.5	Liquid	27.9, 31.45, 35.2
Nitric acid	HNO_3	63.0	63.0	Liquid	50 to 70
Sulfuric acid	H_2SO_4	98.1	49.0	Liquid	77.7 (60° Be[b]) 97 (66° Be[b])

[a]Adapted in part from Eckenfelder (2000).
[b]Baumé scale.

 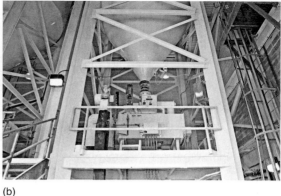

(a) (b)

Figure 9-12
Views of facilities used to prepare lime for addition to RO product water to adjust the Langelier saturation index (LSI): (a) lime saturator (foreground) for hydrated lime [Ca(OH)$_2$] (facilities under construction at the Orange County Water District, Fountain Valley, CA) and (b) lime slaker for the dissolution of quicklime CaO. Storage bin for lime granules is located directly above slaker.

Stability Indexes

Over the years, several indexes have been developed to assess whether a calcium film will be deposited or removed, including the Langelier saturation index (LSI) (Langelier, 1936), the Ryznar stability index (RSI) (Ryznar, 1944), and the Stiff and Davis index (SDI) (Stiff and Davis, 1952). The commonly used LSI is discussed below. The Langelier and Ryznar indexes are also considered in Sec. 19-2 in Chap. 19.

Langelier Saturation Index (LSI) The tendency of treated effluent to develop or to remove CaCO$_3$ scale can be approximated by calculating the LSI of the treated effluent (Langelier, 1936; Larson and Buswell, 1942). The LSI is given by the following expression:

$$\text{LSI} = \text{pH} - \text{pH}_s \tag{9-11}$$

where pH = measured pH in treated water sample
pH$_s$ = saturation pH for CaCO$_3$ at the existing solution concentrations for calcium and bicarbonate [see Eq. (9-13)]

The scaling criteria for the LSI are:

LSI > 0   Water is supersaturated with respect to CaCO$_3$ and scaling may occur

LSI < 0   Water is undersaturated with respect to calcium carbonate. Undersaturated water has a tendency to remove existing calcium carbonate protective coatings in pipelines and equipment which may lead to metallic corrosion

LSI = 0   Water is considered to be neutral (i.e., neither scale forming nor scale removing)

It should be noted that water with LSI < 0 is also sometimes referred to as *corrosive*, but use of the term corrosive is incorrect, as the LSI index only applies to the presence or absence of a calcium carbonate scale.

Computation of the LSI The saturation pH_s in Eq. (9-11) can be computed using the following expression:

$$pH_s = -\log\left\{\frac{K_{a2}(\gamma_{Ca^{2+}})[Ca^{2+}](\gamma_{Alk})[Alk]}{K_{sp}}\right\} \quad (9\text{-}12)$$

where K_{a2} = equilibrium constant for the dissociation of bicarbonate
$\gamma_{Ca^{2+}}$ = activity coefficient for calcium
$[Ca^{2+}]$ = concentration of calcium, mole/L
γ_{Alk} = activity coefficient for alkalinity
$[Alk]$ = alkalinity (typically bicarbonate, HCO_3^- concentration, pH range 6.5 to 9.0), mole/L
K_{sp} = solubility product constant for dissociation of calcium carbonate

The activity coefficient can be estimated using Eq. (9-13).

$$\log \gamma_i = -0.5\,(Z_i)^2\left(\frac{\sqrt{I}}{1+\sqrt{I}} - 0.3I\right) \quad (9\text{-}13)$$

where Z_i = charge on ionic species i
I = ionic strength, mole/L

The ionic strength of a solution can be estimated using the following expression:

$$I = \frac{1}{2}\sum C_i(Z_i)^2 \quad (9\text{-}14)$$

where C_i = concentration of the i-th species, mole/L
Z_i = the valance (or oxidation) number of the i-th species

The ionic strength can also be estimated using Eq. (9-15).

$$I = 2.5 \times 10^{-5} \times TDS \quad (9\text{-}15)$$

The transformed version of Eq. (9-12) often reported in the literature is given by:

$$pH_s = pK_{a2} - pK_{sp} + p[Ca^{2+}] + p[Alk] - \log \gamma_{Ca^{2+}} - \log \gamma_{Alk} \quad (9\text{-}16)$$

Values of K_{a2} and K_{sp} for the carbonate system are given in Table 9-13 as function of temperature. The application of the above equations is illustrated in Example 9-3. Other methods that can be used to calculate calcium carbonate equilibrium conditions can be found in Benefield et al. (1982), Snoeyink and Jenkins (1980), Pankow (1991), and Standard Methods (2005).

Table 9-13
Carbonate equilibrium constants as function of temperature[a]

Temperature, °C	Equilibrium constant[b]		
	$K_{a1} \times 10^7$	$K_{a2} \times 10^{11}$	$K_{sp} \times 10^9$
5	3.020	2.754	8.128
10	3.467	3.236	7.080
15	3.802	3.715	6.02
20	4.169	4.169	5.248
25	4.467	4.477	4.571
30	4.677	5.129	4.078
40	5.012	6.026	3.090

[a]Adapted from Snoeyink and Jenkins (1980) and Pankow (1991).
[b]The reported values have been multiplied by the indicated exponents. Thus, the value K_{a2} at 20°C is equal to 4.169×10^{-11}.

EXAMPLE 9-3. Analysis of Scaling Potential of Water from RO Facility.

Estimate the scaling potential using the Langelier index for the effluent from a reverse osmosis facility with the water quality characteristics given below. If chemical stabilization is required, determine the chemical dosage using lime and/or carbon dioxide (see Table 9-12). Also prepare a summary table.

Constituent	Concentration	
	Unit	Value
Ca^{2+}	mg/L	5.0
Alkalinity	mg/L as $CaCO_3$	12.5
TDS	mg/L	30.5
pH	unitless	6.5
Temperature	°C	20

Solution—Part 1 Assess stability with respect to calcium carbonate

1. Estimate the ionic strength of the treated water using Eq. (9-15).

    $I = 2.5 \times 10^{-5} \times TDS$

    $I = 2.5 \times 10^{-5} \times 30.5 = 76.3 \times 10^{-5}$

2. Determine the activity coefficients for calcium and bicarbonate using Eq. (9-13).

a. For calcium (i.e., divalent ion)

$$\log \gamma_{Ca^{2+}} = -0.5(Z_i)^2 \left(\frac{\sqrt{I}}{1+\sqrt{I}} - 0.3I \right)$$

$$= -0.5(2)^2 \left(\frac{\sqrt{76.3 \times 10^{-5}}}{1+\sqrt{76.3 \times 10^{-5}}} - 0.3 \times 76.3 \times 10^{-5} \right)$$

$$= -0.053$$

$$\gamma_{Ca^{2+}} = 0.885$$

b. For the given pH essentially all of the alkalinity will be in the form of bicarbonate (i.e., monovalent ion)

$$\log \gamma_{HCO_3^-} = -0.5(Z_i)^2 \left(\frac{\sqrt{I}}{1+\sqrt{I}} - 0.3I \right)$$

$$= -0.5(1)^2 \left(\frac{\sqrt{76.3 \times 10^{-5}}}{1+\sqrt{76.3 \times 10^{-5}}} - 0.3 \times 76.3 \times 10^{-5} \right)$$

$$= -0.0133$$

$$\gamma_{HCO_3^-} = 0.970$$

3. Compute the molar concentrations of calcium and alkalinity.

   a. For calcium

   $$[Ca^{2+}] = \frac{(0.005 \text{ g/L})}{(40 \text{ g/mole})} = 0.000125 \text{ mole/L}$$

   b. For alkalinity

   $$[Alk] = \frac{(0.0125 \text{ g/L})}{(50 \text{ g/mole})} = 0.00025 \text{ mole/L}$$

4. Determine the saturation pH_s using Eq. (9-12).

   $$pH_s = -\log \frac{K_{a2}(\gamma_{Ca^{2+}})[Ca^{2+}](\gamma_{Alk})[Alk]}{K_{sp}}$$

   $$pH_s = -\log \left[\frac{(4.17 \times 10^{-11})(0.885)(0.125 \times 10^{-3})(0.970)(0.25 \times 10^{-3})}{5.25 \times 10^{-9}} \right]$$

   $$pH_s = -\log(2.13 \times 10^{-10}) = 9.67$$

5. Determine the Langelier Saturation Index using Eq. (9-11).

   $$LSI = pH - pHs = 6.5 - 9.67 = -3.17$$

   $LSI < 0$ (Water is undersaturated with respect to calcium carbonate)

Solution—Part 2 Determine lime addition needed to stabilize water

1. Compute the lime dose required for a final alkalinity of 40 mg/L as $CaCO_3$.

   For a final alkalinity of 40 mg/L as $CaCO_3$, a lime dose of 27.5 mg/L as $CaCO_3$ (40 − 12.5) is required.

2. Compute the calcium and alkalinity concentrations from the lime addition.
   a. Lime addition =
   27.5 mg/L as $CaCO_3 \times$ [(37 g $Ca(OH)_2$/eq)/(50 g $CaCO_3$/eq)] = 20.35 mg/L
   b. Ca^{2+} = (20.35 mg $Ca(OH)_2$/L) (40 g Ca/mole)/(74 g $Ca(OH)_2$/mole) = 11.0 mg/L = 0.000275 g/mole
   c. Alk = (0.04 g/L)/(50 g/mole) = 0.0008 mole/L

3. Determine the solution pH following lime addition.

   There are a number of methods available to find the equilibrium pH, including computer models, Caldwell-Lawrence diagrams, and spreadsheet simulations. For this system, with a total calcium concentration of 0.0004 mole/L (0.000125 + 0.000275), the solution pH was found to be approximately 10.3.

4. Estimate the dose of CO_2 required to lower the effluent pH to 8.75.

   Again, the methods identified in Step 3 above can also be used to compute the dose of CO_2 required to attain a desired pH. In this case, the CO_2 dose required was found to be 15.5 mg/L.

5. Determine the pH_s for the system following lime addition and recarbonation (note activity coefficients have been recomputed to reflect the increased TDS).

$$pH_s = -\log\left[\frac{(4.17 \times 10^{-11})(0.855)(0.274 \times 10^{-3})(0.962)(0.8 \times 10^{-3})}{5.25 \times 10^{-9}}\right]$$

$$pH_s = -\log(2.09 \times 10^{-9}) = 8.68$$

6. Compute the new LSI.

$$LSI = pH - pHs = 8.75 - 8.68 = 0.07$$

   LSI ≈ 0 (Water is essentially neutral with respect to calcium carbonate)

7. Prepare summary table.

			Value	
Constituent	Unit		Initial	After stabilization
Ca^{2+}	mg/L		5.0	16.0
Alkalinity	mg/L as $CaCO_3$		12.5	40
TDS	mg/L		30.5	50.9
pH	unitless		6.5	8.75
Temperature	°C		20	20

Comment
Depending on the removal characteristics of the RO membranes, the stabilization of the treated water is to some extent arbitrary (e.g., final alkalinity 10 to 50 mg/L), depending on the water chemistry and the management objectives for the distribution system. Some general recommendations for stabilization can be found in Merrill and Sanks (1977a and b, 1978).

Deaeration and Aeration
Depending on the application, it may be necessary to remove or add gases from or to the effluent from the RO process. For example, if the RO process effluent is to be used for boiler makeup water it will be necessary to remove oxygen, carbon dioxide, and other noncondensable gases (e.g., ammonia) from the feedwater to limit corrosion. The presence of these gases is the principal cause of corrosion in boilers, especially high pressure boilers. Corrosion issues are considered further in Chap. 19. Also, because little oxygen remains after NF and RO treatment, aeration may be required in some reuse applications.

Disinfection
Disinfection in most cases will be required to ensure the microbial integrity of the product water and to prevent bacterial regrowth in storage and distribution systems (see Chaps. 11 and 14). However, as higher levels of constituent removal are achieved, the chlorine demand will be much lower.

9-5 PILOT-SCALE STUDIES FOR NANOFILTRATION AND REVERSE OSMOSIS

Because every wastewater is unique with respect to its constituent characteristics, it is difficult to predict *a priori* how a given membrane process will perform. As a result, the selection of the best membrane for a given application is usually based on the results of pilot studies. Membrane fouling indexes (see Table 9-9) can be used to assess the need for pretreatment. In some situations, manufacturers of membranes will provide a testing service to identify the most appropriate membrane for a specific feedwater. Typical pilot-scale facilities used to evaluate the performance of NF and RO treatment processes are shown on Fig. 9-13.

The elements that comprise a pilot plant include: (1) the pretreatment system; (2) tankage for flow equalization and cleaning; (3) pumps for pressurizing the membrane, recirculation, and backflushing with appropriate controls; (4) the membrane test module; (5) facilities for monitoring the performance of the test module; and (6) an appropriate membrane backflushing system. Typical membrane operating parameters and water quality measurements are presented in Table 9-14. Additional specific parameters selected for evaluation will depend on the final use of the product water.

(a) (b)

Figure 9-13
Typical pilot-scale facilities used for the evaluation of a reverse osmosis process (a) test module with six elements. Front view of control panel and instrumentation located on backside, and (b) skid mounted test module with four elements. Cartridge filter and control panel shown in the foreground (Courtesy of Austep r.s.l., Italy).

Table 9-14

Typical operating parameters and water quality measurements used for pilot-testing membrane facilities[a]

Membrane operating parameters
Pretreatment requirements including chemical dosages
Transmembrane flux rate correlated to operating time
Transmembrane pressure
Recovery
Washwater requirements
Recirculation ratio
Cleaning frequency including protocol and chemical requirements
Posttreatment requirements

Typical water quality measurements	
Turbidity	Heterotrophic plate count
Particle counts	Other bacterial indicators
Total organic carbon	Specific constituents that can limit recovery such as silica, barium, calcium, fluoride, strontium, and sulfate
Nutrients	
Heavy metals	
Organic priority pollutants	
Total dissolved solids	Biotoxicity
pH	Fouling indexes
Temperature	

[a]Adapted from Tchobanoglous et al. (2003).

9-6 ELECTRODIALYSIS

Electrodialysis (ED) is an electrochemical separation process in which mineral salts and other ionic species are transported through ion-selective membranes from one solution to another under the driving force of a direct current (DC) electric potential. As compared to NF and RO, which transports pure water through the membrane leaving the salts behind, with ED salt is gradually stripped from solution leaving a dilute solution behind containing particulate matter and neutral species not removed by the ED process. The salt transferred through the membrane then forms the concentrate.

Description of the Electrodialysis Process

The key to the ED process is the ion selective membranes that are essentially ion exchange resins cast in sheet form. Ion exchange membranes that allow passage of positively charged ions such as sodium and potassium are called cation membranes. Membranes that allow passage of negatively charged ions such as chloride and phosphate are called anion membranes. To demineralize a solution using ED, cation and anion membranes are arranged alternately between plastic spacers in a stacked configuration with a positive electrode (anode) at one end and a negative electrode (cathode) at the other (see Fig. 9-14). When a DC voltage is applied, the electrical potential created becomes the driving force to move ions, with the membranes forming barriers to the ions of opposite charge. Therefore, anions attempting to migrate to the anode will pass through the adjacent anion membrane but will be stopped by the first cation membrane

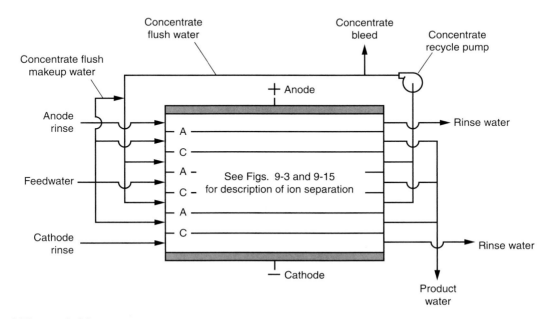

Figure 9-14

Schematic of conventional electrodialysis membrane stack with anode and cathode rinse. The conventional electrodialysis process has been replaced by the electrodialysis reversal (EDR) process (see Fig. 9-15).

they encounter. Cations trying to migrate to the cathode will pass through the cation membrane but will be stopped by the anion membrane. The membranes, therefore, form ion diluting compartments and ion concentrating compartments (www.ionics.com).

An ED assembly, known as a *stack*, consists of multiple cell pairs located between an anode and a cathode. A set of adjacent components consisting of a diluting compartment spacer, an anion membrane, a concentrating compartment spacer, and a cation membrane is called a *cell pair*. Electrolysis stacks can contain as many as 600 cell pairs. Feedwater (filtered wastewater) is pumped through the stack assembly. Typical flux rates are from 0.8 to 1.0 $m^3/m^2 \cdot d$. Dissolved solids removals vary with the (1) wastewater temperature, (2) amounts of electric current passed, (3) type and amount of ions, (4) permeability/selectivity of the membrane, (5) fouling and scaling potential of the feedwater, (6) feedwater flowrates, and (7) number and configuration of stages.

Electrodialysis Reversal

In the early 1970s, the EDR process was introduced. An EDR unit operates on the same principle as ED technology, except that both the product and concentrate channels are identical in construction (see Fig. 9-15). The same membranes are used to provide a continuous self-cleaning ED process that uses periodic reversal of the DC polarity to allow systems to run at high recovery rates. Polarity reversal causes the concentrating and diluting flow streams to switch after every cycle. Any fouling or scaling constituents are removed when the process reverses, sending fresh product water through the compartments filled previously with concentrated waste streams. The reversal

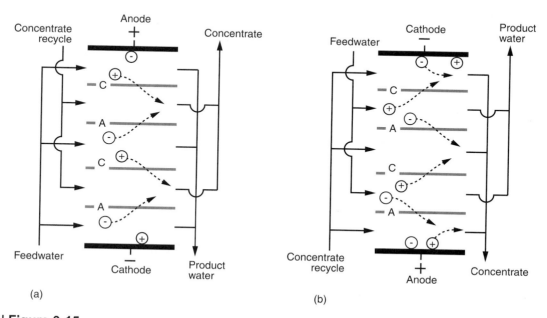

Figure 9-15

Schematic of electrodialysis reversal (EDR) process: (a) negative polarity and (b) positive polarity. Because the polarity is reversed, the anode and cathode rinse shown on Fig. 9-14 is not needed.

process is useful in breaking up and flushing out scales, slimes, and other deposits in the cells before they buildup. Product water is not collected during a short interval of time following reversal.

Electrodialysis reversal systems are able to reduce dissolved ions in process streams containing 10,000 to 12,000 mg/L of total dissolved solids, however, because of energy requirements, EDR is better suited for the treatment of brackish water in the range from 1000 to 5000 mg/L. As a rule of thumb, it takes about 1 to 1.2 kWh/m^3 to remove a kilogram of salt. Typical removal rates can range from 50 to 94 percent removal (www.ionics.com). A view of an EDR installation and an exposed membrane stack are shown on Fig. 9-16. The EDR facility shown on Fig. 9-16 is used to remove TDS from a portion (sidestream) of the reclaimed water produced at the North City plant in San Diego, California. The treated water with a reduced TDS concentration is blended back into the main flow, which has a TDS concentration that varies from 1200 to 1300 mg/L, to produce a final reclaimed water with a TDS equal to or less than 1000 mg/L to meet contractual agreements with the users of the reclaimed water.

The ED/EDR process uses electric power to transfer ions through the membranes and to pump water through the system. Two, or sometimes three, pumping stages are used typically.

Power Consumption

Power Requirements for Ion Transfer

The current required for ED can be estimated using Faraday's laws of electrolysis. Because one Faraday of electricity will cause one gram equivalent of a substance to migrate from one electrode to another, the number of gram equivalents removed per unit time is given by:

$$\text{Gram eq/unit time} = Q(N_{inf} - N_{eff}) = Q\Delta N \quad (9\text{-}17)$$

where gram/eq $= \dfrac{\text{Mass of solute, g}}{\text{Equivalent weight of solute}}$

Q = product flowrate, L/s
N_{inf} = normality of influent (feed), g-eq/L
N_{eff} = normality of effluent (product), g-eq/L
ΔN = change in normality between the influent and effluent, g-eq/L

The corresponding expression for the current for a stack of membranes is given by:

$$i = \frac{FQ(N_{inf} - N_{eff})}{nE_c} = \frac{FQ\Delta N}{nE_c} \quad (9\text{-}18)$$

Where i = current, A
F = Faraday's constant = 96,485 A·s/g-eq = 26.80 A·h/eq
n = number of cell pairs in the stack
E_c = current efficiency, % (expressed as a decimal)

In the analysis of the ED process, it has been found that the capacity of the membrane to pass an electrical current is related to the current density (CD) and the normality (N)

Figure 9-16

Electrodialysis reversal process used to remove TDS from reclaimed water at the North City plant in San Diego, CA: (a) view of cartridge filters used to pretreat the feedstream to the electrodialysis units, (b) view of electrodialysis membrane stack with cover removed, and (c) view of full-scale electrodialysis facility.

of the feed solution. Current density is defined as the current in milliamperes that flows through a square centimeter of membrane perpendicular to the current direction. Normality expresses the concentration of a solution based on the number of gram equivalent weights of a solute per liter of solution. A solution containing one gram of equivalent weight per liter is referred to as normal. The relationship between current density and the solution normality is known as the *current density to normality* (CD/N) ratio.

High values of the CD/N ratio are indicative that there is insufficient charge to carry the current. When high ratios exist, a localized deficiency of ions may occur on the surface of the membrane, causing a condition called *polarization*. Polarization should be avoided as it results in high electrical resistance leading to excessive power consumption. In practice, CD/N ratios will vary from 500 to 800 when the current density is expressed as mA/cm^2. The resistance of an ED unit used to treat a particular water must be determined experimentally. Once the resistance, R, and the current flow, i, are known, the power required can be computed using Ohm's law as follows:

$$P = E \times i = R(i)^2 \qquad (9\text{-}19)$$

Where P = power, W
E = voltage, V
   = R × i
R = resistance, Ω
i = current, A

The application of the above relationships is considered in Example 9-4.

EXAMPLE 9-4. Determine Power Requirements and Membrane Area for ED Treatment of Reclaimed Water.

Determine the power and area required to reduce the TDS content of 4000 m^3/d of a reclaimed water to be used for industrial cooling water. Assume the following data apply.

1. Use an ED unit comprised of 500 cell pairs
2. Influent TDS concentration = 2500 mg/L (~ 0.05 g·eq/L)
3. TDS removal efficiency = 50%
4. Product water flowrate = 90% of feed stream
5. Current efficiency = 90 percent
6. CD/N ratio = (500 mA/cm^2)/(g·eq/L)
7. Resistance = 5.0 Ω

Solution
1. Calculate the current using Eq. (9-18).

$$i = \frac{FQ\Delta N}{nE_c}$$

Q = (4000 m^3/d) × (1000 L/m^3)/(86,400 s/d) = 46.3 L/s

$$i = \frac{(96{,}485 \text{ A} \cdot \text{s/eq})(46.3 \text{ L/s})(0.9)(0.05 \text{ eq/L})(0.5)}{500 \times 0.90}$$

$$i = 223 \text{ A}$$

2. Determine the power required using Eq. (9-19).

$$P = R(i)^2$$

$$P = (5.0 \text{ }\Omega)(223 \text{ A})^2 = 248{,}645 \text{ W} = 249 \text{ kW}$$

3. Determine the power requirement per m³ of treated water.

$$\text{Power consumption} = \frac{(249 \text{ kW})(24 \text{ h/d})}{(4000 \text{ m}^3/\text{d})(0.9)} = 1.66 \text{ kWh/m}^3$$

4. Determine the required surface area per cell pair.
   a. Determine the current density:

   $$CD = (500 \text{ mA/cm}^2)/(\text{g eq/L}) \times 0.05 \text{ g} \cdot \text{eq/L} = 25 \text{ mA/cm}^2$$

   b. The required area is:

   $$\text{Area} = \frac{(223 \text{ A})(1000 \text{ mA/A})}{(25 \text{ mA/cm}^2)} = 8920 \text{ cm}^2 = 0.89 \text{ m}^2$$

Comment

The actual performance will have to be determined from pilot tests. The computed value for the power required per unit volume, 1.66 kWh/m³, is within the range of values reported in Table 9-1 (1.1 to 2.6 kWh/m³) for water with 1000 to 2500 mg/L TDS).

Power Requirements for Pumping

For pumping, the power requirements depend on the concentrate recirculation rate, the need for both product and waste pumping for discharge, and the efficiency of the pumping equipment (USBR, 2003).

Design and Operating Considerations

The ED process may be operated in either a continuous or a batch mode. The units can be arranged either in parallel to provide the necessary hydraulic capacity or in series to obtain the desired degree of demineralization. A typical three-stage, two-line ED flow diagram is shown on Fig. 9-17. The ED process should be protected from particulate fouling by a 10 μm cartridge filter (see Figs. 9-4c and d and 9-7).

A single electrodialysis stack can remove from 25 to 60 percent of the TDS, depending on the feedwater characteristics. Further desalting requires that two or more stacks be used in series (USBR, 2003). A portion of the resulting concentrate is recycled to

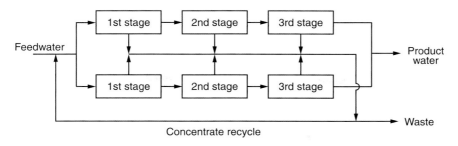

Figure 9-17
Schematic diagram for a three-stage, two-line electrodialysis process.

improve system performance. Makeup water, usually about 10 percent of the feed volume, is required to wash the membranes continuously. A portion of the concentrate stream is recycled to maintain nearly equal flow rates and pressures on both sides of each membrane. Typical operating parameters for the electrodialysis process are reported in Table 9-15.

Problems associated with the ED process for wastewater reclamation include chemical precipitation of salts with low solubility on the membrane surface and clogging of the membrane by the residual colloidal organic matter in wastewater treatment plant effluents. To reduce membrane fouling, some form of filtration may be necessary. With a properly designed plant, membrane cleaning should be infrequent. However, for both ED and EDR systems, clean-in-place systems are provided normally to circulate either hydrochloric acid solution for mineral scale resolution or sodium chloride solution with pH adjustment for organics removal (USBR, 2003).

Membrane and Electrode Life

Membranes for ED and EDR applications have a life of about 10 yr before they are replaced. Effective and timely cleaning in place extends the membrane life and improves product quality and power consumption. Cation membranes typically last longer than anion membranes because anion membranes are particularly susceptible to oxidation by chlorine and other strong oxidants (USBR, 2003). With the development

Table 9-15
Typical operating parameters for electrodialysis units

Parameter	Unit	Range
Flux rate	m^3/m^2·d	0.8–1.0
CD/N ratio	mA/cm^2	500–800
Membrane resistance, Ω	Ohms	4–8
TDS removal	%	50–94
Current efficiency	%	85–95
Concentrate stream flow	% of feed	10–20
Energy consumption[a]	kWh/m^3	1.5–2.6
Approximate energy per kg of salt removed	kWh/m^3·kg	1–1.2

[a]Based on treating reclaimed water with a TDS concentration in the range from 1000 to 2500 mg/L. Not recommended for TDS concentration values beyond 10,000 to 12,000 mg/L.

of the EDR process and new electrode design, the life of anode and cathode electrodes is typically 2 to 3 yr. Anode life is typically less than cathode life. Electrodes can be reconditioned (USBR, 2003).

Advantages and Disadvantages of Electrodialysis versus Reverse Osmosis

In a recently completed study, two advanced treatment processes were compared to reduce the salinity of reclaimed water from a TDS concentration of 750 ± 50 mg/L to 500 mg/L or less (Adham et al., 2004). The two advanced treatment processes evaluated were (1) MF followed by RO and (2) EDR. The study was conducted for a period of about 6 mo. Based on the results of the side-by-side testing, it was found that the EDR process with cartridge prefiltration was more cost effective than the combined MF/RO process. Some of the advantages and disadvantages cited for each advanced treatment process are reported in Table 9-16. As more potential applications of EDR are currently under investigation, current literature should be consulted.

Table 9-16
Comparison of advantages and disadvantages electrodialysis and reverse osmosis for desalination[a]

Advantages	Disadvantages
Electrodialysis	
• Minimal pretreatment may be required (cartridge filtration is recommended) • Operates at a low pressure • Process is much quieter because high pressure pumps are not required • Antiscalant is not required • Membrane life expectancy is longer because foulants are removed continuously during the reversal process • Requires less maintenance than RO due to reversal process	• Limited to 50 percent salt rejection for a single membrane stack (stage) • Requires larger footprint to produce similar quantity and quality of water if multiple staging is used • Electrical safety requirements • Less experience for wastewater demineralization in the U.S. • Not as effective at removing microorganisms and many anthropogenic organic contaminants
Reverse osmosis	
• RO membranes provide a barrier to microorganisms and many anthropogenic organic contaminants (for the treated portion of the water produced) • More demonstrated experience for wastewater demineralization • RO membranes can remove more than 90 percent of TDS • Source water blending will reduce size of systems • Flexibility to provide higher quality water, if desired	• Requires high pressure to achieve high salt rejection • Requires pretreatment processes to minimize scaling and fouling • Requires chemical addition for MF & RO fouling control • More routine maintenance may be required to maintain performance

[a]Adapted from Adham et al. (2004).

9-7 MANAGEMENT OF MEMBRANE WASTE STREAMS

Retentate (concentrate) is produced when NF and RO membranes are utilized in the treatment process (see Fig. 9-4). As noted previously, these processes produce waste streams that are high in total dissolved solids but low in suspended solids. Concerns with membrane retentates, disposal methods, and concentration methods are considered in the following discussion, which has been adapted in part from Mickley (2001) and Crittenden et al. (2005).

Issues that must be addressed in the management of membrane waste streams include: (1) the volume of retentate, (2) quality of the retentate, and (3) environmental classification and regulations. The management of membrane cleaning solutions is also an important consideration. Factors that affect the management and/or disposal of membrane waste streams are presented in Table 9-17 and discussed below.

Membrane Waste Stream Issues

Volume of Retentate

The management of membrane retentate is often problematic because of the relatively large volume produced. Recovery, the amount of product water resulting from treatment of the process feedwater, typically ranges from 50 to 85 percent for an NF and RO facility. If, for example, a 10,000 m^3/d NF process operates at 85 percent recovery, the resulting waste stream will be 1500 m^3/d. With a waste stream of this magnitude, concentrate disposal problems can be formidable. The determination of the volume and concentration of the retentate from an RO process is illustrated in Example 9-5.

Table 9-17

Factors affecting the management and disposal of waste streams from membrane processes[a]

Issue	Description
Volume	With NF, the retentate stream volume ranges from 10 to 30 percent of the feed stream volume. In RO, the retentate stream volume ranges from 15 to 50 percent of the feed stream volume.
Salinity/toxicity	The high salinity of the retentate stream makes it toxic to many plants and animals, limiting options for land application or surface water discharge and rendering it unusable for recycling and reuse. Many retentate streams are anaerobic, which can be toxic to fish if discharged directly to surface waters. In addition, RO processes used for specific contaminant removal (i.e., arsenic, calcium, radium) may produce retentate streams that can be classified as hazardous materials.
Regulations	Retentate is classified as an industrial waste by the U.S. EPA. Retentate disposal is regulated under several different federal, state, and local laws, and the interaction between these regulations can be complex (Kimes, 1995; Pontius et al., 1996). Regulatory considerations are often as important as cost and technical considerations for determining viable retentate disposal options.

[a] Adapted from Crittenden et al. (2005).

EXAMPLE 9-5. Estimate the Quantity and Quality of the Retentate from a Reverse Osmosis Facility.

An RO facility using reclaimed water as a feed source is to be designed to produce 4000 m³/d of water that will be used for industrial cooling operations. Estimate quantity and quality of the retentate (waste stream), the permeate quality, and the total quantity of recycled water that must be processed. Assume the following data apply:

1. Recovery and rejection rates are equal to 90 percent
2. TDS of the feed steam = 1000 mg/L

Solution

1. Determine the flowrate of the concentrated waste stream and the total amount of water that must be processed.
   a. Determine the retentate stream flowrate by combining Eqs. (9-6) and (9-9). The following expression for the retentate stream flowrate results:

$$Q_r = \frac{Q_p(1 - r)}{r}$$

$$Q_r = \frac{(4000 \text{ m}^3/\text{d})(1 - 0.9)}{0.9} = 444 \text{ m}^3/\text{d}$$

   b. Determine the total amount of water that must processed to produce 4000 m³/d of RO water. Using Eq. (9-9) the required amount of water is:

$$Q_f = Q_p + Q_r = 4000 \text{ m}^3/\text{d} + 444 \text{ m}^3/\text{d} = 4444 \text{ m}^3/\text{d}$$

2. Determine the concentration of the permeate stream. The permeate concentration is obtained by writing Eq. (9-7) as follows:

$$C_p = C_f(1 - R)$$
$$C_p = 1000 \text{ mg/L}(1 - 0.9) = 100 \text{ mg/L}$$

3. Determine the concentration of the retentate by rewriting Eq. (9-10).

$$C_r = \frac{Q_f C_f - Q_p C_p}{Q_r}$$

$$C_r = \frac{(4444 \text{ m}^3/\text{d})(1000 \text{ mg/L}) - (4000 \text{ m}^3/\text{d})(100 \text{ mg/L})}{(444 \text{ m}^3/\text{d})}$$

$$C_r = 9108 \text{ mg/L}$$

Retentate Classification

The disposal of retentates from membrane processes is regulated just the same as other residuals from waste treatment processes. In the Clean Water Act (see Chap. 2), residuals from water treatment plant processes that are similar in character to membrane retentates are classified as an industrial waste. Retentate disposal is regulated under several different federal, state, and local laws, and the interaction between these regulatory requirements can be complex (Kimes, 1995; Pontius et al., 1996). Regulations for the control of toxic and hazardous substances must also be considered because toxic effects may result from the high salinity in the retentate. Regulatory considerations are often as important as cost and technical considerations for determining viable retentate disposal options.

Cleaning Solutions

Although the retentate is by far the most voluminous waste stream, NF and RO plants must also dispose of spent cleaning solutions. Frequently, the cleaning solutions are acidic or basic solutions with added detergents or surfactants. In many cases, the cleaning solution volume is so small compared to the retentate stream that the cleaning solution is diluted into and disposed with the retentate. In some cases, treatment of the cleaning solution may be required prior to disposal, but treatment may consist only of pH neutralization or dechlorination. Detergents and surfactants should be selected with disposal issues in mind.

Thickening and Drying of Waste Streams

Because of the potentially large volume of the retentate stream, several alternative processes have been developed to further reduce the volume requiring disposal. Included among the methods that have been developed are: (1) concentration by multiple-stage membrane arrays, (2) solar evaporation, (3) vapor compression evaporator systems (brine concentrators), (4) crystallizers, and (5) spray dryers.

Concentration by Multiple-Stage Membrane Arrays

Increasing the recovery of product water reduces the retentate volume and increases its salinity. Two- and three-stage RO membrane concentration steps (see Fig. 9-18) have been used to increase the concentration of the retentate to TDS values greater than 35,000 mg/L.

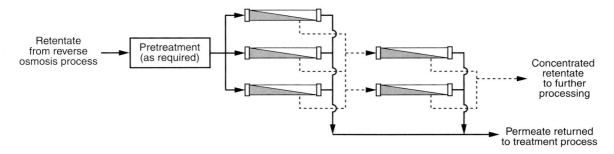

Figure 9-18
Typical schematic diagram for two-stage membrane retentate concentration.

The increase in salinity may limit the number of disposal options available, but the reduced volume will enhance disposal by solar evaporation ponds or deep well injection. The concentrated retentate can also be processed further by a vapor compression evaporator and crystallization.

Solar Evaporation

Where climatic conditions are favorable, the use of evaporation ponds may be feasible. Important factors that affect the performance of evaporation ponds include: relative humidity, wind velocity, barometric pressure, water temperature, and the salt content of the retentate. In some locations, glass-covered solar ponds similar to those used for desalination in many of the dry Mediterranean countries are used to further thicken waste streams and brines by evaporation (see Fig. 9-19).

Falling Film Evaporators

Falling film evaporators, without and with vapor compression, have been used to concentrate retentate solutions including those from NF and RO. In a vapor compression brine evaporator (see Fig. 9-20a), heat released by condensing steam is transferred across a heat exchanger surface to an aqueous solution boiling in the evaporator. The vapor released from the boiling solution is compressed in a vapor compressor. Compression raises the pressure and saturation temperature of the vapor so that it may return to the evaporator. By exchanging heat between the condensed vapors and the feedwater, it is usually possible to operate with little or no makeup heat in addition to the energy needed to drive the vapor compressor. Using the vapor compression approach requires only about 230 kJ of heat energy to evaporate a kilogram (100 BTU/lb) of water. Product water quality is normally less than 10 mg/L TDS. Reject from the evaporator typically ranges between 2 and 10 percent of the

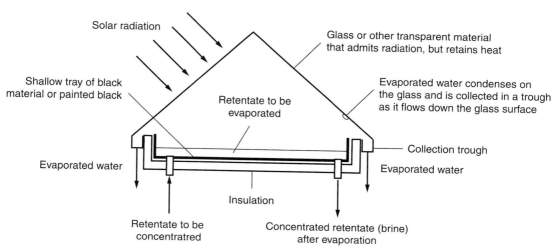

Figure 9-19

Schematic of typical solar evaporation cell for retentate concentration.

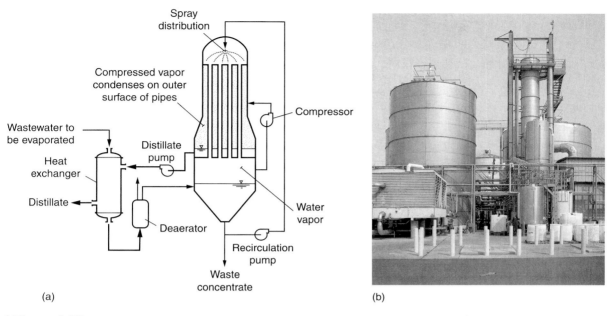

Figure 9-20
Falling film evaporator for membrane retentate: (a) schematic and (b) view of operating falling film evaporator (tall tower on right). Concentrated retentate (brine) is hauled to a disposal site.

feedwater flow, with TDS concentrations ranging as high as 250,000 mg/L. When operated in conjunction with crystallizers (see Fig. 9-21) or spray dryers, zero liquid discharge of RO retentate can be achieved under all climatic conditions (Mickley, 2001). A typical falling film evaporator used for RO retentate is shown on Fig. 9-20b.

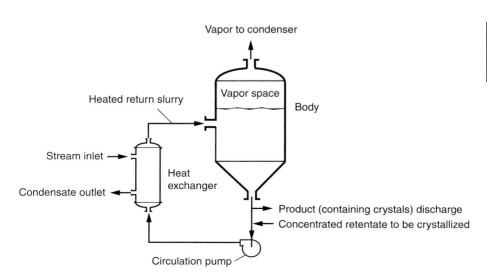

Figure 9-21
Forced circulation crystallizer for concentrated retentate.

Crystallization

Crystallizer technology has been used for many years to concentrate feed streams in industrial processes. More recently, this technology has been applied to retentate streams from desalination processes to reduce wastewater to a transportable form. Crystallizer technology is especially applicable in areas where solar evaporation pond construction cost is high, solar evaporation rates are low, or deep well disposal is not feasible or possible (Mickley, 2001). Crystallizers come in various size ranges and use heat input from either vapor compressors or an available steam supply.

A schematic diagram of a forced-circulation vapor compression crystallizer is shown on Fig. 9-20a. Waste concentrate is discharged to a sump of the crystallizer where it is mixed with recirculating brine. The mixture is pumped to a heat exchanger where it is heated by vapor from a vapor compressor. As water evaporates from the concentrate mixture, crystals form. The crystallizer produces a wet solid that can be readily transported for land disposal. For RO retentate disposal, crystallizers are operated normally in conjunction with a retentate evaporator to reduce the liquid waste to a transportable solid.

Spray Dryers

Spray dryers provide alternatives to crystallizers for reducing waste concentrations. The system consists of a feed tank, vertical spray drying chamber, and a bag filter (see Fig. 9-22). Waste concentrate is transported to the feed tank where it is mixed and pumped to the top of the drying chamber and discharged to the spray drying chamber through a centrifugal brine atomizer. Heated air is introduced into the dryer and drawn through the bag filter. Dry powder is separated in the bag filter and discharged to a receiving station for storage and/or transport to a disposal facility.

Membrane Distillation

Membrane distillation is a new concept for water desalination and concentrating RO retentate and other brine solutions. Schematic diagrams of a membrane distillation

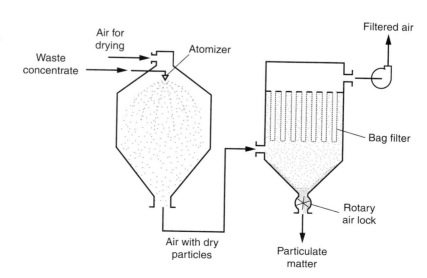

Figure 9-22
Schematic of typical spray dryer for waste concentration and drying.

process are shown on Fig. 9-23. Membrane distillation involves the use of a porous hydrophobic membrane to bring out water vapor generated from a concentrate stream. The hydrophobic membrane allows water vapor to pass through it but not water droplets. As shown on Fig. 9-23a, concentrate with temperatures ranging typically from 50 to 60°C flows along one side of the membrane, water vapor passes through the membrane and is cooled and condensed by cool concentrate flowing through the other side of the membrane (www.takenaka.co.jp). A source of waste heat is required to warm the concentrate such as solar ponds, sludge digester supernatant, or engine jacket water. A pilot-scale installation using membrane distillation in conjunction with NF and RO is being developed in Australia using high TDS treated wastewater (Leslie et al., 2005). Pilot-scale work using membrane distillation on groundwater was also conducted at the University of Texas at El Paso and the authors concluded that it may be competitive in treating RO and NF concentrate (Walton et al., 2004).

Conventional methods used for the disposal of membrane retentates are summarized in Table 9-18. Three of the more common methods now employed are (1) discharge to wastewater collection system, (2) disposal to surface waters, and (3) subsurface injection, and are considered further in the following discussion.

Ultimate Disposal Methods for Membrane Waste Streams

Discharge to Wastewater Collection System

Discharge to the wastewater collection system is a viable consideration where the retentate comes from a satellite treatment facility and the volume of retentate is relatively small compared to the total flow to the central treatment plant. Local regulations will cover the discharge of membrane retentates to the wastewater collection system. In the case of RO plant residuals, which in most cases are primarily concentrated inorganic solutes, the biological process provides little treatment for the retentate stream.

Water Quality Issues Discharge to the collection system of concentrate waste streams should consider that the potential dilution effect with the untreated wastewater. Metered discharge *in lieu* of a slug discharge of concentrate will help minimize any

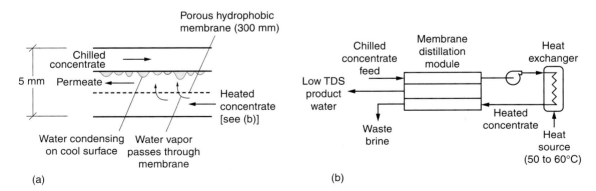

Figure 9-23
Membrane distillation: (a) schematic of basic mechanism and (b) typical process schematic diagram.

Table 9-18
Disposal options for concentrated solutions from membrane processes

Disposal option	Description
Controlled thermal evaporation	Although energy intensive, thermal evaporation may be the only option available in many areas.
Subsurface injection	Depends on whether subsurface aquifer is brackish water or is otherwise unsuitable for domestic uses.
Discharge to wastewater collection system	For very small discharges such that the increase in TDS is not significant (e.g., less that 20 mg/L). For larger discharges, this option is only feasible if the relative volume and TDS do not interfere with operation of treatment facility or effluent disposal or reuse. High retentate concentrations may impact effluent disposal options.
Discharge to treated effluent	Discharge to treatment plant effluent (blend for dilution) for disposal through existing outfalls to surface waters or the ocean will depend on type of discharge permit and discharge location.
Evaporation ponds	Large surface area is required in most areas with the exception of some southern and western states.
Land application	Land application has been used for some low-concentration retentate solutions.
Landfilling	Concentrated retentates may be disposed of in secure hazardous waste landfills. In some cases dried salts may be disposed of in double lined municipal landfills.
Ocean discharge	The disposal option of choice for facilities located in the coastal regions of the United States. Typically, a brine line, with a deep ocean discharge, is used by a number of dischargers. Combined discharge with power plant cooling water has been used in Florida. For inland locations, truck, rail hauling, or a pipeline will be needed for brine transport. Continued ocean disposal of brines from reclaimed wastewater treatment facilities may not be allowed in the future because of concerns over the presence of trace constituents.
Product recovery	Depending on the source of the retentate and the nature of the processing (e.g., crystallization), it may be possible to recover a useful byproduct.
Surface water discharge	Discharge of retentates to surface waters is the most common method of disposal for concentrated solutions. While small amounts of retentate can be tolerated, increased concentrations may not be allowed, depending on the TDS of the receiving water.

potential downstream effects. Pilot-scale investigations and modeling will most likely be required to demonstrate that blending retentate with wastewater will not cause toxicity issues in the treatment process or in the treated effluent thus resulting in violation of the NPDES permit governing wastewater discharge. Further, if effluent from the central treatment plant is used for groundwater recharge, discharge of retentate from a satellite plant to a centralized collection system may be precluded if it causes the effluent TDS to exceed the groundwater TDS limits, for example 500 mg/L (see Chap. 22).

Design Considerations Design of a retentate disposal system should consider providing controlled discharges. Facilities such as an equalization basin and metering pumps can be used to eliminate large slugs of residuals that may upset the wastewater treatment facility. The discharge should be coordinated with the wastewater treatment plant operators so that they may optimize the performance of their process units.

Disposal to Surface Waters

Where permitted, the most cost effective disposal option for RO plants, especially for those plants located in coastal areas, is discharge to brackish or saline receiving waters. The advantage of a surface water disposal system is the relatively low capital and operation and maintenance costs. Disadvantages include (1) the need for demonstrating the retentate will not have any adverse health and environmental effects, (2) the uncertainty of allowing continued practice of surface water discharge in the future, and (3) extensive monitoring of the discharge and the water body to ensure the requirements of the NPDES permit are met.

Water Quality Issues Surface water discharge is dependent on the quality of the retentate. Nominally, the retentate stream is comprised mainly of inorganic solutes from the source stream that have been concentrated. The difference in the TDS content (salinity) between the retentate and the receiving water is an important consideration and most likely would require a well-designed diffuser system to ensure proper dispersion into the marine environment. In addition to TDS, however, it is important to consider the toxicity of individual heavy metals, whose concentration is increased in the same proportion as the TDS. Additionally, many retentate streams are anaerobic, which can be toxic to fish in the receiving water without sufficient dilution. Toxicity can initially be assessed by comparing predicted heavy metal concentrations to regulated limits, but bioassays are often required before permits are issued. Although economic considerations favor discharge to a brackish river or bay near the plant by means of a marine outfall, environmental considerations may influence the feasibility of this option.

Design Considerations Considerations for the design of a surface water disposal system include quality of the retentate, location of the outfall, pumping requirements, flow equalization, and outfall design. Outfall location is also an extremely important concern. The outfall should be located such that it discharges to a point of maximum dispersion. Similarly, the outfall should be designed to disperse the retentate across the well-mixed zone of the water body. The location and design of the outfall significantly impacts the pumping requirements. Consideration should be given to equalizing the residual flow to minimize pump and motor sizing, and to reduce the environmental impact on the receiving water.

Subsurface Injection

Subsurface injection involves pumping the retentate stream into an injection well, typically thousands of meters deep. The injection zone is typically a brackish or saline aquifer, with no potential for use as a potable water supply, which is overlain by thick layers of impermeable rock and sediments that prevent contamination of shallower fresh water aquifers. Deep well injection is employed by about 10 percent of RO plants used in water desalination; its use is becoming more common, particularly in Florida.

Preference for deep well disposal in Florida has arisen because of the existence of a reliable injection zone and public and regulatory resistance to surface water discharge.

Water Quality Issues Discharge of membrane retentate into an aquifer is controlled by federal and local environmental regulations. As a result of the growing concern over the contamination of the nation's groundwater resources from the estimated 300,000 injection wells in the United States, a statutory mandate was included in the Drinking Water Act of 1979 to establish minimum requirements for state programs to protect underground sources of drinking water from contamination by subsurface injection. The Underground Injection Control (UIC) regulations were intended to strengthen state regulations as well as establish minimum federal standards reflecting good engineering practices. Currently 40 states have primacy with regard to the UIC program (Mickley, 2001).

Design Considerations Well construction is governed by regulations for deep well injection of industrial wastes. Wells are constructed of three to four casings, with the space between each casing filled with cement grout (see Fig. 9-24). Each casing ends typically at a different depth. Depending on the local groundwater hydrology, significant potential for groundwater contamination may occur and multiple casings are designed to prevent any leakage from one aquifer to the next. To provide for continuous operation,

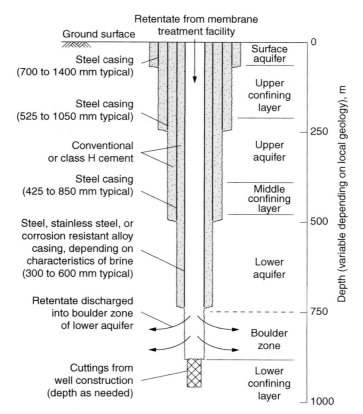

Figure 9-24
Typical injection well for disposal of brine concentrate (From Crittenden et al., 2005).

multiple injection wells must be provided. Deep well injection systems tend to be fairly expensive due to well drilling cost and maintenance costs. The high pressure at the bottom of the injection well and the saline solution tend to enhance the corrosion potential of the well screen and casing. Selection of materials resistant to corrosion under those conditions may prolong the operating life of an injection facility.

PROBLEMS AND DISCUSSION TOPICS

9-1 Four different waters are to be treated by RO using a thin film composite membrane. For water A, B, C, or D (water to be selected by instructor), determine the required membrane area, the rejection rate, and the concentration of the concentrate stream.

Item	Unit	Water A	Water B	Water C	Water D
Feed stream flow	m³/d	4000	5500	20,000	10,000
Feed stream TDS	g/m³	2600	3200	5400	2700
Permeate TDS	g/m³	200	500	400	225
Water mass transfer rate coefficient k_w	m/s	1.5×10^{-6}	1.5×10^{-6}	1.5×10^{-6}	1.5×10^{-6}
Solute mass transfer rate coefficient, k_s	m/s	5.8×10^{-8}	5.8×10^{-8}	5.8×10^{-8}	5.8×10^{-8}
Net operating pressure	bar	28	25	28	30
Recovery	%	88	90	89	86

9-2 Using the data given below, determine the recovery and rejection rates for one of the following a RO units (unit to be selected by instructor).

Item	Unit	Reverse osmosis unit A	B	C	D
Feed stream flow	m³/d	4000	6000	8000	10,000
Retentate flow	m³/d	350	600	7500	9000
Permeate TDS	g/m³	65	88	125	175
Retentate TDS	g/m³	1500	2500	1850	2850

9-3 Using the data given below, determine the water mass flux rate coefficient and the solute mass transfer rate coefficient for one of the following RO units (unit to be selected by instructor).

		Reverse osmosis unit			
Item	Unit	A	B	C	D
Feed stream flow	m³/d	4000	5500	20,000	10,000
Feed stream TDS	g/m³	2500	3300	5300	2700
Permeate TDS	g/m³	20	50	40	23
Net operating pressure, ΔP	bar	28	25	28	30
Membrane area	m²	1600	2000	10,000	5000
Recovery, r	%	85	90	89	86

9-4 Estimate quantity and quality of the retentate stream, and the total quantity of water that must be processed, from an RO facility that is to produce 4000 m³/d of demineralized water. Assume that the recovery and rejection rates are 80 and 85 percent, respectively and that the concentration of total dissolved solids in the feed steam is 1500 mg/L.

9-5 Estimate the SDI for the following filtered wastewater samples. If the water is to be treated with a spiral wound RO, will additional treatment be required?

Test run time, min	Cummulative volume filtered, mL			
	A	B	C	D
2	315	480	180	500
5	575	895	395	700
10	905	1435	710	890
20	1425	2300	1280	1150

9-6 Calculate the modified fouling index (MFI) for the effluent from an MF (water sample to be selected by instructor) using the following experimental data:

Time, min	Volume filtered, L Sample		Time, min	Volume filtered, L Sample	
	A	B		A	B
0			3.5	6.78	7.17
0.5	1.50	1.50	4.0	7.48	8.03
1.0	2.50	2.50	4.5	8.08	9.80
1.5	3.45	3.48	5.0	8.57	10.60
2.0	4.36	4.40	5.5		11.27
2.5	5.22	5.37	6.0		11.90
3.0	6.03	6.28	6.5		12.40

9-7 Estimate the scaling potential for one of the RO-treated water samples with the following chemical characteristics (to be selected by instructor) using both the Langelier and Ryzner (see Sec. 19-2 in Chap. 19) indexes.

		Water sample			
Constituent	Unit	A	B	C	D
Ca^{2+}	mg/L as $CaCO_3$	5	12	15	10
HCO_3^-	mg/L as $CaCO_3$	7	9	16	12
TDS	mg/L	20	40	50	25
pH	unitless	6.5	8.0	6.8	7.8

9-8 Estimate the power cost (based on the current price of electricity) to treat a flow of 2500 m³/d with a TDS concentration of 2000 g/m³ (0.04 g eq/L) to remove 50 percent of the TDS by an electrodialysis EDR unit. Assume the membrane stack will contain 400, 500, and 600 membrane pairs.

9-9 Estimate the power cost (based on the current price of electricity) to treat 2500, 4000, or 6000 m³/d of reclaimed water with the chemical characteristics given below (water and flowrate to be selected by instructor). Assume it is desired to remove 60 percent of the Ca^{2+} and Mg^{2+} and the membrane stack will contain 500 membrane pairs.

	Concentration, mg/L				Concentration, mg/L		
Cation	A	B	C	Anion	A	B	C
Ca^{2+}	610	800	130	HCO_3^-	660	1,220	410
Mg^{2+}	180	380	180	SO_4^{2-}	260	224	50
Na^+	55	1,100	640	Cl^-	80	595	110
K^+	60	230	360	NO_3^-	80	470	850

9-10 Using the data given below for an RO facility producing reclaimed water, determine the quantity and quality of the retentate that must disposed (RO facility to be selected by instructor).

		RO facility			
Item	Unit	A	B	C	D
Feed stream flow	m³/d	4000	6000	8000	10,000
Feed stream TDS	g/m³	1500	2000	1800	2500
Recovery, r	%	88	92	85	90
Rejection, R	%	95	90	92	96

9-11 Review and cite three current articles (within the last 5 yr) dealing with the disposal of NF, RO, and ED brine. What types of processes or process combinations are being used? What are the critical issues in brine treatment and disposal that stand out in your mind?

REFERENCES

Adham, S., T. Gillogly, G. Lehman, E. Rosenblum, and E. Hansen, (2004) *Comparison of Advanced treatment Methods for Partial Desalting of Tertiary Effluents, Desalination and Water Purification Research and Development*, Report No. 97, Agreement No. 99-FC-81-0189, U.S. Department of the Interior, Bureau of Reclamation, Denver, CO.

Adham, S. (2006) Personal communication.

Ahn, K. H., J. H. Y. Song Cha, K. G. Song, and H. Koo (1998) "Application of Tubular Ceramic Membranes for Building Wastewater Reuse," 137, in *Proceedings IAWQ 19th International Conference*, Vancouver, Canada.

ASTM (2002) *D4189−95 Standard Test Method for Silt Density Index (SDI) of Water*, American Society for Testing and Materials, Philadelphia, PA.

AWWA (2004) "Committee Report: Current Perspectives on Residuals Management for Desalting Membranes," AWWA Membrane Residuals Management Committee, *J. AWWA*, **96**, 12, 73–87.

Beck, R. W. (2002) *Technical Memorandum B.7, Demineralization Treatment Technologies*, St. Johns Water Management District, Palatka, FL.

Benefield, L. D., J. F. Judkins, Jr., and B. L. Weand (1982) *Process Chemistry for Water and Wastewater Treatment*, Prentice-Hall, Englewood Cliffs, NJ.

Bourgeous, K., G. Tchobanoglous, and J. Darby (1999) "Performance Evaluation of the Koch Ultrafiltration (UF) Membrane System for Wastewater Reclamation," Center for Environmental and Water Resources Engineering, Report 99-2, Department of Civil and Environmental Engineering, University of California, Davis, CA.

Celenza, G. (2000) *Specialized Treatment Systems, Industrial Wastewater Process Engineering, Vol. 3,* Technomic Publishing Co., Inc., Lancaster, PA.

Childs, W. D., and A. E. Dabiri (1988) *VARI-RO Desalting Pilot Plant Testing and Evaluation*, Water Treatment Technology Program Report No. 30, U.S. Department of the Interior, Bureau of Reclamation, Denver, CO.

Crittenden, J. C., R. R. Trussell, D. W. Hand, K. J. Howe, and G. Tchobanoglous (2005) *Water Treatment: Principles and Design*, 2nd ed., John Wiley & Sons, Inc., New York.

Crozes, G. F., J. G. Jacangelo, C. Anselme, and J. M. Laine (1997) "Impact of Ultrafiltration on Membrane Irreversible Fouling," *J. Membr. Sci.* **124**, 1, 63–76.

Dupont (1977) "Determination of the Silt Density Index," *Technical Bulletin No. 491*, Dupont de Nemours and Co., Wilmington, DE.

Eckenfelder, W. W., Jr., (2000) *Industrial Water Pollution Control*, 3rd ed., McGraw-Hill, Boston, MA.

EPRI (1999) *The Desalting and Water Treatment Manual: A Guide to Membranes for Municipal Water Treatment,* TR-112644, Electric Power Research Institute, Palo Alto, CA.

Freeman, S., G. Leitner, J. Crook, and W. Vernon (2002) "A Clear Advantage—Membrane Filtration is Gaining Acceptance in the Water Quality Field," *Water Environ. Technol.*, **14**, 1, 16–21.

Gagliardo, P., R. P. Merlo, S. Adham, S. Trussell, and R. Trussell (2000) "Application of Reverse Osmosis Membranes for Water Reclamation with Various Pretreatment Processes," *Proceedings of WEFTEC 2000, Water Environment Federation*, Alexandria, VA.

IAEA (2004) *Desalination Economic Evaluation Program (DEEP): User's Manual*. International Atomic Energy Agency, Vienna, Austria.

Kimes, J. K. (1995) "The Regulation of Concentrate Disposal in Florida," *Desalination*, **102**, 1–3, 87–92.

Langelier, W. (1936) "The Analytical Control of Anti-Corrosion Water Treatment," *J. AWWA*, **28**, 10, 1500–1521.

Larson, T., and A. Buswell (1942) "Calcium Carbonate Saturation Index and Alkalinity Interpretations," *J. AWWA*, **34**, 11, 1667–1684.

Lehman, G., T. Gillogly, S. Adham, E. Rosenblum, and E. Hansen (2004) "Comparison of Advanced Water Treatment Methods for Partial Desalting of Tertiary Effluents," *Proceedings of WEFTEC '04*, Water Environment Federation, Alexandria, VA.

Leslie, G., D. Stevens, and S. Wilson (2005) "Designer Reclaimed Water," *Water*, **32**, 6, 75–79.

Lien, L. (1998) "Using Membrane Technology to Minimize Wastewater," *Pollut. Eng.*, **30**, 5, 44–47.

Merrill, D. T., and R. L. Sanks (1977a) "Corrosion Control by Deposition of $CaCO_3$ Films: Part I," *J. AWWA*, **69**, 11, 592–599.

Merrill, D. T., and R. L. Sanks (1977b) "Corrosion Control by Deposition of $CaCO_3$ Films: Part II," *J. AWWA*, **69**, 12, 634–640.

Merrill, D. T., and R. L. Sanks (1978) "Corrosion Control by Deposition of $CaCO_3$ Films: Part III," *J. AWWA*, **70**, 1, 12–18.

Mickley, M. C. (2001) *Membrane Concentrate Disposal: Practices and Regulation*, Desalination and Water Purification Research and Development Program Report No. 69, U.S. Department of the Interior, Bureau of Reclamation.

Pankow, J. F. (1991) *Aquatic Chemistry Concepts*, Lewis Publishers, Chelesa, MI.

Paranjape, S., R. Reardon, and X. Foussereau (2003) "Pretreatment Technology for Reverse Osmosis Membrane Used in Wastewater Reclamation Application—Past, Present, and Future—A Literature Review," *Proceedings of the 76th Annual Technical Exhibition & Conference*, Water Environment Federation, Alexandria, VA.

Pearce, G., M. Wilf, S. Alt, and J. Reverter (2001) "Application of UF Combined with a Novel Low Fouling RO Membrane for Reclamation of Municipal Wastewater," CIWEM Wastewater Conference, Edinborough, UK.

Pontius, F. W., E. Kawczynski, and S. J. Koorse (1996) "Regulations Governing Membrane Concentrate Disposal," *J. AWWA*, **88**, 5, 44–52.

Richard, A., W. Surrat, H. Winters, and D. Kree (2001) "Solving Membrane Fouling for the Boca Raton 40 Mgal/d Membrane Water Treatment Plant: The Interaction of Humic Acids, pH, and Antiscalant with Membrane Surfaces," *Membrane Practices for Water Treatment*, American Water Works Association.

Ryznar, J. W. (1944) "A New Index for Determining the Amount of Calcium Carbonate Scale Formed by Water," *J. AWWA*, **36**, 472–484.

Schippers, J. C., and J. Verdouw (1980) "The Modified Fouling Index, a Method for Determining the Fouling Characterstics of Water," *Desalination*, **32**, 137–148.

Snoeyink, V. L., and D. Jenkins (1980) *Water Chemistry*, John Wiley & Sons,Inc, New York.

Standard Methods (2005) *Standard Methods for the Examination of Water and Waste Water*, 21st ed., American Public Health Association, Washington, DC.

Stephenson, T., S. Judd, B. Jefferson, and K. Brindle (2000) *Membrane Bioreactors for Wastewater Treatment*, IWA Publishing, London.

Stiff, H. A., Jr., and L. E. Davis (1952) "A Method For Predicting The Tendency of Oil Field Water to Deposit Calcium Carbonate," *Pet. Trans. AIME*, **195**, 213–216.

Taylor, J. S., and E. P. Jacobs (1996) "Reverse Osmosis and Nanofiltration," Chap. 9, in J. Mallevialle, P. E. Odendaal, and M R. Wiesner (eds.) *Water Treatment Membrane Processes*, American Water Works Association, published by McGraw-Hill, New York.

Taylor, J. S., and M. Wiesner (1999) "Membranes," Chap. 11, in R. D. Letterman (ed.) *Water Quality and Treatment*, 5th ed., McGraw-Hill, Inc., New York.

Tchobanoglous, G., F. L. Burton, and H. D. Stensel (2003) *Wastewater Engineering: Treatment and Reuse*, 4th ed., eds., McGraw-Hill, New York.

USBR (2003) *Desalting Handbook for Planners*, 3rd ed., Desalination Research and Development Program Report No. 72, United States Department of the Interior, Bureau of Reclamation.

Walton, J., H. Lu, C. Turner, S. Solis, and H. Hein (2004) *Solar and Waste Heat Desalination by Membrane Distillation*, Desalination and Water Purification Research and Development Program Report No. 81, United States Department of the Interior, Bureau of Reclamation.

WCPH (1996a) Total *Resource Recovery Project, Final Report*, Prepared for City of San Diego, Western Consortium for Public Health, Water Utilities Department, San Diego, CA.

WCPH (1996b) *Total Resource Recovery Project Aqua III San Pasqual Health Effects Study Final Summary Report*, Prepared for City of San Diego, Western Consortium for Public Health, Water Utilities Department, San Diego, CA.

Wilf, M. (1998) "Reverse Osmosis Membranes for Wastewater Reclamation," in T. Asano (ed.) *Wastewater Reclamation and Reuse*, Chap. 7, pp. 263–344, Water Quality Management Library, **10**, CRC Press, Boca Raton, FL.

Wong, J. (2003) "A Survey of Advanced Membrane Technologies and Their Applications in Water Reuse Projects," *Proceedings of the 76th Annual Technical Exhibition & Conference*, Water Environment Federation, Alexandria, VA.

10
Removal of Residual Trace Constituents

WORKING TERMINOLOGY   526

10-1 INTRODUCTION TO TECHNOLOGIES USED FOR THE REMOVAL OF TRACE CONSTITUENTS   528
  Separation Processes Based on Mass Transfer   528
  Chemical and Biological Transformation Processes   531

10-2 ADSORPTION   532
  Applications for Adsorption   532
  Types of Adsorbents   533
  Basic Considerations for Adsorption Processes   536
  Adsorption Process Limitations   551

10-3 ION EXCHANGE   551
  Applications for Ion Exchange   552
  Ion Exchange Materials   554
  Basic Considerations for Ion Exchange Processes   555
  Ion Exchange Process Limitations   559

10-4 DISTILLATION   560
  Applications for Distillation   560
  Distillation Processes   560
  Basic Considerations for Distillation Processes   562
  Distillation Process Limitations   563

10-5 CHEMICAL OXIDATION   563
  Applications for Conventional Chemical Oxidation   563
  Oxidants Used in Chemical Oxidation Processes   563
  Basic Considerations for Chemical Oxidation Processes   566
  Chemical Oxidation Process Limitations   567

10-6 ADVANCED OXIDATION   567
  Applications for Advanced Oxidation   568
  Processes for Advanced Oxidation   569
  Basic Considerations for Advanced Oxidation Processes   574
  Advanced Oxidation Process Limitations   577

10-7 PHOTOLYSIS   578
  Applications for Photolysis   578
  Photolysis Processes   579
  Basic Considerations for Photolysis Processes   579
  Photolysis Process Limitations   586

10-8 ADVANCED BIOLOGICAL TRANSFORMATIONS 586
　　　Basic Considerations for Advanced Biological Treatment Processes 587
　　　Advanced Biological Treatment Processes 588
　　　Limitations of Advanced Biological Transformation Processes 590
　　　PROBLEMS AND DISCUSSION TOPICS 591
　　　REFERENCES 594

WORKING TERMINOLOGY

Term	Definition
Absorption	The transfer of a gas phase into a liquid phase, for example, the process of recarbonation consists of bubbling carbon dioxide into water to reduce the pH after treatment with lime.
Activated carbon	A substance used commonly in adsorption processes for the removal of trace constituents from water and odor compounds from air. Activated carbon is derived from an organic base material, prepared (activated) using a high temperature and pressure pyrolysis process, resulting in properties conducive to mass transfer.
Adsorption	The process of accumulating substances that are in a gas or liquid phase onto a suitable surface. Substances deposit on the solid phase due to a number of physical attraction and chemical bonding forces.
Adsorbate	The compound in a gas or liquid phase suspension that is deposited onto the adsorbent.
Adsorbent	The solid, liquid, or gas material onto which adsorption is taking place.
Advanced oxidation	A chemical oxidation process that relies upon the hydroxyl radical (HO·) for the destruction of trace organic constituents found in water. Several of processes have been identified that are able to produce HO·.
Desorption	The release of a volatile gas phase from a liquid phase as in gas stripping or the release of a previously adsorbed compound from an adsorbent material.
Electrical efficiency per log order (EE/O)	The electrical energy (in kWh) required to reduce the concentration of a constituent by one log order per unit water volume.
Electrical potential	The driving force for the exchange of electrons between constituents during a redox reaction, reported in volts with respect to the standard hydrogen electrode.
Extinction coefficient	The fractional amount of ultraviolet (UV) radiation attenuated as the UV light passes through water that contains dissolved substances which absorb energy. The extinction coefficient is also known as the molar absorptivity.
Gas stripping	A process used to remove a volatile constituent from a liquid phase, such as in the removal of ammonia from water in a packed column using air as the gas phase.
Ion exchange	A process used for the removal of dissolved ionic constituents where ions of a given species are displaced from a solid phase material by ions of a different species from solution.
Isotherm	A function used to relate the amount of a given constituent adsorbed from water per concentration of adsorbent at a given temperature.

Mineralization	The complete oxidation and conversion of organic substances into inorganic forms, such as carbon dioxide, water, and mineral acids, through the action of chemically and biologically mediated redox reactions.
Natural organic matter (NOM)	Dissolved and particulate organic constituents that are typically derived from three sources: (1) the terrestrial environment (mostly humic materials), (2) the aquatic environment (algae and other aquatic species and their byproducts), and (3) the microorganisms in the biological treatment process. Typically quantified as total organic carbon (TOC).
Oxidation reactions	A redox reaction involving the loss of electrons. The oxidation reaction can be obtained from tabulated half reactions (by convention, half reactions are reported as reduction reactions) by reversing the direction of the half reaction and multiplying the electrical potential by -1.
Photolysis	A process used for the treatment of trace constituents, where a UV light source is used to supply photons that are absorbed by the constituent which subsequently become unstable and reacts or splits apart.
Quantum yield	A quantity used to describe the frequency at which photon absorption results in a photolysis reaction and is specific to the type of compound and the wavelength.
Quenching	The use of physical or chemical means to stop a chemical reaction.
Redox reaction	The overall reaction resulting from the combination of a reduction and oxidation reaction.
Reduction reactions	A redox reaction involving the gain of electrons. The reduction reaction can be obtained directly from tabulated half reactions.
Reactivation	The desorption of constituents from an adsorbent material followed by the combustion of the remaining sorbed constituents, resulting in restoration of the adsorptive capacity.
Regeneration	The desorption of constituents from an adsorbent material for the partial restoration of the adsorptive capacity.
Reverse osmosis (RO)	The rejection of dissolved constituents by preferential diffusion using a pressure-driven, semipermeable membrane (see Chap. 9).
Scavengers	In advanced oxidation systems, substances that preferentially react with the oxidant and radical species, typically reducing the degradation rate for the compound of interest and overall efficiency of the process.
Separation processes	Physical and chemical processes used in water reclamation that bring about treatment by the isolation of particular constituents. The isolated constituents are concentrated into a waste stream that must be managed (see also Table 10-2).
Sorption	A term used to describe the attachment of organic material (adsorbate) to an adsorbent where it is difficult to differentiate between chemical and physical adsorption.
Synthetic organic compounds (SOCs)	Compounds of synthetic origin used extensively in industrial processes and contained in numerous manufactured consumer products. The presence of SOCs in drinking water as well as reclaimed water is of concern due to toxicity and unknown effects.
Trace constituent	A diverse classification of constituents found at low concentrations in untreated wastewater and not readily removed using conventional secondary treatment. Trace constituents are of concern due to known or suspected toxicity associated with many of these compounds, which may need to be removed during water reclamation depending on the reuse requirements.
Water matrix	The term used to refer to a given water and all of its constituents that result in the physical, chemical, and biological properties of the solution.

Depending on the particular wastewater to be reclaimed, a specific constituent or group of constituents beyond those removed in conventional secondary treatment may need to be removed to meet water quality standards for a given water reuse application. The constituents most commonly encountered in treated municipal wastewater that may need to be removed during water reclamation are the organic and inorganic trace constituents. As described in Chap. 3, trace constituents in reclaimed water are of concern because of issues related to the known or suspected toxicity of these compounds. While the membrane processes described in Chap. 9 are able to remove most of these constituents, the application of alternate treatment processes, individually or in conjunction with membrane treatment, may be more economical or efficient in some applications. In this chapter, which deals with the removal of specific constituents, the following subjects are considered: (1) introduction to technologies used for the removal of trace constituents, (2) adsorption, (3) ion exchange, (4) distillation, (5) chemical oxidation, (6) advanced oxidation, (7) photolysis, and (8) advanced biological treatment.

10-1 INTRODUCTION TO TECHNOLOGIES USED FOR THE REMOVAL OF TRACE CONSTITUENTS

The removal of trace constituents from reclaimed water depends on the characteristics and concentrations of the compounds to be removed. As described in Table 7-12, trace constituents are typically present in reclaimed water at aggregate concentrations ranging from 250 to 1000 µg/L, and may be greater in some cases. While most of the trace constituents present in reclaimed water are added during domestic, commercial, or industrial use, a portion of the residual trace constituents in wastewater may originate from the drinking water supply. A monitoring program should be implemented to identify the presence and concentrations of trace constituents and background water chemistry. Monitoring data is then used during the treatment system selection process to identify constituents that need to be removed and processes that are appropriate for pilot testing.

Conventional and advanced treatment processes used for the removal of trace constituents are presented in Table 10-1. Several water reclamation process flow diagrams that incorporate the processes presented in Table 10-1 are shown on Fig. 10-1. It should be noted that many of these processes have been used historically for the treatment of drinking water, including coagulation, precipitation, and ion exchange. However, as the length of the water cycle is reduced by increasing wastewater discharges coupled with greater water extractions, various technological interventions are needed and many processes developed for drinking water are now being used in water reclamation applications. The two primary types of processes used for the removal of trace constituents from water are based on (1) mass transfer separation and (2) chemical and biological transformation or destruction.

Separation Processes Based on Mass Transfer

The removal of constituents from wastewater by the transfer of mass from one phase to another or by the concentration of mass within a phase is accomplished with various separation processes. The principal separation processes that may be used in water reclamation for the removal of trace constituents are summarized in Table 10-2. It is

Table 10-1
Summary of treatment process effectiveness for selected trace constituents[a]

Selected constituent	Unit	Example drinking water standard	Aeration, stripping	Coagulation, sedimentation, filtration	Lime softening	Adsorption[c,d] GAC	PAC	Activated alumina	Ion exchange Anion	Cation	Membrane filtration Reverse osmosis	Ultra-filtration	Advanced oxidation[e]	Photolysis[e]
Inorganic constituents														
Arsenic (+3)	mg/L	0.010	P	G-E	F-E	F-G	P-F	F-E	G-E	P	E	P		
Arsenic (+5)	mg/L	0.010	P	G-E	F-E	F-G	P-F	F-E	G-E	P	E	F		
Barium	mg/L	2.0	P	P-F	G-E	P	P	P	P	E	E	—		
Chromium (+3)	mg/L	0.10	P	G-E	G-E	F-G	F	P	P	E	E	—		
Chromium (+6)	mg/L	0.10	P	P	P	F-G	F	P	E	P	E	—		
Copper	mg/L	1.3	P	G	G-E	F-G	P	P	P	F-G	E	—		
Fluoride	mg/L	4.0	P	F-G	P-F	G-E	P	E	P-F	P	E	—		
Hardness	mg/L	—	P	P	E	P	P	P	P	E	E	—		
Iron	mg/L	0.30	F-G	F-E	E	P	P	F-G	P	G-E	G-E	—		
Lead	mg/L	0.0	P	E	E	F-G	P-F	P	P	F-G	G-E	—		
Manganese	mg/L	0.05	P-G	F-E	E	F-E	P-F	P	P	G-E	G-E	—		
Mercury	mg/L	0.002	P	F-G	F-G	F-G	P	P	P	F-G	G-E	—		
Nitrate	mg/L	10.0	P	P	P	P	P	P	G-E	P	G	—		
Perchlorate	mg/L	0.018	P	—	—	F-G	—	—	G-E	P	G-E	—		
Radium	pCi/L	5.0	P	P-F	G-E	P-F	P	P-F	P	E	E	—		
Uranium	pCi/L	0.030	P	G-E	G-E	F	P-F	G-E	E	G-E	E	—		
Organic constituents														
VOCs	mg/L	—	G-E	P	P-F	F-E	P-G	P	P	P	F-E	F-E	G-E	G-E
SOCs	mg/L	—	P-F	P-G	P-F	F-E	P-E	P-F	P	P	F-E	F-E	G-E	F-G
Color	CFU	15	P	F-G	F-G	E	G-E	P	P-G	—	—	—	E	P-G
TTHMs	mg/L	0.080	G-E	P	P	F-E	P-F	P	P	P	F-G	F-G	P-F	P
MTBE	mg/L	0.020	G-E	P	P	F-E	P-E	—	P	P	F-E	F-E	G-E	P-F
NDMA	mg/L	0.02	P	—	—	—	—	—	—	—	—	—	G-E	E

[a] Adapted from Crittenden et al., 2005.
[b] Codes for process effectiveness: P—poor (0–20 percent removal); F—fair (20–60 percent removal); G—good (60–90 percent removal); E—excellent (90–100 percent removal); "—"—not applicable/insufficient data.
[c] GAC = granular activated carbon, PAC = powder activated carbon
[d] Granular ferric hydroxide adsorption processes (not shown) are used for high removal of Arsenic (+3) and Arsenic (+5) but are less effective for other constituents.
[e] Advanced oxidation and photolysis are used primarily for the removal of trace organic constituents for selected water reuse applications. The effectiveness of AOPs (i.e., UV alone, UV/H_2O_2, UV/O_3, and O_3/H_2O_2) depends on the indirect reactions that may occur with individual reactants used to generate the hydroxyl radical species.

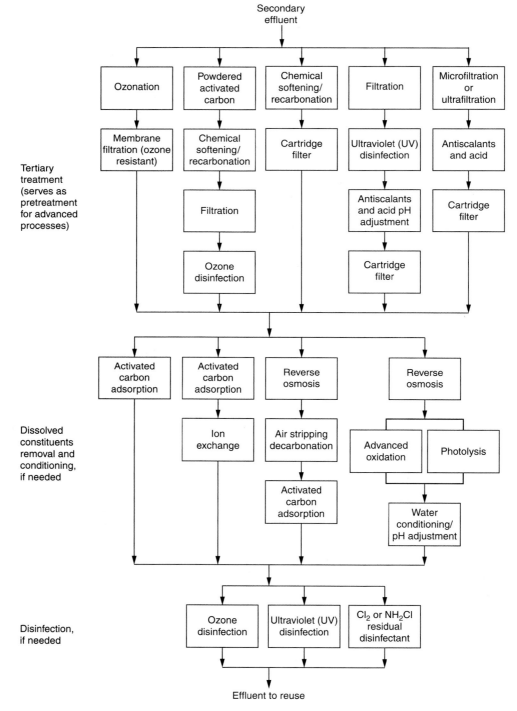

Figure 10-1
Examples of flow diagrams used for advanced treatment for water reclamation.

Table 10-2
Separation processes that are used in wastewater treatment and water reclamation[a]

Term	Phase	Process(es)
Absorption	Gas → liquid	Aeration, O_2 transfer, SO_2 scrubbing, chlorination, chlorine dioxide and ammonia addition, ozonation
Adsorption	Gas → solid Liquid → solid	Removal of inorganic and organic compounds using activated carbon, activated alumina, granular ferric hydroxide, or other adsorbent material
Distillation	Liquid → gas	Demineralization of water, concentrating of waste brines
Gas stripping	Liquid → gas	Removal of NH_3 and other volatile inorganic and organic chemicals
Ion exchange	Liquid → solid	Demineralization of water, removal of specific constituents, softening
Media filtration	Liquid → solid	Particle removal
Microfiltration, ultrafiltration	Liquid → liquid	Particle and colloidal species removal
Nanofiltration, electrodialysis, reverse osmosis	Liquid → liquid	Dissolved and colloidal species removal; softening
Precipitation	Liquid → solid	Softening and dissolved species removal
Sedimentation, flotation	Liquid → solid	Particle and dissolved species removal

[a] Adapted from Crittenden et al., 2005.

important to note that a key characteristic of all separation processes is the generation of a waste stream that will require subsequent management (e.g., processing and disposal). The type and characteristics of the waste stream generated depends on the type and effectiveness of the separation process used. For example, reverse osmosis (RO), described in Chap. 9, generates a liquid waste containing concentrated rejected constituents, adsorption results in a media saturated with trace constituents, and chemical precipitation produces a sludge containing both the precipitate compounds and the chemicals added to cause the precipitation. In many cases, the management of waste streams resulting from separation processes can present a significant technological challenge and cost. Separation processes that have been used for the removal of trace constituents in water reuse applications, introduced in this chapter, include adsorption (Sec. 10-2), ion exchange (Sec. 10-3), and distillation (Sec. 10-4). Additional details on the management of residual waste from separation processes may be found in Chap. 9 and Tchobanoglous et al. (2003).

Transformation of specific constituents from one form to another by chemical and biological conversion is the identifiable characteristic of the second group of processes used for the treatment of trace constituents. Processes that make use of reactions to transform or destroy trace constituents in water typically utilize oxidation and reduction

Chemical and Biological Transformation Processes

reactions. Conventional chemical oxidants that have been used for water reclamation include hydrogen peroxide, ozone, chlorine, chlorine dioxide, and potassium permanganate. Chemical oxidation processes that utilize hydroxyl radical species, referred to as advanced oxidation processes (AOPs), are particularly effective for the transformation and destruction of trace constituents, often resulting in the complete mineralization of trace constituents to carbon dioxide and mineral acids.

Photolysis processes result in both oxidation and reduction reactions. Under the proper conditions, biological processes can be used to treat a wide variety of chemical compounds found in postsecondary effluent. In contrast to the separation processes described above, transformation processes, particularly AOPs, have the potential to remove trace constituents without the formation of any residual waste streams that requires further processing or disposal. Transformation processes introduced in this chapter include chemical oxidation (Sec. 10-5), advanced oxidation (Sec. 10-6), photolysis (Sec. 10-7), and advanced biological transformations (Sec. 10-8).

10-2 ADSORPTION

In water reclamation, adsorption is used for the removal of substances that are in solution by accumulating them on a solid phase. Adsorption is considered to be a mass transfer operation as a constituent is transferred from a liquid phase to a solid phase (see Table 10-2). The *adsorbate* is the substance that is being removed from the liquid or gas phase at the interface. The *adsorbent* is the solid, liquid, or gas phase onto which the adsorbate accumulates. Although adsorption is used at the air–liquid interface in the flotation process (see Sec. 8-6 in Chap. 8), only the case of adsorption at the liquid–solid interface is considered in this section. Activated carbon is the primary adsorbent used in adsorption processes. The basic concepts of adsorption are presented in this section along with elements of design and limitations of the adsorption process in the water reclamation process.

Applications for Adsorption

Adsorption treatment of reclaimed water is usually thought of as a polishing process for water that has already received normal biological treatment. Adsorption has been used for the removal of refractory organic constituents; residual inorganic constituents such as nitrogen, sulfides, and heavy metals; and odor compounds. Under optimum conditions, adsorption can be used to reduce the effluent COD to less than 10 mg/L.

Removal of Trace Organics
Adsorption is used for two principal water reclamation applications, the continuous removal of organics and as a barrier against the breakthrough of organics from other unit processes. In some cases adsorption is used for the control of precursors that may form toxic compounds during disinfection. The capacity to adsorb various compounds onto activated carbon is summarized in Table 10-3. As shown in Table 10-3, activated carbon is known to have a low adsorption affinity for low molecular weight polar organic compounds. If biological activity is low in the carbon contactor (see Sec. 10-8) or in

Table 10-3
Readily and poorly adsorbed organics on activated carbon[a]

Readily adsorbed organics	Poorly adsorbed organics

Aromatic solvents
  Benzene
  Toluene
  Nitrobenzenes
Chlorinated aromatics
  PCBs
  Chlorophenols
Polynuclear aromatics
  Acenaphthene
  Benzopyrenes
Pesticides and herbicides
  DDT
  Aldrin
  Chlordane
  Atrazine
Chlorinated nonaromatics
  Carbon tetrachloride
  Chloroalkyl ethers
  Trichloroethene
  Chloroform
  Bromoform
High molecular weight hydrocarbons
  Dyes
  Gasoline
  Amines
  Humics
	Low-molecular weight ketones, acids, and aldehydes
	Sugars and starches
	Very-high-molecular weight or colloidal organics
	Low-molecular weight aliphatics

[a]From Froelich (1978).

other biological unit processes, low molecular weight polar organic compounds may be difficult to remove with activated carbon.

Removal of Metals

It is possible to remove some metals from wastewater, particularly industrial wastewater, using adsorption. The type of base material used and the method of activation have a strong influence on the potential of an adsorbent material to remove metals effectively. For example, adsorbents of coke, a byproduct derived from high sulfur coal and activated with phosphoric acid, were found to have a high affinity for mercury and silver (Zamora et al., 2000).

Types of Adsorbents

Treatment with adsorbent materials involves either (1) passing a liquid to be treated through a bed of adsorbent material held in a reactor/contactor or (2) blending the adsorbent material into a unit process followed by sedimentation or filtration for removal of the spent adsorbent. The principal types of adsorbents include activated carbon, granular ferric hydroxide (GFH), and activated alumina. Carbon-based adsorbents are used most commonly for reclaimed water adsorption because of their relatively low cost.

Table 10-4

Comparison of various adsorbent materials[a]

Parameter	Unit	Activated carbon Granular (GAC)	Activated carbon Powdered (PAC)	Activated alumina	Granular ferric hydroxide
Total surface area	m²/g	700–1300	800–1800	300–350	250–300
Bulk density	kg/m³	400–500	360–740	0.641–0.960	1.22–1.29
Particle density, wetted in water	kg/L	1.0–1.5	1.3–1.4	3.97	1.59
Particle size range	μm	100–2400	5–50	290–500	320–2000
Effective size	mm	0.6–0.9	na		
Uniformity coefficient	UC	≤1.9	na		
Mean pore radius	Å	16–30	20–40		
Iodine number		600–1100	800–1200		
Abrasion number	Minimum value	75–85	70–80		
Ash	%	≤ 8	≤ 6		
Moisture as packed	%	2–8	3–10		

[a]Specific values will depend on the source material used for the production of the activated carbon.

Other adsorbents that may prove to be effective with further research include manganese greensand, manganese dioxide, hydrous iron oxide particles, and iron oxide coated sand. Regardless of the adsorbent selected for a particular application, pilot testing is necessary for determination of process performance and design parameters. The characteristics of materials used for adsorption are summarized in Table 10-4.

Activated Carbon

Activated carbon is derived by subjecting an organic base material, such as wood, coal, almond, coconut, or walnut hulls to a pyrolysis process followed with activation by exposure to oxidizing gases such as steam and CO_2 at high temperatures. The resulting carbon structure is porous, as illustrated on Fig. 10-2, with a large internal surface area. The resulting pore sizes are defined as follows:

    Macropores >500 Å

    Mesopores >20 Å and <500 Å

    Micropores <20 Å

The surface properties, pore size distribution, and regeneration characteristics that result are a function of both the initial material used and the preparation procedure, therefore many variations are possible. The two size classifications of activated carbon are powdered activated carbon (PAC), which typically has a diameter of less than 0.074 mm (200 sieve) and is added directly to the activated sludge process or solids

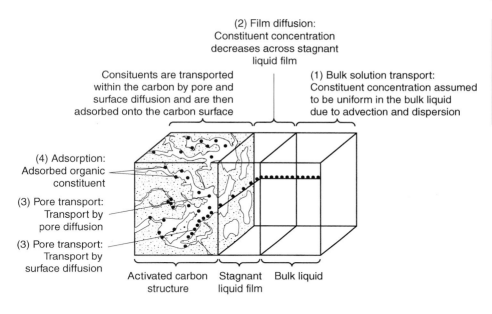

Figure 10-2
Definition sketch for the adsorption of an organic constituent onto an activated carbon particle.

contact processes, and granular activated carbon (GAC), which has a diameter greater than 0.1 mm (~140 sieve) and is used in pressure or gravity filtration.

Granular Ferric Hydroxide

Granular ferric hydroxide (GFH) is manufactured from a ferric chloride solution by neutralization and precipitation with sodium hydroxide. The adsorption capacity of GFH depends on water quality parameters, including pH, temperature, and other constituents in the water. Constituents that have been removed using GFH include arsenic, chromium, selenium, copper, and other metals. Process performance is reduced by suspended solids, and precipitated iron and manganese, and by constituents that compete for adsorption sites including organic matter and other ions (e.g., phosphate, silicate, sulfate). While GFH adsorbents can be effective from a performance standpoint for removal of specific constituents (e.g., arsenic), the cost associated with the process is often prohibitive for large systems. The adsorption capacity of GFH media is reduced significantly following regeneration, thus after reaching capacity, GFH adsorbents are typically disposed of in a landfill and replaced with new media. However, because GFH is not regenerated, the costs associated with management of the waste regenerant can be avoided, making the process viable in some situations, especially where the waste regenerant must be handled as a hazardous waste.

Activated Alumina

Activated alumina is derived from a naturally occurring mineral processed from bauxite that has been treated to remove molecules of water from its crystalline structure. Activated alumina is used in drinking water treatment for the removal of arsenic and fluoride (Clifford, 1999), and may have application in water reclamation for specific constituents. Activated alumina can be regenerated with a strong base followed by a strong acid. The regeneration of activated alumina and subsequent waste management issues result in significant operation and maintenance costs. As mentioned for GFH, pH (best performance at pH of 5.5 to 6), temperature, and competing constituents affect the performance of activated alumina adsorption. The use of powdered activated alumina coupled with membranes (microfiltration and ultrafiltration) is also a promising treatment process that is being investigated currently (2006).

Basic Considerations for Adsorption Processes

The adsorption process, as illustrated on Fig. 10-2, takes place in four more or less definable steps: (1) bulk solution transport, (2) film diffusion transport, (3) pore and surface transport, and (4) adsorption (or sorption). The adsorption step involves the attachment of the material to be adsorbed to the adsorbent at an available adsorption site (Snoeyink and Summers, 1999). Additional details on the physical and chemical forces involved in the adsorption process may be found in Crittenden et al. (2005). Adsorption can occur on the outer surface of the adsorbent and in the macropores, mesopores, micropores, and submicropores, but the surface area of the macro and mesopores is small compared with the surface area of the micropores and submicropores and the amount of material adsorbed there is usually considered negligible.

Because the adsorption process occurs in a series of steps, the slowest step in the series is identified as the rate-limiting step. When the rate of adsorption equals the rate of desorption, equilibrium has been achieved and the capacity of the adsorbent has been reached. The theoretical adsorption capacity for a given adsorbent for a particular contaminant can be determined by developing adsorption isotherms, as described below. Because activated carbon is the most common adsorbent used in advanced water reclamation applications, the focus of the following discussion is on activated carbon.

Adsorption Isotherms

The quantity of adsorbate that can be taken up by an adsorbent is a function of both the characteristics and concentration of adsorbate and the temperature. The characteristics of the adsorbate that are of importance include solubility, molecular structure, molecular weight, polarity, and hydrocarbon saturation. Generally, the amount of material adsorbed is determined as a function of the concentration at a constant temperature, and the resulting function is called an adsorption isotherm.

Adsorption isotherms are developed by exposing a given amount of adsorbate in a fixed volume of liquid to varying amounts of activated carbon. Typically, ten or more containers are used, each containing a different mass of activated carbon. The minimum time allowed for the samples to equilibrate, where powdered activated carbon is used, is 7 d. If GAC is used, it is usually powdered to minimize adsorption times. At the end of the test period, the amount of adsorbate remaining in solution is measured. The adsorbent phase concentration values after equilibrium, which are computed using Eq. (10-1), are used to develop adsorption isotherms, as described below.

$$q_e = \frac{(C_o - C_e)\,V}{m} \tag{10-1}$$

where q_e = adsorbent phase concentration after equilibrium, mg adsorbate/g adsorbent
C_o = initial concentration of adsorbate, mg/L
C_e = equilibrium concentration of adsorbate after absorption has occurred, mg/L
V = volume of liquid in the reactor, L
m = mass of adsorbent, g

Equations often used to describe the experimental isotherm adsorption data were developed by Freundlich; Langmuir; and Brunauer, Emmet, and Teller (BET). The two primary methods used for predicting the adsorption capacity of a given material are known as the Freundlich and Langmuir isotherms (Shaw, 1966).

Freundlich Isotherm The Freundlich isotherm is used most commonly to describe the adsorption characteristics of the activated carbon used in water, wastewater, and reclaimed water treatment. Derived empirically in 1912, the Freundlich isotherm is defined as follows:

$$\frac{x}{m} = q_e = K_f C_e^{1/n} \tag{10-2}$$

where x/m = mass of adsorbate adsorbed per unit mass of adsorbent after equilibrium, mg adsorbate/g activated carbon
K_f = Freundlich capacity factor, (mg absorbate/g activated carbon) × (L water/mg adsorbate)$^{1/n}$
1/n = Freundlich intensity parameter
Other terms as defined previously.

The constants in the Freundlich isotherm can be determined by plotting log (x/m) versus log C_e and making use of the linearized form of Eq. (10-2) rewritten as:

$$\log\left(\frac{x}{m}\right) = \log K_f + \frac{1}{n} \log C_e \tag{10-3}$$

Adsorption isotherms have been developed for a variety of organic compounds, some of which are presented in Table 10-5. As shown in Table 10-5, the variation in the Freundlich capacity factor for the various compounds is extremely wide (e.g., 14,000 for PCB to 6.8×10^{-5} for N-dimethylnitrosamine). Because of the variation, the Freundlich capacity factor must be determined for each new compound. Application of the Freundlich adsorption isotherm is illustrated in Example 10-1, following discussion of the Langmuir isotherm.

Langmuir Isotherm Derived from rational considerations, the Langmuir adsorption isotherm is defined as:

$$\frac{x}{m} = q_e = \frac{abC_e}{1 + bC_e} \tag{10-4}$$

where a, b = empirical constants
Other terms as defined previously.

Table 10-5
Freundlich adsorption isotherm constants for selected organic compounds[a,b]

Compound	pH	K_f (mg/g)(L/mg)$^{1/n}$	$1/n$
Benzene	5.3	1.0	1.6–2.9
Bromoform	5.3	19.6	0.52
Carbon tetrachloride	5.3	11	0.83
Chlorobenzene	7.4	91	0.99
Chloroethane	5.3	0.59	0.95
Chloroform	5.3	2.6	0.73
DDT	5.3	322	0.50
Dibromochloromethane	5.3	4.8	0.34
Dichlorobromomethane	5.3	7.9	0.61
1, 2-Dichloroethane	5.3	3.6	0.83
Ethylbenzene	7.3	53	0.79
Heptachlor	5.3	1220	0.95
Hexachloroethane	5.3	96.5	0.38
Methylene chloride	5.3	1.3	1.16
N-Dimethylnitrosamine	na	6.8×10^{-5}	6.60
N-Nitrosodi-n-propylamine	na	24	0.26
N-Nitrosodiphenylamine	3–9	220	0.37
PCB	5.3	14,100	1.03
PCB 1221	5.3	242	0.70
PCB 1232	5.3	630	0.73
Phenol	3–9	21	0.54
Tetrachloroethylene	5.3	51	0.56
Toluene	5.3	26.1	0.44
1, 1, 1-Trichloroethane	5.3	2–2.48	0.34
Trichloroethylene	5.3	28	0.62

[a]Adapted from Dobbs and Cohen (1980) and LaGrega et al. (2001).
[b]The adsorption isotherm constants reported in this table are meant to be illustrative of the wide range of values that are encountered for various organic compounds. It is important to note that the characteristics of the activated carbon and the analytical technique used for the analysis of the residual concentrations of the individual compounds has a significant effect on the coefficient values obtained for specific organic compounds.

The Langmuir adsorption isotherm was developed by assuming: (1) a fixed number of accessible sites are available on the adsorbent surface, all of which have the same energy, and (2) adsorption is reversible. Equilibrium is reached when the rate of adsorption of molecules onto the surface is the same as the rate of desorption of molecules from the surface. The rate at which adsorption proceeds is proportional to the driving force, which is the difference between the amount adsorbed at a particular concentration and the amount that can be adsorbed at that concentration. At the equilibrium concentration, this difference is zero.

Correspondence of experimental data to the Langmuir equation does not mean that the stated assumptions are valid for the particular system being studied because departures from the assumptions can have a canceling effect. The constants in the Langmuir isotherm can be determined by plotting $C_e/(x/m)$ versus C_e and making use of the linearized form of Eq. (10-4) rewritten as:

$$\frac{C_e}{(x/m)} = \frac{1}{ab} + \frac{1}{a}C_e \qquad (10\text{-}5)$$

Application of the Langmuir adsorption isotherm is illustrated in Example 10-1.

EXAMPLE 10-1. Analysis of Activated-Carbon Adsorption Data.

Determine which isotherm equation (i.e., Freundlich or Langmuir) best fits the isotherm coefficients for the following GAC adsorption test data. Also determine the corresponding coefficients for the isotherm equation. The liquid volume used in the batch adsorption tests was 1 L. The initial concentration, C_o, of the adsorbate in solution was 3.37 mg/L. Equilibrium was obtained after 7 d.

Mass of GAC, m, g	Equilibium concentration of adsorbate in solution, C_e, mg/L
0.0	3.37
0.001	3.27
0.010	2.77
0.100	1.86
0.500	1.33

Solution

1. Derive the values needed to plot the Freundlich and Langmuir adsorption isotherms using the batch adsorption test data.

Adsorbate concentration, mg/L			m, g	x/m,[a] mg/g	$C_e/(x/m)$
C_o	C_e	$C_o - C_e$			
3.37	3.37	0.00	0.000	—	—
3.37	3.27	0.10	0.001	100	0.0327
3.37	2.77	0.60	0.010	60	0.0462
3.37	1.86	1.51	0.100	15.1	0.1232
3.37	1.33	2.04	0.500	4.08	0.3260

[a] $\dfrac{x}{m} = \dfrac{(C_o - C_e)V}{m}$

2. Plot the Freundlich and Langmuir adsorption isotherms using the data developed in Step 1 and determine which isotherm best fits the data.
   a. The required plots are given below.

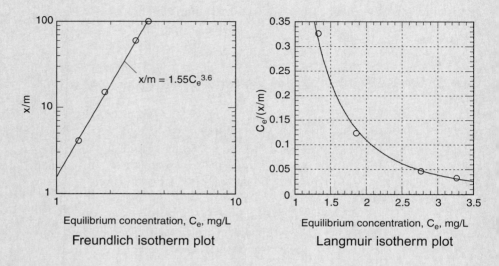

Freundlich isotherm plot

Langmuir isotherm plot

   b. From the above plots, the experimental data are best represented by the Freundlich isotherm. Because the plot for the Langmuir isotherm is curvilinear, use of the Langmuir adsorption isotherm is inappropriate.
3. Determine the Freundlich adsorption isotherm coefficients.
   a. When x/m versus C_e is plotted on log-log paper, the intercept on the x/m axis when $C_e = 1.0$ is the value of K_f and the slope of the line is equal to $1/n$. Thus, $x/m = 1.55$, and $K_f = 1.55$.
   b. When $x/m = 1.0$, $C_e = 0.89$, and $1/n = 3.6$.
   c. The form of the resulting isotherm is:

$$\frac{X}{m} = 1.55 C_e^{3.6}.$$

   d. The Freundlich adsorption isotherm equation may also be determined using a power-type best fit through the data.

Adsorption of Mixtures

In the application of adsorption in water reclamation, mixtures of organic compounds in reclaimed water are always encountered. Typically, there is a depression of the adsorptive capacity of any individual compound in a solution of many compounds, but the total adsorptive capacity of the adsorbent may be larger than the adsorptive capacity with a single compound. The amount of inhibition due to competing compounds is related to the size of the molecules being adsorbed, their adsorptive affinities, and their relative concentrations. It is important to note that adsorption isotherms can be determined for a heterogeneous mixture of compounds including total organic carbon (TOC), dissolved organic carbon (DOC), chemical oxygen demand (COD), dissolved organic halogen (DOH), UV absorbance, and fluorescence (Snoeyink and Summers, 1999). The adsorption from mixtures is considered further in Crittenden et al. (1987a, 1987b, 1987c, 1985) and Sontheimer and Crittenden (1988).

Adsorption Capacity

The adsorptive capacity of a given adsorbent is estimated from isotherm data as follows. If isotherm data are plotted, the resulting isotherm is as shown on Step 2 of Example 10-1. As shown on Fig. 10-3, the adsorptive capacity of the carbon can be estimated by extending a vertical line from the point on the horizontal axis corresponding to the initial concentration C_o, and extrapolating the isotherm to intersect this line. The $q_e = (x/m)_{c_o}$ value at the point of intersection can be read from the vertical axis. The $(q_e)_{c_o}$ value represents the amount of constituent adsorbed per unit weight of carbon when the carbon is at equilibrium with the initial concentration of constituent, C_o. The equilibrium condition generally exists in the upper section of a carbon bed during column treatment, and it therefore represents the ultimate capacity of the carbon for a particular reclaimed water. The value of the breakthrough adsorption capacity $(x/m)_b$ can be determined using the small-scale column test described later in this section. Several equations have been developed to describe the breakthrough curve including those by Bohart and Adams (1920) and Crittenden et al. (1987a).

Mass Transfer Zone

The area of the GAC bed in which sorption is occurring is called the mass transfer zone (MTZ), as shown on Fig. 10-4. After the water containing the constituent to be removed passes through a region of the bed whose depth is equal to the MTZ, the concentration of the contaminant in the water is reduced to its minimum value. No further adsorption occurs within the bed below the MTZ. As the top layers of carbon granules become saturated with organic material, the MTZ will move down in the bed until breakthrough occurs. Typically, breakthrough is said to have occurred when the effluent concentration

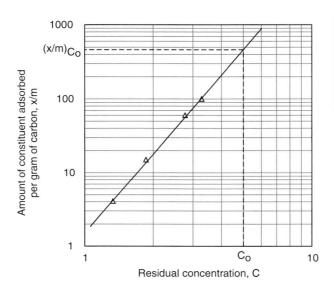

Figure 10-3

Plot of Freundlich isotherm used for determination of breakthrough adsorption capacity.

Figure 10-4
Typical breakthrough curve for activated carbon showing movement of mass transfer zone (MTZ) with throughput volume.

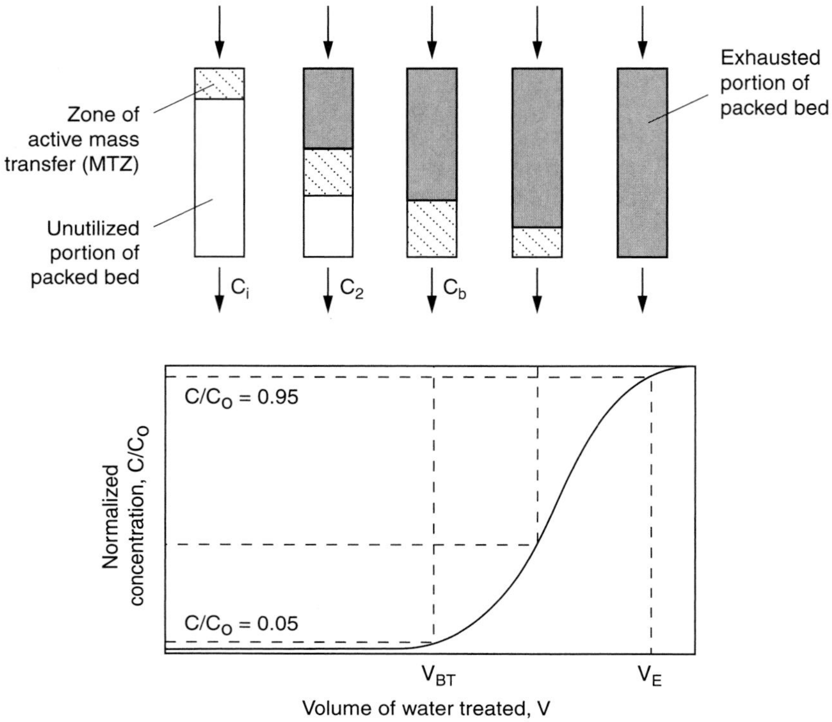

reaches five percent of the influent value. Exhaustion of the adsorption bed is assumed to have occurred when the effluent concentration is equal to 95 percent of the influent concentration. The volume of a given water processed until breakthrough and exhaustion is designated as V_{BT} and V_E, respectively, as shown on Fig. 10-4. The length of the MTZ is typically a function of the hydraulic loading rate applied to the column and the characteristics of the activated carbon. In the extreme, if the loading rate is too great, the length of the MTZ may be larger than the GAC bed depth, and the adsorbable constituents will not be removed completely by the carbon. At complete exhaustion, the effluent concentration is equal to the influent concentration.

In addition to the applied hydraulic loading rate, the shape of the breakthrough curve also depends on whether the applied liquid contains nonadsorbable and biodegradable constituents. The impact of the presence of nonadsorbable and biodegradable organic constituents on the shape of the breakthrough curve is illustrated on Fig. 10-5. As shown on Fig. 10-5, if the liquid contains nonadsorbable constituents, the nonadsorbable constituents appear in the effluent as soon as the carbon column is put into operation. If adsorbable and biodegradable constituents are present in the applied liquid, the breakthrough curve does not reach a C/C_o value of 1.0, but is depressed, and the observed C/C_o value depends on the biodegradability of the influent constituents, because biological activity continues even though the adsorption capacity has been utilized.

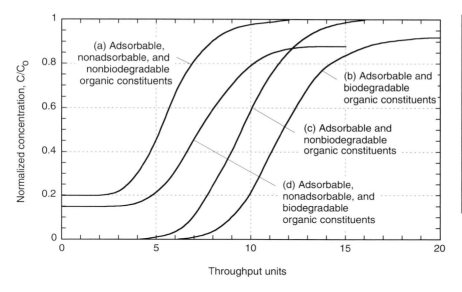

Figure 10-5
Impact of the presence of adsorbable, nonadsorbable, and biodegradable organic constituents on the shape of the activated carbon breakthrough curve. (Adapted from Snoeyink and Summers, 1999.)

If the liquid contains nonadsorbable and biodegradable constituents, the observed breakthough curve does not start at zero and does not terminate at a value of 1.0 (Snoeyink and Summers, 1999). The above effects are observed commonly in reclaimed water adsorption applications, especially with respect to the removal of COD.

In practice, the only way to use the capacity at the bottom of the carbon adsorption column is to have two or more columns in series and switch them as they are exhausted, or to use multiple columns in parallel so that breakthrough in one column does not effect effluent quality. The arrangement of adsorption columns in series and parallel configurations is shown on Figs. 10-6a and 10-6b, respectively. A minimum of two parallel or series carbon contactors is recommended for design. Multiple units permit one or more units to remain in operation while one unit is taken out of service for removal and regeneration of spent carbon, or for maintenance. The optimum flowrate and bed depth, and the operating capacity of the carbon, must be established to determine the dimensions and the number of columns necessary for continuous treatment. These parameters can be determined from dynamic column tests, as discussed below.

Adsorption Contactors

Several types of activated carbon contactors are used for trace constituent removal, including fixed and expanded beds, addition of PAC to activated sludge process, and separate mixed systems with subsequent carbon separation, as summarized in Table 10-6. A typical pressurized, downflow carbon contactor is shown on Fig. 10-7.

The sizing of carbon contactors is based on a number of factors, as summarized in Table 10-7 for a downflow packed bed contactor. For the case where the mass transfer

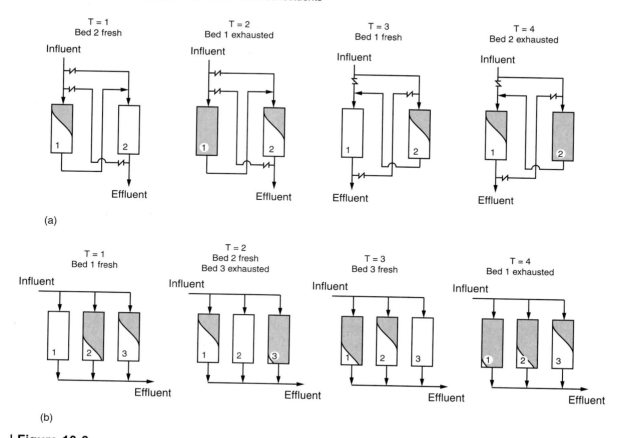

Figure 10-6
Activated carbon contactor configurations: (a) series and (b) parallel operation.

rate is fast and the mass transfer zone is a sharp wave front, a steady-state mass balance around a fixed bed carbon contactor may be written as:

$$\text{Accumulation} = \text{inflow} - \text{outflow} - \text{amount adsorbed}$$
$$0 = QC_o t - QC_e t - m_{GAC} q_e \qquad (10\text{-}6)$$

where Q = volumetric flowrate, L/h
C_o = initial concentration of adsorbate, mg/L
t = time, h
C_e = final equilibrium concentration of adsorbate, mg/L
m_{GAC} = mass of adsorbent, g
q_e = adsorbent phase concentration after equilibrium, mg adsorbate/g adsorbent

From Eq. (10-6), the adsorbent usage rate is defined as:

$$\frac{m_{GAC}}{Qt} = \frac{C_o - C_e}{q_e} \qquad (10\text{-}7)$$

Table 10-6
Application of activated carbon for the removal of trace constituents from wastewater

Configuration	Description
Fixed bed GAC column 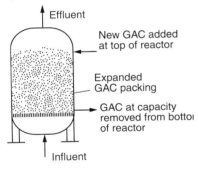	Fixed-bed downflow columns can be operated singly, in series, or in parallel (see Fig. 10-6). Granular-medium filters are used commonly upstream of the activated carbon contactors to remove the organics associated with the suspended solids present in secondary effluent; however, the adsorption of organics and filtration of suspended solids can also be accomplished in a single step. In the downflow design, the water to be treated is applied to the top of the column and withdrawn at the bottom. The carbon is held in place with an underdrain system at the bottom of the column. Provision for backwashing and surface washing is often provided in wastewater applications to limit the headloss buildup due to the removal of particulate suspended solids within the carbon column. Unfortunately, backwashing has the effect of destroying the adsorption front. Although upflow fixed-bed reactors have been used, downflow beds are used more commonly to lessen the chance of accumulating particulate material in the bottom of the bed, where the particulate material would be difficult to remove by backwashing.
Expanded bed GAC (upflow) 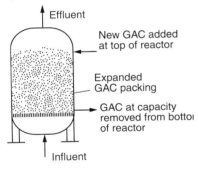	In the expanded (or fluidized) bed system, the influent is introduced at the bottom of the column and the activated carbon is allowed to expand, much as a filter bed expands during backwash. When the adsorptive capacity of the carbon at the bottom of the column is exhausted, the bottom portion of carbon is removed, and an equivalent amount of regenerated or virgin carbon is added to the top of the column. In such a system, headloss does not build up with time after the operating point has been reached. In general, expanded-bed upflow contactors may have more carbon fines in the effluent than downflow contactors because bed expansion leads to the creation of fines as the carbon particles collide and abrade, and allows the fines to escape through passageways created by the expanded bed. While not used commonly, continuous backwash moving-bed and pulsed-bed carbon contactors have been used (see Table 8-4 for filter configurations).
Activated sludge with PAC addition 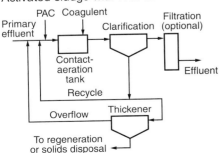	The use of powdered activated carbon (PAC) with the activated sludge process, where activated carbon is added directly to the aeration tank, results in simultaneous biological oxidation and physical adsorption. A feature of this process is that it can be integrated into existing activated sludge systems at nominal capital cost. The addition of PAC has several process advantages, including: (1) system stability during shock loads, (2) reduction of refractory priority pollutants, (3) color and ammonia removal, and (4) improved sludge settleability. In some industrial waste applications where nitrification is inhibited by toxic organics, the application

(Continued)

Table 10-6

Application of activated carbon for the removal of trace constituents from wastewater (*Continued*)

Configuration	Description
Activated sludge with PAC addition (*Continued*)	of PAC may reduce or limit this inhibition. Carbon dosages typically range from 20 to 200 mg/L. With higher solids retention time (SRT) values, the organic removal per unit of carbon is enhanced, thereby improving the process efficiency. Reasons cited for this phenomenon include: (1) additional biodegradation due to decreased toxicity, (2) degradation of normally nondegradable substances due to increased exposure time to the biomass through adsorption on the carbon, and (3) replacement of low molecular weight compounds with high molecular weight compounds, resulting in improved adsorption efficiency and lower toxicity.
Mixed PAC contactor with gravity separation 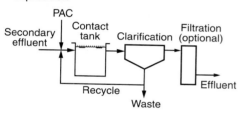	Powdered activated carbon can be applied to the effluent from biological treatment processes in a separate contacting basin. The contactor can operate in a batch or continuous flow mode. In the batch mode, after a specified amount of time for contact, the carbon is allowed to settle to the bottom of the tank, and the treated water is then removed from the tank. The continuous flow operation consists of a basin divided for contacting and settling. The settled carbon may be recycled to the contact tank. Because carbon is very fine, a coagulant, such as a polyelectrolyte, may be needed to aid in the removal of the carbon particles, or filtration through rapid sand filters may be required. In some treatment processes, PAC is used in conjunction with chemicals for the precipitation of specific constituents.
Mixed PAC contactor with membrane separation 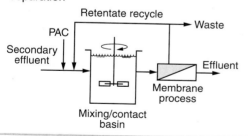	The removal of trace constituents in a complete mix or plug flow contactor may be combined with separation by MF or UF membranes. The PAC is added to the secondary effluent by continuous or pulse addition, followed by concentration of the PAC on the membrane. When the headloss across the membrane reaches a given value, a backwash cycle is initiated. The backwash containing the PAC retentate may be wasted or recycled to the contact basin. A number of full-scale plants have used this process (Snoeyink et al., 2000, Anselme et al., 1997).

If it is assumed that the mass of the adsorbate in the pore space is small compared to the amount adsorbed, then the term $QC_e t$ in Eq. 10-7 can be neglected without serious error and the adsorbent usage rate is given by:

$$\frac{m_{GAC}}{Qt} \approx \frac{C_o}{q_e} \tag{10-8}$$

To quantify the operational performance of GAC contactors, the following terms have been developed and are used commonly.

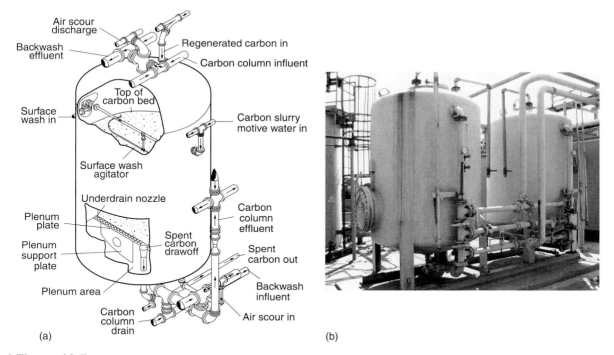

Figure 10-7
Activated carbon contactors: (a) illustration of typical pressure vessel contactor and (b) view of typical granular activated carbon contactors operated in parallel, used for the treatment of filtered secondary effluent.

Table 10-7
Typical design values for GAC contactors[a]

Parameter	Symbol	Unit	Value
Volumetric flow rate	V	m³/h	50–400
Bed volume	V_b	m³	10–50
Cross-sectional area	A_b	m²	5–30
Carbon depth	D	m	1.8–4
Void fraction	α	m³/m³	0.38–0.42
GAC density	ρ	kg/m³	350–550
Approach velocity	V_f	m/h	5–15
Effective contact time	t	min	2–10
Empty bed contact time	EBCT	min	5–30
Operation time	t	d	100–600
Throughput volume	V_L	m³	10–100
Specific throughput	V_{sp}	m³/kg	50–200
Bed volumes	BV	m³/m³	2000–20,000

[a] Adapted from Sontheimer et al. (1988).

1. Empty bed contact time (EBCT)

$$\text{EBCT} = \frac{V_b}{Q} = \frac{A_b D}{v_f A_b} = \frac{D}{v_f} \qquad (10\text{-}9)$$

where EBCT = empty bed contact time, h
V_b = volume of contactor occupied by GAC, m³
Q = volumetric flowrate, m³/h
A_b = cross-sectional area of GAC filter bed, m²
D = length of GAC in contactor, m
v_f = linear approach velocity, m/h

2. Activated carbon density

The density of the activated carbon is defined as:

$$\rho_{GAC} = \frac{m_{GAC}}{V_b} \qquad (10\text{-}10)$$

where ρ_{GAC} = density of GAC, g/L
m_{GAC} = mass of GAC, g
V_b = volume of contactor occupied by GAC, L

3. Specific throughput, expressed as m³ of water treated per gram of carbon:

$$\text{Specific throughput, m}^3/\text{g} = \frac{Qt}{m_{GAC}} = \frac{V_b t}{\text{EBCT} \times m_{GAC}} \qquad (10\text{-}11)$$

Using Eq. (10-10), Eq. (10-11) can be written as:

$$\text{Specific throughput} = \frac{V_b t}{\text{EBCT}(\rho_{GAC} \times V_b)} = \frac{t}{\text{EBCT} \times \rho_{GAC}} \qquad (10\text{-}12)$$

4. Carbon usage rate (CUR) expressed as gram of carbon per m³ of water treated:

$$\text{CUR, g/m}^3 = \frac{m_{GAC}}{Qt} = \frac{1}{\text{Specific throughput}} \qquad (10\text{-}13)$$

5. Volume of water treated for a given EBCT, expressed in liters, L:

$$\text{Volume of water treated, L} = \frac{\text{Mass of GAC for given EBCT}}{\text{GAC usage rate}} \qquad (10\text{-}14)$$

6. Bed life, expressed in days, d:

$$\text{Bed life, d} = \frac{\text{Volume of water treated for given EBCT}}{Q} \qquad (10\text{-}15)$$

The application of these terms is illustrated in Example 10-2.

EXAMPLE 10-2. Estimation of Activated-Carbon Adsorption Breakthrough Time.

A fixed-bed activated carbon adsorber has a fast mass transfer rate and the mass transfer zone is essentially a sharp wave front. Assuming the following data apply, determine the carbon requirements to treat a flow of 1000 L/min, and the corresponding bed life.

1. Compound to be treated = trichloroethylene (TCE)
2. Initial concentration, C_o = 1.0 mg/L
3. Final concentration C_e = 0.005 mg/L
4. GAC density = 450 g/L
5. Freundlich capacity factor, K_f = 28 (mg/g)(L/mg)$^{1/n}$ (see Table 10-5)
6. Freundlich intensity parameter, $1/n$ = 0.62 (see Table 10-5)
7. EBCT = 10 min

Ignore the effects of biological activity within the column.

Solution

1. Estimate the GAC usage rate for TCE. The GAC usage rate is estimated using Eq. (10-7) and Eq. (10-2).

$$\frac{m_{GAC}}{Qt} = \frac{C_o - C_e}{q_e} = \frac{C_o - C_e}{K_f C_o^{1/n}}$$

$$= \frac{(1.0 \text{ mg/L})}{[28 \text{ (mg/g)(L/mg)}^{0.62}](1.0 \text{ mg/L})^{0.62}}$$

$$= 0.036 \text{ g GAC/L}$$

2. Determine the mass of carbon required for a 10 min EBCT.

The mass of GAC in the bed = $V_b \rho_{GAC}$ = EBCT × Q × ρ_{GAC}

Carbon required = 10 min (1000 L/min) (450 g/L) = 4.5 × 10^6 g

3. Determine the volume of water treated using a 10 min EBCT.

$$\text{Volume of water treated} = \frac{\text{Mass of GAC for given EBCT}}{\text{GAC usage rate}}$$

$$\text{Volume of water treated} = \frac{4.5 \times 10^6 \text{ g}}{(0.036 \text{ g GAC/L})} = 1.26 \times 10^8 \text{ L}$$

4. Determine the bed life.

$$\text{Bed life} = \frac{\text{Volume of water treated for given EBCT}}{Q}$$

$$\text{Bed life} = \frac{1.26 \times 10^8 \text{ L}}{(1000 \text{ L/min})(1440 \text{ min/d})} = 87.5 \text{ d}$$

Comment
In this example, the full capacity of the carbon in the contactor was utilized based on the assumption that two columns in series are used. If a single column is used, then a breakthrough curve must be used to determine the bed life.

Bench Scale Tests

Over the years, several bench scale tests have been developed to simulate the results obtained with full scale reactors. One of the early column tests was the high-pressure minicolumn (HPMC) technique developed by Rosene et al. (1983), and later modified by Bilello and Beaudet (1983). In the HPMC test procedure, a high-pressure liquid chromatography (HPLC) column loaded with activated carbon is used. Typically the HPMC test procedure is used to determine the capacity of activated carbon for the adsorption of volatile organic compounds. The principal advantage of the HPMC test procedure is that it allows for the rapid determination of the GAC adsorptive capacity under conditions similar to those encountered in the field.

An alternative procedure known as the rapid small-scale column test (RSSCT) has been developed by Crittenden et al. (1991). The test procedure allows for the scaling of data obtained from small columns (see Fig. 10-8) to predict the performance of pilot- or full-scale carbon columns. In applying the procedure, mathematical models are used to define the relationships between the breakthrough curve for small and large columns for

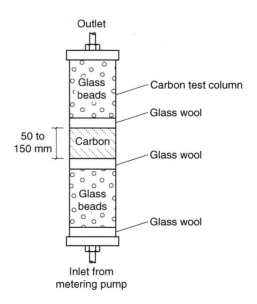

Figure 10-8
Schematic of column used for rapid small-scale column testing (RSSCT) to develop data for pilot- or full-scale carbon columns (Adapted from Crittenden et al., 1991).

process scale-up. Even with HPMC or RSSCT evaluation, pilot-scale testing is still recommended for determination of process design criteria under actual operating conditions.

Carbon Regeneration
In many situations, the economical application of activated carbon depends on an efficient means of regenerating and reactivating the carbon after its adsorptive capacity has been reached. Regeneration is the term used to describe all of the processes that are used to recover the adsorptive capacity of the spent carbon, exclusive of reactivation. Typically, some of the adsorptive capacity of the carbon (about 4 to 10 percent) is lost in the regeneration process, while a loss of 2 to 5 percent is expected during the reactivation process, and a 4 to 8 percent loss of carbon is assumed due to attrition, abrasion, and mishandling. In general, regenerated activated carbon is not used in reclaimed water applications because of the potential for residual constituents, not removed in the regeneration process to desorb and contaminate the reclaimed water. Additional details on carbon reactivation and regeneration may be found in Sontheimer and Crittenden (1988).

Adsorption Process Limitations

The adsorption process in water reuse applications is limited by: (1) the logistics involved with transport of large volumes of adsorbent materials, (2) the area requirements for the carbon contactors, and (3) the production of waste adsorbent that can be difficult to regenerate and may need to be disposed of as hazardous waste due to the presence of toxic constituents. Further, the regeneration of some adsorbents is not feasible, resulting in potentially high media replacement costs. Process monitoring and control is essential, as the performance of carbon contactors is affected by variations in pH, temperature, and flowrate.

10-3 ION EXCHANGE

In water reclamation applications, ion exchange involves the replacement of an ion in the aqueous phase for an ion in a solid phase. The solid phase ion exchange material is insoluble and can be of natural origin such as kaolinite and montmorillonite minerals, or a synthetic material such as a polymeric resin. The exchange materials have fixed charged functional groups located on their external and/or internal surface, and associated with these groups are ions of opposite charge called "counter ions" (See Fig. 10-9). The mobile counter ions are associated by electrostatic attraction to each of the charged functional groups to satisfy the criterion that electroneutrality is maintained at all times within the exchange material as well as in the bulk aqueous solution. Depending on the charge of the functional group on the exchanger, the counter ion can either be a cation if the functional group is negative or an anion if the functional group is positive and can exchange with another counter ion in the aqueous phase. Further, ion exchange resins have an affinity or selectivity for certain counter ions in water, which affects the process performance. In general, a higher selectivity is exhibited for counter ions with the higher charge. However, as with most advanced treatment processes, modeling has limited value and bench and pilot studies should be conducted to determine the actual design and operational parameters.

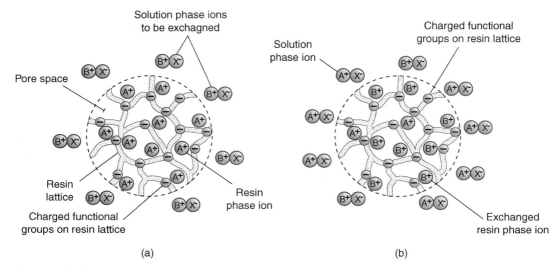

Figure 10-9
Schematic framework of a functional cation exchange resin: (a) resin initially immersed in an aqueous solution containing B⁺ cations and X⁻ anions and (b) cation exchange resin in equilibrium with the aqueous solution of B⁺ cations and X⁻ anions. (From Crittenden et al., 2005.)

Applications for Ion Exchange

The most widespread use of the ion exchange process is in domestic water softening, where sodium ions from a cationic exchange resin are exchanged for the calcium and magnesium ions in the water to be treated, thus reducing the hardness. Ion exchange has been used in water reclamation applications for the removal of nitrogen, heavy metals, and total dissolved solids. In water reclamation applications, ion exchange may be used for: (1) the removal of specified ionic constituents, such as Na^+, Cl^-, SO_4^{2-}, NH_4^+, and NO_3^-; (2) water softening, e.g., removal of Ca^{2+} and Mg^{2+}; or (3) demineralization. Other applications include the removal of specific constituents such as barium, radium, arsenic, perchlorate, chromate, and potentially other constituents. The reactor configurations used for ion exchange are similar to those shown previously on Fig. 10-6 for adsorption.

Nitrogen Control

For nitrogen control, the ions typically removed from the waste stream are ammonium, NH_4^+, and nitrate, NO_3^-. The ion that ammonium displaces varies with the nature of the solution used to regenerate the bed. Although both natural and synthetic ion exchange resins are available, synthetic resins are used more widely because of their durability. Some natural resins (zeolites) have been used for the removal of ammonia from treated effluent for water reuse applications. Clinoptilolite, a naturally occurring zeolite, has proven to be one of the best natural exchange resins. In addition to having a greater affinity for ammonium ions than other ion exchange materials, it is relatively inexpensive when compared to synthetic media. Upon exhaustion, clinoptilolite is regenerated with lime [$Ca(OH)_2$] and the ammonium ion removed from the zeolite is converted to ammonia gas because of the high pH and is removed subsequently by air stripping.

The regenerant solution, stripped of ammonia gas, is collected in a storage tank for subsequent reuse. Excessive calcium carbonate precipitates may form within the zeolite exchange bed and in the stripping tower and piping appurtenances, and may require removal. The zeolite bed is equipped with backwash facilities to remove the carbonate deposits that form within the filter.

When using conventional synthetic ion exchange resins for the removal of nitrate, two problems are encountered. First, while most resins have a greater affinity for nitrate over chloride or bicarbonate, they have a significantly lower affinity for nitrate as compared to sulfate, which limits the useful capacity of the resin for the removal of nitrate. Second, because of the lower affinity for nitrate over sulfate, a phenomenon known as nitrate dumping can occur. Nitrate leaching (dumping) occurs when an ion exchange column is operated beyond the nitrate breakthrough point, at which time sulfate in the feed water will displace the nitrate on the resin, causing a release of nitrate into column effluent.

To overcome the problems associated with low affinity and nitrate breakthrough, new types of resins have been developed in which the affinities for nitrate and sulfate have been reversed. When significant amounts of sulfate are present (i.e., typically greater than 25 percent of the total of the sum of the sulfate and nitrate expressed in meq/L), the use of nitrate selective resins is advantageous. Because the performance of nitrate selective resins varies with the composition of the reclaimed water, pilot testing will usually be required (McGarvey et al., 1989; Dimotsis and McGarvey, 1995). Typical ion exchange test columns used to study the removal of nitrate from water, which has been processed with RO, are shown on Fig. 10-10.

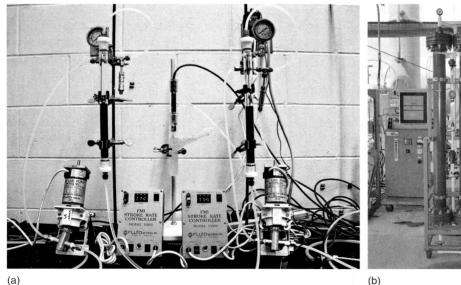

(a) (b)

Figure 10-10

Typical ion exchange test columns used to study the removal of specific constituents from reclaimed water: (a) bench scale (Courtesy of D. Hand) and (b) pilot scale.

Removal of Heavy Metals

Ion exchange is one of the most common forms of treatment used for the removal of metals. Materials used for the exchange of metals include zeolites, weak and strong anion and cation resins, chelating resins, and microbial and plant biomass (Ouki and Kavanaugh, 1999). Chelating resins, such as aminophosphonic and iminodiacetic resins have been manufactured to have a high selectivity for specific metals, such as Cu, Ni, Cd, and Zn. Ion exchange processes are highly pH dependent, as solution pH has a significant impact on the metal species present and the interaction between exchanging ions and the resin. Most metals bind better at higher pH, due to less competition from protons for exchange sites.

Removal of Total Dissolved Solids

For the reduction of the total dissolved solids (TDS), both anionic and cationic exchange resins must be used sequentially. The water to be demineralized is first passed through a cation exchanger where the positively charged ions are exchanged with the hydrogen ions on the resin. The effluent from the cation exchanger is then passed through an anionic exchange resin where the anions are exchanged with the hydroxide ions on the resin. Thus, the anions and cations are replaced by hydrogen and hydroxide ions, which react to form water molecules.

Total dissolved solids removal can take place in separate exchange columns arranged in series, or both resins can be mixed in a single reactor. In reclaimed water applications, rates range from 0.20 to 0.40 m/min (5 to 10 gal/ft^2·min). Typical bed depths are 0.75 to 2.0 m (2.5 to 6.5 ft). In water reuse applications, treatment of a portion of the reclaimed water by ion exchange, followed by blending with reclaimed water not treated by ion exchange, can be used to reduce the TDS to acceptable levels. In some cases, ion exchange may be as cost effective, if not more so, than RO, particularly where one or more few specific constituents need to be removed.

Removal of Organic Matter

Much of the organic matter in reclaimed water is highly ionized and can therefore be removed using ion exchange, primarily using anion exchange resins. The extent of removal is a function of several water quality and resin-specific parameters. Depending on the values of these parameters, a TOC reduction of 50 percent is typical with run lengths ranging from <500 to >5000 bed volumes [V_b from Eq. (10-9)] between regenerations.

Ion Exchange Materials

Important properties of ion exchange resins include exchange capacity, particle size, and stability. The exchange capacity of a resin is defined as the quantity of an exchangeable ion that can be taken up. The exchange capacity of resins is expressed as eq/L or eq/kg (meq/L or meq/kg). The particle size of a resin is important with respect to the hydraulics of the ion exchange column and the kinetics of ion exchange. In general, the rate of exchange is inversely proportional to the square of the particle diameter (i.e., surface area). The stability of a resin is important to the long-term performance of the resin. Excessive osmotic swelling and shrinking, chemical degradation, and structural changes in the resin caused by physical stresses are important factors that may limit the useful life of a resin.

Natural Ion Exchange Materials

Naturally occurring ion exchange materials, known as zeolites, are used for water softening and ammonium ion removal. Zeolites used for water softening are complex aluminosilicates with sodium as the mobile ion. Ammonium ion removal is often accomplished using a naturally occurring zeolite, clinoptilolite.

Synthetic Ion Exchange Materials

Most synthetic ion exchange materials are resins or phenolic polymers. Five types of synthetic ion exchange resins are in use: (1) strong-acid cation, (2) weak-acid cation, (3) strong-base anion, (4) weak-base anion, and (5) heavy-metal-selective chelating resins. Most synthetic ion exchange resins are manufactured by a process in which styrene and divinylbenzene are copolymerized. The styrene serves as the basic matrix of the resin, and divinylbenzene is used to cross link the polymers to produce an insoluble tough resin. Unlike zeolites, synthetic resin materials are highly resistant to regenerant solutions of mineral acids.

While much attention has been placed on the use of conventional synthetic ion exchange resins, research is ongoing to develop specialty resins that are selective for some contaminant ions, such as nitrate. Other developments in resin technology include the use of resins in a mixed slurry contact mode. The MIEX resin, developed in Australia for use in water treatment, contains a magnetized component such that the resin beads act as weak individual magnets (Hammann et al., 2004). In a sedimentation tank the magnetized resin beads readily aggregate, settle rapidly, and can be recovered for regeneration.

Basic Considerations for Ion Exchange Processes

Typical ion exchange reactions for natural and synthetic ion exchange materials are given in Table 10-8. Ion exchange processes can be operated in a batch or continuous mode. In a batch process, the resin is stirred with the water to be treated in a reactor until the reaction is complete. The spent resin is removed by settling, followed by batch regeneration, and is then reused. In a continuous process, the exchange material is placed in a bed or a packed column, and the water to be treated is passed through it. Continuous ion exchangers are usually of the downflow, packed-bed column type. Reclaimed water enters the top of the column under pressure, passes downward through the resin bed, and is removed at the bottom. When the resin capacity is exhausted, the column is backwashed to remove trapped solids and is then regenerated. An example of a full-scale ion exchange process, with ion exchange columns on a rotating platform to facilitate regeneration, is shown on Fig. 10-11. Additional details on principles of ion exchange may be found in Slater (1991).

Reported exchange capacities vary with the type and concentration of regenerant used to restore the resin. Typical synthetic resin exchange capacities are in the range of 1 to 5 meq/mL of resin. Zeolite cation exchangers have exchange capacities of 0.05 to 0.1 meq/mL. Exchange capacity is measured by placing the resin in a known form. For example, a cationic resin could be washed with a strong acid to place all of the exchange sites on the resin in the H^+ form or washed with a strong NaCl brine to place all of the exchange sites in the Na^+ form. A solution of known concentration of an exchangeable ion (e.g., Ca^{2+}) can then be added until exchange is complete and the

Table 10-8
Characteristics of ion exchange resins used in wastewater treatment processes[a]

Resin type	Acronym	Fundamental reaction[b]	Regenerant ions (X)	pK	Exchange capacity, meq/mL	Constituents removed
Strong acid cation	SAC	$n[RSO_3^-]X + M^{+n} \rightleftarrows n[RSO_3^-]M^{+n} + nX$	H^+ or Na^+	<0	1.7 to 2.1	H^+ form: any cation; Na^+ form: divalent cations
Weak acid cation	WAC	$n[RCOO^-]X + M^{+n} \rightleftarrows n[RCOO^-]M^{+n} + nX$	H^+	4 to 5	4 to 4.5	Divalent cations first, then monovalent cations until alkalinity is consumed
Strong base anion (type 1)	SBA-1[c]	$n[R(CH_3)_3 N^+]X + A^{-n} \rightleftarrows n[R(CH_3)_3 N^+]A^{-n} + nX$	OH^- or Cl^-	>13	1 to 1.4	OH^- form: any anion; Cl^- form: sulfate, nitrate, perchlorate, etc.
Stong base anion (type 2)	SBA-2[d]	$n[R(CH_3)_2 (CH_3CH_2OH)N^+]X + A^{-n} \rightleftarrows n[R(CH_3)_2 (CH_3CH_2OH)N^+]A^{-n} + nX$	OH^- or Cl^-	>13	2 to 2.5	OH^- form: any anion; Cl^- form: sulfate, nitrate, perchlorate, etc.
Weak base anion	WBA	$[R(CH_3)_2 N]HX + HA \rightleftarrows [R(CH_3)_2 N]HA + HX$	OH^-	5.7 to 7.3	2 to 3	Divalent anions first, then monovalent anions until strong acid is consumed

[a] From Crittenden et al. (2005)
[b] Term within the brackets represents the solid phase of the resin.
[c] Greater regeneration efficiency and capacity than SBA-2.
[d] Greater chemical stability than SBA-1.

Figure 10-11
Example of full-scale anion and cation exchange columns.

amount of exchange capacity can be measured or in the acid case, the resin is titrated with a strong base. Determination of the capacity of an ion resin by titration is illustrated in Example 10-3. Views of bench scale ion exchange columns used to evaluate resin capacity are shown on Fig. 10-10. Ion exchange processes are known to preferentially remove certain constituents (Anderson 1975, 1979), they will therefore, require pilot-scale testing to determine applicability.

Exchange capacities for resins often are expressed in milequivalents per milliliter (meq/mL), equivalents per cubic meter (eq/m^3), or in terms of grams CaCO$_3$ per cubic meter of resin (g/m^3). Conversion between exchange capacities is accomplished using the following expression:

$$\frac{1 \text{ meq}}{\text{mL}} = \frac{1 \text{ eq}}{\text{m}^3} = \frac{(1 \text{ eq})(50 \text{ g CaCO}_3/\text{eq})}{\text{m}^3} = 50 \text{ g CaCO}_3/\text{m}^3 \quad (10\text{-}16)$$

Calculation of the required resin volume for an ion-exchange process is also illustrated in Example 10-3.

EXAMPLE 10-3. Determination of Ion Exchange Capacity for a New Resin.

A column study was conducted to determine the capacity of a cation exchange resin. In conducting the study, 10 g of resin was washed with NaCl until the resin was in the R-Na form. The column was then washed with distilled water to remove the chloride ion (Cl$^-$) from the interstices of the resin. The resin was then titrated with a solution of calcium chloride (CaCl$_2$), and the concentrations of chloride and calcium were measured at various throughput volumes. The measured concentrations of Cl⁻ and Ca²⁺ and the corresponding throughput volumes are as given below. Using the data given below, determine the exchange capacity of the resin and the mass and volume of a resin required to treat 4000 m³/d of water containing 18 mg/L of ammonium ion NH_4^+. Assume the density of the resin is 700 kg/m³.

Throughput volume, L	Constituent, mg/L		Constituent, normalized concentration, C/C_o	
	Cl⁻	Ca²⁺	Cl⁻	Ca²⁺
2	0	0	0	0
3	trace	0	~0	0
5	7	0	0.099	0
6	18	0	0.253	0
10	65	0	0.915	0
12	71	trace	1	~0
20	71	13	1	0.325
26	71	32	1	0.8
28	71	38	1	0.95
32	$C_o = 71$	$C_o = 40$	1	1

Solution

1. Prepare a plot of the normalized concentrations of Cl⁻ and Ca²⁺ as a function of the throughput volume. The required plot is given below.

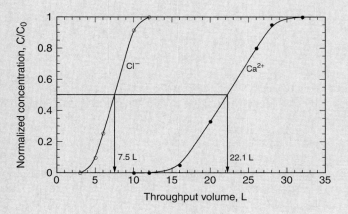

2. Determine the exchange capacity.

   The exchange capacity (EC) of the resin in meq/kg is:

   $$EC = \frac{(22.1 \text{ L} - 7.5 \text{ L})\left[\frac{(40 \text{ mg/L})}{(20 \text{ mg/meq})}\right]}{10 \text{ g of resin}} = 2.92 \text{ meq/g of resin}$$

3. Determine the mass and volume of resin required to treat 4000 m³ of water containing 18 mg/liter of ammonium ion NH_4^+.

   a. Determine the meq of NH_4^+.

   $$NH_4^+, \text{ meq/L} = \frac{(18 \text{ mg/L as } NH_4^+)}{(18 \text{ mg/meq})} = 1 \text{ meq/L}$$

   b. The required exchange capacity is equal to

   $$(1.0 \text{ meq/L})(4000 \text{ m}^3)(10^3 \text{ L/m}^3) = 4 \times 10^6 \text{ meq}$$

   c. The required mass of resin is:

   $$R_{mass}, \text{ kg} = \frac{4 \times 10^6 \text{ meq } (1 \text{ kg}/1000 \text{ g})}{(2.92 \text{ meq/g of resin})} = 1370 \text{ kg of resin}$$

   d. The required volume of resin is:

   $$R_{vol}, \text{ m}^3 = \frac{1370 \text{ kg of resin}}{(700 \text{ kg/m}^3)} = 1.96 \text{ m}^3 \text{ of resin}$$

Comment

In practice, because of leakage and other operational and design limitations, the required volume of resin will usually be about 1.1 to 1.4 times that computed on the basis of exchange capacity. Also, the above computation is based on the assumption that the entire capacity of the resin is utilized, which does not typically occur in full-scale applications.

Ion Exchange Process Limitations

The performance of the ion exchange process is impacted severely by the presence of particulate and colloidal matter, solvents, and organic polymers. To date, ion exchange has been used primarily in water reuse applications to remove specific constituents and for demineralization. However, because of the variability of the chemical composition of the reclaimed water matrix of reclaimed water, it is be necessary to conduct pilot-scale studies with the water to be treated to establish design criteria for full-scale applications.

Pretreatment Requirements

Residual organic matter found in biological treatment effluents can cause blinding of the ion exchange beds, resulting in high headloss and inefficient operation, where ion exchange is used following filtration, for example. Therefore, some form of chemical treatment and clarification is required before ion exchange treatment. Colloidal solids in secondary effluent that would otherwise accumulate in an ion exchange bed can also be removed using microfiltration or by using sacrificial exchange resins before application to the exchange column. As noted above, when ion exchange is used to remove specific constituents, extensive pretreatment is required to optimize process performance.

Regeneration

The key issue with resin regeneration is the potential for irreversible fouling (fouling that can not be removed by regeneration). To make ion exchange economical for advanced wastewater treatment, it would be desirable to use regenerants and restorants that would remove both the inorganic anions and cations and the organic material from the spent resin. Chemical and physical restorants found to be successful in the removal of organic material from resins include sodium hydroxide, hydrochloric acid, methanol, and bentonite. The quantity and quality of regenerate produced and subsequently requiring management must also be considered when selecting regenerants.

Brine Management

The management of regeneration brines will usually involve pH neutralization and brine concentration or dilution by blending with effluent, where acceptable before disposal. Concentrating methods, discussed previously in Chap. 9 for management of RO retentate, are also used for brines. The principal methods used for the disposal of brines are discharge to brackish or saline receiving waters, blending with effluent, and deep well injection. The disposal of the high-TDS regeneration brine remains the primary obstacle to the wider use of ion exchange technology.

10-4 DISTILLATION

Distillation is a unit operation in which the components of a liquid solution are separated by vaporization and condensation. Specially designed reactors are used to vaporize the water undergoing treatment, leaving behind a waste brine that must be disposed of.

Applications for Distillation

Along with RO, electrodialysis, and ion exchange, distillation can be used to control the buildup of salts in critical water reuse applications. Because distillation is expensive, its application is generally limited to applications where: (1) a high degree of treatment is required, (2) contaminants cannot be removed by other methods, and (3) inexpensive heat is available. The basic concepts involved in distillation are introduced in this section. As the use of distillation for water reclamation is a recent development, the current literature must be consulted for the results of ongoing studies and more recent applications.

Distillation Processes

Over the past 30 yr, a variety of distillation processes employing several evaporator types and methods of using and transferring heat energy have been evaluated or used. The principal distillation processes are: (1) boiling with submerged tube heating surface, (2) multiple-effect evaporation by boiling with long-tube vertical evaporator, (3) multistage flash evaporation, (4) forced circulation with vapor compression, (5) solar evaporation, (6) rotating-surface evaporation, (7) wiped-surface evaporation, (8) vapor reheating process, (9) direct heat transfer using an immiscible liquid, and (10) condensing-vapor-heat transfer by vapor other than steam. Of these types of distillation processes, multiple-effect evaporation, multistage flash evaporation, and the vapor-compression distillation are the most feasible processes for water reclamation applications.

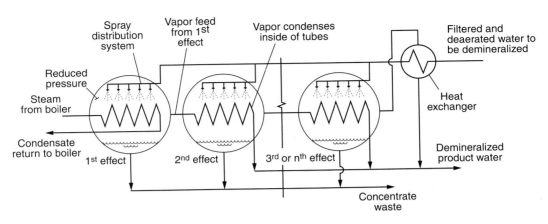

Figure 10-12
Schematic of multiple effect evaporation distillation process.

Multiple-Effect Evaporation

In multiple-effect evaporation (MEE) or distillation (MED) several evaporators (boilers) are arranged in series, each operating at a lower pressure than the proceeding one. For example, in a three-effect (stage) evaporator (see Fig. 10-12), the influent water is first passed through a heat exchanger, where it is preheated (with vapor from the last effect). Preheated feedwater enters the first effect, where it is evaporated by spraying onto heat exchanger tubes heated with steam from a boiler. The resulting vapor from the first effect enters the heat exchanger tubes of the second effect, where it is condensed by spraying preheated feedwater over the tubes. Demineralized product water is formed by condensation within the tubes. Similarly, vapor from the second effect is condensed in the third effect. Vapor from the third effect is used to preheat the feedwater. If entrainment is kept low, almost all of the nonvolatile contaminants can be removed in a single evaporation step. Volatile contaminants, such as ammonia gas and low-molecular weight organic acids, may be removed in a preliminary evaporation step, but if their concentration is so small that their presence in the final product is not objectionable, this step with its added cost can be eliminated. Typical water quality performance data for a multiple-effect distillation process have been reported for a pilot-scale unit by Rose et al. (1999).

Multistage Flash Evaporation

Multistage flash (MSF) evaporation systems have been used commercially in desalination for many years. In the multistage flash process (see Fig. 10-13), the influent feed water is first treated to remove excess TSS and deaerated before being pumped through heat transfer units in the several stages of the distillation system, each of which is maintained at a reduced pressure. Vapor generation or boiling caused by reduction in pressure is known as flashing. As the water enters each stage through a pressure-reducing nozzle, a portion of the water is flashed to form a vapor. In turn, the flashed water vapor condenses on the outside of the condenser tubes and is collected in trays. As the vapor condenses, its latent heat is used to preheat the water that is sent to the main heater where it receives additional heat before being introduced to the first flashing stage. When the concentrated brine reaches the lowest pressure stage, it is pumped out.

Figure 10-13
Schematic of multistage flash evaporation distillation process.

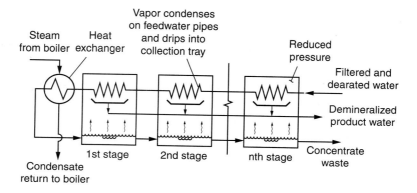

Thermodynamically, multistage flash evaporation is less efficient than ordinary evaporation. However, by combining a number of stages in a single reactor, external piping is eliminated and construction costs are reduced.

Vapor Compression Distillation

In the vapor compression process, an increase in pressure of the vapor is used to establish the temperature difference for the transfer of heat. The basic schematic of a vapor compression distillation unit is shown on Fig. 10-14. After initial heating of the water, the vapor compressor is operated so that the vapor under higher pressure can condense in the condenser tubes, at the same time causing the release of an equivalent amount of vapor from the concentrated solution. Heat exchangers can conserve heat from both the condensate and the waste brine. The only energy input required during operation is the mechanical energy for the vapor compressor. Hot concentrated brine must be discharged at intervals to prevent the buildup of excessive concentrations of salt in the boiler. Vapor compression distillation is also used in the concentration of brines (see Chap. 9).

Basic Considerations for Distillation Processes

The theoretical thermodynamic minimum energy required to raise the temperature of water and to provide the latent heat of vaporization is about 2280 kJ/kg. Unfortunately, because of the many irreversibilities in an actual distillation processes, the thermodynamic minimum energy requirement is of little relevance in the practical evaluation of

Figure 10-14
Schematic of vapor compression distillation process.

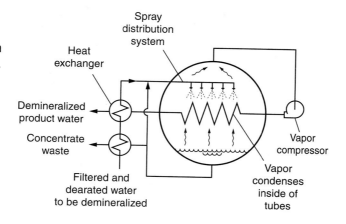

distillation processes. Typically, about 1.25 to 1.35 times the latent heat of vaporization is required. A readily available supply of steam from another process will improve the feasibility of distillation type processes.

Distillation Process Limitations

The principal issues with the application of the distillation processes for water reclamation are the high energy input required to evaporate the feedwater, carryover of volatile constituents found in treated reclaimed water, and the degree of subsequent cooling and treatment that may be required to renovate the distilled water. In addition, all distillation processes reject part of the influent feedwater and produce a concentrated waste stream that must be managed.

The most common operating problems encountered include scaling and corrosion. Due to temperature increases, inorganic salts come out of solution and precipitate on the inside walls of pipes and equipment. The control of scaling due to calcium carbonate, calcium sulfate, and magnesium hydroxide is one of the most important design and operational considerations in distillation desalination processes. Controlling the pH minimizes carbonate and hydroxide scales. Corrosion is controlled by the use of specific materials, including cupronickel alloys, aluminum, titanium, and monel.

10-5 CHEMICAL OXIDATION

Chemical oxidation is used for various applications in water reclamation, including the destruction of anthropogenic or synthetic toxic organics. Processes in which oxidizing chemicals are added to water that directly react with the constituents in water are known as conventional oxidation processes. Processes in which oxidizing chemicals are added to water to cause the formation of hydroxyl radicals (HO·), which then react with the constituents in water, are known as advanced oxidation processes (AOPs). A dot is added after the radical species to show that there is an unpaired electron in the outer orbital. The key difference between conventional oxidation processes and AOPs is the power of the oxidizing agent, known as the oxidation potential. Conventional chemical oxidation processes are described in this section and AOPs that utilize hydroxyl radical species are the subject of Sec. 10-6.

Applications for Conventional Chemical Oxidation

The principal applications of conventional chemical oxidation in water reclamation are for: (1) odor control, (2) hydrogen sulfide control, (3) color removal, (4) iron and manganese removal, (5) disinfection, (6) control of biofilm growth and biofouling in treatment processes and distribution system components, and (7) oxidation of selected trace organic constituents. Each of these applications, with the exception of disinfection, is summarized briefly in Table 10-9. The oxidants listed in Table 10-9 are ordered according to their electrical potential. Because of the importance of disinfection in water reclamation and reuse, a separate chapter (Chap. 11) is devoted to disinfection. Details on the applications of conventional oxidation may be found in Rakness (2005), Crittenden et al. (2005), Tchobanoglous et al. (2003), U.S. EPA (1999), and White (1999).

Oxidants Used in Chemical Oxidation Processes

Oxidants that are frequently used in water reclamation and listed in Table 10-9 are: (1) hydroxyl radical, (2) chlorine, (3) ozone, (4) chlorine dioxide, (5) permanganate, and (6) hydrogen peroxide. The oxidation kinetics of oxygen are usually too slow to be

Table 10-9
Oxidants, forms, and applications methods

Oxidant	Forms	Applications[a]	Application methods	Electrical potential, V
Fluorine	Not used	—	Not used	3.06
Hydroxyl radical	Generated in specially designed reactors at the moment of use due to short life	A, B, C, D	See Sec. 10-6	2.80
Ozone	Ozone is a gas that is generated on site by passing dry compressed air or pure oxygen across a high voltage electrode	B, C, D	Ozone is applied to water as a gas. Because mass transfer is an important issue, special attention is required for contractor design (see Chap. 11)	2.08
Peracetic acid	Stabilized liquid solution	A, D	Concentrated solution mixed with water to be treated	—
Hydrogen peroxide	Liquid solution	A	Concentrated solution mixed with water to be treated	1.78
Permanganate	Available in bulk as granules	A, B	Added as a dry chemical using a feeder or as a concentrated solution (no more than five percent by weight due to its limited solubility)	1.67
Chlorine, free	Chlorine gas, NaOCl solution	A, D	Gas eductors and various diffuser designs	1.49
Chlorine, combined (chloramines)	Addition of ammonia: anhydrous ammonia gas, ammonium sulfate, aqua ammonia (20 to 30 percent ammonia solution)	A, D	Gas eductors, dry chemical feeders, spray jets	—
Chlorine dioxide	Chlorine dioxide gas is produced on-site using a 25 percent sodium chlorite solution. The sodium chlorite solution reacts with the following constituents to form $ClO_{2(g)}$: (1) gaseous chlorine (Cl_2), (2) aqueous chlorine (HOCl), or acid (usually hydrochloric acid, HCl)	A	Gas eductors	1.27
Oxygen	Gas and liquid	—	Pure oxygen or the oxygen in air applied using diffusers or other devices	1.23

[a] A = Oxidation of reduced inorganic species such as soluble metals, complexed metal species, and destruction of odor causing compounds.
B = Oxidation of trace organic compounds, color, and odor causing compounds, and NOM.
C = To improve coagulation.
D = Used as a biocide to control algae in reservoirs and basins, for disinfection, and to control growth in distributions system.

of practical use beyond secondary biological treatment. Chemical oxidants are usually added at specific points during tertiary treatment (e.g., to control odors or membrane fouling) or at the final stage of treatment before distribution (e.g., disinfection). The rate of oxidation typically follows the trend given below; however, there will be exceptions depending on the characteristics of the solution (e.g., pH) and type of compound that is to be oxidized.

$$HO\cdot > O_3 > H_2O_2 > HOCl > ClO_2 > MnO_4^- > O_2 > OCl^- \qquad (10\text{-}17)$$

The behavior of the hydroxyl radical, $HO\cdot$, discussed in detail in Sec. 10-6, is introduced here briefly as related to its formation from ozonation. Of the conventional chemical oxidants, ozonation is effective for the destruction of organic compounds (denoted as R), either by direct reactions with O_3, or indirect reactions with $HO\cdot$ as shown in Eq. (10-18).

$$O_3 \rightarrow \begin{bmatrix} \xrightarrow[O_3]{\text{Direct Pathway}} O_3 + R \rightarrow \text{Product 1} \\ \xrightarrow[\text{NOM}]{\text{Indirect Pathway}} HO\cdot + R \rightarrow \text{Product 2} \end{bmatrix} \qquad (10\text{-}18)$$

The reaction of ozone with natural organic matter (NOM) to produce $HO\cdot$, is among the most important mechanisms used to destroy target compounds (Elovitz and von Gunten, 1999; Westerhoff et al., 1999). However, substantial removals of residual pharmaceuticals in low DOC reclaimed water are possible by direct ozonation, even at low ozone dosages (Huber et al., 2005). Views of ozone contactors used for bench and pilot scale evaluation are shown on Fig. 10-15. Transformation of trace constituents such as

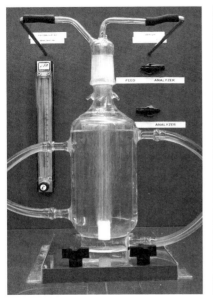

Figure 10-15

Contactors used for evaluation of chemical oxidation using ozone: (a) bench scale and (b) pilot scale.

(a)   (b)

pharmaceuticals is also possible during chlorine disinfection; however, the effectiveness depends on the structure of the chemical compound, the form of chlorine, the contact time, and application of dechlorination (Pinkston and Sedlak, 2004).

Basic Considerations for Chemical Oxidation Processes

In water reclamation, redox reactions take place when an oxidant is added to water and electrons are transferred from the reductant to the oxidant. The constituent that gains electrons (oxidant) is *reduced* and is referred to as the *oxidizing agent*. The constituent that looses electrons (reductant) is *oxidized*, and is referred to as the *reducing agent*. The driving force for the exchange of electrons between an oxidant and a reductant is the difference in their electrical potentials (McMurray and Fay, 2003).

The gain or loss of electrons from redox reactions can be evaluated by the standard electrode potentials for the oxidation and reduction half reactions. Every oxidation or reduction half reaction can be characterized by the electrical potential or electromotive force (emf). This potential is called the standard electrode potential, E^o_{Rxn}, or redox potential for the reaction, and is measured in volts. Many references contain tables with the standard electrode potentials for common reactions in water and wastewater treatment that can be applied to water reclamation (Sawyer et al., 2003; Tchobanoglous et al., 2003).

By convention, half reactions are written as reduction reactions. To obtain the oxidation reaction, the direction of the reduction reaction is reversed and the reduction potential is multiplied by a factor of -1. Addition of the electrode potentials from all relevant reactions results in the overall electrode potential. The value of the redox potential can be determined by adding the reduction and oxidation potentials together, after accounting for the sign convention and normalizing the number of electrons transferred in the reaction. The potentials for the reduction of O_2 and oxidation of H_2 are as follows:

$$O_2 + 4H^+ + 4e^- \rightleftarrows 2H_2O \quad \text{(reduction)} \quad E^o_{red} = 1.23 \text{ V} \quad (10\text{-}19)$$

$$H_2 \rightleftarrows 2H^+ + 2e^- \quad \text{(oxidation)} \quad E^o_{ox} = 0 \text{ V} \quad (10\text{-}20)$$

where E^o_{red} = standard electrode potential for the reduction reaction, V.
E^o_{ox} = standard electrode potential for the oxidation reaction, V.

The overall redox reaction can be obtained by multiplying the terms of Eq. (10-20) by a factor of 2, summing the terms of Eqs. (10-19) and (10-20), and eliminating electrons and H^+ from both sides of the equations. The final overall equation for the formation of water is shown in Eq. (10-21).

$$O_2 + 2H_2 \rightleftarrows 2H_2O \quad \text{(overall)} \quad E^o_{Rxn} = 1.23 \quad (10\text{-}21)$$

where E^o_{Rxn} = standard electrode potential for the overall redox reaction, V.

The value of E^o_{Rxn} is determined by summing the terms E^o_{red} and E^o_{ox}, after multiplying by the respective factors. If the value of E^o_{Rxn} is positive, then, theoretically, the reaction will proceed as written. However, the outcome of a particular reaction will depend on both the electrical potential and free energy of the reaction at the expected solution concentrations. Additional details on analyzing chemical reactions may be found in Crittenden et al. (2005).

Chemical Oxidation Process Limitations

Aside from the expense of chemical addition, the primary concern with any chemical oxidation process is the potential for the formation of toxic byproducts due to incomplete oxidation. While the hydroxyl radical oxidation processes discussed in Sec. 10-6 are able to achieve complete mineralization of many constituents under optimal conditions, conventional chemical oxidation is typically not powerful enough to reach this endpoint. Therefore, subsequent treatment processes are needed to remove the oxidation byproducts, such as adsorption. Further, chemical oxidation increases the biodegradability of some constituents, potentially requiring the use of a biological process to remove residual biodegradable organic material (see Sec. 10-8). Byproduct formation may be controlled by removal of the byproduct precursors before application of the oxidant and careful control of the oxidant dose.

The properties of chemical oxidants that justify their use in water reclamation also contribute to their potential to be corrosive under certain conditions. Thus, careful control of oxidant dosage and the use of compatible materials are important factors to prevent corrosion of facilities and equipment. A number of methods are available to evaluate the potential, type, and rate of corrosion by chemical oxidants with a given material under certain conditions, including thermodynamics, electrokinetics (mixed-potential models), and experimental testing.

10-6 ADVANCED OXIDATION

Advanced oxidation processes can be used to destroy trace constituents that cannot be oxidized completely by conventional oxidants, including constituents that are known to affect the endocrine system (Rosenfeldt and Linden, 2004). Reclaimed water following tertiary treatment (see Fig. 10-1) typically still contains a variety of natural and synthetic organic chemicals at low concentrations that may need to be removed or destroyed to protect public health and the environment, especially in indirect potable reuse applications. The conventional oxidants described in Sec. 10-5 are able to remove some of the constituents of interest, however, there is uncertainty regarding the formation of toxic byproducts following conventional chemical oxidation. Moreover, some trace constituents may be found in the permeate from RO treatment.

The advantage of advanced oxidation is the ability to generate elevated concentrations of hydroxyl radical (HO·), a strong oxidant capable of the complete oxidation of most organic compounds into carbon dioxide, water, and mineral acids (e.g., HCl). As noted previously, the dot is added after the radical species to denote that an unpaired electron is present in the outer orbital. Because of the unpaired electron, hydroxyl radicals are reactive electrophiles (electron-loving) that react rapidly with nearly all electron-rich organic compounds. The reactions with hydroxyl radical are second order because the reactions depend on the concentration of the constituent that is oxidized and the concentration of the hydroxyl radical species. The second order hydroxyl radical rate constants for many dissolved organic compounds are on the order of 10^8 to 10^9 L/mol·s (Buxton and Greenstock, 1988), three to four orders of magnitude greater than second order rate constants for other oxidants.

Applications for Advanced Oxidation

The relative oxidizing power of the hydroxyl radical expressed as electrochemical oxidation potential, along with other common oxidants, is summarized in Table 10-9. As shown, with the exception of fluorine, the hydroxyl radical is one of the most active oxidants known. Advanced oxidation processes differ from the other treatment processes discussed (such as adsorption, ion exchange, or stripping) because organic compounds in water are degraded rather than concentrated or transferred into a different phase. Further, compounds that are not adsorbable or only partially adsorbable may be destroyed by reaction with hydroxyl radical. Because a secondary waste stream is not generated, there is no additional cost to dispose of or regenerate materials. Hydroxyl radicals are capable of oxidizing almost all reduced materials present without restriction to specific classes or groups of compounds, as compared to other oxidants. In addition to being nonselective, many AOPs operate at normal temperature and pressures. Other processes that can generate hydroxyl radicals, but require elevated temperature and/or pressure, include catalytic oxidation, gas-phase combustion, supercritical oxidation, and wet oxidation processes. Additional details on AOPs may be found in Singer and Reckhow (1999) and Crittenden et al. (2005).

Degree of Degradation

Depending on the application it may not be necessary to completely oxidize a given compound or group of compounds as partial oxidation may be sufficient to render specific compounds more amenable to subsequent biological treatment or to reduce their toxicity. The oxidation of specific compounds may be characterized by the extent of degradation of the final oxidation products as follows (Rice, 1996):

1. *Primary degradation.* A structural change in the parent compound.
2. *Acceptable degradation (defusing).* A structural change in the parent compound to the extent that toxicity is reduced.
3. *Ultimate degradation (mineralization).* Conversion of organic carbon to inorganic CO_2.
4. *Unacceptable degradation (fusing).* A structural change in the parent compound resulting in increased toxicity.

Oxidation of Refractory Organic Compounds

Hydroxyl radicals are used most commonly for the oxidation of trace amounts of refractory organic compounds found in highly treated effluents (e.g., following treatment by RO). The hydroxyl radicals, once generated, can attack organic molecules by: (1) radical addition, (2) hydrogen abstraction, (3) electron transfer, and (4) radical combination (SES, 1994) as described below.

1. By radical addition

    The addition of the hydroxyl radical to an unsaturated aliphatic or aromatic organic compound (e.g., C_6H_6) results in the production of a radical organic compound that can be oxidized further by compounds such as oxygen or ferrous iron to produce stable oxidized end products. Radical addition is much more rapid than hydrogen abstraction. In the following reactions the abbreviation R is used to denote the reacting organic compound.

$$R + HO\cdot \rightarrow ROH \qquad (10\text{-}22)$$

2. By hydrogen abstraction

   The hydroxyl radical can be used to remove a hydrogen atom from organic compounds. The removal of a hydrogen atom results in the formation of a radical organic compound, initiating a chain reaction where the radical organic compound reacts with oxygen, producing a peroxyl radical, which can react with another organic compound.

$$R + HO\cdot \rightarrow R\cdot + H_2O \quad (10\text{-}23)$$

3. By electron transfer

   Electron transfer results in the formation of ions of a higher valence. Oxidation of a monovalent negative ion will result in the formation of an atom or a free radical. In the following reaction, n is used to denote the charge on the reacting organic compound R.

$$R^n + HO\cdot \rightarrow R^{n-1} + OH^- \quad (10\text{-}24)$$

4. By radical combination

   Two radicals can combine to form a stable product.

$$HO\cdot + HO\cdot \rightarrow H_2O_2 \quad (10\text{-}25)$$

The reactions of HO· with organic compounds by radical addition reactions with double bonds and hydrogen abstraction are among the most common processes. In general, the reaction of hydroxyl radicals with organic compounds at completion will produce water, carbon dioxide, and mineral acids and salts; this process is also known as *mineralization*.

Disinfection

Because it was recognized that free radicals generated from ozone were more powerful oxidants than ozone alone, it was reasoned that the hydroxyl free radicals could be used effectively to oxidize microorganisms in reclaimed water. Unfortunately, because the half-life of the hydroxyl free radicals is short, on the order of microseconds, it is not possible to develop high concentrations. With extremely low concentrations, the required detention times for microorganism disinfection, based on the C_Rt concept (see Chap. 11), are prohibitive. However, AOPs that incorporate high dosages of UV energy (1000 to 2000 mJ/cm^2) to initiate photolysis reactions may be of sufficient intensity to accomplish significant levels of disinfection. Challenge testing using pilot- or full-scale installations may be used to determine the actual level of disinfection accomplished.

Processes for Advanced Oxidation

Based on numerous studies, it has been found that AOPs are more effective than any of the individual agents (e.g., ozone, UV, hydrogen peroxide). Several technologies are available to produce HO· in the aqueous phase (U.S. EPA 1998). Selected technologies are summarized in Table 10-10. In water reclamation, AOPs are usually applied to low COD reclaimed waters (typically following treatment by RO) because of the cost of ozone and/or H_2O_2 required to generate the hydroxyl radicals. Of the technologies reported in Table 10-10, the commercially available AOPs in the United States for water reclamation are ozone/UV, ozone/hydrogen peroxide, and hydrogen peroxide/UV. It should be noted that some countries do not permit the use of some chemicals, such as hydrogen peroxide, thus some ADPs are not appropriate for use in all parts of the world.

Table 10-10
Advantages and disadvantages of various oxidation processes that produce hydroxyl radicals[a]

Advanced oxidation process	Advantages	Disadvantages
Commercially available AOPs for water reclamation		
Hydrogen peroxide/ultraviolet light	H_2O_2 is fairly stable and can be stored on site temporarily prior to use	H_2O_2 has very poor UV absorption characteristics and if the water matrix absorbs a lot of UV light energy then most of the light input to the reactor will be wasted Special reactors which are designed for UV illumination are required Residual H_2O_2 must be addressed Potential for UV lamp fouling
Hydrogen peroxide/ozone	Waters with poor UV-light transmission may be treated Special reactors designed for UV illumination are not required Volatile organics will be stripped from the ozone contactor and may require treatment	Production of O_3 can be an expensive and inefficient process Gaseous ozone which is present in the off gas of the ozone contactor must be removed Maintaining and determining the proper O_3/H_2O_2 dosages may be difficult Low pH is detrimental to the process
Ozone/UV	Easier to control dosage of O_3 Residual oxidant will degrade rapidly (typical half life of O_3 is 7 min) Ozone absorbs more UV light than an equivalent dosage of hydrogen peroxide (~200 times more at 254 nm) Volatile compounds will be stripped from the process and may require treatment	Using O_3 and UV light to produce H_2O_2 is inefficient compared to just adding H_2O_2. Special reactors which are designed for UV illumination are required Ozone off-gas must be removed Potential for UV lamp fouling
Other selected AOPs		
Ozone/UV/H_2O_2	Commercial processes that utilize the technology are available. H_2O_2 promotes ozone mass transfer Volatile compounds will be stripped from the process and may require treatment	Special reactors that are designed for UV illumination are required Ozone off-gas must be removed Potential for UV lamp fouling
Fenton's reactions (Fe/ hydrogen peroxide, photo-Fenton's or Fe/ozone)	Some effluents may contain sufficient Fe to drive the Fenton's reaction Commercial processes are available that utilize the technology	Process requires low pH
Titanium dioxide/UV	Activated with UV light; consequently greater light transmission is achievable	Fouling of the catalyst may occur. When used as a slurry, the TiO_2 must be recovered. Potential for UV lamp fouling
Ozone at elevated pH (8 to >10)	Does not require the addition of UV light or hydrogen peroxide	Ozone off-gas must be removed pH adjustment may not be practical Process does not yield an appreciable destruction for contaminants for reason provided in Sec. 8-6

[a] Adapted from Crittenden et al. (2005).

The major advantages and disadvantages of various AOPs are also provided in Table 10-10. It should be noted that following oxidation, constituents that were previously resistant to degradation may be transformed into biodegradable compounds that may require further biological treatment (e.g., biologically active filration).

Ozone/UV

Production of the free radical HO· with UV light can be illustrated by the following reactions for the photolysis of ozone (Glaze et al., 1987; Glaze and Kang, 1990). The first step of the ozone/ultraviolet light (O_3/UV) process is the formation of H_2O_2 by photolysis of ozone.

$$O_3 + H_2O + UV(\lambda < 310 \text{ nm}) \rightarrow O_2 + HO\cdot + HO\cdot \rightarrow O_2 + H_2O_2 \quad (10\text{-}26)$$

As shown in Eq. (10-26), the photolysis of ozone in wet air results in the formation of hydroxyl radicals. In water, the photolysis of ozone leads to the formation of hydrogen peroxide, which is subsequently photolyzed or reacted with O_3 to form hydroxyl radicals. The ozone/UV process can degrade compounds through direct ozonation, photolysis, or reaction with the hydroxyl radical, resulting in a process that is more effective when the compounds of interest can be degraded through the absorption of the UV irradiation and through the reaction with the hydroxyl radicals. Basic components of the ozone/UV process include ozone gas generation, ozone injection facilities, and UV photolysis reactors. A schematic flow diagram and view of a typical ozone/UV oxidation process is illustrated on Fig. 10-16.

While it is possible for UV light to split H_2O_2 into HO·, the extinction coefficient for O_3 is greater than that for H_2O_2 at 254 nm. Thus, using ozone to produce H_2O_2, which in turn reacts with O_3 to produce HO·, may not be the most efficient way to produce HO·

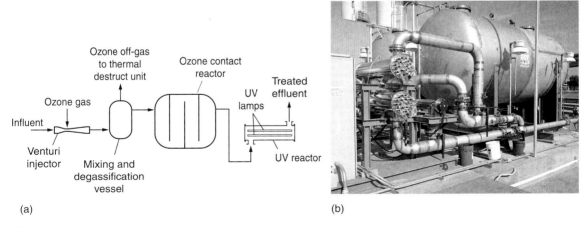

Figure 10-16

Advanced oxidation process involving the use of ozone and UV radiation: (a) schematic representation (ozone contactor shown without chimneys (see Fig. 11-24 in Chap. 11) and (b) view of full scale installation.

radicals because of the large amount of energy required to form ozone onsite. Processes involving ozone and UV dosages in the range of 16 to 24 mg/L and 810 to 1610 mJ/cm^2, respectively, have been found to have a significant impact on TOC concentrations and disinfection byproduct formation compared to the use of either UV or ozone alone (Chin and Bérubé, 2005). As with all UV processes, fouling of the UV lamp sleeve, lamp replacement costs, and energy consumption are important considerations.

Ozone/Hydrogen Peroxide

For compounds that do not adsorb UV or where the transmittance of the water to be treated inhibits photolysis, AOPs involving ozone/H_2O_2 may be more effective than ozone/UV. Processes using O_3/H_2O_2 have been used to reduce the concentration of assorted VOCs, petroleum compounds, industrial solvents, and pesticides in water (Karimi et al.; 1997, Mahar et al., 2004; Chen et al., 2006). The overall reaction for the production of hydroxyl radicals using hydrogen peroxide and ozone is as follows:

$$H_2O_2 + 2O_3 \rightarrow HO\cdot + HO\cdot + 3O_2 \qquad (10\text{-}27)$$

According to Eq. (10-27), 0.5 mols of H_2O_2 are needed for every mol of O_3 or a mass ratio of 0.354 kg of H_2O_2 is needed for every kg of O_3. However, there are several issues that impact the proper dosages of H_2O_2 and O_3. First, O_3 tends to be more reactive with background organic matter and inorganic species than H_2O_2. As a result, the required O_3 dosage will be higher than estimated from stoichiometry. Typical ozone and hydrogen peroxide concentrations range from 5 to 30 mg/L and 5 to 15 mg/L, respectively. Pilot studies are usually conducted to determine the chemical dosage required for a given level of trace constituent removal. However, an excess O_3 dose has the potential of wasting O_3, forming oxidation byproducts (e.g., bromate), and quenching $HO\cdot$ radicals via the following reaction:

$$O_3 + HO\cdot \rightarrow HO_2\cdot + O_2 \qquad (10\text{-}28)$$

The $HO_2\cdot$ radical, formed by Eq. (10-28), may react to produce additional $HO\cdot$. To overcome the problem of byproduct formation and quenching of $HO\cdot$, new reactor designs have incorporated the addition of H_2O_2 or O_3 at multiple points in a single reactor and by using multiple reactors in series. A schematic flow diagram and view of a reactor used for reacting hydrogen peroxide and ozone is shown on Fig. 10-17. Excess H_2O_2 is also detrimental to the H_2O_2/O_3 AOP because it may scavenge $HO\cdot$. Further, the H_2O_2 residual can be more problematic than ozone because hydrogen peroxide is more stable than ozone and, in some applications, it may be necessary to remove residual hydrogen peroxide before reuse. Hydrogen peroxide reacts quickly with hypochlorite to form water, oxygen, and chloride ion.

Hydrogen Peroxide/UV

Hydroxyl radicals are also formed when water containing H_2O_2 is exposed to UV light (200 to 280 nm). The following reaction can be used to describe the photolysis of H_2O_2:

$$H_2O_2 + UV \text{ (or } h\nu, \lambda \approx 200 \text{ to } 280 \text{ nm)} \rightarrow HO\cdot + HO\cdot \qquad (10\text{-}29)$$

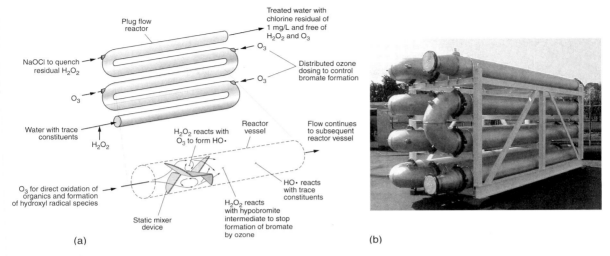

Figure 10-17
Advanced oxidation process involving the use of ozone and hydrogen peroxide: (a) schematic of HiPOx reactor and (b) view of reactor described in (a) (Courtesy of Applied Process Technology, Inc.)

In some cases the use of the hydrogen peroxide/UV process has not been feasible because H_2O_2 has a small molar extinction coefficient, requiring high concentrations of H_2O_2, and high UV dosages. A schematic flow diagram and a typical installation of the hydrogen peroxide/UV process are shown on Fig. 10-18.

The basic elements of the hydrogen peroxide/ultraviolet light (H_2O_2/UV) process includes hydrogen peroxide injection and mixing followed by a reactor that is equipped with UV lights. As shown on Fig. 10-18, typical H_2O_2/UV reactors configurations include inline stainless steel reactors with low pressure (low and high intensity) or medium pressure UV lamps arranged parallel to the flow, perpendicular to the flow, or in upflow columns with crisscrossing lamps oriented perpendicular to the direction of flow.

The H_2O_2/UV process has not been used commonly for potable water treatment because it normally results in high effluent H_2O_2 concentrations. However, the residual hydrogen peroxide is not a concern in water reclamation. High effluent H_2O_2 concentrations occur because high initial dosages of H_2O_2 are required to efficiently utilize the UV light and produce hydroxyl radical. The residual H_2O_2 consumes chlorine and interfere with disinfection. In some instances, where high UV doses are required, as in the photolyisis of N-nitrosodimethylamine (NDMA) (see Chap. 3), H_2O_2 may be added to achieve advanced oxidation of other constituents that are resistant to photolysis alone (Linden et al., 2004). This method of operation is now being used in a number of water reclamation applications. The details required for modeling the H_2O_2/UV process can be found in Crittenden et al. (1999). As discussed in Chap. 11, UV processes are subject to fouling of the UV lamp sleeve, lamp replacement costs, and high energy consumption.

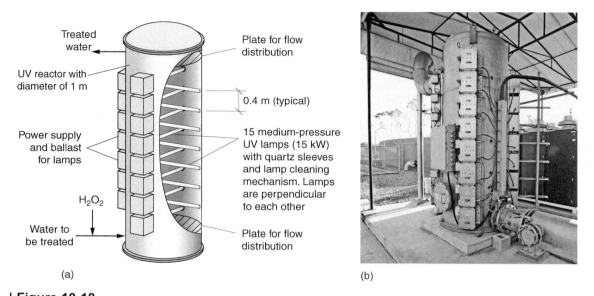

Figure 10-18
Hydrogen peroxide and UV radiation advanced oxidation process: (a) schematic diagram (from Crittenden et al., 2005) and (b) photograph of typical vertical flow UV reactor.

Other Processes

Other reactions which yield HO· include the reactions of H_2O_2 and UV with Fenton's reagent and the adsorption of UV by semiconductor metal oxides such as TiO_2 suspended in water, which act as catalysts. Other processes are currently under development.

Basic Considerations for Advanced Oxidation Processes

The engineering of advanced oxidation processes (AOPs) consists of the selection of a process to generate hydroxyl radicals, estimation of the reaction kinetics with the compounds of interest, and design of a reactor in which the reaction will take place. In addition, the presence of background organic and inorganic matter that reacts with the HO· will reduce the process efficiency for the target compound. Therefore, pilot studies are always necessary to determine process efficiency for a given water matrix.

Commercially available AOPs are rated for a given hydroxyl radical output. Reported field HO· concentrations range from 10^{-11} to 10^{-9} mole/L (Glaze et al., 1987; Glaze and Kang, 1990). The second order hydroxyl radical rate constants for several compounds of interest are presented in Table 10-11. As mentioned previously, the reactions are second order because they depend on the concentration of the hydroxyl radical and the compound undergoing oxidation. The reaction between HO· and an organic compound of interest, R, is represented as follows:

$$\text{HO·} + \text{R} \rightarrow \text{byproducts} \tag{10-30}$$

The second order rate law, r_R, corresponding to the reaction shown in Eq. 10-30, is given by the following expression:

$$r_R = -K_R C_{HO} \cdot C_R \tag{10-31}$$

Table 10-11
Hydroxyl rate constants for commonly occurring trace organic constituents[a]

Compound name	HO• rate constant, L/mole·s	Compound name	HO• rate constant, L/mole·s
Ammonia	9.00×10^7	Hypobromous acid	2.0×10^9
Arsenic trioxide	1.0×10^9	Hypoiodous acid	5.6×10^4
Bromide ion	1.10×10^{10}	Iodide ion	1.10×10^{10}
Carbon tetrachloride	2.0×10^6	Iodine	1.10×10^{10}
Chlorate ion	1.00×10^6	Iron	3.2×10^8
Chloride ion	4.30×10^9	Methyl tertiary butyl ether (MTBE)	1.6×10^9
Chloroform	5×10^6	Nitrite ion	1.10×10^{10}
CN^-	7.6×10^9	N-Dimethylnitrosamine (NDMA)	4×10^8
CO_3^{2-}	3.9×10^8	Ozone	1.1×10^8
Dibromochloropropane	1.5×10^8	p-Dioxane	2.8×10^9
1,1-Dichloroethane	1.8×10^8	Tetrachloroethylene	2.6×10^9
1,2-Dichloroethane	2.0×10^8	Tetrachloroethylene	1.0×10^7
H_2O_2	2.7×10^7	Tribromomethane	1.8×10^8
HCN	6.0×10^7	Trichloroethylene	4.2×10^9
HCO_3^-	8.5×10^6	Trichloromethane	5.0×10^6
Hydrogen sulfide	1.5×10^{10}	Vinyl chloride	1.2×10^{10}

[a]Adapted from Crittenden et al., 2005.

where r_R = second order rate law, mole/L·s
k_R = second order rate constant for the destruction of R with HO• radicals, mole/L·s
$C_{HO•}$ = concentration of hydroxyl radical, mole/L
C_R = concentration of the target organic R, mole/L

The half-life of the target organic compounds may be calculated assuming that the HO• is constant and equal to a typical field value or manufacturers specification. The expression for the half-life of an organic compound is obtained by substituting the rate law into a mass balance on a batch reactor whose contents are mixed completely, and solving and rearranging the result, as follows:

$$\frac{dC_R}{dt} = -k_R C_{HO•} C_R \quad (10\text{-}32)$$

$$t_{1/2} = \frac{\ln(2)}{k_R C_{HO•}} \quad (10\text{-}33)$$

where $t_{1/2}$ = the half life of the organic compound R, s

The use of Eqs. (10-32) and (10-33) is shown in the following example.

EXAMPLE 10-4. Advanced Oxidation Process for Removal of NDMA.

N-nitrosodimethylamine (NDMA) ($C_2H_6N_2O$) is a compound of concern present in many secondary and tertiary treated wastewater effluents. From Table 10-11, the second order rate constant of HO· for NDMA is 4×10^8 L/mole·s. Compute the time required to lower the concentration of NDMA from 200 μg/L to 20 μg/L for a HO· concentration of 10^{-9} mole/L using an ideal plug flow reactor. Assume that the residence time for an ideal plug flow reactor is equivalent to the residence time in a completely mixed batch reactor.

Solution

1. Develop an expression of the concentration of NDMA as a function of time in a completely mixed batch reactor (CMBR).
   a. Using Eq. (10-32), the rate expression for a CMBR, where C_R represents the concentration of NDMA is given by:

   $$r_R = \frac{dC_R}{dt} = -k_R C_{HO·} C_R = -k' C_R$$

   where $k' = k_R C_{HO·}$.

   b. The integrated form of the rate expression for a CMBR is:

   $$\int_{C_{R0}}^{C_R} \frac{dC_R}{C_R} = -\int_0^t k' t$$

   $$C_R = C_{R_0} e^{-k't}$$

2. Calculate the time it would take to achieve a concentration of 20 μg/L using the equation developed in Step 1.
   a. Rearrange the above equation to solve for t.

   $$t = \frac{1}{k'} \ln \frac{C_{R_0}}{C_R}$$

   b. Solve for t.

   The value of k' from Step 1 is

   $$k' = k_R C_{HO·} = (4 \times 10^8 \text{ L/mole·s})(10^{-9} \text{ mole/L}) = 0.4 \text{ 1/s}$$

   $$t = \frac{1}{(0.4)} \ln \left(\frac{200}{20}\right) = 5.8 \text{ s}$$

Comment

Advanced oxidation of NDMA appears to be feasible given the short contact time required for the reaction. Because some organic compounds of interest react more slowly with HO·, these compounds will require longer reaction times and/or high concentrations of HO·. The presence of background organic matter, carbonate, bicarbonate, and pH will also reduce the efficiency of the AOP and must be considered during process design. As discussed in Chap. 6, pilot testing will be required to determine site specific process design and operational parameters.

Advanced Oxidation Process Limitations

The feasibility and efficiency of AOPs are determined by a number of factors, including interferences and the production of byproducts, as described below. Means of overcoming most process limitations in water reuse applications are also considered.

Byproducts of Advanced Oxidation Processes

Advanced oxidation processes (and processes that use ozone), have been found to produce brominated byproducts and bromate (BrO_3^-) in waters containing bromide ion. The concentration of bromide ion, TOC concentration, and pH determine the quantity of brominated byproducts formed. Some AOPs have been designed to minimize bromate formation by pH control or ammonia addition.

Both hydrogen abstraction and radical addition produce reactive organic radicals. The organic radicals undergo subsequent oxidation and may combine with dissolved oxygen to form peroxy organic radicals ($ROO\cdot$), which subsequently undergo radical chain reactions that produce a variety of oxygenated byproducts. A general pattern of oxidation is presented in Eq. (10-34) (Bolton and Cater, 1994).

$$\text{Organic compound} \rightarrow \text{Aldehydes} \rightarrow \text{Carboxylic acids} \rightarrow \text{Carbon dioxide and mineral acids} \quad (10\text{-}34)$$

Carboxylic acids are of particular concern as the second order rate constants for these compounds are much lower than for most other organics, which may inhibit degradation of these constituents. Other byproducts that may be of concern are the halogenated acetic acids, formed from the oxidation of halogenated alkenes such as TCE (Crittenden et al., 2005).

Impact of Bicarbonate and Carbonate

High concentrations of carbonate and bicarbonate in some reclaimed water can react with $HO\cdot$ and reduce the efficiency of AOPs. Bicarbonate and carbonate ions are known scavengers of $HO\cdot$ radicals and reduce the rate of organics destruction significantly. Unfortunately, the concentrations of HCO_3^- and CO_3^{2-} are often three orders of magnitude higher than the organic pollutants targeted for destruction. Even low alkalinities (50 mg/L) reduce the rate of TCE destruction by a factor of 10 at a pH of 7 (Crittenden et al., 2005). However, at high pH a given alkalinity is more detrimental because the second order rate constant with CO_3^{2-} is much larger than HCO_3^-. Reclaimed water with high pH and alkalinity are more difficult to treat using AOPs. To overcome these difficulties and improve the effectiveness of AOPs, pretreatment processes such as softening or RO are used to remove the alkalinity.

Impact of pH

The pH affects AOP performance because it determines the distribution of the carbonate species, HCO_3^- and CO_3^{2-}, as discussed above. The pH will also control the concentration of HO_2^- (H_2O_2 has a pK_a of 11.6), which is important in H_2O_2 type AOPs. For example, in the H_2O_2/UV process, HO_2^- has about 10 times the UV absorbance at 254 nm (228 L/mol·cm) than does H_2O_2; consequently, H_2O_2/UV may be more effective at higher pH especially if the background water matrix absorbs a significant amount of UV light.

Raising the pH to improve the process performance would only be practical if the pH was raised for other purposes such as softening. Finally, pH affects the charge on the organic compounds if they are weak acids or bases. The reactivity and light absorption properties of the compound can be affected by its charge, an observation to be considered in the design of AOPs (Crittenden et al., 2005).

Impact of Metal Ions
Metal ions in reduced oxidation states, such as Fe(II) and Mn(II), can consume a significant quantity of chemical oxidants as well as scavenge HO· radicals. Consequently, the concentration of reduced metal ions should be measured as part of any treatability study and the dosage of oxidants needed should include the COD of the reduced metal species.

Impact of Other Factors
Other factors that also affect the treatment process include suspended material (which affects light transmission) and the type and nature of the residual TOC and COD. For example, NOM, which reacts with hydroxyl radicals, can have a great influence on the reaction rate. Because the chemistry of the water matrix is different for each reclaimed water, pilot testing is almost always required to test the technical feasibility, to obtain usable design data and information, and to obtain operating experience with a specific AOP.

Means of Overcoming Process Limitations
To overcome the problems noted above, AOPs are typically applied following treatment by RO. Further, if adequate reaction time is provided, greater than 99 percent of the organic constituents (as measured by a TOC mass balance) are mineralized (Stefan and Bolton, 1998; Stefan et al., 2000). Flowsheets demonstrating the application of AOPs are shown on Fig. 10-1.

10-7 PHOTOLYSIS

Photolysis is a process by which constituents are broken down by exposure and absorption of photons from a light source. In natural systems, sunlight is the light source for photolysis reactions; however, in engineered systems, UV lamps are used to produce the photonic energy. The photons that are absorbed cause the electrons in the outer orbital of some compounds to become unstable and split or become reactive. The effectiveness of the photolysis process depends, in part, on the characteristics of the reclaimed water, structure of the compounds, design of the photolysis reactor, and dose and wavelength of the applied light. The photolysis rate can be estimated from the rate at which the compound absorbs light and the photonic efficiency of the reaction (quantum yield).

Applications for Photolysis

Photolysis may be used for the removal of various compounds, such as NDMA (see Chap. 3) and other trace organic constituents. It should be noted that many compounds are not removed using photolysis alone, and that the addition of hydrogen peroxide can enhance the degradation of these constituents, however, the addition of hydrogen peroxide may actually reduce the photolysis of some compounds, such as NDMA (Linden et al., 2004). As described in Sec. 10-8, the photolysis of hydrogen peroxide, resulting in the formation of hydroxyl radicals, is an AOP known for its effective destruction of most organic compounds.

10-7 Photolysis

Photolysis Processes

Engineered photolysis reactions are conducted in specially designed reactors optimized with respect to UV dose. Photolysis reactors are typically comprised of a stainless steel column or pipe containing UV lamps arranged parallel to the flow, perpendicular to the flow, or in a crisscrossing pattern perpendicular to the direction of flow. An example of a reactor used for photolysis is shown on Fig. 10-19. Fouling that may occur on the outside of the protective quartz sleeve may be managed using an automatic cleaning system consisting of a collar that periodically moves along the lamp to remove precipitates and intercepted matter. When used in conjunction with RO pretreatment, the chemicals added to reduce the pH for scale control in the membranes will also reduce the potential for precipitation on the UV lamps.

Photolysis reactions are caused by light emission in the UV range (200 to 400 nm, see also Fig. 11-26 in Chap. 11). Three types of UV lamps used for photolysis processes: (1) low-pressure low-intensity, (2) low-pressure high-intensity, and (3) medium-pressure high-intensity lamps. Low-pressure lamps emit much of their energy at a wavelength of 254 nm while medium pressure lamps emit energy at multiple wavelengths (see Fig. 11-26 in Chap. 11). The type of lamp used and reactor configuration depends on the constituent to be removed the water matrix and site-specific conditions.

Basic Considerations for Photolysis Processes

Photolysis occurs when an electron in the outer orbital of a constituent molecule absorbs a photon and forms an unstable compound that splits apart or becomes reactive. Many nontarget constituents may be present in reclaimed water, depending on the prior treatment applied, that absorb light during the photolysis process. As an introduction to the photolysis process, the concepts involved in photolysis are presented in the following discussion for the case of a single absorbing solute. The fundamentals of photolysis

Figure 10-19
View of photolysis reactors for removal of NDMA from reclaimed water.

consist of: (1) absorption of UV light by a compound in water, (2) rate of photolysis, (3) electrical efficiency, and (4) photolysis process limitations.

Absorption of UV light

The absorption of light by a compound in water or other aqueous solution can be described using the Beer-Lambert Law. The absorbance of a solution is a measure of the amount of light absorbed by constituents in the solution using a spectrophotometer at a specified wavelength and over a fixed path length.

$$A(\lambda) = -\log\left(\frac{I}{I_o}\right) = \varepsilon(\lambda)\,Cx = k(\lambda)x \tag{10-35}$$

where $A(\lambda)$ = absorbance, dimensionless
I = light intensity after passing through solution containing constituents of interest at wavelength, λ, einsteins/cm$^2 \cdot$s (note: an einstein is equal to one mole of photons)
I_o = light intensity after passing through a blank solution (i.e., distilled water) of known depth (typically 1.0 cm) at wavelength, λ, einstein/cm$^2 \cdot$s
$\varepsilon(\lambda)$ = base 10 extinction coefficient or molar absorptivity of light-absorbing solute at wavelength λ, L/mole·cm
λ = wavelength, nm
C = concentration of light absorbing solute, mole/L
x = length of light path, cm
$k(\lambda)$ = absorptivity (base 10), 1/cm

The extinction coefficient is a function of wavelength because as the wavelength decreases more energetic photons are absorbed and the absorptivity of a light-absorbing compound increases. Values of the extinction coefficients for several compounds at various wavelengths are given in Table 10-12. The use of Eq. (10-35) is presented in Example 10-5.

Table 10-12
Selected quantum yields and extinction coefficients for compounds commonly found in water[a]

Compound	Primary quantum yield in aqueous phase, mole/einstein	Extinction coefficient at 253.7 nm, L/mole·cm
NO_3^-	—	3.8
HOCl (at 330 nm)	0.23	15
OCl$^-$	0.23	190
HOCl	—	53.4
OCl$^-$	0.52	155
O_3	0.5	3300
ClO_2	0.44	108
Sodium chlorite	0.72	—
TCE	0.54	9
PCE	0.29	205
NDMA	0.3	1974
Water	—	0.0000061

[a]Adapted from Crittenden et al., 2005.

EXAMPLE 10-5. UV Absorbance by NDMA.

The chemical compound NDMA (see Chap. 3) is commonly found at low concentrations in reclaimed water, even after treatment by RO. Estimate the absorptivity of NDMA ($C_2H_6N_2O$) at a wavelength of 254 nm, assuming NDMA is present at a concentration of 30 ng/L.

Solution

1. Convert the mass concentration of NDMA in solution to mole/L.

    Using the periodic table from the back inside cover of this textbook, the molecular weight of NDMA is 74.09 g/mole. The concentration is determined as follows:

$$C = \frac{(30 \text{ ng/L})}{(74.09 \text{ g/mole})} (1\text{g}/10^9 \text{ng}) = 4.05 \times 10^{-10} \text{ mole/L}$$

2. Compute the absorptivity of NDMA using Eq. (10-35).
    a. Determine the extinction coefficient, $\varepsilon(\lambda)$, of NDMA from Table 10-12. The extinction coefficient of NDMA at a wavelength of 254 nm is 1974 L/mole·cm.
    b. The absorptivity, $k(\lambda)$, of NDMA is:

$$k(\lambda) = \varepsilon(254)C = (1974 \text{ L/mole} \cdot \text{cm})(4.05 \times 10^{-10} \text{ mole/L})$$
$$= 8.0 \times 10^{-7} \text{cm}^{-1}$$

Comment

Because of the low concentration of NDMA in the water, the absorptivity is also low. When photolysis is to be used for removal of NDMA, other constituents in the water matrix that will absorb photons and the background absorbance of the water matrix need to be considered.

Light adsorption by a single compound in water was described in the previous analysis. In practice, however, a number of absorbing compounds will be present in solution. The absorption of light as it passes through a solution containing several different compounds may be determined by summing the absorption that would result from each individual compound as shown in the following expression:

$$\ln\left(\frac{I}{I_0}\right) = -\left[\sum \varepsilon'(\lambda)_i C_i\right] x \quad (10\text{-}36)$$

where $\varepsilon'(\lambda)_i$ = extinction coefficient of compound i at wavelength λ (base e), L/(mole·cm); note $\varepsilon'(\lambda)_i = 2.303 \, \varepsilon(\lambda)_i$

C_i = concentration of compound i, mole/L

Other terms as defined previously.

The relationship shown in Eq. (10-36) is based on a single incident wavelength, such as with low-pressure UV lamps. Multiple wavelength situations, such as with the use of medium-pressure UV lamps, can be determined using a similar approach, i.e., summing adsorption of each compound for each wavelength.

Energy Input for Photolysis

The lamp output and reactor size can be used to estimate the energy input for the photolysis reaction. The theoretical maximum photonic energy input per unit volume of the reactor can be determined using the following expression:

$$P_R = \frac{P \times \eta}{N_P \times V \times h\nu} \tag{10-37}$$

where P_R = photonic energy input per unit volume of the reactor, einstein/L·s
 P = lamp power, J/s (W)
 η = output efficiency at the wavelength of interest (as a fraction)
 N_P = number photons per mole (as einstein), 6.023×10^{23} 1/einstein
 V = reactor volume, L
 h = Planck's constant, 6.62×10^{-34} J·s
 $\nu = \frac{c}{\lambda}$ = frequency of light 1/s
 c = the speed of light, 3.00×10^8 m/s
 λ = wavelength of the light, m

While the above analysis is satisfactory for a theoretical assessment, the actual performance of a photoreactor is expected to be lower than computed using Eq. (10-37) due to light being absorbed by the reactor walls or blocked by the precipitate that forms on the lamp sleeve. While a safety factor specific to a particular system could be applied to compensate for these inefficiencies, pilot studies are used to obtain more reliable design criteria.

Rate of Photolysis

The rate at which a compound is photolyzed depends on the rate and frequency of photon absorption. The volumetric photon adsorption rate, derived from Eq. (10-36), is:

$$I_V = -\frac{dI}{dx} = \varepsilon'(\lambda) \cdot C \cdot I_o \cdot e^{-\varepsilon'(\lambda)Cx} \tag{10-38}$$

where I_V = rate that photons are absorbed per volume of solution at a particular point, einstein/cm³·s
 $\varepsilon'(\lambda)$ = base e extinction coefficient or molar absorptivity of light absorbing solute at wavelength $\lambda = 2.303\varepsilon(\lambda)$, L/mole·cm
Other terms as defined previously.

The quantum yield is a quantity used to describe the frequency at which photon absorption results in a photolysis reactions and is specific to the type of compound and the wavelength. The quantum yield, $\phi(\lambda)$, is defined as being equal to the number of photolysis reactions divided by the number of photons absorbed by the molecule as follows:

$$\phi(\lambda) = \frac{-r_R}{I_V} = \frac{\text{Reaction rate}}{\text{Rate of photon absorption}} \qquad (10\text{-}39)$$

where $\phi(\lambda)$ = quantum yield at wavelength λ, mol/einstein
r_R = photolysis rate, mole/(cm$^3 \cdot$s)

As a general rule, the quantum yield increases as wavelength decreases (increasing photonic energy). Selected quantum yields at a wavelength 254 nm are summarized in Table 10-12.

Typically, the light absorption by the component that is targeted for removal is minor as compared to the light absorption by the background water matrix (Crittenden et al., 2005). The pseudo first order rate law for the photolysis reaction is:

$$r_{avg} = \left[\phi(\lambda)P_R\frac{\varepsilon'(\lambda)}{k'(\lambda)}\right]C_i = kC_i \qquad (10\text{-}40)$$

where r_{avg} = overall average photolysis rate of the constituent in the reactor, mole/L$\cdot$s
$k'(\lambda)$ = measured absorptivity of the water matrix at wavelength (base e) λ, 1/cm
k = pseudo-first order rate coefficient, 1/s
Other terms as defined previously.

After obtaining the rate law, r_{avg}, an appropriate reactor model may be used for determination of the expected performance.

Electrical Efficiency

The electrical energy requirement for photolytic reactions is significant due to the process inefficiencies. Consequently, it is important to compare process efficiency on the basis of electrical usage per amount of compound destruction. One such measure is the electrical efficiency per log order (EE/O) of compound destruction (Bolton and Cater, 1994). The definition of EE/O is the electrical energy (in kWh) required to reduce the concentration of a constituent by one log order per unit volume of water.

$$\text{EE/O} = \frac{P \times t}{V \times \log\left[\frac{C_i}{C_f}\right]} \quad \text{(for batch systems)} \qquad (10\text{-}41)$$

$$\text{EE/O} = \frac{P}{Q \times \log\left[\frac{C_i}{C_f}\right]} \quad \text{(for continuous flow systems)} \qquad (10\text{-}42)$$

where EE/O = electrical efficiency per log order reduction, kWh/m³.
P = lamp power output, kW
t = irradiation time, h
V = reactor volume, m³
C_i = initial concentration, mg/L
C_f = final concentration, mg/L
Q = water flowrate, m³/h

For a flowthrough system, the power input can be divided by the EE/O to obtain an estimate of the flow rate that can be treated in a given reaction and achieve one order of magnitude reduction in concentration. Consequently, EE/O is a convenient measure because it can be used to estimate the energy that is required to reduce the contaminant concentration by one order of magnitude.

Based on currently available technology (2006), the required EE/O value for a one log order of reduction (i.e., 100 to 10) of NDMA using UV is on the order of 21 to 265 kWh/10³ m³·log order (0.08 to 1.0 kWh/10³ gal·log order) with a 5 to 6 mg/L dose of H_2O_2, although it does not appear that the peroxide is necessary (Soroushian et al., 2001).

EXAMPLE 10-6. Design of Direct Photolysis Process for NDMA.

A water reclamation plant produces 1.9×10^4 m³/d (5 Mgal/d) of RO effluent containing 50 ng/L of NDMA. Determine the number of photolysis reactors needed to reduce the NDMA concentration of the RO effluent to 1 ng/L prior to groundwater injection. The photolysis reactors under evaluation are 0.5 m in diameter and 1.5 m long with an effective water volume of 242 L. Each reactor has 72 lamps rated at 200 W per lamp and an output efficiency of 30 percent at 254 nm. Assume that the hydraulic detention time, τ, of the reactor can be described using the tanks in series model, $\tau = n[(C_e/C_o)^{1/n} - 1]/k$, where k is the reaction rate constant and n is the number of tanks in series. Use three tanks in series and neglect all other losses. The RO water has an absorptivity measured at a wavelength of 254 nm of $k'(\lambda) = 0.02$ cm^{-1}. Calculate the EE/O and daily energy usage for the photolysis process.

Solution
1. Calculate the photonic energy input per unit volume of the reactor.
   a. Calculate the total lamp power:
   $$P = (72 \text{ lamps} \times 200 \text{ W/lamp}) = 14{,}400 \text{ W} = 14{,}400 \text{ J/s}$$
   b. Calculate the photonic energy input for the reactor using Eq. (10-37)
   $$P_R = \frac{(14{,}400 \text{ J/s})(0.3)(254 \times 10^{-9} \text{ m})}{(6.023 \times 10^{23} \text{ 1/einstein})(6.62 \times 10^{-34} \text{ J·s})(3.0 \times 10^8 \text{ m/s})(242 \text{ L})}$$
   $$= 3.80 \times 10^{-5} \text{ einstein/L·s}$$

2. Calculate the rate constant for NDMA.
   a. The extinction coefficient of NDMA at 254 nm can be obtained from Table 10-12.

   $$\varepsilon(254) = 1974 \text{ L/mole·cm}$$
   $$\varepsilon'(254) = 2.303\ \varepsilon(254) = 2.303 \times 1974 = 4546 \text{ L/mole·cm}$$

   b. The quantum yield for NDMA can be obtained from Table 10-12.

   $$\phi(\lambda)_{NDMA} = 0.3 \text{ mole/einstein}$$

   c. Compute k_{NDMA} using Eq. (10-40).

   $$k_{NDMA} = \phi(\lambda)_{NDMA} P_R \frac{\varepsilon'(\lambda)_{NDMA}}{k'(\lambda)}$$

   $$= (0.3 \text{ mole/einstein})(3.80 \times 10^{-5} \text{ einstein/L·s}) \left[\frac{(4546 \text{ L/mole·cm})}{(0.01/\text{cm})} \right]$$

   $$= 2.59 \text{ 1/s}$$

3. Calculate the flowrate that can be treated per reactor.
   a. Calculate hydraulic detention time for the reactor.

   $$\tau = \frac{n[(C_{NDMA,o}/C_{NDMA,e})^{1/n} - 1]}{k_{NDMA}} = \frac{3[(50/1)^{1/3} - 1]}{(2.59 \text{ 1/s})} = 3.11 \text{ s}$$

   b. Calculate the flowrate that can be processed by one reactor.

   $$Q = \frac{V}{\tau} = \frac{242 \text{ L}}{3.11 \text{ s}} = 77.7 \text{ L/s}$$

4. Determine the number of reactors needed to treat the full flow.
   a. The total flow to be treated is $1.9 \times 10^4 \text{ m}^3/\text{d} = 219 \text{ L/s}$.
   b. The number of reactors needed is $(219 \text{ L/s})/(77.7 \text{ L/s}) = 2.8$ (use 3).
   c. The actual number of reactors needed will be greater than the computed value to compensate for lamp failure, fouling, and maintenance; and for peak flow conditions. It should be noted that the extra reactors will not be in continuous operation, but will only be used when needed or in a service rotation to reduce costs.

5. Calculate the EE/O for the photolysis process using Eq. (10-42).

   $$\text{EE/O} = \frac{P}{Q \times \log\left[\frac{C_i}{C_f}\right]}$$

   $$= \frac{14.4 \text{ kW} \times (10^3 \text{ L/m}^3)}{(77.7 \text{ L/s}) \times \left[\log\left(\frac{50 \text{ ng/L}}{1 \text{ ng/L}}\right)\right] \times (3600 \text{ s/h})} = 0.0303 \text{ kWh/m}^3$$

The computed EE/O value is low compared to the typical range for ground and surface waters because of the high quality effluent from the RO process. Reverse osmosis removes or reduces many of the constituents that would interfere with photolysis of specific constituents and can produce effluent with low absorbance, improving the efficiency of the photolysis process.

6. Estimate the overall daily energy usage for the process.

   For the two operational reactors, the estimated energy usage is:

   $$3 \text{ reactors} \times 14.4 \text{ kW} \times 24 \text{ h/d} = 1037 \text{ kWh/d}$$

Comment

The photolysis reactors sized in the design example represent a minimum size and do not include correction factors for nonideal flow, variability in lamp output, and other inefficiencies. Pilot studies are always required to determine actual design parameters.

Photolysis Process Limitations

The efficiency of the photolysis process depends in part on the characteristics of the water matrix and compounds targeted for degradation. For example, the extinction coefficient for residual organic matter varies over a wide range and may interfere with the photolysis of other compounds. In addition, the light energy input may be absorbed by other constituents, there may be photon losses upon reflection off the reactor wall, and the precipitate that builds up on the exterior surface of sleeves that cover the lamps due to the elevated temperature will reduce or block light transmission. For some constituents, the performance of direct photolysis processes have been improved by the addition of hydrogen peroxide (Linden et al., 2004), as described in Sec. 10-6. To overcome the limitations associated with photolysis processes related to absorbance of UV energy by nontarget constituents in water reuse, pretreatment using RO should be used to remove most of the interfering compounds and improve the overall process performance. Pilot studies should be conducted to characterize the expected efficiency of photolysis and the rate and characteristics of fouling of the lamp sleeve.

10-8 ADVANCED BIOLOGICAL TRANSFORMATIONS

Microbial systems, particularly bacteria, are capable of performing a wide diversity of reactions that can result in the transformation and degradation of many compounds in water to be reclaimed. Naturally occurring organisms are used in the activated sludge process for the oxidation of nonspecific organic compounds that constitute BOD and some reduced nutrients, such as ammonia. However, some constituents present in water to be reclaimed are only partially or not affected by activated sludge treatment for a number of reasons, including: (1) the compounds are not easily biodegradable,

(2) the organisms required for degradation of certain constituents are not present in sufficient numbers, and (3) an environmental condition is prohibiting biodegradation. Advanced biological transformation processes are now available and new processes are being developed that can be used to treat residual constituents of concern.

Microorganisms have the ability to carry out a number of reactions that affect the type and concentration of reclaimed water constituents. The organism may carry out these reactions for a variety of reasons, such as to obtain energy (catabolism) or to synthesize new cell mass (anabolism). In addition, some reactions may be carried out to detoxify the cell environment. A summary of potential microbial processes is presented in Table 10-13. The importance of energetics of constituent degradation, bioaugmentation, and biostimulation in the development of advanced biological treatment processes is considered briefly in the following discussion.

Basic Considerations for Advanced Biological Treatment Processes

Table 10-13

Microbial processes during wastewater treatment that affect the fate of trace constituents

Processes	Description
Accumulation	The storage of a compound within or on the surface of a microbial cell. Chemical substances can also be stored in external cell polymers.
Cometabolism	The inadvertent reaction of a microbial enzyme with a compound for which the organism cannot utilize. Other organisms may be able to utilize the breakdown products.
Electron acceptor utilization	The utilization of compounds participating in redox reactions. Requires the presence of a viable electron donor for reaction to proceed.
Immobilization	Change in the oxidation state or chemical structure of a compound in solution that results in the binding or fixing of the compound.
Mineralization	Conversion of an organic substance to CO_2 and mineral acids resulting from biodegradation.
Mobilization	Change in the oxidation state or chemical structure of an immobilized compound that results in the release into solution.
Nutrient utilization	The uptake of nutrients resulting from primary substrate utilization.
Primary substrate utilization	Degradation of substrate that serves as main source of carbon and energy.
Secondary substrate utilization	Degradation of substrate that serves as source of carbon and energy, but requires presence of primary substrate as well.
Transformation	Partial modification of a compound resulting from biodegradation. Transformations can also affect the toxicity of a compound (increase, no change, or decrease in toxicity) or the effectiveness of subsequent treatment processes.

Energetics of Constituent Degradation

Naturally, microorganisms will carry out reactions that are energetically most favorable. Energetically, oxygen is the most favorable electron acceptor and, when coupled with the glucose as the electron donor, yields a favorable reaction. Not all organisms can utilize oxygen and, therefore, these organisms may be outcompeted under aerobic conditions. However, if the oxygen is depleted, organisms that can utilize the next most energetically favorable electron acceptor may become dominant. Further, as the readily available electron donors are depleted, new electron donors must be utilized for cell energy and growth. As the electron donors are gradually depleted, even the most recalcitrant compounds can be degraded given favorable environmental conditions. As these reactions may yield only little energy to the cell, growth rates and reaction kinetics are reduced. Further, it may require time for an organism that is adapted to degrade the compound to reach sufficient population, or even for new organisms to evolve to utilize the compound.

Bioaugmentation

In some cases, microorganisms needed to conduct a given reaction may not be available in sufficient numbers or may be absent completely. Research is being conducted to determine the feasibility of inoculating a given treatment process with organisms capable of degrading a compound of interest (Maier et al., 2000). As more information becomes available regarding the capacity and growth requirements for certain organisms, the role of bioaugmentation in water reclamation treatment will be expanded in the future. Further, some researchers have identified pathways for the exchange of genetic material among bacteria, and are investigating the potential to genetically modify treatment organisms for the conversion of toxic constituents.

Biostimulation

In some environments, the conditions that occur during water reclamation treatment may limit the removal of a given constituent biologically. For example, some chlorinated hydrocarbons can be biodegraded through utilization as an electron acceptor under anaerobic conditions, but persist under aerobic conditions. Biostimulation is an approach to manipulate the environmental conditions to facilitate the breakdown or transformation of a certain constituent. In addition to the presence or absence of electron donors and acceptors, microbial processes can be affected by temperature, the presence of macro- and micronutrients, toxins, and bioavailability. While nutrients can be added directly as needed, toxins may need to be removed by adsorption. An understanding of cell requirements and of the physical cell environment is necessary to diagnose what types of biostimulation may be necessary to improve biodegradation.

Advanced Biological Treatment Processes

The two processes described below, biological activated carbon and membrane biofilm reactor, are under development for post secondary treatment and water reclamation. These processes may be representative of future progress in water reclamation, making use of current technologies but applying in new ways. Further, as microbial communities are quantified and understood more fully, future treatment processes may include systems composed of highly specialized microbial communities developed for a particular purpose.

Biological Activated Carbon

Biological activated carbon (BAC) filtration, used commonly for water treatment, is GAC in which biological activity is encouraged and used to treat organic matter often found in surface waters and groundwaters along with providing for filtration. The organic matter in surface waters is usually comprised of a complex mixture of compounds formed from the breakdown of plant and animal material. The mixture of compounds is collectively identified as NOM. The organic matter in groundwater, typically derived from the decay of humic materials, contributes to color and is quite common in groundwaters drawn from deep aquifers. Because the NOM found in surface and groundwater potable water supplies is stable and difficult to degrade, these compounds along with those produced during biological treatment will also be found in reclaimed water, and may need to be removed for specific applications. When ozone or chlorine is applied to disinfect water containing these organic compounds, disinfection byproducts (DBPs) will form and they need to be removed. Treatment by BAC has been found to be effective in the removal of DBPs (Wu and Xie, 2005; Wobma et al., 2000).

Many of the microorganisms that grow on the activated carbon can utilize the organic matter in water as a substrate for growth. However, because many of the compounds that comprise NOM are not easily converted biologically, a pretreatment step, such as ozonation or advanced oxidation in which larger organic molecules are converted to simpler compounds, is used to enhance the performance of BACs. For example, ozone dosages in the range from 1 to 2 mg/L have been used in a pretreatment step to break down and improve the biodegradability of NOM. Typical pilot facilities used to develop design parameters and to assess changing operating conditions is shown on Fig. 10-20.

(a) (b)

Figure 10-20

Typical pilot facility used at treatment plant to develop design parameters and to assess ongoing operation of the biological activated carbon (BAC) process: (a) ozone pretreatment system and (b) BAC columns.

The assimilable organic carbon (AOC) test is used to assess whether the NOM can be processed without pretreatment (Standard Methods, 2005). The type of treatment applied following BAC will depend in part on requirements of the reclaimed water application. Where ozone is used as a pretreatment step, it can also serve as the primary disinfectant. Chloramines are generally used following BAC filtration to maintain a chlorine residual in the water distribution system although free chlorine may be used when NOM is show.

Although the BAC process is reasonably stable, the organisms that grow on the carbon are sensitive to water quality parameters and temperature. In some cases, there may be nutrient deficiencies. To maintain the performance of the BAC filter, it has been found, based on practical experience, that the filters should be backwashed once per day, typically without a residual disinfectant. Backwashing once per day keeps the microorganisms on the carbon in an active growth condition and eliminates the growth of secondary forms, such as protozoa which graze on the bacteria. Grazing of the bacteria will reduce the observed performance of the filters. If secondary forms develop, the filters should be backwashed with chlorinated water and restarted. Longer periods between backwashing are possible, depending on the concentration of the constituents to be removed and the degree of pretreatment (e.g., prefiltration). Based on the experience with existing treatment facilities, the carbon bed may have to be replaced every 3 to 5 yr (Bonné et al., 2002).

Membrane Biofilm Reactor

The membrane biofilm reactor (MBfR) is a biological process that has been developed for the removal of specific constituents from water when the limiting factor for biological metabolism or transformation is an electron acceptor (e.g., oxygen gas) or donor (hydrogen gas, methane gas). The technology is used to supply the electron donor or acceptor through a hollow-fiber membrane to the treatment organism, which grows as a biofilm on the outside of the membrane. Constituents that have been evaluated for removal by microbially mediated reduction using an MBfR include perchlorate, chlorinated solvents (TCE), bromate, selenate, heavy metals (chromate), and radionuclides (Nerenberg, 2005).

For example, in one autotrophic denitrification process, denitrifying bacteria are supplied with hydrogen gas that diffuses outward in the radial direction through the membrane, providing dissolved hydrogen gas to the biofilm. A typical pilot-scale MBfR process used for the denitrification of groundwater is shown on Fig. 10-21 (Lee and Rittmann, 1999). Biological denitrification is carried out by the autotrophic biofilm as NO_3^- is used as a terminal electron acceptor for respiration while hydrogen is used as the electron donor under anoxic conditions. As the thickness of the biofilm increases, the transfer of the electron donor through the membrane becomes less efficient and the biofilm must be scoured off to maintain removal rates.

Limitations of Advanced Biological Transformation Processes

The efficiency of advanced biological processes is limited by a number of potentially complicating factors. While biostimulation, as discussed above, may be used to provide for some of the cell requirements, other issues should also be considered. Microbial systems are inherently complex and difficult to model accurately. Further, sensitivity to

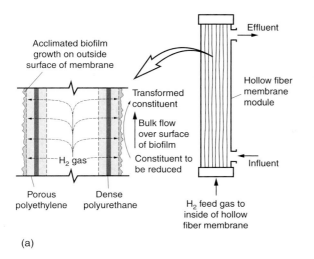 

Figure 10-21
Membrane biofilm reactor process: (a) schematic of the bench scale model and (b) view of pilot scale MBfR reactor used for the removal of nitrate. (Courtesy of Applied Process Technology, Inc.)

upstream process upsets, changes in water quality, and microbial acclimation to the constituents to be removed can affect the reliability of a biological process. Therefore, precautions should be taken to ensure the stability of biological processes.

PROBLEMS AND DISCUSSION TOPICS

10-1 Given the following list of common water reuse applications, select one or more specific constituents or constituent categories that may need to be removed, if applicable, from tertiary treated effluent prior to reuse. Develop a hypothetical treatment process flow diagram for each of the applications.

Vehicle washing

Landscape irrigation

Fodder and fiber crops

Food crop for uncooked human consumption

Groundwater recharge for indirect potable reuse

Wetlands for wildlife habitat

10-2 Select a trace constituent of concern in water reuse and review the current literature to determine the range of concentration values detected in reclaimed water. Comment on the range and distribution of the values and possible reasons for the variation, if any.

10-3 Using the following isotherm test data, determine the type of model that best describes the data and the corresponding model parameters. Assume that a 1 L sample volume was used for each of the isotherm experiments.

Mass of GAC, m, g	Equilibrium concentration of adsorbate in solution, C_e, µg/L			
	A	B	C	D
0	5.8	26	158.2	25.3
0.001	3.9	10.2	26.4	15.89
0.01	0.97	4.33	6.8	13.02
0.1	0.12	2.76	1.33	6.15
0.5	0.022	0.75	0.5	2.1

10-4 Using the results from Problem 10-3, determine the amount of activated carbon that would be required to treat a flow of 4800 m³/d to a final COD concentration of 2 mg/L if the COD concentration after secondary treatment is equal to 30 mg/L.

10-5 Design a fixed-bed activated carbon process using the following data. Determine the number of contactors, mode of operation, carbon requirements, and corresponding bed life. Ignore the effects of biological activity within the column.

Parameter	Unit	System			
		A	B	C	D
Compound		Chloroform	Heptachlor	Methylene chloride	NDMA
Flowrate	m³/d	8×10^3	4×10^3	0.5×10^3	16×10^3
C_o	ng/L	500	50	2000	200
C_e	ng/L	50	10	10	10
GAC density	g/L	450	450	450	450
EBCT	min	10	10	10	10

10-6 Referring to the data presented in Table 10-5, prepare a list of the top five most and least readily adsorbable substances.

10-7 The following normalized test data were obtained using a 10 g (Resin A) and 15 g (Resin B) sample of resin. The concentration of calcium chloride solution used for the titration was 100 mg/L. The resin density is 690 kg/m³ for Resin A is 720 kg/m³. Determine the cation exhange capacity and the amount of resin (A or B to be selected by instructor) that would be required to treat a flowrate of 2500 m³/d to reduce the concentration of Cr^{6+} from 500 to 50 ng/L?

	Resin A		Resin B	
Throughput volume, L	Cl^-	Ca^{2+}	Cl^-	Ca^{2+}
0	0	0	0	0
5	0.02	0	0	0
10	0.2	0.04	0	0
15	0.56	0.18	0	0
20	0.9	0.36	0.02	0
25	0.99	0.56	0.24	0
30	1	0.74	0.48	0.02
35	1	0.88	0.71	0.04
40	1	0.98	0.88	0.07
45		1	0.98	0.12
50		1	1	0.24
55		1	1	0.47
60			1	0.75
65				0.94
70				1
75				1
80				1

10-8 Determine the exchange capacity for one of the resins given in Problem 10-7 (resin to be selected by instructor). How much resin would be required to treat a flowrate of 4800 m³/d to reduce the concentration of calcium Mg^{2+} from 115 to 15 mg/L? Size the exchange process, including the number of reactors and mode of operation.

10-9 Comment on the application of ion exchange versus RO (see Chap. 9) for the preparation of reclaimed water for indirect potable reuse, including the water quality requirements, facility requirements, and overall process advantages and disadvantages.

10-10 Determine the hydroxyl radical concentration required to remove each of the following compounds using an advanced oxidation process with a contact time of 10 s. Comment on the feasibility of removing each of the compounds under the given conditions.

	Concentration, µg/L			
	Water A		Water B	
Compound	Influent	Effluent	Influent	Effluent
Chlorobenzene	100	5	120	7
Chloroethene	100	5	150	5
TCE	100	5	180	10
Toluene	100	5	200	15

10-11 Design an AOP to achieve a 95% reduction of one of the following compounds for a flowrate of 3800 m³/d. Specify reactor dimensions and hydroxyl radical concentration required for the process.

Compound	Initial concentration, µg/L
A	25
B	10
C	100
D	75

10-12 A water reclamation plant produces 1×10^5 m³/d of effluent containing 100 ng/L of NDMA. Determine the number of photolysis reactors needed to reduce the NDMA concentration of the RO effluent to 10 ng/L prior to indirect potable reuse using absorptivity values of $k'(\lambda) = 0.01, 0.05$, and 0.1 cm^{-1} (measured at a wavelength of 254 nm). The photolysis reactors under evaluation are 0.5 m in diameter and 1.5 m long with an effective water volume of 250 L. Each reactor has 25 lamps rated at 500 W per lamp and an output efficiency of 30 percent at 254 nm. Assume that the reactors operate as four mixed tanks in series and neglect all other losses, lamp fouling, and process inefficiencies. Calculate EE/O and daily energy usage for the photolysis process. Comment on the importance of absorptivity and recommend an appropriate pretreatment process.

10-13 Estimate the electricity cost (based on the current price of electricity) to treat a flow of 3800 m³/d with a NDMA concentration of 100 ng/L to 10 ng/L using a photolysis unit.

10-14 For the following list of compounds, determine the most suitable method of treatment to reduce the concentration from 100 µg/L to 10 µg/L using the advanced wastewater treatment processes discussed in this and Chap. 9.

Benzene

Chloroform

Dieldrin

Heptachlor

N-Nitrosodimethylamine

Trichloroethylene (TCE)

Vinyl chloride

10-15 Compare and contrast the differences and similarities of the MBfR discussed in this chapter, and the MBR discussed in Chap. 7.

REFERENCES

Anderson, R. E. (1975) "Estimation of Ion Exchange Process Limits by Selectivity Calculations," in I. Zwiebel and N. H. Sneed (eds.) *Adsorption and Ion Exchange*, AIChE Symposium Series, **71**, 152, 236.

Anderson, R. E. (1979) "Ion Exchange Separations," in P. A. Scheitzer (ed.), *Handbook of Separation Techniques For Chemical Engineers*, McGraw-Hill, New York.

Bilello, L. J., and B. A. Beaudet (1983) "Evaluation of Activated Carbon by the Dynamic Minicolumn Adsorption Technique," in M. J. McGuire and I. H. Suffet (eds.) *Treatment of Water by Granular Activated Carbon*, American Chemical Society, Washington, DC.

Bohart, G. S., and E. Q. Adams (1920) "Some Aspects of the Behavior of Charcoal with Respect to Chlorine," *J. Am. Chem. Soc.* **42**, 523–544.

Bolton, J. R., and S. R. Cater (1994) "Homogeneous Photodegradation of Pollutants in Contaminated Water: An Introduction," in Helz, G. R. (ed.) *Aquatic and Surface Photochemistry*, CRC Press, Boca Raton, FL.

Bonné, P. A. C., J. A. M. H. Hofman, and J. P. van der Hoek (2002) "Long Term Capacity of Biological Activated Carbon Filtration for Organics Removal," *Water Sci. Technol.: Water Supply*, **2**, 1, 139–146.

Buxton, G. V., and C. L. Greenstock (1988) "Critical Review of Rate Constants for Reactions of Hydrated Electrons, Hydrogen Atoms and Hydroxyl Radicals in Aqueous Solution," *J. Phys. Chem. Ref. Data*, **17**, 2, 513–886.

Chen, W. R., C. M. Sharpless, K. G. Linden, and I. H. Suffet. (2006) "Treatment of Volatile Organic Chemicals on the EPA Contaminant Candidate List Using Ozonation and O_3/H_2O_2 Advanced Oxidation Process," *Environ. Sci. Technol.*, **40**, 8, 2734–2739.

Chin, A., and P. R. Bérubé (2005) "Removal of Disinfection By-Product Precursors with Ozone-UV Advanced Oxidation Process," *Water Res.*, **39**, 2136–2144.

Clifford, D. A. (1999) "Ion Exchange and Inorganic Adsorption," Chap. 9, in R. D. Letterman (ed.), *Water Quality and Treatment: A Handbook of Community Water Supplies*, 5th ed., AWWA, McGraw-Hill, New York.

Crittenden, J. C., P. Luft, D. W. Hand, J. L. Oravitz, S. W. Loper, and M. Art (1985) "Prediction of Multicomponent Adsorption Equilibria Using Ideal Adsorption Solution Theory," *Environ. Sci. Technol.*, **19**, 11, 1037–1043.

Crittenden, J. C., D. W. Hand, H. Arora, and B. W. Lykins, Jr. (1987a) "Design Considerations for GAC Treatment of Organic Chemicals," *J. AWWA*, **79**, 1, 74–82.

Crittenden, J. C., T. F. Speth, D. W. Hand, P. J. Luft, and B. W. Lykins, Jr. (1987b) "Multicomponent Competition in Fixed Beds," *J. Environ. Eng. Div. ASCE,* 113, EE6, 1364–1375.

Crittenden, J. C., P. J. Luft, and D. W. Hand (1987c) "Prediction of Fixed-Bed Adsorber Removal of Organics in Unknown Mixtures," *J. Environ. Eng. Div.*, ASCE, 113, EE3, 486–498.

Crittenden, J. C., P. S. Reddy, H. Arora, J. Trynoski, D. W. Hand, D. L. Perram, and R. S. Summers (1991) "Predicting GAC Performance with Rapid Small-Scale Column Tests," *J. AWWA*, **83**, 1, 77–87.

Crittenden, J., S. Hu, D. Hand, and S. Green (1999) "A Kinetic Model for H_2O_2/UV Process in a Completely Mixed Batch Reactor," *Water Res.*, **33**, 10, 2315–2328.

Crittenden, J. C., R. R. Trussell, D. W. Hand, K. J. Howe, and G. Tchobanoglous (2005) *Water Treatment: Principles and Design*, 2nd ed., John Wiley & Sons, Hoboken, NJ.

Dimotsis, G. L., and F. McGarvey (1995) "A Comparison of a Selective Resin with a Conventional Resin for Nitrate Removal," *IWC*, No. 2.

Dobbs, R. A., and J. M. Cohen (1980) *Carbon Adsorption Isotherms for Toxic Organics*, EPA-600/8-80-023, U.S. Environmental Protection Agency, Washington, DC.

Eckenfelder, W. W., Jr. (2000) *Industrial Water Pollution Control*, 3rd ed., McGraw-Hill, Boston, MA.

Elovitz, M. S., and U. von Gunten, (1999) "Hydroxyl Radical/Ozone Ratios during Ozonation Processes," *Ozone Sci. Eng.*, **21**, 239–260.

Froelich, E. M. (1978) "Control of Synthetic Organic Chemicals by Granular Activated Carbon: Theory, Application and Reactivation Alternatives," Presented at the Seminar on Control of Organic Chemical Contaminants in Drinking Water, Cincinnati, OH.

Glaze, W. H., J. W. Kang, and D. H. Chapin (1987) "The Chemistry of Water Treatment Processes Involving Ozone, Hydrogen Peroxide, and Ultraviolet Radiation," *Ozone Sci. Eng.*, **9**, 4, 335–342.

Glaze, W. H., and J. W. Kang, (1990) "Chemical Models of Advanced Oxidation Processes," In *Proceedings Symposium on Advanced Oxidation Processes,* Wastewater Technology Centre Environment Canada, Burlington, Ontario, Canada.

Hammann, D., M. Bourke, and C. Topham (2004) "Evaluation of a Magnetic Ion Exchange Resin to Meet DBP Regs at the Village of Palm Springs," *J. AWWA*, **96**, 2, 46–50.

Huber, M. M., A. Göbel, A. Joss, N. Hermann, D. Löffler, C. S. Mcardell, A. Ried, H. Siegrist, T. A. Ternes, and U. von Gunten (2005) "Oxidation of Pharmaceuticals during Ozonation of Municipal Wastewater Effluents: A Pilot Study," *Environ. Sci. Technol.* **39**, 11, 4290–4299.

Karimi, A. A., J. A. Redman, W. H. Glaze, and G. F. Stolarik (1997) "Evaluating an AOP for TCE and OPCE Removal," *J. AWWA*, **89**, 8, 41–53.

Kawamura, S. (2000) *Integrated Design and Operation of Water Treatment Facilities*, 2nd ed., John Wiley & Sons, New York.

LaGrega, M. D., P. L. Buckingham, and J. C. Evans (2001) *Hazardous Waste Management*, McGraw-Hill Book Company, Boston, MA.

Lee, K. C., and B. E. Rittmann (1999) "A Novel Hollow-Fiber Membrane Biofilm Reactor for Autohydrogenotrophic Denitrification of Drinking Water," *Water Sci. Technol.*, **41**, 4, 219–226.

Linden, K. G., C. M. Sharpless, S. A. Andrews, K. Z. Atasi, V. Korategere, M. Stefan, and I. H. M. Suffet (2004) "Innovative UV Technologies to Oxidize Organic and Organoleptic Chemicals," *AWWA Research Foundation*, Denver, CO.

Mahar, E., A. Salveson, N. Pozos, S. Ferron, and C. Borg (2004) "Peroxide and Ozone: A New Choice For Water Reclamation and Potable Reuse," In *Proceedings of WateReuse Assocation's 9th Annual WateReuse Symposium*, September 19–22, 2004, Phoenix, AZ.

Maier, R. M., I. L. Pepper, and C. P. Gerba (2000) *Environmental Microbiology*. Academic Press, San Diego, CA.

McGarvey, F., B. Bachs, and S. Ziarkowski (1989) "Removal of Nitrates from Natural Water Supplies," Presented at the Amer. Chem. Soc. Meeting, Dallas, TX.

McMurry, J., and R. C. Fay, (1998) *Chemistry*, 2nd ed., Prentice-Hall, New York.

Nerenberg, R. (2005) "Membrane Biofilm Reactors for Water and Wastewater Treatment," In *Proceedings of 2005 Borchardt Conference: A Seminar on Advances in Water and Wastewater Treatment*, February 23–25, Ann Arbor, MI.

Ouki, S. K., and M. Kavanaugh (1999) "Treatment of Metals-Contaminated Wastewaters by use of Natural Zeolites," *Water Sci. Technol.*, **39**, 10–11, 115–122.

Pinkston, K. E., and D. L. Sedlak (2004) "Transformation of Aromatic Ether- and Amine-Containing Pharmaceuticals during Chlorine Disinfection," *Environ. Sci. Technol.*, **38**, 14, 4019–4025.

Rakness, K L. (2005) *Ozone in Drinking Water Treatment: Process Design, Operation and Optimization*, American Water Works Association, Denver, CO.

Rice, R. G. (1996) Ozone Reference Guide, Prepared for the Electric Power Research Institute, Community Environment Center, St. Louis, MO.

Rose, J., P. Hauch, D. Friedman, and T. Whalen (1999) "The Boiling Effect: Innovation for Achieving Sustainable Clean Water," *Water* **21**, 9–10, 16.

Rosene, M. R., R. T. Derthorn, J. R. Lutchko, and N. J. Wagner (1983) "High pressure Technique for Rapid Screening of Activated Carbons for Use in Potable Water," in M. J. McGuire and I. H. Suffet (eds.) *Treatment of Water by Granular Activated Carbon*, Amer. Chem. Soc., Washington, DC.

Rosenfeldt, E. J., and K. G. Linden (2004) "Degradation of Endocrine Disrupting Chemicals Bisphenol A, Ethinyl Estradiol, and Estradiol During UV Photolysis and Advanced Oxidation Processes," *Environ. Sci. Technol.*, **38**, 20, 5476–5483.

Sawyer, C. N., P. L. McCarty, and G. F. Parkin (2003) *Chemistry for Environmental Engineering*, 5th ed., McGraw-Hill, New York.

SES (1994) *The UV/Oxidation Handbook*, Solarchem Environmental Systems, Markham, Ontario, Canada.

Shaw, D. J. (1996) *Introduction to Colloid and Surface Chemistry*, Buttermorth, London, England.

Singer, P. C., and D. A. Reckhow (1999) "Chemical Oxidation," Chap. 12, in R. D. Letterman, (ed.), *Water Quality And Treatment: A Handbook of Community Water Supplies*, 5th ed., AWWA, McGraw-Hill, New York, NY.

Slater, M. J. (1991) *Principles of Ion Exchange Technology*, Butterworth Heinemann, New York.

Snoeyink, V. L., and R. S. Summers (1999) "Adsorption of Organic Compounds," Chap. 13, in R. D. Letterman, ed., *Water Quality and Treatment: A Handbook of Community Water Supplies*, 5th ed., *AWWA*, McGraw-Hill, New York.

Snoeyink, V. L., C. Campos, and B. J. Marinas (2000) "Design and Performance of Powered Activated Carbon/Ultrafiltration Systems," *Water Sci. Technol.*, **42**, 12, 1–10.

Sontheimer, H., J. C. Crittenden, and R. S. Summers (1988) *Activated Carbon For Water Treatment*, 2nd ed., in English, DVGW-Forschungsstelle, Engler-Bunte-Institut, Universitat Karlsruhe, Germany.

Sontheimer, H., and C. Hubele (1987). "The Use of Ozone and Granulated Activated Carbon in Drinking Water Treatment," in P. M. Huck and P. Toft (eds.) *Treatment of Drinking Water for Organic Contaminants*, Pergamon Press, City.

Soroushian, F., Y. Shen, M. Patel, and M. Wehner (2001) "Evaluation and Pilot Testing of Advanced Treatment Processes for NDMA Removal and Reformation," in *Proceedings of the AWWA Annual Conference*, AWWA, Washington, DC.

Standard Methods (2005) *Standard Methods for the Examination of Water and Wastewater*, 21st ed., American Public Health Association, Washington, DC.

Stefan, M. I., and Bolton, J. R. (1998) "Mechanism of the Degradation of 1, 4-Dioxane in Dilute Aqueous Solution Using the UV/Hydrogen Peroxide Process," *Environ. Sci. Technol.*, **32**, 11, 1588–1595.

Stefan, M. I., J. Mack, and J. R. Bolton (2000) "Degradation Pathways during the Treatment of Methyl Tert-Butyl Ether by the UV/H_2O_2 Process," *Environ. Sci. Technol.*, **34**, 4, 650–658.

Tchobanoglous, G., F. L. Burton, and H. D. Stensel (2003) *Wastewater Engineering: Treatment, and Reuse*, 4th ed, McGraw-Hill, New York.

U.S. EPA (1998) *Advanced Photochemical Oxidation Processes*, EPA 625-R-98-004, Office of Research and Development, U.S. Environmental Protection Agency, Washington, DC.

U.S. EPA (1999) *Alternative Disinfectants and Oxidants Guidance Manual*, EPA 815-R-99-014, U.S. Environmental Protection Agency, Cincinnati, OH.

Westerhoff, P., G. Aiken, G. Amy, and J. Debroux (1999) "Relationships between the Structure of Natural Organic Matter and Its Reactivity Towards Molecular Ozone and Hydroxyl Radicals," *Water Res.*, **33**, 10, 2265–2276.

White, G. C. (1999) *Handbook of Chlorination and Alternative Disinfectants*, 4th ed., John Wiley & Sons, New York.

Wobma, P., D. Pernitsky, B. Bellamy, K. Kjartanson, and K. Sears (2000) "Biological Filtration for Ozone and Chlorine DBP Removal," *Ozone Sci. Eng.*, **22**, 4, 393–413.

Wu, H. W., and Y. F. F. Xie (2005) "Effects of EBCT and Water Temperature on HAA Removal Using BAC," *J. AWWA*, **97**, 11, 94–101.

Zamora, R., R. M., R. Schouwenaars, A. Durán Moreno, and G. Buitrón Méndez, (2000) "Production of Activated Carbon from Petroleum Coke and its Application in Water Treatment for the Removal of Metals and Phenol," *Water Sci. Technol.*, **42**, 5–6, 119–126.

11 Disinfection Processes for Water Reuse Applications[a]

WORKING TERMINOLOGY 600

11-1 DISINFECTION TECHNOLOGIES USED FOR WATER RECLAMATION 602
Characteristics for an Ideal Disinfectant 602
Disinfection Agents and Methods in Water Reclamation 602
Mechanisms Used to Explain Action of Disinfectants 604
Comparison of Reclaimed Water Disinfectants 605

11-2 PRACTICAL CONSIDERATIONS AND ISSUES FOR DISINFECTION 606
Physical Facilities Used for Disinfection 606
Factors Affecting Performance 609
Development of the $C_R t$ Concept for Predicting Disinfection Performance 616
Application of the $C_R t$ Concept for Reclaimed Water Disinfection 617
Performance Comparison of Disinfection Technologies 618
Advantages and Disadvantages of Alternative Disinfection Technologies 618

11-3 DISINFECTION WITH CHLORINE 622
Characteristics of Chlorine Compounds 622
Chemistry of Chlorine Compounds 624
Breakpoint Reaction with Chlorine 626
Measurement and Reporting of Disinfection Process Variables 631
Germicidal Efficiency of Chlorine and Various Chlorine Compounds in Clean Water 631
Form of Residual Chlorine and Contact Time 631
Factors that Affect Disinfection of Reclaimed Water with Chlorine 633
Chemical Characteristics of the Reclaimed Water 635
Modeling the Chlorine Disinfection Process 639
Required Chlorine Dosages for Disinfection 641
Assessing the Hydraulic Performance of Chlorine Contact Basins 644
Formation and Control of Disinfection Byproducts 650
Environmental Impacts 654

11-4 DISINFECTION WITH CHLORINE DIOXIDE 654
Characteristics of Chlorine Dioxide 655
Chlorine Dioxide Chemistry 655
Effectiveness of Chlorine Dioxide as a Disinfectant 655
Byproduct Formation and Control 656
Environmental Impacts 657

[a]Adapted in part from Tchobanoglous et al., 2003.

11-5 DECHLORINATION 657
Dechlorination of Reclaimed Water Treated with Chlorine and Chlorine Compounds 657
Dechlorination of Chlorine Dioxide with Sulfur Dioxide 660

11-6 DISINFECTION WITH OZONE 660
Ozone Properties 660
Ozone Chemistry 661
Ozone Disinfection Systems Components 662
Effectiveness of Ozone as a Disinfectant 666
Modeling the Ozone Disinfection Process 666
Required Ozone Dosages for Disinfection 669
Byproduct Formation and Control 670
Environmental Impacts of Using Ozone 671
Other Benefits of Using Ozone 671

11-7 OTHER CHEMICAL DISINFECTION METHODS 671
Peracetic Acid 671
Combined Chemical Disinfection Processes 672

11-8 DISINFECTION WITH ULTRAVIOLET RADIATION 674
Source of UV Radiation 674
Types of UV Lamps 674
UV Disinfection System Configurations 678
Mechanism of Inactivation by UV Irradiation 682
Factors Affecting Germicidal Effectiveness of UV Irradiation 684
Modeling the UV Disinfection Process 690
Estimating UV Dose 691
Ultraviolet Disinfection Guidelines 700
Analysis of a UV Disinfection System 708
Operational Issues with UV Disinfection Systems 708
Environmental Impacts of UV Irradiation 711

PROBLEMS AND DISCUSSION TOPICS 712

REFERENCES 718

WORKING TERMINOLOGY

Term	Definition
Absorbance	A measure of the amount of light of a specified wavelength that is absorbed by a solution and the constituents in the solution.
Breakpoint chlorination	A process whereby enough chlorine is added to react with all oxidizable substances in reclaimed water, such that if additional chlorine is added, it will remain as free chlorine.
Chlorine residual, total	The concentration of free or combined chlorine in reclaimed water, measured after a specified time period following addition. Chlorine residual is measured most commonly amperometrically.

Combined chlorine	Chlorine combined with other compounds [e.g., monochloramine (NH_2Cl), dichloramine ($NHCl_2$), and nitrogen trichloride (NCl_3), among others]. Combined chlorine is measured most commonly amperometrically.
Combined chlorine residual	Chlorine residual comprised of combined chlorine compounds [e.g., monochloramine (NH_2Cl), dichloramine ($NHCl_2$), and nitrogen trichloride (NCl_3) and others].
$C_R t$	The product of chlorine residual expresses in mg/L and contact time expressed in min. The term $C_R t$ is used to assess the effectiveness of the disinfection process.
Dechlorination	The removal of residual chlorine and chlorine compounds from solution by a reducing agent such as sulfur dioxide or by reacting it with activated carbon.
Disinfectant	A chemical (e.g., chlorine) or physical agent (e.g., UV radiation) that inactivates or destroys pathogens.
Disinfection	The partial destruction and inactivation of disease-causing organisms from exposure to chemical agents (e.g., chlorine) or physical agents (e.g., UV radiation).
Disinfection byproducts (DBPs)	Chemicals that are formed with the residual organic matter found in reclaimed water as a result of the addition of a strong oxidant (e.g., chlorine or ozone) for the purpose of disinfection.
Dose	As used in disinfection practice, dose is defined as the concentration or intensity of a disinfecting agent times the exposure time.
Dose response curve	The relationship between the degree of microorganism inactivation and the dose of the disinfectant.
Free chlorine	The total quantity of hypochlorous acid (HOCl) and hypochlorite ion (OCl^-) in solution.
Inactivation	Rendering microorganisms incapable of reproducing, and thus their ability to cause disease.
Natural organic matter (NOM)	Dissolved and particulate organic constituents that are derived typically from three sources: (1) the terrestrial environment (mostly humic materials), (2) the aquatic environment (algae and other aquatic species and their byproducts), and (3) the microorganisms in the biological treatment process.
Pathogens	Microorganisms capable of causing diseases of varying severity.
Photoreactivation/dark repair	The ability of microorganisms to repair the damage caused by exposure to UV radiation.
Reduced equivalent dose (RED)	The inactivation observed through the UV disinfection system as compared to the UV dose response derived from a collimated beam dose response study.
Sterilization	The total destruction of disease-causing and other organisms.
Total chlorine	The sum of the free and combined chlorine.
Transmittance	The ability of a solution to transmit light. Transmittance is related to absorbance.
Ultraviolet (UV) light	Electromagnetic radiation with a wavelength less than that of visible light in the range from 100 to 400 nm.
Ultraviolet (UV) irradiation	A disinfection process in which exposure to UV radiation (or light) is used to inactivate microorganisms.

Because of the critical importance of the disinfection process in water reuse applications, the purpose of this chapter is to introduce the reader to the important issues that must be considered in the disinfection of reclaimed water with various disinfectants to render it safe for reuse in a variety of applications. The four categories of human enteric organisms found in reclaimed water that are of the greatest consequence in producing disease are bacteria, protozoan oocysts and cysts, helminths, and viruses. Diseases caused by these waterborne microorganisms have been discussed previously in Chap. 3. Disinfection, the primary subject of this chapter, is the process used to achieve a given level of destruction or inactivation of pathogenic organisms. Because not all the organisms present are destroyed during the process, the term *disinfection* is differentiated from the term *sterilization*, which is the destruction of all organisms. Dechlorination is a process used to remove residual chlorine from reclaimed water for the purpose of protecting natural resources (e.g., fish) that may be adversely impacted by chlorine. Dechlorination is also discussed in this chapter as it is used in concert with chlorine disinfection.

To delineate the issues involved in disinfection, the following topics are considered: (1) an overview of disinfection technologies used for water reuse applications, (2) practical considerations and issues in disinfection of reclaimed water, (3) disinfection with chlorine and related compounds, (4) disinfection with chlorine dioxide, (5) dechlorination, (6) disinfection with ozone, (7) disinfection with other chemicals, and (8) disinfection with UV radiation. Design details for disinfection systems may be found in the companion textbook *Wastewater Engineering* (Tchobanoglous et al., 2003).

11-1 DISINFECTION TECHNOLOGIES USED FOR WATER RECLAMATION

Before discussing the practical aspects of disinfection and the individual disinfection technologies that follow in detail, it is appropriate to consider the characteristics of an ideal disinfectant, the major types of disinfection agents used for water reclamation, and to provide a general comparison between disinfectants.

Characteristics for an Ideal Disinfectant

To provide a perspective on the disinfection of reclaimed water, it is useful to consider the characteristics of an ideal disinfectant as given in Table 11-1. As reported, an ideal disinfectant would have to possess a wide range of characteristics such as being safe to handle and apply, stable in storage, toxic to microorganisms, nontoxic to higher forms of life, and soluble in water or cell tissue. It is also important that the strength or concentration of the disinfectant be measurable in reclaimed water. The latter consideration is an issue with the use of ozone, where little or no residual remains after disinfection, and UV disinfection where no residual is measurable.

Disinfection Agents and Methods in Water Reclamation

Disinfection in water reuse applications is accomplished most commonly by the use of chemical agents and irradiation. Each of these techniques is considered briefly in the following discussion. Other methods of disinfection are mentioned for completeness.

Chemical Agents

Chlorine and its compounds and ozone are the principal chemical compounds employed for the disinfection of reclaimed water. Other chemical agents that have been

Table 11-1 Characteristics of an ideal chemical disinfectant[a]

Characteristic	Properties/response
Alteration of solution characteristics	Should be effective with minimum alteration of the solution characteristics such as increasing the total dissolved solids (TDS)
Availability	Should be available in large quantities and reasonably priced
Deodorizing ability	Should deodorize while disinfecting
Homogeneity	Must be uniform in composition
Interaction with extraneous material	Should not be absorbed by organic matter other than bacterial cells
Noncorrosive and nonstaining	Should not disfigure metals or stain clothing
Nontoxic to higher forms of life	Should be toxic to microorganisms and nontoxic to humans and other animals
Penetration	Should have the capacity to penetrate through particle surfaces
Safety	Should be safe to transport, store, handle, and use
Solubility	Must be soluble in water or cell tissue
Stability	Should have low loss of germicidal action with time on standing
Toxicity to microorganisms	Should be effective at high dilutions
Toxicity at ambient temperatures	Should be effective in ambient temperature range

[a] Adapted from Tchobanoglous et al. (2003).

used as disinfectants in different applications include: (1) bromine, (2) iodine, (3) phenol and phenolic compounds, (4) alcohols, (5) heavy metals and related compounds, (6) dyes, (7) soaps and synthetic detergents, (8) quaternary ammonium compounds, (9) hydrogen peroxide, (10) peracetic acid, (11) various alkalies, and (12) various acids. Disinfection with chemical agents is accomplished by mixing thoroughly the diluted disinfecting agent with the liquid to be disinfected (treated effluent in water reclamation applications) and allowing sufficient time for the disinfectant to react with the microorganisms that may be present in the liquid.

Radiation

The major types of radiation are electromagnetic, acoustic, and particle. For example, the decay of microorganisms observed in oxidation ponds is due, in part, to their exposure to the ultraviolet (UV) light component of sunlight of the electromagnetic spectrum. Special lamps developed to emit UV light have been used successfully to disinfect reclaimed water. Disinfection with UV light is accomplished by exposing the microorganisms in the liquid to UV light.

Other Disinfectants

Other means of disinfection that have been applied include the use of physical agents. The removal of microorganisms by mechanical and biological means is considered in

Mechanisms Used to Explain Action of Disinfectants

the following section. Heat and sound waves are physical agents that can be used to disinfect reclaimed water. Heating water to the boiling point, for example, destroys the major disease producing nonspore-forming bacteria. Heat is used commonly in the beverage and dairy industry, but it is not a feasible means of disinfecting large quantities of reclaimed water because of the high cost for the amount of energy required. However, pasteurization of sludge is used extensively in Europe.

The five principal mechanisms that have been proposed to explain the action of disinfectants are: (1) damage to the cell wall, (2) alteration of cell permeability, (3) alteration of the colloidal nature of the protoplasm, (4) alteration of the organism DNA or RNA, and (5) inhibition of enzyme activity. A comparison of the mechanisms of disinfection using chlorine, ozone, and UV radiation, is presented in Table 11-2. To a large extent, observed performance differences for the various disinfectants can be explained on the basis of the operative inactivation mechanisms.

Damage, destruction, or alteration of the cell wall by oxidizing chemicals, such as chlorine and ozone, results in cell lysis and death. Oxidizing chemicals can also alter the chemical arrangement of enzymes and inactivate the enzymes. Some oxidants can inhibit the synthesis of the bacterial cell wall. Exposure to UV radiation can cause the formation of double bonds in the DNA of microorganisms as well as rupturing some DNA strands. When UV photons are absorbed by the DNA in bacteria and protozoa and the DNA and RNA in viruses, covalent dimers can be formed from adjacent thymines in DNA or uracils in RNA. The formation of double bonds disrupts the replication process so that the organism can no longer reproduce and is thus inactivated.

Table 11-2 Mechanisms of disinfection using chlorine, ozone, and UV radiation

Chlorine	Ozone	UV radiation
1. Oxidation
2. Reactions with available chlorine
3. Protein precipitation
4. Modification of cell wall permeability
5. Hydrolysis and mechanical disruption	1. Direct oxidation/destruction of cell wall with leakage of cellular constituents outside of cell
2. Reactions with radical byproducts of ozone decomposition
3. Damage to the constituents of the nucleic acids (purines and pyrimidines)
4. Breakage of carbon-nitrogen bonds, leading to depolymerization	1. Photochemical damage to RNA and DNA (e.g., formation of double bonds) within the cells of an organism
2. The nucleic acids in microorganisms are the most important absorbers of the energy of light in the wavelength range of 240–280 nm
3. Because DNA and RNA carry genetic information for reproduction, damage of these substances can effectively inactivate the cell

Comparison of Reclaimed Water Disinfectants

Using the criteria defined in Table 11-1 as a frame of reference and the issues discussed above, the disinfectants that have been used in water reclamation applications are compared in Table 11-3. Additional details on the relative performance of the various disinfection technologies are presented in the following section and in Table 11-4. In reviewing Table 11-3, important comparisons that should be noted include safety

Table 11-3
Comparison of commonly used disinfectants in water reclamation[a]

Characteristic[b]	Chlorine gas[c]	Sodium hypochlorite[c]	Combined chlorine	Chlorine dioxide	Ozone	UV radiation
Deodorizing ability	High	Moderate	Moderate	High	High	na[d]
Interaction with organic matter	Oxidizes organic matter	Oxidizes organic matter	Oxidizes organic matter	Oxidizes organic matter	Oxidizes organic matter	Absorbance of UV irradiation
Corrosiveness	Highly corrosive	Corrosive	Corrosive	Highly corrosive	Highly Corrosive	na
Toxic to higher forms of life	Highly toxic	Highly toxic	Toxic	Toxic	Toxic	Toxic
Penetration into particles	High	High	Moderate	High	High	Moderate
Safety concern	High	Moderate to low	High to moderate[e]	High	Moderate	Low
Solubility	Moderate	High	High	High	High	na
Stability	Stable	Slightly unstable	Slightly unstable	Unstable[f]	Unstable[f]	na
Effectiveness as disinfectant						
Bacteria	Excellent	Excellent	Good	Excellent	Excellent	Good
Protozoa	Fair to poor	Fair to poor	Poor	Good	Good	Excellent
Viruses	Excellent	Excellent	Fair	Excellent	Excellent	Good
Byproduct formation	THMs and HAAs[g]	THMs and HAAs	Traces of THMs and HAAs, cyanogens, NDMA	Chlorite and chlorate	Bromate	None known in measurable concentrations
Increases TDS	Yes	Yes	Yes	Yes	No	No
Use as a disinfectant	Common	Common	Common	Occasional	Occasional	Increasing rapidly

[a] Adapted in part from Tchobanoglous et al. (2003) and Crittenden et al. (2005).
[b] See Table 11-1 for a description of the characteristics.
[c] Free chlorine (HOCl and OCl$^-$).
[d] na = not applicable.
[e] Depends on whether chlorine gas or sodium hypochlorite is used to combine with nitrogenous compounds.
[f] Must be generated as used.
[g] THMs = trihalomethanes and HAAs = haloacetic acids.

Table 11-4
Removal or destruction of total coliform by different treatment processes

Process	Log removal
Coarse screens	0–0.7
Fine screens	1.0–1.3
Grit chambers	1.0–1.4
Plain sedimentation	1.4–2.0
Chemical precipitation	1.6–1.9
Trickling filters	1.9–2.0
Activated sludge	1.9–3.0
Depth filtration	0–1.0
Microfiltration	2–>4
Reverse osmosis	>4–7
Disinfection with chlorine	4–6

(e.g., chlorine gas vs. sodium hypochlorite) and the increase in TDS (e.g., chlorine gas vs. UV irradiation). These issues are also addressed in the subsequent sections.

11-2 PRACTICAL CONSIDERATIONS AND ISSUES FOR DISINFECTION

The purpose of this section is to introduce the practical considerations and issues involved in the disinfection process. The background material presented is intended to serve as a basis for the discussion of individual disinfectants considered in the following sections. Topics to be discussed include: (1) an introduction to the physical facilities used for disinfection, (2) the factors that affect the performance of the disinfection process, (3) development of the $C_R t$ values for predicting disinfection performance, (4) application of the $C_R t$ values for reclaimed water disinfection, (5) a comparison of the performance of alternative disinfection technologies, and (6) a review of the advantages and disadvantages of each disinfection technology. Costs, both capital and operation and maintenance, have not been provided other than in a general context. Costs are influenced by many site-specific factors and must be evaluated on a case-by-case basis.

Physical Facilities Used for Disinfection

In general, the disinfection of reclaimed water is accomplished as a separate unit process in specially designed reactors. The purpose of the reactors is to maximize contact between the disinfecting agent and the liquid to be disinfected. The specific design of the reactor depends on the nature and action of the disinfecting agent. The types of reactors used are illustrated on Figs. 11-1 and 11-2 and described below briefly.

Chlorine and Related Compounds

As shown on Figs. 11-1*a* and 11-1*b* baffled serpentine contact chambers or long pipelines are used for the application of diluted chlorine and related compounds. Both

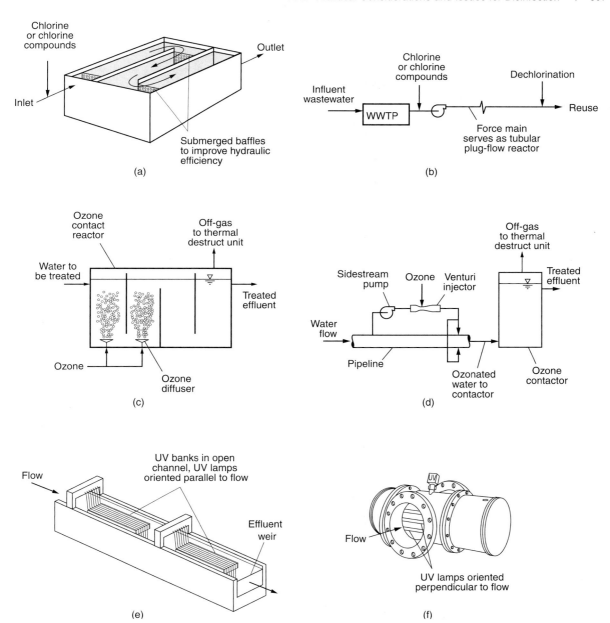

Figure 11-1

Types of reactors used to accomplish disinfection process: (a) plug-flow reactor in back-and-forth configuration, (b) force main which serves as a tubular plug-flow reactor, (c) multiple chamber in-line ozone contactor, (d) sidestream ozone injection system, (e) UV irradiation in an open channel with two UV banks with flow parallel to UV lamps, and (f) UV irradiation in a closed reactor with flow perpendicular to UV lamps.

Figure 11-2

Views of reactors used for disinfection: (a) serpentine plug-flow chlorine contact basin with end deflectors, (b) serpentine plug-flow chlorine contact basin with rounded corners and flow deflection baffles, (c) typical ozone generator, (d) ozone contactors used in conjunction with sidestream ozone injection, (e) open channel UV reactor, and (f) closed in-line UV reactor.

of these contact chambers are designed to perform as ideal plug-flow reactors. As will be discussed later, the efficacy of disinfection is affected by the degree to which the flow in these chambers is less than ideal. Views of full-scale chlorine contact basins are shown on Figs. 11-2a and 11-2b.

Ozone
Ozone is typically applied by bubbling ozone gas through the liquid to be disinfected in a contact chamber (see Figs 11-1c) or in a sidestream (see 11-1d). Fine bubble diffusers are used to improve ozone transfer to the liquid. Venturi injectors are used in sidestream designs. To limit the amount of short circuiting that can occur in a single contact chamber, a series of baffled chambers are used (see Fig. 11-1c). An ozone generator and contactor are shown on Figs. 11-2c and 11-2d, respectively.

Ultraviolet Light (UV)
Both open and closed contact chambers (reactors) are used for UV disinfection (see Figs. 11-1e and 11-1f). Open channel reactors are used commonly for low-pressure low-intensity and low-pressure high-intensity UV lamps. Closed proprietary reactors are used for low-pressure high-intensity and medium pressure high-intensity UV lamps. Because the contact time is short in UV reactors (seconds), the design of open channel and closed reactors is of critical importance. An open channel and a closed UV reactor are shown on Figs. 11-2e and 11-2f, respectively.

Factors Affecting Performance

In applying disinfection agents or physical processes, the following factors must be considered: (1) contact time and hydraulic efficiency of contact chambers, (2) concentration of the disinfectant, (3) intensity and nature of physical agent or means, (4) temperature, (5) types of organisms, (6) nature of suspending liquid (i.e., reclaimed water quality), and (7) the upstream treatment processes.

Contact Time
Perhaps one of the most important factors in the disinfection process is contact time. Once the disinfectant has been added, the time of contact before the effluent is to be reused is of paramount importance. As shown on Fig. 11-1, disinfection reactors that have special physical features are provided to ensure that an adequate contact time is provided.

Working in England in the early 1900s, Harriet Chick observed that for a given concentration of disinfectant, the longer the contact time, the greater the kill (see Fig. 11-3). This observation was first reported in the literature in 1908 (Chick, 1908). In differential form, Chick's law is:

$$\frac{dN_t}{dt} = -kN_t \qquad (11\text{-}1)$$

where dN_t/dt = the rate of change in the number (concentration) of organisms with time
k = inactivation rate constant, T^{-1}
N_t = number of organisms at time t
t = time

Figure 11-3
Log inactivation of dispersed microorganisms as a function of time in a batch reactor using increasing disinfectant dosages.

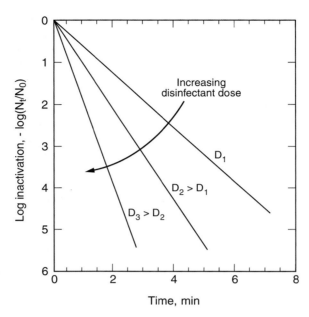

If N_0 is the number of organisms when t equals 0, Eq. (11-1) can be integrated to:

$$\ln \frac{N_t}{N_0} = -kt \tag{11-2}$$

The value of the inactivation rate constant, k, in Eq. (11-2) can be obtained by plotting $-\ln (N_t/N_0)$ versus the contact time t, where k is the slope of the resulting line of best fit.

Concentration of Chemical Disinfectant

Also working in England in the early 1900s, Herbert Watson reported that the inactivation rate constant was related to the concentration as follows (Watson, 1908):

$$k = \Lambda C^n \tag{11-3}$$

where k = inactivation rate constant, T^{-1}
Λ = coefficient of specific lethality, units vary with the value of n
C = concentration of disinfectant, mg/L
n = empirical constant related to dilution, dimensionless

The following explanation has been offered for various values of the dilution constant n:

n = 1 (both the concentration and time are equally important.)
n > 1 (concentration is more important than time.)
n < 1 (time is more important than concentration.)

The value of n can be obtained by plotting C versus t on log-log paper for a given level of inactivation. When n is equal to one, the data are plotted on log-arithmetic paper.

Combining the expressions proposed by Chick and Watson in differential form yields (Haas and Karra, 1984a, b):

$$\frac{dN_t}{dt} = -\Lambda C^n N_t \tag{11-4}$$

The integrated form of Eq. (11-4) is:

$$\ln \frac{N_t}{N_0} = \Lambda_{base\ e} C^n t \text{ or } \log \frac{N_t}{N_0} = -\Lambda_{base\ 10} C^n t \tag{11-5}$$

If n is equal to 1, a reasonable assumption based on past experience (Hall, 1973), Eq. (11-5) can be written as follows:

$$\frac{1}{\Lambda_{base\ e}} \ln \frac{N_t}{N_0} = Ct = D \tag{11-6}$$

where D = *germicidal dose* for a given degree of inactivation, mg·min/L.

The concept of dose (concentration times time) is significant as the performance of the disinfectants, as discussed subsequently, is based on the concept (Morris, 1975). This concept has also been adopted by the U.S. EPA (1986, 2003a) in establishing guidelines for disinfection (see "Development of the $C_R t$ Concept for Predicting Disinfection Performance" later in the chapter).

EXAMPLE 11-1. Determination of the Coefficient of Specific Lethality Based on the Chick-Watson Expression.

Using the data given below, determine the coefficient of specific lethality using Eq. (11-6).

C, mg/L	Time, min	Number of organisms, No./100 mL
0	0	1.00×10^8
4.0	2	1.59×10^7
4.0	4.5	1.58×10^6
4.0	8	2.01×10^4
4.0	11	3.16×10^3

Solution
1. To determine the coefficient of lethality, prepare a plot of log(N/N₀) as a function of Ct and fit a linear trend line through the data.

a. Determine the values of log(N/N₀) and Ct. The required data table is shown below.

C, mg/L	Time, min	Number of organisms, No./100 mL	C·t, mg·min/L	log(N/N₀)
0	0	1.00×10^8	0	0
4.0	2	1.59×10^7	8	−0.8
4.0	4.5	1.58×10^6	18	−1.8
4.0	8	2.01×10^4	32	−3.7
4.0	11	3.16×10^3	46	−4.5

b. Prepare a plot of log(N/N₀) as a function of Ct. The required plot is shown below.

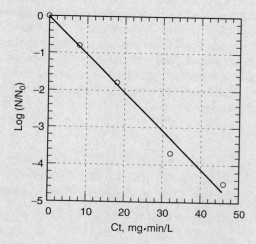

2. Determine the coefficient of specific lethality. The slope of the line in the above plot corresponds to the coefficient of specific lethality, Λ_{CW} (base 10). From the plot,

$$-\Lambda_{CW} = \frac{-5 - 0}{49 - 0}$$

$$\Lambda_{CW} = 0.102 \text{ L/mg·min}$$

Check, when Ct = 46,

$$\log \frac{N_t}{N_0} = -\Lambda_{base10} Ct = -0.102(46) = -4.69 \text{ versus } 4.5, \text{ OK}$$

Temperature

The effect of temperature on the rate of kill with chemical disinfectants can be represented by a form of the van't Hoff-Arrhenius relationship. Increasing the temperature results in a more rapid kill. In terms of the coefficient of specific lethality, Λ, the effect temperature is given by the following relationship:

$$\ln \frac{\Lambda_1}{\Lambda_2} = \frac{E(T_2 - T_1)}{RT_1T_2} \tag{11-7}$$

where Λ_1, Λ_2 = coefficient of specific lethality at temperatures T_1 and T_2, respectively
E = activation energy, J/mole
R = gas constant, 8.3144 J/mole·K

Typical values for the activation energy for various chlorine compounds at different pH values are given in Sec. 11-3. The effect of temperature is considered in Example 11-2.

EXAMPLE 11-2. Effect of Temperature on Disinfection Times.

Estimate the time required for a 99.9 percent kill for a chlorine dosage of 0.05 mg/L at a temperature of 20°C. Assume the activation energy is equal to 26,800 J/mole (from Table 11-12 in Sec. 11-3). The following coefficients were developed for Eq. (11-5) at 5°C using a batch reactor:

$$\Lambda = 10.5 \text{ L/mg·min}$$
$$n = 1$$

Solution

1. Estimate the time required for a 99.9 percent kill using Eq. (11-5).

$$\ln \frac{N_t}{N_0} = -10.5 Ct$$

$$\ln \frac{0.10}{100} = -10.5 \text{ L/mg·min}(0.05 \text{ mg/L})t$$

$$t = \frac{-6.91}{(-10.5)(0.05)} = 13.2 \text{ min at } 5°C$$

2. Estimate the time required at 20°C using the form of the van't Hoff-Arrhenius equation given in Eq. (11-7).

$$\ln \frac{\Lambda_1}{\Lambda_2} = \frac{E(T_2 - T_1)}{RT_1T_2}$$

$$\ln \frac{10.5}{\Lambda_2} = \frac{26,800 \text{ J/mol } (278 - 293)K}{(8.3144 \text{ J/mol·K})(293)(278)}$$

$$\ln \frac{10.5}{\Lambda_2} = -0.594$$

$$\frac{10.5}{\Lambda_2} = e^{-0.594} = 0.552$$

$$\Lambda_2 = 19.0 \text{ L/mg} \cdot \text{min}$$

$$t = \frac{-6.91}{(-19.0)(0.05)} = 5.43 \text{ min at } 20°C$$

Intensity and Nature of Physical Agent

As noted earlier, ultraviolet light (UV) is used commonly for the disinfection of reclaimed water. It has been found that the effectiveness of UV disinfection is a function of the average UV intensity, expressed as milliwatts per square centimeter (mW/cm^2). When the exposure time is considered, the dose of UV radiation to which the microorganisms in the liquid are exposed to given by the following expression.

$$D = I_{avg} \times t \tag{11-8}$$

where D = UV dose, mJ/cm^2 (Note: mJ/cm^2 = mW·s/cm^2)
I_{avg} = average UV intensity, mW/cm^2
t = exposure time, s

The UV dose is expressed in mW·s/cm^2 which is equivalent to mJ/cm^2 (millijoule per square centimeter). Thus, the concept of dose can also be used to define the effectiveness of UV light in a manner analogous to that used for chemical disinfectants.

Types of Organisms

The effectiveness of various disinfectants is influenced by the type, nature, and condition of the microorganisms. For example, viable growing (vegetative) bacterial cells are often killed more easily than older cells that have developed a slime (polymer) coating. Bacteria that are able to form spores enter this protective state when a stress, such as increased temperature or a toxic agent, is applied. Bacterial spores are extremely resistant, and many of the chemical disinfectants normally used have little or no effect on them. Similarly, many of the viruses and protozoa of concern respond differently to each of the chemical disinfectants. In some cases, other disinfecting agents, such as heat or UV radiation, may have to be used for effective disinfection. The inactivation of different types of microorganism groups is considered in the following sections.

Nature of Suspending Liquid

In reviewing the development of the relationships developed by Chick and Watson for the inactivation of microorganisms, as cited above, it is important to note that most of the tests were conducted in batch reactors using distilled or buffered water, under laboratory conditions. In practice, the nature of the suspending liquid must be evaluated carefully. Two constituents of reclaimed water are significant: natural organic material (NOM) and suspended material. The NOM found in reclaimed water reacts with most oxidizing disinfectants and reduces their effectiveness or results in greater dosages to effect disinfection. The NOM found in water reclamation plants is derived from three sources: (1) the terrestrial environment (mostly humic materials), (2) the aquatic environment (algae and other aquatic species and their byproducts), and (3) the microorganisms in the biological treatment process. The presence of suspended matter also reduces the effectiveness of disinfectants by absorption of the disinfectant and by shielding entrapped bacteria.

Because of the interactions that can occur between the disinfecting agent and the reclaimed water properties, departures from the Chick-Watson rate law [Eq. (11-5) or (11-6)] are common as shown on Fig. 11-4. As shown on Fig. 11-4a, there can be a lag or shoulder effect in which constituents in the suspending liquid (treated reclaimed water) react initially with the disinfectant rendering the disinfectant ineffective. The tailing effect in which large particles shield the organisms to be disinfected is shown on Fig. 11-4b. The combined effects of lag and tailing are illustrated on Fig. 11-4c. In general, Eq. (11-5) as applied to reclaimed water fails to account for the variable, heterogeneous characteristics of reclaimed water.

Effect of Upstream Treatment Processes

The extent to which upstream processes remove NOM and suspended matter greatly influences the disinfection process. Bacteria and other organisms are also removed by

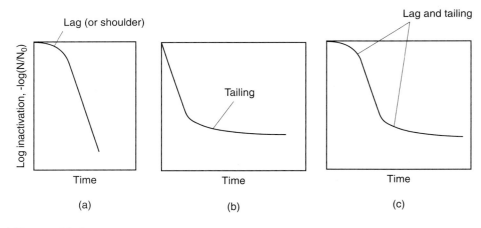

Figure 11-4
Departures observed from the Chicks' law: (a) lag or shoulder effect in which the disinfectant reacts first with constituents in the suspending liquid after which the response is log-linear, (b) log-linear response followed by tailing effect in which large particles shield the organisms to be disinfected following the inactivation of dispersed organisms, and (c) combined lag, log-linear, and tailing effects.

mechanical and biological means during wastewater treatment. Typical removal efficiencies for various treatment operations and processes are reported in Table 11-4. The first and last four operations listed are essentially physical. The removals accomplished are byproducts of the primary function of the process.

Another factor that impacts the performance of both chlorine and UV disinfection for unfiltered effluents (especially when coliform bacteria are used as the regulatory indicator) is the number of particles with associated coliform bacteria. It has been observed that for activated sludge plants the number of particles with associated coliform organisms is a function of the solids retention time (SRT). The relationship between the fraction of wastewater particles with one or more associated coliform organisms and the SRT is illustrated on Fig. 11-5. As illustrated, longer SRTs result in a decrease in the fraction of particles containing coliform bacteria. The use of deep final clarifiers will also reduce the number of large particles that may shield bacteria (see Fig. 7-7*b* in Chap. 7). In general, without some form of filtration, it is difficult to achieve extremely low coliform concentrations in the settled effluent from activated sludge plants operated at low SRT values (e.g., 0.75 to 2 d).

Development of the $C_R t$ Concept for Predicting Disinfection Performance

Although the disinfection models discussed above are useful for analyzing disinfection data, they are difficult to use to predict disinfection performance over a wide range of operating conditions. In the water treatment field, before the adoption of the Surface Water Treatment Rule (SWTR) (circa 1989) and the importance of *Cryptosporidium* as a causative agent in waterborne disease outbreaks was recognized, meeting water quality requirements was quite straightforward. Chlorine and its compounds were generally used to inactivate coliform bacteria to meet the drinking water standards in effect at that time.

In developing the rationale for the first SWTR, the U.S. EPA needed some way to ensure the safety of public water supplies that were unfiltered (e.g., New York City, San Francisco,

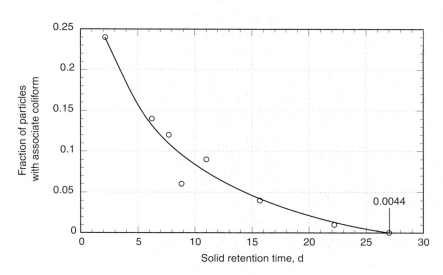

Figure 11-5
Fraction of particles in settled wastewater with one or more associated coliform organisms as function of the solids retention time. (From Emerick et al., 1999.)

Seattle). Based on ongoing research, the U.S. EPA determined that four logs of virus and three logs of *Giardia* reduction would be required by means of disinfection. Recognizing that guidance was required on how to achieve adequate disinfection, the U.S. EPA undertook an evaluation of the most commonly used disinfectants for the disinfection of viruses and *Giardia* cysts. In conducting their evaluation, the U.S. EPA adopted the C_Rt concept (residual disinfectant concentration in mg/L times the contact time in minutes), derived from the simplified Chick-Watson model [see Eq. (11-6)], as a measure of performance. The C_Rt values obtained, typically in laboratory bench scale studies, are used as a surrogate measure of disinfection effectiveness. Thus, if a given C_Rt value were achieved, it could be assumed generally that disinfection requirements had been met.

Although *Cryptosporidium* had been identified at the time the SWTR was adopted in 1989, C_Rt values for *Cryptosporidium* were not included because it would have delayed adoption of the SWTR. It has since been found that many pathogens including *Cryptosporidium* can exist in treated drinking water in the presence of concentrations of chlorine and its compounds sufficient to kill most other pathogens. Based on ongoing work, the U.S. EPA has now published extensive tables of C_Rt values for a variety of disinfectants, microorganisms, and operating conditions (U.S. EPA, 2003a). In addition, corresponding UV dose values have also been published for *Cryptosporidium*, *Giardia*, and virus. From a practical standpoint, the utility of the C_Rt or UV dose approach can be appreciated as it is relatively easy to measure the residual concentration of the disinfectant or the UV intensity and the exposure contact time. With respect to the contact time, the t_{10} value (the contact time during which no more than 10 percent of the influent water has passed through the process—see discussion in Sec. 11-3) is used commonly in the field of water treatment for disinfectants other than UV irradiation.

Application of the C_Rt Concept for Reclaimed Water Disinfection

Use of the C_Rt concept to control the disinfection process is now becoming more common in the water reclamation field. In some states, the C_Rt value and the chlorine contact time are specified in regulatory requirements. For example, the State of California requires a minimum C_Rt value of 450 mg·min/L (based on combined chlorine residual) and a modal contact time of 90 min at peak flow for certain water reclamation applications. It is assumed, based on past testing, that a minimum C_Rt value of 450 mg·min/L produces a four-log inactivation of poliovirus. As the use of the C_Rt concept becomes more common in the water reclamation field, there are a number of limitations that must be considered in the application of this concept for regulatory purposes. Most of the C_Rt values reported in the literature are obtained using: (1) complete-mix batch reactors (i.e., ideal flow conditions) in a laboratory setting under controlled conditions, (2) discrete organisms grown in the laboratory in pure culture, (3) a buffered fluid for the suspension of the discrete organisms, and (4) an absence of particulate matter.

Further, many of the C_Rt values reported in the literature were based on older analytical techniques. As a consequence, C_Rt values used for regulatory purposes often do not match what is observed in the field. Referring to Fig. 11-6, it can be seen that in the tailing region, the residual concentration of microorganisms is essentially independent of the C_Rt value. In addition, some compounds present in reclaimed water will (1) react

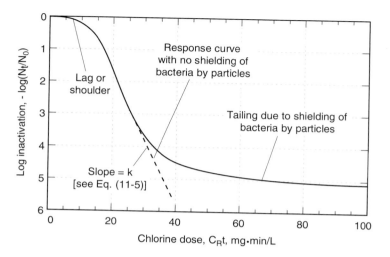

Figure 11-6
Typical disinfection curve obtained with wastewater containing oxidizable constituents and suspended solids. Both lag and tailing effects are evident.

with the chlorine and its compounds, (2) be measured as combined residual, and (3) have no disinfection properties (see Sec. 11-3). In a similar manner, dissolved constituents such as metals and humic acids reduce the effectiveness of UV disinfection. Thus, it is difficult to develop standardized $C_R t$ or UV dose values suitable for all conditions in reclaimed water treatment. Clearly, as discussed subsequently, site-specific testing is required to establish the appropriate disinfectant dose.

Performance Comparison of Disinfection Technologies

A general comparison of the germicidal effectiveness of the disinfection technologies based on Eq. (11-6), by classes of organisms is presented in Table 11-5. Additional information is presented in the sections dealing with the individual technologies. It is important to note that the values given in these tables are only meant to serve as a guide in assessing the effectiveness of these technologies. The $C_R t$ values also vary with both temperature and pH. Because the characteristics of each reclaimed water and the degree of treatment will significantly impact the effectiveness of the various disinfection technologies, site-specific testing must be conducted to evaluate the effectiveness of alternative disinfection technologies and to establish appropriate dosing ranges.

Advantages and Disadvantages of Alternative Disinfection Technologies

The general advantages and disadvantages of using chlorine, chlorine dioxide, UV, and ozone for the disinfection of reclaimed water are summarized in Table 11-6. In most water reuse applications, the choice of disinfectant has usually been between chlorine and UV. Recently, however, with increased awareness of trace constituents of concern, a renewed interest has developed in the use of ozone. Deciding factors in the selection of a disinfectant are commonly (1) economic evaluation, (2) public and operator safety, (3) environmental effects, and (4) ease of operation (Hanzon et al., 2006). Other treatment objectives are important in selecting a disinfectant for reclaimed water. Potential concerns with pesticides, trace constituents of concern, endocrine disruptors, and similar compounds may influence the choice of disinfectants. Each disinfectant offers varying treatment performance with regard to these potential concerns.

Table 11-5
Typical $C_R t$ values for various levels of inactivation of bacteria, viruses, and protozoan oocysts and cysts[a]

Disinfectant	Unit	Inactivation			
		1-log	2-log	3-log	4-log
Bacteria					
Chlorine (free)	mg·min/L	0.1–0.2	0.4–0.6	3–4	8–10
Chloramine	mg·min/L	4–6	10–12	20–40	70–90
Chlorine dioxide	mg·min/L	2–4	8–10	20–30	50–70
Ozone	mg·min/L		3–4		
UV irradiation	mJ/cm^2		30–60	60–80	80–100
Viruses					
Chlorine (free)	mg·min/L		1–4	8–16	20–40
Chloramine	mg·min/L		600–700	900–1100	1400–1600
Chlorine dioxide	mg·min/L		4–6	10–14	20–30
Ozone	mg·min/L		0.4–0.6	0.7–0.9	0.9–1.0
UV irradiation	mJ/cm^2		30–40	50–70	70–90
Protozoa[b]					
Chlorine (free)	mg·min/L	30–40	60–70	90–110	
Chloramine	mg·min/L	600–650	1200–1400	1800–2000	
Chlorine dioxide	mg·min/L	7–9	14–16	20–25	
Ozone	mg·min/L	0.4–0.6	0.9–1.2	1.4–1.6	
UV irradiation[c]	mJ/cm^2	5–10	10–20	20–30	

[a]Adapted in part from Montgomery (1985), U.S. EPA (1999b). Test data obtained using batch reactors with dispersed organisms in buffered clean water (pH ~ 7–8.5, ~20°C).
[b]Protozoan cysts and oocysts will, in general, require higher values.
[c]Based on the results of infectivity studies.

Note: Because there is such a wide variability in the susceptibility of different microorganisms to the different disinfection technologies, a wide range of dosage values has been reported in the literature. Thus, the data presented in this table are only meant to serve as a general guide to the relative effectiveness of the different disinfection technologies and are not for a specific microorganism.

Table 11-6
Advantages and disadvantages of chlorine, chlorine dioxide, ozone, and UV radiation for the disinfection of reclaimed water[a]

Advantages	Disadvantages
Chlorine	
1. Well-established technology 2. Effective disinfectant 3. Chlorine residual can be monitored and maintained 4. Combined chlorine residual can also be provided by adding ammonia 5. Germicidal chlorine residual can be maintained in long transmission lines 6. Availability of chemical system for auxiliary uses such as odor control, dosing RAS, and disinfecting plant water systems 7. Oxidizes sulfides 8. Capital cost is relatively inexpensive, but cost increases considerably if conformance to Uniform Fire Code regulations is required 9. Available as calcium and sodium hypochlorite that are considered to be safer than chlorine gas 10. Can be generated on-site	1. Hazardous chemical that can be a threat to plant workers and the public; thus, strict safety measures must be employed especially in light of the Uniform Fire Code 2. Relatively long contact time required as compared to other disinfectants 3. Combined chlorine is less effective in inactivating some viruses, spores, and cysts at low dosages used for coliform organisms 4. Residual toxicity of treated effluent must be reduced through dechlorination 5. Forms trihalomethanes and other DBPs,[b] including NDMA (see Table 11-15) 6. Releases volatile organic compounds from chlorine contact basins 7. Oxidizes iron, magnesium, and other inorganic compounds (consumes disinfectant) 8. Oxidizes a variety of organic compounds (consumes disinfectant) 9. Increases TDS level of treated effluent 10. Increases chloride content of treated effluent 11. Acid generation; pH of the wastewater can be reduced if alkalinity is insufficient 12. Chemical scrubbing facilities may be required to meet Uniform Fire Code regulations 13. Formal risk management plan may be required 14. Not effective disinfectant for *cryptosporidium*
Chlorine dioxide	
1. Effective disinfectant for bacteria, *Giardia* and viruses 2. More effective than chlorine in inactivating most viruses, spores, cysts, and oocysts 3. Biocidal properties not influenced by pH 4. Under proper generation conditions, halogen-substituted DBPs are not formed 5. Oxidizes sulfides 6. Provides residuals	1. Unstable, must be produced on-site 2. Oxidizes iron, magnesium, and other inorganic compounds (consumes disinfectant) 3. Oxidizes a variety of organic compounds 4. Forms DBPs (i.e., chlorite and chlorate), limiting applied dose 5. Potential for the formation of halogen-substituted DBPs 6. Decomposes in sunlight 7. Can lead to the formation of odors 8. Increases TDS level of treated effluent 9. Operating costs can be high (e.g., must test for chlorite and chlorate)

Table 11-6

Advantages and disadvantages of chlorine, chlorine dioxide, ozone, and UV for the disinfection of reclaimed water[a] (Continued)

Advantages	Disadvantages
Ozone	
1. Effective disinfectant
2. More effective than chlorine in inactivating most viruses, spores, cysts, and oocysts
3. Biocidal properties not influenced by pH
4. Shorter contact time than chlorine
5. Oxidizes sulfides
6. Requires less space
7. Contributes dissolved oxygen
8. At higher dosages than required for disinfection, ozone reduces the concentration of trace organic constituents	1. Ozone residual monitoring and recording requires more operator time than chlorine residual monitoring and recording
2. No residual effect
3. Less effective in inactivating some viruses, spores, cysts at low dosages used for coliform organisms
4. Forms DBPs (see Table 11-15)
5. Oxidizes iron, magnesium, and other inorganic compounds (consumes disinfectant)
6. Oxidizes a variety of organic compounds (consumes disinfectant)
7. Off gas requires treatment
8. Safety concerns
9. Highly corrosive and toxic
10. Energy intensive
11. Relatively expensive
12. Highly operational and maintenance sensitive
13. Has been shown to control the growth of filamentous microorganisms, but more expensive than chlorine
UV radiation	
1. Effective disinfectant
2. Requires no hazardous chemicals
3. No residual toxicity
4. More effective than chlorine in inactivating most viruses, spores, and cysts
5. No formation of DBPs at dosages used for disinfection
6. Does not increase TDS level of treated effluent
7. Effective in the destruction of resistant organic constituents such as NDMA
8. Improved safety
9. Requires less space than chlorine disinfection
10. At higher UV dosages than required for disinfection, UV irradiation can be used to reduce the concentration of trace organic constituents of concern such as NDMA (see Sec. 10-8 in Chap. 10)	1. No immediate measure of whether disinfection was successful
2. No residual effect
3. Less effective in inactivating some viruses, spores, and cysts at low dosages used for coliform organisms
4. Energy intensive
5. Hydraulic design of UV system is critical
6. Capital cost is relatively expensive, but price is coming down as new and improved technology is brought to the market
7. Large number of UV lamps required where low-pressure low-intensity systems are used
8. Low-pressure low-intensity lamps require acid washing to remove scale
9. Lacks a chemical system that can be adapted for auxiliary uses such as odor control, dosing RAS, and disinfecting plant water systems
10. Fouling of UV lamps
11. Lamps require routine periodic replacement
12. Lamp disposal is problematic due to presence of mercury

[a]Adapted in part from Crites and Tchobanoglous (1998), U.S. EPA (1999b), and Hanzon et al. (2006).
[b]DBPs = disinfection byproducts.

11-3 DISINFECTION WITH CHLORINE

Of all the chemical disinfectants, chlorine is the one used most commonly throughout the world. Specific topics considered in this section include a brief description of the characteristics of the various chlorine compounds, a review of chlorine chemistry and breakpoint chlorination, an analysis of the performance of chlorine as a disinfectant and the factors that may influence the effectiveness of the chlorination process, a discussion of the formation of disinfection byproducts (DBPs), and a consideration of the potential impacts of the discharge of DBPs to the environment. Disinfection with chlorine dioxide and dechlorination are considered in the following two sections, respectively.

Characteristics of Chlorine Compounds

The principal chlorine compounds used at water reclamation plants are chlorine (Cl_2), sodium hypochlorite (NaOCl), and chlorine dioxide (ClO_2). Calcium hypochlorite [$Ca(OCl)_2$], another chlorine compound is used in small treatment plants because of its ease of handling. Many large cities have switched from chlorine gas to sodium hypochlorite because of the safety concerns and regulatory requirements related to the handling and storage of liquid chlorine (see Table 11-3). The characteristics of Cl_2, NaOCl, and $Ca(OCl)_2$ are considered below. The characteristics of chlorine dioxide and its use as a disinfectant are discussed in the following section.

Chlorine

Chlorine (Cl_2) can be present as a gas or a liquid. Chlorine gas is greenish yellow in color and about 2.48 times as heavy as air. Liquid chlorine is amber colored and about 1.44 times as heavy as water. Unconfined liquid chlorine vaporizes rapidly to a gas at standard temperature and pressure with 1 L of liquid yielding about 450 L of gas. Chlorine is moderately soluble in water, with a maximum solubility of about 1 percent at 10°C (50°F). The general properties of chlorine are summarized in Table 11-7.

Although the use of chlorine for the disinfection of both potable water supplies and reclaimed water has been of great significance from a public health perspective, serious concerns have been raised about its continued use. Important concerns include:

1. Chlorine is a highly toxic substance that is transported by rail and truck, both of which are prone to accidents.
2. Chlorine potentially poses health risks to treatment plant operators, and the general public, if released by accident.
3. Because chlorine is a highly toxic substance, stringent requirements for containment and neutralization must be implemented as specified in the Uniform Fire Code (UFC).
4. Chlorine reacts with the organic constituents in reclaimed water to produce odorous compounds.
5. Chlorine reacts with the organic constituents in reclaimed water to produce byproducts, many of which are known to be carcinogenic and/or mutagenic.
6. Residual chlorine in reclaimed water is toxic to aquatic life.
7. The discharge of chloro-organic compounds, whose long-term effects are not known, maybe detrimental to the environment.

Table 11-7
Properties of chlorine, chlorine dioxide, and sulfur dioxide[a]

Property	Unit	Chlorine (Cl_2)	Chlorine dioxide (ClO_2)	Sulfur dioxide (SO_2)
Molecular weight	g	70.91	67.45	64.06
Boiling point (liquid)	°C	−33.97	11	
Melting point	°C	−100.98	−59	
Latent heat of vaporization at °C	kJ/kg	253.6	27.28	376.0
Liquid density at 15.5°C	kg/m³	1422.4	1640[b]	1396.8
Solubility in water at 15.5°C	g/L	7.0	70.0[b]	120
Specific gravity of liquid at °C (water = 1)	s.g.	1.468		1.486
Vapor density at 0°C and 1 atm	kg/m³	3.213	2.4	2.927
Vapor density compared to dry air at 0°C and 1 atm	unitless	2.486	1.856	2.927
Specific volume of vapor at 0°C and 1 atm	m³/kg	0.3112	0.417	0.342
Critical temperature	°C	143.9	153	157.0
Critical pressure	kPa	7811.8		7973.1

[a]Adapted in part from U.S. EPA (1986); White (1999).
[b]At 20°C.

Sodium Hypochlorite

Sodium hypochlorite (NaOCl) (i.e., liquid bleach) is only available as a liquid and usually contains 12.5 to 17 percent available chlorine at the time it is manufactured. Sodium hypochlorite can be purchased in bulk or manufactured onsite; however, the solution decomposes more readily at high concentrations and is affected by exposure to light and heat. A 16.7 percent solution stored at 26.7°C (80°F) will lose 10 percent of its strength in 10 d, 20 percent in 25 d, and 30 percent in 43 d. It must, therefore, be stored in a cool location in a corrosion-resistant tank. Another disadvantage of sodium hypochlorite is the chemical cost. The purchase price may range from 150 to 200 percent of the cost of liquid chlorine. The handling of sodium hypochlorite requires special design considerations because of its corrosiveness, the presence of chlorine fumes, and gas binding in chemical feed lines. Several proprietary systems are available for the generation of sodium hypochlorite from sodium chloride (NaCl) or seawater. These systems are electric power intensive and, in the case of seawater, result in a very dilute solution, a maximum of 0.8 percent hypochlorite. On-site generation systems have been used only on a limited basis, typically at relatively large plants, due to their complexity and high power cost.

Calcium Hypochlorite

Calcium hypochlorite [$Ca(OCl)_2$] is available commercially in either dry or wet form. In dry form it is available as an off-white powder or as granules, compressed tablets, or pellets. Calcium hypochlorite granules or pellets are readily soluble in water, varying from about 21.5 g/100 mL at 0°C (32°F) to 23.4 g/100 mL at 40°C (104°F). Because of its oxidizing potential, calcium hypochlorite should be stored in a cool, dry

Chemistry of Chlorine Compounds

The reactions of chlorine in water and the reaction of chlorine with ammonia are as follows:

Chlorine Reactions in Water

When chlorine in the form of Cl_2 gas is added to water, two reactions take place: *hydrolysis* and *ionization*.

Hydrolysis may be defined as the reaction in which chlorine gas combines with water to form hypochlorous acid (HOCl).

$$Cl_2 + H_2O \rightarrow HOCl + H^+ + Cl^- \tag{11-9}$$

The equilibrium constant, K_H, for this reaction is

$$K_H = \frac{[HOCl][[H^+][Cl^-]}{[Cl_2]} = 4.5 \times 10^{-4} \text{ at } 25°C \tag{11-10}$$

Because of the magnitude of the equilibrium constant, large quantities of chlorine can be dissolved in water.

Ionization of HOCl to hypochlorite ion (OCl$^-$) may be defined as:

$$HOCl \rightleftarrows H^+ + OCl^- \tag{11-11}$$

The ionization constant, K_i, for this reaction is

$$K_i = \frac{[H^+][OCl^-]}{[HOCl]} = 3 \times 10^{-8} \text{ at } 25°C \tag{11-12}$$

The variation in the value of K_i with temperature is reported in Table 11-8.

Table 11-8
Values of the ionization constant of hypochlorous acid at different temperatures[a]

Temperature, °C	$K_i \times 10^8$, mole/L
0	1.50
5	1.76
10	2.04
15	2.23
20	2.62
25	2.90
30	3.18
35	3.44

[a] Computed using equation from Morris (1966).

The total quantity of HOCl and OCl⁻ present in water is called the *free chlorine*. The relative distribution of these two species (see Fig. 11-7) is very important because the killing efficiency of HOCl is many times that of OCl⁻. The percentage distribution of HOCl at various temperatures can be computed using Eq. (11-13) and the data in Table 11-8.

$$\frac{[HOCl]}{[HOCl] + [OCl]} = \frac{1}{1 + [OCl]/[HOCl]}$$

$$= \frac{1}{1 + K_i/[H]} = \frac{1}{1 + K_i 10^{pH}} \qquad (11\text{-}13)$$

Hypochlorite Reactions in Water

Free chlorine can also be added to water in the form of hypochlorite salts. Both sodium and calcium hypochlorite hydrolyze to form hypochlorous acid (HOCl) as follows:

$$NaOCl + H_2O \rightarrow HOCl + NaOH \qquad (11\text{-}14)$$

$$Ca(OCl)_2 + 2H_2O \rightarrow 2HOCl + Ca(OH)_2 \qquad (11\text{-}15)$$

The ionization of hypochlorous acid was discussed previously [see Eq. (11-11)].

Chlorine Reactions with Ammonia

Untreated wastewater contains nitrogen in the form of ammonia and various combined organic forms. The effluent from most water reclamation plants also contains significant amounts of nitrogen, usually in the form of ammonia, or nitrate if the plant is designed to achieve nitrification. Because hypochlorous acid is a very active oxidizing agent, it reacts readily with ammonia in reclaimed water to form three types of chloramines in successive reactions:

$$NH_3 + HOCl \rightarrow NH_2Cl \text{ (monochloramine)} + H_2O \qquad (11\text{-}16)$$

$$NH_2Cl + HOCl \rightarrow NHCl_2 \text{ (dichloramine)} + H_2O \qquad (11\text{-}17)$$

$$NHCl_2 + HOCl \rightarrow NCl_3 \text{ (nitrogen trichloride)} + H_2O \qquad (11\text{-}18)$$

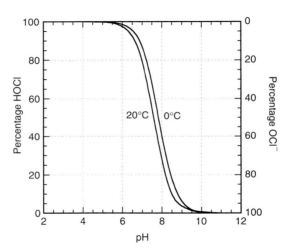

Figure 11-7
Distribution of hypochlorous acid and hypochlorite in water as a function of pH at 0 and 20°C.

These reactions are dependent on the pH, temperature, and contact time, and on the ratio of chlorine to ammonia (White, 1999). The two species that predominate, in most cases, are monochloramine (NH_2Cl) and dichloramine ($NHCl_2$). The ratio of dichloramine to monochloramine as a function of the ratio of chlorine to ammonia at various pH values is presented in Table 11-9. The amount of nitrogen trichloride present is negligible up to chlorine-to-nitrogen ratio of 2.0. As will be discussed subsequently, chloramines also serve as disinfectants, although they are slow-reacting. When chloramines are the only disinfectants, the measured residual chlorine is defined as *combined chlorine residual* as opposed to free chlorine in the form of hypochlorous acid and hypochlorite ion.

Breakpoint Reaction with Chlorine

The maintenance of a residual (free or combined) for the purpose of reclaimed water disinfection is complicated because free chlorine not only reacts with ammonia, as noted previously, but also is a strong oxidizing agent. The term *breakpoint chlorination* is the term applied to the process whereby enough chlorine is added to react with all oxidizable substances such that if additional chlorine is added, it will remain as free chlorine. The main reason for adding enough chlorine to obtain a free chlorine residual is that effective disinfection can usually then be assured. The amount of chlorine that must be added to reach a desired level of residual is called the *chlorine demand*. Breakpoint chlorination chemistry, acid generation, and the buildup of dissolved solids are considered in the following discussion.

Breakpoint Chlorination Chemistry

The stepwise phenomena that result when chlorine is added to reclaimed water containing oxidizable substances and ammonia can be explained by referring to Fig. 11-8. As chlorine is added, readily oxidizable substances, such as Fe^{2+}, Mn^{2+}, H_2S, and organic matter, react with the chlorine and reduce most of it to the chloride ion (point A on Fig. 11-8). After meeting this immediate demand, the added chlorine continues to react with the ammonia to form chloramines between points A and B, as discussed above. For mole ratios of chlorine to ammonia less than 1, monochloramine and dichloramine are formed. At point B, the mole ratio of chlorine (Cl_2) to ammonia (NH_4^+) is equal to 1.

Table 11-9
Ratio of dichloramine to monochloramine under equilibrium conditions as a function of pH and applied molar dose ratio of chlorine to ammonia[a]

Molar Ratio $Cl_2:NH_4$	pH 6	pH 7	pH 8	pH 9
0.1	0.13	0.014	1E-03	0.000
0.3	0.389	0.053	5E-03	0.000
0.5	0.668	0.114	0.013	1E-03
0.7	0.992	0.213	0.029	3E-03
0.9	1.392	0.386	0.082	0.011
1.1	1.924	0.694	0.323	0.236
1.3	2.700	1.254	0.911	0.862
1.5	4.006	2.343	2.039	2.004
1.7	6.875	4.972	4.698	4.669
1.9	20.485	18.287	18.028	18.002

[a]From U.S. EPA (1986).

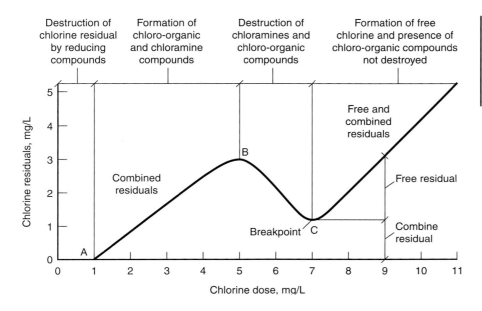

Figure 11-8
Generalized curve obtained during breakpoint chlorination of wastewater.

The distribution of these two forms is governed by their rates of formation, which are dependent on the pH and temperature. Between point B and the breakpoint, some chloramines are converted to nitrogen trichloride [see Eq. (11-18)], the remaining chloramines are oxidized to nitrous oxide (N_2O) and nitrogen (N_2), and the chlorine is reduced to chloride ion. With continued addition of chlorine, most of the chloramines will be oxidized at the breakpoint. Continued addition of chlorine past the breakpoint C, as shown on Fig. 11-8, will result in a directly proportional increase in the free chlorine. Theoretically, the weight ratio of chlorine to ammonia nitrogen at the breakpoint is 7.6 to 1 (see Example 11-3) and the mole ratio is equal to 1.5 to 1.

Possible reactions to account for the appearance of N_2 and N_2O and the disappearance of chloramines during breakpoint chlorination are as follows (Saunier, 1976; Saunier and Selleck, 1976):

$$NH_4^+ + HOCl \rightarrow NH_2Cl + H_2O + H^+ \quad (11\text{-}19)$$

$$NH_2Cl + HOCl \rightarrow NHCl_2 + H_2O \quad (11\text{-}20)$$

$$NHCl_2 + H_2O \rightarrow NOH + 2HCl \quad (11\text{-}21)$$

$$NHCl_2 + NOH \rightarrow N_2 + HOCl + HCl \quad (11\text{-}22)$$

The overall reaction, obtained by summing Eqs. (11-19) through (11-22), is given as:

$$2NH_4^+ + 3HOCl \rightarrow N_2 + 3H_2O + 3HCl + 2H^+ \quad (11\text{-}23)$$

Occasionally, serious odor problems have developed during breakpoint-chlorination operations because of the formation of nitrogen trichloride and related compounds. The presence of additional compounds that react with chlorine, such as organic nitrogen,

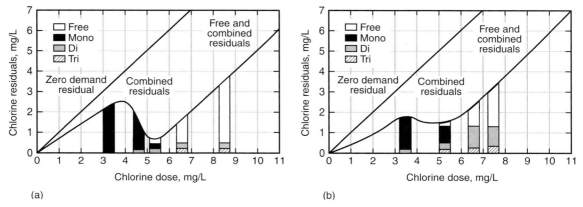

Figure 11-9
Curves of chlorine residual versus chlorine dosage for wastewater: (a) for wastewater containing ammonia nitrogen and (b) for wastewater containing nitrogen in the form of ammonia and organic nitrogen. (Adapted from White, 1999.)

may greatly alter the shape of the breakpoint curve, as shown on Fig. 11-9. The formation of disinfection byproducts is considered later in this section.

Acid Generation

When chlorine is added to water, the hydrolysis reaction results in the formation of HOCl as given by Eq. (11-9). The reaction of HOCl with ammonia also results in the formation of acid as given by Eq. (11-23). The total moles of hydrogen that must be neutralized can be determined by combining Eq. (11-9) with Eq. (11-23), which results in the following expression:

$$2NH_4^+ + 3Cl_2 \rightarrow N_2 + 6HCl + 2H^+ \tag{11-24}$$

In practice, the hydrochloric acid formed during chlorination [see Eq. (11-23)] reacts with the alkalinity of the reclaimed water, and under most circumstances, there is a slight pH drop. Stoichiometrically, 14.3 mg/L of alkalinity, expressed as $CaCO_3$, are required for each 1.0 mg/L of ammonia nitrogen that is oxidized in the breakpoint-chlorination process (see Example 11-3).

Buildup of Total Dissolved Solids (TDS)

In addition to the formation of hydrochloric acid, the chemicals added to achieve the breakpoint reaction also contribute an incremental increase in the TDS. As shown in Eq. (11-24), 6 moles of HCl and 2 moles of H^+ are formed, while 2 moles of NH_4^+ are removed from solution. In situations where the level of TDS may be critical with respect to water reuse applications, this incremental buildup from breakpoint chlorination should always be checked. The TDS contribution for each of several chemicals that may be used in the breakpoint reaction is summarized in Table 11-10. The magnitude of the possible buildup of TDS is illustrated in Example 11-3 in which the use of breakpoint chlorination is considered for the seasonal control of nitrogen.

Table 11-10
Effects of chemical addition on total dissolved solids in breakpoint chlorination[a]

Chemical addition	Increase in total dissolved solids per unit of NH_4^+ consumed
Breakpoint with chlorine gas	6.2:1
Breakpoint with sodium hypochlorite	7.1:1
Breakpoint with chlorine gas—neutralization of all acidity with lime (CaO)	12.2:1
Breakpoint with chlorine gas—neutralization of all acidity with sodium hydroxide (NaOH)	14.8:1

[a] From U.S. EPA (1986).

EXAMPLE 11-3. Analysis of Breakpoint Chlorination Process Used for Seasonal Control of Nitrogen.

Estimate the daily required chlorine dosage, the required alkalinity, if alkalinity needs to be added, and the resulting buildup of TDS when breakpoint chlorination is used for the seasonal control of nitrogen. Assume that the following data apply to this problem:

1. Plant flowrate = 3800 m³/d
2. Reclaimed water characteristics
   a. BOD = 20 mg/L
   b. TSS = 25 mg/L
   c. NH_4^+–N = 5 mg/L
   d. Alkalinity = 150 mg/L as $CaCO_3$
3. Required effluent NH_4^+–N concentration = 1.0 mg/L
4. Any alkalinity added is in the form of lime (CaO)

Solution

1. Determine the molecular weight ratio of hypochlorous acid (HOCl), expressed as Cl_2, to ammonia (NH_4^+), expressed as N, using the overall reaction for the breakpoint reactions given by Eq. (11-23).

$$2NH_4^+ + 3HOCl \rightarrow N_2 + 3H_2O + 3HCl + 2H^+$$

2(18)   3(52.45)

2(14)   3(2 × 35.45)

$$\text{Molecular ratio} = \frac{Cl_2}{NH_4^+ - N} = \frac{3(2 \times 35.45)}{2(14)} = 7.60$$

2. Estimate the required Cl_2 dosage using the molecular ratio developed in Step 1.

$$\text{kg } Cl_2/d = (3800 \text{ m}^3/d)[(5-1) \text{ g/m}^3](7.60 \text{ g/g})(1 \text{ kg}/10^3 \text{ g}) = 115.5 \text{ kg/d}$$

3. Determine the alkalinity required.
   a. The total number of moles of H^+ that must be neutralized per mole of NH^+ oxidized is given by Eq. (11-24), which has been divided by 2.

   $$NH_4^+ + 1.5Cl_2 \rightarrow 0.5N_2 + 3HCl + H^+$$

   b. When using lime to neutralize the acidity, the required alkalinity ratio is computed as follows:

   $$2CaO + 2H_2O \rightarrow 2Ca^{2+} + 4OH^-$$

   $$\text{Required alkalinity ratio} = \frac{2(100 \text{ g/mole of } CaCO_3)}{14 \text{ g/mole of } NH_4^+ \text{ as N}} = 14.3$$

   c. The required alkalinity is

   $$Alk = \frac{[(14.3 \text{ mg/L alk})/(\text{mg/L } NH_4^+)][(5-1) \text{ mg/L } NH_4^+](3800 \text{ m}^3/d)}{(10^3 \text{ g/kg})}$$

   $$Alk = 217.4 \text{ mg/L as } CaCO_3$$

4. Determine whether sufficient alkalinity is available to neutralize the acid during breakpoint chlorination.

   Because the available alkalinity (150 mg/L) is less than the required alkalinity (217.4 mg/L), alkalinity will have to be added to complete the reaction.

5. Determine the increment of TDS added to the reclaimed water. Using the data reported in Table 11-10, the TDS increase per mg/L of ammonia consumed when CaO is used to neutralize the acid formed is equal to 12.2 to 1.

   $$\text{TDS increment} = 12.2(5-1) \text{ mg/L} = 48.8 \text{ mg/L}$$

Comment

The ratio computed in Step 1 will vary somewhat, depending on the actual reactions involved. In practice, the actual ratio typically has been found to vary from 8:1 to 10:1. Similarly, in Step 3, the stoichiometric coefficients will also depend on the actual reactions involved. In practice, it has been found that about 15 mg/L of alkalinity are required because of the hydrolysis of chlorine. In Step 5, it should be noted that although breakpoint chlorination can be used to control nitrogen, it may be counterproductive if in the process the treated effluent is rendered unusable for other applications because of the buildup of TDS and the formation of disinfection byproducts.

Measurement and Reporting of Disinfection Process Variables

To provide a framework in which to consider the effectiveness of disinfection and the factors that affect the disinfection of reclaimed water, it is appropriate to consider how the effectiveness of the chlorination process is now assessed and how the results are analyzed. When using chlorine for the disinfection of reclaimed water, the principal parameters that can be measured, apart from environmental variables such as pH and temperature, are the number of organisms and the chlorine residual remaining after a specified period of time.

Number of Organisms Remaining

The number of coliform group bacteria remaining can be determined by multiple tube fermentation (MTF), membrane filter (MF) technique, or enzymatic substrate test (Standard Methods, 2005). The most probable number (MPN), based on a statistical analysis, is used to quantify results from MTF and enzymatic substrate testing, while direct counting is used for the MF technique. Coliform concentrations are reported typically as number per 100 mL. Use of coliform group bacteria as indicator organisms is also discussed in Chap. 3.

Measurement of Chlorine Residual

The chlorine residual (free and combined) is typically measured using the amperometric method, which has proved to be the most consistently reliable method now available. Also, because almost all the commercial analyzers of residual chlorine use the amperometric method, its adoption allows the results of independent studies to be compared directly.

Reporting of Results

Disinfection process results are reported in terms of the number of organisms and the chlorine residual remaining after a specified period of time. When the results are plotted, it is common practice to plot the logs of removal versus the corresponding $C_R t$ value as shown previously on Fig. 11-6.

Germicidal Efficiency of Chlorine and Various Chlorine Compounds in Clean Water

Numerous tests have shown that when all the physical parameters controlling the chlorination process are held constant, the germicidal efficiency of disinfection, as measured by the survival of "discrete bacteria," depends primarily on the form of the residual chlorine concentration, C_R, the contact time, t, and temperature.

Form of Residual Chlorine and Contact Time

A comparison of the relative germicidal efficiency of HOCl, OCl⁻, and NH_2Cl is presented on Fig. 11-10. For a given contact time or chlorine residual, the germicidal efficiency of HOCl, in terms of either time or residual, is significantly greater than that of either the OCl⁻ or NH_2Cl. For a given contact time, the germicidal efficiency of HOCL is from 100 to 200 times more effective than OCl⁻. Thus, because of the equilibrium between HOCl and the OCl⁻, maintenance of the proper pH is extremely important if effective disinfection is to be achieved. It should be noted, however, that given adequate contact time, NH_2Cl is nearly as effective as free chlorine in achieving disinfection. In addition to the data for the chlorine compounds, $C_R t$ values have been added for the purpose of comparison. As shown, the disinfection data presented on Fig. 11-10 can be represented quite well with the $C_R t$ relationship.

Referring to Fig. 11-10, it is clear that HOCl offers the most positive way of achieving disinfection. For this reason, with proper mixing, the formation of HOCl following the

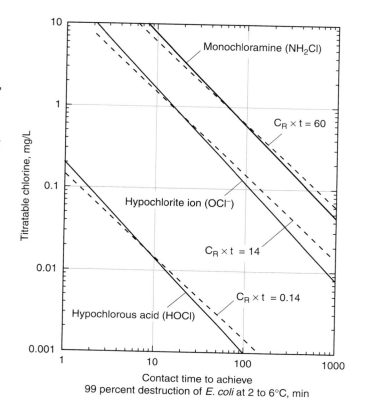

Figure 11-10
Comparison of the germicidal efficiency of hypochlorous acid, hypochlorite ion, and monochloramine for 99 percent destruction of *E. coli* at 2 to 6°C with $C_R t$ values added for the purpose of comparison. (From Butterfield et al., 1943.)

breakpoint is most effective in achieving reclaimed water chlorination. However, when free chlorine is present, the formation of disinfection byproducts (DBPs) is enhanced, as discussed later in this section. If sufficient chlorine cannot be added to achieve the breakpoint reaction, great care must be taken to ensure that the proper contact time is maintained to ensure effective disinfection.

Effect of Temperature in Clean Water

The importance of temperature on the disinfection process with chlorine and chloramines was investigated by Butterfield and his associates in 1943 (Butterfield et al., 1943). Based on these published results, Fair and Geyer (1954) determined the activation energy values reported in Table 11-11 for the disinfection of *E. coli* in clean water. Reviewing the data in Table 11-11, it is important to note the magnitude of the activation energy as a function of pH. As the pH increases, the value of the activation energy increases which corresponds to a reduced effectiveness that is consistent with the data presented in Fig. 11-10.

Relative Germicidal Effectiveness of Chlorine and Chlorine Compounds

In view of the growing interest in public health, environmental water quality, and water reclamation, the effectiveness of the chlorination process is of great concern. Generalized data on the relative germicidal effectiveness of chlorine for the disinfection

Table 11-11
Activation energies for aqueous chlorine and chloramines at normal temperatures[a]

Compound	pH	E, cal/mole	E, J/mole
Aqueous chlorine	8.5	6400	26,800
	9.8	12,000	50,250
	10.7	15,000	62,810
Chloramines	7.0	12,000	50,250
	8.5	14,000	58,630
	9.5	20,000	83,750

[a]Adapted from Fair et al. (1948) who developed the reported values using the data developed by Butterfield et al. (1943).

of different microorganisms are presented in Table 11-5. It is important to note that the data presented in Table 11-5 were derived primarily using batch reactors operated under controlled conditions, and, as such, are of limited use other than for the purpose of illustrating the relative differences in the effectiveness of the different disinfectants for different organism groups. As shown, there are significant differences in the effectiveness of the various disinfectants for each organism group. Unfortunately, similar data on the disinfection for reclaimed water are not available, because of the differences observed in the disinfection response of different reclaimed waters.

Factors that Affect Disinfection of Reclaimed Water with Chlorine

The purpose of the following discussion is to explore the important factors that affect the disinfection efficiency of chlorine compounds in water reclamation applications. These include

1. Initial mixing
2. The chemical characteristics of the reclaimed water
3. The NOM content
4. The impact of particles and particle-associated microorganisms
5. The characteristics of the microorganisms
6. Time of contact

Each of these factors are discussed in more detail below.

Initial Mixing

The importance of initial mixing on the disinfection process cannot be overstressed. It has been shown that the application of chlorine in a highly turbulent regime ($N_R \geq 10^4$) results in kills two orders of magnitude greater than when chlorine is added separately to a conventional rapid-mix reactor under similar conditions. Although the importance of initial mixing is well delineated, the optimum level of turbulence is not known. Examples of mixing facilities designed to achieve the rapid mixing of chlorine with the water to be disinfected are illustrated on Fig. 11-11.

Based on recent findings, questions have now been raised about the form in which the chlorine compounds are added. In some plants where chlorine injectors are used, there

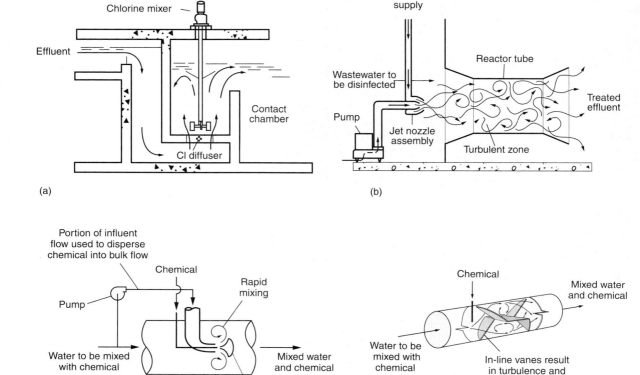

Figure 11-11
Typical mixers for the addition of chlorine: (a) in-line turbine mixer, (b) injector pump type (from Pentech-Houdaille), (c) pumped flash mixing (for large flows multiple stacked units can be used), and (d) in-line static mixer.

is concern over the practice of using chlorinated wastewater for the chlorine injection water. The concern is that if nitrogenous compounds are present in the wastewater, a portion of the chlorine that is added reacts with these compounds, and by the time the chlorine solution is injected, it is in the form of monochloramine or dichloramine. The formation of chloramines can be a problem if adequate retention time is not available in the chlorine contact basin as combined chlorine requires a longer contact time. Again, it should be remembered that HOCl and NH_2Cl are equally effective as disinfecting compounds; only the required contact time is different (see Fig. 11-10).

The formation of DBPs is another major concern with the use of free chlorine, in which molecular chlorine is added directly to the reclaimed water by means of eductors. When reclaimed water is exposed to free chlorine, competing reactions such as the formation of chloramines (free chlorine and ammonia), the formation of DBPs, and the formation of *N*-nitrosodimethylamine (NDMA) (free chlorine, nitrite, and amines) can occur. The predominant reaction depends on the applicable kinetic rates for the various reactions. The formation and control of DBPs is discussed later in this section.

Chemical Characteristics of the Reclaimed Water

It has often been observed that, for treatment plants of similar design with exactly the same effluent characteristics measured in terms of biochemical oxygen demand (BOD), chemical oxygen demand (COD), and nitrogen, the effectiveness of the chlorination process varies significantly from plant to plant. To investigate the reasons for this observed phenomenon, and to assess the effects of the compounds present in the chlorination process, Sung (1974) studied the characteristics of the compounds in untreated and reclaimed water. Among the more important conclusions derived from Sung's study are the following:

1. In the presence of interfering organic compounds, the total chlorine residual cannot be used as a reliable measure for assessing the bactericidal efficiency of chlorine.
2. The degree of interference of the compounds studied depended on their functional groups and their chemical structure.
3. Saturated compounds and carbohydrates exert little or no chlorine demand and do not appear to interfere with the chlorination process.
4. Organic compounds with unsaturated bonds may exert an immediate chlorine demand, depending on their functional groups. In some cases, the resulting compounds may titrate as chlorine residual and yet may possess little or no disinfection potential.
5. Compounds with polycyclic rings containing hydroxyl groups and compounds containing sulfur groups react readily with chlorine to form compounds that have little or no bactericidal potential, but which still titrate as chlorine residual.
6. To achieve low bacterial counts in the presence of interfering organic compounds, additional chlorine and longer contact times are required.

From the results of Sung's work, it is easy to see why the efficiency of chlorination at plants with the same general effluent characteristics can be quite different. Clearly, it is not the value of the BOD or COD that is significant, but the nature of the compounds that make up the measured values. Thus, the nature of the treatment process used in any plant also has an effect on the chlorination process. The impact of reclaimed water characteristics on chlorine disinfection is presented in Table 11-12. The presence of oxidizable compounds such as humics and iron causes the inactivation curve to have a lag or shoulder effect as shown on Fig. 11-6. In effect, the added chlorine is being utilized in the oxidization of these substances and is not available for the inactivation of microorganisms.

Because more water reclamation plants are now removing nitrogen, operational problems with chlorine disinfection are now reported less frequently. In treatment plants where the effluent is nitrified completely, the chlorine added to reclaimed water is present as free chlorine, after satisfying any immediate chlorine demand. In general, the presence of free chlorine reduces significantly the required chlorine dosage. However,

Table 11-12
Impact of wastewater constituents on the use of chlorine for wastewater disinfection

Constituent	Effect
BOD, COD, and TOC	Organic compounds that comprise the BOD and COD can exert a chlorine demand. The degree of interference depends on their functional groups and their chemical structure
NOM (natural organic matter)	Reduces effectiveness of chlorine by forming chlorinated organic compounds that are measured as chlorine residual, but are not effective for disinfection
Oil and grease	Can exert a chlorine demand
TSS	Shield embedded bacteria
Alkalinity	No or minor effect
Hardness	No or minor effect
Ammonia	Combines with chlorine to form chloramines
Nitrite	Oxidized by chlorine, formation of N-nitrosodimethylamine (NDMA)
Nitrate	Chlorine dose is reduced because chloramines are not formed. Complete nitrification may lead to the formation of NDMA due the presence of free chlorine. Partial nitrification may lead to difficulties in establishing the proper chlorine dose
Iron	Oxidized by chlorine
Manganese	Oxidized by chlorine
pH	Affects distribution between hypochlorous acid and hypochlorite ion
Industrial discharges	Depending on the constituents, may lead to a diurnal and seasonal variations in the chlorine demand

the presence of free chlorine may lead to the formation of NDMA, an undesirable disinfection byproduct. In treatment plants that do not nitrify completely, or partially nitrify, control of the chlorination process is especially difficult because of the variation in the effectiveness of the chlorine compounds. Some of the chlorine is used to satisfy the demand of the residual nitrite and/or ammonia. Because of the uncertainties involved in knowing to what degree the plant is nitrifying at any point in time, the chlorine dosage that is added is based on the dosage required if the disinfection of the reclaimed water is to be accomplished by combined chlorine compounds resulting in excessive chlorine use.

Impact of Particles Found in Reclaimed Water

Another factor that must be considered is the presence of suspended solids in the reclaimed water to be disinfected. As shown previously on Fig. 11-6, when suspended solids are present, the disinfection process is controlled by two different mechanisms. The large bacterial inactivation that is observed initially, after the shoulder effect, is of individual free swimming bacteria and bacteria in small clumps. The straight line portion of the bacterial inactivation can be described using Eq. (11-6). In the curved portion of the curve, the bacterial kill is controlled by the presence of suspended solids. The slope of the curved portion of the curve is a function of (1) the particle size distribution and

(2) the number of particles with associated coliform organisms. Further, as noted previously, if particles contain significant numbers of organisms, the organisms can provide protection to other organisms embedded within the particle by limiting the penetration of chlorine through diffusion. Unfortunately, the observed variability caused by the presence of particles often is masked by the addition of excess chlorine to overcome both chemical and particle effects.

Characteristics of the Microorganisms

Other important variables in the chlorination process are the type, characteristics, and age of the microorganisms. For a young bacterial culture (1 d old or less) with a free chlorine dosage of 2 mg/L, only 1 min was needed to reach a low bacterial number. When the bacterial culture was 10 d old or more, approximately 30 min were required to achieve a comparable reduction for the same applied chlorine dosage. It is likely that the resistance offered by the polysaccharide sheath, which microorganisms develop as they age, accounts for this observation. In the activated sludge treatment process, the operating SRT, which to some extent is related to the age of the bacterial cells in the system, will, as discussed previously, affect the performance of the chlorination process. Some recent data on the disinfection of bacteriophage MS2 and poliovirus are shown on Fig. 11-12. As shown on Fig. 11-12, it is clear that a $C_R t$ value of 450 mg·min/L, as used by the state of California, does not result in a four-log reduction of virus, when the measured residual chlorine is combined chlorine (i.e., mono- and dichloramine). Clearly, site-specific testing is required to establish the appropriate chlorine dose.

Some representative data on the effectiveness of chlorine for the inactivation of *E. coli* and three enteric viruses are reported on Fig. 11-13. Again, because of newer analytical techniques that have been developed, the data presented on Fig 11-13 are only meant to illustrate the differences in the resistances of different organisms. From the available evidence on the viricidal effectiveness of the chlorination process, it appears that chlorination beyond the breakpoint to obtain free chlorine is required to inactivate many of the viruses of concern. Where breakpoint chlorination is used, it is necessary to dechlorinate the treated reclaimed water before reuse in sensitive applications to

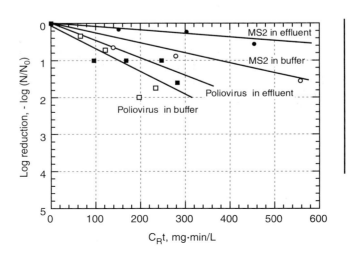

Figure 11-12

Inactivation of MS2 coliphage and poliovirus in a buffer solution and treated wastewater effluent with combined chlorine. (From Cooper et al., 2000.)

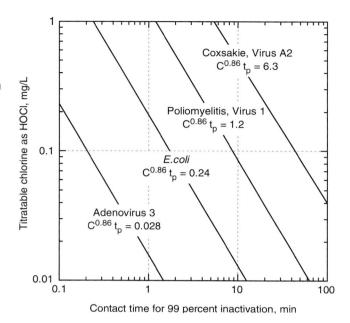

Figure 11-13
Concentration of chlorine as HOCl required for 99 percent inactivation of *E. coli* and three enteric viruses at 0 to 6°C. (From Butterfield et al., 1943.)

reduce any residual toxicity that may remain after chlorination. Recently, based on the use of integrated cell culture-polymerase chain reaction (PCR) techniques, it has been reported that the inactivation of poliovirus may require five times more chlorine than thought previously (Blackmer et al., 2000).

Contact Time

Along with the residual concentration of the disinfectant, contact time is of critical importance in the design and operation of chlorination facilities. The principal design objective for chlorine contact basins is to ensure that some defined percentage of the flow remains in the chlorine contact basin for the design contact time to ensure effective disinfection. The mean contact time is usually specified by the regulatory agency and may range from 30 to 120 min; periods of 15 to 90 min at peak flow are common. To be assured that a given percentage of the flow remains in the chlorine contact basin for a given period of time, the most common approach is to use long plug-flow, around-the-end type of contact basins (see Fig. 11-14). For example, for water reuse applications the Department of Health Services of the state of California requires a $C_R t$ value of 450 mg·min/L based on a modal contact time of 90 min at peak flow. In other states the t_{10} time, which corresponds to the time that 10 percent of the flow has passed through the contact basin, is the time used in the $C_R t$ relationship (see subsequent discussion on assessing the performance of chlorine contact basins).

Issues related to the design of chlorine contact basins not included in this chapter include (1) basin configuration, (2) the use of baffles and guide vanes, (3) number of chlorine contact basins, (4) precipitation of solids in chlorine contact basins, (5) solids transport velocity, and (6) a procedure for predicting disinfection performance. These subjects are considered in detail in Tchobanoglous et al. (2003).

Figure 11-14
Views of chlorine contact basins: (a) and (b) serpentine plug-flow chlorine contact basins with flow deflection baffle(s), (c) spiral plug-flow chlorine contact basin, and (d) plug-flow basin with rounded corners.

Modeling the Chlorine Disinfection Process

When considering the disinfection of reclaimed water, both the lag or shoulder effect and the effect of the residual particles (see Fig. 11-6) must be considered. As noted previously, depending on the constituents in the reclaimed water, a shoulder region may be observed in which there is no reduction in the number of organisms as the result of the addition of a disinfectant. As additional chlorine is added beyond some limiting value, a log linear reduction in the number of organisms is observed with increased chlorine dosages. If particles (typically greater than 20 μm) are present, the disinfection curve starts to diverge from the log linear form and a tailing region is observed due to particle shielding of the microorganisms. The tailing region is of importance as more restrictive standards are to be achieved (e.g., 2.2 MPN/100 mL). It is interesting that the tailing region was identified in an early report on the chlorination of treated wastewater (Enslow, 1938). Further, because large particles have little effect on turbidity, effluents with low measured turbidity values can still be difficult to disinfect due to the presence of undetected large particles (Ekster, 2001, see also the discussion of turbidity in Chap. 8).

The Collins-Selleck Model

In the early 1970s, Collins conducted extensive experiments on the disinfection of various wastewaters (Collins, 1970; Collins and Selleck, 1972). Using the batch reactor whose contents were well stirred, Collins and Selleck found that the reduction of coliform organisms in a chlorinated primary treated effluent followed a linear relationship when plotted on log-log paper (see Fig. 11-15). The equation developed to describe the observed results is:

$$\frac{N}{N_0} = \frac{1}{(1 + 0.23 \, C_R t)^3} \tag{11-25}$$

Note that the form of the equation developed by Collins accounts for the shoulder effect and for tailing. A number of other models have been proposed including an empirical model proposed by Gard (1957), Hom (1972), which was subsequently rationalized by Haas and Joffe (1994), and Rennecker et al. (1999, 2001).

The Refined Collins-Selleck Model

A refinement of the original Collins model for the disinfection of secondary effluent in which a shoulder effect and tailing is observed, as proposed by White (1999), is:

$$N/N_0 = 1 \quad \text{for } C_R t < b \tag{11-26}$$

$$N/N_0 = [(C_R t)/b]^{-n} \quad \text{for } C_R t > b \tag{11-27}$$

where C_R = residual concentration of chemical agent at the end of time t, mg/L
 t = contact time, min
 n = slope of inactivation curve
 b = value of x-intercept when $N/N_0 = 1$ or $\log(N/N_0) = 0$ (see Fig. 11-16)

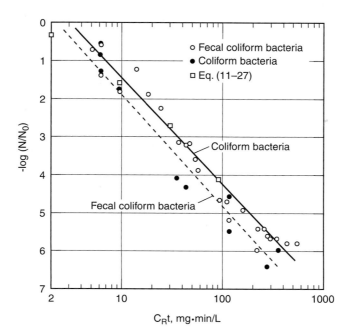

Figure 11-15
Coliform survival in a batch reactor as a function or amperometric chlorine residual and contact time (temperature range 11.5 to 18°C). (From Collins, 1970; Collins and Selleck, 1972.)

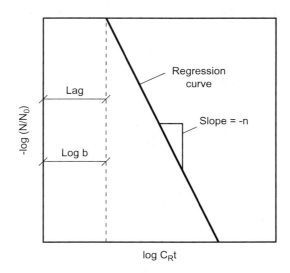

Figure 11-16
Definition sketch for the application of Eq. (11-27).

Typical values for the coefficients n and b for secondary effluent for coliform and fecal coliform organisms are 2.8 and 4.0 and 2.8 and 3.0, respectively (Roberts et al., 1980; White, 1999). However, because of the variability of the chemical composition of the reclaimed water and the variable particle size distribution, it is recommended that the constants be determined for the reclaimed water in question.

Effluent from Membrane Processes

The most important characteristic of effluents from membrane processes is that they do not contain particles that can shield microorganisms. Depending on the type of membrane process used (microfiltration, ultrafiltration, nanofiltration, or reverse osmosis), moderate to significant reductions in the number of microorganisms present will also be observed (see discussion in Chaps. 8 and 9). For these effluents, the Chick-Watson model, as given by Eq. (11-6) or, if a shoulder exists, the Collins-Selleck relationship can be used to model the disinfection process with chlorine. Typically, the shoulder effect is reduced considerably, especially with reverse osmosis effluent.

The required chemical dosage for disinfection can be estimated by considering (1) the initial chlorine demand of the reclaimed water, (2) the allowance needed for decay during the chlorine contact time, and (3) the required chlorine residual concentration determined using Eq. (11-27) for the organism under consideration (e.g., bacteria, virus, or protozoan oocysts and cysts). The chlorine dosages required to meet the initial demand depends on the constituents in the reclaimed water. It is important to remember that the chlorine added to meet the initial demand due to inorganic compounds is reduced to the chloride ion and will not be measured as chlorine residual.

Required Chlorine Dosages for Disinfection

Chlorine that combines with humic materials is effective as a disinfectant, but is nevertheless measured as a chlorine residual contributing to the lag term, b, in Eq. (11-27). Typical decay values for chlorine residual are on the order of 2 to 4 mg/L for contact time of about 1 h. To reduce the decay due to UV oxidation observed in open channel chlorine contact basins, several types of floating and fixed covers have been added to existing

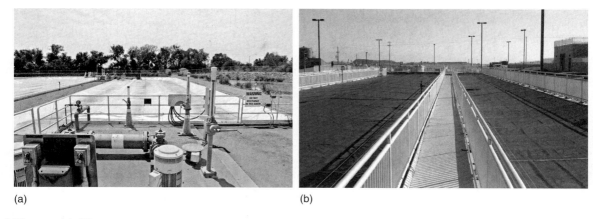

(a) (b)

Figure 11-17
Typical examples of chlorine contact basins covered to limit the oxidation of chlorine by sunlight: (a) basins covered with inexpensive floating trap and (b) basins covered with specially designed polypropylene cover.

contact basins (see Fig. 11-17). Typical chlorine dosage values for various reclaimed waters for total coliform, based on a contact time of 30 min, are reported in Table 11-13. It should be noted that the dosage values given in Table 11-13 are only meant to serve as a guide for the initial estimation of the required chlorine dose. As noted above, site-specific testing is required to establish the appropriate chlorine dose. Estimation of the required chlorine dose is illustrated in Example 11-4.

Table 11-13
Typical chlorine dosages, based on combined chlorine unless otherwise indicated, required to achieve different total coliform disinfection standards for various wastewaters based on a 60-min contact time

Type of wastewater	Initial coliform count, MPN/100 mL	Chlorine dose, mg/L Effluent standard, MPN/100 mL			
		1000	200	23	≤2.2
Raw wastewater	10^7–10^9	5–15			
Primary effluent	10^7–10^9	5–10	6–15		
Trickling filter effluent	10^5–10^6	1–2	2.5–5	16–22	
Activated sludge effluent	10^5–10^6	1–2	2.5–5	16–20	
Filtered activated sludge effluent	10^4–10^6	0.25–0.5	0.5–1.5	1.8–7	7–25
Nitrified effluent[a]	10^4–10^6	0.1–0.2	0.3–0.5	0.9–1.4	3–5
Filtered nitrified effluent[a]	10^4–10^6	0.1–0.2	0.3–0.5	0.9–1.4	3–4
Microfiltration effluent	10^1–10^3		0.1–0.15	0.15–0.2	0.2–0.5
Reverse osmosis[a]	~0	0	0	0	0–0.3
Septic tank effluent	10^7–10^9	5–10	6–15		
Intermittent sand filter effluent	10^2–10^4		0.02–0.05	0.1–0.16	0.4–0.5

[a]Based on free chlorine.

EXAMPLE 11-4. Estimate the Required Chlorine Dose for a Typical Secondary Effluent.

Estimate the chlorine dose needed to disinfect a reclaimed water (filtered secondary effluent) assuming a shoulder effect exists and that the following conditions apply:

1. Effluent total coliform count before disinfection = 10^7/100 mL
2. Required summer effluent total coliform count = 23/100 mL
3. Required winter effluent total coliform count = 240/100 mL
4. Initial effluent chlorine demand = 4 mg/L
5. Demand due to decay during chlorine contact = 2.5 mg/L
6. Required chlorine contact time = 60 min
7. Use the typical values given in the above discussion for the coefficients.
   b = 4.0
   n = 2.8

Solution

1. Estimate the required chlorine residual using the refined Collins-Selleck Model, Eq. (11-27), and the given coefficients.

$$N/N_0 = (C_R t/b)^{-n}$$

a. Summer

$$23/10^7 = (C_R t/4.0)^{-2.8}$$
$$(23/10^7)^{-1/2.8} = (C_R t/4.0)$$
$$(234.3)4 = C_R(60)$$
$$C_R = 15.6 \text{ mg/L}$$

b. Winter

$$240/10^7 = (C_R t/4.0)^{-2.8}$$
$$C_R = 3.0 \text{ mg/L}$$

2. The required chlorine dosage is
   a. Summer
   Chlorine dosage = 4.0 mg/L + 2.5 mg/L + 15.6 mg/L = 22.1 mg/L
   b. Winter
   Chlorine dosage = 4.0 mg/L + 2.5 mg/L + 3.0 mg/L = 9.5 mg/L

Comment

The chlorine dosage increases significantly as the effluent standards become more stringent. In the above computation, it was assumed that the reclaimed water to be disinfected remained in the chlorine contact tank for the full 60 min. Thus, it is clear that the proper design of a plug-flow chlorine contact basin is critical to the effective use of chlorine as disinfectant. The design of chlorine contact basins is discussed in Tchobanoglous et al. (2003).

Assessing the Hydraulic Performance of Chlorine Contact Basins

To be assured that a chlorine contact basin performs properly, most regulatory agencies request that tracer studies be conducted to determine the hydraulic characteristics of the chlorine contact basin. The types of tracers that have been used, the conduct of tracer tests, and analysis of tracer data are reviewed briefly below.

Types of Tracers

Tracers of various types are used commonly to assess the hydraulic performance of reactors used for reclaimed water disinfection. Dyes and chemicals that have been used successfully in tracer studies include Congo red, fluorescein, fluosilicic acid (H_2SiF_6), hexafluoride gas (SF_6), lithium chloride (LiCl), Pontacyl Brilliant Pink B, potassium, potassium permanganate, rhodamine WT, and sodium chloride (NaCl). Pontacyl Brilliant Pink B (the acid form of rhodamine WT) is especially useful in the conduct of dispersion studies because it is not readily adsorbed onto surfaces. Because fluorescein, rhodamine WT, and Pontacyl Brilliant Pink B can be detected at low concentrations using a fluorometer, they are the dye tracers used most commonly in the evaluation of the performance of wastewater treatment facilities.

Conduct of Tracer Tests

In tracer studies, typically a tracer (i.e., a dye, most commonly) is introduced into the influent end of the reactor or basin to be studied (see Fig. 11-18). The time of its arrival at the effluent end is determined by collecting a series of grab samples for a given period of time or by measuring the arrival of a tracer using instrumental methods (see Fig. 11-18). The method used to introduce the tracer controls the type of response observed at the downstream end. Two types of dye input are used, the choice depending on the influent and effluent configurations. The first method involves the injection of a quantity of dye

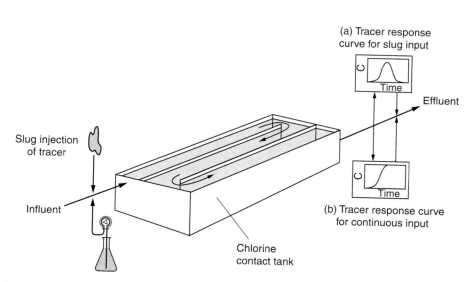

Figure 11-18
Schematic of setup for the conduct of a tracer study of a plug-flow chlorine contact basin using either a slug of tracer added to flow or a continuous input of tracer. The tracer response curve is measured continuously.

(sometimes referred to a pulse or slug of dye) over a short period of time. Initial mixing is usually accomplished with a static mixer or an auxiliary mixer. With the slug injection method, it is important to keep the initial mixing time short relative to the detention time of the reactor being measured. The measured output is as described on Fig. 11-18a. In the second method, a continuous step input of dye is introduced until the effluent concentration matches the influent concentration. The measured response is as shown on Figs. 11-18b. It should also be noted that another response curve can be measured after the dye injection has ceased and the dye in the reactor is flushed out.

Analysis of Tracer Response Curves

Tracer response curves, measured using a slug or continuous injection of a tracer, are known as C (concentration versus time) and F (fraction of tracer remaining in the reactor versus time) curves, respectively. The fraction remaining is based on the volume of water displaced from the reactor by the step input of tracer. The generalized results of three different dye tracer tests are shown on Fig. 11-19. As shown on Fig. 11-19, each of the three basins is subject to differing amounts of short circuiting. Length-to-width ratios (L/W) of at least 20 to 1 (preferably 40 to 1) and the use of baffles and guide vanes helps to minimize short circuiting. In some small plants, chlorine contact basins have been constructed of large diameter sewer pipe. The beneficial effect of using submerged baffles to improve the hydraulic efficiency of serpentine chlorine contact basins is illustrated on Fig. 11-20.

Tracer curves, such as shown on Figs. 11-19 and 11-20, are used to assess the hydraulic efficiency of chlorine contact basins. Parameters used to assess the hydraulic efficiency of chlorine contact basins are summarized in Table 11-14 and are illustrated on Fig. 11-21. As discussed previously, the mean, modal, and t_{10} times have been used to define the contact time in the $C_R t$ relationship. The analysis of a tracer response curve is illustrated in Example 11-5. Additional details on the analysis of tracer response curves may be found in Tchobanoglous et al. (2003); Crittenden et al. (2005).

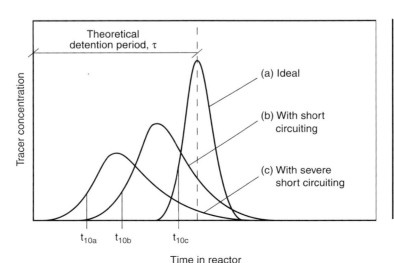

Figure 11-19

Typical chlorine contact basin tracer response curves for three different basins with the same hydraulic detention time. The degree of short circuiting is illustrated clearly by the shape of the tracer curve.

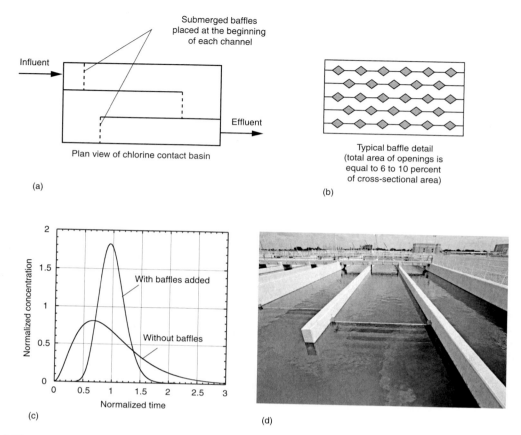

Figure 11-20
Baffling in chlorine contact basins: (a) placement of baffles in chlorine contact tank at the beginning of each channel (or pass) is critical (adapted from Crittenden et al., 2005), (b) typical submerged baffle detail (adapted from Kawamura, 2000), (c) effect of the use of baffles in chlorine contact basins (adapted from Hart, 1979), and (d) view of chlorine contact tank with submerged wooden baffles placed at the beginning and end of each channel.

Figure 11-21
Definition sketch for the parameters used in the analysis of concentration versus time tracer response curves.

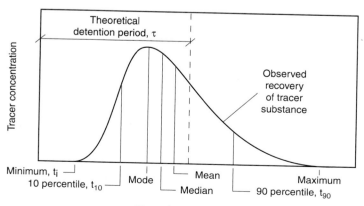

Table 11-14
Various terms used to describe the hydraulic performance of chlorine contact basins[a]

Term	Definition
τ[b]	Theoretical hydraulic residence time (V, volume/Q, flowrate).
t_i	Time at which tracer first appears.
t_p	Time at which the peak concentration of the tracer is observed (mode).
t_m	Mean hydraulic residence time, which corresponds to the time to reach centroid of the residence time distribution (RTD) curve.
t_{10}, t_{50}, t_{90}	Time at which 10, 50, and 90 percent of the tracer had passed through the reactor.
t_{90}/t_{10}	Morrill Dispersion Index, MDI (Morrill, 1932).
1/MDI	Volumetric efficiency as defined by Morrill (1932).
t_i/τ	Index of short circuiting. In an ideal plug-flow reactor, the ratio is 1, and approaches 0 with increased mixing.
t_p/τ	Index of modal retention time. Ratio will approach 1 in a plug-flow reactor, and 0 in a complete-mix reactor. For values of the ratio greater than or less than 1, the flow distribution in the reactor is not uniform.
t_m/τ	Index of average retention time. A value of 1 would indicate that full use is being made of the volume. A value of the ratio greater than or less than 1.0 indicates the flow distribution is not uniform.
t_{50}/τ	Index of mean retention time. The ratio t_{50}/τ is a measure of the skew of the RTD curve. A value of t_{50}/τ of less than 1 corresponds to an RTD curve that is skewed to the left. Similarly, for values greater than 1.0 the RTD curve is skewed to the right.
$t_m \approx \dfrac{\sum t_i C_i \Delta t_i}{\sum C_i \Delta t_i}$	Expression used to determine the mean hydraulic residence time, t_m, if the concentration versus time tracer response curve is defined by a series of discrete time step measurements, where t_i is time at ith measurement, C_i is concentration at ith measurement, and Δt_i is time increment about C_i.
$\sigma_t \approx \dfrac{\sum t_i^2 C_i \Delta t_i}{\sum C_i \Delta t_i} - (t_m)^2$	Expression used to determine variance for a concentration versus time tracer response curve, which is defined by a series of discrete time step measurements.

[a] Adapted from Morrill (1932), Fair and Geyer (1954), and U.S. EPA (1986).
[b] The symbols θ and θ_h have also been used for the theoretical hydraulic residence time.

EXAMPLE 11-5. Analysis of Tracer Data for a Chlorine Contact Basin.

The following tracer data were gathered during a tracer test of a chlorine contact basin. During the tracer test, the chlorine residual measured at the tank outlet was 4.0 mg/L. Using these data, determine the mean hydraulic residence time, t_m, the variance, σ_t, and the t_{10} time. Determine the $C_R t$ values corresponding to the t_m and t_{10} times. To further assess the performance of the chlorine contact basin, determine the Morrill Dispersion Index (MDI) and the corresponding volume efficiency (1/MDI) as defined in Table 11-14.

Time, min	Tracer concentration, μg/L	Time, min	Tracer concentration, μg/L
0.0	0.000	144	9.333
16	0.000	152	16.167
40	0.000	160	20.778
56	0.000	168	19.944
72	0.000	176	14.111
88	0.000	184	8.056
96	0.056	192	4.333
104	0.333	200	1.556
112	0.556	208	0.889
120	0.833	216	0.278
128	1.278	224	0.000
136	3.722		

Solution

1. Determine the mean hydraulic residence time and variance for the tracer response data using equations given in Table 11-14.
   a. Set up the required computation table. In setting up the computation table given below, the Δt value was omitted as it appears in both the numerator and in the denominator of the equations used to compute the residence time and the corresponding variance.

Time, t, min	Conc., C, μg/L	$t \times C$	$t^2 \times C$	Cumulative conc.	Cumulative percentage
88	0.000	0.000	0		
96	0.056	5.338	512.41	0.05	0.05
104	0.333	34.663	3604.97	0.39[a]	0.38[b]
112	0.556	62.227	6969.45	0.94	0.92
120	0.833	99.996	11,999.52	1.78	1.74
128	1.278	163.558	20,935.48	3.06	2.99
136	3.722	506.219	68,845.81	6.78	6.63
144	9.333	1343.995	193,535.31	16.11	15.75
152	16.167	2457.384	373,522.37	32.28	31.58
160	20.778	3324.480	531,916.80	53.06	51.91
168	19.944	3350.592	562,899.46	73.00	71.41
176	14.111	2483.536	437,102.34	87.11	85.22
184	8.056	1482.230	272,730.39	95.17	93.10
192	4.333	831.994	159,742.77	99.50	97.34
200	1.556	311.120	62,224.00	101.06	98.87
208	0.889	184.891	38,457.37	101.94	99.73
216	0.278	60.005	12,961.04	102.22	100.00
224	0.000	0.000			
Total	102.222	16,702.229	2,757,959.48		

[a] $0.056 + 0.333 = 0.39$
[b] $(0.39/102.222) \times 100 = 0.38$

b. Determine the mean hydraulic residence time.
$$t_m \approx \frac{\Sigma t_i C_i \Delta t_i}{\Sigma C_i \Delta t_i} = \frac{16{,}702.23}{102.22} = 163.4 \text{ min} = 2.7 \text{ h}$$

c. Determine the variance.
$$\sigma_t \approx \frac{\Sigma t_i^2 C_i \Delta t_i}{\Sigma C_i \Delta t_i} - (t_m)^2 = \frac{2{,}757{,}959.48}{102.22} - (163.4)^2 = 280.5 \text{ min}^2$$

$$\sigma_t = 16.7 \text{ min}$$

d. Determine the t_{10} time using the cumulative percentage values. Because of the short time interval, a linear interpolation method can be used.

$$(15.75\% - 6.63\%)/(144 \text{ min} - 136 \text{ min}) = 1.14\%/\text{min}$$

$$136 \text{ min} + (10\% - 6.63\%)/(1.14\%/\text{min}) = 139.0 \text{ min}$$

e. Identify the t_m and t_{10} times on the tracer curve.

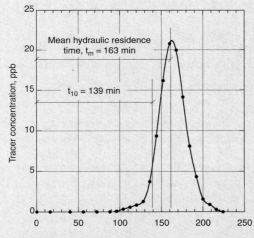

2. Another technique that can be used to obtain the above times is to plot the cumulative concentration data on log-probability paper. Such a plot is also useful for determining the MDI. The required plot is given below.

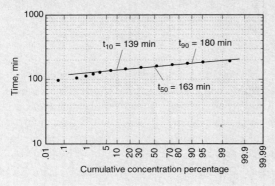

The mean hydraulic retention and t_{10} times are read directly from the above plot.

$$t_{50} = 163 \text{ min}$$
$$t_{10} = 139 \text{ min}$$

3. Determine the corresponding $C_R t$ values for the t_m and t_{10} time determined above in Step 1.

$$C_R t\,(t_m) = (4.0 \text{ mg/L})(163.4 \text{ min}) = 654 \text{ mg} \cdot \text{min/L}$$
$$C_R t\,(t_{10}) = (4.0 \text{ mg/L})(139 \text{ min}) = 556 \text{ mg} \cdot \text{min/L}$$

4. Determine the MDI and the corresponding volume efficiency using the expressions given in Table 11-14 and the values from the plot given in Step 2 above.
   a. The Morrill Dispersion Index is:

   $$\text{Morrill Dispersion Index, MDI} = \frac{t_{90}}{t_{10}} = \frac{180}{139} = 1.30$$

   b. The corresponding volumetric efficiency for the chlorine contact basin is:

   $$\text{Volumetric efficiency, \%} = \frac{1}{\text{MDI}} = \frac{1}{1.3} \times 100 = 77\%$$

Comment

The variance computed in Step 1 is useful in assessing the dispersion in the chlorine contact basin (Tchobanoglous et al., 2003; Crittenden et al., 2005).

The $C_R t$ values, based on the modal and t_{10} times, exceed the $C_R t$ value of 450 mg·min/L required in California. It is important to note that if the tracer curve is very skewed, it may not be possible to achieve effective disinfection, especially if the t_{10} value is used. Thus, the design of a chlorine contact basin to achieve near plug flow is of critical importance.

The MDI value (1.30) is characteristic of a chlorine contact basin with low dispersion. A MDI value below 2.0 has been established by the U.S. EPA as an effective design (U.S. EPA, 1986). Similarly, the volumetric efficiency is high, signifying near-ideal plug flow with a small amount of axial dispersion.

Formation and Control of Disinfection Byproducts

In the early 1970s, it was found that the use of oxidants, such as chlorine and ozone, in water treatment plants for disinfection; for taste, odor, and color removal; and other in-plant uses resulted in the production of undesirable disinfection byproducts (DBPs) (Rook, 1974; Bellar and Lichtenberg, 1974). The DBPs occurring most frequently and with the highest concentration are trihalomethanes (THMs) and haloacetic acids (HAAs) that result from chlorination. In addition to THMs and HAAs, other DBPs are also produced. The principal DBPs that have been identified are reported in Table 11-15.

Table 11-15

Known byproducts of chlorine, chloramine, ozone, and chlorine dioxide application during drinking water treatment

Class	Byproduct	Chemical agent	Molecular formula
Trihalomethanes	Chloroform	Chlorine	$CHCl_3$
	Bromodichloromethane	Chlorine	$CHBrCl_2$
	Dibromochloromethane	Chlorine	$CHBr_2Cl$
	Bromoform	Chlorine, ozone	$CHBr_3$
	Dichloroiodomethane	Chlorine	$CHICl_2$
	Chlorodiiodomethane	Chlorine	CHI_2Cl
	Bromochloroiodomethane	Chlorine	$CHBrICl$
	Dibromoiodomethane	Chlorine	$CHBr_2I$
	Bromodiiodomethane	Chlorine	$CHBrI_2$
	Triiodomethane	Chlorine	CHI_3
Haloacetic acids	Monochloroacetic acid	Chlorine	$CH_2ClCOOH$
	Dichloroacetic acid	Chlorine	$CHCl_2COOH$
	Trichloroacetic acid	Chlorine	CCl_3COOH
	Bromochloroacetic acid	Chlorine	$CHBrClCOOH$
	Bromodichloroacetic acid	Chlorine	$CBrCl_2COOH$
	Dibromochloroacetic acid	Chlorine	$CBr_2ClCOOH$
	Monobromoacetic acid	Chlorine	$CH_2BrCOOH$
	Dibromoacetic acid	Chlorine	$CHBr_2COOH$
	Tribromoacetic acid	Chlorine	CBr_3COOH
Haloacetonitriles	Trichloroacetonitrile	Chlorine	$CCl_3C{\equiv}N$
	Dichloroacetonitrile	Chlorine	$CHCl_2C{\equiv}N$
	Bromochloroacetonitrile	Chlorine	$CHBrClC{\equiv}N$
	Dibromoacetonitrile	Chlorine	$CHBr_2C{\equiv}N$
Haloketones	1,1-Dichloroacetone	Chlorine	$CHCl_2COCH_3$
	1,1,1-Trichloroacetone	Chlorine	CCl_3COCH_3
Aldehydes	Formaldehyde	Ozone, chlorine	$HCHO$
	Acetaldehyde	Ozone, chlorine	CH_3CHO
	Glyoxal	Ozone, chlorine	$OHCCHO$
	Methyl glyoxal	Ozone, chlorine	CH_3COCHO
Aldoketoacids	Glyoxylic acid	Ozone	$OHCCOOH$
	Pyruvic acid	Ozone	$CH_3COCOOH$
	Ketomalonic acid	Ozone	$HOOCCOCOOH$
Carboxylic acids	Formate	Ozone	$HCOO^-$
	Acetate	Ozone	CH_3COO^-
	Oxalate	Ozone	$OOCCOO^{2-}$
Oxyhalides	Chlorite	Chlorine dioxide	ClO_2^-
	Chlorate	Chlorine dioxide	ClO_3^-
	Bromate	Ozone	BrO_3^-
Nitrosamines	N-nitrosodimethylamine	Chloramines	$(CH_3)_2NNO$
Cyanogen Halides	Cyanogen chloride	Chloramines	$ClCN$
	Cyanogen bromide	Chloramines	$BrCH$
Misc.	Chloral hydrate	Chlorine	$CCl_3CH(OH)_2$
Trihalonitromethanes	Trichloronitromethane (Chloropicrin)	Chlorine	CCl_3NO_2
	Bromodichloronitromethane	Chlorine	$CBrCl_2NO_2$
	Dibromochloronitromethane	Chlorine	CBr_2ClNO_2
	Tribromonitromethane	Chlorine	CBr_3NO_2

Adapted from Krasner (1999), Krasner et al. (2001), and Thibaud et al. (1987).

Many of these compounds have also been identified in reclaimed water that has been disinfected using chlorine, chloramines, chlorine dioxide, and ozone.

Formation of DBPs is of great concern in indirect and direct potable reuse because of the potential impact of these compounds on public health and the environment. Chloroform, for example, is a well-known animal carcinogen and many of the haloforms are also thought to be animal carcinogens. In addition, many of these compounds have been classified as probable human carcinogens. Still others of these compounds are known to cause chromosomal aberrations and sperm abnormalities. Recognizing the many unknowns and the potential public health and environmental risks associated with these compounds, the U.S. EPA has moved aggressively to control their formation in drinking water.

Formation of DBPs When Using Chlorine for Disinfection

Trihalomethanes and other DBPs are formed as a result of a series of complex reactions between free chlorine and a group of organic acids known collectively as humic acids. The reactions lead to the formation of single carbon molecules that are often designated as HCX_3, where X is either a chlorine (Cl^-) or bromine (Br^-) atom. For example, the chemical formula for chloroform is $HCCl_3$.

The rate of formation of DBPs is dependent on a number of factors including:

- Presence of organic precursors
- Free chlorine concentration
- Bromide concentration
- pH
- Temperature
- Time

The type and concentration of the organic precursor affects both the rate of the reaction and extent to which the reaction is completed.

The presence of free chlorine was thought to be necessary for the THM formation reaction to proceed, but it appears that THMs can form in the presence of combined chlorine (chloramines), but at a very reduced rate. It is important to note that initial mixing can affect the formation of THMs because of the competing reactions between chlorine and ammonia and chlorine and humic acids. If bromide is present, it can be oxidized to bromine by free chlorine. In turn, the bromine ion can combine with the organic precursors to form THMs, including bromodichoromethane, dibromochoromethane, and bromoform. The rate of formation of THMs has been observed to increase with both pH and temperature. Additional details on the formation of THMs may be found in U.S. EPA (1999b).

Although chloramines, as discussed above, produce THMs at reduced rates, they can, nevertheless, produce other DBPs that are of concern. Other DBPs that are produced when reclaimed water is disinfected with chloramines include NDMA, a member of a class of compounds known as nitrosamines: cyanogen chloride and cyanogen bromide (see Table 11-15). As a class of compounds, nitrosamines are among the most powerful carcinogens known (Snyder, 1995). The compounds in this class have been found to produce cancer in every species of laboratory animal tested.

One pathway leading to the formation of NDMA can be illustrated with the following two reactions:

$$\underset{\text{nitrite ion}}{NO_2^-} + \underset{\text{hydrochloric acid}}{HCl} \rightarrow \underset{\text{nitrous acid}}{HNO_2} + \underset{\text{chloride ion}}{Cl^-} \quad (11\text{-}28)$$

$$\underset{\text{nitrous acid}}{HNO_2} + \underset{\text{dimethylamine}}{CH_3-NH-CH_3} \rightarrow \underset{\text{N-nitrosodimethylamine}}{CH_3-\overset{\overset{\displaystyle NO}{|}}{N}-CH_3} \quad (11\text{-}29)$$

The concern in biological wastewater treatment is that some nitrite may leak through the process. While the concentration of nitrite may be too low to measure by conventional means, concentrations of NDMA as low as 1 or 2 ng/L are being measured and the California DHS notification level for groundwater recharge is 10 ng/L. Based on a limited number of test locations, it has been observed that the concentrations of NDMA in the incoming wastewater can be quite variable, with concentrations as high as 14,000 ng/L being measured.

In addition to the formation of NDMA as outlined above, it appears the addition of chloramines for disinfection can serve to amplify the concentration of any NDMA precursors that may be present in the treated effluent before disinfection. In a series of studies conducted by the Los Angeles County Sanitation Districts (Jalali et al., 2005), it was found that chloramination increased the concentration of NDMA in treated effluent following disinfection by tenfold.

Other DBPs resulting from the use of chloramines as disinfectants in reclaimed water include cyanogen chloride and cyanogen bromide, where bromides are present (see Table 11-15). Cyanogen chloride is used in tear gas, in fumigant gases, and as a reagent in the formation of other compounds. In the body, cyanogen chloride is metabolized rapidly to cyanide. Because there is limited information on the toxicity of cyanogen chloride, proposed guidelines are based on cyanide. The cyanogens compounds are of concern and they are now beginning to be regulated in effluent discharge permits. Current NPDES permit limits for cyanide are 5 mg/L.

Control of DBP Formation When Using Chlorine for Disinfection

The principal means of controlling the formation of THMs and other related DBPs in reclaimed water is to avoid the direct addition of free chlorine. Based on the evidence to date, it appears that the use of chloramines generally does not lead to the formation of THMs in amounts that would be of concern relative to current standards. As discussed previously, other DBPs may be produced which are of equal concern, but for other reasons (see following discussion). It is important to note that if chloramines are to be used for disinfection, the chloramine solution must be prepared with a potable water supply containing little or no ammonia (i.e., *treated plant effluent should not be used*). If the formation of DBPs is of concern due to the presence of specific organic precursors (i.e., humic materials), the practice of breakpoint chlorination cannot be used. Further, if humic materials are present consistently, it may be appropriate to investigate alternative means of disinfection such as UV irradiation.

The control of DBPs produced when chloramines are used as the disinfectant is more challenging. With respect to NDMA, it appears that with proper control and operation of the biological treatment process, the potential for the formation or amplification of this compound can be reduced. Removals of 50 to 70 percent have been reported for NDMA when using reverse osmosis employing thin film composite membranes (see Chap. 9). The use of UV irradiation has also proven to be effective in the control of NDMA. Where the formation of NDMA and cyanogen chloride is a persistent concern, several wastewater agencies have switched to UV irradiation for disinfection. In the study cited above (Jalali et al., 2005), it was also found that there was no net change in the total cyanide (CN^-) concentration in the treated effluent due to UV irradiation.

Environmental Impacts

The environmental impacts associated with the use of chlorine and chlorine compounds as a reclaimed water disinfectant include the discharge of DBPs in the reclaimed water and the regrowth of microorganisms.

Discharge of DBPs

It has been shown that many of the DBPs can cause environmental impacts at very low concentrations. The occurrence of DBPs and compounds such as NDMA raises serious questions about the continued use of free chlorine for reclaimed water disinfection.

Regrowth of Microorganisms

In many locations, a regrowth of microorganisms has been observed in receiving water bodies and in long transmission pipelines following dechlorination of reclaimed water disinfected with chlorine. The regrowth of microorganisms is not unexpected as it is well known that many microorganisms survive the disinfection process. It has been hypothesized that regrowth (also known as aftergrowth) results, in part, because (1) the amount of organic matter and available nutrients in reclaimed water is sufficient to sustain the limited number of organisms remaining after disinfection, (2) predators such as protozoa are absent, (3) temperatures are favorable, and (4) disinfectant residuals are ineffective. Because regrowth is an important issue in transmission lines used for the transport of reclaimed water, a suitable combined chlorine residual (on the order of 1 to 2 mg/L, depending on local conditions) should be maintained in the pipeline to control regrowth (a common practice in water distribution systems). In long pipelines, it may be necessary to add additional chlorine at intermediate points along the length of the pipeline.

11-4 DISINFECTION WITH CHLORINE DIOXIDE

Chloride dioxide (ClO_2), another bactericide, is equal to or greater than chlorine in disinfecting power. Chlorine dioxide has proven to be an effective viricide, being more effective in achieving inactivation of viruses than chlorine. A possible explanation is that because ClO_2 is adsorbed by peptone (a protein), and that viruses have a protein coat, adsorption of ClO_2 onto this coating could cause inactivation of the virus. In the past, ClO_2 did not receive much consideration as a reclaimed water disinfectant due to its high costs; sodium chlorite is about 10 times as expensive as chlorine on a weight basis.

Characteristics of Chlorine Dioxide

Chlorine dioxide (ClO_2) is, under atmospheric conditions, a yellow to red unpleasant smelling irritating unstable gas with a high specific gravity. Because chlorine dioxide is unstable and decomposes rapidly, it is usually generated onsite before its application. Chlorine dioxide is generated by mixing and reacting a chlorine solution in water with a solution of sodium chlorite ($NaClO_2$) according to the following reaction:

$$2NaClO_2 + Cl_2 \rightarrow 2ClO_2 + 2NaCl \quad (11\text{-}30)$$

Based on Eq. (11-30), 1.34 mg sodium chlorite reacts with 0.5 mg chlorine to yield 1.0 mg chlorine dioxide. Because technical grade sodium chlorite is only about 80 percent pure, about 1.68 mg of the technical grade sodium chlorite is required to produce 1.0 mg of chlorine dioxide. Sodium chlorite may be purchased and stored as a liquid (generally a 25 percent solution) in refrigerated storage facilities. The properties of chlorine dioxide were presented previously in Tables 11-3 and 11-6.

Chlorine Dioxide Chemistry

The active disinfecting agent in a chlorine dioxide system is free dissolved chlorine dioxide (ClO_2). At the present time, the complete chemistry of chlorine dioxide in an aqueous environment is not understood completely. Because ClO_2 does not hydrolyze in a manner similar to the chlorine compounds discussed in the previous section, the oxidizing power of ClO_2 is often referred to as *equivalent available chlorine*. The definition of the term equivalent available chlorine is based on a consideration of the following oxidation half reaction for ClO_2.

$$ClO_2 + 5e^- + 4H^+ \rightarrow Cl^- + 2H_2O \quad (11\text{-}31)$$

As shown in Eq. (11-31), the chlorine atom undergoes a 5 electron change in its conversion from chlorine dioxide to the chloride ion. Because the weight of chlorine in ClO_2 is 52.6 percent and there is a 5 electron change, the equivalent available chlorine content is equal to 263 percent as compared to chlorine. Thus, ClO_2 has 2.63 times the oxidizing power of chlorine. The concentration of ClO_2 is usually expressed in g/m^3. On a molar basis, one mole of ClO_2 is equal to 67.45 g, which is equivalent to 177.5 g (5×35.45) of chlorine. Thus, 1 g/m^3 of ClO_2 is equivalent to 2.63 g/m^3 of chlorine.

Effectiveness of Chlorine Dioxide as a Disinfectant

Chlorine dioxide has an extremely high oxidation potential which probably accounts for its potent germicidal powers. Because of its extremely high oxidizing potential, possible bactericidal mechanisms may include inactivation of critical enzyme systems or disruption of protein synthesis. It should be noted, however, that when ClO_2 is added to reclaimed water it is often reduced to chlorite (ClO_2^-) according to the following reaction:

$$ClO_2 + e^- \rightarrow ClO_2^- \quad (11\text{-}32)$$

Equation (11-32) may help to explain the variability that is sometimes observed in the performance of ClO_2 as a disinfectant.

Modeling the Chlorine Dioxide Disinfection Process
As discussed previously in Sec. 11-3, the models that have been developed to describe the disinfection process with chlorine can also be used with appropriate caution for chlorine dioxide. As with chlorine, the shoulder effect and the effect of the residual particles must be considered. Further, the differences between (1) secondary and filtered secondary effluent and (2) microfiltration and reverse osmosis effluent must also be considered.

Required Chlorine Dioxide Dosages for Disinfection
The required chlorine dioxide dosage will depend on the pH and the specific organism of concern. Relative $C_R t$ values for chlorine dioxide are given in Table 11-5, presented previously in Sec. 11-2. In general, the effectiveness of chlorine dioxide is similar to that of combined chlorine for bacteria. However, there is a significant difference in the effectiveness of chlorine dioxide for the disinfection of viruses, which is essentially the same as that for free chlorine. Chlorine dioxide appears to be more effective than free chlorine in the inactivation of protozoan cysts. Because the data on chlorine dioxide in the literature are limited, site-specific testing is recommended to establish appropriate dosage ranges although the $C_R t$ values given in Table 11-5 can be used as a starting point.

Byproduct Formation and Control

The formation of DBPs is of great concern with the use of chlorine dioxide. The formation and control of DBPs with chlorine dioxide is considered in the following discussion.

Formation of DBPs Using Chlorine Dioxide for Disinfection
The principal DBPs formed when chlorine dioxide is used as a disinfectant are ClO_2^- and chlorate (Cl_2O_2), both of which are potentially toxic. The principal sources of the ClO_2^- are from the process used to generate the chlorine dioxide and from the reduction of chlorine dioxide. As given by Eq. (11-30), all of the $NaClO_2$ reacts with chlorine to form chlorine dioxide. Unfortunately, on occasion some unreacted chlorite ion can escape from the reactor where the chlorine dioxide is being generated and find its way into the reclaimed water that is being treated. The second source of chlorite is from the reduction of chlorine dioxide as discussed above [see Eq. (11-32)]. The chlorate ion can be derived from the oxidation of chlorine dioxide, from the impurities in the sodium chlorite feedstock, and from the photolytic decomposition of chlorine dioxide.

The chlorine dioxide residuals and other end products are believed to degrade more quickly than chlorine residuals, and, therefore, may not pose as serious a threat to aquatic life as chlorine residuals. An advantage in using chlorine dioxide is that it does not react with ammonia to form the potentially toxic chlorinated DBPs. It has also been reported that halogenated organic compounds are not produced to any appreciable extent. This finding is of importance with respect to the formation of chloroform, which is a suspected carcinogenic substance.

Control of DBP Formation Using Chlorine Dioxide for Disinfection
The formation of ClO_2^- can be controlled by careful management of the feedstock or by increasing the chlorine dose beyond the stoichiometric amount. Treatment methods for the removal of the chlorite ion involve reducing the chlorite ion to the chloride ion using either ferrous iron or sulfite. Granular activated carbon can also be used to absorb

trace amounts of chlorite. At the present time, there are no cost-effective methods for the removal of the ClO_2^-. The control of the chlorate ion depends primarily on the effective management of the facilities used for the production of chlorine dioxide (White, 1999).

Environmental Impacts

The environmental impacts associated with the use of chlorine dioxide as a reclaimed water disinfectant are not well known. It has been reported that the impacts are less adverse than those associated with chlorination. Chlorine dioxide does not dissociate or react with water as does chlorine. However, because chlorine dioxide is normally produced from chlorine and sodium chlorite, free chlorine may remain in the resultant chlorine dioxide solution (depending on the process) and impact the receiving aquatic environment, as does chlorine and its byproducts. However, chlorine dioxide has been found to be less harmful to aquatic life than chlorine.

11-5 DECHLORINATION

Chlorination is one of the most commonly used methods for the destruction of pathogenic and other harmful organisms that may endanger human health. As noted in the previous sections, however, certain organic constituents in reclaimed water interfere with the chlorination process. Many of these organic compounds may react with chlorine to form toxic compounds that can have long-term adverse effects on the beneficial uses of the waters to which they are discharged. To minimize the effects of free and combined chlorine, and other potentially toxic compounds containing chlorine on the environment, dechlorination of treated effluent and reclaimed water is necessary. Dechlorination may be accomplished by reacting the residual chlorine with a reducing agent such as sulfur dioxide (SO_2) or sodium bisulfite ($NaHSO_3$), or by adsorption on and reaction with activated carbon.

Dechlorination of Reclaimed Water Treated with Chlorine and Chlorine Compounds

Where effluent toxicity requirements are applicable, or where dechlorination is used as a polishing step following the breakpoint chlorination process for the removal of ammonia nitrogen, SO_2 is used most commonly for dechlorination. Other chemicals that have been used are sodium sulfite (Na_2SO_3), sodium bisulfite ($NaHSO_3$), sodium metabisulfite ($Na_2S_2O_5$), and sodium thiosulfate ($Na_2S_2O_3$). Activated carbon has also been used for dechlorination. The use of these chemicals for dechlorination is discussed below.

Dechlorination with Sulfur Dioxide

Sulfur dioxide is available commercially as a liquefied gas under pressure in steel containers. Sulfur dioxide is handled in equipment very similar to standard chlorine systems. When added to water, SO_2 reacts to form sulfurous acid ($H_2SO_3^-$), a strong reducing agent. In turn, the sulfurous acid dissociates to form hydrogen sulfite (HSO_3^-) that will react with free and combined chlorine, resulting in formation of chloride and sulfate ions. Sulfur dioxide gas successively removes free chlorine, monochloramine, dichloramine, nitrogen trichloride, and poly-n-chlor compounds as illustrated in Eqs. (11-33) through (11-38).

Reactions between sulfur dioxide and free chlorine are:

$$SO_2 + H_2 \rightarrow HSO_3^- + H^+ \quad (11\text{-}33)$$

$$\underline{HOCl + HSO_3^- \rightarrow Cl^- + SO_4^{2-} + 2H^+} \quad (11\text{-}34)$$

$$SO_2 + HOCl + H_2O \rightarrow Cl^- + SO_4^{2-} + 3H^+ \quad (11\text{-}35)$$

Reactions between sulfur dioxide and monochloramine, dichloramine, and nitrogen trichloride are:

$$SO_2 + NH_2Cl + 2H_2O \rightarrow Cl^- + SO_4^{2-} + NH_4^+ + 2H^+ \quad (11\text{-}36)$$

$$SO_2 + NHCl_2 + 2H_2O \rightarrow 2Cl^- + SO_4^{2-} + NH_3 + 2H^+ \quad (11\text{-}37)$$

$$SO_2 + NCl_3 + 3H_2O \rightarrow 3Cl^- + SO_4^{2-} + NH_4^+ + 2H^+ \quad (11\text{-}38)$$

For the overall reaction between sulfur dioxide and chlorine [Eq. (11-35)], the stoichiometric weight ratio of sulfur dioxide to chlorine is 0.903:1 (see Table 11-16). In practice, it has been found that about 1.0 to 1.2 mg/L of sulfur dioxide is required for the dechlorination of 1.0 mg/L of chlorine residual (expressed as Cl_2). Because the reactions of sulfur dioxide with chlorine and chloramines are nearly instantaneous, contact time is not usually a factor and contact chambers are not used, but rapid and positive mixing at the point of application is an absolute requirement.

The ratio of free chlorine to the total combined chlorine residual before dechlorination determines whether the dechlorination process is partial or proceeds to completion. If the ratio is less than 85 percent, it can be assumed that significant organic nitrogen is present and that it interferes with the free residual chlorine process.

In most situations, sulfur dioxide dechlorination is a very reliable unit process, provided that the precision of the combined chlorine residual monitoring device is adequate. Excess sulfur dioxide dosages should be avoided, not only because of the chemical wastage, but also because of the oxygen demand exerted by the excess sulfur dioxide. The relatively slow reaction between excess sulfur dioxide and dissolved oxygen is given by the following expression:

$$HSO_3^- + 0.5O_2 \rightarrow SO_4^{2-} + H^+ \quad (11\text{-}39)$$

Table 11-16

Typical information on the quantity of dechlorinating compound required for each mg/L of residual chlorine

Dechlorinating compound			Quantity, mg/(mg/L) residual	
Name	Formula	Molecular weight	Stoichiometric amount	Range in use
Sulfur dioxide	SO_2	64.09	0.903	1.0–1.2
Sodium sulfite	Na_2SO_3	126.04	1.775	1.8–2.0
Sodium bisulfite	$NaHSO_3$	104.06	1.465	1.5–1.7
Sodium metabisulfite	$Na_2S_2O_5$	190.10	1.338	1.4–1.6
Sodium thiosulfate	$Na_2S_2O_3$	112.12	0.556	0.6–0.9

The result of this reaction is a reduction in the dissolved oxygen contained in reclaimed water, a corresponding increase in the measured BOD and COD, and a possible drop in the pH. All these effects can be eliminated by proper control of the dechlorination system.

Dechlorination with Sulfite Compounds

When Na_2SO_3, $NaHSO_3$, and $Na_2S_2O_5$ are used for dechlorination, the following reactions occur. The stoichiometric weight ratios of these compounds needed per mg/L of residual chlorine are given in Table 11-16.

Reactions between Na_2SO_3 and free chlorine residual and combined chlorine residual, as represented by NH_2Cl:

$$Na_2SO_3 + Cl_2 + H_2O \rightarrow Na_2SO_4 + 2HCl \quad (11\text{-}40)$$

$$Na_2SO_3 + NH_2Cl + H_2O \rightarrow Na_2SO_4 + Cl^- + NH_4^+ \quad (11\text{-}41)$$

Reactions between $NaHSO_3$ and free chlorine residual and combined chlorine residual, as represented by NH_2Cl:

$$NaHSO_3 + Cl_2 + H_2O \rightarrow NaHSO_4 + 2HCl \quad (11\text{-}42)$$

$$NaHSO_3 + NH_2Cl + H_2O \rightarrow NaHSO_4 + Cl^- + NH_4^+ \quad (11\text{-}43)$$

Reactions between $Na_2S_2O_5$ and free chlorine residual and combined chlorine residual, as represented by NH_2Cl:

$$Na_2S_2O_5 + Cl_2 + 3H_2O \rightarrow 2NaHSO_4 + 4HCl \quad (11\text{-}44)$$

$$Na_2S_2O_5 + 2NH_2Cl + 3H_2O \rightarrow Na_2SO_4 + H_2SO_4 + 2Cl^- + 2NH_4^+ \quad (11\text{-}45)$$

Dechlorination with Sodium Thiosulfate and Related Compounds

Often used as a dechlorinating agent in analytical laboratories, the use of $Na_2S_2O_3$ in full-scale water reclamation treatment plants is limited for the following reasons. It appears that the reaction of $Na_2S_2O_3$ with residual chlorine is stepwise, creating a problem with uniform mixing. The ability of sodium thiosulfate to remove residual chlorine is a function of the pH (White, 1999). The reaction with residual chlorine is only stoichiometric at pH = 2, making prediction of the required dose impossible in reclaimed water applications. As reported in Table 11-16, the stoichiometric weight ratio of sodium thiosulfate per mg/L of residual chlorine is 0.556. Although not in common use, calcium thiosulfate (CaS_2O_3), ascorbic acid ($C_6H_8O_6$), and sodium ascorbate ($C_6H_7NaO_6$) have all been used at full scale for dechlorination.

Dechlorination with Activated Carbon

Both combined and free residual chlorine can be removed by means of adsorption on and reaction with activated carbon. When activated carbon is used for dechlorination, the following reactions occur once chlorine or chlorine compounds have been adsorbed.

Reactions with free chlorine residual:

$$C + 2Cl_2 + 2H_2O \rightarrow 4HCl + CO_2 \qquad (11\text{-}46)$$

Reactions with combined chlorine residual as represented by mono- and dichloramine:

$$C + 2NH_2Cl + 2H_2O \rightarrow CO_2 + 2NH_4^+ + 2Cl^- \qquad (11\text{-}47)$$

$$C + 4NHCl_2 + 2H_2O \rightarrow CO_2 + 2N_2 + 8H^+ + 2Cl^- \qquad (11\text{-}48)$$

Granular activated carbon (GAC) is used in either a gravity or pressure filter bed. If carbon is to be used solely for dechlorination, it must be preceded by an activated carbon process for the removal of other constituents susceptible to removal by activated carbon. In treatment plants where granular activated carbon is used to remove organics, either the same or separate beds can also be used for dechlorination.

Because GAC in column applications has proved to be effective and reliable, activated carbon should be considered where dechlorination is required. However, this method is quite expensive. The primary application of activated carbon for dechlorination is in situations where high levels of organic removal are also required.

Dechlorination of Chlorine Dioxide with Sulfur Dioxide

Where reclaimed water is disinfected with chlorine dioxide, dechlorination can be achieved using SO_2. The reaction that takes place in the ClO_2 solution can be expressed as:

$$SO_2 + H_2O \rightarrow H_2SO_3 \qquad (11\text{-}49)$$

$$5H_2SO_3 + 2ClO_2 + H_2O \rightarrow 5H_2SO_4 + 2HCl \qquad (11\text{-}50)$$

Based on Eq. (11-50), it can be seen that 2.5 mg of sulfur dioxide is required for each mg of chlorine dioxide residual (expressed as ClO_2). In practice, 2.7 mg SO_2/mg ClO_2 is normally used.

11-6 DISINFECTION WITH OZONE

Although historically used primarily for the disinfection of water, recent advances in ozone (O_3) generation and solution technology have made the use of ozone economically more competitive for the disinfection of reclaimed water. Further, interest in the use of O_3 for reclaimed water disinfection has also been renewed because of its ability to reduce or eliminate trace constituents. Ozone can also be used in water reuse applications for the removal of soluble refractory organics, in lieu of the carbon-adsorption process. The characteristics of O_3, the chemistry of O_3, the generation of O_3, an analysis of the performance of O_3 as a disinfectant, and the application of the ozonation process are considered in the following discussion.

Ozone Properties

Ozone is an unstable gas produced when oxygen molecules dissociate into atomic oxygen. Ozone can be produced by electrolysis, photochemical reaction, and radiochemical reaction by electrical discharge. Ozone is often produced by ultraviolet light and lightning

Property	Unit	Value
Molecular weight	g	48.0
Boiling point	°C	−111.9 ± 0.3
Freezing point	°C	−192.5 ± 0.4
Latent heat of vaporization at 111.9°C	kJ/kg	14.90
Liquid density at −183°C	kg/m³	1574
Gaseous density at 0°C and 1 atm	g/mL	2.154
Solubility in water at 20.0°C	mg/L	12.07
Vapor pressure at −183°C	kPa	11
Vapor density compared to dry air at 0°C and 1 atm	unitless	1.666
Specific volume of vapor at 0°C and 1 atm	m³/kg	0.464
Critical temperature	°C	−12.1
Critical pressure	kPa	5532.3

Table 11-17 Properties of ozone[a]

[a]Adapted in part from Rice (1996), U.S. EPA (1986), White (1999).

during a thunderstorm. The electrical discharge method is used for the generation of ozone in water and reclaimed water disinfection applications. Ozone is a blue gas at normal room temperatures, and has a distinct odor. Ozone can be detected at concentrations of 2×10^{-5} to 1×10^{-4} g/m³ (0.01 to 0.05 ppm$_v$, by volume). Because ozone has an odor, it can usually be detected before health concerns develop. The stability of O_3 in air is greater than it is in water, but in both cases is on the order of minutes. Gaseous O_3 is explosive when the concentration reaches about 240 g/m³ (20 percent weight in air). The properties of ozone are summarized in Table 11-17. The solubility of ozone in water is governed by Henry's law. Typical values of Henry's constant for ozone are presented in Table 11-18.

Ozone Chemistry

Some of the chemical properties displayed by O_3 may be described by its decomposition reactions which are thought to proceed as follows:

$$O_3 + H_2O \rightarrow HO_3^+ + OH^- \tag{11-51}$$

$$HO_3^+ + OH^- \rightarrow 2HO_2 \tag{11-52}$$

Temperature, °C	Henry's constant, atm/mole fraction
0	1940
5	2180
10	2480
15	2880
20	3760
25	4570
30	5980

Table 11-18 Values of Henry's constant for ozone[a]

[a]U.S. EPA (1986).

$$O_3 + HO_2 \rightarrow HO\cdot + 2O_2 \quad (11\text{-}53)$$

$$HO\cdot + HO_2 \rightarrow H_2O + O_2 \quad (11\text{-}54)$$

The dot ($\cdot$) that appears next to the hydroxyl (HO$\cdot$) and other radicals is used to denote the fact that these species have an unpaired electron. The free radicals formed, HO_2 and HO$\cdot$, have great oxidizing powers and are probably the active participants in the disinfection process. These free radicals also possess the oxidizing power to react with other impurities in aqueous solutions (see Chap. 10).

Ozone Disinfection Systems Components

A complete O_3 disinfection system, as illustrated on Fig. 11-22, is comprised of the following components: (1) facilities for the preparation of the feed gas, (2) power supply, (3) the O_3 generation facilities, (4) two types of facilities for contacting the O_3 with the liquid to be disinfected (in-line or sidestream), and (5) facilities for the destruction of the off gas (Rice, 1996; Rakness, 2005). Additional details on the design of O_3 systems and related components may be found in Rakness (2005).

Preparation of Feed Gas

Ozone can be generated using air, high-purity oxygen, or oxygen-enriched air. If air is used for O_3 generation, it must be conditioned by removing the moisture and particulate matter before being introduced into the O_3 generator. The following steps are involved in conditioning the air: (1) gas compression, (2) air cooling and drying, and (3) air filtration. If high-purity oxygen is used, the conditioning steps are not required. The liquid oxygen (LOX) supply is stored onsite and trucked in as needed. In the oxygen-enriched air system, high-purity oxygen is generated on-site with a vacuum swing adsorption (VSA)

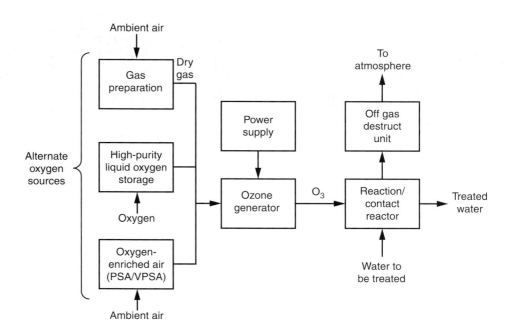

Figure 11-22
Schematic flow diagram for complete ozone disinfection system. (Adapted from U.S. EPA, 1986.)

Table 11-19
Typical energy requirements for the application of ozone

Component	kWh/lb ozone	kWh/kg ozone
Air preparation (compressor and dryers)	2–3	4.4–6.6
Ozone generation		
Air feed	6–9	13.2–19.8
Pure oxygen	3–6	6.6–13.2
Ozone contacting	1–3	2.2–6.6
All other uses	0.5–1	1.2–2.2
Typical total system–Air	10–12	22–26
Typical total system–Oxygen	3.5–5.5[a]	7.7–12.1

[a]Must add the cost of oxygen when estimating total operational cost.

system or pressure swing adsorption (PSA) system. Typically, VSA is used for larger plants and PSA is used for smaller treatment plants. Both oxygen generation systems have facilities for adsorbing moisture, which can damage the ozone generator dielectrics, and for the removal of hydrocarbons and nitrogen to enhance the purity of the oxygen. The choice of feed gas is influenced by the local cost of high-purity oxygen.

Power Supply
The major requirement for power is for the production of O_3 from oxygen. Additional power is required for preparation of the feed gas, contacting the O_3, destroying the residual O_3, and for the controls, instrumentation, and monitoring facilities. The energy requirements for the major components are reported in Table 11-19.

Ozone Generation
Because O_3 is chemically unstable, it decomposes to oxygen very rapidly after generation, and thus must be generated onsite. The most efficient method of producing ozone today is by electrical discharge. Ozone is generated either from air or high-purity oxygen when a high voltage is applied across the gap of narrowly spaced electrodes (see Fig. 11-23). The high-energy corona created by this arrangement dissociates one oxygen molecule, which reforms with two other oxygen molecules to create two

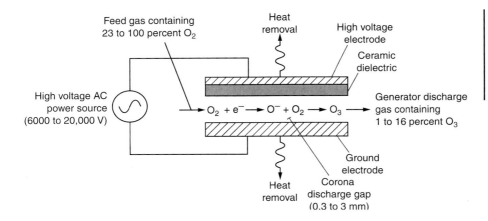

Figure 11-23
Schematic detail of the generation of ozone. (Adapted from U.S. EPA, 1986.)

ozone molecules. The gas stream generated by this process from air contains about 1 to 3 percent ozone by weight, and from pure oxygen about 8 to 12 percent O_3. Ozone concentrations up to 12 percent are now being generated with medium frequency ozone generators.

In-Line Ozone Contact/Reaction Reactors

The concentration of O_3 generated from either air or pure oxygen is so low that the transfer efficiency to the liquid phase is an extremely important economic consideration. To optimize O_3 dissolution, deep and covered contact chambers are normally used. Two 4-compartment O_3 contact reactors are shown schematically on Fig. 11-24 without and with chimneys. The chimneys shown on Fig. 11-24b are used to enhance the countercurrent flow within the reactor. The chimneys also provide locations for O_3 residual sampling.

Ozone is introduced by means of porous diffusers or injectors into the bottom of the first and second, and in some cases, the third chamber. Fast ozone reactions occur in the first chamber. The combined water-ozone mixture then enters the second chamber

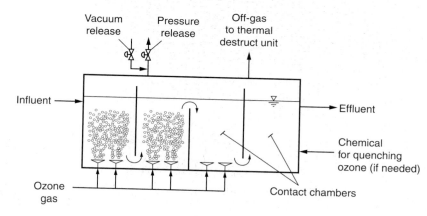

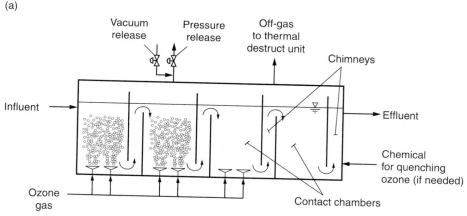

Figure 11-24
Schematic of typical four compartment ozone contactors: (a) without chimneys and (b) with chimneys. The chimneys in (b) are used to enhance the counter current flow through the reactor. (Adapted in part from Crittenden et al., 2005.)

where slower reactions occur. Disinfection generally occurs in the second chamber. The third and fourth chambers are used to complete the slow reactions and to allow the ozone to decompose. The first and second chambers are identified as the *reaction* chambers. The third and fourth chambers, without ozone addition, are known as the *contact* chambers. The number of chambers used depends on the treatment objectives.

Sidestream Ozone Contact/Reaction System

With the ability to generate higher concentrations of ozone (e.g., 10 to 12 percent), sidestream injection of O_3 (see Fig. 11-25) is now a viable alternative to the use of porous diffusers in deep tanks as described above. As shown on Fig. 11-25a, the O_3 injection

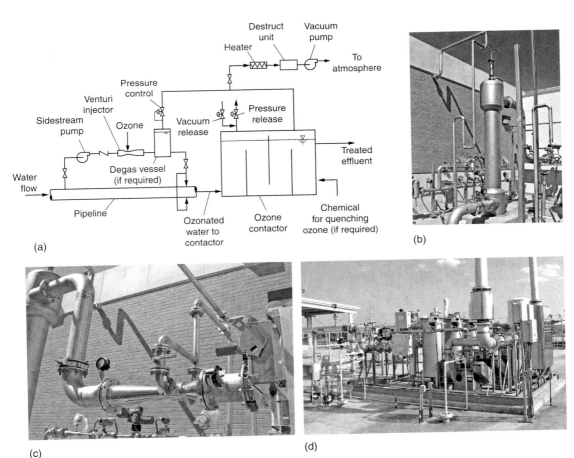

Figure 11-25

Sidestream ozone injection for disinfection: (a) typical schematic for sidestream injection system (Adapted from Rakness, 2005), (b) view of degas vessel, (Venturi injector located on back right), (c) Venturi injector used in conjunction with degas vessel shown in (b) [Photos (b) and (c) courtesy of G. Hunter, Process Applications, Inc.], and (d) view of sidestream injection system located above enclosed ozone contactor, including Venturi injectors (left side), degas vessels (center), and destruct units (right).

system is independent of the O_3 contactor. The O_3 is injected under pressure through a Venturi injector. Two sidestream configurations are used: (1) one with the inclusion of a degas vessel and (2) one without. The purpose of the degas vessel is (1) to minimize the DO level in the water which has been ozonated and (2) to minimize the number of gas bubbles in the downstream pipe which serves as a reactor. The pipeline into which the ozonated water is injected also serves as a reactor prior to the discharge into the contactor (Rakness, 2005).

Destruction of Off Gases and Residual Ozone Quenching

The off gases from the contact chamber and the degas vessel must be treated to destroy any remaining O_3 as it is an extremely irritating and toxic gas. Off gas is destroyed to a concentration of <0.1 ppm_v. The product formed by destruction of the remaining ozone is pure oxygen which can be recycled if pure oxygen is being used to generate the ozone.

If the treatment facility is enclosed and manned, ozone residual quenching is still required to meet U.S. Occupational Safety and Health Administration (OSHA) indoor ambient air quality standards. Ozone quenching is also required to prevent or limit the corrosion of downstream piping and equipment. Where required, hydrogen peroxide, sodium bisulfate, and calcium thiosulfate have been used to quench residual ozone. Where O_3 quenching is required, the chemicals used to quench the residual O_3 are added to in the 4th chamber (see Fig. 11-24).

Effectiveness of Ozone as a Disinfectant

Ozone is an extremely reactive oxidant and it is generally believed that bacterial kill through ozonation occurs directly because of cell wall disintegration (cell lysis). The impact of the reclaimed water characteristics on O_3 disinfection is reported in Table 11-20. The presence of oxidizable compounds causes the O_3 inactivation curve to have a shoulder effect as discussed previously for chlorine (see Fig. 11-6).

Ozone is also a very effective viricide and is generally believed to be more effective than chlorine. (The relative germicidal effectiveness of O_3 for the disinfection of different microorganisms was presented previously in Table 11-5.) Ozonation does not produce dissolved solids and its effectiveness is not affected by the ammonium ion or the influent pH. For these reasons, ozonation is considered as an alternative to either chlorination or hypochlorination, especially where dechlorination may be required and high-purity oxygen facilities are available at the treatment plant.

Modeling the Ozone Disinfection Process

As discussed previously in Sec. 11-3, the mathematical relationships that have been developed to describe the disinfection process with chlorine have also been adapted for ozone. Equations (11-26) and (11-27) have been modified as follows (Finch and Smith, 1989, 1990; U.S. EPA, 1986).

$$N/N_0 = 1 \quad \text{for } U < q \tag{11-55}$$

$$N/N_0 = [(U)/q]^{-n} \quad \text{for } U > q \tag{11-56}$$

Table 11-20
Impact of wastewater constituents on the use of ozone for wastewater disinfection

Constituent	Effect
BOD, COD, TOC, etc.	Organic compounds that comprise the BOD and COD can exert an ozone demand. The degree of interference depends on their functional groups and their chemical structure
NOM (natural organic matter)	Affects the rate of ozone decomposition and the ozone demand
Oil and grease	Can exert an ozone demand
TSS	Increase ozone demand and shielding of embedded bacteria
Alkalinity	No or minor effect
Hardness	No or minor effect
Ammonia	No or minor effect, can react at high pH
Nitrite	Oxidized by ozone
Nitrate	Can reduce effectiveness of ozone
Iron	Oxidized by ozone
Manganese	Oxidized by ozone
pH	Effects the rate of ozone decomposition
Industrial discharges	Depending on the constituents, may lead to a diurnal and seasonal variations in the ozone demand
Temperature	Effects the rate of ozone decomposition

where N = number of organisms remaining after disinfection at time t
N_0 = number of organisms present before disinfection
U = utilized (or transferred) O_3 dose, mg/L
n = slope of dose response curve
q = value of x intercept when $N/N_0 = 1$ or $\log(N/N_0) = 0$ (assumed to be to equal to the initial ozone demand)

The required O_3 dosage must be increased to account for the transfer of the applied O_3 to the liquid. The required dosage can be computed with the following expression:

$$D = U\left(\frac{100}{TE}\right) \quad (11\text{-}57)$$

where D = the total required O_3 dosage, mg/L
U = utilized (or transferred) O_3 dose, mg/L
TE = O_3 transfer efficiency, percent

Typical O_3 transfer efficiencies vary from about 80 to 90 percent. Application of the above equation is illustrated in Example 11-6.

EXAMPLE 11-6. Estimate the Required Ozone Dose for a Typical Reclaimed Water (Secondary Effluent Plus Filtration).

Estimate the O_3 dose needed to disinfect a reclaimed water to an MPN value of 240/100 mL using the following disinfection data obtained from pilot scale installation. Assume the starting coliform concentration is 1×10^6/100 mL and that the ozone transfer efficiency is 95 percent.

Test number	Initial coliform count, N_0, MPN/100 mL	Ozone transferred, mg/L	Final coliform count, MPN/100 mL	$-\log(N/N_0)$
1	95,000	1	1500	1.80
2	470,000	2	1200	2.59
3	3,500,000	5	730	3.68
4	820,000	7	77	4.03
5	9,200,000	14	92	5.00

Solution

1. Determine the coefficients in Eq. (11-56) using the pilot plant data.
   a. Linearize Eq. (11-56) and plot the log inactivation data versus the O_3 dose on log-log paper to determine the constants in:

   $$N/N_0 = [(U)/q]^{-n}$$

   $$\log(N/N_0) = -n[\log(U) - \log(q)]$$

   b. The required log-log plot is given below.

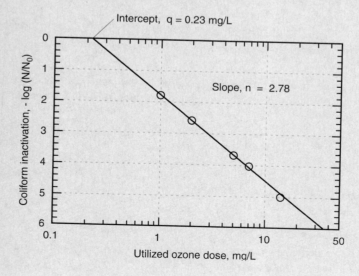

c. The required coefficients are:
   q = 0.23 mg/L
   n = 2.78
2. Determine the O_3 dose required to achieve an effluent coliform concentration of 240 MPN/100 mL.
   a. Rearrange Eq. (11-56) to solve for U.
   $$U = q(N/N_0)^{-1/n}$$
   b. Solve for U.
   $$U = q(N/N_0)^{-1/n} = (0.23 \text{ mg/L})(240/10^6)^{-1/2.78} = 4.61 \text{ mg/L}$$
3. Determine the O_3 dose that must be applied using Eq (11-57), for a transfer efficiency of 95 percent.
   $$D = U\left(\frac{100}{TE}\right)(4.61 \text{ mg/L})\left(\frac{100}{95}\right) = 4.85 \text{ mg/L}$$

Required Ozone Dosages for Disinfection

The required O_3 dosage for disinfection can be estimated by considering (1) the initial O_3 demand of the reclaimed water and (2) the required O_3 dose using Eqs. (11-56) and (11-57). The O_3 dosages required to meet the initial demand depends on the constituents in the reclaimed water. Typical values for the O_3 demand for the disinfection of coliform organisms for various wastewaters based on a contact time of 15 min are reported in Table 11-21. It should be noted that the dosage values given in Table 11-21 are only meant to serve as a guide for the initial estimation of the required O_3 dose. In most cases, bench and pilot scale studies (see Fig. 10-15 in Chap. 10) will need to be conducted to establish the required dosage ranges.

Table 11-21 Typical ozone dosages required to achieve different coliform disinfection standards for various wastewaters based on a 15- to 30-min contact time[a,b]

Type of wastewater	Initial coliform count, MPN/100 mL	Ozone dose, mg/L			
		Effluent standard, MPN/100 mL			
		1000	200	23	≤2.2
Raw wastewater	10^7–10^9	15–30			
Primary effluent	10^7–10^9	10–25			
Trickling filter effluent	10^5–10^6	4–8			
Activated sludge effluent	10^5–10^6	3–5	5–7	12–16	20–30
Filtered activated sludge effluent	10^4–10^6	3–5	5–7	10–14	16–24
Nitrified effluent	10^4–10^6	2–5	4–6	8–10	16–20
Filtered nitified effluent	10^4–10^6	2–4	3–5	5–7	10–16
Microfiltration effluent	10^1–10^3		2–3	3–5	6–8
Reverse osmosis	nil				1–2
Septic tank effluent	10^7–10^9	15–30			
Intermittent sand filter effluent	10^2–10^4	2–4	4–6	8–10	16–20

[a] Adapted in part from WEF (1996); White (1999).
[b] The amount of ozone absorbed depends on the characteristics of the wastewater.

Byproduct Formation and Control

As with chlorine, the formation of unwanted byproducts is one of the problems associated with the use of O_3 as a disinfectant. The formation and control of DBPs when using O_3 are considered in the following discussion.

Formation of DBPs Using Ozone for Disinfection

One advantage of ozone is that it does not form chlorinated DBPs such as THMs and HAAs (see Table 11-15). Ozone does, however, form other DBPs (see Table 11-22) including aldehydes, various acids, and aldo- and keto-acids when significant amounts of bromide are not present. In the presence of bromide, the following DBPs may also be produced: inorganic bromate ion, bromoform, brominated acetic acid, bromopicrin, brominated acetonitriles, cyanogen bromide, and bromate (see Table 11-15) (Haag and Hoigné, 1983; Kim et al., 1999). On occasion, hydrogen peroxide can also be generated. The specific amounts and the relative distribution of compounds depend on the nature of the precursor compounds that are present. Because the chemical characteristics of reclaimed water vary from location to location, pilot testing will be required to assess the effectiveness of ozone as a disinfectant and the formation of DBPs.

Control of DBP Formation Using Ozone for Disinfection

Because the nonbrominated compounds appear to be readily biodegradable, they can be removed by passage through a biologically active filter or carbon column or other biologically active process. The nonbrominated compounds can also be removed by soil application. The removal of DBPs formed when bromine is present is more complex. If brominated DBPs are going to be a problem, it may be appropriate to investigate alternative means of disinfection such as UV irradiation.

Table 11-22
Representative disinfection byproducts resulting from the ozonation of wastewater containing organic and selected inorganic constituents[a]

Class	Representative compounds
Acids	Acetic acids
	Formic acid
	Oxalic acid
	Succinic acid
Aldehydes	Acetaldehyde
	Formaldehyde
	Glyoxal
	Methyl glyoxal
Aldo- and ketoacids	Pyruvic acid
Brominated byproducts[b]	Bromate ion
	Bromoform
	Brominated acetic acids
	Bromopicrin
	Brominated acetonitriles
	Cyanogen bromide
Other	Hydrogen peroxide

[a] Adapted, in part, from U.S. EPA (1999b, 2002).
[b] The bromide ion must be present to form brominated byproducts.

It has been reported that ozone residuals can be acutely toxic to aquatic life (Ward and DeGraeve, 1976). Several investigators have reported that ozonation can produce some toxic mutagenic and/or carcinogenic compounds. These compounds are usually unstable, however, and are present only for a matter of minutes in the ozonated water. White (1999) has reported that ozone destroys certain harmful refractory organic substances such as humic acid (precursor of trihalomethane formation) and malathion. Whether toxic intermediates are formed during ozonation depends on the ozone dose, the contact time, and the nature of the precursor compounds. White (1999) has also reported that ozone treatment ahead of chlorination for disinfection purposes reduces the likelihood for the formation of THMs. **Environmental Impacts of Using Ozone**

An additional benefit associated with the use of ozone for disinfection is that the dissolved oxygen concentration of the effluent will be elevated to near saturation levels as ozone rapidly decomposes to oxygen after application. The increase in oxygen concentration may eliminate the need for reaeration of the effluent to meet required dissolved oxygen water quality standards. **Other Benefits of Using Ozone**

11-7 OTHER CHEMICAL DISINFECTION METHODS

Because of the concerns over the effectiveness of disinfection processes and concern over the formation of DBPs, ongoing research is continuing into the evaluation of alternative disinfection methods. The use of peracetic acid and combined disinfection processes are introduced and considered briefly in this section. Because research on these and other disinfection methods is ongoing, current literature and conference proceedings must be consulted for the latest findings.

In the late 1980s, the use of peracetic acid (PAA, CH_3CO_3H) was proposed as a wastewater disinfectant. Peracetic acid, made up of acetic acid and hydrogen peroxide, has been used for many years as a disinfectant and sterilizing agent in hospitals. Peracetic acid is also used as a bactericide and fungicide, especially in food processing. Interest in the use of PAA as a reclaimed water disinfectant arises from considerations of safety and the possibility that its use will not result in the formation of DBPs. The use of PAA is considered briefly in this section as an example of the continuing search for alternative disinfectants to replace chlorine. **Peracetic Acid**

Peracetic Acid Chemistry and Properties
Commercially available PAA, also known as ethane peroxide acid, peroxyacetic acid, or acetyl hydroxide, is only available as a quaternary equilibrium solution containing acetic acid, hydrogen peroxide, peracetic acid, and water. The pertinent reaction is as follows:

$$CH_3CO_2H \ + \ H_2O_2 \ \rightleftarrows \ CH_3CO_3H + H_2O \quad\quad (11\text{-}58)$$

$$\text{Acetic acid} \quad\quad \text{Hydrogen peroxide} \quad\quad \text{Peracetic acid}$$

The undissociated PAA is considered to be the biocidal form in the equilibrium mixture; however, hydrogen peroxide may also contribute to the disinfection process.

Table 11-23
Properties of various peracetic acid (PAA) formulations[a]

Property	Unit	PAA, % 1.0	PAA, % 5	PAA, % 15
Weight PAA	%	0.8–1.5	4.5–5.4	14–17
Weight hydrogen peroxide	%	min 6	19–22	13.5–16
Weight acetic acid	%	9	10	28
Weight available oxygen	wt, %	3–3.1	9.9–11.5	9.3–11.1
Stabilizers	Yes/no	Yes	Yes	Yes
Specific gravity		1.10	1.10	1.12

[a] Adapted from Solvay Interox (1997).

Hydrogen peroxide is also more stable than PAA. The properties of PAA are summarized in Table 11-23.

Effectiveness of Peracetic Acid as a Disinfectant

The effectiveness of PAA has been studied by Lefevre et al. (1992), Lazarova et al. (1998), Liberti et al. (1999), Gehr (2000, 2006), Wagner et al. (2002), and Gehr et al. (2003) among others. An additional review was published by Kitis (2004). The findings to date are mixed concerning the bactericidal effectiveness of PAA, as well as the impact of reclaimed water characteristics on the effectiveness of PAA, especially when used alone. When combined with UV the effectiveness of PAA appears to be enhanced significantly (see discussion of combined disinfectants presented later). It has been hypothesized that the principal means by which disinfection is accomplished by PAA may be by the release of hydroxyl radicals (HO·) and the active oxygen resulting from secondary reactions (Caretti and Lubello, 2003). The current literature must be consulted for more information on the application of PAA.

In a report by the U.S. EPA (1999a), PAA was included among a total of five possible disinfectants for use on combined sewer overflows (CSOs). Based on data for disinfection of secondary treatment plant effluents, it was suggested that PAA should be considered for CSO disinfection. Among the desirable attributes listed are absence of persistent residuals and byproducts, not affected by pH, short contact time, and high effectiveness as a bactericide and virucide.

Formation of Disinfection Byproducts

Based on the limited data available, the principal end products identified were CH_3COOH (acetic acid or vinegar), O_2, CH_4, CO_2, and H_2O, none of which are considered toxic in the concentrations typically encountered.

Combined Chemical Disinfection Processes

Interest in the sequential or simultaneous use of two or more disinfectants has increased within the last few years, especially in the water supply field. Reasons for the increased interest in the use of multiple disinfectants include (U.S. EPA, 1999b):

- The use of less-reactive disinfectants, such as chloramines, has proven to be quite effective in reducing the formation of DBPs, and more effective for controlling biofilms in the distribution system.
- Regulatory and consumer pressure to produce water that has been disinfected to achieve high levels of inactivation for various pathogens has forced both the water

and reclaimed water industry to search for more effective disinfectants. To meet more stringent disinfection standards, higher disinfectant doses have been used which, unfortunately, has resulted in the production of increased levels of DBPs.

- Based on the results of recent research, it has been shown that the application of sequential disinfectants is more effective than the additive effect of the individual disinfectants. When two (or more) disinfectants are used to produce a synergistic effect by either simultaneous or sequential application to achieve more effective pathogen inactivation, the process is referred to as interactive disinfection (U.S. EPA, 1999b).

Currently, extensive research is being conducted on these processes. Some examples for the use of combined and/or sequential application of disinfectants are presented in Table 11-24. Because the application of multiple disinfectants is, at present, site specific,

Table 11-24

Effectiveness of combined disinfectants and processes for water and wastewater treatment[a,b]

Combined disinfectants	Response	Reference
In water treatment		
Ozone (O_3), UV, and chloramines replaced chlorination	Increase in Ct credits by as much as 3 log	Malley (2005)
Ozone, UV, and chloramines replaced chlorination	Increase in Ct credits by as much as 5 log	Malley (2005)
UV, O_3, and chloramines replaced chlorination	Increase in Ct credits by as much as 3 log	Malley (2005)
Sequential sonification and chlorine	Increase in effectiveness over use of sonification or chlorine alone	Plummer and Long (2005)
Sequential UV and chlorine for inactivation of adenoviruses	Increase in effectiveness over use of UV or chlorine alone	Sirikanchana et al. (2005)
In wastewater treatment		
Peracetic acid (PAA) and UV	Increase in effectiveness over use of UV or PAA alone	Chen et al. (2005); Lubello et al. (2002)
PAA and UV and PAA and ozone	Increase in effectiveness over use of PAA and UV alone	Caretti and Lubello (2005)
PAA and hydrogen peroxide (H_2O_2), H_2O_2 and UV, and H_2O_2 and O_3	No improved effectiveness	Caretti and Lubello (2005); Lubello et al. (2002)
Ozone, PAA, H_2O_2 and copper (Cu)	PAA and H_2O_2 alone had no effect, addition of 1 mg/L Cu had a dramatic effect	Orta de Velasquez et al. (2005)
PAA/UV and H_2O_2/UV	PAA/UV had synergistic effects, whereas H_2O_2/UV did not	Koivunen (2005)
Ultrasound and UV	Increase in effectiveness over use of UV alone	Blume et al. (2002); see also Blume and Neis (2004)

[a] Adapted from Gehr (2006).
[b] Additional combinations are reviewed in U.S. EPA (1999b).

depending on the microorganism, the disinfection technologies employed, and other nondisinfection process objectives, the current literature must be reviewed to assess the suitability and effectiveness of combined disinfection technologies.

11-8 DISINFECTION WITH ULTRAVIOLET RADIATION

The germicidal properties of ultraviolet (UV) radiation sources have been used in a wide variety of applications since the use of UV was pioneered in the early 1900s, having been discovered first in the 1880s. First used on high-quality water supplies, the use of UV light as a reclaimed water disinfectant evolved during the 1990s with the development of new lamps, ballasts, and ancillary equipment. With the proper dosage, UV irradiation has proven to be an effective disinfectant for bacteria, protozoa, and virus in reclaimed water, while not contributing to the formation of toxic byproducts. To develop an understanding of the application of UV for the disinfection of reclaimed water for reuse, the following topics are considered in this section: (1) source of UV radiation, (2) UV system configurations, (3) the germicidal effectiveness of UV irradiation, (4) modeling the UV disinfection process, (5) estimating the UV dose, (6) ultraviolet disinfection guidelines, (7) analysis of a UV disinfection system, (8) operational issues with UV systems, and (9) the environmental impacts of disinfection with UV irradiation.

Source of UV Radiation

The portion of the electromagnetic spectrum in which UV radiation occurs, as shown on Fig. 11-26a, is between 100 and 400 nm. The UV radiation range is characterized further according to wavelength as longwave (UV-A), also known as near-UV radiation; middlewave (UV-B); and shortwave (UV-C), also known as far-UV radiation (see Fig. 11-26b). The germicidal portion of the UV radiation band is between about 220 and 320 nm, principally in the UV-C range. The UV wavelengths between 255 to 265 nm are considered to be most effective for microbial inactivation (see Fig. 11-26c). Most commonly, UV radiation is produced by striking an electric arc between two electrodes in specially designed lamps containing mercury vapor, as well as other gas mixtures. The energy generated by the excitation of the mercury vapor contained in the lamp results in the emission of UV light.

When used for reclaimed water disinfection, quartz sleeves are used to isolate the UV lamps from direct water contact and to control the lamp wall temperature by buffering the effluent temperature extremes to which the UV lamps are exposed, thereby maintaining a fairly uniform UV lamp output. The output of UV disinfection systems also decreases with time due to a reduction in the electron pool within the UV lamp, deterioration of the electrodes, and the aging of the quartz sleeve. Lamps with other gas mixtures and without electrodes, as described below, are also used to generate UV light.

Types of UV Lamps

The principal electrode-type lamps used to produce UV radiation (or light) fall into three categories based on the internal operating parameters: *low-pressure low-intensity, low-pressure high-intensity,* and *medium-pressure high-intensity* systems. Comparative information on the operational characteristics of these three types of UV lamps is presented in Table 11-25. In the brief discussion of these types of UV lamps presented below, it is important to note that UV lamp technology is changing rapidly. It is, therefore, imperative

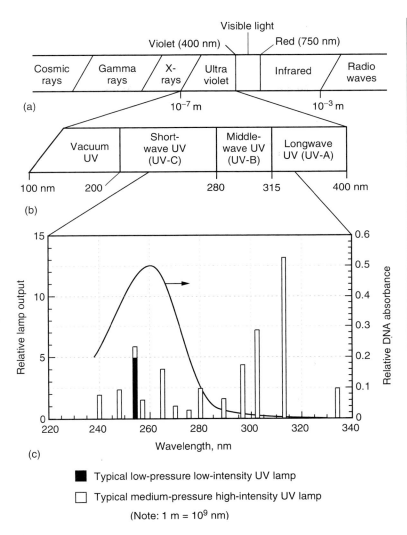

Figure 11-26
Definition sketch for ultraviolet (UV) radiation: (a) identification of the ultraviolet radiation portion of the electromagnetic spectrum, (b) identification of the germicidal portion of the UV radiation spectrum, and (c) UV radiation spectra for both low-pressure low-intensity and medium-pressure high-intensity UV lamps and the relative UV adsorption for DNA superimposed over spectra of the UV lamps.

that current manufacturers' literature be consulted when designing a UV disinfection facility. The ballasts used in conjunction with UV lamps are also discussed briefly.

Low-Pressure Low-Intensity UV Lamps

Low-pressure low-intensity mercury-argon electrode type UV lamps (see Fig. 11-27a) are used to generate a broad spectrum of essentially monochromatic radiation in the UV-C region with an intense peak at a wavelength of 253.7 nm (essentially 254 nm) and a lesser peak at about 184.9 nm. The peak at 254 nm is close to the 260 nm wavelength considered to be most effective for microbial inactivation. Approximately 85 to 88 percent of the lamp output is monochromatic at 254 nm, making it an efficient choice for disinfection processes. Because there is an excess of liquid mercury in the low-pressure low-intensity UV lamp, the mercury vapor pressure is controlled by the coolest part of the lamp wall. If the lamp wall does not remain relatively near the optimum temperature of 40°C, some

Table 11-25
Typical operational characteristics for UV lamps

Item	Unit	Low-pressure low-intensity	Low-pressure high-intensity	Medium-pressure high-intensity
Power consumption	W	40–100	200–500[a]	1000 to 10,000
Lamp current	ma	350–550	Variable	Variable
Lamp voltage	V	220	Variable	Variable
Germicidal output/input	%	30 to 40	25 to 35	10 to 15[b]
Lamp output at 254 nm	W	25–27	60–400	
Lamp operating temperature	°C	35–50	60–100	600–900
Pressure	mm Hg	0.007	0.01–0.8	10^2–10^4
Lamp length	m	0.75–1.5	Variable	Variable
Lamp diameter	mm	15–20	Variable	Variable
Sleeve life	yr	4 to 6	4 to 6	1 to 3
Ballast life	yr	10 to 15	10 to 15	1 to 3
Estimated lamp life	h	8000 to 12,000	7000 to 10,000	3000 to 8000

[a]Up to 1200 W in very high output lamp.
[b]Output in the most effective germicidal range (~ 255–265 nm, see Fig. 11-26).

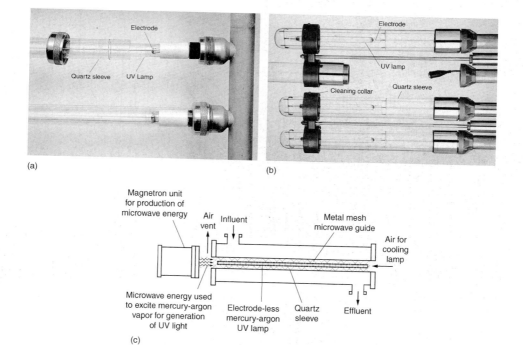

Figure 11-27
Typical examples of UV lamps: (a) low-pressure low-intensity with quartz sleeve removed from socket to expose UV lamp (Courtesy of M. Fan), (b) medium-pressure high-intensity lamps with cleaning device (Courtesy of Trojan Technologies, Inc.), and (c) schematic illustration of the electrode-less microwave driven UV lamp (Adapted from Quay Technologies, Ltd.) (see also Fig. 11-29d).

of the mercury will condense back to its liquid state, thereby decreasing the number of mercury atoms available to release photons of UV; hence, UV output declines.

Low-Pressure High-Intensity UV Lamps

Low-pressure high-intensity UV lamps are similar to the low-pressure low-intensity lamps (see Fig.11-27a) with the exception that a mercury-indium amalgam is used in place of mercury. Use of the mercury amalgam allows greater UV-C output, typically from two to four times the output of conventional low-intensity lamps. One manufacturer offers a lamp that is said to have 20 times the output at 254 nm. The amalgam in the low-pressure high-intensity UV lamps is used to maintain a constant level of mercury atoms, and, thus, provides greater stability over a broad temperature range, and greater lamp life (25 percent greater than other low-pressure lamps). Current manufacturer's literature should be reviewed for lamp specifications as new low-pressure high-intensity lamps are being developed continuously.

Medium-Pressure High-Intensity UV Lamps

Several medium-pressure high-intensity UV lamps have been developed over the last decade. Medium-pressure high-intensity UV lamps, which operate at temperatures of 600 to 800°C and pressures of 10^2 to 10^4 mm Hg, generate polychromatic irradiation (see Fig. 11-26c). Medium-pressure high-intensity UV lamps (see Fig. 11-27b) generate approximately 50 to 100 times the total UV-C output of the conventional low-pressure low-intensity UV lamp. Their use is limited primarily to higher reclaimed water flows, stormwater overflows or on space-limited sites because fewer lamps are required and the footprint of the disinfection system is greatly reduced (i.e., contact time is reduced).

Because the high-intensity UV lamp operates at temperatures at which all the mercury is vaporized, the UV output can be modulated across a range of power settings (typically 60 to 100 percent) without significantly changing the spectral distribution of the lamp. The ability to modulate the power is significant with respect to total power usage. Further, because of the high operating temperature, mechanical wiping of the quartz sleeve is essential to avoid the formation of an opaque film on the surface of the sleeve. Although there are a number of manufacturers of high intensity UV lamps, most of the lamp manufacturers do not market complete UV disinfection systems. The particular UV lamp selected by UV system manufacturers is chosen on the basis of an integrated design approach in which the UV lamp, ballast, and reactor design are interdependent.

Emerging UV Lamp Technologies

New technologies are being developed that may have applications for reclaimed water disinfection. Some examples of the types of lamps that are being developed and applied to reclaimed water include (1) the pulsed energy broadband xenon lamp (pulsed UV), (2) the narrowband excimer UV lamp, and (3) the mercury-argon electrode-less microwave powered high-intensity UV lamp.

The pulsed UV lamp produces polychromatic light at high levels of radiation. It is estimated that the radiation produced by the pulsed UV lamp is 20,000 times as intense as sunlight at sea level (EPRI, 1996; O'Brien et al., 1996). Narrowband excimer lamps produce essentially monochromatic light in three wavelength: 172, 222, and 308 nm depending on the gas used in the lamp. Gases that have been used for the purpose include xenon (Xe), xenon chloride (XeCl), krypton (Kr), and krypton chloride (KrCl). In the microwave-powered UV lamp, UV light is generated by striking a mercury-argon filled electrode-less UV lamp with microwave energy generated with a magnetron (see Fig. 11-27c). Because the lamp does not contain electrodes, longer lamp life is claimed.

Again, as noted above, because developments in UV technology are occurring at such a rapid pace, it is essential that the current literature be consulted when designing UV disinfection systems. Note that in most cases, emerging technologies do not have a proven track record of cost-effective, reliable performance.

Ballasts For UV Lamps

A ballast is a type of transformer that is used to limit the current to a lamp. Because UV lamps are arc discharge devices, the more current in the arc, the lower the resistance becomes. Without a ballast to limit current, the lamp would destroy itself. Thus, matching the lamp and ballast is of critical importance in the design of UV disinfection systems. Three types of ballasts are used: (1) standard (core coil), (2) energy efficient (core coil), and electronic (solid-state). In general, electronic ballasts are about 10 percent more energy efficient than magnetic ballasts. Electronic ballasts are now used most commonly for controlling the UV lamps used for disinfection.

UV Disinfection System Configurations

In addition to the type of lamp used, UV systems for the disinfection of reclaimed water can also be classified according to whether the flow occurs in open or closed channels. Each of these system configurations is described below.

Open Channel Disinfection Systems

The principal components of low-pressure low- and high-intensity open channel UV systems used for the disinfection of reclaimed water are illustrated on Fig. 11-28. As shown, lamp placement can be either horizontal and parallel to the flow (see Fig. 11-28a) or vertical and perpendicular to the flow (see Fig. 11-28b). Each module contains a specified number of UV lamps encased in quartz sleeves. The total number of lamps is specific to each application, but the number of lamps in each module depends on the channel configuration and lamp manufacturer. A spacing of 75 mm (3 in.) between the centers of UV lamps is currently the most frequently used by UV manufacturers. A weighted flap gate, an extended sharp crested weir, or automatic level controller is used to control the depth of flow through each disinfection channel. Level control is essential to maintain submergence of the lamps at all times. Each channel typically contains two or more banks of UV lamps in series, and each bank is comprised of a number of modules (or racks of UV lamps). It is important to note that a standby bank or channel is normally provided for system reliability. The design flowrate is usually divided equally among a number of open channels. Typical examples of horizontal and vertical low-pressure low-intensity UV disinfection systems for reclaimed water are shown on Fig. 11-28c through 11-28f, respectively.

To overcome the effect of fouling, which reduces the intensity of light in the liquid medium, the lamps must either be removed occasionally from the flow channel and cleaned or equipped with a mechanical cleaning system. Mechanical cleaning systems are always used with low- and medium-pressure high-intensity systems to avoid fouling of

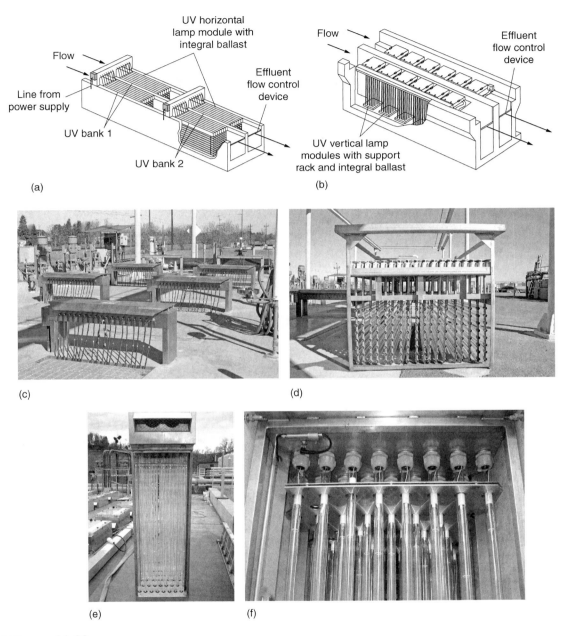

Figure 11-28
Isometric cut-away views of typical open channel UV disinfection systems: (a) horizontal lamp system parallel to flow (adapted from Trojan Technologies, Inc.), (b) vertical lamp system perpendicular to flow (adapted from Infilco Degremont, Inc.), (c) view of system with three UV banks per channel with horizontal lamp placement, (d) view of one UV bank removed for cleaning, (e) vertical lamp module removed from channel for cleaning, and (f) close-up view of mechanical cleaning device. [Photos (e) and (f) Courtesy of P. Friedlander and C. LeBlanc.]

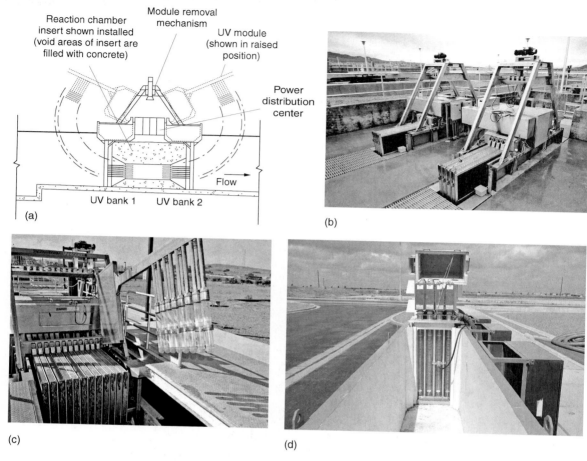

Figure 11-29
Typical examples of medium-pressure and microwave open channel UV disinfection systems: (a) schematic view through UV reactor (Adapted from Trojan Technologies, Inc.), (b) typical medium pressure UV system installed in open channel, (c) medium pressure UV system with one lamp module out of the reactor, and (d) microwave UV lamps with magnetron located above lamps (see also Fig. 11-27c) in vertical orientation in open channel.

the quartz sleeves. Low-pressure high-intensity UV disinfection systems are similar in appearance to those shown on Fig. 11-28a through 11-28f. A typical medium-pressure UV disinfection system is shown on Figs. 11-29a and 11-29b. The lamps are arranged in modules and are positioned in a reactor with a fixed geometry (see Fig. 11-29c). The lamp cleaning sleeves can be seen in Fig. 11-29c. Vertical mercury-argon electrode-less microwave-powered high-intensity UV lamps are shown on Fig. 11-29d.

Closed Channel Disinfection Systems
A number of low- and medium-pressure high-intensity UV disinfection systems are designed to operate in closed channels. In most design configurations, the direction of flow is perpendicular to the placement of the lamps, as shown on Fig. 11-30a. There

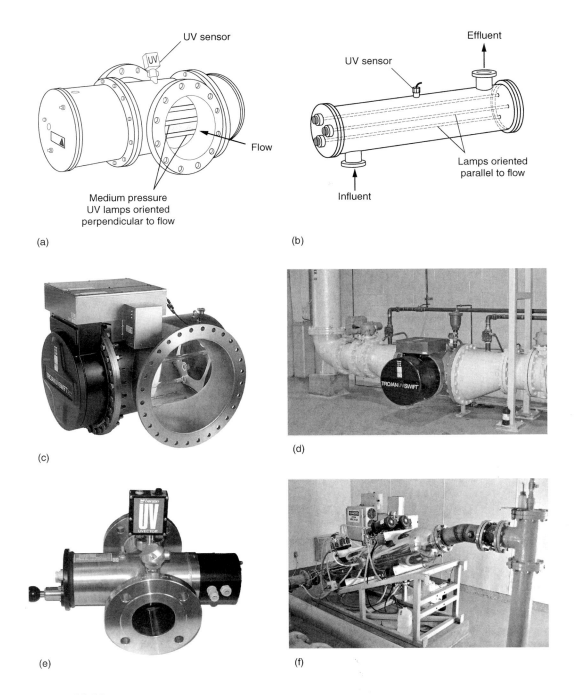

Figure 11-30
Views of medium-pressure high-intensity closed in-line UV disinfection systems: (a) schematic of close reactor with flow perpendicular to UV lamps, (b) schematic of close reactor with flow parallel to UV lamps, (c) view through in-line UV reactor (Courtesy of Trojan Technologies, Inc.), (d) view of installed UV system, (e) close up of small in-line UV system with manual cleaning device, and (f) view of pulsed UV reactor.

are, however, design configurations in which the direction of flow is parallel to the UV lamps (see Fig. 11-30b). Because high-intensity UV lamps operate at a lamp wall temperature of between 600 to 800°C, the UV output of these lamps is unaffected by the effluent temperature. A typical medium-pressure UV disinfection reactor is shown on Figs. 11-30c and 11-30d. Essentially all of the closed or fixed geometry systems used for the disinfection of reclaimed water incorporate some form of mechanical wiping of the quartz sleeves to maintain performance. Some small closed channel UV systems have mechanical cleaning devices that are operated manually (see Fig. 11-30e). A closed system pulsed UV reactor is shown on Fig. 11-30f.

Mechanism of Inactivation by UV Irradiation

Ultraviolet light is a physical rather than a chemical disinfecting agent; the mechanism of inactivation and photoreactivation are of importance in considering the use of UV irradiation for reclaimed water disinfection.

Inactivation Mechanisms

UV radiation penetrates the cell wall of the microorganism and is absorbed by the nucleic acids, which guide the development of all living organisms. Damage to the nucleic acid interferes with normal cell processes such as cell synthesis and cell division. Deoxyribonucleic acid (DNA) controls the structure, while ribonucleic acid (RNA) controls the metabolic processes. Typically, DNA is a double-stranded helical structure with four nucleotides: adenine, guanine, thymine, and cytosine except in some viruses that contain single stranded DNA. In contrast, RNA is a single-stranded structure with the nucleotides adenine, guanine, uracil, and cytosine.

Exposure to UV radiation damages DNA by dimerizing adjacent thymine molecules as illustrated on Fig. 11-31. Cytosine-cytosine and cytosine-thymine dimers can also be formed. Thus, organisms rich in thymine such as *Cryptosporidium parvum (C. parvum)* and *Giardia lamblia (G. lamblia)* tend to be more sensitive to UV irradiation (see Table 11-30). Viruses contain either DNA or RNA, which is either single or double stranded. Adenovirus contains double-stranded DNA, which is considered as a possible explanation for its high sensitivity to UV light (Sommer et al., 2001). Exposure to UV radiation can also cause more severe damage such as breaking chains, cross-linking DNA with itself, cross-linking DNA with other proteins, and forming other byproducts

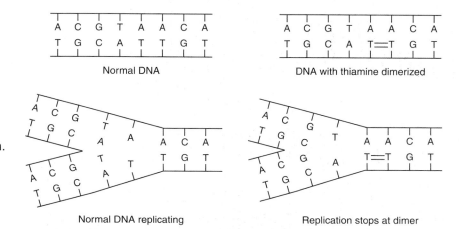

Figure 11-31
Formation of double bonds in microorganisms exposed to ultraviolet radiation. Formation of double bonds inhibits replication.

(Crittenden et al., 2005). In summary, exposure to UV radiation must result in formation of a significant number of bonds or other damage to the cell to be effective.

Microbial Repair Following UV Irradiation

Because some organisms are able to maintain some metabolic activities after being exposed to UV radiation, they may be able to repair the damage caused by the exposure. Many organisms in nature have evolved mechanisms for reversing UV damage. Two different types of mechanisms are used to reverse UV damage: (1) photoreactivation and (2) dark repair.

Photoreactivation Photoreactivation involves specific enzymes which can repair sections of damaged DNA after being energized by exposure to light. The mechanism of photoreactivation, first discovered in 1949 for *Streptomyces griseus* by Kelner (1949) and for bacteriophage by Dulbecco (1949), was demonstrated to be enzyme-catalyzed (Rupert, 1960). The enzyme responsible for DNA repair is named *photolyase*. Photoreactivation can be described as the two-step enzymatic reaction between photolyase and its substrate, pyrimidine dimers (Friedberg et al., 1995). The first step is for photolyase to recognize any dimers (see Fig. 11-31) and specifically bind them to form an enzyme-substrate complex. The first step is light-independent and, therefore, can occur even under dark conditions. The enzyme-dimer complex is stable and goes through the second repair step in which the dimers are broken utilizing the energy of light at wavelengths between 310 and 490 nm. The second step is dependent only on light input.

For example, the *E. coli* photolyase has a round shape with a hole inside, which recognizes and structurally binds to the pyrimidine dimers sticking out from the genome DNA. Once the pyrimidine dimers are repaired (i.e., broken) and the structure is changed, the bond is loosened and the enzyme leaves the dimer (Friedberg et al., 1995). In the case of pathogenic parasites, the effects of photoreactivation are unclear. Based on infectivity studies, it was reported that the oocysts of *C. parvum* did not undergo photoreactivation (Shin et al., 2001). In another study, it was reported that repair of the pyrimidine dimers did occur in oocysts of *C. parvum* (Oguma et al., 2001). What appears to be happening is that the repair of DNA following UV irradiation may not be sufficient for the organism to regain its infectivity. Although the necessary enzymes needed for repair are missing in viral DNA, the enzymes of the host cell can be used to accomplish the repair.

It should also be noted that the ability for an organism to repair itself appears to depend on a number of factors including UV dose (the effect is diminished at higher UV doses), UV wavelength, UV light intensity, and exposure time to photoreactivating light (Martin and Gehr, 2005). *Escherichia coli* exposed to monochromatic low-pressure UV light were able to repair themselves whereas *E. coli* exposed to polychromatic medium-pressure UV light were unable to repair themselves (Zimmer and Slawson, 2002; Oguma et al., 2002). However, *Legionella pneumophila* exhibited very high photoreactivation ability after exposure to either low-pressure or medium-pressure UV light (Oguma et al., 2004). From a review of some recent published findings, it appears that if reclaimed water that has undergone UV disinfection is subsequently kept in the dark for approximately 3 h, the regrowth potential is reduced significantly (Martin and Gehr, 2005). Clearly, more research needs to be done to understand what is causing the effect observed with medium-pressure UV light.

Dark Repair In the early 1960s, it was found that UV radiation-induced DNA damage could be repaired without light (Hanawalt et al., 1979). Dark repair appears to be accomplished by two mechanisms: (1) excision repair and (2) recombination repair. In excision repair enzymes remove the damaged section of DNA, and in recombination repair the damaged DNA is regenerated using a complimentary strand of DNA. Although the necessary enzymes needed for repair are missing in viral DNA, the enzymes of the host cell can be used to accomplish the repair. As with photoreactivation, although the necessary enzymes needed for dark repair are missing in viral DNA, the enzymes of the host cell can be used to accomplish the repair. Contrary to photoreactivation, with high specificity to pyrimidine dimers, dark repair can act on various kinds of damage in the genome. Dark repair is a rather slow process compared to photoreactivation.

Factors Affecting Germicidal Effectiveness of UV Irradiation

The overall effectiveness of the UV disinfection process depends on a number of factors including (1) the chemical characteristics of the reclaimed water, (2) the presence of particles, (3) the characteristics of the microorganisms, and (4) the physical characteristics of the UV disinfection system. Before considering these subjects, it is appropriate to consider the definition of UV dose to provide a frame of reference for the discussion of the factors affecting UV disinfection. The material presented below will also be useful in assessing the modeling of the UV process, which is considered subsequently.

Definition of UV Dose

The effectiveness of UV disinfection is based on the UV dose to which the microorganisms are exposed. The UV dose, D, as defined previously, is given by Eq. (11-8), which is repeated here for convenience.

$$D = I_{avg} \times t$$

where D = UV dose, mJ/cm^2 (note mJ/cm^2 = mW·s/cm^2)
I_{avg} = average UV intensity, mW/cm^2
t = exposure time, s

Note that the UV dose term is analogous to the dose term used for chemical disinfectants (i.e., $C_R t$). As given by Eq. (11-8), the UV dose can be varied by changing either the intensity or exposure time. Because the UV intensity is attenuated with distance from the quartz sleeve as defined by the Beers-Lambert Law, the average UV intensity within a UV disinfection system is often computed mathematically. The Beers-Lambert Law is:

$$\log\left(\frac{I}{I_0}\right) = -\varepsilon(\lambda)Cx \quad (11\text{-}59)$$

where I = light intensity at distance x from the light source, mW/cm^2
I_0 = light intensity at light source, mW/cm^2
$\varepsilon(\lambda)$ = molar absorptivity (also known as the extinction coefficient) of the light-absorbing solute at wavelength λ, L/mole-cm
C = concentration of light-absorbing solute, mole/L
x = light path length, cm

When the left-hand side of Eq. (11-59) is expressed as a natural logarithm, the right-hand side of the equation must be multiplied by 2.303 because the absorbance coefficient is determined in base 10. The term on the right-hand side of Eq. (11-59) is defined as the absorbance, $A(\lambda)$, which is unitless, but is often reported in units of cm^{-1}, which corresponds to absorptivity $kA(\lambda)$. If the length of the light path is 1 cm, absorptivity is equal to the absorbance.

$$k(\lambda) = \varepsilon(\lambda)C = A(\lambda)/x \qquad (11\text{-}60)$$

where $k(\lambda)$ = the absorptivity, cm^{-1}
$A(\lambda)$ = absorbance, dimensionless

The absorptivity of reclaimed water is an important aspect of UV reactor design. Reclaimed waters with higher absorptivity absorb more UV light and need a higher energy input for an equivalent level of disinfection. Absorbance is measured using a spectrophotometer typically using a fixed sample path length of 1.0 cm. The absorbance of water is typically measured at a wavelength of 254 nm. Typical absorbance and transmittance values for wastewater after several different treatment processes are presented in Table 11-26.

The transmittance of a solution $T(\lambda)$ is defined as:

$$\text{Transmittance, } T(\lambda), \% = \frac{I}{I_0} \times 100 \qquad (11\text{-}61)$$

The transmittance at a given wavelength can also be derived from absorbance measurements using the following relationship:

$$T(\lambda) = 10^{-A(\lambda)} \qquad (11\text{-}62)$$

The term percent transmittance, commonly used in the literature is:

$$T(\lambda), \% = 10^{-A(\lambda)} \times 100 \qquad (11\text{-}63)$$

Thus, for a perfectly transparent solution $A(\lambda) = 0$, $T(\lambda) = 1$, and for a perfectly opaque solution $A(\lambda) \to \infty$, $T(\lambda) = 0$.

The principal water characteristics that affect the percent transmittance include inorganic compounds (e.g., copper, iron), organic compounds (e.g., organic dyes, humic

Table 11-26
Absorbance and transmittance values for various wastewaters

Type of wastewater	Absorbance, a.u./cm	Transmittance, %
Primary	0.55 to 0.30	28 to 50
Secondary	0.35 to 0.15	45 to 70
Nitrified secondary	0.25 to 0.10	56 to 79
Filtered secondary	0.25 to 0.10	56 to 79
Microfiltration	0.10 to 0.04	79 to 91
Reverse osmosis	0.05 to 0.01	89 to 98

substances, and aromatic compounds such as benzene and toluene), and small colloidal particles (≤0.45 μm). As will be discussed later, the use of mathematical modeling has not yet proven to be satisfactory for the design on UV disinfection systems, given the many variables that can affect performance (Tang et al., 2006).

Effect of Chemical Constituents in Reclaimed Water

The effect of reclaimed water constituents on UV disinfection is presented in Table 11-27. Dissolved constituents impact UV disinfection either directly via absorbance (increasing absorbance serves to attenuate UV light to a larger degree) or via fouling of UV lamps such that a reduced intensity is applied to the bulk liquid medium. One of the most perplexing problems encountered in the application of UV disinfection for reclaimed water disinfection is the variation typically observed in the absorbance (or transmittance) within a treatment plant. Often, the variations in transmittance are caused by industrial discharges, which can lead to diurnal as well as seasonal variations in disinfection effectiveness.

Table 11-27
Impact of wastewater constituents on the use of UV irradiation for wastewater disinfection

Constituent	Effect
BOD, COD, and TOC	No or minor effect, unless humic materials comprise a large portion of the BOD
NOM (natural organic matter)	Strong absorbers of UV radiation
Oil and grease	Can accumulate on quartz sleeves of UV lamps, can absorb UV radiation
TSS	Absorption of UV radiation, can shield embedded bacteria
Alkalinity	Can impact scaling potential. Also effects solubility of metals that may absorb UV light
Hardness	Calcium, magnesium and other salts can form mineral deposits on quartz tubes, especially at elevated temperatures
Ammonia	No or minor effect
Nitrite	No or minor effect
Nitrate	No or minor effect
Iron	Strong absorber of UV radiation, can precipitate on quartz tubes, can adsorb on suspended solids and shield bacteria by adsorption
Manganese	Strong absorber of UV radiation
pH	Can affect solubility of metals and carbonates
TDS	Can impact scaling potential and the formation of mineral deposits
Industrial discharges	Depending on the constituents (e.g., dyes), may lead to a diurnal and seasonal variations in the transmittance
Stormwater inflow	Depending on the constituents, may lead to short term as-well-as seasonal variations in the transmittance

Common industrial impacts are related to the discharge of inorganic and organic dyes, wastes containing metals, and complex organic compounds. Of the inorganic compounds that affect transmittance, iron is considered to be the most important with respect to UV light absorbance because dissolved iron can absorb UV light directly. Organic compounds containing double bonds and aromatic functional groups can also absorb UV light. Absorbance values for a variety of compounds found in wastewater are given in Table 11-28. From a review of the information presented in Table 11-28, it is clear that the presence of iron in reclaimed water can have a significant impact on the use of UV. If iron salts are used within the treatment process, it may be necessary to switch to another chemical if UV disinfection is to be used.

It is also important to note that stormwater inflows can cause wide variations, especially when humic materials from terrestrial sources are present. In either case, the solution to the problem of varying transmittance requires monitoring of industrial discharges, the implementation of source control programs, and correcting sources of infiltration. In some cases, biological treatment will mitigate the influent variations. In some extreme situations, the conclusion may be that UV disinfection does not work.

Where the use of UV disinfection is being assessed, it is useful to install online transmittance monitoring equipment to document the variations that occur in the transmittance with time. The scaling potential of reclaimed water, as defined by the Langelier Saturation Index (see Sec. 9-4 in Chap. 9), should also be checked to assess whether scaling may be a problem. The scaling potential is especially important when the feasibility of using high-intensity UV lamps is being assessed.

Effect of Particles in Reclaimed Water

The presence of particles in reclaimed water can also impact the effectiveness of UV disinfection (Qualls et al, 1983; Parker and Darby, 1995; Emerick et al., 1999). The manner in which particles can affect UV performance is illustrated on Fig. 11-32. Many organisms of interest in wastewater (e.g., coliform bacteria) occur both in a disperse state (i.e., not bound to other objects) and a particle-associated state (i.e., bound to other objects such as other bacteria or cellular debris). Coliform bacteria are of particular

Table 11-28
UV Absorbance of water and common chemicals found in wastewater

Compound	Form or designation	Molar absorption coefficient, L/mol·cm	Threshold concentration, mg/L
Ferric iron	Fe[III]	3069	0.057
Ferrous iron	Fe[II]	466	9.6
Hypochlorite ion	ClO$^-$	29.5	8.4
N-nitrosodimethylamine	NDMA	1974	
Nitrate	NO$_3^-$	3.4	
Natural organic matter	NOM	80 to 350	
Ozone	O$_3$	3250	0.071
Zinc	Zn^{2+}	1.7	187
Water	H$_2$O	6.1×10^{-06}	

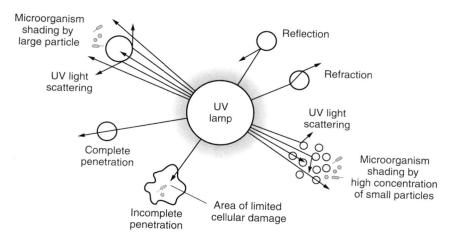

Figure 11-32

Particle interactions that affect the effectiveness of UV disinfection, including microorganism shading; light scattering, reflection, and refraction; and incomplete penetration.

importance because of the central role they play in discharge permits (i.e., coliform bacteria are used as indicators for the presence of other pathogenic organisms and their inactivation is assumed to correlate with the inactivation of other pathogenic organisms). Disperse coliform bacteria are inactivated readily because they are exposed fully to the average UV light intensity as compared to particle embedded microorganisms (see Fig. 11-32). Treatment process-related disinfection problems, when disinfecting unfiltered effluent, usually result from the influence of particle-associated organisms (see also Fig. 11-5). In fact, coliform bacteria can associate with particles to such a degree that they are completely shielded from UV light, resulting in a residual coliform bacteria concentration.

It has been hypothesized that a minimum particle size (reclaimed water specific, but on the order of 10 μm) governs the ability to shield coliform bacteria from UV light (Emerick et al., 2000). Due to the inherent porous nature of the particles in reclaimed water, particles smaller than that critical size are unable to reduce the applied intensity, and thus embedded organisms are inactivated in a manner similar to dispersed organisms. Particles greater than the critical size can shield coliform bacteria similarly. Particle size does not appear to be a governing factor once the critical size is exceeded because coliform bacteria are located randomly within particles and are not typically located in the most shielded regions within particles.

Characteristics of the Microorganisms

The effectiveness of the UV disinfection process depends on the characteristics of the microorganisms. Typical values for the disinfection of coliform organisms with UV light for various wastewaters are reported in Table 11-29. Note that the dosage values given in Table 11-29 are only meant to serve as a guide for the initial estimation of the required UV dose. The range of the reported values reflects the variable nature of wastewater. The relative effectiveness of UV radiation for disinfection of representative microorganisms of concern in reclaimed water is reported in Table 11-30. As with the values given in Table 11-5, the values given in Table 11-30 are only meant to serve

Table 11-29 Typical UV dosages required to achieve different effluent total coliform disinfection standards for various wastewaters

Type of wastewater	Initial coliform count, MPN/100 mL	UV dose, mJ/cm² Effluent standard, MPN/100 mL			
		1000	200	23	≤2.2
Raw wastewater	10^7–10^9	90–130			
Primary effluent	10^7–10^9	90–130			
Trickling filter effluent	10^5–10^6	40–50	50–70	70–90	90–110
Activated sludge effluent	10^5–10^6	40–50	50–70	70–90	90–110
Filtered activated sludge effluent	10^4–10^6	35–45	50–60	70–80	80–100
Nitrified effluent	10^4–10^6	35–45	50–60	70–80	80–100
Filtered nitrified effluent	10^4–10^6	30–40	50–70	70–80	80–100
Microfiltration effluent	10^1–10^3		25–35	30–40	40–50
Reverse osmosis	~ 0	–	–	–	5–10
Septic tank effluent	10^7–10^9	90–130			
Intermittent sand filter effluent	10^2–10^4		10–15	30–40	50–60

Table 11-30 Estimated relative effectiveness of UV irradiation for the disinfection of representative microorganisms of concern in wastewater[a]

Organism	Dosage relative to total coliform dosage[b]
Bacteria	
Escherichia coli (*E coli*)	1.0
Fecal coliform	0.9–1.0
Pseudomonas aeruginosa	1.5–1.8
Salmonella typhi	1.2–1.5
Staphylococcus aureus	1.0–1.5
Vibrio cholerae	0.6–0.8
Viruses	
Adenovirus	6–9
Coxsackie A2	0.8–1.0
F specific bacteriophage	0.8–1.0
Hepatitis A	3.5–4.5
Polio type 1	0.6–0.9
MS-2 bacteriophage	0.8–1.0
Norwalk	0.8–1.0
Rotavirus	4–6
Protozoa	
Acanthamoeba	6–8
Cryptosporidium parvum oocysts	0.1–0.3
Giardia lamblia cysts	0.1–0.2

[a] Adapted in part from Wright and Sakamoto (1999), U.S. EPA (2003b), Hijnen et al. (2006).
[b] Relative doses based on discrete nonclumped single organisms in suspension. If the organisms are clumped or particle-associated, the relative dosages are not applicable.

as a guide in assessing the relative UV dose required for different microorganisms. Knowledge concerning the required UV dose for specific pathogen inactivation is changing continuously as improved methods of analysis are applied. For example, before infectivity studies were conducted, it was thought that UV irradiation at reasonable dosage values (i.e., less than 200 mJ/cm^2) was not effective for the inactivation of *C. parvum* and *G. lamblia*. However, based on infectivity studies, it has been found that both of these protozoans are inactivated with extremely low UV dosage values (typically in the range of 5 to 15 mJ/cm^2) (Linden et al., 2001). The current literature should be consulted to obtain the most contemporary information regarding required UV dosages for the inactivation of specific microorganisms.

Impact of System Characteristics

Problems with the application of Eq. (11-8) for use in the design of UV disinfection reactors are associated with (1) inaccurate knowledge of the average UV intensity and (2) the exposure time associated with all of the pathogens passing through a UV disinfection system. In practice, field scale UV disinfection reactors have dose distributions resulting from both the internal intensity profiles and exposure time distribution. The internal intensity profiles are a reflection of the nonhomogeneous placement of lamps within the system, lack of ideal radial mixing within the system, the scattering/absorbing effects of particulate material, and the absorbance of the liquid medium. The distribution associated with exposure time is a reflection of nonideal hydraulics leading to longitudinal mixing.

One of the most serious problems encountered with UV disinfection systems in open channels is achieving a uniform velocity field in the approach and exit channel. Achieving a uniform velocity field is especially difficult when UV systems are retrofitted into existing open channels, such as converted chlorine contact basins, a practice that is not recommended if the performance of the UV disinfection system is to be optimized.

Modeling the UV Disinfection Process

Although, in general, it is believed that the concentration of suspended solids has a deleterious impact on UV disinfection performance, Emerick et al. (1999) reported that among different treatment processes there is no correlation between the total suspended solids concentration and the number of particles containing coliform bacteria. This lack of correlation underlies the need for inactivation models based on more fundamental water quality parameters.

The use of series-event or multi-hit kinetics has been suggested to describe the initial resistance that homogeneous populations of organisms tend to exhibit to UV light in addition to the subsequent log-linear inactivation behavior (Severin et al., 1983). However, the measurement of the overall response of a mixed population of bacteria (e.g., coliform bacteria) in reclaimed water tends to mask the initial resistances of specific bacterial species/strains. For UV doses greater than 10 mJ/cm^2 (i.e., as is typically applied for reclaimed water disinfection), the following equation can be used for modeling the log-linear inactivation of disperse coliform bacteria in a batch system (Jagger, 1967; Oliver and Cosgrove, 1975; Qualls and Johnson, 1985):

$$N_D(t) = N_D(0)e^{-kD} \qquad (11\text{-}64)$$

where $N_D(t)$ = total number of surviving disperse coliform bacteria at time t
$N_D(0)$ = total number of disperse coliform bacteria prior to UV light application (at time t = 0)
k = inactivation rate coefficient, cm²/mW·s
D = average UV dose, $I_{avg} \times t$, mJ/cm²
I_{avg} = average intensity of UV light in bulk solution, mW/cm²
t = exposure time, s

The fundamental difference between disperse coliform bacteria and particle-associated coliform bacteria is the UV intensity reaching the organism. The above equation is applicable only to disperse organisms because all members of that group receive the same intensity of UV light (assuming perfectly mixed conditions). An organism embedded within a particle receives a reduced UV light intensity relative to that applied to the bulk solution. Knowledge of the distribution of applied intensities allows a model, analogous to that presented above, to be developed to describe the inactivation of both disperse and particle-associated coliform bacteria. Emerick et al. (2000) demonstrated the applicability of the following modeling equation for describing the inactivation of both disperse and particle-associated coliform bacteria (see Fig. 11-6, presented previously) when knowledge of the applied intensity to the bulk liquid medium is known.

$$N(t) = N_D(0)e^{-kD} + \frac{N_P(0)}{kD}(1 - e^{-kD}) \qquad (11\text{-}65)$$

where $N_P(0)$ = total number of particles containing at least one coliform bacteria at time t = 0
other terms as defined above

Equation (11-65) is best used to describe the underlying constraints to UV disinfection performance. The numbers of particles containing coliform bacteria, the inactivation rate coefficient, and the applied UV dose (product of intensity and exposure time) have fundamental impacts on UV disinfection performance. From experience it has been found that it is more convenient to design disinfection systems using collimated beam inactivation data and validated UV disinfection equipment, as discussed below.

Estimating UV Dose

The first step in assessing the performance of a UV disinfection system is to determine the UV dose needed to inactivate the challenge microorganism to a level that is protective of public health. Three methods have been used to estimate the UV dose. In the first method, an average UV dose is determined by assuming an average system UV intensity and exposure time. The average UV intensity is estimated using a computational procedure known as the point source summation (PSS) method (U.S. EPA, 1992). The PSS method is currently used less frequently by designers due to its dependence on system-specific hydraulics (i.e., pilot study results are a function of the pilot unit used during the course of study).

The second method involves the use of computational fluid dynamics (CFD) to integrate both the distribution of UV intensities and velocity profiles within the reactor to obtain a distribution of UV doses within a system (Batchley et al., 1995). Although the CFD method is promising, its use is limited at the present time (2006) because (1) the

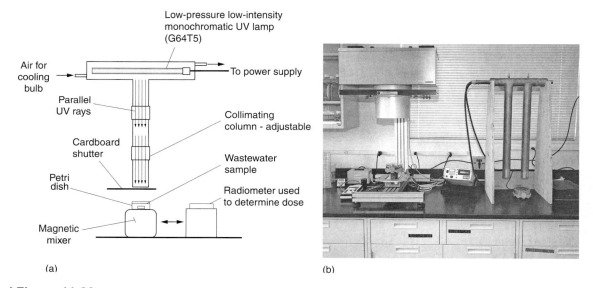

Figure 11-33
Collimated beam device used to develop dose-response curves for UV disinfection: (a) schematic and (b) view of two different types of collimated beam devices. The collimated beam on the left is of European design; the collimated beam on the right is of the type shown schematically in (a).

methodology is not standardized, (2) the methodology has not been validated thoroughly over a range of disinfection systems, and (3) the reporting of a distribution of UV doses, though accurate, is problematic for UV disinfection system specification. In the third, and most widely used, method the UV dose is determined using a collimated beam bioassay. Use of the bioassay approach in designing UV disinfection systems is discussed below.

Determination of UV Dose by Collimated Beam Bioassay

The most common procedure for determining the required UV dose for the inactivation of challenge microorganism involves the use of a collimated beam and a small reactor (i.e., a Petri dish) to which a known UV dose is applied. Typical collimated beam devices are shown on Fig. 11-33. Use of a monochromatic low-pressure low-intensity lamp in the collimated beam apparatus allows for accurate characterization of the applied UV intensity. Use of a batch reactor allows for accurate determination of exposure time. The applied UV dose, as defined by Eq. (11-8), can be controlled either by varying the UV intensity or the exposure time. Because the geometry is fixed, the depth-averaged UV intensity within the Petri dish sample (i.e., the batch reactor) can be computed using the following relationship:

$$\begin{aligned} D &= I_m t(1-R)P_f \left[\frac{(1-10^{-\alpha d})}{2.303(\alpha d)}\right]\left(\frac{L}{L+d}\right) \\ &= I_m t(1-R)P_f \left[\frac{(1-e^{-2.303\alpha d})}{2.303(\alpha d)}\right]\left(\frac{L}{L+d}\right) \end{aligned} \quad (11\text{-}66)$$

where D = average UV dose, mW/cm²
I_m = incident UV intensity at the center of the surface of the sample, mW/cm²
t = exposure time, s
R = reflectance at the air water interface at 254 nm
P_f = Petri dish factor
α = absorbance of sample, absorbance units per centimeter, a.u./cm (base 10)
d = depth of sample, cm
L = distance from lamp centerline to liquid surface, cm

The term (1 − R) on the right hand side of Eq. (11-66) accounts for the reflectance at the air water interface. The value of R is typically about 2.5 percent. The term P_f accounts for the fact that the UV intensity may not be uniform over the entire area of the Petri dish. The value of P_f is typically greater than 0.9. The term within the brackets is the depth-averaged UV intensity within the Petri dish and is based on the Beer-Lambert Law. The final term is a correction factor for the height of the UV light source above the sample. The application of Eq. (11-66) is illustrated in Example 11-7.

The uncertainty of the computed UV dose, D, can be estimated using the sum of the variances as given by either of the following expressions:

Maximum uncertainty

$$U_E = \sum_{n=1}^{n=n} \left| U_{V_n} \frac{\partial D}{\partial V_n} \right| \qquad (11\text{-}67)$$

Best estimate of uncertainty

$$U_E = \left[\sum_{n=1}^{n=n} \left(U_{V_n} \frac{\partial D}{\partial V_n} \right)^2 \right]^{1/2} \qquad (11\text{-}68)$$

where U_E = uncertainty of UV dose value, %
U_{V_n} = uncertainty or error in variable n
V_n = variable n
$\partial D/\partial V_n$ = partial derivative of the expression for D with respect to the variable V_n

The maximum estimate of uncertainty as given by Eq. (11-67) represents the condition where every error will be a maximum value. The best estimate of uncertainty, as given by Eq. (11-68), is used most commonly because it is unlikely that every error will be a maximum at the same time and that some errors may cancel each other.

Knowledge of the average UV intensity and exposure time allows calculation of the average applied UV dose using Eq. (11-8). The UV dose is then correlated to the microorganism inactivation results as discussed below.

EXAMPLE 11-7. Determination of UV Dose Delivered in Collimated Beam Test.

The following measurements were made to establish the UV dose using a collimated beam. Using these data, determine the average UV dose delivered to the sample and best estimate of the uncertainty associated with the measurement.

$I_m = 5 \pm 0.35$ mW/cm² (accuracy of meter $\pm 7\%$)
$t = 60 \pm 1$ s
$R = 0.025$ (assumed to be the correct value)
$P_f = 0.94 \pm 0.02$
$\alpha = 0.065 \pm 0.005$ cm^{-1}
$d = 1 \pm 0.05$ cm
$L = 40 \pm 0.5$ cm

Solution

1. Using Eq. (11-66) estimate the delivered dose.

$$D = I_m t(1 - R)P_f \left[\frac{(1 - 10^{-\alpha d})}{2.303(\alpha d)}\right]\left(\frac{L}{L + d}\right)$$

$$D = (5 \times 60)(1 - 0.025)(0.94)\left[\frac{(1 - 10^{-0.065 \times 1})}{2.303(0.065 \times 1)}\right]\left(\frac{40}{40 + 1}\right)$$

$$D = (300)(0.975)(0.94)(0.928)(0.976) = 249.1 \text{ mJ/cm}^2$$

2. Determine the best estimate of uncertainty for the computed UV dose. The uncertainty of the computed dose can be estimated using Eq. (11-68). The procedure is illustrated for one of the variables and summarized for the remaining variables.

   a. Consider the variability in the measured time, t. The partial derivative of the expression used in Step 1 with respect to t is

   $$U_t = t_e \frac{\partial D}{\partial t} = t_e \left\{I_m(1 - R)P_f\left[\frac{(1 - 10^{-\alpha d})}{2.303(\alpha d)}\right]\left(\frac{L}{L + d}\right)\right\}$$

   $$U_t = 1.0\left\{(5)(1 - 0.025)(0.94)\left[\frac{(1 - 10^{-0.065 \times 1})}{2.303(0.065 \times 1)}\right]\left(\frac{40}{40 + 1}\right)\right\}$$

   $$U_t = 4.15 \text{ mJ/cm}^2$$

   Percent = $100\, U_t/D = (100 \times 4.15)/249.1 = 1.67\%$

   b. Similarly for the remaining variables the corresponding values of the partial derivatives are as given below.

   $U_{I_m} = 17.44$ mJ/cm² and 7.0%

   $U_{P_f} = 5.30$ mJ/cm² and 2.13%

   $U_\alpha = -1.40$ mJ/cm² and -0.56%

$$U_d = -1.21 \text{ mJ/cm}^2 \text{ and } -0.49\%$$
$$U_L = 0.076 \text{ mJ/cm}^2 \text{ and } 0.03\%$$

   c. The best estimate of uncertainty using Eq. (11-68) is

$$U = [(4.15)^2 + (17.44)^2 + (5.30)^2 + (-1.40)^2 + (-1.21)^2 + (0.076)^2]^{1/2}$$
$$U = 18.79 \text{ mJ/cm}^2$$

$$\text{Percent} = (100 \times 18.79)/249.1 = 7.54 \text{ percent}$$

3. Based on the above uncertainty computation the best estimate of the UV dose is

$$249.1 \pm 18.8 \text{ mJ/cm}^2$$

Comment
Thus, a conservative estimate of the UV dose that can be delivered consistently is 230.3 mJ/cm² (249.1 − 18.8). An alternative approach to uncertainty is given in Example 11-9.

Bioassay Testing

To assess the degree of inactivation that can be achieved at a given UV dose, the concentration of microorganisms is determined before and after exposure in a collimated beam apparatus (see Fig. 11-33). Microorganism inactivation is measured using an MPN procedure for bacteria, a plaque count procedure for viruses, or an animal infectivity procedure for protozoa. To verify the accuracy of the laboratory collimated beam dose-response test data, the collimated beam test must be repeated to obtain statistical significance. To be assured that the stock solution of the challenge microorganisms is monodispersed, the laboratory inactivation test data must fall within an accepted set of quality control limits. Quality control limits proposed by the National Water Research Institute (NWRI, 2003) and the U.S. EPA (2003b) for Bacteriophage MS2 spores are as follows:

NWRI

$$\text{Upper bound: } -\log_{10}(N/N_0) = 0.040 \times D + 0.64 \quad (11\text{-}69a)$$

$$\text{Lower bound: } -\log_{10}(N/N_0) = 0.033 \times D + 0.20 \quad (11\text{-}69b)$$

U.S. EPA

$$\text{Upper bound: } -\log_{10}(N/N_0) = -9.6 \times 10^{-5} \times D^2 + 4.5 \times 10^{-2} \times D \quad (11\text{-}70a)$$

$$\text{Lower bound: } -\log_{10}(N/N_0) = -1.4 \times 10^{-4} \times D^2 + 7.6 \times 10^{-2} \times D \quad (11\text{-}70b)$$

where D = UV dose, mJ/cm²

As will be illustrated in Example 11-8, the bounds proposed by the U.S. EPA are more lenient as compared to those used by NWRI. Similar bounding curves have been proposed for *B. subtilus* (U.S. EPA, 2003b). The NWRI guidelines are used for water reuse applications in California.

EXAMPLE 11-8. Verification of Laboratory Procedures for Bacteriophage MS2 Response.

The following collimated beam test results were obtained for a stock solution of bacteriophage MS2 which is to be used to test a UV reactor. Verify that the laboratory test results are acceptable.

Dose, mJ/cm^2	Surviving concentration, phage/mL	Log survival, log (phage/mL)	Log inactivation
0	1.00×10^7	7.0	0.0
20	1.12×10^6	6.05	0.95[a]
40	6.76×10^4	4.83	2.17
60	1.95×10^4	4.29	2.71
80	4.37×10^3	3.64	3.36
100	1.20×10^3	3.08	3.92
120	7.08×10^1	1.85	5.15
140	1.48×10^1	1.17	5.83

[a]Sample calculation: $-$Log inactivation $= 7.00 - 6.05 = 0.95$.

Solution

1. Plot the collimated beam test results and compare to the quality control range expressions provided in the NWRI [Eqs. (11-69a) and (11-69b)] and U.S. EPA [Eqs. (11-70a) and (11-70b)] UV Guidelines. The results are plotted in the figure given below.

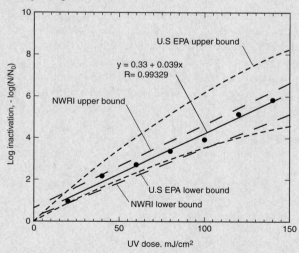

2. As shown in the above, plot all of data points fall within the acceptable range.

Comment

As shown in the above plot there is a considerable difference in the upper quality control limit between the NWRI and the proposed U.S. EPA UV guidelines. Also note that the U.S. EPA guidelines are curvilinear, whereas the NWRI guidelines are linear. Clearly, the NWRI guidelines are more restrictive.

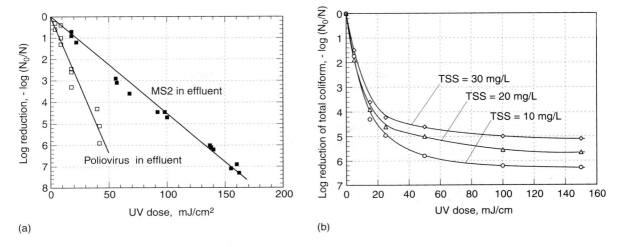

Figure 11-34
Typical dose response curves for UV disinfection developed from data obtained using a collimated beam device: (a) for dispersed discrete microorganisms (Cooper et al., 2000) and (b) wastewater containing varying concentrations of TSS.

Reporting and Using Bioassay Collimated Beam Test Results

The results of collimated beam bioassays are reported in the form of dose response curves (see Fig. 11-34). The inactivation curve shown on Fig. 11-34a is for dispersed discrete organisms exposed to UV light whereas the curve shown on Fig. 11-34b is for reclaimed water containing particulate material. The analysis and use of collimated beam test results in establishing the UV dose required for the inactivation of coliform organisms is illustrated in Example 11-9.

EXAMPLE 11-9. Determination of UV Dose Requirement for Total Coliform Bacteria Using Results from Collimated Beam Testing.

The following dose response data were obtained by conducting collimated beam tests once per month over a 12-mo period for a given reclaimed water. Using these data, determine (1) the mean, standard deviation, and confidence interval associated with the surviving number of total coliform bacteria at each UV dose investigated and (2) the dose required (site-specific) to comply with a permit limitation of 23 total coliform bacteria per 100 mL (30-d median).

Test number	Survival at applied UV dose, mJ/cm²						
	0	20	40	60	80	100	120
1	3,500,000	280	43	6.8	5.5	5.4	6.0
2	79,000	920	23	6.8	5.5	36	22
3	920,000	58	17	13	10	1.8	1.8
4	430,000	540	110	24	430	14	8.1
5	9,200,000	2800	540	24	46	1.8	21
6	210,000	54,000	9200	920	110	2.0	5.5
7	16,000,000	36	23	13	5.5	17	5.5
8	1,700,000	180	46	4.0	4.0	69	4.5
9	920,000	540	49	21	1.8	3.6	5.5
10	5,600,000	2400	31	69	19	24	1.8
11	79,000	920	280	280	81	12	1.8
12	4,400,000	110	9.1	84	22	54	95

Solution

1. Determine the mean, standard deviation, and confidence interval for the monthly dose response data. Because biological UV dose response data are generally log-normally distributed, log-transform the observed survival data to enable use of student-t statistics (student-t statistics must be used because there are not enough data to apply normal statistics, typically 30 samples are required).

   a. Log-transform the observed number of surviving total coliform bacteria. For example, for test 1, the log-transform data point associated with a UV dose of 40 is log (43) = 1.63.

   b. Determine the average and standard deviation for the log transformed data for each investigated UV dose.
   For the UV dose of 20, the average is 2.75.
   For the UV dose of 20, the standard deviation is 0.86.
   The observed mean and standard deviation for each UV dose is provided in the following table:

Test number	Log survival at applied UV dose, mJ/cm²						
	0	20	40	60	80	100	120
1	6.54	2.45	1.63	0.83	0.74	0.73	0.78
2	4.90	2.96	1.36	0.83	0.74	1.56	1.34
3	5.96	1.76	1.23	1.11	1.00	0.26	0.26
4	5.63	2.73	2.04	1.38	2.63	1.15	0.91
5	6.96	3.45	2.73	1.38	1.66	0.26	1.32
6	5.32	4.73	3.96	2.96	2.04	0.30	0.74

(Continued)

Test number	Log survival at applied UV dose, mJ/cm^2						
	0	20	40	60	80	100	120
7	7.20	1.56	1.36	1.11	0.74	1.23	0.74
8	6.23	2.26	1.66	0.60	0.60	1.84	0.65
9	5.96	2.73	1.69	1.32	0.26	0.56	0.74
10	6.75	3.38	1.49	1.84	1.28	1.38	0.26
11	4.90	2.96	2.45	2.45	1.91	1.08	0.26
12	6.64	2.04	0.96	1.92	1.34	1.73	1.98
Average	6.08	2.75	1.88	1.48	1.25	1.01	0.83
Stand. dev.	0.78	0.86	0.83	0.70	0.70	0.58	0.51

c. Determine an adequate confidence interval. Because the permit is based on a 30-d median value, designing based on the mean survival risks occasional permit violations. The 75% confidence interval is often used to ensure compliance with a median permit limit.

  i. For a dose of 60 mJ/cm^2, the 75% confidence interval is calculated using the following expression (Larson and Faber, 2000):

$$75\% \text{ confidence limit} = \bar{x} \pm t_{0.125}\left(\frac{s}{\sqrt{n}}\right)$$

where $\bar{x}$ = mean survival at a specific UV dose = 1.48

$t_{0.125}$ = student t value associated with a 75% level of confidence
    = 1.214 (obtained from statistical tables, Larson and Faber, 2000)

Note that the degrees of freedom are
    $n - 1 = 12 - 1 = 11$
    n = number of replicates = 12
    s = sample standard deviation = 0.70

$$75\% \text{ confidence limit} = 1.48 \pm 1.214\left(\frac{0.70}{\sqrt{12}}\right) = 1.48 \pm 0.245$$

  ii. Transform the mean and confidence interval back to base 10. The mean and confidence interval associated with each investigated UV dose is provided in the table given below.

UV dose, mJ/cm^2	Surviving total coliform per 100 mL		
	Average	Lower 75% CI	Upper 75% CI
0	1,200,000	623,000	2,320,000
20	560	280	1200
40	76	38	150
60	30	17	54
80	18	10	32
100	10	6	17
120	7	4	10

2. Estimate the required UV dose.

   Based on the upper 75% confidence intervals in the above table, it can be concluded that a design UV dose of 100 mJ/cm² is adequate to obtain a 30-d median survival of 23 total coliform bacteria per 100 mL.

   Comment

   The variability in the data reported in Step 1 is representative of what is observed in practice based on limited testing. To gain a better understanding of the variability associated with the reclaimed water of interest, it is recommended that replicate tests be conducted (a minimum of three tests is recommended).

Ultraviolet Disinfection Guidelines

The National Water Research Institute and the American Water Works Association Research Foundation published "Ultraviolet Disinfection Guidelines for Drinking Water and Wastewater Reclamation" (NWRI, 1993; NWRI and AWWARF, 2000; NWRI, 2003). The following elements are considered in the UV guidelines: (1) reactor design, (2) reliability design, (3) monitoring and alarm design, (4) the field commissioning test, (5) performance monitoring, and (6) an engineering report for unrestricted effluent reuse applications. Some of the items may not be applicable when utilizing UV disinfection for less demanding uses.

Application of UV Guidelines

The guidelines that cover reclaimed water are similar to those that cover drinking water systems. The primary difference is that recommended doses are provided for reclaimed water systems whereas there is no mention of recommended doses for drinking water systems. For reclaimed water systems, the recommended design UV doses are 100 mJ/cm² for granular medium filtration effluent, 80 mJ/cm² for membrane filtration effluent, and 50 mJ/cm² for reverse osmosis effluent. The different dose requirements reflect the different virus density concentrations expected within each type of process effluent. The dosages selected are intended to provide 4 log of poliovirus inactivation with a factor of safety of about 2.

In addition to differing dose recommendations as a function of effluent quality, there are differing design transmittance recommendations. For granular medium, microfiltration, and reverse osmosis effluents, the design transmittances are 55, 65, and 90 percent, respectively. The differing transmittance values are based on field observations made to date. All UV disinfection systems installed for either drinking water or unrestricted reuse applications must undergo validation testing prior to their installation. Although the guidelines do not apply to the disinfection of secondary effluent, the general design issues addressed are applicable.

Relationship of UV Guidelines to UV System Design

The design of a UV disinfection system requires three general steps: (1) determination of the UV dose required, based on bioassay testing, for adequate inactivation of the challenge (target) microorganism(s); (2) validation of manufacturer-specific UV disinfection system performance; and (3) determination of an optimal UV system configuration (e.g., the number of lamps per module, modules per bank, banks per channel, and the overall number of channels). The first two issues are addressed directly in the guidelines, and general guidance is provided on design aspects. Determination of the UV dose required to comply with a permit limitation was discussed previously and will be illustrated in Example 11-10. Specific details on the culture of the microorganisms and the conduct of the test are given in the guidelines.

Test Protocol for UV System Performance Validation in Reclaimed Water

Validation testing consists of quantifying the inactivation of a virus surrogate (e.g., Bacteriophage MS2) as a function of flowrate through the UV disinfection system. To quantify the inactivation achieved through the UV disinfection system, the UV dose response of the challenge microorganism to be used is determined using a collimated beam device, as illustrated in Example 11-10. The inactivation observed through the UV disinfection system is compared to the UV dose response to establish a term called the *delivered dose, validated dose, or reduction equivalent dose* (RED) which corresponds to the UV dose delivered by the UV disinfection system. The determination of the delivered dose is illustrated on Fig. 11-35. It should be noted that a variety of correction factors are given in the U.S. EPA Guidance Manual (2003b), primarily for water

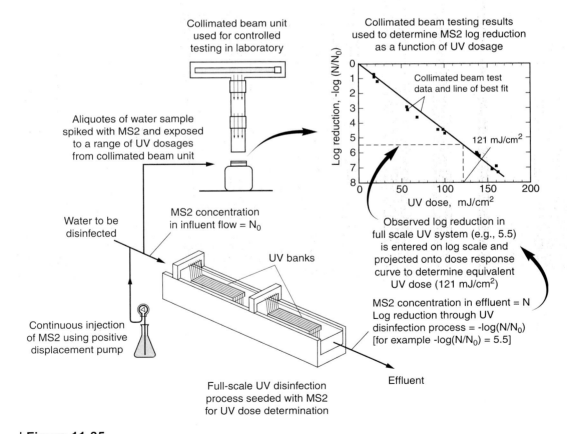

Figure 11-35

Schematic illustration of the application of biodosimetry as used to determine the performance of a test or full scale UV reactor (Adapted in part from Crittenden et al., 2005).

treatment, for (1) RED bias—correcting the dose for different microorganisms, (2) polychromatic bias—account for spectral differences in light output, (3) uncertainty factor—accounts for uncertainty in measurement taken during validation testing.

Validation Testing Based on NWRI UV Guidelines

Validation testing is important because the test results can be used to compare competing UV disinfection technologies and eliminates the need to make choices based on manufacturers' claims, often not verified by an independent third party. The process flow diagram used for testing both open and closed UV reactors is illustrated on Fig. 11-36.

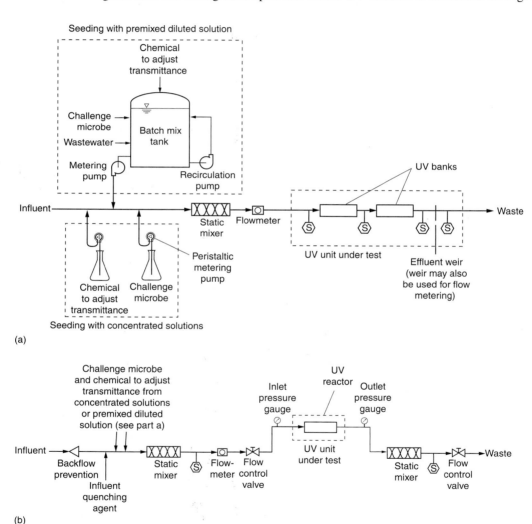

Figure 11-36

Schematic of the experimental setup used for the validation testing of UV reactors using seeded microorganisms and UV transmittance adjustment: (a) seeding with premixed diluted solution from batch tank or from concentrated solutions in open channel systems and (b) seeding with premixed solution for closed channel systems.

Validation testing of UV disinfection equipment, using the setup shown in Fig. 11-36, consists of the following steps:

1. Selection of a representative test water for the use in the validation testing of the disinfection system.
2. Selection of the configuration of the UV disinfection system to be tested (for low-pressure low-intensity UV systems, a minimum of two banks must be tested, typically more are used. If the power to the UV lamps cannot be turned down to simulate the end-of-life lamp performance, then aged UV lamps must be used in the test.
3. Testing of the hydraulic performance of the UV disinfection system. Hydraulic testing is done to verify the uniformity of the approach and exit velocities.
4. Quantification of the inactivation of the viral indicator as a function of hydraulic loading rate through the UV test reactor (see Fig. 11-37). Typical dosing arrangements for both open channel and closed UV systems are illustrated on Fig. 11-36.
5. Simultaneously conduct a collimated beam test on the test water to determine the inactivation of the viral indicator as a function of applied UV dose.
6. Verify the accuracy of the laboratory collimated beam dose-response test data. The laboratory test data must fall within the area bound by Eqs. (11-69a) and (11-69b) given previously.
7. Assign UV doses to the pilot reactor based on the measured inactivation observed during the collimated beam test as a function of applied UV dose.

The steps required in conducting a validation test are illustrated in Example 11-10.

Figure 11-37
Large closed UV reactor undergoing validation testing.

EXAMPLE 11-10. Analysis of Pilot Test Results Used to Validate UV Disinfection System Performance.

A manufacturer has supplied a pilot scale UV disinfection system to be tested for the assignment of UV doses as a function of lamp hydraulic loading rate. For this test, the manufacturer chose to make use of a four-lamp per bank pilot facility with three banks in series to achieve the total applied dose. Each bank of lamps is hydraulically independent of subsequent banks. Therefore, the results can be applied to full-scale reactors up to 40 lamps per bank (i.e., full-scale facility can utilize up to 10 times as many lamps per bank). Aged lamps were placed in the pilot facility to simulate the performance of the UV lamps at the end of their warranted life.

The testing was conducted on tertiary effluent from a local water reclamation facility. Normal transmittance of the tertiary effluent is 75 percent. A transmittance reducing agent (e.g., instant coffee) was injected into the effluent stream to lower the transmittance to 55 percent. The manufacturer has specified that the UV disinfection system should be tested for hydraulic loading rates ranging from 20 to 80 L/min·lamp. Because the titer of the virus indicator (i.e., Bacteriophage MS2) to be used for performance testing was approximately 1×10^{11} phage/mL, it was decided to test the system under the conditions outlined in the following table:

Hydraulic loading rate, L/min·lamp (1)	Process flow, L/min (2)	Virus titer concentration, phage/mL (3)	Virus titer injection flowrate, L/min (4)	Approximate resulting virus concentration in process flow, phage/mL (5)
20	240	1×10^{11}	0.024	1×10^{7}
40	480	1×10^{11}	0.048	1×10^{7}
60	720	1×10^{11}	0.072	1×10^{7}
80	960	1×10^{11}	0.096	1×10^{7}

Notes on column entries:
(1) Desired range to be tested as specified by the manufacturer.
(2) The pilot system contained three banks with 12 lamps total. Therefore, at a hydraulic loading rate of 20 L/min·lamp, the process flowrate needs to be (12 lamps)(20 L/min·lamp) = 240 L/min.
(3) Provided by the laboratory.
(4) It was desired to obtain a virus titer in the process flow of about 1×10^{7} phage/mL. Therefore, at 240 L/min, the solution containing the virus had to be injected at a rate of 0.024 L/min to obtain the desired initial titer.

In conducting the test, each flowrate was tested randomly with respect to order. Three distinct replicate samples were collected per flowrate. An inlet and outlet sample (i.e., that containing the concentration of phage prior to any inactivation) was collected with each process replicate. The inlet test results are as follows:

Flowrate, L/min	Replicate	Inlet concentration, phage/mL	Log-transformed inlet conc., log(phage/mL)	Average log-transformed inlet conc., log(phage/mL)
240	1	5.25×10^6	6.72	
240	2	1.00×10^7	7.00	6.93
240	3	1.15×10^7	7.06	
480	1	1.00×10^7	7.00	
480	2	1.23×10^7	7.09	7.07
480	3	1.29×10^7	7.11	
720	1	1.23×10^7	7.09	
720	2	1.05×10^7	7.02	7.03
720	3	9.55×10^6	6.98	
960	1	1.23×10^7	7.09	
960	2	1.20×10^7	7.08	7.02
960	3	7.94×10^6	6.90	

The outlet results are as follows:

Flowrate (L/min)	Replicate	Number of operational banks[a]	Outlet concentration, phage/mL	Log-transformed outlet conc., log(phage/mL)
240	1	2	2.09×10^2	2.32
240	2	2	1.44×10^2	2.16
240	3	2	1.66×10^2	2.22
480	1	3	3.80×10^2	2.58
480	2	3	3.31×10^2	2.52
480	3	3	3.09×10^2	2.49
720	1	3	1.32×10^4	4.12
720	2	3	6.03×10^3	3.78
720	3	3	4.27×10^3	3.63
960	1	3	4.79×10^4	4.68
960	2	3	1.86×10^5	5.27
960	3	3	6.61×10^4	4.82

[a]Notice that at the low flowrate investigated (240 L/min), only two operational banks were investigated rather than three. Only two banks were tested because three operational banks resulted in no detectable viruses in the effluent. Because the banks were hydraulically independent, it is allowed under the UV guidelines to investigate the inactivation for only two banks and extrapolate to performance expected for additional banks of lamps.

Because the UV disinfection system was tested with filtered secondary effluent, determine the range of flows expressed as L/min·lamp over which the UV disinfection system will deliver a dose of 100 mJ/cm². Assume the MS2 UV dose response curve given in Example 11-8 will be used for the analysis of the test results.

Solution

1. Determine the 75% level of confidence for the degree of inactivation achieved for each flowrate that was evaluated. The results of the analysis are presented in the following table. Note that the 75% level of confidence is determined using the student-t distribution (a minimum of 30 samples are required for use of the normal distribution).

Flowrate, L/min	Replicate	Log inactivation	Average log-inactivation	Sample standard deviation	75% confidence log-inactivation
240	1	4.61[a]			
240	2	4.77	4.69	0.08	4.63
240	3	4.71			
480	1	4.49			
480	2	4.55	4.54	0.05	4.50[b]
480	3	4.58			
720	1	2.91			
720	2	3.25	3.19	0.25	2.95
720	3	3.40			
960	1	2.34			
960	2	1.75	2.10	0.31	1.81
960	3	2.20			

[a] Sample calculation. From the previous table, the average inlet log concentration was observed to be 6.93. Therefore, the log inactivation for replicate 1 is $6.93 - 2.32 = 4.61$.
[b] Sample calculation. For the flowrate of 480 L/min, the 75% level of confidence occurs at 4.50 as shown below.

$$75\% \text{ confidence limit} = \bar{x} \pm t_{0.125}\left(\frac{s}{\sqrt{n}}\right)$$

$$= 4.54 - 1.214 \frac{0.05}{\sqrt{3}} = 4.54 - 0.04 = 4.50$$

2. Assign UV dosages to the investigated hydraulic loading rates, and present results graphically.
   a. From Example 11-8, the equation of the linear regression used to determine the required dose as a function of log MS2 inactivation is given on the following figure:

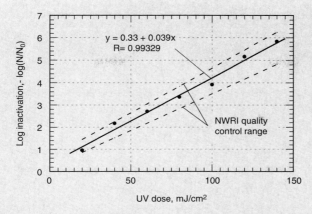

b. The calculated UV dosages are presented in the following table:

Flowrate, L/min	Hydraulic loading rate, L/min·lamp	75% confidence log-inactivation	Equivalent UV dose[b], mJ/cm²
240	20	(1.5)(4.63) = 6.95[a]	170
480	40	4.50	107
720	60	2.95	67.2
960	80	1.81	37.9

[a] The inactivation for this flowrate was extrapolated from the two-bank results. Because the system is a three-bank system, the inactivation for three banks is 150 percent greater than the inactivation observed with two operational banks.

[b] Sample calculation. Using the linear regression expression derived from the collimated beam test, the equivalent UV dose at a flowrate of 480 L/min is:

$$\text{Dose, mJ/cm}^2 = \frac{\text{log inactivation} - 0.33}{0.039} = \frac{4.50 - 0.33}{0.039} = 107 \text{ mJ/cm}^2$$

c. Plot the UV dosages determined in the previous step. The results are plotted on the following figure:

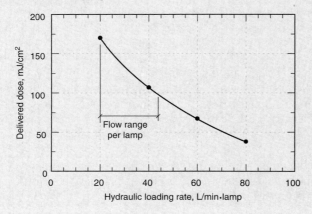

3. Determine the flow per lamp over which the system will deliver 100 mJ/cm². From the plot given above, the system is capable of delivering a dose of 100 mJ/cm² within the range of 20 to 43 L/min·lamp.

Comment

Because aged lamps were used and the transmittance value was adjusted to 55 percent, the test results represent the performance of the UV disinfection system under the worst possible conditions, which provides a factor of safety under typical operating conditions. When the lamps are new, it may not be necessary to operate all three banks, depending on the test results. The curve given above in Step 3 is then used to determine the optimal configuration of the full-scale UV disinfection system.

Analysis of a UV Disinfection System

Factors that affect the minimum number of UV lamps necessary for disinfection are: (1) the hydraulic loading rate determined in the equipment validation test as outlined in the previous example, (2) the aging and fouling characteristics of the UV lamp/quartz sleeve assembly, (3) reclaimed water quality and its variability, and (4) the nature of the discharge permit itself and the level of confidence desired in meeting that permit. Hydraulic behavior has a significant impact on field reactor performance. The flow per lamp determined using the collimated beam bioassay has a corresponding velocity that maintains that inactivation performance. The process configuration must maintain adequate system velocity to ensure that the bioassay results are applicable to the field installation. Although beyond the scope of this textbook, the UV guidelines cited above should be reviewed carefully before undertaking the design of a UV system. For the selection and sizing of a UV system, a current reference such as Tchobanoglous et al. (2003) may be consulted.

Operational Issues with UV Disinfection Systems

Operational issues associated with UV disinfection are related primarily to the inability to achieve permit conditions. Some issues that must be considered when diagnosing problems associated with UV disinfection systems are discussed below.

UV Disinfection System Hydraulics

Perhaps one of the most serious problems encountered in the field is erratic or reduced inactivation performance due to poor system hydraulics. The most common hydraulic problems are related to: (1) the creation of density currents which can cause the incoming reclaimed water to move along the bottom or top of the UV lamp banks resulting in short circuiting, (2) inappropriate entry and exit conditions which can lead to the formation of eddy currents which ultimately create uneven velocity profiles that induce short circuiting, (3) the creation of dead spaces or zones within the reactor resulting in short circuiting. The occurrence of short circuiting or dead zones reduces the average contact time, leading to ineffective use of the UV system.

The principal hydraulic design features that can be used to improve system hydraulics in open channels include the use of: (1) submerged perforated diffusers, (2) corner fillets in rectangular open channel systems with horizontal lamp placement, (3) flow deflectors in open channel systems with vertical lamp placement, and (4) serpentine effluent overflow weirs used in conjunction with perforated diffusers. In some cases, power input to mix the incoming flow may be necessary. Some of these corrective measures for open channel UV disinfection systems are illustrated on Fig. 11-38. Submerged perforated baffles should have an open area of about 4 to 6 percent of the cross-sectional area of the flow channel. In closed UV disinfection systems, the use of perforated plates is typically not required when the units are plumbed correctly. The use of computational fluid dynamics is of great value in studying the effect of various physical interventions in bringing about a more uniform approach velocity flow field (Blatchley et al., 1995).

Biofilms on Walls of UV Channels and on UV Equipment

Another serious problem encountered with UV disinfection systems is the development of biofilms on the exposed surfaces of the UV reactor. The problem is especially serious in open channel systems covered with standard grating. It has been found that if the UV channels are exposed to any light, even dim light, biofilms (typically fungal

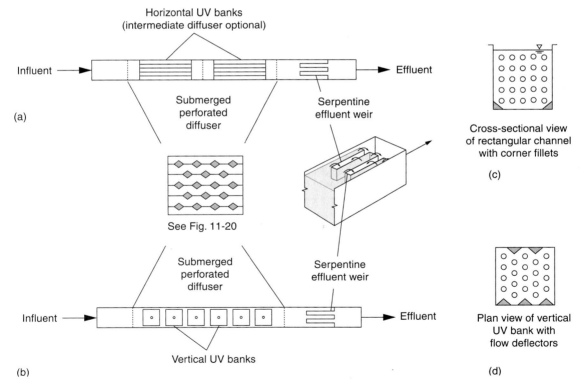

Figure 11-38
Typical examples of physical features that can be used to improve the hydraulic performance of open channel UV reactors: (a) and (b) use of diffusers, (c) use of corner fillets in reactors with horizontal UV lamps, and (d) use of flow deflectors in reactors with vertical UV lamps. For both the horizontal and vertical lamp arrangements, wall effects are reduced as the number of lamp modules is increased.

and filamentous bacteria) develop on the exposed surfaces. The problem with biofilms is that they can harbor and effectively shield bacteria. When the clumped biofilms break away from the attachment surface, bacteria can be shielded as the clumps pass through the disinfection system. The best control measure is to completely cover the UV channels. In addition, the channels can be occasionally cleaned and disinfected using hypochlorite, paracetic acid (see Sec. 11-8), or another suitable cleaning agent/disinfectant.

It should be noted that biofilm development can also occur in closed UV systems, but the severity is usually less, with the exception of UV systems in which medium-pressure high-intensity UV lamps are employed. Because medium-pressure high-intensity UV lamps emit some light in or near the visible light range (see Fig. 11-26), they can stimulate the growth of microorganisms on exposed surfaces. In some cases, growths approaching 300 mm in length have been found attached to the lamp support structure. The amount of light emitted in the visible light range will vary with each type of lamp (i.e., manufacturer). Removal of these growths with a suitable disinfectant must be conducted on a periodic basis.

Overcoming the Impact of Particles by Increasing UV Intensity

It was thought at one time that the impact of particles on the performance of UV disinfection systems could be overcome by increasing the UV intensity. Unfortunately, it has been found that increasing the UV intensity tenfold has little effect on reducing the number of surviving particle-associated coliform bacteria because the absorption of UV radiation by particles in reclaimed water is typically 10,000 times or more greater than the bulk liquid medium. Particles essentially block the transmission of UV light. Particles larger than some critical size (a function of the size of the target organism) effectively shield the embedded microorganisms (Emerick et al., 1999; Emerick et al., 2000). Because the effectiveness of UV disinfection is governed primarily by the number of particles containing coliform bacteria, to improve the performance of a UV disinfection system either the number of particles with associated coliform bacteria must be reduced (e.g., by selecting an appropriate upstream treatment process), or the particles themselves must be removed (e.g., by improved clarifier design or use of some form of filtration). Currently, to meet the stringent total coliform bacteria requirements for body contact water reuse applications (i.e., equal to or less than 2.2 MPN/100 mL), some form of effluent filtration is required.

Effect of Upstream Treatment Processes on UV Performance

The number of particles with associated coliform bacteria is another factor that impacts the performance of a UV disinfection system. As noted previously in Sec. 11-2, it has been observed that for activated sludge plants the number of particles with associated coliform organisms is a function of the SRT (see Fig. 11-5). Thus, both the mode of operation of the biological process and the design and operation of the secondary sedimentation facilities must be evaluated carefully, especially where an unfiltered effluent is to be disinfected. Even with effluent filtration, attention must be focused on the distribution of particle sizes in the filtered effluent (Darby et al., 1999; Emerick et al., 1999).

Environmental Impacts of UV Irradiation

The environmental impacts associated with the use of UV radiation as a reclaimed water disinfectant include the discharge of altered chemical compounds and regrowth of microorganisms.

Discharge of UV Altered Compounds

Because ultraviolet light is not a chemical agent, no toxic residuals are produced. However, certain chemical compounds may be altered by the ultraviolet irradiation. On the basis of the evidence to date, it appears that the compounds formed are harmless or are broken down into more innocuous forms at the dosages used for the disinfection of reclaimed water (80 to 200 mJ/cm^2). Photooxidation, which does alter the structure of compounds, occurs in the kJ/cm^2 (kilojoule) range. Thus, the disinfection of reclaimed water with ultraviolet light is not considered to have any adverse environmental impacts. The impacts associated with some of the new very high-energy lamps, which may operate in the kilojoule range, is not known at present (2006).

Regrowth of Microorganisms

Because, as discussed previously, microorganisms have enzymes that are capable of repairing damage to DNA following exposure to UV light, regrowth is a factor that should be considered where UV is used for water reuse applications, especially where reclaimed water is to be transported over great distances in pipelines.

Assessment of Regrowth Potential Because the capacity of microorganisms to repair themselves differs greatly among species and strains of microorganisms, it is, therefore, important to investigate the repair of specific pathogenic and indicator microorganisms, individually. The question that must be answered is once UV-induced damage is repaired either by photoreactivation or dark repair, can the organisms regain their ability to replicate themselves, resulting in regrowth? And, if regrowth does occur, is the organism capable of causing disease? Differences between monochromatic low-pressure UV and polychromatic medium-pressure UV light with respect to photoreactivation must also be considered.

Control Measures Based on the results of studies reported previously in Sec. 11-7, it appears that the regrowth of microorganisms due to photoreactivation following UV disinfection may be reduced significantly if the disinfected effluent is kept in the dark for about 3 h following exposure to UV light (Martin and Gehr, 2005). To be assured that little or no regrowth will occur due to dark repair, it may be necessary to increase the UV dose used for disinfection. If UV-disinfected reclaimed water is to be pumped to a distant water reuse location, the addition of small amounts of disinfectant (e.g., chlorine, peracetic acid) may be necessary for the control of slime growths that develop in long pipelines, regardless of the operational care devoted to eliminating them. Where UV-disinfected effluent is discharged to an open water body, the addition of small amounts of chlorine or other disinfectant may be necessary to limit regrowth. Where reclaimed water has undergone UV-based advanced oxidation, the UV dose needed for oxidation may be sufficient to limit regrowth.

PROBLEMS AND DISCUSSION TOPICS

11-1 Determine the inactivation rate constant for total coliform for one of the following four treated effluents (sample to be selected by instructor) assuming Chick's law applies. The effluent temperature was 20°C.

Log of organisms remaining	Time, min Sample			
	A	B	C	D
7	0.0	0.0	0.0	0.0
6	12	4.0	6	8
5	24	8.0	11.8	16.6
4	36	13	17.2	25.2
3	47.8	17	25	35
2	60.2	20	30.4	40
Combined chlorine residual, mg/L	5	8	10	7

11-2 Using the rate constant developed in Problem 11-1, determine the chlorine dose required to achieve a 99.99 percent inactivation of total coliform in 60 min at 15 and 25°C.

11-3 Estimate the daily required chlorine dosage, the required alkalinity, if alkalinity will have to be added, and the resulting buildup of total dissolved solids when breakpoint chlorination is used for the seasonal control of nitrogen. Assume that the following data apply to this problem:
1. Plant flowrate = 4800 m³/d
2. Effluent characteristics
   a. BOD = 15 mg/L
   b. Total suspended solids = 15 mg/L
   c. NH_4^+–N = 4 mg/L
   d. Alkalinity = 125, 145, or 165 mg/L as $CaCO_3$ (value to be selected by instructor)
3. Required effluent NH_4^+–N concentration = 1.0 mg/L

11-4 The chlorine residuals measured when various dosages of chlorine were added to four different reclaimed waters are given below. Determine (reclaimed water to be selected by instructor) (a) the breakpoint dosage and (b) the design dosage to obtain a residual of 1, 2, 3.5, or 5 mg/L (residual to be selected by instructor) free chlorine.

Dosage, mg/L	Residual, mg/L			
	A	B	C	D
0	0.0	0.0	0.0	0.0
1	0.6	1.0	0.95	1.0
2	0.2	2.0	1.7	1.98
3	1.0	2.98	2.3	2.9
4	2.0	3.95	1.2	3.4
5	3.0	4.3	0.9	2.7
6		3.6	1.7	1.2
7		2.3	2.7	1.2
8		0.7	3.7	2.1
9		0.7		3.1
10		1.7		4.1
11		2.8		
12				

11-5 Review the current literature and prepare an assessment of the use of chlorine gas versus sodium hypochlorite for the disinfection of reclaimed water. A minimum of 4 articles and/or reports dating back to 1997 should be cited in your assessment.

11-6 The following data were obtained for several treated effluents. Using these data estimate the coefficients for the refined Collins-Selleck model [Eq. (11-27)] (sample to be selected by instructor).

$-\log(N/N_0)$	Time, min			
	Sample			
	A	B	C	D
1	3.8	7	7	9
2	5.8	15	12	17
3	9.7	36	22	31
4	16	80	37	59
5	27.5	190	66	110
6	45	430	115	200

11-7 A consultant has proposed using chorine dosages of 20 and 10 mg/L during the summer and winter, respectively, for effluent disinfection. If the effluent total coliform count before disinfection is $10^8/100$ mL, estimate the final total coliform counts that can be achieved during the summer and winter with these dosages.

1. Initial effluent chlorine demand = 5 mg/L
2. Demand due to decay during chlorine contact = 2.0 mg/L
3. Required chlorine contact time = 45 min
4. Use the typical values given above for the coefficients
    b = 4.0
    n = 2.8

11-8 The following data were obtained from dye tracer studies of five different chlorine contact basins. Using these data, determine the mean hydraulic residence time and the corresponding variance, the t_{10} time, and the Morrill Dispersion Index and the volumetric efficiency for one of the basins (to be selected by instructor). How would the performance basin selected for analysis be classified according to the U.S. EPA guidelines?

Time, min	Concentration, ppb Basin				
	A	B	C	D	E
0	0.0	0.0	0.0	0.0	0.0
10	0.0	0.0	0.0	0.0	0.0
20	3.5	0.1	0.1	0.0	0.0
30	7.6	2.1	2.1	0.0	0.7
40	7.8	7.5	10.0	0.3	4.0
50	6.9	10.1	12.0	1.8	9.0
60	5.9	10.2	10.2	4.5	12.5
70	4.8	9.7	8.0	8.0	11.5
80	3.8	8.1	6.0	11.0	8.8
90	3.0	6.0	4.3	11.0	5.5
100	2.4	4.4	3.0	9.0	3.0
110	1.9	3.0	2.1	4.3	1.8
120	1.5	1.9	1.5	2.0	0.8
130	1.0	1.0	1.0	1.0	0.4
140	0.6	0.4	0.5	0.2	0.1
150	0.3	0.1	0.1	0.0	0.0
160	0.1	0.0	0.0	0.0	0.0
170	0.0	0.0	0.0	0.0	0.0

11-9 Using the following dose response data for an enteric virus and the tracer data for four different chlorine contact basins, determine for one of the basins (to be selected by instructor) the expected effluent microorganism concentration based on the t_{10} and mean hydraulic residence times. Also, estimate the chlorine residual that would be required to achieve 4 logs of removal with the existing basins.

Dose response data for enteric viruses

C_Rt, mg/L·min[a]	Log number of organisms remaining
0	10^7
100	$10^{6.2}$
200	$10^{5.4}$
400	$10^{3.8}$
600	$10^{2.1}$
800	$10^{0.6}$
1000	10^{-1}

[a] Combined chlorine residual = 6.0

Tracer data for chlorine contact basins

	Tracer concentration, mg/L			
	Chlorine contact basin			
Time, min	A	B	C	D
0	0.0	0.0	0.0	0.0
10	0.0	0.0	0.0	0.0
20	20.0	0.0	0.0	0.0
30	0.1	0.0	0.0	0.0
40	2.0	0.0	0.0	0.0
50	7.3	1.1	0.1	0.0
60	7.0	7.0	1.3	0.1
70	5.2	7.3	8.0	1.5
80	3.3	5.7	8.5	7.5
90	1.7	4.2	6.2	8.0
100	0.7	2.9	2.9	5.5
110	0.2	1.7	1.3	3.5
120	0.0	0.9	0.4	1.8
130		0.3	0.0	0.9
140		0.1		0.3
150		0.0		0.1
160				0.0
τ, min	80	85	90	100

11-10 Determine the amount of sulfur dioxide (SO_2), sodium sulfite (Na_2SO_3), sodium bisulfite ($NaHSO_3$), sodium metabisulfite ($Na_2S_2O_5$), and activated carbon (C) that would be required per year to dechlorinate treated effluent containing a chlorine residual of 5.0, 6.5, 8.0, or 7.7 mg/L as Cl_2 (residual to be selected by instructor) from a plant with an average flowrate of 1400, 3800, 4500, or 7600 m^3/d (flowrate to be selected by instructor).

11-11 Estimate the ozone dose needed to disinfect a filtered secondary effluent to an MPN value of 240/100 mL using the following disinfection data obtained from pilot scale installation. Assume the starting coliform concentration will be $1 \times 10^6/100$ mL and that the ozone transfer efficiency is 80 percent.

Test number	Initial coliform count, N_0 MPN/100 mL	Ozone transferred, mg/L	Final coliform count, MPN/100 mL	$-\log(N/N_0)$
1	95,000	3.1	1500	1.80
2	470,000	4.0	1200	2.59
3	3,500,000	4.5	730	3.68
4	820,000	5.0	77	4.03
5	9,200,000	6.5	92	5.00

11-12 Review the current literature and prepare an assessment of the use of ozone for the disinfection of reclaimed water. A minimum of four articles and/or reports dating back to 1995 should be cited in your assessment.

11-13 Review the current literature and prepare an assessment of the use of peracetic acid alone or in combination with other disinfectants. A minimum of three articles and/or reports dating back to 1997 should be cited in your assessment.

11-14 Given the following measurements and data, determine the average UV dose delivered to the sample and best estimate of the uncertainty associated with the measurement.

$I_m = 10 \pm 0.5$ mW/cm² (accuracy of meter $\pm 7\%$)
$t = 30 \pm 1$ s
$R = 0.025$ (assumed to be the correct value)
$P_f = 0.94 \pm 0.02$
$\alpha = 0.065 \pm 0.005$ cm^{-1}
$d = 1 \pm 0.05$ cm
$L = 48 \pm 0.5$ cm

11-15 If the intensity of the UV radiation measured at the water surface in a Petri dish is 10 mW/cm², determine the average UV intensity to which a sample will be exposed if the depth of water in the Petri dish is 10, 22, 14, 15, or 16 mm (water depth to be selected by instructor).

11-16 Assume the intensity of UV radiation measured at the water surface in a Petri dish in Problem 11-12 is 5 mW/cm², and that the computed UV dose was based on a water depth of 10 mm. What would be the effect if the actual water depth in the Petri dish were 20 mm? For one of the test runs (to be selected by instructor) plot the actual test results reported in Problem 11-12 and the corrected values using a water depth of 20 mm.

11-17 Determine the mean, standard deviation, and confidence interval for following MS2 bacteriophage inactivation data, obtained using a collimate beam device. What UV dose would be required to achieve a 4-log inactivation of MS2 with a confidence interval of 75 percent.

Log reduction, $-\log(N/N_0)$	Applied UV dose, mJ/cm² Test				
	1	2	3	4	5
1	17	21	26	24	20
2	37	43	51	47	40
3	56	66	80	70	60
4	75	89	105	94	80
5	94	110	131	120	100
6	114	133	160	143	121
7	131	155	185	170	142

11-18 A UV reactor comprised of two banks with 4 lamps per bank was tested on two different reclaimed waters (A and B) at four flowrates using MS2 bacteriophage as the test organism. The hydraulic loading rates were varied from 25 to 100 L/min·lamp. In conducting the test, each flowrate was tested randomly with respect to order. The measured inlet and outlet bacteriophage concentration are as follows:

Flowrate, L/min	Replicate	Water A, phase/mL		Water B, phase/mL	
		Inlet	Outlet	Inlet	Outlet
200	1	9.65×10^6	1.88×10^2	1.05×10^7	2.19×10^2
200	2	1.00×10^7	1.54×10^2	6.98×10^6	1.54×10^2
200	3	1.15×10^7	1.68×10^2	1.15×10^7	1.70×10^2
400	1	1.00×10^7	3.65×10^2	1.00×10^7	3.75×10^2
400	2	1.29×10^7	3.39×10^2	1.23×10^7	3.62×10^2
400	3	9.55×10^6	3.29×10^2	1.12×10^7	3.08×10^2
600	1	1.23×10^7	1.12×10^4	1.20×10^7	1.32×10^4
600	2	1.05×10^7	9.03×10^3	1.05×10^7	1.05×10^4
600	3	1.25×10^6	8.56×10^3	9.55×10^6	9.95×10^3
800	1	1.13×10^7	4.79×10^4	1.03×10^7	5.95×10^4
800	2	1.08×10^7	8.35×10^4	1.19×10^7	1.00×10^5
800	3	8.95×10^6	6.61×10^4	1.11×10^7	7.68×10^4

Using the given data, determine for water A or B (water to be selected by instructor) the range of flows expressed as L/min·lamp over which the UV disinfection system will deliver a dose of 80 mJ/cm^2. Assume the MS2 UV dose response curve given in Example 11-8 will be used for the analysis of the test results.

11-19 Review the current literature and prepare an assessment of the use of low-pressure low-intensity versus low-pressure high-intensity UV disinfection systems for the disinfection of reclaimed water. A minimum of three articles and/or reports dating back to 1995 should be cited in your assessment.

11-20 Review the current literature and prepare an assessment of the use of medium-pressure high-intensity UV disinfection systems for the disinfection of reclaimed water. A minimum of three articles and/or reports dating back to 1995 should be cited in your assessment.

REFERENCES

Bellar, T. A., and J. J. Lichtenberg (1974) "Determining Volatile Organics at Microgram-per-Litre Levels by Gas Chromatography," *J. AWWA*, **66**, 12, 739–744.

Blackmer, F., K. A. Reynolds, C. P. Gerba, and I. L. Pepper (2000) "Use of Integrated Cell Culture-PCR to Evaluate the Effectiveness of Poliovirus Inactivation by Chlorine," *Appl. Environ. Microbiol.*, **66**, 5, 2267–2268.

Blatchley, E. R. et al. (1995) "UV Pilot Testing: Intensity Distributions and Hydrodynamics," *J. Environ. Eng. ASCE*, **121**, 3, 258–262.

Blume, T., I. Martinez, and U. Neis (2002) "Wastewater Disinfection Using Ultrasound and UV Light," in U. Neis (ed.) *Ultrasound in Environmental Engineering II*, TUHH Reports on Sanitary Engineering, **35**, Hamburg.

Blume, T., and U. Neis (2004) "Combined Acoustical-Chemical Method for the Disinfection of Wastewater," 127–135, *Chemical Water and Wastewater Treatment, Vol. VIII*, Proceedings of the 11th Gothenburg Symposium, Orlando, FL.

Butterfield, C. T., E. Wattie, S. Megregian, and C. W. Chambers (1943) "Influence of pH and Temperature on the Survival of Coliforms and Enteric pathogens When Exposed to Free Chlorine," *U.S. Public Health Service Report*, **58**, 51, 1837–1866.

Caretti C., and C. Lubello (2003) "Wastewater Disinfection with PAA and UV Combined Treatment: a Pilot Plant Study," *Water Res.*, **37**, 2365–2371.

Chen D., X. Dong, and R. Gehr (2005) "Alternative Disinfection Mechanisms for Wastewaters Using Combined PAA/UV Processes," in *Proceedings of WEF, IWA and Arizona Water Pollution Control Association Conference "Disinfection 2005,"* Mesa, AZ.

Chick, H. (1908) "Investigation of the Laws of Disinfection," *J. Hygiene*, British, **8**, 92–158.

Collins, H. F. (1970) "Effects of Initial Mixing and Residence Time Distribution on the Efficiency of the Wastewater Chlorination Process," paper presented at the California State Department of Health Annual Symposium, Berkeley and Los Angeles, CA, May 1970.

Collins, H. F., and R. E. Selleck (1972) "Process Kinetics of Wastewater Chlorination," *SERL Report* 72–75, Sanitary Engineering Research Laboratory, University of California, Berkeley, CA.

Cooper, R. C., A. T. Salveson, R. Sakaji, G. Tchobanoglous, D. A. Requa, and R. Whitley (2000) "Comparison of the Resistance of MS2 and Poliovirus to UV and Chlorine Disinfection," presented at the California Water Reclamation Meeting, Santa Rosa, CA.

Crites, R., and G. Tchobanoglous (1998) *Small and Decentralized Wastewater Management Systems*, McGraw-Hill, New York.

Crittenden, J. C., R. R. Trussell, D. W. Hand, K. J. Howe, and G. Tchobanoglous (2005) *Water Treatment: Principles and Design*, 2nd ed., John Wiley & Sons, New York.

Darby, J., R. Emerick, F. Loge, and G. Tchobanoglous (1999) "The Effect of Upstream Treatment Processes on UV Disinfection Performance," Project 96-CTS-3, Water Environment Research Foundation, Washington, DC.

Dulbecco, R. (1949) "Reactivation of Ultraviolet Inactivated Bacteriophage by Visible Light," *Nature,* **163**, 949–950.

Ekster, A. (2001) Personal Communication.

Emerick, R. W., F. J. Loge, D. Thompson, and J. L. Darby (1999) "Factors Influencing Ultraviolet Disinfection Performance Part II: Association of Coliform Bacteria with Wastewater Particles," *Water Environ. Res.*, **71**, 6, 1178–1187.

Emerick, R. W., F. Loge, T. Ginn, and J. L. Darby (2000) "Modeling the Inactivation of Particle-Associated Coliform Bacteria," *Water Environ. Res.*, **72**, 4, 432–438.

Enslow, L. H. (1938) "Chlorine in Sewage Treatment Practice," Chap. 8, in L. Pearse (ed.) *Modern Sewage Disposal*, Federation of Sewage Works Associations, New York.

EPRI (1996) *UV Disinfection for Water and Wastewater Treatment*, Report CR-105252, Electric Power Research Institute, Inc., report prepared by Black & Veatch, Kansas City, MO.

Fair, G. M., J. C. Morris, S. L. Chang, I. Weil, and R. P Burden (1948) "The Behavior of Chlorine as a Water Disinfectant," *J. AWWA,* **40**, 1051–1056.

Fair, G. M., and J. C. Geyer (1954) *Water Supply and Waste-Water Disposal*, John Wiley & Sons, New York.

Finch, G. R., and D. W. Smith (1989) "Ozone Dose-Response of Escherichia coli in Activate Sludge Effluent," *Water Res.*, **23**, 8, 1017–1025.

Finch, G. R., and D. W. Smith (1990) "Evaluation of Empirical Process Design Relationships for Ozone Disinfection of Water and Wastewater," *Ozone Sci. Eng.*, **12**, 2, 157–175.

Friedberg, E. R., G. C. Walker, and W. Siede (1995) *DNA Repair and Mutagenesis,* WSM Press, Washington.

Gard, S. (1957) "Chemical Inactivation of Viruses," in CIBA Foundation, *Symposium on the Nature of Viruses*, Little Brown & Company, Boston, MA.

Gehr, R. (2000) Seminar Lecture Notes, Universidad Autonoma Metrpolitana, Mexico City, Mexico.

Gehr, R. (2006) Seminar Notes, Presented at III Simposio Internacional en Ingenieria y Ciencias para la Sustentabilidad Ambiental, Universidad Autonoma Metrpolitana, Mexico City, Mexico.

Gehr, R., M. Wagner, P. Veerasubramanian, and P. Payment (2003) "Disinfection Efficiency of Peracetic Acid, UV and Ozone after Enhanced Primary Treatment of Municipal Wastewater," *Water Res.*, **37**, 19, 4573–4586.

Haag, W. R., and J. Hoigné (1983) "Ozonation of Bromide-Containing Waters: Kinetics of Formation of Hypobromous Acid and Bromate," *Environ. Sci. Technol.*, **17**, 5, 261–267.

Haas, C., and S. Karra, (1984a) "Kinetics of Microbial Inactivation by Chlorine—I. Review of Result in Demand-Free Systems," *Water Res.,* **18**, 11, 1443–1449.

Haas, C., and S. Karra (1984b) "Kinetics of Microbial Inactivation by Chlorine—II. Review of Results in Systems with Chlorine Demand," *Water Res.*, **18,** 11, 1451–1454.

Haas, C., and S. Karra (1984c) "Kinetics of Wastewater Chlorine Demand Exertion," *J. WPCF*, **56**, 2, 170–182.

Haas, C. N., and J. Joffe (1994) "Disinfection Under Dynamic Conditions: Modification of Hom's Model for Decay," *Environ. Sci. Technol.*, **28**, 7, 1367–1369.

Hall, E. L. (1973) "Quantitative Assessment of Disinfection Interferences," *Water Treat. Exam.*, **22**, 153–174.

Hanawalt, P. C., P. K. Cooper, A. K. Ganesan, and C. A. Smith (1979) "DNA Repair in Bacteria and Mammalian Cells," *Annu. Rev. Biochem.*, **48**, 783–836.

Hanzon, B., J. Hartfelder, S. O'Connell, and D. Murray (2006) "Disinfection Deliberation," *Water Environ. Technol.,* **18**, 2, 57–62.

Hart, F. L. (1979) "Improved Hydraulic Performance of Chlorine Contact Chambers," *J. WPCF*, **51**, 12, 2868–2875.

Hijnen, W. A. M., E. F. Beerendonk, and G. J. Medema (2006) "Inactivation Credit of UV Radiation for Viruses, Bacteria, and Protozoan (oo)cysts in Water: A Review," *Water Res.*, **40**, 1, 3–22.

Hom, L. W. (1972) "Kinetics of Chlorine Disinfection in an Eco-System," *J. Environ. Eng. Div.* ASCE, **98**, SA1, 183–194.

Jagger, J. H. (1967) *Introduction to Research in UV Photobiology*, Prentice-Hall, Englewood Cliffs, NJ.

Jalali, Y., S. J. Huitric, J. Kuo, C. C. Tang, S. Thompson, J. F. Stahl (2005) "UV Disinfection of Tertiary Effluent and Effect on NDMA and Cyanide," paper presented at WEF Technology 2005, San Francisco, CA.

Kawamura, S. (2000) *Integrated Design and Operation of Water Treatment Facilities*, 2nd ed., Wiley Interscience, New York.

Kelner, A. (1949) "Effect of Visible Light on the Recovery of Streptomyces Griseus Conidia from Ultra-violet Irradiation Injury," *Proc. Nat. Acad. Sci.*, **35**, 73–79.

Kim, J., M. Urban, S. Echigo, R. Minear, and B. Marinas, (1999) "Integrated Optimization of Bromate Formation and *Cryptosporidium Parvum* Oocyst Control in Batch and Flow-Through Ozone Contactors," *Proc. 1999 American Water Works Association Water Quality Technology Conference*, on CD, Tampa, FL.

Kitis, M. (2004) "Disinfection of Wastewater With Peracetic Acid: A Review," *Environ Int.*, **30**, 1, 47–55.

Koivunen, J. (2005) "Inactivation of Enteric Microorganisms with Chemical Disinfectants, UV Irradiation and Combined Chemical/UV Treatments," *Water Res.*, **39**, 8, 1519–1526.

Krasner, S. W. (1999) "Chemistry of Disinfection By-Product Formation," in P. C. Singer, (ed.) *Formation and Control of Disinfection By-Products in Drinking Water*, AWWA, Denver, CO.

Krasner, S. W., S. Pastor, R. Chinn, M. J. Sclimenti, H. S. Wienberg, S. D. Richardson, and A. D. Thruston, Jr., (2001) "The Occurrence of a New Generation of DBPs (Beyond the ICR)," paper presented at the AWWA Water Quality Technology Conference, Nashville, TN.

Larson, R., and B. Faber (2000) *Elementary Statistics*, Prentice-Hall, Upper Saddle River, NJ.

Lazarova, V., M. L. Janex, L. Fiksdal, C. Oberg, I. Barcina, and M Ponimepuy (1998) "Advanced Wastewater Disinfection Technologies: Short and Long Term Efficiency," *Water Sci. Technol.*, **38**, 12, 109–117.

Lefevre, F., J. M. Audic, and F. Ferrand (1992) "Peracetic Acid Disinfection of Secondary Effluents Discharged Off Coastal Seawater," *Water Sci. Technol.*, **25**, 12, 155–164.

Liberti, L., A. Lopez, and M. Notarnicola (1999) "Disinfection with Peracetic Acid for Domestic Sewage Re-Use in Agriculture," *J. Water Environ. Mgmt.* (Canadian), **13**, 8, 262–269.

Linden, K., G. Shin, and M. Sobsey (2001) "Comparative Effectiveness of UV Wavelengths For the Inactivation of *Cryptosporidium Parvum* Oocysts in Water," *Water Sci. Technol.*, **43**, 12, 171–174.

Lubello C., C. Caretti, and R. Gori (2002) "Comparison Between PAA/UV and H_2O_2/UV Disinfection for Wastewater Reuse," *Water Sci. Technol.: Water Supply*, **2**, 1, 205–212.

Malley, J. P. (2005) "A New Paradigm for Drinking Water Disinfection," presented at the 17th World Ozone Congress, International Ozone Association, Strasbourg, Germany.

Martin, N., and R. Gehr (2005) "Photoreactivation Following Combined Peracetic Acid-UV Disinfection of a Physicochemical Effluent," presented at the Third International Congress on Ultraviolet Technologies, IUVA, Whistler, BC, Canada.

Montgomery, J. M., Consulting Engineers, Inc. (1985) *Water Treatment Principles and Design*, A Wiley-Interscience Publication, John Wiley & Sons, New York.

Morrill, A. B. (1932) "Sedimentation Basin Research and Design," *J. AWWA*, **24**, 9, 1442–1458.

Morris, J. C. (1966) "The Acid Ionization Constant of HOCL from 5°C to 35°C," *J. Phys. Chem.* **70**, 12, 3798–3806.

Morris, J. C. (1975) "Aspects of the Quantitative Assessment of Germicidal Efficiency," Chap. 1, in J. D. Johnson (ed.) *Disinfection: Water and Wastewater*, Ann Arbor Science Publishers, Inc., Ann Arbor, MI.

NWRI (1993) *UV Disinfection Guidelines for Wastewater Reclamation in California and UV Disinfection Research Needs Identification,* National Water Research Institute, prepared for the California Department of Health Services. Sacramento, CA.

NWRI and AWWARF (2000) *Ultraviolet Disinfection Guidelines for Drinking Water and Wastewater Reclamation,* NWRI-00-03, National Water Research Institute and American Water Works Association Research Foundation, Fountain Valley, CA.

NWRI (2003) *Ultraviolet Disinfection Guidelines for Drinking Water and Water Reuse,* 2nd ed., National Water Research Institute, Fountain Valley, CA.

O'Brien, W. J., G. L. Hunter, J. J. Rosson, R. A. Hulsey, and K. E. Carns (1996) Ultraviolet System Design: Past, Present, and Future, Proceedings Disinfecting Wastewater for Discharge & Reuse, Water Environment Federation*,* Alexandria, VA.

Oguma, K., H. Katayama, H. Mitani, S. Morita, T. Hirata, and S. Ohgaki (2001) "Determination of Pyrimidine Dimmers in *Escherichia Coli* and *Cryptosporidium Parvum* During Ultraviolet Light Inactivation, Photoreactivation and Dark Repair," *Appl. Environ. Microbiol.*, **67**, 4630–4637.

Oguma, K., H. Katayama, and S. Ohgaki (2002) "Photoreactivation of *Escherichia Coli* after Low- or Medium-Pressure UV Disinfection Determined by an Endonuclease Sensitive Site Assay," *Appl. Environ. Microbiol.*, **68**, 12, 6029–6035.

Oguma, K., H. Katayama, and S. Ohgaki. (2004) "Photoreactivation of *Legionella Pneumophila* after Inactivation by Low- or Medium-Pressure Ultraviolet Lamp," *Water Res.*, **38**, 11, 2757–2763.

Oliver, B. G., and E. G. Cosgrove (1975) "The Disinfection of Sewage Treatment Plant Effluents Using Ultraviolet Light," *J. Chem. Eng.*, (Canadian), **53**, 4, 170–174.

Orta de Velasquez, M. T., I. Yanez-Noguez, N. M. Rojas-Valencia, and C. l. Lagona-Limon (2005) "Ozone in the Disinfection of Municipal Wastewater Compared with Peracetic Acid, Hydrogen Peroxide, and Copper after Advanced Primary Treatment," presented at the 17th International Ozone Association World Congress & Exhibition, Strasburg, France.

Parker, J. A., and J. L. Darby (1995) "Particle-Associated Coliform in Secondary Effluents: Shielding from Ultraviolet Light Disinfection," *Water Environ. Res.*, **67**, 7, 1065–1075.

Plummer, J. D., and S. C. Long (2005) "Enhancement of Chlorine Inactivation with Chemical Free Sonication," presented at the Water Quality Technology Conference, Quebec City, Canada.

Qualls, R. G., M. P. Flynn, and J. D. Johnson (1983) "The Role of Suspended Particles in Ultraviolet Disinfection," *J. WPCF*, **55**, 10, 1280–1285.

Qualls, R. G., and J. D. Johnson (1985) "Modeling and Efficiency of Ultraviolet Disinfection Systems," *Water Res.,* **19**, 8, 1039–1046.

Rakness, K. L. (2005) *Ozone in Drinking Water Treatment: Process Design, Operation and Optimization*, American Water Works Association, Denver, CO.

Rennecker, J., B. Marinas, J. Owens, and E. Rice (1999) "Inactivation of Cryptosporidium Parvum Oocysts with Ozone," *Water Res.,* **33**, 11, 2481–2488.

Rennecker, J., J. Kim, B. Corona-Vasquez, and B. Marinas (2001) "Role of Disinfectant Concentration and pH in the Inactivation Kinetics of *Cryptosporidium Parvum* Oocysts with Ozone and Monochloramine," *Environmental Sci. Technol.*, **35**, 13, 2752–2757.

Rice, R. G. (1996) *Ozone Reference Guide*, prepared for the Electric Power Research Institute, Community Environmental Center, St. Louis, MO.

Roberts, P. V., E. M. Aieta, J. D. Berg, and B. M. Chow (1980) "Chlorine Dioxide for Wastewater Disinfection: A Feasibility Evaluation," Technical Report No. 21, Civil Engineering Department, Stanford University, Stanford, CA.

Rook, J. J. (1974) "Formation of Haloforms During the Chlorination of Natural Water," *Water Treat. Exam.*, **23**, 2, 234–243.

Rupert, C. S. (1960) "Photoreactivation of Transforming DNA by an Enzyme from Baker's Yeast," *J. Gen. Physiol.*, **43**, 573–595.

Saunier, B. M. (1976) "Kinetics of Breakpoint Chlorination and Disinfection," Ph.D. Thesis, University of California, Berkeley, CA.

Saunier, B. M., and R. E. Selleck (1976) "The Kinetics of Breakpoint Chlorination in Continuous Flow Systems," Paper presented at the American Water Works Association Annual Conference, New Orleans, LA.

Severin, B. F, M. T. Suidan, and R. S. Engelbrecht (1983) "Kinetic Modeling of UV Disinfection of Water," *Water Res.*, British, **17**, 11, 1669–1678.

Shin, G. A., K. G. Linden, M. J. Arrowood, and M. D. Sobsey (2001) "Low-Pressure UV Inactivation and DNA Repair Potential of *Cryptosporidium Parvum* Oocysts," *Appl. Environ. Microbiol.*, **67**, 3029–3032.

Sirikanchana, K., J. Shisler, and B. Marinas (2005) "Sequential Inactivation of Adenoviruses by UV and Chlorine," presented at 2005 Water Quality Technology Conference and Exposition, American Water Works Association, Denver, CO.

Snyder, C. H. (1995) *The Extraordinary Chemistry of Ordinary Things*, 2nd ed., John Wiley & Sons, New York.

Solvay Interox (1997) Proxitane™ Peracetic Acid, Company Brouchure.

Sommer, R., W. Pribil, S. Appelt, P. Gehringer, H. Eschweiler, H. Leth, A. Cabal, and T. Haider (2001) "Inactivation of Bacteriophages in Water by Means of Non-Ionizing (UV-253.7 nm) and Ionizing (Gamma) Radiation: A Comparative Approach," *Water Res.*, **35**, 13, 3109–3116.

Sung, R. D. (1974) "Effects of Organic Constituents in Wastewater on the Chlorination Process," Ph.D. Thesis, Department of Civil Engineering, University of California, Davis, CA.

Tang, C-C., J. Kuo, S-J. Huitric, Y. Jalali, R. W. Horvath and J. F. Stahl (2006) "UV Systems for Reclaimed Water Disinfection—From Equipment Validation to Operation" presented at *WEFTEC 06*, Water Environment Foundation, Washington, DC.

Tchobanoglous, G., F. L. Burton, and H. D. Stensel (2003) *Wastewater Engineering: Treatment and Reuse*, 4th ed., Metcalf and Eddy, Inc., McGraw-Hill, New York.

Thibaud, H, J. De Laat, and M. Dore (1987) "Chlorination of Surface Waters: Effect of Bromide Concentration on the Chloropicrin Formation Potential," paper presented at the Sixth Conference on Water Chlorination: Environmental Impact and Health Effects, Oak Ridge, TN.

U.S. EPA (1986) *Design Manual, Municipal Wastewater Disinfection*, U.S. Environmental Protection Agency, EPA/625/1-86/021, Cincinnati, OH.

U.S. EPA (1992) *User's Manual for UVDIS, Version 3.1, UV Disinfection Process Design Manual*, U.S. Environmental Protection Agency, EPA G0703, Risk Reduction Engineering Laboratory, Cincinnati, OH.

U.S. EPA (1999a) *Combined Sewer Overflow Technology Fact Sheet, Alternative Disinfection Methods*, U.S. Environmental Protection Agency, EPA 832-F-99-033, Cincinnati, OH.

U.S. EPA (1999b) *Alternative Disinfectants and Oxidants Guidance Manual*, U.S. Environmental Protection Agency, EPA 815-R-99-014, Cincinnati, OH.

U.S. EPA (2002) *National Primary Drinking Water Regulations: Long Term 2 Enhanced Surface Water Treatment Rule*, (LT2SWTR) Proposed Rule, *Federal Register*, **68**, 154, 47640.

U.S. EPA (2003a) *EPA Guidance Manual*, Appendix B. Ct Tables, LT1ESWTR Disinfection Profiling and Benchmarking, U.S. Environmental Protection Agency, Washington, DC.

U.S. EPA (2003b) *Ultraviolet Disinfection Guidance Manual*, Draft, US. Environmental Protection Agency, Office of Water, Washington, DC.

Wagner, M., D. Brumelis, and R. Gehr (2002) "Disinfection of Wastewater by Hydrogen Peroxide or Peracetic Acid: Development of Procedures for Measurement of Residual Disinfectant and Application to a Physicochemically Treated Municipal Effluent," *Water Environ. Res.*, **74**, 33, 33–50.

Ward, R. W., and G. M. DeGraeve (1976) *Disinfection Efficiency and Residual Toxicity of Several Wastewater Disinfectants*, EPA-600/2-76-156, U.S. Environmental Protection Agency, Cincinnati, OH.

Watson, H. E. (1908) "A Note on the Variation of the Rate of Disinfection with Change in the Concentration of the Disinfectant," *J. Hygiene* (British), **8**, 536.

Wattie, E., and C. T. Butterfield (1944) "Relative Resistance of *Escherichia Coli* and Eber-thella Typhosa to Chlorine and Chloramines," *U.S. Public Health Service Report*, **59**, 52, 1661–1671.

WEF (1996) *Wastewater Disinfection, Manual of Practice FD-10*, Water Environment Federation, Alexandria, VA.

White, G. C. (1999) *Handbook of Chlorination and Alternative Disinfectants*, 4th ed., John Wiley & Sons, New York.

Wright, H. B., and G. Sakamoto (1999) "UV Dose Required to Achieve Incremental Log Inactivation of Bacteria, Virus, and Protozoa," Trojan Technologies, Inc., London, Ontario, Canada.

Zimmer, J. L., and R. M. Slawson (2002) "Potential Repair of *Escherichia Coli* DNA following Exposure to UV Radiation from Both Medium- and Low-Pressure UV Sources Used in Drinking Water Treatment," *Appl. Environ. Microbiol.*, **68**, 7, 3293–3299.

12 Satellite Treatment Systems for Water Reuse Applications

WORKING TERMINOLOGY   726

12-1   INTRODUCTION TO SATELLITE SYSTEMS   727
Types of Satellite Treatment Systems   728
Important Factors in Selecting the Use of Satellite Systems   730

12-2   PLANNING CONSIDERATIONS FOR SATELLITE SYSTEMS   730
Identification of Near-Term and Future Reclaimed Water Needs   730
Integration with Existing Facilities   731
Siting Considerations   731
Public Perception, Legal Aspects, and Institutional Issues   734
Economic Considerations   735
Environmental Considerations   735
Governing Regulations   735

12-3   SATELLITE SYSTEMS FOR NONAGRICULTURAL WATER REUSE APPLICATIONS   735
Reuse in Buildings   736
Landscape Irrigation   736
Lakes and Recreational Enhancement   736
Groundwater Recharge   736
Industrial Applications   737

12-4   COLLECTION SYSTEM REQUIREMENTS   738
Interception Type Satellite System   738
Extraction Type Satellite System   738
Upstream Type Satellite System   739

12-5   WASTEWATER CHARACTERISTICS   739
Interception Type Satellite System   740
Extraction Type Satellite System   740
Upstream Type Satellite System   741

12-6   INFRASTRUCTURE FACILITIES FOR SATELLITE TREATMENT SYSTEMS   741
Diversion and Junction Structures   741
Flow Equalization and Storage   744
Pumping, Transmission, and Distribution of Reclaimed Water   745

12-7   TREATMENT TECHNOLOGIES FOR SATELLITE SYSTEMS   745
Conventional Technologies   745
Membrane Bioreactors   746
Sequencing Batch Reactor   746

12-8	INTEGRATION WITH EXISTING FACILITIES 748	
12-9	CASE STUDY 1: SOLAIRE BUILDING, NEW YORK, NEW YORK 751	
	Setting 751	
	Water Management Issues 751	
	Implementation 752	
	Lessons Learned 753	
12-10	CASE STUDY 2: WATER RECLAMATION AND REUSE IN TOKYO, JAPAN 755	
	Setting 755	
	Water Management Issues 755	
	Implementation 756	
	Lessons Learned 758	
12-11	CASE STUDY 3: CITY OF UPLAND, CALIFORNIA 760	
	Setting 760	
	Water Management Issues 760	
	Implementation 760	
	Lessons Learned 761	
	DISCUSSION TOPICS AND PROBLEMS 761	
	REFERENCES 762	

WORKING TERMINOLOGY

Term	Definition
Blowdown	A portion of the circulating water in a cooling tower that is bled off and replaced with low-salt makeup water. Blowdown water contains concentrated minerals resulting from evaporation losses in the cooling tower and chemical conditioning agents added to the feedwater.
Cycles of concentration	The ratio of the concentration of salt in the blowdown to its concentration in the makeup water. Cycles of concentration are used in determining the amount of cooling tower blowdown or the amount of makeup water required.
Extraction type satellite system	Wastewater to be reclaimed is extracted (mined) from a collection system main, trunk, or interceptor sewer.
Flow equalization	The damping of flowrate variations to obtain a constant or nearly constant flowrate, usually by means of a storage (equalization) basin.
Interception type satellite system	Wastewater to be reclaimed is intercepted before it reaches the collection system.
Membrane bioreactor (MBR)	A process that combines a suspended growth biological reactor with a membrane separation system; membrane separation is accomplished by either microfiltration or ultrafiltration membranes.

Residuals	Waste streams produced by the water reclamation processes. Residual streams include waste sludge, waste washwater from the backwash process, concentrate, and chemical cleaning wastes.
Satellite treatment system	System used for the treatment of wastewater in water reclamation plants located close to the point of reuse. Satellite treatment plants generally do not have solids processing facilities; solids are returned to the collection system for processing in a central treatment plant located downstream. Three types of satellite systems are identified: (1) interception type, (2) extraction type, and (3) upstream type (see separate definitions).
Sequencing batch reactor (SBR)	A fill-and-draw type of reactor system involving a single complete-mix reactor in which all steps of the activated sludge process occur.
Sewer mining	The withdrawal of all or a portion of the wastewater from a wastewater collection system for localized treatment and reuse.
Upstream type satellite system	Wastewater reclamation facilities are used to reclaim water from developments located at the extremities of a centralized collection system.

In most collection and treatment systems, wastewater is transported through the collection system to a centralized treatment plant located at the downstream end of the collection system near the point of disposal. Oftentimes opportunities for instituting water reuse applications, especially for agricultural and landscape irrigation or groundwater recharge, are limited as the points of use are located remotely from the wastewater treatment facilities. The infrastructure costs for storing and transporting reclaimed water to the points of use are often prohibitive, thus making reuse uneconomic. An alternative to the conventional approach of transporting reclaimed water from a central treatment plant is the concept of satellite treatment at upstream locations with localized reuse. Residuals generated by satellite treatment process are discharged to the collection system for processing downstream at the central treatment plant.

Subjects presented in this chapter include: (1) an introduction to satellite treatment and reuse systems, (2) planning considerations for satellite systems, (3) satellite systems for nonagricultural water reuse applications, (4) assessing collection system requirements, (5) wastewater characteristics, (6) infrastructure facilities, (7) treatment technologies, and (8) integration of satellite systems with existing facilities. Three case studies of satellite systems are presented in Secs. 12-9 through 12-11. Decentralized systems, which are used for small community applications, are discussed in Chap. 13.

12-1 INTRODUCTION TO SATELLITE SYSTEMS

The concept of satellite treatment was introduced in the early 1960s in a major way when the Sanitation Districts of Los Angeles County (SDLAC) placed into operation the Whittier Narrows Water Reclamation Plant. The plant was located in an upstream reach of the collection system and remotely from SDLAC's central treatment plant. Wastewater was diverted from the collection system for treatment and for recharge of the local groundwater.

The Whittier Narrows facility employed conventional activated sludge treatment, and the residuals generated by the process were returned to the collection system. Subsequently, other satellite plants, seven in all, were constructed by SDLAC for similar purposes, but with higher levels of treatment for residual solids removal and disinfection.

In addition to the need to recycle reclaimed water, the focus of this textbook, satellite facilities have also been utilized (1) to reduce the flow to a centralized facility because of limitations of capacity in the collection system and treatment facilities, (2) as means of eliminating discharges to impacted receiving water bodies, and (3) as a means for reducing discharges to impacted water bodies. The first situation is occurring in a number of large cities that have continued to expand. The second situation was a major driving force in the development of the reclamation and reuse program in St Petersburg, FL, for eliminating discharges to Tampa Bay.

In this section, the types of satellite systems that are now used and important factors that must be considered in planning satellite systems are introduced. These subjects are explored in greater detail in subsequent sections.

Types of Satellite Treatment Systems

Satellite treatment systems fall generally into three categories: (1) interception type, (2) extraction type, and (3) upstream type. Each of these types of satellite systems is illustrated on Fig. 12-1 and described further below. The distinction between satellite types is made because the characteristics of the wastewater to be treated, the treatment technologies that will be used, and the infrastructure needed to implement them are somewhat different, and, in some cases, quite different.

Interception Type
In the interception type, as illustrated on Fig. 12-1a, the wastewater to be reclaimed is intercepted before it reaches the collection system. Typical applications for this type of satellite system are for water reuse in high-rise commercial and residential buildings. The quantity of flow to be intercepted and reclaimed will depend on the local and seasonal water reuse requirements. Typically, all of the flow from an individual building will be intercepted for reuse. In some cases, it may be necessary to supplement the intercepted flow with potable water. Should excess flow occur, it would be discharged to the collection system.

Extraction Type
In the extraction type, as illustrated on Fig. 12-1b, the wastewater to be reclaimed is extracted (mined) from a collection system main, trunk, or interceptor sewer. Typical applications for this type of satellite system are for reuse in landscape, park, and greenbelt irrigation; for reuse in nearby high-rise commercial and residential buildings; and for commercial and industrial cooling tower applications. The quantity of flow to be extracted and reclaimed will depend on the local and seasonal water reuse demands, especially so for landscape irrigation applications.

Upstream Type
In upstream type, as illustrated on Fig. 12-1c, the wastewater reclamation facilities are used to reclaim water from developments located at the extremities of a centralized collection

12-1 Introduction to Satellite Systems   729

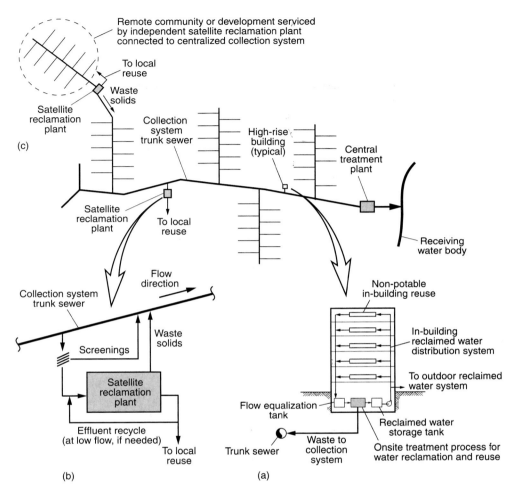

Figure 12-1
Schematic illustration of three types of satellite water reclamation systems: (a) interception type where wastewater to be reclaimed and recycled is intercepted before discharge to a centralized collection system, (b) extraction type (i.e., sewer mining) in which wastewater is extracted (i.e., pumped) from a centralized collection system for local reuse, and (c) upstream type for treatment and reuse for a remote community or development with solids discharged to a centralized collection system. Note: Remote upstream treatment facilities without discharge to a collection system are covered in Chap. 13, which deals with decentralized systems.

system and where opportunities for water reuse (e.g., golf course and roadway median strip irrigation) are available and the capacity of the collection system is limited. Typical applications for this type of satellite system are for new housing developments and remote commercial centers and research parks. Reclamation systems for new housing developments and commercial centers that are not connected to a wastewater collection system are considered in Chap. 13 which deals with decentralized wastewater management systems. The quantity of flow to be intercepted and reclaimed in upstream satellite systems will depend on the local and seasonal reclaimed water demands. In general, all of the flow from a housing development will be intercepted for reuse. In some cases, however, it may be necessary to divert some of the flow directly to the centralized collection system, before or after treatment.

Important Factors in Selecting the Use of Satellite Systems

To implement satellite treatment in connection with a centralized system, a number of factors will have to be considered in planning and siting the facilities. Some of the key factors that must be considered include:

- What is the need for reclaimed water? How much is needed, for what purpose, at what quality, when, and where?
- What facilities will be needed and how will these interface with the existing collection and treatment system?
- What type of delivery system will be needed to convey the reclaimed water to the site(s) of intended use?
- What is the availability of suitable sites for the treatment facilities and supporting infrastructure?
- What will be the system cost and how will it be paid for? What is the potential for revenue generation?
- What environmental factors have to be considered?
- What are the applicable codes and regulations that apply to the implementation of satellite systems?
- What jurisdictional issues, such as the implementing authority, will have to be considered?

These factors are considered further in the subsequent sections. Dual plumbing is considered separately in Chap. 15.

12-2 PLANNING CONSIDERATIONS FOR SATELLITE SYSTEMS

The principal planning considerations for satellite systems are related to (1) identification of localized near-term and future reclaimed water needs; (2) integration with existing facilities; (3) siting considerations; (4) public perception, legal aspects, and institutional issues; (5) economic considerations; (6) environmental considerations; and (7) governing regulations.

Identification of Near-Term and Future Reclaimed Water Needs

In developing or developed areas, there are many opportunities for the use of reclaimed water. Typical uses include landscape irrigation for parks, golf courses, cemeteries, and roadway medians; cooling water for industrial use; and recreational and environmental enhancement such as the creation of artificial lakes and wetlands. Examples of localized reuse where reclaimed water is used for golf course irrigation and artificial lakes are shown on Fig. 12-2. In identifying potential near-term and future reuse applications,

(a) (b)

Figure 12-2
Examples of localized reuse: (a) golf course irrigation and (b) artificial lake created with reclaimed water.

factors that need to be considered include the types of reuse applications; the amount of water required; flowrate requirements, i.e., diurnal, daily, or seasonal variations in demand; water quality requirements; and proximity to potential satellite treatment sites.

Water quality requirements, the amount of reclaimed water required, and whether use is continuous or intermittent must be defined where different reuse applications are to be served from a single facility. The demand variations are particularly important when matching use with the production of reclaimed water in determining storage and the delivery system requirements. Issues related to the storage and distribution of reclaimed water are discussed in Chap. 14. Flow projections must be developed to select sources of wastewater that can accommodate the near-term and future uses of reclaimed water. The potential for service area growth and how it affects future needs must also be evaluated.

Integration with Existing Facilities

The elements of a satellite system may include: (1) the portion of the collection system tributary to the satellite treatment plant, (2) diversion structure (if needed), (3) treatment facilities, (4) reclaimed water delivery system to the point(s) of reuse, (5) return sewer to the collection system for discharge of residuals removed in satellite treatment, and (6) site-specific ancillary facilities. Each of these elements is discussed in later sections of this chapter. Important elements in satellite system planning include determining how those facilities, (i.e., the diversion and return of wastewater) can be integrated into the existing collection system and what effect the residuals return will have on the wastewater characteristics and the operation of the central treatment plant. These topics are discussed further in Sec. 12-8.

Siting Considerations

The selection of a suitable site is critical in implementing any wastewater treatment facility including satellite water reclamation facilities. Siting issues deal largely with finding acceptable locations of sufficient size for the various facilities required. Many of the factors that affect site selection are listed in Table 12-1 for the different types of

Table 12-1
Important factors that must be considered in site selection for a satellite treatment plant

Factor	Specific consideration
Site location	Distance from trunk sewer and point(s) of reuse Elevation relative to trunk sewer
Compatibility with land use	Current land use Proposed future land use Zoning and adjacent land use Proximity to current or planned developed areas Expansion potential
Topography	Ground slope Flood potential
Environmental constraints	Proximity to residential areas Wind direction Presence of rare or endangered species Traffic impacts
Potential changes in water quality	May affect planned or future uses of reclaimed water

satellite systems. Factors associated with the selection of sites for satellite facilities in areas with varying degrees of development are considered in the following discussion. The location of collection systems is also considered.

Sites in Densely Populated Areas

Densely populated areas are defined as those with multiple dwelling units such as multistory apartment buildings or condominiums and supporting commercial development. An illustration of potential sites and applications for satellite facilities in densely populated areas is shown on Fig. 12-3. Selection of sites in densely populated areas can be daunting because of the limited availability of suitable sites at reasonable cost. Thus, site selection may be confined to public lands such as parks and parking lots where a portion of the property can be used for the construction of belowground structures.

The size and configuration of storage facilities, however, are limited in most cases because of insufficient space, costly construction, and the potential aesthetic unacceptability of aboveground structures. The installation of pipelines in city streets where other utilities have been installed may also be a costly undertaking. If the points of use are located in nearby areas such as for landscape irrigation, toilet flushing in commercial buildings, and ornamental ponds, the pipe sizes may be relatively small and the piping would be easier to install. If the reclaimed water is to be used for fire protection, the supporting infrastructure (storage, pumping, and pipelines) will be significantly greater in size to meet fire flow requirements.

Sites in Urban Areas

Urban areas in the context of this discussion may include low-rise residential structures, public buildings such as governmental centers, shopping malls, and industrial parks with

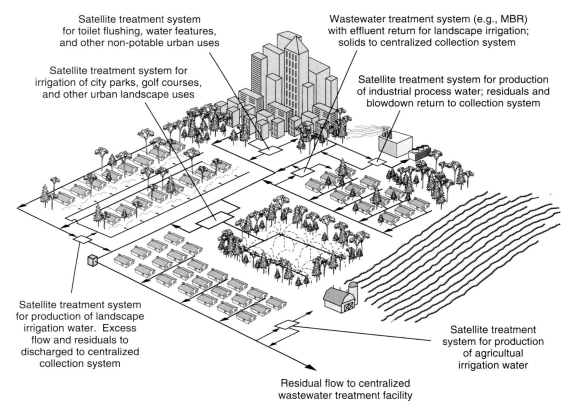

Figure 12-3
Illustration of potential sites and applications for satellite reclamation facilities in densely populated areas.

some light industry. Probable reuse applications might include landscape irrigation, industrial process or cooling water, or ornamental fountains or ponds. Sites in urban areas may be easier to find than in highly developed or densely populated areas but environmental impacts and aesthetic considerations may be overriding issues in site selection. Use of aboveground structures for purposes such as surface storage or enclosing treatment facilities might be possible depending on the adjacent land use, e.g., industrial or commercial development. Architectural designs compatible with surrounding structures should also be considered.

Sites in Suburban and Rural Areas

Typically, suburban or rural areas are largely residential with some supporting commercial facilities such as shopping centers. Greater opportunities for water reuse may be available and consist of agricultural and landscape irrigation, including parks and golf courses, recreational lakes, wetlands enhancement, and groundwater recharge. Because these areas are of low density with more open space, greater opportunities for site acquisition might be available. Environmental and aesthetic considerations for siting the facilities, with particular emphasis on appearance and noise and odor abatement, may not have to be addressed as extensively as they would in higher density areas.

Collection System Locations

Along with the siting issues discussed above, the location of major trunk sewers is also important. Where interception and remote-type satellite systems are used, sufficient collection system capacity is necessary for discharge of excess wastewater (beyond what is needed for reuse) and solids resulting from treatment. In the case of extraction systems, locations of potential connections to the collection system are of critical importance to: (1) obtain wastewater of sufficient quantity and quality and (2) return solids generated in the treatment process. The wastewater flowrate variations in the collection system also need to be determined to (1) ensure that adequate wastewater flow is available at all times for treatment to meet the reuse applications, and (2) size the treatment, storage, and ancillary facilities required (see Secs. 12-6 and 12-7).

The trunk sewer location also affects other facilities as the satellite treatment plant should be located ideally in close proximity to the point of diversion of flow from the collection system, otherwise pumping facilities and lengthy pipelines might be required at added cost to transfer the diverted wastewater to the treatment facilities.

Public Perception, Legal Aspects, and Institutional Issues

For each type of water use, there are significant perception, legal, and institutional issues related to using reclaimed water. These issues are magnified when uses imply the possibility of human contact and the perception of a threat to public health.

Perception Issues

The first obstacle to the use of reclaimed water is that of public acceptance. Because reclaimed water originates from wastewater, a psychological objection to close contact with reclaimed water may occur where a health threat is perceived. Although landscape irrigation and industrial use of reclaimed water is accepted generally, the possibility of public contact is not. Thus, the safe use of reclaimed water for an application such as toilet flushing may require demonstrated assurances and education to ensure acceptance. Additionally, depending on the type of sprinkler system used in landscape irrigation, aerosols can be created that, under windy conditions, can be carried long distances. Safeguards need to be provided in monitoring water quality and in the design of the irrigation system to minimize the impacts of aerosols.

Legal Issues

Legal obstacles focus primarily on liability. In a society that is quick to seek legal remedy for real and/or perceived damages, agencies considering implementation of a water reuse program need to consider potential legal ramifications. Explicit disclosure of the use of reclaimed water and potential health effects has helped many agencies head off any lawsuits based on fears rather than facts (Asano, 1998). Thus, a water reuse program must not only be based on sound engineering for the development of a safe and reliable system, it must be forthcoming in supplying factual educational information to the public, describing the benefits and potential adverse effects of the proposed water reuse applications.

Institutional Issues

Institutional issues may involve possible disagreements between agencies that deliver potable versus nonpotable water and conflicts in building and plumbing codes, public health regulations, financing and rate-setting authority, and utility and user responsibilities. Each of these issues has to be explored carefully.

Economic Considerations

Of major significance in the planning and implementation of a satellite treatment system are the economic considerations. Although the need for supplemental water may be the driving force in initiating a water reuse plan, the affordability issue of a water reclamation system may be a crucial factor in implementation. Monetary benefits that accrue through the use of reclaimed water should be considered in the economic analysis. Examples of monetary benefits include: (1) savings in the cost of producing, storing, and distributing potable water, (2) savings in fertilizer cost where reclaimed water is used for landscape and other irrigation, and (3) the savings over the cost of developing an alternative potable water supply to cover any supply shortfall. One of the advantages of a satellite system is that a less extensive infrastructure is needed to deliver reclaimed water to the point(s) of use, thus improving the affordability of the project. An additional economic factor is the potential for revenue generation through the sale of reclaimed water.

Environmental Considerations

The environmental impacts of the proposed satellite facility are as important, if not more so, as economic considerations. If environmental concerns are not addressed adequately, a "fatal flaw" might occur, thus impeding or stopping implementation of the project. Because of their location, special attention must be devoted to the environmental issues associated with the use of satellite systems.

Governing Regulations

Systems for the use of reclaimed water for commercial and residential buildings are covered generally by various regulations including plumbing codes (see Sec. 15-3 in Chap. 15). As discussed in Chap. 15, the regulations for dual plumbing in commercial and residential buildings are designed for the protection of public health and for the safe installation of the piping system and ancillary devices. For other applications, existing reuse requirements, as discussed in Chap. 4, will apply in most situations. In some cases, depending on the application, it may be necessary to implement more stringent requirements.

12-3 SATELLITE SYSTEMS FOR NONAGRICULTURAL WATER REUSE APPLICATIONS

The initial task in planning a water reuse system is to determine the amount of water needed for specific applications. In this section, some of the factors that need to be considered in projecting water use in nonagricultural applications are addressed. Topics discussed are: (1) reuse in buildings, (2) landscape irrigation, (3) lakes and recreational enhancement, (4) groundwater recharge, and (5) industrial applications. Although satellite treatment can be used for agricultural applications, the economics of producing agricultural water on a large scale by satellite treatment should be investigated carefully. Agricultural uses of reclaimed water are discussed in Chap. 17.

Reuse in Buildings

Commercial and public office buildings offer opportunities for nonpotable uses such as air conditioning and toilet and urinal flushing. These uses constitute up to 80 to 90 percent of the total water use for large commercial centers. High-rise office buildings have the greatest potential for flushing water use because toilet facilities are centrally located in almost always the same location on each floor. These configurations allow for a common riser and short piping runs. Water demand for high-rise office buildings is a function of the number of employees, number of toilet fixtures, and type of fixture (AWWA, 1994). For these uses, installation of a dual plumbing system is required; one system for potable water and one system for reclaimed water. Many cities have plumbing codes for the installation of dual plumbing systems. Installation of a dual system should be done during building construction because the cost of retrofitting an existing plumbing system to install dual plumbing is prohibitive. Dual plumbing systems are considered in Chap. 15.

Landscape Irrigation

By far, the greatest potential for urban use of reclaimed water is for landscape irrigation. Potential places of use include parks, playgrounds, cemeteries, street and freeway medians, golf courses, plant nurseries, and building grounds. Irrigation demands vary and depend on rainfall, runoff, evapotranspiration, soils, geohydrology, vegetation, and local practices. The time and duration of irrigation varies widely and depends on local conditions. Typically, parks, playgrounds, cemeteries, and street medians are irrigated at night or in the early morning hours. Golf courses are often irrigated at dusk. Because the daily water requirement is applied in only a few hours, the rate of demand may be many times the average daily requirement. Landscape irrigation is discussed in Chap. 18.

Lakes and Recreational Enhancement

The amount of water needed for lakes and recreational enhancement varies seasonally and depends on rainfall-runoff conditions, temperature, and amount of sunshine. The quantity of water for open water bodies such as lakes and ponds is needed to offset evaporation. The amount needed can be estimated from pan evaporation data available from the weather service, although these data do not consider all potential water losses. Shallow lakes and ponds are susceptible to algal growth, especially from reclaimed water containing nutrients and runoff containing fertilizer. Problems can be minimized by maintaining good circulation in the ponds, maximizing side water depth and slope, adding artificial aeration, and minimizing detention times; 7 to 10 d are suggested (AWWA, 1994). Water quality issues with open storage reservoirs are reviewed in Chap. 14; environmental and recreational uses of reclaimed water are discussed in Chap. 20.

Groundwater Recharge

Groundwater recharge of reclaimed water has been used (1) in coastal areas of California and Florida to prevent saltwater intrusion into freshwater aquifers (see Chap. 2), (2) to augment potable and nonpotable aquifers, (3) to provide additional treatment for future reuse, and (4) to provide storage and subsurface transmission of reclaimed water. Groundwater recharge may be done by surface spreading or injection wells (see Fig. 12-4). Groundwater recharge is site-specific and the feasibility of recharge depends on a number of factors including water quality characteristics, geologic and hydrogeologic conditions at the proposed sites, and uses of the underlying groundwater. Considerations for groundwater recharge are presented in Chap. 22.

(a) (b)

Figure 12-4
Facilities for groundwater recharge with reclaimed water (a) surface spreading basin (Coordinates: 28.491 N, 81.628 W, view at altitude 4 km) and (b) injection well.

Industrial Applications

Potential industrial applications include cooling water makeup, process water or boiler feed. The amount and water quality requirements vary depending on the type of industry. The use of reclaimed water for industrial applications is discussed in greater detail in Chap. 19.

Cooling Water Makeup
Cooling water makeup is one of the largest water uses at a typical industrial facility. Cooling water use is site-specific and depends on a number of factors including climate, water chemistry, and the type of industrial facility. Inorganic constituents in the reclaimed water affect the operation of cooling water systems as they can decrease the number of cycles of concentration (see Working Terminology) and increase corrosion and chemical costs. Typically, when the cycles of concentration are on the order of 3 to 7, some of the dissolved solids in the circulating water, such as calcium, phosphorus, and silica, can exceed their solubility limits and precipitate, causing scale formation in pipes and coolers (Tchobanoglous et al., 2003). Ammonia also reduces the effectiveness of copper and brass heat exchanger surfaces. Dissolved organics and nutrients in reclaimed water increase the demand for biocides to control algal growth.

Process Water
The use of reclaimed water for process water is also site-specific. For example, reclaimed water is used commonly for cleaning process equipment and facilities. Reclaimed water is also used in fabric dyeing operations and in the manufacture of paper.

Boiler Feedwater
Using reclaimed water for boiler feedwater is also site-specific and a much higher quality water is required than that used for cooling water. Typically for use as boiler feedwater, reclaimed water has to undergo demineralization because minerals will deposit on the hot metal surfaces and eventually affect boiler performance. Higher mineral

content in the feedwater also results in more frequent blowdown, thus causing heat loss that negatively affects boiler efficiency.

12-4 COLLECTION SYSTEM REQUIREMENTS

Because satellite systems are connected to collection systems for the treatment of residuals, it is important to consider the collection system requirements for each type of satellite system. The characteristics of the wastewater are considered in the following section.

Interception Type Satellite System

In general, the flow of residuals from an interception type satellite system will be much smaller than the flow in the main waste water collection system. Nevertheless, the principal collection system requirement for interception type satellite systems is sufficient capacity to handle the total flow in the event the satellite system is offline.

Extraction Type Satellite System

A major issue in developing an extraction type satellite facility is the selection of a wastewater source of supply. Important factors in source selection are the availability of sufficient flow in the tributary collection system and its variability.

Availability of Flow in the Collection System

Sufficient flow must be available in the collection system to meet the projected needs of the reclaimed water system. The amount of wastewater available may be determined by direct and indirect means, but the most reliable method is by flow monitoring. Flow monitoring devices can be installed in access ports at the crown of the sewer for ultrasonic depth measurement or at the sewer invert to measure movement of particle velocity by Doppler measurement. Flow measurement should be done over a period of time, preferably for at least a month, to measure diurnal and daily flow variation, as discussed below. Sampling for wastewater characteristics should also be coordinated with flow monitoring.

Where flow measurements cannot be made, various methods of estimating flowrates can be used such as: (1) water consumption, (2) flowrate characteristics from similar municipalities, and (3) estimates of flowrates from typical residential, commercial, and other sources. Information on estimating flowrates from various sources can be found in AWWARF (1999), and Tchobanoglous et al. (2003). Flowrate estimating, however, lacks the accuracy of data obtained from actual flow monitoring.

Variability of Flowrate

Hourly, daily, and seasonal flowrate variations in the collection system are affected by many factors, including wet weather conditions when inflow or infiltration into the system can occur. An example of how flowrates can vary during the day for a small community of 61 homes is shown on Fig. 12-5. The occurrence and duration of minimum flowrates, especially during dry periods, are of particular importance as they will affect operation of the treatment facilities, the need for storage, and possible contractual arrangements for supplying reclaimed water. Wastewater generation rates in resort areas and municipalities with large seasonal industries can also vary greatly from month to month. Seasonal demand variations for certain types of reuse, such as irrigation, are generally large and these variations require substantial seasonal storage volumes or

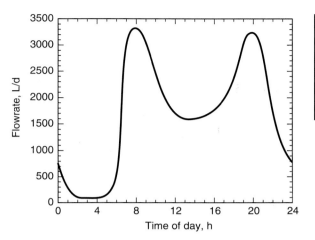

Figure 12-5
Typical daily flowrate variation in the collection system serving a small residential community.

large increases in plant production capacities. Seasonal storage in some cases may be accomplished in a groundwater basin.

Upstream Type Satellite System

For upstream type satellite systems, the collection system requirements vary, depending on the characteristics of the collection system tributary to the satellite plant, and the purpose and use of the satellite system. It should be noted that an upstream satellite system may be served by a variety of wastewater collection systems, including

1. Conventional gravity collection systems
2. Effluent septic tank effluent gravity (STEG) sewers
3. Effluent septic tank effluent pump (STEP) sewers
4. Pressure sewers with grinder pumps
5. Vacuum collection systems

Each of these types of collection systems is considered further in Chap. 13. Also, the differences in the wastewater characteristics to be expected where the above types of collection systems are used are discussed in the following section.

If the purpose of the satellite system is for seasonal or intermittent use, there should be sufficient capacity in the downstream collection system leading to the centralized treatment facility to handle the total flow when the plant is offline. If one of the purposes of the satellite system is to reduce the flow and load contributory to the centralized system, then the capacity of the downstream collection system may only have to be sufficient to handle residuals from the satellite plant and any excess wet weather flow from the satellite area served.

12-5 WASTEWATER CHARACTERISTICS

The wastewater characteristics for each type of satellite system are considered separately in the following discussion, and the pretreatment requirements will change with the characteristics of the wastewater.

Interception Type Satellite System

The sources of wastewater for interception type satellite systems are typically from toilet and urinal flushing, washing and bathing facilities, and food preparation facilities. Because the wastewater that reaches the reclamation facility has not undergone much transformation, different types of screening will be required, especially if membrane bioreactors are used. For example, toilet paper which has not disintegrated completely must be removed by fine screens (see Fig. 12-6). Again, because little transformation has occurred, a relatively large fraction of the particulate organic matter will be removed by fine screening.

Other wastewater characteristics that are different will be the nutrient, especially ammonia, and TDS concentrations which have not been diluted by extraneous inflows. Similarly, intercepted wastewater will often have higher concentrations of residual drugs and medicines. Because the concentration of wastewater constituents will vary, a sampling program should be conducted over a period of at least 2 wk at a similar type of facility to assess the wastewater characteristics.

Extraction Type Satellite System

A wastewater source that is potentially available to an extraction type satellite reclamation system may have high organic, heavy metal, or dissolved solids concentrations depending on the nature and type of contributors to the collection system. The nature of the wastewater needs to be determined not only to characterize the constituents but also to identify substances, such as salts from the regeneration of water softeners, which may interfere with potential reuse applications. The organic strength of the wastewater is important as it affects bioreactor design. A sampling program conducted over a period of at least 2 wk should produce sufficient representative data on the wastewater characteristics.

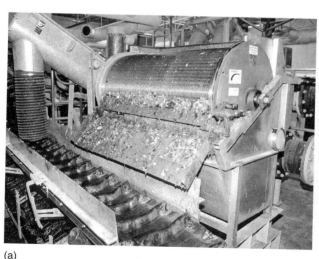

(a) (b)

Figure 12-6

Fine screens used to remove particulate matter from raw wastewater (a) wedge-wire screen with 3 mm (0.125 in.) openings and (b) disk screen with 250 μm openings used to remove undisintegrated toilet paper and hair from screened wastewater. The removal of toilet paper and hair is especially important where membrane bioreactor systems are used.

Upstream Type Satellite System

The relative characteristics of wastewater from upstream sources will vary depending on the type of collection system used to serve the development. In general, the presence of constituents associated with heavy commercial and industrial activities will be absent.

1. Where conventional gravity collection systems are used, the wastewater characteristics will be similar to those of conventional wastewater with the exception that the constituent concentrations will be somewhat higher due to a lack of dilution, especially in the beginning when infiltration will be limited due to improved construction materials and practices.
2. Where septic tank effluent gravity (STEG) sewers are used, the strength of the wastewater, as compared to conventional wastewater, will be reduced for settleable constituents that are present in both dissolved and particulate forms (e.g., BOD and TSS). Other dissolved constituents that do not settle or are adsorbed on solids will be higher due to lack of dilution.
3. Where septic tank effluent pump (STEP) sewers are used, the wastewater characteristics are similar to wastewater from the STEG system.
4. Where pressure sewers with grinder pumps are used, the strength of the wastewater will be increased due to increased solubilization of the wastewater solids.
5. Where vacuum collection systems are used, the strength of the wastewater will be higher due to lack of dilution as compared to conventional wastewater.

12-6 INFRASTRUCTURE FACILITIES FOR SATELLITE TREATMENT SYSTEMS

Of equal importance to the treatment process used for the satellite system is the infrastructure required to deliver the wastewater to the treatment facilities and the reclaimed water to the point(s) of reuse. The principal infrastructure elements are the diversion and junction structures, flow equalization and storage, and pumping and transmission system. Special screens are often required for the interception of solids that affect the treatment process. The infrastructure elements required for each of the three types of satellite systems are summarized in Table 12-2 and discussed in the following paragraphs.

Diversion and Junction Structures

The design and construction of diversion and junction structures vary in most instances due to the physical features or constraints of the affected reach of the collection system. For installation in an existing collection system, important considerations include: (1) the method of flow diversion to the satellite facility, (2) the method of reintroducing the residuals return flow to the collection system, and (3) implementation of construction with a minimum of interruption of existing wastewater flows.

Flow Diversion
Flow can be diverted from the collection system by gravity or by pumping. For gravity diversion, a dam or a weir can be constructed in an existing access port (manhole) with a pipe leading to the satellite plant (see Fig. 12-7a) or a special diversion structure can be designed similar to that shown on Fig. 12-7b. The diversion structures can be similar in design to those used in combined sewer systems for handling wet and dry weather

Table 12-2

Infrastructure requirements for various types of satellite systems

Type of satellite system	Infrastructure facilities required[a]
Interception	• Screen for untreated wastewater • Bypass line to collection system for untreated wastewater (for times when treatment unit is off-line) • Residuals return line to collection system • Reclaimed water storage tank for flow equalization • Reclaimed water pumping and distribution system
Extraction	• Flow diversion structure • Screen for untreated wastewater • Untreated wastewater pumping station (depending on system hydraulic conditions) • Residuals return line to collection system • Reclaimed water storage tank (where required to meet peak demand) • Reclaimed water pumping, transmission, and distribution system
Upstream	• Junction structure (if needed) • In-line headworks for screening and grit removal • Residuals and excess flow return line to collection system • Reclaimed water storage tank • Reclaimed water pumping, transmission, and distribution system

[a] In addition to treatment facilities.

flows. A wetwell can also be constructed near the point of diversion to allow for the installation of submersible pumps for pumping to the satellite plant. A typical wet pit type pumping station is shown on Fig. 12-8. Where possible, the flow diversion structure should be designed to minimize the quantity of screenings intercepted.

Residuals Return

Residuals from the satellite treatment operation may consist of screenings; primary sludge and scum; waste washwater from surface, media, or membrane filters; and waste biological solids. Screenings intercepted in the influent wastewater flow should be: (1) removed from the flow and disposed of separately, especially if there are large quantities of screenings or (2) removed from the flow and ground (to ensure easy of handling) and returned to the residuals discharge line. In most cases, and especially where membrane treatment is used, in-stream screenings grinders should not be used in the feed stream to the treatment unit as the ground solids will clog the membranes and reduce performance. As most satellite treatment plants do not contain solids processing facilities, waste solids from the biological treatment process will typically constitute the largest residual stream. Residuals returned to the collection system contain mostly liquid and fine solids and typically do not interfere with the normal wastewater flow. A pipe connection at a downstream access port should suffice in most instances for returning residual streams.

12-6 Infrastructure Facilities for Satellite Treatment Systems | 743

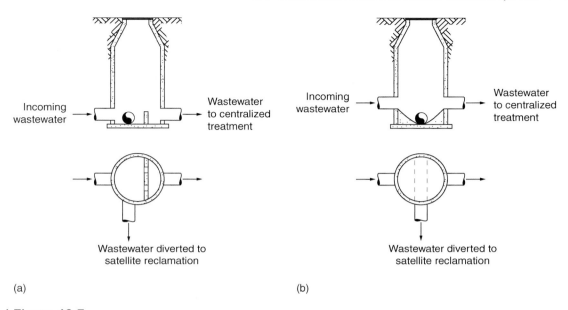

Figure 12-7
Typical devices used for diversion of wastewater: (a) a diversion dam constructed in an access port and (b) special diversion structure.

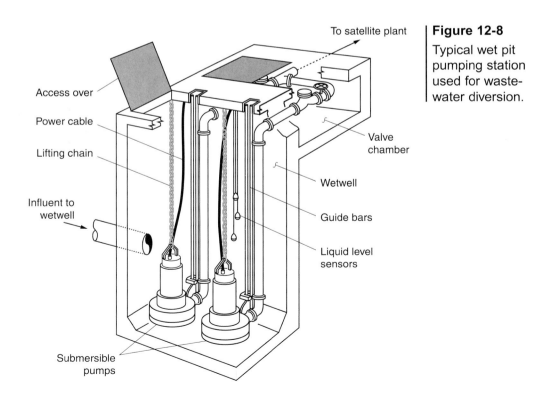

Figure 12-8
Typical wet pit pumping station used for wastewater diversion.

Flow Equalization and Storage

Construction Considerations

One of the principal construction concerns for diversion structures is the handling of wastewater flows if the construction entails interrupting the wastewater flows to construct new or modify existing facilities, i.e., pipelines, junction boxes, and access ports. Special provisions will need to be made to intercept wastewater flows and route them around the work site(s) temporarily. Such provisions are not unusual for construction contractors engaged in modifications and additions to existing collection systems.

Equalization of incoming flows to the satellite treatment facilities may be required, especially if there are wide flowrate variations. Some attenuation of flowrates can be achieved by increasing the wetwell volume in the influent pumping station serving an extraction type satellite plant or by using a sequencing batch reactor treatment system described in Sec. 12-7. In most systems, however, some form of storage is required for product water, especially when product water usage is intermittent (such as for landscape irrigation). Storage can consist of aboveground or underground tanks, open reservoirs or ponds, or aquifer storage. An alternative to storage is to operate the treatment facilities at a constant rate equal to or in excess of the product demand and return excess reclaimed water to the collection system. In this latter case, the return line to the collection system must have sufficient capacity to handle the maximum flow condition.

Flow Equalization

The principal reasons to consider flow equalization are to (1) dampen peak dry weather flows and loads that occur in interception type systems, or (2) regulate wet weather flows from sanitary or combined collection systems in upstream satellite systems experiencing inflow and infiltration. In most cases, equalization of untreated wastewater should be kept to a minimum to limit the space required for equalization, and, most importantly, to reduce the potential for odor generation and need for operator attention. In ideal circumstances, the wastewater flow in the collection system is sufficient to meet the needs for plant operation and water reuse, and the treatment plant has sufficient capacity to meet variations in organic strength. Peak flow attenuation is not required in extraction (i.e., sewer mining) systems applications as only a sufficient amount of wastewater is withdrawn from the collection system as needed. The balance of the peak flow in the collection system continues to flow to the central system. During minimum flow periods, the flow in the collection system may be insufficient to sustain continuous operation of the treatment plant, thus requiring installation of an effluent recirculation system. The amount of storage required for a recirculation system in this case would be minimal.

Reclaimed Water Storage

Two types of reclaimed water storage may be required: operational storage and longer-term seasonal storage. The principal need for operational storage is to balance the daily rate of reclaimed water production and the rate of reclaimed water demand. Operational storage after treatment can be provided by a belowground clearwell, on-site or off-site covered aboveground storage reservoirs, or a groundwater recharge basin. In some cases, such as for landscape irrigation in a hot, summer dry spell or for a nonpotable fire flow system, large volumes of daily or seasonal storage may be necessary to maximize use of the reclaimed water. Some form of disinfection may have to be considered to prevent the growth of microorganisms in the storage and distribution system. Types of reservoirs and reservoir operations are discussed in Chap. 14.

The design of a nonpotable system for the pumping, transmission, and distribution of reclaimed water is similar to the design of a potable water system. The reliability of a nonpotable water system, however, may not be required to be as great as that for a potable water system. Most nonpotable reuse applications such as landscape irrigation and groundwater recharge can withstand short-term outages without major consequences. Systems that supply water for industrial applications, however, may require redundant systems to ensure reliable delivery. Pumping and distribution systems are also discussed in Chap. 14.

Pumping, Transmission, and Distribution of Reclaimed Water

12-7 TREATMENT TECHNOLOGIES FOR SATELLITE SYSTEMS

Several different types of treatment technologies can be used for satellite systems depending upon (1) the type of satellite system, (2) the amount of reclaimed water to be produced, (3) the quality of the reclaimed water required, (4) site and environmental constraints, and (5) compatibility with the existing collection and treatment system. The general categories of technologies that can be used are conventional secondary treatment, compact treatment plants, including membrane bioreactors, and sequencing batch reactors. Conventional biological treatment technologies that are discussed in Chap. 7 can be used in large upstream satellite facilities. It is anticipated that compact technologies such as membrane bioreactors and sequencing batch reactors will be the treatment system of choice for most interception and extraction type satellite treatment applications.

Conventional technologies used customarily in satellite treatment applications are activated sludge, attached growth, or combination processes either with or without nutrient removal. The removal of nitrogen or nitrogen and phosphorus may be required for groundwater recharge, discharge to recreational lakes, or other reuse applications. In any case, nutrient removal can be either an integral part of the biological treatment process or an add-on process. A typical process flow diagram for a conventional activated sludge treatment process is shown on Fig. 12-9.

Conventional Technologies

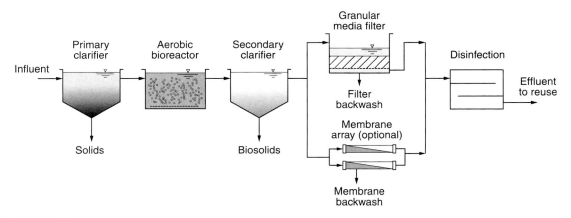

Figure 12-9
Schematic flow diagram for a conventional activated sludge treatment with alternative filtration processes.

Membrane Bioreactors

Membrane bioreactors (MBRs) are particularly well suited for application in satellite treatment as they are compact and can fit into small sites. As described in Sec. 7-5 in Chap. 7, MBRs combine biological treatment with a membrane system to provide enhanced organics and suspended solids removal. The membranes function to replace sedimentation and depth filtration for separating the biomass in suspended growth systems from the treated water. By coupling a biological reactor with a membrane system, overall space requirements and treatment costs can be reduced. The advantages of membranes over conventional systems using clarification and media filtration are that: (1) the facilities are much smaller, (2) higher quality product water can be obtained, (3) return sludge systems can be eliminated or greatly reduced, and (4) the system is simpler to operate. Pretreatment is required generally, usually by a fine screen or a cloth-media filter, to prevent clogging of the membranes and improve performance. A typical process flow diagram of a MBR used in a satellite treatment application is shown on Fig. 12-10.

In addition to their compact size and high levels of performance, MBRs are particularly well suited for satellite systems because (1) the process is stable and less susceptible to upset; (2) as a result of the low turbidity and effective reduction of particle size, high levels of disinfection can be accomplished that make the product water suitable for many local reuse applications as discussed in Sec. 12-2; (3) the potential for odor generation—of especial concern in urbanized and residential areas—is minimal and can be mitigated by facilities enclosure and odor management; and (4) the solids retention time (SRT) normally ranges from 12 to 20 d; thus, the biological solids are well stabilized and do not undergo rapid deterioration and generate odorous and corrosive gases when returned to the collection system.

Sequencing Batch Reactor

The sequencing batch reactor (SBR), described in Sec. 7-3 in Chap. 7, is a variation of the activated sludge process that minimizes space requirements by performing multiple

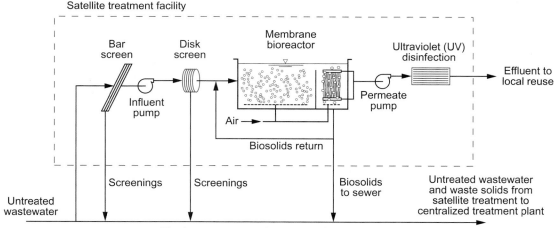

Figure 12-10
Typical process flow diagram for membrane bioreactor used in a satellite treatment application.

treatment steps in a single tank (see Fig. 12-11). The SBR process uses a fill-and-draw reactor with complete mixing during the batch reaction step (after filling); the subsequent steps of aeration and clarification occur in the same tank. As shown on Fig. 12-11a for BOD removal and nitrification, the SBR operation encompasses four sequential steps as follows: (1) fill, (2) react (aeration), (3) settle (sedimentation/clarification), and (4) decant (removal of clarified water). The SBR process is versatile as it can be used, in addition to BOD removal and nitrification, for nitrate removal and nitrogen and phosphorus removal, mainly by altering the length or operation of the aeration cycle, thus enabling anaerobic, anoxic, or aerobic conditions to occur in accordance with the treatment objectives.

For nitrogen removal, preanoxic denitrification occurs after filling as shown on Fig. 12-11b. Mixing is used during the fill period to contact the mixed liquor with the influent wastewater and to remove the nitrate remaining from the settle and decant steps. Separate mixing provides operating flexibility and is useful for anoxic operation during the aeration period, as well as anaerobic or anoxic contacting during the fill period.

For phosphorus removal, if sufficient nitrate is removed during the SBR operation, an anaerobic reaction period can be developed during and after the fill period, as shown on

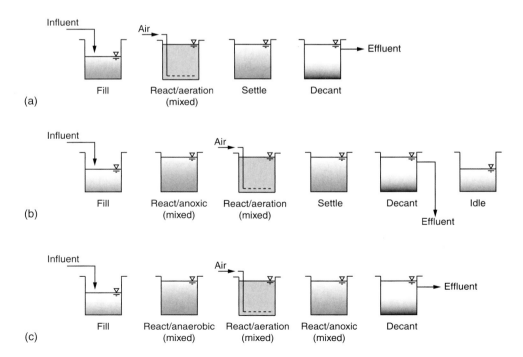

Figure 12-11

Sequencing batch reactor configurations: (a) for BOD removal and nitrification; (b) for preanoxic denitrification; and (c) for phosphorus removal.

Fig. 12-11c. An anoxic operating period is used after a sufficient aerobic time elapses for nitrification and nitrate production. Depending on the water quality requirements for reuse, SBRs can be followed by enhanced solids separation devices such as cloth filters and/or membranes.

12-8 INTEGRATION WITH EXISTING FACILITIES

Although the concept of satellite treatment is relatively simple—intercept or extract wastewater from the collection system, treat it, and return residuals to the collection system—in actuality the effects of introducing satellite treatment on the existing collection and treatment system need to be evaluated to ensure successful integration. The principal effect is the removal of wastewater from the collection system and the return of residuals. In most instances, the interception or extraction of wastewater has a negligible impact on operation of the collection system, and, if wastewater is removed or extracted on a continuous basis, it is in fact beneficial as greater capacity is available for future downstream flows. The effect of residuals return on the downstream system, especially the central treatment plant, may have some impact (such as toxic agents added to the cooling water for algae control) and should be evaluated carefully. An illustration of the effect of residuals return on wastewater characteristics is given in Example 12-1.

EXAMPLE 12-1. Determine the Effects of Satellite Treatment on the Wastewater Loading to a Centralized Treatment System.

A satellite treatment plant and water reuse system is to be added at an upstream location in a centralized system. A schematic of the system, comprised of Trunk Sewer 1 and Trunk Sewer 2, is shown on the figure given below. Flow diversion will occur in Trunk Sewer 1 before it joins with Trunk Sewer 2. Flow and wastewater characteristics of the existing system are also shown below.

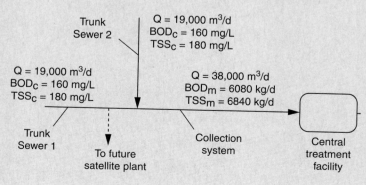

Based on the given information on the above figure and presented below, prepare a solids balance for a satellite wastewater reclamation plant that uses a treatment process consisting of a fine screen and membrane bioreactor. Waste solids from the MBR are to be returned to the sewer for processing at the downstream central treatment plant. Screenings are to be removed and landfilled. Effluent from the satellite plant is used for agricultural irrigation.

1. Definition of terms
   a. CWWTP = central wastewater treatment plant
   b. SWRP = satellite wastewater reclamation plant
   c. BOD_c = BOD expressed as a concentration, g/m³ (mg/L)
   d. BOD_m = BOD expressed as a mass, kg/d
   e. TSS_c = TSS expressed as a concentration, g/m³ (mg/L)
   f. TSS_m = TSS expressed as a mass, kg/d
2. Wastewater flowrates
   a. Total tributary flow to CWWTP = 38,000 m³/d
   b. Total flow in sewer tributary to SWRP = 19,000 m³/d
   c. Total flow to be treated by SWRP = 7600 m³/d
3. Untreated wastewater characteristics
   a. BOD_c = 160 mg/L
   b. TSS_c = 180 mg/L
   c. VSS/TSS = 0.83
4. Reclaimed water characteristics
   a. BOD = 1 mg/L
   b. TSS = 1 mg/L
5. Fine screen performance
   a. BOD removal = 25%
   b. TSS removal = 30%
   c. Note: neglect screenings moisture in computing influent flow to MBR
6. MBR characteristics and performance
   a. MLVSS/MLSS = 0.8
   b. Recovery = 95%

Solution

1. Convert given constituent quantities to daily mass values.
   a. BOD_m in Trunk Sewer 1:
   $$BOD_m = (19{,}000 \text{ m}^3/\text{d})(160 \text{ g/m}^3)/(10^3 \text{ g/kg}) = 3040 \text{ kg/d}$$
   b. TSS_m in Trunk Sewer 1:
   $$TSS_m = (19{,}000 \text{ m}^3/\text{d})(180 \text{ g/m}^3)/(10^3 \text{ g/kg}) = 3420 \text{ kg/d}$$
   c. BOD_m in SWRP influent:
   $$BOD_m = (7600 \text{ m}^3/\text{d})(160 \text{ g/m}^3)/(10^3 \text{ g/kg}) = 1216 \text{ kg/d}$$

d. TSS_m in SWRP influent:

$$TSS_m = (7600 \text{ m}^3/\text{d})(180 \text{ g/m}^3)/(10^3 \text{ g/kg}) = 1368 \text{ kg/d}$$

e. BOD_m after fine screening (influent to MBR):

$$BOD_m = (1216 \text{ kg/d})(0.75) = 912 \text{ kg/d}$$

f. TSS_m after fine screening (influent to MBR):

$$TSS_m = (1368 \text{ kg/d})(0.70) = 957.6 \text{ kg/d}$$

g. BOD_m in screenings:

$$BOD_m = (1216 \text{ kg/d})(0.25) = 304 \text{ kg/d}$$

h. TSS_m in screenings:

$$TSS_m = (1368 \text{ kg/d})(0.30) = 410.4 \text{ kg/d}$$

i. BOD_m in MBR effluent:

$$BOD_m = (7600 \text{ m}^3/\text{d})(0.95)(1 \text{ g/m}^3)/(10^3 \text{ g/kg}) = 7.2 \text{ kg/d}$$

j. TSS_m in MBR effluent:

$$TSS_m = (7600 \text{ m}^3/\text{d})(0.95)(1 \text{ g/m}^3)/(10^3 \text{ g/kg}) = 7.2 \text{ kg/d}$$

2. Determine solids production in MBR that must be wasted.
   a. Operating parameters:

   $$\text{Volatile fraction of MLSS} = 0.80 \text{ TSS}_c$$
   $$\text{Yield coefficient } Y_{obs} = 0.3125$$

   b. Estimate mass of volatile solids produced using Eq. (7-20). Assume effluent BOD is nonbiodegradable.

   $$P_{x,vss} = Y_{obs} Q(S_o - S)/(1 \text{ kg}/10^3 \text{ g})$$
   $$= (0.3125)[(912 - 7.2) \text{ kg/d}] = 282.8 \text{ kg/d}$$

   c. Estimate TSS_m that must be wasted.

   $$TSS_m = (282.8 \text{ kg/d})/0.80 = 353.5 \text{ kg/d}$$

3. Determine total mass of solids returned to sewer.
   a. Mass of solids to be returned

   Total solids = TSS_m after fine screening + TSS_m from bioreactor − TSS_m in effluent

   $$= (957.6 + 353.5 - 7.2) \text{ kg/d} = 1303.9 \text{ kg/d}$$

   b. Flowrate of solids to be returned = $0.05 \times 7600 \text{ m}^3/\text{d} = 380 \text{ m}^3/\text{d}$

4. Compute reclaimed water production.

   $$(7600 - 380) \text{ m}^3/\text{d} = 7220 \text{ m}^3/\text{d}$$

The flows and loads are summarized on the following figure:

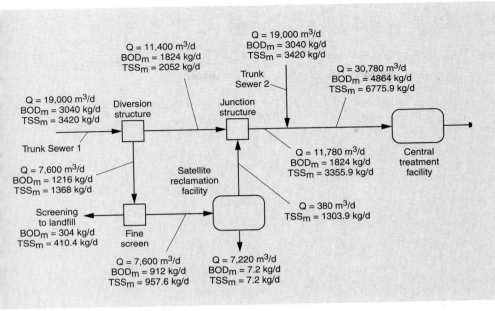

12-9 CASE STUDY 1: SOLAIRE BUILDING, NEW YORK, NEW YORK

The Solaire is a residential high-rise building with a range of environmentally responsible features including an on-site wastewater treatment and reuse system that exemplifies the interception type satellite concept illustrated on Fig. 12-1a. The building received a LEED (Leadership in Energy and Environmental Design) Gold Certification from the U.S. Green Building Council. The Solaire is the first urban, residential water reuse application permitted in the United States. The following discussion is adapted from Zavoda (2005) and www.usgbc.org.

Setting

The Solaire is a 27-story residential building that holds 293 rental apartments (see Fig. 12-12) and is located in Battery Park City in southwest Manhattan adjacent to the site of the former World Trade Center. Battery Park City is a master-planned mixed-use urban development owned by the Hugh L. Carey Battery Park City Authority (BPCA). When fully developed, the 37 ha (92 ac) development site will include 14,000 residential units, 55 ha (6×10^6 ft²) in total floor area of commercial space, over 10 ha (25 ac) of parks, plazas, and the esplanade along the Hudson River. The Solaire is the first building designed in accordance with environmental guidelines instituted in 2000 by the BPCA.

Water Management Issues

The developer sought to achieve a high-level LEED certification for the entire development. The Solaire, the first building to be designed for the LEED program, was designed to require 50 percent less potable water than a conventional residential high-rise building. The building included a wastewater treatment and recycling system to supply

Figure 12-12
The Solaire is a residential high-rise building in New York that has an onsite wastewater system for treating wastewater for toilet flushing and cooling water (coordinates: 40.717 N, 74.016 W).

reclaimed water for toilet flushing and cooling tower makeup. Building features also include water conserving fixtures and appliances such as front-load laundries and rainwater harvesting. The building also qualified for a financial incentive offered by the New York City Department of Environmental Protection Comprehensive Water Reuse Program to reduce water and sewer charges for the building by 25 percent.

Implementation

A process flow diagram of the Solaire water reclamation system is shown on Fig. 12-13 and includes the following elements:

- Aerated influent feed tank
- Trash trap to intercept nonbiodegradable solids
- Three-stage membrane bioreactor (MBR) consisting of an anoxic mix tank, aerobic digestion tank, filter tank containing ultrafiltration membrane units, and recirculation of the mixed liquor to the anoxic tank
- Ozone oxidation for color removal
- Ultraviolet (UV) disinfection
- Finished water storage tanks
- Booster pumping system and reclaimed water distribution piping

Because the water reclamation system was to be installed in the basement level of the building, the footprint of the system was a key to the design. All the required equipment

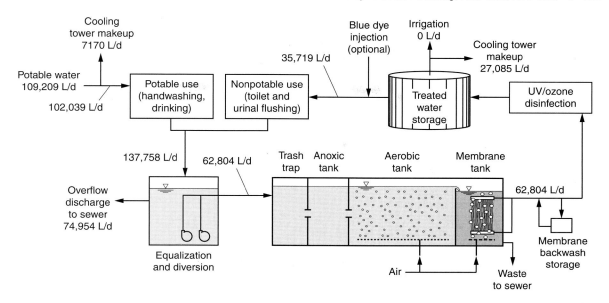

Figure 12-13
Schematic flow diagram of the Solaire water reclamation and reuse system showing average daily flows for the first 12 months of operation (Adapted from Zavoda, 2005).

had to be installed in a 197 m² (2120 ft²) floor space. A membrane bioreactor based wastewater treatment and recycling system was selected to meet the space criteria. Because reclaimed water is used for toilet flushing, ozone treatment was included for the removal of color and odor. Treated reclaimed water is disinfected with UV and stored in two fiberglass storage tanks with a total capacity of 56 m³ (14,700 gal). Stored reclaimed water is recirculated through the UV system to prevent regrowth of microorganisms.

Reclaimed water is pumped from the storage tanks for toilet flushing and cooling tower make-up. Blue dye can be added to the reclaimed water before it is sent to toilet flushing to distinguish reclaimed water from potable water and to obscure any remaining color. The distribution system is also set up to supply reclaimed water for landscape irrigation in Teardrop Park located within Battery Park City.

Lessons Learned

After system installation, the reclaimed water was tested to determine if additional treatment was needed to meet quality requirements for all reclaimed water users. A summary of the reclaimed water quality analysis is shown in Table 12-3. Initially, reclaimed water was used as a primary source of the cooling system. After about 6 mo of operation, however, the total dissolved solids (TDS) level increased significantly because the blowdown (see Chap. 19) from the cooling system was returned to the influent of the on-site water reclamation system. To lower the TDS level, potable water was blended into the makeup water. In addition, to prevent scaling in the cooling tower (see Chap. 19), a system was installed to add an aluminum compound to precipitate phosphorous in the MBR.

Table 12-3
Reclaimed water quality monitoring data from the water reclamation process in the Solaire[a]

		Sample date			
Constituents	Unit	2/5/04	4/13/04	8/25/04	2/16/05
Electrical conductivity	dS/m			0.898	0.405
TDS	mg/L	448	994		242
Calcium	mg/L	11.5		48.9	11.3
Magnesium	mg/L	3.62		33.4	
Total phosphorous	mg/L	6.74			8.0
Orthophosphate	mg/L			21.8	
Sodium	mg/L	124			48.5
Potassium	mg/L	17.4			8.5
Iron	mg/L	0.033		0.051	
Copper	mg/L	0.046		0.031	
Silica	mg/L	10.1		20.07	
Zinc	mg/L			0.045	
Sulfate	mg/L	34			40
Chloride	mg/L	48		158.7	43
Nitrate-N	mg/L	21.3			11.4
Ammonia-N	mg/L	0.16			
Bicarbonate	mg/L as $CaCO_3$	120			
M-Alkalinity	mg/L as $CaCO_3$			97.6	
Total alkalinity	mg/L as $CaCO_3$		290		24
Calcium hardness	mg/L as $CaCO_3$		50		
Langlier index			−0.73		
TOC	mg/L	6.62			
COD	mg/L	22			

[a]Data from Zavoda (2005).

For irrigation water, park horticulturists were concerned about the salinity level, which is less than a level of concern for most plants (see Chap. 18), and required that the TDS level be reduced to less than 100 mg/L. To achieve a lower TDS, a 20 L/min (5 gal/min) reverse osmosis system was installed. A small amount of reclaimed water without RO treatment is blended with the RO treated water to control the level of TDS.

Paper products are present in the wastewater and do not disintegrate before they enter the feed tank because the travel time from the toilets is very short. A device to remove materials that can clog the feed pump, e.g., a horizontal trash basket, is recommended to be installed in the feed tank and cleaned regularly by the operator.

As part of an environmentally-responsible design, water reuse was incorporated in the entire development. Because the tenants were made aware of the design concept, the use of reclaimed water for toilet flushing has been accepted with few complaints.

12-10 CASE STUDY 2: WATER RECLAMATION AND REUSE IN TOKYO, JAPAN

Patterns of water reuse in Japan are unique in that agricultural and landscape irrigation is only a relatively minor use. Most water reuse occurs in urban areas for environmental uses such as streamflow augmentation, for snow melting, and toilet flushing. Of the total reclaimed water use (200×10^6 m^3/yr in 2001), environmental uses comprise more than 60 percent, followed by snow melting (16 percent), and toilet flushing (about 3 percent). The percentage used for toilet flushing is much higher as compared to other countries (JSWA, 2005). Water reuse in Tokyo, Japan, is described in this case study with an emphasis on the reclaimed water use for toilet and urinal flushing. Reclaimed water is produced by regional water reclamation plants and onsite satellite treatment facilities located in high-rise buildings.

Setting

Tokyo is one of the largest cities in the world, with a population of over 12 million, or about 10 percent of Japan's total population, and a population density of over 5500 inhabitants/km^2 (TMG, 2005). Although Japan is in a temperate monsoon climate with above 1400 mm of average annual precipitation, the densely populated metropolitan Tokyo area has suffered from water shortages since the rapid economic and population growth in the 1960s. Water pollution from untreated domestic wastewater and industrial wastes has caused severe deterioration of water quality in the receiving waters as well as in Tokyo Bay. The Japanese government has invested heavily in the construction of drainage and wastewater collection systems starting around the same time as rapid growth began. By 1995, the entire central part of Tokyo (23 Wards) was served by the municipal sewer system (TMG, 2000).

Water Management Issues

A rapid development of water reuse in the Tokyo area started during the period of economic growth in the 1960s. There was increasing concern on how the future water demand could be met for the rapidly growing metropolitan area. To maximize the use of limited water resource, the Tokyo Metropolitan Government (TMG) established an ordinance in 1984 to require all newly constructed large buildings, generally within area greater than 3000 to 5000 m^2 and/or buildings with installed water supply pipe diameters of greater than 50 mm, be equipped with dual plumbing systems and use reclaimed water for toilet and urinal flushing (Suzuki et al., 2002; Yamagata et al., 2002). National subsidies are granted to water reuse projects as "reclaimed water utilization sewerage works." Half of the construction cost is covered by the subsidies. Maintenance and operation costs, however, are not covered by any national subsidies (Maeda et al., 1996).

Reclaimed water quality criteria evolved through several revisions. The original criterion for total coliform was 1000 organisms per 100 mL, much higher than the requirements set forth by California and most other states in the United States (Asano et al., 1996). The most recent revision of Japanese guidelines for water reuse was made in 2005 (see Table 12-4) in which the goal for coliform concentration is "nondetect" with 10 per 100 mL as a maximum (MLIT, 2005). Prior to the revision of Japanese guidelines, the TMG revised their quality criteria to lower the allowable total coliform concentration at less than 1 per 100 mL in 1 wk moving average.

Table 12-4
Reclaimed water quality guidelines for urban uses in Japan[a]

	Unit	Toilet/urinal flushing	Spraying on street and ground	Recreational uses and water features
Total coliform	No./100mL	no detect	no detect	no detect
Turbidity	NTU	2	2	2
pH	pH unit	5.8–8.6	5.8–8.6	5.8–8.6
Appearance	—	not unpleasant	not unpleasant	not unpleasant
Color[b]	CU			<10
Odor[b]	—	not unpleasant	not unpleasant	not unpleasant
Chlorine residual	mg/L	0.1 (free), 0.4 (combined)	0.1 (free), 0.4 (combined)	0.1 (free), 0.4 (combined)
Treatment requirements		Sand filtration or equivalent	Sand filtration or equivalent	Coagulation, sedimentation, and filtration, or equivalent

[a]Adapted from MLIT (2005).
[b]To be adjusted on a case-by-case basis to meet the user's demand.
Note: Water quality is measured at the outflow of the water reclamation plant.

Implementation

Three major wastewater treatment plants are capable of providing reclaimed water for toilet flushing and other miscellaneous nonpotable urban water reuse applications in the central Tokyo area; the plants are: (1) Ochiai, (2) Ariake, and (3) Shibaura. These three plants provide reclaimed water through an area-wide distribution system. In addition, interception type onsite treatment and reuse systems were developed, and reclaimed water has been used for toilet and urinal flushing in high-rise buildings. Ozone is used for color removal, followed by filtration using an ozone resistant membrane, as shown on Fig. 12-14. In 1999, there were over 300 buildings reported to have water reuse systems, either onsite or via area-wide systems (Yamagata et al., 2002). As of 2003, 122 buildings were receiving reclaimed water from the three plants.

One of the largest water reuse systems for toilet flushing is a commercial and business center in the Shinjuku district of Tokyo, designed with dual systems to distribute reclaimed water for toilet flushing. An average flow of 2700 m^3/d with a maximum flow of 4300 m^3/d was delivered to the buildings within the project area in 1993 (Asano et al., 1996; Maeda et al., 1996). A schematic of the reclaimed water system in Shinjuku is shown on Fig. 12-15. The system provides water for toilet flushing at 19 high-rise buildings. Wastewater is tertiary treated with rapid sand filtration at the Ochiai Municipal Wastewater Treatment Plant and pumped to the Shinjuku Recycling Center, located in the basement of the Tokyo Hilton Hotel. Reclaimed water is stored in the distribution reservoir in the Recycling Center before it is disinfected with chlorine and distributed to each building through reclaimed waterlines. The TMG building (see Fig. 12-16) is one of the buildings utilizing reclaimed water for toilet flushing.

Figure 12-14
Ozone resistant membranes used for color removal from reclaimed water for toilet and urinal flushing in a high-rise building in Japan.

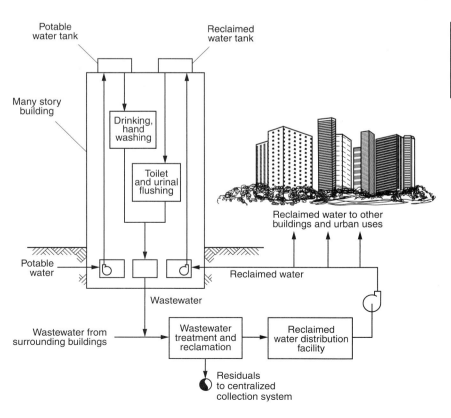

Figure 12-15
Schematic diagram of an area-wide recycling system in Shinjuku, Tokyo.

Figure 12-16

Tokyo Metropolitan Government (TMG) building in which reclaimed water is used for toilet flushing. (Coordinates: 35.689 N, 139.692 E)

The system was expanded and, in 2003, approximately 8000 m³/d of reclaimed water was used for toilet flushing in the project area.

One of the earliest water reuse projects for toilet and urinal flushing was commenced in Sinjuku district. The following water quality criteria were established: (1) less than 1000 total coliforms per 100 mL, (2) a combined chlorine residual, (3) no unpleasant appearance and odor, and (4) pH range between 5.8 and 8.6. The reclaimed water quality constantly met these criteria, and no problems with odor, appearance, or clogging of plumbing were reported in fiscal year 1994. Actual water quality data between April 1994 and January 1995 are shown in Table 12-5.

At the Shibaura wastewater treatment plant, one of the oldest wastewater treatment plants in Japan, the treatment process needed upgrades to meet the water quality criteria for toilet flushing. The original reclamation process included the filtration process with the addition of hypochlorous acid (HClO) before and after filtration. The added treatment included preozone treatment, membrane bioreactor, and second ozone treatment followed by ozone-resistant microfiltration membranes (see Fig. 12-14). The ozone treatment was necessary for color removal. An example of water quality data is shown in Table 12-6. The cost of reclaimed water production with the new system was about $1.50/m³.

Lessons Learned

In recent years in the metropolitan Tokyo area, water shortages have not been as critical as they were a few decades ago due to improvements in the infrastructure and water

Table 12-5
Reclaimed water quality between April 1994 and January 1995 at Shinjuku, Tokyo[a]

Constituent	Unit	Month									
		4	5	6	7	8	9	10	11	12	1
pH	—	7.2	7.2	7.2	7.0	7.3	7.2	7.1	6.8	7.1	7.5
Total coliform	no./mL	0	0	0	0	0	0	0	0	0	0
Combined chlorine residual	mg/L	0.2	1.5	1.0	0.8	0.2	0.2	0.3	0.2	0.4	1.5
Appearance[b]	—	SY	C	C	C	C	C	C	SY	C	C
Odor[c]	—	WM	WM	WS	WS	WM	WS	SS	MS	MS	SS
BOD	mg/L	2.8	2.5	1.0	2.3	2.0	1.4	1.5	3.0	1.9	4.3
Turbidity	NTU	1	1	1	1	1	0	1	0	1	1
Suspended solids	mg/L	0	—	—	0	—	—	0	—	—	0
Total nitrogen	mg/L	13.0	—	—	9.7	—	—	9.6	—	—	16.0
Total phosphorous	mg/L	0.48	—	—	0.70	—	—	1.00	—	—	0.65

[a] Adapted from Maeda et al. (1996).
[b] SY = slightly yellow; C = clear.
[c] WM = weak mold odor; WS = weak sewage odor; MS = moderate sewage odor; SS = strong sewage odor.

conservation efforts. Water reuse projects are implemented mostly to comply with the regulations, and also to demonstrate that the municipalities are making their best effort to conserve water and protect the environment. By requiring dual plumbing systems for reclaimed water reuse in high-rise buildings, water reuse is incorporated at the design phase of a new development, thus, reducing infrastructure costs.

Table 12-6
Comparison of water quality at the Shibaura wastewater treatment plant before and after the installation of an ozone-resistant microfiltration membrane treatment system

Constituents	Unit	Secondary effluent	Reclaimed water after membrane treatment
Suspended solids	mg/L	10.3	0.0
Turbidity	NTU	9.4	<0.1
Color	CU	40	3
Odor	—	slight moldy smell	no smell detected

12-11 CASE STUDY 3: CITY OF UPLAND, CALIFORNIA

An extraction type satellite water reclamation plant in Upland, California, has been in operation for 25 years, producing disinfected tertiary effluent for golf course irrigation. The plant diverts domestic wastewater generated by upstream neighborhoods and returns residual solids generated by the wastewater treatment processes to the downstream collection system (Ripley, 2005).

Setting

The Upland Hills Water Reclamation Plant (UHWRP) is located at the Upland Hills Country Club Golf Course in the City of Upland. The plant, constructed in 1981 with a capacity of 760 m³/d (0.2 Mgal/d), produces Title 22 water which is supplied to the Upland Hills Country Club for golf course irrigation. Treated effluent is stored in the golf course water features before being pumped to the irrigation system.

Water Management Issues

The project was initially conceived and built by a residential developer who needed a golf course to improve the marketability and value of the residential property. Constraints to the project included a limited groundwater supply and limited sewer interceptor capacity—the sewer extraction (mining) concept solved both issues simultaneously and allowed the project to proceed through the entitlement process.

Implementation

Implementation of the project was enhanced by the concept of reclaiming wastewater for a beneficial use in a water-short area. Treatment processes at the plant include influent screening, primary sedimentation, flow equalization, and three aerobic-anaerobic fixed-film reactors in series. Effluent from the third stage reactor is fed to multimedia pressure filters and then is disinfected with chlorine. The plant is enclosed entirely in a residential-type building (see Fig. 12-17) and is provided with odor control facilities. The design features have proved that a wastewater facility can be constructed and operated in a residential-type setting without nuisance.

Figure 12-17

View of satellite treatment plant located in a housing development in Upland, CA. Reclaimed water from the satellite plant is used for golf course irrigation (Coordinates: 34.125 N, 117.641 W.) (Courtesy of D. Ripley, Ripley Pacific Company.)

The UHWRP has performed very well from a water management standpoint. Because the influent wastewater is of domestic origin and the influent has a TDS of about 485 mg/L, the effluent water quality is well suited for golf course irrigation. Application rates can be controlled to meet seasonal requirements of the golf course turf by (1) regulating withdrawals of wastewater from the collection system and (2) providing effluent storage for flow equalization in the golf course ponds.

Lessons Learned

The satellite reclamation project has been successful over the 25 years of operation, but the facilities are in need of upgrading to preserve their integrity. Lack of adequate local financing for improvements is an obstacle in implementing necessary upgrades. The golf course has historically paid about $105/10^3$ m^3 ($130/ac-ft) for reclaimed water, far less than the value of effluent typical in urban southern California of about $650/10^3$ m^3 ($800/ac-ft) (2006 dollars). The lack of adequate payment for the value of the reclaimed water has resulted in unfunded replacement reserves that would have facilitated needed rehabilitation and upgrades to ensure continued operations without full system replacement.

PROBLEMS AND DISCUSSION TOPICS

12-1 Two growing communities are considering expansion of their wastewater system because of proposed residential and commercial development in the upper reaches of their service area. One community anticipates a 10 percent capacity expansion and the other community expects a 25 percent system expansion. What factors should each city consider in determining the type of expansion (satellite or centralized) appropriate for their needs?

12-2 In Example 12-1, if the screenings are returned to the collection system, what effect will they have on the BOD and TSS mass loadings discharged to the centralized treatment plant?

12-3 An upstream satellite treatment facility is being considered for producing reclaimed water for groundwater recharge. The setting is in a suburban area. Prepare two process flow diagrams using alternative treatment technologies and cite the advantages and disadvantages of each for the proposed application.

12-4 Conduct a literature and/or internet search for examples of satellite systems and prepare summaries of at least two case studies. Identify the types of satellite systems used and list factors that were important in their selection and application.

12-5 A satellite water reclamation plant is planned to be located in an urban area to provide reclaimed water for toilet flushing and air conditioning cooling water in an office complex. What are some of the aesthetic and environmental considerations that have to be evaluated in locating and designing the facilities?

12-6 For a satellite water reclamation plant to be located in a suburban area for landscape and golf course irrigation, two alternative systems are being considered: (1) an integrated submerged membrane bioreactor (see Fig. 7-12*b*) and (2) a system using a sequencing batch reactor, cloth filter, and microfiltration (see Fig. 7-17). What are the advantages and disadvantages of each type of system and, based on noneconomic factors, what system would you recommend? State the reasons for your recommendation.

REFERENCES

Asano, T. (ed.) (1998) *Wastewater Reclamation and Reuse*, Water Quality Management, **10**, CRC Press, Boca Raton, FL.

Asano, T., M. Maeda, and M. Takaki (1996) "Wastewater Reclamation and Reuse in Japan: Overview and Implementation Examples," *Water Sci Technol*, **34**, 11, 219–226.

AWWA (1994) *Dual Water Systems*, AWWA Manual M24, American Water Works Association, Denver, CO.

AWWARF (1999) *Residential End Uses of Water*, American Water Works Association Research Foundation, Denver, CO.

JSWA (2005) *Sewage Works in Japan 2005: Wastewater Reuse*, Japan Sewage Works Association, Tokyo, Japan.

Maeda, M., K. Nakada, K. Kawamoto, and M. Ikeda (1996) "Area-Wide Use of Reclaimed Water in Tokyo, Japan," *Water Sci Technol*, **33**, 10–11, 51–57.

MLIT (2005) *Manual on Water Quality for Reuse of Treated Municipal Wastewater*, Japanese Ministry of Land, Infrastructure, and Transport, Tokyo, Japan.

Ripley, G. (2005) Personal Communication.

Suzuki, Y., M. Ogoshi, H. Yamagata, M. Ozaki, and T. Asano (2002) "Large-Area and On-Site Water Reuse in Japan," presented at International Seminar on Sustainable Development of Water Resource Pluralizing Technology, held at Korean National Institute of Environmental Research, Seoul, Korea.

Tchobanoglous, G., F. L. Burton, and H. D. Stensel (2003) *Wastewater Engineering: Treatment and Reuse*, 4th ed., McGraw-Hill, New York.

TMG (2005) website: http://www.metro.tokyo.jp/ENGLISH/

TMG (2000) *Sewerage in Tokyo*, The Bureau of Sewage, Tokyo Metropolitan Government, Tokyo, Japan.

Yamagata, H., M. Ogoshi, Y. Suzuki, M, Ozaki, and T. Asano (2002) "On-Site Insight into Reuse in Japan," *Water 21*, August, 26–28.

Zavoda, M. (2005) "The Solaire, a High-Rise Residential Reuse Case Study," *Proceedings, 20th Annual WateReuse Symposium*, WateReuse Association, Alexandria, VA.

13
Onsite and Decentralized Systems for Water Reuse

WORKING TERMINOLOGY   764

13-1 INTRODUCTION TO DECENTRALIZED SYSTEMS   766
  Definition of Decentralized Systems   766
  Importance of Decentralized Systems   767
  Integration with Centralized Systems   770

13-2 TYPES OF DECENTRALIZED SYSTEMS   770
  Individual Onsite Systems   771
  Cluster Systems   771
  Housing Development and Small Community Systems   772

13-3 WASTEWATER FLOWRATES AND CHARACTERISTICS   774
  Wastewater Flowrates   774
  Wastewater Constituent Concentrations   778

13-4 TREATMENT TECHNOLOGIES   785
  Source Separating Systems   786
  In-building Pretreatment   788
  Primary Treatment   788
  Secondary Treatment   792
  Nutrient Removal   797
  Disinfection Processes   802
  Performance   804
  Reliability   804
  Maintenance Needs   804

13-5 TECHNOLOGIES FOR HOUSING DEVELOPMENTS AND SMALL COMMUNITY SYSTEMS   806
  Collection Systems   807
  Treatment Technologies   815

13-6 DECENTRALIZED WATER REUSE OPPORTUNITIES   816
  Landscape Irrigation Systems   816
  Irrigation with Greywater   818
  Groundwater Recharge   818
  Self-Contained Recycle Systems   821
  Habitat Development   821

13-7 **MANAGEMENT AND MONITORING OF DECENTRALIZED SYSTEMS** 821
Types of Management Structures 821
Monitoring and Control Equipment 824

PROBLEMS AND DISCUSSION TOPICS 826

REFERENCES 827

WORKING TERMINOLOGY

Term	Definition
Aerobic treatment unit (ATU)	A mechanical wastewater treatment system used for secondary treatment of septic tank effluent. System typically consists of diffused bubble aeration in a tank and integrated solids separation. Similar in operation to the activated sludge process, however, with minimal process control. Fixed and suspended packing materials are often used to improve process stability.
Alternative collection systems	Systems for transporting wastewater by gravity, pressure, or vacuum and utilizing small diameter, watertight piping.
Biofilter	A biological filtration process consisting of wastewater distribution onto a fixed packing material. As wastewater moves by gravity through the filter, the microbial community of the packing material (i.e., biofilm) adsorbs and transforms the wastewater constituents.
Blackwater	Wastewater consisting of only toilet flush water and kitchen wastewater containing food waste. Typically higher in organic matter, nutrients, and pathogens.
Cluster system	A wastewater management system used to treat wastewater from a collection of buildings. Typically, the buildings are located adjacent to each other to reduce wastewater transport distance. Combining treatment into one system may improve the maintainability and performance of a treatment system.
Community system	A wastewater management system used to treat wastewater from a community. A watertight collection system is used for transport of septic tank effluent or untreated wastewater.
Composting toilet	A self-contained waterless toilet used for the collection and composting of human waste. Most models incorporate mechanical mixing and/or aeration. Diversion of human waste from wastewater flow may change the nature of treatment processes required.
Constructed wetland	An artificial wetland system design used for water quality improvement, for example, secondary wastewater treatment. Constructed wetland systems require more space than mechanized treatment processes, but use little or no energy for operation.
Decentralized wastewater management (DWM)	The collection, treatment, and reuse of wastewater at or near the point of generation. Decentralized wastewater management systems are used commonly for treating individual onsite and small community-scale wastewater flows from dispersed facilities.
Distributed systems	A term often used to describe satellite and decentralized wastewater management systems.
Drip irrigation	Distribution of irrigation water using a network of tubing and low-flow emitters. Drip irrigation is used for the subsurface dispersal of effluent directly into the root zone to maximize plant uptake (see Chaps. 17 and 18).

Evapotranspiration	The amount of water lost to the atmosphere by evaporation from soil and plant surfaces, and by transpiration from plant tissue. Agricultural extension and government climate agencies typically provide evapotranspiration rates for different areas for estimation of irrigation demand.
Greywater	Wastewater from bathing and washing facilities that does not contain concentrated human waste (i.e., flush water from toilets) or food waste (i.e., kitchen sink, food waste grinders). Examples include bath and shower water, hand wash water, and laundry washwater. Greywater typically contains high concentration of salts and minerals from detergents and soaps. If powered laundry detergents and brine type water softeners are used, increased concentrations of sodium can be expected in the greywater.
Grinder pump	A component used in conjunction with small diameter collection systems that grinds the household wastewater solids and pumps the homogenized wastewater under pressure. Grinder pumps are typically stationed near the wastewater source, and the use of a grinder pump eliminates the need for onsite primary treatment for solids removal.
Hybrid collection system	Wastewater collection system composed of various combinations of gravity, pressure, and vacuum components to increase the efficiency and adaptability of the overall system.
Imhoff tank	A process used for clarification of untreated wastewater. Flow is directed through an upper chamber, and particles that settle out of the wastewater flow are deposited into a lower chamber. Solids in the lower chamber are left to digest and are removed periodically. Baffles are used to deflect rising gas bubbles and particles from entering the top chamber.
Infiltration	The movement of water into the soil. Examples of infiltration include precipitation entering the soil, subsurface effluent discharge systems, and discharge from the bottom of surface water impoundments.
Onsite system	A wastewater management system used at the immediate site of wastewater generation. Onsite systems are designed to accommodate the variability in wastewater generation expected from individual residences or applications. Because the flowrates are low, effluent may be processed further by soil infiltration, or recycled for a given application.
Package plant	A preengineered wastewater treatment process that usually consists of a single unit and can be delivered and installed with minimal effort. An ATU is a common example.
Packed bed filter	A treatment process that makes use of biofilm microbial communities attached to fixed packing materials. As wastewater is distributed on the surface of the packing materials and flows by gravity, it comes in contact with the biofilm followed by adsorption and transformation of wastewater constituents.
Percolation	The movement of water, following infiltration, through the soil vadose zone. When operated properly, high levels of treatment are attained by effluent percolation, comparable to advanced treatment effluent.
Septage	The contents pumped from septic tanks or holding tanks. Septage is typically pumped from tanks using specially designed trucks and hauled to wastewater treatment or other septage handling facilities.
Septic tank	An enclosed water-tight tank used to receive wastewater from a variety of sources such as individual residences, commercial and institutional facilities, and small housing developments. Partial treatment of the wastewater occurs within a septic tank by gravity separation and biological activity. Effluent from the septic tank is dispersed most commonly, with or without further treatment, by subsurface land application or reuse.
Septic tank effluent gravity (STEG) collection system	A wastewater collection system in which small diameter collection pipes are used for the transport of septic tank effluent by gravity to a common treatment facility.

Septic tank effluent pump (STEP) collection system	A wastewater collection system in which small diameter collection pipes are used for the transport of pumped septic tank effluent to a common treatment facility.
Vacuum collection system	A wastewater collection system in which a vacuum pump is used to facilitate transport of wastewater from individual wastewater generation sites to a common treatment facility.
Vadose zone	The unsaturated soil region between the soil surface and saturate zone.

The use of centralized or regional wastewater collection and treatment facilities for the production of reclaimed water is practiced extensively in developed urban regions and other densely populated areas. However, when a centralized collection system is not available, or it is desirable to have independent treatment facilities, decentralized wastewater systems may be an option. Decentralized wastewater reclamation systems have been used widely for landscape irrigation in suburban areas, thereby reducing demand on potable supplies in addition to other benefits discussed in this chapter. In areas located adjacent to a centralized collection system, satellite facilities may also be used to meet some of the reclaimed water demand (see Chap. 12). While satellite facilities share some common characteristics with the decentralized systems described in this chapter, satellite systems are different because they have a direct connection to a centralized wastewater collection system and, therefore, do not have to store or manage solids onsite.

13-1 INTRODUCTION TO DECENTRALIZED SYSTEMS

Decentralized wastewater management (DWM) systems are used most commonly in semiurban, rural, and remote areas, where installation of a centralized sewer system is not technically, politically, environmentally, or economically feasible (see Fig. 13-1). In some areas, decentralized systems are used instead of centralized collection systems to limit and control the type of development in a given area. However, decentralized treatment systems present a significant challenge for the design engineer due to the need for high quality reliable performance in light of a number of constraints, including long periods of time between maintenance activities, lack of redundant systems, high variability in flowrate and constituent concentrations, and site-specific factors.

Decentralized systems are an integral component of smart-growth community design initiatives in unsewered areas (Joubert et al., 2004) and an element of sustainable development because of the potential for low impact wastewater management and other advantages presented below. Further, due to practical and economic limitations, it is recognized that it is not possible or desirable to install centralized sewers to service all areas in the United States. Therefore, DWM systems are necessary for the protection of public health and the environment and for the development of long-term strategies for the management of water resources.

Definition of Decentralized Systems

Decentralized wastewater management is defined as the collection, treatment, and reuse of wastewater at or near the point of waste generation (Crites and Tchobanoglous, 1998). Decentralized facilities may be used for wastewater management from individual

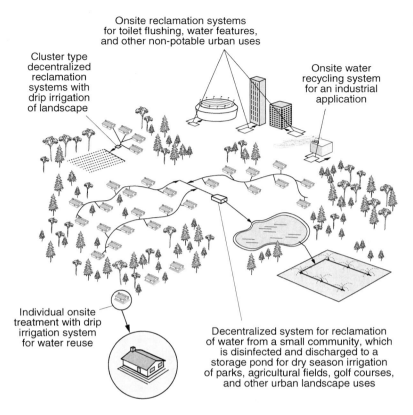

Figure 13-1
Definition sketch for decentralized systems for wastewater collection, treatment, and reuse.

homes, clusters of homes, subdivisions, and isolated commercial, industrial, and agricultural facilities. The wastewater flowrate, quality, and flow distribution is dependent on the types of activities taking place as well as the scale of the application.

At the present time (2006), more than 60 million people in the United States live in homes where decentralized systems are used for wastewater management. Further, the U.S. EPA estimated that about 40 percent of the new homes built in the 1990s were served with decentralized systems for wastewater management (U.S. EPA, 1997). While conventional septic tank systems used for the subsurface leaching of wastewater are onsite decentralized systems, the focus of this chapter is on systems designed to utilize all or a portion of effluent for beneficial local reuse applications. Considerations for the use of decentralized wastewater systems are given below.

Importance of Decentralized Systems

Customized Treatment Processes

The application of decentralized wastewater systems allows for the use of customized treatment processes, specifically designed for the wastewater to be treated. In large, regional wastewater systems, the discharge of substances from anonymous sources and industrial wastewater results in problematic constituents, such as metals, salts, and hazardous trace organic compounds. The commingling of domestic wastewater with commercial and industrial wastewater may result in the presence of constituents, that, in some cases, require advanced treatment processes for removal. In contrast, decentralized systems receive a more homogeneous wastewater from a well defined source. When wastewater from domestic sources is segregated from commercial and industrial sources, it may be treated efficiently onsite using processes adapted to the particular waste stream.

Reduced Infrastructure Needs

A number of benefits related to reduced infrastructure may result from the use of decentralized systems. Managing wastewater locally reduces the size and extent of collection pipes required and the high cost associated with extension and maintenance of conventional collection systems. The cost of expanding and extending centralized collection systems is dependent on several factors, including the capacity of the collection system, development in the area where the collection system is placed, and topographic and geologic limitations. Characteristics of the collection system that are important to overall cost include the length of the collection system laterals and the use of pumping and lift stations. The amount of development that has occurred in an area may also affect the cost, for example, where roads, sidewalks, and property need to be disturbed. Challenging topographic and geologic features include steep slopes, subsurface bedrock formations, and shallow soils and/or water table.

Most original wastewater collection and treatment systems were designed for build out capacity of the urban core areas, while residential development in the peripheral areas of the collection system was not necessarily accounted for or anticipated. In some areas, the useful life of collection facilities and centralized treatment facilities that are already at or near capacity may be extended through reduced loading. Wastewater treatment plants that are operating at capacity may be limited or do not allow additional connections, particularly for outlying areas. Further, most treatment plants have not been designed to accommodate new discharge requirements and may require the addition of advanced treatment processes. The cost to redistribute treated effluent for reuse applications should also be considered, as reuse sites are typically located in remote areas compared to the reclamation plant site. Decentralized facilities may eliminate the need for extensive recycled water transmission networks in some areas where the reuse site coincides with the point of wastewater generation and treatment.

Reliability Issues

The impacts resulting from treatment plant process upsets and other events associated with reliability are less severe for decentralized wastewater systems as compared to large centralized systems because application is on a smaller scale. For example, if a problem develops in a vacuum collection sump serving a cluster of several homes, such as a power outage or valve failure, a limited number of people may be inconvenienced while the system is repaired. Decentralized systems typically have capacity for one or more days of operation, and some gravity flow systems may not be affected at all. Alternatively, if there is a disruption with a centralized collection or treatment system that requires the systems to be shut down for a period of time, thousands of people could be affected, or more likely, partially treated or untreated wastewater may be discharged directly to surface waters. As most decentralized systems are designed with soil dispersal or irrigation for the fate of effluent, surface water discharge is unlikely with these systems.

Watershed Considerations

The nature of large centralized wastewater systems requires point source discharge of effluent, typically to adjacent surface waters. In contrast, decentralized systems that result ultimately in the deep percolation of water may benefit some water systems by retaining water within the local watershed. For example, a reclaimed water irrigation system for a landscape or agricultural application may result in evapotranspiration of all or most of the reclaimed water during the summer growing season. However, during the winter when evapotranspiration demand is low, reclaimed water may percolate down to the groundwater table. The overall effects of distributed infiltration systems is dependent on the local hydrologic conditions, but may help to offset the effects of drought conditions and declining water tables in some areas. However, consideration should be given to effluent quality to ensure that groundwater quality is not impacted by inadequately treated water or by improper dispersal.

Watertight Systems

In addition to the high installation cost of centralized collection systems, issues with nonwatertight joints and damaged sections result in potentially high volumes of inflow and infiltration, or exfiltration in the collection system. Infiltration can more than double the flowrate and dilute wastewater constituent concentrations arriving at treatment facilities in extreme cases. Long-term infiltration into a collection system can also lower groundwater levels. Exfiltration from collection systems may result in groundwater or surface water contamination. While large centralized collection systems are not intended to leak, the nature of large rigid pipes buried in various soils results in more leaks and damage to pipe sections over time. Further, it is costly to identify and repair sections of damaged underground collection system, especially when located below roads and buildings in developed urban areas. Piping used for decentralized facilities is mostly small diameter flexible plastic pipes, typically of polyvinyl chloride (PVC) with solvent welded joints or medium density polyethylene (MDPE) with compression joints, which can be designed for high pressures or vacuum where alternative collection systems are used. Flexible plastic piping is much less likely to leak under normal bedding conditions. These pipes can be installed easily in narrow trenches or by directional drilling that results in minimal disturbance to property and roads.

Treatment Performance

The added flow resulting from expansion of a collection system may change the performance of a given treatment process due to changes in the hydraulics and constituent loading. Given the current and projected future regulations controlling the discharge of effluent constituents into surface waters and the realization that most biological treatment processes are not designed for the removal of many trace constituents of concern, increased loading to existing treatment facilities may require the implementation of advanced treatment processes.

Soil based treatment systems have been found to remove even the most refractory compounds contained in wastewater (see Chap. 22), many of which pass through conventional treatment processes with little attenuation. The complex nature of the soil environment facilitates the removal of pathogens and trace constituents. Decentralized and onsite treatment systems take advantage of the natural purification and assimilative capacity in the soil. By comparison, little assimilative capacity is available by the point discharge of effluent to a surface water body as is common with many centralized systems. Surface water discharges do benefit from solar photolysis, however, which does not occur when effluent is used for irrigation. However, the extent of solar photolysis on degrading trace constituents is difficult to predict due to a number of factors (see Chap. 23). Centralized facilities would need to be equipped with advanced treatment processes to produce a quality effluent that is comparable to the high level of treatment that occurs naturally in the soil. The use of effluent from a DWM system for irrigation fulfills the goals of agricultural and landscape irrigation and high quality treatment in the soil for water that percolates out of the root zone.

Integration with Centralized Systems

Decentralized systems may complement centralized treatment by meeting wastewater management needs in areas limited by practical constraints. It may also be necessary to segregate certain wastewater streams from the bulk flow because of the presence of difficult-to-manage constituents or properties. For example, interception or diversion of flow from an industrial or commercial facility for treatment using a satellite system may be used to alleviate flow or constituent overloading to the centralized facilities.

Many decentralized treatment systems are designed with adequate capacity for extended sludge holding times and solids digestion. For example, primary treatment in septic tanks reduces the overall volume of solids that must be managed (see computation in Example 13-3). However, solids must be removed periodically for processing at a centralized treatment facility, applied to land, or collected in specialized holding facilities for subsequent treatment. Some decentralized facilities make use of thermal combustion, aerobic composting, or subsurface soil treatment for residual solids and therefore do not require any additional solids management.

Data collected from remote decentralized systems, such as pump operation, liquid and solids levels, temperature, alarm status, UV lamp output, and individual constituent concentrations may be consolidated and processed at centralized facilities. A number of systems are available that facilitate the transfer of data over phone lines from remote facilities. Many systems utilize Internet based applications for organizing and viewing data and possibly performing some maintenance functions. The use of centralized facilities for management and monitoring of decentralized facilities is among the most promising approaches to an integrated and comprehensive wastewater management plan.

13-2 TYPES OF DECENTRALIZED SYSTEMS

Decentralized treatment may be applied at different scales and for various applications. Several different decentralized water reclamation and reuse systems were illustrated previously on Fig. 13-1. In many cases, the scale and characteristics of the application are used for process selection. For example, in applications where the service frequency is reduced, such as in remote areas, systems that do not generate large volumes of waste solids are utilized. In areas with high population density, systems that are particularly compact may be used. Other site characteristics that require special design considerations include cold climate regions, areas where effluent discharge is not acceptable, and applications with challenging wastewater characteristics (Crites and Tchobanoglous, 1998). The types of DWM systems used can vary from individual onsite wastewater management systems to satellite treatment systems integrated with centralized systems. Depending on the site characteristics, system design, and type of the application, different approaches are used to implement management and maintenance, as described in Sec. 13-7. The principal categories of DWM systems are reported in Table 13-1.

Individual Onsite Systems

Treatment of wastewater for reuse from an individual building is typically among the most challenging applications. The treatment system must be sized to accommodate the potentially high variability in flowrate and constituent concentrations. Individual systems for wastewater treatment and reuse applications are typically comprised of a septic tank or similar device for primary treatment, an aeration process for the removal of organics and nutrients, and a distribution system for reuse. Typical process flow diagrams and views of onsite wastewater systems are shown on Fig. 13-2. In some reuse applications it may also be necessary to disinfect the treated effluent.

Cluster Systems

The collection of wastewater from several adjacent buildings and processing in a common treatment system is known as cluster type decentralized wastewater treatment. Cluster systems are used commonly for groups of 2 to 12 adjacent buildings. Cluster systems have the advantage of achieving an economy of scale that facilitates the use of monitoring and management systems. A typical flow diagram and view of a cluster

Table 13-1
Types of decentralized wastewater management systems

Type of system	Comment
Indvidual residential	Systems can vary from a septic tank and a gravity fed leachfield to a system comprised of a septic tank, intermittent sand filter, and drip irrigation system.
Clustered residential	Two or more homes are grouped together to form a cluster system for improved wastewater management.
Housing development and subdivisions	Isolated housing developments can be grouped together to achieve wastewater management objectives.
Community	Entire communities may be serviced using alternate collection systems in conjunction with treatment and reuse facilities.
Remote outdoor	Systems located in remote areas often without power and/or running water, such as campgrounds, parks, or other outdoor facilities.
Agricultural	Used for the management of water used in dairy, food processing, and animal housing operations. Typically seasonal and high concentrations of organic matter and nutrients are experienced, depending on the specific application.
Commercial and institutional facilities	Wastewater from individual commercial buildings, buildings, apartments, and institutional and recreational facilities can be managed with complete recycle systems.

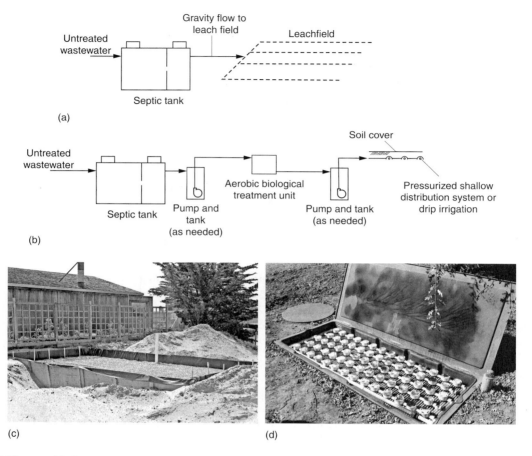

Figure 13-2
Typical onsite wastewater management systems: (a) conventional septic tank system comprised of a septic tank and leachfield leading to *de facto* indirect potable as a result of groundwater recharge, (b) onsite system comprised of septic tank, biological treatment, and effluent reuse, (c) view of sand filter (under construction) used for an individual home, and (d) view of nonsubmerged synthetic media biofilter used for an individual home.

Housing Development and Small Community Systems

system used for wastewater treatment for a number of homes are shown on Fig. 13-3. The treatment processes used for cluster applications is often similar in structure and function to that used for individual onsite systems described previously. Therefore, individual onsite and cluster systems are considered together in the subsequent discussion. Cluster systems are differentiated from systems used for housing developments and small communities by the type and extent of the collection system used.

The collection and treatment of wastewater from housing developments and small communities can be accomplished using alternative collection systems and small treatment facilities (see Sec. 13-5). Small diameter, watertight piping is used for meeting wastewater collection and transport needs. In addition, several options are available that allow for alternative collection systems to be used. For example, wastewater solids may be retained

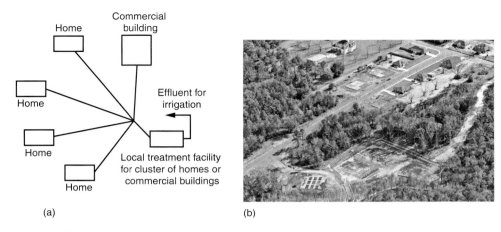

Figure 13-3

Typical example of cluster system: (a) schematic of cluster system and (b) treatment system (lower left), comprised of nine units of the type shown in (Fig. 13-2d), for a cluster of adjacent buildings; treated effluent is used for drip irrigation. (Courtesy of Orenco Systems, Inc.)

and processed in an onsite primary treatment tank, while the liquid portion of the wastewater is discharged to the collection systems and treated downstream near the point of reuse. Other advantages of community treatment systems are economy of scale, the use of more sophisticated treatment processes, and the capacity to have dedicated operations and maintenance personnel. A process flow diagram and example of a small treatment facility used for wastewater treatment at a development are shown on Fig. 13-4.

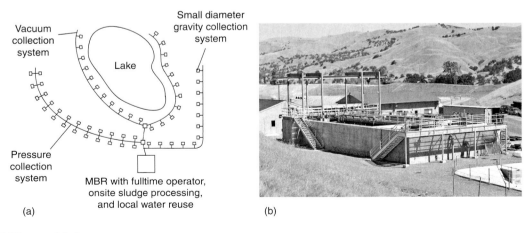

Figure 13-4

Typical example of system for housing development or small community: (a) schematic illustrating different types of collection systems and (b) small membrane bioreactor facility for a commercial development with effluent used for golf course irrigation and toilet flushing.

13-3 WASTEWATER FLOWRATES AND CHARACTERISTICS

To properly design facilities for water reuse, the flowrates and characteristics of the wastewater source must be defined clearly. The variability in flow and constituent mass loading affects the type and size of the processes selected. The flowrates and characteristics of wastewater from an individual building are dependent on the activities occurring inside of the building, the types of fixtures and appliances present, and the habits of the individuals within the building.

Wastewater Flowrates

The volume of water used to carry away waste from residential and commercial sources impacts the size of the treatment system and the concentration of the wastewater constituents. Many processes require a minimum contact or residence time to perform treatment. Variations in flowrate also need to be accommodated. Unlike centralized collection systems, infiltration and inflow to decentralized systems can be minimized because the systems are designed to be watertight. The wastewater flow from an individual home can be estimated using the following equation:

$$\text{Flow (L/home·d)} = 150 \text{ L/home·d} + (130 \text{ L/person·d})(\text{persons/home}) \quad (13\text{-}1)$$

Application of Eq. (13-1) to a home with two, three, and four persons results in average flowrates per person of 205, 180, and 168 L, respectively. In some areas, wastewater flow from a residence is conservatively estimated to be 570 L/d per bedroom, which is assumed to include peaking-factors. However, estimating the flow per bedroom may result in high and low estimates of flow given the variability in water use patterns and the actual number of inhabitants. Further, wastewater treatment processes designed for a particular flowrate and loading regime may not function properly when lightly loaded due to overdesign. Changes in water use such as water conservation practices or greywater diversion reduce the wastewater flowrate, while leaking fixtures increase the wastewater flowrate.

Daily Flow Variations

In decentralized systems serving residential developments, the daily flow variation depends primarily on the schedule of domestic activities. An analysis of flow variation in typical decentralized systems is shown on Fig. 13-5. As shown on Fig. 13-5, the flow distribution is similar for individual residences and a number of residences, with the primary flow occurring in the morning and evening. Changes in lifestyle patterns, such as working from home, may result in modified flow distributions from individual residences. The variation in flowrate resulting from a family leaving home for a vacation would be greater for an individual system than for a collective system receiving wastewater from a number of homes.

The level of flow variations depends on the time frame under consideration. Over short time intervals, such as hours or days, large variations are expected. Similarly, the flow variation is larger for small applications, such as an individual residence, compared to an entire community or city. Typical values of peaking factors which may be used when other estimates are not available are shown in Table 13-2. The use of peaking factors is shown in Example 13-1.

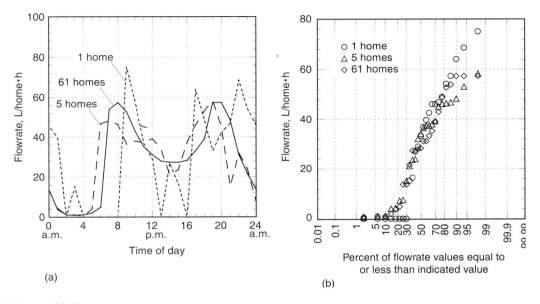

Figure 13-5
Example of variation in flowrate from an individual home, 5 homes, and 61 homes, presented as a (a) function of time of day and (b) probability distribution.

Table 13-2
Peaking factors for wastewater flows from individual residences, small commercial establishments, and small communities[a]

Peaking factor[b]	Individual residence		Small commercial establishment		Small community	
	Range	Typical[c]	Range	Typical	Range	Typical
Peak hour	4–10	6	6–10	7.5	3–6	4
Peak day	2–5	2.5	2–6	3.0	2–4	2.5
Peak week	1.25–4	2.0	2–6	2.5	1.5–3	1.75
Peak month	1.15–3	1.5	1.25–4	1.5	1.2–2	1.25

[a] Adapted from Crites and Tchobanoglous (1998).
[b] Ratio of peak flow to average flow.
[c] Higher values are often reported, but the given values are suitable for sizing onsite wastewater management facilities.

EXAMPLE 13-1. Determination of Design Flowrates from Individual Residences.

Compare the design flowrate based on a per capita allowance and peaking factors presented in Table 13-2 to the design flow based on a per bedroom allowance of 570 L/d.

Solution

1. Determine the per capita design flowrate.
   a. Compute the flowrate for one person using Eq. (13-1)

   $$\text{Flow} = 150 \text{ L/home·d} + (130 \text{ gal/person·d})(1 \text{ person/home})$$
   $$= 280 \text{ L/home·d}$$

   b. Select a peaking factor using the typical value for a peak day presented in Table 13-2; a peaking factor of 2.5 may be used for a single residence. The design flow for a one person residence is:

   $$\text{Design flow} = (2.5)(280 \text{ L/home·d}) = 700 \text{ L/home·d}$$

2. Determine the per bedroom design flowrate.

   For a one-bedroom residence, the design flowrate is:

   $$\text{Design flow} = (570 \text{ L/bedroom·d})(1 \text{ bedroom}) = 570 \text{ L/d}$$

3. The design flowrate for one-, two-, three-, and four-bedroom residences are summarized in the following table:

Number of bedrooms	Number of persons	Flowrate[a], L/capita·d	Peaking factor	Design flow based on peak per capita flow, L/home·d	Design flow based on per bedroom allowance, L/home·d
1	1	280	2.5	700	570
1	2	205	2.5	1025	570
2	3	180	2.5	1350	1140
3	4	168	2.5	1680	1710
4	5	160	2.5	2000	2280

[a] Computed using Eq. (13-1).

Comment

Because the design flowrate affects the overall size and operation of a treatment process, it is important to consider factors such as number of persons using the system and peaking factors. It is also important to note the discrepancy between design flowrate based on number of bedrooms and number of occupants.

Domestic Water Use Practices

Domestic water use includes indoor activities that contribute to wastewater flows, such as toilet flushing, automatic and manual dishwashing, clothes washing, and bathing. Outdoor water usage, such as irrigation and vehicle washing, does not add to wastewater flows. The flowrates from specific domestic activities are summarized in Table 13-3. To reduce domestic water use, water conservation practices may be implemented, which also have the effect of reducing wastewater generation while proportionally increasing the concentration of wastewater constituents. Typical reductions in indoor water use resulting from water conservation practices are shown in Table 13-4. The effect of water conservation on wastewater constituents will be demonstrated in the following discussion.

Greywater Separation

The water from bathing, hand washing, and clothes washing (not including soiled diapers), collectively known as greywater, is sometimes managed separately from human waste because it is relatively free of pathogens, organic matter, and trace constituents. When greywater is separated, wastewater from kitchen sinks, automatic dishwashers, and food waste grinders is discharged typically with toilet flushing water, collectively known as blackwater (note that drainage from kitchen sinks is included in household greywater in Australia). Separated greywater may be treated and reused more easily than combined greywater and blackwater. Some system designs incorporate direct drainage of greywater to mulch basins for tree irrigation, thereby eliminating treatment or storage and greatly reducing the system cost and maintenance needs (Ludwig, 2000). Separated blackwater may be treated separately or discharged to a collection system.

Table 13-3
Typical rates of water use for various devices and appliances in the United States[a]

	US customary units		SI units	
Device or appliance	Units	Range	Units	Range
Automatic home-type washing machine:				
Top loading	gal/load	34–57	L/load	130–216
Front loading	gal/load	12–15	L/load	45–60
Automatic home-type dishwasher	gal/load	9.5–15.5	L/load	36–60
Manual dishwashing basin	gal/use	3–6	L/use	11–23
Bathtub	gal/use	30	L/use	114
Kitchen food waste grinder	gal/d	1–2	L/d	4–8
Shower	gal/min·use	2.5–3	L/min·use	9–11
Toilet, tank, conservation type	gal/use	1.6–3.5	L/use	6–13
Toilet, tank type, standard	gal/use	4–6	L/use	15–23
Washbasin	gal/min·use	2–3	L/min·use	8–11

[a] Adapted from Salvato (1992); and Crites and Tchobanoglous (1998).

Table 13-4
Examples of flow reduction resulting from use of water conservation devices and appliances[a]

Device/appliance	Description of water conservation measures	Water usage, L/capita·d Without water conservation	Water usage, L/capita·d With water conservation
Faucets	Aerators increases the rinsing power of water by adding air and concentrating flow, thus reducing the amount of wash water used.	27	16
Bathing/showering	Pressurized showers mix compressed air with water to create the sensation of conventional shower. Flow-limiting shower heads restrict and concentrate water passage by means of orifices that limit and divert shower flow for optimum use by the bather.	55	35
Toilets	Toilet leak detectors consist of tablets that dissolve in the toilet tank and release dye to indicate leakage of the flush valve. Toilet dams partition the toilet tank to reduce the amount of water used per flush. Low-flush toilets reduce the discharge of water per flush.	80	25
Dishwashing	Water efficient dishwasher reduces the amount of water used to wash dishes.	19	11
Clothes washing	Water efficient clothes washer reduces the amount of water used to wash clothes.	53	15
Total		234	102

[a] Adapted from Crites and Tchobanoglous (1998).

The greywater and blackwater flows may be estimated from the data provided in Table 13-4. Greywater systems are usually expensive to retrofit into a building, and therefore should be included, if possible, during building planning and construction. In some areas the use of greywater for irrigation and toilet flushing is recommended during periods of water shortages. Management of greywater systems may present challenges if there is insufficient planning.

Wastewater Constituent Concentrations

The concentration of wastewater constituents varies on a daily basis depending on factors such as water conservation practices, dietary choices, use of household cleaning products, and water softeners. To estimate the expected effluent quality accurately, calculations based on typical values and specific loading parameters are used. For existing systems, water quality may be determined from analysis of data obtained by repeated sampling events or by composite sampling. Typical constituent values for common water use practices are described below.

Quantities of Waste Discharged by Individuals

Typical values of waste discharged by individuals in the United States are presented in Table 13-5. As shown, the use of food waste grinders can increase the discharge of BOD

Table 13-5
Quantity of waste discharged by individuals on a dry weight basis[a]

Constituent (1)	Range (2)	Value, lb/capita·d Typical without ground up kitchen waste (3)	Value, lb/capita·d Typical with ground up kitchen waste (4)	Range (5)	Value, g/capita·d Typical without ground up kitchen waste (6)	Value, g/capita·d Typical with ground up kitchen waste (7)
BOD	0.11–0.26	0.180	0.220	50–120	80	100
COD	0.30–0.65	0.420	0.480	110–295	190	220
TSS	0.13–0.33	0.200	0.250	60–150	90	110
NH_4^+ as N	0.011–0.026	0.017	0.019	5–12	7.6	8.4
Org. N as N	0.009–0.022	0.012	0.013	4–10	5.4	5.9
TKN as N	0.020–0.058	0.029	0.032	9–21.7	13	14.3
Org. P as P	0.002–0.004	0.0026	0.0028	0.9–1.8	1.2	1.3
Inorg. P as P	0.004–0.006	0.0044	0.0048	1.8–2.7	2.0	2.2
Total P as P	0.006–0.010	0.0070	0.0076	2.7–4.5	3.2	3.5
Oil and grease	0.022–0.088	0.0661	0.075	10–40	30	34

[a]Adapted from Crites and Tchobanoglous (1998).

and TSS by 20 percent. The presence and concentration of pathogenic microorganisms is dependent on individual infection and shedding of microorganisms with human waste. The conversion of mass loading rates into wastewater concentrations is related to the amount of dilution that occurs with waste discharge.

Household Products Discharged with Wastewater

The spectrum of household products used on a daily basis increases the overall salinity of the resulting wastewater. Other chemicals or compounds discharged with wastewater may be toxic to treatment organisms or plants irrigated with the treated effluent. Because the removal of salts and toxic constituents is beyond the scope of most small wastewater treatment applications, source control or dilution may be required for some irrigation reuse applications.

A potential advantage of decentralized treatment systems is that individuals who use the system have direct control of the problematic constituents entering the wastewater stream. Fortunately, the concentration of salts in the water is typically low enough not to be of concern for most applications. However, if the discharge of brine from regenerating water softeners or the use of toxic chemicals is not compatible with a particular process or reuse application, these issues can be discussed with the system users, who also have an interest in proper operation of the system. Examples of substances which have been implicated in adverse impacts to wastewater treatment processes include strong disinfectants, fabric softeners, chemical sanitizers for holding tanks, chemotherapy medications, high amounts of oils or grease, and brine from water softeners. In larger systems, a degree of anonymity exists that makes it difficult to identify the particular source of an offending discharge, and there is less individual responsibility for performance and operational matters, as these activities become the responsibility of the municipality or agency having operational responsibility.

Composition of Domestic Greywater

Ions commonly added to wastewater from domestic water use that contribute to salinity include the cationic species sodium, calcium, magnesium, and potassium, and anionic species bicarbonate, carbonate, chloride, fluoride, and sulfate, as shown in Table 13-6. Of the constituents included in Table 13-6, bicarbonate, sodium, chloride, and sulfate are among the most ubiquitous. An analysis of the source of the constituents, including boron (a plant toxin at high concentrations), is shown in Table 13-7. While

Table 13-6 Typical mineral increase from domestic water use[a]

	Increment range, mg/L[b]	
Constituent	In septic tank effluent	In municipal wastewater
Major anions		
Bicarbonate (HCO_3)	100–200	50–100
Carbonate (CO_3)	2–20	0–10
Chloride (Cl)	40–100	20–50[c]
Sulfate (SO_4)	30–60	15–30
Major cations		
Calcium (Ca)	10–20	6–16
Magnesium (Mg)	8–16	4–10
Potassium (K)	10–20	7–15
Sodium (Na)	60–100[d]	40–70[d]
Aggregate measurements		
Total dissolved solids (TDS)	200–400	150–380
Total alkalinity (as $CaCO_3$)	60–120	60–120
Other minor constituents		
Aluminum (Al)	0.2–0.3	0.1–0.2
Boron (B)	0.1–0.4	0.1–0.4
Fluoride (F)	0.2–0.4	0.2–0.4
Manganese (Mn)	0.2–0.4	0.2–0.4
Silica (SiO_2)	2–10	2–10

[a] Adapted in part from Tchobanoglous et al. (2003).
[b] Based on 450 L/capita·d.
[c] Reported values do not include commercial and industrial additions.
[d] Excluding the addition from domestic water softeners.

Table 13-7
Example amounts of constituents found in household chemicals and products and estimated usage rates.

Household chemical or product	Estimated product usage		Example constituent concentration in product		
	Unit	Value	Unit	Constituent	Value
Liquid bleach	L/capita·d	0.05	g/L	Na^+	23.6
			g/L	Cl^-	36.3
Powdered bleach	kg/capita·d	0.05	g/kg	Na^+	47.7
			g/kg	B	22.4
Liquid laundry detergent	L/capita·d	0.05	g/L	Na^+	40.7
Powdered laundry detergent	kg/capita·d	0.10	g/kg	Na^+	400.5
			g/kg	SO_4^{2-}	119.2
			g/kg	HCO_3^-	565.9
Borax	kg/capita·d	0.02	g/kg	B	113.4
Liquid automatic dishwasher detergent	L/capita·d	0.001	g/L	Na^+	60.2
			g/L	P	45
Powdered automatic dishwasher detergent	kg/capita·d	0.005	g/kg	Na^+	257.0
			g/kg	SO_4^{2-}	119.2
			g/kg	B	5.65
Hand dishwashing soap	L/capita·d	0.025	g/L	Na^+	18
			g/L	P	0.13
Water softener (sodium based)	kg/capita·d	0.3	g/kg	Na^+	393.3
			g/kg	Cl^-	606.6
Water softener (potassium based)	kg/capita·d	0.3	g/kg	K^+	524.4
			g/kg	Cl^-	475.6

not listed explicitly in Table 13-7, many of the cleaning products and detergents available increase the alkalinity of the resulting wastewater and may inhibit growth if used directly to irrigate plants that prefer acidic conditions. The expected increase in salt concentration resulting from domestic use is assessed in Example 13-2.

EXAMPLE 13-2. Addition of Salts Resulting from Typical Domestic Water Use.

Calculate the expected increase in salt concentration after domestic use using the data given in Table 13-7. Compare the results to values cited in Table 17-5 as related to salinity, soil permeability, and plant toxicity. Assume a family of three with an actual average total daily water usage of 570 L/d.

Source water quality data:

Constituent	Concentration, mg/L	Constituent	Concentration, mg/L
Na^+	80	Cl^-	56
Mg^{2+}	63	SO_4^{2-}	152
Ca^{2+}	37	HCO_3^-	350
K^+	1.1	F^-	0.2
B	0.75		

Solution

1. Estimate the salinity of the source water.
   a. Salinity (TDS) can be estimated as the sum of the ionic species.

   $$\text{Salinity} = [Na^+] + [Mg^{2+}] + [Ca^{2+}] + [K^+] + [Cl^-] + [SO_4^{2-}] + [HCO_3^-] + [F^-] = 739.3 \text{ mg/L}$$

2. Compute the sodium adsorption ration (SAR) of the source water using Eq. (17-3) from Chap. 17.

   $$\text{SAR} = \frac{[Na^+]}{\sqrt{([Ca^{2+}] + [Mg^{2+}])/2}} = \frac{[80/23]}{\sqrt{([37/20] + [63/12.15])/2}} = 1.86$$

3. Evaluate the source water for irrigation purposes.
   a. Salinity concentration is of slight concern, may impact sensitive plants
   b. SAR is not a concern, no impact on soil permeability
   c. Specific ion toxicity, slight concern for sodium and boron concentrations; some sensitive plants may be impacted by sodium or boron; and chloride concentration is in a safe range.

4. Estimate the change in constituent concentration resulting from typical domestic water use.
   a. Prepare a table of mass increases of salinity constituents expected to be present in the wastewater using the data in Table 13-7.

Household chemical or product	Estimated product usage		Example constituent concentration in product			Mass increase, g/home·d
	Unit	Value	Unit	Constituent	Value	
Liquid bleach	L/capita·d	0.05	g/L	Na^+	23.6	3.5
			g/L	Cl^-	36.3	5.5
Liquid laundry detergent	L/capita·d	0.05	g/L	Na^+	40.7	6.1

(Continued)

Household chemical or product	Estimated product usage		Example constituent concentration in product			Mass increase, g/home·d
	Unit	Value	Unit	Constituent	Value	
Borax	kg/capita·d	0.02	g/kg	B	113.4	6.8
Hand dish-washing soap	L/capita·d	0.025	g/L	Na^+	18	1.4
			g/L	P	0.13	0.01
Water softener (sodium based)	kg/capita·d	0.3	g/kg	Na^+	393.3	354.0
			g/kg	Cl^-	606.6	545.9

b. Estimate the resulting constituent concentration that contribute to salinity based on three people per home and an average total water usage of 570 L/d, using the data from the table in Step 4a.

For sodium, the computation is as follows:

Sodium = 80 mg/L

$$+ \frac{(3.5 \text{ g/d} + 6.1 \text{ g/d} + 1.4 \text{ g/d} + 354.0 \text{ g/d})(1000 \text{ mg/g})}{(570 \text{ L/d})}$$

= 723.7 mg/L

c. A summary of the increased salinity constituent concentration are shown in the following table:

Constituent	Concentration, mg/L	Constituent	Concentration, mg/L
Na^+	723.7	Cl^-	1028
Mg^{2+}	63	SO_4^{2-}	152
Ca^{2+}	37	HCO_3^-	350
K^+	1.1	F^-	0.2
B	12.75		

5. Estimate the salinity of the resulting wastewater by adding the concentration of each constituent resulting from use to the concentration present in the source water.

$$\text{Salinity} = [Na^+] + [Mg^{2+}] + [Ca^{2+}] + [K^+] + [Cl^-] + [SO_4^{2-}] + [HCO_3^-] + [F^-] = 2368 \text{ mg/L}$$

6. Compute the SAR of the resulting wastewater.

$$\text{SAR} = \frac{[Na^+]}{\sqrt{([Ca^{2+}] + [Mg^{2+}])/2}} = \frac{[723.7/23]}{\sqrt{([37/20] + [63/12.15])/2}} = 16.8$$

7. Evaluate the resulting wastewater for irrigation purposes.
   a. Salinity concentration severely restricts water use to only most salt tolerant plants
   b. SAR is in the range expected to severely impact soil permeability
   c. Specific ion toxicity is of severe concern for sodium, chloride, and boron, many plants are expected to be impacted adversely
8. Estimate the change in constituent concentrations resulting from typical domestic water use if a nonsalt based water softener is used and borax is no longer used in the home.

   Repeat Step 4 without water softener discharges and without boron, the resulting values are shown in the following table:

Constituent	Concentration, mg/L	Constituent	Concentration, mg/L
Na^+	99.4	Cl^-	65.6
Mg^{2+}	63	SO_4^{2-}	152
Ca^{2+}	37	HCO_3^-	350
K^+	1.1	F^-	0.2
B	0.75		

9. Estimate the salinity of the resulting wastewater by adding the concentration of each constituent resulting from domestic water use to the concentration present in the source water.

   $$\text{Salinity} = [Na^+] + [Mg^{2+}] + [Ca^{2+}] + [K^+] + [Cl^-] + [SO_4^{2-}] + [HCO_3^-] + [F^-] = 769 \text{ mg/L}$$

10. Compute the SAR of the resulting wastewater.

    $$\text{SAR} = \frac{[Na^+]}{\sqrt{([Ca^{2+}] + [Mg^{2+}])/2}} = \frac{[99.4/23]}{\sqrt{([37/20] + [63/12.15])/2}} = 2.3$$

11. Evaluate the resulting wastewater for irrigation purposes. The water reuse quality has not been adversely affected from domestic water use, following the removal of the salt-based water softener discharge and borax, as compared to the source water.

Comment

Domestic greywater is also known to contain coliform bacteria and other microorganisms, for example, total and fecal coliform concentrations of 10^5 and 10^4 orgamisms/100 mL, respectively, should be expected.

Table 13-8
Typical unit loading factors and expected wastewater constituent concentrations from individual residences in the United States[a]

Constituent	Unit	Typical value[b]	Concentration, mg/L	
			Dilution volume, L/capita·d (gal/capita·d)	
			190 (50)	460 (120)
BOD	g/capita·d	85	450	187
COD	g/capita·d	198	1050	436
TSS	g/capita·d	95	503	209
NH_4^+ as N	g/capita·d	7.8	41.2	17.2
Org. N as N	g/capita·d	5.5	29.1	12.1
TKN as N	g/capita·d	13.3	70.4	29.3
Org. P as P	g/capita·d	1.23	6.5	2.7
Inorg. P as P	g/capita·d	2.05	10.8	4.5
Total P as P	g/capita·d	3.28	17.3	7.2
Oil and grease	g/capita·d	31	164	68

[a]Adapted from Crites and Tchobanoglous (1998).
[b]Data from Table 3-5, Columns 6 and 7, assuming 25 percent of the homes have kitchen waste food grinders.

Composition of Composite Domestic Wastewater

The concentration of wastewater constituents can be determined by considering the amount of waste discharged and the amount of dilution water used to transport the waste. Using the waste quantities shown in Table 13-5, the wastewater composition may be estimated for different water use patterns. The computed values of constituent concentration in domestic wastewater are presented in Table 13-8.

13-4 TREATMENT TECHNOLOGIES

The challenges in the design of many decentralized systems is to provide the required level of treatment subject to high variability in the flowrate and concentration, and subject to site specific limitations and prohibitive economic constraints, especially in situations where water may not yet be valued economically. In addition, the technologies used for individual onsite and cluster systems must also be able to operate for extended periods of time with low maintenance needs, be fundamentally easy to operate, and be designed to accommodate the level of flow and constituent concentration fluctuations described in Sec. 13-3.

The type of treatment system used for decentralized applications depends on the constraints of the project under consideration and the experience of the system designer. For example, some local site conditions may preclude the use of conventional septic-tank soil absorption disposal fields, including shallow soil cover; percolation rates that are considered too low or too rapid; high groundwater table; proximity of wells and water bodies; steepness of slope; and limited area. Thus, the design engineer must select processes that are expected to meet given reliability and performance criteria. Because there are hundreds of processes, including proprietary and nonproprietary systems (Leverenz et al., 2001; Gaulke, 2006), a general knowledge of design and operational characteristics of decentralized systems is needed. Factors that may be considered during the design and selection of decentralized treatment processes are summarized in Table 13-9.

Source separating systems

Source separating systems include facilities that are used to separate solid and liquid wastes without commingling with the bulk wastewater stream. Human waste can be

Table 13-9
Factors to be considered during design and selection of onsite treatment systems

Issue	Description and Examples
Aesthetics	Odor control (e.g., gas tight lids, carbon filters at air release points and vents)
	Above ground components (e.g., tank covers, air pumps, control panels)
	Noise emissions (e.g., pumps, aerators)
Flowrate	Acceptable variability in flow and constituent loading
Maintenance needs	Frequency (e.g., solids removal frequency, outlet filter cleaning, media packing replacement, cleaning emitters and spray nozzles)
	Responsible party (e.g., system manufacturer, third party, owner)
	Costs and fees associated with maintenance
	Time and skill are required for maintenance activities
Monitoring	Capacity for remote monitoring (e.g., pump on/off cycles, pump run time, tank liquid levels, alarm condition, constituent concentrations, UV lamp status)
	Capacity for remote control (e.g., pump settings, alarm reset)
Performance and reliability	Overall performance and reliability (e.g., nutrient reduction, pathogens)
	During power outage (e.g., short periods, >24 h, extended periods)
	Following extended periods of no flow (e.g., during family vacation)
	After exposure to slug dosing of toxic chemicals (e.g., chlorine bleach)
	Startup time required (e.g., hours, days, weeks)
Power usage	Power may be used for pumping, disinfection, control systems, monitoring, and telemetry equipment
Scalability and retrofitting	Ability to expand or upgrade process to accommodate increased hydraulic or constituent loadings
	Ability to utilize components of existing system, if applicable
Service life	Warranties for process components
	Life span for pumps, electrical components, tankage, packing media, etc.
System owner	System leased to building owner
	System owned by building owner
	Owner and user responsibilities
Tank construction	Noncorrosive, lids watertight, lids lockable, aboveground materials UV resistant
Type of process	See Table 13-12
Volume	Total volume of system and hydraulic retention time for emergency storage in case of power failure or clogging

segregated, with or without the use of water, with composting systems and waste incinerating systems. Collection and processing of human waste (and food waste in the case of in-sink food waste grinders) with a composting toilet or separate wet-composting system can reduce the size of downstream wastewater management systems and produce a compost material that can be used for landscaping purposes (Del Porto and Steinfeld, 1999). An incinerating system uses energy to convert human waste into ash using gas or electric powered combustion.

Source separation can also be used for liquid wastes, including urine diversion and greywater separation. Because of the high nutrient value of urine, toilets have been developed that divert urine to a separate holding tank for reuse in agriculture (Ecosan, 2003). An example of a urine diverting toilet is shown on Fig. 13-6a. Similarly, greywater is often considered for reuse due to the reduced presence of pathogens and organic matter. The level of maintenance and user participation required for source separating systems should be considered carefully when selecting these systems, as many of these processes have failed to work adequately in the field. However, in some areas, where limiting conditions exist, source separating systems may be a preferred alternative.

(a)

(b)

Figure 13-6
In building facilities for source separation and pretreatment: (a) view of a urine diverting toilet for source separation of human waste and potential recovery of nutrients in urine (Photo courtesy of C. Etnier) and (b) view of filter used to remove nonbiodegradable fibers from clothes wash water, potentially improving the operation of some onsite treatment systems. The bag and cartridge filters are rated to remove up to 95 and 99+ percent, respectively, from laundry wash water (Photo courtesy of Septic Protector).

In-building Pretreatment

In-building pretreatment systems are used to manipulate the wastewater characteristics after mixing but before discharge to an external treatment system. Examples of in-building pretreatment include solid-liquid separators, grease separation devices, and filters to remove specific constituents. Solid-liquid separators have been developed that use a vortex action to remove solids from toilet wastewater. The separator is located typically in the basement and uses gravity or centrifugal force to separate and deposit solids into a composting process while the liquid portion of the flow continues to further treatment. Grease separation is important for systems that receive high concentrations of fats, oils, and grease, such as from restaurants. The use of more exotic cooking oils and emulsifying detergents can inhibit the performance of grease separators (septic tanks in series have also been used for grease and oil interception). Screens and filters used for clothes washing wastewater, shown on Fig. 13-6b, reduce the amount of lint and nonbiodegradable fibers discharged with the wastewater. Pretreatment devices that prolong the period of operation of the primary treatment system can positively influence the cost and maintenance requirements and/or performance of subsequent treatment processes.

Primary Treatment

The removal of settleable and floatable particulate materials from wastewater is accomplished using a primary treatment system, typically a septic tank or Imhoff tank. Septic tanks have a large capacity to store intercepted solids, which generally have a high content of biodegradable organic matter and are broken down anaerobically in the tank (see Fig. 13-7). The liquefaction and digestion of intercepted solids results in reduced sludge production compared to primary sedimentation process. Values used to compute solids production from various wastewater treatment processes are shown in Table 13-10. Sludge accumulation in septic tanks can be estimated from data shown on Fig. 13-8.

Septic tanks similar to the models used in current practice were first developed by Louis M. Mouras of Vesoul, France in about 1860 and were referred to as *fosse Mouras* (Dunbar, 1908). Due to the enhanced solids digestion capacity, septic tanks are capable of operating for long periods of time (i.e., several years) with minimal maintenance and without any need for energy input. The efficacy of primary treatment is dependent, in part, upon the primary treatment tank specifications and installation. A variety of technologies have also been developed to improve the operation of primary treatment tanks, including pretreatment devices (described previously), liners and sealants, and outlet filters. An example of a typical septic tank system is shown on Fig. 13-7. Liners and sealants are used to improve the water tightness of existing and new tanks. Leaking tanks result in the unintended release of untreated wastewater, possibly resulting in groundwater contamination. Leaking tanks also affect the ability of a septic tank to retain solids, for example, as the water level fluctuates solids suspended or floating in the tank can bypass baffles used for solids retention. Effluent filters are typically placed on the outlet of the primary treatment tank and act as a screen to capture particles that might otherwise pass out of the tank. Effluent or outlet filters that limit the discharge of particulate solids may improve the performance of subsequent treatment processes, particularly biological treatment processes.

As described in the previous section, food waste grinders contribute to the BOD and TSS mass loading. The composition of septic tank effluent with and without food waste grinders compared to untreated domestic wastewater is shown in Table 13-11. An evaluation of solids digestion in a septic tank compared to a primary sedimentation process is shown in Example 13-3.

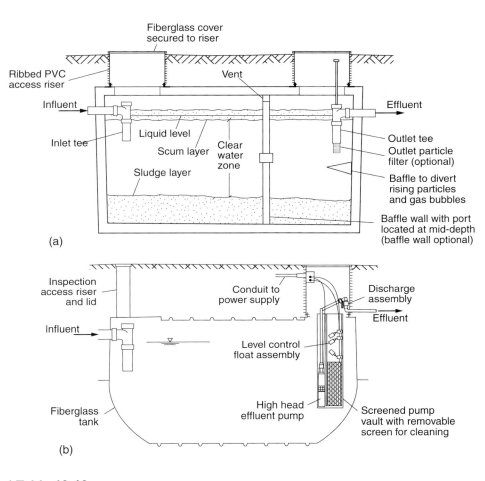

Figure 13-7
Schematic diagrams of septic tanks: (a) two-compartment concrete tank equipped with effluent filter in second compartment and (b) single-compartment plastic septic tank with effluent filter and pump vault. (Adapted from Orenco Systems, Inc.)

Table 13-10

Characteristics of solids from various treatment processes[a]

Process or operation	Total solids, % dry solids		Dry sludge, kg/1000 L	
	Range	Typical	Range	Typical
Primary sedimentation sludge	5–9	6	0.11–0.17	0.15
Waste activated sludge with primary sedimentation	0.5–1.5	0.8	0.07–0.10	0.8
Waste activated sludge from extended aeration[b]	0.8–2.5	1.3	0.08–0.12	0.1
Trickling filter	1–3	1.5	0.06–0.1	0.07
Rotating biological contactor	1–3	1.5	0.06–0.1	0.07
Septic tank solids[c]	4–9	8	0.02–0.1	0.05[d]

[a]Adapted from Tchobanoglous et al. (2003).
[b]Without primary sedimentation.
[c]Based on data from U.S. PHS (1955).
[d]Assuming average accumulation and 8 yr pump out interval (0.13 L sludge/capita·d), sludge specific gravity of 1.02, and 8 percent solids.

Figure 13-8
Analysis of sludge accumulation rates in residential septic tanks.

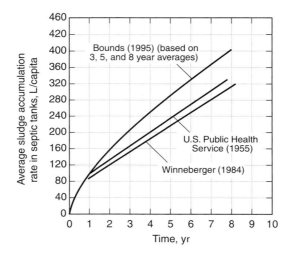

Table 13-11
Typical data on the expected effluent wastewater characteristics from a residential septic tank without and with an effluent filter vault[a]

		Concentration, mg/L						
			Without effluent filter			With effluent filter		
Constituent	Typical complete mix value[b], mg/L	Range	Typical without ground up kitchen waste	Typical with ground up kitchen waste	Range	Typical without ground up kitchen waste	Typical with ground up kitchen waste
(1)	(2)	(3)	(4)	(5)	(6)	(7)	(8)
BOD	450	150–250	180	190	100–140	130	140
COD	1050	250–500	345	400	160–300	250	300
TSS	503	40–140	80	85	20–55	30	30
NH_4^+ as N	41.2	30–50	40	44	30–50	40	44
Org. N as N	29.1	20–40	28	31	20–40	28	31
TKN as N	70.4	50–90	68	75	50–90	68	75
Org. P as P	6.5	4–8	6	6	4–8	6	6
Inorg. P as P	10.8	8–12	10	10	8–12	10	10
Total P as P	17.3	12–20	16	16	12–20	16	16
Oil and grease	164	20–50	25	30	10–20	15	20

[a] Adapted from Crites and Tchobanoglous (1998); Bounds (1997).
[b] Data from Table 13-8, Column 4. Concentration if untreated waste constituents were mixed completely prior to treatment in septic tank.

EXAMPLE 13-3. Rate of Solids Digestion in a Septic Tank System.

Compare the sludge production from a septic tank system to the conventional activated sludge process with primary sedimentation using the data given in Table 13-10. Assume the septic tank is used by two people and has an 8-yr pump out interval. The solids content of the sludge is estimated to be eight percent and the specific weight is 1020 kg/m^3. Because septic tank systems may be operated in conjunction with secondary treatment processes that do not generate waste sludge (i.e., endogenous operation), such as intermittently dosed sand or geotextile packed bed filtration, assume that all waste sludge in the system is accumulated in the septic tank.

Solution
1. Estimate the sludge production from the septic tank system.
   a. Using the data shown on Fig. 13-6, the amount of sludge produced in a septic tank on an 8 yr clean out schedule is 380 L/capita (0.00013 m^3/capita·d).
   b. The mass of sludge produced can be determined using the solids content and specific weight.

   Mass of sludge = (0.00013 m^3/capita·d)(1020 kg/m^3)(0.08) = 0.0105 kg/d

   For two people, the sludge production in the septic tank is 0.021 kg/d.
   c. For two people, the average flowrate is estimated to be 420 L/d.
   d. The sludge production in the septic tank system, normalized for flowrate, is estimated to be 0.05 kg/1000 L.
2. Compare the sludge production in a septic tank to that in an activated sludge process with primary sedimentation using the data given in Table 13-10. The sludge production in a primary clarifier is estimated to be 0.15 kg/1000 L process, with an additional 0.084 kg/1000 L of sludge from the activated sludge process. Therefore, the sludge production from the conventional activated sludge process is higher by a factor of 4.6.

Comment
Solids digestion in a septic tank is relatively passive and because of the long digestion time, solids production is substantially lower as compared to high rate activated sludge processes. Given the expense associated with sludge disposal, the role of onsite solids digestion may improve the economics of wastewater treatment, while the transport of clarified effluent in small diameter, watertight piping will reduce the cost of collection systems. It should also be noted that actual pump out intervals will depend on system design and usage. A typical pump out interval of 3 to 5 yr may be used for design purposes.

In some cases, primary treatment alone may be sufficient for certain types of subsurface irrigation reuse, such as the watering of trees or other point irrigation loads. Some subsurface drip irrigation system components, such as tubing and emitters, are coated or impregnated with biocidal chemical compounds to inhibit biofilm growth and allow for the distribution of primary effluent for subsurface lawn and landscape watering.

Secondary Treatment

To overcome the dispersed nature of onsite and small treatment facilities, decentralized systems are designed for reduced maintenance needs and long-term stable operation with little need for adjustment. The types of secondary treatment systems available are similar in operation to systems used for larger scale applications, these systems (shown on Fig. 13-9) include nonsubmerged attached growth systems (Fig. 13-9a), sequencing

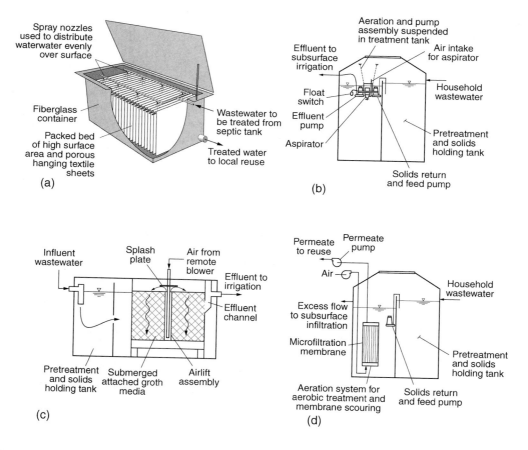

Figure 13-9
Typical examples of treatment systems used for housing development or small community: (a) nonsubmerged synthetic media biofilter (Adapted from Orenco Systems, Inc.), (b) sequencing batch rector (SBR) activated sludge processes (Adapted from ABT Umwelttechnologies GmbH), (c) attached growth submerged treatment process (Adapted from Biomicrobics, Inc.), and (d) membrane bioreactor (MBR) activated sludge process (Adapted from ABT Umwelttechnologies GmbH).

batch reactor suspended growth systems (Fig. 13-9b), and hybrid suspended and attached growth systems (Fig. 13-9c). Membrane based wastewater management systems (e.g., membrane bioreactors, Fig. 13-9d) are also in use and under development for decentralized reuse applications. Because of the need to minimize sludge production, most individual onsite and cluster systems are operated at low hydraulic and organic loading rates, resulting in long solids retention times (SRTs). Descriptions and typical design considerations for these and other decentralized technologies are shown in Table 13-12. The SRT for a typical decentralized treatment process is computed in Example 13-4.

EXAMPLE 13-4. Computation of SRT in an Onsite Aeration Process Treatment System.

Many onsite treatment processes that make use of extended aeration operate at a MLSS of 750 mg/L. The primary mechanism of biomass loss is solids in the effluent flow, with the average effluent solids content of 15 mg/L. Typical aeration tank volume is 4000 L for a flowrate of 2000 L/d. Estimate the SRT for this treatment process. Note that while sludge wasting may be conducted annually as a batch operation, it is not considered because it does not affect the SRT during normal operation.

Solution

1. Assuming that no biomass is entering with the influent flow, the biomass leaving the tank can be determined as the biomass leaving in the effluent and the waste sludge.

    Biomass leaving in the effluent is :

    Effluent biomass = (2000 L/d)(15 mg/L)/(1000 mg/g) = 30 g/d

2. The average biomass in the system is the product of the average MLSS and tank volume.

    Biomass in aeration process = (4000 L)(750 mg/L)/(1000 mg/g) = 2906 g

3. The estimated SRT is determined by dividing the biomass in the aeration process by the biomass leaving the system.

    SRT = 2906 g/(30 g/d) = 97 d

Comment

The SRT in the above aeration process, 97 d, is long compared to the 20 to 30 d SRT used for most activated sludge extended aeration processes. In addition, most sand filtration and other packed bed filtration processes operate reliably for more than 20 yr with effluent TSS concentrations below 5 mg/L and never require solids removal, suggesting much higher SRT values.

Table 13-12
Summary of secondary biological processes used for decentralized wastewater treatment for onsite and cluster systems

Technology	Description	Design considerations
(a) Single-pass packed bed filter	The fundamental components of the single-pass filter are (1) a medium upon which a microbial biofilm community develops, (2) a container or [lined] excavated pit to house the medium, (3) a system for applying the water to be treated to the medium, and (4) a system for collection and discharge of the treated water. Several media with varying properties have been utilized (e.g., sand, glass, peat, rock, foam, geotextiles) and ultimately determine the size of the filter required. The water to be treated is applied, periodically or intermittently, in small doses to the medium to facilitate a high amount of oxygen transfer.	• Organic or hydraulic overloading may result in clogging • Hydraulic loading rate = 40 to 50 L/m²·d • Depth ranges from 0.6 to 0.9 m • Dosing frequency >24 dose/d • Orifice cleaning required on periodic basis • Cost of filter medium may be prohibitive
(b) Multi-pass packed bed filter	Similar in construction to the single-pass packed bed filter, except for the recirculation of the filter effluent through the filter bed several times before discharge. After a dose of the water to be treated passes through the biofilter, a portion (return flow) is returned to the septic tank or intermediate storage for reapplication, and the remainder (effluent) is discharged to subsequent treatment. The return flow is combined with the process influent water (typically effluent from a septic tank) before reapplication to the biofilter. Because of the need for repeated application (recirculation) of the wastewater, pumping and control system needs may be increased, as compared to single-pass systems.	• Medium may need to be replaced or cleaned if performance decreases • Effluent recirculation ratio = 2 to 8 • Organic or hydraulic overloading may result in clogging • Hydraulic loading rate = 200 to 1000 L/m²·d • Depth ranges from 0.6 to 0.9 m • Dosing frequency >24 dose/d
(c) Compact activated sludge	Several types of single tank activated sludge processes have been used for small applications. In general, influent wastewater from a septic tank is mixed with treatment organisms and aerated. Flow moves through the system by hydraulic displacement. Mixed liquor is discharged into sedimentation compartment where settleable solids are returned to the aeration compartment by duct under baffle. Most activated sludge systems used for decentralized application operate under an extended aeration mode for reducing sludge volumes and increased transformation and removal of wastewater constituents.	• Variations in organic loading may result in filamentous growth and solids carryover • Variable performance in unattended applications • MLSS ranges from 400 to 2000 mg/L • HRT = 2 to 4 d • Periodic sludge removal will be required

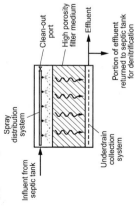

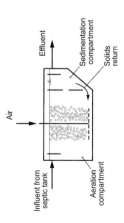

(d) Hybrid activated sludge with fixed or suspended packing

Similar in action to the compact activated sludge process with the addition of fixed internal packing or suspended internal packing. The fixed internal packing is generally a plastic matrix material designed to maximize fixed film microbial growth and contact with wastewater to be treated. In many cases, and air lift pump is used to distribute water on top of the fixed packing. The suspended packing is generally circulated in the aeration tank by currents induced by the aeration device. Wastewater moves through the system by hydraulic displacement, i.e., as water is discharged into the system, an equal volume flows out of the system.

- In general, having a fixed film component has a buffering effect in the event of a process disturbance.
- Other design factors similar to compact activated sludge process given above

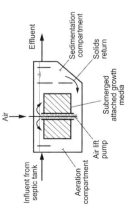

(e) Sequencing batch reactor with or without internal packing

In a sequencing batch reactor (SBR) wastewater is stored in the equalization or aeration chamber of the treatment system until a sufficient volume (a batch) is collected, at which point the treatment process begins. The batch of wastewater is seeded with treatment organisms (activated sludge) and aerated for the treatment period. After the reaction is complete, aeration and mixing are stopped, and the flocculated bacteria and other solid particles settle out. The clear layer (supernatant) is discharged from the reaction chamber, and the next batch of wastewater to be treated begins to flow into the reaction chamber. The fill, react, settle, discharge cycle is repeated continuously. Sequencing batch reactors have been used with and without internal packing materials, however, packing material is recommended for onsite systems to improve process stability.

- Requires additional valves, pumps, and controls compared to other treatment systems
- Potential for high quality treatment for water reuse applications.
- MLSS ranges from 3 to 6 g/L
- HRT ranges from 8 to 14 h
- Sludge handling facilities will be required

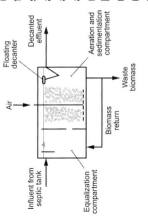

(f) Rotating biological contactor

Rotating biological contactors (RBCs) consist of stacks of rotating disks mounted on a horizontal shaft that are partially submerged and rotated in a reactor as wastewater flows through. The attached microbial community is alternately exposed to the atmosphere and wastewater with each revolution of the disk. The process is optimized by adjusting the speed of rotation and the depth of submergence. For decentralized applications, RBCs have been found to be particularly effective.

- Excellent performance and reliability in several demonstration projects
- Maintenance may include replacing motor, servicing bearings, and cleaning attached growth media as needed
- Process is relatively silent compared to dosing pumps for aeration

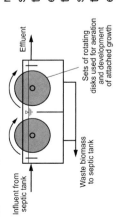

(Continued)

Table 13-12

Summary of secondary biological processes used for decentralized wastewater treatment for onsite and cluster systems (*Continued*)

Technology	Description	Design considerations
(g) Membrane bioreactor	Membrane bioreactors similar to the systems described in Chaps. 7 and 12 have been developed for individual onsite and cluster systems. The system consists of an activated sludge process with a submerged fine pore membrane and a system for maintaining flux through the membrane. Membrane bioreactors are used for wastewater treatment and reclamation at communities of various sizes.	• Variations in organic loading may result in filamentous growth and solids carryover • MLSS ranges from 8000 to 12,000 mg/L • Periodic sludge removal will be required • Systems are available to accommodate individual homes as well as whole communities
(h) Constructed wetland or other natural treatment system	Natural systems for onsite treatment are an alternative to mechanized treatment systems. Natural systems have the advantage of reduced operation and maintenance costs, but are often more difficult to control because their operation may be affected by weather, climate, and seasonal factors and patterns. The principal types of natural treatment processes include surface flow constructed wetlands, subsurface flow constructed wetlands, ecological systems, evapotranspiration systems, and lagoons. Both horizontal and vertical flow wetlands have been used, with better performance with vertical flow.	• Climate and seasonal conditions may affect performance more than other systems • Long HRTs possible • Surfacing effluent should be avoided to reduce public exposure and mosquito growth • Low energy and potential for beneficial habitat • Aesthetic benefit

The systems described in Table 13-12 are able to meet secondary treatment goals and may be used for preparation of effluent for some types of direct reuse, disinfection, or tertiary treatment. In most applications, the treatment systems used for individual onsite and cluster systems are located below grade to take advantage of gravity flow and to make the system unobtrusive.

A number of nutrient related discharge standards have been applied to decentralized wastewater systems because of the concern for nitrate in groundwater and eutrophication in surface waters. As shown in Fig. 13-10, several alternatives have been implemented to reduce nitrogen and phosphorus in wastewater from decentralized treatment systems. However, due to the variable nature of flow and constituent concentrations, reliable nutrient removal may be difficult to implement for individual onsite systems. Cluster systems and larger systems may be better suited to nutrient removal as chemical-dosing facilities and process control may be added to ensure reliable nutrient removal. It should be noted that in small systems several factors limit the effectiveness of nitrification and denitrification. With respect to nitrification, low alkalinity, a characteristic of many surface waters, limits the degree of nitrification. Similarly, the characteristics of the wastewater, particularly, the carbon to nitrogen ratio, limits the degree of denitrification that can be achieved. Both of these subjects are reviewed in the following discussion.

Nutrient Removal

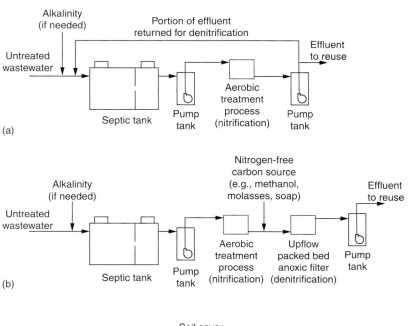

Figure 13-10
Process flow diagrams for nutrient removal in onsite systems: (a) denitrification using internal carbon source (wastewater), (b) denitrification using external carbon source, and (c) aerobic/anaerobic process for phosphorus removal.

Nitrogen Reduction

The basic process used for nitrogen removal consists of two stages, nitrification and denitrification. Other alternative processes have also been developed for nitrogen removal, but are not commonly applied in decentralized systems because of the increased process control required for effective performance. A review of alternate methods of nitrogen removal is presented by Schmidt et al. (2003).

Nitrification is the biological oxidation of ammonia nitrogen in a two-step aerobic process by autotrophic bacteria (i.e., fix and reduce inorganic carbon, such as CO_2 and HCO_3^-) collectively known as nitrifiers. The first step is the oxidation of ammonia to nitrite by the nitroso-group of bacteria (e.g., Nitrosomonas, Nitrosopira, Nitrosococcus). The second step is the conversion of nitrite to nitrate by nitro-group bacteria (e.g., Nitrobacter, Nitrococcus, Nitrospira). Because the organisms responsible for nitrification are autotrophic, rely on aerobic conditions, and utilize nitrogen for energy, they are both slow growing and sensitive to environmental conditions (e.g., oxygen concentration and temperature in the treatment reactor). Nitrifiers are also sensitive to pH, with recommended pH values ranging from 6.5 to 8.0 (U.S. EPA, 1993a). Equations used to represent the nitrification reactions are shown in Table 13-13. As shown in the reactions for the oxidation of ammonia, hydrogen ion is produced.

Table 13-13
Summary of nitrification and denitrification stoichiometric relationships

Nitrification reactions	Stoichiometric relationship
Oxidation of ammonia for energy	$NH_4^+ + 1.5O_2 \rightarrow NO_2^- + H_2O + 2H^+$
Oxidation of nitrite for energy	$NO_2^- + 0.5O_2 \rightarrow NO_3^-$
Overall ammonia oxidation to nitrate	$NH_4^+ + 2O_2 \rightarrow NO_3^- + H_2O + 2H^+$
Ammonia oxidation with cell synthesis based on observed yield	$NH_4^+ + 1.44O_2 + 0.0496CO_2 \rightarrow$ $0.99NO_2^- + 0.01C_5H_7O_2N + 0.97H_2O + 1.99H^+$
Nitrite oxidation with cell synthesis based on observed yield	$NO_2^- + 0.00619NH_4^+ + 0.50O_2 + 0.031\,CO_2 + 0.0124H_2O \rightarrow$ $NO_3^- + 0.00619C_5H_7O_2N + 0.00619H^+$
Overall nitrification with cell synthesis based on observed yield	$NH_4^+ + 1.92O_2 + 0.08CO_2 \rightarrow$ $0.98NO_3^- + 0.016C_5H_7O_2N + 0.95H_2O + 1.98H^+$
Alkalinity utilization for buffering pH changes	$6H^+ + 6HCO_3^- \rightarrow H_2CO_3 + 5CO_2 + 5H_2O$
Denitrification reactions	
Nitrate reduction with wastewater[a] as carbon source based on observed yield	$NO_3^- + 0.17C_{10}H_{19}O_3N + H^+ \rightarrow 0.4N_2 + 0.15C_5H_7O_2N + 1.13H_2O + 0.91CO_2 + 0.17NH_4^+ + 0.17OH^-$
Nitrate reduction with methanol as carbon source based on observed yield	$NO_3^- + 1.08CH_3OH + 0.073H_2CO_3 \rightarrow$ $0.47N_2 + 0.056\,C_5H_7O_2N + 1.51H_2O + 0.83CO_2 + OH^-$
Alkalinity production from hydroxide	$H_2CO_3 + 5CO_2 + 6OH^- \rightarrow H_2O + 6HCO_3^-$

[a] The chemical formula $C_{10}H_{19}O_3N$ is used to represent organic matter in wastewater that exerts a BOD demand.

If sufficient alkalinity is not present to neutralize the hydrogen ion, the pH will drop and potentially inhibit nitrification. An analysis of nitrification reactions is shown in Example 13-5.

EXAMPLE 13-5. Analysis of Nitrification Process.

Nitrification occurs in many decentralized wastewater treatment processes due to the high SRT values. However, in systems with low alkalinity, nitrification is inhibited by low pH conditions. Using the stoichiometric equations given in Table 13-13, estimate the amount of dissolved oxygen and alkalinity required for nitrification of septic tank effluent with an ammonium nitrogen concentration of 50 mg–N/L.

Solution

1. Estimate the amount of oxygen required to oxidize ammonia to nitrate.
    a. Write the stoichiometric equation for nitrification using the overall nitrification reaction with cell synthesis based on observed yield from Table 13-13.
    $$NH_4^+ + 1.92O_2 + 0.08CO_2 \rightarrow 0.98NO_3^- + 0.016C_5H_7O_2N + 0.95H_2O + 1.98H^+$$
    b. Determine the mass of oxygen and nitrogen in the reaction.
       From the equation given in Step 1a, 1.92 mole of oxygen are used for each mole of ammonia nitrogen oxidized.
       The mass of oxygen = (32 g/mole)(1.92 mole) = 61.4 g
       The mass of nitrogen = (14 g/mole)(1 mole) = 14 g
    c. Estimate the oxygen consumed in the reaction per gram of ammonia.
       The amount of oxygen required = (61.4 g)/(14 g) = 4.39 g O_2/g N
    d. Compute the amount of oxygen required to oxidize the ammonia nitrogen.
       Oxygen required = (4.39 mg O_2/mg N)(50 mg–N/L) = 219.5 mg O_2/L

2. Estimate the amount of alkalinity required to neutralize the hydrogen ion, and thus maintain the pH within a range that does not inhibit nitrification.
    a. Determine the mass of hydrogen ion and nitrogen in the reaction.
       From the equation given in Step 1a, 1.98 mole of hydrogen are produced for each mole of ammonia nitrogen oxidized.
       The mass of hydrogen = (1 g/mole)(1.98 mole) = 1.98 g
       The mass of nitrogen = (14 g/mole)(1 mole) = 14 g
    b. Estimate the hydrogen produced in the reaction per gram of ammonia.
       The amount of oxygen required = (1.98 g)/(14 g) = 0.14 g H^+/g N
    c. Determine the amount of alkalinity required to neutralize the hydrogen using the following equation from Table 13-13.
    $$6H^+ + 6HCO_3^- \rightarrow H_2CO_3 + 5CO_2 + 5H_2O$$
    As shown in the above equation, 1 mol of bicarbonate is consumed for each mole of hydrogen ion neutralized, or 8.63 g HCO_3^-/g N oxidized.

d. The stoichiometric alkalinity requirement, expressed as $CaCO_3$, is computed as follows.

$$\text{Alkalinity} = \left(\frac{8.63 \text{ g } HCO_3^-}{\text{g } NH_4^+ - N}\right)\left(\frac{50 \text{ g } CaCO_3}{\text{equivalent}}\right)\left(\frac{\text{equivalent}}{61 \text{ g } HCO_3^-}\right)$$

$$= \frac{7.07 \text{ g } CaCO_3}{\text{g } NH_4^+ - N}$$

e. Compute the amount of alkalinity, as $CaCO_3$, required to buffer the pH change during nitrification of 50 mg/L of ammonium nitrogen.

$$\text{Alkalinity} = (50 \text{ mg–N/L})(7.07 \text{ mg } CaCO_3/\text{mg N}) = 353.5 \text{ mg/L}$$

Comment

As shown in the above example, a substantial amount of oxygen and alkalinity is required for nitrification of septic tank effluent. Systems for individual residences may have difficulty in maintaining reliable nitrification, while larger systems can be maintained and operated under optimized conditions.

The final step of nitrogen removal, denitrification, occurs under anoxic conditions (i.e., without free dissolved oxygen present), requires nitrate or nitrite (the alternate electron acceptor), and a supply of biodegradable organic carbon or alternate electron donor. The bacteria that convert nitrate or nitrite to nitrogen gas are known as facultative aerobes, which are able to complete respiration reactions using oxidized nitrogen only when dissolved oxygen is limited. Both heterotrophic and autotrophic bacteria are able to denitrify (Rittmann and McCarty, 2001). The heterotrophic bacteria (e.g., *Pseudomonas, Bacillus*) use organic compounds such as those present as BOD in wastewater, or other organic compounds added when BOD is insufficient (e.g., methanol, acetate). The autotrophic denitrifiers use an inorganic electron donor (e.g., hydrogen gas or sulfer), which is typically from an external supply.

It is common to use the BOD present in the influent wastewater for denitrification in individual onsite and small-cluster systems because of the expense of using an alternate electron donor, such as methanol or hydrogen, as described above (see Fig. 13-10a). For denitrification with wastewater as the carbon source, nitrified effluent is blended with wastewater that has not been treated aerobically. For individual and small applications, nitrified effluent is discharged to the septic tank for denitrification. It should be noted that the overall nitrogen removal possible when using the BOD in wastewater as the carbon source is further limited by the carbon to nitrogen ration (i.e., BOD/TKN) and amount of effluent recirculation, as a portion of the wastewater applied in the aeration step is discharged without recirculation for denitrification. It has been observed that total nitrogen removal is limited to 50 to 70 percent when BOD in wastewater is used as the carbon source. In some cases, higher nitrogen removals have been obtained than predicted using stoichiometric relationships. The discrepancy is attributed to the use of BOD_5 measurements for characterization of the carbon source, as the BOD_5 test may underestimate the actual amount of the carbon source available. An analysis of denitrification based on stoichiometric relationships is shown in Example 13-6.

EXAMPLE 13-6. Analysis of Denitrification Process.

Denitrification occurs when nitrified effluent, such as from a sand filter, is returned to the inlet of a septic tank and blended with the influent raw wastewater. In this case, denitrification may be limited by the amount of BOD present. Using the stoichiometric equations given in Table 13-13, estimate the amount of BOD required and the amount of alkalinity recovered by denitrification of wastewater with a nitrate concentration of 50 mg–N/L. Assume the composition of the BOD is represented as $C_{10}H_{19}O_3N$ and that only 68 percent of the ultimate BOD is available for denitrification.

Solution
1. Estimate the amount of BOD required to reduce nitrate to nitrogen gas.
   a. The stoichiometric equation for denitrification (nitrate reduction with wastewater as a carbon source based on observed yield from Table 13-13).

   $$NO_3^- + 0.17C_{10}H_{19}O_3N + H^+ \rightarrow 0.4N_2 + 0.15C_5H_7O_2N + 1.13H_2O$$
   $$+ 0.91CO_2 + 0.17NH_4^+ + 0.17OH^-$$

   b. Estimate the ultimate BOD (uBOD) of wastewater assuming complete carbonaceous oxidation to carbon dioxide and ammonia of $C_{10}H_{19}O_3N$. The balanced equation is:

   $$C_{10}H_{19}O_3N + 12.5O_2 \rightarrow NH_3 + 10CO_2 + 8H_2O$$

   From the balanced equation, 12.5 mole O_2 are theoretically required per mole of $C_{10}H_{19}O_3N$, or 1.99 g O_2/g $C_{10}H_{19}O_3N$
   Assume that only 68 percent of the uBOD, which corresponds to the 5 d BOD test, is available for the denitrification reaction. The available BOD is estimated to be:

   $$BOD = (1.99)(0.68) = 1.35 \text{ g } O_2/\text{g } C_{10}H_{19}O_3N$$

   c. Estimate the mass of nitrate and BOD consumed from the equation given in Step 1a.
   The mass of nitrate nitrogen = (14 g/mole)(1 mole) = 14 g
   The mass of BOD = (201 g/mole)(0.17 mole)(1.35 g O_2/g) = 46 g O_2
   d. Estimate the BOD consumed in the reaction per gram of nitrate.
   The amount of oxygen required = (46 g)/(14 g) = 3.3 g O_2/g N
   e. Compute the amount of BOD required to reduce nitrate nitrogen.

   $$BOD \text{ required} = (3.3 \text{ mg } O_2/\text{mg N})(50 \text{ mg-N/L}) = 165 \text{ mg BOD/L}$$

2. Estimate the amount of alkalinity that is produced.
   a. Determine the mass of hydroxide ion and nitrogen in the reaction.
   From the equation given in Step 1a, 0.17 mole of hydroxide ion are produced for each mole of nitrate nitrogen reduced.
   The mass of hydroxide = (17 g/mole)(0.17 mole) = 2.89 g
   The mass of nitrogen = (14 g/mole)(1 mole) = 14 g

b. Estimate the hydrogen produced in the reaction per gram of ammonia. The amount of oxygen required = (2.89 g)/(14 g) = 0.21 g OH$^-$/g N
c. Estimate the increase in alkalinity using the alkalinity production equation from Table 13-13.

$$H_2CO_3 + 5CO_2 + 6OH^- \rightarrow H_2O + 6HCO_3^-$$

As shown in the above equation, 1 mole of bicarbonate is produced for each mole of hydroxide ion, or 4.69 g HCO$_3^-$/g N oxidized.

d. The stoichiometric alkalinity requirement, expressed as CaCO$_3$, is computed as follows:

$$\text{Alkalinity} = \left(\frac{4.69 \text{ g HCO}_3^-}{\text{g NH}_4^+ - \text{N}}\right)\left(\frac{50 \text{ g CaCO}_3}{\text{equivalent}}\right)\left(\frac{\text{equivalent}}{61 \text{ g HCO}_3^-}\right)$$

$$= \frac{3.85 \text{ g CaCO}_3}{\text{g NH}_4^+ - \text{N}}$$

Comment

As shown in the above example, a substantial amount of BOD is required for denitrification. In some applications, the aeration requirements can be reduced to compensate for the reduced BOD concentration. Also, the production of alkalinity can, in part, compensate for the alkalinity utilized during nitrification, as shown in Example 13-5.

Chemical dosing systems have also been used to add an alternate carbon source (that does not contain nitrogen) and these systems are able to achieve complete nitrogen removal (see Fig. 13-10b). Nitrogen removal systems that use an alternate carbon source and other nitrogen removal processes are introduced in Chap. 7. Another type of process that has been used successfully for near complete nitrogen removal utilizes a passive filter bed of wood substrate to filter nitrified wastewater (Robertson et al., 2005).

Phosphorus Reduction

As discussed in Chap. 7, phosphorus may be removed biologically by controlling the redox conditions to facilitate cell accumulation (Tchobanoglous et al., 2003). The biomass with concentrated phosphorus may then be wasted, thus lowering the amount of phosphorus in the water. However, in individual and small cluster systems, it is not practical to waste biomass frequently enough to control phosphorus. Therefore, phosphorus removal from wastewater is accomplished typically by chemical precipitation or adsorption. Dosing treated effluent with ferric chloride followed by filtration of the precipitate has been used successfully. Dosing of alum directly into a septic tank has also been used to control phosphorus (Jowett, 2001). Phosphorus may be removed by contacting with filter media such as lightweight expanded clay and shale aggregates, crushed red brick, and slag (Anderson et al., 1998; Baker et al., 1998; Johansson, 1997; Zhu et al., 1997). Additional information on occurrence and fate of phosphorus in onsite wastewater systems may be found in Lombardo (2006). A flow diagram for phosphorus removal from septic tank effluent is shown on Fig. 13-10c.

Disinfection Processes

When effluent is used for subsurface irrigation, disinfection is not required typically. Disinfection may improve the performance of subsurface drip irrigation where biofilm clogging of emitters is a concern. Varying degrees of disinfection also occur naturally as wastewater percolates through soil, however, most soil infiltration systems have not been designed to take advantage of this effect. Decentralized treatment systems that include a disinfection process ensure protection of groundwater resources and public health.

A summary of disinfection processes that may be utilized for small wastewater flows is presented in Table 13-14. Of the disinfectants reported in Table 13-14, calcium hypochlorite and UV are used most commonly for small systems (U.S. EPA, 2002, 1980). Ozonation has also been used but the cost of an effective system may be prohibitive for small applications. Other processes that are identified in Table 13-14 that may be applied for wastewater disinfection include biological filtration (Gross and Jones, 1999; Emerick et al., 1997) and peracetic acid (Kitis, 2004). Disinfection with chlorine gas and chlorine dioxide is typically not used for decentralized applications as these processes present hazards for small facilities associated with storage, handling, and application. All of the processes identified in Table 13-14 can be used to disinfect wastewater; however, each process has inherent constraints that may limit general application and should be considered (Leverenz et al., 2005; U.S. EPA, 2002). A typical UV disinfection unit and process flow diagram is shown on Fig. 13-11.

Table 13-14
Summary of disinfectants used for disinfection of small wastewater flows

Disinfectant	Formula	Form	Constraints or concerns for application to small flows
Sodium hypochlorite	NaOCl	Liquid	Corrosive, toxic, formation of carcinogenic byproducts; requires chemical feed system; effectiveness may depend on water quality
Calcium hypochlorite	$Ca(OCl)_2$	Solid tablet	Corrosive, toxic, formation of carcinogenic byproducts; requires tablet feed system; effectiveness may depend on water quality; nonuniform tablet erosion can result in variable chlorine dose
Ozone	O_3	Gas	Corrosive, toxic; requires a feed gas preparation unit and a pump for injection of ozone; effectiveness may depend on water quality; usually not cost effective for small systems
Peracetic acid	CH_3CO_3H	Liquid	Corrosive, toxic; not commercially available; requires a chemical feed system; effectiveness may depend on water quality
Ultraviolet (UV) light	—	Electromagnetic radiation	Requires periodic lamp maintenance or replacement; fouling can reduce effectiveness; performance sensitive to water quality
Biological filtration	—	Enzymatic activity, predation	Size of filter may be a limitation; expense of obtaining appropriate media

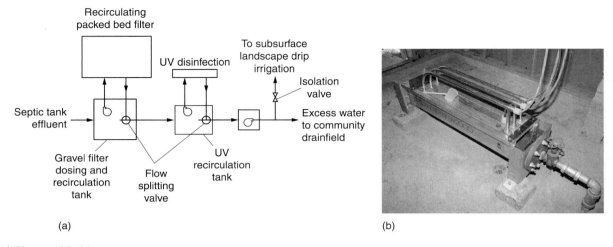

Figure 13-11
Disinfection with UV for landscape irrigation in remote developments: (a) schematic flow diagram of treatment system and (b) view of UV system. (Crites et al., 1997.)

Performance

Performance achievements of the various processes used in decentralized wastewater treatment are described in Table 13-15. The values shown in Table 13-15 are based on monitoring from processes that are maintained properly. Proper maintenance is essential to prevent processes from performing poorly or failing. As shown in Table 13-15, a number of processes may be used for nutrient removal; however, removal of wastewater nutrients may not be required or desirable in all cases. It should also be noted that the performance values given in Table 13-15 are specified for favorable conditions and that site specific limitations could affect performance in terms of individual constituents, e.g., an alkalinity limitation may impact nitrification and therefore denitrification.

Reliability

For accurate determination of the feasibility and overall cost associated with onsite and decentralized water reuse systems, an understanding of reliability is needed. Because of the high peak-to-mean variability experienced in small systems (see Table 13-2), as compared to large systems, onsite and decentralized systems should be designed to accommodate large variations, as well as some amount of neglect or misuse. The reliability of a decentralized system depends on the size and type of process, the system design, and the degree of mechanization and control involved. A number of tools have been used for the analysis of decentralized systems including the development of failure curves, GIS based tools, failure modes and effects analysis, and probability analysis (Etnier et al., 2005).

Maintenance Needs

Because decentralized systems are expected to operate for long periods of time with little or no maintenance or operational adjustments, consideration of process characteristics are critical to proper design and operation. To make decentralized treatment economical,

Table 13-15
Reported treatment levels achievable using various combinations of unit operations and processes used for decentralized wastewater treatment[a]

Process	Typical achievable effluent quality, mg/L, except turbidity, NTU						
	TSS	BOD	COD	TN	NH_4^+-N	PO_4-P	Turbidity
Septic tank without effluent filter[b]	40–140	150–250	250–500	50–90	30–50	8–12	15–30
Septic tank with effluent filter[b]	20–55	100–140	160–300	50–90	30–50	8–12	10–20
Septic tank + intermittent single-pass sand filter	0–5	0–5	10–40	<30	1–5[c]	6–10	0.01–2
Septic tank + intermittent multi-pass media filter	0.5–15	5–10	20–40	7–20	1–3[c]	6–10	0.1–2
Septic tank + compact activated sludge process	10–30	20–60	40–120	20–40	1–5	6–10	
Septic tank + hybrid activated sludge with fixed or suspended packing	5–30	10–40	20–80	20–40	1–5	6–10	
Septic tank + rotating biological contactor	1–15	2–20	10–50	5–30	1–5	6–10	
Septic tank + constructed wetland	10–20	10–20	25–50	5–20	1–10	4–8	
Septic tank + intermittent filter + phosphorus removal	0–5	0–5	10–40	<30	0–5[c]	<0.5	0.01–2
Septic tank + intermittent sand filter + nitrogen removal	0.5–15	10–30	20–60	0.5–5	1–4	6–10	
Septic tank + intermittent sand filter + nitrogen removal + phosphorus removal	0.5–15	10–30	20–60	0.5–5	1–4	<0.5	
Septic tank effluent collection system + membrane bioreactor + phosphorus removal	<1	<5	<30	<10	<1	<0.1	<0.1
Septic tank effluent collection system + extended aeration activated sludge + cloth filter + microfiltration	0	<2	<15	<10	<1	<6	<0.1
Septic tank effluent collection system + sequencing batch reactor	<5	<5	<30	<5	<1	4–8	<1

[a] Adapted from Leverenz et al. (2001) based on literature review and, therefore, data are not standardized.
[b] Values from Table 13-11.
[c] It is assumed that the values reported in the literature were based on processes with sufficient alkalinity for complete nitrification to occur. Residual effluent ammonium values will vary depending on the available alkalinity.

several design features are included typically to compensate for the lost economy of scale (Katehis, 2004) that will affect the maintenance needs:

- Efficient construction practices and compact designs (e.g., common walls, integrated control systems, multipurpose pumps)
- Minimized process control, self-regulating system designs
- No chemical addition, dry-chemical feed systems used when needed
- Minimized mechanical maintenance
- Minimized solids production, no onsite solids processing
- Implementation of telemetry systems for remote monitoring

Proper performance requires familiarity with the wastewater flow to be treated, including flowrate and constituent concentrations, which must be known or estimated in advance as these parameters affect operation.

A typical water reuse system for an individual residence or cluster system may consist of only a septic tank, pump and control system, and drip irrigation system. With proper design and maintenance, this type of system will operate reliably for an extended period of time. For example, the septic tank, if adequately sized, may need to have solids removed every 5 to 10 yr. Similarly, many pumps used for onsite systems are expected to function for about 10 yr. Properly designed drip irrigation systems, with continuous removal of particulate matter and line flushing, are expected to operate without clogging for long periods of time. However, if there is a lapse in maintenance or the septic tank is under designed, solids may carryover into the drip irrigation system and result in emitter clogging that may be difficult to rectify. Therefore, proper design and maintenance of onsite and decentralized systems is of critical importance for long-term operation and performance.

Maintenance needs for onsite treatment processes depend on the type of system, as well as the overall system design and use. For example, onsite treatment systems for sensitive applications that utilize advanced treatment or disinfection processes should be monitored on a quarterly basis or equipped with a telemetry system for continuous remote monitoring. For less sensitive applications, the maintenance needs may be reduced as appropriate. The manufacturer or designer of a treatment system should be consulted for specific maintenance intervals and activities.

13-5 TECHNOLOGIES FOR HOUSING DEVELOPMENTS AND SMALL COMMUNITY SYSTEMS

For housing development and small community applications, wastewater is collected using a network of pipes and appurtenances, and treated using small treatment facilities. Community treatment facilities are used when there is sufficient wastewater flow, sensitive or prohibitive environmental circumstances (e.g., residences adjacent to lakes), or where the arrangement and proximity of buildings in a housing development or small community allow for the use of wastewater collection systems. The types of collection systems typically used in decentralized applications and the commonly used treatment technologies are described below.

Collection Systems

Where cluster or community systems are to be used, a wastewater collection system is required. Although the use of large gravity-flow wastewater collection systems for the collection and transport of wastewater from urban areas continues to be the accepted norm for wastewater practice in the United States, conventional gravity-flow wastewater collection systems may prove to be counterproductive in some areas. For example, the use of conventional gravity-flow wastewater collection systems may not be economically feasible for reasons of proximity, topography, high water table, structurally unstable soils, and rocky conditions. Further, in small communities without wastewater collection systems, the cost of installing conventional gravity-flow wastewater collection systems may be prohibitive, especially if the density of development is low.

To overcome the difficulties posed by conventional gravity-flow wastewater collection systems, a number of alternative types of collection systems have been developed, including:

- Onsite primary treatment and small-diameter variable-grade gravity effluent collection
- Pressure collection with onsite primary treatment or onsite sump and grinder pump
- Vacuum collection with onsite collection tank and controls
- Hybrid collection systems

A comparison of collection systems is shown on Fig. 13-12. Collection systems for decentralized systems are designed to be watertight, so a design allowance for infiltration and inflow is not required. The potential for soil and water contamination from conventional wastewater collection systems through exfiltration is also eliminated through the use of alternative collection systems.

Alternative collection systems used for small communities, neighborhoods, and housing developments are typically based on the number of contributing equivalent dwelling units (EDUs). An EDU is assumed to be the flowrate from an average residence in the community (e.g., household size of 3.5 persons). For design purposes, the peak flowrate is often taken to be 1.3 to 1.9 L/min·EDU. An additional flow of 38 to 76 L/min should be added as a peaking allowance when designing for a small number of EDUs.

Small Diameter Variable Grade Gravity Collection System

Where sufficient slope exists, septic tank effluent can be collected using gravity flow in small diameter piping, known as small diameter gravity wastewater collection systems or *septic tank effluent gravity* (STEG) systems. The basis of the system consists of an onsite septic tank used for solids interception with gravity or pumped flow via service lateral to a small diameter collection system. An analysis of a STEG collection system is shown on Fig. 13-13. The upstream removal of solids permits the use of small diameter pipes for collection and reduces the flowrate required for solids flushing. Where necessary an effluent pumping station can be used. Additional information on the design and operation of STEG systems can be obtained from U.S. EPA (1991), Bounds (1996), and Crites and Tchobanoglous (1998).

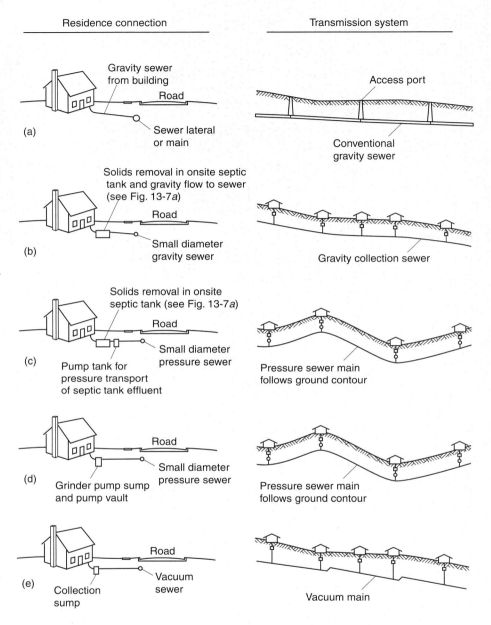

Figure 13-12 Comparison of collection systems: (a) conventional gravity flow system, (b) STEG (septic tank effluent gravity) collection system, also known as small-diameter variable-grade gravity or effluent drainage collection system, (c) STEP (septic tank effluent pump) pressure collection system, (d) grinder pump pressure collection system, and (e) vacuum collection system.

Pressure Collection System

Where there is insufficient slope or other limiting factors that preclude movement of wastewater by gravity, pumping facilities may be used. Septic tank effluent may be pumped directly from the septic tank using a submerged filter and pump assembly or collected in an external sump with a pump. The pump is activated with a water level sensor, and wastewater is discharged into the collection system. This type of system is

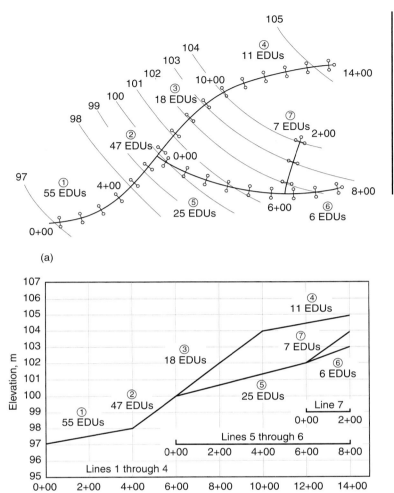

Figure 13-13
Analysis of STEG system: (a) layout showing location of EDUs and (b) a profile diagram of the collection system used for estimating flowrates, flow velocity, and pipe diameter for each section.

generally referred to as a *septic tank effluent pump* (STEP) system. A typical tank used in STEP systems in shown on Fig. 13-7b. An analysis of a STEP system is also shown on Figs. 13-14a and b. A schematic flow diagram of a STEP system used for effluent reuse is shown on Fig. 13-15. Alternatively, a grinder pump can be used to process the entire wastewater stream without the use of onsite septic tank treatment. With all pressure collection systems, a pump is required at each inlet point to the pressure main. Additional information on the design and operation of STEP and grinder pump systems can be obtained from U.S. EPA (1991), Bounds (1996), and Crites and Tchobanoglous (1998).

Figure 13-14
Analysis of STEP system: (a) layout showing location of EDUs, and (b) a profile diagram of the collection system used for determination of pumping requirements and sizing the system.

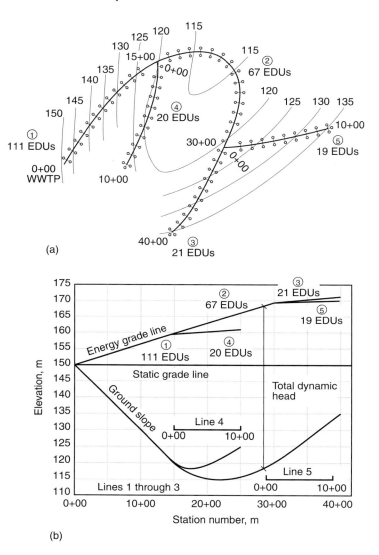

Vacuum Collection System

An alternative to the pressurized sewer is the use of a vacuum system and valves to control the flow of wastewater. In these systems, wastewater from an individual building flows by gravity to the location of a vacuum sump and control system. A valve in the vacuum sump seals the line leading to the main so that a vacuum can be maintained in the main. When a given amount of wastewater accumulates in the sump, the valve automatically opens to allow the wastewater to enter the main under vacuum as a plug. Vacuum pumps are housed at a central vacuum station, usually near the treatment facility or in a convenient location. The efficiency of the vacuum system may be improved where a single vacuum system is used to collect wastewater from a number

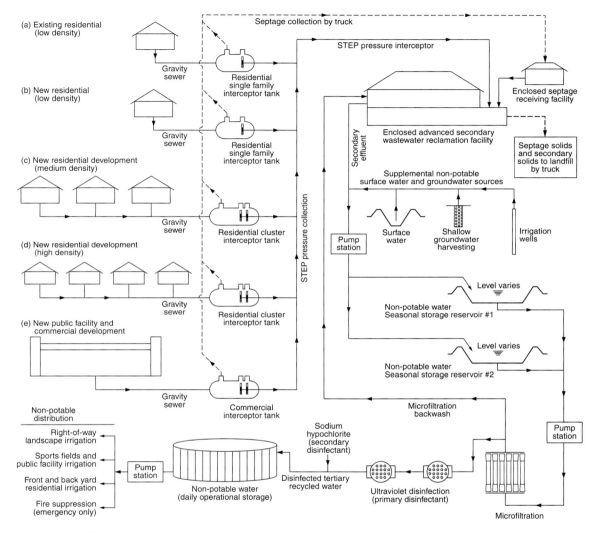

Figure 13-15
Schematic flow diagram of comprehensive water reclamation and reuse plan incorporating STEP systems for low-, medium-, and high-density developments. (Courtesy of D. Ripley, Ripley Pacific Company.)

of holding tanks. An analysis of a vacuum sewer system is shown on Fig. 13-16. Additional information on the design and operation of STEP systems can be obtained from AIRVAC (1989), U.S. EPA (1991), and Crites and Tchobanoglous (1998).

Hybrid Collection Systems
The use of a combination of two or more collection technologies is known as a hybrid collection system. For most applications where alternative collection systems are used, a combination of technologies may prove to be the most efficient design. Typically, a

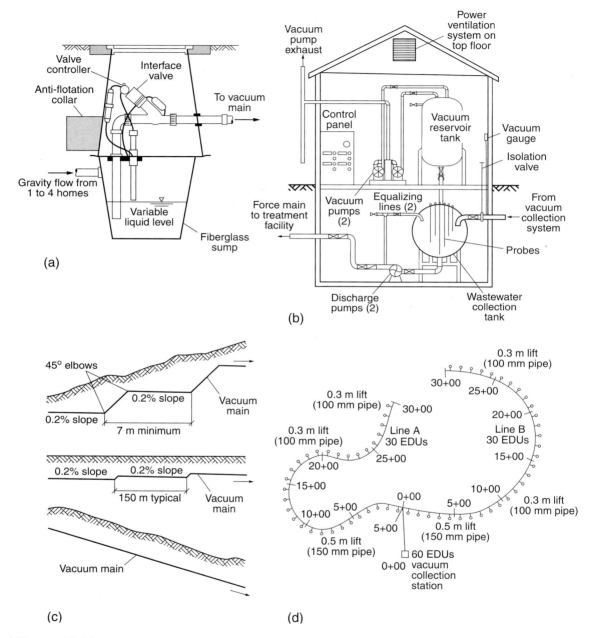

Figure 13-16
Vacuum system for the collection of wastewater: (a) diagram of collection tank and control valve system located at each dwelling unit, (b) vacuum collection station, (c) typical installation of vacuum collection pipes for different grades, and (d) layout showing location of EDUs used for system design.

vacuum system is used in conjunction with grinder pump stations; however, STEP and STEG systems have also been used together. An example of a hybrid collection system for the City of Provincetown, Massachusetts, is shown on Fig. 13-17. In the Provincetown system, failing onsite systems in the coastal areas were impacting the water quality. The hybrid collection system gives residents the option to hookup or maintain their onsite system. The new hybrid system made use of gravity and low pressure collection for 15 percent of the homes, with the remainder of hookups on vacuum collection. The energy usage for each connection is monitored separately; residents pay a flat monthly fee for the collection service.

Comparison of Collection Systems

The type of collection system selected depends on a number of factors, including topography and other constraining factors. In many cases, piping for alternative collection can be installed at depths around 1 m (or below the depth of freezing) using readily available trenching technology. A comparison of conventional and alternative collection systems used for decentralized applications is shown in Table 13-16. In general, systems that utilize onsite solids removal result in costs to the user due to septic tank pumping and maintenance, however, gravity flow systems have the lowest operation cost. Pressure collection systems with grinder pumps have increased maintenance due to the onsite grinder pumps. Vacuum collection systems have been found to have lower energy usage (150 versus 400 kWh/yr) and maintenance needs than grinder pump systems.

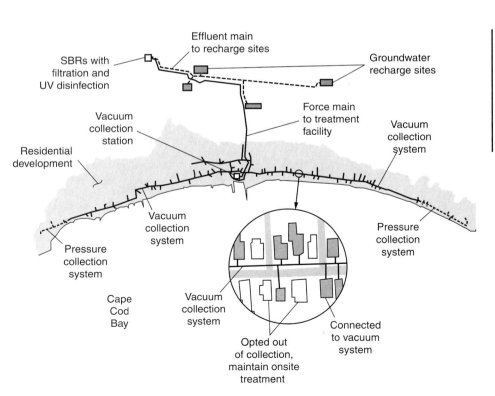

Figure 13-17

Example of hybrid collection system used in Provincetown, MA. (Adapted from Katehis et al., 2004.)

Table 13-16
Comparison of conventional and alternative wastewater collection systems for subdivisions, housing developments, and small communities[a]

Design parameter	Conventional gravity sewer	Alternative decentralized collection systems			
		With onsite primary treatment		With no onsite primary treatment	
		Septic tank effluent pump (STEP) sewer	Septic tank effluent gravity (STEG) sewer	Pressure sewer with grinder pump (GP)	Vacuum sewer
Ideal topography	Downhill only	Undulating, follows topography	Downhill, variable	Uphill	Flat
Ease of construction	Deep and wide trenches are slow to construct and disrupt traffic	Narrow and shallow trenches install quickly with minimal traffic disruption	Narrow and shallow trenches install quickly with minimal traffic disruption	Narrow and shallow trenches install quickly with minimal traffic disruption	Narrow and shallow trenches install quickly with minimal traffic disruption
Construction cost	High	Low	Moderate	Low	Low
Estimated energy use, kWh/user·yr	0	150 to 300	0	300 to 400	100 to 200
Minimum slope or velocity requirement	Yes	No	No	Yes	Yes
Infiltration and exfiltration	Usually	No	No	No	No
Minimum diameter	150 to 200 mm	50 mm	50 mm	50 mm	50 mm
Access to clean main lines	Access points regularly spaced and located at transitions	Cleanouts and pigging ports located at each connection	Cleanouts and pigging ports located at each connection	Cleanouts and pigging ports located at each connection	Cleanouts and pigging ports located at each connection
Trench depth	Minimum depth 5 to 7 m	Maintain minimum depth	Minimum depth to 2 m	Minimum depth	Minimum depth to 1.5 m
Remote pump stations	Needed for areas where downhill slopes can not be maintained	Present at each service point	Needed for areas where downhill slopes can not be maintained	Present at each service point	Not needed, but vacuum station is required
Conflicts with buried utilities	May require some grade adjustment or alignment	Avoided	May require some grade adjustment or alignment	Avoided	Avoided

[a] Adapted from Crites and Tchobanoglous (1998).

Treatment Technologies

The types of technologies used for larger decentralized systems, with flows typically greater than 40,000 L/d range from scaled up versions of the onsite and cluster treatment systems to scaled down versions of the centralized processes described in Chaps. 7, 8, and 9. However, several secondary treatment processes have been found to be well suited to the requirements of decentralized operation, including:

- Sequencing batch reactor
- Oxidation ditch
- Hybrid and conventional attached growth
- Natural treatment using wetland systems
- Membrane bioreactors

The level of maintenance needed for small treatment facilities depends on the type of treatment process used. For example, a membrane bioreactor system with onsite sludge processing may require two or more full time operators, while one part time operator will be sufficient for a large constructed wetland system. The degree of maintenance required should be factored into the selection process when developing options for wastewater reuse.

Multiple Quality Reclaimed Water

Two strategies are used most commonly for the treatment and distribution of reclaimed water to meet multiple quality needs. One strategy is to size the facility to treat all of the flow for a given effluent standard. In this case, the effluent quality standard will be based on the most stringent reuse application due to the cost of implementing multiple distribution networks. In another approach, reclaimed water is produced and distributed to meet the quality requirements of the largest reclaimed water user. Where a higher quality of water is needed than that supplied in the distribution system, local point of use treatment facilities can be used to obtain the required water quality. Thus, the cost of separate distribution pipelines is avoided through application of point of use treatment.

Decentralized systems can also be used to produce reclaimed water to meet a variety of water quality requirements. Customizing treatment processes for a number of reuse applications is known as the multiple quality concept (MQC) (Tchobanoglous et al., 1999). The purpose of the MQC for water reuse is to maximize the beneficial use of wastewater by utilizing different levels of treatment. In the applications shown on Fig. 13-18, three different water qualities are used with respect to biochemical oxygen demand (BOD), total suspended solids (TSS), nutrients, and the presence of pathogenic microorganisms: (a) high quality water for potable use is obtained from a local supply, (b) domestic wastewater is treated for the removal of BOD and TSS and then used for subsurface landscape irrigation, and (c) a portion of the treated effluent is treated for nutrient removal and disinfection before indoor nonpotable reuse, i.e., toilet flushing and clothes washing.

Treatment Requirements

The level of treatment required will depend on the reuse application under consideration; however, the range of options is different from large systems. For example, as

Figure 13-18
Diagram of the multiple quality concept for water reuse at an individual residence.

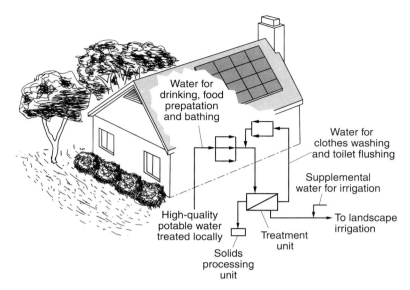

described above, in some cases treated wastewater for reuse does not have to meet recycled water regulations because it will be used directly for subsurface irrigation without the need for reclaimed water distribution piping. Similarly, the treatment process can be suited specifically to the given reuse application. The type of treatment required for a given application needs to be reviewed carefully with an understanding of the water quality in question and local rules and regulations.

For unrestricted water reuse applications, the secondary effluent is processed typically by filtration using granular media, cloth media, or microfiltration, followed by UV disinfection. Membrane bioreactors are followed directly by UV disinfection as additional filtration is not required. Flow diagrams for several characteristic wastewater treatment and reclamation processes incorporating the above processes are shown on Fig. 13-19.

13-6 DECENTRALIZED WATER REUSE OPPORTUNITIES

Given the large number of treatment technologies identified in the preceding sections, it can be reasoned that decentralized systems can be applied to virtually any water reuse design. The most common decentralized water reuse projects include systems for landscape irrigation with drip emitters, groundwater recharge, nonpotable indoor reuse, and habitat development.

Landscape Irrigation Systems

Most decentralized wastewater treatment systems make use of the local soil for dispersal of effluent. These soil infiltration systems are not designed typically to make use of plant uptake of water or nutrients found in wastewater. Therefore, percolation of wastewater to groundwater and plant uptake of wastewater in soil infiltration systems may be

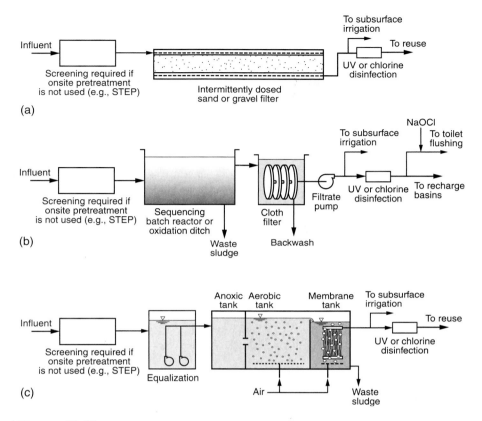

Figure 13-19
Typical treatment process flow diagrams used for decentralized wastewater reclamation facilities for developments and small communities, often used in conjunction with alternative collection systems. The processes are arranged in order of increasing level of sophistication: (a) sand or gravel filter, (b) sequencing batch reactor or oxidation ditch, and (c) membrane bioreactor.

considered unplanned water reuse. The use of planned decentralized water reuse systems for the irrigation of landscapes is practiced in many locations because it reduces water demands, is considered to be safe, and the point of reuse is often near the point of reclaimed water generation. Decentralized wastewater systems designed for water reuse by landscape irrigation are used to apply reclaimed water and nutrients at rates appropriate for plant uptake and typically make use of shallow effluent distribution methods, such as drip irrigation. Typically, wastewater from an apartment building, commercial facility, or a cluster of homes is retained in a large watertight septic tank or other solids separation unit and all or a portion of the flow may be used for tree and median strip watering using subsurface drip irrigation. Because the nutrients in the wastewater are beneficial, the effluent from the septic tank is applied after it has passed an effluent filter to remove coarse solids larger than about 2 to 3 mm. Landscape irrigation systems are described in detail in Chap. 18.

Irrigation with Greywater

The wastewater discharged from bathing and washing facilities that does not contain concentrated human waste or food waste is collectively known as greywater. In some cases, greywater is used for subsurface irrigation without treatment, however, for some reuse applications greywater may need to be treated before reuse. The treatment systems used for greywater include biological treatment, particle removal, and disinfection. If untreated greywater is stored in a holding or equalization tank for any length of time, anaerobic conditions may develop, similar to what occurs in a septic tank. In some cases, greywater can be distributed without treatment for tree irrigation using subsurface dispersal systems (Ludwig, 2003), as shown on Fig. 13-20. As noted previously, the concentration of pathogens, organic matter, trace constituents, and nutrients are expected to be significantly lower than those found in combined wastewater, however, the concentration of sodium and other minerals may be increased depending on the type of detergents and cleaning agents used in the home, as well as the presence of a salt-based water softener. A greywater recycle system is shown later on Fig. 13-23b.

Groundwater Recharge

The nature of most decentralized systems allows for the infiltration of wastewater, a portion of which inevitably reaches the local water table. In areas where adequate pretreatment systems have not been implemented, it is common to find elevated concentrations of nitrate in receiving aquifers or in locally drawn well water. The extensive use of decentralized treatment systems, utilizing appropriate treatment processes, may be used to offset the depletion of groundwater resources in some areas. Where

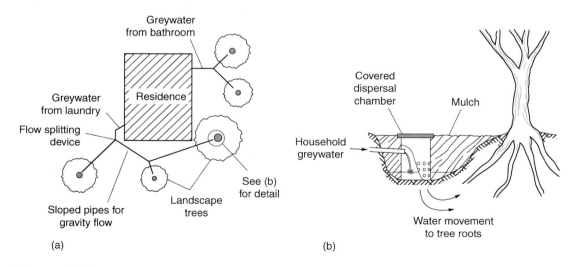

Figure 13-20

Illustration of greywater reuse for tree irrigation: (a) overview diagram showing distribution of greywater to landscape and (b) detail of mulch basin for delivering water to tree roots. (Adapted from Ludwig, 2003.)

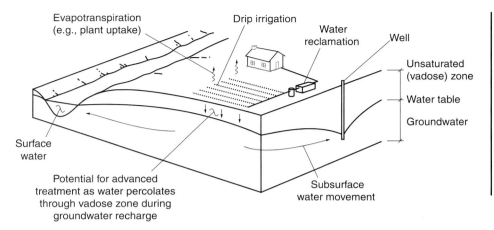

Figure 13-21
Definition sketch of subsurface water and movement of reclaimed water from decentralized systems when soil loading rate exceeds evapotranspiration rate (see Chaps. 17 and 18).

nitrate loading is a concern, wastewater systems should incorporate a nitrogen reduction process.

For many decentralized treatment systems, the flowrate is sufficiently small to allow for soil infiltration of residual effluent. In general, water that is not converted to vapor through physical or biological processes ultimately reaches groundwater. Therefore, soil infiltration of effluent should be conducted in a manner to allow for maximum treatment in the soil. Advanced treatment by soil infiltration can be optimized by using low loading rates, as with drip irrigation. Maintaining an aerobic environment in the unsaturated zone and maximizing soil contact time results in the destruction of trace chemicals found in reclaimed water. Additional details on soil treatment may be found in Chap. 22. A definition sketch of water movement to groundwater is shown on Fig. 13-21. An analysis of water movement by evapotranspiration and deep percolation is given in Example 13-7.

EXAMPLE 13-7. Evaluation of the Potential for Groundwater Recharge from an Onsite Wastewater Reuse System.

When a wastewater system is designed for landscape irrigation, but without the storage capacity needed for equalization of season demand, a portion of flow will infiltrate through the vadose zone to the water table. Using the following water balance analysis, estimate the amount of water that will be infiltrated compared to loss through evapotranspiration. The table is based on a soil effluent loading rate (LW) of 6.1 mm/d. See Chaps. 17 and 18 for additional information on design of effluent irrigation systems using water balance type analysis.

Month	Time, d/mo	Precipitation (Pr), mm/mo	Evaporation (ET_L), mm/mo	Effluent hydraulic loading rate, (L_W), mm/mo	Effluent infiltration, mm/mo
Jan	31	98	20	191	191
Feb	28	90	35	172	172
Mar	31	71	69	191	191
Apr	30	26	111	184	99
May	31	13	140	191	64
Jun	30	5	165	184	24
Jul	31	1	172	191	19
Aug	31	2	152	191	40
Sep	30	9	118	184	76
Oct	31	23	86	191	127
Nov	30	56	41	184	184
Dec	31	62	24	191	191
Total	365	455	1133	2243	1377

Solution

1. Compute the percent of the total wastewater flow that is expected to infiltrate to the water table.

$$\text{Percolation, \%} = \frac{(1377 \text{ mm/yr})}{(2243 \text{ mm/yr})} = 61.4$$

2. Repeat the above analysis for several loading rates.
   A summary of the results are shown on the following plot.

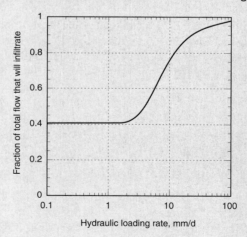

Comment

As shown in the plot, the minimum infiltration occurs below a hydraulic loading rate of 2.5 mm/d (41%). As the loading rate increases the amount of infiltration also increases. For a loading rate of 6.1 mm/d (used commonly for drip irrigation systems), about 61% of the flow is expected to reach the water table. The highest levels of treatment are expected at low hydraulic loading rates (e.g., less than 10 mm/d).

Self-Contained Recycle Systems

Some communities have elected to collect and treat wastewater with a decentralized system, including disinfection, and return the treated wastewater to each residence using a separate water system for the flushing of toilets. As toilet flushing accounts for a large percentage of the overall water needs, the reduced freshwater usage is a primary benefit of this arrangement. Other nonpotable reuse opportunities include landscape irrigation, vehicle washing, fire fighting, recreational uses, and industrial applications.

Several self-contained recycle systems have been developed to take greywater or combined wastewater from buildings or communities, and following appropriate treatment, return the reclaimed water for reuse in toilet and urinal flushing, clothes washing, and nonpotable outdoor uses. Although such processes are expensive, they have been used for apartment and office buildings located in areas without wastewater collection systems and where water for domestic use is in short supply. Diagrams of self-contained water treatment and recycle systems are shown on Fig. 13-22.

Habitat Development

Wastewater treatment processes that make use of plants and other ecological components are usually referred to as natural treatment systems. Natural treatment systems may support the growth of beneficial plants and trees that are used as sanctuary for various organisms. In addition, reduced accessibility to wastewater treatment and reuse areas reduces disturbance to wildlife in these areas. An example system design that makes use of both advanced treatment and natural treatment systems is shown on Fig. 13-23. Because natural processes typically rely on solar energy and gravity flow, the cost of operation and maintenance may be significantly lower than a high rate process, but the treatment processes operate at much lower rates and are more sensitive to environmental and seasonal changes. Additional information on the design of natural treatment systems for wastewater treatment and habitat may be found in Kadlec and Knight (1996), U.S. EPA (1993b), and Crites and Tchobanoglous (1998). Further consideration of water reuse for the development of habitat and other environmental purposes is provided in Chap. 21.

13-7 MANAGEMENT AND MONITORING OF DECENTRALIZED SYSTEMS

The inherent need for management and the dispersed nature of small wastewater systems presents logistical challenges for ensuring proper operation and performance. The purposeful management of DWM systems must be undertaken (1) to improve the performance and reliability of decentralized technologies, (2) to overcome the historical stigma of failed onsite systems, (3) to allow for cost savings using many recently developed technologies, (4) to allow for the development and testing of new technologies, and (5) to allow for the orderly development of areas without wastewater collection systems in the context of a sustainable environment. Topics related to the types of management structures and monitoring and control equipment for DWM systems are presented below. A summary of responsibility for various phases of implementation of decentralized systems is given in Table 13-17.

Types of Management Structures

In developing management strategies, it must be recognized that many different management arrangements are possible; however, the challenge is to find the most suitable

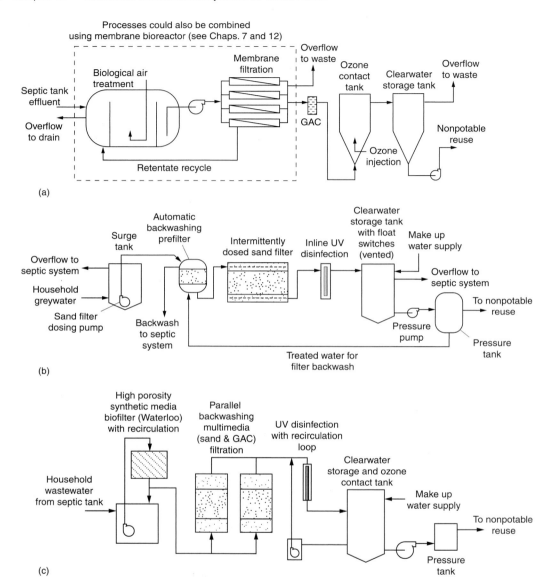

Figure 13-22

Process flow diagrams for self-contained onsite wastewater treatment and recycling systems incorporating different filtration processes for production of reclaimed water: (a) membrane filtration (Adapted from Thetford Systems, Inc.), (b) sand filtration for greywater recycling (Courtesy of www.ecoseeds.org), and (c) granular multimedia filtration (Adapted from Canada Mortgage and Housing Corporation).

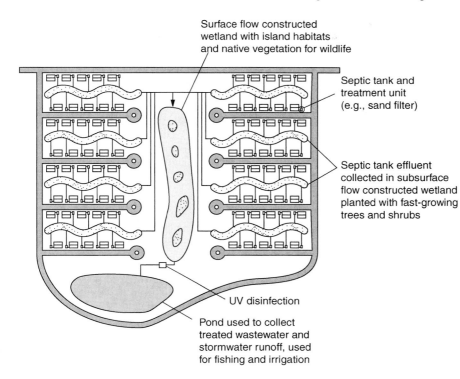

Figure 13-23
Conceptual diagram of decentralized water reclamation system designed to make use of natural and low energy input treatment processes and potentially to provide habitat and sanctuary for wildlife.

Table 13-17
Responsible entities for decentralized wastewater management systems

Issue	Responsible entity[a]
Planning	CE, LG
Funding	PO, LG, SRG
Land acquisition	PO, LG[b]
Permitting	LG, SRG
Design and engineering	CE, M
Construction	CR
Operation	CR, PU
Monitoring	CR, LG, PU
Enforcement	LG, SRG

[a] PO = Property owners
CE = Consultant/engineer
CR = Contractor
M = Technology manufacturer
PU = Public utility
LG = Local government agency
SRG = State or regional government agency

[b] Local government may be in involved in land acquisition for larger systems.

management structure for the situation at hand. To ensure that individual decentralized systems function properly, especially in densely developed areas or where cluster or community type systems are used, it is necessary to organize a maintenance district or contract with a public or private operating agency to conduct periodic inspections and any necessary maintenance. A community served entirely by onsite wastewater systems and with an onsite wastewater management district is shown on Fig. 13-24. Large-scale DWM projects should be allowed only if a responsible management agency has been designated prior to construction. Flexible management arrangements must be developed to deal with the many different types of DWM systems that may be proposed. Additional information on the design and selection of management structures of decentralized systems can be found in U.S. EPA (2005).

Monitoring and Control Equipment

To manage onsite wastewater treatment systems properly, monitoring of process operation and performance is necessary. Because of the increasing number and complexity of onsite wastewater treatment systems, automated monitoring and control systems have become a key component to onsite treatment process management. System controls are necessary for controlling pumps, alarms, and other process equipment. Monitoring equipment can be used to monitor pump on/off cycles, run time, liquid levels, UV lamp operation, and alarm status.

Most manufacturers of onsite wastewater treatment systems provide basic control and alarm systems to alert the system owner of a malfunction. However, remote monitoring using telemetry systems is becoming a more feasible option for onsite applications. Telemetry is the science and technology of automatic measurement and transmission of data by wire, radio, or other means from remote sources to receiving stations for recording and analysis. The centralized management of onsite treatment systems is possible

Figure 13-24

Views of Stinson Beach, a coastal community located north of San Francisco, CA, taken from Highway 1 looking north. The entire community is served with onsite systems under the control of an onsite wastewater management district.

through the use of these automated monitoring and control devices. Several Internet-based software applications have been developed for real-time data management and process control.

Sensors can be digital or analog type output devices. Sensors are used to make a measurement of a physical activity and provide a signal to a monitoring or control device. Float switches, used for monitoring water levels, are among the most common digital sensors used in onsite treatment systems. Analog sensors can be used for measuring water level (using a pressure transducer), pump run time, sludge and scum depth, and constituent concentrations.

Telemetry systems make remote data acquisition and control possible and greatly extend that range of options for management of onsite treatment systems. A typical telemetry system used for monitoring remote wastewater treatment and reuse systems is shown on Fig. 13-25. Data are acquired from system devices and sensors and transmitted by modem or broadband to a specified location. Web-based telemetry systems transmit data to a server, where the data is put into a database that is accessible from any computer with Internet access capabilities. In addition, logic controllers can be used to diagnose and correct system problems before a failure occurs. Remote monitoring systems make it possible to economically and reliably operate and maintain a large number of decentralized systems. Further, online monitoring systems allow management organizations to diagnose and repair problems before the system user is aware of any problem.

(a) (b)

Figure 13-25

Views of a control panels used with telemetry systems for remote monitoring of decentralized treatment systems: (a) control panel for individual residence and (b) control panel for recirculating sand filter.

PROBLEMS AND DISCUSSION TOPICS

13-1 A decentralized system using a STEP system for wastewater collection is to be used to service a housing development with a total of 150 homes. Estimate the mean flowrate, the peak flowrates, and the design flowrate for the treatment process to be used. Estimate the expected annual sludge production and anticipated maintenance needs.

13-2 Estimate your home daily water usage, constituent loading rates, and resulting constituent concentrations. Compute new values if some of the water conservation practices outlined in Table 13-4 are implemented. Comment on how water conservation practices could affect the design of new wastewater treatment facilities and the operation of existing facilities.

13-3 Evaluate the suitability of your local water supply for irrigation (a) as supplied and (b) after domestic use without water conservation practices and (c) after domestic use with water conservation practices.

13-4 Compare the advantages and disadvantages of individual onsite systems with cluster and small community systems. Discuss the factors that you would consider in the application of each type of system.

13-5 In some areas, it is necessary to remove nitrogen, phosphorus, TDS, and/or pathogenic organisms to protect groundwater. Comment on the approaches that might be implemented in decentralized applications and potential limiting factors that might need to be overcome.

13-6 Elaborate on the situations where an alternative collection system might be more appropriate than a conventional centralized collection system.

13-7 Using evapotranspiration and precipitation data for your local climate (or as provided by the instructor), determine a soil loading rate that will (a) minimize the area required for infiltration, (b) provide for irrigation demand during the peak evapotranspiration season, and (c) maximize nitrogen uptake. List all assumptions.

13-8 Based on Fig. 13-5, compute the daily peaking factor for 1, 5, and 61 homes. Estimate the tank size required for flow equalization to obtain a peaking factor of 1.5.

13-9 Discuss the features that can be used for decentralized treatment systems to improve system performance and reliability. Given the technologies available, would you recommend use of decentralized wastewater management instead of a large centralized system? Discuss the rationale for your reponse.

13-10 Develop a process flow diagram for treating the wastewater generated by the housing development in Problem 13-1 assuming water reuse by landscape irrigation.

REFERENCES

AIRVAC (1989) *Design Manual, Vacuum Sewerage Systems*, AIRVAC Company, Rochester, IN.

Anderson, D. L., M. B. Tyl, R. J. Otis, T. G. Mayer, and K. M. Sherman (1998) "Onsite Wastewater Nutrient Reduction Systems (Owners) For Nutrient Sensitive," in *Proceedings of the 8th National Symposium of Individual and Small Community Sewage Systems—2001*, American Society of Agricultural Engineers, 235–244, St. Joseph, MI.

Baker M. J., D. W. Blowes, and C. J. Ptacek (1998) "Laboratory Development of Permeable Reactive Mixtures for the Removal of Phosphorus from Onsite Wastewater Disposal Systems," *Environ. Sci. Technol.*, **32**, 15, 2308–2316.

Bounds, T. R. (1995) "Septic Tank Septage Pumping Intervals," in R. W. Seabloom (ed.) *Proceedings 8th Northwest On-site Wastewater Treatment Short Course and Equipment Exhibition*, University of Washington, Seattle.

Bounds, T. R. (1996) *Alternative Sewer Designs*, Orenco Systems Inc., Roseburg, OR.

Bounds, T. R. (1997) "Design and Performance of Septic Tanks," *Conference of the American Society for Testing and Materials*, Philadelphia, PA.

Crites, R., and G. Tchobanoglous (1998) *Small and Decentralized Wastewater Management Systems*, McGraw-Hill, Boston.

Crites, R., C. Lekven, S. Wert, and G. Tchobanoglous (1997) "A Decentralized Wastewater System for a Small Residential Development in California," *The Small Flows J.*, **3**, 1, Morgantown, WV.

Del Porto, D., and C. Steinfeld (1999) *The Composting Toilet System Book: A Practicle Guide to Choosing, Planning, and Maintaining Composting Toilet Systems, an Alternative to Sewer and Septic Systems*, The Center for Ecological Pollution Prevention, Concord, MA.

Dunbar, Professor, Dr. (1908) *Principles of Sewage Treatment*, Charles Griffen, London.

Ecosan (2003) "Ecosan-Closing the Loop," *Proceedings of The 2nd International Symposium on Ecological Sanitation*, GTZ Publishers, Lubeck, Germany.

Emerick, R., R. M. Test, G. Tchobanoglous, and J. Darby (1997) "Shallow Intermittent Sand Filtration: Microorganism Removal," *Small Flows J.*, **3**, 1, 12–22.

Etnier, C., J. Willetts, C. A. Mitchell, S. Fane, and D. S. Johnstone (2005) *Decentralized Wastewater System Reliability Analysis Handbook*, Project No. WU-HT-03-57, Prepared for the National Decentralized Water Resources Capacity Development Project, Washington University, St. Louis, MO, by Stone Environmental Inc., Montpelier, VT.

Gaulke, L. S. (2006) "Johkasou: On-site Wastewater Treatment and Reuses in Japan," *Proceedings of the Institute of Civil Engineers—Water Management*, **159**, WM2, 103–109.

Gross, M., and S. Jones (1999) "Stratified Intermittent Sand Filter and Ozonation for Water Reuse," in *NOWRA Proceedings of the 8th Annual Conference and Exhibit*, Jekyll Island, GA.

Johansson, L. (1997) "The use of LECA (Light Expanded Clay Aggregates) for the Removal of Phosphorus from Wastewater," *Water Sci. Technol.*, **35**, 5, 87–93.

Joubert, L., P. Flinker, G. Loomis, D. Dow, A. Gold, D. Brennan, and J. Jobin (2004) *Creative Community Design and Wastewater Management*, Project No. WU-HT-00-30, Prepared for the National Decentralized Water Resources Capacity Development Project, Washington University, St. Louis, MO, by University of Rhode Island Cooperative Extension, Kingston, RI.

Jowett, E. C., H. Millar, and K. Pataky (2001) "Four Golf Resorts are Reusing Treated Sewage for Irrigation," *Environ. Sci. Eng.*, **14**, 66–68.

Kadlec, R., and R. Knight (1996) *Treatment Wetlands*, Lewis Publishers, Boca Raton, FL.

Katehis, D., P. Mantovani, and L. Henthorne (2004) "Developing a Successful Decentralized Wastewater Treatment and Reuse Strategy," Presented at the *International Water Demand Management Conference*, May 30 to June 3, 2004, Dead Sea, Jordan.

Kitis, M. (2004) "Disinfection of Wastewater with Peracetic Acid: A Review," *Environ. Int.*, **30**, 1, 47–55.

Leverenz, H., J. Darby, and G. Tchobanoglous (2001) *Review of Systems for the Onsite Treatment of Wastewater*, Center for Environmental and Water Resources Engineering, University of California, Davis, CA. Available online at http://www.waterboards.ca.gov/ab885/technosite.html

Leverenz, H., J. Darby, and G. Tchobanoglous (2005) *Evaluation of Disinfection Units for Onsite Wastewater Treatment Systems*, Center for Environmental and Water Resources Engineering, University of California, Davis, CA.

Lombardo, P. (2006) *Phosphorus Geochemistry in Septic Tanks, Soil Absorption Systems, and Groundwater*, Prepared by Lombardo Associates, Inc., Newton, MA.

Ludwig, A. (2000) *Create an Oasis with Greywater*, 4th ed., Oasis Design, Santa Barbara, CA.

Rittmann, B., and P. McCarty (2001) *Environmental Biotechnology: Principles and Applications*, McGraw-Hill, Boston.

Robertson, W. D., G. I. Ford, and P. S. Lombardo (2005) "Wood-Based Filter for Nitrate Removal In Septic Tank Systems," *Transactions of the ASAE, American Society of Agricultural Engineers*, **48**, 1, 121–128.

Salvato, J. A. (1992) *Environmental Engineering and Sanitation*, 4th ed., Wiley Interscience Publishers, New York.

Schmidt, I., O. Sliekers, M. Schmid, E. Bock, J. Fuerst, J. Gij, S. Kuenen, M. Jetten, and M. Strous (2003) "New Concepts of Microbial Treatment Processes for the Nitrogen Removal in Wastewater," *FEMS Micro. Rev.*, **27**, 4, 481–492.

Tchobanoglous, G., L. Ruppe, H. Leverenz, and J. Darby (1999) "Decentralized Wastewater Management Challenges and Opportunities for the Twenty-First Century," in *10th Onsite Wastewater Treatment Proceedings*, Seattle, Washington, DC.

Tchobanoglous, G., F. Burton, and H. D. Stensel (2003) *Wastewater Engineering: Treatment and Reuse*, 4th ed., McGraw-Hill, Boston, MA.

U.S. EPA (1980) *Onsite Wastewater Treatment and Disposal System—Design Manual*, EPA 625-180-012, Municipal Environmental Research Laboratory, U.S. Environmental Protection Agency, Cincinnati, OH.

U.S. EPA (1991) *Alternative Wastewater Collection Systems*, EPA 625/1-91/024, Center for Environmental Research Information, U.S. Environmental Protection Agency, Cincinnati, OH.

U.S. EPA (1993a) *Nitrogen Control*, EPA/625-93-1-000, U.S. Environmental Protection Agency, Washington, DC.

U.S. EPA (1993b) *Constructed Wetlands for Wastewater Treatment and Wildlife Habitat*, EPA/832-R-93-005, Municipal Technology Branch, Washington, DC.

U.S. EPA (1997) *Response to Congress on Use of Decentralized Wastewater Treatment Systems*, EPA 832-R-97-001b, U.S. Environmental Protection Agency, Washington, DC.

U.S. EPA (2002) *Onsite Wastewater Treatment Systems Manual*, EPA 625-R-00-008, Office of Water, US. Environmental Protection Agency, Washington, DC.

U.S. EPA (2005) *Handbook for Managing of Onsite and Clustered (Decentralized) Wastewater Treatment Systems: An Introduction to Management Tools and Information for Implementing EPA's Management Guidelines*, EPA 832-B-05-001, Office of Water, U.S. Environmental Protection Agency, Washington, DC.

U.S. PHS (1955) *Studies on Household Sewage Disposal Systems, Part III*, U.S. Public Health Service Publication No. 397.

Winneberger, J. H. T. (1984) *Septic-Tank Systems, A Consultant's Toolkit, Volume II the Septic Tank*, Butterworth Publishers/Ann Arbor Science, Ann Arbor, MI.

Zhu, T., P. D. Jennsen, T. Maehlum, and T. Krogstad (1997) "Phosphorus Sorption and Chemical Characteristics of Lightweight Aggregates (Lwa)—Potential Filter Media In Treatment Wetlands," *Water Sci. Technol.*, **35**, 5, 103–108.

14 Distribution and Storage of Reclaimed Water

WORKING TERMINOLOGY   830

14-1   ISSUES IN THE PLANNING PROCESS   831
Type, Size, and Location of Facilities   831
Individual Reclaimed Water System versus Dual Distribution System   832
Public Concerns and Involvement   833

14-2   PLANNING AND CONCEPTUAL DESIGN OF DISTRIBUTION AND STORAGE FACILITIES   833
Location of Reclaimed Water Supply, Major Users, and Demands   834
Quantities and Pressure Requirements for Major Demands   834
Distribution System Network   836
Facility Design Criteria   841
Distribution System Analysis   845
Optimization of Distribution System   847

14-3   PIPELINE DESIGN   856
Location of Reclaimed Water Pipelines   856
Design Criteria for Reclaimed Water Pipelines   858
Pipeline Materials   858
Joints and Connections   860
Corrosion Protection   861
Pipe Identification   862
Distribution System Valves   863
Distribution System Appurtenances   863

14-4   PUMPING SYSTEMS   866
Pumping Station Location and Site Layout   866
Pump Types   867
Pumping Station Performance   870
Constant versus Variable Speed Operation   870
Valves   871
Equipment and Piping Layout   872
Emergency Power   872
Effect of Pump Operating Schedule on System Design   875

14-5   DESIGN OF RECLAIMED WATER STORAGE FACILITIES   877
Location of Reclaimed Water Reservoirs   878
Facility and Site Layout for Reservoirs, Piping, and Appurtenances   879
Materials of Construction   881
Protective Coatings—Interior and Exterior   881

830 | Chapter 14 Distribution and Storage of Reclaimed Water

14-6		**OPERATION AND MAINTENANCE OF DISTRIBUTION FACILITIES** 882
		Pipelines 883
		Pumping Stations 884
14-7		**WATER QUALITY MANAGEMENT ISSUES IN RECLAIMED WATER DISTRIBUTION AND STORAGE** 884
		Water Quality Issues 885
		Impact of Water Quality Issues 887
		The Effect of Storage on Water Quality Changes 887
		Strategies for Managing Water Quality in Open and Enclosed Reservoirs 889
		PROBLEMS AND DISCUSSION TOPICS 892
		REFERENCES 898

WORKING TERMINOLOGY

Term	Definition
Demand	The amount of water required to meet a stated use, e.g., landscape irrigation.
Distribution system	The piping network required to deliver water from a transmission pipeline to the points of connection to users' plumbing systems. Pumping stations are often included as part of the distribution system. In small systems, the distribution system serves functions of both transmission and distribution.
Dual distribution system	Two independent piping systems that are used to deliver potable and reclaimed water.
Dual plumbing system	Piping systems that supply potable and recycled water to users from the point of connection to the distribution main to the points of use.
Easement	A right-of-use giving persons other than the property owner permission to use a property for a specific purpose, e.g., installation of a pipeline.
Emergency storage	Storage capacity that is reserved for emergency use, e.g., when reclaimed water is used for fire protection.
Pressure zone	A portion of a service area, especially in hilly areas, where the pressure in the water mains is maintained within a relatively narrow range.
Right-of-way	A form of an easement granted by a property owner to others for reasonable use of the land, such as access, that does not impair the owner's use of the land. Also, publicly owned land used for roadways, utilities, and other public uses.
Shutoff head	The pressure that occurs in a centrifugal pump at zero discharge flow.
Total dynamic head	The total head (energy) added to the reclaimed water by a pump. Total dynamic head (TDH) is the sum of the static head (elevation difference between source and discharge), friction losses and fitting, and exit losses in the suction and discharge piping.
Transmission line	Pipeline that carries water from the point of production to the distribution system.
Turnout	A connection to a transmission pipeline that provides a supply of reclaimed water to a service area, e.g., the transmission pipeline may be owned by a reclaimed water wholesaler and the service area is served by a reclaimed water retailer.

Water demand	The amount of water needed to meet customers use.
Water hammer	Rapid pressure and flow changes in pipelines caused by pump start-up, pump shutdown, or power failure. Sometimes the transient pressure conditions are accompanied by a hammering-type noise.
Working storage	Storage used to meet peak flow demand in excess of the maximum day demand.

The facilities to store and distribute the reclaimed water to potential users can be planned and designed once the source of reclaimed water and the location and nature of the water reuse areas and demands are known. In most respects, facilities for the storage and distribution of reclaimed water are similar to those for potable water. Because of the characteristics of reclaimed water and the potential changes in water quality that may occur over time (see Sec. 14-7), care must be taken during the planning, design, and operation of distribution and storage facilities to prevent or mitigate any effects. The purpose of this chapter is to introduce the basic issues and concepts involved in the planning and design of reclaimed water storage and distribution systems for centralized and satellite systems. Facilities needed for decentralized and on-site systems are covered in Chap. 13. Dual plumbing systems that deliver water from the distribution system to the points of use are discussed in Chap. 15.

Important issues and factors, typical to most reclaimed water distribution and storage projects, that are addressed in this chapter are: (1) planning and implementation issues, (2) planning and conceptual design of distribution and storage facilities, (3) design of pipelines, (4) design of pumping facilities, (5) operation and maintenance of pipelines and pumping stations, (6) design of storage facilities, and (7) operational issues in reclaimed water storage.

14-1 ISSUES IN THE PLANNING PROCESS

Planning and implementation issues that must be addressed when considering storage and distribution facilities for a reclaimed water project include:

- The type, size, and location of physical facilities.
- The interrelationship between the potable and reclaimed water systems, i.e., is the reclaimed water system being installed in an area where an existing potable water system exists or is a dual distribution system (for potable and reclaimed water) needed?
- Involvement of the public during the planning and implementation process. The public may be affected directly by facilities siting and construction.

The effects of each of these issues are discussed in the following paragraphs.

Type, Size, and Location of Facilities

The primary factors governing distribution and storage facilities include the location of the reclaimed water treatment plant and the location and demand requirements of the reclaimed water users. The principal facilities needed for the delivery of reclaimed water are storage tanks, pumping stations, and transmission and distribution pipelines, and depend on the overall type of reclaimed water system as shown in Table 14-1. The

Table 14-1
Facilities required for alternative wastewater management systems

Facility	Wastewater management system		
	Centralized	Satellite	Decentralized
Distribution	X	X	
Storage	X	X	a
Pumps	X	X	X

[a]May be needed on cluster systems serving multihome developments.

distribution and storage facilities can be classified in several ways as shown in Table 14-2. A conceptualized diagram that includes many of the facilities that may be required in a centralized or satellite system for reclaimed water storage and distribution is shown on Fig. 14-1. Each of the elements of a reclaimed water delivery system is discussed in the following sections in this chapter.

Individual Reclaimed Water System versus Dual Distribution System

A reclaimed water system may be planned, designed, and installed as a system totally separate from the potable water system or planned as part of a dual distribution system that provides both reclaimed and potable water to the service area. The distinction between an individual versus a dual system may at first appear to be obscure, but the integrated planning, design, and construction of a dual system offers advantages in both water resource management and cost savings, as discussed below. The design of the system components in either case meets the general criteria described in Sec. 14-2. The origin and use of dual distribution systems are described in AWWA (1994) and Okun (2005).

Substituting reclaimed water for potable water is one of the primary purposes of dual distribution systems. As stated earlier in this text, the use of reclaimed water for nonpotable purposes serves to conserve the potable water supply for use where drinking water quality is needed. In the planning of a dual distribution system, if the reclaimed water is used for firefighting in lieu of potable water, the potable water pipelines and storage can be sized for delivery of domestic flows and not fire flows. Potable water quality benefits accrue because pipeline and storage sizes are reduced, which in turn reduces the residence time in the potable water system. Long residence times can result in the loss of disinfectant residual and may promote the regrowth of microorganisms, which can affect bacterial quality, tastes, and odors.

Table 14-2
Classifications of distribution and storage facilities

General classification	Types of facilities
Distribution system	Separate reclaimed water system
	Dual distribution systems for reclaimed and potable water
Storage facilities	Long-term storage
	Short-term storage
	Open reservoirs
	Enclosed reservoirs
	Above- and belowground storage tanks
	Aquifer storage

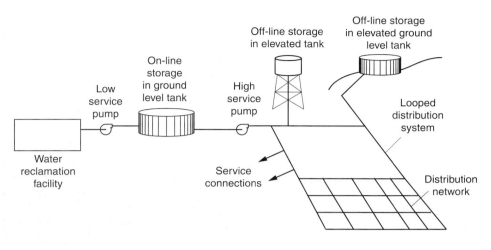

Figure 14-1
Elements of typical reclaimed water storage and distribution system.

Public concerns in planning a water reuse project may involve a range of issues, including:

- Why is water reuse needed?
- What are the public health impacts?
- What facilities are needed and where will they be located?
- How will property values be affected, if at all?
- Will the project stimulate growth?
- What are the construction and operating impacts?

Most of these concerns can be resolved by involving those who are primarily affected and those who may be affected indirectly through various stages of project conception and implementation. A successful program that involves the public and gains their support can also help avoid controversies and delays and accelerate implementation. Various approaches in conducting public involvement and information programs are discussed in Chaps. 25 and 26.

Public Concerns and Involvement

14-2 PLANNING AND CONCEPTUAL DESIGN OF DISTRIBUTION AND STORAGE FACILITIES

The planning process for reclaimed water projects consists of the following tasks:

- Identification of the reclaimed water users and their corresponding water use demands
- Determination of water quantities and pressure requirements for major demands
- Distribution system layout, including pipeline routing, preliminary diameters, and pumping station and storage reservoir locations
- Distribution system analysis
- Optimization of distribution system layout

Location of Reclaimed Water Supply, Major Users, and Demands

These tasks are discussed in the following paragraphs.

One of the initial planning tasks is to determine how the location and production quantities of the reclaimed water supply match the location and demands of the major users. Reclaimed water may be supplied from a centralized water reclamation plant or a satellite reclamation plant located near a major demand area. In most cases, an upstream satellite plant will produce reclaimed water only, and the solids produced during the treatment process will be discharged to the collection system for later treatment at the wastewater treatment plant (see Chap. 12). Occasionally, the reclaimed water supply will be a turnout from a regional reclaimed water supply main.

Important characteristics of the major demands include: (1) the quantity of reclaimed water needed; (2) the physical location and elevation of the points of use within the service area; (3) operating schedule, i.e., time and duration of use; and (4) the required operating pressure at the reuse sites. In addition to agricultural irrigation, other uses will be for irrigation of golf courses, parks, landscape, freeway and expressway median strips, and residential, commercial, and industrial areas where dual distribution piping has been provided. Dual distribution systems will most likely be found in new office areas and subdivisions where separate reclaimed water piping is constructed at the same time as the potable water piping and other infrastructures are installed. Many jurisdictions in the semiarid U.S. southwest have adopted ordinances requiring the construction of dual distribution systems in new developments.

Quantities and Pressure Requirements for Major Demands

Methods for determining the quantities of reclaimed water required for agricultural use and landscape irrigation are presented in Chaps. 17 and 18. For each major demand, the average daily flow, maximum daily flow, and peak hourly flow must be determined. Average and maximum daily flows are important only where there is continuous use such as supply to an industrial facility. Peak hourly demand is the most important flowrate criterion for sizing distribution facilities used in intermittent applications such as agriculture and landscape irrigation. Determination of flowrates is illustrated in Example 14-1.

EXAMPLE 14-1. Estimating Maximum Daily and Peak Hourly Flowrates.

A community park requires an average of 6 mm of irrigation water per day during the maximum demand month of August. However, during high temperature periods lasting a few days at a time, the irrigation demand increases to 12 mm/d. The park turf area is 4 ha in size. The park is open from 8 a.m. to 10 p.m. and sunrise occurs at about 6 a.m. Calculate the average day, maximum day, and peak hourly demands.

Solution

1. Determine the average day demand.

$$\text{Average day demand} = (6 \text{ mm/d})(1 \text{ m}/10^3 \text{ mm})(4 \text{ ha})(10^4 \text{ m}^2/\text{ha})$$
$$= 240 \text{ m}^3/\text{d}$$

2. Determine the maximum day demand.

$$\text{Maximum day demand} = (12 \text{ mm}/6 \text{ mm}) \times \text{average day demand}$$
$$= 2 \times 240 \text{ m}^3/\text{d}$$
$$= 480 \text{ m}^3/\text{d}$$

3. Because the park is occupied during daylight hours, irrigation must occur during the nighttime. In general, irrigation begins 1 or 2 h after park closing, and needs to be completed 1 to 2 h before sunrise to allow some drying before people enter the park. Therefore, the full demand must be delivered in about 4 h.

4. Determine the peak hourly demand.

The peak hourly demand is calculated as follows:

$$\text{Peak hourly demand} = (480 \text{ m}^3/\text{d})/(4 \text{ h/d})$$
$$= 120 \text{ m}^3/\text{h}$$

Comment

If the demand could be met over a 24-h period, the average design flowrate would be 10 m^3/h. Thus, as illustrated in this example, the operating schedule can have a major impact on the design flowrates to be used to size the reclaimed water pumping facilities and distribution system piping. Reduction of the pumping rate would require the addition of storage facilities (see Example 14-2).

In addition to the reclaimed water quantities needed, the required pressure in the reclaimed water distribution system must be determined. The system pressure should be sufficient to operate the onsite facilities, such as irrigation distribution systems and sprinkler fixtures. For example, irrigation sprinkler heads generally require a minimum of 140 kPa (20 lb/in.^{3}) at the sprinkler head to operate properly. Therefore, the pressure in the reclaimed water distribution system should be at least 210 kPa (30 lb/in.^{3}) to allow for a 70 kPa (10 lb/in.^{3}) pressure drop (i.e., headloss) across the onsite irrigation system (assuming a level site). Any variations in elevation at the irrigation site should also be considered in determining the operating pressure requirements. Preferably, the distribution system should deliver reclaimed water to major demand areas with a pressure of at least 280 to 350 kPa (40 to 50 lb/in.^{3}). Pressures greater than 560 kPa (80 lb/in.^{3}) will usually require pressure-reducing valves at the service connection to prevent excessive pressure surges and water hammer in the onsite facilities.

Distribution System Network

The reclaimed water distribution system network consists of all the pipeline routes; the locations, sizes, and type of storage reservoirs; and the locations and capacities of pumping stations. If the service area includes significant elevation changes, it may be necessary to divide the distribution system into two or more pressure zones. Each pressure zone should include sufficient duplication of facilities to ensure reliable reclaimed water delivery during maximum demand periods. The level of reliability depends on whether an uninterruptible supply is required. A conceptual diagram of various distribution system components is shown on Fig. 14-1. The principal elements of the distribution network are discussed in the following paragraphs.

Piping Network

Three types of distribution systems may be used: grid, loop, or tree as described in Table 14-3 and shown on Fig. 14-2. With a grid or loop system, each major reuse area is served from more than one direction, thereby ensuring that all demands will be satisfied even if there is a disruption in a portion of the distribution system. A tree system is less reliable as a failure in the main supply line will shut down service to all or a portion of the users. In general, a tree system is not recommended for use in the distribution of reclaimed water because of the potential for the development of odors in the dead end outlets that are used intermittently. A typical loop system used in St. Petersburg, Florida, is shown on Fig. 14-3.

Table 14-3 Types of reclaimed water distribution systems

System type	Description	Comment
Loop	Large feeder mains surround the areas to be served with smaller crossfeed lines connected to the main loop	Very reliable system as reclaimed water is supplied to the major reuse area from two directions. The headloss across a looped distribution system will generally be less when compared to a tree system. Finally, looping eliminates "dead ends" in the distribution system, thereby reducing the potential for stagnation and deterioration of water quality
Grid	The piping is laid out in a checkerboard fashion, with the piping usually decreasing in size as the distance increases from the source	Advantages are similar to loop system. Reduction in pipe sizes will result in some cost savings for piping materials
Tree	A single main is used that reduces in size with increasing distance from the source. Branch lines emanate from the main	Generally used for systems where a higher degree of reliability afforded by loop and grid systems is not necessary. Provisions for periodic line flushing are required to remove deposits in dead ends

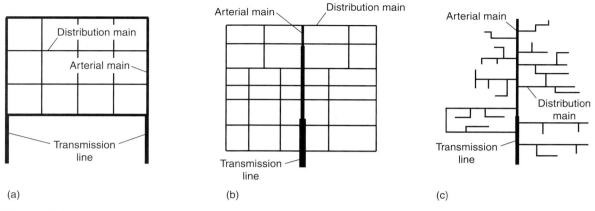

Figure 14-2
Typical water distribution system configurations: (a) loop, (b) grid, and (c) tree. Loop systems are used most commonly for reclaimed water distribution systems. Grid and tree configurations are used in some smaller systems and at the extremities of loop systems. (Adapted from AWWA 2003b.)

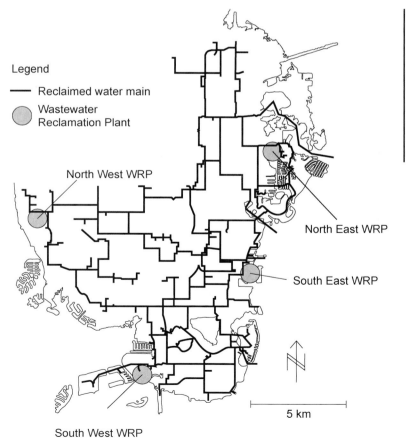

Figure 14-3
Loop distribution system used to interconnect the four water reclamation plants in St. Petersburg, FL. (Adapted from City of St. Petersburg, FL.)

Distribution Storage

Short-term storage is needed in most reclaimed water distribution systems to smooth out the variations between the production and demand of the reclaimed water. Demands for reclaimed water are often seasonal in nature, particularly for crop and landscape irrigation and sometimes for cooling water. Even during a daily or weekly period, demands vary significantly. For example, golf course and other landscape irrigation is often carried out during the night and early morning hours before sunrise. The result is an imbalance between the production rate and reuse of the reclaimed water that must be equalized by providing storage facilities.

Storage requirements are classified as either short-term or long-term. Short-term storage is used to store the variations in the reclaimed water supply and demand over a relatively short time period, usually 1 d or perhaps, 1 wk. Long-term storage is needed when the seasonal differences between production and demand must be contained for an extended period of months or even years in some cases.

Short-term storage is usually provided by steel or concrete tanks, similar to those used for potable water storage. The principal types of short-term storage reservoirs are: (1) ground level with auxiliary pumping (Fig. 14-4*a*), (2) belowground-level storage with auxiliary pumping (Fig. 14-4*b*), (3) elevated storage with and without auxiliary pumping (Fig. 14-4*c*), and (4) small ponds used at golf courses (Fig. 14-4*d*). Elevated storage can be provided by elevated steel tanks or by ground-level reservoirs located at an elevated site.

Long-term storage is usually provided in reservoirs and lakes (see Fig. 14-5). Reservoirs and lakes are common for seasonal storage when the topography of the area is suitable for the construction of dams and large embankments. For smaller projects, or where the topography is flat, earthen ponds or lagoons can be used to provide long-term storage. The sizing of long-term storage reservoirs is considered in Chap. 17.

In semiarid areas, such as the southwestern United States, stream flows are low or nonexistent during the dry summer months. Discharge of effluent from the wastewater treatment plant often is prohibited by regulatory agencies, regardless of the degree of treatment provided. In many cases, irrigation demands are highest during this period and reclamation of the wastewater is common. In this situation, storage is needed to contain the reclaimed water during the fall months after the demand for irrigation water is reduced and before the stream flows increase following the resumption of the winter rains.

In the distribution system, storage reservoirs are usually sized to provide working storage (see "Working Terminology"), as well as an emergency reserve when needed. By providing working storage, reclaimed water demands greater than the maximum day production rate can be served from the storage reservoir, and the capacity of the distribution system and pumping facilities can be reduced. In addition, where an elevated storage reservoir is provided, the supply pumps are not required to run constantly to maintain pressure in the distribution system.

Emergency storage can increase the reliability of the reclaimed water service by providing a source of reclaimed water during periods when other facilities, such as a pumping

Figure 14-4
Typical examples of short-term storage facilities: (a) ground level concrete storage tank, (b) belowground level storage tank with recreational sports field on top (Courtesy of City of San Diego, photo by Jeran-Aero Graphics), (c) elevated steel storage tank, and (d) pond used at golf course.

Figure 14-5
Typical examples of long-term storage facilities: (a) storage reservoir with dam (on right) and (b) lake.

station or pipeline, are out of service for maintenance or repair. Care should be taken not to oversize emergency storage because long residence times can result in water quality deterioration (see Sec. 14-7). In general, emergency storage requirements in reclaimed water distribution systems are not as critical as the requirements for potable systems where fire protection is a major objective.

The location of storage reservoirs within the distribution system needs to be coordinated with the location of the reclaimed water supply and the major reclaimed water reuse areas. Where possible, reservoirs should be located remotely from the supply so that the peak demands can be served from two directions—from the supply pumps and the reservoirs. Hillside locations need to be carefully selected to minimize the visual impact of the reservoir while still providing reliable and cost-effective service.

Pressure Zones

Where the service area includes a large difference in elevation, usually it is good practice to divide the distribution into "pressure zones." If multiple pressure zones are required, critical decisions must be made regarding the "depth" of the pressure zones and the number of pumping stations and storage reservoirs to provide reliable service. The depth of a pressure zone is the pressure differential between the top and bottom of the zone. If pressure zones are too shallow, the number of zones together with their associated pumping and storage facilities will result in excessive costs. On the other hand, if the zones are too deep, i.e., the pressure differential is too large, the static pressure at the bottom of the zone will be excessive and pressure reducers will be required on many of the services to avoid problems with water hammer and pipe failures in onsite irrigation systems. Recommended pressure zone criteria are summarized in Table 14-4.

Pumping Stations

Pumping stations also need to be carefully located. In most cases, a reclaimed water pumping station will be located at the water reclamation plant. Distribution pumping stations, especially those serving a higher pressure zone, often are located at the site of the reservoir serving a lower zone.

Pumping stations used in conjunction with storage facilities should be sized to pump the maximum day demand for the tributary service area. Care must be taken to include operating schedules for the various major reclaimed water demands. In addition, consideration should be given to selecting the capacity and operation of the pumping station to take advantage of lower cost off-peak electric power rates. Off-peak power rates

Table 14-4
Recommended pressure zone criteria

Parameter	Unit	Minimum	Maximum
Pressure differential within zone	kPa ($lb_f/in.^2$)	210 (30)	500 (70)
Static pressure at:			
Top of zone	kPa ($lb_f/in.^2$)	210 (30)	350 (50)
Bottom of zone	kPa ($lb_f/in.^2$)	560 (80)	700 (100)

are available when the demand for electrical energy is low, usually during the nighttime or early morning hours.

After the major facilities of the reclaimed water distribution system have been located, the next step in the planning and design process is to determine the sizes of all the pipelines, pumping stations, and reservoirs. Preliminary design criteria used initially to size the various components of the reclaimed water distribution system are presented in this section; detailed design criteria for pipelines, pumping stations, and reservoirs are presented in Secs. 14-3, 14-4, and 14-5, respectively. After the location and preliminary sizes of all the facilities are known, the system analysis step in the planning and design process can be carried out, as discussed in the next section.

Facility Design Criteria

Piping System Sizing

When designing reclaimed water distribution systems, the size (and cost) of the pipelines must be balanced against the maximum velocity in the pipelines and the total headloss across the distribution system. If the pipelines are too small, headlosses and the pumping requirements and energy costs will be greater. Finally, the operating pressures and the maximum pressures due to surges and water hammer must be known to properly select the pressure class of the pipelines.

Pumping Station Design Factors

The capacity of a pumping station is defined by the peak pumping rate and the maximum operating pressure of the pumps. In general, pumping stations must deliver reclaimed water to the distribution system at a rate sufficient to satisfy the demands of the reclaimed water users. Therefore, the peak pumping rate is equal to the maximum day demand, adjusted to take into account peaking factors for the service area demands, user operating schedules, and energy conservation measures. Where there is little or no system storage, the peak pumping rate will correspond to the maximum hourly demand. Typical peaking factors based on experience for broad categories of reuse such as agricultural and landscape irrigation and cooling water makeup are included in Table 14-5.

Operating schedule considerations include the need to deliver reclaimed water to the reuse sites at times appropriate for the intended use. For example, a city park is often irrigated between the hours of midnight and 4 a.m. to avoid interfering with the normal use of the park (see Example 14-1). In this case, the peaking factor is equal to 24/4, or 6 times the maximum day demand for the park.

Table 14-5
Typical flow peaking factors for various reclaimed water uses

Reclaimed water use	Peaking factors	
	Maximum day/ average day	Peak hour/ maximum day
Agricultural irrigation	1.5–2	2–3
Landscape irrigation	1–1.5	4–6
Cooling water makeup	1–1.5	1–2

Energy conservation considerations may warrant providing a pumping station with a higher capacity so that all pumping can be accomplished during off-peak periods when lower power rates are available. For example, if the local power company has established an off-peak energy rate from 8 p.m. to 6 a.m., the pumping station capacity could be 2.4 times (24 h/10 h) the maximum day demand to fill the reservoir during the off-peak period. Rate schedules should be obtained from the electrical utility supplying energy to the facility when evaluating designs, operating conditions, and costs.

Storage Reservoir Capacity

In a typical reclaimed water distribution system design with elevated short-term storage, much of the peak flow will be provided from the storage reservoirs and not directly from pumping. Storage reservoir capacity is defined by the size needed to provide peak day and hourly demands, together with any emergency storage requirements that have been established for the service area. Storage used to provide peak flow demands above the maximum day demand is called *working storage*.

Working storage can be determined by several methods. One method is to assume that the working storage is equal to a percentage of the maximum daily flow (demand) tributary to the reservoir. An allowance of 25 to 50 percent of the maximum day demand is commonly used. Another method is to perform a cumulative mass balance analysis for the system as shown in Example 14-2. In this method, the inflows (supply) and outflows (demands) are accumulated over the design 24-h period, usually the maximum daily demand period, and plotted as two lines on a graph. The working storage is equal to the volume when the divergence between the cumulative inflow line and the cumulative outflow line is the greatest.

EXAMPLE 14-2. Estimate Working Storage by the Cumulative Mass Balance Diagram Method.

A small town has the hourly distribution of reclaimed water demand during the design maximum day shown in the following table. The town wishes to minimize the pump size by pumping at constant rate. Determine the pump flowrate and the corresponding storage needed to achieve a constant pumping rate.

Period	Time at start of period	Demand, m^3/h
1	12:00 a.m.	680
2	1:00	680
3	2:00	680
4	3:00	710
5	4:00	710
6	5:00	710

(Continued)

14-2 Planning and Conceptual Design of Distribution and Storage Facilities

Period	Time at start of period	Demand, m^3/h
7	6:00	110
8	7:00	110
9	8:00	110
10	9:00	110
11	10:00	410
12	11:00	410
13	12:00 p.m.	410
14	1:00	410
15	2:00	410
16	3:00	410
17	4:00	410
18	5:00	410
19	6:00	110
20	7:00	80
21	8:00	80
22	9:00	80
23	10:00	680
24	11:00	680
	Sum	9600 m^3/d

Solution

1. Determine the constant hourly flowrate.

    The required hourly pumping rate is obtained by dividing the total daily demand by 24.

    Hourly pumping rate = (9600 m^3/d)/(24 h/d) = 400 m^3/h

2. Determine the required storage to achieve the constant flowrate determined in Step 1.

    a. Prepare a computation table that includes the cumulative inflows and outflows. In the last column, calculate the absolute value of the difference between the cumulative inflow and the cumulative outflow to find the maximum value which corresponds to the required reservoir volumetric capacity.

Period	Time at start of period	Inflow, m^3/h Hourly	Inflow, m^3/h Cumulative	Outflow, m^3/h Hourly	Outflow, m^3/h Cumulative	Cumulative difference, m^3 (absolute value)
0			0		0	0
1	12:00 a.m.	400	400	680	680	280
2	1:00	400	800	680	1360	560
3	2:00	400	1200	680	2040	840

(Continued)

Period	Time at start of period	Inflow, m³/h Hourly	Inflow, m³/h Cumulative	Outflow, m³/h Hourly	Outflow, m³/h Cumulative	Cumulative difference, m³ (absolute value)
4	3:00	400	1600	710	2750	1150
5	4:00	400	2000	710	3460	1460
6	5:00	400	2400	710	4170	1770
7	6:00	400	2800	110	4280	1480
8	7:00	400	3200	110	4390	1190
9	8:00	400	3600	110	4500	900
10	9:00	400	4000	110	4610	610
11	10:00	400	4400	410	5020	620
12	11:00	400	4800	410	5430	630
13	12:00 p.m.	400	5200	410	5840	640
14	1:00	400	5600	410	6250	650
15	2:00	400	6000	410	6660	660
16	3:00	400	6400	410	7070	670
17	4:00	400	6800	410	7480	680
18	5:00	400	7200	410	7890	690
19	6:00	400	7600	110	8000	400
20	7:00	400	8000	80	8080	80
21	8:00	400	8400	80	8160	240
22	9:00	400	8800	80	8240	560
23	10:00	400	9200	680	8920	280
24	11:00	400	9600	680	9600	0
	Sum		9600 m³/d		9600 m³/d	

The data in the computation table are summarized on the following figure:

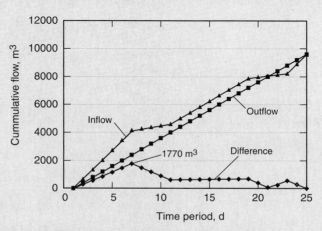

b. The largest cumulative difference occurred at 5 a.m. (time period 6); therefore, the required volumetric capacity of the storage reservoir is 1770 m³.

Finally, and most accurately, the working storage can be determined by performing an extended period analysis using a computerized pipe network analysis program. In this method, a pipe network analysis program is run multiple times using variable values for inputs and outflows over the simulation period, usually the 24 h maximum daily demand period. The pipe diameters of the distribution network and the reservoir sizes are adjusted, as needed, until all the reservoirs can satisfy the maximum day demands and refill before the end of the 24-h simulation period. Additional discussion is provided under "Extended Period System Analysis" in the following subsection.

Distribution System Analysis

Once the reclaimed water distribution system layout is completed, including all pipelines, pumping stations, and reservoirs, a computerized analysis of the distribution system can be done. Several computerized network analysis programs are available. Commercially available programs include *WaterCAD* by Haestad Methods part of Bentley Systems, *H₂ONet* and *H₂OMap* by MWH Soft, *InfoWorks WS* by Wallingford Software, and *KYPipe/Pipe 2000* by KYPIPE Software Center at the University of Kentucky. Public domain network analysis programs are also available, including *EPANET2* developed by U.S. Environmental Protection Agency (EPA), *Branch and Loop* developed by the World Bank, *W-Piper* developed by the Construction Engineering Research Laboratory at the U.S. Corps of Engineers, and *NETIS* by the Water Development Research Unit at South Bank University in London.

Types of Network Analyses

Usually two types of network analyses are carried out for the reclaimed water distribution system—static and extended period. A static analysis is used to examine the distribution system at a single point in time, usually maximum day or maximum hour, and includes all supplies and demands, pumping stations, and reservoirs. The purpose of the static analysis is to determine if the pipeline sizes assumed in the preliminary system layout are adequate to maintain required pressures in the system under peak demands, without excessive velocities. An extended period analysis is used to model the distribution system over an extended period of time, usually 24 h. The 24-h period is usually the maximum day, and the various demands and supplies are varied throughout the period to accurately reflect actual conditions. The purpose of the extended-period analysis is to determine if the distribution system can satisfy pressure and flow requirements throughout the extended period. The analysis will also determine whether the reservoirs will refill before the 24 h peak demand period has elapsed.

Model of Distribution System

The initial step in a system analysis is creation of a model of the distribution system, including all pipes, pumping stations, and reservoirs. The distribution system is depicted by a series of lines (pipes) and nodes (junctions). A simplified example of a hydraulic system model is shown on Fig. 14-6. Data for each junction node include (1) the pipes entering and leaving the node, (2) demand, (3) ground elevation, and (4) depth to pipe centerline. Demand for each node is estimated as the cumulative demand for the portion of the service area tributary to the node. Data for each pipe in

Figure 14-6

Example of a simplified distribution system model illustrating (a) pipelines, (b) junctions, and (c) other appurtenances.

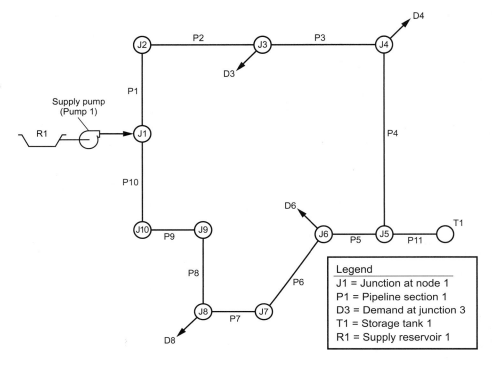

the system include (1) the nodes at each end of the pipe, (2) diameter, (3) length, and (4) friction factor (i.e., Chezy "C," Hazen-Williams "C," Darcy-Weisbach "f," or Manning "n"). Data for pump nodes include (1) ground elevation, (2) pump centerline elevation, and (3) pump characteristic curve. The pump characteristic curve usually consists of three or more points along the combined Q-H (flowrate vs. total dynamic head) curve for the pumps in the station. Reservoir nodes need data on the (1) tank diameter (or side dimensions, if tank is not circular), (2) water depth to overflow, (3) bottom elevation, and (4) ground elevation.

Based on the continuity principle, generalized equations are created by the program for each pipe and node in the distribution network. For pipe junction nodes, the sum of the inflows and outflows (including the demand) must equal zero. Also, the pressure in each of the pipes entering or leaving the node must be the same. Reservoir nodes are similar to junction nodes except that the continuity equation includes a "change in storage volume" term. Nodes for pumping stations can be used to determine the flow through the node using the Q-H curve and the calculated pressures in the entering and exiting pipelines. Flow equations are created for each pipeline using the Chezy, Hazen-Williams, Darcy-Weisbach, or the Manning equation, available in standard hydraulic engineering texts and reference books. The result is a matrix of N equations with N unknowns. Using iterative numerical methods, the values for the unknown pressures and flows at each of the nodes are determined by the network analysis program.

Static System Analysis

In the static system analysis, the individual demands are assigned to each of the nodes in the model for the flow conditions of interest. Pipe diameters and friction factors are assigned to each of the pipelines in the network. The analysis program starts by assuming pressures for each of the nodes and computing the resulting flow in each of the pipes. The flows at each of the nodes are summed to see if the continuity equation is satisfied. If not, a second iteration is carried out using the computed pressures and flows from the first iteration and the results checked for continuity. This process continues until the differences between the beginning and computed pressures and flows at each of the nodes are within an acceptable tolerance, at which time the iterative process is completed. Final output from a static model run will include the demand and pressure at each of the nodes and the flow and velocity in each of the pipelines. Data for pump nodes will include the flow through the pumps and the inlet and outlet pressures (i.e., the total dynamic head). Reservoir nodes will include the net inflow (or outflow) rate for the modeled conditions. A static system analysis is illustrated as part of Example 14-3.

Extended Period System Analysis

An extended period analysis consists of a series of static analyses over a period of time, usually 24 h, where the demands at the nodes are varied hourly to simulate actual conditions during the simulation period. For example, the peak irrigation flowrate used for a park may be the 4-h period from midnight to 4 a.m. Another example would be limiting pumping to refill the reclaimed water reservoirs to the overnight off-peak period from 8 p.m. to 6 a.m. to reduce power costs. Results from extended period analyses will indicate whether localized low pressure conditions will occur during peak flow periods, or whether a reservoir will not completely refill overnight before the start of the next day. An extended period analysis is illustrated in Example 14-3.

Optimization of Distribution System

Optimization of the distribution system is an extension of the analysis phase. As deficiencies in the distribution system are identified from the model runs, alternative pipe sizes, and reservoir and pumping station locations and capacities can be modeled. Through this process, the reclaimed water distribution system can be optimized for performance and cost.

Example 14-3. Design of Reclaimed Water Distribution and Storage System Using Computerized Pipe Network Analysis Program.

The small town from Example 14-2 has a total maximum daily reclaimed water demand of 9600 m³/d. The total demand is distributed among four major users as shown in the following table:

Period	Time at start of period	Total demand, m³/h	Demand, m³/h			
			User 1	User 2	User 3	User 4
1	12:00 a.m.	680	0	0	600	80
2	1:00	680	0	0	600	80
3	2:00	680	0	0	600	80
4	3:00	710	30	0	600	80
5	4:00	710	30	0	600	80
6	5:00	710	30	0	600	80
7	6:00	110	30	0	0	80
8	7:00	110	30	0	0	80
9	8:00	110	30	0	0	80
10	9:00	110	30	0	0	80
11	10:00	410	30	300	0	80
12	11:00	410	30	300	0	80
13	12:00 p.m.	410	30	300	0	80
14	1:00	410	30	300	0	80
15	2:00	410	30	300	0	80
16	3:00	410	30	300	0	80
17	4:00	410	30	300	0	80
18	5:00	410	30	300	0	80
19	6:00	110	30	0	0	80
20	7:00	80	0	0	0	80
21	8:00	80	0	0	0	80
22	9:00	80	0	0	0	80
23	10:00	680	0	0	600	80
24	11:00	680	0	0	600	80
	Totals, m³/d	9600	480	2400	4800	1920

User 1 is a small manufacturing plant with a constant demand during two shifts each day. User 2 is a warehouse that uses evaporative cooling during the heat of the day (10 a.m. to 6 p.m.). User 3 is a large golf course and park complex that irrigates overnight between 10 p.m. and 6 a.m. Finally, User 4 is a small power plant with a constant 24-h demand for makeup cooling water. The user demands are as follows:

User no.	Located at junction no.
1	J3
2	J4
3	J6
4	J8

Determine the required pipeline sizes and pumping and storage requirements for the reclaimed water distribution system using static and extended period network analyses.

The following conditions and assumptions are to be used for the analyses:
- Use the Hazen-Williams hydraulic formula with a "C" value = 130
- Ground elevations of all junction points (except storage tank) = 30 m
- Ground elevation at hillside storage tank site = 60 m
- Storage tank height to overflow = 10 m
- Working storage volume is approximately 1800 m^3 (from Example 14-2)
- Emergency storage allowance = 30 percent of total
- Reclaimed water pump operates at a constant flowrate for 24 h at maximum day demand conditions = 400 m^3/h
- Nominal pumping head (TDH) of the supply pump = 40 m of water
- Demand at all junction points, except junctions J3, J4, J6, and J8 = 0 m^3/s
- Standard pipe diameters, mm = 200, 250, 300, 350, 400, 450, etc.
- Initial (default) pipeline diameter = 200 mm
- Pipeline lengths for the distribution network are as follows:

Pipe no.	Length, m
P1	600
P2	600
P3	450
P4	600
P5	600
P6	250
P7	300
P8	550
P9	150
P10	600

The following design criteria shall apply to the analyses:
- Minimum pressure at all demand junctions, m of water = 30
- Maximum pipeline velocity, m/s = 1.5
- Minimum pipe diameter, mm = 200

Solution

Note: The network analysis program EPANET2 is used in this example because it is available at no cost from the U.S. EPA at www.epa.gov/ORD/NRMRL/epanet.html.

1. First, set up the project by entering default values in EPANET2, including labels for the network map and output tables; units for flow, length, head and pressure; and the headloss formula to be used (i.e., Hazen-Williams, Darcy-Weisbach, etc.).
   a. Network nodes and links are labeled with J (junction) and P (pipe), respectively. Pumps, supply reservoirs, and storage tanks are labeled Pump, R, and T.
   [Note: By convention EPANET2 assumes that supply reservoirs are infinite sources of reclaimed water with a constant water level (head), and storage tanks are nodes that provide storage by varying their water depth (head) and volume over time.]
   b. Select metric units of flow (i.e., flowrate, m^3/h; velocity, m/s; pipe diameter, mm; head and pressure, m of water).
   c. Select the Hazen-Williams headloss equation with a default roughness "C" value of 130.
2. Next, draw the pipeline network with all junctions, pipelines, supply reservoir, pumping station, and storage tank located on an adjacent hillside near the town as shown. By EPANET2 convention, add the objects to the map in the following order: (1) supply reservoir, (2) junctions, (3) storage tank, (4) pipes, and finally (5) the pump.

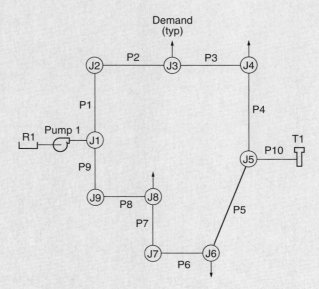

3. Set the properties for all of the objects in the network map.
   a. As each object is added to the network, default values are assigned if they are available. If necessary, the properties for any object are changed using the Property Editor.
   b. For each junction, enter the ground elevation and demand; for each pipeline, enter the length, diameter, and roughness.
   c. In this example, set the ground elevations of all junctions (except the storage tank) to 30 m. Set the ground elevation of the storage tank to 60 m.
   d. Set the demands for all junctions to 0 m^3/h, except for junctions J3, J4, J6, and J8.
   e. Note that two static runs are needed in this example because the maximum demands at junctions J4 and J6 do not occur simultaneously. If a single static run were made with both maximum demands active, portions of the pipe network would be oversized. The demands at junctions J3, J4, J6, and J8 for the separate static runs are shown in the following table:

Junction no.	Demand, m^3/h	
	Static run 1	Static run 2
J3	30	30
J4	0	300
J6	600	0
J8	80	80

   f. Using the assumed working storage of approximately 1800 m^3 and an emergency allowance of 30 percent, the total volume for the storage tank is approximately 2400 m^3. Therefore, with the assumed height to the tank overflow of 10 m, use an initial tank diameter of 18 m.

4. Perform an initial run to solve the pipe network for the maximum demand conditions defined above as static run 1.
   a. The results of the initial run will include negative pressures at several of the junctions. This is because the default pipe diameter of 200 mm is too small in the vicinity of the maximum demand junction, J6. Increase the diameter of the pipes between junction J6 and both the supply pump and the storage tank until an acceptable solution is achieved. In addition, the elevation of the storage tank and the operating curve of the pump can be adjusted to achieve the necessary minimum pressures and maximum pipeline velocities in the network.
   b. After several iterations, the junction pressures and pipe diameters are as shown in the junction and pipe tables.

Pressures at junctions:

Junction no.	Elevation, m	Demand, m³/h	Total head, m	Line pressure, m
J1	30	0	68.3	38.3
J2	30	0	66.7	36.7
J3	30	30	65.0	35.0
J4	30	0	64.4	34.4
J5	30	0	63.8	33.8
J6	30	600	60.1	30.1
J7	30	0	62.1	32.1
J8	30	80	64.5	34.5
J9	30	0	67.5	37.5
R1	30	−451.0	30.0	0.0
T1	63	−259.0	66.5	3.5

By convention, negative demands at reservoirs and tanks are indicated as flows out of the node in EPANET2. Note that the ground elevation of the storage tank needed to be raised to 63 m to provide the minimum pressure of 30 m of water at junction J6.

Hydraulic characteristics of pipelines:

Pipe no.	Length, m	Diameter, mm	Flow, m³/h	Velocity, m/s	Unit headloss, m/km
P1	600	200	79.3	0.70	2.81
P2	600	200	79.3	0.70	2.81
P3	450	200	49.3	0.44	1.16
P4	600	200	49.3	0.44	1.16
P5	600	300	349.3	1.37	6.08
P6	250	250	−250.7	1.42	7.99
P7	300	250	−250.7	1.42	7.99
P8	550	300	−330.7	1.30	5.49
P9	150	300	−330.7	1.30	5.49
P10	600	300	−300.0	1.18	4.59
Pump 1	—	—	410.0	—	−38.3

In EPANET2, flows with negative values are in a direction opposite from the way the pipeline was drawn initially. Similarly, by convention, negative headloss through a pump indicates that the head is added to the head at the supply reservoir (or node). In this example, velocities are maintained less than 1.5 m/s in all pipelines.

5. Perform static run 2 to size the pipelines serving the maximum demand at junction J4.
   a. Modify the junction demands in the Property Editor for the new demand conditions. Keep the pipe diameters from static run 1.
   b. Make the initial run for static run 2. In this initial run, the diameter of one of the pipelines serving J4 must be increased to provide a pressure greater than 30 m as shown in the junction and pipe tables.

Pressures at junctions:

Junction no.	Elevation, m	Demand, m³/h	Total head, m	Line pressure, m
J1	30	0	71.4	41.4
J2	30	0	67.9	37.9
J3	30	30	64.4	34.4
J4	30	300	62.9	32.9
J5	30	0	66.4	36.4
J6	30	0	67.3	37.3
J7	30	0	68.2	38.2
J8	30	80	69.3	39.3
J9	30	0	71.0	41.0
R1	30	−381.6	30.0	0.00
T1	63	−28.4	66.5	3.50

Note that most of the demand at junction J4 is provided by the pump, not the storage tank.

Hydraulic characteristics of pipelines:

Pipe no.	Length, m	Diameter, mm	Flow, m³/h	Velocity, m/s	Unit headloss, m/km
P1	600	200	117.8	1.04	6.85
P2	600	200	117.8	1.04	6.85
P3	450	200	87.8	0.78	4.18
P4	600	250*	−212.2	1.20	8.93
P5	600	300	−162.8	0.64	1.66
P6	250	250	−162.8	0.92	2.54
P7	300	250	−162.8	0.92	2.54
P8	550	300	−242.8	0.95	3.35
P9	150	300	−242.8	0.95	3.35
P10	600	300	−49.4	0.19	0.06
Pump 1	—	—	360.6	—	−41.4

*Increased diameter.

Comment: For static run 2, pipe velocities are less than the design criterion of 1.5 m/s.

6. Extended period analysis

After the static analyses are completed and the network has been defined for the peak demand conditions, run an extended period analysis to determine if the network functions properly over the 24 h peak day period.

a. Use the Pattern Editor to enter the time patterns for the junction demand peaking factors over the 24-h period for the junctions J3, J4, J6, and J8 as shown in the table.

Time patterns for junction demands:

Hour start time	Total demand, m³/h	Demand at junction, m³/h (demand factor)			
		J3	J4	J6	J8
12 a.m.	680	0 (0)	0 (0)	600 (1.0)	80 (1.0)
1 a.m.	680	0 (0)	0 (0)	600 (1.0)	80 (1.0)
2 a.m.	680	0 (0)	0 (0)	600 (1.0)	80 (1.0)
3 a.m.	710	30 (1.0)	0 (0)	600 (1.0)	80 (1.0)
4 a.m.	710	30 (1.0)	0 (0)	600 (1.0)	80 (1.0)
5 a.m.	710	30 (1.0)	0 (0)	600 (1.0)	80 (1.0)
6 a.m.	110	30 (1.0)	0 (0)	0 (0)	80 (1.0)
7 a.m.	110	30 (1.0)	0 (0)	0 (0)	80 (1.0)
8 a.m.	110	30 (1.0)	0 (0)	0 (0)	80 (1.0)
9 a.m.	110	30 (1.0)	0 (0)	0 (0)	80 (1.0)
10 a.m.	410	30 (1.0)	300 (1.0)	0 (0)	80 (1.0)
11 a.m.	410	30 (1.0)	300 (1.0)	0 (0)	80 (1.0)
12 p.m.	410	30 (1.0)	300 (1.0)	0 (0)	80 (1.0)
1 p.m.	410	30 (1.0)	300 (1.0)	0 (0)	80 (1.0)
2 p.m.	410	30 (1.0)	300 (1.0)	0 (0)	80 (1.0)
3 p.m.	410	30 (1.0)	300 (1.0)	0 (0)	80 (1.0)
4 p.m.	410	30 (1.0)	300 (1.0)	0 (0)	80 (1.0)
5 p.m.	410	30 (1.0)	300 (1.0)	0 (0)	80 (1.0)
6 p.m.	110	30 (1.0)	0 (0)	0 (0)	80 (1.0)
7 p.m.	80	0 (0)	0 (0)	0 (0)	80 (1.0)
8 p.m.	80	0 (0)	0 (0)	0 (0)	80 (1.0)
9 p.m.	80	0 (0)	0 (0)	0 (0)	80 (1.0)
10 p.m.	680	0 (0)	0 (0)	600 (1.0)	80 (1.0)
11 p.m.	680	0 (0)	0 (0)	600 (1.0)	80 (1.0)
Totals, m³/d	9600	480	2400	4800	1920

b. Create a time pattern for the supply pump in the Pattern Editor; for this extended period, the pump will run continuously.

c. Finally, enter the time options for the extended period analysis in the Times Options dialog box. For this example, the time duration for the run will be 24 h and the time step will be 1 h.

d. Make the initial run for the extended period run and determine if all of the following conditions are met:

- Pressure is 30 m of water or greater at all junctions over the 24-h period
- Velocity is 1.5 m/s or less in all pipelines
- Water level in the storage tank does not overflow or fall below the minimum level during the extended period; further, the storage tank must refill to its initial level at the end of the analysis period

e. For this example, the diameters of the pipelines serving junction J6 need to be increased to provide the minimum pressure at J6 as summarized in the following table:

	Diameter, mm	
Pipe no.	Static run 2	Extended period run
P1	200	200
P2	200	200
P3	200	200
P4	250	250
P5	300	300
P6	250	300*
P7	250	300*
P8	300	300
P9	300	300
P10	300	300

*Increased diameter.

In addition, the diameter of the storage tank needs to be increased from the assumed 18 to 20 m to provide the emergency storage allowance of 30 percent of the total storage. On the other hand, the extended period analysis indicates that the elevation of the tank base can be lowered to 61 m (from 63 m) and still provide the minimum pressure at all the junctions.

Comment

Sometimes the facilities determined by the static network analysis must be adjusted when an extended period analysis is completed. Often these changes will result in larger facilities to satisfy the design criteria over the extended time period. In this example, both the storage tank and the pipelines serving the pumping station and the storage tank needed to be enlarged from the sizes determined in the static network analysis.

14-3 PIPELINE DESIGN

After planning for the reclaimed water distribution system is completed, design of the pipelines, pumping stations, and reservoirs can proceed. Major design considerations for reclaimed water pipeline systems are presented in this section and include the following:

- Location of pipelines, including separation requirements from potable water pipelines and other utilities
- Pipeline design criteria, including maximum velocities, pressures, and earth loadings
- Pipeline materials, including fittings, joints, and corrosion protection
- Valves
- Appurtenances

The design of reclaimed water pumping stations and storage reservoirs are discussed in Secs. 14-4 and 14-5.

Location of Reclaimed Water Pipelines

Whenever possible, reclaimed water pipelines should be located in public rights-of-way, including streets and roadways, because public rights-of-way provide relatively easy access for constructing and maintaining the pipelines while minimizing project costs. Easement acquisition is necessary to construct pipelines on private property or other public lands that are not rights-of-way, such as parks and playing fields, especially if the agency constructing the pipeline does not own the land. It is important that easements provide for continued access to the pipeline for maintenance and, in addition, include restrictions against construction of permanent improvements within the easement. Separation requirements for some public health jurisdictions in the United States are presented in Table 14-6 and are discussed below.

Separation of Reclaimed Water Pipelines from Potable Water Pipelines

One of the most important factors in locating reclaimed water pipelines is the separation of the reclaimed water pipeline from potable water pipelines and other utilities. The separation of pipelines is especially important where a dual distribution system is used to supply potable and reclaimed water to common users. Most public health jurisdictions have adopted regulations governing pipeline separation that are designed to prevent cross-contamination of potable water with reclaimed water. These regulations generally require minimum horizontal and vertical separation between reclaimed water pipelines and potable water pipelines and other utilities. For example, a minimum horizontal separation of 3 m (10 ft) may be needed between reclaimed water and potable water pipelines without special construction considerations. In addition, it is usually necessary to install reclaimed water pipelines at least 0.3 m (1 ft) deeper than potable water pipelines. If it is necessary to construct a reclaimed water pipeline crossing over a potable water pipeline without the required minimum separation, the reclaimed water pipeline may need to be placed in a steel pipe casing at least 6 m (20 ft) long and centered on the potable pipeline. In this way, reclaimed water that may leak from a reclaimed water pipeline would have to travel at least 3 m (10 ft) through soil before reaching a potable water pipeline. Further, an effective means of minimizing cross-contamination between potable and nonpotable systems is to operate the nonpotable system at a lower pressure [e.g., 69 kPa (10 $lb_f/in.^2$)] within the same service area (AWWA, 1994).

Table 14-6
Pipeline separation requirements for some public health jurisdictions in the United States[a]

State	Drinking water-sanitary sewer	Drinking water-reclaimed water	Reclaimed water-sanitary sewer	Source of standard	Notes
Utah	3 m (10 ft) horizontal	3 m (10 ft) horizontal	3 m (10 ft) horizontal	Utah Administrative Code	If reclaimed water is below or above sewer
Massachusetts	3 m (10 ft) horizontal		Not addressed	2001 Guidelines and Policies for Public Water Systems	Reclaimed water not specifically addressed
Oklahoma	3 m (10 ft) horizontal	1.5 m (5 ft) horizontal	Not addressed	Oklahoma Regulations for Public Water Systems; Water Pollution Control Facility Construction	Sewer and water line cannot occupy same trench
California	3 m (10 ft) horizontal; 0.3 m (1 ft) vertical	Reference to Cal-Nevada AWWA Guidelines for Distribution of Non-Potable Water	Reference to Cal-Nevada AWWA Guidelines for Distribution of Non-Potable Water	California Safe Drinking Water Act	If unable to meet separation; separation as far as possible in separate trenches
Georgia	3 m (10 ft) horizontal; not in same trench with sewer	0.9 m (3 ft) outside to outside of pipe, 460 mm (18 in.) from bottom of water to top of reuse	0.9 m (3 ft) outside to outside of pipe	Georgia Guidelines for Water Reclamation and Urban Water Reuse & Minimum Standards for Public Water Systems	Maximum obtainable separation possible; water-sewer separations less than 3 m (10 ft)—case by case review
Texas	2.7 m (9 ft) outside to outside in all directions	Not specifically addressed	Not specifically addressed	Texas Administrative Code, Title 30, Part 1, Chapter 290.44	Parallel installations require separate trenches
Texas Special Conditions	Nonpressure sewers: determination of no leaks; water 0.6 m (2 ft) above, minimum 1.2 m (4 ft) horizontal New water line: minimum 1,034 kPa (150 $lb_f/in.^2$) pressure rated pipe; water 0.6 m (2 ft) above, minimum 1.2 m (4 ft) horizontal Crossings: water 0.6 m (2 ft) above sewer; if sewer leaking—replace 2.7 m (9 ft) either side 5.5 m (18 ft) total with 1034 kPa (150 $lb_f/in.^2$) rated pipe; new water line installation above sewer-segment centered over sewer 2.7 m (9 ft) to join both directions; new water over existing nonpressure sewer-sewer to have minimum pipe stiffness of 1034 kPa (150 $lb_f/in.^2$) at 5% deflection, sewer embedded in cement stabilized sand (2 bags cement per m^3 of mixture) 150 mm (6 in.) above and 100 mm (4 in.) below sewer				

[a]Adapted from WSWRW (2005).

Separation from Utilities
Separation regulations usually apply to other utilities, such as sanitary sewers and storm drains, as well. For example, it is commonly required that reclaimed water pipelines (as well as potable water pipelines) must be constructed at least 3 m (10 ft) horizontally and 0.3 m (1 ft) above a sanitary sewer or storm drain.

Design Criteria for Reclaimed Water Pipelines

The principal design criteria for reclaimed water pipelines are maximum velocities and working and maximum pressures. In general, maximum velocities between 1.5 and 2.1 m/s (5 and 7 ft/s) in a pipeline will result in the most economical combination of construction cost and ongoing operating costs for pumping. Maximum velocities on the order of 3 m/s (10 ft/s) can be used for relatively short duration pumping during off-peak hours.

Pipeline Materials

Many pipeline materials are available for the distribution of reclaimed water. The most commonly used pipe materials are ductile iron (DI), steel, polyvinylchloride (PVC), and high-density polyethylene (HDPE). The features and limitations of each type of material are summarized in Table 14-7.

Ductile Iron
Ductile iron (DI) pipe has evolved from the original cast iron pipe that has been used to convey liquids for over a century. The manufacture of DI pipe is similar to cast iron pipe, except that DI is stronger and less brittle than cast iron, and it is less susceptible to breakage under normal loadings. Because of its ductility, ductile iron is classified as a flexible pipe and special bedding conditions are required to provide side support for the pipe to help resist vertical trench loadings (AWWA, 2003*a*). Ductile iron pipe is available in several classes, or pipe wall thicknesses, and the designer can select the optimum thickness to resist the combined pressure and vertical trench loadings. Because of its ferrous nature, DI pipe must be protected from both internal and external corrosion. Corrosion protection for DI and other pipe materials are discussed in a later section.

Steel
Steel pipe is also used to convey reclaimed water, especially in larger sizes or in special high-pressure applications. Steel pipe is stronger than ductile iron pipe, but it is also more susceptible to corrosion. Steel pipe is also classified as a flexible pipe for trench bedding purposes, especially if cement mortar is used as an interior lining or exterior coating (AWWA, 2004b). Because cement mortar is brittle, it will break easily if the pipeline deforms more than a minor amount due to external loading. In sizes greater than 1050 to 1200 mm (42 to 48 in.), steel pipe is the most commonly used pipeline material.

Polyvinylchloride
Polyvinylchloride (PVC) pipe is one of the most widely used pipeline materials for conveying reclaimed water, especially in sizes up to 450 mm (18 in.). Polyvinylchloride pipe is manufactured under several industry standard specifications, but the most commonly used PVC pipe for municipal water and reclaimed water conveyance systems is manufactured under the American Water Works Association (AWWA) Specifications C-900 [300 mm, (12 in.) and under] and C-905 [350 to 750 mm, (14 to 30 in.)]. Under these specifications, the dimensions of the pipe and fittings are compatible with ductile iron pipe and can be interconnected without adaptors. The AWWA C-900 pipe is available in three pressure classes: 700, 1050, and 1400 kPa (100, 150, and 200 $lb_f/in.^2$).

Table 14-7
Characteristics of pipe materials used typically in reclaimed water systems

Type	Special features	Applicable AWWA specifications[a]
Ductile iron (DI)	• Strong and flexible • Requires special bedding conditions to limit deflection • Requires protection against internal and external corrosion • Available in three pressure classes • Available in push-on, mechanical joint (MJ), and flanged joints • Special joint restraints required at changes in direction, except for flanged joints	AWWA C151, American National Standard for Ductile-Iron Pipe
Steel	• Stronger than DI and flexible • Requires special bedding conditions to limit deflection • Requires protection against internal and external corrosion • Available in a wide range of wall thickness and pressure ratings • Available in push-on, flanged, and welded joints • Special joint restraints required at changes in direction, except for flanges or welded joints	AWWA C200, Steel Water Pipe—150 mm (6 in.) and Larger
Polyvinyl chloride (PVC)	• Light weight for ease of installation • Flexible, requires special bedding conditions to limit deflection • Special corrosion protection not required • Available in three pressure classes • Allowance for surge pressures not included in rated pressure, may have to limit reclaimed water velocity to reduce surge pressures • Available in push-on joints only • Special joint restraints required at changes in direction	AWWA C900, Polyvinyl Chloride Pressure Pipe, and Fabricated Fittings, 100 through 300 mm (4–12 in.) AWWA C905, PVC Pressure Pipe and Fittings, 350 through 1200 mm (14–48 in.)
High-density polyethylene (HDPE)	• Light weight for ease of installation • Flexible, requires special bedding conditions to limit deflection • Special corrosion protection not required • Available in four pressure classes • Allowance for surge pressures is included in rated pressure • Available in welded joints only • Special joint restraints may not be required at changes in direction	AWWA C906, Polyethylene Pressure pipe and Fittings, 100 through 1575 mm (4–63 in.)

[a]Specifications published by the American Water Works Association, Denver, CO.

Classes for AWWA C-905 pipe are 700, 1140, and 1620 kPa (100, 165, and 235 $lb_f/in.^2$). The nominal pressure class of PVC pipe does not include an allowance for water hammer and surge pressures caused by pump stoppages and rapid valve closures. Therefore, the estimated surge pressure must be added to the maximum operating pressure before selecting the required pressure class of PVC pipe. Some designers of PVC pipelines limit the maximum velocity to prevent over-pressurizing the pipe. And due to its greater elasticity, PVC pipe is even more flexible than ductile iron and steel pipe. Trench design procedures for PVC and other flexible pipe materials are covered in detail in AWWA (2002) and in textbooks and design manuals.

High-Density Polyethylene

High-density polyethylene (HDPE) is a relatively new pipe material for the conveyance of reclaimed water. This material has many similarities to PVC—it is light and easy to handle and install in the field. It is corrosion-resistant and does not require special coatings or linings for the conveyance of reclaimed water. HDPE pipe sections are connected by heat-welded joints using a special joining machine. HDPE pipe is supplied in four standard pressure classes—550, 700, 860, 1140 kPa (80, 100, 125, 165 $lb_f/in.^2$) (PPI, 2001). Unlike PVC, the pressure classes for HDPE pipe are based on the maximum working pressure and include an allowance of 50 percent of the maximum working pressure for water hammer and short-term pressure surges. HDPE is also a flexible material and the pipe requires special bedding conditions similar to PVC pipe to prevent pipe deformation from trench loads.

Joints and Connections

Reclaimed water pipelines can be joined using a variety of connection methods depending on the pipe material and service conditions. Joining methods include push-on joints, mechanical joints, flanged joints, welded joints, and flexible joints (restrained and unrestrained).

Push-on Joints

Push-on joints are available for ductile iron, steel, and PVC pipelines and consist of shaped bell and spigot pipe ends fitted with a compressible elastomeric gasket. Gaskets are available in natural rubber, neoprene, Buna-N, EPDM, and Viton. The shapes of the various bell and spigot pipe ends are similar, but proprietary to the pipe manufacturer. Assembly of push-on joints is achieved by pushing the spigot end of one pipe into the bell end of the adjoining pipe. A typical push-on joint is shown on Fig. 14-7a.

Mechanical Joints

Mechanical joints are similar to push-on joints except that a flanged follower ring is fitted on the spigot end of the pipe that, when tightened with bolts through flange-like ears on the bell, compresses the elastomeric gasket to provide a leak-proof joint. A mechanical joint is shown on Fig. 14-7b. Mechanical joints are commonly used with ductile iron pipe.

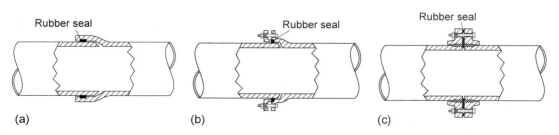

Figure 14-7

Types of joints for various pipe materials: (a) push-on for ductile iron, steel, and PVC pipe, (b) mechanical for ductile iron pipe, and (c) flanged joint for ductile iron and steel pipe.

Flanged Joints

Flanged joints, shown on Fig. 14-7c, are extremely strong and are commonly used on above-ground ductile iron and steel pipelines. Flanges can also be used on buried DI and steel pipes. Because flanged joints are bolted, the pipelines are positively restrained from pulling apart, unlike push-on and mechanical joints that rely on friction between the pipe wall and gasket to prevent separation.

Welded Joints

Welded joints are used for HDPE pipe and often for steel pipes at higher pressures, especially if the pipeline is buried. HDPE pipe joints are heat-welded using a special joining machine as shown on Fig. 14-8. Welded joints are restrained and cannot pull apart under pressure.

Pipe Joint Restraints

Pipe joint restraints are required for pressurized reclaimed water pipelines at most fittings and at all locations where changes in direction occur. The restraints are required to resist the uneven forces caused by the pressure in the fittings or pipeline acting in different directions. Joint restraint can be provided using concrete thrust blocks, restraining rods, mechanical restraining systems, and special restraining joints.

Corrosion Protection

All metallic pipelines in reclaimed water service, either buried or exposed, need to be protected from corrosion. A variety of materials are available to provide this protection, namely, linings and coatings, encasement in plastic sleeves, and cathodic protection. By convention, material applied to the inside of the pipeline is called a *lining* and material applied to the outside is called a *coating*. Common lining and coating materials for buried reclaimed water pipelines include cement mortar, epoxy, and polyurethane. The exterior of aboveground pipelines may also be painted with alkyd, acrylic, epoxy, and polyurethane paints. Cement mortar linings are normally applied by centrifugal force inside a spinning pipe section. Cement mortar coatings are usually sprayed and may include a reinforcing wire mesh or a helically wrapped wire or rod, depending on the

(a) (b)

Figure 14-8

Working with high-density polyethylene (HDPE) pipe: (a) machine used to butt-weld large diameter pipe and (b) welding of tee connection to transmission main. (Courtesy: A. Wilkes, Chevron Phillips Chemical Company, Woodlands, TX.)

pipe diameter. Epoxy linings and coatings may be fusion-bonded or sprayed. Urethane linings and coatings are normally spray applied. Interior lining materials should conform to the requirements of the National Sanitation Foundation standard specifications (NSF Standard 61) for use in potable water service.

Protection of metallic pipe, DI pipe in particular, in a corrosive soil environment can be achieved by encasing the pipe in a tube or sheet of loose polyethylene immediately prior to installation. The polyethylene encasement acts to prevent direct contact of the pipe with corrosive soil and effectively reduces the space between the pipe and the loose wrap to limit the electrolyte available to support corrosion activity. The advantage of polyethylene encasement is its relatively low cost of installation (Stroud, 1988).

Cathodic protection is used more commonly on buried steel pipelines that are installed in a corrosive environment or subject to stray current effects. Cathodic protection is accomplished by making the entire pipeline the cathode of a galvanic or electrolytic corrosion cell. The two basic types of cathodic protection systems, sacrificial anode and impressed current, involve installing anodes that are intended to corrode. Both systems require electrical continuity of the pipeline to ensure complete protection. Additionally, the protected pipeline is typically coated with, or encased in, a dielectric material or bonded coating to reduce the amount of impressed current required. Cathodic protection systems are typically more expensive to install than polyethylene encasement (Stroud, 1988). For a discussion of cathodic protection systems, refer to AWWA (2004c).

Pipe Identification

As a part of the public health regulations designed to prevent accidental cross-connections between potable and reclaimed water supplies, reclaimed water pipelines must be identified clearly when they are manufactured and installed. The most common method, especially with PVC and HDPE reclaimed water pipelines, is to use pipe with a purple color. When the pipe material cannot be colored, purple warning tapes are installed in the trench above the pipe during construction. All reclaimed water pipelines must be marked during manufacture with its diameter, pressure rating, and service. Examples of reclaimed water pipeline identification are shown on Fig. 14-9.

Figure 14-9
View of pipes used for reclaimed water with purple marking tape.

Distribution System Valves

Valves are provided throughout a reclaimed water distribution system to control flows and allow for maintenance of the system. Strategically placed valves will allow sections of the distribution system to be isolated for leaks, repair, and replacement. In some cases, it is necessary to use valves to allow portions of the distribution system to be drained for cleaning. Two types of valves are typically used in reclaimed water distribution systems for shutoff and isolation service—gate valves and butterfly valves.

Distribution System Appurtenances

Appurtenances commonly found in reclaimed water piping distribution systems include the following:

- Services and flowmeters
- Backflow preventers
- Blowoffs and air release valves
- Hydrants

Reclaimed Water Service Requirements

A service is required at every location where reclaimed water is used. A typical reclaimed water service consists of a connection to the reclaimed water distribution main, a service pipeline to the reuse area, a shutoff valve, and a flowmeter. In some jurisdictions, backflow preventers are required to positively protect the reclaimed water supply from contamination by backflow from the users' facilities. Because most non-residential water reuse applications involve large quantities of water, reclaimed water services are larger than a typical household potable water service. The capacities of the various sizes of reclaimed water services are summarized in Table 14-8. Service shutoff valves are typically gate valves or butterfly valves in the sizes greater than 75 mm (3 in.). Smaller services typically will use plug valves for shutoff. Several types of flowmeters are used. Smaller services will typically use displacement-type meters similar to potable water services (see Fig. 15-12c). Larger services will use turbine meters, propeller meters, and magnetic flowmeters.

Backflow Preventers

Backflow preventers are a critical component of any distribution system where reclaimed water and potable water are used. These valves are designed to prevent backflow of reclaimed water from the reuse area to the potable water distribution system

Table 14-8 Capacities of various reclaimed water services

Size of service		Capacity range[a]	
mm	in.	m³/h	gal/min
38	1.5	0.9–45	4–200
50	2	0.9–57	4–250
75	3	1.1–125	5–550
100	4	2.3–284	10–1250
150	6	4.5–528	20–2500
200	8	6.8–1022	30–4500

[a] Lower capacity values are based on the minimum turndown of the service flowmeter.

in the event of a cross-connection or a loss of pressure in the potable water distribution system. Backflow preventers are also used where a potable water system is used as a backup supply for a reclaimed water system. In this way contamination of the potable water supply can be prevented positively. Backflow preventers are almost entirely of the reduced-pressure type, although properly designed double check valve systems can be used in some circumstances. Reduced-pressure backflow preventers consist of two automatically pressure-actuated shutoff valves in series with a third valve located in the space between the two main valves that opens and evacuates any trapped or leaked reclaimed water to the atmosphere. Backflow preventers are described in Chap. 15 and illustrated on Fig. 15-3. Applications are described in Table 15-3.

Blowoffs and Air Release Valves

Blowoffs and air release valves are common components of any reclaimed water distribution system. Blowoffs are small pipe connections with valves located at dead ends and low spots in the distribution system to allow accumulated sediment to be cleaned by flushing and the pipeline drained. Typical blowoff assemblies are shown on Fig. 14-10.

Air release valves are located on all high points in the distribution system to allow trapped air (and other gases) to be removed from the pressurized pipeline without the loss of water. Trapped air, if allowed to accumulate at a high point, acts as a restriction and reduces the capacity of the pipeline by increasing the headloss. A combination air release-vacuum relief valve is similar, except that it will also relieve a vacuum that occurs when a pipeline is drained by allowing atmospheric air to enter the pipeline. A combination air release/vacuum relief valve is shown on Fig. 14-11. Air release and combination air/vacuum valves should be located as shown on Fig. 14-12.

Hydrants

Hydrants are not commonly found in reclaimed water distribution systems, but they can be useful in circumstances where temporary access is needed to the reclaimed water supply or where the reclaimed water system is designed to provide structural fire protection. Typical reclaimed water hydrants are shown on Fig. 14-13. Reclaimed water hydrants are distinguished from potable water hydrants by their distinctive purple color.

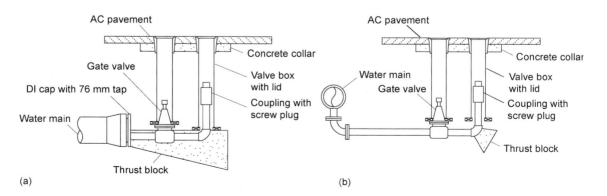

Figure 14-10
Typical blowoff valve assemblies: (a) at dead end and (b) at low point in pipeline.

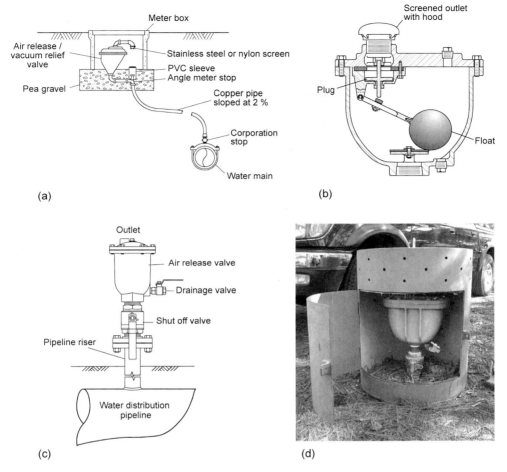

Figure 14-11
Typical air release/vacuum relief details: (a) schematic of belowground type, (b) schematic view of internal valve mechanism (Adapted from Val-Matic Valve and Manufacturing Corp.), (c) schematic of aboveground type, and (d) aboveground type located in protective enclosure.

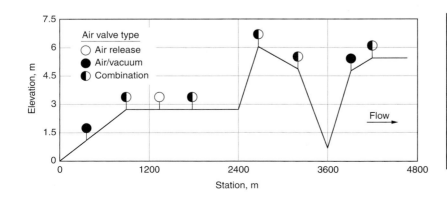

Figure 14-12
Schematic illustrating placement of air/vacuum and air release valves on distribution system. (Adapted from Val-Matic Valve and Manufacturing Corp.)

Figure 14-13
Typical fire hydrants (a) with appropriate signage and (b) with specialized valve operating nut.

(a)   (b)

14-4 PUMPING SYSTEMS

With few exceptions, reclaimed water must be pumped to the storage reservoirs and major reuse areas served by the distribution system. Therefore, at least one pumping station will be needed in a typical reclaimed water distribution system, and often several will be required. Major design considerations for reclaimed water pumping stations are presented in this section and include the following:

- Location and site layout of pumping stations
- Types of pumps for reclaimed water service
- Constant versus variable speed drives for pumps
- Valves and appurtenances
- Emergency power
- Equipment and piping layout
- Effect of pump operating schedule on system design

Pumping Station Location and Site Layout

In general, pumping stations are located near the source of the reclaimed water to be pumped, usually the site of the water reclamation plant. Pumping stations may be stand-alone or may be integrated with the treatment plant facilities. A common location for an integrated reclaimed water pumping station is the outlet end of the chlorine contact tank or the ultraviolet (UV) disinfection system. Booster pumping stations may be needed to fill a distant storage reservoir and may be located at a site along the transmission pipeline. Reclaimed water distribution pumping stations used to supply higher-pressure zones generally are located at the site of the supply storage reservoir.

Table 14-9
Typical features of a pumping station site and appurtenant facilities

Feature	Function/description
General access	Access should be provided for equipment used in the repair and replacement of equipment and general maintenance and operation.
Parking	Depending on the size of the facility, space should be provided for at least one vehicle.
Security	Security should be provided if the structure is unoccupied. Site security can be enhanced by fencing, or the structure can be "hardened" by using concrete or masonry walls, steel doors, and small or high windows.
Valve and metering structures	If flow metering or control valves are located outside the pumping station structure, a secure vault should be provided.
Lighting	Area lighting should be adequate for surveillance of the site from the street. Motion detectors should also be considered.
Utility services	Reliable electric power and communication services should be included. Potable water service and drainage should be considered if sanitary service is needed for operation and maintenance personnel.

Aboveground pumping stations are preferred for ease of access for operation and maintenance, although underground pumping stations can be used. Underground pumping stations tend to be smaller in size or located in areas where an aboveground site is not available. The site layout for reclaimed water pumping stations must provide several features as described in Table 14-9. An example of a reclaimed water pumping station site layout is shown on Fig. 14-14.

Pump Types

Centrifugal pumps are used most commonly for reclaimed water service, including end-suction centrifugal pumps, horizontal split-case centrifugal pumps, and vertical turbine pumps.

End-Suction Centrifugal Pumps

End-suction pumps are available in horizontal and vertical configurations. Examples of horizontal and vertical end-suction centrifugal pumps are shown on Figs. 14-15a and b. An advantage of horizontal end-suction centrifugal pumps is that both the pump and electric motor are located at floor level for easy maintenance. However, the horizontal configuration of the pumping unit and associated piping requires a larger floor area. Vertical end-suction centrifugal pumps require less space for the equipment and piping, but the electric motor is located above the floor and maintenance is less convenient.

Horizontal Split-Case Centrifugal Pumps

Horizontal split-case centrifugal pumps are especially suitable for large-capacity, high-head applications because multiple impellers can be mounted on the horizontal pump shaft. Examples of horizontal split-case centrifugal pumps are shown on Figs. 14-15c and d. In this design, the pump casing is split horizontally and the top portion of the casing can be removed for service or replacement of the impellers and other internal parts. Horizontal split-case centrifugal pumps have the same advantages and disadvantages as

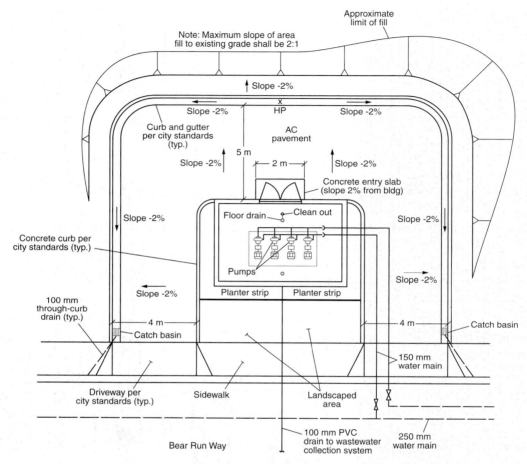

Figure 14-14
Example of a reclaimed water pumping station site layout.

the horizontal end-suction centrifugal pumps, especially the larger area required for suction and discharge piping.

Vertical Turbine Pumps

Vertical turbine pumps are used in small- to medium-capacity reclaimed water pumping stations where available floor space is limited. The turbine pump unit is normally mounted in a suction vessel, called a "can," buried beneath the pumping station floor. Because of its vertical orientation, the space for the pump and piping is relatively small compared to horizontal pumps. Often the piping for the suction can is also buried beneath the floor, resulting is a minimum footprint. Typical vertical turbine pumps are shown on Figs. 14-15*e* and *f*.

Materials of Construction

Special construction materials for reclaimed water pumps generally are not required. General service pumps with cast iron casings and impellers, steel or stainless steel

Figure 14-15
Typical types of pumps used in reclaimed water distribution systems: (a) horizontal end-suction centrifugal pump, (b) vertical end-suction centrifugal pump, (c) small horizontal split-case centrifugal pump (used for high-head applications), (d) large horizontal split-case centrifugal pump (used for high-head applications), (e) vertical turbine pump with aboveground discharge, and (f) vertical turbine pump with belowground discharge.

shafts, and bronze or stainless steel internals are suitable. An exception to this general rule is when reverse osmosis (RO) product water that has not been chemically stabilized is being pumped. In this case, the aggressive nature of RO product water dictates that stainless steel pumps should be used.

Pumping Station Performance

The performance of centrifugal pumps is defined by their head, H, versus flowrate, Q, curves. Typical H-Q curves for end-suction and split-case centrifugal pumps and vertical turbine pumps are shown on Fig. 14-16. The shape of the H-Q curve for each of the pump types can be varied to match the needs of the application but, in general, the slope of end-suction and split-case centrifugal pump curves will be flatter than the curves for similar-sized vertical turbine pumps.

Constant versus Variable Speed Operation

The capacity range of a centrifugal pump operating at constant speed is relatively small with capacity variations depending primarily on the slope of the H-Q curve and the size of the static component of the total pumping head. When the pump is driven by a constant-speed electric motor, the pump capacity can be changed only by throttling the discharge valve to increase the pumping head. By adding a variable-speed drive, the pump can operate over a range of flows within the maximum capacity at full speed, as illustrated on Fig. 14-17. Variable-speed operation of centrifugal pumps is achieved most often using variable frequency drives (VFD) for the electric motors. Other less common methods for achieving variable speed pump operation is through the use of magnetic drives, hydraulic drives, specially designed high-slip squirrel cage electric motors, or wound-rotor electric motors with special electrical control systems. Two-speed electric motors are also used in special applications.

Variable-speed operation is useful in situations where the pump output rate needs to match its input rate. An example would be pumping reclaimed water into a building's plumbing system, with limited storage volume, where the demand for the water being pumped varies over time. In a reclaimed water distribution system that includes adequate

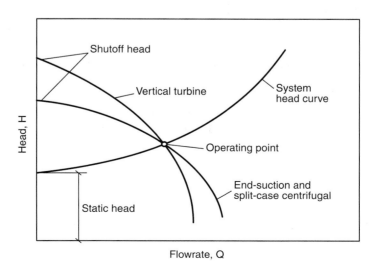

Figure 14-16
Typical head-capacity curves for end-suction and split-case centrifugal pumps and vertical turbine pump.

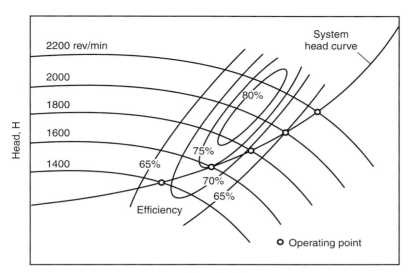

Figure 14-17
Performance of centrifugal pumps at variable speeds.

storage, variable-speed operation may not be necessary and simple start-stop operation of the pump is suitable. In some cases, two-speed motors can be used to provide reduced capacity operation.

Valves

Valves in pumping stations are used for shutoff service, reverse-flow prevention, pump control, pressure reduction, and surge control.

Shutoff, Isolation, and Check Valves
Shutoff and pump isolation functions are usually achieved by using gate valves or butterfly valves. At a minimum, each pump should have shutoff valves on its suction and discharge piping to permit the pump to be removed from service for maintenance. Check valves must be installed on the discharge of each pump to prevent the backflow of reclaimed water when the pump is not operating.

Special Valves for Pump Operation and Surge Control
In many instances, especially in larger pumping stations, special valves are used to control the starting and stopping of the pumps to prevent excessive pressure surges. Under normal operating circumstances, pressure surges can occur both when a pump starts up and when it stops. A surge occurs at pump start-up because the instantaneous flowrate at the time the pump starts is zero, and it increases over a short time to its normal pumping capacity. As a result, for a short period the pump is operating at higher pressures ranging from "shutoff" head to the operating point on the H-Q curve of the centrifugal pump. Shutoff head is the pressure that occurs at zero pump flow. Surges of this type generally are neither excessive nor beyond the pressure rating of the pump and the piping. Of more serious concern are the surges that can occur when a pump stops, either intentionally or due to a power outage. Under extreme conditions (if separation of the water column in the discharge piping occurs), a pressure wave, called a "water hammer," can occur which can damage the piping and pump.

Special valves are available to prevent the development of excessive surge pressures upon pump start-up or shutoff, and to dissipate these if they do occur. These special valves limit the surge in one of two ways: (1) slowly open and close so that the surge is dissipated by flowing through the partially open valve, or (2) open quickly on a bypass pipeline so the surge is relieved to a low-pressure location, typically the supply storage tank. In some instances, it is necessary to use a pressure-reducing valve on the pump discharge to prevent high pressures in the reclaimed water distribution system at low-flow periods. This situation can occur if the shutoff head of the reclaimed water pumps is very high. High pressures in the distribution system during low demand periods can lead to failures in onsite irrigation systems because, often, these low-pressure piping systems are not protected by a pressure-reducing valve.

Equipment and Piping Layout

Many equipment and piping layouts for reclaimed water pumping stations are possible. Some general guidelines for pumping station layouts are summarized as follows:

- Provide adequate access to the building, pumping equipment, and piping and electrical control panels to facilitate initial installation and ongoing maintenance
- Suggested minimum clearances for various purposes include:

    Main building entrance and equipment access area—2 m (6 ft) for small pumps, greater [up to 3 m (10 ft)] for larger pumps

    Around pumps, piping, and special valves—1 m (3 ft) clear space

    Adjacent to electrical control panels—1 m (3 ft) clear space (per National Electrical Code)

- Provide built-in methods for lifting heavy equipment (pumps, motors), such as lifting eyes in the roof or beams, hoists, and trolleys or removable skylights above the equipment
- Provide adequate parking, off-street if possible, for maintenance vehicles, including space for the largest truck needed to transport equipment for repairs

Three examples of reclaimed water pumping station layouts are shown on Fig. 14-18 and Fig. 14-19. The three layouts are for (1) horizontal split-case centrifugal, (2) horizontal end-suction centrifugal, and (3) vertical turbine pumps. The first layout (Fig. 14-18a) for horizontal split-case centrifugal pumps is larger than the other two layouts; the vertical turbine pump layout, shown on Fig. 14-19, has a smaller footprint than the horizontal pump configurations, because of its vertical configuration.

Emergency Power

The decision to provide an emergency electrical power source for a reclaimed water pumping station depends on several factors. Chief among these factors is the need to provide continuous pumping service during a power outage. If adequate storage is available in the pressure zone, or the reclaimed water service is interruptible for short periods, it may not be necessary to provide emergency power unless extended outages that adversely affect delivery and use of the reclaimed water are common. Standby power may consist of a separate utility power source or an engine generator. Further, depending on the customer demand for the reclaimed water, continuous service may not be necessary. For example, in most cases golf course or parkland irrigation may be interrupted during a power outage without adverse impact. On the other hand, if

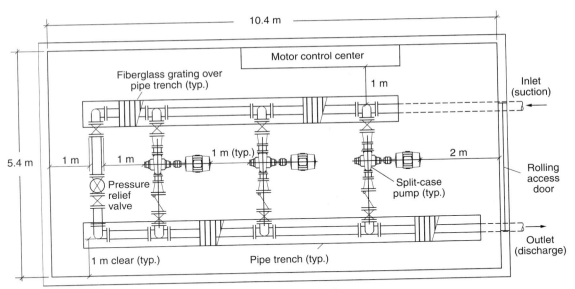

(a)

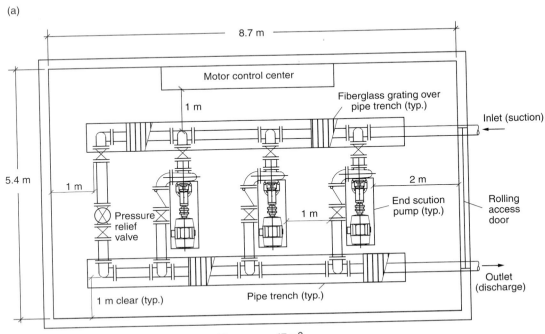

(b)

Figure 14-18
Typical pumping station layouts: (a) split-case centrifugal pumps and (b) end-suction centrifugal pumps.

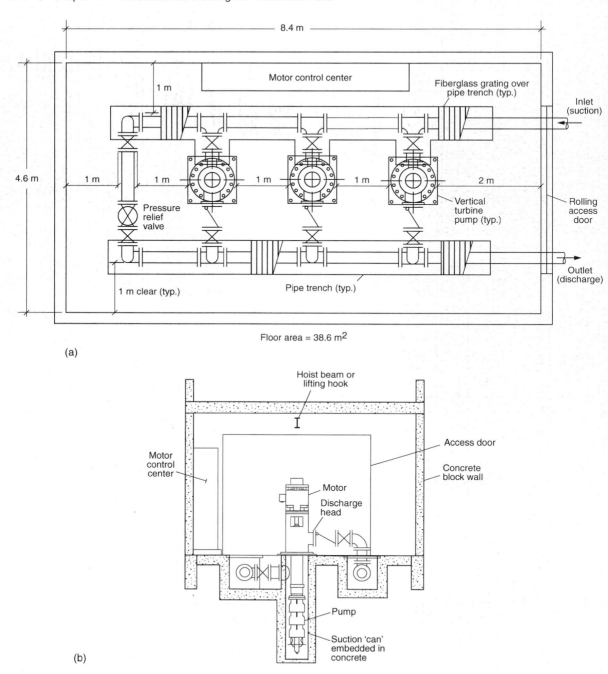

Figure 14-19
Typical pumping station with vertical turbine pumps: (a) plan view and (b) typical cross-section through pump station.

reclaimed water is being used as industrial process water or to provide structural fire protection in dual-plumbed high-rise buildings, emergency power will be an important consideration in the pumping station design. An example of an emergency power system is shown on Fig. 6-10b.

Effect of Pump Operating Schedule on System Design

One of the ways a reclaimed water agency can reduce its operating costs is to design the pumping facilities to operate only during off-peak hours. Local electrical utilities generally offer lower power rates during the evening and nighttime hours when other electrical demands are low. While this practice may be beneficial, the impact of the shorter operating time on the distribution system and storage requirements must be carefully considered before a final decision is made. In many cases, the pipelines serving the pumping station and storage reservoirs must be sized to handle the higher flowrates needed for the shorter operating period. This situation is illustrated in Example 14-4.

EXAMPLE 14-4. Determine the Effect of Pump Operating Schedule on the Design of a Reclaimed Water Distribution and Storage System.

The small town from Examples 14-2 and 14-3 wants to know what changes to the distribution system would be needed if the supply pumping station operates only during off-peak hours, between 8 p.m. to 6 a.m.

Using an extended-period network analysis, determine what changes to the pipe sizes, pumping, and storage facilities will be needed for the shorter pump-operating schedule.

In addition to the assumptions from Example 14-3, the following assumptions are used in this analysis:

- Reclaimed water pump operates at a constant flowrate for 8 h at maximum day demand conditions = 24/8 × 400 m³/h = 1200 m³/h
- Nominal pumping head (TDH) of the supply pump = 40 m of water

Use the same design criteria as Example 14-3 for this analysis, i.e.:

- Minimum pressure at all demand junctions, m of water = 30
- Maximum pipeline velocity, m/s = 1.5
- Minimum pipe diameter, mm = 200

Solution

1. First, use the results of the extended period analysis from Example 14-3 as the starting point for this analysis.

a. Using EPANET2, set the initial pipeline diameters from Example 14-3 as shown in the following table:

Pipe no.	Initial diameter, mm
P1	200
P2	200
P3	200
P4	250
P5	300
P6	300
P7	300
P8	300
P9	300
P10	300

b. Set the initial diameter of the storage tank to 20 m and its ground elevation at 61 m as determined in Example 14-3.

2. Modify the supply pump operating schedule to the desired off-peak power period from 8 p.m. to 6 a.m.
   a. Modify the supply pump run pattern in the Pattern Editor by inserting zero values for the pump off hours between 6 a.m. and 8 p.m.
   b. The hours between 8 p.m. and 6 a.m. should have a value of 1.

3. Extended period analysis
   a. Make the initial extended period run and determine if the design operating requirements can be met.
   b. The results of the initial run include negative pressures at several of the junctions. This is because the larger supply pump requires larger pipelines to carry the higher flow within the maximum velocity allowance. In addition, the pipeline serving the storage tank is too small to handle the higher flowrates.
   c. Increase the diameter of the pipelines serving junction J6 from the supply pump and the storage tank until an acceptable solution is achieved. In addition, adjust the ground elevation of the storage tank and the operating curve for the pump, if necessary, to achieve the minimum pressures and maximum pipeline velocities in the network.
   d. After several iterations, the pipe diameters needed to maintain adequate junction pressures under 24-h and off-peak operations are determined and compared in the following table:

	Pipe diameter, mm	
Pipe no.	24-h pump operation	Off-peak pump operation
P1	200	200
P2	200	200
P3	200	200
P4	250	300*
P5	300	300
P6	300	300
P7	300	300
P8	300	300
P9	300	450*
P10	300	450*

*Increased diameters.

e. More importantly, the diameter of the reclaimed water storage tank would need to be increased from 20 to 28 m to supply all of the reclaimed water demands during those hours when the supply pump is not operating.

Comment

Limiting the pump operating time to off-peak hours can save power costs, but it may also have a significant impact on the size and cost of the other portions of the distribution and storage system. In this example, the pipelines serving the pumping station and the storage tank had to be increased in diameter and the storage tank had to be increased substantially from 20 to 28 m diameter.

14-5 DESIGN OF RECLAIMED WATER STORAGE FACILITIES

The general location and size of the required storage facilities were determined during the planning phase of the reclaimed water project described in Sec. 14-2. The design of short-term storage facilities is the focus of this section.

Short-term storage is usually provided by steel or concrete tanks, similar to those used for potable water storage. Steel tanks are usually of welded construction, although prefabricated bolted steel tanks can also be used. Concrete tanks can be constructed above ground or below ground. Examples of short-term storage facilities are described in Sec. 14-2.

The design of reclaimed water storage facilities involves the following tasks:

- Determining the final location and type of reclaimed water reservoirs
- Preparing facility and site layouts for reservoirs, piping, and appurtenances
- Determining materials of construction for reservoirs and appurtenances
- Selecting protective coatings, if applicable, for reservoirs

These tasks are discussed in this section.

Location of Reclaimed Water Reservoirs

The general location of a storage reservoir is determined in the planning phase. The final location and type of the reservoir will depend on several factors, including:

- Availability of sites at the proper elevation(s)
- Site geology and topography
- Site access
- Visual impacts

Site Elevation and Availability

The availability of potential reservoir sites at the proper elevation will generally limit the options to be evaluated. Further, if more than one reservoir is located in a pressure zone, the high water elevation should be the same for all reservoirs to simplify reservoir operations. If this is not possible, special level control valves must be installed on the inlet piping of the lower reservoir to prevent overfilling and overflow. These valves, called altitude valves, sense the water depth in the reservoir and close automatically when the reservoir is full. However, in some cases, water will not flow from a lower-elevation reservoir during low demand periods due to high pressures in the surrounding distribution system. As a result, stagnation of the reservoir contents can occur. If it is necessary to locate a reservoir at a lower elevation due to the lack of a suitable site, special considerations need to be taken into account, especially the hydraulic head in the distribution system in the vicinity of the reservoir. For example, a separate inlet/outlet pipe to the reservoir, isolated from the local distribution system, can be provided to allow higher pressures to be maintained in the vicinity of the reservoir. Alternatively, a separate outlet pipe from the reservoir to a lower-pressure portion of the distribution system can be provided.

Site Geology and Topography

Site geology and topography can be the deciding factors in the final selection of a reclaimed water reservoir site. Since a reservoir site needs to be leveled, the topography must allow for the needed excavation. Hillsides that are too steep are generally not feasible. Site geology and soils characteristics are also important, often overriding other considerations. Rocky sites may preclude the needed excavation. Weak, friable rock or soils may not provide sufficient strength to support the weight of the tank and its contents. Sites that are subject to land creep or sliding are likewise not suitable. Reservoir sites that are located on or near earthquake fault lines must have a full geotechnical and seismic survey conducted, and if possible, should be avoided. The need for a thorough geotechnical investigation when selecting a reclaimed water reservoir site cannot be overemphasized.

Site Access
Access to the reservoir site for construction and normal operation and maintenance is essential. Access roads need to be adequate to accommodate vehicles and trucks large enough for reservoir maintenance, painting, and equipment repair or replacement.

Visual Impacts
The visual impact of the tank on the surrounding vicinity is an important factor in selecting reclaimed water reservoir sites. Where possible, reservoir sites should be located to minimize the visual impact of the tank and its appurtenances. Landscaping and mounding of soil around the tank can be used for this purpose. Architectural treatment of the exposed exterior surfaces of the reservoir can also help in mitigating visual impacts.

Facility and Site Layout for Reservoirs, Piping, and Appurtenances

The facility and site layout for reservoirs, piping, and appurtenances involves the following considerations:

- Site layout and facility access, parking, drainage, and security
- Site piping, including inlet, outlet, and drainage piping
- Reservoir piping, including inlet, outlet, overflow, and drain piping
- Appurtenances such as tank access manholes and hatches, for metering, mixing, disinfectant addition, and telemetry

Site Layout
After the final location of the reservoir is selected, a site plan must be developed. An access road must be designed to accommodate the largest construction and maintenance vehicle to be used. Design considerations include roadway width, maximum slope, minimum radius of the curves, and pavement materials. The area around the tank should be paved, if possible, to provide access for maintenance vehicles and painting equipment.

Chain-link fencing topped with barbed wire should be provided to increase security and minimize vandalism at the site, which is often located in remote areas away from visual observation.

Site Piping
Site piping includes the inlet and outlet pipelines, the tank overflow and drain pipelines, and the site drainage piping. Either single or separate inlet/outlet pipelines may be used as discussed below. Drainage piping for the reservoir site, including the sloped areas and pavement, should be sized to include the maximum reservoir overflow and drain flows in addition to the runoff from the design rainfall. Inlet and outlet piping and valve and metering vaults should be located to facilitate access for operation and maintenance.

Reservoir Piping
The inlet/outlet piping may be either a single pipeline or two separate pipelines. Using separate pipelines is preferred as it can facilitate mixing of the reservoir contents if the inlet and outlet are located at opposite sides of the reservoir. If separate inlet and outlet pipelines are used, check valves are needed in both pipes to prevent backflows.

Isolation valves are required on all pipelines entering or leaving the reservoir, except the overflow, to allow for maintenance and to prevent reclaimed water from spilling after an earthquake or accident. In active seismic areas, the inlet and outlet pipelines also may be equipped with flexible connections designed to prevent breakage during an earthquake. As discussed previously, altitude valves may be necessary to prevent overfilling of reservoirs located beneath the hydraulic grade line of the surrounding distribution system. All valves should be located in belowground utility vaults to provide protection from freezing and any unauthorized access.

Reservoir Appurtenances

Appurtenances for reservoirs may include any of the following:

- Tank access manholes and hatches
- Water depth measurement and flow metering
- Mixing facilities
- Disinfectant addition facilities
- Telemetry equipment

Each reservoir must be provided with a minimum of two access points to facilitate construction, inspection, maintenance, and painting. For steel tanks, access is usually provided by side manholes located at ground level and hatches located on the reservoir roof. For concrete tanks and reservoirs, access is usually provided by hatches located on the roof. All access points must be lockable to provide security for the reclaimed water in the reservoir.

Water depth measurement and flow metering are common reclaimed water reservoir appurtenances. Water depth is the primary operating parameter for reclaimed water reservoirs and is used to control the pumps serving the reservoir. Although flow metering is not essential to the operation of reclaimed water reservoirs, historical flow data can be important when planning for new storage facilities in the service area. Flowmeters located in single inlet/outlet pipelines must be able to operate with the flow in either direction, such as magnetic flowmeters.

Mixing facilities are becoming more important in storage reservoirs, especially for reclaimed water reservoirs where water quality changes can be significant (see Sec. 14-7). Mixing can be accomplished in several ways, but mechanical mixers and air-mixing are the most common. Mechanical mixers can be roof-mounted or side-mounted with the electric drive motor located outside the reservoir. Recently, submersible mechanical mixers located inside the reservoir have become available. Air-mixing is accomplished by using an air compressor connected to a pipe manifold system installed inside the reservoir. The complexity of the inside manifold system depends primarily on the size of the tank. The air compressor can be permanently installed at the reservoir, or it can be a portable unit brought to the site when needed.

The ability to add disinfectant to remote reclaimed water reservoirs is also becoming more important. Disinfectant addition can be as simple as manually pouring hypochlorite solution through the roof access hatch or as complicated as providing a permanent onsite hypochlorination station. Because addition of disinfectant is usually an intermittent need, especially during the summer, a portable, trailer-mounted hypochlorination feed unit generally will be adequate for most agencies.

Telemetry is another important consideration for reclaimed water reservoirs. Telemetry is used to transfer water depth information from the remote reservoir site to the pumping station and operations center for the control of the feed pumps and for monitoring of the system operation. Telemetry can be transmitted over dedicated wires, including telephone lines. Recently, many telemetry systems transmit information wirelessly, including radio or microwave frequencies. If the reservoir security system includes intrusion alarms, notice of unauthorized entry can also be transmitted to the receiving station.

Materials of Construction

Reclaimed water reservoirs are constructed most commonly of steel or reinforced concrete. Although the basic functions of steel and concrete reservoirs are identical, each material has specific characteristics that must be addressed during the design and construction of the reservoir. Steel tanks for water and reclaimed water service are usually of welded construction, although, prefabricated bolted steel tanks can also be used. Because steel tanks are located aboveground, they must be protected from corrosion with interior and exterior paint or other coating material. Cathodic protection is also employed commonly to prevent electrochemical corrosion of steel. Steel tanks range in size up to 30 to 45 m (100 to 150 ft) in diameter and up to 10 to 15 m (30 to 45 ft) in height, although larger tanks have been used (AWWA, 1998). Elevated steel tanks are available in standard sizes from 190 to 11,360 m^3 (50,000 to 3,000,000 gal) (AWWA, 1996).

Concrete tanks are usually cast-in-place and are reinforced with steel bars or are post-tensioned (wire-wrapped) after concrete placement. The dimensions for concrete tanks are similar to steel tanks, except the diameter can range up to 60 to 70 m (200 to 230 ft). Concrete tanks can be located above or below grade, although below-grade concrete tanks are used when it is important to reduce the visual impact of a large storage structure. The costs of belowground construction generally will be higher than that of an aboveground tank due largely to the cost of earthwork. Coating requirements for concrete construction are generally less severe than steel tanks for water or reclaimed water service. Advantages and disadvantages of steel and reinforced concrete tank construction are given in Table 14-10.

Protective Coatings—Interior and Exterior

The requirements for protective coatings for steel tanks are different for interior and exterior service as described in the following paragraphs. Concrete tanks generally do not require interior or exterior coatings, although they can be provided if desired.

Interior Coatings
Because the interior coating is in constant contact with reclaimed water, it must be suitable for immersion service in chlorinated water and must not dissolve organic or other components into the reclaimed water. The National Sanitation Foundation (NSF) has adopted national voluntary standards for interior coatings for potable and reclaimed water tanks that are widely accepted in the United States and other locations. These standards are included by reference in the American Water Works Association (AWWA)

Table 14-10

Advantages and disadvantages of steel and reinforced concrete construction for enclosed reservoirs

Steel construction	
Advantages	Disadvantages
Welded steel construction is free of leaks	A cathodic protection system is needed to prevent corrosion
Structural integrity is unaffected by thermal expansion and contraction	A regular schedule of maintenance painting is required
Tanks that are coated regularly have an economic life exceeding 50 yr	Tanks are suited only for aboveground construction
Depending on size and configuration, tanks may have a lower initial cost than reinforced concrete	

Reinforced concrete	
Advantages	Disadvantages
Lower maintenance cost than steel construction	May be subject to cracks due to improper design or construction
Tanks can be constructed either aboveground or belowground	Effects of thermal expansion and contraction must be incorporated in design
Special architectural treatment can be more easily incorporated in construction	Initial cost may be higher than steel construction

Standards for Coating Steel Water-Storage Tanks (AWWA D102-03). The most commonly used interior coating materials (in 2006) include epoxy and polyurethane paints. Adequate surface preparation of the steel and concrete is extremely important for the long-term success of an interior coating system.

Exterior Coatings

Service requirements for exterior coatings, including epoxy, urethane, and acrylic paints, are generally not as severe as for interior coatings, primarily because exterior coatings are not subject to continuous immersion. Of greater concern for exterior reservoir coatings is exposure to ultraviolet radiation and the adverse effects it can have on the coating materials.

14-6 OPERATION AND MAINTENANCE OF DISTRIBUTION FACILITIES

Distribution facilities, like treatment facilities, require proper operation and maintenance (O&M) to maintain reliable service to the reclaimed water users. Pipeline O&M tasks include periodic flushing of the pipes to maintain water quality, regular checking of disinfection residuals throughout the system to prevent bacterial regrowth, valve and hydrant exercising, and a rigorous program of cross-connection control. Operation and maintenance tasks for pumping stations include regular and preventative maintenance of the pumping equipment, valves, and electrical equipment and pump controls. Pump controls and reservoir instrumentation need to be checked for proper operation and calibration periodically. Distribution system O&M requirements are discussed in this section. Storage reservoirs are discussed in Sec. 14-7.

Pipelines

Once the disinfected reclaimed water enters the distribution system, changes begin to occur that can adversely affect the quality of the water. First, the residual concentration of the disinfectant (e.g., chlorine, chloramine, chlorine dioxide) will slowly decrease due to chemical reactions with other constituents in the water and with microbial films growing on the inside surface of the pipeline. Constituents in the water can include dissolved oxygen, reduced inorganic compounds, and trace amounts of residual organic matter. In addition, slime layers can build up on the inside surface of the pipeline over time even in the presence of a residual disinfectant. These reactions proceed slowly in the pipeline and generally are more evident as the reclaimed water moves farther from the source. Residual disinfectant decay generally is greater in reservoirs where the reclaimed water is exposed to the atmosphere and, especially in open reservoirs, where it is also exposed to direct sunlight. Residual disinfectant decay is especially serious in the dead ends of distribution pipelines where odors can develop. Whenever possible, dead ends should be avoided in distribution systems.

Periodic Flushing
An important method of maintaining water quality is to institute a program of periodic flushing of the reclaimed water distribution mains. Generally, flushing can be accomplished using hydrants and blowoffs at low points and dead ends in the system to flush the mains. Flushing water can be discharged to the wastewater collection system or, with appropriate permits, discharged to the storm water system or local waterways. If permits cannot be obtained, flushing water may have to be discharged to a tank truck and transported to the wastewater system for disposal. If the flushing water is discharged to the storm sewer system or local waterways, some regulatory agencies require the removal of any residual disinfectant. In some cases, it may be necessary to clean the pipelines with pipe-cleaning pigs (see Fig. 14-20). For large distribution systems, it may be necessary to add disinfectant to remote pipelines or storage reservoirs to increase the disinfection residual to acceptable levels.

Cross-Connection Control
Cross-connection control is probably the most important and serious operational responsibility of any reclaimed water or potable water agency. Only with a well-conceived and executed cross-connection control program can the public health of the consumers be fully protected. Cross-connection protection must be provided whenever a potable source of water is used as a backup or supplemental supply of water for the reclaimed water system. Cross-connection control is discussed in Chap. 15, and various methods used in the prevention and monitoring of cross-connections are described in U.S. EPA (2003) and AWWA (2004a).

Exercising of Valves
Another important operating and maintenance function for reclaimed water distributions systems is a program of regular exercising of the valves and hydrants in the system. Valves and hydrants should be operated at least once annually to ensure their proper operation.

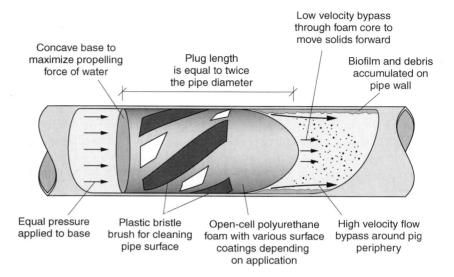

Figure 14-20
Typical polyurethane open-cell foam device for cleaning pipelines, commonly known as a *pig* (adapted from Girard Industries). In general, four types of pigs are used: polyurethane open and closed foam, mandrel or mechanical, solid cast urethane, and other articulated types made of composite materials.

Pumping Stations

Depending on their design, reclaimed water pumping stations may operate nearly continuously during peak demand periods or only during off-peak power periods. If continuous operation is required during peak demand periods, the importance of scheduled preventative maintenance is evident and some redundancy in the pumping capacity should be provided. Maintenance requirements for pumps include regular lubrication of the pumps and motors, checking for leakage from shaft seals and replacement of the packing, if necessary, and checking for proper operation of automatic and manual valves and the pump control system and instrumentation.

14-7 WATER QUALITY MANAGEMENT ISSUES IN RECLAIMED WATER DISTRIBUTION AND STORAGE

Degradation of water quality during distribution and storage is always a concern for the suppliers of reclaimed water, but it is of particular concern with reclaimed water stored in open and enclosed storage reservoirs. Potable water is subject to relatively minor changes in water quality during distribution and storage, usually in the form of reduced disinfectant residual concentrations and bacterial regrowth and slime formation in the distribution piping. These changes are magnified in reclaimed water because the concentrations of dissolved nutrients and residual organic matter usually are higher than for potable water. The following discussion on water quality issues, problems, and management strategies to mitigate these problems is adapted in part from Tchobanoglous and Schroeder (1985), Tchobanoglous et al. (2003), and Miller et al. (2003).

Water Quality Issues

Water quality degradation in reclaimed water can be classified in the following general categories (adapted from Kirmeyer et. al., 2001):

- Physical—temperature, turbidity, suspended solids (sediment)
- Chemical—pH and alkalinity changes, reduced disinfection residuals
- Biological—bacteria (regrowth and birds/animals), algae, low dissolved oxygen, nitrification
- Aesthetic—odors (e.g., hydrogen sulfide), color, turbidity

Although these water quality issues are listed separately, they are interrelated as discussed in this section.

Physical Water Quality

Physical water quality issues in the distribution and storage of reclaimed water generally involve changes in water temperature and increases in turbidity and suspended solids (sediment). Increases in water temperature can be significant when the reclaimed water is stored in open reservoirs or aboveground storage tanks, particularly in areas with hot and arid climates. While elevated temperatures generally will not affect the uses of reclaimed water directly, it does accelerate other changes in the reclaimed water quality. The primary mechanism of change is the dependence of chemical and biological reactions to temperature. For example, bacterial regrowth in distribution pipelines, algae growth in open reservoirs, and chlorine residual reduction, all proceed faster at elevated temperatures.

Turbidity and suspended solids concentrations in reclaimed water can increase significantly when stored in open reservoirs due to the natural growth of plankton and algae and local runoff from the watershed. In extreme cases, it may be necessary to control algae growth by the use of approved herbicides. State-of-the-art methods should be used to control erosion within the watershed to prevent turbidity and suspended solids increases and, in the extreme, it may be necessary to divert some or all of the local runoff away from the reservoir.

Chemical Water Quality

Chemical water quality issues in the distribution and storage of reclaimed water generally involve changes in pH and alkalinity and depletion of the disinfectant residual. Several natural processes in distribution pipelines and storage facilities can change the balance between pH and alkalinity in reclaimed water. Biological oxidation of residual organic matter in reclaimed water releases carbon dioxide (CO_2) as a natural byproduct, lowering the pH of the reclaimed water. In addition, biological nitrification of ammonia to nitrate releases hydrogen ions that lower the pH.

In general, disinfectant concentrations decrease as the residence time of the reclaimed water increases in the distribution piping and storage facilities. The reductions in disinfectant concentration are caused by the presence of low concentrations of residual organic matter in the reclaimed water, slime buildup on distribution piping and reservoir surfaces, and exposure to the atmosphere in enclosed and open reservoirs. Bacterial regrowth can occur if the residual disinfectant levels are allowed to fall to low levels in the distribution system.

Biological Water Quality

Biological water quality issues in reclaimed water are associated primarily with the growth of algae and the growth or regrowth of bacteria and other pathogens in the distribution system and storage facilities. Biological nitrification can also lead to pH and alkalinity changes and low dissolved oxygen (DO) concentrations in the reclaimed water.

The presence of algae in open reservoirs causes diurnal changes in pH, alkalinity, and DO. Algae are phytoplankton that use energy from sunlight and carbon dioxide (CO_2) in the water to reproduce; the production of DO is a natural byproduct of these biological reactions. When sunlight is not available, however, algae use DO in the water for respiration, which releases more CO_2 to the water. The resulting pH and DO variations can be significant if large concentrations of algae are present in the reclaimed water. In addition, excessive algae in reclaimed water stored in open ponds or reservoirs can lead to operation and maintenance problems with irrigation sprinkler heads and drip irrigation equipment. Finally, the presence of algae in reclaimed water is aesthetically unacceptable for many reuse applications.

Bacterial growth and regrowth is a significant water quality problem in most water reuse applications. Bacterial growth will also result from the presence of animals and birds in open storage facilities. Ammonia is a natural component of municipal wastewater and, unless the water reclamation plant includes a biological nitrification step or nitrification/denitrification, significant concentrations of ammonia will be found in reclaimed water. Even though ammonia has value as fertilizer in irrigation applications, it is a major contributor to several of the water quality problems discussed above, especially those related to algae growth and low DO concentrations.

Aesthetic Issues

Odors in reclaimed water can develop over time due to the presence of ammonia and hydrogen sulfide. Ammonia gas has a noticeable odor as it volatilizes from reclaimed water. Because nitrification of ammonia proceeds naturally in the presence of dissolved oxygen, under certain conditions the dissolved oxygen concentration of the reclaimed water can be depleted and anaerobic conditions occur. Loss of oxygen and the formation of odors can happen under stagnant conditions in (1) dead-end pipelines, (2) underutilized enclosed reservoirs, (3) oversized pipelines, and (4) lower levels of stratified open reservoirs where anaerobic conditions occur.

Under anaerobic conditions, dissolved sulfates in the reclaimed water are biologically reduced to hydrogen sulfide by facultative bacteria. In addition, black sulfide precipitates can form in irrigated turf when the natural oxidation of ammonia to nitrate depletes oxygen in the soil profile. When anaerobic conditions exist, generally due to poor soil drainage and excessive irrigation, odors will be released when the turf is disturbed (e.g., divots on golf course).

Impact of Water Quality Issues

The water quality issues discussed above, if left unresolved, can lead to aesthetic, operating, and maintenance problems with regard to the beneficial reuse of reclaimed water. Aesthetic problems include odors and visual impacts. Odors emanating from storage ponds at a park or golf course can seriously impact the acceptance of reclaimed water for reuse. Similarly, heavy algae growth in a decorative pond is a serious aesthetic problem. Excessive regrowth of bacteria is of greater concern because it can lead to public health and regulatory problems. Therefore, it is important to prevent these adverse water quality conditions before they occur.

Operating and maintenance problems include plugging of sprinkler heads and drip irrigation systems and excessive pipeline flushing and storage pond maintenance. Algae and other particles in reclaimed water can cause serious problems in the maintenance requirements of sprinkler heads and drip irrigation emitters. Potential solutions for controlling regrowth problems in distribution systems are presented in Table 14-11. Pipeline flushing, a common method of controlling water quality in distribution systems, is a special concern because the discharge of flushing water to storm drains may not be permitted, and special equipment or arrangements may be required for the capture and disposal of flushing water.

The Effect of Storage on Water Quality Changes

As discussed above, the residence time in reservoirs can have an impact on the quality of the reclaimed water. The effects of storage in open and enclosed storage and mitigation measures to minimize water quality effects are discussed below.

Storage in Open Reservoirs

The most common problems associated with the storage of reclaimed water in open reservoirs are listed in Table 14-12. The principal problems are:

- Release of odors, principally hydrogen sulfide
- Temperature stratification
- Low dissolved oxygen resulting in odors and fish kills
- Excessive growth of algae and phytoplankton
- High levels of turbidity and color
- Regrowth of microorganisms
- Water quality degradation due to bird and rodent populations

Table 14-11
Alternative solutions for the control of bacteria regrowth in distribution systems[a]

Category	Solution
Monitoring	Take water samples and analyze for heterotrophic plate count and chlorine residuals. Correlate bacteriological results with disinfectant residuals. Monitor also for turbidity, temperature, and nutrient levels
Operations management	Increase disinfectant residual, decrease residence time, increase turnover in storage facilities, and isolate and disinfect problem area in pipelines
Maintenance	Conduct periodic unidirectional or zone flushing to remove sediment; scour biofilm from pipe walls with a pig (see Fig. 14-20). Check presence of sediment in storage facilities

[a]Adapted in part from Kirmeyer et al. (2001).

Table 14-12 Problems encountered in the operation of open reservoirs used for the storage of reclaimed water[a]

Reservoir problem	Description
Physical/aesthetic	
Color	The presence of color can affect the aesthetic acceptance of the water. Often caused by the presence of humic materials and fine slits and clays in runoff and the presence of color in the reclaimed water
Odors (primarily H_2S)	One of most common problems encountered with the storage of reclaimed water. In addition to causing odors, H_2S has a chlorine demand
Temperature	Water may be unusable during certain times of the year
Temperature stratification	Usually occurs once or twice a year depending on the latitude
Turbidity	The presence of turbidity can affect the aesthetic acceptance of the water. Turbidity can be caused by runoff containing silt and clay and by algal growth
Chemical	
Chlorine	Chlorine and compounds containing chlorine may be toxic to aquatic life in open reservoirs
Dissolved oxygen	Low DO can cause fish kills and allow the release of odors in open reservoirs
Nitrogen	Nutrient capable of stimulating phytoplankton
Phosphorus	Nutrient capable of stimulating phytoplankton
Biological	
Algae	Presence of excess algae can cause odors, increase turbidity, and clog filters
Aquatic fowl	The presence of excessive numbers of aquatic birds can degrade the water quality of the stored water
Bacteria	Regrowth is a common occurrence in open storage reservoirs. May affect possible applications
Nitrification	Microbial process that oxidizes ammonia nitrogen to nitrite and nitrate nitrogen
Chlorophyll	Presence of excess algae and plant matter
Helminths	May affect possible reuse applications
Insects (mosquitoes)	May require spraying of insecticides
Phytoplankton	Presence of excess algae can cause odors, increase turbidity, and clog filters
Protozoa	May affect possible reuse applications
Viruses	May affect possible reuse applications

[a] Adapted from Tchobanoglous et al. (2003).

Storage in Enclosed Reservoirs

The most common water quality problems encountered with enclosed reservoirs are listed in Table 14-13 and include (Tchobanoglous et al., 2003):

- Stagnation
- Release of odors, principally hydrogen sulfide
- Loss of chlorine residual
- Regrowth of microorganisms

Management strategies that have been used to overcome the problems cited in Tables 14-12 and 14-13 are summarized in Table 14-14. The strategies for open and enclosed reservoirs are as follows:

Open Reservoirs

Although all of the strategies listed in Table 14-14 have been used, the most effective strategy for open reclaimed water storage reservoirs has been the use of aeration to both provide oxygen and destratification. Several types of aeration systems have been used to provide oxygen and to eliminate stratification, including surface aerators with

Strategies for Managing Water Quality in Open and Enclosed Reservoirs

Table 14-13

Problems encountered in the operation of enclosed reservoirs used for the storage of reclaimed water[a]

Reservoir problem	Description
Physical/aesthetic	
Color	Often caused by the presence of humic materials in reclaimed water
Odors (primarily hydrogen sulfide—H_2S)	One of most common problems encountered with the storage of reclaimed water. In addition to causing odors, H_2S has a chlorine demand
Turbidity	The presence of turbidity can affect the aesthetic acceptance of the water
Chemical	
Chlorine	Chlorine and compounds containing chlorine may cause odors. Chlorine is used commonly to control biological growths
Dissolved oxygen	Lack of oxygen can lead to the release of odors in enclosed reservoirs
Biological	
Bacteria	Regrowth has occurred in enclosed storage reservoirs. May affect possible applications
Insects (mosquitoes)	Insects can enter improperly sealed reservoirs. May require spraying of insecticides
Viruses	May affect possible reuse applications

[a]Adapted from Tchobanoglous et al. (2003).

Table 14-14
Management strategies for open and enclosed reservoirs used for the storage of reclaimed water[a]

Management strategies	Comments
Open storage reservoirs	
Aeration/destratification	Installation of aeration facilities can be used to maintain aerobic conditions and eliminate thermal stratification. May result in release of phosphorus from bottom sediments
Alum precipitation	Alum precipitation has been used to remove suspended solids and phosphorus. Can be used to stop release of phosphorus from sediments
Biomanipulation	Control of microorganism growth rates
Copper sulfate addition	Copper sulfate is applied to control the growth of algae. The use of copper may be prohibited because of toxicity concerns over accumulation of copper. Some agencies have banned the use of copper sulfate
Destratification (including recirculation)	Submerged or aspirating mixers can be used to eliminate thermal stratification. Recirculating pumps can also be used. May result in release of phosphorus from bottom sediments
Dilution	Water from other sources can be blended with water from the storage reservoir to manage the water quality
Dredging	Accumulated sediment can be removed annually to limit the formation of deposits and the generation of hydrogen sulfide
Filtration	Water from the storage reservoir can be filtered through a rock filter, a slow sand filter, or a disk type filter to remove algae and to improve the clarity of the water
Natural microorganism decay	The effectiveness of natural decay will depend on the operation of the reservoir and the detention time
Photooxidation	With proper mixing, advantage can be taken the beneficial effects of exposing the water to sunlight
Wetlands treatment	Water from the storage reservoir can be passed through a constructed wetland to improve the clarity of the effluent and to remove algae
Withdrawal from selected depths	Varying water quality can be obtained by drawing off water at selected depths within the reservoir
Enclosed storage reservoirs	
Aeration	Residual level of DO is maintained to eliminate the formation of odors
Chlorination	Used to control the growth of microorganisms
Recirculation	Adequate recirculation can limit the growth of microorganisms and the formation of odors

[a] Adapted from Tchobanoglous et al. (2003), Miller et al. (2003).

high pumping capacity, brush aerators, static tube aerators, and diffused aeration systems. The Irvine Ranch Water District in Irvine, California, uses a system of an air compressor and diffusers to circulate the top and lower layers of the Sand Canyon reclaimed water reservoir to prevent stratification and to increase dissolved oxygen levels (IRWD, 2001).

A relatively new aeration device, the Speece cone (see Fig. 14-21), has a very high oxygen transfer rate and is superior in mixing the contents of a reservoir. The power input required for aeration and destratification is on the order of 0.30 to 0.50 kW/1000 m^3. Obviously, the actual power requirement will vary with the physical characteristics of the reservoir, including surface area, aspect ratio, depth of the reservoir, and temperature.

The growth of plankton in reservoirs can be controlled using copper sulfate or more selective algaecides. The effective use of these chemicals, such as copper sulfate, requires microscopic examination of the water to determine the number and type of organisms involved. Some agencies have banned the use of copper sulfate. Ideally, the control chemicals should be applied just at the time when the number of organisms starts to increase rapidly. The use of chlorine in open reservoirs is not recommended as a control measure. Under certain circumstances, chlorine can combine with odor-causing compounds present in the reservoir to intensify odors.

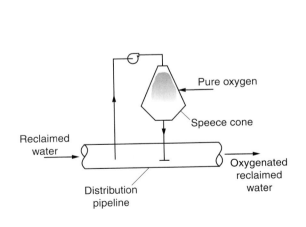

Figure 14-21
Speece cone aeration device for pipeline aeration and/or reservoir recirculation and aeration: (a) schematic and (b) view of typical unit.

Enclosed Reservoirs

In enclosed reservoirs, effective management strategies include providing facilities to (1) recirculate the contents of the storage basin and (2) add additional chlorine to maintain a residual. In addition to using aeration devices and pumps to promote circulation, the inlet and withdrawal piping can be configured to promote circulation. Generally, the addition of chlorine will be limited to small off-line storage reservoirs typically used for landscape irrigation and some industrial reuse applications.

PROBLEMS AND DISCUSSION TOPICS

14-1 A reclaimed water supplier has the supply and demand factors summarized in the following table:

Reclaimed water	Average annual flow	Maximum monthly flow	Maximum daily flow	Peak hourly flow
Supply, m^3/h	2000	2100	2200	2200
Demand, m^3/h	500	1500	2000	2500

a. From these flow rates, determine if distribution system storage is required. Explain the rationale for your recommendation.

b. If distribution storage is required, use the rule of thumb method to estimate the amount of working storage needed to serve the reclaimed water demands.

c. One of the reclaimed water users in the service area is a high-value manufacturing operation with a critical need for a highly reliable, continuous reclaimed water service of 500 m^3/h. Assuming a worst-case scenario, the reclaimed water supply could be disrupted for a total of 2 d. Estimate the emergency storage allowance that would be appropriate for the storage reservoir.

d. Would it be more cost-effective for the critical reclaimed water user to provide its own emergency storage on-site? Assume that the cost of steel storage tanks can be described by the following equation:

$$\text{Cost} = a \times V^b$$

where Cost = dollars (US)
 a = constant = 637
 b = constant = 0.65
 V = tank volume, m^3

14-2 The maximum day reclaimed water demand for a town is shown in the table. The reclaimed water treatment plant has a capacity of 2000 m³/d during the day and evening shifts when the full staff is on duty, and a reduced capacity of 1000 m³/d during the 12 a.m. to 8 a.m. shift when only two operators are on duty.

Period no.	From	To	Reclaimed water supply rate, m³/d	Reclaimed water demand rate, m³/d
1	12 a.m.	1 a.m.	1000	0
2	1	2	1000	0
3	2	3	1000	0
4	3	4	1000	1000
5	4	5	1000	1000
6	5	6	1000	2000
7	6	7	2000	2000
8	7	8	2000	1000
9	8	9	2000	1000
10	9	10	2000	2000
11	10	11	2000	3000
12	11	12 p.m.	2000	3000
13	12 p.m.	1	2000	3000
14	1	2	2000	4000
15	2	3	2000	4000
16	3	4	2000	4000
17	4	5	2000	3000
18	5	6	2000	2000
19	6	7	2000	1000
20	7	8	2000	1000
21	8	9	2000	1000
22	9	10	2000	1000
23	10	11	1000	1000
24	11	12 p.m.	1000	1000

a. Using the cumulative mass diagram method, determine the maximum working storage required for these maximum day demand conditions.
b. Could the maximum working storage be reduced if the town fully staffed the reclaimed water treatment plant for all three shifts? If so, what would the reduced working storage be?

14-3 A small town is implementing a new reclaimed water distribution system as shown in the following figure:

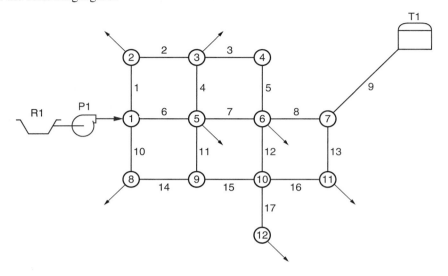

Assume that the distribution system will consist of the following elements:

	Junctions				Pipes	
No.	Elev., m	Maximum day demand, m³/d	Minimum required pressure, m	No.	Length, m	Roughness, mm
1	100	—	—	1	700	0.259
2	110	100	14	2	500	0.259
3	110	500	21	3	600	0.259
4	120	—	—	4	700	0.259
5	110	200	14	5	700	0.259
6	110	300	14	6	500	0.259
7	130	—	—	7	600	0.259
8	110	100	14	8	900	0.259
9	120	—	—	9	1500	0.259
10	130	—	—	10	600	0.259
11	120	600	14	11	600	0.259
12	150	1000	28	12	600	0.259
				13	600	0.259
				14	500	0.259
				15	600	0.259
				16	900	0.259
				17	300	0.259

Total max. day demand	2800

Supply reservoir, R1	Value, m	Elevated storage tank, T1	Value, m
Size	5 × 5	Diameter	10
Base elevation	99	Base elev.	180
Maximum elevation	103	Maximum elev.	186
Minimum elevation	100	Minimum elev.	181

Supply pump, P1	Unit	
Centerline elevation	m	98
Operating point	m³/d @ 90 m	1400

a. Perform a static analysis of the proposed facilities using the EPANET2 network analysis program. Make reasonable assumptions for the initial pipeline diameters. Use standard pipeline diameters from 150 to 450 mm in the analysis. Maximum pipeline velocity shall be 2.0 m/s. Submit the results of your initial analysis in the form of tables and figures showing pipeline diameters, flow rates, velocities, and unit head losses; junction elevations, hydraulic heads, and pressures; final elevated storage tank diameter, base elevation, and hydraulic head; and supply pump operating characteristics. Adjust the distribution system components as needed to balance the inflow to the network between the supply pump and the elevated reservoir. Report the final network configuration with tables and figures similar to the initial analysis.

b. Assuming the supply pump cannot operate during the peak demand period, would the pipeline network and/or elevated tank need to be modified to supply the maximum day demands? Present a brief summary of the changes required and why they are needed.

c. Could any of the pipelines in the network be eliminated and still satisfy all of the users requirements? If so, which pipelines can be eliminated and what would be the effect on the other network elements?

14-4 A new pumping station is needed to convey reclaimed water to a new storage reservoir located on a hill 600 m from the water reclamation plant. Because the reclaimed water supply is variable, the city wishes to use variable speed drives for the needed pumping station. The reclaimed water flowrates range between 30 L/s (minimum) to 45 L/s (average) to 65 L/s (maximum). The water level in the supply sump ranges between 105 and 106 m elevation. The water level in the new reservoir will range between 130 and 133 m. The centerline elevation of the pump is 100 m. The suction and discharge piping consists of the following elements:

Element	Length/number
Suction pipe, 200 mm dia.	6 m
Discharge piping, 200 mm dia.	600 m
Gate valves, 200 mm	2 each
Swing check valve, 200 mm	1 each

a. Assuming the Darcy-Wiesbach friction factor, f, is 0.017, compute the system head curve for the pump piping and force main. Prepare a sketch of the system head curve. (Hint: Do not forget entrance and exit losses.)
b. Show the operating points on the system head curve for the maximum, average, and minimum pump capacities.
c. Two pumps are available with the following operating characteristics:

	Pump 1: vertical turbine	Pump 2: horizontal split-case centrifugal
Shutoff head, m	90	60
Head @ 20 L/s, m	83	56
Head @ 40 L/s, m	68	50
Head @ 60 L/s, m	44	41
Head @ maximum capacity	30 m @ 70 L/s	28 m @ 80 L/s
Head @ maximum efficiency	57 m @ 50 L/s	50 m @ 40 L/s

Prepare a list of the advantages and disadvantages of each pump. Which of these pumps would you recommend for the new pumping station? Why?

14-5 You are the engineering manager of a midsized water/wastewater agency charged with implementing a new reclaimed water program. Among your initial duties is to oversee the planning of a new distribution system to deliver reclaimed water to users in the agency's service area.

The reclaimed water service area is located in a broad, north/south trending valley with hills on either side. The valley floor is approximately 5 km long by 3 km wide and slopes from 110 m above sea level at the high end to 100 m at the outlet. The hills on the west crest at 325 m and the eastern hills crest at 300 m. Approximately 70 percent of the total reclaimed water demand is located on the valley floor and the remaining demand is divided between the western and eastern hills at 10 percent and 20 percent, respectively.

a. Using the recommended pressure zone criteria from Table 14-4, make recommendations for the number of pressure zones needed, their boundary elevations and the associated reservoir overflow elevations for each zone. Explain the rationale for your recommendations.
b. One of the ways to minimize the potential for accidental cross-connections between the potable and reclaimed water systems is to maintain the reclaimed water system at a lower operating pressure than the potable water system. Discuss the impact on your original zone boundary recommendations if your agency decides to adopt a policy of operating the reclaimed water system at a pressure 70 kPa lower than the potable water system.

14-6 You are the design engineer for a new reclaimed water distribution system. A major downtown intersection has the following existing utilities:

Service	Symbol	Pipe sizes, mm				Invert elev., m	Location
		N	E	S	W		
Potable water	PW	250	250	350	300	98.7	3 m clear of SS
Sanitary sewer	SS	300	375	600	375	96.0	Centerline (C/L) of roadway
Storm sewer	SW	300	300	525	300	97.0	1.6 m clear of SS opposite from PW

As shown on the figure, the existing pavement is 12 m wide and has 1.5 m sidewalk/utility rights of way on both sides of the pavement. The pavement has a crown elevation of 100 m at the intersection and slopes 2 percent to the gutters.

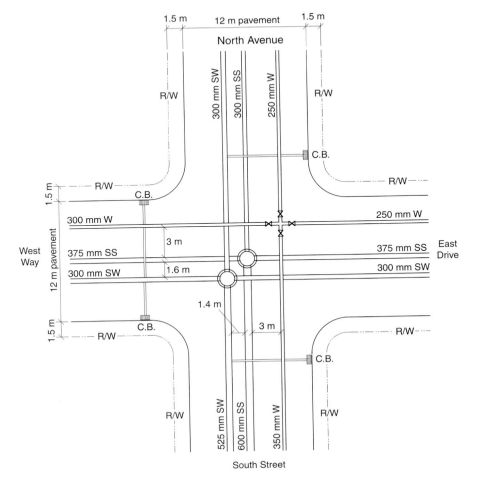

The state health department has adopted the following separation criteria for new reclaimed water (RW) pipelines and new or existing utilities:

Pipelines	Horizontal separation, m	Vertical separation, m	Notes
RW/PW	3	0.3	Less than tertiary quality, PW pipeline above
	1.25	0.3	Tertiary quality only, RW pipeline above
RW/SS	3	0.3	RW pipeline above
RW/SW	1.25	0.3	RW pipeline above

Note: Separations are measured from outside diameter of all pipelines

In addition, the city requires that all new pipelines in the pavement must be at least 0.45 m clear of the face of curb.
  a. Using the above separation criteria, prepare a sketch showing the possible horizontal and vertical location options of the new reclaimed water pipelines. Assume that the reclaimed water may not be of tertiary quality. Show both plan and section views.
  b. If the reclaimed water will be tertiary quality, are there additional options for the location of the new reclaimed water pipelines? If so, show the additional location(s) on the sketch.

REFERENCES

AWWA (1994) *Dual Distribution Systems*, AWWA Manual M24, 2nd ed., American Water Works Association, Denver, CO.

AWWA (1996) *Welded Steel Tanks for Water Storage*, AWWA Standard D100-96, American Water Works Association, Denver, CO.

AWWA (1998) *Steel Water-Storage Tanks (M42)*, American Water Works Association, Denver, CO.

AWWA (2002) *PVC Pipe—Design and Installation (M23)*, 2nd ed., American Water Works Association, Denver, CO.

AWWA (2003a) *Ductile-Iron Pipe and Fittings (M41)*, 2nd ed., American Water Works Association, Denver, CO.

AWWA (2003b) *Principles and Practices of Water Supply Operations*, 3rd ed., Part 3, Water Transmission and Distribution, American Water Works Association, Denver, CO.

AWWA (2004a) *Recommended Practice for Backflow Prevention and Cross-Connection Control*, AWWA Manual M14, American Water Works Association, Denver, CO.

AWWA (2004b) *Steel Water Pipe: A Guide for Design and Installation (M11)*, 4th ed., American Water Works Association, Denver, CO.

AWWA (2004c) *External Corrosion: Introduction to Chemistry and Control (M27)*, 2nd ed., American Water Works Association, Denver, CO.

IRWD (2001) *Reclaimed Water News Briefs*, Irvine Ranch Water District, Irvine, CA.

Kirmeyer, G. J., M. Friedman, K. D. Martel, P. F. Noran, and D. Smith (2001) "Practical Guidelines for Maintaining Distribution System Water Quality," *J. AWWA*, **93**, 7, 62–67.

Miller, G., R. Bosch, A. Horne, A. Lazenby, E. Quinian, and C. Spangenberg (2003) *Impact of Surface Storage on Reclaimed Water: Seasonal and Long Term*, Water Environment Research Foundation, Alexandria, VA, also copublished by IWA Publishing, London.

NFPA (2005) *National Electric Code, 2005, (NFPA 70)*, National Fire Protection Association, Boston, MA.

Okun, D. A. (2005) "Dual Systems to Conserve Water While Improving Drinking Water Quality," 20th Annual WateReuse Symposium, Denver, CO.

PPI (2001) *Handbook of Polyethylene Pipe*, The Plastic Pipe Institute, Washington, DC.

Stroud, T. F. (1988) "Polyethylene Encasement versus Cathodic Protection: A View on Corrosion Protection," *Ductile Iron Pipe News,* Spring/Summer issue, Ductile Iron Pipe Research Association, Birmingham, AL.

Tchobanoglous, G., F. L. Burton, and H. D. Stensel (2003) *Wastewater Engineering: Treatment and Reuse*, 4th ed., McGraw-Hill, New York.

Tchobanoglous, G., and E. D. Schroeder (1985) *Water Quality: Characteristics, Modeling, and Modification*, Addison-Wesley, Reading, MA.

U.S. EPA (2003) *Cross-Connection Control Manual*, EPA 816-R-03-002, U.S. Environmental Protection Agency, Office of Water, Washington, DC.

WSWRW (2005) "Pipeline Separation: Design & Installation Reference Guide," Washington State Water Reuse Workgroup, 20th Annual WateReuse Symposium, Denver, CO.

15 Dual Plumbing Systems

	WORKING TERMINOLOGY 902
15-1	OVERVIEW OF DUAL PLUMBING SYSTEMS 902
	Rationale for Dual Plumbing Systems 902
	Applications for Dual Plumbing Systems 903
15-2	PLANNING CONSIDERATIONS FOR DUAL PLUMBING SYSTEMS 907
	Applications for Dual Plumbing Systems 907
	Regulations and Codes Governing Dual Plumbing Systems 908
	Applicable Health and Safety Regulations 908
15-3	DESIGN CONSIDERATIONS FOR DUAL DISTRIBUTION SYSTEMS 908
	Plumbing Codes 908
	Safeguards 908
15-4	INSPECTION AND OPERATING CONSIDERATIONS 913
15-5	CASE STUDY: IRVINE RANCH WATER DISTRICT, ORANGE COUNTY, CALIFORNIA 915
	Setting 915
	Water Management Issues 915
	Implementation 916
	Operational Issues 918
	Lessons Learned 919
15-6	CASE STUDY: ROUSE HILL RECYCLED WATER AREA PROJECT (AUSTRALIA) 919
	Setting 919
	Water Management Issues 920
	Implementation 920
	Lessons Learned 920
15-7	CASE STUDY: SERRANO, CALIFORNIA 921
	Setting 922
	Water Management Issues 922
	Implementation 923
	Lessons Learned 925
	PROBLEMS AND DISCUSSION TOPICS 925
	REFERENCES 926

WORKING TERMINOLOGY

Term	Description
Air gap	An unobstructed separation between a source of potable water and any potential nonpotable source of supply (including reclaimed water).
Backflow	Any reversal of flow of water from its intended direction.
Backflow prevention device	Any of several devices used to prevent backflow including air gap separation, pressure-type vacuum breaker, atmospheric-type vacuum breaker, double check valve assembly, and reduced pressure zone backflow preventer.
Cross-connection	Any physical connection between a potable water system and any potential source of contamination not protected by an approved device specifically designed to prevent flow between the two.
Dual plumbing system	Separate plumbing systems used to supply potable and reclaimed water from the points of connection from the respective distribution systems to the points of use.
Nonpotable water	Water intended for uses other than potable purposes.
Potable water	Water deemed safe for human consumption, food preparation, and bathing.
Uniform plumbing code (UPC)	Design standards to plumb safely nonresidential buildings with both potable and reclaimed water systems.

Distribution systems for the delivery of reclaimed water to potential users were discussed previously in Chaps. 12 and 14. The plumbing systems required at the point of connection to the distribution main to the points of use in commercial and residential applications are discussed in this chapter. Dual plumbing systems provide: (1) potable water for uses such as drinking, cooking, and bathing and (2) nonpotable water for landscape irrigation and for toilet and urinal flushing. Because of the possibility of cross-connection of potable and reclaimed water supplies and potential contamination of potable water, strict requirements are established for reclaimed water quality and for the design, installation, and operation of dual systems. Subjects considered in this chapter include (1) an overview of dual plumbing systems, (2) planning considerations, (3) design considerations, and (4) operating considerations. Three case studies at the end of the chapter illustrate the successful implementation of dual plumbing systems. Greywater recycling (the recycling of bath and shower water, hand washwater, and laundry washwater) is not covered in this chapter, but is discussed in Chap. 13.

15-1 OVERVIEW OF DUAL PLUMBING SYSTEMS

The rationale for the use of dual plumbing systems and applications for dual plumbing systems are examined in this section. Design details are presented in subsequent sections.

Rationale for Use of Dual Plumbing Systems

The general rationale for implementing water reclamation and reuse from a water resources management perspective was articulated in Sec. 1-1 in Chap. 1. The rationale for using satellite systems that would supply reclaimed water for dual distribution systems was delineated in Chap. 12. For many of the uses of water, potable water is not necessary and appropriately treated and disinfected reclaimed water can be used safely. Reclaimed water can replace potable water for nonpotable applications for exterior use such as landscape irrigation and interior use including fire protection and toilet flushing. Thus, potable water can be conserved and used for essential human health purposes such as drinking, cooking, and bathing, both for present use and the future. Two independent plumbing systems are required, however: one for distributing potable water and one for reclaimed water.

Installation of dual plumbing systems offers opportunities for expanded use of reclaimed water in commercial and residential settings. Dual plumbing systems can be used (1) for commercial buildings such as office buildings and apartment complexes where reclaimed water is furnished for toilet and urinal flushing, fire protection, landscape irrigation, and decorative fountains and ponds or (2) at residential developments for landscape irrigation. These applications are considered further in the following discussion.

Applications for Dual Plumbing Systems

Use in Commercial Buildings

Depending on state and local regulations, reclaimed water can be used for interior use in high-rise and other commercial buildings. In Florida, reclaimed water may be used for (1) fire protection sprinkler systems located in commercial buildings or for industrial facilities and (2) fire protection sprinkler systems and toilet flushing in motels, hotels, apartment buildings, and condominiums where the individual guests or residents do not have access to the plumbing system for repairs and modifications (Godman and Kuyk, 1997). In the Irvine Ranch Water District in California, discussed in Sec. 15-5, reclaimed water is used in high-rise office buildings for flushing toilets and urinals and for priming floor drain traps. As an indication of the potential savings possible through the use of reclaimed water in a typical commercial office building setting, approximately 75 percent of the total water used could be supplied from the reclaimed water system (Holliman, 1998).

The amount of water used for toilet flushing depends on whether water conservation devices are used. Without water conservation, typical water use for toilet flushing is 76.1 L/capita·d (see Table 1-2 in Chap. 1) with an average usage of 13.2 L/flush based on a study of residential water use by the American Water Works Research Foundation (AWWARF, 1999). By using water-conserving low-flow toilets, typical water use for toilet flushing is reduced to 36.3 L/capita·d (Table 1-2).

If in the future reclaimed water is permitted to be used for laundry purposes in commercial buildings, additional savings in water use might be realized. Based on data from the above cited AWWARF study, the mean per capita water use for clothes washing ranged from 71 to 47 L/capita·d for household sizes of one to eight residents, respectively (see Table 15-1). With water-efficient clothes washers, water use may be further reduced by 25 to 30 percent (see Table 1-2). For apartment complexes, the above water use numbers may be used as a guide in estimating daily demand, although hourly demands can vary widely.

Depending on the type of use of the building, i.e., residential or commercial, number of individual units, and number of floors, the demands for reclaimed water from the supply system can be met by (1) feeding directly from the supply main, (2) feeding directly with booster pumping, (3) feeding from an elevated storage tank, or (3) feeding from a hydropneumatic tank system (see Fig. 15-1). The sizing of a reuse system for an apartment building is illustrated in Example 15-1.

Table 15-1 Household water use for clothes washing[a]

Household size (number of residents)	Mean consumption for clothes washing, L/capita·d	Standard deviation, L/capita·d
1	71	55
2	62	40
3	56	38
4	47	23
5	49	24
6	49	21
7	53	20
8	48	17

[a]AWWARF (1999).

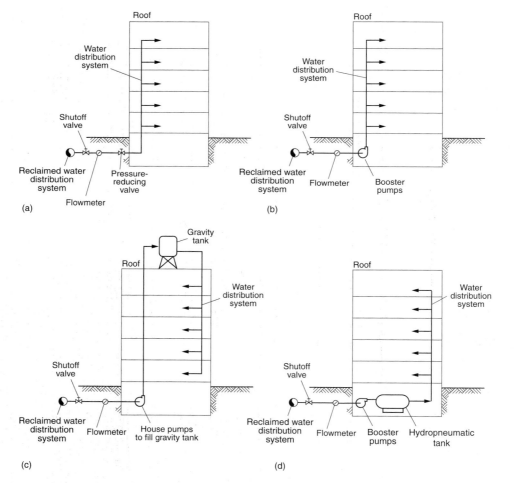

Figure 15-1

Reclaimed water plumbing system for commercial buildings served with reclaimed water from a distribution system from an external wastewater reclamation facility (either centralized or satellite): (a) direct feed; (b) direct feed with booster pump; (c) booster pumping to a gravity tank; and (d) direct feed from hydropneumatic tank.

EXAMPLE 15-1. Estimate the Reclaimed Water Demand for an Apartment Building.

A new apartment building being planned is considering the use of reclaimed water for toilet flushing. Pending approval from local health authorities, reclaimed water is also being considered for use in a centralized laundry facility for clothes washing.

1. Determine the average daily and maximum hourly demand for reclaimed water for the building having the following characteristics:

   Number of one-bedroom units = 30

   Number of two-bedroom units = 36

   Number of three-bedroom units = 12

   Number of floors = 8

2. Determine the percentage in average daily use of potable water use that can be achieved by instituting water reuse for toilet flushing and clothes washing.

Design conditions and assumptions:

1. Average apartment occupancy is 1.5 persons per one-bedroom unit, 2.5 persons per two-bedroom unit, and 4 persons per three-bedroom unit.
2. Amount of water used for toilet flushing is 65 L/capita·d.
3. Amount of water used for clothes washing—interpolate from Table 15-1.
4. A peaking factor of 6 times the average daily demand is assumed to compute maximum hourly demand. (Note: Although no data are available on peaking factors for residential water use, peak wastewater flowrates from individual residences range from 4 to 6 times the average flowrate (Crites and Tchobanoglous, 1998).
5. For all other water uses in the building, use typical water use data without water conservation listed in Table 1-2.

Solution: Part 1

1. Determine the average daily demand.
   a. Set up a computation table to determine the average daily demand for toilet flushing and clothes washing.

Number of bedrooms	Number of units	Number of persons/unit	Total persons	Toilet flushing L/capita·d	Toilet flushing L/d	Clothes washing L/capita·d	Clothes washing L/d
One	30	1.5	45	65	2925	66.5	2993
Two	36	2.5	90	65	5850	59	5310
Three	12	4	48	65	3120	47	2256
Totals			183		11,895		10,559

b. The total for toilet flushing and clothes washing = 22,454 L/d.

2. Determine maximum hourly demand.

$$\text{Maximum demand} = (22{,}454 \text{ L/d}) (1 \text{ d}/24 \text{ h}) (6) = 5614 \text{ L/h}$$

Solution: Part 2

1. Determine total water use in the building by setting up a computation table.

Use	Unit water use, L/capita·d	Units, no. of persons	Total water use, L/d
Toilets	See Part 1		11,895
Clothes washing	See Part 1		10,559
Showers	47.7	183	8729
Faucets	42.0	183	7686
Leaks	37.9	183	6936
Other domestic	5.7	183	1043
Baths	4.5	183	824
Dishwashers	3.8	183	695
Total			48,367

2. Determine potable water savings if reclaimed water is used for toilet flushing and clothes washing.

$$\text{Savings} = \frac{(22{,}454 \text{ L/d})}{(48{,}367 \text{ L/d})} \times 100 = 46.4\%$$

Comment

Under the conditions given in the problem statement, the amount of potable water that can be saved by using reclaimed water for clothes washing is nearly as much as that saved for toilet flushing. In the aggregate, almost half of the potable water use in the building can be reduced by using reclaimed water for nonpotable purposes. Thus, there may be future opportunities for expanded interior use of reclaimed water as its use is proven to meet regulatory standards and public acceptance.

Use at Residential Buildings

Use of reclaimed water at residences in the United States is limited normally to exterior use for landscape irrigation. Depending on the season and the climate characteristics, substantial savings in potable water can be achieved by the use of reclaimed water for landscape irrigation. For example, on an average daily basis, outdoor use amounts to about 7 percent of the residential demand in Pennsylvania and 44 percent in California (AWWA, 1994). In Denver, Colorado, during July and August, approximately 80 percent of all potable water is used for lawn irrigation (U.S. EPA, 1992).

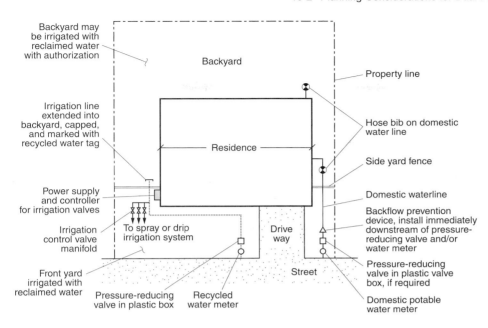

Figure 15-2
Typical layout of a dual plumbing system for a residence. (Adapted from Serrano, 2001.)

Interior use for toilet flushing in individual residences is restricted generally because modifications of the plumbing system may result in possible cross-connections between potable and nonpotable water systems. A layout of a typical dual plumbing system serving a residence is illustrated on Fig. 15-2. Views of a typical residential system where dual plumbing is being installed are presented on Fig. 15-12, discussed in Sec. 15-7.

15-2 PLANNING CONSIDERATIONS FOR DUAL PLUMBING SYSTEMS

Once the need for dual plumbing systems has been established, planning considerations for the implementation of dual plumbing systems are typically related to (1) the type of application (e.g., municipal, commercial, or residential), (2) developing a layout and plan that complies with local and state building codes, and (3) ensuring that applicable health and safety regulations have been met.

Applications for Dual Plumbing Systems

New commercial, residential, and industrial projects where utility systems are in the planning stage provide the greatest opportunities for using dual systems. Retrofitting existing buildings and facilities, however, is more difficult and more costly because of the need to revise interior plumbing systems. Individual residences, in most cases, are limited to the use of reclaimed water for landscape irrigation; only in rare circumstances is reclaimed water permitted for interior use such as toilet flushing. In Tokyo and other large cities in Japan, however, the most important use is for toilet flushing in residential, commercial, and industrial properties (Okun, 2000). With the production of increasingly higher quality reclaimed water and growing applications, other interior uses of reclaimed water may find greater acceptance in the United States. It should be noted that dual piping systems can be supplied with reclaimed water from a centralized wastewater treatment facility or from an interception, extraction, or upstream type satellite system as discussed in Chap. 12. The design features of distribution systems are found in Chap. 14.

Regulations and Codes Governing Dual Plumbing Systems

The use of dual plumbing systems in the United States is controlled by the plumbing codes adopted by the governmental agencies having jurisdiction. Piping systems for the use of reclaimed water for commercial and residential buildings are covered generally by various regulations including plumbing codes. The regulations are designed for the protection of public health and for the safe installation of the piping system and ancillary devices. One of the basic considerations in the development of plumbing codes is to minimize the risk of interconnecting the potable and reclaimed water supply piping, either accidentally or intentionally. To that extent, safeguards are engineered into the system design and operation and backed by rigorous inspection and monitoring during and following construction to meet regulatory requirements.

Applicable Health and Safety Regulations

As discussed in Chap. 4, many state and local regulatory authorities that permit use of reclaimed water establish strict water quality requirements based on type and point of use. Where there is the possibility of human contact with reclaimed water, public health issues are of primary concern. Plumbing systems have to be designed to provide safeguards for the protection of public health. In the future, however, as treatment technologies improve and regulatory requirements are revised, it may not be necessary to use dual plumbing systems if reclaimed water is returned to water supply reservoirs for blending with other surface water and long-term storage, followed by water treatment (see Chap. 23 for further discussion of indirect potable use).

15-3 DESIGN CONSIDERATIONS FOR DUAL DISTRIBUTION SYSTEMS

Design considerations for reclaimed water systems are similar to those for conventional water systems with two important differences: (1) special requirements in plumbing codes and local regulations are established that apply strictly to the reclaimed water system and (2) additional safeguards need to be provided to ensure there is no misuse, either intentional or unintentional, of the reclaimed water occasioned by improper design and operation.

Plumbing Codes

In the United States, a plumbing standard used by many states and localities is the Uniform Plumbing Code (UPC) (IAPMO, 2003). Appendix J of the UPC provides design standards to plumb safely nonresidential buildings with both potable and reclaimed water systems. The general provisions of UPC Appendix J are summarized in Table 15-2.

Safeguards

Where potable and reclaimed water systems are used in close proximity to each other, it is possible that cross-connections can occur accidentally or intentionally. Potable water systems, therefore, must be evaluated at specific points of use to protect against potential sources of contamination. Many safeguards can be implemented to reduce the risk of cross-connection. Safeguards can take the form of backflow prevention devices; pipe separation; color coding of piping, tanks, and appurtenances; reduced pressure of reclaimed water systems; and signage.

General category	Provisions
Pipe material/pipe identification	All metallic reclaimed water pipe and fittings shall be wrapped continuously with purple-colored Mylar tape. The tape shall be imprinted with black uppercase letters with the words "CAUTION: RECLAIMED WATER, DO NOT DRINK." For buried polyvinyl (PVC) pipe, the pipe shall be manufactured with the purple color integral with the plastic material and marked on opposite sides with the same lettering as described above.
Equipment identification	All mechanical equipment appurtenant to the reclaimed water system shall be painted purple to match the purple Mylar wrapping tape.
Installation provisions	• Hose bibs are not allowed on reclaimed water piping systems. • Appurtenances such as valves shall be provided to allow drainage and deactivation of the reclaimed water and potable water systems within buildings. • Reclaimed water pipes shall not be run or laid in the same trench as potable water pipes. A 3 m horizontal separation shall be maintained between pressurized, buried, reclaimed, and potable water piping. Buried potable water pipes crossing pressurized reclaimed water pipes shall be laid a minimum of 300 mm above the reclaimed water pipes. • Each valve or appurtenance shall be sealed with either a crimped lead wire seal or a plastic breakaway seal which, if broken, is evidence that the reclaimed water has been accessed. • To the extent permitted by structural conditions: (a) reclaimed water risers within the toilet room shall be installed in the opposite end of the room containing fixtures served by the potable water system and (b) reclaimed water headers and branches off risers shall not be run in the same wall or ceiling cavity of the toilet room where potable water piping is run.
Signs	All installations using reclaimed water for toilets and/or urinals, valve doors, and reclaimed water equipment rooms shall be provided with prescribed signs indicating that reclaimed water is being used.

Table 15-2

General design provisions of Appendix J of the Uniform Plumbing Code for reclaimed water systems in nonresidential buildings[a]

[a] Adapted in part from IAPMO (2003) and AWWA (1994).

Backflow Prevention

There should be no permanent connection between potable and reclaimed water systems. All temporary connections to an emergency backup, e.g., for fire protection, should be equipped with a backflow prevention device to prevent contamination of the potable water supply from cross-connections and back-siphonage. There are five basic methods of backflow prevention which involve the use of: (1) air gap separation,

(2) pressure-type vacuum breaker, (3) atmospheric-type vacuum breaker, (4) double check valve assembly, and (5) reduced pressure zone backflow preventer. Each type of device is described in Table 15-3 and shown on Fig. 15-3.

Typically, a backflow prevention device is installed on the potable water supply at a location close to the building water meter. It is common practice in high-rise structures, regardless of whether reclaimed water is used, to provide some type of backflow prevention device on potable water systems to protect against back-siphonage of contaminated water. For example, the City of San Francisco requires that an approved backflow prevention device be used on the potable water supply in all buildings four stories and higher (San Francisco, 2001). Backflow prevention devices as illustrated on Figure 15-4, are also used on reclaimed water distribution systems where a connection is made to a potable water supply for make-up water and to prevent back siphoning occurances.

Table 15-3

Backflow prevention devices used to safeguard potable water supplies from cross-connections and back-siphonage[a]

Type	Description/application	See Fig.
Air gap	A typical air gap installation consists of a tank into which water flows from the supply line; water from the tank is then pumped to the points of use. A physical separation (air gap), 50 mm or more, is maintained between the outlet of the building water supply line and the water surface of the tank. This device is suitable for severe hazards.	15-2a
Pressure-type vacuum breaker	A mechanical device designed to prevent backflow caused only by back pressure conditions. It is designed to operate under continuous pressure on both sides of the device. This device is suitable for only minor hazards.	15-2b
Atmospheric-type vacuum breaker	A mechanical device designed to prevent backflow caused only by back-siphonage conditions. It is designed to operate with pressure on only one side of the device. This device is suitable for only minor hazards.	15-2c and d
Double check valve backflow preventer	Two independently operated swing check valves installed in series. This device is suitable for minor hazards.	15-2e
Reduced pressure zone backflow preventer	A mechanical device consisting of two independently operated, spring-operated swing check valves with a pressure-regulated relief valve in between. The relief valve opens under backflow conditions and discharges upstream water to waste until the conditions are corrected. This device is suitable for severe hazards.	15-2f

[a]Adapted from Nayyer (2000) and U.S. EPA (2003).

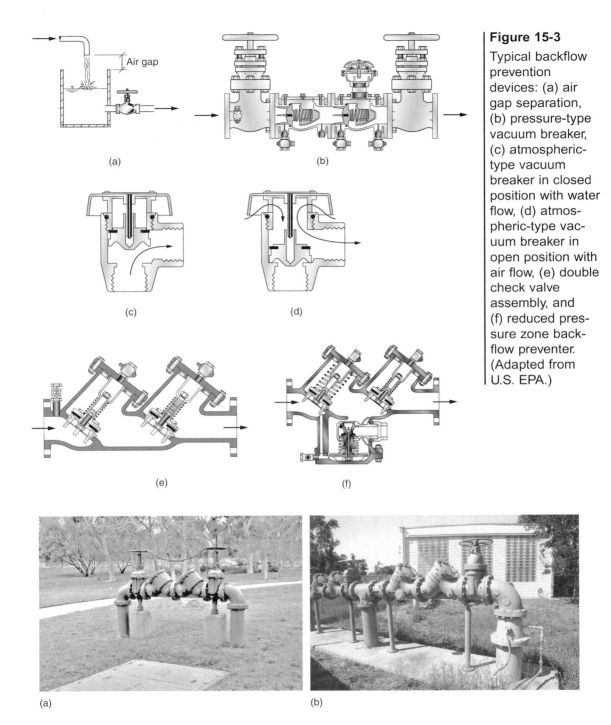

Figure 15-3
Typical backflow prevention devices: (a) air gap separation, (b) pressure-type vacuum breaker, (c) atmospheric-type vacuum breaker in closed position with water flow, (d) atmospheric-type vacuum breaker in open position with air flow, (e) double check valve assembly, and (f) reduced pressure zone backflow preventer. (Adapted from U.S. EPA.)

Figure 15-4
Typical examples of aboveground backflow prevention installations for reclaimed water in (a) California and (b) Florida. The installations are above ground for ease of maintenance and because of the high groundwater table in Florida (typically 0.3 m below the ground surface).

Pipe Separation

If buried potable and reclaimed water pipes are to be installed in the same vicinity, the pipes should be separated to ensure that leakage from the reclaimed waterline does not find its way into the potable waterline resulting in contamination of the potable water. Recommended horizontal and vertical separation distances are given in Table 15-2. The entry piping for potable and reclaimed water to a commercial office building is shown on Fig. 15-5.

Piping System Identification

As described in Table 15-2, provisions of the UPC call for marking and color coding of piping, tanks, hydrants, and appurtenances. Marking includes stenciling the pipe with a warning notice, applying warning tapes, painting with distinctive color, or using materials manufactured with a distinctive integral color such as plastic pipe and fittings. In the United States, purple is the color designated in the UPC for reclaimed water piping; in Australia, the color preference is lilac (Sydney Water, 2003).

Reduced Pressure

An effective means of minimizing cross-connection and contamination between potable and reclaimed water systems is to operate the recycled water system at a lower pressure, on the order of 70 kPa lower, than the pressure in the potable waterline (AWWA, 1994).

Signage

At any outlet or location where reclaimed water is used, appropriate signs should be provided to warn potential users that the facilities contain reclaimed water and it is not to be used for drinking. Where appropriate, other relevant information should be included such as the purpose of use for the reclaimed water (e.g., "nonpotable water for irrigation only"). Examples of warning signs are shown on Fig. 15-6. The signs should be displayed prominently and secured firmly to prevent removal.

Figure 15-5

Separate services for potable water (left side) and reclaimed water (right side) provided for a commercial office building. Note the 3 m (10 ft) separation distance between services cited in Table 15-2.

Figure 15-6
Typical warning signs regarding the use of reclaimed water: (a) sign located in restroom, (b) removable warning sign covering in wall reclaimed water control valves, and (c) and (d) signs used for landscape irrigation.

15-4 INSPECTION AND OPERATING CONSIDERATIONS

A cross-connection can occur during initial construction, when a potable water system is retrofitted to reclaimed water use and potable water connections are overlooked, or when modifications and repairs are made to expand the system or to increase pressure. When installation of the dual system is complete, it is required to be inspected and tested in accordance with the applicable plumbing code. Provisions for inspecting and testing in accordance with the UPC are presented in Table 15-4. In addition to inspecting and testing the system on completion, follow-up inspection and testing should be done subsequently on a basis consistent with regulations to ensure the systems continue to be safe and meet their intended use.

In addition to the pressure testing program for cross-connections described in Table 15-4, other methods—color testing and valve seals—have been suggested to determine the

Table 15-4
Procedure for inspection and testing of dual plumbing systems[a]

Procedure/test	Description
Visual inspection	Prior to cross-connection testing, the following visual inspection of the dual system shall be made, as appropriate: • Meter locations of reclaimed and potable waterlines shall be checked to verify that no modifications were made and no cross-connections are visible. • All pumps and equipment, equipment room signs, and exposed piping in equipment room shall be checked. • All valves shall be checked to ensure valve lock seals are in place and intact. All valve control doors shall be checked to verify no signs have been removed.
Cross-connection test	The following test shall be conducted: • The potable system shall be activated and pressurized. The reclaimed waterline shall be shut down and drained. • The potable water system shall be pressurized for a minimum period specified by the regulating authority, generally not less than 1 h, while the reclaimed water system is dry. • The reclaimed water drainage system shall be checked for flow during and after the test period. • The reclaimed water system shall then be pressurized. • The reclaimed water system shall be pressurized for a minimum period specified by the regulating authority, generally not less than 1 h, while the potable water system is dry. • All potable and reclaimed water fixtures shall be tested and inspected for flow. Flow from any potable water system outlet shall indicate the presence of a cross connection. • The drain on the potable water system shall be checked for flow during the test and at the end of the period. • If there is no flow detected in any of the fixtures which would have indicated a cross-connection, the potable water system can be repressurized.
Follow-up to cross-connection test	In the event that a cross-connection is discovered, the flowing procedures shall be implemented: • Reclaimed water to the building shall be shut down at the meter and the reclaimed water riser drained. • Potable water piping to the building shall be shut down at the meter. • The cross-connection shall be located and disconnected. • The dual system shall be retested in accordance with the procedures of the visual inspection and cross-connection test. • The potable water system shall be chlorinated with 50 mg/L chlorine for 24 h. • After 24 h, the potable water system shall be flushed and a standard bacteriological test performed. If the test results are acceptable, the potable water system can be recharged.
Annual and other inspections	Annual visual inspection and cross-connection tests shall be conducted unless site conditions do not require a cross-connection test. In no case shall a cross-connection test occur less than once in 4 yr.

[a]Adapted from IAPMO (2003).

presence or possible presence of a cross-connection. The color test consists of putting dye into the reclaimed water system and checking for the dye in the potable system. The dye test is not as reliable as the pressure test because small cross-connections may not be detected (AWWA, 1994). Because the pressure and dye tests disrupt water service and may not be acceptable to some users, a system of using unbreakable seals on valves is suggested by the State of California to detect when plumbing work has been done. The master reclaimed water shutoff valve and/or the reclaimed water meter curb cock and each valve within a wall would be sealed to prevent operation of the valve after the reclaimed water system has been approved. The seals, if broken, would provide evidence that the reclaimed water system has been accessed. Log books for recording the type of plumbing work done would also be required (State of California, 2003).

15-5 CASE STUDY: IRVINE RANCH WATER DISTRICT, ORANGE COUNTY, CALIFORNIA

The Irvine Ranch Water District (IRWD) is a full service water and sewer agency located in southern Orange County, California. Since 1967, IRWD has provided wastewater collection and water reclamation services with the defined objective of producing reclaimed water for nonpotable applications. The following discussion is adapted from Crook (2004), Holliman (1998), WEF (1998), and IRWD (2005).

Setting

The IRWD service area encompasses some 46,360 ha (114,560 ac) or approximately 460 km^2 (179 mi^2) of southern Orange County, California. IRWD serves all of the City of Irvine and portions of Tustin, Newport Beach, Costa Mesa, Orange, and Lake Forest. IRWD extends from the Pacific Coast to the mountain foothills, with elevations ranging from sea level to about 1000 m (3300 ft). The IRWD region is semiarid with a mild climate and an annual rainfall of 300 to 330 mm (12 to 13 in.).

Water Management Issues

Water management issues in what is now Orange County as well as other parts of southern California date back hundreds of years, as the area is basically a desert. The critical importance of water became clear during the drought of 1863–1864, when thousands of cattle died and, as a result, much of the land changed ownership due to insolvency. Since that time, the area has been plagued by water problems, principally periodic water shortages. Fast forward to the 1950s and 1960s when what was known as the Irvine Ranch was being transformed into the master planned urban community of Irvine, California, and the University of California at Irvine came into being. Recognizing that water, a limited resource, had to be managed if growth was to continue in a sustainable manner, the IRWD was formed in 1961, under the provisions of the State of California Water Code, to provide irrigation and domestic water for the expanding community. Water management issues encountered by IRWD are related to the management of a limited resource.

Approximately 35 percent of IRWD's drinking water is purchased from the Metropolitan Water District of Southern California. Imported water comes from the

Colorado River via the Colorado River Aqueduct and from northern California via the State Water Project. The remaining 65 percent of the supply comes from local wells.

Implementation

As the IRWD evolved, in 1967 it began serving reclaimed water to agricultural users. The use of reclaimed water next expanded into landscape irrigation, including parks, golf courses, school grounds, play fields, community associations, open space areas, and greenbelts. Water reuse eventually expanded to include large estate-sized lots for front and backyard watering for large commercial buildings, carpet dying, construction dust control, and cooling tower applications. The two water distribution systems—one for potable water and one for reclaimed water—were installed as the service area grew, making the process more economical than building a second distribution system separately. The reclaimed water consists of treated wastewater produced by the Michelson Water Reclamation Plant. Treatment is by activated sludge with nitrification/denitrification, filtration, and chlorine disinfection. Reclaimed water makes up over 25 percent of the water used in the IRWD service area. The reclaimed water used meets Title 22 requirements of the California Department of Health Services (see Chap. 4). Typical views of landscape areas irrigated with reclaimed water are illustrated on Fig. 15-7.

Current Status of Reuse Program

As of July 2005, the population served by the district was 316,287. There are 88,423 domestic water connections. The number of reclaimed water connections was 3812 distributed as follows: commercial—13, industrial—2, landscape irrigation—3742, and agricultural—55. The total amount of water delivered was distributed as follows: treated (potable)—66.08×10^6 m³/yr (53,573 ac-ft/yr), untreated (nonpotable)—7.77×10^6 m³/yr (6301 ac-ft/yr), and reclaimed 27.67×10^6 m³/yr (22,434 ac-ft/yr) for a total of 101.52×10^6 m³/yr (82,307 ac-ft/yr). At the present time (2006), 90 percent of the landscape accounts are served with reclaimed water. The dual distribution system used for reclaimed water is comprised of more than 483 km (300 mi) of pipelines, 13 storage reservoirs, and 15 pumping stations.

To inform people in the service area and to instill a conservation ethic, IRWD provides a weekly facilities tour program for residents, an in-school education program, and newsletters and brochures. Mini-grants are also awarded to local teachers to instruct students in the benefits of conservation and wise water use.

Use of Reclaimed Water in Commercial Buildings

In 1987, IRWD began investigating the feasibility of using reclaimed water in commercial buildings for uses that do not require potable water quality. In 1991, after working closely with health and municipal government officials and local developers and builders, IRWD facilitated construction of six dual-plumbed high-rise office buildings. The first building to be put in service was the 20 story Jamboree Tower 2C (see Fig. 15-8a). Reclaimed water is used for toilet and urinal flushing and for priming floor drains. Restroom facilities in Jamboree Tower 2C consist of a men's and women's restroom for each of 19 floors. The men's restroom contains typically three sinks, two urinals, one toilet, and one floor drain. The women's restroom contains typically two sinks, three toilets, and one floor drain.

Figure 15-7
Typical views of landscape areas irrigated with reclaimed water in Irvine, CA: (a) one of many golf courses, (b) a shopping center, (c) a residential development, and (d) typical residential estate-sized lot for front and backyard watering.

Figure 15-8
Buildings with dual plumbing: (a) Jamboree Tower 2C (on left) in Irvine, CA, in which reclaimed water is used for toilet and urinal flushing (floor drains are also plumbed to use reclaimed water), and (b) new commercial buildings under construction with dual plumbing systems (Coordinates: 33.677 N, 117.838 W).

Important features of the dual distribution system for buildings include:

- All recycled water piping is wrapped with purple Mylar warning tape.
- With the exception of portions of the piping protruding through the walls and connecting directly to the plumbing fixtures, no other access to the piping is available from the restrooms.
- An approved reduced pressure principle backflow prevention device was installed on the potable waterline servicing the building.
- In each restroom, the utility room, equipment room, and valve access doors, signs were posted identifying the use of reclaimed water.
- All reclaimed water valves were locked in the open position with locking valve seals.

A major reason why IRWD considered the use of reclaimed water in nonresidential buildings appropriate was the high level of attention paid to the prevention of cross-connections. IRWD has two full-time employees assigned to the exclusive task of investigating and identifying potential and actual cross-connection occurrences. To complement these efforts, IRWD has a fully equipped water quality laboratory to monitor the quality of the reclaimed water. IRWD also has an on-site water systems group for monitoring construction and inspecting the facilities for cross-connections during and after construction.

Operational Issues

The principal operational problems are related to the buildup of salinity and the need for winter storage.

Salinity

Salinity issues deal with: (1) the increased salinity of the source water (Colorado River), (2) the closed-loop water reclamation system that results in a gradual buildup in minerals, and (3) the use of self-regenerating water softeners that can add large amounts of salt to the collection system. To control water softener discharges, IRWD enacted rules and regulations to prohibit the use of self-generating water softeners within the district boundaries. Although the ban on water softeners was overturned in court, IRWD continues to work legislatively toward restoring the ability of water recycling agencies to control salinity.

Seasonal Storage

Seasonal storage is required because it is difficult to synchronize the production of reclaimed water with the demand, especially for irrigation purposes (see Fig. 15-9). Most of the rainfall occurs in the winter months when the irrigation demand and the needs for reclaimed water are low and wastewater production is at its peak. In the summer months, the irrigation demand is high and can exceed reclaimed water production. Balancing the requirements for seasonal storage is challenging because finding land in an urban setting for reservoir construction or siting open reservoirs or lakes is very difficult. Because of the nutrient content in the reclaimed water, algae growth in the open reservoirs can be a problem. To minimize the need for pretreatment, covered storage tanks have been provided near the open reservoirs for storing diurnal excess flows.

(a)

(b)

Figure 15-9
View of reservoirs used for seasonal storage of reclaimed water at Irvine, CA: (a) reservoir constructed specifically for reclaimed water and (b) concrete-lined potable water supply reservoir converted for use with reclaimed water.

Lessons Learned

Reclaimed water systems require more maintenance than potable water systems because reservoirs need more frequent cleaning; control valves are susceptible to corrosion due to higher residual chlorine levels, and cross-connection control requires ongoing inspection as discussed previously. Additionally, any leaks or spills have to be reported to the county health agency, which requires more personnel time to handle the paperwork. Finally, the success of a water reuse program depends on continued and effective vigilance and effective community educational outreach programs.

15-6 CASE STUDY: ROUSE HILL RECYCLED WATER AREA PROJECT (AUSTRALIA)

The Rouse Hill Development Area (RHDA) located in Sydney's north west, comprising a total area of 13,000 ha (32,124 ac), is one of the largest planned communities in the world using reclaimed water. The following discussion is adapted from Cooper (2003) and www.sydneywater.com.au.

Setting

In 1989, the New South Wales government accepted a proposal by a consortium of landowners to fund, design, and construct water, sewerage, and drainage facilities for the RHDA. Because of environmental concerns, i.e., the impact of treated wastewater discharge on the Hawkesbury-Nepean River, Sydney Water, the implementing agency, proposed to design the new wastewater treatment plant to return high-quality effluent to homes for domestic applications such as garden irrigation, toilet flushing, and car washing. The result would be a reduction of waste loads to the river and conservation of potable water. The project included the construction of the Rouse Hill Recycled Water Plant and a dual water system.

Water Management Issues

The source of wastewater to be reclaimed is almost all domestic, with a small fraction of wastewater coming from a commercial area. The quality of wastewater from the commercial area is strictly controlled to manage the risk of contamination with hazardous chemicals.

Implementation

In November 2001, the first phase of the reclaimed water system was placed into operation and began serving 4000 dwellings. More than 15,000 properties are connected currently to reclaimed water. The next stage of the project was announced in August 2004 and will service an additional 10,000 lots with reclaimed water and potable water. Final completion of the infrastructure is scheduled for 2006. When finished, reclaimed water will be used for garden watering, washing cars, toilet flushing, park and golf course irrigation, and industry in the area. Potable water will continue to be used for most household uses, including drinking, cooking, and bathing.

Wastewater Treatment

The wastewater is treated in a biological nutrient removal wastewater treatment plant at an average dry weather flow of approximately 5.6×10^3 m^3/d. A portion of the treated effluent (3×10^3 m^3/d) is fed to a reclaimed water plant (see Fig. 15-10). The wastewater treatment plant consists of 3 mm fine screens; grit removal; primary sedimentation; a biological reactor (that is divided into five reaction zones) where nitrification, denitrification, and biological phosphorus removal take place; flocculation; clarification; and filtration. The reclaimed water plant feed stream undergoes further treatment by ozonation for virus inactivation, continuous microfiltration (CMF), and disinfection using sodium hypochlorite. Reclaimed water is stored near the areas of use in three 2×10^3 m^3 reservoirs. If the filtrate from the CMF process exceeds 0.5 NTU, the product water is returned to the ozonation basin for reprocessing. Typical water quality is shown in Table 15-5 for treated effluent from the sewage treatment plant and product water from the recycled water plant.

Distribution of Reclaimed Water

Reclaimed water is distributed by gravity flow to the residential areas through about 34 km (21 mi) of pipelines. The reclaimed water taps, piping, and plumbing fittings are colored lilac to distinguish them from the drinking water system and labeled "RECYCLED WATER—DO NOT DRINK," similar to the identifying signs shown on Fig. 15-6. Backflow prevention devices are required on potable water services to prevent contamination from cross-connections. Rechlorination facilities are provided at each storage reservoir for chlorine residual maintenance.

Lessons Learned

The Rouse Hill water recycling scheme has been successful in keeping up with the growing demand. Since 2001, when reclaimed water was made available, the demand for drinking water has been reduced by about 35 percent on average. The cost of installing and operating the dual distribution system was, however, more than anticipated originally. Project implementation did not proceed as rapidly as anticipated, due to slow growth and operational problems at the water recycling plant. The demand for reclaimed water exceeded the production rate and the reclaimed water supply is being supplemented by 2×10^3 m^3/d of potable water. The reclaimed water system is planned to be expanded to 5.2×10^3 m^3/d (as of 2003).

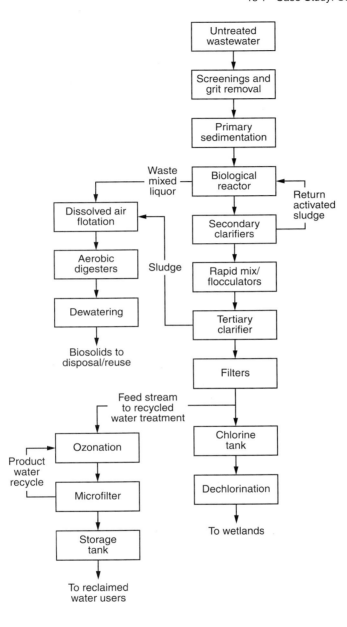

Figure 15-10
Schematic flow diagram of the Rouse Hill wastewater reclamation plant.

15-7 CASE STUDY: SERRANO, CALIFORNIA

Serrano is one of the largest master planned communities in the State of California. As part of the community water supply master plan, reclaimed water is used for landscape and golf course irrigation. Homes are equipped with dual plumbing: potable water for interior use and reclaimed water for landscape irrigation.

Table 15-5
Typical performance data for the Rouse Hill sewage treatment plant and recycled water plant[a]

		Sewage treatment plant			
		EPA license limit		Concentrations achieved	
Constituent	Unit	50 percentile	90 percentile	50 percentile	90 percentile
BOD	mg/L	4	5	<2	<2
TSS	mg/L	5	8	1	4
Ammonia	mg/L	1	2	<0.01	0.03
Total nitrogen	mg/L	10	15	5.2	7.6
Total phosphorus	mg/L	0.2	0.4	0.08	0.18
Chlorine	mg/L	—	0.5	—	0.07
Fecal coliforms	CFU/100 mL	200[b]		35[b]	

	Recycled water plant		
Constituent	Units	RWCC guidelines[c]	Concentrations achieved
Fecal coliforms	CFU/100 mL	<1	<1
Coliforms	CFU/100 mL	<10[d]	<1
Virus	In 50 L	<2	<1
Parasites	In 50 L	<1	<1
Turbidity	NTU	<2[e]	0.01
Color	TCU	<15	4
pH		6.5-8.0	6.9-7.8

[a]Adapted from Cooper (2003). Data are for 2000/2001.
[b]80 percentile.
[c]NSW guidelines for urban and residential use of reclaimed water by NWS Recycled Water Coordination Committee.
[d]In 95 percent of samples.
[e]Geometric mean.

Setting

Serrano is located in El Dorado County, the foothills of the Sierra Nevada Mountains about 50 km (31 mi) from the state capitol, Sacramento. Serrano encompasses about 1400 ha (3500 ac) including a 81 ha (200 ac) golf course and 400 ha (1000 ac) of greenbelt. The terrain is gently rolling and a portion of the site is wooded. Home sites average about 0.13 ha (0.33 ac) in size.

Summer days are very warm but nights are cool. The highest average temperature in the hottest month, July, is in the mid-30°C (90°F) range. The coldest month is January when the temperature ranges from 9 to 12°C (48 to 54°F). Three-fourths of the annual rainfall of 380 to 510 mm (15 to 20 in) occurs between November and March.

Water Management Issues

El Dorado County has adequate potable water but the supply is vulnerable to interruption by natural forces such as a prolonged drought. New supplies are becoming scarcer and demand continues to increase. By using reclaimed water for nonpotable use such as landscape and golf course irrigation (see Fig. 15-11), potable water can be preserved for the primary uses of drinking, cooling, bathing, and washing.

Implementation

The Serrano El Dorado Owners Association made an agreement with the El Dorado Irrigation District (EID), one of the principal water utilities in El Dorado County, to supply reclaimed water from the district's wastewater treatment plants for irrigation purposes. The reclaimed water is treated biologically, filtered, and disinfected and meets the requirements of Title 22 Regulations (see Chap. 4).

Residential Landscape Irrigation

In 1999 Serrano expanded the use of reclaimed water from golf course, greenbelt, parks, and play field applications to irrigation of front and backyards of all new subdivisions where the infrastructure is available. A dual water distribution system was installed and dual plumbing provided for each home. Only potable water is provided to

Figure 15-11
Views of the Serrano development: (a) view of typical home sites, (b) recreational fields, (c) golf course, and (d) typical signage (coordinates: 38.678N, 121.054W; view at attitude 5 km).

about 700 homes that were built prior to the 1999 expansion. Design, inspection, and maintenance of the landscaping in the front yard is the responsibility of the Serrano El Dorado Owners' Association, but design and construction of landscaping and the irrigation system in the backyard are the responsibility of the homeowner (Serrano, 2001). The reclaimed water irrigation system is installed in accordance with the provisions of Uniform Plumbing Code listed in Table 15-2. Views of outside residential plumbing being installed are shown on Fig. 15-12.

The reclaimed water to each home is provided via a service connection from the reclaimed water distribution system. A meter and valve are installed near the sidewalk. To prevent incidental connection to reclaimed water, no hose bibs are allowed on the reclaimed water plumbing. The gate valve on the backyard service line is locked in

Figure 15-12

Example installation of a reclaimed water plumbing system at a residence: (a) view of residence (note that water service is located on the right side of the driveway), (b) reclaimed water piping with flowmeter and check valve from distribution system (on left side of driveway), (c) water meter with plastic body parts, purple in color with warning label on meter face, and (d) view of side yard with connection for backyard and drainage system leading to the street.

accordance with the regulations of the Owners' Association until the backyard irrigation system is designed and approved. The potable waterline is installed at least 3 m (10 ft) from the reclaimed waterline with a meter, valve, and backflow prevention device (see Fig. 15-12). Where a potable waterline and reclaimed waterline cross, the reclaimed waterline must be installed at least 15 cm (6 in.) below the potable waterline in a purple-colored PVC sleeve that extends a minimum of 1.5 m (5 ft) on either side of the potable waterline.

Homeowners Association

New homeowners are supplied with information on the operating and maintenance of the reclaimed water system. In addition, the Serrano El Dorado Owners' Association holds quarterly workshops to educate homeowners and landscape contractors. Only landscape contractors who have attended the workshop are authorized to design, install, or modify irrigation systems on lots with dual plumbing (Klein and Bone, 2005).

Lessons Learned

Although Serrano initially bore the cost of implementing a reclaimed water system in El Dorado County, the program has become increasingly important to EID. Based on the success of Serrano, EID mandated in 2005 that all new developments use reclaimed water where feasible. The district further estimates that using reclaimed water could save up to $100 million in treatment costs through year 2025 by eliminating discharges into area streams (Klein and Bone, 2005).

PROBLEMS AND DISCUSSION TOPICS

15-1 For Example 15-1, discuss the internal water supply infrastructure needed for the apartment complex.

15-2 Obtain a copy of your community's local plumbing code and describe features that apply to the use of reclaimed water. If no provisions exist, describe what purposes and system features you would recommend to allow the safe use of reclaimed water.

15-3 In a warehouse complex, reclaimed water that is treated by secondary treatment, chemical coagulation, depth filtration, and chlorination is proposed to be used for fire protection (sprinklers) and firefighting. What are the hazards associated with using reclaimed water for these purposes? How might some of the hazards be mitigated?

15-4 Obtain at least two examples of warning signs associated with the use of reclaimed water from sources other than this textbook. Do these signs provide a clear and adequate warning to the users of reclaimed water and the affected public? What would you recommend to make the warning more effective?

15-5 Download the U.S. EPA Cross-Connection Control Manual from the Internet (http://www.epa.gov/safewater/crossconnection.html) and summarize the various methods and devices used for the prevention of backflow and back-siphonage.

15-6 Conduct a literature search for a case study of a community that has a dual water system and summarize the features of the system.

REFERENCES

AWWA (1994) *Dual Water Systems*, Manual of Water Supply Practices AWWA M24, American Water Works Association, Denver, CO.

AWWARF (1999) *Residential End Uses of Water*, American Water Works Association Research Foundation, Denver, CO.

Crites, R., and G. Tchobanoglous (1998) *Small and Decentralized Wastewater Management Systems,* McGraw-Hill, New York.

Cooper, E. (2003) "Rouse Hill and Picton Reuse Schemes: Innovative Approaches to Large-Scale Reuse," *Water Sci. Technol.: Water Supply*, **3**, 3, 49–54.

Crook, J. (2004) *Innovative Applications in Water Reuse: Ten Case Studies*, WateReuse Association, Alexandria, VA.

Godman, R. R., and D. D. Kuyk (1997) "A Dual Water System for Cape Coral," *J. AWWA*, **89**, 7, 45–53.

Holliman, T. R. (1998) "Reclaimed Water Distribution and Storage," Chap. 9, in T. Asano (ed.), *Water Reclamation and Reuse*, Water Quality Management Library, **10**, CRC Press.

IAPMO (2003) *Uniform Plumbing Code*, International Association of Plumbing and Mechanical Officials, Ontario, CA.

IRWD (2005) *2005 Urban Water Management Plan,* Irvine Ranch Water District, 15600 Sand Canyon Avenue, Irvine, CA.

Klein, P., and K. Bone (2005) "Recycled Water Benefits New Development," *Source,* Fall 2005 issue, CA-NV AWWA, 26–27, Rancho Cucamonga, CA.

Nayyar, M. L. (2000) *Piping Handbook*, 7th ed., McGraw-Hill, New York.

Okun, D.A. (2000) "Water Reclamation and Unrestricted Nonpotable Reuse: A New Tool in Urban Water Management," *Ann. Rev. Public Health*, **21**, 1, 223–245.

San Francisco (2001) *2001 Building, Electrical, Housing, Mechanical and Plumbing Codes*, v6, Updated Through 9-25-2004, City and County of San Francisco.

Serrano (2001) *Recycled Water User's Manual for Dual Plumbed Homes in Serrano*, Serrano El Dorado Owners Association, El Dorado Hills, CA.

State of California (2003) *Water Recycling 2030: Recommendations of California's Recycled Water Task Force*, Department of Water Resources, Sacramento, CA.

Sydney Water (2003) "Recycled/Reclaimed Water Pipes," *Plumbing Policies Standards and Regulations*, City of Sydney, Australia.

U.S. EPA (1992) *Guidelines for Water Reuse*, EPA/625/R-92/004, U.S. Environmental Protection Agency, Cincinnati, OH.

U.S. EPA (2003) *Cross-Connection Control Manual,* EPA 816-R-03-002, U.S. Environmental Protection Agency, Office of Water.

WEF (1998) *Using Reclaimed Water to Augment Potable Water Resources*, Water Environment Federation, Alexandria, VA.

Part 4
WATER REUSE APPLICATIONS

With advancements in water reclamation technologies, it is technically possible to produce reclaimed water of virtually any quality as described in Part 3. The question then is what level of treatment is necessary and satisfactory for a specific water reuse application? Reclaimed water quality requirements depend not only on the relevant regulations and guidelines, but also on specific applications for which reclaimed water is to be used. Demand and supply balance and needs for infrastructure also vary with various applications. Infrastructure must have a capacity to supply reclaimed water safely, sufficiently, and reliably. Thus, a water reuse project cannot be planned without identifying primary and potential users of the reclaimed water and understanding how reclaimed water is to be used in each application.

In Part 4, various water reuse applications are discussed. An overview of water reuse applications is provided in Chap. 16. Because water quality requirements and infrastructure requirements vary greatly with each specific application, major water reuse applications are discussed in individual chapters. Various nonpotable water reuse applications, including agricultural uses, landscape irrigation, industrial uses, environmental and recreational uses, and urban nonpotable and commercial uses, are discussed in Chaps. 17 through 21, respectively. Groundwater recharge, described in Chap. 22, can be considered as part of indirect potable reuse if the recharged aquifer is connected to potable water production wells. Indirect and direct potable reuse applications are discussed in Chaps. 23 and 24, respectively.

16 Water Reuse Applications: An Overview

WORKING TERMINOLOGY   930

16-1 WATER REUSE APPLICATIONS   930
Agricultural Irrigation   931
Landscape Irrigation   931
Industrial Uses   931
Urban Nonirrigation Uses   933
Environmental and Recreational Uses   933
Groundwater Recharge   933
Indirect Potable Reuse through Surface Water Augmentation   933
Direct Potable Reuse   934
Water Reuse Applications in Other Parts of the World   934

16-2 ISSUES IN WATER REUSE   934
Resource Sustainability   934
Water Resource Opportunities   935
Reliability of Water Supply   935
Economic Considerations   935
Public Policy   935
Regulations   936
Issues and Constraints for Specific Applications   937

16-3 IMPORTANT FACTORS IN THE SELECTION OF WATER REUSE APPLICATIONS   937
Water Quality Considerations   937
Types of Technology   939
Matching Supply and Demand   939
Infrastructure Requirements   939
Economic Feasibility (Affordability)   940
Environmental Considerations   941

16-4 FUTURE TRENDS IN WATER REUSE APPLICATIONS   941
Changes in Regulations   942
Water Supply Augmentation   942
Decentralized and Satellite Systems   942
New Treatment Technologies   942
Issues Associated with Potable Reuse   944

PROBLEMS AND DISCUSSION TOPICS   944

REFERENCES   944

WORKING TERMINOLOGY

Term	Definition
Cross-connection	An inadvertent physical connection between a potable water system and any source containing nonpotable water through which potable water could be contaminated.
Demand, water	The diurnal and seasonal variations in the amount of water needed for a specific application.
Economic feasibility	The economic feasibility of a water reuse project is assessed using one of the following methods of analysis: present worth, total annual cost, or life cycle cost.
Sustainability	The principle of optimizing the benefits of the present water supply system without diminishing the capacity for similar benefits in the future.
Urban water reuse	Types of water reuse applications categorized as urban water reuse in the U.S. EPA water reuse guidelines include landscape irrigation in urban settings, air conditioning, fire protection, toilet and urinal flushing, water features, commercial car washing and laundries, and dust control at construction sites.
Water reuse applications	The methods by which reclaimed water is used.

As communities continue to urbanize and stress existing water supplies, water reclamation and reuse will play an increasingly important role in water resource management. More and more, water reuse will be looked upon as a way to meet growing demands, improve water supply reliability, and preserve existing resources. In a report by BCC Research, the amount of water used for all water reuse applications in the United States is anticipated to grow at an average annual rate of 11.8 percent from 2006 through 2010 (AWWA, 2006; Castelazo, 2006). Agricultural and landscape irrigation will continue to be the largest users of reclaimed water, but the revenue growth of technologies and materials for industrial water reuse, a tangible measure of growth potential, is expected to increase rapidly at an average annual rate of 14.2 percent. Because of the emergence of reclaimed water as an important water resource, an understanding of the types of water reuse applications and the benefits and constraints in their use is increasingly important in developing water reuse programs.

To introduce the water reuse applications discussed in Part 4 of this textbook, the purpose of this chapter is to identify and describe: (1) the types of applications used most commonly, (2) issues in water reuse, (3) important factors in the selection and implementation of water reuse applications, and (4) future trends in water reuse.

16-1 WATER REUSE APPLICATIONS

The various water reuse applications that are used most commonly were identified in Table 1-6 in Chap. 1 and are reviewed in this section for convenience of reference. The general categories of reuse that are practiced in the United States are:

- Agricultural irrigation
- Landscape irrigation
- Industrial uses
- Urban nonirrigation uses
- Environmental and recreational uses
- Groundwater recharge
- Indirect potable reuse

Examples of the first six of these categories are illustrated pictorially on Fig. 16-1. Each category is further defined in this section and discussed in detail in subsequent chapters in Part 4.

Agricultural Irrigation

Agricultural irrigation evolved from the early practice of sewage farming when untreated municipal wastewater (sewage) was directly applied to crops. In some parts of the world that practice is still used, in spite of adverse health and environmental impacts. Agricultural irrigation is the largest user of reclaimed water in the United States and in most of the world. Depending on the use of the crop for food or nonfood purposes and how the irrigation water is to be applied, i.e., by ridge and furrows, sprinklers, or surface/subsurface drip irrigation (see Fig. 16-1a), the degree of the required treatment will vary. In almost all irrigation applications in the United States, a minimum of secondary treatment is required. For spray application to food crops, if allowed, higher levels of treatment including disinfection are mandatory. Other considerations in agricultural irrigation include the type of crop; topography and soil characteristics; effect of water quality, especially dissolved solids, on soils, crops, and the underlying groundwater; runoff and drainage water management; and infrastructure requirements. Agricultural irrigation is discussed in Chap. 17.

Landscape Irrigation

Landscape irrigation is the second largest user of reclaimed water in the United States and is being used increasingly for various locations including golf courses, parks, residential areas, roadway medians and roadside plantings (see Fig. 16-1b), and cemeteries. Because public contact with the applied water is perceived as a potential health hazard, reclaimed water has to meet higher water quality levels for suspended solids and microbial concentrations, as compared to some agricultural applications. Many of the other physical and chemical characteristics of the reclaimed water in landscape applications are similar to those for agricultural use. In irrigation system operation, considerations include limiting the formation and dispersion of aerosols, managing application rates to avoid ponding and runoff, and controlling chlorine residuals to maintain proper disinfection. Landscape irrigation is discussed in Chap. 18.

Industrial Uses

Major industrial users of reclaimed water are power plants, oil refineries, and manufacturing facilities where water is required principally for cooling purposes (see Fig. 16-1c). Water quality, especially total dissolved solids, chlorides, and dissolved oxygen, is of specific concern because of potential scaling or corrosion in piping systems and heat exchangers. Residual organic matter may also contribute to biological growths in heat exchangers and cooling towers. Additional treatment may be necessary at the point

Figure 16-1

Typical reclaimed water reuse applications: (a) agricultural irrigation by means of subsurface drip (coordinates: 38.383 N, 122.770 W), (b) landscape irrigation (highway median strip), (c) industrial reuse for cooling and process water, (d) commercial reuse for car washing, (e) recreational lake (Courtesy of Padre Dam Municipal Water District Staff), and (f) empty rapid infiltration basin (RIB) used for groundwater recharge at Conserve II, Orlando, FL (Coordinates: 28.493 N, 81.620 W). Orange groves are shown in background.

of use depending on the water quality requirements for the specific industrial process. Other important considerations include matching supply with demand, system reliability, and disposal of cooling tower blowdown. Industrial uses are discussed in Chap. 19.

Urban Nonirrigation Uses

Urban nonirrigation uses cover a wide variety of applications including air conditioning cooling water, fire protection, toilet and urinal flushing, ornamental water features, and road care and maintenance. Commercial uses of reclaimed water such as a car washing (see Fig. 16-1*d*) and commercial laundries are practiced typically in urban areas, and they are considered as part of urban nonirrigation uses as described in Chap. 20. Urban use is limited generally to high-density development such as office buildings and apartments where there is economic justification for installing a dual distribution system, one system for potable water and one for reclaimed water. Water needs for most urban nonirrigation water reuse applications are small, and generally, multiple water reuse applications are implemented including landscape irrigation. Principal concerns with dual distribution systems are the high cost of infrastructure and the prevention of cross-connection between the two water supplies. High quality and well-disinfected reclaimed water must also be maintained to ensure public health protection. Urban nonirrigation uses are considered in Chap. 20.

Environmental and Recreational Uses

Environmental and recreational uses include such applications as wildlife habitat maintenance and enhancement in wetlands, low flow augmentation in rivers, and creation of recreational lakes and ponds (see Fig. 16-1*e*). The level of treatment of reclaimed water in most cases depends on the type of waterbody to which reclaimed water is released, and the degree of public contact or the health hazard associated with its use. The need for nutrient and enhanced suspended solids removal is also a consideration. In addition to water quality considerations, important factors include continuous versus intermittent use and matching supply with demand. Environmental and recreational uses are discussed in Chap. 21.

Groundwater Recharge

Groundwater recharge has been used to: (1) reduce, stop, or even reverse declines of groundwater levels (see Fig. 16-1*f*); (2) protect underground freshwater in coastal aquifers against saltwater and brackish water intrusion; and (3) store surface water, including flood or other surplus water and reclaimed water for future use. Water quality requirements may include nitrogen removal or reduction and the control of specific organic and inorganic contaminants. Groundwater recharge with reclaimed water is an approach to water reuse that results in the planned augmentation of potable water supplies (Asano, 1985 and 1998). Groundwater recharge with reclaimed water is discussed in Chap. 22.

Indirect Potable Reuse through Surface Water Augmentation

Planned indirect potable use is a careful and deliberate process to augment water resources while maintaining health and environmental safeguards. Most planned indirect potable reuse is linked to groundwater recharge. However, most indirect potable reuse in practice, whether it is planned or unplanned, occurs through blending with surface water. Discharge of treated wastewater to rivers and streams used as drinking water supplies is in effect *de facto* indirect potable reuse (see Chap. 3), but in practice it is a waste disposal option and not a *planned* indirect potable reuse approach. Indirect potable reuse through surface water augmentation is discussed in Chap. 23.

Direct Potable Reuse

There are no documented direct potable reuse installations in the United States. The only documented case where direct potable reuse has been implemented is in Windhoek, Namibia. A potable reclamation plant was completed in 1968 and has been upgraded on several occasions to meet severe water shortages (see Table E-2 in App. E for a description of the treatment processes). In Windhoek, the reclaimed water is blended with conventionally treated surface water (see the case study in Sec. 24-3). There is no documented direct potable reuse where delivered potable water is 100 percent reclaimed water. Limited discussion on direct potable reuse is presented in Chap. 24.

Water Reuse Applications in Other Parts of the World

In a report by the U.S. Environmental Protection Agency and U.S. Agency for International Development (U.S. EPA, 2004), it is noted that "Reuse of water for agricultural irrigation is practiced today in almost all arid regions of the world." In assessing characteristics of water use in the world, it is further noted that "the best water reuse projects, in terms of economic viability and public acceptance are those that substitute reclaimed water in lieu of potable water for the use in irrigation, environmental restoration, cleaning, toilet flushing, and industrial uses." The benefits cited for these projects are conservation of water resources and pollution reduction. Thus, the water reuse applications described above for the United States apply globally in developed as well as developing countries.

Each of the water reuse applications described above has been used in various countries world-wide depending on specific local needs. Some of the specific projects are described in Table E-2 in App. E, which also includes descriptions of treatment processes employed to produce reclaimed water. In some countries, the use of greywater is widely practiced, but only minor amounts of greywater have been reused in the United States.

16-2 ISSUES IN WATER REUSE

Water reuse projects are implemented for many different reasons, but often the most compelling reason is to mitigate water shortages brought on by drought conditions, increased demand due to growth, or overuse of existing resources. Another major reason is increasingly stringent waste discharge requirements that require significant additions and improvements to wastewater treatment processes. In the broad view, the driving force in resource management should be to preserve and augment existing water supplies to meet not only present needs but also future needs, i.e., resource sustainability. Issues in water reuse may involve improving water supply reliability, reuse opportunities, economic considerations, public policy, and regulatory factors. Other local issues that are specific to particular applications also need to be considered carefully. Each of these issues is examined in this section.

Resource Sustainability

As discussed in Chap. 1, the challenge of sustainable development is to devise and implement integrated and adaptable systems while optimizing water use efficiency. In areas with limited freshwater supplies and increasing water demands, the options for resource sustainability are limited. Possible options include (1) instituting water conservation, (2) developing local water supplies including those of marginal quantity and quality, (3) importing water from a remote location, or (4) a combination of two or more of the above. An additional option is reclaiming treated municipal wastewater for nonpotable uses, thereby reserving freshwater for potable purposes. Thus, water reclamation and reuse becomes an integral part of a long-term plan for conserving, extending, and managing community water resources. In some cases, the use of reclaimed water allows the community to grow beyond the limit which conventional water supply sources can reliably support. Water reuse, however, may be controversial if limited growth of the community is an issue.

Water Resource Opportunities

Historically, water reuse for applications such as agricultural irrigation and industrial use evolved because of geography; the sites of reuse have been in areas close to existing municipal wastewater treatment plants. Under those conditions, water reuse may be more opportunistic rather than the result of a well-planned water resource management program. Too often implementation of a reclaimed water program failed because of the lack of opportunity, i.e., the source of reclaimed water was located too far from potential areas of reuse to be justified economically. In systems where treatment plant locations are diversified (e.g., satellite or decentralized) and are located closer to the points of potential reuse, the opportunities for implementing local reuse are enhanced significantly.

Reliability of Water Supply

During the time of a water shortage, such as a drought, the customary response by water agencies is to curtail use of water. Agricultural and industrial customers are impacted most severely as a reduction in water supply directly affects their operations. Residential customers are also inconvenienced by diminished availability of water for potable use and landscaping may be damaged by the lack of water. By integrating reclaimed water into the water supply system, the reliability of the overall system is enhanced by: (1) reducing the stress on the supply used for potable purposes and (2) providing a locally available alternative source of water for nonpotable use.

Economic Considerations

The economics of water reuse are highly dependent on the local conditions such as the location of the water reclamation plant, locations and types of potential water reuse, water quality considerations, the need for additional treatment, and the cost of competing alternative sources. When considering water supply alternatives, factors to be evaluated include the cost of water acquisition (including water rights), capital cost of facilities, operating costs, and financing costs (see Chap. 25). For water reuse applications, additional factors that should be considered include the incremental treatment costs necessary to meet water quality requirements and the infrastructure costs necessary to provide delivery of reclaimed water to the customers. Factors that help offset costs include: (1) loan and grant programs, where available; (2) revenue generation through reclaimed water sales; (3) savings in potable water treatment costs; and (4) potential savings by avoiding development of another potable water source. Some wastewater treatment cost savings might also be realized if water quality requirements for water reuse are less restrictive than those for discharge of treated effluent. For example, it is less costly to produce reclaimed water suitable for agricultural and landscape irrigation than to provide a higher level of treatment, such as nutrient removal, necessary for discharge into ecologically sensitive surface waters.

Public Policy

Public policy in support of water reclamation projects is a key factor in successful implementation. The advocacy of elected officials and key citizen groups is very important in promoting it. In some water-short regions, state, regional, and local regulatory agencies are mandating water reclamation and reuse for golf course irrigation (see Chap. 4).

Regulations

Besides the water reuse criteria and regulations discussed in Chap. 4, various legal and regulatory issues at various governmental levels must be considered prior to the implementation of water reuse. The rules and regulations at the federal level in the United States that are pertinent to water reuse applications are reported in Table 16-1. The most important component of the Clean Water Act (CWA) is the National Pollution Discharge Elimination System (NPDES). Wastewater treatment plants discharging wastewater effluent to a receiving water body are required to obtain NPDES permits. Water reclamation plants which use all reclaimed water for reuse applications and do not discharge excess water are exempt from the NPDES requirements. Often, the NPDES requirements are more stringent than the quality requirements for water reuse,

Table 16-1

Federal regulations which affect water reuse applications

Key regulations	Description
The Safe Drinking Water Act	Designed to protect the nation's drinking water by establishing standards for drinking water quality. These standards may directly or indirectly constrain the use of reclaimed water for indirect potable reuse (see Chap. 2).
The Clean Water Act (CWA)	Directs the U.S. EPA to assist states in implementing groundwater, surface water, and wetland protection strategies (see Chap. 2). Relevant components of the CWA for water reuse include NPDES permits, total maximum daily load (TMDL) program (see Chaps. 17 and 21), and 404 Permits which regulate the discharge of dredge or fill material into waters including wetlands (see Chap. 21).
The National Environmental Protection Act (NEPA)	Requires an environmental evaluation for every project that requires federal action or relies on federal funding and might significantly affect the environment. In general, water projects fall under the auspices of this act.
The Endangered Species Act	Seeks to conserve threatened and endangered species. Federal agencies must carry out programs for listed species and take action to ensure that projects they authorize, fund, or undertake are not likely to jeopardize the continued existence of threatened and endangered species. Water projects frequently fall under the auspices of this act.
U.S. EPA Surface Water Treatment Rule	Requires that utility systems dependent on surface water sources provide filtration treatment and adequate disinfection to inactivate viruses and protozoan cysts. This rule establishes standards that could directly or indirectly affect the use of reclaimed water for potable purpose.
U.S. EPA Disinfection Byproducts Rule	Provides regulation of trihalomethanes and other potentially carcinogenic organic compounds in drinking water supplies. The concentrations of such compounds depend in part on the concentration of organics in the source water that might constrain water reclamation processes suitable for a particular situation or the amount of reuse possible (see Chap. 3).

which can lead wastewater dischargers to consider water reuse options in lieu of wastewater discharge. Some components of CWA that affect implementation of specific water reuse applications are discussed in subsequent chapters.

The Endangered Species Act (ESA) is another major federal regulation that affects water reuse. Water reuse changes the water balance downstream of the wastewater discharge point, whereas environmental uses of reclaimed water are intended to preserve and enhance wildlife habitats including endangered species. Other important federal regulations not listed in Table 16-1 include the Fish and Wildlife Coordination Act, Coastal Zone Management Act, Wild and Scenic Rivers Act, and National Historic Preservation Act. Parallel regulations to most of these federal rules and regulations likely exist at the state level. In some cases, state regulations may be more restrictive than those enacted at the federal level (Getches et al., 1991).

Issues and Constraints for Specific Applications

In evaluating specific applications using reclaimed water, the issues and constraints associated with each application have to be considered carefully. Typical issues and constraints for each type of application are presented in Table 16-2.

16-3 IMPORTANT FACTORS IN THE SELECTION OF WATER REUSE APPLICATIONS

In selecting water reuse applications, factors that affect selection of water reuse applications include: (1) water quality, (2) types of technology, (3) matching supply with demand, (4) infrastructure, (5) affordability (economic considerations), and (6) environmental mitigation. Each of these factors is discussed in this section.

Water Quality Considerations

As discussed in Chaps. 3 and 5, the acceptability of reclaimed water is dependent on the physical, chemical, and microbiological quality of the water. The effects of physical parameters such as pH, color, temperature, and particulate matter, and chemical constituents such as chlorides, sodium, heavy metals, and trace organics, on turf, other vegetation, soil, and groundwater are well known, and recommended limits have been established for many constituents (see Chaps. 17 and 18). In contrast to the agronomic considerations associated with chemical constituents that may be present in wastewater, microbiological constituents present health considerations for the distribution and use of reclaimed water.

Industrial source control programs can limit the input of chemical and microbiological constituents that may present health, environmental, or irrigation concerns or that may adversely affect treatment processes and subsequent acceptability of the reclaimed water for specific uses. In some arid and semiarid regions, the level of total dissolved solids is a major quality concern, and source control measures are considered for domestic water users such as restrictions on water softener use. Assurance of treatment reliability, as discussed in Chaps. 7, 8, and 9, is an obvious, yet sometimes overlooked, quality control measure.

As discussed in Parts 2 and 3, water quality considerations in water reuse applications are extremely important especially where health and environmental issues are of concern.

Table 16-2
Issues and constraints for municipal wastewater reuse applications[a]

General category	Application	Issues/constraints
Agricultural irrigation	Commercial nurseries Food, fodder, fiber, and seed crops Frost protection Silviculture Sod farms	Buffer zone requirements Marketing of crops Public health concerns Runoff and aerosol control Water quality impacts on soils, crops, and groundwater
Landscape irrigation	Cemeteries Golf courses and greenbelts Industrial parks Public parks and school yards Residential and other lawns Roadway medians and roadside plantings	Controlling residual disinfectants Public acceptance Public health concerns Runoff and aerosol control
Industrial	Boiler feedwater Cooling water Equipment washdown Fire protection Heavy construction (dust control, concrete curing, fill compaction, and cleanup) Process water	Blowdown disposal Cooling tower aerosols Cross-connection with potable water Scaling, corrosion, fouling, and biological growths
Nonpotable urban uses	Air conditioning cooling water Commercial car wash Commercial laundries Decorative fountains and other water features Driveway and tennis court washdown Fire protection Sewer flushing Snow melting Toilet and urinal flushing	Cross-connection with potable water Public acceptance Public health concerns Scaling, corrosion, fouling, and biological growths
Recreation/environmental uses	Artificial lakes and ponds Fisheries Snowmaking Stream flow augmentation Wetlands enhancement	Eutrophication Public health concerns Toxicity to aquatic life
Groundwater recharge	Barrier against brackish or seawater intrusion Groundwater replenishment Ground subsidence control	Availability of suitable sites Groundwater contamination Salt and mineral buildup Toxicological effects of organic chemicals
Indirect potable use	Blending with public water supplies Surface water augmentation	Public acceptance Public health concerns
Direct potable use	None identified at this time	Public acceptance Regulatory agency approval

[a] Adapted in part from AWWA (1994), Tchobanoglous et al. (2003), Crook et al. (2005).

Unless the product water is of sufficient quality to meet the required criteria and regulations for the intended reuse, acceptance by the potential users or beneficiaries will not occur. By the same token, overtreatment that is excessive for its intended use is a waste of resources in terms of energy, labor, equipment, and money.

Types of Technology

As discussed in Part 3, technologies that follow secondary treatment and are suitable for most water reuse applications include depth and surface filters, membrane filtration (pressure or vacuum), carbon adsorption, reverse osmosis, disinfection with ultraviolet radiation, and advanced oxidation. Membrane bioreactors maybe used in place of secondary treatment and membrane filtration.

Membranes represent the most significant development as several new products are now available for a number of water and wastewater treatment and water reuse applications. Membranes had been limited previously to water softening and desalination, but they are now being used increasingly for wastewater applications to produce high-quality reclaimed water suitable for reuse. Treatment trains that incorporate membrane filtration are capable of producing several grades of product water that can serve a range of water reuse applications. Reclaimed water may also be demineralized by means of reverse osmosis and electrodialysis. Increased levels of contaminant removal not only enhance the product water for reuse, but also lessen health risks. Further, the cost of producing high-quality reclaimed water has decreased considerably, largely due to the development of low-pressure membranes and the entrance of a number of suppliers in the competitive marketplace.

Chlorination remains as the most widely used disinfection technology and its effectiveness is vastly improved by improved reclaimed water quality. Increased removal of particulate matter and the development of ultraviolet disinfection technology also improve the applicability of reclaimed water for many more applications. Advanced oxidation is also an important technology for reducing or removing trace constituents and emerging contaminants to safe levels, especially for indirect potable water reuse applications.

Matching Supply and Demand

An important part of analyzing infrastructure requirements is the development of a water demand profile as a basis of matching the supply with demand. A demand profile is the characterization of water use requirements over time. The purpose of a demand profile is to (1) determine the water demand variations, both diurnal and seasonal; (2) show how the demand can be accommodated with the estimated reclaimed water production and operation schedules; and (3) define the type and extent of infrastructure. Important variables that need to be considered in a water demand analysis are presented in Table 16-3. Where reclaimed water use is intermittent or highly variable, the size and location of storage facilities are important in determining system economic and functional feasibility. Storage of reclaimed water for agricultural irrigation is discussed in Chap. 17.

Infrastructure Requirements

The infrastructure (principally pipelines, reservoirs, and pumping stations) is particularly important in evaluating the feasibility of a water reuse project because it can have a significant influence on the economic affordability. Where the points of use are remotely

Table 16-3
Important variables to consider in water demand analysis

Water use application	Consideration
Agricultural and landscape irrigation	Seasonal use Climatic conditions Frequency and duration of use Hydraulic application rates Diurnal variations Maximum and minimum flowrates Types of irrigation methods
Industrial and commercial	Continuous vs. intermittent use Maximum and minimum flowrates Frequency and duration of plant or process shutdowns
Environmental and recreational	Seasonal use Weather effects Frequency and duration of use
Nonpotable urban use	Types of reuse applications including landscape irrigation Extent of irrigated areas within the water reuse system Seasonal use Frequency and duration of use Diurnal variations Fire flow and pressure requirements
Groundwater recharge	Seasonal use Weather effects
Indirect potable use	Types of land use, e.g., residential, commercial Interior vs. exterior use (residential and commercial users) Frequency and duration of use Diurnal variations Maximum and minimum flowrates
Direct potable use	Blending restrictions or requirements Maximum and minimum flowrates

located from the point of reclaimed water generation, the capital and operating costs for the delivery system may be prohibitive, thus rendering the project uneconomic. In the cases of satellite and decentralized systems discussed in Chaps. 12 and 13, respectively, the infrastructure requirements may be significantly less thereby making water reclamation and reuse more affordable.

Most reclaimed water infrastructure facilities are similar to those for drinking water. Descriptions of system components and factors used in the design of storage and distribution systems are discussed in Chap. 14. Ancillary facilities for irrigation systems are discussed in Chaps. 17 and 18.

Economic Feasibility (Affordability)

Project costs—not only the initial construction cost but also annual operation and maintenance costs—are of major significance in the selection of water reuse applications. In some cases, loans and grants from governmental agencies may be available to assist in funding part of the project cost, however, operation and maintenance costs are usually the total responsibility of the operating agency. A feasibility analysis may be developed on the basis of present worth, total annual costs, or life cycle costs (see Chap. 25). In a present worth analysis, all future expenditures are converted to a present worth analysis at the beginning of the planning period. A discount rate is used in the analysis and represents the time value of money (the ability of money to earn interest). In a total annual cost comparison, the capital costs are amortized based on probable interest rates for bonds and the duration of the bond issue. The annual fixed (amortized) cost is added to the annual operating and maintenance costs. Life cycle costs are used to determine the total cost of a facility over its total useful life.

As compared to a strictly wastewater utility that has a product requiring disposal, a water reclamation utility, much like a water utility, has a product of value which has the capability of producing revenue. In the feasibility analysis of a water reclamation application, the potential generation of revenue needs to be assessed. A benchmark for establishing value of reclaimed water is the existing rate structure used for potable water. Determining a pricing structure can be daunting, however, as the prices in actuality may vary. A precursor to the feasibility analysis is a market survey in which potential customers are identified, their usage of reclaimed water is projected, and what economic benefits might be derived, both to the customer and the supplier. Additional discussion on the economics of planning for water reuse is provided in Chap. 25.

Environmental Considerations

The environmental considerations of a proposed water reuse project are sometimes more important than cost considerations. Environmental regulations ensure that probable environmental effects are identified, that a reasonable number of alternative actions and their environmental impacts are considered, and that the public and governmental agencies participate in the decision process. In controversial projects, the environmental process is sometimes used as a means to defeat implementation of a project. For that reason, public participation and support is an important ingredient in the successful implementation of a water reuse project. Implementation issues related to water reclamation and reuse projects are discussed in Chap. 26.

16-4 FUTURE TRENDS IN WATER REUSE APPLICATIONS

The mid-1980s marked the beginning of significant changes in the waterworks industry in the United States. The driving force for many of the changes has been the public attitude toward what constitutes "clean" or "safe" drinking water. These attitudes have affected the future approach to watershed management and water sources, water quality standards, water treatment, water conservation, water reclamation and reuse, and the cost consumers pay for water. In the meantime, aging wastewater treatment plants and increasingly stringent waste discharge requirements necessitate retrofitting many existing wastewater systems, thus providing wastewater authorities with an opportunity to consider potential water reuse applications.

The overriding factor in the water reuse planning will be the continued identification and development of water sources which offer sufficient volume and/or economic advantage for such water reuse efforts. Most of the readily available water sources have already been integrated into water management plans. Other sources become more attractive as the cost of water continues to rise or as the technological advances in treatment techniques mature and become more viable economically. In the long term, other sources may be on the fringes of research and development efforts. Such avenues as water management in space, conducted by the National Aeronautics and Space Administration (NASA), may potentially provide technological breakthroughs, which will allow the utilization of previously untapped water sources.

In the next 10 to 20 yr, nine developments or trends have been identified that will likely be important factors in water reuse applications:

- Changes in water reuse regulations
- Indirect potable reuse through surface water augmentation
- Indirect potable reuse through groundwater recharge
- Improved aquifer storage and recovery
- Exchange of reclaimed water for freshwater for nonpotable purposes
- Decentralized treatment systems
- Satellite treatment systems
- New treatment technologies
- Removal of total dissolved solids

Each of these developments is discussed briefly below and the benefits are given in Table 16-4.

Changes in Regulations

Changes in water reuse regulations are inevitable as water reuse increases and more information is developed about the impacts of trace and emerging constituents. It is also possible that federal regulations will be developed that will supersede state or local regulations and provide encouragement for the implementation of more reuse projects.

Water Supply Augmentation

Indirect potable reuse through surface water augmentation, indirect potable reuse through groundwater recharge, improved aquifer storage and recovery, and exchange of reclaimed water for freshwater for nonpotable purposes represent opportunities to conserve and augment water supplies. More emphasis will be placed on better water resource management instead of providing advanced wastewater treatment for the purposes of effluent disposal only. Utilizing reclaimed water to preserve freshwater for future potable purposes not only extends existing supplies but makes more efficient use of reclaimed water for nonpotable purposes.

Decentralized and Satellite Systems

Decentralized and satellite systems represent new directions in planning wastewater and reclaimed water systems. By developing dispersed systems, the effects of population growth on existing and aging infrastructure can be eased while at the same time creating water reuse opportunities nearer to potential areas of reuse.

New Treatment Technologies

As a result of the developments and improvements in treatment technology, many of the economic and environmental barriers to water reuse have been reduced significantly

Table 16-4
Developments, trends, and benefits in water reuse applications

Development/trend	Benefit
Changes in water reuse regulations	Eliminates some conflicts between state regulations. Establishes requirements for specific reuse applications. Recognizes benefits of water reuse as compared to effluent disposal. Institution of treatment plant reliability standards provides protection against inadequate treatment.
Indirect potable reuse through surface water augmentation	Augments surface water supply. Stabilizes receiving water flowrate during periods of low stream flow. Allows withdrawal of freshwater for water supply without reducing downstream flows. Provides some environmental enhancement.
Indirect potable reuse through groundwater recharge	Augments groundwater supply. Provides some water quality improvement. Serves as a barrier against seawater intrusion in coastal communities.
Aquifer storage and recovery	Transforms water quality during subsurface transport. Improves efficiency of storage and withdrawal.
Exchange of reclaimed water for freshwater	Improves efficiency of use, i.e., lessens demand for potable water for nonpotable use. Allows institution of a "water credit" program. Reduces groundwater pumping.
Decentralized treatment system	Provides water reuse opportunities for small communities. Eliminates wastewater flows and loads from centralized system. Allows use of comparatively "low-tech" treatment processes.
Satellite treatment system	Provides opportunities for a dispersed treatment and reuse system. Improves opportunities for localized reuse by providing reclaimed water near points of use. Reduces hydraulic loads on centralized collection system. Alleviates need to expand existing centralized treatment processes. Reduces impact of treated wastewater discharges to receiving waters. Does not require the same redundant equipment as centralized system. Provides some peak flow reduction.
New treatment technologies Membrane processes (microfiltration, reverse osmosis, electrodialysis) Advanced oxidation Enhanced disinfection	Provide superior water quality for expanded water reuse. Allow production of multiple grades of product water. Provide multiple barriers for protection of public health. Effective in removing trace constituents.
Removal of total dissolved solids (TDS)	Enhances use for a variety of purposes including irrigation, groundwater recharge, and industrial use. Reduces scaling potential of reclaimed water.

and several new opportunities for water reuse applications are possible. Many of these opportunities are associated with membrane treatment and UV disinfection. Health and environmental concerns associated with the use of treated wastewater effluent can be mitigated significantly, thus improving the acceptability of reclaimed water for defined uses such as landscape irrigation and groundwater recharge. As more membrane facilities are installed and operated, and their ability to produce consistently high-quality water becomes more widely known, public acceptance of reclaimed water increases significantly.

Removal of total dissolved solids provides another level of quality for reclaimed water as the suitability for irrigation, industrial use, and aquifer recharge is enhanced significantly. Control of total dissolved solids is the key component, especially in closed loop systems.

Issues Associated with Potable Reuse

The development of water sources is an issue of very broad concern, particularly as it applies to potable reuse projects. In the establishment of drinking water systems, and in implementing the corresponding water quality standards that are inherent to such processes, it is fundamentally understood that "the water supply should be obtained from the most desirable source which is feasible, efforts should be made to prevent and control pollution of the source" (Okun, 1998). This statement in the 1962 U.S. Public Health Service Drinking Water Standards emphasized the identification of sources of pollution, rather than monitoring and analysis of water.

Drinking water standards were promulgated to establish acceptable quality for source water originating from a protected watershed, and presumed not to be applicable to reclaimed water settings. However, unplanned indirect (*de facto*) potable reuse affects many accepted public drinking water supplies. Examples of surface waters used for drinking water, although greatly impacted by wastewater discharges, and the occurrence of *de facto* potable reuse is discussed in Sec. 3-1 in Chap. 3. It is important to note that the predominant form of treated wastewater discharges present in these natural water sources for accepted public drinking water supplies is secondary effluent, not the highly treated reclaimed water normally applied for groundwater recharge or discharge to municipal water supply reservoirs for planned indirect potable reuse.

As pristine water resources become scarcer, and as incentives for water reuse become stronger, the reality and practicality of water reclamation and reuse has to be examined continually. At this time, the primary course of action with regard to water reclamation and reuse is to develop planned nonpotable and indirect potable reuse applications. Indirect potable reuse is already occurring in several water-short regions of the United States as well as in the world (WPCF, 1989; see also Table E-2 in App. E, and Chaps. 22 to 24). Clearly, the augmentation of drinking water supplies (directly or indirectly) with reclaimed water represents the most rigorous and demanding application in water reuse.

PROBLEMS AND DISCUSSION TOPICS

16-1 Conduct a literature or Internet search for regulations governing water reuse applications in your state, province, or country. Summarize the principal water quality criteria for the seven applications discussed in Sec. 16-1.

16-2 Compare the water quality and treatment requirements for the irrigation of food crops and unrestricted urban use for Florida (see Chaps. 17 and 18). Comment on any observed similarities or differences. What requirements does California have that are different from Florida?

16-3 What are the differences in the design of irrigation systems for agriculture as compared to landscape?

16-4 Reclaimed water is being considered as a freshwater source for cooling tower makeup water at a nearby power plant (see Table 19-4). Compare the water quality requirements for cooling water to typical treated water effluent quality given in Table 7-7 and recommend a method of treatment for reclaimed water. State your reasons. Will any additional treatment be required for the removal of specific constituents, and, if so, what options should be considered?

16-5 In your area, identify potential reclaimed water applications that would be candidates for implementation. What are some of the obstacles that would have to be overcome for a successful project?

16-6 Review the case study for indirect potable reuse at the Upper Occoquan Sewage Authority in Sec. 23-8 in Chap. 23 and the case study for direct potable reuse in Windhoek, Namibia, in Sec. 24-3 in Chap. 24. What lessons are learned from these two case studies that can be applied to future water reuse applications?

REFERENCES

Asano, T. (ed.) (1985) *Artificial Recharge of Groundwater*, Butterworth Publishers, Boston, MA.

Asano, T. (ed.) (1998) *Wastewater Reclamation and Reuse*, Water Quality Management Library, **10**, CRC Press, Boca Raton, FL.

AWWA (1994) *Dual Water System*, 2nd ed., AWWA Manual, M24, American Water Works Association, Denver, CO.

AWWA (2006) "Recycling and Reuse to Grow at an Annual Rate of 8%," *J. AWWA*, **98**, 4, 70–71.

Castelazo, M. (2006) *Water Recycling and Reuse: Technologies and Materials*, Report No. GB-331, BCC Research, Norwalk, CT.

Crook, J., J. Mosher, and J. M. Casteline (2005) *Status and Role of Water Reuse: An International View*, Global Water Research Coalition, London, UK.

Getches, D. H., L. J. MacDonnell, and T. A. Rice (1991) *Controlling Water Use: The Unfinished Business of Water Quality Protection*. Natural Resources Law Center, University of Colorado, School of Law, Boulder, CO.

Okun, D. A. (1998) "Philosophy of the Safe Drinking Water Act and Potable Reuse," *Environ. Sci. Technol.*, **14**, 11, 1298–1303.

Tchobanoglous, G., F. L. Burton, and H. D. Stensel (2003) *Wastewater Engineering: Treatment and Reuse*, 4th ed., McGraw-Hill, New York.

U.S. EPA (2004) *Guidelines for Water Reuse*, EPA/625/R-04/108, U.S. Environmental Protection Agency and U.S. Agency for International Development, Washington, DC.

WPCF (1989) *Water Reuse*, 2nd ed., Manual of Practice SM-3, Water Pollution Control Federation, Alexandria, VA.

17
Agricultural Uses of Reclaimed Water

WORKING TERMINOLOGY   948

17-1   AGRICULTURAL IRRIGATION WITH RECLAIMED WATER: AN OVERVIEW   949
Reclaimed Water Irrigation for Agriculture in the United States   950
Reclaimed Water Irrigation for Agriculture in the World   952
Regulations and Guidelines Related to Agricultural Irrigation with Reclaimed Water   953

17-2   AGRONOMICS AND WATER QUALITY CONSIDERATIONS   954
Soil Characteristics   955
Suspended Solids   958
Salinity, Sodicity, and Specific Ion Toxicity   959
Trace Elements and Nutrients   966
Crop Selection   971

17-3   ELEMENTS FOR THE DESIGN OF RECLAIMED WATER IRRIGATION SYSTEMS   971
Water Reclamation and Reclaimed Water Quantity and Quality   977
Selection of the Type of Irrigation System   977
Leaching Requirements   986
Estimation of Water Application Rate   989
Field Area Requirements   997
Drainage Systems   998
Drainage Water Management and Disposal   1003
Storage System   1003
Irrigation Scheduling   1008

17-4   OPERATION AND MAINTENANCE OF RECLAIMED WATER IRRIGATION SYSTEMS   1008
Demand-Supply Management   1009
Nutrient Management   1009
Public Health Protection   1011
Effects of Reclaimed Water Irrigation on Soils and Crops   1011
Monitoring Requirements   1014

17-5   CASE STUDY: MONTEREY WASTEWATER RECLAMATION STUDY FOR AGRICULTURE—MONTEREY, CALIFORNIA   1015
Setting   1016
Water Management Issues   1016
Implementation   1016
Study Results   1017
Subsequent Projects   1021
Recycled Water Food Safety Study   1021
Lessons Learned   1021

17-6   CASE STUDY: WATER CONSERV II, FLORIDA   1022
        Setting   1023
        Water Management Issues   1023
        Implementation   1023
        Importance of Water Conserv II   1027
        Lessons Learned   1027

17-7   CASE STUDY: THE VIRGINIA PIPELINE SCHEME, SOUTH AUSTRALIA—
        SEASONAL ASR OF RECLAIMED WATER FOR IRRIGATION   1028
        Setting   1028
        Water Management Issues   1029
        Regulatory Requirements   1029
        Technology Issues   1029
        Implementation   1030
        Performance and Operations   1032
        Lessons Learned   1035

PROBLEMS AND DISCUSSION TOPICS   1035

REFERENCES   1038

WORKING TERMINOLOGY

Term	Definition
Aquifer storage and recovery (ASR)	The use of a confined aquifer for storage of reclaimed water. Water can be infiltrated into the aquifer or directly injected. Water is extracted for irrigation during the growing season.
Chlorosis	Yellowing of leaf tissue, usually due to lack of essential elements such as iron.
Crop coefficient (K_c)	Ratio of crop evapotranspiration (ET_c) to the reference evapotranspiration (ET_o).
Crop evapotranspiration (ET_c)	Evapotranspiration of a specific crop, ET_c, expressed as depth.
Deep percolation	The percolation of irrigation water below the root zone. Deep percolation occurs when soil has a high hydraulic conductivity value and more water is applied than the crops can utilize. The water then becomes unavailable to the crops.
Drainage coefficient	The rate water is removed from a unit area by drainage, mm/h.
Evapotranspiration (ET)	The amount of water lost to the atmosphere by evaporation from soil and plant surfaces, and by transpiration from plant tissue.
Irrigation efficiency	The percentage of water applied to the field that is used beneficially.
Leaching	Removal of soluble material or other permeable material from the root zone by the application of excess water.
Lysimeter	A device that measures or collects water percolated through soil. A lysimeter is used to measure evapotranspiration. It can also be used to measure quality of water percolated through soil.

Macronutrients	Six nutrients that are required in fairly large amounts for plant growth: nitrogen, phosphorous, potassium, magnesium, calcium, and sulfur.
Reference evapotranspiration (ET_o)	Standardized evapotranspiration from a well irrigated reference surface, expressed as the depth of water applied. The reference surface is usually grass or alfalfa.
Root zone	Depth of soil that plant roots readily penetrate and where predominant root activity occurs.
Salinity	A parameter referring to the presence of soluble salts in waters, or in soils, usually measured as electrical conductivity in dS/m or mmho/cm.
Sodicity	A condition in which sodium is the dominant composition of the salt in the soil solution and irrigation water within the crop root zone. Sodicity is usually expressed in terms of sodium adsorption ratio (SAR).
Sodium adsorption ratio (SAR)	A chemical measure used to assess potential infiltration problems when water is applied to a soil. The SAR is equal to $Na/[(Ca + Mg)/2]^{1/2}$ where the concentration of sodium (Na), calcium (Ca), and magnesium (Mg) are expressed in milliequivalents per liter (meq/L).
Soil permeability	A characteristic of soil representing the ease with which a gas or liquid penetrates or passes through the soil. It is expressed as the rate at which gas or liquid flow through the soil.
Water table	The upper surface of a saturated zone that is located below the soil surface.

In the United States, 190×10^9 m³/yr (137,000 Mgal/d) or about 40 percent of total freshwater withdrawals, are utilized for irrigation (Huston et al., 2004). By comparison, on a worldwide basis, agriculture remains the largest water user consuming about 70 percent of the world's freshwater supply. Given such large water demands, agricultural irrigation offers significant opportunities for the use of reclaimed water.

The purpose of this chapter is to provide practical information for the design, operation, and management of agricultural irrigation using reclaimed water. Topics considered in this chapter include: (1) an overview of agricultural irrigation with reclaimed water, (2) water quality and agronomic considerations, (3) design considerations for reclaimed water irrigation systems, (4) operation and maintenance issues, and (5) case studies to illustrate how actual agricultural irrigation projects have been implemented. The use of reclaimed water for landscape irrigation is considered in Chap. 18.

17-1 AGRICULTURAL IRRIGATION WITH RECLAIMED WATER: AN OVERVIEW

Various agricultural crops can be irrigated with reclaimed water. To ensure public health protection, appropriate water reclamation processes must be adopted, depending on the crop type and irrigation method. The types of agricultural crops that can be irrigated with reclaimed water are listed in Table 17-1, and examples of reclaimed water agricultural irrigation are shown on Fig. 17-1. As an introduction to the sections dealing with the practical aspects of irrigation, it is useful to consider briefly (1) the use of reclaimed water for

Table 17-1
Types of agricultural crops that can be irrigated with reclaimed water[a]

Types	Example of crops	Treatment requirements[b]
Field crops	Barley, corn, oats	Secondary, disinfection
Fiber and seed crops	Cotton flax	Secondary, disinfection
Vegetable crops that can be consumed raw	Avocado, cabbage, lettuce, strawberry	Secondary, filtration, disinfection
Vegetable crops that will be processed before consumption	Artichoke, sugar beet, sugarcane	Secondary, disinfection
Fodder crops	Alfalfa, barley, cowpea	Secondary, disinfection
Orchards and vineyards	Apricot, orange, peach, plum, grapevines	Secondary, disinfection
Nurseries	Flowers	Secondary, disinfection
Commercial woodlands	Timber, poplar	Secondary, disinfection

[a]Adapted from Lazarova and Asano (2004).
[b]Minimum treatment based on the recommendations in the U.S. EPA Guidelines (2004). Requirements may vary in each state.

agriculture irrigation in the United States and in other countries and (2) regulatory requirements associated with reclaimed water irrigation for agriculture.

Reclaimed Water Irrigation for Agriculture in the United States

In arid and semiarid regions, reclaimed water is used commonly for irrigation purposes to reduce the consumption of water from limited water supply sources including surface water and groundwater. The practice has also gained acceptance in more temperate and humid zones, where the regulations for effluent discharges to surface water bodies are more stringent, and where precipitation is irregular or where soils cannot retain water for optimum plant growth.

In 2002, about 300×10^6 m^3/yr (220 Mgal/d), or 46 percent of the total reclaimed water produced in California, was used for agricultural irrigation. The use of reclaimed water for agricultural irrigation in California by crop types is shown on Fig. 17-2. As of 2002, reclaimed water contributes about 0.7 percent of the overall irrigation water use in the state (State of California, 2003, 2004). Even though the current contribution to agricultural irrigation is small, reclaimed water is becoming a vital source of water for irrigation in some regions. For example, about 70 percent of artichokes produced in the United States are grown in Monterey County, where nearly 95 percent of the farmers use reclaimed water as part of their irrigation water supply.

Arizona, Florida, Hawaii, Nevada, Texas, and Washington are among the other major states where reclaimed water is utilized for agricultural irrigation. Although Florida is the largest user of reclaimed water in the United States, agricultural irrigation comprises only about 16 percent of the state's total reclaimed water use, and in 2003 about 130×10^6 m^3/yr (95 Mgal/d) was used for agricultural irrigation (State of Florida, 2004). As of 2004, 40 states had guidelines or regulations (see Chap. 4) for the application of reclaimed water to food crops and nonfood crops (U.S. EPA, 2004).

Figure 17-1

Various crops are grown with reclaimed water: (a) fodder crop (alfalfa), Jordan; (b) date palms, Aquaba, Jordan (Coordinates: 29.563 N, 34.988 E); (c) squash, Santa Rosa, CA (Coordinates: 38.383 N, 122.770 W); and (d) commercial flowers, Monterey, CA (Coordinates: 36.720 N, 121.778 W).

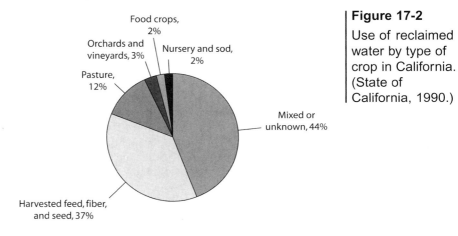

Figure 17-2

Use of reclaimed water by type of crop in California. (State of California, 1990.)

Reclaimed Water Irrigation for Agriculture in the World

Many other countries throughout the world also utilize reclaimed water irrigation in a variety of forms. Examples of reclaimed water irrigation for agriculture in selected countries are shown in Table 17-2. Untreated wastewater also is used for agriculture in parts of the world, but it is not considered in this chapter.

Table 17-2
Examples of agricultural irrigation with reclaimed water in selected countries[a]

Location	Description
Argentina (Mendoza)	Over 160,000 m^3/d of urban wastewater (1 million inhabitants) is treated at the Campo Espejo wastewater treatment plant with a 290 ha lagoon system to meet WHO guidelines for unrestricted irrigation. Reclaimed water is used for irrigation of forest, vineyards, olives, alfalfa, fruit trees, and other crops over 3640 ha.
Australia (Adelaide, Bolivar/Virginia Pipeline Scheme)	The largest reclaimed water system in Australia. Approximately 280×10^6 m^3/yr of reclaimed water from the Bolivar Wastewater Treatment Plant is transported through 150 km of pipelines for irrigation of 200 km^2 of vegetable crop farmland.
China	Gaobeidian sewage treatment plant provides over 500,000 m^3/d of secondary effluent for reuse in industry and agriculture.
France (Clermont-Ferrand)	Over 10,000 m^3/d of tertiary treated urban wastewater is reused for irrigation of 700 ha of maize.
Greece	Several wastewater reclamation projects are under way in Crete, Thessaloniki, Chalkida, and other regions to irrigate crops such as vineyards, sugar beets, tomatoes, and flowers. Hydroponic cultivation is also investigated for reclaimed water.
Israel (Tel Aviv, The Dan Region Project)	About 130×10^6 m^3/yr of reclaimed water after secondary treatment with nutrient removal and soil aquifer treatment (SAT) is used for unrestricted irrigation.
Italy	About 450,000 m^3/yr of reclaimed water from Basso Rubicone fertirrigation plant in Emilia Romagna is used to irrigate 125 ha of farmland.
Jordan	About 70×10^6 m^3/yr of wastewater, treated primarily with stabilization ponds or activated sludge processes, comprising about 10 percent of the nation's water supply, is used for agricultural irrigation.
Kuwait	Three major tertiary treatment facilities, Ardiya, Rikka, and Jahra, produced about 357×10^6 m^3/yr of reclaimed water in 1997, and about 125×10^6 m^3/yr was used for irrigation purposes, primarily agricultural irrigation.
Spain	In Vitoria, about 8×10^6 m^3/yr of wastewater treated with advanced treatment and disinfection is used for unrestricted irrigation. In the Canary Islands, reclaimed water from the electrodialysis reversal (EDR) process has been tested to irrigate several types of banana trees.
Tunisia	Reclaimed water from La Cherguia-Tunis wastewater treatment plant has been used for agricultural irrigation since 1965. The 600 ha (1480 ac) area of La Soukra is irrigated with reclaimed water for citrus and olive trees. Irrigation of vegetables eaten either raw or cooked is prohibited. The actual irrigated area covers 7000 ha.

[a] Adapted from Angelakis et al. (1999), Gotor et al. (2001), Kracman et al. (2001), Ickson-Tal et al. (2003), Hamoda et al. (2004), Lazarova and Asano (2004).

Within the past decade, Australia has become one of the most active countries in promoting water reuse (see Chap. 2). By far, agricultural irrigation is the dominant user of reclaimed water in Australia, utilizing about 420×10^6 m^3 (111×10^3 Mgal), or 82 percent of their total reclaimed water supply in 2000 (Australian Bureau of Statistics, 2004).

Arid countries such as Israel, Jordan, Kuwait, Tunisia, and the United Arab Emirates utilize reclaimed water extensively for irrigation. In Israel, for example it is estimated that 230, partially interconnected, water reuse systems are in operation, that recycle about 72 percent of the municipal wastewater produced, primarily for agricultural irrigation. The water supply from reclaimed water comprises about 15 percent of Israel's water resources (Lazarova and Asano, 2004; Weber and Juanico, 2004). Reclaimed water use for irrigation is becoming an integral component of sanitation and water supply projects along the semiarid Mediterranean and the Middle-East countries.

Regulations and Guidelines Related to Agricultural Irrigation with Reclaimed Water

Water reuse guidelines, criteria, and related regulations are discussed in detail in Chap. 4. Generally, where the effluent is discharged to navigable waters, municipal wastewater treatment plants must obtain permits for the National Pollutant Discharge Elimination System (NPDES) and other waste discharge requirements imposed by each state. Where all wastewater is reused and not discharged to navigable waters, the reclaimed water is exempt from the NPDES permit. Instead, the reclaimed water system must meet the treatment and/or quality criteria set forth by each state. Some of the related regulatory issues for reclaimed water irrigation are discussed below.

Waste Discharge Requirements for Irrigated Land
Discharge of the return flow or runoff water from agricultural land is usually exempt from the waste discharge permits, but because of recent concerns about nonpoint source pollution, more stringent discharge requirements for agricultural water discharge are expected in the near future. For example, the Central Valley Regional Water Quality Control Board of California adopted the Irrigated Lands Conditional Waiver Program in 2003. The program requires the farmers to work collectively or individually to comply with the *California Water Code* and other plans and policies to maintain their status of waiver from the waste discharge requirements (State of California, 2003). These requirements are imposed on farmers, but water planners also need to be aware of the related requirements.

Total Maximum Daily Load
Many constituents in streams are subject to the U.S. EPA Total Maximum Daily Load (TMDL) restrictions. The TMDL is a calculation of the maximum amount of a specific pollutant that a waterbody can receive from all contributing point and nonpoint sources and still maintain the water quality standard. Allocation of allowable pollutant discharge quantities is determined for all contributing point and nonpoint sources.

Regulations and Guidelines for Reclaimed Water Irrigation
As of 2002, 21 states have either regulations or guidelines for the reclaimed water irrigation for food crops, and 40 states for nonfood crops (U.S. EPA, 2004). The treatment and quality criteria in seven selected states for food crops and nonfood crops are summarized in Table 17-3 and 17-4, respectively. Some states use regulations based on the

Table 17-3
Reclaimed water quality and treatment requirements for food crops[a]

Item	Arizona	California	Florida	Hawaii	Nevada	Texas	Washington
Treatment	Secondary treatment, filtration, and disinfection	Oxidized[b], coagulated, filtered, and disinfected	Secondary treatment, filtration, and high-level disinfection	Oxidized, filtered, and disinfected	Secondary treatment and disinfection	NS[c]	Oxidized, coagulated, filtered and disinfected
BOD	NS	NS	20 mg/L CBOD	NS	30 mg/L	5 mg/L	30 mg/L
TSS	NS	NS	5 mg/L	NS	NS	NS	30 mg/L
Turbidity	2 NTU (avg) 5 NTU (max)	2 NTU (avg) 5 NTU (max)	NS	2 NTU (max)	NS	3 NTU	2 NTU (avg) 5 NTU (max)
Coliform	Fecal None detectable[d] 23/100 mL (max)	Total 2.2/100 mL (med)[e] 23/100 mL (max in 30 d)	Fecal 75% of samples below detection 25/100 mL (max)	Fecal 2.2/100 mL (med)[e] 23/100 mL (max in 30 d)	Fecal 200/100 mL (avg) 400/100 mL (max)	Total 20/100 mL (gm)[f] 75/100 mL (max)	Total 2.2/100 mL (avg) 23/100 mL (max)

[a]Adapted from U.S. EPA (2004).
[b]Oxidized means the wastewater is treated with a biological process. The term is used in lieu of "secondary treatment" to avoid specification of the process used to achieve the quality criteria.
[c]NS—not specified by state regulations.
[d]Not detectable in 4 of last 7 daily samples.
[e]Seven-day median.
[f]gm = geometric mean.

California Water Recycling Criteria, whereas other states have developed different regulations, based on their own studies of health effects of reclaimed water uses.

WHO Guidelines for Reclaimed Water Irrigation

While the regulations and guidelines in the United States call for the best available technology and highest levels of public health protection feasible, such regulations may not be feasible in some other countries. The guidelines developed by the World Health Organization (WHO), therefore, take a different approach from the guidelines and regulations in the United States. In the WHO guidelines, consideration is given to the feasibility of adopting costly treatment processes, and the relative health risk of reclaimed water irrigation and other causes of diseases (Blumenthal et al., 2000). World Health Organization guidelines for the safe use of wastewater in agriculture are discussed in Chap. 4, Sec. 4-8.

17-2 AGRONOMICS AND WATER QUALITY CONSIDERATIONS

The sustainability of irrigation schemes depends on the proper management of potential impacts of the available water on soils, crops, and the environment. Knowledge of agronomics is used for all stages of a reclaimed water irrigation project, from the feasibility analysis through the design, construction, and management of the constructed system. Relevant topics in agronomics and water quality for the irrigation with reclaimed water are

Table 17-4
Reclaimed water quality and treatment requirements for nonfood crops[a]

Item	Arizona	California	Florida	Hawaii	Nevada	Texas	Washington
Treatment	Secondary treatment and disinfection	Oxidized[b] and disinfected	Secondary treatment, basic disinfection	Oxidized, filtered, and disinfected	Secondary treatment and disinfection	NS[c]	Oxidized and disinfected
BOD	NS	NS	20 mg/L CBOD	NS	30 mg/L	5 mg/L	30 mg/L
TSS	NS	NS	5 mg/L	NS	NS	NS	30 mg/L
Turbidity	NS	NS	NS	2 NTU (max)	NS	3 NTU	2 NTU (avg) 5 NTU (max)
Coliform	Fecal 200/100 mL[d] 800/100 mL (max)	Total 23/100 mL (med) 240/100 mL (max in 30 d)	Fecal 200/100 mL 800/100 mL (max)	Fecal 2.2/100 mL (avg) 23/100 mL (max)	Fecal 200/100 mL (avg) 400/100 mL (max)	Total 20/100 mL (avg) 75/100 mL (max)	Total 23/100 mL (avg) 240/100 mL (max)

[a]Adapted from U.S. EPA (2004).
[b]Oxidized means the wastewater is treated with a biological process. The term is used in lieu of "secondary treatment" to avoid specification of the process used to achieve the quality criteria.
[c]NS—not specified by state regulations.
[d]Less than 200/100 mL in four of last seven samples.

introduced in this section including (1) soil characteristics; (2) suspended solids; (3) salinity, sodicity; and specific ion toxicity; (4) trace elements and nutrients; and (5) considerations for crop selection.

General guidelines for interpreting water quality parameters and evaluating suitability of reclaimed water for irrigation are summarized in Table 17-5, and each parameter is discussed later in this section. It should be noted that specific water quality requirements depend on soil characteristics, climate, plants to be irrigated, irrigation method, and other local conditions. The most important characteristics of irrigation water for determining its quality are: (1) concentration of soluble salts, (2) relative proportion of sodium to other cations (magnesium, calcium, and potassium), and (3) concentration of boron and other elements that may be toxic to plants.

Soil Characteristics

Applicability of irrigation, crop selection, and selection of the irrigation method are greatly affected by the soil characteristics. References such as the "Soil Survey Manual," available from the Natural Resources Conservation Service under the U.S. Department of Agriculture (USDA, 1993b), should be consulted for detailed information on various soil characteristics, such as soil texture, soil structure, soil depth, soil profile, and chemical properties.

Soil Texture
Soil textural classes, used to classify the physical properties of soil, are based on the relative percentages of sand, silt, and clay. The three classes of soil, i.e., sand, silt, and clay, are determined by the fraction of particles that are smaller than 2 mm in diameter. The classes of soil texture are illustrated on Fig. 17-3. For example, if the

Table 17-5
Guidelines for interpretation of water quality for irrigation[a]

Potential irrigation problem	Units	Degree of restriction on use		
		None	Slight to moderate	Severe
Salinity				
EC_w[b]	dS/m	<0.7	0.7–3.0	>3.0
TDS	mg/L	<450	450–2000	>2000
Sodicity[c]				
SAR, 0–3		and $EC_w \geq 0.7$	0.7–0.2	<0.2
3–6		≥ 1.2	1.2–0.3	<0.3
6–12		≥ 1.9	1.9–0.5	<0.5
12–20		≥ 2.9	2.9–1.3	<1.3
20–40		≥ 5.0	5.0–2.9	<2.9
Specific ion toxicity				
Sodium (Na)[d,e]				
Surface irrigation	SAR	<3	3–9	>9
Sprinkler irrigation	mg/L	<70	>70	
Chloride (Cl)[d,e]				
Surface irrigation	mg/L	<140	140–350	>350
Sprinkler irrigation	mg/L	<100	>100	
Boron (B)	mg/L	<0.7	0.7–3.0	>3.0
Miscellaneous effects				
Nitrogen (Total N)[f]	mg/L	<5	5–30	>30
Bicarbonate (HCO_3) (overhead sprinkling only)	mg/L	<90	90–500	>500
pH	unitless		Normal range 6.5–8.4	
Residual chlorine (overhead sprinkling only)	mg/L	<1.0	1.0–5.0	>5.0

[a]Adapted from University of California Committee of Consultants (1974); and Ayers and Westcot (1985).
[b]EC_w = electrical conductivity of the irrigation water.
[c]SAR = sodium adsorption ratio. See this section for details.
[d]Most tree crops and woody ornamentals are sensitive to sodium and chloride; use the values shown in this table. Most annual crops are not as sensitive; see the salinity tolerance tables (Table 17-12).
[e]With overhead sprinkler irrigation and low humidity (<30%), sodium or chloride greater than 70 or 100 mg/L, respectively, have resulted in excessive leaf adsorption and crop damage to sensitive crops.
[f]Total nitrogen should include nitrate-nitrogen, ammonia-nitrogen, and organic-nitrogen. Although forms of nitrogen in wastewater vary, the plant responds to the total nitrogen.

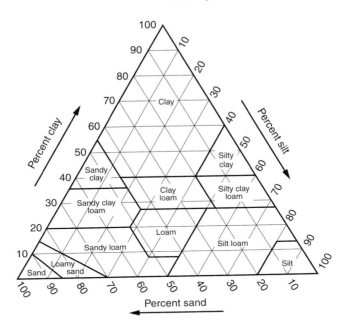

Figure 17-3
Chart showing the percentages of clay, silt, and sand in the basic textural classes. (From USDA, 1993b.)

soil contains 60 percent silt, 30 percent sand, and 10 percent clay, the soil is classified as *silt loam*. The class of the soil can be determined from the particle size distribution or estimated by soil scientists in the field (Crites, 1985). Soils that exhibit the best balance for reclaimed water irrigation are medium textured; very fine sandy loam, loam, silt loam, and silt.

The presence of soil organic matter enhances the soil physical properties, density, water infiltration rate, plant and root development, oxygen availability, biological activity, nutrient availability, and the capacity to hold water (USDA, 1997). Organic matter in soil includes plant and animal residue at various stages of decomposition, and cells, tissues of soil organisms, and stable organic matter also referred to as humus. The percentage of organic matter in organic soils will vary from 20 to 95 percent. The terms used commonly for soils with high organic matter content include muck, peat, and mucky peat.

Physical Structure

Soil particles are aggregated by chemical and biological processes to form natural structures. The structure of soil affects the rate at which air and water move through it. It also affects root development and the nutrient supply to plants (USDA, 1991). The types of soil structure and their effects on the movement of water are illustrated on Fig. 17-4. Soils with prismatic, blocky, and granular structures have better water permeability than the platy or massive (structureless) soils. The prismatic and blocky structures, however, induce peripheral flow of irrigation water around the soil structure, and restrict root development. Therefore, a more granular structure is preferred for growing crops. The structure of soil can be "changed" by specific cultivation practices, including deep tillage, addition of organic matter, and addition of inorganic chemicals such as gypsum (calcium sulfate).

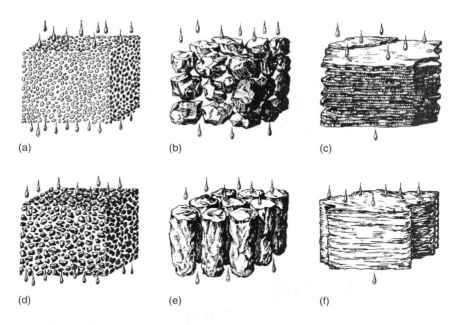

Figure 17-4
Soil structure: (a) single grain, (b) blocky, (c) platy, (d) granular, (e) prismatic, and (f) massive. (From USDA, 1991.)

Soil Depth
The term *soil depth* refers to the depth from the soil surface to a root growth restrictive layer such as bedrock, hardpan, or a water table. The depth of a soil affects its capacity to hold water that a plant can use, and, thus, affects the interval between irrigation periods. The presence of an impermeable layer of soil or shallow groundwater table restricts the downward movement of water and root penetration. The soil profile, i.e., the change in soil characteristics with depth, also affects the downward movement of water.

Soil Chemical Characteristics
Soil is formed primarily from decomposed rocks. The chemical and mineral composition of soil varies with the extent of oxidation, reduction, hydration, hydrolysis, and carbonation that has occurred. Soils composed predominantly of minerals are called mineral soils. In mineral soils, the content of organic matter and oxygen, and the temperature are generally highest at the surface. Microbiological activity is also greatest near the soil surface. The ability of soil to retain cations including nutrients is affected by the soil chemical characteristics and organic content. Soils with limited ability to retain cations may require more frequent application of fertilizers.

Suspended Solids

Suspended solids in reclaimed water could potentially clog the irrigation water distribution lines and emitters. Irrigation systems with small emitter openings and those used at low water velocities are susceptible to emitter clogging problems. Typically, secondary effluent contains 5 to 25 mg/L of total suspended solids (TSS). The suspended solids level is lower in tertiary effluent, typically less than 10 mg/L (see Table 3-14). The TSS concentration of less than 30 mg/L is generally considered suitable for most irrigation systems. However, other factors such as temperature, sunlight, emitter types, and flowrate also affect the clogging potential. Management of irrigation systems and clogging issues in reclaimed water irrigation are further discussed in Sec. 17-4.

Salinity, Sodicity, and Specific Ion Toxicity

As discussed in previous chapters, dissolved solids, added from various sources during domestic and other water uses (see Table 3-11) are generally not reduced during the wastewater treatment and reclamation processes. As a result, reclaimed water generally has higher dissolved solids than potable water and other freshwaters. The impacts of increased dissolved constituents on irrigation are assessed generally in terms of salinity, sodicity, and specific ion toxicity as discussed below.

Salinity Impacts

Salinity is a quantitative measure of the soluble salts in water or soil. The salinity of water is a measure of its total dissolved solids (TDS). The electrical conductivity of water, EC_w, expressed in decisiemens per meter (dS/m), milli-ohms per centimeter (mmho/cm) or micro ohms per centimeter (μmho/cm), is used as a surrogate measure of the TDS concentration. The value of electrical conductivity is affected by temperature. The standard temperature for measuring electrical conductivity is 25°C. For most agricultural irrigation purposes, the values for EC_w and TDS are related to each other and can be converted within an accuracy of about 10 percent using Eq. (17-1):

$$\text{for } EC_w < 5 \text{ (dS/m): TDS (mg/L)} \approx EC_w \text{ (dS/m)} \times 640$$
$$\text{for } EC_w > 5 \text{ (dS/m): TDS (mg/L)} \approx EC_w \text{ (dS/m)} \times 800 \quad (17\text{-}1)$$

Equation (17-1) should be used as a first approximation because it is also dependent upon the ion composition of the water. As salinity increases, the osmotic gradient between soil water and root cells decreases, In turn, the plants need to use more energy to concentrate solutions in root cells allowing them to take up water from the soil, resulting in plant growth reduction and in the development of symptoms similar in appearance to those of drought conditions (see Fig. 17-5).

Salt tolerance ratings of crops are illustrated on Fig. 17-6. Above the salinity threshold level, the crop yield will decrease with an increase in salinity. Definitions of salinity measurements for water and soil are summarized in Table 17-6. The salinity level in soil is stabilized when the amount of salt leached out becomes equal to the amount of salt applied with irrigation water. Because the amount of salt assimilated by plants generally is negligible as compared with the amount of salt applied through irrigation water, leaching is the key to control salinity problems, where reclaimed water is used for irrigation.

Sodicity Impacts

Sodicity is a condition where the sodium is the dominant cation in the soil solution and irrigation water. Under sodic conditions, soil particles disperse and clays swell (see Fig. 17-7), with a result that the soil particles plug large pore spaces in the soil matrix and reduce the rate water and air enter the soil. Water ponding at the soil surface (waterlogging) is observed when the infiltration of water is restricted.

Sodicity is usually expressed by the SAR, which is the ratio of sodium to the calcium and magnesium cations. The SAR is a calculated from laboratory measurements made on a water sample or water extracted from a soil using the following equation.

$$SAR = \frac{[Na^+]}{\sqrt{([Ca^{2+}] + [Mg^{2+}])/2}} \quad (17\text{-}2)$$

Figure 17-5

Effect of salinity on plant growth. The three corn plants shown on this figure were grown from seeds hydroponically in a 1 mM (millimolar) NaCl solution for 14 d. The plants were transferred to solutions containing: (a) 1 mM NaCl, (b) 75 mM NaCl, (c) 100 mM NaCl and grown for an additional 7 d. The plant in the 1 mM NaCl solution served as the control. (Courtesy of Malcolm Drew Agricultural Research Council, Letcombe Laboratory, United Kingdom).

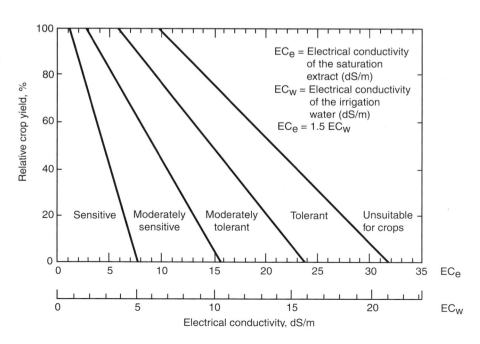

Figure 17-6

Relative salt tolerance ratings of agricultural crops. (From Ayers and Westcot, 1985.)

Table 17-6
Definitions for salinity measurements

Salinity type	Symbol	Definition
Water salinity	EC_w	Electrical conductivity of water
Irrigation water salinity	EC_{iw}	Electrical conductivity of water applied for irrigation
Soil salinity	EC_e	Electrical conductivity for the water contained in the soil sample and extracted for measurement. It also is called the electrical conductivity of saturation extract.
Soil water salinity	EC_{sw}	Electrical conductivity of the soil water. It is difficult to measure soil water salinity, and soil salinity, EC_e, is usually measured. Generally EC_e is approximately half of the soil water salinity, EC_{sw}.
Drainage water salinity	EC_{dw}	Electrical conductivity of water drained from the root zone or a defined layer where percolated water is drained out from the soil.

where $[Na^+]$ = concentration of sodium ion, meq/L
$[Ca^{2+}]$ = concentration of calcium ion, meq/L
$[Mg^{2+}]$ = concentration of magnesium ion, meq/L.

When using Eq. (17-2), it is assumed that the ratio of HCO_3^- to Ca^{2+} is at equilibrium for the given soil-water conditions. When considering reclaimed water irrigation, the calcium concentration $[Ca^{2+}]$ value may need to be adjusted to better estimate the concentration that is expected to remain in the soil water after irrigation water reaches equilibrium with the soil. The adjusted values of calcium $[Ca_x^{2+}]$ can be obtained

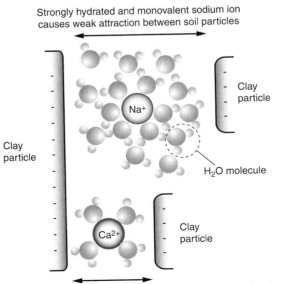

Figure 17-7
The effect of sodium and calcium ions on soil particles. (Adapted from McKenzie, 1998.)

Table 17-7

Values of Ca$_x$ used to compute the adjusted sodium adsorption ratio[a,b,c]

Ratio HCO$_3$/Ca	Salinity of applied water, EC$_w$, dS/m											
	0.1	0.2	0.3	0.5	0.7	1.0	1.5	2.0	3.0	4.0	6.0	8.0
0.05	13.20	13.61	13.92	14.40	14.79	15.26	15.91	16.43	17.28	17.97	19.07	19.94
0.10	8.31	8.57	8.77	9.07	9.31	9.62	10.02	10.35	10.89	11.32	12.01	12.56
0.15	6.34	6.54	6.69	6.92	7.11	7.34	7.65	7.90	8.31	8.64	9.17	9.58
0.20	5.24	5.40	5.52	5.71	5.87	6.06	6.31	6.52	6.86	7.13	7.57	7.91
0.25	4.51	4.65	4.76	4.92	5.06	5.22	5.44	5.62	5.91	6.15	6.52	6.82
0.30	4.00	4.12	4.21	4.36	4.48	4.62	4.82	4.98	5.24	5.44	5.77	6.04
0.35	3.61	3.72	3.80	3.94	4.04	4.17	4.35	4.49	4.72	4.91	5.21	5.45
0.40	3.30	3.40	3.48	3.60	3.70	3.82	3.98	4.11	4.32	4.49	4.77	4.98
0.45	3.05	3.14	3.22	3.33	3.42	3.53	3.68	3.80	4.00	4.15	4.41	4.61
0.50	2.84	2.93	3.00	3.10	3.19	3.29	3.43	3.54	3.72	.3.87	4.11	4.30
0.75	2.17	0.24	2.29	2.37	2.43	2.51	2.62	2.70	2.84	2.95	3.14	3.28
1.00	1.79	1.85	1.89	1.96	2.01	2.09	2.16	2.23	2.35	2.44	2.59	2.71
1.25	1.54	1.59	1.63	1.68	1.73	1.78	1.86	1.92	2.02	2.10	2.23	2.33
1.50	1.37	1.41	1.44	1.49	1.53	1.58	1.65	1.70	1.79	1.86	1.97	2.07
1.75	1.23	1.27	1.30	1.35	1.38	1.43	1.49	1.54	1.62	1.68	1.78	1.86
2.00	1.13	1.16	1.19	1.23	1.26	1.31	1.36	1.40	1.48	1.54	1.63	1.70
2.25	1.04	1.08	1.10	1.14	1.17	1.21	1.26	1.30	1.37	1.42	1.51	1.58
2.50	0.97	1.00	1.02	1.06	1.09	1.12	1.17	1.21	1.27	1.32	1.40	1.47
3.00	0.85	0.89	0.91	0.94	0.96	1.00	1.04	1.07	1.13	1.17	1.24	1.30
3.50	0.78	0.80	0.82	0.85	0.87	0.90	0.94	0.97	1.02	1.06	1.12	1.17
4.00	0.71	0.73	0.75	0.78	0.80	0.82	0.86	0.88	0.93	0.97	1.03	1.07
4.50	0.66	0.68	0.69	0.72	0.74	0.76	0.79	0.82	0.86	0.90	0.95	0.99
5.00	0.61	0.63	0.65	0.67	0.69	0.71	0.74	0.76	0.80	0.83	0.88	0.93
7.00	0.49	0.50	0.52	0.53	0.55	0.57	0.59	0.61	0.64	0.67	0.71	0.74
10.00	0.39	0.40	0.41	0.42	0.43	0.45	0.47	0.48	0.51	0.53	0.56	0.58
20.00	0.24	0.25	0.26	0.26	0.27	0.28	0.29	0.30	0.32	0.33	0.35	0.37

[a] Adapted from Suarez (1981).
[b] The adjusted sodium adsorption ratio (SAR$_{adj}$) is a modification of the SAR procedure. It has long been recognized that calcium in the soil-water is not constant. The calcium concentration at equilibrium depends on both the concentration in the applied water and also the dissolution from soil-calcium or precipitation from soil-water. The effect is to raise or lower the relative sodium content in the soil water. The calcium in solution at equilibrium is influenced by soil-water salinity and the concentration of calcium, bicarbonate, and dissolved carbon dioxide. The effects are reflected in the Ca$_x$ value.
[c] The adjusted sodium adsorption ratio includes the effects of the factors noted in the above footnote and more correctly predicts the sodium hazard and potential infiltration problem caused by water quality. The adjusted sodium adsorption ratio (SAR$_{adj}$) may be substituted for the SAR value when evaluating the potential infiltration problem.
[d] Ratio of bicarbonate to calcium in mEq/L.

from Table 17-7. The values in Table 17-7 are for near-surface soil-water at various water salinities, and are based on the assumption that no precipitation of magnesium is occurring and the partial pressure of CO_2 is 0.071 kPa (0.0007 atm) near the soil surface. The value of the adjusted sodium adsorption ratio, SAR_{adj}, can be calculated from Eq. (17-3).

$$SAR_{adj} = \frac{[Na^+]}{\sqrt{([Ca_x^{2+}] + [Mg^{2+}])/2}} \qquad (17\text{-}3)$$

where $[Ca_x^{2+}]$ is the adjusted concentration of calcium ion from Table 17-7, meq/L

In practice, the difference between the SAR_{adj} and SAR values usually is not significant, and the SAR value is used more commonly. The adjusted value should be used when the water quality and the soil chemical characteristics are likely to affect the equilibrium concentration of calcium significantly. Water with high alkalinity will affect the equilibrium concentration and result in higher SAR_{adj} (see Example 17-1).

The likelihood of an infiltration problem associated with sodicity increases as the electrical conductivity of infiltrating water decreases. The effects of irrigation water SAR and the electrical conductivity (EC) on infiltration are illustrated on Fig. 17-8. By knowing both the EC and SAR, the likelihood of having a water infiltration problem can be predicted. Typically, the calcium level is greater than the magnesium concentration in reclaimed water and in soil. When the reverse occurs, potential sodium problems may be increased slightly because magnesium is hydrated at a higher degree than calcium and held less strongly to the soil particles (Ayers and Westcot, 1985).

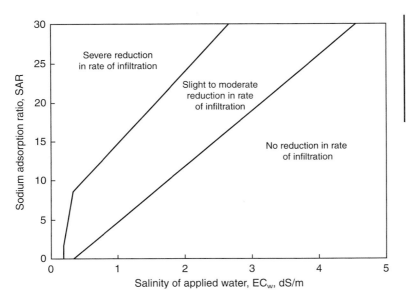

Figure 17-8

Combined effects of irrigation water SAR and EC on soil infiltration rate. (From Hanson et al., 1999.)

EXAMPLE 17-1. Calculation of Sodium Adsorption Ratio.

Estimate the SAR and SAR_{adj} values of a reclaimed water with the following chemical characteristics.

Constituent	Unit	Value
pH	unitless	6.6
Sodium	mg/L	129
Calcium	mg/L	49
Magnesium	mg/L	18
Chloride	mg/L	137
Sulfate	mg/L	163
Alkalinity	mg/L as $CaCO_3$	127
TDS	mg/L	680

Solution

1. Determine the concentrations of sodium, calcium, magnesium, and bicarbonate in meq/L.
   a. Sodium

   Molecular weight = 23.0, $[Na^+]$ = 129/23.0 = 5.6 meq/L

   b. Calcium

   Molecular weight = 40.1, $[Ca^{2+}]$ = 49/(40.1/2) = 2.4 meq/L

   c. Magnesium

   Molecular weight = 24.3, $[Mg^{2+}]$ = 18/(24.3/2) = 1.5 meq/L

   d. Bicarbonate
   At pH = 6.6, alkalinity is primarily from bicarbonate.
   Alkalinity = 127 mg/L as $CaCO_3$

   Milliequivalent concentration = 127/(100.1/2) = 2.54 meq/L

2. Determine the SAR using Eq. (17-2).

$$SAR = \frac{[Na^+]}{\sqrt{([Ca^{2+}] + [Mg^{2+}])/2}}$$

$$SAR = \frac{5.6}{\sqrt{(2.4 + 1.5)/2}} = 4.0$$

3. Determine $[Ca_x]$ using the data given in Table 17-7.
   a. Determine the ratio of HCO_3^-/Ca^{2+}.

$$[HCO_3^-]/[Ca^{2+}] = 2.54/2.4 = 1.06$$

b. Estimate EC_w from the TDS using Eq. (17-1).

$$EC_w \approx TDS/640 = 680/640 = 1.06 \text{ dS/m}$$

c. Determine $[Ca_x]$ using Table 17-7.

$$[Ca_x^{2+}] \approx 2.02$$

4. Determine SAR_{adj} using Eq. (17-3).

$$SAR_{adj} = \frac{[Na^+]}{\sqrt{([Ca_x^{2+}] + [Mg^{2+}])/2}}$$

$$SAR_{adj} = \frac{5.6}{\sqrt{(2.02 + 1.5)/2}} = 4.2$$

Comment

According to the data presented in Table 17-5, with a reclaimed water salinity of $EC_w = 1$ dS/m, and SAR and SAR_{adj} values of 4.0 and 4.2 respectively, there would be no restriction for use of the water for irrigation. If the alkalinity in reclaimed water was 50 mg/L, the SAR_{adj} would be 3.5. If the alkalinity was 200 mg/L, the SAR_{adj} would be 4.5. In both cases, the difference between SAR and SAR_{adj} is not large enough to change the degree of restriction on reclaimed water use for agricultural irrigation.

The soil characteristics affecting the sodicity can be measured in terms of exchangeable sodium percentage (ESP), defined as the percentage of the cation exchange capacity of the soil occupied by sodium ions. The cation exchange capacity is the measure of the capacity of the soil to absorb and hold cations such as calcium, magnesium, sodium, and potassium. The values of SAR and ESP are known to have a linear correlation for a specific soil. Generally, dispersion of soil particles (i.e., soil permeability problem) is expected to occur when the ESP is greater than 15 percent and the soil salinity is less than 4 dS/m. Soils in such condition are called sodic (or alkali) soils.

Calcium sulfate ($CaSO_4$), commonly known as gypsum, is often used to restore soil permeability of irrigated farmland. The amount of gypsum that must be added to lower the SAR enough to ensure acceptable soil permeability depends on the cation exchange capacity and the ESP. Gypsum may be applied directly to the soil or dissolved in the irrigation water. Acidifying amendments such as sulfuric acid are often applied to soils with a high lime ($CaCO_3$) content to release calcium from the soil and, thus, reduce its sodicity.

Specific Ion Toxicity

Many ions present in reclaimed water are beneficial or harmless at low concentrations, but, at higher concentrations, may accumulate and injure plant tissue. Sodium, chloride, and boron are three major chemicals that may cause specific ion toxicity from reclaimed water irrigation. The effects of excessive sodium include leaf burn, chlorosis (yellowing and blanching), and twig dieback. Sodium toxicity has been observed in avocado, citrus, and stone fruit trees such as apricot, cherry, peach, and plum. Low soil concentrations of calcium will affect selectivity in root membranes and will result in uptake and accumulation of sodium in plants. Chloride can cause similar effects in plants, but only for woody plant species. Vegetable, grain, forage, or fiber crops are not affected by either sodium or chloride at typical concentrations in reclaimed water, provided that the SAR values are not extremely high.

Boron can become toxic at levels only slightly greater than that required for good plant growth. Symptoms include leaf burn, leaf cupping, chlorosis, anthocyanin (blue and red leaves), rosette spotting, premature leaf drops, branch dieback, and reduced growth (Westcot and Ayers, 1985). The plant organ affected by boron depends on the plant's internal ability to remobilize boron. It should be noted that boron tolerance varies depending on climate soil conditions and the type of crop.

Trace Elements and Nutrients

In addition to the reclaimed water chemical constituents discussed above, trace levels of metals and inorganic compounds, as well as organic compounds and other water quality parameters such as pH and temperature, can also affect the suitability of reclaimed water for irrigation. Toxic effects of trace elements and the effects of nutrients in reclaimed water are discussed below.

Trace Elements

The term *trace elements* is used for the chemical elements that exist at low concentrations in the natural environment. Some trace elements are essential for plant growth at low concentrations but exhibit plant toxicity at higher concentrations. Plant toxicity varies with the concentration of the elements and the plant species. Trace elements of concern and generally recommended maximum concentrations in irrigation water are reported in Table 17-8. Toxicity such as leaf damage and reduced yield occurs when the trace elements are taken up by the plant roots and then accumulate in leaves and other parts of the plant tissues.

The concentrations of trace elements in reclaimed water vary with the source of the wastewater typically on the order of micrograms to milligrams per liter (μg/L to mg/L) where the source is predominantly domestic wastewater. Higher concentrations have been observed where industrial wastewater is discharged to the municipal wastewater collection system. Typically, the concentrations of the trace elements in reclaimed water are in the range where adverse effects are not likely to occur in the short term. However, long-term application of water containing the trace elements may result in accumulation of the trace elements in soil and may potentially cause plant toxicity and groundwater contamination. Crops may accumulate trace elements in edible parts of the plant, which may pose health risks to humans and animals if concentrations become high enough. Typical concentrations of trace elements in secondary effluent, examples of water quality for tertiary and reverse osmosis processes, and the levels recommended for irrigation by U.S. EPA are presented in Table 17-9.

Table 17-8
Recommended maximum concentrations of trace elements in irrigation waters[a]

Element	Recommended maximum concentration[b], mg/L	Remarks
Al (aluminum)	5.0	Can cause nonproductivity in acid solids (pH < 5.5), but more alkaline soils at pH > 5.5 will precipitate the ion and eliminate any toxicity.
As (arsenic)	0.10	Toxicity to plants varies widely, ranging from 12 mg/L for Sudan grass to less than 0.05 mg/L for rice.
Be (beryllium)	0.10	Toxicity to plants varies widely, ranging from 5 mg/L for kale to 0.5 mg/L for bush beans.
Cd (cadmium)	0.010	Toxic to beans, beets, and turnips at concentration as low as 0.1 mg/L in nutrient solutions. Conservative limits are recommended because of the potential for cadmium to accumulate in plants and soils to concentrations that may be harmful to humans.
Co (cobalt)	0.050	Toxic to tomato plants at 0.1 mg/L in nutrient solution. Tends to be inactivated by neutral and alkaline soils.
Cr (chromium)	0.10	Not generally recognized as an essential growth element. Conservative limits recommended because of lack of knowledge of toxicity to plants.
Cu (copper)	0.20	Toxic to a number of plants at 0.1 to 1.0 mg/L in nutrient solutions.
F (fluoride)	1.0	Inactivated by neutral and alkaline soils.
Fe (iron)	5.0	Not toxic to plants in aerated soils but can contribute to soil acidification and loss of reduced availability of essential phosphorous and molybdenum. Overhead sprinkling may result in unsightly deposits on plants, equipment, and buildings.
Li (lithium)	2.5	Tolerated by most crops up to 5 mg/L; mobile in soil. Toxic to citrus at low levels (>0.075 mg/L). Acts similar to boron.
Mn (manganese)	0.20	Toxic to a number of crops at a few tenths to a few mg/L, but usually only in acid soils.
Mo (molybdenum)	0.010	Not toxic to plants at normal concentrations in soil and water. Can be toxic to livestock if forage is grown in soils with high levels of available molybdenum.
Ni (nickel)	0.20	Toxic to a number of plants at 0.5 to 1.0 mg/L; reduced toxicity at neutral or alkaline pH.
Pb (lead)	5.0	Can inhibit plant cell growth at very high concentrations.
Se (selenium)	0.020	Toxic to plants at concentrations as low as 0.025 mg/L and toxic to livestock if forage is grown in soils with relatively high levels of added selenium. An essential element for animals but in very low concentrations.
Sn (tin)	—	Effectively excluded by plants; specific tolerance unknown.
Ti (titanium)	—	(See remark for tin)
W (tungsten)	—	(See remark for tin)
V (vanadium)	0.10	Toxic to many plants at relatively low concentrations.
Zn (zinc)	2.0	Toxic to many plants at widely varying concentrations; reduced toxicity at pH > 6.0 and in fine-textured or organic soils.

[a] Adapted from Ayers and Westcot (1985) and NRC (1973).
[b] The maximum concentration is based on a water application rate of 1.25 m/yr (4 ft/yr).

Table 17-9

Typical concentrations of trace elements in reclaimed water[a]

Element	Secondary effluent, mg/L		After tertiary treatment[b]	After reverse osmosis	EPA recommended levels for irrigation[c]	
	Range	Median			Long-term	Short-term
As	<0.005–0.023	<0.005	<0.001	0.00045	0.10	10.0
B	<0.1–2.5	0.7	0.3	0.17	0.75	2.0
Cd	<0.005–0.15	<0.005	<0.0004	0.0001	0.01	0.05
Cr	<0.005–1.2	0.02	<0.01	0.0003	0.10	20.0
Cu	<0.005–1.3	0.04	<0.01	0.015	0.20	5.0
Hg	<0.0002–0.001	0.0005	0.0001	—	—	—
Mo	0.001–0.018	0.007	—	—	0.01	0.05
Ni	0.003–0.6	0.004	<0.02	0.002	0.2	2.0
Pb	0.003–0.35	0.008	<0.002	0.002	5.0	20.0
Se	<0.005–0.02	<0.005	<0.001	0.0007	0.02	0.05
Zn	0.004–1.2	0.04	0.05	0.05	2.0	10.0

[a]Adapted from Page and Chang (1985).
[b]Activated sludge treatment, followed by filtration and disinfection.
[c]Adapted from U.S. EPA (2004).

Assuming that the average irrigation rate is 1.2 m/yr, the annual input of a trace element is calculated as:

$$\text{Annual input (kg/ha} \cdot \text{yr)}$$
$$= 1.2 \, (\text{m/yr}) \times 10^4 \, (\text{m}^3/\text{ha} \cdot \text{m}) \times C \, (\text{mg/L}) \quad (17\text{-}4)$$
$$\times 10^3 \, (\text{L/m}^3) \times 10^{-6} \, (\text{kg/mg}) = 12 \times C \, (\text{kg/ha} \cdot \text{yr})$$

where C = concentration of the element in irrigated water (mg/L).

Estimates of the numbers of years for a farmland to reach heavy-metal-loading limits when tertiary treated reclaimed water is used for irrigation are reported in Table 17-10. Because the uptake by plants or the removal from the root zone through leaching is not accounted for, the computed estimates are conservative. Some of the trace elements, however, will accumulate in the soil and may be taken up by the plant tissue. Therefore, it is important to monitor the concentration of trace elements in soil as well as in the plant tissue. The concentration of selected trace elements normally found in soil and plant tissue, and their impact on plant growth are shown in Table 17-11.

Nutrients

Reclaimed water contains nutrients considered to be beneficial for irrigation. Three main macronutrients found in reclaimed water are nitrogen, phosphorus, and potassium. Nitrogen and phosphorous may be present in reclaimed water at significant levels and will have an effect on plant growth. The concentration of potassium in reclaimed water is usually much less and its effect on plant growth is less. Calcium, magnesium,

Table 17-10
Calculated length of time for agricultural soils with reclaimed water irrigation to reach heavy-metal loading limits[a]

Element	Concentration in tertiary effluent, mg/L[b]	Annual input, kg/ha·yr[c]	Suggested loading at soil CEC[d], kg/ha			Time to reach soil loading limit at CEC[d], yr		
			<5	5–15	>15	<5	5–15	>15
Cd	0.0004	0.0048	5	10	20	1042	2083	4167
Cu	0.01	0.12	125	250	500	1042	2083	4167
Ni	0.02	0.24	125	250	500	521	1042	2083
Zn	0.05	0.6	250	500	1000	417	833	1667
Pb	0.002	0.024	500	1000	2000	20,833	41,667	83,333

[a] Adapted from Page and Chang (1985).
[b] Typical concentration, data compiled from tertiary treatment plants in California.
[c] CEC = Cation exchange capacity, expressed in meq/100 g soil.

and sulfate are three other macronutrients needed for plant growth. Typical concentrations of nitrogen and phosphorous in reclaimed water are reported in Table 17-12. Levels of other macronutrients in untreated wastewater vary greatly depending on the source water and the chemicals used for treatment, and, in general, are not reduced

Table 17-11
Concentrations of selected trace elements normally found in soil and plant tissues and their impact on plant growth[a]

Element	Soil concentration, µg/g		Typical concentration, (range) in plant tissue, µg/g	Impact on plant growth[b]
	Range	Typical		
As	0.1–40	6	0.1–5	Not required
B	2–200	10	5–30	Required, wide species differences, toxic at higher concentrations
Be	1–40	6	—	Not required: toxic
Cd	0.01–7	0.06	0.2–0.8	Not required: toxic
Cr	5–3000	100	0.2–1.0	Not required: low toxicity
Co	1–40	8	0.05–0.15	Required by legume at <0.2 µg/g
Cu	2–100	20	2–15	Required at 2–4 µg/g: toxic at >20 µg/g
Pb	2–200	10	0.1–10	Not required: low toxicity
Mn	250–1700	600	15–100	Required: toxicity depends on Fe/Mn ratio
Mo	0.2–5	2	1–100	Required at <0.1 µg/g: low toxicity
Ni	10–1000	40	1–10	Not required: toxic at >50 µg/g
Se	0.1–0.2	0.5	0.02–2.0	Not required: toxic at >50 µg/g
V	20–500	100	0.1–10	Required by some algae: toxic at >10 µg/g
Zn	10–300	50	15–200	Required: toxic at >200 µg/g

[a] Adapted from Allaway (1994); Bowen (1979); Chapman (1965); Lisk (1972), Page (1974); and Chang and Page (1994).
[b] Concentration in plant tissues on a dry-weight (70°C) basis.

Table 17-12
Typical nutrient levels in reclaimed water[a]

		Range of nutrient levels in reclaimed water					
	Unit	Untreated wastewater	Conventional activated sludge	Conventional activated sludge with BNR[b]	Activated sludge with BNR, depth filtration and disinfection	Membrane bioreactor[c]	Secondary with BNR plus MF[d], RO[e], and disinfection
Total nitrogen	mg N/L	20–70	15–35	2–12	2–12	7–18	<1
Nitrate-N	mg N/L	0–trace	10–30	1–10	1–10	5–11	<1
Total phosphorous	mg P/L	4–12	4–10	1–2	<2	0.3–5	<0.05

[a]Adapted from Tchobanoglous et al. (2003).
[b]BNR = biological nutrient removal.
[c]Without BNR.
[d]MF = microfiltration.
[e]RO = reverse osmosis.

significantly by conventional secondary and tertiary treatment processes. Because nutrient requirements also vary greatly with the crop and the stages of plant growth, fertilizer application rates must be adjusted according to the reclaimed water quality and nutrient requirements of the crop. The nutrients in reclaimed water also affect the potential of algal growth in a storage reservoir. If the level of nutrients in reclaimed water is higher than the desired level, reclaimed water quality may have to be altered by nutrient removal, or by blending reclaimed water with water from other sources. Management of nutrients is described further in Sec. 17-4.

Nitrogen The nitrogen demand of crops varies significantly with the crop type. Plants utilize nitrogen in the forms of exchangeable and water-soluble ammonium (NH_4^+) and nitrate (NO_3^-). The organic form of nitrogen is not utilized until it is converted into ammonia or nitrate. Generally, nitrogen concentration in specific plant parts is analyzed to determine the amount of fertilizer to be applied.

Ammonium is the primary form of nitrogen in reclaimed water from conventional secondary treatment without nitrification whereas nitrate is the primary form of nitrogen in a nitrified reclaimed water. The form and range of nitrogen values in reclaimed water depends on the treatment process (see Table 17-12). Due to restrictive nitrogen requirements in many wastewater discharge permits, nitrogen removal is often required for wastewater treatment plants that discharge part or all of the treated effluent to sensitive water bodies. Many water reclamation plants also employ nutrient removal to meet water quality criteria for various water reuse applications (see Chap. 7).

Phosphorous The concentration of phosphorous in reclaimed water ranges from 0.1 mg/L to 15 mg/L, depending on the treatment process (see Table 17-12). Phosphorous added to the soil by reclaimed water irrigation may be taken up by the crop, accumulated in the soil, or lost by leaching, runoff or erosion. The phosphorous in reclaimed water is predominantly in inorganic forms, which can undergo a complex sorption process binding it to the soil. Apparent sorption of phosphorous may be a combination of several processes including a fast reversible sorption on soil particle surfaces, plus various slower time-dependent processes, some of which lead to deposition of phosphorous in the pore space of the soil particles (McGechan and Lewis, 2002).

Crop Selection

Crop selection is a core part of the design process for a reclaimed water irrigation system. The selection of crops is influenced by the reclaimed water quality, climate, economics, management skill, labor and equipment availability, and the regional traditions (George et al., 1985).

Among the important agronomic factors, resistance to salt is the primary plant characteristic considered followed by boron tolerance, when selecting crops. Salt resistance is the ability to withstand salt stress by salt avoidance and/or salt tolerance. Salt avoidance is the mechanism by which plants avoid or delay the onset of salt stress (Carrow, 1994). Salt tolerance is the mechanism that allow plants to tolerate high salt stress. The relative salt tolerance of agricultural crops is shown in Table 17-13. Relative boron tolerance of agricultural crops is shown in Table 17-14. These tables may be useful for a brief review of the salt and boron tolerance of various crops. To further assess the effect of salinity, relative crop yield at various salt levels should be estimated. Generally, there is a threshold salinity below which no adverse effect is observed. Above the threshold salinity level, relative crop yield declines as salinity increases. The relative crop yield can be estimated using Eq. (17-5):

$$Y_r = 100 - b(EC_e - a) \qquad (17-5)$$

where Y_r = relative crop yield, percent
 a = salinity threshold, dS/m
 b = crop-specific incremental decrease (slope) in yield per dS/m, percent
 EC_e = the mean soil salinity in the root zone, dS/m

The values of soil electrical conductivity for various crop yields are shown in Table 17-15. The values reported in Table 17-15 should be used only as a guideline to relative salt tolerance among the crops, because the absolute salt tolerance depends on climate, soil conditions, and cultural practices (Maas and Grattan, 1999).

17-3 ELEMENTS FOR THE DESIGN OF RECLAIMED WATER IRRIGATION SYSTEMS

Design of a reclaimed water irrigation system typically involves consideration of the following topics:

- Selection of wastewater reclamation processes and evaluation of reclaimed water quantity and quality
- Selection of the type of irrigation system
- Estimation of leaching requirement
- Estimation of water application rate
- Field area requirement
- Drainage system
- Storage system
- Irrigation schedule
- Operation and management

Table 17-13
Relative salt tolerance of agricultural crops[a,b]

Common name	Latin name	Tolerant	Moderately tolerant	Moderately sensitive	Sensitive
Field Crops					
Barley	Hordeum vulgare	√			
Bean	Phaseolus vulgaris				√
Broadbean	Vicia faba			√	
Corn (maize)	Zea mays			√	
Cotton	Gossypium hirsutum	√			
Cowpea	Vigna unguiculata			√	
Flax	Linum usitatissimum			√	
Jojoba	Simmondsia chinensis	√			
Oats	Avena sativa		√		
Rice (paddy)	Oriza sativa			√	
Rye	Secale cereale		√		
Sugarbeet	Beta vulgaris	√			
Sugarcane	Saccharum officinarum			√	
Sorghum	Sorghum bicolor		√		
Soybean	Glycine max		√		
Wheat	Triticum aestivum		√		
Vegetable Crops					
Artichoke	Helianthus tuberosus		√		
Asparagus	Asparagus officinalis	√			
Beet, red	Beta vulgaris		√		
Cabbage	Brassica oleracea capitata			√	
Carrot	Daucus carota				√
Celery	Apium graveolens			√	
Cucumber	Cucumis sativus			√	
Lettuce	Lactuca sativa			√	
Onion	Allium cepa				√
Potato	Solanum tuberosum			√	
Spinach	Spinacia oleracea			√	
Squash, zucchini	Cucurbita pepo melopepo		√		
Sweet potato	Ipomoea batatas			√	
Tomato	Lycopersicon esculentum			√	
Turnip	Brassica rapa			√	

Table 17-13
Relative salt tolerance of agricultural crops[a,b] (*Continued*)

Common name	Latin name	Tolerant	Moderately tolerant	Moderately sensitive	Sensitive
		Field Crops			
Alfalfa	*Medicago sativa*			√	
Bermuda grass	*Cynodon dactylon*	√			
Clover, ladino	*Trifolium repens*			√	
Clover, red	*Trifolium pratense*			√	
Fescue, tall	*Festuca elatior*		√		
Foxtail, meadow	*Alopecurus pratensis*			√	
Harding grass	*Phalaris tuberosa*		√		
Lovegrass	*Eragrostis* sp.			√	
Orchard grass	*Dactylis glomerata*			√	
Sesbania	*Sesbania exaltata*			√	
Sphaerophysa	*Sphaerophysa salsula*			√	
Sudan grass	*Sorghum sudanense*		√		
Trefoil, big	*Lotus uliginosus*			√	
Vetch, common	*Vicia angustifolia*			√	
Wheatgrass, fairway crested	*Agropyron cristatum*	√			
Wheatgrass, standard crested	*Agropyron sibiricum*		√		
Wheatgrass, tall	*Agropyron elongatum*	√			
Wildrye, beardless	*Elymus triticoides*		√		
		Fruit Crops			
Almond	*Prunus dulcis*				√
Apricot	*Prunus armeniaca*				√
Blackberry	*Rubus* sp.				√
Date palm	*Phoenix dactylifera*	√			
Grape	*Vitus* sp.			√	
Orange	*Citrus sinensis*				√
Peach	*Prunus persica*				√
Plum, prune	*Prunus domestica*				√
Strawberry	*Fragaria* sp.				√

[a] Adapted from Westcot and Ayers (1985).
[b] These data serve only as a guideline to the relative tolerances among crops. Absolute tolerances vary with climate, soil conditions, and cultural practices.
[c] Sensitivity ratings are defined by the boundaries in Fig. 17-6.

Table 17-14
Relative boron tolerance of agricultural crops[a]

Common name	Latin name	Tolerant	Moderately tolerant	Moderately sensitive	Sensitive
Field Crops					
Barley	Hordeum vulgare		√		
Bean, kidney	Phaseolus vulgaris				√
Corn (maize)	Zea mays		√		
Cotton	Gossypium hirsutum	√			
Groundnut (Peanut)	Arachis hypogaea				√
Oats	Avena sativa		√		
Sorghum	Sorghum bicolor	√			
Sugarbeet	Beta vulgaris	√			
Wheat	Triticum aestivum				√
Vegetable Crops					
Artichoke	Helianthus tuberosus		√		
Asparagus	Asparagus officinalis	√			
Beet, red	Beta vulgaris	√			
Cabbage	Brassica oleracea capitata		√		
Carrot	Daucus carota			√	
Celery	Apium graveolens		√		
Cucumber	Cucumis sativus			√	
Lettuce	Lactuca sativa		√		
Onion	Allium cepa				√
Potato	Solanum tuberosum			√	
Sweet potato	Ipomoea batatas				√
Tomato	Lycopersicon esculentum	√			
Turnip	Brassica rapa		√		
Fodder Crops					
Alfalfa	Medicago sativa	√			
Barley (forage)	Hordeum vulgare				√
Cowpea (forage)	Vigna unguiculata				√
Fruit Crops					
Apricot	Prunus armeniaca				√
Blackberry	Rubus sp.				√
Grape	Vitus sp.				√
Grapefruit	Citrus paradisi				√
Orange	Citrus sinensis				√
Peach	Prunus persica				√
Plum, prune	Prunus domestica				√

[a]Adapted from Maas (1986).
[b]Sensitivity ratings are defined by maximum concentrations of boron in soil water without yield or vegetative growth reductions: Sensitive (<0.5–1.0 mg/L), Moderately sensitive (1.0–2.0 mg/L), Moderately tolerant (2.0–4.0 mg/L), Tolerant (>4.0 mg/L).

Table 17-15
Effects of salinity on agricultural crops: threshold EC_e levels and incremental yield reduction[a,b]

Common (Latin) name	Observed parameter[c]	Threshold EC_e, dS/m[d]	Slope, %/(dS/m)[e]	Rating[f]
	Field Crops			
Barley (*Hordeum vulgare*)[g]	Grain yield	8.0	5.0	T
Bean (*Phaseolus vulgaris*)	Seed yield	1.0	19	S
Broadbean (*Vicia faba*)	Shoot DW	1.6	9.6	MS
Corn (maize) (*Zea mays*)	Ear FW	1.7	12	MS
Cotton (*Gossypium hirsutum*)	Seed cotton yield	7.7	5.2	T
Cowpea (*Vigna unguiculata*)	Seed yield	4.9	12	MT
Flax (*Linum usitatissimum*)	Seed yield	1.7	12	MS
Oats (*Avena sativa*)	Grain yield	—	—	T
Rice (paddy) (*Oriza sativa*)[g]	Grain yield	3.0	12	S
Rye (*Secale cereale*)	Grain yield	11.4	10.8	T
Sugarbeet (*Beta vulgaris*)[h]	Storage root	7.0	5.9	T
Sugarcane (*Saccharum officinarum*)	Shoot DW	1.7	5.9	MS
Sorghum (*Sorghum bicolor*)	Grain yield	6.8	16	MT
Soybean (*Glycine max*)	Seed yield	5.0	20	MT
Wheat (*Triticum aestivum*)	Grain yield	6.0	7.1	MT
Wheat, durum (*Triticum turgidum*)	Grain yield	5.9	3.8	T
	Vegetable Crops			
Artichoke (*Helianthus tuberosus*)	Tuber yield	0.4	9.6	MS
Artichoke (*Cynara scolymus*)	Bud yield	6.1	11.5	MT
Asparagus (*Asparagus officinalis*)	Spear yield	4.1	2.0	T
Beet, red (*Beta vulgaris*)	Storage root	4.0	9.0	MT
Broccoli (*Brassica oleracea botrytis*)	Shoot FW	2.8	9.2	MS
Cabbage (*Brassica oleracea capitata*)	Head FW	1.8	9.7	MS
Carrot (*Daucus carota*)	Storage root	1.0	14	S
Celery (*Apium graveolens*)	Petiole FW	1.8	6.2	MS
Cucumber (*Cucumis sativus*)	Fruit yield	2.5	13	MS
Lettuce (*Lactuca sativa*)	Top FW	1.3	13	MS
Onion (*Allium cepa*)	Bulb yield	1.2	16	S
Potato (*Solanum tuberosum*)	Tuber yield	1.7	12	MS
Radish (*Raphanus sativus*)	Storage root	1.2	13	MS
Spinach (*Spinacia oleracea*)	Top FW	2.0	7.6	MS
Squash, zucchini (*Cucurbita pepo melopepo*)	Fruit yield	4.9	10.5	MT
Sweet potato (*Ipomoea batatas*)	Fleshy root	1.5	11	MS
Tomato (*Lycopersicon esculentum*)	Fruit yield	2.5	9.9	MS
Turnip (*Brassica rapa*)	Storage root	0.9	9.0	MS

(*Continued*)

Table 17-15

Effects of salinity on agricultural crops: threshold EC_e levels and incremental yield reduction[a,b] (Continued)

Common (Latin) name	Observed parameter[c]	Threshold EC_e, dS/m[c]	Slope, % per dS/m[d]	Rating[e]
Fodder Crops				
Alfalfa (Medicago sativa)	Shoot DW	2.0	7.3	MS
Bermuda grass (Cynodon dactylon)[i]	Shoot DW	6.9	6.4	T
Clover, ladino (Trifolium repens)	Shoot DW	1.5	12	MS
Clover, red (Trifolium pratense)	Shoot DW	1.5	12	MS
Foxtail, meadow (Alopecurus pratensis)	Shoot DW	1.5	9.6	MS
Harding grass (Phalaris tuberosa)	Shoot DW	4.6	7.6	MT
Lovegrass (Eragrostis sp.)[i]	Shoot DW	2.0	8.4	MS
Orchard grass (Dactylis glomerata)	Shoot DW	1.5	6.2	MS
Sesbania (Sesbania exaltata)	Shoot DW	2.3	7.0	MS
Sphaerophysa (Sphaerophysa salsula)	Shoot DW	2.2	7.0	MS
Sudan grass (Sorghum sudanense)	Shoot DW	2.8	4.3	MT
Trefoil, big (Lotus uliginosus)	Shoot DW	2.3	19	MS
Vetch, common (Vicia angustifolia)	Shoot DW	3.0	11	MS
Wheatgrass, fairway crested (Agropyron cristatum)	Shoot DW	7.5	6.9	T
Wheatgrass, standard crested (Agropyron sibiricum)	Shoot DW	3.5	4.0	MT
Wheatgrass, tall (Agropyron elongatum)	Shoot DW	7.5	4.2	T
Wildrye, beardless (Elymus triticoides)	Shoot DW	2.7	6.0	MT
Fruit Crops				
Almond (Prunus dulcis)	Shoot growth	1.5	19	S
Apricot (Prunus armeniaca)	Shoot growth	1.6	24	S
Blackberry (Rubus sp.)	Fruit yield	1.5	22	S
Date palm (Phoenix dactylifera)	Fruit yield	4.0	3.6	T
Grape (Vitus sp.)	Shoot growth	1.5	9.6	MS
Orange (Citrus sinensis)	Fruit yield	1.3	13.1	S
Peach (Prunus persica)	Shoot growth, fruit yield	1.7	21	S
Plum, prune (Prunus domestica)	Fruit yield	2.6	31	S
Strawberry (Fragaria sp.)	Fruit yield	1.0	33	S

[a] Adapted from Maas and Grattan (1999).
[b] Presented data serves only as a guideline to relative salt tolerance among crops. Absolute salt tolerance for each crop depends on climate, soil conditions, and cultural practices.
[c] DW = dry weight, FW = fresh weight.
[d] The level of electrical conductivity below which no effect was observed on the specified parameter [see Eq. (17-5)]. In soils containing at least one percent of gypsum (gypsiferous soils), plants will tolerate salinity levels about 2 dS/m higher than indicated.
[e] Reduction in relative yield expressed in percentage per dS/m [see Eq. (17-5)].
[f] Salt tolerance ratings, determined based on Fig. 17-6. S = sensitive, MS = moderately sensitive, MT = moderately tolerant, T = tolerant. Also see Table 17-13.
[g] Less tolerant during seedling stage.
[h] Sensitive during germination and emergence, EC_e should not exceed 3 dS/m.
[i] Average of several varieties.

Each of these topics is considered in the following discussion. Although the discussion follows the order presented above, it should be noted that in practice the sequence in which each of the above topics is addressed will vary with each project and local conditions. In addition to the above topics, the location of the water reclamation facility, the location of the irrigation site, and the needs for auxiliary facilities such as pumping stations, distribution pipelines and storage reservoirs, which will affect the feasibility of reclaimed water irrigation systems must also be considered. These topics are covered in Chap. 14.

Water Reclamation and Reclaimed Water Quantity and Quality

A water reuse project for agricultural irrigation may be planned with an existing wastewater treatment facility, or with a new facility designed specifically for the intended reclaimed water users. With the existing facilities, the feasibility of water reuse for agricultural irrigation depends on the location of the water reclamation facilities relative to the site of use, the quantity and quality of water required for irrigation, and the cost of providing reclaimed water. Depending on the crops to be irrigated, the existing wastewater treatment facilities may have to be upgraded to meet the reclaimed water quality criteria. When planning a new treatment facility to provide reclaimed water for agricultural reuse, it is prudent to locate the facility near reuse opportunities to minimize infrastructure needed to convey reclaimed water. Treatment processes should be selected to produce the quality of water required for existing and future applications (see Part 3: Technologies and Systems for Water Reclamation and Reuse). Reclaimed water quantity and quality information required to assess the feasibility of using reclaimed water for agricultural irrigation is listed in Table 17-16.

Selection of the Type of Irrigation System

Irrigation systems are selected based on the crop types, water quality and quantity requirements, site characteristics, and management costs and skilled labor requirements. The types of irrigation systems and the factors affecting the selection of an irrigation system are discussed below.

Types of Irrigation Systems
Irrigation systems can be classified as: (1) gravity surface flow, (2) gravity subsurface flow and wicking, (3) pressurized surface application systems, and (4) pressurized subsurface systems. Examples of irrigation systems that are used for reclaimed water are shown on Fig. 17-9. Basic features and evaluation of selected irrigation systems are summarized in Tables 17-17 and 17-18. The conditions of use for selected irrigation system are shown in Table 17-19.

Gravity Surface Flow Surface irrigation systems include flood, border, and ridge and furrow irrigation. Flood and border irrigation are generally used for fodder crops whereas ridge and furrow irrigation (see Fig. 17-9a) is used most commonly for food crops.

Gravity Subsurface Flow Several new gravity subsurface flow systems have been developed including manifold and wicking systems. In general, the use of subsurface gravity flow systems is limited to large scale agricultural irrigation applications and are not considered further in this text.

Pressurized Surface Application Pressurized surface irrigation can include the use of water guns (see Fig. 17-9b), wheel roll and center pivot overhead sprinklers (see Fig. 1-6a

Table 17-16
Information required to assess feasibility of using reclaimed water for agricultural irrigation[a]

Information	Decision on irrigation management
Reclaimed water quantity	
The total amount of reclaimed water available during the crop growing season.	Total area that could be irrigated.
Seasonal variability of demand and supply.	Storage requirements and possible use of reclaimed water for other purposes.
The rate of delivery either as m^3 per day or liters per second.	Area that could be irrigated at any given time, layout of fields and facilities, and irrigation system.
Type of delivery: continuous or intermittent, or on demand.	Layout of fields and facilities, irrigation system, and irrigation scheduling.
Mode of supply: delivered to the point of use, or available in a storage reservoir to be pumped by the user.	The need to install pumps and pipes to transport reclaimed water and irrigation system requirements.
Availability of water from other sources	Blending of water to supplement reclaimed water supply, and to control water quality.
Reclaimed water quality	
Microbial quality	Selection of crop types and irrigation methods. The need for additional treatment.
Total salt concentration and/or electrical conductivity of the effluent.	Selection of crops, irrigation method, leaching, and other management requirements.
Concentrations of cations, such as Ca^{2+}, Mg^{2+}, and Na^+.	Assessment of sodium hazard and need to take appropriate mitigating measures.
Concentration of toxic ions, such as heavy metals, Boron, and Cl^-.	Assessment of toxicities that are likely to be caused by reclaimed water irrigation and need for appropriate measures.
Concentration of trace elements (particularly those which are suspected of being phytotoxic).	Assessment of toxicities that are likely to be caused by reclaimed water irrigation and need for appropriate mitigating measures.
Concentration of nutrients, particularly nitrate-N.	Fertilization requirements and crop selection. The need for nutrient removal at the treatment plant.
Suspended solids.	Irrigation system selection and measures to prevent clogging. The need for additional treatment for solids removal.

[a] Adapted from Pescod (1992).

in Chap. 1), fixed (solid set) and movable impact sprinklers (see Fig. 17-9c), conventional sprinkler systems (e.g., lawn type), and aboveground drip irrigation (see Fig. 17-9d). Water guns and overhead sprinklers are used most commonly for fodder crops. Fixed and movable impact sprinklers are used commonly with food crops. Surface drip irrigation systems are used for food crops and for watering grape vines and trees (see Figs. 2-8c and 2-9b).

Figure 17-9

Examples of irrigation methods used with reclaimed water: (a) furrow irrigation, (b) water gun spray irrigation, (c) movable fixed head sprinklers in Watsonville/Monterey, CA (Coordinates: 36.760 N, 121.780 W), and (d) drip irrigation (Courtesy of G. Oron).

Pressurized Subsurface Systems Pressurized subsurface irrigation systems typically involve the use of drip emitters. Drip emitters may be pressure-compensating or tortuous-path type depending on the system design. Drip irrigation systems with reclaimed water are used for playgrounds, athletic fields, median strips, and landscaping (see Chap. 18). An example of a large scale drip irrigation system installed recently is in Forsyth County, Georgia. Reclaimed water, from a new 9.3×10^3 m^3/d (2.5 Mgal/d) ultrafiltration membrane bioreactor plant, is pumped 16.7 km (10 mi) through a 500 mm (20 in.) pipeline to a 73 ha (180 ac) drip irrigation system. The layout of the drip irrigation system which includes 610 km (380 mi) of drip line arranged in seven field areas with 17 operating zones is shown on Fig. 17-10.

Considerations for Irrigation System Selection

When reclaimed water is used for irrigation, special attention must be made to ensure public health protection. Other considerations for irrigation system selection include irrigation efficiency, and the prevention of clogging. Each of these elements is described below.

Table 17-17
Basic features of commonly used irrigation systems[a]

Irrigation method	Factors affecting choice	Special measures for irrigation with reclaimed water
Flood irrigation	Lower cost Exact leveling not required Low irrigation efficiency Low level of health protection	Thorough protection of field workers, crop handlers and consumers needed. Not used commonly for reclaimed water irrigation in U.S.
Furrow irrigation	Low cost leveling may be needed Low irrigation efficiency Medium level of health protection	Low level of wastewater treatment necessary. Protection of field workers, possibly of crop handlers and consumers required. Appropriate crop selection necessary.
Border irrigation	Relatively low cost Leveling required Low irrigation efficiency Medium level of health protection	Low level of wastewater treatment required. Protection of field workers, possibly of crop handlers and consumers required. Crop restriction necessary.
Sprinkler irrigation	Medium to high cost Medium irrigation efficiency Leveling not required Low level of health protection, especially with aerosol generation	Minimum distance (setback distance) from drinking water supply wells, houses and roads required. Water quality restrictions. Anaerobic wastes should not be used due to odor nuisance.
Subsurface and drip irrigation	High cost High irrigation efficiency Higher yields Highest level of health protection	No special protection measures required. Water quality restrictions for the prevention of emitter clogging. Appropriate management to avoid exposure to reclaimed water.

[a] Adapted from WHO (1989).

Public Health Protection The selection of the types of irrigation systems greatly affects the likelihood of human exposure to reclaimed water. Water reuse guidelines and regulations generally dictate the irrigation methods according to the reclaimed water quality and the type of crops to be irrigated. Requirements for irrigation timing and the setback distance from public area are also specified to minimize exposure of field workers and neighbors to reclaimed water (see also Sec. 17-5). The likelihood of exposure to reclaimed water is high for the field workers when surface irrigation methods are used. Aerosols generated from spray irrigation systems and residual irrigation water on the crop pose potential risks for the field workers, neighbors, and workers handling the crops. Spray irrigation is usually restricted to tertiary-treated reclaimed water, and in some states, spray irrigation cannot be used for the food crops that are consumed unprocessed. Drip irrigation is often preferred from the public health standpoint, because of low to negligible human exposure.

Table 17-18
Evaluation of common irrigation methods in relation to the use of reclaimed water[a]

Parameters of evaluation	Furrow irrigation	Border irrigation	Sprinkler irrigation	Drip and subsurface irrigation
Foliar wetting and consequent leaf damage resulting in poor yield	No foliar injury as the crop is planted on the ridge	Some bottom leaves may be affected but the damage is not so serious as to reduce yield	Severe leaf damage can occur resulting in significant yield loss	No foliar injury occurs under this method of irrigation
Salt accumulation in the root zone with repeated applications	Salts tend to accumulate in the ridge which could harm the crop	Salts move vertically downwards and are not likely to accumulate in the root zone	Salt movement is downwards and root zone is not likely to accumulate salts	Salt movement is radial along the direction of water movement, thus a salt wedge is formed between drip points
Ability to maintain high soil water potential	Plants may be subject to water stress between irrigations	Plants may be subject to water stress between irrigations	Not possible to maintain high soil water potential throughout the growing season	Possible to maintain high soil-water potential throughout the growing season and minimize the effect of salinity
Suitability to handle brackish reclaimed water without significant yield loss	Fair to medium; with good management and drainage, acceptable yields are possible	Fair to medium; good irrigation and drainage, practices can produce acceptable levels of yield	Poor to fair; most crops suffer from leaf damage and yield is low	Excellent to good; almost all crops can be grown with very little reduction in yield

[a] Adapted from Pescod (1992).

Irrigation Efficiency Irrigation systems need to be designed and maintained to maximize the efficiency of irrigation. Irrigation efficiency is defined as the percentage of water applied to the field that is used beneficially and is calculated using Eq. (17-6):

$$E_i = \frac{F_b}{F_f} \times 100 \tag{17-6}$$

where E_i = water application efficiency, percent
F_b = water used beneficially, mm/unit time
F_f = water applied to field, mm/unit time

Water used beneficially is potentially accessible for crop evapotranspiration, crop cooling, crop quality control, and leaching of salts from the root zone. Water that is lost by wind drift, runoff, or excess application (i.e., deep percolation) is not counted as beneficially used water.

Table 17-19
Irrigation systems and conditions of use[a]

Irrigation system	Suitability and conditions of use[b]				Irrigation efficiency[c], %
	Crops	Topography	Soil	Water	
		Surface systems			
Straight furrows[d]	Vegetables, row crops, orchards, vineyards	Max grade: 3% Cross slope: 10% (erosion hazard)	IR: 2.5 mm/h NR: if length of furrow is adjusted Depth: sufficient for required grading	Quantity: moderate flows required	70–85
Graded contour furrows[d]	Vegetables, row crops, orchards, vineyards	Max grade: 8% undulating Cross slope: 10% (erosion hazard)	IR: 2.5 mm/h NR: if length of furrow is adjusted Non-cracking soil required	Quantity: moderate flows required	70–85
Narrow graded border up to 4.6 m (15 ft) wide	Pasture, grain, alfalfa, vineyards	Max grade: 7% Cross slope: 0.2%	IR: 7.6–150 mm/h	Quantity: moderate flows required	65–85
Wide graded border up to 30 m (100 ft) wide	Pasture, grain, alfalfa, vineyards	Max grade: 0.5–1% Cross slope: 0.2%	IR: 7.6–150 mm/h Depth: sufficient for required grading	Quantity: large flows required	65–85
Level border	Grain, field crops, rice, orchards	Max grade: level Cross slope: 0.2%	IR: 2.5–150 mm/h Depth: sufficient for required grading	Quantity: moderate flows required	75–90

					Sprinkler systems	
Portable, hand moved	Orchards, pasture grain, alfalfa, vineyards, low-growing vegetable and field crops	Max grade: 20%	Min. IR: 2.5 mm/h WHC: 76 mm	Quantity: NR Quality: high TDS water can cause leaf burn	70–80
Wheel roll	All crops less than 0.9 m (3 ft) high	Max grade: 15%	Min. IR: 2.5 mm/h WHC: 76 mm	Quantity: NR Quality: see above	70–80
Solid set	NR	NR	Min. IR: 1.3 mm/h	Quantity: NR Quality: see above	70–80
Center pivot or traveling lateral	All crops except trees	Max grade: 15%	Min. IR: 7.6 mm/h WHC: 51 mm	Quantity: large flows required Quality: see above	70–80
Traveling gun	Pasture, grain, alfalfa, field crops, vegetables	Max grade: 15%	Min. IR: 7.6 mm/h WHC: 51 mm	Quantity: 380–3800 L/min (100–1000 gal/min·unit) Quality: see above	70–80
					Drip and subsurface systems	
Drip and subsurface	Orchards, vineyards, vegetables, nursery plants	NR	Min. IR: 0.51 mm/h	Quantity: NR	70–90

[a] Adapted from Smith et al. (1985).
[b] IR = infiltration rate, NR = no restriction, WHC = water-holding capacity. WHC is the capability of soils to store and release water to plants.
[c] Based on good management and return of runoff water for surface systems.
[d] Furrow length must be adjusted according to the water intake and infiltration rate.

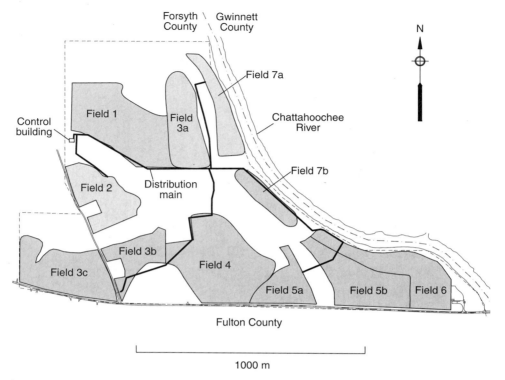

Figure 17-10
Layout of large 73 ha (180 ac) drip irrigation system. The irrigation system includes 610 km (380 mi) of drip line arranged in 7 fields and 17 operating zones, depending on soil and loading characteristics. Each field is divided into two or three operating zones to manage the reclaimed water application rate. (Coordinates: 34.055 N, 84.110 W). (Courtesy of Waste Water Systems, Ellijay, GA.)

Uniformity of irrigation water application is a major factor in determining the irrigation efficiency. Water losses and the uniformity of irrigation are affected not only by the irrigation method used, but also the condition of the irrigation system, soil characteristics, crop type and spacing, irrigation timing and amount of water applied, water management skill, and environmental conditions at the time water is applied (USDA, 1993b). Generally, irrigation efficiency is determined on a monthly basis. With a properly designed and managed irrigation system, the overall irrigation efficiency can be over 80 percent. Typical ranges of irrigation efficiency for various irrigation systems are reported in Table 17-19.

Clogging Prevention When untreated freshwater with high suspended solids is used, a screening filter and/or sand separator are used commonly to remove the solids that are likely to cause emitter clogging. Reclaimed water used for agricultural irrigation is treated typically at secondary or tertiary levels. Suspended solids in treated wastewater are

Figure 17-11
Drip irrigation emitter partially clogged with microbial growth. (Test emitters courtesy of M. Tajrishy.)

mostly biological flocs, and the solids concentration generally is low enough for most irrigation systems. However, reclaimed water irrigation systems with low water velocities, such as drip irrigation systems, are prone to clogging by biological growth and chemical precipitation (see Fig. 17-11). Water quality and clogging potential in drip irrigation systems are reported in Table 17-20. Measures to prevent irrigation system clogging include water quality monitoring, selection of appropriate emitters, control of flowrates, filtration, and maintenance of the irrigation system with periodic chlorination and flushing. Typical filtration devices used for agricultural

Table 17-20
Water quality and clogging potential in drip irrigation systems[a]

Potential problem	Units	Degree of restriction on use		
		None	Slight to moderate	Severe
Physical				
Suspended solids	mg/L	<50	50–100	>100
Chemical				
pH	pH unit	<7.0	7.0–8.0	>8.0
Dissolved solids	mg/L	<500	500–2000	>2000
Manganese	mg/L	<0.1	0.1–1.5	>1.5
Iron	mg/L	<0.1	0.1–1.5	>1.5
Hydrogen sulfide	mg/L	<0.5	0.5–2.0	>2.0
Biological				
Bacterial populations[b]	number/L	<10,000	10,000–50,000	>50,000

[a]Adapted from Pescod (1992); Gilbert et al. (1982); Lazarova et al. (2004).
[b]Heterotrophic plate count.

Figure 17-12

Filtration devices for reclaimed water used for agricultural irrigation: (a) filter used with microspray irrigation system for orange trees, Conserve II, Orange County, FL and (b) filter used with subsurface drip irrigation system, Santa Rosa, CA.

irrigation are shown on Fig. 17-12. Chlorination to a free chlorine residual of 0.5 mg/L at the end of the each irrigation cycle has been reported to be an effective clogging-prevention measure (Cararo et al., 2006).

Leaching Requirements

Salt levels, as described previously, need to be controlled by leaching a portion of the applied water from the plant root zone to prevent problems with salinity, sodicity, and specific ion toxicity. The leaching requirement is defined as the fraction of the water entering the soil that must pass through the root zone to prevent soil salinity from exceeding a specific value. The leaching requirements can be expressed by the leaching fraction (LF) defined as the ratio of the depth of water leached below the root zone to the depth of water applied at the surface, as given by Eq. (17-7).

$$LF = \frac{D_{dw}}{D_{iw}} \qquad (17\text{-}7)$$

where LF = leaching fraction, dimensionless
D_{dw} = depth of water leached out of the root zone (drainage water), mm
D_{iw} = depth of water applied at the surface (irrigation water), mm

At steady-state, the mass of salt in the applied water is equal to the mass of salt in the drained water:

$$D_{dw}EC_{dw} = D_{iw}EC_{iw} \qquad (17\text{-}8)$$

where EC_{dw} = salinity of the drainage water, dS/m
EC_{iw} = salinity of the irrigation water, dS/m.

From Eqs. (17-7) and (17-8), the salinity of the drainage water, EC_{dw}, is calculated as:

$$EC_{dw} = (EC_{iw})/LF \qquad (17\text{-}9)$$

Determination of the LF is illustrated in Example 17-2.

EXAMPLE 17-2. Determination of the Leaching Fraction.

A crop is irrigated with reclaimed water that has a salinity concentration of 1 dS/m. Determine the LF to achieve an average root zone soil-water salinity of 3 dS/m. Assume that the following conditions apply: (1) 40 percent of the total water consumption occurs in the upper quarter of the root zone, (2) 30, 20, and 10 percent of the water consumption occurs in the subsequent quarters, and (3) there is no addition of salt from the soil.

Solution
1. Determine the amount of water consumed at each quarter.

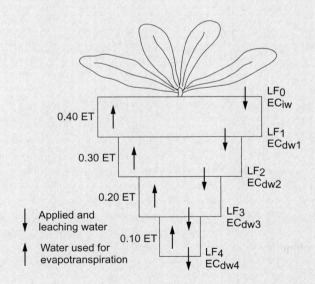

If the LF is x and the applied reclaimed water is A mm/d, the amount of water consumed in each quarter of the root zone and the electrical conductivity of the soil water leaching out from the bottom of each layer are calculated using Eqs. (17-7) and (17-9) as illustrated in the following table.

Layer	Percent ET consumed in each layer	Amount of water consumed	Electrical conductivity of water leached from each layer (EC_{dw}), dS/m	
Soil surface	—	—	1	
First	40	$0.4(1 - x)A$	$1 \times \dfrac{A}{A - 0.4(1 - x)A}$	$= \dfrac{1}{0.6 + 0.4x}$
Second	30	$0.3(1 - x)A$	$1 \times \dfrac{A}{A - (0.4 + 0.3)(1 - x)A}$	$= \dfrac{1}{0.3 + 0.7x}$
Third	20	$0.2(1 - x)A$	$1 \times \dfrac{A}{A - (0.4 + 0.3 + 0.2)(1 - x)A}$	$= \dfrac{1}{0.1 + 0.9x}$
Fourth	10	$0.1(1 - x)A$	$1 \times \dfrac{A}{A - (0.4 + 0.3 + 0.2 + 0.1)(1 - x)A}$	$= \dfrac{1}{x}$

2. Determine the electrical conductivity of the soil water in the root zone.
   a. The average drainage water electrical conductivity in the root zone can be calculated as:

$$\text{Average } EC_{dw} = \dfrac{1 + \dfrac{1}{0.6 + 0.4x} + \dfrac{1}{0.3 + 0.7x} + \dfrac{1}{0.1 + 0.9x} + \dfrac{1}{x}}{5}$$

   b. Using a spreadsheet program, (such as "Solver" in Excel) determine the value of x that corresponds to an average EC_{dw} of 3 dS/m:

$$x = 0.165$$

   Thus, the LF needed to achieve an average EC_{dw} of 3 dS/m is 16.5 percent (0.165×100)

Comment

In some cases, the calculation of LF using the average electrical conductivity may underestimate the effect of salinity on plants, because the salinity at the bottom of the root zone will be higher than the salinity close to the soil surface. The irrigation frequency will also affect the salt buildup between the irrigation periods, especially when the evaporation rate is high and the irrigation water has a high salinity.

The leaching requirements can also be estimated using Fig. 17-12. For example, if reclaimed water containing a salinity level of 1 dS/m is used, and the LF is set to 0.1, the soil salinity (EC_e) will be about 2.1 dS/m and soil water salinity will be about 4.2 (see Table 17-7). At an LF of 0.1, moderately sensitive crops can be irrigated with water containing a salinity level of 1 dS/m. It should be noted that the diagram on Fig. 17-13 was developed for the condition where $EC_e = 1.5 \times EC_w$, assuming a 15 to 20 percent LF and 40–30–20–10 percent water use pattern as described in Example 17-2.

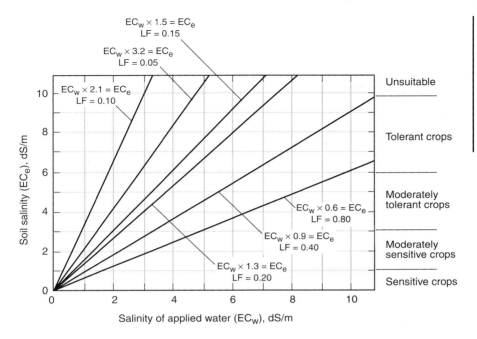

Figure 17-13
Effect of applied water salinity (EC_w) upon root zone soil salinity (EC_e) at various leaching fractions (LF). (From Ayers and Westcot, 1985.)

Estimation of Water Application Rate

Once the crops and the irrigation method(s) are identified, the next step is to estimate the quantity of water to apply. The basic concept for the calculation of irrigation water requirements is the water balance between the applied water plus precipitation and the water loss through evapotranspiration and deep percolation. Major components of the water balance are illustrated on Fig. 17-14. Other factors that affect water requirements include losses through surface runoff, deep percolation, conveyance and distribution;

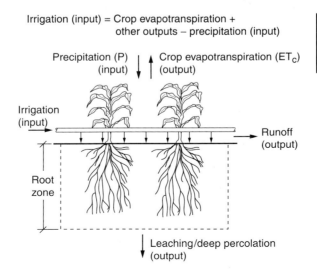

Figure 17-14
Major components of water balance for field irrigation.

uniformity of irrigation; soil characteristics; seed germination; climate control; frost protection; and fertilizer or chemical application. Evapotranspiration (defined below), the net irrigation requirement, the hydraulic loading rate, and the hydraulic loading rate based on nitrogen loading limits, are considered in the following discussion.

Evapotranspiration

The amount of water lost to the atmosphere by evaporation from soil and plant surfaces and by transpiration from plants is called evapotranspiration (ET). By definition, consumptive use of water in irrigation is the sum of ET and the water assimilated into the plant. However, more than 99 percent of applied water is consumed by evapotranspiration (Pescod, 1992) and therefore ET is often assumed equal to the consumptive use. Values of ET depend on climatic conditions, plant ground coverage, type of plants, and soil characteristics. The most precise method to determine ET of a particular site is through direct measurement using devices such as lysimeters. Lysimeters are used to measure the changes in soil water content by irrigation, precipitation, and ET. Lysimeters also can be used to monitor the quality of leaching water below the root zone. Schematic diagrams of three types of lysimeters are shown on Fig. 17-15.

The reference evapotranspiration (ET_o) is the evapotranspiration of a well-irrigated reference surface. It is a parameter determined by the atmosphere at a specific location and time of the year and does not consider the factors affected by the crop and soil. Typically a green grass ground cover is used as the reference surface. In practice, direct measurements of ET are costly and time consuming. Thus, empirical equations are used commonly to estimate the ET_o for the site of interest, and the ET of specific vegetation is estimated from the ET_o value. Data for ET_o are available from various sources including the Food and Agricultural Organization of the United Nations (FAO: national and regional level data), the U.S. Geological Survey (USGS), and state and local government agencies (for example: CIMIS in the State of California). If the data on ET_o are not available for the irrigation site, empirical models can be used to estimate the ET_o values. The Penman-Monteith equation is the most accurate, but is complex. Other methods for estimating ET_o include the radiation method, the temperature method, and the evaporation pan method (Doorenbos and Pruitt, 1977). Details of these methods can be found in references on irrigation such as USDA (1993a).

The values of ET_o can be translated into crop evapotranspiration values, (ET_c) values, using the crop coefficient, K_c, with Eq. (17-10).

$$ET_c = K_c \times ET_o \qquad (17\text{-}10)$$

where ET_c = crop evapotranspiration, mm/unit time
K_c = crop coefficient
ET_o = reference evapotranspiration, mm/unit time

Crop coefficients change with the growth of the plant and development of a plant canopy. Evaporation of water from soil surface is significant in the beginning of a growing

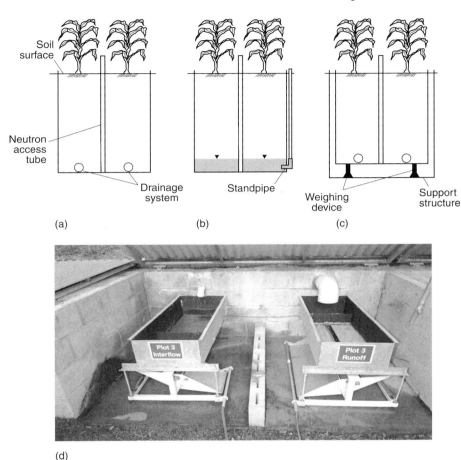

Figure 17-15

Schematic diagrams and view of four types of lysimeters: (a) Nonweighing lysimeter, using a neutron access tube to measure the change in water content within the lysimeter, is designed to prevent deep percolation of irrigated water; (b) the water table lysimeter with neutron access tube, used commonly in humid regions; (c) a weighting lysimeter in which a weighing device is added to the lysimeter diagrams; and (d) tipping-bucket lysimeter used to measure volume of percolate. [Diagrams (a), (b), and (c) adapted from USDA, 1993a.]

season. As the plant develops, transpiration from plant tissue becomes dominant in ET. Crop coefficients for selected crop types are shown in Table 17-21. For precise estimates of crop coefficients, growth stages of the plant and climatic conditions must be considered. A hypothetical crop coefficient curve for typical field and row crops is illustrated on Fig. 17-16. Detailed procedures for the estimation of crop coefficients can be found in irrigation guidelines and handbooks such as USDA (2001) and Snyder et al. (2002).

Table 17-21

Crop coefficient for selected crop types and turfgrass[a]

Crop type or name	K_c values Low[d]	K_c values High[d]
Deciduous orchard[b]	0.50	0.97
Deciduous orchard with cover crop[c]	0.98	1.27
Grape	0.06	0.80
Olive	0.58	0.80
Pistachio	0.04	1.12
Citrus	0.65	0.65
Turfgrass		
Cool season species	0.8	0.8
Warm season species	0.6	0.6

[a] Adapted from University of California and State of California (2000).
[b] Deciduous orchard includes apples, cherries, and walnuts.
[c] When an active cover crop is present, K_c may increase by 25 to 80 percent.
[d] Low values are for early season (March–April) or late season (September–October) and high values for midseason (May–August).

Net Irrigation Requirement

The water requirement calculated based on the simple mass balance of ET, precipitation, and leaching is termed the net irrigation requirement, NR (mm/unit time), and is calculated using Eq. (17-11):

$$NR = (ET_c - P)\left(1 + \frac{LR}{100}\right) \qquad (17\text{-}11)$$

where NR = net irrigation requirement, mm/unit time
P = precipitation, mm/unit time
LR = leaching requirement, percent
Other terms as defined previously.

Figure 17-16

Hypothetical crop coefficient curve for typical field and row crops showing the growth stages and percentages of the season from planting to critical growth dates.

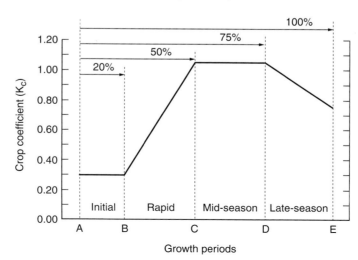

The values of $(ET_c - P)$ are usually calculated on a monthly basis (mm/mo), using the 90 percent exceedance level (i.e., actual values are equal or higher than the given number 90 percent of the time). The 90 percent exceedance level values are used to provide a conservative estimate of the irrigation area. The leaching requirement, as defined previously, is the percent of applied water that percolates below the root zone required to maintain the salt concentration below the level where plants exhibit adverse effects.

Hydraulic Loading Rate

In practice, not all irrigation water is consumed through ET and leaching. Consideration for the nonuniform application of water, water losses in the distribution system, and on the soil surface must be included when calculating hydraulic loading rates. The hydraulic loading rate, $L_{w(1)}$, is most commonly calculated using Eq. (17-12):

$$L_{w(1)} = \frac{NR}{E_i/100} = (ET_c - P) \times \left(1 + \frac{LR}{100}\right) \times \left(\frac{100}{E_i}\right) \quad (17\text{-}12)$$

where $L_{w(1)}$ = irrigation hydraulic loading rate, mm/unit time
  E_i = irrigation efficiency, accounting for surface runoff and wind drift losses, percent
  Other terms as defined previously.

Typical values of the irrigation efficiency for various irrigation systems are shown in Table 17-19. The application of Eq. (17-12) is illustrated in Example 17-3.

When reclaimed water irrigation is considered as the disposal of treated wastewater (land treatment and disposal), available land area may become the limiting constraint. In this case, applied water in excess of the available water capacity of the soil has to percolate under the root zone. The allowable percolation rate, W_p, is the maximum rate at which water can percolate below the root zone. A detailed description of the determination of W_p can be found in the *U.S. EPA Process Design Manual* (U.S. EPA, 1981). The hydraulic loading rate for this condition, $L_{w(2)}$, is calculated using Eq. (17-13a) or (17-13b):

$$L_{w(2)} = (ET_c - P) + W_p + W_r + W_d \quad (17\text{-}13a)$$

or

$$L_{w(2)} = \left(\frac{ET_c}{E_i/100} - P\right) + W_p \quad (17\text{-}13b)$$

where $L_{w(2)}$ = land treatment hydraulic loading rate, mm/unit time
  W_p = allowable percolation rate, mm/unit time
  W_r = water loss from surface runoff, mm/unit time
  W_d = water loss from wind drift, mm/unit time
  Other terms as defined previously.

EXAMPLE 17-3. Calculation of the Hydraulic Loading Rate.

Determine the hydraulic loading rate for an irrigation operation in northern California, subject to the following conditions:

1. Ninety percent exceedance values of ($ET_c - P$) for each month are listed in the following table:

Month	($ET_c - P)_{90}$, mm
Jan	−93.7
Feb	−65.8
Mar	−46.2
Apr	34.0
May	25.9
Jun	120.4
Jul	217.4
Aug	169.7
Sep	52.1
Oct	26.9
Nov	−53.3
Dec	−75.7
Total	311.7

Adapted from Smith et al. (1985).

2. The leaching requirement is 11 percent when irrigation is necessary.
3. No irrigation is necessary when precipitation is greater than ET.
4. Irrigation efficiency is 80 percent throughout the year.

Solution
1. Determine the values of the terms in Eq. (17-12):

$$L_{w(1)} = \frac{NR}{E_i/100} = (ET_c - P) \times \left(1 + \frac{LR}{100}\right) \times \left(\frac{100}{E_i}\right)$$

a. The term $\left(1 + \frac{LR}{100}\right)$ with a leaching requirement of 11 percent is:

$$1 + \frac{LR}{100} = 1 + \frac{11}{100} = 1.11$$

b. The term $\left(\dfrac{100}{E_i}\right)$ with an irrigation efficiency of 80 percent is:

$$\left(\dfrac{100}{E_i}\right) = \dfrac{100}{80} = 1.25$$

2. Determine the hydraulic loading rate for each month using Eq. (17-12). For example, the hydraulic loading rate for June is 120.4 mm × 1.11 × 1.25 = 167.1 mm. The results are summarized in the table below.

Month	$(ET_c - P)_{90}$, mm	$1 + \dfrac{LR}{100}$	$\dfrac{100}{E_i}$	Hydraulic loading rate, $L_{w(1)}$, mm
Jan	−93.7			
Feb	−65.8			
Mar	−46.2			
Apr	34.0	1.11	1.25	47.1
May	25.9	1.11	1.25	35.9
Jun	120.4	1.11	1.25	167.1
Jul	217.4	1.11	1.25	301.6
Aug	169.7	1.11	1.25	235.5
Sep	52.1	1.11	1.25	72.3
Oct	26.9	1.11	1.25	37.3
Nov	−53.3			
Dec	−75.7			

Comment

When using the 90 percent exceedance levels, the estimated hydraulic loading rate meets the water requirement 90 percent of the time. Lower percentage may be used for lower-value crops.

Hydraulic Loading Rate from Nitrogen Loading Limits

Nitrogen in water leaching from the root zone may affect the underlying groundwater or the surface water where drained irrigation water is discharged. The hydraulic loading rate based on the allowable nitrate concentration in percolating water is calculated using Eq. (17-14):

$$L_{w(n)} = \dfrac{(C_p)(P - ET) + 10^2(U)}{(1 - f)(C_n) - C_p} \tag{17-14}$$

where $L_{w(n)}$ = allowable annual hydraulic loading rate based on nitrogen limits, mm/yr
C_p = allowable nitrate concentration in percolating water, mg–N/L
(P − ET) = 90 percentile precipitation minus ET, mm/yr
10^2 = conversion factor, kg/ha·yr to mg·m/L·yr
U = nitrogen uptake by crop, kg N/ha·yr
C_n = nitrogen concentration in applied reclaimed water, mg–N/L
f = fraction of applied nitrogen removed by denitrification and volatilization (0.20 is used typically)

A value 10 mg/L has been used for C_p, based on the U.S. EPA's primary drinking water quality standards (Maximum Contaminant Level, MCL).

The hydraulic loading rate (either $L_{w(1)}$ or $L_{w(2)}$, depending on the limiting constraint) and the allowable hydraulic loading rate calculated from nitrogen loading $L_{w(n)}$ are compared and the smaller value of the two is used for the system design as shown in Example 17-4.

EXAMPLE 17-4. Calculation of the Hydraulic Loading Rate from Nitrogen Loading Limits.

Determine the allowable hydraulic loading rate for an irrigation operation based on the nitrogen loading limit and the following conditions:

1. Ninety percent exceedance values (ET_c − P) for each month are listed in the following table:

Month	$(ET_c - P)_{90}$, mm
Jan	−93.7
Feb	−65.8
Mar	−46.2
Apr	34.0
May	25.9
Jun	120.4
Jul	217.4
Aug	169.7
Sep	52.1
Oct	26.9
Nov	−53.3
Dec	−75.7
Total	311.7

2. Irrigation is not necessary when precipitation is greater than evapotranspiration.

3. Allowable nitrate concentration in percolating water is: C_p = 10 mg–N/L.
4. Nitrogen uptake by crop to be irrigated is: U = 0.03 kg/ha·yr.
5. Nitrogen concentration in applied reclaimed water is: C_n = 15 mg–N/L.
6. Fraction of applied nitrogen removed by denitrification and volatilization is f = 0.20.

Solution

1. Calculate allowable annual hydraulic loading rate using Eq. (17-14):

$$L_{w(n)} = \frac{(C_p)(P - ET) + 10^2(U)}{[(1 - f)(C_n) - C_p]}$$

$$L_{w(n)} = \frac{10 \times 311.7 + 10^2 \times 0.03}{[(1 - 0.20) \times 15 - 10]} = 1560 \text{ mm}$$

2. Compare $L_{w(n)}$ with annual $L_{w(1)}$ calculated in Example 17-3.
   $L_{w(n)}$ = 1560 mm
   $L_{w(1)}$ = 896.8 mm
   $L_{w(n)}$ is greater than $L_{w(1)}$. Therefore, $L_{w(1)}$ is used for design.

Comment

If the nitrogen concentration in the reclaimed water was 20 mg-N/L instead of 15 mg-N/L, the allowable hydraulic loading rate when calculated using the nitrogen limit equation [Eq. (17-14)] would be 520 mm, which is lower than the $L_{w(1)}$ calculated in Example 17-3. Then the value of $L_{w(n)}$ would be used for design and the leaching requirement of 11 percent could not be met. When encountering a situation such as this, one solution that can be used to meet the leaching requirement while avoiding groundwater contamination is to add a nitrification/denitrification process at the water reclamation plant (see Chap. 7). Another solution is to install a subsurface drainage system, as described later in this section. It should be noted that the drainage system will work only when groundwater table is shallow, or there is an impermeable layer right below the root zone.

Field Area Requirements

Based on the design annual hydraulic loading rate calculated above and the average daily available reclaimed water flow, the land area that may be irrigated with reclaimed water, A_w, can be calculated using Eq. (17-15):

$$A_w = \frac{(Q_i)(365 \text{ d/yr}) + \Delta V_s}{L_w \times 10^{-3}} \quad (17\text{-}15)$$

where A_w = irrigated field area, m²
  Q_i = average daily reclaimed water flow (annual average), m³/d
  ΔV_s = net loss or gain in stored reclaimed water volume due to precipitation, evaporation, and seepage at the storage reservoir, m³/yr
  L_w = design annual hydraulic loading rate, mm/yr
  10^{-3} = conversion factor, m/mm

When the area of irrigated land is already fixed and the reclaimed water is to be delivered to meet the design hydraulic loading rate, the average daily reclaimed water flow, Q_i, is determined using Eq. (17-15).

EXAMPLE 17-5. Determination of Field Area Requirements.

Determine the field area needed for reclaimed water irrigation with following conditions:

1. Use the annual hydraulic loading rate as determined in Examples 17-3 and 17-4:

$$L_{w(1)} = 896.8 \text{ mm/yr} = 896.8 \times 10^{-3} \text{ m/yr}$$

2. The average daily flowrate of reclaimed water, Q_i, is 380 m³/d after adjustment for conveyance efficiency.

3. Assume that the net gain/loss in stored reclaimed water is zero.

Solution

1. Calculate the required field area using Eq. (17-15).

$$A_w = \frac{(Q_i)(365 \text{ d/yr}) + \Delta V_s}{L_w \times 10^{-3}}$$

$$A_w = \frac{(380 \text{ m}^3/\text{d})(365 \text{ d/yr}) + 0}{(896.8 \text{ mm/yr})(10^{-3} \text{ m/mm})} = 1.55 \times 10^5 \text{ m}^2 = 15.5 \text{ ha}$$

Comment

For an open storage system, the net gain or loss of reclaimed water is usually not negligible and must be considered in the analysis. Determination of field requirements and storage volume requirements are discussed later in this section.

Drainage Systems

The determination of the leaching requirement and hydraulic loading rate presumes that the irrigated soil can sustain an adequate infiltration rate and the net downward flux of water is maintained. Depending on soil texture, depth to the groundwater table, root depth, and climatic condition, however, the downward movement of water may be inadequate, causing problems with excess water and salinity. To ensure the net downward flux of water, an artificial drainage system may need to be installed on the irrigated land. A drainage system may also be used to prevent excess salts and other pollutants, such as nitrate, from reaching underlying groundwater. However, such a drainage system can be functional only if a layer of low permeability soil exists between the root zone and the groundwater table, and the water percolated through the root zone is built up (Feigin et al., 1991).

Drainage systems can be classified as: (1) surface drainage, (2) subsurface drainage, and (3) interception drainage. Surface drainage is used to remove excess water from the surface of soil with low water permeability. Normally, shallow ditches are used to collect water from the field and the ditches are connected to larger and deeper collector drains. The field is artificially sloped to facilitate the flow of excess water. Subsurface drainage systems are used to remove excess water from the root zone and ensure downward flux of irrigated water. Subsurface drainage systems are either (1) deep open drains or (2) subsurface pipe drains. Deep open drains occupy the land that is otherwise used for growing crops. They also require high maintenance and restrict the use of machines on the field. Thus, in general, pipe drains are preferred. The major factors affecting the design of drainage systems: (1) water table depth, (2) hydraulic properties of the soil, (3) drainage coefficient, and (4) drainage spacing, are discussed briefly below.

Water Table Depth

The depth of the underlying groundwater table is critical to control the upward salt movement and to maintain an aerobic condition in the root zone. When the groundwater is saline, the water table needs to be deep enough to avoid upward movement of the groundwater. When the irrigation interval is relatively long, a water table depth of 1.0 to 2.0 m (3.3 to 6.6 ft) may be required depending on the crop type and soil characteristics. For sprinkler or drip irrigation systems, the watering interval is usually short enough to maintain the downward flux of water. In such cases, aeration of the root zone becomes the limiting factor for determining the water table depth (Feigin et al., 1991). If an area-wide water table exists, existing flow patterns should be evaluated by installing a network of groundwater monitoring devices such as piezometers and observation wells prior to irrigation system design. Recommended minimum depths to the water table for arid areas are shown in Table 17-22.

Hydraulic Properties

Hydraulic conductivity, K, and drainable porosity, p, are the primary parameters to be considered in drainage design. Hydraulic conductivity is the measurement of the ease at which water moves through the soil. Hydraulic conductivity is determined, most commonly, by a field measurement using the auger-hole method or by a laboratory measurement using the undisturbed core sample method. Drainable porosity is defined as the pore volume that drains out of the soil when it is no longer saturated, i.e., the water table is lowered.

Table 17-22
Recommended minimum depth, m, to water table for arid areas[a]

Crop	Fine-textured soil (permeable)	Medium-textured soil	Light-textured soil
Field	0.9	1.2	0.9
Vegetable	0.9	1.1	0.9
Tree	1.4	1.4	1.1

[a]Adapted from Booher (1974).
Note: During fallow periods, the water table should be controlled at a depth of 1.4 m for light- and fine- textured soils and a depth of 1.5 to 1.8 m for medium-textured soils.

Drainage Coefficient

The drainage coefficient is defined as the depth of water the drainage system can remove from the drainage area per unit time; typical units are mm/h. Because the drainage coefficient is site specific, local guidelines and other information should be used to estimate the drainage coefficient. In arid areas, the drain coefficient, q, can be estimated using Eq. (17-16):

$$q = \frac{[(P + C)/100](i)}{24\,F} \qquad (17\text{-}16)$$

where q = drainage coefficient, mm/h
P = deep percolation including leaching requirement, percent
C = field canal loses, percent
i = irrigation application, mm
24 = conversion factor, 24 h/d
F = frequency of application, d

The required drainage coefficient can be estimated from the leaching requirement, water application rate, and irrigation interval as shown in Example 17-6.

EXAMPLE 17-6. Calculation of the Drainage Coefficient.

Calculate the drainage coefficient under the following conditions.
1. The irrigation application is 100 mm.
2. The irrigation interval is two weeks = 14 d.
3. The deep percolation is estimated at 18 percent from consumptive use studies.
4. Field canal loss is estimated at 8 percent of the applied irrigation water.

Solution

1. Using Eq. (17-16), the drainage coefficient is calculated as shown below.

$$q = \frac{[(P + C)/100](i)}{24\,F}$$

$$q = \frac{[(18 + 8)/100]\,(100\text{ mm})}{(24\text{ h/d}) \times (14\text{ d})} = 18\text{ mm/h}$$

Drain Depth and Spacing

The most common subsurface drainage system is comprised of a series of parallel drain pipes located parallel to the groundwater flow underneath the surface of the irrigated field. This type of drainage system is called a relief drain. The capacity handled by the parallel relief drains, Q_r, can be calculated using Eq. (17-17):

$$Q_r = 2.4 \times 10^{-2} \times qS(L + S/2) \qquad (17\text{-}17)$$

where Q_r = relief drain discharge capacity, m³/d
  q = drainage coefficient, mm/h
  S = drain spacing, m
  L = drain length, m
 2.4×10^{-2} = unit conversion factor

The depth and spacing of a drainage system can be calculated using such methods as the Hooghoudt equation and the Donnan formula (Van Schilfgaarde, 1972; USBR, 1993b; USDA, 2001). In the Hooghoudt equation it is assumed that the annual discharge and recharge to the groundwater are about equal. Such a condition is defined as *dynamic equilibrium* (USBR, 1993) and the drain spacing, S, can be calculated using Eq. (17-18).

$$S^2 = \frac{8K_1 d y_o}{q} + \frac{4K_2 y_o^2}{q} \qquad (17\text{-}18)$$

where S = drain spacing, m
  K_1 = hydraulic conductivity above the drain, m/d
  K_2 = hydraulic conductivity below the drain, m/d
  d = the thickness of the "equivalent layer" which takes into account the convergence of flow below the drain, m
  y_o = the water table height above the drain at the beginning of each drain-out period, m
  q = drainage discharge rate, m/d

The Hooghoudt equation is best suited for determining drain spacing in wet regions, but it can be used for semiarid and arid land as well, although the results are not as accurate.

The other commonly used method is the transient flow method developed by the U.S. Bureau of Reclamation. The transient flow method is based on the dynamic equilibrium of water flow in arid areas. Drain spacing can be determined using following parameters and Figs. 17-17 and 17-18:

1. y_o and H: the water table height above the drain, m, at the beginning of each drain-out period. The values y_o and H are for drains above and on the barrier, respectively (see Fig. 17-18). The height is taken at the midpoint between the drains.
2. y and Z: the water table height, m, above the drain at the end of each drain-out period. The values y and Z are for drains above and on the barrier, respectively.
3. Hydraulic conductivity, K_1: the hydraulic conductivity in the flow zone between drains, m/d. The K_1 value is obtained by averaging the result from hydraulic conductivity tests at different locations in the drained area.
4. Specific yield, Y: the amount of groundwater that will drain out of a saturated soil under the force of gravity. The general relationship between hydraulic conductivity and specific yield is shown on Fig. 17-17.

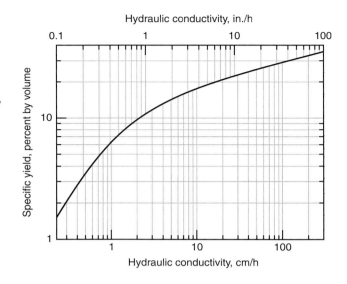

Figure 17-17
General relationship between specific yield and hydraulic conductivity. (From USDA, 2001.)

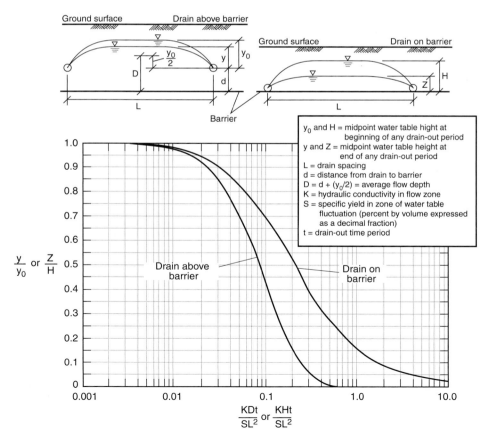

Figure 17-18
Curves showing relationship of parameters needed for drain spacing calculations using transient-flow theory. (From USDA, 2001.)

y_0 and H = midpoint water table hight at beginning of any drain-out period
y and Z = midpoint water table height at end of any drain-out period
L = drain spacing
d = distance from drain to barrier
$D = d + (y_0/2)$ = average flow depth
K = hydraulic conductivity in flow zone
S = specific yield in zone of water table fluctuation (percent by volume expressed as a decimal fraction)
t = drain-out time period

5. Time, t: the drain-out time between irrigations or at specified intervals during the nonirrigation season, day.
6. Flow depth, D: the average flow depth transmitting water to the drain, m. The flow depth is calculated as $D = d + y_o/2$.
7. Drain spacing, S: the distance between parallel drains, m.

From the above parameters, y/y_o or Z/H is calculated and the intersection with the curve on Fig. 17-18 is found. The corresponding value of K_1Dt/YS^2 or K_1Ht/YS^2 is found on the x axis and the drain spacing, S, is derived using other known parameters with successive trials. Further discussion on these types of drainage systems can be found in various guidelines and handbooks such as Tanji and Kielen (2002), USDA (2001), and USBR (1993).

Drainage Water Management and Disposal

Drainage water has a higher salt content than irrigation water. If the salinity of the reclaimed water is low enough to allow irrigation of salt-sensitive crops, the drainage water may be reused for the irrigation of crops with higher salt tolerance. The scheme of irrigating increasingly salt-tolerant crops in a series of irrigated land is an effective way to reuse the drainage water. However, salt accumulated in the drainage water must eventually be disposed in an appropriate manner.

A method or approach to drainage water disposal should be determined in accordance with regulations and guidelines before a reclaimed water irrigation drainage system is designed. Possible methods of drainage water disposal include discharging into the irrigation water conveyance system, streams, or channels; recirculating the water back to the irrigated land; discharging to a marsh; or using evaporation ponds and mechanical evaporation processes. A salt crystallization technology, which has been used in industrial processes more commonly (Bostjancic and Ludlum, 1996), has advanced to the level that it may be used in full-scale operations, making it a viable alternative to disposal of salt-laden water. Even though the high cost of crystallization is still prohibitively expensive for most drainage disposal, crystallization could become a viable option as the problems with brine and high-salt waste become more critical.

Storage System

Irrigation water demand varies significantly with climatic conditions compared to the relatively constant production rate of reclaimed water. To maximize the use of reclaimed water and meet water application requirements throughout the irrigation season, reclaimed water storage is necessary. Storage facilities may be lakes, ponds, or tanks. Confined aquifers may be used for storage and later recovery (ASR) of reclaimed water (Martin et al., 2002; see also Chap. 22). When reclaimed water production exceeds the irrigation rate and the storage capacity, excess reclaimed water may be discharged to receiving waters. It should be noted that the discharge of reclaimed water to surface waters will be subject to NPDES permit requirements.

Reclaimed water storage requirements are reviewed here and a more detailed discussion can be found in Chap. 14. Reclaimed water storage requirements are estimated using the following steps:

1. Calculate the monthly hydraulic loading rate using the evapotranspiration and precipitation data and Eqs. (17-12), (17-13), or (17-14), depending on the site-specific conditions.
2. Determine or estimate the available volume of reclaimed water.

3. Estimate the initial field area using Eq. (17-15), ignoring the net water gain or loss from the reservoir. From the estimated field area and hydraulic loading, calculate the monthly volume requirement.
4. Compute the net change in storage each month by subtracting the required water volume from the available volume.
5. Compute the cumulative storage volume. The computation should start with an empty reservoir at the beginning of the season to store water (i.e., end of the irrigation season).
6. Determine the adjusted field area taking into account the net loss or gain of water from the reservoir due to precipitation, evaporation, and seepage.
7. Determine the adjusted storage volume requirements using Eq. (17-15) by taking into account the net loss or gain of water from the reservoir due to precipitation, evaporation, and seepage.

The application of the above steps is illustrated in Example 17-7.

EXAMPLE 17-7. Estimation of the Volume and Area Requirements for an Open Storage Reservoir.

Determine the volume and area for an open storage reservoir for a reclaimed water irrigation system with the following characteristics:

1. The system design is limited by the amount of available reclaimed water [i.e., use Eq. (17-12) to calculate the hydraulic loading rate].
2. Leaching requirement and irrigation efficiency are estimated at 15 and 85 percent, respectively.
3. The monthly water balance values ET_o, P, $(ET_c - P)_{90}$, and the estimated monthly availability of the reclaimed water, Q_i, are given in the following table (Data from Smith et al., 1985):

Month	ET_o, mm/mo	P, mm/mo	$(ET_c - P)_{90}$[a], mm/mo	$Q_i \times 10^3$, m³/mo
Jan	26.4	98.6	−91.7	93.0
Feb	46.7	70.9	−65.8	90.5
Mar	80.5	49.5	−46.2	116.9
Apr	119.9	38.1	34.0	113.2
May	164.3	13.0	25.9	117.2
Jun	195.1	4.1	120.4	135.9
Jul	207.8	0.25	217.4	135.9
Aug	176.8	0.76	169.7	135.9
Sep	137.9	4.1	52.1	135.9
Oct	93.0	26.4	26.9	118.9
Nov	41.9	51.8	−53.3	90.5
Dec	24.9	81.5	−75.7	94.0

[a] 90% exceedance levels estimated based on historical data. See also Example 17-3.

4. Assume a storage reservoir depth, d_s, of 3.7 m and seepage is negligible.

Solution

1. Determine the monthly hydraulic loading rate, $L_{w(1)}$, using Eq. (17-12).
   For months when the value of $(ET_c - P)$ is negative, $L_{w(1)}$ is assigned a value of zero. The required computation for April is given below.

   $$L_{w(1)} = (ET_c - P) \times \left(1 + \frac{LR}{100}\right) \times \left(\frac{100}{E_i}\right)$$

   $$= (34 \text{ mm/mo}) \times \left(1 + \frac{15}{100}\right) \times \left(\frac{100}{85}\right) = 46 \text{ mm/mo}$$

   The computed hydraulic loading rates are given in column 3 in the following table:

Month (1)	$(ET_c - P)_{90}$, mm/mo (2)	$L_{w(1)}$, mm/mo (3)	$Q_i \times 10^3$, m³/mo (4)
Jan	−91.7	0	93.0
Feb	−65.8	0	90.5
Mar	−46.2	0	116.9
Apr	34.0	46.0	113.2
May	25.9	35.0	117.2
Jun	120.4	162.9	135.9
Jul	217.4	294.1	135.9
Aug	169.7	229.5	135.9
Sep	52.1	70.5	135.9
Oct	26.9	36.4	118.9
Nov	−53.3	0	90.5
Dec	−75.7	0	94.0
Annual		874.5	1377.8

2. Determine the available volume of reclaimed water available for irrigation. A summation of the monthly reclaimed water volumes, Q_i, that are given in the problem statement is given in column (4) in the table presented in Step 1.

3. Estimate the required field area and the monthly volume requirement.
   a. From the annual hydraulic loading rate and annual available volume of reclaimed water, the field area is estimated using Eq. (17-15). For this first estimate, the term for net loss or gain in the reservoir, ΔV_s, due to precipitation, evaporation, and seepage is neglected.

   $$A_w = \frac{(Q) + \Delta V_s}{(L_w \times 10^{-3})} = \frac{(1377.8 \times 10^3 \text{ m}^3 + 0)}{(874.5 \text{ mm})(10^{-3} \text{ m/mm})}$$

   $$= 1.576 \times 10^6 \text{ m}^2 = 157.6 \text{ ha}$$

b. Determine the volume of water irrigated, V_w, for each month. For the month of April, V_w is calculated as:

$$V_w = (A_w)[L_{w(1)}] = (1.576 \times 10^6 \text{ m}^2) \cdot [46 \text{ mm } (10^{-3} \text{ m/mm})] = 72.5 \times 10^3 \text{ m}^3$$

The values of V_w for each month are presented in column (2) of the following table:

Month (1)	Irrigation volume, V_w, $\times 10^3$ m³ (2)	Available reclaimed water, Q_i, $\times 10^3$ m³/mo (3)	Change in storage, ΔS, $\times 10^3$ m³/mo (4)	Cumulative storage, $\Sigma \Delta S$, $\times 10^3$ m³ (5)
Sep	111.1	135.9	24.8	0
Oct	57.3	118.9	61.6	24.8
Nov	0	90.5	90.5	86.4
Dec	0	94.0	94.0	176.9
Jan	0	93.0	93.0	270.9
Feb	0	90.5	90.5	363.9
Mar	0	116.9	116.9	454.4
Apr	72.5	113.2	40.7	571.3
May	55.2	117.2	62.0	612.0
Jun	256.6	135.9	−120.7	**674.0**[a]
Jul	463.4	135.9	−327.5	553.3
Aug	361.7	135.9	−225.8	225.8
Annual		1377.8		

[a] Storage volume requirement, $V_{s(est)}$.

4. Estimate the net change in storage volume, ΔS. The net change is obtained by subtracting V_w (column 2) from Q_i (column 3). The calculated values are shown in column (4) in the computation table prepared in Step. 3. Note that the starting month in the table is rearranged such that the month when ΔS turned from negative to positive is listed first.

5. Calculate the cumulative storage volume and the corresponding surface area of the storage reservoir. In this example, the cumulative volume in September is set to zero, which represents an empty reservoir at the end of the irrigation season.
    a. The cumulative storage volumes are shown in column 5 in the computation table prepared in Step. 3. The maximum value of the cumulative storage volume is the estimated storage volume requirement, $V_{s(est)}$, for the reclaimed water irrigation system. In this example, $V_{s(est)} = 674.0 \times 10^3$ m³.
    b. Determine surface area of the storage reservoir, A_s. Using the assumed storage reservoir depth, d_s, of 3.7 m, the surface area for the storage reservoir can be calculated as:

$$A_s = \frac{V_{s(est)}}{d_s} = \frac{674.0 \times 10^3 \text{ m}^3}{3.7 \text{ m}} = 165.4 \times 10^3 \text{ m}^2 = 16.5 \text{ ha}$$

6. Determine the adjusted field area by considering the net monthly gain or loss in storage.

   a. Using the required surface area, determined in Step 5, the net gain or loss in storage volume, ΔV_s, due to precipitation (P), evaporation (ET_{res}), and seepage can be estimated. The value for ET_{res} can be estimated using the ET_o data multiplied by the crop coefficient for a free water surface (generally 1.05 to 1.15). In this example, ET_{res} was computed as $ET_{res} = 1.1 \times ET_o$. The monthly gain or loss from the storage reservoir for April can be computed as follows:

$$\Delta V_s = (P - ET_{res} - \text{seepage})(10^{-3} \text{ m/mm})(A_s)$$
$$= [38.1 \text{ mm/mo} - 1.1 \times 119.9 \text{ mm/mo}) - 0](10^{-3} \text{ m/mm})(165.4 \times 10^3 \text{ m}^2)$$
$$= -15.5 \times 10^3 \text{ m}^3/\text{mo}$$

The computed values of ΔV_s for each month are shown in column (3) of the following table.

Month (1)	ET_{res}, mm/mo (2)	Net gain or loss in storage, $\Delta V_s \times 10^3$ m³/mo (3)	Adjusted volume, $V_w \times 10^3$ m³/mo (4)	Change in storage, $\Delta S \times 10^3$ m³/mo (5)	Cumulative storage, $\Sigma \Delta S \times 10^3$ m³/mo (6)
Sep	151.7	−24.4	97.6	−13.9	−8.0[a]
Oct	102.3	−12.5	50.4	56.0	13.9
Nov	46.1	0.9	0	91.4	69.8
Dec	27.4	9.0	0	95.0	161.3
Jan	29.1	11.5	0	104.5	256.3
Feb	51.4	3.2	0	93.7	360.8
Mar	88.6	−6.4	0	110.4	454.5
Apr	131.9	−15.5	63.7	34.0	564.9
May	180.8	−27.8	48.5	40.9	598.9
Jun	214.6	−34.8	225.6	−124.5	**639.8**[b]
Jul	228.5	−37.8	407.3	−309.2	515.3
Aug	194.5	−32.0	317.9	−214.1	206.1
Annual		−166.7	1211.1		

[a]Error due to adjustment. Assume zero.
[b]Maximum design storage volume.

   b. Determine the adjusted field area required. Using the values of net gain or loss of water, the adjusted storage area requirement can be estimated using Eq. (17-15). The adjusted field area, A'_w is:

$$A'_w = \frac{Q + \Delta V_s}{L_{w(1)} \times 10^{-3}} = \frac{(1377.8 \times 10^3 \text{ m}^3) - (166.7 \times 10^3 \text{ m}^3)}{(874.5 \text{ mm}) \times (10^{-3} \text{ m/mm})}$$
$$= 1.385 \times 10^6 \text{ m}^2 = 138.5 \text{ ha}$$

Based on the above calculation, the final area required for irrigation is reduced from 157.6 ha to 138.5 ha.

7. Determine the adjusted storage volume required.
   a. The adjusted volume of water irrigated each month, V_w, is calculated using the irrigation area determined in Step 6b. The computation for April is as follows:

   $$V_w = (A_w)[L_{w(1)}] = (1.385 \times 10^6 \text{ m}^2) \cdot [46 \text{ mm } (10^{-3} \text{ m/mm})] = 63.7 \times 10^3 \text{ m}^3$$

   The values of V_w for each month are presented in column (4) of the table presented in Step 6.

   b. The changes in storage, ΔS is a calculated as $Q_i + \Delta V_s - V_w$. The computation for April is as follows:

   $$\Delta S = Q_i + \Delta V_s - V_w = 113.2 \text{ m}^3 + (-15.5 \times 10^3 \text{ m}^3) - 63.7 \text{ m}^3 = 34 \text{ m}^3$$

   The values of ΔS for each month are shown in column (5) of the table presented in Step 6.

   c. The final design storage volume is determined to be $639.8 \times 10^3 \text{ m}^3$ by summing the ΔS values for each month, as shown on column (6) of the table presented in Step 6.

Comment

Theoretically, the adjusted field area, A'_w, could be used to recalculate the irrigation water volume to further refine the design volume. However, the computed values are a conservative estimate, and an additional adjustment of irrigation area and storage volume is usually not necessary.

Irrigation Scheduling

The amount of water applied during each irrigation period, and the timing and frequency of irrigation are determined by the ability of the soil to hold water in the root zone, the allowed water deficit in the root zone between irrigation periods, and the ET of the irrigated area. Because irrigation scheduling is site-specific, involving consideration of a number of complex issues, a detailed discussion is beyond the scope of this textbook. Information relevant to irrigation scheduling may be found in guidance manuals and handbooks such as Pettygrove and Asano (1985), U.S. EPA (1981), and USDA (1997).

17-4 OPERATION AND MAINTENANCE OF RECLAIMED WATER IRRIGATION SYSTEMS

An operation plan for the irrigation systems should be prepared by the designer at the time the construction plans and documents are prepared. A list of information to be included in the operation plan is presented in Table 17-23. Operational issues discussed in this section include management of (1) demand and supply, (2) nutrients, (3) crop and soil, and (4) public health protection. Maintenance issues including monitoring and irrigation system maintenance are also discussed.

Table 17-23
Information to be included in an operation plan[a]

1. A map of the irrigation area showing the following information:
   a. Field or plot numbers, area, and crop
   b. Irrigation system layout and controls
   c. Drainage system layout and controls
   d. Other pertinent information
2. Soil profile information:
   a. Textural changes with depth
   b. Available water capacity
   c. Management-allowed deficiency before irrigation is scheduled
3. Crop information:
   a. How to establish the crop
   b. Crop rotations if necessary
   c. Rooting depth
   d. Critical growth periods
4. Irrigation water to be used:
   a. Source (reclaimed water or blend of reclaimed and fresh water)
   b. Irrigation water quality constituents
   c. Flowrates and time available for irrigation
   d. Operating pressure
   e. Control of flowrate or pressure
5. Schedule irrigation periods
6. Procedure to stop irrigating
7. Determining the number of fields to be irrigated at the same time
8. The order of fields to be irrigated
9. Operating sequence for starting the irrigation system
10. Operating sequence for stopping the irrigation system
11. Safety checks
12. Maintenance procedures and frequency
13. Monitoring schedule required by regulatory agencies and/or for crop management
14. As-built plans of the system (prepared after construction and added to the operation plan)

[a]Adapted from Smith et al. (1985).

Demand-Supply Management

Blending reclaimed water with water from other sources is a common practice where reclaimed water production during peak irrigation period is less than irrigation demand. Blending of different waters is beneficial for (1) increasing irrigation water supply reliability, and (2) improving irrigation water quality. Water from multiple sources can be blended in the reclaimed water storage facility or added to the irrigation system through the use of approved connections, such as air-gap separation.

Nutrient Management

The rate of fertilization varies greatly with crop type and local conditions. As an example, common fertilization rates in California for several crops are shown in Table 17-24. By appropriate irrigation and fertilization management, the amount of fertilizer applied

Table 17-24

Typical fertilization rates in California[a]

Crop or crop category	Common application rate, kg/ha		
	Nitrogen (N)	Phosphorous (P)	Potassium (K)
Citrus and subtropical	137	110	78
Field crops	124	58	116
Fruits and nuts	141	78	253
Pasture	62	35	14
Turf	523	124	247
Vegetables[b]	50–300	50–200	0–300
Grapes	54	27	126

[a]Adapted from Rauschkolb and Mikkelsen (1978).
[b]Application rate is dependent upon the vegetable type.

when irrigating with reclaimed water can be reduced significantly, or even eliminated in some cases. It should be noted that crops may not exhibit an observable difference in response to nutrients in reclaimed water during the first year of irrigation with reclaimed water, especially tree crops (Edraki et al., 2004).

Nitrogen

Nitrogen will be beneficial in the early stages of crop growth, but it is much less beneficial toward maturity. In some cases, however, application of nitrogen in the maturity stage causes excessive vegetative growth, delay in maturity, or reduction in crop quality (Ayers and Westcot, 1985). Generally, a concentration in the reclaimed water below 5 mg N/L will have little or no adverse effect on crops. Sensitive crops may exhibit adverse effects such as reduced yield, late maturing, or poor crop quality, above 5 mg N/L, and above 30 mg N/L for most other crops (see Table 17-5). Excessive application of nitrogen to pastures may result in accumulation of nitrogen in forage and cause adverse health effects on ruminant mammals (Ayers and Westcot, 1985).

Blending reclaimed water with water containing a lower level of nitrogen or changing the water source during the later stage of growth may be helpful for the sensitive crops. Use of reclaimed water for crops that are less sensitive to nitrogen concentrations throughout the growth stages is another option. Many water reclamation plants constructed recently have incorporated nitrification and denitrification to comply with the waste discharge permits (i.e., NPDES), and nitrogen concentrations in the reclaimed water are often less than 5 mg/L.

Phosphorous

If the concentration of phosphorous in reclaimed water is 5 mg/L, 1.0 m of irrigation with reclaimed water per season will provide 50 kg-P/ha of phosphorous to the irrigated land. Depending on the types of the irrigated crops, the amount of phosphorous removed with the harvested crop can be less than what is added by reclaimed water irrigation (see Table 17-24). Excess phosphorous will therefore be accumulated in the soil. Even though accumulated phosphorous may not cause immediate adverse effects for most crops, some plant species are known to be sensitive to high phosphorous concentrations.

Public Health Protection

Exposure to reclaimed water is controlled through regulations and guidelines to minimize potential health risk to the public and farmers. The requirements imposed at the site of reclaimed water irrigation include irrigation methods, setback distance from the irrigated land, and the timing of irrigation. Generally the requirements are specified according to the crop to be irrigated and the method of irrigation. It is important to note that water reuse for agricultural irrigation has been practiced widely in the United States, and other countries. No evidence has been reported that irrigation with reclaimed water has caused adverse health effects in 30 to 40 yr of experience.

Setback Distance from Irrigated Land
Requirements for the setback distance from the area irrigated by reclaimed water are specified in many state regulations and guidelines for the management of public exposure to reclaimed water. Setback distance requirements of selected states are given in Table 17-25.

Potential Health Effect of Trace Organic Compounds
Adverse effects of trace organic compounds through nonpotable water reuse applications are considered to be minimal because water containing these chemicals will not be ingested by humans. To date, limited information is available on the uptake of refractory trace organic contaminants by food crops via reclaimed water irrigation, or on associated human health effects from consumption of crops irrigated with reclaimed water. Various impacts of trace organics have been reported in several studies, but little or no controlled experiments have been conducted.

Effects of Reclaimed Water Irrigation on Soils and Crops

Short-term issues with the use of reclaimed water have been studied extensively and design considerations of reclaimed water irrigation systems are well established. Control of salt is the primary issue in agronomic requirements, whereas public health protection is the basis for the regulations and guidelines of reclaimed water use. Long-term effects of reclaimed water irrigation have not been studied as extensively as the short-term effects, but the effects are not considered to be significantly different from those with conventional irrigation waters.

Source Water Considerations
The quality of source water and reclaimed water in southern California is shown in Table 17-26. As reported, both the salinity and sodicity of the reclaimed water expressed in term of TDS and SAR_{adj}, are both higher than the three major potable water sources. Although the salinity and sodicity of the reclaimed water are within acceptable ranges for most crops and plants, long-term salt accumulation in the root zone and leaching to groundwater should be monitored to ensure sustainability of the water reuse system.

In Israel, where about 70 percent of wastewater is reused for irrigation, the accumulation of salt is becoming a critical issue. Salinity is managed by strict source control, such as prohibition of brine discharge to the wastewater system, changes in water softening agents and detergents, and strict discharge requirements for industries (Weber and Juanico, 2004). Desalination of high salinity wastewater has also been proposed (Rebhum, 2004) to reduce salt in Israel's water recycling system.

Table 17-25
Setback distance requirements for controlling human exposure to reclaimed water in food crop irrigation[a]

State	Secondary effluent without disinfection			Secondary effluent with disinfection			Tertiary or higher treatment with disinfection			Restrictions to irrigation methods and timing
	From domestic water supply well	From impoundments to domestic water well	From public access area	From domestic water supply well	From impoundments to domestic water well	From public access area	From domestic water supply well	From impoundments to domestic water well	From public access area	
California	NA[b]	NA	NA	30 (100)[c]	45 (150)[c]	30 (100)[c]	15 (50)	30 (100)	NS[d]	Irrigation methods specified depending on reclaimed water quality and irrigated crops.
Florida	NA	NA	NA	NA	NA	NA	23 (75)[e]	61 (200)	Low trajectory nozzles required within 30 m (100 ft)	Irrigation methods specified depending on reclaimed water quality and irrigated crops.

Hawaii	45 (150)	305 (1000)	NS	30 (100)	91 (300)	152 (500)	15 (50)	30 (100)	NS	Irrigation methods specified depending on reclaimed water quality and irrigated crops. Irrigation timing specified.
New Jersey	NA	NA	NA	NA	NA	NA	23 (75)[e]	NS	30 (100)	Irrigation methods specified depending on irrigated crops.
Washington	NA	NA	NA	30 (100)		15 (50)	15 (50)	NS	NS	Effluent quality requirements for processed food determined on a case-by-case basis.

[a] Adapted from U.S. EPA (2004).
[b] NA = not allowed.
[c] For spray irrigation.
[d] NS = not specified.
[e] From irrigation site and reclaimed water transmission facility to public water supply well.

Table 17-26
Water quality data from a water district in southern California

Constituent	Unit	Source A (Surface water)[a]		Source B (Groundwater)		Source C (Groundwater)		Reclaimed water	
		Range	Average	Range	Average	Range	Average	Range	Average
pH	—	8.9–8.2	8.2	6.8–9.0	8.0	7.7–7.8	7.8	6.5–6.8	6.6
Sodium	mg/L	55–87	68	34–141	57	97–130	116	116–142	129
Calcium	mg/L	24–56	37	2.8–65	31	42–100	75	37–68	49
Magnesium	mg/L	12–23.5	17.5	ND–14	6.2	11–32	19	11–26	18
Chloride	mg/L	67–105	81	12–35	19	51–71	62	102–183	137
Sulfate	mg/L	41–177	109	4.7–131	47	110–400	267	110–248	163
Alkalinity	mg/L as CaCO$_3$	73–112	89	109–269	153	179–193	186	101–150	127
TDS	mg/L	278–528	384	188–432	267	450–850	670	566–812	680
SAR$_{adj}$	—		2.2		2.7		3.7		4.2
% of total supply	%		45		50		5		

[a] 26–100% from State Water Project (water from northern California), 0–74% Colorado River water.
ND = not detected.

Monitoring Requirements

Reclaimed water quality should be monitored to ensure public heath protection and healthy plant growth. Generally, 24-hr composite or grab samples are taken to monitor various water quality parameters. Sampling frequency for the different water quality parameters and the list of the parameters vary in state regulations, and many of them are to be identified on a case-by-case basis. The sampling frequency should be decided taking into account health risks associated with the reclaimed water applications, the size of the project, and the population exposed. Typical minimum monitoring requirements and sampling frequency in reclaimed water irrigation systems are shown in Table 17-27. Results should be checked against the numerical limits set for different reuse applications.

Two control points should be considered in the water quality monitoring: (1) the point where reclaimed water leaves the reclamation system (treatment plant plus storage, if the storage is included in the treatment process) and (2) the final point of use. Current regulations and guidelines generally require water quality monitoring at the point where reclaimed water is produced, and the monitoring at the final point of use is conducted most commonly by reclaimed water purveyors on a voluntary basis. An approved laboratory should be used to analyze the samples and the results submitted to the appropriate regulatory agency.

When a potable unconfined aquifer exists below agricultural sites irrigated with reclaimed water, a groundwater monitoring program should be conducted. The monitoring based on a set of wells and piezometers has to be defined on a case-by-case basis depending on the reclaimed water quality and the hydrogeological context.

Table 17-27
Typical minimum monitoring requirements and sampling frequency in water reuse systems for irrigation[a]

Parameters	Raw wastewater and reclaimed water	Receiving soils	Groundwater Shallow aquifers	Groundwater Deep aquifers
Coliforms[b]	Weekly to montly	—	Bi-annual	Annual
Turbidity	On-line for unrestricted irrigation	—	—	—
Chlorine residual	On-line for unrestricted irrigation	—	—	—
Volume	Monthly	—	—	—
Water level	—	—	Bi-annual	—
pH	Monthly	Annual	Bi-annual	Annual
Suspended solids	Monthly	—	—	—
Total dissolved solids	Monthly	—	Bi-annual	Annual
Electrical conductivity	Monthly	Bi-annual (EC_e)	Bi-annual	Annual
BOD	Monthly	—	—	—
Ammonia	Monthly	—	Bi-annual	Annual
Nitrites	Monthly	—	Bi-annual	Annual
Nitrates	Monthly	Annual	Bi-annual	Annual
Total nitrogen	Monthly	Bi-annual	Bi-annual	Annual
Total phosphorous	Monthly	Bi-annual (extractable P)	Bi-annual	Annual
Phosphates (soluble)	Monthly	Bi-annual	Bi-annual	Annual
Major solutes (Na, Ca, Mg, K, Cl, SO_4, HCO_3, CO_3)	Quarterly			
Exchangeable cations (Na, Ca, Mg, K, Al)	—	Annual	—	—
Trace elements	Annual	—	—	—

[a] Adapted from Lazarova et al. (2004).
[b] Unrestricted irrigation of landscape and food crops may require higher sampling frequency and additional monitoring parameters.

17-5 CASE STUDY: MONTEREY WASTEWATER RECLAMATION STUDY FOR AGRICULTURE—MONTEREY, CALIFORNIA

Monterey Wastewater Reclamation Study for Agriculture (MWRSA) was the first large scale study designed to investigate the risk and effects of irrigation with reclaimed water on food crops that included raw-eaten vegetables. The MWRSA started in 1976 and the final report of MWRSA was published in 1987. The report has been used as a standard for study design of agricultural irrigation with reclaimed water not only within the United States but also in different countries. A brief overview of the MWRSA is described in this section.

Setting

The area around Castroville in Monterey County, California, is a national center for artichoke production. The area also is a major production site for various food crops including broccoli, asparagus, carrots, cauliflower, celery, spinach, several varieties of lettuce, and more recently, strawberries. Agriculture is a major business in Monterey County, generating almost $3 billion/yr as of 2004.

Water Management Issues

Until the 1980s, groundwater was the primary source of irrigation water in Monterey County. Intensive groundwater withdrawal resulted in depletion of groundwater levels that resulted in seawater intrusion, rendering some well water unsuitable for irrigation. Meanwhile, expansion of the wastewater treatment facilities was required because the existing facilities in the region were reaching full capacity. The Water Quality Management Plan by the California Central Coast Regional Water Quality Control Board recommended that water reclamation be proven safe before regional implementation could be considered. This recommendation became the incentive for conducting the 10-yr MWRSA project to assess the safety and feasibility of agricultural irrigation with reclaimed water (Sheikh et al., 1990).

Implementation

The ultimate objective of the MWRSA was to demonstrate the overall feasibility of wastewater reclamation in northern Monterey County. The three primary concerns were:

1. Wastewater constituents
2. Agronomic concerns
3. Feasibility and public acceptance

Various federal, state and local agencies, as well as local farmers, participated in the MWRSA, with the Monterey Regional Water Pollution Control Agency as the leading agency.

Wastewater Constituents

Before Monterey County started the development of water recycling projects in 1980s, the Castroville Wastewater Treatment Plant was the main treatment plant, treating wastewater at a capacity of 1.5×10^3 m³/d (0.4 mgal/d). The treatment plant was modified and upgraded as a field-scale pilot plant with the process specified in the original California Wastewater Reclamation Criteria (Title-22 process), and a less extensive filtration process (FE process):

- Title-22 process: coagulation, flocculation, sedimentation, filtration, and chlorination
- FE process: coagulation, flocculation, filtration, and chlorination

Dechlorination of the final effluent was practiced for the first 3 yr of the study but discontinued thereafter to prevent microbial regrowth and to ascertain the effects of chlorine residual on crops (Sheikh et al., 1990).

The parameters monitored for the study were:

- Inorganic and organic chemical constituents including heavy metals
- Microbial quality including viruses, bacteria and parasites

Groundwater quality data were collected over the 5-yr study period. Microbial quality of aerosols generated by sprinklers was studied in the early stage of the operation.

Agronomic Concerns

The 5-yr field study began in June 1980, after construction of the pilot treatment plant described above. The 12-ha (30-ac) field was divided into two parts: demonstration fields and experimental fields. Three water types (Title-22 water, FE water and well water), and four fertilization rates (no fertilizer, 1/3, 2/3 and 3/3 of full local fertilizer rate) were tested in the study. Three separate irrigation systems consisting of an underground distribution system with portable aluminum pipes were constructed to supply different water types to each experimental plot. The parameters studied were:

- Survival of viruses and coliform bacteria, and occurrence of other selected pathogens, in irrigation water and on vegetables
- Accumulation of metals in soils and plant tissues
- Soil salinity and sodicity, soil permeability
- Crop yield and crop quality

Plot Design and Crop Rotation

A split-plot design (Little and Hills, 1978) was used to assign randomly various water types and fertilization rates to the experimental plots. This experimental design allowed comparison of both irrigation with different water types and the comparison of the effect of varying fertilization rates at the same time. The rates of fertilizer application varied with crop and year, but they were always based on standard practice in the region. The three types of water were applied either by sprinkler or a furrow irrigation system. The plot design is illustrated on Fig. 17-19. Artichokes were grown on the half of the experimental plots according to normal farming practice in the region. Other vegetables including broccoli, cauliflower, lettuce, and celery were grown on the other half of the plots according to a rotation schedule. The crop rotation schedule is shown on Fig. 17-20. Local farming practices were followed throughout the project.

Feasibility and Public Acceptance

Demonstration of feasibility and public acceptance were primary objectives of the study. Two 5-ha (12-ac) plots in the vicinity of the experimental site were dedicated as a demonstration field to investigate large-scale feasibility. Crops in the demonstration fields were grown using reclaimed water with normal local farming practices and the crops were observed for appearance and vigor. Field observation days were held to show the ongoing activities to local growers and news media. Feedback was obtained regarding their perceptions, questions, and concerns.

Study Results

The pilot tertiary treatment facilities were operated nearly continuously during the period of field study. The field studies began in 1980, and they were completed in 1985.

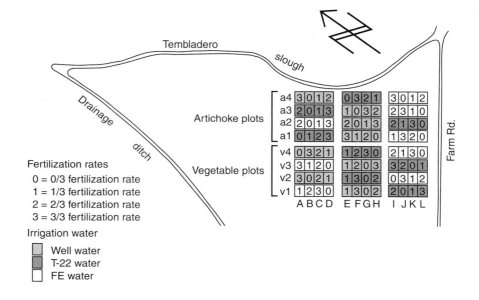

Figure 17-19
Experimental design for test plots. (Adapted from Engineering Science, 1987.)

The farm-scale feasibility study in the demonstration fields was discontinued in years four and five because adequate data were obtained in the first 3 yr. Based on the study results, it was demonstrated that reclaimed water (secondary effluent plus tertiary treatment consisting of coagulation, flocculation, sedimentation, filtration, and disinfection) was safe to use for food crop irrigation. It was also demonstrated that secondary effluent plus filtration and disinfection is sufficient for food crop irrigation (Engineering-Science, 1987; Sheikh et al., 1990; Asano and Levine, 1996; Sheikh et al., 1998).

Wastewater Constituents

The chemical constituents in the irrigation waters used on the experimental plots are shown in Table 17-28. The levels of heavy metals were within the range highly suitable for irrigation water.

Microorganisms lavels in the aerosol from the reclaimed water sprinklers were not significantly different from those in the aerosols from the well water sprinklers. Further, there was no apparent evidence of the application of recycled water in the quality of the shallow shallow groundwater.

Analysis of Results

Analysis of variance (ANOVA) was used to determine the significance of differences in soil and plant characteristics among the plots receiving different water types and fertilization rates. Although the SAR_{adj} values for Title-22 and FE process waters were within the range that could potentially cause problems, the relatively high levels of TDS helped to countract the high SAR values (see Fig. 17-8) such that the waters fell within the favorable range for irrigation (Sheikh et al., 1990). No viruses were found on samples of crops from the experimental plots irrigated with reclaimed water.

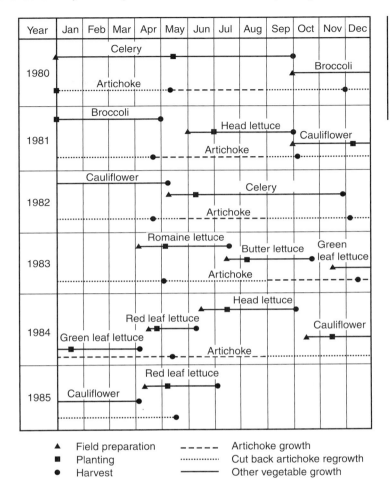

Figure 17-20
Experimental design for crop rotation schedule. (Adapted from Engineering Science, 1987.)

Levels of naturally occurring bacteria were not significantly different between well-water-irrigated crops and reclaimed-water-irrigated crops. The levels of heavy metals in soil were affected by the fertilization rates, but no measurable effect was observed with different water types.

Feasibility and Acceptance

No adverse health effects from exposure of reclaimed water constituents to farmers during conventional farming practices were detected. The quality, yield, appearance, and shelf-life longevity of all the crops irrigated with reclaimed water were equal to or better than those of the crops grown with well water. In the study report, it was concluded that there would be no adverse economic effect. There was no regulatory requirement to label or separate the reclaimed water-grown products, and as long as the products were not labeled, the marketability of the product did not seem to be diminished (Sheikh et al., 1990).

Table 17-28
Chemical constituents of irrigation waters used in the experimental fields[a]

Parameter	Unit	Well water		Tertiary effluent		Filtered effluent	
		Range	Median	Range	Median	Range	Median
pH	unitless	6.9–8.1	7.8	6.6–8.0	7.2	6.8–7.9	7.3
Electrical conductivity	dS/m	400–1344	700	517–2452	1256	484–2650	1400
Calcium	mg/L	18–71	48.0	17–61.1	52.0	21–66.8	53
Magnesium	mg/L	12.6–36	18.8	16.2–40	20.9	13.2–57	22
Sodium	mg/L	29.5–75.3	60.0	77.5–415	166	82.5–526	192
Potassium	mg/L	1.6–5.2	2.8	5.4–26.3	15.2	13–31.2	18
Carbonate	mg/L[b]	0.0–0.0	0.0	0.0–0.0	0.0	0.0–0.0	0.0
Bicarbonate	mg/L[b]	136–316	167	56.1–248	159	129–337	200
Hardness	mg/L[b]	154–246	203	187–416	218	171–435	227
Nitrate	mg/L[c]	0.085–0.64	0.44	0.18–61.55	8.0	0.08–20.6	6.5
Ammonia	mg/L[c]	ND[e]–1.04	—	0.02–30.8	1.2	0.02–32.7	4.3
Total phosphorus	mg/L	ND–0.6	0.02	0.2–6.11	2.7	3.8–14.6	8.0
Chloride	mg/L	52.2–140	104	145.7–841	221	145.7–620	250
Sulfate	mg/L	6.4–55	16.1	30–256	107	55–216.7	84.8
Boron	mg/L	ND–9	0.08	ND–0.81	0.36	0.11–0.9	0.4
Total dissolved solids	mg/L	244–570	413	643–1547	778	611–1621	842
BOD	mg/L	ND–33	1.35	ND–102	13.9	ND–315	19
Adjusted SAR	unitless	1.5–4.2	3.1	3.1–18.7	8.0	3.9–24.5	9.9
MBAS[d]	mg/L	—	—	0.095–0.25	0.14	0.05–0.585	0.15
Cadmium	mg/L	ND–0.1	ND	ND–0.1	ND	ND–0.1	ND
Zinc	mg/L	ND–0.6	0.02	0.07–6.2	0.33	ND–2.08	0.20
Iron	mg/L	ND–0.66	0.1	ND–2.3	0.05	ND–0.25	0.06
Manganese	mg/L	ND–0.07	ND	ND–0.11	0.05	ND–0.11	0.05
Copper	mg/L	ND–0.05	0.02	ND–0.05	ND	ND–0.04	ND
Nickel	mg/L	0.001–0.20	0.04	0.002–0.18	0.04	0.004–0.20	0.04
Cobalt	mg/L	ND–0.057	ND	0.001–0.062	0.002	ND–0.115	0.05
Chromium	mg/L	ND–0.055	ND	ND	ND	ND	ND
Lead	mg/L	ND	ND	ND	ND	0.001–0.70	0.023

[a]Adapted from Sheikh et al. (1990).
[b]As $CaCO_3$.
[c]As N.
[d]Methylene-blue active substance (MBAS).
[e]ND = Chemical concentration below detection limit. Detection limits are as follows: NH_3–N = 0.02 mg/L; P = 0.01 mg/L; B = 0.02 mg/L; BOD = 1 mg/L; and MBAS = 0.05 mg/L; Cd = 0.01–0.1 mg/L; Zn = 0.02–0.5 mg/L; Fe = 0.03 mg/L; Mn = 0.05 mg/L; Cu = 0.001–0.02 mg/L; Co = 0.001–0.1 mg/L; Cr = 0.04–0.2 mg/L; Pb = 0.001–0.2 mg/L.

17-5 Case Study: Monterey Wastewater Reclamation Study for Agriculture—Monterey, California

Table 17-29 Estimated costs of reclaimed water for various tertiary treatment processes[a,b]

Treatment Process	Estimated cost, $/m^3
Filtered effluent	0.05
Filtered effluent with flocculation (FE-F)	0.06
Tertiary with 50 mg/L alum	0.09
Tertiary with 200 mg/L alum	0.13

[a] Assumptions: Plant design flow of 114×10^3 m^3/d; 28×10^6 m^3/d of reclaimed water will be delivered for irrigation; and for FE-F process, estimated capital cost is $11,170,000 and estimated annual O&M cost is $376,000 (in 1990, Engineering News Records Construction Cost Index, ENRCCI = 5200).
[b] Adapted from Sheikh et al. (1990).

Cost of Irrigation with Reclaimed Water

Estimated costs of irrigation with reclaimed water were determined in the MWRSA project. The present worth of 20 yr of operation was estimated for three treatment alternatives: (1) original full treatment including coagulation, flocculation, sedimentation and filtration, (2) filtered effluent, and (3) filtered effluent with flocculation. The estimated costs are shown in Table 17-29.

Subsequent Projects

The Monterey County Water Recycling Project (MCWRP), comprised of the Salinas Valley Reclamation Project (SVRP) and the Castroville Seawater Intrusion Project (CSIP), emerged as a result of the MWRSA (Crites, 2002). The MCWRP involved construction of the Regional Treatment Plant, pumping stations, storage facilities, pipelines and other distribution systems, and environmental mitigation for the water reclamation system. The Regional Treatment Plant began operation in 1988 and treats about 80×10^3 m^3/d (21 Mgal/d) of wastewater with a capacity of about 110×10^3 m^3/d (29.6 Mgal/d) (Monterey County Website). At capacity, approximately 25×10^6 m^3/yr (20,000 ac-ft/yr) of disinfected tertiary recycled water is delivered for irrigation of about 4700 ha (12,000 ac) of food crops (Sheikh et al., 1999).

Recycled Water Food Safety Study

The Recycled Water Food Safety Study was conducted in 1997 to determine if pathogens were present in disinfected tertiary-treated water produced at the Regional Treatment Plant. The efficacy of treatment processes for pathogen removal was also assessed in the study. The hygienic evaluation demonstrated that no *Salmonella, Cyclospora*, or *E. Coli* O157:H7 were detected in tertiary recycled water from the Monterey County Water Recycling Project (Sheikh et al., 1999). The microbial and chemical quality of disinfected tertiary-recycled water is shown in Table 17-30.

Lessons Learned

Currently, reclaimed water quality, including the occurrence of pathogens, is monitored routinely and reported on MRWPCA's website. MRWPCA has worked to reduce salt levels by using more efficient water softeners and replacing sodium chloride with potassium chloride for softener regeneration (Crites, 2002). No harmful effect of salinity has been observed in the sampling program started in 1999.

Table 17-30

Microbial and chemical quality of disinfected tertiary recycled water for the Monterey water reclamation study[a]

Sample	E.Coli O157:H7, CFU/100 mL	Legionella, CFU/100 mL	Salmonella, CFU/100 mL	Giardia, No./L	Crypto-sporidium, No./L	Cyclo-spora, No./L	Fecal Coliform, MPN/100 mL	Turbidity, NTU	Chlorine Residual, mg/L
1	ND[b]	ND	—	—	ND	—	ND	1.9	14
2	—	ND	—	—	—	—	ND	1.7	6.2
3	ND	ND	—	ND	ND	ND	ND	2.7	—
4	ND	ND	ND	0.03	ND	ND	ND	1.2	—
5	ND	ND	—	0.08	ND	ND	ND	2.3	14
6	ND	ND	ND	0.09	ND	ND	ND	1.6	12
7	ND	ND	ND	0.05	ND	ND	ND	1.5	14
Average	ND	ND	ND	0.06	ND	ND	ND	1.8	12
Range	—	—	—	ND–0.09	—	—	—	1.2–2.7	6.2–14

[a] Adapted from Sheikh et al. (1999).
[b] ND = Not detect.

Important lessons learned in the implementation of the MWRSA are:

- The study design for the experimental plots set the standard for the investigation of the effect of reclaimed water on crop irrigation.
- Food crop irrigation with reclaimed water treated with original full treatment required by the *California Wastewater Reclamation Criteria*, as well as the filtered effluent, did not pose any detectable public health hazard in terms of pathogens or heavy metal exposure. The study results led to the modification of the criteria to allow filtered effluent for food crop irrigation.
- The marketability of the product irrigated with reclaimed water did not seem to be diminished.

17-6 CASE STUDY: WATER CONSERV II, FLORIDA

Water Conserv II is the first project in Florida to use reclaimed water to irrigate crops for human consumption. Primary purposes of Water Conserv II were wastewater discharge abatement, agricultural (predominantly citrus) irrigation, and groundwater recharge. Two water reclamation facilities, City of Orlando Water Conserv II Water Reclamation Facility, and Orange County South Regional Water Reclamation Facility, are providing reclaimed water in the project area. Owner agencies for the Water Conserv II are the City of Orlando and Orange County (the City of Orlando is located within the Orange County). A map of the project area is shown on Fig. 17-21. A brief overview of Water Conserv II is presented in this section.

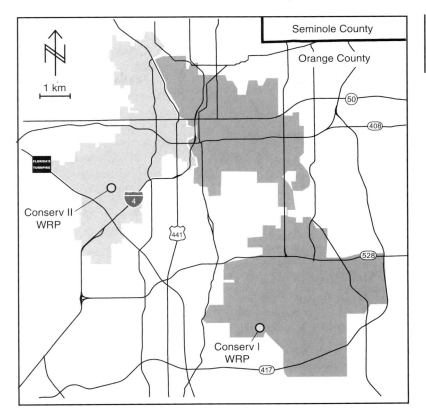

Figure 17-21
Schematic of the Conserv II project area.

Setting

Agriculture in the region is predominantly citrus farming. In 1979, a group called Save Our Lake took legal proceedings against the City of Orlando and Orange County, calling for termination of wastewater effluent discharge from the two wastewater treatment facilities (McLeod Road Wastewater Treatment Facility and Sand Lake Road Wastewater Treatment Facility) into Shingle Creek. The court issued an injunction against the city and county to cease effluent discharge into the creek by 1988 (Cross et al., 2000). To maximize federal funding, the city and county decided to initiate a joint project: the Water Conserv II Water Reclamation Project.

Water Management Issues

After the court decision to cease effluent discharge, the city and county commissioned a federally funded regional wastewater plan, the Southwest Orange County 201 Facilities Plan. The planners investigated different alternatives, and found that a combination of agricultural irrigation and rapid infiltration basins (RIBs) would be the most viable and cost-effective alternative. The recommended alternative was expected to reduce demand on the Floridan aquifer by eliminating the need for well water for irrigation, replenishing the aquifer, and stabilizing area lake levels (City of Orlando, 2006).

Implementation

Construction of facilities continued until late 1986, and operation began in December, 1986. Agriculture and commercial customers use 60 percent of the reclaimed water, and the remaining 40 percent is recharged to groundwater through RIBs (Water Conserv II, 2006).

In 1998, the project supplied reclaimed water to citrus growers, landscape and foliage nurseries, tree farms, landfills (one of which includes a soil cement production facility), an animal shelter, Mid-Florida Citrus Foundation (MFCF, for research on long-term effects), a golf course, and RIBs (Cross et al., 1998).

Water Reclamation Facilities and Transmission Pumping Stations

Water Conserv II utilizes reclaimed water from two reclamation facilities: City of Orlando Water Conserv II Water Reclamation Facility, and Orange County South Regional Water Reclamation Facility. The average flows from the two reclamation plants varies from 114 to 132×10^3 m^3/d (30 to 35 Mgal/d). The permitted annual average daily flow (AADF) at build-out is 260×10^3 m^3/d (68.3 Mgal/d) of which the Permitted Public Access Irrigation is about 180×10^3 m^3/d (46.4 Mgal/d) and the permitted RIB flow is 83×10^3 m^3/d (21.9 Mgal/d).

The Water Conserv II Reclamation Facility has two identical treatment process flow diagrams that consist of screenings and grit removal, primary sedimentation, activated sludge with fine bubble aeration, and secondary clarification. The effluent is filtered and chlorinated prior to being pumped to the distribution center for reuse.

A transmission pumping station was placed at each water reclamation facility. Peak pumping capacity is about 140×10^3 m^3/d (37.5 Mgal/d) per pumping station, with a total capacity of 280×10^3 m^3/d (75 Mgal/d) (Water Conserv II, 2001).

Transmission Pipeline, Distribution Center and Distribution Network

Reclaimed water from two water reclamation facilities is sent to the distribution center and RIB sites by 34 km (21 mi) of transmission pipeline. The transmission pipeline has two surge facilities for surge protection. The distribution center consists of a distribution pumping station, four 3.8×10^3 m^3 (5 Mgal) storage reservoirs, a central control station computer, and the operations and maintenance buildings (see Fig. 17-22). Reclaimed water is then distributed to 76 agricultural and commercial customers, or to the RIB sites through a 79 km (49 mi) pipeline network (Water Conserv II, 2001). The major user of reclaimed water is citrus growers in the service area. Reclaimed water is transmitted through the distribution network and filtered before irrigation (see Fig. 17-12).

Supplemental Water Wells

Groundwater is used in the Water Conserv II project to meet peak demands, notably freeze protection needs of the citrus (York and Wadsworth, 1998). Twenty five supplemental water wells are strategically located on the distribution network to supplement water supply. Peak supplemental water supply capacity is about 212 m^3/min (56,000 gal/min).

Turnouts

A turnout is a point of delivery from city- or county-owned facilities to private customer operations and functions to monitor, record, and regulate flow to the customer. The turnout is kept locked, accessed only by a contract operator. Customers turn on and off according to their needs and have access to the flowmeter for monitoring and record-keeping. Normal system line pressure is 550 to 830 kPa (80 to 120 lb/in^2). Water leaves the distribution center at 380–500 kPa (55–72 lb/in^2) (Water Conserv II, 2001).

17-6 Case Study: Water Conserv II, Florida 1025

Figure 17-22
Views of Conserv II water reclamation facility: (a) signage for Conserv II, (b) control building with radio tower (top cutoff) for control of reuse application facilities, (c) central pump station at distribution center, (d) reclaimed water storage tanks (Coordinates: 28.473 N, 81.647 W), (e) typical rapid infiltration basin (Coordinates: 28.493 N, 81.620 W), and (f) orange trees irrigated with reclaimed water (Coordinates: 28.474 N, 81.658 W).

Rapid Infiltration Basins (RIBs)

During the investigation of alternatives, land about 16 to 24 km (10 to 15 mi) west of McLeod Road and Sand Lake Road facilities was identified as an appropriate site for the RIBs because of conductive geological features. As of 2004, there were seven RIB sites in Orange County (see Fig. 17-22). The total number of RIBs is 66, and each RIB has 1 to 5 cells (total 135 cells). The total RIB bottom percolation area is approximately 79 ha (195 ac), and the capacity infiltration rate is 83×10^3 m^3/d (21.9 Mgal/d). Total site area is about 1170 ha (2900 ac) (Water Conserv II, 2001). A computerized management system, called Groundwater Operational Control System (GOCS) is used to control the flow of the RISBs. The system provides the capability to forecast the impact of RIBs on the regional groundwater system.

Public Acceptance

Initially, the project encountered strong resistance from citrus growers and residents (Cross et al., 2000). Growers were not convinced of the benefits of using reclaimed water for irrigation. Residents mounted opposition by joining forces with the NIMBY (Not in My Back Yard) group (Cross et al., 1998).

The citrus growers accepted the project after the city and county provided research data by R. C. J. Koo, a leading authority on citrus irrigation at the University of Florida's Lake Alfred Citrus Research and Education Center, on the effects of reclaimed water on citrus production and fruit quality. The city and county also agreed to provide funding for research on the long-term effects of the irrigation with reclaimed water. The city and county provided two incentives: (1) reclaimed water would be provided to growers free for the first 20 yr at pressures suitable for microsprinkler irrigation, and (2) water would be provided for enhanced cold protection (Cross et al., 1998).

The area residents accepted the project cautiously after the city and county provided assurances to address and be sensitive to concerns of the residents. The concerns focused on the safety, health, and welfare of the residents and the need to minimize potential adverse environmental impacts.

Study of Long-Term Effects of Irrigation with Reclaimed Water

Mid-Florida Citrus Foundation (MFCF) is a nonprofit organization which conducts research on long-term effects of irrigating citrus with reclaimed water. The MFCF has also conducted research on the use of reclaimed water for other purposes, including different crops and golf course irrigation.

New Options for Reclaimed Water Uses

The city and county realized the importance of diversification of their customer base (Cross et al., 1998). A golf course was constructed as an alternative user of reclaimed water (see Fig. 17-23). Various crops are being investigated for suitability of irrigation with reclaimed water. Residential and commercial development in western Orange County seems inevitable. A new development of the "village" land use was adopted in 1995. As of 2000, the construction was expected to start soon. The village will use

Figure 17-23
Golf course irrigated with reclaimed water in Conserv II, Orange County, FL: (a) view of golf course looking toward club house (coordinates: 28.442 N, 81.626 W) and (b) rapid infiltration basin located at golf course.

reclaimed water for landscape irrigation as well as commercial and light industrial uses (Cross et al., 2000).

The importance of Water Conserv II project for the City of Orlando and Orange County is as follows:

- The discharge of wastewater effluent to surface waters has been eliminated.
- The RIB sites have provided a preserve for endangered and threatened species for plants and animals, as officially cited by city and county decree.
- The Floridan aquifer has been replenished through the discharge of reclaimed water to the RIBs. The demand on the aquifer has also been reduced by eliminating the need for well water for irrigation.
- Reclaimed water use applications have been expanded successfully to meet the zero-discharge requirement.

Importance of Water Conserv II

Important lessons learned in the implementation of Water Conserv II are as follows:

- Extensive scientific studies to demonstrate safety and benefits of reclaimed water irrigation were needed to gain growers' acceptance.
- Distribution of reclaimed water for agricultural customers requiring freeze protection water was not feasible because of the high peak flows needed for freeze protection and the high cost of operation and maintenance.
- Systematic upgrading and expansion of the project, including purchasing of additional RIB sites, was necessary to handle increasing population and development.

Lessons Learned

17-7 CASE STUDY: THE VIRGINIA PIPELINE SCHEME, SOUTH AUSTRALIA—SEASONAL ASR OF RECLAIMED WATER FOR IRRIGATION

The Virginia Pipeline Scheme is a large-scale water reuse project that utilizes seasonal aquifer storage and recovery. In this case study, the use of aquifer storage and recovery for agricultural irrigation and the study on water quality changes in aquifer are described.

Setting

Located in Adelaide, South Australia, the Bolivar Wastewater Treatment Plant (WWTP) has historically discharged 40×10^6 m³/yr (29 Mgal/d) of secondary effluent into the sensitive waters of the Gulf of St. Vincent. The location of the plant relative to the City of Adelaide and the agricultural hub of the Virginia Triangle are shown in Fig. 17-24. In this dry agricultural coastal region (rainfall 600 mm/yr, evaporation 2000 mm/yr), water availability is a limiting factor for crop production, and groundwater resources have been overdrawn for irrigation needs (Kracman et al., 2001). Consistent with a South Australia policy issued in 1993 to encourage sustainable water reuse, and the 1995 Environmental Protection Act further promoting and regulating water reuse, the City of Adelaide considered reclamation and reuse of the Bolivar WWTP effluent to satisfy some seasonal irrigation demands, and to reduce adverse ecological effects caused by nutrients discharged in the marine environment.

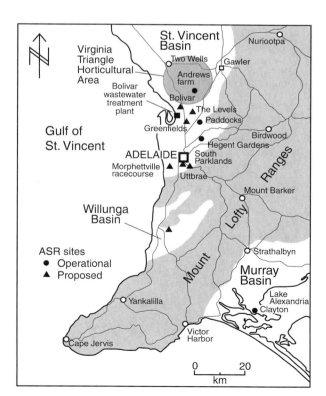

Figure 17-24

Virginia Triangle Horticultural Area and aquifer storage and recovery sites. (Adapted from Barnett et al., 2000.)

Beside aiming at providing reclaimed water for agricultural use during peak demand in the summer time, and minimizing year-round nutrient loads to Gulf of St. Vincent, the water reuse project also presented an opportunity for generating economic benefits in the region, using taxpayer funds to both improve coastal water quality and promote agricultural production, rather than simply building a nonrevenue-generating nutrient removal upgrade for the Bolivar plant. To maximize the economic goals, planners determined that reclaimed water should be stored during the low demand season, thus increasing availability of reclaimed water for the summertime peak irrigation season. **Water Management Issues**

Due to concerns over vast land requirements, recontamination risks, evaporative losses, and waterlogging of surrounding land, surface storage was ruled out (Barnett et al., 2000). Aquifer Storage and Recovery (ASR) had been used recently with success in the region for drinking water applications, and was suggested as the preferred method for reclaimed water storage. The specific constraints posed by reclaimed water ASR justified the launch of a dedicated research program to assess water quality and treatment requirements, to satisfy both public health and irrigation requirements, as well as sustainable long-term ASR wellfield operation.

The existing treatment process at the Bolivar WWTP included primary sedimentation, secondary treatment using biological trickling filters and stabilization lagoons. The effluent was discharged to the marine environment.

Toward safe use of reclaimed water for unrestricted irrigation, South Australian regulations imposed: **Regulatory Requirements**

- Turbidity less than 10 NTU (mean), 15 NTU (max)
- Fecal coliforms less than 10 FCU/100 mL (median)
- Pathogens, less than 1/50L (objective zero)

The additional treatment steps needed to achieve this improved effluent quality would also have to improve the effluent to minimize physical, chemical, and biological processes from occurring in the aquifer and in the injection and recovery wells. A consortium of several governmental and private entities undertook a 3 yr research project to determine the technical feasibility, environmental sustainability, and economic viability of ASR. Research confirmed the viability of dissolved air flotation and filtration (DAF/F) followed by disinfection, as most effective method for polishing the lagoon effluent. The recommended process flow diagram is illustrated on Fig. 17-25.

A 120×10^3 m^3/d (31.7 Mgal/d) DAF/F facility was constructed at the Bolivar site, along with a disinfection contact tank, balancing storage reservoir, and finished water pumping station. As shown on Fig. 17-25, coagulant is added to the algae laden lagoon effluent, prior to dissolved air flotation. The treated effluent from the flotation process is then passed through a granular polishing filter, before undergoing chlorination and operational storage. **Technology Issues**

Seasonal storage of reclaimed water during winter months is achieved by injection into brackish limestone aquifers within the Port Willunga Formation. Because the background

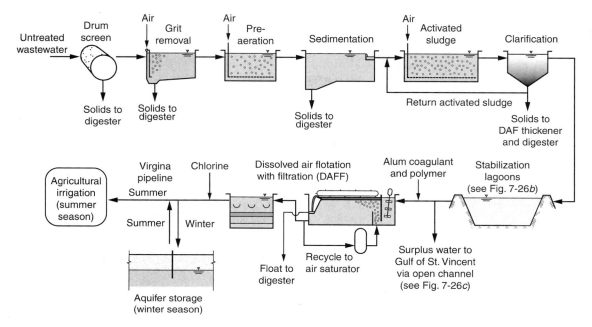

Figure 17-25
Virginia Pipeline Scheme tertiary treatment process for unrestricted crop irrigation and ASR (Coordinates: 35.174 S, 138.595 E, view at altitude 100 km).

aquifer salinity prior to injection was 2100 mg/L, the groundwater was unsuitable for irrigation. Creating a sufficient lens of fresher reclaimed water in the aquifer was key to limit mixing and diffusion phenomena and, thus, recover water suitable for irrigation (Vanderzalm et al., 2002). Views of some of the facilities of the Virginia Pipeline Scheme are shown on Fig. 17-26.

During the irrigation season, produced and extracted reclaimed water is distributed to about 250 client sites through the Virginia Pipeline, a network of 150 km of PVC pipe. A contractor was selected to implement the pipeline under a concession scheme, which includes responsibility for reclaimed water sales. The general layout of the Virginia Pipeline is shown on Fig. 17-27.

Implementation The Bolivar/Virginia Pipeline project is the largest water reuse project in Australia, and the largest ASR project in the world for irrigation quality reclaimed water. It is also one of the first ASR projects to inject lower quality water into a deep confined aquifer to recharge brackish groundwater (Barnett et al., 2000). Key to the success of the project was the 3-yr research, education, and training program intended to gain insight into the sustainability of the project and lead to modernization of the practices involved (Kracman et al., 2001). Monitoring was targeted strategically for cost effectiveness and earliest possible warning of operational incidents and clogging phenomena.

Figure 17-26

Views of Virginia Pipeline Scheme: (a) sign at Bolivar wastewater treatment plant (Coordinates: 34.770 S, 138.583 E), (b) view of one of six large stabilization lagoons used for the further treatment of the secondary treated effluent (Coordinates: 34.756 S, 138.569 E), (c) canal for transporting excess reclaimed water from the lagoons to the Gulf of St. Vincent through mangrove trees, and (d) project trailer at the aquifer storage and recovery trial.

Public acceptance and education on water reuse was considered extremely important to this project. A primary goal was to encourage customers to see reclaimed water as preferable to the continued use of groundwater, to make the project economically feasible and sustainable (Kracman et al., 2001). Reclaimed water samples were displayed at public meetings so that potential customers could understand what the water would look like after treatment. Assurances were given by the South Australian Health Commission that the treatment process would produce a water quality suitable for irrigation of several crops with minimal restrictions.

Over a period of 3 yr, public perception changed to accept reclaimed water as a good alternative to groundwater, rather than as an inferior product. A community liaison program was also created to educate the community about the aquifer, and to consult with

Figure 17-27
General layout of the Virginia Pipeline Scheme with potentiometer surface contours for the confined aquifer. (From Barnett et al., 2000.)

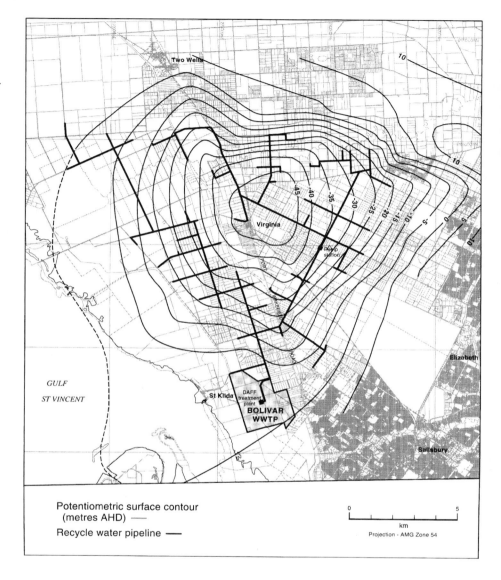

Performance and Operations

the community on the risks and potential benefits of the ASR project. This program was also intended to provide information on use of the aquifer to assure that there was no contamination of the drinking water supply.

The DAF/F plant was monitored during the early commissioning period and results showed that the plant was performing as expected (Kracman et al., 2001). In 24-h composite samples, total suspended solids were reduced from 100–150 mg/L in the feed water to an average of 11 mg/L. The 90th percentile value for fecal coliform was 38 coliforms/100 mL.

The quality of the reclaimed water recovered from ASR was a key monitoring parameter, not only to satisfy stringent requirements for reuse, but because any degradation of aquifer quality would indicate that the process was unsustainable. The main concerns associated with injection and mixing of recycled water with ambient groundwater include redox processes, mineral dissolution, precipitation reactions, ion exchange, clogging, and dissolution processes (Vanderzalm et al., 2002).

Water quality data collected during field trials at Bolivar indicated positive results for both reclaimed water quality and aquifer stability. The injection of a total of 250×10^3 m^3 (66.0 Mgal) of reclaimed water occurred between October 1999 and April 2001 (Vanderzalm et al., 2002). After approximately 16 wk, 150×10^3 m^3 (39.6 Mgal) of water was recovered from the aquifer. The concentration variations observed for the main constituents of interest are shown in Table 17-31. As expected, there were dramatic water quality variations in the first 10^3 m^3 water extracted, which are not reflected in Table 17-31.

The major changes in the recovered water in the field trial included decreases in dissolved oxygen, nitrate, and organic matter, as well as some buffering of pH, calcium, and bicarbonate. Based on the data obtained from the field trials, it was possible to assess environmental concerns about the fate of disinfection-by-products in the reclaimed water injected into the aquifer, particularly trihalomethanes (THMs) and haloacetic acids (HAAs). The concentrations of the major trihalomethanes of concern, including chloroform (CF), bromodichloromethane (BDCM), dibromochloromethane

Table 17-31
Water quality in ambient groundwater, injectant and recovered water during the ASR trial, Bolivar/Virginia Pipeline project, Australia[a]

Parameter	Unit	Ambient (n = 5)[b]	Injectant (n = 15)	Recovered (n = 8)
pH	unitless	7.2–7.3	6.4–7.8	7.0–7.3
EC	dS/m	2.9–3.9	1.8–2.6	1.9–2.5
DO	mmole/L	<0.02	<0.02–0.33	<0.02
Cl$^-$	mmole/L	21–28	10–15	11–16
SO$_4^{2-}$	mmole/L	2.0–3.2	2.0–2.4	2.3–2.7
HCO$_3^-$	mmole/L	3.5–4.9	2.6–6.7	4–5
Ca^{2+}	mmole/L	3.3–3.9	1.0–1.8	1.5–1.9
Na$^+$	mmole/L	16–25	11–15	12–16
Mg^{2+}	mmole/L	2.5–3.7	1.2–1.7	1.2–2.0
K$^+$	mmole/L	0.24–0.38	1.1–1.5	1.0–1.3
Fe-total	mmole/L	0.015–0.024	<0.0005–0.37	0.007–0.12
Sr-total	mmole/L	0.011–0.013	0.0018–0.0046	0.0031–0.0061
TOC	mmole/L	<0.025–0.04	1.1–2.0	0.9–1.2
DOC	mmole/L	<0.025–0.04	1.0–1.9	0.9–1.2
NH$_4^+$	mmole/L	0.003–0.02	0.004–2.1	0.1–1.0
NO$_3^-$	mmole/L	<0.0004	<0.0004–0.34	<0.0004–0.007
LSI^c$_{Calcite}$		0.10–0.19	−1.44–0.13	−0.39−−0.07

[a]Adapted from Vanderzalm et al. (2002).
[b]Number of samples.
[c]Langelier Saturation Index (see Chaps. 9 and 19).

Table 17-32

THM and chloride data in an SAR recharge well during a storage period. Bolivar/Virginia Pipeline project, Australia[a]

Day	\multicolumn{5}{c	}{THM Concentration, mg/L}	Chloride conc., mg/L
	CF	BDCM	DBCM	BF	Total	
0	33	8	46	58	145	415
7	71	20	10	<1	101	382
12	46	6	3	<1	55	360
28	12	2	3	<1	15	358
69	4	<1	<1	<1	4	387
82	2	<1	<1	<1	2	363
109	<1	<1	<1	<1	<4	370

[a]Adapted from Nicholson et al. (2002).
[b]CF = chloroform, BDCM = bromodichloromethane, DBCM = dibromochloromethane, BF = bromoform.
[c]Day 0 represents the recharge water THMs on the last day of recharge.

(DBCM), and bromoform (BF), over time at the ASR well and at an observation well located 4 m (13 ft) from the ASR well are shown in Tables 17-32 and 17-33, respectively, as reported by Nicholson et al., (2002).

Initially, the concentrations of some compounds increased due to the continued formation of THMs as a result of the reaction between residual chlorine in the recharge water and its organic matter content. A decrease of all THMs is observed over time. Because degradation of chloroform can only occur under methanogenic conditions, it appears that such conditions prevail at the ASR well and at the observation well. The lower rate

Table 17-33

THM and chloride data in an observation well located 4 m from the SAR well during a storage period, Bolivar/Virginia Pipeline project. Australia[a]

Day	\multicolumn{5}{c	}{THM Concentration[b], µg/L}	Chloride conc., mg/L
	CF	BDCM	DBCM	BF	Total	
0	41	56	40	6	143	394
7	47	57	38	5	147	370
12	35	41	26	3	105	360
28	33	27	12	1	73	381
69	—	—	—	—	—	—
82	19	9	5	1	34	379
109	14	2	<1	<1	16	357

[a]Adapted from Nicholson et al. (2002).
[b]CF = chloroform, BDCM = bromodichloromethane, DBCM = dibromochloromethane, BF = bromoform.
[c]Day 0 represents the recharge water THMs on the last day of recharge. Travel time between the recharge and observation well is approximately 1 d.

of degradation at the observation well suggests a lower biomass available to carry out the reaction. Haloacetic acids, which are degraded under both aerobic and anaerobic conditions were found to be attenuated rapidly in the aquifer (Nicholson et al., 2002).

Lessons Learned

Important lessons learned in the implementation of the Virginia Pipeline Scheme are as follows:

- A key to the success of the project was an extensive research, education, and training program. Customer confidence was a vital element for the reclaimed water to be marketable.
- Aquifer storage and recovery (ASR) has the potential to create a sustainable water resource cycle, particularly in areas such as the State of South Australia, where a major source of water is groundwater.
- The ASR can be used to provide multiple beneficial effects including protection of the sensitive environment, freshwater recharge of brackish aquifers, the prevention of unsustainable use of freshwater, and attenuation of reclaimed water constituents.

PROBLEMS AND DISCUSSION TOPICS

17-1 Referring to Fig. 17-3, classify soils containing (a) 30 percent silt, 60 percent sand, and 10 percent clay, and (b) 55 percent silt, 15 percent sand, and 30 percent clay. Will the composition of these two soils impact irrigation with reclaimed water?

17-2 Estimate the sodium adsorption ratio, SAR, and the adjusted sodium adsorption ratio, SAR_{adj}, of reclaimed water with the following chemical characteristics.

Constituent	Unit	Value
pH	pH-unit	7.2
Sodium	mg/L	143
Calcium	mg/L	58
Magnesium	mg/L	13
Chloride	mg/L	157
Sulfate	mg/L	123
Alkalinity	mg/L as $CaCO_3$	132
TDS	mg/L	754

17-3 Using data given below and the crop coefficients given in Table 17-21, estimate the agronomic water requirements for olive trees. Assume the leaching requirement is 10 percent when irrigation is necessary, and irrigation efficiency is 75 percent throughout a year. No irrigation is necessary when precipitation is greater than crop evapotranspiration.

Month	Time, d/mo	Precipitation (P), mm/mo	Reference evapotranspiration (ET_o), mm/mo
Jan	31	97.5	25.0
Feb	28	90.0	44.0
Mar	31	71.1	85.8
Apr	30	25.9	139.1
May	31	13.5	175.0
Jun	30	5.1	206.4
Jul	31	1.3	215.6
Aug	31	1.5	189.9
Sep	30	9.1	147.0
Oct	31	22.6	107.6
Nov	30	55.6	51.8
Dec	31	62.2	29.5

17-4 Estimate the expected yield reduction due to salinity for orange trees irrigated with reclaimed water containing 650 mg/L total dissolved solids with a leaching fraction of 15 percent.

17-5 A crop is irrigated with reclaimed water whose salinity, measured by total dissolved solids, is 850 mg/L. If a crop is irrigated to achieve a leaching fraction of 15 percent, estimate (a) the salinity of water leached out of the root zone, and (b) the appropriate leaching fraction to maintain a crop yield above 80 percent. The crop is known to have a threshold salinity level of 4.6 dS/m and the slope of yield reduction curve is 7.6 percent per dS/m.

17-6 If the crop in Problem 17-5 is grown in a climatic condition provided in Problem 17-3, calculate the agronomic water requirements assuming the crop coefficient is 0.8 throughout a year.

17-7 If the crop discussed in Problems 17-5 and 17-6 is grown on a 10 ha field, calculate the amount of salt leached from the irrigated land each year. Discuss the long-term ramification of using reclaimed water for irrigation in terms of salinity, and the measures that can be used to mitigate salinity issues.

17-8 Two methods of treatment are being considered for agricultural irrigation with (a) conventional activated sludge, and (b) activated sludge with biological nutrient removal, followed by filtration. Typical total nitrogen and nitrate levels can be found in Table 17-12. Estimate the amount of nitrogen, in kg/ha, that will be applied to the irrigated field from the above two types of reclaimed water when reclaimed water is applied at an average of 150 mm/mo for the first 3 mo of the growing season.

17-9 Reclaimed water with an average TDS of 1200 mg/L is to be used to irrigate tomatoes. Evapotranspiration for the peak period is 12 mm/d, and the maximum water infiltration rate below the root zone is 8 mm/d. Estimate the maximum crop yield that can be obtained without causing a rise in the water table. Use Table 17-15 to estimate the yield reduction.

17-10 Reclaimed water with the following characteristics is to be used for agricultural irrigation. Estimate the sodium concentration based on the reported water quality data. Also, discuss what should be assessed for crop selection and the irrigation method.

Constituent	Unit	Value
pH	unitless	7.2
Calcium	mg/L	74.4
Magnesium	mg/L	14.8
Ammonia-N	mg–N/L	6.35
Total nitrogen	mg–N/L	10.86
Suspended solids	mg/L	2.6
Total phosphorus	mg–P/L	3.5
Alkalinity	mg/L as $CaCO_3$	191
TDS	mg/L	1067
SAR	–	4.05

17-11 An irrigation water with an EC of 1.8 mmho/cm is to be applied to production of lettuce. Seasonal evapotranspiration is 650 mm, seasonal time of irrigation is to be 165 h, and the water application rate, which is less than the average infiltration rate, is 9 mm/h. Assuming the maximum yield is 450 kg/ha, estimate the expected crop yield.

17-12 Determine the allowable hydraulic loading rate for an irrigation operation based on the nitrogen loading limit. Assume allowable nitrate concentration in percolating water is: $C_p = 10$ mg–N/L, and nitrogen uptake by crop is 0.04 kg/m²·yr. An average total nitrogen concentration in reclaimed water is 25 mg–N/L, and the fraction of applied nitrogen removed by denitrification and volatilization is 0.24. Assume all of the nitrogen leaching out from the root zone is in the form of nitrate. The fraction of applied nitrogen removed by denitrification and volatilization is: $f = 0.20$. The values of $(ET_c - P)$ are shown below.

Month	$(ET_c - P)$, mm
Jan	−20.5
Feb	−2.5
Mar	35.2
Apr	65.0
May	73.4
Jun	−48.8
Jul	−42.5
Aug	−67.9
Sep	−50.5
Oct	20.9
Nov	3.8
Dec	−22.6

17-13 For the evapotranspiration data in Problem 17-12, calculate the annual hydraulic loading rate based on the maximum infiltration rate using Eq. (17-13a) with maximum allowable percolation rate of 13 mm/d, and neglect runoff and drift losses. Compare the results with the hydraulic loading calculated in Problem 17-12. Which hydraulic loading should be used? Provide the reasons for your decision.

17-14 Determine the field area required for reclaimed water irrigation with the following conditions:

- The annual hydraulic loading rate = 655.8 mm/yr.
- The average daily flow of reclaimed water is 1000 m^3/d.
- Conveyance efficiency is 90 percent.
- Neglect the loss or gain of stored reclaimed water.

17-15 In Problem 17-12, the hydraulic loading rate was determined assuming an allowable nitrate in the leaching water is 10 mg–N/L. Discuss the ramification of leaching 10 mg–N/L of nitrate from the root zone in terms of underlying groundwater quality.

17-16 Determine the area of an open storage reservoir for a reclaimed water irrigation system with the water balance data below. Use Eq. (17-12) to calculate the hydraulic loading rate. Leaching fraction and irrigation efficiency are 15 percent and 90 percent, respectively. The storage reservoir will have an average depth of 5 m. Neglect the water loss by seepage.

Month	$(ET_c - P)$, mm
Jan	−84
Feb	−47
Mar	−22
Apr	50
May	132
Jun	183
Jul	219
Aug	183
Sep	98
Oct	36
Nov	−21
Dec	−47

REFERENCES

Allaway, W. H. (1994) "Agronomic Controls over the Environmental Cycling of Trace Elements," *Adv. Agron.*, **20**, 236.

Angelakis, A. N., M. H. F. M. do Monte, L. Bontoux, and T. Asano (1999) "The Status of Wastewater Reuse Practice in the Mediterranean Basin: Need for Guidelines," *Water Res.*, **33**, 10, 2201–2217.

Asano, T., and A. D. Levine (1996) "Wastewater Reclamation, Recycling and Reuse: Past, Present and Future," *Water Sci. Technol.*, **33**, 1–14.

Australian Bureau of Statistics (2004) 4610.0 *Water Account, Australia*, Australian Bureau of Statistics, accessed online: http://www.abs.gov.au/.

Ayers, R. S., and D. W. Westcot (1985) *Water Quality for Agriculture*, FAO Irrigation and Drainage Paper No. 29, Food and Agriculture Organization of the United Nations (FAO), Rome.

Barnett, S. R., S. R. Howles, R. R. Martin, and N. Z. Gerges, (2000) "Aquifer Storage and Recharge: Innovation in Water Resources Management," *Aust. J. Earth. Sci.*, **47**, 1, 13–19.

Blumenthal, U. J., D. D. Mara, A. Peasey, G. Ruiz-Palacios, and R. Stott (2000) "Guidelines for the Microbiological Quality of Treated Wastewater Used in Agriculture: Recommendations for Revising WHO Guidelines, *Bull. World Health Organ.*, **78**, 9, 1104–1116.

Booher, L. J. (1974) *Surface Irrigation*, FAO Agricultural Development Paper No. 95, Food and Agricultural Organization of the United Nations, Rome.

Bostjancic, J., and R. Ludlum (1996) "Getting to Zero Discharge: How to Recycle that Last Bit of Really Bad Wastewater," in *Proceedings, the 57th Annual International Water Conference*, Engineer's Society of Western Pennsylvania, Bellevue, WA.

Bowen, H. (ed.) (1979) *Environmental Chemistry of the Elements*, Academic Press, New York.

Cararo, D. C., T. A. Botrel, D. J. Hills, and H. L. Leverenz (2006) "Analysis of Clogging in Drip Emitters During Wastewater Irrigation," *Appl. Eng. Agric.*, **22**, 2, 1–7.

Carrow, R. N. (1994) "A Look at Turfgrass Water Conservation," in United States Golf Association (ed.), *Wastewater Reuse for Golf Course Irrigation*, Lewis Publishers, Inc., Chelsea, MI, 24–43.

Chang, A. C., and A. L. Page (1994) "The Role of Biochemical Data in Assessing the Ecological and Health Effects of Trace Elements," in *Proceedings of 15th World Congress of Soil Science, Acapulco, Mexico*, International Society of Soil Science, Vienna.

Chapman, H. D. (ed.) (1965) *Diagnostic Criteria for Plants and Soils*, Quality Printing Co., Abilene, TX.

City of Orlando (2006) *Water Conserv II Water Reclamation Program*, accessed online: http://www.cityoforlando.net/public_works/wastewater/reclaim.htm.

Crites, R. W. (1985) "Site Characteristics," Chap. 4, in G. S. Pettygrove and T. Asano (eds.), *Irrigation with Reclaimed Municipal Wastewater: A Guidance Manual*, Report No. 84-1 wr, Lewis Publishers, Chelsea, MI.

Crites, R. (2002) "Agricultural Water Reuse in California," Presented at the 2002 Hawaii Water Reuse Conference, Kauai, Hawaii.

Cross, P., G. Delneky, and T. Lothrop (1998) "Water Conserv II: Past, Present and Future," 281–289, in *98 Water Reuse Conference Proceedings, February 1–4, 1998, Lake Buena Vista, FL*, American Water Works Association, Denver, CO.

Cross, P., G. Delneky, and T. Lothrop (2000) "Worth Its Weight in Oranges," *Water Environ. Technol.*, **1**, 26–30.

Doorenbos, J., and W. O. Pruitt (1977) Guidelines for Predicting Crop Water Requirements, *Irrigation and Drain*, Paper No. 24, 2nd ed., Food and Agricultural Organization of the United Nations (FAO), Rome.

Edraki, M., H. B. So, and E. A. Gardner (2004) "Water Balance of Swamp Mahogany and Rhodes Grass Irrigated with Treated Sewage Effluent," *Agric. Water Mgnt.*, **67**, 3, 157–171.

Engineering Science (1987) *Monterey Wastewater Reclamation Study for Agriculture: Final Report*, Prepared for Monterey Regional Water Pollution Control Agency, Engineering Science, Berkeley, CA.

Feigin, A., I. Ravina, and J. Shalhevet (1991) *Irrigation with Treated Sewage Effluent: Management for Environmental Protection*, Advanced Series in Agricultural Sciences 17, Springer-Verlag, Berlin.

George, M. R., G. S. Pettygrove, and W. B. Davis (1985) "Crop Selection and Management," Chap. 6, in G. S. Pettygrove, and T. Asano (eds.), *Irrigation with Reclaimed Municipal Wastewater: A Guidance Manual*, Report No. 84-1 wr, Lewis Publishers, Chelsea, MI.

Gilbert, R. G., F. S. Nakayama, D. A. Bucks, O. F. French, K. C. Adamson, and R. M. Johnson (1982) "Trickle Irrigation—Predominant Bacteria in Treated Colorado River Water and Biologically Clogged Emitters," *Irrig. Sci.*, **3**, 2, 123–132.

Gotor, A. G., S. O. P. Baez, C. A. Espinoza, and S. I. Bachir (2001) "Membrane Process for the Recovery and Reuse of Wastewater in Agriculture," *Desalination*, **137**, 3, 187–192.

Hamoda, M. F., I. Al-Ghusain, and N. Z. Al-Mutairi (2004) "Sand Filtration of Wastewater for Tertiary Treatment and Water Reuse," *Desalination*, **164**, 3, 203–211.

Hanson, R. B., S. R. Grattan, and A. Fulton (1999) *Agricultural Salinity and Drainage*, University of California Irrigation Program, University of California, Davis, Davis, CA.

Huston, S. S., N. L. Barber, J. F. Kenny, D. S. Lumia, and M. A. Maupin (2004) *Estimated Use of Water in the United States in 2000*, U.S. Geological Survey, Circular 1268, Reston, Virginia.

Icekson-Tal, N., O. Avraham, J. Sack, and H. Cikurel (2001) "Water Reuse in Israel—the Dan Region Project: Evaluation of Water Quality and Reliability of Plant's Operation," *Water Sci. Technol.: Water Supply*, **3**, 4, 231–237.

Kracman, B., R. Martin, and P. Sztajnbok (2001) "The Virginia Pipeline: Australia's largest water recycling project," *Water Sci. Technol.*, **43**, 10, 35–42.

Lazarova, V., and T. Asano (2004) "Challenges of Sustainable Irrigation with Recycled Water," 1-30, in V. Lazarova and A. Bahri (eds.), *Water Reuse for Irrigation: Agriculture, Landscapes, and Turf Grass*, CRC Press, Boca Raton, FL.

Lazarova, V., I. Papadopoulos, and A. Bahri (2004) "Code of Successful Agronomic Practice," 103–150, in V. Lazarova and A. Bahri (eds.) *Water Reuse for Irrigation: Agriculture, Landscapes, and Turf Grass*, CRC Press, Boca Raton, FL.

Lisk, D. J. (1972) "Trace Metals in Soils, Plants and Animals," *Adv. Agron.*, **24**, 267.

Little, T. M., and F. J. Hills (1978) *Agricultural Experimentation: Design and Analysis*, John Wiley & Sons, New York.

Maas, E. V., (1986) "Salt Tolerance of Plants," *Appl. Agr. Res.*, **1**, 1, 12–26.

Maas, E. V., and S. R. Grattan (1999) "Crop Yields as Affected by Salinity," 55–108, in R. W. Skaggs and J. van Schilfgaarde (eds.), *Agricultural Drainage*, Agronomy Monograph No. 38, American Society of Agronomy, Crop Science Society of America, Soil Science Society of America, Madison, WI.

Martin, R., D. Clarke, K. Dennis, J. Graham, P. Dillon, P. Pavelic, and K. Barry (2002) *Boliver Water Reuse Project*, Report DWLBC 2002/02, The Department of Water, Land and Biodiversity Conservation, The Government of South Australia, Adelaide, Australia.

McGechan, M. B., and D. R. Lewis (2002) "Sorption of Phosphorus by Soil, Part 1: Principles, Equations and Models," *Biosyst. Eng.*, **82**, 1–24.

McKenzie, D. C. (ed.) (1998) *SOILpak for Cotton Growers*, 3rd ed., NSW Agriculture, NSW Department of Primary Industries, New South Wales, Australia.

NRC (National Research Council) (1973) *Water Quality Criteria, 1972: A Report of the Committee on Water Quality Criteria*, EPA-R3-73-033, U.S. Environmental Protection Agency, Washington, DC.

Nicholson, B. C., P. J. Dillon, and P. Pavelic (2002) "Fate of Disinfection By-Products During Aquifer Storage and Recovery," *Management of Aquifer Recharge for Sustainability*, Swets and Zeitlinger, Lisse, The Netherlands.

Page, A. L. (1974) *Fate and Effects of Trace Elements in Sewage Sludge when Applied to Agricultural Soils: A Literature Review Study*, EPA-670/2-74-005, U.S. Environmental Protection Agency, Cincinnati, OH.

Page, A. L., and A. C. Chang (1985) "Fate of Wastewater Constituents in Soil and Groundwater: Trace Organics," in G.S. Pettygrove and T. Asano (eds.) *Irrigation with Reclaimed Municipal Wastewater: A Guidance Manual*, Report No. 84-1 wr, Chap. 15, Lewis Publishers, Chelsea, MI.

Pescod, M. B. (1992) "Wastewater Treatment and Reuse in Agriculture," FAO Irrigation and Drainage Paper No. 47, Food and Agriculture Organization of the United Nations (FAO), Rome.

Pettygrove, G. S., and T. Asano (eds.) (1985) *Irrigation with Reclaimed Municipal Wastewater: A Guidance Manual*, Report No. 84-1 wr, Lewis Publishers, Chelsea, MI.

Rauschkolb, R. S., and D. S. Mikkelsen (1978) "Survey of Fertilizer Use in California," University of California Division of Agricultural Science, U.S. Department of Agriculture Bulletin **1887**, 27.

Rebhum, M. (2004) "Desalination of Reclaimed Wastewater to Prevent Salinization of Soils and Groundwater," *Desalination*, **160**, 2, 143–149.

Sheikh, B., R. P. Cort, W. R. Kirkpatrick, R. S. Jaques, and T. Asano (1990) "Monterey Wastewater Reclamation Study for Agriculture," *J. WPCF*, **62**, 3, 216–226.

Sheikh, B., R. P. Cort, R. C. Cooper, and R. S. Jaques (1998) "Tertiary-Treated Reclaimed Water for Irrigation of Raw-Eaten Vegetables," 779–825, in T. Asano (ed.), *Wastewater Reclamation and Reuse*, Water Quality Management Library, Vol. **10**, CRC Press, Boca Raton, FL.

Sheikh, B., R. C. Cooper, and K. E. Israel (1999) "Hygienic Evaluation of Reclaimed Water used to Irrigate Food Crops—A Case Study," *Water Sci. Technol.*, **40**, 4–5, 261–267.

Smith, R. G., J. L. Meyer, G. L. Dickey, and B. R. Hanson (1985) "Irrigation System Design," in G.S. Pettygrove and T. Asano (eds.) *Irrigation with Reclaimed Municipal Wastewater: A Guidance Manual*, Report No. 84-1 wr, Chap. 8, Lewis Publishers, Chelsea, MI.

Snyder, R. L., M. Orang, S. Matyac, and S. Eching (2002) *Crop Coefficients, Quick Answer ET 002*, University of California, accessed online: http://lawr.ucdavis.edu/.

State of California (1990) *California Municipal Wastewater Reclamation in 1987*, California State Water Resources Control Board, Office of Water Recycling, Sacramento, CA.

State of California (2003) *Water Recycling 2030: Recommendations of California's Recycled Water Task Force*, Department of Water Resources, Sacramento, CA.

State of California (2004) *California Irrigation Management Information System*, Department of Water Resources, Office of Water Use Efficiency, accessed online: http://wwwcimis.water.ca.gov/cimis/welcome.jsp.

State of Florida (2004) *2003 Reuse Inventory Report*, Florida Department of Environmental Protection, Tallahassee, FL, accessed online: http://www.dep.state.fl.us/water/reuse/.

Suarez, D. L. (1981) "Relation between pHc and Sodium Adsorption Ratio (SAR) and Alternative Method of Estimating SAR of Soil or Drainage Waters," *Soil Sci. Soc. Am. J.*, **45**, 3, 469–475.

Tanji, K. K., and N. C. Kielen (2002) *Agricultural Drainage Water Management in Arid and Semi-Arid Areas*, FAO Irrigation and Drainage Paper 61, Food and Agricultural Organization of the United Nations (FAO), Rome.

Tchobanoglous, G. F. L. Burton, and H. D. Stensel (2003) *Wastewater Engineering: Treatment and Reuse*, 4th ed., McGraw-Hill, New York.

University of California Committee of Consultants (1974) *Guidelines for Interpretation of Water Quality for Agriculture*, Memo Report, University of California, Cooperative Extension.

University of California and State of California (2000) *A Guide to Estimating Irrigation Water Needs of Landscaping Plantings in California*, University of California Cooperative Extension, Department of Water Resources, Sacramento, CA.

USBR (1993) *Drainage Manual*, 2nd ed., U.S. Department of Interior, U.S. Bureau of Reclamation, Denver Federal Center, Denver, CO.

USDA (1991) "Soil-Plant-Water Relationships," Chap. 1, in U.S. Department of Agriculture, *Irrigation*, National Engineering Handbook, Sec. 15, Soil Conservation Service, U.S. Department of Agriculture, Washington, DC.

USDA (1993a) "Irrigation Water Requirements," Chap. 2, in U.S. Department of Agriculture, *Irrigation*, National Engineering Handbook, Sec. 15, Soil Conservation Service, U.S. Department of Agriculture, Washington, DC.

USDA (1993b) *Soil Survey Manual*, Chap. 3, Soil Conservation Service, U.S. Department of Agriculture Handbook 18, accessed online: http://soils.usda.gov/technical/manual/.

USDA (1997) "Irrigation Guide," Part 652, *National Engineering Handbook*, Natural Resources Conservation Service, U.S. Department of Agriculture, Washington, DC.

USDA (2001) "Drainage," Chap. 14, *Engineering Field Handbook*, Natural Resources Conservation Service, U.S. Department of Agriculture, Washington, DC.

U.S. EPA (1981) *Process Design Manual for Land Treatment of Municipal Wastewater*, EPA 625/1-81-013, U.S. Environmental Protection Agency, Cincinnati, OH.

U.S. EPA (2004) *Guidelines for Water Reuse*, EPA/625/R-04/108, September 2004, U.S. Environmental Protection Agency (U.S. EPA) and U.S. Agency for International Development (U.S. AID), Washington, DC.

Van Schilfgaarde, J. (ed.) (1972) *Drainage for Agriculture*, Number 17 in the series Agronomy, American Society of Agronomy, Madison, WI.

Vanderzalm, J. L., C. Le Gal La Salle, J. L. Huston, and P. J. Dillon (2002) "Water Quality Changes during Aquifer Storage and Recovery at Bolivar, South Australia," 83–88, in P. Dillon (ed.), *Management of Aquifer Recharge for Sustainability*, A. A. Balkema Publishers, Lisse, The Netherlands.

Water Conserv II (2001) *Water Conserv II: A Cooperative Water Reuse Program by the City of Orlando, Orange County, and the Agricultural Community*, Water Conserv II, accessed online: http://www.waterconservii.com/.

Water Conserv II (2006), accessed online: http://www.waterconservii.com/.

Weber, B., and M. Juanico (2004) "Salt Reduction in Municipal Sewage Allocated for Reuse: The Outcome of a New Policy in Israel," *Water Sci. Technol.*, **50**, 2, 17–22.

Westcot, D. W., and R. S. Ayers (1985) "Irrigation Water Quality Criteria," in G. S. Pettygrove and T. Asano (eds.), *Irrigation with Reclaimed Municipal Wastewater: A Guidance Manual*, Report No. 84-1 wr, Chap. 3, Lewis Publishers, Chelsea, MI.

WHO (World Health Organization) (1989) *Health Guidelines for the Use of Wastewater in Agriculture and Aquaculture*, Report of a WHO Scientific Group, Technical Report Series 778, World Health Organization, Geneva.

York, D. W., and L. Wadsworth (1998) "Reuse in Florida: Moving Toward the 21st Century," *Florida Water Res. J.*, **11**, 31–33.

18 Landscape Irrigation with Reclaimed Water

WORKING TERMINOLOGY 1044

18-1 LANDSCAPE IRRIGATION: AN OVERVIEW 1045
Definition of Landscape Irrigation 1045
Reclaimed Water Use for Landscape Irrigation in the United States 1046

18-2 DESIGN AND OPERATIONAL CONSIDERATIONS FOR RECLAIMED WATER LANDSCAPE IRRIGATION SYSTEMS 1047
Water Quality Requirements 1047
Landscape Plant Selection 1050
Irrigation Systems 1054
Estimation of Water Needs 1054
Application Rate and Irrigation Schedule 1065
Management of Demand-Supply Balance 1065
Operation and Maintenance Issues 1066

18-3 GOLF COURSE IRRIGATION WITH RECLAIMED WATER 1070
Water Quality and Agronomic Considerations 1070
Reclaimed Water Supply and Storage 1072
Distribution System Design Considerations 1075
Leaching, Drainage, and Runoff 1076
Other Considerations 1076

18-4 IRRIGATION OF PUBLIC AREAS WITH RECLAIMED WATER 1076
Irrigation of Public Areas 1078
Reclaimed Water Treatment and Water Quality 1079
Conveyance and Distribution System 1079
Aesthetics and Public Acceptance 1079
Operation and Maintenance Issues 1080

18-5 RESIDENTIAL LANDSCAPE IRRIGATION WITH RECLAIMED WATER 1080
Residential Landscape Irrigation Systems 1080
Reclaimed Water Treatment and Water Quality 1081
Conveyance and Distribution System 1081
Operation and Maintenance Issues 1082

18-6	**LANDSCAPE IRRIGATION WITH DECENTRALIZED TREATMENT AND SUBSURFACE IRRIGATION SYSTEMS** 1082 *Subsurface Drip Irrigation for Individual On-site and Cluster Systems* 1082 *Irrigation for Residential Areas* 1086
18-7	**CASE STUDY: LANDSCAPE IRRIGATION IN ST. PETERSBURG, FLORIDA** 1086 *Setting* 1087 *Water Management Issues* 1087 *Implementation* 1087 *Project Greenleaf and Resource Management* 1089 *Landscape Irrigation in the City of St. Petersburg* 1091 *Lessons Learned* 1093
18-8	**CASE STUDY: RESIDENTIAL IRRIGATION IN EL DORADO HILLS, CALIFORNIA** 1093 *Water Management Issues* 1094 *Implementation* 1094 *Education Program* 1096 *Lessons Learned* 1096
	PROBLEMS AND DISCUSSION TOPICS 1097
	REFERENCES 1099

WORKING TERMINOLOGY

Term	Definition
Foliar damage	Damage to leaves of landscape plants. Reclaimed water constituents such as chloride may cause damage to leaves of landscape plants as a result of sprinkler irrigation.
Landscape coefficient	Ratio of evapotranspiration of a landscaping site (ET_L) to the reference evapotranspiration (ET_o).
Restricted access area	Area where public access is limited, such as highway medians, cemeteries, and inside of industrial areas.
Unrestricted access area	Area where public access is not limited, such as golf courses, parks, school yards, commercial areas, and residential areas.
Urban uses of water	Major urban uses of water include landscape irrigation, toilet flushing, air conditioning, street washing, fire hydrants, and some commercial uses such as car washing.
Xeriscape	Landscaping with plants that require little or no water.

Because of its typical location of use, landscape irrigation with reclaimed water is often categorized as an urban water reuse application (U.S. EPA, 2004). Water quality and other agronomic considerations for landscape irrigation, however, follow the same principles used for agricultural irrigation that have been discussed in Chap. 17. Along with an overview of landscape irrigation with reclaimed water, special design and operational considerations for landscape irrigation are discussed in this chapter. In addition, the following landscape irrigation applications are discussed in detail: (1) golf courses, (2) public areas, (3) residential landscape, and (4) landscape irrigation utilizing effluent from decentralized and onsite wastewater treatment systems in rural areas. Two case studies are also presented to illustrate the use of reclaimed water for landscape irrigation. Other urban nonpotable uses of reclaimed water are described in Chap. 20.

18-1 LANDSCAPE IRRIGATION: AN OVERVIEW

Landscape plants provide various functions such as creating an aesthetically pleasing property; creating a buffer between streets, parking lots, and noncommercial areas; and providing vertical and horizontal dimensions to a site. Ornamental plants also maintain moisture and may be used to mitigate the effects of heat in urban areas.

Although the water demand for landscape irrigation varies greatly by geographical location, season, and the types of plants, approximately one-third of residential water use is for landscape irrigation, with significantly higher usage in arid urban areas (U.S. EPA, 1992). For example, the Irvine Ranch Water District in southern California estimates that more than 70 percent of their total water use is for landscape irrigation.

Components of landscape irrigation systems that should be considered include:

- A landscape design and selection of plants that require less water
- Use of irrigation methods that have a high irrigation efficiency
- Use of nonpotable water including reclaimed water

Landscape irrigation with reclaimed water is a viable option to reduce potable water demand, and also as an option to reduce or eliminate wastewater discharge to aquatic environment. Factors motivating many local governments to consider the use of reclaimed water include: (1) the high water demand for landscaping, (2) increasing cost of acquiring additional water in urban areas, and (3) stringent wastewater discharge requirements (see Chap. 2).

Definition of Landscape Irrigation

Landscape irrigation, as used in this textbook, includes irrigation of restricted and unrestricted areas. The definitions of "restricted" and "unrestricted" vary in different state regulations and guidelines, but generally apply to the following applications:

- Landscape irrigation with unrestricted access areas such as:
  - Public parks
  - Playgrounds, school yards, and athletic fields
  - Public and commercial facilities
  - Individual and multifamily residences
  - Golf courses associated with residential properties

- Landscape irrigation with limited or restricted access areas such as:
  - Cemeteries
  - Highway medians and shoulders
  - Landscaping within industrial areas
  - Golf courses not associated with a residential community

The above categorization is based on California's regulations. In Florida, the term "Public Access Areas," applies to cemeteries, highway medians, and golf courses not associated with a residential property (State of Florida, 1999). Examples of landscape areas irrigated with reclaimed water are shown on Fig. 18-1.

Reclaimed Water Use for Landscape Irrigation in the United States

The two largest users of reclaimed water for landscape irrigation in the United States are Florida and California. Florida, the largest user, accounted for approximately 3.8×10^8 m^3 (3.1×10^5 ac-ft) of reclaimed water in 2004, as compared to 1.4×10^8 m^3 (1.1×10^5 ac-ft) in California. A comparison of the use of reclaimed water for landscape irrigation in California and Florida is illustrated on Fig. 18-2. In Florida, over 40 percent of the

Figure 18-1

Examples of landscape areas irrigated with reclaimed water: (a) golf course, Orlando, FL; (b) playground, Marin County, CA; (c) street median strip, Irvine, CA; and (d) residential homes, El Dorado Hills, CA.

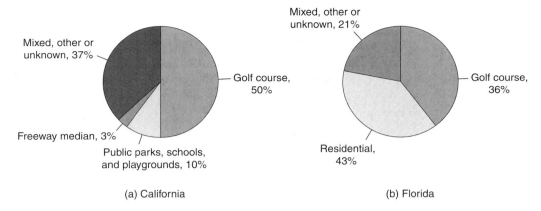

Figure 18-2
Landscape irrigation in (a) California and (b) Florida. (Data from State of California, 1990; State of Florida, 2004.)

reclaimed water is used for landscape irrigation in residential areas, a use not yet popular in other states including California. Golf course irrigation is another major use for reclaimed water, comprising 50 and 36 percent of the total landscape irrigation use in California and Florida, respectively (see Fig. 18-2). Other states that are major users of reclaimed water for landscape irrigation include Arizona, Colorado, Hawaii, Nevada, New Mexico, Texas, and Utah, most of which are located in arid regions. In recent years, however, water-rich regions on the East Coast are using reclaimed water increasingly for landscape irrigation, partly due to stringent waste discharge requirements, and partly due to localized water shortages in densely populated areas. Selected examples of landscape irrigation with reclaimed water in the United States are shown in Table 18-1.

18-2 DESIGN AND OPERATIONAL CONSIDERATIONS FOR RECLAIMED WATER LANDSCAPE IRRIGATION SYSTEMS

Design considerations for urban landscape irrigation systems using reclaimed water are summarized in Table 18-2. For many urban landscape irrigation systems, either the reclaimed water system or the landscape area already exists. To convert these systems to urban landscape irrigation systems using reclaimed water, either the existing irrigation systems need to be retrofitted for conveying reclaimed water, or new landscape areas need to be established adjacent to existing reclaimed water systems. In this section, factors affecting design and operation of landscape irrigation with reclaimed water are described.

Water Quality Requirements

Agronomic water quality requirements were described in Chap. 17 (see Table 17-5). Generally, tertiary treatment or an equivalent level of treatment is required for the purpose of public health protection in irrigation of landscape plants. Treatment processes used to meet the water quality criteria for landscape irrigation are discussed in Chaps. 7 and 8. It should be noted that onsite treatment systems for subsurface irrigation of landscaping plants do not require the same water quality criteria (see Sec. 18-6). When establishing

Table 18-1
Select examples of landscape irrigation with reclaimed water

State	Location	Types of landscape irrigation	Annual flow, $\times 10^6$ m³/yr	Remarks
Arizona	Mesa, the Southeast Water Reclamation Plant (SEWRP)	Golf course, residential	11[a]	Also used for pond replenishment and agricultural irrigation
Arizona	Scottsdale Water Campus	Golf course irrigation	17	Excess water is further treated with RO for vadose zone recharge to groundwater
California	Irvine Ranch Water District	Area-wide reuse system: parks, golf courses, school playfields, athletic fields, and common areas maintained by homeowner associations	11	Also used for agricultural irrigation, industrial water, and toilet flushing
California	The Vallecitos Water District and the Leucadia County Water District (for City of Carlsbad)	Hotels and resort venues, parks, median strips, shopping areas, freeway landscaping, and common areas maintained by homeowner associations	3.5	Approximately 42 km of reclaimed water distribution system, supplying about 60 irrigation sites
Colorado	Denver	Parks, schools, golf courses	41[a]	Also used for industrial cooling water and environmental purposes
Florida	City of St. Petersburg	Residential and other public access area	50	One of the oldest and largest urban irrigation systems in the United States
Georgia	Forsyth County	Park irrigation using subsurface drip irrigation	3.5	Membrane bioreactor is used to treat wastewater
Hawaii	Kihei wastewater reclamation facility, Maui	Golf courses, landscaping of parks, residential areas, community center, schools, and public buildings	2.2–2.8	Also used for agricultural uses, dust control, composting, toilet flushing
Nevada	Clark County Water Reclamation District	Golf course	1.2	Reclaimed water is blended with potable water
Texas	Northwest Wastewater Treatment Plant, El Paso	11 schools, 12 parks, 3 golf courses, cemetery, zoo, residential area, small community	24[a]	Four water reclamation plants in the area provide reclaimed water. Also used for groundwater recharge and industrial water
Texas	San Antonio	Golf courses, schools, commercial sites, cemetery	43	Other uses include industrial cooling and stream augmentation
Utah	Tooele City	Golf course, county recreation property	3.1	Plans to irrigate residential landscape

[a]Flow capacity.

Table 18-2
Typical design considerations for urban landscape irrigation systems

System	Specific consideration	References
Treatment processes	Selection of treatment system to meet quality requirements for landscape irrigation • Pathogens (evaluated by indicator organisms) • Nutrients • Suspended solids	Part 3
Landscape area	Water quality requirements	Chaps. 17, 18
	Plant selection • Salt tolerance • Boron tolerance • Water needs	Chap. 18
	Irrigation method • Required pressure • Irrigation efficiency • Exposure control	
	Leaching requirements	Chap. 17
	Water application rates	Chaps. 17, 18
	Operation and maintenance • Irrigation timing • Irrigation area restriction • Soil conditioning • Sprinkler and emitter clogging control • Monitoring	Chap. 18
Distribution and storage systems	Area-wide distribution main • Flow rate • Pumping requirements • Peaking factor	Chap. 14
	Demand and supply balance • Storage requirements • Blending with other water sources • Multipurpose use of reclaimed water	Chaps. 17, 18
	Cross-connection control • Spacing between reclaimed water and potable waterlines • Pressure difference • Backflow prevention (for potable system) • Coloring of reclaimed water pipes	Chaps. 14, 15

water quality goals, the impact of water quality on the irrigation system also needs to be considered. The impacts of water quality on operation and maintenance of irrigation systems are discussed at the end of this section. A comprehensive salt management guide for landscape irrigation with reclaimed water has been prepared by Tanji et al. (2006).

Agronomic water quality requirements depend on the tolerance of plants to reclaimed water constituents, such as sodium, chloride, and boron, and the effects of salinity and sodicity on irrigated land and landscape plants. Long-term effects of reclaimed water constituents and requirements for leaching and drainage must also be considered when determining water quality requirements. Nutrients are usually considered beneficial, but excess nutrients may cause biofilm growth in the reclaimed water distribution lines and algal growth in the open storage reservoirs. Because of stringent water quality requirements contained in wastewater discharge permits and the need to meet quality requirements of various water reuse applications, it is becoming more common for water reclamation plants to provide nutrient removal.

Aesthetic Quality Considerations

Some aesthetic water quality parameters are also important but they are often not regulated. For example, odor control is important for public acceptance in reclaimed water irrigation but the odor level is not specified in most regulations. Typically, reclaimed water from a tertiary or equivalent treatment has no or only a slight musty odor, unnoticeable when the reclaimed water is used for irrigation. Odors, however, may be generated in the distribution system when the reclaimed water becomes stagnant (see Chap. 14). The development of odors, principally hydrogen sulfide, is of critical concern where the concentration of the sulfate (SO_4^{2-}) is greater than 50 mg/L and the chemical oxygen demand (COD) of the treated effluent is above 20 mg/L. Thus, water quality parameters that are not regulated specifically may be as important as those that are regulated.

Public Health Considerations

Public health considerations are presented in Chap. 4. Two specific concerns relevant to landscape irrigation with reclaimed water are: (1) the health risk associated with potential cross-connection and subsequent contamination of potable water systems, and (2) human exposure to reclaimed water and its constituents during and after irrigation. As of 2003, 28 states have either regulations or guidelines for irrigation of unrestricted access areas, and 34 states have them for irrigation of restricted areas (U.S. EPA, 2004). Typically, the criteria include: (1) minimum treatment levels (2) requirements for disinfection, chemical and microbial water quality, and monitoring, and (3) exposure control measures such as setback distance and irrigation timing. The basis for regulations and guidelines is to minimize the risk of exposure associated with reclaimed water use.

Treatment and Water Quality Requirements

Treatment requirements are specified in most states including Arizona, California, Florida, Hawaii, Nevada, and Washington. Tertiary treatment including filtration and disinfection is required usually for unrestricted uses. In restricted access areas, human exposure to reclaimed water can be controlled more easily; thus, the quality and treatment requirements are typically less stringent than those required for unrestricted use areas. Treatment and water quality criteria in selected states are summarized in Table 18-3.

Landscape Plant Selection

Selection of landscape plants is usually a landscape designer's task, but it also affects the estimation of water demand and quality requirements. As described in Chap. 17, salt tolerance is the most important parameter in plant selection. Salt tolerance of select landscape plants is shown in Table 18-4. Other parameters to be considered are:

- Tolerance to boron and other reclaimed water constituents
- Water needs, drought tolerance
- Native/nonnative to the region

Table 18-3
Various state water quality and treatment requirements for unrestricted and restricted urban uses[a]

Parameter	Arizona	California	Florida	Hawaii	Nevada	Texas	Washington
Unrestricted urban uses							
Treatment	Secondary treatment, filtration, and disinfection	Oxidized,[b] coagulated, filtered, and disinfected	Secondary treatment, filtration, and high-level disinfection	Oxidized, filtered, and disinfected	Secondary treatment and disinfection	ns[c]	Oxidized, coagulated, filtered, and disinfected
BOD, mg/L	ns	ns	20[d]	ns	30	5	30
TSS, mg/L	ns	ns	5.0	ns	ns	ns	30
Turbidity, NTU	2 (avg) 5 (max)	2 (avg) 5 (max)	ns	2 (max)	ns	3	2 (avg) 5 (max)
Coliform, MPN/100 mL	Fecal nondetectable[e] 23 (max)	Total 2.2 (med)[f] 23 (max in 30 d)	Fecal 75% of samples below detection 25 (max)	Fecal 2.2 (med)[f] 23 (max in 30 d)	Fecal 2.2 (avg) 23 (max)	Fecal 20 (avg) 75 (max)	Total 2.2 (avg) 23 (max)
Restricted urban uses							
Treatment	Secondary treatment and disinfection	Oxidized and disinfected	Secondary treatment, filtration, and high-level disinfection	Oxidized and disinfected	Secondary treatment and disinfection	ns	Oxidized and disinfected
BOD, mg/L	ns	ns	20[d]	ns	30	20	30
TSS, mg/L	ns	ns	5	ns	ns	ns	30
Turbidity, NTU	ns	ns	ns	2 (max)	ns	3	2 (avg) 5 (max)
Coliform, MPN/100 mL	Fecal 200 (avg) 800 (max)	Total 23 (med)[f] 240 (max in 30 d)	Fecal 75% of samples below detection 25 (max)	Fecal 23 (med)[f] 200 (max)	Fecal 23 (avg) 240 (max)	Fecal 200 (avg) 800 (max)	Total 23 (avg) 240 (max)

[a] Adapted from U.S. EPA (2004).
[b] *Oxidized wastewater* is wastewater that is treated to oxidize and stabilize organic compounds, and contains dissolved oxygen. The term "oxidized wastewater" is used to avoid specification of treatment processes.
[c] ns = not specified.
[d] CBOD
[e] Not detectable in four of last seven daily samples.
[f] Seven-day median.

Table 18-4
Relative salt tolerance of landscape plants[a]

Common name	Botanical name	Very sensitive	Sensitive	Moderately sensitive	Moderately tolerant	Tolerant	Very tolerant
Aleppo pine	*Pinus halepensis*				√		
Algerian ivy	*Hedera canariensis*		√[c]				
Blue dracaena	*Cordyline indivisa*				√		
Bougainvillea	*Bougainvillea spectabilis*					√	
Brush cherry	*Syzygium paniculatum*					√	
Ceniza	*Leucophyllum frutescens*					√	
Cherry plum	*Prunes cerasifera*			√			
Chinese hibiscus	*Hibiscus Rosa-sinensis*		√				
Chinese holly, cv. Burford	*Ilex cornuta*	√					
Crape myrtle	*Lagerstroemia indica*		√				
Croceum ice plant	*Hymenocyclus croceus*						√
Dodonaea, cv. Atropurpurea	*Dodonaea Viscosa*			√			
European fan palm	*Chamaerops humilis*				√		
Evergreen pear	*Pyrus kawakamii*					√	
Glossy abelia	*Abelia x grandiflora*		√				
Glossy privet	*Ligustrum lucidum*			√			
Heavenly bamboo	*Nandina domestica*		√				
Indian hawthorn	*Raphiolepis indica*			√			
Italian stone pine	*Pinus pinea*			√			
Japanese black pine	*Pinus Thunbergiana*			√			
Japanese boxwood	*Buxus microphylla var. japonica*			√			
Japanese pittosporum	*Pittosporum Tobira*		√				
Laurustinus, cv. Robustum	*Viburnum Tinus*		√				
Natal plum	*Carissa grandiflora*					√	
Orchid tree	*Bauhinia purpurea*			√			
Oleander	*Nerium oleander*				√		

Common name	Scientific name	Very sensitive	Sensitive	Moderately sensitive	Moderately tolerant	Tolerant	Very tolerant
Oregon grape	Mahonia Aquifolium		✓				
Oriental arborvitae	Platycladus orientalis		✓				
Photinia	Photinia x Fraseri			✓			
Pineapple guava	Feijoa Sellowiana			✓			
Purple ice plant	Lampranthus productus						✓
Pyracantha, cv. Graberi	Pyracantha Fortuneana			✓			
Pyrenees cotoneaster	Cotoneaster congestus			✓			
Rose, cv. Grenoble	Rosa sp.			✓			
Rosea ice plant	Drosanthemum hispidum						✓
Rosemary	Rosarinus officinalis				✓		
Southern magnolia	Magnolia grandiflora			✓			
Southern yew	Podocarpus macrophyllus				✓		
Spindle tree, cv. Grandiflora	Euonymus japonica				✓		
Spreading juniper	Juniperus chinensis				✓		
Star jasmine	Trachelospermum jasminoides			✓			
Strawberry tree, cv. Compact	Arbutus Unedo			✓			
Sweet gum	Liquidambar Styraciflua				✓		
Thorny elaeagnus	Elaeagnus pungens				✓		
Tulip tree	Liriodendron Tulipifera			✓			
Weeping bottlebrush	Callistemon viminalis				✓		
White ice plant	Delosperma alba						✓
Xylosma	Xylosma congestum					✓	
Yellow sage	Lantana camara					✓	

[a] Adapted from Maas (1986).
[b] Very sensitive: Max. $EC_w = 0.7 – 1.4$ dS/m
Sensitive: Max. $EC_w = 1.4 – 2.7$ dS/m
Moderately sensitive: Max. $EC_w = 2.7 – 4.0$ dS/m
Moderately tolerant: Max. $EC_w = 4.0 – 5.5$ dS/m
Tolerant: Max. $EC_w = 5.5 – 6.8$ dS/m
Very tolerant: Max. $EC_w > 6.8$ dS/m
[c] In Florida, Algerian ivy is categorized as a salt-tolerant plant.

Note: EC_w = electrical conductivity of the irrigation water. Salinities exceeding the maximum permissible water salinity (Max. EC_w) may cause leaf burn, loss of leaves, and/or excessive stunting. The maximum values shown were derived from maximum permissible EC_e data by a factor of $EC_e = 1.5\ EC_w$.

Generally, plants with low water requirements and high salt tolerance are preferred for use in landscape areas that are irrigated with reclaimed water. Xeriscape, the method of landscaping with plants that require little or no water, is often recommended in arid and semiarid regions to conserve water.

Plant species preferred for a specific area vary with climatic, geological, and cultural conditions. Guidelines and recommendations for plants are usually available from local nurseries, university extensions, and city agencies. In St. Petersburg, Florida, for example, large-scale research projects were carried out during the 1980s to investigate the effects of reclaimed water on landscape plants. In "Project Greenleaf," a total of 203 plant species were examined for the tolerance to reclaimed water irrigation (Parnell, 1988). A summary of the findings from Project Greenleaf is presented in Sec. 18-7.

Irrigation Systems

Landscape irrigation systems with reclaimed water consist of a water reclamation process and a distribution system, including pumps, flowmeters, distribution piping and tubing, and the sprinklers/emitters. A summary of the typical components of a landscape irrigation system is shown in Table 18-5.

In many states, irrigation methods are specified along with the reclaimed water quality requirements. As an example, the requirements for reclaimed water quality and the irrigation methods in California are shown in Table 18-6. Various irrigation methods for landscape irrigation using reclaimed water are summarized in Table 18-7, and further discussion is provided in Chap. 17. Surface sprinklers are used most commonly for turf irrigation. Microsprinklers and drip systems are becoming increasingly popular for landscape irrigation because of high irrigation efficiency and low risk of human exposure to reclaimed water. A subsurface irrigation system practically eliminates human exposure to reclaimed water. Examples of sprinklers and emitters used commonly for landscape irrigation are also shown in Table 18-5.

Estimation of Water Needs

The two essential parameters used to estimate the agronomic water needs for plants are the landscaped area requiring irrigation, and evapotranspiration. The estimation of agronomic water needs is similar to the method used to estimate water needs for agricultural irrigation (see Chap. 17). In this section, only the concepts specific to landscape irrigation are discussed.

Landscape Evapotranspiration

The evapotranspiration that occurs in a landscape area is affected by (1) the plant species, (2) density of vegetation, and (3) microclimate of the landscape site (University of California and State of California, 2000). The crop coefficient (K_c) is used to account for these effects for agricultural irrigation. For landscape irrigation, a landscape coefficient, K_L, is used in lieu of the crop coefficient. The landscape coefficient is defined in Eq. (18-1) as:

$$K_L = k_s \times k_d \times k_{mc} \quad (18\text{-}1)$$

where K_L = landscape coefficient
k_s = species factor
k_d = density factor
k_{mc} = microclimate factor

Table 18-5
Summary of components used for landscape irrigation systems

Landscape irrigation system component	Description
Dripline  Inline emitter	Dripline, constructed of UV-stabilized polyethelene plastic, is used for drip irrigation systems. Emitters may be embedded inline, plugged into outside of tubing, or embedded in the wall of tubing. Some manufacturers produce tubing with biocide coating to inhibit biofilm growth. For use with reclaimed water, driplines should be either colored purple or have a purple strip. Driplines are typically installed at a depth of 150 to 300 mm (6 to 12 in.), with in-line emitters spaced at 450 to 600 mm (18 to 24 in.), and driplines installed in parallel and are separated by 200 to 600 mm (8 to 24 in.)
Inline tortuous path emitter  Flowrate controlled by turbulence in labyrinth type flow path	A tortuous path emitter controls the flowrate, based on turbulent flow through a restricted labyrinth pathway. The flowrate will vary with changes in pressure due to changes in elevation or pump output. Some emitters are coated with herbicide to reduce root intrusion or biocide to reduce biofilm growth. Emitters are operated at pressures ranging from 0.7 to 3 bar (10 to 45 lb/in.^{2})
Inline pressure compensating emitter  Flowrate controlled by flexible diaphragm which covers orifice and adjusts flow path width proportional to pressure	Pressure compensating emitters with an internal diaphragm produce flow that does not vary with changes in line pressure. They are used typically in areas where changes in elevation would cause variable flowrate in nonpressure compensating emitters. Emitters may be less sensitive to biofilm-type clogging, due to flushing action of diaphragm. Root intrusion may depress the diaphragm and reduce the pressure compensating feature. Some models are coated with herbicide to minimize root intrusion or biocide for biofilm control. Emitters are operated at pressures ranging from 0.7 to 3 bar (10 to 45 lb/in.^{2})
Microsprinkler Microsprinkler head	Microsprinklers, sprayers, and jets can be used with flower beds, ground cover, and orchards. Full-, half-, or quarter-circle sprayers can be used at pressures ranging from 0.7 to 2 bar (10 to 30 lb/in.^{2}). Flowrate and the sprinkling radius will vary with the operating pressure, with typical flowrates ranging up to 110 L/h (30 gal/h)
Pop-up sprinkler Pop-up rotor	Pop-up sprinklers can be used for a variety of plants including grass, ground cover, flowerbeds, and shrubs. The rotor type, commonly used for lawns, can be operated at a radius of 3, 4, and 5 m (10, 12, and 15 ft). Full-, half-, or quarter-circle sprinklers are available. For uniform distribution of water, pop-up sprinkler heads should rise above the height of the plants (such as grass) to be irrigated. Operating pressures range from 2 to 2.8 bar (30 to 40 lb/in.^{2})

(Continued)

Table 18-5

Summary of components used for landscape irrigation systems (*Continued*)

Landscape irrigation system component	Description
Screen filter	An in-line filter, also known as a spin or vortex filter, consists of a plastic housing used to enclose a stainless steel mesh screen. Water to be filtered is applied to the interior portion of the cylindrical screen; filtered water is transported through the screen. Particulate matter, deposited on the inside of the screen, gradually accumulates at the bottom of the housing where it can be discharged by opening a valve
Disk filter	The disk filter is similar in action to the vortex filter but consists of a tightly packed stack of plastic disks with grooves for filtration. The grooves cut into the plastic disks allow for water to filter through while oversized solids are removed. On some models, a valve is located at the bottom of the housing to allow for flushing of the accumulated solids
Control panel	To automate dosing of a drip irrigation field, control panels are used to control pump operation, actuate solenoid valves for zone dosing and flushing, and backwashing of filtration devices. Pump run-time and number of dosing cycles may also be recorded to assist in system monitoring. A programmable logic controller is used to set timer operation during normal and abnormal flow events. System alarm functions are used to activate an audible/visual alarm, or, if configured with a modem, to dial a service provider
Float switches	Float switches, typically in groups of three or four, are connected to control panel or used directly to control pumping. Float switches, mechanical or mercury immersion type, are normally open or closed, depending on system needs. A low-level float switch is used to prevent pumping if insufficient water is available; a midrange float is for normal operation, and a high float for high water operation. A fourth float may be used above the high water float to signal an alarm condition
Telemetry system	Moisture sensors can be installed at the irrigated area to monitor moisture content of the ground. Data from the sensors are sent to the control center, where irrigation rates are adjusted to maximize the irrigation efficiency

Table 18-5

Summary of components used for landscape irrigation systems (*Continued*)

Landscape irrigation system component	Description
Automatic zone valve	An automatic multizone valve is used to sequentially direct water to different irrigation zones. Hydraulic pressure is used to advance an internal spring-loaded cam for discharge to the zones in sequence. An automatic valve may be used to reduce the number of solenoid valves, and thus simplify some of the system electronics
Solenoid valve	Solenoid valves, used for effluent irrigation, are operated automatically using a control system. Solenoid valves are opened and closed using low-voltage signals to dose-specific zones or to flush filters and driplines
Air relief valve	Air relief and combination air/vacuum relief valves are located at all local high elevation points of each zone and on both the supply and return manifold to purge air from the system at the start of dosing and to prevent the formation of a vacuum at the end of a dosing cycle. The valve should be effective over the entire range of maximum and minimum operating pressures and must be located at elevations above the dripline laterals. Formation of vacuum in the dripline may cause soil particles to enter the emitter and cause premature fouling
Pressure regulator	Pressure regulators may be of the fixed or variable pressure type. A compression spring located inside of the fitting is used to provide a semiconstant pressure on the downstream side. Pressure regulators must be able to accommodate the maximum pressure from the system pump and provide sufficient pressure to the drip emitters. The use of pressure regulators is usually not necessary with pressure-compensating emitters
Check valves	Check valves are installed in return manifolds to prevent backflow into the dripline. Check valves are also used in the supply manifold when pumping water to a higher elevation to prevent backflow into the pump basin

(*Continued*)

Table 18-5

Summary of components used for landscape irrigation systems (*Continued*)

Landscape irrigation system component	Description
Dripline installation 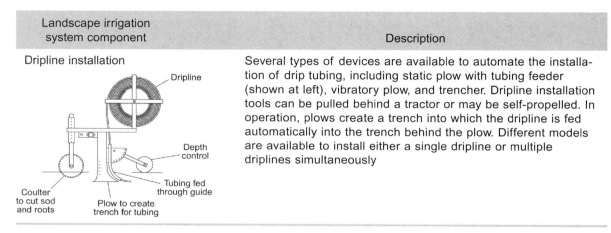	Several types of devices are available to automate the installation of drip tubing, including static plow with tubing feeder (shown at left), vibratory plow, and trencher. Dripline installation tools can be pulled behind a tractor or may be self-propelled. In operation, plows create a trench into which the dripline is fed automatically into the trench behind the plow. Different models are available to install either a single dripline or multiple driplines simultaneously

Typical ranges of each coefficient factor are summarized in Table 18-8. For turf irrigation, a conventional crop coefficient can be used (see Table 17-21 in Chap. 17). Unlike an agricultural field where all the plants are of the same species, a mixture of several plants is usually used for a landscape area, making it difficult to estimate precisely the landscape coefficient. The method for estimating the landscape coefficient in Eq. (18-1) is approximate and judgment by experienced professionals may be necessary to more accurately establish this coefficient.

Table 18-6

Reclaimed water uses for landscape irrigation and irrigation methods in California[a]

	Reclaimed water conditions in which use is allowed			
Uses	Disinfected tertiary	Disinfected secondary with 2.2 total coliform/100 mL	Disinfected secondary with 23 total coliform/100 mL	Undisinfected secondary
Parks, playgrounds, school yards, residential yards, and golf courses associated with residences	Spray, drip or surface	Not allowed	Not allowed	Not allowed
Restricted access golf courses, cemeteries, freeway landscapes	Spray, drip or surface	Spray, drip or surface	Spray, drip or surface	Not allowed
Ornamental plants for commercial use	Spray, drip or surface	Spray, drip or surface	Spray, drip or surface	Not allowed

[a] Adapted from Tchobanoglous et al. (2003).

Table 18-7
Irrigation methods used for landscape irrigation

Irrigation method	Likelihood of exposure	Major causes of exposure	Measures for exposure control	Other remarks
Sprinkler	Moderate	Accidental ingestion of water Inhalation of aerosol Indirect exposure by touching irrigated surface	Timing of irrigation Buffer zone from public access Signs indicating reclaimed water irrigation	Relatively low cost Less problem of clogging than other methods High pressure required
Drip	Low	Indirect exposure by touching irrigated surface	Signs indicating reclaimed water irrigation	Low pressure required Clogging
Subsurface drip	Negligible	Similar exposure as drip system due to inappropriate installation of irrigation system	Signs indicating reclaimed water irrigation	Low pressure required Clogging Relatively high installation and maintenance costs Soil crack and shallow installation may cause water to be exposed to surface

The landscape evapotranspiration, ET_L (mm/d), is then defined as:

$$ET_L = K_L \times ET_o \qquad (18\text{-}2)$$

where K_L = landscape coefficient
ET_o = reference evapotranspiration (mm/d)

Estimation of the landscape coefficient is illustrated in Example 18-1.

Table 18-8
Typical values for landscape coefficient factors[a]

Value	Species factor,[b] k_s	Density factor, k_d	Microclimate factor, k_{mc}
High	0.7 – 0.9	1.1 – 1.3	1.1 – 1.4
Moderate	0.4 – 0.6	1.0	1.0
Low	0.1 – 0.3	0.5 – 0.9	0.5 – 0.9
Very low	<0.1		

[a]Adapted from University of California and State of California (2000).
[b]Species factor values may change during the year, particularly for deciduous species.

EXAMPLE 18-1. Estimation of Landscape Coefficient.

Estimate the landscape coefficient for the mixed planting of bougainvillea, coyote brush, oleander, purple hopseed, and olive trees in a park in the coastal area of southern California. The landscape plants cover the whole ground and are exposed to the sun all day. The site typically has moderate wind.

Solution

1. Select the value of the species factor, k_s.

   All species are classified as plants with low water requirements. From Table 18-8, the value of the *low* species factor ranges from 0.1 to 0.3. In this example, the middle value 0.2 is selected for k_s as no additional information is available about the plants.

2. Select the value of the density factor, k_d.

   The density factor is in the *high* range as there are various plants to form layers of vegetation and the ground is fully covered. An average k_d value of 1.2 is selected from Table 18-8.

3. Select the value of microclimate factor, k_{mc}.

   The microclimate factor describes the specific climate conditions at the landscape area. This site is temperate; neither cold nor hot; dry nor wet. Therefore, the value of 1.0 for the microclimate factor is selected from Table 18-8.

4. Calculate landscape coefficient, K_L.

   The landscape coefficient, K_L, is calculated using Eq. (18-1) as:

   $$K_L = k_s \times k_d \times k_{mc} = 0.2 \times 1.2 \times 1.0 = 0.24$$

Comment

Estimation of water requirements is not as accurate as for agricultural irrigation because (1) the yield of the plant is not critical and (2) it is difficult to estimate water requirements for mixed vegetation.

Irrigation Efficiency

Water sprayed on plant surfaces or excess water on soil surfaces may be lost by evaporation before it is used beneficially. Water may also be lost by runoff from the surface, and watering outside of the landscape area. As defined in Chap. 17, irrigation efficiency is the percentage of water applied to the field that is used beneficially [see Eq. (17-6)]. The factors affecting irrigation efficiency are:

- Uniformity of water application
- Water loss due to the irrigation method
- Water loss in the conveyance system
- Application rate and timing

Table 18-9
Factors affecting water losses during water application in landscape irrigation

Irrigation method	Improper water management[a]	Evaporation from plant surface	Evaporation from soil surface	Evaporation from water surface[b]	Runoff and deep percolation	Leakage from conveyance system
Sprinkler	√	√	√	√	√	√
Microirrigation	√	√	√		√	√
Subsurface	√				√	√

[a] Improper water management includes applying water where it is not needed or in excessive amounts, and the water table is maintained too high or too low.
[b] Evaporation from the water surface will occur when water is applied in excessive amounts resulting in ponding.

An irrigation system needs to be designed and maintained to maximize the efficiency of irrigation, but the unavoidable loss of water must be accounted for in the estimation of water needs. Water losses to be considered for various landscape irrigation methods are shown in Table 18-9. It is difficult to estimate the irrigation efficiency for landscape areas and, to date, a standard method to estimate efficiency is not available. In practice, the irrigation efficiency may be estimated based on an assessment of the design and performance of the irrigation system, or by setting design and management goals. If a similar functioning landscape site is located nearby, the irrigation efficiency may be measured from the flowmeter readings of the existing site and an estimate of evapotranspiration. An irrigation efficiency of 80 to 90 percent can be achieved in a well-designed and maintained system. The range of typical irrigation efficiency for landscape areas is between 65 and 90 percent (University of California and State of California, 2000).

Given an estimated value for irrigation efficiency, and rearranging the terms from Eq. (17-6), the total water applied, F_f, is calculated as follows:

$$F_f = \frac{F_b}{E_i} \times 100 \quad (18\text{-}3)$$

where F_f = agronomic water requirement, mm/unit time
F_b = beneficially used water, mm/unit time
E_i = irrigation efficiency, %

The agronomic water requirement for the landscape area is estimated on a monthly basis using Eq. (18-3). The term, *beneficially used water*, in Eq. (18-3) corresponds to the landscape evapotranspiration, ET_L, determined using Eq. (18-2). The procedure for determining the agronomic water requirement is illustrated in Example 18-2.

EXAMPLE 18-2. Determination of Agronomic Water Requirements.

The office park in Example 18-1 has a total irrigation area of 4 ha. Estimate the total water to be applied using the following monthly evapotranspiration and precipitation data obtained from a public database, and under following conditions:

1. Leaching requirement (LR) is 11 percent when irrigation is necessary.
2. No irrigation is necessary when precipitation is greater than evapotranspiration.
3. The unit application efficiency is 80 percent throughout the year.
4. Use the landscape coefficient determined in Example 18-1 ($K_L = 0.24$).

Month	Reference evapotranspiration (ET_o), mm/mo	Precipitation (P), mm/mo
January	510	110
February	610	750
March	910	180
April	1090	18
May	1470	3.0
June	1220	0.0
July	1520	0.0
August	1440	1.0
September	1100	0.0
October	780	38
November	560	75
December	390	250

Solution

1. Setup a computation table and calculate the estimated evapotranspiration for the irrigation area using the landscape coefficient from Example 18-1 and Eq. (18-2). For example, evapotranspiration for January is estimated as:

$$ET_L = K_L \times ET_o$$

$$ET_L = 0.24 \times (510 \text{ mm/mo}) = 120 \text{ mm/mo}$$

18-2 Design and Operational Considerations for Reclaimed Water Landscape Irrigation Systems

Month	Reference evapotranspiration (ET_o), mm/mo (a)	Evapotranspiration (ET_L) for the irrigated area, mm/mo (b)	Precipitation (P), mm/mo (c)	ET_L-P, mm/mo (d)	Minimum water requirements, mm/mo (e)
January	510	120	110	10	10
February	610	150	750	−600	0
March	910	220	180	40	40
April	1090	260	18	242	242
May	1470	350	3.0	347	347
June	1220	290	0.0	290	290
July	1520	360	0.0	360	360
August	1440	350	1.0	349	349
September	1100	260	0.0	260	260
October	780	190	38	152	152
November	560	130	75	65	65
December	390	94	250	−156	0

2. Subtract precipitation [column (c)] from the calculated evapotranspiration [column (b)] to obtain water budget for each month [column (d)].
3. Determine the minimum irrigation water requirement for each month [column (e)]. The irrigation water requirement is zero when precipitation is more than the evapotranspiration.
4. Determine the hydraulic loading rate, $L_{w(1)}$, defined by the following equation [see Chap. 17, Eq. (17-12)]:

$$L_{w(1)} = \frac{NR}{E_i/100} = (ET_L - P) \times \left(1 + \frac{LR}{100}\right) \times \left(\frac{100}{E_i}\right)$$

where NR = net irrigation requirement, mm/mo
E_i = irrigation efficiency, %
ET_L = landscape evapotranspiration, mm/mo
P = precipitation, mm/mo
L_R = leaching requirement, %

a. Determine the value of the term $\left(1 + \frac{LR}{100}\right)$.

A leaching requirement of 11 percent was given in the problem statement:

$$1 + \frac{LR}{100} = 1 + \frac{11}{100} = 1.11$$

b. Determine the value of the term $\left(\dfrac{100}{E_i}\right)$.

A unit application efficiency of 80 percent was given in the problem statement: $\left(\dfrac{100}{E_i}\right) = \dfrac{100}{80} = 1.25$

c. Using the above values, determine the hydraulic loading rate for each month using the Eq. (17-12). For example, hydraulic loading rate for June is

$$L_{w(1)} = (ET_L - P) \times \left(1 + \dfrac{LR}{100}\right) \times \left(\dfrac{100}{E_i}\right)$$
$$= (290 \text{ mm} - 0) \times 1.11 \times 1.25$$
$$= 402 \text{ mm}$$

The results are shown in column (f) below.

5. Calculate the volume of water required each month to irrigate the landscape area. The required volume for each month is calculated by multiplying the irrigation rate by the irrigation area. For example, the required volume, V, in June is

$$V = (\text{irrigation rate}) \times (\text{the irrigation area})$$
$$= 402 \text{ mm} \times 4 \text{ ha}$$
$$= (402 \times 10^{-3} \text{ m}) \times (4 \times 10^4 \text{ m}^2)$$
$$= 16.1 \times 10^3 \text{ m}^3$$

The results are shown in column (g) below:

Month	Minimum water requirements, mm/mo (e)	Irrigation water requirements, mm/mo (f)	Total volume of water/mo, 10^3 m^3 (g)
January	10	14	0.6
February	0	0	0.0
March	40	56	2.2
April	242	336	13.4
May	347	481	19.2
June	290	402	16.1
July	360	500	20.0
August	349	484	19.4
September	260	361	14.4
October	152	211	8.4
November	65	90	3.6
December	0	0	0
Total	—	2935	117.3

Application Rate and Irrigation Schedule

The primary considerations in determining the rate and timing of irrigation using reclaimed water are:

- Agronomic water requirements
- Peak flowrate and the capacity of the distribution system
- Public exposure to reclaimed water

Many reclaimed water distributors limit pressurized surface irrigation (see Sec. 17-3) of urban landscape areas to the late night and early morning hours (e.g., 9 p.m. to 6 a.m.) to minimize the exposure of the public to reclaimed water. These watering times result in a high peak flow during the nighttime and low (or no) flow for the rest of the day. If the reclaimed water distribution system cannot sustain the required residual pressure at simultaneous peak flow conditions for short durations, the peak irrigation water demand may be attenuated by scheduling irrigation at each site at different times. The timing of irrigation of parks and playgrounds is affected greatly by customer acceptance of wet areas in the morning after irrigation (Young et al., 1998). Because many urban landscape sites are connected to the reclaimed water distribution main, peak flow capacity of the reclaimed water distribution system, the need for storage, and the demand of reclaimed water for other uses must be considered when deciding the irrigation schedule. Details of the reclaimed distribution system are described in Chap. 14.

Management of Demand-Supply Balance

Generally, the flow of wastewater is fairly constant throughout the year and so is the supply of reclaimed water. The demand for reclaimed water for landscape irrigation, however, fluctuates greatly with the climatic conditions of the irrigated area. The demand-supply balance of reclaimed water is managed by using storage or flow equalization, augmenting the supply using other water sources, applying the excess reclaimed water for other purposes (e.g., groundwater recharge), or discharging excess water to receiving waters. When discharge to receiving waters is restricted and excess reclaimed water needs to be applied to the irrigated area, the irrigation rates are determined based on the water infiltration rate (see Eq. 17-13).

Storage Facilities
Storage facilities may be located at the water reclamation plant or near the site of landscape irrigation. Ponds are used commonly for storing irrigation water at golf courses. Covered and underground storage facilities are more costly options than ponds but they are becoming increasingly popular. The advantages of covered and underground storage include: (1) no evaporation loss, (2) lower potential for algal growth, (3) reduction or elimination of odor emission, (4) low visibility especially for the underground storage, (5) no public access to reclaimed water, and (6) no influence of runoff from rainfall. Detailed considerations for storage facilities are discussed in Chaps. 14 and 17.

Other Demand-Supply Issues
The use of water from other sources (e.g., potable water) during peak irrigation periods, discharge of excess reclaimed water, and the utilization of excess reclaimed water for other purposes are considered when the storage system is not sufficient for the water demand-supply balance. For example, excess reclaimed water in Scottsdale Water Campus, Arizona, is further treated with microfiltration and reverse osmosis, and used

to recharge the aquifer through vadose zone injection (see Chap. 22). In an irrigation supply using multiple sources, the quality of water from other sources needs to be monitored as it may affect salinity and sodicity of the irrigation water as well as the fertilizing requirements of the landscape plants (see Chap. 17).

Operation and Maintenance Issues

Proper operation and maintenance (O&M) of a reclaimed water landscape irrigation system is essential to ensure reliable delivery of reclaimed water. Irrigation systems that deliver reclaimed water have specific O&M issues which include:

- Preventing clogging of irrigation devices
- Controlling runoff
- Managing the short-term effects of reclaimed water constituents
- Mitigating the potential long-term effects on soil, plant, and environment caused by reclaimed water constituents

The short-term effects of reclaimed water constituents include salinity and sodicity problems, specific ion toxicity, foliar damage by sprinkler irrigation, and the impacts of nutrients on groundwater. In some cases, the odor of the reclaimed water has resulted in complaints from reclaimed water users and the general public. The long-term effects of reclaimed water constituents include the accumulation of salts and other constituents in the soil and underlying groundwater. Depending on the geological conditions, continuous irrigation may result in a change in the depth of the water table.

Management of Emitter Clogging

Because of its high water use efficiency, microsprinkler and drip irrigation systems are becoming increasingly popular in both agriculture and landscape irrigation. These systems, however, are more susceptible to emitter clogging than high pressure, high volume sprinkler systems. Emitter clogging must be prevented and managed through the control of reclaimed water quality, filtration, emitter design, and appropriate operation and maintenance (Tajrishy et al., 1994). The physical, chemical, and biological factors involved in emitter clogging are reported in Table 18-10. The hazard rating of drip emitter clogging is shown in Table 18-11. Various filtration devices are available for landscape irrigation systems, which can be installed at the point of use. Examples of commonly used filters are shown on Fig. 18-3. The filtration devices used for landscape irrigation are typically smaller than those used for agricultural irrigation because of relatively small flowrates (see Chap. 17). Three major causes of clogging are described below.

Calcium and Magnesium When the bicarbonate concentration is higher than 2.0 milliequivalents per liter (meq/L) and the pH is above 7.5, calcium, as well as, iron can be precipitated out of solution. Precipitation may occur in between irrigation periods, when water remaining on an emitter opening has evaporated and thus the minerals are concentrated. The addition of ammonia for fertilization raises the pH and may encourage precipitation of calcium and magnesium (Keller and Karmeli, 1975).

Bacteria and Algae Biological growth is a critical problem for the maintenance of reclaimed water distribution and storage systems. Chlorine residual in the reclaimed water distribution lines must be monitored and maintained, but the level of chlorine

Table 18-10
Physical, chemical, and biological factors involved in emitter clogging[a]

Cause of clogging	Management strategies
Physical	

Inorganic materials
  Sand (50 – 250 μm)
  Silt (2 – 50 μm)
  Clay (<2 μm)
	Use filter for particle removal

Organic materials
  Aquatic plants
    Phytoplankton, algae
  Aquatic animals
    Zooplankton, snail
  Bacteria
	Periodic flushing of the distribution system, also periodic chlorination

Plastic cutting, lubricant residue	Inspection and flushing before use
Chemical	

Alkaline earths, heavy metal cations
  Calcium
  Magnesium
  Iron
  Manganese
Anions
  Carbonate
  Hydroxide
  Silicate
  Sulfide
Fertilizer sources
  Aqueous ammonia
  Iron
  Copper
  Zinc
  Manganese
  Phosphate
	Maintain pH between 5.5 and 7.0
	Apply sodium hypochlorite
	Aerate the irrigation water and keep it in a reservoir until equilibrium is reached
	Precipitate or chelate iron
	Precipitation by chlorine (Cl_2): 0.64 × ferrous iron concentration
	Manganese precipitation by Cl_2: 1.3 × Mn concentration
	Nutrient removal at reclamation plant

Biological	

Algae
Bacteria
  Filament
  Slime
Microbial depositions
  Iron
  Manganese
  Sulfur
	Chlorination
	Algae: 0.5 – 1.0 mg/L continuously or 20 mg/L for 20 min in each irrigation cycle
	Iron oxidizing bacteria: 1 mg/L more than Fe concentration
	Maintain 1 mg/L free Cl residual
	Periodic flushing of distribution system

[a] Adapted from Bucks et al. (1979), USDA (1987).

Table 18-11

Water quality classification relative to its potential for drip emitter clogging[a]

Clogging factors	Hazard rating		
	Minor	Moderate	Severe
Physical			
Total suspended solids, mg/L	<50	50–100	>100
Chemical			
pH	<7.0	7.0–8.0	>8.0
Total dissolved solids, mg/L	<500	500–2000	>2000
Manganese, mg/L	<0.1	0.1–0.5	>0.5
Total iron, mg/L	<0.2	0.2–1.5	>1.5
Hydrogen sulfide, mg/L	<0.2	0.2–1.5	>1.5
Biological			
Bacterial number, count/mL	<10,000	10,000–50,000	>50,000

[a]Adapted from Nakayama and Bucks (1991).

should be monitored and controlled to avoid adverse effects on landscape plants (see Table 17-5) and other water reuse applications. Reclaimed water distribution lines may be flushed periodically by shock chlorination to eliminate biological growth. Algae can grow extensively when reclaimed water is stored in an open reservoir. By incorporating nutrient removal at the water reclamation process, the potential for algal growth can be reduced significantly. Algal growth can also be prevented by using a covered reservoir and black tubing. Copper sulfate can be used to prevent the algal growth in open reservoirs such as ponds in golf courses. Due to the toxicity of the copper, however, the use of copper sulfate is declining, and herbicides use is increasing.

Iron-Oxidizing Bacteria The soluble form of iron, ferrous ion (Fe^{2+}), is an energy source for iron-oxidizing bacteria. When ferrous iron is exposed to oxygen or iron oxidizing bacteria, it is oxidized to ferric iron (Fe^{3+}) and precipitates as ferric hydroxide,

(a) (b)

Figure 18-3

Filtration devices for reclaimed water used for landscape irrigation: (a) filter used for large-scale irrigation such as golf courses, and (b) disk filter installed in reclaimed water distribution line for commercial properties and individual homes.

$Fe(OH)_3$. The growth of iron-oxidizing bacteria and the resulting precipitated ferric hydroxide may cause clogging problems when the iron concentration is higher than 0.2 mg/L under typical pH conditions (4.0 to 8.5) (Nakayama and Bucks, 1991).

Control of Runoff

Runoff from landscaped areas needs to be minimized to avoid human contact and environmental pollution, and to minimize water-wasting. Runoff is usually considered as a waste discharge and potentially subject to discharge permits. Selection of appropriate irrigation methods and controlling irrigation rates are two ways to eliminate runoff. Using part circle sprinklers can prevent irrigation water from spraying outside of the landscaping boundary. Drip and subsurface irrigation systems can also be used to eliminate off-property spraying. The irrigation rate should be established so it does not exceed the infiltration rate of the irrigated soil. When the infiltration rate varies within the irrigated area, the irrigation rate needs to be established based on the lowest infiltration rate. Drainage systems can be used to collect excess irrigation water to control runoff water.

Salinity and Sodicity Problems

Salt, namely sodium chloride, is not consumed in significant quantities by plants or lost by evaporation. Accumulated salt can cause salinity and sodicity problems, as described in Chap. 17. The management options commonly practiced for reclaimed water-landscape irrigation systems include (USDA, 1993):

- Blend reclaimed water with low salinity water such as stream water, groundwater, and potable water
- Use salt-tolerant plants
- Leach salt out of the root zone by applying excess water
- Modify soil chemical characteristics by using chemicals such as gypsum
- Install drainage systems to artificially remove water of high salt content

Modification of the soil profile with chemicals is commonly used for a reclaimed water irrigation site to improve soil permeability. Gypsum (calcium sulfate, $CaSO_4$) is the chemical used the most for soil permeability management because of its cost, ease of use, and effectiveness. Chemical treatment will be effective where the irrigation water is low in salinity or where the sodium adsorption ratio (SAR) is high (see Chap. 17).

Specific Ion Toxicity

As described in Chap. 17, the most common ions that can accumulate in soil and exhibit toxicity are sodium, chloride, and boron. Some heavy metals such as copper, nickel, zinc, and cadmium may occur in reclaimed water. Trace levels of these heavy metals in reclaimed water may cause toxicity to plants in the long term, but unlike agricultural irrigation, accumulation of heavy metals in plant tissue is not a public health concern.

Management of Foliar Damage

Foliar damage of landscape plants is caused by elevated salinity levels in sprinkler-irrigated water, and by deficiencies of essential minerals in irrigated water, such as iron (iron chlorosis). The foliar damage from sprinkler irrigation is often more noticeable with reclaimed water than with municipal water. Among the ornamental trees commonly planted in the arid southwestern part of the United States, olive, mesquite, aleppo pine, Mondell pine, African sumac, Stone pine, and Raywood ash are relatively tolerant to foliar damage from sprinkler irrigation with reclaimed water. The plants sensitive to salinity-caused foliar damage, such as Modesto ash and Chinese pistache, should be avoided when reclaimed water is used in sprinkler irrigation systems (Jordan et al., 2002). The use of low-profile sprinklers, microsprinklers, and drip irrigation systems can reduce the leaves' contact with reclaimed water, thereby reducing the foliar damage on shrubs and trees.

18-3 GOLF COURSE IRRIGATION WITH RECLAIMED WATER

Golf course irrigation uses the largest amount of reclaimed water among landscape applications. The average area of a golf course in the United States is about 61 ha (150 ac), of which typically 32 to 40 ha (80 to 100 ac) needs to be irrigated. Water demand for golf course irrigation varies greatly with climatic conditions, but typical annual use is 190 to 230 $\times$ 10^3 m^3 (50 to 60 Mgal) for irrigation in the East Coast and 300 to 380 $\times$ 10^3 m^3 (80 to 100 Mgal) for irrigation in the southwest. Because of the large volume of water necessary for golf course irrigation, use of reclaimed water is mandated in some states where reclaimed water is available and is safe to use.

Topics discussed in this section include: (1) water quality and agronomic considerations; (2) reclaimed water supply and storage; (3) distribution system design; (4) leaching, drainage, and runoff; and (5) other issues considered in the planning and designing of golf course irrigation systems.

Water Quality and Agronomic Considerations

Major components of a typical golf course irrigation system with reclaimed water are illustrated on Fig. 18-4. Agronomic considerations for the planning and design of reclaimed water irrigation systems for golf courses include: (1) water quality, (2) turf selection, and (3) irrigation method selection. These issues can be addressed more easily if they are considered during the irrigation system design phase of a new golf course rather than after it has been constructed.

Water Quality

Reclaimed water quality must meet the agronomic and regulatory requirements described earlier in this chapter (see also Chap. 17). If the golf course is the primary user of reclaimed water user, the water reclamation process may be selected and/or adjusted to meet the quality requirements specifically for the golf course. As an alternative, the water quality may be modified by an onsite treatment system. Common onsite conditioning of reclaimed water includes pH adjustment (typically 6.5 to 8.4), corrosion prevention, and SAR adjustment. For example, at the water reclamation plant in Daly City, California, gypsum is added to the reclaimed water to adjust the SAR before delivery to an equalization reservoir for the nearby golf course. Fertilizer may be added to reclaimed water at the beginning of the reclaimed water distribution line (see Fig. 18-5), but it is more common to apply fertilizer directly to the turf and other landscape plants as needed. Fertilizer application rates should be adjusted where the reclaimed water contains significant amount of nutrients, but the value of nutrients in reclaimed water is often ignored, resulting in an overapplication of fertilizers (Tanji et al., 2006; see also Chap. 17).

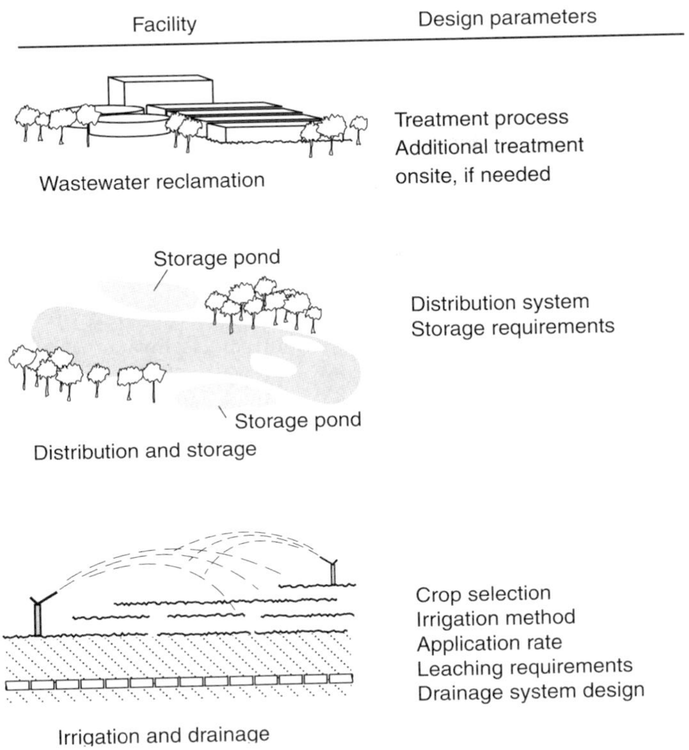

Facility	Design parameters
Wastewater reclamation	Treatment process Additional treatment onsite, if needed
Distribution and storage (Storage pond)	Distribution system Storage requirements
Irrigation and drainage	Crop selection Irrigation method Application rate Leaching requirements Drainage system design

Figure 18-4
Major components of golf course irrigation system.

Figure 18-5
A fertilizer feeding line connected to a reclaimed water distribution line.

In arid regions, it is sometimes difficult to control leaching fractions to maintain salinity levels in root zone, resulting in a buildup of salts at the end of a dry season. Seasonal rainfall is usually sufficient to lower the salinity level in the root zone; however, additional management measures may be necessary if the salinity level is elevated over years of irrigation with reclaimed water. Water from other sources that contains lower salinity may be blended with reclaimed water to lower the salinity level throughout the irrigation season, or water containing low salinity may be used periodically to flush excess salts from the root zone.

Turf Selection

The most critical parameter for turf selection is the salinity level of the reclaimed water. Most turfgrass is not affected significantly by soil water salinity that is less than 3 dS/m (TDS ≈ 1920 mg/L). However, the suitability of reclaimed water for irrigation must be evaluated in light of site-specific conditions (Harivandi et al., 1992). Salt tolerance of turfgrass is reported in Table 18-12. The types of turfgrass that are sensitive to salt, such as annual bluegrass (*Poa annua*), should be avoided when reclaimed water is used for turf irrigation. Salt tolerance of other typical landscape plants was discussed previously in Sec. 18-2 and salt tolerance ratings are given in Table 18-4. Turfgrasses on greens and tees are sensitive to irrigation water quality because they are stressed by high traffic and are cut low to the ground. The use of potable water, or blending reclaimed water with potable water, may be needed if adverse effects of reclaimed water constituents are observed on the turfgrass at greens and tees.

Irrigation Method Selection

The irrigation methods used commonly at golf courses include sprinklers, and surface and subsurface drip systems. The most common method of turf irrigation with reclaimed water is by sprinkler application. Subsurface irrigation is becoming popular, especially for tees and greens, because of high irrigation efficiency and very limited human exposure to reclaimed water.

Clogging of sprinkler heads by particles is not a significant issue when the suspended solids are at the levels observed typically in tertiary-treated reclaimed water and the maximum particle size is controlled. As described in the previous section, however, appropriate management measures need to be taken to prevent emitter clogging of surface and subsurface drip irrigation systems (also see Sec. 17-4).

The entire irrigation system within a golf course is often controlled through a telemetry system and a centralized control station (see Table 18-5). Moisture sensors are placed adjacent to the control valves, and the timing and amount of irrigation are controlled to optimize the irrigation rates. Irrigation is typically restricted to nighttime and early morning hours when no golfers are present, and the turf should be dry by the time golfers are ready to play.

Reclaimed Water Supply and Storage

The agronomic application rate of reclaimed water is determined by ET_o and leaching requirements, as described in Sec. 18-2 and in greater detail in Chap. 17. The capacity of the reclaimed water delivery systems is designed based on the required peak flow, as discussed in Chap. 14. Depending on the relative locations of a water reclamation plant and a golf course, it may be feasible to install a reclaimed water distribution line

Table 18-12
Estimated salt tolerance of common turf grasses[a,b]

Common name (botanical name)	Tolerance[c]			
	Sensitive	Moderately sensitive	Moderately tolerant	Tolerant
Cool-season turfgrass				
Alkaligrass (*Puccinellia* spp.)				√
Annual bluegrass (*Poa annua* L.)	√			
Annual ryegrass (*Lolium multiflorum* Lam.)		√		
Chewings fescue (*Festuca rubra* L. spp. commutate Gaud.)		√		
Colonial bent grass (*Agrostis tenuis* Sibth.)	√			
Creeping bent grass (*Agrostis palustris* Huds.)		√		
Creeping bent grass cv. Seaside			√	
Creeping red fescue (*Festuca rubra* L. spp. *rubra*)		√		
Fairway wheatgrass [*Agropyron cristatum* (L.) Gaertn.]			√	
Hard Fescue (*Festuca longifolia* Thuill.)		√		
Kentucky bluegrass (*Poa pratensis* L.)	√			
Perennial ryegrass (*Lolium perenne* L.)			√	
Rough bluegrass (*Poa trivialis* L.)	√			
Slender creeping red fescue cv. Dawson (*Festuca ruba* L. spp. *trichophylla*)			√	
Tall fescue (*Festuca arundinacea* Schreb.)			√	
Western wheatgrass (*Agropyron smithii* Rydb.)			√	
Warm-season turfgrass				
Bahiagrass (*Paspalum notatum* fluegge)		√		
Bermudagrass (*Cynodon* spp.)				√
Blue grama [*Bouteloua gracilis* (H.B.K.) Lag. ex steud.]			√	
Buffalo grass [*Buchloe dactyloides* (Nutt.) Engelm.]			√	
Centipedegrass [*Eremochloa ophiuroides* (Munro) Hackel]	√			
Seashore paspalum (*Paspalum vaginatum* Swartz.)				√
St. Augustine grass [*Stenotaphrum secundatum* (Walter) Kuntze]				√
Zoysiagrass (*Zoysia* spp.)			√	

[a]Adapted from Harivandi et al. (1992).

[b]The ratings are only to indicate general difficulty in establishing and maintaining the turfgrass. The grasses may tolerate higher salt levels with good growing conditions and optimum care.

[c]Based on soil water electrical conductivity (EC_e). Sensitive ≤3 dS/m, moderately sensitive 3–6 dS/m, moderately tolerant 6–10 dS/m, tolerant ≤10 dS/m.

dedicated to the golf course. Reclaimed water storage facilities may be needed to ensure that the reclaimed water supply meets the daily and seasonal uses.

Lakes and ponds can be used for reclaimed water storage and integrated into the golf course to provide challenging obstacles for golfers. If a golf course uses reclaimed water for irrigation and is adjacent to surface waters that cannot be used for the discharge of excess reclaimed water, additional storage may be necessary to ensure all the reclaimed water produced is used by the golf course (Terrey, 1994). In some golf courses, lakes and ponds also collect stormwater runoff. In the event of heavy rain, incidental overflow of the ponds containing reclaimed water may occur. For a reclaimed water storage pond which also serves as a stormwater collection system, the golf course may need to obtain a National Pollutant Discharge Elimination System (NPDES) permit, and the ponds may have to be designed and/or modified to increase their capacity for stormwater (see Fig. 18-6). Estimating storage requirements based on irrigation water demand is discussed in Chaps. 14 and 17.

Figure 18-6

A reclaimed water storage pond in a golf course deepened to hold runoff from a 100 year storm: (a) deepened pond under construction; (b) excavated pond being lined to minimize percolation losses; (c) view of completed pond; and (d) view of golf course irrigated with reclaimed water. (Photos courtesy of B. Buchanan, City of Roseville.)

Table 18-13
Algae and weed control for reservoir lakes in golf courses[a]

Design/operation tasks	Control option	Remarks
Water reclamation process	Conventional treatment technologies augmented by nitrification, denitrification, or phosphorous removal Advanced treatment	Well established, and adopted at many water reclamation plants
Design of lakes	Narrow and deep lakes	Control by limiting sunlight Poor design and lack of circulation may induce an anoxic condition
Control of algae in the lakes	Aeration by fountains, falls and streams, or mechanical aeration devices	Aesthetically appealing to golfers
	Chemical addition	Copper sulfate and/or aquatic herbicides. Due to copper toxicity, the use of copper sulfate is declining
	Aquatic herbivores (fish)	Low maintenance cost

[a] Adapted from Terrey (1994).

Control of algae and weed growth is a concern in operation and maintenance of golf course lakes and ponds, and should be considered during the design phase of a golf course. Measures to control algae growth in the lakes are discussed in Table 18-13.

Distribution System Design Considerations

The irrigation distribution system conveys the reclaimed water from the transmission system to the application points on the golf course. A distribution system should be designed to minimize dead ends, which can be accomplished by establishing a loop in the distribution system (see Chap. 14). Valves or hydrants for flushing the pipelines should be provided at all dead ends and low points.

Automated systems typically control the time, duration, and application rate of reclaimed water to the turf areas. System settings are established based on course hours of operation and the agronomic water requirements [see Eq. (18-4) and Example 18-2]. Pipes and sprinkler heads for the reclaimed water irrigation system should be easily identifiable; pipes are typically colored purple plastic or marked with identification tape. For a retrofit project, however, typically only the replaced portion of buried pipe is specifically identified as a reclaimed water pipe (Steinburgs, 1994).

In arid regions, dual distribution and plumbing systems with reclaimed water and another source of water such as potable water may be installed for specific areas such as tees and greens, if the effects of salinity by reclaimed water irrigation need to be mitigated using water of low salinity. A dual distribution system may also be considered upstream of the storage ponds to blend low salinity water and reclaimed water for the entire irrigation system. Where potable water is used as the low salinity water, it is important to take appropriate measures to avoid cross-connection. Backflow prevention valves and/or air-gap backflow prevention should be installed on the potable water system, and colored pipes or tapes (normally purple) must be used for the reclaimed water distribution lines (see Chaps. 14 and 15).

Leaching, Drainage, and Runoff

Because water requirements for turfgrass tend to be higher than other types of landscaping plants such as trees and shrubs, and golf courses need to maintain the turf in good condition for golfers, water application rates for golf courses are generally higher than most other landscape irrigation applications. Intense application of water and the need to maintain a good turf condition necessitate careful consideration of leaching and drainage requirements.

Leaching and Drainage

As described in Chap. 17, a drainage system needs to be installed when the groundwater table is not deep enough to avoid the accumulation of salt in the root zone. When a new golf course is constructed with a reclaimed water irrigation system, soil under tees and greens can be modified (e.g., use of a sand bed) to ensure high permeability. It is fairly common to install drainage systems on the greens and tees, where the quality of turf is most crucial, but drainage systems are also important in fairways and roughs (Terrey, 1994). Water of lower salinity, such as potable water, may be used periodically to flush accumulated salts if management of leaching fraction and natural precipitation is not sufficient to control the root zone salinity.

Runoff and Leachate

Runoff and leachate from the irrigation system generally contain higher levels of salt and other chemicals including pesticides and fertilizers. In some cases, with a special permit, the water collected through the drainage system is discharged back to the wastewater collection system. Pretreatment of the collected water may be necessary when the levels of the chemicals are high. Other options include discharge of the runoff and leachate back to the irrigation water reservoir (Terrey, 1994).

Other Considerations

Regulations and guidelines are primarily for public health protection, but some states are requiring the use of reclaimed water to conserve water resources. From an economic perspective, capital costs for constructing reclaimed water systems and all additional operation and maintenance costs associated with reclaimed water use have to be considered. A checklist for planning and designing reclaimed water irrigation systems for golf courses is given in Table 18-14.

18-4 IRRIGATION OF PUBLIC AREAS WITH RECLAIMED WATER

Typically, reclaimed water irrigation systems for public areas are connected to area-wide reclaimed water distribution systems. Most area-wide distribution systems are connected to multiple reclaimed water users, including landscape irrigation in public areas, fire hydrants, cooling systems, and nonpotable in-building uses such as toilet flushing (see Chap. 20). Due to high construction costs of reclaimed water distribution systems in already developed urban areas, satellite and decentralized systems are increasing being

Table 18-14
Checklist for reclaimed water use for golf course irrigation[a]

Soil samples
1. Soil samples should be taken from different sections of the golf course, including tees, greens, fairways, and rough. Sampling should be done in advance of conversion from potable to reclaimed water to allow for tracking of changes due to use of reclaimed water
2. Sample soils on a quarterly basis. Monitoring soils will allow for adjustments in the water irrigation schedule and any necessary mitigating measures

Water quality
1. Establish initial analysis to determine if further treatment is necessary. Water should be analyzed periodically, along with the soil
2. Verify the source of reclaimed water. Reclaimed water from industrial sources can have a greater amount of undesirable elements as compared to reclaimed water originating from residential areas
3. Verify the level of treatment. Determine if filtration is required for the removal of particulate matter
4. Negotiate with the supplying entity to establish maximum levels of biological oxygen demand (BOD), total dissolved solids (TDS), and total suspended solids (TSS). Where possible, it is much easier for the end user to have the treatment plant control undesirable elements

Pumping and water storage
1. If possible, draw water directly from a pressurized source pipeline and apply directly to the golf course
2. If insufficient line pressure exists, install a booster pumping station to increase pressure to meet the golf course sprinkler requirements
3. If water must be stored, the first choice is an enclosed tank, eliminating exposure to sunlight and reducing the formation of algae. The second and most likely choice is storage in lakes. It is preferable to minimize the number of lakes to reduce the problem of algae control
4. The deeper the lake, the better. A deep lake helps reduce sunlight penetration, the water stays cooler, and algae control systems are more effective. Consider the addition of a mechanical aeration device to circulate the contents and maintain aerobic conditions
5. As an alternative to aeration, blending wastewater with freshwater in a pond is usually helpful in controlling BOD, TDS, and TSS. Blending requires separation of the sources to prevent the reclaimed water from contaminating a freshwater source
6. A dual irrigation system is preferred if freshwater is available. Freshwater should be applied to greens, tees, ornamental lakes, and other sensitive plantings

Miscellaneous
1. Drinking fountains need to have self-closing covers. Check with the plumbing code and local authorities
2. Signs should indicate "reclaimed water used to irrigate turf"
3. If reclaimed water is not used currently and there is a possibility that it will be used in the future, consider designing the irrigation system to facilitate easy conversion in the future
   a. Use warning tape or colored pipe depending on local codes
   b. Design for the proper separation of domestic pipelines from reclaimed pipelines (see Chap. 15)
4. Negotiate the hours of operation with the reclaimed water purveyor. Some facilities lower the supply line pressure during daylight hours and increase the pressure at night allowing for direct irrigation without additional pumping
5. Where appropriate, install pressure sensors to shut down the system in the event of operating failures
6. Use an automated system to provide simplified and consistent operation
7. Consider using a backup system, at least for greens and tees. Some reclaimed water treatment plants may shut down periodically for maintenance of their systems

[a]Adapted from Gill and Rainville (1994).

used (see Chaps. 12 and 13). The unique characteristics of, and specific considerations for, public area irrigation with reclaimed water are considered in this section. Irrigation of landscaped areas with minimally treated wastewater using subsurface drip irrigation is an alternative system for the areas where a centralized wastewater collection and treatment system is not available, and is described in Sec. 18-6.

Irrigation of Public Areas

Public areas where reclaimed water can be used for landscape irrigation include street and highway medians; around parking lots, commercial, and business building areas; cemeteries; parks; playgrounds; and school yards (see Fig. 18-7). Sports stadiums are also recognized as potential sites for reclaimed water irrigation. Generally, public areas are categorized as restricted and unrestricted access areas, depending on the likelihood of human contact with irrigation water (see Sec. 18-1). Feasibility of reclaimed water use for landscape irrigation depends on the size of the irrigated area, and proximity of the irrigated area to the reclaimed water distribution system and/or to the water reclamation plant.

Figure 18-7

Examples of public areas irrigated with reclaimed water: (a) ballpark, Dunedin, FL (Coordinates: 28.003 N, 82.787 W); (b) landscape at parking area at a shopping mall, St. Petersburg, FL; (c) public park, Los Angeles, CA; and (d) commercial area, San Diego, CA.

Reclaimed Water Treatment and Water Quality

Both reclaimed water quality and irrigation methods are specified in most water reuse regulations and guidelines to minimize potential human contact with reclaimed water. Typically, tertiary or equivalent treatment is required for unrestricted areas. Although a lower level of treatment is suitable for restricted access areas, reclaimed water landscape irrigation systems are often connected to area-wide reclaimed water distribution systems. Thus, reclaimed water that meets criteria for irrigation of unrestricted areas is used for many restricted areas. In areas where there are multiple users of reclaimed water, a level of quality must be provided that meets the requirements of the majority of users. If a small user requires higher quality water, it may be more cost effective to provide a point-of-use process to meet their needs. The decision about the treatment level must be made based on the existing and potential reclaimed water market and the total costs to serve the market (see Chap. 25).

In most cases, the selection of water reclamation processes depends largely on the existing wastewater treatment facilities and how they can be utilized for water reclamation. As discussed in Chaps. 7 and 8, in the past the most common practice has been to upgrade an existing conventional secondary treatment facility with filtration and improved disinfection to meet the required reclaimed water quality criteria for unrestricted nonpotable uses. With advances in membrane technologies, membrane bioreactors (MBRs) are being used increasingly for water reuse applications, especially where new water reclamation facilities needs to be constructed.

Conveyance and Distribution System

Many landscape sites in urban areas are small, and therefore the reclaimed water demand is limited at each site. Large landscaped areas, such as parks, school yards, and golf courses, and other urban reclaimed-water users (see Chap. 20) may need to be included in an area-wide water reuse system to make the system economically feasible. Another approach is to consider a decentralized or satellite water reclamation and distribution system for a specific landscaped area. In some arid regions where the salinity of the reclaimed water is high throughout a year, blending reclaimed water with low-salinity water should be considered during the planning and design of reclaimed water storage, conveyance, and distribution systems.

Aesthetics and Public Acceptance

It is important to maintain the landscaped areas irrigated with reclaimed water in a condition that is aesthetically appealing as public perception of the use of reclaimed water can be affected greatly by appearance. Reclaimed water may be slightly colored and contain chemical constituents that may stain objects sprayed with the irrigated reclaimed water. The use of surface and/or subsurface drip irrigation can avoid such a problem. Odors can become evident if the reclaimed water distribution system is poorly designed and/or maintained, resulting in an anoxic condition in the reclaimed water distribution system and the generation of hydrogen sulfide (see Sec. 18-2). Periodical flushing, chlorination, and reclaimed water quality monitoring throughout the reclaimed water distribution systems can minimize problems associated with the changes in reclaimed water quality.

Because of the potential risk of accidental ingestion of pathogens of reclaimed water origin, concerns have been raised about the irrigation of school yards and public parks. In theory, ingestion of reclaimed water could occur when children fall or touch the grass and then have hand-to-mouth contact (e.g., eating food without prior hand washing). Although the probability of human exposure to pathogens of reclaimed water origin has been estimated to be extremely low, health concerns remain, namely for children; thus, irrigation of playgrounds and school yards with reclaimed water still could be controversial. As an example, after years of dispute, Redwood City, California, decided to exclude school yards and playgrounds from irrigation with reclaimed water (see Chap. 26).

Operation and Maintenance Issues

Typically, the operation and maintenance of landscape irrigation for public areas is a responsibility of the reclaimed water supplier (e.g., water district, public utilities department, or environmental protection department). Important operation and maintenance issues include: (1) inspection of the construction of new or repaired landscaped areas, (2) cross connection inspection, (3) operation and maintenance of automated irrigation systems and filtration devices, and (4) customer service such as connection and disconnection of the service, metering, and emergency response. An automated irrigation system that is controlled and monitored from a centralized control station through a telemetry system is being used increasingly for landscape irrigation systems in urban areas. Each landscape area typically has a moisture sensor and the amount of water applied to each site is adjusted automatically. A telemetry system is often used for golf course irrigation.

18-5 RESIDENTIAL LANDSCAPE IRRIGATION WITH RECLAIMED WATER

Residential landscape irrigation is a major reclaimed water reuse application in many states and is the focus of this section. Although greywater has been used for irrigation of residential landscaped areas, the primary focus in this section is on the reuse of reclaimed water from municipal sources. Reclaimed water for residential landscape irrigation is treated to meet the criteria for use for the irrigation of unrestricted areas. Examples of a residential landscapes irrigated with reclaimed water are shown on Fig. 18-8.

Residential Landscape Irrigation Systems

Irrigation water for residential areas can be delivered either from the reclaimed water distribution main of an area-wide system or from a satellite or decentralized water reclamation and reuse system. Residential landscape areas include front- and backyards

(a)

(b)

Figure 18-8

Typical examples of residential homes with landscaping irrigated with reclaimed water: (a) El Dorado Hills, CA and (b) St. Petersburg, FL.

of residential homes and public spaces in residential areas such as street median strips. Design criteria for residential landscape irrigation systems with reclaimed water vary, and local regulations and guidelines must be consulted. Typically, sprinkler irrigation is allowed for turf irrigation, but low-profile microsprinklers and drip irrigation are preferred for shrubs. Subsurface drip irrigation can be used to eliminate human contact with reclaimed water.

Reclaimed Water Treatment and Water Quality

Except when a small and decentralized system is used and treated effluent is disposed of by means of irrigation (see Chap. 13), tertiary or equivalent treatment is required for the use of reclaimed water for residential landscaped areas. The reclaimed water distribution systems are typically connected to multiple users, and the water quality requirements are dictated by the principal users, as described in Sec. 18-2.

Conveyance and Distribution System

Where reclaimed water is used for the irrigation of residential landscaping, dual distribution and plumbing systems need to be provided in accordance with the local plumbing code. For example, separation of the potable and reclaimed waterlines is usually specified in the plumbing code (see Chaps. 14 and 15). Where a residential area is to be served by a dual system, extra care is needed to prevent cross-connection of potable and reclaimed waterlines. An example of dual system installation in an individual home is shown on Fig. 18-9. It should be noted that irrigation of the backyard with reclaimed water may not be allowed, depending on local restrictions. Depending on the terms of the service provider, reclaimed water

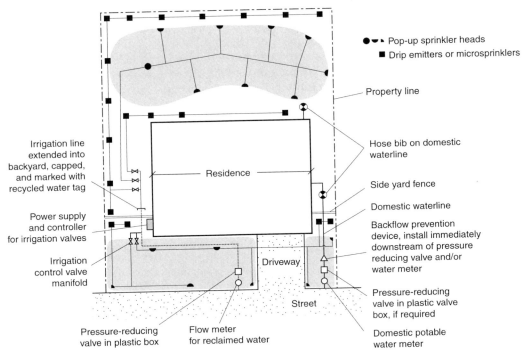

Figure 18-9
Typical design of residential irrigation system using reclaimed water.

Operation and Maintenance Issues

service may be required for all houses connected to the reclaimed water system, or it may be provided only to the houses with a contract, such as in the City of St. Petersburg, Florida.

Typically, operation and management of landscape irrigation for residential areas is a responsibility of the reclaimed water supplier (e.g., water district, public utilities department, and environmental protection department). In some cases where reclaimed water is used in a privately developed community, the homeowners' association may play a significant role in the operation and maintenance of the landscape irrigation system (see Sec. 18-7). Important O&M activities may include: (1) inspection of construction of new residential landscaping, (2) cross connection inspection, (3) operation and maintenance of automated irrigation systems and filtration devices, and (4) customer service such as connection and disconnection of the service, metering, and emergency response. Homeowners are usually responsible for maintaining the plants, turf, and irrigation devices such as sprinklers and drip emitters, but any changes involving the modification of the reclaimed water distribution line must be approved by the reclaimed water purveyor or the homeowners' association.

18-6 LANDSCAPE IRRIGATION WITH DECENTRALIZED TREATMENT AND SUBSURFACE IRRIGATION SYSTEMS

As discussed in Chap. 13, one of the viable options for beneficial use of reclaimed water in decentralized and onsite wastewater treatment and management systems is for landscape irrigation using subsurface irrigation systems. Subsurface irrigation is particularly suitable when the effluent of septic tank is used, as human exposure to reclaimed water can be avoided completely. Because more than 60 million people in the United States live in homes where decentralized systems are used for wastewater management (see Chap. 13), it is useful to discuss the use of reclaimed water for landscape irrigation, with a focus on subsurface drip irrigation systems.

Subsurface Drip Irrigation for Individual Onsite and Cluster Systems

Subsurface drip irrigation can be a reliable and effective method for the distribution of wastewater effluent from onsite wastewater treatment systems to the surrounding environment. In some cases, reclaimed water may be disinfected and distributed using overhead spray irrigation or surface drip irrigation; approaches that may be less costly where the installation of a subsurface system is not feasible. Measures to assure the protection of health and minimization of nuisance conditions such as ponding and runoff, however, must be carefully considered.

Subsurface drip irrigation systems for the use of effluent from an onsite system consist of a treatment process, a pump, and the drip tubing with emitters (see Table 18-5). The treatment process needed depends on the manufacturer of the drip system, but typically consists of a secondary/advanced treatment system. Some drip systems, however, are designed for use with screened septic tank effluent (undisinfected primary effluent). The reclaimed water is collected in a sump and pumped through the drip tubing. Antisiphon valves are placed at the high points in the drip irrigation network to prevent backflow of soil particles into the emitters.

The design of a landscape irrigation system using subsurface drip irrigation system is illustrated in Example 18-3.

EXAMPLE 18-3. Design of a Landscape Irrigation System Utilizing a Cluster Wastewater Treatment System and Subsurface Drip Irrigation.

Using the water balance data presented below, design a landscape irrigation system including the peak irrigation demand, the area to be irrigated, the layout of the drip system, and the pumping requirements for 12 homes, with an average daily flow of 6.8 m³/d (1800 gal/d). The local soil is a clay loam with a recommended maximum reclaimed water application rate of 10.2 mm/d (0.4 in./d). Assume that an irrigation efficiency, E_i, of 90 percent is used.

Month	Time, d/mo	Precipitation (P), mm/mo	Evaporation (ET_L), mm/mo	$ET_L - P$, mm/mo
Jan	31	97.5	20.0	−77.5
Feb	28	90.0	35.0	−55.0
Mar	31	71.1	68.6	−2.5
Apr	30	25.9	111.3	85.4
May	31	13.5	140.0	126.5
Jun	30	5.1	165.1	160.0
Jul	31	1.3	172.5	171.2
Aug	31	1.5	151.9	150.4
Sep	30	9.1	117.6	108.5
Oct	31	22.6	86.1	63.5
Nov	30	55.6	41.4	−14.2
Dec	31	62.2	23.6	−38.6

Note that decentralized reclaimed water applications using drip irrigation are typically sized to meet the maximum irrigation demand. However, as the reclaimed water supply to be dispersed is relatively constant and independent of the irrigation demand, the amount of water applied under constant loading conditions will exceed the irrigation demand during periods of reduced evapotranspiration and is assumed to percolate into the soil.

Solution

1. Select an irrigation loading rate.

   The irrigation loading rate is determined using the water balance presented in Chap. 17. In this example, reclaimed water is applied based on the maximum irrigation demand, i.e., Eq. (17-13b) should be used.

$$L_w = \left(\frac{ET_L}{E_i/100} - P\right) + W_p$$

   Using the value of ET_L, and accounting for the efficiency of the irrigation system used, the first term of Eq. (17-13b), which represents the estimated irrigation demand of the month, is calculated. The peak irrigation demand occurs in July, and the water to be applied with the irrigation system is calculated as

$$\frac{ET_L}{E_i/100} - P = (172.5/0.9) - 1.3 = 190.4 \text{ mm/mo}$$

2. Determine the amount of water expected to infiltrate into the soil for each month using the peak irrigation demand calculated in Step 1 as a uniform hydraulic loading rate, L_w. The amount of water expected to infiltrate is determined from a water balance, computed for January with a correction for irrigation efficiency, as follows:

$$W_p = L_w - \left(\frac{ET_L}{E_i/100} - P\right) = 190.4 - [(20.1/0.9) - 97.5] = 265.7 \text{ mm/mo}$$

All values are summarized in the following table:

Month	Time, d/mo	$\frac{ET_L}{E_i/100} - P$, mm/mo	W_p,[a] mm/mo
Jan	31	−75.2	265.7
Feb	28	−50.8	241.3
Mar	31	5.1	185.4
Apr	30	97.5	92.7
May	31	142.0	48.3
Jun	30	178.3	12.2
Jul	31	190.4	0.00
Aug	31	167.4	23.1
Sep	30	121.7	68.8
Oct	31	73.2	117.3
Nov	30	−9.7	199.9
Dec	31	−36.1	226.6

[a] Based on L_w of 190.4 mm/mo.

3. Check that the maximum rate of infiltration expected is less than the recommended soil loading rate.

The maximum rate of soil infiltration that occurs in January is

$$(265.7 \text{ mm/mo})/(31 \text{ d}) = 8.6 \text{ mm/d}$$

The maximum expected infiltration rate of 8.6 mm/d is less than the recommended maximum application rate for this soil of 10.2 mm/d.

4. Determine the area needed for the irrigation system.

$$A = (6.8 \text{ m}^3/\text{d})/(8.6 \text{ mm/d})(1 \text{ m}/1000 \text{ mm}) = 790.7 \text{ m}^2$$

5. Layout land area for placement of drip irrigation system.

   Choose rectangular land area with width of 26 m and length of 30 m.

6. Select a drip line spacing and emitter spacing.
   a. Drip line spacing is 0.6 m (24 in.)
   b. Emitter spacing is 0.6 m (24 in.)

7. Calculate the length of drip line and number of emitters needed.
   a. Length of drip line = 30 m × (26/0.6) = 1300 m
   b. Number of emitters = (26/0.6) × (30/0.6) = 2166

8. Select an emitter flowrate and determine system flowrate.
   a. Choose an emitter with flowrate of 3.8 L/h
   b. System flowrate = 2166 × (3.8 L/h)/(60 min/h) = 137.2 L/min

9. Select dosing schedule (frequency and duration).

   Choose dose frequency to obtain hydraulic application rate (HAR) less than 2.5 mm/dose.

   A dose frequency of 3 dose/d and a daily application rate of 7.2 mm/d will result in an application rate of

   $$HAR = (7.2 \text{ mm/d})/3 = 2.4 \text{ mm/dose}$$

   The dose duration will be

   $$\text{Dose duration} = (6.8 \text{ m}^3/\text{d})/[(137.2 \text{ L/min})(3 \text{ dose/d})(1 \text{ m}^3/1000 \text{ L})]$$
   $$= 16.5 \text{ min}$$

   The dose volume will be

   $$\text{Dose volume} = (137.2 \text{ L/min})(16.5 \text{ min}) = 2263.8 \text{ L}$$

10. Select a pump to supply the drip irrigation system.

    The pump must be able to supply 137.2 L/min at the calculated total head. Many drip irrigation systems are designed to operate at a pressure of 1.4 to 2.0 bar (20 to 30 lb/in.^{2}) supplied to the drip field. The total head will consist of the static head and friction headloss through the manifold piping, fittings, filters, and valves. The friction headloss can be estimated to be 0.69 bar (10 lb/in.^{2}) for many applications. Assuming a static headloss of 0.28 bar (4 lb/in.^{2}), a pressure of 1.4 bar (20 lb/in.^{2}) required at the drip field, and a friction headloss of 0.69 bar (10 lb/in.^{2}), the pump must supply 137.2 L/min at an estimated total head of 2.37 bar (34 lb/in.^{2}).

Comment

A larger area may be irrigated if makeup water is added. In some cases, drip systems are used with chemicals to reduce biofouling and root intrusion; or biocidal agents are impregnated into the plastic tubing and emitters. As mentioned previously, adequate pretreatment is necessary to ensure that the most problematic constituents are removed in advance of drip irrigation.

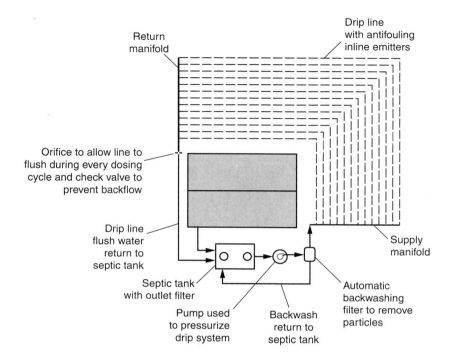

Figure 18-10
Typical design of subsurface irrigation system at a residential home.

Irrigation for Residential Areas

Wastewater collected from a housing development or small community using an alternative collection system may be treated for drip irrigation or other reuse applications. For example, in some developments treated wastewater is recycled for community landscape irrigation. A typical flow diagram of an irrigation system for a residential development is shown on Fig. 18-10. In many areas, the reclaimed water flow during the growing season may not be sufficient to meet landscape irrigation needs and may have to be supplemented or the extent of the landscaping reduced.

18-7 CASE STUDY: LANDSCAPE IRRIGATION IN ST. PETERSBURG, FLORIDA

The reclaimed water distribution system in the City of St. Petersburg, Florida, is the largest urban water reclamation system in the United States. Development of the reclaimed water system began in response to Florida legislature's adoption of the Wilson-Grizzle Act in 1972 (Johnson and Parnell, 1998). The primary objective of the Wilson-Grizzle Act was to reduce wastewater discharge to surface waters. With continuing urban growth and the subsequent shortage of freshwater supplies, the main focus has shifted away from the reduction of wastewater discharge to conservation of potable water. The development and implementation of the reclaimed water system in St. Petersburg is summarized in this section with an emphasis on landscape irrigation applications.

Setting

The City of St. Petersburg is located at the tip of the Pinellas County Peninsula, surrounded by the Gulf of Mexico, Tampa Bay, and Old Tampa Bay on three sides. As of 2003, the population of the city was about 250,000, the fourth largest in the State of Florida.

Water Management Issues

The City of St. Petersburg has been concerned with finding new water sources for more than a century. In the late 1800s, Reservoir Lake (known as Mirror Lake today) was tapped as the city's first major water source. From the early 1900s to the 1970s, the city explored new water sources not only within the city, but also in adjacent areas in Pinellas County as well as in Hillsborough and Pasco Counties by purchasing land and drilling well fields to keep up with increasing water demand.

Water Shortage in St. Petersburg

Continued population growth in the area along with a decrease in rainfall in the early 1970s necessitated St. Petersburg and the three surrounding counties to initiate intergovernmental cooperation to address water issues. Towards the end of the twentieth century, the well fields became highly stressed and St. Petersburg was declared a water-short area. The city prohibited new customers from using potable water to irrigate parcels of land larger than single residential lots, but the population of the area has continued to grow and meeting the area's future water needs is still of great concern (Johnson and Parnell, 1998).

Wilson-Grizzle Act

In 1972, the Florida legislature adopted the Wilson-Grizzle Act. The Act stated, "No facilities for sanitary sewage disposal constructed after the effective date of this act (March 15, 1972) shall dispose of any wastes into Old Tampa Bay, Tampa Bay, Hillsborough Bay, Sarasota Bay, Boca Ciega Bay, St. Joseph Sound, Clearwater Bay, Lemon Bay, and Punta Gorda Bay or any bay, bayou, or sound tributary thereto without providing advanced waste treatment approved by the department of pollution control." The need to abate wastewater discharge led to the development of the water reclamation and reuse system.

Development of Reclaimed Water System

In response to the Florida legislature's adoption of the Wilson-Grizzle Act, the City Council of St. Petersburg decided to implement a water recycling and deep well injection system to achieve zero-discharge. The decision was supported by the results of a pilot study authorized in 1971, in which the feasibility of using reclaimed water for an urban irrigation system was determined. The historical development of St. Petersburg's system between 1970 and 1990 is summarized in Table 18-15. The inception of the first reclaimed water system was in 1977. By 1990, four wastewater reclamation plants (WRPs), Southwest, Northeast, Northwest, and Albert Witted plants, were upgraded to achieve zero-discharge to Tampa Bay. In 2003, the reclaimed water distribution system served over 10,400 active customers with about 460 km (290 mi) of pipelines with a capacity of 136×10^3 m³/d (36 Mgal/d).

Implementation

The unique features of the reclaimed water system in St. Petersburg include: (1) a distribution loop that connects the four WRPs located at the four corners of the city, (2) a deep-well

Table 18-15
Earlier development of St. Petersburg's reclaimed water system[a]

Year	Development
1971	Pilot study for reclaimed water spray irrigation system
1972	Wilson-Grizzle Act
	St. Petersburg City Council decides to implement a recycling and deep well injection system
1974	Federal grant by U.S. EPA ($14 million) to upgrade Southwest plant, design and construct irrigation, and deep-well injection systems
1975	Federal grant by U.S. EPA ($18.8 million) to upgrade the Northeast plant, construct a distribution system, and deep-well injection system
1977	Southwest plant begins providing reclaimed water to the system
	Financial incentive by U.S. EPA for wastewater reuse projects
1978	Federal funds for the 201-Facilities Plan: a $33.4 million federal grant to upgrade the Northwest plant
1980	Expansion of the Northeast plant completed
1981	Construction of over 120 km (76 mi) of major transmission loops is completed
	Reclaimed Water System Master Plan to study expansion of the reclaimed water system into selected residential areas is completed
	City Council adopts policies and regulations for the provision of reclaimed water service
1983	Northwest plant starts to provide reclaimed water to the system
1984	Four critical water quality areas are identified by the city and construction of reclaimed water irrigation system begins [160 km (100 mi) of additional pipeline]
1985	Injection of high chloride effluent to deep-well system at the Northwest plant starts
1986	Project Greenleaf starts
1987	Albert Witted plant upgrade is completed and discharge of its high quality, low chloride effluent to the system begins
	Wilson-Grizzle Act is replaced by the Grizzle-Figg Bill for all wastewater plants to achieve 5 mg/L $CBOD_5$, 5 mg/L TSS, 3 mg/L nitrogen, and 1 mg/L phosphorous by October 1, 1990
	Resources Management Project for landscape irrigation application (1987–1988) is begun
1989	Albert Witted deep wells are permitted for continuous injection and permanently stopping discharge to Tampa Bay

[a] Adapted from Johnson and Parnell (1998).

injection system to discharge excess reclaimed water and the water that does not meet reclaimed water quality standards, and (3) extensive research on the use of reclaimed water for urban landscape irrigation.

Distribution Loop

The four WRPs and the reclaimed water distribution system are shown on Fig. 14-3 in Chap. 14. The reclaimed water distribution system was constructed to form a loop to connect four WRPs at the four corners of the city. The configuration enabled the use of small diameter pipes, thereby significantly reducing the construction cost. The largest pipe used in the distribution main is about 1100 mm (42 in.) in diameter, and most other pipes were less than 900 mm (36 in.).

Deep-Well Injection System
A deep-well injection system was constructed at each WRP as an alternative effluent disposal method to maintain zero-discharge to surface waters. The well systems are designed to dispose of excess reclaimed water and water that does not meet the reclaimed water standard (City of St. Petersburg, 1994).

Research Projects for Landscape Irrigation
Two research projects, Project Greenleaf, discussed below, and Resource Management, were conducted to evaluate the suitability of reclaimed water for landscape irrigation. The list of plants suitable for reclaimed water irrigation can be obtained from the City of St. Petersburg.

Project Greenleaf and Resource Management

Throughout 1985, the City of St. Petersburg received a significant increase in the number of complaints regarding damage of ornamental plants and trees irrigated with reclaimed water. To address the effects of reclaimed water on the plants, a research fund was approved by the city, and the study entitled "Project Greenleaf" was initiated in 1986.

Scope of Project Greenleaf
The objectives of the project were to (Parnell, 1988; Johnson and Parnell, 1998):

- Determine the quality and quantity fluctuations of the influent flow and assess the resulting reclaimed water quality fluctuations and availability
- Assess the effects of reclaimed water on the growth and development of ornamental plants and trees commonly used for landscaping and decorative purposes within the area
- Establish guidelines for the successful use of reclaimed water on ornamental plants
- List necessary management practices for more delicate plant species
- Produce literature both for public education and scientific publications
- Make necessary recommendations to ensure the successful operation of the reclaimed water irrigation system in the future

Findings of Project Greenleaf
From the analysis of wastewater quality data, it was found that saline groundwater and storm water had infiltrated into the sewer lines and affected influent qualities. The effluent quality of the four WRPs was compared with data published previously on the effects of chemical constituents in irrigation water on landscape plants. Eight commonly used landscape ornamental plant species and two saplings were used to investigate the effects of reclaimed water on plant growth and maturation. Three variables; chloride concentration, irrigation method, and open air/indoor conditions, were investigated in the study. The plant growth index was determined by summing plant height and average width, and the condition of each plant was recorded at monthly intervals. Growth data were analyzed to determine if there were differences between the conditions investigated. The results of the plant growth studies are summarized in Table 18-16, in which the plant species are ranked from high to low according to the tolerance to irrigation with reclaimed water. In summary, Boston fern exhibited significantly better growth when irrigated with reclaimed water, and higher chloride concentration also resulted in better growth. Burford holly, Sweet viburnum, and Hibiscus grew better with reclaimed water than potable water, but to a lesser degree with the higher chloride concentration. The last four plants listed in Table 18-16 exhibited high salt sensitivity in both

Table 18-16
Growth difference of 10 ornamental plant species under 10 experimental irrigation conditions[a]

	Open air		Greenhouse		Open air		Greenhouse		Open air		Damage with 180 mg/L chlorides	Damage with 400 mg/L chlorides
Plot location												
Chloride conc., mg/L	15		15		180+		180+		400+			
Water type	Potable		Potable		Reclaimed		Reclaimed		Reclaimed			
Application method	Overhead	Drip	Overhead	Drip	Overhead	Drip	Overhead	Drip	Overhead	Drip		
Response of plant species												
Boston fern	I[b],S[c]	I	NSD[d]		I,S	I		NSD	NSD		None	None
Burford holly	NSD		NSD		NSD		I	I,S	NSD		None	None
Sweet viburnum	NSD		NSD	I,S	NSD		I	I,S	NSD		None	Slight
Hibiscus	NSD			I,T[e]	NSD		I	I,S	NSD		None	Slight
Crepe myrtle		I,T		I,L	I,L	I,S	I,L[f]	I,S	L		Slight	Medium
Avocado	I	I,T		I,L	I,L	I,S	I,L	I,S	I,L	I,S	Slight	Medium
Dwarf azalea	I	I,T		L	L		I,L	I,S	L		High	V. high
Chinese privet	NSD	I,T		D,L	D[g],L	S	D,L	S	D,L	S	High	V. high
Formosa azalea	I	I,T		D,L	D,L	S	D,L	S	D,L	S	High	V. high
Laurel oak	I	I,T		D,L	D,L	S	D,L	S	D,L	S	High	V. high

[a]Adapted from Parnell (1988).
[b]I = Plant size increased significantly.
[c]S = Significantly greater growth than the other water application method.
[d]NSD = No significant difference in growth between treatments
[e]T = Tended to grow greater than the other water application method.
[f]L = Significant leave burn observed.
[g]D = Plant size decreased.

overhead and drip irrigation methods. In addition to the controlled experiment, a total of 203 ornamental plant species were observed in 56 private residences in 1986. The plants were categorized according to the tolerance to reclaimed water irrigation as follows:

Category A: *Plant species that are completely tolerant to reclaimed water irrigation regardless of its chloride concentration*

Category B: Plant species that are tolerant to reclaimed water when the chloride concentration is less than 400 mg/L

Category C: Plant species that may need extra maintenance if the chloride concentration is more than 100 mg/L in the reclaimed water

Category D: Plant species that are not recommended for use with reclaimed water

Plant species in each of the above categories are listed in Table 18-17.

Project Resource Management

The growth experiments in Project Greenleaf were conducted in plastic pots with a constant irrigation rate. Therefore, additional research was carried out to determine the exact amount of irrigation water necessary for optimum plant growth in field conditions. Results of the field trial led to an irrigation rate recommendation of 38 mm/wk with reclaimed water for optimum growth of ornamental plants between March and November. The irrigation rate could be reduced during the low growth season between December and February (Johnson and Parnell, 1998).

Landscape Irrigation in the City of St. Petersburg

In 1993, about 43×10^6 m³/yr of reclaimed water was produced and 37×10^6 m³/yr was used for residential and commercial landscape irrigation (City of St. Petersburg, 1994). A residential home and a golf course irrigated with reclaimed water in St. Petersburg are shown on Fig. 18-11. In March 2002, the City Council of St. Petersburg adopted the Landscaping, Irrigation, Vegetation Maintenance and Tree and Mangrove Protection Ordinance (City Code Chapter 16, Article 15 of the Land Development Code) (City of St. Petersburg, 2002). Design and implementation of landscape irrigation with reclaimed water is controlled under the same code.

Residential homes in the areas where reclaimed water is available can choose whether or not to use reclaimed water. Reclaimed water can be used according to the rule: even addresses on Tuesday, Thursday, and/or Saturday, odd addresses on Wednesday, Friday, and/or Sunday. As of 2006, a monthly fee of $13 is charged for reclaimed water use. If the irrigation system is connected to potable supply, the homeowner is allowed to irrigate their landscaping plants only once a week.

Irrigation of landscaped areas in large commercial properties, multifamily buildings, schools, and government buildings are limited to 3 d/wk, and watering days are specified by the city to control the peak flow. Golf courses, cemeteries, and athletic or recreational facilities are required to submit irrigation plans to the city.

Fire hydrants are connected to the reclaimed water distribution system only for nonresidential areas. The hydrants using reclaimed water are not considered as continuous supply sources; backup hydrants are connected to a potable water distribution system (Crook, 2005). Hydrants in residential areas are all connected to potable water. Reclaimed water hydrants have a cap on the valve to prevent accidental use.

Table 18-17

List of plant species according to the tolerance to irrigation with reclaimed water[a]

Tolerance to reclaimed water irrigation[b]	Common plant name
Category A	*Tree:* Monkey puzzle tree, Norfolk Island pine, Black olive, Australian Pine, Sea grape, Silk oak, Dahoon holly, Torulosa juniper, Cajeput (punk) tree, Wax myrtle, Live oak, Brazilian pepper
	Palm: Pindo palm, European fan palm, Chinese fan palm, Canary Island date palm, Scrub palmetto, Cabbage palm, Washingtonia palm
	Shrub: Century plant, Agave, Asparagus fern, Begonia, Variegated bougainvillea, Bougainvillea, Carissa boxwood, Crown of thorns, African daisy, Hibiscus, Variegated hibiscus, Burford holly, Yaupon holly, Dwarf yaupon holly, Lantana, Lisianthus, Boston fern, Oleander, Dwarf oleander, Pittosporum, Plumbago, Yew podocarpus, Indian hawthorn, Cape honeysuckle, Sweet viburnum, Spanish bayonet, Spineless yucca
	Ground cover: Aloe, Dwarf carissa boxwood, Hottentot fig, Creeping juniper, Mondo grass, Dwarf pittosporum, Purple queen, Confederate jasmine, Wedelia, Coontie
Category B	*Tree:* Carambola, Schefflera, Lemon bottlebrush, Weeping bottlebrush, Pecan, Lime, Lemon, Grapefruit, Tangerine, Orange, Carrotwood, Italian cypress, Italian rosewood, Black sapote, Loquat, Weeping fig, Fig, Indian rubber tree, Laurel fig (Cuban laurel), Golden rain tree, Lingustrum (Privet), Sweetgum, Southern magnolia, Apple, Sapodilla, Chinaberry, Banana, Florida slash pine, Oriental arbor vitae, Cherry laurel, Peach, Guava, Pomegranate, Water oak, Drake elm, Chinese elm
	Palm: Paurotis palm, Queen palm, Ponytail palm, Fishtail palm, Areca palm, Senegal date palm, Pygmy date palm, Lady palm
	Vine: Allamanda, Purple allamanda, Coral vine, Trumpet vine, Night blooming cereus
	Shrub: Copper leaf, Maiden hair fern, Silver queen, Allamanda, Joseph's coat, Zebra plant, Bamboo, Japanese boxwood, Canna lily, Papaya, Periwinkle, Orange cestrum, Chrysanthemum, Croton, Ti plant, Pampas grass, Palay rubber vine, Heather, Queen sago, King sago, Sedge, African iris, Dracena, Crape jasmine, Gerbera daisy, Heliconia, Coral plant, Monstera, Iris, Opuntia cactus, Yellow shrimp plant, Petunia, Philodendron, Red firethorn, Travellers tree, Firecracker plant, Dwarf schefflera, Bird of paradise, Sandankwa viburnum
	Ground cover: Bromeliads, Spider plant, Potos, English ivy, Day-lily, Juniper, Kalanchoe, Liriope, Purslane, Oyster plant, Rosemary, Society garlic, Wandering jew
Category C	*Tree:* Red maple, Silver maple, Cherimoya, Orchid tree, Camphor tree, Persimmon, Longan, Kumquat, Jacaranda, Crape myrtle, Lychee, Mango, Avocado, Laurel oak, Weeping willow
	Vine: Bleeding heart, Passion flower
	Shrub: Caladium, Powder puff tree, Camellia, False aralia, Surinam cherry, Poinsettia, Gardenia, Hydrangea, Ixora, Downy Jasmine, Shrimp plant, Orange jasmine, Geranium, Pentas, Red leaf photinia, Aralia, Rose, Blue sage, Marigold, Verbena
	Ground cover: Bugle weed, Peperomia
Category D	*Shrub:* Chinese privet, Dwarf azalea, Formosa azalea

[a] Adapted from Johnson and Parnell (1998). Original source: Parnell (1988).
[b] Category A: very tolerant regardless of its chloride levels; category B: tolerant when the chloride concentration is less than 400 mg/L; category C: moderately sensitive, extra maintenance may be necessary for reclaimed water with more than 100 mg/L chloride; category D: sensitive and not recommended for use with reclaimed water.

(a)

(b)

Figure 18-11

Landscape irrigated with reclaimed water in St. Petersburg, FL: (a) residential home and (b) golf course.

Lessons Learned

Important lessons learned from the implementation of the reclaimed water landscape irrigation system in St. Petersburg are as follows:

- Even though the water reuse project originally focused on eliminating wastewater discharge, the reclaimed water system has become an essential element of the city's water supply for landscape irrigation and other nonpotable uses in the region facing population growth and water shortage.
- A looped reclaimed water distribution system enabled the use of small diameter pipes, thereby significantly reducing the construction cost.
- The Southwest Florida Water Management District (SWFWMD) adopted rules that mandated reclaimed water use if it is technically and economically feasible (City of St. Petersburg, 1994), and it led to area-wide use of reclaimed water.
- The research project, Project Greenleaf, helped establishing scientific knowledge on plants suitable for reclaimed water irrigation, as well as optimum watering practice with reclaimed water, resulting in the continued success and growth of landscaped areas that were irrigated with reclaimed water.

18-8 CASE STUDY: RESIDENTIAL IRRIGATION IN EL DORADO HILLS, CALIFORNIA

Located in an eastern suburb of the City of Sacramento, the population of the El Dorado Hills, California (18,016 persons in 2000), has been growing rapidly since the early 1990s. In 1990, Serrano became the first development in northern California to include a dual plumbing system for irrigation of both front and backyards with reclaimed water. The community encompasses over 1400 ha (3500 ac), which includes over 400 ha (1000 ac) of greenbelt, two golf courses, and approximately 4500 homes that are expected to be built when fully developed. The dual plumbing system in the Serrano development is discussed in Chap. 15 (see Sec. 15-7). In this case study, the use of reclaimed water for residential landscape irrigation is discussed.

Water Management Issues

The development area is located in the foothill region of the Sierra Nevada Mountains where only limited water is available. As development occurred, it became evident that the area would be short of a reliable water source. The Serrano development was first approved in 1988 with a secure potable water supply for 2000 residential units. The secured water was not sufficient for landscape irrigation of parks, greenbelts, and golf courses. Without additional water, the developer was not able to proceed with construction of the planned community. To accommodate the entire community as initially planned (6100 residential units), the developer needed to seek alternative sources of water by gaining additional water rights, which would have required a new water contract. Because the water was not going to be available in time for the Serrano development, the developer then sought to use reclaimed water for irrigation purposes.

The use of reclaimed water allowed the development to proceed. The El Dorado Irrigation District (EID), the local water agency, also recognized the importance of water reclamation and reuse for their service area, as reclaimed water provides a secure source of water for nonpotable uses and reduces the amount of treated wastewater discharged into streams. As a result, the EID developed a recycled water master plan in 1998 and mandated the use of reclaimed water for all new development where feasible (HDR, 2002).

Implementation

Historical development of the recycled water system in El Dorado Hills is summarized in Table 18-18. The EID operates and maintains two water reclamation facilities, four

Table 18-18

Historical development of Serrano community recycled water system in El Dorado Hills, CA

Year	Events
1979	El Dorado Hills wastewater treatment plant (WWTP) sends secondary treated reclaimed water to El Dorado Hills Golf Course and Wetsel Oviatt
1988	The Serrano development is approved for 2000 homes
1989	El Dorado Hills Development Company purchases a 1400 ha (3500 ac) property
1990	California Department of Health Services (DHS) restricts the use of secondary reclaimed water to nighttime only
1994	The Deer Creek recycled water system is completed
1996	El Dorado Irrigation District (EID) issues a recycled water master plan
	El Dorado Hills WWTP is upgraded to tertiary treatment and the flow capacity increased from 7.6×10^3 to 11.4×10^3 m^3/d (2 to 3 Mgal/d)
	Deer Creek WWTP is upgraded from 9.4×10^3 to 13.6×10^3 m^3/d (2.5 to 3.6 Mgal/d)
1998	Serrano to consider reclaimed water for landscape irrigation
	A 46 cm (18 in.) pipeline for Silva Valley is installed
1999	Serrano starts construction of dual plumbing systems
2001	Village "C" tank is constructed for additional reclaimed water storage
2002	The EID mandates use of reclaimed water for landscape irrigation where feasible

reclaimed water storage tanks, an open reservoir replenished with potable water for use as a backup supply, three booster pumping stations, and reclaimed water distribution lines throughout the service area (see also Sec. 15-7). The average use of reclaimed water is about 5.8×10^3 m³/d (1.5 Mgal/d). The price of reclaimed water is 80 percent of the potable water rate.

Water Reclamation Facilities

Wastewater, mostly of domestic origin, is processed at two water reclamation facilities: Deer Creek Wastewater Treatment Plant (DCWWTP) and El Dorado Hills Wastewater Treatment Plant (EDHWWTP). The DCWWTP was built in 1990 with secondary treatment and a capacity of 9.4×10^3 m³/d (2.5 Mgal/d), and was upgraded to tertiary treatment (direct filtration) with maximum capacity of 13.6×10^3 m³/d (3.6 Mgal/d) in 1996. The EDHWWTP has a capacity of 11.3×10^3 m³/d (3.0 Mgal/d), with an average flow of 6.4×10^3 m³/d (1.7 Mgal/d). Secondary effluent from a BNR-activated sludge process (see Chap. 7) is stored in a 280×10^3 m³ (73 Mgal) open reservoir. Effluent is processed using dissolved air flotation (DAF), filtered and disinfected with chlorine before being pumped out to the reclaimed water distribution system. Reclaimed water from both treatment plants meets the criteria for unrestricted urban reuse specified in the *California Code of Regulations* (Title 22, Disinfected Tertiary Recycled Water). Excess reclaimed water is discharged to Carson Creek according to the NPDES permits issued by the California State Water Quality Control Board.

Landscape Areas and Golf Courses

The Serrano development features two golf courses, public parks, schools, and extensive greenbelts throughout the community (see Fig. 18-12). Landscaping of public areas is designed, inspected, constructed, operated, and maintained by the Owners' Association. The EID inspects the reclaimed water facilities to insure full conformance to all regulations. Golf course landscaping is managed by the golf club. The irrigation system for the golf course uses a remote sensing system to help determine evaporation rates.

(a)
(b)

Figure 18-12
Landscape areas irrigated with reclaimed water in the Serrano development, El Dorado Hills, CA: (a) public area with appropriate signage and (b) golf course. (Coordinates: 38.679 N, 121.051 W, view at altitude 8 km.)

Landscaping for Residential Homes

As of 2005, 600 to 900 new homes are being built annually in the Serrano development. All front yards of residences in the Serrano development are irrigated with reclaimed water. Residential landscape design guidelines are issued by the homeowners' association, and the landscape irrigation system design must follow the guidelines. Potable and reclaimed waterlines are introduced to each property from the distribution mains on the opposite sides of the house (see Fig. 15-12). For each property, both potable and reclaimed water pipes are installed with water meters. A backflow prevention device is installed only on the potable waterline. To avoid an accidental use of reclaimed water for a potable purpose, hose bibs cannot be connected to reclaimed waterlines. The guidelines specify that overhead sprinklers should be used for the irrigation of turf only. Ultra-low volume sprinklers and/or drip emitters should be used for shrubs. Design, inspection, and maintenance of the front yard landscaping are the responsibility of the Serrano El Dorado Owners' Association, but design and construction of landscaping and irrigation in the backyard are the responsibility of each homeowner. The use of reclaimed water for the backyard irrigation is up to the homeowners, and the homeowner is required to submit a landscape and irrigation design to the homeowners' association before the construction of the reclaimed water irrigation system. The plan is reviewed by the irrigation specialist to ensure that it meets the design standards. During the initial installation, a stub-out from the front yard irrigation system to the backyard is placed with a locked valve to prevent unapproved backyard irrigation. The association inspects all plumbing on a dual plumbed lot before the pipe is covered. Additionally, annual inspections are made to ensure that changes have not occurred without following authorization and that the irrigation system is being properly maintained.

Education Program

The homeowners' association has an extensive education program. Workshops are provided on a quarterly basis to builders, landscape designers, contractors, and homeowners at no cost. The purpose of the education program is to familiarize the attendees with the community's design guidelines and the reclaimed water users manual to ensure that their irrigation system is designed and installed according to the guidelines. A package including landscape design guidelines and relevant information is provided at the workshop. Landscape contractors are required to renew their certificate every 18 mo.

Lessons Learned

Important lessons learned from the implementation of the reclaimed water landscape irrigation system in the Serrano development are as follows:

- The use of reclaimed water allowed the water-short area to develop with a minimal increase in freshwater demand.
- Large-scale residential landscape irrigation with reclaimed water, which was yet to be a common water reuse application in northern California, was demonstrated successfully as a viable water reuse option.
- Cooperation between the private and public sectors helped with the introduction of reclaimed water use in an upscale development, and water reuse was recognized as an asset for the community (Hazbun, 2003).
- The EID, the water agency of the area, benefited from water reclamation and reuse by securing a reliable source of nonpotable water and reducing wastewater discharge to the receiving streams.

- An extensive education program in the Serrano community helped the residents to understand benefits of using reclaimed water, to ensure appropriate use of reclaimed water, and to accept reclaimed water as an asset for the community.
- Because the reclaimed water system was installed in a new development and the buyers of the houses were informed about the reclaimed water use in the community, those who decided to live in the community are essentially those who acceptd the use of reclaimed water.

PROBLEMS AND DISCUSSION TOPICS

18-1 A reclaimed water distribution system is connected to a residential area and a golf course for landscape irrigation. Due to high water demands at the golf course, the distribution system is unable to maintain the pressure throughout the distribution system at the peak flow time. Discuss the measures that can be taken to maintain the flowrate and required pressure.

18-2 According to most state regulations and guidelines, the highway median can be irrigated with reclaimed water suitable for restricted access areas. Reclaimed water quality requirements for irrigation of restricted areas are generally less strict than the requirements for irrigation of public access areas. In what kind of a situation is it practical to irrigate a highway median with reclaimed water for restricted areas?

18-3 A reclaimed water purveyor was contacted by a landscape superintendent about odor complaints at a public park irrigated with reclaimed water. The odor is like *rotten eggs* and is prevalent especially in the beginning of each irrigation time. Discuss the possible causes of the odor and measures to alleviate the odor problem.

18-4 Consider a golf course in southern California that has 40 ha (100 ac) of turf, 4 ha (10 ac) of shrubs, and 2 ha (5 ac) of trees that need to be irrigated. Find the precipitation and evapotranspiration data from the California Irrigation Management Information System (CIMIS) and estimate the monthly water demand.

18-5 A water district is considering the feasibility of using reclaimed water for landscape irrigation for a new shopping center to be located in a suburb of a city. The district already has a water reclamation plant that produces tertiary-treated reclaimed water. Discuss the issues that must be addressed to assess the feasibility of using reclaimed water.

18-6 A 700×10^3 m^3/yr (185 Mgal/yr) wastewater treatment plant in Georgia is considering water reuse to abate effluent discharge to the receiving water. The potential reclaimed water user is a public park, and a subsurface drip irrigation system has been proposed by the landscape designer. The recommended maximum application rate for the local soil is 15 mm/d. Calculate the area required to use all of the reclaimed water for irrigation using the precipitation and evapotranspiration data shown below. Determine if a storage system is necessary. Discuss other factors affecting the feasibility of the project.

Month	Time, d/mo	Precipitation (P), mm/mo	Evapotranspiration (ET_L), mm/mo
Jan	31	113.5	6.1
Feb	28	111.5	11.4
Mar	31	130.6	32.2
Apr	30	93.0	57.1
May	31	100.3	96.5
Jun	30	106.7	133.6
Jul	31	126.7	157.0
Aug	31	90.2	142.2
Sep	30	94.0	99.8
Oct	31	77.0	53.6
Nov	30	93.2	24.9
Dec	31	97.5	9.9

18-7 A golf course is planning to use reclaimed water for irrigation. One pond on the golf course will be used for reclaimed water operational storage to alleviate the effect of high water demand on the reclaimed water distribution system. For the conditions given below, calculate the required volume for the reclaimed water operational storage.

- Total area to be irrigated is 24 ha (59 ac).
- Allowed irrigation time is between 10 p.m. to 4 a.m.
- The irrigated area is divided equally and each area is irrigated every other day.
- During the peak water demand month, 12 mm of water is applied in a 1.5 h duration.
- The pond should hold enough water to irrigate the golf course for 2 d without reclaimed water supply.

18-8 The designer of the golf course in Problem 18-7 must design the storage pond to avoid incidental overflow and runoff of the water from the pond. Discuss design options for the pond and the advantages/disadvantages of each option.

18-9 A 1660×10^3 m³/yr (439 Mgal/yr) water reclamation plant in Florida will provide reclaimed water for golf course irrigation. Excess reclaimed water will be diverted to a rapid infiltration basin (RIB) adjacent to the golf course. Under the conditions below, (1) determine if sufficient reclaimed water can be provided without a long-term storage facility, and (2) estimate the amount of water to be diverted to the RIB.

- Area to be irrigated: 55 ha (136 ac).
- Maximum recommended infiltration rate: 18 mm/d.
- Precipitation and evapotranspiration are given in the following table:

Month	Time, d/mo	Precipitation (P), mm/mo	Evapotranspiration (ET_L), mm/mo
Jan	31	85.1	66.0
Feb	28	79.2	76.2
Mar	31	86.6	111.8
Apr	30	56.6	127.0
May	31	81.5	154.9
Jun	30	200.0	137.2
Jul	31	184.7	138.2
Aug	31	207.3	134.6
Sep	30	183.6	111.8
Oct	31	68.8	88.9
Nov	30	69.1	71.1
Dec	31	83.3	58.4

18-10 In planning a landscape irrigation system with reclaimed water in your community, list and explain necessary procedures to complete the planning process, including: (1) water needs, rate, and timing; (2) water quality requirements; (3) landscape plant selection; (4) demand/supply balance; (5) operation and maintenance requirements; and (6) necessary permits.

18-11 A family bought a house in a community where all residential landscape areas are irrigated with reclaimed water. They contacted the homeowner's association and asked if they could plant an apricot tree from which they are hoping to harvest the fruit for consumption. Provide your advice and supporting reasons.

18-12 Compare the similarity and differences of two case studies, St. Petersburg, Florida, and El Dorado Hills, California, given in this chapter in terms of: (1) motivating factors for instituting water reuse, (2) water management issues, (3) implementation and management, and (4) lessons learned.

18-13 In your opinion, how will water reuse for landscape irrigation be viewed in the future, and what factors will influence its future development?

REFERENCES

Bucks, D. A., F. S. Nakayama, and R. G. Gilbert (1979) "Trickle Irrigation Water Quality and Preventive Maintenance," *Agr. Water Mgnt.*, **2**, 149–162.
City of St. Petersburg (1994) *Reclaimed Water Report to City Council 1994*, St. Petersburg, FL.
City of St. Petersburg (2002) *Landscaping and Irrigation Requirements*, December 2002 Revision, City of St. Petersburg, Development Review Services Division, St. Petersburg, FL.

Crook, J. (2005) "St. Petersburg, Florida, Dual Water System: A Case Study," 175–186, in *Proceedings of an Iranian-American Workshop: Water Conservation, Reuse, and Recycling*, National Academies Press, Washington, DC.

Gill, G., and D. Rainville (1994) "Effluent for Irrigation," in United States Golf Association (USGA), *Wastewater Reuse for Golf Course Irrigation*, Lewis Publishers, Boca Raton, FL.

HDR (2002) *Recycled Water Master Plan, Draft*, El Dorado Irrigation District, Prepared by HDR Engineering, accessed online: http://www.eid.org/.

Harivandi, M. A., J. D. Butler, and L. Wu (1992) "Salinity and Turfgrass Culture," 207–229, in D. V. Waddington, R. N. Carrow, and R. C. Shearman (eds.) *Turfgrass*, Agronomic Monograph, No. 32, ASA-C55A-SSSA, American Society of Agriculture, Madison, WI.

Hazbun, A. E. (2003) "Recycled Water for Residential Irrigation: It Works and Is Economically Feasible," in *Proceedings, First Annual Regional Water Reuse Conference, December 7–9, 2003, Amman, Jordan*.

Johnson, W. D., and J. R. Parnell (1998) "Wastewater Reclamation and Reuse in the City of St. Petersburg, Florida," 1037–1104, in T. Asano (ed.) *Wastewater Reclamation and Reuse*, Water Quality Management Library, **10**, CRC Press, Boca Raton, FL.

Jordan, L. A., D. A. Devitt, R. L. Morris, and D. S. Neuman (2002) "Foliar Damage to Ornamental Trees Sprinkler-Irrigated with Reuse Water," *Irrig. Sci.*, **21**, 1, 17–25.

Keller, J., and D. Karmeli (1975) *Trickle Irrigation Design*, Rainbird Sprinkler Manufacturing Corporation, Glendora, CA.

Maas, E. V. (1986) "Salt Tolerance of Plants," *Appl. Agric. Res.*, **1**, 1, 12–26.

Nakayama, F. S., and D. A. Bucks (1991) "Water Quality in Drip/trickle Irrigation: A Review," *Irrig. Sci.*, **12**, 187–192.

Parnell, J. R. (1988) *Project Greenleaf Final Report*, City of St. Petersburg, Public Utilities Department, St. Petersburg, FL.

State of California (1990) *California Municipal Wastewater Reclamation in 1987*, California State Water Resources Control Board, Office of Water Recycling, Sacramento, CA.

State of Florida (1999) *Florida Administrative Code (F.A.C.), Chapter 62-610, Reuse of Reclaimed Water and Land Application*, Florida Department of Environmental Protection, Tallahassee, FL, accessed online: http://www.dep.state.fl.us/.

State of Florida (2004) *2003 Reuse Inventory*, Florida Department of Environmental Protection, Tallahassee, FL, accessed online: http://www.dep.state.fl.us/.

Steinburgs, C. Z. (1994) "Retrofitting a Golf Course for Recycled Water: An Engineer's Perspective," in United States Golf Association (USGA), *Wastewater Reuse for Golf Course Irrigation*, Lewis Publishers, Boca Raton, FL.

Tajrishy, M. A., D. J. Hills, and G. Tchobanoglous (1994) "Pretreatment of Secondary Effluent for Drip Irrigation," *J. Irrig. Drain. Eng., ASCE*, **120**, 4, 716–731.

Tanji, K., S. Grattan, C. Grieve, A. Harivandi, L. Rollins, D. Shaw, B. Sheikh, and L. Wu (2006) *Salt Management Guide for Landscape Irrigation with Reclaimed Water: A Literature Review*, WateReuse Foundation, Alexandria, VA.

Tchobanoglous, G., F. L. Burton, and H. D. Stensel (2003) *Wastewater Engineering: Treatment and Reuse*, 4th ed., McGraw-Hill, New York.

Terrey, A. (1994) "Design of Delivery Systems for Wastewater Irrigation," 130–142, in United States Golf Association (USGA), *Wastewater Reuse for Golf Course Irrigation*, Lewis Publishers, Boca Raton, FL.

University of California and State of California (2000) *A Guide to Estimating Irrigation Water Needs of Landscaping Plantings in California*, University of California Cooperative Extension, California Department of Water Resources, Sacramento, CA.

USDA (1993) "Irrigation Water Requirements," Sec. 15, Chap. 2, in D.S. Department of Agriculture, *Irrigation, National Engineering Handbook*, Soil Conservation Service, U.S. Department of Agriculture, Washington, DC.

USDA (1987) "Trickle Irrigation," Chap. 7, in U.S. Department of Agriculture, *Irrigation, National Engineering Handbook,* Sec. 15, Soil Conservation Service, U.S. Department of Agriculture, Washington, DC.

U.S. EPA (1992) *Manual—Guidelines for Water Reuse*, EPA/625/R-92/004, U.S. Environmental Protection Agency and U.S. Agency for International Development, Washington, DC.

U.S. EPA (2004) *Guidelines for Water Reuse*, EPA/625/R-04/108, September 2004, U.S. Environmental Protection Agency and U.S. Agency for International Development, Washington, DC.

Young, R. E., K. A. Thompson, R. R. McVicker, R. A. Diamond, M. B. Gingras, D. Ferguson, J. Johannessen, G. K. Herr, J. J. Parsons, V. Seyde, E. Akiyoshi, J. Hyde, C. Kinner, and L. Lodewage (1998) "Irvine Ranch Water District's Reuse Today Meets Tomorrow's Conservation Needs," 941–1036, in T. Asano (ed.) *Wastewater Reclamation and Reuse*, **10**, CRC Press, Boca Raton, FL.

19 Industrial Uses of Reclaimed Water

WORKING TERMINOLOGY   1104

19-1   INDUSTRIAL USES OF RECLAIMED WATER: AN OVERVIEW   1105
Status of Water Use for Industrial Applications
in the United States   1105
Water Management in Industries   1107
Factors Affecting the Use of Reclaimed Water
for Industrial Applications   1108

19-2   WATER QUALITY ISSUES FOR INDUSTRIAL USES
OF RECLAIMED WATER   1109
General Water Quality Considerations   1110
Corrosion Issues   1110
Indexes for Assessing Effects of Reclaimed Water Quality
on Reuse Systems   1115
Corrosion Management Options   1126
Scaling Issues   1127
Accumulation of Dissolved Constituents   1129

19-3   COOLING WATER SYSTEMS   1132
System Description   1132
Water Quality Considerations   1132
Design and Operational Considerations   1135
Management Issues   1138

19-4   OTHER INDUSTRIAL WATER REUSE APPLICATIONS   1141
Boilers   1141
Pulp and Paper Industry   1147
Textile Industry   1150
Other Industrial Applications   1154

19-5   CASE STUDY: COOLING TOWER AT A THERMAL POWER
GENERATION PLANT, DENVER, COLORADO   1155
Setting   1155
Water Management Issues   1156
Implementation   1158
Lessons Learned   1158

19-6 CASE STUDY: INDUSTRIAL USES OF RECLAIMED WATER IN WEST BASIN MUNICIPAL WATER DISTRICT, CALIFORNIA   1158
   Setting   1158
   Water Management Issues   1158
   Implementation   1159
   Lessons Learned   1161

   PROBLEMS AND DISCUSSION TOPICS   1161

   REFERENCES   1165

WORKING TERMINOLOGY

Term	Definition
Biofilm	An aggregation of microorganisms that produce polymeric substances for community formation, surface attachment, and protection. Biofilms occur on surfaces in contact with water and may promote localized corrosion and can affect the performance of some industrial processes, such as heat exchangers.
Blowdown (purge) water	Water that is removed from recirculating cooling water to control water quality. Blowdown water is removed either continuously or intermittently. The blowdown water is a waste stream from the cooling process and requires treatment either in a municipal treatment system or onsite.
Boiler	A closed vessel used to heat process water under pressure.
Cooling tower	A facility used to remove heat from process cooling water by evaporation with forced air.
Corrosion	Electro-chemical reaction on metallic surfaces due to external or internal localized conditions. Corrosion causes changes in the properties of the metal that can lead to failure.
Cycles of concentration	The number of times cooling water is recirculated based on the ratio of the blowdown water salt concentration to the salt concentration in the makeup water.
Drift	Water that escapes from evaporative cooling towers due to wind dispersion.
Feedwater	A combination of makeup water and recirculating water that is used for heat exchange in cooling towers and boilers.
Galvanic corrosion	Corrosion induced by electrical potential differences between two dissimilar metals.
Internal water recycling	Recycling of water within an industry for a specific industrial water use.
Makeup water	Water added to the cooling system to replace water lost by evaporation, drift, blowdown, and/or leakage.
Scaling	Formation of oxides, carbonates, hydroxides, and other mineral deposits on the inner surfaces of pipelines, heat exchangers, and tanks. Microbial biofilm can occur along with mineral scaling.
Saturation pH	The pH at which calcium and alkalinity in water are at equilibrium with calcium carbonate at ambient temperature.
Water-pinch analysis	An optimization technique that can be used to determine minimum water requirements for industrial processes.
Zero liquid discharge (ZLD)	Elimination of liquid waste discharges from industrial facilities.

Industrial use of water encompasses quantity and quality requirements that range from the use of large volumes of low-quality water for cleaning applications to the use of high quality process water for manufacturing or boiler feedwater. For industrial activities that have relatively high water usage rates, reclaimed water can serve to either substitute for or augment other water sources, depending on water quality requirements. For some industries, reclaimed water can be used directly; in other cases, additional treatment may be needed to meet more stringent water quality limitations. Examples of industrial applications where reclaimed water can be used directly include cooling water, material conveyance, rinse water, equipment and facility cleaning, and onsite landscape irrigation (see Chap. 18). Supplemental treatment may be needed to modify water quality if reclaimed water is used for manufacturing, boiler feedwater, or other industrial applications with specific water quality requirements. The feasibility of using municipal reclaimed water for industrial applications depends on its relative cost, availability, reliability, and sustainability compared to other water sources. In general, the potential for industrial use of reclaimed water tends to be industry-specific and the extent to which it is used is strongly influenced by localized cost and environmental considerations.

In this textbook, the term *industrial water reuse* refers to the use of reclaimed water in industrial applications such as cooling water, boiler feedwater, and manufacturing processes. In contrast, the term *internal water recycling* refers to the reuse of water within the boundaries of specific industrial facilities. In industrial processes, water used for one process may be *reused* with or without supplemental treatment for another process. Wastewater generated through industrial activities can be treated onsite for *recycling, reuse*, or discharge. Because reclaimed water is used widely for cooling water applications, water quality, design, and operational and maintenance issues are discussed in detail in this chapter. Other industrial water reuse applications are also discussed briefly.

19-1 INDUSTRIAL USES OF RECLAIMED WATER: AN OVERVIEW

Many industrial processes depend on the ready availability of reliable sources of large quantities of water. Examples of water-intensive industries include thermoelectric power generation, pulp and paper manufacturing, textile production, food processing, chemical manufacturing processes, oil refineries, and ore extraction (mining). In general, industries tend to use water at a relatively constant rate throughout the year as compared to the seasonal and geographic variations associated with water use for agricultural and landscape irrigation (see Chaps. 17 and 18). Therefore, industries provide a unique opportunity for year-round use of reclaimed water. As an introduction to industrial reuse, it is useful to review water use patterns in industries, industrial water reuse and recycling, and constraints associated with the use of reclaimed water.

Status of Water Use for Industrial Applications in the United States

A comparison of industrial freshwater withdrawals in 1975, 1995, and 2000 is shown in Table 19-1. In 1975, manufacturing activities consumed about 190×10^9 m³/yr (137 Bgal/d) of which about 100×10^9 m³/yr (75 Bgal/d) was supplied by internal water recycling and reuse. Freshwater withdrawal for manufacturing was about

Table 19-1

Industrial freshwater withdrawals, in 10^6 m³/yr, in 1975 and 2000[a]

Category	1975	1995	2000[b]
Electric power generation	120,000	192,000	188,000
Manufacturing	71,000	35,000	21,000
Total	191,000	227,000	209,000

[a]Data from U.S. Water Resources Council (1979), Solley et al. (1998), and Huston et al. (2004).
[b]Does not include deliveries from public supply, about 0.3 and 12 percent of which was delivered to thermoelectric and industrial water supply, respectively, in 1995. Public supply water withdrawal in 2000 was 67,000 × 10^6 m³/yr (additional 67 × 10^6 and 8000 × 10^6 m³/yr, respectively, if the same percentage was delivered to each water user).

71×10^9 m³/yr (53 Bgal/d) (U.S. Water Resources Council, 1979). In contrast, 1995 water consumption for manufacturing decreased to about half of the 1975 level and has continued to decline. The reduction in manufacturing water use is due to shifting manufacturing practices, water conservation, limits on industrial wastewater discharges, and increased globalization and outsourcing of industrial activities (Huston et al., 2004).

Sources of water used for manufacturing in 1975 and 1995 are compared in Fig. 19-1. To date, the use of reclaimed water for manufacturing applications has been minimal. Based on 1995 data, about 150×10^6 m³/yr (40×10^3 Mgal/yr) of reclaimed water, or 0.4 percent of the total industrial water, was used for industrial purposes (Solley et al., 1998). However, there is growing interest in industrial use of reclaimed water where it is practicable. Selected examples of reclaimed water use in industries are shown in Table 19-2. The extent of use depends on the local water infrastructure, the proximity

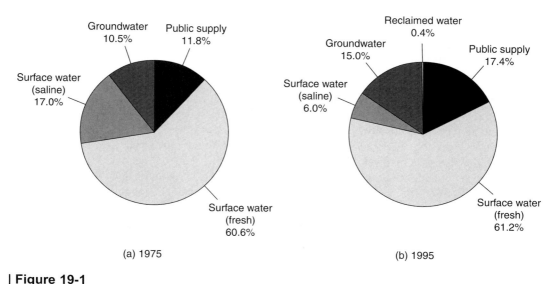

Figure 19-1

Water sources for manufacturing in (a) 1975 and (b) 1995. (Adapted from U.S. Water Resources Council, 1979 and Solley et al., 1998.)

Table 19-2 Selected examples of reclaimed water use in industrial processes

Location	Type of industrial process	Reclamation treatment
Flagstaff, AZ	Pulp and paper (tissue)	Tertiary
Palo Verde, AZ	Nuclear power generation	Tertiary
Los Angeles, CA	Pulp and paper (newspaper)	Tertiary
Los Angeles, CA	Textile (carpet dyeing)	Tertiary
Richmond, CA	Oil refinery	
Santa Rosa, CA	Geyser thermoelectric power generation	Tertiary
West Basin, CA	High-pressure boiler (feedwater)	RO-RO
West Basin, CA	Low-pressure boiler (feedwater)	RO
West Basin, CA	Cooling tower at oil refineries	Tertiary
Denver, CO	Cooling tower at a thermoelectric power generation plant	Tertiary
Pinellas County, FL	Cooling tower and boiler feedwater at waste incineration plants	Tertiary, MF-RO
Vero Beach, FL	Thermoelectric power generation	
Harlingen, TX	Textile	Tertiary-RO

of reclaimed water transmission lines, water quality considerations, and the relative cost and availability of other water sources. In some cases, incentive programs have been developed on a regional or statewide basis to encourage the use of reclaimed water. In other cases, the relative cost of reclaimed water compared to other sources provides an economic incentive. In general, the use of reclaimed water to replace or supplement other water sources for industrial activities has become more prevalent due to the increased availability of reclaimed water and associated cost considerations.

Water Management in Industries

Since the 1990s, industries increasingly have been required to meet more stringent environmental goals, particularly for wastewater discharges. A long-term goal is to eliminate all waste discharges to receiving waters by applying concepts of zero liquid discharge (ZLD). The challenges of meeting ZLD goals have spurred industry into implementing more efficient water use practices and adopting internal water recycling, where practicable.

One step towards realizing ZLD goals is to reduce industrial water use by increasing water use efficiency. Water-pinch analysis is an optimization technique that is used to identify methods of using water more efficiently (Baetens, 2002; Mann and Liu, 1999). Through pinch-analysis techniques, minimum industrial process water requirements can be assessed by considering flowrate requirements in conjunction with water quality factors. The feasibility of internal water recycling and the use of reclaimed water to augment or substitute for existing water sources can be integrated into the pinch-analysis. Conceptual flow diagrams for water reclamation, reuse, and recycling within an industrial facility are illustrated on Fig. 19-2.

Figure 19-2
Water reuse options in industrial processes: (a) without reuse or recycle, (b) with reuse, (c) with reclamation and reuse, and (d) with reclamation and recycle. (Adapted from Mann and Liu, 1999.)

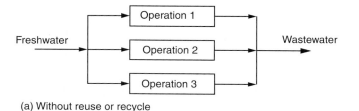

(a) Without reuse or recycle

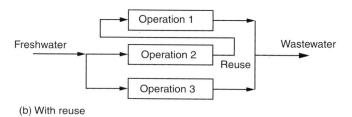

(b) With reuse

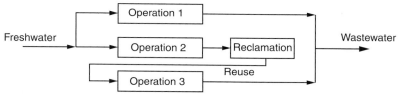

(c) With reclamation and reuse

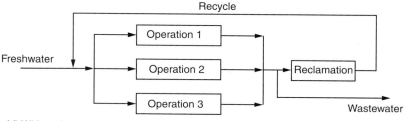

(d) With reclamation and recycle

Factors Affecting the Use of Reclaimed Water for Industrial Applications

As stated previously, a major benefit of using reclaimed water for industrial applications is that water use requirements are relatively consistent year-round in contrast to the seasonal nature of other nonpotable water reuse applications such as irrigation. Thus, the inclusion of an industrial water reuse program in a comprehensive water management plan can obviate the need for storage and/or disposal of excess reclaimed water during periods of low irrigation water demand (see Chaps. 17 and 18). Conversely, industries may be deterred from using reclaimed water due to the lack of readily available transmission systems used to convey reclaimed water from the water reclamation plant to the site of use. In other cases, the costs of implementing supplemental treatment to meet specific industrial process water quality requirements may outweigh the cost-savings associated with the use of reclaimed water. Water quality variables, specific to reclaimed water, may also compromise the ability to use reclaimed water for some industrial purposes. For example, the accumulation of salts and micropollutants may interfere with the operation of closed-loop continuous recirculation cooling water systems (see Sec. 19-2).

Another important consideration is the economic feasibility of using reclaimed water for industrial applications. System components and related major cost elements that must be examined are listed below (Treweek, 1982):

1. Reclaimed water supply—additional cost for supplemental onsite treatment
2. Distribution system—cost for constructing industry-owned pipelines to convey reclaimed water from treatment facilities to industrial sites
3. Onsite re-piping—retrofitting of the existing piping system to accommodate reclaimed water
4. Engineering analysis of water quality and supplemental treatment alternatives
5. Pretreatment capital and operating and maintenance (O&M) costs
6. Internal treatment needs
7. Management of process residuals including sludges and brines (treatment, beneficial reuse, and/or disposal)
8. Institutional, legal, and administrative activities including: (a) identification and coordination among agencies having jurisdiction, (b) meeting all appropriate regulatory requirements, (c) contract negotiations with water supply agencies, (d) permit acquisition, and (e) if appropriate, coordination with other participants in subregional or regional projects

The major costs for industrial reclaimed water systems include: (1) purchase costs of reclaimed water, (2) capital costs associated with construction and installation of offsite and onsite facilities, (3) O&M costs, and (4) financing and administrative costs. Depending on the proximity of the industry to the water reclamation infrastructure, the transmission and distribution system costs may outweigh the economic benefits of using reclaimed water (Treweek, 1982).

19-2 WATER QUALITY ISSUES FOR INDUSTRIAL USES OF RECLAIMED WATER

While specific water quality requirements vary among industries, major water quality issues are associated with the prevention of corrosion, scaling, and biological fouling of equipment and distribution systems. Where there is potential for human contact with the reclaimed water or aerosols generated from reclaimed water, control of pathogenic organisms is particularly important. In this section, general water quality considerations for industrial water reuse applications are discussed. Specific considerations for each application are discussed in the following sections.

General Water Quality Considerations

Reclaimed water quality issues for industrial uses are summarized in Table 19-3 and typical water quality requirements for various industrial processes are given in Table 19-4. Key water quality variables include pH, alkalinity, organics, nutrients, and those that affect corrosion and scaling. It should be noted that water quality requirements are specific for each industrial process and Table 19-4 is intended to provide qualitative information for comparing different water use scenarios.

Corrosion Issues

In the United States, utilities spend over $19 billion per year combating corrosion problems in drinking water distribution systems (Bell and Aranda, 2005). To avoid similar problems in reclaimed water systems, methods for identification, control, and management of corrosion issues are needed, particularly in situations where water temperatures and pressures can fluctuate. Corrosion results from physicochemical reactions between a metal and its environment that lead to changes in the properties of the metal (ISO, 1999). In some cases, electrochemical reactions can act to initiate oxidation and dissolution of metals from piping and other materials that contact the reclaimed water.

In general, corrosion can only occur in the presence of a corrosion cell consisting of four elements: (1) anode, (2) cathode, (3) electrolyte, and (4) a conductor connecting the cathode with the anode, as illustrated on Fig. 19-3. Various types of corrosion and their characteristics, water quality considerations, and corrosion management options appropriate to reclaimed water systems are described below.

Table 19-3
Reclaimed water quality issues for industrial processes[a]

Parameter	Issues
Alkalinity	Effects pH stability
Ammonia	Interferes with formation of free chlorine residual, causes stress corrosion in copper-based alloys, stimulates microbial growth
Calcium and magnesium	Scale formation
Hydrogen sulfide	Corrosion, odors
Iron	Scale formation, staining
Microbiological water quality	Potential for biofouling
Nitrate	Stimulates microbial growth, interferes with dyeing
pH	May affect chemical reactions, solubility of constituents
Phosphorus	Scale formation, stimulates microbial growth
Residual organics	Microbial growth, slime and scale formation, foaming in boilers
Silica	Scale formation
Sulfate	Corrosion
Suspended solids	Deposition, "seed" for microbial growth

[a] Adapted from WPCF (1989), Asano and Levine (1996).

Table 19-4
Typical reclaimed water quality requirements for various industrial processes[a]

Parameters	Unit	Boiler feedwater (bar)[b] 0–10	10–12	48–103	103–344	Cooling water Once-through Fresh	Brackish	Makeup for recirculation Fresh	Brackish
Silica (SiO$_2$)	mg/L	30	10	0.1	0.01	50	25	50	25
Aluminum (Al)	mg/L	5	0.1	0.01	0.01			0.1	
Iron (Fe)	mg/L	1	0.3	0.05	0.01			0.5	
Manganese (Mn)	mg/L	0.3	0.1	0.01				0.5	
Copper (Cu)	mg/L	0.5	0.05	0.05	0.01				
Calcium (Ca)	mg/L		0	0	–[c]	200	520	50	420
Magnesium (Mg)	mg/L		0	0	–[c]				
Sodium (Na)	mg/L								
Ammonia (NH$_3$)	mg/L	0.1	0.1	0.1	0.7				
Bicarbonate (HCO$_3$)	mg/L	170	120	50	–[c]	600		25	
Sulfate (SO$_4$)	mg/L					680	2700	200	2700
Chloride (Cl)	mg/L					600		500	
Fluoride (F)	mg/L					600	19,000	500	19,000
Nitrate (NO$_3$)	mg/L								
Phosphate (PO$_4$)	mg/L								
Dissolved solids	mg/L	700	500	200	0.5	1000	35,000	500	35,000
Suspended solids	mg/L	10	5	0	0	5000	2500	100	100
Hardness	mg/L as CaCO$_3$	20	1.0	0.1	0.07	850	6250	130	6250
Alkalinity	mg/L as CaCO$_3$	140	100	40	0	500	115	20	115
Acidity	mg/L as CaCO$_3$								
pH	unitless	8–10	8–10	8.2–9.2	8.2–9.2	5.0–8.3			
Color	color units								
COD	mg/L	5	5	0.5	0	75	75	75	75
Dissolved oxygen	mg/L	<0.03	<0.03	<0.03	<0.005				
Temperature	°C	49	49	49	49	38	49	38	49
Turbidity	NTU	10	5	0.5	0.05	5000	100		

(Continued)

Table 19-4

Typical reclaimed water quality requirements for various industrial processes[a] (Continued)

		Process water by industry					
Parameters	Unit	Textile	Pulp and paper	Chemical	Petroleum and coal products	Primary metal	Tanning
Silica (SiO_2)	mg/L	25[d]	50	50	60		
Aluminum (Al)	mg/L	8[e]					
Iron (Fe)	mg/L	0.1–0.3	0.3	0.1	1.0		50
Manganese (Mn)	mg/L	0.01–0.05	0.1	0.1			0.2
Copper (Cu)	mg/L	0.01–5					
Calcium (Ca)	mg/L		20	70	75		60
Magnesium (Mg)	mg/L		12	20	30		
Sodium (Na)	mg/L				230		
Ammonia (NH_3)	mg/L				40		
Bicarbonate (HCO_3)	mg/L			130	480		
Sulfate (SO_4)	mg/L	100		100	600		250
Chloride (Cl)	mg/L		200	500	300	500	250
Fluoride (F)	mg/L			5	1.2		
Nitrate (NO_3)	mg/L				10		
Phosphate (PO_4)	mg/L						
Dissolved solids	mg/L	100–200	100	1000	1000	1500	
Suspended solids	mg/L	0–5	10	5	10	3000	
Hardness	mg/L as $CaCO_3$	0–50	475	250	350	1000	150
Alkalinity	mg/L as $CaCO_3$			125	500	200	
Acidity	mg/L as $CaCO_3$					75	
pH	unitless	6–8	4.6–9.4	5.5–9.0	6–9	5–9	6–8
Color	color units	0–5	10	20	25		5
COD	mg/L						
Dissolved oxygen	mg/L						
Temperature	°C		38			38	
Turbidity	NTU	0.3–5					0

[a] Adapted from State of California (1963), U.S. EPA (1973).
[b] 1 bar = 10^5 Pa ≈ 14.5 lb/in.^{2}
[c] Determined by treatment of other constituents.
[d] As SiO_3.
[e] As aluminum oxide, Al_2O_3.

Note: specific quality requirements may vary greatly with each industrial process.

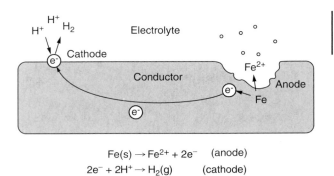

Figure 19-3
Corrosion cell with iron.

$$Fe(s) \rightarrow Fe^{2+} + 2e^- \quad \text{(anode)}$$
$$2e^- + 2H^+ \rightarrow H_2(g) \quad \text{(cathode)}$$
$$Fe(s) + 2H^+ \rightarrow Fe^{2+} + H_2(g) \quad \text{(net)}$$

Types of Corrosion

Corrosion may be categorized into: (1) corrosion induced by electrochemical reactions and (2) corrosion induced by materials and other physical factors. Examples of corrosion induced by electrochemical reactions, and/or by materials and other physical factors, are summarized in Table 19-5. The presence of aggressive (i.e., highly reactive) ions in reclaimed water, such as chloride, sulfide, or sulfate, have the potential to result in localized corrosion such as pitting, crevice, or stress corrosion cracking, depending on the pH, alkalinity, dissolved oxygen content, degree of microbial activity and biofouling, temperature, and other water quality variables. Solubility and stability indexes, and ratios used to assess water corrosivity, are described in Table 19-6.

Examples of corrosion induced by materials coupled with other physical factors include galvanic corrosion and stress corrosion cracking. While galvanic corrosion can result from differences in the electrical potential between two dissimilar metals, stress corrosion cracking is induced by tensile stress and aggressive water quality. The synergistic effects of corrosion and stress propagate the stress corrosion cracks. Even though these types of corrosion are induced primarily by physical factors, chemical and microbiological reactions can play a significant role.

Water Quality Considerations

Water quality and hydraulic factors influence the potential for corrosion to occur in industrial vessels and conveyance structures. For example, pH changes have been correlated to increases in corrosion rates (Stumm, 1956). High levels of alkalinity (over 150 mg/L as $CaCO_3$) have been observed to cause increased corrosion of lead, copper, and zinc (Vik et al., 1996). Neutral salts such as chloride and sulfate can act to increase corrosion rates, particularly at low oxygen concentrations (Clement et al., 2002). A comparison of water quality variables that influence corrosion is given in Table 19-7. Dissolved oxygen can induce and promote corrosion reactions, particularly under elevated temperatures and pressures.

Table 19-5
Types, occurrence, and important factors in the development corrosion[a]

Types of corrosion	Description/occurrence	Important factors
Corrosion induced primarily by electrochemical reactions		
Crevice corrosion	Localized corrosion occurring at narrow openings or spaces between two material surfaces.	Changes in local chemistry within crevice leading to depletion of inhibitor, depletion of oxygen, the creation of acid conditions, and the accumulation of aggressive ions, principally chloride.
Microbiologically induced	Enhance corrosion kinetics by accelerating the rate of redox reactions, most commonly in conjunction with the other types of corrosion cited in this table.	Bacteria and other microorganisms from micro zones on the pipe surface which contain high acidity or concentrations of corrosive species.
Pitting corrosion	Localized corrosion resulting in cavities in the metal. A small portion of the pipe surface becomes a permanent anode and the surrounding pipe serves as a cathode.	High concentrations of aggressive ions principally chloride, acidity, and low oxygen concentrations are major chemical factors leading to pitting corrosion. Other factors include manufacturing defects, condition of metal surface in service.
Tuberculation	Occurs on steel and cast iron when surfaces are exposed to water in presence of oxygen. Oxidized metal will form a tubercle on the surface and corrosion proceeds underneath the tubercle.	The tubercle structure of microbiologically induced tubercles is stronger than tubercles without microorganisms. The outside of the tubercle becomes cathodic, and the inside becomes highly anodic.
Uniform corrosion	Metal conduits exposed to water react with constituents in the water to corrode or become oxidized uniformly.	Oxidizing potential of the water, the presence of aggressive ions, principally chloride.
Corrosion induced primarily by the materials and other physical factors		
Electrolytic	Corrosion occurs where stray direct electrical current from an outside source enters a pipe or other metal structure (either internally or externally) and then leaves to return to the source. In the process, metal is removed at the anode.	Strength of electrical current, internal and external conditions, water pH, conductivity, and velocity; the amount of oxygen reaching the bare metal surface.
Erosion	Corrosion of a metal caused or accelerated by the motion of a corrosive fluid, especially one containing abrasive material. The two most common types are impingement and cavitation.	Velocity of fluid, presence of abrasive particulate matter such as grit in wastewater, presence of imperfections in metal surface, presence of other types of corrosion. Very low pressure resulting in the formation of vapor bubbles.
Galvanic corrosion	Galvanic corrosion occurs when two dissimilar metals are placed in contact with each other.	Difference in the electrical potential between the two metals, difference in the relative areas of the two metals, conductivity of water
Stress corrosion cracking	Induced by tensile stress and a corrosive environment.	Residual stresses caused by welding, improper alignment, repeated cycling coupled with a corrosive environment (principally the presence of chloride).

[a]Adapted from Crittenden et al. (2005).

Table 19-6
Various indexes, parameters, and ratios used to assess the corrosivity of water

Parameter	Rationale	Optimum range
Aggressiveness index (AI)	Empirical relationship of pH, alkalinity and hardness developed for assessing corrosion of asbestos/cement pipes.	>12
Buffer capacity	Measure of the ability of water to resist a change in pH. Higher buffer capacity allows for more consistent pH at the pipe wall and in the bulk water, thus preventing localized metal release.	>0.5 meg/L·pH unit
Calcium carbonate precipitation potential (CCPP)	Estimate of the mass of calcite ($CaCO_3$) that can deposit or dissolve from pipe walls.	4–10 mg/L as $CaCO_3$
Langelier saturation index (LSI)	Estimate of the potential for calcite ($CaCO_3$) to either deposit or dissolve from pipe walls (no information on mass available).	0 to ±0.2 pH unit
Ryznar stability index (RSI)	Empirical variation on the LSI that gives more weighting to the saturation pH.	>6.2 pH unit
Alkalinity to chloride + sulfate ratio	Comparison of availability of anions to react with metals.	>1 eq/eq
Chloride to bicarbonate ratio	Comparison of the potential for chloride and bicarbonate to form metal complexes. Increases in steel corrosion have been observed at values over 0.3.	>0.2 eq/eq
Chloride to sulfate ratio	Measure of potential competitive reactions between anions and released metals.	<0.6 mg/mg

In addition to chemical water quality, microorganisms in reclaimed water may directly or indirectly impact corrosion through the development of biofilms on wetted surfaces of pipelines, condensers, and process tanks. Microbial activity within biofilms produces acid-forming gases such as carbon dioxide and hydrogen sulfide that can lead to localized corrosion (Adams et al., 1980). In addition, the presence of biofilms results in hydraulic discontinuities in pipelines and reaction tanks and provides additional surface area for chemical reaction to occur. Biofilms can also act to shield pathogenic organisms from the actions of disinfectants. Microorganisms implicated in microbiologically induced corrosion are listed in Table 19-8.

Indexes for Assessing Effects of Reclaimed Water Quality on Reuse Systems

One approach for evaluating the potential for water quality induced corrosion and scaling is to use water quality variables to compute corrosion indexes. Examples of indexes used to assess corrosivity are given in Table 19-6. These indexes represent various approaches to quantifying interrelationships between pH, alkalinity, and dissolved inorganic carbon. Although used widely, these indexes are also criticized widely in terms of their applicability to corrosion and scaling control. Details of the indexes are given below. Of these indexes, the Langelier saturation index is used widely to assess the need for post-RO treatment as discussed in Chap. 9.

Table 19-7
Comparison of water quality variables that influence corrosion

Parameter	Optimum Range	Rationale
Alkalinity	Consistency	Variations in alkalinity can influence formation of metal-carbonate complexes and result in increased metal release.
Ammonia	Not present	Ammonia can form complexes with metals, resulting in increased solubility and metal release. Can promote biological growth and microbiologically induced corrosion. Corrosive to Cu and Cu alloys.
Calcium	Consistency	Calcium can form calcite on pipe walls, possibly preventing release of copper. Variations in water quality will promote dissolution of pipe film and promote metal release.
Chloride	Consistency	Chloride can increase corrosion rates under conditions of low dissolved oxygen. Can form complexes with metals increasing solubility and metal release.
Dissolved oxygen	Consistency	Dissolved oxygen can serve as an electron acceptor for corrosion reactions; however, it also reacts to form protective oxide layers that prevent corrosion. Localized differences in dissolved oxygen can promote corrosion.
Magnesium	Consistency	Magnesium can interfere with the deposition of calcium complexes on the pipe wall.
Total organic carbon (TOC)	Consistency	Organic carbon can coat pipe surfaces and prevent metal release. It can also form metal complexes and increase metal release. Can promote biological growth and microbiologically induced corrosion.
Orthophosphate	0.5–5 mg/L as PO_4	Orthophosphate can react with copper to form copper phosphate complexes that coat the interior pipe wall
pH	7.3 to 7.8	pH influences metal solubility and reactions with carbonates. pH may be locally higher at the pipe surface due to OH^- generation.
Sulfate	Consistency	Sulfate can interfere with formation of cupric hydroxide scales, thus increasing the potential for metal release.
Temperature	Consistency	Corrosion rates increase with increasing temperature. Temperature influences solubility, rates of microbiological activity, water density, and associated mixing efficiency.

For calculation of the indexes introduced in this section, it is important to evaluate the effect of ionic strength and activity of ions (i.e. effective concentration) in solution. The ionic strength, I, of a solution is defined by Lewis and Randall (Lewis and Randall, 1921) as:

$$I = \frac{1}{2}\sum C_i (Z_i)^2 \tag{19-1}$$

where I = ionic strength, mole/L
C_i = concentration of the i-th ion, mole/L
Z_i = charge of the i-th ion

Many industries monitor routinely either conductivity or total dissolved solids (TDS). The ionic strength can also be estimated from conductivity or TDS measurements

Table 19-8
Microorganisms most commonly implicated in microbiologically induced corrosion[a]

Genus or species	pH range	Temp. range, °C	Oxygen requirement	Metals affected	Effects of the microorganisms
Bacteria					
Desulfovibrio (most common: D. desulfuricans)	4–8	10–40	Anaerobic	Iron and steel, stainless steels, aluminum, zinc, copper alloys	Utilize hydrogen in reducing SO_4^{2-} to S^{2-} and H_2S; promote formation of sulfide films
Desulfotomaculum (most common: D. nigrificans, a.k.a. Clostridium)	6–8	10–40 (some 45–75)	Anaerobic	Iron and steel, stainless steels	Reduce SO_4^{2-} to S^{2-} and H_2S; (spore formers)
Desulfomonas		10–40	Anaerobic	Iron and steel	Reduce SO_4^{2-} to S^{2-} and H_2S
Thiobacillus thioosidans	0.5–8	10–40	Aerobic	Iron and steel, copper alloys, concrete	Oxidizes sulfur and sulfides to form H_2SO_4; damages protective coatings
Thiobacillus ferroosidans	1–7	10–40	Aerobic	Iron and steel	Oxidizes Fe^{2+} to Fe^{3+}
Gallionella	7–10	20–40	Aerobic	Iron and steels, stainless steels	Oxidizes Fe^{2+} to Fe^{3+} (also Mn^{2+} to Mn^{3+}); promotes tubercle formation
Sphaerotilus	7–10	20–40	Aerobic	Iron and steels, stainless steels	Oxidizes Fe^{2+} to Fe^{3+} (also Mn^{2+} to Mn^{3+}); promotes tubercle formation
Sphaerotilus natans				Aluminum alloys	Oxidizes Fe^{2+} to Fe^{3+} (also Mn^{2+} to Mn^{3+}); promotes tubercle formation
Pseudomonas	4–9	20–40	Aerobic	Iron and steel, stainless steels	Some strains can reduce Fe^{3+} to Fe^{2+}
Pseudomonas aeruginosa	4–8	20–40	Aerobic	Aluminum alloys	Some strains can reduce Fe^{3+} to Fe^{2+}
Fungi					
Cladosporium resinae	3–7	10–45		Aluminum alloys	Produces organic acids in metabolizing certain fuel constituents

[a] Adapted from Davis (2000).

using the Russell or the Langelier approximations, respectively (Langelier, 1936; Russell, 1976):

$$I = 1.6 \times 10^{-5} \times \text{specific conductance, } \mu\text{mho/cm} \tag{19-2}$$

$$I = 2.5 \times 10^{-5} \times \text{TDS, mg/L} \tag{19-3}$$

In using these relationships, it is assumed that the TDS is proportional to the specific conductance where TDS (mg/L) $\approx 0.64 \times$ specific conductance (μmho/cm). Depending on the water quality matrix, this proportionality constant may not be accurate and, if possible, should be verified using actual operating data.

The activity coefficient of an ion A, γ_A, a measure used to determine the effective concentration of an ion in a real solution as compared to an ideal solution, may be estimated using the Davies relationship, a modification of the DeBye-Hückel relationship, as calculated by Eq. (19-4).

$$\log \gamma_A = -0.5(Z_A)^2 \left(\frac{\sqrt{I}}{1 + \sqrt{I}} - 0.3I \right) \tag{19-4}$$

where Z_A is the charge of the ion A.

Generally, the Davies relationship is considered to be a good approximation for solutions with ionic strength of less than 0.5 M, which is the case for most wastewater and reclaimed water (Sawyer et al., 2003).

Given the activity coefficient of an ion A, the activity of A, {A} can be calculated from the molar concentration using Eq. (19-5):

$$\{A\} = \gamma_A \times [A] \tag{19-5}$$

where [A] is the molar concentration of A, mole/L.

In the following discussion on various indexes, activities (e.g., {A}, {B}) are used in equations. In practice, it is often assumed that the activity coefficient, γ, is close enough to 1 (i.e., {A} = [A]), but in some cases, particularly where reclaimed water contains TDS levels over 500 mg/L, activities are significantly less than the measured concentration impacting the solubility and equilibrium distribution of dissolved constituents.

Aggressiveness Index

The aggressiveness index (AI), an empirical relationship, has been used to evaluate the potential of a water to cause corrosion when transmitted through asbestos-concrete (A/C) pipes. The AI is defined as

$$AI = pH + \log (AH) \tag{19-6}$$

where AI = agressiveness index, unitless
pH = actual (measured) pH, unitless
A = total alkalinity, mg/L as $CaCO_3$
H = calcium hardness, mg/L as $CaCO_3$

Developed as a simplified version of the Langelier saturation index, discussed later in this chapter, the AI does not incorporate the effects of temperature or ionic strength (NAS, 1982). Water is assumed to be not aggressive or noncorrosive, if the AI value is 12 or greater. In many cases, particularly cooling water applications, water temperatures may exceed 50°C impacting the solubility of minerals and associated corrosivity or aggressiveness, and limiting the application of the AI relationship.

Buffer Capacity

One of the most important water quality parameters is the buffer capacity, also called buffer intensity. Buffer capacity is a measure of the role of buffering agents, such as carbonates, in controlling the pH stability, defined by Eq. (19-7):

$$\beta = \frac{\Delta\beta}{\Delta pH} \qquad (19-7)$$

where β = buffer capacity, meq/(L·pH unit)
$\Delta\beta$ = number of moles of acid or base added per unit volume of the solution, meq/L
ΔpH = change in pH

The buffer capacity, defined by the slope of the titration curve, represents the amount of base or acid needed to change the solution pH by one unit. Buffer capacity varies with pH, alkalinity, and temperature. Increases in corrosion rates have been correlated with decreases in buffer capacity (Clement et al., 2002; Pisigan and Singley, 1987). Optimizing the buffer capacity of reclaimed water may help to prevent pH changes at pipe/water interfaces, thereby reducing the potential for localized corrosion.

A comparison of the influence of pH on buffer capacity is shown on Fig. 19-4 for alkalinity concentrations ranging from 100 to 250 mg/L as $CaCO_3$. As shown, a minimum buffer capacity exists around pH 8.3. At pH values less than 8, alkalinity has a significant effect on the buffer capacity.

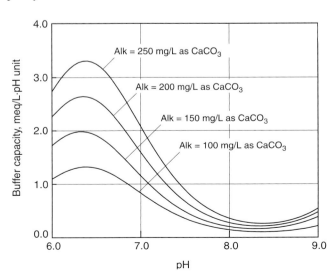

Figure 19-4

A comparison of the influence of pH and alkalinity on buffer capacity.

Figure 19-5
Simplified diagram of carbonate system chemical reactions occuring in concrete pipe: (a) in the presence of CO_3^{2-} and (b) in the presence of H_2CO_3 (Adapted from Snoeyink and Jenkins, 1980).

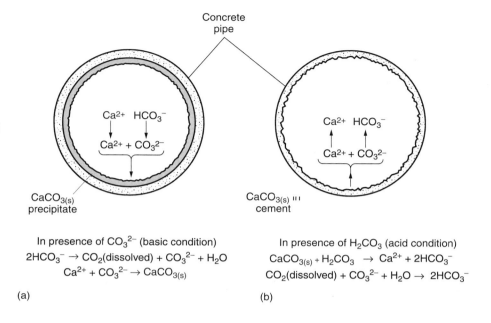

Calcium Carbonate Precipitation Potential (CCPP)

The calcium carbonate precipitation potential (CCPP) is an estimate of the concentration of calcium carbonate that could precipitate from a reclaimed water source under equilibrium conditions (see Fig. 19-5). A negative value of CCPP can be interpreted as the potential to solubilize calcium carbonate precipitates that may be present as a protective film on metallic surfaces, such as pipelines and condensers.

The CCPP index is based on the assumption that only one form of calcium carbonate is likely to form deposits and there are no kinetic barriers to deposition. However, magnesium, copper, zinc, orthophosphate, or polyphosphate may inhibit the formation of calcite and these variables are not considered typically in the computation of the index.

First, to help in understanding the calculation of CCPP, terms used in the carbonate system are defined as follows:

$$C_T = \{H_2CO_3^*\} + \{HCO_3^-\} + \{CO_3^{2-}\} \tag{19-8}$$

$$K_{a1} = \frac{\{H^+\}\{HCO_3^-\}}{\{H_2CO_3^*\}} \tag{19-9}$$

$$K_{a2} = \frac{\{H^+\}\{CO_3^{2-}\}}{\{HCO_3^+\}} \tag{19-10}$$

$$\alpha_0 = \frac{\{H_2CO_3^*\}}{C_T} = \frac{\{H\}^2}{\{H^+\}^2 + K_{a1}\{H^+\} + K_{a1}K_{a2}} \tag{19-11}$$

$$\alpha_1 = \frac{\{HCO_3^-\}}{C_T} = \frac{K_{a1}\{H^+\}}{\{H^+\}^2 + K_{a1}\{H^+\} + K_{a1}K_{a2}} \quad (19\text{-}12)$$

$$\alpha_2 = \frac{\{CO_3^{2-}\}}{C_T} = \frac{K_{a1}K_{a2}}{\{H^+\}^2 + K_{a1}\{H^+\} + K_{a1}K_{a2}} \quad (19\text{-}13)$$

where $\{H_2CO_3^*\}$ = carbonic acid activity at equilibrium, mole/L
$\{HCO_3^-\}$ = bicarbonate activity at equilibrium, mole/L
$\{CO_3^{2-}\}$ = carbonate activity at equilibrium, mole/L
K_{a1} = equilibrium constant for carbonic acid and bicarbonate
K_{a2} = equilibrium constant for bicarbonate and carbonate
$\{H^+\}$ = proton activity at equilibrium, mole/L
$\alpha_0, \alpha_1, \alpha_2$ = ionization fractions for carbonate system, unitless

Assuming that X mole/L of $CaCO_3$ will precipitate (or dissolve) from a reclaimed water, the equilibrium condition can be expressed as:

$$\{Ca^{2+}\}\{CO_3^{2-}\} = K_{so} = (\{Ca^{2+}\}_{initial} - X)(C_T - X)\alpha_2 \quad (19\text{-}14)$$

where $\{Ca^{2+}\}_{initial}$ = calcium activity before equilibrium, mole/L
$\{Ca^{2+}\}$ = calcium activity at equilibrium, mole/L
$C_T = \{H_2CO_3^*\}_{initial} + \{HCO_3^-\}_{initial} + \{CO_3^{2-}\}_{initial}$
K_{so} = solubility product of $CaCO_3 = \{Ca^{2+}\}\{CO_3^{2-}\}$
X = amount of $CaCO_3$ that will precipitate (or dissolve) at an equilibrium, mole/L

Alkalinity in eq/L will be reduced by 2X at the equilibrium condition, and it can be expressed as:

$$Alk_{initial} - 2X \approx (C_T - X)(\alpha_1 + 2\alpha_2) + \frac{K_w}{\{H^+\}} - \{H^+\} \quad (19\text{-}15)$$

where $Alk_{initial}$ = initial alkalinity, eq/L
K_w = equilibrium constant for water

The constants K_{so}, K_{a1}, K_{a2}, and K_w, depend on temperature. Typical values are provided in Table 9-13. To find the value of X, a spreadsheet program can be used as follows (Trussell, 1998):

1. Select an initial pH
2. Solve Eq. (19-9) for X using the initial pH
3. Solve Eq. (19-8) for $\{H^+\}$ and observe the magnitude of the error from the initial pH
4. Repeat Steps 1 through 3 using increasingly smaller pH increments until error in pH is less than 0.01 pH unit.

Target goals for CCPP are between 4 and 10 mg/L as $CaCO_3$. In general, the Langelier saturation index (LSI, see below) and CCPP have been used widely for scale control for waters with high levels of hardness and alkalinity to develop guidelines to avoid excess calcium carbonate deposition (Ferguson et al., 1996; Kirmeyer et al., 2000; Reiber et al., 1996).

EXAMPLE 19-1. Calculation of Calcium Carbonate Precipitation Potential (CCPP).

Calculate the calcium carbonate precipitation potential for a reclaimed water with the characteristics shown below. Neglect ionic strength effects (i.e., $\{A\} = [A]$), calcium complexation, and the effects of other ions such as magnesium, copper, zinc, orthophosphate, or polyphosphate.

$[Ca^{2+}] = 1 \times 10^{-3}$ mole/L (40 mg/L)

pH = 7.5

Alkalinity = 100 mg/L as $CaCO_3$ ($[HCO_3^-] = 2 \times 10^{-3}$ mole/L)

Temperature = 15°C

Equilibrium constants at 15°C from Table 9-13 are given below:

$$K_w = [H^+][OH^-] = 10^{-14.35}$$

$$K_{a1} = \frac{[H^+][HCO_3^-]}{[H_2CO_3^*]} = 10^{-6.42}$$

$$K_{a2} = \frac{[H^+][CO_3^{2-}]}{[HCO_3^-]} = 10^{-10.43}$$

$$K_{so} = [Ca^{2+}][CO_3^{2-}] = 10^{-8.22}$$

Solution

1. Compute the total carbonate concentration, C_T, by rearranging Eq. (19-12) using values for the given water.

$$C_T = \frac{\{HCO_3^-\}(\{H^+\}^2 + K_{a1}\{H^+\} + K_{a1}K_{a2})}{K_{a1}\{H^+\}}$$

$$= \frac{0.002[(10^{-7.5})^2 + (10^{-6.42} \times 10^{-7.5}) + (10^{-6.42} \times 10^{-10.43})]}{(10^{-6.42} \times 10^{-7.5})} = 0.002163$$

2. Use the steps outlined above (Trussell, 1998) to find the value of X. Note a spreadsheet program may be used to automate the solution procedure. As a first trial, choose a pH value of 7.

3. Solve Eq. (19-15) for X using the pH value selected in Step 2.
   a. Compute the values of α_1 and α_2.

$$\alpha_1 = \frac{K_{a1}\{H^+\}}{\{H^+\}^2 + K_{a1}\{H^+\} + K_{a1}K_{a2}}$$

$$= \frac{(10^{-6.42} \times 10^{-7})}{(10^{-7})^2 + (10^{-6.42} \times 10^{-7}) + (10^{-6.42} \times 10^{-10.43})}$$

$$= 0.792$$

$$\alpha_2 = \frac{K_{a1}K_{a2}}{\{H^+\}^2 + K_{a1}\{H^+\} + K_{a1}K_{a2}}$$

$$= \frac{(10^{-6.42} \times 10^{-10.43})}{(10^{-7})^2 + (10^{-6.42} \times 10^{-7}) + (10^{-6.42} \times 10^{-10.43})}$$

$$= 0.000294$$

b. Rearrange Eq. (19-15) to solve for X.

$$X \approx \frac{C_T(\{H^+\} + 2\alpha_2) + \dfrac{K_w}{\{H^+\}} - \{H^+\} - \text{Alk}_{\text{initial}}}{(\alpha_1 + 2\alpha_2 - 2)}$$

$$\approx \frac{0.002163[10^{-7} + (2)(0.000294)] + \dfrac{10^{-14.35}}{10^{-7}} - 10^{-7} - 0.002}{[0.792 + (2)(0.000294) - 2]}$$

$$\approx 0.000237$$

c. Solve Eq. (19-14) for K_{so} and observe the magnitude of the error from the initial K_{so}.

$$K_{so} = ([Ca^{2+}] - X)(C_T - X)\alpha_2$$

$$= (0.001 - 0.000237)(0.002163 - 0.000237)0.000294$$

$$= 4.32 \times 10^{-10}$$

The computed K_{so} value is lower than the target K_{so} value ($10^{-8.22}$), therefore the equilibrium pH value should be increased. Try pH = 8 for the next trial.

4. Repeat Step 3 using increasingly smaller pH increments until the K_{so} value converges.
   a. Recompute the values of α_1 and α_2 using a pH value of 8.

$$\alpha_1 = \frac{K_{a1}\{H^+\}}{\{H^+\}^2 + K_{a1}\{H^+\} + K_{a1}K_{a2}}$$

$$= \frac{(10^{-6.42} \times 10^{-8})}{(10^{-8})^2 + (10^{-6.42} \times 10^{-8}) + (10^{-6.42} \times 10^{-10.43})}$$

$$= 0.971$$

$$\alpha_2 = \frac{K_{a1}K_{a2}}{\{H^+\}^2 + K_{a1}\{H^+\} + K_{a1}K_{a2}}$$

$$= \frac{(10^{-6.42} \times 10^{-10.43})}{(10^{-8})^2 + (10^{-6.42} \times 10^{-8}) + (10^{-6.42} \times 10^{-10.43})}$$

$$= 0.00361$$

b. Rearrange Eq. (19-15) to solve for X.

$$X \approx \frac{C_T(\{H^+\} + 2\alpha_2) + \frac{K_w}{\{H^+\}} - \{H^+\} - Alk_{initial}}{(\alpha_1 + 2\alpha_2 - 2)}$$

$$\approx \frac{0.002163[10^{-8} + (2)(0.00361)] + \frac{10^{-14.35}}{10^{-8}} - 10^{-8} - 0.002}{[0.971 + (2)(0.00361) - 2]}$$

$$\approx -0.000114$$

c. Solve Eq. (19-14) for K_{so} and observe the magnitude of the error from the initial K_{so}.

$$K_{so} = ([Ca^{2+}] - X)(C_T - X)\alpha_2$$
$$= (0.001 + 0.000114)(0.002163 + 0.000114)\,0.00361$$
$$= 9.15 \times 10^{-9}$$

The computed K_{so} value is greater than the target K_{so} value (6.03×10^{-9}), therefore the equilibrium pH value should be decreased. Try pH = 7.75 for the next trial.

5. After successive trials, the K_{so} values converge at a pH value of 7.84, which corresponded to X = −0.0000869 ($CaCO_3$ = −8.69 mg/L).
Thus, CCPP = X = −8.69 mg/L.

Comment

The negative equilibrium value of $CaCO_3$ is representative of an aggressive water. For more accurate calculations, ions that contribute to alkalinity and other ion pairs besides HCO_3^-, CO_3^{2-}, OH^-, and H^+ should be considered. The simplified calculation as presented in this example may underestimate the amount of calcium carbonate that can be dissolved, and overestimate the amount of calcium carbonate precipitation that can occur (Clescerl et al., 1999).

Langelier Saturation Index

The Langelier saturation index (LSI) is the index used most commonly for predicting the aggressiveness of water (Langelier, 1936; Black & Veatch, 1992; Edwards et al., 1996; Faust, 1998; Rothberg, 2000). The LSI is calculated as the difference between the saturation pH and the measured pH.

$$LSI = pH - pH_s \tag{19-16}$$

where pH = measured (or actual) pH of the water, unitless
pH_s = saturation pH (The pH at which calcium and alkalinity in water are in equilibrium with solid calcium and carbonate.), unitless

For pH values between 7.0 and 9.5, the saturation pH (pH_s) is calculated as:

$$pH_s = (pK_{a2} - pK_{so}) + pCa^{2+} + pAlk \tag{19-17}$$

where pK_{a2} = negative $\log_{10}$ of K_{a2}
pK_{so} = negative $\log_{10}$ of K_{so}
pCa^{2+} = negative $\log_{10}$ of the active concentration of calcium
$pAlk$ = negative $\log_{10}$ of alkalinity in eq/L, assuming $\{Alk\} = \{HCO_3^-\}$

The LSI provides an estimate of the thermodynamic driving force for either precipitation or dissolution of calcium carbonate; however, it cannot be used to estimate the concentration of calcium carbonate available for precipitation. The LSI is based on the assumption that a specific form of calcium carbonate is present (calcite, valerite, or aragonite). A general interpretation of the LSI values is presented in Sec. 9-4. For corrosion control, the goal is to have an LSI that is slightly positive to promote coating of the pipe wall by calcium carbonate. In many cases, deposition of calcium carbonate has been observed even when values of the LSI are negative because of localized pH variations at the pipe wall generated by cathodic reduction of oxygen. In addition, the calculation of LSI does not take into account the use of corrosion inhibitors.

Ryznar Stability Index

The Ryznar stability index (RSI) is an empirical variation of the LSI that gives more weighting to the pH at which the water is saturated with calcium carbonate (pH_s). However, its appropriateness for specific metals, such as copper, may need to be verified on a case-by-case basis (Ryznar, 1944; Ferguson et al., 1996; Reiber et al., 1996; Kirmeyer et al., 2000). The RSI is defined as Eq. (19-18).

$$RSI = 2pH_s - pH \qquad (19-18)$$

where pH_s = saturation pH, at which the water is saturated with calcium carbonate
pH = pH of the actual sample

The values of the RSI are generally interpreted as follows:

RSI > 8.5	water is very undersaturated: mild steel corrosion may occur
8.5 > RSI > 6.8	water is undersaturated: tends to dissolve any existing solid $CaCO_3$
6.8 > RSI > 6.2	protective film of $CaCO_3$ may or may not be developed
6.2 > RSI > 5.5	water is oversaturated: tends to be scale forming
5.5 > RSI	water is very oversaturated: heavy scale will form

It should be noted that the value for saturation equilibrium changes with the value of pH_s, and therefore the interpretation of the values must be adjusted with the pH_s.

Other Indexes

The Stiff and Davis stability index (SDI) was developed to predict the tendency of oil field water deposit of calcium carbonate scale (Stiff and Davis, 1952). Puckorius and Brooke (1991) have also proposed a new index for calcium carbonate scaling, developed over a ten year period. Several ratios have been developed to evaluate the relative role of dissolved anions in promoting corrosion (see Table 19-6). The ratio of chloride to bicarbonate, or Larson's Ratio, was originally derived for evaluation of steel and cast iron corrosion (Larson and Skold, 1958).

Corrosion Management Options

Corrosion control options include selection of appropriate (noncorrosive) materials combined with the selective use of coatings, corrosion inhibitors, cathodic/anodic protection, and rational process design. As described above, oxygen can induce corrosion reactions in industrial systems that use high quality water, such as boilers. The use of reducing agents to quench residual oxygen and oxidants can help to protect boiler systems. Ideally, dissolved oxygen levels in feedwater should be below 0.03 mg/L, and preferably less than 0.005 mg/L for high-pressure boilers (McCoy, 1981). In boiler systems, removal of dissolved oxygen is the most common measure for corrosion prevention (Pagliaro, 1997).

Coatings are another practicable approach for protection of metal surfaces from corrosion. Organic or inorganic surface coatings act to prevent metallic corrosion by three mechanisms (Davis, 2000):

- Barrier protection: isolate the substrate from the corrosive environment
- Chemical inhibition: addition of inhibitive pigments to paints
- Galvanic (sacrificial) protection: coat the substrate with a more active metal, by which the substrate becomes the cathode in the corrosion cell

Corrosion inhibitors include anodic inhibitors, cathodic inhibitors, ohmic inhibitors, precipitation inhibitors, and vapor-phase inhibitors (Davis, 2000). Anodic inhibitors are oxidizing agents including chromate, nitrite, nitrate, and nonoxidizing chemicals such as phosphates, tungstates, and molybdates. The inhibitors act to reduce the oxidation potential of iron surfaces so that the iron is less soluble and it is less likely to be oxidized by hydrogen ions (McCoy, 1974). Cathodic inhibitors shift the cathodic reduction processes towards a less negative state, resulting in the slowing of the anodic corrosion reaction that must be balanced. Corrosion inhibitors tend to be more effective under slightly alkaline conditions and their reliability can be impacted by temperature and pH. Selection of inhibitors and dosing should be determined carefully following manufacturer's instructions because an inhibitor that reduces corrosion of one type of metal may accelerate corrosion of other metals or alloys. It is good practice to conduct pilot-scale tests to verify the effectiveness of corrosion inhibitors.

As an alternative to the use of chemical inhibitors, cathodic and/or anodic protection can be used. These control strategies work by shifting the electrical potential of the metal to be protected. The electrical potential can be shifted by applying a direct current from a power supply or by connecting to dissimilar metals to be sacrificed *in lieu* of the piping material (Davis, 2000).

In situ corrosion testing and monitoring are important to optimize corrosion control strategies. Detailed information on corrosion testing and monitoring can be obtained from various sources such as the American Society for Testing Materials (ASTM), the International Organization for Standardization (ISO), the National Association of Corrosion Engineers (NACE) International, and the Materials Technology Institute of the Chemical Process Industries (MTI).

Water quality management can also be a valuable tool for controlling corrosion. The addition of chemicals for the control of pH or alkalinity and or biocides can prevent or mitigate water quality-induced corrosion. When reclaimed water is used in industrial applications, water from sources with varying water quality may be used in a common metal pipeline or tank. It is important to note that the varying water quality may introduce transient conditions that can lead to corrosion. Incremental changes in water quality or off-line blending of water sources may help to alleviate these issues (Videla, 1996).

Scaling Issues

Scaling is the formation of oxides, carbonates, and/or other deposits on the surfaces of pipelines, heat exchangers, tanks, and other surfaces contacted by water. While corrosion is typically associated with dissolution of solid constituents (iron, copper, lead, etc.), scaling represents the precipitation of dissolved constituents onto solid surfaces. A thin film of oxides or carbonates on metal surfaces can be beneficial as it provides a barrier that restricts contact between the liquid and the metal surface thus limiting the potential for corrosion reactions. A thick film of precipitates, however, can adversely affect the system by occluding openings and interfering with water flow and heat exchange. Scaling on a boiler system, for example, can reduce the efficiency of heat exchange, and excessive scaling may restrict flow and eventually block pipes.

Types of Scaling

The most commonly observed scaling associated with reclaimed water is deposition of calcium carbonate (see Fig. 19-6). Calcium carbonate scaling can be controlled through pH optimization and routine monitoring. In some cases, scale inhibitors are used to prevent deposition. Generally, the solubility of dissolved constituents in water increases with temperature. However, in some cases, such as calcium phosphate and calcium sulfate, solubility decreases with an increase in temperature. The precipitation reactions result in the formation of scales on heat exchangers, reducing the heat conductivity. Deposition of calcium phosphate $[Ca_3(PO_4)_2]$ or hydroxyapatite $[Ca_{10}(PO_4)_6(OH)_2]$ can

Figure 19-6

Calcium carbonate deposition inside of a pipe. The reference scale is in centimeters. (Courtesy of R. R. Trussell.)

be problematic in industrial cooling processes because of the relatively low solubility of these compounds. If reclaimed water has not been subjected to nutrient removal processes, the phosphate levels in the treated water may be high enough to result in the formation of phosphate-based scales. In addition to calcium phosphate and hydroxyapatite, other potential scales include magnesium phosphate, silicates such as magnesium silicate, acmite ($Na_2O \cdot Fe_2O_3 \cdot 4SiO_2$), and analcite ($Na_2O \cdot Al_2O_3 \cdot 4SiO_2 \cdot 2H_2O$), and occasionally calcium carbonate (McCoy, 1981).

Scaling Management Options

The best way to prevent scaling is to control chemicals that cause scaling, principally calcium, magnesium, carbonate, phosphate, and organic matter. Dissolved constituents may be controlled by using onsite treatment (see Chap. 9); however, the use of supplemental treatment systems increases capital and operating expenses and requires disposal of waste products. Lime softening and ion exchange have been used, particularly in small-scale processes. Membrane technologies, such as nanofiltration and reverse osmosis (RO), provide alternative approaches for removal of dissolved minerals and can also be used for removal of dissolved organics and control of pathogens (see Chap. 9).

The process often referred to as *cold lime treatment* is effective for removal of phosphates from reclaimed water. By removing the phosphates upstream of the industrial process, the potential for deposition in pipelines, on pumps, and within equipment is reduced. In the cold lime process, phosphate is removed by chemical precipitation as shown in Eq. (19-19).

$$10\ Ca^{2+} + 6\ PO_4^{3-} + 2\ OH^- \rightleftarrows Ca_{10}(PO_4)_6(OH)_2 \qquad (19\text{-}19)$$

In the cold lime process the pH is raised above 11. The high pH results in the precipitation of hydroxyapatite, magnesium, and silica. A byproduct of cold lime treatment is an increase in the calcium concentration. To remove the excess calcium, the water needs to be recarbonated to pH 10 and treated with soda ash (Na_2CO_3) as shown below:

$$Ca^{2+} + CO_3^{2-} \rightleftarrows CaCO_3(s) \qquad (19\text{-}20)$$

$$Ca^{2+} + Na_2CO_3 \rightleftarrows CaCO_3(s) + 2Na^+ \qquad (19\text{-}21)$$

Antiscalants are also used commonly in industrial processes either to prevent scale formation or to mobilize accumulated scale. Many of the antiscalants are phosphate-based compounds, chelants [e.g., ethylene-dinitrilo tetraacetate (EDTA) and nitrilotriacetate (NTA)], and polymers. When phosphate is used for internal treatment of calcium, tan-colored scales of ferric phosphate may be found in heat exchangers (McCoy, 1974). Frequently, details on product formulations are considered to be proprietary information by manufacturers, therefore, the effectiveness of antiscalants needs to be evaluated on a case-by-case basis. Mechanical cleaning of scaled tubes can be effective if the structure of the scale is not rigid. Cleaning devices, commonly called *pigs* (see Fig. 14-20 in Chap. 14), can be used to clean scaling and biological growth from the inner surfaces of pipes and tubing. The pigs are slightly larger in diameter than the tubes and abrade the deposits as they move through the tubes using pressurized water jets. Pigs are especially effective when used in combination with chemical cleaning reagents.

Accumulation of Dissolved Constituents

Water quality can change as a result of industrial use, recycling, and reuse. Chemical compounds may be added to the water for cleaning, corrosion or scale control, or manufacturing purposes, and water may exit as vapor depending on temperature conditions. For example, dissolved solids in cooling water are concentrated due to the loss of water through evaporation. When industrial process water is returned to a municipal wastewater treatment plant, and reclaimed water is repeatedly used for industrial purposes, the constituents in the reclaimed water that are not removed by treatment processes may accumulate to the point where the effluent quality does not meet discharge requirements, or the reclaimed water may become unsuitable for water reuse (McIntyre et al., 2002). For example, most wastewater treatment systems are not designed to remove chlorides, sodium, potassium, or other dissolved constituents, particularly hydrophilic compounds. A mass balance on a water reuse system may be necessary to evaluate the impact of constituents that have potential to accumulate in reclaimed water.

If some reclaimed water is used for irrigation while a larger portion is used for industrial purposes and is returned to the wastewater collection system, the salt content in the reclaimed water may become unsuitable for irrigation even if it can still be used to meet the water quality requirements of the industrial application. Treated wastewater that is not reused must be discharged to receiving waters under National Pollutant Discharge Elimination System (NPDES) permit limits. Depending on the receiving water characteristics, the level of TDS may impact compliance with NPDES discharge requirements when multiple closed-loop water reuse applications are established and high TDS wastewater is returned to the collection system.

A simplified example of a looped water reuse system is illustrated on Fig. 19-7. The mass balance for a constituent of concern is expressed as follows:

For water reclamation plant: $C_o Q_o + C_r Q_r = C_e Q_e + C_e Q_1$ (19-22)

For industrial process: $C_e Q_1 + C_A Q_A = C_L Q_L + C_r Q_r$ (19-23)

where C_o, Q_o = municipal wastewater influent concentration and flowrate
C_e = reclaimed water concentration
C_r, Q_r = industrial wastewater concentration and flowrate
C_A, Q_A = concentration in the water added during the industrial process and flowrate
C_L, Q_L = concentration in the water lost during the industrial process and flowrate
Q_e = net effluent flowrate
Q_1 = flowrate to industrial process

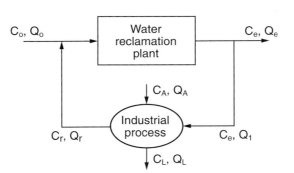

Figure 19-7
Simplified flow diagram of a looped water reuse system.

If reclaimed water is used for industrial cooling and concentrated water (blowdown) is returned to the wastewater collection system, TDS is not added or lost in the cooling tower: $C_A Q_A = M_A = 0$, and $C_L Q_L = M_L = 0$ where M_A and M_L are the corresponding mass values. Combining equations and solving for C_e, the concentration of the constituent in reclaimed water is:

$$C_e = C_o \frac{Q_o}{Q_e} \tag{19-24}$$

Therefore, if reclaimed water is used for an industrial application that concentrates specific constituents, the quality of water in the looped system depends on how much water is returned to the wastewater collection system. In practice, the inputs and outputs of chemical constituents in the looped water reclamation and reuse systems are much more complex, and a mass balance model should be established to examine the accumulation of dissolved constituents in the loop.

EXAMPLE 19-2. Accumulation of Persistent Wastewater Constituents.

Assume two different industries are within the service area of a wastewater reclamation facility as illustrated below. In industrial process A, reclaimed water is used for industrial cooling purposes and the blowdown (concentrated water) is returned to the wastewater collection system. In industrial process B, potable water is used and the industrial wastewater is discharged to the wastewater collection system. Calculate the TDS concentration of (1) the discharged effluent, C_e, from the water reclamation plant and (2) the return flow, C_r, from industrial process A under steady-state conditions.

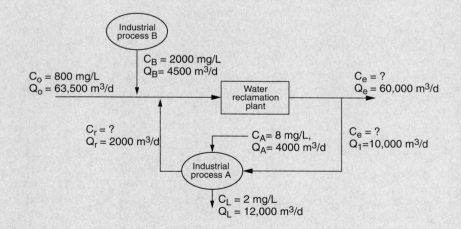

Solution

1. Write a mass balance around the water reclamation plant and industrial process A:

$$C_o Q_o + C_B Q_B + C_r Q_r = C_e Q_e + C_e Q_1$$

Substituting the given numbers yields:

$$800 \times 63{,}500 + 2000 \times 4500 + C_r \times 2000$$
$$= C_e \times 60{,}000 + C_e \times 10{,}000$$

or

$$59{,}800{,}000 + 2000 C_r = 70{,}000 C_e$$

2. Write a mass balance for the industrial process:

$$C_e Q_1 + C_A Q_A = C_L Q_L + C_r Q_r$$

Substituting the given numbers yields:

$$C_e \times 10{,}000 + 8 \times 4000 = 2 \times 12{,}000 + C_r \times 2000$$

Rearrange to obtain:

$$C_r = 5 C_e + 4$$

3. Combine the equations from Steps 1 and 2 and solve for C_e and C_r:

$$59{,}800{,}000 + 2000(5 C_e + 4) = 70{,}000 C_e$$
$$C_e = 996.8 \text{ mg/L}$$

And by substitution

$$C_r = 4988 \text{ mg/L}$$

Comment

The TDS level entering the water reclamation plant before reclaimed water is used for the industrial process A (combined flow from the wastewater collection system and the effluent from the industrial process B) was 879 mg/L. By including the looped system for the Industrial process A, the TDS level increased about 11 percent to 997 mg/L. As the fraction of reclaimed water used for the industrial process increases, the level of dissolved solids in the closed-loop system will increase. Mass balances on salt (TDS) must be evaluated when the water reclamation and reuse system includes a closed-loop application. Measures to control the salinity level include discharge of industrial process water outside of the loop, treatment of high-salt wastewater, and the use of a smaller fraction of reclaimed water.

19-3 COOLING WATER SYSTEMS

The thermoelectric power generation industry is the largest water user in the United States, accounting for nearly 40 percent (188×10^9 m^3/yr) of the total freshwater withdrawals (477×10^9 m^3/yr) (Huston et al., 2004). After electric power generation, cooling systems are the largest water users in the manufacturing, coal and petroleum industries (Wijesinghe et al., 1996), and for commercial air conditioning systems. The constant and high demand for water coupled with the ability to meet water quality constraints, render reclaimed water an attractive water source for cooling water systems. In fact, the use of reclaimed water for cooling purposes represents the largest municipal water reuse application by industry.

System Description

Generally, cooling water systems can be divided into three categories: once-through non-contact cooling, recirculating non-contact cooling, and direct contact cooling (State of California, 1980). In once-through non-contact systems, process heat is transferred to the cooling water in a heat exchanger, which is then discharged from the cooling system. Once-through systems are generally used at locations where water for cooling is abundant and readily accessible; reclaimed water is usually not considered in such cases. In recirculating non-contact systems, warmed water, from a cooling operation or heat exchanger, is cooled by transferring its heat to air through evaporation in a cooling tower. The cooled water is then recirculated back through the heat exchanger and reused to absorb heat. The majority of cooling water systems that use reclaimed water are of the recirculating non-contact type. A simplified diagram of a non-contact recirculating system is shown on Fig. 19-8a, and an example of a cooling tower utilizing reclaimed water is shown on Fig. 19-8b. As shown on Fig. 19-8a, warm water from process cooling is sprayed on the top of the internal packing, used to break up the water through spray into droplets to enhance air/water contact. Cool, dry outside air is pulled up through the cooling tower by a large rotating fan to cool the warm water through evaporation. Typical air to water ratios are on the order of 600 to 800.

Direct contact cooling systems are used for some industrial processes where the water used for cooling is applied directly to the process equipment or product to be cooled. Reclaimed water has been used in a direct contact spray and cascade type cooling systems since 1942 at the Sparrows Point, Maryland, steel plant (formerly Bethlehem Steel). Originally, secondary-treated, disinfected water from Baltimore's Back River wastewater treatment plant was applied directly to cool the surfaces of the blast furnaces, the doors of the open hearth furnaces, the skid pipes, and the steel products themselves (State of California, 1980). Currently the Back River treatment plant is providing reclaimed water from a tertiary treatment facility.

Water Quality Considerations

Reclaimed water can be used successfully as a cooling water source by maintaining the appropriate operational conditions and by controlling the quality of the water. Primary considerations for cooling systems are prevention of corrosion, scaling, and biological fouling. Reclaimed water constituents of concern in cooling tower applications are summarized in Table 19-9. Water quality must be controlled either at the

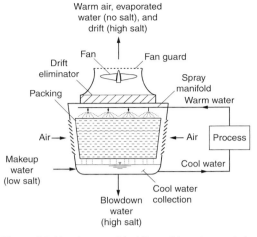

(a)   (b)

Figure 19-8
Cooling water system: (a) a simplified operational flow diagram for a cooling tower and (b) an example of cooling tower utilizing reclaimed water, Largo, FL.

Table 19-9
Reclaimed water constituents of concern for cooling tower applications

Constituents of concern	Potential problems	Control options	Point of control[a]	Reference chapters
Ammonia	Biological fouling	Nitrification, stripping	R/S	7
Carbonate, bicarbonate	Corrosion, scaling	pH control, use of antiscalants	S	9, 19
Calcium	Scaling	Nanofiltration, RO, ion exchange, EDR	R/S	9, 10
Magnesium	Scaling	Nanofiltration, RO, ion exchange, EDR	R/S	9, 10
Microorganisms	Biological fouling	Disinfection, shock chlorination, mechanical cleaning	R/S	11
Organic compounds	Biological fouling	Biological treatment	R	7
Phosphates	Scaling	Lowering pH, biological nutrient removal, cold lime treatment (precipitation)	R/S	7, 8
Silica	Scaling			
Total dissolved solids	Corrosion, scaling	Blowdown	S	19

[a] R = water reclamation plant; S = onsite.

water reclamation plant or by onsite treatment. Some of the specific quality considerations are discussed below.

Ammonia
Ammonia can result in biological growth and can also cause corrosion by forming complexes with metals. Copper or copper alloys are susceptible to corrosion by ammonia (see Table 19-7). If the biological process in the existing wastewater treatment process does not undergo nitrification, a nitrification process may need to be added to the treatment system (see Chap. 7). An advantage of nitrification is that nitrate can act to inhibit calcium carbonate and calcium phosphate scaling.

Carbonate and Bicarbonate (Alkalinity)
The alkalinity of the cooling water can impact the corrosion or scale-forming potential in cooling systems. High alkalinity provides carbonate and bicarbonate ions that can lead to scaling in the presence of calcium, whereas extremely low alkalinity may increase the corrosivity of the water. For makeup water in recirculating cooling systems, alkalinity levels below 20 mg/L as $CaCO_3$ are suggested (see Table 19-4).

Calcium
Calcium scales such as calcium carbonate, calcium sulfate, and calcium phosphate are the principal cause of cooling tower scaling problems (Adams et al., 1980). As described previously, calcium carbonate can be controlled by pH adjustment and is not a major issue. Phosphate control may also be necessary to prevent calcium precipitation.

Phosphate
Removal of orthophosphate and organics can help to reduce the potential for scaling, as described in the previous section. Biological phosphorus removal in the water reclamation process generally results in phosphorus concentrations below the level of concern (see Table 19-4).

Other Chemical Constituents
Other constituents of importance include magnesium, silica, and dissolved organics. Magnesium scales (magnesium carbonate and phosphate) may form, depending on the relative concentrations of magnesium, phosphate, and alkalinity. Silica deposits may also form and can be particularly difficult to remove from heat exchanger surfaces; however, most waters contain relatively small quantities of silica. Organic compounds in reclaimed water along with nutrients can promote the growth of microorganisms in the recirculating water. Metabolites of microbial activity include organic acids, carbon dioxide, soluble organic compounds, and biomass. Depending on the alkalinity, microbial growth can lead to localized changes in pH thereby promoting corrosion reactions.

Microorganisms
In the presence of bioavailable organic compounds, microbial growth in the reclaimed water distribution system and the cooling system can occur. Biological fouling is one of the major concerns in reclaimed water use for cooling systems. Microbial growth must be controlled to prevent fouling of the cooling systems. In using reclaimed water for cooling tower applications, it is important to recognize that pathogenic organisms

may be present. The potential for exposure to microorganisms in cooling water is related to the generation and transport of wind-borne aerosols through drift. Management options for control of pathogenic microorganisms and biological fouling are further discussed below.

The best way to prevent corrosion is to design systems with corrosion-resistant materials, depending on cost and availability factors. Once a cooling system is in operation, effective programs for water quality management are essential to prevent corrosion and scaling problems and to manage public health risks. Material selection for cooling systems used with various water sources is shown in Table 19-10.

A key water quality issue for the operation of cooling water systems is controlling the accumulation of dissolved minerals and organic constituents. During the cooling process, a portion of the cooling water evaporates, resulting in a net increase in the concentration of nonvolatile constituents in the recirculating water. Control of the dissolved solids concentration is accomplished by balancing the addition of supplemental water (makeup water) to replace the evaporated water with removal of higher TDS water through the blowdown process. Because the blowdown water is a waste product of the process, it is important to evaluate the quantity and quality of blowdown water generated to develop appropriate treatment and disposal options.

Blowdown

The water deliberately removed from a cooling system, as shown on Fig. 19-8a, is called *blowdown* (also called *purge*). Small amounts of water are also lost by drift. Using a simple mass balance approach, the flowrate of makeup water should equal the sum of evaporation loss, water loss due to drift, and the flowrate of blowdown:

$$Q_m = Q_e + Q_d + Q_b \quad (19\text{-}25)$$

where Q_m = flowrate for makeup water, L/min
Q_e = water loss rate from evaporation, L/min
Q_d = water loss rate from drift, L/min
Q_b = blowdown flowrate, L/min

Design and Operational Considerations

Table 19-10
Material selection for cooling systems[a]

Water type[b]	Alloy selection
Fresh water	Steel, Copper, Aluminum
Corrosive fresh water	Copper (admiralty)
Brackish water and seawater (low to moderate velocity)	70% Cu–30% Ni, 90% Cu–10% Ni
Unpolluted and polluted seawater (high velocity)	Stainless steel (type 316), ferritic molybdenum stainless steels (Fe-Cr-Mo), titanium
Polluted seawater (low velocity)	Ferritic molybdenum stainless steels (Fe-Cr-Mo), titanium

[a]Adapted from Davis (2000).
[b]Listed in order of increasing corrosivity.

Cycles of Concentration

In a well-operated recirculating cooling system, evaporation water loss is about 1 to 2 percent, and drift loss consists of about 0.1 to 0.001 percent of the recirculation flowrate. The percentage of water wasted as blowdown is determined by the allowable salt concentration in the cooling water system (Asano et al., 1988; McCoy, 1974).

The concentration cycle, or cycles of concentration, is defined as the ratio of dissolved solids concentration in the cooling tower water to the dissolved solids concentration in the makeup water (Cheremisnoff and Cheremisnoff, 1981):

$$C = \frac{X_b}{X_m} \tag{19-26}$$

where C = cycles of concentration
X_b = dissolved solids concentration in blowdown, mg/L
X_m = dissolved solids concentration in makeup water, mg/L

Note that dissolved solids are not lost via evaporation and the solids concentration in drift water is the same as the concentration in the blowdown. By mass balance, the mass of dissolved solids added from the makeup water is equal to the mass of dissolved solids wasted via blowdown and drift.

$$(Q_d + Q_b) \times X_b = Q_m \times X_m \tag{19-27}$$

From Eqs. (19-25) through (19-27), the cycles of concentration can be expressed in terms of flowrates as:

$$C = \frac{X_b}{X_m} = \frac{Q_m}{Q_d + Q_b} \tag{19-28}$$

Eq. (19-28) can be rearranged as

$$Q_d + Q_b = \frac{Q_m}{C} \tag{19-29}$$

From Eqs. (19-25) and (19-29), the makeup water flowrate is:

$$Q_m = Q_e + \frac{Q_m}{C} \tag{19-30}$$

The rate of evaporation can be predicted by assuming that only the latent heat of vaporization is responsible for the cooling. For estimating purposes, approximately 2300 kJ of heat is lost to evaporate 1 kg (1 L) of water. About 4.2 kJ of heat needs to be removed to cool 1 kg (1 L) of water by 1°C. Therefore, the removed heat from the cooling system can be expressed as follows:

$$2300 \times Q_e = 4.2 \times Q_c \times \Delta T \tag{19-31}$$

where Q_c = circulation rate, L/min
ΔT = temperature change, °C

From Eqs. (19-30) and (19-31)

$$\frac{Q_m}{Q_c} = \frac{4.2 \times \Delta T}{2300} + \frac{1}{C} \times \frac{Q_m}{Q_c}$$

or

$$M = \frac{Q_m}{Q_c} \times 100 = \frac{C}{C-1} \times \frac{4.2\Delta T}{2300} \times 100 \approx \frac{\Delta T \times C}{548(C-1)} \times 100 \quad (19\text{-}32)$$

where M = makeup water requirement, percent of the circulation rate

Also from Eq. (19-29)

$$B = \frac{Q_b}{Q_c} \times 100 = \left[\frac{\Delta T}{548(C-1)} - \frac{Q_d}{Q_c}\right] \times 100 \quad (19\text{-}33)$$

where B = blowdown requirement, percent of the circulation rate

Makeup water blowdown requirements derived from Eqs. (19-32) and (19-33) are plotted on Figs. 19-9 and 19-10, respectively; these plots can be used to estimate the required makeup water and blowdown as a function of the cycles of concentration. The cycles of concentration are determined based on the makeup (reclaimed) water quality and allowable dissolved solids concentration (Cheremisinoff and Cheremisinoff, 1981). Typically, the cycles of concentration of a cooling system using reclaimed water range from two to five.

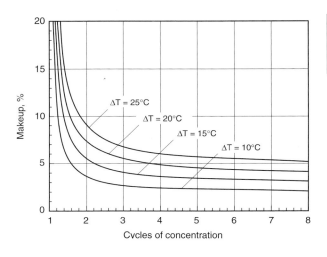

Figure 19-9
Makeup water requirements for a cooling tower.

Figure 19-10

Blowdown requirements for a cooling tower. Assumption: water loss from drift = 0.1 percent.

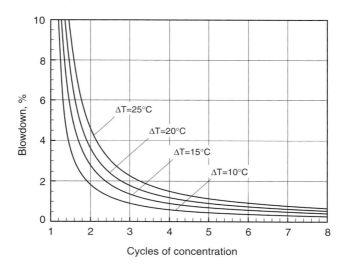

Management Issues

Proper management of cooling water systems is necessary to prevent a proliferation of microorganisms and to control corrosion, scale, and fouling of internal surfaces. Microbial contamination can pose a public health risk if aerosols that are generated through the cooling process are released into the ambient environment and transported through wind. The water that exits cooling systems through wind action is known as *drift*. Many facilities are designed with drift eliminators that are intended to minimize escape of entrained water droplets (Cooling Technology Institute, 2006). It is important to note that the water vapor or steam that is often visible at cooling towers is evaporated water and not drift and it does not contain dissolved solids, microorganisms, or other nonvolatile constituents in cooling water.

Public Health Protection

The potential for human exposure to microorganisms in cooling water is related to generation of aerosols through drift. While reclaimed water is disinfected prior to reuse, some microorganisms persist through treatment. One waterborne disease that has been linked to cooling water systems is legionellosis, a severe and potentially fatal form of pneumonia which is caused by a rod-shaped heterotrophic bacterium, *Legionella*. *Legionella* is also associated with Pontiac fever, which has less severe flu-like symptoms (CDC, 2006). *Legionella* spp. are able to survive in cooling tower environments due to their capacity to survive under a wide range of temperature and dissolved oxygen conditions, and have the ability to proliferate in the interstices of biofilms and within protozoa or algal cells. A summary of the characteristics of *Legionella* is given in Table 19-11.

In general, the survival of *Legionella* within cooling water systems is related to the water temperature, the degree of water stagnation, and the accumulation of sediments and biofilms. An example of the heat tolerance of *Legionella* is shown in Fig. 19-11. The ability to increase the temperature to over 60°C can help to control proliferation of *Legionella*. In addition to temperature, other control strategies include routine water quality monitoring coupled with prudent management practices.

Table 19-11
Characteristics of *Legionella* relevant to growth in cooling towers

Parameter	Description
Size	0.3–0.9 μm by 1.3 μm; can form filaments up to 20 μm
Temperature requirements	15–43°C
Oxygen requirements	Aerobic or microaerophillic (0.2 mg/L)
Generation time	99 min under optimal conditions
Other water quality requirements	Iron
Specific habitats	
Symbiotic microorganisms	
Protozoa	*Acanthamoeba, Hartmanella, Naegleria, Echinamoeba, Vahlkampfia,* and *Tetrahymena*
Cyanobacteria	*Fischerella, Phormidium,* and *Oscillatoria*
Green algae	*Scenedesmus, Chlorella,* and *Gleocystis*
Biofilms	Microorganisms in biofilms provide protection and nutrients
Other locations	Sediment, sludge, scale and organic materials, stagnant water

Direct monitoring of *Legionella* is complicated because the bacteria may be associated with biofilms, sediments, protozoa, algae, or other microorganisms. In addition, the turnaround time for detection of viable microorganisms is fairly long, using cell-culture techniques (1–2 wk). Techniques for the detection and enumeration of *Legionella* using polymerase chain reaction (PCR) are being developed but the specificity and sensitivity of the test results needs to be verified for each sample (Joly et al., 2006). Indirect monitoring by visual inspection for biofilms and turbidity coupled with microscopic analysis of protozoa and algae can be used to identify conditions that might promote the growth of *Legionella*. Direct monitoring is recommended in response to a potential outbreak of legionellosis (Cooling Technology Institute, 2006).

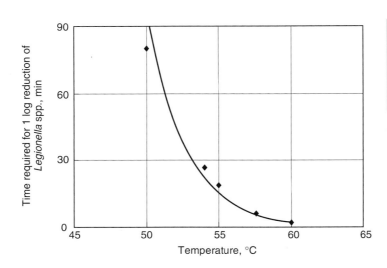

Figure 19-11
Effect of temperature on the time required to achieve 1 log reduction of *Legionella*.

Control of *Legionella* and other pathogenic organisms is best accomplished by prevention and use of prudent management practices. Recommendations from the Cooling Technology Institute (2006) and the American Society of Heating, Refrigerating and Air-Conditioning Engineers, Inc (ASHRAE, 2000) are listed below.

- Minimize water stagnation.
- Minimize process leaks into the cooling system that provide nutrients for bacteria.
- Maintain overall system cleanliness to avoid buildup of sediments that can harbor or provide nutrients for bacteria and other organisms.
- Apply scale and corrosion inhibitors as appropriate.
- Use high-efficiency drift eliminators on cooling towers.
- Maintain good control of the microbiological population.
- Try to allow a physical separation between cooling towers and fresh air intakes.

Routine disinfection of cooling systems can be accomplished using thermal treatment or chemical disinfectants. A summary of chemical disinfectants that are effective for control of *Legionella* is given in Table 19-12. Chlorine is the most widely used biocide. It should be noted that high concentrations of chlorine can be corrosive to metallic surfaces. In addition, chlorination can result in the formation of volatile chlorinated organics that can pose an additional health risk. The efficacy of UV disinfection for control of *Legionella* is uncertain due to its high photoreactivation ability (Oguma et al., 2004; see also Chap. 11).

If an outbreak occurs, the entire cooling system must be taken off-line and disinfected to eradicate the pathogens. Sequential hyperchlorination using two to three applications of at least 10 mg/L free residual chlorine for a 24 h contact time have been recommended (Cooling Technology Institute, 2006). The cleaning operation should be repeated until there is no evidence of biofilm or turbidity in the system.

Corrosion

Corrosion control can be accomplished using two approaches: removal of constituents that cause corrosion and addition of chemicals that inhibit corrosion (see Sec. 19-2). Common corrosion inhibitors for cooling water systems include orthophosphates and polyphosphates that sequester the dissolved minerals (Davis, 2000; Levine and Asano, 2002). Ferrous sulfate ($FeSO_4$) can be used for inhibition of chloride-induced corrosion (Bogaerts and van Haute, 1985).

Table 19-12
Disinfection requirements for control of *Legionella* in cooling water systems

Disinfectant	Typical dose, mg/L	Contact time, min
Ozone	0.3	20
Chlorination		
Continuous	3–5	20
Intermittent	5–10	60 (once per day)
Hyper	10–30	360–1440

Scaling

As described earlier, calcium phosphate is the primary cause of scaling in cooling systems (see Sec. 19-2). Management options for the prevention of scaling include pretreatment of reclaimed water to lower concentrations of calcium and phosphate, application of antiscalants, and periodic cleaning.

Biological Fouling

Biofilm growth in cooling water systems is a major concern when reclaimed water containing biodegradable organics is used. Biofilms can increase the thickness of insulative materials, and eventually plug the recirculating water system (Adams et al., 1980). Either chemical or mechanical means can be used to remove biological growth in a cooling system. Chlorine, sodium hypochlorite (NaOCl), or bromine chloride (BrCl) are used commonly for cleaning cooling towers. In practice, shock chlorination, a short-term high-dose (>10 mg/L) chlorination of the system, can be used periodically to remove biofilm growth in cooling systems. The required chlorination dose and interval depend on the quality of reclaimed water and the degree of biofilm development. Mechanical cleaning in combination with periodic chlorination is usually satisfactory for biofilm removal in most cooling system piping. It should be noted, however, that chlorination of cooling water may result in the production of halogenated organic compounds (Assink and van Deventer, 1995).

19-4 OTHER INDUSTRIAL WATER REUSE APPLICATIONS

Since the 1990s, increased attention has been directed at improving water use efficiency and implementing internal water recycling for industrial facilities. To date, industrial use of reclaimed water from municipal wastewater has been limited by costs, water quality, and the availability of reclaimed water. In some cases, the cost of constructing and maintaining reclaimed water conveyance systems is more expensive than using other water sources. Examples of specific industries that have successfully implemented reclaimed water for industrial applications are described in this section.

Boilers

Many industrial processes involve heating of various materials, necessitating the use of boilers. Boilers produce hot water and/or steam that is used to move turbines and heat other materials. Thermoelectric power generation plants rely on combustible energy sources to heat water to generate steam to operate turbines. In fact, most industrial manufacturing plants need hotwater or steam produced in boilers, for heating, and/or cooling in their production processes.

Reclaimed water can be an economically attractive water source for boiler feedwater for the following reasons: (1) it provides a reliable water supply, (2) boiler feedwater generally goes through an extensive pretreatment process even when using conventional water sources such as surface and groundwater, and (3) industries are often located on the periphery of a large city, where wastewater treatment plants are also located.

Generally, water quality requirements for boiler systems are more stringent than the requirements for cooling water. Where reclaimed water is used, advanced treatment such as RO (see Chap. 9) must be employed. A generic description of a boiler system, water quality considerations, and examples of reclaimed water use in boiler systems are introduced below.

System Description

Industrial boilers are closed vessels in which water is heated or steam is generated from a combustion process driven by a fuel source. The characteristics of the most common types of boilers are summarized in Table 19-13. Boiler systems can be categorized into fire-tube boilers, water-tube boilers, and electric boilers. In fire-tube boilers, heated gas passes through the boiler tubes submerged in water. Fire-tube boilers are generally operated at pressures below 20 bar (300 lb/in.^{2}). In water-tube boilers, water flowing inside the boiler tubes is heated by combustion gases. Water-tube boilers are suitable for operation at higher pressures than fire-tube boilers.

Boiler systems can also be categorized as high-pressure or low-pressure systems. According to the boiler and pressure vessel code by the American Society of Mechanical Engineers (ASME), a low-pressure boiler operates at pressures not exceeding 1 bar (15 lb/in.^{2}), or not exceeding 11 bar (160 lb/in.^{2}) in a hot water heating boiler with temperatures not exceeding 121°C (250°F). The maximum safe temperature range for tube metal is 480 to 540°C (900 to 1000°F), and high-pressure boilers constructed of carbon steel can operate at a maximum temperatures ranging from 480 to 730°C (900 to 1350°F) (Pagliaro, 1997).

Table 19-13
Types, characteristics, and applications of boilers[a]

Type	Efficiency	Pressure range[b]	Typical size, kW[c]	Typical applications
Cast iron	Low	HW, LPS	up to 1960	Heating, process
Electric boiler	High	HW, LPS, HPS to 62 bar	up to 2940	Heating, process
Firebox	Med	HW, LPS	up to 2940	Heating
Fire-tube	High	HW, LPS, HPS to 24 bar	up to 14,700	Heating, process
Fire-tube, vertical	Low/med	HW, LPS, HPS to 10 bar	up to 980	Heating, process
Water-tube, flexible	Med	HW, LPS	400–2450	Heating
Water-tube, industrial	Med	High temp. HW, HPS to 69 bar	—	Process
Water-tube, membrane	Med	HW, LPS, HPS to 41 bar	up to 2450	Heating, process

[a] Data compiled from Cleaver-Brooks (2005).
[b] HW = hot water, LPS = low pressure steam, HPS = high pressure steam, 1 bar = 10^5 Pa ≈ 14.5 lb/in.^{2}.
[c] 1 kW ≈ 0.102 bhp (boiler horsepower); 1 bhp = 33,475 BTU/h, or equivalent to 15.6 kg (34.5 lb) of steam per hour at 100°C.

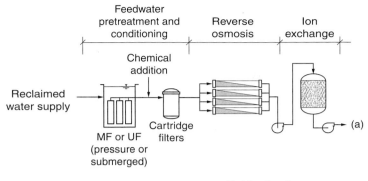
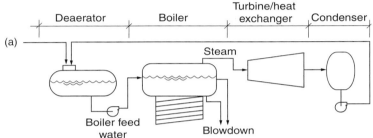

Figure 19-12
A typical flow diagram for a boiler system utilizing reclaimed water.

A simplified diagram of a boiler system utilizing reclaimed water is illustrated on Fig. 19-12. In a steam-generating boiler system, the steam condensate contains low levels of dissolved constituents and can be returned to the boiler feedwater. The returned condensate must be augmented with supplemental makeup water because the amount of condensate available for feedwater is generally less than the required feedwater flow. Typically, makeup water, produced from reclaimed water, is pretreated and conditioned, as discussed below. Makeup water and the returned condensate comprise the boiler feedwater, which is deaerated before being sent to the boiler. The deaeration system is an essential element of a boiler system as dissolved oxygen and other gases can cause corrosion (see Sec. 19-2).

Water Quality Considerations

Boiler feedwater must be of high quality because water impurities can seriously affect the system performance. Scaling reduces heat transfer efficiency and can ultimately lead to system failure, whereas corrosion can result in a system breakdown. Corrosion in boilers is caused principally by low pH, stresses, or excessive amounts of dissolved oxygen and carbon dioxide in the water. Low pH and high water temperatures, conditions often observed in boiler systems, exacerbate corrosion when excessive dissolved oxygen is present in the feedwater (Pagliaro, 1997). Carryover is a phenomenon in which boiler water is found where only steam should be present. The primary cause of carryover is foaming. Constituents in the carryover water can degrade the efficiency of the system and may lead to system failure. Constituents of concern for boiler systems are listed in Table 19-14, and recommended boiler feedwater quality is shown in

Table 19-14

Constituents of concern in boiler feedwater[a]

Constituents	Concerns	Removal methods	Remarks
Soluble gases			
Carbon dioxide (CO_2)	Corrosive; forms carbonic acid in condensate, scale formation with minerals.	Deaeration, neutralization with alkalis.	Filming, neutralizing amines used to prevent condensate line corrosion.
Hydrogen sulfide (H_2S)	Odor, bad tastes, and corrosive to most metals.	Aeration, filtration, and chlorination.	Found mainly in groundwater and polluted streams.
Oxygen (O_2)	Corrosion and pitting of boiler tubes.	Deaeration and chemical treatment with sodium sulfite or hydrazine.	Pitting of boiler tubes and turbine blades, failure of steam lines and fittings.
Suspended solids			
Organic matter	Carryover, foaming, and deposits can clog piping, and cause corrosion. Lower boiler feedwater pH, leading to corrosion	Clarification, filtration, and chemical treatment.	Includes diatoms, molds, bacterial slimes, iron/manganese bacteria. Suspended particles on the surface of the water in the boiler prevent the liberation of steam bubbles rising to the surface.
Sediment and turbidity	Sludge and scale carryover.	Clarification and filtration.	Tolerance of approx. 5 mg/L max. for most applications; 10 mg/L for potable water.
Dissolved and colloidal solids			
Calcium (Ca) and magnesium (Mg)	Scale deposits in boiler, inhibits heat transfer and thermal efficiency. In severe cases can lead to boiler tube failure.	Softening, plus internal treatment in boiler.	Forms are bicarbonates, sulfates, chlorides, and nitrates, in that order. Some calcium salts are reversibly soluble. Magnesium reacts with carbonates to form compounds of low solubility.
Chloride (Cl)	Uneven delivery of steam from the boiler (priming) and carryover in steam lowers steam efficiency and can deposit as salts on superheaters and turbine blades. Foaming if present in large amounts.	Deionization.	Priming, or the passage of steam from a boiler in belches, is caused by sodium carbonate, sodium sulfate, or sodium chloride in solution. Chloride is found in many reclaimed waters in the United States, especially in coastal areas.

Table 19-14

Constituents of concern in boiler feedwater[a] (Continued)

Constituents	Concerns	Removal methods	Remarks
Dissolved and colloidal solids			
Iron (Fe) and manganese (Mn)	Deposits in boiler in large amounts can inhibit heat transfer.	Aeration, filtration, ion exchange.	Most common form of iron in boiler feedwater is ferrous bicarbonate.
Oil and grease	Foaming, deposits in boiler.	Coagulation and filtration.	Enters boiler with condensate
Silica (Si)	Hard scale in boilers and cooling systems; turbine blade deposits.	Deionization, lime soda process, hot-lime-zeolite treatment.	Silica combines with many elements to produce silicates. Silicates form tenacious deposits on boiler tubing. Difficult to remove, often only by fluoric acids. Most critical consideration is volatile carryover to turbine components.
Sodium, alkalinity, NaOH, NaHCO$_3$, Na$_2$CO$_3$ and cannot be removed by	Foaming, carbonates form carbonic acid in steam, return line and steam trap corrosion, can cause embrittlement.	Deaeration of make-up water and causes condensate in Ion exchange, deionization, acid treatment of make-up water.	Sodium salts are found in most waters. They are very soluble condensate return. chemical precipitation. Foaming can also be caused by carbonates in solution which form a light flocculent precipitate on the surface of the water
Sulfate (SO$_4$)	Hard scale if calcium is present	Deionization.	Tolerance limits are about 100 to 300 mg/L as CaCO$_3$

[a] Adapted from EnerCon Consultancy Services (2003).

Table 19-15. It should be noted that individual boiler systems may have specific quality requirements, which may be more or less stringent than the recommended values presented in Table 19-15.

Reclaimed water can be treated for boiler feedwater either at the reclamation plant or at the industrial facility (i.e., point of use). Filtration with various membranes including microfiltration (MF), ultrafiltration (UF), nanofiltration (NF), and reverse osmosis (RO) is used increasingly at water reclamation plants where high-quality reclaimed water is produced (see Fig. 19-12; detailed discussion is given in Chaps. 8 through 10). For high-pressure boilers, water treated with RO is desirable, as the maintenance cost of point-of-use deionization (DI) systems can be reduced substantially. The use of RO

Table 19-15

Recommended industrial boiler feedwater quality criteria (unit in mg/L unless otherwise noted)[a]

Parameter	Low pressure (<10 bar)	Intermediate pressure (10–50 bar)	High pressure (>50 bar)
Aluminum	5	0.1	0.01
Ammonia	0.1	0.1	0.1
Bicarbonate	170	120	48
Calcium	As received	0.4	0.01
Chloride	As received	As received	As received
Copper	0.5	0.05	0.05
Hydrogen sulfide	As received	As received	As received
Iron	1	0.3	0.05
Manganese	0.3	0.1	0.01
Magnesium	As received	0.25	0.01
Silica	30	10	0.7
Sulfate	As received	As received	As received
Zinc	0.01	0.01	—
Alkalinity	350	100	40
Carbon tetrachloride extract	1	1	0.5
Chemical oxygen demand	5	5	1.0
Dissolved oxygen	2.5	0.007	0.0007
Dissolved solids	700	500	200
Suspended solids	10	5	0.5
Hardness	350	1.0	0.07
Methylene blue active substances	1	1	0.5
pH, unitless	7.0–10.0	8.2–10.0	8.2–9.0
Temperature, °C	As received	As received	As received

[a]Adapted from U.S. EPA (1980).
Note: 1 bar = 10^5 Pa.

is also beneficial to industries as it helps to reduce their chemical use and improve the quality of their wastewater discharge.

Other treatment techniques used in conjuction with boilers include deaeration, non-exchange deionization, and precipitation. Deaeration is the most commonly adopted pretreatment method for gaseous constituents known to cause corrosion, such as oxygen and carbon dioxide, from the feedwater. Deionization with anion and cation resins has been used as an effective pretreatment for high-pressure boilers. An ion exchange process called split streaming is popular in treating feedwater. In split streaming, the effluent from a cation exchange softener operating on a standard sodium cycle and the effluent of a cation softener operating on a hydrogen cycle are blended to reduce the alkalinity of the boiler feedwater (Pagliaro, 1997). Phosphate is the most prevalent chemical used to remove calcium hardness at the point of reclaimed water use. However, improper use of phosphate can result in calcium phosphate scale formation in boiler systems (Schroeder, 1991).

Because evaporation results in a net increase in dissolved solids concentrations in boiler water, some water is removed as blowdown to control water quality, similar to the concepts presented in Sec. 19-3 for cooling water systems. Typically, blowdown water is extracted from near the upper water surface for a continuous blowdown, and from the bottom for a manual blowdown. Manual blowdown from the bottom of the boiler allows for removal of precipitated solids.

A Unique Application
In Santa Rosa, California, tertiary-treated reclaimed water is used to supplement the available groundwater used to produce steam at a geothermal power plant facility (Geyser Spring Power Plant, Coordinates: 38.779 N, 122.754 W, view at altitude 4.5 km). Unlike power generation plants that use a boiler to generate steam, the reclaimed water, which is injected into a geyser, is heated along with the groundwater by geothermal energy. Because the power generation system does not involve the use of a boiler to produce steam, the quality of tertiary-treated reclaimed water is sufficient. The plant has been operating with reclaimed water since 2003.

Pulp and Paper Industry

As of 2005, about 550 mills, many of which produce both pulp and paper, were operating in the United States (Center for Paper Business and Industry Studies, 2005). The overall production of paper products has been relatively constant over the last decade. Water use for pulp and paper production at the beginning of the 20th century was estimated at over 600 m^3/ton (72 gal/lb) of pulp produced. In contrast, current approaches for pulp production consume between 20 and 70 m^3 of water per ton of pulp produced, due to improvements in water use efficiency and internal water recycling (Brongers and Mierzwa, 2001; U.S. EPA, 2000, 2004). Even though pulp and paper manufacturing is a water intensive industry, the use of municipal reclaimed water is not a common practice. Instead, much of the water recycling occurs within the manufacturing processes itself. About a dozen pulp and paper mills use municipal reclaimed water in the United States (U.S. EPA, 2004). A brief description of typical pulp and paper production processes and key elements in the use of municipal reclaimed water are presented below.

System Description
The primary processes in the pulp and paper industry are:

1. Pulp production
2. Pulp processing and chemical recovery
3. Pulp bleaching
4. Stock preparation
5. Paper/paperboard production

A typical flow diagram for pulp and paper production including the flow of water and materials is illustrated on Fig. 19-13. In the pulp production process, fibers from wood or other materials such as rags, linters (the fuzz of short fibers that adheres to cottonseed

Figure 19-13

A typical pulp and paper production flow diagram. (Adapted from Brongers and Mierzwa, 2001.)

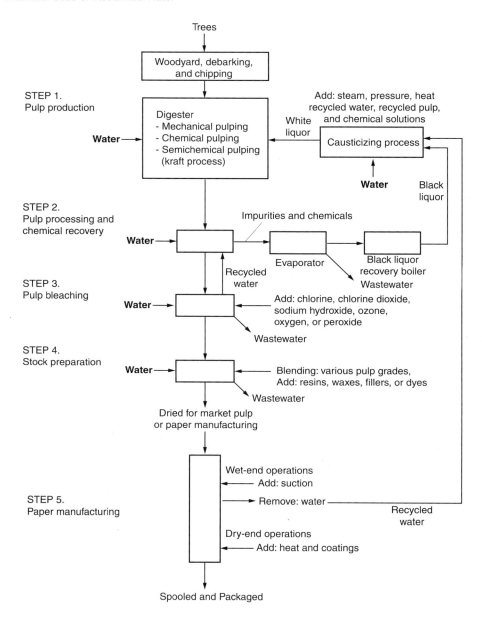

after ginning), wastepaper, and straw, are separated to create pulp. Mechanical, chemical, or a combination of mechanical and chemical means, are used for pulp production. According to the Lockwood-Post's Directory (Paperloop, 2003), approximately 80 percent of the wood pulp is produced using kraft processes, which use sodium sulfate in a digester under high temperature and pressure. Chemical bonds in lignin (glue-like substance) are selectively destroyed through the kraft process to release fibers from the raw materials.

The pulp is cleaned through a series of washing steps to remove impurities and to recycle chemicals from the cooking liquor, known as black liquor. The cleaned pulp is then bleached, combined with various pulps and chemicals including resins, waxes, fillers, and dyes, and dried for market pulp or sent directly to paper manufacturing. Additional additives may be applied after the sheet-making step. The fibers bond together as they are carried through a series of presses and heated rollers to form paper. Water consumption in pulp and paper mills varies with each process, but typically bleaching and paper production require the largest amounts of water (see Fig. 19-14).

Pulp and paper production systems generate wastewater that contains suspended and dissolved organic and/or inorganic materials. Most mills have their own wastewater treatment facilities and internal water recycling systems to meet wastewater discharge requirements and to reduce the amount of waste discharge. The use of reclaimed water is a viable option depending on the availability of transmission lines and the relative cost. The feasibility of reclaimed water use for pulp and paper production depends on the reclaimed water quality and the level of pretreatment necessary to improve the reclaimed water quality for the purpose of specific processes.

Water Quality Considerations

Water quality requirements for pulp and paper production depend on the grade of paper. Generally, lower quality water can be used for the *brown* grade papers such as asphalt or tar-saturated papers, linerboard, low-brightness carton board, packaging and insulating board. The *white* grade papers require higher water quality, and high brightness fine papers require the highest quality process water (Rommelmann et al., 2003). Typical water quality requirements for various papers are shown in Table 19-16. Phosphates, surfactants, and metal ions may affect the efficiency of resins in the stock preparation process. Iron, manganese, and microbial contamination can cause discoloring. Suspended solids also affect the brightness of the paper and the color of reclaimed water may be an issue for production of higher grade paper. In Pomona, California, for example, activated carbon filters were used for color removal to meet the quality criteria for paper manufacturing (U.S. EPA, 2004). Biological growth can affect the texture

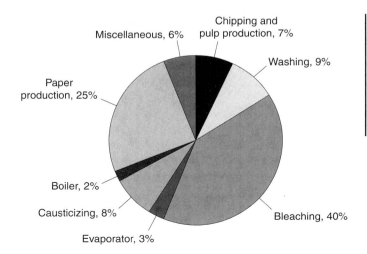

Figure 19-14
Water consumption in pulp and paper production processes. (Adapted from Panchapakesan, 2001.)

Table 19-16
Typical water quality recommended for process water of various paper products[a]

		Brown grades		White grades		
Parameter	Unit	Groundwood (tar) paper	Soda and sulfate (kraft) pulps	Kraft paper, bleached	Unbleached paper	Bleached paper
Turbidity	NTU	70	35	40	140	14–56
Color	c.u.	30	5	25	30–100	5–25
Total dissolved solids	mg/L	250–1000	250–1000	300	75–650	75–650
Total suspended solids	mg/L	40	10	10	10–30	10–30
Alkalinity	mg/L as $CaCO_3$	75–150	75–150	75	150	75–125
Hardness	mg/L as $CaCO_3$	100–200	100–200	100	200	100
Chloride	mg/L	75	75	200	200	200
Iron	mg/L	0.3	0.1	0.2	1.0	0.1
Manganese	mg/L	0.1	0.05	0.1	0.5	0.03
Silica	mg/L as SiO_2	50	20	50	100	9–20
Sulfate	mg/L	trace	—	—	—	100–300
Temperature	°C	<55	27	—	—	15–27

[a]Compiled from U.S. EPA (1973); State of California (1963); WPCF (1989); and Rommelmann et al. (2003).
Note: Water quality requirements for each process may vary.

and uniformity of the paper (Rommelmann et al., 2003). Biological growth is typically controlled by maintaining a chlorine residual, but high chlorine residuals may lead to corrosion problems and formation of chlorinated byproducts.

Reclaimed Water Use in Pulp and Paper Mills
One example of reclaimed water use at a paper mill is a newspaper manufacturing plant in Los Angeles, California, where about 5.3×10^6 m³/yr (1.4 Bgal/yr) of tertiary-treated reclaimed water is being used. Another paper mill in Los Angeles uses about 0.8×10^6 m³/yr (0.21 Bgal/yr) reclaimed water. In Flagstaff, Arizona, a paper mill has been using reclaimed water for most of their paper production processes since 2005. The water reclamation process at the City's Rio de Flag plant includes the Bardenpho process (see Tchobanoglous et al., 2003) for nutrient removal, filtration, and UV disinfection. Industrial wastewater from the mill is returned to the water reclamation plant, forming a closed-loop system. Reclaimed water from the water reclamation plant is also used for irrigation at public schools, parks, cemeteries, a golf course, and residential areas.

Textile Industry
Textile production in the United States has declined steadily in the last decade due to inexpensive import products (U.S. Department of Commerce, 2005). According to the 2002 Economic Census (U.S. Department of Commerce, 2005), there were 3932 textile mills (i.e., fiber, yarn, and thread mills, fabric mills, textile and fabric finishing and fabric coating mills), and 7304 textile product mills (i.e., textile furnishings mills such as carpet and curtain mills, and other textile product mills) in the United States. Because textile production processes require large amounts of water, opportunities for reclaimed

water use exist. In the following discussion, the textile production processes and water quality requirements are described briefly, and the use of reclaimed water in the textile industry is introduced.

System Description

A simplified diagram of a textile production process is shown on Fig. 19-15. A generalized textile production process consists of four steps: (1) yarn formation, (2) fabric formation, (3) wet processing, and (4) fabrication. Generally, water use in the yarn and fabric formation and fabrication processes is negligible. Wet processing is the most important part of the textile industry in terms of water use and waste generation (U.S. EPA, 1997). Wet processing involves: (1) fabric preparation, (2) dyeing, and/or (3) printing (see Fig. 19-15b). For most wool products and some synthetic and cotton products, the yarn is dyed before fabric formation so that patterns are woven into the fabric. A brief summary of wet processing is shown in Table 19-17.

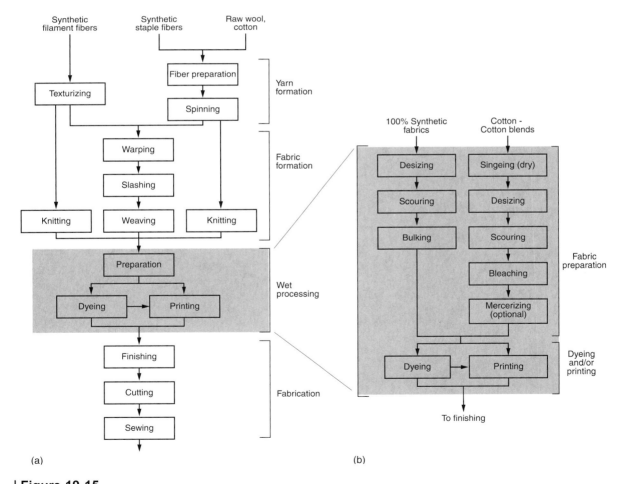

Figure 19-15

A typical textile production processes: (a) entire process and (b) wet processing. (Adapted from U.S. EPA, 1997.)

Table 19-17

Summary of textile production process

Unit process	Description	Remarks
Fabric preparation		
Singeing	Fabrics are passed over a frame or heated copper plates to make a smooth finish. It is useful for the fabrics that are to be printed or a smooth surface is desired.	Singeing is a dry process.
Desizing	Sizing materials added during the fabric formation (weaving) process are removed using hot water and chemicals. It is important to remove the starch before scouring because it may cause color change when exposed to sodium hydroxide in the scouring process.	Enzymes are used for starches applied commonly for natural fibers whereas most synthetic sizing materials can be removed by hot water.
Scouring	Impurities from fibers are removed. Typically alkaline solutions are used to break down and remove natural oils, surfactants, lubricants, dirt and other natural and synthetic materials.	Typically, scouring wastes contribute a large portion organic loadings from of preparation processes. Desizing and scouring operations may be combined.
Bleaching	Unwanted color is removed from textile fibers, yarns, or cloth. The bleaching process typically involves: (1) saturation with the bleaching agent and other necessary chemicals, (2) bleaching action in a raised temperature, and (3) washing and drying.	Typical bleaching agents include hydrogen peroxide, sodium hypochlorite, sodium chlorite, and sulfur dioxide gas. Hydrogen peroxide bleaching is used most commonly (NPI, 1999). The selection of bleaching agents depends on the type of yarns and subsequent processes.
Mercerizing	The fabric is passed through a cold alkali solution, and then stretched out on a tender frame where hot-water sprays remove most of the caustic solution. The fiber increases its strength and affinity for dyes.	The caustic is further removed by several washes under tension. Remaining caustic may be neutralized with a cold acid treatment followed by several more rinses to remove the acid.
Dyeing and printing		
Dyeing	Textiles are dyed using a wide range of chemicals and dyestuffs, techniques, and equipment. Either batch or continuous system is used. Dyeing processes typically consist of dye application, dye fixation with chemicals or heat, and washing.	Continuous dyeing accounts for about 60 percent of total volume of product dyed in the industry (NPI, 1999).
Printing	A decorative pattern or design is applied to constructed fabric using pigments or dyes. Major printing methods include rotary screen, roller, and flat screen.	Pigments are used most commonly.
Finishing		
Mechanical finishes	Mechanical finishes include brushing, ironing, or other physical treatments.	
Chemical finishes	Chemical finishes modify the property of the fabric, such as permanent-press, soil-release, and stain resistant finishing finishes. Chemical finishes are usually followed by drying, curing, and cooling steps.	Chemical finishes are often applied along with mechanical steps.

Water needs in textile production vary greatly with the type of textile produced. Typical quantities of water needed to produce 1 kg of textile for selected fabric types are shown in Table 19-18. In terms of individual unit processes, wet processes have the highest potential for reclaimed water use because they are more water intensive. Typical water demands for textile production processes are shown in Table 19-19.

Water Quality Considerations

The major water quality issues for textile production are to ensure that the dyes react properly and that discoloration or staining does not occur. Turbidity, color, iron, and manganese have the potential to cause staining of fabric during production. Hardness adversely affects soaps used in various cleaning processes and can cause curd-like deposits on the textile. Usually soaps are not deposited evenly with hard water, resulting in dyeing irregularities. Hardness may cause precipitation of some dyes and increase the breakage of silk during reeling and throwing operations (Treweek, 1982). Typical water quality requirements for the textile industry are reported in Table 19-4.

Reclaimed Water Use in Textile Processing

Reclaimed water can be used in various textile production processes including carpet dyeing and cotton fabrication. A few carpet mills in southern California use reclaimed water for their carpet dying process. Examples include Tuftex Carpets in Santa Fe Springs, California, which receives reclaimed water from Central Basin Municipal Water District (CBMWD), and Royalty Carpet Mills, which receives reclaimed water from Irvine Ranch Water District (IRWD). Pilot scale and demonstration projects were conducted at General Dyeing and Finishing, Inc., using reclaimed water from the CBMWD, to determine the suitability of using reclaimed water for fabric dyeing process. It was demonstrated that the quality of reclaimed water received from the CBMWD was comparable to potable water quality with less variability, and was acceptable to meet all process water needs (see Table 19-20) (U.S. EPA, 2004; Water 3 Engineering, Inc., 2005). In Harlingen, Texas, reclaimed water that is treated with secondary treatment, filtration, and RO is sent to the Fruit of the Loom Corporation for their textile processes. The process wastewater is then returned to the municipal wastewater system (Gerston et al., 2002).

Table 19-18
Distribution of water use in textile processing by fabric type[a]

Fabric types	Water use, L/kg production		
	Minimum	Median	Maximum
Wool	111	285	658
Woven	5.0	113	508
Knit	20	83	377
Carpet	8.3	47	163
Stock/Yarn	3.3	100	558
Nonwoven	2.5	40	83
Felted fabric	33	213	933

[a]Adapted from U.S. EPA (1996).

Table 19-19
Typical water demand in textile production, organized by unit process[a]

Processing subcategory	Water consumption, L/kg production
Yarn and fabric forming	0
Slashing	0.5–7.8
Preparation	
Singeing	0
Desizing	2.5–20
Scouring	19–43
Continuous bleaching	2.5–120
Mercerizing	1.0
Dyeing	
Beam	170
Beck	230
Jet	200
Jig	100
Paddle	290
Skein	250
Stock	170
Pad-batch	17
Package	180
Continuous bleaching	170
Indigo dyeing	8.3–50
Printing	25
Print afterwashing	110
Finishing	
Chemical	5.0
Mechanical	0

[a] Adapted from U.S. EPA (1996).

Other Industrial Applications

Oil refineries, industrial chemical manufacturing process, semiconductor industries, and solid waste incineration plants are among other industrial processes that use reclaimed water. Most industrial processes involve heating and cooling, thus reclaimed water can be used for cooling tower and boiler feedwater, depending on the location and the cost of treatment to meet water quality criteria. Reclaimed water can also be used for cleaning and dust control of industrial facilities, concrete mixing, and soil compaction. The practicality of using reclaimed water for other onsite applications depends on the facility's use of reclaimed water for larger applications. Use of reclaimed water for multiple purposes can be an important element in the successful implementation of industrial water reuse. For example, a municipal solid waste incineration plant in Hillsborough County, Florida, uses reclaimed water for cooling tower and facility cleaning, then the used reclaimed water is further reused for ash quenching. Reclaimed water is also used for landscape irrigation (see Fig. 19-16).

Table 19-20
Comparison of water quality values between potable water and reclaimed water in CBMWD, CA[a]

| Parameter | Unit | Potable water | | | Reclaimed water[e] | Typical quality requirements for textile industries[f] |
		State project water[b]	Colorado River water blend[c]	Groundwater[d]		
Alkalinity	mg/L as $CaCO_3$	97	112	181	198	
Aluminum	mg/L	0	0.14	<0.05	NA	8
Chloride	mg/L	64	76	45	116	
Chlorine	mg/L	2	2	na	3.7	
Color	c.u.	2	1.2	4.8	na	0–5
Copper	mg/L	<0.05	<0.05	<0.05	0.03	0.01–5
Heavy metals	mg/L	0	0.002	na	na	
Iron	mg/L	<0.1	<0.1	0.36	0.08	0.1–0.3
Manganese	mg/L	<0.02	<0.02	0.09	0.02	0.01–0.05
pH	pH unit	8	8.1	7.7	7.0	6-8
Sulfate	mg/L	68	174	110	101	100
TSS	mg/L	na	na	na	1.7	0–5
TDS	mg/L	305	487	419	569	100–200
Total hardness	mg/L as $CaCO_3$	147	241	227	217	0–50

[a] Adapted from Water 3 Engineering, Inc. (2005).
[b] Average values from 1999 to 2002, from water quality reports for Metropolitan Water District Jensen plant.
[c] Average values from 1999 to 2002, from water quality reports for Metropolitan Water District Weymouth and Diemer plants (61 to 82 percent Colorado River water, blended with the State project water).
[d] Average values from 1998 to 2000, reported by CBMWD.
[e] Average values from 1998 to 2000, reported by San Jose Water Reclamation Plant.
[f] From Table 19-4.
Note: na = not available.

19-5 CASE STUDY: COOLING TOWER AT A THERMAL POWER GENERATION PLANT, DENVER, COLORADO

Reclaimed water has been used for cooling towers at a thermal power generation plant in Denver, Colorado. Water reuse is facilitated because the power plant is located adjacent to the water reclamation plant.

Setting

The Xcel Energy Cherokee Station in Denver, Colorado, generates electricity by burning low-sulfur coal from western Colorado mines. The plant receives water for boilers and cooling towers from various sources including the South Platte River, Clear Creek, potable water, and reclaimed water from the Denver Water Recycling Plant ("recycling plant"). The water from these sources is stored in the Cherokee Northwest Reservoir and is used for cooling towers and for bottom ash sluicing to the ash ponds.

Figure 19-16
At a waste incineration plant in Tampa, FL, reclaimed water is used for multiple purposes: (a) overview of the plant where reclaimed water is used for facility cleaning, (b) cooling tower using reclaimed water, (c) ash quenching with reclaimed water, and (d) landscape area irrigated with reclaimed water.

Water Management Issues

The recycling plant receives secondary effluent from the Metro Wastewater Treatment Plant where the water reclamation processes used include biological aerated filtration (for nitrification), coagulation, flocculation, sedimentation, granular medium filtration, and chlorine disinfection. The recycling plant produces about 170×10^3 m³/d (45 Mgal/d) of reclaimed water. The reclaimed water is stored in a 42×10^3 m³ (11 Mgal) underground storage reservoir before it is distributed to various sites including irrigation of parks, golf courses, schools, and a local zoo, and to the Xcel Energy's Cherokee Generating Station. Typical effluent water quality is shown in Table 19-21. The cooling tower operated with reclaimed water is shown on Fig. 19-17.

Table 19-21
Reclaimed water quality data at the Denver Water Recycling Plant[a]

Parameter	Unit	Average	Minimum	Maximum
Alkalinity, total as $CaCO_3$	mg/L	78	64	104
Ammonia as N	mg/L	<0.2	<0.2	0.9
Boron	µg/L	291	240	380
Calcium	mg/L	49	44	60
Chloride	mg/L	86	77	102
Chlorine, total	mg/L	1.89	0.63	4.2
Iron	mg/L	0.34	0.18	0.59
Magnesium	mg/L	10.8	8.6	13.4
Manganese	µg/L	23	14	45
Nitrate-N	mg/L	15.2	11.6	17.7
Nitrite-N	mg/L	<0.01	<0.01	0.01
pH	unitless	7.1	6.8	7.4
Phosphorous, total	mg/L	0.172	0.073	0.308
Potassium	mg/L	12	11	14
Sodium	mg/L	117	100	140
Specific conductance	µmhos/cm	891	801	1000
Sulfate	mg/L	139	118	160

[a]Data from Denver Water: samples taken from Oct. 21, 2004, to Sept. 9, 2005.

Figure 19-17
Cooling tower at Xcel Energy Cherokee Generating Station, Denver, CO. (Coordinates: 39.806 N, 104.961 W, view at altitude 2.7 km).

Implementation

Use of reclaimed water from the recycling plant at the Xcel Energy Cherokee Station began in 2004. Prior to the use of reclaimed water, the cooling tower water was treated with antiscalant to avoid scale formation of calcium phosphate. Because reclaimed water has lower concentrations of phosphorus than the original water source, formation of calcium phosphate in the cooling system is controlled by routine cleaning practices (see Table 19-21). Biological growth is controlled through periodic shock chlorination (on alternate days) along with mechanical cleaning with a pig (see Fig. 14-20 in Chap. 14).

Lessons Learned

The recycling plant is located adjacent to the Xcel Energy Cherokee Station, allowing the power plant ready access to the source of reclaimed water without requiring an extensive transmission and distribution system. The recycling plant also benefits by having a constant demand for reclaimed water from the adjacent power plant. In addition, treatment at the recycling plant was designed to meet water quality requirements of Xcel Energy Cherokee Station, one of its major reclaimed water users. For example, to meet the stringent ammonia requirements, a biological aerated filter with polystyrene beads is used to facilitate nitrification. Using reclaimed water, scaling with calcium phosphate is reduced significantly, thereby eliminating the use of antiscalants.

19-6 CASE STUDY: INDUSTRIAL USES OF RECLAIMED WATER IN WEST BASIN MUNICIPAL WATER DISTRICT, CALIFORNIA

The West Basin Municipal Water District (WBMWD), located in southern California, is unique among water reclamation facilities in that six different qualities of reclaimed water are produced. These different qualities of reclaimed water, often termed "designer" by WBMWD, are used for various water reuse applications, the majority of which are industrial. The development of the WBMWD reclamation operation is described in this case study.

Setting

The WBMWD is a water wholesaler for a 480 km^2 (185 mi^2) area of southwest Los Angeles County, serving a population of about 900,000 people. The District purchases secondary treated wastewater from the Hyperion Wastewater Treatment Plant (Hyperion WWTP), one of the largest wastewater treatment plants in the United States with an average flow of 1.4×10^6 m^3/d (362 Mgal/d). Numerous industries are located in the District's service area, including petroleum refineries and manufacturing facilities.

Water Management Issues

After California's severe drought period between the late 1980s and early 1990s, WBMWD initiated plans for transforming itself from a wholesaler of potable water to a purveyor of water from various sources, including reclaimed water. In its Drought-Proof 2000 campaign, WBMWD sought to develop local water resources and diversify the portfolio of water supply sources to reduce the region's dependence on imported water from the Metropolitan Water District (MWD) of Southern California. The MWD's imported water sources consist mainly of water from northern California by way of the California Aqueduct, and from the Colorado River by way of the Colorado River Aqueduct. In 1992, WBMWD received state and federal funding to pursue its water recycling program, including construction of a water reclamation plant in the City of El Segundo. Goals of the recycling program are as follows (Miller, 2003; WBMWD, 2006):

- Reduce the region's dependence on imported water by 50 percent
- Provide an alternative drought-proof local water source to meet present and future water demands.
- Reduce the volume of secondary effluent discharged to the Santa Monica Bay by 25 percent.
- Prevent seawater intrusion into the groundwater basin by injecting reclaimed water into the West Coast Basin Barrier (WCBBP).

The West Basin Water Reclamation Plant (WBWRP), built in 1995, receives secondary effluent from the Hyperion WWTP and subsequently provides treatment and distribution. Various qualities of reclaimed water are used for parks, sports fields, manufacturing processes, and oil refineries. About 7.7 percent of the secondary effluent from the Hyperion WWTP is reclaimed by WBWRP.

Implementation

A unique feature of the water reuse system in the WBMWD is that industrial applications are the major users of reclaimed water, consisting of over 70 percent of total reclaimed water use (see Fig. 19-18). To meet the quality requirements for various industrial applications, the WBMWD is capable of producing six different qualities of reclaimed water to meet specialized water reuse. All reclaimed waters meet the requirements specified in the *California Water Recycling Criteria* (Title 22) (State of California, 2003). The characteristics of the six reclaimed waters are as follows:

- Tertiary treated water for industrial and irrigation uses: secondary effluent from the Hyperion WWTP is coagulated, flocculated, filtered, and disinfected for a wide variety of industrial and irrigation uses.
- Amended tertiary water is used for turf irrigation, augmented to adjust the SAR for soil permeability (see Chap. 17).
- Nitrified water is used for industrial cooling towers.

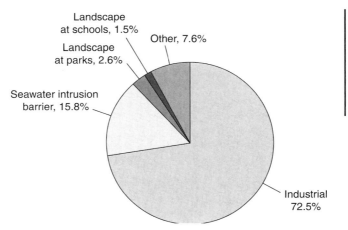

Figure 19-18
Major users of reclaimed water in West/Central Basin, CA. (Adapted from WBMWD, 2006.)

- Softened RO water is used for groundwater recharge: secondary effluent is pretreated by MF, followed by RO and post treatment including postdecarbonation, UV disinfection and stabilization, and peroxide addition for groundwater recharge.
- Water treated by RO is used for low-pressure boiler feedwater.
- Double RO treated water is used for high-pressure boiler feedwater.

Tertiary treated reclaimed water is used for landscape irrigation of parks, playgrounds, and commercial areas, including landscaping at Home Depot National Training Center, Toyota, and Goodyear (see Fig. 19-19). Tertiary treated reclaimed water is also sent to three advanced treatment facilities; the Exxon-Mobil Water Recycling Plant, the Chevron Nitrification Plant, and Carson Regional Water Recycling Plant, for further

(a)

(b)

(c)

(d)

Figure 19-19

Examples of reclaimed water uses in West/Central Basin, CA: (a) Goodyear sports field, (b) landscape irrigation at Home Depot National Training Center, (c) landscape irrigation, toilet and urinal flushing, and cooling towers for air conditioning at Toyota Motor Sales, USA, and (d) cooling towers and boiler feed at Chevron refinery. (Photos courtesy U. Daniel, West Basin Municipal Water District.)

treatment for specific industrial purposes. Highly treated reclaimed water is used for industrial purposes at refineries including Exxon-Mobil, Chevron, and BP (Cearley and Cook, 2003). Reclaimed water that is treated with RO at the WBWRF is injected into the South Bay's groundwater basin to prevent seawater intrusion.

To meet the region's water demand, the WBWRP has been expanded to increase its production of high-quality recycled water. During 2003 to 2004, the WBWRP produced more than 33×10^6 m^3 (8.8 Bgal) of reclaimed water. After two successful expansion projects, WBMWD is undergoing its $52-million Phase IV Expansion Project. The expansion will ultimately increase its reclaimed water supply for groundwater recharge by 18,900 m^3/d (5 Mgal/d), and tertiary treated reclaimed water by 37,900 m^3/d (10 Mgal/d) (WBMWD, 2006). Two other expansion projects include Madrona/Palos Verdes lateral extension and Harbor/South Bay water recycling project. These projects will expand the use of tertiary treated reclaimed water for nonpotable purposes, mostly landscape irrigation.

Lessons Learned

Lessons learned from the water reuse systems in WBMWD are summarized as follows (Miller, 2003):

- Reclaimed water has proved to be a reliable source of new supply.
- The WBMWD uses only about 7.7 percent of effluent from the Hyperion WWTP, and the volume of reclaimed water is independent of climate variability. Wastewater base flow, which is currently much higher than reclaimed water production, can be firmly targeted as a reliable supply source independent of hydrology patterns or cycles.
- Constant and high demand for reclaimed water from industries justified the WBMWD's decision to produce reclaimed water for each customer's quality needs.
- Water reclamation plants are located in the middle of industrial areas, allowing the industries to use reclaimed water without installing extensive reclaimed water distribution lines. The expanded use of reclaimed water enhances the benefits of the investments already made to existing wastewater facilities, conveyance facilities, and ocean outfall facilities.
- Unlike imported or local groundwater supplies, reclaimed water is not subject to legal water rights implications because it is otherwise disposed to the ocean. Further, the use of reclaimed water is not subject to the ecological pressures that are frequently imposed on the development of new water supplies.

PROBLEMS AND DISCUSSION TOPICS

19-1 Consider the three industrial water use options (a), (b), and (c), depicted below. Calculate freshwater needs using a water balance for the scenarios (b) and (c). Discuss advantages and challenges in incorporating scenarios (b) and (c).

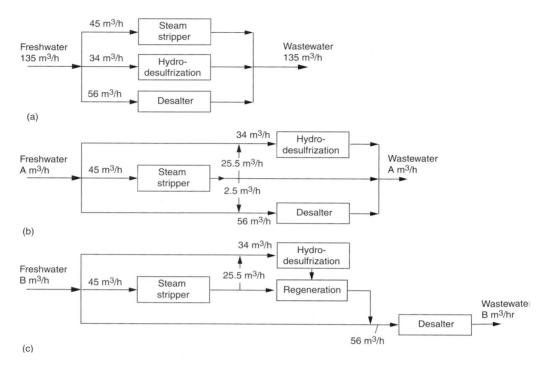

19-2 Discuss the factors affecting the feasibility of using municipal reclaimed water for industrial purposes. Compare the option of using reclaimed water from municipal wastewater with reclaimed water from industrial wastewater generated within the same industrial site.

19-3 In Example 19-1, the effect of ionic strength was assumed to be negligible. Recalculate the calcium carbonate precipitation potential (CCPP) assuming the TDS of the solution is 650 mg/L and taking the effect of ionic strength into account. Discuss how significant the effect of ionic strength is for reclaimed water when the CCPP is estimated.

19-4 Calculate the Langelier saturation index (LSI) and Ryzner stability index (RSI) for a reclaimed water with the characteristics given below. Consider the effects of ionic strength in calculating the indexes.

Parameter	Unit	Concentration
Temperature	°C	20
Total dissolved solids	mg/L	880
Ca^{2+}	mg/L	60
Mg^{2+}	mg/L	33
Total alkalinity	mg/L as $CaCO_3$	90

19-5 Calculate the Langelier saturation index (LSI) and the CCPP for reclaimed water with the characteristics given below at temperatures of 10, 20, 40, 60, and 80°C. Consider the effect of ionic strength in calculating the indices. What is the implication of the calculated results if the water is to be used in an industrial process involving heating and cooling cycles?

Parameter	Unit	Concentration
Total dissolved solids	mg/L	840
Calcium hardness	mg/L as $CaCO_3$	288
Total hardness	mg/L as $CaCO_3$	347
pH	unitless	6.9
Total alkalinity	mg/L as $CaCO_3$	114

19-6 Reclaimed water with an annual average TDS concentration of 375 mg/L is used for a cooling tower at a thermal power generation plant. Calculate the average cycles of concentration to maintain the average total dissolved solids concentration in the blowdown below 1500 mg/L.

19-7 The use of reclaimed water for cooling towers in an oil refinery was proposed. During the feasibility study, effluent quality of the existing wastewater treatment plant was compared with the water quality requirements for the refinery's cooling towers. The water quality comparison is shown below. Discuss additional treatment needs to use the reclaimed water for cooling towers.

Parameter	Unit	Effluent quality	Cooling tower water quality requirements
Alkalinity	mg/L as $CaCO_3$	138	350
Alminum	mg/L	0.2	0.1
Ammonia	mg-N/L	5.6	N/A
Bicarbonate	mg/L	N/A	24
Calcium	mg/L	36.5	125
Chemical oxygen demand	mg/L	19	75
Chloride	mg/L	225	300
Coliforms	No./100 mL	20–40	<2.2
Hardness	mg/L as $CaCO_3$	180	350
Iron	mg/L	0.21	0.5
pH	unitless	6.8	6–9
Manganese	mg/L	0.13	0.5
MBAs (surfactants)	mg/L	0.19	1
Phosphate	mg/L	29.2	<5
Silica (SiO_2)	mg/L	14.4	50
Sulfate	mg/L	124	200
Suspended solids	mg/L	37	100
Total dissolved solids	mg/L	780	500
Total organic carbon	mg/L	12	0.1

19-8 Reclaimed water with the quality characteristics given below is to be used for a cooling tower. Calculate the composition of the blowdown if the cycles of concentration are maintained at 4. Assume that the temperature of water entering the cooling tower is 50°C and the solubility of $CaSO_4$ is 2200 mg/L at 50°C.

Parameter	Unit	Concentration
Calcium hardness	mg/L as $CaCO_3$	86
Total hardness	mg/L as $CaCO_3$	122
Total alkalinity	mg/L as $CaCO_3$	102
SO_4^{2-}	mg/L	18
Cl^-	mg/L	14
SiO_2	mg/L	2.4

19-9 Cooling towers in a waste incineration plant have been operated with reclaimed water. During the inspection of the cooling towers, biofilm buildup in the recirculating water system was observed. Discuss appropriate measures to clean up the system.

19-10 A thermal power generation plant is considering using reclaimed water from the adjacent water reclamation plant. Based on the water quality analysis, target cycles of concentration were set at 4. Given the following conditions, determine (a) the blowdown water flow rate, (b) the drift loss, and (c) the makeup water requirement.

- Total recirculating water flow in the cooling tower = 330 m³/min
- Drift loss = 0.1 percent of the total recirculating flow
- Water temperature drop in the cooling tower = 20°C

19-11 A thermal power generation plant is using reclaimed water for cooling towers. The total dissolved solids concentration of reclaimed water is 680 mg/L. The cycles of concentration for the cooling towers are maintained at 3.5. Discuss the disposal options for the blowdown. What would be the potential impacts if the blowdown is to be discharged to municipal collection system?

19-12 An oil refinery is considering using reclaimed water for boiler feedwater. A water reclamation plant is located adjacent to the refinery and currently is producing tertiary treated reclaimed water for landscape irrigation. Discuss the information required to determine (a) what additional treatment requirements will be needed and (b) where the additional treatment process should be located.

19-13 Typical TDS removal achieved with reverse osmosis treatment is in the range of 90 to 98 percent (see Chap. 9, Table 9-6). High-pressure boilers typically require a TDS level of less than 0.5 mg/L (see Table 19-4). Secondary effluent from a local wastewater treatment plant contains about 600 to 800 mg/L of dissolved solids. Discuss the reclaimed water treatment options to meet the required TDS level for the high-pressure boilers. List which constituents may affect the use of reclaimed water for a high-pressure boiler, and assess the expected performance of the treatment option discussed above.

REFERENCES

Adams, A. P., M. Garbertt, and H. B. Rees (1980) "Bacterial Aerosols Produced from a Cooling Tower Using Wastewater Effluent as Makeup Water," *J. WPCF*, **52**, 3, 498–501.

Asano, T., and A. D. Levine (1996) "Wastewater Reclamation, Recycling and Reuse: Past, Present, and Future," *Water Sci. Technol.*, **33**, 10–11, 1–14.

Asano T., R. Mujeriego, and D. Parker (1988) "Evaluation of Industrial Cooling Systems Using Reclaimed Municipal Wastewater," *Water Sci. Technol.*, **20**, 10, 163–174.

ASHRAE (2000) *ASHRAE Guideline 12-200 Minimizing the Risk of Legionellosis Associated with Building Water*, American Society of Heating, Refrigerating, and Air Conditioning Engineers.

Assink, J. W., and H. C. van Deventer (1995) "Cooling Water Systems: Options and Recommendations for Reducing Environmental Impact," *Euro. Water Pollut. Contrl.*, **5**, 1, 39–45.

Baetens, D. (2002) "Water Pinch Analysis: Minimisation of Water and Wastewater in the Process Industry," 205–228, in P. Lens, L. H. Pol, P. Wilderer, and T. Asano (eds.), *Water Recycling and Resource Recovery in Industry: Analysis, Technologies and Implementation*, IWA Publishing, London, UK.

Bell, G. E. C., and J. Aranda (2005) *Corrosion Engineering Training*, Hawaii Water Environment Association, Corrosion Workshop, Feb. 16, 2005, Honolulu, HI.

Black & Veatch, and Malcolm Pirnie, Inc. (1992) *Lead and Copper Rule Guidance Manual*, Vol. 2, Corrosion Control Treatment, American Water Works Association, Denver, CO.

Bogaerts, W. F., and A. A. Van Haute (1985) "Chloride Pitting and Water Chemistry Control in Cooling or Boiler Circuits," *Corros. Sci.*, **25**, 12, 1149–1161.

Brongers, M. P. H., and A. J. Mierzwa (2001) "Pulp and Paper: Summary and Analysis of Results, Corrosion Control and Prevention," Appendix W, in G. H. Koch, M. P. H. Brongers, N. G. Thompson, Y. P. Virmani, and J. H. Payer (eds.) *Corrosion Cost and Preventive Strategies in the United States*, Department of Transportation, Washington, DC.

CDC (2006) "Legionellosis: Legionnaires' Disease (LD) and Pontiac Fever," in *Disease Listing*, Division of Bacterial and Mycotic Diseases, Centers for Disease Control and Prevention (CDC), accessed online: http://www.cdc.gov/.

Cearley, D., and P. Cook (2003) "Customizing for Industrial Reuse," in *Proceedings, Annual WateReuse Symposium*, San Antonio, TX, WateReuse Association, Alexandria, VA.

Center for Paper Business and Industry Studies (2005) *Pulp Mills, Pulp and Paper Mills, Paper Mills in USA*, Center for Paper Business and Industry Studies (CPBIS), Georgia Institute of Technology, accessed online: http://www.cpbis.gatech.edu/.

Cheremisnoff, P. N., and N. P. Cheremisnoff (1981) *Cooling Towers: Selection, Design and Practice*, Ann Arbor Science Publishers, Ann Arbor, MI.

Cleaver-Brooks (2005) *The Boiler Book*, Cleaver-Brooks, Milwaukee, WI.

Clement, J., M. Hayes, P. Sarin, W. Kriven, J. Bebee, K. Jim, M. Beckett, V. Snoeyink, G. Kirmeyer, and G. Pierson (2002) *Development of Red Water Control Strategies*, American Water Works Association Research Foundation, Denver, CO.

Clescerl, L. S., A. E. Greenberg, and A. D. Eaton (eds.) (1999) *Standard Methods for Examination of Water and Wastewater*, 20th ed., 2330C: "Indices Predicting the Quantity of $CaCO_3$ that Can Be Precipitated or Dissolved," American Public Health Association, Washington, DC.

Cooling Technology Institute (2006) "Legionellosis Guideline: Best Practices for Control of Legionella," *CTI Guidelines WTP-148* (06).

Davis, J. R. (ed.) (2000) *Corrosion: Understanding the Basics*, ASM International, Materials Park, OH.

Edwards, M., Schock, M.R., and Meyer, T. E. (1996). "Alkalinity, pH, and Copper Corrosion Byproduct Release," *J. AWWA*, **88**, 3, 81–94.

EnerCon Consultancy Services (2003) *Boiler Feed-Water Treatment*, accessed online: http://energyconcepts.tripod.com/energyconcepts/.

Faust, S. D., and O. M. Aly (1998) *Chemistry of Water Treatment*, Ann Arbor Press, Chelsea, MI.

Ferguson, J., O. van Franqué, and M. Schock (1996) "Corrosion of Copper in Potable Water Systems," in V. Snoeyink and I. Wagner (eds.), *Internal Corrosion of Water Distribution Systems*, 2nd ed., American Water Works Association Research Foundation and DVGW, Denver, CO.

Gerston, J., M. MacLeod, and C. A. Jones (2002) *Efficient Water Use for Texas: Politics, Tools, and Management Strategies*, Texas Agricultural Experiment Station, Texas A&M University, College Station, TX.

Huston, S. S., N. L. Barber, J. F. Kenny, D. S. Lumia, and M. A. Maupin (2004) *Estimated Use of Water in the United States in 2000*, Circular 1268, U.S. Geological Survey, Reston, VA.

ISO (1999) *ISO 8044: Corrosion of Metals and Alloys—Basic Terms and Definitions*, International Organization for Standardization, Geneva.

Joly, P., P. A. Falconnet, J. Andre, N. Weill, M. Reyrolle, F. Vandenesch, M. Maurin, J. Etienne, and S. Jarraud (2006) "Quantitative Real-Time *Legionella* PCR for Environmental Water Samples: Data Interpretation," *Appl. Envir. Microbiol.*, **72**, 4, 2801–2808.

Kirmeyer, G. J., G. Pierson, J. Clement, A. Sandvig, V. Snoeyink, W. Kriven, and A. Camper (2000) *Distribution System Water Quality Changes Following Corrosion Control Strategies*, AWWA Research Foundation, Denver, CO.

Langelier, W. (1936) "The Analytical Control of Anti-Corrosion Water Treatment," *J. AWWA*, **28**, 10, 1500–1521.

Larson, T. E., and R. V. Skold (1958) "Laboratory Studies Relating Mineral Quality of Water to Corrosion of Steel and Cast Iron," *Corrosion*, **14**, 6, 285–288.

Levine, A. D. (2003) "Use of Reclaimed Wastewater for Cooling Tower Applications, Chap. 3 in *Membranes for Industrial Wastewater Recovery and Reuse*, IWA Publishing, London," UK

Levine, A. D., and T. Asano (2002) "Water Reclamation, Recycling and Reuse in Industry," in P. Lens, L. Hulshoff Pol, P. Wilderer, T. Asano (eds.) *Water Recycling and Resource Recovery in Industry: Analysis, Technologies and Implementation*, IWA Publishing, London, UK.

Lewis, G. H., and Randall, M. J. (1921) "The Activity Coefficient of Strong Electrolytes," *J. Am. Chem. Soc.*, **43**, 1112–1154.

Mann, J. G., and Y. A. Liu (1999) *Industrial Water Reuse and Wastewater Minimization*, McGraw-Hill, New York.

McCoy, J. (1974) *The Chemical Treatment of Cooling Water*, Chemical Publishing, Co., New York.

McCoy, J. (1981) *The Chemical Treatment of Boiler Water*, Chemical Publishing Co., New York.

McIntyre, R., E. Kobylinski, G. Hunter, and S. Howerton (2002) "Returned Reuse Water—Impacts and Approaches of Pollutant Concentration Buildup," in *Proceedings Water Sources Conference*, Las Vegas, NV, American Water Works Association, Denver, CO.

Middlebrook, E. J. (1982) "Municipal Wastewater Reuse in Power Plant Cooling Systems," in E. J. Middlebrooks (ed.) *Water Reuse*, Ann Arbor Science Publishers, Ann Arbor, MI.

Miller, D. G. (2003) "West Basin Municipal Water District: 5 Designer (Recycled) Waters to Meet Customer's Needs," in *Proceedings, Annual WateReuse Symposium*, San Antonio, TX, WateReuse Association, Alexandria, VA.

NAS (1982) Drinking Water and Health, Chap. III, 21–23, The National Academies Press, Washington, DC.

NPI (1999) *Emission Estimation Technique Manual for Textile and Clothing Industry*, National Pollutant Inventory, Environment Australia, Canberra, Australia.

Oguma, K., H. Katayama, and S. Ohgaki (2004) "Photoreactivation of *Legionella Pneumophila* after Inactivation by Low- or Medium-Pressure Ultraviolet Lamp," *Water Res.*, **38**, 2757–2763.

Pagliaro, T. (1997) "Water Treatment Key to Maintaining Boiler Efficiency: DI and RO Work in Tandem to Reduce Dissolved Solids, Hardness," *Water Technology Magazine*, National Trade Publications Inc., accessed online: http://www.watertechonline.com/.

Panchapakesan, B. (2001) "Optimizing White Water System Design," *Paper Age*, **117**, 2, 56–63.

Paperloop (2003) *Lockwood-Post's Directory of the Pulp, Paper and Allied Trades*, 2003 ed., Paperloop, San Francisco, CA.

Pisigan, R. A., and Singley, J. E. (1987). "Influence of Buffer Capacity, Chlorine Residual, and Flow Rate on Corrosion of Mild Steel and Copper," *J. AWWA*, **79**, 2, 62–70.

Puckorius, P. R., and J. M. Brooke (1991) "A New Practical Index for Calcium Carbonate Scale Producing in Cooling Tower Systems," *Corrosion*, **47**, 4, 280–284.

Reiber, S., R. A. Ryder, and I. Wagner (1996) "Corrosion Assessment Technologies," in AWWARF, *Internal Corrosion of Water Distribution Systems*, AWWA Research Foundation, Denver, CO.

Rommelmann, D. W., S. J. Duranceau, M. W. Stahl, C. Kamnikar, and R. M. Gonzales (2003) *Industrial Water Quality Requirements for Reclaimed Water*, AWWA Research Foundation, Denver, CO.

Rothberg, Tamburini, and Windsor, Inc., (2000) *The Rothberg, Tamburini and Windsor Model for Water Process and Corrosion Chemistry*, American Water Works Association, Denver, CO.

Russell, L. L. (1976) *Chemical Aspects of Groundwater Recharge with Wastewaters*, Ph.D Thesis, University of California, Berkeley, CA.

Ryznar, J. W. (1944) "A New Index For Determining The Amount Of Calcium Carbonate Scale Formed By Water," *J. AWWA*, **36**, 472–484.

Sawyer, C. N., P. L. McCarty, and G. F. Parkin (2003) *Chemistry for Environmental Engineering and Science*, 5th ed., McGraw-Hill, New York.

Schroeder, C. D. (1991) *Solutions to Boiler and Cooling Water Problems*, 2nd ed., Van Nostrand Reinhold, New York.

Smith, B. (1986) *Identification and Reduction of Pollution Sources in Textile Wet Processing*, North Carolina Department of Environment, Health, and Natural Resources, Office of Waste Reduction, Raleigh, NC.

Snoeyink, V. L., and D. Jenkins (1980) *Water Chemistry*, John Wiley & Sons, New York.

Solley, W. B., R. R. Pierce, and H. A. Perlman (1998) *Estimated Use of Water in the United States in 1995*, Circular 1200, U.S. Geological Survey, Reston, VA.

State of California (1963) *Water Quality Criteria*, 2nd ed., McKee, J. E. and H. W. Wolf (eds.), State Water Resources Control Board, Sacramento, CA.

State of California (1980) *Evaluation of Industrial Cooling Systems Using Reclaimed Municipal Wastewater: Applications for Potential Users*, Office of Water Recycling, California State Water Resources Control Board, Sacramento, CA.

State of California (2003) *Water Recycling 2030: Recommendations of California's Recycled Water Task Force*, Department of Water Resources, Sacramento, CA.

Stiff, H. A., Jr., and L. E. Davis (1952) "A Method for Predicting the Tendency of Oil Field Water to Deposit Calcium Carbonate," *Pet. Trans. AIME* **195**, 213–216.

Stumm, W (1956) "Calcium Carbonate Deposition at Iron Surfaces," *J. AWWA*, **48**, 300–310.

TMI (1997) *Textiles: America's First Industry*, American Textiles Manufacturers Institute, Washington, DC.

Treweek, G. P. (1982) "Industrial Reuse of Wastewater: Quantity, Quality and Cost," Chap. 23, 521–548, in E. J. Middlebrooks (ed.) *Water Reuse*, Ann Arbor Science Publishers, Ann Arbor, MI.

Trussell, R. R. (1998) "Spreadsheet Water Conditioning," *J. AWWA*, **90**, 6, 70–81.

U.S. Department of Commerce (2005) *2002 Economic Census*, U.S. Department of Commerce, U.S. Census Bureau, accessed online: http://www.census.gov/econ/census02/.

U.S. Department of Commerce (2005) *Trade Data—U.S. Imports and Exports of Textiles and Apparel*, accessed online: http://otexa.ita.doc.gov/.

U.S. EPA (1973) *Water Quality Criteria 1972*, EPA-R-73-033, Washington, DC.

U.S. EPA (1980) *Guidelines for Water Reuse*, EPA-600/8-80-036, Municipal Environmental Research Laboratory, U.S. Environmental Protection Agency, Cincinnati, OH.

U.S. EPA (1996) *Manual, Best Management Practices for Pollution Prevention in the Textile Industry*, EPA/625/R-96/004, Office of Research and Development. National Risk Management Research Laboratory, Center for Environmental Research Information, U.S. Environmental Protection Agency, Cincinnati, OH.

U.S. EPA (1997) *Profile of the Textile Industry*, Sector Notebook Report, EPA/310-R-97-009, Office of Compliance, Office of Enforcement and Compliance Assurance, U.S. Environmental Protection Agency, Washington, DC.

U.S. EPA (2000) *Profile of the Pulp and Paper Industry*, 2nd ed., Sector Notebook Report, EPA/310-R-02-002, Office of Compliance, Office of Enforcement and Compliance Assurance, U.S. Environmental Protection Agency, Washington, DC.

U.S. EPA, and U.S. AID (2004) *2004 Guidelines for Water Reuse*, EPA/625/R-04/108, U.S. Environmental Protection Agency and U.S. Agency for International Development, Washington, DC.

U.S. Water Resources Council (1979) "The Nation's Water Resources *1975–2000*," Vol. 2, *Water Quantity, Quality, and Related Land Considerations*, U.S. Water Resources Council, Washington, DC.

Videla, H. A. (1996) *Manual of Biocorrosion*, CRC Press, Boca Raton, FL.

Vik, E. A., R. A. Ryder, I. Wagner, and J. F. Ferguson (1996) "Mitigation of Corrosion Effects," in AWWARF, *Internal Corrosion of Water Distribution Systems*, AWWA Research Foundation, Denver, CO.

Water 3 Engineering, Inc. (2005) *Analysis of the Potential Benefits of Recycled Water Use in Dye Houses, Final,* Prepared for Central Basin Municipal Water District and California Department of Water Resources, accessed online: http://www.owue.water.ca.gov/recycle/

WBMWD (2006) *2005 West Basin Water Management Plan*, West Basin Municipal Water District, Carson, CA.

Wijesinghe, B., R. B. Kaye, and C. J. D. Fell (1996) "Reuse of Treated Sewage Effluent for Cooling Water Make Up: A Feasibility Study and a Pilot Plant Study," *Water Sci. Technol.*, **33**, 10–11, 363–369.

WPCF (1989) *Water Reuse: Manual of Practice*, SM-3, Water Pollution Control Federation, Alexandria, VA.

20 Urban Nonirrigation Water Reuse Applications

WORKING TERMINOLOGY   1170

20-1   URBAN WATER USE AND WATER REUSE APPLICATIONS: AN OVERVIEW   1171
    Domestic Potable Water Use in the United States   1171
    Commercial Water Use in the United States   1172
    Urban Nonirrigation Water Reuse in the United States   1172
    Urban Nonirrigation Water Reuse in Other Countries   1172

20-2   FACTORS AFFECTING THE USE OF RECLAIMED WATER FOR URBAN NONIRRIGATION REUSE APPLICATIONS   1175
    Infrastructure Issues   1175
    Water Quality and Supply Issues   1176
    Acceptance Issues   1179

20-3   AIR CONDITIONING   1179
    Description of Air Conditioning Systems   1179
    Utilizing Reclaimed Water for Air Conditioning Systems   1181
    Water Quality Considerations   1181
    Management Issues   1183

20-4   FIRE PROTECTION   1183
    Types of Applications   1186
    Water Quality Considerations   1187
    Implementation Issues   1187
    Management Issues   1188

20-5   TOILET AND URINAL FLUSHING   1188
    Types of Applications   1188
    Water Quality Considerations   1188
    Implementation Issues   1192
    Satellite and Decentralized Systems   1193
    Management Issues   1193

20-6   COMMERCIAL APPLICATONS   1195
    Car and Other Vehicle Washing   1195
    Laundries   1196

20-7   PUBLIC WATER FEATURES   1197
    Fountains and Waterfalls   1197
    Reflecting Pools   1197
    Ponds and Lakes in Public Parks   1198

20-8 ROAD CARE AND MAINTENANCE   1198
Dust Control and Street Cleaning   1199
Snow Melting   1199

PROBLEMS AND DISCUSSION TOPICS   1200

REFERENCES   1201

WORKING TERMINOLOGY

Term	Definition
Black water	House wastewater from toilets and bidets that contains human waste.
Blowdown	The water that is removed from recirculating cooling water to control dissolved solids levels. Blowdown is removed either continuously or intermittently (see Chap. 19).
Dual distribution system	Two independent piping systems that are used to deliver potable and reclaimed water (see Chap. 14).
Dual plumbing system	Separate plumbing systems used to supply potable and reclaimed water from the points of connection from the respective distribution systems to the points of use (see Chap. 15).
Greywater	Water from bathing and washing facilities that does not contain concentrated human waste (i.e., toilets) or food waste (i.e., kitchen sink, food waste grinders). Examples include bath and shower water, hand wash water, and laundry water. Greywater typically contains high concentration of salts and minerals from detergents and soaps. If brine water softeners are used, increased concentrations of sodium can be expected in the greywater.
Hydropneumatic system	A hydropneumatic system is comprised of a water storage tank containing water with a pressurized air space above the water (known as a hydropneumatic tank), an external air compressor, and appropriate valves and switches. The hydropneumatic tank is used to minimize excessive pump cycles by maintaining the pressure within a specified operating range while water is withdrawn from the tank.
Legionnaires disease	A disease caused by a type of bacteria called *Legionella*. Symptoms are like other types of pneumonia, and can be fatal to a susceptible population. *Legionella* can be found naturally and sometimes at high concentrations in poorly maintained cooling water systems, hot water tanks, and hot water taps.
Nonpotable urban water reuse	Nonpotable water reuse occurring in urban areas, including irrigation of landscaping plants, toilet flushing, fire fighting, air conditioning, dust control, street cleaning, snow melting, car washing, commercial laundry use, and ornamental uses such as fountains and ponds.
Peaking factor	Ratio of design maximum flow to average daily flow rate. For the design of a reclaimed water distribution system, generally hourly flow rates are used.
Uniform Plumbing Code, Appendix J	The plumbing code by the International Association of Plumbing and Mechanical Codes (IAPMC), specifying the criteria for dual plumbing systems with reclaimed water for nonresidential buildings (see Chap. 15).

Water reuse in urban areas encompasses a wide variety of applications that do not require water potable water. While landscape irrigation is a primary use of reclaimed water in urban areas, reclaimed water that meets applicable water reuse criteria is considered safe from a public health standpoint for a variety of nonirrigation urban uses. The principal types of urban nonirrigation water reuse applications considered in this chapter include: (1) air conditioning, (2) fire protection, (3) toilet and urinal flushing, (4) commercial applications, (5) public water features, and (6) street cleaning and maintenance. However, before discussing these applications in detail, it will be useful to examine, in general terms, the potential nonirrigation urban water reuse applications and the factors affecting the reuse of reclaimed water in urban applications. Reclaimed water distribution and dual plumbing systems, key components of an urban water reuse system, are described briefly in this chapter, but are discussed in greater detail in Chaps. 14 and 15. Indirect potable reuse, another water reuse application that can be implemented in urban areas, is discussed in Chaps. 22 and 23.

20-1 URBAN WATER USE AND WATER REUSE APPLICATIONS: AN OVERVIEW

To understand the potential for urban nonirrigation water reuse applications, it will be useful to review the patterns of potable water use for domestic and commercial purposes in the United States and in other countries.

Domestic Potable Water Use in the United States

Domestic potable water use in the United States during the year 2000 was estimated at about 10.4×10^6 m^3/d (27.6×10^3 Mgal/d). On a national average, the per capita domestic water use in the year 2000 was about 372 L/capita $\cdot$ d (98.3 gal/capita $\cdot$ d), declining by almost 10 percent since 1990 (Solley et al., 1993; Solley et al., 1998; Hutson et al., 2004). Of the total amount of domestic water used, an average of about 65 to 70 percent is used indoors, while 30 to 35 percent is used outdoors. Toilet flushing accounts for about 40 percent of the indoor use (U.S. EPA, 1992; Vickers, 2001). Thus, nonpotable applications such as outdoor uses and toilet flushing comprise more than half of total domestic water use. Patterns of domestic water use, however, vary greatly from region to region due to climatic conditions and water conservation measures. For example, outdoor water use is about 44 percent of domestic water use in California, in contrast to about seven percent in Pennsylvania (U.S. EPA, 1992).

In arid regions where the need for landscape irrigation is high, reclaimed water can supply a major part of the domestic water needs. The percentage of potential reclaimed water use is even higher in highly urbanized areas occupied by offices and other commercial and public buildings. In the Irvine Ranch Water District, California, for example, approximately 90 percent of water used in a high-rise commercial building was used for nonpotable purposes such as air conditioning and toilet and urinal flushing (see Chap. 15).

Commercial Water Use in the United States

Commercial water use in the United States in the year 2000 was about 39.8×10^6 m^3/d (10.5×10^3 Mgal/d), which includes water for motels, hotels, restaurants, office buildings, other commercial facilities, and civilian and military institutions. Public-supply deliveries to golf courses, and for some states offstream water use for fish hatcheries, are included in the estimate (Solley et al., 1998; Hutson et al., 2004). While a significant opportunity exists for the use of reclaimed water in commercial applications, the dispersed nature of the potential users makes it more difficult economically to supply reclaimed water through a centralized storage and distribution system. The development of satellite and decentralized water reclamation systems (discussed in Chaps. 12 and 13) can improve opportunities for reuse in some areas.

Urban Nonirrigation Water Reuse in the United States

Selected examples of urban nonirrigation water reuse applications in the United States are presented in Table 20-1 and on Fig. 20-1. The examples presented are illustrative of the many different types of urban nonirrigation uses that have been developed and continue to be developed. In California, 10 municipal reclaimed water providers (water districts or the cities' public utility) supply reclaimed water for a variety of urban applications including toilet flushing, cooling tower for buildings, car washing, street cleaning, commercial laundries, and dust control at construction sites (State of California, 2003). In Florida, there are five water reuse systems for toilet flushing, one system for fire protection, and nine systems for other urban applications such as air conditioning and street cleaning (State of Florida, 2005). Urban water reuse applications in the eastern part of the United States are generally small scale, and satellite or decentralized water reclamation and reuse systems are being utilized increasingly. In the future, it is anticipated that water reuse in urban areas will have greater importance due to urban population growth, difficulties in obtaining additional water supply, and increasing concerns about environmental effects of wastewater discharges (see Chap. 1).

Urban Nonirrigation Water Reuse in Other Countries

Water reuse in urban settings, especially the applications discussed in this chapter, is observed mostly in well-developed countries. The number of water reuse projects has been increasing rapidly in Australia, due partly to the severe drought from 2001 to 2003 (Radcliffe, 2004). Water reuse in Japan is unique because agricultural and landscape irrigation is not the dominant application; instead, urban uses such as for environmental features (e.g., ornamental lakes, wetlands, and stream flow augmentation), snow melting, and toilet flushing, are more common (JSWA, 2005). Selected examples of urban water reuse outside the United States are presented in Table 20-2.

Many urban water reuse applications in European countries are combined with small and decentralized or satellite water reclamation systems. Greywater, identified in Chap. 13, has been reclaimed in many European countries including the United Kingdom, France, and Germany. In Spain, both greywater and municipal wastewater have been reclaimed and reused for urban water reuse applications. For example, in Portbou, Costa Brava, Spain, tertiary-treated reclaimed water is used for landscape irrigation, street cleaning, and fire protection. The municipality is expanding the reclaimed water distribution system for boat cleaning, at a newly constructed sport marina (Sala, 2003).

Table 20-1
Typical examples of urban water reuse in the United States[a]

Application	Location	Description/remarks
Air conditioning cooling tower	Irvine Ranch Water District, CA	Started operating cooling system with reclaimed water for Opus Center Irvine II in 2002. Reclaimed water is also used for toilet and urinal flushing.
	St. Petersburg, FL	Reclaimed water is used for air conditioning at commercial buildings, including a sports stadium.
	The Solaire, New York	High-rise residential building that has an onsite treatment system. Reclaimed water is also used for toilet and urinal flushing.
Fire protection	Altamonte Springs, FL (Project APRICOT)	75 fire hydrants connected to reclaimed water distribution system.
	Livermore, CA	Sprinklers were installed for commercial buildings.
	St. Petersburg, FL	308 hydrants connected to over 460 km (290 mi) of reclaimed water distribution pipelines.
Toilet/urinal flushing	Irvine Ranch Water District, CA	See above.
	Marin County, CA	Toilet flushing for 330-bed jail facility, parks.
	Patriots Stadium, Foxborough, MA	Onsite treatment facility with 950,000 L/d (250,000 gal/d) capacity.
	San Jose, CA	New city hall and a public library are dual plumbed for reclaimed water use. Several commercial complexes are using reclaimed water for toilet/urinal flushing.
	The Solaire, New York	See above.
Commercial car washing	Largo, FL	Reclaimed water is used for main wash. Potable water treated with RO is used for final rinse.
	Marin County, CA	Part of internally-recycled car wash water is RO-treated and used for the final rinse to prevent spotting.
Commercial laundries	Marin County, CA	Nazareth House uses reclaimed water for laundry. Washing is at 70°C (160°F) for effective cleaning and less detergent usage.
Water features	Tillman Water Reclamation Plant, Los Angeles, CA	Reclaimed water is used for a reflecting pool in the famous Japanese garden.
Street cleaning	Altamonte Springs, FL (Project APRICOT)	Reclaimed water is also used for fire protection and car wash.
	Inglewood, Los Angeles, CA	Both reclaimed and potable water are used.

[a] There are numerous urban water reuse projects in the United States; the above examples are presented only to illustrate opportunities for various urban water reuse applications.

Figure 20-1

Examples of urban nonirrigation water reuse applications: (a) air conditioning for a sports stadium; (b) fire hydrant connected to reclaimed water distribution system; (c) toilet using reclaimed water for flushing; (d) commercial vehicle washing; (e) ornamental pond using reclaimed water; and (f) reclaimed water sprayed on pavement for dust and heat control.

Table 20-2
Examples of urban nonirrigation water reuse in other parts of the world

Country	Description	Major source of reclaimed water
Australia	Reclaimed water is used for toilet flushing in some places such as Rouse Hill, Sydney, Taronga Zoo, Springfield, and Peredenya (Byron Shire Council, 2005). In the Sydney area, reclaimed water is also used for other urban nonpotable purposes such as fire fighting (except in the service area of Sydney Olympic Park Authority), construction purposes, and ornamental features (Radcliffe, 2004).	Municipal wastewater, greywater
France	Greywater is treated and used for toilet flushing on a demonstration scale in Annecy, France (Lazarova, 2005).	Greywater
Japan	More than half of reclaimed water is used to augment stream flow in urban areas. Water reuse for snow melting, which utilizes the warm temperature of the reclaimed water, is a unique application. Toilet flushing is a major use, comprising about three percent of the total reclaimed water use in 2001 (JSWA, 2005).	Municipal wastewater, greywater, rainwater
Spain	In Portbou, Spain, tertiary-treated reclaimed water is being used for landscape irrigation, street cleaning, and fire protection. The municipality is expanding the reclaimed water distribution line for boat cleaning at a newly constructed sport marina (Sala, 2003).	Greywater, municipal wastewater
United Kingdom	Greywater is used more commonly than wastewater (blackwater) for urban applications, most of which are demonstration projects. Onsite treatment and reuse of greywater for toilet flushing in Millennium Dome was a landmark project in UK (Lazarova et al., 2002).	Greywater, rainwater

20-2 FACTORS AFFECTING THE USE OF RECLAIMED WATER FOR URBAN NONIRRIGATION REUSE APPLICATIONS

Important factors affecting the use of reclaimed water for urban applications are related to (1) infrastructure, (2) water quality and supply reliability, and (3) public acceptance. Each of these issues is considered in the following discussion.

Infrastructure Issues

Major infrastructure components in the development of water reuse systems in urban area include (1) dual distribution and storage systems and (2) dual plumbing systems. Both dual distribution systems and dual plumbing systems are easier to install in a new development than in an existing urban area. Retrofit of an developed area to include reclaimed water systems may impose additional issues such as excessive costs for implementation and the risk of cross-connection. Important considerations for dual distribution and plumbing systems are described below.

Dual Distribution Systems and Storage

To convey reclaimed water for urban uses through an area-wide distribution system, dual distribution systems of potable and reclaimed waters must be developed. The reclaimed water demand must be large enough to justify the installation of the necessary infrastructure. Depending on the extent of existing development, the cost associated with the installation and maintenance of dual distribution systems for urban water reuse may be prohibitive, unless the site of water reuse is adjacent to the water reclamation facility. For example, construction of a new area-wide reclaimed water distribution system with more than 15 to 25 km (10 to 15 mi) of pipelines is typically considered infeasible for many urban areas. Satellite and decentralized systems are more viable alternatives for supplying reclaimed water to remote areas (see Chaps. 12 and 13).

Important factors in the design of dual distribution systems include: (1) an area-wide market assessment of reclaimed water demand (see Chap. 25), (2) the potential growth of water reuse within the project area including the identification of large users, and (3) the expandability of the system (Young et al., 1998). Turnouts should be installed where future network connections are anticipated (AWWA, 1994). As discussed in Chap. 14, the availability of adequate storage capacity for reclaimed water is another factor that will affect the feasibility of reclaimed water use in urban areas. Daily variability of reclaimed water demand can be mitigated by installing a short-term storage reservoir, which will also increase the reliability of the reclaimed water supply. Larger long-term storage facilities will be necessary to mitigate seasonal variability of reclaimed water demand. Factors considered in sizing storage facilities are described in Chap. 14.

Dual Plumbing Systems

If reclaimed water is to be used for toilet and urinal flushing and other interior building uses, dual plumbing systems (see Chap. 15) should be installed at the time of building construction. Retrofitting existing buildings and commercial complexes is much more difficult due to the high cost of revising existing plumbing systems and installing the dual system for reclaimed water. Standards for the installation of dual plumbing systems in nonresidential buildings are set forth in the *Uniform Plumbing Code (UPC)*, Appendix J (IAPMO, 2003), or in state and city plumbing codes.

Water Quality and Supply Issues

Consistency of water quality and reliability of water supply are important issues in urban water reuse applications. Water quality requirements are related to public health protection and the design of reclaimed water facilities, as well as acceptance of reclaimed water by users. For example, water supply reliability is a critical issue when reclaimed water is used for fire protection.

Consistency of Water Quality

A summary of water quality requirements for unrestricted urban uses in selected states in the United States is provided in Table 20-3 (see also Chap. 4 for other water quality regulations). In California, filtered and disinfected reclaimed water meeting a total coliform concentration of 2.2/100 mL or less and a turbidity of 2 NTU, generally referred to as Title 22 water, is allowed for most urban nonirrigation water reuse applications. Other states generally have similar criteria for nonirrigation urban water reuse applications.

Table 20-3 Reclaimed water quality and treatment requirements for unrestricted urban uses for selected states in the United States[a]

Parameter	Water quality requirement for indicated state				
	Arizona	California	Florida	Hawaii	Texas
Treatment	Secondary treatment, filtration and disinfection	Oxidized[b], coagulated, filtered and disinfected	Secondary treatment, filtration and high-level disinfection	Oxidized, filtered and disinfected	ns[c]
TSS	ns	ns	5.0 mg/L	ns	ns
Turbidity	2 NTU (avg) 5 NTU (max)	2 NTU (avg) 5 NTU (max)	ns	2 NTU (max)	3 NTU
pH	ns	ns	6.0–8.5	ns	ns
Coliform	Fecal nondetectable in 4 of last 7 daily samples 23/100 mL (max)	Total 2.2/100 mL (7-day median) 23/100 mL (max in 30 d)	Fecal 75% of samples below detection 25/100 mL (max)	Fecal 2.2/100 mL (7-day median) 23/100 mL (max in 30 d)	Fecal 20/100 mL (geometric mean) 75/100 mL (max)

[a] Adapted from U.S. EPA (2004).
[b] *Oxidized wastewater* is wastewater that is treated to oxidize and stabilize organic compounds, and contains dissolved oxygen. The term *oxidized wastewater* is used as an alternative to *secondary effluent* to avoid specifying treatment processes.
[c] ns—Not specified.

Specific Water Quality Requirements Some reclaimed water constituents that are not considered important from a public health standpoint may need to be controlled for nonirrigation urban water reuse applications. For example, constituents that can potentially generate odorous compounds in the distribution system should be monitored and controlled. Other examples include biofilm growth resulting from the presence of residual nutrients, scaling from mineral concentrations in excess of their solubility limits, and corrosion caused by demineralized water that is not buffered sufficiently.

Some water quality requirements for specific water reuse applications may be more stringent than those needed for public health protection. When reclaimed water is used for car washing, it is not used for the final water rinse because the presence of dissolved solids can cause spotting on the washed vehicles. In most modern car wash facilities, water from a reverse osmosis system is used for the final rinse. Reclaimed water quality requirements for specific water reuse applications and the treatment requirements at the point of use must be addressed on a case-by-case basis. Where necessary, pilot testing should be conducted to determine the effectiveness and feasibility of the selected treatment process (also see Part 3).

For multiple water reuse applications, the economic question that must be addressed is whether it is more cost effective to (1) produce multiple grades of reclaimed water to meet the quality criteria for all users, (2) produce reclaimed water of a single quality that meets all criteria, or (3) produce a single grade of quality that meet most criteria and provide treatment at or near the point of use for specific applications. Typically in a water reuse system that involves multiple users and a single quality of product water, reclaimed water quality requirements are determined by a major user that requires the highest quality. For example, if a reclaimed water distribution system is to provide water for landscape irrigation, high-rise building toilet and urinal flushing, and industrial cooling towers, the microbial requirements for toilet and urinal flushing will be critical, whereas industrial cooling tower usage may require nitrogen and phosphorous removal to control biological growth, scaling, and corrosion (see Chap. 19). The aesthetic quality of water is also important for nonirrigation urban water reuse applications, especially for such uses as toilet flushing and urban water features.

Typical Regulations and Guidelines Typical regulations for various nonpotable water reuse applications are given in Table 4-5 in Chap. 4. Typical guidelines used in Japan for various urban uses are presented in Table 20-4.

Reliability of Water Supply

The requirements for water service reliability depends on the need for uninterruptible service such as for fire fighting. Adequate flow and pressure of the reclaimed water supply must be assured at all times when reclaimed water is used for fire fighting. Where

Table 20-4 Reclaimed water quality guidelines for urban uses in Japan[a]

Parameter	Unit	Water quality requirement for indicated use[b]		
		Toilet/urinal flushing	Spraying on street and ground	Recreational uses and water features
Total coliform	No./100mL	non detectable	non detectable	non detectable
Turbidity	NTU	2	2	2
pH	pH unit	5.8–8.6	5.8–8.6	5.8–8.6
Appearance	—	not unpleasant	not unpleasant	not unpleasant
Color[c]	CU[d]			<10
Odor[c]	—	not unpleasant	not unpleasant	not unpleasant
Chlorine residual	mg/L	0.1 (free), 0.4 (combined)	0.1 (free), 0.4 (combined)	0.1 (free), 0.4 (combined)
Treatment requirements		Sand filtration or equivalent	Sand filtration or equivalent	Coagulation, sedimentation, and filtration, or equivalent

[a]Adapted from MLIT (2005).
[b]Water quality is to be measured at the outflow of the water reclamation plant.
[c]To be adjusted on a case-by-case basis to meet the user's demand.
[d]CU = color unit.

reclaimed water service is considered interruptible, a backup water system for fire fighting must be supplied, primarily by the potable water service. Requirements for multiple backup systems may result in a significant increase in the cost of infrastructure (U.S. EPA, 2004). Reliability is less critical when the reclaimed water supply can be interrupted for short periods of time.

The distribution system for reclaimed water can be designed to provide unrestricted, on-demand service, or the reclaimed water can be provided on a restricted schedule. Because the principal use of reclaimed water in urban areas is for landscape irrigation, which is applied generally during the night time hours to minimize human contact and evaporation loss, unrestricted service may result in a high peak flow demand. The peak demand may be several times higher than the daily average flowrate available for producing reclaimed water (U.S. EPA, 2004). In such situations storage reservoirs are needed to meet maximum hourly demands. Where reclaimed water is used for fire fighting, emergency storage can serve as a backup for the distribution system and when pumping stations or pipelines are out of service for maintenance or repair.

Acceptance Issues

Acceptance by the public and the potential users of reclaimed water is an important issue. Both public and specific users of reclaimed water must be protected from potential health hazards associated with the use of reclaimed water. If issues of concern are not addressed satisfactorily, a proposed water reuse application may not be implemented. For example, firefighters in Sydney Olympic Park, Australia, rejected the use of reclaimed water for fire fighting because it was claimed that the firefighters would be exposed to a health risks if they came in contact with reclaimed water and received minor cuts in the performance of their work. To resolve the issue, the City of Sydney decided to use potable water for fire fighting.

The aesthetic quality of reclaimed water must also be assured. Even though reclaimed water generally is treated appropriately by biological treatment, filtration, and disinfection and meets the criteria for unrestricted reuse, it may retain slight color and odor. If the color and odor are noticeable, especially during the toilet and urinal flushing, the users may perceive the water is unsafe. For ornamental uses, residual surfactants in reclaimed water may cause foaming in fountains or other water features. Additional treatment may be needed, therefore, to resolve aesthetic issues and gain public acceptance.

20-3 AIR CONDITIONING

Description of Air Conditioning Systems

Most high-rise commercial and residential buildings have central cooling and heating systems. Water used for air conditioning systems represents one of the major water demands in commercial buildings. Air conditioning systems in commercial applications are considered in this section. More detailed information on cooling towers in industrial applications is given in Chap. 19.

To cool indoor air for buildings and commercial facilities, air conditioning equipment needs a coolant. The coolant can be either water or refrigerants such as liquefied propane gas or hydrofluorocarbons (HFCs). In water-cooled air conditioning systems, a water

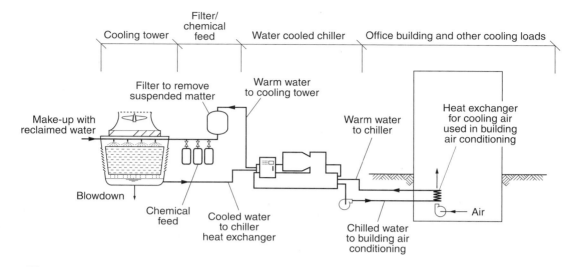

Figure 20-2
Typical schematic of an open-circuit water-cooled chilled water air conditioning system.

chiller is used to chill the coolant water (i.e., chilled water cooling system). Water chillers are air-cooled or water-cooled in releasing heat from the heat exchanger. For a single building, air-cooled chillers are used most commonly. For larger buildings and commercial sites, water-cooled chillers are used. In a water-cooled chiller system, a cooling tower is used to cool the heat exchanger cooling water. Circulating cooling water can be cooled directly with a cooling tower (open-circuit type), or a separate cooling tower water can be used indirectly for cooling (closed-circuit type). A typical diagram of an open-circuit water-cooled chilled water air conditioning system is shown on Fig. 20-2. Views of a typical cooling tower and a water-cooled chiller are presented on Fig. 20-3.

(a)   (b)

Figure 20-3
Air conditioning system using reclaimed water: (a) stainless steel cooling tower and (b) water-cooled centrifugal water chiller.

Utilizing Reclaimed Water for Air Conditioning Systems

Cooling towers used for high-rise buildings and commercial sites are usually located on the top of or behind the building where public access is limited. When reclaimed water is used for cooling tower systems, dual plumbing systems for potable and reclaimed water supplies must be installed. As noted previously, installation of a dual plumbing system usually is limited to new construction as it is not feasible economically to retrofit most existing buildings. An example of a building that uses reclaimed water for air conditioning purposes is the 14-story Opus Center Irvine II building, the first commercial building in Orange County, California, to use reclaimed water for air conditioning (IRWD, 2003). Water used for air conditioning cooling is about 75 percent of total reclaimed water use for the building. The balance of the reclaimed water is used for toilet and urinal flushing. Examples of cooling towers used for high-rise buildings and commercial sites are shown on Fig. 20-4.

Water Quality Considerations

Generally, reclaimed water that meets quality criteria for unrestricted urban uses is acceptable for cooling water, with some specific requirements to prevent scaling, corrosion, and biofouling problems (see also Chap. 19). A summary of the water quality considerations for air conditioning systems is presented in Table 20-5. High ammonia levels in reclaimed water are a special concern as they will cause biological growth and corrosion; thus, denitrified reclaimed water is preferred for cooling systems. Special precautions must be taken when reclaimed water is used for the cooling towers on buildings in urban areas because of the potential exposure of the public to aerosols emitted from the cooling towers.

Another important water quality issue is prevention of biological growth. Extra caution must be taken to prevent the growth of *Legionella spp.*, a type of bacteria that causes respiratory diseases called Legionnaires' disease and Pontiac fever, in cooling towers

(a)

(b)

Figure 20-4
Examples of cooling towers: (a) high-rise buildings and (b) commercial sites.

Table 20-5
Water quality issues to be considered when using reclaimed water for air conditioning systems

Issue	Factors that must be considered	Control/remarks
Biofilm growth	The presence of residual organic matter and ammonia in reclaimed water. With time, even resistant organic compounds can serve as a carbon source for microorganisms in biofilms.	In some cases, it may be necessary to carry a disinfectant residual. Drain and clean on a regular schedule (e.g., quarterly).
Corrosion	Metallic corrosion is caused by dissolved constituents in reclaimed water including ammonia, chlorides, sulfate, and gases. Acid formed by microorganisms can also lead to metallic corrosion.	Use of plastic piping will control metallic corrosion. Where metallic elements are present, rust inhibitors are used.
Drift of water vapor from cooling tower	Safety of water, especially with respect to Legionnaires' disease bacteria (LDB).	Enclose cooling system to limit offsite vapor drift.
Fouling, particulate[a]	Total suspended, colloidal, and dissolved solids in reclaimed water; the presence of microorganisms in reclaimed water; dust blowing into cooling tower.	Nanofiltration or reverse osmosis can be used to reduce the dissolved solids. Ozone is often used to control both biological agents and scale at the same time. The addition of antifouling agents has been used to limit the degree of fouling.
Legionnaires' disease bacteria (LDB)	Measurement of surrogates such as total bacterial counts, total dissolved solids, and pH have not proven to be good indicators of LDB in cooling towers.	Chemical biocides are added to the water to control LDB. Drain and clean on a regular schedule (e.g., quarterly).
Microbial quality	Water used in cooling tower applications must meet Title 22 or equivalent.	In some cases, it may be necessary to carry a disinfectant residual. Drain and clean on a regular schedule (e.g. quarterly).
Public health	Health issues are not of concern as water treated to meet Title 22 requirements is deemed to be safe.	In some cases, it may be necessary to carry a disinfectant residual.
Scaling (precipitation)	Alkalinity, pH, calcium, and phosphorus are of concern. The scaling potential can be assessed using the Langelier saturation index or Ryzner stability index.	Limit the cycles of concentration. Nanofiltration can be used to remove calcium and bicarbonate concentrations to reduce scaling. The addition of an antiscalant has been used to limit the degree of scaling.
Total dissolved solids (TDS)	High TDS concentrations exacerbate fouling and scaling and limit the effectiveness of biocides added to the water to control LDB.	Limit the cycles of concentration. Nanofiltration can be used to remove calcium and bicarbonate concentrations to reduce scaling.

[a] Undesirable deposits that form within a closed loop system that may be due to the formation of crystallized minerals.

(See Chap. 19). Periodic flushing of the cooling system and superchlorination (flushing with high concentration chlorine solution: typically >10 mg/L although the concentration varies greatly depending on the site specific conditions) is generally satisfactory for prevention of biological growth. A comparison of specific water quality requirements for cooling water in San Diego, California with typical reclaimed water and the impacts of using reclaimed water air conditioning systems is presented in Table 20-6 (Gagliardo et al., 2005).

Management Issues

The most critical parameter in the operation and management of a cooling tower system is the buildup of salts, expressed most often as the cycles of concentration. Cycles of concentration is a term used to express the ratio of dissolved solids concentration in the water recirculating in the cooling tower to the dissolved solids concentration in makeup water (see Fig. 19-10; also Eq. 19-26). As water evaporates in the cooling system, the concentration of dissolved solids in the recirculating water increases and eventually reaches a level detrimental to the operation of the cooling system. Blowdown, water removed from recirculating cooling water to control buildup of dissolved solids, is replaced with fresh water. Typically, the cycles of concentration are limited to about three to five when reclaimed water is used because, at higher cycles of concentration, some of the salts in the recirculating water exceed their solubility limits and precipitate. In an unique exception, a synthetic polymer is applied to the recirculating water in cooling towers in Pinellas County, Florida, to prevent scaling and corrosion. With polymer addition, the cycles of concentration have been increased to between 20 and 30, thus reducing the volume of blowdown significantly (West, 2005).

A granular-medium filter can be added to the cooling water system to remove continuously biological flocs and precipitates (see Fig. 20-5). Periodic backwash of the filter should be included as part of the blowdown of the recirculating water. Depending on the frequency of backwash and allowable cycles of concentration, the amount of backwash water wasted may be sufficient to maintain a desirable level of salt content in the recirculating water. Blowdown and backwash water are usually discharged to the wastewater collection system.

Cooling water systems are often provided with chemical conditioning systems for pH adjustment and for the addition of other chemicals such as antiscalant, antifoaming agent, and disinfectant. Periodic superchlorination of the cooling system is also a common practice for cooling systems to control biofilm formation. The frequency of superchlorination depends on the reclaimed water quality and the climatic conditions, but typically once a month is considered sufficient.

20-4 FIRE PROTECTION

The use of potable water for fire protection requires an appropriately sized distribution system capable of supplying a high flowrate for a limited and irregular time period. When a potable water pipeline is used for multiple purposes including fire protection, the average flowrate is much less than the pipe capacity, resulting in long residence times in the pipeline with accompanying biofilm growth and water quality degradation. The use

Table 20-6
A comparison of general water quality requirements for cooling water, typical reclaimed water quality in San Diego, CA, and cooling tower impact[a]

Parameter	Unit	Manufacturer's recommendations[b]			Typical value in reclaimed water	Impact of reclaimed water quality on tower operation/possible remedy
		A	B	C		
Physical parameters						
pH	pH units	6.5–9.0	7.0–9.0	6.5–8.0	8.65	No adverse impact
Temperature	°C	52 (max)	—	—	35	No adverse impact
Indices/surrogate parameters						
LSI[c]	—	0.0–1.0	—	—	2.12	Scaling/increase inhibitor
Total alkalinity	mg/L as $CaCO_3$	100–500	500 (max)	50–300	500	Scaling/increase inhibitor
TDS	mg/L	5000 (max)	1000 (max)	10,000 (max)	3596	Affects thermal performance/reduce cycles of concentration
Chemical parameters						
Ammonia (for corrosion control)	mg/L	50	—	—	0.4	No adverse impact
Ammonia (for biogrowth control)	mg/L	10–25	—	—	0.4	No adverse impact
Chlorine	mg/L	1 (shock residual) or 0.4 (continuous)	—	—	16	Very high concentrations cause wood delignification/need to reduce chlorine residual or use alternative materials

Calcium	mg/L as CaCO$_3$	800	30–500	50–300	364	No adverse impact
Chlorides (for galvanized steel)	mg/L as Cl$^-$	455	125	200	784	Corrosion/add inhibitor
Chlorides (for stainless steel)	mg/L as Cl$^-$	910	—	400	784	No adverse impact/monitor for corrosion
Nitrates	mg/L as NO$_3$	300	—	—	70	Biogrowth/increase biocide use
Silica	mg/L as SiO$_2$	150 (max)	—	—	62	No adverse impact
Sulfates	mg/L as CaCO$_3$	800	125 (max)	—	812	Possible scale with moderate calcium, corrosion of concrete basins/monitoring

[a]Adapted from Gagliardo et al. (2005).
[b]Recommendations from different manufacturers of air conditioning equipment.
[c]LSI = Langlier Saturation Index (see Chap. 19).

Figure 20-5

Sand filters used to remove residual biological flocs and precipitates following pretreatment of reclaimed feedwater for a cooling water system.

of reclaimed water for fire protection allows the potable water pipeline to be designed for much lower peak flowrates and shorter residence times (AWWA, 1994; Okun, 1997).

While the use of reclaimed water for fire protection can reduce the maximum design flow for a potable water system, additional considerations are required in the design, operation, and maintenance of the reclaimed water distribution system. Brief descriptions of fire protection systems, water quality considerations, and issues with design, operation and maintenance of reclaimed water fire protection systems are presented below.

Types of Applications

The types of fire protection systems are either (1) outdoor systems or (2) indoor systems (see Fig. 20-6). Typically, fire hydrants are the principal appurtenances used in outdoor systems. Hydrants are installed directly on the reclaimed water distribution system. Sprinkler systems are used for indoor fire protection.

(a)

(b)

Figure 20-6

Examples of fire protection systems: (a) fire hydrant and (b) reclaimed water plumbing for indoor sprinkler system.

Outdoor System with Fire Hydrants

The two principal types of fire hydrants used are: (1) dry-barrel and (2) wet-barrel (AWWA, 1989). The dry-barrel hydrant is the only one which has a drain mechanism and is used commonly where freezing weather occurs. Wet-barrel hydrants are used commonly in temperate regions. Typical hydrants used for reclaimed water systems are discussed in Chap. 14 (see also Fig. 14-13). Fire hydrants are also used for flushing of the distribution system and can be used to access reclaimed water for dust control at construction sites.

Indoor Sprinkler System

Reclaimed water is rarely used in indoor sprinkler systems because (1) dual plumbing is necessary, (2) there is much greater risk of human exposure, and (3) little or no potable water is saved. The only opportunity that reclaimed water could be used for sprinkler systems is where potable water is not readily available. Situations where reclaimed water is used for sprinkler systems are: (1) at a water reclamation facility where access of general public is restricted and reclaimed water is readily available and (2) for special situations for commercial buildings. An example of the latter case is Livermore, California. The City of Livermore considered the use of reclaimed water for fire protection at a commercial development because the existing potable water distribution system did not have sufficient pressure and flow capacity to meet the fireflow requirements for the sprinkler system. A water reclamation plant is located nearby, which provides sufficient pressure and flow to meet the fire protection needs (Johnson and Crook, 1998). However, the city is not planning to add additional reclaimed water sprinkler systems, due mostly to cost and difficulties in acquiring permits from the California Department of Health Services. Livermore had about 50 fire hydrants available for use with reclaimed water in 2003. For the design of fire suppression systems, Nayyar (2000) may be consulted.

Water Quality Considerations

In the United States, reclaimed water use for fire protection is categorized typically as *unrestricted* nonpotable use (see Table 20-3). Changes in reclaimed water quality that occur in the distribution system pose operational and maintenance concerns for urban water reuse applications. To prevent reclaimed water from becoming stagnant and to maintain the water quality in the distribution system, the use of a looped distribution system and a periodic flushing program are important design and operational considerations. The management of water quality in reclaimed water distribution and storage systems is discussed in Chap. 14.

Implementation Issues

The use of reclaimed water for fire protection poses distinct challenges. Reclaimed water systems must be as reliable for fire protection as potable water systems. Because water for fire protection must be available at all times, the reclaimed water distribution system must be designed and designated as an uninterruptible supply, as prescribed by the insurance rating industry. The storage volume must be sufficient to meet the peak day and hourly demands (see Chap. 14). To meet fire flow requirements pipes having diameters of 150 mm (6 in.) and larger are required generally. The feasibility of providing fire protection with reclaimed water depends on the cost of the necessary infrastructure.

Redundancy and emergency power with an associated increase in cost will be required to achieve high reliability, thus making reclaimed water systems less economical than a potable water system (U.S. EPA, 2004). In St. Petersburg, Florida, reclaimed water is used for fire protection but not as the primary supply because it is considered to be an interruptible source (Crook, 2005).

Management Issues

Management issues include the maintenance of the reclaimed water distribution system, prevention of cross connections, and assurance of adequate flow and pressure. Accidental use of reclaimed water hydrants must be avoided. The reclaimed water hydrants may be modified so that regular tools cannot be used to operate the hydrants (see Fig. 14-13, Chap. 14).

Fire hydrants are flushed periodically to ensure proper functioning and to remove solids that have accumulated in the system. It is a good practice to flush the reclaimed water distribution network periodically because reclaimed water systems are more likely to foster the build up of biofilm and particulate matter as compared to potable water systems. Flushed reclaimed water can be discharged to the wastewater collection system, or with appropriate waste discharge permits, discharged to the storm water system or local waterways. Operation and maintenance issues for distribution facilities are discussed in Chap. 14.

20-5 TOILET AND URINAL FLUSHING

Water demand in high-rise commercial buildings is mostly for toilet and urinal flushing and cooling water for air conditioning; uses that do not require potable water quality. The use of reclaimed water for toilet and urinal flushing in commercial and other buildings can, therefore, reduce the potable water demand. The feasibility of using reclaimed water for toilet and urinal flushing depends primarily on the plumbing and related infrastructure costs. The use of reclaimed water for residential homes and buildings can be more challenging, due to the concerns about potential cross-connections and accidental exposure of reclaimed water, especially to small children.

Types of Applications

Reclaimed water toilet and urinal flushing systems may be installed in high-rise commercial and public buildings, residential buildings, and restrooms in public parks (see Fig. 20-7). Toilet and urinal flushing systems may be designed as part of a mixed urban water reuse plan with a centralized reclaimed water distribution system, or as part of a small scale, onsite treatment and reuse scheme. With respect to toilets and urinals, the same types of fixtures used with potable water can be used with reclaimed water. A simplified diagram of a toilet and urinal flushing system in a high-rise building with dual plumbing is shown on Fig. 20-8.

Water Quality Considerations

Reclaimed water used for toilet flushing must be odorless and colorless for aesthetic reasons, and highly disinfected for public health protection in case of accidental human contact. Water quality issues that should be considered when using reclaimed water for toilet and urinal flushing are presented in Table 20-7. In California, tertiary treated

(a)　　　　　　　　(b)

Figure 20-7
Examples of reclaimed water use for toilet and urinal flushing: (a) commercial buildings and (b) residential buildings.

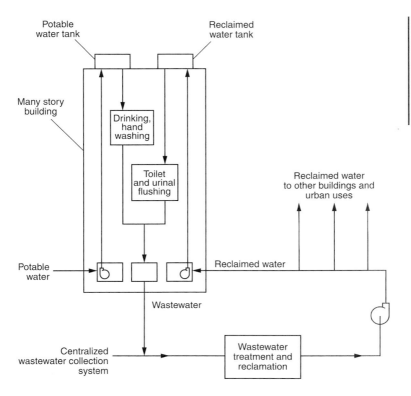

Figure 20-8
Simplified diagram of a toilet and urinal flushing system in a dual-plumbed high-rise building.

Table 20-7
Water quality issues when using reclaimed water for toilet and urinal flushing

Issue	Factors that must be considered	Control/remarks
Biofilm growth	The presence of residual organic matter in reclaimed water. With time, even resistant organic compounds can serve as a carbon source for microorganisms in biofilms.	Maintain circulation in distribution system. Periodic flushing with disinfectant. In some cases, it may be necessary to maintain a disinfectant residual.
Color	The presence of residual organic matter in reclaimed water including indole type compounds found in urine and feces and other organic dyes.	Color has been destroyed with ozone. Ozone-resistant reverse osmosis membranes are being used in Japan (see Chap. 8).
Corrosion, metallic	Because of the constituents present in reclaimed water, metallic corrosion can occur, especially in buried piping systems. Acid formed by microorganisms can also lead to metallic corrosion.	Use of plastic piping will control metallic corrosion. Where metallic elements are present, rust inhibitors can be used.
Fouling, particulate[a]	Total suspended, colloidal, and dissolved solids in reclaimed water. The presence of metallic corrosion products.	Nanofiltration or reverse osmosis can be used to reduce the dissolved solids. The addition of antifouling agents has been used to limit the degree of fouling.
Microbial quality	Water used in urban water reuse applications must meet Title 22 of California regulations or equivalent.	Maintain circulation in distribution system. Periodic flushing with disinfectant.
Odor	The presence of residual organic matter and sulfate at concentrations greater than 50 mg/L, can promote anaerobic conditions, resulting in formation of hydrogen sulfide and other odor causing compounds such as mercaptans.	Maintain circulation in distribution system. Periodic flushing with disinfectant. Avoid dead ends in the distribution system.
Public health	Health issues are not normally of concern as water treated to meet Title 22 requirements is deemed to be safe.	In some cases, it may be necessary to maintain a disinfectant residual.
Scaling (precipitation)	Alkalinity, pH, calcium, and phosphorus are all of concern. Other salts of limited solubility are also of concern. The scaling potential can be assessed using the Langelier saturation index or Ryzner stability index.	Nanofiltration can be used to remove calcium and bicarbonate concentrations to reduce scaling. The addition of an antiscalant has been used to limit the degree of scaling.

[a] Undesirable deposits that form within a closed loop system may be due to the formation of crystallized minerals.

reclaimed water is deemed safe for toilet flushing (see Chap. 4). Most other states that permit the use of reclaimed water for toilet and urinal flushing have adopted similar regulations or guidelines. In Japan, where toilet flushing is one of the most common water reuse applications, criteria for odor and color are included in the water quality requirements. A standard for the allowable pH range is relevant for the design of reclaimed water distribution and plumbing systems as it relates to the potential for scaling and corrosion (see Chap. 19). In the guidelines for New South Wales (NSW), Australia, virus and parasite concentrations are specified, as well as a limit for color. A comparison of water quality requirements in California, Japan, and NSW is presented in Table 20-8.

Color and odor are two major water quality issues for toilet and urinal flushing. Organic compounds (mostly humic substances) and inorganic compounds (such as iron) can cause coloration of reclaimed water. Oxidizing agents such as chlorine, ozone, and hydrogen peroxide can be used for the removal of color and odor. In one of the area-wide water reuse systems in Tokyo, reclaimed water was slightly colored and the color limit could not be met without additional treatment. To meet the color limit, an ozone treatment unit and an ozone-resistant membrane filtration system were added at the Shibaura Water Reclamation Plant (see Fig. 20-9).

When chemical oxidants are used for the control of odors and color and there is a high concentration of iron in the water, ferric hydroxide [$Fe(OH)_3$] will precipitate (Crittenden et al., 2005). Precipitated iron compounds may require removal by filtration before use to avoid staining of fixtures. Chlorine is added commonly to the reclaimed water to ensure a chlorine residual is maintained at the point of use.

High-pressure membranes such as nanofiltration and reverse osmosis can also be used to remove chemical constituents that cause color and odor, but low-pressure membranes such as microfiltration and ultrafiltration are ineffective in removing color and odor (see Chap. 8). In some cases, a blue dye is added to mask the remaining color of reclaimed water. The dye can also help in distinguishing reclaimed water from potable water (Zavoda, 2005).

Table 20-8 Comparison of water quality criteria for toilet flushing for California; Tokyo, Japan; and New South Wales (NSW), Australia

Parameter	California	Tokyo[a]	NSW[a]
Turbidity, NTU	2	2	2 (geometric mean)
Coliforms, No./100 mL	Total <2.2 (median), <23 (max)	Fecal not detected (median), 10 (max)	Total <2.5 (geometric mean), <25 (95% percentile)
pH	NS[b]	5.8–8.6	6.5–8.0 (7.0–7.5 desired)
Chlorine residual, mg/L	5[c]	0.4/0.1 (point of use)[d]	1 (after 30 min)
Other parameters		Odor, color	Virus, parasites, color

[a]Guidelines.
[b]NS = not specified.
[c]Not specified in storage and distribution system. Reclamation plant to maintain 5 mg/L at the overflow to the storage tank.
[d]Combined chlorine = 0.4 mg/L, free chlorine = 0.1 mg/L at the point of use.

Figure 20-9
Ozone-resistant membrane filtration system used at the Shibaura Water Reclamation Plant following ozone treatment for odor and color removal.

Implementation Issues

As discussed in Chap. 15, dual plumbing for potable and reclaimed water is necessary for the use of reclaimed water in buildings (see Fig. 15-1). Booster pumps are used to maintain a pressure and to pump the reclaimed water up to the highest floor in high-rise buildings. Hydropneumatic systems may also be used. For low-rise buildings, the pressure from the distribution main may be sufficient to distribute reclaimed water. Where dual plumbing systems are used, some type of backflow prevention device is required to be installed on the potable water service (see Chap. 15).

The plumbing fixtures used for potable water systems can be used with reclaimed water, although slightly brownish colored ceramic toilet and urinals are generally recommended because reclaimed water, unless highly treated, may be slightly colored as described above. Signage is required for public restrooms to indicate reclaimed water is being used for flushing (see Fig. 20-10). Signage and color coding for dual plumbing systems are discussed in detail in Chap. 15.

Figure 20-10
Signage indicating that reclaimed water is being used for flushing water at a public restroom.

Satellite and Decentralized Systems

Satellite and decentralized water reclamation and reuse systems are becoming increasingly popular, aided by advances in water reclamation technologies that produce a high quality effluent and have a small footprint (see Chaps. 12 and 13). Other types of satellite and decentralized systems utilizing greywater for toilet flushing are found more commonly in European countries and Japan (Asano et al., 1996; Maeda et al., 1996; Suzuki et al., 2002). For example, in Annecy, France a full-scale greywater recycling system for toilet flushing was installed to use reclaimed water in 40 apartments with approximately 120 residents (Lazarova et al., 2003). In Tokyo, a greywater treatment and recycling system is installed in a high-rise commercial building to produce reclaimed water for toilet flushing (see Fig. 20-11). In this system, wastewater (blackwater) from toilet flushing is not reused but is discharged to the wastewater collection system.

Large scale onsite water reclamation and reuse systems are also used commonly for toilet flushing in high population density metropolitan areas of Japan. In 1996, about 2100 buildings either had onsite water recycling systems or were connected to area-wide water recycling and reuse systems. The number of buildings where reclaimed water is used has increased steadily since then. For example, about 130 onsite water recycling systems were installed in 2002 (Gaulke, 2006; Yamagata et al., 2002).

Management Issues

Water quality monitoring is important to ensure that public health is protected. Generally, reclaimed water quality monitoring is conducted routinely at the water reclamation facility before reclaimed water is distributed. Typical monitoring requirements

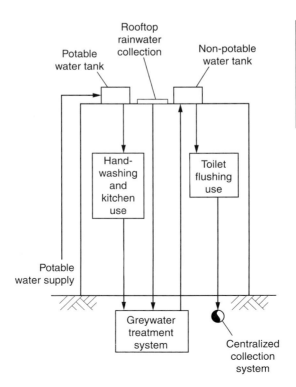

Figure 20-11
Greywater treatment and recycling system installed in a high-rise commercial building for toilet flushing.

Table 20-9

Reclaimed water quality monitoring requirements for unrestricted urban uses in selected states[a]

Parameter	Arizona	California	Florida	Hawaii	Texas
TSS	Case by case	NS[b]	Based on system capacity	Weekly	NS
Turbidity	Case by case	Continuous, following filtration	Continuous	Continuous	Twice/week
pH	Case by case	NS	Based on system capacity	NS	NS
Coliform	Case by case	At least once/day from disinfected effluent	Based on system capacity	Once/day	Twice/week, periodical monitoring in the distribution system may be necessary
Chlorine residual	Case by case	Continuous, at the overflow to the storage tank	Continuous	Continuous	NS
Other parameters	Case by case	NS	*Giardia* and *Cryptosporidium* (once in 2–5 yr based on system capacity), primary and secondary drinking water standards [for facilities >3800 m^3/d (>1 Mgal/d)], monthly to quarterly	BOD (weekly)	BOD or CBOD (twice/week)

[a]Adapted from U.S. EPA (2004).
[b]NS—Not specified.

of reclaimed water for unrestricted uses are reported in Table 20-9. Unfortunately, many reclaimed water purveyors do not have well-established monitoring and maintenance programs for regional water distribution systems (Daniel, 2005). With an inadequate monitoring program, undetected water quality changes can occur in the distribution system that may affect the intended use. Changes that can occur in water quality throughout the distribution system, based on limited sampling within the West Basin Municipal Water District, California, are reported in Table 20-10.

Odor and color problems are sometimes reported by the users of toilet/urinal facilities. Superchlorination (flushing of the system with above 10 mg/L of chlorine solution) is used most commonly when problems with biological fouling and odor are reported. As described previously, the reclaimed water treatment process may need to be modified if the color problem persists.

Table 20-10
Residual chlorine and pH of reclaimed water in distribution system, West Basin Municipal Water District, CA[a]

Location[b]	Estimated age of reclaimed water in pipe, h	Chlorine residual, mg/L	pH
A	<1	5.0	7.0
B	1.7	2.75	8.0
C	4.3	2.38	8.2
D	9.2	0.08	7.9
E	13.3	0.06	7.6
F	13.7	0.16	8.2
G	16.8	ND[c]	8.2
H	21.8	1.84	7.7
I	22.6	0.03	7.9
J	23.7	0.14	7.7
K	>24	0.02	7.8
L	>24	ND	7.7
M	>24	0.21	7.8
N	>24	2.26	8.0
O	>24	ND	7.8
P	>24	0.03	8.1
Q	>24	ND	8.2

[a]Adapted from Daniel (2005). The number of samples was limited and the data do not necessarily represent typical quality changes.
[b]A = Water reclamation plant. B to Q are sampling points along the reclaimed water distribution pipeline.
[c]ND = Not detected.

20-6 COMMERCIAL APPLICATONS

The primary commercial uses of reclaimed water are vehicle washing and commercial laundries. Because the water demand for these uses is relatively small, each project must be handled on a case-by-case basis. While customer satisfaction and the protection of public health are implicit goals, the compatibility of water quality with the application must also be evaluated. Where concerns about water quality are at issue, and where justified by the size of the installation and volume of water to be used, pilot testing should be considered.

Car and Other Vehicle Washing

Several opportunities exist for car and vehicle cleaning, including personal and fleet cars, cargo and freight trucks, trains, and planes. Typical car washing and train cleaning facilities are shown on Fig. 20-12. Due to relatively high dissolved solids in the reclaimed

Figure 20-12
Examples of vehicle washing facilities and train cleaning apparatus: (a) car wash in Marin, CA; (b) car wash in Largo, FL; (c) bus washing facility in Orlando, FL; and (d) train washing apparatus in Tokyo.

water, many car washing facilities that utilize reclaimed water use a separate source of water for the final rinse. Either potable or reclaimed water, treated by reverse osmosis or ion exchange, is used to eliminate spotting and to improve air drying of the washed vehicle.

Laundries

Clothes washing at commercial laundry facilities represents a relatively small use and should, therefore, be used as part of a larger reclaimed water system. An example of a commercial laundry facility is in Marin County, California, where reclaimed water is used in commercial-size washing machines (see Fig. 20-13). Pilot testing was conducted

(a) (b)

Figure 20-13
A laundry facility in Marin County, CA, utilizing reclaimed water with commercial size washing machines: (a) facility building with a notice of reclaimed water use on the door and (b) laundry machine.

prior to the implementation to ensure that laundry could operate properly with reclaimed water, and the test results were satisfactory.

20-7 PUBLIC WATER FEATURES

Public water features include fountains, ponds, and small pools in garden areas. While these uses of reclaimed water do not contribute significantly to the demand for reclaimed water, they can offer a unique educational experience through explanatory signage. By using reclaimed water, water features may be kept operating during drought years without consuming potable water. Water quality should be managed carefully, however, to prevent nuisance conditions such as odors, mosquitoes, and algae.

Fountains and Waterfalls

Reclaimed water can be used for ornamental purposes in urban area, such as fountains and waterfalls. In San Diego, California, an industrial park includes water features supplied by reclaimed water adjacent to commercial buildings (see Fig. 20-14a).

Reflecting Pools

Reflecting pools are aesthetically pleasing in nearly any setting. An example of reclaimed water reflecting pool in a Japanese garden located at the Tillman Water Reclamation Plant in Los Angeles, California, is shown on Fig. 20-14b.

Figure 20-14
Reclaimed water used for ornamental purposes in urban areas: (a) San Diego, CA, commercial complex, (b) Japanese Garden at the Tillman Water Reclamation Plant, Los Angeles, CA (Coordinates: 34.183 N, 118.480 W), (c) recreational lake at Santee, CA. (Coordinates: 32.850 N, 117.006 W), and (d) remains of a moat filled with reclaimed water surrounding an old castle, Osaka, Japan (Coordinates: 34.687 N, 135.526 E.)

Ponds and Lakes in Public Parks

Where reclaimed water is used for landscape irrigation in public parks, it can also be used for the development of lakes and ponds. Signage, as noted previously, must be in place to indicate that reclaimed water is being used for these water features. An example of a recreational lake in Santee, California, is shown on Fig. 20-14c. A moat filled with reclaimed water in Osaka, Japan, is shown on Fig. 20-14d. The use of reclaimed water for environmental and recreational uses is further discussed in Chap. 21.

20-8 ROAD CARE AND MAINTENANCE

Road care and maintenance does not use a significant quantity of reclaimed water; however, there are specific quality requirements that must be taken into consideration. Reclaimed water that is odorous, corrosive, or colored will not be acceptable under

(a) (b)

Figure 20-15
Reclaimed water use for street cleaning and dust control: (a) reclaimed water filling station for street cleaning vehicle and (b) water truck spraying reclaimed water.

normal circumstances. Opportunities for water reuse in road care and maintenance applications include (1) street cleaning, and (2) snow melting.

Dust Control and Street Cleaning

Reclaimed water for dust control and street cleaning is transferred to a truck at a fill station typically located at the water reclamation plant (see Fig. 20-15), or from a reclaimed water hydrant. Generally, fill stations are located in areas not accessible to the public.

Snow Melting

Reclaimed water is utilized for snow melting in some northern regions in Japan (see Fig. 20-16). The snow melting process utilizes the thermal energy of reclaimed water. The capacity to melt snow must be calculated by thermal balance and the volume of reclaimed water to be used (Funamizu et al., 2001). Two methods are used for snow melting:

- Snow is collected by trucks and transported to centralized snow melting tanks in the vicinity of water reclamation plant to fully utilize the heat generated at the plant, or
- A covered channel is constructed under the street. Reclaimed water flows in the channel, and residents dump snow through "throw-in" openings on the side of the road. Debris in the used reclaimed water (mixed with melted snow water) is removed prior to discharge to waterways.

For snow melting applications in Japan, effluent from secondary treatment plants is considered adequate because the risk of human exposure to reclaimed water in the snow melting channel is low (JSWA, 2005).

(a) (b)

Figure 20-16
Snow melting facility in Japan: (a) snow is dumped into a reclaimed water snowmelting facility and (b) inside view of the channel containing reclaimed water into which the snow is discharged. (Courtesy of City of Sapporo, Japan.)

PROBLEMS AND DISCUSSION TOPICS

20-1 Summarize the typical water conservation measures used in your area. Discuss how water conservation and water reuse has been incorporated into the urban water management program.

20-2 Obtain wastewater effluent quality data from a local wastewater treatment plant. If the water is to be reclaimed for use in an air conditioning system for a commercial building, what additional treatment would be necessary? Specify important water quality parameters for public health protection and system operation.

20-3 Cooling towers for air conditioning systems can be open-circuit or close-circuit. When reclaimed water is used for cooling towers, discuss the advantages and disadvantages of each system.

20-4 Discuss the issues in using reclaimed water for toilet flushing in homes. Summarize precautions to be taken during (a) design, (b) operation and maintenance, and (c) user education.

20-5 Reclaimed water is used for toilet and urinal flushing in a commercial building with an onsite treatment and reuse system. Customers noticed that the color of the flush water was slightly yellow. Is there a health issue? What can be done to resolve the color problem? Discuss the advantages and disadvantages of the various options.

20-6 Referring to Table 20-2, expand on the discussion of urban nonirrigation water reuse by adding examples from two other regions, such as Africa, the Middle East, the Mediterranean, and Asia.

20-7 Aesthetic parameters for reclaimed water, such as appearance and odor are identified in Table 20-4. How can these be quantified for regulatory purposes for water reuse?

20-8 What are the underlying statistical assumptions for each of the different coliform criteria identified in Table 20-8 for California, Tokyo, and New South Wales? Which of these criteria are more stringent? What is the meaning and limitation of the criteria "not detected"?

REFERENCES

AWWA (1989) *Installation, Field Testing, and Maintenance of Fire Hydrants*, 3rd ed., AWWA M17, American Water Works Association, Denver, CO.

AWWA (1994) *Dual Water Systems*, AWWA M24, American Water Works Association, Denver, CO.

Asano, T., M. Maeda, and M. Takaki (1996) "Wastewater Reclamation and Reuse in Japan: Overview and Implementation Examples," *Water Sci. Technol.*, **34**, 11, 219–226.

Crittenden, J. C., R. R. Trussell, D. W. Hand, K. J. Howe, and G. Tchobanoglous (2004) *Water Treatment: Principles and Design*, 2nd ed., John Wiley & Sons, Hoboken, NJ.

Crook, J. (2005) "St. Petersburg, Florida, Dual Water System: A Case Study," in National Research Council, *Water Conservation, Reuse, and Recycling: Proceedings of an Iranian-American Workshop*, National Academy Press, Washington, DC.

Daniel, U. (2005) "Reclaimed Water Distribution Maintenance," in *Proceedings, 20th WateReuse Symposium, Denver, CO.*, WateReuse Association, Alexandria, VA.

Funamizu, N., M. Iida, Y. Sakakura, and T. Takakuwa (2001) "Reuse of Heat Energy in Wastewater: Implementation Examples in Japan," *Water Sci. Technol.*, **43**, 10, 277–285.

Gagliardo, P., B. Pearce, and K. Mortensen (2005) "Recycled Water Use in Cooling Towers: Issues and Solutions," in *Proceedings, WateReuse Association California Section Conference, San Diego, CA*, WateReuse Association, Alexandria, VA.

Gaulke, L. S. (2006) "Joukasou: On-Site Wastewater Treatment and Reuses in Japan," in *Proceedings, the Institute of Civil Engineers—Water Management*, **159**, WMZ, 103–109.

Hutson, S. S., N. L. Barber, J. F. Kenny, D. S. Lumia, and M. A. Maupin (2004) *Estimated Use of Water in the United States in 2000*, Circular 1268, U.S. Geological Survey, Reston, VA.

IAPMO (2003) *Uniform Plumbing Code and Uniform Mechanical Code*, International Association for Plumbing and Mechanical Officials (IAPMO), Ontario, CA.

IRWD (2003) *Pipelines: The Newsletter of Irvine Ranch Water District*, Public Affairs Dept., Irvine Ranch Water District, **1**, 1–4.

Johnson, L. J., and J. Crook (1998) "Use of Reclaimed Water in Buildings for Fire Suppression," in *Proceedings, Water Environment Federation 71st Annual Conference and Exposition*, Orlando, FL.

JSWA (Japan Sewage Works Association) (2005) *Sewage Works in Japan 2005: Wastewater Reuse*, Japan Sewage Works Association, Tokyo.

Lazarova, V., S. Hills, and R. Birks (2003) "Using Recycled Water for Non-potable, Urban Uses: a Review with Particular Reference to Toilet Flushing," *Water Sci. Technol.: Water Supply*, **3**, 4, 69–77.

Maeda, M., K. Nakada, K. Kawamoto, and M. Ikeda (1996) "Area-Wide Use of Reclaimed Water in Tokyo, Japan," *Water Sci. Technol.*, **33**, 10–11, 51–57.

MLIT (2005) *Manual on Water Quality for Reuse of Treated Municipal Wastewater*, Japanese Ministry of Land, Infrastructure, and Transportation, Tokyo.

Nayyar, M. L. (2000) *Piping Handbook,* 7th ed., McGraw-Hill, New York.

Okun, D. A. (1997) "Distributing Reclaimed Water Through Dual Systems," *J. AWWA*, **89**, 11, 52–64.

Radcliffe, J. (2004) *Water Reuse in Australia*, Australian Academy of Technological Science and Engineering (ATSE), Ian McLennan House, Victoria, Australia.

Sala, L. (2003) *Water Reuse in Costa Brava*, accessed online: http://www.med-reunet.com/.

Solley, W. B., R. R. Perce, and H. A. Perlman (1993) *Estimated Use of Water in the United States in 1990*, U.S. Geological Survey, Circular 1008, U.S. Department of the Interior, U.S. Geological Survey, Denver, CO.

Solley, W. B., R. R. Pierce, and H. A. Perlman (1998) *Estimated Use of Water in the United States*, U.S. Geological Survey Circular 1200, U.S. Department of the Interior, U.S. Geological Survey, Denver, CO.

State of California (2003) *2002 Statewide Recycled Water Survey*, California State Water Resources Control Board, Office of Water Recycling, Sacramento, CA.

State of Florida (2005) *2004 Reuse Inventory*, Division of Water Resources Management, Florida Department of Environmental Protection, Tallahassee, FL.

Suzuki, Y., M. Ogoshi, H. Yamagata, M. Ozaki, and T. Asano (2002) "Large-Area and On-Site Water Reuse in Japan," in *Proceedings, International Seminar on Sustainable Development of Water Resource Pluralizing Technology*, Korean National Institute of Environmental Research, March 27–29.

U.S. EPA (1992) Manual: *Guidelines for Water Reuse*, EPA/625/R-92/004, Office of Water, U.S. Environmental Protection Agency, Washington, DC.

U.S. EPA (2004) *Guidelines for Water Reuse*, EPA/625/R-04/108, U.S. Environmental Protection Agency and U.S. Agency for International Development (U.S. AID), Washington, DC.

Vickers, A. (2001) *Handbook of Water Use and Conservation*, WaterPlow Press, Amherst, MA.

West, W. (2005) "Innovative Approaches and Sustainable Efficiencies for Cooling Towers Using Reclaimed Water," in *Proceedings 20th Annual WateReuse Symposium, Denver, CO.*, WateReuse Association, Alexandria, VA.

Yamagata, H., M. Ogoshi, Y. Suzuki, M. Ozaki, and T. Asano (2002) "On-Site Insight into Reuse in Japan," *Water 21*, **7**, 26–28.

Young, R. E., K. A. Thompson, R. R. McVicker, R. A. Diamond, M. B. Gingras, D. Ferguson, J. Johannessen, G. K. Herr, J. J. Parsons, V. Seyde, E. Akiyoshi, J. Hyde, C. Kinner, and L. Oldewage (1998) "Irvine Ranch Water District's Reuse Today Meets Tomorrow's Conservation Needs," 941–1036, in T. Asano (ed.), *Wastewater Reclamation and Reuse*, Water Quality Management Library, **10**, CRC Press, Boca Raton, FL.

Zavoda, M. (2005) "The Soloaire, a High Rise Residential Reuse Case Study," *Proceedings 20th Annual WateReuse Symposium, Denver, CO.*, WateReuse Association, Alexandria, VA.

21 Environmental and Recreational Uses of Reclaimed Water

WORKING TERMINOLOGY   1204

21-1   OVERVIEW OF ENVIRONMENTAL AND RECREATIONAL USES
OF RECLAIMED WATER   1205
Types of Environmental and Recreational Uses   1206
Important Factors Influencing Environmental and Recreational Uses
of Reclaimed Water   1207

21-2   WETLANDS   1210
Types of Wetlands   1210
Development of Wetlands with Reclaimed Water   1213
Water Quality Considerations   1216
Operations and Maintenance   1216

21-3   STREAM FLOW AUGMENTATION   1222
Aquatic and Riparian Habitat Enhancement with Reclaimed Water   1222
Recreational Uses of Streams Augmented with Reclaimed Water   1224
Reclaimed Water Quality Requirements   1224
Stream Flow Requirements   1226
Operation and Maintenance   1226

21-4   PONDS AND LAKES   1228
Water Quality Requirements   1228
Operation and Maintenance   1230
Other Considerations   1230

21-5   OTHER USES   1231
Snowmaking   1231
Animal Viewing Parks   1231

21-6   CASE STUDY: ARCATA, CALIFORNIA   1231
Setting   1232
Water Management Issues   1232
Implementation   1232
Lessons Learned   1233

21-7   CASE STUDY: SAN LUIS OBISPO, CALIFORNIA   1234
Setting   1234
Water Management Issues   1235
Implementation   1235
Lessons Learned   1238

21-8   CASE STUDY: SANTEE LAKES, SAN DIEGO, CALIFORNIA   1238
Setting   *1239*
Water Management Issues   *1239*
Implementation   *1239*
Lessons Learned   *1241*

PROBLEMS AND DISCUSSION TOPICS   1242

REFERENCES   1242

WORKING TERMINOLOGY

Term	Definition
Aquatic life	Fish, animal, plants, and other organisms occurring in a waterbody such as wetlands, rivers and streams, lakes and ponds.
Base flow	The portion of stream flow from seepage of groundwater. The base flow comprises the primary source of water during dry weather.
Best management practices (BMPs)	Structural or nonstructural approaches to address point and nonpoint source pollution, and other components of aquatic environment to control, prevent, remove, or reduce pollution.
Clean Water Act (CWA)	Federal regulation that directs the U.S. EPA to assist states in implementing groundwater, surface water, and wetland protection strategies (see Chap. 2).
Constructed wetland	Wetlands intentionally developed in nonwetland areas to duplicate the processes occurring in natural wetlands, but generally for the purpose of improving water quality and with more control.
Emergent plants	Rooted aquatic plants that extend above the water surface.
Endangered Species Act (ESA)	Federal regulation that seeks to conserve threatened and endangered species of fish, wildlife, and plants. Federal agencies must carry out programs for listed species and take action to ensure that projects they authorize, fund, or undertake are not likely to jeopardize the continued existence of threatened and endangered species.
Eutrophication	The aging of lakes and other water bodies resulting from the addition of nutrients that stimulate excessive aquatic plant growth which accumulates and from the addition of debris and sediment.
Floating aquatic plants	Aquatic plants that float on and extend above the water surface.
Impoundment	A surface water or reclaimed water storage facility that is sometimes used for recreational and aesthetic functions.
Minimum flow requirements (or in-stream flow requirements)	The amount of water flow required to sustain, rehabilitate, or restore the functions of a stream including the protection and enhancement of fish and wildlife habitat, recreation activities, navigation, hydropower generation, and water quality control.
Mitigation wetlands	Wetlands established (e.g., constructed or restored) for the purpose of mitigating (replacing) the loss of natural wetlands resulting from development activity.

National Pollutant Discharge Elimination System (NPDES)	A permit program under the CWA to control water pollution by regulating point sources that discharge pollutants into waters of the state.
Natural wetland	Those areas that are inundated or saturated by surface water or groundwater at a frequency and duration sufficient to support a prevalence of vegetation (hydrophytes) typically adapted for life in saturated soil conditions (hydric soils). Wetlands generally include swamps, marshes, bogs, and similar areas.
Riparian habitat	Plants and animals occurring in the thin strip area adjacent to a river or stream.
Vector	An organism that itself does not cause disease, but which spreads infection by harboring disease-causing organisms and transferring them from one host to another.
Waters of the state	Navigable waters including streams, rivers, lakes, and wetlands, as defined in the Clean Water Act (CWA).
Wetlands restoration	Restoring the processes that occur in natural wetlands where they have been altered or destroyed by activities such as draining and filling.

Continued population growth, urbanization, and increased water demands have often caused the reduction of stream flows and waters sustaining wetlands and resulted in deterioration of the water environment, both in terms of water quality and quantity. In recent years, the requirement to maintain minimum water flows in rivers, streams, and wetlands for environmental purposes is becoming one of the major challenges in water resources management. Reclaimed water can be used to restore dry and neglected urban streams as well as augment surface water for various beneficial purposes. The environmental and recreational uses of reclaimed water discussed in this chapter are those projects that are planned and implemented with a clear intention to preserve and enhance the aquatic environment and recreational opportunities.

Topics discussed in this chapter include: (1) types of environmental and recreational water reuse; (2) general considerations for the use of reclaimed water for environmental and recreational purposes; and (3) specific considerations for wetlands, stream flow augmentation, lakes and ponds, and other types of environmental and recreational water reuse applications. Three case studies are also presented to illustrate how some of the environmental and recreational water reuse projects are planned and implemented.

21-1 OVERVIEW OF ENVIRONMENTAL AND RECREATIONAL USES OF RECLAIMED WATER

In recent years, many wastewater treatment plants have been upgraded or expanded to meet increasingly stringent waste discharge requirements. The resulting high-quality effluents being discharged to receiving water are, in fact, contributing to the preservation and enhancement of the aquatic environment in many locations.

Types of Environmental and Recreational Uses

Many environmental uses of reclaimed water have originated historically from the discharge of treated wastewater. Later, when the wastewater systems were upgraded to meet requirements for higher levels of constituent removal, the secondary benefit of releasing higher quality water for environmental enhancement also gained recognition. In many cases, the distinction between discharge of highly treated wastewater (disposal) and planned environmental use of reclaimed water (reuse) is difficult to differentiate.

Recreational uses and water features are generally planned and implemented more deliberately than most environmental uses. Recreational uses often originate from more practical purposes, such as demonstration projects where the benefits of reusing high-quality reclaimed water and reclaimed water storage are demonstrated. Even though some states allow highly treated reclaimed water to be used for impoundments where full body contact can occur, full body contact such as swimming is seldom allowed. The types of environmental and recreational uses are summarized in Table 21-1.

Table 21-1
Types of environmental and recreational water reuse

Type of application	Remarks	
	Environmental	Recreational
Wetlands (natural or constructed)	Reclaimed water may be used to support the development or restoration of wetlands for wildlife habitat and water quality improvement. Wetlands are known to attract many species of birds. In some cases, wetlands may be used for mitigation. Restoration or enhancement of existing or disturbed wetlands is also an environmental benefit.	Wetlands with public access provide opportunities for visitors to view wildlife. School groups may use wetland facilities for outdoor and environmental education programs. Key features include boardwalks, interpretive center, and descriptive signage. For bird watching, spotting scopes and blinds are often used.
Stream flow augmentation	The principal environmental objective of using reclaimed water to augment stream flow is to maintain minimum flows required to protect fish habitat. Smaller streams/creeks may depend on reclaimed water to keep from desiccating during the dry season.	Streams, especially in urban areas, are used for riverside walkways, in some cases for wading and fishing. Swimming and wading in reclaimed water streams are usually restricted to locations where only the highest quality of reclaimed water is used.
Lakes and ponds	Lakes and ponds are used by a number of fish and bird species; however, lakes and ponds using reclaimed water are typically not developed solely for fish and wildlife habitat.	Boating and fishing are the principal recreational uses. The use of reclaimed water for impoundments where people swim is allowed by some states, but it is not common. Reflecting pools are used as aesthetic features in many areas.
Snowmaking	In cold regions, snowmaking is sometimes used as a form of winter water storage. Snowmelt in spring contributes water to streams and other surface water bodies.	Artificial snow may be used to supplement natural snow for winter sports, mainly skiing and snowboarding.

Important Factors Influencing Environmental and Recreational Uses of Reclaimed Water

Compared to other water reuse applications, when reclaimed water is used for environmental purposes, more attention is required for ecosystem protection while recognizing that public health protection is still of primary importance. Reclaimed water quality requirements depend on the type of water body to which reclaimed water is released, and the likelihood of human exposure to the reclaimed water. Regulatory considerations are of paramount importance when environmental and recreational water reuse applications are considered. Important factors influencing environmental and recreational uses of reclaimed water are summarized in Table 21-2; some of these factors are discussed in greater detail in subsequent sections.

Key components of the federal regulations that affect water reuse are shown in Table 21-3. In some states, regulations and guidelines for water reuse also include requirements for environmental and recreational uses. Recreational uses often involve human contact with the reclaimed water and are typically subject to stringent water quality requirements. Reclaimed water regulations for environmental and recreational uses in selected states are summarized in Table 21-4. The use of reclaimed water for surface water augmentation to achieve indirect potable reuse is considered in Chap. 23.

Table 21-2
Factors influencing the use of reclaimed water for environmental and recreational purposes

Factor	Remarks
Public health concerns	The degree of public access and the potential for human contact with reclaimed water are important considerations for environmental and recreational reuse applications. The primary health concern is related to disease resulting from incidental ingestion of pathogenic microorganisms (see Chap. 11).
Ecosystem protection	For environmental uses of reclaimed water, ecosystem protection is a primary consideration. Thus, several regulations and rules pertinent to surface waters apply to the use of reclaimed water for environmental purposes. Depending on the application, some constituents commonly found in reclaimed water, such as residual chlorine, ammonia, and trace constituents, may need to be removed due to fish and wildlife toxicity concerns.
Water quality issues	Reclaimed water treatment and quality requirements for environmental and recreational uses vary with the type of water body to which reclaimed water is released. Discharges to some water bodies may need to comply with surface water quality standards.
Regulatory considerations	The Clean Water Act (CWA) and Endangered Species Act (ESA) are the most important federal regulations affecting the use of reclaimed water for environmental and recreational purposes. Other important regulatory requirements include the environmental impact assessment required under the National Environmental Policy Act (NEPA) and the equivalent state level regulations.
Seasonal reclaimed water availability	The major uses of reclaimed water in the United States have been agricultural and landscape irrigation, which are seasonal, and the demand fluctuates throughout a year. In many regions, summer is dry and reclaimed water demand is high for irrigation purposes, which coincides with periods when stream flows decline. In water-short areas, reclaimed water may need to be diverted for environmental uses to support an established habitat, even at the expense of reducing the water for irrigation or other applications.

Table 21-3

Components of federal regulations affecting water reuse on environmental and recreational applications[a]

Regulation	Remarks
Clean Water Act (CWA)	
National Pollutant Discharge Elimination System (NPDES) permits	Unlike many water reuse applications from which reclaimed water is not returned to receiving waters, environmental uses of reclaimed water are subject to the NPDES requirements because reclaimed water is released to natural waterways. Therefore, a water reclamation plant utilizing its reclaimed water for environmental and recreational uses is required to obtain an NPDES permit. The point where regulations are applied depends on the types of environmental/recreational use. For example, if reclaimed water is released to a constructed wetland, the discharge from the wetland to a surface water body is regulated, but not the reclaimed water discharged into the wetland.
Section 404 Permits	In Section 404 of the CWA, discharge of dredged or fill material to navigable waters including wetlands (defined as waters of the state) is regulated and permits are issued through U.S. Army Corps of Engineers. Any construction projects affecting rivers, streams, and wetlands in any state must obtain the Section 404 permits. Many environmental and recreational water reuse projects are subject to Section 404 of CWA. Related to the Section 404 permits, projects that may result in a discharge of material to waters of the state may be required to obtain Section 401 water quality certifications, which are generally authorized by the state environmental agencies on behalf of the U.S. EPA.
Total maximum daily load (TMDL)	Total maximum daily load is an approach initiated in the CWA. The TMDL approach is intended for the reduction of specific pollutants from all contributing point and nonpoint sources to maintain the water quality standard of a waterbody. Allocation of allowable pollutant discharge is determined for all contributing point and nonpoint sources.
Endangered Species Act (ESA)	
Safe Harbor Agreements	Safe Harbor Agreements are voluntary agreements between the U.S. Fish and Wildlife Service (FWS) or National Oceanic and Atmospheric Administration (NOAA) Fisheries and cooperating non-Federal landowners to promote voluntary management for wildlife species on the FWS endangered species list. Under the agreements, the FWS and the landowners determine the baseline condition for specific wildlife species, and landowners make efforts to preserve and enhance habitat within the property for an agreed period of time; in return, additional future restrictions are not imposed on the landowners. At the end of the term, the landowners are allowed to use the property in any otherwise legal manner to the extent the baseline conditions is preserved.
National Environmental Policy Act (NEPA)	
Environmental Impact Statement (EIS)	Under NEPA, federal agencies are required to assess the environmental impact due to proposed actions. Reasonable alternatives and mitigation measures should be considered in the statement.

[a] Also see Table 16-1 for federal regulations and rules relevant to water reuse projects.

Table 21-4
Water quality and treatment requirements in selected states for environmental and recreational uses[a]

State	California	Florida	Texas	Washington
Environmental Uses				
Treatment		Secondary with nitrification (treatment wetlands)		Oxidized[b] and disinfected
BOD, mg/L		20/5[c]		20[d]
TSS, mg/L		20/5[c]		20[d]
Nitrogen, mg/L		2 (ammonia)/3 (total)[e]		3 (total)
Phosphorous, mg/L		1 (total)		1 (total)
Coliform, MPN/100 mL		NS		23 (avg) 240 (max)
Unrestricted access recreational uses				
Treatment	Oxidized, coagulated, filtered and disinfected	Secondary treatment, filtration and high-level disinfection	NS	Oxidized, coagulated, filtered and disinfected
BOD, mg/L	NS	20[c]	5	30
TSS, mg/L	NS	5.0	NS	30
Turbidity, NTU	2 (avg) 5 (max)	NS	3	2 (avg) 5 (max)
Coliform, MPN/100 mL	Total 2.2 (med)[f] 23 (max in 30 d)	Fecal 75% of samples below detection 25 (max)	Fecal 20 (avg) 75 (max)	Total 2.2 (avg) 23 (max)
Restricted access recreational uses				
Treatment	Oxidized and disinfected	Secondary treatment, filtration and high-level disinfection	NS	Oxidized and disinfected
BOD, mg/L	NS	20[c]	20	30
TSS, mg/L	NS	5	NS	30
Turbidity, NTU	NS	NS	3	2 (avg) 5 (max)
Coliform, MPN/100 mL	Total 23 (med)[f] 240 (max in 30 d)	Fecal 75% of samples below detection 25 (max)	Fecal 200 (avg) 800 (max)	Total 23 (avg) 240 (max)

[a]Adapted from U.S. EPA (2004).
[b]*Oxidized wastewater* is wastewater that is treated to oxidize and stabilize organic compounds and contains dissolved oxygen. The term oxidized wastewater is used to avoid specification of treatment processes.
[c]Requirements, in CBOD. Requirements for wetlands are shown as (treatment wetlands)/(receiving wetlands). Treatment wetlands are considered as part of a treatment process flow system. Receiving wetlands are not considered as a treatment process.
[d]Annual average.
[e]Total nitrogen limit for receiving wetlands only.
[f]Seven-day median.
NS—Not specified.

Water reuse can contribute directly or indirectly to environmental enhancement and preservation of water bodies. Environmental and recreational uses include use of reclaimed water in wetlands, streams, and lakes and ponds. Reclaimed water may be used for nonpotable purposes, thereby reducing the potable water demand. The decision to use reclaimed water to augment existing stream flows, or for other non-potable purposes, is site specific and depends on local conditions.

Planning and implementation of environmental and recreational water reuse requires slightly different approaches than most other water reuse applications. Environmental and recreational uses of reclaimed water are closely related to discharge of treated wastewater into a receiving waterbody and may not be viewed as reuse. Therefore, the permitting process may be complex. For various reasons, abandoned or degraded sites, such as historic wetland sites with altered hydrology, abandoned mining sites, and dry creek beds are used often for the development of environmental or recreational water reuse projects. Considerations for the planning and implementation of water reuse for environmental and recreational purposes are summarized in Table 21-5.

21-2 WETLANDS

Wetlands are defined by the U.S. Army Corp of Engineers and U.S. EPA as areas that are inundated or saturated by surface water or groundwater at a frequency and duration sufficient to support, and that under normal circumstances do support a prevalence of vegetation and a diverse ecological community adapted for life in saturated soil conditions. Wetlands generally include swamps, marshes, bogs, and similar areas found at the interface between aquatic and terrestrial ecosystems. Wetlands also fulfill a number of important functions such as flood control, recharging groundwater, improving water quality as water passes through wetland systems, and providing habitat for wildlife, including numerous threatened plant and animal species (Mitsch and Gosselink, 2000). Unfortunately, many natural wetlands have been drained or altered for agriculture and other human interests. The augmentation of natural and artificial wetlands with reclaimed water for the development of habitat and recreational purposes is considered in this section.

Types of Wetlands

Many classifications, including those based on historical precedent, hydrological conditions, and vegetation characteristics, are used for describing wetland systems. In determining whether to support or augment the flow to wetlands, the nature of the wetland is an important consideration as the introduction of reclaimed water may induce ecological changes. Therefore, a naturally existing wetland system may have limitations that affect the application of reclaimed water if the natural ecology and biodiversity of the system are to be preserved. Whereas, a constructed, or artificial, wetland system designed for use with reclaimed water, and subsequently providing habitat and other ecosystem benefits, will offer more flexibility in terms of water quality requirements and acceptable hydraulic loading rates. Other wetlands types include natural wetlands that

Table 21-5
Checklist for planning and implementation of water reuse for environmental and recreational purposes

Assessment of existing conditions (pre-reuse)

Point of wastewater discharge
Discharge flowrate
Quality of discharged effluent
Status of aquatic and riparian habitat
Identification of endangered species

Identification of planning objectives and conceptual reuse alternatives

Purposes of reuse
Point(s) of reclaimed water discharge
Reclaimed water flowrate and compatibility with seasonal needs of the habitat
Reclaimed water flowrate for recreational purposes
Infrastructure needs; modification or creation of aquatic environment for the specific purpose
Recreational facilities to be constructed
Mitigation measures

Preparation of planning documents necessary to launch the project

Identification of water reuse alternatives
Identification of reclaimed water quality requirements
Identification of federal and state regulations, guidelines
Applications for funding
Preparation of water management and water reuse master plan

Compliance to regulatory requirements

EIS/EIR
NPDES/waste discharge permits
Section 404 Permits
Habitat monitoring/biological opinions
Water reuse regulations/guidelines
Other requirements

Identification of specific water reclamation process

Additional treatment needs
Disinfection/dechlorination

Identification of operational and management requirements

Public health protection measures
Water quality monitoring
Habitat monitoring

Note: See Chap. 25 for more detailed discussion of water reuse planning.

have experienced modifications in their hydrology, typically as a result of human activity, and may be classified as *hydrologically altered wetlands*. The inundation of altered and natural wetlands with reclaimed water requires additional considerations to ensure that the area is not adversely affected from an environmental standpoint.

Constructed Wetlands

Constructed wetlands are artificial wetlands, designed to utilize natural aquatic plants and organisms to improve water quality, retain stormwater for flood control during heavy rain events, and provide wildlife habitat. Treatment of stormwater and wastewater occurs in constructed wetlands by a number of processes. A constructed wetland can also serve as habitat for wildlife, and potentially as a recreational site if it is designed to maintain its principal functions while safeguarding public health. The major types of constructed wetlands are:

- *Free-water-surface (FWS) constructed wetlands.* Wetland systems with open water areas containing submerged, floating, and emergent vegetation.
- *Subsurface-flow (SF) constructed wetlands.* Wetland systems composed of bed of gravel or other granular support packing containing emergent plants. The water to be treated flows through the packing and plant roots during operation.
- *Floating aquatic plant systems.* Engineered wetland system consisting of a channel with floating plants with high surface area roots of varying lengths.
- *Combination systems.* Various arrangements of the systems described above.

Because constructed wetlands are engineered systems, the earthwork for these systems includes the construction of berms to contain the wetland area and an impermeable clay or plastic liner to prevent water exchange with groundwater. Design considerations such as influent and effluent structures, grading, loading rates, and vegetation as well as guidelines for operation and maintenance may be found in Crites et al. (2006) and Kadlec and Knight (1996). Flow structures should be included to facilitate the transfer of water between wetland cells, as may be required to meet water quality limits, as discussed later. When constructed wetlands are located in areas that do not have existing natural wetlands, importing of seed material, such as vegetation, biologically active sediments, fish, and other species may be necessary for establishment of the ecosystem. Examples of constructed wetlands are shown on Fig. 21-1.

Natural Wetlands

Natural wetlands pre-exist in natural environments and are not manmade or significantly altered. The use of reclaimed water in natural wetlands is not practiced commonly because of concerns related to the altering of a natural ecosystem. Specific water quality requirements are applied where reclaimed water is used for augmenting the flow in natural wetlands. For example, the following limits are specified in the Florida Administrative Code (F.A.C. Chap. 62-611): BOD and TSS of ≤ 5 mg/L, total nitrogen ≤ 3 mg/L, and total phosphorus ≤ 1 mg/L. In addition, limits are also specified in F.A.C. Chap. 62-611 for hydraulic loading rates, constituent mass loading rates, and maintaining the natural hydroperiod of the system.

(a) (b)

Figure 21-1
Examples of constructed wetlands using reclaimed water for environmental enhancement in (a) Empuriabrave constructed wetland, Costa Brava, Spain (Courtesy of L. Sala) (Coordinates: 42.243 N, 3.103 E) and (b) Lakeland wetland treatment system, Lakeland, FL (Coordinates: 27.910 N, 81.950 W, view at altitude 7 km).

Development of Wetlands with Reclaimed Water

Hydrologically Altered Wetlands

When the hydrology of a natural wetland system is impacted due to human activity, typically by lowering of the water table by groundwater pumping or artificial drainage, the surrounding terrestrial habitat will gradually encroach on the affected area. Displacement of natural and indigenous wetland plants and animal species is of concern due to the increased rate of species extinction resulting from habitat loss. The restoration of hydrologically altered wetlands using reclaimed water may be desirable in some areas, but the quality and flow of reclaimed water will be subject to the requirements and objectives of the restoration project. The State of Florida specifies requirements for discharging reclaimed water to hydrologically altered wetlands.

Reclaimed water may be discharged to wetland systems for a number of reasons, including wildlife habitat enhancement, habitat restoration, erosion control, and flood control. For any given wetland water reuse application, multiple objectives associated with discharge of reclaimed water must be considered.

Water Quality Improvement

The potential for water quality improvement in wetland systems is well established (U.S. EPA, 1993). Constructed wetlands intended primarily to achieve water quality goals have several notable design features, including densely planted areas for interception of particulate matter and development of biofilm growth, open water areas for reaeration and photolysis reactions, and deep anoxic pools for denitrification. Wetlands designed for effluent polishing typically receive disinfected secondary effluent and rely on the wetland system to accomplish the equivalent of tertiary treatment.

Wetlands can reduce nitrogen levels, but the removal efficiency depends on the type and design of the wetland, and climatic conditions. Nitrogen removal is accomplished primarily by microbial nitrification and denitrification processes. For free water surface constructed wetlands, plant uptake accounts for about 10 percent of the nitrogen removal (Crites and Tchobanoglous, 1998). Typical nitrogen removal efficiency for subsurface flow wetlands ranges from 20 to 70 percent, even though some constructed wetlands are reported to achieve over 80 percent nitrogen removal.

Habitat Value

Wetlands can be developed to provide habitat for aquatic and terrestrial plants and animals, some of which may be listed as threatened or endangered species. The design of wetlands used for wildlife habitat is different from the wetlands designed for treatment, although wetlands used for habitat typically result in water quality improvement. Wetlands developed for habitat purposes may include features such as cells with independent inlet and outlet flow control, undulating bottom topography and varying water depth, basin configuration incorporating protected nesting sites such as islands, and a range of wetland vegetation.

Alternate Dispersal of Reclaimed Water

Wetland systems are sometimes used for reclaimed water dispersal when there is insufficient demand for reclaimed water for other applications. For example, a water reuse system used primarily for summer irrigation applications will need an alternate reuse application during the wet season. It may be difficult to maintain the proper seasonal flow distribution and hydraulic loading rate to a natural wetland where the water needs of the wetland system are independent of the supply of reclaimed water. Potential conflicts may occur when: (1) a wetland system is acclimated to a natural dry cycle, but the supply of reclaimed water is constant and (2) reclaimed water must be diverted to a wetland system, even when other competing applications for reclaimed water exist. Most natural wetlands experience seasonal flow variations to which plant and animal species adapt; however, changes in vegetation and wildlife have been observed in natural wetlands that have been converted to treatment wetlands (U.S. EPA, 1993). Another factor to be considered is that a minimum flow may be required during dry weather to support established plant and animal communities. For example, a reuse system that provides water for irrigation and habitat may need to maintain minimum flows to the wetland, even if there is an insufficient supply of reclaimed water for irrigation.

Restoration and Mitigation

In some cases, reclaimed water has been used to establish or enhance wetlands for environmental or legal purposes. Traditional wetland areas that have been drained or otherwise degraded have been restored using reclaimed water. Wetlands restoration may require seasonal or continuous flooding of the area to be restored, in addition to other restoration activities. In a related practice, wetlands have been developed to compensate for the destruction of wetlands in another area, a practice known as *wetlands mitigation*. For example, a developer may agree to establish or enhance a wetland area in another location in exchange for permission to drain, fill, and build on an existing natural wetland. Unfortunately, mitigation wetlands may not be as successful as the original site due to the artificial nature of their development.

Recreation

Wetlands can be opened to the public for recreational and educational purposes. Many constructed wetland sites, especially those designed for the creation of wildlife habitat, provide accommodations for visitors. Common features include interpretive signage for wildlife identification and water cycle education, elevated boardwalks over water and pathways along berms for walking, and visitor centers (see Fig. 21-2). Wetland areas receiving reclaimed water are used for class fieldtrips, wetland education, and university research. With appropriate facilities and personnel, students can be introduced to issues in water management, wildlife habitat, and local environmental issues. Constructed and natural wetlands may also function as wildlife refuges as well as sites for migratory birds, making these locations popular areas for viewing wildlife. Requirements for reclaimed water quality, signage, and other precautions may be specified by the relevant health department for the desired level of public access and potential water contact (see Table 21-4).

Figure 21-2
Recreational features of the Arcata marsh, Arcata, CA: (a) signage at entrance to marsh, (b) visitor and interpretative center, (c) typical rest area with benches and informational sign, and (d) typical pathway for walking, located on marsh berms.

Water Quality Considerations

The quality of the reclaimed water discharged to a wetland system will depend on a number of factors, including the type of wetland (e.g., natural or constructed), the intended functions of the wetland, potential wildlife issues, and the level of public access. Water quality requirements are usually specified for the influent to the wetland area and the discharge from the wetland; however, requirements at other points in the wetland may also be applied to protect wildlife. Pretreatment requirements usually specify disinfected secondary effluent as a minimum before reuse in a wetland. Typical monitoring considerations for constructed wetland systems are presented in Table 21-6.

Effects on Wildlife

In general, reclaimed water discharged to wetlands for environmental and habitat purposes should comply with relevant surface water quality criteria for organic and inorganic trace constituents, including metals. Treatment requirements for environmental uses differ from many other water reuse applications. For most water reuse applications, a chlorine residual is required to be maintained to prevent biological growth in the reclaimed water distribution system. For environmental reuse, however, residual chlorine must be removed prior to discharge due to the toxic effects of chlorine on aquatic organisms. A point-of-use dechlorination device is shown on Fig. 21-3. The dechlorination process is described in Sec. 11-5 in Chap. 11. For recreational uses of reclaimed water, chlorine residual requirements are determined based on the type of water body to which reclaimed water is released and potential human contact with the reclaimed water. Nonionized ammonia is also a concern for aquatic organisms and should not be present in reclaimed water used for environmental purposes.

Trace organic constituents and metals are also a concern, as there is a potential for these constituents to concentrate at high levels in the food chain, such as birds. For example, methyl mercury (i.e., the toxic form of mercury) is known to form under the anoxic conditions that exist in wetlands. Changes that occur as a result of the introduction of reclaimed water to a wetland, such as shifts in TDS, nutrients, and vegetation, can also affect wildlife. Periodic evaluations of plant and animal distribution within the wetland should be conducted to characterize any changes. Constituents that are anticipated or found to cause problems within a wetland should be removed as part of the water reclamation process.

Effects on Groundwater

Water in natural wetlands is often connected to groundwater. Natural wetlands and unlined constructed wetlands can also recharge groundwater through subsurface infiltration. Therefore, if water quality in the constructed wetland affects groundwater adversely, the wetland needs to be lined to prevent seepage of reclaimed water into underlying groundwater. Where groundwater augmentation is anticipated, the reclaimed water should meet the appropriate treatment standard to ensure that the groundwater quality is not degraded.

Operation and Maintenance

The success of reclaimed water wetland systems in providing safe and enjoyable environmental and recreational opportunities depends, to a large extent, on operation and maintenance considerations for wetland systems. These considerations include management of system hydrology, control of mosquito and nuisance species, odor control, vegetation management, and security issues.

Table 21-6
Summary of suggested monitoring parameters for constructed wetlands[a]

Parameter	Project phase (Pre- or post construction or ongoing)	Location Influent (in), effluent (out), or other	Frequency of collection
Water quality[b,c]			
Dissolved oxygen	Ongoing	In, out, along profile	Weekly
Hourly dissolved oxygen	Ongoing	Selected locations	Quarterly
Temperature	Pre, ongoing	In, out, along profile	Daily/weekly
Conductivity	Pre, ongoing	In, out	Weekly
pH	Pre, ongoing	In, out	Weekly
BOD	Pre, ongoing	In, out, along profile	Weekly
SS	Pre, ongoing	In, out, along profile	Weekly
Nutrients	Pre, ongoing	In, out, along profile	Weekly
Chlorophyll A	Ongoing	Within wetland, along profile	Annually
Metals (Cd, Cr, Cu, Pb, Zn)	Pre, ongoing	In, out, along profile	Quarterly
Bacteria (total and fecal coliform)	Pre, ongoing	In, out	Monthly
EPA priority pollutants	Pre, ongoing	In, out, along profile	Annually
Other organics	Pre, ongoing	In, out, along profile	Annually
Biotoxicity	Pre, ongoing	In, out	Semiannually
Sediments			
Redox potential	Pre, ongoing	In, out, along transects	Quarterly
Salinity	Pre, ongoing	In, out, along transects	Quarterly
pH	Pre, ongoing	In, out, along transects	Quarterly
Organic matter	Pre, post	In, out, along transects	Quarterly

(Continued)

Table 21-6
Summary of suggested monitoring parameters for constructed wetlands[a] (Continued)

Parameter	Project phase (Pre- or post construction or ongoing)	Location Influent (in), effluent (out), or other	Frequency of collection
Vegetation			
Plant coverage	Ongoing	Within wetland, along transects	Quarterly
Identification of plant species	Ongoing	Within wetland, along transects	Annually
Plant health	Ongoing	Within wetland	Observe weekly
Biota			
Plankton (zooplankton tow)	Ongoing	Within wetland, along transects	Quarterly
Invertebrates	Ongoing	Within wetland, along transects	Annually
Fish	Ongoing	Within wetland, along transects	Annually
Birds	Pre, ongoing	Within wetland, along transects	Quarterly
Endangered species	Pre, during, ongoing	Within wetland, along transects	Quarterly
Mosquitoes	Pre, during, ongoing	Within wetland, selected locations	Weekly during critical months
Wetland development			
Flowrate	Ongoing	In, out	Continuous
Flowrate distribution	Ongoing	Within wetland	Annually
Water surface elevations	Ongoing	Within wetland	Semiannually
Marsh surface elevations	Ongoing	Within wetland	Quarterly

[a] Adapted from Tchobanoglous (1993).
[b] Water quality for preconstruction and during construction refers to the wastewater that is to be applied to the wetland.
[c] Permitting agencies may not require all parameters to be tested, nor to be tested at the same frequency.

System Hydrology

As described previously, the flowrate of reclaimed water may need to be adjusted to accommodate natural seasonal changes that affect the growth and life cycle of some species. Some wetland plants are not able to sustain extended periods of inundation or may require an annual dry period. Thus, if the proper conditions are not created, adapted or invasive plant communities will replace these plant species and the specific habitat that they support. Constructed wetlands used for treatment are usually flooded year round and only support vegetation that

Figure 21-3
Small point-of-use dechlorination device for removal of chlorine from reclaimed water before discharge to pond. Sulfur dioxide for dechlorination is generated by burning sulfur.

can tolerate this environment. Inlet and outlet structures should be designed so that water levels and flowrates can be adjusted as necessary. Bypass or transfer structures should also be provided to transfer water to different areas of the wetland as needed (see Fig. 21-4).

System hydrology may also need to be adjusted to meet discharge requirements. For example, as water is lost due to evapotranspiration, the concentration of some constituents will be higher at the outflow of the wetland as compared to the inlet to the wetland. Because NPDES and state waste discharge requirements are imposed on the effluent from the wetland, it may be necessary to design the wetland to adjust the TDS level by diverting portions of reclaimed water flow through the wetland or blending reclaimed water with other low TDS water before discharge.

Figure 21-4
Flow structures used for water level control in natural and constructed wetlands.

Mosquito Control

The control of mosquito populations may be an important component of maintaining wetlands, especially in areas that do not experience mosquito problems and where mosquito-related diseases are a concern. Subsurface flow wetlands are not subject to mosquito growth issues because there is no free water surface. Mosquito growth typically occurs in stagnant flow areas containing water with high organic matter concentration, such as around emergent vegetation in shallow areas, along the perimeter and interface areas, and near inlet and outlet flow structures where standing pockets of water develop. Successful mosquito control in constructed wetlands has been accomplished by implementing the following design and management control measures (Williams et al., 1996; Crites and Tchobanoglous, 1998):

- Stocking open water areas with mosquito fish (*Gambusia*) and developing habitat for fish survival during winter season. Mosquito fish are a robust species that consumes mosquito larvae present in the water column. Mosquito fish may require annual restocking in cold regions.
- Daily sampling and monitoring for mosquito larvae during local mosquito seasons (e.g., April through October).
- Sprinkling water over the wetland during times when mosquitoes would be depositing eggs (i.e., 8 p.m. to 6 a.m.).
- Applications of biological control agents as needed for control of larvae. A popular biological control product is an extract of toxins produced by the bacteria *Bacillus thuringensis israelensis* (Bti) that does not harm fish and other aquatic organisms. Several commercial products are available, some formulations may be more effective than others.
- Use of chemical control agents (e.g., larvicide such as Golden Bear Oil 1111)
- Aeration to disrupt the still water surface.
- Oxidation of ammonia nitrogen (ammonia is considered to be an attractant for mosquitoes).
- Vegetation management to maintain open water and pathways for mosquito fish to gain access to the mosquito larvae. Also clearing vegetation as needed to minimize stagnant water areas.
- Steep grading of banks and interface areas to increase flow and water movement in those areas.

If possible, background mosquito populations should be assessed in the wetland area before beginning and during the construction project to evaluate the impact of the wetland development. Mosquito populations are determined using traps placed strategically and accepted counting techniques. Mosquito surveys may need to be conducted for several seasons to quantify the population accurately.

Nuisance Species

Undesirable species in wetlands can include nonnative and invasive plants and algal blooms that impair water quality. Water hyacinth, an invasive plant, is used sometimes in aquatic treatment systems, but can be problematic in some areas if released to the environment. Where water hyacinths can be contained, the high surface area of its roots can be beneficial for treating water. Where there is danger and risk involved with water hyacinths entering the surrounding environment, they should not be used.

The occurrence of algae in wetlands is natural due to elevated nutrient concentrations in the reclaimed water. However, the presence of algae in the effluent from the wetland can exceed water quality limits. Strategies for controlling algae include using emergent (e.g., cattails) or floating (e.g., duckweed, water hyacinth) dense vegetation around outlet structures to intercept and create a settling zone for algae, as well as to block sunlight required for algae growth. Floating mats and balls have also been used to intercept sunlight. Additional information on algae control can be found in the subsequent section on lakes and ponds.

Other species that inhabit wetlands and may pose a management challenge include, depending on the region, animals that can (1) damage levees and berms (e.g., nutria, muskrats, bores), (2) cause human and pet safety concerns (e.g., bears, alligators, poisonous snakes), (3) damage vegetation (e.g., muskrats, geese, insects), or (4) cause wildlife health problems (e.g., avian influenza). These issues are best addressed with the help of local wildlife biologists and wetland specialists.

Odors
Odor issues are usually the results of anoxic conditions and the release of hydrogen sulfide (H_2S), which is an odorous gas emitted from stagnant water. Odors can be controlled by reducing the organic content of the reclaimed water, providing multiple inlets to the wetland, and by applying a well-oxidized reclaimed water. Techniques used to control odors within the wetland include flow structures that cascade and aerate water where sufficient head exists and submerged aeration devices to mix the water and increase the dissolved oxygen concentration.

Vegetation Management
Vegetation management can include the periodic removal of vegetation to (1) remove carbon and nutrients from the system, (2) prevent the growth of mosquitoes, and (3) improve hydraulics. Removing vegetation for nutrient control is generally not recommended due to negative water quality effects, unless the cell containing the vegetation to be harvested can be taken out of service and the flow diverted. Channelized flows develop naturally, even through densely planted areas, and result in short-circuiting of flow. Managing vegetation to improve the hydraulics is particularly important when the wetland is used for improving water quality. Some control over vegetation patterns can be achieved by varying the water depths, as water depth is a primary factor in determining which species of vegetation can grow.

Security Issues
Wetland facilities that are open to public access may be subject to vandalism, especially when flow diversion structures, signage, and other facilities are accessible. Care should be taken to secure all facilities that are associated with process operation and performance.

21-3 STREAM FLOW AUGMENTATION

Stream flow augmentation with treated wastewater occurrs in every river that receives discharged wastewater effluent. Nevertheless, the stream flow augmentation discussed in this section is a designed and engineered augmentation of stream flow for environmental and recreational purposes. In some rivers in arid regions, reclaimed water (or discharged wastewater) is the primary source of water to maintain the base flow (see Fig. 21-5). Reclaimed water is a reliable water source that can be supplied constantly for aquatic and riparian habitat enhancement, and for aesthetic and ornamental purposes.

Aquatic and Riparian Habitat Enhancement with Reclaimed Water

To evaluate the effect of reclaimed water on aquatic and riparian habitat, baseline conditions, i.e., conditions upstream of the discharge point, and/or before reclaimed water is released, have to be determined first. The baseline conditions and potential effects of reclaimed water are examined typically in compliance with the environmental impact assessment required by the federal and state regulations. If endangered or threatened species are found, mitigation measures to protect these species must be considered. In many cases, the use of reclaimed water for stream flow augmentation is in fact considered as a mitigation measure to protect these species. One recent example of the use of reclaimed water for habitat enhancement in a stream has been implemented in San Luis Obispo, California, described in Sec. 21-7.

Generally, the U.S. Fish and Wildlife Service (U.S. FWS) and the equivalent state agency must be involved when any activity is likely to alter the existing stream and riparian habitat. In some states, there are habitat restoration guidelines that specify the approach and techniques for aquatic and riparian habitat protection and restoration. A summary of techniques suggested by the State of Washington in their stream habitat

(a)   (b)

Figure 21-5

Santa Ana River, Santa Ana, CA: (a) view of effluent dominated stream and (b) typical treated wastewater discharge. (Coordinates: From 33.881 N, 117.738 W to 33.837 N, 117.864 W to 33.630 N 117.957 W)

Table 21-7 Techniques suggested for aquatic habitat preservation and restoration in the State of Washington Stream Habitat Restoration Guidelines[a]

Specified in the aquatic habitat guidelines
Dedicating land and water
Channel modification
Levee modification and removal
Side channel/off-channel habitat restoration
Riparian restoration and management
Fish passage restoration
Nutrient supplementation
Beaver reintroduction
Salmonid spawning gravel cleaning and placement
Instream structures
General design and selection considerations for instream structures
Boulder clusters
Large wood and log jams
Drop structures
Porous weirs
Bank protection construction, modification, and removal
Instream sediment detention basins

Not specified in the aquatic habitat guidelines
Best management practices (BMPs)
Techniques that treat the symptoms of habitat degradation
Techniques that have been used but not successful
Techniques that may be appropriate but not demonstrated to date
Land use planning and establishment of protective regulations
Estuary restoration

[a]Adapted from Saldi-Caromile et al. (2004).

restoration guidelines is shown in Table 21-7. The use of best management practices (BMPs), although not specified in aquatic habitat guidelines, is becoming a common approach in water pollution control. Various techniques, including the use of reclaimed water, should be considered and implemented using the BMP approach.

Aquatic Habitat

Aquatic habitat includes fish and animals occurring in a stream or river. In some cases, the aquatic habitat in the stream includes endangered or threatened species listed by the Endangered Species Act (ESA) and/or in state equivalent regulations. If diadromous fish, such as steelhead (rainbow trout) and salmon, and other fish species listed as endangered species are observed in the stream, the stream flow must be designed such that seasonal flow conditions, water quality, temperature, and other parameters are suitable for those species. The U.S. FWS and state agencies routinely monitor aquatic habitat of major rivers, streams, and other aquatic environment. Therefore, it is likely that there is some record of wildlife occurring in the affected area when a water reuse project is planned.

Riparian Habitat

Riparian habitat consists of plants and animals occurring in a land area adjacent to a stream or river, between the wet and dry zones. Vegetation in the riparian areas is affected greatly by the flow pattern of, and therefore availability of water from, the stream. Continuous supply of reclaimed water throughout a year in a stream that would otherwise be dry may alter the existing condition of riparian environment. Riparian habitat often consists of phreatophyte, a type of plant whose roots extend downward to the water table. Many of these plant species also contribute to improving water quality by hosting microorganisms and taking up nutrients. Riparian habitat also includes a variety of wildlife, some of which may be listed as endangered or threatened species.

Recreational Uses of Streams Augmented with Reclaimed Water

Recreational uses of streams augmented with reclaimed water are limited mostly to an ornamental purpose, even though some streams augmented with reclaimed water for aquatic habitat restoration may become a fishing venue. One of the most prominent examples is the San Antonio River Walk in San Antonio, Texas, where natural stream flow is augmented with groundwater and reclaimed water (see Fig. 21-6*a*). A rare example is a small stream (known as *seseragi* in Japanese) in downtown Tokyo that is supplied with highly treated reclaimed water treated by microfiltration (MF), reverse osmosis (RO), and low-dose chlorination. In this demonstration stream, contact with reclaimed water is allowed and even encouraged as part of environmental education (see Fig. 21-6*b*).

Reclaimed Water Quality Requirements

Because stream flow augmentation with reclaimed water is considered practically the same as discharge of treated effluent to the receiving water, waste discharge regulations must be met. Often water reuse regulations or guidelines are referred to for the microbial quality requirements where human contact to the receiving water is likely (see Table 21-4). Other important parameters influencing the use of reclaimed water

(a)

(b)

Figure 21-6

Examples of recreational use of reclaimed water: (a) San Antonio River Walk, San Antonio, TX (Coordinates: 29.424 N, 98.493 W and (b) stream in Tokyo supplied with MF-RO treated reclaimed water.

Table 21-8
Important water quality parameters for the use of reclaimed water for stream flow augmentation

Parameter	Typical recommended range	Remarks
Chlorine residual	Total Cl_2 < 0.1 mg/L (dechlorinated)	The toxic dose varies but typically 0.1 to 1 mg/L of chlorine is enough to exhibit toxicity in many fish species. Reclaimed water used for many other applications is required to maintain a chlorine residual to prevent microbial growth in the distribution system, but the chlorine residual may have to be removed for environmental uses. UV disinfection of reclaimed water may be a suitable alternative for those applications.
Total dissolved solids	In compliance with NPDES permit	May be toxic to some aquatic organisms and riparian plants. Some municipalities have difficulties complying with TDS requirements due to high TDS in the source water.
Dissolved oxygen	DO ≥ 5 mg/L	Dissolved oxygen is required for maintenance of aquatic species. Reduced concentration of DO can cause stress and death of sensitive fish. The DO requirements should be based on the most sensitive species that is to be protected and may exceed the minimum value of 5 mg/L.
Organic matter	BOD < 20 mg/L	Dissolved oxygen can be depleted by the degradation of organic matter. Depending on the quality and flow conditions, modification to the stream to enhance aeration may be necessary.
Nutrients	Nitrogen <3 mg/L Phosphorous <1 mg/L	Eutrophication must be controlled. The increase in flowrate helps reduce stagnant water, thereby reducing algal growth.
Temperature	±2.8°C (±5°F) of the ambient stream water temperature	An important parameter to control when fish species sensitive to temperature changes are present and need to be protected. Riparian vegetation can be utilized to shade the stream and to lower the water temperature.

for stream flow augmentation include organic matter, nutrients, total dissolved solids, and temperature (see Table 21-8). Where reclaimed water is used for environmental purposes, the effects of reclaimed water constituents on aquatic and riparian habitat are the key considerations in determining quality requirements.

For recreational and ornamental uses, aesthetic quality of reclaimed water must also be managed carefully. The parameters of concern include color, odor, and nutrients that can cause eutrophication. If there is no aquatic organism to protect, chlorine residual helps in preventing microbial and algal growth, and in controlling the generation of odors. In streams affected by human activities, a large number of trace organic constituents have been detected (Kolpin et al., 2002). Several reports have been published on the effects of wastewater constituents on aquatic life, such as fish, in which production of organs or hormones in opposite sex are reported (Jobling et al., 1998; Rodgers-Gray et al., 2000; van Aerle et al., 2001). The cause-effect relationship of trace organic compounds on aquatic habitat is still uncertain, and further study is necessary.

Where reclaimed water is discharged into a river with significant flow, the dilution may be taken into account in determining the quality requirements for discharge. If dilution is to be considered, then the base flow, i.e. minimum flow of a stream and its ambient quality must be assessed to determine the discharge requirements for reclaimed water. When reclaimed water is the primary source of water flow, it must meet a quality standard that provides safety to both humans and the affected habitat. Unlike wetlands that can be considered as part of a treatment system, natural attenuation of water quality in a stream or river is not counted because the point of discharge is considered as the point of use.

Stream Flow Requirements

When considering stream flow augmentation with reclaimed water for environmental purposes, existing flow conditions must be assessed. In some cases, the discharge of wastewater effluent creates an artificial base flow and an aquatic habitat that depends on the effluent flow for its maintenance. For example, during the summer months the base flow in the Santa Ana River is essentially effluent from wastewater treatment plants (see Fig. 21-5).

The required flowrate for an intended environmental purpose is determined through an instream and groundwater flow study, hydrological modeling, and biological resource assessment and analysis of the impact expected from the proposed project. It is also necessary to assess the effects of increased water flow on downstream water quality. For example, maintaining a constant stream flow prevents the development of stagnant areas in the stream, and helps in reducing algal growth and the generation of odors. Stream flow and reclaimed water must be monitored to ensure the objectives of stream flow augmentation are met, and the use is also in compliance with relevant regulations (see Table 21-9).

The stream flow requirements may affect water rights issues for the affected stream/river. In planning and designing the reclaimed water flow dedicated to a stream, it is crucial to clarify the rights to the reclaimed water. Generally, the reclaimed water purveyor possesses the right to the reclaimed water until it is released. However, if the reclaimed water is diverted to a location different from the previous wastewater discharge point, the water rights that pertain to the previous discharge location may become an issue. In some cases, water rights are nullified for the flow dedicated to wildlife enhancement. A further discussion of water rights is included in Chap. 25.

Operation and Maintenance

When using reclaimed water in streams for environmental and recreational purposes, the key issues are management of flowrate and monitoring. Both water quality and habitat must be monitored as required to ensure the environmental protection and the protection of public health. The monitoring program should also include physical characteristics of the stream that is augmented with reclaimed water, such as water depth, water velocity, bed material composition, channel profile, and rate of bank or bed erosion. For recreational use, surveys of visual improvement, recreational use, and community participation to the project should be conducted (Kondolf and Micheli, 1995). The control of stream flow is based on the assessment and implementation of flow requirements discussed above. A thorough monitoring program should continue for a minimum of 3 yr to assess the effects of the project. After the effects of the program have been evaluated, an ongoing monitoring program should continue, but in less frequency for some parameters that are not critical.

Water Quality Monitoring

Water quality monitoring requirements are specified typically in the NPDES permits; water reuse regulations may also require water quality monitoring. Typical water quality monitoring requirements for reclaimed water used for environmental and recreational purposes are shown in Table 21-9. Typically, more frequent monitoring is required in the first year(s) of the project; subsequently, the sampling frequency may be lowered for parameters that do not pose immediate concerns for public health and ecosystem protection. If new aquatic and riparian species are found in the stream after implementation of the project, water quality requirements and monitoring parameters may be adjusted based on the results from the habitat monitoring.

Habitat Monitoring

Aquatic organisms in the stream should be monitored on an ongoing basis to assess the project performance. The monitoring strategies must be developed as part of the project plan. The monitoring during implementation should assess whether the project performs as planned. The monitoring program after implementation is used to determine if the desired effect on habitat or stream conditions was achieved and

Table 21-9 Typical water quality monitoring requirements for environmental and recreational uses of reclaimed water

Parameter	Frequency	Remarks
BOD	Weekly	Weekly composite sampling.
Chlorine residual	Continuous	Usually total residual chlorine. Grab sampling.
Coliform	Daily	Grab sampling. Total or fecal, depending on the requirements.
Dissolved solids	Weekly	
Metals	Twice/week or weekly	Composite sampling.
Nitrogen	Weekly	Total Kjeldahl, ammonia, nitrate, and nitrite, composite sampling.
Oil and grease	Weekly	Grab sampling.
pH	Continuous/daily	Continuous or daily monitoring.
Phosphorous	Weekly	Total- and ortho-phosphorous, composite sampling.
Suspended solids	Daily	Daily composite sampling.
Temperature	Continuous	

Toxicity
  Acute
  Chronic
	Monthly	Organisms to be used for the bioassay depend on the environment of the receiving water. Composite samples can be used.

Trace organics
  PAHs
  Pesticides
  VOCs
	Biweekly	Typically only required in the first year of the project, and in a reduced frequency thereafter. Volatile chemicals must be measured on a grab sample.

Turbidity	Continuous	

resulted in the intended benefit to the environment (Saldi-Caromile et al., 2004). The results of monitoring may lead to further changes to the quality and flow requirements. The parameters to be monitored are case specific, but generally include vegetation, plankton, invertebrates, fish, and birds, some of which may be endangered species (see Table 21-6).

21-4 PONDS AND LAKES

Ponds and lakes supplied with reclaimed water may be used for several purposes, including ornamental and recreational purposes, wildlife habitat, storage of irrigation water, and, in some cases, even as a source of drinking water supply. Some reclaimed water ponds are used as recreational features on golf courses, and concurrently for the storage of irrigation water. The use of lakes and ponds for the storage of irrigation water is discussed in Chaps. 17 and 18. Use of reclaimed water for augmenting surface water supplies is covered in Chap. 23. The following discussion is limited to the use of ponds and lakes supplied with reclaimed water for recreational and habitat purposes.

Water Quality Requirements

The primary water quality concerns related to the use of reclaimed water for the establishment of recreational and habitat ponds and lakes can be categorized as aesthetic and wildlife health considerations. It should be noted that flows to ponds and lakes may also include stormwater runoff and leachate from failed septic systems, which may contain BOD, nutrients, pathogens, and trace constituents. All point and nonpoint source inputs to a pond or lake should be evaluated for potential impact on water quality.

Aesthetic Considerations
When properly designed and maintained, ponds and lakes should not result in nuisance conditions, conditions that could occur include eutrophication and the generation of objectionable odors. Eutrophication is the natural aging process in which a pond or lake becomes organically enriched, leading to increasing domination by aquatic weeds and transformation to marsh land and eventually to dry land. Eutrophication can be accelerated by the input of nutrients. Die-off and settling of plant growth results in sediment oxygen demand, which tends to decrease dissolved oxygen levels. The effects of eutrophication, which may be detrimental to aquatic life, are compounded by large day-night excursions in dissolved oxygen due to photosynthesis and respiration. The process of eutrophication and its relationship to nutrient inputs is complex. In lakes and reservoirs, phosphorus is typically the limiting nutrient, although the presence of nitrogen is also important. A simplified but historic criterion is that algal blooms tend to occur if the concentrations of inorganic nitrogen and phosphorus exceed respective values of 0.3 and 0.01 mg/L (Sawyer, 1947). In ponds and lakes that stratify (depth greater than 5 m), early signs of eutrophication are low dissolved oxygen levels in the hypolimnion, which does not receive any direct reaeration.

The generation of odors may result from the development of anoxic and anaerobic conditions in a pond or lake system. Anoxic conditions are created when dissolved oxygen is depleted from the water column due to aerobic respiration and the rate of reaeration, either natural or artificial, is not sufficient to maintain aerobic conditions. Anaerobic conditions occur when oxygen containing substances have been reduced and anaerobic respiration is occurring. Anoxic and anaerobic conditions result in the formation of numerous odor-causing compounds, including sulfides and mercaptans. In most cases, odor-causing compounds are formed in the sediments or in the hypolimnion of a stratified system, thus aeration of the surface layer may be sufficient to control the release of odors. A surface aeration device is shown on Fig. 21-7.

Wildlife Health Considerations

The main threats to the health and well-being of wildlife in ponds and lakes supplied with reclaimed water are excessive oxygen demand resulting in low dissolved oxygen (DO) concentrations and the potential for toxicity related to inorganic and organic trace constituents. Dissolved oxygen is important to aquatic life, because detrimental effects can occur when DO levels drop below 4 to 5 mg/L, depending on the aquatic species. Ambient DO levels can be affected by the growth of algae (phytoplankton, primary producers) and weeds (macrophytes) feeding on ammonia and nitrate. Algae and weeds constitute an oxygen source during daylight hours due to photosynthesis and a continuous oxygen sink at nighttime due to respiration. Higher enrichment levels, however, lead to high productivity (see eutrophication above), with potentially strong effects on DO fluctuations. Diurnal fluctuations can develop with supersaturated DO levels during daylight hours due to photosynthesis and very low DO levels at night due to respiration. Longer term fluctuations result from photosynthesis/respiration imbalances during high biomass growth and decay periods.

Toxic chemicals include a range of compounds that, at specific concentrations, have detrimental effects on aquatic life or on humans (i.e., upon ingestion of water and/or fish and shellfish). Toxic effects on aquatic life are characterized as acute if they occur after a short exposure (on the order of a few hours) to the toxic agent or chronic if effects require a longer term exposure (on the order of a few days). If chemical toxicity is a concern, supplemental treatment processes should be implemented (see Chaps. 9 and 10), or source control can be used to reduce the concentrations of the constituent. It should also be noted that because ponds and lakes are surface water bodies,

Figure 21-7

Surface aeration device used at a pond augmented with reclaimed water.

they are subject to evaporation and will tend to concentrate water constituents if there is no mechanism to dilute or replace the contents. Using ponds or lakes in a pass-through mode, so that there is continuous flushing, is one strategy to manage the effects of evaporation.

Operation and Maintenance

Operation and maintenance for pond and lake systems include the control of nuisance species and replenishing the contents.

Nuisance Species

The primary nuisance species present in pond and lake systems are algae and aquatic weeds. These growths occur typically in response to high nutrient loading, such as nitrogen. In addition to nitrogen, other nutrients are needed for biomass growth, notably phosphorus and silica. The average molar ratios of nitrogen to phosphorus to carbon in algal protoplasm (Redfield ratios) are approximately $N:P:C = 15:1:105$. If one of these nutrients is available in a smaller proportion to the others than these ratios, it tends to limit growth and any addition of this nutrient will result in a direct increase of biomass. For example, in lakes, phosphorus is typically the limiting nutrient so that addition of phosphorus will spur growth, but addition of nitrogen will have minimal effects. While aquatic herbicide may be applied for temporary control of algae and aquatic weeds, these chemicals will not solve the overall problem, and the long-term solutions should be sought.

Water Replenishment

Replenishment of pond and lake water is intended to control the buildup of salts and other conservative constituents. Rapid buildup of constituents can occur in ponds and lakes in areas with high levels of evaporation and no or limited outflow and fresh replenishment water for dilution in the system. A water and constituent mass balance can be used to determine a preliminary estimate of the steady state concentrations for constituents in the pond or lake. Successful examples for pond and lake systems have been pass-through systems where the reclaimed water is discharged to a subsequent use following temporary storage in the impoundment.

Other Considerations

Pond and lake systems provide numerous opportunities for environmental education and recreation activities.

Environmental Education

The educational potential of pond and lake systems is increased by the use of visitor centers, descriptive signs, and outdoor and environmental education programs. Educational programs involving reclaimed water typically focus on the water cycle. Bird watching and pond ecology laboratories have also been popular educational opportunities (see Fig. 21-2).

Recreation

Recreational opportunities associated with ponds and lakes supplied with reclaimed water may include fishing, boating, and other minimal water contact activities. Wading and swimming should not be allowed unless the reclaimed water is of the highest quality and meets pertinent regulations for body contact and incidental ingestion.

21-5 OTHER USES

A few innovative water reuse options have been implemented for environmental and recreational purposes in demonstration or full scale, such as snowmaking and the reclaimed water use at zoos and other parks that rear animals. Even though these uses are generally minor, it is worthwhile to consider the broad opportunities of beneficial uses of reclaimed water.

Snowmaking

Some ski areas depend on artificial snow to extend the ski season, especially when snowfall is unpredictable or below average. Ski resorts are often located in areas where it is difficult to secure a reliable water supply for artificial snowmaking. Reclaimed water is often a reliable source of additional water, and withdrawal of water from environmentally sensitive sources can be reduced. Snowmaking with reclaimed water can also be considered as seasonal storage and environmental reuse of water because snow will melt gradually and flow into natural streams in warm seasons. The foremost importance is public health protection. Even though the reclaimed water is in the form of snow, the general public will have access to contact snow in ski areas, and incidental ingestion of snow is possible. Typically, the quality requirements for unrestricted access to nonpotable water reuse applications such as irrigation of public areas are required for snowmaking. Because of public health concerns, higher levels of treatment such as ultrafiltration and reliable disinfection should be considered (Tonkovic and Jeffcoat, 2002). The impacts of the snowmelt on receiving streams should be considered when determining water quality requirements because fish and wildlife sensitive to contamination may be present.

A few cases of snowmaking using reclaimed water are found in Australia, including Mt. Buller and Mt. Hotham. In Flagstaff, Arizona, the use of reclaimed water for snowmaking has been planned and approved by the U.S. EPA and Arizona Department of Environmental Quality (ADEQ). However, environmental and health concerns have led to persistent opposition, and it has not been implemented as of 2006.

Animal Viewing Parks

Reclaimed water may be used for various applications associated with animal viewing parks. Examples include the creation of animal habitats, filling of aquariums, and cleaning within animal facilities. Habitats may include small streams and ponds for animal use. A key advantage is that human contact is not possible as the water features within the animal habitat are not accessible. Aquariums and underwater animal features can consume and recycle large amounts of water, and reclaimed water could be adapted to meet part of this demand. For sensitive species, reverse osmosis or equivalent treatment may be required. Cleaning of animal facilities with reclaimed water is usually conducted after business hours, and wash water is collected and directed to a treatment facility.

21-6 CASE STUDY: ARCATA, CALIFORNIA

The Arcata Marsh and Wildlife Sanctuary is one of the most well known examples of using wetlands for tertiary treatment of reclaimed water and establishment of wildlife habitat. In this case study, the evolution of wastewater management for the City of

Setting

Arcata and the development of the Arcata Marsh are reviewed. The key references used for this case study report are U.S. EPA (2000, 1993).

The City of Arcata is located along the Pacific Northcoast, about 450 km (280 mi) north of San Francisco, with a population of approximately 17,000. The city is adjacent to Humboldt Bay, which is one of the major oyster farming sites in California. The average wastewater flowrate from the City of Arcata is about 9.6×10^3 m³/d (2.5 Mgal/d), with a monthly flowrate peaking factor of 2.5. The design flow to the system was 8.7×10^3 m³/d (2.3 Mgal/d) with a wet weather flow of 22×10^3 m³/d (5.9 Mgal/d).

Water Management Issues

The City of Arcata ultimately discharges treated wastewater to Humboldt Bay. The wastewater treatment system for the City of Arcata has evolved from providing primary treatment only in the early 1950s to adding oxidation ponds in 1957, followed by chlorine disinfection in 1966. To protect the oyster farming industry in the bay, it was necessary to develop additional treatment processes prior to discharge. A task force was assembled to evaluate low-cost treatment options as an alternative to the state-recommended regional treatment facility. In response to the task force findings, a research study was conducted using pilot scale wetlands for treatment of oxidation pond effluent. The alternative system was required to meet the state's Bays and Estuary Policy if it was to discharge to Humboldt Bay. The pilot study results were used as justification that wetlands could be used to provide the required treatment. Construction of the wetland system was completed in 1986.

Implementation

The final design of the wetland system was unique because it consisted of two distinct and separate processes, a treatment wetland and a reclaimed water habitat marsh. The first process is a conventional wetland composed of three parallel free water surface treatment cells designed to meet the secondary treatment requirements (BOD, TSS <30 mg/L), followed by chlorine disinfection. Disinfected secondary effluent (i.e., reclaimed water) is discharged to a marsh system that is designed to provide wildlife habitat, nutrient removal, and a public recreation area. An NPDES permit is issued for the discharge from the wetland to ensure reclaimed water quality and public safety in the marsh areas open to public access.

The reclaimed water marsh system (see Fig. 21-8) has three distinct cells with a total area of 12.5 ha (31 ac). The three marsh cells have an average depth of 0.6 m (2 ft), with mixed open areas and stands of emergent plants. Reclaimed water is pumped to the marsh system, which is located on Humboldt Bay. Heavy clay soils that underlie the marsh preclude the need for extensive geotechnical work or a liner system. The marsh construction included earthen berms for directing flow, inlet and outlet flow control structures for each of the marsh cells, and islands for nesting birds. Operation and maintenance activities consist of balancing flow in the marsh cells by adjusting the inlet and outlet weirs. Emergent vegetation is not harvested from the marsh cells. Floating vegetation is harvested periodically from the wetland to maintain open water for habitat and to assist in vector control for mosquitoes.

Effluent from the reclaimed water marsh is pumped back to the treatment facility for disinfection and dechlorination before discharge to the bay. A second NPDES permit is

Figure 21-8
Aerial view of Arcata marsh and wildlife sanctuary, Arcata, CA, developed for public access and educational benefits (Coordinates: 40.859 N, 124.095 W). (Courtesy of R. Gearheart.)

issued for the bay discharge. The mean residence time in the reclaimed water marsh system is about 9 d. On average, effluent BOD, TSS, and TN from the treatment wetland is reduced from 28 to 3.3 mg/L, 21 to 3 mg/L, and 30 to 3 mg/L, respectively, after passage through the habitat wetland. The high-quality effluent from the marsh is blended with effluent from the treatment wetland to meet permit discharge requirements. Nitrogen removal takes place by oxidation in the open water areas followed by denitrification in the sediment and other anoxic zones.

The beneficial use of the habitat marsh is maximized by allowing public access and recreational opportunities around the habitat wetlands. The reclaimed water marshes, and an additional 30 ha of salt and freshwater marshes, brackish ponds, estuaries, and sloughs that do not use reclaimed water, collectively form the Arcata Marsh and Wildlife Sanctuary. The Arcata Marsh and Wildlife Sanctuary includes an interpretive center for wildlife education, trails between and around the marsh cells, and trailside explanatory signs (see Figs. 21-2 and 21-9). The area is a well-known and popular bird watching area and community asset.

Lessons Learned

The Arcata Marsh and Wildlife Sanctuary is an excellent example of a multipurpose facility that was developed with low construction, operating, and maintenance costs. Beneficial uses include a public recreational area, wildlife habitat, and water quality improvement that made a discharge to Humboldt Bay feasible. The marsh has been used for conducting university research and for environmental education programs.

A number of observations have been made related to water quality and facility operation. Observed short-circuiting of flow that occurs in the wetlands could be improved by flow distribution at the inlet and outlet structures for each marsh cell. While fecal coliform detected in the marsh waters is attributed to wildlife, the dense growth of emergent vegetation around the outlets reduces the discharge of bacteria by limiting

Figure 21-9
Views of Arcata marsh: (a) wooden bridge over influent channel to marsh and (b) view of open area of marsh covered with duckweed.

access of birds to the open water areas and acting as a natural filtration and sedimentation area. The water depth is used to control the growth of emergent vegetation, as the shallow areas facilitate growth of emergent plants and deep water areas are used to maintain open water areas.

Discharge permits (mass loadings and concentration limits) for land-based treatment system, such as Arcata's pond/wetland system, need to consider the impact of high individual rainfall events and sustained wet periods. For example, at the Arcata Marsh the effluent flow exceeds the influent flow during major stormwater events. The normal consideration of daily, weekly, and monthly permit requirements do not factor in both the catchment of rainwater, which results in increased effluent flows, and fluctuating hydraulic retention times.

21-7 CASE STUDY: SAN LUIS OBISPO, CALIFORNIA

The water reuse project in the City of San Luis Obispo, California, represents a beneficial use of reclaimed water for multiple purposes, including enhancement of in-stream habitats. In this case study, the use of reclaimed water for environmental enhancement through stream flow augmentation is described.

Setting

San Luis Obispo, California, is located about 320 km (200 mi) northwest of Los Angeles, along California's central coast. The City of San Luis Obispo has a population of about 44,000 and is a home to the California Polytechnic State University, San Luis Obispo. San Luis Obispo Creek, a relatively short river extending about 24 km (15 mi), collects waters from 11 tributaries, runs through the City of San Luis Obispo, part of which flows underground, and flows into the Pacific Ocean. The San Luis Obispo Creek watershed area is about 220 km^2 (84 mi^2).

Water Management Issues

The potable water supply of the City of San Luis Obispo is from imported water from outside of the San Luis Obispo Creek watershed. The flow of the San Luis Obispo Creek varies depending on the season of the year and is affected by the discharge of treated wastewater into the creek. During the dry summer season, the flow in the creek is predominantly treated wastewater. An artificial habitat has been established and depends on the amount of discharged flow to the creek.

The city started considering water reuse in 1988, when the city council decided to upgrade the existing secondary treatment wastewater plant to a tertiary treatment process to comply with the waste discharge requirements imposed by the Central Coast Regional Water Quality Control Board (RWQCB). The water reclamation facility (WRF) improvements for tertiary treatment were completed in 1994. The treatment process consists of primary treatment, secondary treatment with nitrification, and tertiary treatment with filtration and chlorination. The initial plan was to treat wastewater to a tertiary level and use it for landscape irrigation and industrial processes. Part of reclaimed water was to be dechlorinated for the discharge to the creek, but the primary objective was to reduce significantly the discharge into the creek. The temperature of reclaimed water is reduced before discharge, using a cooling tower, to meet the temperature requirement for the fish habitat [within 2.8°C (5°F) of the receiving water temperature]. The WRF is producing approximately 13.6×10^3 m³/d (3.6 Mgal/d) of tertiary treated reclaimed water.

Once the plant upgrade was completed in 1994, the quality of water discharged into the creek improved along with improvements in the creek habitat. These improvements were recognized by various agencies including the Department of Fish and Game (DFG) (DiSimone, 2006) and led to a new challenge to the city's water reuse project, i.e., maintaining stream flow to support aquatic habitat.

Some species of aquatic life found in the creek required special attention, including the southern steelhead (federally listed threatened species) and the tidewater goby in the southern most reaches (federally listed endangered species and state listed species of special concern) (DiSimone, 2006). These species were recognized prior to the upgrade of the WRF.

Implementation

To implement water reuse, the city petitioned the California State Water Resources Control Board (SWRCB) for a change in discharge location, i.e., part of the previously discharged water was to be diverted to the water reuse system. With the petition, the city was required to prepare an environmental impact report (EIR). Six protests were received against the petition, from downstream property owners, Central Coast Salmon Enhancement, and the DFG. The protests were primarily about the reduced flow in the creek and subsequent potential impacts to the downstream environment. The city needed to resolve the protests through the EIR, and a memorandum of understanding (MOU) with the DFG.

As part of the EIR, three studies were conducted to evaluate the impact of reduced flow in the creek by the water reuse project: an instream flow study, a modeling of the hydrology and groundwater of the creek downstream of the reclaimed water outfall, and biological sources assessment and impact analysis. In the EIR, the primary use of reclaimed water identified was irrigation, and about 70 percent of the reclaimed water produced at the WRF was to be used during summer. The rest was to be discharged to maintain the stream flow in the creek. In the final EIR submitted in 1997, the minimum daily average discharge of 4.32×10^3 m^3/d (1.1 Mgal/d) was set to minimize adverse impacts on the aquatic species of concern and was not likely to jeopardize the continued existence of the creek's species (City of San Luis Obispo, 1997).

Institutional Arrangements

Recognizing the benefit of keeping the discharge for instream habitat, various authorities came into play. Issues the city faced during the water reuse project planning and the agencies involved in the process are summarized in Table 21-10. One of the biggest challenges was the compliance to the Endangered Species Act. The city obtained a State Revolving Fund (SRF) loan and a Water Recycling Construction Program (WRCP) grant. Because

Table 21-10

Issues with the water reuse project, involved regulatory agencies, and resolutions; San Luis Obispo, CA[a]

Issues	Resolution	Involved agencies
NPDES permits for discharged wastewater/reclaimed water	Meeting water quality requirements by upgrading to tertiary treatment	RWQCB
Proposal of the water reuse project, including petition for the changes in discharge locations	Submit Environmental Impact Report under the California Environmental Quality Act (CEQA). Minimum discharge of 0.05 m^3/s (1.7 ft^3/s) was suggested for environmental use	SWRCB
Protest on city's petition for the changes in discharge locations	Most protests were withdrawn after completion of EIR. The remaining protest was resolved through a Memorandum of Understanding (MOU), which required the City to take mitigation measures.	DFG
Application to State Revolving Fund and Water Recycling Construction Program (WRCP) grant	Compliance to state and federal requirements. U.S. EPA to request a formal consultation with appropriate agencies	SWRCB, U.S. EPA
Compliance to Endangered Species Act, and issuance of Biological Opinions	Formal consultation to investigate the impact of the project on aquatic species. The requirement of additional discharge was suggested	U.S. FWS, NMFS
Compliance to the California water recycling criteria	Recycling water quality requirements are met by the plant upgrade	California DHS
Minimum discharge under California Water Code Section 1212	Reclaimed water appropriated for the minimum discharge requirement cannot be used for other purpose	SWRCB
Response to the formal consultations	Raised the minimum discharge from 0.05 to 0.07 m^3/s (1.7 to 2.5 ft^3/s)	U.S. FWS, NMFS

[a]Compiled from DiSimone (2006) and City of San Luis Obispo (1997).

the SRF funding source was federal, the city was mandated to comply with federal laws including the Endangered Species Act, and a formal consultation with the appropriate federal agencies was required (DiSimone, 2006). National Marine Fisheries Service (NMFS), the federal agency under National Oceanic and Atmospheric Administration (NOAA), conducted the consultation and issued a Biological Opinion for steelhead. Prior to the project's initiation, the city performed several steelhead surveys for abundance and emigration monitoring as part of compliance with the Biological Opinion. In particular, a 1993 abundance survey of the entire watershed found significant numbers and also a significant proportion of the steelhead population were residing in San Luis Obispo Creek downgradient of the city's discharge point. Based on these data, NMFS indicated its desire for U.S. EPA to request a reinitiated formal consultation to address the findings. The reinitiated formal consultation resulted in an increase in the minimum daily average discharge of reclaimed water to the creek from 4.32×10^3 m³/d (1.1 Mgal/d) to 6.0×10^3 m³/d (1.6 Mgal/d).

Water Reuse Options

After a series of studies and reinitiated formal consultation with the NMFS, the city agreed to maintain 6.0×10^3 m³/d (1.6 Mgal/d), minimum daily average discharge. The San Luis Obispo Creek augmented with reclaimed water is shown on Fig. 21-10. The decision impacted the potential water reuse applications greatly. In conformance with the California Water Code 1212, the reclaimed water discharged under the minimum discharge requirements is appropriated fully for the in-stream uses and, therefore it cannot be used for any other purpose.

The minimum discharge requirement resulted in the reduction of available reclaimed water to maximum of 7.5×10^3 m³/d (2.0 Mgal/d) for other originally planned uses such as irrigation and industrial applications. Because the irrigation water demand is seasonal,

Figure 21-10

San Luis Obispo Creek, San Luis Obispo, CA, augmented with reclaimed water. (Coordinates: 35.252 N, 120.676 W) (Courtesy of K. DiSimone, City of San Luis Obispo.)

the year round use of reclaimed water would be restricted by the maximum demand of irrigation water during summer. The irrigation use is also subject to high peak flow demand as landscape irrigation of public areas is restricted to nighttime. To ensure the feasibility of the water reuse project, the city is in the process of identifying potential year round and daytime reclaimed water users.

Lessons Learned

An upgrade of the wastewater treatment process to comply with the stringent waste discharge requirements resulted in recognition of the benefits from reclaimed water, rather than just discharging treated effluent into the San Luis Obispo Creek. The discovery of endangered and rare aquatic species further shifted the focus from water reuse for conventional purposes such as irrigation to in-stream environmental enhancement. The city worked with various regulatory agencies to find the best solutions to balance the water resources needs and environmental restoration and enhancement through the water reuse project.

21-8 CASE STUDY: SANTEE LAKES, SAN DIEGO, CALIFORNIA

The Santee Recreational Lakes project developed initially as an economical alternative to wastewater disposal in the Pacific Ocean. The series of lakes, known as the Santee Lakes Recreation Preserve, are supplied with reclaimed water and used for various recreational activities (see Fig. 21-11). Currently, the Santee Lakes Recreation Preserve

Figure 21-11
Overhead view of the seven lakes that comprise the Santee Lakes Recreation Preserve, Santee, CA (Coordinates: 32.850 N, 117.006 W.) (Courtesy of Padre Dam Municipal Water District staff.)

encompasses about 77 ha (33 ha of open water) and hosts more than 550,000 visitors annually. The following case study presents a review of the development and success of the Santee Lakes system. Historical background material on the Santee Lakes projects may be found in Stevens (1971).

Setting

The City of Santee is located about 28 km (16.8 mi) northeast of San Diego, California, in Eastern San Diego County. The Santee area was mostly agricultural land for the first half of the twentieth century, but was quickly developed as the population increased in Southern California. The current population of Santee is about 54,500.

Water Management Issues

Historically, wastewater from the City of Santee was discharged to Sycamore Creek, a tributary to the San Diego River that flows to the Pacific Ocean. However, in the late 1950s, the California State Water Resources Control Board (CSWRCB) established discharge standards prohibiting the creek discharge. As an alternative to the relatively expensive options of pipeline transport with ocean disposal or tertiary treatment and creek discharge, a reclaimed water system was envisioned. Subsequent updates of regional water quality plans have imposed more stringent requirements on many water quality parameters.

Implementation

The implementation consists of three phases. The first was a primary sedimentation system with sludge digestion and pond treatment with effluent placed in two constructed lakes. Decommissioned sand and gravel mining pits located in the dry streambed of Sycamore Canyon were reconfigured by dikes to form surface impoundments. The water quality in the storage basins was determined to be of acceptable quality to allow public recreational access to the Santee Lakes site starting in 1961. The second phase occurred due to public pressure and desire to demonstrate the safety and use of the storage basins for recreational purposes, therefore, supplemental treatment was developed. An activated sludge plant with denitrification capability and a capacity of 14.4×10^3 m^3/d (3.8 Mgal/d) was installed in the early 1970s. Filtration and phosphorous removal were accomplished by applying effluent to infiltration basins (see Chap. 22) in cobbly, silty, sandy soil underlain at a depth of 4 to 5 m (13 to 16 ft) by a clay layer. Percolate is collected subsequently for filling the basins (lakes). The lake system was expanded to seven lakes with a total surface area of 24 ha (60 ac). The hydraulic flow is sequential with the overflow from one lake to the subsequent lake (see Fig. 21-11). The last lake discharged to surface waters. Because of operational problems with the nitrification-denitrification facilities and permit restrictions, the plant was operated at 3.6×10^3 m^3/d (0.95 Mgal/d) during the 1980s into the 1990s.

The third phase was completed in 1995 with the construction of a 7.2×10^3 m^3/d (1.9 Mgal/d) Bardenpho plant and a tertiary plant consisting of a coagulation and flocculation system using alum and a lamella settler for turbidity and excess phosphorous removal followed by a denitrification filter using methanol as a carbon source. Chlorine disinfection with a Ct of 450 mg·min/L (see Chap. 11) was applied to all effluent. Effluent to be used in the lake system is dechlorinated with sulphur dioxide. This system met the California water quality regulations for full body contact, for discharge to Sycamore Creek, and for distribution to irrigation and industrial customers in the vicinity.

The treatment process provides multiple barriers for nitrogen, phosphorous, and to some extent, pathogens. Turbidity is monitored continuously before and after filtration and in the final effluent. Chlorine concentration is monitored downstream of the inlet to the contact basin and at the outlet with feed rates controlled by a multiple feedback loop system that also responds to flow. A bypass system is provided, allowing diversion of any noncompliant flows to a downstream regional treatment system.

Recreational activities that occur at Santee Lakes include fishing, boating, and camping (see Fig. 21-12). The fishing program evolved from a catch-and-release only program to fish harvesting for eating and catch and release, depending on species and an aquaculture program. The lakes are stocked all year with catfish and

Figure 21-12
Views of the Santee Lakes Recreation Preserve: (a) information center, (b) typical signage, (c) picnic area located along shore of lake number 4, and (c) full service camp site. (Courtesy of Padre Dam Municipal Water District staff.)

a few other popular species and in winter with trout. A swimming pool facility supplied with reclaimed water was developed and determined to be safe based on the results of microbiological testing and careful observation of swimmers in a public health study. However, it was abandoned because the sand bottom of the pool contributed to turbidity, which potentially shielded pathogens from disinfection. Swimming and wading is now prohibited in the lakes and streams at Santee. Kayaks, canoes, and pedal boats are all available for rental. The recreation area includes a campground with accommodations for 300 full hookup campsites. Many miles of paved trails for walking, running, and biking have been installed around the lakes. Other features include areas for picnicking, playgrounds, a general store, a clubhouse, and swimming pools. The park is operated in two sections, based on the type of use with approximately one-half for camping and one-half for day use. Some lakes are stocked more than others. Tube wading is allowed in 5 lake(s). One lake is used to raise bass.

Lessons Learned

An interesting aspect of the Santee Lakes Recreation Preserve is that the facility is completely self-supported through user fees. The recreation area has become a popular destination in the San Diego area. It is also interesting to note a high level of public acceptance that has been achieved and safety of the reclaimed water for water-related recreation activities has been demonstrated.

Recycled water in the lakes has a total dissolved solids content of between 800 and 900 mg/L. The phosphorous content is very low at about 0.05 mg/L and nitrogen content is typically about 2 mg/L. Algae control is achieved by application of a chelated copper sulfate solution. Aquatic weed control is achieved using a weed harvesting barge that cuts weeds about 1.5 m below the surface and removes them from the water column.

For several years prior to the installation of the Bardenpho and tertiary facilities, the infiltration system served to remove phosphorous as well as turbidity. However, no mechanism was available to remove the adsorbed phosphorous, and the soil column became saturated. An interim treatment step to remove phosphorous consisting of alum addition with sedimentation in a shallow basin was used until the Bardenpho upgrade was installed.

Operation of the infiltration basins contributed to the clogging of the soil and decreased the infiltration rate. The basins were continuously flooded instead of intermittently flooded and dried on a weekly, more or less, basis. When the infiltration rate decreased the ponds were dried and the surface ripped with a bulldozer. Ripping served to drive the organic material deeper into the soil as much as to break up the material to improve porosity.

The system has provided high-quality water to the lakes for many years. There have been no known cases of illness from the incidental contact of fishing and boating. Bacterial water quality degrades slightly through the system because of the bird populations and in winter runoff from lake margins. Monthly monitoring for over a year for *Cryptosporidium* and *Giardia* did not indicate any viable organisms.

PROBLEMS AND DISCUSSION TOPICS

21-1 From a city planning perspective, who is responsible for deciding to implement recreational water features in urban areas? What is the justification for creating these environmental and recreational areas? What are the implications of utilizing reclaimed water for these applications?

21-2 Environmental and recreational uses of reclaimed water are related closely to the discharge of treated wastewater into the receiving water body. What factors differentiate environmental water reuse from a wastewater discharge?

21-3 Discuss the reasons why mitigation wetlands using reclaimed water may not be as successful as natural wetlands. Cite a minimum of three articles from the literature to support your reasons.

21-4 Obtain a copy of the California and Florida guidelines pertaining to the use of reclaimed water in natural, altered, and constructed wetlands. What is the basis for any differences between these two sets of regulations?

21-5 What management options could be used to minimize the impact of heavy metals in reclaimed water on the health of wildlife and the aquatic environment? Cite a minimum of two articles from the literature where different management options were employed.

21-6 What are the potential health risks associated with consuming fish from a lake filled with reclaimed water? Cite a minimum of two references related to health risks.

21-7 Describe the typical diurnal interrelationship between dissolved oxygen, phytoplankton, and pH in a pond system filled with reclaimed water. What effect does the degree of treatment have on your answer?

21-8 It has been found that BOD_5 and TSS in the effluent from the Arcata Marsh increases following storm events. What explanation(s) can you offer to account for the increase? What studies should be conducted to confirm your explanation(s)?

REFERENCES

City of San Luis Obispo (1997) *Water Reuse Project, Final Environmental Impact Report*, SCH No. 92031048, City of San Luis Obispo Utilities Department, San Luis Obispo, CA.

Crites, R. W., E. J. Middlebrooks, and S. C. Reed (2006) *Natural Wastewater Treatment Systems*, Taylor and Francis, Boca Raton, FL.

Crites, R. W., and G. Tchobanoglous (1998) *Small and Decentralized Wastewater Management Systems*, WCB/McGraw-Hill, Boston, MA.

DiSimone, K. (2006) "Dedicated Recycled Water Discharge for Steelhead Trout," in *Proceedings, WateReuse Association California Section Annual Conference*, WateReuse Association, Alexandria, VA.

Jobling, S., M. Nolan, C. R. Tyler, G. Brighty, and J. P. Sumpter (1998) "Widespread Sexual Disruption in Wild Fish," *Env. Sci. Technol.*, **32**, 17, 2498–2506.

Kadlec, R. H., and R. L. Knight (1996) *Treatment Wetlands*, Lewis Publishers, Boca Raton, FL.

Kolpin, D. W., E. T. Furlong, M. T. Meyer, E. M. Thurman, S. D. Zaugg, L. B. Barber, and H. T. Buxton (2002) "Pharmaceuticals, Hormones, and Other Organic Wastewater Contaminants in U.S. Streams, 1999-2000: A National Reconnaissance," *Env. Sci. Technol.*, **36**, 6, 1202–1211.

Kondolf, G. M., and E. R. Micheli (1995) "Evaluating Stream Restoration Projects," *Environ. Mgmt.*, **19**, 1, 1–15.

Mitsch, W. J., and J. G. Gosselink (2000) *Wetlands*, Van Nostrand Reinhold Co., New York.

Rodgers-Gray, T. P., S. Jobling, S. Morris, C. Kelly, S. Kirby, A. Janbakhsh, J. E.Harries, M. J. Waldock, J. P. Sumpter, and C. R. Tyler (2000) "Long-Term Temporal Changes in the Estrogenic Composition of Treated Sewage Effluent and its Biological Effects on Fish," *Env. Sci. Technol.*, **34**, 8, 1521–1528.

Saldi-Caromile, K., K. Bates, P. Skidmore, J. Barenti, and D. Pineo (2004) *Stream Habitat Restoration Guidelines*, Final Draft, Washington Departments of Fish and Wildlife and Ecology, the U.S. Fish and Wildlife Service, Olympia, WA.

Sawyer, C. N. (1947) "Fertilization of Lakes by Agricultural and Urban Drainage," *J. New Engl. Water Works Assoc.*, **61**, 109–127.

Stevens, L. (1971) *The Town that Launders its Water*, Coward, McCann, & Geoghegan, Inc., New York.

Tchobanoglous, G. (1993) "Constructed Wetlands and Aquatic Plant Systems: Research, Design, Operational, and Monitoring Issues," in G. A. Moshiri (ed.), *Constructed Wetlands for Water Quality Improvement*, 23–34, Lewis Publishers, Boca Raton, FL.

Tonkovic, Z., and S. Jeffcoat (2002) "Wastewater Reclamation for Use in Snow-Making within an Alpine Resort in Australia—Resource Rather Than Waste," *Water Sci. Technol.*, **46**, 6–7, 297–302.

U.S. EPA (1993) *Constructed Wetlands for Wastewater Treatment and Wildlife Habitat: 17 Case Studies,* 832-R-93-005, U.S. Environmental Protection Agency, Office of Water, Washington, DC.

U.S. EPA (2000) *Design Manual Constructed Wetlands for Treatment of Municipal Wastewater,* 625-R-99-010, U.S. Environmental Protection Agency, Office of Research and Development, Cincinnati, OH.

U.S. EPA (2004) *Guidelines for Water Reuse*, EPA/625/R-04/108, U.S. Environmental Protection Agency and U.S. Agency for International Development, Washington, DC.

van Aerle, R, M. Nolan, S. Jobling, L. B. Christiansen, J. P. Sumpter, and C. R. Tyler (2001) "Sexual Disruption in a Second Species of Wild Cyprinid Fish (the Gudgeon, *Gobio Gobio*) in United Kingdom Freshwaters," *Env. Tox. Chem.*, **20**, 12, 2841–2847.

Williams, C. R., R. D. Jones, and S. A. Wright (1996) "Mosquito Control in a Constructed Wetland," in *Proceedings, WEFTEC' 96*, Water Environment Federation, Dallas, TX.

22 Groundwater Recharge with Reclaimed Water

    WORKING TERMINOLOGY   1246

22-1   PLANNED GROUNDWATER RECHARGE WITH RECLAIMED WATER   1248
    Advantages of Subsurface Storage   1248
    Types of Groundwater Recharge   1249
    Components of a Groundwater Recharge System   1250
    Technologies for Groundwater Recharge   1251
    Selection of Recharge System   1253
    Recovery of Recharge Water   1254

22-2   WATER QUALITY REQUIREMENTS   1255
    Water Quality Challenges for Groundwater Recharge   1255
    Degree of Pretreatment Required   1255

22-3   RECHARGE USING SURFACE SPREADING BASINS   1256
    Description   1256
    Pretreatment Needs   1257
    Hydraulic Analysis   1259
    Operation and Maintenance Issues   1268
    Performance of Recharge Basins   1271
    Examples of Full-Scale Surface Spreading Facilities   1280

22-4   RECHARGE USING VADOSE ZONE INJECTION WELLS   1282
    Description   1282
    Pretreatment Needs   1283
    Hydraulic Analysis   1285
    Operation and Maintenance Issues   1285
    Performance of Vadose Zone Injection Wells   1286
    Examples of Operational Full-Scale Vadose Zone Injection Facilities   1286

22-5   RECHARGE USING DIRECT INJECTION WELLS   1287
    Description   1287
    Pretreatment Needs   1288
    Hydraulic Analysis   1288
    Operation and Maintenance Issues   1290
    Performance of Direct Injection Wells   1291
    Examples of Full-Scale Direct Aquifer Injection Facilities   1292

22-6	**OTHER METHODS USED FOR GROUNDWATER RECHARGE** 1293	
	Aquifer Storage and Recovery (ASR) 1293	
	Riverbank and Dune Filtration 1294	
	Enhanced River Recharge 1295	
	Groundwater Recharge Using Subsurface Facilities 1296	
22-7	**CASE STUDY: ORANGE COUNTY WATER DISTRICT GROUNDWATER REPLENISHMENT SYSTEM** 1296	
	Setting 1297	
	The GWR System 1297	
	Implementation 1297	
	Lessons Learned 1298	
	PROBLEMS AND DISCUSSION TOPICS 1299	
	REFERENCES 1300	

WORKING TERMINOLOGY

Term	Definition
Anaerobic ammonia oxidation (ANAMMOX)	A form of autotrophic biological nitrogen removal. In the ANAMMOX reaction ammonia serves as the electron donor and nitrite serves as the electron acceptor in the absence of molecular oxygen.
Aquifer	An underground water bearing stratum with sufficient permeability to transmit and yield water in usable quantities.
Aquifer storage and recovery (ASR)	Use of an aquifer for the storage of reclaimed water; water from the aquifer is pumped out subsequently for use when needed.
Assimilable organic carbon (AOC)	AOC is determined by plating out a sample solution incubated with *Pseudomonas fluorescens* and counting the resulting bacterial colonies. The AOC is expressed in terms of the carbon concentration of a standard acetate solution producing the same bacterial growth response.
Confined aquifer	A water bearing stratum (aquifer) restricted by two relatively impermeable layers (e.g., clay lenses).
Disinfection byproduct formation potential (DBPFP)	A test used to assess the potential to form disinfection byproducts when oxidizing chemicals used for disinfection are added to wastewater.
Dissolved organic carbon (DOC)	Concentration of organic matter which passes through a filter of a given size (typically 0.45 μm).
Effluent organic matter (EfOM)	Organic matter remaining in reclaimed water after treatment.
Indirect potable reuse	The planned incorporation of reclaimed water into a raw water supply, such as in potable water storage reservoirs or a groundwater aquifer, resulting in mixing, dilution, and assimilation, thus providing an environmental buffer.
Infiltration rate	The rate at which water enters the soil.
Karst limestone	Porous limestone formation containing deep fissures and sinkholes.

Meelop filter index (MLFI)	An index used to evaluate clogging of native aquifer materials by solids in the recharge water. The MLFI is determined by passing water through columns filled with native aquifer material at flow rates higher than those through the aquifer around the well.
Membrane filtration index (MFI)	An index used to assess the suspended solids content in the water. The MFI is determined by plotting the infiltration rate as a function of volume passing through a membrane filter.
Mounding	The localized increase in elevation of the water table below a recharge basin.
National pollutant discharge elimination system (NPDES)	As authorized by the Clean Water Act, the NPDES is a permit program that controls water pollution by regulating point sources that discharge pollutants into waters of the United States.
Natural organic matter (NOM)	Dissolved and particulate organic constituents that are derived typically from three sources: (1) terrestrial environment (mostly humic materials), (2) aquatic environment (algae and other aquatic species and their byproducts), and (3) microorganisms in the biological treatment process.
Perched aquifer	A water bearing stratum that is located over an impermeable layer (typically, a clay lens).
Personal care products and pharmaceuticals (PCPP)	A term used to describe a wide range of generic and brand name products that contain synthetic organic chemicals and hormonally active agents. When released into the environment, these compounds may have endocrine disrupting effects.
Recharge basins	Shallow earth basins used for groundwater recharge by surface spreading. Also known as rapid infiltration basins (RIBs).
Riverbank filtration (RBF)	A natural filtration system where the river bottom and bank serve as the interface between the surface water and the aquifer being recharged.
Saturated zone	An underground water bearing stratum in which all of the pore spaces are filled with water, corresponding typically to the groundwater.
Soil aquifer treatment (SAT)	The treatment achieved as reclaimed water passes through the soil vadose zone to an aquifer, and subsequently as it flows through the aquifer.
Soluble microbial product (SMP)	Organic compounds produced and released as a result of metabolic activity.
Transmissivity	A measure of rate at which an aquifer can transmit water. More specifically, transmissivity is the volume of water flowing through a unit cross-sectional area of an aquifer under a unit hydraulic gradient in a given amount of time.
Unconfined aquifer	An aquifer that is not confined by an impermeable layer on the top so that the groundwater level can rise or fall.
Unsaturated zone	The zone between the ground surface and the groundwater table including the capillary fringe (use of the term vadose zone is preferred).
Vadose zone	The zone between the ground surface and the groundwater table.

The natural replenishment of underground water occurs very slowly; therefore, excessive continued exploitation of groundwater at a rate greater than this replenishment causes declining groundwater levels in the long term and, if not corrected, can result in mining (depletion) of groundwater resources. To increase the supply of groundwater, artificial recharge of groundwater basins is increasingly important in groundwater management and particularly in situations where the conjunctive use of surface water and groundwater resources is being considered.

Groundwater recharge with reclaimed water, the focus of this chapter, is an approach to water reuse that results in the planned augmentation of groundwater for various beneficial uses. The principal beneficial uses of groundwater include municipal water supply, agricultural irrigation, and industrial water supply. The purposes of artificial recharge of groundwater, identified by Bouwer (1978), Todd (1980), and Asano (1985), are:

- To reduce, stop, or even reverse the decline of groundwater levels
- To protect underground freshwater in coastal aquifers against saltwater intrusion
- To store surface water, including flood or other surplus water and reclaimed water, for future use
- To negate potential problems of land subsidence

Unplanned groundwater recharge with wastewater occurs in a number of situations. For example, groundwater recharge occurs in land treatment of wastewater and in irrigation systems where water is applied at rates exceeding the evapotranspiration demand. Other examples of unplanned recharge with wastewater include leaking from wastewater collection systems and unlined storage basins. In some areas, the disposal of municipal and industrial wastewater via percolation and infiltration is used where surface discharge is not acceptable, or to eliminate the need for an NPDES permit. Unfortunately, these forms of unplanned recharge may compromise groundwater quality and do not necessarily reflect best practice, and therefore, are not considered in this chapter, which deals with planned groundwater recharge.

22-1 PLANNED GROUNDWATER RECHARGE WITH RECLAIMED WATER

Since the 1960s, groundwater recharge with reclaimed water has been practiced for both nonpotable and indirect potable reuse applications in the United States. One of the major purposes associated with groundwater recharge systems is to provide long-term storage; however, water quality improvements through recharge often provide important secondary benefits. Water quality improvements are an essential part of groundwater recharge systems with reclaimed water when the withdrawn water is used for potable purposes.

Advantages of Subsurface Storage

When compared to surface water storage, groundwater recharge has several important advantages, as listed in Table 22-1. The major advantage is the potential improvement in water quality that occurs during the groundwater recharge process. Surface storage with reclaimed water can result in a significant deterioration of water quality from secondary contamination and from algal blooms (see Sec. 14-7 in Chap. 14). Evaporation losses can also result in increased salinity, further adding to accruing salinity problems in arid areas.

Water quality improvements during subsurface transport are mediated by several mechanisms. The principal mechanisms include filtration, biotransformation, adsorption, and hydrolysis. As a result of biodegradation, the transformation of organic compounds and nitrogen may be sustained indefinitely. The most important removal mechanisms are associated with surface mediated reactions as most microorganisms in the subsurface are attached. Therefore, surface area contact during subsurface flow is an important factor.

Table 22-1 The advantages and disadvantages of subsurface storage as compared to surface storage

Advantages
Replenishment of depleted aquifers
Prevent land subsidence from overdraft
Reduce energy costs for pumping from deeper aquifers
Negligible evaporation losses
Low risk of secondary contamination
No algal blooms
Soil aquifer treatment may improve water quality, depending on quality of reclaimed water
Prevent negative environmental impacts from the construction of a surface storage facility
Suitable sites for surface water reservoirs may not be available or environmentally acceptable
Storing water underground can avoid the adverse environmental impacts of building dams
The construction, operation, and maintenance cost of artificial recharge may be less than the cost of equivalent surface water reservoirs
The aquifer serves as an eventual distribution system and may eliminate the need for transmission pipelines or canals for surface water

Disadvantages
Complex geochemical reactions can occur
Absorbed material can be desorbed, depending on characteristics of reclaimed water
Reverse osmosis effluent may have to be stabilized chemically

For subsurface transport through porous media in sand and gravel aquifers, ample surface area is available to support important mechanisms for water quality improvement. For flow through fractured media such as limestone, water quality improvements may not occur because of the limited surface area contact. In the following discussion, it is assumed that suitable aquifer material is present to support transformations.

Types of Groundwater Recharge

Based on practices that have evolved in the United States, the types of groundwater recharge now practiced may be classified as (1) groundwater augmentation for indirect potable reuse, (2) control of seawater and brackish water intrusion into freshwater aquifers, and (3) aquifer storage and recovery. Each of these types of recharge is introduced in the following discussion and examined in greater detail in separate sections.

Groundwater Augmentation for Indirect Potable Reuse

Groundwater recharge with reclaimed water that has undergone advanced treatment and will undergo further treatment as it moves through the underground aquifer is now used to augment existing groundwater resources for potable purposes. With respect to potable reuse, groundwater recharge has several advantages in terms of social issues and public perception. In addition, reclaimed water that has undergone natural treatment is well accepted by the public, when recovered water is used for potable purposes.

Control of Seawater and Brackish Water Intrusion

Seawater intrusion into groundwater basins may be a problem in areas adjacent to the ocean. As groundwater pumping occurs, the freshwater piezometric head may be reduced to below sea level allowing for seawater to enter an aquifer. Injection wells are used to prevent seawater intrusion by creating a hydraulic barrier between the groundwater basin and the ocean. In areas where groundwater withdrawals have exceeded natural recharge such as Los Angeles and Orange counties in southern California, the need to control saltwater and brackish water intrusion led to the development of an extensive network of injection wells. Some wells can inject water into several different aquifers thereby providing protection for both near surface aquifers and deep aquifers. Because the water may be recovered for potable purposes and maintenance of deep injection wells is costly, most of the water is subjected to advanced treatment prior to injection.

Aquifer Storage and Recovery

Storage, as discussed in Chap. 14, is a necessary component of almost all reclaimed water systems because the supply of reclaimed water is relatively constant and the demand may vary both diurnally and seasonally. A classic example of seasonal demand occurs when the primary use of the reclaimed water is for irrigation. Peak demand for irrigation water is in the summer when evapotranspiration rates are greatest while the supply of reclaimed water remains relatively constant through the year. Consequently, the excess reclaimed water produced during winter months must be stored when demand is less than supply. The stored water then becomes available to meet peak demand in the summer. In Arizona, the seasonal disparity between supply and demand has led to the development of decentralized water reclamation, storage, and reuse systems to meet the water supply requirements of continued population growth. As new development occurs, more wastewater is being reclaimed. The reclaimed water is used primarily for irrigation in the summer months and is recharged into the ground and stored during winter months, when irrigation demand is low. The water may then be recovered for either potable or nonpotable purposes as needed. This decentralized strategy for water reuse also eliminates the need for NDPES permits.

Components of a Groundwater Recharge System

A groundwater recharge system may be viewed as a multicomponent system containing both above ground and below ground components as illustrated on Fig. 22-1. The above ground components consist of the reclaimed water source and associated treatment processes prior to groundwater recharge and subsequent extraction for use. The reclaimed water is delivered to the groundwater recharge site where the method of recharge depends on whether the receiving aquifer is confined (saturated) or unconfined (with saturated and unsaturated zones). Following infiltration or injection, reclaimed

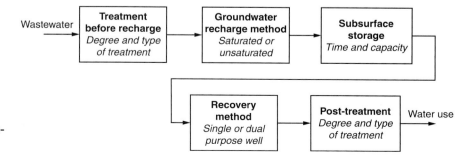

Figure 22-1
The five major components of a groundwater recharge system. The bold words identify the major components. The words in italics describe major factors for design.

water is then stored underground. Water quality improvements are often a function of the storage time as the water moves through the underground strata. The injected water is withdrawn later for use or stored underground for future recovery.

Three types of groundwater recharge are commonly used with reclaimed water: (1) surface spreading, (2) injection into the vadose zone, and (3) direct injection into aquifer. These three methods are illustrated on Fig. 22-2.

Technologies for Groundwater Recharge

Surface Spreading

Surface spreading is the simplest, oldest, and most widely applied method of artificial recharge (Todd, 1980). In surface spreading, reclaimed water from the spreading basin infiltrates and percolates through the vadose (unsaturated) zone. Views of large surface spreading basins in Israel are shown on Fig. 22-3. Surface spreading using recharge basins is the most favored method of groundwater recharge. Recharge basins are favored because space is utilized effectively and they require only simple maintenance. In general, infiltration rates are highest where soil and vegetation are undisturbed.

Where hydrogeological conditions are favorable for groundwater recharge with spreading basins, water reclamation can be accomplished relatively simply by the soil aquifer treatment (SAT) process. The necessary treatment can be obtained as the wastewater percolates through the vadose zone, down to the groundwater and then moves some distance through the aquifer. Pretreatment requirements for SAT systems are considered in Sec. 22-2.

The advantages of groundwater recharge by surface spreading are:

- Groundwater supplies may be replenished in the vicinity of metropolitan and agricultural areas where groundwater overdraft is severe.
- Surface spreading provides the added benefits of treatment as the water percolates through the vadose zone and subsequently through the aquifer.

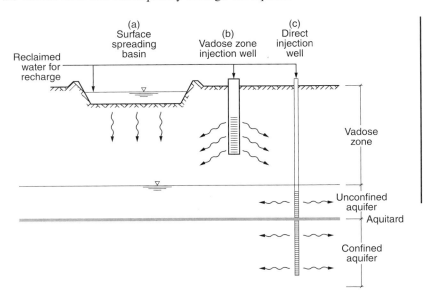

Figure 22-2

Principal methods for groundwater recharge: (a) surface spreading using recharge basins, (b) injection wells in vadose zone, and (c) direct injection wells into an aquifer.

Figure 22-3
Overhead view of large reclaimed water spreading basins in Israel (Coordinates: 31.850 N, 34.710 E). (Courtesy of MEKOROT, Israel National Water Company.)

Injection Wells into Vadose Zone

Vadose zone injection wells, the newest technology developed for groundwater recharge, require the presence of an unsaturated zone. Vadose zone injection wells were developed as a consequence of increasing land costs in urbanized areas. As land prices increased, the need for an alternative cost-effective injection technology to recharge basins for unsaturated aquifers became apparent. The major costs associated with surface spreading are for land acquisition and the distribution system necessary to deliver water to the recharge basins. Because vadose zone injection wells and direct injection wells (discussed below) are not land intensive, they may be located at different locations throughout a distribution system.

Direct Injection Wells into an Aquifer

Direct subsurface recharge is achieved when water is conveyed and placed directly into an aquifer. In direct injection, generally, highly treated reclaimed water is pumped directly into the groundwater zone, usually into a well-confined aquifer. Groundwater recharge by direct injection is practiced:

- Where groundwater is deep or where the topography or existing land use makes surface spreading impractical or too expensive.
- When direct injection is particularly effective in creating freshwater barriers in coastal aquifers against intrusion of saltwater.

Both in surface spreading and direct injection, locating the extraction wells as great a distance as possible from the spreading basins or the injection wells increases the flow path length and residence time of the recharged water. These separations in space and time contribute to the mixing of the recharged water and the aquifer contents, and the loss of identity of the recharged water that originated from municipal wastewater.

The latter is an important consideration in facilitating public acceptance for a successful water reuse project.

Selection of Recharge System

Surface spreading using recharge basins is the most common method for groundwater recharge, but the use of this method is limited to unconfined aquifers with a vadose zone. The use of injection wells in the vadose zone in unconfined aquifers, a recent development, is less common. Direct injection wells may be used with either confined or unconfined aquifers; they may also be designed to inject into several different aquifers at different depths. In addition, direct injection wells may be designed as dual purpose wells capable of both injecting and recovering water. The major factors affecting the selection and design of different recharge methods are listed in Table 22-2.

The two most important considerations for selection of a recharge method are the type of aquifer and the availability of land. If only saturated conditions (i.e., no vadose zone) exist, then direct injection wells are required. Where an unconfined aquifer with a vadose zone is present and land is readily available, recharge basins are a logical choice; if land is not readily available then vadose zone injection wells or direct injection wells may be used. Vadose zone injection wells are most economical when the aquifer is deep and extensive drilling is required if direct injection wells are used. For both vadose zone and direct injection wells pretreatment is required to remove solids and prevent clogging from biological growth. Because the flow cannot be reversed in vadose zone injection wells, clogging may be irreversible and the life span of vadose zone injection wells is uncertain. The advantages and disadvantages of various recharge methods are summarized in Table 22-3.

Table 22-2 Characteristics of principal aquifer recharge methodologies

Characteristic	Recharge basins	Vadose zone injection wells	Direct injection wells
Location where treatment occurs	Vadose zone and saturated zone	Vadose zone and saturated zone	Saturated zone
Aquifer type	Unconfined	Unconfined	Unconfined or confined
Pretreatment requirements	Secondary treatment[a]	Secondary treatment plus filtration[a]	Advanced treatment
Capacity	1000–20,000 $m^3/ha \cdot d$	1000–3000 $m^3/well \cdot d$	2000–6000 $m^3/well \cdot d$
Maintenance requirements	Drying and scraping	Drying and disinfection	Disinfection and flow reversal
Estimated life cycle	>100 yr	5–20 yr	25–50 yr
Estimated major capital costs,[b] US$	Land and distribution system[c]	$100,000–150,000 per well	$100,000–1,500,000 per well

[a] Additional treatment may be required if the recovered recharge water is to be used for potable purposes.
[b] ENRCC Index = 7800.
[c] Costs for recharge basins are principally for land acquisition and distribution system and can vary widely.

Table 22-3
Advantages and disadvantages of various groundwater recharge methods

Advantages	Disadvantages
Surface spreading	
• Relatively easy to construct and operate • Primary or secondary levels of pretreatment may be satisfactory	• Large land area required • Limited availability of suitable sites; soil characteristics are very important in site selection • Wetting and drying cycles required to maintain infiltration rates as well as vector control • Periodic bed maintenance required • Some evaporation losses from open water surface • Algae growth may affect clogging
Vadose zone injection	
• Relatively small site required • Negligible evaporation losses • Less expensive technology than direct injection wells • Greater potential for water quality improvement as compared to direct injection wells	• Relatively new technology • Only short-term life cycle data available • Soil characterization required • Special design and construction of well is necessary • Extensive pretreatment of wastewater is necessary to prevent clogging with solids and development of microbial growth • No effective method available to redevelop clogged well
Direct injection	
• Relatively small site required • May be used for both injection and extraction of reclaimed water • High rate of reclaimed water injection • Flow in well can be reversed for maintenance and redevelopment of well • Can be designed to recharge multiple aquifers	• Relatively expensive to construct • Energy intensive; high pressure pumping required for reclaimed water injection • Design and construction requires greater expertise than vadose zone injection wells • Extensive pretreatment of wastewater is necessary to prevent clogging with solids and development of microbial growth; may require a higher level of treatment than vadose zone injection wells • Limited additional improvement in water quality can be expected

Recovery of Recharge Water

After storage, water is recovered for final use using dedicated recovery wells or dual purpose aquifer storage and recovery wells. Posttreatment of the recovered water may be required for the final use. The location of groundwater recharge sites and recovery wells is often critical as distribution systems to deliver the water to a recharge site and to recover the water near a demand site can be a major expense. Therefore, the use of floodplains for groundwater recharge located near water reclamation plants is often practiced.

22-2 WATER QUALITY REQUIREMENTS

Water quality improvement is an essential part of a groundwater recharge system when the water is recovered for potable purposes, often referred to as *indirect potable reuse* (see Chap. 23). When water is injected directly into the aquifer, extensive treatment such as RO and advanced oxidation eliminates most water quality concerns prior to injection. When recharge basins are used, disinfected secondary or tertiary effluent may be applied to the recharge basins. When recovering the water for potable purposes, concerns over organic carbon, nitrogen, and pathogens must be eliminated. Natural attenuation processes are often sufficient to eliminate these concerns without significant posttreatment requirements. The processes are analogous to attenuation processes that naturally purify groundwater that was once surface water.

Water Quality Challenges for Groundwater Recharge

Groundwater recharge with reclaimed water presents a wide spectrum of technical and health challenges that must be evaluated carefully. Some basic questions related to water quality that need to be addressed include (Asano and Wassermann, 1980; Roberts, 1980; Crook et al. 1990; NRC, 1994):

- What treatment processes are available for producing water suitable for groundwater recharge?
- How do these processes perform in practice?
- How does water quality change during infiltration-percolation and in the groundwater zone?
- What do infiltration-percolation and groundwater passage contribute to the overall treatment system performance and reliability?
- What are the important health issues?
- How do these issues influence groundwater recharge regulations at the points of recharge and extraction?
- What benefits and problems have been experienced in practice?

With respect to the above concerns, water quality factors that are particularly significant in groundwater recharge with reclaimed water are: (1) microbiological quality, (2) total mineral content (total dissolved solids), (3) constituents prone to precipitation such as phosphates, (4) toxic constituents such as heavy metals, (5) nutrients, and (6) trace organic constituents.

Degree of Pretreatment Required

Pretreatment requirements vary considerably, depending on the purpose of groundwater recharge, sources of reclaimed water, recharge methods, and location. Although the surface spreading method of groundwater recharge is in itself an effective form of wastewater treatment, a certain degree of pretreatment must be provided to untreated municipal wastewater before it can be used for groundwater recharge. Pretreatment processes that leave high algal concentrations in the recharge water should be avoided, as algae can severely clog the soil of infiltration basins.

Due to concerns over the presence of trace constituents and scrutiny of indirect potable reuse projects, advanced treatment incorporating RO and advanced oxidation have been applied prior to recharge by surface spreading and direct injection. The demineralized product water is stabilized typically using lime and by blending flow to achieve a water that is not aggressive (see Chap. 9). The decision to apply advanced treatment will depend on several site specific factors, such as the quality of the natural groundwater and level of concern for trace constituents. Pretreatment requirements for the various recharge methods are discussed in their respective sections.

22-3 RECHARGE USING SURFACE SPREADING BASINS

Surface spreading using recharge basins [also known as rapid infiltration basins (RIBs)] is one of the most common and oldest methods for groundwater recharge. Recharge basins are often located in, or adjacent to, floodplains where permeable soils are present (see Fig. 22-23a). Vadose zones without restrictive layers that cause buildup of an excessive groundwater mound and unconfined aquifers with high transmissivity for lateral flow through the aquifer are necessary to prevent excessive mounding.

Description

Recharge basins may vary in size from 0.4 ha (1 ac) to greater than 4 ha (10 ac). Excavation is usually necessary to remove surface soils of low permeability and the excavated soil may be used to construct berms around individual recharge basins (see Fig. 22-4). To prevent erosion, which may introduce fine soil particles that will reduce infiltration rates, the basins are graded to allow sheet flow across the basin when water is first introduced at the bottom of the basin (see Fig. 22-4b).

General Operation

The general operation of recharge basins using reclaimed water requires the use of wetting and drying periods to maintain infiltration rates. As water is applied to the basins,

(a) (b)

Figure 22-4

Typical recharge basins: (a) empty basin at Conserve II in Florida and (b) reclaimed water applied to dry basin moves across the surface in sheet flow. (Coordinates: 28.493 N, 81.620 W.)

existing solids in the reclaimed water are removed in the upper layer of soil. In addition, biological activity increases the accumulation of organic matter in the upper layer of soil. In areas with high solar incidence, algae growth may be the major factor contributing to a reduction in infiltration rates with time. Infiltration rates will continue to decrease with time until the application of water is stopped. As the recharge basin is drained and allowed to dry, the organic material on the surface of the soil will desiccate allowing for the recovery of infiltration rates. If no drying cycle is used, infiltration rates will become unacceptably low unless a submerged cleaning device is used to remove the clogging materials.

Dual Function Basins

Recharge basins may be designed to support wildlife and maintain a population of fish. This dual function is accomplished by excavating a trough at the deep end of the recharge basin. During the drying cycle, fish and other aquatic life can find refuge in the water retained in the trough. When water is reapplied to the basins, the fish and aquatic life can use the full basin area for habitat.

Pretreatment Needs

Because recharge basins can be cleaned to recover infiltration rates and SAT provides robust improvements in water quality under most circumstances, wide ranges of water qualities have been used. Primary, secondary, and tertiary effluents have all been applied to recharge basins. The impact of various wastewater constituents is reported in Table 22-4. Regulatory constraints often require the use of tertiary effluent where the production of potable water is the end purpose. Primary effluent has only been used in test projects and is not presently used in any recharge sites. The Orange County Water District (OCWD) in Fountain Valley, California, is applying reclaimed water, which has been treated with RO and advanced oxidation, in groundwater recharge basins to supplement its potable water supply. The RO and advanced oxidation processes remove essentially all of the constituents of concern before the reclaimed water is applied to

Table 22-4

Impacts of wastewater constituents on groundwater recharge systems

Constituent of concern	Water quality impact	Infiltration impact
Organic carbon	Increased levels lower redox potential	Primary effluent will increase clogging from biofilm development
Trace organic compounds	Effect is negligible Several recalcitrant compounds persist	No impact
Nitrogen	Removal of ammonia occurs with wet/dry cycles Nitrified effluent should not be used	Reduced nitrogen concentrations do not limit algae growth
Suspended solids	Effect is negligible	Reduced suspended solids enhance infiltration rates
Pathogens	Effect is negligible	Effect is negligible

surface spreading basins. Additional details on the OCWD Groundwater Replenishment (GWR) system are presented in Sec. 22-7.

Impact on Water Quality

Water quality concerns associated with reclaimed water used for groundwater recharge include changes in organic constituents, oxidation-reduction (redox) potential, nitrogen, and pathogens.

Organic Matter The removal of organic matter is surprisingly independent of the level of pretreatment when water is recharged from spreading basins (AWWARF, 2001). The reason is the easily biodegradable material removed during wastewater treatment is also removed readily during groundwater recharge. In general, the subsurface environment may be viewed as a large biofilter and the level of pretreatment for the removal of biodegradable organics does not impact significantly the biological removal capacity of the subsurface system when infiltration and storage time scales are on the order of months or years.

Redox Potential As the easily biodegradable organics are removed near the soil/water interface, the oxygen in the water is consumed and the water entering the aquifer is often depleted of oxygen. This oxygen depletion can create an anoxic plume of recharged water that might have adverse interactions with native aquifer materials. Low redox potentials may lead to the solubilization of reduced iron, manganese, and arsenic from native aquifer materials. While solubilization has been observed historically with bank filtration systems in Europe, it has not been a major problem for recharge basins in the United States, although reduced conditions do develop in many aquifers.

Nitrogen Nitrogen removal during above ground treatment is often practiced and it eliminates concerns over potential nitrate contamination of groundwater. When comparing nitrified/denitrified effluents with secondary effluents containing reduced nitrogen, the level of oxygen demand is different. An effluent containing 20 mg/L of ammonium (NH_4^+-N) will have a nitrogenous oxygen demand in excess of 80 mg/L, while a nitrified/denitrified effluent with an NH_4^+-N concentration of 2 mg/L will have a nitrogenous oxygen demand of less than 10 mg/L. Ammonium is often removed by cation exchange onto soil particles during wetting cycles and the adsorbed ammonium is nitrified during drying cycles. The increased oxygen demand associated with ammonia can be a major factor in maintaining low redox potentials in both the vadose zone and the saturated zone.

The use of wet/dry cycles provides an opportunity for nitrogen removal to be sustained when NH_4^+-N is applied as aerobic/anoxic cycling occurs in the vadose zone. Nitrogen removal has been documented at many sites with variable soil conditions and effluent pretreatment. In Arizona, recharge basins may be permitted with nitrogen concentrations exceeding 10 mg NH_4-N/L if appropriate wet/dry cycles are used. Nitrate (NO_3^-) is not adsorbed to soils and there is usually insufficient organic carbon to both create anoxic conditions and support efficient denitrification. To remove 10 mg/L of NO_3^--N, approximately 40 mg/L of carbonaceous BOD (CBOD) is required. Nitrified effluents with nitrate concentrations in excess of 10 mg NO_3-N/L should not be used for groundwater recharge. Nitrified effluents may be applied to combined wetland/recharge basins where plants provide a source of organic carbon to stimulate denitrification.

Pathogens The removal of pathogens prior to wastewater discharge to infiltration basins is primarily for the safety of the public and workers and is analogous to other water reuse projects. The combination of filtration and biodegradation is effective at removing bacteria and parasites during subsurface transport. The survival of viruses is of primary concern during subsurface transport. If public access to recharge basins is limited, then extensive disinfection may not be necessary. Many recharge basins are operated with reclaimed water that is not disinfected prior to application.

Impact of Pretreatment on Hydraulic Capacity

Effluent pretreatment prior to discharge to recharge basins affects the development of clogging layers by controlling the levels of biodegradable organics, suspended solids, and nutrients. Secondary effluent typically contains higher levels of suspended solids as compared to tertiary effluents, resulting in more clogging from suspended solids. However, the suspended solids can limit sunlight penetration and reduce algae growth when solar incidence is high. The levels of easily biodegradable organics are not significantly different between secondary and tertiary effluents, therefore the development of a clogging layer from bacterial growth at the soil/water interface is not a strong function of pretreatment. Although nitrified/denitrified effluents have much lower levels of nitrogen as compared to conventional effluents, the concentrations of nutrients in nitrified/denitrified effluents do not limit the rate of algal growth. Because there are rarely concerns over eutrophication with the final use of recharged water, phosphorous removal to limit algal growth in recharge basins has not been practiced. Phosphorous is removed effectively by calcareous soils during groundwater recharge.

Hydraulic Analysis

The hydraulic capacity of infiltration basins is primarily a function of the vertical hydraulic conductivity of the soils and the rate at which a clogging layer develops (see Fig. 22-5). In addition, the relative time of wetting versus drying is an important factor as water cannot be added to the basins during drying cycles. If the groundwater

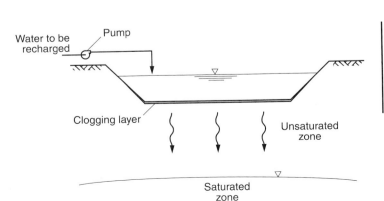

Figure 22-5

Processes related to the infiltration of water from a surface infiltration basin.

level is near the surface or perched water layers develop under the infiltration basin, then horizontal flow to dissipate the recharge water away from the basins may dominate the basin hydraulics (Morel-Seytoux, 1985). The depth of the water in the recharge basins may also affect the hydraulic capacity of the system; however, it is typically not an important factor.

Surface Water Infiltration

Surface water infiltration equations, developed first by Green and Ampt (1911) and subsequently modified by Bouwer (1966) and Neuman (1976), can be used to predict the rate of infiltration. A common form of the equation used to predict infiltration rates is:

$$v_i = K \frac{H_w + L_f - H_{cr}}{L_f} \quad (22\text{-}1)$$

where v_i = infiltration rate, m/d
K = hydraulic conductivity of the wetted zone, m/d
H_w = depth of water above the soil, m
L_f = depth of wetting front, m
H_{cr} = critical pressure head of soil for wetting, m (of water)

Vertical flow, uniform water content, and constant hydraulic conductivity in the wetted zone are assumed in the development of Eq. (22-1). The wetting front is the interface between wetted and nonwetted material. As water enters the recharge basin and begins to infiltrate, the wetting front develops below the basin and it is assumed that the wetting front moves uniformly towards the groundwater. The surface infiltration rate of the water, v_i, corresponds to the Darcy velocity. The actual velocity of the wetting front is greater than the surface infiltration rate as only the pore space is occupied by water. The actual K value of the wetted zone is normally less than the saturated hydraulic conductivity (K_{sat}) value as entrapped air prevents complete saturation of the soils below the recharge basin. In most applications, the K value in Eq. (22-1) is approximately one-half the K_{sat}. The water depth H_w, may range from zero when infiltration begins to more than several meters, for recharge basins and rivers. The pressure head at the wetting front, H_{cr}, is a function of the soil type. Values of H_{cr} range from -0.1 m or greater for coarse soils to -1 m or less for fine soils. As given by Eq. (22-1), the value of L_f increases and the wetting front moves deeper, v_i decreases until it reaches a constant value of K when L_f becomes much greater than $H_w - H_{cr}$. Therefore, the final infiltration rate for a recharge basin with a deep vadose zone is K and the effect of water depth on infiltration is not important.

Movement of the Wetting Front

Calculating the rate of advance of the wetting front dL_f/dt is important to estimate when the wetting front will reach the aquifer and begin to influence mounding. The rate of movement of the wetting front is equal to v_i/f, where f is the fillable porosity (difference between volumetric water content before and after wetting). Equation (22-1) may be rearranged and integrated to solve for t as follows:

$$t = \frac{f}{K}\left\{L_f - (H_w - H_{cr})\left[\ln\left(\frac{H_w + L_f - H_{cr}}{H_w - H_{cr}}\right)\right]\right\} \quad (22\text{-}2)$$

where t = the time since the start of infiltration
Other terms are as defined previously.

The total depth of water infiltrated into the ground, I_t, is equal to $f \times L_f$. Normally, f decreases with depth because the water content of the soils increases with depth which affects the results obtained using Eqs. (22-1) and (22-2). The soil profile must be divided into a number of layers of thickness, Δz, each with its own f value. Equation (22-2) may be used to calculate the time required to travel through each layer. The final relationship between I_t and t may be found by summing the results for $f\Delta z$ and Δt for each layer. Furthermore, this tabular procedure may be extended to nonuniform soils with layers of different hydraulic conductivity. However, a layer of soil with low hydraulic conductivity can limit the infiltration rate of the entire system. The K value of the layer with the lowest hydraulic conductivity will become the limiting infiltration rate as the wetting front passes through that layer.

EXAMPLE 22-1. Analysis of Infiltration in a Recharge Basin.

Reclaimed water is be infiltrated using a recharge basin. Assuming the parameters given below apply for the recharge basin, (a) develop a plot of the depth of the wetting front, L_f, for a distance of 10 m as a function of time, (b) develop a plot of the infiltration rate, v_i, as a function of time, and (c) determine the amount of water infiltrated through a 100 m² recharge basin.

H_w = height of water above ground = 0.7 m
K = hydraulic conductivity = 1 m/d (assume constant)
f = fillable porosity = 0.35, dimensionless
H_{cr} = −0.5 m

Solution
1. Determine the depth of the wetting front as a function of time for depths varying from 0.01 to 10 m using 0.5 m steps.
   a. Using Eq. (22-2) for a depth of 0.5 m, the time is computed as follows:

$$t = \frac{f}{K}\left\{L_f - (H_w - H_{cr})\left[\ln\left(\frac{H_w + L_f - H_{cr}}{H_w - H_{cr}}\right)\right]\right\}$$

$$t = \frac{0.35}{(1 \text{ m/d})}\left\{0.5 \text{ m} - [0.7 \text{ m} - (-0.5 \text{ m})]\left[\ln\frac{0.7 \text{ m} + 0.5 \text{ m} - (-0.5 \text{ m})}{0.7 \text{ m} - (-0.5 \text{ m})}\right]\right\}$$

$$= 0.029 \text{ d}$$

b. Repeating the calculation for depths up to 10 m yields the depth of the wetting front as a function of time, as presented in the figure below:

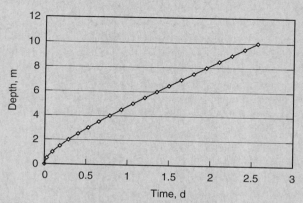

2. Determine the infiltration rate as a function of time
   a. For a wetting front depth of 0.5 m, determine the infiltration rate using Eq. (22-1):

$$v_i = K \frac{H_w + L_f - H_{cr}}{L_f}$$

$$v_i = (1 \text{ m/d}) \left[\frac{0.7 \text{ m} + 0.5 \text{ m} - (-0.5 \text{ m})}{0.5 \text{ m}} \right] = 3.4 \text{ m/d}$$

b. Therefore at a time of 0.029 d, the infiltration rate is 3.4 m/d. Repeating for wetting front depths up to 10 m yields the following figure. Note that after 1 d, the infiltration rate approaches the hydraulic conductivity of 1 m/d.

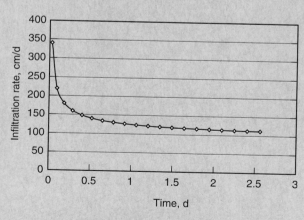

3. Determine the total quantity of water infiltrated.
   a. Estimate the total quantity of water infiltrated for a 100 m² basin. The required information is the depth of the wetting front. When the depth of

the wetting front is 10 m, the volume of water contained in the pore volume of the soil is the volume of water infiltrated.

b. Volume of water infiltrated, V_{inf}, is:

$$V_{inf} = L_f \times f \times area = (10 \text{ m})(0.35)(100 \text{ m}^2)$$
$$= 350 \text{ m}^3$$

Infiltration through the Clogged Layer

In the development of Eq. (22-1) for clean water, two important factors that apply to the operation of infiltration basins were not considered. The first factor is that wetting fronts are not uniform and water contents and hydraulic conductivities tend to increase with time in the wetting front. However, any increase in infiltration from the increase in water content is observed rarely because of the development of clogging layers near the surface. The second factor is that instead of reaching a steady infiltration rate, infiltration rates tend to decline with time as a clogging layer develops. The development of a clogging layer also affects the hydraulics of the basin as the layer with the limiting hydraulic conductivity is at the surface. The saturated wetting front exists in the clogging layer, and unsaturated flow develops below the clogging layer where larger K values exist.

The infiltration rate through a clogging layer is analogous to determining the infiltration rate through an earth liner with a saturated hydraulic conductivity of K_c. The infiltration rate, v_i, may be described as follows:

$$v_i = K_c \frac{H_w + L_c - H_i}{L_c} \qquad (22\text{-}3)$$

where H_w = water depth above the liner, m
L_c = thickness of the earth lining, m
H_i = pressure head of water at bottom of the liner, m
Other terms as defined previously.

The geometry associated with Eq. (22-3) is presented on Fig. 22-6. While actually solving Eq. (22-3) is difficult as most of the parameters such as K_c and L_c will vary with time; an important result can be realized by examining Eq. (22-3). Unlike Eq. (22-1),

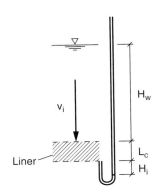

Figure 22-6
Geometry and symbols for Eq. (22-3).

increasing the value of H_w can potentially have a major impact on infiltration rates as the value of L_c should not increase significantly relative to the potential increases in H_w. Therefore, under these conditions, it can be shown theoretically that increasing the water depth in a basin will have a beneficial effect on infiltration rates.

Equation (22-3) cannot be used to predict how K_c will change as the water depth is increased. Bouwer and Rice (1984) analyzed the effect of water depth in groundwater recharge basins on infiltration. It was found that increasing water depth results in compression of the sediments and organic matter that cause clogging. Consolidation theory is used to explain how the compression occurs in the clogging layer and the results were supported by both field and laboratory studies. The general recommendation from these studies was to use the minimum practical water depth when reclaimed waters are the source waters for recharge. The reason is that reclaimed waters contain sufficient nutrients to promote both bacterial and algal growth that contribute to the development of a clogging layer. Therefore, increasing the water depth often results in a temporary increase in infiltration rates followed by a reduction to rates lower than observed previously. Total infiltration rates that include both wetting and drying cycles may also be reduced further when water depths are increased. As a clogging layer becomes more restrictive and water depths are increased, the time required to drain a basin increases as more water must move through a layer of low hydraulic conductivity. The basin must be drained completely before drying and recovery of infiltration rates may occur. When water depths are maintained at the minimum practical level, the clogging layer does not become compressed and less than 0.6 m (2 ft) of water must be drained from the basin, which requires less than a day in most cases.

Impact of Algal Blooms

Algal blooms can greatly exacerbate problems with clogging and provide another reason to maintain lower water depths. Because there are essentially unlimited nutrients available for algal growth in reclaimed water that has only undergone secondary treatment, solar incidence is the primary factor that limits algal growth. Once algae begin to grow, exponential growth can occur and an algal bloom can be the primary reason for clogging. Algae growth may be reduced by decreasing the time water is maintained in a recharge basin. At the Underground Storage and Recovery Facility in Tucson, Arizona, the length of wetting cycles is often determined by the beginning of algal growth. When operators notice that the basins are turning green, flow to the basins is terminated and the basins are drained and dried before an algal bloom develops. Once an algal bloom develops, the algae can dominate the microbial community in a recharge basin. During photosynthesis, the assimilation of carbon dioxide by algae raises the pH of the water. When algal mats are located on the basin surface, localized pH increases can result in calcium carbonate precipitation in the clogging layer. This type of inorganic precipitate clogging may become irreversible over time.

Impact of Mound Development

When the depth to groundwater is shallow, the infiltration rates may become limited by the horizontal flow that allows infiltrating water to move laterally away from the infiltration basin. If a mound develops below the recharge basin (see Fig. 22-7) or if a perched water layer develops, infiltration rates may also be limited by horizontal flow. Equations that can be used to predict the rise and fall of groundwater mounds using horizontal flow theory have been developed by Hantush (1967). The rise in the mound

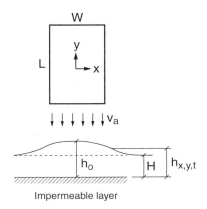

Figure 22-7
Geometry for rectangular infiltration basin with an underlying groundwater mound in an unconfined aquifer.

in unconfined aquifers located below rectangular recharge basins can be predicted using the following equation.

$$h_{x,y,t} - H = \frac{v_a t}{4f} \left\{ \begin{array}{l} F[(W/2 + x)n, (L/2 + y)n] \\ + F[(W/2 + x)n, (L/2 - y)n] \\ + F[(W/2 - x)n, (L/2 + y)n] \\ + F[(W/2 - x)n, (L/2 - y)n] \end{array} \right\} \qquad (22\text{-}4)$$

where $h_{x,y,t}$ = height of water table above an impermeable layer at x, y and time t (see Fig. 22-7), m
  H = original height of water table above impermeable layer, m
  v_a = arrival rate at water table of water from infiltration basin, m/yr
  t = time since start of recharge, d
  f = fillable porosity, dimensionless
  L = length of recharge basin in y direction, m
  W = width of recharge basin in x direction, m
  n = $(4tT/f)^{-1/2}$
  T = transmissivity, m²/d

$$F(\alpha, \beta) = \text{error function} = \int_0^1 [\text{erf}(\alpha \tau^{-(1/2)}) \times \text{erf}(\beta \tau^{-1/2})] d\tau$$

  $\alpha = (W/2 + x)n$ or $(W/2 - x)n$
  $\beta = (L/2 + y)n$ or $(L/2 - y)n$

Values for the error function have been tabulated by Hantush and an abbreviated table of values is presented Table G-1 in App. G.

The transmissivity, T, in the term n is estimated as $K(H + h_{x,y,t})/2$ to compensate for the increase in aquifer thickness as the mound develops. Because the final mound height is not known, T may be estimated initially as K(H) to provide the first approximation at $h_{x,y,t}$. The first calculated $h_{x,y,t}$ value is then used to compute T in the second iteration until $h_{x,y,t}$ is defined accurately. If $h_{x,y,t}$ is greater than 0.5 H, Eq. (22-4) is not valid. The term t should be considered to be the time when the wetted front reaches the water table, typically several days after infiltration is initiated. At the center of the mound, x and y are zero and the sum of the F functions in Eq. (22-4) simplifies to 4F[Wn/2, Ln/2].

Equation (22-5) may be used to estimate the decay of a groundwater mound after infiltration ceases (Hantush, 1967).

$$h_{x,y,t} - H = Z(x, y, t) - Z(x, y, t - t_s) \qquad (22\text{-}5)$$

Where t_s is the time since the infiltrating water has stopped arriving at the water table and the other terms are as defined previously. The terms $Z(x, y, t)$ and $Z(x, y, t - t_s)$ represent the right hand side of Eq. (22-4) with t and $t - t_s$ as the time factors. Because the wetted zone above the mound will continue to drain into the aquifer, the time t_s should be estimated as several days after the water has stopped reaching the water table. In addition, when calculating mound recession, f should be considered for draining, which tends to be less than the original fillable porosity.

EXAMPLE 22-2. Determination of Mound Development and Potential Impacts.

Using the parameters given below for a recharge basin located above an unconfined aquifer, determine (a) the maximum groundwater rise in the mound after 1 yr of recharge basin operation, (b) if the recharge basin will be affected by groundwater mounding, (c) the validity of Eq. (22-4) for these conditions, and (d) the increase in the groundwater level at x = 350 m, y = 350 m.

T = transmissivity = 0.005 m²/s
H = height of groundwater above an impermeable layer (aquifer thickness)
  = 30 m
L = 100 m
W = 100 m
f = fillable porosity = 0.25
v_a = 100 m/yr
Depth to groundwater = 20 m

Solution
1. Determine the maximum rise in the groundwater mound.
    a. The maximum groundwater rise will occur at x = 0, y = 0. Therefore, Eq. (22-4) may be used to find the maximum groundwater rise.

$$h_{x,y,t} - H = \frac{v_a t}{4f} \left\{ \begin{array}{l} F[(W/2 + x)n, (L/2 + y)n] \\ + F[(W/2 + x)n, (L/2 - y)n] \\ + F[(W/2 - x)n, (L/2 + y)n] \\ + F[(W/2 - x)n, (L/2 - y)n] \end{array} \right\}$$

With x = 0 and y = 0, Eq. (22-4) simplifies to

$$h_{x,y,t} - H = \frac{v_a t}{4f}\{4 \times F[(W/2)n, (L/2)n]\}$$

b. Calculate the value of n as defined in Eq. (22-4):

$$n = (4tT/f)^{-1/2} = [(4)(365 \text{ d})(0.005 \text{ m}^2/\text{s})(86{,}400 \text{ s/d})/0.25]^{-1/2}$$
$$= 0.00063 \text{ m}$$

c. Determine the value of the term $F(\alpha, \beta)$. The value of $F(\alpha, \beta)$ is calculated as follows. Values of $F(\alpha, \beta)$ are tabulated in Table G-1 in App. G.

$$\{F[(W/2)n, (L/2)n]\} = F[(100 \text{ m}/2)(0.00063), (100 \text{ m}/2)(0.00063)]$$
$$= F[(0.0315), (0.0315)] = 0.00877$$

d. The groundwater rise can then be calculated as:

$$h_{x,y,t} - H = \frac{(100 \text{ m/yr}) \times 1 \text{ yr}}{4 \times 0.25}(4 \times 0.00877) = 3.51 \text{ m}$$

2. Will groundwater rise impact infiltration?

A groundwater rise of 3.51 m in 1 yr will decrease the depth to groundwater from 20 m to 16.49 m. This decrease should not have a major impact on the recharge basin; however, if the mounding continues for many years or the infiltration rate increases, the groundwater rise may affect infiltration rates.

3. Check whether the use of Eq. (22-4) is valid.
   a. The validity of Eq. (22-4) may be checked based the ratio of groundwater rise to the aquifer thickness (H).

$$\frac{h_{(0,0,1)} - H}{H} = \frac{33.51 \text{ m} - 30 \text{ m}}{30 \text{ m}} = 0.11 < 0.5$$

   b. Because the ratio is less than 0.5, the use of Eq. (22-4) is valid.

4. Estimate the increase in the groundwater elevation at x = 350 m and y = 350 m.
   a. Find the F values in Eq. (22-4) for the coordinates of (350m, 350m). The n value is the same as Step 1a for a time of 1 yr.

$$F[(W/2 + x)n, (L/2 + y)n] = F[(100 \text{ m}/2 + 350 \text{ m})n, (100 \text{ m}/2 + 350 \text{ m})n] = 0.24$$
$$F[(W/2 + x)n, (L/2 - y)n] = F[(100 \text{ m}/2 + 350 \text{ m})n, (100 \text{ m}/2 - 350 \text{ m})n] = -0.195$$
$$F[(W/2 - x)n, (L/2 + y)n] = F[(100 \text{ m}/2 - 350 \text{ m})n, (100 \text{ m}/2 + 350 \text{ m})n] = -0.195$$
$$F[(W/2 - x)n, (L/2 - y)n] = F[(100 \text{ m}/2 - 350 \text{ m})n, (100 \text{ m}/2 - 350 \text{ m})n] = 0.16$$

   b. The groundwater rise at (350 m, 350 m, 1 yr) may be calculated with Eq. (22-4).

$$h_{(350 \text{ m}, 350 \text{ m}, 1 \text{ yr})} = \frac{(100 \text{ m/yr})(1 \text{ yr})}{4 \times 0.25}(0.24 - 0.195 - 0.195 + 0.16)$$
$$= 1.0 \text{ m}$$

Estimation of Clean Water Infiltration Rates

The first step in estimating infiltration rates for surface spreading basins is to determine the clean water infiltration rate using an infiltrometer or by conducting pilot-scale tests. If an infiltrometer is used, it must be either a double ring infiltrometer or of sufficient size to minimize the effects of horizontal flow (Bouwer, 1986). When a double ring infiltrometer is used, the measurements are made in the inner ring where almost all water travels vertically through the soil (see Fig. 22-8). The outer ring of the infiltrometer serves to separate flow patterns where vertically infiltrating water can move horizontally. Equation (22-2) may be used to estimate the hydraulic conductivity of the soil, providing that an accurate estimate of H_{cr} is made. Equation (22-1) may be used to estimate the steady infiltration rates provided that the wetted front is sufficiently greater than the head maintained in the infiltrometer. It is important to conduct infiltrometer tests at several locations in a recharge basin as the surface hydraulic conductivities may vary significantly within a single basin. Therefore, if adequate resources are available, pilot testing with large-scale test equipment with clean water will provide the most accurate estimates of infiltration rates. In some cases pilot testing is impractical because of the large quantities water required.

Operation and Maintenance Issues

After an estimate of the infiltration rate with clean water has been made through testing, the actual infiltration rate must be adjusted for the use of wet/dry cycles and the development of clogging layers. The total time a recharge basin is in operation must include: (1) the time for application of water, (2) the time for water to drain from the basin, and (3) the time for drying. When reclaimed water is used in arid climates, the total application time (TA) is often less than 50 percent of the total time (TT). The remaining time is used for drainage and drying. In addition, the development of clogging

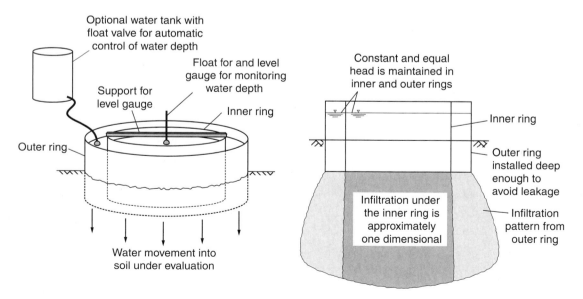

Figure 22-8
A double ring infiltrometer used to measure surface infiltration rates.

layers will reduce infiltration rates significantly as compared to clean water tests. Therefore, the actual infiltration rate must be adjusted for the application time by a factor of TA/TT. The development of clogging layers usually reduces the infiltration rate by a factor of 50 percent. If the measured infiltration rates are very low, less than 0.15 m/d (0.5 ft/d) as observed at the Mesa, Arizona, Northwest Water Reclamation Plant, the development of a clogging layer might not reduce infiltration rates if the subsurface soils have a hydraulic conductivity less than the clogging rate. Because clogging reduces the infiltration rate by up to 50 percent, a system with a measured clean water infiltration rate of 2.5 m/d (8 ft/d) and a projected application time of 50 percent of the total time should be designed based on an average infiltration rate of 0.6 m/d (2 ft/d).

Wet-Dry Cycles

Wetting and drying cycles are an integral part of the operation of recharge basins receiving reclaimed waters. Drying cycles are necessary to stop the accumulation of clogging materials at the soil/water interface. During drying cycles, the accumulated organic material on the soil surface has an opportunity to desiccate. As this organic material desiccates, it tends to shrink in size and separate from soil particles. Furthermore, the introduction of air provides aeration and the organic material may be biodegraded more effectively during the drying cycle. Dry organic material has a very low density and may also be dissipated by wind action. Wetting and drying cycles, as reported in Table 22-5, vary considerably between different recharge sites. Bouwer et al. (1980, 1991) evaluated several different wetting and drying cycles at demonstration sites in Arizona. These sites had favorable hydraulic conditions and the variation in

Table 22-5
Recharge site wetting and drying cycle times and potential impacts on water quality

Recharge site	Reclaimed water	Cycle times	Comments
23rd Avenue, Phoenix, AZ	Secondary effluent	2 wk wet 2 wk dry	69 percent nitrogen removal
Flushing Meadows Phoenix, AZ	Secondary effluent	2–4 d wet 5–10 d dry	Negligible nitrogen removal[a]
Dan Region, Israel	Nitrified-denitrified	1 d wet 2–3 d dry	45 percent nitrogen removal
Tucson Underground Storage and Recovery Facility, AZ	Trickling filter tertiary	2–7 d wet 5–10 d dry	Cycle times controlled by growth of algae, >50 percent nitrogen removal
Mesa Northwest Water Reclamation Plant, AZ	Tertiary nitrified-denitrified	1 wk wet 3 wk dry	Clay lenses limit infiltration, hydraulic connection between basins requires extra time for drainage
Anaheim Forebay, Orange County, CA	Santa Ana River water	Month of wet Periodic drying	Deep basins cleaned with submarine cleaning device
Montebello Forebay, Los Angeles County, CA	Tertiary effluent, stormwater, and surface water	3–4 wk wet 2–4 wk dry	Water availability a major factor in cycle time

[a] Because of the combined effect of a short wetting period and a long drying period, the oil remained aerobic and little or no denitrification occurred.

wetting and drying times was examined to evaluate impacts on water quality. Wetting times may be as short as 1 d in the Dan Region Project in Israel to as long as continuous wetting in the Anaheim Forebay in Orange County, California. The combination of algal blooms and water availability often controls the choice of wetting and drying cycles. At the Mesa, Arizona, Northwest Water Reclamation Plant, clay lenses limit infiltration rates and extensive horizontal flow creates hydraulic connections between the basins. Therefore, long drying periods are necessary to drain the basins effectively and eliminate adverse impacts on the adjacent basins.

The primary concern over operating with wetting and drying cycles is the maintenance of infiltration rates, however, the length of wetting and drying cycles also has an impact on water quality transformations. During wetting, the majority of oxygen is often consumed near the soil water interface where biodegradation of organic carbon occurs. Drying allows for the introduction of air containing oxygen back into the soil. As the wetting front moves below the soil surface, air is drawn back into the soil. On a mass basis, air contains approximately 30 times the amount of oxygen found in water in equilibrium with air. Therefore, drying cycles are effective at aerating the soil and stimulating aerobic degradation of residual oxygen demanding materials. Furthermore, wetting and drying cycles can result in cyclic anoxic/aerobic conditions in the vadose zone where nitrogen transformations and other redox-dependent microbial reactions occur. Consequently, as reported subsequently in Table 22-6, nitrogen removal has been observed at many recharge basins receiving reclaimed water.

Use of Ridges and Furrows

One modification to the design of recharge basins that can be used to minimize maintenance and allow for continuous infiltration is the use of ridges and furrows combined with wave action (Dillon, 2002; Peyton, 2002). Ridges of permeable material are constructed on the basin floor as pictured on Fig. 22-9a. Sediments are washed effectively from the ridges and the ridges maintain high infiltration rates as sediments accumulate

(a)   (b)

Figure 22-9

Methods used to enhance infiltration: (a) ridges and furrows to enhance infiltration without the use of wetting and drying cycles and (b) disking the basin surface, the basin inlet is shown in the foreground.

on the basin floor. The water levels may be varied to allow for drying of portions of the ridges periodically to eliminate any organic accumulation on the ridges and to add oxygen to the infiltrating water. For systems with high sediment loading, infiltration volumes have been increased over 100 percent by eliminating the need for long drying times and cleaning of the basins.

Scraping

As clogging materials accumulate over time on the bottom and banks of infiltration basins, infiltration rates will be reduced and the use of drying cycles alone will not be sufficient to maintain infiltration rates over long operational periods. When reclaimed water is used with a moderate suspended solids loading of less than 10 mg/L, the clogging materials must be removed over a 12 to 24 mo operating period. If reclaimed water or blends of water with higher suspended solids concentrations are used, the frequency of maintenance for the removal of clogging materials must be increased. Front-end loaders or other devices capable of scraping off the clogging layer and removing the materials are most effective. When the clogging layer is disked or plowed, infiltration rates recover, but the effect is short term. Disking moves clogging materials deeper into the soil where they will accumulate. Ultimately, the upper layer of the soil will have to be removed to recover infiltration rates. A view of disking for the recovery of infiltration rates is shown on Fig. 22-9b.

Performance of Recharge Basins

When recharge basins are used for infiltration, transformations may occur in both the vadose zone and the saturated zone. The typical residence time in the vadose zone is less than 2 wk while the residence time in the saturated zone is often months. Therefore, transformations that occur in both zones are important with respect to the improvement of water quality. Under many circumstances, the vertical flow to the basins approaches saturation, as described earlier in this section. Unsaturated conditions exist primarily after a clogging layer develops. Furthermore, clay lenses often exist in alluvial deposits resulting in perched water and saturated zones in the vadose zone. Researchers have demonstrated that most transformations occur under both saturated and unsaturated conditions (AWWARF, 2001). Based on the history of bank filtration, it is known that a vadose zone is not necessary for water quality transformations. However, a vadose zone is necessary to support nitrogen transformations for which cyclic anoxic/oxic conditions are required.

Impact of Intermixing

As infiltrating water reaches the aquifer, a mound is formed and the water flows downgradient from the point of recharge. Because the horizontal flow is typically orders of magnitude greater than vertical flow, the majority of recharged water remains on the surface of the aquifer and also remains on the surface as it moves downgradient. Mixing in the subsurface environment is limited by dispersion, and actually mixing of native groundwater with recharge water is a slow process. For the situation illustrated on Fig. 22-10, the upper alluvial unit of the aquifer became dominated by recharge water after several years of recharge. The upper alluvial unit has a higher hydraulic conductivity than the middle alluvial unit. Monitoring wells located in the upper portion of the middle alluvial unit underwent some mixing of groundwater recharge water, while monitoring wells located in the lower portion of the middle

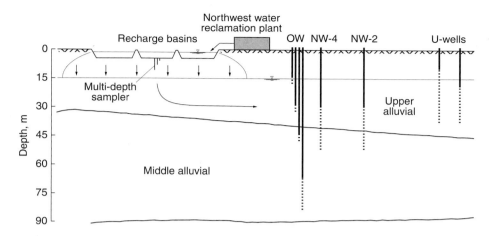

Figure 22-10
Recharge basin in Mesa, AZ, with flow through the vadose zone and saturated flow to downgradient monitoring and recovery wells. (Coordinates: 33.435 N, 111.884 W.)

alluvial unit contained mostly native groundwater. A production well creates drawdown resulting in more vertical movement from the upper layer of the aquifer. Effective dilution of recharge water occurs under most cases because production wells have screens over a hundred meters in length and only the upper portion of the aquifer contains recharge water. The Water Replenishment District of Southern California conducted tracer studies to demonstrate that a well located 15 m (50 ft) horizontally from a basin had a travel time of greater than 6 mo. This length of time was possible as the well was screened such that over 30 m (100 ft) of vertical flow was required before recharge water was recovered by the well.

Impact on Redox Conditions

When water entering a recharge basin is close to equilibrium with the atmosphere, it is nearly saturated with oxygen. The majority of oxygen is consumed as water passes through the soil/water interface where the easily biodegradable carbon is consumed and the ammonia nitrified. The easily biodegradable carbon concentration is approximately equivalent to the ultimate CBOD of the applied water. Most waters have sufficient CBOD to remove the majority of oxygen and the recharge water becomes anoxic as it passes through the vadose zone to the saturated zone. Because there is no mechanism to reaerate the water in the saturated zone, a plume of anoxic recharge water will develop in the aquifer. Further, because reclaimed waters contain nitrogen in either a reduced form or as nitrate, some reduced nitrogen will be nitrified. Concentrations of nitrate of 1 mg/L are sufficient to maintain anoxic conditions and prevent the plume from becoming anaerobic (resulting in sulfate reduction and methanogenisis). Pockets of anaerobic conditions may develop in the vadose zone resulting in sulfate reduction and other potential side reactions. In the Montebello Forebay in Los Angeles County, California, where tertiary treated reclaimed water is used for recharge, along with stormwater and surface waters, the addition of oxygen from the stormwater and surface waters maintains aerobic conditions in the aquifer.

The redox conditions in the vadose zone may vary between anoxic and aerobic as a function of the wetting and drying cycles used. During wetting, anoxic conditions develop in the vadose zone as oxygen is consumed at the soil/water interface. During drying, air reenters the soils and creates aerobic conditions in the upper portion of the vadose zone. The penetration of air into the vadose zone during drying depends on the length of the drying cycle and the dissipation of the groundwater mound below the basin. An important factor that must be considered is the total oxygen demand of the water including both the nitrogenous and carbonaceous demand. Secondary effluents with greater than 20 mg NH_4^+–N/L may have an oxygen demand in excess of 100 mg/L. Ammonium is removed primarily by adsorption onto clays during wetting, and adsorbed ammonium may be nitrified during drying. Because conversion of ammonium to nitrate can consume the majority of oxygen entering the soil during drying, aerobic conditions might only develop in the upper vadose zone. Under these conditions, anoxic conditions are maintained in both the lower portion of the vadose zone and the aquifer.

Transformations of Organic Compounds

Organic compounds may be transformed during subsurface transport indefinitely by microbially mediated reactions. Aerobic and anoxic conditions occur during subsurface transport; however, the majority of microbially mediated reactions are similar under both sets of conditions. If anaerobic conditions develop, the pattern of microbially mediated reactions will change.

Organic Compounds of Concern The concerns over organic compounds come from potential health effects and uncertainties over the composition of effluent organic matter (EfOM). The bulk organic matter may be measured as dissolved organic carbon (DOC) and characterized by a number of methods that have been used to characterize natural organic matter (NOM). These methods include molecular weight fractionation, hydrophobicity, spectroscopic analysis, elemental analysis, and reactivity. Reactivity is an important characteristic as the formation of disinfection byproducts is of concern and the relative reactivity of EfOM as compared to NOM is a major concern for indirect potable reuse. Concerns exist over specific organic compounds that are known to persist after conventional wastewater treatment. These compounds include personal care products and pharmaceuticals (PCPPs), some of which are recognized as endocrine disrupting compounds. Most of these compounds are used by people at high doses and considerable uncertainty exists over their potential health effects at the μg/L or ng/L concentrations often identified in reclaimed waters. The DOC concentrations in reclaimed waters are measured in mg/L and specific organic compounds are often referred to as trace organic constituents, as they represent a small fraction of the DOC present.

Bulk Organic Transformations The bulk organic carbon present in reclaimed water may be divided into four major categories including: (1) natural organic matter (NOM), (2) soluble microbial products (SMPs), (3) easily biodegradable organic compounds, and (4) synthetic organic compounds. The concentration of NOM in reclaimed water depends on the original source of drinking water. The NOM in the water supply is persistent and is not removed by the drinking water treatment processes. Because most NOM is not easily biodegradable, the NOM will persist through water use and wastewater treatment (Drewes and Fox, 2001). During wastewater treatment, the microorganisms produce SMPs that have characteristics similar to NOM in terms of their molecular weight and functional groups. Because the easily biodegradable carbon is removed rapidly during groundwater recharge and synthetic organic carbon compounds make up a small percentage of the organic carbon, the majority of persistent organic carbon is composed of NOM and SMPs.

As groundwater recharge occurs, the transformation of organic compounds may be divided up into different regimes defined as: (1) short-term soil aquifer treatment (SAT) where relatively fast reactions occur and (2) long-term SAT where recalcitrant compounds continue to transform at slower rates over time. Short-term SAT is typically defined as less than 30 d and the majority of easily biodegradable carbon is consumed during this time period. Data comparing a field SAT site with soil column studies completed under aerobic and anoxic conditions are presented on Fig. 22-11. After a period of 20 d, the final DOC concentrations are similar under all conditions. Under aerobic conditions, the majority of easily biodegradable DOC was removed after several days, while the complete time period was required under anoxic conditions. Because the time scales used for most groundwater recharge systems are in the order of months, the removal observed under aerobic or anoxic conditions is similar. The NOM of the drinking water source for these experiments was approximately 2 mg/L and the persistent SMPs contributed approximately 1 mg/L, resulting in a DOC concentration of 3 mg/L after short-term SAT.

As water passes through the saturated zone over longer time scales as compared to short-term SAT, transformations of organic carbon continue. These transformations are similar to those that occur as the natural recharge of surface waters into aquifers results in water quality improvements. The DOC transformations as a function of distance for the Mesa Northwest Water Reclamation Plant (NWWRP) are presented on Fig. 22-12. The plume of reclaimed water at the Mesa NWWRP is anoxic. Each 300 m (1000 ft) of travel is equivalent to approximately 6 mo of travel time. At the monitoring wells closest to the basin, the DOC concentration has actually been reduced to a concentration

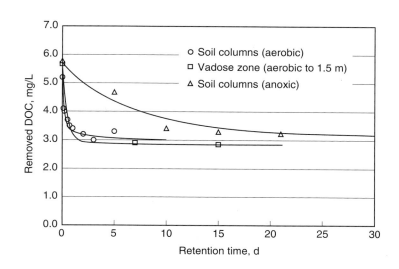

Figure 22-11

Dissolved organic carbon transformations during subsurface transport.

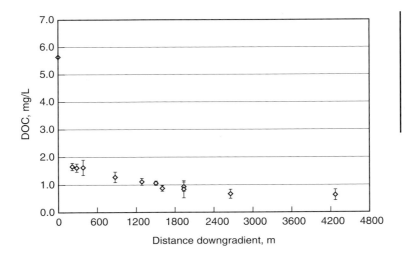

Figure 22-12
Dissolved organic carbon concentrations as a function of distance for the Mesa Northwest Water Reclamation Plant.

lower than the original drinking water DOC concentration. After several years of travel time, the DOC concentrations are less than 1 mg/L, approaching the background concentrations of the aquifer.

The EfOM was characterized in reclaimed water before groundwater recharge, after short-term SAT, and after long-term SAT. The purpose was to compare the EfOM in the final product of a SAT system with the NOM present in the original drinking water source. Using spectroscopic characterization by C^{13}-Nuclear Magnetic Resonance and Fourier Transform Infra-Red, no significant differences were found in the major functional groups. After wastewater treatment, major differences were observed in the organic nitrogen content of the EfOM as compared to NOM, because of the contribution of SMPs. The differences were also verified by fluorescence spectroscopy. However, after long-term SAT, the elemental composition and fluorescence resembled NOM. The majority of differences between EfOM and NOM were eliminated during short-term SAT. Based on the state-of-the-art techniques used to characterize NOM, the bulk organic matter in SAT product water could not be distinguished from NOM (Drewes and Fox, 2001).

Disinfection By-Product Formation Potential The reactivity of the organic carbon after groundwater recharge has been assessed using disinfection byproduct formation potential (DBPFP) tests. The DBPFP of the organic carbon in the recharged groundwater is similar to NOM in native groundwater with a value of 60 µg/mg DOC for trihalomethanes. This similarity is consistent with the extensive analysis that identified organic carbon after SAT as structurally similar to the organic carbon in NOM. If both pools of organic carbon consist of the same types of molecules, their reactivity with chlorine should also be similar. Because SAT product waters should have low DOC concentrations, the total formation potential, when the product water is chlorinated, should be similar to the DBPFP of a native groundwater. Reclaimed waters can contain elevated concentrations of bromide (anthropogenic bromide is added during water use). Because bromide will not be removed during SAT, the bromide will affect the distribution of disinfection byproducts when the product water is chlorinated. The presence of elevated bromide concentrations could be a concern as most brominated DBPs are more toxic than chlorinated DBPs.

Trace Organic Compounds Trace organic compounds are synthetic organic compounds that persist after conventional wastewater treatment and are present at concentrations of μg/L or ng/L. Because these concentrations are too low to support microbial growth directly, the removal of these compounds during subsurface transport is most probably by cometabolism. Cometabolic reactions are mediated by microorganisms, but the microorganisms, do not benefit directly from the reaction. Because many of the organic transformations that occur during subsurface transport involve high molecular weight compounds, the microorganisms produce enzymes to hydrolyze the high molecular weight compounds into small compounds that can be utilized directly. Therefore, the potential for cometabolic transformations of many compounds exists in a biologically active aquifer. These transformations also require long time scales as the reactions are mediated indirectly by microorganisms. Several trace organic compounds have been studied in bank filtration systems in Europe and surface infiltration basins using reclaimed water in the Southwestern United States. The compounds studied include a broad spectrum of PCPPs, organic halides, and detergent residues such as alkylphenol ethoxycarboxylates (APECs).

The fate of ethylenediamine tetraacetate (EDTA) and APECs during long-term SAT at the Mesa Northwest Water Reclamation is presented on Fig. 22-13. The APECs and EDTA are removed to detection limits after approximately 1 yr of travel time. The fate of adsorbable organic halides (AOX) and adsorbable organic iodine (AOI) are also presented on Fig. 22-13. After long-term SAT the AOX concentration is essentially equal to the AOI concentration, implying that chlorinated and brominated compounds were removed and the persistent halogenated compounds are iodated. This result has also been observed in bank filtration systems in Europe (Drewes and Jekel, 1998). Several iodated x-ray contrast agents have been identified as the source of AOI that can persist under conditions that are ideal for biotransformations. Numerous compounds have been

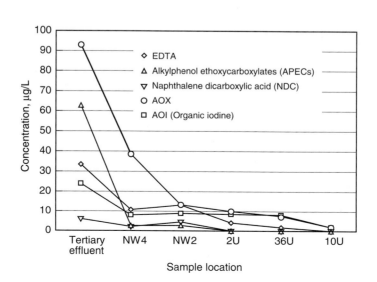

Figure 22-13
The fate of several trace organic compounds during long-term SAT. The adsorbable organic iodine (AOI) persists while other compounds are removed.

identified that are similar to the AOI in terms of their persistence during subsurface transport and these compounds have been found in both Europe and the United States. Common characteristics of these compounds are that they are hydrophilic and have structural features that prevent enzymatic attack. The PCPPs include the antidepressant drugs carbamazepine and primodone, the fire retardant Tri (2-chloroethyl) phosphate, and the mosquito repellant DEET. The persistence of PCPPs in the environment presents a general problem with uncertain health and environmental consequences. The majority of these compounds are susceptible to oxidation and chlorination, in some cases, chemical disinfection may be sufficient to transform these compounds.

Endocrine Disrupting Activity Endocrine disrupting compounds (EDCs) of concern in reclaimed waters include the hormones and detergent residues (APECs). The natural estrogenic hormone 17-β estradiol and the synthetic estrogenic hormone ethinylestradiol contribute a large portion of the estrogenic activity in reclaimed waters and these compounds can affect aquatic life at concentrations of 1 ng/L. The detergent residues are present at μg/L concentrations in reclaimed waters, however, these compounds only exhibit estrogenic activity at elevated concentrations. Estrogenic activity has been demonstrated to be efficiently removed during short-term SAT. Because most EDCs of concern are very hydrophobic and are biodegradable during wastewater treatment, their subsurface transport should be limited (Heberer, 2002). Because these compounds are very hydrophobic, it is possible they might accumulate on soils during subsurface transport. In studies of the near surface soils from recharge basins, it has been found that hydrophobic EDCs are adsorbed onto the soils; however, the compounds are biodegrading with time and no net accumulation or transport to the aquifer is occurring.

Transformation of Nitrogen Compounds

The transformation of nitrogen compounds when reclaimed water is applied to recharge basins can be quite complex and difficult to predict in advance. Where nitrogen control is critical, nitrogen should be removed prior to recharge. Where nitrogen is to be removed in the soil during the recharge process, two nitrogen transformation and removal pathways are possible: (1) aerobic ammonium oxidation, resulting in the production of nitrate (nitrification) followed by heterotrophic denitrification under anoxic conditions where an organic carbon source serves as the electron donor and the nitrate is utilized as the electron acceptor and (2) partial aerobic ammonium oxidation to nitrite (nitritation) followed by autotrophic oxidation of ammonium ion (electron donor) utilizing the nitrite as the electron acceptor under anoxic conditions. The first process is equivalent to conventional denitrification used commonly in biological nitrogen removal. The second process known as anaerobic ammonium oxidation (ANAMMOX) (Mulder et al., 1995), is given by the following reaction.

$$NH_4^+ + NO_2^- \to N_2 + 2H_2O \tag{22-6}$$

The nitrogen transformation and removal processes that take place depend on the water quality, characteristics of the soil environment, and system operation. In many cases, it is difficult to identify the processes that are responsible for the nitrogen transformation and removal. For example, during recharge a substantial amount of the organic carbon is removed at the soil/water interface, resulting in nonideal conditions for conventional denitrification. Thus, if nitrate is present in the water to be recharged (e.g., from a nitrification

Table 22-6

Summary of nitrogen removal at applicable SAT sites

Site	Years of operation	Pretreatment	Average NH$_4^+$ conc., mg N/L	Average percent nitrogen removal[a]
Flushing Meadows,[b] Phoenix, AZ	1967–1978	Secondary activated sludge, no chlorination	21	65
23rd Avenue,[c] Phoenix, AZ	1974–1983	Secondary activated sludge, no chlorination prior to 1980, chlorination after 1980	18	69
TTSA,[d] Tahoe-Truckee, CA	1978–current	Advanced treatment, ion exchange for NH$_4^+$ removal, chlorination	7	70–90
Sweetwater,[e] Tucson, AZ	1986–current	Secondary treatment, chlorination	20	75

[a] Nitrogen removal at a recharge site will vary, depending on the characteristics of the soil, the quality of the applied water, and the operation of the system (see also Table 22-5).
[b] Bouwer and Rice (1974).
[c] Bouwer and Rice (1984).
[d] Woods et al. (1999).
[e] Wilson et al. (1995).

process applied prior to recharge) sufficient organic carbon may not be available to support heterotrophic denitrification in the anoxic zones below. However, when secondary effluent is applied to the recharge basins efficient removal of nitrogen has been demonstrated at several sites (see Table 22-6). While none of the sites had sufficient biodegradable organic carbon in the effluents to support more than 30 percent nitrogen removal by conventional denitrification, removal efficiencies of up to 90 percent have been observed. Explanations for the high levels of nitrogen removal in recharge systems include: (1) organic compounds (measured as COD) which are not measured by the conventional BOD test and are not removed under wastewater treatment conditions may serve as a carbon source in the soil environment, (2) the presence of microzones within the porous medium and biofilm that are favorable for nitrogen removal, e.g., an anoxic zone contained within an aerobic biofilm, and (3) novel microbial transformations, such as ANAMMOX.

The ANAMMOX process was investigated for groundwater recharge systems by Gable and Fox (2003). A sequence of nitrogen transformations during wetting and drying cycles in the vadose zone was observed, as presented on Fig. 22-14. During the initial wetting cycle, the primary removal mechanism for ammonium is adsorption. During the subsequent drying cycle adsorbed ammonium near the surface can be converted to nitrite and nitrate by nitrification. The next wetting cycle mobilizes the produced nitrite and nitrate and moves it deeper into the vadose zone where anoxic conditions exist. If anoxic conditions exist in the deep vadose zone and ammonium ions are adsorbed onto those soils, the conditions are favorable for the ANAMMOX reaction. Significant ANAMMOX

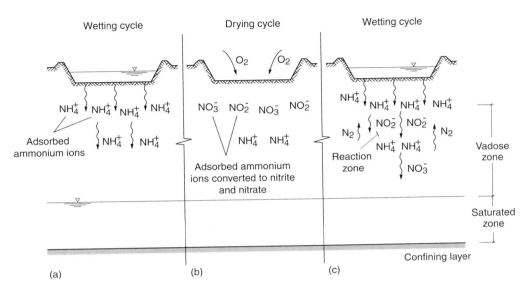

Figure 22-14
Nitrogen transformations in the vadose zone during wetting and drying cycles: (a) wetting cycle-infiltration occurs, ammonium ions are adsorbed onto soil; (b) drying cycle-oxygen enters soil, ammonium is oxidized to nitrite and nitrate; and (c) wetting cycle-infiltration occurs, ammonia and nitrate at surface and possibly at deeper depths.

activity has been found in the soils taken from the Sweetwater Tucson site during sustained nitrogen removal laboratory studies for several years with no addition of organic carbon. The reduction of nitrate to nitrite, which is directly utilized by the *Planctomycetes* responsible for the ANAMMOX reaction, was also observed (Shah and Fox, 2005).

Appropriate wet/dry cycles are important operational parameters that impact nitrogen removal (Bouwer and Rice, 1984) and the State of Arizona uses these wet/dry cycles as guidelines for permitting recharge facilities. The drying cycles are necessary to introduce oxygen into the soil where partial or complete nitrification may occur, as required for the nitrogen transformation and removal reactions described previously. However, when ammonium is applied to the soil, the vadose zone should have sufficient adsorptive capacity to prevent breakthrough of ammonium to the aquifer.

Pathogens

Potential concerns over pathogens during the artificial recharge of groundwater include the fate and transport of parasites, bacteria, and viruses. Studies have been conducted on many types of pathogens during subsurface transport. Because there are no known hosts for pathogenic microorganisms in the subsurface, the growth of pathogens is not a concern. Therefore, the major concern is the ability of pathogens to be transported in the subsurface to a point where the water may be recovered for subsequent use. Because bacteria and parasites are too large to be transported effectively during subsurface flow, the majority of research has focused on the transport and survival of viruses in the subsurface environment. In addition, regulatory criteria in several states have also been based upon ability of viruses to survive.

Soil type and composition, pH, moisture content, and virus strain all interact to affect the adsorptive capacity and virus die-off rate in soil (Goyal and Gerba, 1979; Powelson et al., 1993). Several research findings, however, have been consistent through a number of laboratory studies. The first finding is that decay rates increase with temperature (Nasser and Oman, 1999). Recent findings (AWWARF, 2005) clearly indicate that as temperature increases microbial activity, the decay rate of pathogens increases. The second key finding is that F-specific RNA bacteriophage (male-specific coliphage such as MS2 and f2: FRNA phage) adsorb poorly to soil particles and survive relatively well in groundwater as compared to enteric viruses (Goyal and Gerba, 1979; Powelson et al., 1993; Powelson and Gerba, 1994; Yates et al., 1985). As a result, the FRNA phage has been recommended as a conservative model or indicator of human viruses in certain situations (Havelaar et al., 1993; Havelaar 1993). Reviewing the use of bacteriophage as a model of enteric viruses in the environment, Havelaar (1993) concluded that FRNA phage represent a "worst case" virus model for virus transport in soil.

Field-scale experiments at a specific site under actual recharge conditions with reclaimed water using bacteriophage were also conducted in Los Angeles County as part of the SAT project. A tracer study was done where bacteriophage were seeded along with an inert tracer to a full-scale recharge basin. From an extrapolation of the tracer study data, a 7-log reduction of bacteriophage should occur within approximately 30 m (100 ft) of travel through the subsurface at the Montebello Forebay site in Los Angeles County. Taken together from the monitoring and tracer bacteriophage data, virus contamination of the deeper potable wells located in the recharge area is unlikely to occur under the conditions encountered during this study.

Although efficient virus removal is expected under most groundwater recharge scenarios, it is very difficult to demonstrate the U.S. EPA 10^{-4} risk factor for waterborne disease when indirect potable reuse is desired. Although viruses might be removed efficiently prior to recharge, a minimum of 8 logs of removal must be demonstrated to meet the 10^{-4} risk factor, and it is technically infeasible to conduct performance tests in full-scale systems.

Regulatory criteria have been established based upon the ability of a virus to survive in the environment. In the Netherlands and Germany, travel times of 70 and 50 d, respectively, are used for bank filtration systems. If the travel times exceed the regulatory criteria, no posttreatment for pathogens is necessary and the water may be distributed without disinfection. The City of Berlin uses bank filtration for a large percentage of its water supply and disinfection of the product water is not practiced. In California, the established regulatory criteria require a minimum of 6 mo of subsurface travel time as a safety factor for the potential survival of viruses. The regulatory criteria were designed to provide protection from viruses, however, they also provide sufficient time for the biotransformation of trace organic pollutants by cometabolic reactions (see Chap. 4 and App. F).

Examples of Full-Scale Surface Spreading Facilities

Several types of surface spreading operations have been used in the United States and in other parts of the world, most notably in Israel's Dan Region Project (see Fig. 22-3). Notable surface spreading facilities currently in operation in the United States are reported in Table 22-7. Of the facilities identified in Table 22-7, the largest spreading basins are located in the Central Basin, the main body of groundwater underlying the greater Los Angeles metropolitan area in California (see Fig. 2-3 in Chap. 2 and also

Table 22-7
Examples of full-scale surface spreading and recharge basins

Location	Description
Mesa, AZ (Coordinates: 33.435 N, 111.884 W)	The City of Mesa has two water reclamation plants. Both plants reclaim the water for reuse on golf courses, crop irrigation, industrial uses, freeway landscape watering, and for groundwater recharge. The Northwest Water Reclamation Plant is a state-of-the-art reclamation facility, with a treatment capacity of 68×10^3 m^3/d (18 Mgal/d). This facility has treatment that includes secondary treatment with nutrient removal, filtration, clarification, and disinfection. Reclaimed water from the NWWRP is discharged to two recharge sites and to the Salt River, which also recharges the aquifer. In the near future reclaimed water will also be used for freeway irrigation, on the Riverview Golf Course, and at the Granite Reef Underground Storage Project for recharge purposes. The Southeast Water Reclamation Plant is also a state-of-the-art facility that has a 30×10^3 m^3/d (8 Mgal/d) treatment capacity. The reclaimed water from this plant is used for golf course landscape irrigation, pond replenishment, and agricultural irrigation. Not all groundwater wells are used for drinking water; many are used for crop irrigation, golf course irrigation, and urban lakes. Recharge is an integral part of the City of Mesa's 100-yr water supply requirement for continued development.
Montebello Forebay Groundwater Recharge Project, Los Angeles County, CA (Coordinates: 33.994 N, 118.103 W)	The Montebello Forebay Groundwater Recharge Project, located in southeastern Los Angeles County, is the primary source of replenishment for the Central Basin, the main body of water underlying the greater Los Angeles metropolitan area. During fiscal year 2002–2003, 194×10^3 m^3/d (51.14 Mgal/d, 57,307 ac-ft/yr) of tertiary-treated reclaimed water from the San Jose Creek and Whittier Narrows Water Reclamation Plants was used for groundwater replenishment. In addition, another 8×10^3 m^3/d (2.04 Mgal/d, 2285 ac-ft/yr) of effluent river discharge from the Pomona Water Reclamation Plant was credited toward indirect groundwater recharge. The reclaimed water from these water reclamation plants discharge to rivers or creeks (i.e., flood control channels) that can convey the water by gravity to existing off-stream recharge basins. These basins and the unlined portions of the rivers and creeks permit large volumes of reclaimed water to percolate by gravity into the aquifer. Reclaimed water used in this way incurs no additional capital costs, related operation and maintenance (O&M) costs, or any energy consumption for pumping (SDLAC, 2005).
Water Conserv II, Winter Garden, FL (Coordinates: 28.493 N, 81.620 W)	Water Conserv II is the one of the largest water reuse projects with a combination of agricultural irrigation and rapid infiltration basins (RIBs). Jointly owned by the City of Orlando and Orange County, the system encompasses two water reclamation facilities connected by 34 km (21 mi) of transmission pipeline to a distribution center. From the distribution center, a 78 km (48 mi) pipeline network distributes reclaimed water to 76 agricultural and commercial customers. The reclaimed water that is not used for irrigation is distributed to RIBs. The RIB network contains seven sites with 74 RIBs for a total area of 809 ha (2000 ac). Both the distribution network and RIB site network are monitored and controlled from a central computerized control system (see Table E-1 in App. E).

Table E-1 in App. E). As noted in the above citations, many studies have been conducted on the performance of spreading basins. In planning for the use of spreading and recharge basins, site visits are recommended to obtain the latest information on their performance including operation and maintenance. The OCWD Groundwater Replenishment (GWR) system, currently under construction, is described in Sec. 22-7.

22-4 RECHARGE USING VADOSE ZONE INJECTION WELLS

The use of vadose zone injection wells for reclaimed water began in the 1990s and has been an effective alternative to spreading basins or direct injection wells. At the Scottsdale Water Campus, Scottsdale, Arizona, 27 vadose zone injection wells have been installed to recharge 40,000 m^3/d (10 Mgal/d) of reclaimed water after RO treatment. In Scottsdale, the depth to groundwater is approximately 150 m (500 ft) and land prices are prohibitively high. Even though the life cycle of a vadose zone injection well is uncertain, a life cycle as short as 5 yr makes vadose zone injection wells the most economical choice for recharge at the Scottsdale Water Campus. The actual life cycle has exceeded 5 yr, and the life cycle of the vadose zone injection wells is now projected to be at least 20 yr with the high quality effluent that is being applied.

Description

Vadose zone injection wells are essentially an extension of dry wells that are designed specifically to inject water continuously into the vadose zone. Dry wells are boreholes in the vadose zone, typically 10 to 50 m (30 to 160 ft) deep and approximately 1 to 2 m (3 to 6 ft) in diameter. Dry wells have been used traditionally to aid in the drainage of storm runoff in arid areas that do not have storm sewers or combined sewers. When groundwater depths exceed 100 m (300 ft) and economical land is not available for recharge basins, vadose zone injection wells may be cost effective. A single dry well that is 50 m (160 ft) deep, 1.5 m (5 ft) in diameter and is in soil with a hydraulic conductivity of 1 m/d (3 ft/d) may have an infiltration rate of 3800 m^3/d (1 Mgal/d).

The potentially serious problem with vadose zone injection wells is clogging as the flow cannot be reversed and there is no effective method to redevelop the well. Therefore, clogging must be prevented or minimized through effective pretreatment. The water in the vadose zone injection well must be protected from sloughing of clay from the vadose zone that will cause clogging of the surrounding aquifer. Well protection can be accomplished by using a screen and filling the well with sand or other highly permeable backfill material such as gravel. Vadose zone injection wells used at the Scottsdale Water Campus are illustrated on Fig. 22-15. The vadose zone injection wells should be designed to maintain infiltration rates by allowing the borehole to fill with water without causing air entrainment in the surrounding soils. A perforated pipe is used to introduce water at the bottom of the well and allow the well to fill from the bottom. As the water fills the well, air must be allowed to escape through an air vent at the top of the well to avoid air entrainment.

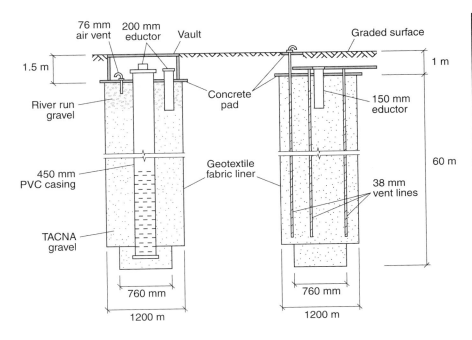

Figure 22-15

Vadose zone injection wells used at the Scottsdale Water Campus. The well on the left is the primary design and contains a 450 mm (18 in.) PVC casing for injection. The well on the right is for emergency use when the supply exceeds the capacity of the main system.

The Scottsdale Water Campus was the first facility to use vadose zone injection wells on a large scale with a capacity in excess of 40,000 m³/d (10 Mgal/d). The Scottsdale Water Campus (see Fig. 2-1a) reclaims water using RO treatment and has not experienced clogging problems in over 10 yr of operation. The Scottsdale Water Campus also uses the same vadose zone injection wells for the injection of microfiltered surface water. Vadose zone injection wells have also been used for recharge with reclaimed water with chlorine residuals in excess of 2 mg/L to prevent microbial growth. Care must be taken to maintain a chlorine residual in the distribution system as microbial growth is sufficient to cause clogging of a vadose zone injection well. Once clogged by microbial growth, attempts to redevelop the well by the addition of chlorine and other cleaning agents have not been effective.

Pretreatment Needs

The pretreatment requirements for vadose zone injection wells are similar to those for surface spreading, where the primary goal of pretreatment is to maintain infiltration rates. A greater potential for water quality improvement can be achieved with vadose zone injection wells as compared to direct injection wells. With the exception of nitrogen transformations, which can require cyclic aerobic/anoxic conditions, water quality transformations for organics and pathogens are similar for most groundwater recharge systems because the sustainable biofiltration mechanisms are similar.

A minimum requirement of tertiary treated and disinfected effluent is necessary for vadose zone injection wells to limit the accumulation of suspended solids at the borehole/soil interface. Therefore, a membrane bioreactor would be an appropriate pretreatment

Hydraulic Analysis

technology (see Chap. 7). Unless an advanced treatment process such as RO is used, the assimilable organic carbon (AOC) concentrations of tertiary effluent will be too high and a chlorine residual must be maintained during injection.

Vadose zone injection wells are analogous to recharge pits or recharge shafts that have been used for the recharge of groundwater with stormwater. The vadose zone injection wells should penetrate permeable layers to enhance infiltration rates. Equations, developed by Zangar (Bouwer and Jackson, 1974), can be used to estimate recharge rates as a function of the hydraulic conductivity of vadose zone soils. The equations presented below were developed originally to estimate the hydraulic conductivity of soils using infiltration test results. The equation used to estimate the hydraulic conductivity is:

$$K = \frac{Q}{2\pi L_w^2} \left\{ \ln\left[\frac{L_w}{r_w} + \sqrt{\left(\frac{L_w^2}{r_w^2} - 1\right)}\right] - 1 \right\} \tag{22-7}$$

where K = hydraulic conductivity, m/s
 Q = volumetric flowrate, m³/s
 L_w = depth of water in the screened interval of the injection well, m
 r_w = radius of the injection well, m

The geometry of the wetted zone that surrounds a vadose zone injection well relative to the depth of an impermeable layer is presented on Fig. 22-16. If the depth from the bottom of the well to an impermeable layer, S_i, is shallow and $S_i < 2L_w$, Eq. (22-7) can be simplified as follows:

$$K = \frac{3Q \ln(L_w/r_w)}{\pi L_w(3L_w + 2S_i)} \tag{22-8}$$

Equation (22-7) is used for most applications of vadose zone injection wells as these wells are most economical when $S_i \gg L_w$.

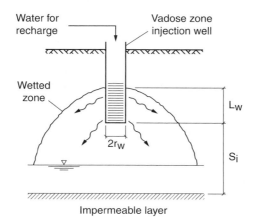

Figure 22-16
Schematic of vadose zone injection well wetted zone.

EXAMPLE 22-3. Estimate Vadose Zone Volumetric Injection Rate.

Estimate the volumetric injection rate, Q, for a vadose zone injection well with the following design parameters:

$K = 0.0005$ m/s
$L_w = 20$ m
$S_i = 60$ m
$r_w = 1$ m

Solution

1. Check the ratio of S_i to L_w:

$$S_i/2L_w = 60/20 = 3$$

Because $S_i > 2L_w$, Eq. (22-7) may be used to estimate Q.

2. Rearrange the terms in Eq. (22-7) and solve for Q.

$$Q = \frac{K 2\pi L_w^2}{\left\{\ln\left[\frac{L_w}{r_w} + \sqrt{\left(\frac{L_w^2}{r_w^2} - 1\right)}\right] - 1\right\}}$$

$$= \frac{(0.0005 \text{ m/s})(2\pi)(20 \text{ m}^2)}{\left\{\ln\left[\frac{20 \text{ m}}{1 \text{ m}} + \sqrt{\left(\frac{20 \text{ m}^2}{1 \text{ m}^2} - 1\right)}\right] - 1\right\}} = 0.029 \text{ m}^3/\text{s}$$

Operation and Maintenance Issues

The prevention of clogging is essential in maintaining infiltration rates in vadose zone injection wells. The principal causes of well clogging are air entrainment, solids accumulation, and biological growths and each will result in a different rate at which clogging develops (see Fig. 22-17). It should be noted that the following discussion of clogging in vadose zone injection wells also applies to direct injection wells considered in Sec. 22-5.

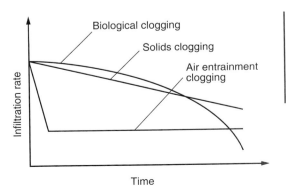

Figure 22-17
Reduction in infiltration rates for injection wells for different types of clogging.

Clogging Due to Air Entrainment

Air entrainment occurs when air becomes trapped in soil pores and effectively blocks the flow of water through the soil. A rapid reduction in infiltration rates is often associated with air entrainment, and recovery of infiltration rates will require an extensive drying period followed by careful reintroduction of water. Care must be taken to fill the vadose zone injection well from the bottom up, using an eductor pipe and air must be vented from the well to the atmosphere.

Clogging Due to Solids Accumulation

A linear reduction in infiltration rates with time is often associated with clogging from suspended solids in the applied water. If the injection rate and suspended solids concentrations are constant, then the loading of solids is constant and infiltration rates will steadily decline as solids accumulate at the interface of the borehole and the soil. For vadose zone injection wells, solids must be removed prior to injection. Clogging from solids might not occur in high permeability zones where solids do not accumulate at the borehole/soil interface. If the solids pass beyond the borehole/soil interface, they will be distributed over a large volume of the aquifer and might not cause clogging.

Biological Clogging

A logarithmic decrease in infiltration rates over time is often the result of biological clogging at the borehole/soil interface. As microorganisms accumulate in the well, their growth rate accelerates causing a reduction in infiltration rates. Because reclaimed water contains relatively high concentrations of biodegradable materials, a high degree of treatment such as RO is necessary to create a biologically stable water that will not result in microbial clogging. Biological growth in vadose zone injection wells may be inhibited by the addition of a disinfectant such as chlorine. If sufficient chlorine is added to prevent growth at the borehole/soil interface, biological growth may occur in the vadose zone away from the borehole where it will not cause clogging. Chlorine residuals of 2 to 5 mg/L have been found to be effective at preventing biological fouling.

Performance of Vadose Zone Injection Wells

Reclaimed water entering vadose zone injection wells may undergo water quality improvements similar to those described for spreading basins. Only limited data have been collected on vadose zone injection wells receiving reclaimed water, however, the mechanisms for removal described for spreading basins should apply except for nitrogen removal mechanisms. While cyclic operation of vadose zone injection wells may be practiced, it is not known if drying cycles can stimulate aeration of soils similar to recharge basins. Another potential concern is the fate of disinfection byproducts during subsurface transport. The requirement for a chlorine residual can increase greatly the concentration of disinfection byproducts injected into the wells.

Examples of Operational Full-Scale Vadose Zone Injection Facilities

Because areas with significant vadose zones do not exist in many parts of the country, groundwater recharge via vadose zone injection is not common. As discussed previously, the principal vadose zone injection facilities are located in Scottsdale, Arizona, where this method of groundwater recharge was pioneered (see Table E-1 in App. E). In other states where potential sites have been identified, questions concerning long-term sustainability have limited the use of this technology.

22-5 RECHARGE USING DIRECT INJECTION WELLS

Direct injection wells are a versatile tool for groundwater recharge with reclaimed water because they may be used in both saturated and unsaturated aquifers and the flow may be reversed, thereby allowing for periodic maintenance and cleaning. Direct injection wells may also be used as aquifer storage and recovery wells where the same well serves for both the injection and recovery. The primary drawback to direct injection wells is the cost for construction when deep aquifers are used for storage. In addition, energy costs for injecting the water to create a sufficient hydraulic gradient to accomplish reasonable infiltration rates can also be significant.

Description

Direct injection wells are constructed like regular pumping wells. In unconsolidated aquifers, a casing, screen, gravel pack, grouting, and a pipe to apply water to the well for infiltration into the aquifer are used (see Fig. 22-18). In consolidated aquifers (fractured rock, limestone and sandstone), the section of the well in the rock is completed as an open borehole without a screen. Recharge wells may be designed to recharge several confined aquifers by using several different injection pipes that inject into screened intervals within each confined aquifer. The injection well system operated by the Orange County Water District (OCWD) in Orange County, California, is an important example of how direct injection into several confined aquifers can be used to prevent salt water intrusion. Typical direct injection wells used for aquifer recharge are shown on Fig. 22-19.

Similar to vadose zone injection wells, the major problem with direct injection wells is clogging. Clogging occurs at the edge of the borehole, usually at the interface between the gravel envelope and the aquifer. Because infiltration rates into the aquifer at the borehole

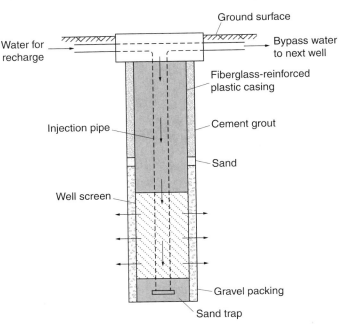

Figure 22-18
Typical schematic of a direct injection well for reclaimed water in a soil matrix.

Figure 22-19
Views of direct injection wells: (a) Florida and (b) Orange County, CA.

are much greater than the infiltration rates in spreading basins, recharge wells are more vulnerable to clogging. Recharge wells may be pumped periodically to reverse the flow of the well and remove clogging materials. The pumped water is often of poor quality and must be treated as wastewater. The best strategy for avoiding problems with the clogging of injection wells is to treat the water extensively prior to injection.

Pretreatment Needs

The pretreatment requirements for vadose zone injection wells apply generally to direct injection wells. For very deep direct injection wells, a higher level of pretreatment is often cost effective, as the cost of redeveloping clogged injection wells increases with depth. When water is injected directly into an aquifer, antidegradation laws regarding the aquifer water quality may apply (see Appendix F). In general, injected water must be equivalent to or better than the existing water quality of the aquifer to comply with most regulatory anti degradation provisions. The restrictions associated with antidegradation of an aquifer may, in some cases, require RO for pretreatment. The assumption that no water quality improvement will occur when water is injected directly into an aquifer is not valid. When water is injected directly into karst or fractured geological formations, however, the potential for water quality improvements decreases. When subsurface transport occurs, a significant surface area is available to support biofiltration reactions.

Hydraulic Analysis

Infiltration rates of injection wells are comparable to production wells except that a reduction in the hydraulic conductivity surrounding the borehole often occurs when reclaimed water is used. The same general equations that describe radial flow to a production well in a confined aquifer can also be used to describe radial flow from an injection well. The Theis solution (1935) that is often used to relate the production of water from a well as a function of the drawdown in hydraulic gradient may be used to estimate injection rates if the aquifer characteristics are known. Many techniques are used to characterize aquifers and it is recommended that aquifer characterization be done by a qualified hydrogeologist. The Theis solution for a direct injection recharge well is:

$$h_o - h(r, t) = \frac{Q}{4\pi T} W(u) \qquad (22\text{-}9)$$

$$u = \frac{r^2 S}{4Tt} \qquad (22\text{-}10)$$

where h_o = pressure used for injection, m (of water)
$h(r, t)$ = pressure at a radial distance r from the centerline of the well at a time t, m (of water)
r = radial distance from the centerline of the well, m
t = time after the start of injection, s
Q = volumetric flowrate of injected water, m³/s
T = transmissivity of the aquifer (hydraulic conductivity × aquifer depth), m²/s
$W(u)$ = well function, dimensionless
u = time parameter, dimensionless
S = storativity of the aquifer, dimensionless

The well function $W(u)$ has been tabulated and is available in groundwater software packages (see also Table G-2 in App. G). Providing that T and S are known, the injection rate may be calculated as a function of the applied pressure used for injection. As the time of injection increases, the pressure in the aquifer will increase and the injection pressure must be increased to maintain the same infiltration rates. Recovery of water either by a recovery well or by the injection well will reduce the pressure in the aquifer allowing for infiltration rates to be maintained without increasing the injection pressure.

EXAMPLE 22-4. Determine the Operating Pressure for a Direct Injection Well.

A direct injection well will be used for the recharge of an aquifer with a transmissivity of 0.02 m²/s and a storativity of 0.0001. The desired volumetric injection rate is 0.03 m³/s and the borehole radius is 0.3 m. Using the Theis equation [Eq. (22-9)], determine the pressure above the peizometric head necessary to maintain this infiltration rate after 10 d of injection.

Solution
1. Rewrite the Theis equation for an injection well.

    The Theis equation is commonly used to calculate drawdown where $h(r, t)$ is less than h_o, however, the equation may be applied for injection by multiplying by a negative sign as follows:

    $$h(r, t) - h_o = \frac{Q}{4\pi T} W(u)$$

2. Determine the well function $W(u)$.
    a. To determine the well function $W(u)$, u must be first determined using Eq. (22-10) as follows:

    $$u = \frac{r^2 S}{4Tt} = \frac{(0.3 \text{ m})^2 \times 0.0001}{(4)(0.02 \text{ m}^2/\text{s})(10 \text{ d})(86{,}400 \text{ s/d})} = 1.3 \times 10^{-10}$$

b. Determine the value of the well function W(u).
   From Table G-2 in App. G, for a u value of 1.3×10^{-10}, the value of W(u) is 22.2

3. Determine the required pressure to achieve the desired flowrate using Eq. (22-9).

$$h(r, t) - h_o = \frac{Q}{4\pi T} W(u) = \frac{(0.03 \text{ m}^3/\text{s})}{4\pi (0.02 \text{ m}^2/\text{s})} \times 22.2 = 2.65 \text{ m}$$

Comment
In applying the Theis equation, the aquifer is assumed to be homogeneous with no effect from aquitards (impermeable layers) or from other pumping or injection wells.

Operation and Maintenance Issues

The general methods for the development of clogging were described in Sec. 22-4 for vadose zone injection wells. While air entrainment can be avoided with direct injection wells, care must be taken to avoid the release of dissolved gases during injection, which may occur when high injection pressures are used and the pressure is dissipated rapidly in the aquifer. As the pressure is reduced, dissolved gases can be released from solution resulting in gas binding of the aquifer. Preventing the dissolution of gases at high pressure or avoiding rapid pressure drops are useful techniques in preventing the release of dissolved gases in the aquifer. Temperature fluctuations may also cause air binding in the aquifer when the injected water temperature increases in the aquifer; dissolved gases will then come out of solution.

Measures of Clogging Potential

Considerable research has been conducted on clogging of research wells, and several parameters have been developed to evaluate the clogging potential due to biological activity and suspended solids (Peters and Castell-Exner, 1993). The membrane filtration index (MFI) has been used to assess the suspended solids content of the water. The MFI is determined by plotting the infiltration rate as a function of the volume filtered using a membrane filter such as a 0.45 μm Millipore filter (see Chap. 8). The slope of the straight line portion of the curve is used to determine the MFI with units of t/vol². The Meeloop filter index (MLFI) has also been used to evaluate clogging from solids with native aquifer materials. The MLFI is determined by passing water through columns filled with native aquifer material at flowrates higher than the corresponding flowrates through the aquifer around the well. The MLFI is also useful for determining if any adverse geochemical interactions occur that will result in clogging. It should be noted that work is underway to develop a more reliable measure that will reflect the presence of fine colloidal material.

Assimilable organic carbon (AOC) is determined by plating out a sample solution incubated with *Pseudomonas fluorescens* and counting the bacterial colonies. The results are expressed in terms of the carbon concentration of an acetate solution producing the same bacterial growth response. If the AOC concentrations in recharge water are less than 10 µg/L, serious clogging of the well should not occur, even if chlorine is not added. As described for vadose zone injection wells, increasing chlorine concentrations can be used to prevent clogging from increasing AOC concentrations.

Need for Full-Scale Testing
While the MFI, MLFI, and AOC parameters can be used to evaluate the relative clogging potential of different waters, they cannot be used as absolute predictors of clogging in actual injection wells. Full-scale studies of injection wells are necessary to determine design and operational criteria under most circumstances. Other practical aspects include variations in flowrates and concentrations of suspended solids that must be considered. The formation of biofilms during periods of low flow and the sloughing of biofilms during periods of high flow can also be very important. During periods of low flow or no injection, it is often useful to maintain a trickle flow with a residual chlorine concentration to prevent biofilm formation.

Because the majority of removal mechanisms described for recharge basins are applicable to both unsaturated and saturated flow, reclaimed water that is injected directly into an aquifer should also be subjected to removal mechanisms similar to those vadose zone injection wells. The injection of RO treated water, however, requires special consideration.

Performance of Direct Injection Wells

While concerns over constituents in RO treated water should be minimal as almost all constituents are removed, the use of RO treated water may limit biological activity which contributes to water quality improvement during subsurface transport. Reverse osmosis removes almost all constituents including biodegradable organic carbon and nutrients that create a biologically active system. Because microbial activity sustains most removal mechanisms during subsurface transport, the effectiveness of an aquifer as a treatment barrier is reduced with RO water. Nonpolar low molecular weight compounds such as chloroform and N-nitrosodimethylamine (NDMA) can pass through a RO membrane and the cometabolic reactions that can remove these compounds during subsurface transport are not stimulated. The compound NDMA is a disinfection byproduct with a health based guideline level of 0.7 ng/L for drinking water, which is below the analytical detection limit. Also, unlike the majority of PCPPs that have been identified in reclaimed waters, the toxicity of NDMA has been evaluated. Because NDMA is photosensitive and may be biotransformed, NDMA has not been a problem with surface spreading basins.

At OCWD where RO water is injected directly into the aquifer to prevent salt water intrusion, NDMA has been demonstrated to persist and has been found in drinking water production wells. When RO water is injected directly into the ground, there is no exposure to light and the microbial activity in the aquifer is limited, thereby creating an ideal situation for the persistence of NDMA. In response to the NDMA problem, OCWD has added an advanced oxidation step to remove NDMA prior to injection. The compound NDMA has also been identified in monitoring wells of the Scottsdale Water Campus where vadose zone injection wells are used to inject RO treated water. Over 90 m (300 ft) of vadose zone exists between the wells and the aquifer.

Examples of Full-Scale Direct Aquifer Injection Facilities

A variety of direct aquifer injection facilities have been used, notably in California and Florida. Examples of direct aquifer injection facilities currently in operation are reported in Table 22-8. Of the facilities identified in Table 22-8, the largest direct injection facilities using reclaimed water are located in Orange County, California, for the control of sea water intrusion in the coastal aquifers as well as inland groundwater replenishment.

Table 22-8
Examples of full-scale direct injection facilities for groundwater recharge

Location	Description
El Paso, TX (El Paso Water Utilities)	Arid El Paso needed a plan to reuse reclaimed water in the face of dwindling groundwater supplies from its aquifer. At the Fred Hervey Water Reclamation Plant, primary effluent enters a two-stage biophysical process which combines activated sludge with powdered activated carbon adsorption. This step of treatment is designed for organics removal, nitrification, and denitrification. Methanol is added to the second stage to provide a carbon source for denitrification. A lime treatment step follows to remove phosphorus and heavy metals, to inactivate viruses, and to soften the reclaimed water. Turbidity removal is provided by sand filters, and disinfection is provided by ozonation. The final product water is passed through a granular activated carbon filter for final polishing before release to storage. A prototype injection project has been in operation in El Paso since 1985, supplying more than 38×10^3 m³/d (10 Mgal/d) of reclaimed water into the Hueco Bolson aquifer. Ultimately, the reclaimed water returns to the city's potable water system after an estimated 2- to 6-yr travel time in the underground. While the reclaimed water currently recharged represents a small percentage of the total aquifer volume, the long-term goal is to provide 25 percent of El Paso's future water needs.
Orange County Water District, Fountain Valley, CA	Orange County Water District's (OCWD) Groundwater Replenishment (GWR) system is a new, planned water purification project unlike previous projects because of its high level of water purification including MF, RO, and hydrogen peroxide/UV advanced oxidation. The GWR system, scheduled to produce water in 2007, is part of an overall plan to help prevent future water shortages in Orange County. The GWR system is being built, and replaces more than 25 yr of successful water reclamation operations at Water Factory 21, a project (design capacity of 57×10^3 m³/d, 15 Mgal/d) built and operated by OCWD. Water Factory 21 was the first project in California to purify municipal wastewater to drinking water quality to be used as a barrier against the intrusion of seawater into a groundwater basin. Although some of the injected water flows toward the ocean forming the seawater barrier, the majority of the water flows into the groundwater basin to augment the potable groundwater supply.
West Basin Municipal Water District, Carson, CA	In 1992, West Basin received state and federal funding to pursue its water recycling program, which consisted of constructing a water reclamation facility in the City of El Segundo. The future expansion will ultimately increase production of reclaimed water for the West Coast groundwater basin by 19×10^3 m³/d (5 Mgal/d) and also increase the production of reclaimed water meeting the Title 22 regulations by 38×10^3 m³/d (10 Mgal/d). Upgrades to the existing barrier water production system will also be installed, improving the efficiency of the treatment process and increasing the quality of the seawater intrusion barrier product water.

22-6 OTHER METHODS USED FOR GROUNDWATER RECHARGE

Other methods that have been used for groundwater recharge include (1) aquifer storage and recovery, (2) riverbank and dune filtration, (3) enhanced river recharge, and (4) subsurface facilities. Each of these methods is described briefly below.

Aquifer Storage and Recovery (ASR)

Direct injection wells may also be used as ASR wells where the wells serve as both recovery and injection wells. Dual use has become a popular method of integrating groundwater recharge into domestic water supplies and is becoming an alternative for reclaimed water systems as well. Another advantage of ASR systems is that they may be used to store water in nonpotable aquifers such as brackish aquifers (Pyne, 1995; Dillon et al., 2006; see also Sec. 17-7 in Chap. 17).

Operational Features of ASR Systems

The major difference between ASR and other groundwater recharge systems is the flow path of the water during storage. As injection occurs in an ASR well, a zone of recharged water is created around the injection well (see Fig. 22-20). A buffer zone exists where the recharged water blends with native groundwater and complex geochemical reactions may occur in this zone. To avoid undesirable geochemical interactions, a large quantity of recharge water may be injected initially to create a large buffer zone surrounding the storage zone. Because water is injected during periods of excess supply and recovered during periods of high demand, the water may be injected and stored for a few or several months before recovery. Because the storage time in ASR systems is variable, the effects on water quality transformations may also be variable. The water at the outside of the storage zone

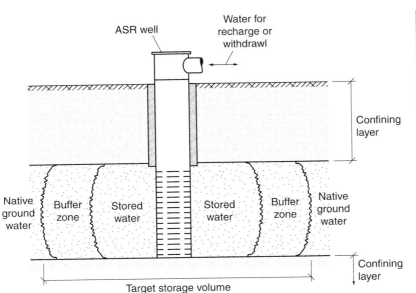

Figure 22-20

Aquifer storage recovery (ASR) well with the target storage volume surrounding the borehole.

Figure 22-21
Reclaimed water injection and recovery wells (a) well being drilled, Scottsdale, Arizona (Courtesy of P. Fox) and (b) operating well system, Australia (Courtesy of P. Dillon, Commonwealth Scientific and Industrial Research Organization, CSIRO, Australia).

is the first water to be injected in and will be the last water to be recovered. A typical operational ASR well is shown on Fig. 22-21*b*.

Water Quality Issues

The use of ASR wells also presents several unique problems with respect to water quality. The first issue is the travel path which requires that the first water recovered has the shortest travel time and vice versa. Therefore, the water quality transformations can vary significantly between the time recovery starts and ends. The effect can be especially problematic for disinfection byproducts as a chlorine residual is often required to prevent biological clogging during injection. Disinfection byproduct formation can continue in the aquifer near the well, and if the water is not stored for a long time, the DBPs will be present in the initial water recovered.

The use of ASR wells with reclaimed water has been limited primarily to recovery for irrigation. As a result, serious concerns over water quality changes during aquifer storage have not been expressed. Because of the flow paths associated with ASR systems, their use for indirect potable reuse is limited to the injection of very high quality water such as RO treated water. Many water quality changes observed in ASR wells are associated with geochemical interactions from the introduction of aerated water into an anoxic aquifer. To avoid problems with these geochemical interactions, a large quantity of water should be injected initially to create a buffer zone where geochemical interactions will not affect recovered water quality.

Riverbank and Dune Filtration

Bank filtration systems use wells to withdraw water from rivers indirectly through the subsurface environment (see Fig. 22-22). Bank filtration systems are used for the purpose of improving water quality and do not provide storage. Basically, the river bank is used as a filter, as the detention time to the extraction wells is short. Riverbank and dune

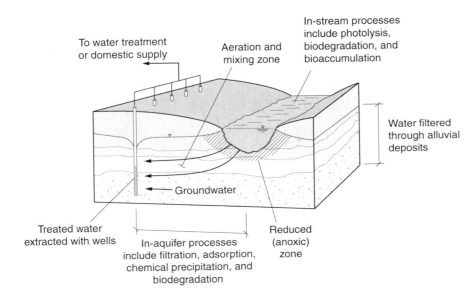

Figure 22-22
Riverbank filtration for the treatment of river waters.

filtration (RBF) systems have been used from ancient times to provide potable water from surface waters that are heavily influenced by wastewater discharges.

Operational Features of Riverbank Filtration Systems

Riverbank filtration uses the river bottom as the interface between the surface water and the aquifer for groundwater recharge. Pumping of a well adjacent to the river draws water from the river through the subsurface to the recovery point (see Fig. 22-22). Natural scour is an important method for maintaining recharge rates in bank filtration systems, as wetting and drying cycles do not exist in most natural river systems. Because RBF systems have been in existence for a long period of time, they have been studied extensively and the knowledge gained may be used for groundwater recharge systems using reclaimed water.

Water Quality Issues

The primary goal of RBF systems is an improvement in water quality. Riverbank and dune filtration systems may also include recharge basins located adjacent to the river to enhance the flow of water through the subsurface to the point of recovery.

Enhanced River Recharge

Natural recharge that occurs as water flows along a river can be enhanced by using a system of weirs, dams, or levees to spread the water over a floodplain. The river recharge system on the Santa Ana River in Orange County, California, is an outstanding example of enhanced river recharge. A set of levees has been developed to enhance the recharge (see Fig. 22-23a). Water from the river is diverted using an inflatable dam (see Fig. 22-23b). The diverted water must flow through a tortuous path around the levees ensuring that the water is spread over a large area, providing ample opportunity for recharge (see Fig. 22-23c). The flow paths are adjusted periodically to allow for drying cycles of portions of the floodplain. During the summer, the flow in the river is essentially treated wastewater from upstream dischargers.

Figure 22-23

In channel infiltration system using T-levees in the Santa Ana River, Santa Ana, CA: (a) aerial view of spreading basins (Courtesy of Orange County Water District, Fountain Valley, CA) (Coordinates: 33.856 N, 117.845, view at altitude 4 km), (b) inflatable rubber dam used to divert water from river to off-river spreading basins, and (c) ground level view of spreading basins in river bed.

Groundwater Recharge Using Subsurface Facilities

Other methods for groundwater recharge include the use of subsurface drains, troughs, and/or trenches. The objective of these methods for groundwater recharge is to increase surface recharge rates without using valuable land. Subsurface drains are similar to the drainage systems used for disposal of effluent from septic tanks, and they may be buried completely allowing the land to be utilized for alternative purposes. Troughs and trenches can be used to increase the surface area for recharge. Excavation often provides access to soils of higher permeability relative to surface soils.

22-7 CASE STUDY: ORANGE COUNTY WATER DISTRICT GROUNDWATER REPLENISHMENT SYSTEM

In closing this chapter, it is appropriate to consider the Orange County, California, Groundwater Replenishment (GWR) system scheduled to go into operation in 2007. When operational, this project will be one of the largest water reclamation facilities of its kind in the world employing the latest advanced treatment technologies including RO and advanced oxidation. This case study was adapted from OCWD (2006) and Daugherty et al. (2005).

Setting

Orange County is located on the southern California coast between Los Angeles County and San Diego County. The Orange County Water District (OCWD) manages the groundwater basin that serves approximately 2.3 million people. Currently, the estimated groundwater usage rate is 333×10^6 m^3/yr (270,000 ac-ft/yr). Both the population served by OCWD and the water demand are projected to increase by 20 to 40 percent by 2030. The primary water supply for the basin is the Santa Ana River, which is recharged into the groundwater basin by means of spreading basins (see Fig. 22-23a). Supplemental water sources for groundwater recharge include the Colorado River and the Sacramento River, delivered to southern California by the California State Water Project. Santa Ana River water, along with the imported water, is also recharged using deep recharge basins (see Fig. 2-5 in Chap. 2). The recharged water is subsequently pumped and serves as the water supply for a large portion of the county population. To augment existing water supplies, OCWD has undertaken the development of the GWR system using highly treated reclaimed water. The GWR system is a project funded jointly by the OCWD and Orange County Sanitation District (OCSD).

The GWR System

When implemented fully the GWR system will produce approximately 1.73×10^8 m^3/yr (140,000 ac-ft/yr) of advanced treated reclaimed water. The GWR project has been implemented to:

1. Protect the groundwater basin from overdraft and seawater intrusion
2. Reduce the amount of treated wastewater discharged to the ocean from OCSD
3. Reduce reliance on other water sources (i.e., imported water: the Colorado and Sacramento Rivers)
4. Provide locally controlled water (i.e., reclaimed water)
5. Help meet the state of California's statewide water objectives
6. Help reduce the mineral buildup in the Orange County groundwater

Implementation

Under construction since 2003, the GWR system is being built at an estimated total program budget of $487 million (2003 estimate). The first phase of the project will be online in 2007 and will supply approximately 88×10^6 m^3/yr (72,000 ac-ft/yr) of water and provide the backbone facilities for future expansion. The GWR system consists of three major components: (1) the Advanced Water Treatment Facility (AWTF) and pumping stations, (2) a 21 km (13 mi) pipeline connecting the AWTF to OCWD's existing groundwater recharge basins, and (3) the expansion of the existing seawater intrusion barriers with additional injection and monitoring wells (OCWD, 2006).

The AWTF process flow diagram, shown on Fig. 22-24, includes microfiltration, cartridge filtration, RO, lime addition, and UV/H$_2$O$_2$ advanced oxidation treatment. Views of the unit processes may be found in other chapters: microfiltration (see Fig. 8-32d), cartridge filters (see Fig. 9-7a), RO (see Figs. 9-2a and b), lime saturator (see Fig. 9-12a), and UV photolysis reactors (see Fig. 10-19). The product water will

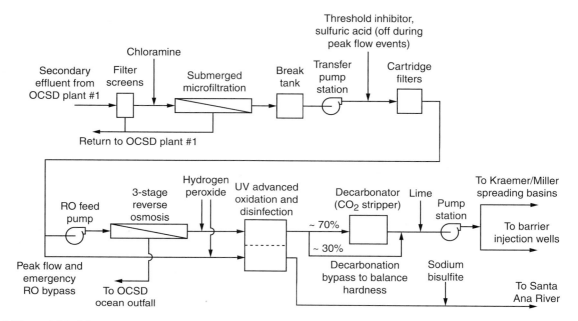

Figure 22-24
Schematic flow diagram for 265 × 10³ m³/d (70 Mgal/d) advanced water treatment facility (currently under construction, 2006), at the Orange County Water District, Fountain Valley, CA, (Coordinates: 33.692 N, 117.942 W). (Adapted from Orange County Water District, Fountain Valley, CA.)

then be introduced into the existing surface spreading basins along with water from other sources. The blended water will be percolated into the groundwater aquifers, where it eventually becomes part of Orange County's drinking water supply. A portion of the reclaimed water, as shown on Fig. 22-24, is to be injected into salt water intrusion barrier wells along the Pacific coastline.

While the GWR system is being constructed, a 19 × 10³ m³/d (5 Mgal/d) facility employing microfiltration, RO, and advanced oxidation has been built and operated to provide water for the injection wells used to control seawater intrusion. Known as the GWR System Phase 1 facility, this facility has also been used to obtain valuable design and operating experience and to allow for hands on training for the operators of the full-scale system.

Lessons Learned

The product water produced from the GWR System Phase 1 facility meets all of the California DHS and Regional Water Quality Control Board requirements for indirect potable reuse. The performance of the Phase 1 facility has validated the effectiveness of the process flow diagram shown on Fig. 22-24 for the full-scale facility. Because of initial concerns for public acceptance and safety, an extensive public outreach program has been conducted to demonstrate the safety of GWR product water and the improvement in groundwater quality.

PROBLEMS AND DISCUSSION TOPICS

22-1 Two locations have been identified for underground storage and recovery of surplus reclaimed water. At location 1, the majority of storage is in an unconfined aquifer. At location 2, the majority of storage is in confined aquifers. Describe the options for groundwater recharge at location 1. What are the options for groundwater recharge in location 2? If both locations are an equal distance from the water reclamation plant, which location would you choose? Explain your choice.

22-2 A recharge basin was evaluated for groundwater recharge and following parameters were determined.
  K = hydraulic conductivity = 1.4 m/d (assume constant)
  H_w = height of water above ground = 0.5 m
  f = fillable porosity = 0.15
  H_{cr} = −0.6 m
Estimate the average infiltration rate for a 7 d wetting period using the Green-Ampt equation. You must assume a value for L_f and check your assumption. Based on the average infiltration rate, estimate the surface area requirements necessary to recharge 120,000 m³/d.

22-3 A recharge basin is operated with a height of water above the ground of 0.5 m. The hydraulic conductivity without clogging is 0.7 m/d. The fillable porosity is 0.15 and the critical pressure head of the soil for wetting is −0.4 m.
  a. Use the Green-Ampt equation to determine infiltration rate, v_i, and depth of the wetted front, L_f, as a function of time. Complete the calculations to a value of L_f of 30 m.
  b. Use the data from Part a to estimate the total quantity of water infiltrated for a 200 m² basin.

22-4 The following parameters have been determined for a recharge basin and the unconfined aquifer below the recharge basin.
  T = transmissivity = 0.003 m²/s
  H = height of groundwater above an impermeable layer (aquifer thickness) = 20 m
  L = 20 m
  W = 120 m
  f = fillable porosity = 0.15
  V_a = 80 m/yr
  Depth to groundwater = 30 m
  a. Calculate the maximum groundwater rise in the mound after 6 mo of recharge basin operation.
  b. Based on the results of Part a, will the groundwater rise affect infiltration?
  c. A landfill is located near the groundwater recharge basin and the bottom of the landfill is 5 m above the groundwater. Estimate the groundwater rise at x = 350 m and y = 350 m where the landfill is located. Will the groundwater rise impact the landfill?

22-5 Estimate the number of vadose zone injection wells necessary to infiltrate 120,000 m³/d of reclaimed water.

The well parameters are

K = hydraulic conductivity = 0.002 m/s
S_i = 100 m
L_w = 25 m
r_w = 1.5 m

Would the solution to the problem change if S_i = 20 m? If so, how many wells would be required?

22-6 Why are wetting and drying cycles used in the operation of percolation basins? Explain in terms of both quantity and quality.

22-7 An unconfined aquifer with K = 2 m/d and H = 5.33 m is recharged from a 20 × 200 m basin with an infiltration rate of 0.2 m/d. How much will the center of the groundwater mound have risen 6 d after the infiltrating water reaches the groundwater table (assume vertical flow directly to the vadose zone)?

If infiltration is stopped 6 d after it started, what will be the height of the mound above the original water table 6 d later?

22-8 Estimate the volumetric injection rate (Q) for a vadose zone well that is 40 m deep and has a radius of 1.5 m. The hydraulic conductivity of the soil is 0.0001 m/s and the depth from the bottom of the well to an impermeable layer is 80 m.

22-9 A direct injection well will be used to recharge an aquifer with a transmissivity of 0.01 m²/s and a storativity of 0.0002. The desired volumetric injection rate is 0.02 m³/s and the borehole radius is 0.2 m. Determine the pressure above the peizometric head necessary to maintain this infiltration rate after 7 d of injection.

22-10 The Groundwater Ambient Monitoring and Assessment (GAMA) Program is a statewide effort to characterize a broad suite of chemicals at low detection limits in groundwater in California. Download and review the results of the GAMA program related to the recharge basins in the Los Angeles and Orange County Focus Area. What is the significance of the findings of this study?

REFERENCES

Asano, T., and K. L. Wassermann (1980) "Groundwater Recharge Operations in California," *J. AWWA*, **72**, 7, 380–385.

Asano, T. (ed.) (1985) *Artificial Recharge of Groundwater*, Butterworth Publishers, Boston, MA.

AWWARF (2001) *Soil Aquifer Treatment for Sustainable Water Reuse*, American Water Works Association Research Foundation, Denver, CO.

AWWARF (2005) *Water Quality Improvements during Aquifer Storage and Recovery*, American Water Works Association Research Foundation, Denver, CO.

Bouwer, H. (1966) "Rapid Field Measurement of Air Entry Value and Hydraulic Conductivity of Soil as Significant Parameters in Flow System Analysis," *Water Res.*, **2**, 729–738.

Bouwer, H., and R. D. Jackson (1974) "Determining Soil Properties," 611–672, in J. van Schilgaarde (ed.) *Drainage for Agriculture*, Agronomy Monograph No. 17, American Society of Agronomy, Madison, WI.

Bouwer, H., and R. C. Rice (1974) "High-rate Land Treatment I: Infiltration and Hydraulic Aspects of the Flushing Meadows Project," *J. WPCF*, **46**, 834–843.

Bouwer, H. (1978) *Groundwater Hydrology*, McGraw-Hill, New York.

Bouwer, H., R. C. Rice, J. C. Lancc, and R. G. Gilbert (1980) "Rapid-Infiltration Research—the Flushing Meadows Project, Arizona," *J. WPCF*, **52**, 10, 2457–2470.

Bouwer, H., and R. C. Rice, (1984) "Effect of Water Depth in Groundwater Recharge Basins on Infiltration," *J. Irrig. Drain. Eng.* **115**, 4, 556–567.

Bouwer, H. (1986) "Intake Rate: Cylinder Infiltrometer," 825–844, in A. Klute (ed.) *Methods of Soil Analysis: Part 1 Physical and Mineralogical Methods*, 2nd ed., American Society of Agronomy and Soil Science Society of America, Madison, WI.

Bouwer, H. (1991) "Role of Groundwater Recharge in Treatment and Storage of Wastewater for Reuse," *Water Sci. Technol.* **24**, 9, 295–302.

Crook, J., T. Asano, and M. Nellor (1990) "Groundwater Recharge with Reclaimed Water in California," *Water Environ. Technol.*, **2**, 4, 42–49.

Daugherty, J. l., S. S. Deshmukh, M. V. Patel, and M. R. Markus (2005) "Employing Advanced Technology for Water Reuse in Orange County," Presented at the California Section of the WateRuse Association Annual Meeting, San Diego, CA.

Dillon, P. J. (ed.) (2002) *Management of Aquifer Recharge for Sustainability*, A. A. Balkema Publishers, Lisse, The Netherlands.

Dillon, P., P. Pavelic, S. Toze, S. Rinck-Pfeiffer, R. Martin, A. Knapton, and D. Pidsley (2006) "Role of Aquifer Storage in Water Reuse," *Desalination*, **188**, 123–134.

Drewes, J. E., and P. Fox (2001) "Effect of Drinking Water Sources on Reclaimed Water Quality in Water Reuse Systems," *Water Environ. Res.*, **72**, 3, 353–362.

Drewes, J. E., and M. Jekel (1998) "Behavior of DOC and AOX Using Advanced Treated Wastewater for Groundwater Recharge," *Water Res.* **32**, 10, 3125–3133.

Freeze, R. A., and J. A. Cherry (1979) *Groundwater*, Prentice-Hall, Englewood Cliffs, NJ.

Gable, J., and P. Fox, (2003) "Sustainable Nitrogen Removal by Anaerobic Ammonia Oxidation," *Proceedings of the 76th Annual Water Environment Federation Conference*, Los Angeles, CA.

Goyal, S. M., and C. P. Gerba (1979) "Comparative Adsorption of Human Enteroviruses, Simian Rotavirus and Selected Bacteriophages to Soils," *Appl. Environ. Microbiol.*, **38**, 242.

Green, W. H., and G. Ampt, (1911) "Studies of Soils Physics, Part I—The Flow of Air and Water through Soils," *J. Agr. Sci.*, **4**, 1–24.

Hantush, M. S. (1967) "Growth and Decay of Groundwater-Mounds in Response to Uniform Percolation," *Water Resour. Res.*, **3**, 1, 227–234.

Havelaar, A. H. (ed.) (1991) "Bacteriophages as Model Viruses in Water Quality Control," IAWPRC Study Group on Health Related Water Microbiology, *Water Res.*, **25**, 529–545.

Havelaar, A. H. (1993) "Bacteriophages as Models of Human Enteric Viruses in the Environment," *ASM News*, **59**, 614–619.

Havelaar, A. H., M. van Olphen, and Y. C. Drost (1993) "F-Specific RNA Bacteriophages are Adequate Model Organisms for Enteric Viruses in Fresh Water," *Appl. Environ. Microbiol.*, **59**, 2956–2962.

Heberer, T. (2002) "Occurrence, Fate, and Removal of Pharmaceutical Residues in the Aquatic Environment: A Review of Recent Research Data," *Toxicol. Lett.*, **131**, 1–2, 5–17.

Morel-Seytoux, H. J. (1985) "Conjunctive Use of Surface and Ground Waters," Chap. 3, in T. Asano (ed.) *Artificial Recharge of Groundwater*, Butterworth Publishers, Boston, MA.

Mulder, A., A. A. van de Graaf, L. A. Robertson, and J. G. Kuenen (1995) "Anaerobic Ammonium Oxidation Discovered in a Denitrifying Fluidized Bed Reactor." *FEMS Microbiol. Ecol.*, **16**, 3, 177–183.

Nasser, A. M., and S. D. Oman (1999) "Quantitative Assessment of the Inactivation of Pathogenic and Indicator Viruses in Natural Water Sources," *Water Res.*, **33**, 7, 1748–1752.

Neuman, S. P. (1976) "Wetting Front Pressure Head in the Infiltration Model of Green and Ampt." *Water Res.*, **12**, 564–566.

NRC (1994) *Ground Water Recharge Using Waters of Impaired Quality*, National Research Council, National Academy Press, Washington, DC.

OCWD (2006) *Groundwater Recharge System*, Accessed at: http://www.gwrsystem.com/about/overview.html.

Peters, J. H., and C. Castell-Exner (eds.) (1993) *Proc. Dutch-German Workshop on Artificial Recharge of Groundwater*, September, 1993, Castricum, The Netherlands. Published by KIWA, P.O. Box 1072, Nieuwegein 3430 BB, The Netherlands.

Peyton, D. E. (2002) "Modified Recharge Basin Floors Control Sediment and Maximize Infiltration Efficiency," 215–220, in P. Dillon (ed.) *Management of Aquifer Recharge for Sustainability*, A. A. Balkema Publishers, Lisse, The Netherlands.

Powelson, D. K., C. P. Gerba, and M. T. Yahya (1993) "Virus Transport and Removal in Wastewater During Aquifer Recharge," *Water Res.*, **27**, 4, 583–590.

Powelson, D. K., and C. P. Gerba (1994) "Virus Removal From Sewage Effluents During Saturated and Unsaturated Flow Through Soil Columns," *Water Res.*, **28**, 10, 2175–2181.

Pyne, R. D. G. (1995) *Groundwater Recharge and Wells, A Guide to Aquifer Storage Recovery*, CRC Press, Inc., Boca Raton, FL.

Roberts, P. V. (1980) "Water Reuse for Groundwater Recharge: An Overview," *J. AWWA*, **72**, 7, 375–379.

SDLAC (2005) Water Reuse Partnership, Sanitation Districts of Los Angeles County, Accessed at: http://www.lacsd.org

Shah, S., and P. Fox (2005) "The Conversion of Nitrate to Nitrate as a Rate Limiting Step for Anaerobic Ammonia Oxidation in Soil Systems," *Proceedings of the 78th Annual Water Environment Federation Conference*, Washington, DC.

Theis, C. V. (1935) "The Relation between the Lowering of Peizometric Surface and the Rate and duration of a Well Using Groundwater Storage," *Trans. Am. Geophys. Union*, **2**, 519–524.

Todd, D. K. (1980) *Groundwater Hydrology*, John Wiley & Sons, New York.

Wilson, L. G., G. L. Amy, C. P. Gerba, H. Gordon, B. Johnson, and J. Miller (1995) "Water Quality Changes during Soil Aquifer Treatment of Tertiary Effluent," *Water Environ. Res.*, **67**, 3, 371–376.

Woods, C., H. Bouwer, R. Svetich, S. Smith, and R. Prettyman (1999) "Study Finds Biological Nitrogen Removal in Soil Aquifer Treatment System Offers Substantial Advantages," *Proceedings of the 72nd Annual Water Environment Federation Conference*, New Orleans, LA.

Yates, M. V., C. P. Gerba, and L. M. Kelly (1985) "Virus Persistence in Groundwater," *Appl. Environ. Microbiol.*, **49**, 4, 778–781.

23 Indirect Potable Reuse through Surface-Water Augmentation

WORKING TERMINOLOGY   1304

23-1   OVERVIEW OF INDIRECT POTABLE REUSE   1305
De Facto Indirect Potable Reuse   1305
Strategies for Indirect Potable Reuse through Surface-Water Augmentation   1307
Public Acceptance   1308

23-2   HEALTH AND RISK CONSIDERATIONS   1308
Pathogens and Trace Constituents   1308
System Reliability   1309
Use of Multiple Barriers   1309

23-3   PLANNING FOR INDIRECT POTABLE REUSE   1309
Characteristics of the Watershed   1310
Quantity of Reclaimed Water to be Blended   1311
Water and Wastewater Treatment Requirements   1312
Institutional Considerations   1312
Cost Considerations   1313

23-4   TECHNICAL CONSIDERATIONS FOR SURFACE-WATER AUGMENTATION IN LAKES AND RESERVOIRS   1314
Characteristics of Water Supply Reservoirs   1314
Modeling of Lakes and Reservoirs   1319
Strategies for Augmenting Water Supply Reservoirs   1320

23-5   CASE STUDY: IMPLEMENTING INDIRECT POTABLE REUSE AT THE UPPER OCCOQUAN SEWAGE AUTHORITY   1323
Setting   1323
Water Management Issues   1323
Description of Treatment Components   1323
Future Treatment Process Directions   1326
Water Quality of the Occoquan Reservoir   1327
Water Treatment   1328
Lessons Learned   1329

23-6   CASE STUDY: CITY OF SAN DIEGO WATER REPURIFICATION PROJECT AND WATER REUSE STUDY 2005   1329
Setting   1330
Water Management Issues   1330
Wastewater Treatment Mandates   1330
Water Repurification Project   1331
2000 Updated Water Reclamation Master Plan   1332

City of San Diego Water Reuse Study 2005 1332
Lessons Learned 1334

23-7 CASE STUDY: SINGAPORE'S NEWATER FOR INDIRECT POTABLE REUSE 1334
Setting 1335
Water Management Issues 1335
NEWater Factory and NEWater 1335
Implementation 1335
NEWater Demonstration Plant Performance 1336
Project Milestones 1336
Lessons Learned 1337

23-8 OBSERVATIONS ON INDIRECT POTABLE REUSE 1340

PROBLEMS AND DISCUSSION TOPICS 1341

REFERENCES 1342

WORKING TERMINOLOGY

Term	Definition
De facto indirect potable reuse	Many cities withdraw drinking water from rivers or surface water reservoirs that contain varying amount of wastewater discharges from upstream cities, industries, and agricultural areas, thus practicing unplanned (*de facto*) indirect potable reuse.
Direct potable reuse	The planned introduction of highly treated reclaimed water either directly into the potable water supply distribution system downstream of a water treatment plant, or into the raw water supply immediately upstream of a water treatment plant.
Indirect potable reuse	The planned incorporation of reclaimed water into a raw water supply, such as in potable water storage reservoirs or a groundwater aquifer, resulting in mixing, dilution, and assimilation, thus providing an environmental buffer.
Multiple barriers	Planned indirect potable reuse system in which several safety measures beyond those normally included in conventional water systems are incorporated to increase the overall system reliability. Multiple barriers include wastewater treatment, dilution, and natural attenuation in the water body, storage in reservoirs, effective drinking water treatment, and extensive raw and treated water monitoring to ensure high quality drinking water.
NEWater	The term adopted by the Singapore Public Utilities Board to describe the product from water reclamation using advanced treatment processes of membrane filtration, reverse osmosis, and ultraviolet disinfection and used for industrial uses and indirect potable reuse after blending in surface water storage reservoirs.
Repurified water	A term used by the City of San Diego to describe the product of a sophisticated treatment regime which is suitable for indirect potable reuse. The treatment steps include the use of several advanced treatment technologies, including membrane filtration; reverse osmosis; ion exchange; advanced oxidation using ozone; and disinfection.

Indirect potable reuse, refers to the planned introduction of reclaimed water into a raw water supply, such as a potable water storage reservoir, the subject of this chapter, or a groundwater aquifer, resulting in mixing and assimilation, thus providing dilution and an environmental buffer. Groundwater recharge with reclaimed water for the purpose of groundwater replenishment and extraction as a potable water supply is discussed in Chap. 22. Indirect potable reuse is motivated by the need to develop additional sustainable water supplies, as discussed in Chap. 1, as well as by recent advances in water reclamation technologies, as discussed in Part 3 of this textbook. In this and the following chapter, indirect potable reuse through surface water augmentation, and direct potable reuse of reclaimed water are examined, respectively. Following an introduction to indirect potable reuse, health and risk considerations, planning for indirect potable reuse, and technical considerations for surface water augmentation in lakes and reservoirs are addressed in the following three sections. Some of the notable planned or existing indirect potable reuse examples are discussed in three case studies (see Secs. 23-5 through 23-7). Observations on indirect potable reuse in the United States are presented in Sec. 23-8, following the case studies.

23-1 OVERVIEW OF INDIRECT POTABLE REUSE

When considering potable reuse as an option for public water supplies, critical distinctions must be made between indirect and direct potable reuse. Currently, in the United States, direct uses of reclaimed water for human consumption are not viable options (see Chap. 24). However, a small but growing number of communities are planning and implementing indirect potable reuse through surface-water augmentation with the added protection provided by advanced water reclamation technologies prior to blending in a water course or in a water supply reservoir. The blended water is then withdrawn after undergoing further water quality improvements by natural processes in the environment. Where a highly treated reclaimed water is blended in a water supply reservoir, it is possible that the quality of the reclaimed water will be degraded, an argument put forth in support of direct potable reuse. *De facto* indirect potable reuse, strategies for planned indirect potable reuse through surface-water augmentation, and public acceptance as related to indirect potable reuse are considered in the following discussion.

De Facto Indirect Potable Reuse

Many communities currently use surface water sources of varying quality for their drinking water supply, including sources that are subject to a significant number of upstream discharges of treated wastewater. For example, more than two dozen major water utilities in the United States use water from rivers that receive wastewater discharges that comprise up to 50 percent or more of the stream flow during low-flow conditions (Swayne et al., 1980). Wastewater discharges, both regulated and unregulated, may include various amounts of treated wastewater effluent, agricultural runoff and return flows, and stormwater runoff and overflows. The use of a water source containing wastewater discharges is referred to as *de facto* potable reuse in this textbook (see also Chap. 3). Although most of the water systems using such water sources meet current drinking water standards, many of the concerns about planned indirect and direct potable reuse of reclaimed water, as discussed in this and Chap. 24, also apply to these conventional drinking water supply systems.

In the case of southern California, both the Colorado and Sacramento River systems and their tributaries, as shown in Fig. 23-1, serve as receiving waters for numerous wastewater discharges before becoming the water supply sources for Los Angeles and San Diego, as well as several other smaller communities in southern California. For example, there are more than 450 permitted wastewater discharges on the Colorado River and it tributaries, although not all are currently active. A major concern with this water supply source is that the level of wastewater treatment and other discharge requirements are not uniform throughout the tributary region. Although dilution and natural attenuation occur in river systems (Gurr and Reinhard, 2006), planned indirect potable reuse cannot be considered in isolation from more general drinking water issues related to water sources containing varying amounts of wastewater.

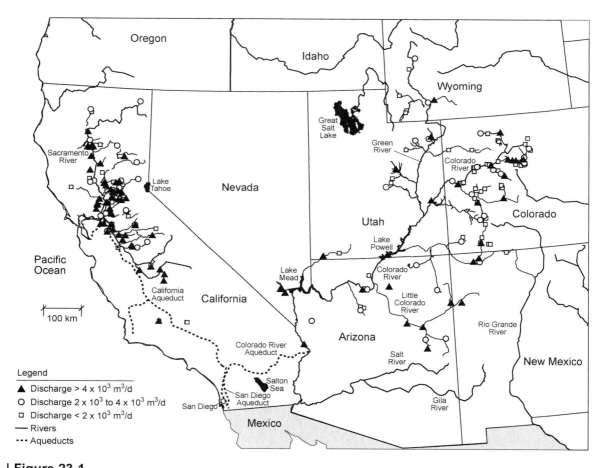

Figure 23-1
Municipal wastewater facilities with NPDES permitted discharges to the Sacramento and Colorado Rivers and their tributaries that serve as water supply sources for southern California. (Adapted from the Water Department, City of San Diego, CA.)

Indirect potable reuse through surface-water augmentation, a planned activity, can occur when treated reclaimed water is introduced into an intervening stream, followed by withdrawal for municipal water supply or direct discharge to a raw water storage reservoir. Schematic diagrams of unplanned, incidental (*de facto*) indirect potable reuse; planned indirect potable reuse; and direct potable reuse of reclaimed water are shown on Fig. 23-2.

Strategies for Indirect Potable Reuse through Surface-Water Augmentation

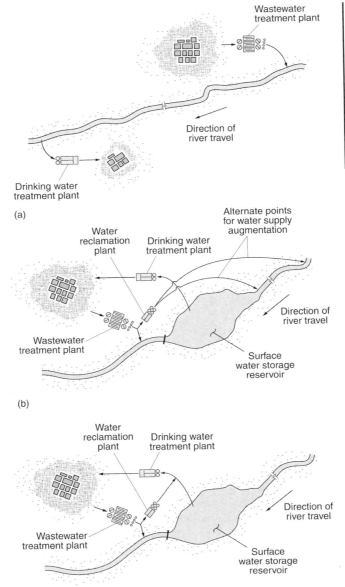

Figure 23-2
Schematic diagrams of water reuse schemes: (a) incidental (*de facto*) indirect potable reuse, (b) planned indirect potable reuse, and (c) direct potable reuse.

In indirect potable reuse, reclaimed water is treated twice prior to its ultimate use for potable purposes. First, it is treated prior to discharge to surface water. Following introduction to and mixing, blending, and assimilation with raw water in the environment, reclaimed water is again treated prior to delivery to the potable water system. For planned indirect potable water reuse systems, advanced processes may be used for the treatment of both wastewater and raw water. Careful evaluations must be made in the application of advanced treatment processes such as reverse osmosis (RO). For example, would it be more advantageous to treat the potable water supply with RO instead of reclaimed water in indirect potable reuse? Some alternative advanced treatment process flow diagrams are shown on Fig. 6-2 in Chap. 6.

Public Acceptance

The ultimate success of any water reuse program is determined by its level of public acceptance. Gaining public acceptance requires a well-conceived program of public involvement, information, and education. An essential component of gaining the support of the public is to inform people of the need to include indirect potable reuse in the overall water supply plan (see Chap. 26). Subjects that should be communicated include:

- Need for additional water supply
- Concept of conservation as it relates to water reclamation and reuse
- Information on successful indirect potable reuse projects
- Actual water quality data, water quality criteria, and technology used in water reclamation plants
- Availability and costs of alternative water supplies

Notwithstanding the fact that some proposed, high profile, indirect potable reuse projects have not been accepted in recent years due to public or political opposition to perceived health concerns, environmental justice, or growth concerns, indirect potable reuse is likely to increase in the future.

23-2 HEALTH AND RISK CONSIDERATIONS

Health concerns about indirect potable reuse are related to the potential presence of pathogens and trace constituents in reclaimed water. Because of the importance of public perception regarding these health concerns, careful consideration must be given to the selection and use of appropriate treatment technologies for the removal of pathogens and trace constituents, system reliability, and multiple barriers. Each of these factors is considered in the following discussion.

Pathogens and Trace Constituents

Recognizing the concerns over pathogens and trace constituents where indirect potable reuse is being considered, various forms of advanced treatment are used to assure that there are essentially no health impacts, although 100 percent assurance can never be provided. Currently, where indirect potable reuse is planned, the advanced treatment process flow diagram includes typically: microfiltration (MF) or ultrafiltration (UF) followed by RO, an advanced oxidation process (AOP), and disinfection. Reclaimed water from such a combination of treatment processes will be free of pathogenic organisms, and most trace constituent concentration levels will be below detection limits. Reclaimed water treated to this extent is of higher quality than most raw water supply sources, especially where the source of raw water contains wastewater discharges (see Fig. 23-1). As mentioned previously, the quality of the reclaimed water may be degraded by blending it with the raw water source.

System Reliability

A critical aspect of using highly treated reclaimed water in an indirect potable reuse application is system reliability, defined as the probability of adequate performance for a specified period of time under specified conditions. System reliability will depend on (1) the variability of the influent wastewater characteristics, (2) the inherent variability of biological treatment processes, (3) the inherent variability of advanced treatment processes, (4) the reliability of the mechanical treatment plant components, and (5) the effectiveness of monitoring. Great strides have been made in the design of treatment facilities taking into account reliability by making processes more robust and incorporating redundancy into design.

Use of Multiple Barriers

Fundamental to the practice of planned indirect potable reuse is the use of multiple barriers to ensure safety of the reclaimed water. In drinking water systems, the types of barriers used traditionally include: (1) source protection, (2) natural attenuation, (3) effective treatment, (4) distribution system integrity, (5) monitoring programs, and (6) responses to adverse conditions. Additional information on these barriers is presented in Table 23-1. As reported in Table 23-1, the concept of multiple barriers is based on redundancy with independent modes of failure.

The multiple barriers used for reclaimed water in indirect potable reuse applications include: (1) source control of discharges to the wastewater collection system, (2) robust and redundant conventional treatment processes, (3) robust and redundant advanced treatment, (4) incorporation into a natural system (e.g., a water supply storage reservoir), (5) water treatment before distribution in the potable water system, and (6) monitoring at various points within the system. Various steps and combinations of multiple barriers applicable to indirect potable reuse are depicted on Fig. 23-3. Although each barrier offers protection, no single barrier is perfect. Thus, an over-reliance on only one barrier at the expense of another may increase the risk of contamination. Independent failure modes should be established (i.e., barriers should be selected so that a failure of one barrier does not result in the failure of all). However, the combination of many independent barriers results in an overall high level of reliability.

23-3 PLANNING FOR INDIRECT POTABLE REUSE

Because planned indirect potable reuse through surface water augmentation is a controversial water reuse application, planning must be done with great care. Securing adequate water supplies for future growth and treating wastewater to preserve water quality in receiving water bodies are cited as impetus for indirect potable reuse applications. Some of the recent examples of indirect potable reuse planning are found in the Metropolitan Atlanta area and in Gwinnett County, Georgia (Yari, 2005; Scarbrough, 2005).

Table 23-1
Examples of barriers, hazards addressed, and risk management options for water treatment operations[a]

Barrier	Hazard addressed	Typical risk management approach
Source protection	Pathogens Chemical contaminants Radionuclides	Watershed protection plan Upgraded wastewater treatment Choice of water source
Treatment	Pathogens Disinfection byproducts Chemical contaminants	Water quality standards Chemically assisted filtration Disinfection
Distribution system	Infiltration Pathogen regrowth	Chlorine residual System pressure Capital maintenance plan
Natural attenuation (dilution, die-off, and decay)	Trace and unregulated contaminants	Natural attenuation and dilution in water body Storage in reservoir
Monitoring	Undetected system failures	Automatic monitors Alarm and shut-offs Log books, trend analysis
Response to adverse condition	Failure to act promptly on system failure Failure to communicate promptly with health authorities and the public	Emergency response plans Boil water advisories or orders

[a] Adapted, in part, from Ontario Ministry of the Attorney General, 2002.

Important factors influencing the consideration of indirect potable reuse are summarized in Table 23-2. In planning indirect potable reuse, the following factors must be evaluated: (1) characteristics of the watershed (2) quantity of reclaimed water to be reused, (3) water and wastewater treatment requirements, (4) institutional considerations, and (5) cost considerations.

Characteristics of the Watershed

Before indirect potable reuse by surface water augmentation is undertaken, the watershed should be evaluated to develop a baseline condition that can be used to assess any long-term impacts on water quality. For those watersheds influenced by other wastewater discharges (unplanned indirect potable water reuse) or non-point sources of pollution, including urban and agricultural runoff, the cumulative effect of all potential sources of pollution must be evaluated.

In addition, it is important that water quality monitoring be performed to characterize adequately the background conditions (i.e., the conditions before planned indirect potable reuse is implemented). Microbiological, organic, and inorganic parameters should be measured at locations and frequencies for use in establishing existing water quality. Monitoring programs should also be designed to reflect seasonal and human-induced

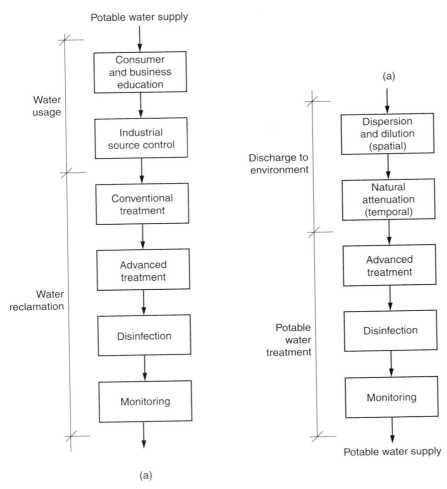

Figure 23-3
Various steps and combinations of multiple barriers in indirect potable reuse.

hydrologic variations that may occur (WEF and AWWA, 1998). A comprehensive set of quantitative molecular microbial source tracking assays can be applied to watershed analysis. Correct identification and quantification of non-point sources of fecal pollution is most important in the watershed being considered for indirect potable water reuse (Thompson et al., 2006).

Quantity of Reclaimed Water to be Blended

The volume of reclaimed water to be introduced to the water supply is a fundamental consideration for planned indirect potable reuse systems. The amount of reclaimed water introduced to the receiving stream, or reservoir, relative to the "native" water determines the percentage of reclaimed water in the water supply. The fraction of reclaimed water that will be in the water supply should be determined and used for planning and communicating with the public and regulatory agencies. Depending on the specific system, seasonal drawdown of reservoirs can create an extreme condition that may require either structural or operational modifications to allow reclaimed water to be mixed fully with reservoir water, as discussed in Sec. 23-4.

Table 23-2
Factors favoring indirect potable reuse[a]

Factor	Description
High cost, environmental impacts, and difficulty in permitting conventional water resources development	Increased infrastructure cost of developing new dams and reservoirs in remote areas and the cost to permit such projects are becoming prohibitive. Indirect potable reuse, in many cases, offers a water supply alternative more tractable than remote conventional water supply development.
Wastewater disposal standards are becoming more stringent	Protection of coastal waters and the inland ecosystem is, for example, resulting in extremely low nutrient limits as well as aquatic life metals criteria. These criteria often require intense chemical treatment and/or membrane treatment for discharge compliance. For this reason, it may be more advantageous to reclaim water from municipal wastewater for indirect potable reuse than to discharge to the aquatic system.
Economically viable, nonpotable reuse opportunities are exhausted	It is often more cost effective to implement more treatment and use the existing potable water supply, treatment, storage, and distribution network than to implement a dual distribution system for nonpotable reuse. Urban demand for nonpotable water fluctuates both diurnally and seasonally which often precludes full usage of the reclaimed water supply, or requires costly storage systems that require near-drinking water quality to be stored.
Flow of reclaimed water is increasing in many watersheds	Due to population growth and urbanization, the flow of treated wastewater discharged to receiving waters is increasing. Higher quality reclaimed water can be introduced into water supply reservoirs.

[a] Adapted, in part, from McEwen, 1998.

Water and Wastewater Treatment Requirements

Conventional water treatment plants that include treatment processes such as chemical coagulation, flocculation, sedimentation, filtration, and disinfection remove suspended solids and inactivate pathogens occurring in the water (see Chaps. 7 and 8). However, conventional water and wastewater treatment processes may not remove many of the trace constituents of concern that might be found in municipal wastewater. For instance, wastewater may contain organic compounds suspected of being carcinogenic that are only partially removed through conventional water and wastewater treatment. Also, certain microorganisms are not removed effectively through conventional treatment. Thus, in a planned indirect potable reuse system, some form of advanced treatment is required to remove constituents of concern (see Chaps. 9 and 10).

Institutional Considerations

Institutional considerations that must be factored into the planning of an indirect potable reuse project include regulatory issues and water rights.

Regulatory Issues

Legal and regulatory issues at both the state and federal levels present many challenges to the implementation of indirect potable reuse. Some of the key rules and regulations that affect water reclamation and reuse applications are discussed in Chaps. 4, 16, and 25. Several parallel rules and regulations exist, usually at the state level, and sometimes state-level requirements are more stringent than the federal requirements.

Water Rights

Water rights issues can constrain water reclamation and reuse projects by imposing restrictions and requirements on the reuse and return of water. Because of the complexities of water rights issues and differences in state water rights laws, expert guidance is advised during the early planning stages of an indirect potable reuse project (see also Chap. 25).

Cost Considerations

In some situations, indirect potable reuse may be the best alternative to make beneficial use of resources, but project implementation may hinge on affordable cost. The lack of infrastructure for nonpotable water reuse (e.g., reclaimed water distribution lines and pumping stations) may be too expensive to implement in a timely manner, particularly, in urban areas (U.S. EPA, 2004). In a cost analysis, all cost factors including indirect costs need to be considered in making a fair comparison. A detailed analysis of direct and indirect costs used in comparing water supply alternatives is presented in Chap. 25. An example of cost estimates for various water reuse applications is shown in Table 23-3.

Table 23-3
Comparative cost estimates for different water reuse options as a function of facility size[a,b]

	Size of facility			
Reuse option	3.8×10^3 m³/d (1 Mgal/d)	18.9×10^3 m³/d (5 Mgal/d)	37.8×10^3 m³/d (10 Mgal/d)	94.6×10^3 m³/d (25 Mgal/d)
Direct aquifer injection, $/m³	5.28	2.94	2.93	2.93
Indirect potable reuse,[c] $/m³	4.91	2.99	2.46	2.30
Irrigation,[d] $/m³	3.72	3.88	3.88	3.88
Rapid infiltration basin,[e] $/m³	1.40	1.77	2.22	3.54
Wetlands,[f] $/m³	0.95	0.95	na[g]	na

[a]Adapted from Beverly et al., 2001.
[b]All costs were based on a 30-yr lifespan and eight percent interest. Both the indirect potable reuse option and the direct aquifer injection option include costs for treating the water with membranes (a dual train of RO or NF combined with MF or UF as a pretreatment) prior to discharge. The direct injection option includes the cost of injection wells and pumps.
[c]Assumes gravity discharge and no permitting issues.
[d]Average cost quotes from municipalities offering public access irrigation (costs do not include wet weather back-up costs).
[e]For areas highly suitable for rapid infiltration basins.
[f]Average cost quotes from municipalities, covers wetlands construction and operation and maintenance only where available. The highest capacity quoted was 18.9×10^3 m³/d (5 Mgal/d). Land costs (which could significantly affect this number) were not included, due to widely varying real estate costs.
[g]na = not available.

23-4 TECHNICAL CONSIDERATIONS FOR SURFACE-WATER AUGMENTATION IN LAKES AND RESERVOIRS

Large lakes and reservoirs provide extensive dilution and mixing capacity for reclaimed water. When introducing reclaimed water into a potable water supply reservoir, the fate of residual constituents, including nutrients, pathogens, and trace constituents, is a primary concern. However, modern water reclamation facilities that employ treatment processes such as microfiltration, reverse osmosis and advanced oxidation are capable of removing virtually all constituents. In such circumstances, the reclaimed water, when blended with the water in the reservoir or lake, may actually improve the overall water quality. Given the development in environmental and drinking water regulations, public involvement, and technological advancements, only treatment processes that produce the highest quality reclaimed water and with the highest degree of reliability are acceptable for planned indirect potable reuse.

Nevertheless, the fate of constituents in the water column and bottom sediments depends on a number of physical, chemical, and biological factors, which can be divided into transport and transformation processes. Transport and transformation processes are both affected by the hydraulic regime of the system under consideration. However, as discussed in this section, the hydraulics of most lake and reservoir systems are inherently complex, and field measurements along with computer models are used to develop an understanding of the transport and transformation processes that take place in a particular system.

Characteristics of Water Supply Reservoirs

Because lakes and reservoirs used for water supply are open systems, various external activities and forces can impact the hydraulics and water quality. In many cases, reservoirs serve several functions and are subject to changes in the environment, including droughts and flooding. Thus, the modeling and management of water quality and hydraulics in a reservoir are interrelated and have some inherent complexities. The general processes that control the fate of the water quality constituents and important process considerations are summarized in Table 23-4.

Multiple Use Facilities

Reservoirs and lakes are used for a variety of functions that will affect their management. In addition to use as municipal and industrial water supplies, lakes and reservoirs are often used for storing flood waters, agricultural irrigation water supply, hydroelectric power generation, navigation, recreation, and habitat for aquatic life. Numerous operations are carried out to maintain water quality and quantity. Optimization models are used typically for simulation of the routing of water between the various competing users. Thus, it is important to recognize that lakes and reservoirs have many functions that may affect augmentation with reclaimed water for indirect potable reuse.

Water Quality Impacts

Water quality impacts originate from both anthropogenic and natural sources. For example, human development in the contributing watershed can result in increased stormwater runoff containing fertilizers and pesticides. Measures to control stormwater runoff

Table 23-4
Constituent transformation and removal processes in water and wastewater treatment and the environment[a]

Process	Comments	Constituent or parameter affected
Adsorption/desorption	Many chemical constituents tend to attach or sorb onto solids. The implication for wastewater discharges is that a substantial fraction of some toxic chemicals are associated with the suspended solids in the effluent. Adsorption, combined with solids settling, results in the removal from the water column of constituents that might not otherwise decay.	Metals, trace organics, NH_4^+, PO_4^{3-}
Algal synthesis	The synthesis of algal cell tissue using the nutrients found in wastewater.	NH_4^+, NO_3^-, PO_4^{3-}, pH
Biodegradation	Bacterial conversion (both aerobic and anaerobic) is the most important process in the transformation of constituents released to the environment. The exertion of BOD and NOD are the most common examples of bacterial conversion encountered in water-quality management. The depletion of oxygen in the aerobic conversion of organic wastes is also known as deoxygenation. Solids discharged with treated wastewater are partly organic. Upon settling to the bottom, they decompose bacterially, either anaerobically or aerobically, depending on local conditions. The bacterial transformation of toxic organic compounds is also of great significance.	BOD, nitrification, denitrification, sulfate reduction, anaerobic fermentation (in bottom sediments), conversion of priority organic pollutants
Bioaccumulation	The uptake of organic and inorganic constituents through the food chain, usually followed by concentration in organisms located highest in the food chain. Bioaccumulation and bioconcentration of toxicants is detrimental to some species.	Inorganic and organic constituents
Chemical reactions	Important chemical reactions that occur in the environment include hydrolysis, photochemical, and oxidation-reduction reactions. Hydrolysis reactions occur between contaminants and water.	Organic compounds
Filtration	Removal of suspended and colloidal solids by straining (mechanical and chance contact), sedimentation, interception, impaction, and adsorption.	TSS, colloidal particles
Flocculation	Flocculation is the term used to describe the aggregation of smaller particles into larger particles that can be removed by sedimentation and filtration. Flocculation is brought about by Brownian motion, differential velocity gradients, and differential settling in which large particles overtake smaller particles and form larger particles.	Colloidal and small particles
Gas absorption/desorption	The process whereby a gas is taken up by a liquid is known as absorption. For example, when the dissolved oxygen concentration in a body of water with a free surface is below the saturation concentration in the water, a net transfer of oxygen occurs from the atmosphere to the water. The rate	O_2, CO_2, CH_4, NH_3, H_2S

(Continued)

Table 23-4

Constituent transformation and removal processes in water and wastewater treatment and the environment[a] (Continued)

Process	Comments	Constituent or parameter affected
Gas absorption/ desorption (*continued*)	of transfer (mass per unit time per unit surface area) is proportional to the amount by which the dissolved oxygen is below saturation. The addition of oxygen to water is also known as reaeration. Desorption occurs when the concentration of the gas in the liquid exceeds the saturation value, and there is a transfer from the liquid to the atmosphere.	
Natural decay	In nature, contaminants will decay for a variety of reasons, including mortality in the case of bacteria, and photooxidation for certain organic constituents. Natural and radioactive decay usually follow first-order kinetics.	Plants, animals, algae, fungi, protozoa, bacteria, viruses, radioactive substances
Photochemical reactions	Solar radiation is known to trigger a number of chemical reactions. Radiation in the near-ultraviolet (UV) and visible range is known to cause the breakdown of a variety of organic compounds.	Organic compounds and microorganisms
Photosynthesis/ respiration	During the day, algal cells in water bodies produce oxygen by means of photosynthesis. Dissolved oxygen concentrations as high as 30 to 40 mg/L have been measured. During the evening hours algal respiration consumes oxygen. Where heavy growths of algae are present, oxygen depletion has been observed during the evening hours.	Algae, duckweed, submerged macrophytes, NH_4^+, PO_4^{3-}, pH, etc.
Sedimentation	The suspended solids discharged with treated wastewater ultimately settle to the bottom of the receiving water body. This settling is enhanced by flocculation and hindered by ambient turbulence. In rivers and coastal areas, turbulence is often sufficient to distribute the suspended solids over the entire water depth.	TSS
Sediment oxygen demand	The residual solids discharged with treated wastewater will, in time, settle to the bottom of streams and rivers. Because the particles are partly organic, they can be decomposed anaerobically as well as aerobically, depending on conditions. Algae, which settle to the bottom, will also decompose, but much more slowly. The oxygen consumed in the aerobic decomposition of material in the sediment represents another dissolved oxygen demand in the water body.	O_2, particulate BOD
Volatilization	Volatilization is the process whereby liquids and solids vaporize and escape to the atmosphere. Organic compounds that readily volatilize are known as VOCs (volatile organic compounds). The physics of this phenomenon are very similar to gas absorption, except that the net flux is out of the water surface.	VOCs, NH_3, CH_4, H_2S, other gases

[a]Adapted from Crites and Tchobanoglous (1998).

have included stormwater treatment facilities, stormwater catchments and infiltration basins, and programs for educating residents on best practices for landscaping and home chemical usage within the watershed area. Untreated or partially treated wastewater can result from leaking wastewater collection systems and failed onsite wastewater treatment systems. The use of watertight piping and conveyance systems and proper maintenance of the wastewater infrastructure are essential to reducing pollution associated with wastewater. Developed areas are also known to have increased atmospheric deposition of aerosols and particulate matter that can impact water quality. Constituents found in vehicle emissions include fuel additives, hydrocarbons, and particulates. Conventional agricultural practices and construction activity may also result in higher silt and sediment loading caused by soil erosion processes.

Stratification in Lakes and Reservoirs

Almost all lakes and reservoirs with a depth of 5 m or more stratify during a substantial part of the year. The exception is run-of-the-river reservoirs with a residence time of a month or less. Stratification is significant because, during the summer when a thermocline develops, there is little mixing between the stratified layers. If reclaimed water is used to augment a reservoir, knowledge of the lake or reservoir stratification conditions will be useful for selecting the depth and location of water-supply withdrawal to maximize water quality. The typical development of stratification in a lake is shown on Fig. 23-4.

Density Currents

Other phenomena associated with stratification are the formation of density currents in the form of plumes. In the initial mixing region, warmer reclaimed water forms a buoyant plume, rising rapidly in the water column. This plume entrains large amounts of ambient water, thereby diluting the introduced reclaimed water. When the water column is stratified, the ambient water which is first entrained is deep, denser water, which reduces the plume buoyancy as it rises into less dense ambient water. At some point in this ascent, the plume density may become equal to that of the ambient water and further rise is impeded. The plume reaches an intermediate equilibrium height of rise. When the water column is weakly stratified or not stratified, as in the winter, the plume rises to the water surface. Beyond the initial mixing region, the reclaimed water disperses by ambient currents and is further diluted by diffusion.

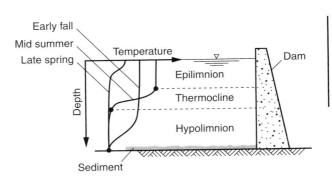

Figure 23-4

Seasonal development of temperature stratification in lakes and reservoirs.

Density current inflow to a lake or reservoir at a particular depth will spread laterally due to hydraulic forces. A consequence is the rapid distribution of inflows over the entire lake area at a particular depth interval. Selective withdrawal is accomplished using structures with multiple inlets, designed to withdraw water from multiple levels. Selective withdrawal is off importance for downstream releases from reservoirs. For example, water withdrawn from deep outlets tends to be cold and may be depleted of oxygen if the hypolimnion is anoxic. Similarly, withdrawals from the epilimnion may contain unacceptable levels of algae at certain times of the year.

While rapid changes of temperature occur over the depth of stratified lakes and reservoirs, surprisingly high horizontal uniformity often exists, even though horizontal distances are usually many times the depth. This horizontal uniformity is in part due to density currents, which are driven by and eventually eliminate any horizontal non-uniformity. As a result, one-dimensional modeling is often appropriate, as discussed later.

Mixing Processes

Mixing in reservoirs and lakes can occur at the discharge point from a water reclamation plant, at the intake from a water supply facility, by natural inflows and outflows, and by natural forces. Inflow mixing, as described above, occurs due to differences in fluid density and kinetic energy of the inflow. Outlet mixing occurs as a zone of influence develops around the outlet point, and depends, in part, on the depth of the outlet. Natural forces that cause mixing in the epilimnion are related to the action of wind shear across the surface, the formation of surface waves, convective mixing from daily heating and cooling cycles, and the Coriolis effect (Martin and McCutcheon, 1999). It should also be noted that under some conditions reservoirs may turn over, due to temperature and density changes in the epilimnion, causing the entire reservoir to mix, occuring typically in the fall and spring.

Residence Time

A residence time distribution (RTD) for a given lake or reservoir systems should be determined using an appropriate hydraulic model. The RTD relates the statistical duration of time any given molecule or packet of water is retained in the system. The RTD can also be used to determine the mean hydraulic residence time (HRT) and to estimate the amount of short-circuiting that occurs in the system. Short-circuiting is the tendency for some of the water to exit the lake or reservoir in a time shorter than the mean HRT and can occur due to changes in the flow regime.

Conditions that lead to short-circuiting can include water column stratification, volume fluctuations, and wind-driven water currents. For example, under extreme drought conditions, when lake and reservoir water volumes are substantially reduced, reclaimed water flow may constitute the entire flow into a reservoir (i.e., no dilution capacity). Therefore, inlet and outlet locations, mixing patterns, and density currents need to be considered carefully to minimize flow short-circuiting under different flow and operating regimes. Flow short-circuiting is of concern when intermediate quality (e.g., tertiary-treated) reclaimed water is used for augmentation. Longer residence times will result in greater attenuation of the residual constituents in the lake or reservoir. However, as discussed previously, the use of the highest quality water (e.g., reverse osmosis with advanced oxidation treatment) for surface-water augmentation, as planned in California and elsewhere effectively makes HRT, short-circuiting of flow, and dilution capacity insignificant from a water quality and risk management perspective. Public acceptance issues will still need to be considered, even with the use of advanced treatment, during conditions when the HRT is short (e.g., less than 6 mo) or the dilution capacity is low (e.g., reclaimed water is greater than 5 to 10 percent of the flow).

Modeling of Lakes and Reservoirs

Modeling of lakes and reservoirs is sufficiently complex that computer models and field measurements are required for most applications. While the level of modeling required and the accuracy of predicting the behavior of a lake or reservoir system are beyond the scope of this textbook, model development and types of models available are introduced briefly to familiarize the reader with basic concepts in lake and reservoir modeling. Following a description of the conservation of mass equation, two types of model analysis are introduced below, for a stratified system and for a complete-mix representation.

Conservation of Mass

The conservation of mass equation, which involves accounting for the fate of constituents in the environment, is the basis for many of the models that have been developed to describe the hydraulics of reservoir and lake systems. Mass balance analysis is based on bookkeeping of the mass of any water quality constituent in a stationary volume of fixed dimensions, known as a *control volume*. The general form of the constituent mass balance can be expressed as follows:

$$\begin{bmatrix} \text{Rate of mass} \\ \text{increase in} \\ \text{control volume} \end{bmatrix} = \begin{bmatrix} \text{Rate of mass} \\ \text{entering} \\ \text{control volume} \end{bmatrix} - \begin{bmatrix} \text{Rate of mass} \\ \text{leaving} \\ \text{control volume} \end{bmatrix} \qquad (23\text{-}1)$$

$$+ \begin{bmatrix} \text{Rate of mass} \\ \text{generated within} \\ \text{control volume} \end{bmatrix} - \begin{bmatrix} \text{Rate of mass} \\ \text{lost within} \\ \text{control volume} \end{bmatrix}$$

Each of the terms in Eq. (23-1) has units of mass per unit time, MT^{-1}. The mass conservation equation is applicable whether the discharge is in a river or lake. However, the specific characteristics of the system under consideration will require different approaches and assumptions to solve for constituent concentrations.

The terms in Eq. (23-1) covering mass entering and leaving the system are associated with transport processes. Three types of transport processes are addressed typically in mass balance models: *advection*, *diffusion*, and *dispersion*. Advection is the movement of a constituent along with the bulk flow, as the movement of a dissolved constituent in the current of a river. Diffusion is movement of a constituent due to (1) a concentration gradient, or (2) turbulent eddy-mixing. Dispersion is associated with turbulence resulting from a velocity gradient.

Stratified System Analysis

The development of a one-dimensional model for a reservoir or lake is based on dividing the system into a number of equal horizontal slices, or elements, as shown on Fig. 23-5.

All of the elements have a constant volume (except for the top element, which has a variable volume); are able to accept and transmit advected and diffused flow from the adjacent layer; can receive laterally advected flow; and can discharge. Heat and mass can also pass through or be transferred to the horizontal elements by advection and diffusion. Outflows are assigned to specific withdrawal points or form the surface element for natural outflows.

Complete Mix Analysis

Small and shallow lakes and reservoirs tend to remain well-mixed due to wind-induced turbulence. Deeper lakes generally stratify during the summer; but for many of these, occurs twice a year, mixing upper and lower strata. Thus, from a long-term point of view, a fully mixed analysis may also be justified for stratifying lakes. In the fully mixed approach, constituent concentrations are assumed to be uniform in the lake or reservoir.

For input into the conservation-of-mass equation, rates of mass gain or loss within the reservoir or lake volume are required. Many constituents are subject to transformation processes which follow a first-order of decay. When these transformation processes are independent, their effects are additive, and the corresponding transformation rates can be simply summed in the conservation-of-mass equation.

Computer Simulations

Water quality models can be static (steady-state) or dynamic. A summary of models categorized by dimensions is presented in Table 23-5. In steady-state models, the input parameters are constant and the equilibrium condition is determined. Dynamic or unsteady-state models are used to analyze systems in which there are fluctuations in the input parameters, and model output over time. Models are also categorized according to dimensions based on simplifications and assumptions that are made regarding the movement and interchange of energy and mass within the system. For example, the stratified system analysis shown on Fig. 23-5 is considered to be a one-dimensional model as exchange only occurs between vertical layer elements. The complete-mix model is considered to be zero-dimensional because it is simply an input-output model with no hydraulic considerations.

Strategies for Augmenting Water Supply Reservoirs

Several approaches have been applied to ensure water quality in lake and reservoir systems. These techniques include augmenting with high quality reclaimed water, providing mixing and aeration to improve water quality in the water body or to improve hydraulics, and the use of selective withdrawals.

Mixing and Aeration

It may be desirable, in some cases, to mix all or part of the contents of a reservoir. For example, in some reservoirs the hypolimnion may become anoxic during certain periods of the year, resulting in deterioration in water quality. Anoxic conditions may form due to the decay of plant material and other organic matter, and in some cases has resulted in the formation of hydrogen sulfide that is harmful to aquatic organisms. Aeration can also be used to eliminate stratification to improve hydraulics through the reservoir. One approach to accomplish this mixing is the use of aeration devices in the bottom of the reservoir, as shown on Fig. 23-6.

Table 23-5
Summary of water quality models used for lakes and reservoirs[a]

Model type	General characteristics	Examples
Water quality	Used to predict the transport and fate of toxicants and microorganisms. Typically couples with hydrodynamic models for specific applications.	WASP (Wool et al., 2001) AQUATOX (Park et al., 2004)
Zero dimension	Control volume or box-models of lake or reservoir systems for which an assumption of complete mixing is applied. Zero-dimension models contain no information of hydrodynamics and therefore have no spatial variation.	Complete-mix reactors and Input-output models (Tchobanoglous et al., 2003; Chapra, 1997)
One dimensional		
Longitudinal	Used for modeling rivers with gradients along river length. May also be used for long and shallow, run-of-the-river reservoirs that are not stratified. Assumes vertical and lateral variations are negligible.	QUAL2K (Chapra et al., 2006) HEC-RAS (Brunner and Bonner, 1994)
Vertical	Stratified lake model composed of a vertical arrangement of horizontal layer elements. Used to compute vertical mixing between stratified layers and model the vertical distribution of energy and water quality parameters.	SELECT (Schneider et al., 2004) CE-QUAL-R1 (WES, 1986a)
Two dimensional		
Laterally averaged	Divides lake or reservoir into a two-dimensional array, layered vertically and along the length. The lateral dimension is assumed to be uniform. Most applicable to long, deep, stratified reservoirs.	BETTER (Bender et al., 1990) COORS (TVA, 1986) CE-QUAL-W2 (WES, 1986b)
Depth averaged	Assumes uniform vertical sections, with interaction and transport between elements along the length and lateral direction. Application to long, wide, and shallow reservoirs and lakes.	RMA4 (King et al., 2005) SWIFT2D (Schaffranek, 2004)
Three dimensional		
	Used to simulate mass transport and flow laterally, along the length, and over the entire depth of a lake or reservoir	HYDRO-3D (Sheng et al., 1990) CH3D-WES (Chapman et al., 1996) HEM3D (Park et al., 1995)

[a] See Martin and McCutcheon (1999) and Wurbs (1994) for a comprehensive discussion of water quality models.

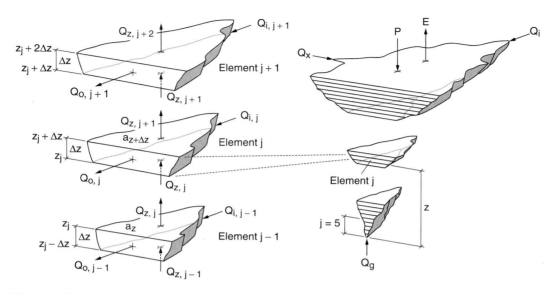

Figure 23-5
Definition sketch of a vertical one-dimensional model for a reservoir system. (After Orlob, 1983.)

Tracer Studies

Tracer studies can be useful for determining the actual hydraulics of a given system. Tracers are conservative substances that can be added to a lake or reservoir and move with the flow of water. Tracer concentrations are quantified by sampling at different points and/or over a period of time to determine the movement of the water. Dyes, such as Rhodamine WT, are used commonly; however, a number of compounds, such as bromide and iodide, which may be present in reclaimed water, may also serve as tracers. The results of tracer studies may be used for model calibration. It should be noted that a large amount of tracer will be required for characterizing flow in a lake or reservoir system, thereby limiting the conduct of tracer studies in water bodies used as water supply. Information on conducting and evaluating tracer data can be found in Tchobanoglous et al. (2003), and Martin and McCutcheon (1999).

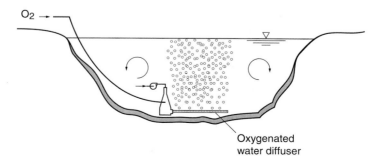

Figure 23-6
Use of a submerged Speece cone aeration device for oxidation of the hypolimnion or destratification of a reservoir.

23-5 CASE STUDY: IMPLEMENTING INDIRECT POTABLE REUSE AT THE UPPER OCCOQUAN SEWAGE AUTHORITY

The Upper Occoquan Sewage Authority (UOSA) plant is a historic (since 1978) full-scale water reclamation facility that has practiced indirect potable reuse via surface-water augmentation in the Occoquan Reservoir. Reclaimed water from the plant currently accounts for about eight percent of the annual average inflow to the reservoir, but during extended droughts, it may account for up to 90 percent of the reservoir inflow. The Occoquan Reservoir is a major raw water supply for more than 1.3 million people in Northern Virginia. The population is projected to reach 1.5 million people in the near future. This case study is prepared from various sources, including WEF and AWWA (1998), NRC (1998), McEwen (1998); and Angelotti et al. (2005).

Setting

The 1475 km² (570 mi²) Occoquan Watershed is located in a water-short area of northern Virginia, with Dulles International Airport on its northern boundary and Washington, DC 24 km (15 mi) to the east. Two tributary subsystems, Bull Run and Occoquan Creek, drain the watershed to the Occoquan Reservoir (capacity: 42×10^6 m³; 11×10^3 Mgal) which is a man-made water supply impoundment.

Water Management Issues

The Occoquan Watershed was largely rural until the 1960s, when the opening of Interstate 66 created a rural/suburban area convenient to people working in Washington, DC. The resulting population growth and land development within the region led to an increase in water demand and a decline in the water quality of the Occoquan Reservoir. The water quality problems in the reservoir included detection of enteric viruses in tributary creeks and the reservoir, frequent and intense algal blooms, frequent taste and odor problems in the water treatment plant, dissolved oxygen depletion and fish kills, and sulfides in the deeper reservoir depths (Robbins and Gunn, 1979).

Based on a comprehensive study of the reservoir in 1969–1970, the major contributors to water quality problems were found to be the effluent discharges from 11 secondary wastewater treatment plants. The Virginia regulatory community recognized that the degraded water quality of the Occoquan Reservoir threatened this valuable water resource for future generations. In 1971, the Virginia State Water Control Board (current Department of Environmental Quality) adopted a comprehensive Occoquan water policy that included the creation of a regional water reclamation agency to build and operate a state-of-the-art treatment system; and an independent agency to monitor water quality in the watershed, and advise state regulatory agencies of measures necessary to protect and preserve the reservoir as a water supply. Thus, the Upper Occoquan Sewage Authority (UOSA) and the Occoquan Watershed Monitoring Program (OWMP) were created as part of a comprehensive policy to manage and protect the Occoquan Watershed. Construction of the UOSA water reclamation system began in 1974, and the system began operation in 1978.

Description of Treatment Components

The current flow diagram of UOSA's 204×10^3 m³/d (54 Mgal/d) capacity Millard H. Robbins, Jr. Water Reclamation Facility is shown on Fig. 23-7. An aerial view of the

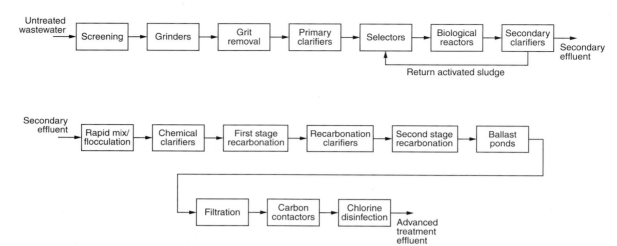

Figure 23-7
Treatment process flow diagram for the Willard H. Robbins, Jr., Water Reclamation Facility, Upper Occoquan Sewage Authority, Contreville, VA. (Adapted from the Upper Occoquan Sewage Authority.)

UOSA plant is shown on Fig. 23-8. The plant is an advanced water reclamation plant comprised of the following processes:

- Advanced biological nutrient removal secondary treatment
- Lime clarification
- Two-stage recarbonation
- Multimedia filtration
- Granulated activated carbon (GAC) adsorption
- Ion exchange and breakpoint chlorination (standby)
- Disinfection and dechlorination

The reclaimed water from the UOSA water reclamation facility is discharged into Bull Run, a tributary of the Occoquan Reservoir. The discharge point is 9.7 km (6 mi) upstream of the headwaters of the reservoir and 32 km (20 mi) upstream of the drinking water supply intake.

System Redundancy

To enhance operational reliability, many fail-safe features are included in the design. Every major electrical and mechanical system has at least one backup unit. Three sources of electrical power serve the plant and the principal pumping stations in the delivery system, with at least one source representing on-site generation. Storage basins at the plant and the principal pumping stations are used for emergencies and when treatment system failures occur. In addition, all main components of the water reclamation plant and delivery system are monitored by a distributed control system, and most of the plant processes are computer-controlled.

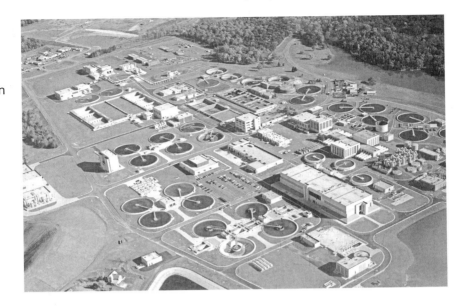

Figure 23-8
Aerial view of the Willard H. Robbins, Jr. Water Reclamation Facility (Coordinates: 38.807 N, 77.463 W) (Courtesy of the Upper Occoquan Sewage Authority, VA).

High-Quality Secondary Effluent

The performance of the advanced water reclamation processes is substantially enhanced by the preceding reliable and high-quality secondary treatment. The secondary treatment results in complete nitrification with good solids separation characteristics. Selectors, either aerobic or anoxic, are installed ahead of the bioreactors to produce good settling solids. Selective wasting of biofilms keeps clarifier solids carryover to a minimum.

Chemical Treatment System

The chemical treatment system includes high-energy mixing, flocculation, chemical clarification, and two-stage recarbonation with intermediate settling. Calcium hydroxide is mixed with secondary effluent to raise the pH to a range of 10.8 to 11.3; whereby the resulting precipitate is flocculated, polymer is added, and the floc is removed by settling. Secondary and chemical treatment processes produce a several-log reduction of viruses, more than 99 percent removal of phosphorus, and substantial reduction in organics, heavy metals, and particulate matter. As a result of physical particle separation steps through flocculation and settling, pathogenic protozoa are also removed.

The high pH of the chemical treatment process effluent is lowered by adding carbon dioxide in a two-stage recarbonation process. In the first stage, the pH is lowered from 9.5 to 10.0, resulting in the formation of precipitable carbonates, which are then removed in recarbonation clarifiers. Second-stage recarbonation reduces the pH to 7.0. The recarbonation effluent is discharged to equalization basins and then pumped at a uniform rate through the remaining treatment processes.

Remaining particulate matter is removed by multimedia pressure filtration. The filters produce an effluent with average suspended solids of approximately 0.3 mg/L and a turbidity of less than 0.3 NTU. Capturing fine particulates further reduces phosphorus and organic matter levels (WEF and AWWA, 1998).

Activated Carbon Adsorption

Because refractory synthetic organic compounds (SOCs) in public water supplies are of increasing concern, GAC is used at UOSA for the removal of trace organic compounds to provide reclaimed water of an acceptable quality for indirect potable reuse. The GAC process targets removal of SOCs and natural organic matter (NOM) and some heavy metals in either chelated or inert forms (Angelotti et al., 2005). The filter effluent is treated by activated carbon adsorption in contactors with a 30-min empty bed detention time. Exhausted carbon is regenerated in a multiple-hearth furnace.

The GAC treatment is another process barrier for the removal of organic matter and is a polishing step employed prior to final filtration, chlorination, and de-chlorination. Laboratory analyses are routinely performed for SOCs in the final effluent. For each of the 114 SOCs analyzed in 2004, the concentrations were below the detection limit (Angelotti et al., 2005). Test results for parameters measured in UOSA's effluent as compared to drinking water limits are shown in Table 23-6. The data in Table 23-6 represent a wide variety of organic compound types and provide a good gauge of the reclaimed water quality with respect to SOCs. In addition, total chemical oxygen demand, total organic carbon, and methylene blue active substances are used as surrogates of the overall water quality with respect to organic constituents and soluble surface-acting agents.

Ion Exchange and Breakpoint Chlorination

The original UOSA plant was designed and constructed with an ion exchange process wherein sodium was exchanged for ammonium ions. The medium was regenerated by purging the ammonia with a concentrated sodium chloride solution. The plant no longer operates the ion exchange process because biological nutrient removal (BNR) is more economical to use. Because nitrification substantially enhances treatment plant performance and is beneficial to the Occoquan Reservoir in most conditions, complete nitrification is the normal mode of operation to meet a total Kjeldahl nitrogen (TKN) limit of 1.0 mg/L. However, during an extreme drought, nitrates in the UOSA discharge could increase the nitrate nitrogen level in the raw water to greater than the drinking water limit of 10 mg/L. Under these conditions, UOSA is required to remove nitrogen to meet its TKN limit of 1.0 and keep the nitrate concentration at the raw water intakes at safe levels. The BNR processes are used for denitrification during such conditions. Nitrogen can also be removed by ion exchange or breakpoint chlorination.

Disinfection

Disinfection is achieved through addition of sodium hypochlorite and by maintaining a high free chlorine residual in the 0.7 to 2.5 mg/L range. Sodium bisulfite is used to remove residual chlorine prior to discharge to the plant's final effluent reservoir. The reservoir serves as a maturation pond and an environmental buffer between the plant discharge and Bull Run, the receiving water stream.

Future Treatment Process Directions

In the future, UOSA will be required to meet nutrient-loading limits to protect the water-quality in Chesapeake Bay. A plant expansion is also planned to increase the treatment capacity to 242×10^3 m^3/d (64 Mgal/d). Expansion plans include membrane bioreactors with chemically enhanced biological nutrient removal followed by deep-bed granular-activated carbon to satisfy the incremental capacity demand and to conform to nutrient-loading limits.

Table 23-6
Synthetic organic compounds detected in the UOSA effluent[a]

Compound category	Compound name	SDWA MCL, μg/L	UOSA effluent conc., μg/L
Aromatic solvents	Benzene	5	<10
	Ethylbenzene	700	<5 to <10
	Toluene	1000	<5 to <10
	Xylenes	10,000	<0.2 to <15
Chlorinated aromatics	Chlorobenzene	100	<10
	Hexachlorobenzene	1	<10
	o-Dichlorobenzene	600	<10
	PCBs	0.5	ND
	p-Dichlorobenzene	75	<10
	Pentachlorophenol	1	<40
Haloacetic acids	Haloacetic acids	60	38
Halogenated alkanes	1,2-Dichloroethane	75	<10
	1,2-Dichloropropane	5	<10
	Trans-1,2-Dichloroethylene	100	<10
	Vinyl chloride	2	<10
Halomethanes	Dichloromethane	5	<10
	Total THMs	100	<10 to 13
Industrial byproducts	Dioxins	0.00003	ND
Organic intermediates	Hexachlorocyclopentadiene	50	<10
Organic solvents	1,1,1-Trichloroethane	200	<10
	1,1,2-Trichloroethane	5	<10
	1,1-Dichloroethylene	7	<10
	Carbon tetrachloride	5	<10
	Tetrachloroethylene	5	<5 to 16
	Trichloroethylene	5	<10
Pesticides	2,4-D	70	<0.2
	2,4,5-TP (Silvex)	50	<0.2
	Chlordane	2	ND
	Endrin	2	<0.05
	Heptachlor	0.4	<0.05
	Lindane	0.2	<0.05
	Methoxychlor	40	<0.05
	Toxaphene	3	ND
Phthalate plasticizers	Bis(2-ethylhexyl)	6	<5 to <10
Polynuclear aromatics	Benzo(a)pyrene	0.2	<10

[a] Adapted from Angelotti et al., 2005.

Water Quality of the Occoquan Reservoir

The Fairfax County Water Authority (FCWA) owns and maintains the Occoquan Reservoir located in Fairfax and Prince William Counties, Virginia. The Reservoir is a major drinking water source serving the Northern Virginia community. To assure a clean, safe supply of water in the reservoir, FCWA maintains the *Occoquan Reservoir Shoreline Easement Policy*. The Policy regulates activity within the shoreline easement such as vegetative cover, structures, storage facilities, and piers/floats. Three developed marinas provide boat rentals, bait, tackle, food, and boat launching facilities (FCWA, 2006).

Due to the urbanization, the Occoquan Reservoir is especially vulnerable to high levels of phosphorous, turbidity, low dissolved oxygen, and copper sulfate (for algal control), and the growing presence of pharmaceuticals. Problems that occur include algal blooms, periodic fish kills, and taste and odor problems, and are directly linked to land uses—including a growing population and difficulties in controlling development in the Occoquan Watershed.

Based on water quality monitoring data from the reservoir and its two primary tributaries, the majority of the nutrient and sediment load to the reservoir comes from nonpoint sources, which are closely tied to land development conditions. Reservoir water quality trends are similar to trends in stream water quality. The tributary in the most urbanized part of the watershed, Bull Run, has been identified as the main contributor of sediment and nutrients to the reservoir (Virginia Tech, 2005).

According to the Prince William Conservation Alliance (2006), development pressures appear overwhelming in Northern Virginia, but there are several potential actions to mitigate the impact on water quality in the area that include: (1) implementing existing rules to protect waterways, (2) considering the direct link between land use and water quality, (3) recognizing the benefits of open space and other natural assets, and (4) developing new ideas in conservation design, low impact development, and other ways to protect natural resources.

Water Treatment

Fairfax Water is Virginia's largest water utility, serving more than 1.3 million people in the Northern Virginia communities of Fairfax, Loudoun, and Prince William Counties and the City of Alexandria. Chartered by the Virginia State Corporation Commission as a public, non-profit water utility, Fairfax Water has operated four water treatment plants with a combined capacity of 992×10^3 m^3/d (262 Mgal/d). The plants include the Corbalis Treatment Plant operating since 1982 on the dependable and free-flowing Potomac River in the northwestern area of Fairfax County, and three water treatment plants located on the impounded Occoquan River in the southeastern area.

The three Occoquan water treatment plants were replaced by a new, state-of-the-art water treatment plant (Frederick P. Griffith, Jr., Water Treatment Plant) which was placed in service on May 4, 2006. The plant can produce up to 454×10^3 m^3/d (120 Mgal/d) of high-quality drinking water to the southern portion of Fairfax County. The plant features advanced drinking water treatment, with ozone and granular activated carbon filters. Ozone enhances the already high quality of treatment by further reducing the production of disinfection byproducts. Carbon filters (with the depth of 1.8 m or 6 ft) provide additional removal of natural organic substances that sometimes cause taste and odors in drinking water. The final product is a high-quality drinking water that serves nearly 1.5 million customers in Northern Virginia. Water is distributed throughout Fairfax Water's service area through more than 5000 km (3100 mi) of water mains. Fairfax Water produces, on average, 560×10^3 m^3/d (148 Mgal/d) of water. Over 235,000 mostly residential accounts in Fairfax County comprise about 60 percent of total water sales. Approximately 40 percent of total water sales is wholesaled to Loudon and Prince William Counties and the City of Alexandria. Revenues in 2004 were $112 million (FCWA, 2006).

The UOSA water reclamation plant has reclaimed water to supplement the Occoquan Reservoir since the system began operation in 1978. UOSA product water is now recognized as the most reliable and best quality source of water in the Occoquan system. As a result, the plant's initial capacity of 57×10^3 m³/d (15 Mgal/d) was subsequently expanded to 102×10^3 m³/d (27 Mgal/d) and then to 204×10^3 m³/d (54 Mgal/d) which includes process modifications and additions as follows (UOSA, 2006):

Lessons Learned

- Modification of the biological reactors to allow full-time nitrification with partial denitrification
- Addition of gravity filters supplementing the existing pressure filters
- Addition of two-stage, upflow/downflow carbon contactors to supplement the single stage upflow, fluidized bed contactors. Both system configurations are still in use today. The GAC unit process is critical to the UOSA mission of protecting the public water supply while augmenting the Occoquan Reservoir's safe yield through successful indirect potable reuse of reclaimed water.

As a result of such success, another expansion of the facility to 242×10^3 m³/d (64 Mgal/d) is in the early stages of design (2006). Upper Occoquan Sewage Authority received unanimous support for these expansions from the Virginia Departments of Health and Environmental Quality, FCWA, and local governments served by the Occoquan Reservoir. Upper Occoquan Sewage Authority's future expansion concept includes low pressure membranes, which will be submerged in a bioreactor. Pilot studies during early conceptual design are intended to solidify process performance and define expected product water quality. In addition to augmenting the supply of potable water, the UOSA project illustrates other recycling measures. For example, anaerobically-digested and waste-activated sludges are blended, dewatered, and dried to produce an exceptional quality biosolid pellet for use as a soil amendment and fertilizer. Digester gas (approximately 70 percent methane and 30 percent carbon dioxide) is used as fuel for the plant boilers, which produce steam to heat digesters and some of the plant's buildings. The boiler, thermal dryer, and GAC regeneration furnace stack gases are the primary source of carbon dioxide used in the recarbonation process. Reclaimed water provides the majority of water needs at the plant and onsite uses include chiller water, process water, fire protection, truck washing, dust suppression, water for construction, decorative water features, a landscape impoundment and turf irrigation (UOSA, 2006).

23-6 CASE STUDY: CITY OF SAN DIEGO WATER REPURIFICATION PROJECT AND WATER REUSE STUDY 2005

At present (2006), the City of San Diego imports 80 to 90 percent of its potable water supply. Recognizing the importance of developing its local water resources, the city undertook a study "To conduct an impartial, balanced, comprehensive, and science-based study of all recycled water opportunities so the City of San Diego can meet current and future water needs" (City of San Diego, 2006a). One of the several alternatives evaluated, as part of the study, was indirect potable reuse through surface-water augmentation. The following case study is developed from various sources, including Trussell et al. (2002), City of San Diego (2006a,b,c).

Setting

San Diego is the oldest and second largest city in California. Located in an arid area in southern California near the border with Mexico, it has an average annual rainfall of approximately 255 mm (10 in.). As of 2006, the population is 1.3 million, which is expected to increase by 50 percent in the next 25 yr. The present potable water use in the City of San Diego averages about 795×10^3 m^3/d (210 Mgal/d). Even with rigorous water conservation measures, the City projects that the demand for potable water, based on the expected population growth, will increase by approximately 25 percent, or an additional 189×10^3 m^3/d (50 Mgal/d).

Water Management Issues

As San Diego and environs have grown, so has the region's reliance on imported water supplies. Historically, up to 90 percent of the city's existing water supply is imported from the Colorado River and the California State Water Project. The city has long recognized the need to develop local water supplies to balance and reduce its dependence on imported water. Many factors outside the city also contribute to the future water needs and the reliability of existing supplies: California's access to surplus water from the Colorado River has been reduced and recurring droughts in both the western United States and the Colorado River watershed have affected imported water supplies. Competing interests statewide between urban users, agricultural uses, and environmental interests are being resolved, but water allocations to each will continue to be adjusted in the future (City of San Diego, 2006a).

Wastewater Treatment Mandates

To understand fully the San Diego context, it is also important to review the City's wastewater situation. Currently, the City of San Diego has a discharge permit waiver, which allows the city to discharge advanced primary effluent to the ocean through a deep outfall. The permit waiver derives from legislation that allows San Diego to avoid secondary treatment of its ocean discharge, provided it constructs 170×10^3 m^3/d (45 Mgal/d) of water reclamation capacity by 2010. Controversial, the legislation was a compromise between environmental interest groups, who preferred secondary treatment and extensive reuse, and taxpayer groups and the local newspaper that did not feel the cost of either was justified (Trussell et al., 2002).

The City began by building separate nonpotable reuse systems. The North City Water Reclamation Plant (NCWRP) became operational in 1997. This plant, which serves the northern service area, can treat up to 114×10^3 m^3/d (30 Mgal/d). The treatment processes at the NCWRP were designed to meet California's Title 22 regulations governing water reclamation for unrestricted reuse. In 2002, a similar facility, the South Bay Water Reclamation Plant (SBWRP), which serves the southern service area, was brought online with a capacity of 57×10^3 m^3/d (15 Mgal/d). The combined design capacity of the two plants meets the waiver requirement. The net production of reclaimed water, at full capacity from the two plants, accounting for treatment plant losses and other onsite uses, is 91×10^3 m^3/d (24 Mgal/d) and 51×10^3 m^3/d (13.5 Mgal/d), respectively.

Water reuse was implemented easily for facilities near the two plants, but the City soon found that the cost of distribution systems to support large flows was prohibitive. Users were situated too far from the plants, located at too high an elevation, and too much construction was required in sensitive areas. In addition, the TDS of reclaimed water was about 1000 to 1100 mg/L, too saline for some applications. Finally, a majority of the city's current customers used reclaimed water for irrigation, primarily for golf courses. Unfortunately, the demand fluctuates according to the weather patterns and season of the year. Due to these varying and seasonal demands, about half of the potential treatment capacity at the water reclamation plants may be unused for part of the year, which is not in accord with the stipulated settlement. Thus, an indirect potable reuse option involving surface-water augmentation, as discussed later, is attractive from the standpoint of efficiency and cost.

Water Repurification Project

Beginning in 1993, the City of San Diego, in cooperation with the San Diego County Water Authority (SDCWA), proposed an indirect potable reuse project called the Water Repurification Project (WRP). The indirect potable reuse project proceeded through various phases of planning, regulatory review, and preliminary design prior to being cancelled by the city council in 1999. Although the WRP was not implemented, the history of this Project is important to any forward-looking evaluation of indirect potable reuse opportunities through surface-water augmentation.

Initially, it was proposed to take reclaimed water from the NCWRP and deliver it to a new, nearby facility for further treatment. The additional treatment steps, which were pilot-tested, would include the use of several advanced treatment technologies including membrane filtration, reverse osmosis, ion exchange, advanced oxidation using ozone, and disinfection. The product water that would be produced from such sophisticated advanced treatment was termed "repurified water." About 25×10^6 m^3/yr or 68×10^3 m^3/d (20×10^3 ac-ft/yr or 18 Mgal/d) of this repurified water was planned to be pumped approximately 32 km (20 mi) to the 111×10^6 m^3 (90×10^3 ac-ft) San Vicente Reservoir, one of the city's potable water sources, where it would be discharged into the reservoir and blended with imported and local water. The repurified water would have been stored in the reservoir for approximately 2 yr, during which time further natural treatment would occur. San Vicente Reservoir water, augmented by repurified water, would then be treated along with other water sources at the city's Alvarado Water Treatment Plant before being distributed to customers.

The California Department of Health Services first granted conditional approval to the project in 1994, and many groups voiced support for the project including the U.S. EPA, the Sierra Club, the San Diego Medical Society, the U.S. Bureau of Reclamation, a citizen's advisory panel, and a variety of business and community interests. Despite this support for the repurification project, public opposition to the project began to emerge. During the 1998 political campaigns, the water repurification project became an issue in several closely contested races. Some members of the public and media began to raise concerns about potable use of reclaimed water, and project opponents began to characterize the project with slogans eliciting a negative reaction from the public (e.g., "toilet to tap"). Another important element was the concern in some stakeholder groups that certain socioeconomic groups were being unfairly targeted to use the repurified water. These factors placed a challenging burden on city policy makers, and subsequently, the city council voted to halt the WRP in January 1999 (City of San Diego, 2006a).

2000 Updated Water Reclamation Master Plan

Because the city is mandated to use the reclaimed water from the reclamation plants, a condition of the U.S. EPA grant and the federal *Ocean Pollution Reduction Act*, the city's water department initiated the Beneficial Reuse Project to assess nonpotable reuse opportunities. It should be noted that the SBWRP was not operational at the time. The Beneficial Reuse Project resulted in the *2000 Updated Water Reclamation Master Plan* and in numerous planned and implemented system improvements to maximize nonpotable use of reclaimed water.

City of San Diego Water Reuse Study 2005

The City, as noted previously, has recognized that it must diversify its sources of water and increase the use of locally produced reclaimed water to assure an adequate and reliable supply for the future. As part of this diversification effort, at the direction of the city council on January 13, 2004, the Water Department undertook a study to evaluate all opportunities for increasing the production and use of reclaimed water. The water reuse study is the response to this direction and forms a master plan component including indirect potable reuse through reservoir augmentation. In addition, the City of San Diego contracted with the National Water Research Institute to organize the Independent Advisory Panel (IAP) to provide technical oversight for the study. During the review process, the IAP offered significant suggestions regarding the reorganization and enhancement of the study as well as the comprehensive science-based projects. Together with the results of a broad public outreach and involvement process, the City will use the findings of this study to determine a future course for the implementation of water reuse projects (City of San Diego, 2006a,b,c).

Reservoir Augmentation Opportunities

Opportunities and constraints of conveying advanced treated water to city-owned, surface water reservoirs have been examined, both for the northern and southern service areas. Regulations require advanced treated reclaimed water to be stored in the reservoir for a minimum of 12 mo to blend with the untreated water within the reservoir and undergo a measure of natural treatment. Consideration was also given to the development of wetlands upstream from the surface water reservoir to provide additional natural treatment processes prior to entering the reservoir. Nine reservoirs were selected as candidate reservoir augmentation concept projects and evaluated in this study, but only three were deemed suitable.

Summary of Indirect Potable Reuse Opportunities that were brought forward for Evaluation

Although many indirect potable reuse opportunities were investigated, not all were brought forward, as previously discussed, for evaluation as components of larger implementation strategies. A summary by service area of the viable opportunities and the facilities required to deliver the reclaimed water for indirect potable uses via surface reservoir augmentation are presented in Table 23-7. The principal findings from the preceding evaluations are as follows (City of San Diego, 2006a):

1. All of the presented alternatives are feasible.
   For both the North City and South Bay service areas there is a range of reuse strategies that are feasible from an engineering, scientific, and regulatory perspective. For the indirect potable reuse strategies, public acceptance will depend on the city's commitment and ability to garner public support through an extensive public involvement program.

Table 23-7
Summary of surface reservoir augmentation opportunities for the City of San Diego[a]

Service area	Augmentation opportunity	Estimated average day demand, m³/d (Mgal/d)	Estimated annual use, m³/d (ac-ft/yr)	Customers served	Facilities required
Northern	Lake Hodges	6.0 × 10³ (1.6)	2.2 × 10⁶ (1800)	Potable water customers—North City and North County San Diego	Phase III recycled water extension 7.7 × 10³ m³/d (2 Mgal/d) advanced water treatment plant, brine disposal pipeline, and connection to Escondido Hale Avenue resource recovery plant
Northern	San Vicente	36.0 × 10³ (9.4)	13.0 × 10⁶ (10,500)	Potable water customers throughout City	60.6 × 10³ m³/d (16 Mgal/d) advanced water treatment plant and 37 km (23 mi) pipeline
Southern	Otay Lakes	19.0 × 10³ (4.9)	6.8 × 10⁶ (5500)	Potable water customers throughout central and southern portions of the City	20.8 × 10³ m³/d (5.5 Mgal/d) advanced water treatment plant and 26 km (16 mi) pipeline

[a]Adapted from City of San Diego, 2006a.

2. **The City faces choices between nonpotable and indirect potable uses.**
   The strategies differ in their type of reuse, specifically, between those that exclusively pursue nonpotable uses and those that include indirect potable reuse. In deciding which strategies to pursue, the City will need to weigh the merits of each type of use.

3. **The City faces choices in deciding how far to pursue a selected strategy.**
   Within each strategy, there are implementation phases that add new units of water reuse, usually at progressively higher incremental costs. In deciding how far along each strategy to advance, the City will need to weigh these costs with the water supply reliability, sustainability, and other values suggested in the report.

Next Steps

The water reuse study assessed the advantages, constraints, and values of the different water reuse opportunities available to the City of San Diego. As directed by the city council, the study does not provide a specific recommended project. This report has been reviewed by the IAP and by the American Assembly group of project stakeholders. The report will be presented to the City's Natural Resources Committee in late 2006, and subsequently to city council for their consideration and direction as to the City's future course of water reuse development.

Lessons Learned

The principal lessons learned are that for any plan involving indirect potable reuse to succeed the following elements are of critical importance: (1) public outreach and involvement, (2) the science conducted in support of the health and safety of indirect potable reuse must be rigorous and defensible, and (3) the engineering documentation must be thorough and defensible.

Public Participation

Learning from the WRP described above, clearly the public's participation is critical. Ensuring the public becomes involved in the study, understands the water reclamation process and helps determine the best use of this water resource is top priority. A wide range of meetings, speaking engagements and communication opportunities are facilitating dialogue and information sharing with city residents and the study team. A significant component of public participation has been conducting two American Assembly-style workshops (City of San Diego, 2006a,b) The workshops brought together stakeholders from around the City to engage in a dialog on reclaimed water issues. Participants were selected by the Mayor's office, city council offices, and from a variety of groups around the city. Workshop participants prepared and adopted position statements at the conclusion of each workshop.

Health and Safety

A major focus of the Water Reuse Study has been on the health and safety aspects of the reclaimed water opportunities. The IAP, comprised of experts from the fields of science, economics, medicine, and education, analyzed the information developed by the City and its consultants and found that, "Water produced with the technologies that have been evaluated, including membrane systems and advanced oxidation, will meet health and safety requirements for any of the water reuse strategies" (City of San Diego, 2006a).

Other Implementation Issues

Other important factors including an assessment of the costs and benefits of the various options, public acceptance, health and safety concerns, and environmental considerations have also been examined (City of San Diego, 2006c). Again quoting from the IAP findings "It is the unanimous conclusion of the panel that appropriate alternative water reuse strategies for the City of San Diego have been identified and these alternatives have been presented clearly so that the citizens of the City of San Diego can make informed choices with respect to water reuse" (City of San Diego, 2006a). It is of particular interest this time to observe how the City will approach indirect potable reuse options and the implementation of the required water reclamation capacity of 142×10^3 m^3/d (37.5 Mgal/d) by 2010.

23-7 CASE STUDY: SINGAPORE'S NEWATER FOR INDIRECT POTABLE REUSE

Water reuse was always an important component of Singapore's water supply, beginning with the Jurong Industrial Water Works project in the early 1970s, which supplied tertiary quality effluent to industries in the southwest of the island. Since 1999, however, the Singapore government, through the Public Utilities Board (PUB), has developed aggressively advanced water reclamation facilities, known as NEWater Factories. Reclaimed water is used to supply the high-tech semiconductor manufacturing sector and to augment the potable water storage reservoirs (Law, 2003; Giap, 2005; Singh, 2005).

23-7 Case Study: Singapore's NEWater for Indirect Potable Reuse

Setting

The Republic of Singapore, a small city-state (land area: 699 km^2), has one of the highest population densities in the world (about 6200 capita/km^2) (Singapore Department of Statistics, 2006). Singapore has developed most of its natural water sources, and buys more than half of its water from Johor, in neighboring Malaysia, under decades-old treaties, which start expiring in 2011. The water trade has sparked occasional disputes between the two nations over pricing and other political and environmental issues.

Water Management Issues

In 1999, the Singapore government launched a strategic initiative to develop alternative and renewable sources of water, in an effort to ensure reliability of supply and consistency of water quality. This initiative, known locally as the *Four Tap Strategy* consisted of: (1) collection and treatment of local surface runoff, (2) importing water from Malaysia, (3) water reuse, and (4) seawater desalination.

NEWater Factory and NEWater

The centerpiece of the strategic initiative is the operation of a 10×10^3 m^3/d (2.6 Mgal/d) dual membrane water reclamation plant called the *NEWater Factory*. The plant is located on a compact site downstream of the Bedok Sewage Treatment Plant, recently renamed the Bedok Water Reclamation Plant (WRP). The NEWater Factory is a membrane-based advanced water reclamation plant consisting of microfiltration (MF), reverse osmosis (RO), and ultraviolet (UV) disinfection. As discussed above, NEWater is the term coined by the PUB to describe the high-quality product water from the membrane- and UV-based advanced treatment processes.

Implementation

The first stage of NEWater program was the decision in 1999 to develop a demonstration project. The success of this demonstration project gave the PUB the confidence to proceed with the program and introduce NEWater to the surface water reservoirs on February 21, 2003.

Design and Process Trains of the NEWater Factory
The design of the NEWater Factory dual-membrane and UV technology process trains are consistent with the recommendations made by the *National Research Council Report* (NRC, 1998). The first design tenet was to ensure rigorous source control of the untreated wastewater. The Bedok WRP was selected as the site of the demonstration plant because it receives more than 95 percent of its wastewater from domestic sources. The second design tenet was the use of multiple physical barriers for the removal of microbial pathogens and chemical contaminants (Seah et al., 2003).

Source (Feedwater) Water
Feedwater to the demonstration plant is a clarified secondary effluent from an activated sludge treatment process that typically contains: 10 mg/L BOD, 15 mg/L TSS, 6.4 mg/L ammonia nitrogen, and 400–1600 mg/L TDS including 12 mg/L of TOC. The secondary effluent is first microscreened (0.3 mm), followed by MF to remove suspended solids, prior to demineralization with RO.

Membrane Processes
The MF process consisted of five self-contained units operating in parallel. The MF units were fitted with polypropylene hollow, fine-fiber membranes, with a nominal pore size of 0.2 μm. The membranes operated in a single-pass mode with a design process water recovery of 90 percent.

Two parallel 5×10^3 m^3/d (1.3 Mgal/d) reverse osmosis processes are provided, each fitted with thin-film aromatic-polyamide-composite membranes configured for 80–85 percent recovery in a three-stage array. As a last step, the RO permeate is disinfected by three UV units in series equipped with broad-spectrum medium-pressure lamps at a minimum design dose of 60 mJ/cm^2 each (Singapore Water Reclamation Study, 2002).

Testing for Robustness and Reliability

The membrane and UV technology was tested over a 2-yr period for robustness and reliability to produce consistently high quality NEWater. NEWater quality and treatment reliability was assessed by a Sampling and Monitoring Program (SAMP), where a suite of physical, chemical and microbiological parameters was measured systematically across the process train to determine the suitability of NEWater as a source of raw water for potable use. The water samples were analyzed for all drinking water parameters listed in the current *U.S. EPA National Primary and Secondary Drinking Water Standards* and *WHO Guidelines for Drinking Water Quality*. Other parameters of potential concern, but not listed in these standards/guidelines, were added to the list of analytes, based on the input of an independent advisory study panel. In total, some 190 physical, chemical and microbiological parameters were monitored and as of April 2002, over 22,000 physical, chemical, and microbiological tests had been performed for the NEWater Study (*Singapore Water Reclamation Study*, 2002; Ong et al., 2004).

This comprehensive monitoring program was considered to be adequate to assess the safety of the NEWater and the reliability of the NEWater process. However, a 2-yr Health Effects Testing Program (HETP) was developed to complement the SAMP, and address the potential health impact of unidentified constituents in the NEWater. The HETP involved a comparative toxicological assessment of NEWater with existing raw potable water sourced from Bedok Reservoir. The NEWater demonstration was the first program to include chronic testing of fish, in parallel with a chronic mice study.

NEWater Demonstration Plant Performance

NEWater Factory was first commissioned in May 2000, and to date plant performance has tracked closely with design specifications. Analytical results indicate that NEWater is of consistently higher quality than those specified under U.S. EPA 1998 and WHO 1993 drinking water regulations and guidelines as shown in Table 23-8.

Project Milestones

The NEWater demonstration study has generated important data to guide the PUB in the development and expansion of the use of reclaimed water in Singapore's overall water management strategy. These milestones in the use of NEWater are chronicled in Table 23-9.

In early 2003, the first full-production NEWater plants went online at Bedok WRP and Kranji WRP with a combined initial capacity of 72×10^3 m^3/d (19 Mgal/d). These plants have provisions to expand to 168×10^3 m^3/d (44 Mgal/d) in the future. The new Bedok and Kranji NEWater plants provide water to the microelectronics industry, thereby saving existing drinking water for domestic use. PUB's goal is to reuse 20 percent of the wastewater generated in Singapore for industrial and potable use by the end of 2010.

Table 23-8
Comparison of the U.S. EPA and WHO *Water Quality Standards* with the NEWater Factory product water[a]

Water quality parameter	Unit	U.S. EPA / WHO[b]	NEWater Factory[c]
Color	Hazen units	15	<5
pH	unitless	6.5–8.5	5.2–6.2
Conductivity	μS/cm	—	39.6–71.1
Alkalinity	mg/L as $CaCO_3$	—	8
Total dissolved solids	mg/L	500	22–41.3
Hardness	mg/L as $CaCO_3$	—	<5
Fluoride	mg/L	1.5	0.18–0.22
Nitrite	mg/L as N	0.91	0.38
Nitrate	mg/L as N	10	0.49–1.65
Ammonia	mg/L	1.5	0.35–0.57
Chloride	mg/L	250	3.6–10.9
Turbidity	NTU	5	<0.1
Aluminum	mg/L	0.2	0.09
Iron	mg/L	0.3	<0.003
Manganese	mg/L	0.05	<0.003
Sulfate	mg/L as SO_4	250	0.16–0.54
Zinc	mg/L	3	<0.004
Silica	mg/L as SiO_2	—	0.21–0.32
Phosphate	mg/L as P	—	0.011–0.044
Sodium	mg/L	200	5.1–9.6
TOC	μg/L	—	60–90
Total coliform	counts/100 mL	ND[d]	<1
Fecal coliform	counts/100 mL	ND[d]	<1
Clostridium perfringens	cfu/100 mL	—	<1

[a] Adapted from *Singapore Water Reclamation Study* (2002).
[b] Lowest limit of either the *U.S. EPA 1998 Surface Water Regulations* or *WHO 1993 Guidelines for Drinking Water*.
[c] Taken from analytical results for the months of June and July, 2000.
[d] ND = Not detectable.

Lessons Learned

To enhance public understanding of NEWater, PUB has embarked on an intensive public education program on NEWater. Advertisements, posters and leaflets were produced. Briefings and exhibitions were held to spread the NEWater message. The NEWater Visitor Centre was also opened in February 2003 for the public to see the use of membrane technology and ultraviolet disinfection in the production of NEWater (see Fig. 23-9).

Besides supplying to industries, NEWater is being introduced to the water supply reservoirs for indirect potable reuse. This involves pumping NEWater into the reservoirs to be mixed and blended with raw water. The mixed water is subject to natural treatment before being treated again in conventional waterworks to produce drinking water. The PUB introduced 13.5×10^3 m³/d (3.6 Mgal/d) of NEWater, about one percent of the amount of water consumed daily, into the raw water reservoir in 2003. The

Table 23-9

Singapore Water Reclamation Milestone Events[a]

Year	Milestone event
1998	Water Reclamation Study (NEWater Study) initiative conceived by the Singapore Public Utilities Board (PUB) and the Ministry of the Environment to determine the suitability of using NEWater as a source of raw water to supplement Singapore's water supply
Jan 1999	Setting up of NEWater Expert Panel to review the findings and to provide independent advice to PUB on the NEWater study.
May 2000	Bedok NEWater Factory Demonstration Plant constructed and commissioned within a 7-mo period. Design capacity of 10×10^3 m^3/d.
Oct 2000	Commencement of the feasiblity study at the Bedok NEWater Factory Demonstration Plant and toxicological assessment on the NEWater: • Sampling and monitoring program for some 190 water quality parameters. • Toxicological assessment using both mice and fish for the first time.
Jan 2001	PUB announces its goal to reclaim 20 percent of secondary-treated effluent for industrial uses.
Jul 2001	Award was made for the engineering design and construction supervision for full-scale Bedok and Kranji NEWater Plants, as well as interactive visitor/public education center at Bedok. Ultimate design capacity of these two plants is 168×10^3 m^3/d.
Jul 2002	Expert Panel recommends the adoption of indirect potable reuse of NEWater to supplement Singapore's existing water supply sources.
Aug 2002	NEWater debuts to wide public acceptance at the National Day Parade and celebrations. Up to 60,000 bottles of NEWater given away at parade day.
Jan 2003	Bedok and Kranji NEWater Plants were inaugurated with an initial capacity of 72×10^3 m^3/d.
Feb 2003	The potable and non-potable use of NEWater is officially launched by the Prime Minister of Singapore at a gala event. At the same time a visitor and public education center was opened to the public.
Nov 2003	Formation of an External Audit Panel to perform independent audit checks on NEWater quality and the robustness of NEWater Plants.
Dec 2003	NEWater Visitor Center received its 100,000th visitor.
Jan 2004	Seletar NEWater Plant commissioned and started supplying 24,000 m^3/d of NEWater.
Jan 2005	Construction of the 4th and largest NEWater Factory at Ulu Pandan commenced. This plant with capacity of 148×10^3 m^3/d NEWater was implemented under a Design-Build-Own-Operate (DBOO) arrangement. PUB will buy NEWater from the private concession company under a 20-yr agreement.

[a] Provided by Singapore Public Utilities Board, 2006.

amount of NEWater will be increased progressively to about two and one-half percent of the total daily water consumption by 2011. The country's target is to obtain 25 percent of its water supply from non-traditional sources: 15 percent from NEWater, five percent from seawater desalination, and the remaining five percent from industrial water, by the year 2012.

Figure 23-9
View of Singapore's NEWater Visitor Center (Courtesy of Singapore Public Utilities Board).

The Expert Panel assembled by the Singapore Government arrived at the following conclusions [*Singapore Water Reclamation Study* (2002)]:

1. NEWater is considered safe for potable use, based on the comprehensive physical, chemical and microbiological analysis of NEWater conducted over a 2-yr period. The quality of NEWater consistently meets the latest requirements of the *U.S. EPA National Primary and Secondary Drinking Water Standards* and *WHO Drinking Water Quality Guidelines*
2. Singapore should adopt the approach of indirect potable reuse based on the following reasons:
   - Blending with reservoir water will provide trace minerals, which have been removed in the reverse osmosis process, necessary for health and taste
   - Storage provides additional safety beyond the advanced technologies used to produce safe, high-quality NEWater
   - Public acceptance. This approach is similar to the precedent practice in the United States with planned indirect potable reuse
3. The Singapore Government should consider the use of NEWater for indirect potable reuse, as it is a safe supplement to the existing water supply
4. A vigilant and continuous monitoring and testing program be carried out.

NEWater was formally launched in 2002 to the public and, on August 9, 2002, at the National Day celebration, 60,000 bottles of NEWater were given away at that occasion (see Table 23-9 and also Fig. 23-9). Furthermore, more than a million bottles of NEWater have been distributed at public events for the purpose of public education, and have received overwhelming public support.

23-8 OBSERVATIONS ON INDIRECT POTABLE REUSE

Because the quantities of treated wastewater discharged into the nation's waters are increasing, much of the research that is focused on drinking water quality from these water sources is becoming equally relevant to planned indirect potable reuse. It may be argued that planned, rather than unplanned, indirect potable reuse exercises more positive engineering control over water quality, and a conscious effort is made to establish multiple barriers to protect public health.

It is essential that the public is educated about drinking water quality issues and the principles and capabilities of water reclamation and reuse technologies and applications. An educated and well-informed public will recognize the need for integrated water resources management in the region and increase the likelihood of full confidence in the integrity of operating and regulatory agencies. Public outreach and education efforts are, thus, essential to successfully achieve acceptance for proposed indirect potable reuse projects.

The following observations are derived from the experiences gained from the existing indirect potable reuse projects:

1. Transparency and public trust in decision-making are of paramount importance.
2. Stakeholder and public acceptance is absolutely essential.
3. It is of critical importance to demonstrate that the water is safe with respect to chemical and microbiological quality.
4. A comprehensive ongoing monitoring program is an essential part of an indirect potable reuse program. The drinking water standards will serve as a benchmark so that safety can be assured.
5. To deal with constituents unregulated by drinking water standards, reasonable precautions should be applied.
6. Safeguards for unregulated compounds should be implemented based on the significance of occurrence, a public health risk assessment, and the feasibility of risk minimization.
7. Treatment process reliability and system redundancies must be incorporated into indirect potable reuse plans.
8. The multiple barrier approach to protect public health and the environment is an essential part of indirect potable reuse.
9. With proper outreach, education, and a sensible message, it is possible to gain public support for indirect potable reuse.

PROBLEMS AND DISCUSSION TOPICS

23-1 Many communities currently use surface water sources of varying quality for their drinking water supply, including sources that contain significant upstream discharges of treated wastewater. Locate one or more river basins in your area where treated wastewater is discharged into the source of the water supply. Discuss the implications for a clean and safe drinking water supply.

23-2 What are your recommendations to manage the increase in *de facto* indirect potable reuse that is occurring due to population growth and urbanization?

23-3 Can technology alone be used to improve water quality to the extent that unplanned and incidental indirect potable reuse is permissible in the future?

23-4 Discuss, based on a review of the literature, the methods used to quantify the attenuation (i.e., removal and transformation) of trace constituents in the environment. Cite a minimum of three references.

23-5 A reservoir is to be used as a drinking water supply for an adjacent community. An upstream community discharges reclaimed water into the same reservoir. Discuss the advantages and disadvantages of locating RO treatment at the water reclamation plant as opposed to locating the RO at the drinking water treatment plant.

23-6 Compare the approach to implementation of water reuse used in the City of San Diego (Case Study 23-9) to that used in City of St. Petersburg (Case Study 26-5). Discuss the successes and limitations of each approach.

23-7 Write an addendum to the City of San Diego Case Study (23-9), based on currently available information.

23-8 In the Singapore Case Study, it is reported that "more than a million bottles of NEWater have been distributed at public events for the purpose of public education resulted in an overwhelming public support." Comment on the reaction that you would expect if a similar campaign was launched in your area and why the reaction in your area may differ from that in Singapore.

23-9 What factors should be assessed to determine how long reclaimed water should be retained in a reservoir before it is safe to use for potable water following water treatment? Discuss approaches that can be used for blending reclaimed water into a reservoir to ensure sufficient residence time.

23-10 Discuss the important characteristics of multiple barriers as related to source control, wastewater treatment and reclamation, environmental buffers, potable water treatment, and monitoring. Describe the advantages and limitations of each barrier discussed above.

REFERENCES

Angelotti, R. W., T. M. Gallagher, M. A. Brooks, and W. Kulik (2005) "Use of Granular Activated Carbon as a Treatment Technology for Implementing Indirect Potable Reuse," *Proceedings of the 20th Annual WateReuse Symposium, Denver, CO.*, WateReuse Association, Alexandria, VA.

Bender, M. D., G. E. Hauser, M. C. Shiao, and W. D. Proctor (1990) *BETTER: A Two-Dimensional Reservoir Water Quality Model: Technical Reference Manual and User's Guide* Report No. WR28-2-590-152, Tennessee Valley Authority Engineering Laboratory, Norris, TN.

Beverly, S. D., W. J. Conlon., and D. F. MacIntyre (2001) *Indirect Potable Reuse and Aquifer Injection of Reclaimed Water,* AWWA, Denver, CO.

Brunner, G. W., and V. R. Bonner (1994) *HEC River Analysis System*, U.S. Army Corps of Engineers, Institute for Water Resources, Hydrologic Engineering Center, Davis, CA.

Chapman, R. S., B. H. Johnson, and S. R. Vemulakonda, (1996) *User's Guide for the Sigma Stretched Version of CH3D-WES; A Three-Dimensional Numerical Hydrodynamic, Salinity, and Temperature Model*, Technical Report HL-96-21, U.S. Army Corps of Engineers Waterways Experiment Station, Vicksburg, MS.

Chapra, S. C. (1997) *Surface Water-Quality Modeling*, McGraw-Hill Series in Water Resources and Environmental Engineering, McGraw-Hill, New York.

Chapra, S. C., G. J. Pelletier, and H. Tao (2006) *QUAL2K: A Modeling Framework for Simulating River and Stream Water Quality*, Version 2.04: *Documentation and Users Manual*, Civil and Environmental Engineering Dept., Tufts University, Medford, MA.

City of San Diego (2006a) *City of San Diego Water Reuse Study 2005: Final Draft Report*, March, 2006, prepared in coordination with Water Department, City of San Diego, PBS&J, and McGuire/Pirnie, San Diego, CA.

City of San Diego (2006b) *Public Participation*, accessed at http://www.sandiego.gov/water/waterreusestudy/involvement/index.shtml#assembly

City of San Diego (2006c) *Study Overview: A Need for More Water*, accessed at http://www.sandiego.gov/water/waterreusestudy/geninfo/overview.shtml

Crites, R., and G. Tchobanoglous (1998) *Small and Decentralized Wastewater Management Systems*, McGraw-Hill, Boston.

FCWA (2006) "Fairfax Water," Fairfax County Water Authority, accessed at http://www.fcwa.org/

Giap, L. C. (2005) "NEWater—Closing the Water Loop," *The Int. Desal. Water Reuse Quar.* **15**, 3, 34–36.

Gurr, C. J., and M. Reinhard (2006) "Harnessing Natural Attenuation of Pharmaceuticals and Hormones in Rivers," *Env. Sci. Technol.*, **40**, 9, 2872–2876.

King, I., J. V. Letter, Jr., B. P. Donnell (2005) *RMA4 WES 4.5*, U.S. Army, Engineer Research and Development Center, Waterways Experiment Station, Coastal and Hydraulics Laboratory, Vicksburg, MS.

Law, I. B. (2003) "Advanced Reuse—From Windhoek to Singapore and Beyond," *Water,* May 2003, 44–50.

Lin, A.Y-C., M. H. Plumlee, and M. Reinhard (2006) "Natural Attenuation of Pharmaceuticals and Alkylphenol Polyethoxylate Metabolites during River Transport: Photochemical and Biological Transformation," *Env. Toxcol. Chem.*, **25**, 6, 1458–1464.

López Ramírez, J. A., J. M. Quiroga Alonso, D. Sales Márquez, and T. Asano (2002) "Indirect Potable Reuse and Reverse Osmosis: Challenging the Course to New Water," *Water 21*, 6, 56–59.

Martin, J. L., and S. C. McCutcheon (1999) *Hydrodynamics and Transport for Water Quality Modeling,* Lewis Publishers, Boca Raton, FL.

McEwen, B. (1998) "Indirect Potable Reuse of Reclaimed Water," Chap. 27, in T. Asano (ed) *Wastewater Reclamation and Reuse*, Water Quality Management Library, **10**, CRC Press, Boca Raton, FL.

NRC (1998) *Issues in Potable Reuse: The Viability of Augmenting Drinking Water Supplies with Reclaimed Water*, 2, National Research Council, National Academy Press, Washington, DC.

Ong, S. L., L. Y. Lee, J. Y. Hu, L. F. Song, H. Y. Ng, and H. Sheah (2004) "Water Reclamation and Reuse—Advances in Treatment Technologies and Emerging Contaminants Detection," Presented at the 10th International Conference in Drinking Water Quality Management and Treatment Technology, held in Taipei, Taiwan.

Ontario Ministry of the Attorney General (2002) *Report of the Walkerton Inquiry: The Events of May 2000 and Related Issues*, Publications Ontario, Toronto, Canada.

Orlob, G. T. (ed.) (1983) *Mathematical Modeling of Water Quality: Streams, Lakes, and Reservoirs*, A Wiley-Interscience Publication, John Wiley & Sons, Chickester, England.

Park, K., A. Y. Kuo, J. Shen, and J. Hamrick, (1995) *A Three-Dimensional Hydrodynamic-Eutrophication Model (HEM3D): Description of Water Quality and Sediment Process Submodels*, Special Report No. 327 in Applied Marine Science and Ocean Engineering, School of Marine Science, Virginia Institute of Marine Science, The College of William and Mary, Goucester Point, VA.

Park, R. A., J. S. Clough, and M. C. Wellman (2004) *AQUATOX (Release 2) Modeling Environmental Fate and Ecological Effects in Aquatic Ecosystems*, Vol. 1 *User's Manual*, U.S. Environmental Protection Agency, Office of Water, Washington, DC.

Prince William Conservation Alliance (2006) Accessed at http://www.pwconserve.org/index.html

Robbins, Jr., M. H., and G. A. Gunn (1979) "Water Reclamation for Reuse in Northern Virginia," **2**, 1311, *Proceedings Water Reuse Symposium*, AWWA, Denver, CO.

Scarbrough, J. H. (2005) "Sustainable Water Resources for Gwinnett County, Georgia," in K. J. Hatcher (ed.), *Proceedings of the 2005 Georgia Water Resources Conference,* Athens, GA.

Schaffranek, R. W. (2004) *Simulation of Surface-Water Integrated Flow and Transport in Two Dimensions: SWIFT2D User's Manual*, U.S. Geological Survey Techniques and Methods, Book 6, Chap. 1, Section B, USGS, Denver, CO.

Schneider, M. L., S. C. Wilhelms, and L. I. Yates (2004) *SELECT Version 1.0 Beta: A One-Dimensional Reservoir Selective Withdrawal Model Spreadsheet*, Coastal and Hydraulics Laboratory, U.S. Army Engineer Research and Development Center, Vicksburg, MS.

Seah, H., J. Poon, G. Leslie, and I. B. Law (2003) "Singapore's NEWater Demonstration Project: Another Milestone in Indirect Potable Reuse," *Water Journal,* 43–46.

Sheng, Y. P., M. Zakikhani, and M. C. McCutcheon (1990) *Three Dimensional Hydrodynamic Model for Stratified Flows in Lakes and Estuaries*, U.S. Environmental Protection Agency, Athens, GA.

Singapore Department of Statistics (2006) "Statistics Singapore-Keystats-Annual Statistics," accessed at http://www.singstat.gov.sg/keystats/annual/indicators.html

Singapore Public Utilities Board (2002) *Singapore Water Reclamation Study—Expert Panel Review and Findings*, Singapore.

Singh S. (2005) "Singapore offers Recipes for Urban Water Management," *Asian Water,* 12, 23–27.

Swayne, M., G. Boone, D. Bauer, and J. Lee (1980) *Wastewater in Receiving Waters at Water Supply Abstraction Points*, EPA-60012-80-044, U.S. Environmental Protection Agency, Washington, DC.

Tchobanoglous, G., F. L. Burton, and H. D. Stensel (2003) *Wastewater Engineering: Treatment and Reuse*, 4th ed., McGraw-Hill, New York.

Thompson, D. E., V. B. Rajal, S. De Batz, and S. Wuertz (2006) "Detection of *Salmonella* spp. in Water using Magnetic Bead Capture Hybridization Combined with PCR or Real-time PCR," *J. Water Health*, **4**, 67–75.

Trussell, R. R., P. Gagliardo, S. Adham, and P. Tennison (2002) "The San Diego Potable Reuse Project: An Overview," *Proceedings of the International Conference on Wastewater*

Management & Technology for Highly Urbanized Coastal Cities, Hong Kong Polytechnical University, Hung Hom, Kowloon, Hong Kong.

TVA (1986) *Hydrodynamic and Water Quality Models of the TVA Engineering Laboratory*, Tennessee Valley Authority, 3, Norris, TN.

UOSA (2006) Personal communication with M. Brooks and R. Angelotti, Upper Occoquan Sewage Authority, Fairfax, VA.

U.S. EPA and U.S. AID (2004) *Guidelines for Water Reuse,* EPA/625/R-04/108, U.S. Environmental Protection Agency and U.S. Agency for International Development, Washington, DC.

Virginia Tech (2005) *An Assessment of the Water Quality Effects of Nitrate in Reclaimed Water Delivered to the Occoquan Reservoir*, Prepared by Occoquan Watershed Monitoring Laboratory, Dept. of Civil and Environmental Engineering, College of Engineering, Virginia Polytechnic Institute and State University (VPI), Manassas, VA.

WEF and AWWA (1998) *Using Reclaimed Water to Augment Potable Water Resources,* Water Environment Federation, Alexandria, VA.

WES (1986a) *CE-QUAL-R1: A Numerical One-Dimensional Model of Reservoir Water Quality; Users Manual, Instruction Report E-82-1 (Revised Edition)*, Environmental Laboratory, U.S. Army Engineer Waterways Experiment Station, Vicksburg, MS.

WES (1986b) *CE-QUAL-W2: A Numerical Two-Dimensional, Laterally Averaged Model of Hydrodynamics and Water Quality; User's Manual*, U.S. Army Corps of Engineers, Waterways Experiment Station, Instruction Report E-86-5, Environmental and Hydraulics Laboratories, Vicksburg, MS.

Wool, T. A., R. B. Ambrose, J. L. Martin, and E. A. Comer (2001) *Water Quality Analysis Simulation Program WASP*, Version 6.0, *User's Manual*, U.S. Environmental Protection Agency, Athens, GA.

Wurbs, R. A. (1994) *Computer Models for Water Resources Planning*, U.S. Army Corp of Engineers, Institute for Water Resources, Alexandria, VA.

Yari, P. F. (2005) "Water Reuse—A Water Supply Option in the Metropolitan Atlanta Area?" in K. J. Hatcher (ed.), *Proceedings of the 2005 Georgia Water Resources Conference, Athens, GA.*

24
Direct Potable Reuse of Reclaimed Water

WORKING TERMINOLOGY   1346

24-1   ISSUES IN DIRECT POTABLE REUSE   1346
Public Perceptions   1347
Health Risk Concerns   1347
Technological Capabilities   1347
Cost Considerations   1348

24-2   CASE STUDY: EMERGENCY POTABLE REUSE IN CHANUTE, KANSAS   1348
Setting   1348
Water Management Issues   1349
Implementation   1349
Efficiency of Sewage Treatment and the Overall Treatment Process   1349
Lessons Learned   1351
Importance of the Chanute Experience   1352

24-3   CASE STUDY: DIRECT POTABLE REUSE IN WINDHOEK, NAMIBIA   1352
Setting   1353
Water Management Issues   1353
Implementation   1354
Lessons Learned   1359

24-4   CASE STUDY: DIRECT POTABLE REUSE DEMONSTRATION PROJECT IN DENVER, COLORADO   1361
Setting   1362
Water Management Issues   1362
Treatment Technologies   1362
Water Quality Testing and Studies   1364
Animal Health Effects Testing   1371
Cost Estimates on the Potable Reuse Advanced Treatment Plant   1372
Public Information Program   1373
Lessons Learned   1374

24-5   OBSERVATIONS ON DIRECT POTABLE REUSE   1375

PROBLEMS AND DISCUSSION TOPICS   1376

REFERENCES   1376

WORKING TERMINOLOGY

Term	Description
Direct potable reuse	The introduction of highly treated reclaimed water either directly into the potable water supply distribution system downstream of a water treatment plant, or into the raw water supply immediately upstream of a water treatment plant. Introduction could either be into a service reservoir or directly into a water pipeline. The water used by consumers could be therefore diluted reclaimed water originating from municipal wastewater.
Indirect potable reuse	The planned incorporation of reclaimed water into a raw water supply, such as in potable water storage reservoirs or a groundwater aquifer, resulting in mixing and assimilation, thus providing an environmental buffer.
Integrated water resources planning	A process that promotes the coordinated development and management of water, land, and related resources to maximize the resultant economic and social welfare in an equitable and sustainable manner.

Direct potable reuse, the subject of this chapter, refers to the introduction of highly treated reclaimed water either directly into the potable water supply distribution system downstream of a water treatment plant, or into the raw water supply immediately upstream of a water treatment plant. As discussed in Chap. 23, indirect potable reuse refers to the planned incorporation of reclaimed water into a raw water supply, such as in potable water storage reservoirs or a groundwater aquifer, resulting in mixing and assimilation, thus providing an environmental buffer.

The key distinction between indirect and direct potable reuse is that direct potable reuse does not include temporal or spatial separation such as natural (environmental) buffers between the reclaimed water introduction and its distribution to the end consumer. In the extreme case, direct potable reuse consists of pipe-to-pipe blending of reclaimed water and potable water. Few direct potable reuse applications have been reported worldwide, although the extraction of water for potable purposes from rivers containing substantial quantities of wastewater effluent is fairly common (*de facto* potable reuse), as discussed in Chap. 3. There are no direct potable reuse applications in the United States.

Following a brief discussion of issues in potable reuse, three case studies are presented to provide a glimpse of different aspects of direct potable reuse: (1) a historical example (1956-57) in Chanute, Kansas, (2) a current international example in Windhoek, Namibia, and (3) a past direct potable reuse demonstration project in the United States. Some observations on direct potable reuse in the United States are presented in Sec. 24-5, following the case studies.

24-1 ISSUES IN DIRECT POTABLE REUSE

Direct potable reuse projects are implemented usually as a result of extreme circumstances, where other potable water alternatives have prohibitive costs or are not available. Depending on the particular conditions, direct potable reuse projects may be temporary, as in the event of a severe drought, or long term, for example where a sufficient local water supply does not exist. In any case, several issues are associated with the practice of direct potable water reuse, including public perceptions, health risk concerns, technological capabilities, and cost considerations.

Public Perceptions

Understandably, direct potable reuse is the most difficult category of water reuse application for a community to accept. A natural resistance exists toward consuming any water that once contained human excreta, regardless of the extent of treatment and subsequent purification to which the reclaimed water is subjected. A reluctance to accept reclaimed water for potable purposes is to be expected, particularly when there is a perception amongst end users that there is no real dilution, natural purification, or loss of identity before the reclaimed water is consumed again. Based on the recent experience in San Diego, California, with proper education and public outreach, support for indirect potable reuse can be achieved (City of San Diego, 2006, 2006a). Additional information on public perception issues can be found in the case studies that follow and in Chap. 26.

Health Risk Concerns

Recent advances in analytical techniques for the measurement of trace inorganic and organic constituents have far exceeded the corresponding knowledge base of the health impacts of these constituents. Thus, it has been argued that it would be prudent to wait until more is known about the health impacts of the many trace constituents that may be found in reclaimed water, before proceeding with direct potable reuse. While it is not possible to be certain that direct potable reuse will be 100 percent safe, it is important to note that an equal amount of unknown risk is associated with traditional water supplies and desalination. Few health and toxicological studies have been undertaken for traditional drinking water sources that are influenced strongly by wastewater discharges from upstream towns and cities (e.g., *de facto* potable reuse, as discussed in Chap. 3). In fact, concern over trace constituents may be unfounded (Snyder et al., 2006). A similar situation exists for desalination of brackish or seawater for drinking water use, which may also be influenced by unknown discharges into estuaries or nearshore waters. There may be similar, but less obvious risks with using so-called "pure" seawater as the feedstock for desalination.

Technological Capabilities

Direct potable reuse implies a confidence in, and reliance on, the applied technology to always produce water that is safe and acceptable to consume, without the opportunity for any natural processes to further improve the water quality. A concern that is often expressed is whether the water agency is competent enough technologically to produce safe reclaimed water consistently for direct potable use, such that it poses no additional health risk over traditional drinking water sources.

To overcome technological concerns, multiple barrier systems, in which sequential and redundant processes are used to remove constituents of concern have been developed, resulting in a high degree of reliability. Also, monitoring technology has improved dramatically, allowing for real-time process monitoring and control. Through scientific studies and technological advances, more robust treatment processes are being developed and implemented, including enhanced membrane systems coupled with advanced oxidation processes, which along with other new technologies are capable of essentially full removal of trace constituents. It is important to note that the technologies that are in place today can be used to produce high-quality potable water which far exceeds current drinking water standards.

Cost Considerations

A significant cost for water reclamation and reuse in urban areas is associated with the need to provide a separate piping and storage system for reclaimed water, as discussed in Chaps. 14 and 15. In any number of cases, the cost of a dual distribution system has been prohibitive and, thus, has limited the implementation of water reuse. Direct potable reuse offers the opportunity to reduce significantly the cost of transporting reclaimed water because the existing water distribution system is used for the transport of blended water to end users. Thus, the continued use of *purple pipe* may in the end prove to be a poor investment. The other significant advantage of both direct and indirect potable reuse is that full reuse can be made of the available reclaimed water, a valuable resource. It can, therefore, be argued that direct or indirect potable reuse can be used to maximize the quantity of water reused, as compared to seasonal irrigation or other intermittent urban water reuse applications.

24-2 CASE STUDY: EMERGENCY POTABLE REUSE IN CHANUTE, KANSAS

As discussed previously, there is no imperative at present for the use of reclaimed water for direct potable reuse in the United States. Nevertheless, the following episode which took place during 1956–57 has historic significance in the chronology of indirect and direct potable reuse of treated wastewater. The following discussion has been adapted from various sources including Metzler et al. (1958); Kasperson and Kasperson (1977); and Dean and Lund (1981).

Setting

During 1956–57 for 150 d (October 14, 1956 to March 14, 1957), emergency use of reclaimed water took place in the city of Chanute, located in eastern Kansas, for the municipal water supply. The reclaimed water system consisted of primary and secondary treatment (trickling filters), a stabilization pond, the water treatment plant, and the water distribution system. One complete cycle from inception of treatment to the time of use required about 20 d.

During the drought of 1952 to 1957, the most severe in Kansas history, the City of Chanute was plagued intermittently by water shortages. Normally, Chanute (population of about 12,000) obtains its municipal water supply from the Neosho River and uses an average of about 5.3×10^3 m^3/d (1.4 Mgal/d) and a maximum of about 7.6×10^3 m^3/d (2 Mgal/d). Local industries account for a substantial part of this water use.

The Neosho River has an average flow of 80×10^3 m^3/d (35 Mgal/d), but the flowrate situation became progressively worse when, in a fifth year of the drought in the summer of 1956, the Neosho River ceased to flow. The river serves not only as a source of water for the City of Emporia and several small neighboring cities and towns upstream of Chanute, but also normally receives diluted treated wastewater from seven upstream communities.

Water Management Issues

As the drought continued, it became evident that additional action was necessary to supply water to the city. City officials investigated and considered several of the following possible courses of action, but none was adopted (Metzler et al., 1958):

- Discontinuance of water service to large industrial water users (a cement plant, an oil refinery, and a wax plant) and conservation of the available supply for domestic use.
- Development of a well water supply, but water from this source was not considered suitable as a permanent supply because of its total hardness of 725 mg/L and its sulfate content of 630 mg/L.
- Hauling of potable water by truck or rail—the cost of hauling was estimated at $1/m^3 ($4/1000 gal). The cost and physical limitations caused its rejection.
- Joining with other cities to pump water from the Smoky Hill River to the upper end of the Neosho River was also considered. If the drought had continued through 1957, this proposal undoubtedly would have received more serious consideration.

Implementation

Recirculation of treated sewage was the final proposal and the one which the City of Chanute officials chose to follow. The Neosho River was dammed below the outfall of the sewage treatment plant and treated effluent backed up to the water intake. "On October 14, 1956, without fanfare, the city opened the valve which permitted mixing of treated sewage with water stored in the river channel behind the water treatment dam" (Metzler et al., 1958). The impounding reservoir for the water treatment plant served very effectively as a waste stabilization pond. In the 17 d retention provided by this intake pool, substantial reductions were observed in BOD, COD, total and ammonia nitrogen, and detergents. Flow diagrams of the water recycling system and treatment processes are shown on Fig. 24-1.

For 5 mo the city reused its treated sewage, circulating it some eight to fifteen times. Because of good treatment including multiple point chlorination, the water met the prevailing public health standards. However, the treated water had a pale yellow color and an unpleasant musty taste and odor. The water also foamed when agitated and contained undesirable quantities of dissolved minerals and organic substances (Metzler et al., 1958).

Efficiency of Sewage Treatment and the Overall Treatment Process

The constituent values and the efficiency of sewage treatment and the overall treatment process during the period of maximum constituent concentrations are shown in Table 24-1. Three weeks' storage in the shallow impoundment resulted in major improvement in water quality (Metzler et al., 1958).

Coliform Organisms

The number of coliform organisms in the raw water during recirculation was considerably less than that found in Neosho river water at Chanute under normal conditions. After chlorination of raw water was started at the water plant on December 20, 1956, water in the first basin had an MPN of less than 3.0/100 mL. The tap water, as judged by standard test, was of satisfactory bacteriological quality during the entire time that water was being reused.

The period of complete direct water reuse ended when heavy rains washed out the temporary dam below the sewage outfall. "Chanute then went back to drinking water that

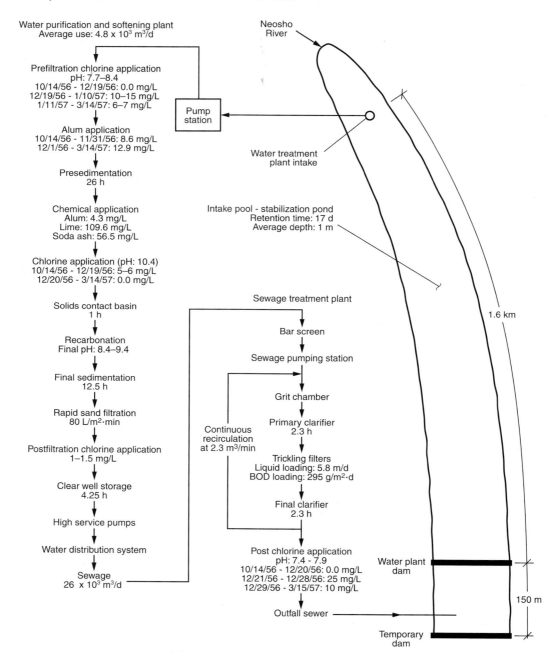

Figure 24-1

Flow diagram of water recycling system and treatment processes at Chanute, Kansas. The system was that used during the 1956–57 period. Adapted from Metzler et al., 1958.

Table 24-1
Selected pollution indicators (average values) during the recirculation period of January 15 to February 13, 1957[a]

Indicator	Raw wastewater, mg/L	Treated wastewater, mg/L	Reduction in wastewater treatment, %	Raw water, mg/L	Treated water, mg/L	Overall reduction (Raw wastewater to treated water), %
Total-N	78	34	56	15	13	83
NH_3-N	47	29	38	12	10	79
COD	656	156	76	56	43	93
BOD	209	29	86	7.3	—	—
ABS[b]	13.6	10.4	25	4.6	4.4	68
Ortho-P as PO_4	22	22	0	12	3.4	88
Total P as PO_4	41	29	29	13	3.9	95
Complex-P as PO_4	19	7	67	1	0.5	97

[a]Adapted from Metzler et al. (1958); Kasperson and Kasperson (1977); Dean and Lund (1981).
[b]ABS = Alkyl benzene sulfonate (hard detergent).

contained Emporia's sewage." Much has been made of the fact that Chanute reused its own municipal wastewater, but little attention has been paid to the fact that Chanute always had some wastewater in its raw water supply and already had a competent water treatment plant before it had to institute 100 percent reuse (Metzler et al., 1958).

Lessons Learned

The water quality met the bacteriological requirements although free chlorine residuals were never present; no viruses were found in the treated water, but there was an abundance of dead and some living organisms including small algae, amoebic cysts, and nematodes. An epidemiological survey showed fewer cases of stomach and intestinal illness during the period of reclaimed water use than in the following winter when Chanute was using river water. All reports agreed that no damage was done, but this type of direct reuse was not recommended in the future (Metzler et al., 1958; Dean and Lund, 1981).

It is not known how much reclaimed water was actually drunk. Bottled-water sales flourished and virtually all grocery stores carried a large stock (Metzler et al., 1958). Thirteen years later, a retrospective telephone survey was made of thirty-nine residents. Results of the survey suggested that most of the population had considered the water to be acceptable (Kasperson and Kasperson, 1977).

The accumulation of nitrogen in reclaimed water makes free residual chlorination impractical, and odor control and taste and color removal difficult. The studies demonstrated the effectiveness of chlorine as a disinfectant, even under very adverse conditions. Bacteriological quality as judged by the coliform test was excellent.

A rapid increase occured in the concentrations of dissolved salts and organic materials, many of which are not amenable to removal by ordinary treatment processes. More information is needed on how effective more elaborate treatment may be in removing some of these pollutants. Many unresolved questions remain concerning the safety of recycled water from a public health standpoint, despite the apparently favorable results obtained in the Chanute study. Standard techniques available to water plant laboratories were not adequate to detect many constituents of health concern (Metzler et al., 1958).

Importance of the Chanute Experience

A historic and comprehensive paper written by Metzler et al. (1958) on the Chanute episode in the *Journal of the American Water Works Association* demonstrated the major differences of the physical, chemical, and biological characteristics of the raw sewage, treated sewage, raw water, and treated water. The paper was co-authored by the leading authorities at that time from the Kansas State Board of Health and the Robert A. Taft Sanitary Engineering Center, U.S. Public Health Service. In addition, discussion was provided by Professor C.H. Connell, University of Texas, Medical Branch, Galveston.

The paper is of historic importance for several reasons: synthetic detergents in wastewater were ABS-based hard (nonbiodegradable) detergent, and froth and foaming were major treatment and aesthetic issues as well as color and odor of the reclaimed water. The frothing of the tap water served as a constant and unpleasant reminder of the source of the city's water supply. Application of large-scale activated carbon technology was still in its infancy then and membrane and advanced oxidation technologies, discussed in Chaps. 9 and 10, were not available until some 10 to 15 yr later. In addition, discoveries related to chlorination of water and its formation of trihalomethanes did not occur until the early 1970s (Rook, 1974). The disinfection processes for water reuse applications are discussed in Chap. 11.

The decision to recirculate treated wastewater was made entirely by the city officials. On December 14, 1956, through published newspaper articles, the reuse of treated wastewater practiced at Chanute came to the attention of the Kansas State Board of Health for the first time. Initial public acceptance of the water was good, probably because the citizens knew that their supply normally received diluted treated sewage from seven upstream communities. No public mention of instituting reuse was made until after recirculation had been started. Public reaction became more adverse when stories appeared in the local newspapers. It is important to note that implementation of water reuse has become much more sophisticated in the last three decades (see Chaps. 25 and 26), and public consultation and acceptance is the key element in any water reuse project, particularly when a indirect or direct potable reuse project is proposed.

24-3 CASE STUDY: DIRECT POTABLE REUSE IN WINDHOEK, NAMIBIA

Since 1968, the City of Windhoek in Namibia has been adding highly-treated reclaimed water to its drinking water supply system. The blending of reclaimed water with potable water takes place directly in the line that feeds its potable water distribution network. Windhoek's Goreangab Reclamation Plant has been a pioneer in direct potable reuse and still today is the only commercial scale operation in existence in the world (du Pisani, 2005). The following case study was prepared from various sources including Harrhoff and Van der Merwe (1996); Odendaal et al. (1998); du Pisani (2005); and Lahnsteiner and Lempert (2005).

Setting

Internationally, the well-known example of direct potable reuse of reclaimed water is at Windhoek in Namibia, located in the southwestern part of Africa bordering the Republic of South Africa. Since 1968, the City of Windhoek has been adding highly-treated reclaimed water to its drinking water supply system. The blending of reclaimed water with potable water takes place directly in the line that feeds its potable water distribution network. Windhoek's Goreangab Reclamation Plant has been a pioneer in direct potable reuse and still today is the only commercial scale operation in existence in the world (du Pisani, 2005).

The City of Windhoek is the capital of Namibia, which is located in the southwestern part of Africa bordering the Republic of South Africa and is the most arid country in sub-Saharan Africa. Namibia encompasses a land area of 825×10^3 km² (318×10^3 mi²) and has a total population of 1.8 million, making it one of the most sparsely populated countries in the world (www.windhoekcc.org.na).

Only ephemeral rivers are in the interior of the country. Perennial rivers are located only on the northern and southern borders of the country, respectively 750 and 900 km (466 to 559 mi) from the capital city. The population of Windhoek is approximately 250,000, and Windhoek is situated almost in the center of the country.

Water Management Issues

Water management issues are concerned with climatic conditions as Namibia is semi-arid and rainfall occurs only a few months of the year. The groundwater water resources are also limited. The following discussion describes some of the constraints in planning for a sustainable water supply system.

Climatic Conditions
The average annual rainfall is 360 mm (14.4 in.) and the annual evaporation amounts to 3400 mm (136 in.). The city relies, for 70 percent of its water, on three surface reservoirs (dams). These reservoirs are built on ephemeral rivers, which run only for a few days after heavy rainfall events. The three dams are located between 70 and 160 km (42 and 96 mi) from the city and are operated by the state-owned water utility, known as NamWater. These dams were built during the period from 1978 to 1993 to supply water to the central areas of Namibia. Windhoek utilizes approximately 90 percent of the water consumed in the central areas.

During the last 10 rainy seasons, only three seasons had yielded above average inflow into these dams. The main "consumer" of water is evaporation, which accounts at times for a volume double that of the water utilized by consumers (du Pisani, 2005). Security of the water supply to the central areas of Namibia and the City of Windhoek is therefore a major challenge, both for the bulk water supplier and the City of Windhoek.

Innovations in an Arid Land
The reason why a settlement originated at Windhoek was the presence of both hot and cold water springs. As the settlement grew, so did exploitation of these sources with the

added digging of wells in the area. The water table subsided as a result and the first municipal borehole was acquired around 1912. Over the period from 1912 to today, some 60 municipal boreholes were developed in an aquifer with a safe assured yield of 1.73×10^6 m³/yr. Groundwater remained the sole source of water for Windhoek until 1933 when the Avis Dam with a capacity of 2.4×10^6 m³ (634 Mgal) was constructed. The dam has a small catchment area and, therefore, had a very small assured yield and often could not supply any water at all. This dam is currently used exclusively for impounding water for recreational purposes. During 1958, a second small surface reservoir, the Goreangab Dam, with a capacity of 3.6×10^6 m³ (951 Mgal), was built and a conventional water treatment plant was constructed to treat water to potable standards (du Pisani, 2005).

Implementation Implementation of the direct water reuse plan from its inception to the present time has evolved. Several issues were addressed including the conversion of the Goreangab Water Reclamation Plant, the 1997 plant upgrade, the building of a new Goreangab Water Reclamation Plant, consideration of a multiple barrier system, process selection, public perception issues, the development of reuse guidelines, and operation and maintenance. These issues are considered briefly in the following discussion.

The Goreangab Water Reclamation Plant
In 1969, the Goreangab water treatment plant was converted to treat not only the water from the Goreangab Dam, but also the final effluent from the city's Gammams Waste Water Treatment Plant. Thus, the Goreangab Water Reclamation Plant was born, with an initial capacity of 4.3×10^3 m³/d (1.1 Mgal/d). The reclaimed water was blended with water from the well field. Because the whole city as well as its informal settlements lies within the catchment area of the Goreangab Dam, the quality of the water from this reservoir is often worse than the treated wastewater and therefore is unfit for water reclamation.

From its inception, one of the cornerstones of water reclamation was that the city separated industrial effluent from domestic effluent and diverted industrial effluents to a separate treatment plant. The effluent used for water reclamation, therefore, originates mostly from wastewaters from domestic and business areas.

Treatment Process Flow Diagrams for the 1997 Upgrade
The initial Goreangab Treatment Plant, now called the "Old" Goreangab Plant, was upgraded several times with the last upgrade undertaken in 1997. The process flow diagram for the 1997 upgrade is shown on Fig. 24-2. The ultimate capacity of this plant was 7.5×10^3 m³/d (2 Mgal/d) of reclaimed water per day.

The New Goreangab Water Reclamation Plant
After independence in 1990 from the Republic of South Africa, the population of Windhoek started growing at a more rapid rate, currently about 5 percent per annum. This growth, together with increased investment and development in the city, placed ever increasing pressure on the supply of water. As the easily accessible natural resources had, to a large extent, been fully exploited and demand management measures implemented successfully, extended water reclamation proved to be the logical choice to augment the water supply. For this purpose, the City of Windhoek obtained a loan from

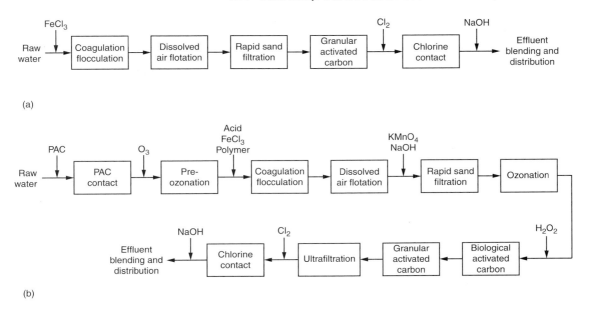

Figure 24-2
Water reclamation process flow diagrams for the Goreangab Water Reclamation Plant in Windhoek, Namibia. (a) 1997 upgrade of the Goreangab process train. (b) The new Goreangab process train. (Adapted from du Pisani, 2005; Lahnsteiner and Lempert, 2005.)

European financial institutions to construct a new 21×10^3 m³/d (5.5 Mgal/d) water reclamation plant on a site adjacent to the old plant. This plant now provides 30 percent of the daily potable water requirements of the city during normal water consumption periods and up to 50 percent during the time of severe droughts (Lahnsteiner and Lempert, 2005).

The Multiple Barrier System

The design philosophy for the new plant is based on a multiple barrier system. In this system, a certain number of safety barriers are set up, depending on the risk associated with a particular substance or contaminant in the water to the end-user. These barriers can be one of three types: (1) treatment, (2) nontreatment, or (3) operational. Treatment barriers are defined as "continually present systems that reduce the undesired substances in the water to an acceptable level" (Lahnsteiner and Lempert, 2005). The nontreatment barriers in the new Goreangab Reclamation Plant include:

- Diversion of industrial effluents to a separate treatment plant
- Complete monitoring at the inlet and outlet of the preceding wastewater treatment plant, allowing corrective action to be taken before the water reaches the water reclamation plant
- Extensive monitoring of drinking water quality
- Blending the water derived from water reclamation with water of different origins, so that at most 30 percent of the drinking water constitutes reclaimed water during times of normal water consumption

Apart from these barriers, operational systems are in action, which take on the role of a further barrier. One example of an operational barrier is the possible addition of powdered activated carbon (PAC) in case the adsorption capacity of the granular activated carbon (GAC) in the process is too low or the organic loading to the plant is too high.

It is clear that a complete removal of impurities from the reclaimed water is practically impossible without reverse osmosis, so the barriers are designed to reduce concentrations of substances to fall within drinking water guidelines. For example, a barrier for turbidity is considered to be a combination of processes such as flocculation, dissolved air flotation (DAF), and dual media filtration, even if turbidity is not completely reduced. However, these three steps are not considered to be a barrier for COD and DOC, but rather only as a step in their partial removal.

The health risks associated with specific constituents vary significantly; for different constituents, the barriers required are as follows:

- Aesthetic parameters, such as turbidity and color, for which there is no direct correlation between them and detrimental effects on health; two barriers
- Microbiological pollutants; three barriers
- Other parameters without health risk, such as calcium carbonate; only one barrier was considered and implemented

The design of the new plant is based on the experience gained over 30 yr of water reclamation and reuse, but also includes new processes such as ozonation and ultrafiltration. The latter two processes were pilot tested onsite over a period of 30 mo, whereby the performance with this specific raw water was thoroughly tested so design decisions could be based on actual recorded results. During these trials and the design process, the City of Windhoek sought the advice of recognized experts in the fields of ozone, membranes, and GAC/BAC (biological activated carbon), as used in the process. The operational protocol of the plant makes provision for the multiple barrier system to be operational at all times.

Selection of the Processes
The new Goreangab Water Reclamation Plant, is made up of a series of unit processes and operations, as shown on Fig. 24-2. An exterior view of the plant is shown on Fig. 24-3.

Public Perception
Without doubt, the most important cornerstone of potable reuse is public acceptance and trust of consumers in the quality of reclaimed water. The most difficult task for anyone who wants to emulate the Goreangab approach in Windhoek would be to overcome the psychological barrier of direct reuse of reclaimed water for potable purposes. Dr. Lucas van Vuuren, a pioneer of water reclamation and reuse in South Africa during the 1970's, coined a phrase: "Water should be judged not by its history, but by its quality" (Odendaal et al., 1998). To gain public acceptance of this viewpoint is, however, not easy and a heavy responsibility is placed onto the City of Windhoek to exercise the required level of control over water quality. For this purpose, the city has over the years invested substantially in laboratory facilities and staffing. The Gammams Laboratory of the city's Scientific

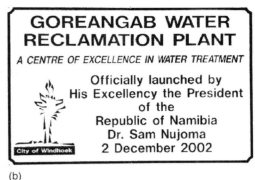

(a)   (b)

Figure 24-3
Goreangab Water Reclamation Plant in Windhoek, Namiba: (a) exterior view of the new Goreangab Water Reclamation Plant (Coordinates: 17.006 E, 22.528 S) (Courtesy of Aqua Services & Engineering, Windhoek, Namibia) and (b) the plate showing official launch on December 2, 2002 (Courtesy of VA Tech Wabag GmbH, Vienna, Austria).

Services Division, therefore boasts state-of-the-art facilities and analytical equipment to provide a level of safety and comfort satisfactory to the customers.

At Goreangab, the history of the feedwater is recognized as treated municipal wastewater and treatment processes were designed accordingly. To retain public confidence, water quality monitoring and control are of utmost importance. Water quality is monitored on an ongoing basis through on-line instrumentation and composite samplers for every major unit process. Should any quality parameter exceed an absolute value, the plant goes into a recycle mode and water is not delivered. The final product water is also monitored continuously and analyzed for the full range of water quality parameters, inclusive of the bacterial pathogens, *Giardia*, and *Cryptosporidium*. Enteric viruses, endocrine disruptors, and other pharmaceutical residues are of concern in direct potable water reuse. Thus, monitoring programs and research studies have been introduced as an ongoing research project to assess the presence of these constituents in the final product water. Only a few enteric virus studies have been conducted on the water reclamation and reuse practices, and much work needs to be devoted to this subject in the world-wide research in the future.

The public of Windhoek is well informed through the local media and municipal newsletters regarding the importance of using water wisely. An ongoing water demand management campaign is aimed at and has been successful in reducing water consumption. To the extent that water consumption is currently estimated, a further reduction of 13 percent in demand would lead to an adverse effect on the economy of the city. Water reclamation is well publicized and the new water reclamation plant was placed into service on December 2, 2002, by the President of the Republic of Namibia with full TV and media coverage (see Fig. 24-3*b*). The plant is often visited by schools and the scientific community, local as well as international. No consumer surveys have been conducted, but over the last 6 yr, no consumer complaints have been lodged about the use of or quality of reclaimed water in Windhoek. The citizens of Windhoek have over time become used to the idea that potable reuse is included in the water supply system and harbor a fair amount of pride that their city in many respects leads the world in direct potable reuse (du Pisani, 2005; Lahnsteiner and Lempert, 2005).

Water Quality Guidelines for Direct Potable Reuse

Because direct potable reuse is not widely practiced, specific water quality guidelines for potable reuse were not readily available. The city therefore compiled water quality standards that were selected from relevant drinking water standards as specifications for reclaimed water for direct potable reuse (du Pisani, 2005; Lahnsteiner and Lempert, 2005). These relevant drinking water standards and guidelines included Namibian Drinking Water Guidelines, U.S. EPA, EU, WHO, and the Rand Water Standards (South Africa).

To ensure optimal performance of the process steps, intermediate treated water criteria were stipulated. These criteria are aimed, for instance, at maximum organic removal through enhanced coagulation to extend carbon life, and the effective removal of iron and manganese to protect the membranes. The criteria consist of target values and absolute values that have to be maintained. Failure to meet intermediate quality criteria at a certain unit process would preclude the delivery of final water and cause the plant to go into recycle mode until conformance to the specified values is achieved. The treated water specifications and the intermediate treated water criteria are shown, respectively, in Tables 24-2 and 24-3.

Operation and Maintenance

The loan conditions of the European Investment Bank required that operation and maintenance (O&M) of the plant be outsourced and include specific internationally recognized operating companies. This requirement exists for the 20 yr term of the loan. Because the construction was already in progress, the options of BOO (build-operate-own), BOT (build-operate-transfer) or BOOT (build-operate-own-transfer) were not considered and the City of Windhoek decided on an O&M model. The city obtained the services of Stallard Burnsbridge, later Katalyst Solutions, to manage the international procurement process for an operator that would satisfy the criteria of the loan agreement.

The successful bidder consisted of a consortium of Veolia Water, Berlinwasser International, and VA Tech Wabag. A contract, the Private Management Agreement (PMA) was concluded, starting in September 2002, for a period of 20 yr. During this period, the operating manager is responsible for the total maintenance of the plant, all scheduled replacements, and specific hand back conditions, which are stipulated.

The PMA was crafted in a way that would provide maximum incentive to the operating manager to reach quality guidelines and to perform according to the requirements of the city. "Payment to the Manager consists of two parts or tolls, being the aggregate of the availability toll and the volumetric toll. Both these tolls are subject to performance failure factors and availability factors" (du Pisani, 2005).

The new Goreangab Water Reclamation Plant went into operation in August 2002, and is being operated by the Windhoek Goreangab Operating Co., Ltd. (WINGOC) under a 20 yr O&M contract. From the Windhoek experience it is evident that highly treated municipal wastewater (reclaimed water) can be reused successfully for potable purposes.

Lessons Learned

Table 24-2
Treated water specification for direct potable reuse in Windhoek, Namibia[a,b]

Parameter	Unit	Target values	Absolute values
Physical and organoleptic constituents			
CCPP[c]	mg/L as $CaCO_3$	N/A[d]	Must be between 0 and 8
Chemical oxygen demand	mg/L	10	15
Color	mg/L Pt	8	10
Dissolved organic carbon	mg/L	3	5
Total dissolved solids	mg/L	Greater of 1000 or 200 above raw water	Greater of 1200 or 250 above raw water
Turbidity	NTU	0.1	0.2
UV_{254}	Abs/cm	N/A	0.06
Macro elements			
Aluminum	mg/L	N/A	0.15
Ammonia	mg/L	N/A	0.10
Chloride	mg/L	Not removed by process	
Iron	mg/L	0.05	0.1
Manganese	mg/L	0.01	0.025
Nitrite and Nitrate	mg-N/L	Not removed by process	
Sulfate	mg/L	Not removed by process	
Microbiological indicators			
Heterotrophic plate counts	count/mL	80	100
Total coliform	count/100 mL	N/A	0
Fecal coliform	count/100 mL	N/A	0
E. Coli	count/100 mL	N/A	0
Coliphage	count/100 mL	N/A	0
Enteric viruses	count/10 L	N/A	Greater of 0 or 4 log removal
Fecal streptococci	count/100 mL	N/A	0
Clostridium spp.	count/100 mL	N/A	0
Clostridium viable cells	count/100 mL	N/A	0
Giardia	count/100 L	Greater of 0 or 6 log removal	Greater of 0 or 5 log removal

(Continued)

Table 24-2
Treated water specification for direct potable reuse in Windhoek, Namibia[a,b] (*Continued*)

Parameter	Unit	Target values	Absolute values
Microbiological indicators			
Cryptosporidium	count/100 L	Greater of 0 or 6 log removal	Greater of 0 or 5 log removal
Chlorophyll A	μg/L	N/A	1
Disinfection byproducts			
Total THMs	μg/L	20	40

[a]Adapted from du Pisani (2005).
[b]Other parameters that are not included in this table will be required to comply with the Rand Water Standards (South Africa) for potable water as valid at the effective date. The treated water will not exceed the lower of the RSA limits or the background concentration for those parameters as found in the raw water.
[c]Calcium carbonate precipitation potential. See Chap. 19, Sec. 19-2.
[d]N/A = not applicable.

Table 24-3
Intermediate treated water criteria[a]

Parameter	Unit	Target values	Target values	Absolute values
After DAF				
Turbidity[b]	NTU	1.5 (exceeded by no more than eight readings in one day) 5.0 (exceeded by no more than four readings in one day)	5.0 (exceeded by no more than four readings in one day)	8.0 (absolute maximum peak reading)
After rapid sand filters				
Turbidity[b]	NTU	0.2 (exceeded by no more than four readings in one day)	0.35 (exceeded by no more than four readings in one day)	0.5 (absolute maximum peak reading)
Manganese	mg/L	0.03	0.05	N/A[c]
Iron	mg/L	0.05	0.05	N/A
After ozonation				
Ozone	mg/L	—	—	0.1 minimum (absolute minimum registered by on-line monitoring)

(*Continued*)

Table 24-3

Intermediate treated water criteria[a] (*Continued*)

Parameter	Unit	Target values	Target values	Absolute values
After ozonation				
COD	mg/L	25	25	N/A
DOC	mg/L	15	15	N/A
Microbiological quality, disinfection byproducts	—	According to treated water specification (see Table 24-2)		
After GAC filters				
DOC	mg/L	5	5	8

[a]Adapted from du Pisani (2005). See Fig. 24-2*b*.
[b]Readings are taken at 15 min intervals.
[c]N/A = not applicable.

In the case of Windhoek, a combination of factors, with the lack of alternatives probably the most notable, makes direct potable reuse a viable option, even in financial terms. It is furthermore evident that the technology exists to produce water reliably that meets all drinking water guidelines and standards and provide the user with an acceptable level of confidence regarding the risk of direct potable reuse.

The old Goreangab plant provides reclaimed water for the irrigation of all sports fields and public parks in the city. In the very near future, all excess reclaimed water will be used to artificially recharge the Windhoek aquifer, albeit under very strong quality constraints.

The Goreangab Water Reclamation Plant is an excellent example of one of the innovations practiced in a country with little resources, both natural and financial. Direct potable reuse has proven in Windhoek that it is possible to overcome public perception and prejudice with persistent and positive marketing. Direct potable reuse in Windhoek is a viable option and fits in well into the concept of regional integrated water resources management.

24-4 CASE STUDY: DIRECT POTABLE REUSE DEMONSTRATION PROJECT IN DENVER, COLORADO

In the period from 1985 to 1992, the City of Denver conducted a direct potable reuse demonstration project. The conduct and findings from this landmark study are presented and reviewed in this case study, which was adapted principally from the work of Lauer and Rogers (1998).

Setting

In 1968, the City of Denver began an evaluation of alternative water supplies to meet the increasing water demands of the Denver metropolitan area, including trans-basin diversions and water reclamation and reuse. The evaluation of alternatives was undertaken as part of a consent decree that allowed water from the Blue River to be diverted for use as Denver metropolitan area water supply. It was recognized that a comprehensive evaluation of all alternatives including trans-basin diversions and various forms of successive water use, would be needed to comply with the consent decree (Works and Hobbs, 1976). In 1970, a pilot-scale advanced treatment facility, as described below, was put into operation to assess a variety of treatment technologies. Based on the results from the pilot study, the Denver Potable Water Demonstration Project was developed and put into operation in 1985. Advanced wastewater treatment processes were used for the demonstration facility to produce water of drinking-water quality from secondary-treated municipal wastewater from the City of Denver. The project included an evaluation of the effectiveness of multiple-barrier treatment processes and a two-year animal health effect study on the toxicology, carcinogenicity, and reproductive toxicity associated with the final product water. The potable reuse project was concluded in 1992.

Water Management Issues

As part of the water resources evaluation program initiated in 1968, known as the "successive use project", a grant was secured from the Federal Water Quality Administration (FWQA) to construct a 19 L/min (5 gal/min) pilot plant to evaluate alternative treatment processes. The pilot plant was in operation from 1970-1979. The Denver Water Department formed a project advisory committee of national experts to review the conduct of the project and advise the department on specific issues. The results of the research conducted at the pilot plant, operated by the Denver Water Department and the University of Colorado's Environmental Engineering Department, were incorporated into the design of the potable reuse demonstration plant described below (Linstedt and Bennett, 1973; CH2M Hill, Inc., 1975). It should be noted that one of the critical conclusions, based on the results of the successive use project, was that direct potable reuse was a viable alternative that should be evaluated further.

In 1979, plans were developed to construct a demonstration facility to study the costs and reliability of potable reuse. The interrelated issues of technical and economic feasibility, product water safety, and public acceptance of direct potable reuse were investigated in this landmark project. An advanced wastewater treatment plant with a capacity of 3,785 m^3/d (1.0 Mgal/d), which served as the main testing facility for the demonstration project, was constructed at a cost in excess of $18 million. The more than $34 million project (1979-1992), of which $7 million was funded by the U.S. EPA, included integral health effects testing (the whole animal health effects studies) as well as comprehensive analytical chemistry studies. A health effects advisory panel was also convened to assess the experimental protocols and results of the studies. In parallel with the scientific investigations, public information programs were conducted. The findings and experience from operation of the demonstration project are important for the development of a direct potable reuse system to meet the water demand for the City of Denver (Rothberg et al., 1979; Lauer and Rogers, 1998).

Treatment Technologies

The advanced treatment processes that were evaluated, the processes used during the health effects testing, and the operation and maintenance issues are described below.

The water quality testing and studies that were undertaken using various treatment technologies are described subsequently.

Description of the Advanced Treatment Processes
The influent to the potable reuse demonstration plant was unchlorinated secondary effluent treated at the Denver Metropolitan Wastewater Reclamation District's regional wastewater treatment facility. The treatment processes at this facility consisted of screening, grit removal, primary sedimentation, activated sludge, secondary sedimentation, and nitrification for part of its influent. However, the portion that fed to the demonstration plant was not nitrified.

The potable reuse demonstration plant employed advanced unit processes and operations to achieve the required high constituent removal. The various processes included high-pH lime treatment, sedimentation, recarbonation, filtration, UV irradiation, carbon adsorption, reverse osmosis (RO), air stripping, ozonation, chloramination, and ultrafiltration (UF) (see Chaps. 9, 10, and 11).

Initial treatment at the potable water demonstration plant consisted of aeration, followed by a high-pH lime treatment, and then by addition of ferric chloride to aid the sedimentation process. Following sedimentation, recarbonation was employed to adjust the pH to approximately 7.8. A tri-media filter system followed the chemical treatment step. The filtration system produced effluent with turbidity to 0.5 NTU. Only 380 m^3/d (0.1 Mgal/d) of the influent was treated further with UV irradiation followed by carbon adsorption and membrane treatment. Other elements included reverse osmosis for TDS rejection; air stripping to remove carbon dioxide and volatile organic chemicals; ozonation as the primary disinfectant; chloramination as the residual disinfectant; and an ultrafiltration sidestream system to compare the difference between ultrafiltration and reverse osmosis systems.

Processes Used during the Health Effects Testing Program
The health effects study was conducted with reclaimed water produced using various advanced treatment technologies, including lime treatment, filtration, UV, carbon adsorption, RO (or, alternately, UF), air stripping, ozonation, and chloramination. A schematic treatment process flow diagram used for the health effects study is shown on Fig. 24-4.

The technologies used in the health effects study were found to be effective for the removal of most constituents from water, with some processes serving as redundant systems from some constituents. Particulate matter and turbidity were removed by chemical lime treatment and filtration. Microbial constituents, such as coliphage virus and coliform bacteria, were also reduced by lime treatment, and practically removed completely by UV irradiation, while RO served as a final barrier to microbial constituents. Dissolved organic constituents were reduced by both chemical lime treatment and adsorption on activated carbon. Dissolved gases, such as carbon dioxide and hydrogen sulfide, were removed primarily by air stripping, which also resulted in an increased pH. Ozone was used as the primary disinfectant and to oxidize any remaining organics. Chlorine was added as a disinfectant residual. The UF system performed effectively the same function as the RO system for the pilot system.

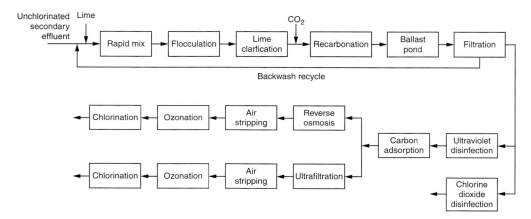

Figure 24-4
Direct potable reuse treatment process used for the Health Effects Testing Program in Denver, CO. (Adapted from Lauer and Rogers, 1998).

Operation and Maintenance

The multiple barrier approach (see also Chap. 23) was implemented in the potable reuse demonstration project as the most effective method to improve the overall process reliability. Redundant and backup systems, put into service when parallel processes were taken off-line for routine maintenance or in the event of process failure, were also used to improve reliability (Work et al., 1980; Lauer and Rogers, 1998). It should be noted that some processes that were not considered to be critical, such as air stripping, did not have a redundant backup.

Water Quality Testing and Studies

A continuous supply and consistent quality of product water was required for the Health Effects Study sampling program for the 2 yr duration of the animal feeding study. To achieve a consistent product water quality for testing, the treatment process (see Fig. 24-4) was operated at steady-state for the duration of the study. The performance of the advanced treatment process was evaluated with a comprehensive sampling and analysis program (Lauer and Work, 1982).

Water Quality Results

Practically every known water constituent of concern at that time was examined in this study and the results are summarized in Table 24-4. Most results were obtained from 24 h composite samples taken with specialized automatic sampling devices from the treatment plant effluent locations. The primary objective was to determine the product water safety; therefore, these sample locations received the most scrutiny. The variability and extent of constituent removal was established based on the plant influent. Constituents were measured following activated carbon adsorption to evaluate the process itself, but also to characterize the additional treatment obtained by UF and RO (Lauer and Rogers, 1998).

Multiple Barrier Approach

The multiple-barrier approach was used to produce a highly reliable process in which no one process is entirely responsible for the removal of a given constituent (see Chap. 23).

Table 24-4
Health effects reuse plant performance[a,b]

Parameter	Unit	Reuse plant influent	Reuse plant product UF permeate	Reuse plant product RO permeate	Denver drinking water
General					
Total alkalinity	mg/L as $CaCO_3$	247	166	3	60
Total hardness	mg/L as $CaCO_3$	203	108	6	107
TSS	mg/L	14.2	d	d	d
TDS	mg/L	583	352	18	174
Specific conductance	µmho/cm	907	648	67	263
pH	unitless	6.9	7.8	6.6	7.8
Turbidity	NTU	9.2	0.2	0.06	0.3
Particle size					
>128 µm	count/50 mL	NA[c]	d[d]	d	d
64–28 µm	count/50 mL	NA	d	d	1
32–64 µm	count/50 mL	NA	18	1.2	18
16–32 µm	count/50 mL	NA	100	58	168
8–16 µm	count/50 mL	NA	448	147	930
4–8 µm	count/50 mL	NA	1290	219	3460
Radiological					
Gross alpha	pCi/L	2.9	<0.1	<0.1	1.3
Gross beta	pCi/L	10.0	5.6	<0.4	2.3
Radium 228	pCi/L	<1	<1	<1	<1
Radium 226	pCi/L	<0.3	<0.3	<0.3	<0.3
Tritium	pCi/L	<100	<100	<100	<100
Radon 222	pCi/L	<20	<20	<20	<20
Plutonium—total	pCi/L	<0.02	<0.02	<0.02	<0.02
Uranium—total	pCi/L	0.004	<0.0006	<0.0006	0.002
Microbiological					
m-HPC	count/mL	1.3×10^6	350[e]	d	3.3
Total coliform	count/100 mL	7.7×10^5	d	d	d
Fecal coliform	count/100 mL	6.3×10^4	d	d	d
Fecal strep	count/100 mL	9.3×10^3	d	d	d
Coliphage B	count/100 mL	1.7×10^4	d	d	d
Coliphage C	count/100 mL	4.8×10^4	d	d	d
Giardia	cysts/L	0.8	d	d	d
Endamoeba coli	cysts/L	0.5	d	d	d
Nematodes	count/L	3.8	d	d	d
Enteric virus	count/L	NA	d	d	d

(Continued)

Table 24-4
Health effects reuse plant performance[a,b] (Continued)

Parameter	Unit	Reuse plant influent	Reuse plant product UF permeate	Reuse plant product RO permeate	Denver drinking water
Microbiological					
Entamoeba histolytica	cysts/L	d	d	d	d
Algae	count/mL	1.1	d	d	1.9
Clostridium perfringens	count/100 mL	8.5×10^3	<0.2	<0.2	<0.2
Shigella	—	Present	Absent	Absent	Absent
Salmonella	—	Present	Absent	Absent	Absent
Campylobacter	—	Present	Absent	Absent	Absent
Legionella	—	Present	Absent	Absent	Absent
Inorganic[f]					
Aluminum	mg/L	0.051	d	d	0.2
Arsenic	mg/L	0.001	d	d	d
Boron	mg/L	0.4	0.3	0.2	0.1
Bromide	mg/L	d	d	d	d
Cadmium	mg/L	d	d	d	d
Calcium	mg/L	77	38	1	27
Chloride	mg/L	98	96	19	22
Chromium	mg/L	0.003	d	d	d
Copper	mg/L	0.024	0.01	0.009	0.006
Cyanide	mg/L	d	d	d	d
Fluoride	mg/L	1.4	0.8	d	0.8
Iron	mg/L	0.025	0.07	0.02	0.03
Potassium	mg/L	13.5	9.1	0.7	2.0
Magnesium	mg/L	12.6	1.8	0.1	7.2
Manganese	mg/L	0.103	d	d	0.008
Mercury	mg/L	0.0001	d	d	d
Molybdenum	mg/L	0.019	0.004	d	0.02
TKN	mg/L	29.5	19	5	0.9
Ammonia-N	mg/L	26.0	19	5	0.6
Nitrate-N	mg/L	0.2	0.3	0.1	0.1
Nitrite-N	mg/L	d	d	d	d
Nickel	mg/L	0.007	d	d	d
Total phosphorus	mg/L	5.6	0.05	0.02	0.01
Selenium	mg/L	d	d	d	d
Silica	mg/L	15.0	8.8	2.0	6.4
Strontium	mg/L	0.44	0.13	d	0.2
Sulfate	mg/L	158	58	1	47
Lead	mg/L	0.002	d	d	d
Uranium	mg/L	0.003	d	d	0.001

(Continued)

Table 24-4
Health effects reuse plant performance[a,b] (Continued)

Parameter	Unit	Reuse plant influent	Reuse plant product UF permeate	Reuse plant product RO permeate	Denver drinking water
Inorganic					
Zinc	mg/L	0.036	0.016	0.006	0.006
Sodium	mg/L	119	78	4.8	18
Lithium	mg/L	0.018	0.014	d	0.007
Titanium	mg/L	0.107	0.035	d	0.005
Barium	mg/L	0.034	d	d	0.04
Silver	mg/L	0.001	d	d	d
Rubidium	mg/L	0.004	0.003	d	0.001
Vanadium	mg/L	0.002	d	d	d
Iodide	mg/L	d	d	d	d
Antimony	mg/L	d	d	d	d
Organic					
Total organic carbon	mg/L	16.5	0.7	d	2.0
Total organic halogens	µg/L	109	23	8	45
Methylene blue active substances	µg/L	400	d	d	d
Total trihalomethanes	µg/L	2.9	d	d	3.9
Methylene chloride	µg/L	17.4	d	d	d
Tetrachloroethene	µg/L	9.6	d	d	d
1,1,1-Trichloroethane	µg/L	2.7	d	d	d
Trichloroethene	µg/L	0.7	d	d	d
1,4-Dichlorobenzene	µg/L	2.1	d	d	d
Formaldehyde	µg/L	d	12.4	d	d
Acetaldehyde	µg/L	9.5	7.2	d	d
Dichloroacetic acid	µg/L	1.0	d	d	3.9
Trichloroacetic acid	µg/L	5.6	d	d	d

[a]From Lauer and Rogers (1998).
[b]Geometric mean values January 9, 1989, to December 20, 1990.
[c]NA = Not analyzed.
[d]d = Below detection limit.
[e]Disinfection considered to be non-optimal at pilot scale.
[f]Additional parameters which were tested but concentrations were below the detection limit or only very limited data are available, include:

Beryllium	Galium	Niobium	Tellurium
Bismuth	Germanium	Osmium	Terbium
Cerium	Gold	Palladium	Thulium
Cesium	Hafnium	Platinum	Tin
Cobalt	Holmium	Praseodymium	Tungsten
Dysprosium	Iridium	Rhodium	Ytterbium
Erbium	Lanthanum	Ruthenium	Yttrium
Europium	Lutetium	Samarium	Zirconium
Gadolinium	Neodymium	Scandium	

The system included redundancies to remove viruses, bacteria, protozoa, metals, inorganics, and organics in reclaimed water. Final effluent from the potable reuse demonstration plant met or exceeded the U.S. EPA Drinking Water Standards (physical, general mineral, microbiological, organic, metals and others) for almost every constituent of concern.

The multiple barrier system was challenge tested using elevated concentrations of several organic compounds. A list of the organic compounds used in this evaluation is presented in Table 24-5. Most of these compounds were removed completely or substantially by lime treatment. The remainder of the compounds was removed by activated carbon. The only exception was chloroform. A very small residual concentration (<1 µg/L) survived reverse osmosis treatment, but was eliminated by air stripping prior to final product water sampling. As with most of these compounds chloroform was added to the treatment plant in an amount 100 times the concentration normally found in the plant influent. The results of the challenge study demonstrated that the multiple-barrier process can remove contaminants to nondetectable levels, even when the given organic compounds are dosed at 100 times the normal concentration (Lauer and Rogers, 1998).

Ultrafiltration/Reverse Osmosis Blended Water Quality

As noted previously, a split process flow diagram, which employed RO and UF, was investigated as part of the health effects study. The advantages of employing split

Table 24-5

Water reuse plant contaminant challenge study: organic compounds cumulative percent removal[a]

Compound	Initial dosage, mg/L	Removal after given treatment phase, percent			Final plant effluent
		Lime	Carbon	Reverse osmosis	
Acetic acid	5054	100	—	—	—
Anisole	23	100	—	—	—
Benzothiazole	86.2	63	100	—	—
Chloroform	229.6	26	99.7	99.9	100
Clofibric acid	17.1	0	100	—	—
Ethyl benzene	25.1	100	—	—	—
Ethyl cinnamate	67.8	100	—	—	—
Methoxychlor	44.6	84	100	—	—
Methylene chloride	230	8	100	—	—
Tributyl phosphate	69.4	51	100	—	—
Gasoline (1st trial)	97.8	100	—	—	—
Gasoline (2nd trial)	2115	—	—	—	—
Toluene	—	25	97	100	—
Benzene	—	40	100	—	—
Ethylbenzene	—	36	100	—	—
Xylene	—	32	100	—	—

[a] From Lauer and Rogers (1998).

treatment were to: (1) eliminate the problems associated with the corrosive water produced from the RO process, if used alone, or alternately to stabilize the product water; (2) achieve lower electricity costs due to the lower operating pressures required by UF; (3) produce various qualities of product water which could be produced and distributed for different beneficial purposes, such as agricultural irrigation; (4) obtain reclaimed water from the activated carbon step that could be used for industrial uses; and (5) produce RO or UF treated water that could be used for special applications requiring higher quality water such as boiler feed water (Lauer and Rogers, 1998).

Various qualities of water could be achieved by blending the product water from the UF and RO processes. The quality of a 50/50 blend of UF and RO water is reported in Table 24-6, along with the quality of the Denver drinking water and the applicable regulatory standards. As reported in Table 24-6, the characteristics of the blended water would be similar to the existing Denver drinking water. The alkalinity, TDS, and specific conductance would be similar and the hardness would be about one half the current value. The microbiological quality of the blended water would meet all regulatory requirements, as disinfection with chlorine would be used in the full-scale facility (Lauer and Rogers, 1998).

According to Rogers (1989), none of the regulated inorganic substances was present in the blended water at concentrations of regulatory significance. The concentrations of calcium and magnesium were lower while the concentrations of sodium and potassium were higher than corresponding values in the Denver drinking water. Higher concentration values, e.g., for nontoxic elements, were well within the WHO guidelines. Although the concentrations for ammonium nitrogen in the blended water were reported to be around 10 mg-N/L, in a full-scale process implementation a biological nitrification step would be used. Thus, it was assumed that the concentration of ammonium nitrogen would be reduced to below the detection limit.

Based on an assessment of the extensive measurements, it was concluded that the concentration of organic constituents in the blended water would be lower as compared to the Denver drinking water. Formaldehyde, formed in low concentrations during ozonation and detectable in the product water, would be near the detection limit in the blended water. While the concentration of formaldehyde at the detection limit would not pose a health concern, the concentration could be lowered further by air stripping after ozonation instead of before air stripping. Thus, while the blended effluent would contain a few inorganic constituents at higher concentrations, the concentration of organic constituents would be lower than the corresponding concentration in the Denver drinking water. Further, the blended water supply would be less corrosive and lower in cost as compared to the use of product water from the reverse osmosis process alone (Lauer and Rogers, 1998).

Summary of Water Quality Studies

As part of the 2 yr study, a comprehensive testing program was initiated to characterize the quality of the unchlorinated effluent from the demonstration plant. To verify plant

Table 24-6 UF and RO blended water quality comparison: mean values for selected parameters[a]

Parameter	Unit	Reuse UF and RO	Denver drinking water	Regulatory standard
General parameters				
Total alkalinity	mg/L as CaCO$_3$	78	64	
Total hardness	mg/L as CaCO$_3$	53	107	
Total dissolved solids	mg/L	180	183	
Specific conductance	mmhos/cm	361	294	
Radiological parameters				
Gross alpha	pCi/L	<0.1	1.3	15
Gross beta	pCi/L	2.8	2.3	50
Microbiological parameters				
m-HPC	count/mL	91	2.8	
Total coliform	count/100 mL	<0.2	<0.2	1
Inorganic parameters				
Aluminum	mg/L	0.011	0.144	
Calcium	mg/L	16.7	25.9	
Chloride	mg/L	55	25	
Fluoride	mg/L	0.4	0.7	4
Iron	mg/L	0.034	0.028	
Magnesium	mg/L	0.9	7.9	
Nitrogen-ammonium	mg/L as N	10.6	0.6	
Potassium	mg/L	4.5	2.0	
Silica	mg/L	5.2	6.1	
Sodium	mg/L	42	19	
Sulfate	mg/L	30	47	
Organic parameters				
Total organic carbon	mg/L	0.7	2.1	
Total organic halide	mg/L	0.015	0.046	
Trihalomethanes	μg/L	<0.5	3.9	100
Formaldehyde	μg/L	6	<5	
Dichloroacetic acid	mg/L	<0.1	3.9	

[a] From Lauer and Rogers (1998).

operation, more than 10,000 samples were analyzed to assess the myriad number of constituents that may be found in water. Regular sampling was used to characterize the relative variability in plant influent and response of the demonstration plant (Lauer and Rogers, 1998).

The reclaimed water from the UF and RO membrane systems satisfied all existing and proposed U.S. EPA, WHO, and EEC standards. Additionally, these product waters compared favorably with Denver's existing drinking water. A 50/50 blend of the two reclaimed waters would also satisfy all standards and provide less corrosive water at a lower cost than reverse osmosis alone. The blended supply would also compare favorably with Denver drinking water. Lower concentrations of calcium, magnesium, and sulfate would be offset by higher (but not in excess of any standard) concentrations of sodium, potassium and chloride making this possible supply relatively soft and quite desirable.

By conducting whole animal health effects studies concurrently with the water quality evaluations, it was possible to establish the physical, chemical, and microbiological characteristics of the product water with far more completeness and for a longer duration than any study of drinking water to date. Based on the results of the testing program, the reclaimed water produced at the potable water demonstration facility was found to meet all health standards and was of equal or better quality than the city drinking water (Lauer and Rogers, 1998).

Animal Health Effects Testing

Animal health testing was conducted to determine the potential health effects of consuming reclaimed water. The studies, adapted from Lauer and Rogers (1998), involved testing with both rats and mice over a 2 yr period. The results of the testing program are discussed below.

Whole Animal Health Effect Testing
As discussed in Chap. 5, it is difficult to apply the results of dose response relationships based on animal feeding studies to an equivalent risk for human for even an individual compound. Animal feeding studies are typically based on exposing an animal to concentration of a compound that are orders or magnitude higher than the anticipated dose that an average human would be exposed to, such as through a lifetime of drinking water ingestion. The presence of multiple compounds adds an additional level of complexity, e.g., synergistic effects. Given these considerations and in response to the position of the National Academy of Sciences (NAS) panel on Quality Criteria for Water Reuse, which stated that the evaluation of potential adverse health effects from reclaimed water should be based on the results from chronic toxicity studies in whole animals (NAS, 1982), the Health Effects Studies Panel recommended animal feeding studies using concentrated constituents present in the demonstration plant product water (Lauer et al., 1990).

Two-Year Chronic Toxicity and Carcinogenicity Rat Study
The objective of the chronic toxicity and carcinogenicity study was to evaluate potential adverse effects on growth and development and potential carcinogenic effects during a two-year (104 week) study.

The rat study tests were conducted with water in which the constituents in RO product water, UF product water, and Denver drinking water were concentrated using assorted adsorbent resins to 500 times the respective concentration. The concentrated samples, fed to F344 rats, did not cause any significant toxicological or carcinogenic effects compared to the control group. Survival ranged from 52 to 70 and 64 to 84 percent in male and female rats, respectively, which were within the ranges normally observed (Lauer and Rogers, 1998).

Two-Year Chronic Toxicity/Carcinogenicity Mouse Study
As with the rat study, a 2 yr study of the health effects associated with ingestion of reclaimed water was also conducted with $B_6C_3F_1$ mice. In the study, constituents in product water from the RO process were concentrated to 500 times the original concentration. No significant differences in toxicity or carcinogenicity were obtained compared to the control group after 104 weeks of feeding, with survival rates ranging from 52 to 70 percent and 64 to 84 percent, for male and female mice, respectively. Further, additional studies were conducted during and at the conclusion of the study to evaluate the clinical, gross, and microscopic pathology. No effects were identified that could be associated with the experimental treatment (Lauer and Rogers, 1998).

Reproductive Toxicity Study
The objective of the reproductive toxicity study was to identify potential adverse effects on reproductive performance, intra-uterine development, and growth and development of the offspring during a two-generation study. A teratology phase was included to identify potential embryo toxicity and teratogenicity of the test article. Fifty male and fifty female Sprague-Dawley rats per 500 times the concentration of the water sample were used for the F_0 generation. The animals were 12 to 15 weeks of age at start of study. The animals were randomly selected to each test group by computer-generated randomization procedures.

Based on the result of the multi-generation reproductive study, no treatment-related effects were observed on the reproductive performance, growth, mating capacity, survival of the offspring or fetal development in any of the treatment groups. Clinical signs or gross tissue alterations, attributed to any of the dose water exposures, were not noted at necropsy in either parental generation. No treatment-related histopathologic findings in parental animals of either generation were observed (Lauer and Rogers, 1998).

Cost Estimates on the Potable Reuse Advanced Treatment Plant

The feasibility of implementing direct potable water reuse depends on several factors: product water safety, regulatory agency approval, public acceptance, and cost. The financial viability of this proposed direct potable reuse scheme depends on the cost to convert secondary treated effluent to potable water quality and the availability and relative cost of alternate traditional water supply development projects. Although accurate estimates from related projects suffer from numerous deficiencies, they can be used to determine relative merit and to support decisions regarding the advisability of proceeding with further studies (see Chap. 25). The pilot plant study was used to establish that the cost of advanced treatment was within acceptable limits; thus, the health effects study proceeded. The following cost estimates utilize the knowledge gained from more than two years of continuous operation of the water reuse demonstration plant used for the Health Effects Testing Program. Also these cost figures have been revised to reflect January 1994 conditions.

U.S. Army Corps of Engineers' Study on Denver Metropolitan Water Supply Alternatives

In 1988, an environmental impact study was completed by the U.S. Army Corps of Engineers on Metropolitan Denver's water supply alternatives (U.S Army Corps of Engineers, 1988). Among the items evaluated with respect to augmentation of metropolitan Denver's water supply was the cost of alternative water supply projects that could be used to meet Denver's projected water demand. Projects were identified that may be needed in less than 20 yr and in 20 to 50 yr in the future. Costs for direct potable reuse were compared to the cost for alternative projects, even though direct potable reuse would not likely be required in the near future. In the final environmental impact study (EIS), long-term project costs ranged from $0.2 to $0.8/m^3 ($250 to $960/ac-ft).

Cost Estimates for a Full-Scale Water Reuse Facility

Based on engineering cost estimates and actual operational data from the demonstration plant, cost estimates for a full-scale water reuse facility were estimated to be in the range from $0.4 to $0.6/m^3 ($534 to $762/ac-ft) (Lauer and Rogers, 1998). The estimated values were similar to those projected for other future water supply augmentation projects and were presented in detail in the final EIS. Thus, cost was not found to be a limiting factor for the implementation of direct potable reuse in Denver.

The amortized capital costs and the anticipated operating costs, derived from the operation experience from the demonstration plant, are presented in Table 24-7 for the construction of the 380 × 10^3 m^3/d (100 mgd) plant. Using the higher operation and maintenance cost estimates based on the demonstration plant along with the capital cost estimates, a conservative value of $0.6/m^3 ($2.34/1000 gal or $762/ac-ft) was projected. Thus, based on the operation of the demonstration plant experience, the cost of a full-scale reclamation facility for production of water for direct potable reuse of $0.4 to $0.6/m^3 ($534 to $762/ac-ft) would be comparable to the City of Denver's other projected water supply augmentation options of $0.2 to $0.8/m^3 ($250 to $960/ac-ft) (Lauer and Rogers, 1998).

Public Information Program

Public awareness of direct potable reuse and the role it might play in meeting Denver's water supply needs was an important objective of the Direct Potable Reuse Demonstration Project.

The public information program was implemented as part of the demonstration project because it was reasoned that a more knowledgeable public would be more receptive in the event that the direct potable reuse program became a reality. A multi-media approach was developed that focused on escorted plant tours with additional information provided in

Table 24-7

Cost estimate for a 380 × 10³ m³/d (100 Mgal/d) reverse osmosis plant[a]

Process	Amortized capital, $/m³	O&M costs, $/m³	Total costs,[b] $/m³
Biological nitrogen removal	0.02	0.01	0.04
High pH lime clarification (including sludge disposal)	0.03	0.08[c]	0.11
Sludge disposal	0.01	0.03	0.04
Filtration	0.01	0.01[c]	0.01
Activated carbon contact and regeneration (including regeneration and replacement)	0.02	0.05[c]	0.07
Reverse osmosis (including brine disposal)	0.12	0.19[c]	0.31
Ozonation	0.002	0.02[c]	0.02
Chloramination	0.0005	0.0008[c]	0.001
Miscellaneous plant	0.002	0.004	0.01
Total[d]			0.6

[a] From Lauer and Rogers (1998).
[b] Some total costs are not additive due to rounding.
[c] O&M cost derived from the Water Reuse Demonstration Plant.
[d] Total treatment cost estimate = $0.6/m³ = $2.34/1000 gal = $762/ac-ft, using demonstration plant operational values; January 1991 cost.

the form of an audio visual presentation and a full color brochure (Lohman, 1988). More than 7,000 visitors, representing more than 40 countries, participated in the educational program.

Other forms of information transfer were also used, including (1) the publication of a newsletter, (2) newspaper and television reports, and (3) a 26 min documentary video ("Pure Water ... Again"). The purpose of the information transfer program was to inform the public about various project milestones and to provide broad education on the general subject of potable water reuse. (Carley, 1972; Lauer and Rogers, 1998).

Lessons Learned

Conducted over a 13 yr period, the results of the demonstration project were used to assess the feasibility of producing potable water from municipal wastewater using advanced treatment technologies. The effectiveness of various advanced treatment processes for the removal of constituents from wastewater was demonstrated conclusively in the 3.8 × 10³ m³/d (1 Mgal/d) pilot-scale facility. Selected project findings and lessons learned are as follows (Lauer and Rogers, 1998):

1. Using advanced treatment processes, it was possible to produce drinking water from municipal wastewater that met all of the then current and proposed U.S. EPA drinking water standards.

2. Toxicity and carcinogenicity studies were conducted over a 2 yr period to evaluate the safety of the reclaimed water. No adverse health effects were detected from lifetime exposure to any of the reclaimed water samples.

3. Reproductive studies were conducted on the reclaimed water. No adverse health effects were detected during a two-generation reproduction study.

4. From comprehensive physical, chemical and microbiological testing, the product water was found to be of extremely high quality, comparable to the existing City of Denver potable water supply.

5. While the uncertainties related to the toxicological and epidemiological testing that were conducted need to be recognized, it should be noted that similar limitations exist for the evaluation of the safety of conventional municipal water supplies.

6. Based on the extensive public education program, a majority of the public was found to be supportive of potable water reuse, if needed and the safety was assured.

24-5 OBSERVATIONS ON DIRECT POTABLE REUSE

At present, there is no imperative for the use of reclaimed water for direct potable reuse in the United States: "Direct use of reclaimed water for human consumption, without the added protection provided by storage in the environment, is not currently a viable option for public water supplies" (NRC, 1998). Nonetheless, direct potable reuse could well be a cost-effective form of water reuse in the long-term. While treatment requirements are clearly greater and public acceptance could be a major obstacle, direct potable reuse would have an advantage of avoiding the unnecessary cost of duplicate water distribution and storage systems. Further, direct potable reuse has the potential to readily utilize all the reclaimed water that could be generated and avoid altogether the need to discharge excess flow to the environment. The pressure to consider reclaimed water as a source of a potable supply must increase in the future as it seems inevitable that, in time, potable reuse in some form will occur (Department of Health and Aged Care, 2001; Law, 2003).

PROBLEMS AND DISCUSSION TOPICS

Some of the following discussion topics are adapted from the Water Recycling Discussion Group website: owner-water-recycling@lists.dnr.qld.gov.au, moderated by the Queensland Water Recycling Strategy project team in Australia.

24-1 Do you believe that water agencies are sufficiently technologically competent to consistently produce reclaimed water for direct drinking water use such that it poses no additional health risk over traditional drinking water sources? State the reasons for your answer.

24-2 Discuss and summarize multiple barrier systems used in direct potable reuse with respect to (1) wastewater treatment barriers, (2) nontreatment related barriers, and (3) operation related barriers.

24-3 The ability to analyze the chemical characteristics of wastewater has exceeded the corresponding ability to assess the toxicology of the chemicals that are measured. Given that, would it not be prudent to wait before facilitating the expansion of another as yet, unknown long incubation period disease as a result of reuse? State the reasons for your answer.

24-4 Direct potable reuse avoids all the reclaimed water distribution obstacles associated with dual pipes for nonpotable reuse. Thus, would it not be better to pursue the potable reuse option, as opposed to the use of a dual pipe system? Discuss pros and cons of this augment.

24-5 What health risk assessment has been undertaken for desalinated seawater for drinking water use given that many wastewater treatment plants discharge into estuaries or near shore waters? Are there similar, but less obvious risks with using "pure" seawater as the feedstock?

24-6 Discuss the findings of studies that have been undertaken for traditional drinking water sources that are strongly influenced by municipal treated wastewater discharges from upstream cities and towns. Cite a minimum of three references.

REFERENCES

Carley, R. L. (1972) "Attitudes and Perceptions of Denver Residents Concerning Reuse of Wastewater," *Proceedings of the 92nd AWWA Annual Conference*, AWWA, Denver, CO.

CH2M Hill, Inc. (1975) "Conceptual Design Report—Potable Water Reuse Plant—Successive Use Program," Report to the Denver Board of Water Commissioners, Denver, CO.

City of San Diego (2006) *City of San Diego Water Reuse Study*, Final Draft Report 2006, prepared in coordination with City of San Diego Water Department, PBS & J, and McGuire/Pirnie, San Diego, CA.

City of San Diego (2006a) Public Participation http://www.sandiego.gov/water/waterreuse-study/involvement/index.shtml#assembly

Dean, R. B., and E. Lund (1981) *Water Reuse: Problems and Solutions*, Academic Press, Inc., London, UK.

Department of Health and Aged Care (2001) *Review of Health Issues Associated with Potable Reuse of Wastewater (RFT200/00, Final Report)*, produced by Gutteridge Haskins & Davey Pth Ltd., Brisbane QLD, Australia.

du Pisani, P. L. (2005) "Direct Reclamation of Potable Water at Windhoek's Goreangab Reclamation Plant," 193-202, in S. J. Khan, A. I. Schäfer, M. H. Muston (eds.) *Integrated Concepts in Water Recycling*, University of Wollongong, NSW, Australia.

Harrhoff, J., and B. Van der Merwe (1996) "Twenty-five Years of Wastewater Reclamation in Windhoek, Namibia," *Water Sci. Technol.,* **33**, 10–11, 25–35.

Kasperson, R. E., and J. X. Kasperson (ed) (1977) *Water Re-Use & the Cities*, The University Press of New England, Hanover, NH.

Lahnsteiner, J., and G. Lempert (2005) "Water Management in Windhoek/Namibia," *Proceedings of the IWA Specialty Conference, Wastewater Reclamation & Reuse for Sustainability*, November 8–11, Jeju, Korea.

Law, I. B. (2003) "Advanced Reuse—From Windhoek to Singapore and Beyond," *Water,* **5**, 44–50.

Lauer, W. C., and S. W. Work (1982) "Denver's Analytical Studies Program," *Proceedings of the Water Reuse Symposium II*, AWWA Research Foundation, Denver, CO.

Lauer, W. C., and S. E. Rogers (1998) "The Demonstration of Direct Potable Water Reuse: Denver's Landmark Project," Chap. 28, in T. Asano (ed.) *Wastewater Reclamation and Reuse,* **10**, Water Quality Management Library, CRC Press, Boca Raton, FL.

Lauer, W. C., F. J. Johns, G. W. Wolfe, B. A. Myers, L. W. Condie, and J. F. Borzelleca (1990) "Comprehensive Health Effects Testing Program for Denver's Potable Water Reuse Demonstration Project," *J. Toxicol. Environ. Health*, **30**, 4, 305–321.

Linstedt, K. D., and E. R. Bennett (1973) "Evaluation of Treatment for Urban Wastewater Reuse," EPA-R2-73-122, U.S. EPA, Cincinnati, OH.

Lohman, L. C., and J. G. Milliken (1985) *Informational/Educational Approaches to Public Attitudes on Potable Reuse of Wastewater*, Denver Research Institute, Denver, CO.

Lohman, L. C. (1988) "Potable Water Reuse Can Win Public Support," *Proceedings of the Water Reuse Symposium IV*, AWWA Research Foundation, Denver, CO.

Metzler, D. F., R. L. Culp, H. A. Stoltenberg, R. L. Woodward, G. Walton, S. L. Chang, N. A. Clarke, C. M. Palmer, and F. M. Middleton (1958) "Emergency Use of Reclaimed Water for Potable Supply at Chanute, Kansas," *J. AWWA*, **50**, 8, 1021–1060.

NAS (1982) *Quality Criteria for Water Reuse*, National Academy of Sciences, National Academy Press, Washington, DC.

NRC (1998) *Issues in Potable Reuse: The Viability of Augmenting Drinking Water Supplies with Reclaimed Water*, National Research Council, National Academy Press, Washington, DC.

Odendaal, P. E., J. L. J. van der Westhuizen, and G. J. Grobler (1998) "Wastewater Reuse in South Africa," Chap. 25, in T. Asano (ed.) *Wastewater Reclamation and Reuse*, **10**, Water Quality Management Library, CRC Press, Boca Raton, FL.

Queensland Government, Australia (2005), accessed online: http://www.epa.qld.gov.au/environmental_management/water/water_recycling_strategy/

Rogers, S. E. (1989) "Biofilm Nitrogen Removal for Potable Water Reuse," Masters Thesis, Department of Civil, Environmental, and Architectural Engineering, University of Colorado, Boulder, CO.

Rook, J. (1974) "Formation of Haloforms During Chlorination of Natural Waters," *Water Treat. Exam.*, **23**, 2, 234–243.

Rothberg, M. R., S. W. Work, K. D. Linstedt, and E. R. Bennett (1979) "Demonstration of Potable Water Reuse Technology, The Denver Project, *Proceedings of the Water Reuse Symposium I*, AWWA Research Foundation, Denver, CO.

Snyder, S., R. Pleus, E. Snyder, J. Hemming, and G. Bruce (2006) "Toxicological Significance of Trace Endocrine Disruptors and Pharmaceuticals in Water," presented at the WateReuse Symposium 2006, San Diego, CA.

U.S. Army Corps of Engineers (1988) "Metropolitan Denver Water Supply Environmental Impact Statement," U. S. Army Corps of Engineers, Omaha District.

Work, S. W., and N. Hobbs (1976) "Management Goals and Successive Water Use," *J. AWWA*, **68**, 2, 86–92.

Work, S. W., M. R. Rothberg, and K. J. Miller (1980) "Denver's Potable Reuse Project: Pathway to Public Acceptance," *J. AWWA,* **72**, 8, 435–440.

Part 5

IMPLEMENTING WATER REUSE

Producing reclaimed water of a specified quality to fulfill multiple water use objectives is now a reality due to the progressive evolution of water reclamation technologies, regulations, and environmental and public health protection. The incentives for a water reuse program make perfect sense to technical experts—a new water source, water conservation, economic advantages, environmental benefits, government support, and the realization that the cost of wastewater treatment makes the product too valuable to "throw away" or dispose. So why has water reuse not been embraced and supported wholeheartedly by the public? An ultimate decision to promote water reclamation and reuse is dependent on necessity and opportunity in terms of economic, regulatory, public policy, and, more importantly, public acceptance, factors reflecting the water demand, safety, and need for a reliable water supply to meet local conditions.

In Part 5 the focus is on planning and implementation for water reuse. Integrated water resources planning including reclaimed water market assessment, and economic and financial analyses is presented in Chap. 25. As technology continues to advance and cost effectiveness and the reliability of water reuse systems are more widely recognized, implementation of water reclamation and reuse plans and facilities will continue to expand as an essential element in sustainable water resources management. Implementation issues in water reclamation and reuse including soliciting and responding to community concerns, gaining public support through educational programs, and avoiding pitfalls that may cause a project to fail are discussed in Chap. 26.

25 Planning for Water Reclamation and Reuse

WORKING TERMINOLOGY 1382

25-1 INTEGRATED WATER RESOURCES PLANNING 1384
  Integrated Water Resources Planning Process 1385
  Clarifying the Problem 1386
  Formulating Objectives 1386
  Gathering Background Information 1386
  Identifying Project Alternatives 1388
  Evaluating and Ranking Alternatives 1389
  Developing Implementation Plans 1389

25-2 ENGINEERING ISSUES IN WATER RECLAMATION AND REUSE PLANNING 1392

25-3 ENVIRONMENTAL ASSESSMENT AND PUBLIC PARTICIPATION 1392
  Environmental Assessment 1393
  Public Participation and Outreach 1393

25-4 LEGAL AND INSTITUTIONAL ASPECTS OF WATER REUSE 1393
  Water Rights Law 1393
  Water Rights and Water Reuse 1395
  Policies and Regulations 1397
  Institutional Coordination 1397

25-5 CASE STUDY: INSTITUTIONAL ARRANGEMENTS AT THE WALNUT VALLEY WATER DISTRICT, CALIFORNIA 1397
  Water Management Issues 1397
  Lessons Learned 1398

25-6 RECLAIMED WATER MARKET ASSESSMENT 1399
  Steps in Data Collection and Analysis 1399
  Comparison of Water Sources 1399
  Comparison with Costs and Revenues 1401
  Market Assurances 1402

25-7 FACTORS AFFECTING MONETARY EVALUATION OF WATER RECLAMATION AND REUSE 1406
  Common Weaknesses in Water Reclamation and Reuse Planning 1407
  Perspectives in Project Analysis 1408
  Planning and Design Time Horizons 1408
  Time Value of Money 1409
  Inflation and Cost Indices 1409

25-8 ECONOMIC ANALYSIS FOR WATER REUSE 1411
 Comparison of Alternatives by Present Worth Analysis 1412
 Measurement of Costs and Inflation 1412
 Measurement of Benefits 1412
 Basic Assumptions of Economic Analyses 1414
 Replacement Costs and Salvage Values 1415
 Computation of Economic Cost 1417
 Project Optimization 1420
 Influence of Subsidies 1421

25-9 FINANCIAL ANALYSIS 1422
 Construction Financing Plans and Revenue Programs 1422
 Cost Allocation 1423
 Influence on Freshwater Rates 1423
 Other Financial Analysis Considerations 1423
 Sources of Revenue and Pricing of Reclaimed Water 1424
 Financial Feasibility Analysis 1425
 Sensitivity Analysis and Conservative Assumptions 1429

PROBLEMS AND DISCUSSION TOPICS 1430

REFERENCES 1432

WORKING TERMINOLOGY

Term	Definition
Actual costs	Costs experienced in the marketplace reflecting inflation.
Appropriative rights	The doctrine of water law that the right of water use is associated with the act of intentionally using water and that the priority of right is based on the principle of first in time is first in right.
Capital cost	The initial cost to construct a project from the inception of planning to completion of construction. While some references confine capital costs to construction of the physical facilities, in this textbook acquisition of land or right-of-way is included in capital cost.
Constant dollars	Costs of materials or services at different points in time adjusted to a common reference point in time, usually through the use of cost indexes.
Cost-effectiveness analysis	An analysis to determine which project alternative will result in the minimum total resources cost over time to meet project objectives. The cost-effective alternative is the project alternative that has the lowest net economic cost (present worth or equivalent annual value) while meeting project objectives, unless nonmonetary costs are overriding.
Debt service	The amount of money required to repay borrowed funds, including the borrowed principal and interest, usually repaid on an installment basis, such as monthly or annually.
Demand management	Management of current and future demands for water, including reclaimed water, by altering net consumption or timing or place of use.
Design life	The period in which the use of a component of facilities to be constructed is expected to reach design capacity.

Discount rate	The interest rate used in computing the present value of future cash flow payments in economic analyses.
Economic analysis	A procedure to determine the total monetary costs and benefits of all resources committed to a project for the purpose of determining whether a project should be built and which alternative has the greatest net benefit.
Externality	Positive or negative impact that results from an action and is experienced outside of the entity performing the action.
Facilities plan	A plan for facilities construction, including presentation of the need or problem being addressed, the feasibility analyses for the alternatives considered, and a detailed implementation plan for the recommended project, incorporating or summarizing results of associated studies such as environmental impact assessments, feasibility reports, market assessment reports, and financial feasibility reports.
Feasibility criteria	The factors used to evaluate the feasibility of a project, including technical, economic, and other factors.
Financial analysis	A monetary analysis used to assess whether a project is feasible financially from the perspectives of all project participants, and, in some cases, external parties that may experience financial effects of a project.
Financing period	The period for meeting debt obligations or required paybacks for undertaking a project. This period may be shorter or longer than the planning period.
Fresh water	Water with low total dissolved solids as compared to seawater, derived from precipitation of atmospheric water vapor, found in surface water bodies and groundwater used as a potable water supply. Only three percent of the water on earth is fresh water.
Inflation	Rise in general price level of goods and services.
Intangible costs or benefits	Costs or benefits that cannot be expressed readily in monetary terms.
Integrated water resources planning (IWRP)	A systematic decision-making process that is used to determine the optimal approach for water resource management, including reclaimed water.
Life cycle cost	Capital and operational costs over the life span of a facility, often expressed as total present worth cost or a unit cost.
Mandatory use ordinance	A rule or law of local governmental bodies, such as water districts, mandating the use of reclaimed water in lieu of alternative water supplies, usually potable water.
Marginal cost	The increment of additional capital and operational costs to provide an additional increment of project output.
Market feasibility	The process of assessing the potential market, i.e., potential uses, use sites, and users, for reclaimed water including issues such as public health, water quality, and user and public acceptance.
National environmental policy act of 1969 (NEPA)	Federal act of the United States, which mandates the assessment and mitigation of environmental impacts caused by federal projects or projects within federal jurisdiction, such as federal funding or regulatory authority.
Participatory planning	Planning based on the principle that those who are affected by decisions or policies should participate or be represented in the policy-making processes.
Planning period	The total period for which facility needs will be assessed and alternatives will be evaluated for cost-effectiveness and long-term implementation.
Present worth analysis	A method of analysis used to compare alternative plans with different cash flows to determine which alternative has the lowest net cost over time.

Prior appropriation	The doctrine which serves as the basis for water law in most western states, based on the principle of first in time, first in right.
Real costs	Costs adjusted to constant dollars, excluding inflation, to reflect actual investment of resources and labor in the production of goods and services.
Retrofit	The conversion of a site from the use of freshwater or potable water to reclaimed water.
Riparian rights	The doctrine of water law that the right to water use is tied to riparian land in contact with the water source.
Study area	An area delineated initially to encompass the geographic scope of the problem being addressed.
Sunk costs	Costs already incurred that cannot be recovered, regardless of future events.
Tangible costs or benefits	Costs or benefits that can be expressed in monetary terms.
Useful life	The estimated period of time during which a facility or component of a facility will be operated before replacement or abandonment.
Water rights law	A legal entitlement allowing the diversion of water from a specified source to be put to beneficial use.

The objectives of this chapter, which deals with the planning and analysis of water reuse projects, are to provide a framework for: (1) water reuse planning and (2) the discussion of issues that will assist in identifying and resolving most of the potential problems associated with planning a water reclamation and reuse project. More specifically, the purpose of the chapter is to: (1) introduce integrated water resources planning, (2) discuss briefly environmental assessment and public participation, (3) introduce legal and institutional aspects of water reclamation and reuse planning, (4) present methods for assessing a reclaimed water market, and (5) describe the elements of economic and financial analyses for water reclamation and reuse.

25-1 INTEGRATED WATER RESOURCES PLANNING

Water systems in the United States historically have been managed in a fragmented manner, reflecting the different geographic and functional scope of local, regional, state, and federal agencies responsible for water supply planning and development. Implementation of water reuse is complicated further by the usual division of responsibilities for water supply and wastewater management into separate agencies. Water reclamation and reuse defies categorization as *either* water supply *or* wastewater management. As described in Chap. 1, water supply and wastewater disposal are interconnected within the hydrologic cycle and affect the quantity and quality of water available to meet societal and ecological needs. In recognition of the integrated nature of water systems and the need for greater sustainability, water agencies are increasingly adopting a whole-system planning model called *integrated water resources planning* (IWRP). In the IWRP process, water reclamation and reuse strategies are evaluated alongside other options for water supply, water demand management (including water conservation), wastewater treatment and disposal, and environmental restoration. The purpose of IWRP is to determine whether any, or all, of these strategies have a place within regional water system management, and how best to implement the selected options.

Integrated Water Resources Planning Process

Integrated water resources planning is an iterative process, cycling through the basic steps of: (1) identifying and clarifying the problem, (2) formulating objectives, (3) gathering background information, (4) identifying alternatives, (5) evaluating and ranking alternatives, and (6) selecting an alternative for implementation. A simplified flow chart of IWRP is illustrated on Fig. 25-1. Ideally, IWRP would be conducted by addressing both water supply and wastewater management problems together and developing an integrated solution. If water reuse is part of the recommended alternative, the next planning iteration would follow each of the IWRP steps, focusing more specifically on water reclamation and reuse feasibility and alternatives. In practice, water reuse is often pursued outside of the context of IWRP; however, IWRP principles can be incorporated into the water reuse planning process (see Fig. 25-1). The initial problem identification addresses the water supply or wastewater management objectives, and alternatives not involving water reclamation and reuse are still addressed, but in a less comprehensive manner than an integrated water resources plan would. Water reuse cannot be justified adequately and evaluated without understanding the broader water resources issues and alternatives under consideration (Beecher, 1995).

Water reuse options are often evaluated in three phases, with subsequent phases reiterating the steps shown on Fig. 25-1. The conceptual or reconnaissance phase involves defining the problem and scoping out potential alternatives for evaluation. The feasibility phase involves a thorough analysis of well-defined alternatives through each of the planning steps. Finally, a recommended alternative will be refined with preliminary design and implementation presented in a facilities plan, incorporating results of all of the phases.

Decision-making in planning is only as good as the foundation upon which it is built. Poor decisions can result from failure to state the problem clearly and accurately, define project objectives, delineate the planning area, or gather adequate background information. Poor decisions can also result from a failure in the early stages of the planning

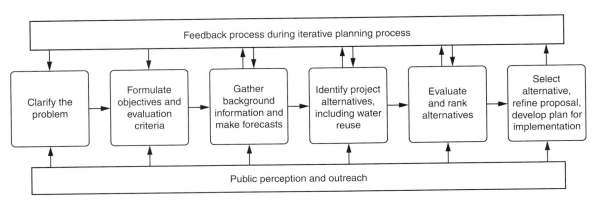

Figure 25-1

Flow diagram of the integrated water resources planning process.

process to assess the public perceptions of needs, objectives, and risks (see Chap. 26). These early steps are often overlooked, but are essential elements of the planning process.

Clarifying the Problem

Water reuse cannot be justified as an end in itself. It is a response or solution to a defined problem or specified need or desire. The stated problem or need is the benchmark for the planning steps. By clarifying the problem a sound foundation is established for identifying measurable objectives and for creating a wide range of alternatives to consider.

Formulating Objectives

To establish project objectives, planners and participating stakeholders begin by creating a list of general concerns. Objectives are drawn from the list and defined more succinctly, often in short, direct phrases such as "minimize economic costs," or "mitigate environmental damage." It is important to include all relevant objectives as part of the decision, not just those that are focused on technical and economic criteria. For example, coordination with programs of other agencies could be an important objective that might be overlooked if not made explicit in the objectives list. The list of objectives is further refined by distinguishing between objectives that are means to an end and those that are ends in themselves. For example, improving water quality is a *means objective* that leads toward the *fundamental objective* of protecting public health. A common mistake in water reuse planning is specifying water reuse as an end in itself, rather than as one way to achieve fundamental objectives such as providing water to meet societal needs (Gregory and Keeney, 2002; Anderson, 2003).

Water reclamation and reuse has the unique potential to meet several water management objectives at once. Water reuse is a means to meet more fundamental objectives, mainly: (1) a reliable water supply, (2) public health protection, (3) environmental protection and restoration, and (4) regional economic development (especially in developing countries). Secondary objectives stemming from these fundamental objectives include: (1) developing a cost-effective means for wastewater treatment and disposal (usually to meet regulatory requirements for achieving public health and environmental protection) and (2) enhancing crop productivity where reclaimed water is used for agricultural irrigation.

Gathering Background Information

Background information needed for water planning includes demographic trends, historic water use, economic indicators, climatic and hydrologic data, current and planned water conservation programs, and conditions and capacities of existing water and wastewater facilities. Background information and forecasts of future conditions are important for clearly understanding the problem to be addressed and for identifying possible solutions. Information on existing infrastructure is necessary for developing delivery alternatives and assessing reclaimed water markets (see Sec. 25-6).

A water demand forecast is needed to establish how much water the planning area will need in the future. Forecasts are performed for normal, dry-year, critical dry-year, and wet-year conditions, as well as for a range of growth projections (AWWA, 2001). Uncertainties in supply and demand forecasts, produced by the range of potential hydrologic and development scenarios, may provide a basis for constructing facilities in phases that correspond to actual conditions. General background information needed for water planning is summarized in Table 25-1. Information required for IWRP may

Table 25-1
Examples of data that may be required in integrated water resources management[a]

Data category	Examples

Demographic	Population
	    Size
	    Distribution
	    Growth rate

Economic	Income levels
	Economic structure
	    Manufacturing
	    Industry
	    Commercial
	    Agriculture

Land use	Residential Industrial Commercial Manufacturing Agricultural
Cultural/historical	Archeological sites Sites of historical interest

Geophysical	Climatological
	Hydrological
	    Surface water (quantity and quality)
	    Groundwater (quantity and quality)
	Geological
	    Rock types and structure
	    Seismic risk

Biological	Ecosystem type and structure Threatened or endangered species Wetlands

Infrastructural	Water and wastewater systems
	    Present facilities
	    Existing flows/flow variations
	    Treatment processes
	    Capacities of treatment plants, transmission and storage facilities
	Plans for new facilities

Legal and Institutional	Water and wastewater entities; e.g., wholesalers, retailers, utilities Regulatory agencies Stakeholder groups Water rights

[a] Adapted from Thompson (1999), Office of Water Recycling (1997).

seem outside the scope of water reuse planning, but it is essential to evaluate water reuse according to the criteria for meeting fundamental objectives, such as increasing water supply or managing wastewater responsibly.

Delineation of Project Study Area

The project study area is delineated initially to encompass the geographic scope of the problem being addressed. However, the project study area may have to include other areas that may contribute to the problem solution or that may be impacted by proposed actions. Water and wastewater planning agencies usually begin with the regions of their primary concern, i.e., the area within their administrative jurisdiction. However, for water reuse planning, the boundaries of an agency are often inadequate for developing and analyzing efficient water reuse projects.

For water reuse planning, the project study area should encompass the locations of all potential reclaimed water sources and where reclaimed water could be delivered and used. The study area also must be sufficiently wide to encompass areas that may benefit or be impacted by water reuse. For example, the most serious impacts from overdrafted groundwater basins may manifest in communities beyond the local area. Implementing water reuse in the project area could, therefore, yield water supply savings for other water users within the groundwater basin. On the other hand, by reducing wastewater effluent discharges, water reuse may deprive downstream users of their source of water supply. In effect, there are multiple study areas: the area for sources of and markets for reclaimed water, and areas that may be impacted positively or negatively by water reuse.

Public Involvement

Public involvement is sought in the early planning stages to properly identify objectives and to establish a collaborative planning effort. Throughout project planning and implementation, water reuse planners should meet with potential users and a broader representation of stakeholders to solicit input, disseminate information, address concerns, and provide access to technical experts to respond to questions. Because of frequent public misconceptions and a lack of trust in public agencies, transparency in the conduct of the study will be critical to gain public support. Public involvement in water reuse planning is discussed further in Chap. 26.

Identifying Project Alternatives

Alternatives are developed to meet the fundamental and secondary objectives of the water reclamation and reuse plan. Alternatives are generated by asking how each individual objective can best be achieved. Public participation can aid planners in identifying alternatives and can lead to innovative solutions (City of San Diego, 2005).

Water reuse is just one option for satisfying the objectives of water planning, and ideally it should be evaluated in parallel with alternative sources of water supply, demand-management options, and wastewater treatment and discharge options. From an analysis of current and future conditions if it is concluded that additional future supplies are necessary, a spectrum of new sources and demand-management approaches must be identified. Sources that may be considered include surface water, groundwater, desalinated, reclaimed, conserved, transferred, and purchased water. In cases of onsite water recycling, such as industrial cooling systems, water reuse may be considered a demand-management option. These conserved or reclaimed "sources" may obviate or delay the need to develop other water supply sources (AWWA, 2001). Without consideration of other alternatives to meet fundamental objectives, justification for implementation of a water reclamation and reuse plan may be weakened or flawed. The water reuse option itself can have several alternatives: different markets, applications, treatment levels, and distribution and storage systems as discussed in other chapters of this textbook. Nonstructural alternatives should be considered alongside structural alternatives. For example, shifting the timing of water demand might reduce or eliminate the need for reclaimed water storage to meet peak demands.

Evaluating and Ranking Alternatives

Alternative solutions are evaluated based on their feasibility and their ability to meet stated objectives. Major criteria considered when evaluating alternatives include: (1) market assessment, (2) engineering factors, (3) economic feasibility, (4) financial feasibility, (5) environmental impact, (6) institutional considerations, and (7) social impact and public acceptance. The first six criteria are discussed in this chapter; public acceptance is considered in Chap. 26.

The last six criteria described above are common to all water resources projects. Market feasibility, the first criterion, takes on particular importance for water reclamation and reuse projects. Securing a reclaimed water market is essential for successful implementation of water reuse projects. Water reuse-related issues such as public health, water quality, and user and public acceptance create an added complexity that generally is not encountered with natural freshwater sources.

The generalized IWRP model shown on Fig. 25-1 is useful in developing and evaluating alternatives for the water reuse plan. Each alternative, including those that do not involve water reuse, is evaluated for its ability to satisfy project objectives, applying the major feasibility criteria. If water reuse compares favorably to other alternatives for accomplishing the stated objectives, the most promising water reuse alternatives are refined and the feasibility criteria are applied with more detail and accuracy.

Depending on the geographic scope and breadth of potential options identified, planning may be an iterative process of alternatives identification, initial application of feasibility criteria to screen the alternatives, refinement of a narrower spectrum of alternatives, and then further application of the feasibility criteria.

Developing Implementation Plans

The facilities planning phase is part evaluation, part implementation. During this phase a thorough cost-effectiveness analysis is conducted for all potential alternatives that have survived the preliminary feasibility screening. The feasibility criteria are applied more rigorously as alternatives are screened and the number of alternatives is narrowed for more detailed analysis. An example of feasibility criteria used in the evaluation of water reuse alternatives is presented in Table 25-2. Potential obstacles are addressed, with a serious attempt to resolve them before committing to a selected alternative and commencing design. All necessary facilities of a recommended project are identified with sufficient detail to develop reliable cost estimates and seek approvals from funding sources and regulatory agencies. Additional refinement of the market assessment for the reclaimed water takes place; the refining process includes informing potential

Table 25-2
Criteria used by the City of San Diego, CA, to evaluate alternative reuse strategies[a]

Criteria	Objective and performance measure
Health and safety	To protect human health and safety with regard to recycled water use.
	Meets or exceeds federal, state, and local regulatory criteria for recycled water uses.
Social value	To maximize beneficial use of recycled water with regard to quality of life and equal service to all socioeconomic groups.
	Comparison of beneficial uses and their effect on human needs and aesthetics, as well as public perception.
Environmental value	To enhance, develop; or improve local habitat or ecosystems and avoid or minimize negative environmental impacts.
	Comparison of environmental impacts and/or enhancements, environmental impacts avoided, and permits required.
Local water reliability	To increase substantially the percentage of water supply that comes from water reuse, thereby offsetting the need for imported water.
	Increases percent of water recycling and improves local reliability.
Water quality	Meets or exceeds level of quality required for the intended use and customer needs.
	Meets all customer quality requirements.
Cost	To minimize total cost to the community.
	Comparison of estimated capital improvement costs, operational costs, and revenues for each reuse opportunity, as well as comparison of estimated avoided costs such as future regional water and wastewater infrastructure costs and costs to develop alternative water supplies (e.g., desalination).
Operational reliability	To maximize ability of facilities to perform under a range of future conditions.
	Level of demand met and opportunities for system interconnections and operational flexibility are addressed.
Ability to implement	To evaluate viability or fatal flaws and assess political and public acceptability
	Level of difficulty in physical, social or regulatory implementation.

[a] From City of San Diego (2005).

reclaimed water users of the conditions of service and the probable price of the reclaimed water. A construction financing plan and revenue program are developed to determine financial feasibility. Institutional feasibility is established by reaching agreement with suppliers, wholesalers, retailers, and users of reclaimed water on legal and operational responsibilities. Market assurances, such as mandatory use ordinances or letters of intent from potential users, are obtained (Mills and Asano, 1998).

Facilities Plan Components

A facilities plan or facilities planning report is the main project documentation that is available to the public, decision-making authorities, regulatory authorities, funding agencies, and potential users. While environmental impact reports, feasibility reports, market assessment reports, or financial feasibility reports provide valuable information, discussion of many important issues and analyses performed during the project planning are omitted because of their narrow scope. Without a single report to tie all of the various analyses together, it is very difficult for independent parties to see the continuity between the various analyses and the final recommended project.

A suggested outline for a facilities plan that includes water reclamation and reuse alternatives is shown in Table 25-3. All of the items shown in Table 25-3 have been relevant at one time or another for water reclamation and reuse projects. Thus, although all of the factors shown do not deserve an in-depth analysis for every project, each item

Table 25-3
Wastewater reclamation and reuse facilities plan outline[a]

Item	Description
1	Study area characteristics: geography, geology, climate, groundwater basins, surface waters, land use, and population growth.
2	Water supply characteristics and facilities: agency jurisdictions, sources and qualities of supplies, description of major facilities and existing capacities, water use trends, future facilities needs, groundwater management and problems, present and future freshwater costs, subsidies, and customer prices.
3	Wastewater characteristics and facilities: agency jurisdictions, description of major facilities, quantity and quality of treated effluent, seasonal and hourly flow and quality variations, future facilities needs, need for source control of constituents affecting reuse, and description of existing reuse (users, quantities, contractual and pricing agreements).
4	Treatment requirements for discharge and reuse and other restrictions: health- and water quality-related requirements, user-specific water quality requirements, and use-area controls.
5	Reclaimed water market assessment: description of market analysis procedures, inventory of potential reclaimed water users, and results of user survey.

6	Project alternative analysis: planning and design assumptions; evaluation of the full array of alternatives to achieve the water supply, pollution control, or other project objectives; preliminary screening of alternatives based on feasibility criteria; selection of limited alternatives for more detailed review, including one or more reclamation alternatives and at least one base alternative that does not involve reclamation for comparison; for each alternative, presentation of capital and operation and maintenance costs, engineering feasibility, economic analyses, financial analyses, energy analysis, water quality effects, public and market acceptance, water rights effects, environmental and social effects; and comparison of alternatives and selection, including consideration of the following alternatives:
	  a. Water reclamation alternatives: levels of treatment, treatment processes, pipeline route alternatives, alternative markets based on different levels of treatment and service areas, storage alternatives
	  b. Freshwater or other water supply alternatives to reclaimed water
	  c. Water pollution control alternatives to water reclamation
	  d. No project alternative

7	Recommended plan: description of proposed facilities, preliminary design criteria, projected cost, list of potential users and commitments, quantity and variation of reclaimed water demand in relation to supply, reliability of supply, and need for supplemental or backup water supply, implementation plan, and operational plan.
8	Construction financing plan and revenue program: sources and timing of funds for design and construction; pricing policy of reclaimed water; cost allocation between water supply benefits and pollution control purpose; projection of future reclaimed water use, freshwater prices, reclamation project costs, unit costs, unit prices, total revenue, subsidies, sunk costs and indebtedness; and analysis of sensitivity to changed conditions.

[a] Adapted from Asano and Mills (1990) and TFWR (1989).

should be considered. The overall level of detail should be commensurate with the size and complexity of the proposed project. Although the emphasis on the wastewater or water supply aspects will vary depending on whether a project is single or multiple purpose, the nature of water reclamation and reuse is such that both aspects must at least be considered.

Facilities Plan Report

All of the basic data, the study procedure, and results of the feasibility analyses are documented in a facilities plan report. The information presented in a structured format is readily accessible for review and thus provides for a careful analysis and evaluation of the project. The facilities plan report provides a vehicle for thorough review by the public as well as the many project participants and funding and regulatory agencies.

25-2 ENGINEERING ISSUES IN WATER RECLAMATION AND REUSE PLANNING

Planning for water reuse projects has many elements related to the planning and design of water supply and wastewater management systems. However, special issues that are unique and have to be addressed are the following:

1. Water quality
2. Public health protection
3. System reliability
4. Existing and new treatment technologies for water reclamation and reuse
5. Application of satellite and decentralized systems for water reuse applications
6. Storage and distribution system siting and design
7. Matching supply and demand for reclaimed water
8. Use site dual plumbing systems and controls
9. Single vs. multiple applications of reclaimed water

The above issues are the subjects of discussion in Parts 2, 3, and 4 of this textbook.

25-3 ENVIRONMENTAL ASSESSMENT AND PUBLIC PARTICIPATION

One of the main feasibility criteria of project evaluation is avoiding or mitigating negative environmental impacts. For federal projects this became a mandate in 1969 when, as an outcome of the environmental movement in the United States, the federal government enacted the National Environmental Policy Act of 1969 (NEPA). This act mandated the assessment and mitigation of environmental impacts caused by federal projects. Public notice and opportunity for public input were also mandated during the environmental assessment process. At least 20 states have passed similar laws that apply to state and local projects. The assessment processes are quite detailed and beyond the scope of this textbook. However, the concepts of public participation, beyond the procedures specified in the laws, are particularly relevant to the success of water reclamation and reuse projects and are addressed in more depth in Chap. 26.

Environmental Assessment

The purposes of NEPA are to: (1) encourage productive and enjoyable harmony between humans and the environment, (2) promote efforts to prevent or eliminate damage to the environment and biosphere and stimulate the health and welfare of humans, (3) enrich the understanding of the ecological systems and natural resources important to the nation, and (4) establish a Council on Environmental Quality (CEQ, 2005). Projects that receive federal funding, such as water reuse projects administered by the U.S. Bureau of Reclamation, are subject to the provisions of NEPA, even if implemented by local agencies.

The required procedures involve a staged environmental analysis first to determine whether a significant environmental impact potentially could occur and whether any significant impacts can be mitigated or reduced to a level less than significant. If the impacts are significant, NEPA requires that federal agencies complete an Environmental Impact Statement (EIS), which is an in-depth analysis of impacts of project alternatives and identification of mitigation measures that will be implemented. State environmental laws are structured similarly (Kontos and Asano, 1996).

Public Participation and Outreach

Participatory planning is based on the principle that those who are affected by decisions or policies should participate or be represented in the policy-making processes. The premise is that both professional planners and community members can make valuable contributions to project planning. The public participation and outreach concept goes beyond merely informing the public of project planning results and seeking comments. The objective is to provide a forum for public input at all stages of planning, including initial identification of project objectives and formulation of alternatives to be analyzed. The challenge for planners is to find the most effective approach to public involvement—who to involve, at what level of influence, at which milestone in the project planning process, and in what form. The level of public involvement in decision-making and the techniques employed will vary depending on the level of public knowledge and interest, potential for controversy, specific concerns, and technical or regulatory constraints (House, 1999; Hartling, 2001; Marks et al., 2003; RWTF, 2003). Approaches to public participation and outreach in water reuse are discussed in Chap. 26.

25-4 LEGAL AND INSTITUTIONAL ASPECTS OF WATER REUSE

Water reuse planning is affected by a variety of laws, policies, rules, and regulations, including water rights laws, water use and wastewater discharge regulations, land use restrictions, and environmental and public health protection laws. Implementation of water reclamation and reuse projects is also influenced by policies on the development of reclaimed water rates and institutional agreements, as well as rules affecting system construction and liability for water reuse (U.S. EPA, 2004).

Water Rights Law

A water right is a right to use water, not the ownership of the water itself. State laws dictating surface water rights are divided generally into two categories: riparian doctrine and prior appropriation doctrine. Groundwater rights are generally considered separately from surface water rights (Getches, 1990). Water rights law and regulations are complex; thus, consulting with skilled legal professionals is recommended.

Riparian Doctrine

The *riparian doctrine* of water rights is the basis for water law in the eastern United States but coexists with other forms of water law in other regions. The basic premise of the riparian doctrine is that the right to use water is a property right—only those who own riparian land have the right to use surface water. *Riparian land* is property in contact with inland water that is inundated at the average high tide, such as streams, rivers, lakes, or bays. The owner of land on the bank of a stream or lake is called a *riparian owner*. The owner has the right to "reasonable use" of the water that flows through the riparian land, subject to some restrictions. Historically, the landowner was required to leave the *natural flow* of the river unchanged to protect the rights of downstream riparian owners. The modern interpretation of the *reasonable use* rule dictates that each riparian owner may use the water, regardless of natural flow, as long as the use does not cause an *unreasonable injury* to any other riparian user (AWWA, 2001). The user is not permitted to store water or to extend the water use to another piece of property. The right to unused portions of water can be held indefinitely and without forfeiture.

There are three derivations of the riparian rights doctrine (AWWA, 2001; Dzurik, 2003):

1. Reasonable use: Upper riparian users can divert water to put to beneficial use as long as it does not interfere with the reasonable use by downstream riparian owners.
2. Correlative rights: Riparian owners are assigned a share of water proportional to land ownership. During times of water shortage, each owner's share is reduced proportionally.
3. Regulated riparian: A direct user must obtain a state permit for water use. Regulated riparian rights overlays a system of government permits and regulation by state agencies on top of the traditional court-made riparian doctrine.

Prior Appropriation

The doctrine of "prior appropriation" is the basis for water law in most of the western states. The doctrine was developed during early European settlement of the western United States as a way to protect the water use of the first settlers, mainly miners, from those who came later. The doctrine was formalized into law through expressed recognition by court decisions, constitutional provisions, and state statutes (AWWA, 2001). The right to use water is allocated based on the "first in time, first in right" principle. The first parties to use the water, *senior appropriators*, have the most senior claims. Later users of the water, *junior appropriators*, are entitled to their water only after senior appropriators have diverted water according to their rights. During times of water shortage, junior appropriators can use water only if there is a surplus after senior appropriators have met their needs (Dzurik, 2003).

Each user is permitted to divert a specific quantity of water, without diverting more water than can be put to beneficial use. Once that water has served the beneficial use, the waste or return flow must be allowed to return to the stream (AWWA, 2001). A senior appropriator may involuntarily forfeit the prior appropriation right if the water is used for a nonbeneficial purpose or if the water is not used during a specified period of time. The interpretation of "beneficial use" has an inherent bias toward economic development, but some western states also include conservation, recreation, and aesthetics as part of beneficial use. The appropriation rights are also linked to a particular use and point of diversion; a change in use or point of diversion may trigger a downgrade to junior rights or may require permission by the state (Dzurik, 2003). A comparison of the riparian and appropriation water rights is shown in Table 25-4. States may operate under a combination of riparian and prior appropriation doctrines.

Groundwater Law

Groundwater law is defined by two legal classifications: underground streams and percolating waters. Underground streams are waters that flow in a known and defined channel. Percolating waters include all other underground waters that do not flow in a definite channel. The rule in most eastern states of the United States dictates that a landowner can withdraw groundwater for reasonable use as long as the withdrawal does not impair the water supply on another property.

Potential liability for harm caused by the recharge of reclaimed water to a groundwater basin is a more difficult matter. There are numerous tort theories that plaintiffs might try to establish liability, including negligence, strict product liability, and strict liability for ultra hazardous activities, warranty, and nuisance theories. Contract provisions might avoid applications of some of these theories, but no entity could entirely insulate itself contractually from liability (Schneider, 1985).

Water Rights and Water Reuse

An important consideration in water reuse planning is who among the discharger, water supplier, other appropriators, or environmental interests owns the right to use reclaimed water (Cologne and MacLaggan, 1998). As reclaimed water becomes a more significant part of the nation's water conservation program, legal disputes are likely to arise. The foreseeable disputes will come from conflict over ownership of the reclaimed water and over ambiguities in contractual obligations. Water reclamation and reuse is a new use of a resource already heavily drawn upon. As water formerly returned to streams after use and treatment is withheld for resale at the treatment site, diminished flow downstream may deprive dependent users of their accustomed supply. Legal actions have been taken to block proposed sales of reclaimed water for this reason. Two examples are described below.

> **People v. City of Roseville, CA** (Civil No. 49608, California Superior Court, Placer County, September 30, 1977). The City of Roseville, CA contracted to sell treated wastewater to certain irrigators in the drought year of 1977. For many years it had released its effluent into Dry Creek after treatment. The California State Water Resources Control Board brought an action to enjoin the sale because the withdrawal of the water would injure other legal users downstream (Richardson, 1985).
>
> **City of Walla Walla, WA.** The municipal corporation of the City of Walla Walla was taken to court by a local irrigation district that wanted the city to continue to discharge wastewater effluent into Mill Creek, a natural channel, for irrigation use. The court decreed on two occasions that the city must discharge all of its wastewater effluent, at all seasons of the year, into the creek (U.S. EPA, 2004).

Many existing contracts between wastewater reclamation agencies and purchasers do not sufficiently clarify the mutual obligations of the parties. As water reclamation projects have expanded, conflicts have arisen concerning the water entitlements of earlier

Table 25-4

Comparison of riparian and appropriation water rights[a]

Issue	Type of water rights	
	Riparian	Appropriation
How are rights acquired?	Riparian rights are acquired by obtaining riparian land, which is defined as property touching the water of a lake or stream. A riparian right is for the use, not the ownership, of the water.	An appropriation right is independent of land ownership. The right to a certain quantity of water may be acquired by applying the water to a beneficial use. The basic principle is that when the supply cannot fulfill the needs of all the perfected appropriations, the last or junior rights are the first to be shut off—first in time is first in right.
What uses may be made?	Water may be used for any reasonable purpose.	Water may be used for any beneficial use as defined by the states' codes. Common beneficial uses may include irrigation, mining, stock watering, manufacturing, municipal uses, domestic uses, and recreational uses.
May water be impounded?	Generally it is unclear whether riparians have the right to impound water at high flow for later use or release. Some riparian states have initiated a permit system for storage for mill or hydroelectric operations.	Impoundment for later use is common.
Where can water be used?	Often riparian water use is limited to riparian lands, but many states permit use on nonriparian lands if other users are not harmed. Additionally, water use may be limited to the watershed of origin.	Water may be used anywhere, and if there is no injury to vested rights, water may be used outside the watershed.
When may water be used?	Whenever it is available.	Often an appropriation right may be limited to a specific time, i.e., day or night, summer or fall.
What is the nature of the right?	Except for domestic uses, riparians on a watercourse are co-sharers and have an equal right to make a reasonable use of the water. No riparian is ever ensured a definite quantity, unless a prescriptive right is obtained.	Appropriation rights are never equal because first in time appropriators are guaranteed an ascertainable amount of water. If an appropriator's needs can be met by use of less water, that appropriator is entitled only to the lesser quantity.
What happens if the water right is not used?	The right does not depend on use. Therefore, it is not lost by nonuse and is not subject to abandonment.	The right is held only as long as proper beneficial use is continued. Appropriation rights are subject to abandonment.

[a] Adapted from AWWA (2001).

versus subsequent water users. Although these incidents have been minor, they demonstrate that the best insurance against breach of contract disputes is to clarify the expectations of the parties at the outset. Recent changes in water codes in several states have attempted to resolve these issues of rights to reclaimed water. In California unless otherwise provided by agreement, the wastewater treatment facility now has exclusive rights to the treated wastewater as against any supplier of raw (untreated) wastewater. These changes eliminate the need to negotiate with any other entity that has contributed to the generation of wastewater. As in any area of the law, however, answers cannot be given with certainty. Until specific legal problems have been addressed by the courts through litigation, or in the legislature through statutes, their solutions can only be stated in probable terms (Richardson, 1985).

Policies and Regulations

Policies and regulations are governed by a number of governmental agencies having jurisdiction. The fundamental principle of U.S. water policy is that the federal government must respect the primary role that individual states have in shaping and controlling their own policies regarding water use and allocation. However, federal legislation, such as environmental regulations, often interferes with state water policy (Cologne and MacLaggan, 1998). For water reuse applications, the principal U.S. federal and state regulations as well as the World Health Organization (WHO) guidelines that have to be considered are described in Chap. 4.

Institutional Coordination

Water reuse projects involve the interaction between, as a minimum, a supplier of reclaimed water and the user. More typically, several entities are involved. Separate entities may be responsible for collection of wastewater, treatment of wastewater, distribution of reclaimed water, wholesale supply of water, and retail distribution of water. All of these entities may have to cooperate and establish areas of responsibility for the successful operation of a water reclamation and reuse project. From a regulatory standpoint, permits or approvals may be necessary from authorities governing water quality, water supply, public health, agricultural affairs, or water rate setting. Regulatory agencies should be involved early in the planning process. Some of the complexities that can occur in implementing a water reclamation and reuse program are described in the following section, which deals with a case study of a California water reuse project.

25-5 CASE STUDY: INSTITUTIONAL ARRANGEMENTS AT THE WALNUT VALLEY WATER DISTRICT, CALIFORNIA

The Walnut Valley Water District, located in southern California, represents a typical situation of a complex organizational arrangement. The District operates a water reclamation project that delivers water not only within its service area, but also within an adjacent retail water district (Bales et al., 1979).

Water Management Issues

To understand the economics and financial feasibility of this project during planning, it was necessary to understand the water supply infrastructure from the sources of supply to the end water user. A chain of water supply agencies are involved. The organizational structure for providing potable water to Walnut Valley Water District is illustrated on Fig. 25-2.

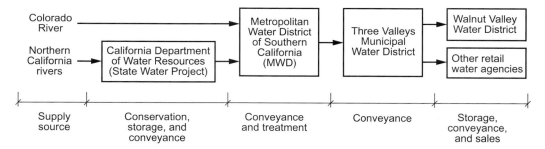

Figure 25-2
Organizational structure governing the delivery of potable water to Walnut Valley Water District.

Reclaimed water reaches the district through another chain of agencies. An agreement between the County Sanitation Districts of Los Angeles County and the City of Pomona entitled the city to all of the effluent from the wastewater treatment plant. An interagency agreement had to be negotiated between Walnut Valley Water District and its supplier of reclaimed water, the City of Pomona. Another agreement was executed between the District and Rowland Water District so users located within Rowland Water District could participate in the project. In this situation, Walnut Valley Water District constructed and operates all facilities for the distribution of reclaimed water. Rowland Water District, while it does not operate the facilities, purchases the reclaimed water from Walnut Valley Water District, reads the water meters, and bills the reclaimed water users within its territory. In establishing wholesale and retail reclaimed water rates, it was necessary to consider the unique wholesale and retail potable water rate structures of each agency. The institutional relationships for Walnut Valley Water District Water Reclamation Project are shown schematically on Fig. 25-3.

Lessons Learned

In developing a viable project, it was necessary to explore jurisdictions beyond any one agency's boundaries for sources of reclaimed water and potential reclaimed water users. Several interagency agreements were needed to secure a source of reclaimed water, to

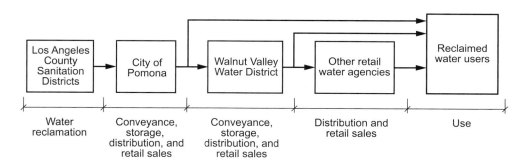

Figure 25-3
Organizational structure governing the delivery of reclaimed water to Walnut Valley Water District.

obtain authority to deliver reclaimed water within several jurisdictions, and to establish cost and revenue sharing among agencies.

25-6 RECLAIMED WATER MARKET ASSESSMENT

A key task in planning a water reclamation project is to find potential users or customers of reclaimed water who are capable and willing to use reclaimed water. The market assessment involves the gathering of background information on the constraints applicable to various categories of reclaimed water use and detailed data on each potential user or use site. The background information and user data will provide a basis for determining if a potential user is capable of using the reclaimed water. This information will also help the potential user to decide whether to use the water. Willingness is also dependent on whether the reclaimed water will be marketed on a voluntary or mandatory basis. Marketing approaches are considered subsequently.

In addition to identifying potential users, the market assessment provides much of the data needed for formulating project alternatives. Information about the users is needed to determine (1) location of facilities and their capacities, (2) design criteria, (3) reclaimed water pricing policy, (4) financial feasibility, (5) the amount and sources of potable or other freshwater displaced, and (6) the institutional framework for the project.

The ability of a user to use reclaimed water will depend on the quality of water, its availability, and its suitability for the type of application involved. Willingness to use reclaimed water will depend on whether the use is voluntary, and, if so, on how well reclaimed water competes with freshwater with respect to cost, quality, and convenience. It is essential to have a thorough knowledge of the water supply context of the users, especially if reclaimed water is to be marketed on a voluntary basis (Asano and Mills, 1990; Mills and Asano, 1998).

Steps in Data Collection and Analysis

The gathering of background information for analysis involves several steps as presented in Table 25-5. Steps 1 through 6 are conducted simultaneously and are addressed, at least on a preliminary basis, before steps 7 and 8. Information required for step 7 is given in Table 25-6. To some extent these steps may be carried out iteratively, returning to them during the course of planning to refine the data or survey additional potential users. The focus of the first phase of data collection is usually on larger water users. The basic information obtained is on the type of application and the expected annual quantity of use. As facilities alternatives take shape, more detailed information will be collected and smaller users will be identified. Potential users may be contacted more than once as planning progresses to gather more information from the users and to provide the users with more information on the potential project. Group presentations with potential users may be useful for disseminating information and providing technical experts to answer questions.

Comparison of Water Sources

If reclaimed water use will displace the existing use of other sources of water or offset the development of alternative new water sources, it is essential to have a thorough understanding of the freshwater sources of supply. This knowledge is necessary to

Table 25-5
Steps in the gathering background information for a reclaimed water market assessment

Step	Description
1	Create an inventory of potential users in the study area and locate them on a map. Group the users by types of use. Cooperation of retail water agencies can be very helpful in this task.
2	Determine public health-related requirements by consulting regulatory agencies. Such requirements will determine the levels of treatment for the various types of use and application requirements that will apply on the sites of use; e.g., backflow prevention devices to protect the potable water supply, irrigation methods that are acceptable, use-area controls to prevent ponding or runoff of reclaimed water, practices to protect workers or the public having contact with the water.
3	Determine water quality regulatory requirements to prevent nuisance or water quality problems, such as restrictions to protect groundwater quality
4	Determine water quality needs of various types of use, such as industrial cooling or irrigation of various crops. University farm advisors may be helpful in this regard.
5	Identify the wholesale and retail water agencies serving the study area. Collect data from them on current and projected freshwater supply prices (rates) that would be applicable to the reclaimed water users. Also, collect data on the quality of freshwater being provided.
6	Identify the sources of the reclaimed water and estimate the probable quality of the reclaimed water after treatment to the level or levels under evaluation. Determine what types of use would be permitted at the various levels of treatment based on public health requirements and requirements suitable for various usages, such as industrial or agricultural uses.
7	Conduct a survey of the identified potential reclaimed water users to obtain detailed and more accurate data for evaluating each user's capability and willingness to use reclaimed water. The types of data that should be collected on each user are shown in Table 25-6. While most of these data must be obtained directly from the user, some of these data may be assessed from the background information obtained from other sources.
8	Inform potential users of applicable regulatory restrictions, probable quality of reclaimed water at various levels of treatment compared to freshwater sources, reliability of the reclaimed water supply, projected reclaimed water and freshwater rates. Determine on a preliminary basis the willingness of the potential user to accept reclaimed water.

obtain acceptance of reclaimed water users and cooperation of affected water suppliers and to determine the net potable water savings and economic and financial feasibility. The information will be collected from the market survey of users as well as by consultation with local and regional water suppliers.

From the viewpoint of the reclaimed water user, reclaimed water will be compared with other sources of water available to the user in terms of quality, quantity, dependability, safety, and cost. The planner must identify the current or future sources of freshwater that each user is or would be using if reclaimed water were not available. It must not be assumed that because a user is located within the boundaries of a retail potable or freshwater supplier, the user purchases water from the supplier. Many users have their own sources of supply, especially wells. The independent well user may see quite different water supply costs than customers of the water supplier.

Table 25-6
Information required for a reclaimed water market survey[a]

Item	Description
1	Potential specific uses of reclaimed water
2	Location of user
3	Recent historic and future quantity needs (because of fluctuations in water demands, at least three years of past use data should be collected)
4	Timing of needs (seasonal, daily, and hourly water demand variations)
5	Water quality needs
6	Water pressure needs
7	Reliability needs regarding availability and quality of reclaimed water, i.e., how susceptible is the user to interruptions in water supply or fluctuations in water quality
8	Needs of the user regarding the disposal of any residual reclaimed water after use
9	Identification of onsite treatment or plumbing retrofit facilities needed to accept reclaimed water
10	Internal capital investment and possible operation and maintenance costs for onsite facilities needed to accept reclaimed water
11	Needed monetary savings on reclaimed water to recover onsite costs or desired payback period and rate of return on onsite investments
12	Present source of water, present water retailer if the water is purchased, cost of present source of water
13	When user would be prepared to begin using reclaimed water
14	Future land use trends that could eliminate reclaimed water use, such as conversion of farm lands to urban development
15	For undeveloped future potential sites, the year in which water demand is expected to begin, current status, and schedule of development
16	After informing user of potential project conditions, a preliminary indication of the willingness of user to accept reclaimed water

[a] Adapted from Mills and Asano (1998).

Comparison with Costs and Revenues

In most cases, the water supplier may be comparing water reclamation costs and revenues to the supplier's current freshwater supplies, future water supply developments, or purchases from its wholesale water suppliers. Wholesale and retail water supply service areas and the potential users associated with each of the suppliers must be identified. The reclaimed water planner must be familiar with the freshwater sources of supply and the major facilities for capturing, treating, and distributing these supplies. With this knowledge, it is possible to evaluate how serving each user with reclaimed water will offset a particular freshwater demand, resulting not only in reducing the cost of providing the freshwater, but also in reducing the associated revenue. In some cases the reduction in revenue will exceed the reduction in costs, resulting in a financial loss to the freshwater purveyor. Costs and benefits must be evaluated from the viewpoint of each wholesale and retail water supplier. With an understanding of how each agency will be impacted, a plan can be developed to share the water reclamation costs and revenues equitably between the water suppliers and wastewater agencies, thus ensuring their full cooperation.

When gathering cost data, it is important to distinguish between fixed and variable costs. The use of reclaimed water will reduce the variable costs of a freshwater supply, but not the fixed costs, such as debt service on existing facilities.

Retail Water Supplies

Retail water suppliers are useful sources of data on water quality, potable or freshwater rates, and water demands of water users. Water suppliers can often easily provide actual water use records of individual users. Obtaining several years of records can help ensure that planners are not misled by data from unusually wet or dry years.

Market Assurances

During the market assessment phase of planning, a list is developed of potential users capable of using reclaimed water. While this list can be used to begin shaping alternative plans to serve these users with reclaimed water, at some point an assessment must be made of the degree of willingness these users have to participate in a project. There are a multitude of reasons that potential reclaimed water users may resist participation. The more frequently encountered reasons are listed in Table 25-7. These concerns can be overcome only by becoming familiar with the user's needs through the market assessment and educating the user about the nature and benefits of the water reclamation project (see Chap. 26).

Measure of the Degree of Willingness of a Potential Reclaimed Water User

There are various means of measuring the degree of willingness of a potential user. During the planning process, the options in order of commitment are usually:

1. Oral interview
2. Letter of interest signed by the potential user, indicating general understanding of the potential project and expressing interest in participation
3. Letter of intent signed by the potential user, incorporating key understanding of the conditions of reclaimed water service, such as quantity of supply, water pressure, responsibilities for making on-site conversions, dates of service, and estimated price of reclaimed water.

Before a commitment is made to commence implementation of a recommended project, a letter of intent is recommended from each user identified with the project.

Table 25-7 Principal customer concerns about use of reclaimed water

Principal concerns
• Price of reclaimed water is too high in relation to freshwater costs
• Inability to finance onsite costs for plumbing conversion
• Water quality and reliability
• Employee exposure to potential health hazards
• Possibility of employee union objections
• Lack of reliable reclaimed water supply
• Water supply costs are too insignificant to tolerate the perceived added inconvenience of using reclaimed water
• Liability and potential lawsuits

The letter should contain many of the details of service and financial responsibilities so that it is clear that the user has been educated fully about the project and, with this understanding, fully intends to participate in the project. However, because these letters are not legally binding, they are insufficient to assure that users will participate in the project. Stronger, legally enforceable contracts are usually called for and recommended before any financial commitments for facilities construction are made.

If a user resists signing a letter of intent during the planning process or a contract before the design phase of a project, several outstanding concerns may not have been resolved. However, if these concerns cannot be overcome before awarding a construction contract, a major investment is at risk. Based on past experience, it has been found that although some potential users will initially express positive interest in using reclaimed water, they will often resist a long-term commitment or refuse to use the reclaimed water after the facilities are built. Negotiations for firm commitments should be completed before project construction. The sometimes laborious process of obtaining commitments can be an educational exercise to win the support and confidence of potential customers and bring out hidden issues much earlier in the project development process (Asano and Mills, 1990).

Without signed contracts, there is limited basis for optimism regarding user participation in water reuse projects. In a survey of 16 operating water reclamation and reuse projects in California, only one quarter of the water supply agencies were delivering the amount of reclaimed water that was planned initially. One quarter of this sample delivered 50 percent or less of the planned amount. As part of the same survey, data on elapsed time for design were collected for 28 projects. One third had not gone to construction after over 2 yr of design, including one having surpassed 5 yr after commencing design. Nearly all of the projects experiencing deliveries below the planned amount and long implementation periods suffered from problems that needed to be addressed in the planning phase of the of water reclamation project. The problems included using unreliable data for estimating water demand quantities, encountering institutional difficulties, and, most frequently, failing to obtain agreements with potential users to purchase reclaimed water (Mills and Asano, 1996). To ensure user commitment, legally enforceable agreements are needed.

Types of Market Assurances

The ultimate measure of success of a water reclamation project is the delivery of reclaimed water in the quantities projected for the life of the project. To ensure that a project will achieve this goal, there must be some assurance that an adequate reclaimed water market exists. Examples of market assurances are listed below:

1. Land ownership: The agency owns the land upon which the reclaimed water will be applied.
2. Land lease: The agency leases the land upon which the reclaimed water will be applied.
3. User contract: The reclaimed water is sold or supplied to a water user under the provisions of a legally enforceable contract between the user and the reclaimed water supplier.

4. **Mandatory reclaimed water use ordinance:** A legal mandate is imposed by the retail water purveyor that water users are obligated to accept reclaimed water in lieu of another source of water available from the purveyor.
5. **Broad water rights authority:** National, provincial, or regional governments have authority to restrict the rights to use fresh water if reclaimed water is available.
6. **Sale by reclaimed water use permit or informal arrangement:** Reclaimed water is sold or supplied to a water user under provisions of a permit or informal arrangement without any long-term obligations.

The mechanisms above are ranked generally in the order of the strongest to weakest form of market assurance. Ordinances and water rights authority have the potential to be very strong, depending on the willingness of the jurisdictional agencies or governments to exercise their authority and the ease with which the authority can be exercised.

If a water reclamation project is intended to meet a water pollution control objective, then the success of the project may be dependent on having users accept all of the effluent from the wastewater treatment plant. A system may be designed such that water reuse is the only means of disposing the effluent, especially if there are regulatory restrictions preventing discharge to surface waters. The facilities plan should identify a reclaimed water market for all present and future projected flows and describe the facilities needed to serve that market, even if the facilities are constructed in phases. Strong market assurances are necessary for the life of the facility to prevent discharge violations. The wastewater agency may have alternative plans to purchase land or obtain long-term leases to exercise direct control over the ability to dispose of effluent to land. Long-term user contracts are also used.

Voluntary or Mandatory Approach to Marketing

The project proponent has to decide whether to take a voluntary or mandatory approach to marketing the reclaimed water. The choice of mechanism depends on the local situation and project purpose. The voluntary approaches can include land ownership if willing sellers are found for the land or land lease if a land owner willingly leases the land to an agency for land application of reclaimed water. If willing landowners are not found, an agency may use the legal powers of eminent domain, taking a mandatory approach. These two mechanisms are most common when the project purpose is for treatment or disposal of treated wastewater.

A user contract is considered a voluntary mechanism in which the reclaimed water purveyor and the user mutually agree upon a long-term, legally binding commitment for the sale or transfer of reclaimed water. The contract should contain provisions specifying the responsibilities, obligations, and conditions of reclaimed water service, and liability of the two parties. A list of desirable provisions for reclaimed water user contracts is given in Table 25-8.

Mandatory Use Ordinance

A *mandatory use ordinance* is a rule or law used by local governmental bodies, such as water districts, to mandate the use of reclaimed water in lieu of alternative water supplies, usually potable water. An ordinance, which may have different names under different

Table 25-8
Desirable provisions of reclaimed water user contracts[a]

Provisions
• Contract duration: terms and conditions for termination.
• Reclaimed water characteristics: source, quality, and pressure.
• Quantity of reclaimed water demand.
• Flow variation of reclaimed water demand (hours and season of use): as needed by user or allowed by supplier.
• Reliability of supply: potential lapses in supply and back-up supply provisions.
• Commencement of use: when user can or will begin use.
• Specific areas and conditions for reuse.
• Financial arrangement: pricing of reclaimed water and payment for facilities, land lease costs.
• Ownership of facilities and rights-of-way.
• Responsibility for operation and maintenance.
• Notification of problems: obligation of user to notify reclaimed water purveyor of violations of regulatory requirements; obligation of purveyor to notify user of water quality violations or supply interruptions.
• Liability: for example, the user contract should indemnify the reuse agency from damages caused by use of reclaimed water due to no fault of the reuse agency.
• Operations plan: operations or practices of the purveyor and the user to ensure monitoring and safe use of reclaimed water; obligation of user to observe regulatory requirements.
• Violations of contract provisions: definition of what constitutes a violation, specification of penalties and remedies.
• Inspection of onsite reuse facilities: the right of entry into the user's premises to inspect the reuse facilities during construction and operation.

[a]Adapted from TFWR (1989).

jurisdictions, has legal effect only if the agency declaring the ordinance is the retail supplier of both the reclaimed water and the alternative supply that a user would have to depend on. If an agency has control over both supplies, it can impose restrictions on the use of freshwater or fines if freshwater is used instead of reclaimed water. A wholesale agency does not have a direct relationship with the water user, so it has limited legal jurisdiction. However, wholesale water suppliers can encourage retail water agencies to mandate the use of reclaimed water by imposing fines or restrictions on the sale of fresh water to retail agencies.

Desirable provisions that should be included in a mandatory use ordinance are listed in Table 25-9. A mandatory use ordinance can apply to existing development that is currently using freshwater. The ordinance can also apply to new development by requiring dual plumbing or dual water distribution systems during new construction. Because reclaimed water service may not be cost effective or feasible in certain geographic areas, the ordinance should specify the service areas governed by the ordinance. There are always special circumstances that prevent the use of reclaimed water, such as sites that would require expensive conversion (retrofit) costs or sites containing salt-sensitive ornamental plants. A mandatory use ordinance should have a formal procedure to apply for, review, and approve a waiver.

Table 25-9
Desirable provisions for mandatory use ordinance[a]

Provisions
• Specification of the types of use of water for which reclaimed water must be used.
• Specification of the conditions under which reclaimed water must be used or new development must be plumbed for future reclaimed water use.
• Procedure for determining which water users are required to either convert to reclaimed water service or be plumbed to accept reclaimed water upon new water service.
• Procedure to provide notice to potential users that they are subject to the ordinance and specification that the notice include information about the project, the responsibilities of the users under the ordinance, the price of the reclaimed water, and description of the on-site retrofit facilities requirements.
• Procedure for request by the users for a waiver.
• A penalty for noncompliance with the ordinance. Example penalties are discontinuance of freshwater service or a freshwater rate surcharge of 50 percent of the freshwater rate.

[a]Adapted from Office of Water Recycling (1997).

Waste and Unreasonable Use of Water

The State of California has a "waste and unreasonable use of water" doctrine incorporated into its constitution. As an extension of this doctrine, the state legislature has created laws that require the use of reclaimed water in lieu of potable water under certain conditions that make allowances for public health protection, reasonable costs, and suitable water quality. However, the exercise of this state authority is cumbersome, requiring a quasi-judicial proceeding. While it is a strong mechanism, it is not in the control of the local water purveyor and has been applied in only a few circumstances (Cologne and MacLaggan, 1998; Mills and Asano, 1998).

25-7 FACTORS AFFECTING MONETARY EVALUATION OF WATER RECLAMATION AND REUSE

Two of the main criteria for project evaluation are economic and financial feasibility. While the two criteria sound similar and the terms are often used interchangeably, they have different roles as defined in classic water resources economics (James and Lee, 1971). Economic feasibility, established through an economic analysis, is a test of whether the total benefits that result with a project exceed those that would accrue without the project by an amount greater than the project cost. A project is considered to be justified economically if the total benefits outweigh the total costs. Procedures have been developed by economists to calculate benefits and costs on a common basis so a comparison can be made between project alternatives. The ability to finance the construction of a project and to raise the revenues for debt service and project operation is based on a financial analysis. Economic justification and financial feasibility do not necessarily go hand in hand. Even though the total monetary benefits would exceed the total costs, the financial mechanisms may not be available to implement a project. Thus, a financial analysis is used to determine whether a project can be implemented rather than to measure the net benefits of a project. Expressed in simpler terms

(Mills and Asano, 1986/87), an economic analysis addresses the question, *should* a project be constructed? A financial analysis addresses the question, *can* a project be constructed?

While there are many textbooks that cover fundamental principles of economic and financial analyses, the distinction between these analyses and the application of the principles to water reclamation and reuse projects are not readily available and are not well understood by engineers and planners. Because economic and financial analyses are often performed by engineers and planners without specialized economics training and because these analyses must be understood by the public and decision-makers, the basic concepts as applied to water reclamation and reuse will be presented here. Common weaknesses in planning that lead to inappropriate calculation of the costs and benefits of water reclamation and reuse projects are addressed below. Issues related to planning perspectives, time horizons, the time value of money, and inflation and cost indexes are also considered. These discussions will provide a basis for Secs. 25-8 and 25-9, devoted separately to economic and financial analyses. For a fuller and more theoretical discussion of the material in this and the following sections, engineering and water resources economics textbooks should be consulted.

Common weaknesses or misconceptions applied to water reclamation and reuse planning that result in omissions of costs and benefits or inappropriate design of projects include: (1) the point of view, (2) lack of consideration of externalities, and (3) comparison of projects on a before- and after-basis.

Common Weaknesses in Water Reclamation and Reuse Planning

Point of View
Costs and benefits are perceived differently depending on particular viewpoints. A common weakness in water reclamation and reuse planning is to take a singular viewpoint. This viewpoint is usually that of the agency proposing the project. The success of a water reclamation and reuse project involves the support of the public at large and the willingness of water users to accept reclaimed water. A project may also involve cooperation of several agencies. Analyses must be conducted with these several viewpoints in mind.

Impact of Externalities
Another common error in planning is to ignore externalities. An externality can be defined as the impact or effect of an action or decision made by an individual, group, or entity on others (individual, group, or entities) who were not considered in the decision making process. An example is the discharge of treated wastewater from a municipality into a river that serves as water supply source for a downstream municipality without considering potential impacts on the downstream user. The failure to consider increased water treatment costs is an example of ignoring externalities. The diversion of treated effluent for local reuse without considering the impacts on a downstream ecological habitat is another example.

Basis of Project Comparisons
In identifying the benefits of a project and its alternatives, the comparison must be *with* and *without* the project rather than *before* and *after* the project (James and Lee, 1971). Some outcomes may occur after implementation of a project that would have occurred without the project in any case. The outcomes that are dependent on implementation of a project must be separated to assess the project costs and benefits. For example, if a project consists of addition of tertiary treatment to expand the use of reclaimed water

already taking place with secondary effluent, the continued delivery of reclaimed water to the existing users cannot be attributed as a benefit of the new tertiary treatment.

Perspectives in Project Analysis

Determining the benefits and costs of a project depends on the perspective from which the analysis takes place: utility, ratepayer, or society perspective. Projects should be examined from each perspective to make sure that no costs or benefits have been overlooked and to provide the most information to decision makers. However, the final analysis of a project must incorporate all perspectives.

Utility Perspective
When an analysis is done from the perspective of a utility, only the costs and benefits that directly impact the utility are included in the analysis. Examples of costs to utilities include payments that the utility makes for land, equipment, supplies, fees, consultants, engineers, contractors, other agencies, operators, financing costs, and utility staff. Private funding or government subsidies reduce the direct utility costs. In general, externalities are missing in the analysis.

Ratepayer Perspective
Analysis from the ratepayer perspective incorporates costs that are passed on to the water user by the utility plus costs or benefits directly experienced by the ratepayer. For example, reclaimed water users will pay fees to the utility for the purchase of reclaimed water. In addition, irrigation users may experience a benefit of lowering their fertilizer costs because of added nutrients in reclaimed water. Industrial users may have to add chemical treatment to reclaimed water to offset the negative effects of constituents in reclaimed water. Potable water users may experience rate increases due to the sale of less potable water to cover potable water system fixed costs.

Society Perspective
For the purpose of determining the optimum alternative considering all project costs and benefits, including external effects, the society perspective is used. For this reason, the society perspective is appropriate for economic analysis.

Planning and Design Time Horizons

There are several time horizons that are important in facilities planning, design, economic evaluation, and financial analyses.

1. *Planning period* is the total period for which facility needs will be assessed and alternatives will be evaluated for cost-effectiveness and long-term implementation.
2. *Design life* is the period in which a phase of a component of facilities to be constructed is expected to reach capacity.
3. *Useful life* is the estimated period of time during which a facility or component of a facility will be operated before replacement or abandonment. The useful life is usually equivalent to the period during which a facility is capable of performing its function. However, in some cases a facility will cease being useful even though it is still functional, in which case its useful life will be shorter than its operable life. The useful life may be shorter or longer than the planning period.
4. *Financing period* is the period for meeting debt obligations or required paybacks for undertaking a project. This period may also be shorter or longer than the planning period.

Alternatives conceived during planning should all meet the planning objectives for the planning period. Each alternative should incorporate an evaluation of phasing of construction during the planning period. While the ultimate design for each alternative should serve the needs for the entire planning period, the initial phase to be constructed may serve only a portion of the planning period.

The optimum period for the design life of each project component is dependent on the degree of certainty for predicting future needs, the useful life of the component, the practical ability to add facility expansions, and the economy of scale related to the component. Present worth cost analyses, discussed in Sec. 25-8, can be used to arrive at the optimum design life by testing different phasing intervals for certain types of facilities. However, if there is relative uncertainty in projecting growth in demand, either in terms of quantities or geographic direction, shorter design lives for the initial phase are desirable. Based on the actual experience for many water reuse projects, projecting future reclaimed water demand is a very uncertain task (Mills and Asano, 1996).

Time Value of Money

It is generally understood that money has a time value; that is, a dollar is worth more to us today than the promise of receiving a dollar a year from now because a dollar earned today can be put to use today by either investing it or spending it for an immediate benefit. This benefit is the basis for charging interest for the use of money. The time value of money presents a dilemma in the evaluation of projects. How can the costs be compared for two different projects that have different streams of capital and operations costs over the lives of the projects? A simple addition of the costs does not reflect the time value of money.

The interest rate is the measure of the time value of money. The rate is influenced by inflation, tax rates, and the risk in an investment. However, even assuming a condition without inflation, taxes, or risk, there is still some expectation of a rate of return on the use of money. Which rate to use in an analysis depends on the purpose of the analysis and will be discussed in Secs. 25-8 and 25-9.

Interest rate factors have been developed that can be used to calculate the value of a monetary transaction made at one time to the value at another time. There are formulas for simple and compound interest factors to calculate interest earned, loan repayments, and equivalent values of funds at different points in time. The formulas with example calculations are shown in Appendix H. The applications of the time value of money in economic and financial analyses are explained in Secs. 25-8 and 25-9.

Inflation and Cost Indexes

Inflation must be considered in estimating costs used for economic and financial analyses. *Inflation* and, less commonly, deflation are increases and decreases, respectively, in the price levels for goods and services. General inflation, as it affects prices in the overall economy, represents changes in the buying power of money. Differential inflation; that is, the difference in the inflation rate for a particular product as compared to the general inflation rate, may represent a change in the scarcity of an item or the efficiency of its production.

Cost estimates used in planning are often based on the actual experience of past projects. Cost indexes are the usual tool for tracking past inflation and adjusting historical costs into current costs. The Consumer Price Index is the common measure of general inflation. Because of the differential inflation, specialized indexes have been developed for many industries or types of construction activities. For public works projects, the *Engineering News-Record* magazine publishes the Construction Cost Index (ENRCCI), which is used commonly for wastewater and water supply projects, including water reclamation and reuse projects. It is, therefore, important to understand the proper use of this index.

As a general rule, a cost at one point in time can be converted to an actual cost at another point in time using the following formula:

$$\text{Cost}_{t_2} = (\text{Cost}_{t_1}) \frac{\text{Index}_{t_2}}{\text{Index}_{t_1}} \tag{25-1}$$

where Cost_{t_2} = adjusted cost at time t_2
Cost_{t_1} = the source cost at time t_1
Index_{t_2} = cost index at time t_2
Index_{t_1} = cost index at time t_1

Because different geographic regions experience different rates of inflation, the ENRCCI is computed for 20 cities in the United States. In addition, these indexes, published monthly, are averaged to produce the U.S. 20-city average. One of the common errors in use of the ENRCCI is to use it to convert the cost in one geographic area to a cost in another. Each city index is tracked independently and can be used to measure cost changes over time but not differentials between cities. The index for each city began at 100 in 1913. Annually in March, *Engineering News-Record* publishes a historical summary of indexes alongwith an explanation of the use of the index (Grogan, 2005).

EXAMPLE 25-1. Use of Cost Index and Interest Factors.

In December 2004, a community proposed to add direct filtration tertiary treatment to its 20,000 m³/d secondary wastewater treatment plant to produce reclaimed water suitable for irrigating school grounds. The capital cost of upgrading treatment was estimated to be $3,275,000 in May 2000 dollars. Construction is expected to occur from February 2007 through December 2007. Estimate the capital cost of the treatment plant construction using the ENRCCI and an assumed inflation rate of 3 percent in the future.

Solution

1. Estimate the capital cost in December 2004 dollars. The historic cost is adjusted to December 2004 dollars using the 20-City Average ENRCCI. Adjusted capital cost through December 2004

$$= (\$3,275,000) \frac{\text{ENRCCI@ Dec 2004}}{\text{ENRCCI@ May 2000}} = (\$3,275,000) \frac{7308}{6233} = \$3,839,836$$

2. Project the construction cost to the midpoint of construction (July, 2007). Because inflation rates are compounded annually, the compound amount factor will be used to adjust the cost from December 2004 through December 2006 and the simple interest factor will be used to complete the adjustment through July 2007. Use Eq. (H-2) in App. H to calculate the compound (F/P) amount factor.

Adjusted cost through July 2007

= ($3,839,836)(F/P, 3%, 2 yr)(Simple interest factor, 3%, 7 mo)

= ($3,839,836)$(1+0.03)^2[1 + (0.03)(7/12)]$ = $4,145,000 (rounded)

Comment

Although published cost indexes used to track historic cost trends are a useful tool for adjusting costs to a common past or current date, judgment is still required in selecting the appropriate index and in assuming inflation rates for future costs.

Inflation rates represent changes in *actual* costs over time, which corresponds to the costs experienced in the marketplace. Cost changes resulting from general inflation do not indicate a change in the actual investment of resources and labor in the production of goods and services. *Real costs* are costs adjusted to constant dollars to exclude inflation to reflect actual investment of resources and labor in the production of goods and services. Planning involves the estimation of costs occurring in the future. Actual costs or real costs can be used in monetary analyses, depending on the purpose of the analysis. The adjustment of future costs for inflation is discussed in later sections on economic and financial analyses.

25-8 ECONOMIC ANALYSIS FOR WATER REUSE

Economic analysis is used to determine the total monetary costs and benefits of all resources committed to a project, regardless of who in society contributes them or who in society receives the benefits. Economic analysis is designed to anticipate and assess the impacts of alternative policies over the longer term and on all affected parties, not only on those affected immediately. Water and the resources required to both exploit and protect it are increasingly scarce; hence, it is in the public interest that economic criteria be applied to water management decisions (Young, 2005). Economic analyses are performed to determine whether project alternatives have a net benefit in monetary terms and to rank alternatives in terms of relative benefit. The perspective of an economic analysis is that of society.

For public works projects, such as water reclamation and reuse projects, an attempt should be made to quantify all costs and benefits associated with a project, not just the costs that will be experienced by the project sponsor. Economic analysis includes costs and benefits to utilities and ratepayers as well as all externalities from other perspectives, such as environmental impact. The objective of an economic analysis is to determine the project alternative that will achieve the highest net benefit to the public as a whole. The traditional approach for performing an economic analysis for resource projects is to perform a benefit–cost analysis. The benefit–cost approach is typically adapted when applied to water reclamation and reuse. Because of difficulties in quantifying benefits in monetary terms, approaches involving least-cost analyses, such as cost-effectiveness, are explained.

Comparison of Alternatives by Present Worth Analysis

A present worth analysis is used to compare two alternative actions with different cash flows to determine which alternative has the highest net benefit or lowest net cost over time. This method of analysis is based on the concept of the time value of money discussed in Sec. 25-7. It involves translating future monetary cost and benefit flow streams into a single present lump sum called the *present worth* or *present value*. Future cash flows are converted to present worth amounts by use of the present worth factor (*P/F*) (see Appendix H). The technique as applied in an economic analysis is illustrated later in this section.

The interest rate used in a present worth analysis is called the *discount rate*. The appropriate discount rate depends on the purpose of the analysis and the time value of money within the context of the purpose. The appropriate discount rate also depends upon whether actual (inflated) costs or real (constant dollar) costs are used in the analysis. The difference between costs is discussed later under the subheading Basic Assumptions of Economic Analyses.

Measurement of Costs and Inflation

For the purpose of an economic analysis, an attempt is made to compare projects on a real-cost basis that will represent the relative investment of material and labor resources in each alternative. Because the rate of inflation represents a change in the value of money rather than a change in the investment of social resources, real costs in constant dollars rather than actual costs are used in estimating future costs for an economic analysis. However, because there is a time value of money, aside from the inflation factor, a present worth analysis is used to compare alternatives on an equivalent basis.

Future costs are estimated in constant dollars, that is, the dollar value of materials and labor at a chosen reference point in time. The reference point is usually a date close to the period of the analysis or the date when a project is expected to be constructed. Cost indexes can be used to adjust costs to a common current date. However, if the reference point is the period of construction, then an assumption of inflation between the present and the future construction date will have to be made and all costs adjusted to that common reference point. Other than adjusting estimated costs to a common reference point, inflation into the future is disregarded. However, if there is a basis for assuming that certain goods are inflating at a different rate from common inflation, it is permissible to adjust costs of those goods by the differential inflation rate, i.e., the difference between the two inflation rates.

Measurement of Benefits

Benefits are measured in terms of the effectiveness of actions or projects in achieving their stated goals (James and Lee, 1971). The two most common goals of water reclamation and reuse projects are the production of a usable water supply and the reduction of pollution in the environment. Benefits can be measured by the market value of project outputs. However, the valuation of the output of a cleaner environment is very difficult. Furthermore, while water has a market value, (i.e., the price of water to customers), the transfer of water does not occur within a free market economic environment. The price of water is often set to recover the costs of production rather than to reflect its worth in the marketplace. The price of water is generally less than the cost of development of new sources of water. The true benefit of an adequate water supply is *intangible*, i.e., its monetary value cannot be measured readily.

Alternative Cost Valuation

An alternative method of measuring project benefits is through the cost of producing the same outputs in an alternative manner, which is the most common approach for public works projects. Within an array of alternatives that achieve the project goals equally effectively, the least-cost alternative is usually the recommended project, unless there are other mitigating circumstances. The benefits are measured as the avoided cost of the second-best alternative. Inherent in this approach is that the second-best alternative would in fact be built if the recommended project were not built. It is always possible to find a more expensive project to compare to, but to provide a valid analysis, the second-best alternative must be a practicable one (Gittinger, 1982; James and Lee, 1971).

Cost-Effectiveness Analysis

Another approach to project analysis when benefits cannot be measured reasonably in monetary terms is the *cost-effectiveness analysis*. A cost-effectiveness analysis is used to determine which project alternative will result in the minimum total resources cost over time to meet project objectives. The premise is that the level of primary benefits is the same for all alternatives under consideration or that all alternatives meet a minimum stated level of benefits. The most cost-effective project alternative is the alternative which the analysis determines to have the lowest net economic cost (present worth or equivalent annual value) unless nonmonetary costs are overriding while meeting minimum project objectives. While a cost-effectiveness analysis relies significantly on the economic analysis, intangible costs and benefits are considered (Gittinger, 1982; Office of the Federal Register, 1991).

Life Cycle Cost

Using the alternative costs as the basis of comparison for projects, the costs of each alternative is determined with the present worth analysis using common assumptions for each alternative, such as an equivalent planning period and discount rate. If the present worth covers the life span of the project or otherwise takes into consideration project life, the present worth is also called the life cycle cost. There are two common derivations from the present worth cost: (1) equivalent annual cost and (2) unit cost.

Equivalent Annual Cost The equivalent annual cost is a uniform annual cost that has a present worth equal to the present worth of the project. It is derived as follows:

$$\text{Equivalent annual cost} = (\text{Present worth})(A/P) \tag{25-2}$$

where the capital recovery factor, A/P [see Eq. (H-5) in App. H], is based on the same discount rate and time period as the present worth.

Unit Cost When there are alternative approaches to achieving a given output, but in different quantities of the output, the unit cost approach based on volume is useful. Alternative water supply projects may not yield the same amounts of water. Comparison on the basis of cost per unit of water yield may be the only means of cost comparison. When comparing unit costs it is important that the units be equivalent. For example, it is invalid to compare dollars per unit of reclaimed water delivered to dollars per unit of potable water delivered if more reclaimed water is needed to serve a purpose than potable water due to water quality differences.

The computation of unit cost involves a computation of the present worth of reclaimed water delivered. A commodity, like money, also has a time value. The delivery of a unit of water today has more value than a promise of its delivery in the future. Therefore, the same concepts of present worth can be applied to a commodity such as water as to money. To obtain a valid unit cost, the time value must be taken into consideration. Thus, the unit cost of reclaimed water delivered

$$= \frac{\text{(Present worth cost, \$)}}{\text{(Equivalent present worth of volume of reclaimed water delivered, m}^3\text{)}} \quad (25\text{-}3)$$

The present worth cost is computed using Eq. (H-3) given in App. H. The equivalent present worth of the reclaimed water is computed using the quantity of water delivered and applying the present worth factor. If the present worth of the volume of water is not computed, the unit cost will be incorrectly underestimated. Another factor leading to underestimation of unit costs is the disregard of buildup of water demand or production in the early project years, by estimating unit costs by dividing annual costs at full project development by the full project deliveries. Applying Eq. 25-3 over the life of the project, the project buildup is incorporated into the calculation. The derivation of the unit cost based on volume and appropriate units of comparison are illustrated by Example 25-3.

Basic Assumptions of Economic Analyses

The basis for expressing costs in an economic analysis is real costs in constant dollars. Thus, the discount rate should be a rate that is not dominated by inflation influences. Market interest rates are influenced by rates of inflation and tend to rise as the rate of inflation rises. Long-term government borrowing rates averaged over several years are often considered a better basis for discount rates in economic analyses. However, the issue is complex and economics references should be consulted for more background (James and Lee, 1971; Riggs and West, 1986; USBR, 1990).

Sunk Costs
Sunk costs are expenditures that occurred in the past. They should not be allowed to influence decisions on future actions. Likewise, continuing debt payments for past investments can generally be considered a sunk cost, because the debt obligation will remain regardless of which future action is taken.

Time Reference Point

The time reference point for economic analyses is usually the beginning of project operation. All costs and benefits for each alternative through the end of the planning period should be identified. The present worth of costs and benefits is computed to this point in time. Design and construction costs occurring before this reference point must be brought forward to this point in time.

Costs Induced by Project Alternatives

All costs induced by project alternatives should be considered in the economic analysis. There is a tendency to ignore costs that may be essential for project implementation but not the responsibility of the entity performing the feasibility study. For example, onsite conversion costs, also called retrofit costs, are necessary for users presently using freshwater to convert the infrastructure to reclaimed water. The installation of dual plumbing or special signage warning against improper use of reclaimed water, are examples. The project proponent may not be intending to pay for these costs. Nevertheless, they are a part of the project and should be calculated in the economic analysis.

Subsidies

Subsidies are not considered in economic analyses because they represent a *transfer of funds*, not an investment of resources. From the society perspective, rebates by a utility to a reclaimed water customer would not be considered. While a rebate is a cost to the utility, it is an equivalent benefit to the consumer, yielding no net cost or benefit to society. The same is true for payments between utilities or government subsidies for utilities. These transactions are all considerations of financial analyses, addressed in Sec. 25-9. Transfers would be subject to consideration of equity, fairness of distribution, and political considerations, which are part of the social impact and public acceptance project evaluation criteria. They are also part of the nonmonetary criteria used along with the economic analysis when applying cost-effectiveness analysis.

Inclusion of Associated Project Costs

If a first phase of a project is dependent upon future phases for fulfillment of reclaimed water delivery projections, then the facilities plan should identify the market, facilities, and associated costs of all dependent phases. It is common, but inappropriate, to attribute an output to a project without identifying all of the facilities and costs that must be added in future to realize this yield.

Replacement Costs and Salvage Values

Facilities that will reach their useful lives during the planning period will need replacement. The replacement costs must be included in the economic and financial analyses. If a useful life exceeds the planning period and if the facility is expected to remain in use, the facility will have a salvage value at the end of the planning period. When the planning period is shorter than the useful life, the life cycle cost can be computed by taking into consideration the salvage value.

There are various formulas for depreciation used for taxing purposes and private sector financing. For the purposes of an economic analysis, straight-line depreciation from the first day of facility operation is assumed for determining salvage value. Computation of salvage value is illustrated in Example 25-2.

EXAMPLE 25-2. Calculation of Replacement Cost and Salvage Values.

A small reclaimed water distribution system will cost $1,000,000. It consists of a pumping station structure costing $200,000, pumps costing $150,000, and pipelines costing $650,000. The useful lives for the structure, pumps, and pipelines are 20, 15, and 50 yr, respectively. What is the salvage value at the end of a 20 yr planning period?

Solution

1. Determine the remaining lives of the facilities at the end of the planning period.
   a. The pumps will be replaced in 15 yr, before the end of the planning period.
   b. The remaining lives of the facilities at the end of the planning period is based on the following relationship.

   Remaining life = useful life − planning period

   Pumping station structure = 20 yr − 20 yr = 0 yr

   Replacement pumps = 15 yr + 15 yr − 20 yr = 10 yr

   Pipelines = 50 yr − 20 yr = 30 yr

2. Determine the salvage value at the end of 20 yr planning period.
   a. Salvage value of facilities is determined using the following relationship:

   Salvage value = (initial cost)(remaining life)/(useful life)

   Pump station structure = ($200,000)(0/20) = $0

   Replacement pumps = ($150,000)(10/15) = $100,000

   Pipelines = ($650,000)(30/50) = $390,000

   b. The total salvage value (TSV) is

   TSV = $100,000 + $390,000 = $490,000

Comment

Salvage values are a means to recognize the different useful lives of project components and to provide credit for facilities that remain useful beyond the planning period.

Application of the various concepts described above for economic analyses is illustrated in Example 25-3.

Computation of Economic Cost

EXAMPLE 25-3. Conduct an Economic Feasibility Analysis for a Water Reclamation and Reuse Project.

A water reclamation and reuse project with a total capital cost of $1.5 million is proposed by a small city. Design and construction are each estimated to take 1 yr, beginning in 2006. An economic analysis is to be conducted using the information given in the following table on the years of occurrence of various project phases, capital costs, useful lives, and salvage values. This example and Example 25-4 are adapted from Mills and Asano (1998).

Item/phase	Years of occurrence	Capital cost, $	Useful life, yr	Salvage value at end of 2027[a], $
I. Construction				
Tertiary treatment addition	2007	600,000	20	0
Pump facilities	2007	50,000	15	33,333[b]
Distribution pipelines	2007	500,000	50	300,000
Onsite plumbing conversions	2007	100,000	50	60,000
Subtotal construction		1,250,000		393,333
II. Design	2006	150,000		0
III. Services during construction	2007	100,000		0
Total project costs		1,500,000		393,333

[a] Assumed straight-line depreciation for 20 yr planning period. Example: For distribution pipelines, ($500,000) × (50 yr − 20 yr)/50 yr = $300,000
[b] Based on salvage value of replacement pumps.

The following factors are to be considered in the analysis:

1. Deliveries of reclaimed water will be 200×10^3 m^3 during the first year of operation, 270×10^3 m^3 during the second year, and 450×10^3 m^3 each year thereafter. Deliveries include 100×10^3 m^3/yr to industrial users for cooling water, beginning in the first year, with the remaining reclaimed water delivery to landscape irrigation users.

2. Because of the increased salinity of the reclaimed water as compared to potable water, more reclaimed water must be applied for landscape irrigation to flush salts from the root zone. Industrial users will have to waste blowdown process water more frequently than would be required for potable water. Thus, each cubic meter of reclaimed water replaces only 0.8 m^3 of potable water.

3. Operation and maintenance (O&M) costs will be $40,000 during the first year, $60,000 the second year, and $85,000 each year thereafter. However, in year 15, assume a pump replacement will be necessary at a cost of $50,000.
4. The fertilizer value of the nutrients in the reclaimed water is $0.04/m³.
5. Consistent with the concept of economic analyses, all costs used in this example are expressed in constant dollars at the midpoint of construction in 2007.

Assuming a 20-yr planning period and an interest rate of six percent, determine the present worth of the project and the unit cost based on volume. Note: Present worth factors, readily available in most economics reference books, are given in Appendix H.

Solution

1. The reclaimed water deliveries and the annual costs, expressed as present worth, are presented in the following table.
2. The total present worth for the project is $2,172,724 (see Column 13).
3. The present worth of the delivered reclaimed water for each year, as given in Column 15, is computed by applying the present worth factor to the annual reclaimed water delivery.
4. The present worth volume after 20 yr is 4,765,000 m³. The unit cost of fresh water delivered, computed using Eq. 25-3, is

$$\frac{\$2{,}172{,}724}{4{,}765{,}000 \text{ m}^3} = \$0.46/\text{m}^3$$

5. Because the amount of reclaimed water deliveries is not equivalent to freshwater replaced, calculations for freshwater replaced are also presented. The unit cost in relation to freshwater replaced would be the appropriate basis of comparison to alternative freshwater projects. Unit cost of fresh water replaced is

$$= \frac{\$2{,}172{,}724}{3{,}812{,}000 \text{ m}^3} = \$0.57/\text{m}^3$$

Comment

1. If the present worth for the project ($2,172,724) is divided by the total volume of water delivered over the 20 yr period (8,570,000 m³), the corresponding unit cost for water would have been $0.25/m³. Because of the large discrepancy in unit cost between the total and present worth volumes of water, the failure to use the discounted volume of water in this analysis would distort the comparative costs when evaluating alternative water projects.
2. When comparing a water reclamation project with a freshwater project, it is important to compare projects on the basis of equivalent freshwater deliveries or freshwater replaced.

	Reclaimed water deliveries, 10^3 m^3			Potable water replaced,[b] 10^3 m^3	Design period costs,[c] $	Construction and related costs,[c] $	Operation, maintenance, and replacement costs,[d] $	Fertilizer credit,[e] $	Salvage value[c]	Total net cost, $	Present worth factor at a 6% discount rate[f]	Present worth of net cost,[g] $	Present worth of potable water replaced,[h] 10^3 m^3	Present worth of reclaimed water deliveries,[i] 10^3 m^3
Year	Industrial use	Landscape irrigation use[a]	Total											
1	2	3	4	5	6	7	8	9	10	11	12	13	14	15
2006					150,000					150,000	1.06000	159,000	0	0
2007						1,350,000				1,350,000	1.00000	1,350,000	0	0
2008	100	100	200	160			40,000	(4,000)		36,000	0.94340	33,962	151	189
2009	100	170	270	216			60,000	(6,800)		53,200	0.89000	47,348	192	240
2010	100	350	450	360			85,000	(14,000)		71,000	0.83962	59,613	302	378
2011	100	350	450	360			85,000	(14,000)		71,000	0.79209	56,239	285	356
2012	100	350	450	360			85,000	(14,000)		71,000	0.74726	53,055	269	336
2013	100	350	450	360			85,000	(14,000)		71,000	0.70496	50,052	254	317
2014	100	350	450	360			85,000	(14,000)		71,000	0.66506	47,219	239	299
2015	100	350	450	360			85,000	(14,000)		71,000	0.62741	44,546	226	282
2016	100	350	450	360			85,000	(14,000)		71,000	0.59190	42,025	213	266
2017	100	350	450	360			85,000	(14,000)		71,000	0.55839	39,646	201	251
2018	100	350	450	360			85,000	(14,000)		71,000	0.52679	37,402	190	237
2019	100	350	450	360			85,000	(14,000)		71,000	0.49697	35,285	179	224
2020	100	350	450	360			85,000	(14,000)		71,000	0.46884	33,288	169	211
2021	100	350	450	360			85,000	(14,000)		71,000	0.44230	31,403	159	199
2022	100	350	450	360			135,000	(14,000)		121,000	0.41727	50,489	150	188
2023	100	350	450	360			85,000	(14,000)		71,000	0.39365	27,949	142	177
2024	100	350	450	360			85,000	(14,000)		71,000	0.37136	26,367	134	167
2025	100	350	450	360			85,000	(14,000)		71,000	0.35034	24,874	126	158
2026	100	350	450	360			85,000	(14,000)		71,000	0.33051	23,466	119	149
2027	100	350	450	360			85,000	(14,000)	(393,333)	(322,333)	0.31180	(100,505)	112	140
Total			8,570		150,000	1,350,000	1,680,000	(262,800)	(393,333)			2,172,724	3,812	4,765

[a] Calculated, Column 4 − Column 2.
[b] Potable water replaced equals 80 percent of reclaimed water deliveries: Column 5 = (Column 4) × 0.8.
[c] Given data
[d] After 15 yr of useful life, the pump replacement cost of $50,000 is included in year 2022.
[e] Calculated, (Column 3) × ($0.04/m^3).
[f] Present worth factor = $1/[(1 + i)^n]$, where i = 0.06, n = years from beginning of operation. All costs or benefits are assumed to occur at end of year.
[g] Calculated, (Column 11) × (Column 12).
[h] Calculated, (Column 5) × (Column 12).
[i] Calculated, (Column 4) × (Column 12).

Note: The values reported in the table, computed using a spreadsheet program without rounding, have been truncated for the purpose of presentation.

Project Optimization

Economic analyses can be used not only to identify the least-cost alternative within a group of distinct alternatives to achieve a project objective, but also to optimize a given alternative with respect to size and features.

Marginal Cost Analysis

Optimization involves the use of marginal cost analysis. The *marginal cost* is the cost associated with adding a particular project feature or adding an increment of size to a project or component of project. It is cost-effective to add the project feature or increase the size if the marginal cost is less than the associated increase in benefits.

Each increment of additional demand added to the reclaimed water system is associated with potential marginal costs of additional wastewater treatment, additional pipelines, additional daily or seasonal storage, and operational costs of treatment and pumping. An example of when marginal cost analysis should be used is in the consideration of whether to upgrade from secondary to tertiary level of treatment to deliver more reclaimed water. A certain market for a hypothetical project may have been identified that could use 1.0×10^6 m³/yr of reclaimed water of secondary treatment quality. Because of fewer restrictions on the use of tertiary treated reclaimed water, the potential market for tertiary quality reclaimed water could be 1.5×10^6 m³/yr. There is a base cost for this project using secondary effluent, consisting of treatment and distribution facilities. To upgrade treatment to tertiary and to make use of this reclaimed water, there are marginal costs of the added treatment level, the expansion of the capacity of the secondary and tertiary treatment to serve additional users, and the expansion of the pipeline distribution system to reach the additional users that could use only tertiary treated reclaimed water. If the basis of project justification is the cost of alternative water supply alternatives, then the upgrade to tertiary treatment would be justified if the marginal costs were less than providing 0.5×10^6 m³/yr from new freshwater development.

Another common example where marginal cost analysis should be applied is in determining the geographic extent of a reclaimed water distribution system. To reach each new user of reclaimed water, there are marginal costs of additional pipeline to the user and expanded capacity of treatment, pumping facilities, and pipelines to serve the additional water demand.

Comparison of Approaches to Cost and Benefit Analysis

A common fallacy in project justification is to use total or average costs and benefits rather than marginal costs and benefits in cost analyses. Total and average costs, as illustrated on Fig. 25-4, tend to mask project components or sizes that are not cost-effective. The hypothetical curves on Fig. 25-4 are used to illustrate typical trends of costs for water reclamation projects. The benefits shown in this illustration are the least-cost alternative freshwater development costs, which are assumed to have a constant marginal unit cost within the range of the reclaimed water project size. The cost curves, shown as smooth curves, in reality are irregular steps representing the addition of individual users or groups of users and the costs associated with each discrete addition.

The marginal cost curve in the lower half of the figure illustrates the initial effect of economy of scale, drawing down the unit cost as the project becomes larger. However,

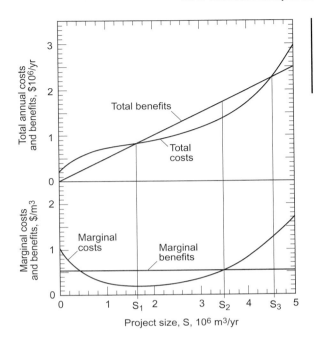

Figure 25-4
Use of marginal costs and benefits for project size optimization.

as the distribution becomes more extensive to reach users progressively farther away from the water reclamation plant or to reach users with progressively smaller water demands, the marginal costs begin to rise to a point of being uneconomical to serve. If total costs alone are compared, as is done in calculating benefit–cost ratios or in comparing average unit costs, any project within the size range between S_1 and S_3 appears to be cost-effective. However, when marginal costs are analyzed, it is seen that any project addition beyond S_2 has a marginal cost greater than the marginal benefit. Thus, it is not cost-effective to build a project larger than S_2. While any project within the range of S_1 and S_2 is justified, the optimum size would be at S_2, where the maximum net benefit is achieved. Making decisions based on benefit–cost ratios or average unit costs may lead to uneconomic, oversized projects.

Influence of Subsidies

The objective of the economic analysis is to identify the project that has the least true cost or the most net benefits. Subsidies, such as grants or rebates received from external sources, should not be incorporated into the analysis. Subsidies would lower the cost curves shown on Fig. 25-4, falsely giving the impression that a given project alternative has a greater net benefit. Subsidies incorporated into analyses for project selection or optimization can result in oversized, if not entirely uneconomic, projects.

Subsidies can come in many forms. For water reclamation and reuse projects, the most common forms are grants or below-market interest rate loans for capital financing and operational rebates based on reclaimed water usage or deliveries. Because subsidies may apply to only certain categories of costs, such as capital costs, they can sometimes influence local decisions contrary to the broader public interest. A high cost alternative characterized by high capital costs but low operational costs may be favored by an agency if there is a subsidy for capital costs but not operational costs. Unless there are overriding non-monetary benefits to the public at large, subsidies should not bias local decisions in favor of more expensive project alternatives.

Subsidies should be a tool to encourage actions that are beneficial to the public at large. After the most cost-effective project is identified considering the economic analysis and other non-monetary factors, subsidies should be used to make this alternative more desirable, affordable, or otherwise capable of being implemented. From a local viewpoint, water reclamation may appear to be financially more expensive than alternative water supplies, even though from an economic analysis it can be shown to be the least-cost water supply. Subsidies, usually from higher levels of government or regional water supply or wastewater agencies, can lower the cost of implementation of the most cost-effective project. Subsidies rightfully belong in the financial feasibility analyses, discussed in Sec. 25-9.

25-9 FINANCIAL ANALYSIS

The *financial analysis* is used to address the question of whether a water reuse project is financially feasible (i.e., whether there is the willingness and capability to pay for the project). Financial feasibility must be addressed from perspectives of all project participants, and, in some cases, external parties that may experience financial effects of a project. There are the viewpoints of the project sponsor, other suppliers and distributors of the reclaimed water, and the reclaimed water users. However, even users that are not participants in a water reuse project may see their water rates rise or fall because of the use of reclaimed water. Each participant or affected party expects to remain the same financially or better off by the use of reclaimed water in lieu of freshwater.

A financial analysis involves the following steps:

1. Identification and estimation of all project monetary costs and benefits at market prices
2. Allocation of costs to project purposes if a project is a multiple-purpose project
3. Allocation of costs between project participants
4. Development of capital financing mechanisms, such as bond sales or loans
5. Design of rate, fee, tax or other revenue producing structures to repay costs
6. Determination of adverse financial impacts on nonparticipants, such as potable water users

Construction Financing Plans and Revenue Programs

Financial analyses for water reuse projects generally fall into two categories: construction financing plans and revenue programs. The options for financing of capital costs are addressed in the construction financing plan. The sources of revenue during the operation of a project to cover debt service and operation and maintenance costs are analyzed in the revenue program. Because capital financing affects annual costs, the two analyses are related.

While the methods of financing were disregarded in the economic analysis, they are an important consideration in financial feasibility analysis. Different sources of capital funds have different interest rates or repayment constraints to consider. As discussed in Sec. 25-8, subsidies of various kinds may be available to make the economically justified alternative more attractive. It is appropriate in financial analyses to incorporate subsidies and actual borrowing interest rates in the computation of debt service and annual costs. The computation of debt service should reflect actual repayment periods, regardless of the planning period or useful life of facilities. Unlike the economic analysis, which uses costs expressed in constant dollars, financial analyses use market prices reflecting projected inflation.

In establishing reclaimed water rates, it may be necessary to consider ongoing costs of facilities existing prior to the project under investigation. Debt service on existing facilities, treated as sunk costs in economic analyses, could be given consideration in financial analyses. Existing costs may have to be melded with the costs of new facilities in establishing water rates. However, care must be taken not to attribute sunk costs as a cost of a new project, because they will be common to all alternatives being compared.

Cost Allocation

Cost allocation is a matter of policy, equity, relative distribution of benefits, and ability to pay. Cost allocation is often a matter of policy and negotiation if different parties are involved. Often, the level of treatment required to meet water quality standards for discharge is allocated to the purpose of water pollution control. Costs for higher levels of treatment needed for reuse of effluent may be considered to serve the purpose of water supply. The costs that are easily associated with a particular purpose are *separable costs*. Where there are *joint costs* that are associated with more than one project purpose, a common calculation procedure is the separable costs–remaining benefits method. Joint costs are divided in proportion to the amount of benefits for each project purpose that exceed the separable costs for each purpose. Detailed procedures can be found in James and Lee (1971) and Gittinger (1982).

Influence on Freshwater Rates

Because reclaimed water use may reduce the use of fresh water, the freshwater rates may be influenced. The existing freshwater infrastructure will have to be financed by a lower base of freshwater sales, which could cause freshwater rates to rise. On the other hand, the use of reclaimed water may offset the purchase of more expensive fresh water. It is best to approach reclaimed water as another component of water supply for the community as a whole, rather than as a separate water system to be financed independently. Reclaimed water rates should be set to reflect the value of the reclaimed water relative to freshwater. Revenues from reclaimed water sales may need to be transferred to offset freshwater revenue losses, or freshwater revenue may be appropriately used to subsidize water reuse to create an equitable sharing of the costs and benefits of the reclaimed water supply.

Other Financial Analysis Considerations

Within the reclaimed water study area, potential reclaimed water users may obtain their water supplies from different sources at different costs. In performing the financial analysis, it is important to consider what users are presently paying for water and would be paying in the future. The use of average water prices for users may be appropriate for preliminary feasibility analyses, but in the final analysis the situation of each potential user must be considered to assess the financial feasibility from the viewpoint of the user.

The costs of onsite conversions from fresh water to reclaimed water use are often ignored or inappropriately assigned. The assumption may be that these costs are to be borne by the users and are of no concern to the project proponent. However, these costs will clearly be an issue of user acceptance of reclaimed water and must be considered early in the marketing analysis and later in the financial analysis. To encourage user participation, the project proponent may elect to pay for onsite costs, to loan the funds to the user at a subsidized rate, or to reduce the reclaimed water rates in the first few years of the project operation to offset the costs of conversion.

Inflation, which is generally ignored in the economic analysis, should be considered in financial analyses insofar as the rate of inflation can be estimated realistically.

Sources of Revenue and Pricing of Reclaimed Water

There are a variety of sources of revenue to cover the costs of a water reuse project. An obvious possible source is payment for the reclaimed water by users. Insofar as water reclamation serves as a means of treatment and disposal of wastewater, sewer use charges may be justified as a source of funds to cover water reclamation costs. As noted earlier, revenues from freshwater sales may be used to offset reclaimed water costs. The primary sources of revenue used to cover debt service and operational costs of reclaimed water systems are summarized in Table 25-10.

Table 25-10
Reclaimed water project revenue sources

Revenue source	Potential rationale
Reclaimed water delivery charges	Reclaimed water users receive a water supply benefit, a more secure supply in droughts.
Property taxes or parcel assessments on reclaimed water users	
Reclaimed water connection fee	There are initial costs for review of new reclaimed water use sites and installation of reclaimed water meters and potable water backflow prevention devices.
	In area of growth constrained by limited water supply, reclaimed water supply allows for new development of use site.
Impact fee on new development	Reclaimed water provides new water supply, freeing potable water for use in new development.
Rebate from regional water supplier for reclaimed water deliveries by retail agency	Reclaimed water use at the local level relieves the need for additional freshwater development by regional water supplier.
Property taxes or parcel assessments on all property in community	The entire community benefits from the use of reclaimed water that otherwise would be served by the community's freshwater supply.
Shared potable or freshwater revenue	The use of reclaimed water reduces wastewater treatment and disposal costs that would have been incurred with discharge of effluent to a surface water.
Shared sewer service charges	

Revenue from reclaimed water users can be collected in various forms, including water service connection fees, fixed monthly or annual charges, and variable charges based on quantities of water use. When other water sources are available, the upper boundary for the reclaimed water rate is the competing freshwater rate. A discount from this upper boundary is used commonly to provide an incentive to use reclaimed water. The discount reflects some added costs for the users resulting from point of use treatment or added maintenance caused by scaling or corrosion. The lower boundary is providing the water for free or even paying users to accept reclaimed water. To the extent that users realize a genuine benefit from the use of reclaimed water, prices should reflect this benefit. Excess revenues can be used to offset other water supply or wastewater management costs to benefit the community as a whole. A useful discussion of capital financing and revenue mechanisms is found in U.S. EPA (2004).

A financial/feasibility analysis for a water reclamation project is illustrated in Example 25-4.

Financial Feasibility Analysis

EXAMPLE 25-4. Conduct Financial Feasibility Analysis for a Water Reclamation Project.

Perform a financial feasibility analysis for the water reclamation and reuse project described in Example 25-3 using the conditions and assumptions listed below. It should be noted that in the economic analysis presented in Example 25-3, all costs are given in constant dollars. In a financial analysis, where actual cash flow is of relevance, inflated costs must be estimated. From the viewpoint of the city, financial feasibility of the project is contingent upon generating sufficient revenue to recover the water reuse project costs, taking into consideration other financial impacts on the city. From the viewpoint of the reclaimed water customers, the expectation is that reclaimed water will cost no more than potable water. For the given conditions, estimate the potential price range for the reclaimed water if the project is to be self-supporting. The following conditions and assumptions apply:

1. Capital costs are financed with a 20-yr loan at 8 percent annual interest.
2. The repayment period for the debt service begins on completion of construction, i.e., the end of 2007. The loan principal is computed as of that date. All costs are assumed to occur at the end of the year, so 1 yr of interest will have accrued for loan disbursements made to cover design costs during 2006.
3. Operation and maintenance costs are predicted to increase with inflation at an annual rate of 4 percent.
4. The costs in the tables for Example 25-3, which is the basis for this example, are adjusted to dollars in midpoint of construction, 2007. Inflation adjustments are made from this year.

5. All of the reclaimed water customers currently purchase potable water from the city proposing the water reclamation project. The retail price charged by the City for potable water is $0.30/m^3$, expressed in 2007 dollars. The price is expected to increase at a rate of 2 percent annually.

6. The city currently obtains potable water from two sources: groundwater pumping at a variable cost of $0.10/m^3$ and purchased imported water at a variable cost of $0.60/m^3$, expressed in 2007 dollars. The imported water cost is expected to increase at a rate of 4 percent annually.

7. The city is responsible for the financing and construction of the treatment and distribution facilities. The city will also reimburse reclaimed water users for making the necessary plumbing changes on their sites to accept reclaimed water.

Solution Part A, Total Project Cost and Net Cost

1. Calculate the loan principal, capital recovery factor, and annual debt service payment.

   a. Loan principal = design cost including interest for 1 yr + construction costs = ($150,000)(1.08) + ($1,350,000) = $1,512,000

   b. Compute the capital recovery factor (A/P) for 20 yr at 8 percent interest using Eq. (H-5) from App. H:

   $$\text{Capital recovery factor} = \frac{i(1+i)^n}{(1+i)^n - 1}$$
   $$= \frac{(0.08)(1+0.08)^{20}}{(1+0.08)^{20} - 1} = 0.101852$$

   c. Annual debt service = loan principal × capital recovery factor
   = ($1,512,000)(0.101852) = $154,001

2. Estimate the net annual cost including operation, maintenance, and replacement costs. Set up a computation table (typically done in a spreadsheet) for the analysis of the annual cash flow and net cost to city. The required computation table is given on the following page. The data in Columns 2, 3, and 5 are from the computation table prepared in Step 1 of Example 25-3. As noted in the problem statement, the operation, maintenance, and replacement costs used in Example 25-3 must be inflated using the assumed 4 percent rate. The inflation factor, shown in Column 6, is based on the compound amount factor. The adjusted operation, maintenance, and replacement costs are given in Column 7. The net project cost, given in Column 14, is $1,867,396.

Solution Part B, Reclaimed Water Price Range

3. Estimate the price range for reclaimed water and the price margin (the spread between the minimum and maximum possible reclaimed water prices). The calculations related to reclaimed water pricing are shown in the table presented following the project cost table. The information in Columns 2, 3, and 4 are from the computation table prepared in Step 1 above. The minimum reclaimed

Computation table Part A, total and net cost

				Operation, maintenance and replacement				Avoided potable water purchases			Lost potable retail water revenue			
Year	Total reclaimed water deliveries,[a] 10^3 m^3	Potable water replaced,[a] 10^3 m^3	Debt service, $	Cost in base year (2007),[a] $	Inflation factor at 4%	Inflation Adjusted,[b] $	Total project cost,[c] $	Adjusted unit cost,[d] $/m^3	Total cost,[e] $	Potable price escalation factor at 2%	Adjusted potable water price,[f] $/m^3	Total lost revenue,[g] $	Net cost,[h] $
1	2	3	4	5	6	7	8	9	10	11	12	13	14
2007													
2008	200	160	154,001	40,000	1.04000	41,600	195,601	0.62	99,840	1.02000	0.31	48,960	144,721
2009	270	216	154,001	60,000	1.08160	64,896	218,897	0.65	140,175	1.04040	0.31	67,418	146,139
2010	450	360	154,001	85,000	1.12486	95,613	249,614	0.67	242,971	1.06121	0.32	114,610	121,254
2011	450	360	154,001	85,000	1.16986	99,438	253,439	0.70	252,689	1.08243	0.32	116,903	117,652
2012	450	360	154,001	85,000	1.21665	103,415	257,416	0.73	262,797	1.10408	0.33	119,241	113,860
2013	450	360	154,001	85,000	1.26532	107,552	261,553	0.76	273,309	1.12616	0.34	121,626	109,869
2014	450	360	154,001	85,000	1.31593	111,854	265,855	0.79	284,241	1.14869	0.34	124,058	105,672
2015	450	360	154,001	85,000	1.36857	116,328	270,329	0.82	295,611	1.17166	0.35	126,539	101,257
2016	450	360	154,001	85,000	1.42331	120,982	274,982	0.85	307,435	1.19509	0.36	129,070	96,617
2017	450	360	154,001	85,000	1.48024	125,821	279,821	0.89	319,733	1.21899	0.37	131,651	91,740
2018	450	360	154,001	85,000	1.53945	130,854	284,854	0.92	332,522	1.24337	0.37	134,284	86,616
2019	450	360	154,001	85,000	1.60103	136,088	290,088	0.96	345,823	1.26824	0.38	136,970	81,235
2020	450	360	154,001	85,000	1.66507	141,531	295,532	1.00	359,656	1.29361	0.39	139,710	75,585
2021	450	360	154,001	85,000	1.73168	147,192	301,193	1.04	374,042	1.31948	0.40	142,504	69,655
2022	450	360	154,001	135,000	1.80094	243,127	397,128	1.08	389,004	1.34587	0.40	145,354	153,478
2023	450	360	154,001	85,000	1.87298	159,203	313,204	1.12	404,564	1.37279	0.41	148,261	56,901
2024	450	360	154,001	85,000	1.94790	165,572	319,572	1.17	420,747	1.40024	0.42	151,226	50,052
2025	450	360	154,001	85,000	2.02582	172,194	326,195	1.22	437,576	1.42825	0.43	154,251	42,869
2026	450	360	154,001	85,000	2.10685	179,082	333,083	1.26	455,079	1.45681	0.44	157,336	35,339
2027	450	360	154,001	85,000	2.19112	186,245	340,246	1.31	473,283	1.48595	0.45	160,482	27,446
Total			3,080,011			2,648,589	5,728,600		6,471,096			2,570,453	1,827,956

[a] Refer to computation table prepared in Step 1 of Example 25-3, columns 4 and 5.
[b] (Column 5) × (Column 6).
[c] (Column 4) + (Column 7).
[d] $0.60/m^3 × (Column 6).
[e] (Column 3) × (Column 9).
[f] $0.30/m^3 × (Column 11).
[g] (Column 3) × (Column 12).
[h] (Column 8) + (Column 10) − (Column 13).

Note: The values reported in the table, computed using a spreadsheet program without rounding, have been truncated for the purpose of presentation.

water price is obtained by dividing the values in Column 3 by the values in Column 2. The maximum reclaimed water price, given in Column 6, is 80 percent of the adjusted potable water price (Column 4). The pricing margin, given in Column 7, is the difference between the maximum (Column 6) and the minimum (Column 5) reclaimed water price.

Computation table Part B, reclaimed water price range

Year	Total reclaimed water deliveries,[a] 10^3 m^3	Net cost,[a] $	Adjusted potable water price,[a] $/m^3	Minimum reclaimed water price,[b] $/m^3	Maximum reclaimed water price,[c] $/m^3	Pricing margin available, $/m^3
1	2	3	4	5	6	7
2006						
2007						
2008	200	144,721	0.31	0.72	0.24	(0.48)
2009	270	146,139	0.31	0.54	0.25	(0.29)
2010	450	121,254	0.32	0.27	0.25	(0.01)
2011	450	117,652	0.32	0.26	0.26	(0.00)
2012	450	113,860	0.33	0.25	0.26	0.01
2013	450	109,869	0.34	0.24	0.27	0.03
2014	450	105,672	0.34	0.23	0.28	0.04
2015	450	101,257	0.35	0.23	0.28	0.06
2016	450	96,617	0.36	0.21	0.29	0.07
2017	450	91,740	0.37	0.20	0.29	0.09
2018	450	86,616	0.37	0.19	0.30	0.11
2019	450	81,235	0.38	0.18	0.30	0.12
2020	450	75,585	0.39	0.17	0.31	0.14
2021	450	69,655	0.40	0.15	0.32	0.16
2022	450	153,478	0.40	0.34	0.32	(0.02)
2023	450	56,901	0.41	0.13	0.33	0.20
2024	450	50,052	0.42	0.11	0.34	0.22
2025	450	42,869	0.43	0.10	0.34	0.25
2026	450	35,339	0.44	0.08	0.35	0.27
2027	450	66,886	0.45	0.15	0.36	0.21
Total		1,867,396				

[a] Refer to table developed in Step 1.
[b] (Column 3)/(Column 2).
[c] Because 1 m^3 of reclaimed water replaces only 0.8 m^3 of potable water, maximum price = (Column 4) × 0.8.
[d] (Column 6) − (Column 5)
Note: 1 m^3 = 8.107 × 10^{-4} ac-ft = 264.17 gal.

Discussion of Findings from Feasibility Analysis

In reviewing the information contained in the two computation tables prepared to assess the financial feasibility for this water reclamation project, it is appropriate to comment on the effect of external impacts, reclaimed water pricing, and pricing margin.

External Impacts As a result of use of reclaimed water, financial impacts external to the project itself need to be taken into consideration. For example, the city will avoid variable costs associated with supplying potable water to the reclaimed water customers. Typically, the city would reduce the most expensive source of supply, the imported water purchase, at $0.60/m^3$ ($2.27/1000 gal), as reflected in Columns 9 and 10 in the computation table prepared in Step 2. However, the city will also lose the revenue from the sale of potable water, as shown in Columns 12 and 13. An external cost possibly borne by the customers is the cost of onsite conversion of plumbing for handling a dual supply. In this example, the onsite conversion costs are assumed to be part of the project cost borne by the city and are financed through the city's debt service. The net cost to the city shown in Column 14 reflects the reduced potable water purchase and potable sales revenue.

Reclaimed Water Pricing The hypothetical minimum price, Column 5 in the computation table prepared in Step 3, is the unit cost to recover the net cost of the project, Column 3. From the perspective of the customers, the maximum reclaimed water price is the price that would have been paid for potable water. If onsite conversion costs were the responsibility of the customers, an allowance for this would have to be considered in setting reclaimed water prices. A further consideration is that it is assumed that customers will have to purchase more reclaimed water than they would have purchased of potable water due to the water quality difference. Additional water could be needed, for example, because of a reduction in the number of cycles of concentration in an industrial cooling tower due to higher salt content or an increase in irrigation flows to leach salts from the plant root zone. The maximum reclaimed water price was set at 80 percent of the potable price to account for this water quality effect.

Pricing Margin As indicated, there is no margin available during the first 3 yr. Unless supplemental income or a subsidy can be obtained to offset a negative cash flow in these years of project operation, this project would not be financially feasible. However, a case can be made to raise additional potable water sales revenue to provide additional income. In the later years, excess reclaimed water revenue can be used to cover potable system costs.

Sensitivity Analysis and Conservative Assumptions

To ensure engineering reliability, design factors may be based on water demands in extreme years, such as dry years; for irrigation. However, for the purpose of water supply planning and cost evaluation, it is important to use estimates typical of average years to have realistic expectations of project yield and realistic estimates of costs and benefits. It is advisable to report estimated project yield as a range and to use appropriate estimates within the range for long-range planning purposes and evaluating economic justification and financial feasibility.

Estimates of reclaimed water deliveries should also reflect the degree of uncertainty that certain users will eventually participate in the project. A sensitivity analysis should be performed testing the effects of high and low project deliveries on economic cost and financial feasibility. Depending on the degree of uncertainty in estimating demands for reclaimed water and the level of market assurances obtained, the worst-case scenario in terms of project cost would be minimum deliveries representing potential water users in existence that are willing to execute a binding agreement to accept reclaimed water. The best case of maximum deliveries would include demands such as future undeveloped land or users located outside the project proponent's boundaries where institutional agreements have not been formalized. It should be made clear that the best-case scenario is associated with the highest degree of risk of failing expectations.

PROBLEMS AND DISCUSSION TOPICS

25-1 An agency is constructing a water reclamation project that will cost $5 million. The agency has three financing options: (1) borrowing funds from a private lender at an interest rate of 8 percent, (2) receiving a federal grant of 25 percent of the cost and borrowing the remainder at an interest rate of 8 percent from a private lender, and (3) receiving a state loan at an interest rate of 4 percent. All borrowing options have a repayment period of 20 yr. Which financing option has the lowest cost? What is the equivalent subsidy of the lowest cost option?

25-2 Barstow Municipal Water District is proposing to construct a reclaimed water distribution system. Reclaimed water will be purchased from Frederick County Sanitation District, which collects and treats municipal wastewater to secondary level and discharges the treated effluent into the Shirley River. To be able to serve the reclaimed water to city parks with unrestricted access, tertiary treatment will be required. Barstow MWD will sell reclaimed water to the Mojave Golf Course and in exchange will receive groundwater pumping rights that the golf course currently uses to obtain its irrigation water. The city of Richard will purchase reclaimed water from Barstow MWD and deliver it to city parks and the privately-owned Whispering Sands Cemetery. The city will take custody of the portion of the reclaimed water distribution system built by Barstow MWD within the city boundaries. Describe the types of agreements, market assurances, or alternative institutional mechanisms needed between the parties involved in the production, distribution, and use of the reclaimed water. Describe the issues and provisions that should be addressed or included in each agreement or other mechanism.

25-3 The City of Sherwood Lake is evaluating the feasibility of serving reclaimed water to the Mathieson Golf Course. The golf course has its own wells to obtain water currently. What data or information are needed in the market assessment phase of the planning study to provide a basis for feasibility analysis according the various planning criteria?

25-4 The Kelseyville Wastewater District has a discharge prohibition for 6 mo of the year to protect the receiving stream during periods of low flow. It is too costly to store all of the effluent during the period of discharge prohibition. What are the possible reuse or land disposal options the district should consider to ensure the ability to dispose of the effluent? What forms of market assurances are needed at the various stages of planning and implementation for each option?

25-5 The Town of Cory operates a wastewater collection system and constructed a new wastewater treatment plant and reclaimed water system in 2007. The current flow at the treatment plant is 8×10^3 m³/d. The 10×10^3 m³/d capacity of the treatment plant is expected to be reached in 2017, at which time it is planned to be expanded to 20×10^3 m³/d. The plant will need to be replaced in 2037. A reclaimed water pumping station has a capacity of 20×10^3 m³/d enough to meet needs through 2017. While the pumping station structure will last until 2037, the pumps must be replaced every 10 yr. The reclaimed water distribution pipeline to the Thompson and Briggs farms will reach capacity by 2027 but is expected to remain in service until 2047. The town has sold bonds to pay for construction of the treatment plant and reclaimed water distribution system that must be repaid in full in 25 yr. Suggest an appropriate planning period that the town would have used in planning the treatment and reclaimed water facilities and explain why. What are the design lives and useful lives of the treatment plant, pumping station, pumps, and distribution pipeline? What is the financing period?

25-6 Reclaimed water pipelines of the same design were constructed in Peniuk Valley near San Francisco in December 1990 and December 1995 at unit costs of $190 and $200/m respectively. Adjusting both costs to December 2003 using the appropriate Engineering News-Record Construction Cost Index, which pipeline was the most expensive in real cost? With the information given, can the cost be estimated for a pipeline of the same size constructed in December 2003 in Kansas City using the ENRCCI index?

25-7 A reclaimed water project will deliver 200×10^3 m³/yr. The estimated costs are $1,000,000 for construction and $50,000/yr for operation and maintenance costs. Assuming a discount rate of 6 percent and useful life of 20 yr, calculate the unit cost in $/m³ using two methods. Assume all costs occur at the end of year incurred. First, use the present worth method to compute the present worth of all costs and of annual deliveries and then derive the unit cost from the present worth values. Second, use the equivalent annual cost method to compute the equivalent annual cost of all costs and then derive the unit cost from the annual costs and deliveries of reclaimed water.

25-8 Compute the salvage value at the end of 20 yr for a project consisting of the following facilities, taking into consideration replacement of equipment during the planning period:

Facility	Construction cost, $	Useful life, yr
Tertiary filtration treatment addition	5,000,000	30
Reverse osmosis treatment	1,000,000	15
Pumping station excluding pumps	800,000	20
Pumps	200,000	10

REFERENCES

Anderson, J. (2003) "The Environmental Benefits of Water Recycling and Reuse," *Water Sci. Technol.: Water Supply*, **3**, 4, 1–10.

Asano, T., and R. A. Mills (1990) "Planning and Analysis for Water Reuse Projects," *J. AWWA*, **82**, 1, 38–47.

AWWA (2001) *Water Resources Planning: Manual of Water Supply Practices, Manual M50*, American Water Works Association, Denver, CO.

Bales, R. C., E. M. Biederman, and G. Arant (1979) "Reclaimed Water Distribution System Planning: Walnut Valley, California," *Proceedings Water Reuse Symposium: Water Reuse—From Research to Application*, American Water Works Association Research Foundation, Denver, CO.

Beecher, J. A. (1995) "Integrated Resource Planning Fundamentals," *J. AWWA*, **87**, 6, 34–48.

CEQ (2005) *National Environmental Protection Act of 1969 as Amended*, Council on Environmental Quality Web site, http://ceq.eh.doe.gov/nepa/regs/nepa/nepaeqia.htm

City of San Diego (2005) *City of San Diego Water Reuse Study 2005*, Draft Report, Nov. 2005, City of San Diego Web site, http://www.sandiego.gov/water/waterreusestudy

Cologne, G., and P. M. MacLaggan (1998) "Legal Aspect of Water Reclamation," 1397–1416, in T. Asano (ed.) *Wastewater Reclamation and Reuse*, Water Quality Management Library, **10**, CRC Press, Boca Raton, FL.

Dzurik, A. (2003) *Water Resources Planning*, 3rd ed., Rowman & Littlefield Publishers, Inc., Lanham, MD.

Getches, D. H. (1990) *Water Law in a Nutshell*, West Publishing Co., St. Paul, MN.

Gittinger, J. P. (1982) *Economic Analysis of Agricultural Projects, Second Edition*, The Johns Hopkins University Press, Baltimore and London, published for the Economic Development Institute of the World Bank.

Gregory, R. S., and R. L. Keeney (2002) "Making Smarter Environmental Management Decisions," *J. Am. Water Res. Assoc.*, **38**, 6, 1601–1611.

Grogan, T. (2005) "How to Use ENR's Cost Indexes," *Engineering News-Record*, **254**, 11, 36–71.

Hartling, E. C. (2001) "Laymanization, An Engineer's Guide to Public Relations," *Water Environ. Technol.*, **13**, 4, 45–48.

House, M. A. (1999) "Citizen Participation in Water Management," *Water Sci. Technol.*, **40**, 10, 125–130.

James, L. D., and R. R. Lee (1971) *Economics of Water Resources Planning*, 161–162, McGraw-Hill, New York.

Kontos, N., and T. Asano, (1996), "Environmental Assessment for Wastewater Reclamation and Reuse Projects," *Water Sci. Technol.*, **33**, 10–11, 473–486.

Marks, J., N. Cromar, H. Fallowfield, and D. Oemcke (2003) "Community Experience and Perceptions of Water Reuse," *Water Sci. Technol.*, **3**, 3, 9–16.

Mills, R. A., and T. Asano (1986/87) "The Economic Benefits of Using Reclaimed Water," *J. Freshwater*, **10**, 14–15.

Mills, R. A., and T. Asano (1996) "A Retrospective Assessment of Water Reclamation Projects," *Water Sci. Technol.*, **33**, 10–11, 59–70.

Mills, R. A., and T. Asano (1998) "Planning and Analysis of Water Reuse Projects," 1143–1161, in T. Asano (ed.) *Wastewater Reclamation and Reuse*, Water Quality Management Library, **10**, CRC Press, Boca Raton, FL.

Office of the Federal Register, National Archives and Records Administration (1991) *Code of Federal Regulations, Protection of Environment*, **40**, Parts 1 to 51, U.S. Government Printing Office, Washington, D.C., Title 40, Part 35, Subpart E, Appendix A, 539–546.

Office of Water Recycling (1997) *Water Reclamation Funding Guidelines*, California State Water Resources Control Board, Sacramento, CA.

Richardson, C. S. (1985) "Legal Aspects of Irrigation with Reclaimed Wastewater in California." in G. S. Pettygrove, and T. Asano (eds.) *Irrigation with Reclaimed Municipal Wastewater—A Guidance Manual,* Lewis Publishers, Chelsea, MI.

Riggs, J. L., and T. M. West (1986) *Engineering Economics,* 3rd ed., McGraw-Hill, New York.

RWTF (2003) *White Paper of the Public Information, Education, and Outreach Workgroup on Better Public Involvement in the Recycled Water Decision Process,* 2002 Recycled Water Task Force, Department of Water Resources, State Water Resources Control Board, Department of Health Services, Sacramento, CA.

Schneider, A. J. (1985) "Groundwater Recharge with Reclaimed Wastewater: Legal Questions in California," Chap. 24, in T. Asano (ed.) *Artificial Recharge of Groundwater,* Butterworth Publishers, Boston, MA.

TFWR (1989) *Water Reuse, Manual of Practice SM-3,* 2nd ed., Task Force on Water Reuse, Water Pollution Control Federation, Alexandria, VA.

Thomas, J. C. (1995) *Public Participation in Public Decisions: New Skills and Strategies for Public Managers,* Jossey-Bass Publisher, San Francisco, CA.

Thompson, S. A. (1999) *Water Use, Management and Planning in the United States,* Academic Press, London.

USBR (1990) "Bureau of Reclamation, Change in Discount Rate for Water Resources Planning," U.S. Bureau of Reclamation, *Federal Register,* **55**, 15, 2265.

U.S. EPA (2004) *Guidelines for Water Reuse,* EPA/625/R-04/108, U.S. Environmental Protection Agency and U.S. Agency for International Development, Washington, D.C.

Young, R. A. (2005) *Determining the Economic Value of Water: Concepts and Methods,* Resources for the Future Press, Washington, D.C.

26 Public Participation and Implementation Issues

WORKING TERMINOLOGY   1436

26-1 HOW IS WATER REUSE PERCEIVED?   1436
  Public Attitude about Water Reuse   1436
  Public Beliefs about Water Reuse Options   1440

26-2 PUBLIC PERSPECTIVES ON WATER REUSE   1440
  Water Quality and Public Health   1441
  Economics   1441
  Water Supply and Growth   1441
  Environmental Justice/Equity Issues   1441
  The "Yuck" Factor   1442
  Other Issues   1442

26-3 PUBLIC PARTICIPATION AND OUTREACH   1443
  Why Involve the Public?   1443
  Legal Mandates for Public Involvement   1443
  Defining the "Public"   1444
  Approaches to Public Involvement   1444
  Techniques for Public Participation and Outreach   1446
  Some Pitfalls in Types of Public Involvement   1448

26-4 CASE STUDY: DIFFICULTIES ENCOUNTERED IN REDWOOD CITY'S LANDSCAPE IRRIGATION PROJECT   1450
  Setting   1450
  Water Management Issues   1450
  Water Reclamation Project Planned   1450
  Lessons Learned   1452

26-5 CASE STUDY: WATER RECLAMATION AND REUSE IN THE CITY OF ST. PETERSBURG, FLORIDA   1453
  Setting   1453
  Water and Wastewater Management Issues   1453
  Development of Reclaimed Water System   1455
  Current Status of Water Reclamation and Reuse   1456
  Lessons Learned   1456
  Access to City's Proactive Water Reclamation and Reuse Information   1459

26-6 OBSERVATIONS ON WATER RECLAMATION AND REUSE   1459
  PROBLEMS AND DISCUSSION TOPICS   1459
  REFERENCES   1460

WORKING TERMINOLOGY

Term	Definition
Environmental justice/equity issues	Imposition of adverse project impacts on disadvantaged communities such as receiving low-quality water and disproportionately burdened with health risks and lower property values, while other communities are serviced with high-quality water.
Growth control measures	Municipalities may use water supply or wastewater management as growth control measures such as water connection bans and strict control of wastewater discharge.
The public; also identified as stakeholders	The public consists of many subsets of society with different motivations, values, and approaches that may be affected either directly or indirectly by the project and can influence it. The community at large includes local ethnic groups; political, social, and economic groups; environmental justice advocates; and environmentalists.
"Yuck" factor	The instinctive aversion of members of the public to coming in contact with or drinking water that was once municipal wastewater.

The WateReuse Association (www.watereuse.org/) estimates that in the United States approximately 11×10^6 m³/d (3×10^3 Mgal/d) of municipal wastewater was reclaimed and reused in 2005 and estimated that water reuse was growing at about 15 percent per year. As in many other large public works projects, water reclamation and reuse projects often have the potential to generate conflict in the community. Despite increasing implementation, water reuse remains a cause of community concern because the initial source of water is municipal wastewater. Because water reuse programs may require a public referendum to approve a financial instrument such as a bond issue for funding capital improvements, diligently soliciting community viewpoints and addressing any concerns early in the planning process are invaluable in garnering support.

The purpose of this chapter is to gain insight into a few issues related to implementation of water reclamation and reuse projects. Several issues related to the water reuse implementation were already discussed in Chap. 1, including: (1) implementation hurdles, (2) public support, (3) acceptance which varies depending on opportunity and necessity, (4) public water supply from polluted water sources, and (5) advances in water reclamation technologies. In this chapter public participation and implementation issues are examined with respect to: (1) how water reuse is perceived, (2) public perspectives on water reuse, (3) public participation and outreach, and (4) two case studies.

26-1 HOW IS WATER REUSE PERCEIVED?

The purpose of the following discussion is to examine: (1) public attitude about water reuse and (2) public evaluation of water reuse options.

Public Attitude about Water Reuse

The importance of public attitude in the implementation of a water reclamation and reuse project has been recognized for some time. Over the past 30 yr, several public surveys have been conducted to assess public attitude toward the use of reclaimed water. The results of some public opinion surveys on various uses of reclaimed water are summarized in Table 26-1. In general, public acceptance of landscape irrigation with reclaimed water ranged from 83 to 98 percent. Landscape irrigation and golf course irrigation with reclaimed water are widespread in the United States (see Chap. 18) and the practices in more than 2000 sites have not experienced any adverse health effects with well-treated reclaimed water. As expected, the majority of respondents are strongly opposed to drinking reclaimed water (direct potable reuse). Too often, however, bias affects both the questions asked and the analysis of survey results. For example, if the questions posed in the surveys reported in Table 26-1 were either: (a) are you opposed to the use of reclaimed water for the following purposes . . . or (b) are you in favor of using reclaimed water for the following purposes? . . . different responses might be obtained. Also, if the survey is based on a "gut response" by the respondents and not accompanied by an information booklet describing what is meant by reclaimed water, its water quality, and the potential uses, other answers might also result. The conclusions from the surveys could be subject to wide interpretations but may be useful in assessing public attitude for the purposes of developing information programs and promoting better understanding of issues being considered in the context of water resources management.

When respondents are given more than a yes or no choice about drinking reclaimed water, the results can be divided into three broad categories of acceptance or nonacceptance as illustrated in Table 26-2. One category is that of nonacceptance, i.e., defined as those who mind a lot about the proposed use of reclaimed water; a second category is for those respondents who have some reservations (minds a little bit); and the third category covers those respondents who generally accept the proposed use of reclaimed water (does not mind). The effect of interpretive bias tends to move respondents with some reservations into the opposed group. As shown in Table 26-2, respondents given a choice tend to support direct potable reuse by more than 70 percent. Respondents, when asked whether they approve or disapprove of drinking reclaimed water without any other options, tend to be more opposed, as reported in Table 26-1.

Direct contact with reclaimed water, as opposed to ingestion, is more broadly accepted. About 15 to 25 percent of those surveyed in Table 26-1 were opposed to swimming in reclaimed water and 14 to 21 percent were opposed to irrigating vegetables with reclaimed water. Little objection to the use of reclaimed water in golf course and landscape irrigation, industrial process, and cooling/air conditioning was noted.

It should be further noted that the U.S. surveys reviewed in Table 26-1 are relatively old. Because nonpotable water reuse has been implemented widely in recent decades, research into public perception has been conducted rarely since the mid-1980s. If surveys were conducted now, it would be in conjunction with a planned indirect potable reuse and the critical factors in obtaining a positive response would be the credibility of the water supply agency and well-informed respondents on water resources needs, water quality, and public health protection.

Table 26-1
Percentage of respondents opposed to various uses of reclaimed water in the general opinion surveys

Item	Bruvold[a] (1972)[e] N = 972[f]	Bruvold[a] (1981) N = 140	Milliken and Lohman[b] (1983) N = 399	Lohman and Milliken[b] (1985) N = 403	Sydney Water[c] (1999) N = not known	ARCWIS[c] (2002) N = 665	City of San Diego[d] (2004) N = 710
1. Drinking water	56	58	63	67	69	74	64
2. Food preparation in restaurants	56						
3. Cooking in the home	55		55	55	62		
4. Preparation of canned vegetables	54						
5. Bathing in the home	37		40	38	43	52	
6. Swimming	24						
7. Pumping down special wells	23						
8. Home laundry	23		24	30	22	30	
9. Commercial laundry	22						
10. Irrigation of dairy pasture	14						
11. Irrigation of vegetable crops	14	21	7	9			31
12. Spreading on sandy areas	13						
13. Vineyard irrigation	13						
14. Orchard irrigation	10						

Use						
15. Hay or alfalfa irrigation	8					
16. Pleasure boating	7					
17. Commercial air conditioning	7					
18. Electronic plant process water	5					18
19. Home toilet flushing	4	3	4			13
20. Golf course hazard lakes	3	8				
21. Residential lawn irrigation	3	5	1	3	4	17
22. Irrigation of recreation parks	3	4		3		15
23. Golf course irrigation	2	4			2	7
24. Irrigation of freeway greenbelts	1					7
25. Road construction	1					
26. Stream or river discharge						
27. Bay or ocean discharge						

[a] Adapted from Bruvold (1972, 1981).
[b] Adapted from Lohman (1987). These studies were in conjunction with the Denver Potable Reuse Demonstration Project in Denver, CO (see Chap. 24).
[c] Adapted from Radcliffe (2004).
[d] Adapted from PBS&J (2005).
[e] Year of survey.
[f] N = number of respondents in survey.

Table 26-2
Respondents' approval ratings for various reuse applications based on their view of potable reuse (1982 and 1985)[a]

Water reuse application	Approval, %					
	Minds a lot		Minds a little bit		Does not mind	
	1982	1985	1982	1985	1982	1985
Drinking	3.2	2.1	21.6	19.9	71.0	74.3
Cooking	8.8	10.4	36.3	33.7	75.6	78.0
Bathing	25.6	30.2	57.6	60.8	86.3	87.2
Laundry	44.0	44.8	76.0	66.9	95.4	91.7
Watering garden	87.2	82.3	88.7	91.2	97.7	96.3
Flushing toilets	92.8	89.6	97.6	96.1	98.5	100.0
Washing cars	92.8	89.6	97.6	97.2	98.5	100.0
Watering lawn	97.6	90.6	98.4	96.7	99.2	100.0

[a] Adapted from Lohman (1988).

Public Beliefs about Water Reuse Options

There are correlations between attitude and beliefs of the public about the need for water supply augmentation, the technical means for water reclamation, and pollution of the present water supply (Bruvold, 1972; 1987; 1988; and 1992). For example, even in the situation of a controversial scenario for the use of reclaimed water for drinking water supply, positive response can be obtained from a respondent, depending on the respondent's beliefs. Respondents are more favorable to the use of reclaimed water for potable purposes who believe: (1) there is a need for water supply augmentation, (2) in the efficacy of technology, and (3) pollution is serious and widespread (see Chaps. 23 and 24).

It should be noted that commercial advertising and marketing influence public beliefs or perception of water supply safety. It has been reported that bottled water advertising and point-of-use treatment systems marketing often utilize the tactic of creating uncertainty about the safety of a public water supply. Any effort to change public or professional evaluation of water reuse must consider these ongoing marketing efforts by pointing out the fallacies in the arguments of product purveyors. In the case of drinking water, it is seldom revealed that most bottled water is not tested fully or certified, or that current tap water meets all drinking water standards.

26-2 PUBLIC PERSPECTIVES ON WATER REUSE

Public and professional acceptance of water reuse varies with the water reuse application. As discussed above, the higher the degree of potential contact with reclaimed water, the more unfavorable both groups are to particular uses (see Tables 26-1 and 26-2). However, when faced with specific proposals for water reuse projects to address community problems, the public weighs a variety of concerns and objectives. For example, respondents in one study were found to favor water reuse options that conserved water, enhanced the environment, protected public health, or held down water and wastewater treatment and distribution costs (Bruvold, 1988). Thus, it is important that project objectives reflect community desires and that public information programs address how alternative solutions compare based on those objectives. The following varied issues and concerns may affect the acceptability of water reuse.

Water Quality and Public Health

Protection of public health is generally the highest concern with the use of reclaimed water. The possible health risks associated with water reuse are related to the adequacy and reliability of the water reclamation system and the extent of exposure to the reclaimed water. Water reuse criteria and regulations are presented in Chap. 4, and health and environmental concerns in water reuse are discussed in detail in Chap. 5.

Economics

The cost of water reuse depends on the application, the level of treatment required, distribution facilities, infrastructure, and onsite adjustments required. It is important that the public understands the water supply and wastewater management context of water reuse (see Chap. 25). A typical basis for comparison of the costs of water reuse will be the avoided costs of alternative water supply or wastewater discharge scenarios. In general, users will expect the rates for reclaimed water to be less than rates for other water supplies. Additional savings in the use of reclaimed water for improved reliability of water supply in drought years should be factored into the equation. The net costs or savings will be shared by ratepayers at large for changes in water and sewer services along with the users of the reclaimed water.

Water Supply and Growth

Municipalities sometimes attempt to use the provision of potable water or wastewater collection as growth control measures: water connection bans or large fees for taps, and strict control of wastewater discharge may both serve to limit growth. Some members of the public fear that water reuse may induce growth by creating a new water resource, resulting in unwanted local development. However, water management can ideally be decoupled from growth by employing local zoning or conservation ordinances or by placing conditions on the use of reclaimed water. For example, some water reuse projects stipulate that the supply of reclaimed water be used only as a replacement for freshwater previously diverted from surface or groundwater supplies, which improves water supply reliability and local control of water supply without increasing overall water availability. This scenario only applies to legally organized communities and does not affect growth outside of community boundaries. In most states, water supply capability is not a prerequisite to development.

Environmental Justice/Equity Issues

A concern that has arisen in several water reuse projects, as with development projects in general, is the imposition of adverse project impacts on disadvantaged communities. Residents of communities in which water reuse projects are sited may fear that they will receive lower-quality water and be disproportionately burdened with health risks and lower property values, while other communities are serviced with high-quality water. To ensure that decisions are made fairly and equitably, and to assuage fears of favoritism, it is important to have open and participatory decision-making along with recognition of the fears of the potentially affected community.

The "Yuck" Factor

Many people who oppose water reuse, or believe that it should be employed only as a last resort, do so because of their instinctive aversion to coming in contact with or drinking water that was once municipal wastewater. This is particularly true for indirect and direct potable reuse options, which have been assigned in the past the derogatory label "toilet-to-tap" in the mass media. This aversion has also been referred to as the "yuck" factor. Many people are unaware that most drinking water contains a percentage of treated wastewater that was discharged to surface water or infiltrated into groundwater (see *de facto* potable reuse in Chap. 3). That is not to say that the perception is invalid or that the lower quality of some drinking water sources should justify potable use. On the contrary, such perceptions should be taken seriously as a reflection of personal or cultural feelings toward water purity and of the public's understanding of hydrologic processes and water management issues (Wegner-Gwidt, 1998; Hartling, 2001). Then the issue can be addressed in a variety of educational or informational venues.

Other Issues

Other important issues that influence public acceptance of water reuse projects include: (1) understanding of local water supply shortages and awareness of reclaimed water as having a place in the overall water supply allocation scheme, (2) understanding of the quality of reclaimed water and how it would be applied and used, and (3) confidence in the adequacy of public health regulations and their enforcement. In recent years, studies on understanding public perception and participation on urban indirect potable reuse has been renewed with respect to the challenge of better integrating water reclamation and reuse systems into urban water supply.

Rapid advances in analytical chemistry for identifying anthropogenic trace contaminants in public water supplies, wastewater, and reclaimed water have caused national and international concern about public and ecological health. References on these topics are found in Kolpin et al. (2002) and Daughton (2005). In addition, guidance and research agendas addressing the social and political complexity of adopting indirect potable reuse as part of a sustainable community strategy have been considered. More research is needed on how humans perceive and calculate risk in relation to reclaimed water projects, and how this research might lead to improvements in relations between water agencies and the public. These research efforts could help in understanding the role of trust in people's decision-making processes to either accept or reject the use of reclaimed water in various situations, and how the limits of scientific knowledge in the field of water use and reuse are perceived (Asano, 1998; Sheikh et al., 1998; Hartley, 2003; Haddad, 2004).

As discussed in Chaps. 23 and 24, indirect or direct potable reuse continues to face intense public scrutiny and conflict in building the necessary public confidence, trust, and support for water reuse projects. While the results of public attitude surveys reported in Tables 26-1 and 26-2 and the prevailing views of water reuse are generally applicable in the implementation of nonpotable water reuse projects, it is not uncommon to find small numbers of individuals who oppose or question the acceptability of such projects. The stated reasons almost always focus on uncertainty relating to public health protection and often insisting on unrealistic "zero risk." There also can be many unrelated or hidden agendas in public debate such as no growth issues in the community. Even an unremarkable landscape irrigation project has been delayed and modified due to opposition by an organized group (see Sec. 26-4).

26-3 PUBLIC PARTICIPATION AND OUTREACH

From the previous discussion, it is clear that perceptions and opinions can make the difference between success or failure. This is a reality that must be recognized in the planning and implementation of every water reuse program. While there are no fool-proof guarantees of success, a sound and proactive communication and education program is essential as discussed in the following sections.

Why Involve the Public?

The public bears part or all of the financial burden of water reclamation projects, experiences possible exposure to the reclaimed water, and may also experience aesthetic or other impacts. While costs and public health have been prominent in public concern, underlying issues of environmental justice, growth, and land development have also been evident. The public has concerns for safe and adequate water supply and environmental protection. Thus, it is important to create awareness of how water reuse can make a positive contribution to all these issues and to form a context that may outweigh other concerns (State of California, 2003).

Public involvement in water reuse planning can have substantial benefits in terms of satisfying community water needs, gaining public support, developing a broad market for reclaimed water, and improving project implementation. Public engagement, a process of two-way communication, serves to inform both the public and planners about issues that may have been overlooked or misunderstood by the other group. For example, information and outreach programs implemented by project planners help to educate the public about scientific, technical, and economic aspects of the water reuse project. Participants can help to broaden the perspective on project objectives, the range of possible solutions, and enhance evaluation of risks, costs, and benefits (see Chap. 25). Potential reclaimed water users provide valuable input regarding the quantity and quality of reclaimed water that is acceptable for their applications.

Legal Mandates for Public Involvement

Public participation in public works projects in the United States was advanced in the late 1960s by federal mandates for public involvement in projects receiving funding from or under the jurisdiction of federal agencies. State and local agencies followed with their own mandates for public involvement. Legislation such as the National Environmental Policy Act of 1969 (NEPA) and the California Environmental Quality Act (CEQA) manifested a growing environmental conscience in the United States and created new demands for technical decision-making that reflects societal values. These federal, state, and local mandates represent the minimum level of public involvement in many public projects.

The Council on Environmental Quality (CEQ) established procedures for implementation of NEPA. For public involvement (CEQ, 2005a; 2005b), agencies are required to:

1. Make diligent efforts to involve the public in preparing and implementing NEPA procedures
2. Provide public notice of NEPA-related hearings, public meetings, and the availability of environmental documents so as to inform those persons and agencies that may be interested or affected

3. Hold or sponsor public hearings or public meetings whenever appropriate or in accordance with statutory requirements applicable to the agency
4. Solicit appropriate information from the public
5. Explain where interested persons can get information or status reports on environmental impact statements and other elements of the NEPA process
6. Make environmental impact statements, the comments received, and any underlying documents available to the public

Despite what appears to be a comprehensive list of requirements, the law allows considerable discretion. The requirements are focused on the environmental assessment process, and the absolute minimum requirements are far short of a fully participatory planning process.

Defining the "Public"

The public consists of many subsets of society with different motivations, values, and approaches who have often been identified as stakeholders. The community at large includes, among others, local ethnic groups; political, social, and economic groups; environmental justice advocates; and environmentalists (State of California, 2003). The primary categories of the public that may have an interest in water reuse projects are listed in Table 26-3. For planning purposes, the public can be divided into two general categories: direct participants (e.g. reclaimed water users) and the broader public that will experience indirect impacts. Public participation strategies should be tailored to the interests and needs of each category. The needs of potential reclaimed water users must be addressed as part of market assessment, which is discussed in Chap. 25.

Approaches to Public Involvement

There are three major approaches to public involvement in water reclamation and reuse planning: (1) heavily involve the public from the earliest stages of planning all the way through to the final stages of adoption, (2) seek public involvement after all planning has been completed and a bond issue vote is needed to ratify the planning and fund actual adoption, and (3) involve the public after planning has identified major options but before public ratification of funding is necessary. The following discussion of the pros and cons of each of the three planning approaches to public involvement has been adapted from Bruvold and Crook (1980).

Table 26-3
Groups representing various public interests in water reuse projects[a]

Citizens in general	Regulatory agencies
Potential reclaimed water users	Resource agencies
Potential users' neighbors	Environmental groups
Homeowners associations	Community and civic organizations
Farmers	Ratepayers
Agricultural agencies	Educational institutions/academic leaders
Water distributors	Political leaders
Food processing industry	Business leaders
Land developers	Community leaders

[a] Adapted from Wegner-Gwidt (1998); U.S. EPA (2004).

Involve the Public from the Earliest Stages of Planning
In support of this approach, it could be argued that the best time for public input into the decision-making on water reclamation and reuse would be early in the process to ensure planning would proceed on a course acceptable to a majority of the stakeholders. Almost all experienced planners and public relations personnel recommend involving the public early in the process by raising issues and avoiding surprises (Kelly, 1998; Wegner-Gwidt, 1998).

It is often noted in the literature that the ordinary voter lacks the technical expertise to formulate options for water reclamation and reuse and then decide, on his or her own, which is best. Further, even if more people are fundamentally capable of understanding the necessary basics of technology involved, they are not interested in nor have the time to develop and assess options. Thus, lack of technological expertise on the part of the voting population can create serious problems for early involvement of voters in the planning process. This scenario might occur when the early public involvement is unguided and the materials released are incomplete or purely technical. Any stage for public involvement requires that informational materials be presented appropriately in lay language.

Seek Public Involvement after Completion of All Planning
In support of this planning approach, public involvement should come much later in the decision-making process where the option selected by technical experts is ratified as in the vote on a local bond issue. This procedure allows technical and professional experts to do the planning and analysis work that requires technical and professional expertise. These individuals select and develop options, decide which of several options is best, and then present the one selected to the voting public for ratification in a yes–no referendum. Here, only one option is presented and the voters have the choice of affirming or denying support to the one choice the technical experts and politicians themselves have chosen.

A major problem with this planning approach is that it may often lead to conflict and failure. In general, any type of local bond issue will likely be hard to pass. Thus, the technical experts and professional planners must be extremely careful to choose water reclamation and reuse projects that can develop sufficient public approval to obtain the support needed to finance construction and operation. Further, a yes-or-no ratification on one option chosen by the experts may weigh too heavily in favor of technocratic expertise and too lightly for public concerns or values. Most voters likely are not expert enough to develop and assess options, but will be able to clearly understand the options and option analyses developed by the technical experts when this information is presented using nontechnical, clear language.

Involve the Public after Major Options Have Been Identified
There are a number of reasons that can be put forth to support this planning approach as one that will more properly balance technical and democratic imperatives while also pragmatically ensuring more success at the polling booth during local elections. This approach requires the selection of a small number of options for water reclamation and reuse that are feasible for the area, which represent a variety of different solutions for water reclamation and reuse. Then, careful comparative analysis of the options can be

developed. Once option development and analyses are completed by the technical experts and planning professionals, the finding can be presented to voters in a straightforward, easy-to-understand language for their reaction and input before going to a formal public ratification. Such a procedure would not require each citizen to become expert in water reclamation and reuse, but it would also not put off public involvement until the last minute. The principal argument against this approach centers on its practicality. Bruvold and Crook (1980) suggested that the technology sector should assess risks and efficiency while the public sector assesses safety and benefits.

Techniques for Public Participation and Outreach

There is a distinction between participation and outreach: participation implies a means for stakeholders to influence the plan, whereas outreach may simply be a way of disseminating or collecting information and educating the public about water reuse planning. Surveys, public information programs, and workshops are public outreach techniques that may be part of an effort to involve citizens in decision-making processes. However, task forces and advisory committees are examples of stakeholder participation in decision-making. Public involvement techniques employed for San Diego's Water Repurification Project are shown in Table 26-4. Techniques and tools that can be used alone or in combination with each other in the public participation and outreach are summarized in Table 26-5.

Table 26-4 Public involvement techniques employed for San Diego's Water Repurification Project[a]

Public Surveys
- One-on-one interviews with city residents
- Telephone interviews
- Focus groups

Citizens' Advisory Committee

Public Information
- Brochure and fact sheets
- Video describing the project
- Slide presentation
- Telephone information line
- Briefings for policy-makers and their staff
- Workshops
- Open houses

Media Coverage
- Milestone news releases
- Media briefings
- TV news stories
- Newspaper articles and editorials

Speaking Engagements
- Enlisted members of advisory committee and area organizations (Sierra Club, County Medical Society, Chamber of Commerce) to speak on behalf of project

Tours of pilot treatment plant, with taste tests of treated water

[a]Adapted from State of California (2003); U.S. EPA (2004); Wegner-Gwidt (1998).

Table 26-5
Techniques for public participation and outreach[a]

Techniques	Descriptions
Surveys	Social surveys, such as one-on-one interviews, questionnaires, and informal discussions, can be used to evaluate community preferences, attitudes and values. Social surveys can help to inform participants and to highlight areas where more information is required by the public. Surveys are also useful in screening potential markets for reclaimed water.
Public information programs	Open, complete access to information can generate trust and good relations with the public. Information should be available in all languages represented in the community and should be presented in various formats, including written materials and radio announcements.
Public meetings/hearings	Public meetings or hearings can be used to: (1) introduce the need for project planning, (2) inform the public of the progress and configuration of plans being developed, (3) obtain public input on planning objectives and alternatives identification for inclusion into the planning process, (4) present alternatives under consideration, (5) seek an understanding by the public as to what is required in the planning of an upgraded wastewater system or a more reliable water supply, and (6) satisfy regulatory requirements for public input in the planning process.
Workshops	Small public workshops provide opportunities for the engineering/planning team to: (1) present technical, environmental, and community impact findings, (2) compare alternatives, and (3) engage in discussion with the public.
Consultation with key contacts	Planners may contact key industry, government, or community leaders, particularly those with expertise and influence, to solicit input. Groups such as engineering firms, land developers, and professional societies can provide insight, data and information, and project support.
Advisory committees	Advisory committees serve to provide in-depth input on the project and to reinforce to the public that their input is important and utilized. Advisory committees are formed generally for an unspecified period of time to provide input continuously. Members of advisory committees should represent a broad range of groups, including private citizens who have no financial interest in the project, representatives of public interest groups, public officials, individuals or representatives of groups with economic interests in the project, potential reclaimed water users and their employees, neighborhood residents, and citizens with specialized expertise in relevant areas (e.g., public health).
Task forces	Task forces, like advisory committees, assist in defining system features and resolving problem areas. Unlike an advisory committee, a task force has clearly defined objectives and disbands once the objectives have been met.

[a] Adapted from House (1999); Umphres et al. (1992), U.S. EPA (2004), Walesh (1999), Wegner-Gwidt (1998), PBS&J (2005), City of San Diego (2006).

Some Pitfalls in Types of Public Involvement

There are many techniques that planners can use to elicit public support without ceding any real influence to those who will be affected by and have an interest in a project. These techniques often leave participants feeling disenfranchised and distrustful, thus making it more difficult to achieve public acceptance of future projects. The following are models or techniques of public involvement and outreach that should be avoided in water reuse planning.

Decide–Announce–Defend

The traditional approach by technical decision-makers is to decide, inform the public, and then justify the decision. This is commonly called decide–announce–defend (DAD). The DAD approach is derived from a belief that technical professionals know best how to address community needs. Professionals define the problem and objectives, identify potential solutions, and select the best alternative, without public involvement except for notification and hearings as legally required.

The DAD approach may be appropriate for engineering projects dominated by technical and regulatory constraints, such as the specifics of water reclamation plant design. But the DAD approach is inadequate to achieve public acceptance of controversial projects, particularly those involving social and environmental impacts (Walesh, 1999).

Controlled Selection of Participants

A common mistake in public involvement programs is to involve only those people or groups who are most inclined to support a project and those individuals who have the greatest influence on whether the plan is adopted. While it appears that this approach will increase the chances of obtaining project approval, excluding other members of the community often backfires when those citizens organize to oppose projects. The problem with choosing only influential people to be in the target groups is that they do not represent the interests of all community members; rather, they may focus on issues of personal interest or high public profile (see Sec. 26-4).

Information Control

Another mistake is presenting only the information that will reflect positively on the project and withholding information that may fuel opposition. Carefully selecting information to be shared with participants—for example, emphasizing benefits while ignoring costs—diminishes their ability to make objective decisions on the merits of the plan. Participants are vulnerable to manipulation if they lack the technical knowledge to detect gaps in the information presented to them. Whether by intention or lack of sensitivity to audience perception, planners and engineers can cause intimidation and confusion by using obscure technical jargon and complicated explanations. This approach will not succeed.

Marketing strategies, if misused, may also be perceived as information control or manipulation. Selling a project by using favorable images may result in a loss of credibility. Appropriate uses of marketing techniques include public surveys and focus groups used to measure public acceptance of and concerns about water reuse for various applications and to identify potential reclaimed water customers. Focus groups also help to identify the best ways to advertise the project; for example, which terms and images will elicit the most supportive response. If marketing strategies are employed during project planning, they should be combined with other participation techniques to avoid being interpreted as manipulation.

Participation without Influence

A final mistake in public involvement programs is soliciting public input without allowing the input to have any influence on the final decision. Planners may provide token concessions to influential individuals or groups, while ignoring concerns and needs of the general public. This approach inevitably generates distrust and opposition. A list of basic principles for dealing with the public from an engineer's point of view is shown in Table 26-6. "If you make the public your partner, everyone can win in the end" (Hartling, 2001).

Table 26-6
An engineer's guide to public relations[a]

Talk in regular English. People won't believe you if they don't understand you. Technobabble alienates them and makes them suspicious. And while you're at it, get to the point and shut up. Rambling isn't popular either.

Talk often. Take every opportunity to spread the word. Rotary Clubs, church groups, city councils—wherever and whenever you can. You can never have too many friends. Taking the time to involve all kinds of people will make you the friends you'll need when last-minute opposition appears.

Know what you're talking about. People will pull all kinds of questions, concerns, and issues out of left field, so make like a Boy Scout and be prepared for anything. Too many "I don't know" answers will not enhance your credibility.

Make yourself "ground zero" and leave no challenge unmet. Some opponents will do and say almost anything—some of it just plain wrong—to derail your project. Take them on directly and forthrightly address everything they say in public meetings, to the media, etc. If you don't, people will assume (consciously or unconsciously) that the opposition statements are true.

Be human, lighten up, and have a sense of humor. People tend to listen and learn more when they're enjoying themselves and if they like you. Think back to that teacher who droned on forever in an unchanging monotone, as if on tranquilizers. Why imitate him or her?

Some opponents just will not accept your project no matter what you do, so don't worry about not getting through to them. There are still people who do not believe we landed on the moon. In any innovation, there will be people who cannot or will not be convinced. In some cases, this group has a hidden agenda for its opposition. They may say the issue is public health, but they're really concerned about selling bottled water or tap water, getting elected, limiting community growth, or other issues that may not garner much public sympathy.

Involve the public in the decision *making* process, not just the decision *accepting* process. People's acceptance comes with their ownership of a project or idea. This is a slow but necessary process.

Have your ego surgically removed, if necessary. So what if you're a big shot engineer? If people don't like your idea, go with the flow and modify it so they do like it. The object is to get the job done, not to get it done your way.

Show people what you're talking about. A picture really is worth 1000 words, so take the people you want to convince on a tour of your facility. It's like giving them a backstage pass.

[a]Adapted from Hartling (2001).

The Need for Transparency and Public Trust

In this day and age with the number of available channels for obtaining information, the opportunity for misinformation is enormous. Ultimately, if the merits of water reclamation and reuse projects are to be considered thoughtfully by the public, two key elements are essential: *transparency* and *public trust*. The planning process for such projects must be transparent with respect to the sources of information and data, the assumptions made, and the methods of analysis used to assess the engineering, economic, and social feasibility. While people may disagree with the findings, the data and methods used to develop the findings must be clear, defensible, and transparent. Coupled with transparency, the other key element of project development is public trust. In general, trust is a natural consequence of promises fulfilled. Project proponents, such as public agencies, must be able to establish, maintain, and increase trust and credibility with the public (stakeholders) including employees, regulatory agencies, citizen groups, the public at large, and the media. Trust is especially important when communicating the risk elements of a project. At the end of the day, in today's world, it is difficult to develop and implement water reuse projects without transparency and public trust. An excellent source of information on environmental risk communication may be found at the following internet site: www.centerforriskcommunication.org.

26-4 CASE STUDY: DIFFICULTIES ENCOUNTERED IN REDWOOD CITY'S LANDSCAPE IRRIGATION PROJECT

This case study is derived from various sources including Ingram et al. (2005) and the Redwood City web site http://www.redwoodcity.org/publicworks/water/recycling/

Setting

Redwood City, California, a community of about 75,000 people located 40 km (25 miles) south of San Francisco, receives 100 percent of its potable water supply from the Hetch Hetchy regional water system operated by the San Francisco Public Utilities Commission (SFPUC). Hetch Hetchy Reservoir is located 64 km (40 miles) from Yosemite Valley via Highway 120 and Evergreen and Hetch Hetchy Roads. Though politically and environmentally controversial, the Hetch Hetchy Reservoir has, since 1923, been very much an integral part of the Yosemite Valley area and the regional water supply system.

Water Management Issues

The city's existing contractual water supply assurance limit is 15×10^6 m^3 per annum (12,243 ac-ft/yr). The city currently consumes 14.8×10^6 m^3/yr (12,000 ac-ft/yr) in excess of the amount contracted from the SFPUC. The excess water is available as a result of other customers not using their full contractual supply. Current regional demand projections indicate that the Hetch Hetchy system will reach full contractual capacity by 2007–2009; thus, eliminating Redwood City's ability to purchase excess water beyond its contractual limit. The shortfall in water supply could increase by approximately 3.3×10^6 m^3/yr (2700 ac-ft/yr) by 2010.

Water Reclamation Project Planned

To bring Redwood City into compliance with its water supply assurance limit, the city determined that water recycling, in conjunction with enhanced water conservation programs, was the only viable near-term option that could be readily implemented within the 2010 timeframe. However, controversy has developed about the plan to use recycled water for irrigation to meet the projected water supply shortfall. In Redwood Shores (a subdivision of the city), many residents objected to the use of reclaimed water in parks and schoolyards, fearing it might make their children sick. Because this is the largest and most complicated project the city has ever undertaken, the people expect and demand a lot of information about the project.

The city initiated the design, permitting, and environmental review for a citywide water reclamation project for landscape irrigation, delivering approximately 2.5×10^6 m^3/yr (2000 ac-ft/yr) in 2002. However, use of reclaimed water for landscape irrigation faced intense local opposition from a small but determined group of citizens who objected to the use of reclaimed water from a public health and safety perspective, particularly in areas where children play. In response to this opposition, the city council eventually formed a community task force and empowered it to help develop a solution to the city's water supply problem. In summary, the use of reclaimed water for landscape irrigation is becoming a more critical issue as the community grows. Each year, the city exceeds the Hetch Hetchy allotment, and the situation is getting worse. Redwood City has successfully managed a pilot recycled water project called *First Step*, and is now expanding its use of reclaimed water for additional landscape irrigation (except at schools and playgrounds where artificial turf is used) as well as aggressively conserving water (see Fig. 26-1).

(a)   (b)

Figure 26-1
Landscape irrigation with reclaimed water in Redwood City, CA: (a) typical residential area and (b) bayside landscaping (Photos courtesy of K. McManus).

Lessons Learned

The lessons learned through the process to create a community-based, consensus-driven solution include developing true community participation, education of policy makers, and prompt and full sharing of information. According to Ingram et al. (2005), the lessons learned are:

1. Assume nothing
If initial informational meetings are poorly attended, do not assume the community is uninterested or "okay" with a proposed project. Based on the city's experience at the first workshop in June 2002, there was no indication that the reclaimed water project should not move forward as planned. The city's First Step Project had been operating successfully for several years, and the public did not seem interested in whether the project continued on a citywide basis.

However, the city learned that people come to understand a project in their own way and time. In Redwood City, a few people stirred a vocal opposition effort from one geographic area of the community, which was then able to influence the schedule, system design, and resulting cost of a project that serves the entire community. Those who are committed to an opposing view can create significant challenges to the process and focus of reclaimed water projects, but they must be heard and respected whenever they choose to participate.

2. Share information early and often
Debunking false information or responding to arguments taken out of context is very difficult; it can put the planner in a defensive rather than leadership position. In Redwood City, individuals seeking to derail the water reclamation project used data and information culled from the Internet to support their position; a heightened reaction was created among citizens and raised the volume if not the substance of the opposition's arguments. Keeping the public educated and informed is critical, as is establishing the city as the reliable and trusted information source. The city learned that an opposition organization with time and resources requires equal or more time and resources on the part of the city to ensure accurate information was disseminated.

3. Begin education at once
Educating policymakers up front can positively affect key decisions made later. Most Redwood City Council members had a limited background in both drinking water and wastewater treatment and regulatory standards, which put them at a "knowledge disadvantage," particularly when the opposition began talking about a wide range of emerging contaminant issues and Internet-gleaned "science." It is important to prepare elected officials for the nature of the debate, and to make sure they are kept informed of the continued development and challenges of the project, so they do not lose continuity with citizens or issues under discussion.

4. Use appropriate tools
Formation of a Task Force was a logical step for Redwood City. The Redwood City staff adopted a core purpose "Build a Great Community Together" and core values "Excellence; Integrity; Service; Creativity" that guide the organization in achieving its core purpose. The water reuse project represented an extraordinary opportunity for the city

to approach the problem at hand in light of its purpose and values. Task Force members committed to a dialog and followed the ground rules that they themselves set, with the city providing full support. This approach fostered an appreciation for the critical nature of the water supply problem and planted the seeds for trust between the various interest groups. It was an engaging and healing process not only for Task Force participants but the entire community and city leadership.

5. Let the citizens speak

It was important that Task Force members deliver their own resolution in their own voice. Two members of the Task Force, who originally held opposing positions, made a unified presentation to the city council of their concerns and recommendations in their own words. The unanimity of voices reflected a positive experience and delivered a powerful and meaningful message. This unanimity enabled the council to enthusiastically adopt the Task Force recommendations and move forward with the project.

6. It all takes time

Building community trust takes time. There was a general sense that the city needed to "slow down to go fast." It became essential to allow time for the Task Force and the community to understand the dimensions of the long-term water supply issue, and how best to approach its resolution. Taking time to review, reflect, learn, and assess enabled the project to ultimately move forward and will continue to affect its future. Developing trust and open communication between water authorities and their customers is an important contributor to the success of some recycled water projects.

26-5 CASE STUDY: WATER RECLAMATION AND REUSE IN THE CITY OF ST. PETERSBURG, FLORIDA

This case study has been developed from various sources including Johnson and Parnell (1989) and the web site http://www.stpete.org/wwwrecla.htm.

Setting

The City of St. Petersburg, located at the tip of the Pinellas County peninsula on Florida's west-central coast, is somewhat unique in that it is surrounded by saltwater on three sides. The city is bordered on its eastern side by Tampa Bay, on its western side by the Gulf of Mexico, on its southern side by the entrance to Tampa Bay from the Gulf, and on its northern side by incorporated communities (see Fig. 26-2). St. Petersburg is Florida's fourth largest city with a resident population of over a quarter of a million persons. In the winter, the population increases by several thousand transient visitors.

Water and Wastewater Management Issues

The supply of drinking water for the ever increasing population and the treatment of wastewater have played dominant roles in the growth and development of this city since its inception in 1880 up to the present time. Many of the water and wastewater issues have been very controversial.

Water Wars

In the early 1970s, continued population growth throughout Pinellas and neighboring Hillsborough and Pasco Counties, combined with an overall decrease in local rainfall,

Figure 26-2
Landsat picture of St. Petersburg, FL. The city is bordered on its eastern side by Tampa Bay, on its western side by the Gulf of Mexico, on its southern side by the entrance to Tampa Bay from the Gulf, and on its northern side by incorporated communities (Coordinates: 27.774 N, 82.674 W, view at altitude 80 km).

placed ever increasing demands on the available groundwater supply. As a result, intergovernmental cooperation was needed to address critical water issues. Because the City of St. Petersburg owned substantial blocks of land in Pinellas, Hillsborough, and Pasco Counties from which they extracted groundwater, the governments of these counties became alarmed when they realized that they might not be able to provide adequate water for their own growing populations. When St. Petersburg and Pinellas County worked jointly to develop another well field in Pasco County, Pasco, Hillsborough, and Hernando Counties joined together to have legislation enacted to block any further water development by municipalities outside of their own jurisdiction.

As a result of these "water wars," in 1974, Hillsborough, Pasco, and Pinellas Counties and the cities of St. Petersburg and Tampa formed a separate governmental entity known as the West Coast Regional Water Supply Authority (WCRWSA) to develop regional water supplies and supply water at wholesale rates to counties and municipalities.

Regional Water Supply

With the formation of the WCRWSA, more well fields have been brought on line in recent years and interconnections between well fields have been constructed. All of the well fields are located within a 777 km^2 (3300 mi^2) area and have a total annual permitted withdrawal of approximately 550×10^3 m^3/d (145 Mgal/d).

The Wilson-Grizzle Act of 1972

In 1969, the U.S. EPA described Tampa Bay as one of the most polluted bodies of water in the nation. As a result, the Florida legislature adopted what was called the "Wilson-Grizzle Act" in 1972. The text of this Act was brief and contained the following statement:

"No facilities for sanitary sewage disposal constructed after the effective date of this act shall dispose of any wastes into Old Tampa Bay, Tampa Bay, Hillsborough Bay, Sarasota Bay, Boca Ciega Bay, St. Joseph Sound, Clearwater Bay, Lemon Bay and Punta Gorda Bay or any bay, bayou or sound tributary thereto without providing advanced waste treatment approved by the department of pollution control." Some confusion over the meaning of "advanced treatment" led to the State Pollution Control Board to set limits of 5 mg/L for BOD and TSS, 1 mg/L for phosphorus and 3 mg/L for total nitrogen and to require minimum treatment efficiency of 90 percent for BOD and TSS removal.

Following the adoption of this Act, in 1972, the City of St. Petersburg evaluated its alternative plans and, based on the cost of constructing and operating advanced wastewater treatment (AWT) facilities and considering the potential potable water supply shortages from the "water wars," the city council selected a plan to upgrade all plants to advanced secondary (i.e., tertiary) treatment and to implement a water reuse and deep injection well program that would result ultimately in zero-discharge to Tampa and Boca Ciega Bays. This bold decision led to the initiation by the city of what has become the largest urban reclaimed water distribution system in the United States.

Development of Reclaimed Water System

In addition to the provisions of the Wilson-Grizzle Act, the 1972 decision by the city council to implement a water reuse and deep injection well system was also influenced by the results of a pilot study that had been authorized in 1971. In this study, designed to determine the feasibility and efficiency of using highly treated wastewater for spray irrigation in an urban environment, it was concluded that spray irrigation using treated wastewater was more feasible and considerably more cost effective than advanced wastewater treatment followed by discharge to Tampa Bay. Also, the construction of a reclaimed water system would benefit the community by reducing the total quantity of water to be imported for potable use.

The approach adopted was to reclaim the wastewater by treating it to a sufficient degree that it would be suitable for the irrigation of parks, schools, and golf courses within the city. The city council also included funds in its 1972 Capital Improvement Program for the upgrading of the regional wastewater treatment facilities to advanced secondary (tertiary) treatment and for the development of plans for a reclaimed water distribution system.

With the passage of the amendments to the Federal Water Pollution Control Act (Clean Water Act of 1972), the door was open for the development of full scale water reuse systems with federal approval and funding. The city began immediately to submit applications for grants to finance upgrading of the four wastewater treatment plants and for the design and installation of the reclaimed water distribution system and the deep-well disposal system.

The initial reclaimed water distribution system, constructed in the late 1970s, was limited to serving golf courses, parks, schools, and large commercial areas. Extensive biological research through the late 1970s and early 1980s resulted in approval by the Florida Department of Environmental Protection (FDEP) and the U.S. EPA for expansion of the reclaimed water system into residential areas. In 1986, a $10 million system expansion was completed to include service to a limited number of residential and commercial sites. Continued expansion of the reclaimed water system has contributed significantly to reducing potable (drinking) water demands.

Current Status of Water Reclamation and Reuse

St. Petersburg's Water Reclamation System is not only the first to be built in the United States, but it remains one of the largest in the world (see Fig. 26-3). The layout of the reclaimed water distribution system and critical water quality areas are discussed previously in Chap. 18 and shown on Fig. 14-3 in Chap. 14.

Some important statistics about the system follow. The system has:

- Over 10,400 active customers
- 470 km (290 mi) of reclaimed water pipelines
- 308 fire hydrants
- 3741 valves
- Four water reclamation facilities
- Reclaimed water production of 136×10^3 m^3/d (36 Mgal/d)

Reclaimed water is not permitted for the following applications:

- Consumption by humans or animals
- Connection to a dwelling for toilet flushing or other internal household use
- Interconnection with another water source
- Sprinkler irrigation of edible crops
- Human bodily contact or water recreation
- Nonreclaimed marked/labeled hose bibs, faucets, quick couplers, and hoses
- Filling of swimming pools, decorative pools, and ponds
- Development of a common reclaimed water service or connection between properties
- Washing equipment such as cars, boats, driveways, and structures

Irrigation with reclaimed water is a smart alternative to using potable water. Like other freshwater sources, a limited amount of reclaimed water is available to water customers for irrigation. In St. Petersburg, the typical residential lawn can require 114 m^3/mo (30,000 gal/mo) of irrigation water during the growing season. The average residential customer discharges 23 m^3/mo (6000 gal/mo) to the wastewater collection system. Therefore, it takes five residential customers to produce enough reclaimed water to supply one residence with irrigation water. As a result, it is not possible at this time to supply all residences in St. Petersburg with reclaimed water. The Water Resources Department's Reclaimed Water Division is currently able to consider in-fill requests for reclaimed water service. All other reclaimed water service requests are put on a waiting list for future consideration when the system is able to accommodate expansion.

Lessons Learned

The reclaimed water system has continued to expand and change in character since its inception, from one of an alternative mode of wastewater effluent disposal to one of a fully operational reclaimed water supply service. The regional water management utility encompasses (1) drinking water treatment and distribution, (2) wastewater collection and treatment, and (3) reclaimed water supply.

Figure 26-3
Various water reuse applications in St. Petersburg, FL: (a) open parkway with reclaimed water used for fire fighting, (b) pond in local park and golf course, (c) cooling towers used in conjunction with air conditioning system at a football stadium (Coordinates: 27.767 N, 82.654 W), and (d) reclaimed water used for cooling towers at a large solid waste processing facility.

Impact of Reclaimed Water System

The growth in the reclaimed water demand since its inception has contributed significantly to the suppression of additional potable water demands. Significant economic and environmental benefits have been derived from the development of a reclaimed water system. Since its inception, the annual demand for potable water has been stabilized while the demand for reclaimed water has increased steadily.

The reclaimed water system has been an economic benefit to all the city's utility customers in that the following potable water system projects have been delayed indefinitely:

- Additional treatment units at the Cosme Water Treatment Plant
- A booster pumping station on the 1200 mm (48 in.) diameter water transmission main in the Safety Harbor area
- South-side booster pumping station and storage facility

The cost avoidance for these projects, i.e., the amount that would have been spent to expand the potable water system, was in the range of $25 to $30 million (ca 1997–1998). In addition, economic savings have also been realized at the wastewater treatment plants by avoiding installation of expensive nutrient removal processes.

Change in Customer Base

The expansion of the distribution system into residential areas also required a change in management policy and attitude. In the early years, the reclaimed water distribution system supplied large land areas such as parks and golf courses where landscape management practices were well established. Thus, the use of this new resource posed no significant problems to these customers and the city received very few complaints. In 1981, only 18.7 percent of the total number of customers was residential users, accounting for less than 0.5 percent of the total land area under irrigation. In 1987, however, residential customers represented 96.2 percent of the total number of users, and the residential area under reclaimed water irrigation was 32 percent of the total area served by the system. Since 1985, the water reuse program has changed from a nonresidential user oriented system to one which now must recognize and respect the needs of the smaller residential homeowner and deal with the more numerous questions and complaints.

Change in Management Policy and Attitude

Together with the growth of the system, the attitude of Public Utilities wastewater treatment personnel towards wastewater effluent has also changed significantly between 1980 and 1987. In St. Petersburg, operators now work in "Water Reclamation Facilities." Signs at plant entrances identify them by this term, and "water production," not "sewage treatment," is the theme of a water reclamation program. Plant operations staff members have a "manufacturing" mentality and not a "treat and dispose" attitude. The value of the reclaimed water in helping St. Petersburg meet its total water needs is recognized by all plant employees.

In the Reclaimed Water System Master Plan Update it was recognized that the system had to continue to expand and change in character from an alternate mode of wastewater disposal to a full operation as a third element of the city's Public Utilities Department. Recommendations on the future development and management of this system were also incorporated in this report.

Achieving Zero Discharge of Wastewater Effluent

The aggressive water reclamation and reuse program has made it possible for St. Petersburg to become the first major municipality in the United Sates to achieve zero discharge of wastewater effluent. This significant achievement represents a remarkable example of what can be accomplished by careful planning and considerable foresight, just as the city fathers had foresight in the 1920s concerning potable water issues.

The development of a water reclamation and reuse program also provides a workable solution to water supply and water pollution problems in coastal areas. The City of St. Petersburg's water reclamation and reuse program serves as a valuable model for many other municipalities now striving to attain similar water reclamation and reuse goals.

Information on the City of St. Petersburg's water reclamation and reuse is easily accessible and contains the wealth of positive and proactive information, at http://www.stpete.org/wwwrecla.htm.

> **Access to City's Proactive Water Reclamation and Reuse Information**

26-6 OBSERVATIONS ON WATER RECLAMATION AND REUSE

The following general observations can be derived from the existing water reclamation and reuse projects:

- Stakeholder and public acceptance is essential.
- It is of critical importance to demonstrate that the reclaimed water is safe with respect to chemical and microbiological quality.
- A comprehensive ongoing monitoring program is an essential part of a water reuse program.
- To deal with unknown constituents, the precautionary principle should be applied, where appropriate.
- Additional safeguards should be implemented, based on risk assessment and risk minimization.
- Treatment process reliability and system redundancies must be incorporated into any water reuse plan.
- In dealing with the public, it is essential that all of the elements of the proposed projects are presented in a transparent manner, regardless of the outcome.
- Project proponents, such as public agencies, must be able to establish, maintain, and increase trust and credibility with the public (stakeholders).
- With diligent outreach programs, it is possible to gain public support for various water reuse projects.

PROBLEMS AND DISCUSSION TOPICS

26-1 Identify possible opportunities for water reuse in your community. What factors promote or prevent water reuse? How can negative factors be overcome so that water reuse can be implemented? Prepare a table listing positive and negative factors along with an evaluation of these factors.

26-2 How can public perspectives on water reuse be analyzed and public participation and outreach be formulated? What would be a role of a water quality specialist in this program?

26-3 Expand the discussion on the need for transparency and public trust in water reclamation and reuse planning and implementation. Cite three recent references on water reuse implementation and discuss how the need for transparency and public trust were dealt with.

26-4 Referring to Table 26-6, how can the merits of a preferred water reuse project be communicated to and ultimately adopted by the public and a city government?

26-5 Discuss the future prospects for water reclamation and reuse with respect to: (1) water resources management, (2) social and economic aspects, (3) environmental protection, (4) public health protection, (5) urban amenities, (6) public acceptance, and (7) sustainability.

26-6 Given that a reclaimed water has been proven beyond a doubt to be safe for direct potable reuse, which of the factors given in Problem 26-5 will be the most difficult to overcome if direct potable reuse is proposed? What is the basis for your answer?

26-7 It may be argued that "necessity and opportunity" are key driving forces for the ultimate use of reclaimed water. Referring to Problem 26-5, which factor has the greatest impact on promoting water reclamation, recycling, and reuse?

REFERENCES

Asano, T (ed.) (1998) *Wastewater Reclamation and Reuse*, Water Quality Management Library, **10**, CRC Press, Boca Raton, FL.

Bruvold, W. H. (1972) *Public Attitude toward Reuse of Reclaimed Water*, Contribution No. 137, University of California Water Resources Center, Los Angeles, CA.

Bruvold, W. H., and J. Crook, (1980) *Public Evaluation of Water Reuse Options*, OWRT/RU-80/2, U.S. Department of the Interior, Office of Water Research and Technology, Washington, DC.

Bruvold, W. H. (1981) "Community Evaluation of Adopted Uses of Reclaimed Water," *Water Resour. Res.*, **17**, 3, 487–490.

Bruvold, W. H. (1987) "Public Evaluation of Salient Water Reuse Options," *Proceedings of Water Reuse Symposium IV*, 1019–1028, AWWA Research Foundation, Denver, CO.

Bruvold, W. H. (1988) "Public Opinion on Water Reuse Options," *J. WPCF*, **60**, 1, 45–49.

Bruvold, W. H. (1992) "Public Evaluation of Municipal Water Reuse Alternatives," *Water Sci. Technol.*, **26**, 7–8, 1537–1543.

CEQ (2005a) *National Environmental Protection Act of 1969 as Amended*, Council on Environmental Quality web site, http://ceq.eh.doe.gov/nepa/regs/nepa/nepaeqia.htm.

CEQ (2005b) *Code of Federal Regulations, Title 40, Part 1506*, Other requirements of NEPA, Council on Environmental Quality web site, http://ceq.eh.doe.gov/nepa/regs/ceq/1506.htm#1506.6.

City of San Diego (2006) *City of San Diego Water Reuse Study 2005: Final Draft Report*, March, 2006, prepared in coordination with Water Department, City of San Diego, PBS&J, and McGuire/Pirnie, San Diego, CA.

Daughton, C. (2005) "Emerging Chemicals as Pollutants in the Environment: A 21st Century Perspective," *Renewable Resour. J.*, **23**, 4, 6–23.

Haddad, B. M. (2004) *Research Needs Assessment Workshop: Human Reaction to Water Reuse*, WateReuse Foundation, Alexandria, VA.

Hartley, T. W. (2003) *Water Reuse: Understanding Public Perception and Participation*, 00–PUM-1, Water Environment Research Foundation, Alexandria, VA.

Hartling, E. C. (2001) "Laymanization, An Engineer's Guide to Public Relations," *Water Environ. Tech.*, **13**, 4, 45–48.

House, M. A. (1999) "Citizen Participation in Water Management." *Water Sci. Technol.*, **40**, 125–130.

Ingram, P. C., V. J. Young, M. Millan, C. Chang, and T. Tabucchi (2005) "From Controversy to Consensus: The Redwood City Recycled Water Experience," presented at the 2005 Annual California WateReuse Conference, San Diego, CA.

Johnson, W. D., and J. R. Parnell (1989) "Wastewater Reclamation and Reuse in the City of St. Petersburg, Florida," 1037–1104, in T. Asano (ed.) *Wastewater Reclamation and Reuse*, Water Quality Management Library, **10**, CRC Press, Boca Raton, FL.

Kelly, L. (1998) "Reuse: Eight Critical Steps to Building Community Support," 1447–1455, in T. Asano (ed.) *Wastewater Reclamation and Reuse*, Water Quality Management Library, **10**, CRC Press, Boca Raton, FL.

Kolpin, D., E. T. Furlong, M. T. Meyer, E. M. Thurman, S. D. Zaugg, L. B. Barber, and H. T. Buxton (2002) "Pharmaceuticals, Hormones, and Other Organic Wastewater Contaminants in U.S. Streams, 1999–2000: National Reconnaissance," *Envi. Sci. Tech.*, **36**, 6, 1202–1211.

Lohman, L. C. (1987) "Potable Wastewater Reuse Can Win Public Support," 1029–1044, *Proceedings of the Water Reuse Symposium IV*, AWWA Research Foundation, Denver, CO.

Lohman, L. C. (1988) "Water Reuse: Professional and Public Perspectives," *Proceedings of the Third North American Chemical Conference*, Toronto, Canada.

PBS&J (2005) "Public Outreach and Involvement," 2005 Water Reuse Study, City of San Diego, Technical Memorandum 8, San Diego, CA.

Radcliffe, J. (2004) "Water Recycling in Australia—Arising Policy Issues," Enviro04 Conference, Australian Water Association, Sydney, Australia.

Sheikh, B., E. Rosenblum, S. Kasower, and E. Hartling (1998) "Accounting for the Benefits of Water Reuse," *Water Reuse Conference Proceedings*, Lake Buena Vista, FL.

State of California (2003) *Water Recycling 2030: Recommendations of California's Recycled Water Task Force*, Department of Water Resources, Sacramento, CA.

Umphres, M. B., F. Stevenson, S. M. Katz, and R. Spear (1992) "Keeping the Public in Public Works Facility Planning," *Proceedings of the Environmental Engineering Sessions, World Water Forum '92*, Baltimore, MD.

U.S. EPA (2004) *Guidelines for Water Reuse*, EPA/625/R-04/108, U.S. Environmental Protection Agency and U.S. Agency for International Development, Washington, DC.

Wegner-Gwidt, J. (1998) "Public Support and Education for Water Reuse," Chap. 31, in T. Asano (ed.) *Wastewater Reclamation and Reuse*, Water Quality Management Library, **10**, CRC Press, Boca Raton, FL.

Walesh, S. G. (1999) "Dad is Out, Pop is In," *J. AWWA*, **35**, 3, 535–544.

Appendix A

Conversion Factors

Table A-1
Unit conversion factors, SI units to U.S. customary units, and U.S. customary units to SI units

SI unit name	Symbol	→	←	Symbol	U.S. customary unit name
Acceleration					
Meters per square second	m/s^2	3.2808	0.3048	ft/s^2	Feet per square second
Meters per square second	m/s^2	39.3701	0.0254	in./s^2	Inches per square second
Area					
Hectare (10,000 m^2)	ha	2.4711	0.4047	ac	Acre
Square centimeter	cm^2	0.1550	6.4516	in.^{2}	Square inch
Square kilometer	km^2	0.3861	2.5900	mi^2	Square mile
Square kilometer	km^2	247.1054	4.047×10^{-2}	ac	Acre
Square meter	m^2	10.7639	9.2903×10^{-2}	ft^2	Square foot
Square meter	m^2	1.1960	0.8361	yd^2	Square yard
Energy					
Kilojoule	kJ	0.9478	1.0551	Btu	British thermal unit
Joule	J	2.7778×10^{-7}	3.6×10^6	kW·h	Kilowatt-hour
Joule	J	0.7376	1.356	ft·lb$_f$	Foot-pound (force)
Joule	J	1.0000	1.0000	W·s	Watt-second
Joule	J	0.2388	4.1876	cal	Calorie
Kilojoule	kJ	2.7778×10^{-4}	3600	kW·h	Kilowatt-hour
Kilojoule	kJ	0.2778	3.600	W·h	Watt-hour
Megajoule	MJ	0.3725	2.6845	hp·h	Horsepower-hour
Force					
Newton	N	0.2248	4.4482	lb$_f$	Pound force
Flowrate					
Cubic meters per day	m^3/d	264.1720	3.785×10^{-3}	gal/d	Gallons per day
Cubic meters per day	m^3/d	2.6417×10^{-4}	3.7854×10^3	Mgal/d	Million gallons per day
Cubic meters per second	m^3/s	35.3147	2.8317×10^{-2}	ft^3/s	Cubic feet per second

(Continued)

Table A-1

Unit conversion factors, SI units to U.S. customary units and U.S. customary units to SI units (*Continued*)

SI unit name	Symbol	→	←	Symbol	U.S. customary unit name
		To convert, multiply in direction shown by arrows			
Flowrate					
Cubic meters per second	m^3/s	22.8245	4.3813×10^{-2}	Mgal/d	Million gallons per day
Cubic meters per second	m^3/s	15850.3	6.3090×10^{-5}	gal/min	Gallons per minute
Liters per second	L/s	22,824.5	4.3813×10^{-2}	gal/d	Gallons per day
Liters per second	L/s	2.2825×10^{-2}	43.8126	Mgal/d	Million gallons per day
Liters per second	L/s	15.8508	6.3090×10^{-2}	gal/min	Gallons per minute
Length					
Centimeter	cm	0.3937	2.540	in.	Inch
Kilometer	km	0.6214	1.6093	mi	Mile
Meter	m	39.3701	2.54×10^{-2}	in.	Inch
Meter	m	3.2808	0.3048	ft	Foot
Meter	m	1.0936	0.9144	yd	Yard
Millimeter	mm	0.03937	25.4	In.	Inch
Mass					
Gram	g	0.0353	28.3495	oz	Ounce
Gram	g	0.0022	4.5359×10^2	lb	Pound
Kilogram	kg	2.2046	0.45359	lb	Pound
Megagram (10^3 kg)	Mg	1.1023	0.9072	ton	Ton (short: 2000 lb)
Megagram (10^3 kg)	Mg	0.9842	1.0160	ton	Ton (long: 2240 lb)
Power					
Kilowatt	kW	0.9478	1.0551	Btu/s	British thermal units per second
Kilowatt	kW	1.3410	0.7457	hp	Horsepower
Watt	W	0.7376	1.3558	$ft \cdot lb_f/s$	Foot-pounds (force) per second
Pressure (force/area)					
Pascal (newtons per square meter)	$Pa(N/m^2)$	1.4504×10^{-4}	6.8948×10^3	$lb_f/in.^2$	Pounds (force) per square inch
Pascal (newtons per square meter)	$Pa(N/m^2)$	2.0885×10^{-2}	47.8803	lb_f/ft^2	Pounds (force) per square foot
Pascal (newtons per square meter)	$Pa(N/m^2)$	2.9613×10^{-4}	3.3768×10^3	in. Hg	Inches of mercury (60°F)

Table A-1
Unit conversion factors, SI units to U.S. customary units and U.S. customary units to SI units (*Continued*)

SI unit name	Symbol	→	←	Symbol	U.S. customary unit name
Pressure (force/area)					
Pascal (newtons per square meter)	Pa (N/m^2)	4.0187×10^{-3}	2.4884×10^2	in. H$_2$O	Inches of water (60°F)
Kilopascal (kilonewtons per square meter)	kPa (kN/m^2)	0.1450	6.8948	lb/in.^{2}	Pounds (force) per square inch
Kilopascal (kilonewtons per square meter)	kPa (kN/m^2)	9.8688×10^{-3}	1.0133×10^2	atm	Atmosphere (standard)
Temperature					
Degree Celsius (centigrade)	°C	1.8(°C) + 32	0.555(°F − 32)	°F	Degree Fahrenheit
Kelvin	K	1.8(K) − 459.67	0.555(°F + 459.67)	°F	Degree Fahrenheit
Velocity					
Meters per second	m/s	2.2369	0.44704	mi/h	Miles per hour
Meters per second	m/s	3.2808	0.3048	ft/s	Feet per second
Volume					
Cubic centimeter	cm^3	0.0610	16.3781	in.^{3}	Cubic inch
Cubic meter	m^3	35.3147	2.8317×10^{-2}	ft^3	Cubic foot
Cubic meter	m^3	1.3079	0.7646	yd^3	Cubic yard
Cubic meter	m^3	264.1720	3.7854×10^{-3}	gal	Gallon
Cubic meter	m^3	8.1071×10^{-4}	1.2335×10^3	ac·ft	Acre-foot
Liter	L	0.2642	3.7854	gal	Gallon
Liter	L	0.0353	28.3168	ft^3	Cubic foot
Liter	L	33.8150	2.9573×10^{-2}	oz	Ounce (U.S. fluid)

Table A-2
Conversion factors for commonly used wastewater treatment plant design parameters

SI units	To convert, multiply in direction shown by arrows →	←	U.S. units
g/m^3	8.3454	0.1198	lb/Mgal
ha	2.4711	0.4047	ac
kg	2.2046	0.4536	lb
kg/ha	0.8922	1.1209	lb/ac
kg/kW·h	1.6440	0.6083	lb/hp·h
kg/m^2	0.2048	4.8824	lb/ft^2
kg/m^3	8345.4	1.1983×10^{-4}	lb/Mgal
kg/m^3·d	62.4280	0.0160	lb/10^3 ft^3·d
kg/m^3·h	0.0624	16.0185	lb/ft^3·h
kJ	0.9478	1.0551	Btu
kJ/kg	0.4303	2.3241	Btu/lb
kPa (gage)	0.1450	6.8948	lbf/in.^{2} (gage)
kPa Hg (60°F)	0.2961	3.3768	in. Hg (60°F)
kW/m^3	5.0763	0.197	hp/10^3 gal
kW/10^3 m^3	0.0380	26.3342	hp/10^3 ft^3
L	0.2642	3.7854	gal
L	0.0353	28.3168	ft^3
L/m^2·d	2.4542×10^{-2}	40.7458	gal/ft^2·d
L/m^2·min	0.0245	40.7458	gal/ft^2·min
L/m^2·min	35.3420	0.0283	gal/ft^2·d
m	3.2808	0.3048	ft
m/h	3.2808	0.3048	ft/h
m/h	0.0547	18.2880	ft/min
m/h	0.4090	2.4448	gal/ft^2·min
m^2/10^{3}m^3·d	0.0025	407.4611	ft^2/Mgal·d
m^3	1.3079	0.7646	yd^3
m^3/capita	35.3147	0.0283	ft^3/capita
m^3/d	264.1720	3.785×10^{-3}	gal/d
m^3/d	2.6417×10^{-4}	3.7854×10^3	Mgal/d
m^3/h	0.5886	1.6990	ft^3/min
m^3/ha·d	106.9064	0.0094	gal/ac·d
m^3/kg	16.0185	0.0624	ft^3/lb
m^3/m·d	80.5196	0.0124	gal/ft·d
m^3/m·min	10.7639	0.0929	ft^3/ft·min
m^3/m^2·d	24.5424	0.0407	gal/ft^2·d
m^3/m^2·d	0.0170	58.6740	gal/ft^2·min
m^3/m^2·d	1.0691	0.9354	Mgal/ac·d

Table A-2
Conversion factors for commonly used wastewater treatment plant design parameters (*Continued*)

SI units	→	←	U.S. units
$m^3/m^2 \cdot h$	3.2808	0.3048	$ft^3/ft^2 \cdot h$
$m^3/m^2 \cdot h$	589.0173	0.0017	$gal/ft^2 \cdot d$
m^3/m^3	0.1337	7.4805	ft^3/gal
$m^3/10^3\ m^3$	133.6805	7.4805×10^{-3}	$ft^3/Mgal$
$m^3/m^3 \cdot min$	133.6805	7.4805×10^{-3}	$ft^3/10^3\ gal \cdot min$
$m^3/m^3 \cdot min$	1,000.0	0.001	$ft^3/10^3\ ft^3 \cdot min$
Mg/ha	0.4461	2.2417	ton/ac
Mm	3.9370×10^{-2}	25.4	in.
ML/d	0.2642	3.785	Mgal/d
ML/d	0.4087	2.4466	ft^3/s

Table A-3
Abbreviations for SI units

Abbreviation	SI unit
°C	Degree Celsius
cm	Centimeter
g	Gram
g/m^2	Gram per square meter
g/m^3	Gram per cubic meter (= mg/L)
h	Hour
ha	Hectare
J	Joule
K	Kelvin
kg	Kilogram
kg/capita·d	Kilogram per capita per day
kg/ha	Kilogram per hectare
kg/m^3	Kilogram per cubic meter
kJ	Kilojoule
kJ/kg	Kilojoule per kilogram
kJ/kW·h	Kilojoule per kilowatt-hour
km	Kilometer
km^2	Square kilometer
km/h	Kilometer per hour
km/L	Kilometer per liter
kN/m^2	Kilonewton per square meter

(*Continued*)

Table A-3
Abbreviations for SI units (*Continued*)

Abbreviation	SI unit
kPa	Kilopascal
ks	Kilosecond
kW	Kilowatt
L	Liter
L/s	Liters per second
m	Meter
m^2	Square meter
m^3	Cubic meter
mm	Millimeter
m/s	Meter per second
mg/L	Milligram per liter ($= g/m^3$)
m^3/s	Cubic meter per second
MJ	Megajoule
N	Newton
N/m^2	Newton per square meter
Pa	Pascal (usually given as kilopascal)
s	Second
W	Watt

Table A-4
Abbreviations for U.S. customary units

Abbreviation	US customary unit
ac	Acre
ac-ft	Acre foot
Bgal	Billion gallons
Btu	British thermal unit
Btu/ft^3	British thermal unit per cubic foot
d	Day
ft	Foot
ft^2	Square foot
ft^3	Cubic foot
ft/min	Feet per minute
ft/s	Feet per second
ft^3/min	Cubic feet per minute
ft^3/s	Cubic feet per second
°F	Degree Fahrenheit
gal	Gallon

Table A-4
Abbreviations for U.S. customary units (*Continued*)

Abbreviation	U.S. customary unit
gal/ft²·d	Gallon per square foot per day
gal/ft²·min	Gallon per square foot per minute
gal/min	Gallon per minute
h	Hour
hp	Horsepower
hp-h	Horsepower-hour
in.	Inch
kWh	Kilowatt-hour
lb_f	Pound (force)
lb_m	Pound (mass)
lb/ac	Pound per acre
lb/ac·d	Pound per acre per day
lb/capita·d	Pound per capita per day
lb/ft²	Pound per square foot
lb/ft³	Pound per cubic foot
lb/in.²	Pound per square inch
lb/yd³	Pound per cubic yard
Mgal/d	Million gallons per day
mi	Mile
mi²	Square mile
mi/h	Mile per hour
mo	Month
ppb	Part per billion
ppm	Part per million
ppt	Part per trillion
s	Second
ton (2000 lb_m)	Ton (2000 pounds mass)
wk	Week
yd	Yard
yd²	Square yard
yd³	Cubic yard
yr	Year

Appendix B

Physical Properties of Selected Gases and the Composition of Air

Table B-1

Molecular weight, specific weight, and density of gases found in water at standard conditions (0°C, 1 atm)[a]

Gas	Formula	Molecular weight, g/mole	Specific weight, lb/ft^3	Density, g/L
Air	—	28.97	0.0808	1.2928
Ammonia	NH_3	17.03	0.0482	0.7708
Carbon dioxide	CO_2	44.00	0.1235	1.9768
Carbon monoxide	CO	28.00	0.0781	1.2501
Hydrogen	H_2	2.016	0.0056	0.0898
Hydrogen sulfide	H_2S	34.08	0.0961	1.5392
Methane	CH_4	16.03	0.0448	0.7167
Nitrogen	N_2	28.02	0.0782	1.2507
Oxygen	O_2	32.00	0.0892	1.4289

[a] Adapted from Perry, R. H., D. W. Green, and J. O. Maloney (eds.) (1984) *Perry's Chemical Engineers' Handbook,* 6th ed., McGraw-Hill, New York.

Table B-2

Composition of dry air at 0°C and 1.0 atmosphere[a]

Gas	Formula	Percent by volume[b,c]	Percent by weight
Nitrogen	N_2	78.03	75.47
Oxygen	O_2	20.99	23.18
Argon	Ar	0.94	1.30
Carbon dioxide	CO_2	0.03	0.05
Other[d]	—	0.01	—

[a] Note: Values reported in the literature vary depending on the standard conditions.
[b] Adapted from *North American Combustion Handbook,* 2nd ed., North American Mfg. Co., Cleveland, OH.
[c] For ordinary purposes air is assumed to be composed of 79 percent N_2 and 21 percent O_2 by volume.
[d] Hydrogen, neon, helium, krypton, xenon.

Note: Molecular weight of air = $(0.7803 \times 28.02) + (0.2099 \times 32.00) + (0.0094 \times 39.95) + (0.0003 \times 44.00) = 28.97$ (see Table B-1).

Appendix B Physical Properties of Selected Gases and the Composition of Air

B-1 DENSITY OF AIR AT OTHER TEMPERATURES

In SI Units

The following relationship can be used to compute the density of air, ρ_a, at other temperatures at atmospheric pressure.

$$\rho_a = \frac{PM}{RT}$$

where P = atmospheric pressure, $1.01325 \times 10^5 \text{ N/m}^2$
M = molecular weight of air (see Table B-1), 28.97 g/mole air
R = universal gas constant, 8314 N·m/(mole air · K)
T = temperature, K ($273.15 + °C$)

For example, at 20°C, the density of air is:

$$\rho_{a,20°C} = \frac{(1.01325 \times 10^5 \text{ N/m}^2)(28.97 \text{ g/mole air})}{[8314 \text{ N}_g \cdot \text{m/(mole air}_g \cdot \text{K)}][(273.15 + 20)\text{K}]}$$

$$= 1.204 \times 10^3 \text{ g/m}^3 = 1.204 \text{ kg/m}^3$$

In U.S. Customary Units

The following relationship can be used to compute the specific weight of air, γ_a, at other temperatures at atmospheric pressure.

$$\gamma_a = \frac{P(144 \text{ in.}^2/\text{ft}^2)M}{RT}$$

where P = atmospheric pressure, 14.7 lb/in.^{2}
M = molecular weight of air (see Table B-1), 28.97 lb/lb mole air
R = universal gas constant, 1544 ft·lb/(lb mole air · °R)
T = temperature, °R ($460 + °F$)

For example, at 68°F, the specific weight of air is:

$$\gamma_a = \frac{(14.7 \text{ lb/in.}^2)(144 \text{ in.}^2/\text{ft}^2)(28.97 \text{ lb/lb mole air})}{[1544 \text{ ft} \cdot \text{lb/(lb mole air} \cdot °\text{R})][(460 + 68)°\text{R}]} = 0.0752 \text{ lb/ft}^3$$

B-2 CHANGE IN ATMOSPHERIC PRESSURE WITH ELEVATION

In SI Units

The following relationship can be used to compute the change in atmospheric pressure with elevation.

$$\frac{P_b}{P_a} = \exp\left[-\frac{gM(z_b - z_a)}{g_c R\, T}\right]$$

where P_b = pressure at elevation z_b, N/m²
P_a = atmospheric pressure at elevation z_a, N/m²
g = acceleration due to gravity, 9.81 m/s²
M = molecular weight of air (see Table B-1), 28.97 g/mole air
z_b = elevation at b, m
z_a = elevation at a, m
R = universal gas constant, 8314 N · m/(mole air · K)
T = temperature, K (273.15 + °C)

The following relationship can be used to compute the change in atmospheric pressure with elevation.

In U.S. Customary Units

$$\frac{P_b}{P_a} = \exp\left[-\frac{gM(z_b - z_a)}{RT}\right]$$

where P_b = pressure at elevation z_b, lb/in.²
P_a = pressure at elevation z_a, lb/in.²
g = acceleration due to gravity, 32.2 ft/s²
M = molecular weight of air (see Table B-1), 28.97 lb_m/lb mole air
z_b = elevation at b, ft
z_a = elevation at a, ft
g_c = 32.2 ft · lb_m/lb · s²
R = universal gas constant, 1544 ft · lb/(lb mole air · °R)
T = temperature, °R(460 + °F)

Appendix C

Physical Properties of Water

The principal physical properties of water are summarized in SI units in Table C-1 and in U.S. customary units in Table C-2. They are described briefly below (Vennard and Street, 1975; Webber, 1971).

C-1 SPECIFIC WEIGHT

The specific weight γ of a fluid is its weight per unit volume. In SI units, it is expressed in kilonewtons per cubic meter (kN/m³). The relationship between γ, ρ, and the acceleration due to gravity g is $\gamma = \rho g$. At normal temperatures γ is 9.81 kN/m³ (62.4 lb_f/ft³).

C-2 DENSITY

The density of ρ of a fluid is its mass per unit volume. In SI units it is expressed in kilograms per cubic meter (kg/m³). For water, ρ is 1000 kg/m³ at 4°C. There is a slight decrease in density with increasing temperature.

C-3 MODULUS OF ELASTICITY

For most practical purposes, liquids may be regarded as incompressible. The bulk modulus of elasticity E is given by

$$E = \frac{\Delta p}{(\Delta V/V)}$$

where Δp is the increase in pressure, which when applied to a volume V results in a decrease in volume ΔV. For water E is approximately 2.17×10^6 kN/m² at normal temperatures and pressures.

C-4 DYNAMIC VISCOSITY

The viscosity of a fluid μ is a measure of its resistance to tangential or shear stress. Viscosity in SI units is expressed in Newton seconds per square meter (N·s/m²).

C-5 KINEMATIC VISCOSITY

In many problems concerning fluid motion, the viscosity appears with the density in the form μ/ρ, and it is convenient to use a single term v, known as the kinematic viscosity and expressed in square meters per second (m²/s) in SI units. The kinematic viscosity of a liquid diminishes with increasing temperature.

C-6 SURFACE TENSION

Surface tension is the physical property that enables a drop of water to be held in suspension at a tap, a glass to be filled with liquid slightly above the brim and yet not spill, or a needle to float on the surface of a liquid. The surface-tension force across any imaginary line at a free surface is proportional to the length of the line and acts in a direction perpendicular to it. The surface tension per unit length σ is expressed in Newtons per meter (N/m) in SI units. There is a slight decrease in surface tension with increasing temperature.

C-7 VAPOR PRESSURE

Liquid molecules that possess sufficient kinetic energy are projected out of the main body of a liquid at its free surface and pass into the vapor. The pressure exerted by this vapor is known as the vapor pressure p_v. In SI units vapor pressure is expressed in kilonewtons per square meter (kN/m²). The vapor pressure of water at 15°C is 1.72 kN/m².

REFERENCES

Vennard, J. K., and R. L. Street (1975) *Elementary Fluid Mechanics*, 5th ed., Wiley, New York.

Webber, N. B. (1971) *Fluid Mechanics for Civil Engineers*, SI ed., Chapman & Hall, London.

Table C-1
Physical properties of water (SI units)[a]

Temperature, °C	Specific weight, γ, kN/m^3	Density[b], ρ, kg/m^3	Modulus of elasticity[b], $E/10^6$, kN/m^2	Dynamic viscosity, $\mu \times 10^3$, N·s/m^2	Kinematic viscosity, $\nu \times 10^6$, m^2/s	Surface tension[c], σ, N/m	Vapor pressure, p_v, kN/m^2
0	9.805	999.8	1.98	1.781	1.785	0.0765	0.61
5	9.807	1000.0	2.05	1.518	1.519	0.0749	0.87
10	9.804	999.7	2.10	1.307	1.306	0.0742	1.23
15	9.798	999.1	2.15	1.139	1.139	0.0735	1.70
20	9.789	998.2	2.17	1.002	1.003	0.0728	2.34
25	9.777	997.0	2.22	0.890	0.893	0.0720	3.17
30	9.764	995.7	2.25	0.798	0.800	0.0712	4.24
40	9.730	992.2	2.28	0.653	0.658	0.0696	7.38
50	9.689	988.0	2.29	0.547	0.553	0.0679	12.33
60	9.642	983.2	2.28	0.466	0.474	0.0662	19.92
70	9.589	977.8	2.25	0.404	0.413	0.0644	31.16
80	9.530	971.8	2.20	0.354	0.364	0.0626	47.34
90	9.466	965.3	2.14	0.315	0.326	0.0608	70.10
100	9.399	958.4	2.07	0.282	0.294	0.0589	101.33

[a] Adapted from Vennard and Street (1975).
[b] At atmospheric pressure.
[c] In contact with the air.

Table C-2
Physical properties of water (U.S. customary units)[a]

Temperature, °F	Specific weight, γ, lb/ft^3	Density[b], ρ, slug/ft^3	Modulus of elasticity[b], $E/10^3$, lb$_f$/in.^{2}	Dynamic viscosity, $\mu \times 10^5$, lb·s/ft^2	Kinematic viscosity, $\nu \times 10^5$, ft^2/s	Surface tension[c], σ, lb/ft	Vapor pressure, p_v, lb$_f$/in.^{2}
32	62.42	1.940	287	3.746	1.931	0.00518	0.09
40	62.43	1.940	296	3.229	1.664	0.00614	0.12
50	62.41	1.940	305	2.735	1.410	0.00509	0.18
60	62.37	1.938	313	2.359	1.217	0.00504	0.26
70	62.30	1.936	319	2.050	1.059	0.00498	0.36
80	62.21	1.934	324	1.799	0.930	0.00492	0.51
90	62.11	1.931	328	1.595	0.826	0.00486	0.70
100	62.00	1.927	331	1.424	0.739	0.00480	0.95
110	61.86	1.923	332	1.284	0.667	0.00473	1.27

(Continued)

Table C-2
Physical properties of water (U.S. customary units)[a] (*Continued*)

Temperature, °F	Specific weight, γ, lb/ft³	Density[b], ρ, slug/ft³	Modulus of elasticity[b], $E/10^3$, lb_f/in.²	Dyamic viscosity, $\mu \times 10^5$, lb·s/ft²	Kinematic viscosity, $\nu \times 10^5$, ft²/s	Surface tension[c], σ, lb/ft	Vapor pressure, p_v, lb_f/in.²
120	61.71	1.918	332	1.168	0.609	0.00467	1.69
130	61.55	1.913	331	1.069	0.558	0.00460	2.22
140	61.38	1.908	330	0.981	0.514	0.00454	2.89
150	61.20	1.902	328	0.905	0.476	0.00447	3.72
160	61.00	1.896	326	0.838	0.442	0.00441	4.74
170	60.80	1.890	322	0.780	0.413	0.00434	5.99
180	60.58	1.883	318	0.726	0.385	0.00427	7.51
190	60.36	1.876	313	0.678	0.362	0.00420	9.34
200	60.12	1.868	308	0.637	0.341	0.00413	11.52
212	59.83	1.860	300	0.593	0.319	0.00404	14.70

[a] Adapted from Vennard and Street (1975).
[b] At atmospheric pressure.
[c] In contact with the air.

Appendix D

Statistical Analysis of Data

The statistical analysis of wastewater flowrate and constituent concentration data involves the determination of statistical parameters used to quantify a series of measurements. Commonly used statistical parameters and graphical techniques for the analysis of wastewater management data are reviewed next.

D-1 COMMON STATISTICAL PARAMETERS

Commonly used statistical measures include the mean, median, mode, standard deviation, and coefficient of variation, based on the assumption that the data are distributed normally. Although the terms just cited are the most commonly used statistical measures, two additional statistical measures are needed to quantify the nature of a given distribution. The two additional measures are the coefficient of skewness and coefficient of kurtosis. If a distribution is highly skewed, as determined by the coefficient of skewness, normal statistics cannot be used. For most wastewater data that are skewed, it has been found that the log of the value is normally distributed. Where the log of the values is normally distributed, the distribution is said to be log normal. The common statistical measures used for the analysis of wastewater management data (Eqs. D-1 through D-9) are summarized in Table D-1.

D-2 GRAPHICAL ANALYSIS OF DATA

Graphical analysis of wastewater management data is used to determine the nature of the distribution. For most practical purposes, the type of the distribution can be determined by plotting the data on both arithmetic- and logarithmic-probability paper and noting whether the data can be fitted with a straight line. The three steps involved in the use of arithmetic- and logarithmic-probability paper are as follows:

1. Arrange the measurements in a data set in order of increasing magnitude and assign a rank serial number.
2. Compute a corresponding plotting position for each data point using Eqs. (D-10) and (D-11).

$$\text{Plotting position (\%)} = \left(\frac{m}{n+1}\right) \times 100 \qquad \text{(D-10)}$$

where m = rank serial number
n = number of observations

The term $(n + 1)$ is used to correct for a small sample bias. The plotting position represents the percent or frequency of observations that are equal to or less than the

Table D-1
Statistical parameters used for the analysis of wastewater management data[a]

Parameter		Definition
Mean value $$\bar{x} = \frac{\sum f_i x_i}{n}$$	(D-1)	$\bar{x}$ = mean value f_i = frequency (for ungrouped data $f_i = 1$) x_i = the mid-point of the ith data range (for ungrouped data x_i = the i-th observation)
Standard deviation $$s = \sqrt{\frac{\sum f_i(x_i - \bar{x})^2}{n - 1}}$$	(D-2)	n = number of observations (note $\sum f_i = n$) s = standard deviation
Coefficient of variation $$C_v = \frac{100\, s}{\bar{x}}$$	(D-3)	C_v = coefficient of variation, percent α_3 = coefficient of skewness α_4 = coefficient of kurtosis M_g = geometric mean
Coefficient of skewness $$\alpha_3 = \frac{\sum f_i(x_i - \bar{x})^3/(n - 1)}{s^3}$$	(D-4)	s_g = geometric standard deviation $\log x_g = \log x_i - \log M_g$ $P_{15.9}$ and $P_{84.1}$ = values from arithmetic or logarithmic probability plots at indicated percent values
Coefficient of kurtosis $$\alpha_4 = \frac{\sum f_i(x_i - \bar{x})^4/(n - 1)}{s^4}$$	(D-5)	**Median value** If a series of observations are arranged in order of increasing value, the middlemost observation, or the arithmetic mean of the two middlemost observations, in a series is known as the median.
Geometric mean $$\log M_g = \frac{\sum f_i (\log x_i)}{n}$$	(D-6)	**Mode** The value occurring with the greatest frequency in a set of observations is known as the mode. If a continuous graph of the frequency distribution is drawn, the mode is the value of the high point, or hump, of the curve. In a symmetrical set of observations, the mean, median, and mode will be the same value.
Geometric standard deviation $$\log S_g = \sqrt{\frac{\sum f_i (\log^2 x_g)}{n - 1}}$$	(D-7)	**Coefficient of skewness** When a frequency distribution is asymmetrical, it is usually defined as being a skewed distribution.
Using probability paper $$s = P_{84.1} - \bar{x} \text{ or } P_{15.9} + \bar{x}$$	(D-8)	**Coefficient of kurtosis** Used to define the peakedness of the distribution. The value of the kurtosis for a normal distribution is 3. A peaked curve will have a value greater than 3 whereas a flatter curve will have a value less than 3.
$$S_g = \frac{P_{84.1}}{M_g} = \frac{M_g}{P_{15.9}}$$	(D-9)	

[a] Adapted from Metcalf & Eddy (1991) and Crites and Tchobanoglous (1998).

indicated value. Another expression often used to define the plotting position is known as Blom's transformation:

$$\text{Plotting position, \%} = \frac{m - 3/8}{n + 1/4} \times 100 \qquad \text{(D-11)}$$

3. Plot the data on arithmetic- and logarithmic-probability paper. The probability scale is labeled as *Percent of values equal to or less than the indicated value.*

If the data, plotted on arithmetic-probability paper, can be fit with a straight line, then the data are assumed to be normally distributed. Significant departure from a straight line can be taken as an indication of skewness. If the data are skewed, logarithmic-probability paper can be used. The implication here is that the logarithm of the observed values is normally distributed. On logarithmic-probability paper, the straight line of best fit passes through the geometric mean, M_g, and through the intersection of $M_g \times s_g$ at a value of 84.1 percent and M_g/s_g at a value of 15.9 percent. The geometric standard deviation, s_g, can be determined using Eq. (D-9) given in Table D-1. The use of arithmetic- and logarithmic-probability paper is illustrated in Example D-1.

EXAMPLE D-1. Statistical Analysis of Wastewater Constituent Concentration Data.

Determine the appropriate statistical parameters for the following set of effluent data from a wastewater treatment plant, collected over a 24-mo period.

Month	Value, g/m³	
	TSS	COD
1	13.50	15.00
2	25.90	11.25
3	28.75	35.35
4	10.75	13.60
5	12.50	15.30
6	9.85	15.75
7	13.90	16.80
8	15.10	15.20
9	23.40	18.75
10	21.90	37.50
11	23.70	27.00
12	18.00	23.30
13	37.00	46.60
14	30.10	36.25
15	21.25	30.00
16	23.50	25.75
17	16.75	17.90

(Continued)

	Value, g/m³	
Month	TSS	COD
18	8.35	11.35
19	18.10	25.20
20	9.25	16.10
21	9.90	16.75
22	8.75	15.80
23	15.50	19.50
24	7.60	9.40

Solution

1. Determine the nature of the distribution by plotting the data on arithmetic- and logarithmic-probability paper.
   a. Determine the plotting position using Eq. (D-10) where n equals 24.

Number	Plotting position, %[a]	Value, g/m³	
		TSS	COD
1	4	7.60	9.40
2	8	8.35	11.25
3	12	8.75	11.35
4	16	9.25	13.60
5	20	9.85	15.00
6	24	9.90	15.20
7	28	10.75	15.30
8	32	12.50	15.75
9	36	13.50	15.80
10	40	13.90	16.10
11	44	15.10	16.75
12	48	15.50	16.80
13	52	16.75	17.90
14	56	18.00	18.75
15	60	18.10	19.50
16	64	21.25	23.30
17	68	21.90	25.20
18	72	23.40	25.75
19	76	23.50	27.00
20	80	23.70	30.00
21	84	25.90	35.35
22	88	28.75	36.25
23	92	30.10	37.50
24	96	37.00	46.60

[a] Plotting position (%) = $\left(\dfrac{m}{n+1}\right) \times 100$

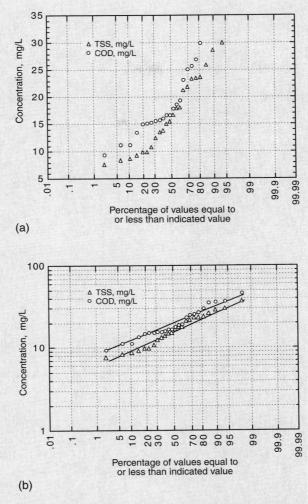

(a)

(b)

b. Plot the above data on both (a) arithmetic- and (b) log-probability paper. As shown in the accompanying plot, both the TSS and COD data are log-normal.

2. Determine the geometric mean for TSS and COD and the corresponding geometric standard deviation using Eq. (D-9).

$$s_g = \frac{P_{84.1}}{M_g} = \frac{M_g}{P_{15.9}}$$

Constituent	M_g	s_g
TSS	17	1.5
COD	20	1.47

Appendix E

Review of Water Reclamation Activities in the United States and in Selected Countries

Table E-1
Milestone water reuse project and research studies in the United States

Year	Description and significant contributions
1912	**Golden Gate Park, City of San Francisco, CA** (Hyde, 1937; Ongerth and Jopling, 1977; Kubick et al., 1992)
	The first beneficial use of reclaimed water in an urban setting was at Golden Gate Park in 1912. First raw sewage and then septic tank effluent was used to fill and maintain a series of ornamental lakes and to irrigate about 100 ha of landscape. An onsite water reclamation plant was constructed and began operation in 1932. Treatment consisted of a bar screen and grit removal, preaeration, primary sedimentation, activated sludge, final sedimentation, and chlorination. The plant is of historic significance because it was constructed for the sole purpose of water reclamation and reuse.
	The Golden Gate Park plant was also the first treatment facility in California that could be operated as a satellite (on-line) water reclamation facility; that is, the source of flow came from sewer mining on an as-needed basis. (As much wastewater as is required is diverted from a sewer and, in periods of plant upset or when the reclaimed water is not needed, the wastewater either is not taken or is returned to the sewer. Sludge treatment or disposal facilities are not needed because all sludge is discharged back to the city sewer.) The current interest and developments in decentralized and satellite systems are discussed in Chaps. 12 and 13.
	Although landscape irrigation from this historic plant was terminated in 1981, the City of San Francisco has adopted a water recycling master plan that envisions the use of approximately 68×10^3 m^3/d by the year 2007. Legislation is in place that requires dual plumbing systems in new and existing landscaping projects of 930 m^2 and larger, and new or remodeled buildings with a total area of 3700 m^2 or greater. In addition, a construction water ordinance requires contractors to use reclaimed water for soil compaction and dust control at construction sites.
1926	**Grand Canyon Village, Grand Canyon National Park, AZ** (Fleming, 1990)
	Tourism at the Grand Canyon began in the 1880s with 80 km trips from Williams, Arizona, via horseback or wagon. The water needs of these early visitors were met by hauling potable water in wagons. In 1924, a plan was developed for providing reclaimed water to Grand Canyon Village. An activated sludge plant with subsequent treatment consisting of rapid sand filtration and chlorine disinfection was constructed

(Continued)

Table E-1

Milestone water reuse project and research studies in the United States (*Continued*)

Year	Description and significant contributions
	and the reclaimed water was pumped 3 km back to the village in a separate distribution system from which it could be withdrawn to serve nonpotable water needs. The reclaimed water was used for toilet flushing in the El Tovar Hotel (Coordinates: 36.057 N, 112.139 W), boiler feed, cooling water for the stationary engines in the electric generating plant, and for makeup water for the steam locomotives. This facility, with a nominal capacity of 493 m^3/d, was constructed in May 1926, and was the first operational water reclamation plant in the United States.
	The practice of water reuse has expanded and continues to expand as a direct result of a water reclamation plant constructed in 1989, at which time the original 63-yr old water reclamation plant was taken off line. In 1990, an estimate of the cost of water production was $0.26/$m^3$ for reclaimed water, and $1.45/$m^3$ for potable water. By comparison, residents of the City of Tusayan paid rates of $4.89/$m^3$ for potable water hauled from Park, and $7.52/$m^3$ for water hauled from Williams.
1972	**City of St. Petersburg, FL** (York and Potts, 1996; Johnson and Parnell, 1998; Walker-Coleman, 1999; See also Chap. 26)
	St. Petersburg (city) has had to deal with water and wastewater issues for more than a century as its population growth mirrored that of the State of Florida. (The population of Florida has grown from 2.77 million in 1950, to 4.95 million in 1960, 9.75 million in 1980 and to about 17.79 million in 2005). Development of the city's reclaimed water system followed the Wilson-Grizzle Act and the 1972 decision by city council to implement a water recycling and deep well injection system to achieve zero-discharge to surface waters. The Wilson-Grizzle Act mandated ". . . wastewater treatment plants discharging to Tampa Bay and its tributaries treat their wastewater to that of drinking water standards or cease discharging to surface waters altogether."
	The conclusions drawn from the results of a pilot study were that spray irrigation using treated wastewater was more feasible and considerably more cost effective than advanced wastewater treatment followed by discharge to Tampa Bay. Also, the construction of a reclaimed water system would benefit the community by reducing the total quantity of water to be imported for potable use. Based on the pilot study results, an approach was adopted to reclaim wastewater by treating it to a sufficient degree that it would be suitable for the irrigation of parks, schools, and golf courses within the city.
	Since the inception of the reclaimed water system, annual demand for potable water has been stabilized while the demand for nutrient rich reclaimed water has steadily increased. The reclaimed water system has been an economic benefit to all the city's utility customers. Several potable water system projects have been delayed indefinitely including additional treatment units at the Cosme Water Treatment Plant, a booster station on the 1.2 m water transmission main in the Safety Harbor area, and the south-side booster station and storage facility. The cost for these projects, should they have been constructed, is estimated to be between $25 and $30 million. Economic savings have also been realized at the wastewater treatment plants by avoiding the need to install expensive nutrient removal wastewater processing systems.

Table E-1
Milestone water reuse project and research studies in the United States (*Continued*)

Year	Description and significant contributions
	The complete implementation of water reclamation and reuse in the city was accomplished over a 15-yr period and St. Petersburg became the first major municipality in the United States to achieve zero discharge of wastewater effluent into surrounding surface waters. The Reclaimed Water System Master Plan Update (1987) recognized that the reclaimed water system had continued to expand and change in character from an alternate mode of wastewater disposal to full operation as a third element (i.e., reclaimed water, water and wastewater) of the City's Public Utilities Department.
1977	**Irvine Ranch Water District, Irvine, CA** (Young et al., 1998; http://www.irwd.com)
	Irvine Ranch Water District (IRWD) promotes the philosophy that water is too valuable to be used just once. Every liter of reclaimed water used to irrigate crops or landscaping means a liter of potable water that can be saved for potable uses.
	In 1991, IRWD became the first water district in the nation to obtain permits for the interior use of reclaimed water from a community system. Reclaimed water is delivered through a completely separate distribution system that includes more than 394 km of pipeline, 8 storage reservoirs, and 12 pump stations.
	The reclaimed water system provides water to 44 agricultural users that have approximately 405 ha of fields and orchards planted with a variety of fruits, vegetables, and nursery products. Over 2818 landscape water meters receive reclaimed water, providing irrigation to approximately 80 percent of all business and community landscaping (e.g., parks, school grounds), which is over 2287 ha of land and includes a few estate-sized residential lots. Many water features such as fountains and the lake at Mason Park are filled with reclaimed water. Reclaimed water is used for toilet flushing in IRWD's facilities as well as in several high rise office buildings constructed with dual piping systems. Potable, or drinking water demands in these buildings have dropped by as much as 75 percent since converting to reclaimed water. In addition, conversion of some local industries to using reclaimed water in their production processes has resulted in additional potable water savings. For example, one carpet mill converted its carpet dyeing process from domestic to reclaimed water, saving 1892 m^3/d to 3785 m^3/d of drinking water.
1977	**Pomona Virus Study County Sanitation Districts of Los Angeles County, Los Angeles, CA** (Sanitation District of Los Angeles County, 1977; State of California, 1978; Dryden et al., 1979; State of California, 1988; Chen et al., 1998; State of California, 2000)
	The Pomona Virus Study was initiated to determine the cost effectiveness of alternative treatment systems that were capable of achieving 5 log (~99.999 percent) removal of waterborne enteric viruses, equivalent to the treatment level typically obtained with a full treatment system. The alternative systems investigated included direct and

(*Continued*)

Year	Description and significant contributions
	contact filtration with low dose (5 mg/L) alum coagulation, followed by disinfection (theoretical chlorine contact time of 2 h with 5 mg/L free and 10 mg/L combined residual Cl_2 and a modal contact time between 90 to 100 min, based on peak dry weather flow). Conclusions drawn from the frequency distributions of various filter effluent turbidities were that an effluent turbidity of 2 NTU or less was achieved approximately 90 percent of the time for the alternative treatment trains, and greater than 99 percent of the time for the full treatment train. In addition, if the filter influent turbidity could be maintained at 4 NTU or less through proper pretreatment, an effluent turbidity of 2 NTU could be achieved more readily. Based on the comparison of turbidity and virus removals, and an analysis of treatment costs, direct filtration was considered the most cost-effective predisinfection treatment alternative to full treatment for producing an essentially pathogen-free disinfected effluent.

A series of specific design and operational requirements followed from the Pomona Virus Study including "Policy Statement for Wastewater Reclamation Plants with Direct Filtration" (State of California, 1988), which was incorporated into the current water recycling criteria (*Code of Regulations, Title 22, Division 4, Chap. 3, Water Recycling Criteria*). |
| 1980–1985 | **Monterey Water Reclamation Study for Agriculture Monterey Regional Water Pollution Control Agency, Monterey, CA** (Kirkpatrick and Asano, 1986; Engineering-Science, 1987; Sheikh et al., 1990; Sheikh et al., 1998)

The Monterey Wastewater Reclamation Study for Agriculture (MWRSA) was a 5-yr field investigation of the safety and feasibility of irrigating artichokes and raw-eaten vegetables such as lettuce, broccoli, celery, and cauliflower with disinfected tertiary reclaimed water. During the early planning stage, Castroville, California was selected as the site for the MWRSA project and an environmental assessment was completed. Next, the existing 1500 m^3/d Castroville treatment plant was upgraded and a field-scale pilot treatment plant for full treatment and direct filtration was designed and constructed, and experimental field plots were established. The 5-yr field study began in 1980 and continued through 1985.

The tertiary treatment processes were designed based on conclusions drawn from the Pomona Virus Study. Virus seeding studies were performed, and conclusions drawn from data gathered during these studies were that both full treatment and direct filtration (which included chlorination with 90-min modal contact time) were capable of reducing the numbers of enteric viruses by 5 log or more; a reconfirmation of the Pomona Virus Study. Thus, the acceptance of direct filtration as an equivalent substitute for the full treatment process train specified in the California Title 22 Water Recycling Criteria was a direct result of the MWRSA pilot field testing. Direct filtration has become a widely practiced technique for tertiary-treated reclaimed water.

Conclusions drawn from data gathered during the 5-yr MWRSA were that food crop irrigation was safe and acceptable with respect to pathogens, heavy metals, and crop quality and yield. No drawbacks were observed in terms of soil or groundwater quality |

Table E-1
Milestone water reuse project and research studies in the United States (*Continued*)

Year	Description and significant contributions
	degradation. Irrigation with tertiary treated and disinfected (with chlorine) reclaimed water produced excellent yields of high-quality produce. Cauliflower and broccoli yields were significantly improved by irrigation with reclaimed water. The MWRSA successfully proved the acceptability of irrigating food crops with reclaimed water from the standpoints of regulatory agencies, farmers, consumers, and wastewater treatment agencies.
1983	**Palo Verde Nuclear Generating Station Palo Verde, Arizona** (Blackson and Moreland 1998; Water Use, 1983)
	Thermal generating plants depend heavily on large quantities of cooling water to condense the steam that drives the plants' turbine generators.
	The Palo Verde Nuclear Generating Station (PVNGS) is the energy cornerstone of the Southwest. It is the largest facility for the peaceful use of nuclear power in the United States. Palo Verde is a standardized triple-unit commercial nuclear power facility. It consists of three identical pressurized water reactors and turbine-generators. Each unit generates 1,270,000 kilowatts of electricity for a total of 3,810,000 kilowatts. The cooling requirement for the three nuclear generating units is approximately 35 m^3/s per unit.
	The location selected for the PVNGS was the arid Sonoran Desert 88 km west of Phoenix, Arizona, far from any natural bodies of water and where the average rainfall is 178 mm/yr. Wastewater effluent from two Phoenix wastewater treatment plants was selected as an economical source of cooling water. The effluents from the 91st Avenue Wastewater Treatment Plant in Phoenix and the Tolleson Wastewater Treatment Plant have proven to be a reliable source of water (approximately 492×10^3 m^3/d) that is treated by the on-site Water Reclamation Facility (WRF). The WRF was designed to produce economical cooling tower makeup water in sufficient quantities and quality to support the operation of the three nuclear power plants at the PVNGS. Continuing efforts are under way to optimize the operation and improve water quality production during the winter months, improve controls of the cooling tower water chemistry, and minimize the blowdown water to the evaporation ponds.
1984	**Health Effects Study at the Montebello Forebay Groundwater Replenishment Project County Sanitation Districts of Los Angeles County, Los Angeles County, CA** (Nellor et al. 1985; Hartling and Nellor 1998; Sanitation Districts of Los Angeles County, 2002)
	The County Sanitation Districts of Los Angeles County (CSDLAC), formed in 1923 by an act of the California State Legislature, has a long history as a pioneer in the field of water reclamation and reuse.
	Reclaimed water, blended with imported river water (Colorado River and State Project water) and local storm water runoff, had been used for groundwater replenishment since 1962. The Health Effects Study was initiated by the CSDLAC in 1978 to evaluate the health effects of using treated wastewater for groundwater recharge.

(*Continued*)

Table E-1

Milestone water reuse project and research studies in the United States (*Continued*)

Year	Description and significant contributions
	The primary goal of the study was to provide information for use by health and regulatory agencies in determining whether the use of reclaimed water in the Montebello Forebay should be maintained at the present level, reduced, or expanded. Specific objectives were: to determine whether the historical level of water reuse had adversely affected groundwater quality or human health and to estimate the relative impact of the different replenishment sources on groundwater quality. Research tasks included water quality characterizations of the replenishment sources and groundwater in terms of their microbiological and chemical content, toxicological and chemical studies of the replenishment sources and groundwater to isolate and identify organic constituents of possible health significance, field studies to evaluate the efficacy of soil for attenuating chemicals in reclaimed water, hydrogeologic studies to determine the movement of reclaimed water through groundwater and the relative contribution of reclaimed water to municipal water supplies, and epidemiologic studies of populations ingesting reclaimed water to determine whether their health characteristics differed significantly from a demographically similar control population.
	No measurable adverse impacts on the area's groundwater or the health of the population ingesting this water were found from an extensive evaluation of the project data. A need for continued monitoring and evaluation of the project with respect to trace organic content of the groundwater and replenishment waters was identified. In addition, investigation of the organic compounds responsible for the mutagenicity detected in bacterial tests was considered prudent in order to determine whether or not any connection could be made to human health.
	Based on the results of the study and recommendations of the Scientific Advisory Panel, authorization was given by the Los Angeles Regional Water Quality Control Board in 1987 to increase the annual quantity of reclaimed water used for replenishment from 40×10^6 m^3 to 62×10^6 m^3. In 1991, the water reclamation requirements for the project were revised to allow for recharge up to 74×10^6 m^3 and 50 percent reclaimed water in any 1 yr as long as the running 3-yr total did not exceed 185×10^6 m^3 or 35 percent reclaimed water. Continued evaluation of the replenishment project is ongoing through an extensive sampling and monitoring program, and by supplemental research projects pertaining to soil aquifer treatment (SAT), chemistry, epidemiology, and microbiology.
1987	**State of California Scientific Advisory Panel on Groundwater Recharge, State of California: State Water Resources Control Board, Department of Water Resources, and Department of Health Services, Sacramento, CA** (State of California, 1987)
	In 1987, three California State agencies—the State Water Resources Control Board, the Department of Water Resources, and the Department of Health Services—commissioned a Scientific Advisory Panel on Groundwater Recharge to review the County Sanitation Districts of Los Angeles County's (CSDLAC's) Health Effects Study and other pertinent information on groundwater recharge with reclaimed water.

Table E-1
Milestone water reuse project and research studies in the United States (*Continued*)

Year	Description and significant contributions
	The purpose of the panel, chaired by Gordon G. Robeck, was to define the health significance of using reclaimed water for groundwater to augment domestic water supply, evaluate the benefits and risks associated with groundwater recharge using reclaimed water, and provide detailed background information needed to establish statewide criteria for groundwater recharge using reclaimed water.
	The panel concluded that the research described in the Health Effects Study final report was thorough and well conducted with state-of-the-art methodology. In addition, limitations of the study were also noted. Limitations included: (1) whether the organic compounds present and of greatest health significance were identified, (2) whether the genotoxicity data were adequate, in the absence of other toxicologic information, to serve as a basis for risk assessment, and (3) whether the ecologic and household survey studies could establish risk because confounding demographic factors could easily overshadow the results. The panel concluded that it was comfortable with the continuation of the Montebello Forebay Groundwater Recharge Project and with the safety of the product water. The panel concluded that the risks, if any, were small and probably not dissimilar from those that could be hypothesized for commonly used surface waters.
1979–1992	**Denver Potable Water Demonstration Project, Denver, CO** (Work et al., 1980; Lauer et al., 1991; Lauer and Rogers, 1998; See also Chap. 24)
	The City of Denver's Direct Potable Water Reuse Demonstration Project was developed to determine whether potable water of comparable quality to Denver's existing water supply could be continuously and reliably produced from secondary treated municipal wastewater. The interrelated issues of product safety, technical and economic feasibility, and public acceptance of direct potable reuse were investigated. The purpose of the project was to provide information necessary to evaluate direct potable reuse as an alternative supply source to meet Denver's future water supply needs. The major objectives for the project were: (1) establish product water safety, (2) demonstrate the dependability of the process, (3) increase public awareness, (4) establish a foundation for future regulatory acceptance, and (5) provide data for a large-scale implementation.
	The key objective upon which all the others depended was the establishment of the safety of the product. Because the health standards thus far established for drinking water were not intended to apply to treated wastewaters or reclaimed water, additional criteria had to be used to establish that the product water was suitable for human consumption.
	Major findings from the project are as follows: • The treatment processes (high pH lime clarification, recarbonation, filtration, activated carbon adsorption, reverse osmosis, or ultrafiltration, air stripping, ozonation, and chloramination) reliably produce a product from secondary treated wastewater that easily satisfied all current (1992) and proposed U.S. EPA drinking water standards. • A complete 2-yr chronic toxicity and carcinogenicity study was conducted for the first time on the reclaimed water and compared to currently used drinking water. No adverse health effects were detected from a lifetime exposure to any of the samples.

(*Continued*)

Table E-1
Milestone water reuse project and research studies in the United States (*Continued*)

Year	Description and significant contributions
	• Reproductive studies were conducted on the reclaimed water and compared to Denver's current drinking water. No adverse health effects were detected during a two-generation reproduction study.
	• Unprecedented physical, chemical, and microbiological testing of the reclaimed water revealed a purity not normally found in domestic water supplies. No compound (organic or inorganic) or organism (bacteria or virus) was found in any of the samples that approached regulatory limits.
	• Public attitudes were found to be cautiously optimistic with the majority expressing a willingness to accept potable water reuse if the need were demonstrated and the safety assured.
	• Although regulatory agency approval was not sought as part of the project, the analyses and health effects test results provide a strong foundation for acceptance.
	• The estimated cost of a full-scale reuse treatment plant, based upon the demonstration plant experience, would compare favorably ($0.43-$0.62/m^3 vs. $0.20-$0.78/m^3) with that of Denver's projected future conventional water supply augmentation projects.
1987	**Water Conserv II, Winter Garden, FL, A cooperative water reuse program by the City of Orlando, Orange County, and the agricultural community** (Cross et al., 2000; See also Chap. 17)
	Water Conserv II is the one of the largest water reuse projects with a combination of agricultural irrigation and rapid infiltration basins (RIBs). It is also the first water reuse project in Florida permitted by the State of Florida Department of Environmental Protection to irrigate crops produced for human consumption, primarily citrus trees, with reclaimed water. Jointly owned by the City of Orlando (city) and Orange County (county), it has taken a liability (effluent previously discharged to surface water bodies) and turned it into an asset (reclaimed water) that benefits the city, the county, and the agricultural community.
	The system encompasses two water reclamation facilities connected by 34 km of transmission pipeline to a distribution center. From the distribution center, a 78 km pipeline network distributes reclaimed water to 76 agricultural and commercial customers. The reclaimed water that is not used for irrigation is distributed to RIBs. The RIB network contains seven sites with 74 RIBs over a total of 809 ha. Both the distribution network and RIB site network are monitored and controlled from a central computerized control system.
	As an incentive to sign on, growers were offered reclaimed water at no cost for 20 yr at pressures suitable for microsprinkler irrigation, eliminating the costs for installation, operation, and maintenance of a pump system, resulting in a savings of $318/ha/yr for citrus growers. In addition, agricultural customers benefited from enhanced freeze protection capabilities, and the project was able to supply enough water to each customer to protect all of their crops.

Table E-1
Milestone water reuse project and research studies in the United States (*Continued*)

Year	Description and significant contributions
	Citrus growers using reclaimed water have realized increased crop yields of 10 to 30 percent and tree growth of up to 400 percent. The increases are not due to the reclaimed water itself, but the availability of water in the soil for the tree to absorb. Because the water is free, growers have maintained higher soil moisture levels. Growers have also benefited from research conducted by the Mid Florida Citrus Foundation (MFCF), another Water Conserv II customer. The MFCF is a nonprofit organization charged with assisting citrus growers in various ways including (1) finding solutions to challenges facing citrus growers, (2) studying the long-term effects of irrigating citrus with reclaimed water, (3) evaluating agricultural crops for economic viability, and (4) studying the effects of golf course irrigation with reclaimed water. • In summary, Water Conserv II has eliminated discharge of wastewater effluent to surface waters. • The RIB sites have provided a preserve for endangered and threatened species for plants and animals, as officially cited by city and county decree. • Water Conserv II replenishes the Floridan aquifer through the discharge of reclaimed water to the RIBs. It also reduced the demand on the aquifer by eliminating the need for well water for irrigation.
Late 1970s–1996	**City of San Diego Total Resources Recovery Project San Diego, CA** (WCPH, 1996; See also Chap. 23.) The City of San Diego imports virtually all of its water supply from other parts of the state and from the Colorado River. New sources of imported water are not readily available, and the city is actively investigating advanced wastewater treatment technologies for utilization of potential resources that may already exist locally. Municipal wastewater, discharged to the Pacific Ocean, is among the potential local water resources that were investigated. The city operated three progressively larger aquatic treatment total resources recovery (TRR) facilities known as Aqua I, II, and III. These were based on water hyacinth pond systems to meet secondary treatment standards, plus tertiary and advanced water treatment (AWT) processes, particularly reverse osmosis (RO), to meet drinking water standards. The TRR project began in the late 1970s and was initially conducted at the 1136 m^3/d pilot-scale facility in Mission Valley (Aqua II). Beginning in October 1994, Aqua III was started up at the 3785 m^3/d full-scale demonstration Aqua III facility in the San Pasqual Valley (3785 m^3/d secondary and 1893 m^3/d AWT). Twelve-mo full-scale operation, monitoring, and health effects studies at the Aqua III TRR facilities were conducted during 1994–1995. The Health Effects Study of the TRR project reconfirmed that the AWT treatment train of coagulation, filtration, reverse osmosis, air stripping, and granular activated carbon (GAC) adsorption can produce water of equivalent or better quality than the existing Miramar raw water supply, which is the main water supply for the city. Based on the study data, GAC adsorption in addition to RO is unnecessary to achieve California's Title 22 requirements for reducing organics of wastewater origin to less than 1 mg/L water in indirect potable reuse reservoir augmentation or direct injection for groundwater recharge.

(*Continued*)

Table E-1

Milestone water reuse project and research studies in the United States (*Continued*)

Year	Description and significant contributions
1998	**City of Scottsdale's Water Campus Scottsdale, Arizona** (Hemken, et al., 1997; Mansfield and Nunez, 1998; Nunez, 2000; See also Chap. 22.)

The Scottsdale Water Campus is a state-of-the-art water and wastewater treatment complex located on a 57 ha site in Scottsdale, Arizona. It includes a 204×10^3 m³/d surface water treatment plant, a 45×10^3 m³/d water reclamation plant (WRP), and a 38×10^3 m³/d advanced water treatment plant (AWT) along with 27 vadose zone injection wells. The Water Campus was build to provide water for a major indirect potable reuse program with an annual groundwater recharge target in excess of 19×10^6 m³.

Tertiary effluent from the WRP is conveyed through a distribution system that delivers reclaimed water to 17 golf courses in north Scottsdale for irrigation. A chlorine contact basin provides disinfection for the WRP effluent prior to flow equalization.

The AWT was designed and constructed to further treat WRP effluent that is not used for golf course irrigation. The AWT includes continuous microfiltration units and three-stage reverse osmosis (low-pressure thin film composite elements) arrays set in a 24-10-5 configuration with a target of 85 percent recovery. The design provides flexibility to operate the entire microfiltration plant with either wastewater effluent or Central Arizona Project (CAP) water. The operational configuration is determined based on wet weather conditions, golf course irrigation demands, and the targeted recharge credits. An enclosed reservoir provides flow equalization storage to allow operation of the AWT facilities at a constant flow rate.

There is a microfiltration system (MCAP), which is operated in parallel with the AWT to treat wet weather flows from the WRP that exceed the capacity of the AWT microfiltration system. During dry weather, surface water from the Colorado River (transported via the CAP Canal) is treated and recharged through the MCAP units.

The Water Campus is a major part of the city's integrated wastewater/water reclamation program to comply with the State of Arizona's 1980 Groundwater Management Act, which requires either natural or artificial recharge to equal groundwater withdrawals.

Year	Description and significant contributions
2001	**Management Practices for Nonpotable Reuse, Water Environment Research Foundation, Alexandria, VA** (Mantovani et al., 2001; Radcliffe, 2004)

The purpose of this study was to document the diverse planning and management experiences by nonpotable water reuse projects based on: (1) a review of approximately 200 water reuse technical publications, covering the past 10 yr, (2) a survey of 40 U.S. projects and 25 projects in 10 other countries that gathered information and experiences on planning and managing nonpotable water reuse systems, and (3) a survey of 20 regulatory agencies, in the United States and abroad, documenting regulations, guidelines, and funding programs for water reuse projects and gathering regulators' perspectives on project planning and management practices. The projects analyzed in this study covered a wide range of system ages, types and reuse applications, system capacities, and reclaimed water customers.

Table E-1
Milestone water reuse project and research studies in the United States (*Continued*)

Year	Description and significant contributions
	Some principal conclusions are summarized as follows: (1) nonpotable water reuse has an excellent and well-documented performance track record, which to date has featured no documented health problems, strong public acceptance, and good regulatory compliance; (2) long-term sustainability of projects are threatened by the difficulties of achieving cost-recovery and/or environmental protection goals, frequently because of shortfalls in reclaimed water deliveries; (3) coordination between water and wastewater agencies as well as between local and state agencies is paramount to ensuring optimal project planning and cost and benefit allocation, and integrated agencies have an inherent advantage in developing projects; (4) while most applications screened options for alternative uses, few examined alternative sites, most projects being constrained as retrofits of preexisting wastewater treatment plants; (5) capital came from widely diverse sources including bonds, government loans and grants, and contributions by stakeholders such as developers and industrial users; (6) water pricing for most projects was driven by the need to provide incentives for potential customers, with very few projects directing their pricing strategies at full capital, operational, and maintenance cost-recovery, and most relied on intra- or inter-agency transfers for financial viability; (7) public outreach and education programs are an essential component of planning a reclaimed water service, and their cost is minor in the overall project cost; and (8) California's Water Recycling Criteria (Title 22) were the basis for many sets of regulations in many states and countries.
2003	**Water Recycling 2030: Recommendations of California's Recycled Water Task Force, California Department of Water Resources** (State of California, 2003b) To meet the needs of California's projected population of 52 million in the year 2030, it was determined that the state's water supply must be augmented and made more efficient. Water conservation, recycling, desalination, trading, and storage of surface and groundwater are the components that will control the state's overall future water supply. The 40-member Recycled Water Task Force was established by California Assembly Bill 331, Chap. 590, Statues of 2001, to evaluate the current framework of state and local rules, regulations, ordinances, and permits to identify the opportunities for and obstacles or disincentives to increasing the safe use of reclaimed water. The key issues and recommendations on water recycling were reported and included: (1) increase funding for water recycling projects, (2) encourage community value-based decision-making model project planning, (3) leadership support for water recycling, (4) develop comprehensive educational curriculum, (5) develop state-sponsored media campaigns, (6) revise the Uniform Plumbing Code appendix, (7) prepare Department of Health Services guidelines on cross-connection control, (8) review local and state health and safety regulations, (9) investigate incidental runoff, (10) create a uniform interpretation of state standards, (11) educate the public on impacts of water softeners, (12) develop uniform analytical methods for economic analysis, (13) provide sustainable state funding for research on recycling water issues, and (14) encourage university academic programs for water recycling.

(*Continued*)

Table E-1
Milestone water reuse project and research studies in the United States (*Continued*)

Year	Description and significant contributions
2004	**Report of the Scientific Advisory Panel—Orange County Water District's Santa Ana River Water Quality and Health Study** (National Water Research Institute, 2004)

The Santa Ana River (SAR) is the primary source of water for recharging the Orange County (OC) groundwater basin, which provides over 2 million residents with about two-thirds of their water supply. As with many densely populated metropolitan areas in the western United States, OC has a limited natural local water supply. Dry summers and mild winters, with an average rainfall of about 330 mm/yr, characterize the region's arid climate. Orange County uses almost twice the amount of water that is currently available in the SAR watershed, and the local water supply must be supplemented by the purchase of imported water from the Colorado River and from Northern California. Population growth projections mean that it is important to maximize local groundwater supplies, especially because adequate supplies of imported water may not be available in the future. Thus, groundwater recharge with SAR water is critical for replenishing groundwater supply.

The Santa Ana River Water Quality and Health Study (SARWQH) was initiated by the Orange County Water District (OCWD) in 1994 to address questions about the use of SAR water for recharging the OC groundwater basin because of the high percentage of treated wastewater in the river's base flow. The study was designed to provide scientific information to help address concerns frequently expressed by the California Department of Health Services (DHS) regarding the use of reclaimed water to recharge groundwater subsequently withdrawn for potable use.

The Scientific Advisory Panel's conclusions presented in the SARWQH study are:

- The recharge of SAR water to the groundwater basin does not currently threaten water quality or public health.
- Water quality in the SAR will continue to change, and these changes may influence OCWD recharge operations.
- Emerging chemical and microbiological constituents of concern (nonregulated and previously unidentified) will require continued surveillance.
- The OCWD should continue to monitor the quality of SAR water and groundwater for chemical and biological constituents of public health concern.
- Groundwater in the SARWQH study area is vulnerable to microbial contamination, as indicated by the occasional presence of bacteriophage in some water samples.
- Utilities using recharged groundwater supplies from vulnerable sources must do more than rely on drinking-water standards and guidelines to ensure safety.
- To minimize any risks that might be associated with the vulnerability of groundwater to fecal contamination, all production wells in the study area that are found positive for bacteriophage should be disinfected.

REFERENCES

Blackson, D. E., and J. L. Moreland (1998) "Wastewater Reclamation and Reuse for Cooling Towers at the Palo Verde Nuclear Generating Station," 1143–1161, in T. Asano (ed.) *Wastewater Reclamation and Reuse*, Water Quality Management Library, **10**, CRC Press, Boca Raton, FL.

Chen, C-L., J. F. Kuo, and J. F. Stahl (1998) "The Role of Filtration for Wastewater Reuse," 219–262, in T. Asano (ed.), *Wastewater Reclamation and Reuse*, Water Quality Management Library, **10**, CRC Press, Boca Raton, FL.

Cross, P., G. Delneky, and T. Lothrop (2000) "Worth Its Weight in Oranges: Florida City and County Turn Wastewater Effluent into Valuable Commodity," *Water Environ. Technol.*, **1**, 26–30.

Dryden, F. D., C-L. Chen, and M. W. Selna (1979) "Virus Removal in Advanced Wastewater Treatment Systems, *J. WPCF,* **51**, 8, 2098–2109.

Engineering-Science (1987) *Monterey Wastewater Reclamation Study for Agriculture Final Report*, Prepared for Monterey Water Pollution Control Agency, Monterey, CA.

Fleming, P. A. (1990) "Water Supply, Reclamation and Reuse at Grand Canyon: A Case Study," *Proceedings of CONSERVE 90*, August 12–16, 1990, 129–133, Phoenix, Arizona, American Society of Civil Engineers, American Water Resources Association, American Water Works Association, and National Water Well Association.

Hartling, E. C., and M. H. Nellor (1998) "Water Recycling in Los Angeles County," 917–940, in T. Asano (ed.) *Wastewater Reclamation and Reuse*, Water Quality Management Library, **10**, CRC Press, Boca Raton, FL.

Hemken, B., M. Craig, and D. Mansfield (1997) "Advanced Reclamation for Scottsdale's Integrated Water Management Strategies," **6**, 229–234. *Proceedings of the Water Environment Federation 70th Annual Conference & Exposition*, Chicago, IL.

Hyde, C. G. (1937) "The Beautification and Irrigation of Golden Gate Park with Activated Sludge Effluent," *Sew. Works J.*, **9**, 6, 929–941.

Johnson, W. D. and J. R. Parnell (1998) "Wastewater Reclamation and Reuse in the City of St. Petersburg, Florida," 1037–1104, in T. Asano (ed.) *Wastewater Reclamation and Reuse*, Water Quality Management Library, **10**, CRC Press, Boca Raton, FL.

Kirkpatrick, W., and T. Asano (1986) "Evaluation of Tertiary Treatment Systems for Wastewater Reclamation and Reuse," *Water Sci. Technol.*, **18**, 10, 83–95.

Kubick, K. S., Ross, R. S., and Pitt, P. (1992) "Up a Dry Creek Without a Paddle: San Francisco's Innovative Solution to Wastewater Reclamation," 189–199, *Proceedings of the Water Environment Federation 65th Annual Conference & Exposition*, **9**. New Orleans, LA.

Lauer, W. C., S. E. Rogers, A. M. Lachance, and M. K. Nealey (1991) "Process Selection for Potable Reuse Health-Effects Studies," *J. AWWA*, **83**, 11, 52–63.

Lauer, W. C., and S. E. Rogers (1998) "The Demonstration of Direct Potable Water Reuse: Denver's Landmark Project," 1269–1334, in T. Asano (ed.) *Wastewater Reclamation and Reuse*, Water Quality Management Library, **10**, CRC Press, Boca Raton, FL.

Mansfield, D. M., and A. A. Nunez (1998) "The City of Scottsdale Water Campus—A Complex State of the Art Indirect Potable Reuse Project," Water Resources Department, and Water and Wastewater Treatment Department, respectively, City of Scottsdale, AZ.

Mantovani, P., T. Asano, A. Chang, and D. A. Okun (2001) *Managing Practices for Nonpotable Water Reuse*, Project 97-IRM-6, Water Environment Research Foundation, Alexandria, VA.

National Water Research Institute (2004) *Report of the Scientific Advisory Panel—Orange County Water District's Santa Ana River Water Quality and Health Study*, Fountain Valley, CA.

Nellor, M. H, R. B Baird, and J. R. Smyth (1985) "Health Effects of Indirect Potable Water Reuse," *J. AWWA*, **77**, 7, 88–96.

Nunez, A. A. (2000) "Scottsdale's Water Campus: Off and Running," *Proceedings Water Reuse Conference*, San Antonio, TX, American Water Works Association, Denver CO.

Ongerth, H. J., and W. F. Jopling (1977) "Water Reuse in California," in H. I. Shuval (ed.), *Water Renovation and Reuse*, Academic Press, Inc., New York.

Radcliffe, J. C. (2004) *Water Recycling in Australia*, Australia Academy of Technological Sciences and Engineering, Parkville, Victoria, Australia.

Sanitation District of Los Angeles County (1977), *Pomona Virus Study Final Report*, Prepared for California State Water Resources Control Board and U.S. Environmental Protection Agency, Sacramento, CA.

Sanitation Districts of Los Angeles County (2002) *Thirteenth Annual Status Report on Reclaimed Water Use*, Fiscal Year 2001–02, Whittier, CA.

Sheikh, B., R. P. Cort, W. R. Kirkpatrick, R. S. Jaques, and T. Asano (1990) "Monterey Wastewater Reclamation Study for Agriculture," *J. WPCF*, **62**, 3, 216–226.

Sheikh, B., R. P. Cort, R. C. Cooper, and R. S. Jaques (1998) "Tertiary-Treated Reclaimed Water for Irrigation of Raw-Eaten Vegetables," 779–825, in T. Asano (ed.) *Wastewater Reclamation and Reuse*, Water Quality Management Library, **10**, CRC Press, Boca Raton, FL.

State of California (1978) *Wastewater Reclamation Criteria, An Excerpt from the California Code of Regulations, Title 22, Div. 4*, Environmental Health, Department of Health Services, Berkeley, CA.

State of California (1987) *Report of the Scientific Advisory Panel on Groundwater Recharge with Reclaimed Wastewater*, prepared for State of California, State Water Resources Control Board, Department of Water Resources, and Department of Health Services, Sacramento, CA.

State of California (1988) *Policy Statement for Wastewater Reclamation Plants with Direct Filtration*, Department of Health Services, Sacramento, CA.

State of California (2000) *Code of Regulations, Title 22, Div. 4, Chap. 3 Water Recycling Criteria*, Secs. 60301 *et seq.*, December 2, 2000.

State of California (2003b) *Water Recycling 2030: Recommendations of California's Recycled Water Task Force*, Department of Water Resources, Sacramento, CA.

Walker-Coleman, L. (1999) "Florida's Reuse Program," in *Southwest Florida Reclaimed Water Guide*, The Southwest Florida Water Management District.

Water Use (1983) A pamphlet, Arizona Nuclear Power Project, Phoenix, AZ.

WCPH (1996) *Total Resource Recovery Project: Final Report,* prepared for City of San Diego Water Utilities Department, prepared by Western Consortium for Public Health in association with EOA, Inc., Oakland, CA.

Work, S. W., M. R. Rothberg, and K. J. Miller (1980) "Denver's Potable Reuse Project: Pathway to Public Acceptance," *J. AWWA*, **72**, 8, 435–440.

York, D. W., and E. A. Potts (1996) "The Evolution of Florida's Water Reuse Program," *Proceedings of Water Reuse 1996*, American Water Works Association and Water Environment Federation, San Diego, CA.

Young, R. E., K. A. Thompson, R. R. McVicker, R. A. Diamond, M. B. Gingras, D. Ferguson, J. Johannessen, G. K. Herr, J. J. Parsons, V. Seyde, E. Akiyoshi, J. Hyde, C. Kinner, and L. Oldewage (1998) "Irvine Ranch Water District's Reuse Today Meets Tomorrow's Conservation Needs," 941–1036, in T. Asano (ed.) *Wastewater Reclamation and Reuse*, Water Quality Management Library, **10**, CRC Press, Boca Raton, FL.

Table E-2
Summary of water reclamation and reuse in select countries of the world

Australia
(Kracman et al., 2001; Dillon, 2001; Radcliffe, 2004; South Australian Water Corp, 2005; Queensland Water Recycling Strategy, 2003)

Australia has a population of approximately 20 million as of 2006. The continent is 80 percent arid and semiarid, with 90 percent of the precipitation falling in the tropical north, where only 10 percent of the population lives. Average annual precipitation is 534 mm, and less than 250 mm/yr falls in the arid and semiarid regions. Australia's water reuse is almost as diverse and as complex as that of the United States. A wide range of water reuse practices is found throughout the country. Agricultural and landscape irrigation is a major use of reclaimed water. On-site water recycling systems are fairly common. One such system, sewer mining, withdraws wastewater from collection interceptors, treats the water to tertiary level by membrane bioreactor, and distributes the reclaimed water for use in the nearby communities. In 2000, 1.6×10^9 m^3 of treated wastewater was generated, with 171×10^6 m^3, or 11 percent, reused.

Virginia Pipeline Project, Adelaide, South Australia Virginia is an agricultural area near Adelaide that provides 35 percent of South Australia's vegetable production. The agricultural sector in the Adelaide region has historically relied on groundwater for irrigation. Deteriorating water quality, over-exploitation of groundwater, and environmental impacts of wastewater discharge prompted water reuse and the construction of the Virginia Pipeline Project (VPP).

The VPP began operation in 1999, and is the largest water reclamation project in Australia. The project consists of secondary treatment at the Bolivar Wastewater Treatment Plant, followed by tertiary treatment and distribution for mainly flower and vegetable irrigation. The Bolivar WTP treats 135,000 m^3/d of domestic and industrial wastewater, 55 percent of which is from the Adelaide metropolitan area. Treatment consists of primary sedimentation, trickling filters, and stabilization lagoons. The VPP further treats the wastewater with a 120,000 m^3/d dissolved air flotation and filtration system followed by disinfection. When fully operational, the VPP will reuse 70 percent of Bolivar wastewater flows and will supply over 20×10^6 m^3 of irrigation water per year.

Rouse Hill Recycled Water Area Project, Sydney, New South Wales The Rouse Hill Project is the first large-scale residential nonpotable water reuse system in Australia. Water reuse was motivated by a need to reduce the impact of new residential development on an adjacent river.

Wastewater is treated in a biological and chemical treatment plant consisting of primary sedimentation, activated sludge operated for nitrogen and phosphorus removal, coagulation, flocculation, clarification, filtration, chlorine disinfection, and pH control. The treated water is distributed to the residences via a dual water supply system. In 2000, the project served 25,000 m^3/d and serviced 15,000 homes. A cross-connection was discovered in 2004 that affected 82 households and was a result of unauthorized plumbing work undertaken in the construction of a household in Rothwell Court, Glenwood.

Canada
(Jackson and Lee, 2000; Exall et al., 2004; Di Pasquale, 2004)

On the whole, Canada has abundant water in most regions. Water reclamation and reuse is on a relatively small scale, involving agricultural and golf course irrigation in British Columbia, Alberta, Manitoba, and Saskatchewan. In April 2000, Alberta Environment published the Guidelines for Municipal Wastewater Irrigation. The guidelines are meant to ensure that reclaimed municipal wastewater is used for irrigation only when environmentally acceptable and agriculturally beneficial. In British Columbia, water reuse is considered for a much broader

(Continued)

Table E-2
Summary of water reclamation and reuse in select countries of the world (*Continued*)

Canada (*Continued*)
(Jackson and Lee, 2000; Exall et al., 2004; Di Pasquale, 2004)

range of applications in the May 2001 documents, "Code of Practice for the Use of Reclaimed Water," a companion document to the Municipal Sewage Regulation.

Because of stringent discharge requirements for nutrients in receiving waters, land applications of treated effluent are common in Canada with a few problems as described below.

The City of Vernon, with a population of 35,000, is located in the semiarid Okanagan Valley of south central British Columbia, 175 km north of the U.S. border. The city is situated at an elevation of 425 m and is surrounded by three lakes. The area is well known for its fruit orchards, vineyards, and as a retirement destination due to warm dry summers, mild winters, and relaxed lifestyle. Vernon's treated effluent is pumped to a 10×10^6 m^3 reservoir and is stored for up to a year before being chlorinated and irrigated on 1000 ha of nearby agricultural, silvicultural, and recreational areas including golf courses, a seed orchard, and a nursery. The program is in keeping with a city policy of not discharging treated effluent into a water course, except on an emergency basis.

The city needs to increase it's irrigation land base to address increasing treated municipal wastewater flows and an increase in precipitation that has occurred over the past few years. Increasing the land base is not readily feasible within the service area of the 10×10^6 m^3 reservoir without capital investment in a transmission system, which cannot be readily offset by end water user charges. There is a new treatment plant (design capacity: 27,000 m^3/d) that can operate in a biological nutrient removal mode and has been designed with final effluent filtration, UV disinfection, and chlorination. The treatment plant will facilitate the ability to meet provincial environment regulations prior to diverting the treated effluent to direct irrigation and bypass the need for detention in the impoundment reservoir. Though this will not preclude the continuing diversion of treated effluent to the reservoir, it will open up opportunities for generating new irrigation land base to address the expansion in municipal wastewater flows.

Israel
(Shelef, 1991; Kanarek and Michail, 1996; Oron, 1998; Friedler, 1999; Ministry of National Infrastructures, 2003)

The population of Israel is approximately 6.3 million. Annual rainfall ranges from around 1000 mm in the north to 30 mm in the south. About 95 percent of the total average annual potential of renewable water is already exploited and used for domestic consumption and irrigation. About 80 percent of the water potential is in the north of the country and only 20 percent in the south.

Reclaimed water is used predominantly for agricultural irrigation. As a result of increasing water demand for domestic and industrial uses, the supply of freshwater to the agricultural sector has decreased from about 77 percent in the 1960s to about 60 percent at present. Of the 410×10^3 m^3 wastewater generated from urban centers, 265×10^3 m^3, or 65 percent, is reused for irrigation. Currently, an estimated 25 percent of the total water supplied for irrigation is reclaimed water. This volume is expected to double by the year 2020.

Dan Region Project, Tel Aviv The Dan Region Project (DRP) is the largest water reclamation scheme in Israel. Domestic and industrial wastewater from the Tel Aviv Metropolitan Region is treated, recharged into a confined aquifer, and subsequently withdrawn and distributed for water reuse.

The DRP treats about 110×10^3 m^3 of wastewater annually. Stage one of the reclamation system has a treatment capacity of 20×10^3 m^3/yr and has been in operation since 1977. Treatment in stage one consists of

Table E-2
Summary of water reclamation and reuse in select countries of the world (*Continued*)

Israel (*Continued*)
(Shelef, 1991; Kanarek and Michail, 1996; Oron, 1998; Friedler, 1999; Ministry of National Infrastructures, 2003)

facultative oxidation ponds with recirculation, and polishing ponds for ammonia stripping and recarbonation. Groundwater recharge in stage one is accomplished via surface spreading at four spreading basins with a total area of 24 ha. Stage two has a capacity of 80×10^3 m³/yr and has been in operation since 1987. Secondary treatment in stage two consists of mechanical-biological treatment by activated sludge operated for nitrification/denitrification. Stage two has two spreading basins with a total area of 42 ha.

After treatment and recharge to the groundwater, reclaimed water is withdrawn through a ring of recovery wells, located between 350 and 1500 m from the nearest recharge basin. The reclaimed water is pumped to southern Israel for use in nonpotable applications. The main application for water reuse is unrestricted agricultural irrigation.

Jeezrael Valley Project   The Jeezrael Vally Project (JVP) reclaims domestic and industrial wastewater from towns and small settlements in and around the Jeezrael Valley for reuse in agricultural irrigation. The valley contains 20,000 ha of intensively cultivated area.

Wastewater from domestic and industrial sources is treated in anaerobic ponds followed by aerated lagoons. Treated wastewater is distributed to storage reservoirs located in rural agricultural areas. The reservoirs are operated in sequencing batch reactor mode, thus providing additional treatment. During the first year of operation, in 1996, 6×10^3 m³ of treated effluent was supplied for irrigation. At full capacity, JVP will supply 13×10^3 m³/yr.

Japan
(Asano et al.,1996; Ogoshi et al., 2001; Yamagata et al., 2002; Food and Agriculture Organization, 2003; Tokyo Metropolitan Government, 2003; Japan Sewage Works Assoc. 2005)

Despite average annual precipitation of 1670 mm, the steep topography and short rivers make large volume water storage difficult in Japan. This is particularly true in the Tokyo metropolitan area because of a very high population density, where 12.5 million people are living in a 2187 km² area. Water reuse in Japan is dominated by nonpotable urban reuses such as in-plant wash water, stream flow augmentation, toilet flushing in large commercial and office buildings, and landscaping. In 1997, the total volume of reclaimed water use in Japan was 206×10^6 m³. Agricultural reuse was only about 16 percent, whereas toilet flushing was about 36 percent. The Tokyo Metropolitan Government Sewerage Bureau has implemented an extensive water reuse program and approximately 9 percent of treated effluent is currently reused in Tokyo.

Makuhari New Center Water Recycling Project in Tokyo   The project uses tertiary treated reclaimed water for toilet flushing, cleansing, and environmental water in a convention center, commercial buildings, hotels, and parks within 0.62 km² of the project area. Operation began in October, 1989. Treatment trains for water reclamation include activated sludge process, chemical coagulation, filtration, ozonation, and chlorination. The supply capacity is 4120 m³/d with an average flow of 2372 m³/d.

Rokko Island Water Recycling Project in Kobe   The project uses reclaimed water for toilet flushing, park irrigation, and cleansing for commercial buildings, schools, and parks within 1.66 km² of the project area. Operation began in April, 1986. The treatment trains for water reclamation include activated sludge, filtration, ozonation, and chlorination. The supply capacity is 2100 m³/d, and the average operating flow is 415 m³/d.

Tokyo Metropolitan Government Building   The project provides 19 high-rise buildings with reclaimed water for toilet flushing. The distribution of reclaimed water began in 1984. Design capacity of the project is 4000 km³/d for a service area of 0.5 km². In 1993, a daily maximum of 4300 m³ and a daily average of 2700 m³

(*Continued*)

Table E-2
Summary of water reclamation and reuse in select countries of the world (*Continued*)

Japan (*Continued*)
(Asano et al.,1996; Ogoshi et al., 2001; Yamagata et al., 2002; Food and Agriculture Organization, 2003; Tokyo Metropolitan Government, 2003; Japan Sewage Works Assoc. 2005)

were delivered to offices and commercial buildings. The tertiary treated wastewater (via sand filtration) is pumped from Ochiai Municipal Wastewater Treatment Plant to the Shinjuku Water Recycling Center, where the water is disinfected with chlorine and distributed to each building for toilet flushing.

Fukuoka City Water Recycling Project The City of Fukuoka is using reclaimed water for toilet flushing, park irrigation, and cleansing for commercial buildings and parks within 7.7 km^2 of the project area. Operation began in 1980. Reclaimed water use is required for buildings with floor space greater than 3000 m^2 or have a water intake pipe diameter larger than 50 mm. Wastewater is reclaimed with activated sludge, filtration, ozonation, and chlorination.

Kuwait
(Al-Attar et al., 1997; von Gottberg and Vaccaro, 2003)

Kuwait has a population of 2.5 million and has experienced rapid growth over the last four decades (<0.8 million people in the 1960s), which in turn has increased the water demand for various activities. It is an arid country with negligible rainfall, and 95 percent of potable water comes from desalination of sea water (multi-stage flash distillation plants). The other major water source is brackish groundwater.

Wastewater is treated in four main municipal treatment plants (Ardiya, Riqqa, Jahra, and Um Al Haimam) by secondary and tertiary processes (extended aeration activated sludge, filtration, and chlorine, plus UV disinfection in the case of Um Al Haimam) with an annual treatment capacity of over 255×10^6 m^3, which is equivalent to 65–80 percent of potable water use. The quantity of wastewater produced was 119×10^6 m^3 in 1994, of which 103×10^6 m^3 were treated.

Water Reuse Projects Reclaimed water from Ardiya treatment plant together with some of the treated effluent from Riqqa is directed to the Data Monitoring Center storage reservoirs (two 170,000 m^3 closed reservoirs) before transfer (36.5×10^6 m^3/yr), for agricultural and landscape irrigation in Sulaibiya. Irrigation methods consist of sprinklers (both center-pivot and sideroll), drip, and furrow systems. The farms produce a variety of agricultural products including animal fodder (which utilizes 75 percent of the total area), horticulture products, and vegetables eaten cooked (e.g., spinach, potatoes, onions, and eggplants). Relatively small quantities of lemons and dates are also grown. Another irrigation project involves reforestation. The charge for treated effluent supplied to private farms is $0.07/m^3. The areas irrigated with tertiary effluent cover around 1680 ha. Proposals for expanding irrigation include 3300 ha. Currently, tertiary treated effluent is also reused for greening the highways and to reduce mobile sand movement. Effluent reuse forms an important part of the program to expand the green areas of Kuwait. A portion of the industrial wastewaters is also recycled after treatment.

The Sulaibiyah Wastewater Reclamation and Reuse Project An ambitious program to upgrade the entire wastewater system has been launched. A new plant was constructed in Sulaibiyah to replace the Ardiya and Jahra plants and to allow unrestricted nonpotable uses. This was a 30-yr concession from the Kuwait government to a consortium to design, build, own, operate, and maintain the facility. The plant will treat 375,000 m^3/d of wastewater in a first stage beginning in 2005 and be expandable to 600,000 m^3/d. The wastewater treatment plant is designed for biological nutrient removal, sludge treatment, and removal of residual constituents, dissolved solids, and pathogens through ultrafiltration and reverse osmosis. Water reuse options include mixing with brackish water and supplying the brackish water system, irrigation, and

Table E-2
Summary of water reclamation and reuse in select countries of the world (*Continued*)

Kuwait (*Continued*)
(Al-Attar et al., 1997; von Gottberg and Vaccaro, 2003)

groundwater recharge. It is estimated that the plant will deliver water of potable water quality at approximately $0.6/m^3$, with $0.4/m^3$ for conventional wastewater treatment and pipeline costs and $0.2/m^3$ for producing potable water from treated effluent.

Namibia
(Haarhoff and van der Merwe, 1996; Odendaal et al., 1998; Du Pisani, 2000; Lahnsteiner and Lempert, 2005)

Namibia is a sparsely populated country of about 1.9 million people. It is the most arid country in sub-Saharan Africa. There are no perennial rivers in the interior of the country, no permanent natural lakes, and the total length of the western coastline is covered by desert. The average annual evaporation in Windhoek is 3400 mm and the rainfall is 370 mm.

Windhoek's Goreangab Water Reclamation Plant   The nation's capital city, Windhoek, has been practicing direct potable reuse since 1968. The Windhoek Water Reclamation Plant serves a population of 220,000. Domestic wastewater from the city is first treated in a conventional biological wastewater treatment plant; the treated wastewater then flows through a series of maturation ponds to a water reclamation plant. The reclamation plant has undergone many reconfigurations and upgrades since 1968, most recently with the construction of the new Goreangab Water Reclamation Plant (WRP) in 2002. Industrial effluents in the city are diverted to a separate sewer and treatment system.

Goreangab WRP has a capacity of 2100×10^3 m^3, and it is internationally renowned as the first in the world to reclaim municipal wastewater to potable water quality as a supplement to Windhoek's very scarce water source. The treatment train consists of dissolved air floatation (DAF), sedimentation, rapid sand filtration, ozonation, carbon adsorption (both granular and powdered), ultrafiltration, and chlorine disinfection. After treatment, reclaimed water is mixed with water from other sources, so that reclaimed water makes up at most 35 percent of the city's drinking water. Potable reuse, despite its potential difficulties elsewhere, is an indispensable element of the Windhoek water system and has proven to be a reliable and sustainable option for over 36 yr.

Singapore
(Public Utilities Board Report, 2003; Law, 2003; Ong et al., 2004; Singh, 2005)

The Republic of Singapore, a small city-state that has one of the highest population density in the world (about $6000/km^2$), now buys more than half of its water from neighboring Malaysia under decades-old treaties, which start expiring in 2011. The water trade has sparked occasional disputes between the two nations over pricing and other political and environmental issues.

The Singapore Water Reclamation Study (NEWater Study) was initiated in 1998 as a joint initiative among the Public Utilities Board (PUB), the Ministry of the Environment, and the National University of Singapore. The primary objective of the joint initiative was to determine the suitability of using NEWater as a source of raw water to supplement Singapore's water supply. NEWater is reclaimed water that has undergone stringent purification and treatment process using advanced dual-membrane (microfiltration and reverse osmosis) and ultraviolet disinfection.

The NEWater Factories at Bedok and Kranji Water Reclamation Plants were commissioned at the end of 2002. Since February 2003, NEWater has been supplied to wafer fabrication plants at Woodlands and Tampines/Pasir Ris and other industries for nonpotable use. In January 2004, another milestone in the

(*Continued*)

Table E-2
Summary of water reclamation and reuse in select countries of the world (*Continued*)

Singapore (*Continued*)
(Public Utilities Board Report, 2003; Law, 2003; Ong et al., 2004; Singh, 2005)

NEWater initiative was accomplished with the commissioning of the third NEWater Factory at Seletar Water Reclamation Plant which began supplying NEWater to the wafer fabrication plants at Ang Mo Kio. The total capacity of the three NEWater factories is 92,000 m^3/d.

Besides supplying to industries, NEWater is also available for indirect potable use. This involves pumping NEWater into the reservoirs to be mixed and blended with raw water. The mixed water undergoes a process of naturalization before being treated again in conventional waterworks to produce drinking water. The PUB introduced 13,500 m^3/d of NEWater, about 1 percent of the amount of water consumed daily, into the raw water reservoir in 2003. The amount of NEWater will be increased progressively to about 2.5 percent of the total daily water consumption by 2011. The country's target is to obtain 25 percent of its water supply from non-traditional sources: 15 percent from NEWater, 5 percent from seawater desalination, and the remaining 5 percent from industrial water, by the year 2012.

South Africa
(Isaäcson et al., 1987; Odendaal et al., 1998; Grobicki and Cohen, 1999; Tredoux et al., 1999)

South Africa has a population of 40 million. The mean annual rainfall is 483 mm for the whole country, ranging from 55 mm/yr on the west coast to 1250 mm/yr in the eastern mountains. The mean evaporation ranges from 1100 mm in the east to 3000 mm in the west.

A study by the Water Research Commission in 1998 showed that less than three percent of available treated wastewater was directly reclaimed (estimated 30×10^3 m^3 out of 1089×10^3 m^3/yr). The major water reuse applications are reuse in the paper industry, cooling in municipal power stations, and aquifer storage and recharge.

Sappi Pulp and Paper Group Enstra Mill is an integrated pulp and fine paper operation near the City of Springs that uses 15,000 m^3/d freshwater and 15,000 m^3/d of reclaimed water as its water supply. The wastewater undergoes secondary treatment at McComb & Ancor works of Springs Municipality, followed by alum coagulation, DAF, and combined chlorine and activated bromine at the water reclamation plant.

Atlantis Water Resource Management Scheme The town of Atlantis, 50 km north of Cape Town, has recharged groundwater aquifers with urban stormwater and treated wastewater for over two decades. Domestic and industrial wastewaters are collected and treated separately for recharge into different portions of the aquifer. Strict regulations are enforced to control the quality of industrial effluent released to the wastewater system. Tertiary treated domestic wastewater is recharged for subsequent withdrawal and reuse. The lower quality reclaimed industrial wastewater is infiltrated through coastal basins to act as a saltwater intrusion barrier. Typical volumes of recharge for indirect reuse and saltwater barrier are 3×10^3 m^3/yr and 2.2×10^3 m^3/yr, respectively.

Spain
(Mujeriego and Asano, 1991, 1999; Herrero, 1998; Sala et al., 2002; López Ramírez et al., 2002; Sala, 2003)

The population of Spain is over 40 million in 2006. The climate in the country ranges from dry on the Mediterranean coast to humid on the Atlantic coast, with accompanying variations in distribution and intensity of precipitation. The average annual precipitation in the Mediterranean region and Atlantic regions are under and over 800 mm, respectively.

Table E-2
Summary of water reclamation and reuse in select countries of the world (*Continued*)

Spain (*Continued*)
(Mujeriego and Asano, 1991, 1999; Herrero, 1998; Sala et al., 2002; López Ramírez et al., 2002; Sala, 2003)

The Consorci de la Costa Brava (CCB, Costa Brava Water Agency) is a public agency that manages drinking water, wastewater treatment, and water reuse in the 27 coastal municipalities of the Girona province, north of Barcelona. The CCB is considered a pioneer of water reuse in Spain. The earliest planned reuse was golf course irrigation with reclaimed water in 1989. The volume of water reused has increased from 0.2×10^6 m^3 in 1989 to 2.6×10^6 m^3 in 2002.

Portbou, Girona The municipality of Portbou, located in northern Costa Brava, has a population of 1600 and is located in a remote mountainous area of the Mediterranean region. Water reuse was motivated by drought conditions that limited nonpotable uses of water, such as landscape irrigation. Wastewater is treated by coagulation, flocculation, direct filtration, and combined UV–chlorine disinfection. The treatment system capacity is 15 m^3/h. Treated water is used for urban nonpotable applications, including landscape irrigation, street cleaning, and fire protection.

Aiguamolls de l'Emporda Nature Preserve (AENP), Girona The AENP is a naturally occurring marsh in northern Costa Brava that was declared a nature preserve in 1984. The Cortalet Lagoon in AENP experienced summer desiccation due to dam construction and urbanization. In 1998, CCB began applying between 500,000 and 750,000 m^3/yr of reclaimed water to the lagoon to keep it from drying up. The wastewater is treated first at the Empuriabrava Wastewater Treatment Plant (WWTP) with extended aeration and polishing lagoons. It then undergoes treatment in a 7-ha constructed wetland designed to reduce the nitrogen content of the reclaimed water.

Integrated Water Resources Management Program in Vitoria Wastewater reclamation and reuse has been the final cornerstone of an ambitious integrated water resources management program for the City of Vitoria (250,000 people, located in the Basque Country, northern Spain) that began in 1995. The wastewater reclamation and reuse project implemented in Vitoria includes a wastewater reclamation facility with a capacity of 35,000 m^3/d (expandable to three additional modules), that satisfies the water quality requirements specified by the Title 22 of California Code of Regulations, and an elaborated pumping, conveyance, and storage system.

The project objectives are: (1) to supply reclaimed water for spray irrigation of 9500 ha during the summer, (2) to provide storage capacity for 7 hm^3 of reclaimed water during the winter for agricultural irrigation in the summer. Ultimately, the project could supply up to 27 hm^3/yr of reclaimed water for stream flow substitution at the discharge base of the drinking water supply reservoir. Construction of the reservoir's earth dam began in 2002 and was completed in June, 2004, becoming one of the largest reclaimed water reservoirs in the western Mediterranean basin. Although agricultural irrigation around the City of Vitoria was the main driving force of the project, the city is planning to use reclaimed water for irrigation of the urban landscape (3×10^6 m^3/yr) and for industrial water supply to a new industrial park. The initial commitment of the project to produce a high quality reclaimed water (suitable for unrestricted irrigation) has been instrumental in its success and its wide acceptance among current and potential users.

Tunisia
(Bahri, 1998; Bahri et al., 2001; Koundi, 2004)

Tunisia has a population of 10 million. Rainfall and evaporation is unevenly distributed across the country. In the north, center, and south of Tunisia, the average annual rainfall is 594 mm, 289 mm, and 156 mm, respectively. Evaporation ranges from 1300 mm in the north to 2500 mm in the south. Only half of the country's 4700×10^6 m^3/yr of available water has a low enough salt content for use without restrictions.

(*Continued*)

Table E-2
Summary of water reclamation and reuse in select countries of the world (*Continued*)

Tunisia (*Continued*)
(Bahri, 1998; Bahri et al., 2001; Koundi, 2004)

In 2003, the volume of wastewater collected was 240×10^6 m³/yr, 187×10^6 m³/yr of which was treated in 70 treatment plants utilizing activated sludge, oxidation ditches, and stabilization ponds. Tunisia has a national water reuse policy launched at the beginning of the 1980s with treatment and reuse coordinated from the planning stage. In 2003, 43×10^6 m³ of treated wastewater was used for agricultural and landscape irrigation, which is approximately four times the volume that was used in 1990.

Agricultural Irrigation Reclaimed water from La Cherguia-Tunis wastewater treatment plant has been used since the early 1960s for agricultural irrigation to reduce the impact of saltwater intrusion due to excessive pumping of groundwater. La Cherguia was constructed in 1958 and provides secondary wastewater treatment. The 600 ha area of La Soukra is located to the northeast of Tunis, and reclaimed water is used to irrigate citrus and olive trees. Irrigation of raw-eaten vegetables is prohibited.

The area currently irrigated with reclaimed water is about 8000 ha, 62 percent of which is located around Tunis and a few other locations near Hammamet, Sousse, Monastir, Sfax, and Kairouan. Farmers pay for the reclaimed water they use to irrigate their fields. The use of secondary treated effluents in Tunisia is for a restrictive irrigation which does not include vegetable crops, whether eaten raw or cooked. The main crops irrigated with treated wastewater are fruit trees (citrus, grapes, olives, peaches, pears, apples, promegranates), fodder (alfalfa, sorghum, berseem), and cereals.

Golf Course Irrigation The eight existing Tunisian golf courses, covering 570 ha, are all irrigated with reclaimed water. Secondary treated wastewater is stored in a series of ponds on the golf courses. Irrigation is conducted at night with low-range sprinklers to prevent public exposure.

REFERENCES

Al-Attar, M. H., N. Al-Awadhi, J. Al-Sulaimi, and G. F. Leitner (1997) "Water Reuse Concepts and Programs in Kuwait Leading to Optimum Use of all Water Resources," *Desal. Water Reuse*, 6, 4, 46–53.

Asano, T., M. Maeda, and M. Takaki (1996) "Wastewater Reclamation and Reuse in Japan: Overview and Implementation Examples," *Water Sci. Technol.*, 34, 11, 219–226.

Bahri, A. (1998) "Wastewater Reclamation and Reuse in Tunisia," 877–916, in T. Asano (ed.), *Wastewater Reclamation and Reuse*, Water Quality Management Library, **10**, CRC Press, Boca Raton, FL.

Bahri, A., C. Basset, F. Oueslati, and F. Brissaud (2001) "Reuse of Reclaimed Wastewater for Golf Course Irrigation in Tunisia," *Water Sci. Technol.*, **43**, 10, 117–124.

Dillon, P. J. (2001) "Water Reuse in Australia: Current, Future and Research," *J. Water (Australia)*, **28**, 3, 18–21.

Di Pasquale, W. J. (2004) Personal communication, Water Reclamation Plant, City of Vernon, BC, Canada.

Du Pisani, P. (2000) "Potable Re-use of Treated Sewage Effluent in Windhoek, Namibia," *Proceedings of 2nd World Water Forum*, The Hague, The Netherlands.

Exall, K., J. Marsalek, and K. Schaefer (2004) "A Review of Water Reuse and Recycling, with Reference to Canadian Practice and Potential: 1. Incentives and Implementation, and 2. Applications," *Water Qual. Res. J. (Canada)*, **39**, 1, 13–28.

Food and Agriculture Organization of the United Nations (FAO) (2003) AQUASTAT, accessed at: http://www.fao.org/

Friedler, E. (1999) "The Jeezrael Valley Project for Wastewater Reclamation and Reuse, Israel," *Water Sci. Technol.*, **30**, 4–5, 347–354.

Grobicki, A. M. W., and B. Cohen (1999) "A Flow Balance Approach to Scenarios for Water Reclamation," *Water (SA)*, **25**, 4, 473–482.

Haarhoff, J., and B. van der Merwe (1996) "Twenty-Five Years of Wastewater Reclamation in Windhoek, Namibia," *Water Sci. Technol.*, **22**, 10–11, 25–35.

Herrero, A. M. (1998) "Water Reuse: National Report Spain," *Water Supply*, **16**, 1–2, 307–310.

Isaäcson, M., A. R. Sayed, and W. Hattingh (1987) *Studies on Health Aspects of Water Reclamation during 1974 to 1983 in Windhoek, South West Africa/Namibia*, Report WRC 38/1/87 to the Water Research Commission, Pretoria, South Africa.

Jackson, E., and J. Lee (2000) "Dual Distribution Planning," *Proceedings of the 2000 Water Reuse Conference*, San Antonio, TX.

Japan Sewage Works Assoc. (2005) *Sewage Works in Japan 2005: Wastewater Reuse*, Tokyo, Japan.

Kanarek, A., and M. Michail (1996) "Groundwater Recharge with Municipal Effluent: Dan Region Reclamation Project, Israel," *Water Sci. Technol.*, **34**, 11, 227–233.

Koundi, A. (2004) Wastewater Management—The Tunisian Experience, Communication at the 1st African Development Bank Water Week, Tunis.

Kracman, B., R. Martin, and P. Sztajnbok (2001) "The Virginia Pipeline: Australia's Largest Water Recycling Project," *Water Sci. Technol.*, **43**, 10, 35–42.

Lahnsteiner, J., and G. Lempert (2005) "Water Management in Windhoek/Namibia," *Proceedings of the IWA Specialty Conference, Wastewater Reclamation & Reuse for Sustainability*, November 8–11, Jeju, Korea.

Law, I. B. (2003) "Advanced Reuse—From Windhoek to NEWater and Beyond," *The Internat. Desal. Water Reuse Quart.*, **13**, 1, 22–28.

López Ramírez, J. A., J. M. Quiroga Alonso, D. Sales Márquez, and T. Asano (2002) "Indirect Potable Reuse and Reverse Osmosis: Challenging the Course to New Water," *Water 21*, **6**, 56–59.

Melbourne Water (2003) *Recycled Water Handbook*, Melbourne Water, Melbourne, Australia, accessed at: http://www.melbournewater.com.au/

Ministry of National Infrastructures, Israel (2003), Accessed at: http://eng.mni.gov.il/english/index.html—once at website enter "wastewater background" in search box.

Mujeriego, R., and T. Asano (eds.) (1991) *Wastewater Reclamation and Reuse*, Pergamon Press, Oxford, UK.

Mujeriego, R., and T. Asano (1999) "The Role of Advanced Treatment in Wastewater Reclamation and Reuse," *Water Sci. Technol.*, **40**, 4–5, 1–9.

Odendaal, P. E., J. L. J. van der Westhuizen, and G. J. Grobler (1998) "Water Reuse in South Africa," 1163–1192, in T. Asano (ed.) *Wastewater Reclamation and Reuse*, Water Quality Management Library, **10**, CRC Press, Boca Raton, FL.

Ogoshi, M., Y. Suzuki, and T. Asano (2001) "Water Reuse in Japan," *Water Sci. Technol.*, **43**, 10, 17–23.

Ong, S. L., L. Y. Lee, L. Y. Hu, L. F. Song, H. Y. Ng, and H. Seah (2004) "Water Reclamation and Reuse—Advances in Treatment Technologies and Emerging Contaminants Detection," Presented at the 10th International Conference on Drinking Water Quality Management and Treatment Technology, June 1–3, 2004 in Taipei, Taiwan.

Oron, G. (1998) "Water Resources Management and Wastewater Reuse for Agriculture in Israel," 757–777, in T. Asano (ed.) *Wastewater Reclamation and Reuse*, Water Quality Management Library, **10**, CRC Press, Boca Raton, FL.

Public Utilities Board Report, Singapore (2003) Accessed at: www.pub.gov.sg

Queensland Water Recycling Strategy (2003) Accessed at: http://www.epa.qld.gov.au/

Radcliffe, J. C. (2004) *Water Recycling in Australia*, Australia Academy of Technological Sciences and Engineering, Parkville, Victoria, Australia.

Sala, L., R. Mujeriego, M. Serra, and T. Asano (2002) "Spain Sets the Example," *Water 21*, **4**, 18–20.

Sala, L. (2003) "Water Reuse in Costa Brava (Girona, Spain)," Mediterranean Network on Wastewater Reclamation and Reuse, Case Studies, Accessed at: http://www.med-reunet.com/05ginfo/05_case.asp.

Shelef, G. (1991) "Wastewater Reclamation and Water Resources Management," *Water Sci. Technol.*, **24**, 9, 251–265.

Singh, S. (2005) Singapore offers Recipes for Urban Water Management, *Asian Water,* December 2005, 23–27.

South Australian Water Corporation (SA Water), Accessed at: http://www.sawater.com.au/SAWater/Education/OurWaterSystems/

Tredoux, G., P. King, and L. Cave (1999) "Managing Urban Wastewater for Maximizing Water Resource Utilization," *Water Sci. Technol.*, **39**, 10–11, 353–356.

Tokyo Metropolitan Government (2003) *Reclamation & Recycling*, published by Bureau of Sewerage, **38**, 6.

von Gottberg, A., and G. Vaccaro (2003) "Kuwait's Giant Membrane Plant Starts to Take Shape," *Desal. Water Reuse*, **13**, 2, 30–34.

Yamagata, H., M. Ogoshi, Y. Suzuki, M. Ozaki, and T. Asano (2002) "On-site Insight into Reuse in Japan," *Water 21*, **4**, 26–28.

Appendix F

Evolution of Nonpotable Reuse Criteria and Groundwater Recharge Regulations in California

The development of water reuse criteria and regulations, as noted in Chap. 4, is of critical importance in the development of water reclamation facilities and reuse applications. The evolution of nonpotable reuse criteria and groundwater recharge regulations in California are presented in Secs. F-1 and F-2, respectively.

F-1 EVOLUTION OF NONPOTABLE WATER REUSE CRITERIA IN CALIFORNIA (Adapted from CROOK, 2002)

The evolution of water reuse criteria began in California, where regulations were first adopted in 1918. As reclaimed water began to be used on an extensive scale for a multitude of applications, criteria had to be revised continually to take into account associated public health concerns, advances in treatment and analytical technology, and other new information from research and demonstration studies. The philosophy behind California's regulations and the rationale supporting decisions on various requirements included in the criteria are indicative of a conservative approach taken to assure public health protection.

Although agricultural irrigation with low-quality wastewater has been practiced for hundreds of years in various parts of the world, there were no significant regulations or guidelines addressing this practice until the early twentieth century. As urban areas began to encroach on sewage farms and as the scientific basis of disease became understood more widely, concern about possible health risks associated with the use of wastewater for irrigation increased among public health officials. This concern led to the establishment of controls on the use of wastewater for agricultural irrigation, which was the first reclaimed water application to be regulated.

History of California Regulations

In California, at least 20 communities were using either raw or settled sewage for agricultural irrigation at the beginning of the twentieth Century. The earliest reference to a public health viewpoint on water quality requirements in California appeared in the California State Board of Health Monthly Bulletin dated February 1906, in which it was noted that "Oxnard is installing a septic tank system of sewage disposal, with an outlet in the ocean. Why not use it for irrigation and save the valuable fertilizing properties in solution, and at the same time completely purify the water? The combination of the septic tank and irrigation seems the most rational, cheap, and effective system for this State." (Ongerth and Jopling, 1977). Thus, the first recommended treatment process for reclaimed water in California was septic tank treatment.

Although the typhoid fever hazard of eating uncooked vegetables irrigated with sewage had been recognized, no particular control was applied to the sewage irrigation of crops until 1907. In the State Board of Health's April 1907 Bulletin, local health authorities were asked to "watch irrigation practices" and not allow use of "sewage in concentrated form and sewage-polluted water, to fertilize and irrigate vegetables which are eaten raw, and strawberries." In 1910, it was reported that there were 35 California communities using sewage for farm irrigation, 11 without any treatment and 24 after septic tank treatment. The March 1910 Monthly Bulletin of the State Board of Health contained the following statement favoring reuse: "In California, where water is so valuable for irrigation, the utilization of sewage for broad irrigation should be carefully considered."

Adoption of First Regulations

The first standards adopted by the State Board of Health in 1918, entitled *Regulation Governing Use of Sewage for Irrigation Practices* (California State Board of Health, 1918), prohibited the use of raw sewage for crop irrigation and limited the use of treated effluent to irrigation of nonfood crops and food crops that were cooked before being eaten or did not come in direct contact with the wastewater. Garden crops of the type that are cooked before being eaten could be irrigated if the application of effluent was not made within 30 d of harvest. The regulations provided several exemptions, such as permitting irrigation of melons if the sewage did not come in contact with the vine or product, and irrigation of tree-bearing fruit or nuts if windfalls or products lying on the ground were not harvested for human consumption.

Revision to Regulations in 1933

The regulations were revised in 1933 and renamed *Regulations on the Use of Sewage for Irrigating Crops* (California Department of Public Health, 1933). These regulations prohibited the use of raw sewage for crop irrigation and prohibited the use of sludge as a fertilizer for growing vegetables, garden truck, or low growing fruits or berries unless the sludge was rendered innocuous. The use of settled or undisinfected sewage effluent for the irrigation of the same type of crops and for the irrigation of orchards or vineyards during seasons in which windfalls or fruit lie on the ground was also prohibited. Irrigation of fodder, fiber, or seed crops with settled or undisinfected sewage was allowed, but milk cows could not be pastured on land that was moist with sewage. The regulations exempted restriction of wastewater for the irrigation of garden truck crops eaten raw if the wastewater was well oxidized, nonputrescible, and reliably disinfected or filtered to meet a bacterial standard approximately the same as the then-current drinking water standard. Disinfection reliability was emphasized in that two or more chlorinators, weighing scales, reserve supply of chlorine, twice daily coliform analyses, and records were required. The 1933 standards marked the first appearance of cross-connection control regulations. Cross connections between wastewater and domestic water supply pipelines were prohibited, and signs warning against drinking the water were specified on pipes and appurtenances that contained wastewater.

The 1933 regulations continued in effect until passage of the Water Pollution Act of 1949 eliminated the permit system that constituted the statutory basis for the regulation. They were reissued without change in 1953 as part of the California Administrative Code; the title was slightly modified to *Regulations Relating to Use of Sewage for Irrigating Crops* (California Department of Public Health, 1953).

F-1 Evolution of Nonpotable Water Reuse Criteria in California (Adapted from Crook, 2002)

Development of Statewide Standards

The number of water reuse projects increased dramatically in the 1960s, and it became necessary to develop water reclamation standards for various types of use. In 1967, a state legislative committee reported that legislation relating to the use of reclaimed wastewater was needed to protect public health and that the State Department of Pubic Health should be required to establish statewide contamination standards. The committee recommended that the Regional Water Quality Control Boards (RWQCBs) establish waste discharge requirements for the use of reclaimed water that are in conformity with statewide contamination standards. These recommendations resulted in revisions to the California Water Code in 1967 (California State Water Resources Control Board, 1967), which gave the Department of Public Health the authority and responsibility to establish reclamation criteria and gave the RWQCBs the responsibility to enforce the criteria. As a result, more comprehensive regulations were enacted in 1968 that were directed mainly at the control of disease agents. These *Statewide Standards for the Safe Direct Use of Reclaimed Water for Irrigation and Impoundments* (State of California, 1968) included treatment and quality requirements intended to assure that the use of reclaimed water for the applications specified in the regulations would not impose undue risks to the public health.

Several studies conducted by the Department of Health in the late 1960s and early 1970s indicated a record of poor reliability at wastewater treatment plants (California Department of Health, 1973; Crook, 1976). At the request of the Department of Health, Sections 13520 and 13521 of the Water Code were revised in 1969 as follows:

> 13520. As used in this article *reclamation criteria* are the levels of constituents of reclaimed water, and means for assurance of reliability under the design concept which will result in reclaimed water safe from the standpoint of public health, for the use to be made.
>
> 13521. The State Department of Health shall establish statewide reclamation criteria for each varying type of reuse of reclaimed water where such use involves the protection of public health (California State Water Resources Control Board, 1967).

The above modification in state law authorized the Department of Health to establish regulations on treatment reliability. The 1968 standards were revised in 1975 to include treatment reliability requirements and were renamed *Wastewater Reclamation Criteria* (State of California, 1978). There have been two subsequent revisions to the criteria, one in 1978 that added general requirements for groundwater recharge and differentiated between different types of landscape irrigation, and one in 2000 that included several changes and additions.

The 1968 Criteria

Most of the rationale supporting the 1968 standards also pertains to subsequent water reuse regulations adopted in California (see Foster and Jopling, 1969). To arrive at a proper balance of realistic and workable standards that would protect public health, an advisory committee was established, and a broad spectrum of interested parties, including waste dischargers, regulatory agencies, and potential users, were able to participate in formulation of the standards via public meetings and other forums. It was determined at that time that the state-of-the-art of potable reuse was not sufficiently advanced to consider that use of reclaimed water.

Inclusion of Descriptive Terms (Treatment Processes)

Both reclaimed water quality limits and wastewater treatment unit processes were included for the following reasons: water quality criteria involving surrogate parameters alone do not adequately characterize reclaimed water quality; a combination of treatment and quality requirements known to produce reclaimed water of acceptable quality obviates the need to monitor the finished water for certain constituents; expensive, time-consuming, and in some cases, questionable monitoring for pathogenic microorganisms is eliminated without compromising health protection; and assurance of treatment reliability is enhanced.

The three types of reclaimed water, i.e., primary effluent, oxidized wastewater, and filtered wastewater, were defined by descriptive terms rather than specific water quality parameters. For example, while the definition of oxidized wastewater, that is, *wastewater in which the organic matter has been stabilized, is nonputrescible, and contains oxygen*, left the definition open to exact interpretation, it was felt that the meaning is clear to those engaged in the wastewater treatment field. Specification of a limit, such as BOD or suspended solids, had the drawback of requiring a monitoring program that could be a significant burden on small water reuse operations, could restrict the development of additional systems, and could cause the abandonment of many existing operations (Ongerth and Jopling, 1977).

Treatment and Quality Limits Requiring primary effluent for the irrigation of fodder, fiber, and seed crops, as well as for the surface irrigation of orchards and vineyards was a continuation of standards that were developed in 1933. Because there were no known adverse health effects at the almost 150 installations where reclaimed water was used for such purposes, there was no obvious reason to change the standard. The allowance of primary effluent for irrigation of processed food crops was added to the standards.

A requirement of oxidation, coagulation, filtration, and disinfection to achieve a total coliform limit of 2.2/100 mL or less for reclaimed water used to irrigate food crops eaten raw was intended to assure the absence of viruses in the reclaimed water and was based on studies carried out at Santee, California, and elsewhere in the state as well as virus research reported in the literature. Helminths and other parasites were not thought to be a major concern in California, and *Giardia* and *Cryptosporidium* were not yet recognized as public health threats. A requirement that wastewater after filtration meet a 10 turbidity unit limit was based on acknowledgment that particulate matter can adversely affect disinfection.

Sampling and Analysis Requirements

The standards required daily sampling for total coliform bacteria, with compliance based on a running median of the last seven samples for which analyses had been completed. Use of a running median was consistent with previous disinfection requirements specified by the Department of Health. Inclusion of a maximum total coliform limit was considered, but rejected based on the fact that short-term unsatisfactory conditions would not be detected, since it takes a minimum of 2 d to obtain bacteriological results using the multiple tube fermentation technique. The bacteriological results were meant to reflect a recent record of the adequacy of the operation rather than a means for promptly determining unsatisfactory conditions (Foster and Jopling, 1969). Provisions to allow an alternative to the coliform standard—implicitly, chlorine residual—also was rejected on the grounds that chlorine residual is an even more indirect measure of microbial destruction than the use of coliform bacteria. The minor cost savings of using chlorine residual *in lieu* of coliform analyses was not deemed sufficient to allow elimination of daily coliform sampling.

The standards reflected the opinion that the risk of disease would increase directly with the expected degree of human contact with the reclaimed water. Primary effluent was deemed acceptable where no expected human contact with the water is expected, e.g., the irrigation of nonfood crops. Where there is a negligible degree of contact, or infrequent contact, such as at golf courses or freeway landscapes, it was determined that the effluent should not contain high numbers of pathogens and that an oxidized wastewater not exceeding a 7-d median total coliform concentration of 23/100 mL was a reasonable limit likely to contain low levels of pathogens. While fecal coliforms were recognized as a better indicator of sewage contamination in surface waters, it was decided to take a conservative approach and use total coliforms as the measure of disinfection effectiveness.

For uses where casual contact with reclaimed water would be expected, such as a recreational impoundment where fishing or boating is allowed, it was felt that an oxidized wastewater not exceeding a 7-d median total coliform concentration of 2.2/100 mL would be unlikely to contain significant levels of pathogens if a high quality secondary effluent was produced prior to disinfection. For the irrigation of food crops eaten raw and nonrestricted recreational impoundments where swimming is allowed, the intent was to require treatment to essentially eliminate viable viruses from the wastewater. The requirements, which included oxidation, chemical coagulation, clarification, and disinfection to achieve a total coliform level not exceeding 2.2/100 mL, were principally predicated on studies conducted at water treatment facilities and wastewater treatment plants.

Use Area Controls
The 1968 standards were based on implementation of proper controls and procedures to protect public health at the individual use areas, and it was decided that a Manual of Good Practices would be developed to describe proper operating controls and practices for each type of reuse included in the standards. In practice, it was several years before guidelines describing use area controls were produced.

The 1975 Criteria

The principal revision in the 1975 *Wastewater Reclamation Criteria* was the addition of reliability requirements. The criteria contained both design and operational requirements to ensure an appropriate level of treatment reliability. The regulations required features such as alarm systems, standby power supplies, treatment process reliability, emergency storage or disposal of inadequately treated wastewater, elimination of means to bypass treatment processes, monitoring devices and automatic controllers, flexibility of design, and certified personnel. Those reliability requirements have remained unchanged in subsequent versions of the criteria.

The turbidity limit was lowered from 10 turbidity units to an average operating turbidity of 2 turbidity units and a maximum of 5 turbidity units at least 95 percent of the time during any 24-h period. The establishment of a more conservative turbidity standard than in the 1968 standards was to ensure that effective coagulation and filtration had taken place. The turbidity standard was predicated on studies to determine the virus removal capability of tertiary treatment processes and on available information indicating that those turbidity levels can be achieved using then-present technology and were being attained at wastewater treatment facilities in California and elsewhere (California Department of Health, 1974).

The 1978 Criteria

The 1975 criteria were revised in 1978, due in part to an influx of proposals for landscape irrigation in urban areas initiated during the 1976–1977 drought and pressure from operating agencies and other regulatory agencies to include criteria addressing groundwater recharge by surface spreading. The revisions increased the treatment and quality requirements for reclaimed water used to irrigate parks, playgrounds, schoolyards, and similar areas and added a section recognizing groundwater recharge by spreading as an acceptable use of reclaimed water (State of California, 1978). Specific criteria for groundwater recharge were not included in the regulations.

The 2000 Criteria

The results of research and demonstration studies conducted in the late 1970s and 1980s, along with advances in treatment technology and a need to include requirements for additional types of reuse, resulted in a protracted effort to revise the 1978 criteria. This effort, begun in 1988, culminated in adoption of a new set of criteria in 2000. These *Water Recycling Criteria* (State of California, 2000a) include requirements for several new applications of reclaimed water, modify some of the treatment and quality requirements, prescribe requirements for dual water systems, include cross-connection control requirements, and include use area requirements that formerly were issued as guidelines. In conformance with terminology in the California Water Code, the word *reclaimed* has been replaced with *recycled*, and *reuse* has been replaced with *recycling* in all regulations. The types of reclaimed water uses covered by the criteria and their attendant treatment and quality criteria are summarized in Table F-1.

Treatment Requirements

Primary effluent is no longer acceptable for the irrigation of fodder, fiber, and seed crops; undisinfected oxidized wastewater is now the minimum treatment allowed. The Department of Health Services (DHS) opined that, while public contact with reclaimed water used for these types of uses is minimal, contact by workers or others is possible and secondary treatment is needed to assure a minimal acceptable quality to protect public health. Undisinfected secondary effluent also is acceptable for irrigation of commercially processed food crops.

The required treatment train for high level uses has been relaxed in the new criteria, to the point where—except for reclaimed water used for nonrestricted recreational impoundments—chemical clarification is no longer required and chemical addition is not required under certain conditions. This change in the criteria is based principally on data produced during the Pomona Virus Study (SDLAC, 1977) that compared the full treatment train included in the 1978 water reuse criteria (oxidation, coagulation,

Table F-1
2000 California water recycling criteria[a]

Type of use	Total coliform limits[b]	Treatment required
Irrigation of fodder, fiber, and seed crops, orchards[c] and vineyards[c], processed food crops[d], nonfood-bearing trees, ornamental nursery stock[e], and sod farms[e]; flushing sanitary sewers	• None required	• Oxidation
Irrigation of pasture for milking animals, landscape areas[f], ornamental nursery stock and sod farms where public access is not restricted; landscape impoundments; industrial or commercial cooling water where no mist is created; nonstructural fire fighting; industrial boiler feed; soil compaction; dust control; cleaning roads, sidewalks, and outdoor areas	• ≤23/100 mL • ≤240/100 mL in more than one sample in any 30-d period	• Oxidation • Disinfection
Irrigation of food crops[c]; restricted recreational impoundments; fish hatcheries	• ≤2.2/100 mL • ≤23/100 mL in more than one sample in any 30-d period	• Oxidation • Disinfection
Irrigation of food crops[g] and open access landscape areas[h]; toilet and urinal flushing; industrial process water; decorative fountains; commercial laundries and car washes; snow-making; structural fire fighting; industrial or commercial cooling where mist is created	• ≤2.2/100 ml • ≤23/100 ml in more than one sample in any 30-d period • 240/100 ml (maximum)	• Oxidation • Coagulation[i] • Filtration[j] • Disinfection
Nonrestricted recreational impoundments	• ≤2.2/100 mL • ≤23/100 mL in more than one sample in any 30-d period • 240/100 mL (maximum)	• Oxidation • Coagulation • Clarification[k] • Filtration[j] • Disinfection
Groundwater recharge by spreading	• Case-by-case evaluation	• Case-by-case evaluation

[a]Adapted from State of California (2000a).
[b]Based on running 7-d median; daily sampling is required.
[c]No contact between reclaimed water and edible portion of crop.
[d]Food crops that undergo commercial pathogen-destroying prior to human consumption.
[e]No irrigation for at least 14 d prior to harvesting, sale, or allowing public access.
[f]Cemeteries, freeway landscaping, restricted access golf courses, and other controlled access areas.
[g]Contact between reclaimed water and edible portion of crop; includes edible root crops.
[h]Parks, playgrounds, schoolyards, residential landscaping, unrestricted access golf courses, and other uncontrolled access irrigation areas.
[i]Not required if the turbidity of influent to the filters is continuously measured, does not exceed 5 NTU for more than 15 min and never exceeds 10 NTU, and there is capability to automatically activate chemical addition or divert wastewater if the filter influent turbidity exceeds 5 NTU for more than 15 min.
[j]The turbidity after filtration through filter media cannot exceed an average of 2 nephelometric turbidity units (NTU) within any 24-h period, 5 NTU more than five percent of the time within a 24-h period, and 10 NTU at any time. The turbidity after filtration through a membrane process cannot exceed 0.2 NTU more than five percent of the time within any 24-h period and 0.5 NTU at any time.
[k]Not required if reclaimed water is monitored for enteric viruses, *Giardia*, and *Cryptosporidium*.

clarification, filtration, and disinfection) to an abbreviated treatment train (oxidation, coagulation, filtration, and disinfection). Based on the results of pilot-scale studies, it was found that both treatment trains were capable of removing approximately 5 logs of seeded virus from the wastewater and met the total coliform and turbidity requirements specified in the criteria. Study conditions included a chlorine residual of 5 mg/L after a modal contact time of approximately 90 min. Another study (Engineering-Science, 1987), conducted during the Monterey Wastewater Reclamation Study for Agriculture, confirmed the virus removal capability of abbreviated treatment train.

The use of reclaimed water for nonrestricted recreational impoundments is the only nonpotable application for which so-called *conventional treatment* is still required. Conventional treatment means a treatment process flow diagram that includes a chemical clarification process between the coagulation and filtration processes. In consideration of the likelihood of ingesting reclaimed water while swimming in nonrestricted recreational impoundments and the paucity of information regarding pathogen removal where a discrete chemical clarification process is omitted, the regulations are more restrictive for this use than other nonpotable uses. Disinfected tertiary reclaimed water may be used *in lieu* of water that has received conventional treatment if the reclaimed water is monitored for enteric viruses, *Giardia*, and *Cryptosporidium*.

Based on a significant body of research and operational data, the criteria now allow membrane processes as a replacement for media filtration. Chemical coagulation preceding a membrane unit process is not required. For uses where direct or indirect human contact with reclaimed water is likely, the turbidity requirements have been modified slightly to include an absolute maximum turbidity limit of 10 NTU. As in the 1975 and 1978 criteria, turbidity must be monitored continuously. If a membrane process is used, the turbidity of the product water cannot exceed 0.2 NTU more than five percent of the time within a 24-h period and cannot exceed 0.5 NTU at any time. The turbidity requirement is based on observed turbidity levels in product water from properly designed and operated microfiltration unit processes having a nominal pore size in the 0.1 μm range and reflects attainability and good engineering practice.

Disinfection Requirements

Where disinfected tertiary treated reclaimed water is required, the criteria require either a chlorine disinfection process that provides a residual chlorine concentration times modal contact time (CT) value of at least 450 mg-min/L at all times with a modal contact time of at least 90 min, or a disinfection process that, when combined with the filtration process, has been demonstrated to reduce the concentration of MS-2 phage or poliovirus by 5 logs, i.e., to 1/100,000 of the concentration in the filter influent. The criteria do not allow modal contact times less than 90 min because of a lack of research demonstrating disinfection effectiveness at modal contact times less than 90 min. These requirements are based on the operating conditions employed during the Pomona Virus Study.

Methods of disinfection other than chlorine are not precluded if demonstrated to the satisfaction of DHS to provide an equivalent degree of disinfection and reliability. For example, if ultraviolet (UV) radiation is used for disinfection, UV facilities must meet design and operation requirements recommended in a document entitled *Ultraviolet Disinfection Guidelines for Drinking Water and Water Reuse* (NWRI, 2003).

Use Area Requirements

The criteria now include use area requirements that previously were used as guidelines by DHS. Guidelines are not enforceable in California and inclusion of previously developed use area guidelines in water reclamation permits issued by the RWQCBs had been inconsistent throughout the state. Reclaimed water use area setback distance requirements include the following: no irrigation or impoundment of undisinfected reclaimed water within 50 m (150 ft) of any domestic water supply well; no irrigation of disinfected secondary-treated reclaimed water within 30 m (100 ft) of any domestic water supply well; no irrigation with tertiary-treated (secondary treatment, filtration, and disinfection) reclaimed water within 15 m (50 ft) of any domestic water supply well unless special conditions are met, and no impoundment of tertiary-treated reclaimed water within 30 m (100 ft) of any domestic water supply well; and only tertiary-treated reclaimed water can be sprayed within 30 m (100 ft) of a residence or places where more than incidental exposure is likely. Other use area controls include confinement of runoff to the reclaimed water use area unless otherwise authorized by the regulatory agency; prohibition of reclaimed water spray, mist, or runoff in dwellings, designated outdoor eating areas, or food handling facilities; protection of drinking water fountains against contact with reclaimed water; signs at sites using reclaimed water that are accessible to the public, although educational programs or other approaches to assure public notification may be acceptable to DHS; and prohibition of hose bibs on reclaimed water piping systems accessible to the public.

The Water Recycling Criteria do not include color-coding requirements. Requirements pertaining to color-coding reclaimed water pipe are included in California's Health & Safety Code (State of California, 2000b), which states, in part, that: "All pipes installed above or below ground, on and after June 1, 1993, that are designed to carry recycled water, shall be colored purple or distinctively wrapped with purple tape." The Health & Safety Code further states that purple pipe or tape is not required for pipes used for water delivered for agricultural use and at municipal or industrial facilities that have established a labeling or marking system for reclaimed water on their premises, as otherwise required by a local agency, that clearly distinguishes reclaimed water from potable water.

F-2 CALIFORNIA DRAFT GROUNDWATER RECHARGE REGULATIONS

The information presented below is based on the most recent draft of the proposed groundwater recharge regulations (California Department of Health Services, 2004); it is likely that substantial changes will be made prior to adoption of the criteria.

The Regulatory Process

While aspects of its regulatory development process have been protracted, California has developed a comprehensive approach to groundwater recharge with reclaimed wastewater. Currently proposed regulations have gone through several iterations and, when finalized and subsequently adopted, will be included in the *Water Recycling Criteria*. The proposed regulations address both surface spreading and injection projects and are focused on potable reuse of the recovered water. The draft regulations, portions of which are summarized in the Table F-2, include requirements for—among other things—source control, water quality, treatment processes, recharge methods, dilution, operational controls, distance to withdrawal, time underground, monitoring wells, and preparation of an engineering report. The criteria are intended to apply only to planned groundwater recharge projects using recycled water, i.e., any water reclamation project planned and operated for the purpose of recharging a groundwater basin designated for use as a domestic drinking water source. They do not apply to wastewater disposal projects.

Filtration Requirements

The draft regulations vary somewhat for surface spreading and injection projects. Disinfected filtered recycled water is the minimum allowed for surface spreading projects, although additional treatment would be required if soil aquifer treatment is ineffective. Additional treatment via reverse osmosis (RO) is required for all injection projects to ensure organics removal, principally unregulated organic compounds (see Table F-2).

Barriers to Microbial Pathogens

As an additional barrier to microbial pathogens, the draft regulations include requirements addressing retention time of the reclaimed underground water and distance between the point of recharge and point of withdrawal. The purpose of establishing these two criteria was to ensure minimal migration of viruses through the soil system and to allow time for the natural die-off or attenuation of viruses to take place. The retention time is longer for recharge by injection than for surface spreading to compensate for the lack of initial soil virus removal and inactivation. The time and space separation between the recharge site and drinking water wells also allows time for corrective action if groundwater monitoring detects water quality problems.

Compliance with State Drinking Water Regulations

The reclaimed water must comply with the following state drinking water regulations: primary maximum contaminant levels (MCL), inorganic chemicals (except nitrogen), MCLs for disinfection byproducts, and action levels for lead and copper. Quarterly monitoring is required, with compliance determined from a running average of the last four samples. The reclaimed water also must be monitored annually for several secondary MCLs. In addition, the reclaimed water must be sampled quarterly for unregulated chemicals, priority toxic pollutants, and chemicals with state notification levels that DHS specifies based on a review of the project. Each year, the reclaimed water must be monitored for endocrine disruptors and pharmaceuticals specified by DHS after reviewing the project.

Nitrogen Limits

The DHS rationale for the nitrogen limits (see Table F-2 for options) is based on the following: (1) the scientific literature indicates that the efficiency of converting ammonia nitrogen to nitrate is about 50 percent, (2) nitrite is formed as an intermediary product, (3) the nitrite MCL is 1 mg/L, (4) it is unlikely that there would be 100 percent conversion to nitrite, and thus, (5) limiting total nitrogen to 5 or 10 mg/L—depending on which option is selected—would not result in nitrite formation greater than 1 mg/L.

Removal of Organics

The proposed regulations specify total organic carbon (TOC) as a surrogate for determining organics removal efficiency. Although TOC is not a measure of specific

Table F-2
California draft groundwater recharge regulations[a]

Contaminant Type	Type of Recharge	
	Surface spreading	Subsurface injection
Pathogenic Microorganisms		
Filtration	≤2 NTU	
Disinfection	5-log virus inactivation[b], ≤2.2 total coliform per 100 mL	
Retention time underground	6 mo	12 mo
Horizontal separation[c]	150 m (500 ft)	600 m (2000 ft)
Regulated Contaminants		
Drinking water standards	Meet all drinking water MCLs (except nitrogen) and new federal and state regulations as they are adopted	
Total nitrogen	• Level specified by DHS for existing project with no RWC increase • ≤5 mg/L for new project or increased RWC at existing project • Or NO_2 and NO_3 consistently met in mound (blending allowed)	
Unregulated Contaminants		
TOC in filtered wastewater	TOC≤16 mg/L in any portion of the filtered wastewater not subjected to RO treatment	
TOC in recycled water	RO treatment as needed to achieve: • TOC level specified by DHS for existing project with no RWC increase • TOC≤(0.5 mg/L)/RWC (new project or increased RWC at existing project) • Compliance point is in recycled water or mound[d] (no blending)	100% RO treatment to achieve: • TOC level specified by DHS for existing project with no RWC increase • TOC≤(0.5 mg/L)/RWC (new project or increased RWC at existing project)
Recycled water contribution (RWC)	≤50 % subject to above requirements 50–100 % subject to additional requirements	

[a] Adapted from California Department of Health Services (2004).
[b] The virus log reduction requirement may be met by a combination of removal and inactivation.
[c] May be reduced upon demonstration via tracer testing that the required detention time will be met at the proposed alternative distance.
[d] If mound monitoring is approved.

organic compounds, it is considered to be a suitable measure of the gross organics content of reclaimed water for the purpose of determining organics removal efficiency. The proposed TOC limit is based on increasing concern over unregulated chemical contaminants and the realization that current technology using membranes can readily reduce TOC to 0.2 mg/L or less. Reductions in TOC concentrations via aboveground treatment processes are less restrictive for surface spreading projects than for injection projects because of additional TOC removal that has been demonstrated to occur in the unsaturated zone at existing surface spreading projects (Asano and Cotruvo, 2004; Crook et al., 1990). The TOC limit applies to TOC of wastewater origin in recharged water. Weekly sampling is required for TOC, with compliance determined monthly from the average of the most recent 20 TOC samples.

Dilution Requirement

The draft criteria require a minimum of 50 percent dilution with water of nonsewage origin, although recharge greater than 50 percent reclaimed water may be considered by DHS if certain conditions are met, such as: annual testing for tentatively identified compounds (TIC), inclusion of an advanced oxidation process (i.e., hydrogen peroxide addition and ultraviolet radiation), and submission of a proposal and report that includes documentation of compliance with all pertinent criteria, the results of any additional studies requested by DHS, and peer review by an independent advisory panel. The reclaimed water contribution must be determined monthly with compliance based on a running 5-yr average.

Groundwater Monitoring

Groundwater monitoring wells must be located within 1 and 3 mo hydraulic travel time from the recharge area to the nearest downgradient domestic public or private water supply well and at additional points. The monitoring wells must be capable of obtaining independent samples from each aquifer that potentially conveys the recharged water. Monitoring wells must be sampled quarterly for TOC, total nitrogen, total coliforms, secondary MCLs, and other constituents specified by DHS that are identified through reclaimed water monitoring.

Required Permits

Any intentional augmentation of drinking water sources with reclaimed water in California requires two state permits. A waste discharge or water recycling permit is required from a RWQCB, which has the authority to impose more restrictive requirements than those recommended by DHS, and a public drinking water system using an impacted source is required to obtain an amended water supply permit from DHS to address changes to the source water.

REFERENCES

Asano T., and J. A. Cotruvo (2004) "Groundwater Recharge with Reclaimed Municipal Wastewater: Health and Regulatory Considerations," *Water Research*, **38**, 8, 1941–1951.

California Department of Health (1973) *Development of Reliability Criteria for Water Reclamation Operations*, California Department of Health Services, Water Sanitation Section, Berkeley, CA.

California Department of Public Health (1933) *Special Bulletin No. 59: Regulations on Use of Sewage for Irrigating Crops*, State of California Department of Public Health, Sacramento, CA.

California Department of Health (1974) *Turbidity Standard for Filtered Wastewater*, State of California Department of Health, Water Sanitation Section, Berkeley, CA.

California Department of Health Services (2004) *Draft Groundwater Recharge Regulations: 12-1-04*. California Department of Health Services, Drinking Water Technical Program Branch, Sacramento, CA.

California Department of Public Health (1953) *Regulations Relating to Use of Sewage for Irrigating Crops*, California Administrative Code, Title 17, Chap. 5, Subchap. 1, Group 7, State of California Department of Public Health, San Francisco, CA.

California State Board of Health (1918) *Regulations Governing Use of Sewage for Irrigation Purposes*, California State Board of Health, Sacramento, CA.

California State Water Resources Control Board (1967) *Wastewater Reclamation and Reuse Law*. California Water Code, Chap. 6, Div. 7, California State Water Resources Control Board, Sacramento, CA.

Crook, J. (1976) *Reliability of Wastewater Reclamation Facilities*, State of California Department of Health, Water Sanitation Section, Berkeley, CA.

Crook, J., T. Asano, and M. Nellor (1990) "Groundwater Recharge with Reclaimed Water in California," *Water Environ. Technol.*, **2**, 4, 42–49.

Crook, J. (2002) "The Ongoing Evolution of Water Reuse Criteria," in *Proceedings of the AWWA/WEF 2002 Water Sources Conf.*, Las Vegas, NV.

Engineering-Science (1987) *Monterey Wastewater Reclamation Study for Agriculture: Final Report*. Prepared for the Monterey Regional Water Pollution Agency by Engineering-Science, Berkeley, CA.

Foster, H. B., and W. F. Jopling (1969) "Rationale of Standards for Use of Reclaimed Water," *J. San. Eng. Div.*, ASCE, **95**, SA3, 503–514.

NWRI (2003) *Ultraviolet Disinfection Guidelines for Drinking Water and Water Reuse*, Report Number NWRI-2003-06, National Water Research Institute, Fountain Valley, CA.

Ongerth, H. J., and W. F. Jopling (1977) "Water Reuse in California," 219–256, in *Water Renovation and Reuse*, Academic Press, Inc., New York.

SDLAC (1977) *Pomona Virus Study: Final Report*, Sanitation Districts of Los Angeles County California State Water Resources Control Board, Sacramento, CA.

State of California (1968) *Statewide Standards for the Safe Direct Use of Reclaimed Water for Irrigation and Recreational Impoundments*, California Administrative Code, Title 17, Group 12, California Department of Public Health, Berkeley, CA.

State of California (1978) *Wastewater Reclamation Criteria*, California Administrative Code, Title 22, Div. 4, California Department of Health Services, Sanitary Engineering Section, Berkeley, CA.

State of California (2000a) *Water Recycling Criteria*, California Code of Regulations, Title 22, Div. 4, Chap. 3, California Department of Health Services, Sacramento, CA.

State of California (2000b) *Water Recycling in Landscaping Act*, Government Code, Title 7, Div. 1, Chap. 3, Sacramento, CA.

Appendix G

Values of the Hantush Function F(α, β) and the Well Function W(u)

Table G-1
Values of the Hantush function F(α, β) for selected values of α and β[a,b]

α	β									
	0.02	0.04	0.06	0.08	0.10	0.3	0.5	0.7	0.9	1.0
0.02	0.0041	0.0073	0.0101	0.0125	0.0146	0.0288	0.0361	0.0401	0.0422	0.0429
0.04	0.0073	0.0135	0.0188	0.0236	0.0278	0.0559	0.0705	0.0785	0.0828	0.0842
0.06	0.0101	0.0188	0.0266	0.0335	0.0398	0.0817	0.1035	0.1154	0.1219	0.1239
0.08	0.0125	0.0236	0.0335	0.0425	0.0508	0.1060	0.1350	0.1509	0.1595	0.1622
0.10	0.0146	0.0278	0.0398	0.0508	0.0608	0.1290	0.1650	0.1849	0.1957	0.1990
0.3	0.0288	0.0559	0.0817	0.10600	0.12900	0.3009	0.3995	0.4553	0.4860	0.4955
0.5	0.0361	0.0705	0.1035	0.1350	0.1650	0.3995	0.5420	0.6254	0.6721	0.6865
0.7	0.0401	0.0785	0.1154	0.1509	0.1849	0.4553	0.6254	0.7272	0.7852	0.8034
0.9	0.0422	0.0828	0.1219	0.1595	0.1957	0.4860	0.6721	0.7852	0.8504	0.8710
1.0	0.0429	0.0842	0.1239	0.1622	0.1990	0.4955	0.6865	0.8034	0.8710	0.8924

[a]Adapted from Hantush, M. S. (1967) "Growth and Decay of Groundwater-mounds in response to Uniform Percolation," Water Resour. Res., **3**, 1, 227–234.
[b]For a complete set of function values refer to the original article or a standard text on groundwater hydrology.

Table G-2

Values of the well function $W(u)$ as a function of the dimensionless parameter u[a]

u	1.0	2.0	3.0	4.0	5.0	6.0	7.0	8.0	9.0
	0.219	0.049	0.013	0.0038	0.0011	0.00036	0.00012	0.000038	0.00
$\times 10^{-1}$	1.82	1.22	0.91	0.70	0.56	0.45	0.37	0.31	0.26
$\times 10^{-2}$	4.04	3.35	2.96	2.68	2.47	2.30	2.15	2.03	1.92
$\times 10^{-3}$	6.33	5.64	5.23	4.95	4.73	4.54	4.39	4.26	4.14
$\times 10^{-4}$	8.63	7.94	7.53	7.25	7.02	6.84	6.69	6.55	6.44
$\times 10^{-5}$	10.94	10.24	9.84	9.55	9.33	9.14	8.99	8.86	8.74
$\times 10^{-6}$	13.24	12.55	12.14	11.85	11.63	11.45	11.29	11.16	11.04
$\times 10^{-7}$	15.54	14.85	14.44	14.15	13.93	13.75	13.60	13.46	13.34
$\times 10^{-8}$	17.84	17.15	16.74	16.46	16.23	16.05	15.90	15.76	15.65
$\times 10^{-9}$	20.15	19.45	19.05	18.76	18.54	18.35	18.20	18.07	17.95
$\times 10^{-10}$	22.45	21.76	21.35	21.06	20.84	20.66	20.50	20.37	20.25
$\times 10^{-11}$	24.75	24.06	23.65	23.36	23.14	22.96	22.81	22.67	22.55
$\times 10^{-12}$	27.05	26.36	25.96	25.67	25.44	25.26	25.11	24.97	24.86
$\times 10^{-13}$	29.36	28.66	28.26	27.97	27.75	27.56	27.41	27.28	27.16
$\times 10^{-14}$	31.66	30.97	30.56	30.27	30.05	29.87	29.71	29.58	29.46
$\times 10^{-15}$	33.96	33.27	32.86	32.58	32.35	32.17	32.02	31.88	31.76

[a]Wenzel, L. K. (1942) "Methods for Determining Permeability of Water-Bearing Materials with Special Reference to Discharging-Well Methods," *U.S. Geological Survey Water-Supply Paper* 887, p. 192, Washington, DC.

Appendix H

Interest Factors and Their Use

Compounding or discounting monetary values to compute future or present worths involve the use of interest factors. While some factors are built into hand calculators or electronic spreadsheets, others are not. Formulas for the nine common factors are given in Table H-1. It is important to understand the formulas the factors are based on to apply them to specific situations. Interest factors incorporate two values: the interest rate per time period, i, and the number of time periods, n. Tables of calculated factors for various interest rates are published in many economics textbooks and other references, such as Gittinger (1984).

Interest rates are expressed as percent per annum. If the time period of compounding is less than a year, as with home mortgage payments, the interest rate per period is the nominal rate per annum divided by the number of periods per year, such as 12 for monthly periods. For water resource projects, analyses are usually conducted for annual periods.

Financial transactions can be assumed to occur at the beginning of time periods or the end of the periods. Formulas used in calculators or in spreadsheets can be set for either assumption. It is important to select the appropriate assumption to derive the correct interest factors. Following the usual convention for economic or financial analyses of water resource projects, the formulas in Table H-1 are based on the assumption that the financial events occur at the end of the period.

The present worth at the beginning of the first time period is designated as P. The future worth at the end of the last, nth, period is designated as F. The annuity, A, is equal to a periodic payment or receipt occurring at the end of each of n periods. For recurring periodic events, such as equal annual operating expenses, the factors are often called *series* factors, such as series present worth factor. A gradient series is a series that increases by the same amount each period. The gradient series formulas in Table H-1 are based on n-year end-of-year payments of 0, G, 2G, 3G, . . . (n–1)G where G is the periodic incremental change.

The interest factors in Table H-1 are based on the assumption that interest is compounded. Simple interest is interest that is accumulated based on the principal or present worth amount regardless of the number of periods. For compound interest, which is more common than simple interest, the interest is calculated for each period on the balance due at the beginning of the period, not on the original principal. Thus, interest is accrued on both the original balance and all previously accrued interest.

Even if compounding is applied, it may be necessary to use a simple interest calculation for time spans of less than the standard period. For simple interest, n can be in fractional parts of the period. To illustrate a simple interest calculation for a period of less than 1 yr, assume that an amount is due after 7 mo on a loan of $1000 borrowed at a rate of five percent per annum.

$$F = \$1000\left[(1 + (0.05)\left(\frac{7 \text{ mo}}{12 \text{ mo}}\right)\right] = \$1029.17$$

Table H-1
Compound interest factors

Factor	Formula	Eq. No.	Use
Simple interest future worth factor (F/P)	$1 + (i)(n)$	(H-1)	To calculate using simple interest the future worth (F/P) at the end of the nth period at interest rate i from a single present amount (P) occurring at the beginning of the initial period. n can be expressed in fractional periods.
Compound amount factor (F/P)	$(1 + i)^n$	(H-2)	To calculate the future worth (F) at the end of the nth period at interest rate i from a single present amount (P) occurring at the beginning of the initial period.
Present worth factor (P/F)	$\dfrac{1}{(1 + i)^n}$	(H-3)	To calculate the present worth (P) at the beginning of the initial period of a future amount (F) occurring at the end of the nth period.
Sinking fund factor (A/F)	$\dfrac{i}{(1 + i)^n - 1}$	(H-4)	To calculate the amount of an equal payment (A) to be made at the end of each of n periods to accumulate a given future amount (F) at the end of the nth period.
Capital recovery factor (A/P)	$\dfrac{i(1 + i)^n}{(1 + i)^n - 1}$	(H-5)	To calculate a uniform payment (A) at the end of each of n periods to recover a present amount (P) by the end of the nth period, as in calculating the annual loan repayments, A, to pay off a loan of amount P at interest rate i.
Series compound amount factor (F/A)	$\dfrac{(1 + i)^n - 1}{i}$	(H-6)	To calculate the future accumulated value (F) at the end of the nth period of equal payments (A) made at the end of each of n periods.
Series present worth factor (P/A)	$\dfrac{(1 + i)^n - 1}{i(1 + i)^n}$	(H-7)	To calculate the present worth (P) at the beginning of the initial period of a series of equal payments (A) made at the end of each of n periods.

Table H-1
Compound interest factors (*Continued*)

Factor	Formula	Eq. No.	Use
Gradient series compound amount factor (F/G)	$\left(\dfrac{1}{i}\right)\left[\dfrac{(1+i)^n - 1}{i} - n\right]$	(H-8)	To calculate the future accumulated value (F) at the end of the nth period of a series of payments that begin at 0 the first period and increase by amount G at the end of each subsequent period for n periods.
Gradient series present worth factor (P/G)	$\left(\dfrac{1}{i}\right)\left[\dfrac{(1+i)^n - 1}{i(1+i)^n} - \dfrac{n}{(1+i)^n}\right]$	(H-9)	To calculate the present worth (P) at the beginning of the initial period of a series of payments that begin at 0 the first period and increase by amount G at the end of each subsequent period for n periods.
Gradient series to uniform series equivalent factor (A/G)	$\left(\dfrac{1}{i}\right)\left[\dfrac{n}{(1+i)^n - 1}\right]$	(H-10)	To convert a gradient series of payments that begin at 0 the first period and increase by amount G at the end of each subsequent period for n periods to a uniform series of payments A occurring at the end of each of n periods.

To illustrate the meaning and use of several of the interest factors, four examples are provided. Additional explanation and examples of usage of these factors, as well as additional factors to fit more specialized situations, can be found in engineering economics textbooks (Gittinger, 1984; Riggs and West, 1986).

EXAMPLE H-1.
A pump having a 10 yr life will cost $10,000 to replace. How much money must be set aside at the end of each year into a savings account earning five percent interest to yield $10,000 in 10 yr?

$$A = (F)(\text{Sinking fund factor}) = (\$10,000)\left[\dfrac{0.05}{(1+0.05)^{10} - 1}\right] = \$795.05$$

EXAMPLE H-2.
A loan is approved for $100,000. It must be repaid in monthly installments for 30 yr at an interest rate of six percent. What must the monthly repayment be? This problem has an added complication in that the interest rate is expressed in a per annum amount, but the period between repayments is only monthly. The interest rate must be converted to a monthly equivalent:

$$i = \text{Monthly interest rate} = \dfrac{(0.06/\text{yr})}{(12 \text{ mo/yr})} = 0.005/\text{mo}$$

$$n = \text{Number of monthly payments} = (12/\text{yr})(30 \text{ yr}) = 360$$

$$A = (P)(\text{Capital recovery factor}) =$$
$$(\$100{,}000)\left[\frac{0.005\,(1 + 0.005)^{360}}{(1 + 0.005)^{360} - 1}\right] = \$599.55$$

EXAMPLE H-3.

A deposit of $100,000 is set aside annually in a sinking fund for future equipment replacement. The savings account earns six percent interest. What will be the account balance in 10 yr?

$$F = (A)(\text{Series compound amount factor}) =$$
$$(\$100{,}000)\left[\frac{(1 + 0.06)^{10} - 1}{0.06}\right] = \$1{,}318{,}079$$

EXAMPLE H-4.

A wastewater reclamation plant operation and maintenance costs are $150,000 per year. The planning period for analysis is 20 yr and the discount rate is six percent. What is the present worth of the operation and maintenance costs assuming that the costs occur at the end of each year?

$$P = (A)(\text{Series present worth factor}) =$$
$$(\$150{,}000)\left[\frac{(1 + 0.06)^{20} - 1}{0.06(1 + 0.06)^{20}}\right] = \$1{,}720{,}488$$

REFERENCES

Gittinger, J. P. (1984) *Compounding and Discounting Tables for Project Analysis: With a Guide to Their Applications,* 2nd ed., published for the Economic Development Institute of the World Bank, The Johns Hopkins University Press, Baltimore, MD.

Riggs, J. L., and T. M. West (1986) *Engineering Economics,* 3rd ed., McGraw-Hill, New York.

Name Index

The following convention was used for the citation of authors of articles or reports in this textbook. Where a single author is involved, the citation is given as (Asano, 2001); where two authors are involved, the citation is given as (Asano and Tchobanoglous, 2006); and where three of more authors are involved, the et al. notation is used (Asano et al., 2001). However, to recognize the contributions of all of the authors of articles or reports, not cited in the text, but appearing in the end of chapter reference lists, they are all listed in this name index. Individuals or organizations that provided photographs and other graphics are also cited. Where both an abbreviation and spelled out name have been used in the text, the full name is spelled out in the name index followed by the abbreviation in parentheses. Where there is a discrepancy in the usage of initials for the same author, the most complete form of the name (known to the writers of this textbook) has been used (e.g., Crites, R. W. is used for Crites, R.).

Abbaszadegan, M., 128
Abe, A., 119, 124
ABT Umwerttechnologies GmbH, 792
Adams, A. P., 1115, 1134, 1141, 1165
Adams, C. D., 119, 124
Adams, D. B., 20, 33
Adams, E. Q., 541, 595
Adamson, K. C., 1040
Addiss, D. G., 127
Adham, S., 334, 371, 372, 481, 508, 522, 523, 1343
Aertgeerts, R., 181, 187
Ahn, K. H., 487, 522
Aieta, E. M., 127, 722
Aiken, G., 597
AIRVAC, 811, 827
Akiyoshi, E., 1101, 1202, 1498
Al-Attar, M. H., 1502, 1506
Alavanji, M., 124, 126
Al-Awadhi, N., 1506
Al-Ghusain, I., 1040
Allaway, W. H., 969, 1038
Al-Mutairi, N. Z., 1040
Al-Sulaimi, J., 1506
Alt, S., 523
Alum, A., 128
Alvarez-Cohen, L., 118, 124, 125, 128
Aly, O. M., 1124, 1166
Amato, T., 459
Ambrose, R. B., 1344
American Society for Testing and Materials (ASTM), 479, 522
American Society of Civil Engineers (ASCE), 9, 33
American Society of Heating, Refrigerating, and Air-Conditioning Engineers (ASHRAE), 1140, 1165

American Water Works Association (AWWA), 60, 67, 86, 124, 389, 448–450, 453, 459, 522, 736, 762, 832, 837, 856, 858, 859, 862, 881–883, 898, 906, 909, 912, 915, 926, 930, 945, 1176, 1186, 1201, 1386, 1389, 1394, 1396, 1432
American Water Works Association Research Foundation (AWWARF), 738, 762, 903, 926, 1257, 1271, 1280, 1300
Ames, B. N., 253
Ampt, G., 1260, 1301
Amy, G. L., 597, 1302
Anabela, D., 371
Andersen, A., 287, 293
Anderson, D. L., 802, 827
Anderson, J. M., 30, 33, 59, 66, 293, 1386, 1432
Anderson, R. E., 557, 594, 595
Ando, T., 125
Andre, J., 1166
Andrews, S. A., 596
Angelakis, A. N., 39, 41, 58, 66, 181, 187, 952, 1038
Angelotti, R. W., 1323, 1326, 1327, 1342
Angermeier, P. L., 34
Anselme, C., 522
Applet, S., 722
Applied Process Technologies, Inc., 573, 591
Aqua Aerobic Systems, Inc., 419, 420, 423
Aqua Services and Engineering, 1357
Aranda, J., 1110, 1165
Arant, G., 1432
Ares-Mazas, M. E., 129
Arizona Nuclear Power Project, Water Use, 1489, 1498
Arora, H., 595

Arrowood, J., 722
Arrowood, M. J., 127
Art, M., 595
Arthur, J., 180, 187
Asano, T., 9, 13, 30, 33, 34, 43, 48, 58, 60, 66, 67, 68, 120, 124, 127, 235, 251, 254, 459, 755, 756, 762, 933, 945, 950, 952, 953, 1008, 1018, 1038, 1039, 1040, 1041, 1110, 1136, 1140, 1165, 1166, 1193, 1201, 1202, 1248, 1255, 1300, 1301, 1342, 1390, 1391, 1393, 1399, 1401, 1403, 1406, 1407, 1409, 1417, 1432, 1442, 1497, 1498, 1460, 1489, 1501, 1504, 1506, 1507, 1508, 1520, 1521
Ashbolt, N. J., 225, 235, 251, 254
Assink, J. W., 1141, 1165
Atasi, K. Z., 596
Audic, J. M., 720
Aulicino, F. A., 124
Austep s.r.l., Italy, 397, 450, 479, 500
Australian Bureau of Statistics, 953, 1039
Avraham, O., 1040
Ayers, R. S., 956, 963, 966, 967, 973, 989, 1010, 1039, 1042
Azov, Y., 58, 68

Babcock, R., Jr., 333, 334, 371
Bachir, S. I., 1040
Backs, B., 596
Badger, P. G., 251, 253
Baetens, D., 1107, 1165
Baez, S. O. P., 1040
Bahri, A., 58, 61, 67, 68, 204, 1040, 1505, 1506
Baird, R. B., 60, 67, 1497
Baker, M. J., 802, 827
Balbus, J., 253
Bales, R. C., 1397, 1432
Barber, L. B., 115, 124, 1243, 1461

1529

Name Index

Barber, N. L., 68, 1040, 1166, 1201
Barcina, I., 720
Barenti, J., 1243
Barhorst, T. A. S., 126
Barker, P. L., 344, 371
Barlaz, M. A., 129
Barnard, J. L., 324, 371
Barnett, S. R., 1028, 1030, 1039
Baron, J. S., 9, 30, 34
Barry, K., 1040
Bartell, S., 253
Bartram, J., 180, 187
Barty-King, H., 39, 41, 43, 67
Barwick, R. S., 84, 85, 124
Basset, C., 1506
Batchley, E. R., 691, 718
Bates, K., 1243
Bauer, D., 1343
Baumann, D. D., 7, 34
Baumann, L. C., 113, 124
Beach, M. J., 124, 126, 127, 129
Beaudet, B. A., 550, 595
Bebee, J., 1165
Beck, M. B., 34
Beck, R. W., 491, 522
Beckett, M., 1165
Beecher, J. A., 1385, 1432
Beek, R. W., 522
Beerendonk, E. F., 720
Bell, G. E. C., 1110, 1165
Bellamy, B., 597
Bellar, T. A., 113, 124, 650, 718
Belosevic, M., 125
Bender, M. D., 1319, 1342
Benefield, L. D., 495, 522
Bennett, E. R., 1362, 1377
Bennett, J. V., 86, 124
Bens, M. S., 127
Bentem, A. G. N., 372
Berg, J. D., 722
Berger, P. S., 125
Bernstein, D. L., 254
Bérubé, P. R., 572, 595
Beverly, S. D., 1313, 1342
Biederman, E. M., 1432
Bilello, L. J., 550, 595
Biomicrobics, Inc., 792
Birks, R., 1201
Bixio, D., 58, 67
Black and Veatch, 1124, 1165
Black, S. E., 128
Blackburn, B. G., 84, 124, 129
Blackmer, F., 638, 718
Blackson, D. E., 1489, 1497
Blair, K. A., 127
Blair, S., 371
Blancq, S. M., 254
Blatchley, E. R., 718
Block, D. E., 253
Blowes, D. W., 827
Blume, T., 673, 718
Blumenthal, U. J., 62, 67, 188, 954, 1039

Bock, E., 828
Bogaerts, W. F., 1140, 1165
Bohart, G. S., 541, 595
Boland, J. J., 34
Bolton, J. R., 125, 577, 578, 583, 595, 597
Bone, K., 925, 926
Bonné, P. A. C., 590, 595
Bonner, V. R., 1319, 1342
Bonnet, L., 189
Bonomo, L., 58, 67
Bontoux, L., 66, 67, 1038
Booher, L. J., 999, 1039
Boone, G., 1343
Borden, R. C., 129
Borenzstajn-Roten, Y., 189
Borg, C., 596
Borzlleca, J. F., 1377
Bosch, R., 899
Bostjancic, J., 1003, 1039
Botrel, T. A., 1039
Bounds, T. R., 790, 807, 809, 827
Bourgeous, K., 315, 363, 406, 436, 437, 459, 459, 489, 522
Bourke, J., 596
Bouwer, H., 7, 34, 117, 124, 1248, 1260, 1264, 1268, 1269, 1278, 1284, 1288, 1301
Bowen, H., 969, 1039
Braden, J. B., 9, 3434
Bradley, D. J., 125
Bradley, P. M., 129
Bratby, J. R., 354, 355, 371
Brennan, D., 827
Bresee, J. S., 125
Brighty, G., 1243
Brindle, K., 372, 523
Brissaud, F., 58, 67, 1506
Brongers, M. P. H., 1147, 1148, 1165
Brooke, J. M., 1125, 1167
Brooks, M. A., 1342
Brown, G. K., 124
Brown, S. K., 129
Bruce, G., 1378
Brumelis, D., 723
Brunner, G. W., 1319, 1342
Brusseau, M. L., 253
Bruvold, W. H., 1439–1441, 1444, 1446, 1460
Buchanan, B., 1074
Buckingham, P. L., 596
Bucks, D. A., 1067, 1068, 1099, 1100, 1040
Buechler, S., 61, 67
Buisson, H., 371
Buitrón Méndez, G., 597
Bukhari, Z., 124
Burden, R. P., 719
Buring, J. E., 207, 252
Burton, F. L., 129, 291, 293, 294, 371, 372, 460, 524, 597, 722, 762, 828, 899, 945, 1041, 1100, 1343
Bush, E. O., 126

Buswell, A., 494, 523
Butler, J. D., 1100
Butterfield, C. T., 632, 633, 638, 718, 723
Buxton, G. V., 567, 595
Buxton, H. T., 1243, 1461

Cabal, A., 722
Caccio, S. M., 87, 124
Cairns, W., 127
Calderon, R. L., 81, 124, 125, 127, 129
California Department of Health Services, 1517, 1519, 1520, 1521
California Department of Health, 165, 187, 1511, 1514, 1520
California Department of Public Health, 1510, 1520, 1521
California State Board of Health, 1521
California State Water Resources Control Board (CSWRCB), 48, 67, 1511, 1521
Caliskaner, O., 395, 459
Camann, D. E., 148, 187, 188
Campbell, A. T., 128
Camper, A., 1166
Campos, C., 597
Canadian Mortgage and Housing Corporation, 822
Canter, K. P., 114, 124, 126, 219, 252
Cararo, D. C., 986, 1039
Caretti, C., 672, 673, 718, 720
Carley, R. L., 1374, 1376
Carnahan, R. P., 128
Carns, K. E., 721
Carr, R. M., 59, 62, 67, 184, 188
Carrow, R. N., 971, 1039
Casson, L. W., 129
Castelazo, M., 930, 945
Casteline, J. M., 945
Castell-Exner, C., 1290, 1302
Cater, S. R., 577, 583, 595
Cave, L., 1508
Cearley, D., 1161, 1165
Celenza, G. J., 441, 442, 444, 459, 468, 473–476, 522
Center for Disease Control and Prevention (CDC), 87, 89, 91, 124, 1165
Center for Paper Business and Industry Studies, 1147, 1165
CH2M Hill, Inc., 112, 125, 1362, 1376
Chakrabarti, C., 69
Chambers, C. W., 718
Chandler, D. P., 82, 129
Chang, A. C., 34, 114, 124, 968, 969, 1039, 1040, 1497
Chang, C., 1460
Chang, M., 460
Chang, S. L., 719, 1377
Chapin, D. H., 596
Chapman, H. D., 969, 1039
Chapman, R. S., 1319, 1342
Chappell, C. L., 253
Chapra, S. C., 1319, 1342

Name Index

Charnley, G., 194, 205, 206, 251
Chauret, C., 87, 125
Chen, C. L., 44, 67, 1488, 1497
Chen, D., 673, 718
Chen, N., 124, 129
Chen, W. R., 572, 595
Cheremisnoff, N. P., 1136, 1137, 1165
Cheremisnoff, P. N., 1136, 1137, 1165
Cherry, J. A., 1288, 1301
Chick, H., 609, 718
Chick, S. E., 253
Chiemchaisri, C., 330, 371
Childs, W. D., 492, 522
Chin, A., 572, 595
Chinn, R., 720
Choi, E., 253
Chow, B. M., 722
Christiansen, L. B., 1243
Chu, K. H., 125
Cikurel, H., 1040
City of Orlando, FL, 1023, 1039
City of Roseville, CA, 1074
City of San Diego, CA, 839, 1306, 1329–1334, 1342, 1347, 1376, 1388, 1390, 1432, 1447, 1460
City of San Francisco, CA, 910, 926
City of San Luis Obispo, CA, 1236, 1237, 1242
City of Sapporo, Japan, 150, 1200
City of St. Petersburg, FL, 837, 1089, 1091, 1093, 1099
Clancy, J. L., 89, 125
Clark, R. M., 46, 67
Clarke, D., 1040
Clarke, N. A., 1377
Cleasby, J. L., 412, 459
Cleaver-Brooks, 1142, 1165
Clement, J., 1113, 1119, 1165, 1166
Clescerl, L., 1124, 1165
Clifford, D. A., 535, 595
Clough, J. S., 1343
Cochran, K. W., 188
Cockerham, L. G., 200, 251
Cohen, B., 1504, 1507
Cohen, J. M., 538, 595
Cohn, P. D., 83, 86, 87, 90, 92, 125
Cole, E., 129
Colford, J. M., 251, 252
Collins, H. F., 640, 718
Cologne, G., 1395, 1397, 1406, 1432
Comer, E. A., 1344
Commonwealth Scientific and Industrial Research Organization (CSIRO), Australia, 1294
Condie, L. W., 1377
Conlon, W. J., 1342
Cook, P., 1161, 1165
Cooling Technology Institute (CTI), 1138–1140, 1165
Cooper, E., 919, 922, 926
Cooper, P. F., 39, 41, 43, 67
Cooper, P. K., 720

Cooper, R. C., 80, 92, 125, 225, 227, 251, 253, 254, 697, 718, 1041, 1498
Cooper, W. J., 113, 125
Cornel, P., 372
Corona-Vasquez, B., 721
Cort, R. P., 1041, 1498
Cosgrove, E. G., 690, 721
Cosgrove, W. J., 15, 18, 34
Cote, P., 353, 371
Cothern, C. R., 205, 251
Cotruvo, J. A., 215, 251, 1520
Council of Environmental Quality (CEQ), 1393, 1432, 1443, 1460
Couto, H. J. B., 450, 459
Cox, M., 125
Crabtree, K. D., 252
Craig, M., 1497
Craik, S. A., 87, 89, 125
Craun, G. F., 81, 86, 87, 91, 124, 125, 126, 127, 129
Crawford, G., 333, 334, 355, 357, 361, 364, 371
Crites, R. W., 99, 125, 284, 289, 293, 387, 437, 459, 621, 719, 766, 771, 775, 777, 779, 785, 790, 804, 807, 809, 811, 814, 821, 827, 905, 926, 957, 1021, 1039, 1212, 1214, 1220, 1242, 1316, 1342, 1480
Crittenden, J. C., 80, 81, 98, 99, 125, 450, 458, 459, 490, 491, 509, 522, 529, 531, 536, 541, 550, 552, 563, 566, 568, 570, 573, 574, 575, 577, 578, 580, 583, 595, 597, 605, 645, 646, 650, 683, 719, 1201
Croman, N., 1432
Crook, J., 48, 67, 97, 101, 125, 140, 143, 188, 254, 522, 915, 926, 938, 945, 1091, 1100, 1186, 1188, 1201, 1255, 1301, 1444, 1446, 1460, 1511, 1520, 1521
Cross, P., 1023, 1024, 1026, 1027, 1039, 1493, 1497
Crozes, G. F., 487, 522
Culp, R. L., 1377
Cunningham, K. M., 194, 252
Cunningham, T., 372
Current, W. L., 88, 125

Dabiri, A. E., 492, 522
Dahm, C. N., 34
Daigger, G., 371
Dalton, D., 459
Daniel, U., 1160, 1194, 1195, 1201
Danielson, R. E., 251, 253, 254
Darby, J. L., 459, 522, 687, 710, 719, 721, 827, 828
Daugherty, J. I., 1297, 1301
Daughton, C. G., 214, 251, 1442, 1460
Davis, D. N., 127
Davis, J. R., 1117, 1126, 1135, 1140, 1165
Davis, L. E., 494, 523, 1125, 1167

Davis, W. B., 1039
Davis, W. Y., 34
De Batz, S., 1343
de Fur, P. O., 126
De Giacomo, M., 124
De Koning, J., 67
De Lat, J., 722
Dean, R. B., 78, 125, 1348, 1351, 1377
DeBacker, E., 128
Debroux, J., 597
DeCarolis, J. F., 334, 371
Deeb, R. A., 120, 125
DeGeorge, J., 254
DeGraeve, G. M., 671, 723
Del Porto, D., 787, 827
Delneky, G., 1497
DeLorenze, G., 129
Dennis, K., 1040
DeOreo, W. B., 34
Department of Health and Aged Care, Australia, 1377
Derthorn, R. T., 596
Deshbhratar, P. B., 69
Deshmukh, S. S., 1301
Deventer, H. C., 1141
Devitt, D. A., 1100
Devol, M., 372
Di Pasquale, W. J., 1499, 1506
Diamadopoulos, E., 66
Diamond, R. A., 1101, 1202, 1498
Dickey, G. L., 1041
Dickson, J., 128
Dillon, P. J., 1040, 1042, 1270, 1293, 1294, 1301, 1499, 1506
Dimotsis, G. L., 553, 595
DiSimone, K., 1235–1237, 1242
Dobbs, R. A., 538, 595
Dold, P. L., 344, 371
Dong, X., 718
Donnell, B. P., 1342
Doorenbos, J., 990, 1039
Dore, M., 722
Dosemeci, M., 124, 126
Dow, D., 827
Drewes, J. E., 1273, 1275, 1276, 1301
Drost, Y. C., 126, 1301
Dryden, F. D., 44, 67, 1488, 1497
du Pisani, P. L., 1353–1355, 1358, 1360, 1361, 1377, 1503, 1506
Dudely, R. H., 227, 251
Dulbecco, R., 683, 719
Dunbar, Professor, Dr., 788, 827
Dupont, 479, 522
Durán Moreno, A., 597
Duranceau, S. J., 1167
Duran-Oreiro, D., 129
Dussert, B. W., 125
Dziegielewski, B., 34
Dzurik, A., 1394, 1395, 1432

Eaton, D. L., 219, 251
Echigo, S., 720

Eching, S., 1041
Eckenfelder, W. W., Jr., 493, 522, 595
Ecosan, 787, 827
Eddy, H. P., 39, 41, 43, 60, 68
Edraki, M., 1010, 1039
Edwards, M., 1124, 1166
Edzwald, J. K., 449, 450, 459
Eisenberg, J. N. S., 199, 229, 231, 239, 249, 251, 252, 253, 254
Ekster, A., 639, 719
Electric Power Research Institute (EPRI), 476, 522, 677, 719
Eliassen, R., 401, 460
Elovitz, M. S., 565, 595
Emerick, R., 459, 616, 687, 688, 690, 691, 710, 719, 803, 827
EnerCon Consultancy Services, 1145, 1166
Engelbrecht, R. S., 722
Engineering-Science, 140, 188, 1018, 1039, 1489, 1497, 1516, 1521
Englehardt, J., 202, 249, 252
Ennever, F. K., 254
Enslow, L. H., 639, 719
EOA Inc., 252
Erdal, Z., 293
Ershow, A. G., 219, 252
Eschweiler, H., 722
Espinoza, C. A., 1040
Esposito, K. M., 287, 293
Etienne, J., 1166
Etnier, C., 787, 804, 827
European Economic Community (EEC), 67
Evans, J. C., 596
Exall, K., 1499, 1506

Faber, B., 699, 720
Fair, G. M., 39, 67, 632, 633, 647, 719
Fairfax County Water Authority (FCWA), VA, 1326, 1328, 1342
Falconnet, P. A., 1166
Falkenmark, M., 19, 34
Fallowfield, H., 1432
Fan, M., 676
Fane, S., 827
Fankhauser, R. L., 90, 125
Fannin, K. F., 148, 188
Farrah, S. R., 128
Faruqui, N., 62, 67
Fattal, B., 189
Faubert, G., 127
Faust, S. D., 1124, 1166
Fay, R. C., 566, 596
Fayer, R. G., 88, 125
Feachem, R. G., 97, 102, 125
Feigin, A., 998, 999, 1039
Fell, C. J. D., 1168
Ferguson, D., 1101, 1202, 1498
Ferguson, F., 1168
Ferguson, J., 1121, 1125, 1166
Fernandez, A., 356, 361, 371

Ferrand, F., 720
Ferron, S., 596
Fewtrell, L, 187, 188
Fiksdal, L., 720
Finch, G. R., 125, 666, 719
Fleischer, E. J, 371
Fleming, P. A., 1485, 1497
Filter Loading Evaluation for Water Reuse (FLEWR), Monterey, CA, 459
Flinker, P., 827
Florida Department of Environmental Protection (FDEP), 165, 168, 188, 264
Flynn, M. P., 721
Foellmer, J., 251
Fonseca, A. N., 371
Food and Agriculture Organization (FAO), 1501, 1506
Ford, G. I., 828
Foster, H. B., 1511, 1513, 1521
Foster, K. R., 15, 34
Foussereau, X., 459, 523
Fox, K. R., 127
Fox, P., 124, 1273, 1275, 1278, 1294, 1301, 1302
Frankel-Conrat, H., 90, 125
Freeman, A. W., 68
Freeman, S., 474, 522
Freeze, R. A., 1288, 1301
French, O. F., 1040
Friedberg, E. R., 683, 719
Friedlander, P., 679
Friedler, E., 1500, 1507
Friedman, D., 596
Friedman, J., 898
Froelich, E. M., 533, 595
Fuerst, J., 828
Fuhs, G. W., 227, 252
Fulton, A., 1040
Funamizu, N., 1199, 1201
Furlong, E. T., 127, 1243, 1461

Gable, J., 1277, 1301
Gagliardo, P., 478, 522, 1183, 1185, 1201, 1343
Gaines, B., 371
Gallagher, T. M., 1342
Ganesan, A. K., 720
Garbertt, M., 1165
Garcia, L. S., 88, 125
Gard, S., 640, 719
Gardner, E. A., 1039
Garelick, H., 125
Gator, A. G., 952
Gaulke, L. S., 786, 827, 1193, 1201
Gavaghan, P. D., 129
Gearheart, R, 1233
Gee, D., 34
Gehr, R., 672, 673, 683, 711, 718, 719, 720, 723
Gehringer, P., 722
Gennaccaro, A. L., 89, 125

George, M. R., 971, 1039
Gerba, C. P., 80, 82, 86, 90, 91, 102, 126, 127, 129, 189, 235, 245, 252, 253, 254, 596, 718, 1280, 1301, 1302
Gerges, N. Z., 1039
Gerston, J., 1153, 1166
Getches, D. H., 937, 945, 1393, 1432
Geyer, J. C., 39, 67, 632, 647, 719
Giap, L. C., 1334, 1342
Giese, T. P., 356, 359, 371
Gij, J., 828
Gilbert, R. G., 985, 1040, 1099, 1301
Gill, G., 1077, 1100
Gillogly, T., 522, 523
Gilmour, R. A., 128
Gingras, M. B., 1101, 1202, 1498
Ginn, T., 719
Girard Industries, 884
Gittinger, J. P., 1413, 1423, 1432, 1525, 1527, 1528
Glass, R. I., 125
Glaze, W. H., 571, 574, 596
Gleick, P. H., 8, 9, 34
Göbel, A., 596
Godman, R. R., 903, 926
Goebel, R. P., 253
Gold, A., 827
Gonzales, R. M., 1167
Gordis, L., 208, 211, 252
Gordon, H., 1302
Gori, R., 720
Gosselink, J. G., 1210, 1243
Gotor, A. G., 1040
Goveia, M., 253
Goyal, S. M., 1280, 1301
Grabow, W. O. K., 126
Gradus, M. S., 127
Grady, C. P. L., Jr., 344, 371
Graham, J., 1040
Grattan, S. R., 971, 976, 1040, 1100
Green, S., 595
Green, W. H., 1260, 1301
Greenstock, C. L., 567, 595
Gregory, R. S., 1386, 1432
Grieve, C., 1100
Grizzard, T. J., 371
Grobicki, A. M. W., 1504, 1507
Grobler, G. J., 68, 1377, 1507
Grogan, T., 1410, 1432
Gross, M., 803, 827
Gujer, W., 371
Gullick, R. Q., 119, 126
Gunn, G. A., 1323, 1343
Gurr, C. J., 1306, 1342

Haag., W. R., 670, 719
Haarhoff, J., 1503
Haas, C. N., 126, 189, 199, 201, 202, 225, 227, 235, 252, 254, 611, 640, 719
Hach Company, 384, 459
Haddad, B. M., 1442, 1460
Haider, T., 722

Hairston, N. G., 34
Hale, R. C., 117, 126
Hall, E. L., 611, 719
Hallenbeck, W. H., 194, 252
Halling-Sorensen, B., 293
Hammann, D., 555, 596
Hamoda, M. F., 952, 1040
Hamrick, J., 1343
Hanawalt, P. C., 684, 720
Hand, D. W., 125, 459, 522, 553, 595, 719, 1201
Hanemann, W. M., 34
Hansen, E., 522, 523
Hanson, R. B., 1040, 1041
Hantush, M. S., 1264, 1301, 1523
Hanzon, B., 618, 621, 720
Harding, H. J., 187
Hargy, T. M., 125
Harivandi, M. A., 1072, 1073, 1100
Harremoës, P., 14, 15, 34
Harrhoff, J., 123, 126, 1353, 1377, 1507
Harries, J. E., 1243
Hart, F. L., 646, 720
Hartfelder, J., 720
Hartley, T. W., 1442, 1460
Hartling, E. C., 1393, 1432, 1442, 1449, 1460, 1461, 1490, 1497
Harvey, E., 126
Hattingh, W., 68, 1507
Hauch, P., 596
Hauser, G. E., 1342
Havelaar, A. H., 93, 126, 181, 189, 200, 226, 236, 254, 1280, 1301
Hayes, M., 1165
Hazbun, A. E., 1096, 1100
HDR Engineering, 236, 252, 1094, 1100
Heberer, T., 1277, 1301
Hein, H., 524
Hekimain, K. K., 251
Heller-Grossman, L., 128
Hemken, B., 1494, 1497
Hemming, J., 1378
Hennekens, C. H., 207, 252
Henze, M. C., 344, 371
Henze, M., 34
Herman, W. H., 129
Hermann, N., 596
Hermanowicz, S. W., 7, 26, 30, 34, 372
Herr, G. K., 1101, 1202, 1498
Herrero, A. M., 1504, 1507
Hertog, W., 67
Herwaldt, B. L., 127
Hiasa, M., 372
Higgins, M. J., 371
Hightower, M. M., 288, 293
Hijfnen, W. A. M., 689, 720
Hildesheim, M. E., 114, 124, 126
Hill, R. R., 119, 126
Hill, V., 124, 129
Hills, D. J., 1100, 1039
Hills, F. J., 1017, 1040
Hills, S., 1201

Hirata, T., 721
Hlavsa, M. C., 87, 88, 126
Hobbs, N., 1361, 1362, 1378
Hodgson, E., 212, 252
Hofman, J. A. M. H., 595
Hoigne, J., 670, 719
Holbrook, R. D., 355, 360, 371
Holliman, T. R., 903, 915, 926
Hom, L. W., 640, 720
Homberg, S. D., 124
Hopkins, B., 129
Hoppin, J., 194, 252
Horne, A., 899
Horvath, R. W., 722
House, M. A., 1393, 1432, 1447, 1460
Howe, K. J., 125, 459, 522, 595, 719, 1201
Howerton, S., 1166
Howles, S. R., 1039
Hoxie, N. J., 127
Hu, J. Y., 1343
Hu, L. Y., 1507
Hu, S., 595
Hubbard, A. H., 252
Hubele, C., 597
Huber Technologies, 337
Huber, M. M., 565, 596
Huesemann, M. W., 7, 34
Huffman, D. E., 81, 90, 91, 117, 126
Huitric, S. J., 720, 722
Hulsey, R. A., 721
Hultquist, R. H., 188
Humphrey, C. D., 125
Hunter, G. L., 721, 1166
Hunter, G., 665
Hunter, P. R., 80, 126, 188
Huston, J. L., 1042
Huston, S. S., 1106, 1132, 1166, 1040, 1042
Hutchinson, S. L., 119, 126
Hutson, S. S., 46, 47, 68, 1171, 1172, 1201
Hyde, C. G., 1485, 1497
Hyde, J., 1101, 1202, 1498
Hydrotech, 419

Icekson-Tal, N., 952, 1040
Iida, M., 1201
Ikeda, M., 762, 1201
Infilco Degremont, Inc., 679
Ingram, P. C., 1450, 1452, 1460
Institute of Medicine (IOM), 206, 252
International Association of Plumbing and Mechanical Officials (IAPMO), 908, 909, 914, 926, 1176, 1201
International Atomic Energy Agency (IAEA), 484, 491, 523
International life Sciences Institute (ILSI), 225, 226, 252
International Organization of Standardization (ISO), 1110, 1166

International Water Management Institute (IWMI), 17, 20, 34, 62
Irvine Ranch Water District (IRWD), CA, 891, 898, 915, 926, 1181, 1201
Isaäcson, M., 59, 68, 1504, 1507
Israel, K. E., 1041

Jacangelo, J. G., 127, 522
Jackson, E., 1499, 1507
Jackson, M. H., 128
Jackson, R. D., 1284, 1301
Jackson, R. B., 34
Jacobs, E. P., 480, 523
Jacquez, G., 253
Jagger, J. H., 690, 720
Jakubowski, W., 129
Jalali, Y., 653, 654, 720, 722
James, L. D., 1406, 1407, 1412–1414, 1423, 1432
Janbakhsh, A., 1243
Janex, M. L., 720
Japan Sewage Works Association (JSWA), 59, 68, 755, 762, 1172, 1175, 1199, 1201, 1501, 1507
Jaques, R. S., 1041, 1498
Jarraud, S., 1166
Jeffcoat, S., 1231, 1243
Jefferson, B., 371, 372
Jefferson, B., 523
Jeffs, G. E., 126
Jekel, M., 1276
Jenkins, D., 372, 495, 496, 523, 1120, 1167
Jennsen, P. D., 828
Jeran-Aero Graphics, 839
Jetten, M., 828
Jim, K., 1165
Jimenez, B., 58, 68, 127
Jobin, J., 827
Jobling, S., 1225, 1243, 1521
Joffe, J., 640, 719
Johannessen, J., 1101, 1202, 1498
Johansson, L., 802, 827
Johns, F. J., 1377
Johns, M. M., 126
Johnson, B. H., 1342
Johnson, B., 1302
Johnson, D. E., 148, 187, 188
Johnson, J. D., 690, 721
Johnson, L. J., 1186, 1201
Johnson, R. M., 1040
Johnson, T. L., 288, 293
Johnson, W. D., 54, 68, 1086–1092, 1100, 1453, 1486, 1461, 1497
Johnston, C. A., 34
Johnston, D. S., 827
Joly, P., 1139, 1166
Jones, C. A., 1166
Jones, R. D., 1243
Jones, S., 803, 827
Jopling, W. F., 41, 43, 45, 68, 1485, 1498, 1509, 1511–1513, 1521

Jordan, L. A., 1070, 1100
Joss, A., 596
Joubert, L., 766, 827
Jowett, E. C., 802, 827
Juanico, M., 953, 1011, 1042
Juby, G., 439, 459
Judd, C., 332, 372
Judd, S., 332, 371, 372, 523
Judkins, J. F., Jr., 522
Juranek, D. D., 88, 126, 127
Juwarkar, A. S., 69

Kadlec, R., 821, 827, 1212, 1243
Kadlec, R. H., 1212
Kamnikar, C., 1167
Kanarek, A., 1500, 1507
Kang, J. W., 571, 596
Kappus, K. K., 87, 126
Karim, M. R., 90, 126
Karimi, A. A., 572, 596
Karmeli, D., 1066, 1100
Karra, S., 611, 719
Kasower, S., 1461
Kasperson, J. X., 1348, 1351, 1377
Kasperson, R. E., 1348, 1351, 1377
Katayama, H., 721, 1166
Katehis, D., 806, 813, 827
Katz, S. M., 1461
Katzenelson, E., 189
Kavanaugh, M. C., 118, 125, 128, 554, 596
Kawamoto, K., 762, 1201
Kawamura, S., 409, 459, 596, 646, 720
Kawczynski, E., 523
Kaye, R. B., 1168
Kazmierczak, J. J., 127
Keck, C. W., 206, 254
Keeney, R. L., 1386, 1432
Keller, J., 1066, 1100
Kelly, C., 1243
Kelly, L. M., 1302
Kelly, L., 1445, 1461
Kelner, A., 683, 720
Kelty, C. A., 129
Kenny, J. F., 68, 1040, 1166, 1201
Keyes, C. G., 288, 293
Keys, J., 34
Kidmi, S., 189
Kiefer, J. C., 34
Kielen, N. C., 1003, 1041
Kim, J., 720, 721
Kimball, D. T., 188
Kimball, P. C., 125
Kimes, J. K., 509, 511, 523
King, I., 1319, 1342
King, P., 1508
Kinner, C., 1101, 1202, 1498
Kirby, S., 1243
Kirkpatrick, W., 1041, 1489, 1497, 1498
Kirmeyer, G. J., 885, 887, 898, 1121, 1125, 1165, 1166
Kitis, M., 672, 720, 803, 827
Kjartanson, K., 597

Klaassen, C. D., 209, 214, 251, 252, 253
Klein, P., 925, 926
Klime, D. E., 115, 126
Knapton, A., 1301
Knight, R., 821, 827, 1212, 1243
Knowlton, D. R., 254
Kobylinski, E., 1166
Koch Industries, Inc., 339
Koivunen, J., 673, 720
Kolluru, R., 193–196, 205, 206, 253
Kolpin, D. W., 115, 117, 127, 1225, 1243, 1442, 1461
Kondolf, G. M., 1226, 1243
Konnan, J. I., 251, 253
Kontos, N., 120, 127, 1393, 1432
Koo, H., 522
Koopman, J. S., 231, 232, 253
Koorse, S. J., 523
Kopp, H., 35
Korategere, V., 596
Korhonen, L. K., 102, 127
Korick, D. G., 89, 127
Koundi, A., 1505, 1507
Kracman, B., 952, 1028, 1030–1032, 1040, 1499, 1507
Kramer, M. H., 84, 85, 88, 127
Krasner, S. W., 113, 127, 651, 720
Krause, S., 372
Kraybill, H. R., 119, 127
Krayer von Krauss, M. M. B. A., 15, 34
Kree, D., 523
Krieger, G. R., 211, 254
Kriven, W., 1165, 1166
Krogstad, T., 828
Kroner, R. C., 124
Kruidenier, L., 254
Kshirsagar, D. G., 69
Kubick, K. S., 1485, 1497
Kubota Corporation, 336
Kuenen, J. G., 1302
Kuenen, S., 828
Kulik, W., 1342
Kuo, A. Y., 1343
Kuo, J. F., 67, 720, 722, 1497
Kurian, J., 62, 68
Kutz, S. M., 102, 127
Kuyk, D. D., 903, 926

La Chance, A. M., 127, 1497
Lagona-Limon, C. I., 721
LaGrega, M. D., 538, 596
Lahnsteiner, J., 1353, 1355, 1358, 1377, 1503, 1507
Laine, A., 371
Laine, J. M., 522
Lampert, G., 1377
Lampert, Y., 189
Lamphiear, D. E., 188
Lance, J. C., 1301
Landmeyer, J. E., 129
Langelier, W., 494, 523, 1118, 1124, 1166
Larsen, M. D., 356, 371

Larson, R., 699, 720
Larson, T. E., 494, 523, 1125, 1166
Lauer, W. C., 112, 127, 1361, 1362, 1364, 1367–1374, 1377, 1491, 1497
Law, I. B., 1334, 1342, 1343, 1377, 1503, 1507
Lawrence, D. P., 372
Lazarova, V., 58, 68, 672, 720, 950, 952, 953, 985, 1015, 1193, 1040, 1201
Lazenby, A., 899
Le Clech, P., 371
Le Gal La Salle, C., 1042
LeBlanc, C., 679
LeChevallier, M. W., 90, 126
Lee, J., 1343, 1499, 1507
Lee, K. C., 590, 596
Lee, L. Y., 1343, 1507
Lee, R. R., 1406, 1407, 1412–1414, 1423, 1432
Lee, S. H., 84, 85, 124, 127, 129
Leenheer, J. A., 124, 142, 188
Lefevre, F., 672, 720
Lehman, G., 467, 522, 523
Leitner, G. F., 522, 1506
Lekven, C., 827
Lempert, G., 1353, 1355, 1358, 1503, 1507
Leong, Y. C., 251
Leslie, G., 471, 515, 523, 1343
Leth, H., 722
Letter, J. V., Jr., 1342
Leverenz, H. L., 786, 803, 805, 828, 1039
Levi, P. E., 212, 252
Levine, A. D., 48, 60, 67, 68, 80, 378, 379, 380, 459, 1018, 1110, 1039, 1140, 1165, 1166
Levy, D. A., 84, 85, 124, 127, 129
Levy, J. A., 125
Lewis, B. L., 252
Lewis, D. R., 971, 1040
Lewis, G. H., 1116, 1166
Lewis, R., 355, 364, 371
Liberti, L., 672, 720
Lichtenberg, J. J., 124, 650, 718
Lien, L., 470, 523
Lim, H. C., 371
Lin, A. Y-C., 115, 127, 1342
Linden, K. G., 87, 127, 567, 573, 578, 586, 595, 596, 690, 720, 722
Linder, S., 125
Lindh, G., 19, 34
Linstedt, K. D., 1362, 1377
Lisk, D. J., 969, 1040
Lisle, J. T., 88, 127
Little, T. M., 1017, 1040
Liu, Y. A., 1107, 1108, 1166
Llamas, M. R., 35
Lodewage, L., 1101
Löffler, D., 596
Logan, B. E., 119, 127
Loge, F., 459, 719
Logsdon, G. S., 412, 459

Lohman, L. C., 1373, 1377, 1439, 1440, 1461
Lombardo, P., 802, 828
Long, S. C., 673, 721
Loomis, G., 827
Loper, S. W., 595
López Ramirez, J. A., 1342, 1504, 1507
Lopez, A. D., 181, 188
Lopez, A., 720
Lorenz, W., 372
Lorenzo-Lorenzo, M. J., 129
Losch, H. J., 372
Lothrop, T., 1497
Loucks, D. P., 9, 34
Love, N. G., 371
Loyer, M., 371
Lozier, J., 371
Lu, H., 524
Lubello, C., 672, 673, 718, 720
Lubin, J., 124, 126
Ludlum, R., 1003, 1039
Ludwig, A., 777, 818, 828
Luft, P., 595
Luiz, F., 371
Lumia, D. S., 68, 1040, 1166, 1201
Lumsden, L. L., 41, 68
Lund, E., 78, 125, 1348, 1351, 1377
Lutchko, J. R., 596
Lykins, B. W., Jr., 595
Lynch, A., 293
Lynch, C. F., 124, 126
Lynch, S. T., 288, 293
Lyon, S. A., 214

Maas, E. V., 971, 974, 976, 1040, 1053, 1100
MacDonnell, L. J., 945
MacGarvin, M., 34
MacIntyre, D. F., 1342
Mack, J., 597
MacKenzie, D. C., 961, 1040
MacKenzie, W. R., 88, 127
Mackun, P. J., 17, 19, 34
MacLaggan, P. M., 1395, 1397, 1406, 1432
MacLeod, M., 1166
Macler, B. A., 185, 188
Maddaus, W. O., 8, 34
Madore, M. S., 88, 127
Maeda, M., 755, 756, 759, 762, 1193, 1201, 1506
Maehlum, T., 828
Maggi, L. J., 459
Magnuson, M. L., 129
Mahar, E., 572, 596
Mahmood, T., 372
Maier, R. M., 596
Maier, R. M., 92, 127, 588
Makri, A., 236, 253
Malcolm Drew Agricultural Research Council, 960
Malcolm Pirnie, Inc., 1165

Malcolm, R. L., 128
Malley, J. P., 673, 720
Manka, J., 128
Mann, J. G., 1107, 1108, 1166
Mansfield, D. M., 1494, 1497
Mantovani, P., 13, 34, 827, 1495, 1497
Mara, D. D., 62, 67, 68, 125, 180, 188, 1039
Marcino, S. A., 287, 293
Marecos do Monte, M. H. F., 58, 66, 68, 1038
Marias, G. v. R., 371
Marinas, B. J., 597, 720, 721, 722
Marks, J., 1393, 1432
Markus, M. R., 1301
Marsalek, J., 1506
Marshall, M. M., 125
Marshall, W. E., 126
Martel, K. D., 898
Martikainen, P. J., 102, 127
Martin, J. L., 1318, 1319, 1322, 1340, 1344
Martin, N., 683, 711, 720
Martin, R. R., 1003, 1039, 1040, 1301, 1507
Martinez, I., 718
Martínez-Cortina, L., 35
Massarani, G., 459
Masschelein, W. J., 128
Matsuo, T., 371, 372
Matyac, S., 1041
Maupin, M. A., 68, 1040, 1166, 1201
Maurin, M., 1166
Maya, C., 89, 127
Mayer, P. W., 8, 34
Mayer, T. G., 827
Mazas, E. A., 128
Mcardell, C. S., 596
McBride, G., 201, 202, 227, 253
McCann, J., 211, 253
McCarty, P. L., 597, 800, 828, 1167
McCoy, J., 1126, 1128, 1136, 1166
McCutcheon, S. C., 126, 1318, 1319, 1322, 1342, 1343
McEwen, B., 1312, 1323, 1343
McFeters, G. A., 102, 128
McGarvey, F., 553, 595, 596
McGauhey, P. H., 136, 188
McGechan, M. B., 971, 1040
McGuire, M. J., 127
McIntyre, R., 1129, 1166
McKay, W. A., 129
McKee, J. E., 134, 188
McKenzie, D. C., 1040
McLaughlin, M. R., 126
McManus, K., 1451
McMurray, J., 566, 596
McQuarrie, J. P., 293
McVicker, R. R., 1101, 1202, 1498
Mead, J. R., 127
Mechalas, B. J., 251
Medema, G. J., 254, 720

Megregian, S., 718
Mekorot Water Company Ltd., 61, 1252
Melbourne Water, 1507
Melin, T., 67
Melo, M. V., 459
Merlo, R. P., 361, 372, 522
Merrill, D. T., 499, 523
Messner, M. J., 202, 253
Metcalf & Eddy, 1480
Metcalf, L., 39, 41, 43, 60, 68
Metzler, D. F., 1348–1352, 1377
Meyer, J. L., 1041
Meyer, M. T., 127, 1243, 1461
Meyer, T. E., 1166
Michail, M., 1500, 1507
Micheli, E. R., 1226, 1243
Mickley, M. C., 509, 513, 514, 518, 523
Middlebrooks, E. J., 1242
Middleton, F. M., 1377
Mierzwa, A. J., 1147, 1148, 1165
Mikkelsen, D. S., 1010, 1041
Millan, M., 1460
Millar, H., 827
Miller, D. G., 1158, 1161, 1166
Miller, G., 899
Miller, J., 1302
Miller, K. J., 1378, 1498
Milliken, J. G., 1377
Mills, R. A., 1390, 1391, 1399, 1401, 1403, 1406, 1407, 1409, 1417, 1432
Miltner, R. S., 129
Minear, R., 720
Ministry of Land, Infrastructure and Transport (MLIT), Japan, 755, 756, 762, 1178, 1202
Ministry of National Infrastructures, Israel, 1500, 1507
Mitani, H., 721
Mitch, W. A., 118, 128
Mitchell, C. A., 827
Mitsch, W. J., 1210, 1243
Mitta, P., 372
Mohr, T. K. G., 119, 128
Monroe, S. S., 125
Montgomery Consulting Engineers, Inc., J. M., 619, 721
Monto, A. S., 188
Moore, L. A., 129
Moreland, J. L., 1489, 1498
Morel-Seytoux, H. J., 1260, 1301
Morita, S., 721
Morrill, A. B., 647, 721
Morris, J. C., 611, 624, 719, 721
Morris, R. L., 1100
Morris, S., 1243
Mortensen, K., 1201
Mosher, J., 945
Mourato, S., 359, 372
Moyer, N. P., 86, 128
Mujeriego, R., 58, 67, 68, 1165, 1504, 1507, 1508
Mulder, A., 1278, 1302

Murray, C. J. L., 181, 188
Murray, D., 720
Murthy, S. N., 371
Myers, B. A., 1377

Nakada, K., 762, 1201
Nakayama, F. S., 1068, 1100, 1040, 1099
Nashikkar, V. J., 69
Nasr, J., 62, 69
Nasser, A. M., 1280, 1302
National Academy of Sciences (NAS), 211, 253, 1166, 1371, 1377
National Fire Protection Association (NFPA), 899
National Institutes of Health (NIH), 214, 253
National Pollutant Inventory (NPI), Australia, 1166
National Research Council (NRC), 80, 81, 87, 93, 94, 97, 115, 117, 128, 150, 156, 179, 188, 189, 196, 197, 199, 226, 253, 967, 1040, 1255, 1302, 1323, 1335, 1343, 1375, 1377
National Water Research Institute (NWRI) and American Water Works Association Research Foundation (AWWARF), 700, 721
National Water Research Institute (NWRI), 144, 189, 695, 700, 721, 1517, 1497, 1521
Nayyar, M. L., 910, 926, 1186, 1202
Nealy, M. K., 127, 1497
Neis, U., 673, 718
Nellor, M. H., 1301, 1490, 1497, 1521
Nelson, J. O., 34
Nelson, K. L., 80, 126, 417
Nerenberg, R., 288, 293, 590, 596
Neuman, D. S., 1100
Neuman, S. P., 1260, 1302
Newsome, S., 251
Ng, H. Y., 1343, 1507
Niang, S., 67
Nicholson, B. C., 1034, 1035, 1040
Noel, J. S., 125
Nolan, M., 1243
Noran, P. F., 898
Notarnicola, M., 720
Novak, J. T., 371
Noyes, T. I., 124
Nunez, A. A., 1494, 1497, 1498
Nurizzo, C., 67

Oberg, C., 720
O'Brien, W. J., 677, 721
O'Connell, S., 720
Odendaal, P. E., 59, 68, 1353, 1356, 1377, 1503, 1504, 1507
Oemcke, D., 1432
Office of the Federal Register, 1413, 1432
Office of Water Recycling, California State Water Resources Control Board, 1387, 1406, 1432

Ogoshi, M., 762, 1202, 1501, 1507, 1508
Oguma, K., 683, 721, 1140, 1166
Ohgaki, S., 460, 721, 1166
Okhuysen, P. O., 253
Okun, D. A., 34, 39, 41, 43, 68, 144, 189, 832, 899, 907, 926, 944, 945, 1186, 1202, 1497
Oldewage, L., 1202, 1498
Oliver, B. G., 690, 721
Olivier, M., 422, 424, 459
Olivieri, A. W., 80, 92, 125, 227, 233, 236, 239, 251, 253, 254
Oman, S. D., 1280, 1302
O'Melia, C. R., 401, 459
Ong, S. L., 1336, 1343, 1503, 1507
Ongerth, H. J., 41, 43, 45, 60, 68, 1485, 1498, 1509, 1512, 1521
Ongerth, J. E., 60, 68
Ontario Ministry of the Attorney General, Canada, 1310, 1343
Opitz, E. M., 34
Orang, M., 1041
Orange County Sanitation District (OCSD), CA, 96, 118, 128
Orange County Water District (OCWD), CA, 50, 1296, 1297, 1298, 1302
Oravitz, J. L., 595
Oregon Health Division, 87, 128
Orenco Systems Inc., 284, 776, 789, 792
Orlob, G. T., 1243
Oron, G., 979, 1500, 1507
Orta de Velasquez, M. T., 673, 721
Otis, R. J., 827
Ott, E. M., 251
Ottson, J., 251
Oueslati, F., 1506
Ouki, S. K., 554, 596
Owens, J., 721
Ozaki, M., 762, 1202, 1508

Padre Dam Municipal Water District (PDMWD) staff, CA, 932, 1238, 1240
Page, A. L., 124, 968, 969, 1039, 1040
Pagliaro, T., 1126, 1142, 1143, 1147, 1167
Palmer, C. M., 1377
Pan, G., 124
Panchapakesan, B., 1149, 1167
Pankow, J. F., 495, 496, 523
Papadopoulos, I., 1040
Paperloop, 1148, 1167
Paranjape, S., 437, 459, 478, 523
Paranychianakis, N. V., 66
Parashar, U. D., 125
Park, K., 1319, 1343
Park, R. A., 1319, 1343
Parker, D., 371, 1165
Parker, J. A., 687, 721
Parkin, G. F., 597, 1167
Parkin, R., 253, 254
Parnell, J. R., 54, 68, 1054, 1086–1092, 1100, 1453, 1461, 1486, 1497

Parsons, F. Z., 125
Parsons, J. J., 1101, 1202, 1498
Pastor, S., 720
Pataky, K., 827
Patania, N. L., 127
Patel, M. V., 1301
Patel, M., 597
Paton, C. A., 128
Paustenbach, D. J., 195, 197, 253
Pavelic, P., 1040, 1301
Payment, P., 719
PBS&J, 1439, 1447, 1461
Pearce, B., 371, 1201
Pearce, G., 474, 523
Pearce, N., 209, 253
Pearson, H., 180, 188
Peasey, A., 67, 1039
Peeters, J. E., 88, 128
Pekkanen, J., 209, 253
Pelletier, G. J., 1342
Penny, J. P., 372
Pepper, I. L., 127, 200, 253, 596, 718
Pereira, W. E., 124
Perlman, H. A., 69, 1202
Pernitsky, D., 597
Perry, J., 459
Perz, J. F., 236, 254
Pescod, M. B., 981, 985, 990, 1040
Peters, J. H., 1290, 1302
Peterson, D. E., 127
Petterson, S. R., 235, 251, 254, 1041
Pettygrove, G. S., 1008, 1039
Peyton, D. E., 1270, 1302
Pezzey, J., 7, 35
Phelps, C., 459
Phillips, P. J., 293
Pidsley, D., 1301
Pierce, R. R., 69, 1202
Pierson, G., 1165, 1166
Pineo, D., 1243
Pinkston, K. E., 566, 596
Pisigan, R. A., 1119, 1167
Pitblado, R., 253
Pitt, P., 1497
Plastic Pipe Institute (PPI), 860, 899
Pleus, R., 1378
Plumlee, M. H., 127, 1342
Plummer, J. D., 673, 721
Poff, N. L., 34
Ponimepuy, M., 720
Pontius, F. W., 509, 511, 523
Poon, J., 1343
Porco, T. C., 252
Postel, S. L., 15, 19, 35
Potts, E. A., 1486, 1498
Powelson, D. K., 1280, 1302
Pozio, E., 124
Pozos, N., 596
Praderie, M., 371
Prettyman, R., 1302
Prevost, R. J., 188
Pribil, W., 722

Prince William Conservation Alliance, VA, 1328, 1343
Process Applications, Inc., 665
Proctor, M. E., 127
Proctor, W. D., 1342
Pruitt, W. O., 990, 1039
Prüss, A., 181, 189
Ptacek, C. J., 827
Public Utilities Board Report, Singapore, 1503, 1507
Puckorius, P. R., 1125, 1167
Pyne, R. D. G., 1293, 1302

Qualls, R. G., 687, 690, 721
Quay Technologies, Ltd., 676
Queensland Government, Australia, 1377
Queensland Water Recycling Strategy, Australia, 13, 35, 1499, 1507
Quinian, E., 899
Quintero-Betancourt, W., 126
Quiroga Alonso, J. M., 1342, 1507

Radcliffe, J. C., 59, 68, 1172, 1175, 1202, 1439, 1461, 1495, 1498, 1499, 1508
Rai, R. P., 69
Rainville, D., 1077, 1100
Rajal, V. B., 1343
Rakness, K. L., 563, 596, 662, 665, 666, 721
Randall, C. W., 372
Randall, M. J., 1116, 1166
Rauschkolb, R. S., 1010, 1041
Ravetz, J., 34
Ravina, I., 1039
Reagan, K. M., 127
Reardon, R., 459, 523
Rebhun, M., 113, 117, 128, 1011, 1041
Reckhow, D. A., 113, 128, 568, 597
Recycled Water Task Force (RWTF), CA, 1393, 1433
Reddy, P. S., 595
Redman, J. A., 596
Redwood, M., 67
Reed, S. C., 1242
Rees, H. B., 1165
Refling, D. R., 372
Regli, S., 185, 188, 189, 225, 227, 235, 252, 254
Reiber, S., 1121, 1125, 1167
Reinhard, M., 127, 1306, 1342
Reiss, J., 422, 459
Rennecker, J., 640, 721
Repacholi, M. H., 34
Requa, D. A., 718
Reverter, J., 523
ReVoir, G. J., II, 359, 372
Reynolds, K. A., 718
Reyrolle, M., 1166
Rice, E., 71
Rice, R. C., 1264, 1278, 1301
Rice, R. G., 568, 596, 661, 662, 721
Rice, T. A., 945

Richard, A., 478, 523
Richardson, C. S., 1395, 1397, 1433
Richardson, S. D., 720
Richter, B. D., 34
Ried, A., 596
Rigby, M. G., 251
Riggs, J. L., 1414, 1433, 1527, 1528
Right, H. B., 723
Rijsberman, F. R., 15, 18, 34
Rinck-Pfieffer, S., 1301
Riolo, C. S., 253
Ripley, D., 760, 762, 811
Rittmann, B., 590, 596, 800, 828
Robbins, M. H., Jr., 1323, 1343
Roberts, D. R., 126
Roberts, J. M., 126
Roberts, P. V., 641, 722, 1255, 1302
Robertson, L. A., 1302
Robertson, L. J., 88, 128
Robertson, W. D., 802, 828
Robertson-Bryan Inc., 252
Rodgers-Gray, T. P., 1225, 1243
Rogers, M. F., 124
Rogers, P. P., 20, 35
Rogers, S. E., 127, 1361, 1362, 1364, 1367–1374, 1377, 1491, 1497
Rogers, S., 372
Rohwer, B., 293
Rojas-Valencia, N. M., 721
Rollins, L., 1100
Rommelmann, D. W., 1149, 1150, 1167
Rook, J. J., 113, 128, 650, 722, 1352, 1377
Rose, J. B., 87, 88, 126, 127, 128, 189, 225, 236, 245, 252, 254, 561
Rose, J., 596
Roseman, J. M., 129
Rosenblum, E., 522, 523, 1461
Rosene, M. R., 550, 596
Rosenfeldt, E. J., 567, 596
Ross, R. S., 1497
Rosson, J. J., 721
Rothberg, M. R., 1362, 1377, 1378, 1498
Rothberg, Tamburini, and Windsor, Inc., 1124, 1167
Ruiz-Palacios, G., 67, 1039
Rupert, C. S., 683, 722
Ruppe, L., 828
Russell, L. L., 1118, 1167
Ryan, T., 253
Ryder, R. A., 1167, 1168
Ryu, H., 89, 128
Ryznar, J. W., 494, 523, 1125, 1167

Sack, J., 1040
Safe Drinking Water Act (SDWA), 221, 254
Sakaji, R. H., 188, 251, 254, 438, 460, 718
Sakakura, Y., 1201
Sakamoto, G., 723

Sala, L., 58, 61, 68, 1172, 1175, 1202, 1213, 1504, 1508
Saldi-Caromile, K., 1223, 1228, 1243
Sales Márquez, D., 1342, 1507
Salvato, J. A., 777, 828
Salveson, A., 596, 718
San Diego County Water Authority, 51, 68
Sandvig, A., 1166
Sanin, F., 361, 372
Sanitation Districts of Los Angeles County (SDLAC), 44, 49, 68, 140, 189, 1302, 1488, 1490, 1498, 1514, 1521
Sanks, R. L., 499, 523
Sarin, P., 1165
Sattar, S., 124
Saunier, B. M., 627, 722
Savic, D., 67
Sawyer, C. N., 566, 597, 1118, 1167, 1228, 1243
Sayed, A. R., 68, 1507
Scanlan, P. A., 124
Scarbrough, J. H., 1309, 1343
Schaefer, K., 1506
Schaffranek, R. W., 1319, 1343
Schiff, G. M., 254
Schippers, J. C., 480, 523
Schmid, M., 828
Schmidt, I., 798, 828
Schneider, A. J., 1395, 1433
Schneider, C., 372
Schneider, M. L., 1319, 1343
Schock, M. R., 1166
Schonning, C., 251
Schott, M. J., 129
Schouwenaars, R., 597
Schroeder, C. D., 1147, 1167
Schroeder, E. D., 35, 86, 128, 254, 413, 460, 885, 899
Schuerch, P., 718
Schwartzbrod, J., 127
Sclimenti, M. J., 720
Scott, J., 252
Scutchfield, F. D., 206, 254
Seah, H., 1335, 1343, 1507
Sears, K., 597
Secrist, N. D., 124
Sedlak, D. L., 118, 124, 128, 566, 596
Selleck, R. E., 627, 640, 718, 722
Selna, M. W., 67, 1497
Selvin, S., 251, 253
Sen, D., 324, 372
Septic Protector, 787
Serra, M., 68, 1508
Serrano Association, LLC, 907, 924, 926
Seto, E. Y. W., 251, 253
Severin, B. F., 690, 722
Seyde, V., 1101, 1202, 1498
Shah, S., 1278, 1302
Shalhevet, J., 1039
Shane, B. S., 200, 251
Shapiro, M. A., 129

Sharp, J. O., 128
Sharpless, C. M., 595, 596
Shaw, A. R., 293
Shaw, D. J., 597, 1100
Sheah, H., 1343
Sheikh, B., 1016, 1018–1022, 1041, 1100, 1442, 1461, 1489, 1498
Shelef, G., 58, 68, 69, 1500, 1508
Shen, J., 1343
Shen, Y., 597
Shende, B. C., 61, 69
Sheng, Y. P., 1319, 1343
Sherman, K. M., 827
Sherwood, J. R., 254
Shiao, M. C., 1342
Shih, T., 125
Shiklomanov, I. A., 15, 35
Shin, G. A., 127, 683, 722
Shin, G., 720
Shisler, J., 722
Shuval, H. I., 185, 189
Siede, W., 719
Siegrist, H., 293, 596
Siemans Water Technologies Corporation, 337
Sikdar, S., 7, 35
Simon, C. P., 253
Simpson, J., 13, 35
Sinclair, N. A., 127
Singapore Department of Statistics, 1335, 1343
Singapore Public Utilities Board, 1336, 1337, 1339, 1343
Singer, P. C., 128, 568, 597
Singh, S. N., 126
Singh, S., 1334, 1343, 1503, 1508
Singley, J. E., 1119, 1167
Sirikenchana, K., 673, 722
Skidmore, P., 1243
Skold, R. V., 1125, 1166
Skopec, M., 127
Slater, M. J., 555, 597
Slawson, R. M., 683, 723
Sliekers, O., 828
Slifker, R., 125
Smit, J., 62, 69
Smith, C. A., 720
Smith, C. L., 126
Smith, D. W., 666, 719
Smith, D., 898
Smith, H. V., 128
Smith, P. G., 128
Smith, R. G., 983, 994, 1005–1007, 1009, 1041
Smith, R. K., 60
Smith, S., 371, 1302
Smyth, J. R., 1497
Snoeyink, V. L., 495, 496, 523, 536, 541, 543, 546, 597, 1120, 1165, 1166, 1167
Snyder, C. H., 653, 722
Snyder, E., 1378

Snyder, R. L., 991, 1041
Snyder, S. A., 117, 128, 1347, 1378
So, H. B., 1039
Sobsey, M. D., 127, 720, 722
Solarchem Environmental Systems (SES), 597
Solis, S., 524
Soller, J. A., 226, 230–233, 236, 239, 240, 249, 252, 253, 254
Solley, W. B., 46, 69, 1106, 1167, 1171, 1172, 1202
Solomon, S. L., 124
Solvay Interox, 672, 722
Sommer, R., 682, 722
Song Cha, J. H. Y., 522
Song, K. G., 522
Song, L. F., 1343, 1507
Sontheimer, H., 547, 597
Sorber, C. A., 129, 187
Soroushian, F., 584, 597
South Australian Water Corporation, 1499, 1508
Spangenberg, C., 899
Spear, R. C., 251, 253, 254, 1461
Speth, T. F., 595
Springthorpe, S., 124
Stahl, J. F., 67, 720, 722, 1497
Stahl, M. W., 1167
Standard Methods, 93, 128, 380, 384, 460, 495, 523, 597, 1165
State of California, 6, 35, 45, 47, 48, 50–53, 69, 152, 189, 264, 294, 915, 926, 950, 951, 953, 1041, 1047, 1100, 1112, 1132, 1150, 1159, 1167, 1172, 1202, 1443, 1444, 1446, 1461, 1488, 1491, 1498, 1511, 1514, 1515, 1517, 1521
State of Florida, 48, 53–58, 69, 294, 1041, 1046, 1047, 1172, 1202
Stefan, M. I., 578, 596, 597
Steinburgs, C. Z., 1075, 1100
Steinfeld, C., 787, 827
Steinman, A. D., 34
Stensel, H. D., 129, 294, 361, 372, 460, 524, 597, 722, 762, 828, 899, 945, 1041, 1100, 1343
Stenstrom, M. K., 113, 124
Stenström, T-A., 187, 251
Stephenson, T., 356, 372, 468, 523
Sterling, C. R., 127
Stevens, A. A., 113, 129
Stevens, D., 523
Stevens, L., 1239, 1243
Stevenson, D. A., 128
Stevenson, F., 1461
Stiff, H. A., Jr., 494, 523, 1125, 1167
Stiles, C. W., 68
Stinson, B. M., 293
Stirling, A., 34
Stolarik, G. F., 596
Stoltenberg, H. A., 1377
Stott, R., 67, 1039

Stowell, R., 387, 460
Straub, T. M., 82, 129
Street, R. L., 1475, 1476, 1477
Stricoff, S., 253
Stroud, T. F., 862, 899
Strous, M., 828
Stumm, W., 401, 459, 1113, 1167
Suarez, D. L., 962, 1041
Suffet, I. H., 125, 595, 596
Suidan, M. T., 722
Sullivan, J. B., 211, 254
Summers, R. S., 46, 67, 536, 541, 543, 597
Sumpter, J. P., 1243
Sung, R. D., 635, 722
Surrat, W., 523
Susarla, S., 126
Suzuki, Y., 755, 762, 1193, 1202, 1507, 1508
Svetich, R., 1302
Swan, S. H., 129
Swartout, 202, 252
Swayne, M., 1305, 1343
Sydney Water, Australia, 912, 926
Sykora, J. L., 87, 129
Sztajnbok, P., 1040, 1507

Tabucchi, T., 1460
Tajrishy, M. A., 985, 1066, 1100
Takaki, M., 762, 1506
Takakuwa, T., 1201
Takizawa, S., 62, 69
Tanaka, H., 235, 245, 254
Tang, C. C., 686, 720, 722
Tanji, K., 1003, 1041, 1049, 1070, 1100
Tao, H., 1342
Task Force on Water Reuse (TFWR), 1391, 1405, 1433
Tate, C. H., 381, 460
Taylor, J. S., 480, 491, 523, 524
Tchobanoglous, G., 66, 89, 95, 96, 99, 107, 120, 125, 129, 253, 254, 263, 278, 284, 289, 293, 294, 298, 304, 307, 310, 312, 313, 319, 323, 328, 340–342, 343–347, 361, 362, 370, 372, 377, 387, 395, 396, 400, 401, 410, 411, 413, 415, 418, 421, 422, 437, 443, 445, 459, 460, 491, 500, 522, 524, 531, 563, 566, 595, 597, 599, 602, 603, 605, 621, 638, 643, 645, 650, 708, 718, 719, 722, 737, 738, 762, 766, 771, 775, 777, 779, 780, 785, 789, 790, 802, 807, 809, 811, 814, 821, 827, 828, 885, 888–890, 899, 905, 926, 938, 945, 1041, 1058, 1100, 1201, 1214, 1218, 1220, 1242, 1243, 1316, 1319, 1322, 1342, 1343, 1480
Teltsch, B., 189
Tennessee Valley Authority (TVA), 1344
Tennison, P., 1343

Ternes, T. A., 214, 251, 293, 596
Terrey, A., 1073–1076, 1100
Terzieva, S. I., 102, 128
Test, R. M., 827
Teunis, P. F. M., 200, 226, 236, 254
Theis, C. V., 1288, 1302
Thetford Systems, Inc., 822
Thibaud, H., 651, 722
Thoeye, C., 67
Thomas, J. C., 1433
Thomas, J. M., 114, 129
Thomas, R. E., 188
Thompson, D. E., 1311, 1343
Thompson, D., 354, 355, 371, 372, 719
Thompson, K., 251, 1101, 1202, 1498
Thompson, S. A., 7, 35, 1387, 1433
Thompson, S., 720
Thruston, A. D., Jr., 720
Thurman, E. M., 1243, 1461
Till, D., 253
Tobiason, J. E., 459
Todd, D. K., 1248, 1251, 1302
Tokyo Metropolitan Government (TMG), 204, 755, 762, 1501, 1508
Toles, C. A., 126
Tonkovic, Z., 1231, 1243
Tooker, N., 438
Topham, C., 596
Toze, S., 1301
Tredoux, G., 1504, 1508
Treweek, G. P., 1109, 1153, 1167
Trojan Technologies, Inc., 676, 679, 680, 681
Trussell, R. R., 125, 128, 372, 381, 459, 460, 522, 595, 719, 1122, 1127, 1167, 1201, 1329, 1330, 1343
Trussell, R. S., 331, 340, 342, 353, 354, 372, 522
Trynoski, J., 595
Tsagarakis, K. P., 66
Tsuchihashi, R., 293
Tubor, C. F., 124
Turner, C., 524
Tyl, M. B., 827
Tyler, C. R., 1243

U.S. Agency for International Development (U.S. AID), 69
U.S. Army Corps of Engineers, 1373, 1378
U.S. Bureau of Reclamation (USBR), 10, 466, 506–508, 524, 1001, 1003, 1041, 1414, 1433
U.S. Census Bureau, 16, 35
U.S. Department of Agriculture (USDA), 955, 957, 958, 984, 990, 991, 1001, 1003, 1008, 1041, 1042, 1069, 1100, 1101
U.S. Department of Commerce, 1150, 1167
U.S. Department of Health, Education, and Welfare, 155, 189
U.S. Department of the Interior, 23, 35
U.S. Environmental Protection Agency (U.S. EPA) and U.S. Agency for International Development (U.S. AID), 43, 69, 80, 157, 158, 160, 161, 164, 169, 177, 179, 184, 187, 189, 1147, 1153, 1313, 1344
U.S. Environmental Protection Agency (U.S. EPA), 46, 69, 119, 129, 148, 150, 155, 165, 180, 184, 194, 216, 226, 243, 254, 263, 294, 563, 569, 597, 619, 621, 623, 626, 629, 647, 650, 652, 661–663, 666, 670, 672, 673, 691, 695, 722, 767, 798, 803, 807, 809, 811, 821, 824, 828, 883, 899, 906, 910, 911, 926, 934, 945, 950, 953–955, 968, 993, 1008, 1013, 1042, 1044, 1045, 1050, 1051, 1101, 1112, 1146, 1147, 1149–1151, 1153, 1154, 1168, 1171, 1177, 1179, 1188, 1194, 1202, 1209, 1213, 1214, 1232, 1243, 1393, 1395, 1425, 1433, 1446, 1447, 1461
U.S. Geological Survey (USGS), 20, 21, 35, 77, 116, 129
U.S. Public Health Service (U.S. PHS), 789, 828
U.S. Water Resources Council, 1106
Umphres, M. B., 1447, 1461
Ungar, B. L. P., 88, 125
United Nations Environment Programme (UNEP) and Global Environment Centre Foundation (GEC), 59, 69
United Nations, 15, 16, 18, 35
University of California and State of California, 992, 1054, 1059, 1061, 1041, 1100
University of California Committee of Consultants, 956, 1041
Upper Occoquan Sewage Authority (UOSA), 1324, 1325, 1329, 1344
Urasse, T., 439, 460
Urban, M., 720
Urbansky, E. T., 120, 129

VA Tech Wabag GmbH, 1357
Vaccaro, G., 1502, 1508
Valentine, R. L., 128
Val-Matic Valve and Manufacturing Corp., 865
van Aerle, R., 1225, 1243
van Asselt, M. M. B. A., 34
van de Graaf, A. A., 1302
Van de Roest, H. F., 331, 332, 339, 372
van der Hoek, J. P., 595
Van der Merwe, B., 123, 126, 1353, 1377, 1503, 1507
van der Westhuizen, J. L. J., 68, 1377, 1507
van Deventer, H. C., 1165
van Franqué, O., 1166
Van Haute, A. A., 1140, 1165
van Ierland, E. C., 9, 34
Van Olphen, M., 126, 1301
Van Schilfgaarde, J., 1001, 1042
Van Veenhuizen, R., 67
Vandenesch, F., 1166
Vanderzalm, J. L., 1030, 1033, 1042
Vaz, S. G., 34
Vecchia, P., 34
Veerasubramanian, P., 719
Vemulakonda, S. R., 1342
Vennard, J. K., 1475, 1476, 1477
Verdouw, J., 480, 523
Vernon, W., 522
Vesilind, P. A., 361, 372
Vickers, A., 8, 35, 1202
Videla, H. A., 1127, 1168
Vigneswaran, S., 371
Vik, E. A., 1113, 1168
Villacorta-Martinez De Maturana, L., 88, 128, 129
Villate, J. T., 125
Virgadamo, O., 363
Virginia Tech, 1328, 1344
Visvanathan, C., 62, 68
von Gottberg, A., 1502, 1508
von Gunten, U., 565, 595, 596

Wadsworth, L., 54, 56, 70, 1024, 1042
Wagenknecht, L. E., 90, 129
Wagner, I., 1167, 1168
Wagner, M., 354, 355, 372, 719, 723
Wagner, N. J., 596
Waldock, M. J., 1243
Walesh, S. G., 1447, 1448, 1461
Walker, G. C., 719
Walker-Coleman, L., 1486, 1498
Waller, K., 114, 129
Wallis-Lage, C., 334, 353, 356, 372
Walton, G., 1377
Walton, J., 515, 524
Ward, R. L., 254
Ward, R. W., 671, 723
Washington State Water Reuse Workgroup (WSWRW), 857, 899
Wasserman, L., 371
Wassermann, K. L., 1255, 1300
Waste Water Systems, 984
Water 3 Engineering, Inc, 1153, 1155, 1168
Water Conserv II, 1023, 1024, 1026, 1042
Water Environment Federation (WEF) and American Water Works Association (AWWA), 1311, 1323, 1325, 1344
Water Environment Federation (WEF), 46, 70, 263, 294, 329, 372, 669, 723, 915, 926
Water Pollution Control Federation (WPCF), 944, 945, 1110, 1150, 1168
Water Resources Institute (WRI), 15, 35
WateReuse Association, 47

Waterways Experiment Station (WES), 1319, 1344
Watkins, J. B., III, 214, 253
Watson, H. E., 610, 723
Watson, J. C., 126
Wattie, E., 718, 723
Weand, B. L., 522
Webber, N. D., 1475, 1476, 1477
Weber, B., 953, 1011, 1042
Wegner-Gwidt, J., 32, 35, 1442, 1443–1447, 1461
Wehner, M. P., 188, 597
Weil, I., 719
Weill, N., 1166
Weldon, D., 125
Wellman, M. C., 1343
Wenzel, L. K., 1525
Wert, S., 827
West Basin Municipal Water District (WBMWD), CA, 1158, 1159, 1160, 1161, 1168
West, T. M., 1414, 1433, 1527, 1528
West, W., 1183, 1202
Westcot, D. W., 956, 963, 966, 967, 973, 989, 1010, 1039, 1042
Westerhoff, P., 124, 128, 565, 597
Western Consortium for Public Health (WCPH), 109, 112, 129, 472, 524, 1498
Westrell, T., 251
Whalen, T., 596
White, G. C., 563, 597, 623, 626, 628, 640, 641, 657, 659, 661, 669, 671, 723
Whitley, R., 718
Widstrand, M., 19, 34
Wienberg, H. S., 720

Wiesner, M., 480, 491, 524
Wijesinghe, B., 1132, 1168
Wilderer, P. A., 7, 35
Wilf, M., 485, 523, 524
Wilhelms, S. C., 1343
Wilkes, A., 861
Willetts, J., 827
Williams, C. R., 1220, 1243
Williams, G., 417
Wilson, L. G., 1278, 1302
Wilson, S. R., 17, 19, 34
Wilson, S., 523
Winneberger, J. H. T., 790, 828
Winters, H., 523
Wintgens, T., 67
Wobma, P., 589, 597
Wolf, H. W., 134, 188
Wolfe, G. W., 1377
Wolfe, N. L., 126
Wong, J., 472, 474, 475, 524
Wood, W. L., 718
Woods, C., 1278, 1302
Woodward, R. L., 1377
Wool, T. A., 1319, 1344
Work, S. W., 1361, 1362, 1364, 1377, 1378, 1491, 1498
World Commission on Environment and Development (WCED), 7, 33, 35
World Health Organization (WHO), 19, 35, 59, 70, 179–184, 187, 189, 190, 194, 205, 206, 225, 254, 980, 1042
Wright, N., 372
Wright, S. A., 1243
Writer, J. H., 124
Wu, H. W., 589, 597
Wu, L., 1100
Wuertz, S., 86, 128, 1343

Wurbs, R. A., 1319, 1344
www.ecoseeds.org, 822
Wynne, B., 34

Xie, Y. F. F., 589, 597

Yahya, M. T., 1302
Yamagata, H., 755, 756, 762, 1193, 1202, 1501, 1508
Yamamoto, K., 330, 371, 372, 460
Yamasaki, E., 253
Yanez-Noguez, I., 721
Yari, P. F., 1309, 1344
Yates, L. I., 1343
Yates, M. V., 80, 129, 1280, 1302
Yoder, J. S., 85, 124, 129
Yoon, Y., 128
York, D. W., 54, 56, 70, 1024, 1042, 1486, 1498
Young, C. E., 254
Young, H. W., 54, 70
Young, R. A., 1411, 1433
Young, R. E., 1065, 1101, 1176, 1487, 1202
Young, V. J., 1460

Zacheis, A., 459
Zakikhani, N., 1343
Zamora, R., 533, 597
Zaugg, S. D., 127, 1243, 1461
Zavoda, M., 751, 753, 754, 762, 1191, 1202
Zenker, M. J., 119, 129
Zenon Membrane Solutions, 335
Zhu, T., 802, 828
Ziarkowski, S., 596
Zimmer, J. L., 683, 723

Subject Index

Because a number of the subjects covered in this text can be referenced (i.e., indexed) under different alphabetical listings, it has been necessary to develop an approach to limit the degree of duplication, yet not affect the utility of the index. The approach used is as follows. Each subject with multiple subentries is indexed in detail under one letter of the alphabet. Where the same subject is indexed under another letter of the alphabet, inclusive page numbers are given and a *See also* citation is given to the location where the subject is indexed in detail. Where an abbreviation is used for a term that is spelled out (e.g., BOD for biochemical oxygen demand) in the same letter of the alphabet, the following convention is used. If the subject entry contains multiple subentries, the abbreviated term BOD is followed by (See Biochemical oxygen demand). If the term that is spelled is followed by a single page listing (Aggressiveness Index, 1118–1119) then the abbreviated term AI is followed by [(Aggressiveness Index), 1118–1119]. To access the number of data tables in the textbook more easily, an index entry followed by the capital letter T in parenthesis [e.g., (T)] is used to denote a data table related to the subject matter. Further, because the are so many individual locations listed which relate to water reclamation and reuse they are all listed together under the heading *Locations*. Similarly, case studies are listed together under the heading *Case studies*.

Abbreviations:
    for SI units, 1467–1468
    for U.S. customary units, 1468–1469
Abiotic reactions, 74, 106
Absorbance:
    definition, 580, 685
    of UV radiation by DNA, 675
    relationship to transmittance, 685
    typical values for wastewater (T), 685
Absorptivity, 580–581, 684–685
Acceptable risk (*See* Tolerable risk)
Acceptance of water reclamation and reuse, 31
Accumulation of dissolved constituents, 1129–1131
Acid generation, 628
Activated alumina, 535–536
Activated carbon adsorption, 534–535, 543–550
    activated sludge with PAC addition, 545–546
    breakthrough time estimation, 549
    expanded bed GAC, 545
    fixed bed GAC column, 545
    GAC contactor design values, 547
    illustration, 547
    mixed PAC contactor with gravity separation, 546
    mixed PAC contactor with membrane separation, 546
    series *vs.* parallel operation, 544
    sizing, 543, 544, 546, 548
    at Upper Occoquan Sewage Authority, 1323–1324
Activated carbon dechlorination, 659–660

Activated sludge processes, 385–386
    advantages and limitations of for BOD removal and nitrification (T), 312
    analysis and design of:
        for biosolids production, 345–347
        kinetic coefficients, 344, 345
        kinetic equations, 342–344
    description of (T), 309–310
    floc strength, 386
    microorganisms, 386
    particle size distribution, 385, 386
    reliability evaluation, 316–318
    total suspended solids and turbidity, 385, 386
Activated sludge with fixed film packing, 324, 325
Activated sludge with PAC addition, 545–546
Adenoviruses, 92
Adsorption, 532–551
    activated alumina, 535–536
    activated carbon, 534–535
    activated carbon contactors, 534–535, 543–550
    bench scale tests, 550–551
    capacity, 541
    carbon regeneration, 551
    Freudlich isotherms, 537, 538, 540, 541
    granular ferric hydroxide, 536
    isotherms, 536–541
    Langmuir isotherms, 537–540
    mass transfer zone, 541–543
    for metals removal, 533
    of mixtures, 540–541
    process limitations, 551
    for trace organics removal, 532–533

Adsorption (*Cont.*):
    types of, 533–536
Advanced oxidation processes (AOPs), 569–578
    applications for, 568–569
    considerations, 574–576
    and degree of degradation, 568
    for disinfection, 569
    hydrogen peroxide/UV, 570, 572–574
    NMDA removal, 576
    for oxidation of refractory organic compounds, 568–569
    ozone/hydrogen peroxide, 570, 572
    ozone/UV, 571–572
    process limitations, 577–578
    processes, 569–575
Advanced wastewater treatment (AWT):
    application of, 108–112, 1494
    definition, 74
    flow diagrams, 530
    technologies for (T), 464, 529, 531
Aeration:
    alpha values in activated sludge processes, 354–355
    following NF and RO treatment, 499
    for membrane bioreactors, 354–355
    of reclaimed water, 891
    of reservoirs, 1320
    with Speece cone, 891, 1320
Aerosols:
    from cooling towers, 1138
    limiting exposure, 148
    pathogen survival, 147, 148
    setback distances, 148–149, 1517
    from spray irrigation systems, 980

Subject Index

Aesthetic issues:
  with ponds and lakes, 1228–1229
  with reclaimed water, 886, 1050, 1188–1190, 1225, 1228
  for water quality, 1050
Affordability, 940–941
Aggregate organic parameters, 103, 115, 378
Aggressiveness Index (AI), 1118–1119
Agricultural irrigation, 149–150, 931, 932, 947–1035
  agronomics and water quality considerations (See Agronomics and water quality considerations)
  crop contamination, 149
  crop processing, 149
  crop selection in, 971
  design elements for irrigation systems (See Agricultural irrigation systems design)
  globally, 952–953
  level of treatment, 150
  Monterey Water Reclamation Study for Agriculture, 1015–1022
  nutrients in, 968–971
  operation and maintenance of irrigation systems (See Agricultural irrigation systems operation and maintenance)
  pathogen concerns, 149
  regulations and guidelines, 953–955
  salinity in, 959, 960
  sodicity in, 959, 961–965
  terminology, 948–949
  trace constituents, 149–150
  in U.S., 950
  Virginia (South Australia) pipeline, 1028–1035
  Water Conserv II, 1022–1027
Agricultural irrigation systems design, 971, 977–1008
  drainage systems, 998–1003
  drainage water management and disposal, 1003
  estimation of water application rate, 989–997
    evapotranspiration, 990, 991
    hydraulic loading rate, 993–995
    hydraulic loading rate from nitrogen loading limits, 995–997
    net irrigation requirement, 992, 993
  field area requirements, 997–998
  irrigation scheduling, 1008
  leaching requirements, 986–989
  selection of type of irrigation system (See Agricultural irrigation systems selection)
  storage system, 1003–1008
  water quantity and quality, 977, 978

Agricultural irrigation systems operation and maintenance, 1008–1015
  demand-supply management, 1009
  monitoring requirements, 1014–1015
  nutrient management, 1009–1010
  public health protection, 1011–1013
  soils and crops, effects of reclaimed water irrigation on, 1011, 1014
Agricultural irrigation systems selection, 977–986
  clogging prevention, 984–986
  efficiency, 981, 982–984
  gravity subsurface flow systems, 977
  gravity surface flow systems, 977
  pressurized subsurface application systems, 979
  pressurized surface application systems, 977–979
  public health protection, 980
  types of systems, 977–984
Agriculture:
  WHO water reuse guidelines, 180, 182–184
  U.S. EPA water reuse guidelines, 171–173
Agronomics and water quality, 954–976
  crop selection, 971–976
  Monterey Water Reclamation Study for Agriculture case study, 1017–1022
  nitrogen, 970
  nutrients, 968–971
  phosphorous, 970–971
  salinity, 959, 960
  sodicity, 959, 961–965
  soil characteristics, 955, 957–958
  specific ion toxicity, 965–966
  suspended solids, 958–959
  trace elements, 966–968
  trace elements and nutrients, 966–971
AI (Aggressiveness Index), 1118–1119
Air, composition and properties, 1471–1473
Air conditioning, 1179–1186
  description of systems, 1179–1180
  management issues, 1183, 1186
  reclaimed water utilized for, 1181
  system description, 1179–1180
  water quality considerations, 1181–1185
Air entrainment, clogging due to, 1286
Air release valves, 864
Air supply (in MBR), 354–355
Algae:
  concentration in pond effluent, 387
  control of in golf course irrigation, 1075
  control of in open reservoirs, 886
  effect on emitter clogging, 1066, 1068
  impact of blooms, 1264
  removal by flotation, 451–452

Alkalinity, impact or importance in:
  advanced oxidation processes, 577
  buffer capacity, 1119
  chlorine disinfection, 628
  cooling towers, 1134
  corrosion, 1113
  dissolved air flotation, 453
  distribution and storage, 885–886
  expressed as calcium carbonate, 800
  Langelier saturation index, 495, 1124
  nitrogen removal, 797–798
  sodium adsorption ratio, 963
Alpha (α) factor in aeration, 354–355
Alternative cost valuation, 1413
Altitude valves for distribution systems, 878, 880
Ammonia:
  chlorine reactions with, 625–626
  in cooling water systems, 1134
Anaerobic ammonia oxidation (ANAMMOX), 1277–1279
  microorganisms for, 1279
  reaction, generalized, 1277
Analysis, statistical (See Statistical analysis)
ANAMMOX (See Anaerobic ammonia oxidation)
Animal health effects testing, 1371–1372
Animal viewing parks, 1231
Anthracite, for depth filtration:
  properties of, 412
  sizes used in depth filters, 410–411
  typical particle size distribution, 409
AOPs (See Advanced oxidation processes)
A/O process, 324–326
A^2/O process, 325–326
Appurtenances:
  in distribution system, 863–866
    air release valves, 864
    backflow preventers, 863, 864
    blowoffs, 864
    hydrants, 864, 866
    reclaimed water service requirements, 863
  reservoir, 880–881
Aquaculture, WHO guidelines for, 180, 182–184
Aquatic habitat, 1223–1224
Aquifer characterization, 154
Aquifer recharge, 154
Aquifer storage and recovery (ASR):
  application in Virginia Pipeline Scheme, 1028–1035
  definition, 1250
  description, 1293–1294
Arithmetic probability distribution, 1479–1483 (See also Statistical analysis)

Subject Index

Ascaris lumbricoides, 89
ASR (*See* Aquifer storage and recovery)
Atlantis Water Resource Management Scheme, 1504
Atmospheric pressure, elevation changes in, 1472–1473
Attached growth processes, description of:
   for BOD removal and nitrification, 308, 313, 314
   for nitrification and denitrification, 321–323
Attitudes about water reuse, 1436–1440
AWT (*See* Advanced wastewater treatment)

BAC (*See* Biological activated carbon)
Backflow prevention:
   devices for (T), 910, 911
   in distribution system, 863, 864
   in dual plumbing systems design, 909–911
Bacteria, 83, 86–87
   Ascaris lumbricoides, 89
   Campylobacter jejuni, 87
   coliform limits, 164
   and emitter clogging, 1066, 1068
   Escherichia coli, 86–87
   Salmonella, 86
   Schistosoma mansoni, 89
   Shigella, 86
   Yersinia enterocolitica, 87
Bacteriophages, 93–94 (*See also* Coliphage)
Ballasts for UV lamps, 678
Base flow of a stream (*See* Stream flow augmentation)
Beers-Lambert Law, 684
Beliefs about water reuse, 1440
Bench-scale studies,
   for adsorption, 550–551
   for dissolved air flotation, 453–454
   for ion exchange, 553
   for ozone, 565
   for UV irradiation, 691–701
Beta-Poisson model, 201–202
Bicarbonate:
   advanced oxidation processes impacted by, 577
   in cooling water systems, 1134
Bioassay testing, for determination of UV dose, 695–697
Bioaugmentation, 588
Biochemical oxygen demand (BOD):
   biological processes used for removal of, 310–311, 313
   typical values in raw and treated wastewater (T), 107, 109–111
   variability in:
      activated sludge process effluent, 315–318
      untreated wastewater, 302–304

Biochemical oxygen demand (BOD) (*Cont.*):
   wastewater treatment requirements, 143–144, 264
Biofilm:
   clogging of injection and recharge wells, 1257, 1291
   definition, 1104
   impacts on corrosion, 1115
   impacts on growth of *legionella,* 1138–1140
   impacts on UV disinfection, 709–710
   in attached growth treatment processes, 308, 321, 324
   in cooling water systems, 1141
   in drip irrigation systems, 1055
   in pipelines and distribution systems, 883, 885–887
   management, 1188
   membrane fouling, 487–489
   stimulation of, 152
   use in membrane biofilm reactors, 288, 590–591
Biofor process, 321, 323
Biological activated carbon (BAC), 589–590
Biological clogging, 1286
Biological fouling, 489, 1141
Biological phosphorus removal, 324–326
Biological solids production (MBR analysis and design), 345–351
   estimation based on published data, 345–346
   estimation using kinetic coefficients, 346–351
Biological transformations, advanced, 586–591
   and bioaugmentation, 588
   biological activated carbon, 589–590
   and biostimulation, 588
   and energetics of constituent degradation, 588
   membrane biofilm reactor, 590
   process limitations, 590–591
   processes, 588–590
Biological water quality, 886
Bioreactor suspended solids concentration, 353
Biosolids processing, 361
Biosolids production and management, 358
Biostimulation, 588
Biostyr process, 321, 323
Biotic reactions, 106
Blending, 492, 494
Blowdown, 1135
Blowoffs for distribution systems, 864
BOD (*See* Biochemical oxygen demand)

Boilers, 1141–1147
   constituents of concern in, 1144–1145
   feedwater for, 737, 1144–1145
   in Santa Rosa, California, 1147
   system description, 1142–1143
   types of, 1142
   unique application of, 1143, 1145–1147
   water quality considerations for, 1143, 1145–1147
Boron tolerance, 966, 974
Brackish water intrusion control, 1248
Breakpoint chlorination chemistry, 626–628
Breakthrough time estimation (of activated-carbon adsorption), 549
Brine (*See also* Membrane waste stream management):
   definition, 462
   from distillation processes, 561–563
   from ion exchange, 560
   from water softeners, 779–780
   surface discharge, 517
   subsurface discharge, 517
   thickening, 511–515
Buffer capacity, 1119
Buildings, reclaimed water used in, 736
Bulk organic transformations, 1273–1275
Byproducts:
   of advanced oxidation processes, 577
   disinfection (*See* Disinfection byproducts)

Calcium:
   in cooling water systems, 1134
   and emitter clogging, 1066
Calcium carbonate precipitation potential (CCPP), 1120–1124
Calcium hypochlorite, 623–624
Caliciviruses, 90, 91
California:
   regulations, 1509–1520
      groundwater recharge, 1517–1520
      indirect potable reuse, 167
      nonpotable water reuse, 1509–1517
      treatment facility reliability, 165
      water reuse and recycling, 51, 52
   type and quantity of water reuse in, 48
   water quality ranges for reuse applications in, 264
   Water Recycling 2030, 1495
Campylobacter jejuni, 87
Capital recovery factor (*See* Economic analysis)
Captor process, 324, 325
Car washing with reclaimed water, 1195–1196
Carbon regeneration, 551
Carbonate:
   advanced oxidation processes impacted by, 577
   in cooling water systems, 1134

Carcinogens and carcinogenicity, 1371–1372
  EPA's qualitative assessment of, 222–223
  identification of compounds likely to be, 221
  risk extrapolation for, 221
  whole animal tests for, 213, 214
Case studies:
  Agricultural irrigation in Monterey, CA, 1015–1022
  Agricultural irrigation at Water Conserv II, FL, 1022–1027
  Agricultural irrigation in South Australia, 1028–1035
  Cooling tower water at Denver, CO, 1155–1158
  Direct potable reuse demonstration project in Denver, CO, 1361–1375
  Direct potable reuse in Chanute, KS, 1348–1352
  Direct potable reuse in Windhoek, Namibia, 1352–1361
  Dual plumbing at Irvine Ranch Water District, CA 915–919
  Dual plumbing at Rouse Hill, Australia, 919–921
  Dual plumbing at Serrano, CA 921–925
  Environmental and recreational uses at Arcata, CA, 1231–1234
  Environmental and recreational uses at San Luis Obispo, CA, 1234–1238
  Environmental and recreational uses at Santee Lakes, CA, 1238–1241
  Groundwater Replenishment at Orange County Water District, CA, 1296–1298
  Indirect potable reuse in Singapore, 1334–1340
  Indirect potable reuse at Upper Occoquan Sewage Authority, VA, 1323–1329
  Industrial uses at West Basin Municipal Water District, CA, 1158–1161
  Institutional arrangements at Walnut Valley Water District, CA, 1397–1399
  Landscape irrigation in Redwood City, CA, 1450–1453
  Landscape irrigation in St. Petersburg, FL, 1086–1093
  Residential landscape irrigation in El Dorado Hills (Serrano), CA, 1093–1097
  Satellite system for the Solaire Building, NY, 751–754
  Satellite system for Tokyo, Japan, 755–759
  Satellite system for the Upland, CA, 760–761
  Water reclamation and reuse in St. Petersburg, FL, 1453–1459

Case studies (Cont.):
  Water repurification project at San Diego, CA, 1329–1334
  Water reuse in California, 47–53
  Water reuse in Florida, 53–58
Cation exchange capacity (CEC), 965
CCPP (Calcium carbonate precipitation potential), 1120–1124
CDI (See Chronic daily intake)
CEC (cation exchange capacity), 965
Centralized (conventional) gravity flow collection system, 728–729, 807, 808
Centralized treatment facilities:
  advantages and disadvantages of (T), 285
  site location for, 282, 283, 285–286
Centrifugal pumps, 867, 868
Check valves for distribution systems, 871
Chemical addition:
  in dissolved air flotation, 453
  NF and RO design/operational considerations, 492, 494
  phosphorus removal by, 327–329
Chemical cleaning, 490
Chemical concentration, as disinfection performance factor, 610–612
Chemical conditioning, 453
Chemical constituents in wastewater, 103–116, 140, 142
  after AWT, 108–112
  after primary treatment, 108–111
  after secondary treatment, 108–111
  after tertiary treatment, 108–111
  in cooling water systems, 1135
  DBP formation, 113–114
  domestic, commercial and industrial additions, 104–107
    in collection systems, 106
    composition of untreated wastewater, 106–107
    minerals, 104–107
    odor/corrosion control additions, 106
    from stormwater, 106
  dose-response assessment, 198, 199
  hazard identification, 198
  impact of constituents remaining after treatment, 113
  and regulations/guidelines, 140, 142
  removal of trace constituents, 113
  surrogate parameters, 115–116
  and treated vs. natural water, 114, 116
  in treated wastewater, 108–113
  in untreated wastewater, 103–104
  and UV disinfection, 686–687
Chemical disinfectants:
  characteristics of an ideal disinfectant (T), 603
  comparison of (T), 605
  properties of:
    chlorine, chlorine dioxide, and sulfur dioxide (T), 623
    ozone (T), 661
  peracetic acid (T), 672

Chemical oxidation, 563–567
  applications for, 563, 564
  considerations for, 566
  oxidants used in, 563–566
  process limitations of, 567
Chemical risk assessment, 215–225
  nonthreshold toxicants, 220–224
    carcinogens, 221
    enforceable national drinking water standards, 224
    three-category approach for setting MCLGs, 223–224
    U.S. drinking water regulations, 221–222
    U.S. EPA's qualitative assessment of carcinogens, 222–223
  and regulations, 215–220
    incremental life risk, 215–217
    noncarcinogenic effects, 218–220
  risk considerations, 224–225
Chemical treatment:
  NF/RO considerations for, 490
  in UOSA case study, 1325
Chemical water quality, 885–886
Chick's Law, for disinfection, 609
Chick-Watson disinfection model
  application, 611–612
  definition, 611
Chloramines:
  byproducts (T), 651
  formation of, 625–627
  germicidal effectiveness, 622–623
Chloride:
  buildup due to chlorination, 628
  buildup due to evaporation, 1129
  effect on plants (T), 1090
  guidelines for irrigation (T), 956
  increase due to domestic usage (T), 105, 780, 781
  increase due to water softening, 781–784
  typical values in wastewater (T), 107
Chlorine:
  characteristics of, 622–623
  dechlorination of, 660
Chlorine contact basins, 644–650
  analysis of tracer response curves, 645–650
  conduct of tracer tests, 644–645
  hydraulic analysis (T), 647
  tracer used, 644
  volumetric efficiency, 647
Chlorine dioxide disinfection, 654–660
  advantages and disadvantages of, 620
  byproduct formation/control, 656–657
  characteristics, 655
  chemistry, 655
  dosage requirements, 656
  effectiveness, 655–656
  environmental impacts, 657
  germicidal efficiency of, 655–656
  modeling, 656

Chlorine disinfection, 622–654
  advantages and disadvantages of, 620
  breakpoint reaction with, 626–630
    acid generation, 628
    chemistry, breakpoint chlorination, 626–628
    TDS buildup, 628–630
  byproduct formation/control, 650–654
  characteristics of chlorine compounds, 622–624
    calcium hypochlorite, 623–624
    chlorine, 622–623
    sodium hypochlorite, 623
  chemical characteristics of reclaimed water, 635–639
    contact time, 638–639
    microorganism characteristics, 637–638
    particles found in reclaimed water, impact of, 636–637
  chemistry of compounds, 624–626
    chlorine reactions in water, 624–625
    chlorine reactions with ammonia, 625–626
    hypochlorite reactions in water, 625
  contact basins, 644–650
  DBPs, discharge of, 654
  dosage requirements, 641–643
  effluent characteristics of, 635–639
  environmental impacts, 654
  factors affecting efficiency of, 633–635
  germicidal efficiency, 631
  hydraulic performance assessment of chlorine contact basins, 644–650
    analysis of tracer response curves, 645–650
    conduct of tracer tests, 644–645
    types of tracers, 644
  models of, 639–641
    Collins-Selleck, 640
    effluent from membrane processes, 641
    refined Collins-Selleck, 640, 641
  physical facilities for, 606–609
  process variable measurement/reporting, 631
  regrowth of microorganisms, 654
  relative germicidal effectiveness, 632–633
  temperature effect in clean water, 632
Chlorine reactions:
  with ammonia, 625–626
  in water, 624–625
Chlorine residual, measurement of, 631
Chronic daily intake (CDI):
  application, 217
  definition, 215
Clean-in-place frequency, 444–445
Clean Water Act (CWA), 45–46, 155–156
Clean water infiltration rates, estimation of, 1268
Cleaning solutions, 511

Clogged layer, infiltration through in surface spreading basins, 1261–1262
Clogging:
  Biological in vadose zone, 1286
  in direct injection wells, 1290–1291
  due to air entrainment, 1286
  due to solids accumulation, 1286
  in irrigation system, 984–986
  in membranes, 355, 445
Clogging prevention, 984–986
Closed channel disinfection systems, 680–682
Cloth-media filter (CMF), 420–422, 424–425
Cluster systems, 1082–1086
CMF (*See* Cloth-media filter)
Codes, in dual plumbing systems design, 908
Coefficient, drainage, 1000
Coefficient of specific lethality:
  application, 611–612
  definition, 610
  effect of temperature on, 613
Cold lime treatment, 1028
Cold weather discharge permits, 44
Coliform bacteria:
  coliform bacterial limits, 164
  fecal coliform, 93
  E. coli, 86–87
  survival in the environment (T), 102
Coliphage:
  description, 93–94
  inactivation by chlorine, comparison to polio virus, 637
  occurrence in wastewater, 97–98
  removal during treatment, 408
  removal in groundwater recharge, 1280
  used as indicator organism, 96, 144
Collimated beam:
  application, 694–695, 697–700
  determination of UV dose with, 691–693, 701
  reporting collimated beam results, 701–703
  schematic, 692
Collection systems, wastewater:
  comparison of collection systems (T), 813–814
  constituents formed as result of abiotic and biotic reactions, 106
  satellite treatment systems, 738–739
  siting considerations for, 734
  types of:
    conventional gravity flow, 728–729, 807, 808
    hybrid, 811, 813
    grinder pump pressure, 808
    septic tank effluent pump (STEP) pressure, 789, 808, 810–811
    septic tank effluent gravity (STEG), 807–808

Collection systems, wastewater, types of (*Cont.*):
  small diameter variable grade gravity, 807–808
  vacuum, 810, 812
Collins-Selleck model, 640
Colloidal particles:
  concentration in untreated and reclaimed waters (T), 110
  impact on membranes, 445, 477, 481
  size range of, 375–379
Colorado:
  cooling tower at thermal power generation plant case study, 1155–1158
  Denver Potable Water Demonstration, 1491–1492
  direct potable reuse of reclaimed water case study, 1361–1375
Combined chemical disinfection processes, 672–674
Commercial applications:
  car and vehicle washing, 1195–1196
  laundries, 1196
Commercial buildings:
  dual plumbing systems in, 903–906
  Irvine Ranch Water District case study, 916, 918
Commercial chemical constituents, 104–107
Commercial water use in U.S., 1172
Compliance point, monitoring of, 145–146
Compound interest, 1525–1528 (*See also* Economic analysis)
Computer simulation models (T), 1321
Concentration, cycles of, 1136–1138
Conductivity, hydraulic (*See* hydraulic conductivity)
Conductivity, of water, electrical:
  definition of, 300
  guidelines for irrigation water, 956, 960–963, (T) 975
  relationship to ionic strength, 1118
  relationship to TDS (salinity), 959
Connections, distribution system, 860–861
  flanged joints, 861
  mechanical joints, 860
  pipe joint restraints, 861
  push-on joints, 860
  welded joints, 861
Conservation of mass, 1319
Constant speed operation:
  in pumping systems, 870–871
  variable speed operation *vs.*, 870–871
Constituent concentration values, 299, 301
Constituent concentration variability, 302
Constituent concentrations in decentralized systems, 778–785
  domestic greywater, 780–785
  household products, 779–780

Subject Index

Constituents, wastewater, 139–142, 299, 300, 525–591
  adsorption, 532–551
  biological transformations, advanced, 586–591
  chemical, 140, 142
  chemical oxidation, 563–567
  distillation, 560–563
  ion exchange, 551–560
  microbial, 139–141
  oxidation, advanced, 567–578
  photolysis, 578–586
  technologies used, 528–532
  untreated municipal wastewater, 260, 261, (T) 261–262
Constructed wetlands, 1212, 1213, 1216–1218
Construction, EPA water reuse guidelines regarding, 174
Construction financing plans, 1422–1423
Construction materials for reservoirs, 881, 882
Contact basins, chlorine (*See* Chlorine contact basins)
Contact time, 609–610
  chlorine disinfection, 638–639
  as disinfection performance factor, 609–610
Contaminants, emerging (municipal wastewater), 117–120
  1, 4-dioxane, 118–119
  endocrine disruptors and pharmaceutically active chemicals, 117
  methyl tertiary-butyl ether and other oxygenates, 120
  n-nitrosodimethylamine, 118
  new and reemerging microorganisms, 120
  perchlorate, 119–120
Control measures, EPA water reuse guidelines for, 178
Conversion factors, 1463–1467
  conversion factors between SI and U.S. customary units (T), 1463–1465
  for environmental engineering computations (T), 1466–1467
Conveyance and distribution system:
  for public areas landscape irrigation, 1079
  for residential landscape irrigation, 1081, 1082
Cooling water systems, 1132–1141
  blowdown, 1135
  cycles of concentration, 1136–1138
  description, 1132, 1133
  design and operational considerations, 1135–1138
  management issues, 1138–1141
    biological fouling, 1141
    corrosion, 1140
    public health protection, 1138–1140
    scaling, 1141

Cooling water systems (*Cont.*):
  system description, 1132
  water makeup in, 737
  water quality considerations, 1132–1135
    ammonia, 1134
    calcium, 1134
    carbonate and bicarbonate (alkalinity), 1134
    chemical constituents, other, 1135
    microorganisms, 1134
    phosphate, 1134
Corrosion, 1110, 1113–1117
  considerations for, 1113, 1115
  in cooling water systems, 1140
  types of, 114, 1113
Corrosion control additions, 106
Corrosion management options, 1126–1127
Corrosion protection:
  in pipeline design, 861, 862
  for steel tanks, 881, 882
Cost-effectiveness analysis, (*See* Economic analysis)
Cost indexes, 1409–1411
Costs:
  allocation of, 1423
  for direct potable reuse, 1348
  for indirect potable reuse, 1313
  induced by project alternatives, 1415
  life cycle, 1413–1414
  measurement of, 1412
  replacement, 1415
  total and average *vs.* marginal, 1420–1421
Criterion (definition), 134–135
Crop rotation, 1017
Crop selection, 971–976
  in agricultural irrigation, 971
  boron tolerance, 971, 974
  effects of salinity on crops, 971, 975–976
  salt tolerance, 971–973
  types, 950
Crops:
  contamination of, 149
  effects of reclaimed water irrigation on, 1011, 1014
  processing of, 149
  relative boron tolerance of, 974
  relative salt tolerance of, 972–973
  salinity affecting, 975–976
  yield, 971, 975–976
Cross-connection control:
  dual distribution systems and in-building uses, 151–152
  inspection and testing for, 913–915
  of pipelines in distribution system, 883
$C_R t$ concept in water reuse, 616–618
Cryptosporidium parvum, 88–89
Crystallization for brine concentration, 513

CWA (Clean Water Act), 45–46, 155–156
Cycles of concentration, 1136–1138

DAD (decide-announce-defend) approach, 1448
DAF (*See* Dissolved air flotation)
DALYs (disability adjusted life years), 180–181
Dark repair:
  definition, 601
  of microorganisms following UV irradiation, 684
  regrowth and control, 711
DBPFP (*See* Disinfection byproduct formation potential)
DBPs (*See* Disinfection byproducts)
DCMF (Diamond cloth-media filter), 421–423
De facto indirect potable reuse:
  definition, 74
  impact on public water supplies, 77–78
  in surface water supplies, 1305–1308
Deaeration:
  In boiler feedwater treatment, 1143–1146
  in reverse osmosis applications, 499
Decentralized systems, 763–825
  advantages and disadvantages of (T), 286
  definition, 766–767
  future of, 289, 942
  housing developments/small community, 806–816
  management and monitoring of, 821–825
  overview of, 766–770
  site location for, 283–286
  for toilet and urinal flushing, 1193
  treatment technologies for, 785–806
  types of, 770–773
  wastewater constituent concentrations in, 778–785
  wastewater flowrates in, 774–778
  water reuse opportunities with, 816–821
Decentralized wastewater reclamation systems, 763–825
  housing developments/small community, 806–816
  management and monitoring of, 821–825
  overview of, 766–770
  treatment technologies for, 785–806
  types of, 770–773
  wastewater constituent concentrations in, 778–785
  wastewater flowrates in, 774–778
Dechlorination, 657–660
  with activated carbon, 659–660
  with sodium thiosulfate and related compounds, 659
  with sulfur compounds, 659
  with sulfur dioxide, 657–659

Decide-announce-defend (DAD) approach, 1448
Deep percolation, 981, 989
Deep-well injection system (*See* Direct injection wells for groundwater recharge)
Degree of degradation in advanced oxidation processes, 568
Demand-supply management, 1009, 1065–1066
Density:
 of air at other temperatures, 1472
 of gases, 1471
 of water, 1475, 1477–1478
Density currents:
 in lakes and reservoirs, 1317–1319
 in secondary sedimentation tanks, 320
Deoxyribose nucleic acid (DNA):
 absorbance of UV radiation, 675
 approximate size range, 376
 damage during chemical disinfection, 604,
 damage by UV radiation, 682
 in viruses, 90
 repair of following UV irradiation, 683–684, 711
Depth filtration, 392–418
 design considerations, 407–415
  filter bed characteristics, 413
  filter-medium characteristics, 408
  selection of filter medium, 409–412
  selection of filtration technology, 413–415
 operational issues, 417, 418
 performance of, 398–408
  microorganism removal, 407, 408
  operational considerations, 398, 399
  particle removal mechanisms, 399–401
  particle size alteration, 405–407
  total suspended solids removal, 402
  turbidity removal, 401–402
  variability in turbidity and total suspended solids removal, 402–405
 pilot-scale studies, 415–417
 for reclaimed water, 388–390
 technologies, 392–397
Design considerations for dual plumbing systems, 908–913
 codes, 908
 safeguards, 908–913
DF (Discfilter), 421–422
Diamond cloth-media filter (DCMF), 421, 423
1, 4-dioxane, 118–119
Direct injection wells for groundwater recharge, 1287–1292
 description, 1287–1288
 examples (T), 1292
 hydraulic analysis, 1288–1290
 operation/maintenance issues, 1290–1291

Direct injection wells for groundwater recharge (*Cont.*):
 performance, 1291
 pretreatment needs, 155, 1288
 using ASR wells, 1293–1294
Direct observation of particle size, 383–384
Direct potable reuse of reclaimed water, 934, 1345–1376
 Chanute, Kansas, case study, 1348–1352
 cost considerations, 1348
 Denver case study, 1361–1375
 health risk concerns, 1347
 issues in, 1344–1346
 public perceptions, 1347
 and technological capabilities, 1347
 technological capabilities for, 1347
 Windhoek (Namibia) case study, 1352–1361
Disability adjusted life years (DALYs), 180–181
Discfilter (DF), 421–422
Discharge:
 to surface waters, 517
 of UV altered compounds, 711
 to wastewater collection system, 515–517
Disease, probability of, 229
Disease transmission, dynamics of, 208
Disinfectants, 602–606
 characteristics of ideal, 602, 603
 chemical, 602, 603
 comparison of, 605–606
 concentration of chemical, 610–611
 mechanisms of, 604
 radiation, 603
Disinfection, 606–621
 advanced oxidation process for, 569
 combined processes for (T), 672–673
 $C_R T$ concept, 617–618
 factors affecting performance, 609–616
 NF and RO design/operational considerations, 499
 physical facilities for, 606–609
 technology comparisons, 618–621
Disinfection by-product formation potential (DBPFP) in groundwater recharge, 1275–1276
Disinfection byproducts (DBPs):
 from chlorine dioxide disinfection, 656–657
 from chlorine disinfection, 650–654
 formation of, 113–114
 from ozone disinfection, 670
 from peracetic acid disinfection, 672
Disinfection facilities, 606–609
 chlorine and related compounds, 606–609
 ozone, 607–609
 ultraviolet light, 607–609

Disinfection processes, 599–711
 (*See also specific process, e.g.:* Chlorine disinfection)
 combined chemicals in, 672–674
 considerations and issues, 606–621
  advantages and disadvantages, 618, 620–621
  application of $C_R t$ concept, 617–618
  development of $C_R t$ for predicting performance, 616, 617
  factors affecting performance, 609–616
  performance comparison, 618, 619
  physical facilities, 606–609
 for decentralized systems, 802–804
 dechlorination, 657–660
 peracetic acid, 671–672
 performance factors, 609–616
  concentration of chemicals, 610–612
  contact time, 609–610
  intensity and nature of physical agent, 614
  nature of suspending liquid, 614–615
  temperature, 613–614
  types of organisms, 614
  upstream treatment processes, effect of, 615, 616
 physical facilities for, 606–609
  chlorine and related compounds, 606–609
  ozone, 607–609
  ultraviolet light, 607–609
 technologies, 271–273, 602–606
  agents and methods in water reclamations, 602–604
  comparison, 605–606
  ideal disinfectant characteristics, 602, 603
  mechanisms, 604
 terminology, 600–601
 U.S. EPA water reuse guidelines, 169–179
 wastewater treatment requirements, 144
Dissolved air flotation (DAF), 445–454
 design considerations for, 448–453
  air-solids ratio, 448, 452, 453
  algae removal by flotation, 451–452
  chemical addition, 453
  high solids concentrations, quantity of air for, 452–453
  low solids concentrations, quantity of air for, 448, 451–452
  principal design factors for (T), 449, 450
 operating considerations for, 453
 performance of, 448
 pilot-scale studies of, 453–454
 process description, 445–448
 for reclaimed water, 389, 390

1548 Subject Index

Dissolved air flotation (DAF) (*Cont.*):
  recycle-flow *vs.* full-flow saturation, 447, 448
  sludge removal, 448
Dissolved constituent removal technologies, 463–469
  electrodialysis, 466–467, 501–508
  membrane separation, 463, 464
  nanofiltration/reverse osmosis, 465–466, 467–500
  osmotic pressure definition, 463–465
  typical process applications/flow diagrams, 467, 469
Dissolved constituent removal with membranes, 461–519
  design of NF and RO systems, 475–499
  electrodialysis for, 501–508
  and management of waste streams, 509–519
  nanofiltration for, 467–473
  pilot-scale studies for NF and RO, 499–500
  reverse osmosis for, 473–475
  technologies used in, 463–469
  terminology, 462–463
Dissolved constituents, accumulation of, 1129–1131
Dissolved organic matter removal, 268–270
Distillation, 560–563
  applications for, 560
  considerations, 562, 563
  membrane, 514
  multiple-effect evaporation, 561
  multistage flash evaporation, 561–562
  process limitations, 563
  processes, 560–562
  vapor compression distillation, 562
Distribution and storage facilities planning and design, 833–855
  distribution system network, 834–841
  location in, 834
  location issues, 834
  optimization of system, 847–855
  preliminary design criteria, 841–845
    piping system sizing, 841
    pumping system, 841–842
    storage reservoir capacity, 842–845
  quantity/pressure requirements in, 834–835
  system analysis, 845–847
    extended period system analysis, 847
    model of distribution system, 845–847
    static system analysis, 847
    types of, 845
  system network, 836–841
    distribution storage, 837, 840
    piping network, 836
    pressure zones, 840
    pumping stations, 840–841

Distribution and storage facilities planning and design (*Cont.*):
  types of:
    description (T), 836
    schematic, 837
Distribution and storage of reclaimed water, 829–892
  enclosed reservoirs, 888, 890, 892
  open reservoirs, 887–891
  operation/maintenance, 882–884
  pipeline design in (*See* Pipelines in distribution systems)
  planning/conceptual design of facilities in (*See* Distribution and storage facility planning and design)
  planning process issues, 831–833
    individual *vs.* dual distribution system, 832
    public concerns and involvement, 833
    type, size, and location of facilities, 831, 832
  pumping systems in, 866–877
    constant *vs.* variable speed operation of, 870, 871
    emergency power, 872, 875
    layout of equipment and piping, 872–874
    location/layout of pumping station, 866, 867
    operating schedule affecting design of, 875–877
    performance of, 870
    types of pumps, 867–870
    valves, 871–872
  quality issues, 884–892
    aesthetic issues, 886
    biological water quality, 886
    chemical water quality, 885–886
    impact of, 887
    management, 889–892
    physical water quality, 885
    storage affecting, 887–889
  storage facility design, 877–882
  terminology, 830–831
Distribution of reclaimed water, 745
Distribution system analysis, 845–855
  commercially available programs, 845
  extended period system analysis, 847
  model of distribution system, 845–847
  static system analysis, 847
  types of, 845
Distribution system, looped, 836–837, 1088
Distribution of reclaimed water, 745
Distribution system network, 836–841
  description of, 836–837
  pressure zones, 840
  pumping stations, 840–841
  storage, 838–840
Diversion structures, 741–744
DNA (*See* Deoxyribose nucleic acid)

Domestic wastewater chemical constituents, 104–107
Dose, of disinfectant:
  definition of, 611
  for chemical disinfectants, 616–618
  for UV radiation, 614
Dose-response assessment, 198–200
  chemical constituents, 198, 199
  microbial constituents, 198, 200
Dose-response models, 200–203
  beta-Poisson model, 201–202
  model coefficients, 202–204
  multistage models, 201
  single-hit models, 200–201
Drainage systems:
  agricultural irrigation systems, 1003
  depth and spacing, drain, 1000–1003
  golf course irrigation, 1076
  hydraulic properties, 999
  water table depth, 999
Drinking water standards:
  enforceable national, 224
  U.S. regulations for, 221–222
Drip irrigation systems, 767, 772, 887, 978–979, 984–985, 1082–1086
Dual distribution systems:
  in-building uses, 151–152
  individual distribution systems *vs.*, 832
  storage, 1176
Dual function recharge basins, 1257
Dual plumbing systems, 901–925, 1176
  applications for, 903–907
  code issues with, 908
  in commercial buildings, 903–906
  design considerations for, 908–913
  health and safety regulations for, 908
  inspection and operating considerations for, 913–915
  Irvine Ranch Water District case study, 915–919
  overview of, 902–907
  planning considerations for, 907–908
  regulations and codes governing, 908
  in residential buildings, 906–907
  Rouse Hill Recycled Water Area Project case study, 919–922
  safeguards in design of, 908–913
    backflow prevention, 909–911
    page separation, 912
    piping system identification, 912
    reduced pressure, 912
    signage, 912, 913
  Serrano, California, case study, 921–925
  terminology, 902
  typical layout for a residence, 907
  use of, 902–903
Ductile iron pipe, 858, (T) 859
Dune filtration systems, 1294–1295
Dust control, with reclaimed water, 1199

Dynamic models for microbial risk assessment, 229–232
  accounting for additional factors, 230
  application, 239–244
  deterministic/stochastic modeling, 231
  epidemiological states, 230–231
  equivalence of static and, 231
  evaluating, 232
  microbial risk characterization model complexity, 231–232
  movement from susceptible to exposed state, 229
Dynamic viscosity of water:
  definition, 475
  SI units (T), 1477
  U.S. customary units (T), 1477–1478

EBCT (See Empty bed contact time)
EC (See Conductivity, of water, electrical)
Economic analysis, 1411–1422
  associated project costs, 1415
  benefits measurement, 1412–1414
    alternative cost valuation, 1413
    cost-effectiveness analysis, 1413
    life cycle cost, 1413–1414
  cost-effectiveness analysis, 1413
  equivalent annual cost, computation of, 1413–1414
  feasibility analysis example, 1417–1419
  interest factors (T), 1526
    applications, 1527–1528
    capital recovery factor, 1526
    interest, compounded, 1526
    other factors, 1526
    present worth factor, 1526
    sinking fund factor, 1526
  measurement of costs and inflation, 1412
  present worth analysis of alternatives, 1412
  project alternatives, costs induced by, 1415
  and project optimization, 1420–1421
    marginal cost analysis, 1420
    total and average vs. marginal costs/benefits, 1420–1421
  replacement costs and salvage values, 1415–1416
  subsidies, 1415
  subsidies, influence of, 1421–1422
  sunk costs, 1414
  time reference point, 1415
  unit cost, 1414
Economics:
  industrial applications, 935
  public perspectives, 1441
  satellite treatment systems, 735
  secondary treatment selection, 367–368
Ecosystems, effects of municipal wastewater on, 121–122
Ecotoxicology, 214
ED (See Electrodialysis)

EDCs (See Endocrine disrupting compounds)
EDR (See Electrodialysis reversal)
EE/O (See Electrical energy per log order reduction),
Effective size (d_{10}):
  application, 411
  definition of, 408
  of granular activated carbon (GAC), 534
  in selection of filter materials, 409, (T) 410–411
Effluent:
  final use of, 362, 454
  membrane process modeling of, 641
  reverse osmosis and stability of, 492, 493
  water quality, 342
Effluent constituent values:
  of biological nutrient removal, 311, 312, 314, 315, 328
  of membrane bioreactor, 340
Effluent constituent variability:
  of biological nutrient removal, 312, 313, 315–318, 328
  of membrane bioreactor, 340
  for nanofiltration, reverse osmosis, and electrodialysis, (T) 472
Effluent quality:
  after removal of residual particulate matter (T), 391
  after secondary treatment, (T) 110, 111, 314
Electrical conductivity (See Conductivity, of water, electrical)
Electrical energy per log order reduction (EE/O):
  application, 584–586
  definition, 526, 583
  typical values for NDMA, 584
Electrodialysis (ED), 466–467, 501–508
  advantages/disadvantages of (T), 508
  applications for (T), 468
  design/operating considerations, 506–507
  dissolved constituent removal, 466–467
  general characteristics (T), 464
  membrane/electrode life, 507–508
  operating parameters (T), 507
  power consumption, 503–506
  process description, 501–502
  process flow diagram for, 469
  removal of dissolved constituents with, 466–467
  reversal, 502–504
  typical applications (T), 468
Electrodialysis reversal (EDR):
  applications for, 468
  power requirements, 503–506
  process description, 502–504
  process flow diagram for, 469
Elevation, atmospheric pressure changes with, 1472–1473

Emergency power, 872–875
Emerging contaminants in municipal wastewater, 117–120
  1, 4-dioxane, 118–119
  endocrine disruptors and pharmaceutically active chemicals, 117
  methyl tertiary-butyl ether and other oxygenates, 120
  n-nitrosodimethylamine, 118
  new and reemerging microorganisms, 120
  perchlorate, 119–120
Emerging technologies, for ultraviolet lamps, 676–678
Emitter clogging, 1066–1069
  algae, 1066, 1068
  bacteria, 1066, 1068
  biological factors affecting, 1067
  calcium, 1066
  chemical factors affecting, 1067
  hazard ratings for, 1068
  iron oxidizing bacteria, 1068–1069
  magnesium, 1066
  physical factors affecting, 1067
Empty bed contact time (EBCT):
  application, 549–550
  definition, 548
  values for GAC adsorption, 547
Enclosed reservoirs:
  managing, 890, 892
  storage affecting, 888
End-suction centrifugal pumps, 867
Endocrine disrupting compounds (EDCs):
  definition, 75
  in reclaimed water, 117
  removal by soil aquifer treatment, 1273, 1277
Energy consumption (See also Power requirements):
  devices to recover energy from RO, 491, 492
  for NF and RO (T), 491
  impacts of various technologies, 290, (T) 291
Enforceable national drinking water standards, 224
Engineering issues in water reclamation and reuse planning, 1392
Engineering News Record Construction Cost Index (ENRCCI):
  description, 1409–1411
  application, 1410–1411
Enhanced river recharge, 1295
Enteric viruses, 90, 97, 101–102, 244–248
  definition, 90
  inactivation by chlorine, 637–638
  monitoring for, 164, 178, 1515
  occurrence in wastewater, 97
  removal during groundwater recharge, 1279–1280

1550 Subject Index

Enteric viruses (*Cont.*):
  removal during treatment, 101
  in risk assessment, 235, 240–248
  survival in the environment, 102
Enteroviruses, 91–92
Environmental and recreational uses of reclaimed water, 1203–1241
  animal viewing parks, 1231
  applications, 932, 933
  Arcata, California, case study, 1231–1234
  checklist for planning/implementation, 1211
  factors influencing, 1207–1210
  ponds and lakes, 1228–1230
    environmental education, 1230
    operation/maintenance, 1230
    recreation, 1230
    water quality requirements, 1228–1230
  San Luis Obispo, California, case study, 1234–1238
  Santee Lakes, California, case study, 1238–1241
  snowmaking, 1231
  stream flow augmentation, 1222–1228
    aquatic/riparian habitat enhancement, 1222–1224
    operation/maintenance, 1226–1228
    recreational uses, 1224
    stream flow requirements, 1226
    water quality requirements, 1224–1226
  types of, 1206
  wetlands, 1210–1221
    development, 1213–1215
    operation/maintenance, 1216, 1218–1221
    types of, 1210, 1212–1213
    water quality considerations, 1216–1219
Environmental assessment, 1392–1393
Environmental epidemiology, 208–209
Environmental ethics, 13–15
Environmental impact(s):
  of chlorine dioxide disinfection, 657
  of chlorine disinfection, 654
  of municipal wastewater, 120–122
  of nonpotable reclaimed water, 154
  of ozone disinfection, 670
  of ultraviolet radiation disinfection, 711
Environmental justice, 1441
Environmental Protection Agency (EPA) water reuse guidelines (*See* U.S. EPA water reuse guidelines)
Environmental public health indicators (EPHIs), 208
EPA water reuse guidelines (*See* U.S. EPA water reuse guidelines)
EPHIs (environmental public health indicators), 208

Epidemiological states, 230–231
Epidemiology, 208–209
  disease transmission, 208
  environmental epidemiology issues, 208–209
  study design, 208–211
Equitable water allocation, 14
Equity issues, 1441
Escherichia coli (E. coli), 86–87
ESP (exchangeable sodium percentage), 965
Ethics, environmental, 13–15
Evaporation, 561–562
Evapotranspiration, 990–992, 1054, 1059
Exchangeable sodium percentage (ESP), 965
Existing facilities, integration with, 731, 748–751
Expanded bed GAC, 545
Expansion of existing plant, new construction *vs.*, 362
Exposure assessment, 204
Extended period system analysis, 847
Exterior protective coatings, 882
Extraction-type satellite treatment systems, 728, 729
  collection system requirements, 738–739
  wastewater characteristics, 740

Facilities plan, 1390–1392
Falling film evaporators, 512, 513
Feasibility analysis, 1425–1429
Fecal contamination indicator organisms, 95
Federal statutes, impact of, 45–46
  Clean Water Act, 45–46
  Safe Water Drinking Water Act, 46
Feedwater:
  boiler, 1144–1145
  in NF and RO, 475–477
  pretreatment of, 490
Field area requirements for agricultural irrigation systems, 997–998 (*See also* Irrigation system)
Field crops:
  boron tolerance of, 974
  salinity affecting, 975
  salt tolerance of, 972
Filter bed, 413
Filter medium:
  characteristics of, 408
  selection of, 409–412
Filtration:
  depth (*See* Depth filtration)
  for irrigation systems, 985–986, 1066–1068
  membrane (*See* Membrane filtration)
  nano- (*See* Nanofiltration)
  surface (*See* Surface filtration)
Final use of effluent, 362, 454

Financial analysis of water reclamation and reuse, 1422–1430
  considerations, 1423–1424
  construction financing plans and revenue programs, 1422–1423
  cost allocation, 1423
  freshwater rates, influence on, 1423
  sensitivity analysis and conservative assumptions, 1429–1430
  sources of revenue and pricing of reclaimed water, 1424–1429
Fire hydrants, 864, 866, 1186–1188
Fire protection, 1183, 1186–1188
  implementation issues, 1187–1188
  indoor sprinkler system, 1186, 1187
  management issues, 1188
  outdoor system with fire hydrants, 1186–1187
  types of applications, 1186–1187
  water quality, 1187
Flanged joints, 861
Floc strength, 386
Florida:
  regulations
    for indirect potable reuse, 167–169
    treatment facility reliability, 165
  type and quantity of water reuse in, 48
  Water Conserv II, 1022–1027, 1492–1493
  water quality ranges for reuse applications in, 264
Florida water reuse case study, 53–58
  current status, 54–56
  experience, 54, 55
  policies and recycling regulations, 56
  potential future uses, 56–58
Flow availability, in collection systems for extraction, 738
Flow diversion, 741–742
Flow equalization, 357, 744
Flowrate:
  variation, 775
  variability in collection system, 738–739
Fluorometer, 644
Flushing, periodic, 883
Flux rates, for membranes, 491
Fodder crops:
  boron tolerance of, 974
  salinity affecting, 976
  salt tolerance of, 973
Foliar damage, 1069–1070
Food crop irrigation, 1012–1013
Fouling indexes, 479–482
  limitations of, 481
  modified fouling index (MFI), 480
  mini-plugging factor index (MPFI), 480, 481
  silt density index (SDI), 479–480, 482
Fountains, with reclaimed water, 1196, 1197
Freudlich adsorption isotherm, 537, 538, 540, 541

Fruit crops:
  boron tolerance of, 974
  salinity affecting, 976
  salt tolerance of, 973
Furrows, use of with ridges in recharge basins, 1270, 1271

GAC (*See* Granular activated carbon)
Garnet, for depth filtration:
  properties of, 412
  sizes used in depth filters, 410–411
Gases, physical properties of (T), 1471
Gastroenteristis outbreaks, 84–85
Geometric standard deviation (s_g) (*See also* Statistical analysis):
  application of, 316–317, 403–405
  definition of, 1480
  relationship to peaking factors, 301
  typical values for treatment processes (T), 315, 398
Germicidal dose, 611
Germicidal efficiency, chlorine compounds, 631–633
GFH (granular ferric hydroxide), 536
Giardia lamblia, 87–88
Global water shortages, potential, 19–20
Golf course irrigation, 1070–1077
  for algae control, 1075
  checklist for reclaimed water use for, 1076, 1077
  distribution system design considerations, 1075–1076
  in El Dorado Hills, California case study, 1095
  leaching, drainage, and runoff, 1076
  method selection, 1072
  reclaimed water supply and storage, 1072, 1074–1075
  regulatory requirements, 1070–1072
  storage, 1072–1075
  turf selection, 1072
  water quality and agronomic considerations, 1070–1073
  for weed control, 1075
Golf Course Irrigation study, 1506
Granular activated carbon (GAC), 534–536, 539, 549
  adsorption isotherms, 536–540
  applications for, 532–533
  contactor design values (T), 547
  definition sketch, 535
  mass transfer zone, 541–542
  modes of operation, 544
  properties of (T), 534
Granular ferric hydroxide (GFH), 536
Graphical analysis of data, 1479, 1481–1483
Grasses, turf, 1072, 1073
Gravity subsurface flow, 977
Gravity surface flow, 977
Greywater:
  composition of, 780–785

Greywater (*Cont.*):
  definition, 765
  examples in various countries (T), 1175
  irrigation, 818
  recycling in office buildings, 1193
  separation of, 777–778, 787
Grinder pump collection systems, 808
Groundwater:
  effect of wetlands on, 1216
  effects of municipal wastewater on, 121
  monitoring of, 146
Groundwater law, 1395, 1397
Groundwater recharge, 736–737, 1245–1298
  aquifer storage and recovery, 1293–1294
  direct injection wells, 1287–1292
    description, 1287–1288
    examples, 1292
    hydraulic analysis, 1288–1290
    operation/maintenance issues, 1290–1291
    performance, 1291
    pretreatment needs, 1288
  enhanced river recharge, 1295, 1296
  Florida regulations for, 168
  Orange County Water District case study, 1296–1298
  planning, 1248–1254
    advantages/disadvantages of subsurface storage, 1248–1249
    components of groundwater recharge system, 1250–1251
    recovery of recharge water, 1254
    selection of recharge system, 1253–1254
    technologies for groundwater recharge, 1251–1253
    types of groundwater recharge, 1249–1250
  regulations and guidelines, 154–155
  riverbank and dune filtration, 1294–1295
  surface spreading basins:
    examples, 1280–1282
    hydraulic analysis, 1259–1268
    operation/maintenance issues, 1268–1271
    pathogens, 1279–1280
    performance of, 1271–1279
    pretreatments needs, 1257–1259
  U.S. EPA water reuse guidelines, 175–176
  vadose zone injection wells, 1282–1286
    description, 1282–1283
    examples, 1286
    hydraulic analysis, 1284–1285
    operation/maintenance issues, 1285–1286
    performance, 1286
    pretreatment needs, 1283–1284
  water quality requirements, 1255–1256
  water reuse applications, 154–155, 932, 933

Groundwater recharge (*Cont.*):
  water reuse opportunities in decentralized systems, 818–820
Guidelines (*See* Regulations and guidelines)

Habitat:
  development, 821
  enhancement by stream flow augmentation, 1222–1224
  monitoring, 1227, 1228
  value, 1214
Half reactions, 566
Hantush function F(α, β):
  definition and use, 1265
  values (T), 1523
HAV (Hepatitis A virus), 90
Hazard identification, 198
Hazard ratings, for emitter clogging, 1068
HDPE (high-density polyethylene) pipe, 859–861
Health and risk concerns, 78–83
  with direct potable reuse, 1347
  with dual plumbing systems, 908
  with indirect potable reuse, 1308–1311
    multiple barriers, use of, 1309–1311
    pathogens/trace constituents, 1308–1309
    system reliability, 1309
  with nonpotable reclaimed water, 154
  with reclaimed municipal wastewater, 78–83
    historical events, 79, 80
    waterborne diseases, 80–83
Health Effects Study at Montebello Forebay Groundwater Replenishment Project, 1489–1490
Health protection measures, WHO agriculture guidelines for, 183–184
Health risk analysis, 194–197
  definitions/concepts, 196–197
  elements of, 194
  historical development of, 194, 195
  objectives/applications of human, 194, 196
Health risk assessment, 191–250
  chemical, 215–225
    considerations, 224–225
    nonthreshold toxicants, risks from potential, 220–224
    safety/risk determination in regulation, 215–220
  definitions/concepts, 196
  dose-response assessment, 198–200
  dose-response models, 200–203
  elements of, 194
  exposure assessment, 204
  hazard identification, 198
  historical development of, 194, 195
  interrelationships of four steps of, 205
  limitations in applying, 249–250

Health risk assessment (*Cont.*):
  microbial, 225–248
    application, 225–248
    dynamic models, 229–232
    infectious disease paradigm for, 225–228
    methods, 227
    selecting models, 232–234
    static models, 227–229
  regulations and guidelines, 157
  risk characterization, 204–205
  and risk communication, 206–207
  and risk management, 205
  tools and methods, 207–214
    ecotoxicology, 214
    epidemiology issues, 208–209
    NTP cancer bioassay, 213, 214
    public health issues, 207–208
    toxicology issues, 209, 211–214
  WHO agriculture guidelines, 182
Heavy metals:
  adsorption, removal by, 533
  concerns, 141, 153, 517, 1069
  ion exchange, removal by, 554
Helminths (worms):
  description, 89, 140
  importance in water reuse, 178, 180, 182–183
  occurrence in wastewater, 97–98
  removal during treatment, 101
  size range, 376
Hepatitis A virus (HAV), 90
High-density polyethylene (HDPE) pipe, 859–861
Historical development of water reuse:
  milestones, 1485–1508
  post 1960, 41–45
  pre 1960, 39–41
Horizontal split-case centrifugal pumps, 867, 868
Housing development decentralized systems, 806–816
  treatment technologies in, 815–816
  waterwater collection in, 807–814
HRT (*See* Hydraulic residence time)
Huber Technology, 334, 337–338, 357
Human exposure to reclaimed water, 1012–1013
Humid climatic regions, 43
Hybrid collection systems, 811, 813
Hybrid process:
  description of, 310
  for decentralized treatment (T), 795
  illustration of (T), 313
Hydrants, fire, 864, 866, 1186–1188
Hydraulic analysis:
  of chlorine contact basins, 644–650
    analysis of tracer response curves, 645–650
    conduct of tracer tests, 644–645
    types of tracers, 644
  of direct injection wells, 1288–1290

Hydraulic analysis (*Cont.*):
  of subsurface spreading basins, 1259–1268
    algal blooms, impact of, 1264
    clean water infiltration rates, estimation of, 1268
    infiltration through clogged layer, 1263–1264
    mound development, impact of, 1264–1267
    movement of wetting front, 1260–1263
    surface water infiltration, 1260
  of vadose zone injection wells, 1284–1285
Hydraulic conductivity:
  of drainage water, 999, 1001–1002
  importance for operation of recharge basin, 1259–1271
  importance for operation of vadose zone injection wells, 1282, 1284–1286
  importance for operation of direct injection wells, 1288–1292
Hydraulic flushing, 490
Hydraulic loading rate:
  for Cloth-Media filtration, 421
  for Discfilter, 422
  for dissolved air flotation system, 449–450
  for irrigation systems, 993–997, (*See also* Irrigation systems)
  for ultraviolet irradiation, 703–708
Hydraulics, in UV disinfection system, 708–709
Hydraulic residence time (HRT):
  definition, 647
  in lakes and reservoirs, 1318–1319
  in MBRs, 353
Hydrogen peroxide:
  in advanced oxidation, 569–574, 1298
  in chemical oxidation, 564
  in combined disinfection processes, 673
  for ozone quenching, 666
  reaction for generation of peracetic acid, 671–672
Hydrologically altered wetlands, 1213
Hydrology, wetlands and system, 1218, 1219
Hypochlorite reactions in water, 625

Identification of compounds likely to be carcinogenic, 221
Illustration, 547
Impoundments, 152–153
In vitro tests, 211
In vivo tests, 212, 213
Inactivation, by UV irradiation, 682–683
Inactivation rate constant, 610
Incremental life risk, 215–217
Indicator organisms, 92–96
  bacteriophages, 93–94

Indicator organisms (*Cont.*):
  characteristics of ideal indicator organism, 92–93
  coliform group bacteria, 93
  definition, 75
  of fecal contamination, 95
  monitoring for in reclaimed water, 144–147, 164
  organisms used as (T), 95
  for performance criteria, 96
  use and limitations of, 144
Indices for assessing effects of reclaimed water quality on reuse systems, 1115, 1116, 1118–1125
Indirect potable reuse, 155–157
  chemical constituents and pathogens, 156–157
  de facto, 1305–1307
  most protected water source, 155
  observations on, 1340
  regulations and guidelines, 155–157
    CWA and SDWA, 155–156
    health risks assessment, 157
    most protected water source, use of, 155
  of states, 167–169
  trace chemical constituents and pathogens, 156–157
  U.S. EPA guidelines, 175–179
  through surface water augmentation (*See* Surface water augmentation, indirect potable reuse through)
Indirect potable reuse planning, 1309–1313
  costs considerations, 1313
  institutional considerations, 1312–1313
  quantity of reclaimed water to be blended, 1311
  water and wastewater treatment requirements, 1312
  watershed characteristics, 1310, 1311
Individual distribution system, dual *vs.*, 832
Indoor sprinkler system, 1186, 1187
Industrial applications, 737–738, 1103–1161
  boiler feedwater, 737–738
  boilers, 1141–1147
    constituents of concern in, 1144–1145
    feedwater of, 1144–1145
    system description, 1142–1143
    types of, 1142
    unique application of, 1143, 1145–1147
    water quality, 1143, 1145–1147
  cooling water systems, 737, 1132–1141
    design and operation, 1135–1138
    management, 1138–1141
    system description, 1132
    water quality, 1132–1135

Subject Index | 1553

Industrial applications (*Cont.*):
  Denver thermal power cooling tower case study, 1155–1158
  factors affecting use of reclaimed water for, 1108–1109
  non-cooling, 166
  process water, 736
  pulp and paper industry, 1147–1150
  at solid waste incinerator plant, 1154, 1156
  state regulations and guidelines, 166
  terminology, 1104
  textile industry, 1150–1155
  U.S. EPA water reuse guidelines, 174
  water management in industries, 1107–1108
  water quality issues, 1109–1131
    Aggressiveness Index (AI), 1118–1119
    buffer capacity, 1119
    calcium carbonate precipitation potential (CCPP), 1120–1124
    corrosion issues, 1110, 1113–1117
    general considerations, 1110–1112
    indices for assessing effects of reclaimed water quality, 1115, 1116, 1118–1125
    Langelier saturation index (LSI), 1124–1125
    Larson's Ratio, 1125
    Ryznar stability index (RSI), 1125
    Stiff-Davis Stability Index (SDI), 1125
  water reuse, 153, 931–933
  West Basin Municipal Water District case study, 1158–1161
Industrial chemical constituents, 104–107
Infection(s):
  probability of, 229
  secondary, 249
Infectious disease paradigm for microbial risk assessment, 225–228
  complexities of person-to-person interactions, 226–228
  risk analysis framework, 226
Infiltration:
  basins, 1254–1280
  problem in irrigation, 963
Infiltrometer, double ring, 1268
Inflation, 1409–1412
Influent flowrates, variability in, 302
Influent wastewater parameters, variability in, 301–304
Influent water quality issues, 341, 342
Information control, 1448–1449
Information needs, 184–185
Infrastructure:
  new concepts and designs in, 290–291
    energy efficiency, 290, 291
    robust treatment processes, 290, 291
    security concerns, 291

Infrastructure (*Cont.*):
  planning issues, 28–30
  for satellite treatment systems, 741–745
    diversion/junction structures, 741–744
    flow equalization/storage, 744
    pumping/transmission/distribution of reclaimed water, 745
  for urban nonirrigation water reuse, 1175–1176
  for water reuse applications, 280, 281, 939, 940
Initial mixing (*See* Mixing)
Injection wells:
  for disposal of brine concentrate, 517–519
  for direct aquifer recharge, 1287–1292 (*See also* Direct injection wells for groundwater recharge)
  for vadose zone aquifer recharge, 1282–1286 (*See also* Vadose zone injection wells)
Institutional issues:
  and satellite treatment siting, 735
  of water reuse, 735, 1397
Integrated fixed film activated sludge process, 324
Integrated Water Resources Management Project in Vitoria, 1505
Integrated water resources planning (IWRP), 24–27, 1384–1392
  background information gathering, 1386–1388
  evaluating and ranking alternatives, 1389
  facilities plan components, 1390–1392
  facilities plan report, 1392
  identifying project alternatives, 1388–1389
  implementation plan development, 1389–1392
  objectives formulation, 1386
  problem clarification, 1386
  process, 1385–1386
  project study area delineation, 1388
  public involvement, 1388
  substituting reclaimed water for nonpotable uses, 26
  water use patterns, 27
Internal water recycling, 1105, 1107, 1141, 1147, 1149
Interception-type satellite treatment systems, 728, 729
  collection system requirements, 738
  wastewater characteristics, 740
Interest, compound (*See* Economic analysis)
Interior protective coatings, 881–882
Intermixing, impact of, 1271–1272
Ion exchange, 551–560, 1326
  applications for, 552–554, 1323–1324
  capacity, 555, 557–559
  considerations for, 555–559

Ion exchange (*Cont.*):
  for heavy metals removal, 554
  materials, 554–555
  natural materials, 555
  for nitrogen control, 552–553
  for organic matter removal, 554
  process limitations, 559–560
  and resin characteristics, 556
  synthetic materials, 555
  for total dissolved solids removal, 554
Ionic strength:
  application, 496–499
  computation of, 495, 1116
  definition, 1116
  estimated based on TDS values, 495, 1118
Ion toxicity, specific, 965–966, 1069
Iron, ductile pipe, 858
Iron oxidizing bacteria, 1068–1069
Irradiation, ultraviolet, 603
Irrigated land, regulations and guidelines for, 953
Irrigation:
  agricultural (*See* Agricultural irrigation)
  efficiency, 981, 984, 1060–1064
  with greywater, 818
  landscape (*See* Landscape irrigation)
  methods of, 981
  reclaimed water, 981
  scheduling, 1008
Irrigation systems, 982–983
  consideration for, 979–985
  field area requirements, 997–998
  hydraulic loading rate, 993–997
  operation and maintenance, 1008–1015
  types of, 977–979
Isolation valves, use in distribution systems, 871
Isotherms, adsorption, 536–541
  Freudlich, 537, 538, 540, 541
  Langmuir, 537–540
IWRP (*See* Integrated water resources planning)

Joints (distribution system piping), 860–861
  flanged joints, 861
  mechanical joints, 860
  pipe joint restraints, 861
  push-on joints, 860
  welded joints, 861
Junction structures, for satellite systems, 741–744

Kaldnes process, 324, 325
Kinematic viscosity of water:
  definition, 475
  SI units (T), 1477
  U.S. customary units (T), 1477–1478
Kinetic coefficients, 344–345
Kinetic equations, 342–344
Koch, 339–340
Kubota, 334–336, 356

Subject Index

Laboratory studies (*See* Bench-scale studies)
Lagoons, 387
Lakes:
 ponds (*See* Ponds and lakes)
 in public parks, 1197
 reclaimed water used with, 736, 1228–1230
 as reservoirs (*See* Reservoirs)
Land use and development, 122
Landscape irrigation, 150, 151, 736, 931, 932, 1043–1097
 with decentralized and subsurface irrigation systems, 1082–1086
 with decentralized systems, 816–817
 definition, 1045–1046
 design considerations (*See* Landscape irrigation design)
 El Dorado Hills, California, case study, 1093–1097
 golf courses, (*See* Golf course irrigation)
 operation/maintenance, 1066–1070
  emitter clogging, 1066–1069
  foliar damage, 1069–1070
  runoff control, 1069
  specific ion toxicity, 1069
 public access, 150–151
 public areas, 1076, 1078–1080
 Redwood City, California, case study, 1450–1453
 residential, 1080–1082
 St. Petersburg, Florida, case study, 1086–1093
 with subsurface irrigation systems, 1082–1086
  for on-site and cluster systems, 1082–1085
  in residential areas, 1086
 terminology, 1044
 trace constituents, 151
 in U.S., 1046–1048
 use area controls, 151
Landscape irrigation design, 1047, 1049–1066
 application rate and irrigation schedule, 1065
 components for, 1055–1058
 demand-supply balance management, 1065–1066
 irrigation systems, 1054–1059
 operation/maintenance, 1069–1070
 plant selection, 1050, 1052–1054
 with reclaimed water, 1048
 urban, 1049, 1051
 water needs estimation, 1054, 1058–1064
  evapotranspiration, 1054, 1058
  irrigation efficiency, 1060–1064
  landscape coefficient, 1054, 1059–1060
 water quality requirements, 1047, 1049–1051

Landscape plants, 1052–1053
Langelier Saturation Index (LSI):
 application, 469–499
 definition of, 494–496, 1124–1125
Langmuir adsorption isotherm, 537–540
Larson's Ratio, 1125
Laundries with reclaimed water, 1196–1197
LC_{50} (*See* Lethal concentration)
LD_{50} (*See* Lethal dose)
Leachate from irrigation systems, 1076
Leaching from root zone in agricultural irrigation:
 fraction (LF), 986
 requirements, 986–989, 1076
Legal aspects of water reuse, 1393–1397
 institutional coordination, 1397
 policies/regulations, 1397
 with satellite treatment systems, 734–735
 water rights law, 1393–1396
Legal mandates for public participation, 1443–1444
Lethal concentration (LC_{50}), 211, 213
Lethal dose (LD_{50}), 211, 213
Life cycle cost, 1413–1414
Linpor process, 324, 325
LOAEL (Lowest observed adverse effect level), 211–213
Locations:
 Arizona:
  Grand Canyon Village, 1485–1486
  Palo Verde Nuclear Generating Station, 1489
  Scottsdale's Water Campus, 1494
 Australia, 1028–1035, 1499
 California:
  Arcata Marsh and Wildlife Sanctuary, 1231–1234
  El Dorado Hills, 1093–1097
  Los Angeles County Sanitation Districts, 1487–1488
  Montebello Forebay Groundwater Replenishment Project, 1489–1490
  Monterey Water Reclamation Study for Agriculture, 1488–1489
  Orange County, 1496
  Redwood City, 1450–1453
  San Diego Total Resources Recovery Project, 1493
  San Diego water repurification and reuse, 1329–1334
  San Francisco Golden Gate Park plant, 1485–1486
  San Luis Obispo, 1234–1238
  Santa Ana River, 1496
  Santa Rosa, 1147
  Santee Lakes, San Diego, 1238–1241
  Serrano, 921–925
  Upland Hills Water Reclamation Plant, 760–761
  Walnut Valley Water District, 1397–1399

Locations (*Cont.*):
 Canada, 1499–1500
 Denver Potable Water Demonstration, 1491–1492
 East Coast, Metropolitan region, 22
 Eastern Midwest region (of U.S.), 22
 Florida:
  St. Petersburg, 1086–1093, 1453–1459, 1486–1487
  Winter Garden, 1492–1493
 Great Lakes region, 22
 Great Plains region, 21–22
 Israel:
  Dan River Project, Tel Aviv, 1500–1501
  Jeezrael Valley Project (JVP), 1501
 Japan:
  Kobe, 1501–1502
  Tokyo Metropolitan Government Building, 1501–1502
 Kansas, 1348–1352
 Kuwait, 1502–1503
 Mid-Atlantic region (of U.S.), 22
 Namibia, 1352–1361, 1503
 New York City, 22
 Rio Grande, 22
 Singapore, 1334–1340, 1503–1504
 South Africa, 1504
 Southeast region (of U.S.), 23
 Spain, 1504–1505
  AENP (Aiguamolls de l'Emporda Nature Preserve), 1505
  Girona, 1505
  Portbou, 1505
  Vitoria, 1505
 Tunisia, 1505–1506
 Western region (of U.S.), 23, 41, 43
Logarithmic (Log-normal) probability distribution, 1479–1483 (*See also* Statistical analysis)
Log removal:
 definition, 83
 values for microorganisms in treatment processes (T), 101, 606
 for MS2 coliphage, 408
Low-pressure high-intensity ultraviolet lamps, 676, 677
Low-pressure low-intensity ultraviolet lamps, 675–677
Lowest observed adverse effect level (LOAEL), 211–213
Lumped parameter, 103, 115, 378
LSI (*See* Langelier Saturation Index)

Magnesium, 1066
Management practices for nonpotable reuse, 1494–1495
Mandatory use ordinance, 1404–1406
Mandatory water reuse, voluntary *vs.*, 165
Marginal cost analysis, 1420
Marginal costs and benefits, 1420–1421

Market assessment for reclaimed water (*See* Reclaimed water market assessment)
Market assurances, 1402–1406
  mandatory use ordinance, 1404–1406
  measure of degree of willingness of potential user, 1402–1403
  types of, 1403–1404
  voluntary *vs.* mandatory approach, 1404
  waste/unreasonable use of water, 1406
Mass balance:
  for accumulation of dissolved constituents during evaporation, 1129–1131
  for cycles of concentration in cooling towers, 1136
  for irrigation requirements, 992
  for lakes and reservoirs, 1319
  for membrane systems, 436, 485–487
  for reservoir storage capacity, 842–845
Mass flux rate, in membrane systems, 483
Mass loadings, variability of, 301–304
  constituent concentrations, 302
  influent flowrates, 302
  influent wastewater parameters, 301–304
Mass transfer zone (MTZ), 541–543
Materials mass balance (*See* Mass balance)
Maximum contaminant level goal (MCLG), 221–225
MBR (*See* Membrane bioreactor)
MCLG (Maximum contaminant level goal), 221–225
MDI (*See* Morrill dispersion index)
Mean cell residence time (MCRT), (*See* Solids retention time)
Measurement of benefits, 1412–1414
  alternative cost valuation, 1413
  cost-effectiveness analysis, 1413
  life cycle cost, 1413–1414
Measurement of costs and inflation, 1412
Measures of disease, 207
Mechanical joints, 860
Medium-pressure high-intensity ultraviolet lamps, 676, 677
Meeloop filter index (MLFI), 1290–1291
Membrane biofilm reactor (MBfR), 590
Membrane bioreactor (MBR), 328, 330–340
  airlift, 339
  analysis/design of (*See* Membrane bioreactor analysis and design)
  characteristics of proprietary systems (T), 334
  description, 330–331
  other membrane systems:
    airlift, 339
    Koch/Puron, 339–340
    sequencing batch reactor/cloth filter/microfiltration process, 338–339

Membrane bioreactor (MBR) (*Cont.*):
  principal proprietary systems, 333–338
    Huber Technology, 334, 337–338
    Kubota, 334–336
    Mitsubishi, 334, 335
    USFilter, 334, 336–337
    Zenon Environmental, 333–335
  process performance:,
    expectations, 340
    range of effluent quality (T), 314
    range of effluent variability (T), 315, 316
  in satellite treatment systems, 746
  suitability for reclaimed water applications, 331–332
  thickening and dewatering of MBR biosolids, 361
  types of, 332–333
Membrane bioreactor (MBR) analysis and design, 340–361
  alpha (α) correction factor for aeration system, 354–355
  biosolids processing, 361
  characteristics of proprietary systems (T), 334
  design considerations, 353–358
    air supply, 354–355
    biosolids production and management, 358
    membrane fouling control and cleaning, 355–357
    peak flow management, 357
    pretreatment, 353–354
  equations commonly used in, 341
  nutrient removal, 358–361
    biosolids processing, 361
    nitrogen, 358–359
    phosphorus, 360–361
  process analysis, 340–353
    biological solids production, 345–351
    bioreactor suspended solids concentration, 353
    coefficients, 344–345
    key wastewater constituents, 341
    kinetic equations, 342–344
    membrane flux rate, 352
    membrane life, 352
    pore size, 352
    process variables, 352
    solids and hydraulic retention times, 353
    temperature, 352
    water quality issues, 341, 342
Membrane distillation, 513, 514
Membrane filtration, 425–445
  classification, 426, 427
  flow patterns in, 426, 428, 429, 431
  micro-/ultrafiltration, 430–442
    design considerations, 441, 442
    membrane performance, 436–441
    operating characteristics/strategies, 436, 437

Membrane filtration, micro-/ultrafiltration (*Cont.*):
    operational modes, 431–434
    pressurized, 430–434
    process analysis, 435–436
    submerged, 431, 433, 434
  operational issues, 443–445
    clean-in-place frequency, 444–445
    membrane life, 443
    membrane performance, 443
    operating efficiency, 443, 444
  pilot-scale studies, 441, 443, 444
  for reclaimed water, 388–390
  terminology, 426
  types, 426, 428–430
Membrane flux:
  MBR analysis and design, 352
  NF and RO design/operational considerations, 482–487
    mass balance equations, 485–487
    mass flux rate, 483
    for proprietary systems, (T) 334
    recovery ratio, 484–485
    rejection efficiency, 485
    water flux rate, 482–483
Membrane fouling:
  biological, 489
  control and cleaning in MBR, 355–357
  in NF and RO, 487–490
  organic, 489
  particulate, 487–489
  scaling, 488, 489
Membrane life, 352, 443
Membrane performance, 436–441
  broken fibers, impact of, 440–441
  as operational issue, 443
  primary effluent filtration, 439
  secondary effluent filtration, 437–441
Membrane separation, 463, 464
Membrane waste stream management, 509–519
  issues in, 509–511
    cleaning solutions, 511
    retentate classification, 511
    volume of retentate, 509, 510
  thickening/drying of waste streams, 511–515
    concentration by multiple-stage membrane arrays, 511–512
    crystallization, 513
    falling film evaporators, 512, 513
    membrane distillation, 513, 514
    solar evaporation, 512
    spray dryers, 513
  ultimate disposal methods, 515–519
    discharge to wastewater collection system, 515–517
    disposal options (T), 516
    disposal to surface waters, 517
    subsurface injection, 517–519
Metal ions (*See* Heavy metals)
Methyl tertiary-butyl ether (MTBE), 120

1556　Subject Index

MF (*See* Microfiltration)
MFI (modified fouling index), 480
Microbial constituents:
　of concern in wastewater, 139–141
　dose-response assessment, 198, 200
　hazard identification, 198
　regulations and guidelines, 139–142
Microbial limits, EPA water reuse guidelines for, 178
Microbial pathogens, 94–102
　in environment, 102
　in primary effluent, 99, 101
　in secondary effluent, 99–101
　survival of, 102
　in tertiary and advance wastewater treatment effluent, 100, 102
　in treated wastewater, 97–102
　in untreated wastewater, 94, 96–98
Microbial repair following UV irradiation, 683–684
　dark repair, 684
　photoreactivation, 683–684
Microbial risk, tolerable, 181–182
Microbial risk assessment (MRA), 225–248
　application of, 234–248
　　with dynamic models, 239–244
　　for enteric viruses, 244–248
　　with static models, 234–238
　dynamic models, 229–232
　　accounting for additional factors, 230
　　application, 239–244
　　deterministic/stochastic modeling, 231
　　epidemiological states, 230–231
　　equivalence of static/dynamic models, 231
　　microbial risk characterization model complexity, 231–232
　　movement from susceptible to exposed state, 229
　for enteric viruses, 244–248
　infectious disease paradigm for, 225–228
　　complexities of person-to-person interactions, 226–228
　　risk analysis framework, 226
　methods, 227
　model selection for, 232–234
　　caveats, 233–234
　　data required, 232–233
　　evaluating static/dynamic models, 232
　　parameters most strongly impacted, 233
　　risk manager's role, 234
　static models, 227–229
　　application, 234–238
　　model states, 227–229
　　probability of infection/disease, 229
　　required health effects information, 229

Microbial risk characterization model complexity, 231–232
Microfiltration (MF), 430–442
　design considerations, 441, 442
　membrane performance, 436–441
　　broken fibers, impact of, 440–441
　　primary effluent filtration, 439
　　secondary effluent filtration, 437–441
　operating characteristics/strategies, 436, 437
　operational modes, 431–434
　pressurized, 430–434
　process analysis, 435–436
　　mass balance, 436
　　operating pressure, 435
　　permeate flow, 435
　　recovery, 435–436
　　rejection, 436
　submerged, 431, 433, 434
Microorganisms (*See also* specific types)
　in activated sludge processes, 386
　and chlorine disinfection, 637–638
　in cooling water systems, 1135
　depth filtration and removal of, 407, 408
　new and reemerging, 120
　regrowth of, 654, 711
　secondary treatment processes, 376, 377
　surface filtration and removal of, 423
　UV disinfection, 688–690
Mineral increase, from domestic water use, 104–105, 778–785
Mini-plugging factor index (MPFI), 480, 481
Mitigation of wetlands, 1214
Mitsubishi, 334, 335, 356
Mixed PAC contactor with gravity separation, 546
Mixed PAC contactor with membrane separation, 546
Mixing:
　in biological treatment, 307, 312, 747
　facilities for mixing chemicals, 634
　impact on formation of disinfection byproducts, 652
　importance of initial mixing for chemical disinfection, 633–635
　in lakes and reservoirs, 880, 1317–1320
　indexes for, 647 1318, 1322
Mixtures, adsorption of, 540–541
MLE (*See* Modified Ludzack-Ettinger process)
Model coefficients, for chemical risk analysis, 202–204
Modified fouling index (MFI), 480
Modified Ludzack-Ettinger (MLE) process, 321, 322
Modulus of elasticity of water, 1475, (T) 1477–1478
Molecular weight of elements (*See* inside back cover)

Monetary evaluation of water reclamation and reuse, 1406–1411
　inflation and cost indexes, 1409–1411
　perspectives on, 1408
　planning weaknesses in, 1407–1408
　time horizons, planning and design, 1408–1409
　time value of money, 1409
Monitoring:
　of agricultural irrigation systems, 1014–1015
　of habitat, 1227, 1228
　of reclaimed water quality, 145–146, 1193–1195, 1308–1309
　of stream water quality, 1227
Monitoring and control equipment, for decentralized systems, 824–825
Montebello Forebay Groundwater Replenishment Project, 1489–1490
Morrill dispersion index (MDI):
　application, 647–650
　definition, 647
　relationship to volumetric efficiency, 647
Mosquito control for constructed wetlands, 1220
Most protected water source, use of, 155
Mound development, groundwater:
　determination of, 1264–1267
　Hantush fuction for (T), 1523
Movement from susceptible to exposed state, 229
Movement of wetting front, 1260–1263
MPFI (*See* Mini-plugging factor index)
MRA (*See* Microbial risk assessment)
MTZ (Mass transfer zone), 541–543
Mukuhari New Center Water Recycling Project, 1501
Multiple barriers:
　concept of, 263, 265
　in Denver potable reuse demonstration project, 1364, 1368
　in Goreangab Water Reclamation project, 1353–1354, 1364, 1368
　with indirect potable reuse, 1309–1311
Multiple-effect evaporation, 561
Multiple-stage membrane arrays, for concentration of brine, 511–512
Multiple tube fermentation (MTF), 79, 164, 631, 1512–1513
Multiple use facilities, 1314
Multistage flash evaporation, 561–562
Multistage models, 201
Municipal wastewater, 73–122
　chemical constituents in, 103–116
　　DBP formation, 113–114
　　domestic, commercial, and industrial additions, 104–107
　　in treated wastewater, 108–113
　　treated wastewater to natural water comparison, 114–116
　　in untreated wastewater, 103–104

Municipal wastewater (*Cont.*):
  de facto potable reuse of, 77–78, 1303–1304
  emerging contaminants in, 117–120
    1, 4-dioxane, 118–119
    endocrine disruptors and pharmaceutically active chemicals, 117
    methyl tertiary-butyl ether and other oxygenates, 120
    n-nitrosodimethylamine, 118
    new and reemerging microorganisms, 120
    perchlorate, 119–120
  environmental issues, 120–122
    development and land use, effects on, 122
    ecosystems, effects on, 121–122
    soils and plants, effects on, 121
    surface water and groundwater, effects on, 121
  health issues, 78–83
    etiology of waterborne diseases, 81, 83
    historical events, 79–80
    waterborne diseases, 80–82
  indicator organisms, 92–96
    bacteriophages, 93–94
    characteristics of ideal indicator organism, 92–93
    coliform group bacteria, 93
    of fecal contamination, 95
    for performance criteria, 96
  microbial pathogens:
    pathogens in environment, 102
    pathogens in treated wastewater, 97–102
    pathogens in untreated wastewater, 94, 96–98
    survival of pathogens, 102
  terminology, 74–76
  U.S. EPA water reuse guidelines, 170–177
  waterborne pathogenic microorganisms, 83–92
    bacteria, 83, 86–87
    helminths, 89
    log removal, 83
    protozoa, 87–89
    terminology conventions for organisms, 83
    viruses, 89–92

N-nitrosodimethylamine (NMDA) removal:
  advanced oxidation, removal by, 573, 575, 576
  description, 118
  concern in groundwater recharge, 1291
  formation of, 113, 653
  photolysis, removal by, 580–581, 584–586
  reaction with chlorine, 653–654
  risk assessment, 217

Nanofiltration (NF), 465, 467–473
  applications for (T), 468, 471
  design and operational considerations, 475–499
    control of membrane fouling, 490
    feedwater considerations, 475–477
    membrane flux/area requirements, 482–487
    membrane fouling, 487–489
    posttreatment, 492–499
    pretreatment, 477–479
    process design consideration (T), 467
    treatability testing, 479–482
  dissolved constituent removal, 465–466
  general characteristics of (T) 464
  membrane types used in, 468, 470
  performance expectations, 471–473
  pilot-scale studies, 499–500
  process flow diagram for, 469
  process operating parameters, 490–492
  rejection rates (T), 472
National pollutant discharge elimination system (NPDES), 953, 1003, 1010, 1074, 1095, 1129, 1208, 1211, 1219, 1225, 1227, 1232, 1236
National Toxicology Program (NTP) cancer bioassay, 213–214
Natural ion exchange materials, 555
Natural water, chemical constituents in treated water *vs.*, 114, 116
Natural wetlands, 1212
NDMA (*See N*-nitrosodimethylamine)
Nephelometric turbidity unit (NTU), 384 (*See also* Turbidity)
Net irrigation requirement, 992, 993
NEWater Study, 1503–1504
NF (*See* Nanofiltration)
Nitrification, definition of, 321 (*See also* Nitrogen removal)
Nitrogen:
  in agricultural irrigation, 970
  ion exchange for control of, 552–553
  in groundwater recharge, 1256–1257
  in reclaimed water, 1010
Nitrogen compounds, transformation of, 1277
Nitrogen control, 320–321
Nitrogen loading limits, 995–997
Nitrogen removal, 321–325
  activated sludge with fixed film packing, 324, 325
  by ANAMMOX reactions, 1277–1279
  application, 799–802
  in groundwater recharge, 1256
  by ion exchange, 552–553
  in membrane bioreactors, 358–359
  process flow diagrams for:
    Biofor, 323
    Biostyr, 323
    Membrane bioreactors, 358
    Modified Ludzack-Etinger, 322

Nitrogen removal, process flow diagrams for (*Cont.*):
  oxidation ditch, 323
  sequencing batch reactor, 322
  step feed, 322
  reaction stoichiometry (T), 798
  submerged attached growth processes, 321, 323, 324
  suspended growth processes, 321–323
No observed adverse effect level (NOAEL), 211–213
NOAEL (*See* No observed adverse effect level), 211–213
Nonagricultural satellite treatment systems, 735–737
  groundwater recharge, 736–737
  industrial applications, 737
  lakes/recreational enhancement, 736
  landscape irrigation, 736
  reuse in buildings, 736
Noncarcinogenic effects, 218–220
  relative source contribution, 219
  safety factors, 219–220
Nonmembrane processes for secondary treatment, 307–329
  nutrient control and removal, 320–329
    nitrogen control, 320–321
    nitrogen removal, 321–325
    phosphorus removal, 324–329
  process descriptions, 308–314
    attached growth processes, 308, 313, 314
    hybrid process, 310, 313
    suspended growth processes, 308–312
  process performance expectations, 310–318, 328
    effluent constituent values, 311, 312, 314, 315, 328
    evaluation of activated sludge process reliability, 316–318
    variability in effluent constituents, 312, 313, 315–318, 328
  secondary sedimentation tank design, 318–320
    modification of sedimentation facilities, 319, 320
    physical factors, 319, 320
    sidewater depth, 319
  suitability for reclaimed water applications, 307–308
Nonpotable uses of reclaimed water, 153–154
  environmental impacts, 154
  health considerations, 154
  state regulations and guidelines, 165–167
Nonthreshold toxicants, risks from potential, 220–224
  enforceable national drinking water standards, 224
  EPA's qualitative assessment of carcinogens, 222–223

Nonthreshold toxicants, risks from potential (*Cont.*):
  identification of compounds likely to be carcinogenic, 221
  risk extrapolation for carcinogens, 221
  three-category approach for setting MCLGs, 223–224
  U.S. drinking water regulations, 221–222
Noroviruses, 90, 91
NPDES (*See* National pollutant discharge elimination system)
NTP cancer bioassay (National Toxicology Program cancer bioassay), 213–214
NTU (*See* Nephelometric turbidity unit), 384, (*See also* Turbidity)
Nuisance species, 1220–1221, 1230
Nutrient control and removal, 320–329
  in MBR, 358–361
  nitrogen control, 320–321
  nitrogen removal, 321–325, 358–359
  phosphorus removal, 324–329, 360–361
Nutrients:
  in agricultural irrigation, 968–971
  in irrigation water, 968–971
  in lakes and ponds, 1228
  management of, 1009–1010
  treatment technologies for removal of, 268–270

Odor control additions, 106
Odor issues:
  in landscape irrigation, 1050
  in toilet and urinal flushing, 1190–1191
  in wetland, 1221
Office buildings:
  dual plumbing systems for, 904, 917
  use of reclaimed water in, 736
O&M (*See* Operation and maintenance)
Onsite systems, 1082–1086
Open channel disinfection systems, 678–680
Open reservoirs:
  managing, 889–891
  storage affecting, 887–888
Operating efficiency, 443, 444
Operating pressure, 435, 491
Operating schedule, 875–877
Operation and maintenance (O&M):
  and emitter clogging, 1066–1069
  of public areas landscape irrigation, 1080
Operation and maintenance issues:
  foliar damage, 1069–1070
  of residential landscape irrigation, 1082
  runoff control, 1069
  salinity problems, 1069
  sodicity problems, 1069
  specific ion toxicity, 1069

Organic compounds, transformation of, 1273–1277
  bulk organic transformations, 1273–1275
  compounds of concern, 1273
  disinfection by-product formation potential, 1275–1276
  endocrine disrupting activity, 1277
  trace organic compounds, 1276–1277
Organic fouling, of membranes, 489
Organic matter, removal of:
  in biological processes, 310–311, 313
  in groundwater recharge, 1256
  in ion exchange, 554
Osmotic pressure, 463–465
Outdoor system with fire hydrants, 1186, 1187
Oxidation, 563–578
  advanced (*See* Advanced oxidation processes)
  applications for advanced, 568–569
  chemical, 563–567
  considerations for advanced, 574–576
  process limitations for advanced, 577–578
Oxidation ditch:
  advantages and limitations of, 312
  description of (T), 310
  for nitrogen removal (T), 323
Oxygenates, 120
Ozone disinfection, 609, 660–671
  advantages and disadvantages of, 621
  benefits, 671
  byproduct formation and control, 670
  chemistry, 661, 662
  dosage requirements, 669
  effectiveness, 666, 667
  environmental impacts, 671
  modeling, 666–669
  physical facilities for, 607–609
  process modeling in, 666–669
  properties, 660–661
  systems components, 662–666
    destruction of off gases and residual ozone quenching, 666
    feed gas preparation, 662, 663
    generation, ozone, 663–664
    in-line ozone contact/reaction reactors, 664–665
    power supply, 663
    sidestream ozone contact/reaction system, 665–666
Ozone/hydrogen peroxide AOP, 572
Ozone/UV AOP, 572–574

PAA (*See* Peracetic acid)
PAC (*See* Powdered activated carbon)
Pipe separation, for potable and reclaimed water lines, 907, 912
Parallel operation (of activated-carbon contactors), 544
Parameters, aggregate and lumped, 103, 115, 378

Particle counters, 380, 381, 416
Particle removal mechanisms, 399–401
Particle size:
  alteration during filtration, 405–407
  analysis of particle size data, 380–384
  analytical techniques for (T), 380
  direct observation of, 383–384
  distribution (PSD) of, 378–384
    in activated sludge processes, 315, 385, 386
    direct observation, 383–384
    electronic particle size analyzers, 378, 380–383
    in secondary clarifiers, 318–319
    serial filtration, 378, 379
    in surface filtration, 422, 424
  particle size data for constituents found in wastewater, 376
  of soil, 957
Particles:
  impact on chlorine disinfection, 636–637
  increasing UV intensity for overcoming impact of, 710
  impact on UV disinfection, 687, 688
Particulate fouling, 487–489
Pathogenic microorganisms, waterborne, 83–92:
  bacteria, 83, 86–87
  helminths, 89
  log removal, 83
  protozoa, 87–89
  terminology conventions for organisms, 83
  viruses, 89–92
Pathogens:
  agricultural irrigation, 149
  groundwater recharge, 1279–1280
  indirect potable reuse, 1308–1309
  limits and monitoring for, 164
  regulations and guidelines, 156–157
  survival of, 102, 147, 148
Peak flow management, 357
Peaking factors:
  application of, 776
  for individual homes and small communities (T), 776
  for pumping station design, (T) 841
Peracetic acid, disinfection:
  chemistry of, 671
  in combined disinfection processes, 673
  effectiveness as a disinfectant, 672
  properties of (T), 672
Perchlorate, 119–120
Periodic flushing, 883
Permeate flow, 435
Person-to-person interactions, complexities of, 226–228
Personnel, 27
pH:
  advanced oxidation processes impacted by, 577–578

pH (*Cont.*):
  chemical commonly used for control of, 493
Pharmaceutically active chemicals, 117
Phoredox, 324, 326
Phosphate, 1134
Phosphorous:
  in agricultural irrigation, 970–971
  in reclaimed water, 1010
Phosphorus removal, 324–329, 1134
  advantages and disadvantages of metal salts and lime addition (T), 329
  biological, 324–326
  by chemical addition, 327–329
  in MBR, 360–361
  process flow diagrams:
    A/O, 326
    A²/O, 326
    MBR, 360
    chemical addition, 327
    Phoredox, 326
Photolysis, 578–586
  absorption of UV light, 580–582
  applications for, 578
  electrical efficiency, 583–586
  energy input for, 582
  process limitations, 586
  processes, 578
  rate of, 582–583
Photoreactivation, 683–684, 711
Physical disinfection agents, 603–604
Physical properties:
  of untreated municipal wastewater, 261–262
  of wastewater, 141, 142
  of water, 1475–1478
Pig, for pipe cleaning, 884
Pilot-scale studies:
  for depth filtration, 415–417
  for dissolved air flotation, 453–454
  for membrane filtration, 441, 443, 444
  for nanofiltration and reverse osmosis, 499–500
  need to conduct, 276
  for residual particulate matter removal, 455
  secondary treatment selection issues, 362, 363
  for surface filtration, 425
Pipe joint restraints, 861
Pipelines in distribution systems, 856–865
  appurtenances, distribution system, 863–866
  corrosion protection, 861, 862
  cross-connection control, 883
  design criteria for, 858–860
  ductile iron, 858
  high-density polyethylene, 860
  identification, 862
  identification of, 862
  joints and connections, 860–861
  location of, 856–858

Pipelines in distribution systems (*Cont.*):
  materials, 858–860
  operation/maintenance, 883–884
  periodic flushing, 883
  pipe identification, 862
  polyvinylchloride, 858, 859
  reclaimed pipeline location, 856–858
  steel, 858
  separation requirements, 856–858
  system appurtenances, 863–865
  valves, 863
Piping system:
  identification, 912
  sizing, 841
Planning for groundwater recharge with reclaimed water, 1248–1254
  advantages/disadvantages of subsurface storage, 1248–1249
  aquifer storage and recovery, 1250
  components of groundwater recharge system, 1250–1251
  control of seawater/brackish water intrusion, 1250
  direct injection wells into aquifer, 1252–1253
  direct injection wells into vadose zone, 1252
  groundwater augmentation for indirect potable reuse, 1249
  recovery of recharge water, 1254
  selection of recharge system, 1253–1254
  surface spreading, 1251, 1252
  technologies for groundwater recharge, 1251–1253
  types of groundwater recharge, 1249–1250
Planning for indirect potable reuse, 1309–1313
  costs considerations, 1313
  institutional considerations, 1312–1313
  quantity of reclaimed water to be blended, 1311
  treatment requirements, 1312
  watershed characteristics, 1310, 1311
Planning for satellite treatment systems, 730–735
  economics, 735
  environment, 735
  institutional issues, 735
  integration with existing facilities, 731
  legal issues, 734–735
  public perception, 734
  regulations, 735
  siting, 731–734
  water needs identification, 730–731
Planning for water reclamation and reuse, 1381–1430
  economic analysis, 1411–1422
    basic assumptions, 1414–1415
    feasibility analysis example, 1417–1419

Planning for water reclamation and reuse, economic analysis (*Cont.*):
    measurement of benefits, 1412–1414
    measurement of costs and inflation, 1412
    present worth analysis of alternatives, 1412
    project optimization, 1420–1421
    replacement costs and salvage values, 1415–1416
    subsidies, influence of, 1421–1422
  engineering issues, 1392
  environmental assessment/public participation, 1392–1393
  financial analysis, 1422–1430
    considerations, 1423–1424
    construction financing plans and revenue programs, 1422–1423
    cost allocation, 1423
    freshwater rates, influence on, 1423
    sensitivity analysis and conservative assumptions, 1429–1430
    sources of revenue and pricing of reclaimed water, 1424–1429
  infrastructure and, 27–29
  integrated, 1384–1392
    background information gathering, 1386–1388
    evaluating/ranking alternatives, 1389
    identifying project alternatives, 1388–1389
    implementation plan development, 1389–1392
    objectives formulation, 1386
    problem clarification, 1386
    process, 1385–1386
  legal/institutional aspects, 1393–1397
    institutional coordination, 1397
    policies/regulations, 1397
    water rights law, 1393–1396
    water rights/water reuse, 1395, 1397
  monetary evaluation factors, 1406–1411
    differing perspectives, 1408
    inflation and cost indexes, 1409–1411
    planning weaknesses, 1407–1408
    time horizons, planning and design, 1408–1409
    time value of money, 1409
  reclaimed water market assessment, 1399–1406
    costs and revenues comparison, 1401–1402
    data collection and analysis, 1399–1401
    market assurances, 1402–1406
    water sources comparison, 1399, 1400
  Walnut Valley Water District case study, 1397–1399
Plants, effects of municipal wastewater on, 121

Subject Index

Polluted water sources, public water supply from, 31
Polyvinylchloride (PVC) pipe, 858, (T) 859
Pomona Virus Study, 1487–1488
Ponds and lakes, 1197, 1228–1230
   environmental education, 1230
   operation/maintenance, 1230
   recreation, 1230
   in urban nonirrigation water reuse applications, 1198
   and water quality, 1228–1230
Pools, reflecting, 1196, 1197
Population growth:
   in humid climatic regions, 43
   in West, 41, 43
Pore size, 352
Posttreatment in NF and RO, 492–499
   blending and/or chemical addition, 492, 494
   deaeration and aeration, 499
   disinfection, 499
   effluent stability, 492, 493
   stability indexes, 494–499
Potable water reuse:
   de facto, 77–78
   direct (*See* Direct potable reuse of reclaimed water)
   indirect (*See* Indirect potable reuse)
   issues with, 944
   U.S. domestic, 1171
Potency factor (PF):
   application of, 217
   definition, 216
   for selected potential carcinogenic constituents (T), 216
Potential global water shortages, 19–20
Potential U.S. regional water shortages (*See* U.S. potential regional water shortages in)
Powdered activated carbon (PAC):
   contactors for, 545–546
   properties of, 534
   use of, 534–535
Power requirements (*See also* Energy consumption):
   for activated sludge processes, 364
   for electrodialysis, 503–506
   importance of in technology selection, 275, 289, 455
   for ozone disinfection (T), 663
   for photolysis, 583
   for pumping, 506
   for UV lamps, 676
Precautionary principle, 14–15
Present worth analysis of alternatives, 1412
Pressure:
   atmospheric, 1472–1473
   operating, 435
   osmotic, 463–465
   reduced, 912
   vapor, 1478

Pressure zones, 840
Pressurized filtration, 430–434
Pressurized subsurface application, 979
Pressurized surface application, 977–979
Pretreatment:
   for direct injection wells, 1288
   feedwater pretreatment, 490
   for groundwater recharge with reclaimed water, 1255–1256
   in-building, 788
   and ion exchange, 559–560
   in MBR, 353–354
   for nanofiltration/reverse osmosis systems, 477–479
   NF and RO design/operational considerations, 477–479, 490
   of pathogens, 1259
   for pathogen removal in surface spreading basins, 1257
   for vadose zone injection wells, 1283–1284
Pricing margin, 1429
Pricing of reclaimed water, 1424–1429
Primary effluent:
   filtration of, 439
   pathogens in, 99, 101
Primary treatment, chemical constituents remaining after, 108–111
Prior appropriation, 1394–1396
Probability of infection or disease, 229
Process variables, in MBR analysis and design, 352
Process water, 736
Project Greenleaf, 1089–1092
Project optimization, 1420–1421
   marginal cost analysis, 1420
   total and average *vs.* marginal costs/benefits, 1420–1421
Project study area, delineation of, 1388
Protective coatings, 881, 882
Protozoa, 80, (T) 82, 87–89, 140
   $C_R t$ values for protozoa, 619
   Cryptosporidium parvum, 88–89
   Disinfection, 605, 689
   Entamoeba histolytica, 89
   Giardia lamblia, 87–88
   life cycle, 88
   occurrence in wastewater, 97, 110
   size range for, 376
PSD (*See* particle size)
Public acceptance of urban nonirrigation water reuse, 1179
Public access, 150–151, 1046
Public areas irrigation, 1076, 1078–1080
   acceptance of, 1079–1080
   aesthetics of, 1079
   background for, 1078
   conveyance and distribution system for, 1079
   operation and maintenance issues, 1080
   treatment and quality, reclaimed water, 1079

Public health:
   agricultural irrigation systems, 1011–1013
   cooling water systems, 1138–1140
   landscape irrigation systems, 1050
   urban nonirrigation water reuse applications, 1182, 1188, 1190
   environmental public health indicators, 208
   measures of disease, 207
   public perspectives of, 1441
   water quality, 1050, 1441
Public information program, 1373, 1374
Public participation, 1435–1459
   approaches to, 1444–1446
   after identification of major options, 1445–1446
   in early stages, 1445
   in post-planning stage, 1445
   attitudes about water reuse, 1436–1440
   beliefs about water reuse, 1440
   defining, 1444
   economic perspective, 1441
   environmental justice/equity issues, 1441
   and growth, 1441
   legal mandates for, 1443–1444
   observations on, 1459
   outreach techniques, 1446–1447
   pitfalls, 1448–1450
   controlled selection of participants, 1448
   decide-announce-defend, 1448
   information control, 1448–1449
   need for transparency and public trust, 1450
   participation without influence, 1449
   planning for water reclamation and reuse, 1388, 1392, 1393
   and public health, 1441
   rationale for, 1443
   Redwood City case study, 1450–1453
   St. Petersburg, Florida case study, 1453–1459
   and water quality, 1441
   "yuck" factor, 1442
Public perceptions:
   of direct potable reuse, 1347, 1356–1358
   satellite treatment siting, 734
   of satellite treatment systems, 734
Public policy, 935–936
Public support for water reclamation and reuse, 31
Public water features, 1197–1198
Public water supply from polluted water sources, 31
Pulp and paper industry, 1147–1150
   reclaimed water use in, 1150
   system description, 1147–1149
   water quality considerations for, 1149, 1150

Pump operation valves, 871–872
Pumping of reclaimed water, 745
Pumping station:
  layouts, 868, 873–874
  operation and maintenance, 884
Pumping system design, 841–842
Pumping systems in distribution and storage of reclaimed water, 866–877
  constant *vs.* variable speed operation in, 870–871
  and emergency power, 872–875
  equipment/layout of, 872
  layout of equipment and piping, 872–874
  and location/site layout, 866–867
  and operating schedule, 875–877
  performance of, 801, 870
  types of pumps, 867–870
  valves, 871–872
Pumps, 867–870
  construction materials for, 868, 869
  end-suction centrifugal pumps, 867
  horizontal split-case centrifugal pumps, 867, 868
  vertical turbine pumps, 868
Puron, 339–340
Push-on joints, 860
PVC (Polyvinylchloride), 858–859

Quality (*See* Water quality)

Radiation disinfectants, 603
Rapid infiltration basins (RIBs), 1026
RBF (*See* Riverbank filtration systems)
Recharge, groundwater (*See* Groundwater recharge)
Recharge basins, surface spreading using (*See* Surface spreading using recharge basins)
Reclaimed air (for air conditioning), 1181
Reclaimed water:
  aerosols and windborne sprays, 147–149
  application rates, 147
  dechlorination of, 657–660
  irrigation methods and, 981
  for irrigation of crops, 950, 951, 953, 954
  landscape irrigation design with, 1048
  nitrogen in, 1010
  phosphorous in, 1010
  in public areas landscape irrigation, 1079
  quality monitoring, 145–147
  regulations and guidelines for, 953, 954
  for residential landscape irrigation, 1081
  storage of, 744
  system development for, 1087
Reclaimed water application rates, 147

Reclaimed water applications
  technologies, 388–390
  depth filtration, 388–390
  dissolved air flotation, 389, 390
  membrane filtration, 388–390
  process flow diagrams, 390
  process performance expectations, 390–391
  suitability, 392
  surface filtration, 388–390
Reclaimed water market assessment, 1399–1406
  costs and revenues comparison, 1401–1402
  data collection and analysis, 1399–1401
  market assurances, 1402–1406
  water sources comparison, 1399, 1400
Reclaimed water service requirements (in distribution system), 863
Reclaimed water distribution and storage, water quality management issues, 884–892
Reclamation facilities, 1095
Recovery ratio, for membrane systems, 484–485
Recreational uses of reclaimed water, 736
  for ponds and lakes, 1230
  for streams augmented with reclaimed water, 1224
  for wetlands, 1215
Recycle systems for water, self contained, 821–822
Redox conditions, impact on, 1272–1277
Redox potential in groundwater recharge, 1257–1258
Redox reations:
  in chemical oxidation, 566
  definition, 527
  in groundwater recharge, 1258, 1272
  half reactions for, 566
Redundancy:
  for indirect potable reuse systems, 1323, 1325
  technology selection, 279–280
Refined Collins-Selleck model, 640, 641
Reflecting pools, 1196, 1197
Refractory organic compounds, oxidation of, 568–569
Regeneration, ion exchange and, 560
Regrowth of microorganisms:
  following chlorine disinfection, 654
  following ultraviolet radiation disinfection, 711
Regulation (definition), 135
Regulations and guidelines, 131–185
  aerosols and windborne sprays, 147–149
  agricultural irrigation, 149–150, 953–955
  in California, 51, 52
  constituents and physical properties of concern, 139–142

Regulations and guidelines (*Cont.*):
  development of, 136–139
  dual distribution systems and in-building uses, 151–152
  for dual plumbing systems, 908
  future changes in, 942
  future directions in, 184–185
  future technologies and new, 287–288
  groundwater recharge, 154–155
  impoundments, 152–153
  indirect potable reuse, 155–157
    CWA and SDWA, 155–156
    health risks assessment, 157
    most protected water source, use of, 155
    of states, 167–169
    trace chemical constituents and pathogens, 156–157
    U.S. EPA guidelines, 175–179
  industrial uses, 153
  landscape irrigation, 150–151
  nonpotable uses, 153–154
  planning for indirect potable reuse, 1312–1313
  planning for water reclamation/reuse, 1397
  reclaimed water application rates, 147
  reclaimed water quality monitoring, 145–146
  research and development of, 44–45
  satellite treatment siting, 735
  specific applications, 149–155
  state water reuse (*See* State water reuse regulations and guidelines)
  storage requirements, 146, 147
  technologies and treatment systems, 287–288
  terminology, 132–135
    criterion, 134–135
    guidelines, 135
    regulation, 135
    standards, 134–135
    water reclamation and reuse, 135
  for U.S. drinking water, 221–222
  U.S. EPA (*See* U.S. EPA water reuse guidelines)
  wastewater treatment and water quality considerations, 142–144
    BOD, TSS, and turbidity requirements, 143–144
    disinfection requirements, 144
    economic considerations, 143
    indicator organisms, use and limitations of, 144
    types, 142–143
  water policy, 1397
  water reuse issues, 936–937
  WHO guidelines (*See* World Health Organization guidelines for water reuse)
Relative source contribution (RSC), 219

Reliability (*See also* Statistical analysis):
  decentralized systems, 804
  process, 276, 278, 279
  system, 1309
  of water supply, 935
Removal mechanisms in depth filtration (T), 400–401
Replacement costs, 1415–1416
Required health effects information, 229
Research and development:
  need for future, 291–292
  of regulations and guidelines, 44–45
Reservoirs:
  aeration of, 889–891, 1320
  appurtenances, 880–881
  characteristics of, 1314–1319
  density currents, 1317–1318
  design of, 877–882
  enclosed, 888, 890, 892
  location of, 878–879
  management strategies for, 889–892
  materials of construction, 881, 882
  mixing processes, 1318
  modeling of, 1319–1320
  multiple use facilities, 1314
  open, 887–891
  piping, 879–880
  plankton control, 891
  protective coatings, 881, 882
  residence time, 1318, 1319
  stratification, 1317
  water quality issues, 1314, 1317
Residence time distribution (RTD), 1318–1319
Residential areas, subsurface irrigation systems in, 1086
Residential buildings, dual plumbing systems in, 906–907
Residential homes, 1096
Residential landscape irrigation, 1080–1082
  conveyance and distribution system for, 1081, 1082
  in Irvine Ranch Water District case study, 923–925
  operation and maintenance issues of, 1082
  systems for, 1080, 1081
  treatment and quality, reclaimed water, 1081
Residual chlorine, 631–633
Residual-dissolved constituent removal, 271
Residual particulate matter removal, 373–456
  constituents/properties of concern, 375–385
    distribution of particle sizes, 378–384
    microorganisms, 376, 377
    suspended particles, 377–378
    turbidity, 384–385
  depth filtration (*See* Depth filtration)

Residual particulate matter removal (*Cont.*):
  dissolved air flotation (*See* Dissolved air flotation)
  membrane filtration (*See* Membrane filtration)
  from secondary effluent, 269, 270
  from secondary treatment processes, 385–388
    activated sludge processes, 385–386
    lagoons, 387
    trickling filters, 387
  selection of technology for, 454–456
    comparative performance, 455
    disinfection process type, 455
    economic considerations, 455–456
    energy considerations, 455
    final use of effluent, 454
    future water quality requirements, 455
    pilot-scale study results, 455
    site constraints, 455
  surface filtration (*See* Surface filtration)
  technologies for, 388–392
    process flow diagrams, 390
    process performance expectations, 390–391
    reclaimed water applications, 388–390
    suitability for reclaimed water applications, 392
Residuals return, 742
Resources management, 6–15
  environmental ethics, 13–15
  sustainability criteria, 7–13
Restoration of wetlands, 1214
Restricted access, reuse facilities, 171, 1046, 1209
Retentate classification, 511
Retentate volume, 509, 510
Retrofitting existing treatment plants, 288
Reuse program, 916
Revenue programs, 1422–1423
Revenue sources, 1424–1429
Reverse osmosis (RO), 465–466, 473–475
  advantages/disadvantages of (T), 508
  applications for, 468, (T) 474
  definition of, 464
  design and operational considerations, 475–499
    control of membrane fouling, 490
    feedwater considerations, 475–477
    membrane flux/area requirements, 482–487
    membrane fouling, 487–489
    posttreatment, 492–499
    pretreatment, 477–479
    process operating parameters, 490–492
    treatability testing, 479–482
  dissolved constituent removal, 465–466

Reverse osmosis (RO) (*Cont.*):
  general characteristics (T), 464
  loose RO, 465, (T) 472
  membrane types used in, 473–474
  performance expectations, 474–475
  pilot-scale studies, 499–500
  pretreatment methods, 477–479
  process design consideration (T), 476
  process flow diagram for, 469
  process operating parameters, 490–492
  rejection rates, 474, (T) 475
RSI (Ryznar stability index), 494, 1125
RIBs (rapid infiltration basins), 1026
Ridges and furrows, use in recharge basins, 1270–1271
Riparian doctrine, 1394, 1396
Riparian habitat, 1224
Risk, tolerable, 181–182
Risk analysis, 193–194, 196–197
Risk analysis framework, 226
Risk assessment (*See also* Health risk assessment)
  historical development of, 194, 195
  interrelationships of four steps of, 205
  objectives/applications of human health, 194, 196
  relative nature of, 249
Risk characterization, 204–205
Risk communication, 206–207
Risk extrapolation for carcinogens, 221
Risk management, 205
Risk manager, 234
Riverbank filtration systems (RBF), 1294–1295
RO (*See* Reverse osmosis)
Road care and maintenance, 1198–1200
  dust control, 1199
  snow melting, 1199–1200
  street cleaning, 1199
  urban nonirrigation water reuse applications for, 1199
Robust treatment processes, 290, 291
Rokko Island Water Recycling Project, 1501
Rotating biological contactor (RBC), for decentralized treatment (T), 795
Rotaviruses, 90–91
Rouse Hill Development Area (RHDA)
  case study, 919–922
  implementation, 920–922
  lessons learned, 920
  setting, 919
  water management issues, 920
Rouse Hill Recycled Water Area Project, 1499
RSC (relative source contribution), 219
RTD (*See* Residence time distribution), 1318–1319
Runoff, golf-course irrigation, 1076
Runoff control, 1069
Ryznar stability index (RSI), 494, 1125

Safe Drinking Water Act (SDWA), 46, 155–156
Safeguards in dual plumbing systems design, 908–913
  backflow prevention, 909–911
  page separation, 912
  piping system identification, 912
  reduced pressure, 912
  signage, 912, 913
Safety factors in risk assessment, 219–220
Safety of manufactured products, 153
Salinity:
  definition, 961
  effects on crops, 975–976
  in agricultural irrigation, 959, 960
  in Irvine Ranch Water District case study, 918
  and landscape irrigation design, 1069
  total dissolved solids (TDS), 959
  water quality considerations, 959, 960
Salmonella, 86
Salt tolerance:
  of crops, 972–973
  of landscape plants, 1052–1053
  of turf grasses, 1073
Salvage values, 1415–1416
Sand, for depth filtration:
  properties of, 412
  sizes used in depth filters, 410–411
  typical particle size distribution, 409
San Diego—Santee Lakes Recreation Preserve (*See* Santee Lakes reclaimed water case study)
San Diego Total Resources Recovery Project, 1493
San Diego Water Reuse Study, 1332–1333
Sappi Pulp and Paper Group, 1504
SAR (*See* Sodium adsorption ratio)
SAT (*See* Soil aquifer treatment)
Satellite treatment systems, 725–761
  advantages and disadvantages of, 285
  case studies, 751–761
  collection system requirements, 738–739
  extraction type, 728, 729, 738–740
  future of, 942
  future technologies in, 289
  infrastructure facilities for, 741–745
    diversion/junction structures, 741–744
    flow equalization/storage, 744
    pumping/transmission/distribution of reclaimed water, 745
  integration with existing facilities, 748–751
  interception type, 728, 729, 738, 740
  nonagricultural water reuse applications, 735–737
    groundwater recharge, 736–737
    industrial applications, 737
    lakes/recreational enhancement, 736

Satellite treatment systems, nonagricultural water reuse applications (*Cont.*):
    landscape irrigation, 736
    reuse in buildings, 736
  planning considerations for, 730–735
    economics, 735
    environment, 735
    instituitional issues, 735
    integration with existing facilities, 731
    legal issues, 734–735
    public perception, 734
    regulations, 735
    siting, 731–734
    water needs identification, 730–731
  selection factors, 730
  site location for, 282–283, 285–286
  site selection factors (T), 732
  siting considerations, 731–734
  Solaire Building, New York, case study, 751–754
  terminology, 726–727
  for toilet and urinal flushing, 1193
  Tokyo, Japan, case study, 755–759
  treatment technologies for, 745–748
    conventional, 745
    membrane bioreactors, 746
    sequencing batch reactor, 746–748
  types of, 728–730
  Upland, California, case study, 760–761
  upstream type, 728–730, 739, 741
  wastewater characteristics, 739–741
SBR (*See* Sequencing batch reactor)
Scaling, 1127–1128
  in cooling water systems, 1141
  management options for, 1128
  NF and RO design/operational considerations, 488, 489
  types of, 1127, 1128
Schistosoma mansoni, 89
Scientific Advisory Panel on Groundwater Recharge, 1490
Scraping, of recharge basins, 1271
SDI (*See* Silt density index)
SDI (Stiff and Davis stability index), 1125
SDWA (*See* Safe Drinking Water Act)
Seasonal storage, 918–919
Seawater control, 1250
Secondary effluent:
  filtration of, 437–441
  high quality, 1325
  pathogens in, 99–101
  residual-particulate-matter removal from, 269, 270
Secondary infections, 249
Secondary treatment:
  chemical constituents remaining after, 108–111
  constituent removal, 295–368
    activated sludge *vs.* MBR, 367
    comparative performance of treatment processes, 362

Secondary treatment, constituent removal, (*Cont.*):
    disinfection process types, 362
    economic considerations, 367–368
    energy considerations, 363–364
    expansion of existing plant *vs.* new construction, 362
    final use of effluent, 362
    future water quality requirements, 363
    MBR analysis/design (*See* MBR analysis and design)
    membrane bioreactor (*See* Membrane bioreactor)
    nonmembrane processes (*See* Nonmembrane processes for secondary treatment)
    pilot-scale studies, 362, 363
    selection issues, 361–368
    site considerations, 364–366
    terminology, 296–298
    untreated wastewater, 299–304
    water reuse applications, 304–307
  removal of dissolved organic matter, suspended solids, and nutrients by, 268–270
  residual particle removal, 385–388
    activated sludge processes, 385–386
    lagoons, 387
    trickling filters, 387
  technologies for water reuse applications, 304–307
Security:
  of facilities, 291
  of wetlands, 1221
Sedimentation tanks:
  modification for improved performance, 319
  particle removal performance in shallow and deep clarifiers, 318
  physical factors that affect performance, 319, 320
  sidewater depth, 319
Self contained recycle, 821
Sensitivity analysis, 1429–1430
Separation processes, 528–531
Septic tank effluent gravity (STEG) collection systems, 807–808
Septic tank effluent pump (STEP) pressure collection systems, 789, 808, 810–811
Sequencing batch reactor (SBR):
  advantages and limitations of (T) 312
  for BOD and nitrification (T), 310
  for decentralized treatment system (T), 795
  for nitrogen removal (T) 322
  used in satellite treatment systems, 746–748
  used with membrane systems, 338
Serial filtration for particle size analysis, 378, 379

Subject Index

Series operation (of activated-carbon contactors), 544
Setback distances, 148–149
Settled sludge removal, 448
Shigella, 86
Shutoff valves, 871
SI units:
    abbreviations for (T), 1467–1468
    conversion to U.S. customary units (T), 1463–1465
Sidewater depth, 319
Sieve analysis for depth filter media, 408–409
Signage, for reclamation and reuse systems, 14, 43, 55, 102, 138, 912, 913, 923, 1031, 1078, 1240, 1357
Silt density index (SDI), 479–480, 482
Single-hit models, 200–201
Site considerations, 281–286
    access to site, 879
    for centralized treatment plants, 282, 283, 285–286
    for collection systems, 734
    for decentralized treatment facilities, 283–286
    in densely populated areas, 732, 733
    elevation and availability, 878
    geology and topography, 878
    for satellite treatment facilities, 282–283, 285–286, 731–734
    and secondary treatment selection, 364–366
    in suburban/rural areas, 733–734
    in urban areas, 732, 733
    visual impacts, 879
    for water reservoirs, 878–879
Site layout, 879–881
Site piping, 879
Slime growths (*See* Biofilm)
Sludge removal, 448
Snow, applications with reclaimed water:
    making, 1231
    melting, 1199–1200
Sodicity, 959, 961–965
    definition, 959
    in agricultural irrigation, 959, 961–965
    in landscape irrigation, 1069
    and water quality, 959, 961–965
Sodium adsorption ratio (SAR), 959–965
    adjusted, 961–963
    calculation of, 963–965
    combined effects of salinity and, 963
Sodium hypochlorite, 623
Sodium thiosulfate and related compounds, 659
Soil(s):
    in agricultural irrigation, 955, 957–958
    characteristics of, 955, 957–958
    conditioning, 965
    depth, 958

Soil(s) (*Cont.*):
    effects of municipal wastewater on, 121
    effects of reclaimed water irrigation on, 1011, 1014
    physical structure, 957–958
    texture, 955–957
Soil aquifer treatment (SAT):
    application, 1296–1298, 1490
    definition, 1247
    description, 154–155, 1251
    pretreatment requirements, 1257–1259
    performance, 1271–1280
Solar evaporation, for brine theckening, 512
Solid waste incinerator plant, 1154, 1156
Solids accumulation, clogging due to, 1286
Solids retention time (SRT):
    application, 348–351, 793
    computation of, 343
    definition, 298
    impact on floc strength, 386
    impact on microorganisms in floc particles, 407, 616
    impact of particle size distribution, 315
    impact on solids production, 346–347
    in MBRs (T), 334, 353
Source control, for constituents of concern, 263, 937, 1011, 1229, 1307, 1333
Specific ion toxicity:
    in agricultural irrigation, 965–966
    in landscape irrigation design, 1069
Specific weight of water:
    definition, 1475
    values of (T), 1477–1478
Speece cone for aeration, 891, 1320
Split-case centrifugal pumps, horizontal, 867, 868
Split treatment:
    in phosphorus removal, 327
    in reverse osmosis, 471, 492
Spray dryers, 513
Sprinkler systems, 983
SRT (*See* Solids retention time)
Stability indexes, 494–499
Standards (definition), 134–135
Standby facilities, for process and power, 279–280
State statutes, impact of, 45
State water reuse regulations and guidelines, 157–169
    California, 51–52
    continuing development of, 184
    Florida, 56
    indirect potable reuse, 167–169
    and nonpotable uses of reclaimed water, 159–160, 162–163, 165–167
    specific reuse applications, 158–165
    status of, 158
    variations among states, 158, 162–165

State water reuse regulations and guidelines (*Cont.*):
    voluntary *vs.* mandatory water reuse, 165
Static models for microbial risk assessment, 227–229
    application, 234–238
    equivalence of dynamic and, 231
    evaluating, 232
    model states, 227–229
    probability of infection/disease, 229
    required health effects information, 229
Statistical analysis:
    application to wastewater data:
        activated sludge, effluent, 316–318
        depth filtration, effluent, 403–405
        flowrate, 775
        plant performance data, 408
        surface filtration, effluent, 424
        variability of secondary treatment processes (T), 315
    common statistical parameters, 1479, (T) 1480
    determination of type, 1481–1483
    graphical, 1479–1483
    statistical distributions:
        arithmetic, 1479
        log-normal, 1479
Steel pipe, 858, (T) 859
Steel tanks, 881
STEG collection systems, 807–808
STEP collection systems, 789, 808, 810–811
Stiff and Davis stability index (SDI), 1125
Stockholm framework, 180
Storage:
    affecting reclaimed water, 887–889
    in agricultural irrigation systems, 1003–1008
    reclaimed water, 744
    requirements for, 146, 147
    water quality management issues, 884–889
Storage facilities, 877–882
    concrete tanks, 881
    elevated, ground level, and surface storage, 838–839
    layout, 879–881
    location of, 878–879
    materials of construction, 881
    protective coatings, 881–882
    reservoir capacity, 842–845
    reservoir location, 878–879
    site layout of, 879–881
Storage tanks (*See* Storage facilities)
Stormwater in combined collection systems and infiltration, 106
Stratification in lakes and reservoirs, 1317
Stratified system analysis in reservoirs, 1319–1321
Stream flow augmentation, 1222–1228
    base flow, 1226

Subject Index | 1565

Stream flow augmentation (*Cont.*):
  habitat enhancement, 1222–1224
  operation/maintenance, 1226–1228
  recreational uses, 1224
  stream flow requirements, 1226
  water quality requirements, 1224–1226
Street cleaning, 1199
Study design in epidemiology, 208–211
Submerged attached growth processes, 321, (T) 323, 324
Submerged (vacuum) filtration, 431, 433, 434
Subsidies in economic analysis, 1415, 1421–1422
Substituting reclaimed water for non-potable uses, 26
Subsurface application, pressurized, 979
Subsurface drip irrigation systems, 1082–1086
Subsurface facilities for groundwater recharge, 1295–1296
Subsurface flow, gravity, 977
Subsurface injection, 517–519
Subsurface irrigation systems, 1082–1086
  for on-site and cluster systems, 1082–1085
  in residential areas, 1086
Subsurface systems, 983
Sulaibiyah Wastewater Reclamation and Reuse Project, 1502–1503
Sulfur compounds, 659
Sulfur dioxide, 657–660
Sunk costs (*See* Economic analysis)
Supply and demand, 939, 940
Surface application, pressurized, 977–979
Surface filtration, 417, 419–425
  cloth-media filter, 420–421
  design considerations, 423, 425
  diamond cloth-media filter, 421, 423
  Discfilter, 421, 422
  performance of, 422–424
  pilot-scale studies, 425
  for reclaimed water, 388–390
  technologies, 419–423
Surface flow, gravity, 977
Surface spreading basins, 1251, 1252, 1256–1282
  description, 1256–1257
  dual function basins, 1257
  examples of, 1280–1282
  hydraulic analysis, 1259–1268
    algal blooms, impact of, 1264
    clean water infiltration rates, estimation of, 1268
    infiltration through clogged layer, 1263–1264
    mound development, impact of, 1264–1267
    movement of wetting front, 1260–1263
    surface water infiltration, 1260

Surface spreading basins (*Cont.*):
  operation/maintenance issues, 1268–1271
    ridges/furrows, use of, 1270, 1271
    scraping, 1271
    wet-dry cycles, 1269–1270
  pathogens, 1279–1280
  performance, 1271–1279
    anaerobic ammonia oxidation, 1277–1279
    intermixing, impact of, 1271–1272
    redox conditions, impact on, 1272–1277
    transformation of nitrogen compounds, 1277
    transformation of organic compounds, 1273–1277
  pretreatments needs, 1257–1259
    hydraulic capacity, impact on, 1259
    water quality, impact on, 1257–1259
Surface systems, 982
Surface tension, of water:
  definition, 1476,
  values (T), 1477–1478
Surface water, effects of municipal wastewater on, 121
Surface water augmentation, indirect potable reuse through, 933, 1303–1340:
  case studies:
    San Diego water repurification project and water reuse case study, 1329–1334
    Singapore's NEWater case study, 1332–1340
    Upper Occoquan Sewage Authority case study, 1322–1329
  de facto indirect potable reuse, 1305–1307
  factors favoring reuse (T), 1312
  future of, 942
  health and risk considerations, 1308–1311
    multiple barriers, use of, 1309–1311
    pathogens/trace constituents, 1308–1309
    system reliability, 1309
  lakes and reservoirs, technical considerations, 1314–1322
    characteristics of water supply reservoirs, 1314–1319
    modeling, 1320–1321
    strategies for augmenting, 1320
  planning for, 1309–1313
    costs considerations, 1313
    institutional considerations, 1312–1313
    quantity of reclaimed water to be blended, 1311
    treatment requirements, 1312
    watershed characteristics, 1310, 1311

Surface water augmentation, indirect potable reuse through (*Cont.*):
  public acceptance, 1308
  strategies for, 1307–1308
  Surface water infiltration, 1260
Surge control valves, 871–872
Surrogate parameters, 115
Suspended growth processes:
  advantages and limitations of for BOD removal and nitrification (T), 312
  for BOD removal and nitrification (T), 309–310
  kinetic coefficients for, 344, 345
  kinetic equations for, 342–344
  for nitrogen removal, 321, (T) 322–323, 325
  for phosphorus removal, 324–326
  types of, 308
Suspended particles, 377–378
Suspended solids, 958
  in agricultural irrigation, 958–959
  monitoring of, 145
  treatment technologies for removal of, 268–270
Suspending liquid, nature of, 614–615
Sustainability, resource:
  challenges, 7
  criteria, 7–13
  definitions, 7
  environmental ethics, 13–15
  issues, 934–935
  principle of, 7
  and water conservation, 8, 10, 11
  and water reclamation/reuse, 10–14
Sustainable engineering, 27, 28
Synthetic ion exchange materials, 555
System analysis, in distribution/storage of reclaimed water, 845–847
  extended period system analysis, 847
  model of distribution system, 845–847
  static system analysis, 847
  types of, 845
System hydrology for constructed wetlands, 1218, 1219
System network in distribution and storage of reclaimed water, 836–841
  distribution storage, 837, 840
  piping network, 836
  pressure zones, 840
  pumping stations, 840–841

Tanks (*See* Storage facilities)
TDS (*See* Total dissolved solids)
Technologies and treatment systems, 257–292
  advances in, 31–32, 184
  decentralized systems (*See* Technologies for decentralized systems)
  for direct potable reuse, 1347
  future of, 184, 286–292, 942–944
    decentralized treatment facilities and systems, 289

1566 Subject Index

Technologies and treatment systems, future of (*Cont.*):
  new infrastructure concepts and designs, 290–291
  new regulations, 287–288
  new treatment plants, 289
  research needs, 291–292
  retrofitting existing plants, 288
  satellite treatment systems, 289
  trace constituents, 287
 issues in, 260, 262–265
  multiple barrier concept, 263, 265
  need for multiple treatment technologies, 265
  potential applications, 262, 263
  water quality requirements, 262–264
 needs relating to, 27–29
 plant location, 281–286
  for centralized treatment plants, 282, 283, 285–286
  for decentralized treatment facilities, 283–286
  for satellite treatment facilities, 282–283, 285–286
 for residual suspended particulate removal, 388–392
 satellite treatment systems (*See* Satellite treatment systems)
 selection factors (*See* Technology selection factors for water reuse)
 terminology, 258–259
 untreated municipal wastewater constituents, 260, 261
  physical properties, 261–262
 for water reclamation applications (*See* Technologies for water reclamation applications)
 for water reuse applications, 304–307
Technologies for decentralized systems, 785–806
 disinfection processes, 802–804
 in-building pretreatments, 788
 maintenance, 804, 806
 nutrient removal, 797–802
 performance, 804, 805
 primary treatments, 788–792
 reliability, 804
 secondary treatments, 792–797
 source separating systems, 786–787
Technologies for water reclamation applications, 265–272
 disinfection processes, 271–273
 residual dissolved constituent, removal of, 271
 residual particulate matter in secondary effluent, removal of, 269, 270
 secondary treatment for removal of dissolved organic matter, suspended solids, and nutrients, 268–270
 trace-constituent removal, 271, 272

Technology selection factors for water reuse, 272–281
 infrastructure needs, 280–282
 multiple water reuse applications, 273–277
 pilot-scale testing, 276, 278, 279
 process reliability, 276, 278, 279
 standby and redundancy, 279–280
 trace constituent removal, 273, 276
 types of technology, 939
Temperature, conversion between SI and U.S. customary units, 1465
Temperature, effects on:
 adsorption, 551
 aquatic life, 1223, 1225
 boiler systems, 1142
 biological treatment kinetics, 343, 348
 carbonate equilibrium constants (T), 496, 1121
 chlorine disinfection, 632, 640
 chlorine ionization, 624
 cooling towers, 1136–1138
 corrosion (T), 1116, 1126
 disinfection kinetics, 613–614
 distribution systems, 885
 formation of disinfection byproducts, 342–346
 lake and reservoir stratification, 1317–1318
 legionella inactivation, 1139
 membrane flux rates, 491
 microorganism survival (T), 102
 physical properties of air, 1472–1473
 physical properties of water, 1477–1478
 MBR design, 352
 UV lamps, 674, 676
Tertiary treatment:
 chemical constituents remaining after, 108–111
 pathogens in effluent of, 100, 102
Textile industry, 1150–1155
 process summary for, 1152
 reclaimed water use in, 1153
 system description, 1151, 1153
 water quality considerations for, 1153
Thickened sludge removal, 448
Three-category approach for setting MCLGs, 223–224
Time horizons, planning and design, 1408–1409
Time reference point, 1415
Time value of money, 1409
TMDL (total maximum daily load), 953
Toilet and urinal flushing, 1188–1195
 decentralized systems, 1193
 implementation issues, 1192
 management issues, 1193–1195
 satellite and decentralized systems, 1193
 satellite systems, 1193
 types of applications, 1188, 1189
 and water quality, 1188, 1190–1192

Tolerable (acceptable) risk, 181–182
Total dissolved solids (TDS):
 chlorine disinfection buildup of, 628–630
 ion exchange for removal of, 554
 in irrigation water (*See* Salinity)
  accumulation of, 1129–1131
Total maximum daily load (TMDL), 953
Total suspended solids (TSS):
 biological processes used for the removal of, 310–311, 313
 depth filtration and removal of, 402–405
 relationship to turbidity, 402
 surface filtration and removal of, 422, 424
 typical values in raw and treated wastewater (T), 107, 109–111
 wastewater treatment requirements, 143–144
 variability in:
  activated sludge process effluent, 315–318
  untreated wastewater (T), 302–304
Toxicity, specific ion, 965–966
Toxicology, 209, 211–214
 LD_{50} and LC_{50}, 211, 213
 NOAEL and LOAEL, 211, 212
 in vitro tests, 211
 in vivo tests, 212, 213
 whole animal tests for carcinogenicity, 213, 214
Trace constituents, 1308–1309
 agricultural irrigation, 149–150
 future water reuse, 287
 health concerns for indirect potable use, 1306–1307
 landscape irrigation, 151
 need to remove, 273, 276
 regulations and guidelines, 156–157
 treatment technologies for removal of, 271, 272
Trace constituents removal, 525–591
 adsorption, 532–551
 advanced biological transformations, 586–591
 advanced oxidation processes, 569–578
 chemical and biological transformation processes, 531–532
 chemical constituents remaining after, 113
 chemical oxidation, 563–567
 distillation, 560–563
 effectiveness of processes, 529
 flow diagrams, 530
 ion exchange, 551–560
 mass transfer-based separation processes, 528, 531
 photolysis, 578–586
 terminology, 526–527
Trace elements, 966–969

Subject Index

Trace organic compounds in groundwater recharge, 1276–1277
Trace organics removal, 532–533
Tracer studies, 1322:
    analysis of data from, 647–650
    characteristics of tracer curve, 646
    indexes used for (T), 647
    in lakes and reservoirs, 1322
    response curve analysis, 645–650
    schematic diagram for conduct of, 644
    types of tracers, 644
    used to detect short circuiting, 645
Transformation of nitrogen compounds, 1277
Transformation of organic compounds in recharge basins, 1273–1277
    bulk organic transformations, 1273–1275
    disinfection byproduct formation (DBPFP), 1275–1276
    endocrine disrupting activity, 1277
    organic compounds of concern, 1273
    trace organic compounds, 1276–1277
Transformation processes, 531–532
Transmission of reclaimed water, 745
Transmission pipeline, 1024
Transmission pumping stations, 1024
Transmittance:
    adjustment for equipment validation studies, 702–708
    definition of, 262, 685
    determination of, 384
    importance in UV disinfection, 686–687
    relationship to absorbance, 685
    typical values for wastewater, 685
Transparency, in public projects, 1450
Treatability testing, 479–482
    limitations of fouling indexes, 481
    mini-plugging factor index, 480, 481
    modified fouling index, 480
    silt density index, 479–480, 482
Treated wastewater:
    chemical constituents in, 108–113
        after AWT, 108–112
        after primary treatment, 108–111
        after secondary treatment, 108–111
        after tertiary treatment, 108–111
        impact of constituents remaining after treatment, 113
        removal of trace constituents, 113
    pathogens in, 97–102
Treated water, chemical constituents in natural water *vs.*, 114, 116
Treatment facilities:
    expansion of existing plant *vs.* new construction, 362
    location of, 281–286
    new, 289
    reliability of, 165
    retrofitting existing, 288

Treatment systems (*See* Technologies and treatment systems)
Trends in water reuse, 63–65
Trickling filters, 308, 313, 314, 387
TSS (*See* Total suspended solids)
Turbidity, 384–385
    in activated sludge processes, 385
    breakthrough in filtration, 399, 418
    guidelines for reuse, 264
    limitations of measurement, 385
    limitations of use as a measure of membrane integrity, 440
    measurement of, 384
    monitoring, 145
    occurrence in wastewater, 110
    probability distribution in filtered effluent, 403
    relationship to TSS, 402
    removal by depth filtration, 401–402, 417
    removal by surface filtration, 422, 424
    variability in removal of, 315, 386, 398, 402–405, 424
    wastewater treatment requirements, 143–144
Turbine pumps, 868
Turf grasses, 1072, 1073
Turnouts, 1024
2006 WHO agriculture guidelines, 182–184
    health-based targets, 182–183
    health protection measures, 183–184
    health risk assessment, 182

UF (*See* Ultrafiltration)
Ultimate disposal methods, 515–519
    discharge to wastewater collection system, 515–517
    disposal to surface waters, 517
    subsurface injection, 517–519
Ultrafiltration (UF), 430–442
    design considerations, 441, 442
    membrane performance, 436–441
        broken fibers, impact of, 440–441
        primary effluent filtration, 439
        secondary effluent filtration, 437–441
    operating characteristics/strategies, 436, 437
    operational modes, 431–434
    pressurized, 430–434
    process analysis, 435–436
        mass balance, 436
        operating pressure, 435
        permeate flow, 435
        recovery, 435–436
        rejection, 436
    submerged, 431, 433, 434
Ultraviolet lamps, 674–678
    ballasts for, 678
    emerging technologies, 676–678
    low-pressure high-intensity, 676, 677
    low-pressure low-intensity, 675–677
    medium-pressure high-intensity, 676, 677

Ultraviolet radiation (UV):
    absorption of, 580–582, 675, 687
    definition, 675
    disinfection with, 609
    sources of, 674
Ultraviolet (UV) disinfection, 674–711
    advantages and disadvantages of, 621
    analysis of system, 708
    closed channel, 680–682
    dosage estimation, 691–700
        bioassay testing, 695–697
        collated beam bioassay, 692–695, 697–700
        reporting results, 697
    environmental impacts, 711
    factors affecting effectiveness, 684–690
        definition of UV dose, 684–686
        effect of chemical constituents in reclaimed water, 686–687
        effect of particles in reclaimed water, 687, 688
        impact of system characteristics, 690
        microorganism characteristics, 688–690
    guidelines, 700–708
        application, 700
        and design, 700–701
        test protocol, 701–702
        validation testing, 702–708
    inactivation, 682–683
    mechanism, 682–684
    microbial repair following, 683–684
    modeling, 690–691
    open channel, 678–680
    operational issues, 708–710
        biofilms, 709–710
        hydraulics, 708–709
        increasing UV intensity for overcoming impact of particles, 710
        upstream treatment effect, 710
    physical facilities for, 607–609
    source of, 674, 675
    system configurations, 678–682
    UV lamps, 674–678
Uniform Plumbing Code, 908, (T) 909
Uniformity coefficient:
    definition of, 408–409
    typical values for depth filters (T), 410–411
Units:
    for environmental engineering computations (T), 1466–1467
    conversion factors between SI and U.S. customary units, 1463–1465
    abbreviations for SI, 1467–1468
    abbreviations for U.S. customary, 1468–1469
Unit cost (*See* Economic analysis)
United States:
    agricultural irrigation in, 950
    commercial water use in, 1172

United States (*Cont.*):
  domestic potable water reuse in, 1171
  drinking water regulations in, 221–222
  evolution of water reclamation and reuse in, 42–43
  industrial applications in, 1105–1107
  landscape irrigation in, 1046–1048
  potential regional water shortages in, 20–23
  water reclamation activities in, 1485–1496
  water reuse status in, 46–47
U.S. EPA water reuse guidelines, 169–179
  agricultural, 171–173
  carcinogens, 222–223
  construction, 174
  control measures, 178
  disinfection requirements, 169–179
  environmental, 175
  groundwater recharge, 175
  indirect potable, 175–179
  industrial, 174
  landscape impoundments, 173
  microbial limits, 178
  municipal wastewater, 170–177
  restricted access area irrigation, 171
  urban reuse, 170
Unrestricted access in landscape irrigation, 1045, 1209
Untreated wastewater, 299–304
  chemical constituents in, 103–104
    natural water, 103–104
    public water supplies, 104
  composition of, 106–107
  constituents and physical properties in, 260–262, 299–304
    concentration values, 299, 301
    variability of mass loadings, 301–304
  pathogens in, 94, 96–98
Upstream treatment processes, effect of:
  as disinfection performance factor, 615, 616
  on UV disinfection performance, 710
Upstream-type satellite treatment systems, 728–730
  collection system requirements, 739
  wastewater characteristics, 741
Urban landscape irrigation design, 1049, 1051
Urban nonirrigation water reuse, infrastructure facilities for, 1176
Urban nonirrigation water reuse applications, 932, 933, 1169–1200
  air conditioning, 1179–1183, (T) 1184–1185
    management issues, 1183
    reclaimed air for, 1181
    system description, 1179–1180
    water quality considerations for, 1181–1183
  for car and vehicle washing, 1195–1196
  commercial applications, 1195–1197

Urban nonirrigation water reuse applications (*Cont.*):
  commercial water use in U.S., 1172
  domestic potable water use in U.S., 1171
  EPA water reuse guidelines, 170
  factors affecting use, 1175–1179
    infrastructure issues, 1175–1176
    water quality, 1176–1179
  fire protection, 1183, 1186–1188
    applications, types of, 1186–1187
    implementation issues, 1187–1188
    management issues, 1188
    water quality considerations, 1187
  globally, 1172, 1175
  for laundries, 1196
  public water features, 1197–1198
  road care and maintenance, 1198–1200
  terminology, 1070–1071
  toilet and urinal flushing, 1188–1195
    decentralized systems, 1193
    implementation issues, 1192
    management issues, 1193–1195
    satellite systems, 1193
    types of applications, 1188
    water quality considerations, 1188–1192
  in U.S., 1172–1174
Urinal flushing (*See* Toilet and urinal flushing)
U.S. customary units:
  abbreviations for (T), 1468–1469
  conversion to SI units (T), 1463–1465
U.S. Filter, 334, 336–337, 356–357
UV (*See* Ultraviolet radiation)
UV absorbing compounds (T), 685
UV disinfection system hydraulics, 708–709
UV radiation sources, 674–677
UV validation testing, 701–708, (*See* also Ultraviolet disinfection)

Vacuum filtration (*See* Submerged filtration)
Vacuum collection system, 810, 812
Vadose zone injection wells, 1282–1286
  description, 1282–1283
  examples, 1286
  hydraulic analysis, 1284–1285
  operation/maintenance issues, 1285–1286
  performance, 1286
  pretreatment needs, 1283–1284
Valves:
  air release, 864
  distribution system, 863
  excercising, 883
  pumping system, 871–872
Vapor compression distillation, 562
Vapor pressure of water:
  definition, 1476
  values (T), 1477–1478
Vector, in the transfer of disease, 208, 1232, 1252

Vegetable crops:
  boron tolerance of, 974
  salinity affecting, 975
  salt tolerance of, 972
Vegetation management in wetlands, 1221
Vehicle washing, 1195–1196
Vertical turbine pumps, 868
Viruses, 89–92
  classification of,
  concern in water reclamation and reuse, 140
  description of, 89–92
  dose response parameters for risk assessment, 203
  infectious dose, 98
  models, risk assessment, 235–236, 237–238, 240–248
  monitoring for, 145, 164
  occurrence in wastewater, 97–98, 314
  removal of:
    by chlorination, 619, 637–638
    comparison of disinfection processes for (T), 605
    by groundwater recharge, 1259, 1279–1280
    by membranes, 437–438
    by ozonation, 619
    by reverse osmosis, 464, 472, 475
    by surface filtration, 423
    by UV irradiation, 604, 619, 682–683, 689, 697
  risk assessment for water reuse from enteric, 244–248
  selected examples (T), 82:
    adenoviruses, 92
    enteroviruses, 91–92
    hepatitis A, 90
    noroviruses and caliciviruses, 90, 91
    rotaviruses, 90
  sizes range, 376
Viscosity of water:
  definition, 475
  SI units (T), 1477
  U.S. customary units (T), 1477–1478
Voluntary marketing approach, 1404
Voluntary water reuse, mandatory *vs.*, 165

Waste and unreasonable use of water, doctrine, 1406
Waste discharge requirements for irrigated land, 953
Wastewater:
  in decentralized systems
    constituent concentrations, 778–785
    flowrates, 774–778
  equations used in characterization of, 341
  and pathogenic microorganisms, 97–102
  physical properties of concern in, 139–142
  in satellite treatment systems, 739–741
  and water quality, 142–144

Wastewater treatment, 142–144
  BOD, TSS, and turbidity requirements, 143–144
  disinfection requirements, 144
  economic considerations with, 143
  indicator organisms, use and limitations of, 144
  types, 142–143
Water, physical properties of, 1475–1478
Water application rate estimation, 989–997
  evapotranspiration, 990–992
  hydraulic loading rate, 993–997
  net irrigation requirement, 992, 993
Water Conserv II case study, 1022–1027, 1492–1493
  implementation, 1023–1027
  importance of, 1027
  lessons learned, 1027
  setting, 1023
  water management issues, 1023
Water conservation, 8, 10, 11, 777–778
Water Environment Research Foundation, 1494–1495
Water flux rate, 482–483
Water hammer, in distribution systems, 871
Water issues, 3–32
  future, 30–32
  reclamation and reuse, 23–32
  resources management, 6–15
  shortages, 15–23
  terminology, 4–6
Water management in industries, 1107–1108
Water needs estimation:
  evapotranspiration, 1054, 1058
  irrigation efficiency, 1060–1064
Water pinch analysis, 1107–1108
Water quality:
  agricultural irrigation systems, 977, 978
  agronomics, 954–976
  as application selection factor, 937, 939
  aquifer storage and recovery wells, 1294
  BOD, TSS, and turbidity requirements, 143–144
  consistency of in urban water reuse applications, 1176–1178
  in cooling water systems, 1132–1135
  direct potable reuse guidelines, 1358–1361
  dual distribution systems, 152
  economic considerations with, 143
  for fire protection, 1187
  for groundwater recharge, 1255–1259
  impoundments, 153
  indicator organisms, use and limitations of, 144
  industrial issues with (*See* Water quality issues for industrial uses)
  management, 884–892

Water quality (*Cont.*):
  MBR analysis and design, 341, 342
  monitoring, 145–147
  ponds/lakes, 1228–1230
  public health, 1441
  in reclaimed distribution systems (*See* Water quality management in reclaimed distribution systems)
  reliability issues of, 1178–1179
  reservoirs, 1314, 1317
  residential landscape irrigation, 1081
  residual particulate matter removal, 455
  riverbank filtration systems, 1295
  secondary treatment selection issues, 363
  standards, basis for, 136
  stream flow augmentation, 1224–1227
  technology issues, 262–264
  urban nonirrigation water reuse, 1176–1179
  wastewater treatment, 142–144
  wetlands, 1213–1214, 1216, 1219
Water quality issues for industrial uses, 1109–1131
  accumulation of dissolved constituents, 1129–1131
  Aggressiveness Index, 1118–1119
  buffer capacity, 1119
  calcium carbonate precipitation potential (CCPP), 1120–1124
  corrosion issues, 1110, 1113–1117
  corrosion management options, 1126–1127
  general considerations, 1110–1112
  indices for assessing effects of reclaimed water quality, 1115, 1116, 1118–1125
  Langelier saturation index (LSI), 1124–1125
  Larson's Ratio, 1125
  Ryznar stability index (RSI), 1125
  scaling issues, 1127–1128
Water quantity, in agricultural irrigation systems, 977, 978
Water reclamation and reuse, 10–13, 23–32
  acceptance depending on opportunity and necessity, 31
  advances in technologies, 31–32
  challenges, 32
  definition of, 135
  evolution of, 39–45
  future of, 30–32
  implementation hurdles, 31
  important role, 23–30
  infrastructure/planning issues, 28–30
  integrated water resources planning, 24–27
  personnel needs/sustainable engineering, 27, 28
  planning for (*See* Planning for water reclamation and reuse)

Water reclamation and reuse (*Cont.*):
  public support, 31
  public water supply from polluted water sources, 31
  and sustainability criteria, 10–14
  treatment/technology needs, 27–29
Water Recycling 2030, 1495
Water replenishment in ponds and lakes, 1230
Water reuse, 37–65
  California case study, 47–53
  case studies, 47–58
  current U. S. status, 46–47
  EPA guidelines, 169–179
  evolution of, 39–45
  regulations, 135–155, 157–184
  state and federal statutes, impact of, 45–46
  and sustainability criteria, 10–14
  trends, 63–65
  types of, 24
  WHO guidelines for, 179–184
  worldwide, 58–63
Water reuse applications, 149–155, 929–945
  agricultural irrigation, 149–150, 931, 932
  decentralized, 816–819
  direct potable reuse, 934
  dual distribution systems, 151–152
  environmental and recreational use, 932, 933
  future trends in, 941–944
  global, 934
  groundwater recharge, 154–155, 932, 933
  impoundments, 152–153
  in-building uses, 151–152
  indirect potable reuse through surface water augmentation, 933
  industrial, 153, 931–933
  issues, 934–938
    economic considerations, 935
    public policy, 935–936
    regulations, 936–937
    reliability of water supply, 935
    resource opportunities, 935
    resource sustainability, 934–935
    specific applications, 937, 938
  landscape irrigation, 150, 151, 931, 932
  nonpotable, 153–154
  selection factors, 937, 939–941
    economic feasibility, 940–941
    environmental considerations, 941
    infrastructure requirements, 939, 940
    supply and demand, 939, 940
    technology types, 939
    water quality considerations, 937, 939
  technologies for, 304–307
  technology selection factors for multiple, 273–277

Water reuse applications (*Cont.*):
  urban nonirrigation uses, 932, 933
  water reuse guidelines, 59, 135–139
Water rights law, 1393–1396
  groundwater law, 1395
  and planning for indirect potable reuse, 1313
  prior appropriation, 1394–1396
  riparian doctrine, 1394, 1396
Water scarcity, 19–20
Water shortages, 15–23
  potential global, 19–20
  potential U.S. regional, 20–23
  in St. Petersburg, 1087
  world population, impact of, 15–19
Water supply:
  augmentation of, 942
  and community growth, 1441
Water table depth, 999
Water use patterns, 27
Water wars, 1453–1454
Water wells, supplemental, 1024
Waterborne diseases, 80–83
Waterborne pathogenic microorganisms, 83–92
  bacteria, 83, 86–87
  helminths, 89
  log removal, 83
  protozoa, 87–89
  terminology conventions, 83
  viruses, 89–92
Waterfalls, 1196, 1197
WBMWD case study (*See* West Basin Municipal Water District case study)
Weed control, 1075
Welded joints, 861

Well function, W(u):
  definition and use, 1289
  values (T), 1524
West Basin Municipal Water District (WBMWD) case study, 1158–1161
  implementation, 1159–1161
  lessons learned, 1161
  setting, 1158
  water management issues, 1158–1159
Wet-dry cycles, 1269–1270
Wetlands, 1210–1221
  for decentralized systems, 796, 821
  development of, 1213–1215
    alternate dispersal of reclaimed water, 1214
    habitat value, 1214
    recreation, 1215
    restoration/mitigation, 1214
    water quality improvement, 1213–1214
  operation/maintenance, 1216, 1218–1221
    mosquito control, 1220
    nuisance species, 1220–1221
    odors, 1221
    security issues, 1221
    system hydrology, 1218, 1219
    vegetation management, 1221
  state regulations and guidelines, 166
  types of, 1210, 1212–1213
  water quality considerations, 1216–1219
Wetting front, movement of, 1260–1263
WHO guidelines for water reuse (*See* World Health Organization (WHO) guidelines for water reuse)

Whole animal tests for carcinogenicity, 213, 214
Wilson-Grizzle Act, 1087, 1454–1455
Windborne sprays, 147–149
  limiting exposure, 148
  pathogen survival, 147, 148
  setback distances, 148–149
World Health Organization (WHO) guidelines for water reuse, 59, 179–184, 954
  agriculture and aquaculture, 180
  disability adjusted life years, 180–181
  Stockholm framework, 180
  tolerable microbial risk in water, 181–182
  tolerable risk concept, 181
  2006 guidelines for safe use of wastewater in agriculture, 182–184
World population, water shortages impacted by, 15–19
  domestic and industrial water uses, 19
  irrigation water use, 17, 18
  urbanization, 16–19
Worldwide water reuse, 58–63
  in developing countries, 59, 61–63
  evolution of, 40–41
  significant developments, 58–60
  WHO guidelines, 59

Xeriscape, 1054

Yersinia enterocolitica, 87
"Yuck" factor, 1442

Zenon Environmental, 333–335, 355–356
Zero liquid discharge (ZLD), 1107

Atomic numbers and atomic masses[a]

Element	Symbol	Z	Mass	Element	Symbol	Z	Mass
Actinium	Ac	89	227.0278	Mercury	Hg	80	200.59
Aluminum	Al	13	26.98154	Molybdenum	Mo	42	95.94
Americium	Am	95	(243)	Neodymium	Nd	60	144.24
Antimony	Sb	51	121.75	Neon	Ne	10	20.179
Argon	Ar	18	39.948	Neptunium	Np	93	237.0482
Arsenic	As	33	74.9216	Nickel	Ni	28	58.70
Astatine	At	85	(210)	Niobium	Nb	41	92.9064
Barium	Ba	56	137.33	Nitrogen	N	7	14.0067
Berkelium	Bk	97	(247)	Nobelium	No	102	(259)
Beryllium	Be	4	9.01218	Osmium	Os	76	190.2
Bismuth	Bi	83	208.9804	Oxygen	O	8	15.9994
Boron	B	5	10.81	Palladium	Pd	46	106.4
Bromine	Br	35	79.904	Phosphorous	P	15	30.97376
Cadmium	Cd	48	112.41	Platinum	Pt	78	195.09
Calcium	Ca	20	40.08	Plutonium	Pu	94	(244)
Californium	Cf	98	(251)	Polonium	Pu	84	(209)
Carbon	C	6	12.011	Potassium	K	19	39.0983
Cerium	Ce	58	140.12	Praseodymium	Pr	59	140.9077
Cesium	Cs	55	132.9054	Promethium	Pm	61	(145)
Chlorine	Cl	17	35.453	Protactinium	Pa	91	231.0389
Chromium	Cr	24	51.996	Radium	Ra	88	226.0254
Cobalt	Co	27	58.9332	Radon	Rn	86	(222)
Copper	Cu	29	63.546	Rhenium	Re	75	186.207
Curium	Cm	96	(247)	Rhodium	Rh	45	102.9055
Dysprosium	Dy	66	162.50	Rubidium	Rb	37	85.4678
Einsteinium	Es	99	(254)	Ruthenium	Ru	44	101.07
Erbium	Er	68	167.26	Samarium	Sm	62	150.4
Europium	Eu	63	151.96	Scandium	Sc	21	44.9559
Fermium	Fm	100	(257)	Selenium	Se	34	78.96
Fluorine	F	9	18.99840	Silicon	Si	14	28.0855
Francium	Fr	87	(223)	Silver	Ag	47	107.868
Gadolinium	Gd	64	157.25	Sodium	Na	11	22.98977
Gallium	Ga	31	69.72	Strontium	Sr	38	87.62
Germanium	Ge	32	72.59	Sulfur	S	16	32.06
Gold	Au	79	196.9665	Tantalum	Ta	73	180.9479
Hafnium	Hf	72	178.49	Technetium	Tc	43	(97)
Helium	He	2	4.00260	Tellurium	Te	52	127.60
Holmium	Ho	67	164.9304	Terbium	Tb	65	158.9254
Hydrogen	H	1	1.0079	Thallium	Tl	81	204.37
Indium	In	49	114.82	Thorium	Th	90	232.0381
Iodine	I	53	126.9045	Thulium	Tm	69	168.9342
Iridium	Ir	77	192.22	Tin	Sn	50	118.69
Iron	Fe	26	55.847	Titanium	Ti	22	47.90
Krypton	Kr	36	83.80	Tungsten	W	74	183.85
Lanthanum	La	57	138.9055	Uranium	U	92	238.029
Lawrencium	Lr	103	(260)	Vanadium	V	23	50.9414
Lead	Pb	82	207.2	Xenon	Xe	54	131.30
Lithium	Li	3	6.941	Ytterbium	Yb	70	173.04
Lutetium	Lu	71	174.97	Yttrium	Y	39	88.9059
Magnesium	Mg	12	24.305	Zinc	Zn	30	65.38
Manganese	Mn	25	54.9380	Zirconium	Zr	40	91.22
Mendelevium	Md	101	(258)				

[a] From *Pure Applied Chemistry.*, vol. 47, p. 75 (1976). A value in parentheses is the mass number of the longest lived isotope of the element.